INDIA

A Comprehensive Geography

List of Books by the same Author

1. **India—A Comprehensive Geography (All Universities) (English)**
2. **Physical Geography (All Universities) (English)**
3. **ISC Geography for Class-XI**
4. **ISC Geography for Class-XII**
5. **India—A Comprehensive Geography (All Universities) (Hindi)**
6. **Geography of India and Practical Geography (Hindi), Rajasthan**
7. **Physical Geography and Practical Geography (Hindi), Rajasthan**
8. **Geography of India and Practical Geography (English)**
   (For all universities in Haryana)
9. **Geography of India and Practical Geography (Hindi)**
   (For all universities in Haryana)
10. **Physical Geography and Practical Geography (Hindi)**
    (For all universities in Haryana)
11. **Physical Geography (Atmosphere and Hydrosphere) and Practical Geography (Hindi)**
    (For all universities in Haryana)
12. **Human Geography and Practical Geography (Hindi)**
    (For all universities in Haryana)
13. **Economic Geography and Practical Geography (Hindi)**
    (For all universities in Haryana)
14. **Remote Sensing and Quantitative Methods (Hindi)**
    (For all universities in Haryana)
15. **Practical Geography (Hindi) for Different Universities**
16. **Human Geography (Hindi) for Different Universities**
17. **Commercial Geography for Andhra Pradesh and Telangana**
18. **Physical Geography (Geomorphology) and Practical Geography (English)**
    (For all universities in Haryana)
19. **Physical Geography (Atmosphere & Hydrosphere) and Practical Geography (English)** (For all universities in Haryana)

INDIA

A Comprehensive Geography

D.R. Khullar
Former Principal
Government Post Graduate College
KALKA

KALYANI PUBLISHERS
LUDHIANA • NEW DELHI • NOIDA (U.P.) • HYDERABAD • CHENNAI
KOLKATA • CUTTACK • GUWAHATI • KOCHI • BENGALURU

KALYANI PUBLISHERS

Head Office

3251/B, Street No. 5, Near Sandhu Tower, Gurdev Nagar, **Ludhiana-**141 001 • Ph : 0161-2760031
E-mail : kalyanibooks@yahoo.co.in **Website :** www.kalyanipublishers.co.in

Administrative Office

4779/23, Ansari Road, Daryaganj, **New Delhi-**110 002 • Ph : 011-23271469, 23274393
E-mail : kalyani_delhi@kalyanipublishers.in

Works

B-16, Sector-8, **NOIDA** (U.P.)

Branches

1, Mahalakshmi Street, T. Nagar, **Chennai-**600 017 • M : 9444117261
Gopabandhu Lane, Behind Govt. Bus Stand, Badambadi,
Cuttack-753 012 (Odisha) • Ph : 0671-2311391
4-3-65/2/A, Behind HVS Public School, Raghunath Bagh
Sultan Bazar, **Hyderabad-**500 095 • Ph : 040-24750368
10/2B, Ramanath Mazumdar Street, **Kolkata-**700 009 • Ph : 033-35571565
Arunalaya, 1st Floor, Saraswati Road, Pan Bazar, **Guwahati-**781 001 • Ph : 0361-2731274
Koratti Parambil House, Convent Road, **Kochi-**682 035 • Ph : 0484-2367189
No. 24 & 25, 1st Floor, Hameed Shah Complex, Cubbonpet Main Road, **Bengaluru-**560 002

Acknowledgement **© Government of India, 2005**

The following pertains to all maps :

1. Based upon Survey of India map.
2. The territorial waters of India extend into the sea to a distance of twelve nautical miles measured from the appropriate base line.
3. The administrative headquarters of Chandigarh, Haryana and Punjab are at Chandigarh.
4. The inter state boundaries between Arunachal Pradesh, Assam and Meghalaya shown on these maps are as interpreted from the North-Eastern Areas (Reorganisation) Act, 1971, but have yet to be verified.
5. The external boundaries and coastlines of India agree with the Record/Master Copy certified by Survey of India.
6. The inter state boundaries between Uttaranchal-Uttar Pradesh, Jharkhand-Bihar and Chhattisgarh-Madhya Pradesh have not been verified by the Governments concerned.
7. The external boundary and coastline of India on the maps agree with the copy certified by the Survey of India, Dehradun.
8. The responsibility for the correctness of internal details rests with the publisher.
9. The information in the maps given inside are for understanding the facts and has got no political value.

U 41426 3 44 + 1088 = 1132 Pages 20" x 30" 141.5 Forms

U-42921 3

First Edition, 2000
Second Edition, 2006 • Reprinted, 2007 - 2013
Third thoroughly Revised and Enlarged Edition, 2014 • Reprinted, 2017, 2018
Fourth thoroughly Revised and Enlarged Edition, 2018 • Reprinted, 2019
Fifth thoroughly Revised Edition, 2020
Sixth thoroughly Revised Edition, 2023
Seventh Revised Edition, 2024

Reprinted, 2025

Typesetting at
Times Printographic

ISBN 978-93-272-9148-3

PRINTED IN INDIA

at Diamond Agencies Pvt. Ltd., B-125, Sec. 63, NOIDA

Dedicated to the everlasting memory of

my wife

Mrs. ASHA KHULLAR

who has been a perpetual source of

inspiration to me throughout life

PREFACE

SEVENTH REVISED EDITION

I am immensely delighted to place the **Seventh Revised Edition** of this book before the readers. The revision of text books, particularly the data based text books, is always an ongoing perennial process. In order to maintain high quality of teaching and general awareness, revision of text books should be accomplished at the earliest possible.

This revision has become necessary in view of the fact that the book has been pirated by some unscrupulous people to have financial gain accruing from the unparalleled popularity which the book has been enjoying for the last several years. The change in the content will help in discouraging the unwanted pirators and in providing latest information to the readers.

I am beholden to the readers for love and affection shown by them to the earlier editions of this book. The old and obsolete matter has been removed and is replaced by new and refreshing matter.

Maps and diagrams comprise the soul of a book on any subject of geography. As such the book is documented with over 326 maps and diagrams which are extensive and designed for their utility, clarity and beauty. This makes the book picturesque, and its reading enjoyable. Further, 305 statistical tables containing latest data have been given to make the subject matter perfectly authentic. These features will surely enhance the utility of the book and the readers will find its reading more rewarding and comfortable.

I am sure that with the above mentioned changes, this edition provides the reader an entirely new orientation to Geography of India, and will serve the purpose for which it is intended.

My foremost obligation is to my readers, who act as a perennial source of inspiration for me for making the book more and more comprehensive. I shall ever remain grateful to them for providing me the requisite stimulus. Encouragement, stimulation, guidance, help and suggestions have continued to pour in, as ever before, from a number of friends and institutions. It would be remiss not to express my sincere thanks to them.

I will be failing in my duty if I do not express my gratefulness to M/s Kalyani Publishers, reputed as one of the leading publishers of India, for bringing out this book in its present form. My special thanks are to Sh. Raj Kumar. But for his keen interest and deep involvement in the project, this edition of the book would not have seen the light of the day.

Suggestions for further improvement of the book are solicited.

July, 2024

1150, Sector 16, Faridabad-121 002
(M) 9311139488

D.R. KHULLAR
Email: khullardr@yahoo.in

PREFACE

FIRST EDITION

The present book entitled "India—A Comprehensive Geography", is the outcome of my long teaching experience spanning over 34 years, supplemented by a long eventful innings of 27 years as an author of several books on different subjects of Geography. The primary object of writing this book is to meet the growing demand for a suitable text book for the university and college students at the post-graduate and under-graduate levels. Besides, the book will be of great value for candidates appearing in the competitive examinations conducted by the Union Public Service Commission (UPSC) and the State Public Service Commissions. It is also designed to cater to the needs of those who are more serious in adding to their existing knowledge of geography of India.

Students and teachers at the university and college level have always been feeling handicapped in the absence of a good book on the geography of India. A few books available in the market do not fully meet their academic needs. Students must be exposed to fundamentals of 'Geography of India' in a manner that they may easily comprehend and grasp it.

The book has been written in simple, lucid, clear and modern style so that the teachers and students of geography could draw maximum benefit. Ample use of maps and diagrams has been made. This has made the subject matter interesting, picturesque and easy to learn. India has made rapid strides in the recent past and drastic changes have taken place in almost all the spheres. To keep pace with ever changing scenario on the geographical horizon of India, up-to-date data and information have been used. This will help the reader to acquaint himself with the latest developments.

The economic prosperity of a country largely depends upon its natural resources and the way in which its inhabitants utilize these resources. Keeping this dictum in view, the entire volume has been divided into the trio of *land, people* and *economy.* Part I on THE LAND deals with the natural set up of the country and includes chapters on Relief, Geology, Drainage, Climate, Soils, Natural Vegetation, Natural Hazards and Disasters and Environment, apart from the Introductory chapter.

India is a huge reservoir of manpower with a vast ocean of humanity and is the world's second most populous country next only to China. If utilized properly, the poverty stricken, much cursed and ever growing population of India could prove to be a boon rather than bane for the national development. Thus, Part II of the book is aptly devoted to THE PEOPLE of the country. This part includes chapters on Growth and Distribution of Population, Composition of Population, Migration Patterns, Rural Settlements and Trends in Urbanization.

Part III highlights the ECONOMIC GROWTH of the country. An effort has been made to analyse the process of economic development and also the factors responsible for slow progress in economic and commercial growth. India is still a developing economy in spite of phenomenal progress made by the country in almost all the spheres, particularly after Independence. The chapters included in this part are on Irrigation, Multipurpose Projects, Livestock, Fisheries, Land Utilization, Agriculture, Green Revolution, Major Crops, Mineral Resources, Energy Resources, Manufacturing Industries, Transport, Foreign Trade, and Tourism.

The above mentioned scheme of dividing the book into three distinct and well defined parts helps in understanding the interplay of natural and human forces. A systematic appraisal of spatial and temporal analysis of the country's natural and human resources makes it an interesting and useful reading.

Written from the standpoint of broadening the horizon and not confining to strict disciplinary bounds, some unconventional chapters have been added to the present work. These include chapters on droughts and floods, natural hazards and disasters, environment, green revolution, tourism, composition of population and migration patterns. This new approach will certainly provide a glimpse of the new developments, problems and issues emerging over the geographical horizon of India and help us to comprehend looming scenario in the twenty-first century.

Selected references have been given at the end of each chapter which authenticate the subject matter given in the book. This will be of great benefit to those who are interested in digging out more details about the concerned chapter and help any serious reader to enrich his knowledge of the subject.

I would like to express my deep sense of gratitude to the University Grants Commission, New Delhi for granting me financial assistance to write this book. Encouragement, stimulation, guidance, help and suggestions have been received from a number of esteemed friends and institutions for completing this work. It would be remiss not to express my sincere thanks to them: The needed moral and academic support for the completion of this book came particularly from Professors M.H. Qureshi of J.N.U., New Delhi, Sudesh Nangia of J.N.U., New Delhi, R.C. Chandna of P.U., Chandigarh, Gopal Krishan of P.U., Chandigarh, Majid Hussain of Jamia Milia Islamia,

New Delhi, S.K. Aggarwal of D.U., Delhi, V.K. Asthana of K.U., Kurukshetra and last but not the least S.C. Gupta of University College, Rohtak. I am particularly beholden to these persons for taking extra interest in this project.

I will be failing in my duty if I do not express my gratefulness to M/s Kalyani Publishers, reputed as one of the leading publishers of India, for bringing out this book in its present form. My special thanks are to Sh. Raj Kumar. But for his keen interest and deep involvement in the project, this book would not have seen the light of the day.

I remain indebted to my wife, Mrs. Asha Khullar, who has shown fortitude in the face of my pre-occupation for writing this book. I have bestowed long hours, often working against the clock, thus unwillingly ignored her on certain occasions. I would like to reassure her that I do care for her in no less measure than my work.

The outcome of a really good book is a continuously dynamic process. Valuable suggestions for improvement of the book are solicited.

August 15, 1999
1150, Sector-16, Faridabad

D.R. KHULLAR

CONTENTS

LIST OF TABLES

LIST OF FIGURES

CHAPTER 1

Introduction

INDIA AS A GEOGRAPHICAL UNIT

India is a country of great geographical extent. It sprawls from the snowy ranges of the Himalayas in the north to the shores of the Indian Ocean in the south. It belongs to Asia which is the largest continent of the world. It forms a part of south Asia and is separated by the Himalayas from the rest of the continent. It encompasses vast areas of diverse landmasses. In the north are the lofty Himalayas, parts of which are permanently ice-covered. To the south of Himalayas is the Great Indo-Gangetic Plain which is well-known for its fertile soils. The western part of this vast plain is the Thar Desert. South of this plain is the Peninsular India comprising of the uneven plateau which is surrounded by Eastern Coastal Plain in the east and Western Coastal Plain in the west. Indian landmass gets an abundance of sunshine from the tropical sun and splashing rains from the monsoons. These are two most important climatic factors for the Indian people. Due to its vastness and diversities, *India is considered to be a subcontinent* as it possesses all the characteristics of a continent.

India extends from 8° 4′ north to 37° 6′ north latitude and 68° 7′ east to 97° 25′ east longitude. Thus, its latitudinal and longitudinal extent is about thirty degrees. Away from the mainland of India, the southernmost point of the country in the Andaman and Nicobar Islands, the Pygmalion Point or Indira Point is located at 6° 45′ north latitude. Its north-south extent from Indira Col in Kashmir to Kanniyakumari is 3,214 km while its east-west width from the Rann of Kachachh to Arunachal Pradesh is 2,933 km (**Fig. 1.1**). The latitudinal extent of India is about one-third the angular distance between the Equator and the North pole and its longitudinal extent is about one-twelfth of the circumference of the Equator. The longitudinal difference between Saurashtra in the west and Arunachal Pradesh in the east is about 30°. The earth moves around its axis through 360° in 24 hours. Thus, a difference of 1° longitude will make a difference of 4 minutes in time. Therefore the difference of local time between Lakhpat near Bhuj in Gujarat and Vijayanagar in Arunachal Pradesh is 30 × 4 = 120 minutes or 2 hours. Since Arunachal Pradesh is towards the east, it will have sunrise about two hours before the sunrise in Saurashtra. Thus, the sun is quite high in the sky in Arunachal Pradesh while Saurashtra still waits for the first ray of the sun. Latitudinal extent also has its own impact. Rainfall, temperature and vegetation vary with latitude. The difference between the longest and the shortest day in Kerala is hardly 45 minutes

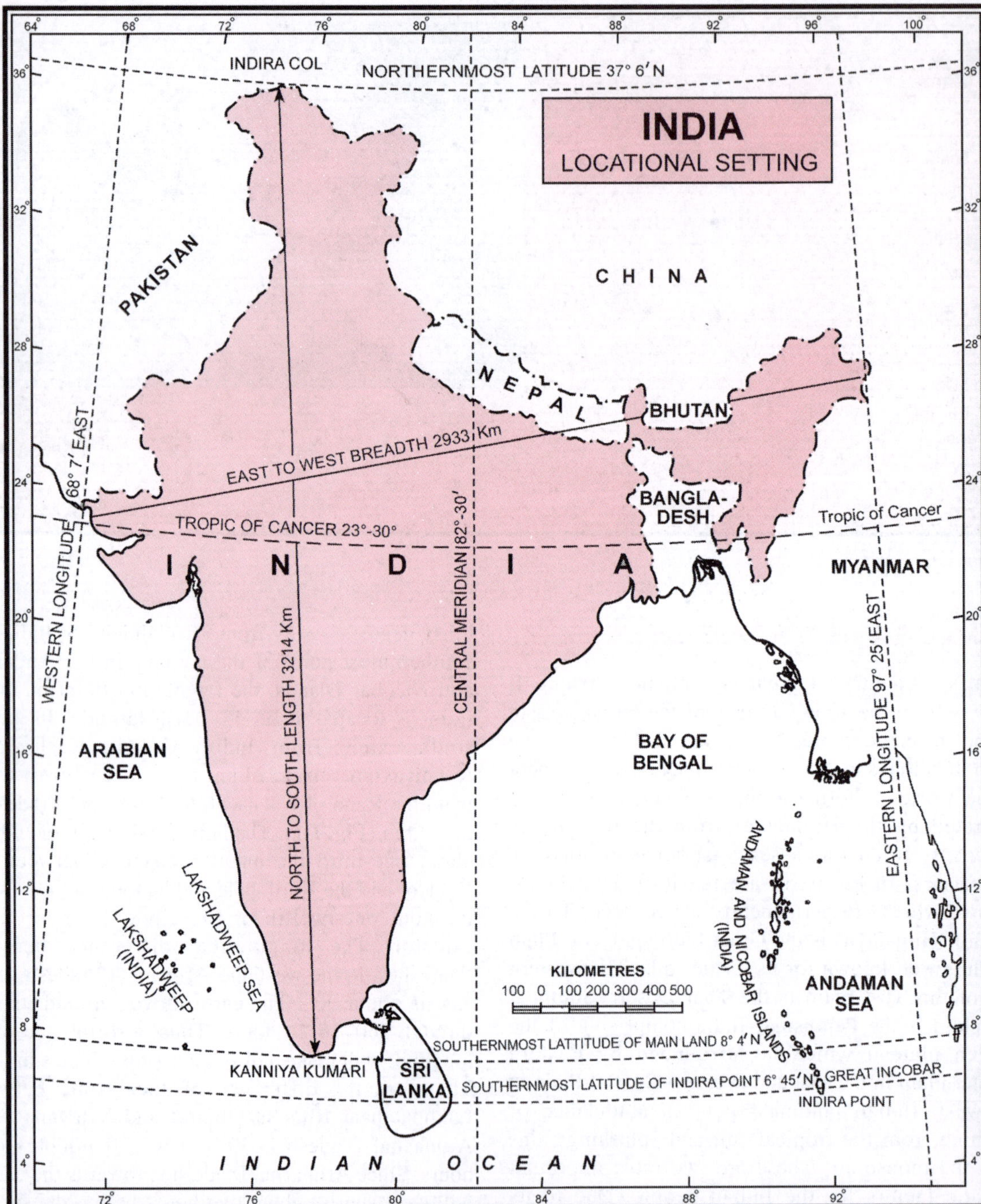

FIG. 1.1. India : Locational Setting

whereas this difference may be as large as 4 hours in Leh and Ladakh. The difference between the longest and the shortest day increases with latitude.

With an area of 32,87,263 sq km, India is the seventh largest country of the world. (**Table 1.1** and **Fig. 1.2**)

TABLE 1.1. Area of seven largest countries of the world

Name of the country	Area in sq km
1. Russian Federation	17,075,200
2. Canada	9,984,670
3. U.S.A.	9,626,091
4. China	9,596,960
5. Brazil	8,511,965
6. Australia	7,686,850
7. India	3,287,263*

Source : Dorling Kindersley World Atlas, 2010, pp. 198-206

* Data for India have been taken from India 2024 : A Reference Annual, p. 1.

Figures regarding area of India include the area under unlawful occupation of Pakistan and China. This area includes 78,114 sq km under illegal occupation of Pakistan, 5,180 sq km illegally handed over by Pakistan to China and 37,555 sq km under illegal occupation of China.

India accounts for about 2.4 per cent of the total land surface area of the world. India is nearly 13 times as large as Great Britain, the country which ruled over us for about two centuries. Many of the Indian states are larger than several countries of the world.

The peninsular tableland juts into the Indian Ocean for a distance of about 1,600 km.

The Tropic of Cancer passes through the middle of the country dividing it into two latitudinal halves being about 15° from either end. But the northern portion is very broad and the area to the north of Tropic of Cancer is nearly twice the area which lies to the south of it. The enormous width of India is often forgotten, being overshadowed by the more popular fact of its length. We are habitual of expressing the dimension of the country as *'from Kashmir to Kanniyakumari'* and not from '*Rann of Kachchh to Arunachal Pradesh*'. The east-west extent of India is almost equal to the combined longitudinal extent of Spain, France, Germany, Belgium, the Netherlands and Poland. South of 22° north latitude, the country tapers off over 1500 km into the Indian Ocean as a peninsula and divides this ocean into two arms—the Arabian Sea in the west and the Bay of Bengal in the east.

It is natural to look upon India as being divided into northern temperate and southern tropical lands by the Tropic of Cancer. Thus the temperate part of the country should be twice as much as its tropical part. But India has always been treated as a tropical country for two widely different reasons. The reasons are those of physical and cultural geography. The country is separated from the rest of Asia by a mountain wall forming an insulated compartment. Its climate is dominated by the tropical monsoons and the temperate air masses are restricted by the mountain chain. Further, although the night temperatures in January at several places in Punjab may come down to the level of those prevailing in temperate lands, yet clear skies and intense insolation raise the day temperatures to a tropical level, so that the entire area south of the Himalayas is essentially tropical from climatic point of view. Outside the Himalayas, almost everywhere agriculture is tropical in type. As per cultural geography, the culture of India is totally different from that of the temperate countries of the western civilisation.

Girdled by the rampart of the young folded mountains in north-west, north and north-east and washed by the Indian Ocean in the south, India along with its neighbouring countries, is a well defined

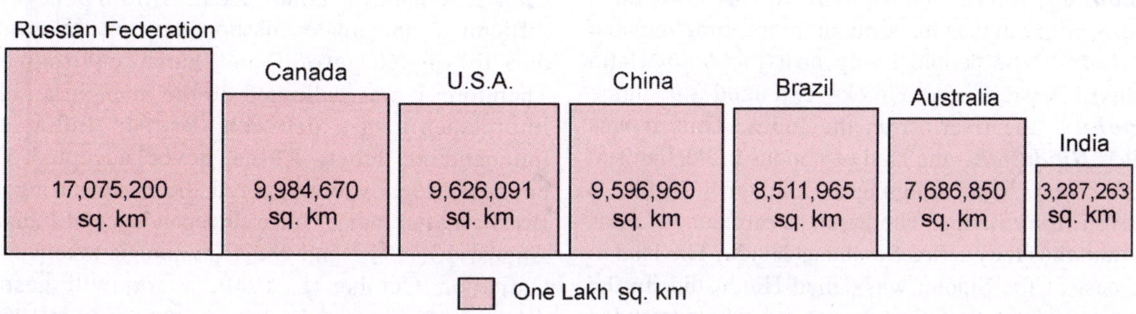

FIG. 1.2. Seven largest countries of the world

geographical unit. South of the mountain chain, India, Pakistan, Bangladesh, Nepal and Bhutan form a definite realm of South Asia which is often referred to as the ***Indian subcontinent.*** India alone accounts for about three-fourths of the total area of the subcontinent. According to **Prof. Chisholm**, 'there is no part of the world better demarcated by nature as a region by itself than the Indian subcontinent'. **G.B. Cressey** strongly advocated that India may be termed as subcontinent because it is a distinct geographical unit with many physical and cultural units. But some Indian geographers are of the view that the use of the term ***'subcontinent'*** for India is a misnomer and is a legacy of the British rule which tended to divide the area on the basis of region and religion. These geographers contend that this expression has never been used for much vaster and more diverse geographical units like China and Russia. It is worth mentioning that the subcontinental theory of the Britishers worked very well in dividing the area into different nations giving credibility to their basic policy of ***'divide and rule'***. Total area of the country before partition was 42,35,201 sq km. The partition of the country on 15th August, 1947 gave birth to a new country of Pakistan. This led to a loss of 8,03,940 sq km area of West Pakistan (now called Pakistan) and 1,43,998 sq km of East Pakistan (the present Bangladesh). Thus the present extent of the country is reduced to about three-fourths of its original size. Consequently, some scholars tend to replace the term Indian subcontinent by South Asia which includes, India, Pakistan, Bangladesh, Nepal, Bhutan, Sri Lanka and sometimes even Afghanistan.

Nomenclature of the Country

This country has been variously named as ***India, Hindustan, Bharat*** and ***Aryavarta.*** The word India has its origin in the Greek literature meaning the land of '*Indoi*', the people living near the *Indos* (latin Indus). Persians and Greeks extended the name ***Sindhu***—'the river' from the Indus. Thus it was called ***Hindustan***—the land of Hindus in Persian and other West Asian languages. The term Hindu is derived from Sindhu. The Persians pronounce 'S' as 'H' and thus they called Sindhu as Hindu. The land to the east of the Sindhu was called Hindustan. In the Hindu literature the subcontinent as a whole is styled as *Bharat* or *Bharat-Varsha*, the land of the legendary King Bharata who visualised the fundamental unity of the country. However, some scholars believe that the name has been derived from the Bharath tribe, who among others inhabited the area. In the European languages it is popularly known as ***India***. The name *Aryavarta* refers to the land of the Aryan race. At present only *India* and *Bharat* are officially recognised, although *Hindustan* is also in common use.

INDIA'S FRONTIERS

Encompassed between the Himalayas in the north, north-west and north-east, the Indian Ocean in the south and marshy Rann of Kachchh, vast desert of Rajasthan and fertile plain of Punjab in the west, India has both land and water frontiers.

Land Frontiers

India shares her 15,200 km long land frontier with Pakistan, in the west and north-west. Afghanistan in the north-west, China, Nepal and Bhutan in the north, and Bangladesh and Myanmar in the east. India's longest border is with Bangladesh while the shortest border is with Afghanistan as is indicated in **Table 1.2. Figure 1.3** shows land frontiers of India.

1. Border with China

India shares 3,917 km long border with China which is over one-fourth of the total land border of the country. This is the second longest border of India, next only to its border with Bangladesh. Four Indian states, namely Himachal Pradesh, Uttarakhand, Sikkim and Arunachal Pradesh and one Union Territory of Ladakh touch the Indian boundary with China. This border is the product of Manchu policy, Chinese Republican policy and the British policy. It is difficult to demarcate this boundary on the ground due to rugged terrain and harsh environment. Therefore, it was delimited on the maps in a rather imprecise form, between British India and independent Tibet. China never accepted this boundary legally but ignored the issue during the British rule in India. Since Independence of India on August 15, 1947 and the Communist takeover of China on October 1, 1949, a goodwill gesture developed between these two countries and in 1954 a general agreement containing the principle *Panch*

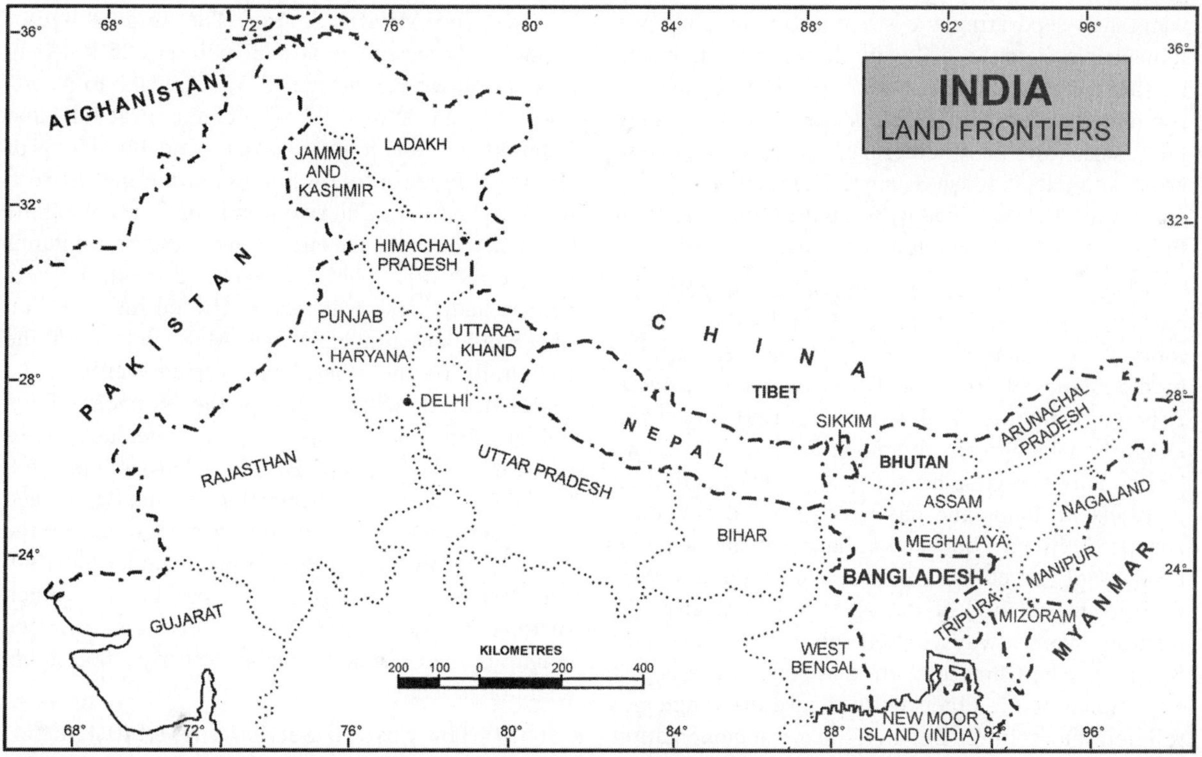

FIG. 1.3. India : Land Frontiers

Sheel was announced. These five principles were (*i*) mutual respect for each others territorial integrity and sovereignty, (*ii*) mutual non-aggression, (*iii*) mutual non-interference in each others internal affairs, (*iv*) equality and mutual benefit, and (*v*) peaceful co-existence. In the same year, India recognised the Chinese sovereignty over Tibet and it was brought under the direct control of China. Thus, the buffer character of Tibet ceased to function and a razor-thin boundary between India and China became operative. However, China attacked India in 1962 to take revenge of the so called Indian interference in Tibetan affairs.

The Sino-Indian border is generally divided into three sectors namely : (*i*) the Western sector, (*ii*) the Middle sector, and (*iii*) the Eastern sector.

(*i*) The Western Sector. This 2,152 km long sector of the Sino-Indian border separates Union Territory of Ladakh of India from the Sinkiang province of China. The boundary in the western sector runs along the Muztaghata Range and the Aghil Mountain across the Karakoram Pass via Quara Tagh pass and along the main Kunlun Range to a point east of 80°E longitude and 40 km north of Hajit Langer. It forms a physical boundary between Gilgit area and Sinkiang following the main Karakoram

TABLE 1.2. Length of India's Border with the Neighbouring Countries

Name of the country	Length of border (in kilometres)	Percentage of total length of border
1. Bangladesh	4,096	26.95
2. China	3,917	25.77
3. Pakistan	3,310	21.78
4. Nepal	1,752	11.53
5. Myanmar	1,458	9.59
6. Bhutan	587	3.86
7. Afghanistan	80	0.52
Total	**15,200**	**100.00**

watershed separating the streams flowing into the Indus basin from those flowing into the Tarim basin. Farther south-east the boundary runs along the watershed across Lanak La, Kone La and Kepsang La, then follows the Chemesang river across Pangong Lake and the Kailash Range. Here the boundary constitutes the watershed between the Indus system in India and the Khotan system in China.

The western sector boundary is largely the outcome of the British policy towards the state of Jammu and Kashmir. These boundaries were defined by the treaties of 1665 and 1686 (known as ***Ladakh-Tibet*** agreements) and were confirmed by 1842 Dogra-Ladakh agreement among Kashmir, Tibet and China. However, this boundary was never delimited precisely on maps and has led to several boundary disputes between China and India. The Chinese claim rests mainly on ethnic grounds, and on the assertion that the wastelands of the Aksai Chin in disputed territory were always linked more with Tibet and Sinkiang. The Chinese claim that Aksai Chin is just an extension of Tibet with regard to language, religion and culture. But in Chinese documentation regarding the actual occupation of the area by Tibet is inconclusive. The Indians, on the other hand, claim that the area has been historically administered by the state of Jammu and Kashmir since 1849, and that the Indo-Tibet Treaties of 1665, 1686 and 1842 confirmed the boundary between Tibet and Ladakh. China claims the Aksai Chin district, the Changmo valley, Pangong Tso and the Sponggar Tso area of north-east Ladakh as well as a strip of about 5,000 sq km down the entire length of eastern Ladakh. China also claims a part of Hunza-Gilgit area in North Kashmir (ceded to it in 1963 by Pakistan), although the whole territory had been effectively under the British sovereignty since 1895.

Since 1954, the Chinese have repeatedly violated the international border between India and China and penetrated deep into the Indian territory in the western sector. China renewed aggression in 1959 and the Line of Actual Control (LoAC) became a series of positions occupied by the Chinese forces rather than a well defined border between the two countries. In 1962, China waged a full scale war and its forces intruded far deeper into the Indian territory. Currently the Chinese occupation line runs 16 to 24 km west of traditional line.

(*ii*) The Middle Sector. The middle sector boundary between China and India is 625 km long and runs along the watershed from Ladakh to Nepal. Two Indian states of Himachal Pradesh and Uttarakhand touch this border. The boundary of Himachal Pradesh follows the water parting between the Spiti and Para Chu rivers and continues along the watershed between the eastern and western tributaries of the Satluj. The Uttarakhand boundary is demarcated by the watershed between the Satluj on one hand and the Kali, the Alaknanda, and the Bhagirathi on the other. This boundary crosses the Satluj near the Shipki La on the Himachal-Tibet border. Thereafter, it runs along the watershed passes of Mana, Niti, Kungri-Bingri, Dharma and Lipu Ladakh. It finally joins trijunction of China, India and Nepal. This part of the border was approved by the Tibetan and the British governments under the 1890 and 1919 treaties. Although there are not much serious territorial problems between the two countries, the Chinese lay claim to nearly 2,000 sq km area in this sector.

(*iii*) The Eastern Sector. The 1,140 km long boundary between India and China runs from the eastern limit of Bhutan to a point near Talu-Pass at the trijunction of India, Tibet and Myanmar. This line is usually referred to as the **Mc Mahon Line** after Sir Henry Mc Mahon, then foreign secretary of British India, who negotiated the boundary agreement between Great Britain and Tibet at Shimla accord in 1913-14. This line normally runs along the crest of the Himalayas between Bhutan in the west and Myanmar in the east. India has stressed that the Mc Mahon Line is the international boundary between Tibet and India as was agreed to between the governments of India, China and Tibet. On the other hand China considers the Mc Mahon line as illegal and unacceptable, claiming that Tibet had no right to sign the 1913-14 convention held in Shimla which delineated the Mc Mahon Line on the map. India challenges such a position, maintaining that Tibet was independent and in fact concluded several independent treaties which were considered valid by all parties, and were in operation for decades.

China declined the validity of the Mc Mahon Line as an international boundary and laid claims to areas south of this line up to the foot of the Himalayan range in the Brahmaputra valley. The whole of Arunachal Pradesh, measuring over 83 thousand sq km has been claimed by China as Chinese territory—calling it ***"South Tibet"***. This area has been administered by India since 1947.

Chinese government never formally questioned the validity of 1913-14 Shimla convention until 1959. India has been striking at the two crucial points. First, Britain and India had been exercising jurisdiction over the area since 1914 and 1947 respectively and second China never disputed the Indian control over the area until 1959. Even in 1956 when the Chinese attention was drawn to certain maps drawn by China which showed these areas to be parts of China, the Chinese government promised to look into the 'cartographic errors' in their maps.

China's relations with India started taking a turn for the worse in 1950s when China started consolidating its position in Tibet and then in Ladakh. By 1956, China started intensifying its intrusions into the Indian territory. India wishfully hoped that the insurmountable barrier of the Himalayas would prevent Chinese to attack Indian territory. Indian apprehensions grew in 1956 when China built a road through Ladakh linking West Tibet with Sinkiang and quickly moved into Aksai Chin and eastern Ladakh. China also moved its forces into the North-East Frontier Agency (NEFA), the present Arunachal Pradesh. Reported small-scale armed clashes between 1959 and 1962 escalated into a full-scale war in October, 1962 when China launched a major offensive in Ladakh in the western sector and NEFA in the eastern sector. Indian armed forces were not well trained to fight a modern war in a rough and rugged mountainous terrain. They were outnumbered and outgunned by the Chinese forces.

Following lessons have been learnt from Chinese aggression over India :

(*i*) The myth that the Himalayas were an effective defence barrier was exploded.

(*ii*) India's naive confidence in China's friendliness had dulled its perception regarding effective security measures in the Indo-China borderlands.

(*iii*) The prompt and positive response of western countries in rushing military supplies to the war zone helped improve the image of the West in Indian eyes.

(*iv*) India realised that posture of "non-alignment" was no substitute for defence preparedness.

Surprisingly, Chinese forces pulled back without annexing the disputed territory in the eastern sector. In the western sector, however, they did not pull out of most of Aksai Chin. Perhaps they did so due to the fear that their advance troops would have been cut off from supply bases in Tibet in winter of 1962 when the high passes in the Himalayas would have been closed by snow. The Chinese justified their withdrawal by stating that they had no further territorial ambitions.

The primary aim of Chinese invasion was to deprive India of its moral leadership in the world especially among the Afro-Asian countries, and to put pressure on it to join the socialist camp. China's collusion with Pakistan is a clear indication of this political ambition. Perhaps China miscalculated its political power strategy. The support of the Western countries to India compelled China to rethink its strategy. India's unity in the face of Chinese aggression was another deterrent for China to achieve political mileage.

An uneasy truce had been prevailing for a long time since the October, 1962 war. Neither side made a serious effort to normalise the situation on the border. The Colombo powers, spearheaded by Ceylon (now Sri Lanka) and the erstwhile Soviet Union failed to bring about a respectable agreement between the two countries. Several countries condemned China as an aggressor. However, Cuba, Albania and Portugal supported the Chinese.

Of late, leaders of both the countries have engaged themselves in improving bilateral relations. Several meetings have been held to resolve the problem of border issues. The first important step to resolve dispute was the signing of Border Peace and Tranquillity Agreement at Beijing on September 7, 1993. The pact consists of confidence-building measures (CBMs). An agreement on CBMs and the military field along the Line of Actual Control (LAC) was signed between the two sides on November 29, 1996. A major development took place in October, 2003 when China made a significant change in its official website. It removed the mention of Sikkim from its list of nations. In May, 2004, China recognised Sikkim as a part of India for the first time.

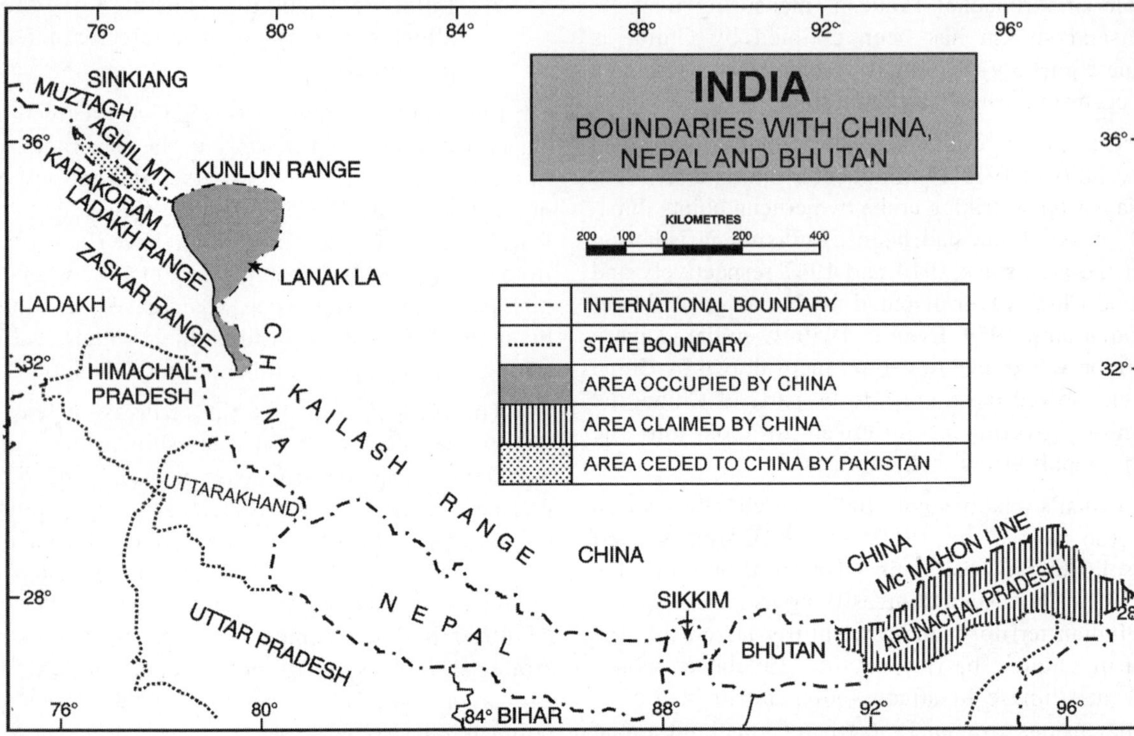

FIG. 1.4. India : Boundaries with China, Nepal and Bhutan

The world map in World Affairs Year Book 2003/2004 does not show Sikkim as a separate country in Asia. It also does not mention Sikkim in its index of countries. The move was significant since it involved recognition of Sikkim-China border which is a part of the Mc Mahon Line. A protocol on modalities for implementing military CBMs was signed on April 11, 2005. If things go as planned, the two nations should be able to work out the guidelines for transforming the Line of Actual Control (LoAC) into a mutually acceptable and internationally recognised boundary. However, China still insists that Arunachal Pradesh is a disputed territory. China has also suggested a 'package deal'. It asked India to accept Chinese domination over Aksai Chin in return of China's acceptance of Mc Mahon Line as the international boundary. India, however, has advocated 'sector-by-sector' approach.

However, China's so called goodwill gestures cannot be trusted so easily. There was Doklam standoff between Indian armed forces and the Peoples' Liberation Army (PLA) in Doklam in June, 2017. Doklam is a small plateau at the trijunction of India, China and Bhutan borders. This put severe pressure on India-China ties though the issue was resolved through diplomatic means. Again in December, 2017, about 1800 Chinese troops have virtually established a permanent presence in Doklam. Further, China is building roads and providing other infrastructural facilities in the area for smooth movement of troops. This can again put strain on Indo-Chinese relations.

China's Aggressive Postures

China has been adopting aggressive postures against India and some other neighbouring countries ever since the Indo-China war of 1962. With its fast growing economy, China is dreaming of becoming a super power like the U.S.A. and one of the main devices to realise this dream is to dominate over its main rivals. From this point of view, India is the main target because India is also progressing rapidly. Some

Chronology of China's Agressive Postures against India and India's efforts to ease the situation.	
Year	**China's aggressive postures against India**
1962	China waged a full scale war against India in both western and eastern sectors in which India suffered heavy losses as India was not prepared for such a situation.
1967	Chinese attempts to enter Indian territory along Sikkim border was thwarted by the Indian armed forces. A total of 80 Indian soldiers and 460 Chinese soldiers were killed in action in Nathu La.
1975	Four Indian solders were killed when their petrol was ambushed by Chinese soldiers at Tulung La in Arunachal Pradesh. Both sides exchanged fire along the LAC.
June 1986 - Aug. 1987	Biggest face-off after 1962 war in Sumdorong Chu valley, north of Tawang in Arunachal Pradesh. Entire infantry brigade was airlifted under 'Operation Falcon' to counter China's aggressive moves.
1988	India, China agreed to set-up a joint working group on boundary issue.
1993	Agreement signed on Maintenance of Peace and Tranquility along the Line of Actual Control, thus recognising the LAC.
1996	India, China signed agreement on Confidence Building Measures along LAC.
1999	Amid Kargil conflict with Pakistan, China built 5 km track right upto Pangong Tso's southern bank.
2000	Chinese soldiers build permanent roads and set up bunkers inside Indian part of LAC in Ladakh's Aksai Chin area.
May-April 2013	Chinese troops intrude 19 km into Indian territory in Despang Valley in Ladakh. 21 day face-off saw two armies pitching tents. Stand off resolved after negotiations. In October, 2013, the two sides sign Border Defence Cooperation Agreement.
2014	Chinese soldiers attempted to build a road in Chumar area of Ladakh.
June-Aug. 2017	73-day confrontation on Bhutanese territory of Doklam near the Sikkim-Bhutan Tibet trijunction after Indian soldiers blocked Chinese troops from extending a road. PLA permanently stationed troops and built helipads in north Doklam since then.
2020	Tension at multiple areas along LAC from Sikkim to Ladakh sees troops build up before violent escalation occurs in Galwan valley in Ladakh. Aggressive postures by Chinese troops in disputed area of Pangong Tso, Galwan Valley, Demchok and Daulat Beg Oldi in eastern Ladakh along LAC. Chinese troops were forced to retreat from finger four to finger eight of Pangong Tso.

of the recent developments are pointers towards the unwanted intentions of China, so far as India is concerned.

1. China's Water Strategy. China has prepared an ambitious plan to harness the huge water potential of the upper Brahmaputra river course in Tibet. The plan includes construction of gigantic 38,000 MW. Motuo dam along the Brahmaputra's ***'great bend'*** when the river curves to flow south into Arunachal Pradesh of India. China considers it to be the word's biggest hydroelectric project which would supply clean energy and would save 200 million tons of carbon emissions per year. It would involve 15-25 km long tunnels attached to downstream pipes. China claims that this dam would be 1.5 times the size of Three Gorges dam—currently world's biggest. In all 28 dams are proposed to be constructed across the main Brahmaputra river and its tributaries in Tibet. These dams are primarily motivated by Beijing's desire to develop larger hydropower projects to power resource extraction, infrastructure development and ultimately for supply to coastal Chinese cities where demands are the highest.

The construction of these 28 dams will put India at a greater risk, because China will gain significant capacity to control the Brahmaputra's flow. China will be able to hold back water in the river when it is most needed in India during spring and summer, and release more water during the monsoon when there is fear of flood in large tracts of Assam and Arunachal Pradesh in the middle course of the river. Thus India will become dependent on China for flow of what is now a ***"free-flowing international river."*** Further, the proposed projects pose greatest risk for India because of seismic activity in the region. Thus water dispute has added a new dimension to the already complex relations between India and China. Limited water resources and rapidly growing demand for water resources on both sides of McMahon line are bothering our security planners. Since India is the lower riparian country, she is likely to suffer severely due to the Chinese plans and a conflict between the two giants of Asia can break at any time. Some defence experts on the Indian side have gone to the extent of calling this plan of China as ***'Chinese Water Bomb.'*** It may be pointed out that the turbulent Tibetan plateau is the source of most of China's water.

According to a report published by Institute for Defence Studies and Analysis (IDSA), Beijing has plans of damming the Yarlung-Tsangpo (Brahmaputra) and controlling the flow of the river water to China's advantage. The IDSA report, "Water Security for India : The External Dynamics", states, "Harnessing the potential of the Yarlung-Tsangpo is critical to China's overall development" plans and is part of its grand design to divert waters from the south to the north. As a lower riparian state, India will be vulnerable to any major storage projects planned on the Yarlung-Tsangpo. The report also held that, "Given the political situation between the two countries it is hard to imagine China playing the role of a responsible upper riparian country by releasing *re-regulated* flows from power houses immediately back into the river." In fact China has advantage of using water resources as a ***military tool.*** Thus there is an urgent need for a regional basin approach by which China, India and Bangladesh decide to treat the water of the river systems of Tibet as a common resource and develop a formula to reasonably distribute it among themselves.

China's rail and road projects. Realising the strategic significances of railways and roads from the defence point of view, China has drawn a gigantic plan to construct railways and roads in the border areas. China has already completed its 3,900 km long Beijing-Lhasa rail link. This is considered as a great achievement by China because of the rough and rocky terrain and adverse climatic conditions prevailing in the Tibet plateau. China is also pushing ahead with several other railroad projects adjoining the Indian border. Two major projects are to connect Lhasa with Shingatse near Sikkim border and Nijingchi near Arunachal Pradesh border.

China has also announced plans to extend rail connectivity to its outpost at Rulli, adjoining its border with Myanmar. Proposals are also afoot to build 5,000 km of rail links, with emphasis on establishing connectivity in the Tibetan autonomous Region. Further, China has also proposed to build a rail network in Nepal.

In contrast, India is very poorly prepared so far as railway construction in border areas is concerned. There is practically no rail service in the whole of the Himalayan region except Jammu-Baramula rail link. Another plan has been drawn to construct 497 km Bilaspur-Manali-Leh rail link stretching from Himachal Pradesh to Ladakh.

China is showing considerable interest in railway development in Pakistan also. Pakistan happens to be China's close ally so far as economic and defence ties are concerned. That country is helping Pakistan to build railway infrastructure very close to Indo-Pak border along the six km long Munabao (in Rajasthan) and Khakrapur (in Pakistan) making India vulnerable to Pakistani attack. Another 1,800 km long railway line has been planned by China. It will connect Kashgar in Xinjiang in China to Gwadar Port of Pakistan, passing through PoK, Islamabad and Karachi. India has objected to this project.

China is also constructing China-Pakistan Economic Corridor (CPEC) at a cost of $46 billion. This 3000 km long road will connect the Gwadar port on the Arabian Sea to Kashgar town on the Chinese border. It will have three alignments namely (*i*) the Eastern Alignment, (*ii*) the Central Alignment and (*iii*) the Western Alignment. The Eastern Alignment will pass through Karachi, Hyderabad, Multan, and Lahore, the Central Alignment will connect Muzaffargarh and the Western Alignment will provide connectivity to Quetta and Peshawar. This will give China a ready access to Indian Ocean through Arabian Sea as China will be able to use Gwadar and Karachi ports. It will also be a best case for One Belt, One Road (OBOR) proposed by China to which India has raised serious objection as it will be against India's political interests.

To counter China's aggressive posture in Pakistan, India is developing Chabahar port in Iran, just 100 km west of Gwadar port of Pakistan (as the crow flies). This port is crucial for the following reasons:

- Sits at the mouth of Strait of Hormuz area.
- Connects three regions: Central Asia, South Asia and West Asia.
- Junction of shipping, oil trade routes.
- About 100,000 ships sail by yearly.
- The region holds two-thirds of world's oil reserves.
- About 17 billion barrels of crude oil passes this way daily.
- Chabahar port route helps connect India with energy rich Azerbaijan, Turkmenistan and other Central Asian countries.

- This route is 40 per cent shorter, 30 per cent less expensive than trade via the Red Sea-Suez Canal-Mediterranean route.

China is also building Karakoram highway, a road of great strategic significance in the Pakistan occupied Kashmir (PoK) (**Fig. 1.7**).

All the above mentioned developments put India to a great risk because India will find herself at a great disadvantage as compared to China and Pakistan regarding the movement of men and materials in the event of a war.

China's Increasing Influence in the Indian Ocean. Although China has its coast on the Pacific Ocean only, that country is increasing its influence in the Indian Ocean also. As part of a new strategy called ***'far sea defence'***, the Chinese military is seeking to project naval power well beyond the Chinese coast, from the oil ports of West Asia to the shipping lanes of the Pacific. According to the Chinese admirals, Chinese warships will escort and defend tankers and other ships from as far as Persian Gulf and Gulf of Aden, sailing across the Indian Ocean to the Strait of Malacca carrying commercially important cargo. China imports most of its mineral oil requirements through this ocean route. According to the U.S. sources, China aims to have a formidable naval presence in the Indian Ocean. According to Robert Kaplan of the centre for a New American security, "The rise of China as a sea power is one of the biggest development of the last one decade. China's ambition of becoming a big power in the Indian ocean is reflected in the plan to build ports at Chittagong in Bangladesh, Hambantota in Sri Lanka and Kyaukpyu in Myanmar." He further remarked that China becoming a sea power across the Indian Ocean and India too, is raising and becoming a sea power— brings China and India into competition for the first time in their histories.

Influence on Nepal. Apart from China's efforts to increase its influence in the Indian Ocean, that country is also trying to influence Nepal by providing facilitates for infrastructural development such as railways and roads and some industries. The latest venture is to develop Lumbini, the birth place of Lord Buddha, as Mecca for Buddhists. A Chinese government backed NGO, has signed a deal to develop Lumbini, a UN world heritage site, with funds worth $ 3 billion. The project is aimed at creating a 'special development zone' in the sleepy town by building airports, convention centres and infrastructural network. The project could be a major security concern for India due to long-term presence of Chinese personnel in the area.

FIG. 1.5. China's CPEC in Pakistan

2. The India-Nepal Boundary

India has common border with Nepal which stretches for 1,752 km from the trijunction of Nepal, Tibet and Uttarakhand state of India in the west to Sikkim state of India in the east. Five states of India, namely Uttarakhand, Uttar Pradesh, Bihar, West Bengal and Sikkim touch the Nepalese border with India. Major portion of Indo-Nepalese border runs in the east-west direction almost along the foothills of the Shiwalik Range. Although there had been numerous border conflicts between British India and Nepal since 1769, the present Indo-Nepal boundary is peaceful and there is no boundary dispute between the two countries.

Nepal is a landlocked country which is sandwiched between two giant countries of China and India. Consequently, Nepal has pursued its foreign policy very carefully and has allowed both India and China to construct roads linking its capital city of Kathmandu with its two big neighbours. Nepal has close contacts with India by treaties of friendship and protection from external aggression. However, under the influence of China, Nepal became a hostile country and laid claim on two Indian territories. One is Kalapani, Limpiyadhura and Lipulek of Uttarakhand and the other is Sirsia in West Champaran district of Bihar. That country prepared a new map to claim these two territories which was outrightly rejected by India.

3. The India-Bhutan Boundary

Bhutan is a 'buffer' state between India and China and shares 587 km long border with India. Bhutan is protected from external invasion, although it became a fully sovereign nation and became a member of the United Nations in 1971. India's border with Bhutan is quite peaceful and there is no boundary dispute between the two countries.

Bhutan's border with China follows the crest of the Great Himalayas—the watershed for the northern part. China rejected the watershed principle and claimed nearly 775 sq km of northern Bhutan as its own territory. Bhutan has friendly relations with India which are based on the 1949 treaty. This treaty provides a sound framework to ensure 'perpetual peace and friendship'. Under this treaty, India has got the rights to protect Bhutan's sovereignty and defend its borders. Accordingly, India formally protested in 1959 against China's claim over Bhutanese territory. China denounced India's rights of protection of Bhutan, and asserted that any border dispute with Bhutan could be settled with Bhutan without Indian interference, India maintained its right to defend Bhutanese borders as per provisions of the 1949 treaty. India has been closely helping Bhutan in its defence efforts. Indian army units are permanently stationed all along the Bhutan-China border. India took a firm stand in favour of Bhutan when China claimed a part of Doklam in June, 2017.

4. The Indo-Pakistan Boundary

The Indo-Pakistan boundary is the result of partition of the country in 1947 under the **Radcliffe** award of which Sir Cyril Radcliffe was the chairman. This boundary runs through varied relief features and is marked by a large number of incongruities, anomalies and irrationalities. It starts from the marshy Rann of Kachchh in Gujarat, and traverses through the sandy deserts of Rajasthan, fertile plain of Punjab, hills and mountains of Jammu and Kashmir inaccessible mountain ranges of Ladakh and reaches right upto the Karakoram Range in its northernmost reach (**Fig. 1.6**).

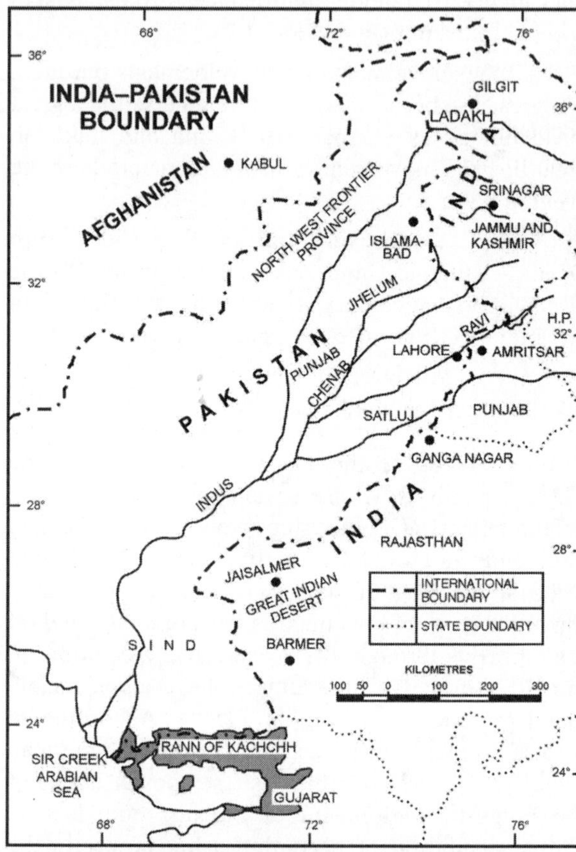

FIG. 1.6. India-Pakistan Boundary

This boundary has created several problems by dividing the fertile erstwhile Punjab, the eastern part remaining with India and the western-part going to Pakistan in the form of West Punjab. Here the boundary line cuts across the canals, roads, railways, towns, villages, fields and even the places of worship. Thus, it is a purely unnatural boundary created by man as a result of partition of the country which is based entirely on communal considerations.

Consequently the boundary has been marked by a number of disputes and hostile activities from both the countries ever since it came into being. Following disputes are worth mentioning.

(*i*) The Rann of Kachchh Dispute. Rann of Kachchh boundary was well defined according to the

Radcliffe award but Pakistan advocated that it was not properly delineated. That country argued that the Rann was not a marsh but a land-locked sea or lake and as such it should be equally divided between India and Pakistan. India countered Pakistani claim by presenting documentary proof that the Rann was a marsh which had always remained a part of the Kachchh state. As this state merged with India after independence, the entire Rann of Kachchh should be given to India. Pakistan claimed about 9,065 sq km out of a total area of 20,720 sq km of the Rann of Kachchh. During the 1965 Indo-Pakistan war, the Pakistani troops invaded this area and reached inside the Indian territory. The matter was referred to a boundary tribunal constituted by the International Court of Justice which awarded 906.5 sq km area of the Kachchh to Pakistan. After a number of protests and representations both India and Pakistan accepted the award and the demarcation of the boundary was completed in June 1969.

***(ii)* The Kashmir Problem.** The former princely state now converted into a Union Territory within the Indian Union is undoubtedly the most complex and serious problem between India and Pakistan and is the major stumbling block in normalising the relations between the two countries. The Kashmir problem has led to bloody wars between the two countries in 1948, 1965, 1971 and 1999. There had been large scale violations of LoC by Pakistan ever since it was demarkated.

The erstwhile state of Jammu and Kashmir was ruled by a Hindu whereas about 77 per cent of the population of the consisted of Muslims before partition of the country. It is an extremely important part of the country from the strategic point of view because of its contiguity to Pakistan, China and Afghanistan and proximity to Tajikistan (formerly a part of the U.S.S.R.). The state was sought by Pakistan on the basis of its Muslim majority status. Moreover, major rivers of Pakistan namely Ravi, Chenab, Jhelum and Indus flow through Kashmir. Unlike most of the rulers of princely states who had acceded to either India or Pakistan before August 15, 1947, ruler of Kashmir did not make up his mind. The Maharaja thought that any decision to accede to India or Pakistan may spark off chain reactions and disturb the peace of the state. Pakistan cut off the communications and stopped the supply of essentials to coerce Kashmir to accede to Pakistan. On October 22, 1947, fully armed tribesmen, supported by Pakistani armed forces, invaded Kashmir. They indulged in large scale killing of Hindus and Muslims and committed large scale looting and arson. This forced the ruler to make a desperate appeal to the Indian government for military help. India promised to help the ruler but only after he decided to accede to the Indian Union. The ruler agreed to this condition and the instrument of accession of Kashmir was signed by Maharaja Hari Singh on October 26, 1947. The accession was accepted by the Governor General of India on October 27, 1947. Thus, Kashmir became legally and constitutionally an integral part of the Indian Union.

Two major considerations have guided India's policy about Kashmir problem. *First,* India is secular state and surrender of Kashmir to Pakistan on religious grounds would amount to denial of the nation's essential principles. *Second,* the strategic location of Kashmir is of great importance to India. China's occupation of Tibet and part of Ladakh and building a road through Aksai Chin to serve Tibet and Sinkiang, and full scale aggression on India in 1962 further lent urgency to the strategic aspect of Kashmir's location. China helped Pakistan in building a new road across its occupied area to Sinkiang and Pakistan, in turn, ceded a large territory of 5,180 sq km to China. This move by Pakistan has been termed as illegal; because India claimed that the entire state including the territory controlled by Pakistan, annexed by China and ceded by Pakistan to China belonged to India. Thus, out of a total area of 2,22,236 sq km, a territory of 78,114 sq km is under illegal occupation of Pakistan, and 5,180 sq km illegally handed over by Pakistan to China and 37,555 sq km under illegal occupation of China in Ladakh (**Fig. 1.7**).

Pakistan has so far made four abortive attempts to conquer Kashmir. Its forces had to trace back in 1947-48 war when Kashmir was legally acceded to India. Pakistan again invaded in 1965 and it was badly mauled by the Indian forces. The 1971 India-Pakistan conflict led to the liberation of East Pakistan and emergence of a new country by the name of Bangladesh. This left Pakistan a truncated country and added much to the already increasing bitterness between India and Pakistan. In 1999,

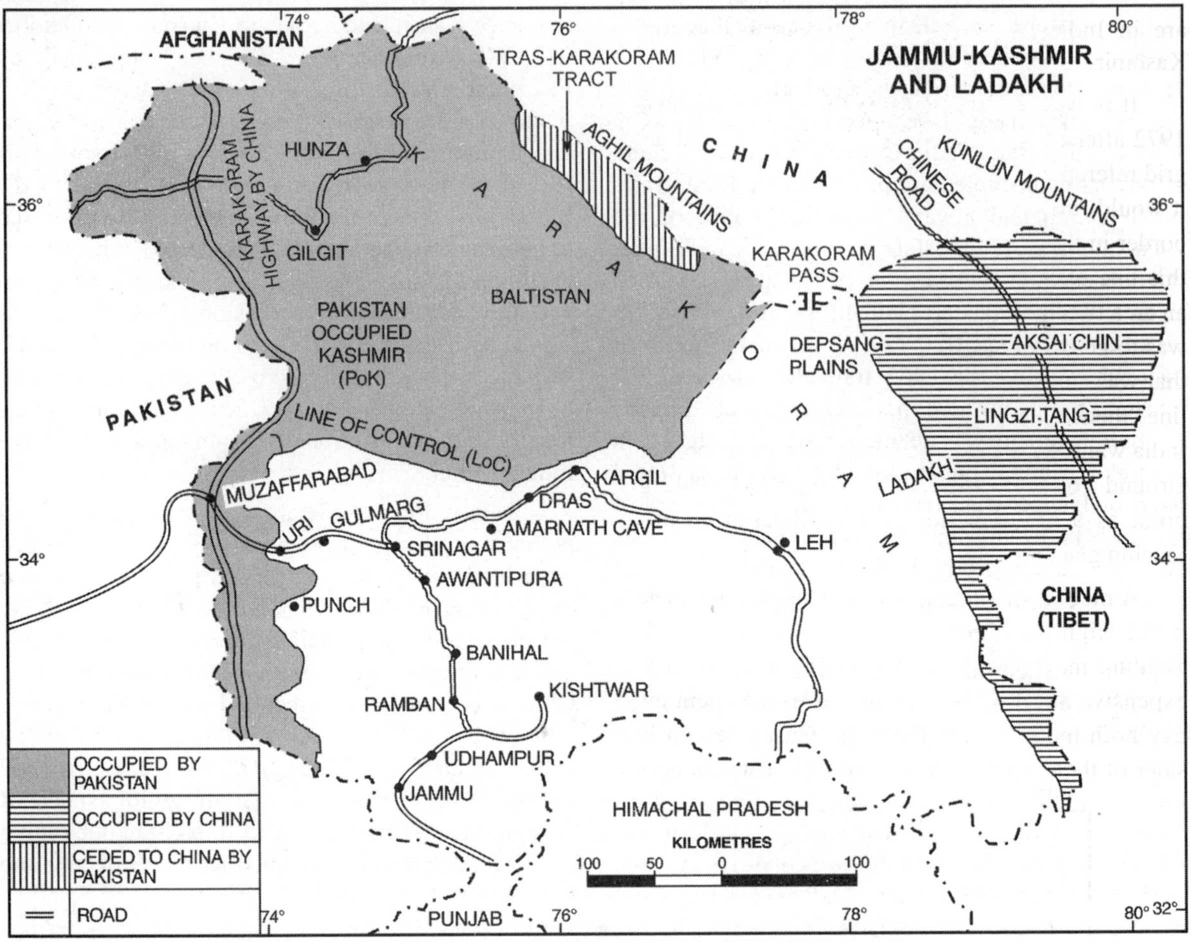

FIG. 1.7. Jammu and Kashmir and Ladakh

Pakistani troops stealthily occupied certain positions in the Kargil sector for which India had to use strong force. Unable to beat India in regular warfare, its forces are launching proxy war in Kashmir and sending foreign mercenaries and terrorists to disturb law and order and incite communal tension in Kashmir. Due to its link with Islamic terrorist organisations and ISI, Pakistan has become a hot-bed of international terrorism and is posing a serious threat to world peace. Even after seven decades of Kashmir's accession to India, there seems no sign of resolving the Kashmir problem as both sides stick to their respective positions. India is taking no chance and has taken corrective measures by scrapping Article 370 which gave special status to the erstwhile state. The central government has also done away with Article 35A, which enamated from it. India has also made a strong claim over Pakistan occupied Kashmir (PoK).

***(iii)* The Siachin Glacier Dispute.** Siachin glacier is about 75 km long and 2 to 8 km wide. This glacier has the distinction of being the largest glacier outside the Polar or sub-Polar regions. It covers an area of about 450 sq km at an altitude of about 5,800 metres above sea level in Ladakh region near Karakoram range. The Karakoram highway between China and Pakistan is very close which enhances its strategic value. India occupies about two-third area of the glacier in its south-eastern part. Here Nubra river emerges from Karakoram glacier and meets the Shyok river which is a tributary of the Indus river. Indian troops use the Nubra valley to reach the glacier. The glacier has four passes. Of these Gasherbrum, Saltoro and Vilafondala are in India and Gyongla is under Pak occupied Kashmir.

It is worth mentioning that LoC demarcated in 1972 after the Shimla agreement stopped dead at the grid reference N.J. 9842, with no indication as to how it would run along the 70-odd km left to the Chinese border in the north. While Pakistan decided to extend this line eastward to Karakoram Pass, thus claiming an area two-thirds the size of Sikkim, India's decision was to go by the wording of 1949 Karachi agreement that was superseded by LoC of 1972 which spoke of line "thence northwards joining the glaciers". Further, India wants Pakistan to officially recognise the Actual Ground Position Line (AGPL) begin N.J. 9842 grid point as a pre-requisite to de-militarisation of the Siachin glacier.

A number of skirmishes have taken place making it the highest battlefield in the world (**Fig. 1.8**). Sending men and material in such an area is a very expensive affair and crores of rupees are spent every day both by India and Pakistan. Casualties on both sides of the border are also very high. Even in normal circumstances, Indian troops suffer one casualty every second day while Pakistani troops suffer one casualty every fourth day. Besides there are psychological disorders, frostbite, high altitude pulmonary and cerebral edema and snow blindness. Such heavy losses can be avoided if both the countries start living in peace as friendly neighbouring countries.

***(iv)* Sir Creek.** Sir Creek forms the boundary between Gujarat state of India and Sind Province of Pakistan. This creek is extremely rich in oil and marine life and both the countries desire to include it in their respective territories (**Fig. 1.9**). Sir Creek has become a contentious issue between the two countries as Pakistan has been claiming a 40 km region inside India's territory. The Sir Creek involves the last portion of our western land border that is in the form of a 100-km estuary that empties into the Arabian sea. Pakistan claims that the border runs along its eastern shore, while India invokes the "Thalweg principle" in international law to say that it lies in the middle of the channel.

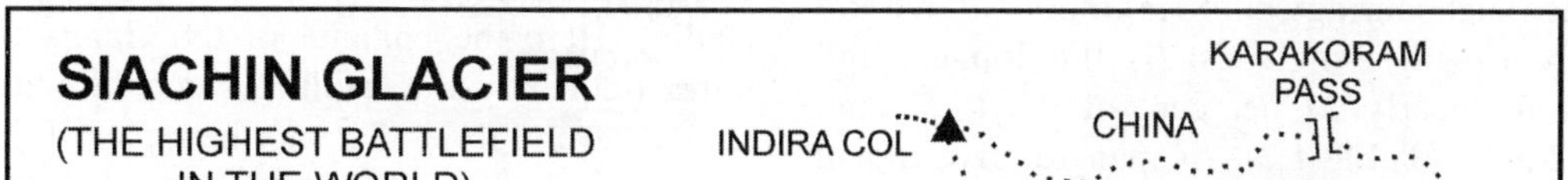

FIG. 1.8. The Siachin Glacier : The Highest Battlefield in the world.

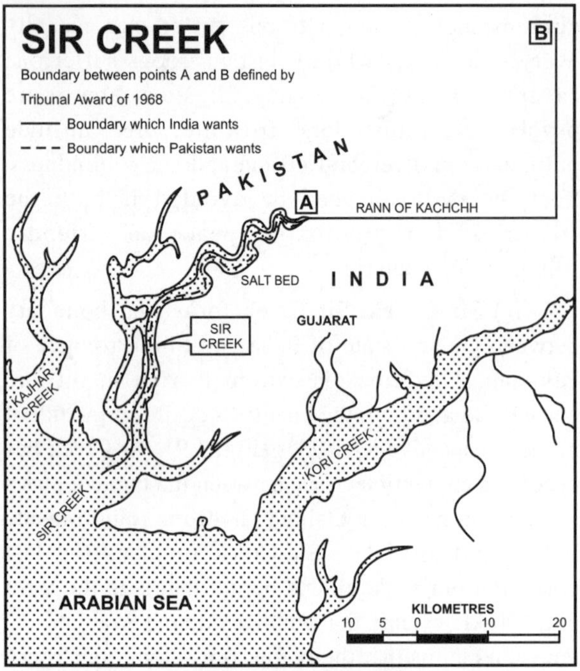

FIG. 1.9. Sir Creek

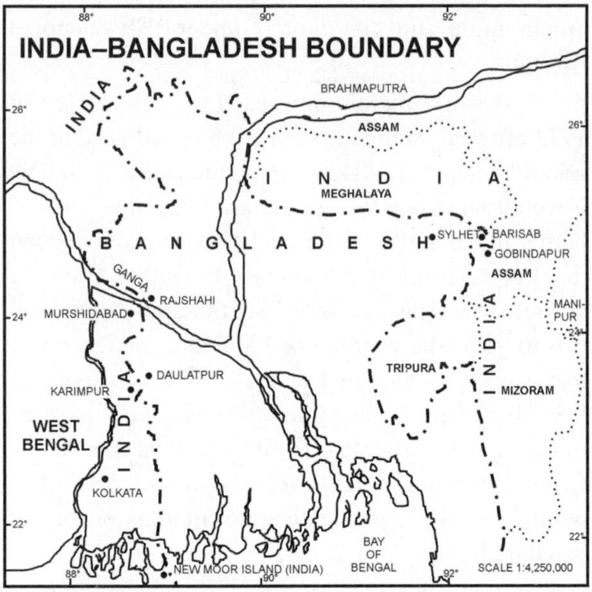

FIG. 1.10. India-Bangladesh Boundary

5. The India-Bangladesh Border

Of all the neighbouring countries, India's 4,096 km long border with Bangladesh is the longest and accounts for nearly 27 per cent of the total land border of India (**Table 1.2**). As many as five Indian states share the international boundary with Bangladesh. These are West Bengal, Assam, Maghalaya, Tripura and Mizoram. The boundary line between India and Bangladesh criss-crosses the vast Ganga-Brahmaputra delta. This boundary runs through an entirely flat country, in which there is not even a small mound or hill which could be used for demarcating the boundary between the two countries. It is, therefore, a very unsatisfactory border because it fails to separate the people living on either side of the frontier. This boundary has been determined under the Radcliffe Award which divided the erstwhile province of Bengal into two parts. The eastern part went to the then East Pakistan which became an independent country, Bangladesh in 1971, and western part remained with India as the state of West Bengal. West Bengal has 2,272 km long border with Bangladesh which is the longest border shared by any Indian state with Bangladesh. Although Indo-Bangladesh border has remained more or less peaceful and there are no serious border problems between the two countries, yet there were four overlapping claims which were referred to the tribunal. The first was between Rajshahi district of Bangladesh and Murshidabad district of West Bengal (India). Here the channel of the Ganga fluctuates rather frequently as a result of which the boundary has shifted several times. The northern part of the district was confirmed as the international boundary. The *second* dispute was between Karimpur district (India) and Daulatpur district (Bangladesh). In this case, Radcliffe award was amended and half way line in the Matabhanga River was treated as the international boundary. Bangladesh gained about 13 sq km as a result of this amendment. The *third* dispute was in Sylhet district of Bangladesh and Garo, Khasi, and Jaintia Hills districts of Meghalaya in India. This is a densely forested area where demarcation is difficult. Here Radcliffe Award was confirmed and India's claim was sustained. The *fourth* dispute existed between Barisari district (Bangladesh) and Gobindpur district (Assam).

Farakka barrage was another major issue between India and the then East Pakistan (now Bangladesh). Immediately after the Indus Water Treaty between India and Pakistan, a dispute arose over the use of the Ganga river water. India planned the construction of a barrage across the Ganga at Farakka. This barrage aimed at rejuvenating the river system of West Bengal, flood control, efficient distribution of water between the northern and southern parts of West Bengal and for desilting the Kolkata harbour. Pakistan feared that construction of Farakka barrage would seriously reduce water supply to East Pakistan. Negotiations failed to produce a settled agreement and the work on the construction of Farakka barrage started in mid 1960s.

During the 1980s a refugee influx of Bangladeshis in Tripura and Assam created problems for India. By 1989 more than six lakh refugees from Bangladesh had entered India. The refugee influx continues even today, creating economic, social, political and demographic problems for India. Currently, almost all the Indian states touching the Bangladesh border are infested by refugees from Bangladesh. India's plan to erect barbed-wire fence along the Bangladesh border to check the illegal migration of refugees from Bangladesh has not been welcomed by that country.

In the beginning, the emergence of an independent and generally friendly country of Bangladesh in place of the initial East Pakistan was viewed positively by India. But the recent growth of anti-India elements backed by ISI of Pakistan is likely to create problems between the two countries.

However, things have slightly changed in the positive direction as India and Bangladesh agreed on 6th September, 2011 to exchange 166 enclaves in each other's territory, settle all issues related to adverse possessions and demarcate the lost stretch of 6.5 km of the 4906 km border. According to the deal, India got back 111 enclaves (17,158 acres and population of 37,369) while 55 (7,110 acres and population 14,215) had gone to Bangladesh. But there is strong resentment in Assam as that state has lost 665 acres of land across Dhubri and Karimganj districts to Bangladesh. Bangladesh initially wanted 793 acres of land from India but was given only 267.5 acres according to the agreement signed by the two countries on 6th September, 2011. Indian states bordering Bangladesh are also sour about the illegal Bangladeshi immigrants to their states. It is estimated that there are over 40 lakh illegal Bangladeshis in Assam alone. Sharing of waters of the river Teesta—including those of Feni, Manu, Muhuri, Khowai, Gumti, Dharla and Dudhkumar still remains the main irritant.

6. India-Myanmar Boundary

The boundary between India and Myanmar is 1,458 km long and runs from India-China-Myanmar trijunction in the north to the southern tip of Mizoram. This boundary runs roughly along the watershed between the Brahmaputra and Ayeyarwaddy. It passes through thickly forested hill country, with Mizo Hills, Manipur and Nagaland on the Indian side and Chin Hills, Naga Hills and Kachin state on the Myanmar side. Although defined under the treaty of Yundaboo in 1926, this boundary was determined precisely by bilateral treaty signed on March 10, 1967. Barring a few minor incidents, the Indo-Myanmar-border has remained peaceful due to the goodwill gesture of give and take on both sides.

In May, 2012, India and Myanmar signed several agreements in Naypyidaw (new capital of Myanmar, over 300 km north of old capital Yangon) to strengthen relations between two neighbouring countries.

However, a new problem has come up as about 40,000 Rohingyas have illegally crossed over to India from Rakhine state of Myanmar. That country has not recognised the Rohingya among its 135 ethnic groups under a 1982 citizenship Act. They are fleeing Myanmar in the face persecution by the army. India wants to deport Rohingyas because they will be burden on our resources and will disrupt the demographic set up. Moreover, they can prove to be a big security threat because many of them are supposed to be linked with terrorist groups.

Water Frontiers

About 6,100 km long coastline of mainland of India is washed on three sides of the country by the Indian Ocean and its two arms namely the Arabian Sea in the west and the Bay of Bengal in the east. If we add to this the coastlines of Andaman and Nicobar Islands in the Bay of Bengal and Lakshadweep Islands in the Arabian Sea, the total coastline stretches to 7,516.6 km. Next to the Himalayas, the Indian Ocean is the most dominant factor which has influenced the destiny of India. It is through the waters of the Indian Ocean that India could establish her trade contacts with the neighbouring countries in the historic past. One of the greatest climatological phenomena, the Monsoon winds, have their origin in the Indian Ocean. Our water frontiers had been considered to be safe in the past. But the recent technological advancements in naval warfare have made these frontiers vulnerable to attack by the enemy forces and this has necessitated greater defence preparedness. "To be secure on land, we must be secure at sea."

7. India-Sri Lanka Boundary

India and Sri Lanka are separated from each other by a narrow and shallow sea called Palk Strait. Dhanushkodi on the Tamil Nadu coast in India is only 32 km away from Talaimanar in Jaffna peninsula in Sri Lanka. These two points are joined by a group of islets forming Adam's Bridge. There have been close and cordial relations between the two countries in the historical times. Every phase of Peninsular Indian culture found its way into the island. Consequently

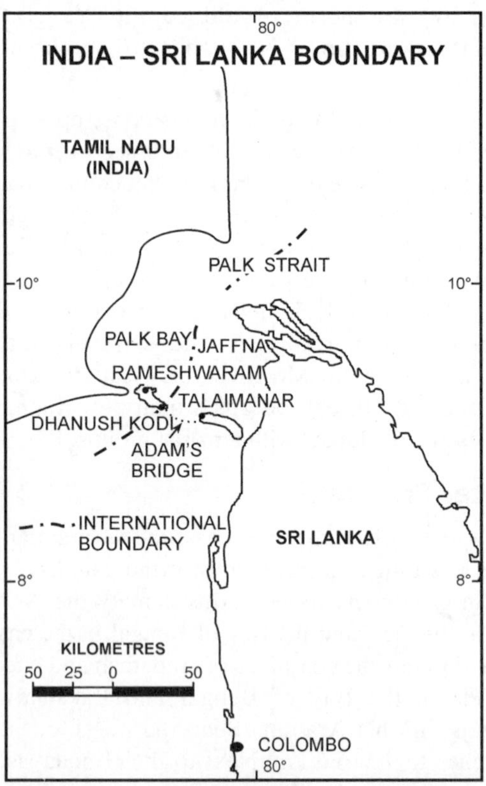

FIG. 1.11. India-Sri Lanka Boundary

more than half of the island had remained under the peninsular influences. The northern and north-eastern parts of the island have large number of Tamilians who migrated from Tamil Nadu to that country.

The maritime boundary between India and Sri Lanka passes through Palk Strait as Palk Bay touching Dhanushkodi (**Fig. 1.11**). This boundary has remained peaceful barring a few minor clashes between the fishermen of the two countries over the fishing rights. Some bitterness was created over the ownership of Kachchitevu Island (area 1.92 sq km) in the Palk Strait about 16 km from the Tamil Nadu coast. The problem was resolved with the demarcation of India-Sri Lanka boundary along the median line. This island was given to Sri Lanka as a result of agreement of 1974. The maritime boundary between India and Sri Lanka became lively in 1980s with the insurgent activity of L.T.T.E. which demanded a separate homeland for Sri Lankan Tamils within the island. Since then the border between the two countries has remained peaceful.

GEOPOLITICS OF THE INDIAN OCEAN

With an estimated area of about 73,442,700 sq km the Indian ocean is the third largest ocean in the world after the Pacific Ocean and the Atlantic Ocean. Although the Indian Ocean is much smaller in size than the Pacific and the Atlantic Oceans, yet it is of immense importance for us because it is located to the south of India. This is the only ocean in the world to be named after the name of a country, *i.e.*, Indian Ocean after India. In a way, it is just half an ocean because it does not open out northwards in the Arctic Ocean. It is bounded by South Asia in the north, Indonesian islands and Australia in the east and by Africa in the west. The meridian of Cape of Tasmania (147°E) forms the boundary line between the Indian and the Pacific Oceans while the meridian of Cape Town (18° 22′E) forms the dividing line between the Atlantic Ocean and the Indian Ocean. In the south, it extends to the Antarctic continent from where it merges with the Atlantic and the Pacific. The International Hydrographic Organisation has recognised the coast of Antarctica as the southern extreme of the Indian Ocean. Tropic of Cancer forms its northernmost limit.

The Indian Ocean has great strategic importance for India. The "landlocked" nature of the Indian

Ocean has given India a commanding position. From the eastern coast of Africa and the shores of the Persian Gulf to the Strait of Malacca, no other country rivals India's dominant location in the Indian Ocean. The strategic importance of this ocean is further enhanced by the fact that it is accessible from the west and the east through narrow straits only. The Red Sea and the Persian Gulf are the narrow outlets in the west while in the east, there are the Strait of Malacca and the Timore and Arafura sea.

The Indian Ocean has limited outlets. Before the opening of the Suez Canal in 1869, the only contact of littoral states of the Indian Ocean with the western countries was via Cape of Good Hope by circum-navigating the whole continent of Africa. On the eastern side there are two outlets—one through the islands of Indonesia and the second is to the south of Australia. The Indian Ocean can be choked any time by controlling these outlets. Since the Indian Ocean and the countries surrounding it are very rich in natural resources, such a possibility has considerably enhanced the geopolitical strategy of this ocean.

In spite of above mentioned geopolitical limitations, the Indian Ocean has never been a barrier between the countries. On the contrary it has served as a great linkage between the countries lying on its coasts and even further beyond. We can reach West Asia, Africa and Europe from the west coast and South East Asia, Far East and Oceania from the eastern coast. The Indian Ocean, thus, bridges a gap between east and west. This ocean is encircled by 46 countries (27 littoral including Australia, 7 island countries and 12 landlocked countries as recognised by the U.N.), with great diversity in almost every respect; shape, size, people, resources, economy, polity, culture, etc. (**Fig. 1.12**)

The Indian Ocean is endowed with rich variety of natural resources of which mineral and power resources as well as food resources are very important. Some of the resources are briefly described as under :

Aggregates

Marine aggregates comprise sand, gravel or shell deposits and are used primarily in construction industry. They are, at present the most important commodities mined offshore, both quantitatively and by value. They are mainly found on the continental shelves. Offshore calcareous deposits are formed by fragmentation of shells by waves and currents. These are used for manufacturing cement.

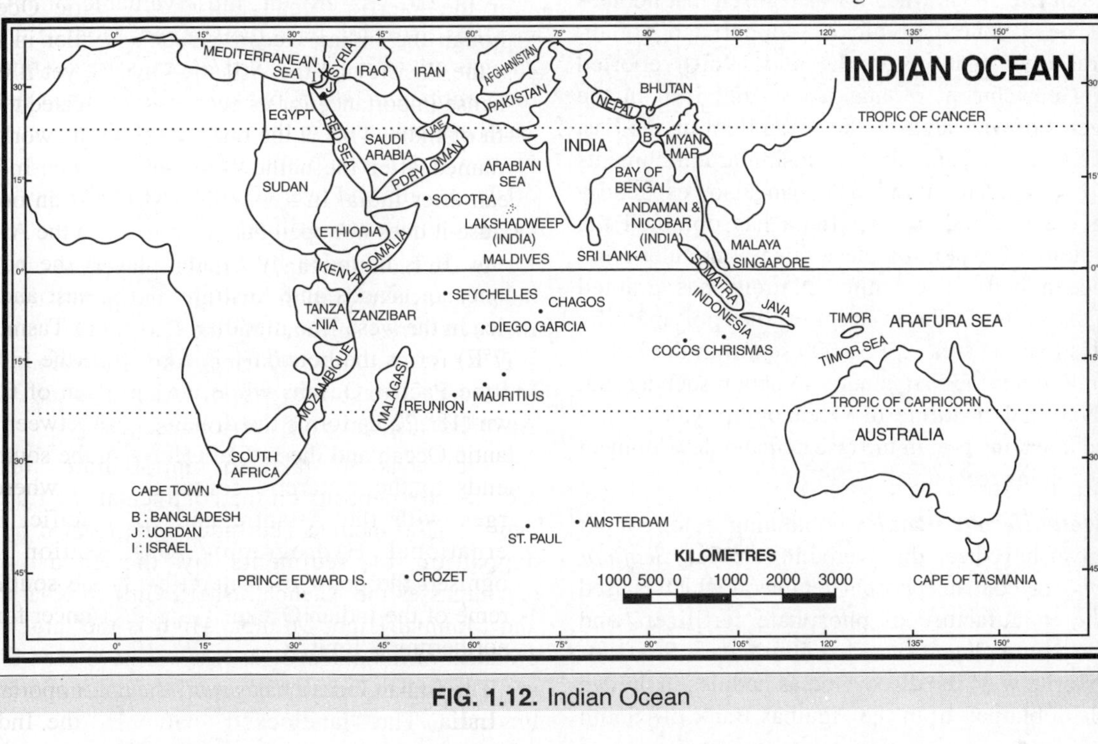

FIG. 1.12. Indian Ocean

Placers

Placer deposits are concentrations of heavy, resilient, and chemically resistant minerals eroded from existing ore bodies by mechanical weathering. Such deposits include native gold, native platinum, tin, titanium, magnetite (iron), zirconium, monozite (thorium) and gemstones. In the Indian ocean, such placers are found mainly along the coasts of Sri Lanka, India, Australia, Malaysia and Indonesia. Sri Lanka, India and Australia have titanium sands, whereas Malaysia and Indonesia have tin deposits. Indonesia is one of the main areas in the world where offshore placers are mined. Primary tin deposits occur in granitic rocks onshore and heavy minerals extend offshore.

Polymetallic Nodules

Polymetallic nodules are those which contain several metals, the important metals being manganese, copper, nickel, cobalt, etc. They occur in many shapes, sizes and forms and are generally friable.

Manganese nodules were first discovered on the 1872-76 scientific voyage of *Challenger,* but systematic exploration and detailed studies only started in late 1950s when it was realised that nodules might be a source of nickel, copper and cobalt. Several investigations in the mid 1960s reported finding enrichment of manganese and iron in the sediments of wide areas of the ocean floor on the East Pacific Rise axis, and since then similar enrichments have been found in Indian ocean also, especially along the mid-Indian rise. India has obtained the technology of exploiting these mineral nodules from the ocean beds. The United Nations has granted permission to India to exploit the polymetallic nodules over an area of 1,50,000 sq km in the Indian Ocean. India is the first country to obtain such a right. The National Insititute of Oceanography, Goa has played a leading role in the research and development of this mining technique.

Metalliferous nodules containing zinc, copper and silver have been discovered in Red Sea. *Seafloor phosphorite* contains phosphate mineral which is used for the manufacture of phosphate fertilizers and certain phosphate based chemicals. Seafloor phosphorite was first discovered as nodules in dredge samples obtained from the Agulhas Bank off South Africa during the *Challenger* Expedition of 1872-76 and have since been found elsewhere. They occur mostly on the continental margins and upper parts of the continental slopes at depths of less than 500 metres. They are normally confined to the zone between 40°N and 40°S latitudes. In the Indian ocean phosphorite deposits occur along the coast of Arabian Peninsula, east coast of India in the Bay of Bengal, near Andaman and Nicobar Islands and to the south of Sri Lanka. Phosphorites occur widely on Agulhas Bank off South Africa. It has become one of the most intensely studied areas of the world. The total reserves are estimated at 450 million tonnes.

Estimates of reserves suggest that recoverable *nickel* and *copper* are of the same order as known land economic resources. But the current level of technology is not competent enough to exploit these resources. Specialised mining systems have to be developed to recover the nodules without lifting the underlying sediment from depths in excess of 4000 metres. The environmental conditions of potential mine sites and possibilities of mineral extraction have been ascertained.

Oil and Gas

Oil and gas are the most valuable of all the minerals extracted from the sea bed. Most of the oil and gas producing areas of the oceans are confined to the *continental shelf*, but oil wells in much deeper seas have been dug in the recent past. At present half of the world's total output of oil and gas comes from offshore wells and over 80 countries of the world are engaged in offshore drilling.

In the Indian ocean, the major players in offshore oil and gas exploration, drilling and production are India, countries surrounding the Persian Gulf and the south-east Asian countries like Indonesia and Malaysia. Coastal areas of Australia (western Australia) also have offshore oil reserves.

India has 3.2 lakh sq km offshore area of sedimentary deposits on the continental shelf upto a depth of 200 metres. The Indus cone formed by the deposition of sediments by the Indus river encompasses the Kachchh Shelf, Gulf of Khambhat and Mumbai High. Mumbai High is the largest oil producer of India. Huge gas reserves have been found in the Krishna-Godavari basin off the coast of Andhra Pradesh.

Oil and gas reserves of the Persian Gulf are simply amazing. But these reserves remained unexploitated for a pretty long time because this gulf has land based reserves also and those reserves were to be exploited first. The offshore potential is very important, especially for smaller states which have restricted land areas. The Persian Gulf has the advantage of being sheltered from the open ocean, is shallow and relatively free of hazards. It is the most rapidly developing part of the industrialising world as its onshore and offshore wealth is re-invested in industry. Kuwait, Saudi Arabia, Bahrain, Qatar, United Arab Emirates, Iran and Iraq are the beneficiary states.

In south-east Asian waters, offshore, oil is becoming increasingly significant for a number of countries in this vast archipelagic area. There is increasing shift from exploitation of land resources to offshore resources. This phenomena is particularly seen in Indonesia and Malaysia where the main producing areas are off southern Sumatra. The South-east Asian waters are considered as one of the most promising for immediate exploitation.

Off the coast of Western Australia there are substantial reserves of oil and gas. These reserves experienced rapid development in the 1970s.

Besides there is great scope for harnessing the non-conventional energy resources in this ocean.

National Institute of Ocean Technology (NIOT) is developing a technology in which a 12 tonne crawler with censors, pumps and crusher will be tested on the sea bed filled with metallic ores such as cobalt, nickel and copper located around 6000 km from the Indian shores. If trials are successful, India will be the third country after China and Korea to develop technology for mining polymetallic nodules. These nodules are a big source of nickel, cobalt, copper, manganese and iron. Estimated polymetallic nodule resource potential is 380 million tonnes containing 4.7 million tonnes of nickel, 4.29 million tonnes of copper, 0.55 million tonnes of cobalt and 92.59 million tonnes of manganese. The exploration will be in a 75,600 sq km region in the Central Indian Ocean Basin.

Scientists from the Geological Survey of India (GSI), discovered the presence of millions of tonnes of precious metals and minerals deep under the waters that surround the peninsular India. The huge presence of resources was first identified off Mangaluru, Chennai, Mannar Basin, Andaman and Nicobar Islands and around Lakshadweep in 2014. The amount of lime mud, phosphate-rich and calcareous sediments, hydrocarbons, metalliferous deposits and micro-nodules that geologists came across was a clear indication that deeper and more extensive exploration could lead to a larger treasure trove. After three years of exploration, GSI has generated 181,025 sq.km. of high-resolution sea-bed morphological data and established the occurrence of more than 10,000 million tonnes of lime mud within the Exclusive Economic Zone (EEZ) of India. It has also confirmed the presence of phosphate sediment off Karwar, Mangaluru, and Chennai coasts, gas hydrate in Channel-levee system of Mannar Basin off the Tamil nadu coast, cobalt-bearing ferro-manganese crust from the Andaman Sea and micro-manganese nodules around Lakshadweep Sea.

The above mentioned and some other natural resources of this ocean can be a great cause of rivalry among the nations of the world.

Historical background. India has used the Indian Ocean for thousands of years. This ocean had been a powerful medium of trade, defence and diffusion of Indian culture particularly in Southeast Asia. The Mauryan kings had established ports on the Bay of Bengal in the fourth century B.C. Kautilya made a mention of a separate administrative division of overseas maritime activities in his famous *Arthsastra*. Large naval kingdoms of the Cholas and the Chalukyas were set up in South India. Sri Vijaya empire, set up by the Indian rulers in South-east Asia from eighth to eleven centuries A.D., maintained strong cultural and commercial ties with south India through the Indian Ocean. The Arab explorers and traders increased their activities in the Arabian Sea from the middle of the 13th century to the beginning of the 16th century. However, they could not exercise much control on this ocean. The European influence in the Indian Ocean started after the landing of Vasco da Gama at Calicut on the western coast of India in 1498. Major European powers were eventually drawn into a long and bloody struggle for power in the Indian Ocean. Out of all the European powers, Britain had the best naval force and gained supremacy over the Indian Ocean and the Indian subcontinent.

During the 18th and 19th centuries, sea ports like Mumbai, Kolkata and Chennai were developed to boost up sea trade and maritime activities. The British became a super power by dint of its naval superiority. According to an Indian historian and diplomat, K.M. Panikkar (1945), the history of British control in Indian Ocean illustrates the basic geopolitical principle that *the power which rules the sea eventually rules the adjoining land*. He observed that the pre-British invasions and land directed conquests of India led to the founding of political dynasties, which in a short period were Indianised. Only the British rulers could remain unassimilated as they could draw their strength from England through the naval supply lines. It is worth noting that although a large portion of north India was conquered many times by foreigners, India was never ruled by a monarch who did not have his capital in India except whom it was under the naval power of the British (Panikkar, 1951). Japanese effort to capture Singapore and Andaman Islands during the World War II also corroborated the above statement.

Most of the former British colonies became independent countries after World War II. This resulted in shrinking of the British empire and reducing of the British influence in the Indian Ocean. By mid 1960s, it lost almost all the colonies and serious economic strains were experienced in Britain. Consequently Britain started withdrawing its military forces from the Indian Ocean region. This withdrawal created "*a power vacuum*" which gave birth to super power rivalry in the Indian Ocean region. The USA immediately jumped into the fray and made efforts to fill the '*power vacuum*' created by the withdrawal of the British forces. The USA purchased Diego Garcia island from Britain to build a military base there. The base is fully equipped with nuclear weapons and provides decisive advantage to the USA to command the areas of the Middle East, South Asia, Central Asia, Russia and China. In addition to Diego Garcia, the USA has set up military bases near Asmara (Ethiopia), Woomera and Harsld E. Holt (Australia) as well as in Bahrain and Mahe. The military bases at Gan Island, Masirah Island, Simonstown (South Africa) and Port Louis (Mauritius) are jointly owned by the US and the UK. France has also set up bases in Diego Suarez and Reunion Island.

The Russian interest in the Indian Ocean region is as intense as that of the USA. The individuality of the ocean makes it possible for both the countries to exercise their power even though they may be far removed from the Indian Ocean. In order to counter the US influence, Russia sent its fleet in 1968. Since 1970, Russia has been steadily building trade and economic connections with India and the countries bordering the Indian Ocean. That country has entered into bilateral treaties with several countries of the region such as India, Sri Lanka, Bangladesh, Iraq, Egypt, Mauritius, Somalia, Seychelles, etc. Russia maintains military bases at Berbera, Masira, Umakas, Dahalak and Socotra islands. It constructed naval radio stations and ammunition depots too at the northern end of the Red Sea. It has also obtained access to port facilities in Somalia, Mauritius and Singapore. Russia is trying to gain a foothold in the oil rich countries of the Middle East also.

In the recent past, China has also shown keen interest in the Indian Ocean. With the former U.S.S.R. having lost its super-power status, China wants to take full advantage of the situation and is trying hard to fill in the vacuum, and ultimately assume the leadership of the Afro-Asian world. China has shored up its presence in the Indian Ocean, investing billions of dollars to build ports of Gwadar in Pakistan and Hambantota in Sri Lanka and plan maritime trade routes as part of its 'One Belt, One Road' (OBOR) initiative to help increase its market reach. Sri Lanka has long been in India's orbit, but that country's relationship with China has strengthened in the last few years.

India could establish her supremacy in the Indian Ocean in 1988 when Sri Lankan mercenaries invaded Maldives and Indian forces had come to its rescue. Maldives has been consistently turning down lucrative offers by big powers to provide facilities for establishing military bases on its soil in order to make the Indian Ocean a zone of peace.

The littoral countries encircling the Indian Ocean are closely associated with this ocean traditionally, culturally and economically. These countries are awfully worried about the geopolitical affairs of the ocean. The concerned countries want to keep the Indian Ocean a zone of peace rather than an area of cold war, great power rivalry, super power confrontation and a battle-field.

Maintaining the environment of peace in the Indian Ocean is very important for the littoral states because almost all the countries surrounding this ocean are at the developing stage. They are in the process of economic reconstruction and development, social change towards modernization and building the democratic political system. As such these countries cannot afford confrontation and war and they want to maintain peace in the Indian Ocean at any cost. Majority of these countries are members of the Non-aligned movement and do not want any external interference in the affairs concerning Indian Ocean. The member countries of the Non-aligned group strongly feel that the socio-economic prosperity in the region will only be possible when there is an atmosphere of economic co-operation and co-ordination among the richly endowed littoral states. This can happen only if the external powers are not allowed to show their presence in the Indian Ocean and are deterred from influencing the littoral states. The conference of the Heads of Non-aligned nations held at Lusaka in 1970 called upon all states to identify the Indian Ocean as a zone of peace and should also be free of nuclear weapons. The UN General Assembly has also passed a resolution to keep the Indian Ocean zone as a '*zone of peace*.'

However, there are several obstacles in maintaining peace in the Indian Ocean. A large number of the littoral states are small, economically poor and politically immature. Some of the littoral states such as India, Australia, Republic of South Africa, Iran and Indonesia are strong enough to influence the political events at the regional and international levels. Environment of politically strained relations between some of the neighbouring countries is also a great hindrance in the way of peaceful co-existence. The situation has become rather complicated with the growing interference of China and Japan. China wants to establish itself as a major power in the Afro-Asian region. Similarly, Japan wants to increase its influence because nearly 50 per cent of the Japanese international trade is carried through the Indian Ocean. The Gulf War made the Indian Ocean the focal point of the world geopolitics. The growing terrorism in the Muslim world has also assumed dangerous posture for the world in general and for the Indian Ocean in particular.

Thus, peace in the Indian Ocean is very fragile and it can be maintained only by the cooperation of the littoral states. Only such cooperation can save this ocean from exploitation by the developed countries and the erstwhile colonial powers.

ROUTES INTO INDIA

It has already been mentioned that India is separated from the rest of Asia by a broad *no-man's land* of mountains, from the north, north-west and north-east. Most of the Himalayan ranges are either ice bound or covered by thick vegetation and offer an almost an impenetrable wall. This makes India an intelligible isolate. However, there are a few passes which provide routes into India.

Three high passes *viz. Muztagh* (Snowy mountain), the *Karakoram* and the *Changchemo* are all over 6,000 metres above sea level and offer little scope for interaction between the two sides of the mountain ranges. Some passes in the high Himalayas such as *Burzil* and *Zoji La* in Ladakh; *Bara Lacha La* and *Shipki La* in Himachal Pradesh; *Thagla La*, *K-Lang, Niti and Lipu Lekh* in Uttarakhand; *Nathu La* and *Jelep La* in Sikkim are used for crossing the main Himalayan range.

Towards the north-west, the frontier formed by the Himalayas is continued by the *Karakoram* range and the *Hindukush*. Further south lie the *Safed Koh, Sulaiman* range and the *Kirthar range*, which separate India from Afghanistan and Baluchistan. Thus the real natural frontier of India lies far away from the present man-made and artificial frontier between India and Pakistan which is the creation of partition of the country in 1947. The mountain ranges here become shallow and narrow down to less than 500 km between Turkestan and the Punjab (undivided) Plains. The *Hindukush* is pierced by numerous passes which are snow covered in winter but can easily be crossed in other seasons. There are a few passes which are less than 1,600 metres above sea level and offer easy access into India. The passes of *Khyber*, *Malakand*, *Tochi*, *Gomal* and *Kohat* have their base in *Badakhshan* province of Afghanistan and provide passage from Afghanistan to the Indian subcontinent. Towards the south, the highways of *Herat* and *Seistan* converge at Quetta. It is worth noting that Quetta lies at the head of the *Bolan pass* and provides one of the most significant gateways into the Indian subcontinent. History reveals that the barrier was at no time an insuperable one, and at all periods, invaders, settlers and traders have found their way over the high and desolate passes into India, while Indians have carried their commerce and culture beyond her frontiers by the same routes. India's isolation has never been complete, and the effect of the mountain wall in developing her unique civilisation has been over emphasised. People after people, literally nations have pressed into the Indian subcontinent through these passes and changed the fate of various dynasties in India. Hordes of invaders have attacked and conquered India through these passes.

Perhaps the earliest immigrants of civilised man into India were the *Dravidians* about whom little evidence is available. It is believed that they migrated into India from the *Tigris Valley* along the *Makran Coast* of present Pakistan. This was followed by many others in the pre-historic times, the most prominent being the *Aryans* who migrated by the Kabul approaches. Later in the historic times, several other people such as *Greeks*, *Parsees*, *Huns*, *Mughals* and *Mohammadans* entered India mostly through these passes.

In the east, the hills forming the boundary between India and Myanmar are comparatively low and rarely exceed 3,000 metres in elevation. But the dense forests, heavy rainfall for half the year, difficult terrain and a large number of swift flowing rivers have acted as effective barriers between the two countries. The only passages are along the courses of rivers Brahmaputra, Mekong, Salween and Ayeyarwadi—the main gaps being the *Ann*, *Teju*, *Manipur, Tulu, Tongap,* etc. Therefore, not many foreigners have crossed through this frontier into India. In fact, reaching Ayeyarwadi valley in Myanmar from India is much easier by sea than by land. The opening of a new trade route in 1995 connecting *More* in Manipur with *Tamu* in *Myanmar* may help in developing trade and cultural relations between the two countries. The bus service between Imphal (Manipur) and Mandalay covering a distance of 579 km has further strengthened ties between India and Myanmar.

It is surprising that until the coming of European seamen, no considerable power was founded in India from the sea; but some were from India. Initiative to the east really came from India. Hindu traders and colonisers took their civilisation by sea to the south-east Asian countries. Buddhism originated in India and spread to the whole of eastern Asia. The Cholas had an empire in the East Indies. Several other Hindu cultural traits are still found in large parts of the Far East.

LAND OF DIVERSITIES

India is a land of great diversities and contrasts. It is but natural that an area as vast as the Indian subcontinent should have considerable physical diversity. *The extremes of physical and human geography of India are extreme indeed.* In the south is the Plateau which is one of the oldest and least disturbed land masses on the earth's surface. Its rocks have never been extensively covered by sea since their formation in the pre-cambrian period over 3,000 million years ago. In contrast to this, the Himalayan mountain ranges and the Great Plains represent the most recent formations. The denudational processes have made these contrasts still sharper. While the mountain ranges in the north have very youthful topography with sharp peaks and steep-sided valleys, the Peninsular plateau shows old and senile topography with gently rising ridges and wide valleys. The Aravalis and the Himalayas are perhaps the oldest and the youngest ranges respectively on the earth's surface. The differences are most striking even in the landforms of recent origin. Within a distance of a few hundred kilometres from north to south, one can reach from the highest peak of the world to the flat, featureless and monotonous plain.

Even the rivers of the Himalayan and the Peninsular regions have contrasting characteristics. The Himalayan rivers have their origin in the snow covered areas. As such, they receive water even in dry season due to snow melt and are perennial. On the other hand, rivers of the Peninsular plateau carry only rain water and the quantity of water carried by them decreases considerably in dry season. They are, therefore, termed as seasonal rivers. Moreover, the rivers of the Peninsular plateau have reached maturity whereas the Himalayan rivers are still in their youthful stage. Most of the Himalayan region is made of sedimentary rocks while the Peninsular plateau has mostly igneous and metamorphic rocks. Hemmed between the Himalayas in the north and the Peninsular plateau in the south, lies one of the largest alluvial plain of the world. This vast plain stretching in a great curve from Arabian Sea to the Bay of Bengal, consists of extremely fertile soils washed down by the streams and rivers for thousands of years.

Climatic contrasts are no less pronounced than the physiographic contrasts. Although a typical monsoonal realm, Indian climate exhibits a wide range of climatic variations. While the mercury may dip to (–) 40°C during winter nights at Dras or Kargil in Ladakh, the temperature may stand at a fairly high level of 20°C to 25°C at Chennai. In summer, the day temperature at Barmer in Rajasthan may soar to 48°C–50°C while the higher reaches of the Himalayas may still remain snow covered. The differences are equally striking in rainfall patterns also. Mawsynram near Cherrapunji receives an annual rainfall of over 1,221 cm as compared to only 12 cm received at Jaisalmer. Several places in Garo hills receive more rainfall in a single day than received by Jaisalmer in a long span of ten years. The people of Mumbai experiencing maritime climate may not have any idea of extremes of climate. But Delhiites living in continental climate have to pass through the entire cycle of seasonal changes. As a result of climatic extremes, the natural vegetation varies from dense tropical evergreen forests of the Western Ghats, North-Eastern states and Andaman and Nicobar islands to the scanty shrub bush vegetation of the Thar Desert and its adjoining areas.

Extremes and diversities of the physical features and climatic conditions have produced cultured heterogeneity of a high order. On one end of the scale are the vast uninhabited areas of Ladakh and the Thar desert and on the other end are the river valleys and deltas accounting for some of the highest population densities in the world and that too based on purely agrarian economy. The length and breadth of the country comprises a rich mosaic of religions, languages, cultures and races and is, thus, heterogeneous in character and spirit. All levels of economic development from that of the *stone age* to the *satellite* and *computer* age are seen in the country. In large parts of the country, purely tribal, agricultural, industrial and commercial economies exist simultaneously. Considering its physical and cultural diversities India is often called the ***'epitome of the world'*.**

UNITY IN DIVERSITY

It must not be forgotten that the above mentioned diversity of India is based on its underlying unity. India has been able to project itself as a single territorial unit in the face of physical, political, social and economic contrasts. The unifying role of the Great Plains between the Himalayan ranges on one hand and the Peninsular India on the other can hardly be ignored. Climatically, the monsoonal rhythm of seasons provides a strong element of uniformity. The concentration of monsoonal rainfall to a few months in a year and the associated agricultural activities are an all India phenomena. Many of our cultural traditions are strongly tied to the monsoons. Our saints have spread the message of universal brotherhood which has helped a great deal in uniting different sections of society in India and making the country a united nation. The *Ramayana* and the *Mahabharata* have provided themes even for tribal dances and music for thousands of years and still continue to do so. Although *Hinduism* is the way of life for majority of Indians, *Hinduism* and *Islam* are intertwined into a composite matrix on the Indian land. It is in this composite matrix that integration of India is strongly rooted. In fact there are almost as many Muslims in India as in Pakistan. Centuries of foreign rule failed to disrupt our cultural ties.

On economic front, the development of inter-regional linkage and the emergence of a national home market have played a significant role in uniting the country. For example, tea from Assam, wheat from Punjab and Haryana, minerals from Chotanagpur plateau and spices from the hill areas of the southern states are used all over the country and in turn these areas obtain items required by them from other areas. This process has been accelerated by rapid development of transportation and communication.

LAND OF POTENTIALITIES

India is blessed with a large variety of natural resources in huge quantities. Many of these still await exploitation. Whereas nature has been bountiful to

India, the inhabitants of this country have not been able to develop requisite technology, in the modern context, to harness these resources. India is, therefore, often referred to as '*poorly developed rich country*' or '*a rich country inhabited by poor people*'. Sometimes it is also called the '*land of the future*' or '*land of potentialities*'.

India's huge mineral wealth encompasses a wide range which is supposed to be sufficient enough for developing a modern industrial base. India is particularly rich in the deposits of high grade iron ore, manganese and chromite. Her reserves are said to be adequate with respect to limestone, bauxite, coal and strategic minerals.

Although India's forest resources are not very large and there is urgent need to increase the area under forests to meet the growing demand for forest products and to maintain the ecological balance, yet they are able to feed many forest based industries like paper, plywood, match box, furniture, resins, lac, etc.

India has vast resources of both surface and ground water which can be used for irrigation, power generation, industries and for drinking purposes. Only a small proportion of this natural resource has actually been utilised offering us enormous potentialities. It is estimated that about 48% of the cropped area is presently irrigated whereas about three-fourths of the cropped area can be irrigated by making optimum use of the water resources available to us. Similarly only one-fourth of the total water power potential in the country is actually used. Indian rivers offer numerous sites suitable for generating hydroelectricity.

India is one of the leading agricultural countries of the world and agriculture is the most important occupation of Indians. But the overall production and the yields per hectare are very low as compared to that in the advanced countries. The total production can be increased only by increasing the yields as almost all the cultivable land has already been brought under plough. This requires intensive use of high yielding varieties of seeds, fertilisers and irrigation facilities.

Of course, the greatest asset of India is its population which is second only to that of China. The poverty stricken, much cursed and ever growing vast ocean of humanity, if properly channelized, could be converted into huge manpower resource and could prove to be a boon, rather than a bane, for national development. The only problem is that not many avenues are open due to backward state of economy in the present day context.

INDIA AND THE ORIENTAL WORLD

The Oriental world includes into its fold the countries of East Africa, South West Asia, South Asia and South East Asia. These countries surround the Indian Ocean which unites them through sea routes. Economically and culturally, India's major contacts with outside world for the last two millennia have been by sea, earning India, the title of ***'Mistress of the Eastern Seas'***. The opening of the Suez Canal has brought the countries of Southern Europe and North Africa very close to India. In ancient days, the boats of the Babylonians, the Egyptians and the Phoenicians used to sail in the Arabian Sea. Hindu and Buddhist cultures and religions spread to far off areas towards south and east Asia. There are a number of temples in Thailand with *slokas* from *Ramayana* engraved on the walls. Bali in Indonesia was the seat of Hindu Kingdom. Mauritius in the Indian Ocean is a '*miniature India*'.

Our relationship through land frontiers are much older than our maritime contacts. Passes, gorges and valleys in the mountain chains offered passages to travellers in the ancient times. Waves of settlers from the north-west were lured by the riches of India in general and that of the Great Plains in particular. Mongols, Turks, Arabs and Iranians entered India as conquerors and settled down here. These settlers got assimilated in the national mainstream with the passage of time which led to a mixture of races, cultures, people and ethnic groups. Some of the invaders brought architectural excellence to India and took back to their homeland the Indian numerals, the decimal system and the ideas of *Upnishads*. *Buddhism* travelled from India to Tibet through land routes and further crossed over to China, Korea and even Japan. Thus, India is an integral part of the oriental world and holds a significant position in the region.

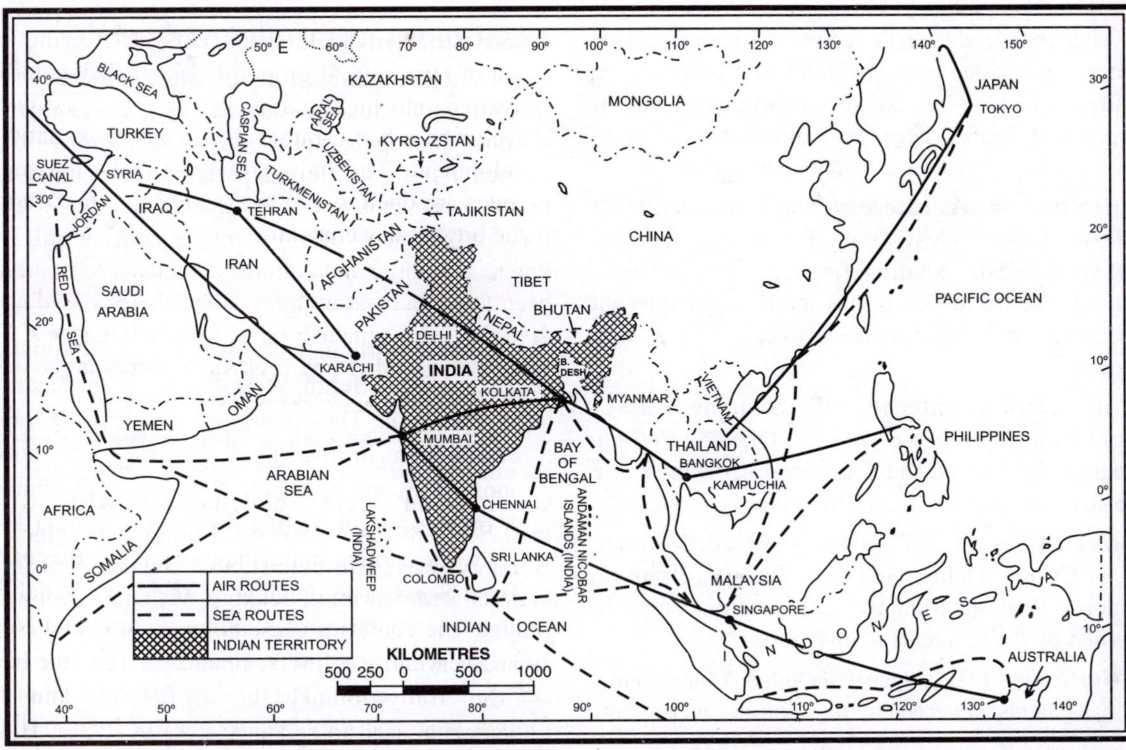

FIG. 1.13. Location of India at the head of Indian Ocean in the Oriental World on the International Highway of Trade and Commerce

STRATEGIC SIGNIFICANCE OF INDIA

India stands at the head of the Indian Ocean at the very centre of the *Eastern Hemisphere* commanding trade routes running in all directions. Its location, size and economic resources have made it the most dominating country among the *littoral states*. Indian Ocean is the only ocean in the world which has been named after a country *i.e*. India. This only proves the great prominence enjoyed by India in early days when these oceans were given their names. The trans-Indian Ocean routes connecting the developed countries of Europe in the west and the developing countries of Asia in the east must skirt the shores of India. No other country has such a long coastline on this ocean as India has. Therefore, *the Indian Ocean is truly Indian Ocean*. Most of the air routes between Europe, West Asia and Africa in the west and East Asia, South East Asia, Japan and Australia in the east also pass through India.

The centrality and the consequent significance of India to broader Indian Ocean geopolitics and economics has been proved beyond doubt. It was India's status as the principal entrepot in a highly profitable oceanic trade structure which attracted the West Asian Muslims, and the European colonial powers towards India. India became the '*jewel in the crown*' of the vast British empire due to its wealth and because of its role as a fulcrum upon which the imperial and commercial system in Asia was balanced.

India thus commands an important strategic position on the globe with respect to trade as well as social and cultural interaction. India is a concept. India is an experiment through ages. India is unique, India is India—no parallel example exists.

INDIA'S POLITICAL DIVISIONS

About two centuries of British rule left India fragmented into nine British provinces and 562 small princely states at the time of independence in 1947. The constitution of India was adopted with the establishment of Democratic Republic of India on 26th Jan., 1950 and India became a Union of States.

After Independence, the country was reorganised by merging various princely states and other states. During the first phase of reorganisation of Independent India, there were four categories of states :

(*i*) **States of 'A' category** which included Uttar Pradesh, Bihar, West Bengal, Assam, Odisha, Madhya Pradesh, Madras (present Tamil Nadu, Andhra Pradesh and Telangana) and Bombay (present Maharashtra and Gujarat). These were governed by a governor.

(*ii*) **States of category 'B'** including PEPSU, Central India, Mysore (present Karnataka), Saurashtra, Rajasthan, Hyderabad and Travancore-Cochin. These were governed by the chief of the states.

(*iii*) **States of 'C'** category included Ajmer, Kutch, Coorg, Delhi, Bilaspur, Bhopal, Tripura, Himachal Pradesh, Manipur and Vindhya Pradesh, governed by the Lieutenant Governors.

(*iv*) **States of 'D'** category included Andaman and Nicobar Islands, governed by the central government.

States Reorganisation Commission

The above classification lacked economic viability or a sustainable administrative machinery and did not represent the unity of the nation. The Indian government devised a mechanism based on such considerations as distinctive regional, linguistic, cultural and economic characteristics. With this background and with growing demand for creation of linguistic states, the first linguistically based Andhra state was created by separating Telugu speaking areas from the composite Madras state on October 1, 1953. Telangana area was added to Andhra state and Andhra Pradesh was formed on November 1, 1956. The formation of Andhra state in 1953, quickly sparked renewed demand of political recognition by other linguistic and cultural groups. This led to the appointment of States Reorganisation Commission in December, 1953. The commission submitted its report on 30th September, 1955, recommending reorganisation of India into 16 states and 3 territories. The government accepted the major part of the report but proposed the reorganisation of India into 15 states and 7 territories. Finally the Parliament passed the States Reorganisation Act, 1956 reorganising India into 14 states and 6 union territories as on 1st November, 1956.

Post-Reorganisation Changes

The Reorganisation of the states as per recommendations of the States Reorganisation Commission could not meet the aspirations of the people living in different parts of the country and pressing demands have been made from time to time for readjustment of politico—territorial units on linguistic, ethnic and cultural grounds. The trend had been towards creating more states thereby leading to division and fragmentation of the existing states. The influence of underlying centrifugal forces in the unity of India was conspicuous in these changes.

Within a short span of five years after the recommendations of the States Reorganisation Commission were implemented with slight modifications, the erstwhile bilingual province of Bombay was bifurcated into Gujarati speaking Gujarat and Marathi speaking Maharashtra on 1st May, 1961. The Portuguese colonies of Goa, Daman and Diu were the last remnants of the European occupation in India. These were liberated from the foreign yoke and they became part of India on 16th December, 1961. This completed the territorial integration of the Indian Union.

Several new states have been carved out in the north-eastern region of the country in response to demands by the tribals. In 1957 the north eastern part of Assam was separated and was designated as North-East Frontier Agency (N.E.F.A.). The Naga hills territory was a centrally administered area and was known as the Naga Hills Tuensang Area. It was renamed as Nagaland in 1961 and was given the status of a state on 1st December, 1963. Manipur became a Union Territory under States Reorganisation Act 1956 and was made a full fledged state of Union in 1972. Meghalaya was created as an autonomous state within the state of Assam on 2nd April, 1970. The full-fledged state of Meghalaya came into existence on January 21, 1972. Tripura became centrally administered territory in 1956. It was granted statehood in 1972. Thus three states in the north-eastern region came into existence in 1972. Sikkim ceased as a protectorate of India in 1974 and became an Associate State of India thereafter. It became an integral part of India as a result of referendum held on April 14, 1975. The North-East Frontier Agency (N.E.F.A.) was given the new name of Arunachal Pradesh and was granted full statehood on February

20, 1987. Mizoram was one of the districts of Assam till 1972 when it became a union territory with the implementation of the North-Eastern Reorganisation Act. It was granted full statehood on February 20, 1987.

Turning again to the south, it is worth mentioning that the Mysore State, created in 1956, was renamed as Karnataka in 1973 on the demand of Kannada speaking majority of the area. Goa which was liberated in 1961 was conferred statehood on May 30, 1987 although Daman & Diu still remain as union territory.

In the north-western part of the country, the bilingual state of Punjab was divided into Punjabi speaking Punjab and Hindi speaking Haryana on

FIG. 1.14. India : Political Divisions at Present

1st November, 1966. Thus the new state of Haryana was born on this date. Certain hill areas of erstwhile Punjab were transferred to Himachal Pradesh. It was granted full statehood on January 25, 1971.

Three new states came into being in November 2000. Chhattisgarh was carved out of Madhya Pradesh on November 1, 2000 as the 26th state of the union to fulfill the long cherished demand of the tribal people. The hill areas of Uttar Pradesh were separated to form a new state of Uttarakhand on November 9, 2000. This is the 27th state of India which shares its international boundaries with China (Tibet) in the north and Nepal in the east. Jharkhand, the 28th state of India was carved out of Bihar on November 15, 2000. This tribal state largely comprises forest tracts of Chotanagpur plateau and Santhal Pargana and has distinct cultural traditions. 29th state of Telangana was carved out of Andhra Pradesh on 2nd June, 2014 thus meeting the long standing demand of people belonging to Telangana region of Andhra Pradesh. So far this is the youngest state of the Indian Union.

There are some small scattered areas in the Union of India which are centrally administered and are known as Union Territories.

The erstwhile state of Jammu and Kashmir enjoyed special status under Article 370 which was a big discrimination against other states and union territories of India. On 5th August, 2019, the Government of India issued a constitutional order scrapping Article 370 and Jammu and Kashmir Reorganisation Act was passed by parliament, enacting the division of the state of Jammu and Kashmir into two union territories to be called Union Territory of Jammu and Kashmir and Union Territory of Ladakh. The reorganisation took place on 31 October, 2019.

On December 3, 2019, the Parliament passed a bill to merge two Union Territories namely Daman and Diu and Dadra and Nagar Haveli which came into effect on 26 January 2020. The new Union Territory is known as Dadra and Nagar Haveli and Daman and Diu (DNHDD). Its capital is at Silvassa. The total area of DNHDD is 603 sq. km. As the things stand today the Indian Union comprises of 28 states and 8 union territories. The boundary between the union territories of Jammu-Kashmir and Ladakh shown in figure 1.14 is as per map released by the Survey of India on Nov. 2, 2019. The Union Territory of Jammu and Kashmir comprises 22 districts including Muzaffarabad and Mirpur in PoK (Pak occupied Kashmir). The Union Territory of Ladakh has two districts of Kargil and Leh.

TABLE 1.3. Political Divisions of India

Sl. No.	States	Area in sq km	Capital	Principal languages	No. of districts as in 2021
1.	Andhra Pradesh	1,62,267	Amaravati	Telugu, Urdu, Hindi, Banjara	26
2.	Arunachal Pradesh	83,743	Itanagar	Monpa, Miji, Aka, Sherdukpen, Nyishi, Apatani, Tagin, Hill Miri, Adi, Digaru-Mismi, Idu-Mishmi, Khamtli, Miju—Mishmi, Nocte, Tangsa and Wancho	25
3.	Assam	78,438	Dispur	Asamiya, Bodo	35
4.	Bihar	94,163	Patna	Hindi, Bhojpuri, Maithili	38
5.	Chhattisgarh	1,35,194	Raipur	Hindi	28
6.	Goa	3,702	Panaji	Konkani, Marathi, Kannada	2
7.	Gujarat	1,96,024	Gandhinagar	Gujarati	33
8.	Haryana	44,212	Chandigarh	Hindi, Haryanvi	22
9.	Himachal Pradesh	55,673	Shimla	Hindi and Pahari	12
10.	Jharkhand	79,714	Ranchi	Hindi	24
11.	Karnataka	1,91,791	Bengaluru	Kannada	31

12.	Kerala	38,863	Thiruvananthapuram	Malayalam	14
13.	Madhya Pradesh	3,08,252	Bhopal	Hindi, Urdu	52
14.	Maharashtra	3,07,713	Mumbai	Marathi	36
15.	Manipur	22,327	Imphal	Manipuri	16
16.	Meghalaya	22,429	Shillong	Khasi, Garo Pnar and English	12
17.	Mizoram	21,081	Aizawl	Mizo, Lushai and English	11
18.	Nagaland	16,579	Kohima	Angami, Ao, Chang, Konyak, Lotha, Sangtam, Sema and Chakhesang	13
19.	Odisha	1,55,707	Bhubaneshwar	Odia	30
20.	Punjab	50,362	Chandigarh	Punjabi	23
21.	Rajasthan	3,42,239	Jaipur	Hindi and Rajasthani	50
22.	Sikkim	7,096	Gangtok	Lepcha, Bhutia, Nepali, Limboo and Sikkimese	6
23.	Tamil Nadu	1,30,058	Chennai	Tamil	38
24.	Telengana	1,12,077	Hyderabad	Telugu, Urdu	33
25.	Tripura	10,491	Agartala	Bengali and Kokborok	8
26.	Uttar Pradesh	2,40,928	Lucknow	Hindi and Urdu	75
27.	Uttarakhand	53,483	Dehra Dun	Hindi, Garhwali, Kumauni	13
28.	West Bengal	88,752	Kolkata	Bangla	25
Union Territories					
1.	Andaman & Nicobar Islands	8,249	Port Blair	Hindi, Nicobarese, Bengali, Malayalam, Tamil and Telugu	3
2.	Chandigarh	114	Chandigarh	Hindi, Punjabi	1
3.	Dadra and Nagar Haveli and Daman and Diu (DNHDD)	603	Silvassa	Konkani, Gujarati, Hindi, English, Portuguese	3
4.	Delhi	1,483	New Delhi	Hindi, Punjabi, Urdu and English	11
5.	Lakshadweep	30	Kavaratti	Malayalam, Dhivehi	1
6.	Puducherry	490	Pondicherry	Tamil, Telugu, Malayalam, English and French	4
7.	Jammu and Kashmir	1,20,356	Srinagar	Urdu, Kashmiri, Dogri, Pahari, Punjabi	20
8.	Ladakh	96,751	Leh	Ladakhi, Tibetan, Hindi	2

Notes :

(*i*) Andhra Pradesh Government has proposed to have three capitals for the balanced development of its three regions. They are (1) Amravati (Legislative Capital), (2) Vishakhapatnam (Administrative/executive Capital) and (3) Kurnool (Judicial Capital).

(*ii*) Jammu and Kashmir has summer capital at Srinagar and winter capital at Jammu.

(*iii*) Uttarakhand has proposed to have its permanent capital at Gairsain in Chamoli district. Dehra Dun is its temporary capital.

(*iv*) Number of districts and area for the new Union Territories of Jammu-Kashmir and Ladakh are according to Survey of India after scrapping of Article 370. These include PoK also.

(*v*) Number of districts for the remaining States and Union Territories is as per Census of India, lgdirectory.gov.in

Sources :

(*i*) Census of India 2011, Provisional Population Totals. (*ii*) Census of India 2011, Administrative Atlas of India

(*iii*) Parliamentary Proceedings on Article 370.

(*iv*) Parliamentary Proceedings on Dadra and Nagar Haveli and Daman and Diu

(*v*) India 2024; A Reference Annual.

It is clear from **Table 1.3** that Rajasthan with an area of 3,42,239 sq km is the largest state of the Indian Union. In fact Rajasthan, Madhya Pradesh and Maharashtra have area of more than three lakh sq km each and these three states account for about thirty per cent of the total area of the country. On the other end, Goa is the smallest state with an area of 3,702 sq km only. With the exception of Arunachal Pradesh and Assam, most of the north-eastern states are of small size. From administrative point of view, Uttar Pradesh has the largest number of 71 districts while Goa has only two districts. Bihar, Chhattisgarh, Haryana, Himachal Pradesh, Jharkhand, Madhya Pradesh, Uttar Pradesh, Uttarakhand, Rajasthan and NCT of Delhi have Hindi as their main language.

REFERENCES

Bains, J.S. (1962) : *India's International Disputes : A Legal Study,* Asian Publishing House, New York.

Bhashan, A.L. (1994) : *The Wonder That Was India,* Rupa and Co., New Delhi, Chapter 1.

Bhagwan Vishnu (1999) : *Indian Constitution* : An Appraisal, Atma Ram & Sons, New Delhi.

Census of India 2011 : Administrative Atlas of India.

Frankel, E.G. (1979) : *India's Political Economy, 1947-77,* Princeton.

Government of India (1955) : *Report of States Organisation Commission,* New Delhi.

Government of India, (2024) India : *A Reference Annual,* Publication Division, New Delhi.

Gupta, Sisir (1966) : *A Study of India Pakistan Relation,* Asia Publishing House, New Delhi.

Kant, Surya (1988) : *Administrative Geography of India,* Rawat Publications, Jaipur.

Kothari, Rajni (1970) : *Politics in India,* Boston.

Krishnaswamy, A. (1965) : *The Indian Union and the States: A Study in Autonomy & Integration,* Pergamon Press, New York.

Kuriyan, George (1970) : *India : A General Survey,* National Book Trust, India, New Delhi (Chap. 1)

Lamb, Alostair (1964) : *The India-China Border, The Origin of the Disputed Boundaries,* Oxford University Press, New York.

Lamb, Alostair (1968) : *Asian Frontiers : Studies in a Continuing Problem,* Frederick A. Praeger, New York.

Maxwell, N. (1971) : *India's China War,* New York.

Menon, V.P. (1956) : *The Story of the Integration of Indian States,* New York.

Morris-Jones, W.H. (1967) : *The Government and Politics of India,* New York.

Muthiah, S. *et al.* (ed.) (1987) : *A Social and Economic Atlas of India,* Oxford University Press, New Delhi.

Nadkarni, M.V. (ed.) (1991) : *India : The Emerging Challenges,* Bangalore.

Nag Prithvish and Sengupta, Smita (1992) : *Geography of India,* Concept Publishing Company, New Delhi, (Chap. 1).

National Atlas and Thermatic Mapping Organisation (NATMO) (1982) : *National Atlas of India.*

Panikkar, K.M. (1951) : *India and the Indian Ocean : An Essay on the Influence of Sea Power on Indian History,* George Allen and Unwin, London.

Panikkar, K.M. (1964) : *A Survey of Indian History,* Bombay.

Rai, Satya M. (1965) : *Partition of the Punjab,* London.

Raina, A.N. (1971) : *Geography of Kashmir,* National Book Trust, India, New Delhi.

Robinson Francis *et al.* (eds.) (1989) : *The Cambridge Encyclopedia of India, Pakistan, Bangladesh, Sri Lanka,* Cambridge University Press, Cambridge.

Schewartzherg, J.E. (ed.) (1978) : *A Historical Atlas of South Asia,* Chicago.

Singh Harpal Mair and Tiwana, D.S. (2004) : *Geography of Punjab*, National Book Trust, N. Delhi.

Singh, R.L., *et al.* (ed.) (1971) : *India : A Regional Geography,* National Geographical Society of India, Varanasi.

Spate, O.H.K. (1947) : *The Partition of Punjab and Bengal,* The Geographical Journal, Vol. 110, Nos. 4-6, pp. 210-11.

Spate, O.H.K. (1948) : *The Partition of India and the Prospects of Pakistan,* The Geographical Review, Vol. 38.

Srivastava, R.P. (1996) : *The Changing Map of India,* in New Frontiers in Indian Geography, Ed. by R.L. Singh, Prof. R.N. Dubey Foundation, Allahabad.

Sukhwal, B.L. (1985) : *Modern Political Geography of India,* Sterling Publishers, New Delhi.

Tayyeb, A. (1966) : *Pakistan : A Political Geography,* Oxford University Press, London.

Tinker, Hugh (1962) : *India and Pakistan,* New York.

Tirath, Ranjit and Krishan Gopal (2001) : *Geography of India,* Rawat Publications, Jaipur, Chaps. 1 & 8.

Van Eckelen, W.F. (1967) : *Indian Foreign Policy and the Border Dispute,* The Hague.

◆◇◆

CHAPTER 2

Geological Structure

INTRODUCTION

The geological structure, which includes the arrangement and deposition of rocks in the earth's crust, plays a dominant role in determining the relief of land and nature of soil. It also helps in knowing about the vast mineral wealth buried beneath the earth's surface. As such, study of geological structure plays a vital role in agricultural and industrial growth and in the economic prosperity of a country. For example, the vast alluvial Indo-Gangetic plain has very fertile soils and is extremely useful for agriculture. But it is almost completely devoid of any mineral deposits worth the name. On the other hand, igneous and metamorphic rocks of the peninsular plateau, especially those of the Chotanagpur plateau, are very rich in mineral resources. Similarly the study of geological structure helps in land use planning, development of transport and communication lines, increasing potentials for irrigation, determining the quality and quantity of ground water resources and understanding disasters like earthquakes, volcanoes, landslides, floods, etc. Therefore, it is necessary for us to study the geology of India before we proceed to learn more about the geography of the country.

GEOLOGICAL REGIONS OF INDIA

Geologically, India is divided into three regions, *viz.*, (1) *the Peninsular region*, including the Meghalaya plateau in the north-east and the Kuchchh-Kathiawar region in the west; (2) *the extra-peninsular region*—the Himalayas and their eastern extensions including Andaman and Nicobar Islands and (3) *the Indo-Gangetic plain*, between Peninsular and exra-peninsular region.

Although the *triple tectonic division* of India, as mentioned above, is generally held valid and is readily accepted by majority of geologists, some scholars recognise only two geological divisions of India *i.e.* the *Peninsular* block and the *Extra-Peninsular* region comprising the Himalayan ranges and the Indo-Gangetic plain. These macro-regions present most striking geological contrasts. The Peninsula is one of the oldest land-masses of the earth and is dominated by open senile topography. The Extra-Peninsula, on the other hand, presents the most youthful relief of the earth in the form of the Himalayas. The alluvium filled Indo-Gangetic plain presents flat, featurless and monotonous topography. The Indian Peninsula has not undergone marine submergence since the *Cambrian period* and is not much affected by the tectonic forces. In contrast, the extra-peninusLar region has its origin in the Tethys Sea and is prone to tectonic forces resulting in devastating earthquakes. It is a weak and flexible portion of the earth's crust which has been folded, faulted and overthrust.

GEOLOGICAL HISTORY OF INDIA

Geologically India represents a monumental assemblage of rocks of different character belonging to different ages, ranging from pre-Cambrian (prior to 600 million years) to the recent times. **Sir T. Holland** of the *Geological Survey* of India has classified the rock systems of the country into following four major divisions :

1. The Archaean Rock System.
2. The Purana Rock System.
3. The Dravidian Rock System.
4. The Aryan Rock System.

The major geological systems of India are shown in **Fig. 2.1**.

1. THE ARCHAEAN ROCK SYSTEM

The word 'Archaean' was first used by **J.D. Dana** for rock structure prior to the Cambrian system. Obviously the Archaean rocks are the oldest in the world. The Archaean rock system includes the following rock groups :

(*a*) The Archaean System-Gneisses and Schists. These are the oldest rocks and were the first to be formed at the time of cooling and solidification of the upper crust of the earth's surface in the pre-Cambrian era (about 4000 million years ago). Although their broad characteristics are well known, the details of their origin continue to arouse considerable speculation. They are all azoic, thoroughly crystalline, extremely contorted and faulted, often formed as plutonic intrusions and generally have a well defined foliated structure. They often underlie strata formed subsequently and the system is generally known by names of the ***'Fundamental Complex'*** or the ***'Basement Complex'***. Their details are complex but recent advances in geochemistry are likely to enhance our understanding of their formation.

The most common Archaean rock covering about two-thirds of the Peninsular surface is gneiss. This is the rock which, in mineral composition, may vary from granite to gabro, possesses a constant, more or less foliated or banded structure. The schists, mostly crystalline, include mica, talc, hornblende, chlorite, epidote sillimanite and graphite schists.

The crystalline metamorphosed sediments and gneissic rocks cover large parts of India. The central and the southern parts of the Peninsula are occupied by this rock system. To the north-east of the Peninsula, they occupy wide areas in Odisha, Meghalaya, Madhya Pradesh, Chhattisgarh and Chotanagpur plateau of Jharkhand. They also cover the whole of Bundelkhand in the north and to the north-west, they are found in a number of isolated outcrops, extending from north of Vadodara to a long distance along the Aravalis. In the extra-peninsula, these rocks are exposed all along the Himalayas, forming the bulk of the high ranges and the backbone of the mountain system (**Fig. 2.2**).

(*b*) The Dharwar System. This system derives its name from the rocks first studied in the Dharwar district of Karnataka where such rocks are found in abundance. The Dharwars include some of the highly metamorphosed rocks of both sedimentary and igneous origin. According to **Wadia**, the Dharwar System is the most ancient metamorphosed sedimentary rock-system of India, as old as, and in some cases older than, the basement gneisses and schists. He further adds that the weathering of the pristine Archaean gneisses and schists yielded the earliest sediments which were deposited on the bed of the sea, and formed the oldest sedimentary strata, known in the geology of India as the Dharwar system. Some of the metamorphosed rocks of the igneous origin are also included in this system. Most rocks of the Dharwar system are so metamorphosed that they are practically indistinguishable from their primitive formations. The major rocks of the Dharwar system are hornblende, schists, quartzites, phyllites, slates, crystalline limestones and dolomites. These rocks were deposited in three major cycles, the earliest one is over 3,100 million years old and the latest one about 2,300 million years. They were metamorphosed around 1,000 million years ago.

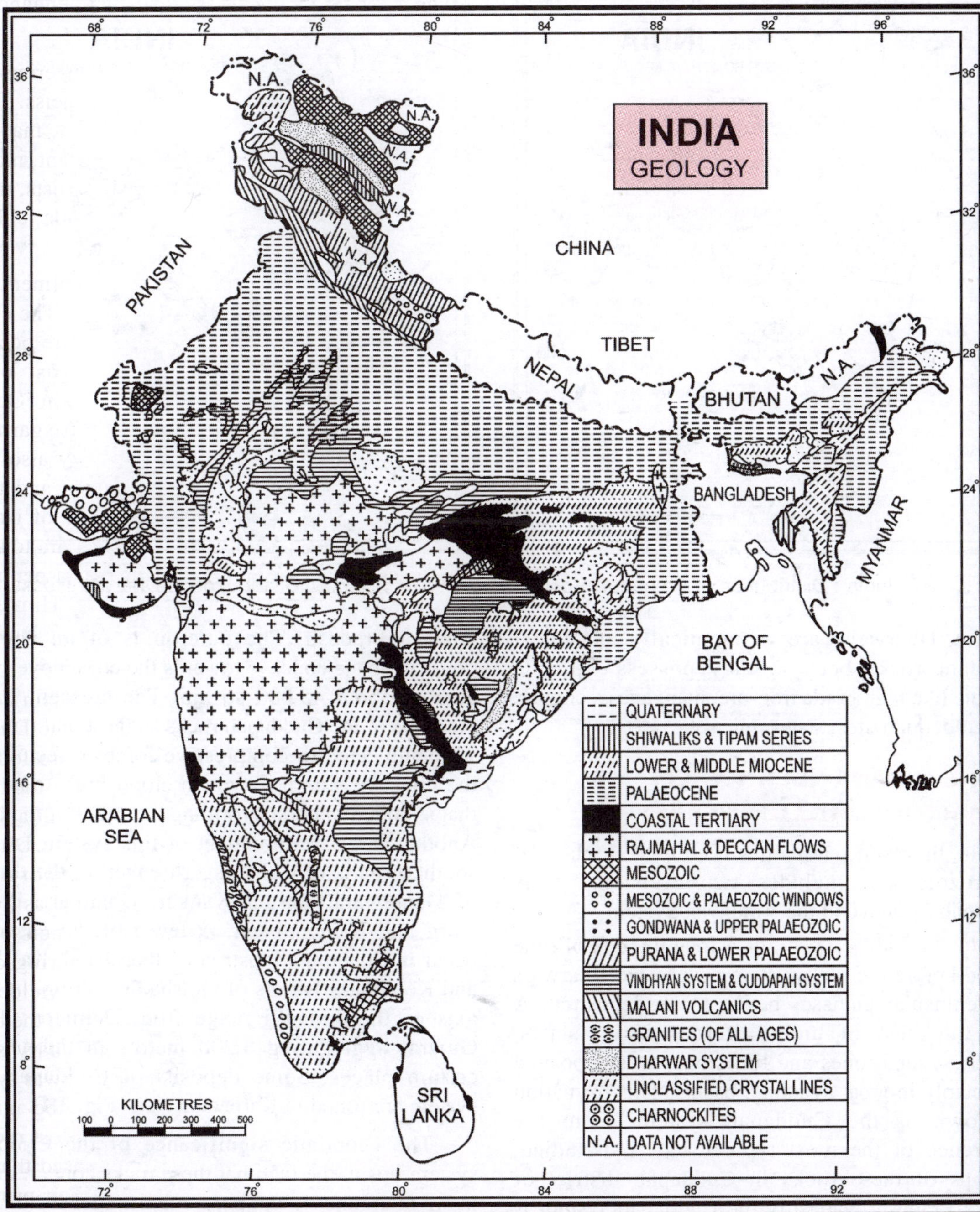

FIG. 2.1. India : Geology

The Dharwar system is very well developed in the Dharwar-Bellary-Mysore belt of Karnataka. It also occurs in Jharkhand (Ranchi, Hazaribagh), Madhya Pradesh (Balaghat, Rewa), Chhatisgarh (Bastar, Dantewara, Kanker), Odisha (Sundergarh, Keonjhar) and in the Aravalis between Jaipur and Palanpur. In the extra-Peninsular region, the Dharwar system is well represented in the Himalayas both in the c entral and northern zones as well as in the Meghalaya plateau (**Fig. 2.3**).

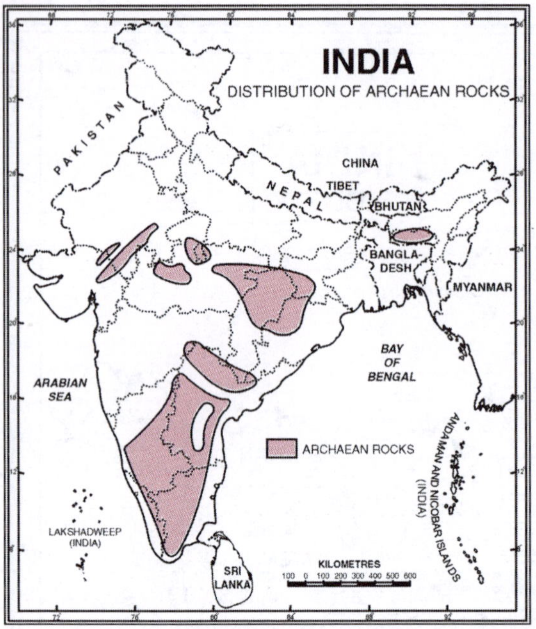

FIG. 2.2. India : Distribution of Archaeon Rocks

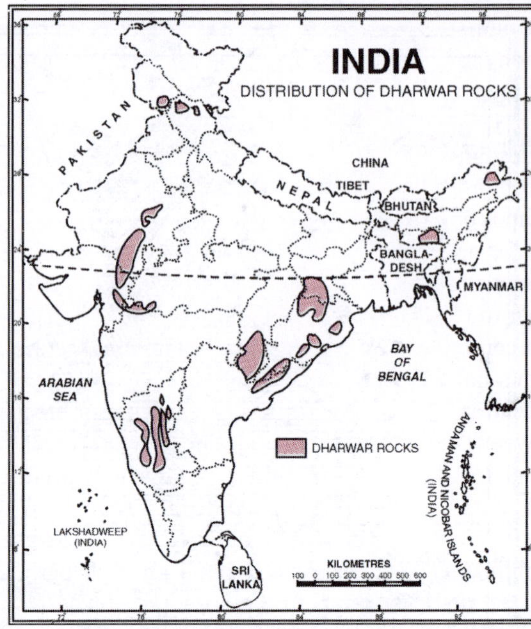

FIG. 2.3. India : Distribution of Dharwar Rocks

The Dharwars are economically the most important rocks because they possess valuable minerals like high grade iron-ore, manganese, copper, lead, gold, quartzites, slates, mica, etc.

2. THE PURANA ROCK SYSTEM (1400–600 MILLION YEARS)

In india, the word purana has been used in place of Proterozoic and includes two divisions : the Cuddapah System and the Vindhyan System.

The Cuddapah System. A long interval of time elapsed before the rock system next to the Dharwars and Peninsular gneisses began to be deposited. A great thickness of unfossiliferous clays, slates, quartzites, sandstones and limestones was deposited presumably in great synclinal basins. This formation is known as the Cuddapah system, from the occurrence of the most typical and first-studied, outcrops of these rocks in Cuddapah district of Andhra Pradesh. Naturally the Cuddapah system is separated from the Dharwar system by a great unconformity. In some parts of southern India, we have the Cuddapah system amounting to 6,100 metres in thickness with several unconformities.

The most extensive occurrence of this system is in Cuddapah district followed by Kurnool district of Andhra Pradesh. The outcrop is of an irregular crescent shape concave towards the coast covering an area of about 45,000 sq km. The crescent shaped outcrops of the Cuddapah rocks suggest that the rocks were subjected to compressive forces directed from the concave side near which stood high mountains that supplied materials forming the rocks of the basin. Another large development of this system is in the southern parts of Chhattisgarh covering the districts of Dantewara, Bastar, Kanker, Dhamtari, Raipur, Durg and Rajnandgaon. A few isolated exposures occur in Singhbhum district of Jharkhand, Kalahandi and Keonjhar districts of Odisha and along the main axis of the Aravalli range from Delhi to Idar in Gujarat aggregating 5,200 metres in thickness at certain places. Some deposits of Cuddapah rock system are found in Karnataka also (**Fig. 2.4**).

The economic significance of the Cuddapah system lies in the fact that these rocks contain ores of iron, manganese, copper, cobalt, nickel, barytes, jasper, asbestos, steatite and cherts. They also contain large deposits of building purpose quartzites and cement grade limestones.

The Vindhyan System (1300-600 million years). This system derives its name from the great Vindhyan mountains although ***Spate*** has tried to distinguish between the Vindhyan rocks and the Vindhyan hills. The system comprises of ancient sedimentary rocks superimposed on the Archaean base. It is a vast stratified formation of sandstones, shales and limestones, often over 4000 m thick. Except a few traces of animal and vegetable life, this group is devoid of any recognisable fossils. Occupying a large area of over 1,00,000 sq km, the Vindhyan system stretches from Sasaram and Rohtas in western Bihar to Chittaurgarh in Rajasthan with the exception of a central tract in Bundelkhand which makes a gap in this belt. Large area of this belt is covered by the Deccan trap. The outcrop has the maximum breadth between Agra and Neemuch. These rocks are also found in Chhattisgarh, Bhima Valley of Karnataka and Kurnool district of Andhra Pradesh. The Vindhyan system has been found to continue to the north under the Gangetic alluvium and it is perhaps buckled down underneath the Himalayas. The rocks of the Vindhyan system comprise two distinct but unequal sets of deposits. The *Lower Vindhyan* (1,300-1,100 million years) is marine in origin, mostly calcareous in nature and shows tectonic deformation by folding movements. This system is well placed in the Son valley, in Chhattisgarh, in the valley of the Bhima and in a separate basin in Mewar. *The upper Vindhayan system* (1000-600 million years), on the other hand, is fluviatile in origin and is gently lying in undisturbed horizontal strata. The Lower and the Upper Vindhyans are separated by an unconformity which is quite prominent in the north but almost disappears in southern areas of Mewar, Chittaurgarh and the Son valley.

The Upper Vindhayan beds enclose two diamond bearing horizons, from which Panna and Golconda diamonds have been mined. The Vindhyan system, on the whole, is devoid of metalliferous minerals but provides large quantities of excellent and durable stones, flagstones, ornamental stones, limestone, pure glass making sand and some coal.

3. THE DRAVIDIAN ROCK SYSTEM (PALAEOZOIC)

The rocks of the Dravidian system came into being about 600-300 million years ago. Most rocks of this system are found in the Extra Peninsular region and they are conspicuous by their absence in the Peninsular India except for one or two small patches of lower Permian age near Umaria. The Cambrian and Devonian rocks containing recognisable fossils are unknown in the Peninsula. The rocks belonging to the Dravidian System contain abundant fossils which help in determining correctly the age of the rocks and make correlation of rocks possible over distant areas.

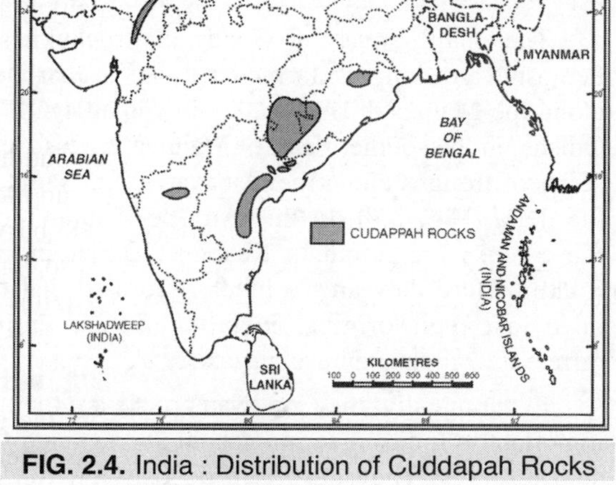

FIG. 2.4. India : Distribution of Cuddapah Rocks

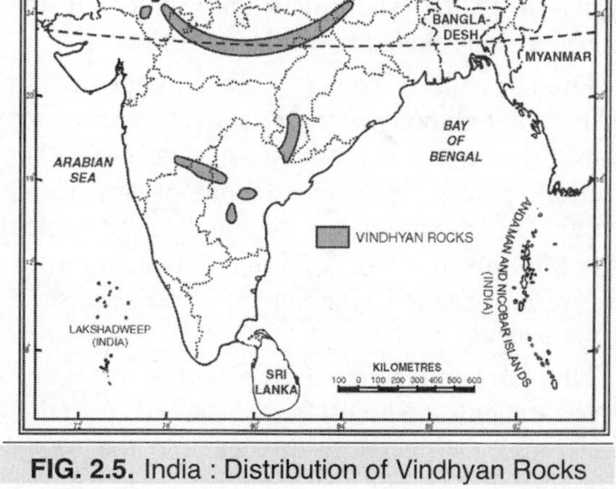

FIG. 2.5. India : Distribution of Vindhyan Rocks

The rocks of Cambrian, Ordovician, Silurian, Devonian and Carboniferous periods are included in the Dravidian system.

The Cambrian rocks (600 million years) named after Cambria, the Latin name for Wales in Great Britain, include slates, clays, quartzites and limestones. They are best developed in the north west Himalayan region. In the Spiti valley of Himachal Pradesh, there is an extensive fauna known as the ***Haimanta System.*** This 1600 metre thick deposit consists of slates, quartzites and dolomites. Similar formations are found north of Kullu and Lahul in Himachal Pradesh as well as in the Baramula district of Jammu and Kashmir. In Kumaon region of Uttarakhand, some slaty and sandy rocks of the Cambrian period are found. In the salt range on the Indo-Pak border, the Cambrian rocks are represented by 900 metres of fossiliferous sandstones, shales and dolomites underlain by salt marl (clayey salt) known as the ***saline series***.

The Ordovician rocks (500 million years) after the Ordovices, former inhabitants of Wales include quartzites, grits, sandstones and limestones. They overlie the Haimanta system in all parts of the Spiti in the form of a thick series underlain by conglomerates. They are also present in the Lidar valley of Kashmir and in the Kumaon region.

The Silurian rocks (440 million years) after the Silures, former inhabitants on the borders of Wales and England. In the Spiti valley, the Silurian rocks are in continuation with the Ordovician rocks. Round the core of the Lidar anticline there runs a thin but continuous band of Silurian strata. The Lahul and the Kullu valleys of Himachal Pradesh also have some Silurian deposits. The limes and shales of the Kumaon region belong to the Silurian period.

The Devonian rocks after Devonshire in England (400 million years) are a great thickness of massive white quartzite reaching a thickness of 900 m at certain places. They are devoid of any fossil remains. These rocks have definitely been identified in the Muth quartzite of Spiti and Kumaon, on the flanks of Lidar anticline and in the Haridwar district of Uttarakhand.

The Carboniferous rocks (350 million years) comprise mainly of limestone, shale and quartzite. These rocks are generally divided into the *Upper Carboniferous, Middle Carboniferous* and *Lower Carboniferous* systems. *The Upper Carboniferous* rocks are made of limestone and dolomite. *Mount Everest* is composed of Upper Carboniferous limestones. *The Middle Carboniferous* has been the age of great upheavels. The rocks of this group are mainly found in the Spiti valley, Kashmir, Shimla and in the eastern Himalayas. In the Lower *Carboniferous group* are included slates of different types, Pir Panjal trap, and some rocks of the Kumaon region. Coal formation started in the Carboniferous age. Carboniferous in geology means *coal bearing*.

4. THE ARYAN ROCK SYSTEM

The Aryan System comprises the rock formations ranging from the *Upper Carboniferous* to the *Recent*. They are fairly preserved in the Peninsular India and are found in a perfect sequence in the Himalayan region along the entire northern border.

The Gondwana System derives its name from the Kingdom of the Gonds, the most primitive people living in Telangana and Andhra Pradesh. It has relevance with the Gond region of Madhya Pradesh also where these rocks were first discovered. The Gondwanas consist of sandstones with some shales and clays. They are of continental origin, fluviatile and lacustrine deposits laid down in geosynclinal troughs on ancient plateau surface. As the sediments accumulated, the loaded troughs subsided which led to thick deposits of fresh water and subaerial sediments into which were embedded the terrestrial plants and animals. These flat sedimentary strata, some 6,000 m thick, were laid down from the start of the Permian period some 250 million years ago. The system shows several climatic changes during its deposition.

The main areas of Gondwana rocks in the Peninsula are along the Damodar Valley in Jharkhand, along the Mahanadi river valley in Chhattisgarh and Odisha, in the southern part of Madhya Pradesh and a series of troughs along the Godavari from Nagpur to the delta (**Fig. 2.6**). In the extra-peninsular region these rocks are found in Kashmir, Darjeeling and Sikkim. Here they are so much deformed that they have lost their original identity and are entirely different from the peninsular rocks.

Economically, the Gondwana rocks are the most important in India containing about 98 per cent of her coal reserves. They have rich deposits of iron ore, copper, uranium and antimony also. Sandstones, slates and conglomerates are used as building materials.

The Triassic System meaning three fold (280-225 million years) is almost unknown in the Peninsula but is found over extensive areas from Hazara in Pakistan to Nepal. Impressive sections of the system are exposed on the south flank of the Great Himalaya range from Kashmir to Byans in Eastern Kumaon. The trias are generally divided into lower, middle and upper divisions. *The Lower Trias* is over 100 m thick in Kashmir, 50 m in Byans and only 12 m in Spiti. *The Middle Trias* is 275 m thick in Kashmir. *The Upper Trias* of Kashmir are 1200-1800 m thick well bedded limestones. In Spiti both these stages are represented by about 1,000 m thick dominantly calcareous strata.

The Jurassic System, after Jura mountains on the borders of France and Switzerland (180 million years) overlies the Triassic, covering wide areas in Tibet, South Ladakh, Spiti, Nepal and Bhutan, where limestone occurs to a depth of 600-900 m. The marine transgression in the latter part of the Jurassic gave rise to thick series of shallow water deposits in Rajasthan and in Kachchh. There is 190 km long and 64 km wide area in Kuchchh which is covered by rocks of the Jurassic system. Coral limestone, oolitic limestone, sandstone, conglomerates and shales occur in Kuchchh. The Jaisalmer area of Rajasthan also has some Jurassic rocks. Another transgression on the east coast of the Peninsula between Guntur and Rajahmundry has resulted in the development of marine Upper Jurassic strata interbedded with the upper Gondwana formation.

The Cretaceous System, from Creta, the Latin name for chalk (135-70 million years), is one of the best developed marine systems of India showing a variety of faces and represented by a wide variety of rocks, deposited on the land, sea estuaries and lakes. No other system is so widely distributed in India as the Cretaceous System is, both in the Peninsular and the Extra-peninsular parts.

The Cretaceous rocks of the Spiti area include sandstones, quartzites, limestones and shales. In the Kumoun region, the important rocks are sandstones and shales. Cretaceous limestones are also found in Rupshu and Burzil areas of Ladakh. The plateau of Meghalaya has sandstones. The upper Cretaceous system occurs in the Puducherry-Tiruchchirapalli belt. Some detached outcrops of marine Cretaceous, known as Bagh Beds in Gwalior are found along the Narmada Valley where they underlie the Deccan Trap. In the Central parts of the Peninsula occur estuarine and lacustrine deposits called the ***Lemetas*.**

The Deccan Trap. From the end of the Cretaceous till the beginning of the Eocene, stupendous volcanic outburst overwhelmed a vast area of the Peninsular India, like the one which is seldom known anywhere else in the world geological history. A vast area of about ten lakh sq km was flooded by the outpourings of extremely mobile basalt lava from fissures and cracks covering fully the pre-existing topography. These volcanic deposits have flat tops and steep sides so that they appear as gigantic steps from a distance, and are therefore called ***'trap'***, the name derived from the Swedish word meaning a '***stair'*** or '***step'***. However, **Spate** feels that the term Deccan Trap is used in old literature and he prefers to use the term ***'Deccan Lava'*** instead. The individual lava flows that make up the Deccan Trap plateau vary greatly from a fraction of a metre to 36 metres in thickness. The hills formed by them are in some places over 1200 metres high.

The process of denudation over a long period has reduced the Deccan Trap to almost half of its original size and the present Deccan Trap covers about 5 lakh sq km mainly in parts of Kachchh, Saurashtra, Maharashtra, the Malwa plateau and northern Karnataka (**Fig. 2.7**). Parts of Telangana, Tamil Nadu, Jharkhand and Uttar Pradesh also have some outliers of the Deccan Trap. Thickness of the Deccan Trap is not uniform everywhere. It is as much as 3,000 metres along the coast of Mumbai which is reduced to 600-800 metres towards the southern limit, 800 metres in Kachchh and only 150 metres at Amarkantak near its eastern limit. The Deccan Trap has been divided into following groups :

(*a*) The Upper Trap, found in Maharashtra and Saurashtra has numerous inter-trappean beds and layers of volcanic ash and has an average thickness of about 450 m.

(*b*) The Middle Trap is about 1200 m thick and is spread over Central India and Malwa. It has numerous ash-beds in the upper portion and is practically devoid of inter-trappeans.

(*c*) The Lower Trap is only 150 m thick with inter trappean beds but rare ash beds. It is found in Central India and Tamil Nadu.

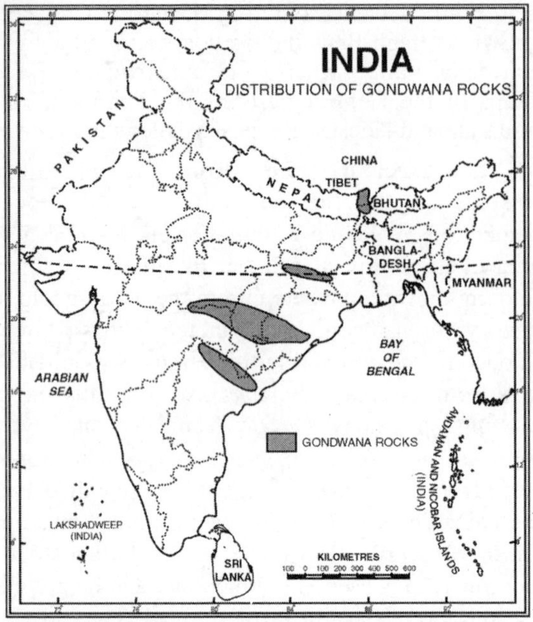

FIG. 2.6. India : Distribution of Gondwana Rocks

FIG. 2.7. India : Deccan Trap

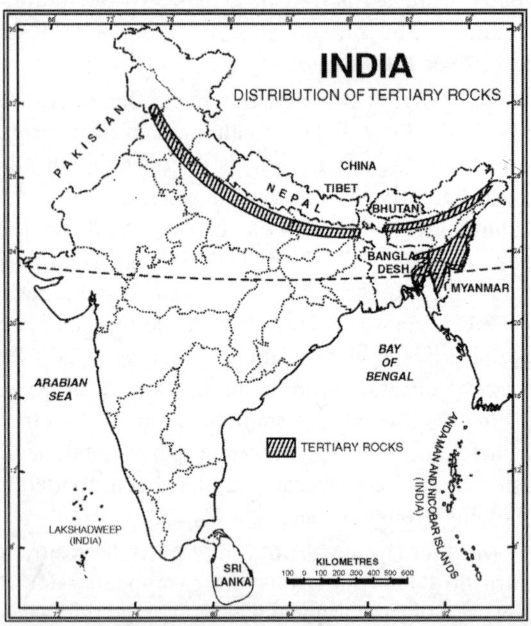

FIG. 2.8. India : Distribution of Tertiary Rocks

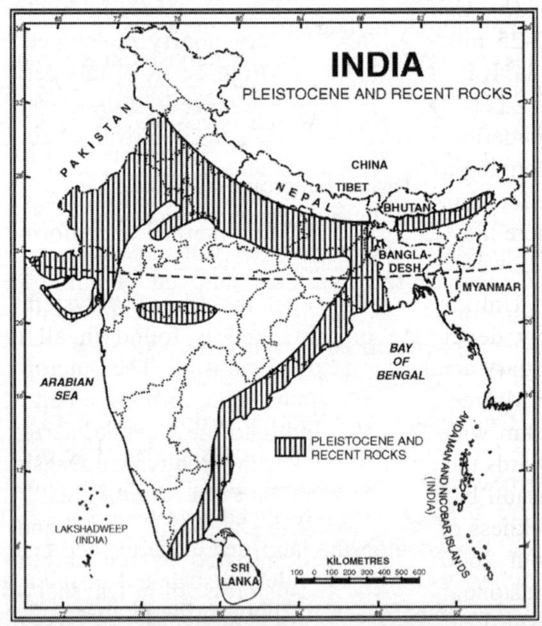

FIG. 2.9. India : Pleistocene and Recent Rocks

Basalt is the main rock but dolorite, rhyolite, gabro and many other rocks are also found. These rocks are a great source of quartz, agate, calcite, building stones and road building material. The weathering of these rocks for a long time has given birth to black cotton soil known as ***'regur'***.

The Tertiary System. The rocks of the Tertiary System were formed from Eocene to Pliocene about 60 to 7 million years ago. The Tertiary is the most significant period in India's geological history because the Himalayas were born and India's present form came into being in this period. The Tertiary has been called the ***Age of the Mammals*** because of the abundance of the fossil remains of these animals in the deposits of this period. The Tertiary System is generally divided into the following three systems.

(*a*) The Eocene System (60 million years ago) is mainly found in Jammu and Kashmir, Himachal Pradesh, Rajasthan, Gujarat and in the north-eastern part of India. The Eocene belt of Jammu goes upto Shimla and further east upto Mussoorie. Such deposits also occur in the southern flank of the Pir Panjal. In Rajasthan, the lignite deposits belong to the Eocene System. The Kuchchh, Surat and Broach areas of Gujarat have deposits of the Eocene System. In the north eastern part of the country, the Eocene is represented by limestones and coal-bearing sandstones in Jaintia series in the southern and eastern parts of the Meghalaya plateau. The Barail series has a wide distribution in Surma Valley and in the Naga Hills. The Tertiary coal deposits of Upper Assam are worth noting. Some Eocene deposits have recently been noticed in Puducherry.

(*b*) Oligocene and Lower Miocene System (40-25 million years) is very poorly developed in India. It is estimated that during a part of this period the Tertiary outcrops suffered considerable denudation which resulted in the removal of rocks belonging to this system. Rocks of the Oligocene age are found in greater part of the Barail series of Assam where they are overlain with a marked unconformity by Lower Miocene rocks.

Unlike the Oligocene, the Miocene system is fully developed in India and is found in all the Tertiary areas of the extra-Peninsula. The outcrop of the Murree series in Jammu hills, forming a belt 25-40 km wide from the Indus to the Chenab, narrows towards the east and merges into the typical Dagshai-Kasauli band (Sirmur series) is of Shimla area. The thickness of the deposits in Dagshai-Kasauli band is about 2600 m and is dominated by red clay and sandstone. No major fossils are seen in Dagshai and Kasauli group. The coal measures of Assam, belonging to the Barail series are overlain by the Surma series. It is composed of sandstones and sandy shales, mudstones and conglomerates. The deposits of Garo Hills have large number of marine fossils. Besides some rocks of Kachchh in Gujarat, Mayurbhanj in Odisha, Durgapur in West Bengal, and Kollam in Kerala also belong to this group.

The Shiwalik System—Middle Miocene to Lower Pleistocene (14 to 0.2 million years ago) takes its name from Shiwalik hills between rivers Ganga and Yamuna, where they were first known to science. It is from here the first palaeontological treasures were obtained which have made this system extremely famous with the geologists all over the world. The Shiwalik strata are found all along the length of the Shiwalik hills. Sandstones, grits, conglomerates, clays, and silts comprise the rocks of this system. They have been deposited in lagoons and fresh water lakes by the rivers of that time upto a thickness of about 5000 m. The great bulk of the Shiwalik formation is nonfossiliferous, but in certain areas some formations are highly fossiliferous. The fauna records the sedimentation from Middle Miocene to Lower Pleistocene and has yielded a variety of fossils showing wide range of environment—from humid forest conditions to aridity. The mammalian fossils were first discovered from the Shiwaliks near Haridwar in 1839.

TABLE 2.1. Division of Shiwalik Strata

Upper Shiwaliks	Boulder Conglomerate beds, Sandstones, thick earthy clays. Richly fossiliferous in Shiwalik hills	Lower Pleistocene to Upper and Lower Pliocene
Middle Shiwaliks	Gravels, sandstones, shales and clays	Upper to Middle Miocene
Lower Shiwaliks	Bright red shales and sandstones	Middle Miocene

The Shiwalik strata are divided into three sections, the passage from one into the other division being quite gradual and transitional.

In Assam the Shiwalik system is almost equivalent to the Tipam, Dupi Tila and Dihing series. In the southern part of Assam, the lower part of the Tipam series consists of coarse ferruginous sandstones and sandy shales, and has some marine fossils. Above the Tipam series lies the Dupi Tila series which consists of sandstone and conglomerates. Its lower beds contain the oil fields of Lakhimpur and Digboi. The Dihing beds overlie the Dupi Tila and correspond to the Upper Shiwaliks.

The Shiwalik rocks are also found in Kachchh and Saurashtra areas of Gujarat. Highly fossiliferous beds have been struck in a well at Karikal on the Thanjavur coast (Tamil Nadu). In Mayurbhanj area of Odisha, there are rocks of Miocene age. Some rocks of Middle Miocene and Lower Pleistocene age are seen in Varkala (Kerala) as well as in the sandstones of Cuddalore and Rajahmundry.

The Shiwalik system is famous for its oil resources, lignite, bauxite and clays of different varieties.

The Pleistocene and the Recent (Quaternary) is a brief period of nearly one million years and is said to have just begun. It has two divisions without a clear-cut boundary between them. The older is the Pleistocene which is marked by cold climate and glaciation. The younger division is called ***Recent***. It started about 12,000 years ago since the withdrawal of the last glaciation. The Pleistocene age has been marked by several advances and retreats of the glaciers. This age is divided into four glacial and three inter-glacial periods by ***de Terra*** and ***Paterson***. During the glacial periods, sea level was lowered by about 100 metres than it is now and this resulted in much larger extension of land.

The Himalayan glaciers were much larger during the glacial period than their present size. They have left indelible impression on the Himalayan topography. The First Glacial and Inter-glacial periods belong to the lower Pleistocene in Kashmir. The Second Glacial and Inter-glacial periods embrace Middle Pleistocene. Heavy deposition of sediment consisting of boulder fans and thick fluvio-glacial material took place in this period. The *Karewas* of Kashmir are supposed to be of the second inter-glacial period. The flat-topped terraces of the Kashmir valley, and on the flanks of the Pir Panjal consisting of clays, sands, silts together with lenticles of conglomerate from old deltaic fans are known as ***Karewas*** in Kashmiri language. The alluvial deposits of the Narmada are also of the same period. The Upper Pleistocene includes the Third and the Fourth glaciations and the intervening Third Inter-Glacial period. Pir Panjal experienced uplift during the First and the Third Glacial periods, and the terrace deposits of Kashmir occurred at the end of the Fourth Glaciation.

The fossiferous clays, sands and gravels of the Upper Satluj and the alluvial deposits in the valleys of Tapi, Godavari and Krishna are also of the Pleistocene age.

The most important, extensive and recent deposits in India are the Indo-Gangetic alluviums filling the great depression between the foot of the Himalayas and the northern edge of the Peninsula. The older alluvium called ***bhangar*** is of the Middle or Upper Pleistocene age. The newer alluvium occupies the lower areas of the river valleys prone to annual floods is called ***khadar*** and belongs to the Upper Pleistocene age.

THE COASTLINE OF INDIA

India has about 6100 km long coast most of which is more or less uniform and regular with a few creeks, inlets, back waters and promontories here and there. *Suess* has opined that the straight and regular coastline of India is the result of disruption and faulting of the Gondwanaland during the Cretaceous period. As such the coast of India does not offer many sites for good natural harbours. The peninsular shape of South India has divided the entire coast into western and eastern parts with Kanniyakumari at the southernmost tip as the dividing point. The Bay of Bengal and the Arabian Sea came into being during the Cretaceous or early Tertiary period after the disintegration of *Gondwanaland*.

The east coast of India extends from the Ganga delta to Kanniyakumari facing the Bay of Bengal. It is marked by deltas of great rivers like the Mahanadi, the Godavari, the Krishna and the Cauvery. There are several lagoons of which the Chilka lake and the Pulicat lagoon are outstanding.

The West Coast runs almost straight in the north-south direction right from the Gujarat plains to Kanniyakumari. However, it is dotted with a large number of coves, creeks and a few estuaries. The estuaries of the Narmada and the Tapi are well known. The Malabar coast in the south has some lakes, lagoons and backwaters, the largest being the Vembanad Lake.

There is evidence to show that large parts of Indian coast have undergone submergence and emergence during the geological times and even during the historical times. The most outstanding example of submergence of the sea coast is the existence of a submerged forest on the eastern side of the island of Mumbai. It is located at depths of 6-12 m below the present sea level where a number of tree-stumps are seen with their roots *in situ* embedded in the old soil. The submergence is further proved by the raised terraces formed of coral reefs or loosely cemented fragmentary shell limestone rocks. A similar submerged forest or old land surface, about half an acre, is seen on the Tinnevelli coast. A thick bed of lignite found at Puducherry 73 m below ground level also provides proof of submergence. Such vegetable debris is also found in the Ganga delta. Sudden increase in the depth of sea about 30 km from the coast of Makran in Pakistan is due to the submergence of a cliff lying at the coast. Marine archaeologists have recently discovered the ancient city of Dwarka lying under shallow waters off the coast of Saurashtra.

Examples of both submergence and emergence have been found in the Rann of Kachchh. As a result of the great earthquake in 1819, an area of 5180 sq km on the western coast of the Rann of Kachchh was suddenly submerged under the sea upto a depth of 3.6 to 4.5 m. The fort of Sindree on the sea coast was completely submerged excepting a single turret which remained above the water level. Simultaneously another area of about 1550 sq km was elevated by several metres. In fact there have been repeated falls in sea level in the Rann of Kachchh area which is evident from the presence of coastal dunes, sea cliffs in the islands of the Gulf, sea caves within the plains and raised beaches of littoral concrete on the fringes of the islands. The total fall in the sea level has been estimated at 26 m. Even during historic times the Rann of Kachchh was a gulf which was silted up by the rivers falling into it. This process was supplemented by the slow elevation of the floor of the gulf and the entire area was converted into a low-lying tract. Before the Rann became dry, the Luni river used to enter the Arabian Sea through the Kori estuary crossing the entire length of the great Rann, and the Western Banas flowed over the Little Rann into the Gulf of Kachchh. Some islands off the Arabian Sea coast came into existence due to submergence of low-lying areas around them. The coral archipelago of the Lakshadweep islands probably marks the site of submerged land.

A large number of evidences of uplift of the western coast have been given by the geologists. Raised beaches are found at altitudes ranging between 30-45 metres at many places, while marine shells are found at many places on land at a height far above the level of tides. The steep face of the Sahyadri parallel to the west coast of India, suggests that the escarpment is the result of a recent elevation of the Ghats from the sea.

Proofs of uplift of the coastal plain of Kerala are numerous and decisive. The occurrence of coral reefs below the alluvium several kilometres from the present coast and the presence of a number of lakes, lagoons, backwaters and spits are a few of them. At least two phases of uplift have been noticed.

Away from the coast in the Bay of Bengal the Andaman and Nicobar Islands were once connected with the Arakan Yoma. They have come into being as a result of submergence of the surrounding low-lying areas in the geological times. However, the shape and size of some of the islands near the Ganga delta change with the change in sea level. Old islands disappear and some new islands emerge above the sea level. The recent emergence of New Moore Island in the Ganga delta is an outstanding example of such changes.

Sri Lanka was a part of India in the geological times, and the two landmasses were connected with each other by Adam's Bridge. This link was drowned in the Quarternary times and is now represented by Pamban Island on Indian side and the Mannar Island on the Sri Lankan side along with several other small islands lying between these two islands.

Ice Ages in India

Following three ice ages are generally recognised in India.

1. The Dharwar Ice Age. Ice age of Dharwar is reflected by conglomerates near Kaldurg in South India.

2. The Gondwana Ice Age. The Talcher series of the Gondwana system in Odisha provide proof of the glaciation during that period. Boulders found here are similar to those existing in Shimla, Hazara and Salt Range.

3. The Pleistocene Ice Age. During the Pleistocene period the effect of ice age was noticed in the Himalayas. This period was not a continuous frigid spell, rather it was marked by cold and warm spells in succession. In the Himalayas, evidence of extensive glaciation is found upto height of 1800 m while glacial drifts and terminal moraines covered hill sides and valley floors down to 1,400 m. In Kashmir region, four or five periods of glaciation with inter-glacial periods have been noticed by **de Terra** and **Peterson.** Large blocks and boulders are normally seen in different parts of the Himalayas. Rock polishing and grooving on lower steps of the Himalayas, the buff coloured sands and luminated clays inter-stratified among the Karewa deposits of Kashmir, accumulations of marine debris found on the tops and sides of many ranges of the middle Himalayas which do not have any glaciers now, ancient moraines found before the snouts of existing glacier at an elevation of about 1,650 m amply prove that this part of India experienced a glacial age in the Pleistocene period. Glaciation also led to the formation of a number of glacial lakes in different parts of the Himalayas. Kailash kund, the Sanasar basin near Batoti, the Gulmarg basin, the Sheshnag, the Kaunsarmag, etc. are some of the examples of this type of lakes. The impact of glaciation also led to sudden and large scale reduction of the Shiwalik mammals. The Peninsular part of India has no evidence of glaciation during the Pleistocene period.

REFERENCES

Ahmad, E. (1962) : *Geomorphic Regions of Peninsular India,* Journal of Ranchi University, Vol. I Nov. 1, pp. 1-29.

Betz, F. Jr. (Ed.) (1975) : *Environmental Geology,* Halsted, Stroulsburg (Pennsylvania).

Bhattacharya, S. (1969) : *Geography and Geology of Assam,* National Book Trust, India, New Delhi.

Burrard, S.G., H.H. Hayden, A.M. Heron (1933) : *A Sketch of Geology and Geography of the Himalaya Mountains and Tibet,* Part IX, Govt. of India Publication, New Delhi.

Coates, D.R. (1981) : *Environmental Geology,* John Wiley, New York.

Chatterjee, S.P. (1941) : *Gneissic Topography of Ranchi Plateau,* Calcutta Geographical Review, No. 1, pp. 45-48.

Chugh, R.S. and J.C. Bhattacharji (1974) : *Study of Isostasy in the Himalayan Region, Himalayan Geology*, Wadia Institute of Himalayan Geology, Delhi, Vol. IV pp. 453-64.

Davis, S.N., Reiton, P.H. and Pestrong, P. (1976) : *Geology and Our Environment,* McGraw Hill, New York.

de Terra, H. (1939) : *Studies on the Ice Age in India,* Carn. Inst. Washington, No. 493.

Dey, A.K. (1968) : *Geology of India,* National Book Trust, India, New Delhi.

Gansser, A. (1964) : *Geology of the Himalaya,* John Wiley and Sons, London.

Geological Survey of India (1976) : '*Geology of the Himalaya Appraisal of Status of Definition of Problems*'. Key Paper to Section I. Himalayan Geology Seminar, p. 48.

Haward, A.D. and Renson, I. (1978) : *Geology in Environmental Planning,* McGraw Hill, New York.

Jhingran, A.G. (1981) : *Geology of the Himalayas,* Oxford University Press, New Delhi.

Krishnan, M.S. (1968) : *Geology of India and Burma*, Higginbothams Ltd. Madras.

Krishnan, M.S. (1982) : *Geology of India and Burma,* CAS Publishers and Distributors, Delhi.

Kumar, R. (1983) : *Historical Geology and Stratigraphy of India,* Wiley Eastern Ltd. New Delhi.

Liverson, Davis (1980) : *Geology and Urban Environment,* Oxford University Press, New York.

Mathur, M.S. (2004) : *Physical Geology of India,* National Book Trust, India, New Delhi.

Mears, B., Jr. (1970) : *Essentials of Geology :* Van Nortrand, New York.

Mahdiratta, R.C. (1962) : *Geology of India, Pakistan and Burma,* Atma Ram and Sons, Delhi.

Mukerji, A.B. (1964), "*Alluvial Morphology of the Upper Ganga Yamuna Doab*", Deccan Geographer, 126, Vol. 2, No. 2, pp. 100-126.

Pardhasaradhi, Y.J. and Vaidhyanathan R. (1974) : Evolution of Landforms over the Nilgiri, South India, *Journal of the Geological Society of India,* Vol. 15 : No. 2, pp. 182-188.

Pascoe, E.H. (1973) : *A Manual of the Geology of India and Burma*, Controller of Publications, New Delhi.

Qureshi, M.N. (1971) : *Geophysical Investigation in the Himalaya,* Himalayan Geology (Delhi), Vol. 1, pp. 165-177.

Robinson Francis (ed.) (1989) : *The Cambridge Encyclopedia of India, Pakistan, Bangladesh, Sri Lanka*, Cambridge University Press, Cambridge, pp. 12-15.

Siddhartha, K. (1989) : *Geological Evolution of India,* Hindustan Times, Patna.

Singh, R.P. (1987) : Tectonics and Surfaces of Peninsular India, *Souvenir International Symposium on Geomorphology and Environmental Management,* Allahabad Geographical Society, Allahabad.

Singh, R.L. (ed.) (1971) : *India : A Regional Geography,* National Geographical Society of India, Varanasi.

Spate, O.H.K. and Learmonth, A.T.A. (1967) : *India and Pakistan: A General and Regional Geography,* Methuen & Co. Ltd., Chap. 1.

Wadia, D.N. (1965) : *Geology* in Chopra P.N. (Ed.). *The Gazetteer of India*, Vol. One, Govt. of India Publication, New Delhi (Chap. III).

Wadia, D.N. Meher (1993) : *Geology of India*, Tata McGraw Hill, New Delhi.

◆◇◆

CHAPTER 3

Physiography

INTRODUCTION

Physiography is that branch of geography which studies the present relief features of the earth's surface or of natural features in their causal relationships. As described in the previous chapter, the present surface features of India owe their formation to various geological events which took place in different geological periods. Both endogenetic and exogenetic forces have been consistently working to shape the present land forms on the surface of the earth. India has a unique personality with regard to physiography. The great diversity of relief features encompassed in the vast dimensions of the country is simply amazing. The physiographic diversity of India embraces lofty young fold mountains, flat plains and one of the oldest plateaus of the world. The Indian islands have their own unique personality. A rough estimate made by the Census Commission in 1951, shows that of the total land area, 10.7 per cent is more than 2135 m above sea level and is mountainous, 18.6 per cent is hilly (305 to 2135 m), 27.7 per cent is plateau (305 to 915 m) and the remaining 43 per cent is plain area.

Due to geological complexities and geomorphological diversities, division of India into physiographic regions is a difficult task. The views expressed by geographers in this regard are as diverse as the diversities of landforms in India. Some scholars follow the *triple tectonic division* viz. (*i*) the Himalayan Mountains (*ii*) the Indo-Gangetic plains and (*iii*) the Indian Peninsula. There are some scholars who feel that the coastal plains have a separate identity and should be treated as such. More recently, it has been felt that the distant islands in the Bay of Bengal and the Arabian Sea should also be treated separately. Thus, to be more realistic and for the sake of convenience, it is preferred to divide India into following five physiographic divisions (**Fig. 3.1**).

1. The Himalayan Mountains.
2. The Great Plain of North India.
3. The Peninsular Plateau.
4. The Coastal Plains.
5. The Islands.

1. THE HIMALAYAN MOUNTAINS

The Himalayan mountains are also known as the ***Himadri***, ***Himavan*** or ***Himachal***. The Himalayas consist of the youngest and the loftiest mountain chains in the world. The magnitude of their size can be estimated from the fact that the central axial range of the main Himalayas alone stretches for a distance

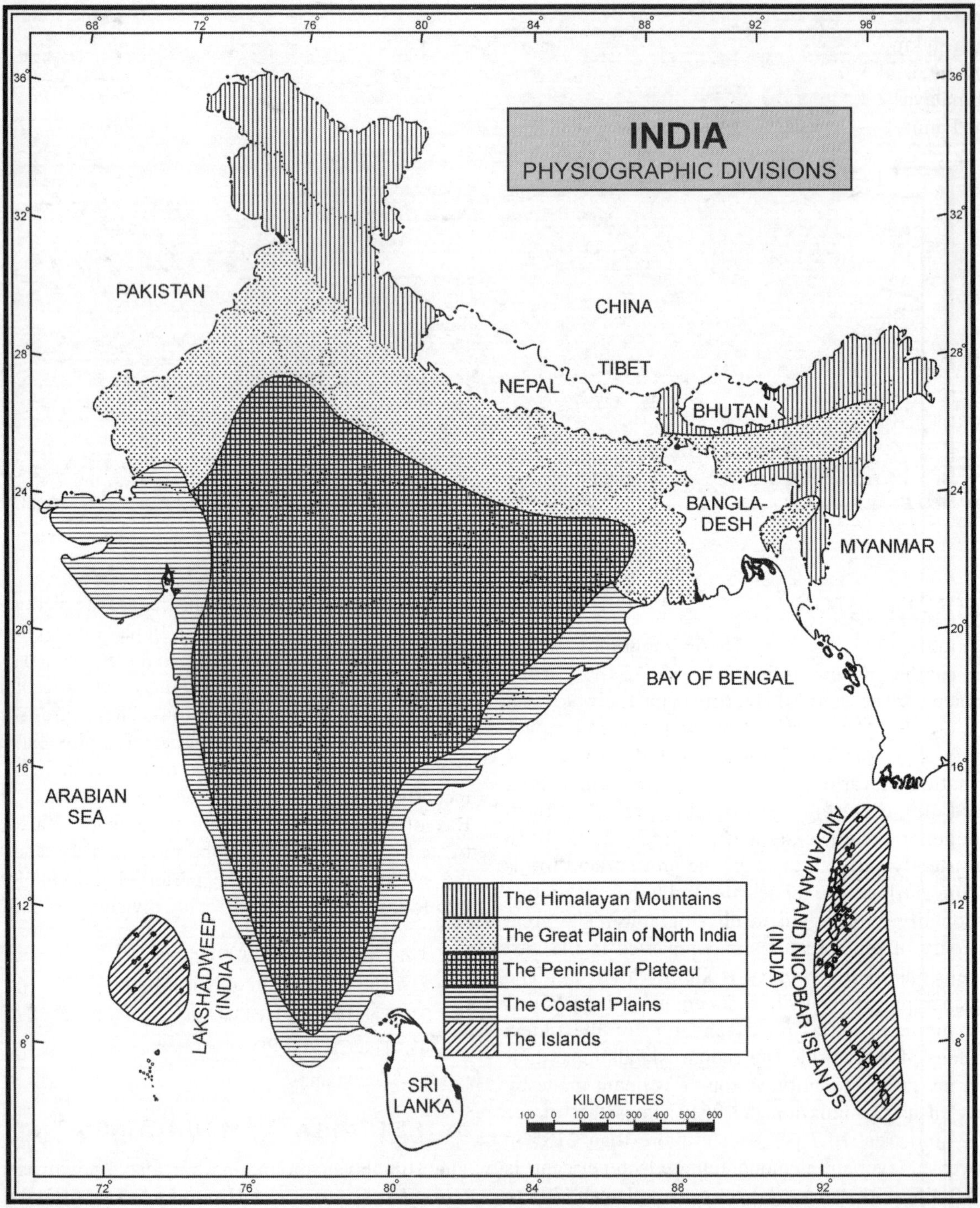

FIG. 3.1. India : Physiographic Divisions

of over 2,400 km (over 22° longitude) from the Indus gorge in the west to the Brahmaputra gorge in the east.

The width of the Himalayas varies from 500 km in Kashmir to 200 km in Arunachal Pradesh. The total area of the Himalayan mountain region is nearly five lakh sq km. The Pamir, popularly known as the ***roof of the world*** is the connecting link between the Himalayas and the high ranges of Central Asia. From the Pamir, the Himalayas extend eastward in the form of an arcuate curve which is convex to the south. The southern boundary of the Himalayas is well defined by the foothills (300 m contour line) but the northern boundary is rather obscure and merges with the edge of the Tibet Plateau.

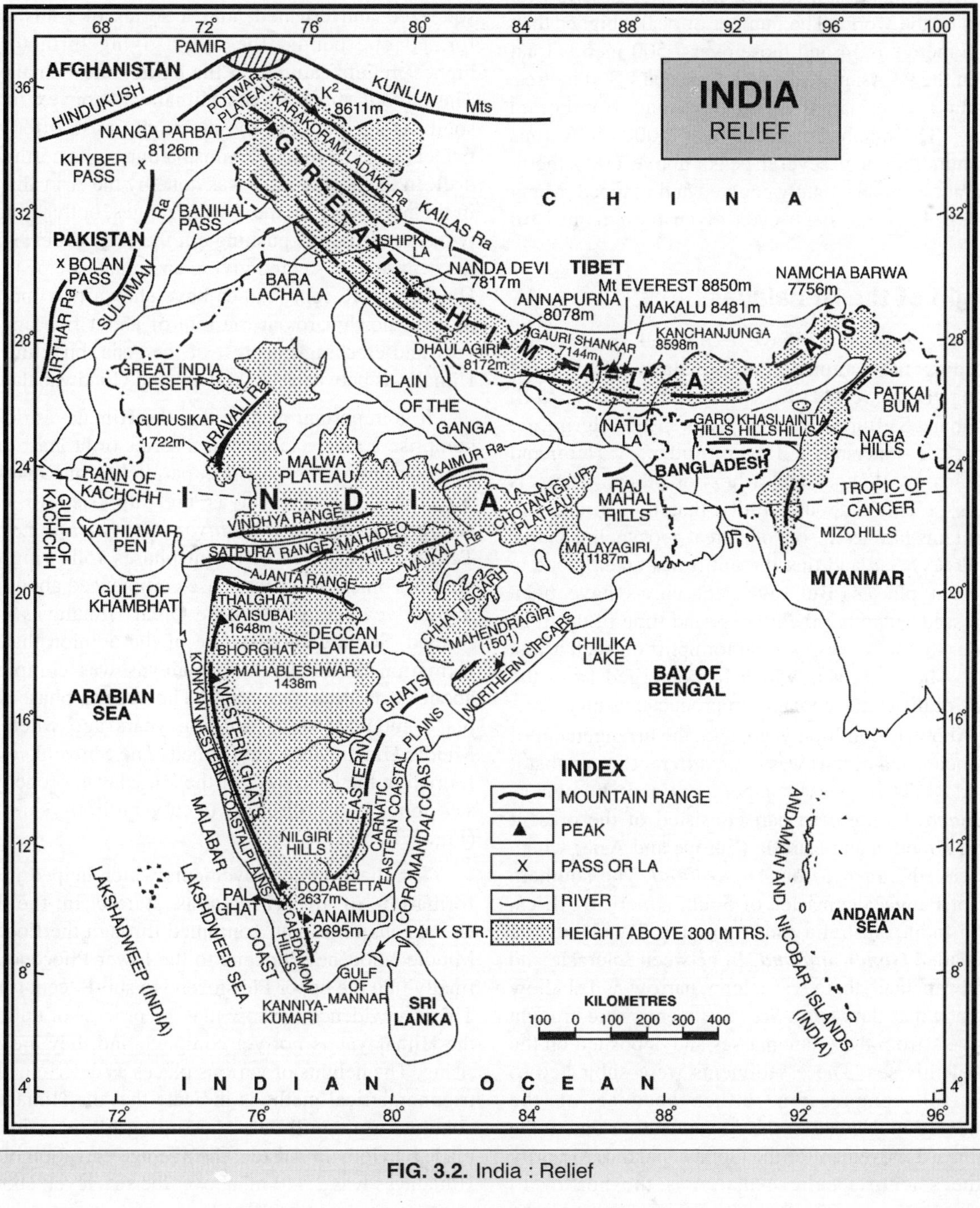

FIG. 3.2. India : Relief

The Himalayas have attained a unique personality owing to their high altitude, steep gradient, snow capped summits, deeply dissected topography, youthful drainage, complex geological structure and rich temperate flora in the subtropical latitudes.

The Himalayan mountain chains have the loftiest peaks of the world. The number of peaks higher than 8,000 metres is 14 and those over 7,500 metres total 20. Of the 94 Asian peaks which exceed 7,300 metres, all but two are in the Himalaya and Karakoram ranges. The number of peaks over 7,000 metre runs into hundreds and several peaks above 6000 metre have not been adequately counted and mapped. Many of them have not been even given their names till now.

Origin of the Himalayas

Several scholars have expressed their views regarding the origin of the Himalayas. Prominent among them are O.H.K. Spate, D.N. Wadia, M.S. Krishnan, S. Burrard, E.H. Pasco, G.E. Pilgrim, de Terra, T.T. Paterson, T. Hagen, Auden, A. Heim and A. Gansser, Wagner and a host of others. There is almost a complete unanimity that the Himalayan mountains have come out of a great geosyncline called the ***Tethys Sea*** and that the uplift has taken place in different phases. But divergent views have been expressed regarding the process and time of uplift as well as the forces responsible for uplift on such a vast scale. The consensus which has emerged from the views of different scholars is reproduced as under :

About 120 million years ago, the arrangement of continents and oceans was quite different from what it is today. There was a super continent known as ***Pangaea***. Its northern part consisted of the present North America and Eurasia (Europe and Asia) which was called ***Laurasia*** or ***Angaraland***. The southern part of Pangaea consisted of South America, Africa, South India, Australia and Antarctica. This landmass was called ***Gondwanaland***. In between Laurasia and Gondwanaland, there was a long, narrow and shallow sea known as the ***Tethys Sea***. Sediments were brought by rivers from these landmasses and deposited on the bed of this sea. These sediments were subjected to powerful compression, either because of the southward movement of the Angaraland or due to the northward movement of the Gondwanaland. Majority of the scientists believe that it is the northward movement of the Gondwanaland which caused compression in the sediments at the floor of the Tethys Sea. In any case, whether Angaraland moved southwards or Gondwanaland moved northwards or both moved towards each other, the net result would be the same; the sediments in the Tethys Sea was squeezed and crushed, and a series of folds were formed one behind the other giving birth to the highest relief features on the earth—the Himalayas. The curved shape of the Himalayas convex to the south, is attributed to the maximum push offered at two ends of the Indian Peninsula during its northward drift. In the northwest it was done by the Aravalis and in the north-east by the Assam ranges, both acting as two extended arms pushing out the extremities, while the central area sagged giving an arcuate shape to the Himalayas. Recent studies have shown that India is moving northwards at the rate of about five cm per year and crashing into rest of the Asia, buckling the Himalayas between Angaraland and Gondwanaland.

It is important to note that the Himalayas do not comprise a single range but a series of at least three ranges running more or less parallel to one another. Therefore, the Himalayas are supposed to have emerged out of the *Himalayan Geosyncline i.e.* the Tethys Sea in three different phases following one after the other. The first phase commenced about 120 million years ago, when the Great Himalayas were formed. Some geologists are of the opinion that the formation of the Great Himalayas was completed about 70 million years ago. The second phase took place about 25 to 30 million years ago when the Middle Himalayas were formed. The Shiwaliks were formed in the last phase of the Himalayan orogeny—say about two million to twenty million years ago (**Fig. 3.3**).

The diastrophic movements which helped in the formation of the Himalayas started in the late Cretaceous times and continued through the Eocene, Middle Miocene, Pliocene to the lower Pliocene and finally into the upper Pleistocene to sub-Recent times. There is evidence to show that the process of uplift of the Himalayas is not yet complete and they are still rising. The heights of various places as determined by trigonometrical methods indicate that the Himalayas continue to rise till date. According to the estimates made by Godwin Austen, the average elevation of the Himalayas was 2,440 m above the sea level about a

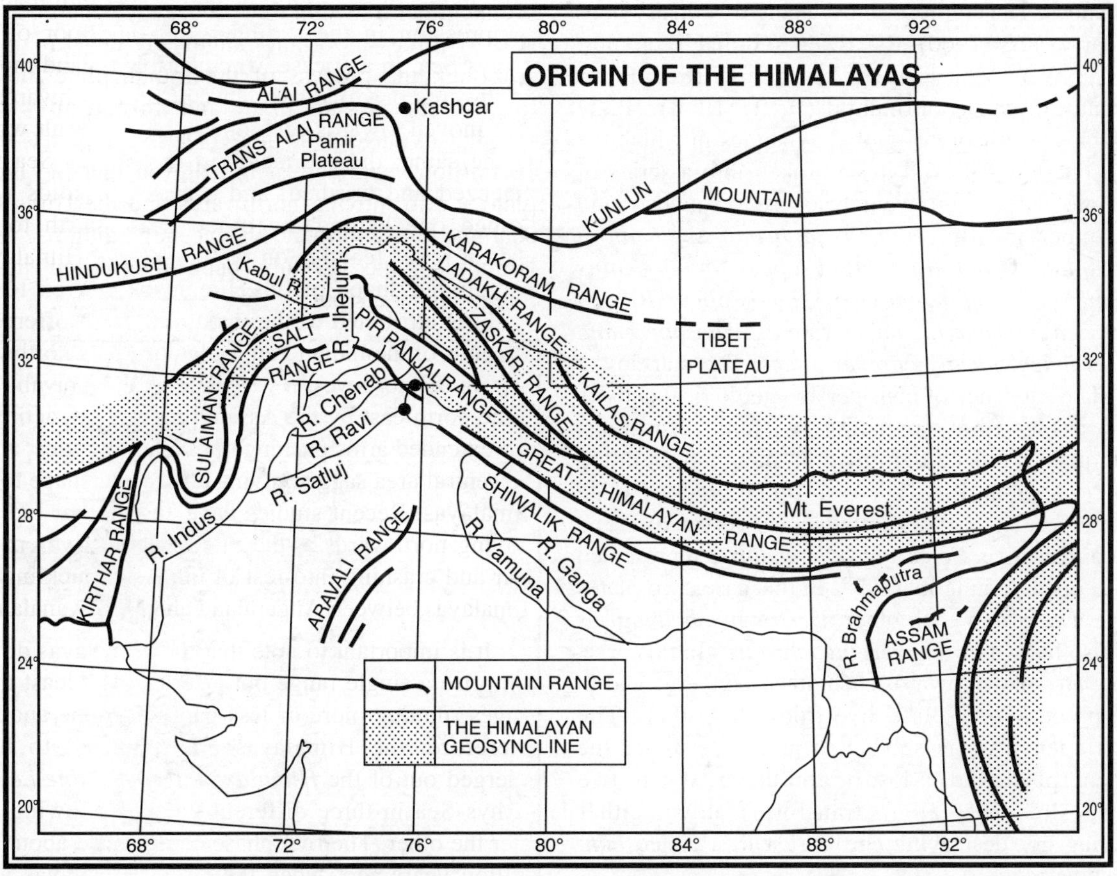

FIG. 3.3. Origin of the Himalayas

million years ago which has now risen to 3,050 m. The Mahabharat range is still in a state of rigorous uplift. Following evidence is cited to prove that the Himalayas are still rising :

(*i*) Some of the fossil formations found in the Shiwalik hills are also available in the Tibet plateau. It indicates that the past climate of the Tibet plateau was somewhat similar to the climate of the Shiwalik hills and that the elevation of Tibet plateau was almost the same as that of the present Shiwalik hills and the plateau has since risen to its present elevation.

(*ii*) Desiccation of lakes of Tibet has been observed within the recent or even in historical times. Surrounding these lakes, the sand and gravel terraces at higher levels, sometimes 60-100 metres above the present water level, are seen which prove that the water stood at a much higher level till recent times. This could be possible only in the event of uplift of the region.

(*iii*) The frequent occurrence of earthquakes in the Himalayan region shows that the Himalayas have not yet attained *isostatic equilibrium* and they still continue to rise further.

(*iv*) The Himalayan rivers are still in their youthful stage and have been rejuvenated in recent times. This is another proof of rising trend in the Himalayas. Terraces found at the valley sides suggest rejuvenation of the valley region due to uplift.

The present rate of uplift of the Himalayas has been calculated at 5 to 10 cm per year.

Plate Tectonics

Plate tectonics is the most recent and widely acclaimed theory which gives most satisfactory answers to intricate and puzzling questions regarding origin of continents and oceans, formation of mountains, occurrence of earthquakes and eruption of volcanoes. The credit for introducing this theory goes to Harry Hess (1960), R.S. Dietz (Global Tectonics, 1961), W.J. Morgan and Le Pichon (Sea Floor Spreading and Continental Drift, 1968). Plate tectonics is a theory of global dynamics in which the lithosphere is believed to be broken into a series of separate plates that move in response to convection in the upper mantle. *Plate is a broad segment of lithosphere (including rigid upper mantle plus oceanic and continental crust) that is about 100 km thick and floats on the underlying asthenosphere and moves independently of other plates.* The margins of the plates are sites of considerable geological activity such as sea-floor spreading, volcanic eruptions, earthquakes, crustal deformation, mountain building and continental drift. There are three types of plate boundaries, *viz*., (*a*) constructive, (*b*) destructive, and (*c*) conservative. *Constructive boundaries* represented zones of 'diverging margins'. In this case, two plates move away from each other. *Destructive boundaries* are also known as '*converging margins*'. In this case, two plates move towards each other, converge and in the process one plate overrides the other. The overridden plate is subducted and goes under the asthenosphere and is lost or consumed. Where two plates slide past along transform faults neither creating nor destroying earth crust it is called ***conservative zone***.

It is the converging boundary of the plates where folded mountains like the Himalayas build up. When two convergent plates composed of continental crusts collide against each other, the denser plate is subducted under the lighter plate. The resultant lateral compression squeezes and folds the sediments deposited on either side of the continental plate margins or sediments of the geosynclines lying between the two.

The Himalayas are the product of such a process on the convergence zone of the Asiatic plate in the north and the Indian plate in the south. Some 70 million years ago, the Indian plate started moving towards the Asian plate and the Tethys sea began to contract due to the movement of Indian and Asian plates towards each other. Since the Indian plate was made up of denser material than the Asian plate, the former began to subduct under the latter causing lateral compression of the marine sediment on the bed of the Tethys (**Fig. 3.4**). Geologists believe that the sediment got folded in three successive phases giving rise to three important ranges of the Himalayas. Although the process of the Indian plate moving towards the Asian plate and formation of the Himalayan ranges was more or less completed about 10 million years ago, it is believed that the Indian plate is still moving northwards and the Himalayas continue to rise further. Evidence regarding rise in the Himalayas has already been mentioned in the foregoing paragraphs.

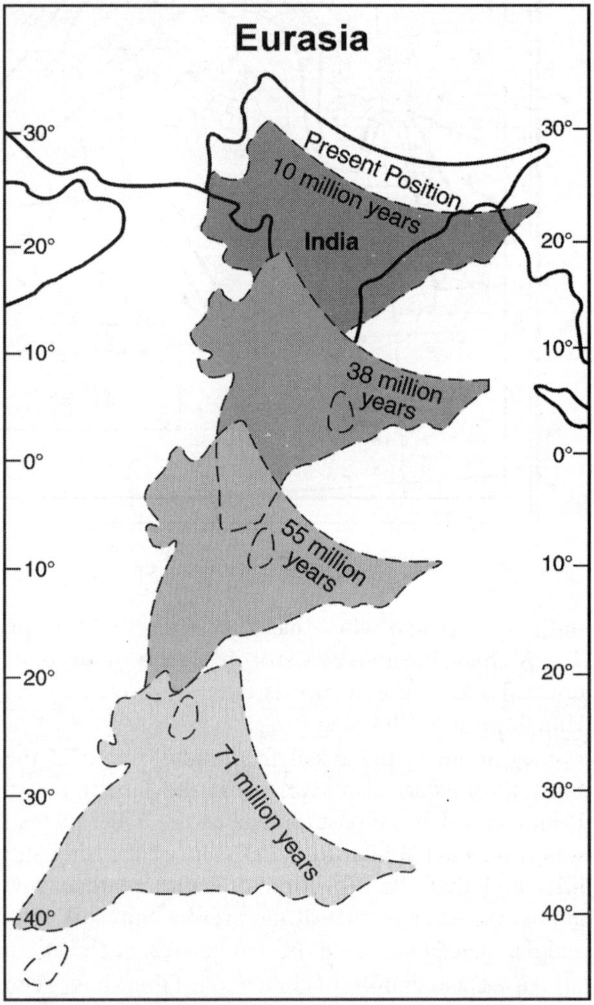

FIG. 3.4. The northerly drift of Indian plate and its collision with Asian plate

Different stages of movement of the Indian plate and the evolution of the Himalayas are illustrated in **Fig. 3.5**.

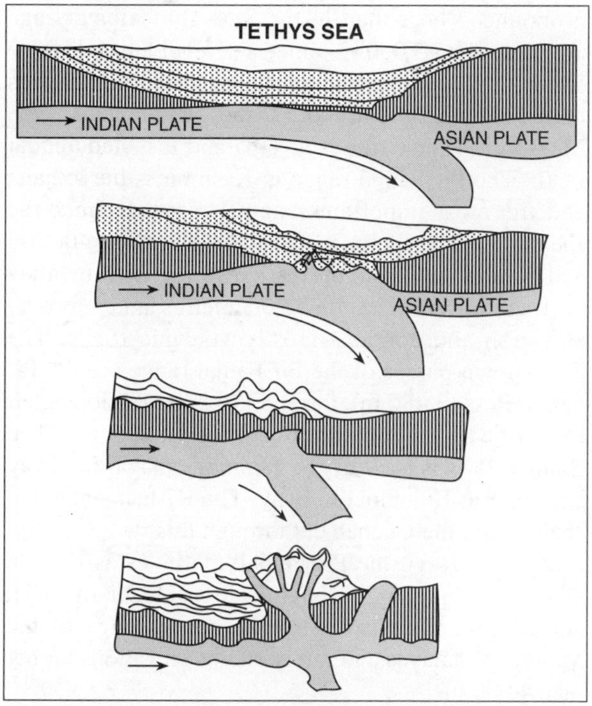

FIG. 3.5. Plate Tectonics and Evolution of the Himalayas

Recent studies have shown that convergence of the Indian plate and the Asian plate has caused a crustal shortening of about 500 km in the Himalayan region. This shortening has been compensated by sea floor spreading along the oceanic ridge in the Indian Ocean. Since the northward movement of the Indian plate is still continuing, the height of the Himalayan ranges is increasing with the passage of time. It has been estimated that the Indian plate is still moving northward at a rate of about 5 cm per year. As a result of this movement, one more fracture has appeared in the outer fringes of the Shiwalik Hills which demarcates these hills from the Ganga Plain. Geologists postulate that a new mountain chain would be formed in the Ganga basin forcing the rivers of this region to migrate further south. This clearly indicates that the process of upheaval of the Himalayas is not yet over and will continue for a pretty long period in the times to come.

Division of the Himalayas

Geographers generally tend to divide the Himalayas on geographical, regional and geological bases. Geographically the entire Himalayan region can be divided into: (*i*) The Himalayan Ranges; (*ii*) The Trans-Himalayas and (*iii*) The Eastern Hills.

(*i*) The Himalayan Ranges

The Himalayas are not a single chain or range of mountains, but a series of several, more or less parallel or converging ranges. These ranges are separated by deep valleys. As in all young fold mountains, we find a densely dissected "*ridge-and-valley-topography*" in the Himalayas. The most outstanding valleys in the Himalayan region are the vale of Kashmir and the Karewas, the Kangra and Kulu valley in Himachal Pradesh; the Dun valley; the Bhagirathi valley (near Gangotri) and the Mandakini valley (near Kedarnath) in Uttarakhand and the Kathmandu valley in Nepal. The individual ranges have very steep gradient towards the south but they present a much gentler slope towards the north. In the eastern section the Himalayas rise abruptly from the plains of Bengal and Oudh and suddenly attain great elevations within a short distance from the foot of the mountains. Thus the peaks of Kanchenjunga and Everest are only a few kilometres from the plains and are clearly visible from there. In contrast, the western Himalayas rise gradually from the plains through a series of ranges. Their peaks of perpetual snow are 150 to 200 km away from the plain areas. Most of the Himalayan ranges fall in India, Nepal and Bhutan, but the northern slopes are partly situated in Tibet while the western extremity lies in Pakistan, Afghanistan and Central Asia. A succession of at least three parallel ranges from the Indo-Gangetic plain to the Tibet plateau may be recognised in the Himalayas.

(*a*) The Shiwalik Range. The Shiwalik comprises the outermost range of the Himalayas and is also known as the ***Outer Himalayas***. With its southern steep slopes, it assumes a ***hogback*** appearance. Flat scarps, anticlinal crests and synclinal valleys are the chief characteristic features of this range. Overlooking the Great Plain, this chain of hills runs almost parallel to the lesser Himalayas for a distance of about 2,400 km from the Potwar Plateau to the Brahmaputra valley. The width of the Shiwaliks varies from 50 km in Himachal Pradesh to less than 15 km in Arunachal Pradesh. It is an almost unbroken succession of low hills except for a gap of 80-90 km which is occupied by the valley of the Tista River.

The gorges of the Tista and the Raidak have jointly formed a gap 80-90 km wide in the Shiwalik range. The altitude varies from 600 to 1500 metres. The Shiwaliks are formed of great thickness of Mio-Pleistocene sands, gravels and conglomerates which have been brought by the rivers flowing from the higher ranges of the Himalayas. These have been folded, faulted and indurated by the earth movements from the Middle Miocene to the lower Pliestocene ages. Obviously the Shiwaliks were formed last of all the ranges. The Shiwaliks are known by different names in different areas. They are called Jammu Hills in Jammu and Dafla, Miri, Abor and Mishmi Hills in Arunachal Pradesh. The Dhang Range, Dundwa Range (Uttarakhand) and the Churia Ghat Hills of Nepal also form part of the Shiwalik range.

As the Shiwalik Hills were formed after the formation of the Himalayas, they obstructed the courses of the rivers draining from the higher reaches of the Himalayas and formed temporary lakes. The debris brought by those rivers were deposited in these lakes. After the rivers had cut their courses through the Shiwalik Range, the lakes were drained away leaving behind plains called ***'duns'*** or ***'doons'*** in the west and ***'duars'*** in the east. Dehra Dun in Uttarakhand is the best example of such a plain which is 75 km long and 15-20 km wide. It is covered with boulder and clay deposits. Kotah, Patli Kothri, Chumbi, Kyarda and Chaukhamba are other duns. In the Jammu hills the extensive picturesque duns of Udhampur and Kotli are quite typical.

The eastern part of the Shiwalik range upto Nepal is covered with thick forests but the forest cover becomes thin in the west. The southern slopes of this range are almost completely devoid of forest cover in Punjab and Himachal Pradesh and are highly dissected by several seasonal streams locally called ***Chos***.

(*b*) The Middle or the Lesser Himalaya. In between the Shiwaliks in the south and the Great Himalaya in the north is the Middle Himalaya running almost parallel to both the ranges. It is also called the ***Himachal*** or ***Lower Himalaya***. It has an intricate system of ranges which are 60-80 km wide having elevations varying from 3,500 to 4,500 m above sea level. Locally linear longitudinal ranges have also developed, with steep, bare southern slopes and more gentle, forest covered northern slopes, they give these ranges typical ***hogback*** look, more pronounced here than in the Shiwalik. Many peaks are more than 5,050 m above sea level and are snow covered throughout the year. The important ranges included are the Pir Panjal, the Dhaola Dhar, the Mussoorie Range, the Nag Tiba and the Mahabharat Lekh. The Pir Panjal range in Kashmir is the longest and the most important range. It extends from the Jhelum river to the upper Beas river for 300-400 km and is separated from the Zaskar Range by the valley of Kashmir. It rises to 5,000 metres and more in elevation and contains mostly volcanic rocks. The best known passes of the Pir Panjal range are the Pir Panjal Pass (3,480 m), the Bidil (4,270 m), Golabghar Pass (3,812 m) and Banihal Pass (2,835 m). The Banihal Pass is used by the Jammu-Srinagar highway and Jammu-Baramula railway. The Kishanganga, the Jhelum and the Chenab cut through this range. South-east of the Ravi, the Pir Panjal continues as Dhaola Dhar range, passing through Dalhousie, Dharmshala, and Shimla. This is the southernmost range of the Middle Himalayas and rarely attains elevations higher than 4,000 metres.

Between the Pir Panjal and the Zaskar Range of the main Himalayas, lies the famous valley of Kashmir running over a distance of about 135 km in a south-east to north-west direction. This valley is about 40 km broad in its middle. Its total area is 4921 sq km and its average elevation is 1,585 m above mean sea level. Geographers and geologists have divergent views regarding the origin of this valley. **Wadia** thinks that the valley of Kashmir is the synclical basin enclosed between the Pir Panjal to the south and an offshoot of the central axial range to the north while **de Terra** feels that it is a recently depressed intermont basin pointing to evidence of faulting on the Himalayan flank. However, it is generally believed that this basin was occupied by a lake in the Pleistocene age. This was filled with sediment and uplifted to form the Kashmir valley. The synclinal basin of the valley is floored with a variety of alluvial deposits, lacustrine, fluvial and fluvioglacial, through which the Jehlum River meanders majestically before entering the deep gorge it has cut through the Pir Panjal. In Himachal Pradesh there is *Kangra Valley*. It is a strike valley and extends from the foot of the Dhaola Dhar Range to the south of Beas. On the other hand, the *Kulu Valley* in the upper course of the Ravi is transverse valley.

Further east, the Middle Himalayas are marked by the Mussoorie and the Nag Tibba ranges. The Mussoorie range has an average elevation of 2,000-2,600 m and runs over a distance of 120 km from Mussoorie to Lansdowne. Mussoorie, Nainital, Chakrata and Ranikhet are important hill stations situated between 1,500 to 2,000 metres above sea level. The Mahabharat Lekh, in southern Nepal is a continuation of the Mussoorie Range. Its summits rise to 3000 m and its average elevation varies from 1,500 to 2,000 m above sea level. The Kathmandu valley to its north is a very important feature of this area. East of the Sun Kosi River, the Sapt Kosi, Sikkim, Bhutan, Miri, Abor and Mishmi hills represent the lower Himalayas.

On the whole, the Middle Himalayan ranges are less hostile and more friendly to human contact. Majority of the Himalayan hill resorts like Shimla, Mussoorie, Ranikhet, Nainital, Almora and Darjeeling, etc. are located here.

(*c*) The Great Himalaya. This is also known as ***Inner Himalaya***, ***Central Himalaya*** or ***Himadri***. This is the northernmost or the innermost of all the Himalayan ranges. With an average elevation of 6,100 m above sea level and an average width of about 25 km, this is the loftiest and the most continuous mountain range of the world. It is about 150 km away from the northern edge of the plains of Northern India. It is mainly formed of the central crystallines (granites and gneisses) overlain by metamorphosed sediments. The folds in this range are asymmetrical with steep south slope and gentle north slope giving ***'hogback'*** topography. This mountain arc, convex to the south, terminates abruptly in the Nanga Parbat in north-west and in the Namcha Barwa in the north-east.

This mountain range boasts of the tallest peaks of the world, most of which remain under perpetual snow. There are several peaks over 8,000 m in altitude. The highest is the Mount Everest which is 8,850 m* above sea level. Its Nepalese name is ***Sagarmatha*** which means *The* ***Goddess of the Sky***. The Tibetans call it ***Chomlungma***. Located at 27° 59′ 16″ N latitude and 86° 55′ 46″ E longitude this sky touching peak was first located by ***Sir George Everest***, the then Surveyor General of India in 1841 and in 1852 it was established as the highest peak of the world by the Great Trigonometrical Survey of India. In 1865 it was named as Mount Everest as a tribute to Sir George Everest. Before that, it was known as Peak XV. However, Chinese want the peak to be named as Chomolungma, meaning the "Mother of the World". China claims that this peak was mapped by Qing dynasty officials in 1717. The other important peaks in descending order of altitude are, Kanchenjunga, Lhotse, Makalu, Dhaulagiri, Manaslu, Cho Oyu, Nanga Parbat, Annapurna, Gosainthan or Shisha Pangma. Some of the important peaks between 7,100 and 8,000 m elevation are Nanda Devi, Kamet, Namcha Barwa, Gurla Mandhata, Badri Nath, Trisul, etc. (*see* **Table 3.1** and **Figs. 3.6** and **3.8**).

This range is so formidable that it cannot be easily crossed even through the passes because they are generally higher than 4,570 m above sea level and are snowbound for most part of the year. The Burzil Pass and the Zoji La in Jammu and Kashmir, the Bara Lacha La and the Shipki La in Himachal Pradesh, the Thaga La, the Niti Pass and the Lipu Lekh pass in Uttarakhand and the Nathu La and the Jelep La in Sikkim are the important passes through the Great Himalaya. The Hindustan-Tibet Road connecting Shimla with Gartok in Western Tibet passes through the Shipki La. Another important trade route connecting Kalimpong (near Darjeeling) with Lhasa in Tibet, passes through Jelep La (4,386 m).

(ii) The Trans Himalayas

The Himalayan ranges immediately north of the Great Himalayan range are called the trans-Himalayas. This part of the Himalayan ranges is also called the ***Tibetan Himalaya*** because most of it lies in Tibet. The Zaskar, the Ladakh, the Kailas and the Karakoram are the main ranges of the Trans-Himalayan system. It stretches for a distance of about 1,000 km in east-west direction and its average elevation is 3000 m above mean sea level. The average width of this region is 40 km at the eastern and western extremities and about 225 km in the central part.

The *Zaskar Range* branches off from the great Himalayan Range near 80° E longitude and runs more or less parallel to it. The Nanga Parbat (8126 m) forms its culmination in the north-west but the adjoining Deosai Mountain may also be included in it. North of the Zaskar Range and running parallel to it is the *Ladakh Range*. It is about 300 km long and its

TABLE 3.1. Some Important Peaks of the Great Himalayas and their Heights

Name of the Peak	Height above sea level (in metres)	Name of the Peak	Height above sea level (in metres)
1. Mount Everest	8,848.86	9. Gosainthan or Shisha Pangma	8,013
2. Kanchenjunga	8,598	10. Nanda Devi	7,817
3. Lhotse I	8,501	11. Kamet	7,756
4. Makalu	8,481	12. Namcha Barwa	7,756
5. Dhaulagiri	8,172	13. Gurla Mandhata	7,728
6. Manaslu	8,156	14. Trisul	7,140
7. Cho Oyu	8,153	15. Badarinath	7,138
8. Annapurna	8,078		

History of the Everest Altitude

- According to popular belief the Everest height was first calculated by a team lead by Sir George Everest who served as a Surveyor General of India from 1830 to 1843.
- However, some Indian scholars believe that Sir George Everest, neither visited the peak nor measured the height of the peak himself. He employed Sh. Radhanath Sikdar of Kolkata (Calcutta) in the Survey of India Department. He was an expert in Spherical Trigonometry and took pains to calculate the height of the peak. Sir George Everest got retirement in 1843 and Col. Andrew Scott Waugh was appointed as Surveyor General of India. He asked Radhanath Sikdar to continue with the process of computing the height of the peak. At that time it was known as peak XV and it is Col. Andrew Scott Waugh who suggested the name of Mount Everest in honour of his predecessor Sir George Everest. In 1852, Sh. Radhanath Sikdar calculated the height of the peak as 29,002 ft. (8840 m). But to be doubly sure, he repeated the process and came to final conclusion that the height of the peak was exactly 29,002 ft. (8,840 m).
- In 1954, the Indian officials computed the height of the Mt. Everest using more modern and accurate instruments and conclude that the height of the peak is 29,028 ft. (8,848 m).
- A survey team from Italy calculated the height at 29,022 ft. (8,845.9 m) in 1992.
- Another team from the U.S.A. in 1999 determined the height to be 29,035 ft. (8,849.8 m).
- In 2005 a Chinese team calculated the height as 29017 ft. (8,844.4 m).
- In 2020, the Chinese and Nepalese surveyors calculated the height of the Mount Everest by using GPS. The most contentious matter was whether to calculate the depth of snow on a mountain peak as a part of its height. The Nepalese surveyors used a ground penetrating radar to actually measure the height of snow standing on top of Everest's rocky crest but in the part it has been debated whether the snow should be part of calculation. By mutual consensus, the Chinese and Nepalese surveyors arrived at the conclusion that the height of Mt. Everest is 29,032 ft. (8,848.86 m).

Chronology of height of Mount Everest

Particulars	Year	Height
1. British India	1,856	29,002 ft. (8,840 m)
2. India	1,954	29,028 ft. (8,848 m)
3. Italy	1992	29,022 ft. (8,845.9 m)
4. U.S.A.	1999	29,035 ft. (8,849.8 m)
5. China	2005	29,017 ft. (8,844.4 m)
6. Nepal-China	2020	29,032 ft. (8,848.86 m)

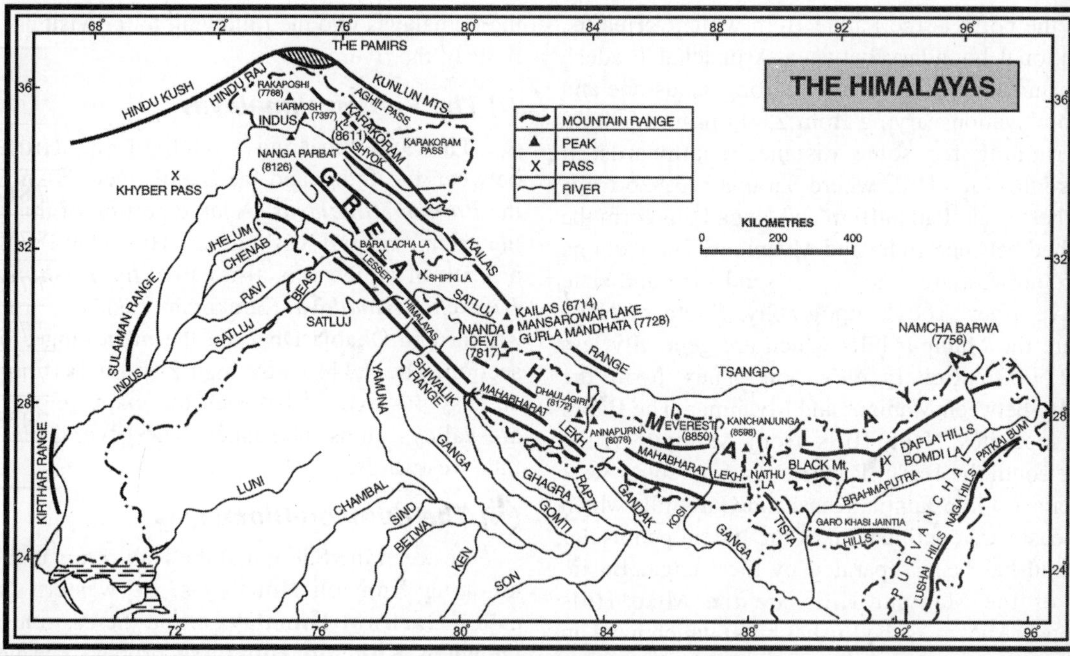

FIG. 3.6. The Himalayas

average elevation is 5800 metres above the sea level. Only a few peaks of this range attain heights of over 6000 metres. The Rakaposhi-Haramosh Ranges beyond the Indus may be treated as extensions of the Ladakh Range to the northwest. The Kailas Range (*Gangdise in Chinese*) in western Tibet is an offshoot of the Ladakh Range. Its average elevation is 5,500-6,000 m above sea level and its average width is 30 km. The highest peak is Mount Kailas (6714 m). River Indus originates from the northern slopes of the Kailas range. The northern most range of the Trans-Himalayan Ranges in India is the Great Karakoram Range also known as the Krishnagiri Range. It forms India's frontier with Afghanistan and China. It extends eastwards from the Pamir for about 800 km. The average width of this range is 120-140 km. It is a range with lofty peaks and its elevation hardly ever falls below 5,500 m. It is the abode of some of the largest glaciers of the world outside the polar regions. Some of the peaks are more than 8,000 metre above sea level. K^2 (8,611 m) is the second highest peak in the world and the highest peak in the Indian Union and rises majestically like a cone. It has been named as ***Mt Godwin Austen*** by the Britishers and ***Qogir*** by the Chinese. The other peaks located in its neighbourhood and rising more than 8,000 m above sea level are the Gasherbrum I or Hidden Peak (8,068 m), Broad Peak (8,047 m) and Gasherbrum II (8,035 m). Another 19 peaks in Karakoram cross the 7,600 m elevation mark and those over 7,000 m have not yet been fully enumerated.

The Ladakh Plateau lies to the north-east of the Karakoram Range. With an average elevation of over five thousand metres above sea level, it is the highest plateau of the Indian Union. It has been dissected into a number of plains and mountains, the most outstanding among them being Soda Plains, Aksai Chin, Lingzi Tang, Depsang Plains and Chang Chenmo.

(iii) The Eastern Hills or The Purvanchal

After crossing the Dihang gorge, the Himalayas take a sudden southward turn and form a series of comparatively low hills running in the shape of a crescent with its convex side pointing towards the west. These hills are collectively called the ***Purvanchal*** because they are located in the eastern part of India. Extending from Arunachal Pradesh in the north to Mizoram in the south, they form India's boundary with Myanmar. Differing markedly from the Himalaya in the scale of their relief and in their morphology, these hill ranges none-the-less stem from the same orogeny.

In the north is the Patkai Bum which forms the international boundary between Arunachal Pradesh and Myanmar. It is made up of strong sandstone and rises to elevation varying from 2,000 m to 3,000 m. After running for some distance southwards, it merges into Naga Hills where Saramati (3,826 m) is the highest peak. Patkai Bum and Naga Hills form the watershed between India and Myanmar. The Kohima hills to the west are made up of sandstone and slate and have a very rough topography. South of Naga Hills are the Manipur hills which are generally less than 2,500 metres in elevation. They form the boundary between Manipur and Myanmar. The Barail range separates Naga Hills from Manipur Hills. Further south the Barail Range swings to south-west and then west into Jaintia, Khasi and Garo hills which are an eastward continuation of the Indian peninsular block and has been separated by the Bengal Basin. South of the Manipur Hills are the Mizo Hills (previously known as the Lushai hills) which have an elevation of less than 1,500 metres. The highest point is the Blue Mountain (2,157 m) in the south. It is obvious that the elevation of the eastern hills decreases as we move from north to south. Although comparatively low, these hill ranges are rather forbidding because of the rough terrain, dense forests and swift streams.

Regional Division of the Himalayas

Sir Sidney Burrard divided the entire length of the Himalayas into the following four divisions on the basis of the river valleys :

(i) The Punjab Himalayas

The 560 km long stretch of the Himalayas between the Indus and the Satluj rivers is known as the ***Punjab Himalayas***. A large portion of this sector lies in Jammu and Kashmir and Himachal Pradesh as a result of which it is also called the ***Kashmir and Himachal Himalaya***. Karakoram, Ladakh, Pir Panjal, Zaskar and Dhaola Dhar are the main ranges of this section. The 3,444 metre high Zoji La pass provides an easy passage. In between the main ranges, there are valleys, duns, and lakes. The general elevation falls westwards.

(ii) The Kumaon Himalayas

Between the Satluj and the Kali rivers is the 320 km long Kumaon Himalayas. Its western part is called **Garhwal Himalaya** while the eastern part is known as **Kumaon Himalaya** proper. The general elevation is higher as compared to Panjab Himalayas. Nanda Devi (7,817 m), Kamet (7,756 m), Trisul (7,140 m), Badrinath (7,138 m), Kedarnath (6,968 m), Gangotri (6,510 m) are important peaks. The sources of sacred rivers like the Ganga and the Yamuna are located in the Kumaon Himalayas. There are several *duns* between the Middle Himalayas and the Shiwalik Hills. Nainital and Bhimtal are important lakes.

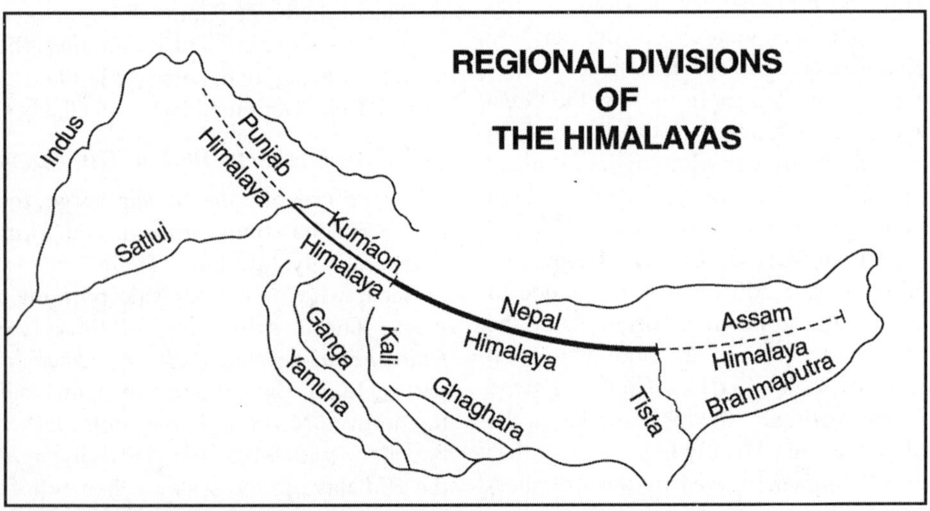

FIG. 3.7. Regional Divisions of the Himalayas

(iii) The Nepal Himalayas

This section of the Himalayas stretches for a distance of 800 km between the Kali and the Tista rivers. Most of it lies in Nepal as a result of which it is called the ***Nepal Himalayas***. This is the tallest section of the Himalayas and is crowned by several peaks of perpetual snow. The Mount Everest (8,850 m) is the tallest peak of the world. The other major peaks are Kanchenjunga (8,598 m), Lhotse I (8,501 m), Makalu (8,481 m), Dhaula Giri (8,172 m), Cho Oyu (8,153 m) and Annapurna (8,078 m). Kathmandu is a famous valley in this region.

(iv) The Assam Himalayas

The Himalayan ranges from Tista to Brahmputra rivers covering a distance of 750 km are called the ***Assam Himalayas***. This part of the Himalayas spreads over large parts of Sikkim, Assam and Arunachal Pradesh and has elevation much lesser than that of the Nepal Himalayas. The southern slopes are very steep but the northern slopes are gentle. The Lesser Himalayas are very narrow and are very close to the Great Himalayas. The improtant peaks of this region are Namcha Barwa (7,756 m), Kula Kangri (7,554 m) and Chomo Lhari (7,327 m).

Apart from Sir Sydney Burrard, some other scholars have also divided the Himalayas in their own way. For example, **Prof. S.P. Chatterjee** (1,964) divided the Himalayan region into three meso physiographic regions. Their names are (1) Western Himalayas (Kashmir, Punjab and Kumaon Himalayas), (2) Central Himalayas (Nepal Himalayas) and (3) Eastern Himalaya—besides the Purvanchal consisting of the north-eastern ranges. **R.L. Singh** (1971) also made three fold subdivision of the Himalayas. His division was slightly different from that made by **S.P. Chatterjee**. **Prof. R.L. Singh's** division comprises (*i*) Western Himalaya (1. Kashmir Himalaya and 2. Himachal Himalaya), (*ii*) Central Himalaya (3. U.P. Himalaya—now Uttarakhand Himalaya, 4. Nepal Himalaya), and (*iii*) Eastern Himalaya (5. Darjeeling—Bhutan—Assam Himalaya and 6. Purvanchal).

From the above discussion, it sounds more logical to divide the Himalayas into the following three broad divisions :

1. The Western Himalayas. Stretching for 880 km between the Indus in the west and the Kali river in the east, the Western Himalayas spread over Jammu and Kashmir, Ladakh, Himachal Pradesh and Uttarakhand. It encompasses three physiographic provinces namely Kashmir Himalaya, Himachal Himalaya and Kumaon Himalaya.

The *Kashmir Himalayas* lie almost entirely in the Union Territories of Jammu and Kashmir and Ladakh. The average height of this region is nearly 3,000 m above sea level. This province is marked by a number of glaciated troughs. The Indus is the most important river which traverses this area diagonally for about 650 km. North of this river lies the imposing Karakoram range. This range has lofty peaks and a large number of big glaciers. The second highest peak of the world Godwin Austin or K^2 (8,611 m) lies in this range. The Ladakh plateau and the Kashmir valley are two important areas of this region. The chief characteristic features of the Kashmir Himalaya are high snow covered peaks, deep valleys, interlocked spurs and high mountain passes.

The *Himachal Himalayas* are mainly confined to Himachal Pradesh. All the three ranges of the Himalaya are represented in this area. The Greater Himalaya is represented by the Zaskar range, lesser Himalaya by Pir Panjal and Dhauladhar ranges and the Outer Himalaya by the Shiwalik range. The main range has some peaks which rise above 6,000 m. The southern slopes are rugged and forested while the northern slopes are bare and show plains with lakes. Zojila La, Rohtang, Bara Lacha La and Shipki La are important passes. This area has beautiful valleys of Kullu, Kangra and Lahul and Spiti.

The *Kumaon Himalayas* lie in Uttarakhand and extend from the Satluj to the Kali river. Its main peaks are Nanda Devi (7,817 m), Kamet (7,756 m), Mana (7,273 m), Trisul (7,140 m), Badrinath (7,138 m), Kedarnath (6,968 m), Jauli or Shivling (6,638 m) and Gayatri (6,515 m). The Gangotri glacier is the source of the sacred Ganga. The region is dotted by about 36 lakes. The Lesser Himalayas in Kumaon Himalaya is represented by the Mussoorie and Nag Tiba ranges. The area has the famous Nainital lake. The Shiwalik in this region is about 900-1000 m high and runs south of the Mussoori range between the Ganga and the Yamuna rivers for a length of about 75 km. The flat valleys between the Lesser Himalaya and the Shiwalik range are called ***'Doons'*** or ***'Duns'*** of which ***Dehra Dun*** is the most famous. The Kumaon

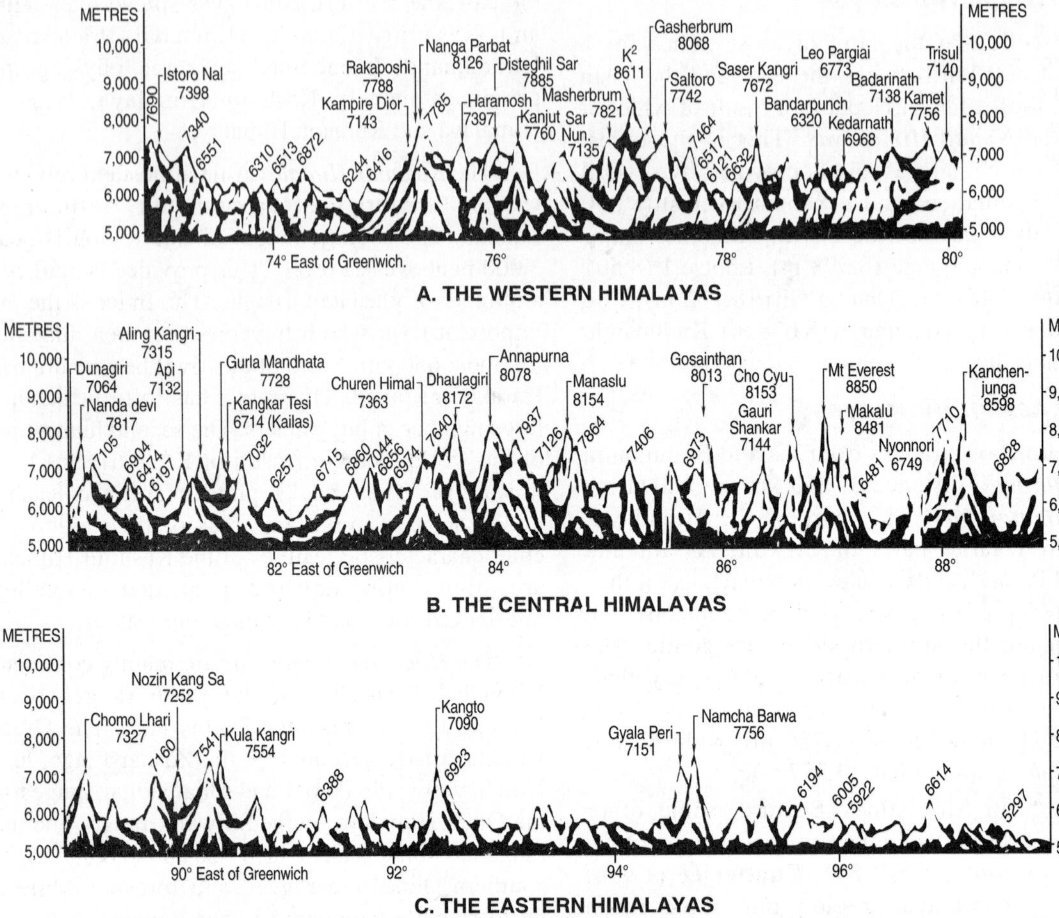

FIG. 3.8. Longitudinal cross section of the Himalayas showing important peaks

Himalaya is connected to Tibet by a number of mountain passes of which Thaya La, Muling La (5,669 m), Mana (5,611 m), Niti (5,068), Marhi La (4,953 m), Kungribingri, Darma and Lipin Laot are important.

2. The Central Himalayas. The Central Himalayas stretch for a distance of about 800 km between river Kali in the west and river Tista in the east. All the three ranges of the Himalayas are present here. The Great Himalaya range attains maximum height in this portion. Some of the world famous peaks Mt. Everest, Kanchenjunga, Makalu, Annapurna, Gosainthan and Dhaulagiri are located here (**Fig. 3.8**). The Lesser Himalaya is known as ***Mahabharat Lekh*** in this region. The range is crossed by rivers like Ghaghara, Gandak, Kosi, etc. In between the Great and the Lesser Himalayas, there are Kathmandu and Pokhra valleys which represent lacustrine deposits. The Shiwalik range comes very close to the lesser Himalaya towards the east and is almost non-existent beyond Narayani (Gandak).

3. The Eastern Himalayas. This part of the Himalayas lies between the Tista river in the west and the mighty Brahmaputra river in the east and stretches for a distance of about 720 km. Also known as the ***Assam Himalayas***, the Eastern Himalayas occupy mainly the areas of Arunachal Pradesh and Bhutan. This part of the Himalayas is known by a number of different names according to the names of the tribal people living in different areas. The prominent peaks of this area are Namcha Barwa (7,756 m), Kula Kangri (7,554 m), Chamo Lhari (7,327 m), Hozin

Kang Sa (7,252 m), Gyalaperi (7,151 m) and Kangto (7,090). A number of mountain passes are found in the Eastern Himalayas. The important passes are Jalep La (4,535 m), Bum La (4,331 m), Tse La (4,740 m), Tunga (5,044 m), Yonggyap (3,962 m), and Kangri Karpo La (5,636 m). The Jelep La in the Chumbi valley of Sikkim and Bum La in Arunachal Pradesh provide passage to the Tibetan Capital Lhasa. The Assam Himalayas show a marked dominance of fluvial erosion due to heavy rainfall.

The Himalayas take a sudden southward turn after the Dihang gorge and the hill ranges running in more or less north-south direction along India's border with Myanmar are collectively known as the ***Purvanchal***. These are known by various local names such as *Patkai Bum*, *Naga hills*, *Kohima hills*, *Manipur hills*, *Mizo hills* (previously known as the Lushai hills), Tripura hills and Barail range. The extension of the Purvanchal Himalaya continues southwards upto Andaman and Nicobar Islands through the Myanmar range and even upto the Indonesian archipelago.

Syntaxial Bends of the Himalayas

It has already been mentioned that the Himalayas extend in the east-west direction from the Indus gorge in the west to the Brahmaputra gorge in the east. This east-west trend of the Himalayas is suddenly terminated at its western and eastern extremities and the Himalayan ranges take sharp southward bends. These bends are called ***syntaxial bends*** of the Himalayas. The bend on the western extremity is called the ***western syntaxial bend*** and that on the eastern extremity is known as the ***eastern syntaxial bend***. The ***western syntaxial bend*** occurs near the Nanga Parbat where the Indus river has cut a deep gorge. It forms a great knee-bend and effects the strike of the ranges probably as far as the foot of the Pamirs. The geological formations take a sharp hairpin bend as if they were bent round pivotal points. The ***eastern syntaxial bend*** occurs at the eastern extremity of the Himalayas where the Himalayan range takes a sharp bend in the southern direction and the eastern trend of the ranges terminates here. The bend is conspicuous in Arunachal Pradesh where the mountain ranges turn southward after crossing the Brahmaputra. The strike of the ranges also has a deep knee-bend towards the south.

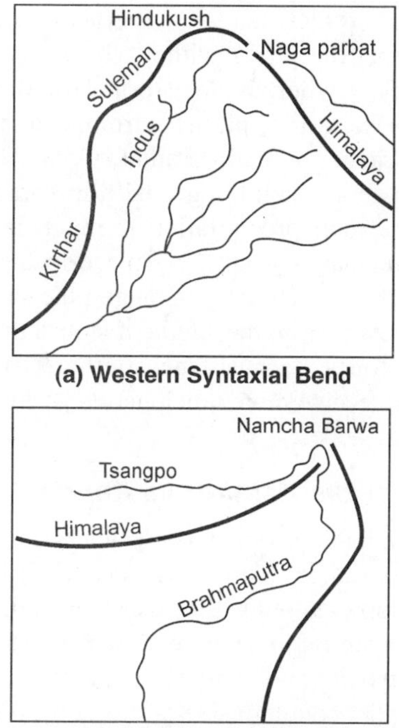

(a) Western Syntaxial Bend

(*b*) Eastern Syntaxial Bend

FIG. 3.9. Syntaxial Bends of the Himalayas

Glaciers of the Himalayas

As the name implies, the Himalayas are the *abode of snow* and are called the ***storehouse of snow***. Out of a total area of five lakh square kilometres of the Himalayas, about 33,000 sq km area is covered by snow. Evidence of large snow areas in the past is available in abundance. The active glaciers and the associated landscape provide an ideal platform for investigation of historic glacial variation. Obviously the Pleistocene ice was far more extensive than that of today. There are about 15,000 glaciers in the Himalayas lying between the two syntaxial bends in the east and west. The *snow line*, *the lowest level of perpetual snow*, varies to a great extent in different parts of the Himalayas depending upon latitude, amount of precipitation and local topography. While the glaciers of Kanchenjunga in Sikkim portion hardly move below 3,965 m, and those of Kumaon and Lahul to 3,660 m, the glaciers of the Kashmir and Ladakh Himalayas may descend to 2,500 m above sea level. **T.D. La Touche** while elaborating the causes of these variations, has suggested that in part it is due to the increase in latitude from 28° N in Kanchenjunga to 36° N in the Karakoram, and in part because of the fact that the eastern Himalayas rise abruptly from the plains without the intervention of high ranges, whereas Western Himalayas rise gradually from the plains through a series of mountain ranges like Pir Panjal, Great Himalaya, Zaskar Range, Ladakh Range and Karakoram Range. Though the total precipitation is much less in the western Himalayas it all takes place in the form of snow. In the Great Himalayan ranges, the snow line is at lower elevation on the southern slopes than on the northern slopes because the southern slopes are steeper and receive more precipitation as compared to the northern slopes.

Glaciers of the Karakoram Range

Maximum development of glaciers occurs in the Karakoram Range. This range accounts for about 16,000 sq km or about half of the snow bound area of the Himalayan region. Some of the largest glaciers outside the polar and sub-polar regions are found in this range. The southern side of this range nourishes a number of gigantic glaciers, some of which are exceeded in size only by the great Humboldt of Greenland. The 75 km long *Siachen Glacier* in Nubra valley has the distinction of being the largest glacier outside the polar and the sub-polar regions. Lolofond and Teram Shehr are its main tributaries. The second largest is the 74 km long *Fedchenko Glacier*. It covers an area of 450 sq km in the north-western Pamir and has a depth of nearly 550 m of ice. Third comes the *Hispar Glacier*. It is 62 km long and occupies a tributary of the Hunza River. It combines with 59 km long *Biafo Glacier* occupying about 65 sq km area of Braldoh valley. Kunyong or Lak (24 km) is an important tributary of the Hispar Glacier. The west flowing *Baltoro Glacier* also flows in the Braldoh valley. The depth of solid ice at the end of the Baltoro Glacier is about 120 m; the thickness in the middle of the glacier could be much greater. The great *Godwin Austen Glacier* drains three sides of K^2 and joins the Baltoro Glacier. The *Punmah Glacier* also flows in the Braldoh valley. This 27 km long glacier has a complex system of branches in its upper reaches. The *Batura Glacier* draining the Hunza is 58 km long and is considered to be the fifth longest glacier, along with Baltoro, outside the polar and sub-polar regions. The sixth largest glacier outside the polar and sub-polar regions is the 50 km long *Chogo Lungma* glacier in the Rakaposhi Range. It terminates at an altitude of 2075 m, the lowest recorded in the

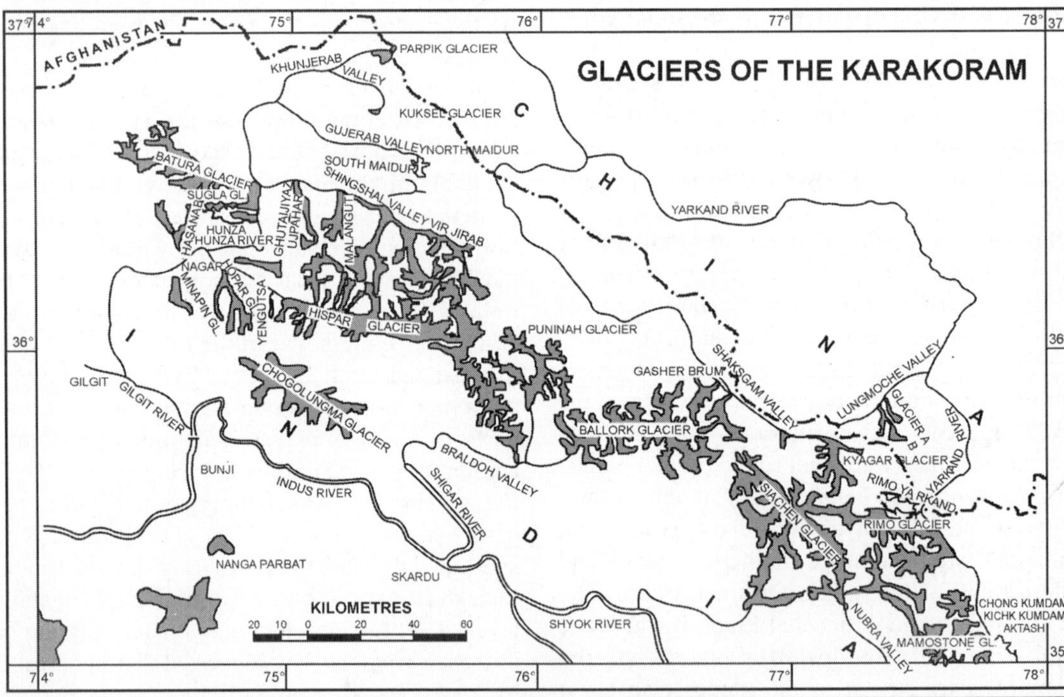

FIG. 3.10. Glaciers of the Karakoram

Himalayas. The other two glaciers draining into the Hunza Valley are the *Pasu* (25 km) and *Sasaini* or *Ghulkin* (18 km). The *Rimo Glacier* occupying the Shyok valley is about 40 km long. It is joined by the 20 km long *Yarkand Rimo* from the north. The other Glaciers in the Shyok valley are the *Chong Kumdan* (21 km), *Kichik Kumdan (*11 km) and *Aktash* (8 km). The *Khurdopla Glacier* in Shingshal valley is 47 km long. The other glaciers of this valley are the *Virjerab Glacier* (40 km), the *Mombil Yaz* (32 km), *Yazghil* (31 km) and the *Malangutti Yaz* (23 km) The *Gasherbrum* and the *Kyagar* in the Shaksgam valley are 21 km long each. The *Urdok Glacier* lying below these two glaciers is nearly 23 km long.

Glaciers of the Pir Panjal Range

The glaciers of the Pir Panjal Range are less numerous and smaller in size as compared to those of the Karakoram Range. The longest *Sonapani Glacier* in the Chandra Valley of Lahul and Spiti region is only 15 km long. Rising at about 4000 m near the Rohtang Pass and flowing in south west direction it has developed a dry lake basin about 2.5 km long and 1.5 km wide, just below the ice cliff. The *Bara Shigri Glacier* also occupies the Chandra Valley. The rough estimates put its length varying from 10 to 20 km. The largest glacier in the Nun Kun massif is the *Gangri Glacier* which is 13 km long. The glaciers of the Nanga Parbat massif are small in size and are moving fast due to steep slope. The *Chungphar, Rakhiot*, *Buzhi* and *Tashan* are important glaciers.

Glaciers of the Kumaon-Garhwal Region

In the Kumaon-Garhwal region of the Himalayas, the largest is the 30 km long *Gangotri Glacier* which is the source of the holy Ganga. It is fed by five tributary glaciers, the largest of which is 24 km long. The second largest glacier of this region is the *Milam Glacier* which occupies the valley of the Gori Ganga River. It is 20 km long and is formed by the union of nine tributary glaciers. The other glaciers of this region are the *Mana Glacier* (18 km), the *Bhagirath-Kharak Glacier* (18 km), the *Satopanch* (16 km) the *Sankalpa Glacier* (12 km) and the *Raikane Glacier* (9 km).

Glaciers of Central Nepal

Central Nepal comprises the mountain ranges between Cho Oyu and Dhaulagiri. Little known

TABLE 3.2. Important Glaciers of the Himalayan Mountains

Name of the Glacier	Length (km)	Location
The Karakoram Range		
1. Siachen	75	Nubra Valley
2. Fedchenko	74	South-Western Pamir
3. Hispar	62	Tributary of Hunza River
4. Biafo	59	Braldoh Valley
5. Batura	58	Hunza
6. Baltoro	58	Braldoh Valley
7. Chogo Lungma	50	Rakaposhi Range
8. Khurdopla	47	Shingshal Valley
9. Lolofond	40	West of Siachen
10. Yarkand Rimo	40	Shyok Valley
11. Mombil Yaz	32	Shingshal Valley
12. Yazhil	31	Shingshal Valley
13. Godwin Austen	30	K_2
14. Pasu	25	Hunza
15. Kunyang	24	Muztagh, Karakoram
16. Chong Kumdon	21	Shyok
17. Gasherbrum	16	Gasherbrum
The Pir Panjal Range		
18. Sonapani	15	Chandra Valley of Labul and Spiti
19. Bara Shigri	10-20	Chandra Valley
20. Rakhiot	15	Nanga Parbat
21. Gangri	13	Nun Kun Massif
22. Chungphar	13	Nanga Parbat
The Kumaon-Garhwal Region		
23. Gangotri	30	Source of the Ganga
24. Milam	20	Gori Ganga
25. Bhagirath Kharak	18	Near Badrinath
26. Mana	18	Mana Valley north of the Gangotri
27. Satopanch	16	Near Badrinath

The Central Nepal Region		
28. Yepokangara	13.5	Gosainthan
29. Lidanda	11	Manaslu
30. Chhuling	11	Manaslu
31. Mayondi	11	Dhaulagiri Himal
The Kanchenjunga-Everest Region		
32. Rongbuk	52	Northern (Tibetan) side of the Mount Everest
33. Zemu	25	Zemu Valley (Forms headwaters of the Tista River)
34. Kanchanjunga	21	North of Kanchanjunga, occupies head of Kangchen river.
35. Khumbu	20	South of Mt. Everest
36. Kangshung	19	East of the Mt. Everest
37. Tolam Bau	19	South-west of the Mt. Everest
38. Barun	15	North-east of Baruntse peak
39. Rambang	10	Kanchanjunga

glaciers of this region are confined to the surroundings of Gosainthan, Manaslu, Annapurna and Dhaulagiri peaks. The 13.5 km long *Yepokangara* glacier is in the Gosainthan mountain. Two 11 km long ones are the *Lidanda* and *Chhuling* glaciers, both on the south-eastern side of the Manaslu. The *Annapurna* glacier is in the Annapurna mountain. Another 11 km long Mayondi glacier is in the north of the Dhaulagiri.

Glaciers of the Kanchenjunga-Everest Region

In the Kanchenjunga-Everest region, there is *Rongbuk Glacier* on the northern side of the Mount Everest. This is 52 km long and is considered to be the largest outside the Karakoram. Another important glacier of the Everest group is the *Khumbu Glacier*, (20 km) lying to the south of the Everest. The 25 km long *Zemu Glacier* flows in an easterly direction at the head of the Zemu River. It is about one kilometre wide with 180 metre thick ice. The 21 km long *Kanchenjunga Glacier* descends from the peak of the same name and occupies the head of Kangchen river. Both the Zemu and the Kanchenjunga glaciers are formed as a result of the union of several branches coming down from the peaks. The *Yalung Glacier* is 16 km long and flows in a south-west direction from the Kanchenjunga peak. The *Talung Glacier* (13 km) also flows southwest wards. This glacier is separated from the *Alukthang Glacier* by a ridge. Though only 5 km long the Alukthang glacier is unique because it is clearly visible from Darjeeling at the foot of the Kanchenjunga in clear weather. The *Kangshung Glacier* is 19 km long and flows east of the Everest. The *Barun Glacier* flowing down the Baruntse peak is about 15 km long.

Main Passes of the Himalayas

Although the mighty Himalayas rise abruptly from the Indus-Ganga-Brahmaputra plain and stand like a great wall providing little opportunity to cross, yet there are some passes across this mountain wall which offer some opportunity to cross their mountain wall. Some of the important passes are briefly described below :

A. Passes of the Western Himalayas

The Western Himalayas are dotted with a number of passes which offer passages through them. Union Territories of Jammu and Kashmir and Ladakh, and states of Himachal Pradesh and Uttarakhand are included in this part of the Himalayas.

I. Jammu and Kashmir and Ladakh

1. Mintaka Pass. It lies near the trijunction of India-China and Afghanistan border and joins Ladakh with China.

2. Parpik Pass. It lies to the east of Mintaka pass on the Indo-China border and joins Ladakh with China.

3. Khunjerab Pass. This pass lies at an altitude of 4934 m on the Indo-China border and gives access between Ladakh and China.

4. Aghil Pass. Lying in the north of K^2 Peak (the highest peak in India and the second highest peak in the world). This pass is situated at an elevation of about 5000 m above sea level and joins the Ladakh region of India with the Xinjiang (Sinkiang) Province of China. Being located at a high altitude and surrounded by lofty mountains, it remains snow covered during the winter season and is closed from November to May.

5. Banihal Pass. It is situated at an elevation of 2832 m across the Pir-Panjal Range. It remains snow covered during winter season and cannot be used as a transport route in that season. To provide round-the-year transport facilities between Jammu in the south and Srinagar in the north, a tunnel named as ***The Jawahar Tunnel*** (after Pandit Jawaharlal Nehru, the first prime minister of India) was inaugurated in December, 1956. Another 11 km long tunnel provides railway link between Banihal and Kazigund. It was thrown open to railway transport in July, 2013.

6. Chang-La. Located at the border between India and China at an altitude of 5360 m, this pass joins Ladakh with Tibet. Road after Chang La is very steep. This has a temple dedicated to Chang-La Baba after whom the temple has been named. This pass remains closed to traffic during the winter season because of heavy snowfall.

7. Khardung La. This pass is situated at an altitude of 5602 m near Leh in the Ladakh range. The world's highest motorable road passes through this pass. However, this road remains closed in winter due to heavy snowfall.

8. Lanak La. Located near the border between India and China at an altitude exceeding five thousand metres in the Akasai-Chin area in north-eastern part of Ladakh, this pass provides passage between Ladakh and Lhasa. A road to connect Xinjiang Province with Tibet has been constructed by the Chinese.

9. Pir-Panjal. Lying across the Pir Panjal range, it had been a traditional pass on the Mughal Road and provides the shortest and the easiest metalled road between Jammu and Kashmir Valley. But this route had to be closed down as a result of partition of the country.

10. Qara Tag La. It is situated on the Indo-China borer across the Karakoram Range. It remains snow bound and closed to traffic during winters because it is located at an elevation of over six thousand metres. It was an offshoot of the Great Silk Route.

11. Imis La. Situated at the Indo-China border at an altitude of over 4500 m, this pass provides passage between Ladakh region of India and Tibet in China. However, the access is not very easy due to difficult terrain and steep slopes. It remains snowbound and closed during the winter season.

12. Pensi La. Situated to the east of the famous Zoji La in the Greater Himalayas at an elevation of over five thousand metres, this pass provides a vital link between the Kashmir Valley and Kargil. It remains closed to traffic from November to mid-May due to heavy snowfall.

13. Zoji La. It is located at an altitude of 3850 m above sea level and provides an important road link between Srinagar on one side and Kargil and Leh on the other side. The road passing through this pass has been designated as National Highway (NH-1D). Border Road Organisation (BRO) is responsible for maintaining the road and clearing it of snow during winter. In spite of all these efforts, the road through this pass remains closed from December to mid-May.

II. Himachal Pradesh

14. Bara Lacha La. This mountain pass is situated at an altitude of 4883 m and provides passage between Himachal Pradesh and Ladakh. National highway connecting Mandi in Himachal Pradesh with Leh in Ladakh passes through this pass. Being situated at high altitude, it remains snow covered in winter and is not used as a transport route.

15. Debsa Pass. Situated at an elevation of 5270 m above sea level in the Greater Himalayas, it provides a link between Kullu and Lahul and Spiti districts. It offers a much easier and shorter alternative route to traditional Pin-Parbati Pass route between Kullu and Spiti.

16. Rohtang Pass. It is located at an altitude of 3979 m and provides road link between Kullu, Lahul and Spiti Valleys. Border Roads Organisation (BRO) is responsible for constructing and maintaining roads in this area. Rohtang pass is a great tourist attraction and traffic jams are very common because this route is widely used by military, public and private vehicles.

17. Shipki La. It is located at the Indo-China border at an altitude of over 5669 m in Kinnaur district and provides a road connection between Himachal Pradesh and Tibet. It remains snow bound for most of the winter season and is not available for transport. River Satluj enters India near this pass.

III. Uttarakhand

18. Lipu Lekh. Situated near the trijunction of Uttarakhand (India), Tibet (China) and Nepal borders, in Pithoragarh district, it provides a link between Uttarakhand and Tibet. This pass is used by pilgrims to Kailash-Mansarowar. Use of this pass becomes difficult due to landslides in the rainy season and avalanches in the winter season.

19. Mana Pass. Situated a little north of the holy place of Badrinath at an elevation of 5610 m near the Indo-China border in the Greater Himalayas, this pass connects Uttarakhand with Tibet. It remains closed for six winter months in the year due to heavy snowfall.

20. Mangsha Dhura Pass. Situated at an altitude of over five thousand metres at the Indo-China border in the Greater Himalayas in Pithoragarh district, this pass connects Uttarakhand with Tibet. It is used by pilgrims going to Kailash-Mansarowar. Landslides during the rainy season and avalanches during the winter season pose great threat to pilgrims using this route.

21. Niti Pass. Located at an altitude of 5068 m at the Indo-China border across the Greater Himalayas, this pass joins Uttarakhand with Tibet. It remains snow covered and hence closed to traffic from November to mid-May.

22. Muling La. It is situated to the north of Gangotri at an elevation of 5669 m in the Great Himalayas. It provides passage between Uttarakhand and Tibet but remains closed during winter season due to heavy snowfall.

B. Passes of the Eastern Himalayas

This part of the Himalayas includes states of Sikkim and Arunachal Pradesh.

IV. Sikkim

23. Nathu La. Situated at an altitude of 4310 m on the Indo-China border, it forms part of an offshoot of the ancient Silk Route. It connects Sikkim with Tibet and is an important trade route between India and China. It was closed after the Chinese aggression on India in 1962 but was reopened in 2006 as the governments of the two countries decided to enhance their trade through land routes. An alternative route to Kailash-Manasarowar was thrown open for pilgrims through this pass in 2015.

24. Jelep La. It lies at the Sikkim-Bhutan border at an altitude of 4538 m and passes through Chumbi Valley. This pass provides an important link between Sikkim and Lhasa.

V. Arunachal Pradesh

25. Bomdi La. Situated at an altitude of 4331 m near the eastern boundary of Bhutan in the Greater Himalayas, this pass connects Arunachal Pradesh with Lhasa. It is snowbound in winter and remains closed for traffic.

26. Dihang Pass. Situated at an elevation of more than 4000 m it provides passage between Arunachal Pradesh and Myanmar.

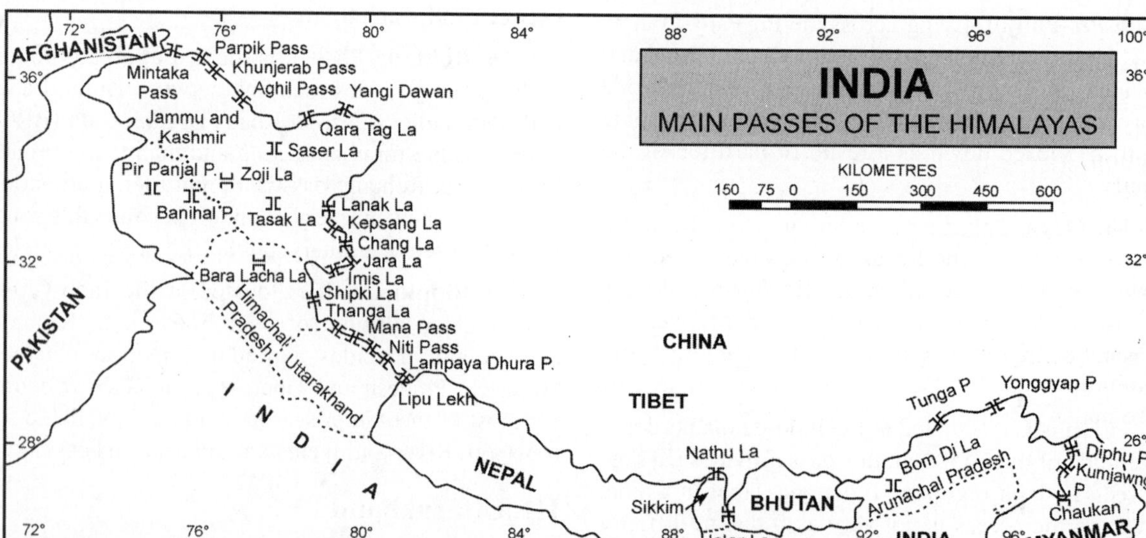

FIG. 3.11. India : Main passes of the Himalayas

27. Yonggyap Pass. It lies at an altitude of 3962 m on the Indo-China border and joins Arunachal Pradesh with Tibet.

28. Diphu Pass. Lying at the trijunction of India, China and Myanmar border at an altitude of 4353 m, it provides an easy access between Arunachal Pradesh and Mandalay in Myanmar. It is an important land trade route between India and Myanmar and remains open throughout the year.

29. Kumjawng Pass. It lies on the Indo-Myanmar border at an altitude of 2929 m and joins Arunachal Pradesh with Myanmar.

30. Hpungan Pass. Lying at an altitude of 3072 m on the Indo-Myanmar border, it provides an important link between India and Myanmar.

31. Chaukan Pass. This pass also lies on the Indo-Myanmar border at an elevation of 2432 m and joins Arunachal Pradesh with Myanmar.

The Significance of the Himalayas

The Himalayas comprise the most dominating geographical feature of India. No other mountain range anywhere in world has influenced the life of people and shaped the destiny of a nation as the Himalayas have in respect of India. *The Himalayas are the body and soul of India*. In a very special measure, the Himalayas constitute India's national mountain system. The following few points will bring out the significance of the Himalayan mountains to India.

1. Climatic Influence. The Himalayas play a very significant role in influencing the climate of India. By virtue of their high altitude, length and direction, they effectively intercept the summer monsoons coming from the Bay of Bengal and Arabian Sea and cause precipitation in the form of rain or snow. Besides, they prevent the cold continental air masses of central Asia from entering into India. Had there been no Himalayas, the whole of India would have been a desert in the absence of precipitation and its winters would have been very severe under the influence of cold airmasses coming from Central Asia. According to the latest meteorological studies, the Himalayas are responsible for splitting the jet stream into two branches and these in turn play an extremely important role in bringing monsoons in India.

2. Defence. The Himalayas have been protecting India from outside invaders since the early times thus serving as a defence barrier. But the Chinese aggression on India in October, 1962 has reduced the defence significance of the Himalayas to a considerable extent. In spite of advancement in modern warfare technology, the defence significance of the Himalayas cannot be ignored altogether.

3. Source of Rivers. All the great rivers of North India have their sources in the Himalayan ranges. Abundant rainfall and vast snow-fields as well as large glaciers are the feeding grounds of the mighty rivers of India. Snow melt in summer provides water to these rivers even during dry season and these are perennial rivers. The Himalayan rivers, along with hundreds of their tributaries, form the very basis of life in the whole of north India.

4. Fertile Soil. The great rivers and their tributaries carry enormous quantities of alluvium while descending from the Himalayas. This is deposited in the Great Plain of North India in the form of fertile soil, making the plain one of the most fertile lands of the world. It has been estimated that the Ganga and the Indus carry 19 and 10 lakh tonnes of silt, per day respectively and the silt carried by the Brahmaputra is even more. It is, therefore, often said that *the great plain of north India is a Gift of the Himalayas*.

5. Hydroelectricity. The Himalayan region offers several sites which can be used for producing hydroelectricity. There are natural waterfalls at certain places while dams can be constructed across rivers at some other places. The vast power potential of the Himalayan rivers still awaits proper utilisation.

6. Forest Wealth. The Himalayan ranges are very rich in forest resources. In their altitude, the Himalayan ranges show a succession of vegetal cover from the tropical to the Alpine. The Himalayan forests provide fuel wood and a large variety of raw materials for forest based industries. Besides many medicinal plants grow in the Himalayan region. Several patches are covered with grass offering rich pastures for grazing animals.

7. Agriculture. The Himalayas do not offer extensive flat lands for agriculture but some of the slopes are terraced for cultivation. Rice is the main crop on the terraced slopes. The other crops are wheat, maize, potatoes, tobacco and ginger. Tea is a unique crop which can be grown on the hill slopes only. A wide variety of fruits such as apples, pears, grapes, mulberry, walnut, cherries, peaches, apricot, etc. are also grown in the Himalayan region.

8. Tourism. By virtue of their scenic beauty and healthy environment, the Himalayan ranges have developed a large number of tourist spots. The hilly areas in the Himalayas offer cool and comfortable climate when the neighbouring plains are reeling under the scorching heat of the summer season. Millions of tourists from different parts of the country as well as from abroad throng the Himalayan tourist centres to enjoy their natural beauty and to escape from the summer heat of the plains. The increasing popularity of winter sports and the craze to enjoy snowfall has increased the rush of tourists in winters also. Srinagar, Dalhousie, Dharamshala, Chamba, Shimla, Kullu, Manali, Mussoorie, Nainital, Ranikhet, Almora, Darjeeling, Mirik, Gangtok, etc. are important tourist centres in the Himalayas.

9. Pilgrimage. Apart from places of tourists interest, the Himalayas are proud of being studded with sanctified shrines which are considered to be the abodes of the Gods. Large number of pilgrims trek through difficult terrain to pay their reverence to these sacred shrines. Kailas, Amarnath, Badrinath, Kedarnath, Vaishnu Devi, Jwalaji, Uttarkashi, Gangotri, Yamunotri, etc. are important places of pilgrimage.

10. Minerals. The Himalayan region contains many valuable minerals. There are vast potentialities of mineral oil in the tertiary rocks. Lignite is found in Kashmir. Copper, lead, zinc, nickel, cobalt, antimony, tungsten, gold, silver, limestone, semi-precious and precious stones, gypsum and magnesite are known to occur at more than 100 localities in the Himalayas. Unfortunately, many of the mineral resources cannot be exploited at the present level of technological advancement due to adverse geographical conditions. Further advancements in modern technology may help in exploiting these resources. So the future possibilities of mineral exploitation in the Himalayas are great.

2. THE GREAT PLAIN OF NORTH INDIA

To the south of the Himalayas and to the north of the Peninsula lies the Great Plain of North India. It is an aggradational plain formed by the depositional work of three major river systems *viz.*, the Indus, the Ganga and the Brahmaputra. This arcuate plain is also known as Indo-Gangetic-Brahmaputra Plain. This is the largest alluvial tract of the world extending for a length of 3,200 km from the mouth of the Indus to the mouth of the Ganga, of which the Indian sector alone accounts for 2,400 km in length. Its average width varies from 150 to 300 km. It is widest in the west where it stretches for about 500 km. Its width decreases in the east. It is about 280 km wide near Pryagraj and 160 km near Rajmahal Hills. It widens to about 460 km in West Bengal but narrows down in Assam where it is only 60-100 km wide. It covers a total area of 7.8 lakh sq km. The northern boundary of this plain is well defined by the foothills of the Shiwaliks but its southern boundary is a wavy irregular line along the northern edge of the Peninsular India.

Rivers flowing through this plain, especially those originating in the Himalayas have deposited a thick layer of alluvium throughout the length and breadth of this plain. Thus it is a classical example of an aggradational plain. However, the thickness of the alluvium deposit varies from place to place and different estimates have been made about it. According to **Oldham**, the maximum depth of the alluvium is about 5,000 m near its southern edge. It has probably maximum depth between Delhi and Rajmahal Hills and is shallow in Rajasthan and Assam. According to recent computations of seismic soundings, the maximum depth of the alluvium upto the basement rocks is about 6,100 m. The depth of alluvial deposits at some important locations are Meerut (1066.8 m), Kalyan (2286.0 m), and Siliguri (5577.8 m). The variation in thickness of the alluvium largely depends upon the morphological processes. The cones of Kosi in the north and those of Son in the south exhibit greater alluvial thickness while the intra-cone areas have relatively shallower deposits.

Extreme horizontality of this monotonous plain is its chief characteristic. Whereas its average elevation is about 200 m above mean sea level, its highest elevation is 291 m above mean sea level between Ambala and Saharanpur. The comparatively higher area near Ambala forms the watershed which divides the drainage system of the Ganga from that of the

Indus. It is, in fact, a low land and is hardly perceptible as one enters Panjab-Haryana plain from the Uttar Pradesh plains. Its average gradient for about 1500 km from Saharanpur to Kolkata is 20 cm per km and it decreases to 15 cm per km from Varanasi to the Ganga delta.

Origin of the Plain

It is almost universally accepted that this vast plain has been formed as a result of filling of a deep depression lying between the Peninsular and the Himalayan regions by the depositional work of the rivers coming from these two landmasses. However, divergent views have been expressed regarding the origin of this great depression and the process of filling it. **Wadia** postulates that these plains were originally a deep depression or furrow lying between the Peninsula and the mountain region. The great Austrian geologist **Edward Suess** has suggested that a "*foredeep*" was formed in front of high crust-waves of the Himalayas as they were checked in their southward advance by inflexible solid landmass of the Peninsula. This foredeep was like a large syncline in which alluvium brought by the Himalayan and the Peninsular rivers was deposited. In due course of time, this was filled with alluvium and the Great Plain of North India was formed. It rests on the hard and crystalline rocks through which the region is connected to the Himalayan and the Peninsular blocks (**Fig. 3.12**). **Sir Sydney Burrard**, on the other hand, thinks that the Indo-Gangetic alluvium conceals a great deep rift, or fracture, in the earth's sub-crust, several thousand metres deep, the hollow being subsequently filled up by detritus. He ascribes to such sub-crustal cracks or rifts a fundamental importance in geotectonics. Such sunken tracts between parallel, vertical dislocations are called ***'Rift Valleys'***. The rift valley between the Himalayan ranges and the Peninsula which gave birth to this plain was about 2,400 km long and hundreds of metres deep. His findings were based on some anomalies in the observations of the deflections of the plumb line and other geodetic considerations. He described some other rift valleys of the Himalayan region as well as the rift valleys of Narmada and Tapi in the Peninsular India. Scholars like **Hayden** and **R.D. Oldham** as well as other geologists of the Geological Survey of India have not accepted Burrard's view of the Indo-Gangetic depression. The main objection to Burrard's views is that there is no trace of a rift valley at the northern edge of the Peninsula and that such a vast rift valley is not possible.

According to the recent views expressed by many geologists and geographers, sediment deposited at the bed of the Tethys Sea was folded and warped due to northward drift of the Peninsula. Consequently the Himalayas and a trough to the south were formed. The origin of this depression or trough, lying at the foot of the mountain, is doubtless intimately connected with the latter. The Great plain represents the infilling of the foredeep warped down between the advancing Peninsular Block and the Himalayas. The infilling has been done by the deposition of the detritus of the mountain brought by the numerous rivers emerging from them during the period of great gradational activity. Geologically most parts of this plain are of the Pleistocene and Recent formations. Thus, the surface deposits of this tract belong to the last chapter of the earth's history and conceal beneath them the older peninsular and other formations.

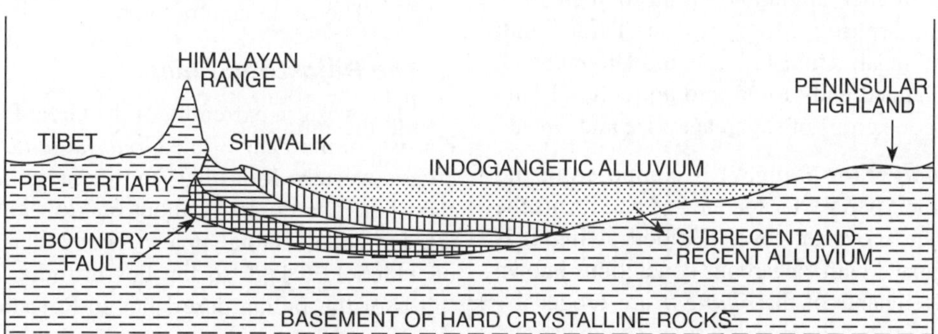

FIG. 3.12. Deposition of alluvium in the Indo-Gangetic trough and formation of the Indo-Gangetic Plain

Geomorphology of the Plain

There is a tendency amongst geographers to treat the Great Plain of North India as a monotonous, flat and featureless plain; but it has its own geomorphological varieties which have their own significance. The following geomorphological features may be noticed depending upon the nature of relief and soil structure.

The Bhabar is a narrow belt about 8-16 km wide running in east-west direction along the foot of the Shiwaliks with a remarkable continuity from the Indus to the Tista. Rivers descending from the Himalayas deposit their load along the foothills in the form of alluvial fans. These fans, consisting of gravel and unassorted sediments, have merged together to build up the bhabar belt which forms the northern boundary of the Great Plain. The porosity of the pebble studded rock beds is so high that most of the streams sink and flow underground. Therefore, the area is marked by dry river courses except in the rainy season. The Bhabar belt is comparatively narrow in the east and extensive in the western and north-western hilly region. The area is not suitable for agriculture and only big trees with large roots thrive in this belt.

The Tarai (Hindi : *'Tar'-wet*) is a 15-30 km wide marshy tract in the south of Bhabar running parallel to it. It is marked by the re-emergence of the underground streams of the Bhabar belt. The re-emerged water converts large areas along the rivers into ill-drained marshy lands of excessive dampness covered with thick forests giving shelter to a variety of wild life. The Tarai is more marked in the eastern part than in the west because the eastern parts of the plain receive higher amount of rainfall than their western counterparts. Most of the Tarai land, especially in Punjab, Uttar Pradesh and Uttarakhand, has been reclaimed and turned into agricultural land which gives good crops of sugarcane, rice and wheat.

The Bhangar **(or Bangar)** is composed of old alluvium of the Middle Pleistocene age and forms the alluvial terrace above the level of flood plains. It is often impregnated with calcareous concretions known as *kankar*. Remnants of the *bhangar* are eroded by every change in the direction of river channels, and are being levelled down by their meandering tendencies. It mostly occupies the Pleistocene terrace dating back to Middle and Upper Pleistocene periods. It also contains fossils of animals like rhinoceros, hippopotamus, elephant, etc.

The Khadar is composed of newer alluvium and forms the flood plains along the river banks. A new layer of alluvium is deposited by river flood almost every year. These deposits are normally confined to the vicinity of the present river channels. The clays have less kankar, and the organic remains entombed in them belong to still living species. The *khadar* imperceptibly merges into the deltaic and other accumulations of prehistoric times.

Reh or ***Kallar*** comprises barren saline efflorescences of drier areas in Uttar Pradesh and Haryana. Reh areas have spread in recent times with increase in irrigation.

Bhur denotes an elevated piece of land situated along the banks of the Ganga river especially in the upper Ganga-Yamuna Doab. This has been formed due to accumulation of wind-blown sands during the hot dry months of the year.

Regional Divisions of the Great Plain of India

Although the Great Plain of North India is treated as a geographical unit with low elevation and gentle slope, this vast area exhibits distinctive fluvial patterns, directions of flow and geomorphology in different parts allowing it to be divided into the following four major regions : (**Fig. 3.13**)

1. The Rajasthan Plain.
2. The Punjab-Haryana Plain.
3. The Ganga Plain.
4. The Brahmaputra Plain.

1. The Rajasthan Plain

The western extremity of the Great Plain of India consists of the *Thar* or the *Great Indian Desert* which covers western Rajasthan and the adjoining parts of Pakistan. The desert is about 650 km long and 250-300 km wide. Its total area is about 2.0 lakh sq km out of which 1.75 lakh sq km lies in India. About two-thirds of the Indian desert lies in Rajasthan, west of the Aravali Range, and the remaining one-third is in the neighbouring states of Haryana, Punjab and Gujarat. Recently, some geomorphic studies by using remote sensing techniques in conjunction with ground truth were undertaken by the Central Arid Zone Research Institute, Jodhpur. This vast desert is an undulating plain whose average elevation is about 325 m above mean sea level. It descends to about 150 m above mean sea level near the Indo-Pak border as well as towards the Indus Valley and the Rann of Kachchh. The desert proper is called ***Marusthali*** and accounts for greater part of the *Marwar* plain. The average elevation of the Marusthali is 200-250 m above mean sea level. It has a vast stretch of sand with a few outcrops of bedrock of gneisses, schists and granites which proves that geologically it is a part of the Peninsular Plateau and it is only at the surface that it looks like an aggradational plain. In general, the eastern part of the Marusthali is rocky while its western part is covered by shifting sand dunes locally known as ***dhrian***.

The eastern part of the Thar Desert, upto the Aravali Range is a semi-arid plain which is known as the ***Rajasthan Bagar***. It runs in a north-east to south-west direction from the edge of the Aravalli in the east to the 25 cm *isohyet* (line joining places of equal rainfall) in the west. It is drained by a number of short seasonal streams originating from the Aravali and supports agriculture in some patches of fertile tracts called ***rohi***. Even the important river, the Luni, is a seasonal stream which flows towards the south-west to the Rann of Kachchh. The tract north of the Luni is known as ***thali*** or sandy plain.

The LANDSAT imageries have been used to study the geomorphological characteristics of this Great Desert including its sand dunes. Based on the morphological, spatial and spectral characteristics, six types of sand dunes, *viz.* obstacle, parabolic, longitudinal, transverse, barchan and shrub-coppice have been identified in this desert. The mean height of the dunes varies from 8 m in Jaipur and Sikar districts to 30 m in Barmer district. The length of the dunes also varies greatly and some may be 5-10 km long.

North of the Luni basin, there is a large area of inland drainage on the eastern edge of the Thar Desert having several saline lakes. They are a great source of common salt and many other salts. The *Sambhar*, the *Didwana*, the *Degana*, the *Kuchaman*, the *Sargol* and

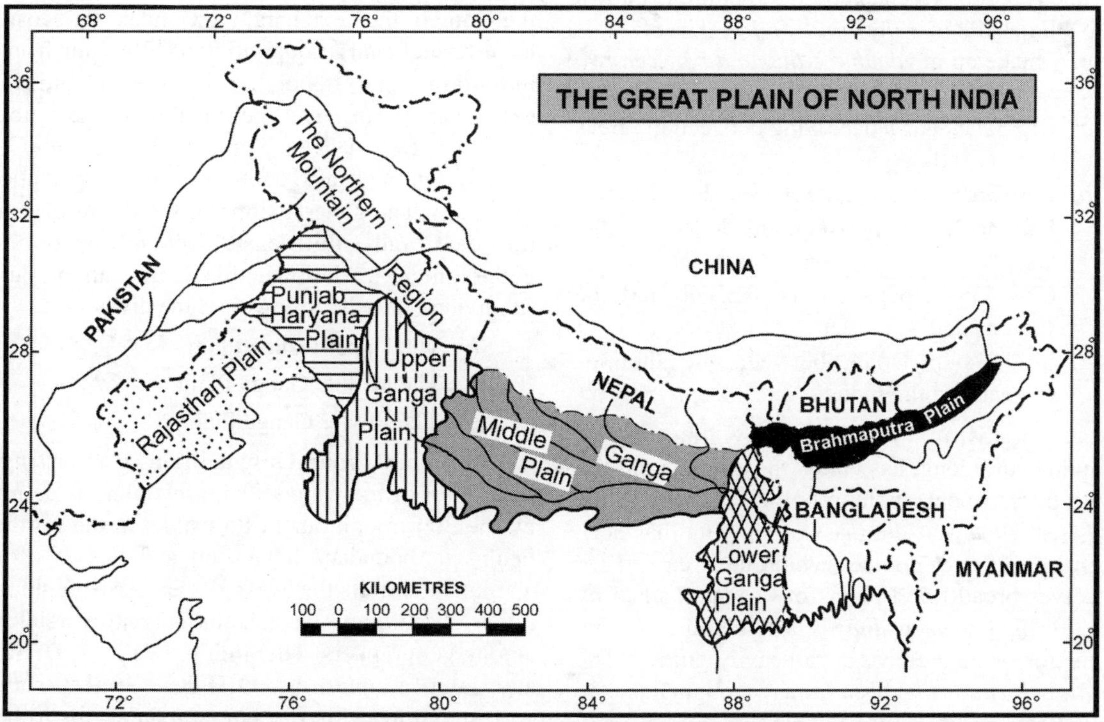

FIG. 3.13. Regional Divisions of the Great Plain of North India

the *Khatu* are some of the important lakes. The largest and the most outstanding is the Sambhar lake about 65 km west of Jaipur. Situated at an altitude of about 360 m in the Aravali terrain, this lake is about 30 km long with an average width of 3-8 km. It occupies an area of about 225 sq km during the rainy season but shrinks considerably in dry season.

2. The Punjab-Haryana Plain

The Great Indian Desert imperceptibly gives way to the fertile plains of the Punjab and Haryana towards the east and north east. The entire plain extends for a length of 640 km in north-west to south-east direction and is about 300 km wide in east-west direction. The total area of this plain is about 1.75 lakh sq km. Its eastern boundary in Haryana is formed by the Yamuna river. The average elevation of the plain is about 250 m above mean sea level. Its northern part is nearly 300 m above mean sea level and it drops to about 200 m in the south-east. Part of the plain shows a flat to slightly convex planation controlled by subsurface Delhi-Aravali ridge.

The part of the plain, formed as a result of alluvial deposits by five rivers, *viz*., the Satluj, the Beas, the Ravi, the Chenab and the Jhelum, is known as the Punjab Plain—***the land of five rivers.*** It is primarily made up of '***doabs***'—*the land between two rivers*. From east to west these doabs are as under :

(*a*) Bist-Jalandhar Doab, lying between the Beas and the Satluj;

(*b*) Bari Doab, between the Beas and the Ravi;

(*c*) Rachna Doab, between the Ravi and the Chenab;

(*d*) Chaj Doab, between the Chenab and the Jhelum; and

(*e*) Sind Sagar Doab, between the Jhelum-Chenab and the Indus.

The depositional process by the rivers, continuing since long, has united these doabs and has given a homogenous geomorphological entity to the entire area. However, the mass of alluvium has been broken by the river courses which have carved for themselves broad flood plains of khadar flanked by bluffs, locally known as ***dhayas***. These bluffs, as high as 3 metres or more, have been heavily gullied. The *khadar* belt, known as ***bet lands***, though liable to flooding, is agriculturally valuable due to fertile soil deposited by rivers during annual floods.

The northern part of this plain adjoining the Shiwalik hills has been intensively eroded by numerous streams called ***Chos***. This has led to enormous gullying. The erosion by the Chos is particularly noticed in Hoshiarpur district of Punjab. In a short stretch of about 130 km nearly a hundred Chos debouch on the plains. To the south of the Satluj river there is *Malwa* plain of Punjab.

The area between the Ghaggar and the Yamuna rivers lies in Haryana and is often termed as ***'Haryana Tract'***. It acts as water-divide between the Yamuna and the Satluj rivers. The only river between the Yamuna and the Satluj is the Ghaggar which is considered to be the present day successor of the legendary **Saraswati River**.

3. The Ganga Plain

This is the largest unit of the Great Plain of North India stretching from Delhi to Kolkata in the states of Uttar Pradesh, Bihar and West Bengal covering an area of about 3.75 lakh sq km. The Ganga is the master river after whose name this plain is named. The Ganga along with its large number of tributaries originating in the Himalayan ranges *viz*., the Yamuna, the Gomati, the Ghaghara, the Gandak, the Kosi, etc. have brought large quantities of alluvium from the mountains and deposited it here to build this extensive plain. The peninsular rivers such as Chambal, Betwa, Ken, Son, etc. joining the Ganga river system have also contributed to the formation of this plain. The general slope of the entire plain is to the east and south-east. Depending upon its geographical variations, this plain can be further subdivided into the following three divisions :

(*a*) The Upper Ganga Plain.

(*b*) The Middle Ganga Plain.

(*c*) The Lower Ganga Plain.

(*a*) The Upper Ganga Plain. Comprising the upper part of the Ganga Plain, this plain is delimited by the 300 m contour in Shiwaliks in the north, the Peninsular boundary in the south and the course of the Yamuna river in the west. Its eastern boundary is rather obscure and has become a controversial topic among geographers. The limit drawn by **L.D. Stamp** and later adopted by **O.H.K. Spate**, roughly corresponding with 100 cm *isohyet* seems to be far from being practical. Physiographically, the 100 m contour (line joining places of equal height) has been accepted by the geographers as the most effective line of demarcation. This plain is about 550 km long in the east-west direction and nearly 380 km wide in north-south direction, covering an approximate area of 1.49 lakh sq km. Its elevation varies from 100 to 300 m above mean sea level. The plain is drained by the Ganga and its tributaries like the Yamuna, the Ram Ganga, the Sarda, the Gomati and the Ghaghara rivers. Almost all the rivers follow NW-SE course concomitant with the lie of the land. The average gradient of the land is about 25 cm per km. The gradient is comparatively steep in the northern part. The rivers flow sluggishly in the plain as the gradient decreases. The monotony of this flat and featureless plain is broken by the *tarai-bhabar* submontane belt and on micro level by the river bluffs, river meanders and oxbow lakes, levees, abandoned river courses, sandy stretches (Bhurs) and the river channels themselves. The western part of this plain consists of comparatively higher Ganga-Yamuna Doab. East of this doab are the low lying Rohilkhand plains which merge into the Avadh plains further east. The Ghaghara is the main stream of the Avadh Plains. The khadar of this river is very wide because the river meanders through this area. Moreover it often changes its course. At places this khadar is 55 km wide. Consequently, there is a constant fear of devastating floods.

(*b*) **The Middle Ganga Plain.** To the east of the Upper Ganga plain is Middle Ganga plain occupying eastern part of Uttar Pradesh and Bihar. It measures about 600 km in east-west and nearly 330 km in north-south direction accounting for a total area of about 1.44 lakh sq km. Its northern and southern boundaries are well defined by the Himalayan foothills and the Peninsular edge respectively. Its western and eastern boundaries are rather ill defined and the region is wide open on both the sides giving it the personality of the *east-west continuum* of the vast isotropic Ganga Plain. There is no physical boundary worth the name and the plain imperceptibly opens up in the west from out of the upper Ganga Plain and so invisibly dies out into the lower Ganga Plain in the east. As such, it is a *transitional region par excellence*, interposed in the enormity of the Ganga Valley. However, several efforts have been made to demarcate the western and the eastern boundaries of this transitional zone. The most accepted boundaries are those made by 100 m contour in the west, 75 m contour in the north-east and by 30 m contour in the south-east. Obviously this is a very low plain, no part of which exceeds 150 m in elevation.

This plain is drained by the Ghaghara, the Gandak and the Kosi rivers, all tributaries of the Ganga coming from the Himalayas. These rivers are responsible for filling up with alluvial deposits of 2,000 metre deep trough at the foot of the Nepal Himalayas. They flow sluggishly in this flat land as a result of which the area is marked by local prominences such as levees, bluffs, oxbow lakes, marshes, *tals*, ravines, etc. The *kankar* formation is comparatively less due to the preponderance of the *khadar*. Almost all the rivers keep on shifting their courses making this area prone to frequent floods. The Kosi river is very notorious in this respect. It used to flow near Purnea in 1736 and now its course is about 110 km west of it. At occasions its water level has risen by 10 metres in a short span of 24 hours. It has long been called the ***'Sorrow of Bihar'***. Strenuous efforts both by India and Nepal are being made to tame this river. The major units of this plain are Ganga-Ghaghara doab, Ghaghara-Gandak doab and Gandak-Kosi doab (Mithila Plain).

Some rivers join the Ganga from the south also, the Son being the most important. Here the gradient is a bit steeper, 45 cm per km, as compared to 9-10 cm per km in east Uttar Pradesh and only 6 cm per km in the Mithila Plain. East of Son river is the Magadh Plain.

(*c*) **The Lower Ganga Plain**. This plain includes the Kishanganj district in Bihar, the whole of West Bengal (excluding the Purulia district and the mountainous parts of Darjeeling district) and most parts of Bangladesh. It measures about 580 km from the foot of the Darjeeling Himalaya in the north to Bay of Bengal in the south and nearly 200 km from the Chotanagpur Highlands in the west to the Bangladesh border in the east. The total area of this plain is about 81 thousand sq km. Its width varies greatly and it narrows down to a mere 16 km between the Rajmahal Hills and the Bangladesh border. The 50 m contour roughly corresponds with its western boundary. The northern part of this plain has been formed by the sediment deposited by the Tista, Jaldhaka and Torsa. Besides, this area is marked by the Duars (Darjeeling *Tarai*) and the Barind plain, a tract of old alluvium between the Kosi-Mahananda corridor in the west and the river Sankosh in the east. The delta formation accounts for about two-thirds of this plain. This is the largest delta in the world. The Ganga river divides itself into several channels in the delta area. The slope of the land here is a mere 2 cm per km. Two thirds of the area is below 30 m above mean sea level. The entire land upto Kolkata would be completely submerged if the sea level rose by only 7 metres. The seaward face of the delta is studded with a large number of estuaries, mud flats, mangrove swamps, sandbanks, islands and forelands. Large part of the coastal delta is covered by thick impenetrable tidal forests. These are called the ***Sunderbans*** because of the predominance of *Sundri* tree here.

4. The Brahmaputra Plain

This is also known as the Brahmaputra valley or Assam Valley or Assam Plain as most of the Brahmaputra valley is situated in Assam. Though often treated as the eastern continuation of the Great Plain of North India, it is a well-demarcated physical unit girdled by the Eastern Himalaya of Arunachal Pradesh in the north, Patkai and Naga Hills in the east and the Garo-Khasi-Jaintia and Mikir Hills in the south. Its western boundary is formed by the Indo-Bangladesh border as well as the boundary of the lower Ganga Plain. Extending from the easternmost end of Assam near the syntaxial bend of the Eastern Himalayas to the west of Dhubri near the Bangladesh border this plain is about 720 km long and its average width is 60-100 km. The entire plain covers an area of about 56 thousand sq km. It is an aggradational plain built up by the depositional work of the Brahmaputra and its tributaries. The Brahmaputra river enters this plain near Sadiya and flows farther to Bangladesh after turning southwards near Dhubri. The general level of the plain varies from 130 m in the east to 30 m in the west. The average gradient of the land is 12 cm per km in the N.E. to S.W. direction. The area is well demarcated by 150 m contour beyond which the surrounding hill terrain dominates the scene. The northern margin has steep slope from the foothills of Arunachal Pradesh but the southern margin is marked by gradual fall from the hill ranges. The innumerable tributaries of the Brahmaputra river coming from the north debouch abruptly upon the main valley and form a number of alluvial fans. Consequently, the tributaries branch out in many channels giving birth to river meandering leading to formation of *bills* and ox-bow lakes. There are large marshy tracts in this area. The alluvial fans formed by the coarse alluvial debris have led to the formation of tarai or semi-tarai conditions. Some southern tributaries also have meandering courses and there are a good number of *bills* and ox-bow lakes.

Significance of the Plain

With its fertile alluvial soils, flat surface, slow moving perennial rivers and favourable climate, the Great Plain of North India is of great significance. It is the home of about half of the Indian population although it accounts for less than one fourth of the total area of the country. The plain supports some of the highest population densities depending purely upon agro-based economy in some of these areas. The extensive use of irrigation has made some parts of this plain, especially Punjab, Haryana and western part of Uttar Pradesh the *granary* of *India*. The entire plain except the Thar Desert, has a close network of roads and railways which has led to large scale industrialisation and urbanization. The development of trade and commerce in this plain is a natural sequel of industrialization and urbanization. There are many religious places along the banks of the sacred rivers like the Ganga and the Yamuna which are very dear to Hindus. Here flourished the religions of Budha and Mahavira and the movements of Bhakti and Sufism. In short, this vast plain is the hearthrob of India and constitutes its very soul.

3. THE PENINSULAR PLATEAU

The Peninsular Plateau of India is roughly triangular in shape with base coinciding with the southern edge of the great plain of North India and its apex is formed by Kanniyakumari in the southern extremity. The northern boundary of the Peninsular block is an irregular line running from Kachchh along the western flank of the Aravali range to near Delhi, and thence roughly parallel to the Yamuna and the Ganga as far as the Rajmahal Hills and the Ganga Delta. It is surrounded by the hill ranges on all the three sides. To its north are the Aravali Range, the Vindhya, the Satpura, the Bharmer and the Rajmahal Hills. To the south of about 22° N latitude, the Western Ghats (The Sahyadris) and the Eastern Ghats form its western and eastern boundaries respectively (**Fig. 3.14**). The entire

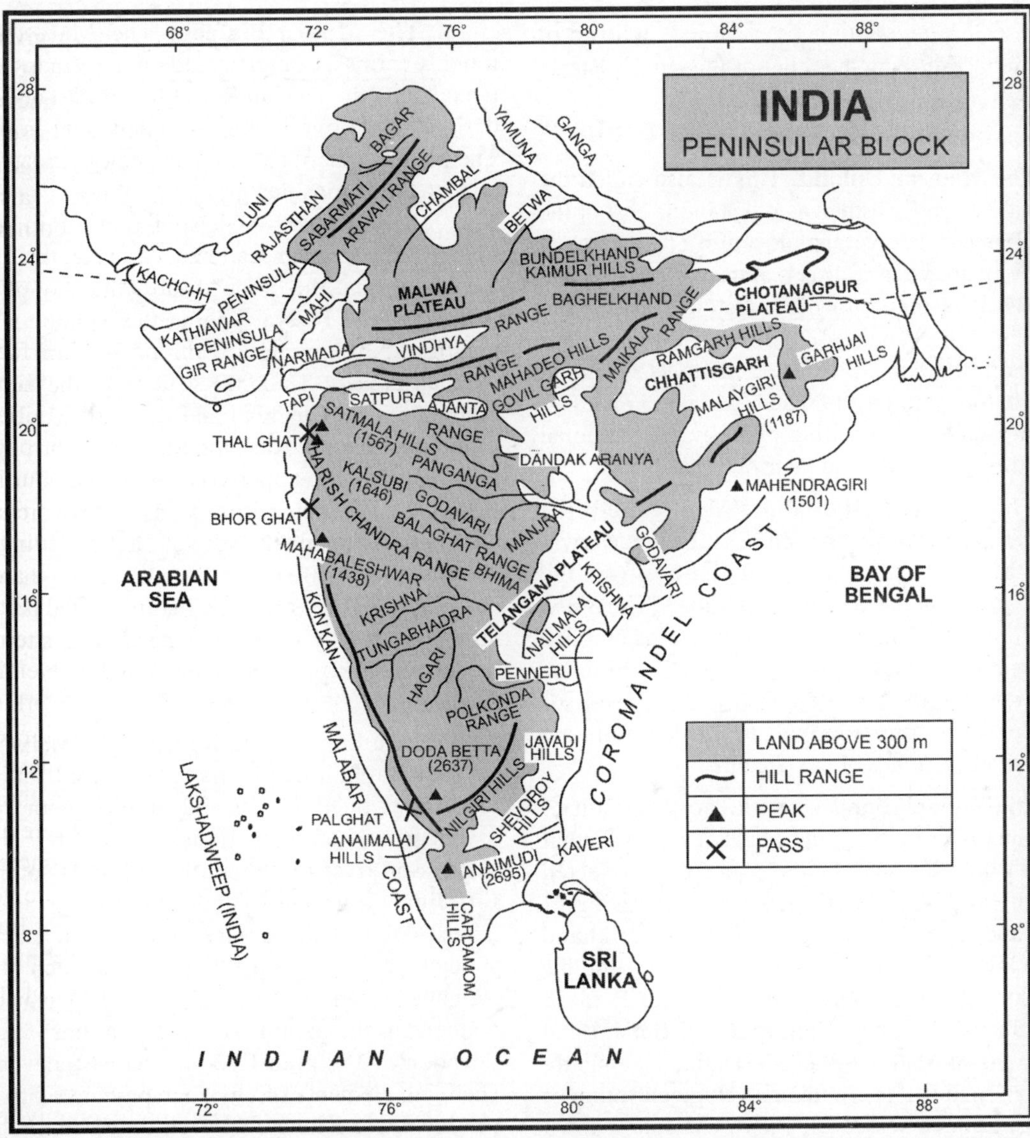

FIG. 3.14. India : Peninsular Block

plateau measures about 1,600 km in north-south and 1,400 km in east-west direction. It covers a total area of about 16 lakh sq km which is about half of the total land area of the country. It is thus the largest physiogrpahic unit of India. The average height of the plateau is 600-900 m above sea level although many parts are well over 1000 m. The general slope of the palteau is from west to east with the exception of Narmada-Tapi rifts which slope westwards.

The Peninsular Plateau is an ancient tabular block composed mostly of the Archaean gneisses and schists. It has been a stable shield which has gone through little structural changes since its formation. Ever since the dawn of geological history the Peninsula has been a land area and has never been submerged beneath the sea except in a few places where marine transgressions have been made and that too locally and temporarily. Undoubtedly, the entire Peninsular Plateau is an aggregation of several smaller plateaus and hill ranges interspersed with river basins and valleys. A brief description of these plateaus, hill ranges and the intervening river valleys will be of great help in describing the relief of this great plateau.

The Plateaus of the Peninsular India

1. The Marwar Upland. This is also called the Upland of eastern Rajasthan as it lies in the east of the Aravali Range. The average elevation of this area is 250-500 m above sea level and it slopes down eastwards. It is made up of sandstone, shales and limestones of the Vindhayan period. The Banas river originates in the Aravali Range and flows for about 400 km before joining the Chambal river. The area has been carved into a rolling plain by the erosional work of the Banas river and its tributaries.

2. The Central Highland. Also called the ***Madhya Bharat Pathar*** is in the east of the Marwar Upland. Most of it comprises of the basin of the Chambal river which flows in a rift valley. The *Sindh* and the *Parbati* are its main tributaries. It is an open rolling plateau made of old rocks which is interspersed with rounded hills composed of sandstone. Thick forests grow here. To the north are the *ravines* or *badlands* of the Chambal river.

3. The Bundelkhand Upland. To the south of the Yamuna river between the Madhya Bharat Pathar and the Vindhyan scarplands is the old dissected upland of the 'Bundelkhand gneiss' comprising of granite and gneiss. This is called Bundelkhand upland. Covering an area of about 54,560 sq km this upland spreads over five districts of Uttar Pradesh (Jalaun, Jhansi, Lalitpur, Hamirpur and Banda) and four districts of Madhya Pradesh (Datia, Tikamgarh, Chhatarpur, and Panna). With an average elevation of 300-600 m above sea level, this area slopes down from the Vindhyan Scarp toward the Yamuna River. The area is recognised by a mass of rounded hummocky hills made of granite and sandstone. In the north-west and north-east, the system is covered by the Ganga-Yamuna alluvium and in the south-west by the Deccan Trap. The erosional work of the rivers flowing here has converted it into an undulating area and rendered it unfit for cultivation. The region is characterised by senile topography. About 67.7 per cent of the area is under 300 m and only 3.6 per cent is above 450 m in altitude. Streams like Betwa, Dhasan and Ken have carved out steep gorges, precipitous rocky banks and waterfalls.

4. The Malwa Plateau. The Malwa Plateau roughly forms a triangle based on the Vindhyan Hills, bounded by the Aravali Range in the west and sharply defined scarp overlooking Bundelkhand in the east. The plateau inherits a complex geology; scarcely any one of the peninsular groups is unrepresented here. This plateau has two systems of drainage; one towards the Arabian sea (The Narmada, the Tapi and the Mahi), and the other towards the Bay of Bengal (Chambal and Betwa, joining the Yamuna). With a length of 530 km and a width of 390 km, it spreads over an area of 1,50,000 sq km. In the north it is drained by the Chambal and many of its right bank tributaries like the Kali, the Sindh and the Parbati. It also includes the upper courses of the Ken and the Betwa. It is composed of extensive lava flow and is covered with black soils. The general height decreases from 600 m in the south to less than 500 m in the north. There are rolling surfaces and flat-topped hills dissected by rivers flowing through the area. In the north, the plateau is marked by the *Chambal ravines*.

5. The Baghelkhand. East of the Maikal Range is the Baghelkhand made of limestones and sandstones on the west and granite in the east. It covers an area of about 1.4 lakh sq km. It is bounded by the Son river on the north, and to its south occur anticlinal highlands and synclinal valleys of sandstones and limestones. The central part of the plateau acts as a water divide between the Son drainage system in the north and the Mahanadi river system in the south. The region has a general elevation of 150 m to 1,200 m and has uneven relief. The main elements of physiography are scarps of the Vindhyan sandstones between the Ganga plain and the Narmada-Son trough. The Bhanrer and Kaimur are located close to the trough-axis. The general horizontality of the strata shows that this area has not undergone any major disturbance.

6. The Chotanagpur Plateau. East of Baghelkhand, the Chotanagpur plateau represents the north-eastern projection of the Indian Peninsula. It covers an area of over 87 thousand sq km mostly in Jharkhand, northern part of Chhatisgarh and Purulia district of West Bengal. The Son river flows in the north-west of the plateau and joins the Ganga. The average elevation of the plateau is 700 m above sea level. This plateau is composed mainly of Gondwana rocks with patches of Archaean granite and gneisses and Deccan Lavas. The Chotanagpur plateau virtually consists of a series of plateaus standing at different levels of elevation. The highest general elevation of about 1,100 m is in the mid-western portion known as the *Pat* lands (high level laterite plateau). From here, the land descends in all directions in a series of steps which are marked by waterfalls across the rivers. The plateau is drained by numerous rivers and streams in different directions and presents a *radial drainage* pattern. Rivers like the Damodar, the Subarnrekaha, the North Koel, the South Koel and the Barakar have developed extensive drainage basins. The Damodar river flows through the middle of this region in a rift valley from west to east. Here are found the Gondwana coal fields which provide bulk of coal in India.

North of the Damodar river is the **Hazaribagh plateau** with an average elevation of 600 m above mean sea level. This plateau has isolated hills; Parasnath in the east rises to 1,366 m. The area is made of granites and gneisses while the hills have quartz rocks. It looks like a peneplain due to large scale erosion.

The **Ranchi Plateau** to the south of the Damodar Valley rises to about 600 m above mean sea level. The maximum height is found in western part where *pats* or high *mesas* capped with laterite rise steeply to an altitude of about 1100 m. The *Netarhat Pat* and *Goru* rise to 1,119 and 1,142 m above sea level respectively. Most of the surface is rolling where the city of Ranchi (661 m) is located. At places it is interruped by monadnocks and conical hills.

The **Rajmahal Hills** forming the north eastern edge of the Chotanagpur Plateau are mostly made of basalt and are covered by lava flows. They run in north-south direction and rise to average elevation of 400 m (highest mount is 567 m). These hills have been dissected into separate plateaus.

7. The Meghalaya Plateau. The rocks of the peninsular plateau of India extend further north-east beyond the Rajmahal hills and form a rectangular block known as the Meghalaya or the Shillong plateau. This plateau has been separated from the main block of the peninsular plateau by a wide gap known as the ***Garo-Rajmahal Gap***. This gap was formed by down-faulting and was later on filled by sediments deposited by the Ganga. Extending over an area of about 35 thousand sq km, this plateau is largely formed of Archaean (Dharwarian) quartzites, shales and schists with granite intrusions and some basic silts. The plateau slopes down to Brahmaputra valley in the north and the Surma and Meghna valleys in the south. Its western boundary more or less coincides with the Bangladesh border. The western, central and the eastern parts of the plateau are known as the ***Garo Hills*** (900 m), the ***Khasi-Jaintia Hills*** (1,500 m) and the ***Mikir Hills*** (700 m). Shillong (1,961 m) is the highest point of the plateau.

8. The Deccan Plateau. This is the largest unit of the Peninsular Plateau of India covering an area of about five lakh sq km. This triangular plateau is bounded by the Satpura and the Vindhya in the north-west, the Mahadev and the Maikal in the north, the Western Ghats in the west and the Eastern Ghats in the east. With an average elevation of 600 m it rises to 1000 m in the south but dips to 500 m in the north. Its general slope is from west to east which is indicated by the flow of its major rivers like the Mahanadi, the Godavari, the Krishna and the Cauvery. These rivers have further subdivided this plateau into a numbmer of smaller plateaus described as under :

(a) The Maharashtra Plateau lies in Maharashtra and forms the northern part of the Deccan Plateau. Much of the region is underlain by basaltic rocks of lava origin. The area looks like a rolling plain due to weathering. The horizontal lava sheets have led to the formation of typical *Deccan Trap* topography. The broad and shallow valleys of the Godavari, the Bhima and the Krishna are flanked by flat-topped steep sided hills and ridges. The Ajanta range lies to the south of the Tapi river. The entire area is covered by black cotton soil known as ***regur***.

(b) The Karnataka Plateau also known as the Mysore plateau lies to the south of the Maharashtra plateau. Made up primarily of the Archaean formations, it is a rolling country with an average elevation of 600-900 m. It is highly dissected by numerous rivers rising from the Western Ghats. It contains the heads of the *Tungabhadra* and the *Cauvery rivers*. The general trend of the hills is either parallel to the Western Ghats or athwart it. The highest peak (1913 m) is at Mulangiri in *Baba Budan Hills* in Chikmagalur district. The plateau is divided into two parts called ***Malnad*** and ***Maidan***. The

Malnad in Kannada means hill country. It is dissected into deep valleys covered with dense forests. The *Maidan* on the other hand is formed of rolling plain with low granite hills. The plateau tapers between the Western Ghats and the Eastern Ghats in the south and merges with the Niligiri hills there.

(*c*) ***The Telangana plateau*** consists of Archaean gneisses at an average elevation of 500-600 m. The southern part is higher than its northern counterpart. The region is drained by three river systems, the Godavari, the Krishna and the Penneru. The entire plateau is divided into two major physiographic regions, namely, the *Ghats* and the *Peneplains*.

9. The Chhattisgarh Plain. The Chhattisgarh plain is the only plain worth the name in the vast stretch of plateaus and hill ranges of the Peninsular plateau. It is a saucer shaped depression drained by the upper Mahanadi. The whole basin lies between the Maikala Range and the Odisha hills. The region was once ruled by *Haithaivanshi Rajputs* from whose thirty six forts (*Chhattis garh*) it derives its name. The basin is laid with nearly horizontal beds of limestone and shales deposited during the Cuddapah age. The general elevation of the plain ranges from 250 m in the east to 330 m in the west.

Hill Ranges of the Peninsular Plateau

The above mentioned plateaus of the Peninsula are separated from one another by hill ranges and river valleys. The hills of the Peninsular India are much lower than the Himalayan ranges. Most of these hills are of the relict type, being remnants of the originally higher mountains but some typical horsts are also seen. Some of the important hill ranges are described as under :

1. The Aravali Range. One of the major physiographic elements of the Peninsular India is the Aravali range running in a north-east to south-west direction for 800 km between Delhi and Palanpur (near Ahmedabad) in Gujarat. The Aravalis represent the relict of the world's oldest mountain formed as a result of folding at the close of the Archaean era. Obviously, their dimensions were much larger and probably even higher than the present Himalayas. Although its north-eastern end is marked by the Delhi ridge, it is supposed to continue upto Haridwar buried under the allivium of the Great Plains. **M.S. Krishnan,** has expressed the view that one Branch of the Aravalis extends to the Lakshadweep Archipelago through the Gulf of Khambhat and the other into Andhra Pradesh and Karnataka. According to **D.N. Wadia**, the Aravalis were a prominent feature in the old Palaeozoic and Mesozoic periods and extended as a continuous chain of lofty mountains from Deccan to possibly beyond Garhwal. What we at present see of them are the eroded remnants of these mountains; their mere stumps laid bare by repeated cycles of erosion.

Prof. S.P. Chatterjee has brought out some interesting features of the Aravali Range. According to the conclusion drawn by him, the Aravali, which occupies the site of an ancient geosyncline and was uplifted and folded into a synclinorium for the first time during the *Algonkian* age, 600 to 700 million years ago. The present Aravali has lost its range character in many parts and is now a relic of what it was in the past when it formed India's main watershed, extending from Kumaon Himalaya to the farthest end of the Peninsular plateau on the south, and perhaps one of its arms reaching eastwards across Central India. In that remote age several of its summits rose above the snow-line and nourished glaciers of stupendous magnitude, which in their turn fed many rivers. This range was later reduced by long continuous erosion almost to sea level. There is evidence to suggest that in the late Mesozoic times, perhaps about 100 million years ago, the peneplaned Aravali was uplifted for the second time by at least 1200 m near Udaipur, and 200 m at its two ends near Delhi and Ahmedabad.

Although its general elevation is only 400-600 m, parts of this range rise well above 1,000 m. The Aravali range has a lower elevation between Delhi and Ajmer where it is characterised by a chain of detached and discontinuous ridges. However, it becomes a continuous range south of Ajmer where it rises to 900 m. At the south-west extremity the range rises to over 1,000 m. Here Mt. Abu (1,158 m), a small hilly block, is separated from the main range by the valley of the Banas. Guru Sikhar (1,722 m), the highest peak, is situated in Mt. Abu. Barr, Pipli Ghat, Dewair and Desuri passes allow movement by road and rail.

2. The Vindhyan Range. The Vindhaya Range rises as an escarpment flanking the northern edge of the Narmada-Son Trough overlooking the Narmada valley. It runs more or less parallel to the Narmada Valley in an east-west direction from Jobat in Gujarat to Sasaram in Bihar for a distance of over 1,200 km. The general elevation of the Vindhyan Range is 300 to 650 m and it rarely goes above 700 m. The northern slope of this range is rather gentle and there are no well marked spurs, steep falls and valleys. Most parts of the Vindhayan Range are composed of horizontally bedded sedimentary rocks of ancient age. The western part of this range is covered with lava. The Vindhayas are continued eastwards as the *Bharner* and *Kaimur* hills. The Vindhya-Kaimur scarp exceeds 610 m at a few places only. This range acts as a watershed between the Ganga system and the river systems of south India and forms the northern boundary of the Deccan. The rivers Chambal, Betwa and Ken rise within 30 km of the Narmada.

3. The Satpura Range. '*Sat*' in Sanskrit means seven and '*pura*' means mountains. Therefore, the Satpura range is a series of seven mountains. It runs in an east-west direction south of the Vindhyas and in between the Narmada and the Tapi, roughly parallel to these rivers. Commencing from the Rajpipla Hills in the west, through the Mahadev Hills to the Maikala Range, it stretches for a distance of about 900 km. These hills appear to be affected by tectonic disturbances. There is evidence that parts of the Satpuras have been folded and upheaved. They are regarded as structural uplift or 'horst'.

No other east-west tectonic mountain of Peninuslar India is as high as the Satpura. Most of the hills rise to elevation of 900 to 1,000 m. Dhupgarh (1,350 m) near Pachmarhi on Mahadev Hills is the highest peak. The other peaks are the Astamba Dongar (1,325 m) and Amarkantak (1,127 m).

4. The Western Ghats (or The Sahyadris). Forming the western edge of the Deccan tableland, the Western Ghats run in north-south direction, parallel and close to the Arabian Sea coast, from the Tapi valley (21° N latitude) to a little north of Kanniyakumari (11° N latitude) for a distance of 1,600 km. As the name implies, the Western Ghats are, down to Malabar, steep-sided, terraced, flat-topped hills or cliffs presenting a mesa-like stepped topography facing the Arabian Sea coast. This is due to the horizontally bedded lavas, which on weathering, have given a characteristic '*landing stair aspect*' to the relief of this mountain chain. South of Malabar, the Nilgiris, Anamalai, etc. present quite different landscape due to the difference in geological structure. The Western Ghats abruptly rise as a sheer wall to an average elevation of 1,000 m from the Western Coastal Plain and appear to be an imposing mountain. But they slope gently on their eastern flank and hardly appear to be a mountain when viewed from the Deccan tableland.

The northern section of the ghats from 21° N to 16° N latitudes *i.e.* from Tapi valley to a little north of Goa is made of horizontal sheets of Deccan lavas forming a formidable wall looking over the Western Coastal Plain. The average height of this section of the Ghats is 1,200 m above mean sea level, but some peaks attain more heights. Kalasubai (1,646 m) near Igatpuri, Salher (1,567 m) about 90 km north of Nashik, Mahabaleshwar (1,438 m) and Harishchandragarh (1,424 m) are important peaks. Thalghat and Bhorghat are important passes which provide passage by road and rail between the Konkan Plains in the west and the Deccan Plateau in the east. LANDSAT imagery shows a large density of faults along the trend of the Ghats.

The Middle Sahyadri runs from 16°N latitude upto Nilgiri hills. This part is made of granites and gneisses and presents rougher topography. Lying at a stone's throw from the Arabian Sea coast, this area is covered with dense forests. The western scarp is considerably dissected by headward erosion of the west flowing streams. The average height is 1200 m but many peaks exceed 1500 m. The Vavul Mala (2,339 m.), the Kudremukh (1,892 m) and Pushpagiri (1,714 m) are important peaks. The Nilgiri Hills which join the Sahyadris near Gudalur rise abruptly to over 2,000 m and mark the junction of the Western Ghats with Eastern Ghats. Doda Betta (2,637 m) and Makurti (2,554 m) are important peaks of this area.

The southern part of the Western Ghats is separated from the main Sahyadri range by Palghat Gap which presents a sudden break in the continuity of this mountain range. The high ranges terminate abruptly on either side of this gap. This 24-30 km wide gap has an elevation ranging from 75 to 300 m while the bordering hills rise to 1500-2000 m above mean sea level. In all probability, it is a rift valley which has been formed by subsidence of the land between two parallel fault lines. This gap is used by a number of roads and railway lines to connect the plains of Tamil Nadu with the coastal plain of Kerala. It is through this gap that moisture-bearing clouds of the south-west monsoon can penetrate some distance inland, bringing rain to parched Mysore region. South of the Palghat Gap there is an intricate system of steep and rugged slopes on both the eastern and western sides of the Ghats. Anai Mudi (2,695 m) is the highest peak in the whole of southern India. This is a nodal point from which three ranges radiate in different directions. These ranges are the Anaimalai (1800-2000 m) to the north, the Palni (900-1,200 m) to the north-east and the Cardamom Hills or the *Ealaimalai* (1000-1500 m) to the south.

5. The Eastern Ghats. Bordering the eastern edge of the Deccan Plateau, the Eastern Ghats run almost parallel to the east coast of India leaving broad plains between their base and the coast. In striking contrast to the continuous eminence of the Western Ghats, the Eastern Ghats are a chain of highly broken and detached hills starting from the Mahanadi in Odisha to the Vagai in Tamil Nadu. The hills constituting the Eastern Ghats have neither structural unity nor physiographic continuity. In fact, they almost disappear between the Godavari and the Krishna. The name Eastern Ghats is, therefore, something of a misnomer and various mountains and hill groups are generally treated as independent units. In view of the heterogenous character, **Spate** has preferred to use the terms ***Eastern Hills*** for the northern, ***Cuddapah Ranges*** for the central, and ***Tamilnad Hills*** for the southern group in place of the collective term of Eastern Ghats. Still many geographers tend to use the term Eastern Ghats to this discontinuous linear assemblage of hills, each group of hills having its separate identity.

Depending upon the relief and structure, the Eastern Ghats can be divided into northern and southern parts, the dividing boundary lying somewhere south of the Godavari valley. It is only in the northern part, between the Mahanadi and the Godavari, that the Eastern Ghats exhibit true mountain character with a width of 200 km in the north and 100 km in the south. This part comprises the Maliya and the Madugula Konda ranges. The peaks and ridges of the Maliya range have a general elevation of 900-1,200 m and Mahendra Giri (1,501 m) is the tallest peak here. The Madugula Konda range has higher elevations ranging from 1,100 m to 1,400 m with several peaks exceeding 1,600 m. Arma Konda (1,680 m), Gali Konda (1,643 m) and Sinkram Gutta (1,620 m) are important peaks.

Between the Godavari and the Krishna rivers, the Eastern Ghats lose their hilly character and the saddle between these two rivers is occupied by Gondwana formations. The Eastern Ghats reappear as more or less a continuous hill range in Cuddapah and Kurnool districts of Andhra Pradesh. Here, the arcuate Nallamalai Range with general elevation of 600-850 m is the most prominent. It is composed of quartz and slate. The southern part of this range is called the Palkodna range. To the south, the hills and plateaus have low altitudes; only Javadi Hills and the Shevroy-Kalrayan Hills form two distinct features of 1,000 m elevation. The Biligiri Rangan Hills in Coimbatore district attain a height of 1,279 m. These hills have steep slope and bold relief because they are made of charnockites. Further south, the Eastern Ghats merge with the Western Ghats.

Significance of the Peninsular Plateau

The Peninsular Plateau of India is the oldest and the most stable landmass of the Indian sub-continent. It contains a rich variety of minerals which occur in large quantities. There are huge deposits of iron, manganese, copper, bauxite, chromium, mica, gold, etc. Above all, 98 per cent of the Gondwana coal deposits of India are found in the Peninsular Plateau. Besides, there are large reserves of slate, shale, sandstones, marbles, etc. A large part of north-west plateau is covered with fertile black lava soil which is extremely useful for growing cotton. Some other areas of the Peninsular Plateau are suitable for the cultivation of tea, coffee, rubber, millets, spices, tobacco and oilseeds. Some low lying areas of the plateau are suitable for growing rice. A variety of tropical fruits is also grown here. The highlands of the plateau are covered with different types of forests which provide a large variety of forest products. The rivers originating in the Western Ghats offer great opportunity for developing hydroelectricity and providing irrigation facilities to the agricultural crops. The plateau is also known for its hill resorts such as Udagamandalam (Ooty), Pachmarhi, Kodaikanal, Mahabaleshwar, Khandala, Matheron, Mount Abu, etc.

4. THE COASTAL PLAINS

The narrow coastal strip between the edges of the Peninsular Plateau and the coastline of India running for a distance of about six thousand kilometres from the Rann of Kachchh in the west to the Ganga-Brahmaputra delta in the east is called the coastal plains. The area between the Western Ghats and the Arabian Sea coast is known as the ***West Coastal Plain*** and that between the Eastern Ghats and the Coast of the Bay of Bengal is called the ***East Coastal Plain.*** The two coastal plains meet each other at the southernmost tip *i.e.* Kanniyakumari.

The West Coastal Plains

Stretching from Rann of Kachchh in the north to Kanniyakumari in the South, there are narrow west coastal plains with an average width of about 65 km. It is quite narrow in the middle and a bit broader in the northern and southern parts. Depending upon relief and structure, it can be divided into following subdivisions.

I. The Kachchh Peninsula was an island surrounded by seas and lagoons. These seas and lagoons were later filled by sediment brought by the Indus river which used to flow through this area. Thus, the island became a part of the mainland and comparatively broad plain was formed. Some scholars do not consider it as part of the west coastal plain and treat it as a separate entity. There are other geographers who consider Kachchh and Kathiawar as part of the Peninsular plateau because Kathiawar is made of the Deccan Lava and there are tertiary rocks in the Kachchh area. The true west coastal plain, according to them, lies between Surat and Kanniyakumari for a distance of 1600 km. But the ground reality is that it is, more or less, a plain area and lies near the west coast of India. Hence it should be treated as an integral part of the West Coastal Plain.

Due to scarcity of rain and flowing surface water, the work of wind is felt everywhere and this has given rise to arid and semi-arid landscape. Coastal sand dunes, sandy plains, interrupted with bare rocky hills are the chief characteristic physiographic features.

All along the north of Kachchh there lies a broad level salt-soaked plain. This is the ***Great Rann***. Its southern continuation, known as the ***Little Rann*** lies on the coast and to the south-east of Kachchh. The flat and unbroken Great Rann is about 320 km long with a maximum width of 160 km, covering an area of about 21,500 sq km. It rises only a few metres above the sea level and is flooded by the Banas and the Luni rivers during the rainy season. Some places are actually below sea level and are inundated during high tides. Most of the area is formed of sun-baked dark silt encrusted with salt. A few patches of high ground are covered with grass and break the monotony of the otherwise flat plain.

II. The Kathiawar Peninsula lies to the south of the Kachchh. It is encircled on the east and north-east by the Little Rann and the Nal Basin. The average elevation is less than 200 m. The central part is a highland of Mandav Hills from which small streams radiate in all directions. Mt. Girnar (1,117 m), the highest point, is supposed to be of volcanic origin. The Gir Range is located in the southern part of the Kathiawar peninsula. It is covered with dense forests and is famous as home of the Gir lion.

III. The Gujarat Plain lies east of Kachchh and Kathiawar and slopes towards the west and south west. It may almost be described as an intrusion of Indo-Gangetic conditions into the Peninsula. Formed by the rivers *Narmada*, *Tapi*, *Mahi* and *Sabarmati*, the plain includes the southern part of Gujarat and the coastal areas of the Gulf of Khambhat. This is a low plain no part of which exceeds 150 m in elevation. The eastern part of this plain is made of sediments and is fertile enough to support agriculture, but the greater part near the coast is covered by wind blown loess which has given rise to semi-arid landscape. A chain of saline marshes near the coast is prone to floods during high tide.

IV. The Konkan Plain lies to the south of the Gujarat plain and extends from Daman to Goa for a distance of about 500 km with its width varying from 50 to 80 km. It has some features of marine erosion including cliffs, shoals, reefs and islands in the Arabian Sea. Mumbai was an island but parts of the sea lying between the mainland and the island have been reclaimed to connect it with the mainland. The Thane creek of the Ulhas around Mumbai is an important embayment which provides an excellent natural harbour on the southern side of the island. South of Mumbai, the Konkan coast has a series of small bays and coves lying between jutting headlands containing beaches of sand. Behind the alluvial coastal belt, there is a series of parallel ridges reaching 450-600 m in which rivers like the Vaitarni, Ulhas and Amba have lower courses more or less parallel to the coast before reaching it transversely. Some lateritic hillocks rise to 100 m above mean sea level.

V. The Karnataka Coastal Plain from Goa to Mangaluru is about 225 km long. It is a narrow plain with an average width of 30-50 km, the maximum being 70 km near Mangalore. The central part of this plain is crossed by numerous spurs projecting from the Ghats. These spurs approach so close to the coast that the breadth of the lowland is reduced to 5-7 km at 14°N latitude where the Ghat's crest is only 13 km away from the sea. Running like ridges, the spurs attain heights of more than 600 m near the Ghats. At some places the streams originating in the Western Ghats descend along steep slopes and make waterfalls. The Sharavati while descending over such a steep slope makes an impressive waterfall known as ***Gersoppa (Jog) Falls*** which is 271 m high. Marine topography is quite marked on the coast.

VI. The Kerala Plain also known as the ***Malabar Plain***, between Mangaluru and Kanniyakumari is about 500 km long. This is much wider than the Karnataka plain and at places it is 96 km wide. It is a low lying plain and at no place its height exceeds 30 m. The existence of lakes, lagoons, backwaters, spits, etc. is a significant characteristic of the Kerala coast. The backwaters, locally known as ***kayals*** are the shallow lagoons or inlets of the sea, lying parallel to the coastline. The largest among these is the Vembanad lake which is about 75 km long and 5-10 km wide and gives rise to a 55 km long spit. Kochi is situated on its opening into the sea. This and several other lagoons have been joined together by canals to provide excellent inland waterways from the mouth of the Ponnani in the north to Thiruvananthapuram in the south.

The East Coastal Plains

Between the Eastern Ghats and the east coast of India are located the East Coastal Plains extending from the Subarnarekha river along the West Bengal-Odisha border to Kanniyakumari. A major part of the plains is formed as a result of the alluvial fillings of the littoral zone by the rivers Mahanadi, Godavari, Krishna and Cauvery comprising some of the largest deltas. Its western boundary is a discontinuous line of the Eastern Ghats, more precisely by the contours of 75 m in Odisha, 100 m in Andhra Pradesh and 150 m in Tamil Nadu. In contrast to the West Coastal Plains, these are extensive plains with an average width of 120 km although these may be as wide as 200 km in the deltaic regions and as narrow as 35 km in-between the deltas. This plain is known as the ***Northern Circars*** between the Mahanadi and the Krishna rivers and ***Carnatic*** between the Krishna and the Cauvery rivers. Depending upon physiographic variations, the entire plain is divided into three regions.

The Utkal Plain comprising coastal areas of Odisha is about 400 km long. It includes the Mahanadi delta with Cuttack at its head. There is thick layer of alluvium covering this delta. The most prominent physiographic feature of this plain is the **Chilka Lake** in the south of the Mahanadi delta. This lagoon on the Odisha coast is about 70 km long and its maximum width in the north-east is nearly 22 km narrowing to about 7 km at its south-western end. It is the biggest lake in the country and its area varies between 780 sq km in winter to 1,144 sq km in the monsoon months. South of Chilka Lake, low hills dot the plain.

The Andhra Plain lies south of the Utkal Plain and extends upto *Pulicat Lake*, some 40 km north of Chennai. This lake has been barred by a long sand spit known as Sriharikota Island, on which is located the satellite launching station of the Indian Space Research Organisation. The lagoon is about 60 km long and about 16 km wide in its widest part. The most significant feature of this plain is the delta formation by the rivers Godavari and Krishna. In fact, the two deltas have merged with each other and formed a single physiographic unit. The combined delta has advanced by about 35 km towards the sea during the recent geological past. This is clear from the present location of the Kolleru lake which was once a lagoon at the shore but now lies far inland. This part of the Andhra plain is quite wide. Andhra plain has a straight coast and badly lacks good harbours with the exception of Vishakhapatnam and Machilipatnam.

The Tamil Nadu Plain stretches for 675 km from Pulicat lake to Kanniyakumari along the coast of Tamil Nadu. Its average width is 100 km. The most important feature of this plain is the Cauvery delta where the plain is 130 km wide. The fertile soil and large scale irrigation facilities have made the Cauvery delta the **granary of South India**.

Significance of the Coastal Plains

Large parts of the coastal plains of India are covered by fertile soils on which different crops are grown. Rice is the main crop of these areas. Coconut trees grow all along the coast. The entire length of the coast is dotted with big and small ports which help in carrying out trade. About 98% of our international trade is carried through these ports. The sedimentary rocks of these plains are said to contain large deposits of mineral oil. The sands of Kerala coast have large quantity of monazite which is used for nuclear power. Fishing is an important occupation of the people living in the coastal areas. Low lying areas of Gujarat are famous for producing salt.

5. THE INDIAN ISLANDS

Apart from the large number of islands in the near proximity of the Indian coast, there are two main groups of islands in the Indian Ocean far away from the coast. One of these is the Andaman and Nicobar Archipelago in the Bay of Bengal and the other is a group of tiny islands known as the Lakshadweep Islands in the Arabian Sea. These islands have gained much importance and their study has become almost indispensable in view of the increasing interest of super powers in the geopolitics of the Indian ocean.

The Andaman and Nicobar group of islands form an arcuate chain, convex to the west, extending from 6° 45′ N to 13° 41′ N and from 92° 12′ E to 93° 57′ E for a distance of about 590 km with a maximum width of about 58 km. This archipelago is composed of 572 big and small islands covering a cumulative area of about 8249 sq km. The entire chain consists of two distinct groups of islands. The Great Andaman group of islands in the north is separated by the *Ten Degree Channel* from the Nicobar group in the south. The Andaman is a closely knit group of islands. It is 260 km long and 30 km wide with a total area of 6596 sq km. This group of islands is divided into three major groups *viz*. North Andaman, Middle Andaman and South Andaman. The capital city of Port Blair is located in South Andaman. Little Andaman is separated from the Great Anadmans by 50 km wide Duncan Passage. Only 36 islands are inhabited, 24 in Andaman and 12 in Nicobar.

The Nicobar group of islands consists of 7 big and 12 small islands together with several tiny islands. They are scattered over a length of 262 km with maximum width of 58 km covering an area of 1,653 sq km. The Great Nicobar, as its name suggests, is the largest island measuring 50 × 25 km. It is the southernmost island and is only 147 km away from Sumatra island of Indonesia. India's southernmost point *'Indira Point'* is located at the southern tip of the Great Nicobar. However, much of it was destroyed by tsunami of 26th December, 2004.

Most of these islands are made of tertiary sandstone, limestone and shale resting on basic and ultrabasic volcanoes. The Barren and Narcondam islands, north of Port Blair, are volcanic islands. Some of the islands are fringed with coral reefs. Many of them are covered with thick forests and some are highly dissected. Most of the islands are mountainous and reach considerable heights. Saddle peak (737 m) in North Andaman is the highest peak.

Names of following three islands were changed in 2018:

1. Ross Island = Netaji Subhash Chandra Bose Dweep
2. Neil Island = Shaheed Dweep
3. Havelock Island = Swaraj Dweep.

The Lakshadweep Islands in the Arabian Sea, though literally mean one lakh islands is only a group of 25 small islands. They are widely scattered over an area of 108.78 sq kms extending from 8° N to 12° 20′ N and 71° 45′ E to 74 °E about 200-500 km south-west of the Kerala coast. The islands north of 11° N are known as *Amindivi Islands* while those south of this latitude are called *Cannanore Islands*. In the extreme south is the Minicoy island. All are tiny islands of coral origin and are surrounded by fringing reefs. The largest and the most advanced is the Minicoy island with an area of 4.53 sq km. Betra has an area of only 0.12 sq km. Most of the islands have low elevation and do not rise more than five metres above sea level. Their topography is flat and relief features such as hills, streams, valleys, etc. are conspicuous by their absence. Shallow lagoons are seen on their western side, while on the eastern seaboard the slopes are steeper.

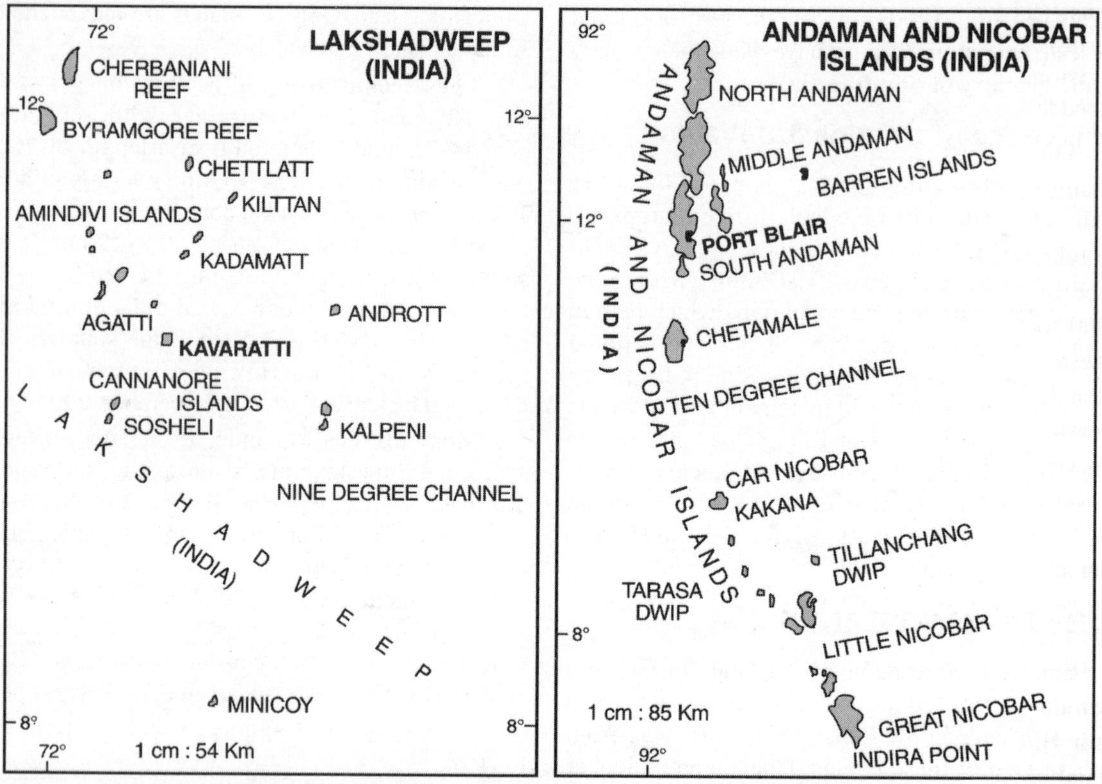

FIG. 3.15. Andaman and Nicobar Islands and Lakshadweep

PHYSIOGRAPHIC REGIONS OF INDIA

The term physiography was conceived as 'a description of nature' or 'of natural features' in their causal relationships. After some time, the term became almost synonymous with physical geography. Gradually, it became limited to the study of landforms, especially in the U.S.A. By some authorities the term has now been superseded by the more scientific term geomorphology. However, other authorities still regard it in wider terms as an integration of geomorphology, plant geography and pedology. We will confine ourselves to the study of land features of India in this context.

L.D. Stamp was the first scholar who made an attempt to classify natural regions or physiographic regions of India in 1922. His scheme was based on the homogeneity of physiography, structure and climate. He divided India into 3 major natural regions and 22 sub-regions. Another attempt was made by **J.N.L. Baker** in 1928. His scheme was based on the work of Wood and was in close agreement with the work of L.D. Stamp. Following the empirical approach adopted by Stamp and Baker, **O.H.K. Spate** presented a detailed scheme of physiographic regions of India in 1954 and slightly revised it at a later stage. In the third edition of his book entitled "India and Pakistan : A General and Regional Geography" published in 1967, he divided India into 3 macro regions which were further divided into 34 regions of first order (excluding islands), 74 regions of second order and 225 subdivisions of these regions. Three macro regions, as suggested by Spate are (*i*) the mountain rim, (*ii*) the Indo-Gangetic plain and (*iii*) the Peninsula.

S.P. Chatterjee presented his detailed study of physiographic divisions of India in The Gazetteer of India Vol. I published by Publication Division of India in 1965. He divided India into seven major physiographic divisions on the basis of topography. They are (1) Northern Mountains which include the Himalayas and mountain ranges in the north-east, (2) Great Plains, (3) Central Highlands, (4) Peninsular

Plateaus, (5) East Coast, (6) West Coast and (7) Bordering Seas and Islands. Further divisions and subdivisions of these major regions are briefly described as under :

1. The Northern Mountains. This chain of mountains include the mighty Himalayas and the mountain ranges which extend in the north-east. The Himalayas extent almost uninterruptedly for a distance of 2500 kms and cover an area of about 5,00,000 sq km. This mountain range is divided into the Western Himalaya and the Eastern Himalaya. Between these two is the Himalayan country of Nepal. The **Western Himalayas** are further divided into four regions *viz.* (*a*) the North Kashmir Himalayas, (*b*) the South Kashmir Himalayas, (*c*) the Punjab Himalayas and (*d*) the Kumaon Himalayas. The **Kashmir Himalayas** are the broadest part which is 700 km from west to east and 500 km from north to south. It spreads over an area of about 3,50,000 sq km. The Great Himalayan Range acts as the dividing line between the North Kashmir Himalayas and the South Kashmir Himalayas. Kashmir Valley lying between Pir Panjal in the south and Zaskar Range in the north is the most densely populated part of the Western Himalayas. Srinagar, the summer capital of Jammu and Kashmir lies in the heart of this valley. It is situated at an elevation of 1893 m above sea level. The **Punjab Himalaya** includes the portion of the Himalaya lying in Himachal Pradesh and Punjab. It covers an area of about 45,000 sq km and is drained by four important rivers namely the Satluj, the Beas, the Ravi and the Chenab. These rivers have their sources in the upper reaches of the Himalayas. The **Kumaon region** lies in the state of Uttarakhand and stretches over an area of about 30,000 sq km. The most important rivers, the Ganga and the Yamuna, originate in the Kumaon Himalayas. The religious places like Haridwar, Rishikesh, Kedarnath and Badrinath as well important places of tourist interest like Mussorie, Nainital, Almora and Ranikhet are located in this part of the Himalayas.

The **Eastern Himalayas** can be divided into (*a*) the Sikkim Himalaya, (*b*) the Darjeeling Himalaya, (*c*) the Bhutan Himalaya and (*d*) The Assam Himalaya.

The **Sikkim Himalaya** is an 'anticlinal valley' and this area is drained by the Tista river (a tributary of the Brahmaputra river). The **Darjeeling Himalaya** mainly consists of two north-south ranges. The Darjeeling Range is known all over the world for its tea gardens and Darjeeling tea is considered to be the best tea available anywhere in the world. The **Bhutan Himalaya** consists of high ranges and deep valleys and covers an area of about 22,500 sq km. The **Assam Himalaya** lies in Assam and Arunachal Pradesh. It spreads over an area of about 67,500 sq km.

The north-eastern mountains are subdivided into three physiographic provinces namely (*a*) Purvanchal, (*b*) the Meghalaya or Shillong Plateau and (*c*) the Assam valley. The **Purvanchal** includes lofty 'mountain ranges' which border India in the north-east, hills and plains in Manipur, Tripura and adjoining districts of Assam. The Purvanchal may be further sub-divided into six sub-divisions. Arunachal Pradesh, the Nagaland, the Manipur Hills, the North Cachar Hills, the Mizo Hills and the Tripura Hills. **The Meghalaya Plateau** contains Chirrapunji, the Khasi, the Jaintia, the Garo and the Mikir Hills. The **Assam Valley** can be divided into two physiographic parts *viz.* the Upper Assam and the Lower Assam. Each is about 75 km wide with low relief, easy accessibility and extremely fertile land.

2. The Great Plains. In between the Himalayan mountain ranges in the north and the peninsular plateau region in the south, lies the Great Plain of India which stretches from the Indus delta in the west to the Ganga delta in the east. A part of it lies in Pakistan but major part is in India. This Great Plain can be divided into following broad divisions.

The **Western plain** consists of arid land of Rajasthan. It has a well defined boundary in the east marked by the Aravalli range. This is a sandy desert with distinct aeolian topography. The eastern part is less sandy, comparatively more humid and is clothed with steppe vegetation. The **Punjab plains** are flat and the general elevation is 200 to 240 (except the Hoshiarpur plains). To the south of the Punjab plains lies the Haryana tract.

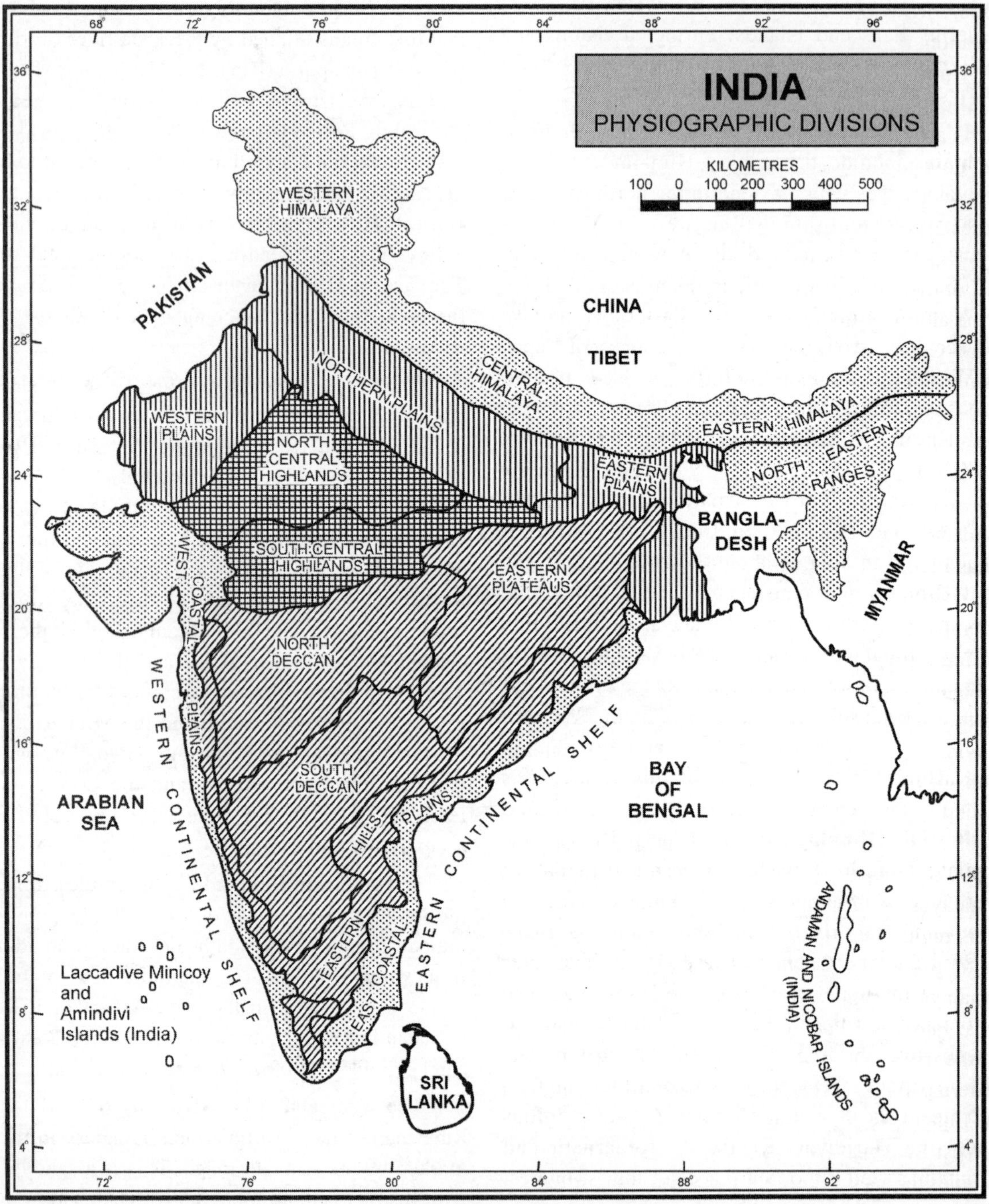

FIG. 3.16. Physiographic Divisions

To the east of the Punjab plains is the **Ganga** plain which occupies about 55% of the total area of the Great Plains and stretches over the states of Uttar Pradesh, Bihar and West Bengal. The region between Ganga and its main tributary Yamuna is known as **Ganga-Yamuna Doab** which can be divided into

Upper, Middle and Lower sections depending on elevation, rainfall and character of older flood plains. The **Upper Doab** extends from Haridwar in the north to Aligarh in the south. This region is irrigated by East Yamuna and Upper Ganga canals and is one of the most fertile areas giving bumper crops. It also receives more rainfall than the Middle and the Lower Doabs. The **Lower doab** is a flat area which ends at Pryagraj at the confluence of Ganga and Yamuna. East of the Ganga-Yamuna Doab, lies the vast stretch of the alluvial plains, from the foot of the Himalayas to the Ganga. These are known as the **Rohilkhand** plains. These plains lie in UP and cover an area of about 35,000 sq km. The **Avadh plains** are other plains of Uttar Pradesh. To the east of the Rohilkhand and the Avadh plains, lie **Bihar plains** covering an approximate area of 88,000 sq km. These plains are divided into the North Bihar Plains and the South Bihar Plains. The North Bihar Plains are drained by the rivers Ganga, Ghaghara, Gandak and Kosi accompanied by a large number of mountain streams passing through these plains. "The combined work of these streams has resulted in a 2000 m deep trough at the foot of the Nepal Himalaya being filled up with alluvial deposits. One of the most extensive alluvial plains of the world, 54,000 sq km in North Bihar was formed." The portion of the Great Plains coming under North Bengal covers an approximate area of 23,000 sq km and extends from the foot of the Eastern Himalayas to the northern limit of the Bengal basin. "This region has evolved from an extensive sheet of waste materials brought down from the Eastern Himalaya by a number of powerful streams like the Tista, Jaldhaka and Torsa." Its eastern part is drained by the rivers joining the Brahmaputra while its western part is drained by the tributaries of the Ganga. The Bengal basin embraces most of the alluvial plains of West Bengal and Bangladesh and the Ganga delta occupies the major portion of this basin.

3. Central Highlands. Lying in the central part of India, the Central Highlands separate the Great Plains in the north from the plateaus and coastal plains of the Deccan. "It forms a compact block of mountains, hills, and plateaus, interspersed with valleys and basins, covering about one sixth of the total area of India." It covers about half of Madhya Pradesh one-third of Rajasthan and a small part of Uttar Pradesh.

The Central Highlands can be divided into North Central Highlands and South Central Highlands. The **North Central Highlands** can further be subdivided into four divisions namely (*a*) the Aravalli Range, (*b*) Eastern Rajasthan upland, (*c*) *Madhya Bharat Pathar* and (*d*) Bundelkhand upland. Similarly, the **South Central** Highlands can be divided into four divisions *viz*. (*a*) Malwa Plateau, (*b*) Vindhya Scarplands, (*c*) Vindhya Range and (*d*) Narmada Valley.

The **Aravalli range** runs for a distance of 800 km from Delhi to Ahmedabad. To the east of the Aravalli range lies **East Rajasthan Upland.** It's elevation varies from 250 to 500 m. It is drained by the Chambal river which enters Rajasthan near the northern end of the Gandhi Sagar and flows in this state for a distance of 360 km. The **Madhya Bharat Pathar** lies to the east of the Chambal river. It is a rocky surface with dense forests. The **Bundelkhand** lies between the Yamuna and the northern Aravali scarp of the Vindhyan plateau.

The southern part of the Central Highlands consists of the Malwa Plateau, the Vindhya Scarplands, the Vindhya Range and the Narmada Valley. The **Malwa Plateau** forms a great triangle based on the Vindhyan Hills and bounded on the north-west by the Great Boundary Fault of Aravallis and on the east by the sharply-defined scarp overlooking Bundelkhand. Most part of this plateau lies in Madhya Pradesh. It has general northward slope, good drainage and black soil. **The Vindhyan** range is really an escarpment which varies in character and height depending on the structure and lithology of the underlying sector. For the first 100 km from its western terminus, Gomanpur peak (554 m) in the Dhar district of Madhya Pradesh, the Vindhya range runs in a curve, its convex side facing the Narmada valley and following a 300 m contour line. It has dense forests but sparse population. For the next 160 km a more open type of country prevails and the escarpment, still built of basalt, becomes more prominent. Near Hoshangabad, the rock type changes; the Vindhya range comes down very close to the Narmada river and presents a terraced slope built of hard sand stones alternating with shales. The strong sand stones of Kaimur, Rewa and Bhander are the principal makers of the Vindhyan scarpland. The **Narmada Valley** lies to the extreme south of the central highlands. It is a rift valley caused by tectonic forces in which the Narmada river flows from east to west and empties itself in the Arabian Sea.

4. The Peninsular Plateaus. The Peninsular Plateaus constitute the largest physiographic division of India facing the Bay of Bengal in the east and the Arabian Sea in the west. Its maximum length from Pachmarhi in the north to Cape Comorin (Kanniyakumari) in the south is 1600 km and its maximum width from the Sahyadri in the west to the Rajmahal hills in the east is 1400 kms. The peninsular plateaus consist of five distinct physiographic subdivisions, namely (*i*) Western hills, (*ii*) North Deccan plateau, (*iii*) South Deccan plateau, (*iv*) Eastern plateaus and (*v*) Eastern hills.

(*i*) The **Western hills** cover all the three sections of the Sahyadris (Western Ghats) namely Northern Sahyadri, Middle Sahyadri and Southern Sahyadri. **(*ii*)** the **North Deccan plateau** comprises Satpura range and the Maharashtra plateau. The Satpura range extends in east-west direction between the Narmada river in the north and the Tapi river in the south. It represents folding and upheaval and is an ancient tectonic range. Its present physiography is that of scarped blocks (on the whole steeper towards the Tapi), largely covered with Deccan Lavas but with some inliers of gneissic plateau country. To the south of Solapur is the Maharashtra plateau, almost whole of which is formed of plateau basalt, which on weathering has given rise to rolling plains with intervening shallow valleys. The west-flowing Tapi river forms its northern boundary. Further south is the Godavari river which enters Maharashtra plateau at Nasik and flows for 650 km eastwards to the farthest end of this region. (*iii*) **The South Deccan** plateau can be divided into Telangana plateau and the Karnataka plateau. The Godavari river, after receiving a number of tributaries, enters the **Telangana plateau** and divides it into two parts. The northern part is very hilly and forested and is sparsely populated. The southern part is dotted with low hills and shallow depressions. One such depression, surrounded by low hills, 130 m high provided the site for the twin cities of Hyderabad-Secunderabad. The **Karnataka plateau** can also be divided into two sections on the basis of altitude. These sections are the northern section and the Mysore plateau. The Krishna and the Tungabhadra flow through the northern section. **The Mysore plateau** is the loftiest and well defined plateau of South India. This plateau is bounded by the Sahyadri in the west, the Eastern Ghats in the east and the Nilgiris in the south. Physiographically the Mysore plateau may be divided into two sections *viz*. Malnad and Maidan. The **Malnad** is a hilly area bordering the Sahyadri. It has a mean elevation of about 1000 m and its average width is 35 km. It is dissected into steep hills and deep valleys and has dense forests. The **Maidan** consists of rolling plains with low granitic hills. Cauvery is the most important river. (*iv*) **The Eastern plateaus** consist mainly of the plateaus of **Chotanagpur** of which Ranchi plateau and Hazaribagh plateau are the major parts. Rounded hills of massive granite and slightly elevated terraces of older flood plains mark the topography of the Ranchi plateau. The Damodar river rises in the hills of Chotanagpur plateau and drains through Ranchi, Hazaribagh, Dhanbad, Santhal Parganas, Bankura and Burdwan. It flows in a rift valley and traverses a distance of 541 km before it joins the Hugli river near Kolkata. To the west of the Chotanagpur plateau is the **Bhagelkhand plateau.** It is bounded by the Ganga plain in the north, Maikala hills in the south and Kaimur hills in the west. Its general slope is in the north and north-west direction as is indicated by the flow of the Son river and its tributaries. The Mahanadi basin lies in the southern part of the Eastern plateau and the central part of the Mahanadi basin is known as the Chhattisgarh basin. To its south lies the **Dandakaranya.** It encompasses the districts of Bastar, Kalahandi and Koraput and some parts of Andhra Pradesh. The region is well demarcated and enjoys the central location between Chhattisgarh basin in the north, Andhra Pradesh in the south,

Maharashtra Plateau in the west and Eastern coastal plains towards the east. It has an undulating topography with well marked elevations and depressions.

5. East Coast. The eastern coastal strip lies between the Eastern Ghats and Bay of Bengal and stretches from the Ganga delta to Cape Comorin (Kanniyakumari). It is marked by a number of deltas such as the deltas of the Mahanadi, the Godavari, the Krishna, and the Cauvery. It is known by different names in different parts. In Odisha, it is known as Utkal plain which includes the Mahanadi delta with Cuttack at its head. From the southern limit of the Utkal plain, stretches the Andhra plain. Rivers Godavari and Krishna flow through this region and form deltas in their lower reaches. To the south of the Andhra plain is the Tamil Nadu plain. This plain is about 675 km long and 100 km wide. The Cauvery delta is the most important physiographic feature of this plain. In Tamil Nadu and Andhra Pradesh, the East coast is called the Payan Ghat which extends from Cape Comorin (Kanniyakumari) northwards to the united deltas of the Krishna and the Godavari for 1100 km with an average width of 100-130 km.

6. West Coast. The west coast strip extends from the Gulf of Cambay (Gulf of Khambhat) in the north to Cape Comorin (Kanniyakumari) in the south and from the Western Ghats in the east to the Arabian sea coast in the west. It is made up of mud brought down by the short streams originating from the Western Ghats. At the northern end are two peninsulas namely Kutch (Kachchh) and Kathiawar and the vast extensive plain of Gujarat. To the south of the Gujarat plain are the coastal plains which are divided into three plains. Starting from north to south, they are (*i*) the Konkan coast, (*ii*) the Karnataka coast and (*iii*) the Kerala coast. The limits of these areas are conforming to the state boundaries.

7. The Islands. India is blessed with a large number of islands in both the Bay of Bengal and the Arabian sea. There are hundreds islands in the Bay of Bengal which are known as the Andaman and Nicobar Islands. The Andaman islands are divided into three main island groups *i.e.* North, Middle and South. Separated from these islands by the Duncan passage is the Little Andaman. The North Andaman Island is 80 km long and has a maximum width of 20 km. The Middle Andaman is much wider and less indented. It is 70 km long and 30 km wide. The South Andaman consists of parallel ranges and valleys. Port Blair, the capital of Andaman Nicobar Islands and the most important port town lies in the South Andaman. Off these main islands lie numerous smaller islands. South of Andamans are the Nicobar islands, consisting of 19 islands 12 of which are inhabited. The Great Nicobar is the largest, occupying 862 sq km. The Car Nicobar is the northernmost island, 229 km from Port Blair.

In the Arabian Sea, there are three groups of islands. (*i*) Amindivi Islands (consisting of six main islands of Amini, Keltan, Chetlat, Kadmat, Bitra and Perumul Par). (*ii*) Laccadive Islands (consisting of five major islands of Androth, Kalpeni, Karavatti, Pitti and Suheli Par) and (*iii*) Minicoy. At present these islands are collectively known as Lakshadweep.

REFERENCES

Ahmed, E. (1962) : Geomorphic Regions of Peninsular India, *Journal of Ranchi University*, Vol 1, No. 1, pp. 1-29.

Ahmed, E. (1972) : *Coastal Geomorphology of India*, Orient Longman, N. Delhi.

Ahmad, E. (1992) : *Geography of the Himalayas*, Kalyani Publishers, Ludhiana.

Bhattacharya, S. (1969) : *Geography and Geology of Assam*, National Book Trust, India, New Delhi.

Burrard, S. (1915) : Origin of the Indo-Gangetic Trough—Commonly called Himalayan Foredeep; Proceedings of Royal Society, NOGIA, London, pp. 22-38

Burrard, S.G. (1933) : *A Sketch of Geology and Geography of the Himalaya Mountains and Tibet*, Part IV, Govt. of India, New Delhi.

Chatterjee, S.P. (1965) : *Physiography* in Chopra P.N. (ed). *The Gazetteer of India*, Vol. One, Ministry of I and B, Govt. of India, New Delhi (Chap. 1).

Chatterjee, S.P. (1999) : *India : A Physical Geography*, Publications Division, Government of India, New Delhi.

Chaudhary, Srivastava and Suba Rao (1974) : Seismotectonic Investigation of the Himalaya, *Himalayan Geology* Vol. 4, Delhi.

Cox, Allan, (1973) : *Plate Tectonics and Geomagnetic Reversals.* W.H. Freeman, California.

Crawford, A.R. (1974) : The Indus Sutur Line, the Himalaya, Tibet and Gondwanaland, *Geological Magazine*, Vol. 5.

Deshpande, C.D. (1992), *India—A regional Interpretation*, Indian Council of Social Science Research and National Book Centre, New Delhi.

Dikshit, K.R. *et al.* (eds.) (1994) : *India : Geomorphological Diversity*, Rawat Publications, Jaipur (Chap. 3, 7, 8, 15 & 20).

Gansser, A. (1964) : *Geology of Himalayas,* Interscience Publishers, John Wiley, N. York.

Geddess, A. (1960) : The Alluvial Morphology of the Indo-Gangetic Plain, *Transaction and Papers* I.B.G., Vol. 28, pp. 262-63.

Johnson, B.L.C. (1969) : *South Asia,* Heinemann Educational Books Ltd., London (Chap. 2).

Kalia, K.L. and Hari Narain (1976) : Evolution of the Himalaya based on Seismotectonics and Deep Seismic Soundings'. Key paper to Section II, Part 2, *Himalayan Geology Seminar*, p. 30.

Kuriyan, George (1970) : *India : A General Survey*, National Book Trust, India, New Delhi.

Lall, J.S. (1982) : *Himalayas Aspects of Change*, Natraj, Dehra Dun.

Mathur, S.M. (2004) : *Physical Geology of India*, National Book Trust, India, New Delhi.

Mukherji, A.R. (1964) : Alluvial Morphology of the Upper Ganga-Yamuna Doab, *Deccan Geographer*, Vol. 2, No. 2, pp. 100-25.

Niyogi, D. and A. Chakravarti (1967) : Applied Geomorphology along Digh Beach, Midnapur District, West Bengal. *Proc. Geomorphological Studies in India*, Saugar University, Sagar.

Pande, J.C. (1975) : *Recent Advances in the Himalayan Geology*, Presidential Address, Indian Science Congress, 62nd Session.

Randhawa, M.S. (1947) : *The Birth of the Himalayas.*

Raina, A.N. (1971) : *Geography of Jammu and Kashmir*, National Book Trust, India, New Delhi.

Ray, S. (1978) : *Geomorphology of India*, Firma K.L.M. Private Ltd., Calcutta.

Robinson, H. (1967) : *Monsoon Asia: A Geographical Survey.*

Saigal, Umesh (1994) : *Lakshadweep*, National Book Trust, New Delhi, India.

Shafi, M. (1968) : Ganga-Yamuna Doab in R.L. Singh (ed.) : *India : Regional Studies*, National Committee for Geography, Calcutta.

Siddhartha, K. (1999) : *India : The Physical Aspects* Centre for Development of Environment and Resources, Patna (Chap. 2).

Singh, L.R. (1965) : *The Tarai Region of U.P. : A Study of Human Geography*, Ram Narain Beni Prasad, Allahabad.

Singh, R.L. *et al.* (eds) (1971) : *India: A Regional Geography*, National Geographical Society of India, Varanasi.

Singh, R.P. (1962) : The Geomorphology of the Peninsular India, *Abstract Indian Science Congress.*

Singh, R.P. (1968) : Geomorphology of the Shillong Plateau of Assam, *Proc, Pre-Cong. Sym. on Meghalaya* II, I.G.U.V, Gauhati.

Singh, R.P. (1969) : Geomorphological Evolution of Chotanagpur Highlands, India, *National Geographical Society of India*, Varanasi, pp. 108-120.

Singh, R.P. (1987) : Tectonics and Surfaces of Peninsular India, *Souvenir*, *International Symposium on Geomorphology and Environmental Management*, Allahabad Geographical Society, Allahabad.

Singh, Savindra (1978) : Physiographic Regions, Landforms and Erosion Surfaces of Ranchi Plateau, *National Geographer*, Vol. 13, No. 1, pp. 43-65.

Singh, Savindra and D.P. Upadhyay (1985) : Physiographic Regions, Landforms and Erosion Surfaces of S.E. Chotanagpur Region in *Facets of Geomorphology*, Ed. by A. Kumar, Thinkers Library, Allahabad, pp. 55-76.

Spate, O.H.K. and Learmonth, A.T.A. (1967) : *India and Pakistan* : *A General and Regional Geography*, Metheun & Co. Ltd., Chap. 1.

Srinivasan, P.B. Raju, K.C.C. and Raju D.V. (1980) : Geomorphology and its influences on Kerala Coast. *Bull O.N.G.C.* 17 (10), pp. 277-84.

Srivastava, L.S.P., and Singh, V.N. Singh (1974) : Tectogenesis and Seismotectonics of the Himalayas, *5th Symposium on Earthquake Engineering*, India.

Tamta, B.R. (1994) : *Andaman and Nicobar Islands*, National Book Trust, New Delhi, India.

Valdaya, K.S. (1977) : Structural set-up of the Kumaon Lesser Himalaya, In : *Himalaya Science de la Terra C.N. R.S. Paris* 268 pp. 448-62.

Wadia D.N., Meher (1993) : *Geology of India*, McGraw Hill, New Delhi.

◆◇◆

CHAPTER 4

Drainage

INTRODUCTION

India is blessed with hundreds of large and small rivers which drain the length and breadth of the country. Rivers constitute the most useful natural resource. They have attracted the attention of planners, economists, geographers, geologists, hydrologists and a host of other specialists from different fields. They are a great source of water for irrigation, industry and domestic purpose and offer innumerable sites for producing hydro-electricity. Almost all the fertile areas of the country have been formed by the depositional work of rivers. Many rivers are used as channels for inland water transport. Fishing has become an important occupation along the river courses.

According to an estimate made by **S.P. Dasgupta**, the annual yield of water in the rivers of the country is 1,858,100 million cubic metres of which more than one third (33.8%) is contributed by the Brahmaputra, followed by the Ganga (25.2%), the Godavari (6.4%), Indus (4.3%), the Mahanadi (3.6%), the Krishna (3.4%), and the Narmada (2.9%). The remaining 20.4% is contributed by other rivers (percentages are calculated for basin areas in Indian territory only).

Classification of Drainage Systems of India

There are several bases for classifying drainage systems of India. The main types of drainage systems are briefly described below :

1. Size of the Catchment area. River systems of India are divided into three categories depending on their size. *Major river basins* have catchment area of 20,000 sq km and above. *Medium river basins* have catchment are between 2,000 and 20,000 sq km and river basins with catchment area less than 2,000 sq km are known as *minor river basins*.

2. Origin. Depending upon the origin of rivers two broad drainage systems of India are generally recognised, *viz*. (*a*) The Himalayan rivers including the Indus, the Ganga, the Brahmaputra and their tributaries and (*b*) The Peninsular rivers which include along with their tributaries, a large number of rivers such as the Mahanadi, the Godavari, the Krishna, the Cauvery, the Narmada and the Tapi.

3. Orientation to the sea. The Indian drainage is divided into two major drainage systems on the basis of orientation to the sea. These include : (*i*) the Bay of Bengal drainage and (*ii*) Arabian sea drainage. About

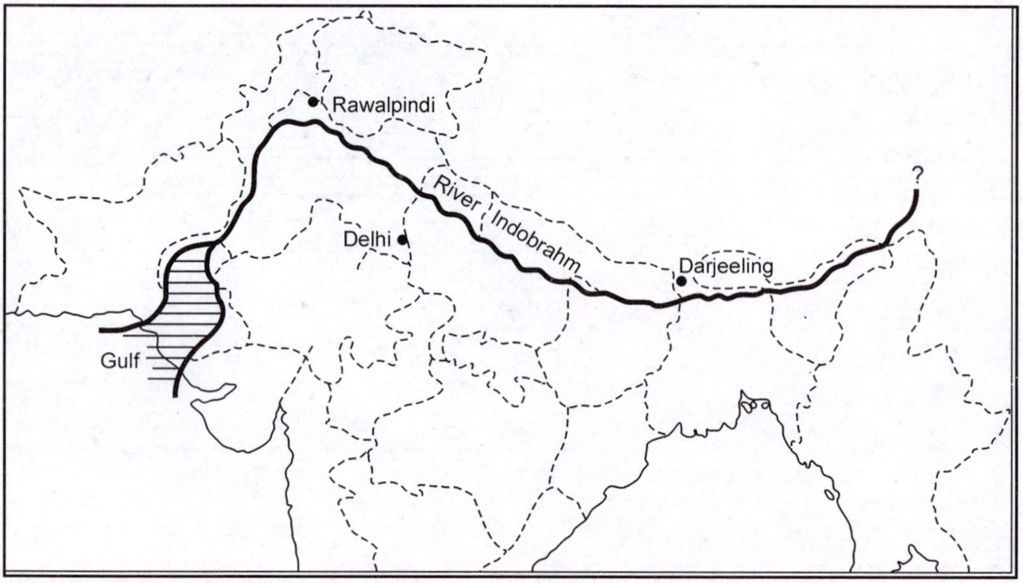

FIG. 4.1. The supposed course of the Indo-Brahm River (According to Pascoe)

77 per cent of the drainage area of the country is oriented towards the Bay of Bengal. It consists of a large number of rivers like the Ganga, the Brahmaputra, the Mahanadi, the Godavari, the Krishna, the Cauvery, the Penneru, the Penneiyar, the Vaigai, etc. The Arabian Sea drainage spreads over 23 per cent of the country's surface flow area which commands river basins like the Indus, the Narmada, the Tapi, the Sabarmati, the Mahi and the large number of swift flowing western coast rivers descending from the Sahyadris.

It may be mentioned here that the area covered by the above mentioned drainage systems differs markedly from the amount of water carried by them.

Over 90 per cent of the water carried by the Indian rivers is housed into the Bay of Bengal; the rest is drained into the Arabian Sea or forms inland drainage. This lop sided distribution is due to the location of the watershed separating the drainage systems falling into the Bay of Bengal and the Arabian Sea. This 2,736 km long watershed runs from Kanniyakumari through the Western Ghats, the Ajanta, the Maikala, the Vindhayas and the Aravalis ranges to the Shiwalik hills near Haridwar (**Fig. 4.3**).

THE HIMALAYAN RIVER SYSTEMS

Three Major river systems *viz.* the Indus, the Ganga and the Brahmaputra comprise the Himalayan river systems. Many of the Himalayan rivers existed even before the Himalayan ranges were uplifted. These rivers originate on the Tibetan side beyond the mountain ranges of the Himalayas. The gorges of the Indus, the Satluj, the Alaknanda, the Gandak, the Kosi and the Brahmaputra clearly indicate that these rivers are older than the mountains themselves. It is believed that they continued to flow all through the building phase of the Himalayas; their banks rising steeply while the beds went lower and lower, thus cutting deep gorges through the rising Himalayan ranges. Thus, many of the Himalayan rivers are typical examples of *antecedent drainage*. The evolution and flow of the Himalayan rivers is described as under :

Evolution of the Himalayan Rivers

1. The Indo-Brahm or Shiwalik River Theory

Some of the unwarranted features of the Himalayan rivers such as the longitudinal courses of the Indus, the Satluj and the Brahmaputra; deep gorges cut by the rivers across the Himalaya; and still westerly flowing tributaries in their upper reaches are a great puzzle and require proper explanation. This has been partly done by assuming a mighty river which flowed from Assam to Punjab and even beyond upto Sind. This hypothetical ancient river was called the ***Indo-Brahm*** by **E.H. Pascoe** who thought that the present day Indus and Brahmaputra were the severed

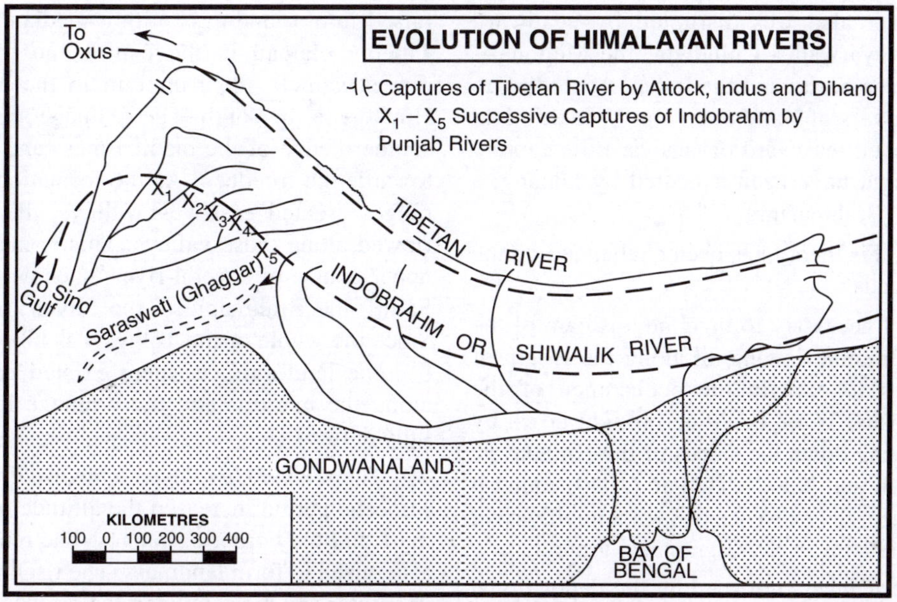

FIG. 4.2. Evolution of the Himalayan Rivers

parts of the original river. However, it was named as the ***Shiwalik River*** by E.G. Pilgrim who considered that the course of the primitive river is occupied by the present day Shiwalik hills. Pascoe and Pilgrim presented an elaborate and comprehensive hypothesis, independently in 1919, on the assumption that the Shiwalik deposition occurred along this great river. The river came into being due to earth movements which took place in *Tertiary* period and is believed to be successor of the *Himalayan sea*. In the *Eocene Epoch* a gulf extended from Sind to Afghanistan and from there extended eastward and south-eastward through Kohat and Punjab to the neighbourhood of Nainital. This gulf gave place to a great river. With its headwater consisting of portions of the Brahmaputra, this master stream flowed along the foot of the Himalayas first westward and then north-westward as far as north-western Punjab where it turned southward more or less along the course of the modern Indus, and emptied itself into the Arabian Sea. Later, this mighty stream got dismembered into the following systems and sub-systems :

(*a*) the Indus,

(*b*) the five tributaries of the Indus in Punjab,

(*c*) the Ganga and its Himalayan tributaries, and

(*d*) the stretch of the Brahmaputra in Assam and its Himalayan tributaries.

The dismemberment was the result of the following two events :

(*i*) Upheavals in the western Himalayas including the Potwar Plateau in the Pleistocene age and

(*ii*) Headward erosion by the tributaries of the Indobrahma river.

As a result of the above mentioned dismemberment of the Indobrahma river, the Indus and its tributaries, the Ganga and its tributaries and the Brahmaputra and its tributaries came into being. It is supposed that the Yamuna was first a tributary of the Indus. During the late Pleistocene and early Holocene ages, it joined with the Sarswati somewhere near Suratgarh and continued to flow as Ghaggar, finally joining the Indus. Some other scholars believe that it reached upto Rann of Kachchh as an independent river. Later on, it changed its course due to tectonic disturbance along the Aravali axis and was annexed by the Ganga to become its tributary.

Pascoe also envisaged a great '*Tibetan River*' flowing north–westwards along the Tsangpo—Manasarovar Lakes—Satluj—Gartang —Indus trough. This river might have emptied itself into the Oxus Lake, or might have debouched on to the plains by one of a number of transverse gaps such as Photu

Pass. This river also was disrupted by headward erosion by Ayeyarwadi—Chindwin, the Meghna—Brahmaputra, the Satluj and the Indus. According to de Terra, the rivers of Karakoram and Ladakh might have flowed south-eastward or eastward. Its capture in the east might have been executed by Dihang, a tributary of the Brahmaputra.

However, this theory has been challenged on the following grounds :

(*i*) It is not necessary to imagine a stream of the size of Indo-Brahma flowing all along the length of the Himalayas to explain the occurrence of the Shiwalik deposits. These could be brought down by the rivers flowing down the slopes of the Himalayas. **M.S. Krishnan** and **N.K.N. Iyngar** (1940) found it difficult to accept the existence of such a mighty river on geological as well as physiographical grounds.

(*ii*) The evidences furnished by the depositional history in the Ganga delta and in Assam does not fit well with the concept of Indobrahma river.

(*iii*) The width, thickness and lithology of the Tipam sandstones of Assam which were deposited in an estuary situated so close to the source of the Indobrahma also speak against this theory.

(*iv*) The upper course and the source of the Satluj river also do not fit in this theory. It is fed by underground water from the Manasarovar lake and it flows in a deep canyon cut in the soft fluviated beds of Nari Khorsum. The upper course of the Satluj is distinctly arid and the river appears to be a *misfit*. This could be explained if it were an old outlet of the '*Tibetan River*'. Pascoe argues that the Satluj captured a part of the '*Tibetan River*' and then lost again to the rejuvenated Tsangpo after the Dihang had cut back into the furrow. This argument has been rejected by many scholars.

E. Ahmad (1965 and 1971) has given his own interpretation of evolution of the Himalayan drainage. He believed that the Tethys remained as a basin of sedimentation from the Cambrian to the Eocene period but the major portion of the Himalayan region was occupied by the Gondwana landmass. During the *first Himalayan upheaval* in Oligocene period, part of the Tethyian geosyncline and probably the Gondwana shield covering the major part of the Himalayan region was uplifted. Most probably, this marks the initiation of the Himalayan drainage. The Tethys was raised into a landmass with a median mass of high Tibetan plateau in the centre and two bordering ranges namely the Kun Lun in the north and the Himadri in the south. The drainage started from the southern edge of the median mass and flowed south towards the foredeep. As the formation of east-west ranges created east-west valleys, the rivers partly flowed along these valleys. This is indicated by the upper course of several rivers such as the Indus, the Satluj, the Brahmaputra, the Shyok, the Arun, etc. Since the whole of the Tethys was not fully raised to become land surface, there existed patches of sea along the margins and the drainage lines were not fully defined.

The *second Himalayan upheaval* during the mid-Miocene period increased the altitude of the medium mass and the bordering ranges. The remnant sea was also raised to form landmass. The rise in land resulted in greater and more invigorated drainage. Along with these changes, the region to the south of the first Himalayan range was raised as Lesser Himalayan range. Earlier streams on the southern margin of the Tibetan plateau cut down deep valleys to maintain their courses. Along the southern slopes of the lesser Himalayas, a number of consequent streams also emerged which drained into the southern foredeep.

The *third Himalayan upheaval* during the Pleistocene period resulted in the folding of the Shiwalik foredeep into hill ranges. Also the height of earlier ranges and the Tibetan plateau was raised. The rise in the Tibetan plateau blocked the streams that had gone northward into the Tibetan sea. These streams were diverted east or west which probably led to the formation of the trans-Himalayan master stream. This master stream was broken into two (the proto-Indus and the proto-Brahmaputra) by the formation of the Kailas Range. The uplift of the Shiwalik range gave rise to the last set of consequents originating on the crest of the range emptying into older streams.

2. Multiple River Theory

An alternative explanation regarding the evolution of the Himalayan drainage has been offered by the Multiple River Theory. The protagonists of this theory find it difficult to accept the existence of a large river like the Indo-Brahm or Shiwalik on geological and physiographical grounds.

This theory postulates that the Eocene sea extended upto Sind, Rajasthan and from Punjab to Jammu and thereafter Lansdowne and Nainital. This was connected to Tethys. Existence of such a sea is evidenced by the presence of shallow water bodies indicative of coast near Landsdowne. This limit also coincides with the eastern continuation of one of the ridges of the Aravali Range which presumably acted as barrier. At the same time, another ridge extended from the Rajmahal Hills to the Meghalaya or Shillong plateau (Rajmahal—Garo gap) which is now occupied by the Ganga-Brahmaputra basin. The sea was broken by the first upheaval of the Himalayas to form an isolated basin in which sediments were deposited. In the next upheaval, a pronounced foredeep all along the southern border of the Himalayas was formed. This foredeep contained numerous lagoons in which flowed streams from the Peninsula and the newly uplifted Himalayas. These streams brought sediments which later came to be known as Shiwalik deposits. The outlet of this foredeep was through the Rajmahal-Garo gap in the Bay of Bengal and the Arabian Sea in the west. Later on the lagoons got dried up and numerous transverse streams flowing from the Himalayan region formed what is now known as the Himalayan drainage.

THE INDUS RIVER SYSTEM

The Indus (Sanskrit—Sindhu, Greek—Sinthos; Latin—Sindus), along with its tributaries forms one of the largest drainage systems of the world. It is from this river that India got her name. The Indus valley has been the cradle of one of the oldest civilizations in the world—'the Indus Valley Civilization'.

The mighty Indus rises near Manasarovar Lake from the glaciers of the Kailas Range in Western Tibet (31°15′ N and 81°40′E) at an elevation of 5,182 metres. Starting from its source, it flows for a distance of 257 km in the north west direction in the trans-Himalaya region under name of *Singge Khabab* until it is joined by the Dhar. A short distance lower down, it enters India at an elevation of 4,206 m and continues to flow in the same north-west direction between the Ladakh and the Zaskar Ranges. The gradient of the river in this course is very gentle (about 30 cm per km). Here it encircles the town of Leh and is joined by the Zaskar river. About 50 km before Skardu, it is joinc 1 by the Shyok at an elevation of about 2,700 m. The *Gilgit*, *Gortang*, *Dras*, *Shigar*, *Hunza* are the other Himalayan tributaries of the Indus. After passing through a 480 km long antecedent and very deep gorge (5,181 metres at Bunji north of Nanga Parbat), it takes a sharp southerly bend and reaches Attock at an altitude of about 610 m. Here it ends its mountainous journey and is joined by the Kabul river from Afghanistan. Thereafter it flows through the Potwar plain and crosses the Salt Range. Some of the important tributaries below Attock include the Kurram, Toch and the Zhob-Gomal. Just above Mithankot, about 805 km from the sea, at an elevation of 79 m the Indus receives from *Panjnad* (*Panchnad*), the accumulated waters of the five eastern tributaries—the Jhelum, the Chenab, the Ravi, the Beas and the Satluj. The river finally empties itself into the Arabian Sea, south of Karachi after forming a big delta. The total length of the Indus river from its source to its mouth is about 2,880 km out of which only 709 km length falls in India. The average annual flow of water in the Indus river is 110,450 million cubic metres at Kalabagh.

The left bank tributaries are big rivers in themselves and deserve a brief mention here. These rivers join together one by one before they ultimately meet the main river.

The Jhelum rises in a spring at Verinag in the south-eastern part of the Kashmir Valley. It flows northwards from its source to Wular Lake and further down south-westwards. The river flows through the Kashmir Valley and forms a 200 m deep defile with almost vertical walls through Pir Panjal Range below Baramula. A number of tributaries notably the Lidar, the Sind and the Pohru, which rise in Kashmir, join the main river. At Muzaffarabad, the river takes a sharp hairpin swing southward and the Kishaganga joins it on its right bank. Thereafter, it forms the India-Pakistan boundary for 170 km and emerges at the Potwar Plateau near Mirpur. After skirting the outlying spurs of the Salt Range, it debouches on the plains near the city of Jhelum about 402 km from its source. About 322 km lower, it joins the Chenab at Trimmu. The river is navigable for about 160 km out of a total length of 724 km. The average annual flow of water in the river at Mangla is 27,890 million cubic metre. The catchment area upto the Indo-Pakistan border is 34,775 sq km.

TABLE 4.1. The Indus Drainage System

Name of the river	Source	Length in km	Area drained (sq km)	Volume of average annual flow (in million cubic metres)
Indus	Near Manasarovar Lake (31°15′N, 81°40′E at 5,182 m elevation	2,880 (709 in India)	1,178,440 (321,284 in India)	1,10,450 at Kalabagh in Pakistan
Jhelum	Verinag	724	34,775 upto Indo-Pak Border	27,890 at Mangla in Pakistan
Chenab	Bara Lacha Pass	1,180	26,155 upto Indo-Pak Border	29,000 at Marala in Pakistan
Ravi	Near Rohtang Pass	725	14,442 (5,957 in India)	8,000 at Madhopur
Beas	Near Rohtang Pass at 4,062 m height	460	20,303	15,800 at Mandi
Satluj	Manasarovar-Rakas Lakes at 4,570 m altitude	1,450 (1,050 in India)	25,900	16,600 at Rupnagar

The Chenab originates from near the Bara Lacha Pass in the Lahul-Spiti part of the Zaskar Range. Two small streams on opposite sides of the pass, namely ***Chandra*** and ***Bhaga***, form its headwaters at an altitude of 4,900 m. The united stream, called the ***Chandrabhaga*** flows in the north-west direction through the Pangi valley, parallel to the Pir Panjal range, and enters Jammu and Kashmir as the Chenab river at an elevation of 1,838 m. Near Kistwar, it cuts a deep gorge, at places 1,000 m deep. Here, it flows for 290 km between steep cliffs of high mountains and turns southwards and flows in this direction for a short distance. Further, it turns to the west and enters the plain area near Akhnur in Jammu and Kashmir after traversing a distance of about 330 km. From here it swings to the south-west through the plains of Pakistani Punjab for a distance of 644 km to reach Panchnad where it joins the Satluj after receiving the waters of Jhelum and Ravi rivers. The total length of the river is 1,180 km. Its catchment area upto the Indo-Pakistan border is 26,155 sq km. Its annual flow at Marala is 29,000 million cubic metres.

The Ravi has its source in Kullu hills near the Rohtang Pass in Himachal Pradesh. Flowing in a northwest direction from its source, it drains the area between the Pir Panjal and the Dhaola Dhar ranges. After crossing Chamba, it takes a south-westerly turn and cuts a deep gorge in the Dhaola Dhar range. It enters Punjab Plains near Madhopur and later enters Pakistan 26 km below Amritsar. It joins the Chenab a little above Rangpur in Pakistani Punjab after flowing for a distance of 725 km from its source. Its total catchment area is 14,442 sq km of which only 5,957 sq km lies in India. The annual flow of water in this river at Madhopur is 8,000 million cubic metres.

The Beas originates near the Rohtang Pass, at a height of 4,062 m above sea level, on the southern end of the Pir Panjal Range, close to the source of the Ravi. Fairly steep in its upper portion (24 metres per km) it crosses the Dhaola Dhar range through a deep gorge 900 m deep from Lorji to Talwara. On meeting the Shiwalik hills the river sweeps sharply northward, then bending round the base of the hills it takes a southerly direction and debouches on the plain. Thereafter, it takes a south-westerly direction and meets the Satluj river at Harike in Punjab. It is a comparatively small river which is only 460 km long but lies entirely within the Indian territory. Its total catchment area is 20,303 sq km. The river carries about 15,800 million cubic metres of water at Mandi.

The Satluj rises from the Manasarovar-Rakas Lakes near Darma Pass in western Tibet at a height of 4,570 m within 80 km of the source of the Indus. Like the Indus, it takes a north-westerly course upto the *Shipki La* on the Tibet-Himachal Pradesh boundary. Here the river flows at an altitude of about 3,000 m above sea level. It cuts deep gorges where it pierces the Great Himalaya and the other Himalayan ranges. In Nari Khorsan province of Tibet, it has created an extraordinary canyon, comparable to the Grand

Conyon of Colorado; the channel here is 900 m deep. Its tributaries in Himachal Pradesh are short in length except the Spiti, which drains a large trans-Himalayan area and joins it at Namgia near the Shipki La. Before entering the Punjab plain, it cuts a gorge in Naina Devi Dhar, where the famous Bhakra dam has been constructed. After entering the plain at Rupnagar (Ropar), it turns westwards and is joined by the Beas at Harike. From near Ferozepur to Fazilka it forms the boundary between India and Pakistan for nearly 120 km. During its onward journey it receives the collective drainage of the Ravi, Chenab and Jhelum rivers and about 70 km further downstream, it joins the Indus a few kilometres above Mithankot. Out of its total length of 1,450 km, it flows for 1,050 km in Indian territory. It drains an area of 25,900 sq kms and its average annual flow at Rupnagar (Ropar) is 16,660 million cubic metres.

The water assets of the Indus river are shared by India and Pakistan according to the Indus Water Treaty signed between the two countries on 19th September, 1960. According to this treaty, India can utilize only 20 per cent of its total discharge of water.

THE SARASWATI—MYTH OR REALITY

A lot has been said about the mythical river Saraswati. A new research—led by Physical Research Laboratory (PRL) Ahmedabad in collaboration with IIT-Bombay has reported unequivocal evidence that the mighty Saraswati flowed majestically in the plains of north-west India roughly along the course of the modern Ghaggar. In fact, Ghaggar is a residue of the Saraswati river which flowed in ancient times. The report published in "Scientific Report" says that Saraswati was a perennial river and had flowed from Higher Himalayas between 7,000 B.C. and 2,.500 B.C. The river had its source in the glaciated region of the Higher Himalayas, similar to the Ganga, the Yamuna and the Satluj. The conclusion has been drawn by determining the depositional ages of the coarse-grained white sand layer that occurred 3-10 metres below modern alluvium of the Ghaggar's flood plain. The dating was done with radiocarbon and optically stimulated luminescence methods. The Saraswati lost its perennial character after Satluj fed channels ran dry due to extreme aridity.

THE GANGA RIVER SYSTEM

The Ganga river system consists of the master river Ganga and a large number of its tributaries. This system drains a very large area comprising the middle part of the Himalayas in the north, the northern part of the Indian Plateau in the south and the Ganga Plain in-between.

The Ganga originates as Bhagirathi from the Gangotri glacier in Uttar Kashi District of Uttarakhand at an elevation of 7,010 m. Alaknanda joins it at Devaprayag. But before Devaprayag is reached, the Pindar, the Mandakini, the Dhauliganga and the Bishenganga rivers pour into the Alaknanda and the Bheling flows into the Bhagirathi. The Pindar river rising from Nanda Devi and East Trisul (6,803 m) joins Alaknanda at Karan Prayag and Mandakini or Kali Ganga meets at Rudra Prayag. The combined water of the Bhagirathi and the Alaknanda flows in the name of the Ganga below Devaprayag. After travelling 280 km from its source, the Ganga reaches Haridwar, debouches from the hills and enters plain area. From here it flows in south and south-east direction for a distance of 770 km to reach Pryagraj. Here it is joined by the Yamuna which is its most important tributary. It sweeps another 300 km eastwards to reach the Bihar plain. Near Rajmahal Hills it turns to the south-east and south of Farraka, it ceases to be known as the Ganga. It bifurcates itself into Bhagirathi-Hugli in West Bengal and Padma-Meghna in Bangladesh. After traversing 220 km further down in Bangladesh, the Brahmaputra (or the Jamuna as it is known here) joins it at Goalundo and after meeting Meghna 100 km downstream the Ganga joins the Bay of Bengal.

Before entering the Bay of Bengal, the Ganga, along with the Brahmputra, forms the largest delta of the world between two arms : the Bhagirathi / Hugli and the Padma / Meghna covering an area of 58,752 sq km. The delta front of the Ganga is a highly indented area of about 400 km length extending from the mouth of the Hugli to the mouth of the Meghna. The delta is made of a web of distributaries and islands and is covered by dense forests called the ***Sunderbans***. A major part of the delta is a low-lying swamp which is flooded by marine water during high tide.

The total length of the Ganga river from its source to its mouth (measured along the Hugli) is 2,525 km, of which 310 km is in Uttarakhand, 1,140 km is in Uttar Pradesh, 445 km is in Bihar and 520 km in is West Bengal. The remaining 110 km stretch of the

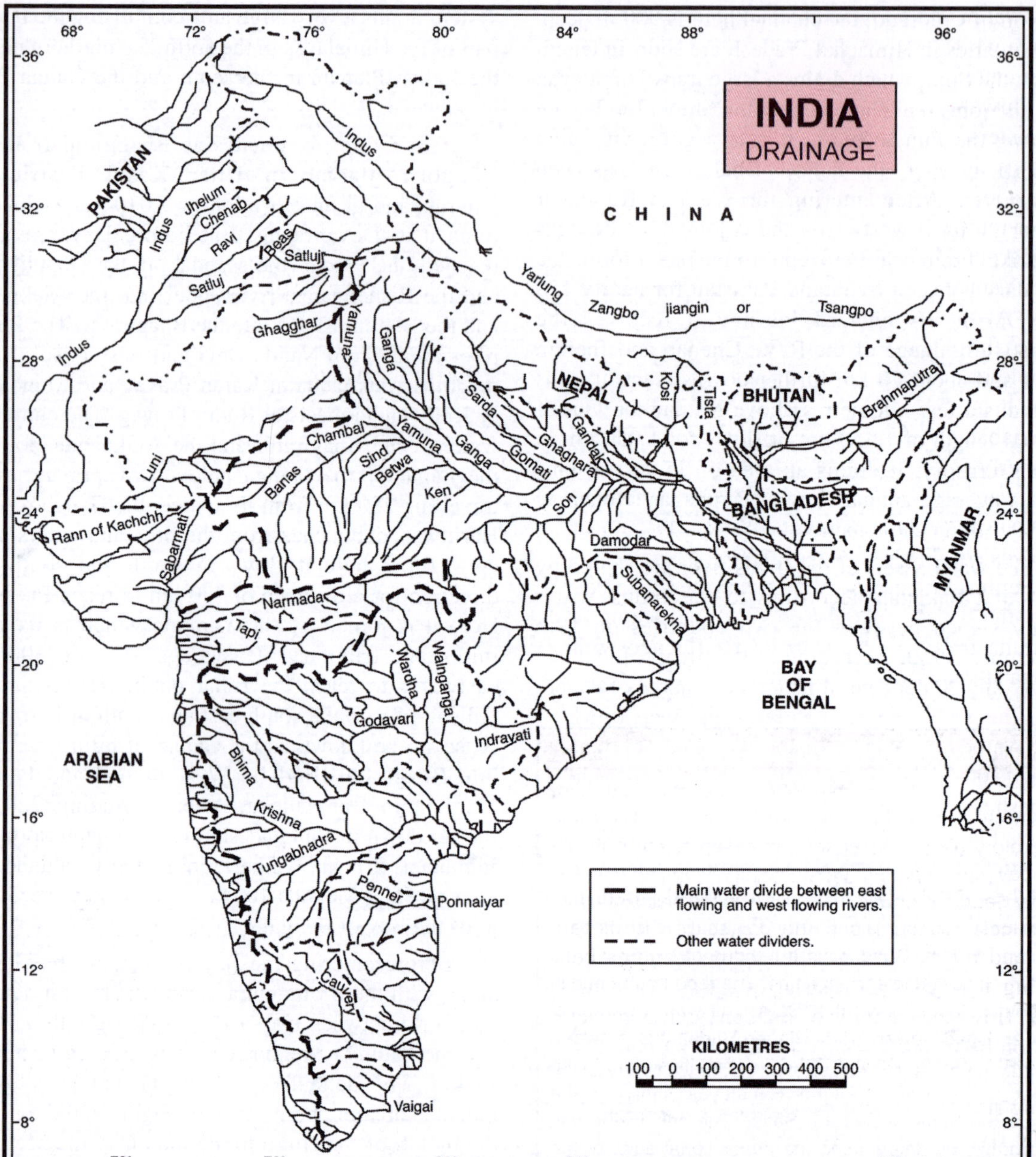

FIG. 4.3. India : Drainage

Ganga forms the boundary between Uttar Pradesh and Bihar. The river flows majestically from Gangotri to Bay of Bengal with an average gradient of 2.77 m per km.

The Ganga is joined by a large number of tributaries both from the left as well as from the right. Majority of them originate in the Himalayan ranges but some of them have their sources in the Peninsular plateau.

The Yamuna is the largest and the most important tributary of the Ganga. It originates from the Yamnotri glacier on the Bandarpunch Peak in

Garhwal in Uttarakhand at an elevation of 6,330 m, a source which is very close to that of the Ganga itself. After cutting across the Nag Tibba, the Mussoorie and the Shiwalik ranges it emerges out of the hilly area and enters plains near Tajewala. Many small streams such as the Rishiganga, the Uma and the Hanuman Ganga join it in the mountains. Its main affluent in the upper reaches is the *Tons* which also rises from the Bandarpunch glacier at an altitude of 3,900 m and joins Yamuna below Kalsi before the latter leaves the hills. At this site, the water carried by the *Tons* is twice the water carried by the Yamuna. After completing a journey of about 300 km in the plain area, it is joined by the 256 km long Hindon near Ghaziabad in Uttar Pradesh. The Yamuna takes a southerly course upto Mathura and south-easterly in its onward journey upto Pryagraj where it unites with the Ganga. Between Agra and Pryagraj, the Yamuna receives some important tributaries originating from the Peninsular Plateau. These are the Chambal, the Sind, the Betwa and the Ken. The total length of the Yamuna from its origin to its confluence with the Ganga at Pryagraj is 1,376 km.

The Chambal rises 15 km south-west of Mhow (Dr. Ambedkar Nagar) in the highlands of Janapao Hills (700 m) in the Vindhya Range in Madhya Pradesh and takes a north-westerly course through the Malwa Plateau. It then enters a gorge at Chaurasigarh, 312 km from its source. The gorge is 96 km long and stretches upto Kota city. Below Kota, it turns to north-east and after reaching Pinahat, it turns to the east and runs nearly parallel to the Yamuna before joining it in Etawah district of Uttar Pradesh. The river flows much below its banks due to severe erosion because of poor rainfall and numerous deep ravines have been formed in the Chambal Valley, giving rise to **badland topography**. The total length of the river is 1,050 km. *The **Banas*** is an important tributary of the Chambal. It originates in the southern part of the Aravali Range, and takes a north-eastern course to join the Chambal near Sawai Madhopur. **The Sind** originates in Vidisha Plateau of Madhya Pradesh at an elevation of 543 m. It flows for a distance of 415 km before it joins the Yamuna. It drains an area of 25,085 sq km. **The Betwa** rises at an elevation of 470 m in Bhopal district and joins the Yamuna near Hamirpur. It has a total length of 590 km and a catchment area of 45,580 sq km. **The Dhasan** is its important tributary. The 360 km long **Ken river** rising from the Barner Range of Madhya Pradesh is another tributary which joins the Yamuna near Chila; some 60 km downstream from the mouth of the Betwa.

The Son river springs from the Amarkantak Plateau, not far from the origin of the Narmada, at an elevation of about 600 m. After flowing for some distance to the north, it meets the Kaimur Range which turns its course to the north-east. It passes over the cascades in the hill reaches and further beyond through the Palamau district of Jharkhand. It joins the Ganga about 16 km upstream of Danapur in Patna district of Bihar after flowing for a distance of 784 km from its source. It has a total catchment area of 71,259 sq km in which there are wide fluctuations of water flow changing with the change of season. The important tributaries of the Son are the *Johilla*, the *Gopat*, the *Rihand*, the *Kanhar* and the North *Koel.* Almost all the tributaries join it on its right bank.

The Damodar river rises in the hills of the Chotanagpur plateau and flows through a rift valley. It used to cause devastating floods as a result of which it earned the dubious name of ***'Sorrow of Bengal'***. After flowing through Bankura and Burdwan districts of West Bengal, it joins the Hugli 48 km below Kolkata. The total length of the river is 541 km and its catchment area is 25,820 sq km.

Still more important tributaries of the Ganga originate in the Himalayas and join it on its left bank. The major tributaries of this category, apart from the Yamuna, are the Ramganga, the Gomati, the Ghaghra, the Gandak, the Burhi Gandak, the Bagmati, and the Kosi.

The Ramganga river rises in the Garhwal district of Uttarakhand at an altitude of 3,110 m and enters the Ganga Plain near Kalagarh. It joins the Ganga at Kannauj after traversing a distance of 596 km. The Khoh, the Gangan, the Aril, the Kosi, and the Deoha (Gorra) are important tributaries which join the Ramganga. Its basin covers 32,493 sq km.

The Ghaghra river originates near the Gurla Mandhata peak, south of Manasarovar in Tibet. Obviously it is a river of the trans-Himalayan origin and carries sufficient water. It is known as *Karnaili* in western Nepal. It first flows south-east and then south-southwest cutting across the Great Himalaya through a succession of steep defiles. Its important tributaries are the *Sarda*, the *Sarju* on the bank of which Ayodhya is located and the *Rapti*. The Ghaghara joins the Ganga a few kilometres downstream of Chhapra in Bihar after a 1,080 km long journey from its source. Soon after reaching the plain area, its stream gets divided into many branches of which, *Koriyab* and *Garwa* are important. These two streams meet again in Bahraich district of Uttar Pradesh. From here, the river bed is sandy and sudden bends start occurring in the stream. Flowing over a wide sandy expanse, this river crosses Gorakhpur, Deoria, Azamgarh and Balia districts of Uttar Pradesh as a divided stream. The total catchment area of the river is 127,950 sq km out of which 45% is in India and the rest lies in Nepal and Tibet. The river has a high flood frequency and has shifted its course several times. Its average annual flow is 94,000 million cubic metres.

The Kali river also rises in the high glaciers of snow covered region of the trans-Himalaya. It has several feeders in the high ranges and forms the boundary between Nepal and Kumaon for most of its transit across the mountains. It is known as the ***Sarda*** or ***Chauka*** after it reaches the plains near Tanakpur. Below this point it takes a south-easterly course and joins the Ghaghara.

The Gandak which originates near the Tibet-Nepal border in the north-west of Dhaulagiri at a height of 7,620 m receives a large number of tributaries in Nepal Himalaya, the most outstanding being the *Kali Gandak*, the *Mayangadi*, the *Bari* and the *Trishuli*. It debouches into the plains at Tribeni and pours into the Ganga at Hajipur in Bihar after flowing for a distance of 425 km in India. Its drainage area is 46,300 sq km out of which 7,620 sq kms, is in India. The average annual flow is 52,200 million cubic metres.

The Burhi Gandak originates from the western slopes of Sumesar hills near the India-Nepal border and flows in south-east direction. It joins the Ganga opposite Monghyr town. Its length is 610 km and its drainage area is 12,200 sq km.

The Kosi river consists of seven streams namely ***Sun Kosi, Tamba Kosi, Talkha, Doodh Kosi, Botia Kosi, Arun*** and ***Tamber*** and is popularly known as ***Saptkaushiki***. These streams flow through a large part of eastern Nepal which is known as the ***Sapt Kaushik*** region. *Arun* is the main stream which rises to the north of Gosainthan. The sources of seven streams of the Kosi are located in snow covered areas which also receive heavy rainfall. Consequently, huge volume of water flows with tremendous speed. Coming out of the Saptakosi region, the streams flow in south-west direction. Seven streams mingle with each other to form three streams named the ***Tumar, Arun*** and ***Sun Kosi***. These three streams contribute respectively 44%, 37% and 19% of the total water flow in the river. They unite at Triveni north of the Mahabharata Range to form the Kosi. The river enters

TABLE 4.2. The Ganga Drainage System

Name of the river	Source	Length (in km)	Area drained (sq km)	Volume of average annual flow (in million cubic metres)
Ganga	Gangotri Glacier at 7,010 m	2,525	861,404	1,52,000 at Pryagraj, 4,59,040 at Patna
Yamuna	Yamnotri Glacier at 6,330 m	1,376	366,223	93,020 at Pryagraj
Chambal	Near Mhow (Madhya Pradesh)	1,050	139,468	30,050 at confluence with the Yamuna
Ramganga	Garhwal district at 3,110 m	596	32,493	15,258
Ghaghara	Near Gurla Mandhota peak south of Manasarovar	1,080	127,950	94,400
Gandak	Tibet-Nepal border at 7,620 m	425 in India	46,300 (7,620 in India)	52,200
Kosi	Sikkim-Nepal-Tibet Himalaya	730 in India	86,900 (21,500 in India)	61,500

the *Tarai* of Nepal after cutting a narrow gorge for 10 km in the Mahabharata Range at Chatra. The Kosi flows for a distance of 730 km in India and joins the Ganga near Kursela. The total drainage area of the river is 86,900 sq km out of which 21,500 sq km lies in India.

Soon after debouching onto the plain the river becomes sluggish and large scale deposition of eroded material takes place. The river channel is braided and it shifts its course frequently. This has resulted in frequent devastating floods and has converted over 10,000 sq km cultivable land into waste land in Bihar. Thus the river is often termed as the ***'Sorrow of Bihar'***. In order to tame this river, a barrage was constructed in 1965 near Hanuman Nagar in Nepal. Embankments for flood control have been constructed as a joint venture of India and Nepal.

THE BRAHMAPUTRA RIVER SYSTEM

The Brahmaputra (meaning the *son of Brahma*) rises in the great Chemayungdung glacier in the Kailas Range of the Himalayas a little south of the Lake Konggyu Tsho at an elevation of about 5,150 m located at 30° 31′ N latitude and 82° 10′ E longitude. This source is about 150 km away from the source of the Indus and only 35 km from the birth place of the Satluj. Both these rivers flow westwards while the Brahmaputra flows eastwards from this region. *Mariam La* separates the source of the Brahmaputra from the Manasarovar Lake. With a total length of 2,900 km the Brahmaputra is one of the longest rivers of the world and passes through Tibet, India and Bangladesh.

Known as the ***Tsangpo*** (meaning 'The Purifier'), in Tibetan and ***Yarlung Zangbo Jiangin*** in Chinese language, the Brahmaputra flows eastwards in Southern Tibet for about 1,800 km and for most part of this journey it passes through the depression formed by the *Indus-Tsangpo Structure Zone* between the Great Himalayas in the south and the Kailas Range in the north. Inspite of the exceptionally high altitude of the land through which the Tsangpo flows, it has a gentle slope and the river is sluggish and has a wide navigable channel for about 640 km from Lhatse Dzong down-stream. This is one of the most remarkable navigable waterways of the world where boats sail at an altitude of 3,000 metres and above. Its bed is 3,525 m high at Tradom (Teladuomu) and 3,300 m at Shigatse. It receives a large number of tributaries in Tibet. The first major tributary is the Raga Tsangpo from the north meeting the Tsangpo near Lhatse Dzong. The river Ngangchu flows through the trade centre of Gyantse in the south and joins the main river near Zhikatse. Further east, the Kyichu (Gya Chhu) from the north runs past the sacred town of Lhasa and joins it near Chushul Dz. At Tsela Dz, the river is joined by the Griamda from the north.

Towards the end of its journey in Tibet, it abruptly turns to north-east and north and then traverses in a succession of rapids between high mountains of Gyala Peri (7,151 m) and Namcha Barwa (7,756 m). Further, it turns to the south and south-west and cuts across the eastern Himalaya through the Dihang or Siang Gorge and emerges from the mountains near Sadiya in the Assam Valley. Here it first flows under the name of Siong and then as the Dihang. The Dihang gorge is only 50 metres wide at places but is 5,485 m below Namcha Barwa. The steepness of its slope can be judged from the fact that its altitude falls suddenly from 2,450 m north of Namcha Barwa to only 135 m at Sadiya. Here it is joined by two important tributaries *viz.* the Dibang (or Sikang) from the north and Lohit from the south forming what looks like a '*delta in reverse*'.

From Sadiya onwards, this mighty river is known as the ***Brahmaputra*** and flows majestically in the Assam Valley for a distance of about 720 km. Here also, a large number of tributaries, both from the north and from the south, pay their tribute to the master river. The main streams merging with the Brahmaputra from the north are, Subansiri, Kameng, Belsiri, Dhansiri (north), Nyera Ama, Manas—Mora—Manas, Champaman, Gangadhar, Raidah, Dharla and Tista. The Tista was a tributary of the Ganga prior to the devastating floods of 1787 after which it diverted its course eastwards to join the Brahmaputra. Those pouring in the main river from the south are the Dibru, Burhi Dihing, Noa Dihing, Dikhu, Dhansiri (south) and Kalang.

The Brahmaputra has a braided channel for most of its passage through Assam. There is a constant shifting of the river channels and the sandy shoals. Its southward shift in the Assam plains has been quite conspicuous in the recent years. It carries a lot of silt and there is excessive meandering. The river is nearly

16 km wide at Dibrugarh and forms many islands, the most important of which is *Majuli*. It is 90 km long and measures 20 km at its widest, having an area of 1,250 sq km. This island has 144 villages with a population of over 1,50,000. With rainfall concentrated during the monsoon months only the river has to carry enormous quantities of water and silt which results in disastrous floods in the rainy season. The floods caused by the Brahmaputra in Assam Valley affect a vast area, and cause huge loss of life and property. The Brahmaputra is thus truly a *River of Sorrow*. However, the river is navigable for a distance of 1,384 km upto Dibrugarh from its mouth and serves as an excellent inland water transport route.

Traversing round the stairs of the Garo Hills, the Brahmaputra bends southwards and enters Bangladesh near Dhubri. It flows for a distance of 270 km under the name of Jamuna river and joins the Ganga at Goalundo. The united stream of the Jamuna and the Ganga flows further under the name of Padma. About 105 km further downstream, the Padma is joined on the left bank by the Meghna, originating in the mountainous region of Assam. From the confluence of Padma and Meghna, the combined river is known as the ***Meghna*** which makes a very broad estuary before pouring into the Bay of Bengal.

THE PENINSULAR DRAINAGE

The Indian Peninsula is traversed by a large number of rivers which have existed for a much longer period than the Himalayan rivers. As such the peninsular rivers have reached mature stage and have almost reached the base level of their erosion. This is characterised by broad, shallow and largely graded valleys. The beds have very subdued gradient except for a limited tract of river where faulting has allowed steepening of the gradient. Almost the entire peninsula presents a senile topography showing features of mature drainage.

The velocity of water in the rivers and the load carrying capacity of the streams is low due to low gradient and the rivers form big deltas at their mouths. This is especially observed in the east flowing rivers pouring into the Bay of Bengal. But the west flowing rivers of Narmada and Tapi as well as those originating from the Western Ghats and falling in the Arabian Sea form estuaries in place of deltas.

However, there are instances of superimposed and rejuvenated drainage represented by waterfalls. The Jog on the Sharvati (289 m), Yenna of Mahabaleshwar (183 m), Sivasamundram on the Cauveri (101 m), Gokak on the Gokak (55 m), Kapildhara (23 m) and Dhuandhar (15 m) on the Narmada are the major waterfalls in the Peninsular India.

Evolution of the Peninsular Drainage

Geologists believe that the Sahyadri-Aravali axis was the main water divide in the geological past. One hypothesis assumes that the existing peninsula is the remaining half of bigger landmass. The Western Ghats were located in the middle of this landmass dividing the entire drainage into two symmetrical parts—one flowing in the east and the other flowing in the west. But the western part of the Peninsula cracked and submerged in the Arabian Sea and disturbed the symmetrical plan of the rivers on either side of the watershed during the early Tertiary period. During the collision of the Indian plate, a second major distortion was introduced in the Peninsular block and it was subjected to subsidence and consequent trough faulting through which now flow main west flowing rivers of the Peninsula, namely the Narmada and the Tapi. This is amply proved by the straight coastline, steep western slope of the Western Ghats, and the absence of delta formations on the western coast.

Another view is put forward keeping in view the exceptional behaviour of the Narmada and the Tapi. It is believed that these two rivers do not flow in the valleys formed by the rivers themselves. Rather they have occupied two fault planes or alluvium filled rifts in rocks running parallel to the Vindhyas. These faults are supposed to be caused by bending or 'sagging' of the northern part of the Peninsula at the time of upheaval of the Himalayas. According to **D.N. Wadia** (1975) the Peninsular block, south of the cracks, tilted slightly eastwards thus giving new orientation to the entire drainage towards the Bay of Bengal. While this line of thought explains the present drainage system of the Peninsular India with greater satisfaction, it still leaves some questions to which there is no satisfactory answer. **R.C. Mehdiratta** (1962) argues that tilting should have increased the gradient of the river valleys and caused some rejuvenation of the rivers. This type of phenomenon is absent in the Peninsula, barring a few exceptions such as waterfalls, as mentioned earlier.

The Peninsular River Systems

Although the general direction of flow of the Peninsular rivers is from west to east, a careful study reveals at least three main directions of flow :

(*i*) The Mahanadi, the Godavari, the Krishna, the Cauvery and several smaller rivers draining south-east into the Bay of Bengal.

(*ii*) The Narmada and the Tapi flowing west as well as several small streams originating from the Western Ghats flow westwards into the Arabian Sea.

(*iii*) Tributaries of the Ganga and the Yamuna such as the Chambal, the Betwa, the Ken, the Son and the Damodar flow in the north-easterly direction.

The East Flowing Rivers of the Peninsula

The Mahanadi (literally meaning big river) is an important river of the Peninsular India. It has its source in the northern foothills of Dandakaranya near Sihawa in Raipur District of Chhattisgarh at an elevation of 442 m. Its upper course lies in the saucer-shaped basin called the 'Chhattisgarh Plain'. This basin is surrounded by hills on the north, west and south as a result of which a large number of tributaries join the main river from these sides. The main tributaries are the *Ib* (251 km), the *Mand* (241 km), the *Hasdo* (333 km) and the *Sheonath* (383 km) on the left bank and the *Ong* (204 km), the *Jonk* (196 km), and the *Tel* (295 km) on the right bank. From its source, the river takes a north easterly course. Beyond Seorinarayan it flows eastwards and after entering Odisha, it turns southwards below the Hirakud Dam. Further below it turns eastwards near Sonepur. On reaching the Eastern Ghats, the river flows through a narrow gorge for 23 km near Tikkarpara Range and finally emerges in a delta at Naraj 11 km west of Cuttack. The Mahanadi finally empties itself in the Bay of Bengal after flowing for a distance of 857 km. The delta of Mahanadi spreads over an area of 9,500 sq km and is over 150 km broad.

The Godavari is the largest river system of the Peninsular India and is next only to the Ganga and the Indus Systems regarding sanctity, picturesqueness and utility and is held in reverence as *Vridha Ganga* or *Dakshina Ganga*. Its total length is 1,465 kilometres and the area drained by this river is 3,12,812 sq km. The source of this river is in the Trimbak Plateau of North Sahyadri near Nashik in Maharashtra which is only 80 km from the shore of the Arabian Sea. From its source it flows eastwards in a narrow rocky bed upto Nashik but the river valley opens out below this point. It receives a large number of tributaries both from the left as well as from the right. But the left bank tributaries are more in number and larger in size than the right bank tributaries. The *Manjra* (724 km) is the only important right bank tributary. It originates from Jamkhed Hill in Central Maharashtra and joins the Godavari near Kondalwadi after passing through the Nizam Sagar. The *Penganga*, the *Wardha*, the *Wainganga*, the *Indravati* and the *Sabari* are important left bank tributaries. The *Penganga* (676 km) rises from the Buldana Range and joins the Wardha River (483 km) near Ghughus. The Wardha in its turn joins the Wainganga. The united Wardha and Wainganga rivers become the short span *Pranhita* river which joins the Godavari below Sironcha. The *Indravati* (535 km) river from the Kondhan Hills of Eastern Ghats joins the master river about 48 km downstream from Sironcha. Below its confluence with Indravati, the Godavari flows in a gorge, 60 km long and 200 m wide through the Eastern Ghats. It is located about 100 km from the mouth of the river and is supposed to be formed due to faulting. Below Rajahmundry, the river divides itself into two main streams, the *Gautami Godavari* on the east and the *Vashishta Godavari* on the west and forms a large delta before it pours into the Bay of Bengal. The current at Rajahmundry is not rapid and varies from 1.2 to 3.3 metres per second. The delta has a front of 120 km and projects about 35 km into the sea. The delta of the Godavari is of *lobate type* with a round bulge and many distributaries. The Godavari causes heavy floods in its lower course below Polawaram. The Godavari is navigable upto a distance of 300 km from its mouth.

The Krishna is the second largest east flowing river of the Peninsula. It rises in the Western Ghats just to the north of Mahabaleshwar, about 64 km from the Arabian Sea and flows for a distance of 1,400 km to the Bay of Bengal in a general easterly direction. The *Koyna*, the *Ghataprabha*, the *Malprabha*, the *Bhima*, the *Tungabhadra*, the *Musi* and the *Muneru* are its important tributaries. **The Koyna** is a small tributary but is very famous for Koyna Dam. This dam was perhaps the main cause of the devastating earthquake in 1967. **The Bhima** originates from the Matheron Hills and joins the Krishna 26 km from Raichur after flowing in the south-easterly direction for a distance of 861 km. Mula, Mutha, Ghod and Nora are its sub-tributaries. **The Tungabhadra** is formed by the unification of the *Tunga* and the *Bhadra* originating from Gangamula in the Central Sahyadri. Its total length is 531 km. At Wazirabad, it receives its last important tributary, the *Musi*, on whose banks the famous city of Hyderabad is located. The Krishna forms a large delta spreading over an area of 4,600 sq km with a shoreline of about 120 km. The Krishna delta appears to merge with that formed by the Godavari and extends about 35 km into the sea.

Kaveri (Cauvery) is the most revered and sacred river of south India and is designated as the '*Dakshina Ganga*' or '*the Ganga of the South*'. The Source of this river lies at Taal Cauvery on the Brahmagiri range of hills in the Western Ghats at 1,341 m elevation (12° 25′ N and 75° 34′ E) situated in the Coorg Plateau (Coorg district of Karnataka). It flows eastwards for a distance of about 800 km before it empties in the Bay of Bengal. This river is unique in the sense that its upper catchment area receives rainfall during summer by the south-west monsoon and the lower catchment area during winter season by the retreating north-east monsoon. It is, therefore, almost a perennial river with comparatively less fluctuations in flow and is very useful for irrigation and hydroelectric power generation. Thus the Cauvery is one of the best regulated rivers and 90 to 95 per cent of its irrigation and power production potential already stands harnessed.

The Cauvery river falls rapidly by 450 m from its source of origin within a short distance of 8 km. The middle course of the river over the 750 m high Mysore Plateau is characterised by broad and sweeping meanders. The river descends from the South Karnataka Plateau to the Tamil Nadu Plains through the Sivasamudram waterfalls (101 m high). The river which was more than one kilometre wide above the Sivasamudram Falls, narrows down considerably below it and enters a long, picturesque gorge, 200 m deep at places, cuts through the Eastern Ghats. The river divides itself into two distinct channels at Srirangam. The northern channel is called Kollidam (Coleroon) while the southern one retains the name Cauvery. The delta formation starts from a point about 16 km away from Tiruchchirappalli. The Cauvery has formed a quadrilateral delta spreading over an area of 8,000 sq km. It has almost a straight front which runs for a distance of 130 km along the coast of Bay of Bengal.

The water flowing in the Cauvery is supplemented by a large number of tributaries which join the master river at different places. The main tributaries are the *Herangi*, the *Hemavati*, the *Lokpavani*, the *Shimsha* and the *Arkavati* from the north and the *Laksmantirtha*, the *Kabani*, the *Suvarnavati*, the *Bhavani* and the *Amravati* from the south.

Among the other east flowing rivers of the Peninsular India are, from north to south, the Subarnarekha, the Brahmani, the Penneru, the Ponnaiyar and the Vaigai. **The Subarnarekha** originates from the Ranchi Plateau in Jharkhand at an elevation of 600 m and flows in south-east direction forming the boundary between West Bengal and Odisha in its lower course. It joins Bay of Bengal forming an estuary between the Ganga and Mahanadi deltas. Its total length is 395 km. **The Brahmani** river comes into existence by the confluence of the *Koel* and the *Sankh* rivers near Rourkela. It has a total length of 800 km. Its main tributaries are the *Kura*, the *Sankhad* and the *Tikra*. Its mouth becomes the northern end of the Mahanadi delta. **The Penneru** river springs from the *Nandi Durg* peak in Karnataka and flows in northward direction. It then enters Andhra Pradesh and takes an easterly course. It forms a narrow estuary before it joins the Bay of Bengal. The total length of the river is 597 km. The principal tributaries of this river are the *Jayamangali*, the *Kunderu*, *Saigileru*, *Chitravari*, *Papagni* and *Cheyyeru*. The *Ponnaiyar* is a small stream which is confined to the coastal area only. South of the Cauvery delta, there are several streams, of which the *Vaigai* is the longest. This flows through dry channels and tends to get lost intermittently and appear again on the surface.

The west flowing rivers of the Peninsula

The west flowing rivers of the Peninsular India are fewer and smaller as compared to their east flowing counterparts. The two major west flowing rivers are the *Narmada* and the *Tapi*. This exceptional behaviour of these two rivers is explained by the supposition that they do not flow in the valleys formed by themselves but have usurped for their channels two fault planes running parallel to the Vindhyas. These faults are supposed to have originated with the bending or '*sagging*' of the northern part of the Peninsula at the time of upheaval of the Himalayas. It is interesting to note that the Peninsular rivers which fall into the Arabian Sea do not form *deltas*, but only *estuaries*. This is due to the fact that the west flowing rivers, especially the Narmada and the Tapi flow through hard rocks and are not able to form distributaries before they enter the sea. The Sabarmati, Mahi and Luni are other rivers of the Peninsular India which flow westwards. Hundreds of small streams originating in the Western Ghats flow swiftly westwards and join the Arabian Sea.

The Narmada is the largest of all the west flowing rivers of the Peninsular India. It rises from the western flanks of the Amarkantak plateau about 22°40′N and 81°45′E at an elevation of 1,057 m in Shahdol district of Madhya Pradesh and flows westwards through a rift valley between the Vindhyan Range on the north and the Satpura Range on the south. Its total length from its source in Amarkantak to its estuary in the Gulf of Khambhat is 1,310 km. For 1,078 km, it flows through Madhya Pradesh. For the next 32 km, it forms the boundary between Madhya Pradesh and Maharashtra and for another 40 km between Maharasthra and Gujarat. The remaining 160 km of its course is in Gujarat. After flowing for 400 km from the source, the river slopes down Jabalpur, where it cascades 15 m into a gorge to form the most spectacular and world famous *Dhuandhar (Cloud of Mist) Falls*. Since the gorge is composed of marble, it is popularly known as the *Marble Rocks*. The average descent of the river in its upper basin is 7-9 metres per kilometre. Below Jabalpur, it flows in a narrow elongated and well defined basin and forms a few rapids. It makes two waterfalls of 12 m each at Mandhar and Dardi. Near Maheshwar the river again descends from another small fall of 8 m, known as the Sahasradhara Falls. Emerging from the hills near Gardeshwar, it meanders through an alluvial plain past Bharuch and makes an estuary before entering the Gulf of Khambhat. There are several islands in the estuary of the Narmada of which Aliabet is the largest. At its confluence with the sea, the mouth of the river is about 28 km wide. The Narmada is navigable upto 112 km from its mouth.

Since the river flows through a narrow valley confined by precipitous hills, it does not have many tributaries. The absence of tributaries is especially noted on the right bank of the river where the *Hiran* is the only exception. It is 188 km long and flows parallel to the Bhanru Range in Jabalpur district before falling into the Narmada, north-east of Chhindwara. The other right bank tributaries are the *Orsang*, the *Barna* and the *Kolar*. A few left bank tributaries drain the northern slopes of the Satpura Range and join the Narmada at different places. The important tributaries joining the Narmada on its left bank are the *Burhner* (177 km), the *Banjar* (184 km), the *Shar* (129 km), the *Shakkar* (161 km) the *Tawa* (172 km) and the *Kundi* (169 km).

The Tapi (also known as the *Tapti*) is the second largest west flowing river of the Peninsular India and is known as ***'the twin'*** or ***'the handmaid'*** of the Narmada. It originates from the sacred tank of Multai on the Satpura Plateau in Betul district of Madhya Pradesh at an elevation of 730 m above sea level. It first traverses on open plain and then plunges into a rocky gorge of the Satpura hills between the Kalibhit Range in Nimar and Chikalda in Berar. At a distance of 192 km from its source, the river enters the Nimar region. After crossing the Nimar region of Madhya Pradesh it enters the Khandesh Plain of Maharashtra, lying between the Satpura and the Ajanta Ranges. On entering the Khandesh Plain, it receives the *Purna* river on its left bank. This is the main tributary of the Tapi which originates in the Gawilgarh Hills and joins the main river near Bhusawal. Further west it crosses the Western Ghats through a deep and narrow valley, passes through the alluvial plain of Surat and makes an estuary before falling into the Gulf of Khambhat.

The total length of the river is 730 km. About 48 km stretch of its course is tidal and the river is navigable for only about 32 km from the sea. The important tributaries of this river apart from the Purna River, are the Betul, Patki, Ganjal, Dathranj, Bokad, Bokar Suki, More, Kanki, Guli, Aner, Arunavati, Gomai and Valer on right bank and Ambhora, Khursi, Khandu, Kapra, Sipra, Garja, Khokri, Utaoli, Mona,

TABLE 4.3. Major Rivers of the Peninsular India

Name of the river	Source	Length in km	Area drained (sq km)	Volume of average annual flow (million cubic metres)	Important tributaries
Mahanadi	Dandakaranya near Sihawa in Raipur district of Chhattisgarh	857	141,600	67,000	Ib, Mand, Hasdo, Sheonath, Ong, Jonk, Tel
Godavari	Trimbak Plateau near Nashik in Maharashtra	1,465	312,812	105,000	Manjra, Penganga, Wardha, Wainganga, Indravati, Sabari, Pranhita
Krishna	Near Mahabaleshwar in Maharashtra	1,400	258,948	67,670	Koyna, Ghatprabha, Malprabha, Bhima, Tungabhadra, Musi, Muneru
Cauvery	Taal Cauvery Western Ghats	800	87,900	20,950	Herangi, Hemavati, Lokpavani, Shimsha, Arkavati, Lakshmanatirtha, Kabani, Suvanavati, Bhavani, Amaravati
Narmada	Amarkantak Plateau	1,310	98,795	40,700	Hiran, Orsang, Barna, Kolar, Burhner, Banjar, Shar, Shakhar, Tawa, Kundi
Tapi	Multai in Betul Distt. of M.P.	730	65,145	17,980	Purna, Betul, Patki, Ganjal, Dathranj, Bokad, Amravati

Vaghur, Girna, Bori, Panjhara, Buray and Amravati on the left bank.

The Sabarmati is the name given to the combined streams the *Sabar* and *Hathmati*. It rises from the hills of Mewar in the Aravali Range and is 320 km long. It flows through a gorge at Dharoi and falls into the Gulf of Khambhat. The important tributaries of this river are the *Sedhi*, the *Wakul*, the *Harnav*, the *Meshwa* and the *Vatrak*. The average annual flow of this river is 3,200 million cubic metres.

The **Mahi** river rises in the Vindhyas at an elevation of 500 m and empties itself into the Gulf of Khambhat after flowing for a distance of 533 km. The main tributaries of this river are the *Som*, the *Anas* and the *Panam*.

The **Luni** or the **Salt River** (*Lonari or Lavanavari* in Sanskrit) is named so because its water is brackish below Balotra. Its source lies to the west of Ajmer in the Aravalli at an elevation of 550 m and it flows in a south-west direction. It is a small stream but has got the distinction of flowing through the Thar Desert. The river is known as the *Sagarmati* in its upper course. It is joined by the Sarsuti after passing Govindgarh and it is from this confluence that the river gets its proper name as the Luni. It traverses for a distance of 482 km and is finally lost in the marshy grounds at the head of the Rann of Kachchh. The river has a low gradient and flows through a wide plain without a well marked valley.

West flowing Rivers of the Sahyadris

About six hundred small streams originate from the Western Ghats and flow westwards to fall into the Arabian Sea. The western slopes of the Western Ghats receive heavy rainfall from the south-west monsoons and are able to feed such a large number of streams. Although only about 3% of the areal extent of the basins of India is drained by these rivers, they contain about 18% of the country's water resources. The important rivers falling in this category are the Mandovi, Zuari and Rachol in Goa; Kalinadi, Gangavalli-Bedti, Sharavati, Tadri and Netravati in Karnataka, and Beypore, Pannam, Bharatapuzha, Periyar and Pamba in Kerala. Most of the streams flow swiftly down the steep slope and some of them make waterfalls. The *Jog* or *Gersoppa Falls* (289 m) made by the Sharavati river is the most famous waterfall of India.

Inland Drainage

Some rivers of India are not able to reach the sea and constitute inland drainage. Large parts of the Rajasthan desert and parts of Aksai Chin in Ladakh have inland drainage. The *Ghaggar* is the most important river of inland drainage. It is a seasonal stream which rises on the lower slopes of the Himalayas and forms boundary between Haryana and Punjab. It gets lost in the dry sands of Rajasthan near Hanumangarh after traversing a distance of 465 km. Earlier, this river was an affluent of the Indus, the dry bed of the old channel is still traceable. Its main tributaries are the Tangri, the Markanda, the Saraswati and the Chaitanya. It contains a lot more water in rainy season when its bed becomes 10 km wide at places. Most of the streams draining western slopes of the Aravalli Range dry up immediately after they enter the sandy arid areas to the west of this range.

Differences between the Himalayan and the Peninsular River Systems

The Himalayan rivers are quite different from the rivers of the Peninsular India from the point of view of the drainage features and hydrological characteristics. A brief summary of these differences is given below :

Basis	The Himalayan River System	The Peninsular River System
1. Name	These rivers originate from the lofty Himalayan ranges and are named as the Himalayan rivers.	These rivers originate in the Peninsular Plateau and are named as Peninsular rivers.
2. Basins	These rivers have large basins and catchment areas. The total basin area of the Indus, the Ganga and the Brahmaputra is 11.78, 8.61 and 5.8 lakh square kilometres respectively.	These rivers have small basins and catchment areas. The Godavari has the largest basin area of 3.12 lakh square kilometres only which is less than one-third the basin area of the Indus.
3. Valleys	The Himalayan rivers flow through deep I-shaped valleys called gorges. These gorges have been carved out by down cutting carried on side by side with the uplift of the Himalayas. These are examples of *antecedent drainage*.	The Peninsular rivers flow in comparatively shallow valleys. These are more or less completely graded valleys. The rivers have little erosional activity to perform. These are examples of *consequent drainage*.
4. Water Flow	The Himalayan rivers are perennial in nature, *i.e.*, water flows throughout the year in these rivers. These rivers receive water both from the monsoons and snow-melt. The perennial nature of these rivers makes them useful for irrigation.	The Peninsular rivers receive water only from rainfall and water flows in these rivers in rainy season only. Therefore, these rivers are seasonal or non-perennial. As such these rivers are much less useful for irrigation.
5. Stage	These rivers flow across the young fold mountains and are still in a youthful stage.	These rivers have been flowing in one of the oldest plateaus of the world and have reached maturity.
6. Meanders	The upper reaches of the Himalayan rivers are highly tortuous. When they enter the plains, there is a sudden reduction in the speed of flow of water. Under these circumstances these rivers form meanders and often shift their beds.	The hard rock surface and non-alluvial character of the plateau permits little scope for the formation of meanders. As such, the rivers of the Peninsular Plateau follow more or less straight courses.
7. Deltas and Estuaries	The Himalayan rivers form big deltas at their mouths. The Ganga-Brahmaputra delta is the largest in the world.	Some of the Peninsular rivers, such as the Narmada and the Tapi form estuaries. Other rivers such as the Mahanadi, the Godavari, the Krishna and the Cauvery form deltas. Several small streams originating from the Western Ghats and flowing towards the west enter the Arabian Sea without forming any delta.

RIVER BASINS OF INDIA

River basin is the area drained by the main river and its tributaries. For example, Ganga basin is the area drained by the main Ganga river and its tributaries like Yamuna, Ghaghara, Gandak, Kosi and their sub-tributaries. In a large country like India, it is natural to have different types of river basins. Depending on the size of their catchment area : **Dr. K.L. Rao** (1977), divided the Indian river basins into following three categories :

(*i*) **Major river basins** with catchment area of 20,000 sq km and above account for 85 per cent of the total run-off of all the rivers of India. They are fourteen in all and are in high rainfall areas (*i.e.* 63 million cubic metres per 100 sq km). There are three river systems to the north of Tropic of Cancer, seven between Tropic of Cancer and 20°N latitude and four in the Peninsular India.

(*ii*) **Medium river basins** with a catchment area between 2,000 sq km and 20,000 sq km account for 7% of the total run off of all the rivers. They are 44 in all and have medium rainfall (*i.e.* 45 million cubic metres per 100 sq km). The medium river basins are divided into three sub-groups depending upon their direction of flow. Of them 19 are west flowing, 21 are east flowing and the remaining 4 flow in other directions.

(*iii*) **Minor river basins** with a catchment area below 2,000 sq km account for 8 per cent of the total run off of all the rivers of India. There are 55 such river basins. They are located either in dry areas or in the coastal areas, especially on west coast where the rivers are much shorter due to narrow coastal plain between the coast and the Western Ghats.

Major River Basins

India has 14 major river basins. They are Indus, Ganga, Brahmaputra, Mahanadi, Godavari, Krishna, Kaveri, Narmada, Tapi, Penneru, Brahmani, Mahi,

TABLE 4.4. Surface Flow in Major River Basins of India

Sl. No.	River basins	Basin area* (km²)	%	Annual water yield (million m³)	%	Rate of flow (m³/km²)	Storage capacity (million m³)
	Major rivers						
1.	Ganga	861,404	26.2	468,700	25.2	442,170	33,467
2.	Indus	321,284	9.8	79,500	4.3	247,441	14,419
3.	Godavari	312,812	9.5	118,000	6.4	377,223	14,859
4.	Krishna	258,948	7.9	62,800	3.4	243,403	29,860
5.	Brahmaputra	258,008	7.8	627,000	33.8	1,081,034	142
6.	Mahanadi	141,589	4.3	66,640	3.6	470,658	7,926
7.	Narmada	98,795	3.0	54,600	2.9	970,658	2,550
8.	Kaveri	87,900	2.7	20,950	1.1	237,770	5,428
9.	Tapi	65,145	2.0	17,982	0.9	267,770	8,140
10.	Penneru	55,213	1.7	3,238	0.2	58,646	1,978
11.	Brahmani	39,033	1.2	18,310	1.0	202,701	3,953
12.	Mahi	34,481	1.0	11,800	0.6	388,681	4,140
13.	Subarnarekha	19,296	0.6	7,940	0.4	411,484	283
14.	Sabarmati	21,895	0.7	3,800	0.2	173,556	1,017
	Medium & minor rivers	711,833	21.6	296.840	16.0	417,008	18,452
	Total (India)	**3,287,782**	**100.0**	**1,858,100**	**100.0**	**565,153**	**146,826**

*Area means basin area in India

Source : S.P. Das Gupta, 1989.

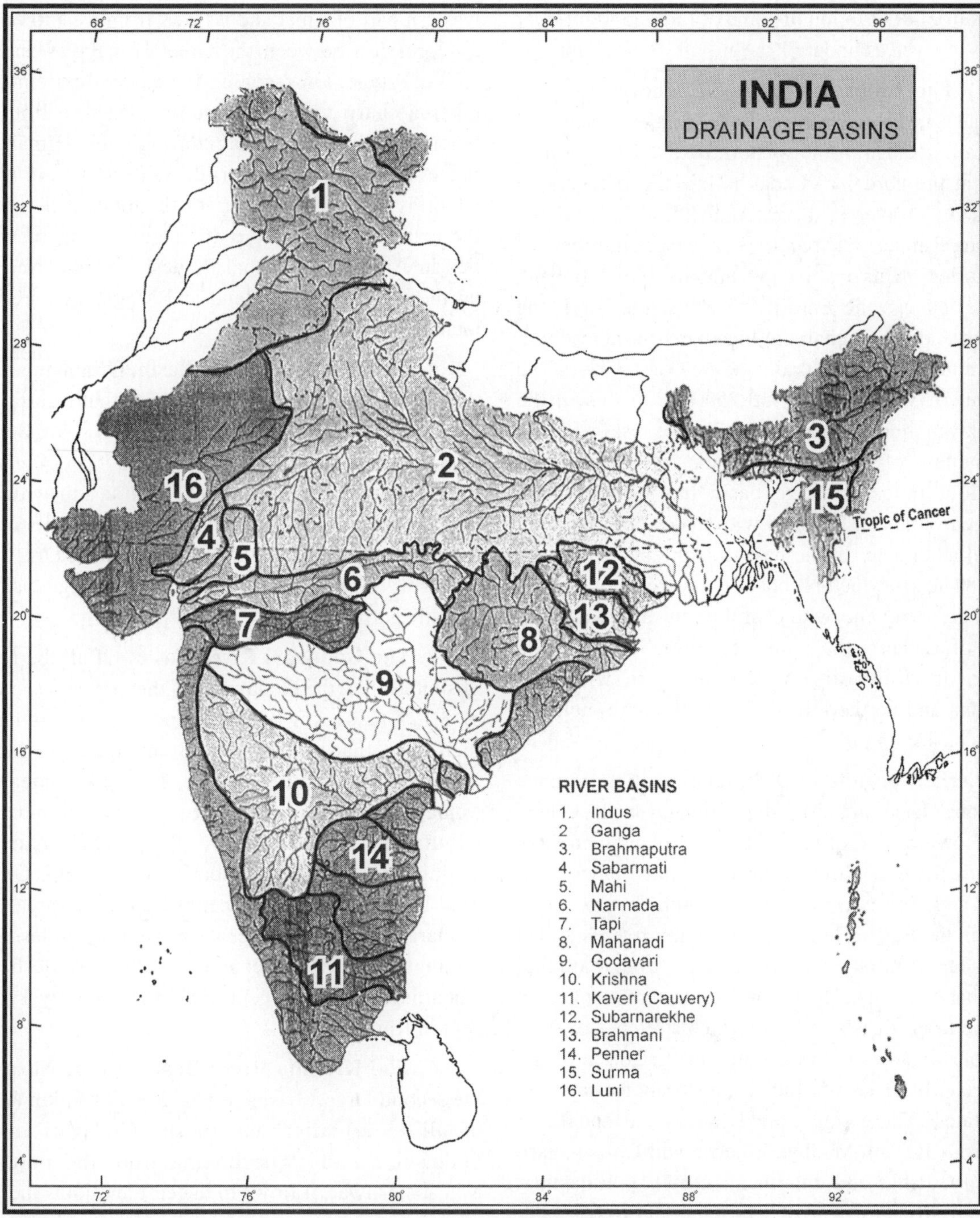

FIG. 4.4. India : Drainage Basins

Sabarmati and Luni. **S.P. Das Gupta** has treated Subarnarekha also as the major river basin although its basin area is less than 20 thousand sq km (**Table 4.4**). Some other geographers tend to treat Surma and Luni also as major river basins as is indicated in **Fig. 4.4**.

A brief description of some of the important river basins is given as under :

1. The Indus River Basin. The Indus river originates near Manasarover Lake from the glaciers of the Kailas Range in the western part of Tibet, enters India at the border of Ladakh Union Territory with China and traverses between the Ladakh and the Zaskar Ranges. It receives a large number of tributaries on its right bank and its five left bank tributaries *viz.* the Satluj, the Beas, the Ravi, the Jhelum and the Chenab are big rivers in themselves. The Indus has a total drainage area of 1,178,440 sq km out of which about 453,250 sq km lies in the Himalayan mountains and foothills, the rest lies in the Indus plain of India and Pakistan. As a result of partition, Indus drainage basin has been divided between India and Pakistan with major portion going to Pakistan and India has to content with only 321,284 sq km which is about 26.2 per cent of its total drainage area. The whole of Jammu and Kashmir, most of Ladakh, the whole of Punjab, almost the whole of Himachal Pradesh, northern fringe of Haryana and northern tips of Rajasthan are included in this basin.

2. The Ganga River Basin. Originating from the Gangotri glacier in the Western Himalayas, the Ganga flows towards the south in Uttarakhand and then turns towards the east, flowing through Uttar Pradesh and Bihar and then turn towards the south-east to empty itself in the Bay of Bengal. This mighty river is joined by a large number of tributaries like Yamuna, Ghaghara, Gandak, Kosi, Son, etc. In fact, the Ganga has the largest basin in India which stretches over an area of 861,404 sq km and amounts to 26.2% of the total river basin area of India. This basin is shared by ten states. These states are Uttarakhand and Uttar Pradesh (34.2%), Madhya Pradesh and Chhattisgarh (23.1%), Bihar and Jharkhand (16.7%), Rajasthan (13.0%), West Bengal (8.3%), Haryana (4.0%) and Himachal Pradesh (0.5%). The Union Territory of Delhi accounts for 0.2% of the total area of the Ganga Basin.

3. The Brahmaputra River Basin. Like the Indus, the Brahmaputra river also has its source in the western part of Tibet and it flows from west to east in a depression between the Great Himalayas and the Kailas Range. On reaching the eastern extremity, it takes a sharp turn in the south-west direction and enters India after cutting through the Himalayan range. Its total basin area is 2,58,008 sq km, in India, which is about 7.8 per cent of the total basin area of the country. Most of the drainage area of this river lies in Assam, Arunachal Pradesh and Sikkim. The northern parts of Nagaland and Meghalaya are also included in this basin.

4. The Mahanadi River Basin. Originating from the foothills of Dandakaranya in Chhattisgarh, this river flows to the east and joins the Bay of Bengal after traversing over a distance of 857 km. The total basin area of this river is 141,589 sq km which is about 4.3% of the total basin area of India. Its major tributaries are Ib, Mand, Hasdo, Seonath, Org, Jonk and Tel. This basin spreads over a large part of Chhattisgarh and middle of Odisha.

5. The Godavari River Basin. Of all the rivers of south India, the Godavari has the largest basin area of 3,12,812 sq km (9.5% of India) which is larger than even that of the great Brahmaputra in Indian territory. Originating in the Trimbak Plateau in Maharashtra, it is joined by a large number of tributaries like Manjra, Penganga, Wardha, Wainganga, Indravati, Sabari and Pranhita. Out of total basin area 48.6 per cent of the basin area is in Maharashtra, 23.8 per cent in Andhra Pradesh and Telangana, 20.7 per cent in Madhya Pradesh and Chhattisgarh, 5.5 per cent in Odisha and only 1.4 per cent in Karnataka.

6. The Krishna River Basin. The Krishna has the second largest basin area of 258,948 sq km (7.9% of all India) after that of the Godavari in the Peninsular India. Originating from the north of Mahabaleshwar, it flows in the east and joins the Bay of Bengal. Its important tributaries are Koyna, Ghataprabha, Malprabha, Bhima, Musi, Tungabhadra and Muneru. Its basin area is quite wide in the west in the states of Maharashtra and Karnataka and narrows down considerably in Telengana and Andhra Pradesh and it is narrowest in the delta region.

7. The Kaveri (Cauvery) River Basin. Originating in the Western Ghats, it traverses 800 km in the eastern direction before joining the Bay of Bengal. The tributaries joining the main river from the north are the Herangi, the Hemavati, the Lakpavani, the Shimsha and the Arkawali while those joining from the south are the Lakshmantirtha, the Kabani, the Suvarnavati, the Bhavani and the Amravati. Its total basin area is 87,900 sq km (2.7% of the India) which is shared by two states of Karnataka and Tamil Nadu.

8. The Narmada River Basin. This river originates from the Amarkantak Plateau and flows westward to join the Arabian sea in the Gulf of Khambhat after forming an estuary. Since it flows through a rift valley, it does not receive any tributary worth the name as a result of which its basin is elongated, quite long in the east and west direction and quite narrow in the north-south direction. The total area of this basin is 98,795 sq km (3.0% of all India), most part of which lies in Madhya Pradesh with small part lying in Gujarat.

9. The Tapi River Basin. South of the Narmada river flows the Tapi river in the east-west direction. It originates from the Multai Tank in Betul district of Madhya Pradesh and traverses a distance of 730 km before entering the Arabian sea. Its total basin area of 65,145 sq km which is about 2.0% of the all India basins. Like the Narmada, the Tapi river also does not have large tributaries and its basin area is elongated with its long side in east-west direction.

10. The Penneru River Basin. The Penneru river originates from the Nandi Durg in Karnataka and enters Andhra Pradesh where its flows in an easterly direction to join the Bay of Bengal after forming a small estuary. This is a comparatively small river basin covering an area of 55,213 sq km (1.7%).

11. The Brahmani River Basin. The Brahmani river takes its birth at the confluence of the Koel and the Sankh rivers near Rourkela. Tributaries like Kura, Sankhad and Tikra join the main river enabling the Brahmani river basin to spread over an area of 39,033 sq km (1.2%) in the state of Odisha.

12. The Mahi River Basin. This 533 km long river originates in the Vindhyas and empties itself in the Gulf of Khambat. The Jakam, the Anas and the Panam are its main tributaries. Its total basin area is 34,481 sq km which is shared by Rajasthan (47%), Gujarat (34%) and Madhya Pradesh (19%).

13. The Sabarmati River Basin. This 320 km long river is the combined name of the Sabar and the Hathmati which originate in the hills of Mewar in the Aravali range and empties itself in the Gulf of Khambhat. Its main tributaries are the Sedhi, the Wakul, the Harnav, the Mehswa and the Vatrak. The total basin area is 21,895 sq km (0.7%), which entirely lies in Gujarat.

14. The Luni River Basin. Part of this river has salty water as a result of which it is known as the Luni river. It originates to the west of Ajmer and flows through the Thar Desert and is lost in the marshy grounds at the head of the Ranu of Kachchh. It flows through a wide plain without a well marked valley. This river has a vast basin which spreads over Rajasthan and Gujarat.

River Regimes

River regime is the seasonal fluctuation in the volume of water in a river. The climatic differences in the sources of the Himalayan and the Peninsular rivers lead to the differences in the drainage patterns of these two areas. The Himalayan rivers are perennial because their regimes depend upon water supply both from rainfall and snow-melt. Thus, the regimes of the Himalayan rivers are both monsoonal as well as glacial. On the other hand, the Peninsular rivers receive water from rainfall only and their regimes are only monsoonal.

Fig. 4.5 shows hydrographs of two Himalayan and two Peninsular rivers depicting their regimes. These graphs show the monthly discharge of water flowing in these rivers. This discharge is measured in *cusecs* (cubic feet per second) or *cumecs* (cubic metres per second).

The Ganga has minimum flow from January to June. The maximum is attained in August or

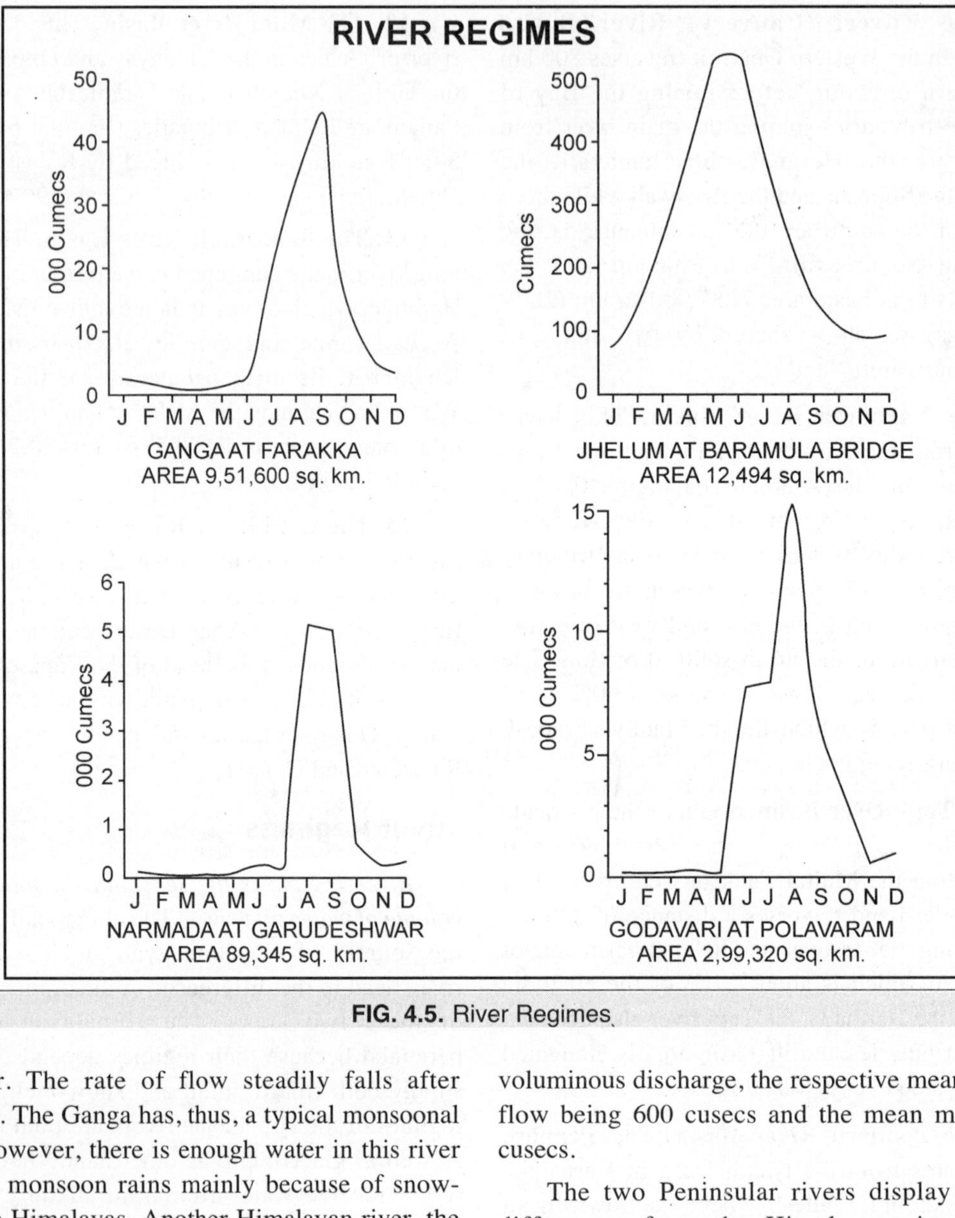

FIG. 4.5. River Regimes.

September. The rate of flow steadily falls after September. The Ganga has, thus, a typical monsoonal regime. However, there is enough water in this river before the monsoon rains mainly because of snow-melt in the Himalayas. Another Himalayan river, the Jhelum attains its maximum flow in June or even in May mainly because of Himalayan snow-melt. An interesting difference between the regimes of the two rivers is seen in the range of variation between the maximum and the minimum flow. This difference is sharper in the case of the Ganga than the Jhelum. The mean maximum discharge of the Ganga at Farakka is about 45,000 cumecs, while the mean minimum is only 1,300 cumecs. The Jhelum, however, has a less voluminous discharge, the respective mean maximum flow being 600 cusecs and the mean minimum 50 cusecs.

The two Peninsular rivers display interesting differences from the Himalayan rivers in their regimes. The Narmada has a very low volume of discharge from January to July which suddenly rises in August when the maximum is attained. The river, however, maintains its fall as spectacular as the rise in August. The Godavari flows at a low level until May. It has a double maxima—one in May-June and other in July-August. After August, there is a sharp fall in discharge, although the volume of flow in October and November is higher than in any of the months from January to May. The mean maximum discharge of the Godavari at *Polauaram* is 3,200 cumecs, while the mean minimum flow is only 50 cumecs. The respective values for the Narmada, as recorded at Garudeshwar, are 2,300 cusecs and 15 cumecs.

Usability of Rivers

Rivers are of great use for us because they comprise a great source of fresh water. In fact most of our requirements of fresh water are met by rivers. The volume of annual precipitation in the country is estimated at about 37,00,400 million cubic metres. A large part of it seeps into the ground and some part is lost by evaporation and transpiration. We use river waters for a variety of purposes such as irrigation, hydro-electric production, navigation, etc. Large quantity of water is supplied to cities and villages for domestic consumption. A large number of industries also depend upon water.

1. Irrigation. The largest amount of river water is used for irrigation. Indian rivers carry about 18,58,100 million cubic metres of water per year. Due to uneven topography and seasonal flow of the rivers, all this amount is not usable. About 5,55,166 million cubic metres of river waters are actually used for irrigation through canals. This amounts to about 30 per cent of the annual flow of the Indian rivers.

2. Hydro-electric Production. Large rivers have great water power potential. The Himalayas in the north, the Vindhyas, the Satpura and the Aravalli in the west, the Maikala and Chhotanagpur in the east, the Meghalaya plateau and Purvachal in the northeast, and Western and the Eastern Ghats of the Deccan plateau offer possibilities of large scale water power development. Sixty per cent of the total river flow is concentrated in the Himalayan rivers, 16 per cent in the Central Indian rivers (the Narmada, the Tapi, the Mahanadi, etc.), and the rest in the rivers of the Deccan plateau. Dependable power generation from the peninsular rivers requires impounding of water during the monsoon months. The Himalayan rivers do not have such problem as their flow is appreciable even during the critical winter months. They, however, have other kind of problems, namely, difficulty in construction of large storage on account of narrow valleys, high seismicity of the region and vast alluvial plain with no variation of relief. The country has an exploitable power potential of about 41 million KW at 60 per cent load factor from these rivers.

3. Navigation. The Ganga and the Brahmaputra in the north and northeastern part of the country, the Mahanadi in Odisha, the Godavari and the Krishna in Andhra and Telangana, the Narmada and the Tapi in Gujarat, and the lakes and tidal creeks in coastal states possess some of the important and useful waterways of the country. In the past they were of great importance, which suffered a great deal with the advent of rail and roads. Withdrawal of large quantities of water for irrigation resulted in dwindling flow of many rivers. The most important navigable risers are the Ganga, the Brahmaputra and the Mahanadi. The Godavari, the Krishna, the Narmada and the Tapi are navigable near their mouths only.

REFERENCES

Bharati, Radhakant (2004) : *Rivers of India* (Translated by Sandeep Sinha) National Book Trust, India, New Delhi.

Chatterjee, S.P. (1999) : *India, A Physical Geography,* Publications Division, Govt. of India.

Chhiber, H.L. (1948) : Drainage Pattern Observed in India and Adjacent Countries, National Geographical Journal of India, Vol. 2(1), pp. 2.5.

Dasgupta, S.P. (1989) : *India's Water resource potential In :* S.P. Das Gupta *(ed.), Man and Ecology,* School of Fundamental Research, Calcutta.

Kuriyan, George (1970) : *India: A General Survey,* National Book Trust, New Delhi, India, Chapt. II.

Mathur, S.M. (2004) : *Physical Geology of India,* National Book Trust, New Delhi, India.

Mehdiratta, R.C. (1962) : *Geology of India, Pakistan and Burma,* Atma Ram & Sons, Delhi.

Misra, S.D. (1970) : *Rivers of India,* National Book Trust, New Delhi, India.

Mithal, R.S. (1968) : 'The Physiographical and Structural Evolution of the Himalaya, *Mountains and Rivers of India,* 21st International Geographical Congress, New Delhi, pp. 287-304.

Pascoe, E.H. (1919) : Early History of the Indus, Brahmaputra and the Ganges, *Qurts Journ. Geol. Soc.* Vol. 75, London, p. 139.

Pichamuthu, C.S. (1969) : *Physical Geography of India,* National Book Trust, New Delhi, India.

Pilgrim, G.E. (1919) : The Siwalik River, *Journ. Assit. Soci. Bengal,* Vol. 15, New Series, pp. 81-99.

Rama (1994) : *Our Water Resources,* National Book Trust, New Delhi, India.

Rao, K.L. (1979) : *India's Water Wealth,* Orient Longman Ltd., New Delhi, India.

Rao, K.L. (1979) : *River Basins of India* in A. Ramesh (ed.), (1984) Resource Geography, Heritage Publishers, New Delhi.

Siddartha, K. (1999) : India, The Physical Aspects, Centre for Development of Environment and Resources, Patna (Chap. 3).

Spate, O.H.K. and Learmonth, A.T.A. (1967) : *India and Pakistan : A General and Regional Geography,* Methuen, London.

Vij, G.K. and Sheonoy, R.C. (1968) : *Hydrology of Indian Rivers in* B. C. Low *(ed.) The Mountains and Rivers of India,* 21st International Geographical Congress, New Delhi.

◆◇◆

CHAPTER 5

Climate

INTRODUCTION

Climate of a country includes the study of temperature, rainfall, atmospheric pressure, as well as the direction and velocity of winds over a long period of time. These elements of climate are largely influenced by latitudinal extent, relief, and areal distribution of land and water. The *Tropic of Cancer* passes through the middle of the country. In the north, the Indian subcontinent is separated from the rest of Asia by the Himalayan ranges as a result of which the cold air masses from Central Asia cannot enter India. During winter, the northern half of India is warmer by 3°C to 8°C than other areas located on these latitudes. Thus the whole of India, south of the Himalayas is climatically treated as a tropical country. According to **L.D. Stamp**, "India is basically a tropical country although its northern part is situated in the temperate belt." In the south, the Indian coasts are washed by the Arabian Sea and the Bay of Bengal branches of the Indian Ocean which give it a typical tropical monsoon climate. "*India is par excellence, a tropical monsoon country.*"

DIVERSITY IN THE UNITY OF INDIAN MONSOON CLIMATE

Notwithstanding its broad climatic unity, the climate of India has many regional variations, expressed in the pattern of winds, temperature and rainfall, rhythm of seasons and the degree of wetness or dryness. These climatic differences are due to location, altitude, distance from sea and relief. The vast size of the country, coupled with its topographical variations is responsible for a great variety of climatic conditions in India. **Blanford** was highly impressed by climatic diversities of India and stated "We may speak of climates but not the climate of India, for the world itself affords, no greater contrast than is to be met with at one and the same time within its limits." ***Marsden*** too believed that India possesses all types of climate found in the world. The main examples of diversities in the monsoonal unity of Indian climate are briefed as under :

1. The day temperature in the month of June may soar to 48°-50°C at Barmer in Rajasthan while it may hardly reach 22°C at Gulmarg or Pahalgam in Kashmir at the same time.
2. On a December night, the temperature may dip to –40°C at Dras or Kargil in Ladakh while it may be as high as 20°C-22°C at Thiruvananthapuram or Chennai at the same time.

3. Mawsynram in Meghalaya receives as high as 1,221 cm of average annual rainfall while at Jaisalmer (Rajasthan) the annual rainfall rarely exceeds 12 cm. Tura, in the Garo Hills, may sometimes receive as much rainfall in a single day as is received by Jaisalmer in 10 years.
4. The Ganga delta and the coastal plains of Odisha are hit by strong rain storms almost every third or fifth day in July and August, while the Coromandel Coast, a thousand kilometres to the south, goes dry during these months.
5. The people of Mumbai and the Konkan Coast do not have to suffer the extremes of climate but these extremes affect the life of people of interior parts of the country such as Delhi and Agra.
6. Places like Thiruvananthapuram, Hyderabad, Bhubaneshwar and Patna get rains by the first quarter of June while the rains are awaited till the end of June or early July at places like Agra, Delhi and Chandigarh.

FACTORS INFLUENCING THE CLIMATE OF INDIA

The climate of India is a complex phenomenon and is influenced by a large number of geographical factors. Some of the important factors are briefly discussed as under :

1. Location and Latitudinal Extent. The mainland of India extends roughly from 8°N to 37°N and the Tropic of Cancer passes through the middle of the country. Areas south of the Tropic of Cancer are closer to the equator and experience high temperature throughout the year. The northern parts on the other hand lie in the warm temperate zone. Hence they experience comparatively lower temperatures. Some places record considerably low temperatures particularly in winter. Water bodies comprising the Arabian sea and the Bay of Bengal surround the peninsular India and make climatic conditions mild along the coastal areas.

2. Distance from the Sea. Areas near the coast have equable or *maritine climate*. On the contrary, interior locations are deprived of the moderating influence of the sea and experience extreme or *continental climate*. For example the annual range of temperature at Kochi does not exceed 3°C whereas it is as high as 20°C at Delhi. Similarly, the amount of annual rainfall at Kolkata is 119 cm which falls to a low of 24 cm at Bikaner.

3. The Northern Mountain Ranges. As mentioned earlier, India is separated from the rest of Asia by the impenetrable wall of the Himalayan mountain ranges. These ranges protect India from the bitterly cold and dry winds of Central Asia during winter. Further, these mountain ranges act as an effective physical barrier for rain bearing south-west monsoon winds to cross the northern frontiers of India. Thus, the Himalayan mountain ranges act as a climatic divide between the Indian Sub-continent and Central Asia.

4. Physiography. Physiography of India has a great bearing on major elements of climate such as temperature, atmospheric pressure, direction of winds and the amount of rainfall. In fact, physical map of India is very closely related to the climatic conditions of the country. Places located at higher altitude have cool climate even though they are located in the peninsular India, *e.g.,* Ooty. Several hill stations in the Himalayan ranges are much cooler than the places located in the Great Plain of North India. The greatest control of physiography in the peninsular India is seen in the distribution of rainfall. The south-west monsoon winds from the Arabian sea strike almost perpendicularly at the Western Ghats and cause copious rainfall in the Western Coastal plain and the western slopes of the Western Ghats. On the contrary, vast areas of Maharashtra, Karnataka, Telangana, Andhra Pradesh and Tamil Nadu lie in *rainishadow* or *leeward* side of the Western Ghats and receive scanty rainfall. The physiographic control of the mighty Himalayas over the climate of the country goes without saying. The monsoon winds from the Bay of Bengal are bifurcated into two branches by physiographic features. One branch goes to the Brahmaputra valley through the Meghalaya plateau. Here the funnel shaped Cherrapunji valley forces the moisture laden monsoon winds to rise along the steep slope and make this area the wettest place in the world. The other branch of monsoons from the Bay of Bengal enters the Ganga valley. Its northward-movement is obstructed by the Himalayan ranges and it advances westwards up the Ganga plain. Initially this branch causes heavy rainfall but the amount of rainfall decreases as the monsoons loose much of the moisture content while advancing westwards.

5. Monsoon Winds. The most dominating factor of the Indian climate is the 'monsoon winds' as a result of which it is often called the ***monsoon climate***. The complete reversal of the monsoon winds brings about a sudden change in the seasons—the harsh summer season suddenly giving way to eagerly awaited monsoon or rainy season. The south-west summer monsoons from the Arabian sea and the Bay of Bengal bring rainfall to the entire country. The north-eastern winter monsoon travel from land to sea and do not cause much rainfall except along the Coromandel coast after getting moisture from the Bay of Bengal.

6. Upper Air Circulation. The changes in the upper air circulation over Indian landmass influences the climate of India to a great extent. Jet streams in the upper air system influence the Indian climate in the following ways :

(i) Westerly Jet Stream. Westerly jet stream blows at a very high speed during winter over the sub-tropical zone. This jet stream is bifurcated by the Himalayan ranges. The northern branch of this jet stream blows along the northern edge of this barrier. The southern branch blows eastwards south of the Himalayan ranges along 25° north latitude. Meteorologists believe that this branch of jet stream exercises a significant influence on the winter weather conditions in India. This jet stream is responsible for bringing western disturbances from the Mediterranean region in to the Indian sub-continent. Winter rain and storms in north-western plains and occasional heavy snowfall in hilly regions are caused by these disturbances. These are generally followed by cold waves in the whole of northern plains.

(ii) Easterly Jet Stream. Reversal in upper air circulation takes place in summer due to the apparent shift of the sun's vertical rays in the northern hemisphere. The westerly jet stream is replaced by the easterly jet stream which owes its origin to the heating of the Tibet plateau. This leads to the development of an easterly cold jet stream centered around 15°N latitude and blowing over peninsular India. This helps in the sudden onset of the south-west monsoons.

7. Tropical Cyclones and Western Disturbances. Tropical cyclones originate in the Bay of Bengal and Arabian Sea and they influence large parts of the peninsular India. Majority of the cyclones originate in the Bay of Bengal and influence the weather conditions during the south-west monsoon season. Some cyclones are born during the retreating monsoon season, *i.e.*, in October and November and influence the weather conditions along the eastern coast of India.

The western disturbances originate over the Mediterranean sea and travel eastward under the influence of westerly jet stream. They influence the winter weather conditions over most of Northern-plains and Western Himalayan region.

8. El-Nino Effect. El-Nino is a narrow warm current which occasionally appears off the coast of Peru in December (See **Fig. 5.9**). It is a temporary replacement of the cold Peru current which normally flows along the coast. This current is responsible for wide spread floods and droughts in the tropical regions of the world. Sometimes it becomes more intense and increases the surface water temperatures of the sea by 10°C. This warming of tropical Pacific waters affects the global pattern of pressure and wind systems including the monsoon winds in the Indian Ocean. Meteorologists believe that the severe droughts of 1987, 2009 and 2014 in India were caused by El-Nino.

9. La Nina. After an El-Nino, weather conditions return to normal. However, sometimes trade winds become so strong that they cause abnormal accumulation of cold water in the central and eastern Pacific region. This event is called ***La Nina***, which in effect is the complete reversal of El Nino. A La Nina also marks an active hurricane season. But in India, the presence of La Nina portends exceptionally good news. It is the harbinger of heavy monsoon showers in India.

La Nina usually follows a strong El Nino. How the two weather pattern anomalies compare :

El Nino the little boy	*La Nina the little girl*
• Trade winds weaken, warm waters move east	• Strong Pacific trade winds blow from surface water westward.
• Pacific jet stream is pulled further south than normal; picks up storms the jet stream could normally miss	• Cold water rises to the surface
• Weakens Indian monsoons	• Strengthens Indian monsoons
• El Nino occurs after 3 to 5 years	• La Nina occurs roughly half as often as El Nino; lasts from 1 to 3 years.

10. Southern Oscillation. There is a strange linkage of meteorological changes often observed between the Indian and the Pacific Oceans. It has been noticed that whenever the surface level pressure is high over the Indian Ocean, there is low pressure over the Pacific Ocean and *vice-versa*. This inter-relation of high and low pressure over the Pacific and the Indian Ocean is called ***Southern Oscillation***. When the winter pressure is high over the Pacific Ocean and low over the Indian ocean, the south-west monsoons in India tend to be stronger. In the reverse case, the monsoons are most likely to be weaker.

THE MONSOON WINDS

The term monsoon has been derived from the Arabic *mausin* or from the Malayan *monsin* meaning '*season*'. Thus the monsoons are seasonal winds which reverse their direction of flow with the change of season. They flow from sea to land during the summer and from land to sea during winter. In other words, the monsoon is a double system of seasonal winds, that is, the sum of summer and winter winds. There seems to be a lack of agreement on a precise definition of the term monsoon and different scholars have tried to define the monsoon winds in different ways.

According to **A.A. Rama Sastry**, "*Monsoons are large scale seasonal wind systems flowing over vast areas of the globe, persistently in the same direction, only to be reversed with the change of season.*" H.G. **Dobby** is of the opinion that "*reversal of wind system is the key-note of the monsoonal climate*". The reversal of the monsoon wind system is fully emphasised by **Conrad**. According to him, "*a true thermal monsoon demands a complete reversal of winds, that is an angle of about 180° between the dominant winds at extreme seasons*". This is further elaborated by **P.A. Menon**, when he expressed the opinion that "*the main criterion used in demarcating the monsoon areas is the reversal of wind systems between summer and winter*". Thus between July and January there should be a shift of nearly 180° in the prevailing wind direction with seasonal wind showing high degree of steadiness. The reversal of monsoon winds takes place in a definite manner keeping rhythm with change of season. Therefore, it is often said that *rhythm is the key-note of the monsoonal climate*. Some scholars tend to treat the monsoon winds as land and sea breezes on a large scale. **Koppen** (1923), **Hann** (1932) and **Angot** (1943) believe that the "*monsoons represent simply a land and sea breezes on a large scale, and that the annual period of the monsoon corresponds to the diurnal period of the breezes*."

While discussing the monsoon winds **C.S. Ramage** (1971) suggested the following four features of monsoon winds :

(*i*) The prevailing wind direction should shift by at least 120° between January and July.

(*ii*) The average frequency of prevailing wind directions in January and July should exceed 40 per cent.

(*iii*) The mean resultant wind velocity in at least one of the months should exceed 3 m/s.

(*iv*) There should be fewer than one cyclone-anticyclone alternation every two years, in either month, over a five degree latitude/longitude grid.

On the basis of above criteria he demarcated the area of the monsoon region as a rectangle roughly extending from 35° N to 25° S latitude and 30° W to 173° E longitude.

Mechanism of the Monsoons

The origin of monsoons is still shrouded in mystery. Several attempts have been made to explain

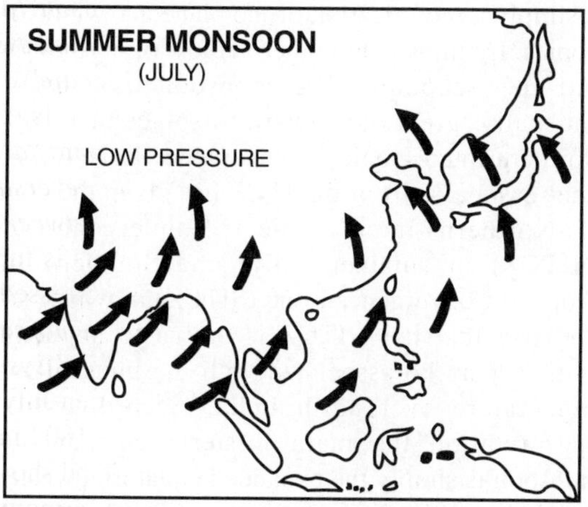

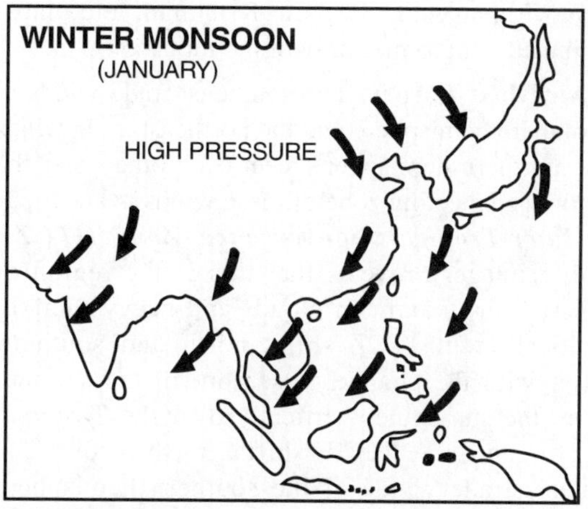

FIG. 5.1. Summer and Winter Monsoon

the mechanism of the monsoons but no satisfactory explanation is available till date. Over the years many mysteries of the monsoons have been unravelled but still much remains to be done. The theories regarding the monsoons are generally divided into following two broad categories :

1. Classical Theory, and
2. Modern Theories.

1. Classical Theory. Although monsoons are mentioned in our old scriptures like the *Rig Veda* and in the writings of several Greek and Buddhist scholars, the credit for first scientific study of the monsoon winds goes to the Arabs. Near about the tenth century, Al Masudi, an Arab explorer from Baghdad, gave an account of the reversal of ocean currents and the monsoon winds over the north Indian Ocean. Date of commencement of monsoons at several places was reported by Sidi Ali in 1554 A.D.

In 1686 the famous Englishman Sir Edmund Halley explained the monsoon as resulting from thermal contrasts between continents and oceans due to their differential heating. Accordingly, Halley conceived summer and winter monsoons depending upon the season.

(a) Summer Monsoon. In summer the sun shines vertically over the *Tropic of Cancer* resulting in high temperature and low pressure in Central Asia while the pressure is still sufficiently high over Arabian Sea and Bay of Bengal. This induces air flow from sea to land and brings heavy rainfall to India and her neighbouring countries.

(b) Winter Monsoon. In winter the sun shines vertically over the Tropic of Capricorn. The north western part of India grows colder than Arabian Sea and Bay of Bengal and the flow of the monsoon is reversed (**Fig. 5.1**).

Halley's ideas are basically the same as those involved in land and sea breeze except that in the case of the monsoon day and night are replaced by summer and winter, and the narrow coastal strip and adjacent sea are replaced by large portions of continents and oceans.

2. Modern Theories. Halley's classical theory based on differential heating of land and water as the main driving force of the monsoon winds dominated the scene for about three centuries. However, the monsoons do not develop equally everywhere and the thermal concept of Halley fails to explain the intricacies of the monsoons. Besides differential heating, the development of monsoon is influenced by the shape of the continents, orography, and the conditions of air circulation in the upper troposphere. Therefore, Halley's theory has lost much of its significance and modern theories based on air masses and jet stream are becoming more relevant. Although Halley's ideas have not yet been outrightly rejected, studies during the last seven decades have thrown much light on the genesis of the monsoons. During these years, Flohn, Thompson, Stephenson, Frost, M.T. Yin, Hwang, Takahashi, E. Palmen, C. Newton and Indian meteorologists including P. Koteswaram, Krishnan, Raman, Ramanathan, Krishna Murti, Rama

Rattan, Ramaswami, Anant Krishnan, etc. have contributed a lot to the study of the monsoon winds.

Air Mass Theory. The southeast trade winds in the southern hemisphere and the northeast trade winds in the northern hemisphere meet each other near the equator. The meeting zone of these winds is known as the ***Inter-Tropical Convergence Zone (ITCZ)***. Satellite imagery reveals that this is the region of ascending air, maximum clouds and heavy rainfall. The location of ITCZ shifts north and south of equator with the change of season. In the summer season, the sun shines vertically over the *Tropic of Cancer* and the ITCZ shifts northwards. The southeast trade winds of the southern hemisphere cross the equator and start blowing from southwest to northeast direction under the influence of *coriolis force* (**Fig. 5.2**). These displaced trade winds are called ***south-west monsoons*** when they blow over the Indian sub-continent. The front where the south-west monsoons meet the north-east trade winds is known as the ***Monsoon Front***.

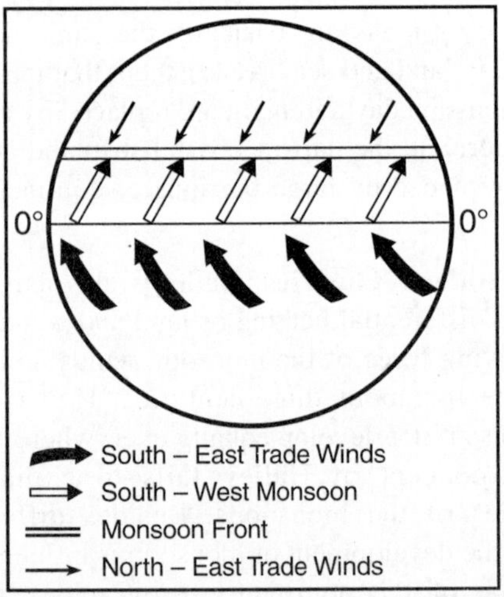

FIG. 5.2. Northward shifting of ITCZ and the wind system in summer season of northern hemisphere.

In the month of July the ITCZ shifts to 20°- 25° N latitude and is located in the Indo-Gangetic Plain and the south-west monsoons blow from the Arabian Sea and the Bay of Bengal (**Fig. 5.3**). The ITCZ in this position is often called the ***Monsoon Trough***.

H. Flohn of the German Weather Bureau, while rejecting the classical theory of origin of monsoons suggested that the tropical monsoon of tropical Asia is simply a modification of the planetary winds of the tropics. He thinks of the thermal low of northern India and the accompanying monsoon as simply an unusually great northward displacement of the Northern Inter-Tropical Convergence Zone (NITCZ). The seasonal shift of the ITCZ has given the concept of Northern Inter-Tropical Convergence Zone (NITCZ) in summer (July) and Southern Inter-Tropical Convergence Zone (SITCZ) in winter (Jan.). The fact that the NITCZ is drawn to about 30°N latitude may be associated with the unusually high temperature over north India. According to this interpretation the main westerly current of the monsoon is simply the expanded equatorial westerlies which lie embedded in the great mass of tropical easterties or the trade winds. NITCZ is the zone of clouds and heavy rainfall.

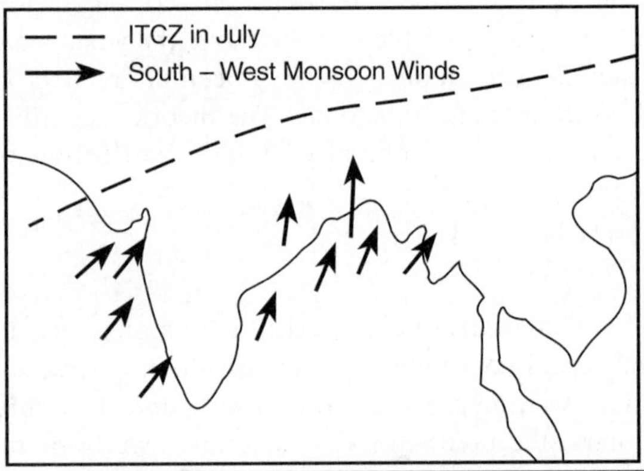

FIG. 5.3. ITCZ and S.W. Monsoon winds

Jet Stream Theory. Jet stream is a band of fast moving air from west to east usually found in the middle latitudes in the upper troposphere at a height of about 12 km. The wind speeds in a westerly jet stream are commonly 150 to 300 km/h with extreme values reaching 400 km/h Jet stream is the latest theory regarding the origin of the monsoons and has earned world wide acceptance and acclaim from the meteorologists.

M.T. Yin (1949), while discussing the origin of the monsoons expressed the opinion that the burst of monsoon depends upon the upper air circulation. The low latitude upper air trough shifts from 90° E to 80° E longitude in response to the northward shift of the western jet stream in summer. The southern jet becomes active and heavy rainfall is caused by south-west monsoons.

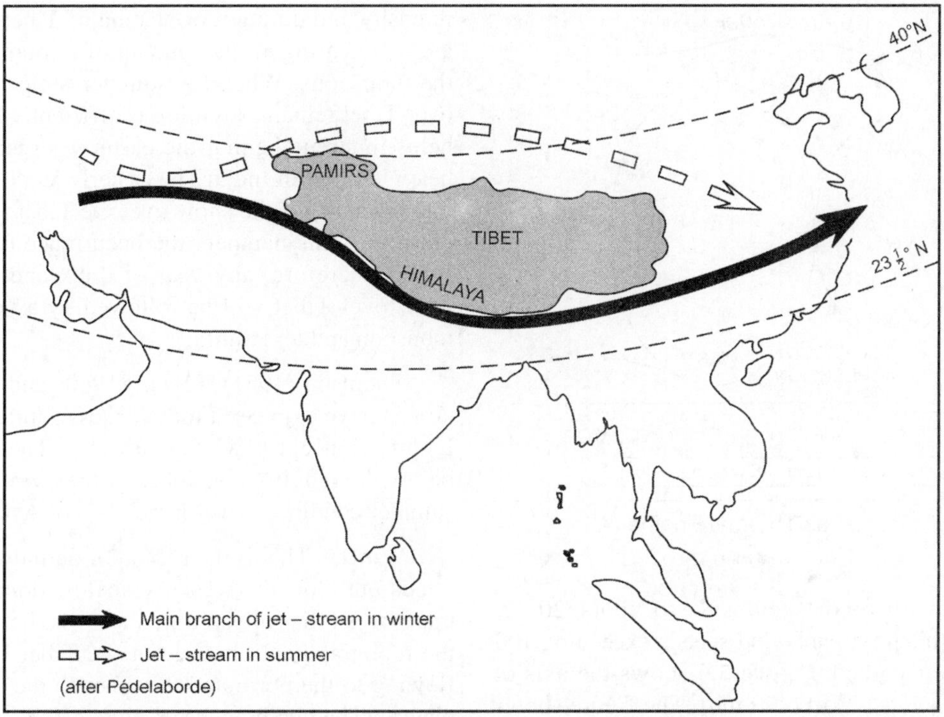

FIG. 5.4. Jet Stream (After Pedelaborde)

Yin's ideas are well recognised by **Pierre Pedelaborde** (1963), in his book entitled ***'The Monsoon'***. The map, showing the seasonal shift of the westerly jet stream, has been reproduced in **figure 5.4**. It shows that in winter the western jet stream flows along the southern slopes of the Himalayas but in summer it shifts northwards, rather dramatically, and flows along the northern edge of the Tibet Plateau. The periodic movements of the Jet stream are often indicators of the onset and subsequent withdrawal of the monsoon.

P. Koteswaram (1952), put forward his ideas about the monsoon winds based on his studies of upper air circulation. He has tried to establish a relationship between the monsoons and the atmospheric conditions prevailing over Tibet Plateau. Tibet is an ellipsoidal plateau at an altitude of about 4,000 m above sea level with an area of about 4.5 million sq km This plateau is surrounded by mountain ranges which rise 6,000-8,000 m above sea level. It gets heated in summer and is 2°C to 3°C warmer than the air over the adjoining regions. Koteswaram, supported by Flohn, feels that because the Tibet Plateau is a source of heat for the atmosphere, it generates an area of rising air. During its ascent the air spreads outwards and gradually sinks over the equatorial part of the Indian Ocean. At this stage, the ascending air is deflected to the right by the earth's rotation and moves in an anti-clockwise direction leading to anticyclonic conditions in the upper troposphere over Tibet around 300-200 mb (9 to 12 km). It finally approaches the west coast of India as a return current from a south-westerly direction and is termed as equatorial westerlies (**Fig. 5.5**). It picks up moisture from the Indian Ocean and causes copious rainfall in India and adjoining countries.

The south-west monsoon in southern Asia is overlain by strong upper easterlies with a pronounced jet at 100 to 200 mb. These easterly winds, which often record speeds exceeding 100 knot are known as the ***Easterly Jet Stream of the tropics***. The Easterly Jet Stream was first inferred by P. Koteswaram and P.R. Krishna in 1952 and aroused considerable interest among tropical meteorologists. A careful study of the jets would suggest that the core of the easterly jet is at 13 km (150 mb) while that of the westerly jet is at 9 km. Over India, the axis of the strongest winds in the easterly jet may extend from

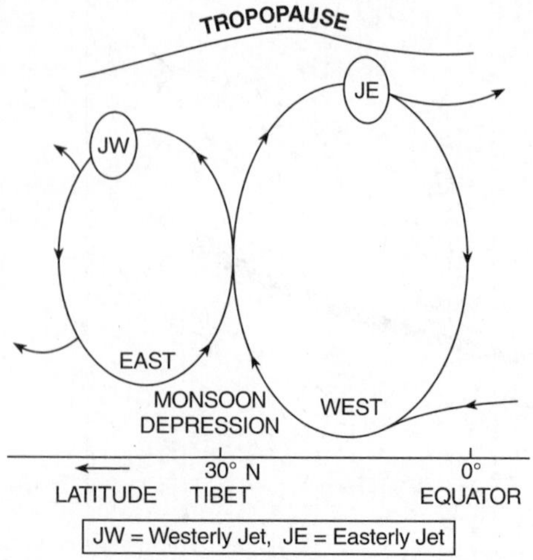

FIG. 5.5. Monsoon as Thermal Engine (After P. Koteswaram)

the southern tip of the peninsula to about 20° N latitude. In this jet stream wind speeds exceeding 100 knot may be recorded. **Figure 5.6** shows the axis of the easterly jet at 12 km (200 mb). The figure shows that there is the subtropical westerly jet to the north of the Himalayas besides the easterly jet over the peninsular India. It has already been made clear in **Fig. 5.4** that the westerly jet stream is located along the southern slopes of the Himalayas in winter but it suddenly shifts to the north with the onset of the monsoon. The periodic movements of the sub-tropical jet stream provide a useful indication of the onset and subsequent withdrawal of the monsoon. In fact, northward movement of the subtropical jet is the first indication of the onset of the monsoon over India.

Recent observations have revealed that the intensity and duration of heating of Tibet Plateau has a direct bearing on the amount of rainfall in India by the monsoons. When the summer temperature of air over Tibet remains high for a sufficiently long time, it helps in strengthening the easterly jet and results in heavy rainfall in India. The easterly jet does not come into existence if the snow over the Tibet Plateau does not melt. This hampers the occurrence of rainfall in India. Therefore, any year of thick and widespread snow over Tibet will be followed by a year of weak monsoon and less rainfall.

Thomson (1951), Flohn, (1960) and Stephenson (1965) have expressed more or less similar views but Flohn's concept is widely accepted. These ideas can be explained by considering the winter and the summer conditions over large parts of Asia.

Winter. This is the season of outblowing surface winds but aloft the westerly airflow dominates. The upper westerlies are split into two distinct currents by the topographical obstacle of the Tibet Plateau, one flowing to the north and the other to the south of the plateau. The two branches reunite off the east coast of China (**Fig. 5.7**). The southern branch over northern India corresponds with a strong latitudinal thermal gradient which, along with other factors, is responsible for the development of southerly jet. The southern branch is stronger, with an average speed of about 240 km p.h. at 200 mb compared with 70 to 90 km p.h. of the northern branch. Air subsiding beneath this upper westerly current gives dry outblowing northerly winds from the subtropical anticyclone over northwestern India and Pakistan. The surface winds blow from northwest over most parts of northern India.

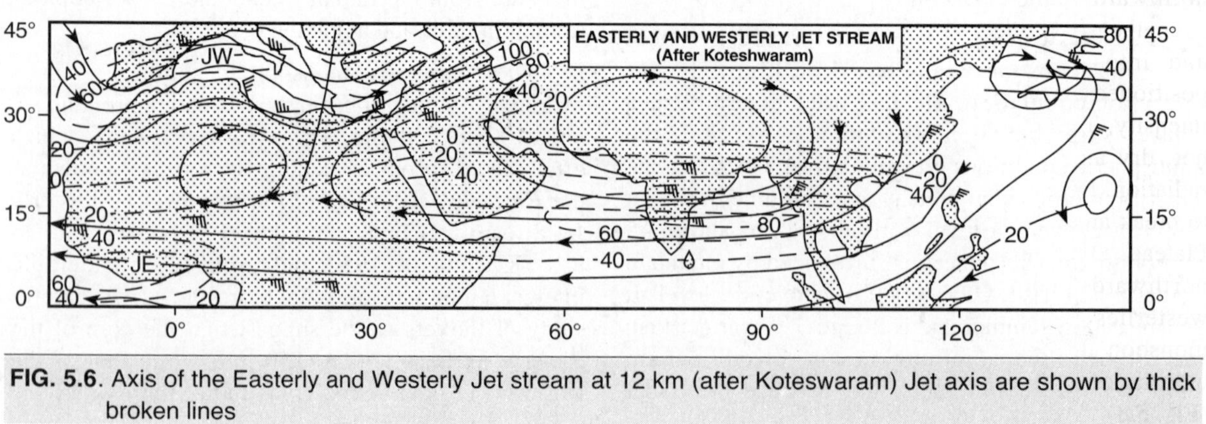

FIG. 5.6. Axis of the Easterly and Westerly Jet stream at 12 km (after Koteswaram) Jet axis are shown by thick broken lines

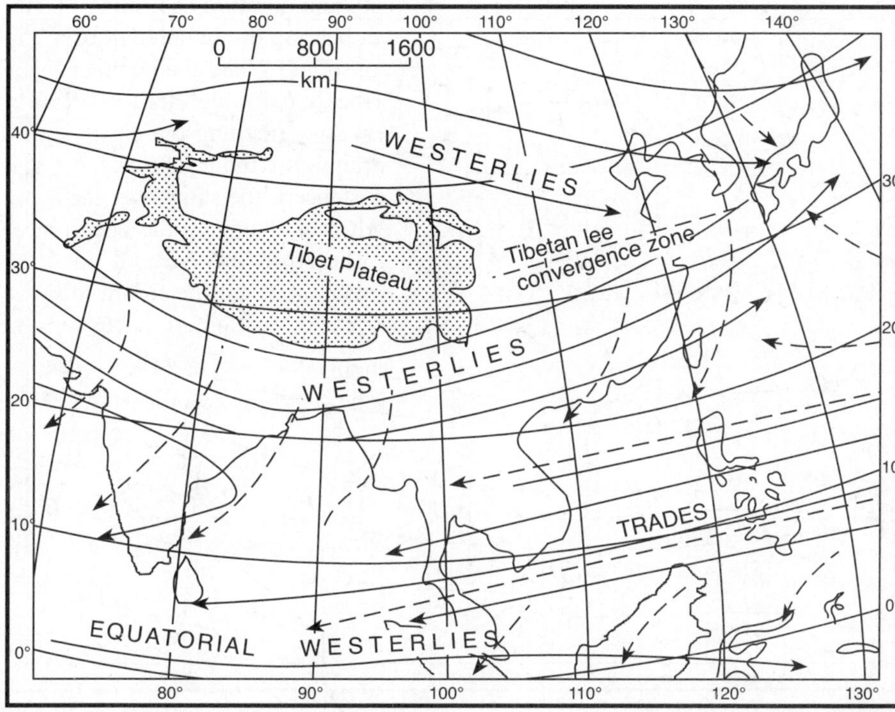

FIG. 5.7. The characteristic air circulation over southern and eastern Asia in winter. Solid lines indicate air flow at about 3000 m and dashed lines that at about 600 m. The names refer to the wind systems aloft.

The upper jet is responsible for steering of the western depressions from the Mediterranean Sea. Some of the depressions continue eastwards, redeveloping in the zone of jet stream confluence about 30° N, 105° E beyond the area of subsidence in the immediate lee of Tibet.

Summer. With the beginning of summer in the month of March, the upper westerlies start their northward march, but whereas the northerly jet strengthens and begins to extend across central China and into Japan, the southerly branch remains positioned south of Tibet, although weakening in intensity. The weather over northern India becomes hot, dry and squally due to larger incoming solar radiation. By the end of May the southern jet begins to break and later it is diverted to the north of Tibet Plateau. Over India, the Equatorial Trough pushes northwards with the weakening of the upper westerlies south of Tibet, but the burst of the monsoon does not take place until the upper-air circulation has switched to its summer pattern (**Fig. 5.8**). The low level changes are related to the high level easterly jet stream over southern Asia about 15° N latitude.

T.N. Krishnamurti used data of the upper atmosphere to calculate the patterns of divergence and convergence at 200 mb for the period of June-August, 1967. He observed an area of strong divergence at 200 mb over northern India and Tibet, which coincides with the upper-level divergence associated with the easterly jet. Similarly he found a northerly component to the flow from this region which represents the upper branch of the *Hadley cell*. These happenings are closely related to the Indian monsoon. **S. Rama Rattan** opined that the development of monsoon winds is deeply connected with the jet stream in addition to the differential heating of land and sea. The upper air circulation in summer has anticyclonic pattern between 40° N and 20° S whereas cyclonic conditions prevail at the surface. Western and eastern jets flow to the north and south of the Himalayas respectively. The eastern jet becomes powerful and is stationed at 15° N latitude. This results in more active south-west monsoon and

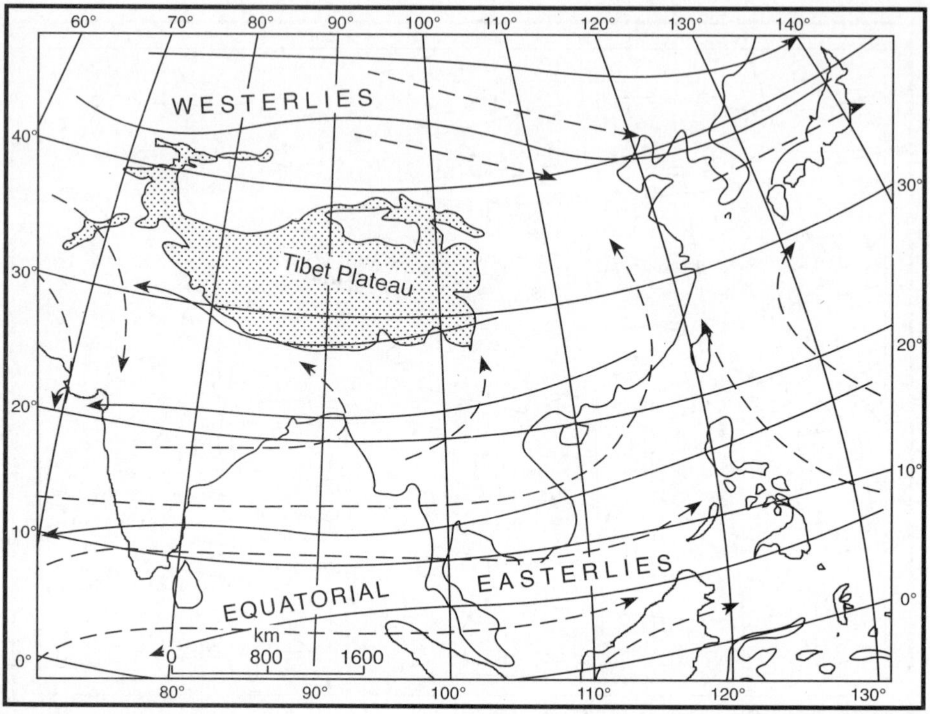

FIG. 5.8. The characteristic air circulation over southern and eastern Asia in summer. Solid lines indicate air flow at about 6,000 m and dashed lines that at about 600 m. Note that the low-level flow is very uniform between about 600 and 3,000 m.

heavy rainfall is caused. **Raman** and **Ramanathan** while discussing the tropical easterly jet stream suggested that the easterly winds become very active in the upper troposphere after the beginning of the rainy season. The latent heat produced due to cloud cover results into *inversion of temperature* and causes rainfall. **Ananth Krishnan** is of the opinion that the south-west monsoons are deeply influenced by the subtropical cyclones in the upper troposphere between 20° and 25° N latitudes. These winds start developing in the beginning of the summer season and shift to 30° N about 5-6 weeks later. Besides intensive heat between 20° and 40° N latitudes gives further strength to the south-west monsoons. **S. Parthasarthy** in his essay on '*Trying to solve the Monsoon Riddle*' expressed the view that the monsoons are influenced by the north-east trade winds. A weak north-east trade wind results in weak monsoon and leads to drought conditions.

The Indian monsoons, particularly the south-west monsoons, have generated a lot of interest among the meteorologists all over the world. Concerted efforts on data collection and of intensive studies of monsoon regimes by various meteorological services and organisations from different nations have been made during the last six decades. Much has been done but much more is yet to be done. The first attempt was made during International India Ocean Expedition (IIOE) from 1962 to 1965. It was organised jointly by the International Council of Scientific Unions. (ICSU), Scientific Committee on Ocean Research (SCOR) and UNESCO with World Meteorological Organisation (WMO) joining the meteorology programme. Special oceonographic and atmospheric studies were carried out with the aid of research vessels, instrumented aircrafts, rockets as well as special upsonde and dropsonde soundings. Two more experiments were conducted, jointly, by India and the former USSR in 1973 and 1977, with limited participation from other countries. These experiments are known as the ***Indo-Soviet Monsoon Experiment (ISMEX)*** and Monsoon-77 respectively. It was observed from these experiments that there is a specific zone off the coast of Kenya where the monsoons from the southern hemisphere crossed the equator on their way to India. It was also observed that the fluctuations in the intensity of low-level across the equator resulted in the fluctuations of rainfall over Maharashtra. Upper air observations over the Bay of Bengal were also made in 1977.

More intensive data collection effort was made under the aegis of another international experiment—the Monsoon Experiment in 1979. It is popularly known as ***MONEX-1979***. It was organised jointly by Global Atmospheric Research Programme (GARP) of the International Council of Scientific Unions (ICSU) and the World Meteorological Organisation (WMO) under their World Weather Watch (WWW) programme. It is so far the largest scientific effort made to extend the frontiers of our knowledge of the monsoons by the international scientific community. As many as 45 countries pooled their talents and resources under the aegis of the United Nations for this great venture. Some idea of the dimensions of this experiment may be had from the fact that in May 1979 as many as 52 research ships were deployed over the tropical oceans between 10° N and 10° S latitudes. Besides 104 aircraft missions were successfully completed over different parts of the Pacific, the Atlantic and the Indian Oceans.

The great MONEX was designed to have three components considering the seasonal characteristics of the monsoons :

(*i*) Winter Monex from 1 December 1978 to 5 March 1979 to cover the eastern Indian Ocean and the Pacific along with the land areas adjoining Malaysia and Indonesia.

(*ii*) Summer Monex from 1 May to 31 August 1979 covering the eastern coast of Africa, the Arabian Sea and the Bay of Bengal along with adjacent landmasses. It also covered the Indian Ocean between 10° N to 10° S latitudes.

(*iii*) A West African Monsoon Experiment (WAMEX) over western and central parts of Africa from 1 May to 31 August 1979.

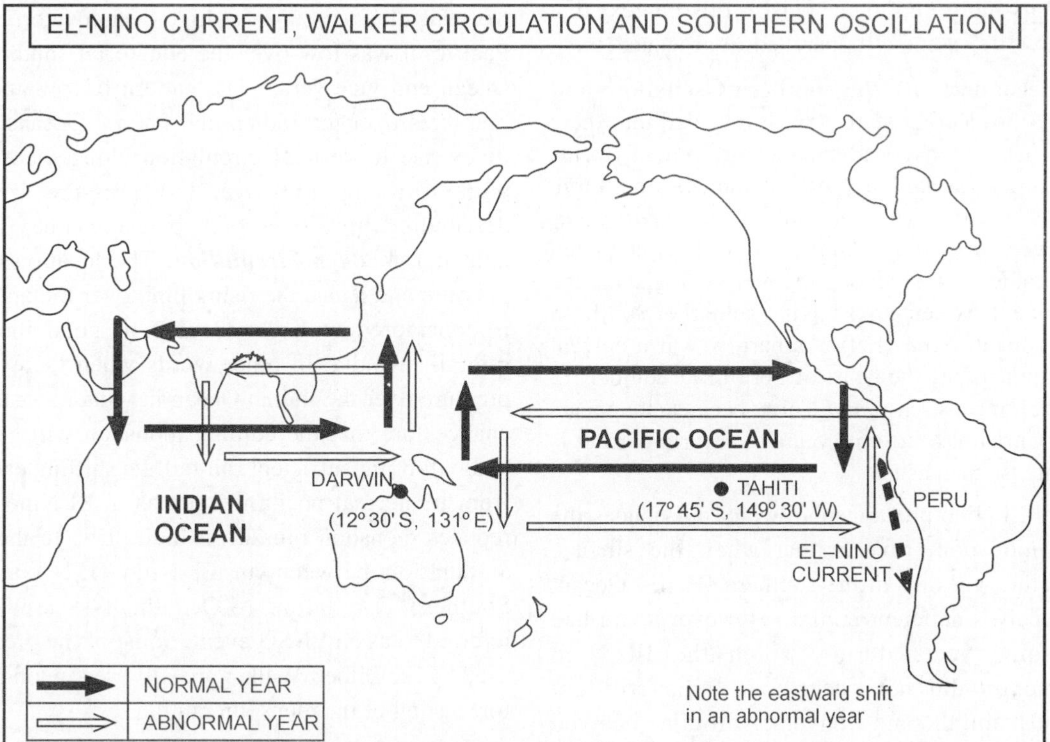

FIG. 5.9. El Nino, Walker Circulation and Southern Oscillation

International MONEX Management Centres (IMMC) were set up in Kuala Lumpur and New Delhi to supervise the winter and summer components of the experiment.

MONEX-1979 suffered some setback due to abnormal behaviour of the monsoons in that year. None of the cold surges was intense in China Sea during the winter MONEX. A strong anticyclone developed in the Arabian Sea in summer of 1979. The southwest monsoon was deflected southwards before touching the Kerala coast under the influence of this anticyclone and started blowing parallel to the coast. Consequently the onset of southwest monsoon over Kerala was delayed by 12 days. Moreover, July was characterised by several weak or break-monsoon occurrences and there was only one monsoon depression. Therefore, 1979 was not a normal monsoon year and MONEX failed to study the normal behaviour of the monsoons. But the vagaries of the monsoons are proverbial and in a scientific and analytical understanding of the monsoons, a study of anomalies is perhaps more important. It is in this context that MONEX-1979 assumes unparalled significance.

Teleconnections, the Southern Oscillation and the El Nino. Recent studies have revealed that there seems to be a link between meteorological events which are separated by long distances and large intervals of time. They are called ***meteorological teleconnections***. The one which has aroused considerable interest among the meteorologists is the difference between an El Nino and the Southern Oscillation. ***El Nino*** (EN) is a narrow warm current which appears off the coast of Peru in December. In Spanish, it means *The Child Christ* because it appears around Christmas. In some years this warm current is more intense than usual.

The El Nino phenomenon, which influences the Indian monsoon, reveals that when the surface temperature goes up in the southern Pacific Ocean, India receives deficient rainfall. However, there had been some years during which the El Nino phenomenon did not occur, but India still got deficient rainfall, and conversely, India received sufficient rainfall during an El Nino year. A study of the one hundred years from 1870 to 1970 of the Indian monsoons shows that out of 43 deficient monsoon years, 19 were associated with an El Nino. On the other hand, there were 6 El Nino years which were also years of good monsoon rain. Analysis of meteorological data for three decades from 1972-73 to 2013-14 shows that out of 14 drought years 9 had been El Nino years and 5 were non El Nino years. Years 1972-73, 1979-80, 1987-88, 2002-03, 2009-10, 2013-14 and 2014-15 were the years of severe drought and witnessed strong El Nino. On the other hand, years 1991-92, 1994-95, 1997-98, 2004-095 and 2006-07 non El Nino years without drought conditions. Thus, although there is a tendency for poor monsoons to be associated with an El Nino, there is no one-to-one correspondence.

Southern Oscillation (S.O.) is the name ascribed to the curious phenomenon of *sea-saw pattern* of meteorological changes observed between the Pacific and Indian oceans. This great discovery was made by Sir Gilbert Walker in 1920. While working as the head of the India Meteorological Service, he noticed that when the pressure was high over equatorial south Pacific, it was low over the equatorial south Indian Ocean and vice versa. The pattern of low and high pressures over the Indian and Pacific Oceans (S.O.) gives rise to vertical circulation along the equator with its rising limb over low pressure area and descending limb over high pressure area. This is known as ***Walker Circulation***. The location of low pressure and hence the rising limb over Indian Ocean is considered to be conducive to good monsoon rainfall in India. In other words when there is low pressure over the Indian Ocean in winter months, the chances are that the coming monsoon will be good and will bring sufficient rainfall. Its shifting eastward from its normal position, such as in El Nino years, reduces monsoon rainfall in India. Due to the close association between an El Nino (E.N.) and the Southern Oscillation (S.O.), the two are jointly referred to as an ENSO event. Some of the predictors used by Sir Gilbert Walker are still used in long-range forecasting of the monsoon rainfall.

The main difficulty with the Southern Oscillation is that its priodicity is not fixed and its period varies from two to five years. Different indices have been used to measure the intensity of the Southern Oscillation, but the most frequently used is the Southern Oscillation Index (SOI). This is the difference in pressure between Tahiti (17°45′S, 149°30′W) in French Polynesia, representing the Pacific Ocean and Port Darwin (12°30′S, 131°E), in northern Australia representing the Indian Ocean. The positive and negative values of the SOI *i.e.* Tahiti minus the Port Darwin pressure are pointers towards good or bad rainfall in India.

Scientists of India Meteorological Department (IMD) joined an international study programme called the Tropical Oceans and Global Atmosphere (TOGA) in 1985. This is an interesting and ambitious programme which investigates both teleconnections effects and the internal variability. As a follow up to TOGA, the climate variability (CLIVAR) was set up in January 1995, to develop an internationally operational climate forecasting system.

Another major programme is the Indian Middle Atmospheric Programme (IMAP) initiated by the Department of Space. This programme has been launched to augment the existing weather prediction scheme.

Western Disturbances: Western disturbances comprise a part of temperate cyclones. Several theories have been put forward regarding the origin of western disturbances but the **Polar Front Theory** propounded by **Vilhelm Bjerknes** and his son **Jakob Bjerknes** the two famous meteorologists of Norway in 1918 is the most popular and is widely accepted. According to this theory, temperature cyclones are formed due to collision of two air masses of contrasting characteristics of temperature and humidity. Air masses of low temperature and humidity from polar areas meet air masses of high temperature and humidity in the temperate zone. They meet each other at 60° latitudes. Here they do not merge with each other, rather they form a front known as **Polar Front**. The lighter air with high temperature and humidity ascends over the heavier air with low temperature and humidity and in the process, temperate cyclones are formed.

Temperate cyclones reaching India originate over the Mediterranean Sea and travel from west to east direction. They enter the north-western part of India after crossing over Iraq, Iran, Afghanistan and Pakistan. They are known as western disturbances since they travel from the west. They are more active in the winter season and deeply impact the weather conditions in north India. They are in the form of **Occlude Front** when they enter India. Although their main impact is felt in Jammu and Kashmir, Punjab, Haryana, Himachal Pradesh and western part of Uttar Pradesh, some times they affect area as far as Bihar, Bengal, Assam and even Arunachal Pradesh. They cause heavy snowfall in the mountainous areas and bring light rainfall in the plain areas. The average amount of rainfall caused by these disturbances is about 60 cm in the Himalayan region, 12 cm in Punjab and Haryana, 5.3 cm in Delhi and 2.5-1.8 cm in Uttar Pradesh and Bihar. Although very small in amount, this winter rainfall is very useful for the rabi crops, especially the wheat crop.

The average life span of the western disturbances is 3-4 days. The temperature falls 5-10°C below normal and cold wave conditions prevail over vast areas . Mist and fog in the morning and evening hours is a common phenomenon. Sometimes 2-3 western disturbances follow each other and continue to influence the weather conditions for 8-10 days. Jet stream plays a significant role in bringing these disturbances to India. Normally, they are active from December to March and 4-6 disturbances visit India per month. According to **Kandrew**, the number of western disturbances visiting India is 2 in November, 4 in December, 5 each in January, February, March and April and 2 in May.

Indian Ocean Dipole (IOD): Unlike El Nino, which is an abnormal warming of equatorial Pacific Ocean, the Indian Ocean Dipole (IOD) is defined by the difference in the sea surface temperatures between the two equatorial regions of the Indian Ocean—the western end near the Arabian Sea and the eastern one off Indonesia. Positive IOD conditions are likely to be favourable for a normal/above normal monsoon. Although there is no fully established correlation between summer monsoon rainfall in India and the IOD, records of India Meteorological Department (IMD) show that IOD was positive in four of the 16 El Nino years during 1951-2017 (Table 5.1) and the rainfall was normal *i.e.* 96-104% of the Long Period Average (LPA) in three of these four years. The year 1997 saw one of the strongest El-Nino of the 20th

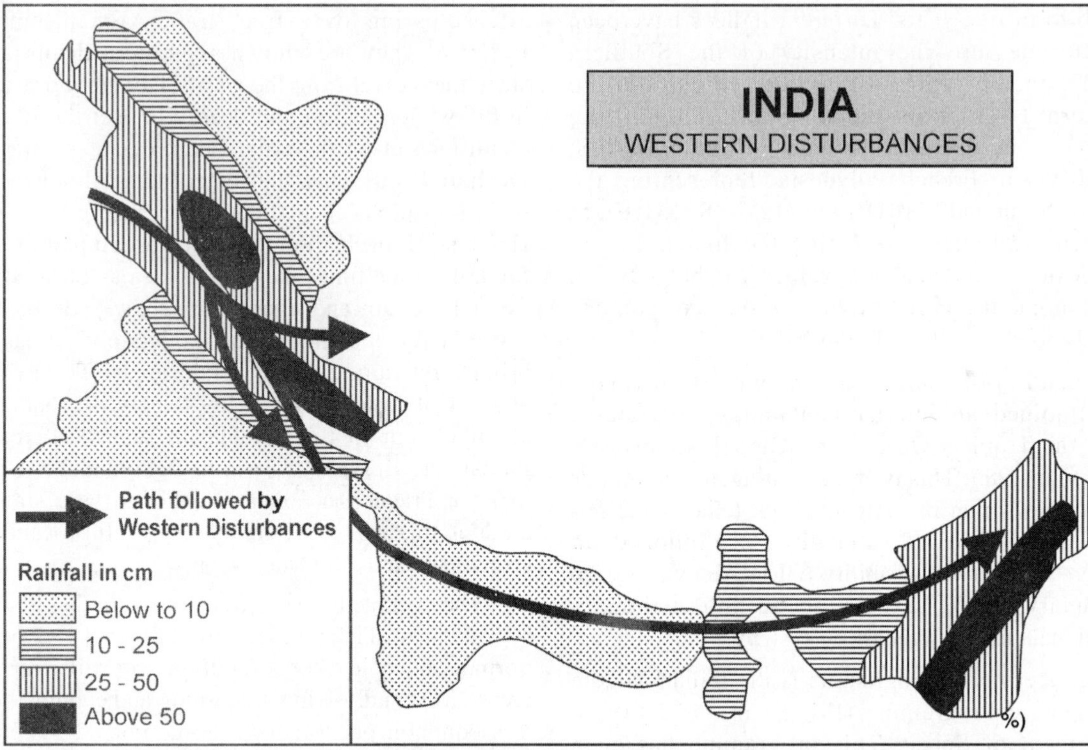

FIG. 5.10. India : Western Disturbances

Century, but India received normal (102% of the LPA) because this was the year of positive IOD.

After the severe drought of 1987, parametric and power regression models have been developed to forecast monsoon rainfall by utilising signals from 15 parameters. Some of the parameters are global while others are regional. These parameters are divided into four broad categories, *viz*. (*a*) temperature, (*b*) pressure (*c*) wind pattern and (*d*) snow cover and are listed below :

(*a*) *Temperature related parameters*

1. El Nino in current year
2. El Nino in previous year
3. Northern India (March)
4. East coast of India (March)
5. Central India (May)
6. Northern hemisphere (Jan. and Feb.)

(*b*) *Wind related parameters*

7. 500 hPa (1 hecta pascal, equals 1 mb) ridge (April)

TABLE 5.1. El Nino, Phase of IOD and rainfall in India from 1951 to 2017

Year	El Nino Strength	Phase of IOD	Rainfall (% of LPA)
1951	Weak	Neutral	81
1953	Weak	Neutral	110
1957	Strong	Positive	98
1963	Moderate	Positive	98
1965	Strong	Neutral	82
1969	Weak	Neutral	100
1972	Strong	Neutral	76
1982	Strong	Neutral	85
1987	Strong	Neutral	75
1991	Moderate	Positive	91
1997	Strong	Positive	102
2002	Moderate	Neutral	81
2004	Weak	Neutral	86
2009	Moderate	Neutral	78
2015	Strong	Neutral	86
2017	Weak	Positive	—

Source: India Meteorological Dept.

TABLE 5.2. Southern Oscillation Index (SOI) and associated weather phenomena

Positive SOI :	Negative SOI :
(*i*) Tahiti pressure greater than that of Port Darwin	(*i*) Port Darwin pressure exceeds that of Tahiti.
(*ii*) Pressure high over east Pacific and low over Indian Ocean.	(*ii*) Pressure high over Indian Ocean and low over east Pacific.
(*iii*) Low rainfall over eastern Pacific and prospects of good monsoon rain over India and Indian Ocean.	(*iii*) Low rainfall or poor monsoon over Indian Ocean and higher than usual rain over east Pacific.

8. 50 hPa ridge-trough extent (Jan. and Feb.)
9. 10 hPa (30 km) westerly wind (Jan.)

(*c*) *Pressure anomaly (SOI)*

10. Tahiti-Darwin (Spring)
11. Darwin (Spring)
12. South America, Argentina (April)
13. Indian Ocean Equatorial (Jan.-May)

(*d*) *Snow cover related parameters*

14. Himalayan (Jan.-March)
15. Eurasian (Previous December)

It was observed in late eighties that whenever more than 50% parameters showed favourable signals, the monsoon rainfall in India was normal and when 70% or more parameters were favourable, the monsoon rainfall was above normal.

Somewhat similar set of predictors for monsoon was suggested by H.N. Srivastava and S.S. Singh in 1994 while discussing long range weather forecasting techniques.

One more parameter, *viz.*, surface pressure anomaly of north-eastern hemisphere was also added later on, thus making a total of 16 parameters. These 16 parameters have been used by the IMD to develop the power regression model. Although this model has been used for forecasting rainfall in India with greater accuracy since 1989, it is far from being an elaborate and foolproof model. A model capable of forecasting area specific rainfall is yet to be built. The study of data flowing from MONEX, TOGA and other experiments is continuing and our meteorologists are hopeful of discovering more parameters which may help in developing better models capable of predicting rainfall more accurately.

In the recent past, the study of satellite imageries and remote sensing has helped a lot to forecast the monsoon rainfall.

SEASONAL RHYTHM

Indian climate is characterised by distinct seasonality. Seasons come and go one after the other with surprising precision. They depict the annual cycle of weather and reflect the changing moods of nature. Each season has its distinct features. India Meteorological Department (IMD) has recognised the following four distinct seasons :

1. The cold weather season or winter season,
2. The hot weather season or summer season,
3. The south-west monsoon season or rainy season, and
4. The season of the retreating monsoon or cool season.

1. The Cold Weather Season or Winter Season

The cold weather season commences in November and continues till March. **Clear sky, pleasant weather, low temperature and humidity, high range of temperature, cool and slow northern winds are the chief characteristics of this season.**

(*a*) **Temperature.** The northern two thirds of the country have mean temperatures below 21°C, with afternoon temperatures of 27°C. January is the coldest month when the temperature in the Ganga plain varies from 12.5° to 17.5°C. The mean minimum temperatures are about 5°C over northwest India and 10°C over the Gangetic plains. However, on individual days the temperatures may fall much below the mean values. The night temperature often falls below freezing point in many hilly areas. Dras in Ladakh is the coldest place in India. Here the minimum temperature often falls below –40°C. The minimum temperature recorded at Dras was –45°C on 28th December, 1908.

The southern one-third has rather warm conditions and does not have a distinctly defined winter weather. The isotherm of 20°C runs in east-west direction, roughly parallel to the Tropic of Cancer and divides India climatically into the northern and southern parts. To the south of this

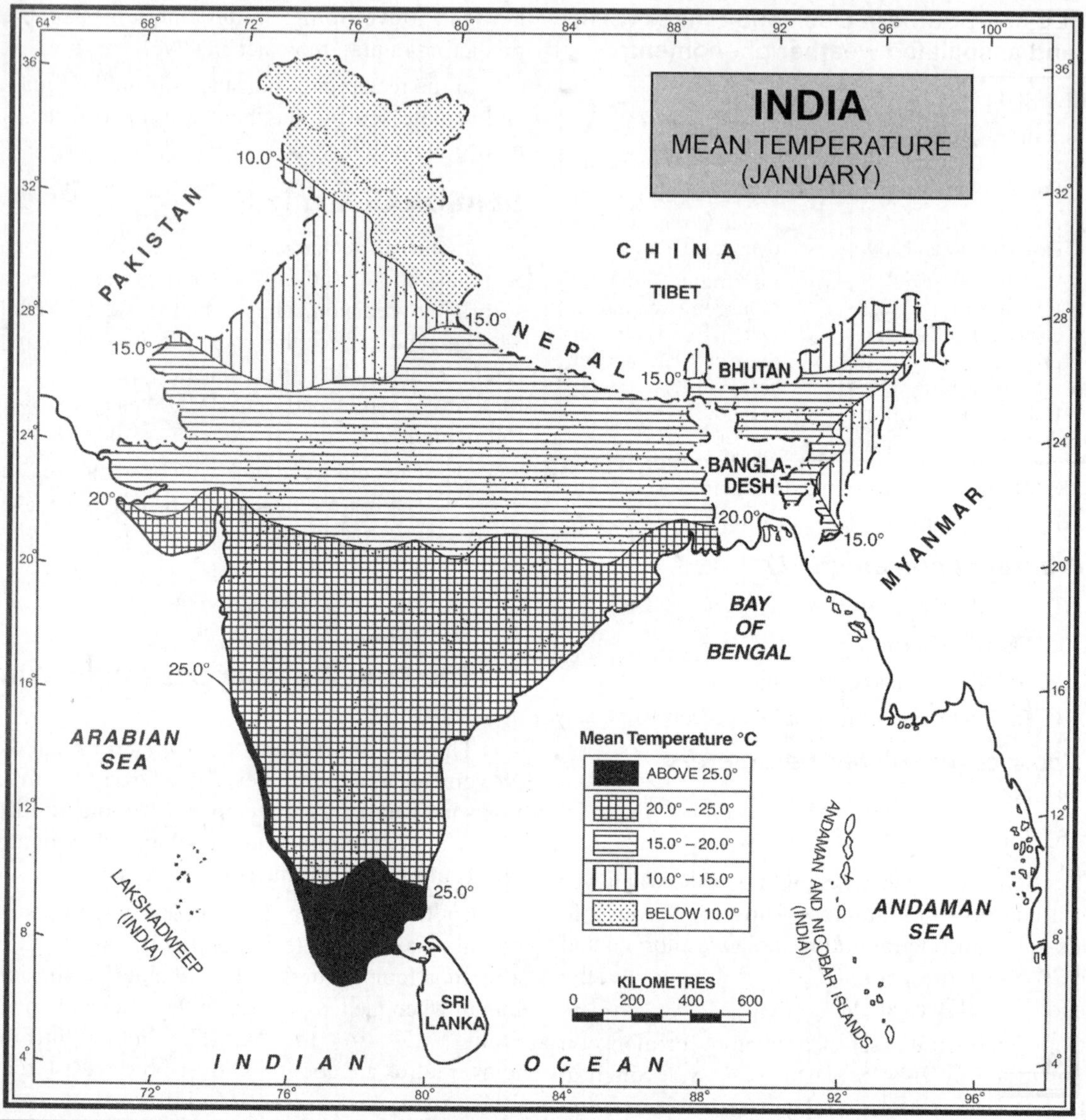

FIG. 5.11. Mean temperature (January)

isotherm the temperatures are invariably above 20°C. In the extreme south the temperature may be well above 25°C (**Fig. 5.11**). January temperature at Thiruvananthapuram is 31°C. The diurnal range of temperature, especially in interior parts of the country, is very high. It may reach 15°C over western India.

(*b*) Pressure and Winds. High air pressure prevails over large parts of north-west India due to low temperature conditions there. Normally the pressure varies from 1015 to 1020 mb. However, pressure is comparatively lower in south India. The isobar of 1019 mb is seen in north-west India while the isobar of 1013 mb touches the southern tip of the country (**Fig. 5.12**). The winds start blowing from high pressure area of north-west to low pressure area of south-east. The wind velocity is low due to low pressure gradient. Depending upon the pressure and physiography, the winds blow from west and north-west in Punjab, Haryana, Uttar Pradesh and Bihar, from north-west in Bengal and from north-east in the Bay of Bengal.

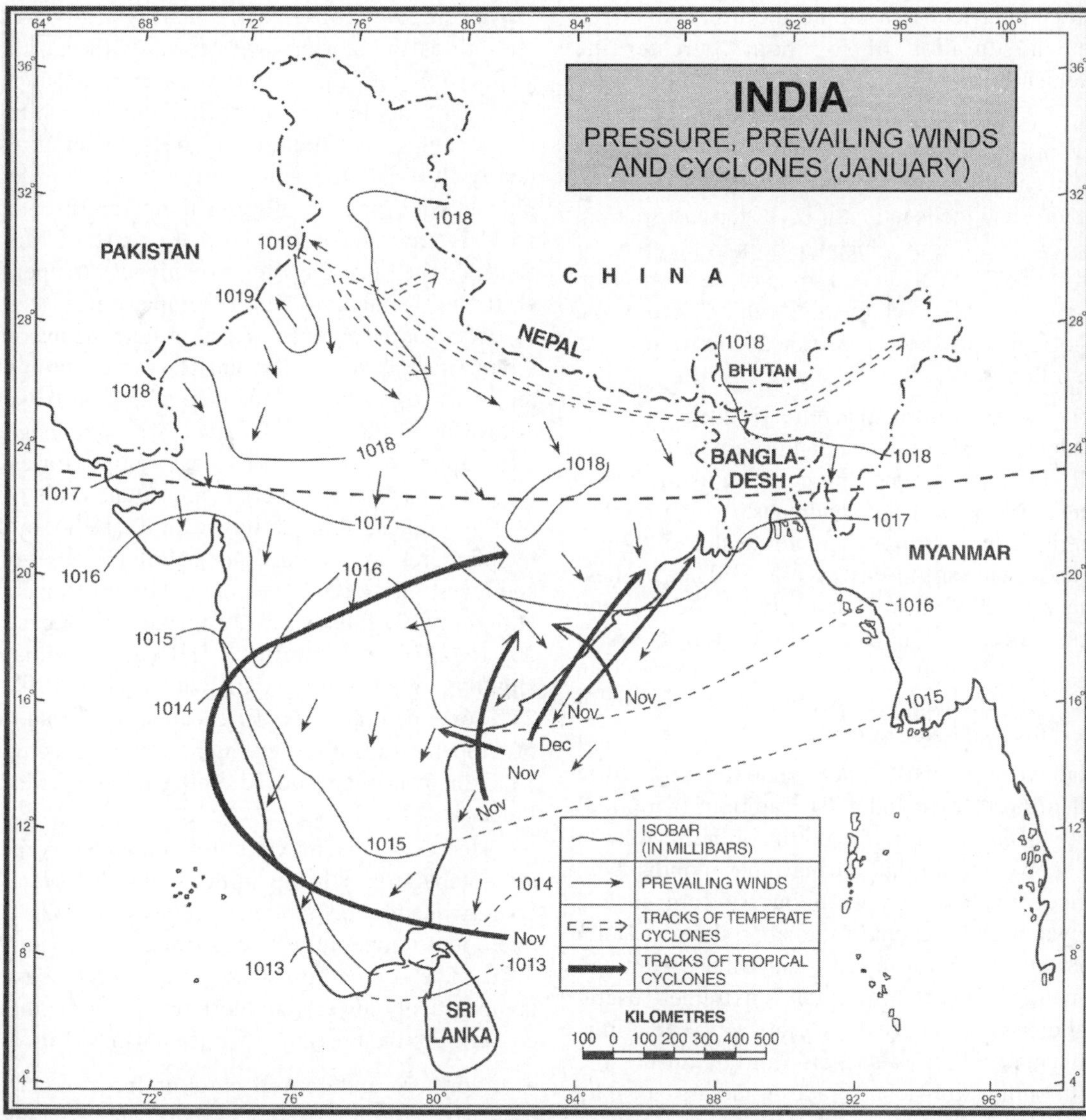

FIG. 5.12. India : Pressure, Prevailing Winds and Cyclones in January

Western Disturbances and Tropical Cyclones. Although sky is generally clear, the spell of fine weather is often broken due to inflow of depressions from the west. These low pressure depressions are called ***western disturbances***. They originate in the Mediterranean Sea and enter India after crossing over Iraq, Iran, Afghanistan and Pakistan. On their way, their moisture content is augmented from the Red sea and the Persian Gulf. They are often in an occluded form when they reach India. They intensify over Rajasthan, Panjab, and Haryana. They move eastwards across the sub-Himalayan belt and reach right upto Arunachal Pradesh (**Fig. 5.12**), causing light rain in the Indus-Ganga plains and snowfall in the Himalayan belt. After the passage of the disturbance, widespread fog and cold waves lowering the minimum temperature by 5° to 10°C below normal are experienced. Haze is common in morning and evening.

Fog occurring in this season causes great inconvenience for transport in general and air transport in particular in the northern part of India. Longer the duration of fog hours, greater the inconvenience for transport.

This is the season of least tropical cyclone activity and the frequency decreases with the advancement of season. This is due to low sea surface temperature and the location of ITCZ farthest south in this season. The storms which are born in the Bay of Bengal strike Tamil Nadu and some of them cross the southern peninsula over to the Arabian Sea. Some storms originate in the Arabian Sea and move towards either north or west.

(*c*) Precipitation. The retreating winter monsoons blow from land to sea and do not cause much rainfall. These winds pick up some moisture while crossing the Bay of Bengal and cause winter rainfall in Tamil Nadu, south Andhra Pradesh, south-east Karnataka and south-east Kerala. The highest seasonal rainfall of about 75 cm between October and December occurs along the south-eastern coast of Tamil Nadu and adjoining parts of Andhra Pradesh. Thereafter, it gradually decreases except for a small area towards south-east Kerala.

The western disturbances also cause a little rainfall in north-west India. The amount of rainfall gradually decreases from the north and north-west to east. The average rainfall during three months from December to February is about 60 cm in the Himalayan region, 12 cm in Punjab, 5.3 cm at Delhi and 1.8 cm to 2.5 cm in U.P. and Bihar. Although very small in amount, this rainfall is extremely useful for rabi crops, especially the wheat crop. The north-eastern part of India also gets rainfall during the winter months. Arunachal Pradesh and Assam may get as much as 50 cm of rainfall during these months. The distribution of winter rainfall in India is shown in **Fig. 5.13**.

2. The Summer Season

Period from March to June is called the ***summer season***. ***High temperature and low humidity are the chief characteristics of this season.*** Hence it is also known as ***hot weather season*** or ***hot dry summer season***. Sometimes, it is referred to as ***pre-monsoon*** period.

(*a*) Temperature. As the season advances, sun's vertical rays move northwards and large parts of the country, south of the Satpura range, are heated up.

There is a progressive northward march of warmth as the sun proceeds towards the north, the southern parts being distinctly warmer in March and April whereas in June, north India has much higher temperatures. In March, the highest temperatures are nearly 40°C in the southern parts of the Deccan plateau; in April the highest temperature of about 45°C is recorded in the northern parts of Madhya Pradesh; in May the scene of highest temperature shifts to Rajasthan where temperatures as high as 48°C may be recorded and in June the maximum temperature is in Punjab and Haryana. The highest temperatures so far recorded are 50.5°C at Alwar on 10th May, 1956 and 50.6°C at Sri Ganganagar on 14th June, 1935. The highest temperatures are recorded in May but in certain areas June is the month of the highest temperatures. In fact the highest temperatures are recorded just before the onset of the southwest monsoons. Even the minimum temperature at night rarely falls below 20°C mark. It ranges from 20° to 25°C over northern and central India and slightly higher in the south with a pocket of over 27°C over central parts of Deccan. The diurnal range of temperature is also very high. It may be as high as 18°C in Rajasthan and adjoining areas of Madhya Pradesh.

However, the maximum summer temperatures are comparatively lower in the southern parts of the country due to moderating effect of the sea. The mean maximum temperature at most places is about 26°–30°C. The temperatures along the west coast are comparatively lower than those prevailing on the east coast due to the prevailing westerly winds (**Fig. 5.14**).

Northern and central parts of India experience *heat waves* in this season. A *heat wave* is generally defined with reference to the normal maximum temperature of a particular region. According to Indian convention, departure of the maximum temperature of the order of 6° to 7°C above normal is termed as ***'moderate'*** and 8°C and more as ***'severe' heat wave***. Most of the heat waves develop over Rajasthan, Punjab and Haryana. This is due to the location of these areas far away from the sea. From here they spread over Uttar Pradesh and Bihar. The strong north westerly winds with a long land trajectory over hot regions check the onward march of the sea breeze over eastern coastal belt and create heat wave conditions over coastal Odisha and Andhra

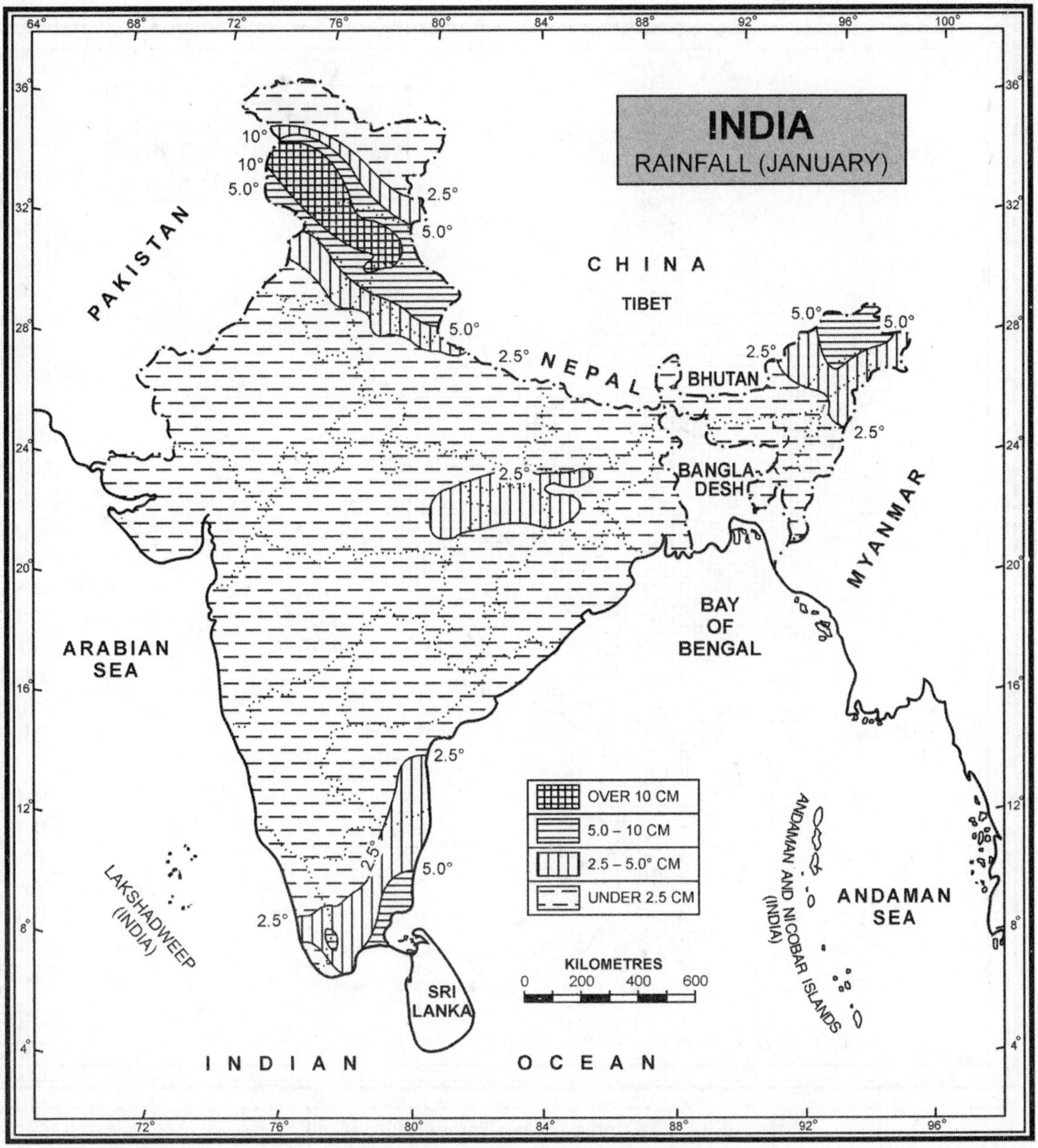

FIG. 5.13. India : Rainfall (January)

Pradesh. The heat waves start striking by the end of April and their maximum occurrence is in May and June till the onset of southwest monsoon. The normal duration of heat waves is 4 to 5 days. However, heat waves are rare over the peninsula south of 13°N latitude due to maritime conditions prevailing there.

(*b*) **Pressure and Winds.** The atmospheric pressure is low all over the country due to high temperature. Unusually low pressure is noticed in north-west India where temperatures are exceptionally high. The overall pressure gradient is low. The difference of pressure between the extreme north and south of the country does not exceed 3-4 mb. The isobars run more or less parallel to the coast indicating differences in pressure conditions over land and sea.

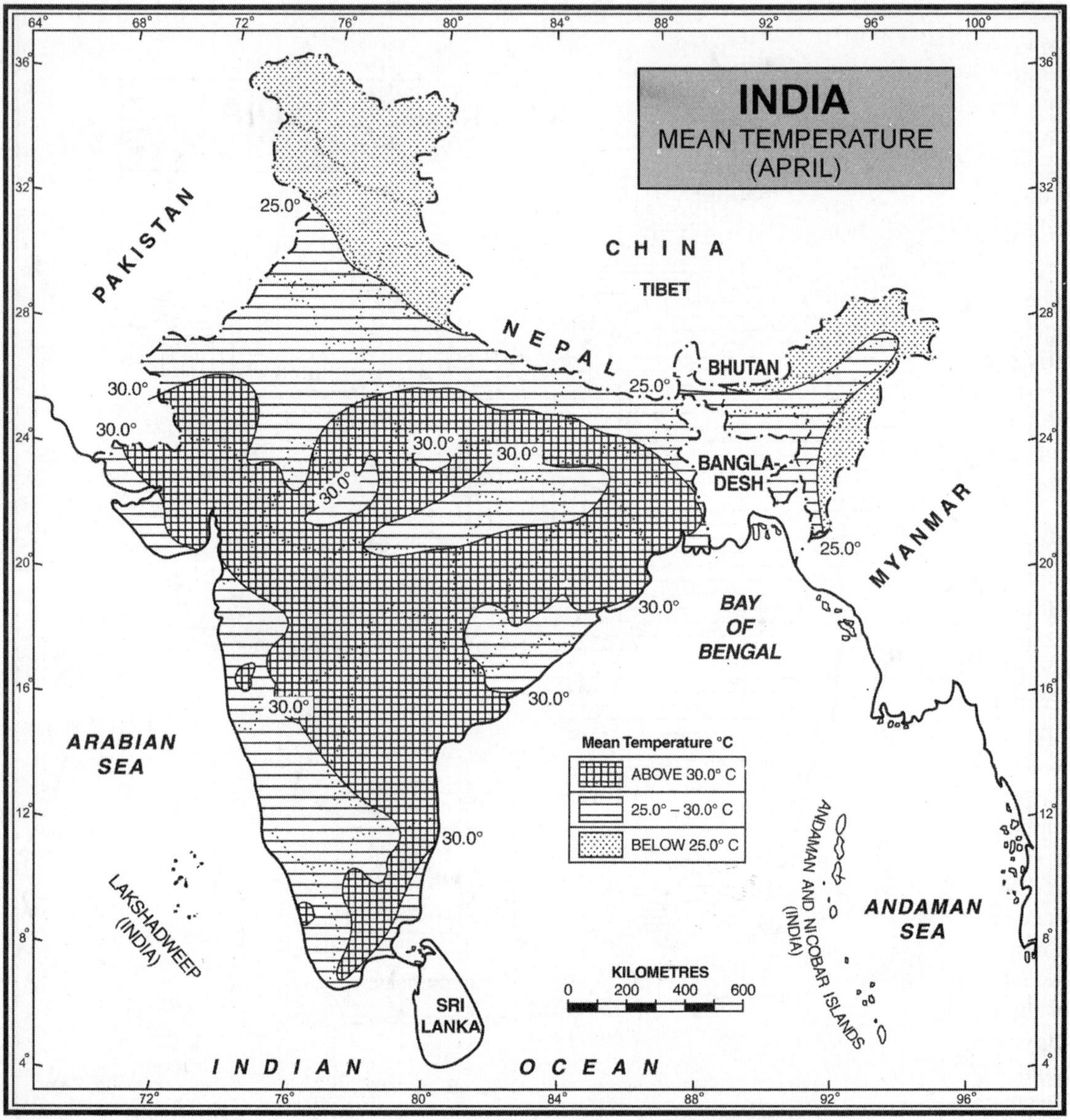

FIG. 5.14. India : Mean Temperature (April)

There is a marked change in the direction and speed of the winds from the winter conditions. The winds are by and large light and variable. However, there are some exceptions. In May and June, high temperature in northwest India builds steep pressure gradient which is often of the order of 1.0 to 1.5 mb per degree of latitude. Under such conditions, hot, dust laden and strong wind known as ***loo*** blows. Loo normally starts blowing by 9.00 A.M., increases gradually and reaches maximum intensity in the afternoon when the temperature is maximum. It blows with an average speed of 30-40 km per hour and persists for days together; often 3 to 10 days at a stretch. The strong dust storms resulting from the convective phenomena are locally known as ***andhis*** which literally mean ***blinding storms***. They are essentially duststorms, which move like a solid wall of dust and sand. The wind velocity often reaches 50-

60 km per hour and the visibility is reduced to a few metres ; sometimes the visibility is nil. Such dust storms are common in Rajasthan, Haryana, Punjab, Jammu region, Delhi, Uttar Pradesh, Bihar and Madhya Pradesh. They are invariably short lived but the squall and showers which follow these storms bring down the temperature sharply and afford much needed relief, although temporary, from the scorching heat.

The strong convectional movements with divergence related to the westerly jet stream or westerly disturbances in the upper troposphere lead to thunderstorms in eastern and north-eastern part of the country. They normally originate over Chota Nagpur plateau and are carried eastwards by westerly winds. The areas with highest incidence of thunderstorms are Assam, Arunachal Pradesh, Nagaland, Mizoram, Manipur, Tripura, Meghalaya, West Bengal and the adjoining areas of Odisha and Jharkhand. In West Bengal and the adjoining areas of Jharkhand, Odisha and Assam, the direction of squalls is mainly from the northwest, and they are called ***norwesters***. They are often very violent with squall speeds of 60 to 80 km per hour. Hailstones sometimes accompany showers and occasionally attain the size of a golf ball. They cause heavy damage to standing crops, trees, buildings, livestock and even lead to loss of human lives. However, they are, some times, useful for tea, jute and rice cultivation. In Assam, these storms are known as ***'Barodoli Chheerha***. The period of maximum occurrence of these storms is the month of *Vaisakh* (mid-March to mid-April) and hence, they are locally known as ***Kalabaisakhis***, the *black storms* or a *mass of dark clouds of Vaiasakha*. In the south the thunderstorms occur in Kerala and adjoining parts of Karnataka and Tamil Nadu, particularly during evenings and nights. The highest occurrences are in the extreme south-western tip with seasonal incidence of 38 days over Thiruvananthapuram. This frequency gradually decreases northwards and is only 14 days at Mangaluru.

Western Disturbances and Tropical Cyclones. The western disturbances which influence large parts of north-west and north India during the winter season still persist with their frequency and intensity gradually decreasing with advancement of this season. Approximately 4, 3 and 2 western disturbances visit north-west India in March, April and May respectively. These disturbances cause changes in weather conditions by bringing cloudiness and convective activity in this part of the country. They also cause snowfall in higher reaches of the Himalayas.

This is also the season of tropical cyclones originating in the Bay of Bengal and the Arabian Sea. A few cyclones are formed in the Bay of Bengal in the month of March but they do not affect the mainland of India. Their frequency rises steeply in April and the number of cyclones originating in May is more than double than those originating in April. About three-fourths of the tropical cyclones are born in the Bay of Bengal and the rest originate in the Arabian Sea. Most of the depressions in April originate to the south of 10°N latitude while those originating in May are born to the north of this latitude. About 60 per cent of the storms of this season initially move west or north-west but later they recurve northeast and strike Bangladesh and the Arakan Coast of Myanmar. About 30 per cent of the storms cross the Indian coast and a few dissipate over the sea itself. The whole of the east coast of India, the coastal areas of Bangladesh and Arakan Coast of Myanmar are liable to be hit by tropical storms in May. Many of them are quite severe and cause heavy damage to life and property. In the Arabian Sea, major storms are formed in May between 7° and 12° N latitudes. About 75 per cent originate to the west of 70° E longitude and move away from the Indian coast in a north-westerly direction and dissipate in the sea. About 25 per cent originate close to the Indian coast. They move towards the north-east and hit somewhere along the west coast of India.

(*c*) Precipitation. As mentioned earlier, this is a dry season. But this is not a totally rainless season although only one per cent of the annual rainfall of India is received during this season. In the northeastern parts of the country, dust storms bring some rainfall. Rajasthan, Gujarat and Madhya Pradesh receive less than 2.5 cm rainfall during this season. The precipitation in Kashmir is mainly in the form of snow or sleet in March and early April and is caused by western disturbances. Higher reaches more than 5,000 metre above sea level experience snowfall even in May. These disturbances move eastwards and cause snowfall or thunderstorms in central and eastern Himalayas till the end of April. The *norwesters* bring

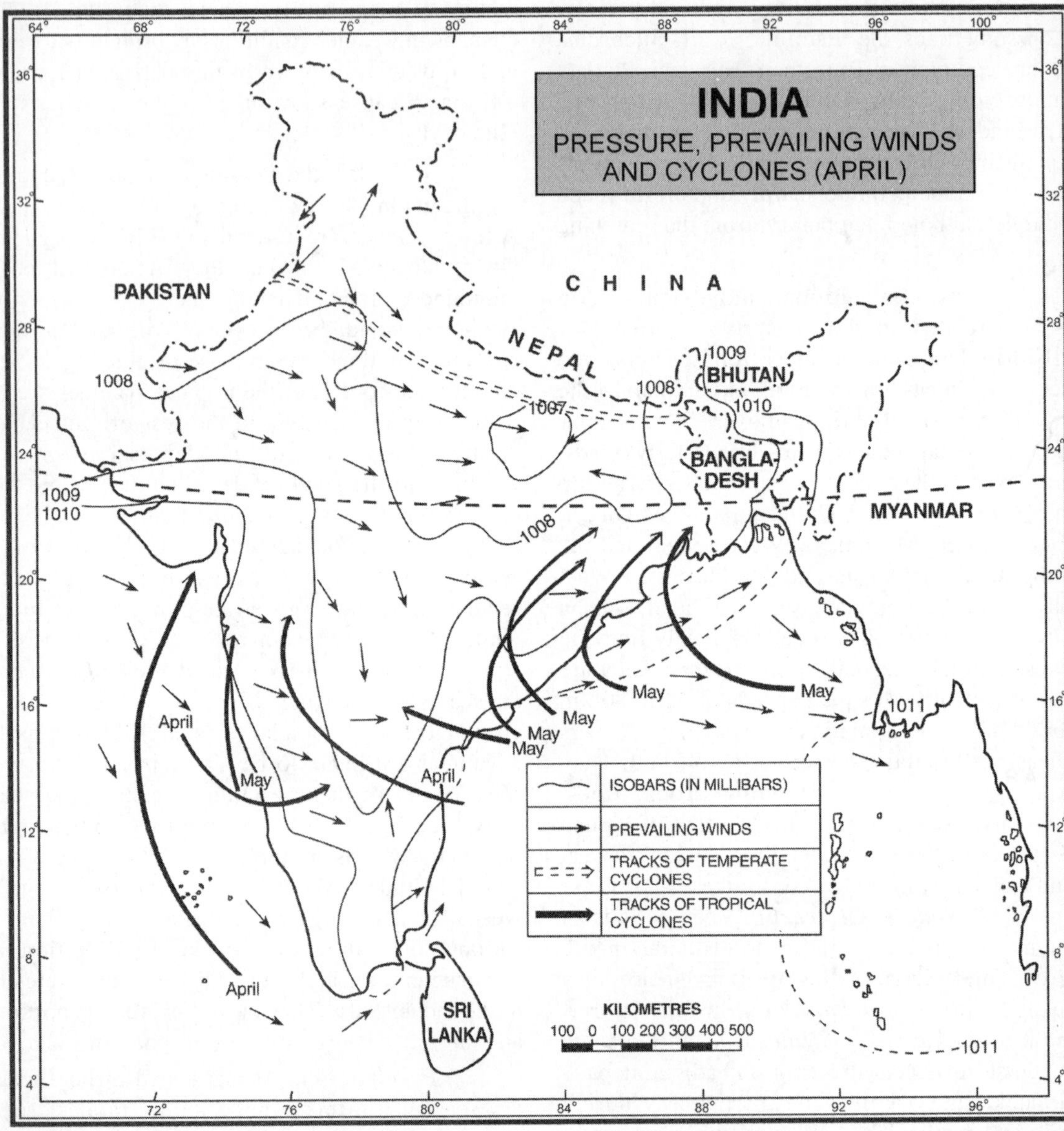

FIG. 5.15. Pressure, Prevailing Winds and Cyclones (April)

about 50 cm of rainfall in Assam and about 10 cm in West Bengal and Odisha. The intensity of rainfall is high. Often as much as 5 cm of rain is recorded in one hour but the rainfall is short lived. The rainfall brought by the norwesters is known as the ***spring storm showers***. In Assam the rainfall received in May is about two-thirds of the rainfall received in June. This small amount of rainfall is very useful for the cultivation of tea, jute and rice and is known as ***tea showers*** in Assam. Coastal areas of Kerala and Karnataka receive about 25 cm of rainfall from thunderstorms. These thunderstorms also cause about 10 cm of rainfall in the interior of south India. Such showers are called ***mango showers*** in Tamil Nadu and Andhra Pradesh because they are very beneficial to mango crop. In Karnataka they are called ***cherry blossoms*** due to their salutary effect on the coffee plantations. The distribution of rainfall in April is shown in **Fig. 5.16**.

FIG. 5.16. Distribution of rainfall (April)

3. The Rainy Season

The rainy season in India starts with the onset of the southwest monsoon in June and continues till the middle of September. This is also called the ***monsoon season***, the ***southwest monsoon season***, the ***wet season*** and the ***hot-wet season***. The weather conditions all over the country change with the onset of the monsoon winds. *High temperature, high humidity, extensive clouding and several spells of moderate to heavy rain with strong surface winds are the chief characteristics of this season.*

(*a*) Temperature. There is a significant fall in temperature with the beginning of rainy season. The June temperature in south India is 3° to 6°C lower than the May temperature. Similarly July temperature in northwest India is 2° to 3°C lower than the June temperature. But once the temperature falls from its dry summer level, it remains more or less uniform throughout the rainy season. However, the temperature rises again in September with the cessation of rains and secondary maximum temperature period is experienced all over the country. Also there is rise in temperature whenever there is break in the monsoons and rainfall does not occur for a number of days. Night temperatures are more uniform than the day temperatures. The diurnal range of temperature is small due to clouds and rains. It ranges between 4°C and 8°C when the monsoon is fully established.

The highest temperatures of over 32°C are experienced in the Thar desert of Rajasthan. At places especially west of the Aravali the temperature may be as high as 38° to 40°C. This is due to lack of clouds and the predominance of continental airmass. The remaining parts of northwest India also have temperatures above 30°C. The temperatures are quite low over the Western Ghats due to high elevation and also due to heavy rainfall, but the rainshadow area is comparatively warmer on account of low elevation and scanty rainfall. The coastal areas of Tamil Nadu and adjoining parts of Andhra Pradesh have temperatures above 30°C because they receive little rainfall during this season. (**Fig. 5.17**)

(*b*) Pressure and Winds. The temperatures in northwest India are still very high as a result of which low pressure conditions prevail there. The most conspicuous feature of the surface pressure distribution during this season is an elongated trough across the Ganga basin right upto the head of Bay of Bengal. This is called the ***monsoon trough***. There are frequent changes in its location and intensity depending upon the weather conditions. The atmospheric pressure in most parts of north India is less than 1,000 mb. It increases steadily southwards where it ranges between 1,008 mb and 1,010 mb. The isobar of 1,009 mb crosses parts of Kerala and Tamil Nadu besides Arabian Sea and the Bay of Bengal.

Under the influence of the above mentioned pressure distribution, winds blow in a southwest to northeast direction from Arabian sea and Bay of Bengal. They maintain this direction throughout peninsular India. But their direction undergoes a change in Indo-Gangetic plain where they move from east to west.

(*c*) Rainfall. India's three fourths of the total annual rainfall is received during this season. In some areas it is much more than this average. For example, the average rainfall over the plains of India in this season is about 92 cm, or about 87 per cent whereas during the remaining part of the year only 14 cm of rainfall occurs. This season has the maximum number of rainy days as a result of which it is called the ***'wet season'***.

Rainfall during this season all over the country is caused by southwest monsoons coming from the Indian Ocean. **Figure 5.19** shows *isolines* of normal dates of arrival of the monsoons in different parts of the country. It is clear from this map that the normal date of the arrival of the monsoon is 20th May in Andaman and Nicobar Islands. It is worth mentioning that the advance of the monsoon is much faster in the Bay of Bengal than in the Arabian Sea. This is evident from the pronounced curve shown by isolines of the monsoon onset in **Fig. 5.19**. The monsoon current advances to nearly 20°N latitude in the Bay of Bengal by the third week of May, while it is still south of Kerala at about 7°N latitude in the Arabian Sea. The normal date of onset of the southwest monsoon over Kerala *i.e.* the first place of entry in the mainland of India is 1st June. The monsoons advance with startling suddenness accompanied with a lot of thunder, lightning and heavy downpour. This sudden onset of rain is termed as ***monsoon burst***. Although the normal date of onset of the southwest monsoon on the southern tip of the peninsula is 1st June, the actual onset may be earlier or later than this date. On 60% occasions, the onset occurs between 29th May and 7th June. The earliest onset was on 11th May in 1918 and 1955, while the most delayed onset was on 18th June in 1972. Satellite imagery is used to identify the advance of the monsoon on a day to day basis.

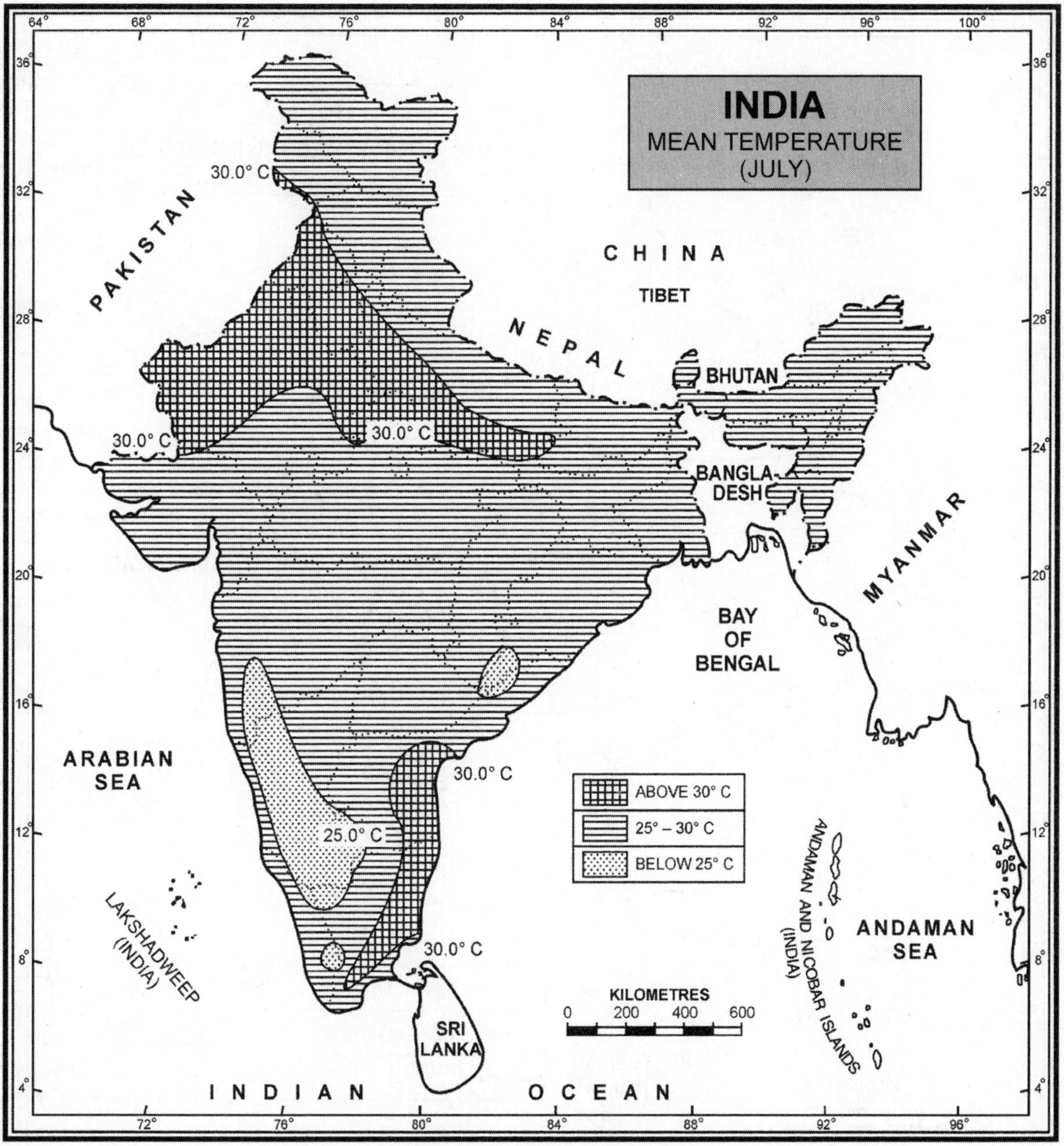

FIG. 5.17. Mean Temperature (July)

The progress of the monsoon winds beyond south Kerala is in the form of two branches *viz.* the Arabian Sea branch and the Bay of Bengal branch. The Arabian Sea branch gradually advances northwards. It reaches Mumbai by 10th June and spreads over Saurashtra-Kachchh and central parts of the country by 15th June. The progress of the Bay of Bengal branch is more spectacular. It spreads rather rapidly over most of Assam. The normal date of its arrival at Kolkata is 7th June. On reaching the foothills of the Himalayas the Bay branch is deflected westward by the Himalayan barrier and it advances up

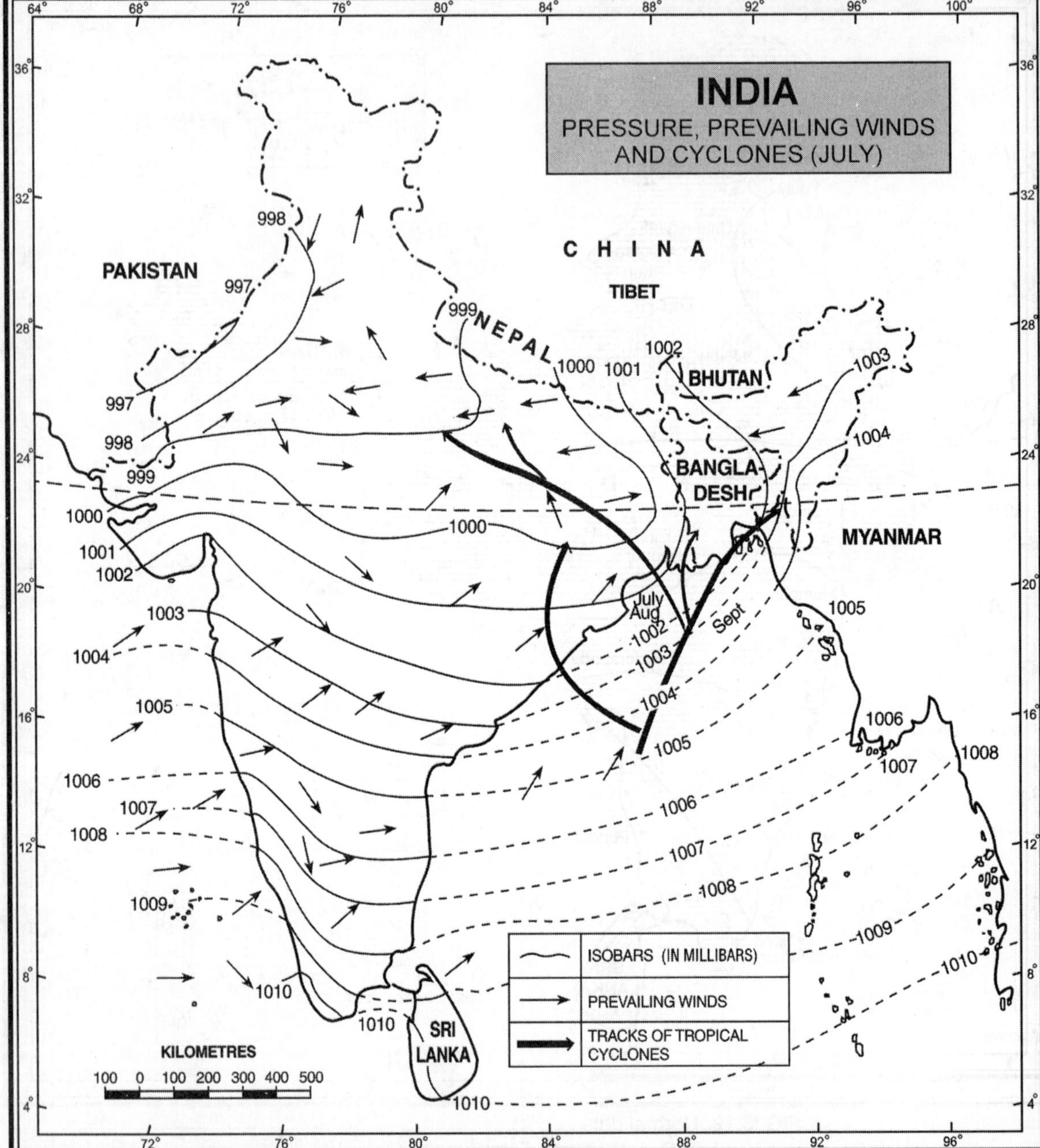

FIG. 5.18. Pressure, Prevailing Winds and Cyclones (July)

the Gangetic plain. The two branches meet roughly along the line running through Agra and Ferozepur and merge with each other to form a single current. The combined current gradually extends to west Uttar Pradesh, Haryana, Punjab, Rajasthan and finally to Himachal Pradesh and Kashmir. By the end of June the monsoon is usually established over most parts of the country. By mid-July, the monsoon

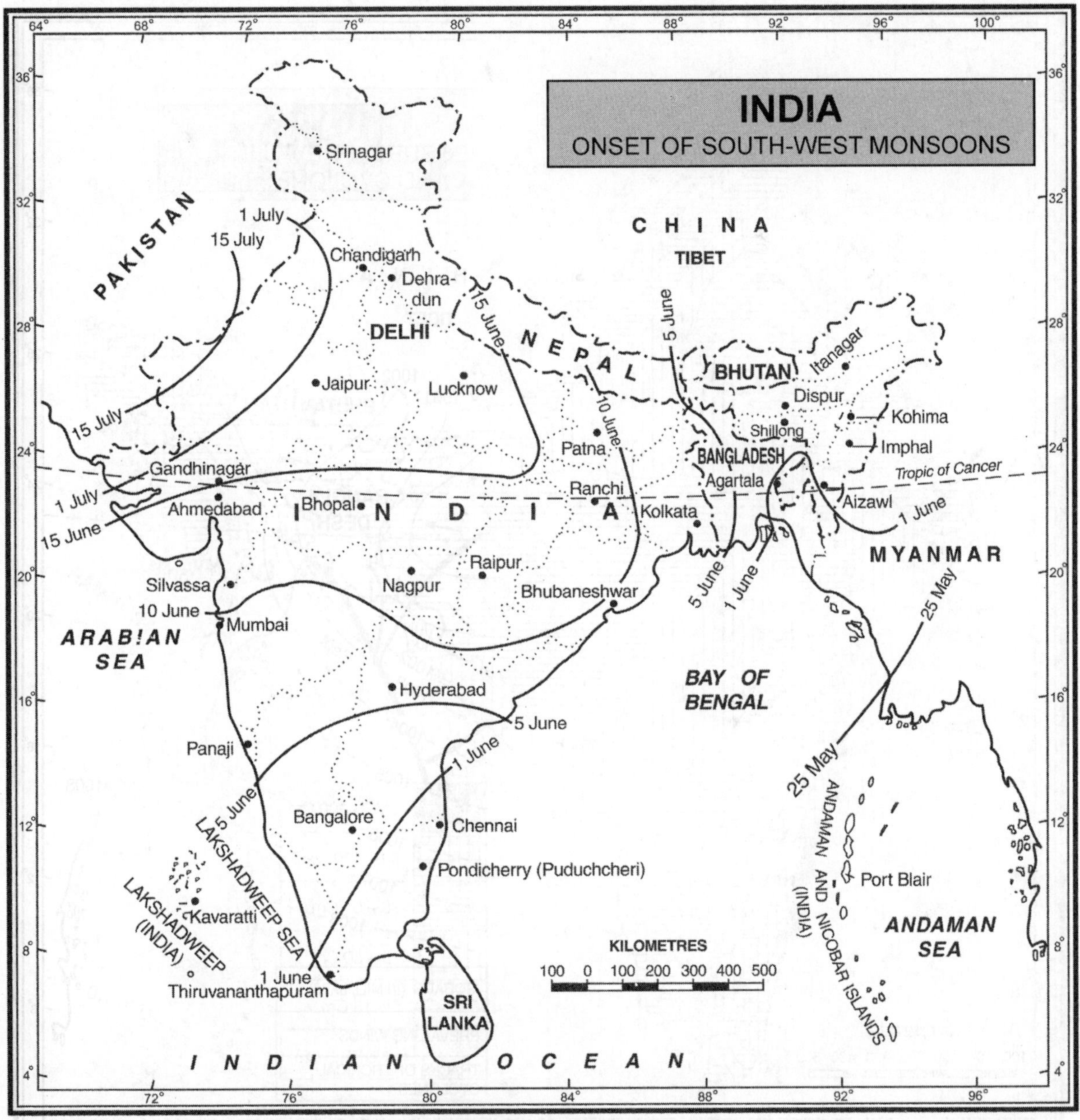

FIG. 5.19. Normal dates of onset of south-west Monsoons

extends into Kashmir and the remaining parts of the country, but only as a feeble current because, by this time, it has shed most of its moisture. It is often difficult to say whether the Arabian Sea branch or the deflected Bay of Bengal branch will be the first to arrive. For example, at a place like Delhi, the first showers are sometimes brought by the Bay of Bengal branch from the east but on a number of other occasions it is the Arabian Sea branch which brings the first monsoon rain from the south. It is interesting to note that the Arabian Sea branch of the monsoon is much more powerful than the Bay of Bengal branch for two reasons : (*i*) The Arabian Sea is larger than the Bay of Bengal, and (*ii*) the entire Arabian

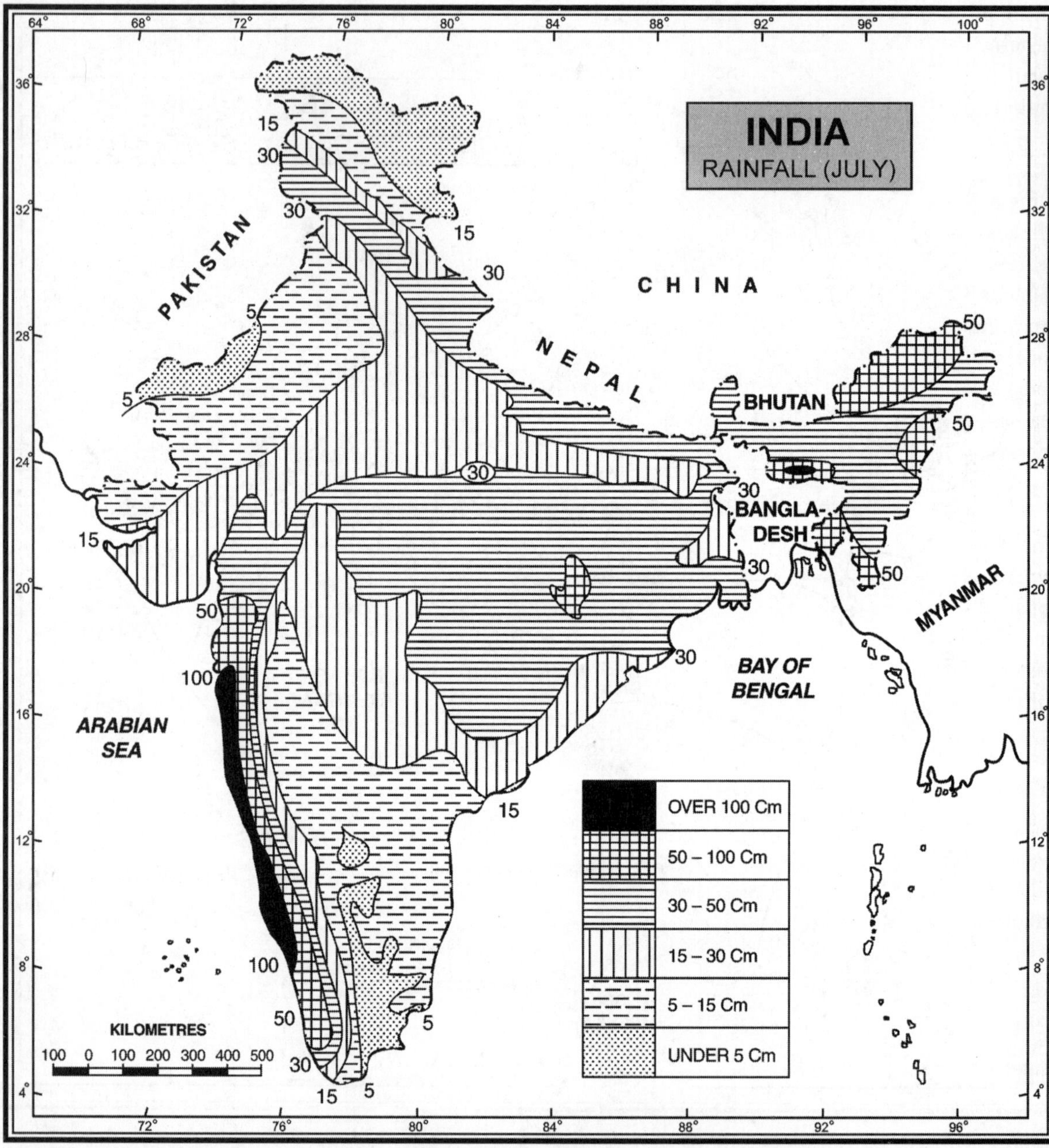

FIG. 5.20. Rainfall (July)

Sea current advances towards India, whereas only a part of the Bay of Bengal current enters India, the remainder proceeding to Myanmar, Thailand and Malaysia.

The Arabian Sea branch of the southwest monsoons is divided into three distinct streams on arriving in the mainland of India.

The first stream impinges on the west coast of India and gives extremely heavy rainfall of over 250 cm particularly between 10° to 20°N. latitudes where the imppinge is perpendicular to the direction of the Western Ghats. The Western Ghats rise abruptly like a wall from the western coastal plains more or less parallel to the coastline. This wall like mountain range forces the moisture laden southwest monsoons from the Arabian Sea to ascend the slope thereby giving heavy rainfall in the west coastal plain and still heavier rainfall on the western slopes of the Western Ghats, *i.e.*, the windward side of the mountain range. But these winds have to descend the slope after crossing the crest of the ghats. In the process their temperature rises and their humidity decreases. Therefore, they cause little rainfall and the area east of the Ghats is called the ***'leeward side'*** or the ***'rain shadow region'***. Thus while Mumbai on the west coast records about 190 cm, Khandala about 50 km east a bit below the crest gets 60 cm and Pune about 160 km away from Mumbai on the leeward side receives only 50 cm rainfall during the monsoon season. This phenomenon is observed almost all along the Western Ghats. The crest of the Western Ghats receives about 400 to 500 cm annual rainfall which is drastically reduced to about 30-50 cm within a distance of 80-100 km leeward from the crest. This speaks volumes of orographic control of the monsoon rainfall. There is a narrow belt of marked aridity on the immediate leeward side of the Western Ghats. But once it is passed, the air starts rising again and the amount of rainfall increases further east.

The second stream enters Narmada–Tapi troughs and reaches central India. It does not cause much rain near the coast due to the absence of any major orographic obstacle there. Nagpur receives about 60 cm rainfall from this stream.

The third stream moves in a north-easterly direction parallel to the Aravali Range. Since the orientation of the Aravali Range is parallel to the direction of the prevailing monsoon winds, it does not offer major obstacle in the way of the winds and these winds move further without causing much rainfall. Consequently a large part of Rajasthan is a desert area. However, some orographic effect is discernible here as the south-eastern edge of the Aravali Range comes in the way of the monsoon winds and receives reasonably good rainfall. Mt. Abu gets about 113 cm rainfall while the surrounding plains have only 60 to 80 cm rainfall.

The Bay of Bengal Branch of the southwest monsoon is divided into two distinct streams :

The first stream crosses the Ganga-Brahmaputra delta and reaches Meghalaya. It is here that the orographic effect on the monsoon winds and the consequent amount of rainfall is most pronounced. Cherrapunji, a small town (25° 15′ N, 91° 44′ E), located at an elevation of 1,313 m above mean sea level receives an annual rainfall of 1,102 cm, major portion of which occurs from June to August. Till recent times, this has been considered as the highest amount of rainfall for any station in the world. But the recently recorded observations have shown that Mawsynram (25° 18′ N, 91° 35′ E) located at 1,329 m above sea level just 16 km to the west of Cherrapunji records higher annual rainfall of 1,221 cm. Both the stations are located on the southern slopes of the Khasi hills at the northern end of a deep valley running from south to north. When the monsoon winds blow from the south, they are trapped within the funnel shaped valley and strike Cherrapunji and Mawsynram in a perpendicular direction and give copious rains. However, the heaviest rainfall occurs when the winds blow directly on the Khasi hills. Cherrapunji and Mawsynram may receive more rainfall in a day than the annual rainfall of many parts of the country. The highest records of rainfall in a day for these two stations are 103.6 cm and 99 cm respectively. The rainfall is well over 200 cm in most parts of the north eastern states. But the amount of rainfall decreases sharply on the leeward side of the Khasis. Guwahati, only 90 km from Cherrapunji gets only 161 cm of rainfall.

The second stream of the Bay of Bengal branch goes to the Himalayan foothills and after reaching there, it is deflected to the west by the size and orientation of the Himalaya and brings widespread rainfall to Ganga plain. The rainfall by this stream is characterised by a steady decline as we move from east to west up the plain. For example, Kolkata gets 119 cm, Patna 105 cm, Pryagraj 91 cm, Delhi 51 cm and Bikaner only 24 cm rainfall during the south-west monsoon period.

FIG. 5.21. Normal dates of withdrawal of South-west Monsoon

The eastern coastal belt, particularly in Tamil Nadu, remains relatively dry during the south-west monsoon period. This is because the Tamil Nadu coast lies in the rainshadow area of the Arabian Sea current and is parallel to the Bay of Bengal current.

Break in the Monsoon. During the rainy season, particularly, in July and August, there are certain periods when the monsoons become weak. The cloud formation decreases and rainfall practically ceases over the country outside the Himalayan belt and southeast peninsula. This is known as ***break in the monsoon***. The latest studies have revealed that the breaks are likely to occur more frequently during the second week of August. The normal duration of the break is about a week but on some occasions this could be longer. The longest breaks have been known to persist for two to three weeks, but such occasions are rare. The breaks are believed to be brought about by the collapse of the Tibetan High. This results in the northward shifting of the *monsoon trough*. The axis of the trough lies at the foothills of the Himalayas during the break period. Even when most parts of the country have to content without rainfall during the break period, heavy rainfall occurs over the sub-Himalayan regions and the southern slopes of the Himalayas. This leads to high flooding of the rivers having their catchment areas in the Himalayas. On an average one or two breaks do occur during the rainy season. Statistical studies of the monsoon show that in 85 out of 100 years there is a break in the monsoons.

Monsoon Depressions. A major part of the monsoon rainfall is generated by depressions originating in the Arabian Sea but more so in the Bay of Bengal. Some depressions develop over land also. About 3-4 depressions are formed per month from June to September. Almost all of them are sucked inward through the deltas of great rivers, the Ganga, the Mahanadi, the Godavari, the Krishna and the Cauvery and cause heavy rain in these areas.

In June the depressions in the Bay of Bengal originate between 16° N and 21° N and west of 92° E. The majority of them move towards the north-west. During July and August they originate north of 18° N and to the west of 90° E and move generally in a west north-westerly direction. In September the Bay storms originate to the north of 15° N and to the west of 90° E. Majority of the cyclones move along the monsoon trough and most of them merge with the seasonal low over north-west India.

In the Arabian Sea the formation of depressions in June is between 17°N and 20°N. They move either in north-west or in northerly direction and may affect Gujarat or Maharashtra. Storms during August and September are rare and are formed close to Maharashtra-Gujarat coast.

Most of the rainfall in central and northern parts of the country is caused by these depressions. The satellite imagery shows thick clouds associated with these depressions. Sometimes they give excessive rainfall, as much as 60 cm in a single day, resulting in heavy loss of life and property. The absence of depressions or a change in their tracks result in deficit or no rain.

Chief Characteristics of Monsoonal Rainfall

(*i*) Rainfall by the southwest monsoon is seasonal in character, the major part of which is received between June and September.

(*ii*) Monsoonal rainfall is largely orographic in its mode of occurrence and is governed by relief. The Himalayas and the Western Ghats are the main rainfall controlling relief features. The Himalayas obstruct the moisture laiden monsoon winds from the Indian Ocean and cause rainfall in the northeastern States and in the Indus-Ganga-Brahamputra plain. Again the windward side of Western Ghats receive more than 250 cm annual rainfall whereas most parts of the leeward side of the Western Ghats receive less than 60 cm annual rainfall.

(*iii*) The amount of rainfall decreases with increasing distance from the sea. For example Kolkata receives 119 cm during the southwest monsoon period. Patna 105 cm, Pryagraj 76 cm and Delhi 56 cm.

(*iv*) The monsoon rains occur in wet spells of a few days interspersed with rainless intervals known as ***'breaks'***.

(*v*) The rainfall by the southwest monsoon comes in the form of heavy downpour which results in large scale run off and soil erosion.

(*vi*) Indian monsoon rains are vital for agrarian economy of the country.

(*vii*) There are large scale spatial variations in the distribution of rainfall. The amount of annual rainfall varies from about 12 cm in western Rajasthan to over 250 cm in the west coastal plains.

(*viii*) Monsoons often fail to keep date. Sometimes the beginning of rains is delayed considerably over the whole of country or a part thereof.

(*ix*) Sometimes the monsoons withdraw before the scheduled time causing considerable damage to the crops.

RAIN CLOUDS ARE THINNING OUT IN WEST AND CENTRAL INDIA

A study of 50 years (1960-2010) by India Meteorological Department has revealed that low-lying clouds (usually upto 2000 m) that normally bring rain in the monsoon season have thinned out and reduced over the years. This means that there will be more number of hotter days, less number of rainy days and more days with higher maximum temperature during day and lower minimum temperature at night leading to higher diurnal range of temperature. The study indicates that most of the Indo-Gangetic plains, north-west India and western Himalayas showed increased low cloud cover by 4-8% per decade. These include Punjab, Haryana, Uttar Pradesh, Himachal Pradesh, Uttarakhand and Bihar. The rest of India, especially central India, south-east peninsula and west coast have recorded 4-6% decline per decade in the low cloud cover during the monsoon season.

Although no specific reason has been assigned to this unwanted trend, increase in particulate matter in the air is said to be the most probable reason.

4. The Cool Season

The cool season starts with the beginning of the withdrawal of southwest monsoon in the middle of September and continues upto November, after which the winter season starts. The monsoons withdraw from the extreme north-west end of the country in September, from the peninsula by October and from the extreme south-eastern tip by December. Due to retreat of the monsoon, this season is also called the ***season of retreating monsoon***. Sometimes it is referred to as the ***post monsoon season***.

A simple comparison of **figure 5.19** with **5.21** will show that the south-west monsoons reach north-west India last of all and withdraw from there first of all. Therefore, the duration of the south-west monsoon period is minimum there. In Punjab, for example, the south-west monsoons reach in the first week of July and withdraw from there in the second week of September, giving only two and a half months of monsoon activity. As against this, the south-west monsoons reach Coromandel coast in the first week of June and withdraw from there only in the middle of December. Thus the south-west monsoons remain active for more than six months over the Coromandel coast. Unlike the *sudden burst* of the advancing monsoons, the withdrawal is rather gradual and takes much longer time.

(*a*) Temperature. With retreat of the monsoons, the clouds disappear and the sky becomes clear. The day temperature rises a little at the initial stage but soon starts falling rather steeply. In October the day temperature over larger parts of Rajasthan is 37°C. It decreases to 33°C in the peninsula and to 30°C in north-east India. Nights are cool with minimum temperature varying from 20° to 25°C. In the Himalayan region the minimum temperatures reach freezing point. The average temperature in most parts of the country varies from 25° to 30°C. **Figure 5.22** shows the distribution of temperature in the month of October. The diurnal range of temperature increases due to lack of cloud cover.

(*b*) Pressure and Winds. As the monsoons retreat, the elongated trough of the low pressure across the Indus-Ganga plains weakens and gradually shifts southward. By October it reaches the Bay of Bengal and moves further southwards as the season advances. The axis of low pressure roughly runs in an east-west direction along 13°N latitude. The surface pressure in most parts of the country varies from 1,010 to 1,012 mb. Consequently the pressure gradient is low.

This is the season when south-west monsoons yield place to the north-east monsoons. Unlike the south-west monsoons, the onset of the north-east monsoon is not clearly defined. In fact, on many occasions, the meteorologists fail to draw a clear demarcation between the withdrawal of the summer monsoon and the onset of winter monsoon over peninsular India. However, the direction of winds over large parts of the country is influenced by the local pressure conditions. (**Fig. 5.23**).

Cyclones. This is the season of the most severe and devastating tropical cyclones originating in the Indian seas especially in the Bay of Bengal. The highest frequency of the cylcones is in the month of October and the first half of November. More cyclones are born in October and then in November and more cylcones originate in the Bay of Bengal than in the Arabian Sea. In October, the Cyclones of the Bay of Bengal originate between 8°N and 14°N. Initially they move in a west or northwesterly direction , but many of them later recurve and move towards the north-east; About 55 per cent of the Bay storms cross or affect the Indian coast. The areas most vulnerable to these storms include the coastal

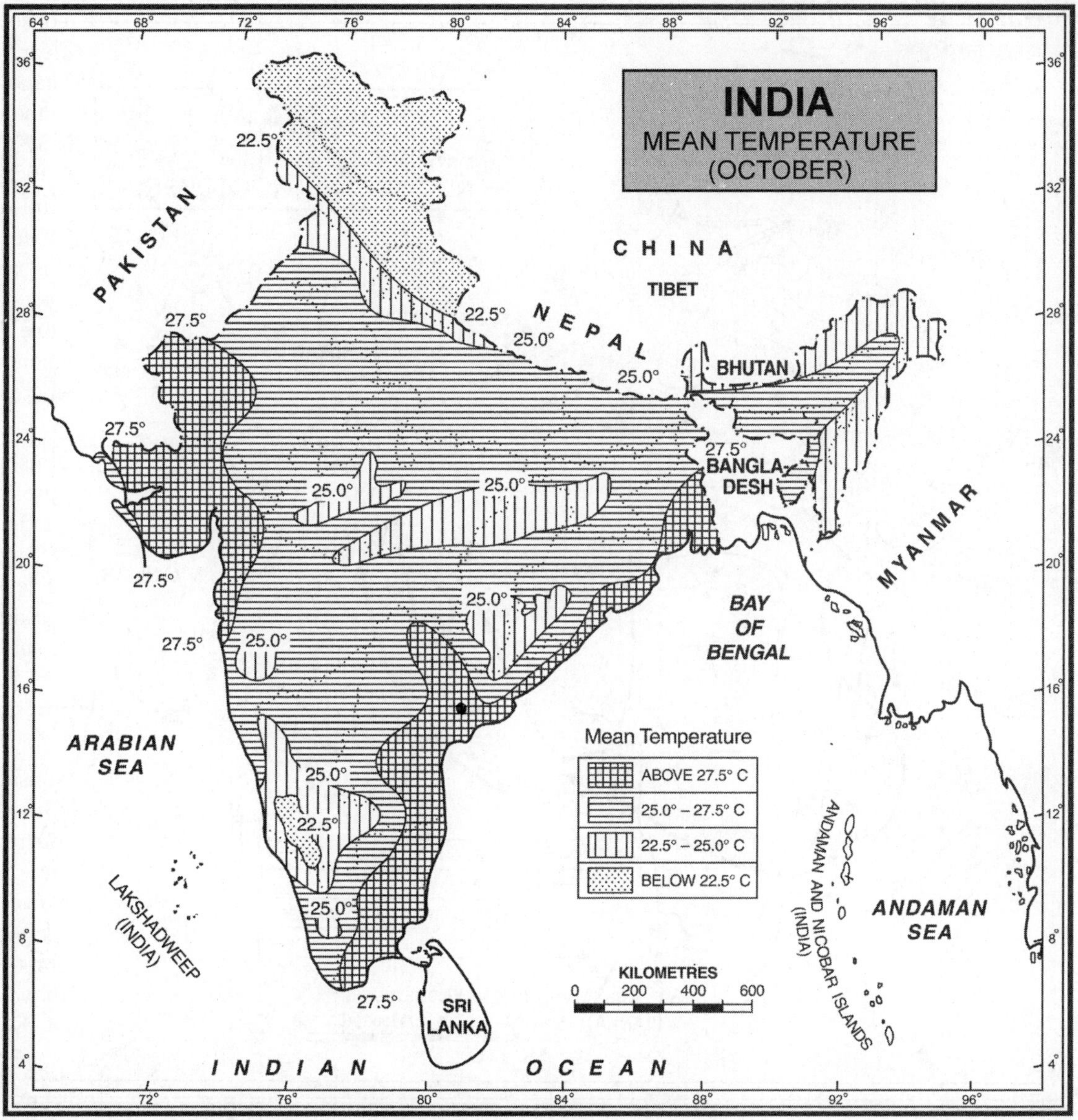

FIG. 5.22. Mean Temperature (October)

belts of Tamil Nadu, Andhra Pradesh, Odisha and West Bengal. Many of the cyclones which strike the eastern coast of India, south of 15°N latitude cross the southern Peninsula and enter Arabian Sea. During this process, they may weaken, but on re-entry over the Arabian sea they reintensify into cyclonic storms. The storms of Arabian sea originate between 12°N and 17°N latitudes in October and between 8° N and 13° N latitudes in November. Generally they move away from the coast in a north-westerly direction. But about 25% of them later recurve towards northeast and strike the Maharashtra or Gujarat coast.

The cyclones originating in the Bay of Bengal and the Arabian Sea are accompanied by strong surface winds, dense clouds, thundering and heavy downpours. A rainfall of 50 cm in a day is not uncommon. The heavy rainfall may continue for 2-3 days.

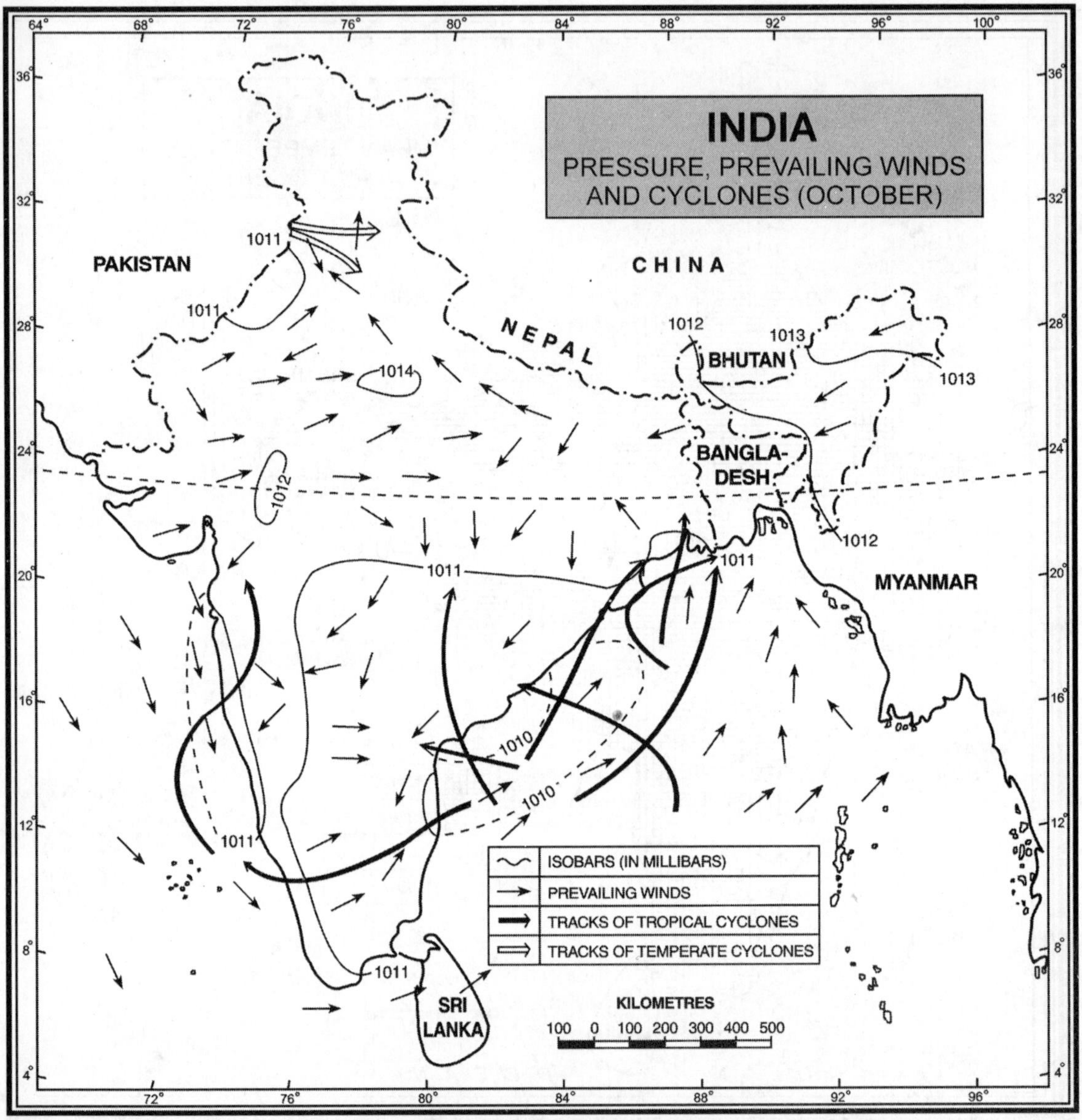

FIG. 5.23. Pressure, Winds and Cyclones (October)

In north-west India the western disturbances produce clouding and light rainfall in the otherwise fine weather. The average incidence of western disturbances is 1 to 3, the frequency showing increasing trend with the advance of the season. The precipitation is in the form of snow in higher reaches of Jammu and Kashmir, Ladakh, Himachal Pradesh and in Kumaon Hills.

(c) Precipitation. The humidity and cloud cover are much reduced with the retreat of the south-west monsoons and most parts of the country remain without much rainfall. However, October-November is the main rainy season in Tamil Nadu and adjoining areas of Andhra Pradesh to the south of the Krishna delta as well as a secondary rainy period for Kerala. In fact Kerala has bi-modal pattern of rainfall with main rains in the south-west monsoon season and subsidiary rains in this season. The retreating monsoons absorb moisture while passing over the Bay of Bengal and cause this rainfall. According to a

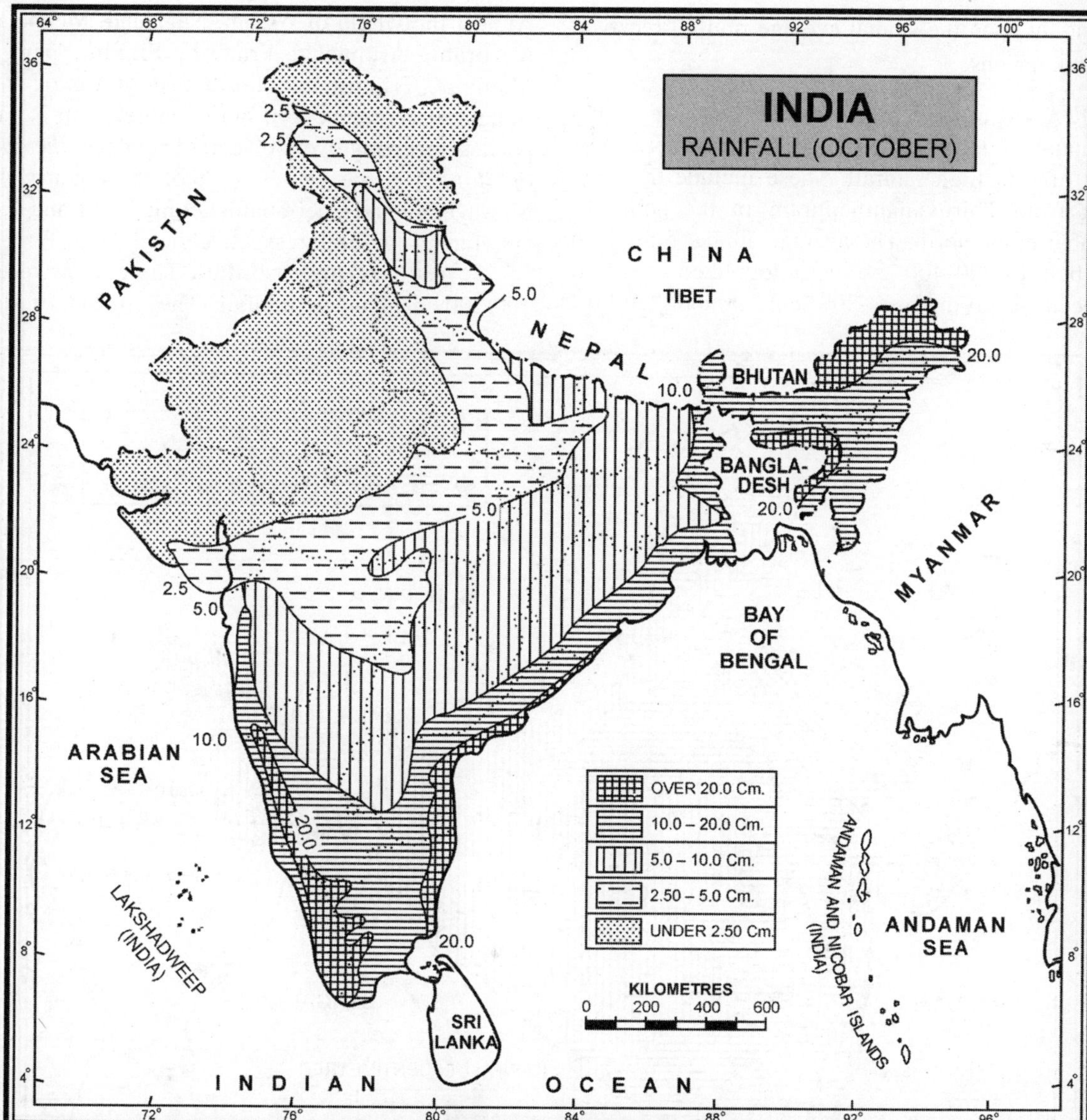

FIG. 5.24. India : Rainfall (October)

study conducted by **Krishna Rao** and **Jagannathan** the average rainfall in Tamil Nadu in October and November is 38.25 cm which is about 38.73 per cent of the annual rainfall received here. However, the coastal areas receive about 50% of their annual rainfall during this season.

Annual Rainfall

Figure 5.25 shows the distribution of annual rainfall in India. The average annual rainfall of India is about 118 cm which is the highest for a country of such vast dimensions anywhere in the world. But there are large spatial and temporal variations. Large parts of Meghalaya receive about 1,000 cm annual rainfall while on the other end of the scale, some districts in south-west Rajasthan hardly receive annual rainfall of 15 cm. In the extreme south, Kanniyakumari and Nellai Kattabomman districts get less than 30 cm during the south-west monsoon, while heavy rains of 200 cm or more, lash the nearby Kerala Coast. **L.D. Stamp** has divided India into following regions depending upon the annual average rainfall received by these regions.

1. Areas of very high rainfall. Areas receiving an annual rainfall of 200 cm and above are termed as areas of very high rainfall. These include the west coast from Thiruvananthapuram in the south to Mumbai in the north. The average annual rainfall in this belt is 200-400 cm with localized areas in between receiving 400-800 cm annual rainfall. Almost the whole of Assam, Nagaland, Meghalaya, Mizoram, Arunachal Pradesh, Sikkim, parts of Manipur, Tripura and north-eastern tip of West Bengal also receive 200 cm or more, with isolated pockets receiving over 400 cm. Meghalaya (*the abode of clouds*) is the wettest part of the country with Mawsynram and Cherrapunji getting 1,221 and 1,102 cm of annual rainfall respectively.

2. Areas of high rainfall. These areas receive 100-200 cm annual rainfall. They include eastern

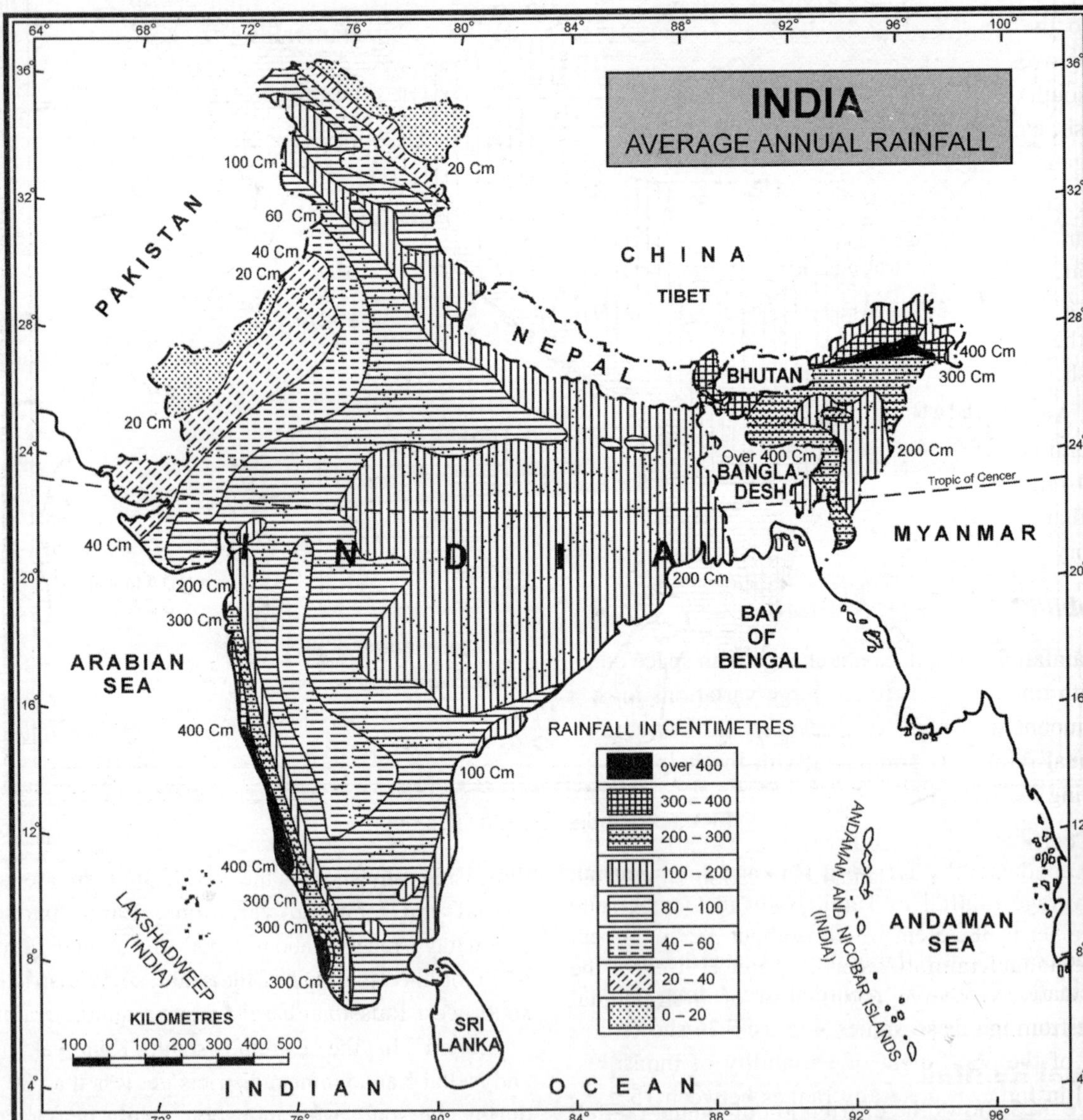

FIG. 5.25. Average Annual Rainfall

slopes of the Western Ghats, major part of the northern plain, Odisha, Madhya Pradesh, Telangana, Andhra Pradesh and Tamil Nadu. The *isohyet* (the line joining places of equal rainfall) of 100 cm rainfall runs southwards from Gujarat coast roughly parallel to the crest of the Western Ghats upto Kanniyakumari. The rainfall to the west of this line is above 100 cm. To the north, the 100 cm isohyet trends eastward passing over Jammu and Kashmir, Himachal Pradesh and northern Uttar Pradesh.

To the east of Pryagraj, it bends to the west passing over Bundelkhand in Uttar Pradesh. Turning west-south-westwards, it runs over western Madhya Pradesh, eastern Maharashtra, northern Andhra and Telangana.

3. Areas of low rainfall. These areas receive 50-100 cm annual rainfall and include large parts of Gujarat, Maharashtra, western Madhya Pradesh, Andhra Pradesh, Telangana, Karnataka, eastern Rajasthan, Punjab, Haryana and parts of Uttar Pradesh.

4. Areas of very low rainfall. These are desert and semi-desert areas receiving less than 50 cm of annual rainfall. They include large areas of western Rajasthan, Kachchh and most of Ladakh Union Territory.

Variability of Annual Rainfall

Rainfall in India does not change with space only but with time also. There are large variations in the total amount of rainfall from year to year. Variability of annual rainfall is computed with the help of the following formula.

$$\text{Coefficient of Variability (CV)} = \frac{\text{Standard Deviation}}{\text{Mean}} \times 100$$

The coefficient of variability indicates the amount of fluctuation in rainfall recorded over a long period of time from the mean values. **Figure 5.26** shows the values of the coefficient of variability of monsoon rainfall in India. It generally ranges between 15 and 35 per cent for the country as a whole. The coefficient of variation in large parts of Rajasthan, Gujarat, and parts of Haryana and Panjab is well over 40 per cent. In extreme western parts of Rajasthan and Gujarat, it exceeds 60% and reaches 80% in the extreme west of Rajasthan. Incidentally these are the areas which receive minimum rainfall. The other areas of low rainfall and high coefficient of variability are Leh and Ladakh in the north and Rayalaseema in the south.

The areas of low coefficient of variability include the west coast, the sub-Himalayan belt including Sikkim, Arunachal Pradesh, Assam and West Bengal and the north-eastern hilly regions of Nagaland, Manipur and Mizoram. The coefficient of variability in these areas ranges between 15 and 20 per cent. These areas receive heavy annual rainfall of about 200 cm. In general the variability increases from west to the interior of the plateau as well as from Odisha and West Bengal towards the north and north-west.

A major conclusion that can be drawn from the above description is that the variability is highest where the amount of rainfall is the lowest. In other words, *the rainfall variability is inverse to total amount of rainfall and reliability of rainfall is inversely proportional to the coefficient of variability*. As such the spatial distribution of rainfall variability has a great bearing on Indian agricultural productivity. Since the areas of high variability are also the areas of low rainfall, they suffer from chronic deficiency of water, frequent droughts and crop failures. The areas with marginal rainfall are the worst sufferers.

However, the study of the coefficient of variation helps us in delineating those parts of the country where long range prediction will be useful. This prediction is not much required in areas of small coefficient of variation. For example, we can safely say that the actual rainfall will be within 10 to 15 per cent of the normal rainfall in Assam and West Bengal. But in Rajasthan and adjoining areas of north-west India the rainfall may vary by as much as thirty to fifty per cent or even more from the normal. Similarly the variability of the annual rainfall is high in large part of the Indian Peninsula. Evidently these are the areas where advance information about seasonal rainfall plays a crucial role.

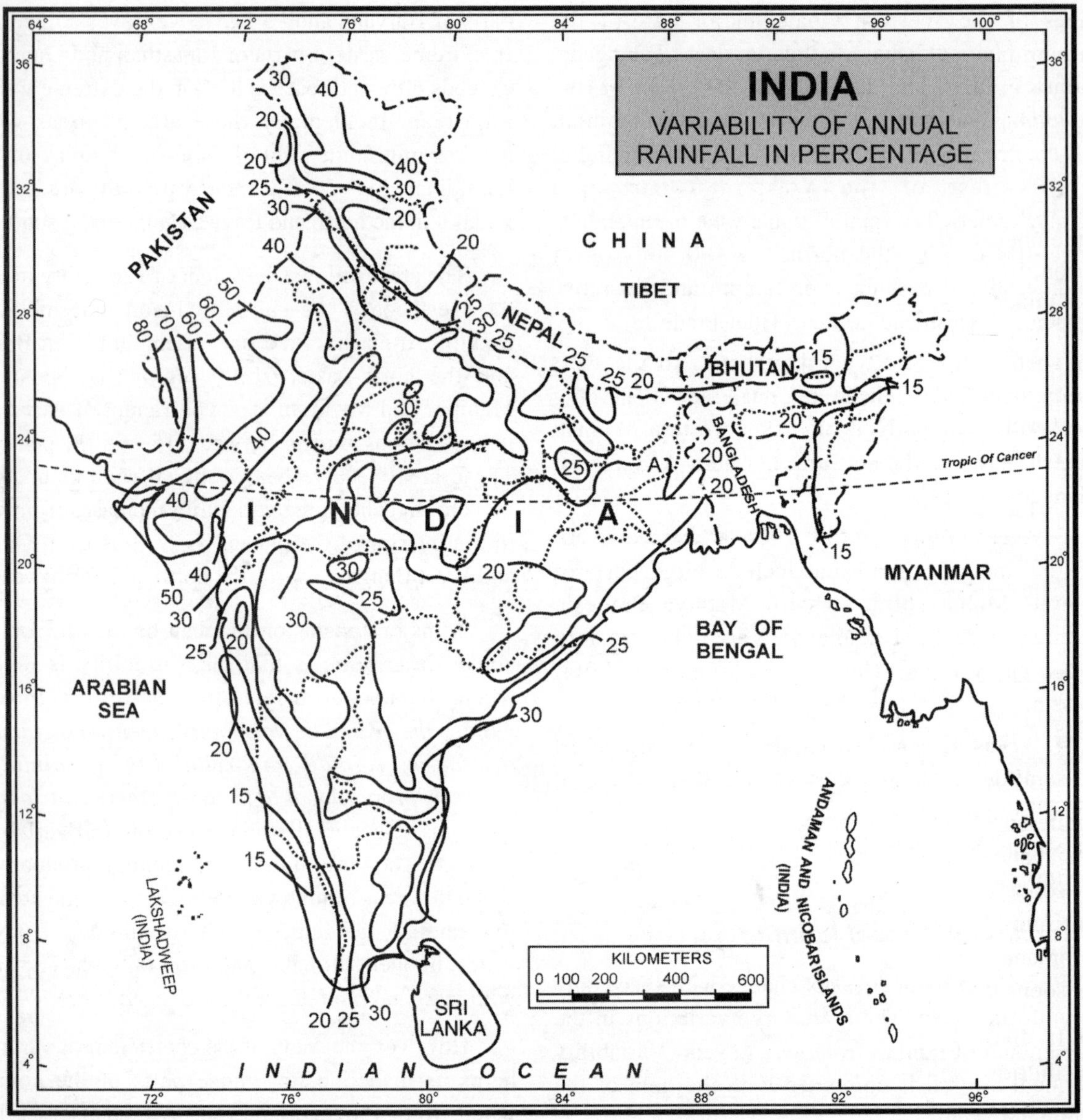

FIG. 5.26. India : Variability of Annual Rainfall

PECULIARITIES AND SIGNIFICANCE OF INDIAN CLIMATE

Indian climate is primarily dominated by south-west monsoons and is peculiar in many ways. The extremes of temperature, rainfall and humidity are well known. The rainfall from the monsoon winds is variable and quite undependable. The monsoon may advance much before its due date or may be considerably delayed. Further, the amount of rainfall may be more than the normal or there may be deficient rains. Some parts of the country may be facing the fury of floods due to heavy rains while other parts may be reeling under drought conditions due to scanty rainfall. The variability of rainfall in time and space plays havoc with agriculture which shatters the very foundation of economy in a predominantly agricultural country like India. It is often said that *Indian budget is a gamble in the monsoon*. In fact *monsoon is the pivot upon which the whole economic life of India swings*. Nowhere else in the world, so many people over so vast a land are so intimately wedded to the monsoon regime as they do in India. Another peculiar feature of the Indian rainfall is that it is concentrated in a few months of the year. Of the country's total rainfall, about 75 per cent is received in the monsoon months from June to September, 13 per cent comes in the post monsoon season, 10 per cent in the pre-monsoon season and the remaining 2 per cent in the winter season. At the same time, it is worth mentioning that one part or the other gets rainfall in each month of the year *i.e.* in no month of the year the whole of India is completely dry. In January and February, north-west India gets rainfall from the western disturbances. In March thunderstorms start influencing Assam and West Bengal and give occasional pours till the arrival of the monsoons in June. Rainfall by south-west monsoons continues till the withdrawal of monsoons. Coromandel coast receives rainfall by the north-east monsoons in the winter season.

TABLE 5.3. Distribution of Rainfall

Percentage of the total land area	Amount of annual rainfall (cm)
11	above 200
21	125 to 200
37	75 to 125
24	35 to 75
7	Below 35

It has already been mentioned that the distribution of rainfall in India is very uneven. According to the calculations made by census of India in 1951 only 11 per cent area of India gets over 200 cms. of annual rainfall while about one third of the total land area of the country has to content with a mere 75 cm annually. **Table 5.3** gives areal distribution of rainfall in India.

Indian rainfall is basically torrential in nature. Much of the rainfall is received in 3-4 months of the rainy season. Even in this season the actual rainy days are 40-45 only. The heaviest downpours occur in association with cyclones which originate in the Bay of Bengal and the Arabian Sea. A rainfall of 50-60 cm in a rainy day is not uncommon. The highest record, as already mentioned, is 103.6 cm in 24 hours at Cherrapunji. This place gets rainfall of 1,102 cm in 180 rainy days. Sri Ganganagar receives 12 cm of rain in 10-12 rainy days. Hence the statement, *it pours, it never rains in India,* is true whether it be Meghalaya or Rajasthan. The sudden heavy downpour results in devastating floods and excessive soil erosion.

Another very important aspect of Indian rainfall is that it is largely controlled by orography. The effects of the Himalayas and the Western Ghats on the amount and distribution of rainfall and the inability of the Aravalis to cause rainfall have already been discussed. The whole of India would have been a vast desert but for the size and lie of the Himalayas and the Western Ghats. Therefore, we can easily say that the *rainfall over the country is primarily orographic*.

CLIMATIC REGIONS OF INDIA

Although India has tropical monsoon climate as a whole, there are large regional variations in important climatic elements such as rainfall and temperature. This is quite natural for a vast country like India having subcontinental dimensions. Variations in rainfall are much more marked than those of temperature. Hence, most geographers have given more importance to rainfall than to temperature. The first attempt to divide India into climatic regions was made by Blanford towards the close of the 19th century. Among the subsequent attempts made to divide India into climatic regions, those made by W.G. Kendrew, L.D. Stamp, Koppen, Thornthwaite, G.T. Trewartha and Johnson are worth mentioning. Among the Indian geographers, attempts made by Subrahmanyam (1955), Bharucha and Shanbhag (1957) and R.L. Singh (1971) deserve appreciation.

Stamp's Classification of Climatic Regions

Dr. L. Dudley Stamp's classification of Indian climate is very much akin to that suggested by **W.G. Kendrew**. This classification is empirical, arbitrary and subjective, but uses quantitative limits to the regions in easily understood units like temperature and rainfall. Stamp's scheme has been presented here with slight modifications here and there. Stamp used

18°C isotherm of mean monthly temperature for January to divide the country into two broad climatic regions, *viz.*, temperate or continental zone in the north and tropical zone in the south. A look at **Fig. 5.27** will show that this line runs roughly across the root of the peninsula, more or less along or parallel to the Tropic of Cancer. The two major climatic regions are further divided into eleven regions depending upon the amount of rainfall and temperature.

A. The Temperate or Continental India has been divided into following five regions :

1. The Himalayan region (heavy rainfall)
2. The north-western region (moderate rainfall)
3. The arid low land
4. The region of moderate rainfall
5. The transitional zone

B. The tropical India has been divided into following six regions :

6. Region of very heavy rainfall
7. Region of heavy rainfall
8. Region of moderate rainfall
9. The Konkan Coast
10. The Malabar Coast
11. Tamil Nadu

A. Climatic regions of temperate or continental India

1. The Himalayan Region. This region embraces the entire Himalayan mountain area which includes from west to east, Union Territories of Jammu and Kashmir and Ladakh, Himachal Pradesh, large parts of Uttarakhand, the north hill region of West Bengal, Sikkim and Arunachal Pradesh. The winter and summer temperatures are 4°-7°C and 13°-18°C respectively. The higher reaches are perpetually under snow and ice. The average annual rainfall exceeds 200 cm in the east but it is much less in the west. Shimla in the west and Darjeeling in the east are its representative cities.

2. The North-western Region. It includes the northern parts of Punjab and southern parts of Jammu and Kashmir. The winter and summer temperatures are 16°C and 24°C respectively. Amritsar is its representative city.

3. The Arid Lowland. This is a vast dry area which includes the Thar desert of Rajasthan, south western part of Haryana and Kachchh of Gujarat. The average temperature in winter varies from 16° to 24°C which may shoot up to 48°C in summer. Jaipur is its representative city. The average annual rainfall does not exceed 40 cm.

4. The region of moderate rainfall. Parts of Punjab, Haryana, western Uttar Pradesh, Union Territory of Delhi, north-west Plateau area of Madhya Pradesh and eastern Rajasthan are areas of average rainfall with an annual rainfall of 40 to 80 cm. Temperatures in January and July are 15°-18°C and 33°-35°C respectively. Most of the rainfall occurs in summer. Delhi represents this region very well.

5. The Transitional Zone. Eastern Uttar Pradesh and Bihar comprise the transitional zone between areas of average rainfall on the west and areas of heavy rainfall in the east. The average annual rainfall in this zone is 100-150 cm. The temperatures in January and July vary between 15°-19°C and 30°-35°C. Patna is the representative city of this zone.

B. Climatic Regions of Tropical India

6. Region of very heavy rainfall. These areas receive more than 200 cm of annual rain and include large parts of Meghalaya, Assam, Tripura, Mizoram and Nagaland. Temperatures remain around 18°C in January and rise to 32°-35°C in July. Cherrapunji and Mawsynram in Meghalaya receive 1,102 cm and 1,221 cm of annual rainfall respectively.

7. Region of heavy rainfall. Chhattisgarh, Jharkhand, Gangetic West Bengal, Odisha and coastal Andhara Pradesh receive 100-200 cm annual rainfall and are termed as areas of heavy rainfall. The rainfall is primarily brought by the monsoon winds coming from the Bay of Bengal. The cyclones originating in the Bay of Bengal also bring some rainfall. The amount of rainfall decreases as we move from east to west and from north to south. The January and July temperatures range from 18°-24°C to 29°-35°C respectively. Kolkata is the representative city of this region.

8. Region of moderate rainfall. It includes mostly those areas between Western and Eastern Ghats which receive annual rainfall of 50-100 cm. Rainfall is comparatively low because this region lies in the rain shadow area of the Western Ghats. The

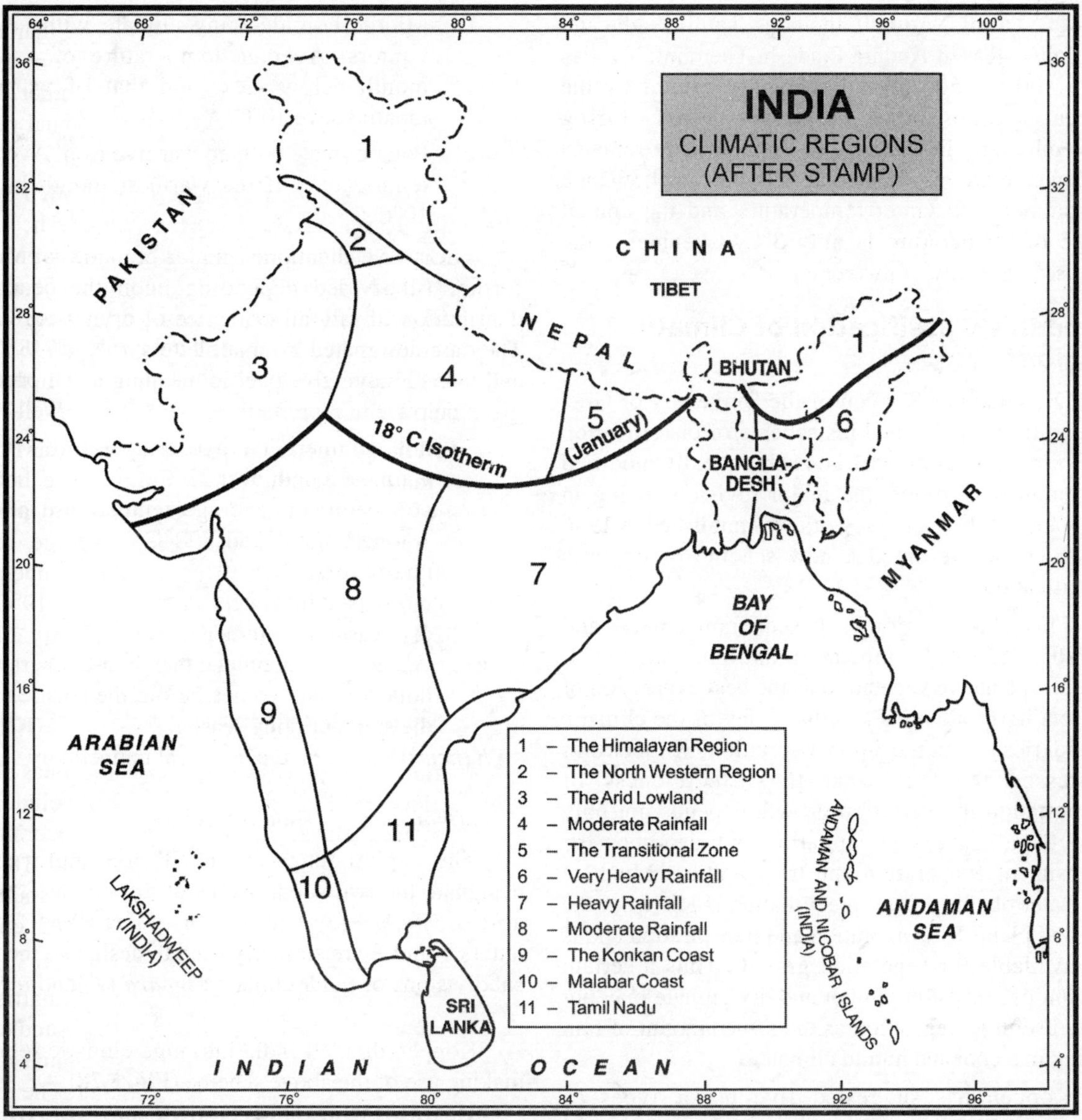

FIG. 5.27. Climatic Regions of India (After Stamp)

average temperature in winter is 18°-24°C which rises to 32°C in summer. This region is represented by Hyderabad in Telangana.

9. The Konkan Coast. Extending from Mumbai in the north to Goa and coastal Karnataka in the south, the Konkan Coast receives over 200 cm annual rainfall brought by the Arabian Sea branch of the south-west monsoons. The temperature remains fairly high and varies from 24°-27°C. Thus the annual range of temperature is very low; to the tune of 3°C only. Mumbai is the representative city of this region.

10. The Malabar Coast. It extends from Konkan coast to Kanniyakumari and receives heavy annual rainfall of over 250 cm. The rainfall is mainly brought by the south-west monsoon winds coming from the Arabian Sea and continues for about six months in a year. The temperature remains in the vicinity of 27°C and the annual range of temperature is only 3°C. This region is represented by Thiruvananthapuram.

11. Tamil Nadu. It includes Tamil Nadu and adjoining areas of Andhra Pradesh. The rainfall varies from 100 to 150 cm and is mainly caused by the retreating monsoons from north-east during November and December. The temperature remains somewhere around 24°C. There is not much change in summer and winter temperature and the annual range of temperature is only 3°C. Chennai is the representative city of this region.

Koppen's Classification of Climatic Regions

Dr. Wladimir Koppen of the University of Graz (Austria) first published his scheme of classification of world climates in 1901 and subsequently modified it a number of times, the major revisions being in 1918 and 1931. His latest work was published in 1936 in which he presented a new scheme of climatic classification.

This classification is based upon annual and monthly means of temperature and precipitation. It accepts the native vegetation as the best expression of the totality of a climate, so that many of the climatic boundaries are based upon vegetation. Koppen has expressed the view that the effectiveness of precipitation in vegetation growth depends not only upon the amount of precipitation, but also upon the intensity of evaporation and transpiration. Much of the water obtained from precipitation is lost from the soil and plants by evaporation and transpiration and is not available for vegetation growth. Thus a certain amount of rain falling in hot and dry climate may not be as useful to vegetation as the same amount of rain falling in a cool and humid climate.

Koppen has suggested five major types of climate which correspond with five principal vegetation groups. Each climatic type is represented by a capital letter explained as under :

A : Tropical rainy climate with no cool season. Temperature of the coolest month above 18°C.

B : Dry climate in which there is an excess of evaporation over precipitation.

C : Middle-latitude rainy climate with mild winters. Average temperature of coldest month below 18°C but above –3°C. Average temperature of warmest month over 10°C.

D : Middle-latitude rainy climate with severe winters. Average temperature of coldest month below –3°C and that of warmest month above 10°C.

E : Polar climate with no warm season. Average temperature of the warmest month below 10°C.

The above mentioned major climatic types are further subdivided depending upon the seasonal distribution of rainfall or degree of dryness or cold. They are designated by small letters *a, c, f, h, m, g, s* and *w* each having a specific meaning as per details given below :

a : hot summer, average temperature of the warmest month over 22°C

c : cool summer, average temperature of the warmest month under 22°C

f : no dry season

w : dry season in winter

s : dry season in summer

g : Ganges type of annual march of temperature; hottest month comes before the solstice and the summer rainy season.

h (heiss) : average annual temperature under 18°C

m (monsoon) : short dry season.

The capital letters *S* and *W* are employed to designate the two subdivisions of dry climate : semi arid or *Steppe (S)* and arid or *desert (W)*. Capital letters *T* and *F* are similarly used to designate the two subdivisions of polar climate : *tundra* (*T*) and *icecap* (*F*).

Koppen divided India into nine climatic regions making use of the above scheme (**Fig. 5.28**).

1. Amw (Monsoon type with short dry winter season). This climate is found in the western coastal region, south of Mumbai. This area receives over 300 cm of annual rainfall in summer from the south-west monsoons.

2. As (Monsoon type with dry season in high sun period). This is the region in which rainfall occurs in winter and summer is dry. Coromandel coast experiences this type of climate. Coastal Tamil Nadu and adjoining areas of Andhra Pradesh are included in it. The amount of rainfall, mostly in winter is 75-100 cm and is received from the retreating monsoons.

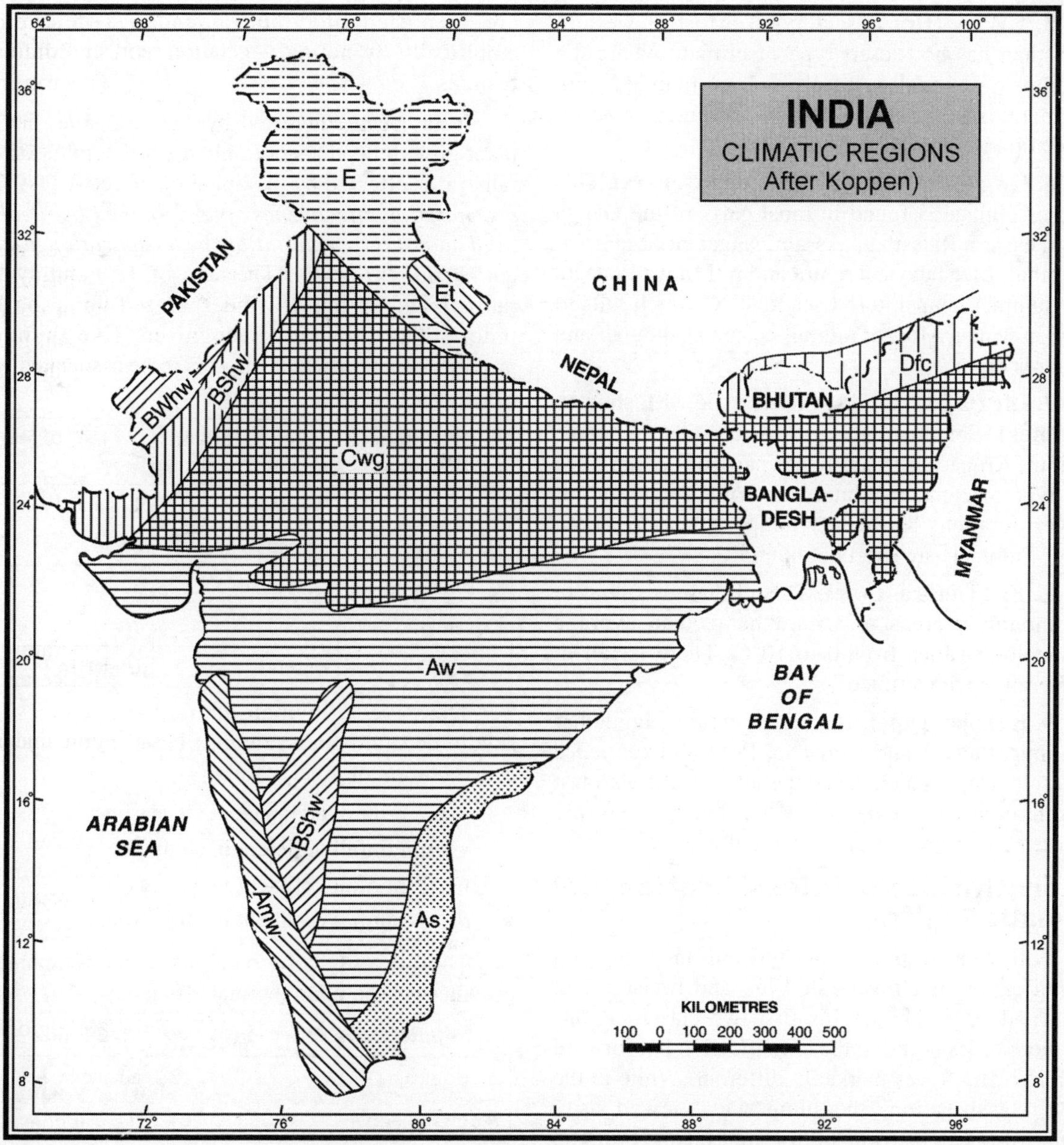

FIG. 5.28. Koppen's Climatic Regions of India

3. Aw (Tropical Savanah type). This climate is found in most parts of the peninsular plateau barring Coromandel and Malabar coastal strips. The northern boundary of this climatic region roughly coincides with the Tropic of Cancer. The average annual rainfall is about 75 cm which is received in summer season from the south west monsoons. Winter season remains dry.

4. BShw (Semi-arid Steppe type). Some rainshadow areas of Western Ghats, large part of Rajasthan and contiguous areas of Haryana and Gujarat have this type of climate. Rainfall varies from 12 to 25 cm and most of it occurs in summer. Winter is completely dry. Some arid steppe vegetation is found here.

5. BWhw (Hot desert type). Most of western Rajasthan has hot desert type of climate where the amount of annual rainfall is less than 12 cm. Temperatures are very high in summer. Natural vegetation is almost absent.

6. Cwg (Monsoon type with dry winters). This type of climate is found in most parts of the Ganga Plain, eastern Rajasthan, Assam, Nagaland, Manipur, Mizoram, Meghalaya, Tripura and in Malwa Plateau. The summer temperature rises to 40°C which falls to 27°C in winter. Most of rainfall occurs in summer and winter is dry.

7. Dfc (Cold, Humid winters type with shorter summer). Some of the north-eastern states such as Sikkim, Arunachal Pradesh and parts of Assam have this type of climate. Winters are cold, humid and of longer duration. The winter temperatures are about 10°C. Summers are short but humid.

8. Et (Tundra Type). This climate is found in the mountain areas of Uttarakhand. The average temperature varies from 0 to 10°C. There is fall in temperature with altitude.

9. E (Polar Type). The higher areas of Jammu & Kashmir, Ladakh and Himachal Pradesh experience polar climate in which the temperature of the warmest month varies from 0° to 10°C. These areas are covered with snow for most part of the year.

Thornthwaite's Classification of Climatic Regions

Following Koppen, Thornthwaite presented his classification of climates in 1931 and revised it in 1933 and 1948. Though his first two classifications are more or less similar, the classification put forward by him in 1948 was markedly different. While in the 1931 classification, the plant was viewed as a meteorological instrument for measuring climatic character, in the 1948 classification, vegetation is regarded as a physical mechanism by means of which water is transported from soil to the atmosphere; *it is the machinery of evaporation as the cloud is the machinery of precipitation*. The combined loss through evaporation from the soil surface and transpiration from plants is called ***evapotranspiration***. His classification is based on the principle that '*the plant is in the nature of a meteorological instrument which is capable of measuring all the integrated climatic elements*'. Under this system climatic types are identified and their boundaries determined empirically by noting vegetation, soil and drainage features.

It has been suggested by Thornthwaite that the plant growth depends not only upon precipitation but rather it is affected by precipitation effectiveness. *The precipitation effectiveness is determined by dividing total monthly precipitation by evaporation and is called the P/E ratio*. The sum of 12 monthly P/E ratios is called the P/E index. Based upon the P/E index, Thornthwaite recognised five humidly provinces each of which appears to be associated with a characteristic vegetation.

Humidity Province	Characteristic Vegetation	P/E Index
A, wet	Rainforest	128 and above
B, humid	Forest	64-127
C, subhumid	Grassland	32-63
D, semiarid	Steppe	16-31
E, arid	Desert	Under 16

The five principal humidity provinces are subdivided into four subtypes based upon seasonal concentration of precipitation.

r = rainfall adequate in all seasons

s = rainfall deficient in summer

w = rainfall deficient in winter

d = rainfall deficient in all seasons.

Thornthwaite also recognised six temperature provinces based upon thermal efficiency :

Temperature Province	T/E Index
A', tropical	128 and above
B', mesothermal	64-127
C', microthermal	32-63
D', taiga	16-31
E', tundra	1-15
F', frost	0

A total of 120 combinations of three elements, precipitation effectiveness, seasonal concentration of rainfall, and thermal efficiency are possible but Thornthwaite recognised only 32 climatic types on the world map of climatic classifications prepared by him. Of these only 12 climatic types are found in India.

1. AA′r is tropical wet climate in which the P/E as well as T/E index is above 128. Rainfall occurs in all the seasons and dense rain forests grow. This climate is found along the west coast and in Tripura and Mizoram.

2. BA′w is tropical humid climate in which thick forests grow. T/E ratio is above 128 but P/E ratio is 64-127. Rainfall is deficient in winter. This climate is found on the eastern slopes of the Western Ghats and in West Bengal.

3. BB′w is temperate humid climate which is mainly found in parts of Assam, Meghalaya Mizoram and Nagaland. P/E index and T/E index vary from 64 to 127. The rainfall is deficient in winter. Such a climate is useful for the growth of vegetation.

4. CA′w covers lerge parts of the peninsular India. The T/E index is 128 and above but the P/E index falls to 32-63. Being a comparatively dry climate, it is suitable for grasslands only.

5. CA′w′ is similar to *C A′w* with the only difference that the rainfall occurs in winter. Most of Tamil Nadu and the neighbouring parts of Andhra Pradesh enjoy this climate.

6. CB′w is temperate subhumid climate which covers most of the northern plain of India. The T/E index is 64 to 127 and the P/E index is still lower ranging between 32 and 63. This climate is suitable for grasslands.

7. DA′w is a tropical semi-arid climate which is mainly found in the northern Gujarat and southern Rajasthan. T/E index is 128 or above while P/E index is 16 to 31. This climate is suitable for steppe type of vegetation.

8. DB′d is temperate semi arid climate in which rainfall is deficient in all the seasons. T/E index is 64-127 while P/E index is 16-31. This climate is found in Jammu and Kashmir and is suitable for steppe vegetation.

9. DB′w is similar to DB′d climate with the only difference that the rainfall is deficient in winter. Parts of Punjab, Haryana and Rajasthan as well as some parts of rainshadow area in the peninsular India experience this climate.

10. D′ is a taiga type climate in which T/E index is 16 to 31. It is found in Jammu and Kashmir, Himachal Pradesh, Punjab, and Uttarakhand in the west and Arunachal Pradesh, Sikkim and parts of upper Assam in the east.

11. E′ is a cold tundra type of climate which is found in higher reaches of the Himalayas.

12. EA′d is extremely hot and dry climate in which T/E index is over 128 and P/E index remains below 16. Rainfall is deficient throughout the year. This type of climate is found in western part of Rajasthan.

Trewartha's Classification of Climatic Regions

Prof. Glenn T. Trewartha of the University of Wisconsin in the U.S.A., modified and simplified Koppen's classification and presented his own classification in his book entitled 'An Introduction to Climate'. He has recognised six major climates of the world. They are :

A. Tropical humid climates
B. Dry climates
C. Humid mesothermal climates
D. Humid microthermal climates
E. Boreal climates
F. Polar climates

In addition, he recognised H. undifferentiated climate of highlands.

Out of the above climatic types four types *viz*. *A*, *B*, *C* and *H* are found in India. These four types have further been divided into seven sub-types. A brief description of each climatic type is given below :

A is tropical humid climate in which temperature does not fall below 18°C. It has two subdivisions.

(i) Am is tropical monsoon type with average annual temperature of 27°C and annual rainfall of over 250 cm. Western Coastal plain and Tripura have this climate.

(ii) Aw is tropical wet and dry (savanna) type. The temperature here also remains at 27°C but the annual rainfall is much less, about 100 cm. The entire rainfall occurs in summer season. A major part of peninsular plateau and north-eastern state of Mizoram experience this climate.

FIG. 5.29. India : Climatic Regions after Thornthwaite classification of climates

B is a dry climate in which the annual rainfall is always less than 100 cm. This climate is divided into following three sub types.

(i) ***Bs*** is semi arid or steppe type which is found in the rain shadow of the Western Ghats. The average temperature is above 27°C but the rainfall is below 100 cm. Rainfall occurs mainly in summer and supports grasslands.

(ii) ***Bsh*** is tropical and subtropical desert climate in which the average annual temperature is above 27°C and the rainfall declines to 50-100 cm. Large parts of Gujarat, Rajasthan and south-west Haryana experience this type of climate.

(iii) ***Bwh*** is middle latitude desert climate which is found in west Rajasthan and Kachchh region of Gujarat. Here the temperatures are sufficiently high but the annual rainfall further declines to a low of 20 cm. This climate supports only thorny bushes.

C is mesothermal or subtropical climate in which winter is dry and cold.

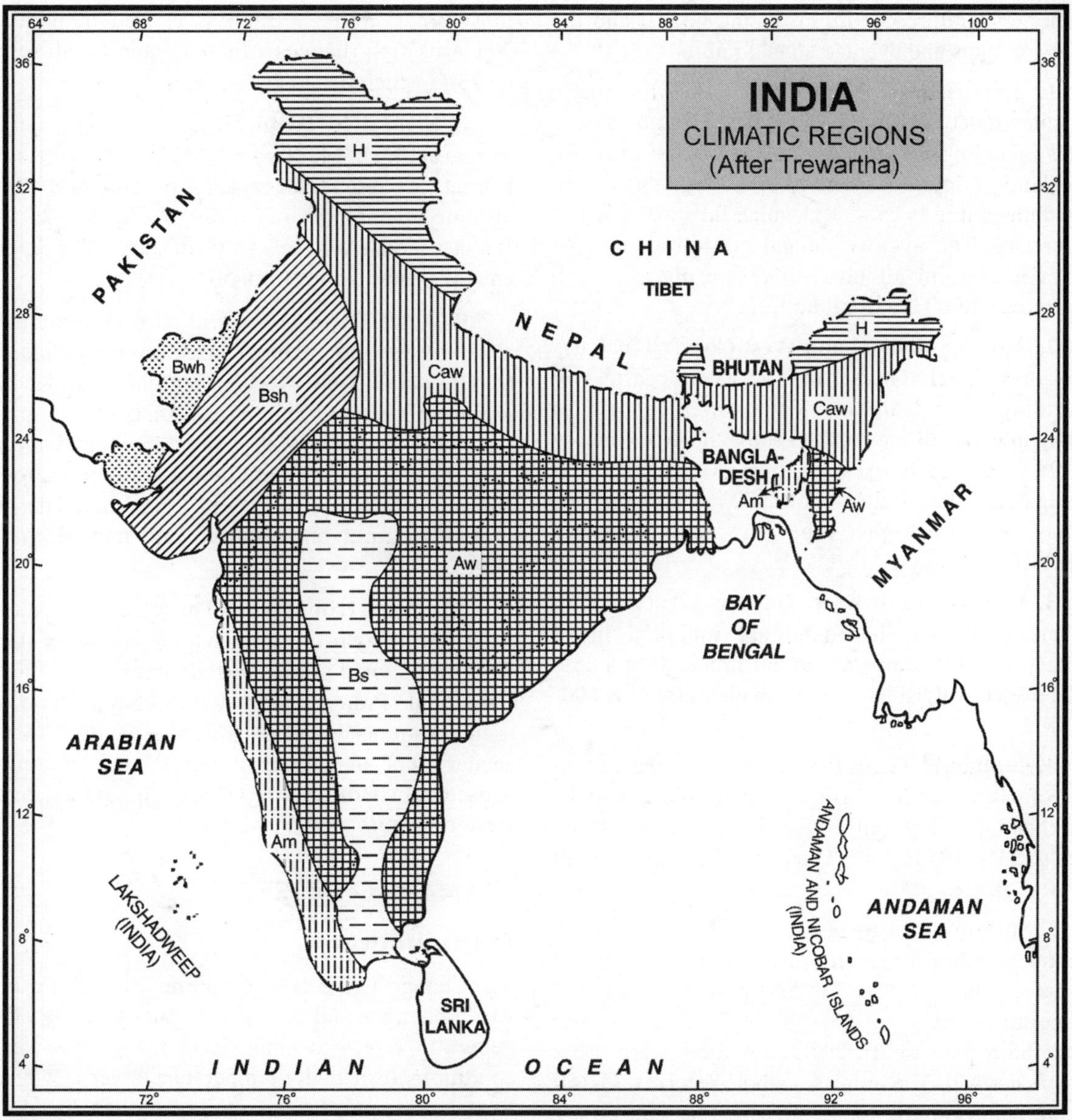

FIG. 5.30. Climatic Regions of India (after Trewartha)

Caw. Most parts of the northern plain of India have this type of climate where winter temperature falls below 18°C. Most of the rainfall occurs in summer and the amount of rainfall decreases from east to west. Western disturbances give light winter rainfall in the western part.

H climate indicates undifferentiated highlands where the temperatures are invariably low. Higher areas have snowfall in winter. Rainfall is caused by the monsoon winds in summer and by western disturbances in winter. Most of the Himalayan region including Union Territories of Jammu and Kashmir and Ladakh, Himachal Pradesh, Uttarakhand, Sikkim and Arunachal Pradesh have this climate.

R.L. Singh's Classification of Climatic Regions

Following Kendrew and Stamp, Dr. R.L. Singh presented his climatic divisions of India in 1971. He divided the country into 10 climatic divisions based on the temperature conditions of the hottest and the coldest months and average annual rainfall.

1. Per Humid North-East. As its name indicates, it covers most parts of the north-eastern states including Sikkim, Assam, Arunachal Pradesh, Nagaland, Tripura, Mizoram, and Meghalaya. The July temperature is 25°-33°C which falls to 11°-24°C in January. The average annual rainfall at most of places is 200 cm although there are places which record over 1,000 cm of rainfall.

2. Humid Sahyadri and West Coast. It includes Sahyadri (Western Ghats) and its western coastal belt extending from Narmada Valley in the north to Kanniyakumari in the south. Temperature in January is 19°-28°C which rises to 26°-32°C in July. The average annual rainfall is about 200 cm but at places it may be much higher especially on the western slopes of the Western Ghats.

3. Humid South-East. Odisha, West Bengal, Chhattisgarh and Jharkhand are included in it. January and July temperatures are 12°-27°C and 26°-34°C respectively. The average annual rainfall is 100-200 cm.

4. Subhumid Transition. It embraces the eastern part of Uttar Pradesh, Bihar and northern part of Jharkhand. January temperature is 9° to 24°C which rises to 24°-41°C in July. The average annual rainfall is 100-200 cm.

5. Subhumid Littorals. Eastern Tamil Nadu and coastal Andhra Pradesh have subhumid littoral climate. May is the hottest month when the temperature rises to 28°-38°C. In January the temperature falls to 20°-29°C. Summers are dry but winters are wet. The area as a whole receives 75-150 cm annual rainfall, most of which is caused by retreating monsoons in November-December.

6. Subhumid Continental. This climate is primarily found in the Ganga plain where January and July temperatures are 7°-23°C and 26°-41°C respectively. The average annual rainfall varies from 75 to 150 cm.

7. Semi arid and Subtropical. This climate prevails in the Satluj-Yamuna water divide which includes Punjab, Haryana, eastern Rajasthan and the Union Territories of Delhi and Chandigarh. The average rainfall is 25 to 100 cm, most of which falls in summer. Some rainfall is caused in winter by western disturbances. January temperatures are 6°-23°C which rise to 26°-41°C in May.

8. Semi arid Tropical. Large parts of Gujarat, Maharashtra, Madhya Pradesh, Chhattisgarh, Karnataka and Telangana have semi arid tropical climate. Temperature varies from 13°-29°C in January and from 26°-42°C in July. The average annual rainfall varies from 50 to 100 cm.

9. Arid. The areas of arid climate comprise the Thar desert and includes western Rajasthan, part of Haryana, south-western Punjab and Kachchh region of Gujarat. This is extremely dry climate in which the annual rainfall is only 25 cm and at places it is as little as 10 cm. January temperatures are 5°-22°C which rise to 20°-40°C in June. The range of temperature, both diurnal and annual, is very large.

10. West Himalayan. This climate is found in west Himalayan region which includes Union Territories of Jammu and Kashmir and Ladakh, Himachal Pradesh and Uttarakhand. The July temperature is 5°-30°C which dips to 0°-4°C in January. The annual rainfall is 150 cm. The rainfall is caused by south-west monsoons in summer and by western disturbances in winter.

CLIMATE CHANGE

Introduction

Change is the law of nature. Everything in the world changes and so does climate. Climate change generally refers to changes in the global weather conditions (especially temperature) over a long-term time scale. It can be caused either by natural processes, such as change in heat-radiation received by the Earth from the Sun or by human activities—in particular manmade greenhouse gases (GHG) emissions that build up in the atmosphere and alter its chemical composition. The study of climate change has become imperative in the present day world because it affects mankind as well as fauna and flora. The growth and decline of the past human civilizations as well as other species of plants and animals and also the future planning for sustainable life largely depend on climate and climate change.

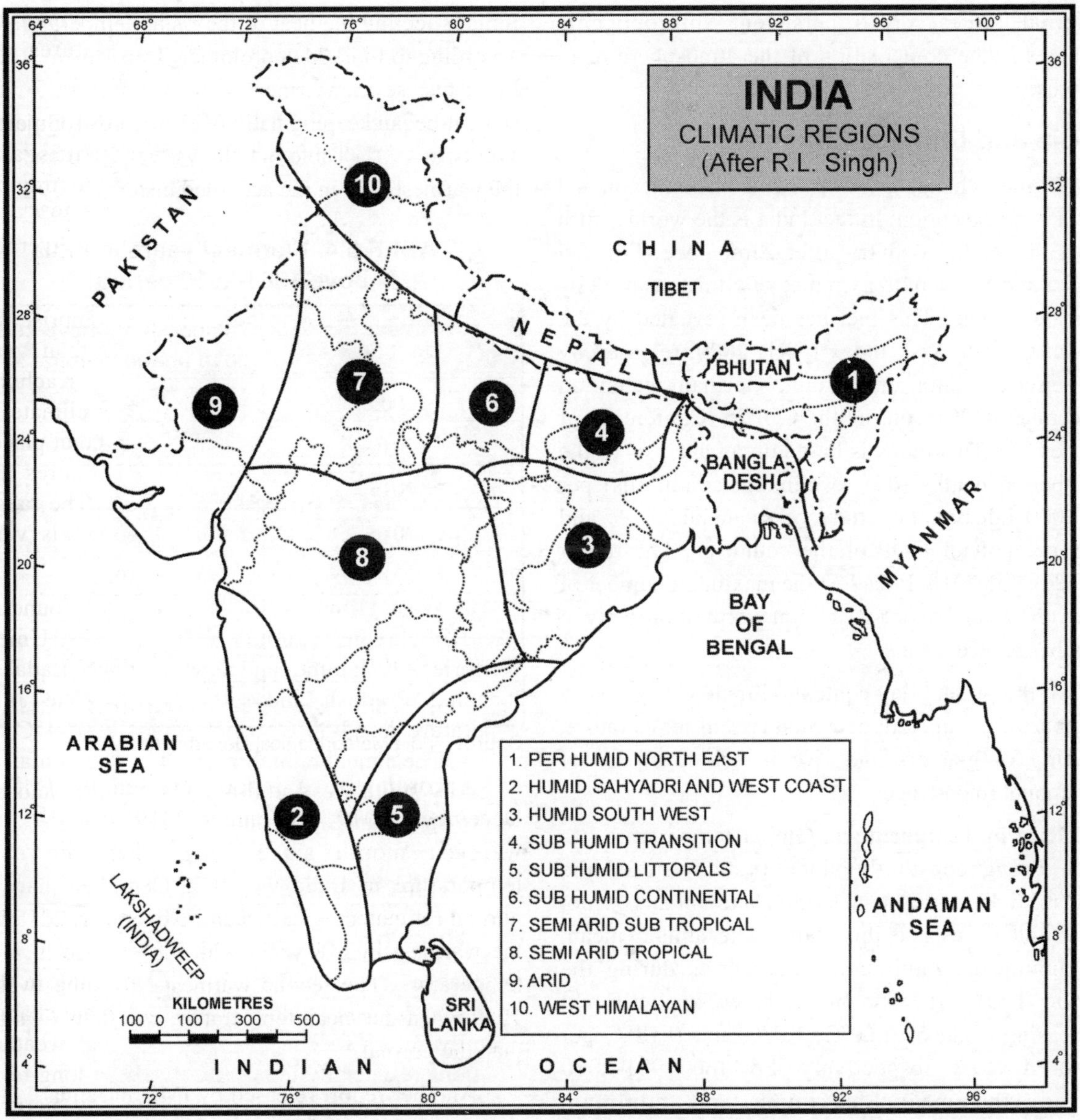

FIG. 5.31. India's Climatic Regions after R.L. Singh

Definitions

The United Nations Framework Convention on Climate Change (UNFCCC) defines 'climate change' as a change of climate which is attributed directly or indirectly to human activity that alters the composition of the global atmosphere and which is in addition to natural climate variability observed over comparable time periods.

(OR)

The Inter-governmental Panel on Climate Change (IPCC) defines climate change as a change in the state of the climate that can be identified (*e.g.*, by using statistical tests) by changes in the mean and/or the variability of its properties, and that persists for an extended period, typically decades or longer. Climate change may be due to natural internal processes or external forces, or to persistent anthropogenic changes in the composition of the atmosphere or in land use.

India and Climate Change

Climate change is a world wide phenomenon and India is no exception. In fact India is the world's fifth most vulnerable country after Zimbabwe, Fiji, Sri Lanka and Vietnam to extremes of climate caused by climate change. This fact has been revealed by the global Climate Risk Index (CRI) 2020, released by the environmental think tank German watch on December 4, 2019 during the COP25 (Conference of Parties 25). This index is based on an analysis of the number of deaths (due to climatic conditions) per 100,000 inhabitants, extent of financial losses and loss per unit of GDP of the countries. The report noted that in 2016, India lost the maximum number of lives (2019) and over $ 21 billion worth of property to weather related events.

In the present day context, climate change has a number of parameters of which rise in temperature, melting of glaciers and rise in sea level are of paramount importance.

Rise in Temperature. Global temperature has been rising constantly since the advent of the Industrial Revolution in Britain and India has not been able to escape this trend. The annual mean, maximum and minimum temperatures during the period 1901-2010 for India, show a significant increasing trend of 0.60°C, 1.0°C and 0.18°C per hundred years, respectively. For the 1981-2010 period, the mean, maximum and minimum temperatures increased almost at an equal rate of around 0.2°C per decade, which is much higher than the trends for 1901-2010.

Human anthropogenic activities are primarily responsible for rise in the global average temperature. As per the recent Intergovernmental Panel on Climate Change (IPCC) special report on Global warming of 1.5°C, human-induced warming reached approximately 1°C (likely between 0.8C and 1.2°C) above the pre-industrial levels in 2017, increasing at 0.2°C (likely between 0.1°C and 0.3°C) per decade. India is feeling the heat of the rising temperature and according to India Meteorological Department (IMD). India had seven warmest years from 2009 to 2019 since it began keeping nationwide records for weather events in 1901. Table 5.4 shows that 2016 had been the warmest year in the accorded history.

TABLE 5.4. Warmest years in India (Above 1981-2010 period)

Year	Temperature above 1981-2010 period average in °C
2009	+0.56
2010	+0.54
2015	+0.42
2016	+0.72
2017	+0.55
2018	+0.41
2019	+0.58
2020	+0.29

Source: India Meteorological Department (IMD)

According to a report released by IMD in December, 2019, November 2019 was the third warmest month since 1901 when the mean temperature in India was 0.72°C higher than the normal (climateological mean). November, 2015 was the warmest in 118 years with 0.87°C rise in mean temperature. The second warmest November was in 1979 when the mean temperature was 0.86°C higher than the normal.

Another report released by IMD in January 2020 showed that 2010-2019 was India's hottest decade with temperature 0.36°C higher than the normal.

Scientists with the National Oceanic and Atmospheric Administration found that January, 2020 was the warmest month ever in 141 years of record keeping. The Global average temperatures in January, 2020 were 1.14°C above average, slightly higher than in January 2016 (the warmest year ever recorded).

Figure 5.32 shows that the maximum temperatures in March had been steadily rising

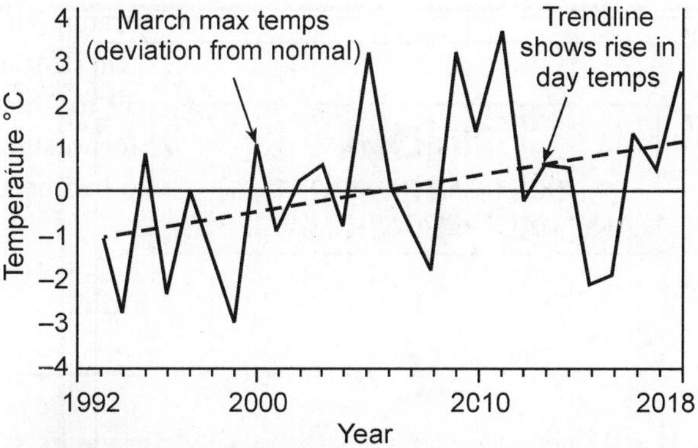

FIG. 5.32. Deviation of maximum temperature and overall rise in temperature in India from 1992 to 2018

between 1992 and 2018, a trend that could be linked to global warming.

Figure 5.33 shows that the average decadal temperature in India has risen by 0.5°C from 25.3°C in 1901-10 to 25.8°C in 2011-18. This is a serious trend and has to be checked at all costs.

Figure 5.34 gives a clear indication of rising trends in temperature anomaly from 1971-2000 average. The temperature anomaly from 1971-2000 average was negative from 1901 till 1980s, after which there had been positive anomaly and this anomaly has been increasing at a sharp rate.

Temperature anomaly is the difference between the mean temperature and actual temperature.

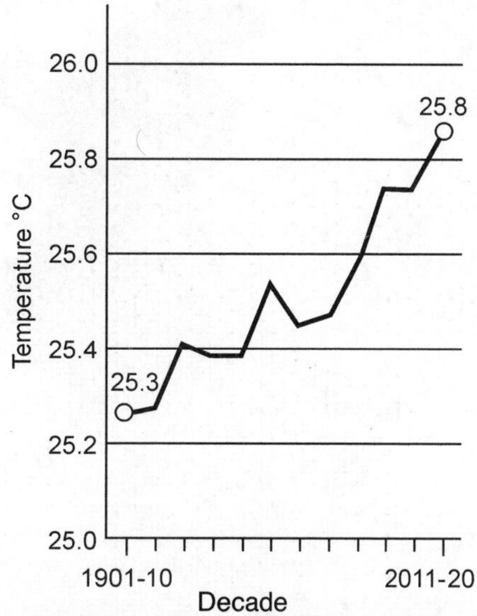

FIG. 5.33. Decadal average temperature (in degree Celsius)

Source: Environment Statistics, 2019

However, temperature rise differed in different parts of the country. Among the four major metropolises of the country Kolkata recorded the

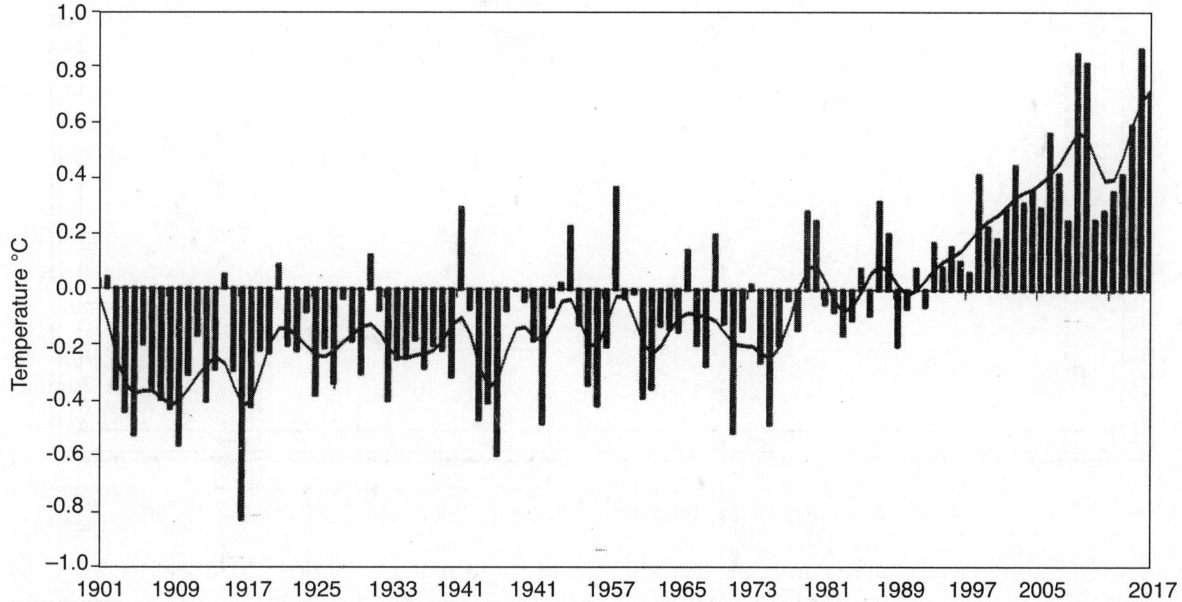

FIG. 5.34. Temperature anomaly from 1971-2000 average

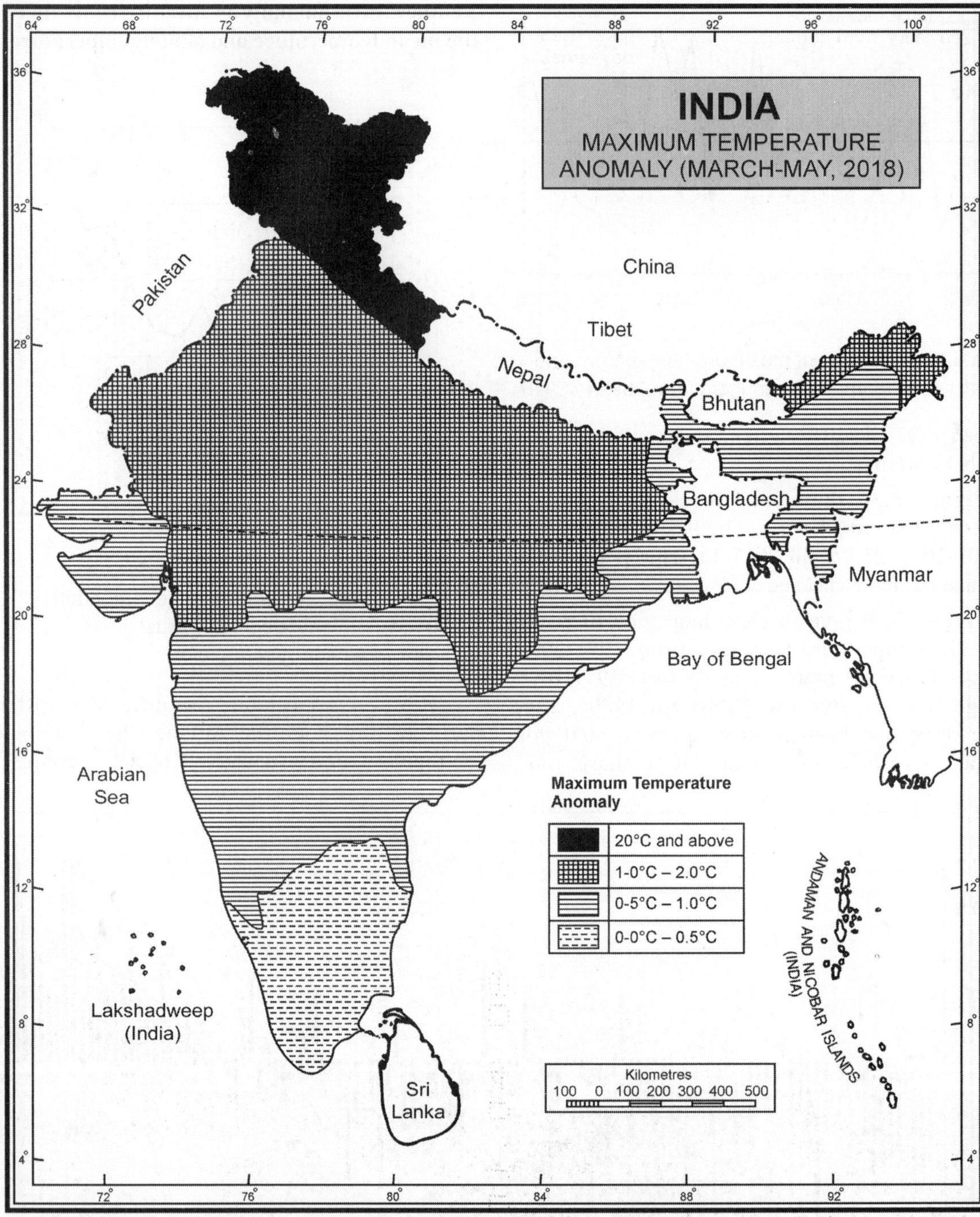

FIG. 5.35. Maximum Temperature Anomaly (March-May, 2018)

highest 1.2°C rise in temperature during the last 150 years. This was followed by Delhi (1.0°C), Mumbai (0.7°C) and Chennai (0.6°C). The overall temperature is likely to rise by 1.5°C by 2050 and by 2.6°C by 2100 even if strenuous efforts are made to control the temperature rise at the global level.

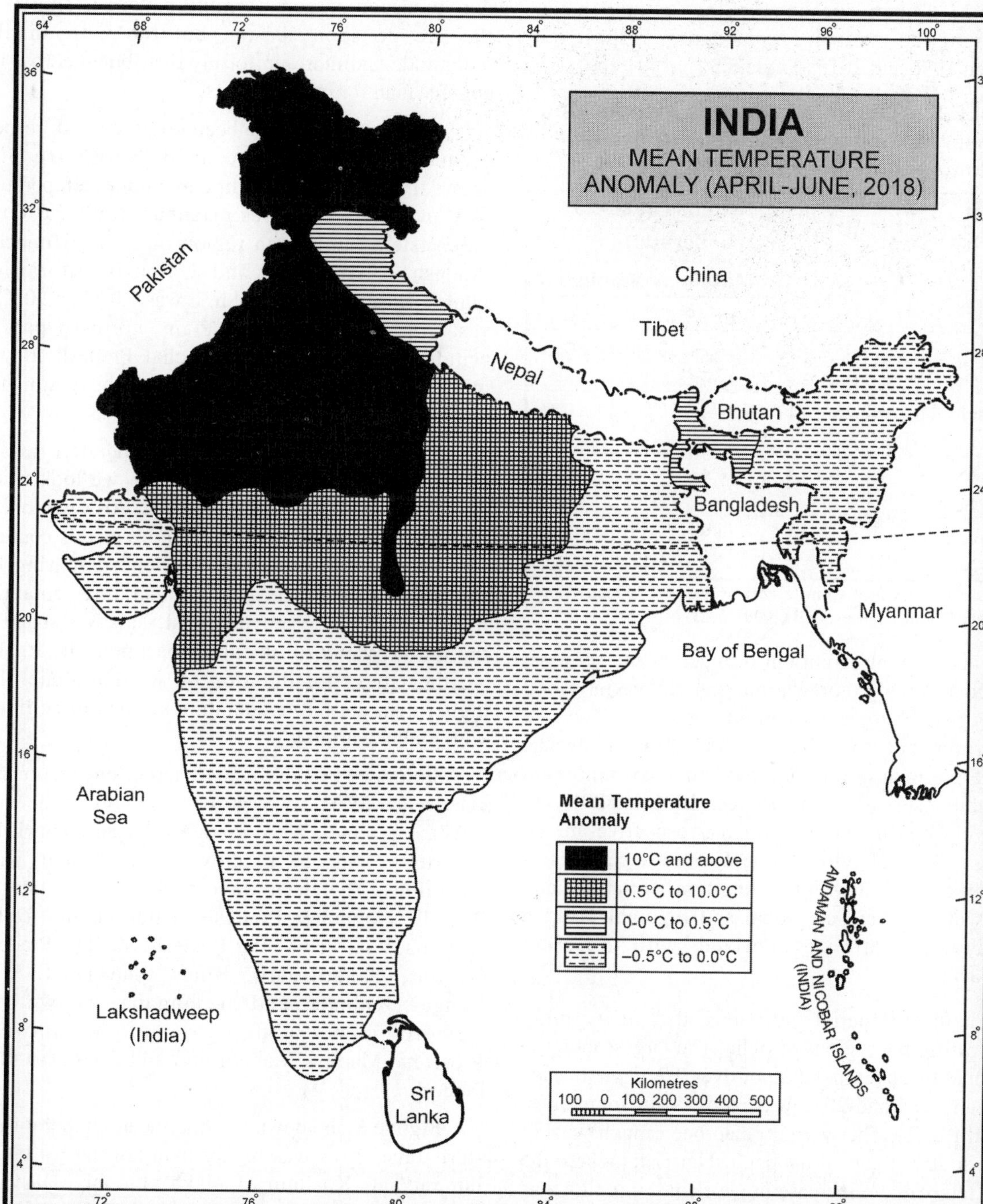

FIG. 5.36. India : Mean Temperature Anomaly (April-June 2018)

According to the first ever "Assessment of Climate Change Over the Indian Region' released by the Ministry of Earth Sciences (MOES), Government of India on 16 June 2020, the mean temperature over India has warmed from the mid-twentieth century with an increased rate of warming of 0.15°C, 0.15°C and 0.13°C per decade for the annual mean, maximum and minimum temperatures respectively between 1986 and 2015 (**Table 5.5**)

TABLE 5.5. Observed changes in India land mean annual and seasonal surface air temperature between 1986 and 2015

Season	Temperature trends 1986-2015 (°C per decade)		
	Mean	**Maximum**	**Minimum**
Annual	0.15 ± 0.09	0.15 ± 0.10	0.13 ± 0.10
Winter (Dec.-Feb.)	0.05 ± 0.16	0.03 ± 0.20	0.07 ± 0.18
Pre Monsoon (March-May)	0.26 ± 0.17	0.29 ± 0.20	0.20 ± 0.16
Monsoon (June-Sept.)	0.11 ± 0.12	0.10 ± 0.17	0.11 ± 0.08
Post-monsoon (Oct.-Nov.)	0.17 ± 0.17	0.14 ± 0.22	0.19 ± 20

Source: Assessment of Climate change over the Indian Region, Ministry of Earth Sciences, GOI (2020), p. 23.

The warming is not uniform across the seasons, with considerably more warming in the pre-monsoon season (March to May) than other seasons. The rate of warming of 0.26°C, 0.29°C and 0.20°C per decade for the pre-monsoon mean, maximum and minimum temperatures, respectively between 1986 and 2015 are relatively higher than that for the respective annual values. This behaviour of temperature changes is broadly consistent with the long-term variations and trends among various global and regional gridded temperature estimates derived with different approaches.

There are spatial variations also in warming among different regions of India. The largest increase in annual mean temperature of over 0.2°C per decade between 1986 and 2015 was recorded in some areas of north India. The warming had been much weaker in the southern part of the peninsular India where the mean temperature increase in west coastal region was less than 0.1°C per decade. The winter warming is limited to peninsular India. The pre-monsoon season shows predominant warming of more than 0.5°C per decade over north India. The summer monsoon season warming is confined to the eastern parts of the Indus-Ganga plains and adjoining central India. The post-monsoon season warming pattern is similar to the pre-monsoon season, but the with smaller magnitude and more uniformly distributed across the country than for other seasons.

Different trends have been observed with respect to maximum temperature anomaly. **Figure 5.35** shows that anomaly regarding maximum temperature in March-May, 2018 was maximum (over 2.0°C) in the Western Himalayan region including Himachal Pradesh, Uttarakhand and Union Territories of Jammu-Kashmir and Ladakh. It was 1.0°C to 2.0°C in vast area covering North Plain, northern part of peninsular plateau and Arunachal Pradesh. It was minimum in the southern part of the peninsular plateau covering Tamil Nadu and Kerala.

A further investigation into temperature studies reflect still a different picture when we look at the mean temperature anomaly. The year 2018 was one of the hottest years since the beginning of this century. Most parts of north India experienced soaring hot summer with average mean temperature remaining more than a degree above normal. This was stated by the India Meteorological Department in its hot weather study for duration of April to June. The higher than normal temperature seemed to be linked with global warming.

Summer season of 2018 experienced unusual occurrence of heat waves over the core 'heat wave (HW) zone'. India Meteorological Department (IMD) reported about 38% above normal heat wave conditions. The 'heat wave zone' covers states of Punjab, Himachal Pradesh, Uttarakhand, Delhi, Haryana, Rajasthan, Uttar Pradesh, Madhya Pradesh, Gujarat, Chhattisgarh, Bihar, Jharkhand, West Bengal, Odisha and Telangana as well as meteorological subdivisions of Marathwada, Vidarbha, Madhya Maharashtra and Coastal Andhra Pradesh.

Figure 5.36 shows that the mean temperature in April-June, 2018 was higher than normal by 1°C in Jammu and Kashmir, Ladakh, Punjab, Haryana, southern parts of Himachal Pradesh and Uttarakhand, Chandigarh, Delhi, large parts of Rajasthan and western Uttar Pradesh. It was 0.5°C to 1.0°C in northern parts of Himachal Pradesh and Uttarakhand, central parts of Uttar Pradesh, parts of Madhya Pradesh, Rajasthan and Gujarat. Northern part of

West Bengal (Darjeeling), Sikkim and adjoining parts of eastern Assam had 0.0°C to 0.5°C higher than normal mean temperature anomaly. The rest of India had –0.5°C to 0.0°C mean temperature anomaly.

Rising Heat Index. Heat index is a measure of the stress placed on humans by increased levels of temperature and moisture. The higher the heat index,the hotter the weather one feels because sweat does not readily evaporate and cools the skin. According to a recent research by India Meteorological Department and Indian Institute of Tropical Meteorology, the average heat index is increasing significantly per decade at the rate of 0.56°C in summer and 0.32°C in the monsoon season.

Of 25 mega cities analysed, 17 including Mumbai, Pune, Kolkata, and Ahmedabad, have long-term average heat index in 'very hot category'. Most mega cities in north India, including Delhi and Amritsar, are in the 'hot category' of heat index during the summer season and in very hot category in monsoon.

The study found that 10 mega cities show significant increasing trends in monsoon heat index.

The main anthropogenic cause of temperature rise is emission of Green House Gases (GHG) like carbon dioxide, methane, nitrous oxide and fluorinated gases. Carbon dioxide is the largest gas accounting for nearly 76% of all GHG gases, followed by methane (16%), nitrous oxide 6% and fluorinated gases (2%). Share of these gases is shown in **figure 5.37**.

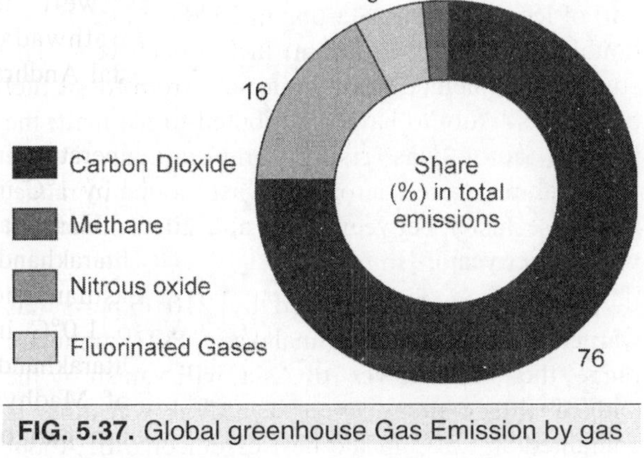

FIG. 5.37. Global greenhouse Gas Emission by gas

Not only the percentage of carbon dioxide is the highest among GHGs, its concentration in the atmosphere is increasing at an alarming rate. Concentration of carbon dioxide increases by burning fossil fuels such as coal, oil, gas and wood and by destroying the forest cover. Increase in concentration of carbon dioxide in the atmosphere has been observed particularly after the Industrial Revolution. It was 294 ppm (parts per million) in 1780. During 120 years between 1860 and 1980, the CO_2 content in the air increased by 13% and it stood at 334 ppm in 1979. **Figure 5.38** shows that CO_2 increased slowly in the beginning and rather rapidly at a later stage. In the year 2000, the amount of CO_2 in the atmosphere stood at 370 ppm. The consumption of fossil fuels is increasing at the rate of 4% annually. At this rate the amount of CO_2 in the atmosphere will double by 2030. Even if burning of fossil fuels is reduced to half of its present rate, the amount of carbon dioxide will be doubled by 2050. According to a research conducted into Antarctic ice cores in 2005, the levels of CO_2 are now 27% higher than at any point of time in the last 6,50,000 years.

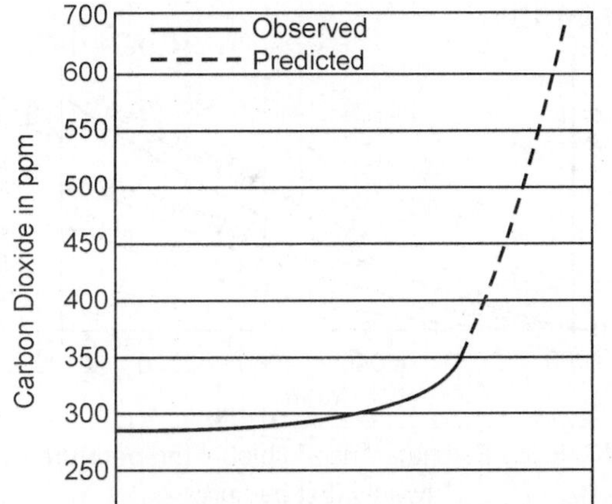

FIG. 5.38. Observed and predicted atmospheric carbon dioxide

Increase in concentration of CO_2 is shown in **Table 5.6**.

Carbon dioxide permits the short wave radiations from the sun to pass through it while long wave radiations reflected by the earth's surfaces are obstructed by it. Thus, the increasing trend in carbon

TABLE 5.6. Lower atmospheric concentration of CO_2

Year	CO_2 Concentration	PPM
1825	0.021	210
1885	0.028	280
1985	0.035	350
2000	0.036	370
2020	0.041	410
2050 (Projected)	0.045	450

dioxide results in increase in heat energy which leads to rise in global temperature. **Figure 5.38** makes it clear that if increase in carbon dioxide is checked to a great extent, then the global temperature will rise by 2°C too by the end of 21st century. But if the present rate of increase in carbon dioxide continues unabated, the global temperature will rise by 4.5°C by the end of 21st century. The greater possibility between these two extremes is that if rise in carbon dioxide is partially changed and reduced to some extent, the global temperature will rise by 3.0°C by 21st century (**Figure 5.39**).

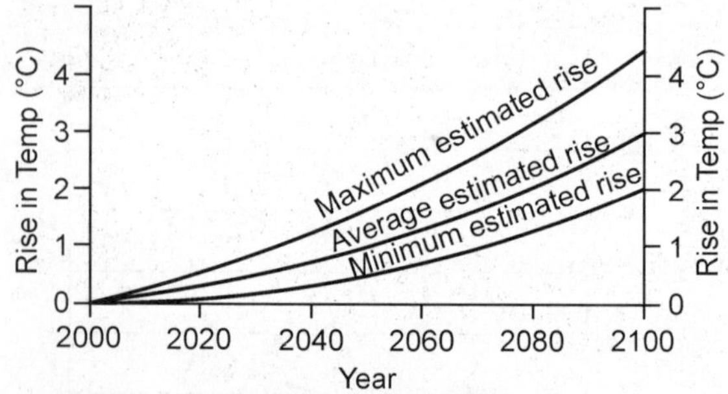

FIG. 5.39. Estimated rise in global temperature by twenty first century

In a report by researchers of IIT Delhi, published in Science Report, a journal of the Nature Group, it has been emphasised the GHGs released by human activities are solely responsible warming climate in India as per the temperature studies from 1956 to 2005. According to the model used by the researchers, the warming contributed by natural factors was just 0.005°C while GHG caused about 1.85°C warming. The study found that the sharpest rise in temperature was in the western Himalayas, specifically Jammu and Kashmir and adjoining areas. Apart from the western Himalayas, the study covered east-coast, west-coast, interior peninsula, north-west India, north-east and north-central India where comparatively less warming was observed.

Consequences of Climate Change

In the present context, climate change largely means rise in temperature in different parts of India due to global warming. Following are the main consequences of climate change in India :

1. Melting of Glaciers

A study published in 'Science Advances' in June, 2019 found that the Himalayan glaciers have lost a foot and a half of ice every year since 2000, melting at a far faster pace than in the previous 25 year period. According to a study based on 40 years of satellite data, climate change is 'eating' the glaciers of the Himalayas, posing a grave threat to hundreds of millions of people, who live downstream. In recent years, the glaciers have lost about eight billion tonnes of water every year. In February, 2019 a report warned that the Himalayas could lose upto a third of their ice by the end of this century even if the world fulfills its goal of keeping global average temperatures from rising only 1.5°C above the preindustrial levels. Scientists fear that the global temperature may rise between 3° to 5°C by the end of this century. Another study published in May 2019 found that the Himalayan glaciers are melting faster in summer than they are being replenished by snow in winter. The researchers at Columbia University relied on analysis of satellite images of 650 glaciers across 2000 km length of the Himalayas. They found that from 1975 to 2000, glaciers lost 10 inches (about 25 cm) of ice every year. Starting in 2000 the rate of loss doubled to about 20 (50 cm) inches each year. The study also concluded that while soot from fossil fuel burning is likely to have contributed to ice melt, the bigger factor was rising temperatures. While temperatures varied across the vast mountain range, they rose faster between 2000 and 2016 compared with earlier years.

Another study compiled by 210 experts has warned the Hindu Kush Himalayan (HKH) mountain range, the ***'water tower'*** of Asia, will vanish by the end of this century even if global warming is contained at 1.5°C in the best care scenario. About

10-20 per cent of this region is covered by 9,000 glaciers storing about 12,000 cubic km of fresh water. The study says that the eastern Himalayas could face near total loss of glaciers, around 95%, if global warming goes unchecked. Even in best scenario (of 1.5°C warming), the region is likely to lose 64% of its glaciers by the year 2100.

The study published in Science Advances Journal, claims the Himalayas lost an average of 4 billion tonnes of ice per annum from 1975 to 2000. After 2000, however, the glaciers started melting twice as fast losing about 8 billion tonnes every year upto 2016.

The glaciers in HKH region, often referred to as '*Asias future bank of water*', spanning over eight Asian countries from Afghanistan to Myanmar, including India, have the biggest store of ice outside the poles. The region sometimes called the '*third pole*' supplies water to nearly 2 billion people.

According to the first ever "Assessment of Climate Change over the Indian Region," released by the Ministry of Earth Sciences (MoES), Government of India, more than 30,000 sq km of the Himalayan region is covered by glaciers which provide 8.6 million cubic metres of water every year. Climate change over the past several decades has influenced the Himalayan mountain glaciers significantly. Glaciers in the Chinese Himalayas are reported to have significantly retreated since the 1950s, with accelerated retreat after 1990s. Almost all the glaciers in the region of Mount Everest have experienced retreat from the late 1990s. Rongbuk glacier which is an important source of fresh water into Tibet, was retreating at the rate of 20 metres per annum. The gangotri glacier in India, which provides a significant source of meltwater for Ganga, has retreated about 30 metres every year during the recent decades. An analysis of 83 glaciers utilising remote sensing and in *situ* observations have reported a loss in glacier extent by 12.6 ± 7.5% for the last 40 years. The maximum loss in the glacier extent has been reported in the Eastern Himalaya, near Tista and Mt. Everest region, followed by Bhutan and Western Himalayas, with minimum loss over the Karakoram Himalayas.

Glacier inventory estimated from satellite image, topographical maps, and aerial photographs provides information about 9040 glaciers covering an area of 18528 sq km in the Hindu Kush Himalayan (HKH) region, shows that the retreat of the Himalayan glaciers ranges from a few metres to almost 61 metres per year, depending on the terrain and meteorological parameters. Mapping of nearly 20,060 out of over 40,000 sq km of glaciated area, distributed in all major climate zones of the Himalayas, suggests a loss in glacier area by about 13% during the last four decades.

2. Floods and Droughts

Melting of the Himalayan glaciers will supply more than normal water to those rivers which originate in these glaciers. This will cause flood conditions to vast areas of the North Plain of India. Currently this plain is the home to over 40% of the Indian population. In short term, melting glaciers leave behind rock debris that creates dams, and if these debris dams burst under hydraulic pressure, it will result in furious floods which will inandate vast areas causing heavy damage to life and property.

After partial or complete melting of glaciers, there will be short supply of water and many areas will face drought conditions. Even otherwise, climate change leads to extremes of weather conditions resulting in high rainfall variability. High frequency and intensity of floods and droughts is the natural sequel. It is worth noting that the number of floods and draughts affecting India has increased during the last few years. A detailed description of floods and droughts is given in chapter 9.

3. Rise in Sea Level

Water generated by melting of glaciers flows through rivers and reaches the sea. The amount of water in the sea increases and there is rise in the sea level. The latest Intergovernmental Panel on Climate Change (IPCC) report shows sea level globally is rising at the rate of 3.6 mm per year and the rate of sea level rise is accelerating with the passage of time. The IPCC report released in Monaco in September, 2019 said that sea levels would rise globally by around 30-60 cm by the end of this century even if GHG emissions are sharply reduced and global warming is limited to well below 2°C. However, the good news is that the annual average sea level rise in India is about 1.7 mm which is about half of the world average. Even this small annual rise can have

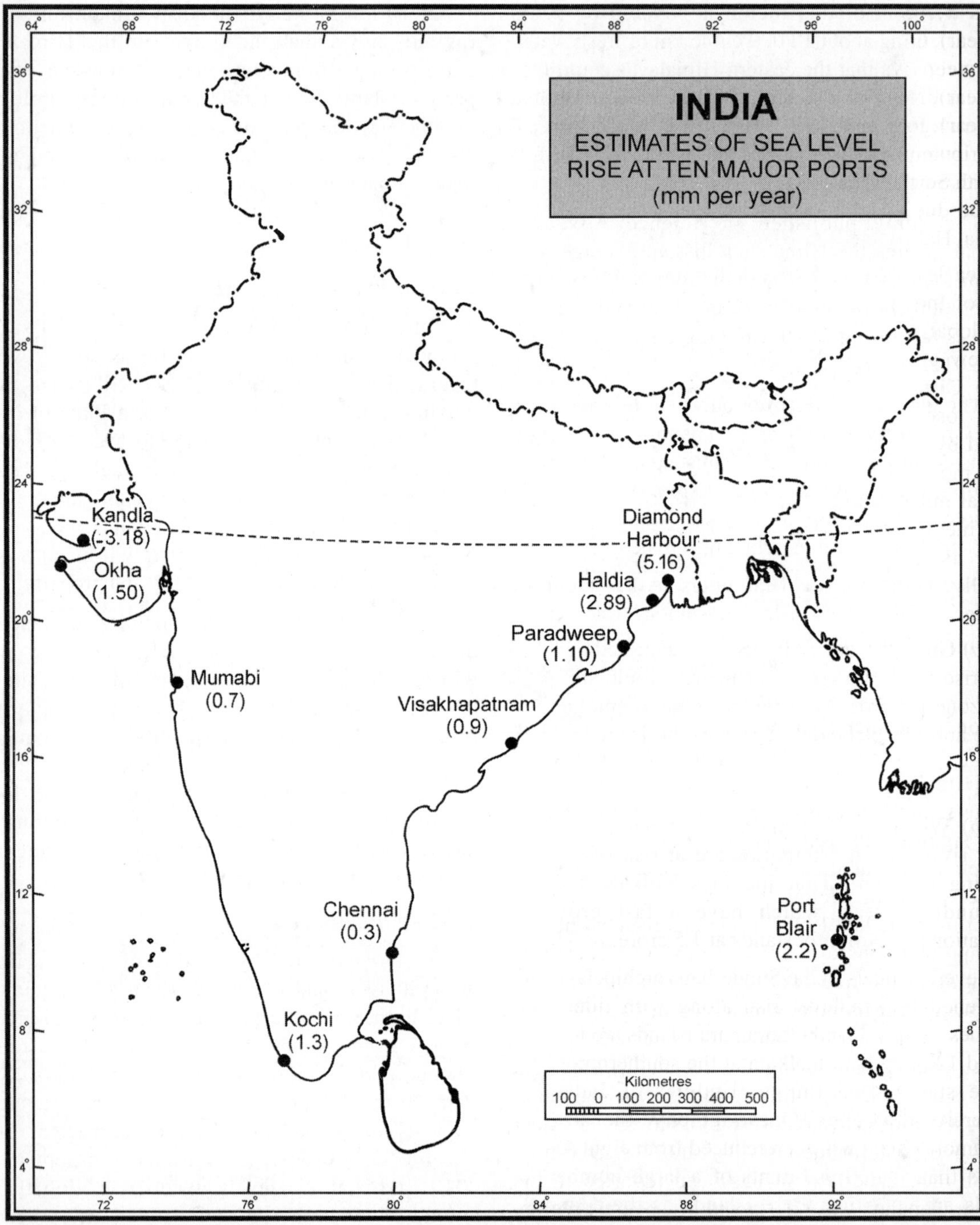

FIG. 5.40. India : Estimates of Sea Level Rise at Ten Major Ports

far reaching consequences because sea levels along the Indian coast have risen by 8.5 mm during the last 50 years. The data collected at 10 Indian ports indicates sea level at Diamond Harbour in West Bengal recorded the highest annual average increase 5.16 mm/year followed by Deendayal or Kandla (3.18

mm/year), Haldia (2.89 mm/year), Port Blair (2.2 mm/year), Okha (1.5 mm/year). Kochi (1.3 mm/year), Paradweep (1.10 mm/year), Visakhapatnam (0.9 mm/year), Mumbai (0.7 mm/year), and Chennai (0.3 mm/year). However, the entire sea level rise may not be attributed to climate change because the Ministry of Earth Sciences has opined that part of sea level rise may be due to land subsidenes as has happened at Kandla, Haldia and Port Blair.

Sea level rise can wreck havoc in the coastal areas of India if the present trend of climate change and global warming is not checked immediately. As per new projections from the U.S. based research group **Climate Central**, about 36 million Indians could lose their homes and livelihoods by 2050 caused by rising sea levels. This is more than seven times more than predicted by the previous models. Several millions more could end up below the high tide line by 2100. After China and Bangladesh, India is the third most threatened country globally. Following few examples will testify the findings.

(*a*) Gujarat. Older projections suggested that sea level rise of even a couple of metres would shift the tidal zone towards Surat which is the diamond and textile hub of the country. The latest projection shows that a vast chunk of the city faces the risk of annual floods.

(*b*) West Bengal. Hundreds of towns along the Hugli river and its tributaries are at risk of annual flooding by 2050. That includes Kolkata and its surrounding areas, which have a fast growing population that currently stands at 1.5 crore.

Several islands in the Sunderbans archipelago are threatened by sea level rise along with tides and cyclones. Mousuni and Ghoramara islands are located around 120 km from Kolkata at the southernmost tip of the state. According to findings of Jadavpur University School of Oceanography, the size of Ghoramora island has been reduced from eight sq km to less than half. Inhabitants of a large number of islands are deserting these islands as their land is swallowed by rising sea level.

(*c*) Maharashtra. Home to about 1.8 crore people, Mumbai is at risk of being virtually swept away by coastal flooding. The city stands on low-lying land reclaimed from the sea and it has been suggested that even if the target of limiting global warming to 1.5°C above the pre-industrial levels is achieved, tropical floods will continue to pose a big threat.

(*d*) Kerala. According to the findings of the new model, large swathes of two districts—Alappuzha and Kottayam—could be in annual coastal flood risk zone by 2050.

(*e*) Tamil Nadu. Districts along the northern coast of Tamil Nadu, including Chennai, Thiruvallur and Kanchipuram, are among the state's most vulnerable regions. Chennai alone is the home to at least 7 million people.

(*f*) Odisha. Communities all along the Odisha coast fall within the risk zone with more than half a million people vulnerable to coastal flooding. Sea level rise may also threaten the deltas of Mahanadi, Godavai, Krishna and Cauvery on the east coast of India. The coastal ground water can turn saline and endanger wetlands.

4. Coral Bleaching

Coral reefs are masses of limestone and dolomite, accumulated by lime-secreting organisms known as coral polyps of different colours like green, yellow, pink, violet, white, etc. Coral reefs are also defined as *cemented and compacted deposits of skeletons of corals in the tropical and subtropical oceans and seas*.

Coral bleaching is the change in colour of the coral because of decay in algae living in their tissues as symbiotic partners. Algae is also the reason that gives corals their distinct colour. Corals can survive in the temperature range of 18°C to 28°C *i.e.* the SST (Sea Surface Temperature) should be above 18°C in winter and below 28°C in summer. The maximum duration of heat stress which corals can tolerate is 28 days. SST has been rising consistently as a result of global warming. As the sea temperature rises above normal summer maxima, the corals are forced to expel the algae and turn white. These blushed corals thereafter get prone to disease and ultimately die.

A study undertaken by ISRO's Space Applications Centre (SAC) research team and the department of geophysics, Kurukshetra University (Haryana) in August, 2019 analysed SST from 1982 and reported increase in number of heat stress days.

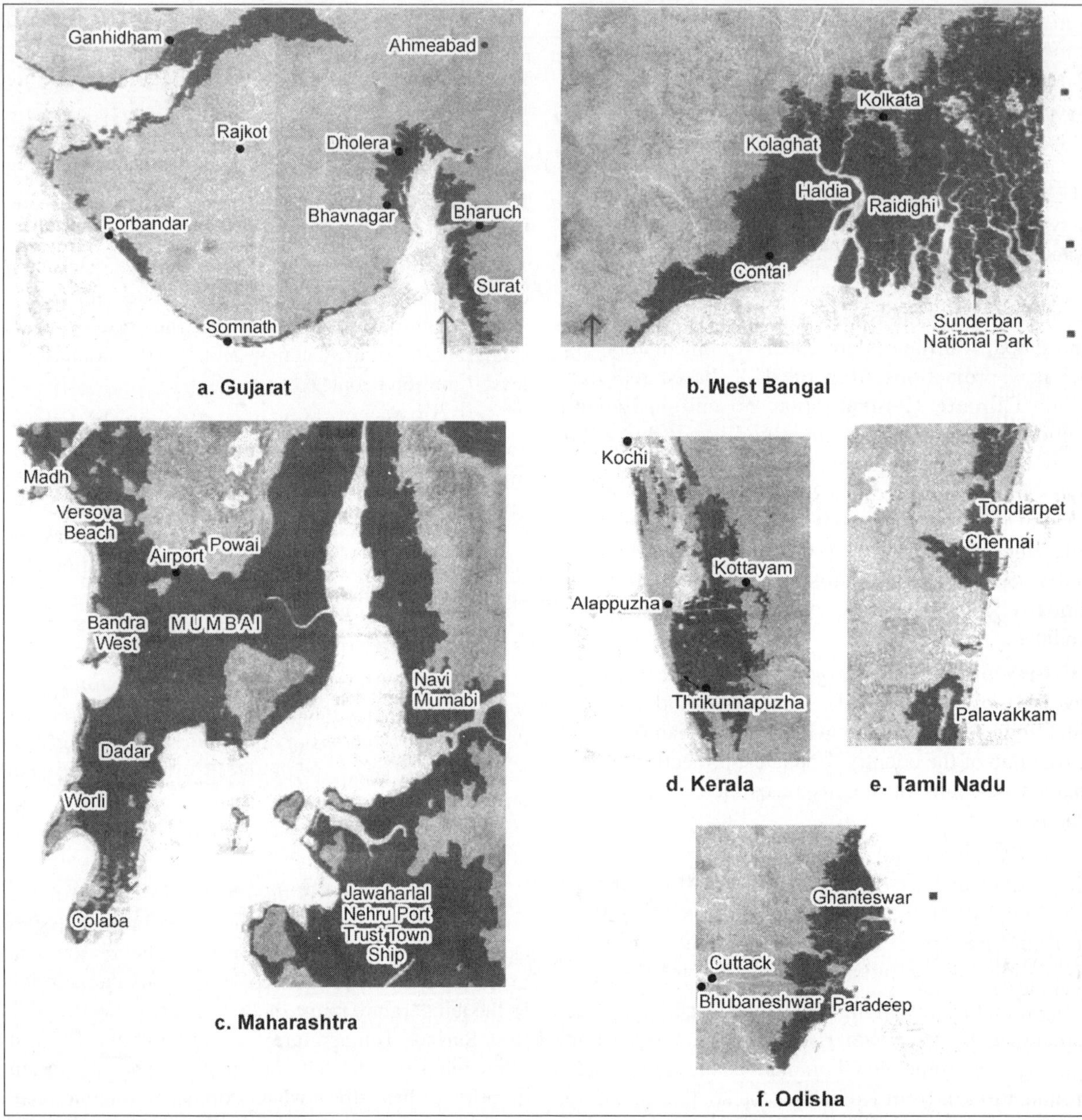

FIG. 5.41. Areas (in dark shade) likely to be threatened by sea level rise

The study includes three distinct years of Mass Coral Bleaching (MCB) instances of 1998, 2010 and 2016 recorded during El Nino Southern Oscillation (ENSO) that caused abrupt rise in temperature (Table 5.7). It has been found that 50 to 70 per cent of the corals have been bleached since 1998.

Normally corals can re-establish themselves (or return to their pre-bleaching state) in a decade or two. But the quick frequency and high intensity of Mass Coral Bleaching (MCB) overwhelms the coral reef's ability to recover from the heat stress episodes.

According to Climate Risk Index (CRI) released on 24th January, 2021, India was the seventh worst hit country in the world due to extreme weather events in 2019, both in terms of fatalities and economic losses. Six countries worse than India were

TABLE 5.7. Mass Coral Bleaching Years in Indian Coral Reef Regions

Coral bleaching indices/year		Andaman	Nicobar	Lakshadweep	Gulf of Mannar	Gulf of Kachchh
1998	Warmest in summer months	30.76°C	30.69°C	30.71°C	30.84°C	29.58°C
	Duration of thermal stress	61 days	67 days	68 days	67 days	37 days
2010	Warmest in summer months	31.43°C	31.26°C	30.77°C	30.87°C	30.44°C
	Duration of thermal stress	87 days	85 days	73 days	48 days	91 days
2016	Warmest in summer months	31.03°C	31.2°C	30.86°C	31.21°C	29.51°C
	Duration of thermal stress	53 days	67 days	91 days	63 days	7 days

(1) Mozambique, (2) Zimbabwe, (3) The Bahamas, (4) Japan, (5) Malawis and (6) Afghanistan. In that year India suffered 2267 fatalities and economic losses of $ 68,812 million.

5. Increase in Extreme Weather Events

As a result of climate change, the frequency of extremes of weather events has increased and that too at an accelerated rate. According to Climate Risk Index (CRI) released on 24th January, 2021, India was the seventh worst hit country in the world due to extreme weather events in 2019, both in terms of fatalities and economic losses. Six countries worse than India were (1) Mozambique, (2) Zimbabwe, (3) The Bahamas, (4) Japan, (5) Malawi, and (6) Afghanistan. In that year India suffered 2267 fatalities and economic losses of $ 68,812 million. Apart from frequency, the intensity and duration of extreme weather events has also increased and reached an alarming stage. The latest data released by the India Meteorological Department in November 2019 reveals that the number of cyclones and severe cyclones in the Arabian Sea and Bay of Bengal has risen by nearly 11% in the last decade with an alarming 32% increase recorded in the last five years. On an average, four cyclones affected India each year during 2010 and 2019, higher than the average of three in the previous decades since 1980. Back-to-back cyclones have caused havoc and disrupted weather in recent past five years in comparison with previous decades could be an alarming pointer to the calamitous effects of global warming. There were seven cyclones each in 2018 and 2019, the highest since 1985. Similarly six severe cyclones each hit India in 2018 and 2019—the biggest since 1976.

Similarly frequency of heat and cold waves has also increased considerably in the recent past. A **heat wave** is a prolonged period of abnormally hot weather that typically lasts from a few days to few weeks. A lethal heat wave occurs when average daily maximum wet-bulb temperature (the lowest to which air can be cooled with evaporating water) reaches a point that can kill a healthy human resting in shade.

In India, a heat wave episode may last for about a week when the temperature is sufficiently high above the normal. Since India is a tropical country, practically no part of India is absolutely free from the impact of the heat waves. However, the Northern Plain of India from Punjab to Bihar is more prone to heat waves. India could face an annual threat of deadly heat waves, like the one in 2015 that killed atleast 2500 people.

A study by the Indian Institute of Tropical Meteorology (IITM) has revealed that the frequency of heat waves and their duration may begin to increase as early as 2020. According to this study El Nino Modoki could be responsible for the increase in heat waves. El Nino Modoki is different from El Nino as it becomes effective when the central part of the Pacific Ocean warms more with cooling in the eastern and western parts. The study says that El Nino Modoki may be connected with climate change.

The study called "Future Projections of Heat Waves over India from CMIP5 Model and published in the international journal "Climate Dynamics" examined nine climatic models to understand how the frequency, intensity and duration of heat waves in India will increase. Its severe and adverse impact on health was also analysed. The models used in the study identified 54 heat waves events in India between 1961 and 2005. The study says that between 2020 and 2064, there will be increase of two heat

TABLE 5.8. Number of heat and cold waves and deaths caused by these waves in India

Year	Heat waves		Cold waves	
	Number of heat waves	Deaths due to heat waves	Number of cold waves	Deaths due to cold waves
2010	21	242	33	450
2011	4	12	24	722
2012	37	729	13	139
2013	14	1433	12	271
2014	30	547	9	58
2015	35	2081	6	18
2016	37	1111	8	42

Source: (*i*) India Meteorological Department in EnviStats India 2018, pp. 245, 46, 50
(*ii*) State of India's Environment 2020, p. 168.

waves and of 12-18 days in heat waves duration in this period.

The impact of heat waves on individuals varies greatly. The elderly people are the most vulnerable because heat puts more stress on weak hearts and bodies.

According to the IPCC report released in September, 2019, marine heat waves have doubled in frequency between 1982 and 2016. Increased frequency of marine heat waves and over all rise in sea surface temperature would lead to fall in the production of sea food.

On the other extreme, India suffers a lot from cold waves. For a cold wave day, the minimum temperature should be 4°C or more notches below normal or less than 4°C. These waves are effective in the northern part of India and their impact is more pronounced in the winter months of December and January. Details of heat and cold waves is given in **Table 5.8**.

Storms, whether accompanied by thunder, dust, hail or lightning have also become more frequent in the past few years. The year 2018 witnessed the worst damage by these storms. In seven of the last ten years, storms have killed close to 2,500 people. Lightning alone has accounted for more than 1,500 of these deaths.

Extremes of rainfall are no less pronounced than those of temperature. Daily rainfall observation during 1901-2004 indicate that the frequency of extreme rainfall events (rain rate > 100 mm/day) has a significant positive trend of 6 per cent per decade.

TABLE 5.9. Year-wise damage due to natural extreme events

Year	Humans killed (numbers)	Cattle lost (in numbers)	Houses damaged (in numbers)	Cropped areas affected (in lakh hectares)
2001-02	834	21,269	3,46,878	18.72
2005-06	2,698	1,10,997	21,20,012	35.52
2010-11	2,310	48,778	13,38,619	46.25
2011-12	1,600	9,126	8,76,168	18.87
2012-13	946	24,293	6,67,319	14.49
2013-14	5677*	10,299	12,10,227	63.75
2014-15	1,674	92,180	7,25,390	26.85
2015-16	1,460	59,057	13,13,371	31.09
2016-17 (P)	1,487	41,965	5,46,518	25.49
2017-18 (P)	2,057	46,488	9,15,878	47.44

* Includes persons missing in the natural disaster.
P = Provisional
Source : State of India's Environment 2019, (In figures), p. 93.

Table 5.9 gives details of losses caused by natural extreme events in India for a few selected years.

6. Adverse effect on GDP and Living Standard

In a first-of-its kind report released by the World Bank in June, 2018, a serious concern has been expressed that unchecked climate change will dent India's GDP by 2.8% and depress the living standards of about 600 million people by 2050. People living in severe 'hotspot' districts of Central India, particularly Vidarbha are likely to suffer over 10% dip in economic consumption. A hotspot is a **'location where gradual changes in average temperature and rainfall patterns will have negative impacts on living standards in future.** This is the scenario even if the 2015 Paris Agreement goal of containing global warming to less than 2°C is achieved. Some of the states are likely to suffer much higher drop in living standards than the 2.8% average for India as a whole (**Table 5.10**).

TABLE 5.10. Drop in living standards as a result of climate change

State	Drop in living standard (in percentages)
1. Chhattisgarh	9.4
2. Madhya Pradesh	9.1
3. Rajasthan	6.4
4. Uttar Pradesh	4.9
5. Maharashtra	4
6. Jharkhand	4.6
7. Haryana	4.3
8. Punjab	3.8
9. Andhra Pradesh	3.4
All India	2.8

Source: World Bank Report, South Asia's Hotspots.

7. Change in Agricultural Patterns

Climate change is definitely going to have a deep impact on the agricultural patterns in India. While area and production of Kharif crops may increase due to rise in temperature, rabi crops may suffer severe loss at the hands of climate change. Increase in temperature and carbon dioxide (CO_2) can increase the production of some crops in certain areas. But to realize these benefits, nutrient-levels, soil moisture, water availability and other conditions must also be met. However, other factors such as changing temperatures, ozone and water and nutrient constraints may counteract these benefits. Indian agriculture is highly vulnerable to vagaries of weather, and the looming threat of climate change may expose its vulnerability further. Guiteras (2009) finds that crop yields will decline 4.5-9.0% from 2010 to 2039 and by a whopping 25% in the long run (2070-2099) in the absence of adaptation by farmers. According to Ministry of Agriculture and Farmers Welfare, climate change can have negative effects on irrigated crop yields across agro-ecological regions both due to temperature rise and changes in water availability. Rainfed agriculture will be primarily impacted due to rainfall variability and reduction in number of rainy days. Analysis of impact of climate change under National Innovations in Climate Resilient Agriculture (NICRA) project has found that climate change is expected to affect yields, particularly in crops like rice, wheat and maize. It is estimated that wheat yield will drop by 5 to 25% by 2050.

8. Effect on Human Health

In the present context, climate change is generally associated with global warming. Rising temperature is often associated with higher level of humidity. Hot and humid climate is more harmful to human health than cold and dry climate. Change in temperature, humidity, rainfall, and sea level rise could all affect the incidence of infectious diseases. Mosquitoes, ticks and fleas are sensitive to subtle temperature and humidity changes. But vector-borne diseases are also dependent on many other interacting factors. A large number of ailments like malaria, dengue, chikungunya, cholera, diarrhoea, cardiovascular failure, asthma, etc. are likely to increase with respect to their frequency and intensity. **Figure 5.42** shows some of the consequences of climate change and their effects on human health.

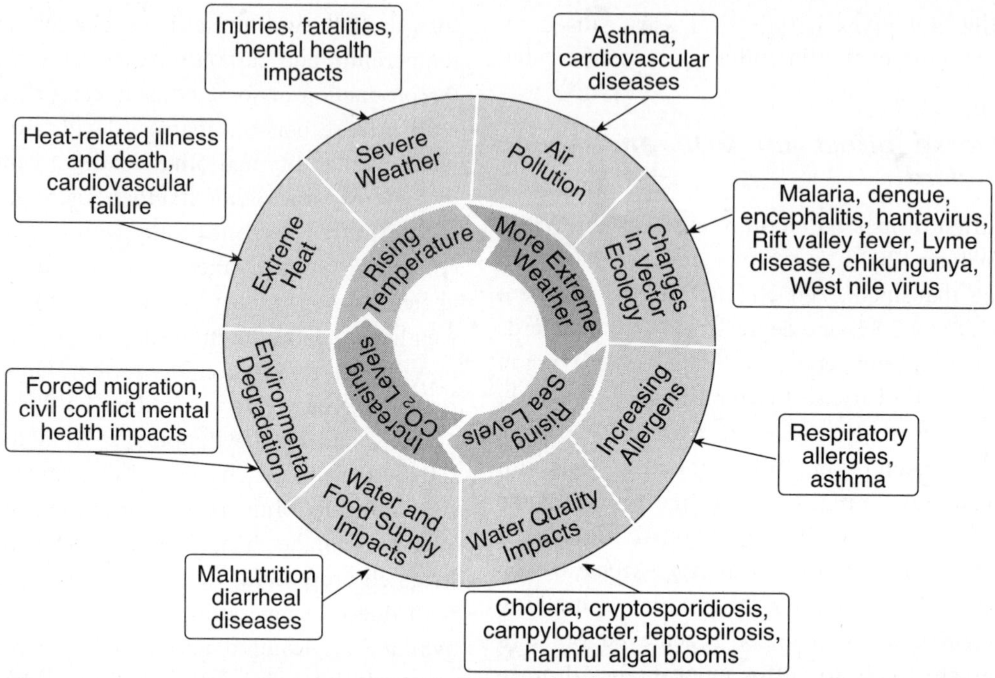

FIG. 5.42. Impact of climate change on human health

Source : Climate change reader for Universities, 2018.

Mitigation

There seem to be two mitigation strategies to combat the effects of climate change. They are : (*i*) restrict the climate change to manageable limit and (*ii*) adopt life style according to climate change.

- According to a special report by the World Health Organisation (WHO), India can gain \$ 3.28 to 8.4 trillion by making efforts to limit the global warming to 1.5°C by the end of this century.
- India has taken up the challenge of climate change seriously and has engaged herself constructively at the multilateral level under the United Nations Framework Convention on Climate Change (UNFCCC). Further India is actively engaged in the efforts towards developing guidelines for effective implementation of the Paris Agreement on climate change.
- India's National Action Plan on Climate Change (NAPCC) was launched in 2008. It has through eight National Missions focussed on advancing the country's climate change related objectives of adaptation, mitigation and preparedness on climate risks. The Government has decided to revise the NAPCC in line with Nationally Determined Contribution (NDC) submitted by India under the Paris Agreement to make it more comprehensive in terms of priority areas. Eight developments under NAPCC are : (1) National Solar Mission (NSM), (2) National Mission for Enhanced Energy Efficiency (NMEEE), (3) National Mission for Green India (GIM), (4) National Mission for Sustainable Habitat (NMSH), (5) National Water Mission (NWM), (6) National Mission for Sustainable Agriculture, (7) National Mission for Sustaining Himalayan Eco-system and (8) National Mission on Strategic Knowledge for Climate Change (NMSKCC).
- As part of the mission on strategic knowledge on climate change, India has established 8 Global Technology Watch

Groups in the areas of Renewable Energy Technology, Advance Coal Technology, Enhanced Energy Efficiency, Green Forest, Sustainable Habitat, Water, Sustainable Agriculture and Manufacturing.

- The broad policy initiatives of the central government—are supplemented by actions at the sub-national levels. 32 States and Union Territories have put in place the State Action Plans on Climate Change attempting to mainstream climate change concerns in their planning process.
- Climate Change Action Plan (CCAP), a Central Sector Scheme, as approved in January 2014 to build and support the scientific and analytical capacity for assessment of climate change in the country, establish appropriate institutional framework and implement climate actions for a duration of five years. Two important components of CCAP scheme are National Carbonaceous Aerosols Program (NCAP) and Long-Term Ecological Observations (LTEO). NCAP is a multi-institutional program being implemented by a consortium of 17 institutions led by IIT Bombay. It was inaugurated in the year 2017-18 and is a five year duration project. LTEO is another project under the CCAP scheme for duration of 5 years and was started in 2019-20. The duration of CCAP was initially from 2017-18 to 2019-20 and this was later extended till 2020-21.
- National Adaptation Fund on Climate Change established in 2015 to support concrete adaptation activities which are not covered under on-going activities through the schemes of State and Central Governments, continues till 31st March 2020.
- India is one of the few countries where, despite ongoing development, forest and tree cover has increased transforming country's forests into a net sink owing to national policies aimed at conservation and sustainable management of forests. India's growth in the forest cover has been in the positive territory from 1991 to 2020 while that for Indonesia and Brazil, which are countries with substantial forest cover, the growth has been in the negative territory during the same period.
- The Brown to Green Report 2019 released by the Climate Transparency noted that India is the only country among G20 nations that is close to 1.5°C temperature 'pathway'. Limiting global temperature increase to 1.5°C instead of 3°C avoids over 70% of the climate related impacts in water, health and agriculture sectors, said the report.
- The outcomes of the above mentioned initiatives are reflected in India's Second Bienninal Update Report (BUR) submitted to UNFCCC (United Nations Framework on Climate Change) in December 2018. The report shows that emission intensity of India's GDP came down by 21 per cent between 2005 and 2014 and its achievement of climate goal for pre 2020 period is on the track. A total of 2.607 billion tonnes of CO_2 equivalent of GHGs were emitted and about 12% of the emissions were offset by carbon sink action of forest land, cropland and settlements.

HORRIFYING PROJECTIONS ABOUT CLIMATE CHANGE IN INDIA

In a first of its kind report on climate change in the Indian Subcontinent, the Ministry of Earth Science (MoES) Govt. of India, released on June 16, 2020, following projections have been made which are quite horrifying :

- The average temperature in India is projected to rise by 4.4°C by the end of 21st century over 1976-2005 average.
- The report "Assessment of climate change over the Indian Region" shows temperatures of the warmest day and the coldest night of the year have risen by about 0.63°C and 0.4°C in the recent 30 year period (1986-2015) and these temperatures are projected to rise approximately by 4.7°C and 5.5°C respectively by the end of the century in business as usual (BAU) conditions. BAU conditions refer to a scenario where very little or no action is taken to reduce the emission of greenhouse gases that have resulted in global warming and consequent climate change.

- India has witnessed an increase in frequency of droughts and floods.
- The humid regions of central India have notably become drought prone regions.
- The flood risks have increased over the east coast, West Bengal, eastern U.P., Gujarat and Konkan regions as well as in majority of urban areas such as Mumbai, Kolkata and Chennai.
- Summer monsoon (June-September) rainfall has declined by 6% from 1951 to 2015. Notable decrease in Indio-Gangetic plains and Western Ghats. Frequency of localised heavy rainfall has increased,. Projected substantial increase in daily extreme rainfall.
- Sea-level rise in North Indian Ocean (NIO) occurred at the rate of 1.06-1.75 mm per year during 1874-2004. It is projected to rise by 300 mm (30 cm) by 2100 over average of 1986-2005. Global mean rise during the period is expected to be 180 mm.
- Frequency of very severe cyclonic storms has increased significantly during the last two decades and is projected to increase further during the current century.

SOME OTHER STARTLING FACTS ABOUT CLIMATE CHANGE IN INDIA

- According to global Climate Risk Index (CRI) released on January 24, 2021, India is seventh worst hit country due to extreme weather events after Mozambique, Zimbabwe, The Balamas, Japan, Malawi and Afghanistan. Even some of the least developed countries like South Sudan, Niger, Bolivia, etc. are better placed than India.
- The first of its kind national and district wise climate vulnerability assessment report released by the **Science and Technology Ministry** on 17th April, 2021 has mentioned that eastern part of the country, including Jharkhand, Mizoram, Odisha, Chhattisgarh, Assam, Bihar, Arunachal Pradesh and West Bengal, is highly vulnerable to climate change. The report titled **'Climate vulnerability Assessment for Adaptation Planning in India using a Common Framework'** noted that all districts or states are vulnerable, but some are more vulnerable than others.
- A study carried out by the University of Chicago and released on 25-3-2021 has found that for every 1°C rise in temperature, industrial plants in Delhi and across the country are showing a drop in productivity by around 2% annually. Productivity has been hit the most in labour intensive plants.
- A study released by the council on Energy, Environment and Water (CEEW) on 10th December, 2020 has pointed out that over 75% of Indian districts, home to 63.8 crore people, are vulnerable to extreme climate events such as cyclones, floods, droughts and both cold and heat waves. The study also found that there had been an increase in the frequency and intensity of these extreme events which rose dramatically 2005 onwards. While India witnessed 250 extreme climate events between 1970 and 2005, it recorded 310 extreme weather events between 2005 and 2020.
- 2011-20 was the warmest decade with warmest six years all being since 2015 on record, say the World Meteorological Organisation (WMO).
- A report released by Mc Kinsey Global Institute (MGI) in November 2020 has mentioned that extreme heat poses $ 250 billion risk to India's GDP by 2030 since 50% of India's GDP is dependent on heat-exposed work like agriculture, mining and construction. It also found that 160 to 200 million people could be living in areas that are vulnerable to deadly heat by 2030.
- According to India Meteorological Department (IMD), extreme weather events claimed 1429 lives in 2020. The event-wise casualties were thunderstorm (569), rain and floods (452), cold wave (149), lightning (148) and cyclone (111).
- A study published in the Climate Change journal by Spring Nature on 29th July, 2020 has warned that Jammu-Kashmir and Ladakh may be hotter by 7°C by the end of this century.

REFERENCES

Anna Mani, (1981) : *The Climate of the Himalaya*, Oxford University Press.

Barry, R.G. and Chorley, R.J. (2003) : *Atmosphere, Weather and Climate,* The English Language Book Society and Methuen & Co. Ltd. London.

Basu, S. (1965) : *Weather and Climate,* in Chopra P.N. (*Ed.*) *The Gazatter of India*, Govt. of India Publication, (Chap II).

Bijerknese, J. (1969) : *Atmospheric Teleconnections from the Equatorial Pacific,* Mon. Wea. Rew., 97, 163-172.

Bunker, A.F. (1965) : *Interaction of summer monsoon air with the Arabian Sea* (Invited lecture), Proc. Synp. On Int. Ind. Ocean Expd. Bombay.

Centre for Science and Environment, New Delhi (2018). Climate Change Now, The Story of Carbon Colonisation,

Centre for Science and Environment, New Delhi (2018); Climate Change Reader for Universities.

Center for Science and Environment (2019, 2020); State of India's Environment.

Centre for Science and Environment (2018); Climate Change Now.

Chang, C.P. and Krishnamurti, T.N. (1987) : *Monsoon Meteorology,* Oxford University Press, Oxford.

Chatterji, A.K. (1992) : *Monsoons, Cyclones and Floods in India,* Birla Economic Research Foundation, Radiant Publishers, New Delhi.

Chatterji, S.B. (1953) : *Indian Climatology, Climostatics, Climate Classification of India with Special reference to the Monsoons,* Calcutta, Pp. 201-317.

Das, P.K. (1988) : *Why Monsoons Fail,* Sc. Reporter, May, 88. Publications and Information Directorate (CSIR), New Delhi.

— — (1990) : *The Ocean Connection*, Sc. Reeboprter, July 1990, Publications and Information Directorate (CSIR), New Delhi.

— — (2004) : *The Monsoons,* National Book Trust, New Delhi, India.

Findlater, J. (1981) : An experiment in monitoring cross-equatorial airflow at low level over Kenya and rainfall of Western India during the northern summer, *Monsoon Dynamics,* Ed. Sir James Lighthill and R.P. Pearec, Cambridge University Press, pp. 309-318.

Flohn, H. (1968) : *Contribution to a meteorology of the Tibetan Highlands,* Atoms. Sci. Paper No. 130, Colorado State University, Fort Collins.

— — (1969) : *Climate and weather*, Weiderfield & Nicolson, London

Ghosh, S.K. *et al.* (1978) : Influence of Arabian Sea on the Indian Summer Monsoon, Tellus, 30, 117-125.

Government of India : Economic Survey (2018-19), Chap 4, India and Climate Change, pp. 74-76.

Government of India : Economic Survey (2019-20), Chap. 5, Climate Change, pp. 119-20.

Government of India : Economic Survey (2020-21), Chap. 6, Climate Change, pp. 207-15.

Government of India : Economic Survey (2021-22), Chap. on Sustainable Development and Climate Change, pp. 220-28.

Government of India : Economic Survey, Chap. 7, pp. 217-42.

Government of India, Ministry of Earth Sciences (2020), Assessment of Climate Change over the Indian Region.

Gowarikar, Vasant (1990) : *Forecasting Monsoon Rain,* Science Reporter, July, 1990, Publications and Information Directorate (CSIR), New Delhi.

Griffith, J.F. (1968) : *Applied Climatology,* Oxford University Press.

Hess Darrel (2011) : *Physical Geography,* PH1 Learning Pvt. Ltd., New Delhi.

Holton, J.R. (1973) : *Introduction to Dynamic Meteorology,* Academic Press.

Jay, S., Fien and Pamela, L. Stephens (1987) (eds.) : *Monsoon,* John Willy & Sons, New York.

Johnson, B.L.C. (1969) : *South Asia,* Heinemann Educational Books

Karnani, Roop (1993) : Charting the Monsoon, *Business India,* June 21-July 1993, pp. 114-115.

Katiyar, V.S. (1990) : *The Indian Monsoon and its Frontiers,* Inter India Publications, New Delhi.

Kelkar, R.R. and Rao, A.V.R.K. (1992) : *Extraction of quantitative meteorological information from INSAT imagery Asian Monsoon,* in M. Murakami (Ed.) Met. Soc. of Japan.

Kendrew, W.G. (1961) : *Climates of Continents,* Oxford University Press.

Koteswaram, P. (1953) : *An Analysis of high tropospheric wind circulation over India in Winter,* Indian Journal of Met. Geophysics.

— — (1958) : *Easterly Jet Stream in the Tropics,* Tellus 10, 43-57.

Koppen, Wladimer (1931) : Grundriss der Klimakund, Walter de Gruyler Company, Berlin.

— — (1936) : Das Geographische System der Klimate, *Handbook der Klimablogie,* Ed. W. Koppen and R. Geiger, Vol. I, Berlin.

Lamb, H.H. (1871) : *Climate—Past, Present and Future,* Methuen & Co. Ltd. London.

Lighthall, M.J. (1969) : *Dynamic Response of the Indian Ocean to the Onset of the Southwest Monsoon,* Phil. Trans. Ray. Soc. London, p. 265.

Lockwood, G. (1983) : *World Climatology, An Environmental Approach,* ELBS, London.

Lorenz, E.N. (1967) : *The nature and theory of general circulation of the Atmosphere,* WNO, Geneva.

Lutgens, Frederick K. and Tarbuck Edward J. (2012) : *The Atmosphere : An Introduction to Meteorology,* PHI Learning Pvt. Ltd., New Delhi.

Menon, P.A. (1993) : *Our Weather,* National Book Trust, New Delhi, India.

Pedelaborde, P. (1963) : *The Monsoon,* Methuen & Co. Ltd., London.

Pichamuthu, C.S. (1967) : *Physical Geography of India,* National Book Trust, New Delhi.

Pisharoty, P.R. (1965) : *Evaporation from the Arabian Sea and the Indian Southwest Monsoon,* Prec. Symp. On 110E, Bombay, 43-54.

Rama Sastry, A.A. (1983) : *Weather and Weather Forecasting,* Publication Division, Ministry of I&B. Govt. of India.

Ramage, C.S. (1971) : *Monsoon Meteorology,* Academic Press, New York & London.

Ramaswamy, C. (1956) : *On the sub-tropical jet stream and its role in the development of large-scale convection,* Tellus 8, 26-60.

— —(1962) : *Breaks in the Indian summer monsoon as a phenomenon of interaction between the easterly and the sub-tropical westerly jet streams,* Tellus 14, 3 37-49.

Rao, Krishna, P.R. and Jagannathan P. (1953) : *A study of the N.E. monsoon rainfall of Tamil Nadu,* Ind. J. Met. Geophys 4.

Rao, Y.P. (1976) : *South-west Monsoon,* Met. Monograph No. 1/1976.

Rao, Y.P. and Ramamurthy. (1968) : *Climate of India,* Forecasting Manual Part I No. 2.

Roy, A.K. (1964) : *Tracks of Storms and Depressions in the Bay of Bengal and Arabian Sea (1877-1960),* India Meteorological Deptt.

Saha, K.R. (1976) : *Asymmetry of Tropical Monsoons;* Proc. Symp. on Tropical Monsoons I.I.T.M., Pune.

Shah, K.R., J. Shukla and F. Sanders, (1981) : *Esterward Propagating Predecessors of Monsoon Depredations,* Mon. We. Rev. U.S.A., pp. 330-343.

Shukla, J. (1975) : *Effects of Arabian Sea—Surface Temperature Anomaly on Indian Summer Monsoon. A numerical experiment with the GFDL Model,* J. Atmos. Sci. U.S.A. pp. 503-511.

Siddhartha, K. (1999) : *India : The Physical Aspects,* Centre for Development of Environment and Resources Patna, Chap. 4.

Singh Narayan *et al.* (eds) (2016) : Climate Change and Environmental Issues. The Energy and Resources Institute (TERI), New Delhi.

Singh, R.L. (ed.) (1971) : *India : A Regional Geography,* National Geog. Society of Indian,

Singh, Savindra, (2009) : *Climatology,* Prayag Pustak Bhandar, Allahabad.

Srinivasan, K. Ramamurthy K. and Nene, Y.R. (1973) : *Norwesters, Andhis and Large Scale Convective activity, etc.* Forecasting Manual Part III No. 2.2.

Srivastava, H.N. and Singh, S.S. (1994) : *Long range weather forecasting techniques in India and using an EOF approach,* Proc. Ind. Nat. Sci. Acad. (Physical Sci.)

Stamp, L.D. (1958) : Asia : Regional and Economic Geography

Tarakanov, G.G. (1982) : *Tropical Meteorology* (translated from Russian by Michael Edelev), Mir Publishers, Moscow.

Thornthwaite, C.W. (1933) : The Climates of the Earth, *Geographical Review,* Vol. 23, No. 3 pp. 433-440.

Thornthwaite, C.W. (1948) : An Appraisal Towards Rational Classification of Climate, *Geographical Review,* Vol. 38, pp. 55-94.

Trewartha Glenn, T. (1954) : *An Introduction to climate,* McGraw Hill, New York.

— — (1961) : *The Earth's Problem Climates,* McGraw Hill, New York.

Trewartha Glenn, T. and Horn Lyle, H. (1980) : *An Introduction to Climate,* McGraw Hill.

Trujillo Alan P. and Truman Harold V. (2012), Essentials of Oceanography PHI Learning Pvt. Ltd., New Delhi.

Webster, P. (1981) : 'Monsoon' *Scientific American,* Vol. 245, No. 2, pp. 108-118.

◆◇◆

CHAPTER 6

Soils

INTRODUCTION

Soil is the thin surface-layer on the earth, comprising of mineral particles formed by the break-down of rocks, decayed organic materials, living organisms, water and air. Soil is formed under specific natural conditions and each element of the natural environment contributes to this complex process, described by the soil scientists as the process of *pedogenesis*.

Soil is a very important natural resource of India because agricultural production is basically dependent on the fertility of soil. The rich, deep fertile soils of the Ganga Plain, especially its delta, and coastal plain of Kerala support high density of population through agricultural prosperity. On the other hand, the shallow and coarse grained soils of Telangana and Rajasthan support only a small population because these soils do not provide a base for prosperous agriculture.

SOIL FORMATION IN INDIAN CONDITIONS

Indian soils, as in other parts of the world, reflect a combination of factors which have contributed to their formation. In Indian conditions, there is wide diversity with respect to geographical conditions such as, physiography, climate and vegetation. Consequently soils display a wide variety of physical and chemical characteristics. In India, soil formation is mainly related to the parent rock material, surface relief, climate and natural vegetation. Animals, insects and man also play an important role in soil formation. Some of the important factors of soil formation are mentioned below :

1. Parent Material. The material for soil formation is mainly derived from the rocks and is termed as the parent material by soil scientists. The parent material determines the colouration of the soil, its mineral composition and texture. India possesses a great variety of parent material which is generally categorised into following six classes :

(*i*) Ancient crystalline and metamorphic rocks

(*ii*) Cuddapah and Vindhyan rocks

(*iii*) Gondwana rocks

(*iv*) Deccan basalts

(*v*) Tertiary and Mesozoic sedimentary rocks of extra peninsular India

(*vi*) Recent and sub-recent rocks.

The surface rocks are exposed to the process of weathering and suffer decay and decomposition. In this process, the rocks are converted into fine grains and provide a base for the soil formation. The *ancient crystalline and metamorphic rocks* constitute greater part of the peninsular region. These rocks are basically granites, gniesses and schists which are rich in ferromagnesian materials. Such rocks give rise to red soils on weathering. The red colour of these soils is largely due to the presence of iron oxide. The *Cuddapah and Vindhyan rocks* have weathered to give calcareous and argillaceous soils. These soils are native. The *Gondwana rocks* give rise to comparatively less mature soils of more or less uniform character but of low fertility. *The Deccan traps* are composed of basalt. Basalts are quite rich in titanium, magnetite, aluminium and magnesium. Consequently the weathering of these rocks has given rise to soils of darker colour. The soil derived from the Deccan trap is fertile with high moisture holding capacity and is popularly known as ***'regur'*** or ***black cotton soil***. The *Tertiary and Mesozoic* sedimentary rocks of extra peninsular India have given rise to soils with high porosity. These are generally immature *recent and sub recent* rocks and result in alluvial soils on weathering.

The soils of the Northern Plain of India have been largely derived from the depositional work of the Himalayan rivers. This depositional work has been continuing for thousands of years. These are alluvial fertile soils consisting of fine silts and clay. These soils have little relation with the original rocks. On the other hand, the soils of peninsular plateau are generally coarse-grained and are closely related to the parent rocks. The peninsular soils are generally less fertile.

2. Relief. Relief influences the process of soil formation in many ways, the most important being the slope of land. Steep slope encourages the swift flow of water and hinders the process of soil formation. There may even be soil erosion in areas of steep slope. Chambal ravines offer an excellent example of soil erosion. The areas of low relief or gentle slope generally experience deposition and have deep soils. Because of this reason, there are thick layers of fertile alluvial soils in the northern plain of India whereas the soils are generally shallow in the plateau area. The exceptions in the plateau are river basins where the soil layers are sufficiently deep. The degree of slope also largely determines the fertility of soil.

3. Climate. Climate is the single most important factor in soil formation. Most important climatic factors affecting soil formation are the amount and seasonal distribution of temperature and rainfall. Climate controls the type and effectiveness of weathering of the parent material, the quantity of water seeping through the soil and the type of micro-organisms present therein. In areas of heavy rainfall and high temperature, the soils are red or lateritic. Torrential rainfall during the rainy season washes the upper soil and leaches the materials into deeper horizon. During the dry summer season the evaporation exceeds precipitation and through capillary action iron and aluminium sesquioxides are transported to the surface making the soil red. In areas of alternate wet and dry climate, the leached material which goes deep down in the horizon is brought up and the blazing sun bakes the top soil so hard that it resembles a brick. Therefore, this soil is called **lateritic which literally means brick**. In arid and semi-arid regions, evaporation always exceeds precipitation. Under such circumstances, two main factors determine the nature and properties of soils. Firstly there is very little vegetation and the soils badly lack humus content. Hence the soils are invariably of light colour. Secondly, the excess of evaporation makes soils lime accumulating. Thus they are bound to be pedocal in nature. Such soils are widely spread in the extreme western part of the country. In cold climates of the Himalayan region, the process of vegetation decay is very slow and the soils formed under such circumstances are acidic in nature. When the climatic control acts for a sufficiently long period, it reduces the differences in the parent materials. Two different parent materials may develop the same soil in the same type of climate. Similarly, the same parent material may produce two different types of soils in two different types of climates. The crystalline granites produce laterite soil *in relatively moist* parts of the monsoonal region and non-laterite soil in drier areas. Hot summer and low rainfall develops black soil as is found in some parts of Tamil Nadu irrespective of the parent rock. In Rajasthan, both granite and sandstone give birth to sandy soil under arid climate. This soil is poor in organic matter.

4. Natural Vegetation. Natural vegetation reflects the combined effects of relief and climate. The formation and development of soil is very much influenced by the growth of vegetation. The decayed leaf material adds much needed **humus** to soil thereby increasing its fertility. The densely forested areas contain some of the best soils in India. There is a close relationship between the vegetation types and soil types in India.

MAJOR SOIL GROUPS OF INDIA

India is a country of vast dimensions with varied conditions of geology, relief, climate and vegetation. Therefore, India has a large variety of soil groups, distinctly different from one another. Different criteria have been applied to classify Indian soils, the outstanding being geology, relief, fertility, chemical composition and physical structure, etc. Any classification based on any one of the aforesaid criteria has its own inherent drawback. Even the most competent pedologist would find it difficult to present an accurate, complete, comprehensive and generalised account of the Indian soils.

During the British rule in India, a vast body of fascinating accounts had emerged in district gazetteers and official reports. These accounts were generally directed towards the assessment of differential soil fertility and land revenue collection, but did not attempt classification of soil types in the country. The earlier studies of Indian soils were made by foreign scholars like *Volckar* (1893), *Leather* (1898), *Schokalskaya* (1932), *Champion* (1936), etc. Indian scholars including *Wadia* (1935), *Basu* (1937), *Vishwanath and Ukil* (1944), *Chatterjee, Krishnan, and Roychaudhary* (1954) made strenuous efforts to classify soils of India. In 1957, The *National Atlas Organisation* (Kolkata) published a soil map of India in which Indian soils were classified into 6 major groups and 11 broad types. The *Irrigation Atlas of India* (1972) and Spate's *India, Pakistan* and *Ceylon* (1976) utilised the 7th approximation soil classification developed by the U.S. Department of Agriculture (USDA). The 7th approximation defines soil classes strictly in terms of their morphology and composition as produced by a set of natural and human forces. The classification is determined by quantifiable criteria.

Geologically, Indian soils can broadly be divided into two main types : (*a*) Soils of peninsular India and (*b*) Soils of extra-peninsular India.

The soils of Peninsular India are those which have been formed by the deomposition of rocks *in situ*, *i.e.* directly from the underlying rocks. They are transported and redeposited to a limited extent and are known as sedentary soils. On the other hand, the soils of the Extra-Peninsula are formed due to the depositional work of rivers and wind. They are mainly found in the river valleys and deltas. They are very deep and constitute some of the most fertile tracts of the country. They are often referred to as *transported or azonal soils*.

The Indian Council of Agricultaral Research (ICAR) set up an All India Soil Survey Committee in 1953 which divided the Indian soils into eight major groups. They are (1) Alluvial soils, (2) Black soils, (3) Red soils, (4) Laterite and Lateritic soils, (5) Forest and Mountain soils, (6) Arid and Desert soils, (7) Saline and Alkaline soils and (8) Peaty and Marshy soils (See **Fig. 6.1**). This is a very logical classification of Indian soils and has gained wide acceptance. A brief account of these eight soil groups is given as under :

1. Alluvial Soils. Alluvial soils are by far the largest and the most important soil group of India. Covering about 15 lakh sq km or about 45.6 per cent of the total land area of the country, these soils contribute the largest share of our agricultural wealth and support the bulk of India's population. Most of the alluvial soils are derived from the sediments deposited by rivers as in the Indo-Gangetic plain although some alluvial soils in the coastal areas have been formed by the sea waves. Thus the parent material of these soils is all of transported origin. The streams bring with them the products of weathering of rocks from the mountains and deposit them in the low-lying areas. The alluvial soils are yet immature and have weak profiles. They differ in consistency from *drift sand to rich loams* and from *silts* to *stiff clays*. A few occasional *kankar* beds are also present. However, pebbly, stony or gravelly soils are rare in this group. The chemical composition of the alluvial soils make this group of soils as one of the most fertile in the world. The proportion of nitrogen is generally low, but potash, phosphoric acid and alkalies are adequate, while iron oxide and lime vary within a wide range. The porosity and texture provide good drainage and other conditions favourable for bumper crops. These soils are easily replenished by the recurrent river floods and support uninterruped crop growth.

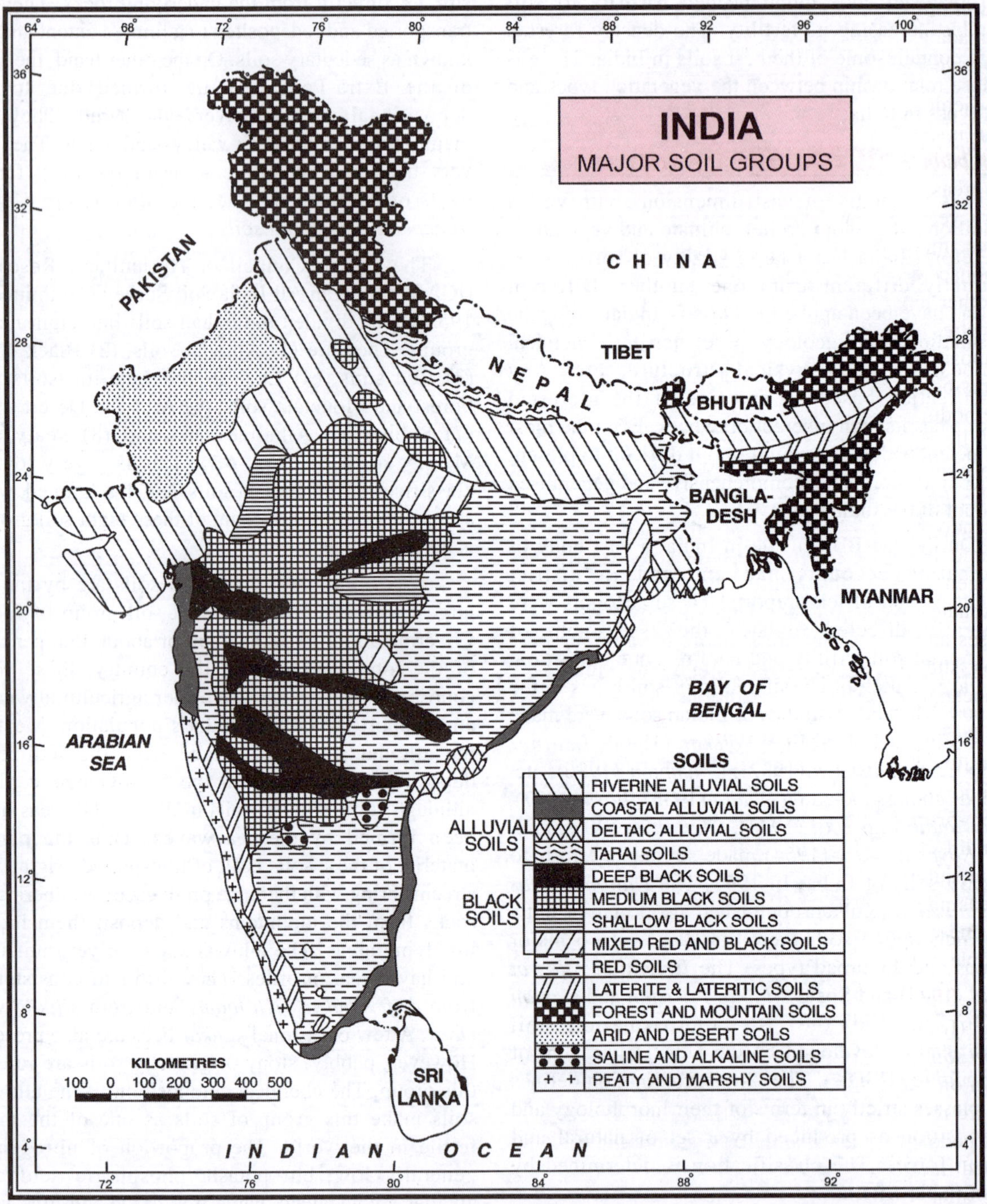

FIG. 6.1. India : Major Soil Groups

The widest occurrence of the alluvial soils is in the Great Indo-Gangetic Plain starting from Punjab in the west to West Bengal and Assam in the east. They also occur in deltas of the Mahanadi, the Godavari, the Krishna and the Cauvery, where they are called ***deltaic alluvium***. Along the coast they are known as

coastal alluvium. Some alluvial soils are found in the Narmada and Tapi valleys. Northern parts of Gujarat also have some cover of alluvial soils.

Geologically, the alluvium of the Great plain of India is divided into newer or younger *khadar* and older *bhangar* soils. The ***khadar*** soils are found in the low areas of valley bottom which are flooded almost every year. They are pale brown, sandy clays and loams, more dry and leached, less calcareous and carbonaceous *i.e.* they are less *kankary*. ***Bhangar***, on the other hand, is found on the higher reaches about 30 metres above the flood level. It is of a more clayey composition and is generally dark coloured. A few metres below the surface of the bhangar are beds of lime nodules known as ***kankar***. Along the Shiwalik foothills, there are alluvial fans having coarse, often pebbly soils. This zone is called ***bhabar***. To the south of the *bhabar* is a long narrow strip of swampy lowland with silty soils. It covers an area of 56,600 sq km and is called ***tarai***. The tarai soils are rich in nitrogen and organic matter but are deficient in phosphate. These soils are generally covered by tall grasses and forests but are suitable for a number of crops such as wheat, rice, sugarcane, jute and soyabean under reclaimed conditions.

Due to the softness of their strata and fertility the alluvial soils are best suited to irrigation and respond well to canal and well/tube-well irrigation. When properly irrigated, the alluvial soils yield splendid crops of rice, wheat, sugarcane, tobacco, cotton, jute, maize, oilseeds, vegetables and fruits. Increasing trend of rice cultivation in Punjab and Haryana with the help of intense irrigation is a living example.

2. Black Soils. The black soils are also called ***regur*** (from the Telugu word Reguda) and ***black cotton soils*** because cotton is the most important crop grown on these soils. Several theories have been put forward regarding the origin of this group of soils but most pedologists believe that these soils have been formed due to the solidifaction of lava spread over large areas during volcanic activity in the Deccan Plateau, thousands of years ago. Most of the black soils are derived from two types of rocks, the Deccan and the Rajmahal trap, and ferruginous gneisses and schists occurring in Tamil Nadu. The former are sufficiently deep while the later are generally shallow.

Geographically, black soils are spread over 5.46 lakh sq km (*i.e.* 16.6 per cent of the total geographical area of the country) encompassed between 15°N to 25°N latitudes and 72°E to 82°E longitudes. This is the region of high temperature and low rainfall. It is, therefore, a soil group of the dry and hot regions of the Peninsula. These soils are mainly found in Maharashtra, Madhya Pradesh, parts of Karnataka, Telangana, Andhra Pradesh, Gujarat and Tamil Nadu.

The black colour of these soils has been attributed by some scientists to the presence of a small proportion of titaniferous magnetite or even to iron and black constituents of the parent rock. The black colour of this soil may even be derived from crystalline schists and basic gneisses such as in Tamil Nadu and parts of Andhra Pradesh. Various tints of the black colour such as deep black, medium black, shallow black or even a mixture of red and black may be found in this group of soils.

The black soil is highly retentive of moisture. It swells greatly and becomes sticky when wet in rainy season. Under such conditions, it is almost impossible to work on such soil because the plough gets stuck in the mud. However, in the hot dry season, the moisture evaporates, the soil shrinks and is seamed with broad and deep cracks, often 10 to 15 cm wide and upto a metre deep. This permits oxygenation of the soil to sufficient depths and the soil has extraordinary fertility. Remarkably "*self-ploughed*" by loosened particles fallen from the ground into the cracks, the soil "*swallows*" itself and retains moisture. This soil has been used for growing a variety of crops for centuries without adding fertilizers and manures, or even fallowing with little or no evidence of exhaustion.

A typical black soil is highly argillaceous with a large clay factor, 62 per cent or more, without gravel or coarse sand. It also contains 10 per cent of alumina, 9-10 per cent of iron oxide and 6-8 per cent of lime and magnesium carbonates. Potash is variable (less than 0.5 per cent) and phosphates, nitrogen and humus are low. The structure is cloddish but occasionally friable. In all *regur* soils in general, and in those derived from ferromagnesian schists in particular, there is a layer rich in *kankar* nodules formed by segregation of calcium carbonate at lower depths. As a general rule, black soils of uplands are of low fertility but they are darker, deeper and richer in the valleys.

Because of their high fertility and retentivity of moisture, the black soils are widely used for producing several important crops. Some of the major crops grown on the black soils are cotton, wheat, jowar, linseed, virginia tobacco, castor, sunflower and millets. Rice and sugarcane are equally important where irrigation facilities are available. Large varieties of vegetables and fruits are also successfully grown on the black soils.

3. Red Soils. This comprehensive term designates the largest soil group of India, comprising several minor types. Most of the red soils have come into existence due to weathering of ancient crystalline and metamorphic rocks. The main parent rocks are acid granites and gneisses, quartzitic and felspathic. The colour of these soils is generally red, often grading into brown, chocolate, yellow, grey or even black. The red colour is due more to the wide diffusion rather than to high percentage of iron content.

The red soils occupy a vast area of about 3.5 lakh sq km which is about 10.6 per cent of the total geographical area of the country. These soils are spread on almost the whole of Tamil Nadu, parts of Karnataka, south-east of Maharashtra, Telangana, Andhra Pradesh, Madhya Pradesh, Chhattisgarh, Odisha and Chota Nagpur in Jharkhand. In the north the red soil area extends in large parts of south Bihar; Birbhum and Bankura districts of West Bengal; Mirzapur, Jhansi, Banda and Hamirpur districts of Uttar Pradesh; Aravallis and the eastern half of Rajasthan, parts of Assam, Nagaland, Manipur, Mizoram, Tripura and Meghalaya.

By and large, the red soils are poor in lime, magnesia, phosphates, nitrogen and humus, but are fairly rich in potash. In their chemical composition they are mainly siliceous and aluminous, with free quartz as sand, the alkali content is fair, some parts being quite rich in potassium. The texture of these soils varies from sand to clay, the majority being loams. On the uplands, the red soils are thin, poor and gravelly, sandy or stony and porous, but in the lower areas they are rich, deep dark and fertile.

The red soils respond well to the proper use of fertilizers and irrigation and give excellent yields of cotton, wheat, rice, pulses, millets, tobacco, oilseeds, potatoes and fruits.

4. Laterite and Lateritic Soils. The word 'laterite' (from Latin word meaning brick) was first applied by Buchanan in 1810 to a clayey rock, hardening on exposure, observed in Malabar. In the opinion of George Kuriyan, "It is probably the end product of decomposition found in regions of heavy rainfall, more than 200 cm. Such climatic conditions promote leaching of soil whereby lime and silica are leached away and a soil rich in oxides of iron and aluminium compounds is left behind. We have numerous varieties of laterite which have bauxite at one end and an indefinite mixture of ferric oxides at the other. Almost all laterite soils are very poor in lime and magnesia and deficient in nitrogen. Sometimes, the phosphate content may be high, probably present in the form of iron phosphate but potash is deficient. At some places, there may be higher content of *humus*.

Laterite and lateritic soils are widely spread in India and cover an area of 2.48 lakh sq km. They are mainly found on the summits of Western Ghats at 1000 to 1500 m above mean sea level, Eastern Ghats, the Rajamahal Hills, Vindhyas, Satpuras and Malwa Plateau. They also occur at lower levels and in valleys in several other parts of the country. They are well developed in south Maharashtra, parts of Karnataka, Telangana, Andhra Pradesh, Odisha, West Bengal, Kerala, Jharkhand, Assam and Meghalaya.

Due to intensive leaching and low base exchange capacity, typical laterite soils generally lack fertility and are of little use for crop production. But when manured and irrigated, some laterites and lateritics are suitable for growing plantation crops like tea, coffee, rubber, cinchona, coconut, arecanut, etc. In low lying areas paddy is also grown. Some of the laterite soils in Kerala, Karnataka, Chota Nagpur region of Jharkhand, Odisha and Assam respond well to the application of fertilizers like nitrogen, phosphorus and potassium. In some areas, these soils support grazing grounds and scrub forests.

Laterite and lateritic soils have a unique distinction of providing valuable building material. These soils can be easily cut with a spade but harden like iron when exposed to air. Because it is the end-product of weathering, it cannot be weatherd much further and is indefinitely durable.

5. Forest and Mountain Soils. Such soils are mainly found on the hill slopes covered by forests. These soils occupy about 2.85 lakh sq km which is about 8.67 per cent of the total land area of India. The formation of these soils is mainly governed by the characteristic deposition of organic matter derived from forest growth. These soils are heterogeneous in nature and their character changes with parent rocks, ground-configuration and climate. Consequently, they differ greatly even if they occur in close proximity to one another. In the Himalayan region, such soils are mainly found in valley basins, depressions, and less steeply inclined slopes. Generally, it is the north facing slopes which support soil cover; the southern slopes being too precipitous and exposed to denudation to be covered with soil. Apart from the Himalayan region, the forest soils occur on Western and Eastern Ghats as well as in some parts of the Peninsular plateau.

The forest soils are very rich in *humus* but are deficient in potash, phosphorus and lime. Therefore, they require good deal of fertilizers for high yields. They are especially suitable for plantations of tea, coffee, spices and tropical fruits in Karnataka, Tamil Nadu and Kerala and wheat, maize, barley and temperate fruits in Jammu and Kashmir, Himachal Pradesh and Uttarakhand.

6. Arid and Desert Soils. A large part of the arid and semi-arid region in Rajasthan and adjoining areas of Punjab and Haryana lying between the Indus and the Aravalis, covering an area of 1.42 lakh sq km (or 4.32% of total area) and receiving less than 50 cm of annual rainfall, is affected by desert conditions. The Rann of Kachchh in Gujarat is an extension of this desert. This area is covered by a mantle of sand which inhibits soil growth. This sand has originated from the mechanical disintegration of the ground rocks or is blown from the Indus basin and the coast by the prevailing south-west monsoon winds. Barren sandy soils without clay factor are also common in coastal regions of Odisha, Tamil Nadu and Kerala. The desert soils consist of aeolian sand (90 to 95 per cent) and clay (5 to 10 per cent). Some of these soils contain high percentages of soluble salts, are alkaline with varying degree of calcium carbonate and are poor in organic matter. Over large parts, the calcium content increases downwards and in certain areas the subsoil has ten times calcium as compared to that of the top soil. The phosphate content of these soils is as high as in normal alluvial soils. Nitrogen is originally low but its deficiency is made up to some extent by the availability of nitrogen in the form of nitrates. Thus, the presence of phosphates and nitrates make them fertile soils wherever moisture is available. There is, therefore, great possibility of reclaiming these soils if proper irrigation facilities are available. The changes in the cropping pattern in the Indira Gandhi Canal Command Area is a living example of the utility of the desert soils. However, in large areas of desert soils, only the drought resistant and salt tolerant crops such as barley, rape, cotton, wheat, millets, maize and pulses are grown. Consequently, these soils support a low density of population.

7. Saline and Alkaline Soils. These soils are found in Andhra Pradesh, Telangana and Karnataka. In the drier parts of Bihar, Uttar Pradesh, Haryana, Punjab, Rajasthan and Maharashtra, there are salt-impregnated or alkaline soils occupying 68,000 sq km of area. These soils are liable to saline and alkaline efflorescences and are known by different names such as ***reh***, ***kallar***, ***usar***, ***thur***, ***rakar***, ***karl*** and ***chopan***. There are many undecomposed rock and mineral fragments which on weathering liberate sodium, magnesium and calcium salts and sulphuric acid. Some of the salts are transported in solution by the rivers, which percolate in the sub-soils of the plains. In canal irrigated areas and in areas of high sub-soil water table, the injurious salts are transferred from below to the top soil by the capillary action as a result of evaporation in dry season. The accumulation of these salts makes the soil infertile and renders it unfit for agriculture. It has been estimated that about 1.25 million hectares of land in Uttar Pradesh and 1.21 million hectares in Punjab has been affected by *usar*. In Gujarat, the area around the Gulf of Khambhat is affected by the sea tides carrying salt-laden deposits. Vast areas comprising the estuaries of the Narmada, the Tapi, the Mahi and the Sabarmati have thus become infertile.

8. Peaty and Marshy Soils. Peaty soils originate in humid regions as a result of accumulation of large amounts of organic matter in the soils. These soils contain considerable amount of soluble salts and 10-40 per cent of organic matter. Soils belonging to this group are found in Kottayam and Alappuzha districts of Kerala where it is called ***kari***. Marshy soils with a high proportion of vegetable matter also occur in the coastal areas of Odisha and Tamil Nadu, Sunderbans of West Bengal, in Bihar and Almora district of Uttarakhand. The peaty soils are black, heavy and highly acidic. They are deficient in potash and phosphate. Most of the peaty soils are under water during the rainy season but as soon as the rains cease, they are put under paddy cultivation.

Characteristics of Indian Soils. Following are the chief characteristics of Indian soils :

1. The Indian soils have been formed under varied geographical conditions and differ widely in their physical properties, chemical composition and fertility level.
2. Most soils are old and mature. Soils of the peninsular plateau are much older than the soils of the great northern plain.
3. Indian soils are largely deficient in mineral salts, humus, organic materials and nutrients. According to a report released by the Ministry of Agriculture and Farmers Welfare released in December, 2019, soil in dozen states including Andhra Pradesh, Chhattisgarh, Gujarat, Madhya Pradesh, Uttar Pradesh and Telangana face nitrogen deficiency while phosphorous deficiency has been found in 14 states/union territories including Assam, Haryana, Delhi, Puduchery, Uttar Pradesh, Odisha and Goa. Potassium deficiency in soil has been found in six states and UTs (five of them in north-east) and these include Arunachal Pradesh, Meghalaya, Tripura and Andaman and Nicobar Islands. Nitrogen, phosphorus and potassium, known as NPK are essential for plant growth.
4. Plains and valleys have thick layers of soils while hilly and plateau areas depict thin soil cover.
5. Some soils like alluvial and black soils are fertile while some other soils such as laterite, desert and alkaline soils lack in fertility and do not yield good harvests.
6. Indian soils have been used for cultivation for hundreds of years and have lost much of their fertility. As such there is urgent need of giving scientific treatment to our soils.
7. Indian climate is characterised by seasonal rainfall and our soils need irrigation during the dry period.
8. Indian soils suffer from soil erosion and other allied problems.

PROBLEMS OF INDIAN SOILS

Indian soils suffer from a number of problems, some of them taking a serious turn. Some of the important problems faced by Indian soils are (*i*) soil erosion, (*ii*) deficiency in fertility, (*iii*) desertification, (*iv*) waterlogging (*v*) salinity and alkalinity (*vi*) wasteland, (*vii*) over exploitation of soils due to increase in population and rise in living standards and (*viii*) encroachment of agricultural land due to urban and transport development.

SOIL EROSION

Soil erosion is the removal of soil by the forces of nature, particularly wind and water, more rapidly than the various soil forming processes can replace it. Soil erosion is a serious menace which is adversely affecting our agricultural productivity and the economy of the country as a whole. Although the process of soil erosion is imperceptibly slow and it can scarcely be detected, the loss over periods of time is indeed very great.

TYPES OF SOIL EROSION

Two natural agents i.e., water and wind, are constantly at work causing soil erosion. Therefore, we can talk of two types of soil erosion *viz*., water erosion and wind erosion.

Water Erosion. During heavy rains, water removes a lot of soil. Rain drops fall with an approximate speed of 10 metres per second and wash away the top soil. The soils most readily detached by raindrop splash erosion are sands and silt. Coarser particles are not shifted about as much because of their greater volume and weight. Run off water is responsible for much soil erosion, moving the soil particles by *surface creep*, *saltation* and *suspension*. Water erosion consisting of *rilling*, *gullying*, *sheet-wash and rain peeling process* is mainly confined to the mountains, hills and upper slopes of the piedmont

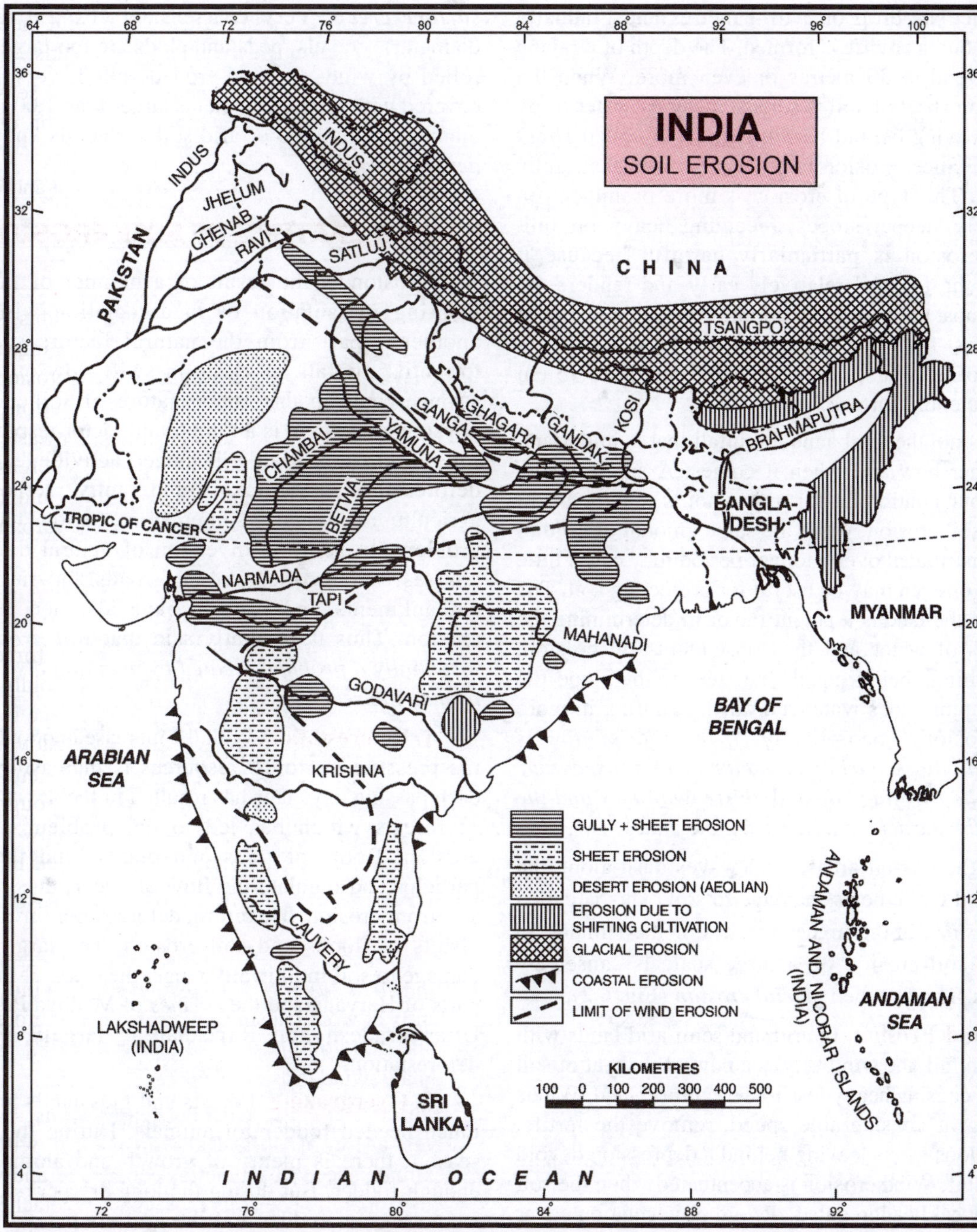

FIG. 6.2. India : Soil Erosion

zone. If erosion continues unchecked for a sufficient time, numerous finger-shaped grooves may develop all over the area due to silt laden run off. The whole pattern resembles that of the twigs, branches and trunk of a tree. This is called ***rill erosion***. With further erosion of the soil, the rills may deepen and become enlarged and are ultimately turned into **gullies**. The cutting of soil goes to immense size and volume and the entire area may be turned into **badland topography**. When a gully bed cuts into the soil with an immediate drop of 3 to 4 metres and gradually flattens out, a **ravine** is formed. The depth of a ravine may extend to 30 metres or even more. When the entire top sheet of soil is carried away by water or by wind, leaving behind barren rock, it is called ***sheet erosion***. Sheet erosion is no less harmful than gully erosion. This type of erosion is more prominent on relatively steeper slopes, receiving heavy rainfall. Sheet erosion is particularly harmful because it attacks the top soil relatively early and renders the land almost unfit for cultivation. If one cubic metre of soil is lost from one hectare of land each week for a period of 30 years, the loss would amount to 15 cm from the entire surface.

It is not the total annual rainfall that is important, but rather how and when it comes. A single heavy downpour continuing for a few hours may result in severe soil erosion, while the same amount of rainfall fairly distributed over a longer period may cause little erosion or even may be useful for protecting soil. The slope of the land is a potent factor in determining the velocity of water and the consequent soil erosion. Other things being equal, the steeper the slope the more rapidly does water run down resulting in more soil erosion. Theoretically, *if the rate of flow is doubled, the scouring capacity is increased four times, the carrying capacity thirty two times and the size of the particles carried sixty four times*.

In the coastal areas, tidal waves dash along the coast and cause heavy damage to soil. This is called ***sea erosion***. In the higher reaches of the Himalayan region, soil erosion on a large scale is caused by glaciers. This is called ***glacial erosion*** (**Fig. 6.2**).

Wind Erosion. In arid and semi arid lands with little rainfall, the wind acts as a powerful agent of soil erosion causing heavy loss to agricultural land. Winds blowing at considerable speed, remove the fertile, arable, loose soils leaving behind a depression devoid of top soil. Wind erosion is accentuated when the soil is dry, weakly aggregated, devoid of vegetation cover along with over grazing and the winds are strong. Even modest wind velocities can keep individual particles of humus, clay and silt in suspension. Very fine, fine and medium sands are moved by wind in a succession of bounds and leaps, known as ***saltation***. Coarse sand is not usually airborne but rather is rolled along the soil surface. This type of erosion is called ***surface creep***. Very coarse sand (1.0-2.0 mm in diameter), gravels, peds and clods are too large to be rolled by wind, so wind-eroded soils have surfaces covered with coarse fragments larger than 1.00 mm in diameter. This kind of arid soil surface is known as ***desert pavement***.

HUMAN FACTORS OF SOIL EROSION

Soil erosion is the result of a number of factors, working in isolation or in association with one another. Apart from the natural factors such as torrential rainfall, resulting in swift flow of water, strong winds in dry areas, nature of soil and the physiography, man is an important factor responsible for soil erosion. Man's ill judged activities such as deforestation, overgrazing and faulty methods of agriculture have made soil erosion a serious national problem. In addition diversion of natural drainage courses, wrong orientation of roads and railways, embankments and bridges have also led to soil erosion. Thus it is rightly said that *soil erosion is essentially a problem created by man and also faced by man himself*.

1. Deforestation. With the increase in population, the pressure on forest resources is increasing with each passing day. This has resulted in reckless cutting of forests which has led to the problem of soil erosion. Roots of trees and plants bind the soil particles and regulate the flow of water, thus saving soil from erosion. Therefore, deforestation invariably results in floods and soil erosion. The large scale damage to soil in Shiwalik range, the *Chos* of Punjab, parts of Haryana and the *ravines* of Madhya Pradesh, Uttar Pradesh and Rajasthan is largely due to deforestation.

2. Overgrazing. Forests and grasslands provide much needed fodder for animals. During the rainy season, there is plenty of growth and animals get enough fodder. But during the long dry period, there is shortage of fodder and the grass is grazed to the ground and torn out by the roots by animals. This leads to loose structure of the soil and the soil is easily washed away by rains. Moreover, soil is pulverised by the hoofs and teeth of animals, especially by sheep and goats and thus proves detrimental to top soil when heavy showers fall on it. Soil erosion due to overgrazing is a common sight in the hilly areas of Himachal Pradesh, Uttarakhand, Jammu and Kashmir and in the dry areas of Rajasthan, Madhya Pradesh, Maharashtra, Karnataka and Andhra Pradesh.

3. Faulty Methods of Agriculture. Much of the soil erosion in India is caused by faulty methods of agriculture. The most outstanding are wrong ploughing, lack of crop rotation and practice of shifting cultivation. If the fields are ploughed along the slope, there is no obstruction to the flow of water and the water washes away the top soil easily. In some parts of the country, the same crop is grown year after year which spoils the chemical balance of the soil. This soil is exhausted and is easily eroded by wind or water.

Another outstanding example of faulty method of agriculture is the *shifting cultivation* practised in some areas in the north-eastern states of Arunachal Pradesh, Assam, Meghalaya, Manipur, Tripura, Mizoram as well as in Odisha. It is sporadically practised in Telangana, Andhra Pradesh, Madhya Pradesh, Chhattisgarh, Maharashtra, Kerala, Karnataka and Tamil Nadu. In this method, a piece of forest land is cleared by felling and burning of trees and crops are grown. The removal of the forest cover leads to the exposure of the soil to rains and sun which results in heavy loss of top soil, especially on the hill slopes. Thus the soil becomes unfit for cultivation and the tribes move to another piece of land after 2-3 years, returning to the earlier one after a gap of 10-15 years. In this way, the whole of the forest area is adversely affected by shifting cultivation resulting in intensive soil erosion in vast areas.

EXTENT OF SOIL EROSION IN INDIA

It has been estimated that an area of over 80 million hectares or about one-fourth of our total area is exposed to wind and water erosion out of which 40 million hectares of land has undergone serious erosion. Ironically the extent of soil erosion is increasing inspite of our efforts to check soil erosion. Experts have estimated that about 40,000 hectares of our land is permanently lost to cultivation and much larger area is rendered less productive every year due to wind and water erosion. About 21 million hectares are subject to severe wind erosion in Rajasthan and adjoining areas of Punjab, Haryana, and Gujarat (**Fig. 6.2**).

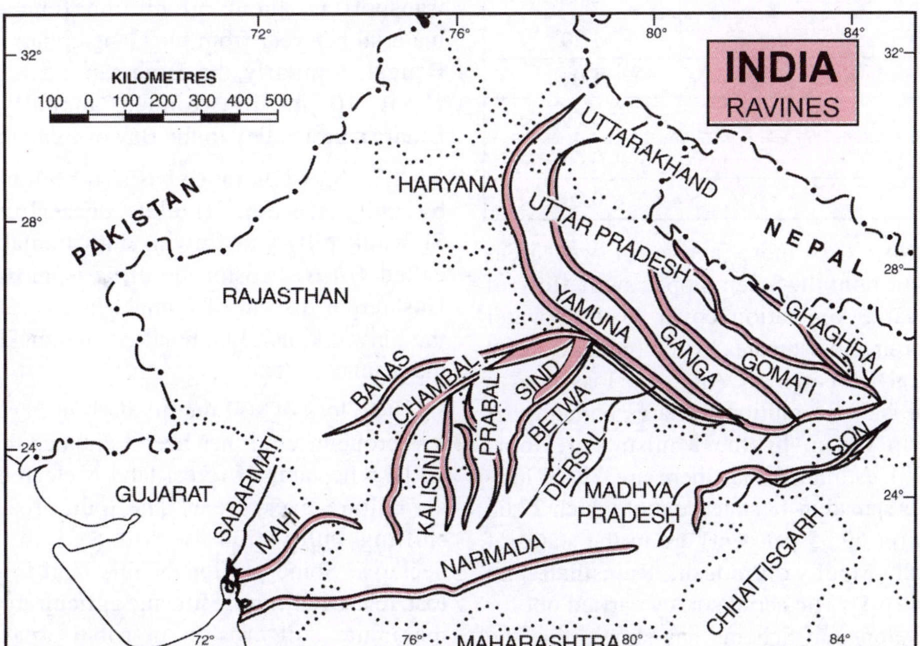

FIG. 6.3. India : Ravines

Wind erosion is a serious problem in arid and semi-arid parts of north west India. About 45 million hectares of land is subject to severe wind erosion in Rajasthan and adjoining areas of Punjab, Haryana, Gujarat and Western Uttar Pradesh (**Fig. 6.2**). It is estimated that 34 lakh tonnes of fertile soil is removed by wind every year in the districts of Jodhpur, Bikaner, Kota, Jaipur, Bharatpur, Kishangarh etc. in Rajasthan. These areas receive scanty rainfall. They are devoid of vegetation cover and have sandy soil. As such these areas are exposed to wind erosion. Faulty farming practices, failure to conserve moisture, lack of management and over grazing make the problem of wind erosion more complicated. According to the latest estimates about 8 million tonnes of soil has been removed from every square kilometre area in desert during the last hundred years.

TABLE 6.1. Area Under Ravines in India (lakh hectares)

State	Area
Uttar Pradesh	12.30
Madhya Pradesh	6.83
Rajasthan	4.52
Gujarat	4.00
Punjab	1.20
West Bengal	1.04
Bihar	0.60
Tamil Nadu	0.60
Maharashtra	0.20

Water erosion is more active in wet areas receiving more rainfall. Steep slope, swift flow of rivers and scarce vegetation cover lead to water erosion. According to estimates by the Indian Council of Agricultural Research (ICAR), the loss due to water erosion is 53.34 million hectares annually. A working group set up by the Ministry of Home Affairs in 1971 estimated that there are 39.75 lakh hectare ravines spread in 18 states, out of which 27.65 lakh hectares (or 69.55 per cent) are in the states of Uttar Pradesh, Madhya Pradesh, Rajasthan and Gujarat (**Table 6.1**). The aerial survey carried out for Chambal Development Scheme has shown that the area covered by ravines upto 4.5-6.0 metres depth is about 50,600 hectares. In Madhya Pradesh about 4 to 8 lakh hectares are affected by deep gullies and ravines along the banks of rivers Chambal and Kali Sindh. Out of this about 2.4 lakh hectares are in the districts of Gwalior, Morena and Bhind. In Uttar Pradesh the ravines are mostly along the banks of the Chambal, the Yamuna, the Gomati, Son and their tributaries (**Fig. 6.3**). Agra, Etawah and Jalaun are the worst affected districts. According to another estimate the Chambal-Yamuna badlands covering a total area of about 32 lakh hectares are the result of 1000 years of soil erosion wherein 0.25 tonnes of soil is being removed every day.

In Timil Nadu, ravines are common in South Tirunelveli, Tirupur, Kanniyakumari, Tiruchchirapalli, Chingleput, Salem and Coimbatore districts. In West Bengal numerous gullies and ravines exist in the upper catchment areas of the Kangsabati river in Purulia district. In Bihar, river courses of the Ganga, the Gandak, the Kosi and the Son are affected by ravines.

The flood plains of the Ganga and its tributaries in Uttar Pradesh and Bihar also suffer from the problem of soil erosion caused by water. These rivers are carving deep furrows and removing fertile top soil. According to one estimate the Ganga river is transporting about 30 million tonnes of eroded material per year from the Ganga plain to the Bay of Bengal. Similarly the Brahmaputra is transporting about 10 million tonnes annually from the Brahmaputra valley to the Bay of Bengal.

The Shiwalik range has also been badly affected by gully erosion. Rivulets descending from the Shiwalik hills and flowing into Punjab are locally called ***Chos***. Erosion by ***chos*** is most marked in Hoshiarpur district of Punjab. In the 130 km belt of the Shiwalik, nearly a hundred streams debouch onto the plains.

The loss of soil due to *shifting agriculture* is no less pronounced. It has been estimated that more than 15 lakh hectares of forest land is cleared for shifting agriculture every year. The total area affected by shifting cultivation is estimated to be 45 lakh hectares. Some portion of this land is permanently lost to agriculture. Shifting agriculture has caused maximum soil erosion in tribal areas of Assam, Meghalaya, Tripura, Nagaland, Mizoram, Kerala, Andhra Pradesh, Telangana, Odisha, Madhya

Pradesh, Chhattisgarh, etc. It is reported that about 207,287 hectares in Assam, 41,963 hectares in Tripura and 21,862 hectares in Manipur are under shifting cultivation. In Odisha about 33,08,502 hectares of land are subjected to shifting cultivation.

In Tamil Nadu the bad effects of misuse of land are illustrated in the Nilgiri Hills where potato is one of the main crops. Here potato cultivation is done on steep slopes, sometimes exceeding 60 per cent. This has led to intense erosion of soil and the yields of potato have gone down by about 50 per cent in spite of heavy application of fertilisers. Land steeper than 1 in 4 should not be open for cultivation but put under grasses and trees.

Coastal erosion is another form of erosion caused by constant pounding of tidal waves and currents on the sea coast. Waves and currents strike the coast with great force and break the hanging cliff rocks. The broken material is carried away by the retreating waves. Coastal erosion is quite pronounced in the season of monsoon winds and during storms and cyclones. Several coastal areas in Gujarat, Maharasthra, Karnataka, Kerala, Tamil Nadu, Andhra Pradesh and Odisha have suffered heavily at the hands of sea erosion. Of the 560 km long coast of Kerala, about 32 km stretch consisting of sandy beaches is subject to severe sea erosion. Erosion of beaches along the Kerala coast is evidenced by uprooting of coconut trees.

Effects of Soil Erosion. The adverse effects of soil erosion are reflected in the following points :

1. Top soil is eroded which leads to loss of soil fertility and fall in agricultural productivity.
2. Flooding and leaching result in loss of mineral nutrients.
3. Ground water level is lowered and there is decrease in soil moisture.
4. Natural vegetation cover dries up and arid lands expand.
5. Frequency and intensity of floods and drought increases.
6. Rivers, canals and tanks are silted and their water holding capacity decreases.
7. The incidence and damaging power of landslides increases.
8. Economy as a whole suffers a great setback.

Soil Salinity and Soil Alkalinity. Soil salinity and soil alkalinity are the results of over irrigation in canal irrigated areas. In canal irrigated areas plenty of the water is available and the farmers indulge in over irrigation of their fields. Under such conditions, the ground water level rises and saline and alkaline efflorescences consisting of salts of sodium, calcium and magnesium appear on the surface as a layer of white salt through capillary action. *Salinity* means the predominance of chlorides and sulphates of sodium, calcium and magnesium in the soils in sufficient quantity to be able to seriously interfere with the growth of most plants. *Alkalinity* implies the dominance of sodium salts, specially sodium carbonate. Although salts of alkali are somewhat different in their chemical properties from the salts of saline soils both soils occur in the same areas. Increasing salinity and alkalinity always indicate extension of waterlogging salt encrustation (saline efflorescence) or ***thur*** tendencies. Sandy soils are more prone to alkalinity and the loamy soils to salinity-alkalinity. Salinity and alkalinity have adverse effect on soil and reduce soil fertility. It is estimated that about 80 lakh hectares of land (2.43% of the country's total area) is affected by the problem of salinity and alkalinity. Vast tracts of canal irrigated areas in Uttar Pradesh, Punjab and Haryana; arid regions of Rajasthan, semi-arid areas of Maharashtra, Gujarat, Andhra Pradesh, Telangana and Karnataka and coastal areas of Odisha, Gujarat and West Bengal are facing this problem (**Table 6.2**). The western part of Uttar Pradesh is one of the worst sufferers at the hands of salinity and alkalinity. In some parts of Uttar Pradesh the internal drainage is greatly restricted and the soils are charaterised by alkalinity. Some of the most fertile soils in Punjab and Haryana have been rendered useless by salinity and alkalinity. In Punjab about 6,000 to 8,000 hectares of good land is becoming barren every year due to salinisation. Although Indira Gandhi canal in Rajasthan has turned the sandy desert into a granary, it has given birth to serious problems of salinity and alkalinity. Alkali soils are met with almost all over the state of Maharashtra. In Gujarat, the area around the Gulf of Khambhat is affected by sea tides carrying salt-laden deposits. Nearly 173,530 sq km comprising estuaries of the Narmada, the Tapi, the Mahi and the Sabarmati have been damaged in this way. Portions of Dharwar

TABLE 6.2. Salinity Affected Areas in India

States/Union territories	Area (lakh hectares)
1. Uttar Pradesh	12.95
2. Punjab	12.25
3. Gujarat	12.14
4. West Bengal	8.50
5. Rajasthan	7.28
6. Maharashtra	5.34
7. Haryana	5.26
8. Karnataka	4.04
9. Odisha	4.04
10. Madhya Pradesh	2.24
11. Andhra Pradesh and Telangana	0.42
12. Delhi	0.11
13. Bihar	0.04
14. Tamil Nadu	0.04

district and of Bijapur taluk are affected by what is locally known as ***karl*** soils which are saline, alkaline and fairly deep and clayey. Salt lands of the Nira valley have developed due to excessive irrigation on deep black soils of the locality.

Effects of salinity and alkalinity

Soil salinity and alkalinity has many adverse effects, some important effects are as under :

(*a*) Soil fertility is reduced which results in crop failure. Cultivation is not possible on saline soils unless they are flushed out with large quantities of irrigation water to leach out the salts.

(*b*) Choice of crops is limited because some crops are sensitive to salinity and alkalinity. Only high salt tolerant crops such as cotton, rape, barley, etc. and medium salt tolerant crops like wheat, rice, linseed, pulses, millets etc. can be grown.

(*c*) Quality of fodder becomes poor.

(*d*) Salinity and alkalinity create difficulties in building and road construction.

(*e*) These cause floods due to reduced infiltration, leading to crop damage in the adjoining areas.

Steps to treat salinity and alkalinity

Following steps are necessary to treat salinity and alkalinity and restore the fertility of soil.

1. Providing outlets for water to drain out excess water and lower water table. Efforts should also be made to seal all points and strips of the leakage from canals, tanks and ponds by lining them.
2. Minimising the use of water. Making judicious use of irrigation facilities.
3. Planting salt tolerant vegetation and crops such as cotton, rape, barley, date palm, linseed, etc. and certain grasses as fodder crops can be helpful.
4. It has been found that crop rotation involving ***dhaincha*** (green fodder)—cotton in the Deccan plateua, ***dhaincha***—rice in the Uttar Pradesh and ***dhaincha***—rice-barseem in Punjab and Haryana have been very helpful.
5. Liberal application of gypsum in upper 15 cm thick soil to convert the alkalies into soluble compounds. Alkali can be removed by adding sulphuric acid or acid forming

TABLE 6.3. Area under Arid and Semi-arid conditions in India (as per cent to the total geographical area)

States	Arid	Semi-arid
Rajasthan	57.42	36.67
Gujarat	33.72	47.50
Haryana	29.32	59.77
Punjab	24.60	75.40
Maharashtra	0.42	61.17
Andhra Pradesh and Telangana	7.18	44.66
Karnataka	4.27	72.60
Ladakh	31.13	6.22
Uttar Pradesh	—	21.73
Madhya Pradesh	—	13.38
Tamil Nadu	—	15.54
Total average	**12.13**	**29.13**

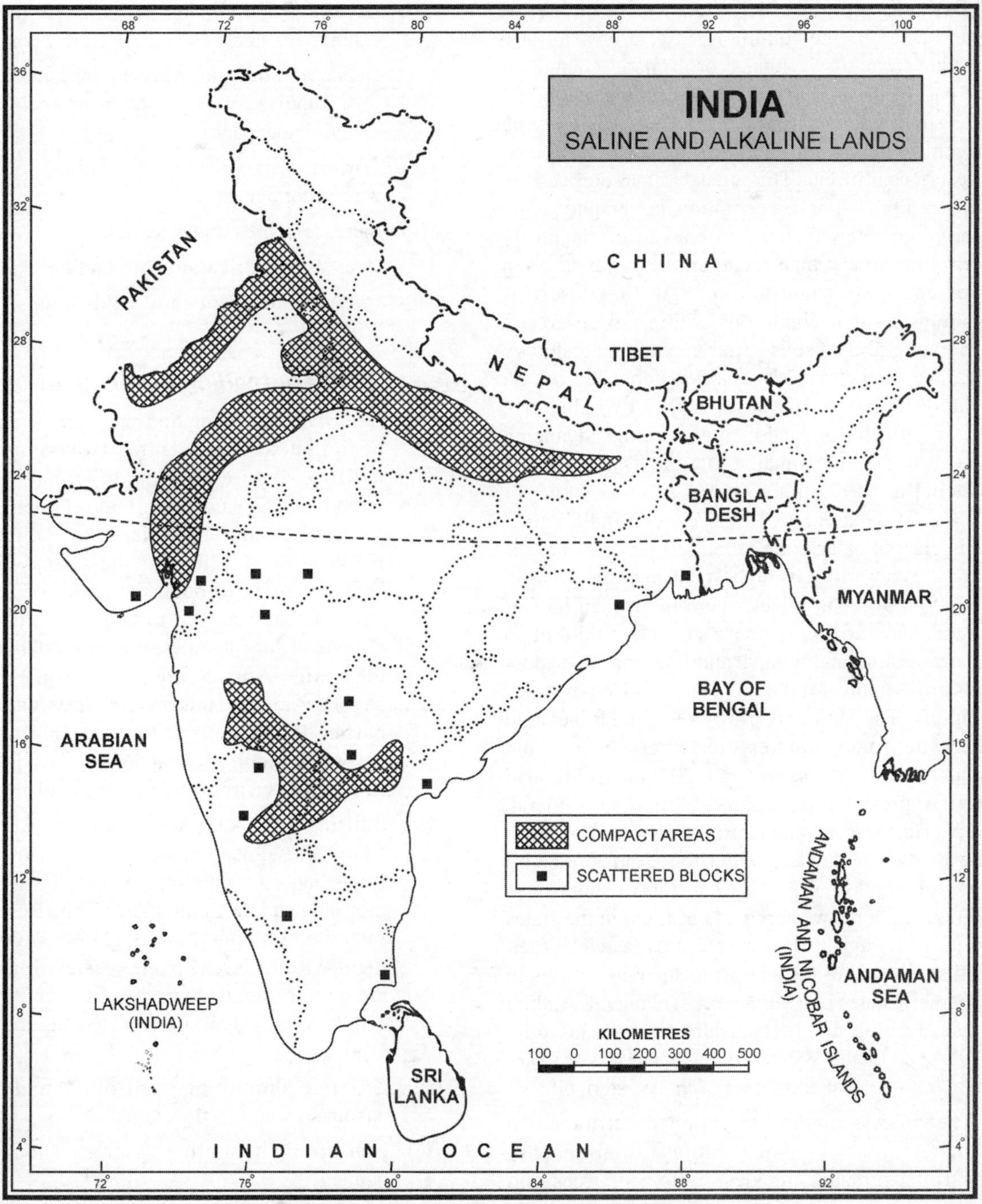

FIG. 6.4. India : Saline and Alkaline Lands

substances like sulphur and pyrite. Also organic residues such as rice husk and rice straw can be added to promote formation of mild acid as a result of their decomposition.

6. Flushing the salt by flooding the fields with excess water. However, this practice can lead to accumulation of saline water in the downstream area.

Desertification. Desertification can be defined as spread of desert like conditions in arid or semi-arid areas due to man's influence or climatic change. A large part of the arid and semi-arid region belonging to Rajasthan and south Punjab and Haryana and lying between the Indus and the Aravali range is affected by desert conditions. This area is characterised by scanty rainfall, sparse vegetation cover and acute scarcity of water. The Rajasthan desert is partly natural but largely man made and the fear is often expressed, quite rightly too, that the desert is spreading and that conditions within the desert are deteriorating. Desert soils suffer maximum erosion by wind. The sand carried by wind is deposited on the adjoining fertile lands whose fertility dwindles and sometimes the fertile land merges with the advancing desert. It has been estimated that the Thar Desert is advancing at an alarming rate of about 0.5 km per year. Currently about 12.13 per cent of India's total area is classed as arid and about 30 per cent area is termed as semi-arid. The highest concentration of arid land is found in Rajasthan where more than half of the total area is arid. Gujarat has about one-third of its total area as arid land. Punjab and Haryana have about one-fourth of the total land characterised by arid conditions. The Union Territory of Ladakh has cold desert where temperatures are extremely low and annual rainfall hardly excess 20 cm. Semi arid conditions prevail in large parts of Rajasthan, Gujarat, Punjab, Haryana, Maharashtra, Telangana, Andhra Pradesh, Karnataka, Ladakh, Uttar Pradesh, Tamil Nadu and Madhya Pradesh (**Table 6.3**). Semi-arid conditions in north-western parts of India in the states of Rajasthan, Punjab, Haryana etc. are created by their long distances from the sea while such conditions in the southern states like Karnataka, Telangana Andhra Pradesh, Tamil Nadu, Maharashtra etc. are caused by the presence of the Western Ghats. These states are located in the rain shadow area of the Western Ghats.

The process of desertification is attributed to various causes of which more important are uncontrolled grazing, reckless felling of trees and growing population. Climate change has also contributed to the spread of deserts.

Desertification has several ecological implications. Some important implications are listed below.

(*i*) Drifting of sand and its accumulation on fertile agricultural land.

(*ii*) Excessive soil erosion by wind and to some extent by water.

(*iii*) Deposition of sand in rivers, lakes and other water bodies thereby decreasing their water containing capacity.

(*iv*) Lowering of water table leading to acute water shortage.

(*v*) Increase in area under wastelands.

(*vi*) Decrease in agricultural production.

(*vii*) Increase in frequency and intensity of droughts.

Measures of Controlling Desertification

(*i*) Intensive tree plantation programme should be initiated. The Central Government is mulling ambitious plan to create about 1400 km long and 5 km wide 'Green Belt' from Porbandar in Gujarat to Panipat in Haryana on the lines of the "Great Green Wall" running through the width of Africa from Dakar (Senegal) to Djibouti to combat climate change and desertification. Covering the entire Aravali range and some areas beyond Aravalis, this green belt will act as a barrier for dust from the west and check eastward march of Thar Desert. It will also restore a large tract of degraded land.

(*ii*) Shifting sand dunes in Bikaner, Barmer, Churu, Jaisalmer and Jhunjhunun districts of Rajasthan cover an approximate area of 74 thousand sq km. Central Arid Zone Research Institute, Jodhpur has suggested mulching them with different plant species. Mulches are put in small squares and serve as an effective physical barrier to the moving sand.

(*iii*) Grazing should be controlled and new pastures should be developed.

(*iv*) Indiscriminate felling of trees should be banned.

(*v*) Alternative sources of fuel can reduce the demand for fuelwood and save the trees from destruction hence checking the onward march of the desert.

(*vi*) Sandy and wastelands should be put to proper use by judicious planning.

Waterlogging. The flat or nearly level surfaces and saucer-like depressions make the movement of surface water sluggish leading to accumulation of rain water which in turn results in waterlogging and flooding of the soil. Waterlogged soils are soaked or saturated with water. Moreover seepage from unlined channels or canal systems also leads to waterlogging in the contiguous arable lands. It has been estimated that extent of waterlogged soils is about 12 million hectares in India; half of which lies along its coast and the other half in the inland area under existing or newly created irrigation command areas (**Fig. 6.5**).

Waterlogging is believed to be one of the chief causes of salinity. The problem of land reclamation under waterlogging conditions is a complex one and must therefore, be tackled with great care as the scheme involves huge expenditure. Proper layout of drainage schemes is the only way to overcome the menace of waterlogging. The basic methods of removing excess water from waterlogged soils are (*a*) surface drainage and (*b*) vertical drainage.

(a) Surface Drainage. Surface drainage involves the disposal of excess water over ground surface through an open drainage system with an adequate outlet. Surface drainage is helpful where (*i*) soils are deep with low infiltration rates, where (*ii*) intensity of rainfall is high, where (*iii*) terrain is level to nearly level and where (*iv*) the water table is high.

(b) Vertical Drainage. Any bore or well from which the underlying water is extracted is defined as vertical drainage. The success of vertical drainage depends upon the presence of favourable aquifer and water table for lifting the ground water on sustainable basis and the favourable quantity of water that could be re-utilised for irrigation purposes. Such conditions prevail in the Indo-Gangetic plain. In the Punjab-Haryana plain the water-table has been successfully lowered through tubewell pumping. As a consequence waterlogged saline lands could be reclaimed.

SOIL CONSERVATION

Soil conservation includes all those measures which help in protecting the soil from erosion and exhaustion. Soil erosion has been continuing over such a large part of India for such a long time that it has assumed alarming proportions. According to Prof. **S.P. Chatterjee,** "*Soil erosion is the greatest single evil to Indian agriculture and animal husbandry*". Soil is our most precious asset and no other gift of nature is so essential to human life as soil. Productive soil alone ensures prosperous agriculture, industrial development, economic betterment and a higher standard of living. A healthy agriculture is bound up with healthy soil. While emphasising the significance of conservation, **G.T. Renner** has said that conservation is defined as "the greatest good to the largest number for the longest time." According to **S.I. Kayastha,** "*With soil conservation people rise and with its destruction they fall.*" It has been estimated that about two thirds of our arable land needs conservation measures. There is, therefore, an urgent need to conserve soil for the sake of prosperity of our masses. Unfortunately, it has not attracted the attention that it deserves. Our peasantry is not fully aware of many benefits of soil conservation and *neglect of soil is like killing the hen that lays the golden eggs*. Following methods are normally adopted for conserving soil :

1. Afforestation. The best way to conserve soil is to increase area under forests. Indiscriminate felling of trees should be stopped and efforts should be made to plant trees in new areas. A minimum area of forest land for the whole country that is considered healthy for soil and water conservation is between 20 to 25 per cent but it was raised to 33 per cent in the second five year plan; the proportion being 20 per cent for the plains and 60 per cent for hilly and mountainous regions.

2. Checking Overgrazing. Overgrazing of forests and grass lands by animals, especially by goats and sheep, should be properly checked. Separate grazing grounds should be earmarked and fodder crops should be grown in larger quantities. Animals freely move about in the fields for grazing and spoil the soil by their hoofs which leads to soil erosion. This should be avoided.

3. Constructing Dams. Much of the soil erosion by river floods can be avoided by constructing dams across the rivers. This checks the speed of water and saves soil from erosion.

4. Changing Agricultural Practices. We can save lot of our valuable soil by bringing about certain

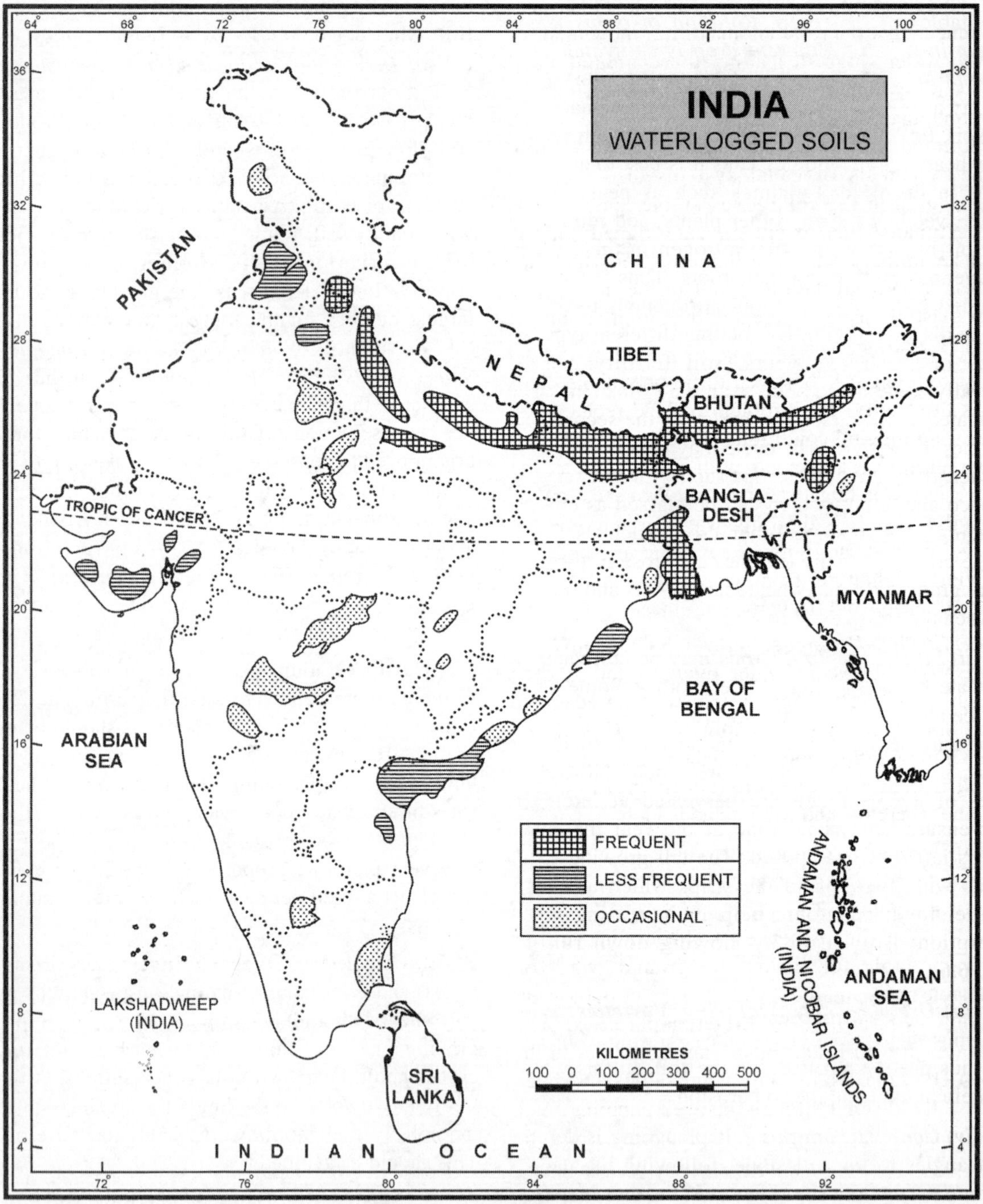

FIG. 6.5. India : Waterlogged Soils

changes in our agricultural practices. Some of the outstanding changes suggested in this context are as under :

(i) ***Crop Rotation.*** In many parts of India, a particular crop is sown in the same field year after year. This practice takes away certain elements from

the soil, making it infertile and exhausted rendering it unsuitable for that crop. *Rotation of crops is the system in which a different crop is cultivated on a piece of land each year*. This helps to conserve soil fertility as different crops make different demands on the soil. For example, potatoes require much potash but wheat requires nitrate. Thus it is best to alternate crops in the field. Legumes such as peas, beans, clover, vetch and many other plants, add nitrates to the soil by converting free nitrogen in the air into nitrogenous nodules on their roots. Thus if they are included in the crop rotation, nitrogenous fertilisers can be dispensed with. By rotating different types of crops in successive years, soil fertility can be naturally maintained. For example, wheat may be cultivated in the first year, barley in the second and legumes in the third. The cycle may then be repeated. Further, there are some crops such as maize, cotton, tobacco and potato which can be classed as erosion inducing, whilst some other crops such as grass, forage crops and many legumes are erosion resisting. Small grain crops like wheat, barley, oats and rice are between these two extremes.

(ii) Strip Cropping. Crops may be cultivated in alternate strips, parallel to one another. Some strips may be allowed to lie fallow while in others different crops may be sown *e.g.,* grains, legumes, small tree crops, grass, etc. Various crops ripen at different times of the year and are harvested at intervals. This ensures that at no time of the year the entire area is left bare or exposed. The tall growing crops act as wind breaks and the strips which are often parallel to the contours help in increasing water absorption by the soil by slowing down run off (**Fig. 6.6**).

(iii) Use of Early Maturing Varieties. Early maturing varieties of crops take less time to mature and thus put less pressure on the soil. In this way it can help in reducing the soil erosion.

(iv) Contour Ploughing. If ploughing is done at right angles to the hill slope, following the natural contours of the hill, the ridges and furrows break the flow of water down the hill. This prevents excessive soil loss as gullies are less likely to develop and also reduce run-off so that plants receive more water. Thus by growing crops in contour pattern, plants can absorb much of the rain water and erosion is minimised. When viewed from above, the field looks like a contour map (**Fig. 6.7**).

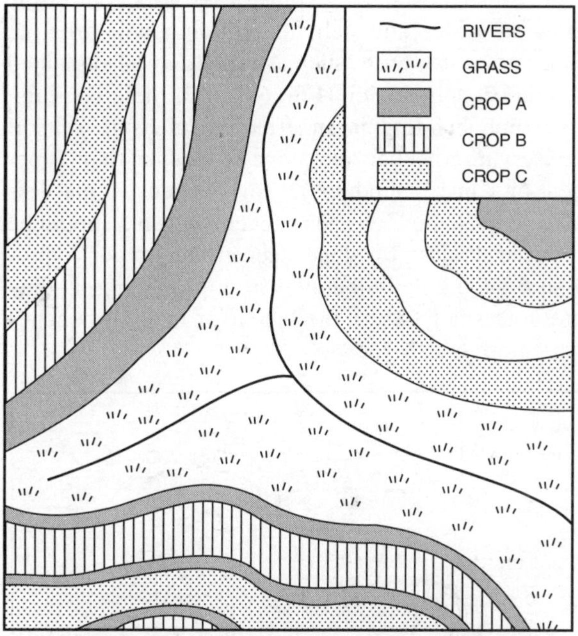

FIG. 6.6. Strip farming following the contour pattern

FIG. 6.7. Contour Ploughing

(v) Terracing and Contour Bunding. Terracing and contour bunding across the hill slopes is a very effective and one of the oldest methods of soil conservation. Hill slope is cut into a number of terraces having horizontal top and steep slopes on the back and front. Contour bunding involves the construction of banks along the contours. Terracing and contour bunding which divide the hill slope into numerous small slopes, check the flow of water, promote absorption of water by soil and save soil from erosion. Retaining walls of terraces control the flow of water and help in reducing soil erosion. Sometimes tree crops such as rubber are also planted to combat soil erosion (**Fig. 6.8**). But there is a limit to which bunding is an effective measure of soil conservation. When the slope is steeper than 8 per cent or 1 in 12, bunding becomes expensive and less effective. Nothing over 20 per cent or 1 in 5 should be terraced. Fields of a slope steeper than 15 per cent or 1 in 6 should be withdrawn from ploughing as they are not usually worth the labour of making benches very close together.

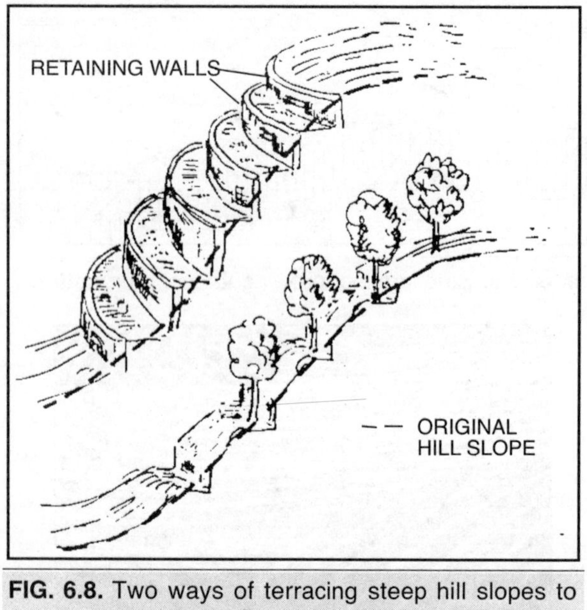

FIG. 6.8. Two ways of terracing steep hill slopes to prevent soil erosion and to produce flat-land conditions artificially.

(*vi*) Checking Shifting Cultivation. Checking and reducing shifting cultivation by persuading the tribal people to switch over to settled agriculture is a very effective method of soil conservation. This can be done by making arrangements for their resettlement which involves the provision of residential accommodation, agricultural implements, seeds, manures, cattle and reclaimed land.

(*vii*) Ploughing the Land in Right Direction. Ploughing the land in a direction pependicular to wind direction also reduces wind velocity and protects the top soil from erosion.

ACHIEVEMENTS IN SOIL CONSERVATION

Soil and water conservation measures have been adopted as one of the essential inputs for increasing agricultural output in the country right from the beginning of the First Five Year Plan. Appreciating the value of soil and the seriousness of the problem of soil erosion, the Central Government as well State Governments have taken a number of steps to conserve soil. The soil and water conservation division set up by Central Government in the Department of Agriculture and Cooperation aims at providing an overall perspective of problems like water and wind erosion, degradation through waterlogging, salinity, ravines, torrents, shifting cultivation and coastal sands.

Another scheme was sponsored by the Central Government in the Third Plan which aimed at preventing premature siltation of multi-purpose reservoirs. Scheme of Flood-Prone Rivers (FPR) was started in the Sixth Five Year Plan. Both the schemes were clubbed together during the Ninth Five Year Plan. Under the programme for the catchment management of River Valley Projects and Flood Prone Rivers, 53 catchments are covered, spread over 27 states. The total catchment area is 96.14 million hectares with Priority Area needing urgent treatment in 26 million hectares.

The *trio* of salinity-alkalinity-acidity leads to deficiency in chemical-biological nutrients and their problem has been acquiring increasing importance. Reclamation of saline and alkaline soils is a difficult task. Removal of the top soil, providing proper drainage to the waterlogged areas, avoiding seepage of water from canals, installing tube-wells in canal irrigated areas, and afforestation are some of the measures which can check the growth of unwanted salts in these soils. Application of farmyard manures, green manures, and addition of gypsum, lime, and sulphur can add to the fertility of these soils. Cultivation of salt-resistant crops like rice, barseem, barley, jowar and sugarcane has been suggested in such soils. A centrally sponsored scheme for reclamation of alkaline (usar) soils initiated in the seventh Five Year Plan is continuing in Haryana, Punjab and Uttar Pradesh. The scheme was extended to Gujarat, Madhya Pradesh and Rajasthan during the Eighth Five Year Plan. During Ninth Plan extension of the scheme to all other states was approved where alkali soil problems exist as per scientific parameters. The scheme aims at improving physical conditions and productivity status of alkali soils for restoring optimum crop production.

Land reclamation in the ravines spreading over 27.65 lakh hectares in Madhya Pradesh, Rajasthan, Uttar Pradesh and Gujarat is a challenging job. A ravine reclamation programme was launched in 1987-88 which was transferred to state sector in 1991-92 as per decision of the National Development Council. Major components of the ravine reclamation programme are peripheral bunding to halt any ingress of ravines, table land treatment in vicinity of upstream of the peripheral bund, afforestation, irrigation of shallow ravines and rehabilitation. About 1.266 lakh hectares of land has been treated so far.

Contour bunding and levelling have been very effective in checking soil erosion and in increasing crop yields. In Maharashtra, Karnataka, Telangana, Andhra Pradesh, Gujarat and Punjab, an increase of 25 per cent in the crop yields has been recorded as a result of contour bunding. This has also increased the water recharging capacity of wells in the adjoining areas. According to the reports of National Commission on Agriculture, the 'Kharif' and 'Rabi' crops raised in semi-arid alluvial plains of Uttar Pradesh have recorded yield increases of 35 per cent due to bunding and 60 per cent due to levelling. By bench terracing in the Kunda catchement in Nilgiris (Tamil Nadu) potato yields have gone up by 15 to 65 per cent.

A scheme for controlling shifting agriculture was implemented in Arunachal Pradesh and Mizoram in 1986-87. A larger scheme was launched in 1987-88 to cover nine states of Arunachal Pradesh, Assam, Meghalaya, Manipur, Tripura, Nagaland, Mizoram, Odisha and Andhra Pradesh. Till 1990-91, 26,532 families were selected for rehabilitation. The scheme involves different activities such as land development, irrigation, agriculture, horticulture, cottage industries, fisheries, social and fuel-wood forestry and minimum housing programmes. During Eighth Plan 67 thousand hectare area was treated through treatment packages. During Ninth Plan upto March 2002, 1.5 lakh hectares had been treated. The new guidelines for the scheme on the basis of new watershed to common approach has been effective from November 2000. During Tenth Plan, an area of 0.4 lakh hectares has been treated upto 2003-04.

All India Soil and Land Survey Organisation is using remote sensing technology to keep itself abreast of the latest trends.

REFERENCES

Agrawal, M.C., Agrawal, R.P. and Chaudhary, R.S. (1974) : Infiltration Characteristics of soils as related to soil physical proposition. *Jour. Indian Soc. Soil Soc.,* 22 : 285-89.

Agarwal, R.R. and Gupta, R.N. (1968) : Saline Alkali Soils in India; ICAR Tech. Bull (Apic. Series).

Anon (1977) : *Desertification and its Control,* Indian Council of Agricultural Research New Delhi.

Battawar, H.B. and Rao, V.P. (1969) : Effectiveness of crop cover for reducing runoff and soil losses *Ind. Jour. soil and Water Conservation,* 17 (3-4) : 39-50.

Bordoloi, B.N. (1976) : The problem of rehabilitation of the Jhumiya families of the hill areas of Assam in Negi S.S. *Environmental Problems of the Himalaya,* Bishan Singh, Mahendrapal Singh Dehradun (1981).

Bruun, P. and Nayak, B.U. (1980) : *Manual on Protection and Control of Coastal Erosion in India,* National Institute of Oce, Panaji (Goa)

Chatterjee, S.P. (1964) : *Fifty years of Science in India—Progress of Geography,* Indian Science Congress Association, Calcutta.

Chinnamani, S. (1977) : *Soil and Water Conservation in Hills of Western Ghats,* Soil Conservation Digest, vol. 5., pp. 25-32.

Das, D.C. Bali, Y.P. and Kaul, R.N. (1981) : Soil conservation in multipurpose river valley catchment : problems, programme, approach and effectiveness; *Indian Jour. Soil and Conservation 9(1) :* 6-20.

Donahue Roy, L. (1963) : *Our soils and Their Management,* First Indian Edition, Asia Publishing House, Bombay.

Donahue Roy, L., Miller, Raymond, W., and Shickluna, John C. (1987) : *Soils: An Introduction to Soils and Plant Growth,* Prentice Hall of India Pvt. Ltd., New Delhi.

Doughlas, I. (1981) : Erosion risk assessment and land classification for soil conservation in R.P.C. Morgan (Ed.) : *Soil Conservation Problems* and *Prospects*, John Wiley, New York pp. 63-73.

Goh Cheng Leong and Gillian C. Morgan (1999) : *Human and Economic Geography.* Oxford University Press, pp. 137-141.

Government of India. (1966) : *Study on Soil Conservation in Catchment Areas Above Dams in River Valley Projects*, Part I, Planning Commission, New Delhi.

Gupta, R.S. (1956) : 'Soil Survey and Soil Classification in India with Special Reference to Forest Conditions.' *Indian Forest,* 89 : 18-38.

India (2024) : A Reference Annual, Publication Division Ministry of I & B, Govt. of India, New Delhi, pp. 384-390.

Indian Council of Agricultural Research, **Hanbook of Agriculture** (1961) : ICAR, New Delhi, pp. 22-27.

Indian Council of Agricultural Research (1983) : *All India Soil and Land Use Survey,* ICAR, New Delhi.

Kayastha, S.I. (1965) : *Some Aspects of Soil Erosion and Conservation in India with Special Reference to Agriculture and Food Production* in A. Ramesh (1984) (ed.) *Resource Geography* (*Contribution to Indian Geography*), Heritage Publishers, New Delhi, pp. 225-33.

Khybri, M.L. (1973) : Ravine reclamation in Kota district, Rajasthan. *Soil Conservation Digest*, 201 : 567-60.

Kuriyan, George (1970) : *India—A General Survey*, National Book Trust, India, New Delhi, pp. 42-45.

Mehta, K.K. (1983) : *Reclamation of Alkali, Soils in India*, Oxford and IBH, New Delhi.

NATMO (1978) : *Atlas of Agricultural Resources of India*, National Atlas and Thematic Mapping Organisation, Calcutta.

Olson, G.K. (1981) : *Soil and Environment*, Chapman and Hall, New York.

Patnaik, N. (1977) : *Soil Conservation Works in River Valley Projects* (*RVP*), Soil Conservation Digest, pp. 53-61.

Raychoudhuri, S.P. (1986) : *Land and Soil*, National Book Trust, India, New Delhi, Chapt. IV & IX.

Rayachaudhri, S.P. *et al.* (1963) : *Soils of India*, Indian Council of Agricultural Research, New Delhi.

Sharma, H.S. (1980) : *Ravine Erosion in India,* Concept Pub. New Delhi.

Siddhartha, K. (1999) : *India : The Physical Aspect,* Centre for Development of Environment and Resources, Patana, Chap. 6

Singh, Jasbir and Dhillon, S.S. (1984) : *Agricultural Geography*, Tata McGraw Hill Publishing Co. Ltd., New Delhi, pp. 88-91.

Spate, O.H.K. (1957) : *India and Pakistan—A General and Regional Geography*, Methuen & Co. Ltd., London, Chapt. III

Wadia, D.N. (1945) : *Soils of India*, Journal of Scientific and Industrial Research, Vol. 3, pp. 359-367.

◆◇◆

CHAPTER 7

Natural Vegetation and Wildlife

INTRODUCTION

Natural vegetation is the primeval plant cover unaffected by man either directly or indirectly. The word forest is derived from Latin 'fores' meaning outside, the reference being to a village boundary or fence and it must have included all uncultivated and uninhabited land. In general, a forest is defined as an area set aside for the production of timber and other forest produce or maintained under woody vegetation for certain benefits which it provides. Ecologically, a forest is a plant community, predominantly of trees and other woody vegetation, usually with a closed canopy.

FACTORS INFLUENCING VEGETATION

The geographical factors which influence natural vegetation include climate, soil and topography. The main climatic factors are rainfall and temperature. Generally speaking, rainfall is more important than temperature except in the Himalayas. The seasonal rainfall regime, the length of dry season and its relation to the march of temperature, are also important. The amount of annual rainfall has a great bearing on the type of vegetation. Areas receiving 200 cm or more rainfall per annum have stands of evergreen rain forests while monsoon deciduous forests dominate in areas with rainfall between 100 and 200 cm. In areas having 50 to 100 cm rainfall there are drier deciduous or tropical savana grading into open thorny scrub while those with less than 50 cm rainfall have only dry thorny scrub and low open bush merging into semi desert. In higher altitudes of the Himalayas and the hills of the Peninsula at elevation of more than 900 metres above sea level, temperature plays a more important role. As the temperature falls with altitude in the Himalayan region the vegetal cover changes from tropical to sub-tropical, temperate and finally alpine. Changes in soil conditions have given birth to different types of vegetation in different parts of the country. Mangrove forests, swamp forests, and beach and sandy coastal forests are some of the outstanding examples of influence of soil on natural vegetation. Topography in the narrow sense is responsible for certain minor types *e.g.* alpine flora, tidal forests, etc.

CLASSIFICATION OF NATURAL VEGETATION

A great variety of vegetation is found in different parts of India due to unequal distribution of rainfall and temperature as well as their seasonal variation, besides varied edaphic and biotic conditions. A generalised classification of Indian forests is, therefore, a difficult job. Various systematic classifications have been made by a number of botanists and ecologists. **H.G. Champion** (1936) distinguished 116 types of vegetation in India, some with further sub-divisions. Champion's classification has been much simplified by **G.S. Puri** (1960), **Legris** (1963), **Champion** and **S.K. Seth** (1968) and **S.S. Negi** (1990). Based on the suggestions of these scholars India's vegetation can be divided into 5 main types and 16 sub-types as per details given below :

A. Moist Tropical Forests

1. Tropical Wet Evergreen
2. Tropical Semi-Evergreen
3. Tropical Moist Deciduous
4. Littoral and Swamp

B. Dry Tropical Forests

5. Tropical Dry Evergreen
6. Tropical Dry Deciduous
7. Tropical Thorn

C. Montane Sub-tropical Forests

8. Sub-tropical broad leaved hill
9. Sub-tropical moist hill (pine)
10. Sub-tropical dry evergreen

D. Montane Temperate Forests

11. Montane Wet Temperate
12. Himalayan Moist Temperate
13. Himalayan Dry Temperate

E. Alpine Forests

14. Sub-Alpine
15. Moist Alpine scrub
16. Dry Alpine scrub

A. MOIST TROPICAL FORESTS

1. Tropical Wet Evergreen Forests. These are typical rain forests which grow in those areas where the annual rainfall exceeds 250 cm, the annual temperature is about 25°-27°C, the average annual humidity exceeds 77 per cent and the dry season is distinctly short. Due to high heat and high humidity, the trees of these forests do not shed their leaves annually, at least not together, and are termed as evergreen forests. These are lofty, very dense multilayered forests with mesosphytic evergreens. The trees often reach 45 metres in height, individual trees exceed 60 metres. The entire morphology looks like a green carpet when viewed from above. The sun light cannot reach the ground and owing to deep shade, the undergrowth is formed mainly of tangled mass of canes, bamboos, ferns, climbers, etc. A variety of orchids flourishes on the trees.

The true evergreen forests are found along the western side of the Western Ghats (between 500 to 1370 metres above sea level) south of Mumbai, in a strip running from north-east to south-west direction across Arunachal Pradesh, Upper Assam, Nagaland, Manipur, Mizoram and Tripura upto a height of 1070 metres and in the Andaman and Nicobar Islands.

The timber of these forests is fine-grained, hard and durable as a result of which it has high commercial value. The important species of these forests are mesua, white cedar, calophyllum, toon, dhup, palaquium, hopea, jamun, canes, etc. in the Western Ghats, and gurjan, chaplasha, jamun, mesua, agar, muli, bamboo etc. in the Himalayan region. Although these forests contain valuable commercial species, they have not been properly exploited due to dense undergrowth, absence of pure stands and lack of transport facilities.

2. Tropical Semi-Evergreen Forests. Bordering the areas of tropical wet evergreen forests, are comparatively drier areas of the tropical semi-evergreen forests. Here the annual rainfall is 200-250 cm, the mean annual temperature varies from 24°C to 27°C and the relative humidity is about 75 per cent. Obviously these forests are found on the Western coast, Assam, lower slopes of the Eastern Himalayas, Odisha and Andamans.

The semi-evergreen forests are less dense but more gregarious (especially in Assam) than the wet evergreen forests. At places, these forests represent a transition from wet evergreen to deciduous forests. Further, these forests are characterised by many species, frequently buttressed trunks, rougher and thicker bark, heavy climbers, less bamboos and abundant epiphytes. The important species of these forests are aini, semul, gutel, mundani, *hopea*, benteak, kadam, irul, laurel, rosewood, mesua, haldu, kanju, byasal, kusum, thorny bamboo, etc. on the Western Ghats, and bonsum, white cedar, Indian chestnut, *litsa*, hollock, champa, mango, etc. in the Himalayan region.

3. Tropical Moist Deciduous Forests. These forests are found in areas of moderate rainfall of 100 to 200 cm per annum, mean annual temperature of about 27°C and the average annual relative humidity of 60 to 75 per cent. Such areas include a belt running along the Western Ghats surrounding the belt of evergreen forests both on the western and the eastern slopes, a strip along the Shiwalik range including terai and bhabar from 77° E to 88° E, Manipur and Mizoram, hills of eastern Madhya Pradesh and Chhattisgarh, Chota Nagpur Plateau, most of Odisha, parts of West Bengal and in the Andaman and Nicobar islands.

The trees of these forests drop their leaves for about 6-8 weeks during the spring and early summer when sufficient moisture for the leaves is not available. The sub-soil water is not enough to enable the trees to retain their leaves throughout the year. The general appearance is burnt up and bare in April-May. These forests again become green when the leaves grow with the onset of the rainy season. Tropical moist deciduous forests present irregular top storey of different species, 25 to 60 metres high, heavily buttressed trees and fairly complete shrubby undergrowth with patches of bamboos, climbers and canes.

These are very useful forests because they yield valuable timber and several other forest products. The main species found in these forests are teak, sal, padauk, laurel, white chuglam, badam, dhup, chikrosi, kokko, haldu, rosewood, mahua, bijasal, lendi, semul, irul, dhaman, amla, kusum, tendu, paula, jamun, bamboo, etc. It is comparatively easy to exploit these forests due to their high degree of gregariousness.

These forests occupy a much larger area than the evergreen forests but large tracts under these forests have been cleared for cultivation.

4. Littoral and Swamp Forests. These forests occur in and around the deltas, estuaries and creeks prone to tidal influences and as such are also known as *delta* or *tidal* forests. While littoral forests occur at several places along the coast, swamp forests are confined to the deltas of the Ganga, the Mahanadi, the Godavari, the Krishna and the Cauvery. The most peculiar feature of these forests is that they can survive and grow both in fresh as well as brackish water. Dense mangroves occur all along the coastline in sheltered estuaries, tidal creeks, backwaters, salt marshes and mudflats covering a total area of 6,740 sq km which is about seven per cent of the world's total mangrove area. It provides useful fuel wood. The most pronounced and the densest is the *Sunderban* in the Ganga delta where the predominant specie *Sundri* (Heriteera) grows abundantly. It provides hard and durable timber which is used for construction and building purposes as well as for making boats.

The important species found in these forests are Sundri, burguiera, sonneratia, agar, bhendi, keora, nipa, amur, bhara, *rhizophora*, *screw pines*, canes and palms, etc.

B. DRY TROPICAL FORESTS

5. Tropical Dry Evergreen Forests. Along the coasts of Tamil Nadu are areas which receive annual rainfall of about 100 cm mostly from the north-east monsoon winds in October-December. Here, the mean annual temperature is about 28°C and the mean humidity is about 75 per cent. These areas are covered by the tropical dry evergreen forests. The growth of evergreen forests in areas of such low rainfall arouses great botanical interest, and the reason for such a phenomena is difficult to explain. It may be due to the seasonal distribution of rainfall. Most of rainfall occurs in winter. The chief characteristics of these forests are short statured trees, upto 12 m high, with complete canopy, mostly of coriaceous leaved trees of short boles, no canopy layer differentiation, bamboos are rare or absent and grasses not conspicuous. The important species are *khirni, jamun, kokko, ritha, tamarind, neem,*

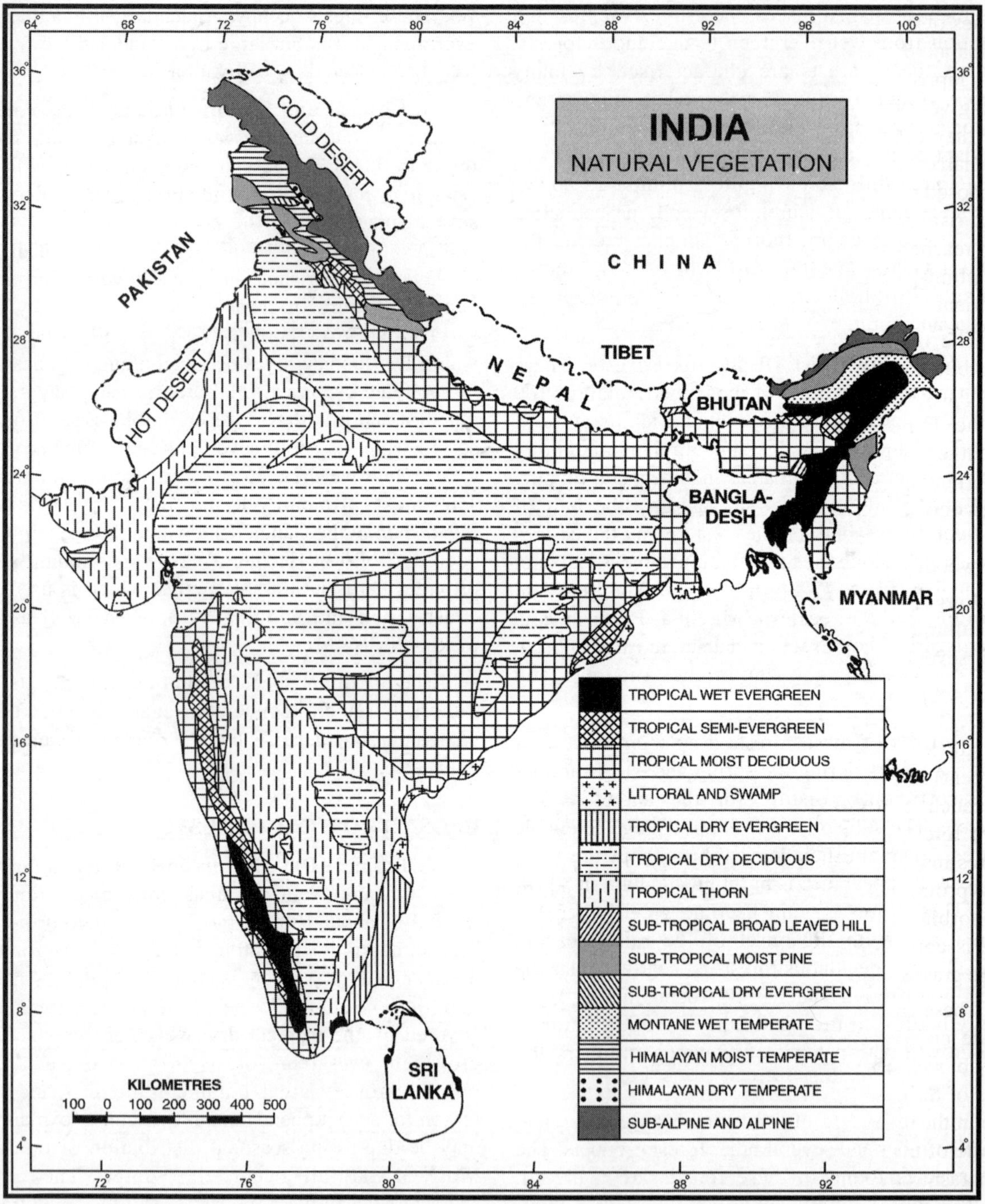

FIG. 7.1. India : Natural Vegetation

machkund, *toddypalm*, *gamari* canes, etc. Most of the land under these forests has been cleared for agriculture or casuarina plantations.

6. Tropical Dry Deciduous Forests. These are similar to moist deciduous forests and shed their leaves in dry season. The major difference is that the species of dry deciduous forests can grow in areas of comparatively less rainfall of 100-150 cm per annum. They represent a transitional type; on the wetter side, they give way to moist deciduous and on the drier side they degenerate into thorn forests. Such forests are characterised by closed and rather uneven canopy, composed of a mixture of a few species of deciduous trees, rising upto a height of 20 metres or so. Enough light reaches the ground to permit the growth of grass and climbers. Bamboos also grow but they are not luxuriant.

The tropical dry deciduous forests are widely distributed over a large area. They occur in an irregular wide strip running north-south from the foot of the Himalayas to Kanniyakumari except in Rajasthan, Western Ghats and West Bengal. The important species are teak, axlewood, tendu, bijasal, rosewood, amaltas, palas, haldu, kasi, bel, lendi, common bamboo, red sanders, anjair, harra, laurel, satinwood, papra, achar, sal, khair, ghont, etc.

Large tracts of this forest have been cleared for agricultural purposes and these forests have suffered from severe biotic factors such as over cutting, over grazing and fire, etc.

7. Tropical Thorn Forests. In areas of low rainfall (less than 75 cm), low humidity (less than 50 per cent) and high temperature (25°-30°C), there is not much scope for thick forests and only tropical thorn forests are found. The trees are low (6 to 10 metres maximum) and widely scattered. Acacias are very prominent, widely and pretty evenly spaced. Euphorbias are also conspicuous. The Indian wild date is common, especially in damper depressions. Some grasses also grow in the rainy season.

These forests are found in the north-western parts of the country including Rajasthan, south-western Panjab, western Haryana, Kachchh and neighbouring parts of Saurashtra. Here they degenerate into desert type in the Thar desert. Such forests also grow on the leeside of the Western Ghats covering large areas of Maharashtra, Karnataka, Telangana, Andhra Pradesh and Tamil Nadu. The important species are khair, reunjha, neem, babul, thor, cactii, khejra, kanju, palas, ak, nirmali, dhaman, etc.

C. MONTANE SUB-TROPICAL FORESTS

8. Sub-tropical Broad-leaved Hill Forests. These forests are found in the eastern Himalayas to the east of 88°E longitude at altitudes varying from 1000 to 2000 m where the mean annual rainfall is 75 cm to 125 cm, average annual temperature is 18°-21°C and the average humidity is 80 per cent. They form luxurious forests of evergreen species, are fairly high (20-30 metres) and dense. Evergreen oaks and chestnuts predominate with some ash and beech. Sals and pines may occur on lower and higher margins respectively. Climbers and epiphytes are common.

These forests are not so distinct in the southern parts of the country due to restricted area of southern hills. They occur in the Nilgiri and Palni hills at 1070-1525 metres above sea level. It is essentially a "stunted rain-forest" and is not so luxuriant as the true tropical evergreen. The higher parts of the Western Ghats such as Mahabaleshwar, the summits of the Satpura and the Maikal ranges, highlands of Bastar in Chhattisgarh and Mt. Abu in the Aravali Range also carry sub-types of these forests. But most of these forests have been destroyed, reduced and changed beyond recognition.

9. Sub-tropical Moist Pine Forests. These forests also occur at the same height as the wet hill forests *i.e.* at 1000 to 2000 metres above sea level but in the western Himalayas between 73°E and 88°E longitudes. Some parts of Arunachal Pradesh, Manipur, Naga Hills and Khasi Hills of Meghalaya are also covered by such forests at similar altitudes. Chir or Chil is the most dominant tree which forms pure stands. It provides valuable timber for furniture, boxes and buildings. It is also used for producing resin and turpentine. There is often a grassy floor with bulbous plants and little undergrowth, except for stunted evergreen oaks in wetter areas.

10. Sub-tropical Dry Evergreen Forests. Like the tropical moist evergreen, this occurs in a restructed area and is found in the *Bhabar*, the Shiwaliks and the western Himalayas upto about 1000 metres above sea level. Here the annual rainfall is 50-100 cm (15 to 25 cm in December-March). The summers are sufficiently hot and winters are cold enough for the occurrence of frequent frost. This is a low, practically scrub forest with small evergreen stunted trees and shrubs, including thorny species, herbs and grasses. Olive, acacia *modesta* and *pistacia* are the most predominant species. Considerable tracts are covered by dwarf creeping palm *Nonnorpops*.

D. MONTANE TEMPERATE FORESTS

11. Montane Wet Temperate Forests. This variety of forests grows at a height of 1800 to 3000 m above sea level in areas where the mean annual rainfall is 150 cm to 300 cm, the mean annual temperature is about 11°C to 14°C and the average relative humidity is over 80 per cent. These are mainly found in the higher hills of Tamil Nadu and Kerala, in the Eastern Himalayan region to the east of 88°E longitude including the hills of West Bengal, Assam, Arunachal Pradesh, Sikkim and Nagaland. These are closed evergreen forests in which the trees are mostly short-boled and branchy attaining a large girth. Leaves are dense and rounded. Branches are clothed with mosses, ferns and other epiphytes. Woody climbers are common. The trees rarely achieve a height of more than 6 metres.

Deodar, Chilauni, Indian chestnut, birch, plum, *machilus*, *cinnamomum*, *litsea*, *magnolia*, blue pine, oak, hemlock, etc. are important species.

12. Himalayan Moist Temperate Forests. Occurring in the temperate zone of the Himalayas between 1500 and 3300 metres where the annual rainfall varies from 150 cm to 250 cm, the Himalayan moist temperate forests cover the entire length of this mountain range in Kashmir, Himachal Pradesh, Uttarakhand, Darjeeling and Sikkim. Such forests are mainly composed of coniferous species, mostly pure, 30 to 50 m high. In the wetter east, the broad leaved evergreens are mixed with the dominant conifirs. Pines, cedars, silver firs, spruce, etc. are most important trees. They form high but fairly open forest with shrubby undergrowth including oaks, rhododen-drons, laurels and some bamboos. In comparatively drier western parts where the rainfall varies from 115 to 180 cm especially to the west of 80°E longitude, deodar dominates the scene and forms pure stands. It provides fine wood which is of much use for construction, timber and railway sleepers.

13. Himalayan Dry Temperate Forests. These are predominantly coniferous forests with xerophytic shrubs in which deodar, chilgoza, oak, ash, maple, olive, celtis, parrotia, etc. are the main trees. Such forests are found in the inner dry ranges of the Himalayas where south-west monsoon is very feeble and the precipitation is below 100 cm, mostly snow. Such areas are in Ladakh, Lahul, Chamba, Kinnaur, Garhwal and Sikkim.

E. ALPINE FORESTS

The Alpine forests occur all along the Himalayas at altitudes ranging between 2,900 to 3,500 m or even upto 3800 m above sea level, depending upon the location and the variety of species. These forests can be divided into : (1) sub-alpine; (2) moist alpine scrub and (3) dry alpine scrub. The sub-alpine forests occur at the upper limit of tree forest adjoining alpine scrub and grasslands and comprise of dense growth of small crooked trees and large shrubs with coniferous overwood. It is a mixture of coniferous and broad-leaved trees in which the coniferous trees attain a height of about 30 m while the broad leaved trees reach only 10 m. Fir, kail, spruce, rhododendron, plum, yew, etc. are important species. *The moist alpine scrub* is a low evergreen dense growth of rhododendron, birch, berberis and honeysuckle which occurs from 3,000 metres and extends upto snowline. *The dry alpine scrub* is the uppermost limit of scrub xerophytic, dwarf shrubs, over 3,500 metres above sea level and found in dry zone. Juniper, honeysuckle, artemesia, potentilla, etc. are important species.

As different forest groups grow under different geographical conditions, the area covered by each group differs greatly. The extent of forests of different groups for the country as a whole, is given in **table 7.1.**

Apart from the major classification of Indian forests described earlier, the Indian forests may also be classified on the basis of statutes, ownership, composition and exploitability.

1. Legal or Administrative Classification. This classification has been done to protect the forests against indiscriminate destruction. From legal or administration point of view, the Indian forests have been divided into : (*i*) Reserved, (*ii*) Protected and (*iii*) Unclassed.

(i) Reserved Forests. It is very harmful to cut these forests and usually nobody is allowed to cut forests belonging to this category. These forests are reserved to avoid soil erosion, for flood control, to check the onward march of deserts and to maintain climate as well as for ecological reasons. Even grazing of animals is prohibited in these forests. Such forests cover about 54% of the total forest area of India.

(ii) Protected Forests. Grazing of animals and cutting of wood is allowed in these forests. But these activities are allowed under strict control so that there is no undue harm to the forests. These forests cover about 29% of the total forest area of the country.

***(iii)* Unclassed Forests.** These forests have not been classified so far. There is no restriction for grazing of animals and cutting of wood in these forests. They account for 17 per cent of the total forest area of the country. This area is largely degraded, unproductive and unprofitable.

2. Classification based on Ownership. Forests are divided into following classes on the basis of ownership.

(i) State Forests. These forests are owned by the Government through different departments such as forest department. State forests account for about 96% of the total forests of the country.

(ii) Corporate Bodies Forests. These forests are owned by local bodies such as municipal corporations and zila parishads, etc. This type of forests cover about 3% of the total forest area of India.

(iii) Individual Forests. These forests are owned by individuals and account for only 1% of the total forest area.

3. Classification according to Composition. Two types of forests are recognised based on composition :

(i) Coniferous Forests. Coniferous forests have cone-shaped trees and are mainly found in the Himalayan region. They cover about 5% of the total forest area in the country. Some species of the coniferous forests are deodar, chir, fir, spruce, etc. Though they contain valuable softwood timber, they are not properly exploited due to their inaccessibility, difficult terrain and lack of transport facilities.

(ii) Broad Leaved Forest. Trees of these forests have broad leaves as a result of which such forests are called broad leaved forests. They are spread over vast areas and cover about 95% of the total forest area of the country. Sal and teak are the most important species and provide valuable timber. They cover about 16.55 per cent and 13.82 per cent of the total area under broad leaved forests respectively. Bamboo covers about 7.48 per cent of the total area under broad leaved forests. The other species of the broad leaved forests are rosewood, Indian laurel, shistram, garyan and benteak.

4. Classification according to Exploitability. From the exploitability point of view, the forests can be classified into : *(i)* exploitable *(ii)* potentially exploitable and *(iii)* others.

(i) Exploitable Forests. These are very useful forests and provide many valuable forest products. Such forests spread over 50% of the total forest area of India.

(ii) Potentially Exploitable Forests. Such forests can be potentially used and are reserved for future use. These forests cover about one-fourth of the total forest area of the country.

Both the exploitable and potentially exploitable forests occur in Assam, Arunachal Pradesh, Tripura, Western Ghats, Satpura, Maikal, Chota Nagpur plateau, Andaman and Nicobar Islands, Odisha and adjoining areas of Andhra Pradesh and Chhattisgarh.

TABLE 7.1. India : Area under Forests

Sl. No.	Forest Group	Area in million hectares	Percentage of total area
1.	Tropical Wet Evergreen	5.120	8.0
2.	Tropical Semi-evergreen	2.624	4.1
3.	Tropical Moist Deciduous	23.680	37.0
4.	Littoral and Swamp	0.384	0.6
5.	Tropical Dry Evergreen	0.128	0.2
6.	Tropical Dry Deciduous	18.304	28.6
7.	Tropical Thorn	1.664	2.6
8.	Sub-tropical Broad-leaved hill	0.256	0.4
9.	Sub-tropical Moist Hill (pine)	4.224	6.6
10.	Sub-tropical Dry Evergreen	1.600	2.5
11.	Montane Wet Temperate	2.304	3.6
12.	Himalayan Moist Temperate	2.176	3.4
13.	Himalayan Dry Temperate	0.192	0.3
14.	Alpine (including sub-alpine, moist and dry alpine scrub)	1.344	2.1
	Total	**64.000**	**100.0**

Source : K.P. Sagreiya, Forests and Forestry (2004), p. 67.

TABLE 7.2. India : Statewise Area under Forest (sq. km.) in 2017

States/Union Territories	Geographical Area	Forest area	Percentage of forest area to geographical area
STATES			
Andhra Pradesh	1,60,200	37,258	23.26
Arunachal Pradesh	83,743	51,407	61.39
Assam	78,438	26,832	34.21
Bihar	94,163	6,877	7.30
Chhattisgarh	1,35,192	59,772	44.21
Goa	3,702	1,225	33.09
Gujarat	1,96,024	21,647	11.04
Haryana	44,212	1,559	3.53
Himachal Pradesh	55,673	37,033	66.52
Jammu & Kashmir	2,22,236	20,230	9.10
Jharkhand	79,716	23,605	29.61
Karnataka	1,91,791	38,284	19.96
Kerala	38,852	11,309	29.11
Madhya Pradesh	3,08,252	94,689	30.71
Maharashtra	3,07,713	61,579	20.01
Manipur	22,327	17,418	78.01
Meghalaya	22,429	9,496	42.34
Mizoram	21,081	5,641	26.76
Nagaland	16,579	8,623	52.01
Odisha	1,55,707	61,204	39.31
Punjab	50,362	3,084	5.19
Rajasthan	3,42,239	32,737	9.56
Sikkim	7,096	5,841	82.31
Tamil Nadu	1,30,060	22,817	17.54
Telangana	1,14,840	26,904	23.43
Tripura	10,486	6,294	60.02
Uttarakhand	53,483	38,000	71.05
Uttar Pradesh	2,36,287	16,582	7.02
West Bengal	88,752	11,879	13.38
UNION TERRITORIES			
A. & N. Islands	8,249	7,171	86.93
Chandigarh	114	35	30.70
D. & N. Haveli	491	204	41.55
Daman & Diu	111	8	7.21
Delhi	1,483	102	6.88
Lakshadweep	30	0	—
Puducherry	490	13	26.53
India	**32,87,263**	**7,67,419**	**23.34**

Source : Data computed from EnviStat India, 2018, pp. 92-93.

(iii) Other Forests. Large proportion of our forests is inaccessible for effective exploitation and are also termed as *non-merchantable*. They lie in the high reaches of the Himalayas in Kashmir and Arunachal Pradesh where they cannot be exploited due to lack of transport facilities.

GEOGRAPHICAL DISTRIBUTION OF FOREST AREA

The total geographical area of the country is 32,87,263 sq km out of which an area of 7,67,419 sq km or 23.34 per cent was under forests in 2017. This is much below the average of 30.4 per cent for the world. The picture is very gloomy when we compare our percentage of forests area with 34.4 per cent of Canada, 36.0 per cent of Germany, 41.07 per cent of the U.S.A., 52.00 of Zaire, 56.10 per cent of Brazil, 57.8 per cent of Sweden and 67.6 per cent of Japan. India with about 17.5 per cent of the world's population has nearly 2 per cent of the total forests, and if the forest growing stock has been correctly estimated, it is one of the lowest per unit area *i.e.,* less than 1 per cent of that of the forests of the world. According to the National Forest Policy, the minimum desired area which is considered safe for a tropical country like India is about 33 per cent. As per broad policy recommendations, about 60 per cent of the area in the Himalayas and the Peninsular hills and 20 per cent in the Great Plains should be under forests. Furthermore, the per capita forest area in India is a bare 0.05 hectares against the world average of 1.08 hectares. The per capita forest cover is 1.8 hectares in the U.S.A., 3.2 hectares in Sweden, 3.5 hectares in Russia, 5.1 hectares in Australia, 8.6 hectares in Brazil and 22.7 hectares in Canada. The production of Indian forests is also very low, being only 0.5 cubic metre per hectare a year as against the world average of 2.1 cubic metres per hectare per year. Thus the useful forest resources are extremely inadequate for a predominantly agricultural country like India.

The forest resources are very unevenly distributed in India. Madhya Pradesh has the largest area of over 94,689 sq km under forests (**Table 7.2**). The other states with considerable area under forests are Maharashtra (61,579 sq km), Odisha (61,204 sq km), Chhattisgarh (59,772 sq km) and Arunachal Pradesh (51,407 sq km). Incidently, most of them are very large states and the total forest area is not a good index of forest prosperity.

Percentage of forest area to total area can serve as a better index of the forest cover. Unfortunately, the forests are conspicuous by their absence where they are needed the most. For example only 5 per cent of the total area in the thickly populated Ganga plain is under forests. In north-west India about 11 per cent of the area is under forests while the forests cover about 20 per cent of the area in the Tarai. As per percentage of forest area to total area, Andaman Nicobar Islands. Sikkim and Manipur are very rich areas. These regions have over 75 per cent of their geographical area under forests. Forest cover is reasonably good in the states of Uttarakhand, Arunachal Pradesh, Himachal Pradesh, Nagaland and Tripura. These states have more than 50% of the total area under forests. States of Assam, Chhattisgarh, Goa, Meghalaya and Odisha, have between one-third to one-half of the total area under forests. States and Union Territories of Jharkhand, Kerala, Madhya Pradesh, Mizoram, Telangana, Chandigarh and Puducherry have percentage of forest area to total area above the national average but below the required average. But there are other states where the percentage of forest area to total area is desperately low. For example, Punjab, Haryana and Rajasthan have very low percentage of their land under forests. Other large and important states like Uttar Pradesh, Bihar, Gujarat and Jammu and Kashmir also have very low percentage of the area under forests. Rajasthan and Gujarat receive scanty rainfall and experience arid and semi-arid climate. Much vegetation cover is not expected under such conditions. In states like Punjab, Haryana, Uttar Pradesh and Bihar vast forest tracts have been cleared to make land available for cultivation. These are very rich areas from agriculture point of view. Among the union territories, Daman and Diu, Delhi and Lakshadweep are poor. In Jammu and Kashmir, the low percentage of the forest area is primarily due to low rainfall, steep barren slopes and snow covered peaks. The semi-arid lands of Maharashtra, Karnataka, Andhra Pradesh, Telangana, and Tamil Nadu in the rain-shadow of the Western Ghats receive much less rainfall and are devoid of thick forest growth. The coastal areas of Tamil Nadu and Andhra Pradesh receive sufficient rainfall in winter and support good forest cover but most of the forests in this region have been cleared for cultivation.

The forest areas do not follow political boundaries, and as such it would be better to study their growth and distribution according to the geographical conditions governing them. The areas more conducive for the forest growth are those receiving heavy rainfall, having sufficiently high temperature and soils as well as relief suitable for forest growth (see **figure 7.2**). **Table 7.3** would help in understanding the geographical distribution of forests in India.

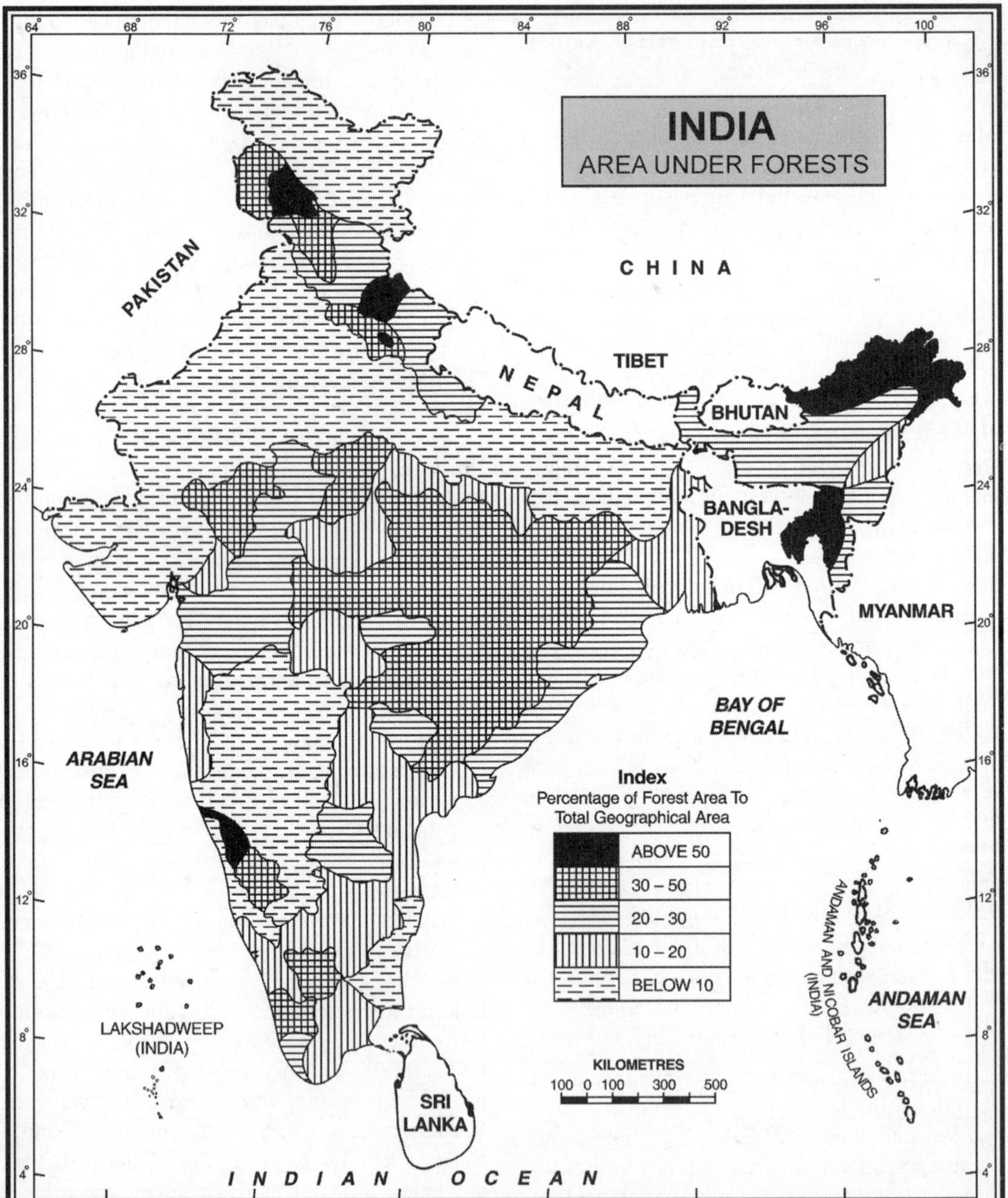

FIG. 7.2. India : Area under Forests

TABLE 7.3. Geographical Distribution of Forests in mainland of India

Sl. No.	Geographical Region	Percentage of total forest of India
1.	Himalayan Region	18.00
2.	The Great Plain of India	5.00
3.	Peninsular plateaus and Hills	57.00
4.	Western Ghats and Western Coastal Plain	10.00
5.	Eastern Ghats and Eastern Coastal plain	10.00
	Total	**100.00**

FOREST PRODUCTS

Forests constitute one of the major natural resource of India. They produce a large variety of woods which are used as fuel, timber and industrial raw material.

They also provide many more things out of which bamboos, canes, herbs, drugs, lac, grasses, leaves, oils, etc. are important. Accordingly the forest products of India are classified into two categories *viz.*, major products and minor products.

Major Forest Products

Major forest products consist of timber, smallwood and fuelwood including charcoal. Indian forests produce about 5,000 species of wood, of which about 450 are commercially valuable. Both hard and soft woods are obtained from Indian forests. *Hard woods* include important species such as teak, mahogany, logwood, iron-wood, ebony, sal, greenheart, kikar, semal, etc. These woods are used for furniture, wagons, tools, etc. *Soft woods* include deodar, poplar, pine, fir, cedar, balsam, etc. They are light, strong, fairly durable and easy to work and as such are very useful as constructional timbers. They also provide useful raw materials for making paper pulp. It is interesting to note that 70 per cent of hard wood is burnt as fuel and only 30 per cent is used in industries while 70 per cent of the soft wood is used in industries and only 30 per cent is burnt as fuel. Forests meet about 40 per cent of energy needs of the country including more than 80 per cent of the rural energy requirements. The current production of wood is about 52 million cubic metres per year out of which about 40 million cubic metres or nearly 77 per cent of total is used as fuel wood. The current production of wood is too short of our present demand which is increasing at an alarming rate. The increase in demand for industrial wood will be much more keeping in view the large scale industrialisation in the country.

Minor Forest Products

Minor forest products include all products obtainable from the forests other than wood and thus comprise products of vegetable and animal origin. Some of the imporant forest products of minor nature are described as under :

1. Grasses, bamboos and canes
2. Tans and Dyes
3. Oils
4. Gums and Resins
5. Fibres and Flosses
6. Leaves
7. Drugs, spices and poisons
8. Edible products
9. Animal products

1. Grasses, Bamboos and Canes. Different types of grasses grow in different parts of the country. Most of the grasses are used as fodder or for thatching, but some grasses are better used for cordage, matting and as an important raw material for manufacturing paper. Grasses like *sabai*, *bhabar* and elephant are used for papermaking. *Sabai* is the most important grass which provides the basic raw material for paper industry. It is a perennial grass which grows on the bare slopes of the sub-Himalayan tract and in Bihar, Odisha, West Bengal, Madhya Pradesh and western part of Himachal Pradesh. Annually over two million tonnes of sabai grass is collected and supplied to paper mills. The roots of *khus* grass are used for making cooling screens. *Munj*, a tall grass is used for making chicks, stools, chairs, etc. and the leaves are twisted into strings.

Bamboo belongs to grass family but grows like a tree. It is woody, perennial and tall. It may attain a height of as much as 30 metres. More than 100 species of bamboo grow in the Indian forests covering an area of over one lakh sq km. Although Madhya Pradesh has the largest bamboo producing area, the bulk of production comes from the north-eastern states. The total production of bamboo is about five million tonnes per year. About 8.6 million people depend on bamboo for their livelihood. India is second only is China in terms of bamboo diversity. Bamboo is called the *poor man's timber* as it provides cheap material for roofing, walling, flooring, matting, basketry, cordage, carthoods and a host of other things. Young tender culms are eaten; the seed is collected and eaten as grain. However, the most significant commercial use of the bamboo is for making pulp for the production of paper and newsprint. Of the total bamboo consumed in India, 32 per cent is for construction, 30 per cent for rural use, 17 per cent for making paper pulp, 7 per cent for packaging and the remaining 14 per cent for other purposes.

Cane grows abundantly in moist forests of Andaman and Nicobar Islands, Karnataka, Madhya Pradesh, Kerala, Maharashtra, Nagaland, Manipur, Arunachal Pradesh and Mizoram. These are major producers of cane in India. Some parts of Assam, West Bengal, Kerala, Tamil Nadu, Jharkhand, Chhattisgarh and Odisha are also suitable for growth of cane. It is mainly used for making strings, ropes, mats, bags, baskets, furniture, walking sticks, umbrella handles, sports goods, etc.

2. Tans and Dyes. Tannins are secretion products of plant tissues. Tanning materials are used in leather industry. The most commonly used tanning materials are mangrove, amla, oak, hemlock, anwal, wattle, myrobalans, ratanjot, flowers of dhawri, babul, avaram, etc.

Some of the important dyes are obtained from red sander (bright red), khair (chocolate), flowers of Palas, fruits of *Mallotus phillipensis*, bark of wattle and roots of *Morinda tinctoria*. About two lakh tonnes of tans and dyes are produced every year in India.

3. Oils. A large number of plants and trees which grow in Indian forests contain several types of oils which are used to manufacture soaps, cosmetics, confectionary, pharmaceutical preparations and many more things. Commercially important oils are those obtained from sandalwood, lemon grass, khus and *euclyptus globulus*.

4. Gums and Resins. Gums are exuded from the stems or other parts of different trees, partly as a natural phenomenon and partly by injury to the bark or wood or blazing the tree. The most important gum is Karaya obtained from *Sterculia urens* or *S. villosa* trees of dry deciduous forests. It is mainly used in textiles, cosmetics, confectionery, medicines, inks, pastes, cigars, etc. Madhya Pradesh is the largest producer of gums in India. This state is closely followed by Maharashtra, Andhra Pradesh (including Telangana), Gujarat and Karnataka are other producers. A large proportion of Indian gums is exported to the USA, the UK and France.

Resin is obtained mainly from Chir pine which grows in the Himalayan region in Arunachal Pradesh, Uttarakhand, Himachal Pradesh, Jammu and Kashmir and some parts of Punjab. The main producers are Arunachal Pradesh, Punjab and Jammu and Kashmir. Some resin is produced in Manipur also. Crude resin consists of two principal constituents; a liquid known as oil of turpentine (25%) and a solid called resin (75%). They are separated after distillation. Turpentine is mainly used as a solvent for paints and varnish, synthetic camphor, pine oil, disinfectants, pharmaceutical preparations, wax, boot polish and industrial perfumes. Resin is an important raw material for several industries of which paper, paint, varnish, soap, rubber, water proofing, linoleum, oils, greases, adhesive tape, phenyl, plastic, etc. are important.

5. Fibres and Flosses. Fibres are obtained from the tissues of some trees. Most of such fibres are coarse and are used for rope making. However, the fibres of Ak (*Calotropis* spp.) is fine, strong and silky which is used for making fishing nets. Flosses are obtained from certain fruits and are used for stuffing pillows, mattresses, etc.

6. Leaves. Different types of leaves are obtained from the trees and are used for different purposes, the most important being the tendu leaves used as wrappers for *bidis*. The tendu tree grows in large numbers in Madhya Pradesh, Chhattisgarh, Andhra Pradesh, Telangana, Bihar, Maharashtra, Gujarat, Rajasthan, Karnataka and Uttar Pradesh. According to Indira Gandhi National Forest Academy, Dehradun, India produces about 375 thousand tonnes of tendu leaf every year. With 123.0 thousand tonnes per year, Madhya Pradesh is the largest producer. This is followed by Chhattisgarh (74.0 thousand tonnes), Odisha (50.0 thousand tonnes), Maharashtra (46.0 thousand tonnes), Andhra Pradesh (39.0 thousand tonnes), Bihar (24.0 thousand tonnes), Rajasthan (6.5

thousand tonnes), Uttar Pradesh (5.0 thousand tonnes), Gujarat (5.0 thousand tonnes), Tamil Nadu (2.0 thousand tonnes) and West Bengal (0.5 thousand tonnes). Tendu leaves and *bidis* are exported to Pakistan, Bangladesh, Sri Lanka and some other Asian and African countries.

Leaves of *Bauhinia vahlii* are converted into plates and leaf cups and are also used as wrappers by vendors of sweets.

7. Drugs, Spices and Poisons. Thousands of drugs are obtained from fruits, flowers, roots, stems and leaves of different types of trees, plants and herbs. Quinine is the most important drug obtained from the Indian forests.

Spices are used to add aroma or pungency to food to flavour certain dishes. The important spices are galangal, cinnamon or *dalchini*, lesser cardamom (*Chhoti Ilayachi*), greater cardamom (*Badi Ilayachi*), etc.

Indian forests produce some poisonous substances which can act as good medicines when taken in small, regular doses. Some outstanding poisons are strychnine, aconite, dhatura, ganja, etc.

8. Edible Products. Fruits, flowers, leaves or roots of various species provide edible products. Mango, *bel*, *ber*, *jamun*, *khirni*, *phalsa*, *sitaphal*, etc. are important fruits obtained from the forests. Among the kernels cashewnut, *akhrot* or walnut, *achar*, *chilgoza* and *kimal* are important. *Amla*, *anar*, *imli*, *karaunda*, *munga*, *kachnar*, *kaith*, *mushroom*, *zimikand*, *guchchi*, etc. are important products used as pickles or vegetables. Palmyrah, palm, mahua, corolla are used to obtain liquor and their seeds are eaten. *Tejpata*, used for flavouring curries are leaves of a small evergreen tree.

9. Animal Products. Lac is the most important animal product obtained from the forests. It is secreted by a minute insect (*Laccifer lacca*) which feeds on the saps of a large variety of trees like palash, peepul, kusum, sissoo, sisir, kul, gular, ber, banyan, jujuba and ghont. These trees grow extensively in the Chota Nagpur plateau of Jharkhand, eastern districts of Madhya Pradesh, Chhattisgarh, western border areas of West Bengal, eastern part of Maharashtra, northern districts of Odisha and to a lesser extent in Assam, Andhra Pradesh, Tamil Nadu, Uttar Pradesh, Karnataka and Punjab. India practically holds a monopoly in the production of lac. At present it is widely used in medicines, plastics, electrical insulation material, dyeing silk, making bangles, paints, sealing wax, leather and wood finishing, ornamental articles, etc.

The other animal products are honey, wax, silk moths, horns and hides of dead animals, ivory, antlers of deer, etc.

The above mentioned major and minor forest products are used for various purposes and form an important sector of Indian economy. About 3.5 million persons are engaged in different forest activities. About two per cent of the government revenue comes from the forests. There has been a steady increase in government revenue from the forests. In some states like Maharashtra, Uttarakhand, Karnataka, Kerala, Odisha and Andhra Pradesh, gross revenue from forests is far greater than the expenditure on forest activities. Some foreign exchange is also earned by exporting forest products.

INDIRECT USES OF FORESTS

Apart from their direct uses described above, the forests are of immense use to man indirectly also. They prevent soil erosion, regulate the flow of rivers and reduce the frequency and intensity of floods, check the spread of deserts, add to soil fertility and ameliorate the extremes of climate.

1. Prevention and control of soil erosion. Forests play a significant role in the prevention and control of soil erosion by water and wind. The destruction of forest cover leads to increased runoff of rain water and its diminished seepage and storage in soil. The structure of the soil suffers, runoff increases and loosens the soil which is carried away to other regions. The fertility of the soil is thus lost, and it becomes barren and unproductive. The reckless destruction of forests in the Shiwalik hills, Western Ghats, Chota Nagpur plateau and in the Deccan table-land has resulted in serious problem of soil erosion. The most effective way to check soil erosion is to stop reckless cutting of forests and to plant more trees.

2. Flood Control. Roots of the trees absorb much of the rain water and use it slowly during the dry season. Thus they regulate the flow of water and help in controlling the floods. The forest cover acts as a **rain-holder** and a **rain banker**. Trees also act like millions of tiny dams and check the flow of water like a barrage. With the increased rate of deforestation, the frequency and the intensity of floods has increased in different parts of the country. In the absence of forests, the increased runoff along with sand and silt, especially after heavy downpour, comes in as a sudden rush and often gets blocked in the silted streams and causes devastating floods.

3. Checks on spread of deserts. Sand particles are blown away by strong winds in the deserts and are carried over long distances, thus resulting in the spread of deserts. The roots of trees and plants bind the sand particles and do not permit their easy transportation by wind. Therefore, the forests are a great instrument to put a check on the spread of deserts. In the long run, the forests add humidity to the atmosphere and further help in checking the spread of deserts. A great success has been achieved in checking the march of the Thar desert by planting a 650 km long and 8 km wide strip of trees on its periphery.

4. Increase of soil fertility. The fallen leaves of trees add humus to soil after their decomposition. Thus forests help in increasing the fertility of soil.

5. Effect on Climate. Forests have a far reaching effect on climate. They ameliorate the extremes of climate by reducing the heat in summer and cold in winter. They also influence the amount of rainfall by lowering the temperature of moisture laden winds and increase the relative humidity of the air through the process of transpiration. They reduce the surface velocity of winds and retard the process of evaporation.

PROBLEMS OF INDIAN FORESTRY

Indian forests face a number of problems which are both natural and manmade. Some of the major problems are briefly discussed as follows :

1. Inadequate Forest Area and Dwindling Forest Cover. The biggest problem of the Indian forests is the inadequate and fast dwindling forest cover. It has already been mentioned that forests cover only 23.34 per cent of the area against the required coverage of 33 per cent. Even this small percentage of forest cover is seriously threatened by the increasing demand for major and minor forest products. These products are badly needed for fuel, building and to feed a large number of forest based industries. Vast forest tracts have been cleared for agriculture. Shifting agriculture in different parts of the country has played havoc with forests. Overgrazing is a big factor which is responsible for serious damage to forests. India possesses a livestock population of about 535.78 million about one-tenth of which graze in the forests. Wherever forests are easily accessible, the livestock entirely depends on grazing in them.

TABLE 7.4. India : Percentage change of Forest Cover in Green Wash Areas between 1999 and 2013

States/Union Territories	Percentage loss between 1999 and 2013
1. Daman and Diu	–100.0
2. Chandigarh	–92.7
3. Delhi	–90.7
4. Dadra and Nagar Haveli	–76.7
5. Punjab	–55.6
6. Haryana	–53.7
7. Nagaland	–36.8
8. Arunachal Pradesh	–36.7
9. Sikkim	–35.1
10. Jammu and Kashmir	–34.3
11. Himachal Pradesh	–33.1
12. Gujarat	–29.4
13. Uttarakhand	–20.6
14. Maharashtra	–19.7
15. Jharkhand	–17.8
16. Uttar Pradesh	–16.7
17. Assam	–16.2
18. Karnataka	–15.2
19. Rajasthan	–14.0
20. West Bengal	–13.6

Source : Forest Survey of India, *State of Forest Reports 1999*, 2009 and 2013

Data released by the Ministry of Environment, Forest and Climate Change in July, 2015 reveals that despite an increase in green cover, the loss of original forest cover has increased exponentially and the country may have lost close to 10.6 million hectares of original forests between 1999 and 2013. Punjab, Haryana, Nagaland, Arunachal Pradesh, Sikkim, Jammu and Kashmir and Himachal Pradesh are the greatest loosers as they have lost over one-third of their original forests. Union territories of Daman and Diu, Chandigarh, Delhi and Dadra and Nagar Haveli have lost more than three-fourths of their original forest cover (**Table 7.4**). The only gainers are Tripura (+15%) and Mizoram (+1.9%).

According to State of Forest Report, 2019 released by the Ministry of Environment Forest and Climate Change, India's total forest cover has registered a 1.5 per cent increase between 2015 and 2019. But seemingly good increase is misleading because the bulk of increase has taken place in open forest category, which includes commercial plantations. This has happened at the cost of moderately dense forest, which is normally the area close to human habitations. Very dense forest occupies only 3% of the total forest cover (**Table 7.5**).

2. Low Productivity. Productivity of Indian forests is very low as compared to some other countries. For example, annual productivity of Indian forest is only 0.5 cubic metre per hectare while it is 1.25 cubic metre per hectare in the USA, 1.8 cubic metre per hectare in Japan and 3.9 cubic metre per hectare in France.

3. Nature of Forests and their Uneconomical Utilisation. The forests are thick, inaccessible, slow growing and lack in gregarious stands in many parts of the country. Some of them are very thin and comprise only of thorny bushes. These factors make their utilization uneconomical because there is a good deal of wastage and this makes it very expensive in spite of the cheap labour available in India.

4. Lack of Transport Facilities. One of the biggest problems faced by the Indian forests is the lack of proper transport facilities. About 16 per cent of the forest land in India is inaccessible and does not have proper transport facilities. It must be remembered that the major product of the forests is timber which is a cheap and bulky commodity. As such it cannot afford high freights charged by the railways and roadways. Therefore, Indian forests cannot be economically exploited without the availability of cheap and efficient transport facilities. Unfortunately, in India, the railways serve thickly populated areas only and are not of much use to forests. All weather roads in the forest areas are badly lacking. Water transport has only limited scope. Considering these facts, we can easily say that transport with reference to forests is inadequate in India.

TABLE 7.5. Category-wise forest area in India

Forest Category	2015		2017		2019	
	Total area (sq km)	Percentage of total area	Total area (sq km)	Percentage of total area	Total area (sq km)	Percentage of total area
1. Very dense forest with tree canopy density of 70% and above	85,904	2.61	98,158	2.99	99,278	3.02
2. Moderately dense forest with tree canopy density of 40 to 70%	3,15,374	9.59	3,08,318	9.38	3,08,472	9.38
3. Open forest with tree canopy densely 10 to 40%	3,00,395	9.14	3,01,797	9.18	3,04,499	9.26
4. Degraded forest lands with canopy densely less than 10%	41,362	1.26	45,979	1.40	46,297	1.41
5. Non forest area	25,44,228	77.40	25,33,217	77.05	25,28,923	76.93

Source: India State of Forest Report, Ministry of Environment, Forest and Climate Change (Quoted in State of India's Environment, 2021, Centre for Science and Environment, p. 424).

TABLE 7.6. Number of incidences of Forest Fire

	States	2015	2016	2017
1.	Madhya Pradesh	294	2,675	4,781
2.	Odisha	1,467	2,763	4,416
3.	Chhattisgarh	1,272	2,808	4,373
4.	Maharashtra	721	1,874	3,487
5.	Assam	1,656	1,766	1,887
6.	Andhra Pradesh	1,075	1,699	1,877
7.	Telangana	1,052	1,154	1,748
8.	Mizoram	2,468	1,318	1,587
9.	Meghalaya	1,373	966	1,454
10.	Karnataka	295	831	1,333
11.	Uttar Pradesh	130	691	1,170
12.	Jharkhand	457	740	1,133
13.	Manipur	1,286	1,105	1,094
	Others	2,391	4,427	5,548
	Total	**15,937**	**24,817**	**35,888**

Note : States have been arranged in descending order of incidences of Forest Fires in 2017.

Source : EnviStats, 2018, p. 100.

5. Forest Fires. Large tracts of vegetal cover are destroyed every year by forest fires. Insufficiency of properly trained personnel to prevent and fight fires is a big handicap.

About 50% of India's forest cover is prone to fires. Most fires occur in dry season from February to June before the onset of monsoons. Incidents of forest fires are increasing rapidly due to mismanagement and laxity of the forest personnel.

The number of incidents of forest fires more than doubled in a short span of two years from 15,937 in 2015 to 35,888 in 2017 (**Table 7.6**). Madhya Pradesh, Odisha and Chhattisgarh account for more than one-third of the total forest for fires in India. As many as 27 states and union territories registered increase in number of forest fires against only 3 which registered a decrease in such incidents. Dadra and Nagar Haveli, Daman and Diu and Lakshadweep did not report any incident of forest fire. The annual economic loss due to forest fires has been estimated at ₹ 1,176 crore at 2016 prices.

The Central Government provides financial assistance to fight forest fires so that damages caused by forest fires is controlled and is kept at the lowest level (**Fig. 7.3**). However, it is felt that funds released by the Government are not enough to control or even to reduce forest fires.

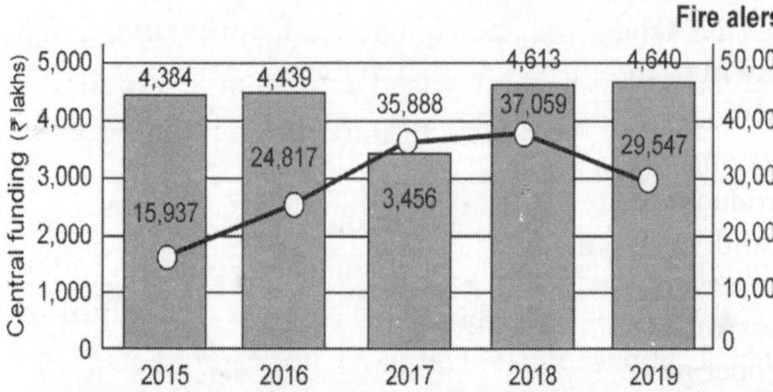

FIG. 7.3. Number of forest-alerts and funds released by the Government for controlling forest fires. Line graph shows fire alerts and bars show central funding.

Main Causes of Fire

- Increasing heat, less rain and thunderstorms/ lightning
- Dry forest floor of twigs, leaves and pine needles
- Invasive weed lantana is a fire risk—a 2020 report found lantana had invaded 40% of India's tiger reserves
- Carelessness of collectors of non-timber forest produce—a beedi or cigarette can start a fire
- Locals often set fire to sal leaves, for mahua. In times of scarcity, a boiled mix of mahua flowers and sal seeds is used as a grain substitute
- Poachers start fires to chase animals into straps
- Slash-and-burn is a trigger in the Northeast.
- Timber mafia
- Corrupt officials.

India's worst forest fires

1995 : 375,000 hectares affected in Uttarakhand

1999 : 86,000 hectares in Ganga-Yamuna watershed

2008 : 10,000 hectares in Maharashtra.

2016 : Uttarakhand 1,900 hectares affected in 1,600 forest fires.

2020 : Uttarakhand, 46 incidents of forest fires.

6. Plant Diseases, Insects and Pests. Large tracts of forest cover suffer from plant diseases, insects and pests which lead to considerable loss of forest wealth. For example, thousands of hectares of sal forests in Madhya Pradesh and Chhattisgarh are being threatened by sal borer for which no remedial measures have been adopted so far. Forest officials are only using the primitive methods of hiring the tribals to catch and kill the insects.

7. Obsolete Methods of Lumbering and Sawing. In most of the Indian forests, obsolete methods of lumbering, sawing, etc. are practised. This system leads to a lot of wastage and low forest productivity. Large quantities of inferior wood which could be put to better use through seasoning and preservation treatment remain unutilised or go waste. Saw mills use old obsolete machinery and do not get proper power supply.

8. Lack of Commercial Forests. In India most of the forests are meant for protective purposes and commercial forests are badly lacking. Growing awareness about environmental degradation has forced us to look at forest wealth as a protective agent for environment rather than treating it as a commercial commodity.

9. Lack of Scientific Techniques. Scientific techniques of growing forests are also lacking in India. Only natural growth of forests takes place in India whereas in many developed countries new scientific techniques are being used through which tree growth is quickened. A large number of trees are malformed or consist of species which are slow growing and poor yielders.

10. Undue Concessions to Tribals and Local People. In vast forest tracts, tribals and local people have been granted customary rights and concessions for free-grazing as well as removing timber fuel and minor forest products. They are also allowed to continue with age-old shifting cultivation. These practices have led to the reduction in forest yield. In addition, there has been encroachment on these forests by the village people inhabiting the peripheral areas.

Remedies

At present there is an urgent need of undertaking silvicultural operations on a large scale. This can be done through the following measures :

1. Intensive development schemes for afforestation should be adopted. High yielding varieties should be planted in suitable areas.
2. Improved techniques of logging and extraction should be used.
3. Proper transport facilities should be provided to remote and inaccessible forest areas.
4. Saw mills should get uninterrupted power supply.
5. Latest techniques of seasoning and preservation are necessary to avoid wastage.
6. Proper arrangements to save forests from fires and plant diseases can go a long way to solve several problems.
7. A thorough inventory of forest resources is necessary to make an accurate assessment of our forest resources and make plans for their proper use.
8. Shifting cultivation should be discouraged and tribals depending on this type of cultivation should be provided with alternate sources of livelihood.
9. People associated with forest protection should be properly trained.

SOCIAL FORESTRY

Social forestry means the management and protection of the forests as well as afforestation of barren lands aimed at helping in environmental, social, and rural development as against the traditional objective of securing revenue. It is aptly described as *forestry of the people, by the people and for the people*. The main thrust of social forestry is to reduce pressure on the traditional forest areas by developing plantations of fuelwood, fodder and grasses. The term social forestry was used for the first time by the National Commission on Agriculture in 1976, to denote tree raising programmes to supply firewood, small timber and minor forest produce to rural population. This commission divided social forestry into; (*i*) agro forestry, (*ii*) expansion forestry, (*iii*) rehabilitation of degraded and low grade forests and (*iv*) recreation forestry. The programme was formally launched in 1978 and it became an integral part of the Sixth Plan in 1980. In the second Forestry Conference (1980) it was decided that social forestry scheme should be given priority over barren and waste land, community land and lands along the roads, railways and canals.

Agro forestry, community forestry, commercial farm forestry, non-commercial farm forestry and urban forestry are the main components of social forestry.

Agro forestry involves the raising of trees and agricultural crops either on the same land or in close association in such a way that all land including the waste patches is put to good use. This enables the farmer to get food, fodder, fuel, fruit and timber from his land. The land gives maximum production and provides employment to rural masses.

Community forestry involves the raising of trees on public or community lands aimed at providing benefit to the community as a whole. Although the plants and seedlings are provided by the forest department, the protection of plantations is primarily the responsibility of the community as a whole. India is second only to China in community forestry. Remarkable success has been achieved in community forestry in some major states like Gujarat, Tamil Nadu, Rajasthan, Kerala, Karnataka, Uttar Pradesh, Haryana, Odisha and Himachal Pradesh.

Commercial farm forestry involves growing of trees in the fields in place of food and other agricultural crops. In this way, the farmers grow trees for direct commercial gain. This type of forestry has become very popular in Haryana, Punjab, Gujarat, Uttar Pradesh, Rajasthan and in some parts of south India. Farmers in these areas grow eucalyptus, poplars, and casuarina on their land and sell them to paper, pulp, rayon and match industries and also use them for making packing cases and fuel. The state governments are encouraging this forestry in a big way.

Non-commercial farm forestry involves tree planting by farmers on their own land for their own use and not for sale. The land mainly used for this purpose includes margins of agricultural fields, wastelands and marginal lands which are not usually cultivated, grasslands and pastures and land around homes and cowsheds. This type of forestry is practised in Gujarat, Haryana, Kerala and Karnataka for obtaining fuel, fodder and fruit.

Urban forestry pertains to raising and management of trees on public and privately owned lands in and around urban centres. It includes green belts, roadside avenues, recreational parks, wildlife parks, etc. Its main objectives are reduction of environmental pollution, recreation and improving aesthetic values.

FOREST CONSERVATION

Forests comprise a unique gift of nature to man and constitute one of the prized assets of a nation. They play a significant role in the national economy of a primarily agricultural and developing country like India. The agricultural and industrial progress of the country is not only stabilized but accelerated by a proper conservation and utilization of forest resource. As mentioned earlier, the uses of forests, both direct and indirect, are so large that they are aptly termed as *an index of prosperity of a nation*. Keeping in view the benefits which we derive from the forests, it is of utmost importance that strong steps should be taken to conserve forests.

Our increased demand for forest products has led to increasing destruction and degradation of our forests which is causing heavy erosion of top soil, erratic rainfall and frequent devastating floods. In short, depletion of forests has a chain reaction in eco-system. Though it is a renewable resource, it takes its own time to regenerate. We have been destroying our forest resources so ruthlessly and so quickly that large forest tracts of yesteryears are now devoid of any forest cover. India's woods, once dark and deep, are now a living example of man's ravage and destruction. The saying that *man finds forests but leaves deserts could not be more true to India*. Over the past four decades, about 25 million hectares of land that originally had tree cover has been laid bare for agriculture and other purposes. The latest reports of the National Remote Sensing Agency (NRSA) indicates that the country is losing about 1.3 million hectares of forest cover every year. This will be detrimental to our national interest. *Nature never forgives the abuse of her gifts*. Hence, the urgent need for conservation of forests.

Forest conservation does not mean the denial of use, but rather the proper use without causing any adverse effect on our economy or environment. But any scheme of conservation of forests on a piecemeal basis will not solve the problem. Conservation of forests is a national problem and should be tackled as such. There should be perfect coordination between the forest department and other departments for an effective conservation of forests. People's participation in any forest conservation is of vital importance. ***Van Mahotsava*** was launched in 1950 to make people aware of the importance of planting trees. ***Chipko movement*** is a living example of general public awareness about forests.

The year 2011 has been declared as the International Year of Forests.

Forests have to be developed and worked for obtaining various raw materials and for providing an effective means of flood control, checking soil erosion, for regulating the flow of water in streams and for conserving moisture in the soil. Therefore, a carefully coordinated scientific policy for conservation of forests should be the first step in any scheme of national planning of the country.

While contribution of forests to the nation's economy, apart from their vital role in environment, can never be underestimated, the investment in forestry sector has been rather low. For example, from the first to the sixth Five Year Plans, the investment in forestry was between 0.39 per cent and 0.71 per cent of the total plan outlay. In the seventh Five Year Plan, the allocation was raised to 1.03 per cent of the total plan outlay. But again, the investment in the forestry sector during the eighth Five Year Plan had fallen to 0.94 per cent of the total plan outlay. At present the total investment in forestry is about ₹ 800 crore against the required investment of about ₹ 3,000 crore. This is too small an investment and unless it is increased, it will not be possible to ensure sustainable supply of goods and services for the huge sector of population dependent on forests.

The Forest (Conservation) Act, 1980 enacted to check indiscriminate deforestation/ diversion of forest lands was amended in 1988 to make it more stringent by prescribing punishment for violations. Guidelines have been prepared for working plans. Some salient features are : (*i*) working plans should be up-to-date and stress conservation; (*ii*) preliminary working plan should have multi-disciplinary approach; (*iii*) tribal rights and concessions should be highlighted along with control mechanism; (*iv*) grazing should be studied in detail and specific prescriptions should cover fodder propagation; (*v*) shifting cultivation and encroachments need to be controlled; (*vi*) clear-felling with artificial regeneration should be avoided as far as possible and clear-felling blocks should not exceed 10 hectares in hills and 25 hectares area in plains and (*vii*) banning all felling above 1000 metre altitude for a few years should be considered to allow these areas to recover. Critical areas in hills and catchment areas prone to landslides, erosion, etc. should be totally protected and quickly afforested.

The Indian Council of Forestry Research and Education (ICFRE) was created in 1987 under the Central Ministry of Environment and Forests. Later on it was constituted into an autonomous body on the pattern of the Indian Council of Agricultural Research with its headquarters at Dehra Dun. Following forestry research institutions are working under this organisation :

(*i*) Forest Research Institute, Dehra Dun

(*ii*) Institute of Arid Zone Forestry Research, Jodhpur

(*iii*) Institute of Rain and Moist Deciduous Forests, Jorhat

(*iv*) Institute of Wood Science and Technology, Bengaluru.

(*v*) Tropical Forestry Research Institute, Jabalpur

(*vi*) Institute of Forest Genetics and Tree Breeding, Coimbatore

(*vii*) Temperate Forest Research Centre, Shimla

(*viii*) Centre for Forest Productivity, Ranchi and

(*ix*) Centre for Social Forestry and Environment, Pryagraj.

The future welfare and prosperity of India would very much depend upon our ability, effort and success in conserving, developing and proper utilisation of our forest resources. It is, therefore, high time that the nation as a whole awakens to this burning problem for the sake of a better future.

Forest Policy and Law

(*i*) Forest Policy of 1894. Forest policy was first declared by the British Government of India on 19 October, 1894 at the recommendation of **Dr. Voelcker**, a German expert who studied the role of forests vis-a-vis agriculture.

(*ii*) Forest Policy of 1952. With the passage of time and with India becoming an Independent country, major changes of economic and political nature took place. The population, both of human beings and livestock, had increased very substaintially, resulting in a heavier pressure on the forests for securing more land for agriculture and pasturage. The reconstruction schemes also leaned heavily on the forest products. The Government of free India thought it desirable to come out with a new

National Forest Policy in 1952. While retaining the major clauses of the earlier forest policy, this policy laid greater emphasis on some other points also. It proposed the classification of forests on a functional basis into *protected forests, national forests and village forests*. The policy recommended the establishment of tree-lands wherever possible. It also stressed upon increasing pastures and timber and made a strong plea against the indiscriminate extension of agriculture by excision of forests. The policy also stressed that the *productive, protective* and *bioaesthetic* roles of forest entitle them to an adequate share of land. It also emphasised the need to protect wildlife. The policy laid stress on, (*i*) weaning the primitive people by persuation, from the baneful practice of shifting cultivation, (*ii*) increasing the efficiency of forest administration by having adequate forest laws, (*iii*) giving requisite training to the staff of all ranks, (*iv*) providing adequate facilities for management of forests and for conducting research in forestry and forest products utilization, (*v*) controlling grazing in the forestry and (*vi*) promoting welfare of the people.

(*iii*) Forest Policy of 1988. The country made rapid strides after Independence under planned development and changes of far reaching consequences occurred in the economic, political, environmental and social set up of the country since the forest policy of 1952 was declared. Therefore, the Government of India came out with a new forest policy in December, 1988. The main plank of the revised forest policy of 1988 is protection, conservation and development of forests. Its aims are : (*i*) maintenance of environmental stability through preservation and restoration of ecological balance (*ii*) conservation of natural heritage ; (*iii*) check on soil erosion and denudation in catchment area of rivers, lakes and reservoirs ; (*iv*) check on extension of sand dunes in desert areas of Rajasthan and along coastal tracts ; (*v*) substantial increase in forest/tree cover through massive afforestation and social forestry programmes ; (*vi*) steps to meet requirements of fuel wood, fodder, minor forest produce and timber of rural and tribal populations; (*vii*) increase in productivity of forest to meet the national needs; (*viii*) encouragement of efficient utilisation of forest produce and optimum substitution of wood and (*ix*) steps to create massive people's movement with involvement of women to achieve the objectives and minimise pressure on existing forests.

(*iv*) Forest Policy of 2018. Released on March 15, 2018, this policy seeks to replace the 1988 Forest Policy. It calls for promoting urban greens, public private partnership models for afforestation, strengthening forest fire prevention measures and plantations in catchment areas to rejuvenate water bodies among several proposals to protect India's green cover. The policy is proposed to synchronize with the country's forestry-related 'Nationally Determined Contribution' targets under the Paris Agreement where India has promised to rapidly increase its forest cover so that an additional carbon sink of 2.5 to 3.0 billion tonnes of carbon dioxide equivalent is created by 2030. The ministry of Environment, Forest and Climate Change is of the opinion that though the National Forest Policy of 1988 had been instrumental in increasing forest and tree cover and also in reducing diversion of forest land for other uses despite compelling demands from increasing population, a new policy was needed in face of challenges of climate change, human-wildlife conflict and deteriorating environment. Following are the salient features of Draft National Forest Policy, 2018.

1. Public-private participation for afforestation in degraded forest areas.
2. Promoting urban green
3. Measures to safeguard ecosystems from forest fire
   - Map the vulnerable areas.
   - Develop and strengthen early warning systems.
4. National Community Forest Management Mission for participatory forest management.
5. Afforestation in catchment areas for river rejuvenation and water recycling.
6. Economic valuation of forests.
7. Forest certification to provide premium on forest produce.
8. Integration of climate change concerns in forest management.
9. Promote agro-forestry and farm forestry.
10. Management of north-eastern forests.

WILDLIFE

Wildlife comprises animals, birds, and insects living in forests. With large regional variations in physiography, climate and edaphic types, Indian forests offer a wide range of habitat types which are responsible for a large variety of wild life in India. India boasts of more than 90,000 species of animals which is about 6.5% of the word's total species. Indian fauna includes about 6,500 invertebrates, 5,000 molluscs, 2,546 species of fishes, 2,000 species of birds, 458 species of reptiles, 4 species of panthers and over 60,000 species of insects.

Elephant is the largest Indian mammal which only a few centuries ago, was found in large numbers in vast forest tracts of India.

India's elephant habitat has expanded with jumbo population stable. The total population of elephants in India was 27,657 to 27,682 in 2007, 29,391 to 30,711 in 2012 and 30,051 in 2017. Decline between 2012 and 2017 has been attributed to difference in counting methods. Most of the elephants are found in southern states of Kerala, Karnataka, Tamil Nadu and Odisha. In the north-eastern part of India, Assam has the largest number of elephants (**Table 7.7**).

TABLE 7.7. Distribution of Elephants in India (2017)

State	Number of Elephants
1. Kerala	6,182
2. Karnataka	6,068
3. Assam	5,620
4. Tamil Nadu	4,015
5. Odisha	1,930
6. Meghalaya	1,811
7. Uttarakhand	1,346
8. Arunachal Pradesh	890
9. Jharkhand	688
10. West Bengal	647
Others	854
All India	**30,051**

Source: Centre of Wildlife and Protection, Ministry of Environment, Forest and Climate Change.

The one-horned ***rhinoceros***, India's second largest mammal was once found throughout the Indo-Gangetic Plain as far west as Rajasthan. The number of this mammal drastically decreased with the passage of time and at one time it was close to extinction. But steps taken to conserve this rare species have borne fruit and at present, India has over 2900 rhinoceros. They are mainly confined to Assam and West Bengal. Some rhinoceros are found in Uttar Pradesh also (**Table 7.8**).

TABLE 7.8. Distribution of Rhinoceros in India (2017)

States/Area		Number of Rhinoceros
Assam	Kaziranga Tiger Reserve	2,401
	Orang Reserve	100
	Manas Reserve	93
	Pobitora Wildlife Sanctuary	28
	Total	2,622
West Bengal	Jaldapara National Park	204
	Gosumara National Park	49
	Darjeeling	2
	Total	255
Uttar Pradesh	Dudhwa	32
Total		**2,909**

Source: Ministry of Environment, Forest and Climate Change (2017).

However, poaching is a big threat to their survival as rhino horn is very valuable and is used in traditional Asian medicines. They survive under strict protection in various areas like Kaziranga and Manas sanctuaries of Assam and Jaldapara sanctuary of West Bengal.

Asiatic Lion is another important wild animal found in Indian forests. There has been a gradual increase in number of lions and area of lions. The total number of lions in 1990 was only 284 which rose to 674 in 2020. The area of lions also increased from 6,600 sq km to 30,000 sq km during the same period. The population estimation of Asiatic lions is conducted at an interval of five years (**Table 7.9**).

TABLE 7.9. Number of Asiatic Lions and their area in India

Year	Number of Asiatic Lions	Lion distribution area (sq km)
1990	284	6,600
1995	304	10,000
2001	327	12,000
2005	359	13,000
2010	411	20,000
2015	523	22,000
2020	674	30,000

This has happened due to use of multiple strategies like people's participation, use of modern technology, wildlife health care, habitat management, man-lion conflict mitigation, etc.

The ***arna*** or wild buffalo is found in Assam and in Bastar district of Chhattisgarh. The ***gaur*** or the Indian bison is one of the largest existing bovine and is found in the forests of Central India. India is home to about 70% ***tigers*** in the world. In 2006, there were 1411 tigers in India. This number increased to 1706 in 2011 and 2226 in 2014. Most tigers of India are found in the forests of eastern Himalayan foothills and in parts of peninsular India. There are about 51 tiger reserves in India.

Leopard is widely found in the forests of a large number of states. The leopard population in India recorded an unprecedented increase of 62% in four years—from 7,910 in 2014 to 12,852 in 2018. According to **'Status of Leopards in India, 2018'** report released on 21st December, 2020, Madhya Pradesh has the highest number of 3,421 leopards. This is followed by Karnataka (1,783), Maharashtra (1,690), Tamil Nadu (868), Chhattisgarh (852) and Uttarakhand (839). These six states account for about 73% of the total leopard population of India. Remaining leopards live in Kerala, Odisha, Andhra Pradesh, Uttar Pradesh, Goa, Bihar and Jharkhand (**Table 7.10**).

However, the grim reality is that a large number of leopards are killed every year in India. The number of leopards killed in 2014 was 331 which rose to 460 in 2018. Thus on an average more than one leopard is killed every day. Apart from natural deaths, poaching, rail and road accidents, attacks by villagers, and actions of forest department are the main causes of deaths of leopards. Poaching occurs for different parts of their bodies, the important being hides, bones and claws. The highest number of 93 deaths in 2018 occurred in Uttarakhand. This was followed by Maharashtra (90), Rajasthan (46), Madhya Pradesh (37) and Uttar Pradesh (27).

TABLE 7.10. Distribution of leopards in India (2018)

State	Total number of leopards	Percentage of total
1. Madhya Pradesh	3,421	26.62
2. Karnataka	1,783	13.87
3. Maharashtra	1,690	13.15
4. Tamil Nadu	868	6.75
5. Chhattisgarh	852	6.63
6. Uttarakhand	839	6.53
Others	3,399	26.45
Total	**12,852**	**100.00**

Source: Leopard Census; Status of Leopards in India Report, 2018.

Note: Actual number will be higher since survey only covered forested habitats in tiger-occupied states.

The *black* ***Panther*** is a widely distributed predator. *Desert* and *jungle cats* live in north western parts of the country. ***Lynx*** lives upto 3,010 m in Ladakh. *Brown, Black* and *Sloth Bear* are found at high altitudes in the north-western and central Himalayas. India was once home of many ***Cheetahs*** but the last of them was killed in 1947. It was declared extinct in India in recorded history.

***Cheetahs*:** India was once home to a large number of cheetah but their number reduced drastically due to large scale hunting. In 1947, only three cheetahs were left and they too were hunted mercilessly in Sal forests. Consequently cheetahs were declared extinct in India in 1952. On 17th September, 2022, eight cheetahs (three males and five females) were flown from Namibia (Africa) and were released in Kuno National Park in Madhya Pradesh. They are fast adjusting themselves to the new environment and it is hoped that their number will increase very soon.

Yak, the ox of snow, is largely found in Ladakh and is tamed to be used as a draught animal. Several species of wild sheep and wild goats are also found in India. The *shapu* or *urial*, *bhoral* the blue sheep, and *nayan,* a huge sheep with curved horns are the main types of wild sheep. *Serow* and *goral* are the goat antelopes of the Himalayas. The Kashmir *markhar*, the *ibex* and the Himalayan *thar* are some of the Himalayan wild goats.

Deer also used to roam widely across the Indian forests, although their number has been drastically reduced. Stag or ***barasingha*** is found in Assam, Chhattisgarh and Madhya Pradesh. The ***Munjac*** or barking deer are found extensively in the lower wooded slopes of the Himalayas and in the forests of southern India. The ***kastura*** or the musk deer, much sought after for its musk pod, live in the birch woods in the higher forests of the Himalayas. *Thamin* is a pretty deer found in Manipur containing *Kasturi*.

Several species of ***monkeys*** are found in almost all the forest areas of India, the two commonest being the Rhesus Macaque and the common ***Langur***. There are wide regional variations in the body structure and behaviour of monkeys in India.

The ***Chinkara*** or the Indian gazelle, the black buck or the Indian antelope, the ***nilgai*** or the blue bull, the mouse deer or the Indian chevrotain, the ***chawsinga*** or the four horned antelope, wild dog, the fox, the jackal, the hyena, the mongoose, shrews, hedgchogs, moles, bats, rodents and squirrels are the other mammals found in the Indian forests.

India also abounds in large variety of reptiles, although many of them are now endangered species. There are more than 200 species or subspecies of snakes, the best known being the ***Cobra***, ***Krait*** and ***Russel's Viper***. These are poisonous snakes while *Dhaman* is a non-poisonous large snake. The length of the King Cobra may be well over five metres which makes it the longest poisonous snake. However, the Rock Python and the Reticulated Python may be seven metres long, weighing over 115 kg. Several snakes living in water are also poisonous, although many of them are non-poisonous. The Blunt Nosed or Marsh Crocodile (the *Magar* or *Mugger*) and the long nosed *Gharial* are important large sized reptiles, although their number has drastically reduced. They are hunted for their skins which fetch handsome prices. The big Estuarine Crocodile is still found from the Ganga to the Mahanadi. The lizards include the well known Chameleon and the monitor lizard or varanus. They are found both in deserts and forests but are endangered species. India has some important breeding beaches for a number of species of turtles. In Odisha about 3,00,000 Olive Ridley Turtles breed while Hawksbill Turtles breed in southern Tamil Nadu.

India is extremely rich in bird life. There are about 2,000 species of birds in India which is about three times the number of species found in Europe. Although most of the birds have their origin in India, a number of them have their source in other areas. Some birds such as ducks, cranes, swallows, and fly-catchers migrate from central Asia to the wetlands of Bharatpur every winter. Recently, some migratory birds have been seen near Mathura.

Indian bird-life has all the varieties of birds including aquatic, gallinaceous and arboreal. Aquatic birds include a large variety of storks, herons, ducks, flamingoes, egrets and cormorants. Among the waders and shore birds are the snipes, iluses, gulls, cranes and the lapwings. The Great Indian Bustard, pea fowl, jungle fowl, quail and partridge are the main ground birds. Babblers, barbits, bulbuls, mynas, pigeons, parakeets, doves, cuckoos, rollers, beaters, fly-catchers, orioles, warblers, wagtails, finchlarks, finches, drongos, hoops, etc. are other important birds. India is the home of a large number of birds of prey, the important being owl, eagle, kite, falcon, kestrel, etc. *Peacock* is the ***national bird of India***. Its magnificent plumage symbolizes the colour and wealth of India's bird life.

PRESERVATION OF WILDLIFE

The limited forest cover in India has adversely affected wildlife in the country. The number of several species has been drastically reduced, some are endangered species, the others are on the verge of extinction, while some of them have already disappeared. This is a very serious trend and will disturb the ecological balance. Therefore, it must be checked and reversed at all costs. This calls for urgent measures for preserving wildlife.

National Board for Wildlife was constituted in 1952. The main purpose of the board was to advise the Government on the means of conservation and protection of wildlife, construction of national parks, sanctuaries and zoological gardens as well as promoting public awareness regarding conservation of wildlife. The wildlife (Protection) Act, 1972 is a comprehensive law which has been adopted by all states. It governs wildlife conservation and protection of endangered species. The Act prohibits trade in rare and endangered species. An Inter-State Committee has been set up to review the Wildlife (Protection) Act, 1972 and other laws. The wildlife (Protection) Amendment Bill 2002 proposes to enhance penalties for violation of the provision of the Act. Under this, export or import of endangered species and their products are governed by the conditions and stipulations laid down therein.

Project Tiger, one of the premier conservation efforts in the country was launched on 1 April 1973. It is a centrally financed scheme under which 51 Tiger Reserves have been set up. The tiger population had slid to a near extinction level before the implementation of Project Tiger. There were only 711 tigers in all the tiger reserves in 1979. This figure rose to 1,141 in 1984, 1,327 in 1989, 1,396 in 1993 and 1,575 in 1995. The number of tigers further increased to 3642 in 2001 after which there was a sharp decline and their number fell to 2226 only in 2014.

However, the number of tigers in India rose to 2967 in 2018, a 33% jump over 2014. Madhya Pradesh had the largest number of 526 tigers, followed by Karnataka at 524. Earlier, Karnataka was having the largest number of tigers but was dislodged by Madhya Pradesh (**Table 7.11**).

Table 7.12 shows that some states have recorded increase in tiger population while other states have lost a large number of tiger population. This loss of tiger population in some states is a matter of serious concern.

TABLE 7.11. Distribution of tigers in India (2018)

State	Number of Tigers
1. Madhya Pradesh	526
2. Karnataka	524
3. Uttarakhand	442
4. Maharashtra	312
5. Tamil Nadu	264
6. Kerala	190
7. Assam	190
8. Uttar Pradesh	173
Others	346
All India	**2,967**

Source: Status of Tigers in India, 2018.

The primary object of the project tiger is "to ensure maintenance of the viable population of tigers in India for scientific, economic, aesthetic, cultural and ecological values, and to preserve for all times, areas of biological importance as a national heritage for the benefit, education and employment of the people". Following are the salient features of project tiger :

- Amendment of the Wild Life (Protection) Act 1972 for providing provisions for constitution of the National Tiger Conservation Authority and the Tiger and other Endangered Species Crime Control Bureau.
- 100 per cent Central Assistance to Tiger Reserves for deployment of Tiger Protection Force, comprising ex-army personnel and local workforce.
- Constitution of the National Tiger Conservation Authority with effect from September 4, 2006.
- Constitution of a multi-disciplinary Tiger and Other Endangered Species Crime Control Bureau (Wildlife Crime Control Bureau) w.e.f. 6-6-2017.
- Approval accorded for declaring eight new Tiger Reserves. Notification for Sahyadri Tiger Reserve in Maharashtra has been issue in January, 2011.
- The revised Project Tiger guidelines have been issued to states for strengthening tiger conservation.
- A scientific methodology for estimating tiger (including co-predators, prey animals and assessment of habitat status) has been evolved and mainstreamed. According to the refined technology, an estimated land of 93,697 sq km has been observed as tiger habitat. An area of 29,284.76 sq km has been notified by 15 Tiger States as core or critical tiger habitat.
- India has a Memorandum of Understanding (MoU) with Nepal on controlling transboundary illegal trade in wildlife and conservation, apart from a protocol on tiger conservation with China.
- A Global Tiger Forum of Tiger Range Countries has been created for addressing international issues related to tiger conservation.

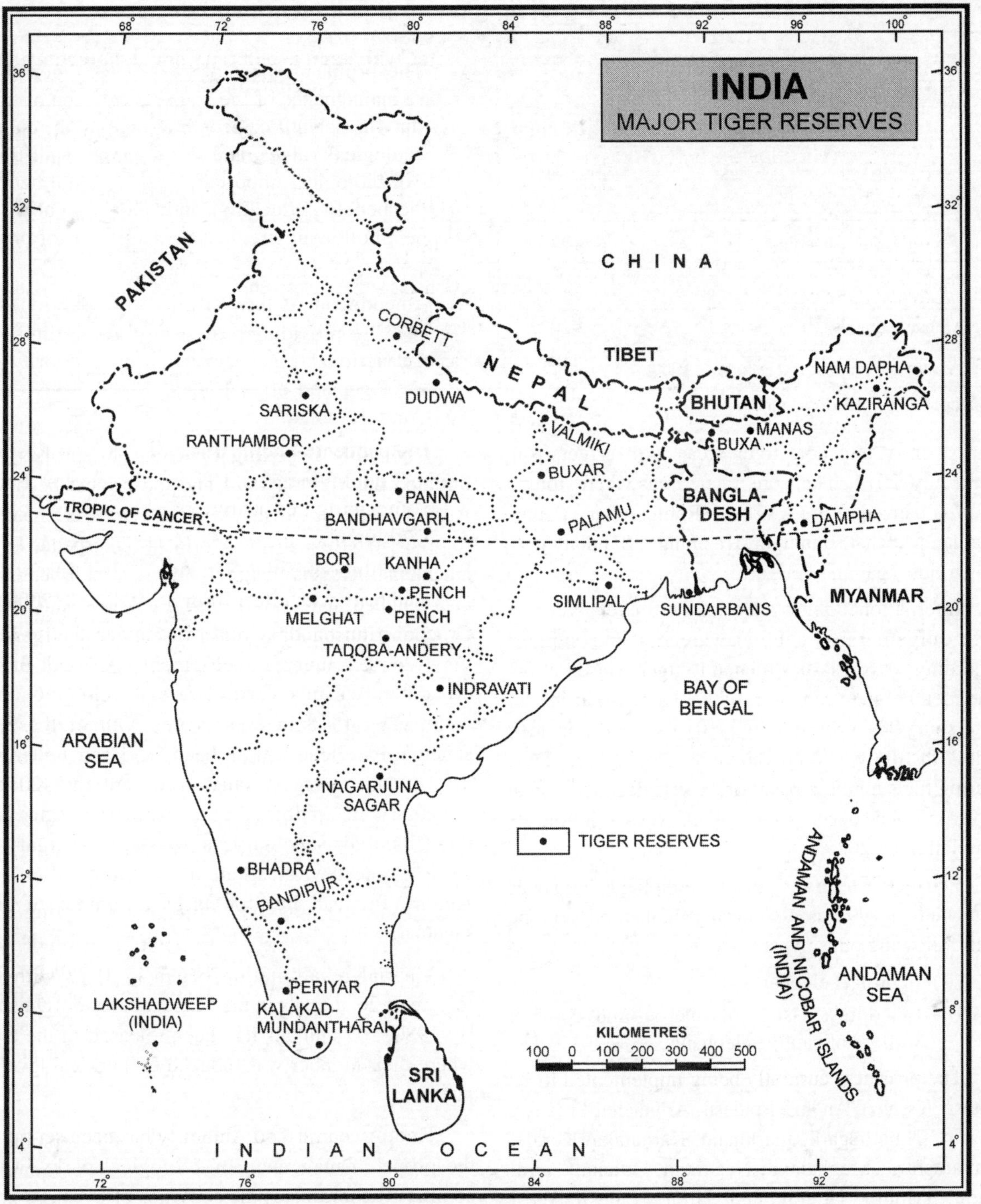

FIG. 7.4. India : Major Tiger Reserves

A tiger crisis cell has also been formed in the Ministry of Environment, Forests and Climate change. Tiger has been declared as our **national animal**. Honourable Supreme Court has banned

TABLE 7.12. States with largest tiger population increase and decrease

States with increase in tiger population between 2006 and 2018				States with decrease of tiger population between 2006 and 2018			
State	**Population in 2006**	**Population in 2018**	**Change in %age**	**State**	**Population in 2016**	**Population in 2018**	**Change in %age**
1. Kerala	46	190	+313	1. Mizoram	6	0	–100
2. Tamil Nadu	76	264	+247.4	2. Odisha	45	28	–37.8
3. Bihar	10	31	+210	3. Chhattisgarh	26	19	–26.9
4. Maharashtra	103	312	+202.9	4. Andhra Pradesh and Telangana	95	74	–22.1
5. Assam	70	190	+171.4				

Source: Status of Tigers in India

tourist entry into core tiger areas with effect from 25th July 2012. The core tiger areas where tourist entry is banned are Corbett, Ranthambar, Panna, Kanha, Melghat, Bandipur, Kaziranga, Bandhavgarh, Nagarajunsagar and Periyar.

International Tiger's Day is held every year on 29th July to raise public awareness and support regarding several issues related to tiger conservation. The idea behind such an initiative came up at the St. Petersburg Tiger Summit in 2010. It was also decided that 29th July would be known as Global Tiger Day. During the summit, a panel of experts declared a goal towards tiger conservation, which was to double its population by 2022.

Project Elephant. Project elephant (PE) was launched in 1992 as a Centrally sponsored scheme with following objectives:

(*i*) to protect elephants.

(*ii*) to address issues of man-animal conflict welfare of captive elephants.

The project is currently being implemented in 22 states/UTs, *viz*., Andhra Pradesh, Arunachal Pradesh, Assam, Chhattisgarh, Jharkhand, Karnataka, Kerala, Maharashtra, Meghalaya, Nagaland, Odisha, Tamil Nadu, Tripura, Uttarakhand, Uttar Pradesh, West Bengal, Rajasthan, Andaman and Nicobar Islands, Bihar, Punjab, Gujarat and Haryana. There are 29 notified and 2 proposed Elephant Reserves in the country.

Crocodile Breeding Project. This project was initiated on April 1, 1974 and the project began on April 1, 1975 in Odisha. Gharial eggs were hatched for the first time in the world at Tikerpada, Distt. Dhankanal, Odisha in June, 1975. A small betal was also hatched at Kurkrail, near Lucknow same year. Crocodile husbandry work was undertaken with a view to sanctuary development. A total of 16 crocodile rearing centres were developed in the country by 1978 in eight states. Out of these 16 sanctuaries, eleven have been declared under the project. Two largest sanctuaries are the Krishna Sanctuary in Andhra Pradesh (3,600 sq km) and the Chambal Sanctuary, a tri-state sanctuary in Uttar Pradesh, Rajasthan and Madhya Pradesh. Andhra Pradesh has the largest number of five sanctuaries.

Gharial rehabilitation began in 1977 with the release of 26 animals in the Mahanadi River (Odisha). By 1980, 107 animals had been released in the river where their number was reduced to a critical level of just five.

Zoo. A central Zoo Authority has been set-up for the proper management of zoological parks in the country. It coordinates the activities of over 200 zoos and also supervises the exchange of animals. A national policy on zoos prepared by the authority provides appropriate directions to the government and other zoo operations.

The National Wildlife Action Plan (NWAP) provides the framework of strategy as well as programmes for conservation of wildlife. The first National Wildlife Action Plan of 1983 has been revised and new Wildlife Action Plan (2002-2016) has been adopted. The Indian Board of Wildlife is the apex advisory body overseeing and guiding the implementation of various schemes for wildlife conservation.

National Park. A national parks is relatively large land or water area which contains representative samples and sites of major natural regions, features, scenery, and/or plant and animal species of national or international significance and are of special scientific, educational and recreational interest. Usually the national parks contain one or several entire ecosystems that are not materially altered by human exploitation or occupation. National parks are protected and managed by the government in a natural or near natural state. Visitors enter under special conditions for inspirational, educational, cultural and recreational purposes.

Wild Life Sanctuary. It is more or less similar to a national park which is dedicated to protect the wildlife and concerned species. A wild life sanctuary is an area constituted by competent authority in which killing and capturing of any form of wild life is prohibited, except with permission and boundaries and character of which are sacrosanct. Entry to the sanctuary is restricted, and allowed only under permission of the chief wild life warden, for research photography, tourism or other lawful business with those residing within the sanctuary. The people who are permitted to live in the sanctuary are authorised officers or persons with right over the immovable property within the limits of the sanctuary. Grazing or movement of livestock is regulated and the livestock permitted to enter the sanctuary have to be immunised against communicable diseases. The chief warden is authorised to allow or disallow entry into the sanctuary or construction of roads, buildings, fences, etc. Hunting is also restricted and strictly regulated. Approval of the government is necessary before the chief warden gives his permission for hunting.

Biosphere Reserves. A biosphere reserve is a unique and representative ecosystem of terrestrial and coastal areas which are internationally recognised within the framework of UNESCO's Man and Biosphere (MAB) programme. It consists of more or less concentric zones with core at the centre which is followed by buffer zone, transitional zone and finally zone of human encroachment (**Fig. 7.5**).

The objectives of Man and Biosphere Programme (MAB) are as follows :

1. Conserve representative samples of all ecosystems and habitat types.
2. Provide long term *in-situ* conservation of genetic diversity.

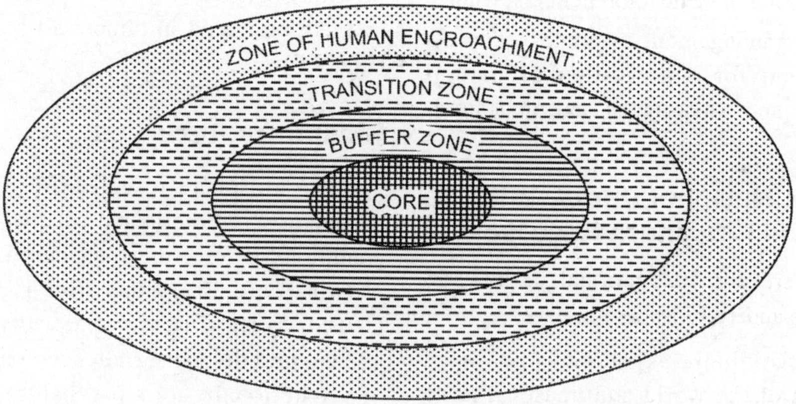

FIG. 7.5. The concept of biosphere reserve

Distinction between National Park, Sanctuary and Biosphere Reserve

National Park	Sanctuary	Biosphere Reserve
1. Hitched to the habitat for particular wild animal species such as tiger, lion, hangul, rhino , etc.	Generally species-oriented such as citrus, pitcher plant, Great Indian Bustard, etc.	Hitched to the whole ecosystem *i.e.* totality of all forms of life *i.e.* ecosystem-oriented.
2. In India, the size of a national park may range from 0.04 to 3162 sq km. The most common average size is 100 to 500 sq km (in about 40% case) and 500 to 1,000 sq km (about 15%)	Size range is 0.61 to 7818 sq km. Most common (in about 40%) is 100 to 500 sq km. In 25% cases the size is 500 to 1,000 sq km.	Size range over 5,670 sq km.
3. Boundaries are fixed by legislation	Boundaries are not sacrosanct	Boundaries are fixed by legislation
4. Except the buffer zone, no biotic interference	Limited biotic interference	Except the buffer zone, no biotic interference.
5. Tourism permissible	Tourism permissible	Tourism normally not permissible.
6. Research and scientific management lacking	Lacking	Managed
7. So far no attention to genopools and conservation	So far no such attention	Attention given

3. Promote and facilitate basic and applied research and monitoring of wild life.
4. Promote mass education, awareness and capacity building.
5. Promote appropriate sustainable management of natural resources.
6. Dissemination of experience so as to promote sustainable development elsewhere.
7. Biosphere management mandates people's participation for conservation of natural resources and shaping of national land-scape.

Distribution of National Parks, Wildlife Sanctuaries and Biosphere Reserves

At present there are 104 national parks, 566 wildlife sanctuaries and 18 biosphere reserves.

National Parks. India is one of the mega diversity countries of the world and houses a large number of wild species found nowhere else in the world. This is the reason that India has currently as many as 104 national parks spread through the length and breadth of the country. Major national parks are shown in **Fig. 7.6**.

Wildlife Sanctuaries. As mentioned earlier, India currently has 566 wildlife sanctuaries spread over an area of 1,22,420 sq km. Like National Parks they provide shelter to a large variety of animal species and can be traced in almost all parts of the country (**Fig. 7.7**).

Biosphere Reserves

The biosphere reserves are required to meet a minimal set of criteria and adhere to a minimal set of conditions before they are admitted to the World Network of Biosphere Reserves designated by UNESCO. These Reserves are very rich in biological and cultural diversity and encompass unique features of exceptionally pristine nature. India has been divided into 10 biogeographic zones and these zones together consist of 25 biogeographic provinces. The aim is to designate one representative site as Biosphere Reserve in each bio-geographic province for long term conservation.

FIG. 7.6. Location of important national parks in India.

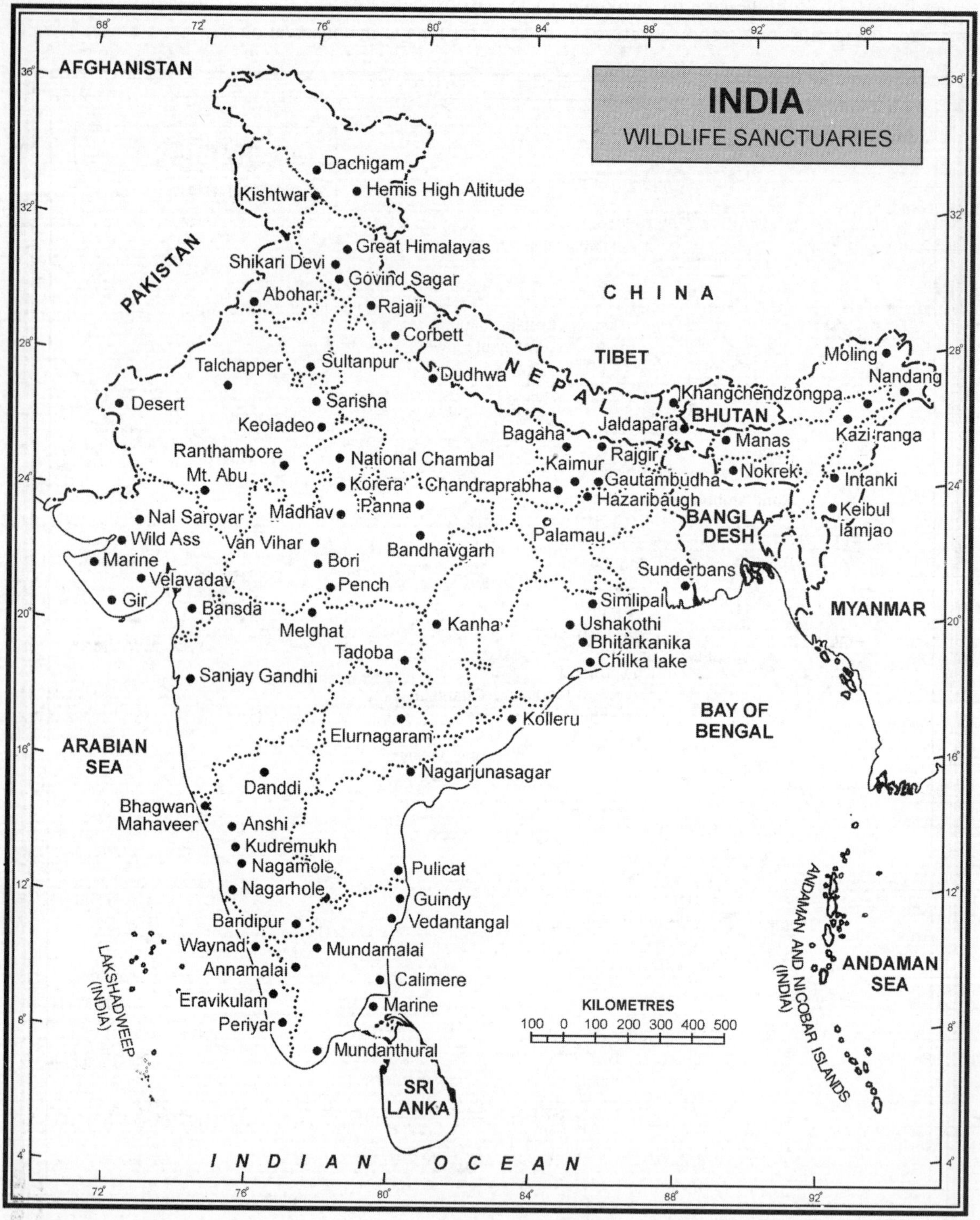

FIG. 7.7. Location of some important wildlife sanctuaries in India.

TABLE 7.13. Biosphere Reserves in India

Sl. No.	Name of Biosphere Reserve	Area (in sq. km.)	Date of Notification	Location (State) and Bio-Geographic Zones
1.	Nilgiri*	5,520.00	01-08-1986	Part of Wyanad, Nagarhole, Bandipur and Mudumalai, Nilambur, Silent Valley and Siruvani hills (Tamil Nadu, Kerala and Karnataka)—Western Ghats.
2.	Nanda Devi*	5,860.69	18-01-1988	Part of Chamoli, Pithoragarh and Almora Districts and valley of flowers (Uttarakhand)—West Himalayas.
3.	Nokrek*	820.00	01-09-1988	Part of Garo Hills (Meghalaya)—East Himalayas.
4.	Manas*	2,837.00	14-03-1989	Part of Kokrajhar, Bongaigaon, Barpeta, Nalbari, Kamrup and Darang districts (Assam)—East Himalayas.
5.	Sunderbans*	9,630.00	29-03-1989	Part of Delta of Ganges and Brahmaputra river system (West Bengal)—Gigantic Delta.
6.	Gulf of Mannar*	10,500.00	18-02-1989	Indian part of Gulf of Mannar between India and Sri Lanka (Tamil Nadu)—Coasts.
7.	Great Nicobar*	885.00	06-01-1989	Southern Most Islands of Andaman and Nicobar (A&N Islands)—Islands
8.	Similipal*	4,374.00	21-06-1994	Part of Mayurbhanj district (Odisha)—Deccan Peninsula.
9.	Dibru-Saikhowa	765.00	28-07-1997	Part of Dibrugarh and Tinsukhia districts (Assam)—East Himalayas.
10.	Dehang Debang	5,111.50	02-09-1998	Part of Siang and Debang Valley in Arunachal Pradesh—East Himalayas.
11.	Pachmarhi*	4,926.00	03-03-1999	Part of Betul, Hoshangabad and Chindwara Districts of Madhya Pradesh—Semi-arid-Gujarat Rajputana.
12.	Kanchanjunga*	2,619.92	07-02-2000	Parts of North and West Sikkim.
13.	Agasgthyamalai*	1,828.00	12-11-2001	Part of Thirunelveli and Kanniyakumari districts in Tamil Nadu and Thiruvananthapuram, Kollam and Pathanamthita of Kerala (Tamil Nadu & Kerala).
14.	Achankumar*-Amarkantak	3,835.51	30-03-2005	Part of Anuppur and Dindori Dist., of Madhya Pradesh, part of Bilaspur Distt. of Chhattisgarh State (Madhya Pradesh & Chhattisgarh).
15.	Kachchh	12,454.00	29-01-2008	Parts of Kachchh, Rajkot, Surendranagar and Patan Civil Districts of Gujarat State.
16.	Cold Desert	7,770.00	28-08-2009	Pin Valley National Park and surroundings; Chandratal and Sarchu and Kibber Wildlife Sanctuary in Himachal Pradesh.
17.	Seshachalam	4,756.00	20-09-2010	Seshachalam hill range in Eastern Ghats encompassing part of Chittoor and Kadapa district in Andhra Pradesh.
18.	Panna	2,998.98	25-08-2011	Part of Panna and Chhattarpur district in Madhya Pradesh.

Source: Ministry of Environment, Forests and Climate Change.

* Biosphere reserves included in World Network of Biosphere Reserves.

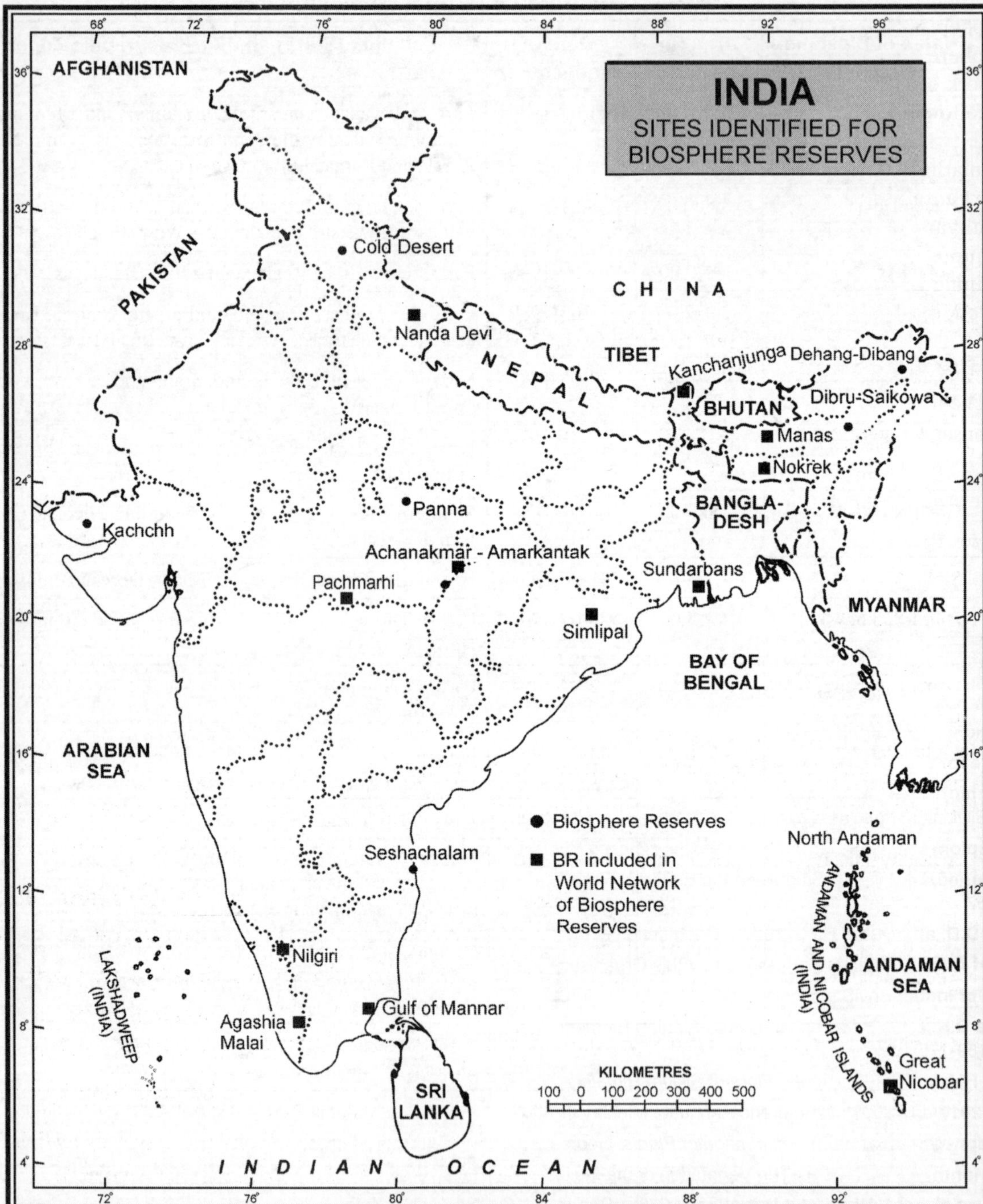

FIG. 7.8. Biosphere Reserves

The programme was initiated in 1986 and till date 18 sites have been designated as Biosphere Reserves in different parts of the country. Out of these 18 Biosphere Reserves, 11 Biosphere Reserves have been included in the World Network of Biosphere Reserves till now. Nilgiri was the first Biosphere Reserve to be included in the World Network of Biosphere Reserve in the year 2000. This was followed by Gulf of Mannar (2001), Sundarbans (2001), Nanda Devi (2004), Nokrek (2009), Pachmarhi (2009), Simlipal (2009), Achanakmar-Amarkantak (2012), Great Nicobar (2013), Agasthyamala (2016) and Kanchanjunga (2018). Strenuous efforts are being made for getting remaining Biosphere Reserves included in the World Network of Biosphere Reserves.

Measures of Conserving Wildlife

The following measures can prove effective tools for conserving wildlife :

(*a*) Ban on hunting should be strictly implemented.

(*b*) Poachers and herdsmen should not be allowed to enter the forests.

(*c*) More national parks and wildlife sanctuaries should be established.

(*d*) Existing national parks and sanctuaries should be further developed and more amenities should be provided in them.

(*e*) Captive breeding of wildlife should be encouraged.

(*f*) Seminars, workshops, exhibitions should be arranged in national parks and sanctuaries to improve the wildlife and general awareness among the public.

(*g*) Adequate medical facilities should be provided in national parks and sanctuaries for the wildlife so that their health is improved.

(*h*) Proper conditions should be created for living and breeding of wild fauna in the national parks and sanctuaries.

REFERENCES

Arora, R.K. (1985) : *Erosion of Plant Genetic Resources*, Natraj, Publishers.

Bharucha, J.P. (1982) : *Vegetation of India*, Oxford India, Bombay.

Centre for Science and Environment; State of Indias Environment (2019, 2020)

Champion, H.G. (1936) : *A Preliminary Survey of Forest Types of India and Burma*, Indian Forest Record, New Series, Silviculture, Vol. 1, New Delhi.

Champion, H.G. and Seth, S.K. (1968) : *A Revised Survey of Forest Types of India*, Manager of Publications, Delhi.

Chatterjee, D. (1939) : "Studies on the Endemic Flora of India and 'Burma'. *J. Roy, Asiat, Soc. Bengal, Calcutta (Sc) 5(3) : 19-67*

Deb, D.B. and Dutta, R.M. (1971) : Contribution to the Flora of Tirap Frontier Division. *J. Bombay Nat. Hist. Soc. 68 : 573-95.*

Forest Atlas of India (1976) : National Atlas Organisation, Calcutta.

Govt. of India : EnviStats India, 2018.

Gupta R.K. (1972) : Boreal and Arcto-Alphine Elements in the Flora of Western Himalaya'. *Vegetation.* The Hague 24 (1-3) : 159-175.

Hara, Hiroshi (1963) : *Spring Flora of Sikkim Himalaya,* Hosp Publishing Co., Osaka.

India 2017 and 2018 : A Reference Annual , Ministry of I&B Govt. of India, New Delhi.

Kingdon-Ward, F. (1960) : *Pilgrimage of Plants*, George G. Hassap, London.

Learmonth, A.T.A. (1964) : The Vegetation of the Indian Sub-Continent, Canberra.

Ministry of Agriculture and Irrigation (1976) : *Report of the National Commission on Agriculture Part IX, Forestry*, Govt. of India, New Delhi.

Maheshwari, P. *et al.* (1965) : Flora in The Gazetters of India, Indian Union, Vol. 1. Country and People (Nashik).

Mathur, V.P. (1968) : *Forest Management*, Jugal Kishore and Co., Dehradun.

Mathiah, S. *et al.* (eds) (1987) : *A Social and Economic Atlas of India,* Oxford University Press, N. Delhi, pp. 85-89.

Negi, S.S. (1990) : *Forest Types of India, PEBA, New Delhi.*

Patel, V.J. (1983) : A *New Strategy of High Density Agroforestry,* Jivanjihai Agrocentre, Kardej (Gujarat).

Pathak, Bindeshwar : *Property does grow on trees,* The Hindustan Times, May 4, 1995.

Prabodh K. Maiti and Paulami Maiti (2011): Biodiversity : Perception, Peril and Preservation, PHI Learning Pvt. Ltd. N. Delhi.

Puri, G.S. (1960) : *Indian Forest Ecology,* Oxford Books & Stationary Co.

Ramakrishnan (2010) : *Primer on Characterising Biodiversity,* National Book Trust, India, N. Delhi.

Rau, M.A. (1974) : *"Vegetation and Phytogeography of the Himayala"* in M.S. Mani, Ecology and Biogeography in India.

Richards, P.W. (1952) : *Tropical Rain Forest : An Ecological Study:* Cambridge University Press, England.

Robinson Francis, *et al.* (ed) (1989) : *The Cambridge Encyclopaedia of India, Pakistan, Bangladesh and Sri Lanka,* Cambridge University Press, p. 19-27.

Sagreiya, K.P. (2004) : *Forests and Forestry,* National Book Trust India, New Delhi.

Sahni, K.C. (1981) : *Botanical Panorama of the Eastern Himalaya,* Oxford University Press.

Shafi, M. (1950) : *Forests in the National Economy of India* in A. Ramesh, *(ed.),* Resource Geography (1984), pp. 119-27.

Sharma, P.D. (2010) : *Ecology and Environment,* Rastogi Publications, Meerut.

Siddartha, K. (1999) : *India : The Phyiscal Aspects,* Center for Development of Environment and Resources, Patna pp. 113-154.

Singh J.S., Singh, S.P. and Gupta S.R. (2006) : Ecology, Environment and Resource Conservation, Anamaya Publishers, N. Delhi.

Singh, J.S. *et al.* (1984) : *Indias Silant Valley and its theratened rainforests ecosystems.* Environment Conservation 11 : 223-33.

Spate, O.H.K. (1967) : *India and Pakistan, A General and Regional Geography*, Methuen & Co., London, Chap. III.

Whyte, R.O. (1956) : *The Grassland and Fodder Resources of India*, Indian Council of Agricultural Research, Scientific Monograph, No. 22, New Delhi.

◆◇◆

CHAPTER 8

Environment

INTRODUCTION AND DEFINITION

The word environment has been derived from the French word *Environir* which means '*to surround*'. Thus environment refers to the sum total of conditions which surround man at a given point of space and time. It is a composite term for the conditions in which organisms live. It includes both biotic and abiotic substances. In other words, environment is the totality of all physical, social, and biological factors, individually as well as collectively, that comprise the natural and man made surroundings.

Considering its basic structure, environment may be divided into two broad types *viz.* physical environment and biotic environment. Physical environment is subdivided into three broad categories *viz.*, (*i*) solid, (*ii*) liquid and (*iii*) gas which represent three basic realms of the earth known as the ***lithosphere***, the ***hydrosphere*** and the ***atmosphere.*** The biotic component of environment consists of plants and animals including man as an important factor. Of all the organisms, man is the most intelligent, skilled and civilized creature. He is both the creator and the moulder of his environment. With his tremendous scientific and technological advancement, man has to a large extent, succeeded in controlling and dominating nature. In this process, he has interfered with nature and disrupted his biotic and abiotic environments.

Pollution. The word pollution has been taken from the Latin word *Pollutionem*, meaning defilement from polluere, to soil or defile (make dirty). Later on Oxford English Dictionary used the word *pollute* with reference to physical contamination of terrestrial or aquatic environments in the nineteenth century. In the twentieth century the word pollution was used with reference to contamination of water, soil and air. According to the report of the Environment Pollution Panel of the **U.S. President's Science Advisory Committee** (1965), environmental pollution may be defined as the unfavourable alteration of our surroundings, wholly or largely as a byproduct of man's actions, through direct or indirect effects, of changes in energy patterns, radiation levels, chemical and physical constitution and abundance of organisms. **R.F. Dasmann** (1975) has defined pollution as "*the accummulation of substances, or forms of energy, in the environment in the quantities, or at rates of flow, which exceed capacity of ecosystems to either neutralize or disperse them to harmless levels*". In simple words, pollution is an undesirable change in the physical, chemical or biological characteristics of our air, land and water that may or will harmfully affect human life or that of desirable species, our industrial processes, living conditions and cultural assets or deteriorate our raw material resources.

Pollutant. According to **Sir Fedrick Warner**, "A substance is normally considered to be a pollutant if it adversely alters the environment by changing the growth rate of species, interferes with the food chain, is toxic or interferes with the health, comfort, amenities or property values of people." Pollutants may be solids, liquids as well as gases. The solid particulate pollutants include aerosols, industrial wastes, such as lead, mercury, asbestos, etc. The liquid pollutants are dissolved solids, ammonia, urea, nitrates, chlorides, fluorides, carbonates, insecticides and pesticides—all in dissolved form, oil and greases, etc. The major gaseous pollutants are carbon dioxide, sulphur dioxide, nitrogen oxide, etc.

ENVIRONMENTAL POLLUTION IN INDIA

There has been large scale pollution of the natural environment in India, especially after Independence, as a result of unprecedented urbanization and industrialization. The most important source of pollution in India is its fast growing population. Most parts of India are now crowded with people resulting in large scale consumption of resources and creation of wastes. Pollution problems increase almost in proportion to the increase in population because the pollutants and wastes also increase proportionately. In India, the problem of environmental pollution is further complicated by widespread poverty of the masses. A large proportion of Indian population lives below the poverty line in slum areas without basic civic amenities. In fact, *poverty and underdevelopment are India's prime pollution problems which are roots of the major ecological imbalances and poor quality of human environment in the country*.

AIR POLLUTION

According to World Health Organization (WHO), Air Pollution is defined as, "*substances put into air by the activity of mankind into concentration, present for a sufficient time and under circumstances which interfere significantly with the comfort, health or welfare of persons or with the full use of enjoyment of his property*."

Engineers Joint Council of the USA defines air pollution as "*the presence in the outdoor atmosphere of one or more contaminants such as dust, fumes, gas, mist, odour, smoke or vapour in quantities, with characteristics, and of duration such as to be injurious to human, plant or animal life or to property or which unreasonably interferes with the comfortable enjoyment of life and property*."

Air is most essential for all types of life in the biosphere because it helps in breathing. *Man can live without food for a few weeks, without water for a few days but he cannot live without air even for a few minutes*. A person breathes 22,000 times a day inhaling about 16 kg of air. It is mostly the oxygen which is consumed while carbon dioxide is released during the breathing process. **Table 8.1** gives the composition of respiratory act.

TABLE 8.1. Composition of Respiratory Air

Name of the gas	Volume %of	
	Inhaled air	Exhaled air
Oxygen	20.95	16.4
Nitrogen	79.01	79.5
Carbon dioxide	0.04	4.1

The atmospheric air is recognised as clean when there are no drastic variations in its natural composition and it is devoid of any solid particulate matter. It must be borne in mind that we cannot find clean or pure air in any part of the world because it is constantly being polluted both by nature and man. When we breath, not only oxygen but some other gases such as sulphur dioxide, hydrogen sulphide, carbon monoxide, dust particles, emissions from volcanoes, etc. also enter our respiratory system. Thus air becomes polluted when its normal composition is disturbed either by nature or by man or by both.

Air Pollutants

Air pollutants are classified as primary or secondary based on their characteristics while they are emitted and physical/chemical changes they undergo while in the atmosphere. The pollutants emitted into the atmospheric directly from the identifiable sources that remain scattered in the atmosphere in the same chemical form as at the time of emission from source are known as ***primary pollutants***. The pollutants which undergo chemical changes in the atmosphere as a result of reactions among two or more pollutants are called ***secondary pollutants***. The pollutants like sulphur dioxide, nitrogen dioxide and particulates are recognised as primary pollutants while several other air pollutants are categorised as secondary pollutants.

Normally speaking, urban air is more polluted than rural air. The routine pollutants in urban air include sulphur dioxide, nitrogen oxides and suspended particulate matter. Besides, there is a severe threat from a range of other air toxins such as carbon monoxide, small particulate emissions, lead, benzene, polycyclic aromatic hydrocarbons (PAH) and ozone. A brief description of various air pollutants commonly found in the urban atmosphere is given below :

1. Sulphur dioxide (SO_2) : Critical Pollutant. Sulphur dioxide is generated from natural resources such as the bacterial decomposition of sulphurs in the soil from the oxidation of hydrogen sulphide produced by the decay of organic matter and from volcanoes, etc. The anthropogenic sources of sulphur dioxide emission arise mainly from combustion of fuels because of trace amount of inorganic and organic sulphur contained in the fossil fuels and ores. It has been estimated that about one-third of the sulphur dioxide emission in the atmosphere arises from the activities of man, chiefly the combustion of fossil fuels. Significant quantities of sulphur dioxide are emitted by oil refineries, automobiles, thermal power plants, acid plants, smelters, incinerators, etc.

2. Nitrogen Oxide (NO_2) : Killer Gas. Generally recognised as sum of nitric oxide (NO) and nitrogen dioxide (NO_2), it is a reddish brown corrosive gas. The automobile exhaust is one of the largest sources of NO_2 emission in the ambient air, as it is formed during combustion as a result of oxidation of atmospheric nitrogen and organic nitrogen. The significant concentration of nitrogen oxides in gaseous emissions occurs from the industrial emissions where nitric acid is produced or is used in chemical reactions. The residence time of NO_2 in the atmosphere is about a few days and is scavenged from the atmosphere through formation of nitric acid, nitrites or nitrates and through their dry dissociation.

The steady upward trend of annual NO_2 levels in Mumbai, Delhi and Chennai indicates that the problem is going to take an even more serious turn in the near future.

Though annual mean concentration of NO_2 in most cities is still within the limits of tolerance, maximum levels in several cities are well above the permissible limits. Gajraula in Uttar Pradesh, Ludhiana, Jalandhar and Patiala in Punjab, Parwanoo in Himachal, Kottayam in Kerala, Puducherry, Mysuru in Karnataka, Haora in West Bengal, Dhanbad, Sindri and Jharia in Jharkhand, Ahmedabad, Surat, Vapi, Rajkot and Ankleshwer in Gujarat, Nagda and Jabalpur in Madhya Pradesh, Jaipur, Alwar and Kota in Rajasthan are some of the towns and cities where NO_2 concentration is already well beyond the standard or it is increasing rapidly.

3. Particulate Matter (P.M.). The particulate matter refers to the solid or liquid particles in its form of dust fumes, mist or smoke and originated either by dispersion of particles from breakdown of solid bulk material or condensation originated, built up from molecular dimension after heating or cooling. The particles have several physico-chemical properties such as size, mass, volume, settling velocity, chemical aerodynamics and optical properties. These properties of the particles play a vital role in atmospheric processes. The size, density and shape of the particles are of prime importance because these factors not only influence their cleansing rate from the environment, but also their effects. The particle size is one of the most important physical characteristics of air borne particulate matter because it controls the residence time of particles in the ambient air. The size of particles may vary from 0.002 μm, (micron) to 500 μm (one micron is equal to one millionth of a metre). Particles larger than 50 μm can be seen with naked eye. Following two types of particulates are recognised depending upon their size.

*(i) **Suspended Particulate Matter (SPM).*** Particles with size less than 100 μm are very small particles and tend to remain suspended in the atmosphere for a long period of time. Hence, they are known as ***suspended particulate matter (SPM)***. In other words, SPM can be defined as solid and liquid particles in the air that are small enough not to settle down onto the earth's surface under the influence of gravity. The main contributors of SPM in ambient air are dust from different sources like construction activities, open fields, mining, stone crushers etc. and

from industrial sources such as thermal power plants and smelters.

(ii) Respirable Suspended Particulate Matter (RSPM). Respirable Suspended Particulate Matter (RSPM or PM_{10}) consists of particulates having diameter less than 10 μm. These are very small particles. As such they can be inhaled and may enter deep into the respiratory tract and pulmonary system of human beings. The sources of RSPM include road dust, wind blown dust and the dust from agriculture, construction and combustion processes.

$PM_{2.5}$. It refers to atmospheric particulate matter (PM) that has a diameter less than 2.5 micrometres which is about 3% the diametre of human hair. Particles of this category are so small that they can only be detected by an electronic microscope. These particles come from various sources like power plants, motor vehicles, airplanes, residential wood burning, forest fires, agricultural burning, volcanic eruptions and dust storms. Since they are very small particles, they tend to stay for a much longer period in the air than heavier particles. This increases the possibility of humans and animals inhaling them into their bodies. Owing to their minute size, they are able to bypass the nose and throat and penetrate deep into the lungs and some may even enter the circulatory system.

Studies have found a close link between exposure to fine particles and premature death from heart and lung disease. These particles are also known to trigger or worsen chronic diseases such as asthma, bronchitis and other respiratory problems. It is estimated that for every 10 micrograms per cubic metre ($\mu g/m^3$) increase in fine particulate air pollution, there is an associated 4%, 6% and 8% increased risk of all-cause cardiopulmonary and lung cancer mortality, respectively.

Exposure to $PM_{2.5}$ exceeds World Health Organisation (WHO) guidelines (10 $\mu g/m^3$) in 99% of districts in India. Delhi-NCR had been the dirtiest in terms of $PM_{2.5}$, while East Kameng (Arunachal Pradesh) is the cleanest (annual exposure is 6.7 $\mu g/m^3$). About 76% of Indian population was exposed to $PM_{2.5}$ (Indian air quality standard 40 $\mu g/m^3$) in 2019.

According to a study published in Nature Communications in June 2021, one fourth of air pollution deaths in India is due to residential emissions where solid biofuel combustion contributes to the ambient fine particulate matter $PM_{2.5}$. It identified 15 polluting sources with residential emissions from biofuels burning (indoor air pollution due to emissions from cooking and heating) resulting in 25.7% of $PM_{2.5}$ share. This is followed by industry (14.8%) and energy (12.5%) in 2019 (**Table 8.2**). The total attributable deaths due to $PM_{2.5}$ in the country are estimated at 866,566 and 953,857 in 2017 and 2019 respectively. The analysis underlined that one-fourth of these deaths could be avoided by eliminating solid biofuel combustion (especially coal) primarily used for residential cooking and heating.

TABLE 8.2. Main contributors to $PM_{2.5}$ in India in 2019

Contributor	Percentage of the total
1. Residential	25.7
2. Industry	14.8
3. Energy	12.5
4. AFCID (Anthropogenic fugitive combustion and industrial dust)	11.5
5. Agriculture	9.4
6. Transport	6.7
7. Waste	4.2
8. Wind blown dust	3.4
9. Other combustion	3.0
10. Commercial	2.0
11. Agricultural waste burning	1.0
12. Other fires	1.0
13. Shipping	0.7
14. Solvents	0.4
15. Others	3.7

Figure 8.1 shows that the Indus-Ganga Plain is the worst affected area. The regions landlocked nature is a factor, but so is the seeping nature of air pollution. Air pollution does not follow the etiquette of borders, it travels with wind affecting widespread areas. The Indus-Ganga plain experiences smog episodes almost every winter. The assessment of annual average level of $PM_{2.5}$ and maximum 24 hour average in cities spread across Indus-Ganga plain—

Sirsa, Jind, Karnal, Bathinda, Meerut, Lucknow, Varanasi, Patna, Gaya and even cities of West Bengal at the fringe of the region—shows that urban areas with much lower annual average levels of $PM_{2.5}$ compared to Delhi can experience significantly high daily smog build up.

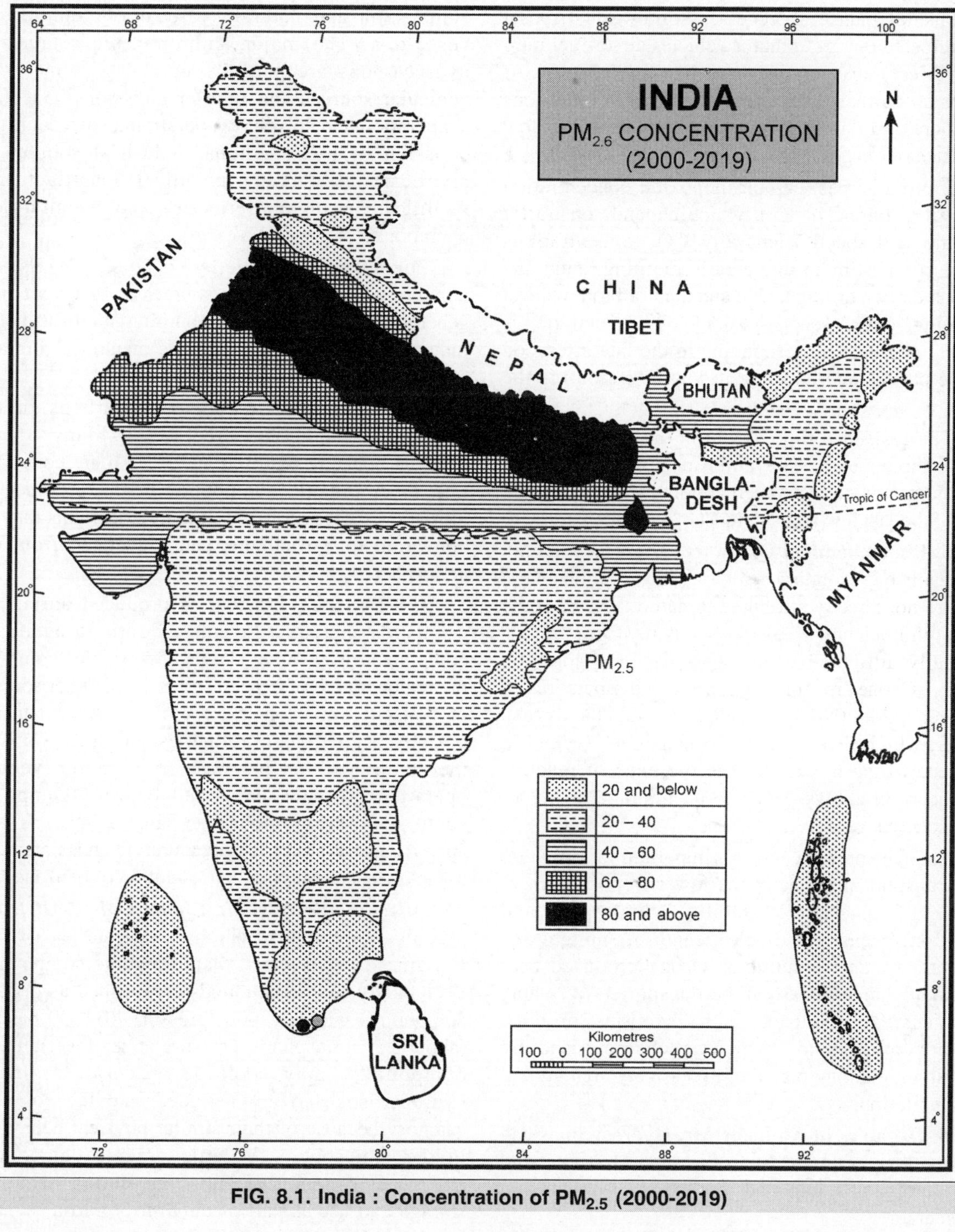

FIG. 8.1. India : Concentration of $PM_{2.5}$ (2000-2019)

4. Carbon Monoxide (CO). Incomplete combustion of carbon results in the production of carbon monoxide (CO). The anthropogenic sources of CO are motor vehicles, coal combustion, fuel oil combustion, industrial processes, solid waste disposal and refuse burning. In motor vehicles air-to-fuel ratio has a direct impact on carbon monoxide emissions. At lower air-to-fuel ratio, carbon monoxide emissions are increased due to incomplete combustion in low presence of oxygen.

In urban areas, carbon monoxide concentration follows a diurnal pattern, which depends on traffic volume and speed. Generally CO concentrations reach to a maximum in the early morning hours due to peak early morning traffic and then fall to low level during the day. A second peak of CO concentration is usually observed corresponding to the late afternoon traffic period, and decrease to low levels during the night.

The residence time of CO in the atmosphere is about three years, The anthropogenic sources of carbon monoxide have been continuously increasing during the last few years in most of the urban areas.

5. Photochemical Oxidants. These are mainly the result of secondary reactions in the atmosphere and are not directly attributed to nature. *Ozone* is the main photochemical oxidant and its formation is normally attributed to nitrogen dioxide photolytic cycle. Ozone is formed through dissociation of nitrogen oxides (NO_2) by sunlight to yield the oxygen atoms, which then react with molecular oxygen to produce ozone molecule. The presence of reactive hydrocarbon allows ozone to accumulate at higher than steady levels.

The meteorological factor important in formation and transport of ozone in the lower troposphere are degree of atmospheric stability, wind speed and direction, intensity and wave length of sunlight and synoptic weather conditions. Ozone produced near pollutant sources/areas can be transported over long distances. A result of such reactions produce photochemical smog which is characterised by grayish haze during period of excessively high ground level pollution.

6. Organics in Ambient Air. Hydrocarbons and Polynuclear aromatic hydrocarbons, Benzene and volatile organic compounds are the chief organics in ambient air.

(i) Hydrocarbons and Polynuclear- aromatic hydrocarbons (PAHs). The ambient hydrocarbon composition includes the unburned hydrocarbon from fuels such as gasoline, species formed during combustion and natural hydrocarbons emitted by vegetation. The major anthropogenic sources of hydrocarbons are partially burned gasoline from vehicular sources and interior emissions. Gasoline evaporation and solvent evaporation also account for emission of hydrocarbons. Industrial sources of hydrocarbons include chemical manufacturing facilities, petroleum refineries and metallurgical operations.

Polynuclear aromatic hydrocarbons (PAHs) are homologs of benzene. The sources of PAHs are heat generation using coal, refuse burning, motor vehicles, industries such as steel and coke manufacturing and carbon-black manufacturing.

The residence time of hydrocarbons in the atmosphere is about three years. Many of the hydrocarbons are oxidised and several are converted into other organic compounds in the presence of nitrogen oxides. Ultimately the hydrocarbons may be converted to particulates and scavenged from the atmosphere.

(ii) Benzene. Benzene is produced during the distillation of crude petroleum and forms a significant component of gasoline. Transport vehicles comprise the major source of benzene emissions. Other sources of benzene emission are chemical manufacturing, coke ovens, petroleum refineries, etc. It is estimated that emissions of benzene from motor vehicle operation are substantial and benzene comprises about 4% of automobile exhaust emissions and approximately 1% of fuel evaporative emissions. The urban areas have higher concentration of benzene.

(iii) Volatile Organic Compounds (VOC). The organic compounds which evaporate easily are recognised together as volatile organic compounds. Technically such compounds are defined as organic compounds with a vapour pressure of 1300 pascals (about 1% atmospheric pressure at sea level). These compounds may include several aromatic hydrocarbons and are included into the common category because of their similar physical behaviour in the atmosphere. Volatile organic compounds emerge as evaporative emissions during handling, storage and use as part of unburnt or partially burnt hydrocarbon mostly along with exhaust gases from vehicles. The volatile organic compounds (VOC) are emitted to the atmosphere from transport, use of industrial solvents and for domestic purposes. These compounds are scavenged with condensation along with water vapour, absorption on surfaces or on particles. Some outstanding examples of volatile organic compounds are aldehydes, ketones, organic acids, alcohols, furans, etc.

7. *Trace Metals in Ambient Air.* Various metals are found in the ambient air in trace and ultra trace concentrations. These include lead, cadmium, zinc, nickel, iron, chromium, etc. They are briefly described as under :

Lead. The airborne soil is the primary source of atmospheric lead. Smelter and refinery processes, as well as incineration of waste containing lead are major point sources of lead. *One of the most important source of lead is release of lead compounds from motor transport using* lead gasoline containing tetra-ethyl or tetra methyl lead. It is estimated that about 75% of lead is emitted into the atmosphere in the form of particles less than 1 μm in size which can stay in atmosphere for long time. Approximately 70-80% lead which is added to the leaded gasoline for increasing its octane number is discharged into the atmosphere. Other major sources of lead emission are burning of lead containing products like paints, batteries, plastics, combustion of coal and oil.

Sources of ***chromium*** include metallurgical and chemical industries, products employing chromate compounds, cement and asbestos. Sources of ***zinc*** include zinc refineries, manufacturing of brass, zinc and galvanising process. Sources of ***cadmium*** include metal industries engaged in extraction, refining, electroplating and welding of cadmium material. It is also emitted as byproduct of refining laced zinc and copper. It is estimated that about 76% of all

TABLE 8.3. Typical Sources of some Air Pollutants in Ambient Air

Sl. No.	Air Pollutants	Major Sources
1.	Sulphur Dioxide (SO_2)	Fuel combustion, power stations, industrial processes, chemical processes, diesel vehicles, solid waste disposal, smelters.
2.	Nitrogen Oxide (NO_2)	Transport (road, rail, passenger and commercial), fuel combustion, power stations, industrial boilers, chemical processes, waste incinerators, smelters.
3.	Particulate Matter (SPM, RSPM-PM_{10}, RSPM-$PM_{2.5}$)	Fuel combustion, power stations, construction activities, industrial processes, diesel vehicle exhaust, re-suspended road dust, domestic refuse burning, domestic wood.
4.	Carbon Monoxide (CO)	Transport, combustion, industrial processes, solid waste disposal, refuse burning.
5.	Ozone (O_3)	Secondary pollutants formed during photochemical reaction.
6.	Organic compounds	Transport, oil based fuel combustion sources, chemical processes, solvent use, waste incinerators, vaporization of fuel.
	Benzene	Petrol combustion products, petrol filling stations, chemical process.
	Polynuclear aromatic hydrocarbons (PAH)	Fuel combustion, industrial emissions.
	Volatile organic compounds (VOC)	Transport, solvents (especially used in industrial and domestic sector).
7.	Trace metals	Fuel combustion, chemical processes, transport, metal production and finishing operations, product manufacture.
	Lead (Pb)	Lead additives in gasoline, soil originated particles.
	Cadmium (Cd)	Fuel combustion, metal production process, transport.

Source : Air Quality Status and Trends in India (2000), Central Pollution Control Board, pp. 17-18.

anthropogenic cadmium emissions come from non-ferrous metal industries, Sources of ***nickel*** include metallurgical plants using nickel, engines burning fuels containing nickel additives, burning of coal and oil, nickel plating facilities and incineration of nickel products. Sources of ***iron*** in ambient air include iron and steel plants, fly ash from combustion of coal and fuel oil, municipal waste incineration and use of welding rods.

Table 8.3 gives a brief summary of typical sources of some air pollutants in ambient air.

Table 8.4 gives an idea of health problems arising from major pollutants. It shows that the most common diseases resulting from air pollution are asthma, lung diseases, bronchitis, respiratory problems, eye irritation, kidney damage, etc.

Effects of Air Pollutants

Air pollutants have varied effects on human life, the most important being their effects on human health. Their other effects are on materials, and vegetation. A brief description of effects of various air pollutants on health is given as under :

1. Health Effects Associated with Air Pollutants

Air pollutants have serious adverse effects on human health. People living in urban and industrial areas are particularly prone to varied types of diseases due to air pollution.

I. *Health Effects of Sulphur Dioxide (SO_2).* Human beings chronically exposed to SO_2 have higher incidence of cough, shortness of breath, bronchitis, colds of long duration and fatigue. Most of the SO_2 in the atmosphere is converted to sulphate salts, which are removed by sedimentation or by washout alongwith precipitation thereby making rain water acidic due to sulphuric acid formation. The most common acute exposure to SO_2 at concentration > = 0.4 ppm (parts per million) is induction of asthmatics after exposure lasting only 5 minutes. Increased prevalence of cough in children with intermittent exposure to SO_2 levels of 1.0 ppm is observed.

II. *Health Effects of Nitrogen Dioxide (NO_2).* The oxides of nitrogen are toxic gases which enter the human body during breathing. High concentration of NO_2 may increase susceptibility to respiratory pathogens and also increases risk of acute respiratory diseases like bronchitis, chronic fibrosis, emphysema and bronchopneumonia. NO_2 exposure can cause decrement in lung functions. It has been established that continuous exposure with as little as 0.1 ppm NO_2 in air over a period of one to three years

TABLE 8.4. Health Effects of Major Pollutants

Pollutant	Short term effect	Long term effect
Carbon Monoxide (CO)	Head ache, nausea, dizziness and anxiety	Reduced mental alertness, cardio-vascular diseases, impaired foetal development
Sulphur Dioxide (SO_2)	Eye irritation, wheezing, sore throat	Asthma, bronchitis, lung damage
Carbon Dioxide (CO_2)	Lower oxygen level in blood, slower reflexes, makes people sleepy	Permanent damage to vital organs
Nitrogen Oxide (NO_2)	Susceptibility to respiratory infections, irritation of lungs	Asthma, chronic bronchitis
Ozone (O_3)	Eye irritation, respiratory tract infections	Bronchitis, asthma, reduction in lung function
Particulate Matter (PM)	Eye and nose irritation, asthma	Can cause severe lung damage, heart disease and cancer
Lead (Pb)	Anaemia and high blood pressure	Brain and kidney damage, cancer, still birth
Mercury (Hg)	Shaking of hands and feet	Memory problems, kidney damage and respiratory failure

Source : All India Institute of Medical Sciences (AIIMS), Delhi.

increases incidence of bronchitis, emphysema, oedema and adversely affects the performance of lungs. US study by Hasselblad *et al.* (1992) indicate that repeated NO_2 exposure increases respiratory illness in children. The epidemiological studies suggest that an increase of 30 $\mu g/m^3$ in NO_2 levels results in about 20 per cent increase in respiratory illness and diseases. Continuous exposure with more than 2.0 ppm NO_2 may cause extensive morphological alteration, lung dispensability and permanent changes in the lung (bronchiolitis).

III. ***Health Effects of Particulate Matter.*** The particles may influence the human body on setting and cause external effects *e.g.,* effects on skin. However, certain groups of particles pass into the blood stream on being inhaled and act as systematic poison. The effect of irritant particles in respiratory tract depends upon the size of the particles, their solubility, penetration, deposition and clearance mechanism in human respiratory tract. Fine particles may cause irritation of broncospasm, pulmonary oedema and allergic alveolitis, while certain moulds of larger particle size cause obstructive lung disease. As the presence of finer particle size increases, the percentage of particles deposited in upper part of respiratory tract decreases, thus particles are inhaled deeper. The soluble nature of particles of chemical origin may cause systematic intoxication in the body. Increased particle presence in ambient air increases the frequency of cough and phlegm. The susceptibility of infection in pulmonary system increases in case the inhaled particles are active particles constituting bacterial, fungal spores or viral strains. The acute effect of particulate air pollution results in changes in respiratory health status and depict several respiratory symptoms. The symptoms are often recorded into upper respiratory symptoms such as stuffy or running nose, sinusitis, sore throat, wet cough, head cold, hay fever and burning or red eyes. The lower respiratory symptoms include wheezing, dry cough, phlegm, shortness of breath, chest discomfort and pain. Cough is the most frequently reported symptom due to continued exposure to high particulate laden ambient air.

Asthma and allergic alveolitis are two main respiratory diseases related to chronic exposure to particles. Studies reveal a direct relationship in asthmatic attack with exposure to particles. Particles containing fungi, viral or bacterial pathogens loading in ambient air may play their role in transmission of infectious diseases. Increased particulate exposure increases the incidence of bronchitis. The bronchitis or pneumonia induced by air pollution in the presence of pre-existing heart problems might precipitate congestive heart failure and cardiovascular mortality.

IV. ***Health Effects of Carbon Monoxide (CO).*** Carbon monoxide is absorbed from the lung tissue in blood stream. Toxic effects of CO are mainly due to its high affinity for Hb which is 240 times greater than oxygen affinity. The COHb in blood of exposed population may be between 3.0 and 5.3 per cent whereas the safe limit is less than 2 per cent.

High dose exposure of CO may affect lung tissue and may lead to acute decrement in lung functions. CO level to about 5 per cent may cause cardiovascular effect in young healthy, non-smoking individuals leading to fatigue and reduced ability to work. The recurrent episodes of exertial angina increase risk of precipitating heart attack, fatal arrhythmia or myocardial damage, increased risk of sudden death with coronary artery disease. Increased concentration of carbon monoxide also leads to stroke, head injury, artherosclerosis, hypertension, etc. High CO concentration has special effect on children and infants. There are strong evidences of reduction in birth weight, cardio megaley, delay in behavioural development and disruption of cognitive function and sometimes even infant death syndrome. The other systematic influence of CO poisoning include effects on lever, kidney, bone, immune capacity and spleen.

V. ***Health Effects of Ozone.*** Variations in ozone levels in urban environment are the main cause of concern, both with regard to health risk associated with concentration exceeding the standard and possible health effects associated with human exposure to somewhat lower ozone concentration over long period of many days. High elevation of ozone causes major problem of human health which includes eye, nose and throat irritation, chest discomfort, cough and headache, Ozone is a respiratory irritant, reacts rapidly with tissues and airways lining of lungs. The acute reversible decrement in lung functions and increased respiratory symptoms occur in individual exposed for 1 to 3 hours with ozone concentration ranging between 235 to 314 $\mu g/m^3$. Acute exposure to ozone may cause lung inflammation within a few hours. Long term ozone exposure has been associated with inflammation of lungs which may be involed in progression from chronic health effects. Ozone also increases susceptibility to pulmonary bacterial infections and it can exacerbate its severity of influenza infection. Acute exposure to low ozone levels decreases activity pattern, may affect immune system leading to potential health risk.

VI. *Health Effects of Benzene.* Benzene is a hazardous air pollutant which accelerates carcinogenicity and human health risk from ambient air. Various studies have provided evidence regarding effect of benzene linked to genetic change, chromosomal aberration, etc. International Agency for Research on Cancer (IARC) has classified benzene as carcinogenic to humans which increases the frequency of cancer in human beings. Exposure to excessively high level of benzene may cause cancer in kidney, testes, brain, pancreas, stomach, lungs, respiratory tract, bladder and uterus. Benzene acts as Leukaemogen in human beings, acting as etiologic agent of aplastic anaemia leading to acute myclogenous leukaemia. World Health Organisation (WHO) estimates a four in one million risk of leukaemia on exposure to benzene to a concentration of 1 mg/m^3. Benzene has also been recognised as causing DNA damage in mammalian cells. Prolonged exposure to even mild form of benzene may cause euphoria followed by giddiness, irregular heart beat, headache, dizziness, nausea and unconsciousness. Breathlessness, nervous irritability and unsteadiness in walking may persist for a long time. Acute benzene poisoning includes extensive petechial haemorrhage in the brain pleurae, pericardium, urinary tract, mucous membrane and skin. Pneumonitis and bronchitis can also be caused by direct action of benzene. The other effects of benzene are disorders of blood, harmful effect on bone marrow, anaemia and reduced ability of blood to clot, damage to immune system and reproductive process.

It has been found that women are more susceptible than men to benzene. Benzene exposure may cause menstrual disorder and retard foetal development.

VII. *Health Effects of Volatile Organic Compounds (VOC).* The volatile compounds have potential carcinogenic effects on human beings and are termed as air toxins. These compounds react with oxides of nitrogen in the presence of sun light and give rise to photochemical smog. This smog is a dense haze which restricts visibility. Hazy fumes cause irritation to eyes and lungs and damage plant life.

VIII. *Health Effects of Lead.* Lead particles from the ambient air can be inhaled, can settle down as dust in neighbouring area, on vegetation and water bodies and may be partly ingested. Out of the total estimated release of lead from the vehicular emission, about 50-70% is released as emission into the environment and the remaining part gets deposited. *Lead is pervasive environmental poison which affects virtually every system in the body.* It can damage the kidneys, the nervous system, the reproductive system and cause high blood pressure.

Children are more prone to lead pollution because they absorb lead more readily than adults. It affects the development of brain of foetuses and young children. Children exposed to lead show lack of intelligence, behavioural problems and decreased ability to concentrate. Blood lead levels as low as 10 μg/decilitre are associated with harmful effects on the learning process of the children. Elevated blood lead level can be more harmful. At extremely high level (70 μg/decilitre or higher), seizure, coma and even death can occur.

Lead is specially harmful to pregnant women and infants. Lead can get accumulated and stored in bones for decades and can be released whenever there is demand for calcium such as during pregnancy and lactation. During lactation, lead crosses placenta and is detected in breast milk. This is the major source of lead in infants which causes neurological problems in the developing child.

Not to speak of women, infants and childrens, even adult males are not spared by lead pollution in the air. Chronic exposure with elevated blood lead levels are associated with hypertension, headache, confusion, irritability and insomnia. Further higher levels cause drowsiness, loss of muscular coordination, kidney damage, fatigue, apathy and susceptibility to infection and anaemia. Higher blood lead level (80 μg/decilitre or more) also leads to gastro-intestinal problems and liver damage.

Deaths and Economic Losses due to Air Pollution. According to Ministry of Health and Family Welfare, air pollution is responsible for about 13 per cent of deaths in India and its burden is highest in the northern region, especially Delhi, Punjab, Uttar Pradesh and Haryana. About 51% of the people who died due to air pollution were younger than 70 years. It is estimated that about 77 per cent of Indians breath

TABLE 8.5. Deaths and economic losses due to air pollution in India

State/Union Territory	Deaths		Economic Loss	
	Total	Percentage to total deaths	US $ Million	Percentage of state GDP
1. Andhra Pradesh	62,808	15.6	1,349.5	1.09
2. Arunachal Pradesh	789	11.0	26.0	0.74
3. Assam	657	1.42	5,130.0	2.15
4. Bihar	1,26,460	18.8	1,552.8	1.95
5. Chhattisgarh	41,519	16.9	690	1.55
6. Delhi	17,248	18.2	1,206.5	1.08
7. Goa	1,396	13.3	79.9	0.72
8. Gujarat	87,811	18.9	2,859.6	1.33
9. Haryana	34,119	19.0	1,566.3	1.49
10. Himachal Pradesh	10,383	18.5	253.8	1.16
11. J&K and Ladakh	15,997	20.2	252.2	1.14
12. Jharkhand	33,136	16.7	543.3	1.24
13. Karnataka	89,184	16.6	2,680.7	1.22
14. Kerala	36,392	14.3	1,090.5	0.98
15. Madhya Pradesh	1,12,009	18.7	1,970.5	1.70
16. Maharashtra	1,39,118	16.7	3,975.4	1.06
17. Manipur	40	1.08	5130.3	2.15
18. Meghalaya	1,870	11.7	39.2	0.8
19. Mizoram	770	11.3	22.4	0.7
20. Nagaland	1,281	12.6	33.6	0.86
21. Odisha	43,409	12.7	806.6	1.14
22. Punjab	41,090	18.8	1,148.9	1.52
23. Rajasthan	113,361	21.2	2,294.3	1.70
24. Sikkim	488	14.4	25.5	0.67
25. Tamil Nadu	84.587	13.8	2,529.1	1.06
26. Telengana	35,364	15.5	1,165.9	0.91
27. Tripura	4,925	19.1	91.1	1.26
28. Uttar Pradesh	3,49,926	19.5	5,130.3	2.15
29. Uttarakhand	16.989	18.6	526.6	1.5
30. West Bengal	1,22,833	20.8	2,125.3	1.26
31. Other Small Union Territories	2,688	13.3	120.3	0.79

Source : Health and economic impact of air pollution in the states of India, the Global Burden of Disease Study 2019, quoted in State of Indias, Environment 2021, A Down to Earth Annual, pp. 370-71.

air that is worse than the levels recommended by the National Ambient Air Quality Standards. In 2017, about 1.24 million deaths were attributed to air pollution. On an average 2802 people per 1,00,000 died due to air pollution in 2017.

In the year 2019, 1.67 million Indians died due to air pollution. Half of these deaths were just in five states—Uttar Pradesh, Maharashtra, Bihar, West Bengal and Rajasthan. The economic cost was US $ 36,803 million, which is equivalent to 1.36 per cent of India's GDP. Table 8.5 gives details of deaths and economic losses caused by air pollution in India.

Reduction of Life Expectancy. According to an assessment by Energy Policy Institute at the University of Chicago, pollution levels in India shave off 5.2 years from the life expectancy of the average Indian and it most acutely hits people living in the Indus-Ganga plains. The assessment says that India is the second most polluted country globally after Bangladesh. The authors of the assessment an air quality life index (AQLI), which takes particulate air pollution and determines the hit it has on life expectancy.

In 2018, Lucknow's (the most polluted district in the country) residents were exposed to an average annual $PM_{2.5}$ concentrations of 114.6 $\mu g/m^3$ which is likely to cut short life expectancy by 10.3 years. In the National Capital Territory of Delhi, the average $PM_{2.5}$ concentration was 166 $\mu g/m^3$ which can lead to loss of 9.4 life years compared to if Delhi had met WHO guidelines for air quality.

Years of life people stand to lose at current pollution levels			
Average Indian	North India residents	Residents of Delhi	Lucknow residents
5.2	8.0	9.4	10.3

The study says that a quarter of India's population is exposed to pollution levels not seen in any other country. Between 1998 and 2018, Indians, on average, lost 1.8 years of their life span. About 248 million residents of north India are likely to lose more than eight years of life in the current scenario. The findings also indicate that in states like Bihar and West Bengal, people can add more than seven years to their life if air quality meets the WHO guidelines while the count will be eight years in Haryana.

2. Effects of Air Pollutants on Materials

Air pollutants produce physical and chemical change in materials which results in their damage and destruction. The natural effects of corrosion and weathering are aggravated when the air is polluted. The most destructive air pollutants to materials are smoke, grit, dust and oxides of sulphur. Sulphur dioxide is the most dangerous air pollutant. It changes to sulfurous and sulphuric acid with moisture and accelerates the rate of corrosion. Amount of moisture in the air determines the rate of corrosion—more the moisture, more the corrosion.

Different types of metals and metallic structures such as iron and steel, aluminium and aluminium alloys, copper and copper alloys are corroded when exposed to polluted air. Building materials are also corroded and disfigured with increasing pollution of air. Smoke, grit and soot deposits disfigure the buildings. During high winds, larger particulates can result in surface erosion. The oxides of sulphur react with limestone to form calcium sulphate. Slow loss of substances from surface occurs during rains which leads to blistering. Effects of various air pollutants on materials are shown in **Table 8.6**.

3. Effect of Air Pollution on Vegetation

In addition to its effects on health and materials, air pollution has damaging effect on vegetation also. The effects of air pollutants on vegetation depends upon their chemical nature, level of concentration and duration of exposure.

The principal air pollutants of prime concern to agriculture and vegetation are sulphur dioxide, SPM and photochemical oxidants. The air pollutants effects on vegetation are demonstrated by plants through following characteristics :

- Qualitative and quantitative changes in solar radiations input on leaf surface and alteration in energy exchange process.
- Decrease in chlorophyll and chloroplast injury.
- Increase in gaseous exchange process.
- Dust induced alteration in physico-chemical parameters.

Table 8.7 shows the impact of major air pollutants *viz.* sulphur dioxide, ozone and suspended particulate matter on vegetation.

TABLE 8.6. Air Pollutants and their Effects on Materials

Sl. No.	Air Pollutants	Effects
1.	Particulate Matter (PM)	— Physical erosion with abrasive action. — Particulate deposits cause streaky appearance. — Corrosion of metallic substances. — Deposition on electric contacts interfere with function, accelerates corrosion. — Soiling to textile, reduced wear life and abrasive effect.
2.	Sulphur dioxide (SO_2) and sulphur trioxide (SO_3)	— Flaking of surfaces due to formation of sulphuric acid from SO_x. — Acceleration of corrosion of steel and other metals. — Embrittlement of paper and leather. — Reduced strength of fibres in textiles.
3.	Oxides of nitrogen (NO_x)	— Corrosion effect on surface and metals. — Fabric discoloration and fading.

Source : Air Quality Status and Trends in India (2000), Central Pollution Control Board, p. 44.

TABLE 8.7. Effects of Air Pollutants on Vegetation

Sl. No.	Air Pollutants	Effects on Vegetation
1.	Sulphur dioxide	— Enters into leaf through stomata. — Excessive exposure causes injury on blade with ivory colour, brown to reddish brown spots, depending on plant and environmental conditions.
2.	Ozone	— High concentration causes dark brown to black lesions on upper surface of leaves.
3.	Suspended Particulate Matter	— Block the stomata through deposition on leaf surface. — Excessive dust deposition retards the growth of plant. — Automobile exhaust smoke damage lower surface of leaves, bronzing and silvering, upper surface shows fleck like marking.

Source : Air Quality Status and Trends in India (2000), Central Pollution Control Board, p. 47.

National Air Monitoring Programme (NAMP)

Central Pollution Control Board initiated National Ambient Air Quality Monitoring (NAAQM) programme in the year 1984. Subsequently, the programme was renamed as National Air Monitoring Programme (NAMP).

Objectives

The objectives of the NAMP are as follows :

- To determine status and trends of ambient air quality.
- To ascertain whether the prescribed ambient air quality standards are violated, and to assess health hazard and damage to materials.
- To continue ongoing process of producing periodic evaluation of air pollution situation in urban and industrial areas of the country.
- To identify non-attainment cities.
- To obtain the knowledge and understanding necessary for developing preventive and corrective measures.
- To understand the natural cleansing process undergoing in the environment through dilution, dispersion, wind based movement, dry deposition, precipitation and chemical transformation of pollutants generated.

INDIAN SCENARIO

According to the report of WHO, released in 2018, in which data for 3000 cities from 103 countries was collected, the Northern Plains of India were more polluted than the other parts of the country. This region contains 13 of 15 most polluted cities of India. The main reasons for such a state are its geographical location, climate, population, urbanisation, industrialisation and agricultural practices as well as fossil fuel emissions. The region is located next to the Himalayas and is a wind convergence zone which transports pollution from other areas to the India-Gangetic plains. The convergence factor shows up the most in winters when the region sees spells of dense fog due to cold winds coming from the west. Fog traps pollutants, leading to sharp deterioration in air quality. Cold winds also lead to 'inversion', wherein pollutants are trapped near the surface.

The northern plains also have among the highest population densities in the world. This translates into higher pollution from the human activities.

The other major source of pollution in the recent years in the practice of crop stubble burning.

Vehicular Pollution

Vehicular pollution has grown at an alarming rate due to growing urbanisation in India. The air pollution from vehicles in urban areas, particularly in big cities, has become a serious problem. The pollution from vehicles has begun to tell through symptoms like cough, headache, nausea, irritation of eyes, various bronchial and visibility problems.

The main pollutants emitted from the automobiles are hydrocarbons, lead/benzene, carbon monoxide, sulphur dioxide, nitrogen dioxide and particulate matter. The main cause of vehicular pollution is the rapidly growing number of vehicles. The other factors of vehicular pollution in the urban areas are 2-stroke engines, poor fuel quality, old vehicles, inadequate maintenance, congested traffic, poor road condition and old automotive technologies and traffic management system. In India, the number of vehicles increased from 306 thousand in 1950-51 to 295772 thousand in 2018-19. About half the vehicles are concentrated in 39 metropolitan cities (cities with population of over one million). The two wheelers are the major contributors of vehicular air pollution followed by four-wheelers (*e.g.*, car, jeep, taxi etc.), trucks and buses in decreasing order of magnitude.

Delhi is a typical example of air pollution by vehicles. Delhi's vehicular population increased from an insignificant of 2.17 lakh in 1971 to 110 lakh in 2019. Over 1,400 vehicles are registered each day in Delhi. Delhi has more vehicles than Mumbai, Kolkata and Chennai put together. Unfortunately, number of vehicles in Delhi outpaces the road length.

Share of vehicular pollution in Delhi has increased from 23% in 1970-71 to 75% in 2018-19. In contrast share of industrial pollution decreased from 56% in 1970-71 to less than 18% in 2018-19.

Much of the vehicular air pollution can be avoided by maintaining proper speed of the vehicles. Vehicles standing on the road crossings or in traffic jams cause more pollution. The quantity of harmful emissions decreases with increasing speed (*See* **Table 8.8**).

Delhi is a major point of intersection in north India, for both passenger and freight traffic. Delhi is connected to five national highways directly (NH 1—GT Karnal road, NH 2 Mathura road, NH 8—Gurgaon road, NH 10—Rohtak road, NH 24—Hapur road) and indirectly to two (NH 24 carries load of NH 58 and NH 91). In order to divert the interstate traffic the 'Peripheral Expressways' have been constructed. The 135 km long Eastern Peripheral Expressway from

TABLE 8.8. Emissions from Vehicles at Varied Speeds (gm/km)

Speed (km/hour)	Auto		Bus	
	Carbon monoxide	Hydrocarbons	Carbon monoxide	Hydrocarbons
10	33.02	4.47	22.6	5.7
25	21.20	2.60	14.6	2.3
50	9.80	1.30	8.2	0.0

Source : Centre for Science and Environment.

Kundli-Gaziabad and Palwal whereas the 135.6 km long Western Peripheral Expressway joins Kundli-Manesar and Palwal. These Expressways have reduced vehicular traffic within the limits of Delhi to a great extent. Besides, air traffic also adds to the air pollution. More than 950 tonnes of pollutants are released by air traffic every day. With the increase in Air Traffic this pollution is bound to increase in future. Even smog (mixture of smoke and fog) is becoming a real threat to Delhi's air environment.

In Mumbai about 52 per cent of the total pollution load is contributed by vehicles. Vehicles contribute about 54 per cent of SPM and 52 per cent of nitrogen dioxide in Mumbai. About 30 per cent of the total pollution load in Kolkata comes from vehicles. Among the metropolitan cities of India. Chennai seems to have clearer air. However, experts warn that SPM in many locations in Chennai is the cause for concern. According to latest studies, high SPM levels in some of the residential and other non-industrial localities is largely due to emissions by vehicles.

According to the World Air Quality Report, 2020 released in March, 2021, Delhi is still the world's most polluted capital although the air quality in the city improved by 15% in 2020 due to lock down caused by Covid-19. Delhi with an average annual $PM_{2.5}$ concentration of 84.1 $\mu g/m^3$ topped the list of the most polluted capitals. It was followed by Dhaka of Bangladesh (77.1 $\mu g/m^3$), Kabul of Afghanistan (46.4 $\mu g/m^3$) and Doha of Qatar (44.3 $\mu g/m^3$). The report said that 22 out of 30 most polluted cities in the world were located in India. Thirteen of 15 most polluted cities in the world were from north India. Nine of these cities are in the National Capital Region. The report adjudged the Chinese city of **Hotan** as the most polluted in the world.

Industrial Pollution

There has been phenomenal growth of industries in India, particularly after independence. Industries emit all sorts of pollutants in the air and result in air pollution to great extent. The major industrial air pollutants are suspended particulates, carbon monoxide, sulphur dioxide, nitrogen oxides and many more.

In Delhi, besides big units there are 70,000 uncontrolled small scale industries, discharging into atmosphere gases, liquids and solid wastes producing health hazards. A recent study conducted by Central Pollution Control Board has revealed that Najafgarh Road, Lawrence Road, Wazirpur, Kirti Nagar, DLF Industrial Area and Moti Nagar are the most polluted areas which suffer at the hands of polluting industries. In Mumbai, belt between Chembur and Trombay is highly industrialised and has 3 to 6 times more pollution than the remaining parts of the city. Tarapur Atomic Power Plant continues to spew out dangerous radiation doses. Kolkata's industrial units have led to precarious situation with regard to air pollution. According to a report submitted by National Environmental Engineering Research Institute (NEERI), Nagpur, the total daily emission from all sources amounts 1,307 tonnes in Kolkata Metropolitan District wherein 900 tonnes of pollutants are generated in the industrial belt of Kolkata and the rest in Haora industrial belt. The major pollutants are suspended particulates (560 tonnes), carbon monoxide (450 tonnes), sulphur dioxide (125 tonnes), hydrocarbons (102 tonnes) and nitrogen oxides (70 tonnes). Surat is another example of air pollution. This industrial city has a very high average of suspended particulate matter (SPM) amounting to about 267 micrograms per cubic metre of air per day. The discharges of sulphur dioxide, nitrogen oxide and hydrogen sulphide are 20.4, 24.6 and 0.5 microgram per cubic metre of the air respectively. Ahmedabad has a large number of textile mills as well as other industries which are contributing to air pollution.

Central Pollution Control Board (CPCB) has classified 241 major industries into following categories based on their air, water and soil polluting potential.

1. Red. 59 types of industries fall in this category (coal based power plants, fertiliser, oil refinery, nuclear power plant, pharma, etc.)

Environmental clearance valid for 5 years as opposed to earlier 3 years.

2. Orange. 93 types of industries fall in this category (synthetic detergent, dry coal processing, marble cutting/polishing, etc.)

Environmental clearance valid for 10 years as opposed to 5 years earlier.

3. Green. 53 types of industries fall in this category (bakery, cotton spinning/weaving, flour mills, etc.)

Environmental clearance valid for life time as opposed to 10 years earlier.

4. White. 36 types of industries fall in this category (bio-fertiliser, bio pesticide, organic manure, wind power and mini hydel of less than 25 MW, etc.)

Not necessary to take any consent for environment clearance.

Thermal Power Plants

Thermal power plants comprise the second most dangerous source of air pollution, next only to vehicles. Most of the thermal power plants use large quantities of coal which produce huge amounts of smoke, ash and other pollutants. Coal based thermal power plants account for 60% of Suspended Particulate Matter (SPM), 50% of SO_2, 30% of NO_x, 80% of mercury emissions and 70% of fresh water withdrawal—of all industrial sectors. The impact of thermal power plants on environment is of growing concern given the anticipated growth of this sector.

Most of the coals used in Indian power plants are of low grade with high ash contents varying from 30% to 45% and low calorific value of 3,500-4,000 kcal/kg. As a result, emissions of fly ash, SPM and SO_2 are of very high order. There had been huge capacity addition to the coal power sector resulting in increasing rate of fly ash generation. Between 2009-10 and 2018-19, the annual coal consumption rose from 367 million tonnes to about 629 million tonnes. The annual fly ash generation increased from over 123 million tonnes to more than 217 million tonnes during the same period which accounts an increase of about 76 per cent (**Table 8.9**).

The major cause of concern is the amount of residual ash which accumulated in large quantities in due course of time due to low utilisation. Huge left over stocks require large tracts of land for disposal of left over ash. According to the estimates made by Centre for Science and Environment the unused ash stock from coal power plans of India from 2009-10 to 2019-20 is 627 million tonnes.

Sulphur dioxide (SO_2) is a significant contributor to air pollution. As a reactive pollutant, SO_2 reacts with other air pollutants to form sulphate particles and makes important part of particulate matter ($PM_{2.5}$) which has the largest public health impact. It is estimated that SO_2 commonly makes up more than 10% of deadly fine particles in India. The Greenpeace Study released in August, 2019 found a number of hotspots with respect to SO_2 emissions (**Fig. 8.2**).

TABLE 8.9. Trends in fly ash utilisation and accumulation of unused ash in India

Year	Fly ash generation (million tonnes)	Fly ash utilisation (million tonnes)	Percentage ash utilisation	Unused and residual ash (million tonnes)
2019-10	123.54	77.33	63	46.21
2010-11	131.09	73.13	56	57.96
2011-12	145.42	85.05	58	60.37
2012-13	163.56	100.37	61	63.19
2013-14	172.87	99.62	58	73.25
2014-15	184.14	102.54	56	81.60
2015-16	176.74	107.77	61	68.97
2016-17	169.25	107.10	63	62.15
2017-18	196.44	131.87	67	64.57
2018-19	217.04	168.40	78	48.64
Total unused ash (2009-19)				627

***Source*:** State of India's Environment (2021) Centre for Science and Environment, p. 200.

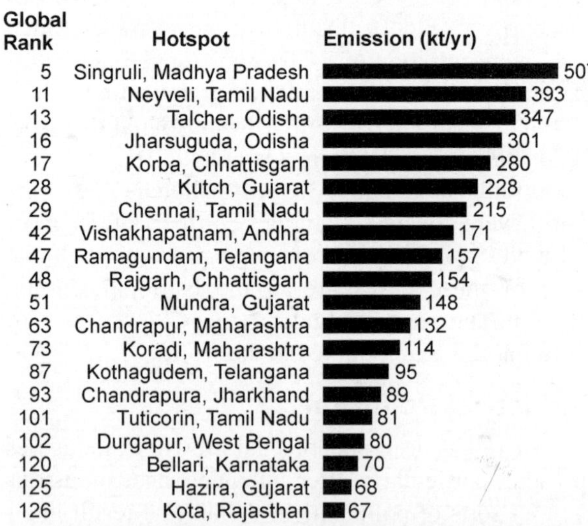

FIG. 8.2. India : Worst emitters of Sulphur Dioxide (Thermal Power Plants based on coal)

Supply of the beneficated coal—coal washed to lower ash content—has emerged as a serious policy issue in the last decade. Pollution control equipment can also function more efficiently if ash load is reduced by using washed coal.

Crop Stubble Burning

In large stretches of Punjab, Haryana, and Uttar Pradesh, crop stubble are burnt after crop is harvested. This is done to prepare the fields for the next crop because the farmers do not find sufficient time to clear the field off the stubble in a short duration of just 20-25 days. Moreover, removing crop stubble is labour and cost intensive. Burning of crop stubble leads to air pollution over a wider area and metros like Delhi have to suffer the most. Farmers are advised not to burn crop stubble but they do not find any viable alternative to practice of burning. There are two crop stubble burning seasons in north India, in October after the paddy harvest and April end to May following wheat harvest. Stubble burning has become a big issue because of the following reasons.

- Burning of one ton of straw release 3 kg particulate matter, 60 kg carbon monoxide (CO), 1460 kg carbon dioxide (CO_2), 199 kg ash and 2 kg sulphur dioxide (SO_2).
- It also accounts for loss of 5.5 kg nitrogen, 2.5 kg potassium, 2.3 kg phasporus and 1.2 kg sulphur.
- Heat from burning penetrates into soil up to 1 cm and ruins soil health.
- Massive depletion of ground water resources.
- Repeated burning in the field permanently diminishes bacterial population by 50 per cent in top 2.5 cm of soil increasing farmers' dependence on fertilisers and pesticides.
- Wheat and paddy harvesters together leave a stubble of about 25 cm, which is difficult to plough back.
- Subsidies offered by the Punjab and Haryana governments have not attracted the farmers because of cumbersome formalities.
- Fine of ₹ 5000 per acre in Haryana and Punjab has not deterred farmers from burning stubble.

Table 8.10 gives details of crop stubble burning in Punjab and Haryana. IIT Delhi has identified a number of districts in these two states which contribute most to Delhi's air pollution in October and November. The study conducted by IIT-Delhi's Centre for Excellence for Research on Clean Air (CERCA) shows that Amritsar, Tarn Taren, Firozepur, Ludhiana, Faridkot, Bathinda, Mansa, Patiala and Sangrur in Punjab, and Hisar, Sirsa and Kaithal in Haryana are the biggest contributors of crop stubble burning and air pollution.

TABLE 8.10. Details of crop stubble burning in Punjab and Haryana in 2019

Particulars	Punjab	Haryana
Estimated paddy crop residue	2 crore tonnes	70 lakh tonnes
Estimated paddy crop residue burnt	98 lakh tonnes	12.4 lakh tonnes
Number of fires	50,738	6,364
Estimated area of stubble burning (acres)	27.7 lakh	5.9 lakh
Percentage of area burnt of total paddy cultivated	37.4	17.7

Note: Data pertain to October and November, 2019

***Source*:** Environment Pollution (Prevention and Control) Authority (EPCA)

It must be noted that the **table 8.10** does not include wheat stubble burning which remains operative in April and May as against October for paddy stubble burning. Number of fire counts recorded between April 1 and May 16 are shown in **Table 8.11.**

TABLE 8.11. Number of fire counts recorded between April 1 and May 16 in Punjab and Haryana

Year	Punjab	Haryana
2019	10,307	5,443
2020	8,172	2,523
2021	12,033	4,825

***Source* :** Council on Energy, Environment and Water (CEEW)

Table 8.11 shows that there was considerable reduction in the incidents of crop stubble fire in 2020 as compared to 2019 but the situation took a bad turn in 2021 when there was pronounced increase in crop stubble burning both in Punjab and Haryana.

Centre for Science and Environment (CSE) analysed the data compiled by System of Air Quality and Weather Forecasting and Research (SAFAR) and divided the daily contribution into four categories over 30%, between 20% and 30%, between 10% and 20% and less than 10%. Till November, 2020 it was found that on seven days, the contribution of stubble burning to Delhi's $PM_{2.5}$ concentration was over 30% in contrast to only three days in 2019 and 2018.

On six days in 2020, the contribution of smoke farm fires was between 20% and 30% compared to just four days in 2019. The contribution ranged between 10% and 20% on 16 days compared to 15 such days in 2019. However, there were 23 days with less than 10% contribution in 2020, down from 30 days in 2019 (**Table 8.12**).

TABLE 8.12. Contribution of stubble burning to Delhi's $PM_{2.5}$ particulate matter

Stubble burning contribution to $PM_{2/5}$ in percentages	Number of days	
	2020	2019
Over 30	7	3
20-30	6	4
10-20	16	15
Less than 10	23	30

Source : SAFAR data analysed by CSE

The above discussion shows that crop burning has a huge impact on the air quality in the National Capital Region (NCR). The impact of crop fires on the air quality is accentuated by hostile meteorological conditions. During the smog episode, for example, ground wind speed is very low and wind direction is usually from north and northwest through Punjab and Haryana where crop stubble is burnt over large areas.

According to India Meteorological Department (IMD), mixing heights (also called boundary conditions)—the levels at which ground level winds mix with higher winds and allow dispersion—are very low, leading to high concentration of pollution particles.

However, there is a ray of hope of solving the problem of crop stubble burning as some farmers have started using straw as an input for future crops. Even as the new wheat crop grows, the old residue sits in the field enriching the soil, conserving water, nourishing the new plants and creating conditions for decreased use of fertilisers. The technology of using paddy residue for 'greener agriculture' existed for years but recent improvements have made it more viable and scalable solution to the vexed problem of stubble burning. The technology is a combination of Happy Seeder and straw spreader (straw management system or SMS) for rice-wheat farming. Developed by Ludhiana's Punjab Agricultural University, The Happy Seeder is a machine that sows seeds without the need to till the field or remove paddy straw. It works best when the straw is spread evenly on the field through SMS device attached to a combine harvester. The combination of SMS and Happy Seeders is a win-win for farmers and the environment. It does away with stubble-burning while increasing farmers' profitability and promoting climate smart agriculture. The spin-off are significant—lower chemical load on soil, more crop per drop, reduction in CO_2 emissions, better soil health, less weeds and sturdier plants with deep roots.

WHY LEAVE CROP RESIDUE IN FIELD

- Residue decomposes, providing nitrogen, carbon and nutrients, improving soil health.
- Retains moisture that saves farmer one round of watering.
- Crops grow longer roots, making them stronger in facing weather hazards.

Farm stubble to Biochar. JNUs School of Environmental Sciences (SES) has suggested that stubble should be converted in biochar. It is fine grained black charcoal like product. It is high in organic carbon that is resistant to decomposition. Biochar is created after heating of biomass in the absence of oxygen. A temperature of 450°C is needed to convert stubble biomass to biochar in a few hours to 24 hours time. According to the Ministry of new Renewable Energy, the annual availability of biomass in India is estimated at 500 million tonnes. Unfortunately, a large portion of unused crop residues is burned in the fields after harvest, causing serious air pollution and producing carbon dioxide that contributes to global warming. It also causes a high loss of carbon feedstock that can be used to improve soil fertility. Following are the advantages of biochar.

- It reduces carbon from the carbon cycle, improves soil fertility and increases agricultural yields.
- It helps in increasing the water holding capacity of soil from about 11 wt (weight percentage) to 22 wt. and increasing its organic carbon.
- Biochar can remove pollutants such as fluoride, phosphate and arsenic from the ground water.
- It can increase microbial activity and helps retention of nutrients in the soil.

Other Causes of Air Pollution

Large amount of garbage, lack of sanitary facilities and sewer wastage are also responsible for air pollution. Cigarette, *biri*, cigar, and *hooka* smoking has become a great pollutant of air both in urban as well as rural areas. The worst affected are the closed areas such as cinema houses, buses and railway compartments, etc. Smoking leads to some of the serious ailments like heart disease and cancer. The smoker not only spoils his own health but affects adversely the health of the silent smoker who is sitting by his side. A total ban on smoking will go a long way in reducing the pollution levels of air.

Large scale use of pesticides and insecticides, in the agricultural field during the last over five decades, has resulted in air pollution in rural India. The life style of the Indian rural masses is closely related to air pollution. Most people in rural India burn wood, cowdung cakes, coal and kerosene oil to cook food. The kitchens are generally housed in unventilated rooms, where unscientific or conventional *chulhas* (ovens) are used. The fuels used in these kitchens emit large quantities of smoke, consisting of carbon monoxide, carbon dioxide, sulphur, etc. thereby polluting the air to a great extent. On an average, about 100 million tonnes of fire wood is burnt in India annually which produces about 3.14 million tonnes of particulate matter, 2.35 million tonnes of hydrocarbons, 1.96 million tonnes of sulphur dioxide, 0.16 million tonnes of carbon monoxide and 0.39 million tonnes of nitric oxides. The world's worst air pollution problem could be the wood smoke inhaled by poor Indian women, especially in rural areas, while cooking. A tonne of particulate matter from household wood stoves may actually lead to more than 500 times the human exposure than a tonne of particulate matter from coal fired power station. About 55 million tonnes of cowdung cakes are burnt in India every year emitting 7.2 lakh tonnes of particulate matter, 5.4 lakh tonnes of hydrocarbons, 4.5 lakh tonnes of sulphur dioxide, 90 thousand tonnes of nitric oxides and 38 thousand tonnes of carbon monoxide.

Air Quality Index (AQI). The AQI is an index for reporting daily air quality. It tells how clean your air is, and what associated health effects might be a concern for you. Environmental Protection Agency (EPA) calculates the AQI for five major air pollutants regulated by Clean Air Act. They are (1) ground level ozone, (2) particle pollution (also known as particulate matter), (3) carbon dioxide, (4) sulphur dioxide, and (5) nitrogen oxide. EPA has established national air quality standards for each of these pollutants in order to protect public health.

How does the AQI work. AQI is a kind of yardstick which runs from 0 to 500. The higher the AQI value, higher the level of pollution and higher the health risk. For example, an AQI value of 50 represents good air quality with little health problem, while an AQI value over 300 represents higher level of air pollution and hazardous air quality.

Understanding the AQI. In order to understand the AQI it is divided into the following six categories. It is an indication of effects of air quality on public health.

TABLE 8.13. Air Quality Index (AQI) and health concerns

AQI range	Air quality conditions : Levels of health concern	Colour
0-50	Good	Green
51-100	Moderate	Yellow
101-150	Unhealthy for sensitive groups	Orange
151-200	Unhealthy	Red
201-300	Very unhealthy	Purple
301-500	Hazardous	Maroon

Each category corresponds to a different health concern :

1. Good AQI is 0 to 50. Air quality is considered satisfactory, and air pollution poses little or no risk.
2. Moderate AQI is 51 to 100. Air quality is acceptable; however, for some pollutants there may be a moderate health concern for a small number of people. For example, people who are usually sensitive to ozone may experience respiratory symptoms.
3. Unhealthy for 'Sensitive Groups' AQI is 101 to 150. Although general pubic is not likely to be affected at this AQI range, people with lung disease, older adults and children are at greater risk from exposure to ozone, whereas persons with heart and lung disease are at greater risk from the particles present in the air.
4. Unhealthy AQI is 151 to 200. Everyone may begin to experience adverse health effect, and members of sensitive groups may experience more serious effects.
5. Unhealthy AQI is 201-300. Everyone may begin to experience adverse health effects, and members of sensitive groups may experience more serious effects.
6. Hazardous AQI is greater than 300. This would trigger health warnings for emergency conditions. The entire population is more likely to be effected.

Delhi-NCR region is the worst affected area, particularly in the winter season with AQI crossing 300 mark at most places for a pretty long time. As a matter of fact the whole of the Satluj-Ganga plain suffers from poor air quality particularly in winters. The southern parts of India have comparatively clean air where AQI varies from 101 to 200. This is because there is no pronounced effect of winter season.

INITIATIVES FOR CONTROL OF AIR POLLUTION

Various measures have been taken to control air pollution from vehicles and industries. These measures are described as under :

Measures taken for Control of Air Pollution from Vehicles

1. Vehicular Emission Norms

The vehicular emission norms in India are detailed below :

Strict emission norms, particularly in metropolitan areas like Delhi, Mumbai, Kolkata, Chennai, etc., have been introduced and enforced from time to time so that air pollution level is controlled to the tolerable limits. Different phases of Euro and Bharat are worth mentioning in this context. However, rapidly increasing number of vehicles in our cities and towns has more or less nullified the impact of vehicular emission norms with respect to air pollution.

2. Fuel Quality Specifications

Diesel and Gasoline fuel quality with respect to environment related parameters had been notified under Environment (Protection) Act during April 1996. The specifications include low leaded gasoline, unleaded gasoline and low sulphur diesel.

***(i)* Unleaded Gasoline.** With the progressive reduction of lead content in petrol (from 0.56 gm/l to 0.15 gm/l and then to 0.013 g/l in unleaded petrol) introduction of unleaded petrol for new passenger cars from April, 1995 and supply of only unleaded petrol for all vehicles from September, 1998, in NCT—Delhi a lethal pollutant from vehicular exhaust has been removed. The lead content in the atmosphere near traffic intersections of Delhi has reduced by more than 60% with the introduction of unleaded petrol.

***(ii)* Benzene Reduction.** The fear of increased emission of benzene and reduced performance of engines by the use of unleaded petrol has also been falsified. The oil refineries were told to combine the benzene content in the unleaded petrol upto 5% (v/v) in 1996 and 3% (v/v) from the year 2000. In addition to phasing out of lead, it is considered necessary to reduce the benzene (to 1% or lower) and aromatics in petrol not only for Delhi but also for other parts of the country.

***(iii)* Sulphur in Diesel.** Sulphur content in diesel supplied in Delhi was reduced to 0.5% in 1996 and it was further reduced to 0.25% from April 1996 onward. The diesel with 0.25% sulphur has been made available throughout the country by September, 1999. Considering the fact that several countries have introduced diesel with much lower sulphur content it is necessary to have low sulphur diesel for meeting the emission norms.

3. Quality Lubricants

Specifications of 2T oil for two stroke engines with respect to smoke emission has been notified under EPA during September 1998 for implementation from 1.4.1999 throughout the country. Pre-mix 2T oil dispenser has been installed at all petrol filling stations in big cities so that excessive oil is not being used by the vehicle owners. Sale of loose 2T oil has been banned from December 1998 in Delhi.

4. Alternate Fuels

(i) Bio-fuels mainly Ethanol and Biodiesel (in B20 form) from Jatropa feedstock are the prospective options.

***(ii)* CNG (Compressed Natural Gas).** CNG is a better and clean fuel providing limited emissions of various toxic gases. All Government vehicles were required to compulsorily fit CNG kit or catalytic converter by December 1996. New CNG taxies are being registered in Delhi as well as there is no restriction on registration of CNG vehicles in National Capital Territory (NCT) as they already comply EURO-II norms. CNG kits have been exempted from customs duty for promotion of installation of CNG kits in vehicles. Emission norms for CNG vehicles have been notified under Motor Vehicles Rules dated 24.4.2001. Besides Delhi and Mumbai the supply of CNG has been extended to a large number of cities.

***(iii)* LPG.** The use of LPG as an alternate fuel in automobiles has been made applicable for which amendment has been made in Motor Vehicles Act to legally permit the use of LPG as automobile fuel Hon'ble Supreme Court permitted dual mode facility (CNG + Petrol) for the vehicles in its order dated 10th May 2000. Emission norms for LPG vehicles were modified on 24.4.2001.

***(iv)* Battery driven vehicles.** Battery driven vehicles have been introduced in few corridors in Delhi and in some other big cities.

***(v)* E-Vehicles.** Electric operated cars is the most recent trend in big cities. These vehicles do not cause any pollution. The government is providing subsidiary to encourage the use of these vehicles so that pollution is kept at low level.

5. Phase out of Grossly Polluting Vehicles

(i) Registration of new auto rickshaws with conventional engine has been banned from May 1996 and registration of Defence Service and Govt. auctioned vehicles has been banned from April 1998 in Delhi.

(ii) Commercial vehicles more than 10 years old have been prohibited from plying in Delhi and other major cities of the country.

(iii) Registration on alternation of vehicles by replacing petrol engine with diesel has been banned from 1.4.1998 in Delhi.

6. Promotion of Comprehensive inspection and Certification

It has been possible to reduce 30-40% pollution loads generated by vehicles through proper periodical inspection and maintenance of vehicles. Such inspection and maintenance of vehicles is being carried on by State Pollution Control Boards, Pollution Control Committees and Transport Directorates in different parts of the country.

7. Traffic Management

(i) Restriction has been imposed on goods vehicles during day time in several big cities.

(ii) Time clocks have been installed in important red lights to enable the drivers to switch off their vehicles depending on the time left in the time clocks.

(iii) More fly-overs and subways have been constructed and T-Junctions have been closed for better traffic flow.

(iv) Almost all big cities are plagued with traffic jams on important roads. Standing vehicles cause much air pollution. Steps should be taken to avoid traffic jams as far as possible.

8. Public Transport System

(*i*) To discourage the use of individual motor vehicles by people, public transport system is augmented from time to time in various urban areas of the country. The number of buses has been increased in big cities.

(*ii*) Private sector has been allowed to operate public transport buses to increase mobility.

(*iii*) Mass Rapid Transport System (MRTS) has been launched. Metro Rail Transport System is making rapid progress in a large number of cities and is likely to reduce pressure on transport system of these cities.

9. Technology

(*i*) Fitment of catalytic converter into new petrol passenger cars has been made compulsory.

(*ii*) Two wheeler scooters with four stroke engines have been introduced in the market from October 1998.

(*iii*) Registration of only rear engine auto rickshaws is being allowed from May 1996 onwards.

(*iv*) Only four stroke two wheelers are being registered.

10. Information Dissemination/Mass Awareness

(*i*) Messages/articles related to vehicular emissions are disseminated through newsletters, pamphlets, newspapers, magazines, Television, Radio, Internet and through Workshops, Summer Courses, Exhibitions, displays, Pollution Control Camps, etc.

(*ii*) Display of ambient air quality data through Electronic Display System as well as dissemination through newspapers, daily news and Internet.

(*iii*) Publishing reports related to vehicular pollution control and dissemination to various organisations.

(*iv*) Regular publication of air quality statistics regarding ambient air quality status in the country.

(*v*) Non-Government Organisations (NGOs) working in the area of Vehicular Pollution Control in different parts of the country are being encouraged for creating mass awareness.

Measures Taken for Controlling Air Pollution from Industries

The measures taken for controlling air pollution from industries are as follows :

1. Emission standards have been notified under the Environment (Protection) Act, 1986 to check pollution.
2. Industries have been directed to install necessary pollution control equipment in a time bound manner and legal action has been initiated against the defaulting units.
3. As many as 24 critically polluted areas have been identified. Action plans have been formulated for restoration of environmental quality in these areas.
4. Environmental guidelines have been evolved for siting industries.
5. Environmental clearance has been made compulsory for 29 categories of development projects involving public hearing/NGO participation as an important component of Environmental Impact Assessment process.
6. The process of Environment Auditing has been initiated in highly polluting industries. The methodology has been standardised and finalised for respective group of industries. Submission of Environmental Statement has been made mandatory.
7. Under Indo-German Bilateral Programme, methodology for zoning, mapping and siting of industries has been developed in various states in collaboration with State Pollution Control Boards in order to identify the existing characteristics of the district, unsuitable zones for the industries, air quality mapping assessment of risk due to siting of air polluting industries and industrial suitability mapping. Based on zoning/siting programme, the site clearance procedure has been streamlined.
8. Minimal National Standards (MINAS) have been presented for highly polluting industries under The Air (Prevention and Control of

Pollution) Act, 1981 and Environment (Protection) Act 1986.

9. Power plants (coal based) located beyond 1000 kms from the pit-head are required to use low ash content coal (not exceeding 34%) with effect from 1.6.2002. Power plants located in the sensitive areas are also required to use low ash coal irrespective of their distance from the pit head.

   It is estimated that about 40 million tonnes of fly ash is generated per annum from coal based thermal power plants and contribute to particulate matter leading to environmental pollution. Fly ash possesses good pozzalinic properties due to presence of active and finely divided silica, alumina and calcium oxide, which provide it with cement like qualities in combination with lime rich material. Thus fly ash emitted by thermal power plants can be used for manufacturing bricks, blocks, aggregates and cement.

10. Use of cleaner technologies is a new dimension emerging rapidly for cleaner production and to increase production efficiency, and at the same time eliminate or atleast minimise emission and waste at their source rather than to treat them at the end of the production chain after they are generated.

11. Industrial wastes like slags, red mud, etc. are generated from iron and steel and during extraction of non-ferrous metals such as aluminium and copper. The slags are dumped in the vicinity of plant while red mud is disposed of as slurry. This slurry becomes air borne after getting dried. In phosphoric fertilizer plants about 4.5 million tonnes of phospho-gypsum (with floride contents 0.7 to 1.5%) are produced. This can be used for cement, gypsum board, partition panels, ceiling tiles, artificial marble and fibre boards. A thrust will have to be made for proper disposal and reutilisation of these waste materials.

The National Clean Air Programme (NCAP)

This programme was launched in January, 2019. This is a step towards creating a national framework for air quality management and compliance. Under this programme, 122 cities have been identified which do not meet the National Ambient Air Quality Standards (NAAQS) for PM_{10} and $PM_{2.5}$. As such they are termed as "non-attainment" cities. About 60% of these towns and cities are in the Indus-Ganga Plain and the remaining 40% are in other six major climatic zones.

The list of "non-attainment" cities includes megacities like Delhi, Mumbai, Kolkata and Bengaluru, as well as a large number of small cities. About 70% of these cities and towns have less than one million population each. In such towns and cities localised action can certainly reduce local exposure to air pollution, but if their larger influence area is neglected, their air cannot be cleaned up beyond a point. It is on account of this reason that there is a need to shift the focus to larger airsheds. *Airsheds are a complex mix of multiple state jurisdiction and governance system*. However, there is no clear mechanism for delineating regional airsheds and creating a common and harmonised framework of action, monitoring and compliance across states.

WATER POLLUTION

Water is the most important element in the biosphere because it sustains all forms of life on the planet earth. The purity and quality of water is of basic concern to mankind since it is directly related to human health. Nature has given us plenty of water and even then we are feeling the shortage of water for drinking, washing, irrigation and industrial purposes because we have misused and polluted our water resources to a great extent. Water pollution may be defined as *alteration in the physical, chemical and biological characteristics of water, which may cause harmful effects on human and aquatic life*. Water pollution is now-a-days considered not only in terms of public health but also in terms of conservation, aesthetics and preservation of natural beauty and resources.

For the sake of simplicity, the water resources are divided into two categories *viz. surface water* and *ground water*. Surface water is found in the form of rivers, canals, lakes and ponds, etc. Some of the rain water percolates in the rocks under the surface and is

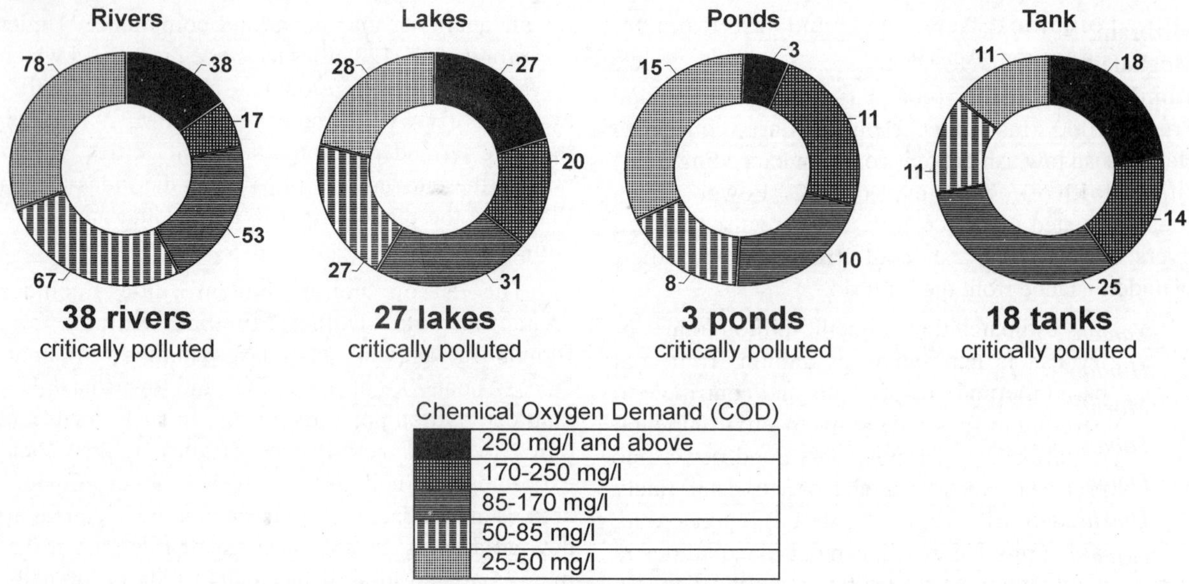

FIG. 8.3. State of surface water pollution

Source: Ministry of Environment, Forest and Climate Change as on March 2018

available as underground water. This is used through hand pumps, wells, tube wells and springs.

In India, water pollution has been taking place on a large scale and for a long time. Both surface and ground water bodies are polluted to a great extent.

Surface Water Pollution

Surface water is available in the form of rivers, lakes, ponds and tanks. According to a report released by the Ministry of Environment, Forest and Climate Change in March, 2018, as many as 86 (17% of the total water bodies, surveyed) are critically polluted having chemical oxygen demand concentration of more than 250 mg/l. Chemical oxygen demand (COD) is a parameter to assess the chemical contamination of water bodies. Figure 8.3 shows that 38 out of 253 rivers, 27 out of 133 lakes, 3 out of 47 ponds and 18 out of 79 tanks are critically polluted.

Maharashtra has the maximum number of 9 critically polluted river stretches. This is followed by Gujarat (5), Uttar Pradesh and Tamil Nadu (4 each), Madhya Pradesh, Assam and Uttarakhand (3 each).

According to a report of NEERI, a staggering 70 per cent of the available water in India is polluted. From Dal Lake in the north to Periyar and Chalyar rivers in the south and from Damodar and Hugli in the east to Thane Creek in the west, the picture of water pollution is uniformly gloomy. Most of the rivers are carrying polluted water. We have been dumping all sorts of waste matter into our rivers—industrial waste, municipal sewage, residues of chemical fertilizers and what not. These wastes are both, solids as well as in dissolved form. Apart from polluting the river, they decrease the depth of the river and raise the level of river beds. Thus the river becomes a garbage can, without depth and flow. Most of the sewage of towns and cities is dumped into water courses without any treatment rendering the natural water course down stream unfit for drinking and even for bathing.

In 2015, the Central Pollution Control Board (CPCB) identified 302 polluted stretches along 275 rivers in 28 states and 6 Union Territories. According to 2018 CPCB report, out of India's 29 states and seven Union Territories, 27 have critically polluted stretches (Daman and Diu, and Dadra and Nagar Haveli have been bracketed by CPCB as a single unit). The report identifies 351 pollutted stretches on 323 rivers in these states and union territories. Maharashtra has the highest 53 polluted river stretches. This is followed by Assam, Madhya Pradesh, Kerala, Gujarat, Odisha, West Bengal, Karnataka, Uttar Pradesh, Goa, Uttarakhand,

Mizoram, Manipur, Jammu & Kashmir, Telangana, Meghalaya, Jharkhand, Himachal Pradesh, Tripura, Tamil Nadu, Nagaland, Bihar, Chhattisgarh, Andhra Pradesh, Sikkim, Punjab, Rajasthan, Puducherry, Haryana and Delhi. There are 370 major towns and cities along the course of these rivers which are constantly discharging their waste water into these rivers. Some of the most polluted river stretches are as under:

Yamuna : Delhi to Agra
Hindon : Saharanpur to Ghaziabad
Morna : Maharashtra
Sabarmati : Ahmedabad
Damanganga : Daman and Diu
Umkhrah : Shillong
Gaggar : Punjab
Cauvery : Tamil Nadu

Main Causes of Water Quality Degradation

There are several causes of water pollution in India. The main causes are briefly described as under :

1. Urbanisation

Rapid urbanisation in India during the recent decades has given rise to a number of environmental problems such as water supply, wastewater generation and its collection, treatment and disposal. Many towns and cities which have come up on the banks of rivers have not given a proper thought to problem of wastewater, sewage, etc. In urban areas, water is tapped for domestic and industrial use from rivers, streams, lakes, ponds, wells, etc. Nearly 80% of the water supplied for domestic use passes out as wastewater. In most cases, this wastewater is let out untreated and causes large scale pollution of the surface water. A part of it percolates into the ground and contaminates the ground water. About 70% of the population of class I cities is provided with sewerage facility. The Ganga river basin contributes about one-third of the total wastewater of India.

Central Pollution Control Board (CPCB) report of 2015 brought out the fact that 61,948 million litres of urban sewage is generated on a daily basis in India. But cities have an installed sewage treatment capacity

TABLE 8.14. Sewage generation and installed treatment capacity and the number of polluted river stretches

States	Sewage generated in urban areas (MLD)	Installed treatment capacity (MLD)	Number of polluted river stretches identified till 2017
1. Andhra Pradesh	2,871	247.27	6
2. Arunachal Pradesh	DNA	DNA	DNA
3. Assam	703	0.51	28
4. Bihar	1,879	124.55	5
5. Chhattisgarh	951	0	5
6. Goa	145	74.56	8
7. Gujarat	4,119	3062.92	20
8. Haryana	1,413	852.7	2
9. Himachal Pradesh	110	114.72	8
10. Jammu & Kashmir	547	264.74	9
11. Jharkhand	1,270	117.24	8
12. Karnataka	3,777	1304.16	15
13. Kerala	2,552	152.97	13
14. Madhya Pradesh	3214	482.23	21
15. Maharashtra	8143	5160.36	49
16. Manipur	132	0	12
17. Meghalaya	95	1	10
18. Mizoram	DNA	DNA	DNA
19. Nagaland	92	0	3
20. Odisha	1,121	385.54	12
21. Punjab	1,664	1,245.45	2
22. Rajasthan	2,736	865.92	8
23. Sikkim	24	31.88	5
24. Tamil Nadu	5,599	1799.2	2
25. Telangana	1,671	685.8	7
26. Tripura	154	0.05	2
27. Uttarakhand	495	152.9	5

28. Uttar Pradesh	7,124	2646.84	13
29. West Bengal	4,667	416.9	17
Union Territories			
1. Andaman & Nicobar Islands	DNA	DNA	DNA
2. Chandigarh	DNA	DNA	DNA
3. Dadra & Nagar Haveli	55	0	1
4. Daman & Diu	—	—	—
5. NCT of Delhi	4,155	2,693.7	1
6. Lakshadweep	DNA	DNA	DNA
7. Puducherry	DNA	DNA	DNA

Notes : (*i*) DNA : Data Not Available
(*ii*) MLD : Million Litres per Day

Source : State of India's Environment, 2018, p. 54.

of only 38% of this. In reality more than this amount goes untreated into the rivers or water bodies as the treatment capacity of major sewage treatment plants (STPs) in the country is around 66 per cent of the installed capacity. As a result more than 38,000 million litres of waste water goes into the major rivers, water bodies and even percolates into the ground every day.

Municipal water treatment facilities in India, at present, do not remove traces of heavy metals. Given the fact that heavily polluted rivers are the major sources of municipal water supply for most towns and cities along their courses, it is believed that every consumer has been, over the years, exposed to unknown quantities of pollutants in water they have consumed. To add to this, Indian towns and cities have grown in an unplanned manner due to rapid population growth.

Facilities for running water have been provided in many towns and even in some villages during the last couple of decades. This has resulted in the use of flush-latrines and much larger use of water in homes for bathing, washing of clothes, utensils, etc., generating large quantities of wastewater. Use of soaps and detergents and amounts of various food materials going to sink have also grown considerably with improved life standards. But sewerage has lagged far behind water supply. According to estimates made by the Central Pollution Control Board, only 22% of the wastewater from class I cities and 14% from class II cities is being collected through sewerage. A large number of cities/towns either do not have any sewerage system or the sewerage system is overloaded or defunct. All this results in large quantity of wastewater remaining uncollected.

Situation in big cities is worsened by migration of poor people from the surrounding rural areas. These people migrate to the cities in search of livelihood. According to an estimate by CPCB, only about 40-50% of the population of the major cities like Delhi, Mumbai, Kolkata, Chennai and Bengaluru are served by sewer systems. Even in areas where sewers exist, they often leak or overflow, releasing their contents to storm-water or other surface drains or percolate in to the soil to reach ground water. Very often uncollected and untreated sewerage water reaches the streams thereby polluting their water.

2. Industries

Most Indian rivers and other sources of fresh water are polluted by industrial wastes or effluents. All these industrial wastes are toxic to life forms that consume this water. The total wastewater generated from all major industrial sources is 83,048 Mld which includes 66,700 Mld of cooling water generated from thermal power plants. Out of remaining 16,348 Mld of wastewater, thermal power plants generate another 7,275 Mld as boiler blow down water and overflow from ash ponds. Engineering industries comprise the second largest generator of wastewater in terms of volume. Under this category the major polluting industries are electroplating units. Other significant contributors of wastewater are paper mills, steel plants, textile and sugar industries. The major contributors of pollution in terms of organic load are distilleries followed by paper mills.

Both large scale industries and small scale industries contribute their share of water pollution. While many large scale industries claim to have installed costly treatment and disposal equipments, these are often not in proper working order. Several examples can be cited, such as oil wastes present in the storm-water channel along Haldia Refinery of ammonia pollution in ground water around a urea factory of Kanpur and a natural spring close to the Zuari Agro Urea plant in Goa.

Small scale and cottage industries cause no less water pollution than the large scale industries. There are about 3 million small scale and cottage industrial units in India. These units neither have, nor can they afford, appropriate sanitation and/or pollutant disposal systems, and yet have not hasistated in adopting highly polluting production technologies such as chrome, tanning of leather, use of azo-dyes in fabrics, use of cadmium in ornaments and silver-ware, electroplating with cyanide baths, production of dye-intermediates and other refractory and toxic chemicals, etc. Their solid wastes and sludges get scattered around or dumped in unlined pits and effluents flow to streams through storm-drains or stagnate in depressions to percolate, leach or get washed-off during the next rainy season. This is the story of many industrial areas and urban centres in the country.

3. Agricultural runoff and improper agricultural practices

Traces of fertilizers and pesticides are wasted into the nearest waterbodies at the onset of the monsoons or whenever there are heavy showers. As the point of entry of such agricultural inputs is diffused throughout the river basin, they are termed as *non-point* sources of pollution. Although irrigation has increased considerably in the country, precious little has been done to tackle the problem of the high salinity return water. This is the situation in Punjab and Haryana. In Haryana, the 40 km long drain No. 8 pours 250,000 kg/day of chlorides into the Yamuna to raise the chloride concentration in the river from 32 mg per litre just upstream of the drain confluence to 150 mg per litre just downstream of it. And most of these chlorides are from agricultural return flows. According to the findings of the CPCB, some of the seepage into the drain contains over 15,000 mg per litre of chlorides. Intensive and ever increasing usage of chemical fertilizers, pesticides, weedicides and other chemicals is adding a new dimension to such pollution. According to A.K. Dikshit, senior scientist with the Indian Agricultural Research Institute (IARI), New Delhi, farmers often indulge in excess usage of fertilizers and pesticides. When these are used more than the recommended doses, they pollute water, land and air.

Flood-plain cultivation is another significant contributor to water pollution. Fertilizers and pesticides used in these tracts of land are bound to be washed into rivers particularly during the monsoons.

4. Withdrawal of Water

Indian rivers, particularly the Himalayan rivers, have plenty of water in their upper courses. They are, however, starved of water when they enter the plain area. Irrigation canals whisk away clean water soon after the rivers reach the plains, denying water to flow in the river downstream. What flows into the river is water trickling in from small insignificant streams and drains carrying untreated sewage and effluents. The river-turned drains flow downstream with little or no fresh water unless a large river augments the depleted flows.

As the quantity of fresh water in the river is negligibly small, pollution—either from urban and rural areas, industries or even natural forms of pollution—cannot get diluted and its ill effects are not reduced. The Yamuna has almost no water at Tajewala in Haryana where the Eastern Yamuna Canal and the Western Yamuna Canal draw large quantities of water for irrigation. Similarly, the Upper Ganga Canal and the Lower Ganga Canal have left the Ganga downstream almost dry. When the Yamuna and the Ganga flow past Delhi and Kanpur respectively, they are turned into stinking sewers. Therefore, it is essential that a minimum level of flow of water must be maintained in the river. This is known as ***minimum flow of rivers.*** According to a report of the Ministry of Jal Shakti on the study of minimum flows in the Ganga, impact on river water quality resulting from discharges of treated or untreated wastewater into the river will depend on the dilution offered by the quantum of flows in the river. Minimum flow in the recipient river will be required to maintain the desired water quality. Further, the study has expressed the view that it is not possible to fix the minimum flow of water in the entire course of the river because it depends on the pollution discharged at different points on the river. For example the existing minimum flow in the Ganga at Kanpur in May is hardly 50 cumecs (cubic metres per second) whereas the required minimum in the same month is 350 cumecs. The study further says that since the water is scarce it is not possible to add further fresh water for dilution. The solution lies in less amounts of pollution entering the river. In view of the increased demand of water for irrigation, the minimum flow is likely to fall further in future. In the words of K.C. Sivaramakrishnan, former director of the Ganga Action Plan (GAP), "maintenance of minimum flow is an important point. In simple terms a non-existent river cannot be cleaned. In case of the Ganga between Bijnor and Kanpur, the river is just a small stream. In case of the Yamuna, from Delhi till the point where the Chambal joins, the river is just a trickle. Other rivers like Sabarmati are almost devoid of water."

The Yamuna is dying a slow death in Delhi. In fact it is a dead river as it flows past Delhi. This river is relatively less polluted when it enters Delhi at Wazirabad barrage, but a mere 100 metres downstream the barrage, the river receives untreated sewage and industrial waste. The committee on minimum flows in the Yamuna indicates that if the minimum flows requirement in Delhi is met, that would suffice for the entire course of the river. According to report of the committee, discharge downstream of Tajewala and Okhla is less than 5 cumecs whereas minimum flow of 10 cumecs is required between Tajewala and confluence of the Yamuna with the Chambal. The committee states that this shortfall can be met either by creating storage facilities in the catchment area or from imports from another river basin. Increasing demand for the Yamuna water for irrigation and for meeting the urban requirements would leave very little freshwater in the river to maintain the minimum flow.

The maintenance of minimum flow, to sustain a river ecology through its course as well as its confluences, is a *recent awakening which requires serious thought*. This policy must be pursued vigorously so that river pollution is kept at certain permissible limit.

5. Religious and Social Practices

Religious faith and social practices also add to pollution of our river waters. Carcasses of cattle and other animals are disposed of in the rivers. Dead bodies are cremated on the river banks. Partially burnt bodies are also flung into the rivers. All this is done as a matter of religious faith and in keeping with ancient rituals. These practices pollute the river water and adversely affect the water quality. Mass bathing in a river during religious festivals is another environmentally harmful practice. Studies have revealed that the biochemical oxygen demand (BOD) goes up drastically when thousands of people simultaneously take a 'holy dip'. Religious practices also demand that offerings from a ***puja*** be immersed in a river. It is now common to see people immersing offerings in plastic bags. Plastic bags are very dangerous and further add to the pollution load of the river. In 2013, Pryagnagar played host to the biggest and the most spectacular gathering of humanity in the world—the Maha Kumbh. A staggering 100 million congregated at the Sangam-up from 70 million in 2001 and the Ganga had to bear the brunt of this massive load of humanity. Over 9.1 million devotees thronged Mahakumbh at Haridwar in 2021. This was a much less gathering due to Covid-19 surge. This Kumbh mela was terminated before time because of large number of devotees got Covid-19 infection.

Effects of Water Pollution

Water pollution adversely affects the health and life of man, animals and plants alike. Polluted water is also harmful for agriculture as it adversely affects the crops and the soil fertility. Pollution of sea water damages the oceanic life.

1. Health Aspects of Water Quality. Consumption of polluted water is a major cause of ill health in India. Polluted water causes some of the deadly diseases like cholera, dysentery, diarrhoea, tuberculosis, jaundice, etc. About 80 per cent of stomach diseases in India are caused by polluted water. Water borne diseases are the single most important factor responsible for nearly 80 per cent of human mortality in India. Children are the worst affected, especially in rural areas and urban slums.

2. Effects of Organic Pollution on Water Quality. All organic materials can be broken down or decomposed by microbial and other biological activity (biodegradation). Organic and some of the inorganic compounds exhibit a biochemical oxygen demand (BOD) because oxygen is used in the degradation process. Oxygen is a basic requirement of almost all aquatic life. Aquatic life is adversely affected if sufficient oxygen is not available in the water. Typical sources of organic pollution are sewage from domestic and animal sources, industrial wastes from food processing, paper mills, tanneries, distilleries, sugar and other agro based industries.

3. Effects of Nutrients on Water Quality. Water supports aquatic life because of the presence of nutrients in it. Here the primary focus is on fertilizing chemicals such as nitrates and phosphates. Although these are important for plant growth, too much of nutrients encourage the over-abundance of plant life and can result in environmental damage called ***"entrophication'***. This can occur at both microscopic level in the form of algae or macroscopic level in the form of aquatic weeds. Nitrates and phosphates are contributed by sewage, agricultural run-off and run-off from un-sewered residential areas.

4. Effects of High Dissolved Solids (HDS) on Water Quality. Water is the best solvent and can dissolve a large variety of substances which come in its contact. The amount of dissolved solid is a very important consideration in determining its suitability for drinking, irrigation and industrial uses. In general, waters with a total dissolved solids of less than 500 mg/litre are most suitable for drinking purposes. Higher amount of dissolved solids may lead to impairment of physiological processes in human body. Dissolved solid is a very important criteria for irrigation. This is due to the fact that dissolved solid accumulates on the ground resulting in salinization of soil. In this way it renders the agricultural land non-productive. Dissolved solids are harmful for industries also because they form scales, cause foaming in boilers, accelerate corrosion and interfere with the colour and taste of many finished products.

5. Effects of Toxic Pollutants on Water Quality. Toxic pollutants mainly consist of heavy metals, pesticides and other individual xenobiotic pollutants. The ability of a water body to support aquatic life, as well as its suitability for other uses depends on many trace elements. Some metals *e.g.,* Mn, Zn and Cu present in trace quantities are important for life as they help regulate many physiological functions of the body. Some metals, however, cause severe toxicological effects on human health and the aquatic ecosystem.

6. Effects of Thermal Discharges on Water Quality. The discharge of cooling water from industrial and commercial operations generally heats up the aquatic environment. Organisms may become physiologically stressed or may even be killed when exposed to heated water. If water heating is supplemented by the summer heat, the impact on aquatic environment can be disastrous. Thermal pollution also causes a decrease in the driving force or oxygenation which may directly kill aquatic life through asphyxiation. If toxic pollutants are present in the aquatic environment, thermal pollution may increase their toxicity to the aquatic life. Bioavailability of many pollutants may also increase due to thermal pollution, which may ultimately adversely affect the aquatic life.

Water Quality Monitoring

Water quality monitoring is defined by the International Organisation for Standardization (ISO) as : "the programmed process of sampling, measurement and subsequent recording or signalling or both, of various water characteristics, often with the aim of assessing conformity to specific objectives." Water quality monitoring is essential to ensure that the water quality is being maintained or restored at desired level. With the introduction of Water (Prevention and Control of Pollution) Act, 1974 and its subsequent amendment in 1981, the prime responsibility has been to maintain or restore the wholesomeness of water. It was therefore imperative to establish a water quality monitoring network to update on the existing water quality *vis-a-vis* the derived quality depending on the designated best use. The first water quality (WQ) stations were chosen on the river Yamuna. Central Pollution Control Board (CPCB) initiated water quality monitoring in 1976-77 with 18 stations on the Yamuna river. The number of stations gradually increased based on available resources and water quality problems. CPCB has been identified as the Government of India's agency to serve as a focal point for carrying water quality monitoring under the Global Environment Monitoring System (GEMS) programme of World Health Organisation (WHO) in India.

Another national programme, known as Monitoring of Indian National Aquatic Resources (MINARS) was started in 1984 with a total of 113 stations spread over 10 river basins. Presently the inland water quality monitoring network is operated under a three-tier programme-GEMS, MINARS and Yamuna Action Plan (YAP). Currently the network consists of 2,500 stations in 28 states and six Union Territories. The monitoring network covers 445 rivers, 154 lakes, 12 tanks, 78 ponds, 41 creeks/sea water, 25 canals, 45 drains, 10 water treatment plants (raw water) and 807 wells. Among the 2,500 stations, a total of 1,275 are on rivers, 190 on lakes, 45 on drains, 41 on canals, 12 on tanks, 41 on creeks/sea water and 79 on ponds.

Water Quality Status

The water quality monitoring results indicate that organic and bacterial pollution are the critical parameters in Indian aquatic resources. This is mainly due to discharge of domestic wastewater, mostly in untreated form, from the fast growing urban centres of the country. By and large the municipal corporations do not have the requisite resources and are not able to treat the increasing load of municipal sewage flowing into the water bodies. Besides the receiving water bodies do not have adequate water for dilution. Therefore, the oxygen demand and bacterial pollution in Indian water bodies is increasing with each possing day. This is the main cause of increasing incidence of water borne diseases.

Most of the surface water in India is found in rivers and there are several river basins in India. It will, therefore, be relevant to study the problem of water pollution on the basis of river basins. The pollution status of major river courses is shown in **Fig. 8.4** and is discussed as under :

1. The Ganga Basin

This is the largest river basin of India. It covers over one-fourth of the country's total geographical area and is the home to about half of her population. The Ganga basin accounts for nearly 50 per cent of Class I and Class II towns of the country and the mode of discharge of the municipal waste is mainly into the river system.

The polluted stretches include large stretches of the Ganga itself, the Yamuna (Delhi to Etawah), the Chambal (downstream of Nagda to downstream of Kota), the whole of the Kali river, the Hindon (Saharanpur to confluence with the Yamuna), Kshipra (around Ujjain and downstream), Damodar (downstream of Dhanbad to Haldia), Gomati (Lucknow to confluence with Ganga), Betwa (along Mandideep and Vidisha), Son, etc.

Pollution of water in some of the major rivers of the Ganga basin is described as under :

GANGA. The Ganga is the most sacred and at the same time one of the most polluted rivers of India. Keeping in view the rapidly falling water quality level of the Ganga and its tributaries, a large number of water quality monitoring stations have been set up in this river basin. Out of its total length of 2,525 km from its source in Gangotri to its mouth in the Bay of Bengal, as much as 600 km or about one-fourth of total length is reported to be badly polluted. The main cause of water pollution in the Ganga is rapid growth of urban and industrial centres all along the length of its banks.

The story of polluting the holy Ganga starts from Rishikesh and continues upto Kolkata. The main polluting cities are Haridwar, Kanpur, Prayagraj, Varanasi, Patna and Kolkata.

There are 97 towns along the Ganga river, which generate 2953 million litres per day (MLD) of sewage, estimates the Central Pollution Control Board (CPCB). Only a fraction of this is treated before being dumped into the river. Between 70 and 80 per cent of the total waste generated comes from domestic sewage. Another 15 per cent is added by industrial affluent. Of the 6440 tonnes of solid waste generated in towns, 55 per cent is dumped into the river unprocessed. High microbial load is also a big concern. In 2018, CPCB monitored 94 locations and found high biochemical oxygen demand (BOD) in long river stretches of Uttar Pradesh and West Bengal. BOD did not meet the bathing standards from Kannauj to Prayagraj and from Mirzapur to Ghazipur in Uttar Pradesh and from Serampore to Uluberia in West Bengal.

High levels of faecal coliform have been found in the stretch from Garhmukhteshwar to Kannauj, in Bithoor in Kanpur, Mirzapur and Varanasi in Uttar Pradesh. The entire stretch of the Ganga in Bihar and West Bengal has faecal coliform too high for bathing.

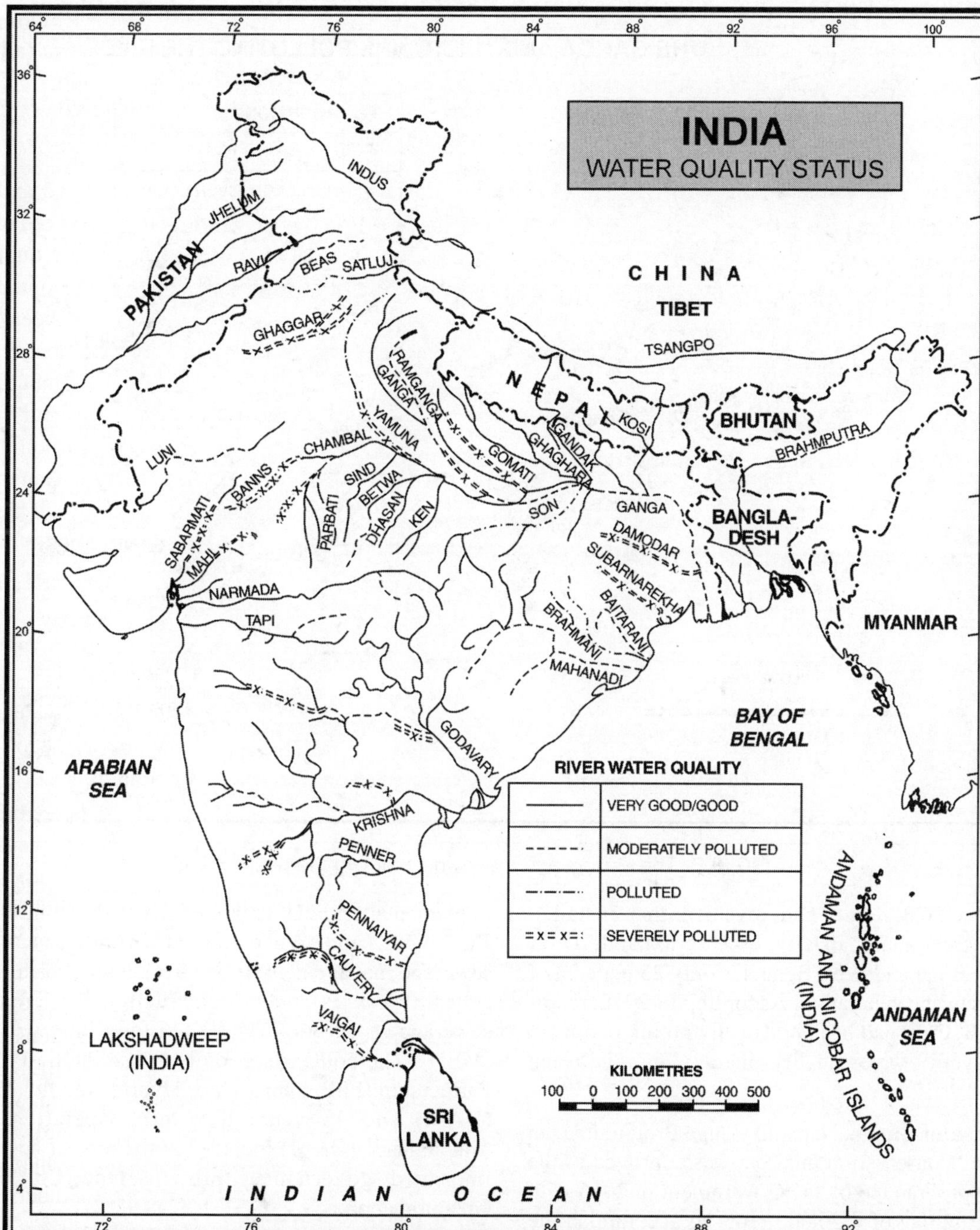

FIG. 8.4. India : Water Quality Status

According to a report published in the international journal '*Nature*' in 2013, the Ganga was ranked as the second-most polluted river in the world after Citarum river of Indonesia. The pollution level of the Ganga was about 3,000 times the safe limit prescribed by the World Health Organisation (WHO) for human use. Only 45% of sewage that flows into the Ganga is treated. People living in the vicinity of the river are vulnerable to cancer.

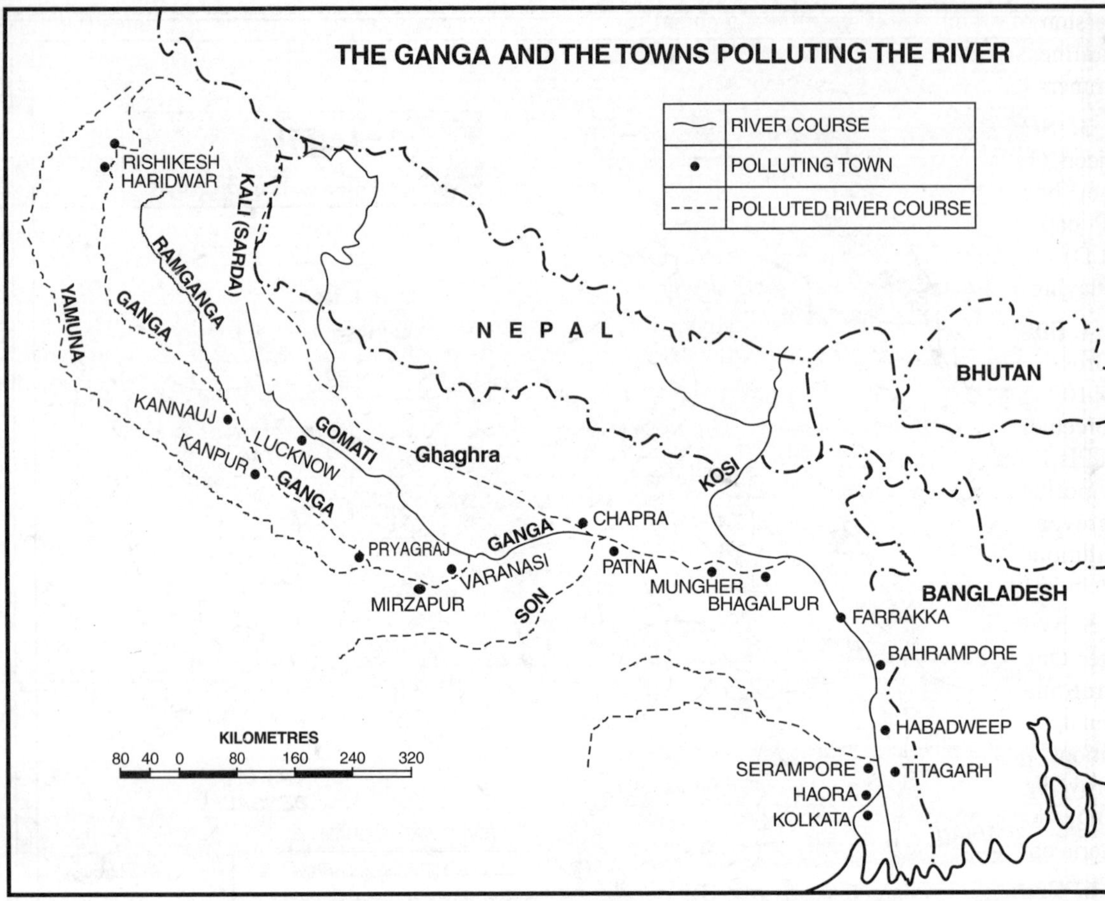

FIG. 8.5. The Ganga and the towns polluting the river

A 2017 CPCB report had revealed that installed sewage treatment capacity in Uttarakhand, Uttar Pradesh, Bihar and West Bengal is only 25 per cent of current sewage generation. According to 2018, report by CPCB, there had been practically no improvement in the overall water quality of the Ganga between 2014 and 2017.

Namami Gange. Namami Gange Programme, an integrated conservation mission, was approved as the flagship programme by the Government in 2015 with a budget outlay of ₹ 20,000 crore to accomplish the twin objects of effective abatement of pollution, conservation and rejuvenation of the Ganga.

Following are the major components of Namani Gange Mission :

1. Sewerage Project Management. As many as 63 sewerage management projects are under implementation in Uttarakhand, Uttar Pradesh, Bihar, Jharkhand, and West Bengal. Another 12 new sewerage management projects are launched in these states. Work is under construction for creating sewerage capacity of 1187.33 Million Litre per Day (MLD). The policy decision is to use Public-Private Partnership (PPP) approach of Hybrid Annuity Mode (HAM) and 15 years long-term Operation and Maintenance (O&M) includes in the project cost and improved governance through 'One City One Operator' approach.

2. Urban Sanitation. According to a report by the Consortium of 7 IITs, about 109 cities are found to contribute more than 60 per cent pollution load in the Ganga. The Mission extended a comprehensive coverage of these cities with construction and rehabilitation of Sewerage Treatment Plants (STPs) for a prospective year of 2030; inception and diversion of drains, solid waste management through cleanliness drives on *ghats* and deployment of skimmers for river surface cleaning.

3. Sewerage Infrastructure. 150 sewerage projects (11 on Ganga stem and 39 in the tributaries) have been approved for creation of new STP capacity of 3729.92 million litres per day (MLD), 1,149.39 MLD rehabilitated STP capacity and laying of about 4,972.35 km sewerage networks.

4. Industrial Effluent Monitoring. The number of Grossly Polluting Industries (GPIs) has been found to 1109 till the end of 2019. Regulation and enforcement through regular and surprise inspections of GPIs is carried out for compliance verification of the pollution norms and process modification, wherever required through third party technical institutions. The compliance rate of GPIs was only 39% in 2017 which increased to 76% in 2018.

5. Water Quality Monitoring. 36 Real Time Water Quality Monitoring Stations (RTWQMS) are operational under Namami Gange Programme. Manual water quality monitoring is carried out at 94 locations. Between 2017-18, Dissolved Oxygen levels improved at 36 locations, Biological Oxygen Demand (BOD) decreased at 42 locations and coliform bacteria count decreased at 47 locations.

6. Creating River-Front Development. Under the Mission, 28 River Front Development Projects and 33 Entry level projects for construction, modernization and renovation of 182 Ghats and 118 crematoria has been initiated.

7. River Surface Cleaning. River surface cleaning for collection of floating solid waste from the surface of the Ghats and River and its disposal are afoot and pushed into service at 11 locations.

8. Biodiversity Conservation. Several Biodiversity Conservation projects have been initiated. Some of the projects are Biodiversity Conservation and Ganga Rejuvenation, Fish and Fishery Conservation in Ganga River, Ganges River Dolphin Conservation, etc. Five bio-diversity centres have been developed at Dehradun, Narora, Prayagraj, Varanasi and Barrackpore.

9. Afforestation. Afforestation along the banks of the Ganga has been taken up scientifically with the help of Forest Research Institute, Dehradun. Local communities have been involved in massive afforestation drive in five Ganga states (Uttarakhand, Uttar Pradesh, Bihar, Jharkhand and West Bengal) with total plantation of 96,46,607 leading to increase in forest area of 8,631 hectares.

10. Rural Sanitation. 4465 villages on the Ganga stem have been declared Open Defacation Free (ODF) with the construction of about 11 lakh independent toilets. Support is also being extended to 1662 Gram Panchayats along the Ganga for solid and liquid waste management.

11. Urban River Management. National Mission for Clean Ganga (NMCG) in partnership with National Institute of Urban Affairs (NIUA) is preparing an Urban River Management Plan to project and enhance the status of river health within the city, to prevent their deterioration and to ensure sustainable use of water resources. A comprehensive survey for generating high-resolution Light Detection and Ranging (LIDAR) maps of the entire Ganga stretch to create a baseline of its spatial status has also been initiated.

THE DAMODAR deserves a special mention with respect to water pollution. Traditionally notorious as the sorrow of Bengal, the river is now the sorrow of both Bengal and Jharkhand. It is one of the most polluted rivers of the country. The high level of pollution of this river has rendered its water unfit for drinking, bathing and even for agriculture. Indiscriminate and uncontrolled discharge of industrial effluents has led to this sad state of affairs. The presence of high concentration of oil and grease (mostly mineral oils), toxic substances, etc., have made the Damodar water harmful and useless. These substances are of industrial origin and have rendered the river a ***'biological desert'*** showing hardly any existence and propagation of aquatic life in the vulnerable stretch of about 300 km between Girdih and Durgapur. Nevertheless, the river is the main source of drinking water for innumerable rural, urban and industrial people along its banks. Raw materials are available in plenty and the requisite infrastructure has been developed. Thus, the region has seen unprecedented industrial growth (**Fig. 8.6**). This has resulted in gross pollution of the river, rampant deforestation, soil erosion and siltation of the river course. Most of the catchment area of the river falls in the mining belt. Overburden from open cast mines are often dumped near the mined areas and are

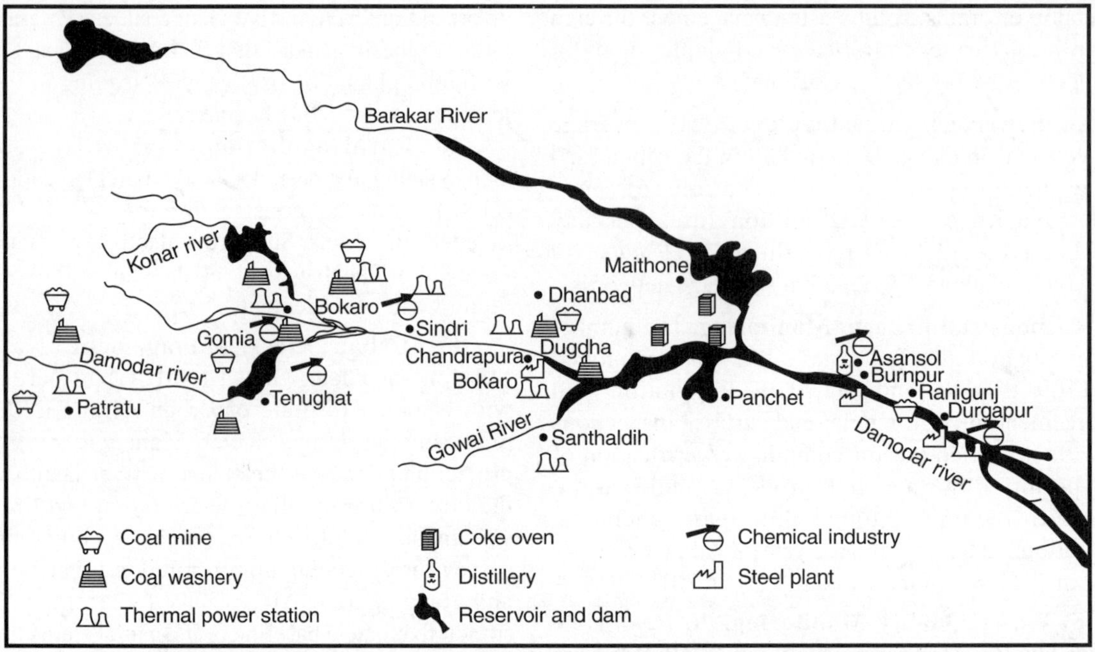

FIG. 8.6. Industries responsible for water pollution in the Damodar River

carried off with surface runoff. At places, this has even blocked the river. At present, many stretches of the Damodar and its tributaries (Konar and Barakar), resemble large drains carrying black, highly turbid water. The total suspended soils (TSS) count at most places is 40 to 50 times higher than the permissible limit. For most parts the river carries a film of oil and grease from industrial effluents. The Damodar basin has about one-third of India's coal reserves and is the most outstanding producer of coal in the country. Consequently over half of the industries of the region comprise coal mining, coal washeries—and coke oven plants. Since coal is easily available, several thermal power plants, steel plants and ancillary industries have come up in the basin.

As many as 50 medium and 100 small scale industries are concentrated in the middle-stretch of the river. A total of about 6,000 Mld of mostly untreated industrial wastewater is discharged into the river. This excludes the wastewater from mine base activities and sewage from towns and cities on the course of the river. Conservative estimates put the daily outfall of pollutants and effluents at 60 tonnes of BOD (Biochemical Oxygen Demand) load, 2 tonnes of non-metallic toxins and 1.2 tonnes of toxic metallic substances. Industrial effluents generally carry high suspended solids in terms of fine coal particles and flyash. According to recent studies, the lowest observed concentration of oil and grease is no less than 2 milligrammes per litre which is 200 times higher than the accepted limit for use as source for public water supply. An estimated 3,100 tonnes of nitrogen, 500 tonnes of phosphorus (from fertilizers) and 30 kg of persistent pesticides reach the river annually, through surface runoff from agriculture.

THE YAMUNA. The Yamuna, the most important and sacred tributary of the Ganga, is also one of the most polluted river of India. Sewage and runoff from industries, towns and vast agricultural tract in Haryana and Uttar Pradesh have been emptied into the Yamuna for the last several years. Still it is the main source of drinking water for millions of inhabitants of Delhi, Faridabad, Mathura, Agra and so many other towns located on its banks. The water quality of the river begins to deteriorate as soon as it enters the plain area after coming down from the mountain area. At every step, fresh water is drawn from the river and large loads of polluted water are poured into it. This results in successive degradation of the river water. The Hindon, Chambal, Sindh, Betwa and Ken join the river at various places in its 1376 km long journey through the plains. From

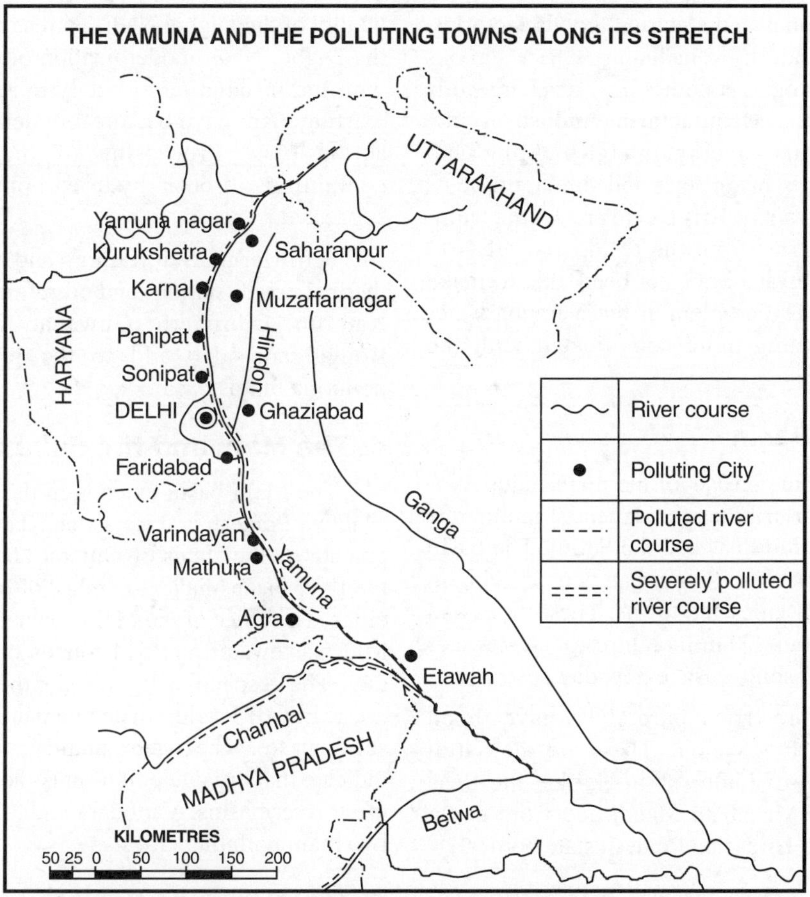

FIG. 8.7. The Yamuna and the polluting towns along its stretch

pollution point of view, the Yamuna's course has been divided into five segments—Himalayan (172 km), Upper (224 km), Delhi (22 km), Eutrophicated (490 km) and Diliated (468 km). Of these, the Upper segment, the Delhi segment and Eutrophicated segment are badly polluted stretches.

Delhi is the largest polluter of the Yamuna river. The Yamuna is comparatively less polluted before entering into Delhi. But as soon as it enters Delhi, the river is polluted at a much faster rate. In Delhi, the Yamuna seems to be *dying a slow death*. Everyday about 3,600 million litres of domestic and 300 million litres of industrial waste water is discharged into the Yamuna through 17 open drains. While Delhi has a capacity to treat 2330 Mld, only 1478 Mld is actually treated. The toxic effluents discharged into the *Yamuna* carry 132 tonnes each of B.O.D. (Biological Oxygen Demand) and suspended solids and 250 tonnes of dissolved solids. Before the Yamuna enters Delhi, 100 millilitres of its water contain only 7,500 disease causing bacteria but after getting polluted by Delhi the same amount of water contains 24 million disease causing bacteria. The Yamuna is rightly described as the ***'Open sewer of Delhi'***. Interestingly, the length and the basin area of the Yamuna in Delhi is only 2 per cent of the total length of the river, but the pollution load contributed by this city is 71 per cent of the total waste water and 55 per cent of total BOD discharged into the river every day. Industrial waste from 20 large, 25 medium and about 93,000 small scale industries located in Delhi also flows through these drains. Though the large and medium industries form only about 0.05 per cent of the total industries located in Delhi, they contributed about 50 per cent of the total industrial waste. These are mostly engineering, textiles, chemicals, electronics and electrical goods factories. Common service industries, including service stations and garages catering to automobiles, discharge oils and other effluents. Manufacturing industries like electroplating units, dyeing, metal-pickling and anodizing units—discharge acids and similar effluents into the open drains. Efforts to re-locate non-conforming industries from the residential areas to alternate site at Bawana have not been able to make much difference. The problem of the Yamuna water in Delhi is becoming more complicated with the passage of time.

2. The Indus Basin

The Indus basin consists of the main Indus river and its five major tributaries *viz.* the Jhelum, the Chenab, the Ravi, the Beas and the Satluj. The basin is the home of over six million urban people living in Class I and Class II cities and towns. These cities and towns generate about 800 million litres of wastewater and 1700 tonnes of solid waste every day.

Three polluted river stretches have been identified in this river system. These are the Satluj from down stream of Ludhiana to Harike, the Beas from upstream of Manali to Mandi and from down stream of Mandi to Himachal Pradesh state border.

3. The Brahmaputra Basin

The Brahmaputra river basin covers large parts of Assam, Arunachal Pradesh, Nagaland, Meghalaya and West Bengal. The basin and catchment area receives heavy rainfall and the Brahmaputra has the highest discharge of all the rivers of India. There are a number of class I and Class II cities in this basin which generate about 233 Mld of wastewater and 414 tonnes of solid waste every day. Guwahati produces the largest amount of wastewater and solid waste of all the towns located in this basin. Although the water quality of the river and its tributaries is not affected much due to enormous dilution available in the river, still there is need for water management.

One of the 22 problem areas identified by CPCB for priority action is Digboi which is located in this basin. It has caused water pollution in the Digboi Nadi which is a sub-tributary of the river Brahmaputra. Oil refinery located at Digboi is primarily responsible for deterioration of the water quality of the Digboi Nadi. To restore water quality of the Digboi Nadi, modernization of the refinery along with the installation of catalytic reforms unit in the existing plant have been recommended as part of the action plan. Processing of oil bearing sludge containing 50-60% hydrocarbons has also been suggested.

Fertilizer units at Kamrup and Namrup as well as thermal power stations and oil refineries in Kokrajhar, Kamrup, Dibrugarh, Guwahati, Numaligarh and Bongaigaon also add to the problem of water pollution in this basin.

4. The Mahi and the Sabarmati Basins

The Mahi basin extends in the states of Gujarat, Madhya Pradesh and Rajasthan. The Sabarmati basin extends in the states of Gujarat and Rajasthan. Both the river basins have a number of class I and class II cities. The cities of the Mahi basin generate about 160 Mld wastewater and 514 tonnes of solid wastes per day. The respective figures for the Sabarmati basin are about 677 Mld wastewater and 1877 tonnes of solid waste. The water quality monitoring results indicate that organic pollution is the major problem in these river basins. Vadodara and Ahmedabad are the two main polluting cities.

5. The Narmada Basin

The Narmada is the largest west flowing river of the peninsular India. The major urban centres in this basin are Jabalpur, Hoshangabad, Dewas and Khandwa in Madhya Pradesh and Bharuch in Gujarat. Class I and Class II cities generate 68.6 Mld of wastewater and about 430 tonnes of solid waste every day. Dhar, Jabalpur, and Bharuch are the main industrial districts. These districts are marked by clusters of pharmaceuticals, pesticides, dyes and distilleries, leather and fertilizer units. In Jabalpur, Khandwa and Hoshangabad, the main industrial activity consists of paper mills.

6. The Tapi Basin

The urban areas of the Tapi basin generate about 293 Mld wastewater and nearly 1650 tonnes of solid wastes per day. The most populous city in the Tapi basin is Surat followed by Amravati and Dhule. Industrialisation has taken place in the East Nimar

(Khandwa) district of Madhya Pradesh and Jalgaon district of Maharashtra. Distillery units contribute the largest share of pollution in Maharashtra whereas textiles occupy the predominant place in Gujarat followed by food and beverages and chemical industries. The Tapi river from Nepanagar to Burhanpur has been identified as a polluted stretch under the National River Action Plan (NRAP).

7. The Mahanadi Basin

The Mahanadi, along with its tributaries like the Seonath, the Kharoon, the Hasdeo, the Mand, the Ib and the Tel forms an important river basin in the north east of the Deccan plateau. Mahanadi river basin is comparatively less polluted. However, certain stretches like downstream portion of river Ib at Brajrajnagar, downstream of Sambalpur and Cuttack have higher degree of pollution. The Ib river is seriously polluted by discharges from the paper mill at Brajrajnagar. A large number of towns on the banks of the main Mahanadi river and its tributaries discharge domestic and industrial wastewater. Korba has been identified as a critically polluted area in this basin. Industrial as well as domestic wastewater is being discharged into the Hasdeo river. Besides iron and steel industry at Bhilai, cement industries at Durg and Raipur, distillery at Raipur, textile industry at Rajnandgaon are the main pollutants.

The Odisha portion of the Mahanadi basin is marked by paper, textiles and thermal power plants at Choudwar, fertilizer and breweries at Paradeep, sugar industry at Nayagarh, Badamba, cement industry at Bargarh, paper industry at Brajrajnagar, coal mining areas of Rampur and Ib valley and an aluminium smelter at Hirakud.

8. The Godavari Basin

The Godavari is the largest river basin of the peninsular India and extends over ten per cent of the total geographical area of the country. It may be mentioned here that inspite of its massive catchment area, the discharge of water is not very impressive due to moderate annual average rainfall in the basin. On account of high concentration of population and industries, towns and cities in this basin discharge large quantities of domestic and industrial waste into the river. It is observed that Class I cities and Class II towns of the basin generate about 761 Mld of wastewater and 2773 tonnes of solid waste every day. Out of the total wastewater only 85.4 Mld is treated.

Amongst the five states which share the Godavari river basin, Odisha is the least industrialised followed by Madhya Pradesh, Chhattisgarh and Karnataka, with Maharashtra having high concentration of industrial pockets. The highest concentration of industries is found in Aurangabad and Nashik in Maharashtra and East and West Godavari districts of Andhra Pradesh. In Maharashtra, sugar and distillery units are in large numbers. This is followed by pharmaceuticals, leather, pulp and paper and pesticide units. In Andhra Pradesh, sugar and distillery units are in large numbers followed by pulp and paper and fertilizer industries. All these industries consume a lot of fresh water and generate large quantities of wastewater.

The river Godavari from downstream of Nashik to Nanded in Maharashtra and upstream of Bhadrachalam at Mancheral and Ramgundam in Andhra Pradesh have been identified as the polluted stretches to be taken up under the National River Action Plan (NRAP). The major sources of pollution in the polluted stretches are from domestic and industrial wastewater generated from cities between Nashik and Nanded cities in Maharashtra and Mancheral, Ramgundam and Bhadrachalam cities in Andhra Pradesh.

9. The Krishna Basin

Though the Krishna is the third largest river of India, yet it has poor water wealth because of low rainfall in its basin. Lying mainly in the Deccan plateau, this basin covers large areas in the states of Maharashtra, Karnataka and Andhra Pradesh. The river has two large tributaries—the Bhima and the Tungabhadra and four smaller tributaries—the Ghataprabha, the Malprabha, the Musi and the Muneru.

The urban areas of this basin generate about 1,404 million litres of wastewater daily (out of which only 204 Mld is treated) and 4,488 tonnes of solid waste every day.

With an annual consumption of about 7,941 tonnes of pesticides, agricultural runoff adds 120 mg/L nitrogen, 26 mg/L phosphorus and 31 mg/L potassium to the river basin. The total domestic pollution load in the Krishna basin is 1,433,084 kg per day—50.6% rural and 49.4% urban. Of the total BOD load generated, Andhra Pradesh accounts for 40.6%, Karnataka 32.7% and Maharashtra 26.7 per cent. Urban BOD load is 707.8 tonnes per day of which domestic is 77.6 per cent and industrial 22.4 per cent.

Several industrial units have come up along the river and the load of industrial waste is quite high in this river. The largest industrial pollution load in Maharashtra part of the basin comes from Pimpri Chinchwada New Corporation part of Pune urban agglomeration. In Karnataka, the Harihar Polyfibres Ltd. at Harihar in Dharwar district abstracts nearly 35,000 cumec of water per day from the Tungabhadra and discharges about 33,000 cumec per day of wastewater into the river. Bhadra river is badly polluted by Mysore Paper Mills and Visveswaraya Iron and Steel Limited.

10. The Cauvery Basin

With almost 95 per cent abstraction of water, the Cauvery is one of its most exploited rivers of India. The Cauvery has watered agricultural lands of Tamil Nadu, Karnataka and Kerala. Sharing of the Cauvery waters has been the bone of contention between the two contending states, *viz.,* Tamil Nadu and Karnataka. High rate of application of chemical fertilizers and insecticides in agricultural field results in discharge of large quantities of chemicals in the river water, thereby polluting the river to a great extent. Coffee production in the districts of Coorg, Hassan and Chikkamagaluru contributes heavily to the BOD level of the river water which ranges from 2,000-4,000 mg/L. Coffee industry contributes 4,730 tonnes of BOD load in each season.

The urban areas in this basin generate about 750 million litres of wastewater and 3,500 tonnes of solid waste every day. Important cities are in Karnataka (Bengaluru, Hassan, Mandya and Mysuru) and in Tamil Nadu (Salem, Erode, Coonoor, Tiruppur, Valparai, Dindigul, Tiruchchirapalli, Kumbakonam, Thanjavur and Coimbatore). The industrial activity is very high in this basin, particularly in the Bengaluru area in Karnataka and Mettur as well as Coimbatore in Tamil Nadu, followed by districts of Mysore and Mandya in Karnataka and Periyar and Salem in Tamil Nadu. A total of 61 industries in Karnataka and 1,139 in Tamil Nadu contribute heavily to the pollution load, especially during the lean flow period of the year. In Tamil Nadu, some water-intensive industries such as textiles, sugar mills, paper mills, chemical units, engineering units, tanneries, etc., have come up which contribute to the pollution load of the river.

Noyyal is one of the most polluted tributaries of the Cauvery river. It is polluted by 800 odd dyeing and bleaching units located at Tiruppur. This 173 km long river carries untreated sewage and industrial effluents from the towns of Coimbatore and Tiruppur for most part of the year.

GROUNDWATER QUALITY

The term 'groundwater' is defined as the water that occurs below the surface of the earth. The source of groundwater is precipitation and infiltration of surface water from runoff and streams. In a country of monsoon climate like India, where rainfall is seasonal, erratic, undependable and variable and where the surface water is always in short supply, groundwater plays a vital role in everyday life of the masses. Most of the drought prone areas largely depend on groundwater.

Sources and Types of Groundwater Contaminants

Groundwater contamination leads to several problems like taste, colour, hardness and foaming. The problem of groundwater contamination becomes serious when pathogenic organisms, flammable or explosive substances or toxic chemicals are present. Once polluted, groundwater may remain unusable or even in hazardous condition for decades or even centuries. It is often difficult to identify the nature and sources of groundwater pollutants and water quality problems. Normally speaking, the quality of groundwater is affected by waste disposal and land use. The water-soluble substances which are dumped, spilled, spread or stored on the land surface eventually may infiltrate into the ground water. The disposal of fluids through wells and sinkholes directly into aquifers also results in contamination of the groundwater. Infiltration of contaminated water also causes groundwater contamination.

Arsenic Contaminated Water

Arsenic contamination of water occurs when water passes through arsenic rich rocks. It may also occur due to mining and industrial activities.

About 20% of India's population across 153 districts in 20 states and Union Territories of India drink water that contains unacceptably high levels of arsenic. In effect, they are being slowly poisoned.

About 65% of Assam's population, is drinking arsenic contaminated water, while it is 60.7% in Bihar, and 44.3% in West Bengal. In terms of absolute numbers, however, Uttar Pradesh has the largest number of people, over 70 million exposed to arsenic-contaminated water. This is followed by Bihar, West Bengal and Assam.

The above description presents a gloomy picture indeed. The World Health Organisation (WHO) has warned that long-term intake of such water leads to arsenic poisoning or arsenicosis, with cancer of skin, bladder, kidney or lung or diseases of skin (colour changes, and hard patches on palms and soles), or blood vessels of legs and feet. Fresh evidence indicates a possible association between intake of contaminated water and onset of diabetes, hypertension, and reproductive disorders states a WHO document.

WHO's guideline value for acceptable levels of arsenic in drinking water is 0.01 mg/litre. In 153 districts of 20 states and UTs, the arsenic level in groundwater is above the benchmark.

SOIL OR LAND POLLUTION

Soil is a very important environmental attribute because it supports all sorts of plant life found on land. Soil becomes polluted due to the *misdeeds of man* or at times the *mischiefs of nature*. The main factors of soil pollution are the high state of soil erosion, excessive use of chemical fertilizers, biocides (pesticides, insecticides and herbicides), polluted liquids and solids from urban and industrial areas, forest fires, water-logging and related capillary processes, leaching, drought, etc. Some of the micro-organisms and unwanted plants enter the soil and result in soil pollution. Some of the air-borne pollutants from the industries are deposited on the land surface and pollute the soil. Solid particles from mining areas pollute the neighbouring land to a great extent rendering it unsuitable for agriculture. Many areas near mica and manganese mines in Jharkhand have fallen prey to this type of pollution. Soils near copper smelting units are so polluted that no plant growth is possible there. Main sources of land pollution are briefly described as under :

1. Chemical Fertilizers and Biocides. The accelerated use of chemical fertilizers and biocides in agriculture is the major cause of soil pollution. They are used to increase the yields and to save the crops from insects, pests and unwanted plant growth. It should be particularly noted that biocides first kill germs and unwanted plants and then degrade the quality of soil. Among the pesticides, the most widely used are the chlorinated hydrocarbons, *e.g.* D.D.T., B.H.C., endrin, aldrin, dieldrin and lindane and organophosphorus compounds such as parathion and malathion. When these are used in excess, their remnants are absorbed by soil particles and contaminate crops grown in such soils. They are further transferred into carnivores through herbivores and finally enter the human bodies in course of food chains. They are responsible for several incurable diseases and even cause death. Biocides are, thus, called as ***creeping deaths.***

The use of biocides gained momentum in India with the commencement of Green Revolution in 1966-67. The introduction of high yielding varieties (HYV) of seeds gave birth to heavy doses of chemical fertilizers and biocides. The overall per hectare consumption of fertilizers rose from a mere 0.55 kg in 1950-51 to 67 kg in 1992-93 and to 141.43 kg in 2018-19.

For reducing the impact of chemical fertilizers, it is suggested that use of organic manures, composts, and agricultural wastes should be encouraged. Composition of NPK in different farmyard manures and composts is given in **Table 8.15**. Organic farming refers to farming which does not use any form of chemical fertilizers or other agro chemicals and is dependent entirely on organic sources of crop nutrition and crop husbandry. Organic farming can also be defined as a system in which the maintenance of soil fertility and the control of pests and diseases are achieved through the enhancement of biological

TABLE 8.15. Composition of NPK in different Farmyard Manure and Compost (in percentage)

Waste	Nitrogen (N)	Phosphorus (P)	Potassium (K)	Total Nutrient
Rural compost	0.75	0.5	0.5	1.75
Urban compost	1.00	1.00	1.00	3.00
Neem cake	5.20	1.00	1.40	7.60
Farmyard Manure	0.60	0.20	0.60	1.40
Poultry droppings	3.00	2.60	1.40	7.00

Source : Environmental Atlas, (June 2001), Central Pollution Control Board, p. 97.

processes and ecological interaction. The major component of organic farming is the maintenance of soil fertility through maximizing nutrient recycling and minimizing losses. Organic farming also helps in improving the physical properties, microbial production and humus content of soil while increasing its water holding capacity.

2. Municipal Solid Waste. Municipal solid waste (MSW) is a heterogeneous mixture of various constituents. According to Scavenging and Cleaning Act, MSW includes :

- dust, ashes, refuse and rubbish.
- trade refuse.
- carcasses of dead animals and other matter.
- sweepings, sand stones, leaves and other dead vegetation.
- wastes from shops and market areas including paper, straw and cardboard packing, decaying fruits and vegetables and other described items; and
- other solid wastes generated from establishments such as hospitals, schools, offices and small cottage industries.

Municipal Solid Waste Generation in India

The quantity of municipal solid waste generated by a city depends upon a number of factors of which size of population, standard of living, food habits, scale of commercial and industrial activities, etc. are of primary importance. As such the per capita generation of solid waste differs greatly from one city to another. It varies from 200 to 500 gm/person/day with an average of 376 gm/person/day. With the passage of time, the level of urbanisation and standard of living of the urban people is likely to improve which will result in increased generation of total solid waste as well as per capita solid waste.

According to Ministry of Environment, Forest and Climate Change (MOEFCC), India generated 53 million tonnes of solid waste in 2016. This is likely to increase to 165 million tonnes in 2030 and 436 million tonnes in 2050. But the World Bank Report of 2018 says that India generated about 277.1 million tonnes of solid waste which is more than five times the estimates made by MOEFCC. This is about 80% of the 334 million tonnes of waste generated across South Asia and about 13% of the global 2.01 billion tonnes of waste generated every year. According to this report the solid waste generation in India may reach 387.8 million tonnes in 2030 and 543.3 million tonnes in 2050 (**Fig. 8.8**).

There are large scale regional variations with respect to solid waste generation in India. According to data released by Statista, 2021. Maharashtra is the largest producer of solid waste followed by Uttar Pradesh, West Bengal, Tamil Nadu, Karnataka and Delhi. Among the states, Sikkim generates the minimum quality of solid waste while among the Union Territories, Lakshadweep produces the least (**Table 8.16**).

Disposal of such large quantities of garbage is a difficult job. The dumping grounds of such garbage are invariably polluted. It is estimated that about 25 human diseases are associated with solid wastes. Rats and flies flourish on garbage heaps. Rats are carriers of insects and other bio-organisms and are responsible for spreading deadly diseases like *plague, cholera*, etc. The flies which carry pathogenic organisms cause diseases such as dysentery, diarrhoea, etc. Studies reveal that an approximate of 70,000 flies can flourish on one cubic foot of garbage.

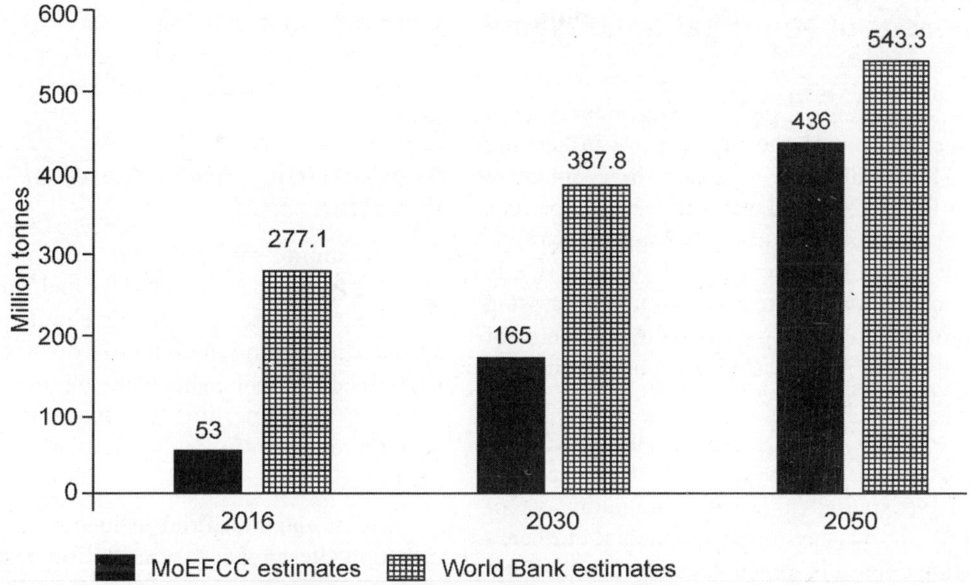

FIG. 8.8. Solid waste generation in India

Source: CSE, 2020: Compiled from MoEFCC and World Bank estimates.

TABLE 8.16. Distribution of Municipal solid waste in India during the financial year 2019

State/Union Territory	Municipal Solid waste generated (Metric tonnes/day)	State/Union Territory	Municipal Solid waste generated (Metric tonnes/day)
1. Maharashtra	23,844.6	19. Uttarakhand	1,527.5
2. Uttar Pradesh	17,377.3	20. Assam	1,293.7
3. West Bengal	14,613.3	21. Puducheri	599.3
4. Tamil Nadu	13,968.0	22. Chandigarh	470.0
5. Karnataka	11,958.0	23. Tripura	445.7
6. Delhi	10,817.0	24. Himachal Pradesh	389.0
7. Telangana	8,497.0	25. Nagaland	339.5
8. Madhya Pradesh	8,000.0	26. Arunachal Pradesh	271.0
9. Rajasthan	6,625.6	27. Mizoram	251.4
10. Andhra Pradesh	6,440.0	28. Goa	236.4
11. Haryana	4,635.8	29. Manipur	218.6
12. Punjab	4,634.5	30. Meghalaya	170.6
13. Kerala	3,903.0	31. Andaman and Nicobar Islands	120.0
14. Odisha	2,564.4	32. Daman Diu and Dadra Nagar Haveli	98.0
15. Bihar	2,272.0		
16. Jharkhand	2,205.0	33. Sikkim	75.1
17. Chhattisgarh	1,650.0	34. Lakshadweep	35.0
18. Jammu-Kashmir	1,530.0		

Source: Statistics 2021.

Management of Municipal Solid Waste (MSW)

In India, recently solid waste management systems are assuming larger dimensions in keeping with the Municipal Solid Wastes (Management & Handling) Rules, 2000. Many of the municipalities are taking appropriate actions to improve various component systems like collection of solid waste from generation areas, its transportation to processing and disposal site(s), utilising the recycling potential of Municipal Solid Waste (MSW) and ultimately disposing of by landfilling.

In view of this, under the sponsorship of Central Pollution Control Board (CPCB), NEERI has carried out extensive studies to assess the pathways of pollution for various environmental media. Further, a site selection criteria has been developed in the form of guidelines to suit Indian conditions in keeping with the findings of the other studies.

The urban solid waste management is an essential municipal service for protection of environment and health of citizens. All the citizens, industries, hospitals and NGOs should co-operate with the Municipal authorities to ensure safe management of urban solid waste. Segregation of inorganic recyclable materials like plastic, glass, metal, paper, etc. at the source should be promoted and every effort should be made to provide collection of these in separate containers or bags in each house. Solid waste should be collected from each house on a daily basis and transported to the disposal site. Direct transfer of garbage from primary collection carts to covered transportation vehicles should be encouraged.

Refuse from vegetable and fruit markets should be collected and transported to the composting facilities. Large restaurants/hotels should be encouraged to develop their own onsite treatment facilities.

Sanitary landfills would be the main option for disposal of urban solid waste. The concerned authorities must make adequate provision of land allocation for scientific landfill sites.

The urban solid wastes of Indian cities have low calorific value and high moisture content with high percentage of non-combustible materials and hence it is unsuitable for thermal technologies. However, application of technologies such as incineration, palletisation and pyrolysis-gasification should be evaluated through research and development and pilot scale studies.

Prevention and Control of Soil Pollution

Maximum soil pollution is done by use of biocides (pesticides, insecticides and herbicides) and chemical fertilizers. Farmers should be properly educated to make judicious use of these chemicals. Immediate restriction should be imposed on the use of DDT. At the same time, efforts should be made to develop less harmful and more useful chemicals for use in agriculture.

Urban and industrial effluents can be used for irrigation after proper treatment. Compostable organic substances such as vegetables, plant leaves, animal and human wastes should be properly composted, so that there is proper disposal of the solid waste and it can also produce organic manure. According to an estimate, a town with one lakh population may produce 20,000 tonnes of garbage and 8,000 tonnes of night soil per day which can be converted into 18,000 tonnes of compost (organic manure). Today, plastics are proving to be a big environmental hazard because they are practically indestructable. It takes hundreds of years for plastics to disintegrate, adding to solid waste buildup, especially in urban and industrial areas. There is an urgent need to restrict the use of plastics and to find out ways and means of recycling them.

Solid Waste Management Campaigns

Swachh Survekshan was launched in 2015 to foster a healthy competition among cities for improving cleanliness standards. Ministry of Housing and Urban Affairs (MAHUA) started "Swachh Survekshan" which ranks cities on cleanliness and other aspects of urban sanitation. In 2016-17, 434 towns and cities were ranked on solid waste management (SWM) and sanitation. Many more cities would be covered under the campaign in time to come. Ranking system has pushed cities to take better initiatives towards segregation and waste management. A "Segregation at Source" campaign was launched in 2017 where 4,041 statutory towns and cities were part of campaign which pushes for segregation at source.

Solid Waste Management Rules, 2016. The Central Government revised the rules for management of solid waste in the municipal areas after 16 years. The new rules are now applicable beyond municipal areas and extend to urban agglomerations; census towns, notified industrial townships; areas under the control of Indian Railways; airports; airbase, ports and harbours, defence establishments; special economic zones; state and central government organisations, places of pilgrimage, religious and historical importance. The segregation at source has a been mandated. Responsibilities of generators have been fixed to segregate the waste into three streams. These streams are : (1) Wet (biodegradable), (2) dry (plastic, paper, metal, wood, etc.) and (3) ***domestic hazardous wastes*** (diapers, napkins, empty containers of cleaning agents, mosquito repellents, etc.)

Plastic Waste

Plastic waste has become a very serious environmental problem because of its non-degradable nature. As per CPCB, India produces close to 25,940 tonnes of plastic waste each day of which nearly 15,000 tonnes is recycled. However, per capita consumption of plastic in India is only 11 kg per annum which is much lower than global consumption of 28 per capita per annum and much lower than 65 kg in European Union and 139 kg in U.S.A. According to the claim by the Central Government, the plastic recycling rate in India is relatively high at 60% as compared to global recycling rate of only 14 per cent. Considering 42 per cent of the total plastic waste as packaging share, 10,894 tonnes of packaging waste is generated every day. With a 60 per cent recycling capacity, it can be inferred that around 4,500 tonnes of plastic waste ends up littered or in landfalls. The uncollected plastic waste causes choking of drainage and river systems, littering of marine ecosystem, soil and waste pollution, ingestion by stray animals, and open air burning leading to adverse effect on human health and environment.

Although India's plastic waste production is not as large as that of some developed/developing countries, India faces a ticklish problem of managing and disposing off this waste. In order to solve this problem Ministry of Environment; Forest and Climate Change introduced *Plastic Waste Management Rules, 2016*. Following are the provisions of Rules, 2016.

(*i*) Increase in minimum thickness of plastic carry bags from 40 to 50 microns.

(*ii*) First time cover and stipulate minimum thickness of 50 microns for plastic sheets being used for packaging and wrapping commodities to facilitate and recycle of plastic waste.

(*iii*) Revamped pricing mechanisms for plastic carry bags by introducing plastic waste management fee to be paid by retailers/street vendors willing to provide carry bags as pre-registration charge.

(*iv*) Ways and means to promote gainful utilization of plastic waste such as energy recovery in road construction, etc.

(*v*) Introducing user charge and spot fines by the local authority.

It is worth noting that one-sixth of the total plastic waste is generated by 60 major cities with Delhi, Chennai, Kolkata, Mumbai and Bengaluru together generating more than 50% of the total contribution from these cities. According to a survey of 60 cities conducted by the CPCB in 2019, it was found that these 60 cities generate a total 4059 tonnes of plastic waste every day. Contribution of ten top cities is depicted in **Fig. 8.9**.

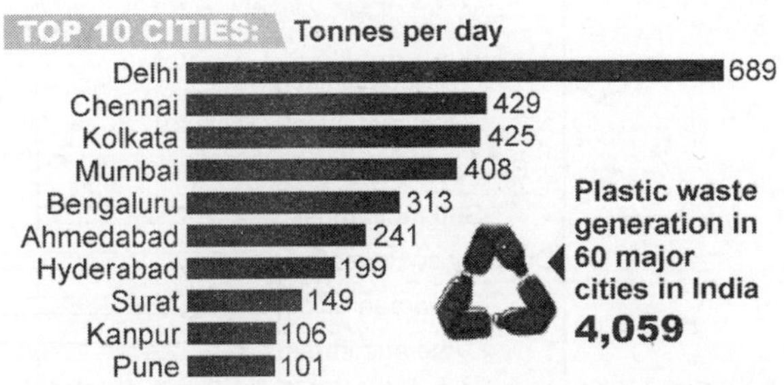

FIG. 8.9. Ten top plastic waste generating cities.

NOISE POLLUTION

Noise is unwanted sound and covers all sounds which can result in hearing impairment or are harmful to health, or otherwise dangerous. According to **K.E.**

Maxwell, "Noise is any sound that is not wanted. It is one of the more common forms of atmospheric pollution." *Noise pollution may be defined as the state of discomfort and restlessness caused to humans by unwanted high intensity sound*. More and more noise is created as the modern civilization moves ahead. It has now become a major environmental pollutant especially in urban areas. In 1972, the UN Environment Conference at Stockholm, noise pollution has been accepted as a problem, which needs proper control.

Measurement of Sound/Noise and Its Intensity

The most popular measure of noise level is the *decibel* measured by an instrument known as ***decibel meter***. A sound between 0 and 1 decibel is about the weakest that the average human can hear. For testing purposes, 0 decibel is considered to be the threshold of hearing. A whisper is about 20 decibels and an average speaking voice is about 60 decibels. The loudest sound that a person can stand without discomfort is about 80 decibels. Automobile horns may reach 90 decibels and a jet airplane may have an intensity of about 140 decibels. The levels of common noise are given in **Table 8.17** and has also been depicted in **Fig. 8.10**.

Actually, loudness alone is not the sole cause of noise problem, although this is the most apparent characteristic of noise that creates environmental problems. Pitch of the noise also influences the

TABLE 8.17. Intensity of Various Sounds

Sound Source	Intensity (in decibels)	Response Criteria
	0 –10	Just audible
Broadcasting studio	20	—
Soft whisper	30	very quiet
Library, bedroom slow radio	30–50	quiet
Light auto traffic, air conditioning unit	50 –70	intrusive
Freeway traffic, freight tram	70–80	annoying
Motor cycle, heavy traffic	80–100	very annoying; hearing damage
Rail noise	110-130	intolerable
Lighting-thunder	120	intolerable
Carrier deck jet operation	140–150	painfully loud

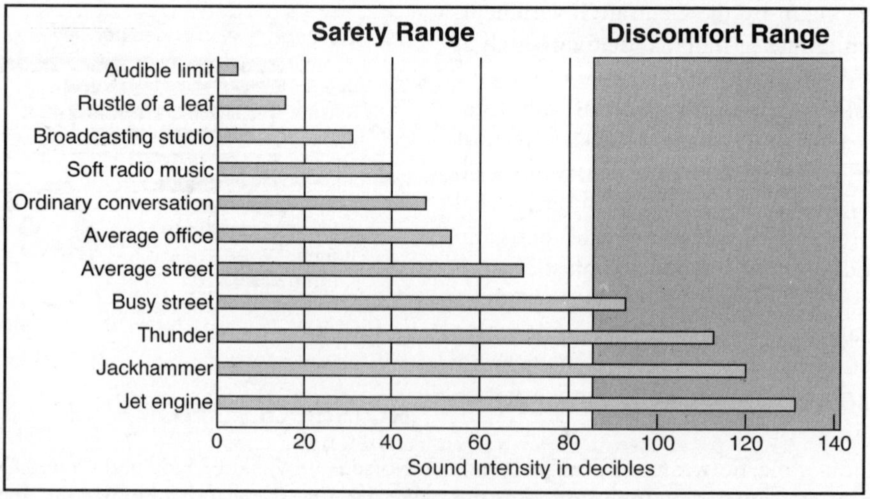

FIG. 8.10. Levels of Common Noise (After T. Berland)

degree of annoyance—the higher the pitch, the greater the annoyance. Similarly, duration of the noise also determines how annoying or irritating noise will be. The degree of noise in an enclosed space or indoors is different from the level of outdoor noise.

TABLE 8.18. Permissible Noise Limit in dB

Zone/Area	Day time (from 6 AM to 10 PM	Night time (from 10 PM to 6 AM)
Silent Zone	50	40
Commercial area	65	55
Residential area	55	44
Industrial area	75	70

Source : World Health Organisation (WHO).

NOISE POLLUTION IN INDIA

Noise pollution in India has increased considerably with the increase in urbanization and industrialization. In most of the Indian cities, early morning tranquillity is shattered by automobiles, factory machines and loudspeakers from religious places. Television and radio sets played at very high pitch, marriages and other processions, festivals, cultural programmes and a host of other sources of noise add to the problem of noise pollution. Most of the big cities of India have noise pollution much higher than the permissible limit of 60-70 dB. Delhi (89 dB), Kolkata (87 dB), Mumbai (85 dB), Chennai (82 dB), Kochi (80 dB), Madurai (75 dB) and Nagpur (75 dB) are some of the examples of cities suffering from noise pollution.

Effects of Noise Pollution

Continuous and prolonged exposure to noise pollution leads to many disorders and ailments. The most immediate and acute is the impairment of hearing. It also results in mental tension, blood pressure, heart diseases and stomach trouble. Noise pollution also causes annoyance, irritation and fatigue which result in low efficiency and high rate of errors. High and sudden noise may lead to abortion in early stages of pregnancy. According to Kudesia, "It is possible that due to disco dance and rock 'n' roll music we are raising a nation of teenagers who will be hard of hearing before they reach the age of 35 years. Those who listen songs from radio on loud pitch fall in the same line". A study to measure the Noise Induced Hearing Loss (NIHL) among the factory workers in the textiles, automobiles, fertilizers and chemical industries in Chennai, Coimbatore, Madurai, Kochi and Thiruvananthapuram revealed that about 25% of the factory workers and 10% people such as traffic constables and pavement vendors suffer from NIHL. About 60% of the students in the age group of 5-10 years living in industrial areas, railway colonies and other areas of high noise level cannot concentrate on their studies.

Prevention and Control of Noise Pollution

It has been observed that noise pollution, like other forms of pollution, has become a serious problem. Since noise pollution is created by man, it can be controlled by adopting certain measures. Some of the measures are as follows :

(*i*) Noise producing industries should be located away from residential areas.

(*ii*) Inside industries, proper arrangements to minimise noise be made by constructing sound proof walls and also to provide such instruments to workers which can protect their ears from noise.

(*iii*) Much of the industrial noise can be reduced by replacement of old noise producing machines with quieter alternatives, proper maintenance, greasing and oiling of the machines.

(*iv*) The automobile horns should be designed in such a way that the noise they produce may not be harmful.

(*v*) Use of horns should be minimum and pressure horn should be banned as has been done in many countries.

(*vi*) Noise pollution by road transport can be reduced by quieter engines, enforcing speed limits and laying the transport routes away from the residential areas such as by-passes. Tree plantation along the road side reduces noise by 10 to 15 dB.

(*vii*) The noise created by railways can be checked by construction of ballastless rail tracks.

(*viii*) Special arrangements be made to check noise near air bases. Control of aircraft noise requires several changes, which should be made.

(*ix*) Intensity of noise can be reduced to a great extent by proper acoustic designing of cinema, dance and community halls, as well as religious places and other buildings of high noise intensity.

(*x*) The noise created by musical instruments, and other indoor equipments can be checked by individuals in their own interest.

(*xi*) Effect of noise on the ear can be reduced by 40 to 50 dB by putting ear-plugs or earmuffs.

REFERENCES

Agarwal, A.L. (1993) : Air Quality Management Goals in India, Paper presented at International Workshop on Urban Air Quality Management, Metropolitan Environmental Improvement Programme, Mumbai, April, *mimeo.*

Agarwal, Anil (ed) (1997) : *Homicide By Pesticides,* Centre for Science and Environment, New Delhi, February.

Agarwal, Anil; Narain, Sunita and Sen, Srabani (eds.) (2002) : *The Citizens' Fifth Report, State of India's Environment,* Centre for Science and Environment, New Delhi.

Ahluwalia, V.K. (2019) : Energy and Environment. The Energy and Resources Institute (TERI), N. Delhi.

Anon (1996) : Air Quality Assessment in *Urban Air Quality Management in Asia, Greater Mumbai Report,* International Bank of Reconstruction and Development, Washington DC.

Anon (1997) : *MIS Report of Programmes under National River Conservation Directorate,* Vol. I & II, Ministry of Environment and Forests, Govt. of India, New Delhi.

Anon (1997) : Polluted River, *Down to Earth,* Society for Environmental Communications, New Delhi, December 31, Vol. 6, No. 15.

Aziz, Abdul (1992) : *Cities and Environment, in* Raza Mehdi (Ed.) *Development and Ecology*, Rawat Publications, Jaipur.

Bagchi, S. (1981) : Dust pollution with special reference to manganese and mica ores. In A.K. Sharma and Archana Sharma (eds.). *Impact of Development of Science and Technology on Environment* Ind. Sci. Cong. Assci. Calcutta, pp. 132-134.

Bandopadhyaya, J.S. (1982) : *India's Environment Crisis and Response,* Natraj, Dehradun.

Banerjee, Brojendra Nath (1989) : *Can Ganga be Cleaned*, B.R. Publishing Corporation, New Delhi.

Barlelmus, Peter (1986) : *Environment and Development*, Allen and Unwin, London.

Bharat, Dogra (1981) : *Protecting Ganga River : Linking reversence to effective Action,* Rural Litigation and Entitlement Kendra, Dehradun.

Botkiri, D.B. and Keller, E.A. (1982) : *Environmental Studies,* C.E. Merrill Publishing Company, A Bell and Howell Company, Columbus.

Buch, M.N. (1985) : *Crisis in Urban Ecosystems and the Lessons of Bhopal,* Natraj Publishers.

Central Pollution Control Board (CPCB), Delhi (1995) : *Action Points for Pollution Control in Problem Areas.*

CPCB Delhi (2000) : *Air Quality Status and Trends in India.*

CPCB Delhi (2000) : *Ambient Noise Level Status in Delhi During 1995 and 1999.*

CPCB Delhi (1995) : *An Inventory of Major Polluting Industries in Ganga Basin and their Pollution Control Status.*

CPCB Delhi (2003) : *Assessment of Industrial Pollution.*

CPCB Delhi (1999) : *Control of Noise Pollution from Diesel Generator Sets.*

CPCB Delhi (2001) : *Criteria for Hazardous Waste Landfills.*

CPCB Delhi (2001) : *Development of Emission Factors for Cement Industries.*

CPCB Delhi (2003) : *Development of Site Selection—Methodology for Landfilling.*

CPCB Delhi (2001) : *Environmental Atlas of India.*

CPCB Delhi (1994) : *Environmental Auditing in Polluting Industries.*

CPCB Delhi (2003) : *Environmental Management in Selected Industrial Sectors.*

CPCB Delhi (1996) : *Groundwater Quality in Problem Areas—A Status Report (Part I-IV).*

CPCB Delhi (1995) : *Implementation Status of the Pollution Control Programme in Major Polluting Industries.*

CPCB Delhi (2003) : *Municipal Sewage Pollution Along Indian Coastal Waters.*

CPCB Delhi (2003) : *Municipal Solid Waste Processing Technologies Review and Guidelines for Local Bodies.*

CPCB Delhi (2004) : *National Ambient Air Quality Status.*

CPCB (2012) : *National Ambient Air Quality Status and Trends in India, 2010.*

CPCB Delhi (2002) : *Oil Pollution and Marine Environment.*

CPCB Delhi (2001) : *Pollution Control in Small Scale Industries.*

CPCB Delhi (1985) : *River Basins Atlas.*

CPCB Delhi (1995) : *Status of Industrial Pollution Control Programme (Phase-1).*

CPCB Delhi (2000) : *Status of Municipal Solid Waste Generation, Collection, Treatment and Disposal in Class-I Cities.*

CPCB Delhi (2000) : *Status of Solid Waste Generation, Collection, Treatment and Disposal in Metrocities.*

CPCB Delhi (2000) : *Status of Solid Waste Generation Collection, Treatment and Disposal in Class-II Towns.*

CPCB Delhi (1987) : *Status of Water Quality of Some Rivers of India Monitored Under Global Environmental Monitoring Systems.*

CPCB (2011) : *Status of Water Quality in India.*

CPCB Delhi (2000) : *Status of Water Supply and Wastewater Generation, Collection, Treatment and Disposal in Class-II Towns.*

CPCB Delhi (2001) : *Transport Fuel Quality for year 2005.*

CPCB Delhi (1983) : *Water Pollution Control—An Overview.*

CPCB Delhi (1994) : *Water Quality Atlas of India.*

CPCB Delhi (2002) : *Water Quality—Criteria and Goals.*

CPCB Delhi (2003) : *Water Quality in India—Status and Trends.*

CPCB Delhi (1997) : *Zoning Atlas for Siting of Industries (Based on Environmental Considerations). The Conceptual Framework.*

Centre for Development of Environment and Resources (CENDER); Patna (1999) : *India's Environment.*

Centre for Science and Environment (2017, 2018, 2019, 2020, 2021) : *State of India's Environment.*

Centre for Science and Environment (2019) State of India's Environment 2019 in Figures.

Chandola, R.P. (1995) : *Environment and Life*, RBSA Publishers, Jaipur.

Chaudhari, A.B. (1993) : *Tree and the Environment : An Indian Scenario*, Ashish Publishing House, New Delhi.

Dasseman, R.F. (1976) : *Environmental Conservation,* John Wiley, New York.

Desh Bandhu (1981) : Environmental Situation in India. In Desh Bandhu (Ed.) *Environmental Management,* Indian Environmental Society, New Delhi, pp. 29-43.

Detweyler, T.R. (1971) : *Man's Impact on Environment,* McGraw Hill, New York.

Dikshit, R.D. (1984) : *Geography and Teaching of the Environment,* Department of Geography, Pune University, pp. 68-83.

Economic Survey 2018-19 : Vol. 2, pp. 108-12.

Futchelly, Laeeq (1994) : *Our Environment,* National Book Trust, India, New Delhi.

Government of India; EnviStats of India 2018

Gregory, K.J. and Willing, D.E. (1981) : *Man and Environmental Processes,* Butterworths, London.

Grossman, L. (1977) : *Man-environment relationship in anthropology and geography,* Annals of the Assoc. American Geog. (AAAG), Vol. 67, pp. 126-44.

India 2022, *A Reference Annual*, Publication Division, Ministry of I&B., Govt. of India, New Delhi.

Jadhav, H.V. (1995) : *Environmental Pollution*, Himalaya Publishing House, Mumbai.

Kassas, M, (1994) : *Global Environment: Tasks for the 21st Century,* in Shafi Mohammad *and* Raza Mehdi (Eds.) *Geography of Environment*, Rawat Publications, Jaipur.

Kayastha, S.L. (1992) : *Environment and Development: Emerging Perspectives and Imperatives* in Raza Mehdi (Ed.) *Development and Ecology*, Rawat Publications, Jaipur.

Khosboo, T.N. and Ahmad K.T. (1981) : Air Pollution and Plants. In A.K. Sharma and Archana Sharma (Eds.) *Impact of the Development of Science and Technology on Environment.* Ind. Sci. Cong. Assoc. Calcutta, pp. 78-94.

Khullar, D.R. and Roa, J.A.C.S. (2016) : *Environment for Civil Services (Prelims & Mains)*, McGraw Hill Education, New Delhi.

Kudesia, V.P. (1995) : *Pollution (Everywhere)*, Pragati Prakashan, Meerut.

Kumra, V.K. (1982) : *Kanur City : A Study in Environmental Pollution,* Tara Book Agency, Varanasi.

Manivasakam, N. (1984) : *Environmental Pollution*, National Book Trust, New Delhi, India.

Miller, G.T. and Armstrong, P. (1982) : *Living in the Environment,* Wordsworth International, Belmont, California.

Misra, R.C. (1985) : Environmental Problems created by exploration of mineral resources in the Himalaya. In Singh, J.S. (Ed.) *Environmental Regeneration in Himalaya,* Cen. Him. Env. Assoc. Nainital, pp. 155-168.

Nadkarni, M.V. (1994) : *Agriculture and Environment,* in Rathore, M.S. (Ed.), *Environment and Development* (1996), Rawat Publications, Jaipur.

Roychaudhary, Anumita (1996) : *Slow Murder : The Deadly Story of Vehicular Pollution in India,* Centre for Science and Environment, New Delhi, November, p. 29.

Saxena, H.M. (2017) : *Environmental Geography,* Rawat Publications, Jaipur.

Shafi Mohammad (1994) : *Sustainable Agriculture and Environmental Security in India,* in Mohammad Shafi *et al.* (eds.) *Geography of Environment*, Rawat Publications, Jaipur.

Singh, H.H., Nag, Kumara, V.K. *et al.* (eds.) (1986) : *Geography and Environment* : *Issues and Challenges*, Concept Publishing Company, New Delhi.

Singh, Savindra (1991) : *Environmental Geography*, Prayag Pustak Bhawan, Allahabad.

Sinha, Upendra Kumar. (1986) : *Ganga : Pollution and Health Hazard*, Inter India Publications, New Delhi.

Strahler, A.N. and Strahler A.H. (1976) : *Geography and Man's Environment,* John Wiley, New York.

Sundaresan, B.B. (1982) : Waste Management, *in State of Environment—Some Aspects,* National Committee on Environmental Planning, Department of Environment, New Delhi, pp. 1-41.

The Hindu : *Survey of Environment,* 2014.

Tolba, M.K. (1983) : *Development without Destruction,* Tycooly, Dublin.

Vyas Vijay, S. and Reddy V. Ratna (1993) : *Environmental Policies and Policy Implementation in India*: *An Assessment,* in Rathor M.S. (1996), *Environment and Development,* Rawat Publications, Jaipur.

WHO—UNEP (1992) : *Urban Air Quality in Megacities of the World,* Backwell Publishers.

Wilkinson, H. (1963) : *Man and Natural Environment,* Department of Geography, University of Hull. Occasional Papers in Geography, No. 1.

◆◇◆

CHAPTER 9

Natural Hazards and Disasters

Man has been facing natural hazards and disasters since the dawn of civilisation. The earth has been undergoing various changes; both slow and catastrophic from the very beginning. It is worth understanding some of the basic terms before we make a detailed study of the natural hazards and disasters and their effects on human beings.

HAZARDS

Changes that affect humans adversely are called ***hazards***. A hazard comprises a dangerous condition or event, natural or manmade, that could cause injury, loss of life or damage to property, livelihood or environment. In other words, hazards are defined as "***phenomena that pose a threat to people, structures or economic assets and which may cause a disaster. They could be either manmade or naturally occurring in our environment.***" (Disaster Preparedness Training Manual, Philippine National Red Cross 1954). A *natural disaster* pertains to a natural phenomenon which occurs in proximity and poses a threat to people, structures and economic assets caused by biological, geological, seismic, hydrological or meteorological conditions or processes in the natural environment.

TYPES OF HAZARDS

The list of hazards is very long and hazards can be classified in a number of ways. Generally hazards are classified depending upon their causes, mitigation strategies or their effects on societies. We shall confine ourselves to classification of disasters depending upon their causes only.

Types of Hazards depending upon their causes

Hazards could be caused by nature or man and are consequently called natural hazards and manmade hazards. A third type of hazards are caused both by nature and man and are called ***socio-natural hazards***.

1. Natural Hazards. These are caused by the forces of nature and man has no role to play in such hazards. Following are the main types of natural hazards :

- Earthquakes
- Volcanic eruptions

- Cyclonic storms
- Tsunamis
- Floods
- Droughts
- Landslides
- Avalanches

2. Man-made Hazards. These are caused by undesirable activities of man. Such hazards include :

- Explosions
- Leakage of toxic waste
- Pollution of air, water and land
- Dam failures
- War and Civil Strife
- Terrorism

3. Socio-natural Hazards. These are caused by the combined effect of natural forces and misdeeds of man. For example, the frequency and intensity of floods and droughts may increase due to indiscriminate felling of trees, particularly in the catchment areas of the rivers. Storm surge hazards may be worsened by the destruction of mangroves. Although landslides are normally caused by natural forces, yet they may be caused and their frequency and impact may be aggravated as a result of building roads in mountainous areas, excavating tunnels and by mining and quarrying.

DISASTER

Extreme form of hazard leads to disaster. Disaster (French *des* meaning 'bad' and *aster* meaning 'star') is a manmade or natural event which results in widespread loss of life and property. Thus disaster is defined as "*A serious disruption of the functioning of a society, causing widespread human, material or environmental losses which exceed the ability of the affected society to cope using its own resources.*" The United Nations defines disaster as "*...the occurrence of a sudden or major misfortune which disrupts the basic fabric and normal functioning of a society (or community)*. The World Bank defines disaster *as an extraordinary event of limited duration which causes serious disruption of the economy of the country*. It is an event or a series of events which give rise to casualties and/or damage or loss of property, infra-structure, essential services or means of livelihood on a scale that is beyond the normal capacity of the affected communities to cope with unaided.

Hazards and disasters are closely related and are sometimes used as synonymous with each other. *Hazard is a threat, while disaster is an event*. The latter is a calamity or tragedy or a consequence of a hazard. *Natural hazards that cause great loss to human life and economy are known as* ***disasters*** *and* ***catastrophes***. The term ***disaster*** is sometimes also used to describe a "*catastrophic situation in which the normal patterns of life (or eco-systems) have been disrupted and extraordinary emergency interventions are required to save and preserve human lives and/or the environment*."

Classification of Disasters

Disasters are generally classified on the basis of their origin as tectonic (earthquakes, volcanoes), topographical (landslides, avalanches), meteorological (hurricanes, cyclones, tornadoes, floods, droughts, frost, loo, heat wave, cold wave), infestic (locust invasion of crops, epidemics) and human (industrial accidents, nuclear bombs).

VULNERABILITY

Hazards and disasters are unevenly distributed with respect to time and space. The terms like ***'earthquake prone'***, ***'drought prone'*** or ***'flood prone'*** areas are used to describe the distributional effect of the concerned hazards. People living in such areas are vulnerable to hazards and disasters of varied types. **Thus vulnerability is the extent to which an individual or a community or an area is exposed to the impact of a hazard.** According to National Institute of Disaster Management (formerly National Centre for Disaster Management), vulnerability is defined as the extent to which a community, structure, service or geographic ***area is likely to be damaged or disrupted by the impact of a particular hazard on account of either nature, construction and proximity to hazardous terrain or disaster prone area."*** Thus, vulnerability implies a measure of risk combined with the level of social and economic ability to cope with the resulting event in order to resist major disruption or loss. In other words, vulnerability is ***"a set of prevailing or consequential conditions composed of physical, socio-economic and/or political factors which increase a community's susceptibility to calamity or which adversely affect its ability to respond to events."***

NATURAL HAZARDS AND DISASTERS IN INDIA

Because of its subcontinental dimensions, geographical situation and behaviour of the monsoon, India is exposed to various natural hazards and disasters like droughts, floods, cyclones, earthquakes, etc. year after year. While all the states and union territories in the country are likely to face one or a combination of disaster situations, 27 states/union territories are more vulnerable to these disasters. Four major disasters which adversely affect different parts of the country are drought, flood, cyclone and earthquake. Only one state (West Bengal), faces all four types of disasters, seven states face three types of disasters, ten face two types of disasters and nine states face one type of disaster. It is not uncommon to experience more than one or two types of disasters affecting different parts of the country at the same time. For example, there may be flood in the Brahmaputra Valley, drought in Rajasthan and cyclonic storm in some coastal area.

Disasters occur with unfailing regularity and despite better preparedness to meet all such contingencies, the economic and social cost on account of losses caused by the natural disasters continues to mount year after year. The main hazards in India are earthquakes, droughts, floods and cyclones. Since the beginning, they have been causing heavy losses of life and property forcing man to ***'learn to live'*** with natural disasters.

The most disturbing factor is that these hazards and disasters are hitting different parts of the country with increased frequency and with greater force.

INDIA'S KEY VULNERABILITIES

- According to UN report, India has been ranked No. 3, after China and U.S.A. in recording highest number of natural disasters between 2000 and 2019.
- 59 per cent of total area is in Seismic Zones III-IV and vulnerable to earthquakes.
- 68 per cent of net sown area is vulnerable to drought.
- 40 million hectares of land is vulnerable to floods.
- 8 per cent of the total land area particularly along the eastern coast and Gujarat coast is vulnerable to tropical cyclones.
- Sub-Himalayan region and Western Ghats are vulnerable to landslides.
- Of the 7,516 km long coastline close to 5,700 km is prone to tropical cyclones and tsunamis.

EARTHQUAKES

Earthquake is a violent tremor in the earth's crust, sending out a series of shock waves in all directions from its place of origin. If you throw a stone in a pond of still water, a series of concentric waves are produced on the surface of water. These waves spread out in all directions from the point where the stone strikes the water. Similarly, any sudden disturbance in the earth's crust may produce vibrations in the crust which travel in all directions from the point of disturbances. Earthquakes constitute one of the worst natural hazards which often turn into disaster causing widespread destruction and loss of human lives.

Magnitude and Intensity of Earthquakes

Magnitude and intensity are two ways of measuring the strength of an earthquake.

Earthquake's magnitude is determined by the use of a seismograph, an instrument which continuously records, ground vibration. It is measured on the Richter scale. This scale was developed by Charles Francis Richter in 1935 and is known after his name. It was modified in 1965 by Richter and his colleague Beno Gutenberg. It is an instrumentally measured scale and is a measure of the amount of force or energy released at the earthquake source. It is calculated from the amplitude measured by a standard Wood-Anderson seismograph and corrected in terms of a seismograph supposed to be kept at a distance of 100 km from the epicentre. The number indicating magnitude or intensity on Richter Scale ranges between 0 and 9 but actually the scale has no upper limit of number because it is an **open ended** and **logarithmic scale**. Each whole number on this scale represents 10-fold increase in the measured wave amplitude. Translated into energy, each whole number demonstrates a 31.5 times increase in amount of energy released. Thus a 3 on the Richter Scale represents 31.5 times more energy than a 2 and 992 times more energy than a 1. The vibration of an earthquake with a magnitude of 2 are 10 times greater in amplitude than those of an earthquake with a magnitude of 1, and the vibrations of an earthquake with a magnitude of 8 are one million times greater in amplitude than those of an earthquake with a magnitude of 2. The largest earthquake ever recorded had a magnitude of 8.9 on the Richter Scale. It is worth noting that earthquakes larger than this magnitude are not likely to occur because rocks are not strong enough to accumulate more energy. A 2 point quake is barely perceptible, a 5 may cause damage to structures, a 7 is severe and an 8 is a violent quake.

Intensity of Earthquake. The intensity or destructive power of an earthquake is an evaluation of the severity of ground motion at a given location. *It is measured in relation to the effects of the earthquake on human life*. Generally destruction is described in terms of damage caused to buildings, dams, bridges and other structures. The intensity of an earthquake is measured by **Modified Mercalli (MM)** scale. This scale was developed by an Italian seismologist Mercalli in 1902 and modified by Wood and Newman in 1931. It expresses the intensity of the effects of earthquakes on people, structures and earth's surface in values from I to XII. (**Table 9.1**).

TABLE 9.1. Modified Mercalli Intensity Scale (Abridged) (*Source* : IS : 1893-1984) and its Correlation with Richter Scale

	Description	Magnitude according to Richter Scale
I.	Not felt except by very few under especially favourable circumstances.	0
II.	Felt only by a few persons at rest, especially on upper floors of buildings. Delicately suspended objects may swing.	3.5
III.	Felt quite noticeably indoors, especially on upper floors of buildings, but many people do not recognise it as an earthquake. Standing motorcars may rock slightly. Vibrations like passing of lorry. Duration estimated.	4.2
IV.	During the day felt indoors by many, outdoors by few. At night some awakened. Dishes, windows, doors disturbed; walls make cracking sound. Sensation like heavy lorry striking building. Standing motor cars rocked noticeably.	4.8
V.	Felt by nearly everyone; many awakened. Some dishes, windows, etc. broken; a few instances of cracked plaster; unstable objects overturned. Disturbance of trees, poles and other all objects sometimes noticed. Pendulum clocks may stop.	4.9 – 5.4
VI.	Felt by all; many frightened and run outdoors. Some heavy furniture moves; a few instances of fallen plaster or damaged chimneys. Damage slight.	5.5 – 6.1
VII.	Every body runs outdoors. Damage negligible in buildings of good design and construction; slight to moderate in well built ordinary structures; considerable in poorly built or badly designed structures; some chimneys broken. Noticed by persons driving motor cars.	6.5
VIII.	Damage slight in specially designed structures; considerable in ordinary substantial buildings with partial collapse; great in poorly built structures. Panel walls thrown out of frame structures. Fall of chimneys, factory stacks, columns, monuments, walls. Heavy furniture overturned. Sand and mud ejected in small amounts. Changes in well water. Disturbs persons driving motorcars.	6.9
IX.	Damage considerable in specially designed structure; well designed frame structures thrown out of plumb; great in substantial buildings with partial collapse. Buildings shifted off foundations. Ground cracked conspicuously. Underground pipes broken.	7 – 7.3
X.	Some well-built wooden structures destroyed; masonry and frame structures and their foundations destroyed; ground badly cracked. Rails bent. Landslides considerable from river banks and steep slopes. Shifted sand and mud. Water splashed over banks of rivers, etc.	7.4 – 8.1
XI.	Few, if any, masonry structures remain standing. Bridges destroyed. Broad fissures in ground. Underground pipelines completely out of service. Earth slumps and landslips in soft ground. Rails bent greatly.	8.2
XII.	Damage total. Wave seen on ground surface. Lines of sight and level distorted. Objects thrown upward in the air.	8.2 and above

Source : U.S. Geological Survey.

The other scales used for expressing the intensity of the earthquake are Rossi-Forel (RF), Medvedev Sponheuer Karnik (MSK) and Japan Meteorological Agency (JMA) scale. All the scales are close-ended stepped scales, RF having ten points (I to X), MM and MSK with 12 points (I to XII) and JMP with 7 points (I to VII). However, Richter scale is the most popular and widely used for expressing the intensity of an earthquake.

Earthquake Hazards in India

India has a very long history of earthquake occurrences. According to the latest seismic map prepared by the National Disaster Management Authority (NDMA) and Building Material and Technology Promotion Council (BMTPC) and released in September, 2016, nearly 59% of India's land area is vulnerable to moderate or severe earthquake. It also shows that 95% of households are vulnerable to earthquakes in different degrees. The most vulnerable areas according to the seismic zone map of India are located in Himalayan and sub-Himalayan regions, Kachchh and the Andaman and Nicobar Islands. Depending on varying degrees of seismicity, the entire country can be divided into the following seismic regions :

1. **Kashmir and Western Himalayas.** This region covers the Union Territories of Jammu and Kashmir and Ladakh, state of Himachal Pradesh and sub-mountainous areas of Punjab.
2. **Central Himalayas.** This region includes the mountainous and sub-mountainous regions of Uttarakhand and Uttar Pradesh.
3. **Northeast India.** This region comprises the whole of Indian territory to the east of northern West Bengal.
4. **Indo-Gangetic Basin and Rajasthan.** This region comprises of Rajasthan, plains of Punjab, Haryana, Uttar Pradesh, Bihar and West Bengal.
5. Khambhat and Rann of Kachchh.
6. Peninsular India including the Islands of Lakshadweep.
7. The Andaman and Nicobar Islands.

The distribution of magnitudes equal to or more than 5.0 in these regions and their average return periods are represented in the **table 9.2**.

Based on intensities of the earthquakes recorded on MM Scale the Indian Standards Institute has published a map of India showing seismic zones. Map given in **Figure 9.1** shows five Seismic zones based on MM Scale.

Zone I — Intensity V or below (Instrumental, feeble, slight, moderate, rather strong)
Zone II — Intensity VI (Strong)
Zone III — Intensity VII (Very strong)
Zone IV — Intensity VIII (Destructive)
Zone V — Intensity IX and above (Ruinous, disastrous, very disastrous, catastrophic)

According to seismologists, one part of India or the other experiences one earthquake almost everyday. But most of them are of low intensity, are

TABLE 9.2. Region-wise Earthquake (M > 5.0) Occurrence in India (1897–2014)

Sl. No.	Seismic Region	No. of Earthquakes of Magnitude				Return Period
		5.0–5.9*	6.0–6.9	7.0–7.9	8.0 +	
1.	Kashmir & Western Himalayas	25	8	2	2	2.5–3 years
2.	Central Himalayas	68	28	4	1	1 Yr.
3.	North East India	200	128	15	4	< 4 months
4.	Indo-Gangetic Basin and Rajasthan	14	6	—	—	5 Yrs
5.	Khambhat and Rann of Kutch	4	4	1	1	20 Yrs
6.	Peninsular India	31	11	—	—	2.5–3 Yrs.
7.	Andaman & Nicobar Islands	90	80	1	1	< 8 months

* Numbers of earthquake under-category 5.0–5.9 are approximate.

harmless and go unnoticed. However, major earthquakes of higher intensity are strong enough to wreck havoc and result in great loss of life and property.

Depending upon the frequency and intensity of the earthquakes, the whole country can be divided into three broad seismological zones.

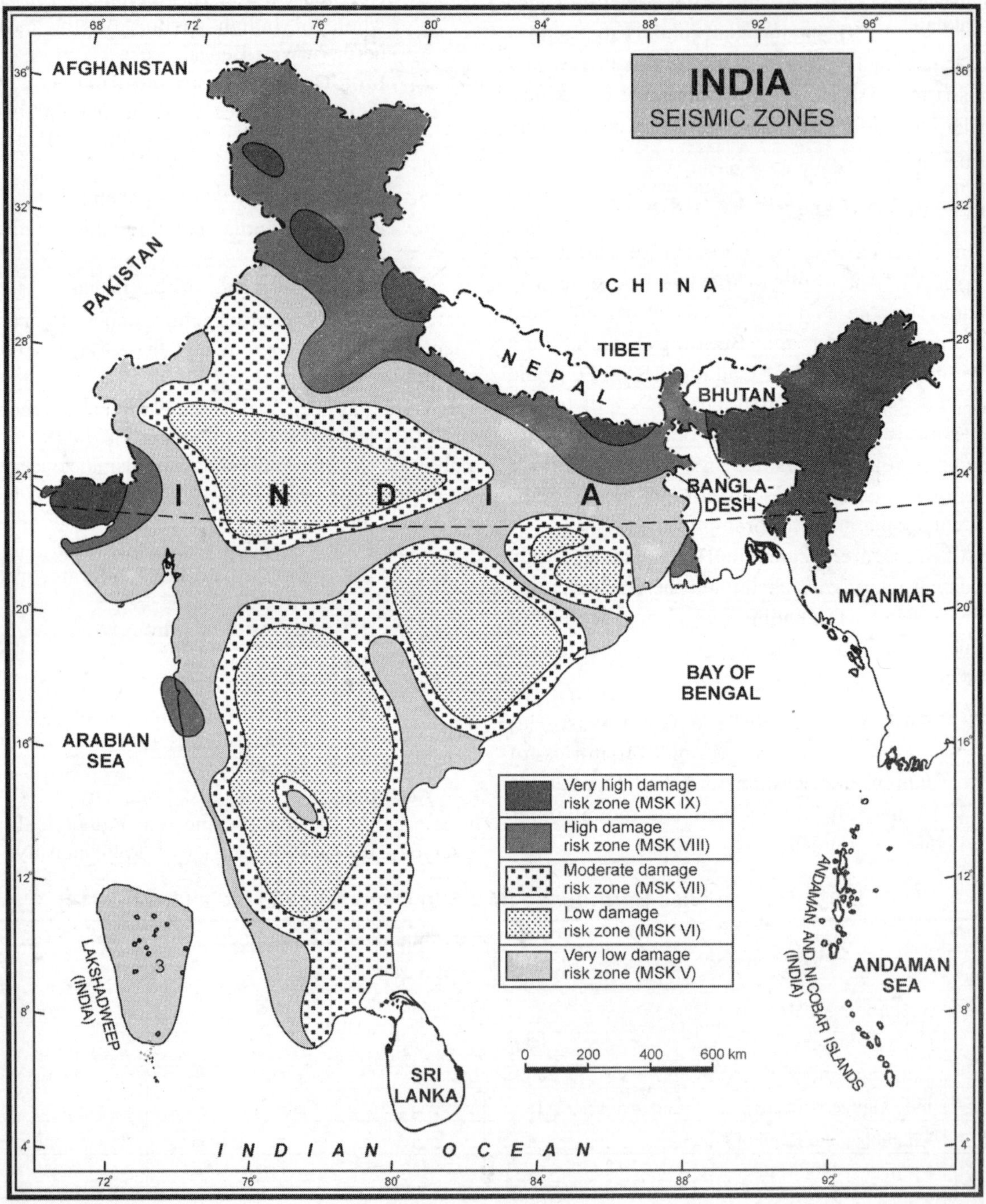

FIG. 9.1. India : Seismic Zones

1. Himalayan Zone. The areas most prone to earthquakes in India are the fold mountain ranges of the Himalayan zone. The Union Territories of Jammu and Kashmir and Ladakh, Himachal Pradesh, Uttarakhand, Bihar, the Bihar-Nepal border and the north eastern states, especially Assam fall in this zone. The earthquakes in this zone are primarily due to plate tectonics. The Indian plate is pushing in the north and north-east direction at an annual rate of 5 cm subducting the Eurasian plate along the

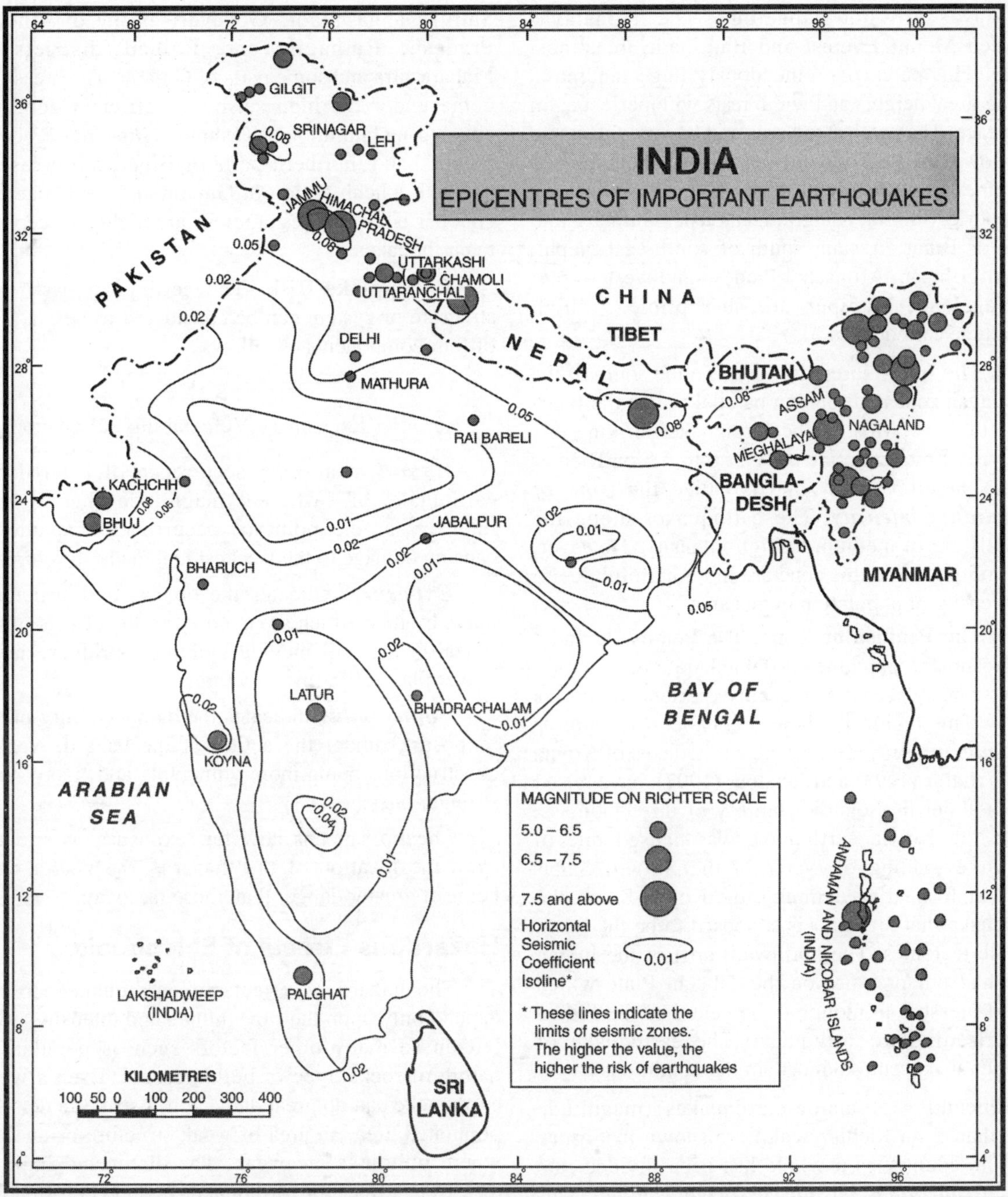

FIG. 9.2. India : Epicentres of Important Earthquakes

Himalayas. The Himalayas have not yet attained *isostatic equilibrium* and are rising. The region along the Himalayas where two plates meet is highly earthquake-prone. This is known as the ***zone of maximum intensity***. The absence of Nepal from the list of earthquake prone areas shows that the whole of Himalayas are not dangerous. The Himalayas between Mount Everest and Badrinath are almost stable. This patch of tremendously huge landmass, having great heights and width rests with perfect calm and peace. The patches towards and beyond this patch are, however, very violent because of their hurried movement. Areas north-east of the arc joining Mussoorie, Shimla, Kangra, Dalhousie, Gulmarg and areas of Bihar, Assam, south of south-east Nepal, Sikkim, Bhutan, Arunachal Pradesh and western part of Nagaland, Manipur are susceptible to high magnitude earthquakes.

2. The Indo-Gangetic Zone. To the south of the Himalayan zone and running parallel to it is the Indo-Gangetic zone. Most of the earthquakes striking this zone are of moderate intensity of 6 to 6.5 on Richter scale. Therefore, this zone is called the ***zone of comparative intensity***. The earthquakes along the foothills are of medium to high intensity. However, the earthquakes of this zone are more harmful due to high density of population in this area.

3. The Peninsular Zone. The Peninsular India has presumably remained a stable landmass and only a few earthquakes have been experienced in this region. This region is, therefore, called the ***zone of minimum intensity***. But severe earthquakes of Koyna (1967), Latur (1993) and Jabalpur (1997), have raised doubts about the seismic stability of this landmass. While the Koyna earthquake was caused due to excessive loading of water in the Shivaji Sagar reservoir formed by damming the Koyna River, the earthquake that hit Latur is supposed to be the result of plate tectonics. The northward drift of the Indian Plate had put pressure on the Tibetan Plate which caused pressure to mount at the centre of the Indian plate, leading to earthquake. The earthquake of Jabalpur also occurred under similar conditions.

Epicentres of major earthquakes (magnitude more than 5 on Richter scale) are shown in **Figure 9.2**. The country as a whole has been divided into zones by ***horizontal seismic coefficient isolines.*** This coefficient varies from 0.01 to 0.08. The low horizontal seismic coefficient zone (0.01–0.02) constitutes the areas that are quite secure from earthquakes. This zone is mainly spread over extensive parts of peninsular India, Madhya Pradesh and Rajasthan. Areas having medium horizontal seismic coefficient (0.04–0.05) are spread over the Satluj-Ganga plain, Godavari basin of Andhra Pradesh, Ratnagiri and Raigad districts of Maharashtra and some parts of Gujarat. The areas that come under the high seismic coefficient zone are situated in Jammu and Kashmir, Himachal Pradesh, Uttarakhand, northern parts of Bihar, north-eastern states, Kachchh region of Gujarat and Andaman and Nicobar Islands. These regions are highly susceptible to earthquakes.

Earthquake Risk. The earthquake risk to a structure or system can be considered to be made up of four components as follows :

Seismic Risk =
Hazard × Exposure × Vulnerability × Location

Hazard means the occurrence of a terrifying earthquake of sufficient magnitude (hence peak intensity at the point of occurrence) capable of causing damage to the weakest manmade structures.

Exposure indicates the objects and structures made by man which are exposed to the effects of the 'hazard' and will include buildings, bridges, dams, power plants, life-line structures, etc.

Vulnerability indicates the damageability of the 'exposure' under the action of the hazard, weaker constructions being more vulnerable and 'risky' than stronger ones.

Location means how far 'exposure' is situated from the location of the 'hazard', the nearer ones being at greater danger than those far away.

Hazardous Effects of Earthquakes

The hazardous effects of earthquakes do not depend only upon their magnitude and intensity alone but on so many other factors such as population, nature of rocks, type of buildings, etc. Even a weak earthquake can do great damage if it strikes a densely populated area, an area of weak structure or an area where buildings are weak. The direct and indirect disastrous effects of earthquakes include deformation of ground surface, damage and destruction to human structures such as buildings, rails, roads, dams, bridges, factories, destruction to towns and cities, loss of human and animal lives, fires, floods, landslides, etc.

Major hazards of earthquakes are briefly described as under :

1. Loss of Life and Property. There is devastating loss of life and property if the intensity of the earthquake is more than 5 on the Richter scale. Buildings, roads, railways, bridges, dams, etc. suffer severe damage when an earthquake strikes them. Several villages, towns and cities are completely ravaged. The towns of Bhuj, Bhachau, Anjar, Gandhidham, and Ratnal were completely destroyed by the earthquake which struck Bhuj on 26th January, 2001. Property worth ₹ 2,000 crore was destroyed. The maximum damage is noticed near the epicentre of the earthquake. This damage is reduced as we move away from the epicentre. An earthquake becomes a hazard or disaster only when it strikes populated area and causes unaccountable loss to life and property.

2. Topographical Changes. The main effects of earthquakes on topographical features are seen in the form of offsets along known faults, fissures, scarps, elevation and depression of coasts, etc. Earthquakes are often followed by landslides in hilly areas. Due to earthquake vibrations, the looser material at or near its maximum static stable angle may become unstable and move along the slope of hill. Cracks and fissures which occur as a result of a severe earthquake facilitate landslides. The severe earthquake which struck Uttarkashi in Uttarakhand (1991) caused many cracks and fissures in *Varunavat Parvat*. This gave birth to landslides which caused heavy damage to Uttarkashi town in the year 2003. Large masses of dry earth and rock, called ***earth avalanches,*** slide over considerable distances. If wet soil slips down the hill side, a ***slump*** occurs. ***Earthflow*** is another phenomenon which is primarily restricted to ground water. The earthquake is either accompanied or succeeded by a sudden burst of water from a locality where it normally appears in springs.

3. Liquefaction. Soil liquefaction is a phenomenon where low density saturated sands of relatively uniform size inside the earth start behaving like a jelly with no strength to hold a building up, and the building just sinks or gets tilted on one side. The phenomenon of liquefaction is particularly important for dams, bridges, underground pipelines and buildings close to river banks, sea shore or large lakes. Vast tracts of plains of Uttar Pradesh and Bihar where soil is generally soft and the water table is high, offer favourable conditions for such effects. The great Bihar-Nepal earthquake of 1934 produced a belt of slumping extending from Bettiah in north-west to Purnea in the south-east (a distance of nearly 320 km), surrounding the epicentral tract, in which all buildings either tilted or sank in soft alluvium. Widespread subsidence over large area was observed. There were innumerable fissures through which large quantities of sand and water were thrown up to the surface, thereby causing large scale destruction to standing crops. The soil became totally unfit for cultivation. Rann of Kachchh also covers a vast area where large scale liquefaction of ground was observed during the Bhuj earthquake of 2001.

Pattern and Causes of Damage

1. Buildings. Great damage is done to buildings by earthquakes particularly if the buildings are constructed with brick, mud or timber. Reinforced concrete buildings suffer least damage, while affected adobes and unreinforced brick buildings are severely damaged by the earthquakes. Projecting cornices, balconies, towers and arcades render the building more vulnerable.

An earthquake causes damages to the buildings by setting up oscillating (backward and forward) motion of the ground. As the part of the earth's crust affected by the earthquake oscillates, the buildings on the ground start responding to oscillation in varying degrees depending upon how these have been designed and constructed. Close to the epicentre, there is also a vertical movement. Due to oscillatory movements caused by the earthquake, the foundations of the building move with the ground, but the inertia of the rest of the building prevents it from moving instantaneously and there is a slight delay before the upper portions start moving. This delay leads to differential stresses and subsequent cracking because the roof tends to separate from the support and the walls tend to be torn apart. The force exerted on the building depends upon the movement of the ground caused by the earthquake and the weight of the building. It is axiomatic *that earthquakes do not kill—collapsed buildings do*. The Muzaffarabad earthquake on 8 October 2005 hit Pakistan and India with a magnitude of 7.4 in Pakistan and 6.8 in India and killed over one lakh persons. Several buildings collapsed like a pack of cards. In contrast a magnitude of 7.1 earthquake in October 2004 in Hokkaido (Japan) caused only 17 injuries. The main reason for the low Japanese casualties was their seismically engineered buildings. Therefore, it is advisable to construct light weight buildings with lighter roofs in earthquake prone regions. Rapid progress in earthquake engineering has standardised techniques for earthquake resistant design of buildings in different seismic zones. India has been classified in five seismic zones based on Modified Mercalli Intensity Scale. In zones III, IV and V destruction and damage to buildings is considerable. Zone II could also have minor damage in rural buildings. but collapse of houses in this zone is not considered probable, while zone I may be taken as non damaging seismically.

The extent of damage is dependent on a number of factors; the intensity of the earthquake, distance from the epicentre, soil condition, type of structure (mass, permissible stress, elasticity, dynamic load response and durability of materials), design of building and quality of construction.

Shape of the building has a great bearing on its resistance to earthquake risk. Geometric shapes such as square or rectangle usually perform better than buildings in the shape of L, T, U, H, +, O or a combination of these.

2. Damage to Transport System. Damage to transport system includes damage to highways, railways, airports, marine and river systems, water supply and sewage, fuel and oil, energy transmission and communication systems. The facilities of transport may be built either on surface or underground. The earthquake affects differently on these two types of structures. The underground structures such as oil or gas pipelines are mainly governed by the strain in the surrounding ground caused by the propagation of earthquake waves, while the surface structures like bridges are governed by vibrating response of the structure to earthquake ground motion. The second category of damage arises due to the blocking of road by jammed cars, fire, uprooted poles and collapsed buildings. Blocking of highways may stop various activities needed for earthquake relief and rescue.

3. Fire. In the event of an earthquake, short circuits, contact with live electrical wires, kitchen fires, damage to blast furnaces and other fire related appliances are the major causes of fire.

4. Floods. Often earthquakes lead to distortion and displacement of the surface rocks which block the flow of a river. This causes floods in the upper course of the river. Often dams and embankments develop cracks and the course of the river lower down the dam site is flooded.

5. Public Health. People suffer multiple injuries and many become permanently disabled. Hazards of disease on pollution of water bodies, breakdown of sewage and sanitary pipes and other unhygenic conditions could lead to epidemics. Many people die of heart attack.

6. Civic Services. Civic services like water pipes, sewers, electric connections, etc. are disrupted. Fire hydrants supply lines, if vulnerable, could hamper service operations.

7. Economic Activities. Economic activities like agriculture, industry, trade, transport and other services may be severely affected by widespread damage to infrastructure in the event of an earthquake. It becomes one of the major causes of civic unrest.

Earthquake Safety Rules

Before an earthquake

1. Keep stock of drinking water, some foodstuffs, first-aid equipment, a crow bar, shovel, pick and rope, electric torch and some candles and a helmet for every member of the family.
2. Ensure that water heaters and other gas appliances are firmly fixed and shut off when not in use, as broken gas pipes or appliances are likely to cause fire hazards.
3. Securely fasten fixtures, refrigerators and shelves to the wall and place large and heavy objects on the lower shelf. Top heavy objects should be braced or anchored.

4. Find out the location of the nearest first-aid post, Warden Post and Fire Station and approach for help if required.
5. Join the Civil Defence Organisation and train yourself and the members of your family in first-aid, rescue fire fighting etc. which will help you, your family and neighbours.
6. Conduct occasional home earthquake drills so that your family has the knowledge to avoid unnecessary injuries and panic in the event of an earthquake.
7. More responsible members of the family may be taught how to turn off electricity, gas and water at the main switches or valves.
8. Be aware of what to do in various situations so that you are prepared in the event of an earthquake, *e.g.*, at home, whilst driving a car, at work in a shop, a public theatre, cinema hall, etc.
9. Avoid the risk of an epidemic which usually follows earthquakes by using safe water and clean food.
10. Evacuate old dilapidated buildings as they are sure to tumble first.

Certain other precautions, which are necessary in case of fire, should also be taken as far as possible :

1. Keep adequate water/dry sand stocked in accessible places.
2. All lofts and attics should be kept free from inflammable material.
3. Keep all inflammable stores, which are required, as much dispersed as possible.

During the earthquake

1. Do not panic. The motion is frightening, but unless it shakes something down on top of you it is harmless. The earth does not yawn open, gulp down a neighbourhood, and close up. So keep calm.
2. In the event, the safest place is an open space away from building. However, when this is not suitable do not try to run from a building during earthquake.
3. If it catches you indoors, take cover under a desk, table, bench or in doorways, halls, and against inside walls or staircase. Stay away from glass.
4. Do not use candles, matches, or other open flames, either during or after the earthquake. Put out all fires.
5. If the earthquake catches you outside, move away from buildings and utility wires. Once in the open stay there until the tremor stops.
6. Do not run through or near buildings. The greatest danger from falling debris is just outside doorways and close to outer walls.
7. If you are in moving car, stop as quickly as safety permits, but stay in the vehicle. A car is an excellent seismometer, and will jiggle on its springs during the earthquake, but it is a good place to stay until the earthquake stops.

After an earthquake

1. Keep calm, switch on the transistor radio and obey any instructions you hear on the radio.
2. Keep away from beaches and low banks of rivers. A huge wave may sweep in.
3. Expect aftershocks.
4. Turn off the water, gas and electricity.
5. Do not smoke and do not light matches or use a cigarette lighter. Do not turn on switches. There may be gas leaks or short-circuits.
6. Use a torch.
7. If there is a fire, try to put it out. If you cannot, call the fire brigade.
8. If people are seriously injured, do not move them unless they are in danger.
9. Immediately clean up any inflammable products that may have spilled (alcohol, paint, etc.)
10. If you know that people have been buried, tell the rescue teams. Do not rush and do not worsen the situation of injured persons or your own situation.
11. Avoid places where there are loose electric wires and do not touch any metal object in contact with them.

12. Do not drink water from open containers without having examined it and filtered it through a sieve, a filter or an ordinary clean cloth.
13. Eat something. You will feel better and more capable of helping others.
14. If your home is badly damaged you will have to leave it. Collect water containers, food, and ordinary and special medicines (for persons with heart complaints, diabetes, etc.).
15. Do not re-enter badly damaged buildings and do not go near damaged structures.
16. Do not walk around the streets to see what has happened. Keep clear of the streets to enable rescue vehicles to pass.

TSUNAMI

Introduction and Definition

Tsunami is derived from Japanese word '***tsu***' meaning harbour and '***nami***' meaning wave. According to Report of High Powered Committee on Disaster Management (2001). "Tsunami is an ocean wave produced by an event at the sea, like an earthquake, landslide or volcanic eruption" (**Fig. 9.3**). A tsunami is not a single wave but a series of waves generated by the geological changes near or below the ocean floor. These waves may reach enormous size and have been known to travel across the oceans.

Formation of Tsunamis

(i) Undersea earthquakes

Although tsunami may be caused by landslides, volcanic eruptions or even by the impact of a large meteorite falling on the ocean, most destructive tsunamis are generated by massive undersea earthquakes, occurring at depth less than 50 km with the epicentre or fault line near or on the ocean floor. A strong undersea earthquake with magnitude greater than 7.5 on the Richter scale tilts and deforms large areas of the sea floor ranging from a few square kilometres to 1000 square kilometres and even more. As the sea floor is tilted or deformed by the tectonic earthquake (earthquake associated with the earth's crustal deformation), the sea water above is displaced from its equilibrium position. Waves are formed as the displaced water attempts to regain its equilibrium under the influence of gravity. It is this vertical

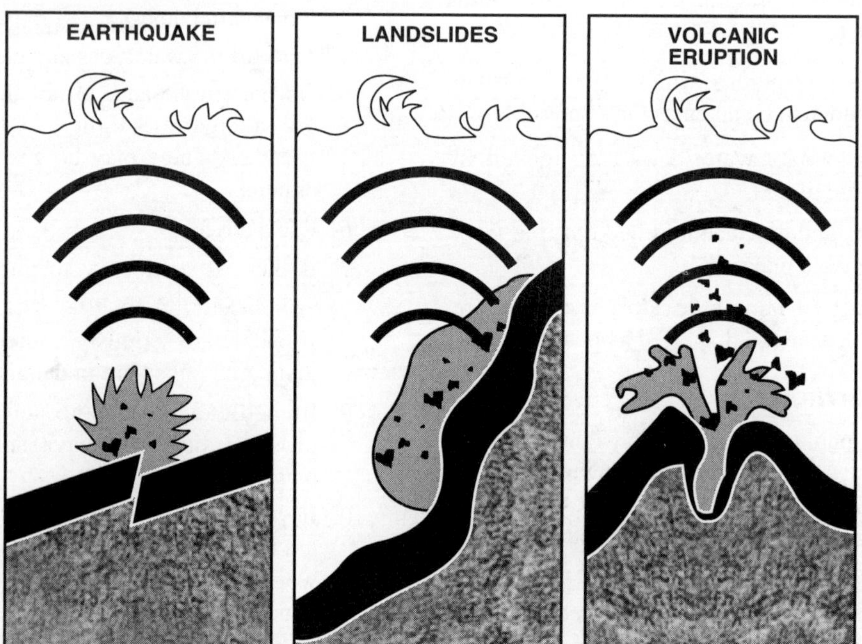

FIG 9.3. Triggering of Tsunami

movement of the entire water column that generates destructive tsunami waves.

The displacement of sea floor, and occurrence of an earthquake and formation of tsunamis can best be explained on the basis of plate tectonics. When two converging lithospheric plates come closer together, then heavier plate is thrust under the lighter plate and displacement of the crust takes place at the subduction zone. A fault is created and an earthquake occurs, giving rise to tsunamis.

It must be noted that a tsunami is usually not generated if the sea floor movement is horizontal. Besides, not all undersea earthquakes create tsunami, as it depends upon the nature and degree of displacement of seawater column. It is only the vertical displacement of the seawater due to abrupt, jerky movements of fault blocks on seabed that gives birth to tsunamis. Once formed, the monstrous waves soon begin their journey towards the nearest coastline, ringing the bells of doom (**Fig. 9.4**).

(ii) Landslides

Tsunami waves are also generated by displacement of seawater resulting from landslides as well as rock falls, icefalls etc. Construction work of an airport runway along the coast of Southern France in the 1980s caused an underwater landslide. This triggered the destructive tsunami waves in the harbour of Thebes. Underwater landslides may also occur when a strong earthquake shakes the sea floor, thus forming tsunamis. These waves rapidly travel away from the source due to dissipation of energy, and create havoc in the nearby coastlines.

(iii) Volcanic Eruptions

Whenever a violent volcanic eruption takes place under the sea, it causes sudden displacement of a large volume of seawater and tsunami waves are formed. Similarly, when the roof of a volcano collapses that has a large empty magma chamber owing to continuous flow of lava, a crater sometimes as large as one kilometre in diameter is formed. As the seawater gushes into this crater, the water column of the sea is disturbed which gives rise to tsunami waves.

One of the largest and the most destructive tsunami ever recorded was generated on August 26, 1883 after the explosion and collapse of the volcano of Krakatoa in Indonesia. This explosion generated waves with a towering height of about 40 m, that wrecked havoc on the coastal areas along the Sunda Strait in both the islands of Java and Sumatra killing more than 36,000 people. It is believed that the destruction of the Minoan civilisation in Greece in 1490 B.C. was caused by tsunamis which were formed by an explosion of the volcano of Santorini in the Aegean Sea.

(iv) Meteorites and Asteroids

There is a potential danger of a tsunami being formed by the fall of meteorites and asteroids in the ocean. Researchers in California have developed a computer simulation depicting the ocean impact of asteroid 1950 DA, a gigantic space rock that would be very close to the earth in 2880. Although the possibility of such an impact is very remote, the computer model definitely gives the researcher an insight into the destructive power of tsunami caused by near-Earth object. Some of the meteorites have been uncomfortably close to the earth and could wreck havoc in different forms including tsunamis.

Propagation of the Tsunamis

Tsunamis consist of a series of very long waves which travel outwards on the surface of the ocean in all directions away from their place of origin. Their movement is just like ripples created by throwing a pebble into a pond of water. In deep sea the tsunamis travel at very very high speed (say 500–800 km per hour), almost as much as the speed of a jet aircraft. Their wavelength is very long which often exceeds 500–700 km. However, the amplitude of tsunamis in deep sea is very low and rarely exceeds 1 metre. Physically they propagate as long waves with speed given by

$$(\text{Water depth} \times \text{gravitation acceleration})^{1/2}$$

Since the tsunamis have very long wavelength and very low amplitude in deep ocean, they cannot be seen or detected from the air. Therefore, passengers on boats cannot feel or see the tsunami waves as the killer waves pass by underneath at high speeds. It may only appear as a gentle rise and fall of the sea. Thus tsunamis are always deceptive and are able to conceal their killing capacity in the deep water of

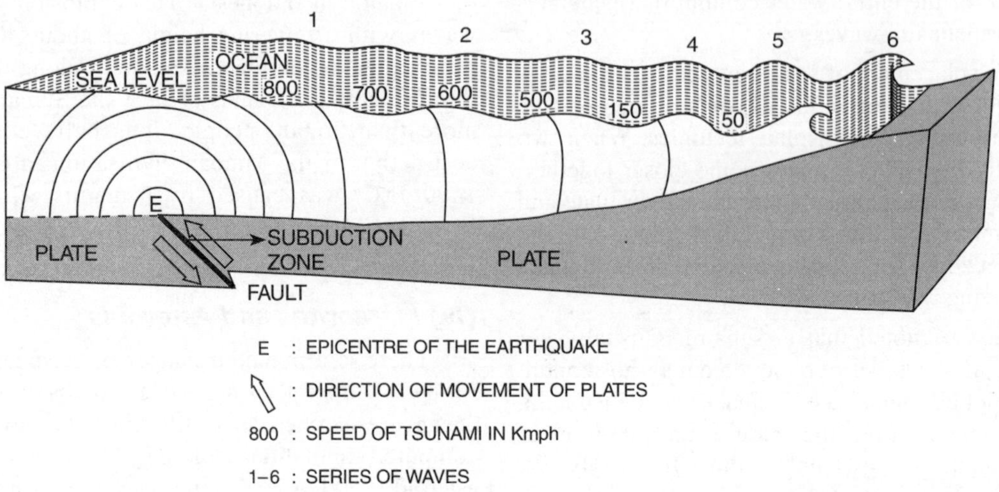

FIG. 9.4. Origin and Propagation of Tsunamis

open area. For example, the Great Sanriku tsunami, which struck Honshu in Japan on June 15, 1896 was completely undetected by fishermen as its deep water height was only about 40 cm. A monster in disguise, this tsunami transformed into huge waves when it arrived on the shore and ravaged 275 km of coastline killing 28,000 people. So from the sky tsunami waves cannot be distinguished from ordinary ocean waves. But beneath, a tremendous amount of energy lurks. Since the rate at which the wave loses its energy is inversely related to its wavelength, tsunamis not only propagate at high speeds, they can also travel great transoceanic distances with limited energy losses.

As the tsunamis leave the deep water of the open ocean and travel towards the shallow water, they are transformed in two ways. Firstly their speed is reduced considerably and secondly they attain enormous height often exceeding 10 metres and occasionally may reach 30 metres. **Figure 9.4** shows the origin of tsunamis and their propagation from deep water of open sea to the coast.

Effects of Tsunamis

Tsunami pose serious danger to the inhabitants of the coastal areas. They attack the sea shore as gigantic waves, moving with great force, appearing without a warning and hitting the coastline like a water bomb. Loaded with enormous energy, the killer waves wreck havoc by flooding hundreds of metres inland, past the typical high water level. They flatten houses and wipe out villages, uproot electric poles, throw cars into swirling waters and toss boats ashore all in mad fury, and finally drag thousands of hapless victims out to sea as they recede. Large rocks weighing several hundred tonnes and other debris can be moved hundreds of metres inland by a tsunami. Tsunamis can even travel up rivers and streams that lead to the ocean.

Tsunamis in the Indian Ocean

Tsunamis have been generated in the Indian Ocean at a number of times. In 1883, eruptions from the Krakatoa fueled tsunamis which killed 36,000 people in Java and Sumatra islands of Indonesia. During the earthquake in 1880 which had its epicentre near the centre of the Bay of Bengal, tsunamis were reported. During the earthquakes of 1819 and 1845 near the Rann of Kachchh, there was rapid movement of water into the sea. Tsunamis 12 to 15 metres high were generated in the Persian Gulf by 1945 Makran earthquake. The estimated height of waves was about 2 metres at Mumbai.

Tsunamis of 26th December 2004 in the Indian Ocean

On 26th December 2004, the Indian Ocean was hit by tsunamis which is considered to be the most catastrophic in the living memory of the inhabitants of the coastal areas of this ocean. It was caused by a severe earthquake which measured 8.9 on the Richter scale. Seismologists at Northwestern University in Illinois later upgraded the earthquake to magnitude 9.3 on Richter scale. This is perhaps the highest magnitude for any earthquake ever recorded anywhere in the world. This earthquake had its epicentre off the coast of Sumatra (Indonesia) at 3.5° north latitude and 95° east longitude. This place happens to be at the tri-junction of the Indian, Australian and Burmese (Myanmarese) plates. The earthquake was triggered by the collision of the Indian plate with Burmese plate. It occurred at the point where the Indian plate subducts below the Burmese plate due to the northward movement of the Indian plate (**Fig. 9.5**). Seismologists have noted a 15 metre slip in the vertical direction along the crack that is about 1000 km long extending upto Andaman and Nicobar Islands in the northern direction.

Damages Caused by Tsunami

This tsunami had been the most damaging in the world history. It had wrecked havoc to as many as 11 countries of south and southeast Asia and East Africa extending from Indonesia to Somalia. Hence it is rightly called the ***tragedy of international dimensions***. Various types of damages caused by the tsunami of 26th December 2004 are briefly described as under.

1. Death Toll

The tsunamis that hit the Indian Ocean on 26th Dec. 2004 claimed over 1.84 lakh lives in different countries of Asia and Africa. The soaring waves killed people of at least 40 nationalities including tourists from various countries of the world.

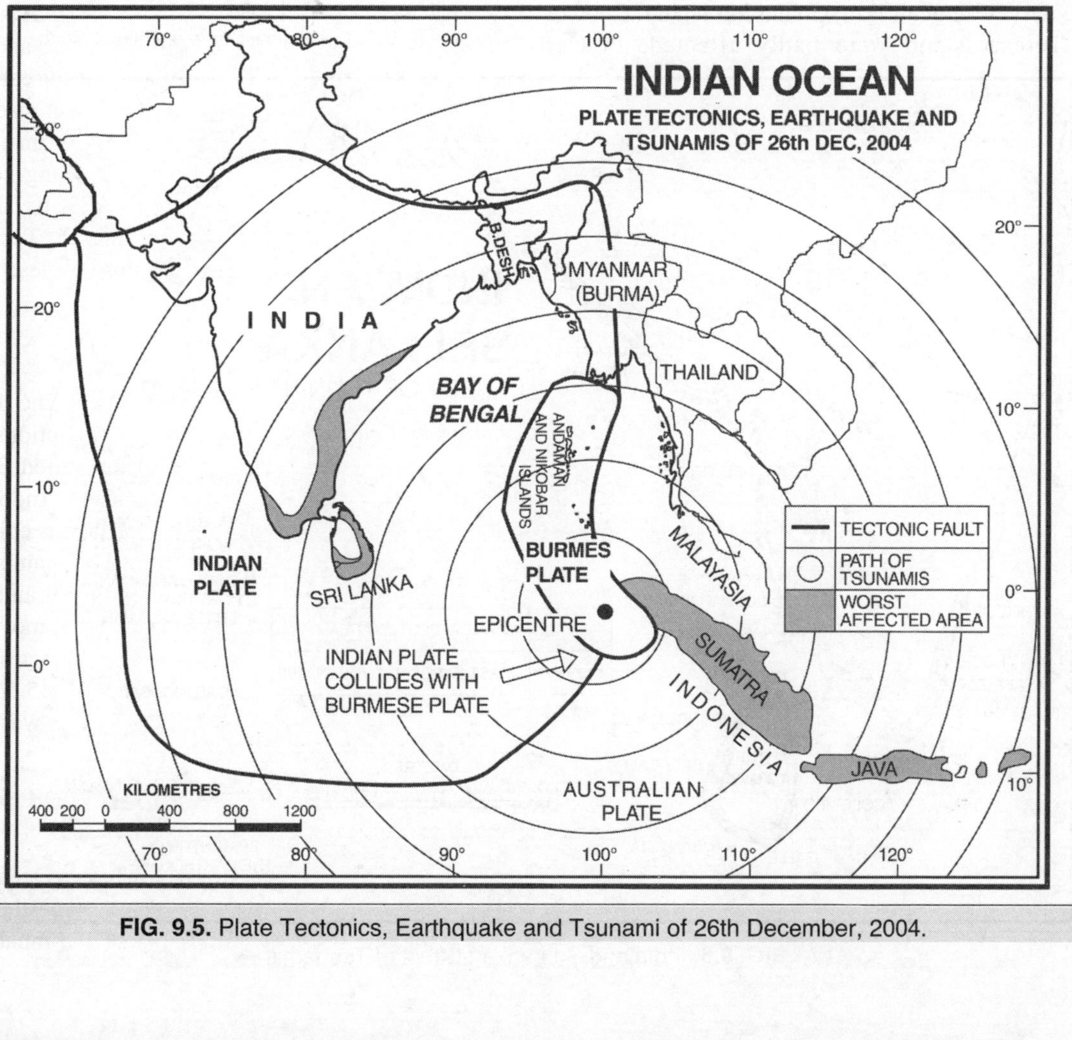

FIG. 9.5. Plate Tectonics, Earthquake and Tsunami of 26th December, 2004.

Table 9.2 shows that Indonesia has been the worst sufferer with death toll of over one lakh persons. Sri Lanka, India and Thailand also suffered heavy tolls. The other countries which suffered loss of life include Myanmar, Bangladesh, Maldives and the distant African countries of Somalia, Kenya, Seychelles and Tanzania.

Surprisingly Malaysia suffered only a fraction of destruction despite its being located so near to the epicentre of the earthquake. This is because of the location of Sumatra which acted as buffer for Malaysia.

Besides over one million people were affected and several others were rendered homeless.

In India, the Andaman and Nicobar Islands were the closest to the epicentre of the earthquake and hence the origin of tsunamis, and were the worst sufferers. Areas like Car Nicobar, Katchal, Nancowry, Campbell Bay, Champion Island, Chowra and Teresa Island were badly affected. In Car Nicobar, half of the total population of about 20 thousand was reportedly missing.

TABLE 9.3. Number of persons dead and untraceable due to Tsunami of 26 Dec., 2004

Country	Dead	Untraceable
1. Indonesia	1,31,934	37,066
2. Sri Lanka	35,222*	—
3. India	12,405	5,640
4. Thailand	595	2,817
5. Maldives	82	26
6. Malaysia	69	5
7. Myanmar	61	—
8. East Africa (Somalia, Kenya, Seychelles, Tanzania, Malagasy	164	139

*Only combined data of dead and untraceable are available with respect to Sri Lanka

Source: (*i*) U.S. Geological Survey, (*ii*) Geoscience Australia.

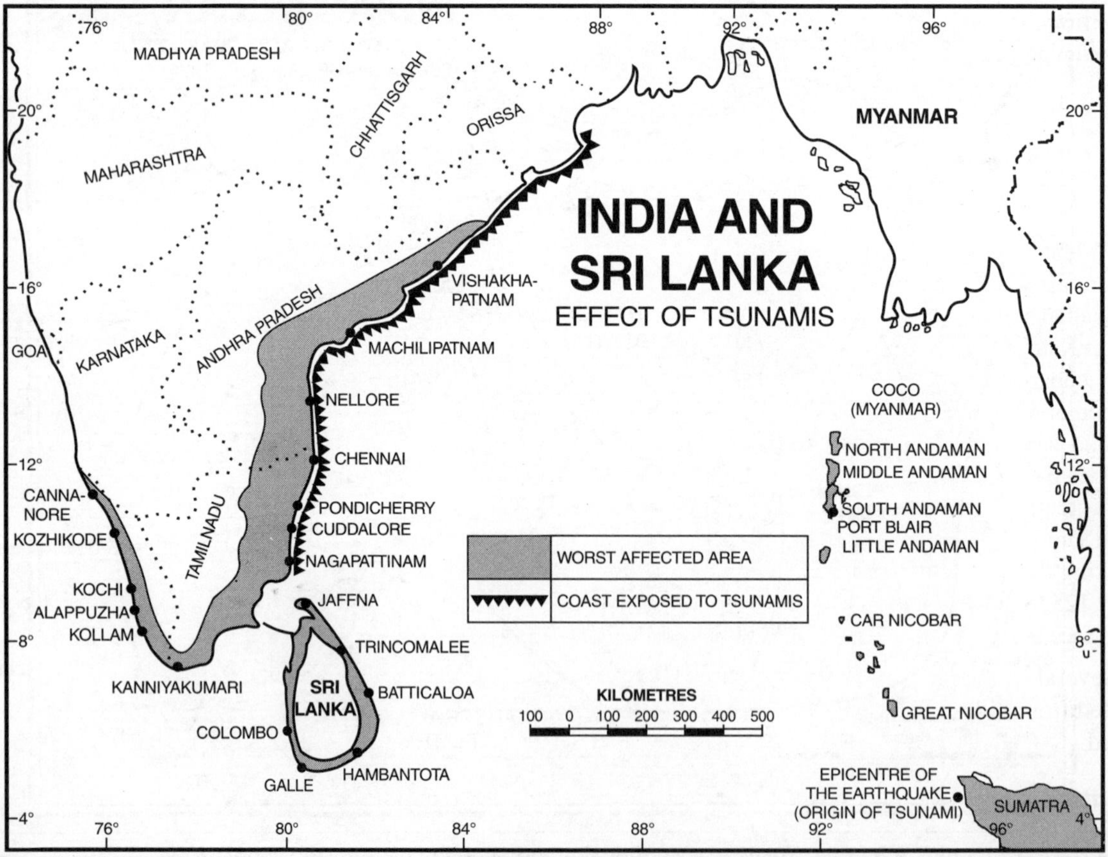

FIG. 9.6. India and Sri Lanka : Effect of Tsunamis

On the mainland of India, the main attack of tsunamis was on the coastal areas of Tamil Nadu, Andhra Pradesh and Kerala and the union territory of Puducherry. Of these three states, Tamil Nadu suffered the most. This was followed by Andhra Pradesh and Kerala.

2. Loss of Property

Property worth crores of rupees has been damaged as a result of attack by tsunamis. Infrastructural elements like houses, public buildings, transport and communication system etc. had been damaged almost beyond repair. Sea water even entered the nuclear power plant at Kalpakkam which was closed for a number of days. In Sri Lanka, rail tracks were twisted near Colombo and a train was derailed in which about 1,000 persons were killed. According to preliminary findings of the government of India, the coastal areas of India, which include large coastal tracts of Tamil Nadu, Andhra Pradesh, Puducherry and Kerala, have suffered financial losses of billions of rupees. The financial loss in the Andaman and Nicobar Islands was pegged at ₹ 2,500 crore.

3. Physiographical Changes

Tsunamis of 26th December 2004 were so strong that they could bring about drastic physiographic changes in different parts of the world. Satellite pictures of the tsunami affected areas show conspicuous changes in Chennai (particularly Adyar river course) as well as Trinket Katchall and Camorta Islands of the Andaman and Nicobar. Water level in many islands had risen, number of beaches in many islands like Campbell Bay had vanished and Trinket Island was split into two. Car Nicobar, which was worst affected sank to some extent. Indira Point, the southern-most tip of the Indian Union, was almost completely washed away, shrinking the coastline inland. These islands are hardly 125 to 200 km from the origin of tsunamis and had to face the worst fury of the killing waves.

Several other physiographic changes had been reported from Indonesia, Maldives and large areas of sea bed.

4. Motion of the Earth

The US Geological Survey expressed the opinion (as expressed by Ken Herdnut) that tremendous energy released by the earthquake made the earth wobble on its axis. According to Richard Gross, a NASA geophysicist, the earthquake might have permanently accelerated the earth's rotation due to shift of mass towards the earth's centre. This had caused the planet to spin 3 microseconds or one millionth of a second faster and to tilt about 2.5 cm on its axis. In other words, day is shortened by about 3 microseconds and the north pole has shifted towards east Siberia by 2.5 cm. Besides earth's oblateness (flattering at the poles and bulging at the equator) decreased by one part in 10 billion.

5. Decline in Soil Fertility and Agricultural Production

Vast low lying coastal areas were submerged under sea water which increased the salinity of the soil and reduced agricultural production. Cuddalore and Nagapattinam districts in Tamil Nadu were the worst affected. Soil profile tests showed high salinity varying from 6.8 to 9.10 pH value (neutral value for pH is 7). This is highly saline condition in which no paddy crop could be cultivated. In Nagapattinam district alone, more than 9,500 hectares of land had been rendered unfit for cultivation by increased salinity. Horticulture also suffered heavy losses. Farmers had been advised to sow plants like cashewnut which are saline-resistant.

6. Effect on Marine Life

The killer tsunamis had badly affected the marine life of the Indian Ocean. A large section of the coral reefs of the Andaman and Nicobar Islands archipelago had been destroyed, while others suffered extensive damages. According to marine biologists, satellite pictures showed that 45 per cent of the fragile coral reefs had been destroyed. The surviving reefs were damaged by the debris washed into the sea from the islands. Experts say it would take at least 700 to 800 years for reefs to re-form. The coral reefs around the Andaman and Nicobar Islands are of fringing type *i.e.,* they lie just off the coastline. Hence they have suffered extensive damage. Some of the species that were unique to the archipelago could have become extinct.

Fishing also suffered heavy losses at the hands of the powerful tsunamis. Sea beaches along the coasts of the Indian Ocean became graveyards of the dead fishes after the tsunami swept across the ocean. Mangrove areas that acted as nursery habitats to fish and shrimp were also damaged. The breeding, feeding and other activities of large sea mammals such as whales, dolphins etc. were also adversely affected.

Tsunami Distaster Management and Mitigation

Tsunamis are so powerful oceanic waves that it is not in man's hands to control or check them. However, following few tips may be helpful if forewarning about advancing tsunamis is received.

1. Get off the beaches and head for higher ground.
2. Stay away from rivers leading into the ocean.
3. Wait for all clear as there can be several giant waves in succession.

Advance Warning about Tsunamis

The best way of escaping the fury of the killing tsunami waves is to give advance warning of impending tsunamis to the vulnerable communities. Following few tips will help the common man to apprehend the danger of tsunamis.

(*i*) If there is a major undersea earthquake, beware of the danger because tsunamis could reach the coast in a few minutes.

(*ii*) Shaking of ground is a sufficient warning for the people living in the coastal areas about the impending tsunami.

(*iii*) A perceptible rapid rise or fall in coastal water can also be taken as a forewarning about an approaching tsunami.

(*iv*) Coastal areas less than eight metres above sea level run the greatest risk of an attack by tsunamis. They should be immediately evacuated if there is danger of an approaching tsunami.

It is possible to predict the onslaught of tsunamis by studying their speed, wavelength and depth of the sea. Tsunamis often take some time to reach the sea coast from their place of origin (**Fig. 9.7**) because the wave length and speed of tsunamis is drastically reduced when they approach the shallow sea. But on reaching the coastal shallow waters, their energy is compressed and they attack the coastal population as lethal waves. Thus a correct warning a few hours before the arrival of tsunamis can save thousands of lives and property worth millions of rupees.

Tsunami Warning System in India

Unfortunately India and other Indian Ocean rim countries did not have tsunami warning system before 26th December 2004 when the deadly tsunamis hit the coasts of Indian Ocean. Had public warnings been issued about the killing tsunami waves across the Indian Ocean, thousands of lives and property worth billions of rupees could be saved as the tsunami which hit the Sumatra Island took about three hours to crash into the Indian Coast. Pacific ocean is often hit by tsunamis and 26 countries located on the Pacific coast had already developed their scientific system of

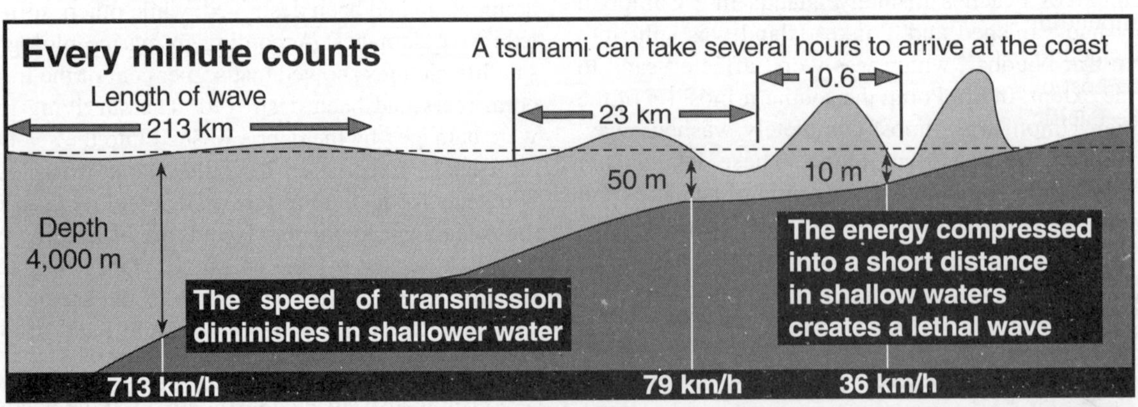

FIG. 9.7. Study of movements of Tsunamis can help in warning the people about the coming waves

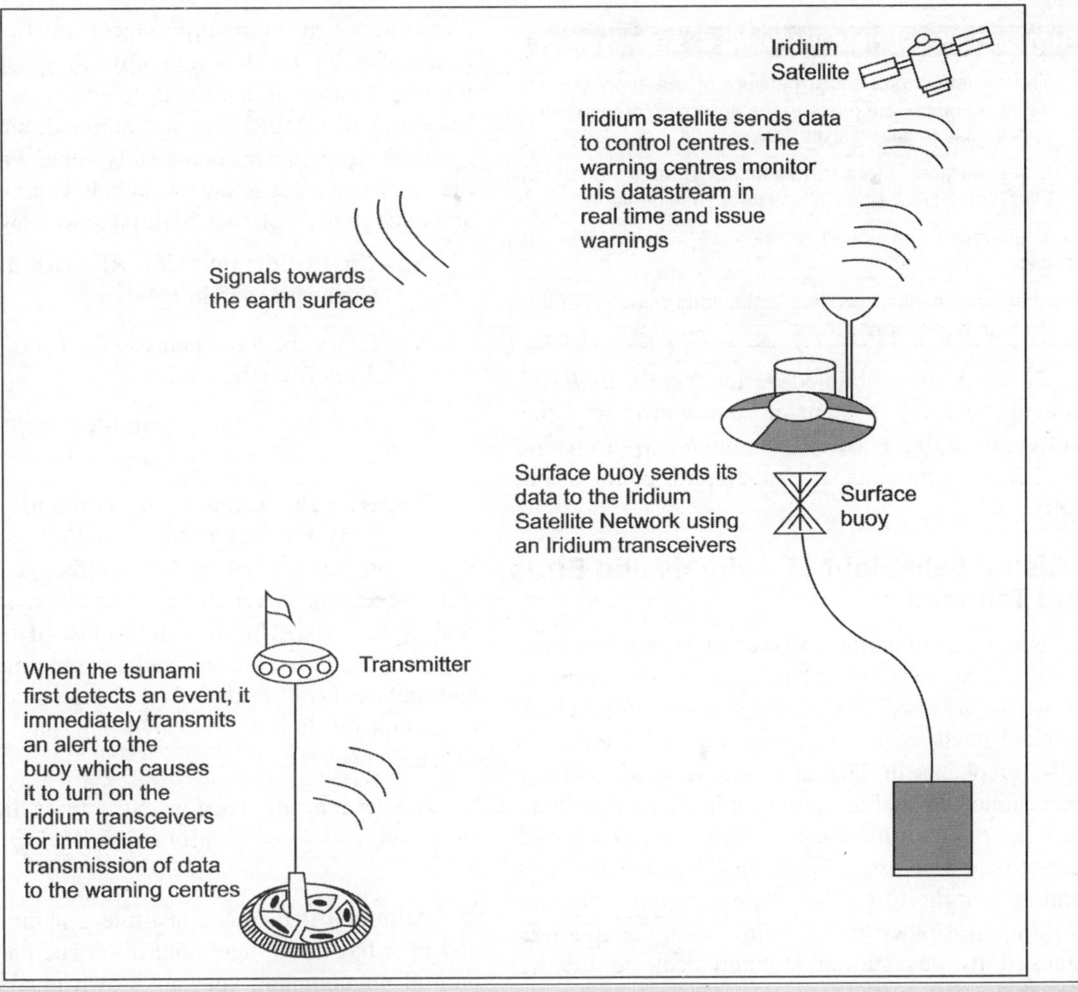

FIG. 9.8. India's Tsunami warning system.

Source : NOAA Centre for Tsunami Research.

impending tsunami. After the catastrophe, India decided to install her own tsunami warning system at the cost of ₹125 crore rather than becoming a part of the Pacific Warning System. It was planned to install the warning system by the end of the year 2007 and was inaugurated on 15th October, 2007. The system comprises a real-time network of seismic stations, bottom pressure recorders (BPR) and the tide gauges to monitor earthquakes in the sea and tsunamis triggered by them. The centre receives the data from the seismic network of the India Meteorological Department and other international set ups.

The system can detect earthquakes above 6 magnitude occurring in the Indian Ocean in less than 20 minutes and alert agencies within 13 minutes. This is done with the help of the BPRs—the key sensors that confirm the triggering of the tsunami. Six BPRs, have been installed in all out of which four are in the Bay of Bengal and two are in the Arabian Sea. The system was successfully tested during the 8.4 magnitude earthquake that hit Indonesia on September 12, 2007.

The centre generates and disseminates timely advisories to the control room of the Home Ministry which issues an alert to the public. To warn the ministry, a satellite based network for disaster management support has been established. This enables the centre to also issue alerts to the state emergency operations centres. The messages are to be sent by telephone, SMS and e-mails to the authorised officials.

HOW INDIA'S TSUNAMI WARNING SYSTEM WORKS

- The system comprises a network of real-time seismic stations and bottom pressure recorders (BPRs) installed in deep sea to detect disturbances.
- It senses all tremors in the Indian Ocean measuring over 6 on Richter scale in less than 20 minutes.
- The system can sound an alert 13 minutes after an undersea quake.
- After equipment sends data to the early warning centre, a computer generates alerts.

It has been acclaimed as the ***"most modern"*** tsunami warning system in the word by **Peter Koltermann**, the head of the tsunami co-ordination unit intergovernmental oceanographic commission, UNESCO.

Unusual Behaviour of Animals and Birds and Tsunamis

As in case of earthquakes, animals start behaving in unusual manner, before tsunamis actually strike the coast. At the time of catastrophic tsunamis which wrecked havoc in Indian Ocean on 26th December 2004, elephants in Thailand started wailing. They soon calmed down, but started wailing again about an hour later. This time they could not be comforted despite their ***mahauts***' efforts. The elephants just kept running for the hill. The elephants that were not working, broke their hefty chains. Soon the area was attacked by devastating tsunamis caused by the earthquake (8.9 on Richter scale) of Sumatra.

Abnormal behaviour of birds and animals was also observed in the coastal belt of Tamil Nadu (the worst hit area in India). Animals in forests were feeling restless. Birds stopped chirping and there was unusual silence. About two months before the tsunamis struck the Indian coast, fishermen of Puducherry found fish with reddish tails, called ***Red Bait***, in their trawler nets. The sighting of this fish had always preceded a natural calamity. Catches of similar type in 1977, 1979 and 1996 were followed by major cyclones. It is a fishermen's instinct that Red Bait portends disaster and it has often come true. According to Central Marine Fisheries Research Institute (CMFRI), the Red Bait (measuring about 12 cm when fully grown and is edible) is a deep water fish which surfaces during 'upwelling' of water (a phenomenon of water at the bottom coming-up). This phenomenon is more pronounced on the Western Coast of India. But this time upwelling happened on the East Coast and that too during the non-upwelling season. Still CMFRI was not alarmed, since unlike cyclones, tsunami was completely unknown. Some of the other instances of abnormal behaviour of animals and birds noticed on that fateful day were as under.

- Crows flew into CMFRI's fish hatchery in Chennai and didn't budge.
- Cows on the Chennai beach ran like mad away from the coast.
- At Point Calimere Wildlife Sanctuary, black bucks and deer fled towards higher ground.

Vegetation. Impact of tsunamis can be considerably reduced by growing sea coast vegetation. Sea vegetation can be effective protection if the vegetation cover along the coast exceeds 70 per cent of the area. Unfortunately most of the natural vegetation has been destroyed for obtaining fodder and fuel. In Tamil Nadu only 110 km stretch out of a total coastal length of 1,076 km has sufficient vegetation cover.

Structural Protection. Structures like walls, ridges etc. can act as protective devices to some extent.

Although tsunamis constitute a natural tragedy and man has hardly any control on such a powerful natural phenomenon yet man's own misdeeds have made the situation very dangerous and helped in aggravating the fury of tsunamis. For example, no construction is allowed within 500 metres of the coast. There is legal obligation in India to leave this stretch of land vacant but all sorts of constructions residential, commercial, recreational etc.) are carried on without caring for the law. People living in this belt are most vulnerable to tsunamis and other sea related hazards.

Mining of sand and other minerals from the sea, as is done in Kanniyakumari, makes the concerned area highly vulnerable to tsunamis. The sand deposited on the sea coast absorbes much energy of the waves, and saves much of the coastal areas from their fury. There is urgent need to check unwanted construction and mining in ecologically sensitive coastal areas.

DROUGHTS

Introduction and Definition

Drought is a temporary reduction in water or moisture availability significantly below the normal or expected amount for a specific period. According to High Powered Committee on Disaster Management Report, "Any lack of water to satisfy the normal needs of agriculture, livestock, industry or human population may be termed as a drought." This condition occurs either due to inadequacy of rainfall, or lack of irrigation facilities, under-exploitation or deficient availability of water for meeting the normal crop requirements in the context of the agro-climatic conditions prevailing in any particular area. This has been scientifically computed as Moisture Index (MI). Drought, in this context, can be defined as adverse MI or adverse water balance which may be attributable not only to a prolonged dry spell due to lack of sufficient rainfall but also due to such other factors as excessive evapo-transpiration losses, high temperature, low soil holding capacity, etc. The inadequacy is with reference to the prevailing agro-climatic conditions in any particular area. Therefore, there is a drought in Jaisalmer (Average annual rainfall 200 mm) if rainfall is not sufficient to grow grass and paltry coarse-grains, whereas in Bolangir or Koraput (Odisha-annual rainfall above 1000 mm) there is a drought if there is not enough rainfall for bringing the paddy crop to maturity.

Drought is a distress situation caused by lack of rainfall. The failure of rains may be reviewed from two aspects. Firstly, the rainfall may be insufficient. Secondly it may be sufficient for the region as a whole but with wide gap, separating two wet spells. Thus both the amount as well as spatial distribution of rainfall are important. In other words, drought is a relative phenomenon. Therefore, the amount of rainfall is not that important as its effectiveness.

Types of Drought

Following main types of drought may be recognised.

1. Meteorological Drought

It describes a situation where there is a reduction in rainfall for a specific period (days, months, seasons or year) below a specific amount (long term average for a specific time). The India Meteorological Department (IMD) has defined drought as a situation occurring in any area when the mean annual rainfall is less than 75% of the normal rainfall. IMD has further classified droughts into two broad categories *viz.* a ***severe drought*** when the deficiency of rainfall exceeds 50% of the normal rainfall and ***moderate drought*** when the deficiency of rainfall is between 25 and 50% of the normal rainfall.

It is worth noting here that the effectiveness of rainfall is more important then the amount of rainfall so far as meteorological drought is concerned. On an average India receives 118 cm annual rainfall which is considered to be highest anywhere in the world for a country of comparable size. But the uncertain, unreliable and erratic nature of rainfall by south-west monsoons creates drought conditions in different parts of the country. The major causes of meteorological drought may be summed up as under :

(*i*) Lean monsoons and below average rainfall due to absence of depressions over India.

(*ii*) Late onset or early withdrawal of monsoons.

(*iii*) Prolonged breaks in monsoon.

(*iv*) Re-establishment of southern branch of jet stream.

2. Hydrological Drought

Hydrological drought is associated with reduction of water. A meteorological drought often leads to hydrological drought. Generally it takes two successive meteorological droughts before the hydrological drought sets in. There are two types of hydrological droughts *viz.*, (*i*) surface water drought and (*ii*) ground water drought.

(i) Surface-water Drought. It is concerned with drying up of surface water resources such as rivers, streams, lakes, ponds, tanks, reservoirs, etc. There are many processes, besides meteorological drought, which lead to surface water drought. Large scale deforestation is the main cause of surface water drought. Some other unwanted human activities have led to the enhancement of flood/drought duo. Important among them are ecologically hazardous mining, indiscriminate road construction, and spread of non-terraced agriculture. In the Doon valley limestone quarrying has drastically changed the surface water flow in the valley turning several perennial rivers into carriers of monsoon floods which go dry after the monsoon.

(*ii*) ***Ground-water Drought.*** Ground-water drought is associated with the fall in the ground water level. This happens due to excessive pumping of ground water without compensatory replenishment and creates more or less irreversible ground water drought even in normal rainfall conditions. Ground water replenishment depends upon the availability of surface water obtained from rainfall and nature of soil. The Northern Plain of India has soft and porous alluvial soils. These soils permit percolation of water and help in ground water replenishment. In contrast, the peninsular plateau area is made up of hard and impervious rocks which hinder the process of ground water replenishment.

3. Agricultural Drought

Agricultural drought is concerned with the impact of meteorological/hydrological drought on crop yields. When soil moisture and rainfall conditions are not adequate enough to support a healthy crop growth to maturity thereby causing extreme moisture stress and wilting of major crop area, it leads to agricultural drought. Agricultural drought may occur even when there is no meteorological drought and *vice-versa*. It is worth noting that agricultural drought is a relative category, depending upon the value of plant and soil. What could be a drought condition for the cultivation of rice, could well be a suitable condition for wheat and a condition of excess soil moisture for dry crops like bajra or jowar. Thus the choice of crops evolve according to variations of climatic and soil conditions. The indigenous dry-crops prove very high-yielding when there is optimum use of water. Howeveı, under extreme condition of soil water drought, no plant can survive and this condition is termed as ***desertification.***

With the onset of green revolution in India in 1960s, High Yielding Varieties (HYV) of seeds have replaced the traditional drought resisting seeds. The use of fertilisers has also increased the water requirements of different crops. The agriculture based on the Green Revolution technology is precariously dependent on irrigation and any delay in supply of water will cause serious agricultural drought. It has been found that the organic matter content of HYV seeds is quite less. Organic matter input to soil increases its water holding capacity and soils rich in organic content don't dry quickly. Hence the use of HYV seeds has increased the risk of agricultural drought to a great extent. Recent studies have revealed that with the addition of manure and organic fertiliser, the water retentivity of the soil increases by 2 to 5 times.

Change in cropping pattern, particularly with the introduction of Green Revolution technology, has also led to soil water drought. The indigenous cropping pattern was in tune with the agroclimatic conditions of the concerned region. With the introduction of an alien cropping pattern, more water demanding crops have gained importance which has led to soil water drought. For example, rice cultivation was practically unknown in Punjab, Haryana and Western Uttar Pradesh before Green Revolution and this region is not suitable for rice cultivation considering the amount of rainfall received by this region. The area receives an average annual rainfall from 50 to 75 cm whereas minimum amount of rainfall for any successful cultivation of rice is 100 cm. This change in the cropping pattern has put heavy strain on meagre water resources leading to acute occasional droughts.

In the drought affected areas of Maharashtra, Karnataka and Andhra Pradesh there is a change in the cropping pattern in which cash crops are given more importance than the traditional crops. Thus while the staple crops are denied water, cash crops like sugarcane and grapes are given priority with respect to irrigation. This leads to a condition of soil water drought which is created not by absolute scarcity of water but by preferential diversion of limited water resources.

4. Soil Moisture Drought

This is a situation of inadequate soil moisture particularly in rainfed areas which may not support crop growth. This happens in the event of a meteorological drought when the water supply to soil is less and water loss by evaporation is more.

5. Socio-Economic Drought

It reflects the reduction of availability of food and income loss on account of crop failures endangering food and social security of the people in the affected areas.

6. Famine

A famine occurs when large scale collapse of access to food occurs which, without intervention, can lead to mass starvation and even deaths.

7. Ecological Drought

Ecological drought takes place when the productivity of a natural eco-system falls significantly as a consequence of distress induced environmental damage.

Causes of Droughts

Droughts in India occur in the event of a weak south-west monsoon. A weak monsoon results in deficient rainfall and droughts occur. Droughts also occur due to late arrival or early withdrawal of the monsoons. Prolonged breaks in the monsoon during rainy season also result in droughts. Although a drought may occur at any time and in any part of the country, most of the drought prone areas are those having marginal rainfall and high variability of rainfall.

Although there can be several reasons for weak monsoon and deficit rainfall, yet El Nino effect is supposed to be the most probable reason for such a situation. However, El Nino does not always lead to deficiency in rainfall by the South-West monsoons. Analysis of data since 1880 show that less than half of El Nino events have caused poor rainfall. But data from 1950 to 2015 show that 10 of the 13 drought years have coincided with El Nino. In 1997, a strong El Nino year, rainfall was normal. Yet, in 2009, an El Nino triggered India's worst drought in three decades.

TABLE 9.4. Rainfall Vagaries and Drought Conditions

1. Deficient rainfall	A drought occurs when the rainfall is less than 75% of the normal.
2. Delay in onset	Maximum of three weeks from the normal date of arrived for a given region
3. Timely onset and sudden break	A maximum break of two weeks after the timely onset of the monsoon
4. Early withdrawal	A drought occurs when the monsoons withdraw from north-west India by the last week of August

Following are indications of a drought with respect to monsoon rainfall (**Table 9.4**).

Effects of Droughts

Droughts have a wide range of effects on the masses in a developing country like India. The impact of droughts is specifically conspicuous in view of the tropical monsoon character of the country. Rainfall by the south-west monsoon is notorious for its vagaries. The impact of droughts in India can be summed up under the following headings :

(*i*) Physical Impact. Meteorological drought adversely affects the recharge of soil moisture, surface runoff and ground water table. Soils dry up, surface runoff is reduced and ground water level is lowered. Rivers, lakes, ponds and reservoirs tend to dry up, wells and tube-wells are rendered unserviceable due to lowering of the ground water table.

(*ii*) Impact on Agriculture. Indian agriculture still largely depends upon monsoon rainfall where over half of the arable land lacks irrigation facilities and is termed as ***rainfed***. The effect is manifested in the shortfalls of agricultural production in drought years. History is replete with examples of serious shortfall in cultivated areas and drop in agricultural productivity. Severe shortage of foodgrains had been felt and the country had to resort to import of foodgrains to save the poor people from hunger and starvation. However, India has been able to build a buffer stock of foodgrains and threat from droughts is not as serious as it used to be before the Green Revolution.

It is worth mentioning here that the shortfall in agricultural production may be the direct impact of meteorological droughts but the succeeding hydrological and agricultural droughts have a long range and far reaching impact on agriculture. This impact may be in the form of changes in the cropping patterns and impoverishment of cattle.

(*iii*) Social and Economic Impact. Social and economic impact of a drought is more severe than the physical and agricultural impacts. A drought is almost invariably associated with famine which has its own social and economic consequences. The impact of drought manifests itself in the following sequence :

1. Decline in cultivated area and fall in agricultural production.
2. Fall in employment in agricultural sector.
3. Fall in purchasing power.
4. Scarcity of drinking water, foodgrains and fodder.
5. Rise in inflation rate.
6. Distress sale of cattle and loss of cattle life.
7. Low intake of food and widespread malnutrition.
8. Ill health and spread of diseases like diarrhoea, dysentery, cholera and opthalmia caused by malnutrition, hunger and starvation.
9. Distress sale and mortgage of land, jewellery and personal property.
10. Migration of people from drought hit areas to other areas in search of livelihood and food.
11. Death due to malnutrition/starvation/diseases.
12. Slowing down of secondary and tertiary activities due to fall in agricultural production and decline in purchasing power.
13. Low morale of the people.
14. Social stress and tension, disruption of social institutions and increase in social crime.
15. Growth of fatalism and belief in supernatural powers and superstitions.

The greatest impact of a drought is seen on the weaker sections of society. These include landless labourers, small and marginal farmers as well as artisans like weavers. Such people live in hand to mouth economy and do not have enough stock to sustain in the event of a drought. Whatever little stock they have, it is quickly exhausted and they are compelled to go in for distress sale or mortgage their belongings to rich landlords. Thus whereas a drought situation brings miseries and sufferings for the poor people, the rich people take undue advantage of the situation and exploit the poor people. Often the poor becomes poorer and the rich becomes richer in a drought situation. A series of bad harvest plunges the small and marginal farmers in a vicious circle of poverty making them landless and penniless. The money-lender charges high rates of interest and the inability of the farmers to repay the loan compells them to forfeit their mortgaged property. In extreme cases, the farmers tend to commit suicide. Cases of suicide by farmers in Andhra Pradesh, Telangana, Karnataka, Odisha, Maharashtra and even in agriculturally rich states of Punjab and Haryana have been reported from time to time. (**Table 9.4**)

Drought Prone Areas

A drought prone area is defined as one in which the probability of a drought year is greater than 20 per cent. A chronic drought prone area is one in which the probability of a drought year is greater than 40 per cent. A drought year occurs when less than 75 per cent of the normal rainfall is received.

In India, the hard core of recurring drought affects 16 per cent of the total area and 12 per cent of the population, although the total average drought prone area may be as much as 10 lakh sq km or about one third of the total land area of the country. The areas regularly haunted by droughts are those receiving low (generally below 75 cm annually) and highly unreliable (variability over 40 per cent) rainfall and with inadequate irrigation facilities. In all 77 districts receiving less than 75 cm of rainfall per annum are drought prone. This accounts for 34 per cent of the net sown area. In addition, there are 22 districts in Maharasthra, Gujarat, Madhya Pradesh, Karnataka, Rajasthan and Uttar Pradesh accounting for 9 per cent of the cultivated area of the country which receives 75 to 85 cm of rainfall per annum. This rainfall is of doubtful efficacy and as such these districts should also be considered vulnerable to drought. However, the severest droughts have occurred in comparatively wet areas such as West Bengal, Bihar and Odisha where rainfall is normally plentiful enough to allow high density of population and where failure of rainfall can affect millions of people.

There are some well defined tracts of drought which are briefly described as under (**Fig. 9.9**).

(*a*) The desert and semi-desert region covering about 6 lakh sq km. This area is demarcated by lines from Ahmedabad to Kanpur on one side and from Kanpur to Jalandhar on the other. Some areas in this

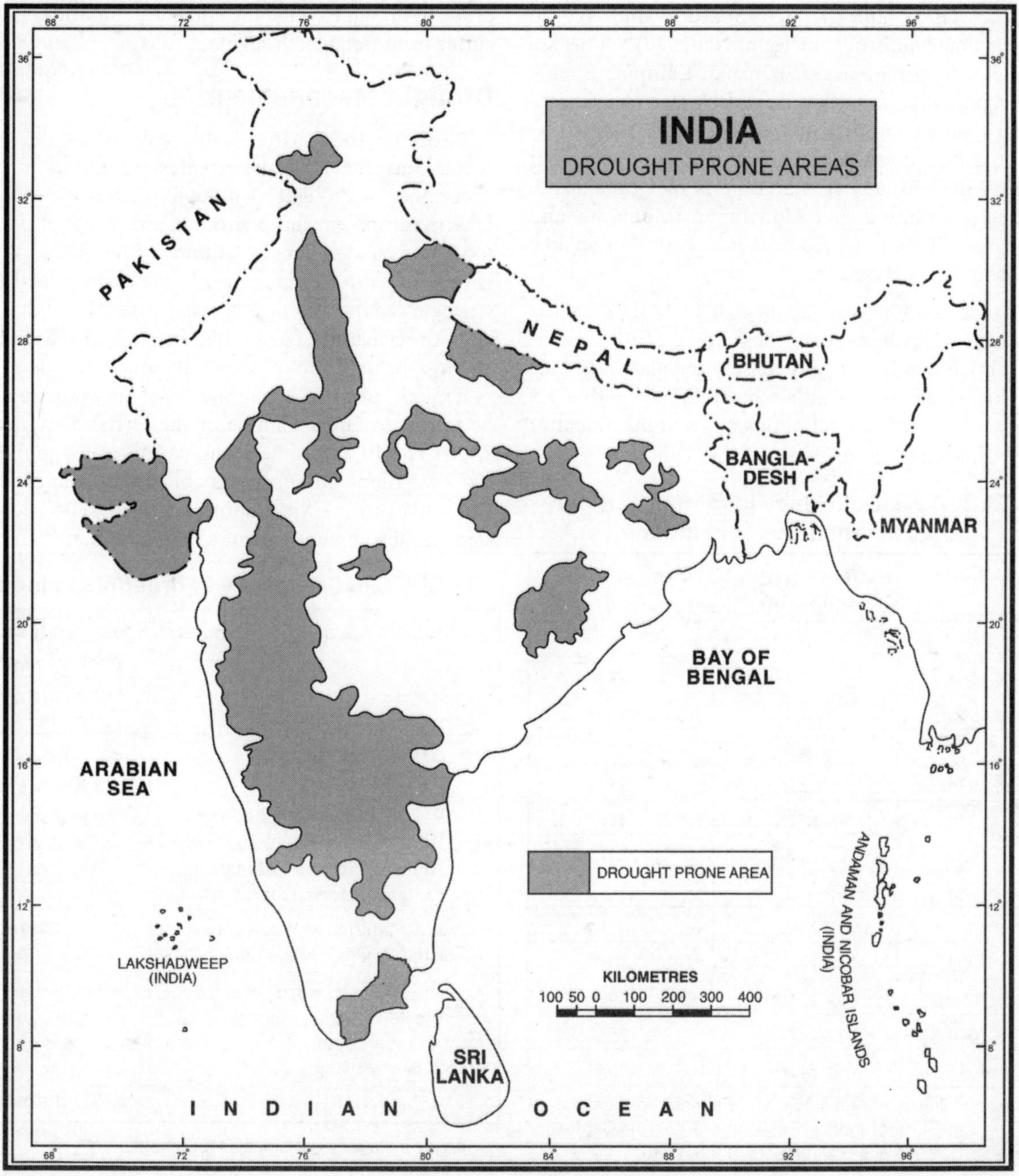

FIG. 9.9. India : Drought Prone Areas

region are without irrigation and comprise the worst famine tracts of the country.

(*b*) The dry region lying in the leeside of the Western Ghats upto a width of 300 km stretching eastwards up to 100 km from the East Coast and reaching right upto the southern tip of the peninsula. It covers an area of about 3.7 lakh sq km.

(*c*) Outside the above mentioned two major regions, there are isolated pockets which experience frequent droughts and are termed as drought prone areas. They are (*i*) Coimbatore and Nellai Kottabomman districts in Tamil Nadu; (*ii*) Saurashtra and Kachchh regions; (*iii*) Jhansi, Lalitpur, Banda, Mirzapur, Pilibhit, Kheri and Bahraich districts of Uttar Pradesh; (*iv*) Palamau plateau of Jharkhand; (*v*) Purulia district of West Bengal and (*vi*) Kalahandi region of Odisha, (*vii*) Large parts of Uttarakhand, and (*viii*) Jammu and Uddampur in Jammu and Kashmir. These scattered pockets account for an area of about one lakh sq km.

In a drought year, about half of India's area is affected by drought conditions. Some of these areas are agriculturally rich and densely populated. Hence, the losses due to drought are much higher. **Table 9.5** gives an idea of rainfall deficiency and the extent of area affected by droughts.

TABLE 9.5. Deficiency in rainfall and the resultant drought in India

Year	Deficiency in rainfall (%)	Area affected (%)
1951	–18.7	34
1965	–18.2	37
1966	–13.2	34
1972	–26.0	40
1974	–12.0	33
1979	–18.0	34.8
1982	–14.5	28.9
1985	–7.1	43.3
1987	–25.0	17.9
2002	–19.0	10
2009	–22.0	59
2014	–12.0	25
2015	–14.0	39

Source : Ministry of Agriculture and Farmers Welfare quoted in State of India's Environment (2016), Centre for Science and Environment, p. 58.

Frequency of Droughts

Frequency of droughts refers to the number times droughts occur in a given duration of time. Different parts of the country experience different frequencies of droughts depending on the amount of rainfall, variability of rainfall and water requirements. **Table 9.5** shows that areas with higher rainfall are less prone to droughts while the arid and semi-arid regions suffer from frequent droughts.

Drought Management

Among the various natural disasters the one which has received the greatest attention is the occurrence of drought. A drought often leads to total loss of crops or sharp drop in the production of foodgrains and creates conditions of famine. History is replete with great famines in the country and the strategies adopted to face the challenges. Drought is a slow onset natural hazard and offers sufficient time and opportunity to mitigate its impact. The first systematic attempt at famine relief measures could be traced to Great Famine or the Orissa Famine of 1866. The first Famine Commission was appointed by the then Government of India under the Chairmanship of Sir George Cambell. The second step was the appointment of a Famine Commission in 1880 following the famines of 1873 and 1876-78. Following the recommendations of the Famine Commission, the administration decided to promulgate Famine Codes from 1883 onwards which ushered in the modern policy of relief administration.

TABLE 9.6. Frequency of droughts in India

Meteorological sub-divisions	Recurrence of the period of highly deficient rainfall
1. Assam, N.E. Region	Very rare, once in 15 years
2. West Bengal, Chhatisgarh, Bihar, Madhya Pradesh, Coastal Andhra Pradesh, Maharashtra, Konkan, Kerala, Odisha	Once in 5 years
3. West interior Karnataka, Eastern Uttar Pradesh, Vidarbha	Once in 4 years
4. Gujarat, Eastern Rajasthan, Western Uttar Pradesh, Tamil Nadu, Kashmir, Rayalseema, Telangana	Once in 3 years
5. Western Rajasthan	Once in 2.5 years.

Source : India Meteorological Department.

Moving from Drought Relief to Drought Management. After Independence, India experienced major droughts in the years 1965-67, 1972-73, 1979-80 and 1985-88. The analysis of India's drought response efforts reveals that the pre-1987 efforts to mitigate droughts were primarily focused to intervene at later stages, particularly at the socio-economic drought stage. The approach was to deal with ultimate consequences of the drought thus preventing the deterioration drought leading to famine conditions, essentially providing employment opportunities to the affected population through relief projects and distribution of foodgrains through public distribution system (PDS). The contingent drought relief expenditure imposed a serious strain on public finances as huge amounts had to be diverted from development for undertaking relief projects. Analysis of the drought relief expenditure incurred by the Rajasthan Government in 1980s reveals that the drought relief outlays exceeded development outlays.

By incorporating contingency crop planning with drought response efforts through relief approaches acquired the status of Drought Management Strategies in early 1970s. The drought management approach differed from drought relief approach with regard to objectives, reliance of early warning indicators and timing of public intervention. The objective of relief approach was to protect entitlements of the affected population by ensuring physical and economic access to food through relief projects and public distribution system of foodgrains. As against this, the drought management approach aimed at ensuring entitlement to produce food so as to obviate the need for taking up ***ad hoc*** relief projects. While drought relief approach relied on socio-economic indicators like crop production data, price rise, migration of the people and increased rate of petty crimes, etc. for drought declaration and intervention, drought management approach relied on hydro-agro indicators like rainfall, water level of reservoirs and progress of cropping pattern to detect early signs of developing drought situation. While drought relief approach enabled the government to intervene only in the month of November/December when the rainy season is over and the kharif crops have been harvested, the drought management approach enabled the government to intervene in the monsoon season itself.

According to report of the High Powered Committee on Disaster Management (2002), "Drought management is generally by focus on employment generation, water conservation and power supply, standing crop saving and public distribution supplies of essential commodities".

Establishment of Crop-weather Watch Group

This watch group was set up during 1979 drought by the Ministry of Agriculture. It consists of representatives from the Department of Agriculture, India Meteorology Department, Indian Council of Agricultural Research, Ministry of Information and Broadcasting, and others. A two pronged strategy was adopted which focused on curative and preventive measures. They were to provide weekly reports of rainfall, agricultural operations, employment and other activities for occurrences affected by drought. The twelve-point programme was created to avert ***Trikal (Akal, Jalkal, Tinkal)***, which means to take care of food, water, and fodder. Although drought affected 240 districts and 220 million people, the severity was not widely felt and no starvation deaths were reported because of the buffer stocks of food grains available with the government.

Integrated Watershed Management

Integrated watershed management as a preventive measure plays a key role in moderating drought conditions. This approach ensures planning on the basis of the total available water resources, conjunctive use of surface and groundwater, allocating priority for rational use of water and also the preparation of a coordinated plan. Thus, watershed management holds the promise of conservation of land and water resources and their optimal utilization in reality. Under the National Watershed Programme for Rainfed Areas, a large number of watersheds have been established in different rainfed regions of the country.

FLOODS

Definition and Introduction

Flood is a state of high water level along a river channel or on sea coast that leads to inundation of land which is normally submerged. Flood is an important component of hydrological cycle of a drainage basin. As a matter of fact droughts and floods are two extreme ends of the hydrological cycle. While droughts occur due to failure of rainfall caused by the southwest monsoons, floods occur in the event of excessive rainfall. Flood is a natural hazard which occurs in response to heavy rainfall and it *becomes a disaster when it inflicts heavy loss to life and property*.

Causes of Floods

Floods are generally caused by one or more unfavourable meteorological and physical factors. A combination of unfavourable meteorological and physical factors working together leads to a serious flood situation resulting in a disaster. In recent times, the impact of meteorological and physical factors has been accentuated by unwanted human activities. Following are the major causes of floods in India.

A. *Meteorological Factors*

1. Heavy and Continuous Rainfall
2. Cloud Burst
3. Tropical Cyclones

B. *Physical Factors*

1. Large Catchment Area
2. Obstacles in the flow of River

C. *Human Factors*

1. Deforestation
2. Siltation
3. Faulty Agricultural Practices
4. Faulty Irrigation Practices
5. Bursting of Dams
6. Accelerated Urbanisation

A brief description of the above mentioned causes is given as under :

A. Meteorological Factors

1. Heavy and Continuous Rainfall. A flood situation arises whenever heavy rainfall by the south-west monsoon continues for a sufficiently long period. India receives annual rainfall of 118 cm which is highest for a country of this size. Areas located in the east of 78°E longitude receive more than 100 cm annual rainfall. About 80% of this rainfall is caused by the south-west monsoons during the rainy season from June to September. Thus rainy season is marked by floods in a large number of river basins while the rest of year remains dry. The west coastal plain and the western slopes of Western Ghats receive more than 250 cm annual rainfall and the danger of flood always looms large there. Some parts of catchment areas of the Brahmaputra and its tributaries receive over 500 cm annual rainfall and that area experiences floods almost every year.

2. Cloudbursts. Cloudburst is an event when torrential rainfall occurs in a short period of time. Cloudburst is a frequent event in the Himalayan region, Odisha, Central and Western India. The Gujarat cloudburst in 1979 caused very heavy rainfall of 35 to 50 cm in a short duration just one day. As a result of heavy and concentrated rainfall Machhu Dam was damaged and the whole of Morvi town was washed away. Cloudburst in the Himalayan region in 1971 caused floods in vast areas.

3. Tropical cyclones. In the coastal areas of India, especially the coast of Andhra Pradesh, Odisha and Gujarat, the greatest cause of floods comprise the tropical cyclones originating in the Bay of Bengal and Arabian Sea. The tropical cyclones of 1982 and 1983 resulted in rise of water level by more than two metres in as many as 27 dams. The Super Cyclone which hit Odisha in 1999 flooded vast areas and caused death to 10,000 people.

B. Physical Factors

1. Large catchment areas. Most rivers of India have large catchment areas which results in excess of water in medium and small rivers. Large catchment area is the main cause of floods in the Ganga and the Godavari rivers.

2. Obstacles in the flow of river. Often the normal flow of the river is obstructed due to landslides, sediment deposition in the river bed, meandering, etc. A sort of dam is built across the river and water starts accumaling behind the dam. The process of water accumulation continues till the dam yields to pressure of increasing water. Once this dam is broken, water gushes downstream and large area is flooded.

C. Human Factors

Apart from meteorological and physical factors as mentioned above, human factors also cause widespread floods in different parts of the country. Unwanted activities of man not only cause floods, they also accentuate the non-human causes of floods. Main human causes are briefly described as under :

1. Irrational Deforestation. There is large scale deforestation by humans for obtaining different forest products. Roots of plants can absorb much of rain water and recharge ground water. This results in limited surface flow and avoids flood situation. In the absence of forests, the surface water flow increases which causes floods over vast area. Large scale deforestation in the Shiwalik Hills. Lesser Himalayas, Chota Nagpur Plateau and Western Ghats has resulted in increased frequency and intensity of floods.

2. Faulty Agricultural Practices. In most parts of the country fields are ploughed along the slope of land. This facilitates the easy flow of water along the slope and causes flood in the lower areas. For avoiding such a situation fields should be ploughed in a direction perpendicular to the slope of land. This is known as **contour ploughing**.

3. Faulty Irrigation Practices. There is a close network of irrigation canals in Punjab, Haryana and Western part of Uttar Pradesh. Canal water percolates into the ground. This results in rise in ground water level and leads to water logging. In such a situation, even a small amount of rainfall can cause floods over vast areas.

4. Other Causes. Siltation, bursting of dams, and accelerated urbanisation are other human causes of floods. Rapidly increasing urbanisation leads to floods. Almost all the urban garbage is thrown into the rivers. This reduces the water carrying capacity of the river and obstructs the natural flow of the river water. In such a situation, water accumulates in the river and floods are caused. Natural flow of river is also obstructed by constructing bridges, roads and dams which causes floods.

Effects of Floods

Floods have multipronged effects on human life. A more frightening fact is that floods are becoming more damaging as their frequency, intensity and magnitude increases with the passage of time. The most damaging effect of floods is the colossal loss of life and property. The other losses include damage to crops, cattle loss, break down of communication, dislocation of transport system and disruption of essential services.

Earlier the impact of floods was not as intense and widespread as it is now. There are several reasons for this deteriorating condition. Floods are becoming more frequent and furious due to ecological degradation such as deforestation. Moreover, the population in India is increasing at a rapid pace and the areas prone to floods, such as flood plains and even river beds are inhabited by the people. The principle '*where a river has the right of way stay out of its way*' is not followed by the people who have little option in setting themselves or locating industrial projects. In fact floods cause more damage than any other single disaster. India is the most flood affected country in the world next only to Bangladesh. The most severely affected are generally the poor people because they live on the periphery of human habitat. **Table 9.7** gives an idea of flood damages in India.

TABLE 9.7. Annual Average Flood Damages (based on data from 1953 to 2000)

Damage Head	Magnitude
1. Land area affected	7.56 million hectare
2. Population affected	32.03 million
3. Human lives lost	1,504 number
4. Livestock lost	96,713
5. Houses damaged	11,683 (₹ 136.615 crore)
6. Crops damaged	₹ 460.07 crore
7. Public utilities damaged	₹ 377.248 crore
Total Losses	**₹ 982.126 crore**

Source : Manual on Natural Disaster Management in India, NDM Division (2001), p. 29.

According to the latest report of the UN office for Disaster Risk Reduction (UNDRR), India is second worst affected by floods after China with 17 floods every year compared to China's 20 floods per year.

Flood Prone Areas

The Rashtriya Barh Ayog (RBA) or National Commission on Floods, set up by the Government of

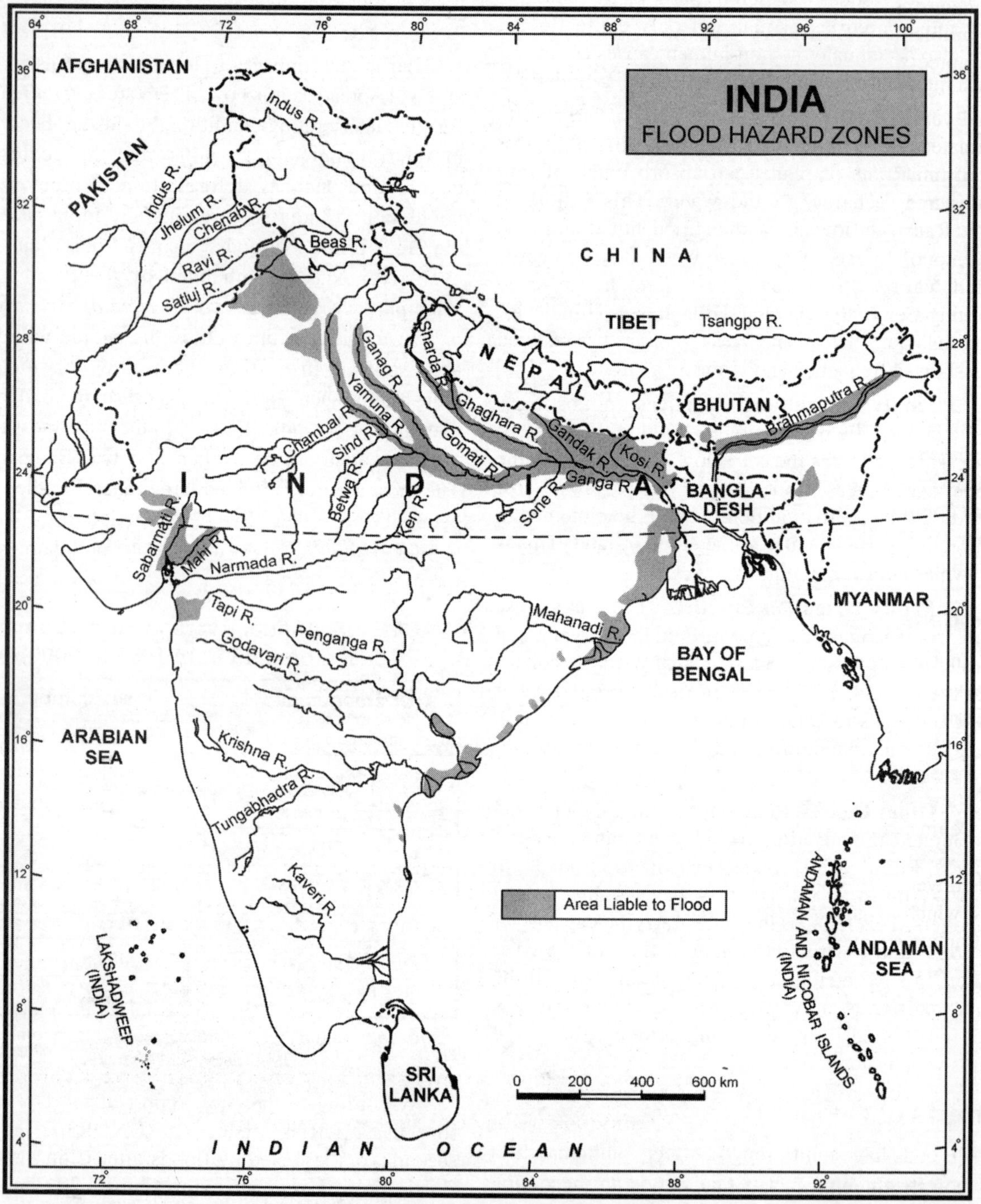

FIG. 9.10. India : Flood Hazard Zones

India in 1976, provided statistical evidence of the flood problem in the country. The commission took the maximum area affected by floods in a state in any one year, as its flood prone area and added up the flood prone areas of all the states to get the flood prone area of the country. This proved to be erroneous method as it underestimates the severity of the problem. This is due to the fact that there is no guarantee that floods in any year will affect only those areas which were affected during the maximum flood year. On several occasions, flood waters have inundated areas which were never flooded before. Yet the commission found that the country's flood prone area increased from 25 mha in 1960s to 34 mha in 1978. Rashtriya Barh Ayog (RBA) had estimated in 1980, total flood prone area in the country as 40 mha which was revised further to 49.815 mha by the Working Group on Flood Management set-up by the Planning Commission for the XII Five Year Plan. This shows that there has been a rapid increase in the flood-prone area of the country. A glance at **Fig. 9.10** will show how the flood prone areas are distributed. It is estimated that over three-fourths of the total damage done to crops and property is in the plains of Northern India comprising of Punjab, Haryana, Uttar Pradesh, Bihar, West Bengal and Assam. The geographical distribution of flood prone areas in India is as under :

The Ganga River Region

In Uttar Pradesh and Bihar, the mighty Ganga receives a large number of tributaries such as the Gomati, the Ghaghra, the Gandak and the Kosi from the left bank as well as the Yamuna and the Son from the right bank. This brings huge quantities of water to these areas both from the Himalayan region and from the Peninsular India resulting in devastating floods. River Kosi often shifts its course flooding new areas and converting once fertile areas into wasteland. The Kosi which means ***kosna*** (curse), brings flood fury to vast area every year and is living upto its name. The Yamuna is an important right bank tributary of the Ganga and floods large areas in Uttar Pradesh and Haryana. The Chambal and the Betwa meet the Yamuna and add to the flooding capacity of the Yamuna.

In *Uttar Pradesh* floods are frequent in the eastern districts, mainly due to spilling of the Rapti, the Sharda, the Ghaghra and the Gandak. The problem of drainage congestion exists in the western parts of Uttar Pradesh, particularly in Agra, Mathura and Meerut districts. Erosion is experienced in some places on the left bank of the Ganga and on the right banks of the Ghaghra and the Gandak.

In *Bihar* the floods are largely confined to the northern part of the state where occurrence of floods is almost an annual feature. The rivers such as the Burhi Gandak, the Bhagmati and the Kamala and other smaller rivers of the Adhwara Group, the Kosi in the lower reaches and the Mahananda spill over their banks causing considerable damage to crops and dislocation of traffic.

Uttar Pradesh and Bihar are the worst flood affected states and account for over one-third of the flood prone area of the country.

In *West Bengal* the southern and central parts are flooded by the Mahananda, the Bhagirathi, the Ajoy, the Damodar, etc. due to an inadequate capacity of river channels and tidal effect. There are occasional floods caused by the Damodar river even after the construction of four dams and a barrage under the Damodar Valley Project. In 1956, about 25,000 sq km of area was flooded in Southern districts of West Bengal by this river. The Ganga delta is often flooded.

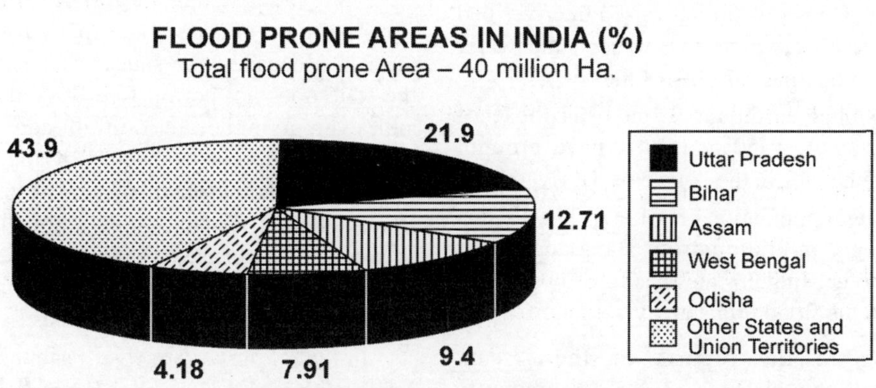

FIG. 9.11. Flood Prone Areas in India (%)

There is also the problem of erosion of banks of some of the rivers and on left and right banks of the Ganga upstream and downstream respectively of the Farakka Barrage.

The Bhrahmaputra River Region

In the Brahmaputra basin, floods are almost an annual feature. The main cause of floods here is heavy rainfall amounting to over 250 cm during the rainy season. Large amount of silt is deposited here by the Brahmaputra and its tributaries which makes the river channel shallow and its capacity to carry large amount of water is reduced. This results in flooding of vast areas in and around the valley. Earthquakes, which occur at frequant intervals, cause change in the level of river course and the flow of water in the river is obstructed. This leads to inundation of large areas in this region. Landslides are very common here. Huge rock material falling as a result of landslides acts as a temporary dam across the river and vast area is submerged under water. Later it gives way under the pressure of water and floods large area downstream. The Assam Valley is considered to be one of the worst flood-affected areas of India. The main causes of floods in the Brahmaputra river system can be summed up as under :

(*i*) There are 34 major tributary rivers of the Brahmaputra. These bring huge quantities of water and silt which cause floods.

(*ii*) Very heavy rainfall exceeding 250 cm per annum.

(*iii*) Narrowness of the Brahmaputra valley with a maximum width of about 81 km surrounded by hills.

(*iv*) Heavy deposit of silt has raised the river bed considerably which has reduced the water accommodating capacity of the river.

(*v*) Occasional earthquakes, such as those of 1897, 1930, 1950 and 1984 have brought about changes in the course of the river.

(*vi*) Very high population pressure, primarily due to illegal migration from Bangladesh and some other Indian states has forced people to live in the flood prone areas.

According to the reports of the Assam government, all the districts of the Brahmaputra valley are inundated almost every year.

Though most of the flood affected areas in Assam are rural in character, yet some urban areas are also affected by floods each year. The worst flood affected area of the Brahmaputra valley is the world's largest river island, Majuli. The urban areas of Assam, namely Dhubri, Guwahati, Dibrugarh, Tezpur etc. ate frequently flooded. It has been estimated that an area of 30 lakh hectares out of 78 lakh hectares *i.e.*, about 45 per cent of Assam's total area is flood prone.

The North West Rivers Region

The flood problem in this region is less serious as compared to the one prevailing in the Ganga and the Brahmaputra river regions. The major problem is that of inadequate surface drainage which causes inundation and waterlogging over vast areas. In the Punjab-Haryana plain rain water in the waterlogged and poorly drained areas inundates large parts. Major and minor rivers like the Satluj, the Beas, the Ghaggar and the Markanda are in spate during the rainy season and bring flood havoc to vast areas.

In *Punjab* floods are an annual feature though intense floods are experienced normally at an interval of 4-5 years. The main reason of floods in Punjab is obstruction of poor natural drainage by man made features. Some of the major canals (the Bhakra System) do not follow the natural flow and create obstacles. Secondly, National Highway No. 1 and the main railway line run almost perpendicular to the natural flow. Cultivation of area near river banks and construction activities in low lying areas, especially in cities like Ludhiana, Patiala, etc. have together created obstacles in the natural flow of water.

The state of *Haryana* experiences severe flood once in a decade. The saucer shaped topography of the state does not permit free flow of surface runoff and even a moderate rainfall can cause floods. Sometimes, floods are caused by the Ghaggar river also. This river used to disappear in the sands of Rajasthan after flowing through Punjab and Haryana. In recent years, besides flooding Punjab and Haryana areas, it has become active in Rajasthan territory also, occasionally submerging large areas.

In the north-western river basin covering Jammu and Kashmir and Himachal Pradesh, the Satluj, the Beas, the Ravi and the Chenab often flood large areas.

Floods occur periodically in Jhelum and its tributaries in the Kashmir Valley causing a rise in level of the Wular Lake thereby submerging marginal areas of the lakes and sometimes threatening Srinagar and other areas along the river banks. Similarly the Chenab and its tributaries like Tawi are often in spate endangering several densely populated areas in Jammu and Akhnoor.

The Central India and the Deccan Region

The southern states of Andhra Pradesh, Telangana, Karnataka, Tamil Nadu, Odisha, Jharkhand, Chhattisgarh, Maharashtra, Gujarat and parts of Madhya Pradesh are included in this region. The floods do not pose a serious problem in this region because most of the rivers have well defined and stable courses. However, the deltas of the Mahanadi, the Godavari, the Krishna and the Cauvery suffer from occasional floods owing to the large scale silting and the consequent change in the river courses. Indiscriminate felling of trees in the catchment areas of major rivers has complicated the flood problem. High tide at the time of flood aggravates the flood situation. Lower courses of the Narmada and the Tapi in Gujarat are also prone to floods.

The small rivers of *Kerala*, originating in the Western Ghats and flowing to the Arabian sea, cause considerable damage when in spate.

The above description of flood prone areas reveals that one or other part of the country is affected by floods and flooding is almost an annual phenomenon.

Urban Floods

Mark Pelling, geographer and climate change expert at King's College, London has mentioned in his book entitled ***The Vulnerability of Cities,*** "urbanisation affects disasters just as profoundly as disasters can affect urbanisation." Urban flooding has emerged as a major problem in the recent past. With global warming and climate change, the weather is becoming more erratic with the passage of time and highly concentrated population in urban areas is facing increased risk of floods. Flooding in urban areas has its own peculiarities. The main reason of urban flooding is surface water runoff, which is the excess water from rain or melt snow that flows over the earth's surface without getting absorbed in urban landscape. It is controlled and managed artificially through drains which flush out the runoff from the city, unlike in rural settings where the runoff is absorbed naturally by farmlands and ponds.

Urban areas have impervious surfaces such as roads, pavements and buildings that decrease a city's capacity to absorb water making it dependent solely on the outflow of surface water runoff. Under such circumstances, even a moderate rainfall can lead to flash floods in low-lying areas of the city. Cities located on the river banks face additional problem if the river flows at a higher level. An example of this situation is Guwahati, a low-lying city on the bank of the Brahmaputra river which has to face the flood fury almost every year during the rainy season when the river is in spate. In coastal cities like Mumbai, a high tide may worsen the flood situation due to reverse flow of water.

Another reason for urban flooding is the lack of proper drainage system in urban areas that overflow even during low rains. In the past few years, flooding in Delhi due to overflow of the city's 18 major drains has become a common phenomenon. Heavy rains in the Yamuna's upstream increases its water level in Delhi, due to which drains in the city experience reverse flow.

A research paper by US De, farmer additional director general of meteorology (research), Pune, analysed flooding in four megacities of India—Delhi, Chennai, Kolkata and Mumbai. The paper demonstrated how rapid and uncontrolled urbanisation was at the root of floods and flood-related damages in these cities. It noted that the mechanism of urban flooding is complex and location specific. Hence each city needs its own flood management practices. The paper notes with deep concern that the flood plains of the Yamuna in Delhi have thousands of illegal colonies which are thickly populated areas. High population density in these colonies demand more infrastructure leading to environmental degradation.

The major urban flood events in India have occurred in Mumbai (2005, 2014, 2017), Bengaluru (2005, 2007, 2015), Chennai (2002, 2004, 2005, 2006, 2007, 2015), Ahmedabad (2017) and Kolkata (2007, 2017). It is worth noting that three major metropolitan Indian cities experienced severe flooding in the same year i.e. Mumbai in July 2005,

TABLE 9.8. List of recent major flood events over India

Year	Region	Cause
2005 (July)	Mumbai flood	Heavy downpour resulted in a huge rainfall as much as 994 mm of rain fell in just 24 h and 684 mm in only 12 h. Resulted in massive flooding of the Mithi river. The impact was further amplified by the inadequate drainage and sewage resulting in massive flooding.
2007 (August)	Bihar flood	Extremely heavy and long-term rainfall flooded various rivers in Bihar and Uttar Pradesh.
2008 (August)	Bihar flood	The flooding of the Kosi river valley in the northern Bihar due to breaking of Kosi embankment
2012 (June)	Brahmaputra floods	Extremely heavy monsoon rainfall resulted in over-flowing of Brahmaputra river and its tributaries.
2013 (June)	North India floods (Uttarakhand)	Notable natural disaster in Uttarakhand. Continuous heavy monsoon rainfall followed by landslides in the hills led to flash flooding. This region has recently experienced frequent flooding and landslides.
2013 (July)	Brahmaputra floods	Similar to the 2012 event, extremely heavy monsoon rainfall resulted in over-flowing of Brahmaputra river and its tributaries.
2014 (September)	Kashmir floods	Continuous rainfall for more than three days resulted in floods and landslides in Jammu and Kashmir after the Jhelum river reached above the dangerous level.
2015 (June and August)	Assam floods	Extremely heavy monsoon rainfall resulted in the bursting of Brahmaputra river and its tributaries causing landslides in the region.
2015 (July)	Gujarat flood	Monsoon deep depression over the Arabian Sea caused intense rainfall and flooding across the coast of Gujarat.
2015 (November)	Chennai floods	The transition of low pressure into a deep depression after crossing the coast resulted in very heavy rainfall. It is likely due to blocking of clouds by the Eastern Ghats which led continuous rainfall and produced massive urban flooding. Also, rampant urban development could have played a vital role.
2016 (July)	Assam floods	Extremely heavy monsoon rainfall mostly during monsoon breaks resulted in the bursting of Brahmaputra river and its tributaries.
2017 (June and July)	North-east India floods	Extremely heavy monsoon rainfall mostly during monsoon breaks resulted in the bursting of Brahmaputra river and its tributaries.
2017 (July)	Bihar flood	The torrential rain in the Nepal region resulted in a sudden increase in the discharge in all the eight rivers in Bihar, which led to massive flooding.
2017 (July)	West Bengal floods	Week-long continuous rainfall due to cyclone Komen during monsoon resulted in dangerous floods in West Bengal and Jharkhand.
2017 (July)	Gujarat flood	Simultaneous occurrence of rainfall due to low pressure systems from Arabian sea as well as Bay of Bengal. It is also likely that the heavy inflow into dams Dharoj and Dantiwada resulted in the massive flooding.

2017 (August)	Mumbai flood	Massive Mumbai flood after 2005. The high tide and the extreme rainfall (468 mm in 12 h) along with inadequate drainage and sewage resulted in massive flooding.
2018 (August)	Kerala floods	The unusual rainfall during the monsoon season has resulted in massive flooding. Other reasons are a sudden discharge of water from the reservoir, land-use changes and landslides.
2019 (July, August and September)	Widespread over Indian regions	A series of devastating floods over areas of several states (such as Maharashtra, Karnataka, Kerala, Gujarat, Rajasthan, Andhra Pradesh, Odisha, Uttarakhand, Madhya-Pradesh, Bihar, Uttar Pradesh, West Bengal, Assam and Punjab) due to persisting monsoonal deluges with excessive rain rates, stream flow and run-off during peak monsoon months extending into September.

***Source*:** Assessment of climate change over the Indian Region, A Report of the Ministry of Earth Sciences, Govt. of India, 2020, p. 128.

Bengaluru and Chennai in October and December respectively.

One of the most serious urban flooding was reported in Mumbai on July 26-27, 2005. During those two days, the city witnessed an unprecedented 944 mm rainfall in 24 hours. The intrusion of sea water into the city resulted in mass inundation due to complex drainage system. Studies also suggest that Mumbai is highly vulnerable to climate change, sea level rise, storm surge and extreme precipitation. In the same year, 10 severe urban floods were reported from across the country. Three-fourths of Chennai was inundated. The flood affected more than 5,00,000 people.

In 2006, as many as 22 cities in India suffered from floods and the number of such cities rose to 35 in 2007. According to US De, "Floods and droughts will become more frequent. One projection shows that the intensity and number of tropical cyclones will increase in the near future." Anil K. Gupta, Director of the Wadia Institute of Himalayan Geology in Dehradun (Uttarakhand), emphasis that the entire Himalayan range is vulnerable to rising temperature. "Each and every valley—be it Kashmir, Kedarnath or Badrinath—faces the threat of increased precipitation." The number of rainy days in the region in 2030s may increase by 5 or 10. The intensity of rainfall is likely to increase by 1-2 mm per day. "What is increasing is sudden precipitation, which happened during the Kashmir floods in 2014."

For the last four major disasters—Mumbai floods of 2005, Leh cloudburst of 2010, Uttarakhand disaster of 2013 and Jammu-Kashmir floods of 2014—India Meteorological Department (IMD) has cited disturbances in the wind current and monsoon as the reason but has failed to assign the reason for high frequency of these events.

Urban floods are a new challenge and the problem of urban floods is becoming more serious with the passage of time. A 2008 study by the National Institute of Disaster Management showed that the annual economic losses from urban flooding are much higher than those incurred from other disasters.

Mumbai was flooded on August 28, 2017 following heavy rains, 468 mm rainfall occurred in past 12 hours. This was the worst flooding since 2005 floods.

Chandigarh was hit by sudden heavy rainfall on the morning of August 21, 2017. By afternoon the city received 112 mm of rainfall, 23 times the city's average daily monsoon rainfall. Similarly on August 15, 2017, Bengaluru received 37 times and in August 11-12, 2017, Agartala received more than 11 times its average daily monsoon of rainfall of the last five years.

Chennai had a series of five "extreme precipitation events" between 4 and 8 December, 2015. Before the city could recover from one event, it was bettered by the next. Economic loss to Chennai from the flood of 2015 was estimated at ₹ 21,381 crore. Chennai is more prone to tropical cyclones, which often leads to flooding of major rivers and clogging of drainage systems.

FLOOD DISASTER MANAGEMENT

Flood disaster management implies not letting the excess runoff water flow suddenly and intensively in the drainage network. After the unprecedented floods of 1954, flood management works were taken up in a planned manner by the State Governments. The main thrust of managing floods in different river basins was to modify the floods through specific structural measures such as reservoirs, embankments, channel improvement, town protection and river training works. The main steps for flood disaster management are briefly discussed below.

1. Flood Forecasting

Flood forecasting involves giving prior information regarding the occurrence of floods. This is essential and is extremely useful for taking timely action to prevent loss of human lives, livestock and movable property. The Central Water Commission (CWC) started flood forecasting in November 1958 when the first forecasting station was established at Old Railway Bridge, Delhi. Since then it has been extended to cover almost all the major inter-state flood prone rivers of the country. At present there are 226 flood forecasting stations in the country which includes 166 river level forecasting and 60 inflow forecasting stations on major dams/barrages. It covers 19 major river systems in the country including 72 river sub-basins. They pertain to 19 states and three Union Territories *viz.,* Andhra Pradesh, Arunachal Pradesh, Assam, Bihar, Chhattisgarh, Gujarat, Haryana, Jharkhand, Karnataka, Madhya Pradesh, Maharashtra, Odisha, Rajasthan, Tamil Nadu, Telangana, Tripura, Uttarakhand, Uttar Pradesh and West Bengal, and Union Territories of Jammu and Kashmir and Dadra and Nagar Haveli as well as National Capital Territory of Delhi. These centres issue daily-flood forecasts and warnings throughout the flood season from May to October.

Normally, these forecasts are issued 6 to 48 hours in advance, depending upon the river terrain, the location of the flood forecasting sites and base station. In addition to conventional flood forecasting methodology, mathematical model forecasting based on rainfall-runoff methodology is also being used in some areas. This has enabled CWC to issue a 3-day advance flood advisory.

On an average, 6,000 forecasts at various places in the country are issued during the monsoon season every year with over 98 per cent accuracy. A forecast is considered accurate if forecast water level is within ±15 cm of actual water level and the inflow forecast (*i.e.,* discharge) is within ± 20% of actual discharge.

A Memorandum of Understanding was signed by India and Nepal in 1988 according to which 45 Hydrological and Hydro-meteorological Stations were to be set up in India and Nepal for issuing flood forecasts to benefit both the countries.

A Memorandum of Understand with China has also been signed according to which China provides hydrological information with respect to Brahmaputra and Satluj rivers during the monsoon season. The MoU is renewed from time to time for regular flow of information. This information is utilised in the formulation of flood forecasts by the Central Water Commission.

2. Reduction of Runoff

Reduction of runoff is one of the very effective methods of flood disaster management. Runoff can be reduced by inducing and increasing infiltration of the surface water into the ground in the catchment area. This can be done by large scale afforestation particularly in the catchment area. Afforestation helps in reduction of runoff in the following ways :

(*i*) The canopy of the forest cover intercepts the falling raindrops and the roots, the leaf litter and humus are capable of holding water.

(*ii*) Together these encourage infiltration and reduce runoff.

(*iii*) Runoff reduction helps in reducing soil erosion which leads to reduced sediment load of the streams.

(*iv*) Reduction in stream sediment load reduces siltation and helps in maintaining the water accommodating capacity of the rivers.

In the Indo-Gangetic plain, the runoff can be reduced by artificially inducing infiltration by digging wells along the beds of ephemeral channels. A series of dug wells helps in storing and channelising the surface water. There is vast scope of using this method in the Northern Plain of India due to soft soil and vastness of aquifers.

3. Reducing Flood Peaks by Volume Reduction (Constructing Dams and Detention Basins)

The flood peaks can be reduced by construction of dams and detention basins. Dams have the capacity of holding huge quantity of water during the flood period and help in reducing flood peak volume of water. Water stored in reservoirs created by constructing dams can be allowed to flow down the stream under controlled conditions depending upon the accommodating capacity of the river downstream the dam. A number of reservoir projects have been completed in India since the launching of the National Flood Control Programme in 1954. These dams have helped in mitigating flood-peak in the down stream reaches. Notable among these are Tilaiya, Konar, Maithon and Panchet Hill dams in the Damodar Valley System, Bhakra Dam on the Satluj, Hirakud Dam on the Mahanadi, Pong Dam on the Beas, Nagarjun Sagar and Tunghbhadra on the Krishna and Ukai Dam on the Tapi. All these dams have afforded reasonable degree of protection to about 13.64 lakh hectares of land.

Apart from dams as described above ponds, tanks and surface storage structures also check flood and help in harvesting water for dry seasons. Other types of detention basins include natural depressions such as marshes in plains and old quarries and mines.

4. Reducing Flood Levels

Flood levels can be reduced in the following ways.

(*i*) Stream Channelisation. A close network of canals reduces flood hazard to a great extent because flood water flowing in the river can be diverted to canals. Canals serve as temporary storage and hold water as its flood waves move downstream. Thus they help in reducing the severity of the flood.

(*ii*) Channel Improvement. Channel improvement is done by deepening, widening, straightening, lining and cleaning out of vegetation and debris from the river channel. These changes in the river channel increase the flood conveyance capacity of the river. Channel improvement is supplemented by bank stabilisation by constructing ripraps, dykes or spurs and planting deep rooted trees on embankments. In a meandering river, meander loops impede drainage and retard disposal of flood water. Whenever, the river meanders become extremely sharp, they can be straightened by artificially cutting individual or a series of bends. This method can be applied to the meandering courses of the rivers like the Gandak, the Gomati, the Rapti, the Kosi, etc.

(*iii*) Flood Diversion. Flood diversion is the process of diverting the flood water in marshes, lakes, depressions and spreading it thinly over paddy fields and desert drylands. One such scheme is the Ghaggar Diversion Scheme which diverts 340 cumecs (cubic metres per second) of water before its entry into Rajasthan into the depressions and the areas between the sand dunes. In this way, discharge of water in the Ghaggar river is kept within the safe limits during the flood period.

5. Protection against Inundation (Construction of Embankments)

Building of embankments was considered to be the only way of controlling floods in 1940s. It is still considered to be one of the very effective devices against inundation of the inhabited areas and agricultural land. Construction of embankments has been taken up on a large scale in India. Between 1954 and 1978, 10,821 km long embankments had been built. By March 2000, more than 33,630 km of new embankments had been constructed. Most of the embankments have been constructed in north India where the Brahmaputra valley of Assam, northern parts of Bihar, Uttar Pradesh (Ganga, Yamuna and Ghaghara) and Punjab (Satluj, Beas and Ravi) are the main beneficiaries. In south India, embankments have been constructed mainly in the deltaic parts of the Mahanadi, the Godavari, the Krishna and along the banks of Penneru (**Fig. 9.12**).

As mentioned earlier, the Brahmaputra valley in Assam is the most frequently and severely flooded part of India. As such it is also the most heavily embanked part of the country. About one third of the total embankments of India have been constructed in Assam alone. The largest construction of embankments has taken place along the Brahmaputra river itself. Efforts have been made to control floods in the Brahmaputra river by constructing embankments at several places. At present the total length of embankments along the Brahmaputra is 934 km and 2,400 km on its various tributaries. These embankments provide protection to an area of 13.27 lakh hectares.

Next to Assam, Bihar is the most heavily embanked state, About 20 per cent of the total embankments of India have been constructed in Bihar. With the increase in flood prone area over the years from 2.5 million hectares in 1952 to 6.89 million hectares in 1994, the length of embankments has also grown from 160 km in 1952 to 3,465 km in 1998 *i.e.*, an increase of almost 22 times. The Kosi, and the Burhi Gandak are the most heavily embanked rivers. The Gandak, the Bhagmati, the Son and the Mahananda also have long stretches of embankments. These embankments have provided considerable protection to large areas particularly in north Bihar.

However, there are severe limitations of embankments as a flood control device. As a matter

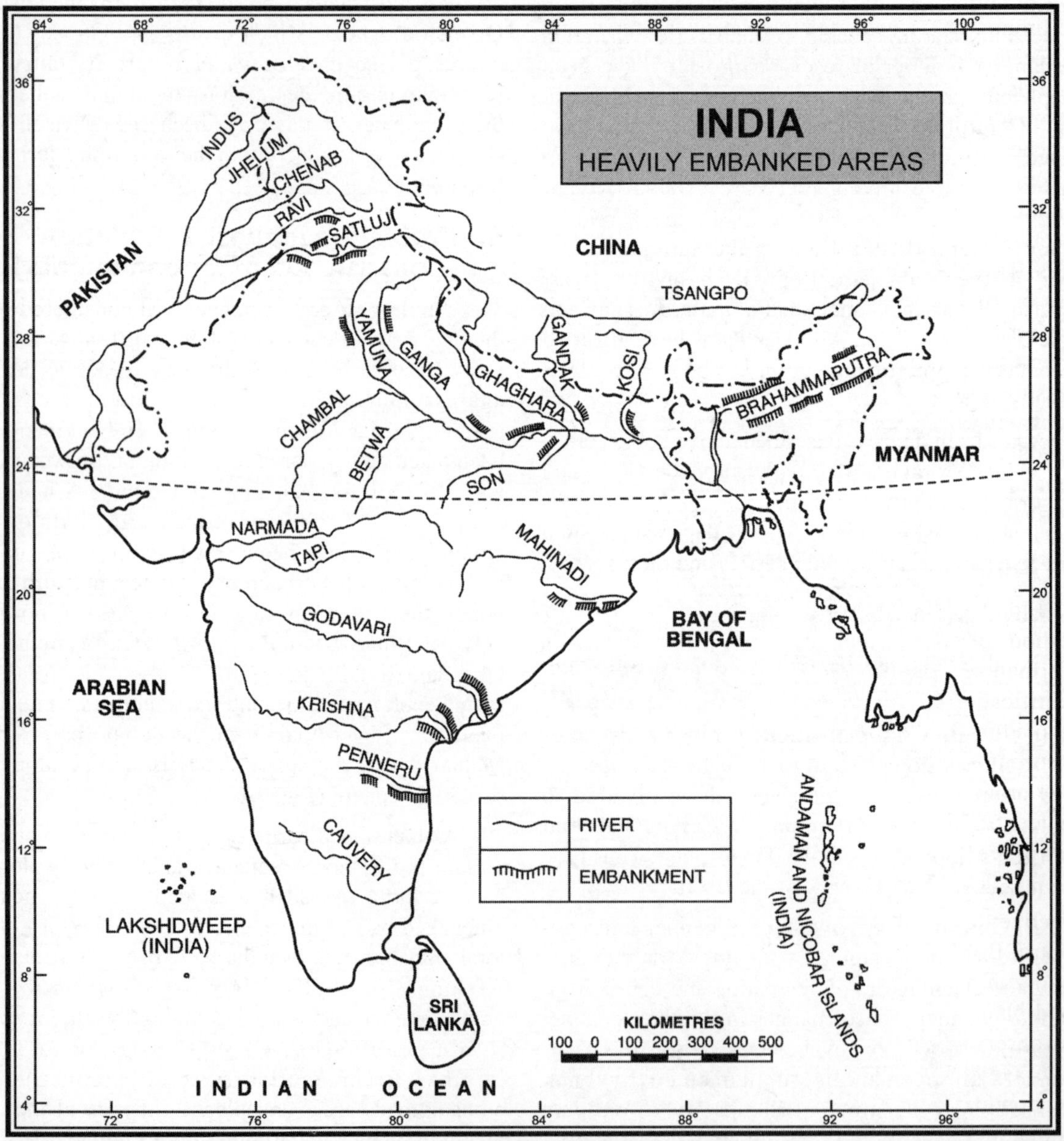

FIG. 9.12. India : Heavily Embanked Areas

of fact, they are not so much a method of flood control as flood transfer. Embankments may protect the neighbouring areas but they often cause floods in the downstream areas. In the event of high flood levels, the embankments may develop breaches and floods cause heavy damage to life and property in the low lying areas near the embankments. Construction of embankments puts limits to the river channel. The sediment which was to be deposited in a much wider area in the absence of embankments, is deposited in the limited river channel. Thus the river bed rises at an accelerated rate and consequently the flood water level rises. Under such circumstances the flood water may overflow the embankments or there may be breaches in the embankments due to intense hydraulic pressure. Under both the circumstances, the flood situation takes a serious turn and causes untold miseries to the people living in the neighbouring areas. Thus whereas constructing embankments is a very useful method of flood control it can lead to a very serious flood situation. The *Rashtriya Barh Ayog* report states. "Embankments are not a feasible measure of flood protection in cases where the country runoff draining into the river is so large as to inundate appreciatively the area protected by the embankments from river spills, during periods when the river is running at high flood stages."

6. Flood Plain Zoning (FPZ)

Flood plain zoning is another very effective method of flood management. It is based on information regarding flood plains, particularly the identification of floodways in relation to land use. Detailed maps of flood prone areas are prepared after a thorough study of flood cycles. Some areas are more prone to floods than the others. Different zones are identified and demarcated. After that necessary control is exercised with respect to land use.

Recognising the significance of FPZ as an effective device for flood control, the Central Flood Control Board mooted the idea in 1957 to demarcate flood zones to prevent indiscriminate settlement in flood plains. The increasing trend in flood damages was observed in India even during the 1970s and the State Governments were requested to give due attention to development of flood plains in a regulated manner. A model bill on Flood Plain Zoning was circulated to State Governments as early as 1975 with a request to enact suitable legislation for restricting the encroachment of flood plains and for their development in a regulated manner. The main features of the model bill were :

(*a*) flood plain zoning authority,
(*b*) delineation of flood plain,
(*c*) notification of limits of flood plains,
(*d*) restrictions on use of flood plains,
(*e*) compensation, and
(*f*) power to remove constructions after prohibition.

FLOOD CONTROL PROGRAMMES AND POLICY

After devastating floods in 1954, the Government of India announced a National Flood Management Programme. The programme was divided into three phases—immediate, short term and long term.

(*a*) **Immediate phase** extending over a period of 2 years was adopted for intensive collection of data and execution of emergent flood protection measures.

(*b*) **Short term measures** covering next 4 to 5 years, included construction of spurs and embankments at selected sites.

(*c*) **Long term measures** included the construction of storages, reservoirs on rivers/tributaries and additional embankments.

It has been estimated that out of 40 million hectares of flood prone area only 32 million hectares can be provided with reasonable degree of flood protection by structural measures.

Since the launching of the National Programme of Flood Management in 1954, 18.22 million hectares of land has been provided with a reasonable degree of protection against floods by construction of embankments, drainage channels, town protection works and by raising villages.

Post Flood Management

Post flood management includes the following :

1. Speedy restoration of transport system particularly roads and railways and postal services.

2. Supply of safe drinking water to the affected areas either by tankers or by fire tenders.
3. Repair of power, telephone and sewerage lines.
4. Proper arrangement for supply of food, shelter and clothing to affected people.
5. Adequate supply of necessities of life.
6. Constitution of a survey team to assess the loss and compensation to be given to the affected people.
7. Assistance for repair/rebuilding of private properties.
8. Desilting and dewatering of inundated areas.
9. Taking up appropriate measures according to Contingency Plan for agricultural sector.

TROPICAL CYCLONES

The World Meteorological Organisation (WMO) uses the term ***"Tropical Cyclone"*** to cover weather systems in which winds exceed 'gale force' (minimum of 34 knots or 63 kph). Over warm water in the tropical ocean, little away from the equator within the belt of 30°N and 30°S, the occurrence of tropical cyclones is almost a worldwide phenomenon. However, their characteristics like frequency, intensity and coastal impact vary from region to region. Tropical cyclones are the most destructive of the seasonally recurring rapid onset natural hazards. Between 80 and 100 tropical cyclones occur around the world each year. But these have been the deadliest when crossing the coast bordering the Bay of Bengal (coastal areas of Tamil Nadu, Andhra Pradesh, Odisha, West Bengal and Bangladesh) mainly because of serious storm surge problem in this area.

Characteristics and Destructive Power of Tropical Cyclones

Tropical cyclones are characterised by destructive winds, storm surges and exceptional levels of rainfall, which may cause flooding. Wind speeds upto 200 km/h, rainfall of upto 50 cm/day for several consecutive days and storm surges of 7 m high are not uncommon. A mature cyclone releases energy equivalent to that of 100 hydrogen bombs. Cyclone is a heat engine whose heater is the oceanic water. The released heat after condensation converts into kinetic energy for the cyclone. Yet it is described as a poor heat engine since it converts only 3% of the heat generated into kinetic energy.

Destructive Winds

The strong winds that blow counter-clockwise in the Northern Hemisphere, while spiralling inward and increasing toward the cyclone centre are highly destructive. Wind speeds progressively increase towards the core. As the eye arrives, winds decrease to become almost calm but rise again just as quickly as the eye passes and are replaced by hurricane force winds from a direction nearly the reverse of those previously blowing.

Wind speeds greater than 120 kmph are characteristic of severe cyclonic storms with a core of hurricane winds. All physical structures are vulnerable to the extreme pressures exerted by winds and thus collapse or are damaged. Wind speeds of cyclonic storms have been incorporated into the building code for coastal areas.

Damages due to wind are not confined to the coastal areas only. Damage can occur well in the interior. For example, a number of structures around Vijayawada (which is about 100 km from the coast) were damaged in a cyclone. Almost the whole east coast of India from Tamil Nadu to West Bengal is vulnerable to wind hazard. The coasts of Kachchh and Kathiawar in Gujarat are also highly vulnerable to wind hazard. Such wind hazards are mostly associated with tropical cyclones. High wind velocity of 50 m/s is not uncommon in these coastal regions.

Storm Surges

One of the peculiar characteristics and having a very high damage potential is storm surge. Storm surges are an abnormal rise of seawater due to tropical cyclone and are greatly amplified where the coastal water is shallow, in the estuarian region and where the shape of the coast is like a funnel.

The major factors include

- A fall in the atmospheric pressure over the sea surface
- The effect of wind
- The influence of the sea bed

- A funnelling effect
- The angle and speed at which the storm approaches the coast
- The tides

The coastal areas of north Bay of Bengal satisfy most of the above mentioned criteria and the storm surge gets enormously amplified there. Due to several favourable factors in these areas, the world's highest storm surge of 41 feet (over 13 metres) was reported from the area in 1876 near Bakerganj. Cyclonic storms are sometimes accompanied by tidal waves with heights of five metres and sometimes hit 20 km inland with wind speed of 150 kmph.

The low-pressure area or "eye" of the cyclone allows the sea level to rise. The high-speed winds surrounding the "eye" drive more water over this rise. The sloping bed of the sea and contours off the shoreline add further to the height. A further contribution to the height of the storm surge is added if the cyclone arrives at high-tide time.

Storm surges are not waves though they may look like them. A storm surge is a mass of water, which will submerge everything in its path, till it recedes back into the sea. It moves at the same speed as that of the cyclone. It travels upto the point where the ground height is equal to the height of the surge. The period of submergence of the land may be as long as 45 minutes or more depending on the depth it reaches inland.

As the leading edge of the surge crashes against the coastline and the water continues to travel inland there will be surface waves created which criss-cross each other and carry much under water turbulence. The destruction caused by the surge is tremendous. Houses are the worst affected. First the speed of the surge places great stress on the walls. The turbulence and currents created destroy the foundations of the structure. The debris like uprooted trees, fences and parts of broken houses act as battering rams, which cause further damage. The sand and gravel carried by the fast moving currents at the bottom of the surge can cause sand papering action on the foundations. The huge volume of water can cause such pressure differences that the house "floats" and once the house is lifted from the foundations water enters the structure and causes collapse of the building.

Damage occurs to every kind of asset built above the ground level due to the above characteristics and crops get affected very badly.

Exceptional rainfall occurrences

The world's highest rainfall spread over one or two days has occurred during tropical cyclones. The very high specific humidity condenses into exceptionally large raindrops and giant cumulus clouds, resulting in high precipitation rates. When a cyclone makes landfall, the rain rapidly saturates the catchment areas and the rapid runoff may extensively flood the usual water sources or create new ones.

Rainfall is generally very heavy and spreads over a large area thus leading to excessive amount of water, which leads to flooding. The size of drops, in a rainfall, increases with increase in rainfall intensity. Soil erosion results on a large scale as raindrops strike the ground with energies substantially greater than those in ordinary rainfall. The heavy rains waterlog the ground and cause softening of the earth due to soaking. This contributes to weakening of tank embankments, the leaning over of utility poles or collapse of pole type structures.

TROPICAL CYCLONES IN INDIA

With about 6 per cent of the world wide cyclones, the Indian subcontinent is one of the worst cyclone affected areas of the world. About 8 per cent of the total land area, particularly along the eastern coast and Gujarat coast is vulnerable to tropical cyclones. In fact, Indian ocean is one of six major cyclone prone regions of the world. On an average, about 5-6 tropical cyclones are formed in the Bay of Bengal and the Arabian Sea every year, out of which 2 or 3 may be severe. More cyclones form in the Bay of Bengal than in the Arabian Sea. As such, the eastern coast is more prone to cyclones and about 80 per cent of the total cyclones generated in the Indian Ocean strike the east coast of India. There are two definite seasons of tropical cyclones in the North Indian Ocean. One is from May to June and the other from mid-September to mid-December. May, June, October and November are known for ***severe storms***. The entire east coast from West Bengal to Tamil Nadu is vulnerable to cyclones with varying frequency and intensity. Along the west coast, the Gujarat and Maharashtra coasts are more vulnerable as compared to the southern part. Most of the cyclones have their origin between 10°N

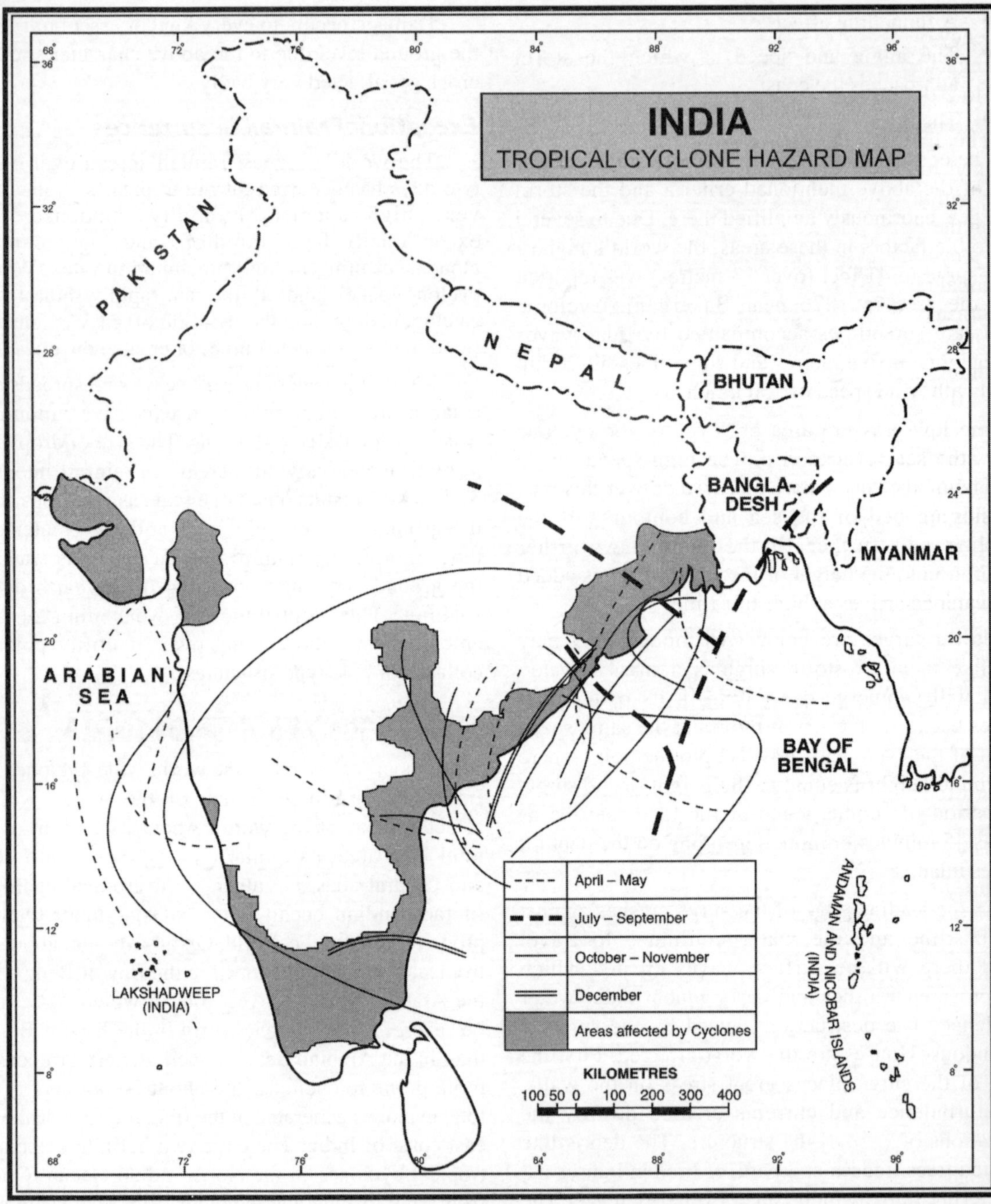

FIG. 9.13. India : Tropical Cyclone Hazard Map

and 15°N during the monsoon season. Almost all storms in the Bay of Bengal have their genesis between 16°N and 21°N and west of 92°E in June. By July, the Bay storms form north of 18°N and west of 90°E. It is also noteworthy that most July storms move along a westerly track. They are generally confined to the region between 20°N and 25°N and their recurving to the Himalayan foothills is comparatively rare. **Figure 9.13** shows the direction and seasonal distribution of cyclonic storms as well as the areas affected by the cyclonic storms. The frequency of cyclones and severe cyclones in the Bay of Bengal and the Arabian Sea during the period from 1897 to 2014 is shown in **Table 9.9**.

The distructive effect of cyclonic storms is confined to coastal districts, maximum destruction being within 100 km from the centre of the cyclones and on the right of the storm track. Principal dangers from a cyclone are : (*i*) Gales and strong winds, (*ii*) Torrential rain, and (*iii*) High tidal waves (also known as 'Storm surges'). Most casualties are caused by coastal inundation by tidal waves and storm surges. Maximum penetration of severe storm surges varies from 10 to 20 km inland from the coast. Heavy rainfall and floods come next in order of devastation. They are often responsible for much loss of life and damage to property. Death and destruction purely due to winds are relatively less. The collapse of buildings, falling trees, flying debris, electrocution, train and aircraft accidents and disease from contaminated food and water in the post-cyclone period also contribute to loss of life and destruction of property. **Figure 9.13** shows that the states of Andhra Pradesh, Odisha on the east coast and Gujarat on the west coast of India are extremely vulnerable to tropical cyclones.

MAIN MITIGATION STRATEGIES

1. Cyclone Shelters

One of the most successful means of reducing loss of human lives during cyclones is the provision of cyclone shelters. In densely populated coastal areas, where large scale evacuation is not always feasible, community buildings and buildings used for gathering of large number of persons, like schools, *dharamshalas*, hospitals, prayer halls, etc. can be used as cyclone shelters. These buildings can be so designed, as to provide a blank facade, with a minimum of apertures in the direction of the prevailing winds. The shorter side of the building should face the storm. Alternately these buildings can be designed on a circular/ellipsoidal plan, so as to impart least wind resistance. Earth berms and green belts can be used in front of these buildings to reduce the impact of the storm. These shelters should be located in relatively elevated areas with provision for community kitchen, water supply and sanitation.

TABLE 9.9. Frequency of Tropical Cyclones (1897-2014)

Month	Bay of Bengal		Arabian Sea	
	Cyclonic storm	Severe cyclonic storm	Cyclonic storm	Severe cyclonic storm
January	4	2	2	0
February	0	1	0	0
March	2	3	0	0
April	11	11	2	4
May	15	34	4	15
June	33	6	6	12
July	33	7	3	0
August	27	3	2	0
September	24	15	5	3
October	44	35	14	11
November	40	54	6	21
December	23	20	5	2
Total	**256**	**191**	**49**	**68**

Source : India Meteorological Department, New Delhi.

2. Engineered Structures

Cyclone proof structures which can withstand the force of winds and torrential rains can be of several types. Such facility is generally lacking due to widespread poverty of the people living in coastal areas. However, following precautions can be taken :

(*i*) *Kachha* house with thatched roofs should be avoided in coastal areas. Instead, loans should be given to vulnerable people for constructing cyclone proof houses and proper guidance should be given to them. As far as possible, houses should be constructed on stilts or on earth mound.

(*ii*) Houses can be strengthened to resist wind and flood damage. All elements holding the structures need to be properly anchored to resist the uplift or flaying off of the objects. For example, avoid large overhangs for roofs and projections should be tied down. Trees planted in a row can act as a shield, because such a provision reduces the energy of the cyclones.

3. Flood Management

Cyclones are accompanied with torrential rainfall which causes flooding of widespread areas. Therefore, flood mitigation measures (as mentioned earlier in the chapter) should also be taken.

4. Coastal Shelter Belts

Rows of strong rooted trees with needle like leaves are planted in the direction facing the wind. The trees in first few rows are provided fenced support to save them from being uprooted. These shelter belts can be grown all over the coastal area. Tidal forest trees and other hardwood trees are also recommended. Shelter belts grown along the coastline is a powerful tool to mitigate the impact of strong cyclonic winds. They also help in checking the soil erosion and inward sand drift, thereby protecting cultivated fields, houses and homesteads near the coast.

CYCLONE WARNING

Giving advance warning about a coming cyclone is one of the most effective methods of reducing the disastrous effects of cyclones. India Meteorological Department is the pioneer institute which issues such warnings.

A cyclone warning division has been set up in New Delhi to coordinate and supervise the totality of cyclone warning programmes in the country.

Cyclone Tracking

Cyclone tracking in India is done with the help of the following :

(*i*) Regular observations from weather network of surface and upper air observation stations

(*ii*) Reports from ships

(*iii*) Cyclone detection Radars

(*iv*) Satellites

(*v*) Reports from commercial aircrafts

Communication Network (Dissemination of Cyclone Warnings)

Cyclone warnings are communicated to Crisis Managers and other concerned organizations by high priority telex, telephones and Police wireless. Cyclone warnings are provided by the IMD from the Area Cyclone Warning Centres (ACWCs) at Kolkata, Chennai and Mumbai and Cyclone Warning Centres (CWCs) at Vishakhapatnam, Bhubaneswar and Ahmedabad. Cyclone warning bulletins for All India Radio/Doordarshan, and cyclone advisories for the north Indian Ocean to Bangladesh, Myanmar, India, Maldives, Pakistan, Sri Lanka and Thailand are being issued from Meteorological Office in New Delhi. This office also issues tropical cyclone advisories for the tropical cyclones in the south west Indian ocean to Mauritius. There is also a Satellite based communication system called the Cyclone Warning Dissemination Systems (CWDS) for transmission of warnings. There are 250 such cyclone-warning sets installed in the cyclone prone areas of east and west coast. The general public, the coastal residents and fishermen, are also warned through the Government machinery and broadcast of warnings through AIR and Television.

Recently India Meteorological Department has developed a system known as ***"Disaster Warning System" (DWS)*** to transmit cyclone warning bulletins through INSAT-DWS to the recipients.

TABLE 9.10. India : Intense Cyclones in the recent years in Arabian Sea and Bay of Bengal

Date and Year	Observed/ Estimated Maximum wind velocity (kmph.)	Damage
Arabian Sea		
May 16-22, 1999	102	Severe damage in Kachchh and Jamnagar districts of Gujarat. Loss of property about ₹ 80 crore. In Rajasthan 5104 cattle heads perished, 5133 houses partially damaged.
May 5-10, 2004	—	Widespread damage in Lakshadweep. 45 houses damaged. 35 boats and small cargo ships lost. Property lost worth ₹ 30 crores.
Nov. 9-12, 2009	83	Damage to lives, crops and property in Goa and Konkan region of Maharashtra. About 1000 houses damaged, 7 persons died, and 44 fishermen missing.
Dec. 10-15, 2009, WARD	83	No damages in India
June 10-14, 2014, **Nanauk**	83	No damages in India
Oct. 25-31, 2014, **Nilofar**	204	No damages in India
Oct. 28, 2015, **Chapla**	115	No damages in India
June 10-13, 2019 **Vayu**	180	2.5 lakh people evacuated, 8 dead. Property worth $ 140,000 damaged. Damage could be much higher but the cyclone changed its course.
May 18-20 2021 Tauktae (Pronounced as TauTe)	155-165 gusting upto 185	Deadliest storm since 1998 to hit Gujarat and Diu. 193 people died in six states. 2.6 lakh people evacuated, 2,240 trees uprooted in Ahmedabad alone. Property worth ₹ 15,000 crores destroyed. Effect of this cyclone was felt all over North-West India including Delhi-NCR. Tidal waves 2-3 metre high inundated many areas. Pappa (P. 305) ship rank in the sea in which 70 officials died and 16 were missing.
Bay of Bengal		
Oct. 25-31, 1999 Super Cyclone (Odisha)	260	9887 people died, 129.22 lakh people affected. Property worth $ 4.44 billion damaged.
Dec. 23-28, 2000	167	Three districts of Tamil Nadu (Ramnathpuram, Tirunelveli and Tuticorin) badly affected. Agricultural land affected and crops destroyed.
Dec. 11-16, 2003	102	81 people died, 1637 buildings destroyed completely, 7453 buildings destroyed partially, 61898 hectres of agricultural land damaged. Property loss worth ₹ 23903.13 lakh.
Nov. 28, Dec. 02, 2005 **Baaz**	83	Nellore, Chittoor and Cuddapah districts of Andhra Pradesh badly affected. 11 persons died.
Oct. 29-30 2006 **OGNI**	185	24 persons and 3,61,533 livestock perished. Loss of crops 1,99,986 acres. 900 villages submerged. Total loss ₹ 216.1 lakhs.
Oct. 25-27, 2008 **Rashmi**	93	Heavy loss to life and property in Meghalaya, Assam and Arunachal Pradesh.

Nov. 25-27, 2008 **Nisha**	83	78 persons lost their lives in Tamil Nadu and Andhra Pradesh. Over 12 lakh hectares of farm land affected and crops damaged.
May 23-26, 2009, **Aila**	112	100 people died, 2.2 million people affected (1,000 houses fully damaged, 1,32,000 houses partially damaged in West Bengal. In Odisha 1,000 acres of cropland affected badly.
Dec. 25-31, 2011, **Thane**	139	46 persons died in Tamil Nadu and Puducherry. 6000 persons shifted to shelters. 73,292 houses fully damaged and 94633 partially damaged.
Oct. 5-14, 2013, **Phailin**	213	Heavy damage in Andaman and Nicobar Islands, Andhra Pradesh and Odisha.
Oct. 7-14, 2014, **Hudhud**	185	It originated near Andaman and Nicobar Islands and affected large parts of Andhra Pradesh, Odisha, Chhattisgarh and Madhya Pradesh. Its effect was felt upto the Himalayas. 124 persons lost their lives and property worth $ 3.4 billion was damaged.
Dec. 6-13, 2016 **Vardah**	155	38 dead in Tamil Nadu and Andaman and Nicobar Islands. Property worth $ 3.37 billion destroyed. Estimated agricultural loss valued at ₹ 35,571 crores.
Nov. 29-Dec. 5, 2017 **Ockhi**	150	35 Persons lost their lives; 199 went missing and 6868 houses damaged in Tamil Nadu. In Kerala 75 persons dead, 137 fisherman reported missing and 3600 houses damaged.
May 3-11, 2019 **Fani**	225	41 dead, 14 lakh evacuated, 1.5 crore people affected in 14 districts of Odisha. 5 lakh houses damaged, 34 lakh livestock dead, and monetary loss of ₹ 5,000 crore.
Nov. 5-12, 2019 **Bulbul**	165	10 people died, 60,000 houses damaged, 4.6 lakh people affected, 8 fishermen went missing in West Bengal
May 16-21, 2020 **Amphan**	160	72 dead, 5.5 lakh evacuated in West Bengal, 1.6 lakh evacuated in Odisha, 1.8 lakh hectares crop destroyed.
26-27 May 2021 **Yaas**	165	About 20 lakh people evacuated and shifted to safer places. Only 16 casualties were reported due to large scale planning and rescue operations. 128 villages marooned in Odisha. In West Bengal, three lakh houses and 134 embankments were damaged. Thousands of tree were uprooted. It affected the states of Odisha, West Bengal, Jharkhand, Bihar, Tamil Nadu and Andhra Pradesh and Union Territory of Andaman and Nicobar Islands.

Source : India Meteorological Department, New Delhi.

Post Disaster Assistance

Post disaster assistance is very essential for rescue and rehabilitation of the people in distress. It includes the following.

- Evacuation
- Emergency shelter
- Search and rescue
- Medical assistance
- Provision of short term food and water
- Water purification
- Epidemiological surveillance
- Provision of temporary lodging
- Reopening of roads
- Re-establishment of communications networks and contact with remote areas
- Debris clearance
- Disaster assessment
- Provision of seeds for replanting

HOW ARE CYCLONES NAMED ?

The World Meteorological Organisation/ Economic and Social Commission for Asia and the Pacific Panel on Tropical Cyclones, at its 27th session held in 2000 in Muscat, Oman, agreed to assign names to the tropical cyclones in the Bay of Bengal and Arabian Sea. The naming of the tropical cyclones over north Indian Ocean Commenced from September, 2004. Eight countries along the Bay of Bengal and Arabian Sea suggested names that are sequentially tested. The nations suggested alphabetically Bangladesh, India, Maldives, Myanmar, Oman, Pakistan, Sri Lanka and Thailand. Some of the names suggested by India are Agni, Jali, Bijli, Akash while Mala, Helen, and Nilofar were suggested by Sri Lanka, Bangladesh and Pakistan respectively. These lists are used sequentially and they are not rotated every few years, unlike the Atlantic and Eastern Pacific lists. If the public wants to suggest the name of a cyclone to be included in the list, the proposed name must meet some fundamental criteria. "The name should be short and readily understood when broadcast. Further, the names must not be culturally sensitive and must not convey some unintended and potentially inflammatory meaning."

The practice of naming the storm first started in the USA. This helped in identifying the storm and also aided the researchers.

The mid 1900s saw the start of practice of using feminine names for storms. Before the end of 1900s, forecasters started using male names for storms forming in the Southern Hemisphere. Since 1953, Atlantic tropical storms have been named from lists originated by National Hurricane Centre. They are now maintained and updated by an international committee of World Meteorological Organisation. The original name lists featured only women's names. In 1979, men's names were introduced and they alternate with women's names. Six lists are used in rotation. Thus the 2019 list will be used again in 2025.

Some times a storm is so deadly or costly that the future use of names becomes inappropriate for reason of sensitivity. That is the only time when there is a change in the list. If that happens, then at an annual meeting by WMO Tropical Cyclone Committees, the offending name is stricken from the list and is replaced another name. The infamous storm names like Mangkhut (Philippines, 2018, Irma and Maria (Caribbean, 2017), Katrina (USA 2005), etc. are examples for this.

The main advantage of giving specific name to a storm is that it is less subject to error than the old more cumbersome latitude-longitude identification methods. This advantage is especially important in exchanging detailed storm information between hundreds of widely scattered stations, coastal bases and ships at sea.

LOCAL SEVERE STORMS

Local severe storms are small scale disturbances which are formed due to strong convective motions in a moist and unstable atmosphere and originate from well grown cumulonimbus clouds. The destructive effects of local severe storms are thunderstorms, strong winds, hail storms, lightening, heavy rains and tornadoes

Thunderstorms

Thunderstorms are a meso scale phenomena covering an area of tens to hundreds of square kilometres and lasting a few minutes to a few hours. They occur when there is instability in the atmosphere under adequate moisture conditions and a mechanism exists to release the instability. Thunderstorms constitute an important form of local severe storms which cause considerable damage to life and property every year. They occur over all parts of India and there is no month when one part or the other of the country is not affected by thunderstorms. The period of least thunderstorm activity is December to March. However, the most widespread thunderstorm activity all over the country occurs during the hot weather period from middle of March to middle of June. This is also known as the ***pre-monsoon period***. The main regions of the pre-monsoon thunderstorm activity are (*i*) Northeast India (*ii*) Northwest India, (*iii*) Central parts of the country, and (*iv*) Southwest Peninsula. The maximum activity is during March-September in northeast India while it is April-May and October-November in southwest peninsula. Severe thunderstorms occur generally during the afternoon and evening hours.

Thunderstorms in West Bengal, Chhotanagpur plateau of Jharkhand and northeast India during March-June are often accompanied by violent destructive squalls reaching speed of 100 km per hour. They are known as ***'Norwesters'*** in meteorological language in India as their modal direction is from northwest. They are also called ***'Kalbaisakhis*** which means *evil or black storms of month of Vaisakh*. They cause widespread damage to crops and kill cattle and at times, people too. During March-June the thunderstorms of northwestern and central India are often preceded by dust storms. There are thundersqualls which raise walls of dust and sand rendering visibility to practically nil and are locally known as ***'andhis'*** (blinding storms). Significant duststorm activity begins in April and reaches maximum in June.

During the post-monsoon (October and November) period thunderstorms occur in association with cyclonic storms and depressions mostly over peninsular India.

Basic parameters leading to the development of thunderstorm are observed through conventional synoptic surface and upper air meteorological observations. Frequency of their observation varies from two to six over the surface and two to four over upper air daily. Hourly observations are taken in some selected areas and are used for forecasting severe thunderstorms. The most reliable tracking of severe storms is through radar surveillance. As soon as the thunderstorm cells or squall lines are observed on the radar screen, their speed and direction of movement can be measured by marking their progress over the radar screen. Indications are also available on the radar screen regarding the severity of the thunderstorm and its potential for producing rain and hail. Such information is used for forecasting local severe storms. Cloud observations from geostationary satellites and polar orbiting satellites are useful in demarcating broadly the areas of potential for the convective development where severe storms may occur.

At present there is provision for issuing warnings based on certain known facts for high speed wind, hail and heavy rain associated with the thunderstorms generally 24 hours in advance to the warnees listed with different meteorological offices in the country. In addition, such warnings are issued on radio, television and other news media from different meteorological offices in India.

Hailstorms

Hail is precipitation in the form of ice pellets. Thunderstorms producing hail are known as ***hailstorms***. The size of hailstones may vary from less than a centimetre to about 5 cm or even more in diameter. Hail is the product of violent convection, formed in a cumulonimbus (Cb) cloud. A hailstone starts its life as an ice crystal. It grows progressively as it comes into contact with ice crystal and water droplets alternately in up and down currents in the Cb cell. This is evident from the study of its structure. A cut section of a hailstone resembles an onion with alternate layers of glaze ice and opaque rim. The glaze ice layers of the hailstone are formed during its downward journey in the region of super-cooled water while the opaque rim layers form at higher levels where smaller drops predominate. Once the hailstones become too large to be supported by the vertical updrafts, they start falling on the ground. They fall with great speed which is often accentuated by the squall. Sometimes the hailstones attain large size (as large as a golf ball) and cause heavy damage to crops, property and life. Hail is a grave hazard to aircraft in flight.

During the period from March to May thunderstorms over northern India, particularly the submountain areas are accompanied by hail. Hailstorms also occur in association with western disturbances during the winter months from December to February. However, hail is rare in the southwest monsoon period. The annual frequency of hail is the highest in extreme northeast of the Brahmaputra valley with nearly 15 occasions. The other high incidence areas are Tripura, Dumka and Pakaur (Santhal Parganas) of Jharkhand, West Bengal and submontane regions of Uttarakhand. However, no region is free from hail.

Tornado

Tornadoes are extremely severe vortices of very small dimensions occurring in association with intense and large cumulonimbus (Cb) clouds or cyclonic storms. They comprise the most destructive meso-scale convective phenomena which builds up in thunderstorm. The visible symptom of a tornado is a small funnel extending down from a Cb cloud with winds reaching several hundred kilometres per hour, revolving tightly around the core. Actual records of wind speed are not available because anemometers cannot withstand such high speeds. From the sound etc., it is presumed that the speed of the wind in a tornado may exceed the speed of sound. Wind speeds of as high as 400 to 500 kmph are not uncommon. Tornadoes travel along narrow paths with lengths from a few km to 100 km or more. The width of the path is small, ranging from 30 to 100 metres. The diameter of the funnel touching the ground may vary from less than a metre to a few tens or hundreds of metres. The entire disturbance moves at speed varying between 100 to 150 kmph.

In view of their severe intensity, tornadoes have high potential for destruction. They cause immense damage and devastation to life and property. Because of vortex, they suck everything in their path. They are also known as ***"twisters"***. The pressure may be lower by 10% inside the vortex than at the periphery. Hence, when the funnel passes over, buildings seem to explode. The debris and even persons may be sucked and carried aloft by the tornado. On 24th March, 1998, tornadoes claimed 200 lives in West Bengal and Odisha. Tornadoes occurring over sea suck water up to the base of the mother cloud. The cloud becomes linked with water body and the phenomena is termed as ***'water spout'***.

The most probable regions of tornado occurrence in India are Assam and adjoining states, West Bengal, Odisha, the Ganga plain, Punjab and Haryana.

Considering the random occurrence of tornadoes, their short life period and very low frequency in India (1 to 2 per year), the warning services for tornadoes are not seriously attempted as yet. In fact, advance warning of tornadoes is a difficult task. The radar comes in quite handy for tornado monitoring and warning. Unless the warning of tornado occurrence is disseminated to public at large, due to its short life, adequate steps cannot be taken by public to get away from the path of tornado. Various state governments in the target region have promulgated building laws which specify design of houses capable of withstanding tornado fury.

Cloudburst

Cloudburst is another extreme natural hazard which causes large scale loss of life and property within a span of a few minutes.

WHAT IS A CLOUDBURST ?

A cloudburst refers to a sudden downpour within a radius of a few kilometres. The area affected by a cloudburst typically does not exceed 20-30 sq km. The downpour is always short lived but is capable of causing heavy loss of life and property and leading to untold miseries to human beings, animals, etc. Over 100 mm of rain in an hour is the basic criteria for a cloud to be called a ***cloudburst***.

Highly concentrated rainfall over a limited area in a short duration of time is the main criteria for any weather event to be termed as cloudburst. Meteorologists believe that cloudbursts occur due to rapid lifting of clouds by the steep orography of the region. The clouds get vertically lifted. Such clouds are generally called ***convective clouds*** which can

TABLE 9.11. Most Recent Cloudbursts

Date	Place/Area	Extent of Loss
August 18, 2014	Pauri district of Uttarakhand	14 persons killed and many houses burried
July 25, 2015	Baltal and Pahalgam routes of Amarnath Yatra	Pilgrimage to Amarnath Cave stopped for some time, 3 killed and 19 injured
August 8, 2015	Poonch, Jammu and Kashmir	19 structures damages, many people injured
July 3, 2016	Tehri, Uttarakhand	20 houses burried, 18 dead
August 17, 2017	Doda, Jammu & Kashmir	6 killed, 11 injured
July 16, 2018	Chamoli, Uttarakhand	15 houses damaged with mud and rubble, 10 vehicles, 3 ropeway bridges and motor bridge washed away.
June 3, 2020	Madurai (Tamil Nadu)	Heavy loss to life and property
May 11, 2021	Devapryag (Uttarakhand)	Many shops and houses were either buried under the debris or were washed away by the rain water. Many people were shifted to safe places. Some bridges and water pipelines were damaged.

extend upto 15 kilometres in the air above the ground. This process is called the ***"cumulonimbus convection condition"*** which results in formation of towering vertical dense clouds. The lifting is generally dynamic which causes thermodynamic instability resulting in rapid condensation and highly concentrated heavy downpours.

HOW A CLOUDBURST COMES INTO BEING ?

1. Warm and humid air is pushed up along the mountain slope.
2. The air mass keeps rising and forms large thunder clouds.
3. Lack of upper-level wind prevents dissipation of the thunder clouds.
4. Concentrated localised rainfall occurs.

Further, it is believed that in the Himalayan region, the clouds which are being lifted up at a fast rate are also accompanied by soil moistened by earlier precipitation. This moistened soil acts as an additional source of moisture and might also have an important role in the frequent cloudbursts in the Himalayan region. Cloudbursts are more frequent in the monsoon season.

The above discussion and **table 9.11** make it clear that most cloudbursts occur in the Himalayan region.

But that does not mean that other parts of the country are free from the fury of cloudbursts. Western ghats, western coastal areas, central and western India are also prone to cloudburst. In these areas, cloudbursts occur rarely but they strike with tremendous force and heavy downpour in a short duration of time. For example cloudburst over Musi river (a tributary of the Krishna river) in 1908 claimed 15,000 lives and destroyed 80,000 houses in Hyderabad. Cloudburst in 1979 was responsible for breaching the Machhu dam following incessant rain which resulted in heavy flooding of the downstream area. Morvi town was almost completely destroyed. 1,800-25,000 people died and property worth ₹ 100 crore destroyed.

On July 26, 2005, Mumbai received 1,448 mm of rain in 10 hours from a cloudburst; throwing entire life of the city out of gear. According to India Meteorological Department, Mumbai had never experienced such high rainfall in such a short duration for about 500 years and it may not happen in the coming 200 years.

LANDSLIDES

Introduction and Definition

Landslide is a rapid movement of rock, soil, and vegetation down the slope under the influence of gravity. Landslides are generally sudden and sporadic. An earthquake and heavy rainfall may cause large landslides in mountainous areas. Wherever mountain slopes are steep, there is a possibility of

A Checklist of Causes of Landslides

1.	***Ground Causes*** 1. Weak, sensitivity, or weathered materials 2. Adverse ground structure (joints, fissures etc.) 3. Physical property variation (permeability, plasticity, etc.)
2.	***Morphological Causes*** 1. Ground uplift (volcanic, tectonic etc.) 2. Erosion (wind, water) 3. Scour 4. Deposition loading in the slope crest 5. Vegetation removal (by forest fire, drought, etc.)
3.	***Physical Causes*** 1. Prolonged precipitation 2. Rapid draw-down 3. Earthquake 4. Volcanic eruption 5. Thawing 6. Shrink and swell 7. Artesian pressure
4.	***Man-made Causes*** 1. Excavation (particularly at the toe of slope) 2. Loading of slope crest 3. Draw-down (of reservoir) 4. Deforestation 5. Irrigation 6. Mining 7. Artificial vibrations 8. Water Impoundment and leakage from utilities.

Source : Manual on Natural Disaster Management in India (2001), National Centre for Disaster Management, p. 63.

large disasterous landslides. They can also result when the base of a slope is oversteepened by excavation or river erosion. Man breaks rocks for constructing roads, railways, buildings, tunnels, etc. In such cases rocks become loose and landslides occur. Mining activity by man is also responsible for landslides. Landslides are very common along the steep slopes of the Himalayas, the Western Ghats and along the river valleys. **Natural removal of soil and rock from slopes is known as mass wasting.** Landslide as a hazard has long been recognised by the people living in the mountains. It becomes specially dangerous when there is heavy rainfall or snowfall on the steep hill sides which makes the rocks to slip and break.

The extent of landslides depends on the steepness of the slope, the bedding plane of rocks, the amount of vegetation cover and the extent of folding and faulting of the rocks. It is the rocks that break and carry with it the soil and debris. A major cause which triggers off a landslide is the weight of the overlying material and the presence of a lubricating material like water. This is known as ***solifluction.*** Freezing and thawing of the rocks on mountain slopes cause them to break and roll down the slopes. The over bearing weight of snow or ice or water which has seeped into the soft permeable rocks also leads to slipping and breakage of hill slopes. Near the sea coast, cliffs are eroded at the base by sea waves and the rocks jutting out on top break off and fall. Landslides occur frequently during the rains. Deforestation as a result of felling of trees for timber and removal of vegetation cover for development activities are also responsible for soil erosion and destabilization of slopes. It is estimated that the construction of just one kilometre long road requires removal of 40,000 to 80,000 cubic metres of debris, which slide down the slopes, killing vegetation and choking mountain streams.

From the above discussion, it can be concluded that landslides can be caused by poor ground conditions, geomorphic phenomena, natural physical forces and quite often due to heavy spells of rainfall coupled with impeded drainage.

Onset-Type and Warning. Generally speaking, landslides occur gradually but often they occur without warning. They may take place in combination with earthquakes, floods and volcanoes. It is difficult to predict the actual occurrence of landslides. The areas of high risk can be determined by using information on geology, hydrology, vegetation cover, post occurrence and consequences in the region.

Elements at Risk. The most common elements at risk are the settlements built on the steep slopes, at the toe, and at the mouths of streams emerging from the mountain valley (**Fig. 9.14**). Buildings constructed without appropriate foundation for a given soil and in sloppy areas are also at risk. Roads, communication lines, and buried utilities are also at risk.

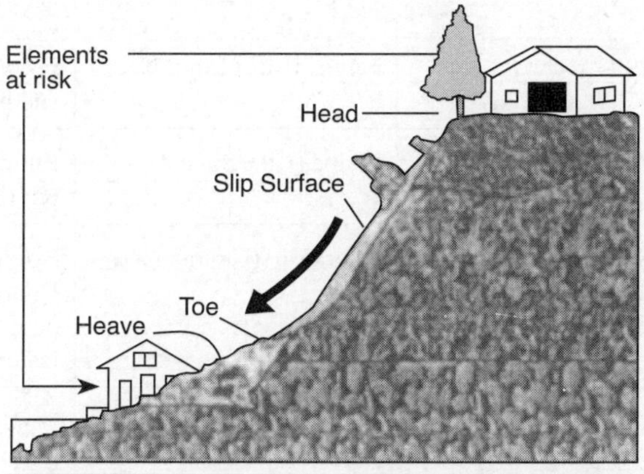

FIG. 9.14. Landslides : Elements at Risk

TABLE 9.12. Incidences of landslides in India

Region	Incidences of landslides
1. Himalayas	High to very high
2. North-eastern Hills	High
3. Western Ghats and Nilgiris	Moderate to high
4. Eastern Ghats	Low
5. Vindhyachal	Low

Source: Manual on Natural Disaster Management in India (2001) p. 62.

Landslide Vulnerability Zones

According to Geological Survey of India, about 12% of Indian land is prone to landslides as climate change increases the risk.

Following landslide vulnerability zones are recognised in India :

***(i)* Very High Vulnerability Zone.** This zone includes young mountains of the Himalayas, Andaman and Nicobar, steep and rainy slopes of the Western Ghats and Nilgiris, the north-eastern regions and areas of intense human activities particularly those related to construction of roads, dams etc.

***(ii)* High Vulnerability Zone.** These areas have geographical conditions similar to those which have very high vulnerability. The only difference is that the intensity and frequency of landslides is less as compared to areas of very high vulnerability. All the Himalayan states and the states from the north-eastern region except the plains of Assam are included in the high vulnerability zones.

***(iii)* Moderate to Low Vulnerability Zone.** These include areas which receive less precipitation such as Trans-Himalayan areas of Laddakh and Spiti, Aravalli hills, rainstorm areas of the Western and

TABLE 9.13. Major Landslides and their effects

Date/Year	Location	Effects
1993	Ratighat (Uttarakhand)	Nainital hill area remained cut off for about a week. It took five days to clear the debris.
1993	Nilgiri Hills (Tamil Nadu)	Forty people died, over 600 families shifted to safer places. Roads and houses destroyed.
August 12-18, 1998	Malpa (Uttarakhand)	Entire village was washed away. About 384 persons including 60 pilgrims going to Lake Mansarovar were killed.
Sept. 24 to Oct. 10, 2003	Uttarkashi (Uttarakhand)	Boulders from Varunavat Parvat razed several parts of the Uttarkashi town. Landslides affected 362 families and nearly 3,000 people were evacuated. Several portions of the national highway leading to Gangotri were damaged.
July 6, 2004	Badrinath (Uttarakhand)	Landslide blocked the pilgrimage route to Badrinath. Over 2,000 pilgrimage were stranded in the mountains near Badrinath. Nearly 1,000 shopkeepers and 2,500 villagers were trapped and remained cut off for 3 days. Roads were blocked. Massive rescue work had to be launched.
June 16, 2013	Uttarakhand	Floods and landslides caused by heavy rains resulted in untold damage to life and property.
July 30, 2014	Malin village 80 km north of Pune	The whole village burried in the debris. 151 people feared dead.
August 13, 2017	Kotrupo Village in Mandi district of H.P.	50 people burried under debris.
June 4, 2018	Lunglawn	10 people died and one pucca house along with occupants burried.
August 7, 2020	Munnar in Idukki district of Kerala	70 killed, mostly workers in tea and coffee plantations. 44 persons remained missing.
October 23, 2020	Sonapati (Manipur)	Conductivity via National Highway-2 cut off.
February 8, 2021	Nanda Devi	More than two dozen people killed and 150 missing.

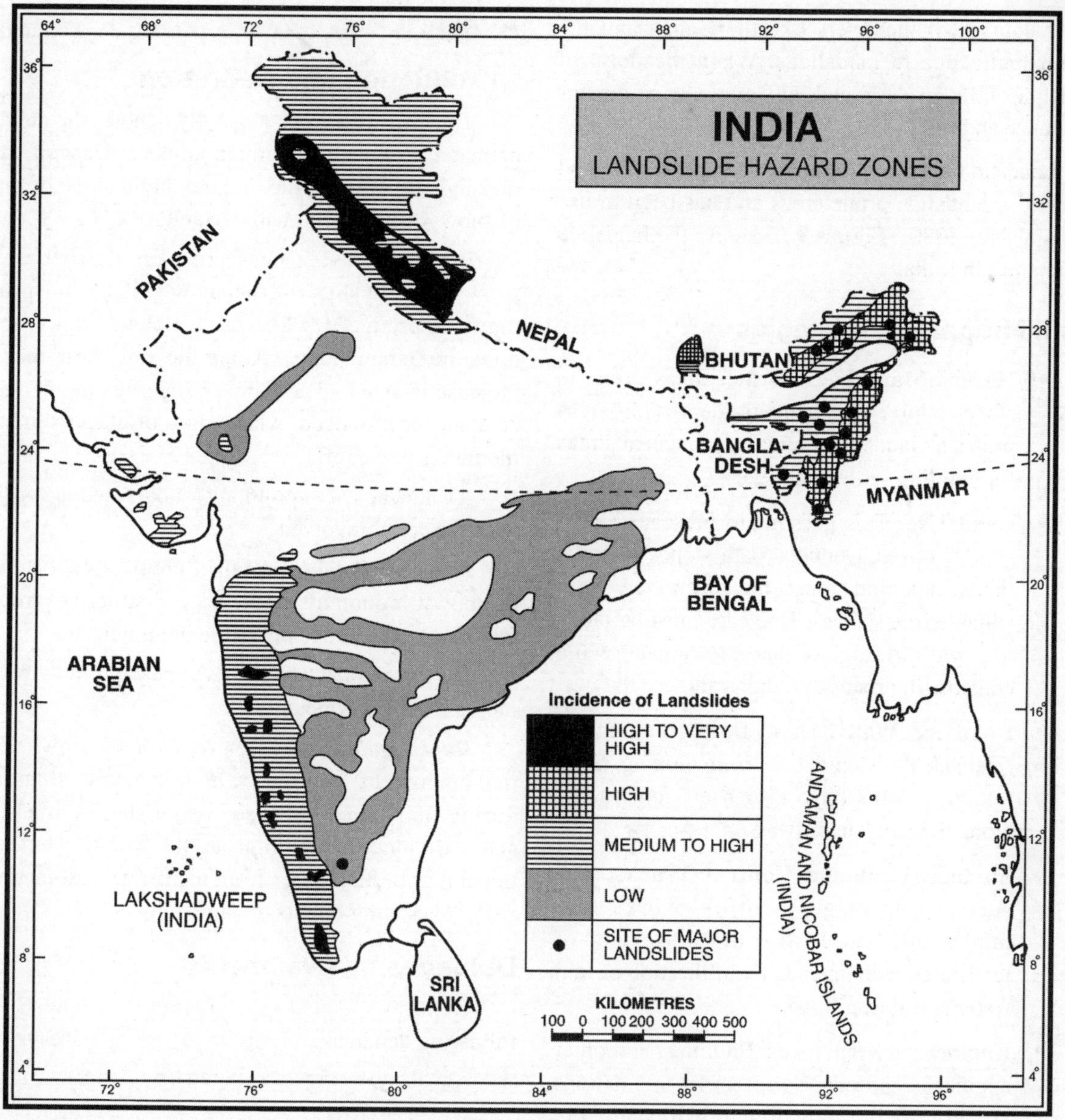

FIG. 9.15. India : Landslide Hazard Zones

Eastern Ghats and Deccan plateau, Landslides due to mining and subsidence are most common in states like Jharkhand, Odisha, Chhattisgarh, Madhya Pradesh, Maharashtra, Andhra Pradesh, Telangana, Karnataka, Tamil Nadu, Goa and Kerala.

***(iv)* Other Areas.** The remaining parts of the country especially states like Rajasthan, Haryana, Uttar Pradesh, Bihar, West Bengal (except district Darjeeling), Assam (except district Karbi Anglong) and Coastal regions of the southern states are safe as far as landslides are concerned.

Consequences of landslides

Landslides have serious consequences at the local level. They obstruct the flow of water in the river and cause flash floods. There are frequent floods in the upper courses of the rivers like the Ganga and the Brahmaputra due to landslides. When the mass of landslides falls on roads and railway tracks, it creates hindrance and rail traffic.

Landslide Zonation Mapping is a modern method to identify landslide prone areas and has been in use in India since 1980s. **Figure 9.15** shows the landslide prone areas in India.

Main Mitigation Strategies

- **Hazard Mapping.** It locates areas prone to slope failure. This helps in identifying areas prone to landslides and avoidance of areas for building settlements.
- **Land Use.** This pertains to preservation of vegetal cover. Denuded path slopes provoke landslides and must be reforested with suitable tree species. Due care must be taken to avoid blockage of natural drainage while constructing roads, buildings and canals.
- **Retaining Walls.** These are built along the road sides to stop debris from slipping. Such retaining walls have also been constructed along the Konkan railway line.
- **Surface Drainage Control Works.** The surface drainage control works are implemented to control the movement of landslides accompanied by infiltration of rain water and spring flows.
- **Engineered Structures.** Buildings and other engineered structures with strong foundations are in a better position to withstand the ground movement forces. Underground installations such as pipes and cables should be made flexible to move in order to withstand forces caused by the landslides.
- **Increasing Vegetation Cover.** This is the cheapest and the most effective way of arresting landslides. This helps in binding the top layer of the soil with layers below, while preventing excessive run-off and soil erosion.

AVALANCHES

Introduction and Definition

The term avalanche generally denotes the descent of material down a mountain slope, but specifically meaning the hurtling down a mountain slope of mass of snow, compounded with ice and rock.

In winter, avalanches are caused when fresh snow falls and slips off the snow surface. In spring, they are produced when partly thawed snow rolls down mountain slopes. Along the way these masses increase in size and acquire a dangerous momentum. Avalanches also occur when edges of glaciers in high mountains break off.

Avalanches are commonly subdivided according to the material involved.

Snow avalanches occur in predictable locations in snowy mountains and create distinctive ground features as they move down the mountain side.

Debris avalanches involve the rapid downslope movement of sediment.

Rock avalanches are very rapid downslope movements of bedrock which become shattered during movement. These avalanches sometimes achieve velocities as high as 400 kmph. They can travel tens of kilometres from their source, sometimes with devastating effects on human life.

Damages by Avalanches

Generally speaking, avalanches cause the following damages :

(*i*) Roads are damaged by snow of the avalanches.

(*ii*) Traffic is blocked when an avalanches falls on a road.

(*iii*) Road structures such as retaining walls are overturned.

(*iv*) Structures coming in the way of an avalanche are damaged.

Preventive Measures

(*i*) Afforestation of avalanche prone area.

(*ii*) Trapping the avalanche by control measures.

TABLE 9.14. Major avalanche disasters in India

Location	Time	Damages
Himachal Pradesh	March, 1978	30 people died. Severe damage to roads and other properties.
Lahul and Spiti	March 1979	About 237 people died. Communication system badly disrupted.
Jammu-Kashmir	December, 1982	Nearly 100 persons lost their lives. Communication system badly damaged.
Ladakh	December, 1984	Avalanche caused by artificial fire took life of 27 soldiers.
Jammu-Kashmir	March, 1988	80 persons died and communication system failed.
Himachal Pradesh	March 1991	Transport system closed for 40 days.
Himachal Pradesh	September, 1995	Huge amount of ice melted and fell below which flooded large areas.
Jammu-Kashmir	January, 2010	Avalanche in Khilanmarge took life of 17 soldiers.
Jammu-Kashmir	April, 2012	Siachin glacier, 135 Pakistani soldiers burred under snow.
Jammu-Kashmir (Gurej Sector)	January 27, 2017	24 including 20 soldiers died. Large number of people reported missing.
Chamoli (Uttarakhand)	April 24, 2021	At least 8 persons died and 384 were rescued. Road access was cut-off at 4-5 locations.

(*iii*) Disposing the avalanche potential snow packs by artificial triggering

(*iv*) Predicting the occurrence of avalanches through stability analysis and issuing warning about impending avalanche.

(*v*) Guiding the residents about the emergency evacuation shelters.

Avalanche Prone Areas in India

The higher reaches of the Himalayan region are prone to attack by avalanches. The snowy region of the Western Himalayas is particularly prone to avalanches. It includes higher areas in Ladakh,

Jammu and Kashmir, Himachal Pradesh and Uttarakhand. The details are as under :

1. Jammu and Kashmir – Higher reaches of Kashmir and Gurez valleys.
2. Kargil and Ladakh and some of the major roads.
3. Himachal Pradesh – Chamba, Kullu-Spiti and Kinnaur are vulnerable areas.
4. Uttarakhand – Parts of Tehri Garl.wal and Chamoli districts are vulnerable areas.

Avalanche Control Structures

Two major types of avalanche control structures are (*a*) Prevention Structures and (*b*) Protection structures.

(*a*) Prevention Structures. These structures are meant for preventing the occurrence of avalanches. Following are the major preventive structures.

(i) Avalanche Prevention Forests**.** These prevent the movement of avalanches by the resistance of tree trunks and branches, increase the stability of snow cover by uniformly distributing it and control quick changes in snow cover.

(ii) Stepped Terraces. Theses help in stabilising the snow cover. Stepped terraces are easy to construct but are not effective in controlling surface layer avalanches.

(iii) Avalanche Control Piles. Avalanche Control Piles are assemblies of single piles driven into slopes in avalanche zones to control surface layer avalanches. Spacing of piles depends upon the type of snow or topographical features. The average spacing is about 5 metres.

(iv) Avalanche Control Fence. Avalanche control fence is installed on slopes of avalanche zones to prevent full depth or surface layer avalanches.

(v) Suspended Fences. Suspended fences are used on steep slopes or in areas where foundations cannot be properly installed because of poor ground conditions. These are useful in small area.

(vi) Snow Cornice Control Structures. These structures are installed at tops of mountain areas to prevent the development of snow cornices that can cause avalanches.

(*b*) Protection Structures. These structures are installed in the path of the avalanche or in snow deposit areas to change direction of flow of avalanches, to reduce their energy to block their flow or to allow their passage. Following are the main protective structures.

(i) Protective Fences. These are installed to block the avalanches and their action is similar to that of retaining walls. They are normally constructed of steel and are used mainly for blocking small avalanches.

(ii) Retaining Walls. Retaining walls are normally installed in snow deposit areas to block the flow of avalanches before they reach the roadside. These walls need a pocket large enough to store snow deposited by avalanches and are not very effective unless they are installed on gentle slopes of 20 degrees or less.

(iii) Deflecting Structures. As the name indicates, these structures are installed to deflect the flow of an avalanche. This is done particularly to avoid interference of the avalanche in road traffic.

(iv) Snow sheds. Snow shed is a roofed structure installed over a road to allow the flow of an avalanche over the roof. This is most reliable of the various avalanche protection structures.

(v) Retarding Structures. These are structures to reduce the flow velocity or the scale of the avalanches. There are various types such as earth mounds, retarding piles, grating crib work and retarding fences.

Other Control Measures. Apart from the above mentioned measures, there are other control measures which are briefly described as under.

(i) Prediction and Forecasting. Prediction and forecasting is a very effective method of reducing the risk from avalanches can not only prevent avalanche disasters but can also make it efficiently dispose off dangerous snow deposits and cornices.

(ii) Disposal of Avalanches Potential Snow Packs. Methods that dispose of snow packs on hazardous slopes include hard labour, mechanical methods that use blasting powder. In general, small avalanches are usually disposed of by blasting.

REFERENCES

Agrawal, Krishan Mohan : *The making of an Earthquake,* The Hindustan Times, November 7, 1993.

Ajay Kanchan : *Rajasthan Drought (2000) :* A Human Made Disaster, Centre for Advocacy and Communications, New Delhi.

All India Symposium on Drought Prone Areas of India (Jan. 4–6, 1978), Sri Venteshwara University, Department of Geography.

All India Symposium on Drought Prone Area of India (1971) Tirupati : Rayalseema Geographical Society sponsored by the U.G.C.

Anandha Kumar, K.J. (2013), Uttarakhand Flash Floods Lessons for Disaster Management in Science Reporter, August 2013 p. 2829.

Anon (1977) : *Desertification and its Control* : Indian Council of Agricultural Research, New Delhi.

Arya, A.S. (1992) : *Possible effects of a major earthquake in Kangra region of Himachal Pradesh,* Current Science, 62.

Arya, A.S. (1999) : *Action Plan for Earthquake Disaster Mitigation* in Vinod K. Sharma (Ed), Disaster Management, National Centre for Disaster Management, New Delhi, pp. 60–73.

Arya, A.S. (1999) : *Case Study in India ; Diagnosis, Repair and strengthening of Damaged Buildings in* Vinod K. Sharma (Ed.). Disaster Management NCDM, New Delhi. pp 139-158.

Ayal, J.R. (1996) : *Precursor Seismicity, foreshocks and aftershocks of Uttarkashi earthquake of October 20, 1991 at Garhwal Himalaya,* Techno Physics.

Bagchi, Kathakali (1991) : *Drought Prone India ; Problems and Prospective,* New Delhi.

Balachandran, C.; Thanavelu and R., Pitchai Muthu : Marappalam Landslide, The Nilgiri District, Tamil Nadu, India, A Case Study in *Proceedings, International Conference of Disaster and Mitigation,* Madras, India 19-22 January 1996, Vol. 1, Theme : An Earthquake and Landslide, pp. A4-21 to A 4-28.

Bansal, R.C., and Mathur, H.N. (1976) : Landslides—The nightmare of roads. *Soil Conservation Digest* 5(1) : 36-37.

Basu, Sreelekha : Floods and Water Management, *Yojana 29 (6, 7),* April 16, 1985.

Benarde, Melvin A. (1972) : *Race against famine,* Orient Longman, Bombay.

Bhandari, R.K. and Gupta, C. (1985) : Problems of Landslides in the Himalaya and future directions in J.S. Singh (Ed), *Environmental Regeneration of Himayala: Strategy and Concept.* Cent, Him, Env. Assoc. Nainital pp. 39-57.

Bhattacharrya S.K. (2013) : *Landslide Risks,* Head the Warning in Science Reporter, August, 2013, pp. 20-27.

Bhatia, B.M. (1967) : *Famines in India,* Asia Publishing House, Delhi.

Biswas, M.R. and Biswas, A.K. (Eds.) (1980) : *Desertification Case Studies,* Pergamon. Oxford.

Caine, N. and Mool, P.K. (1982) : Landslides in the Kolpu Khola drainage; Middle Mountains, Nepal, *Mountain Research and Development,* 2(2) : 157-73.

Carter T.M. Clark J.P. and Leik, R.K. (1979) : *Constraints on organisational and household response to earthquake predictions. Int. synop. Earthq. Pred.* (UNESCO , Paris), Abstracts.

Chatterji, A.K. (1992) : *Monsoons, Cyclones and Floods in India,* Birla Economic Research Foundation, Radiant Publishers, New Delhi.

Chawla Parvinder : The Slayer Tsunamis; *Science Reporter,* February 2005, pp. 10 -15 and 24-25.

Chowdhery, Dandukar M.M and P.S. Pact (1989) : *Variability in drought incident over India*—A statistical approach, Mausum, Vol. 40. No. 2. pp. 207-214.

Devis, I (1984) : The vulnerability and reduction of damage-risk in small houses subjected to natural hazards in K.J. Mathew (Ed), *International Karakoram Expedition,* Vol. 1, Cambridge.

Gangopadhyay, B. (1985) : A case history of Kaliyasauer Landslide at km 147 on Rishikesh Joshimath road. *Proc. Sem. Construction of Roads in Hill Areas,* Ind. Road Congress, New Delhi, v : 21-32.

Gautam, Ashutosh (1994) : *Earthquake* : *A Natural Disaster,* Ashish, New Delhi.

Govt. of India (2020) : Ministry of Earth Sciences, Assessment of Climate Change over the Indian Region.

Govt. of India (1987) : Ministry of Water Resources , Central Water Commission, *Flood Atlas of India.*

Govt of India (1997) : Building Materials & Technology Promotion Council (BMTPC), Ministry of Urban Development, *Vulnerability Atlas of India.*

Govt. of India (2004) Building Material and Technology Promotion Council (BMTPC), Ministry of Urban Development, Landslide Hazard Zonation Atlas of India.

Govt. of India : *India a Reference Annual 2024* : Publication Division, Ministry of Information and Broadcasting.

Govt. of India : *National Disaster Response Plan* : A document prepared by High Powered Committee on Disaster Management (Sept. 2001), Ministry of Agriculture.

Govt. of India : *High Powered Committee on Disaster Management Report* (October 2001), Ministry of Agriculture.

Govt. of India : Department of Science and Technology, *Geological Hazards* (2001).

Govt. of India : IMD, *Water and Climate Related Disasters, 2001.*

Govt. of India : National Institute of Disaster Management (NIDM), Ministry of Home Affairs : *Disaster Management in India : A status Report (August 2004).*

Gupta, J.H.K., Kupel, H.J., Radhakrishna, I. Chedda, R.K. and Grecksch, G. (1997) : *In situ pore pressure studies in the Koyna area, India* 291 IASPEI General Assembly, Thessalmiki, Aug. 18-28, Greece (Abstract).

Gupta H.K., Rastogi, B.K. Chadha R.K., Mandal, P. and Sarma, C.S.P. (1997) : *Enhanced reservoir induced earthquakes in Koyna region, India during 1993-95,* Journal of Seismology, 1.

Gupta, H.K., Singh, V.P. and Mohan, I. (1979) : *Earthquake Prediction research in Assam, north-east India,* Int. Symp. Earth. Pred. (UNESEO, Paris), Abstracts.

Gupta, M.C., Sharma, Vinod K., Gupta, L.C. and Tamini, B.K. (2001) : *Manual on Natural Disaster Management in India,* National Centre for Disaster Management (NCDM) Ministry of Agriculture, Govt. of India.

Hemmady, A.K.R. (1996) : *Earthquakes,* National Book Trust India, New Delhi.

Indian Standards Institute (1976) : *Criteria for the Earthquake Resistant Design of Structures,* Delhi.

Jain, P.C. (1999) : *Flood Mitigation Practices in India* in Vinod K. Sharma (Ed.), Disaster Management (NCDM), New Delhi. pp. 38-45.

Jaiswal, N.K. (1981) : *Development of Drought Prone Areas,* NIRD, Hyderabad, pp. 461-498.

Jolly Asil (2013) How the Hills can Kill Again in India : Today, August, 2013 pp. 36-58.

Kanwar Sain (1979) : Flood Problem of India, *Birla Institute of Scientific Research.*

Kapoor Anu (2009) : Vulnerable India : A Geographical Study of Disasters, Indian Institute of Advanced Study, Shimla, Sage Publications India, Pvt. Ltd. New Delhi.

Kayastha, S.L. (1983) : *Floods in India—A study of their occurrences, causes, forecasting and control,* National Geographical Journal of India, Vol. 29, Nos. 3 and 4, pp. 121-41.

Khathri, K.N. (1987) : *Great earthquakes seismicity gaps and potential for earthquake disaster along the Himalayan plate boundary.* Tectnophysics 138.

Kolvankar, V.G. (2001) : *Earthquakes Sequence of 1991 from Valsa Region Gujarat BARC* / 2001/ E/ 006.

Kumar, G. and Shome, S.K. (1970) : Alaknanda tragedy—a geotechnical review. *In*: *Proc. River Valley Projects in North India,* Roorkee University and Society for Engineering Geology, Calcutta.

Lyngdoh, J.M. : Disaster management, A case study of Kosi Security in North-East Bihar, *Jour. of Rural Development,* (5); Sept. 1988, pp. 519-539.

Mahalanobis, P.C. : *Rain Storms and river floods in Orissa,* Sankhya, Calcutta 5, 1, 120.

Mandal, G.S. (1999) : *Natural Disasters* in Vinod K. Sharma, Disaster Management NCDM, New Delhi, pp. 178-189.

Mason, B.J. (1957) : The *Physics of Clouds,* Clarendon Press. Oxford.

Mehta, Balraj : The Challenge of drought, *Indian Express,* 15 July, 1992.

Misra, Girish K. (1984) : *Strategy and Resource Analysis for Drought Prone Areas in* A. Ramesh (ed.) *Resource Geography (Contributions to Indian Geography),* Heritage Publishers, India, New Delhi.

Misra, R.S. *et al.* (2001) : A *seismotectonic evaluation of Pandhana Swarm activity in Satpura block Khandwa-Khargone Belt M.P.,* Int. Conf. Seismic hazard with particular reference to Bhuj. earthquake of January 26, 2001, Oct. 3-5, New Delhi (Abstract).

Muthiah, S. *et al.* (eds.) : (1987) : *A Social and Economic Atlas of India,* Oxford University Press, New Delhi.

Ni, J. and Barazangi, M. (1984) : *Seismatectonics of the Himalaya collision zone; geometry of the underthrusting Indian Plate beneath the Himalaya. Jr. Geophysics Res. 89.*

Padmanabhan, B.S. : Disaster management and development, *Yojana,* 36 (22) 15 Dec. 1992, p. 15-16.

Pakiser, C.C., Eaton, J.P., Healy, J.H., and Raleigh, C.B. (1969) : Earthquake predictions and control, *Science 166* (3912) : 1467-74.

Pandit, A.K. and Quadri, S.S. Floods threatening Kashmir Wetlands, *Journal of Environmental Management,* 31(4), Dec. 1990, pp. 299-311.

Prakash Surya (2013) : Managing Disaster, Involving Communities in Science Reporter, August, 2013 pp. 32-36.

Press, F. (1975) : Earthquakes prediction. *Scientific American, 232.*

Rajan, K. (1999) : *Natural Disaster Management in National Development—An Indian Perspective* in Vinod K. Sharma, Disaster Management; NCDM., New Delhi, pp. 26-37.

Rana, Madhukar S.J.B. : Disaster Management, Prashant, 21(2), Mar. 90 , p. 97-110.

Rastogi, B.K., Singh, J. and Verma, R.K. (1973) : Earthquake mechanism and tectonics in the Asia-Burma region Tectonophysics, 18.

Saha, S.N. (1988) : *Microearthquake and Earthquake Prediction, Oriented Studies in Northeast India,* Proc. Symp. Earthquake Prediction, Pune.

Sain, Kanwar (1979) : *The Flood Problem India*, Birla Institute of Scientific Research, New Delhi.

Samra, J.S. and Singh, Gurbachan (2002) : Drought Management Strategies, Indian Council of Agricultural Research, New Delhi.

Sarkar, R.P. (1979) : *Droughts in India and their predictability*, Proc. Symp. on Hyd. Aspects of Droughts IIT, Delhi 33-40.

Sanjay K. Jain *et al.* (2013) : *Flash Floods in Scientific Reporter*, Aug., 2013, pp. 12-18.

Scholz, C.H. Sykes, L.R. and Aggarwal, Y.P. (1973) : *Earthquake prediction*, a physical basis, Science 181

Sen, Nandita, Impact of flood on crop damages—A Case study of lower Damodar Valley Region, *Indian Journal of Regional Science* 24 (2) 92 p. 65-72.

Seth, S.L., Das, D.C. and Gupta, G.P. (1982) : *Floods in Arid and Semiarid Areas—Rajasthan,* National Flood Commission Report, Chap. 10.

Sharma, N.K. : Flood Control in India, *Bhagirath*, 332 (1) Jan. 1985, p. 2-9.

Sharma, R.K. : *Role of farm extenion in draught management,* Agricultural Economic Research Centre, University of Delhi, 1990.

Sharma, S.C. (1975) : Landslips in the Deecan Traps with special reference to Western Ghat. *Proc. Sam. Landlides and Toe Erosion Problem with Special Reference to Himalayan Region. Indian Sec. Eng. Geol.* Calcutta, pp. 255-12

Sharma, Vinod K. (Ed.) (1999) : *Disaster Management*, National Centre for Disaster Management (NCDM), New Delhi.

Sharma, Vinod K. *et al.* (2000) : *India : IDNDR & Beyond,* NCDM, New Delhi.

Simizu, M. (1979) : *On the response to earthquake predictions and the organisational structure for earthquake prediction. Int. Sympo. Earth. Pred.* (UNESCO, Paris).

Siddhartha, K. (1999) : *Floods in India : A model of generating factors* CENDER, Patna.

Siddhartha, K. (1999) : India : *The Physical Aspects*, CENDER, Patna, Chap. 7 & 8.

Singh, Ranjit (1987) : *Indian droughts and Quasi Biennial Oscillations of Stratospheric Winds*, Proc. National Symp. on Hydrology, N.I.H. Roorkee.

Singh, S.K. *et al.* (2003) : *Estimation of Ground Motion for Bhuj (26 Jan. 2001; M. 7.8) and for future earthquakes in India*, Bull Seism. Soc. Amer. 93.

Singh, Tapeshwar (1978) : *Drought Prone Areas in India : Aspects of Identification and development,* Peoples Publishing House, New Delhi.

Sinha, Nishi : Flood control measures, *Financial Express,* 20 Oct., 92.

Srivastava, H.N. (2004) : *Earthquake; Forecasting and Mitigation,* National Book Trust, India, New Delhi.

Srivastava, H.N. (2007). *Coastal Hazards, (Cyclones, Tsunamis and other Disasters),* National Book Trust, New Delhi.

Subbiah, A.R. (1999) : *Drought Management through Anticipatory Multidimensional Approaches — A case study* in Vinod K. Sharma (Ed.) Disaster Management, NCDM, pp. 91-107.

Subramaniam, A.R. (1991) : *Agroclimatic Studies in India in* Diddee Jaymala (*ed.*) *Emerging Trends in Indian Geography*, Rawat Publications, Jaipur.

Tandon A.N. *et al.* (2001) : *Interplate vis a vis intraplate seismicity in relation of great Bhuj earthquake, 2001.* Proc. Symp. on Bhuj and Chamoli earthquakes, Roorkee (Vol. 2).

Uppal, J.N. (1984) : *Bengal famine of 1903 : a man-made tragedy,* Atma Ram & Sons, Delhi.

Valdiya, K.S. (1984) : *Aspects of Tectonics : Focus on South-Central Asia,* Tata McGraw Hill, New Delhi.

Virk, H.S. and Singh, B. (1995) : *Correlation of Randon Anomalies with Uttarkashi Earthquake* , Geol. Soc. of India Mem. 30.

Zaruba , Q. and Mencl, V. (1969) : *Landslides and their Control* Elsevier, Amsterdam.

◆◇◆

CHAPTER 10

Population : General Characteristics

SIZE OF INDIAN POPULATION

With a total population of 1,210.2 million according to 2011 census figures, India is the second most populous country of the world, next only to China, a country with a population of 1,341.0 million in 2010. India covers only 2.4 per cent of the land area of the world, but is the home of about 17.5 per cent of the world's population as compared to 19.4 per cent of the world's population living in China whereas that country occupies more than 7 per cent of the world's area. Thus a little more than one out of every six persons in the world is from India. With about 308.7 million population, the USA is the third largest country of the world with respect to population size. However, there is yawning gap of more than 901 million between the population of India and the USA which is about three times the total population of that country. Further it is worth noting that gap between population of India and China is only 130.8 million. Our population is almost equal to the combined population of the USA, Indonesia, Brazil, Pakistan, Bangladesh and Japan put together, the population of these six countries totals 1214.3 million. If we look at the area of the major countries of the world, Russian Federation is more than five times, Canada is over three times, the USA is 2.8 times, Brazil is 2.6 times and Australia is 2.3 times as large as India. But their combined population is only 63 per cent of the total population of India. India's population is a little over twice the population of Latin America and 1.2 times the population of whole of Africa.

CENSUS OF POPULATION

Data regarding population are collected through censuses all over the world. A census count offers a spectrum of population at a particular point in time covering a wide range of demographic, social and economic attributes of population. The year 1872 marked the beginning of census taking in India. Although it marked an auspicious beginning, it was neither a synchronous project nor did it cover the entire country. The first complete and synchronous

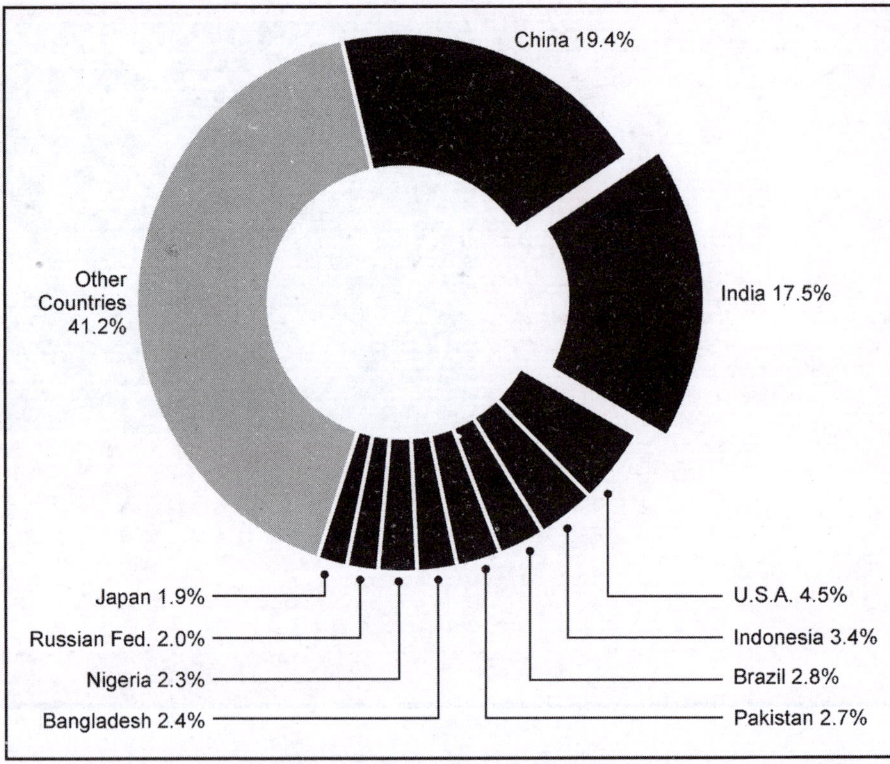

FIG. 10.1. India in world population

census covering the entire country and providing vital demographic data was conducted in 1881. Since then the census in India has been conducted regularly after every ten years. The 2011 census represents the fifteenth census of India as reckoned from 1872 and seventh after Independence. With a view to widen the scope and improve the quality of census data, modifications in its schedule and questionnaire have been introduced from time to time.

India's population is so large that population of some states is larger or almost equal to the total population of several large countries of the world (*see* box).

Comparison of population of some Indian states with that of some large countries of the world. Population in millions (2011)

Indian States	Countries
1. Uttar Pradesh : 199.6	Brazil : 190.7
2. Maharashtra : 112.4	Japan : 128.1
3. Bihar : 103.8	Philippines : 98.0
4. West Bengal : 91.3	Germany : 82.3
5. Madhya Pradesh : 72.6	Egypt : 73.0
6. Gujarat : 60.4	Italy : 59.3, UK : 61.0
7. Odisha : 41.9	Argentina : 41.0
8. Kerala : 33.4	Iraq : 33.0; Afghanistan : 32.9
9. Chhattisgarh : 25.5	Ghana 25.4, Australia : 21.9
10. NCT of Delhi : 16.7	Angola : 16.3

The combined population of Uttar Pradesh and Maharashtra is about 312 million which is substantially greater than the population of U.S.A., third largest country of the world with respect to population.

Growth of Population : Basic Concepts

1. Growth Rate. Growth of population is the change in the number of people living in a particular area between two given points of time. The net change between two points of time is expressed in percentage and is described as the growth rate of population.

2. Natural Growth. The difference between the natural birth-rate and death-rate is called the *natural growth* of population.

3. Migratory Growth. This growth of population is caused by migration of people.

4. Positive Growth. When the population increases between two given points of time, it is called ***positive growth***. It takes place when the birth rate is higher than the death-rate or people migrate from other countries.

5. Negative Growth. The growth of population is called ***negative*** if the population decreases between two given points of time. It takes place if the birth-rate is lower than the death-rate or people migrate to other places.

There has always been positive growth of population in India ever since first census of India was conducted. However, 1921 is an exception when the growth was negative.

Several attempts have been made to estimate the population of India before first census was conducted in 1872. According to an estimate, India's population was just 100 million in 1600 A.D. It rose to 120 million in 1800, 130 million in 1841, and 255 million in 1871. This can be termed as a negligible growth compared to the present trends.

Population Growth Since 1901

Trends in population growth since 1901 have been given in **Table 10.1**. A close look at this table shows that there have been significant demographic divides as far as trends in population growth are concerned. These significant turning points are the census years 1921, 1951 and 1981. Thus the demographic history of India can be charted and classified into the following four distinct phases.

1. Period of Stagnant Population (1901-1921)
2. Period of Steady Growth (1921-1951)
3. Period of Rapid High Growth (1951-1981)
4. Period of High Growth with Definite Signs of Slowing Down (1981-2011)

1. Period of Stagnant Population (1901-1921)

During most of the 19th century India witnessed sporadic, irregular and slow growth of population which drifted into the twentieth century until 1921. Thus the population growth during this period can be termed more or less stagnant when compared to the growth rates observed during the consequent periods. The high birth rate was counterbalanced by high death rate. The progressive growth rate in 1921 over 1901 was only 5.42 per cent. In fact, the census year 1921 registered a negative growth rate of –0.31 per cent which happened only once throughout the demographic history of India. It is because of this decline in place of rise in population that the year 1921 is called the '***demographic divide***'

TABLE 10.1. Decadal Growth Rates in India, 1901-2011

Census Year	Total Population	Growth Rate*	
		Absolute Number	% of Growth
1901	238396327	......	......
1911	252093390	(+) 13697063	(+) 5.75
1921	251321213	(–) 772117	(–) 0.31
1931	278977238	(+) 27656025	(+) 11.00
1941	318660580	(+) 39683342	(+) 14.22
1951	361088090	(+) 42427510	(+) 13.31
1961	439234771	(+) 78146681	(+) 21.64
1971	548159652	(+) 108924881	(+) 24.80
1981	683329097	(+) 135169445	(+) 24.66
1991	846421039	(+) 163091942	(+) 23.87
2001	1028737436	(+) 182316397	(+) 21.54
2011	1210193422	(+) 181455986	(+) 17.64

*Decadal growth rate : $g = \frac{P_2 - P_1}{P_2} \times 100$

Where P_1 = Population of the base year

P_2 = Population of the present year

Source : Census of India, 2011, Provisional Population Totals Paper 1, p. 41.

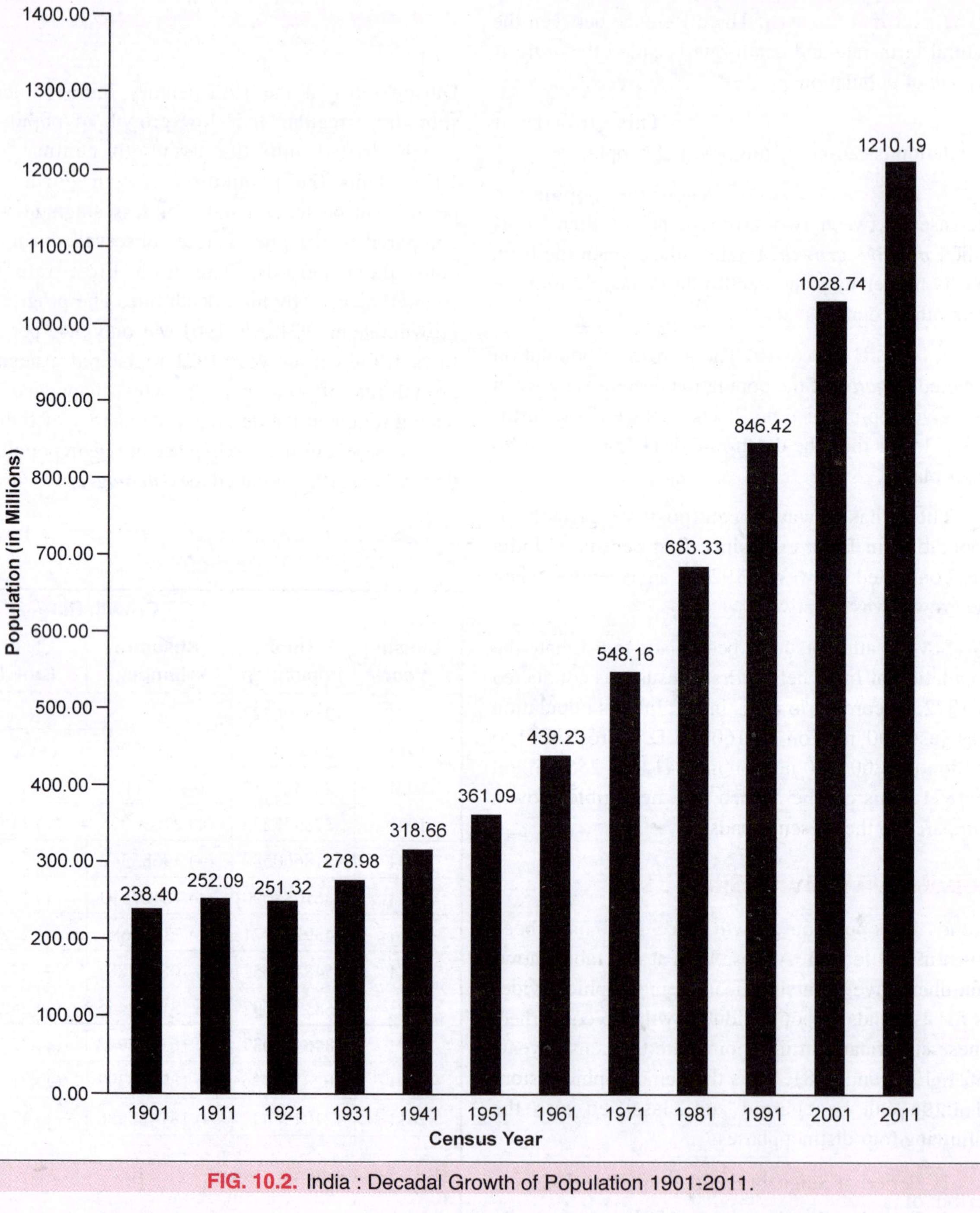

FIG. 10.2. India : Decadal Growth of Population 1901-2011.

in the demographic history of India. The high mortality during this period was the product of large scale abnormal deaths due to epidemics of influenza, plague, small pox, cholera, etc. Influenza alone claimed 12 million lives in 1918. Food shortages caused by severe droughts in 1911, 1913, 1915, 1918 and 1920 claimed their own toll. In addition, thousands of Indian soldiers lost their lives during the

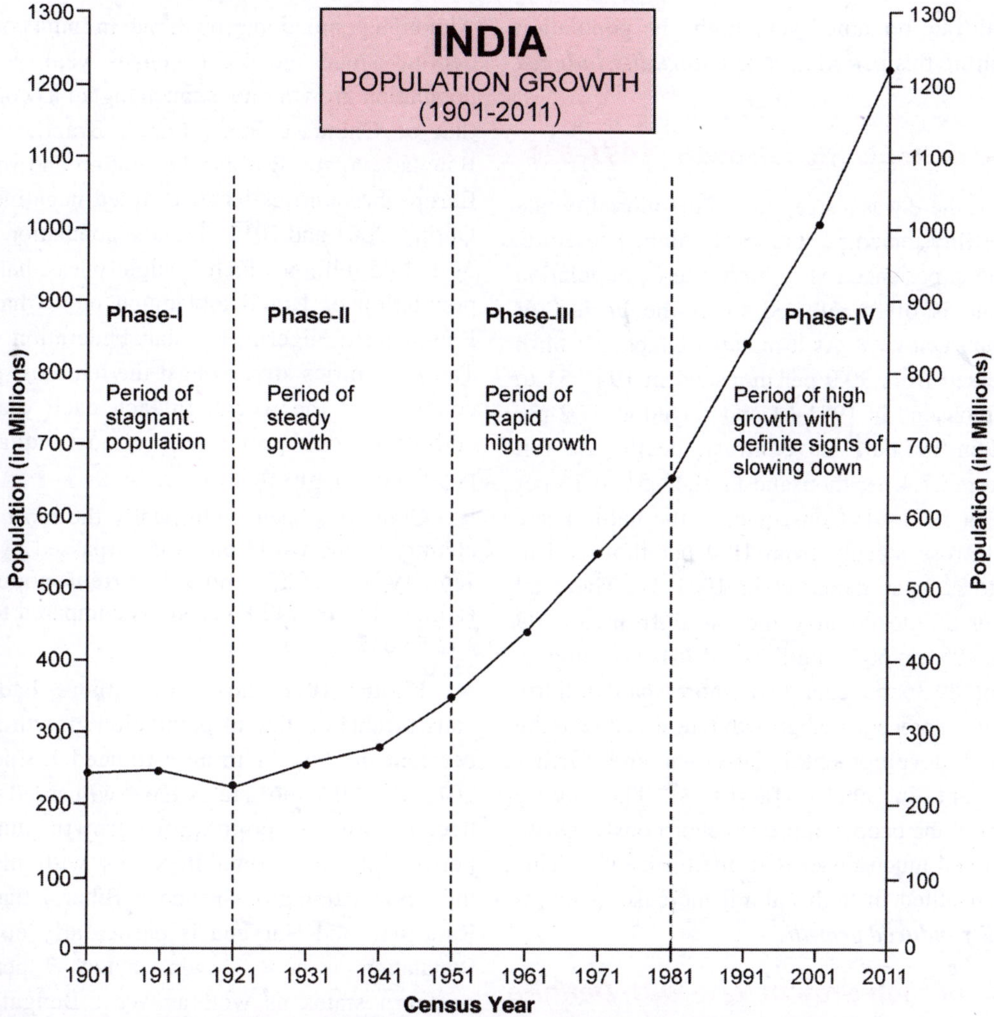

FIG. 10.3. India : Population Growth : 1901-2011

World War I (1914-18). Lakhs of people emigrated to a number of countries in Africa.

2. Period of Steady Growth (1921-51)

During 1921-51, the population of India increased from 251 million to 361 million (**Table 10.1**). This duration of 30 years has thus registered a growth of 47.3 per cent. Therefore, this period is called the period of steady growth rate. The mortality rate started showing downward trend as a result of improvement in general health and sanitation conditions after 1921. These developments helped in controlling epidemics lil-e plague, cholera and malaria. The crude death rate which stood at a high of 47.2 per thousand in 1911-21, declined to 27.4 per thousand in 1941-51. On the contrary, the crude birth rate continued to stay at an abnormally high level and decline only to 39.9 per thousand in 1941-51 as against 48.1 per thousand in 1911-21. Decline in death rate was also achieved partly through the improvement in the distribution system as a result of improved transportation so that timely supplies of food could be made available to drought and famine stricken areas. The combined effect of these factors was that the population started increasing steadily. Since crude death rate declined considerably and crude birth rate remained very high, the population growth during this period is called ***mortality induced*** growth.

3. Period of Rapid High Growth (1951-81)

After 1951, there was a steep fall in the mortality rate but the fertility remained stubbornly high. Therefore, this period experienced very high rate of population growth and is often referred to as the ***period of population explosion***. As a matter of fact, the birth rate increased from 39.9 per thousand in 1941-51 to 41.7 per thousand in 1951-61 and stayed at 37.2 per thousand in 1971-81. In contrast, death rate fell rapidly from 27.4 per thousand in 1941-51 to 15 per thousand in 1971-81. Consequently the natural rate of growth, rose steeply from 18.9 per thousand in 1951-61 to 22.2 per thousand in 1971-81. The total population of the country increased from 361.09 million in 1951 to 683.3 million in 1981 recording an increase of 89.36 per cent in a short span of thirty years. This unprecedented growth rate was due to the accelerated developmental activities and further improvement in health facilities. The living conditions of the people improved enormously. Death rates declined much faster than the birth rates. This situation resulted in high natural increase. Thus, it was ***fertility induced growth***.

4. Period of High Growth Rate with Definite Signs of Slowing Down (1981-2011)

The last phase of 20th century and the early phase of 21st century *i.e.,* the period between census years 1981 and 2011 is known as the period of high growth with definite signs of slowing down. Although the rate of growth was still very high, it started declining after 1981. The highest ever growth rate of 22.2 per cent was recorded in 1971-81 which remained at a high of 19.7 in 1981-91 also. It declined to 14.7 per cent in 2001-11. Thus the growth rate registered the sharpest decline during the decade 2001-2011. Declining trend of death rate continued but at a slower rate. This declining trend is a positive indicator of the official efforts of birth control and people's own inclination to opt for smaller families. Although population growth rate in India continues to decline since the 1971 census year yet India's population growth rate is much higher as compared to that of China, U.S.A., Japan, Brazil, Indonesia, Bangladesh, etc. Russian Federation and some other European countries have recorded negative growth. During 2001 and 2011, India's population increased by 181.45 million which is slightly less than the total population of Brazil and much more than that of Bangladesh, Nigeria, Russian Federation or Japan. These countries are amongst the ten most populated countries of the world. In fact, each year we are adding to our population almost equal to the population of Australia.

China has been traditionally the most populous country of the world but was surpassed by India on January 21, 2023. India's current population is estimated to be 142.86 crore as compared to China's 142.57 crore.

Figure 10.4 shows that there had been a considerable decline of population growth from 2.48 per cent in 1961-71 to an estimated 1.3 per cent in 2011-16. All major states have witnessed a marked deceleration in population growth during this period; the slowdown in states with historically high population growth such as Bihar, Uttar Pradesh, Rajasthan and Haryana is particularly noteworthy. Population is now growing below 1 per cent in southern states as well as West Bengal, Punjab, Maharashtra, Odisha, Assam and Himachal Pradesh.

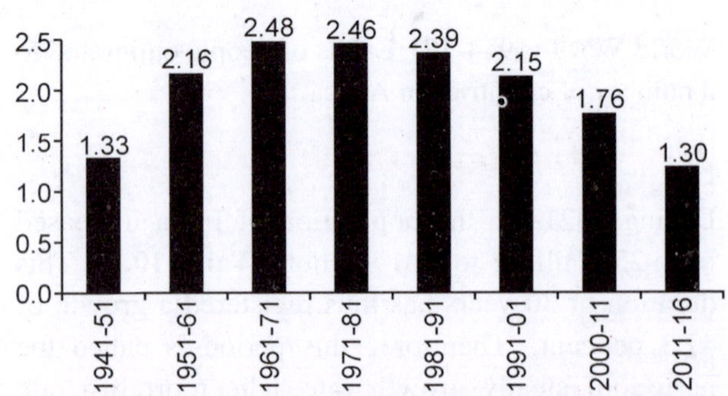

FIG. 10.4. Annual Population Growth Rate in India (per cent)

Source: Census 2011, International Institute for Population Sciences (IIPS) estimates.

A key driver of this trend has been the steady decline in India's total fertility rate (TFR) from 5.9 in 1950-55 to 2.0 in 2020-25 (**Fig. 10.5**). *Total fertility rate refers to the total number of children born or likely to be born to a women of child-bearing age in her lifetime*. Globally, a fertility rate of 2.1 is considered the replacement rate. This means that an average of 2.1 babies per women will keep the present population intact *i.e.* the population exactly replicates itself from one generation to the other without migration. India's TFR may already be close the effective replacement level fertility after accounting for its skewed sex ratio. The 2.1 replacement rate is based on the assumption that the number of male and female babies born is equal. But in India, the sex ratio at berth (SRB)—the average value of sex ratio at birth is around 1.05 *i.e.* 105 boys born per every 100 girls.

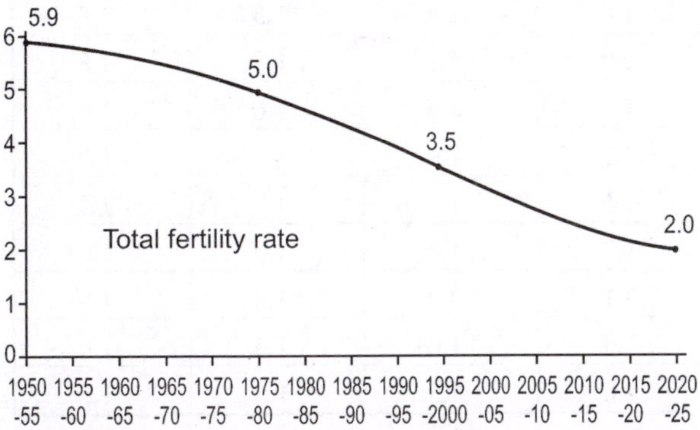

FIG. 10.5. TFR in India

Source: Sample Registration System

It must be noted that there are wide variations at the state level with respect to TFR. As per calculations for 2021, it is now below the replacement level fertility in 21 out of 22 major states and Union Territories. Only Bihar has reported TFR of 2.5 which is slightly more than the replacement level. The projected values for 2021-41 suggest that TFR at the national level will continue to decline rapidly and will lie below replacement level fertility at 1.7 as early as 2031 (**Table 10.2**).

Further, it is interesting to note that all the religious groups have shown a considerable drop in TFR between 2005-06 and 2015-16 (**Table 10.3**).

Table 10.3 shows that people belonging to Buddhist religion had the lowest TFR of 1.39 which is much below the replacement level of 2.1. The other religions with TFR lower than the replacement level are Sikhs, Christians and Buddhists. Hindus have TFR almost equal to the replacement rate. A big surprise is that Muslims have recorded the highest decline of –46.5% TFR between 1992-93 and 2019-21; although they still have the highest TFR among all the religious groups.

Population Growth Trajectory

Demographic projections show that India's population growth will continue to slow rapidly over the next two decades, growing less than 1 per cent during 2021-31 and under 0.5 per cent during 2031-41. Positive population growth in the next two decades will be due to population momentum and the continued rate in life expectancy.

Large scale variations in trajectory of population and population growth are expected at the state level. States ahead of demographic transition will see a continued deceleration in population growth and reach near zero growth rates by 2031-41. Tamil Nadu's population growth will start declining during 2031-41 unless offset by inward migration. Population growth will be close to zero in Andhra Pradesh and as low as 0.1-0.2 per cent in Karnataka, Kerala, Telangana, Himachal Pradesh, West Bengal, Punjab and Maharashtra.

States lagging behind in the demographic transition will also witness a marked slowdown in population growth during 2021-41. Population growth will be half over the next two decades in Uttar Pradesh, Chhattisgarh, Rajasthan and Madhya Pradesh. Only Bihar will have a population growth rate of one per cent.

TABLE 10.2. TFR for India and Major States, 2001-2041

States	2001	2011	2016	2021	2031	2041
INDIA	3.1	2.4	2.3	1.8	1.7	1.7
Andhra Pradesh	2.3	1.8	1.7	1.5	1.5	1.5
Assam	3.0	2.4	2.3	1.8	1.8	1.8
Bihar	4.4	3.6	3.3	2.5	2.0	1.8
Chhattisgarh	3.9	2.7	2.5	1.8	1.8	1.8
Delhi	2.1	1.9	1.6	1.5	1.5	1.5
Gujarat	2.9	2.4	2.2	1.9	1.8	1.8
Haryana	3.1	2.3	2.3	1.8	1.8	1.8
Himachal Pradesh	2.2	1.8	1.7	1.6	1.6	1.6
Jammu & Kashmir	2.5	1.9	1.7	1.5	1.5	1.5
Jharkhand	4.4	2.9	2.6	1.8	1.8	1.8
Karnataka	2.4	1.9	1.8	1.5	1.5	1.5
Kerala	1.8	1.8	1.8	1.8	1.8	1.8
Madhya Pradesh	3.9	3.1	2.8	2.0	1.8	1.8
Maharashtra	2.4	1.8	1.8	1.5	1.5	1.5
Odisha	2.6	2.2	2.0	1.8	1.8	1.8
Punjab	2.4	1.8	1.7	1.6	1.6	1.6
Rajasthan	4.0	3.0	2.7	1.9	1.8	1.8
Tamil Nadu	2.0	1.7	1.6	1.5	1.5	1.5
Telangana	2.3	1.8	1.7	1.6	1.6	1.6
Uttar Pradesh	4.5	3.4	3.1	2.0	1.8	1.8
Uttarakhand	4.5	3.4	1.9	1.6	1.6	1.6
West Bengal	2.4	1.7	1.6	1.5	1.5	1.5

Source: Census, Sample Registration System, IIPS projections.

Note : Projected values for 2021-41 are based on fitted model values from 20-year trends of TFR based on estimates from Sample Registration System. Before 2016, the values of Andhra Pradesh are assigned to Telangana.

TABLE 10.3. Religion-wise Total Fertility Rate

Community	NFHS-5 (2019 to 2021)	NFHS-4 (2015 to 2019)	NFHS-3 (2005 to 2006)	NFHS-2 (1998 to 1999)	NFHS-1 (1992 to 1993)	% Change since NFHS-4	% Change since NFHS-1
1. Hindu	1.94	2.13	2.59	2.78	3.3	–8.9	–41.2
2. Muslim	2.36	2.62	3.4	3.59	4.41	–9.9	–46.5
3. Christian	1.88	1.99	2.34	2.44	2.87	–5.5	–34.5
4. Sikh	1.61	1.58	1.95	2.26	2.43	1.9	–33.7
5. Buddhist/Neo-Buddhist	1.39	1.74	2.25	2.13	N.A.	–20.1	N.A.
6. Jain	1.6	1.2	1.54	1.9	N.A.	33.3	N.A.
India	2	2.2	2.7	2.9	3.4	–9.1	–41.2

Source: NFHS = National Family Health Survey.

TABLE 10.4. Annual Population Growth Rate (per cent) in India

Decade	Growth rate (per cent)
2001-11	1.77
2011-21	1.12
2021-31	0.72
2031-41	0.46

Source: Census, International Institute for Population Sciences (IPS) projections.

In spite of all the above mentioned facts about declining TFR and population growth rates, India's population is likely to increase from 1210.6 million in 2011 to 1510.2 million in 2041 and at that time India will be the most populous country of the world (Table 10.5).

TABLE 10.5. Projected Population in India

Year	Population in millions
2011	1210.6
2016	1286.1
2021	1346.9
2031	1443.2
2041	1510.2

Source: Census, Sample Registration System, IIPS Projections

Demographic Transition. Demographic transition is the process of change in population of a society. It consists of the following four stages :

(*a*) ***Stage 1.*** High death and birth rates, low growth rate.

(*b*) ***Stage 2.*** Rapid decline in death rate, continued high birth rate, very high growth rate.

(*c*) ***Stage 3.*** Rapid decline in birth rate, continued decline in death rate, growth rate begins to decline.

(*d*) ***Stage 4.*** Low death and birth rates, low growth rate.

The story of population growth in India is fairly in tune with the classical theory of demographic transition. During most of the nineteenth century India witnessed a fluctuating but ultimately more or less a stagnant growth of population, which drifted into the twentieth century until 1921. Thereafter, the country passed through successively all the phases of demographic transition and is now widely believed to have entered the final phase which is normally characterised by rapidly declining fertility. It is yet to be seen as to how long will this phase extend and when India will achieve a stable population.

Spatio-Temporal Variations in Population Growth

The average population growth rate of 17.64 per cent during 2001-11 does not give true picture as there are differences in the growth rate with reference to space and time. These are called spatio-temporal variations in population growth. There can be different reasons for differential growth-rate in different parts of India. A study of **Table 10.6** and **Fig. 10.6** gives an idea of regional variations in population growth.

Exactly half of the 20 most populous states, each with a population of ten million or more, have added lesser persons in the decade 2001-2011 compared to the decade 1991-2001. Had these ten states added the same number of persons during 2001-2011 also as they did in the previous decade, India would have added another 9.7 million persons during this decade. The phenomenon of low growth has spread beyond the boundaries of the southern states during 2001-2011, where in addition to Andhra Pradesh, Tamil Nadu and Karnataka in the south, Himachal Pradesh and Punjab in the north, West Bengal and Odisha in the east and Maharashtra in the west have registered growth rate between 11 and 16 per cent in 2001-2011.

Among the smaller states and union territories, Dadar and Nagar Haveli and Daman and Diu registered the highest growth rate of 55.5 and 53.54 per cent respectively. In contrast, Lakshdweep, Andaman Nicobar Islands and Goa have registered low growth rate remaining in single digit only. A glaring down trend in the growth has been observed

TABLE 10.6. Population, percentage decadal growth and average annual exponential growth rates 1991-2001 and 2001-2011

State/ UT Code	India/State/Union Territory	Total Population		Percentage Decadal growth	
		2001	2011	1991-2001	2001-2011
1	2	3	4	5	6
	INDIA	**1,02,87,37,436**	**1,21,01,93,422**	**21.54**	**17.64**
1	Jammu & Kashmir	1,01,43,700	1,25,48,926	29.43	23.71
2	Himachal Pradesh	60,77,900	68,56,509	17.54	12.81
3	Punjab	2,43,58,999	2,77,04.236	20.1	13.73
4	Chandigarh	9,00,635	10,54,686	40.28	17.1
5	Uttarakhand	84,89,349	1,01,16,752	20.41	19.17
6	Haryana	2,11,44,564	2,53,53,081	28.43	19.9
7	NCT of Delhi	1,38,50,507	1,67,53,235	47.02	20.96
8	Rajasthan	5,65,07,188	6,86,21,012	28.41	21.44
9	Uttar Pradesh	16,61,97,921	19,95,81,477	25.85	20.09
10	Bihar	8,29,98,509	10,38,04,637	28.62	25.07
11	Sikkim	5,40,851	6,07,688	33.06	12.36
12	Arunachal Pradesh	10,97,968	13,82.611	27	25.92
13	Nagaland	19,90,036	19,80,602	64.53	–0.47
14	Manipur	22,93,896	27,21,756	24.86	18.65
15	Mizoram	8,88,573	10,91,014	28.82	22.78
16	Tripura	31,99,203	36,71,032	16.03	14.75
17	Meghalaya	23,18,822	29,64,007	30.65	27.82
18	Assam	2,66,55,528	3,11,69,272	18.92	16.93
19	West Bengal	8,01,76,197	9,13,47,736	17.77	13.93
20	Jharkhand	2,69,45,829	3,29,66,238	23.36	22.34
21	Odisha	3,68,04,660	4,19,47,358	16.25	13.97
22	Chhattisgarh	2,08,33,803	2,55,40,196	18.27	22.59
23	Madhya Pradesh	6,03,48,023	7,25,97,565	24.26	20.3
24	Gujarat	5,06,71,017	6,03,83,628	22.66	19.17
25	Daman & Diu	1,58,204	2,42,911	55.73	53.54
26	Dadra & Nagar Haveli	2,20.490	3,42,853	59.22	55.50
27	Maharashtra	9,68,78,627	11,23,72,972	22.73	15.99
28	Andhra Pradesh	7,62,10,007	8,46,65,533	14.59	11.10
29	Karnataka	5,28,50,562	6,11,30,704	17.51	15.67
30	Goa	13,47,668	14,57,723	15.21	8.17
31	Lakshadweep	60,650	64,429	17.3	6.23
32	Kerala	3,18,41,374	3,33,87,677	9.43	4.86
33	Tamil Nadu	6,24,05,679	7,21,38,958	11.72	15.6
34	Puducherry	9,74,345	12,44,464	20.62	27.72
35	Andaman & Nicobar Islands	3,56,152	3,79,944	26.9	6.68

Source: Census of India, 2011; Provisional Population Totals, paper 1 of 2011. Series 1. p. 54.

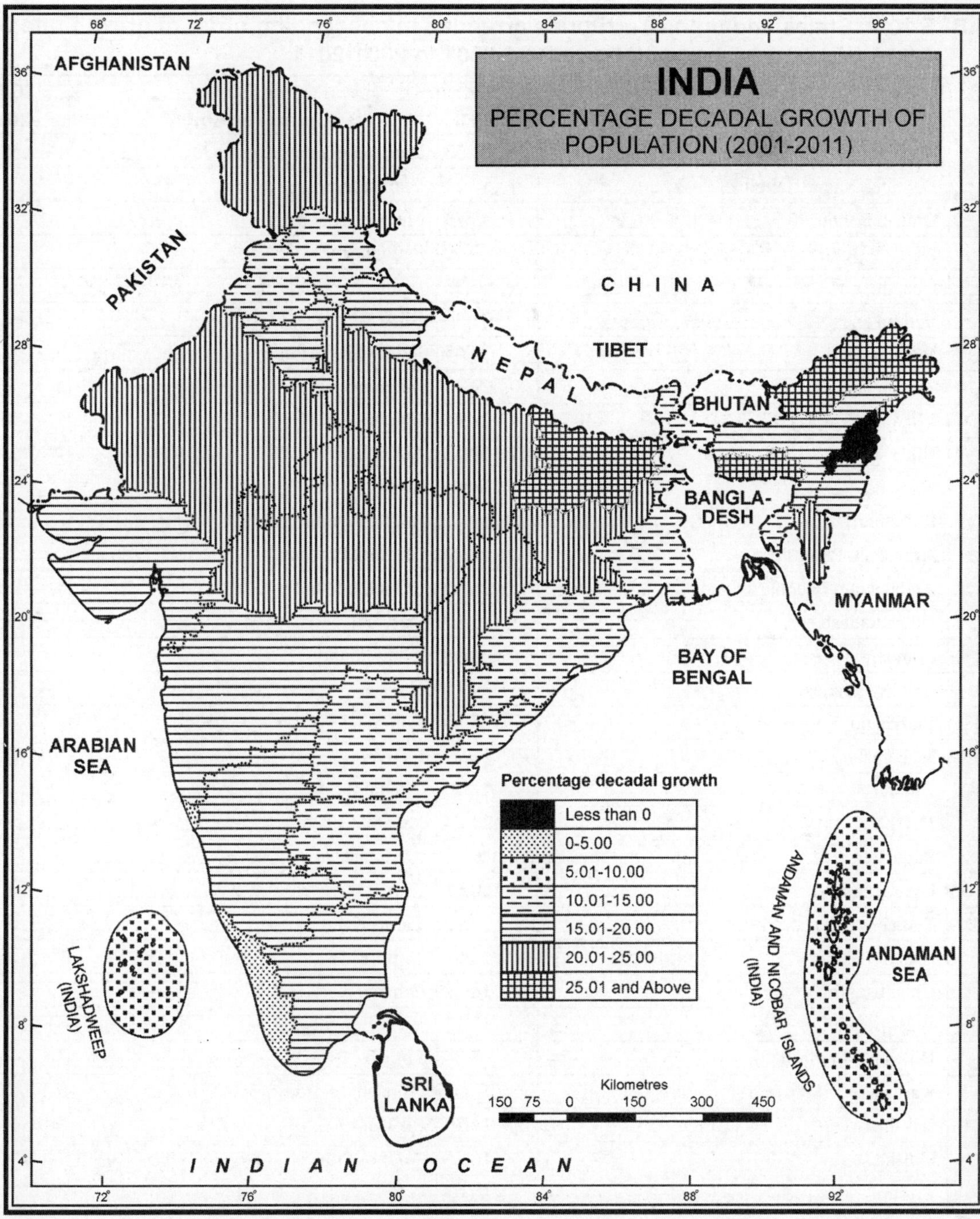

FIG. 10.6. India : Percentage decadal; growth of population (2001-2011)

in Nagaland, where there had been a steep fall in growth rate from 64.53 per cent in 1991-2001 to negative growth rate of –0.47 per cent. The second minimum growth of 4.86 per cent has been recorded by Kerala. This state has reached high level of demographic transition and can be easily compared with the advanced countries of Europe and America. Some of the more populous states have registered

TABLE 10.7. States and union Territories arranged in descending order of growth rate of population : 1991-2001 to 2001-2011

Rank	India/State/ Union Territory#	1991 -2001	India/State/ Union Territory#	2001 -2011
1	Nagaland	64.53	Dadra & Nagar Haveli#	55.50
2	Dadra & Nagar Haveli#	59.22	Daman & Diu#	53.54
3	Daman & Diu#	55.73	Meghalaya	27.82
4	NCT of Delhi#	47.02	Puducherry#	27.72
5	Chandigarh#	40.28	Arunachal Pradesh	25.92
6	Sikkim	33.06	Bihar	25.07
7	Meghalaya	30.65	Jammu & Kashmir	23.71
8	Jammu & Kashmir	29.43	Mizoram	22.78
9	Mizoram	28.82	Chhattisgarh	22.59
10	Bihar	28.62	Jharkhand	22.34
11	Haryana	28.43	Rajasthan	21.44
12	Rajasthan	28.41	NCT of Delhi#	20.96
13	Arunachal Pradesh	27.00	Madhya Pradesh	20.30
14	Andaman & Nicobar Islands#	26.90	Uttar Pradesh	20.09
15	Uttar Pradesh	25.85	Haryana	19.90
16	Manipur	24.86	Uttarakhand	19.17
17	Madhya Pradesh	24.26	Gujarat	19.17
18	Jharkhand	23.36	Manipur	18.65
19	Maharashtra	22.73	**INDIA**	**17.64**
20	Gujarat	22.66	Chandigarh#	17.10
	INDIA	**21.54**	Assam	16.93
21	Puducherry#	20.62	Maharashtra	15.99
22	Uttarakhand	20.41	Karnataka	15.67
23	Punjab	20.10	Tamil Nadu	15.60
24	Assam	18.92	Tripura	14.75
25	Chhattisgarh	18.27	Odisha	13.97
26	West Bengal	17.77	West Bengal	13.93
27	Himachal Pradesh	17.54	Punjab	13.73
28	Karnataka	17.51	Himachal Pradesh	12.81
29	Lakshadweep#	17.30	Sikkim	12.36
30	Odisha	16.25	Andhra Pradesh	11.10
31	Tripura	16.03	Goa	8.17
32	Goa	15.21	Andaman & Nicobar Islands#	6.68
33	Andhra Pradesh	14.59	Lakshadweep#	6.23
34	Tamil Nadu	11.72	Kerala	4.86
35	Kerala	9.43	Nagaland	–0.47

Union Territory

Source : Census of India, Provisional Population Totals, Paper 1 of 2011 Series 1, p. 167.

TABLE 10.8. India : Child Population in the age group 0–6 (2011)

India/State/Union Territory	Person	Male	Female	Decadal Change 2001-11	Percentage proportion of child population to total population
INDIA	**15,87,89,287**	**8,29,52,135**	**7,58,37,152**	**–50,48,108**	13.12
Uttar Pradesh	2,97,28,235	1,56,53,175	1,40,75,060	–18,96,393	14.90
Bihar	185,82,229	96,15,286	89,66,949	17,76,166	17.90
Maharashtra	1,28,48,375	68,22,262	60,26,113	–8,22,751	11.43
Madhya Pradesh	1,05,48,295	55,16,957	50,31,338	–2,33,919	14.50
Rajasthan	1,05,04,916	55,80,212	49,24,704	–1,46,086	15.31
West Bengal	1,01,12,599	51,87,264	49,25,335	–13,01,623	11.07
Andhra Pradesh	86,42,686	44,48,330	41,94,356	–15,29,171	10.21
Gujarat	74,94,176	39,74,286	35,19,890	–38,228	12.41
Tamil Nadu	68,94,821	35,42,351	33,52,470	–3,40,339	9.56
Karnataka	68,55,801	35,27,844	33,27,957	–3,26,299	11.21
Jharkhand	52,37,582	26,95,921	25,41,661	2,80,755	15.89
Odisha	50,35,650	26,03,208	24,32,442	–3,23,160	12.00
Assam	45,11,307	23,05,088	22,06,219	13,232	14.47
Chhattisgarh	35,84,028	18,24,987	17,59,041	29,112	14.03
Kerala	33,22,247	16,95,935	16,26,312	–4,70,899	9.95
Haryana	32,97,724	18,02,047	14,95,677	–37,813	13.01
Punjab	29,41,570	15,93,262	13,48,308	–2,30,259	10.62
Jammu & Kashmir	20,08,642	10,80,662	9,27,980	5,22,839	16.01
NCT of Delhi#	19,70,510	10,55,735	9,14,775	–46,339	11.76
Uttarakhand	13,28,844	7,04,769	6,24,075	–31,188	13.14
Himachal Pradesh	7,63,864	4,00,681	3,63,183	–29,273	11.14
Meghalaya	5,55,822	2.82,189	2,73,633	87,843	18.75
Tripura	4,44,055	2 27,354	2,16,701	7,609	12.1
Manipur	3,53,237	1,82,684	1,70,553	26,871	12.98
Nagaland	2,85,981	1,47,111	1,38,870	–3,697	14.44
Arunachal Pradesh	2,02,759	1,03,430	99,329	–3,112	14.66
Mizoram	1,65,536	83,965	81,571	21,802	15.17
Goa	1,39,495	72,669	66,826	–6,473	9.57
Puducherry#	1,27,610	64,932	62,678	10,451	10.25
Chandigarh#	1,17,953	63,187	54,766	2,340	11.18
Sikkim	61,077	31,418	29,659	–17,118	10.05
Dadra & Nagar Haveli#	49,196	25,575	23,621	8,997	14.90
Andaman & Nicobar Islands#	39,497	20,094	19,403	–5,284	10.40
Daman & Diu#	25,880	13,556	12,324	5,302	10.65
Lakshadweep#	7,088	3,715	3,373	–2,003	11.00

Source : Census of India 2011, Provisional Population Totals, Paper 1 of 2011, Series 1, pp. 64-65.

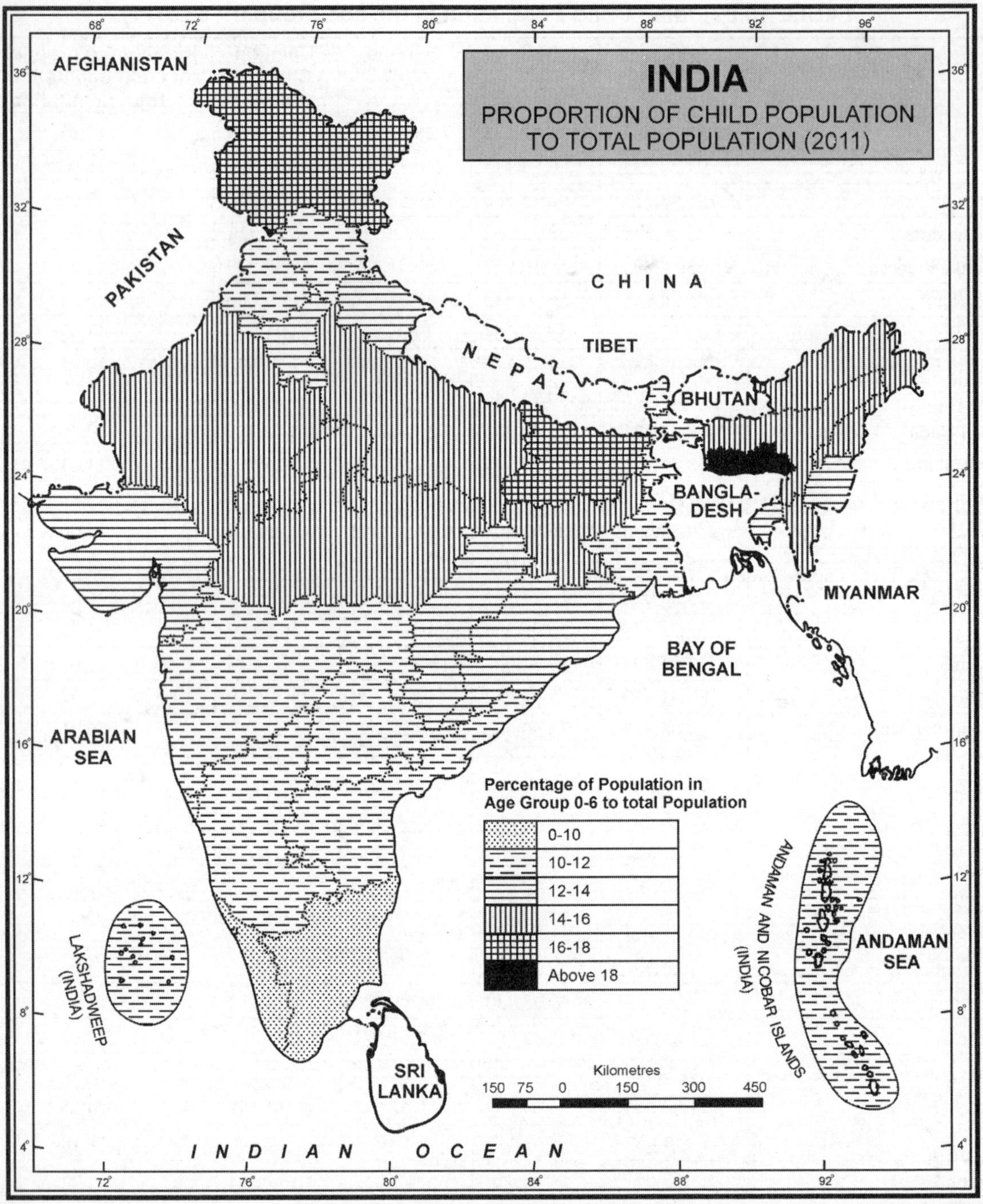

FIG. 10.7 India : Proportion of child population to total population (2011)

very high growth rate of over 20 per cent. Among them Bihar (25.07%), Jammu & Kashmir (23.71%), Chhattisgarh (22.59%) and Jharkhand (22.34) are worth noting. Some other states with small population but higher growth rate are Meghalaya (27.82%) and Arunachal Pradesh (25.92%).

Fifteen states and union territories registered a decline of over five percentage points in decadal growth rate from the previous census decade. These states and union territories are Jammu & Kashmir, Punjab, Haryana, Rajasthan, Uttar Pradesh, Sikkim, Nagaland, Manipur, Mizoram, Maharashtra, Goa, Delhi, Chandigarh, Lakshadweep, and Andaman and Nicobar Islands and account for more than 39% of India's population. Delhi has registered the sharpest drop of 26 percentage points during 2001-2011, followed by Haryana (8.53), Rajasthan (6.97) and Maharashtra (6.74). Seventeen states and union territories have shown 1 to 5 per cent points fall in their growth rates during 2001-11 as compared to 1991-2001. These states and union territories account for more than 52% of total population of India. Thus more than nine out every ten Indians live in states and union territories that have registered declining trend in population growth.

Growth of Child Population

Child population in the age group 0-6 years has special significance in our demographic scene because this segment of population determines the future course of trends in population growth when it reaches the reproductive age. According to 2011 census figures, the total number of children in the age-group 0-6 years is 158.8 million which is about 5 million less than the number recorded in 2001. Reduction in child population in the age group of 0-6 years is an indication of a fall in fertility rate which is a positive sign. However, the total number of children in India is a little more than the total population of Nigeria—the eighth largest country of the world with respect to population. It is important to note that out of the absolute increase of 181 million in India's population during the decade 2001-2011, 88 per cent has been contributed by the Child Population. Five states namely Uttar Pradesh (29.7 million), Bihar (18.6 million), Maharashtra (12.8 million), Madhya Pradesh (10.5 million) and Rajasthan (10.5 million) have the largest number of children constituting 52% of India's child population. On the other hand, Lakshadweep, Daman & Diu, Andaman and Nicobar Islands, Dadra and Nagar Haveli and Sikkim have least number of children. A comparison of figures of census 2011 with those of census 2001 reveals that the maximum decline in absolute numbers of children has been in Uttar Pradesh. This is followed by Andhra Pradesh, West Bengal, Maharashtra, and Kerala. On the other extreme are the states of Bihar, Jammu and Kashmir, Jharkhand, Meghalaya and Chhattisgarh, where maximum increase in child population has been recorded between 2001 and 2011. As per gender composition of child population, it has been found that decline in females (29,91,976) has been much more as compared to decline in males (20,56,132) during the decade 2001-2011. **Table 10.8** shows that 20 states and union territories have over one million children in the age groups 0-6 years. On the other hand there are five states and union territories that are yet to cross one lakh mark.

DISTRIBUTION AND DENSITY OF POPULATION

One of the most important aspects of India's population is its uneven distribution. On one hand the population of India is highly concentrated in some pockets such as highly urbanized and industrialised areas and areas of high agricultural productivity, while on the other hand there are virtually demographic deserts in high mountains, arid lands, thickly forested areas and some remote corners of the country. Such a situation needs some explanation and the explanation is found, to a great extent, by the study of some geographical factors which affect the distribution and density of population in a given area. Besides some social, demographic, political and historical factors play their own role in influencing the distribution and density of population. It may further be emphasised that these factors act in totality and not individually. While some scholars attach more importance to natural factors, Clarke and Zelinsky are of the view that cultural factors are more prominent in determining the concentration of population in an area. According to Clarke, economic conditions, technological development, social organisation, government policy, etc. play a vital role in the distribution of population. Major factors influencing the distribution and density of population are described as under :

1. Terrain. Terrain of land is a potent factor which influences the concentration and growth of population. Normally speaking, plain areas encourage higher density of population as compared to mountain regions. The steep slopes in mountain areas restrict the availability of land for agriculture, development of transport, industries and other economic activities which may tend to discourage concentration of population and its proper growth. It is because of these adverse circumstances that the Himalayan region, though occupies about 13 per cent of India's land area, supports only 1-2 per cent of the country's population. In contrast to this, the Great Plain of North India is a land of extremely gentle slope and offers great opportunities for the growth of agriculture, transport and industries. This results in higher concentration of population. Although the Great Plain of North India covers less than one fourth of the country's land area, it is the home to more than half of India's population.

2. Climate. Climate is as important as terrain in influencing population. Of all the climatic factors, twin elements of rainfall and temperature play the most important role in determining the population of an area. Man cannot go beyond the limits set by climate. Extremes of climate discourage the concentration of population. Such climates include the too cold climate of the Himalayas, and the too hot and dry climate of the Thar Desert. A moderate climate, on the other hand, is favourable for population.

Of the twin factors of rainfall and temperature, rainfall is more effective in determining the distribution of population. It is generally said that *the population map of India follows its rainfall map*. Rainfall supplies sufficient water for agriculture which is the main occupation of Indian masses. As we move from the Ganga-Brahmaputra Delta in the east towards the Thar Desert in the west, the amount of rainfall and consequently the density of population decreases. However, there are a few exceptions to this general observation. The Assam valley in the north-east and the Circars coast on the Bay of Bengal have moderate density of population although these areas receive heavy rainfall. Similarly, southern face of the Himalayas is scarcely populated though this area receives sufficiently high rainfall. Some of the adverse factors such as steep slope, frequent floods, infertile soils and dense forests counterbalance the positive effect of rainfall. Increased use of irrigation facilities in north-west India comprising Punjab, Haryana and western Uttar Pradesh has resulted in higher concentration of population than normally expected considering the amount of rainfall received by this region.

Since India is a tropical country, temperature is fairly high and does not play as important a role as is done by rainfall except in extreme cases. On high altitudes, in the Himalayan region, climate is too cold beyond 2,000 m and population is sparse there. There is practically no population in areas over 3,000 m above sea level.

3. Soil. Soil is an important factor in determining the density of population in an overwhelmingly agricultural country like India. Fertile soil supports higher population density while infertile soil leads to low density. In the northern plain of India, the soil is regularly enriched by annual floods of the great rivers like the Indus, the Ganga and the Brahmaputra and their tributaries. Therefore, this is an area of high population density. The coastal plains also have fertile soils and are areas of high population density. The Black soil of the *Deccan Plateau* also supports high population density. On the other hand, desert soils, mountain soils, laterite soils are infertile soils and are not capable of supporting high population densities. However, new technology in the agricultural field may change the future population scenario to some extent.

4. Water Bodies. Availability of water plays a significant role in determining the population of a given area. Water is the basic necessity for several purposes including irrigation, industries, transport and domestic use. Rivers are the greatest source of fresh potable water. Therefore, most of the population is concentrated in the river valleys.

5. Mineral Resources. Minerals act as great source of attraction for people from different areas, which results in higher density of population. The higher population densities in the Chhota Nagpur Plateau of Jharkhand and in the adjoining areas of Odisha are largely due to the availability of minerals.

6. Industries. Industrial growth offers massive employment opportunities and acts as a great magnet to attract people, particularly from the neighbouring areas. This results in higher population density. Industrial areas are almost invariably associated with areas of high population densities. One hectare of industrial land is capable of supporting several thousand persons, while the most fertile area devoted to agriculture may not support more than a few hundred persons per hectare. One of the major causes of high population density in West Bengal, Bihar, Jharkhand, Odisha, Maharashtra and Gujarat is the phenomenal growth of industries in these states.

7. Transport. Growth of population is directly proportional to the development of transport facilities. The northern plain of India has a dense network of transport routes and is a densely populated region. The peninsular plateau has moderate network of transport routes and is a moderately populated area. The Himalayan region badly lacks transport facilities and is sparsely populated.

8. Urbanization. Urbanization and population concentration go hand-in-hand and are closely related to each other. All the urban centres are marked by high density of population. The minimum density, that an area should have to be designated as urban, is 400 persons per sq km. The highly urbanized districts of Kolkata, Chennai, Greater Mumbai, Hyderabad, Delhi and Chandigarh have population densities of over 6,000 persons per sq km. Delhi has the highest population density of 11,297 persons per sq km as per 2011 census figures.

DISTRIBUTION OF POPULATION

The total population of India according to the 2011 census is 1210.2 million. A casual look at Table 10.9 will reveal that the distribution of India's population is very uneven. The contrasts in population distribution are quite clear at the level of the states, and are further sharpened at the level of districts. These contrasts are due to varying size of the states and wide variations in their resource base. Uttar Pradesh has the largest population of 199.5 million. This is followed by Maharashtra (112.3 million), Bihar (103.8 million). West Bengal (91.3 million) and Andhra Pradesh including Telangana (84.6 millions). These five states account for about half of the country's population. More than one-fourth of our people live in two-states of U.P. and Maharashtra alone. This, however, does not imply that states with large areas have large population also. For example. Rajasthan is the largest state of India accounting for over 10.4 per cent area of the country. But this state supports only 5.67 per cent population of India. Similarly, Madhya Pradesh, the second largest state in terms of area, has only 6.0 per cent of population on 9.38 per cent of area of the country. Contrary to this, Uttar Pradesh supports 16.49 per cent of population on only 7.33 percent of area of the country. In fact, U.P. has more people than the two largest states of Rajasthan and Madhya Pradesh. The three southern states of Kerala, Karnataka and Tamil Nadu together have less population than Uttar Pradesh. Bihar has 9.29 per cent of population on 9.86 per cent of area. In all, in eleven states and six union territories population size is much larger in comparison to the areas. This means that these states have higher pressure of population than the national average. On the other hand, Jammu and Kashmir covers 6.76 per cent area but supports only 1 per cent population of India. Arunachal Pradesh has 0.11 per cent of population on 2.55 per cent of area. Sikkim, a Himalayan mini-state has only 6 lakh population which is only 0.05 per cent of the total population of India. In fact, Sikkim has the smallest population among all the states of India. Delhi with 16.75 million has the largest population among all the union territories. It is a matter of fact that more people live in Delhi than in the state of Jammu and Kashmir or in all the other Union Territories put together.

Causes of Uneven Distribution of Population

The uneven distribution of population described above is the result of several factors of which physical, socio-economic and historical factors are more important.

***(i)* Physical Factors.** Among the physical factors, relief, climate and availability of water are the chief factors which determine the population of an area. It is because of these factors that the North Indian Plains, deltas and coastal plains have higher proportion of population than the interior districts of southern and central Indian states, the Himalayas, and some of the north eastern and the western states. However, development of irrigation facilities by Indira Gandhi Canal in Rajasthan, rich deposits of mineral and energy resources in Chotanagpur plateau of Jharkhand have resulted in moderate to high proportion of population in these areas which were previously very thinly populated.

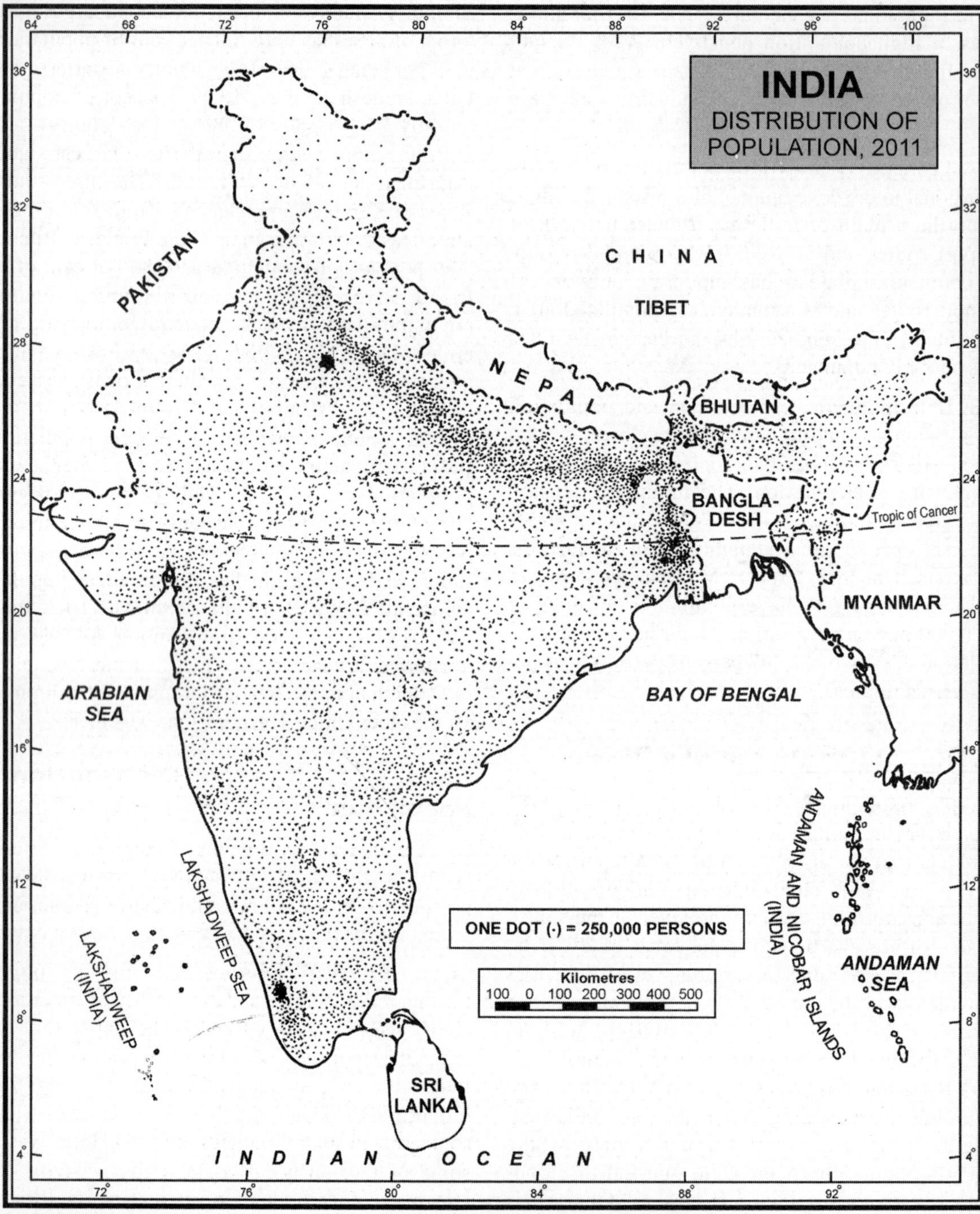

FIG. 10.8. India : Distribution of population, 2011

***(ii)* Socio-economic and Historical Factors.** Evolution of settled agriculture and agricultural development; pattern of human settlement; development of transport network, industrialisation and urbanisation are some of the important socio-economic and historical factors which influence the

TABLE 10.9. Ranking of States and Union Territories by Population : 2001 and 2011

Rank in 2011	India/State/Union Territory	Population 2011	Percent to total population of India	
			2011	2001
	India	**1,21,01,93,422**	**100.00**	**100.00**
1	Uttar Pradesh	19,95,81,477	16.49	16.16
2	Maharashtra	11,23.72,972	9.29	9.42
3	Bihar	10,38,04,637	8.58	8.07
4	West Bengal	9,13,47,736	7.55	7.79
5	Madhya Pradesh	7,25,97,565	6.00	5.87
6	Tamil Nadu	7,21,38,958	5.96	6.07
7	Rajasthan	6,86,21,012	5.67	5.49
8	Karnataka	6,11,30,704	5.05	5.14
9	Gujarat	6,03,83,628	4.99	4.93
10	Andhra Pradesh	4,93,78,776	4.08	—
11	Odisha	4,19,47,358	3.47	3.58
12	Telangana	3,52,86,757	2.91	—
13	Kerala	3,33,87,677	2.76	3.10
14	Jharkhand	3,29,66,238	2.72	2.62
15	Assam	3,11,69,272	2.58	2.59
16	Punjab	2,77,04,236	2.29	2.37
17	Chhattisgarh	2,55,40,196	2.11	2.03
18	Haryana	2,53,53.081	2.09	2.06
19	NCT of Delhi#	1,67,53,235	1.38	1.35
20	Jammu & Kashmir	1,25,48,926	1.04	0.99
21	Uttarakhand	1,01,16,752	0.84	0.83
22	Himachal Pradesh	68,56,509	0.57	0.59
23	Tripura	36,71,032	0.30	0.31
24	Meghalaya	29,64,007	0.24	0.23
25	Manipur	27,21,756	0.22	–0.22
26	Nagaland	19,80,602	0.16	0.19
27	Goa	14,57,723	0.12	0.13
28	Arunachal Pradesh	13,82,611	0.11	0.11
29	Puducherry#	12,44,464	0.10	0.09
30	Mizoram	10,91,014	0.09	0.09
31	Chandigarh#	10,54,686	0.09	0.09
32	Sikkim	6,07,688	0.05	0.05
33	Andaman & Nicobar Islands#	3,79,944	0.03	0.03
34	Dadra & Nagar Haveli#	3,42,853	0.03	0.02
35	Daman & Diu#	2,42,911	0.02	0.02
36	Lakshadweep#	64,429	0.01	0.01

Union Territory

Source : Census of India, 2011 : Provisional Population Totals, Paper 1, Series 1, p. 47.

Note : Population of Andhra Pradesh and Telangana has been calculated on the basis of district level population in both the states.

distribution of population. Generally speaking river plains and coastal areas have larger concentration of population due to early history of human settlement and development of transport network. Urban areas like Delhi, Mumbai, Kolkata, Bengaluru, Pune, Ahmedabad, Chennai and Jaipur have high concentration of population due to high level of industrialisation and urbanisation. People in large numbers migrate from rural to urban areas.

DENSITY OF POPULATION

Density of population is a better measure of understanding the variation in the distribution of population. It is expressed as number of persons per unit area. In other words, it is the ratio of total population to the total area of the country or a part thereof. For example the total population of India according to 2011 census is 1210.1 million living on a total area of 3.17 million square kilometres (excluding the area of Jammu and Kashmir and Ladakh illegally occupied by Pakistan and China). Therefore, the density of population of India in 2011 is :

$$\frac{\text{Total population}}{\text{Total area}} = \frac{1210.1}{3.17}$$
$$= 382 \text{ persons per sq. km)}$$

India's population density of 382 persons per sq. km is much higher than China's 141 persons per sq. km. Among the most populous ten countries of the world, India stands second in density; the first being Bangladesh (1141 persons per sq km). Thus heavy pressure of population on land is one of the serious problems of the country. The main cause of worry is that the population density in India has been consistently increasing since 1921 (**Table 10.10**). There has been almost five fold increase in the density of population between 1921 and 2011. The density of population increased rapidly between 1951 and 2011.

TABLE 10.10. Density of Population, India : 1901-2011

Census Year	Density (per sq km)	Absolute Increase	%age Increase
1	2	3	4
1901	77		
1911	82	5	6.5
1921	81	–1	–1.2
1931	90	9	11.1
1941	103	13	14.4
1951	117	14	13.6
1961	142	25	21.4
1971	177	35	24.6
1981	216	39	22
1991	267	51	23.6
2001	325	58	21.7
2011	382	57	17.5

Source : Census of India 2011, Provisional Population Totals, Paper 1 of 2011 Series 1, p. 138.

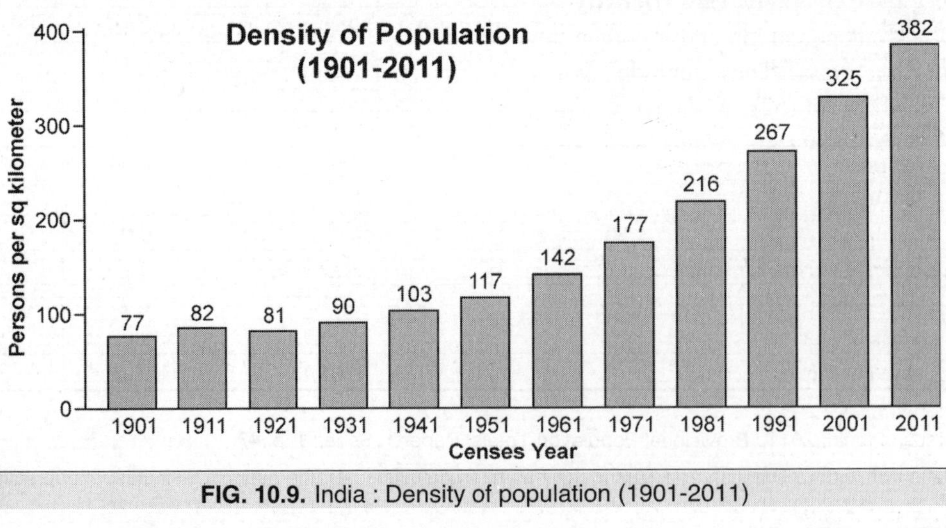

FIG. 10.9. India : Density of population (1901-2011)

It is clear from **table 10.10** and **Figure 10.9** that density of population in India is very high and is increasing at a very fast rate. During the decade 2001-11 alone, India's population density increased by 57 persons per sq km which means each sq km of land has to feed 57 new mouths. This is a matter of great concern as it puts immense pressure on our natural resources. This increase is nearly double the existing density of the USA (30 persons/sq km), seven times the existing density of Russia (8 persons/sq km) and a more than one-third the existing density of China (141 persons/sq km). Hence among the large sized countries of the world, India is not only the most densely populated but is also adding to its density at an alarming rate.

State Level Patterns

Table 10.11 shows that the national average of 382 persons per sq km does not give a clear idea of nature of unevenness of population density as it varies from a minimum of 17 persons per sq km in Arunachal Pradesh to a maximum of 1102 persons per sq km in Bihar. Among the union territories, Delhi is the most thickly populated with 11,297 persons per sq km, while Andaman and Nicobar Islands have the lowest density of only 46 persons per sq km.

For the sake of convenience, the spatial distribution of population density is classified into following categories :

1. Areas of Extremely Low Density. Areas having 100 persons per sq km and less than that are included in this class. They include Arunachal Pradesh (17), Mizoram (52), Andaman and Nicobar Islands (46) and Sikkim (86). Arunachal Pradesh and Mizoram are located in a remote and inaccessible part of north-east India. Sikkim is also a mountainous area with low density of population. Andaman and Nicobar Islands are situated far away from the Indian mainland. Hot and humid climate of these islands is injurious to health and very little economic development has taken place here.

2. Areas of Low Density. Areas having population density of 101 to 250 persons per sq km are included in this class. These states are Nagaland

TABLE 10.11. Ranking of States and Union Territories by density : 2011 and 2001

Rank in 2011	State/Union Territory	Density (per sq km)		Rank in 2001
		2011	2001	
1	2	3	4	5
	India	**382**	**325**	
1	NCT of Delhi	11,297	9,340	1
2	Chandigarh	9,252	7,900	2
3	Puducherry	2,598	2,034	3
4	Daman & Diu	2,169	1,413	5
5	Lakshadweep	2,013	1,895	4
6	Bihar	1,102	881	7
7	West Bengal	1,029	903	6
8	Kerala	859	819	8
9	Uttar Pradesh	828	690	9
10	Dadra & N. Haveli	698	449	13
11	Haryana	573	478	12
12	Tamil Nadu	555	480	11
13	Punjab	550	484	10
14	Jharkhand	414	338	16
15	Assam	397	340	15
16	Goa	394	364	14
17	Maharashtra	365	315	17
18	Tripura	350	305	18
19	Karnataka	319	276	20
20	Andhra Pradesh (+ Telangana)	308	277	19
21	Gujarat	308	258	21
22	Odisha	269	236	22
23	Madhya Pradesh	236	196	23
24	Rajasthan	201	165	24
25	Uttarakhand	189	159	25
26	Chhattisgarh	189	154	26
27	Meghalaya	132	103	29
28	Jammu & Kashmir	124	100	31
29	Himachal Pradesh	123	109	28
30	Manipur	122	103	30
31	Nagaland	119	120	27
32	Sikkim	86	76	32
33	Mizoram	52	42	34
34	Andaman & Nicobar Islands	46	43	33
35	Arunachal Pradesh	17	13	35

Source: Census of India, 2011; Provisional Population Totals, Paper 1, p. 140.

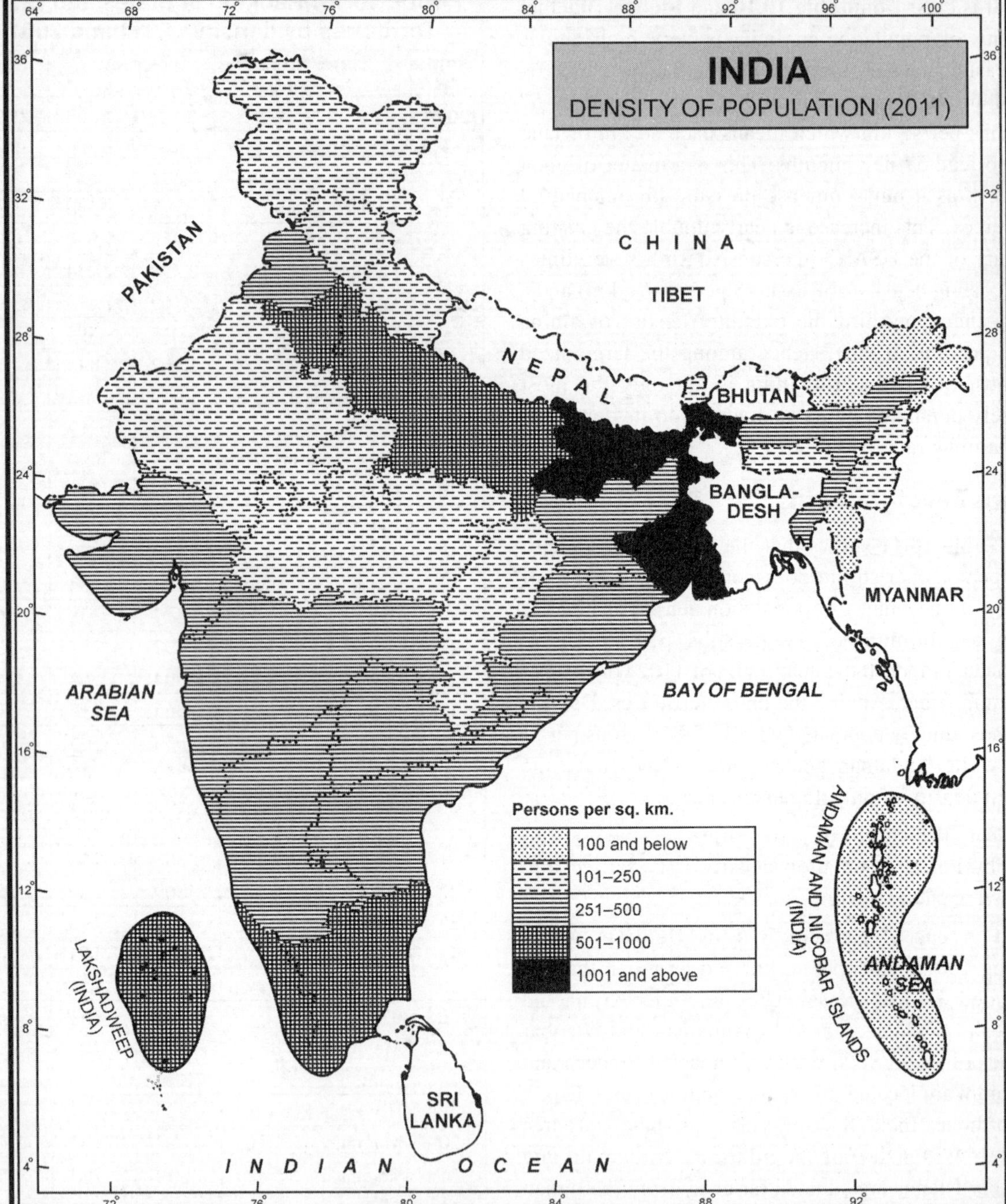

FIG. 10.10. India: Density of Population 2011.

(119), Manipur (122), Himachal Pradesh (123), Jammu and Kashmir (124), Meghalaya (132), Chhattisgarh (189), Uttarakhand (189), Rajasthan 201) and Madhya Pradesh (236). Meghalaya, Manipur, and Nagaland are hilly, forested and dissected areas of north-east India. These areas suffer from almost the same problems as those of Arunachal Pradesh and Mizoram, although to a lesser extent.

Himachal Pradesh and Uttarakhand are parts of the north-western Himalayan region and have very little level land to support high population density. Jammu and Kashmir has vast areas devoid of population. Only some parts of Jammu region and Kashmir valley are thickly populated. Large stretches of Leh (Ladakh) and Kargil have population density less than ten persons per sq km. On the whole Kargil has population density of 10 persons/sq km while Leh (Ladakh) has only 3 persons per sq km. These are dry and cold areas and badly lack the basic amenities of life. Rajasthan is the largest state of India. There are obviously large variations in the density of population in different parts of the state depending upon the local conditions. Most of Rajasthan is a sandy desert lacking in water resources and does not support high population density. Western part of the state is having even less than 50 persons per sq km whereas eastern and north-eastern parts of this state have sufficient resources and have comparatively higher density of population. Madhya Pradesh is a part of the Deccan Plateau and is having rugged topography of hard rocks. Like Madhya Pradesh, Chhattisgarh has rugged topography, is thickly forested and is largely inhabited by the tribal people. As such, the population density in this state also is low.

3. Areas of Moderate Density. This class includes those areas which are having 251 to 500 persons per sq km. The average for whole of India (382 persons per sq km) also falls in this class. Odisha (269), Gujarat (308), Andhra Pradesh including Telangana (308), Karnataka (319), Tripura (350), Maharashtra (365), Goa (394), Assam (397), and Jharkhand (414) are included in this category. These areas are wide apart from one another and there are different reasons for moderate density of population in different areas. For example, Assam has tea estates whereas Andhra Pradesh, Telangana, Odisha, Karnataka and Jharkhand have agricultural and mineral resources. Maharashtra is highly urbanised and industrialised state. The neighbouring state of Gujarat also has urban and industrial growth, although on a scale smaller than that of Maharashtra. Among the north-eastern states, Tripura has sufficient level land which supports moderate population density.

4. Areas of High Density. These are areas having population density of 501 to 1000 per sq km. States and union territories included in this category are Punjab (550), Tamil Nadu (555), Haryana (573), Dadra & Nagar Haveli (698), Uttar Pradesh (828) and Kerala (859). Punjab and Haryana have highly developed agriculture based on heavy inputs in the form of high yielding varieties of seeds, chemical fertilizers and canal and tube-well irrigation. Similarly, Tamil Nadu's population is based on agriculture and industries. The coastal plain of Kerala is also very fertile. However, Kerala has started showing decline in the growth rate of population. Uttar Pradesh is located in the fertile Ganga Plain and supports high population density.

5. Areas of Very High Density. Areas having more than 1000 persons per sq km are termed as areas of very high population density. West Bengal (1029), Bihar (1102), Lakshadweep (2013), Daman & Diu (2169) Puducherry (2548), Chandigarh (9252) and Delhi (11,297) have very high density of population due to different factors operating in different areas. Like Uttar Pradesh, Bihar is located in the fertile plain of Ganga and supports very high population density. It seems that measures to control population growth have not yielded the desired results and Bihar has now surpassed West Bengal as the state with highest density of population among the major states. West Bengal is located in the Ganga delta which is one of the most fertile areas of the world, producing 3-4 crops of rice in a year. In addition, India's biggest industrial cluster is located in the Hugli basin. These factors combine together to make West Bengal the second most densely populated state of India. Among the union territories, Delhi has experienced one of the fastest population growths as a result of which its population density has increased considerably. This growth is primarily due to large scale migration of people from the surrounding areas. People migrate to Delhi in large numbers in search of livelihood, and better amenities of life.

Population Problems

As mentioned earlier, India is the second most populous country of the world, next only to China. Obviously India is facing serious problems regarding population and there are spatial and temporal variations in these problems. Perhaps the most serious problems being faced by India today are in connection with her people. Some of the major population problems are briefly described as under :

1. Rapid Growth of Population. In spite of several steps taken by the government, population of India is growing at a rapid pace. India's population was just 36.1 crores at the time of first census (1951) after Independence which rose to over 121 crores in 2011. Thus our population has increased more than three fold in a short span of half a century. During the last century, the world population grew by three times while India's population grew by four times. Although the growth rate of our population has come down from 24.8 per cent in 1961-71 to 17.64 per cent in 2001-11, it is still vary high as compared to world average growth rate of 12.3 per cent and is much higher than 5.3 per cent of the most populous country of the world—China. During 2001-11 India's population increased by 181.4 million against China's 79.7 million. At this rate our population is likely to exceed the population of China in 2027. The population growth was so high during 1951-81 that it is often referred to as ***'population explosion'***.

2. Uneven distribution of Population. One of the most striking features of India's population is its uneven distribution. On one hand, there are almost empty lands like Arunachal Pradesh where population density is only 17 persons per sq km and large tracts of the Himalayas have less than five persons per sq km. On the other hand Bihar's one sq km has to support as many as 1102 persons on an economy which is heavily dependent on agriculture. Urbanized union territories like Chandigarh and Delhi have unmanageable densities of 9,252 and 11,297 persons per sq km respectively.

Rapid growth and uneven distribution of population have put enormous pressure on our scarce natural resources to varying degrees in different parts of the country. Main problems arising out of such a situation are unemployment, poverty, hunger, malnutrition, environmental degradation and low standard of living. These problems will be discussed in the following paragraphs.

3. Unemployment. In view of rapidly growing population and limited resources, the employment seekers far outnumber and outpack the growth of employment opportunities. Indian economy is heavily dependent on agriculture which the largest employment provider in the country. At present about two-third of India's population is living in rural areas which almost solely depends on agriculture for its livelihood. More than 54% of the population of the country earns its livelihood from agriculture. But in spite of heavy inputs in agriculture in the form of high yielding variety of seeds, chemical fertilizers and machines, agricultural progress has failed to keep pace with the growing demand for employment due to fast increasing population. Moreover, agriculture is a seasonal activity in which labour is required only during the sowing and harvesting seasons and the agricultural labourers are without employment for about six months in a year. The secondary and tertiary sectors have also not grown in proportion to demand for employment. As such, the employment opportunities for unskilled, semi-skilled, and skilled people are very limited. Even the educated and professional technocrats are finding it difficult to get a suitable jobs and they have a tendency to migrate to developed countries resulting in a serious crisis of **brain drain.** According to official figures there are crores of unemployed youths in India. This situation of unemployed youth is very dangerous because the younger people are more liable to be involved in anti-social activities like theft, cheating, drugs, murder, rape, etc. However, the number of unemployed persons has decreased from 3.83 crore in 2017-18 to 2.81 crore in 2019-20 (**Table 10.12**).

4. Hunger and Malnutrition. Situation of hunger and malnutrition arises when the population

TABLE 10.12. Estimates of Labour Force, Employment and Unemployment in cores

Year	2017-18			2018-19			2019-20		
	Male	Female	Total	Male	Female	Total	Male	Female	Total
Labour Force	38.73	12.24	50.97	39.37	12.45	51.82	40.87	15.47	56.34
Employment	36.29	10.85	47.14	37.01	11.77	48.78	38.77	14.76	53.53
Unemployment	2.44	1.39	3.83	2.36	0.68	3.04	2.10	0.71	2.81

Source: Economic Survey 2021-22, p. 371.

growth outpaces the growth of foodgrains and food items become so costly as to be out of reach of the common man.

India's ranking is very poor on Global Hunger Index (GHI) which is designed to comprehensively measure, and track hunger globally, and by country and region. Published annually since 2006, it ranks countries on 0 to 100 point scale calculated by taking into account four indicator parameters. Zero means least score (no hunger) and 100 means worst condition. Values less than 10 reflect low hunger, values from 20 to 34.9 indicate serious hunger, values from 35 to 49.9 alarming and values 50 or more are extremely alarming. India's score was 30.3 putting it in the serious hunger category. Four parameters are (*i*) undernourished population (1/3rd of weight), (*ii*) child wasting (1/6th of weight), (*iii*) child stunting (1/6th weight) and (*iv*) infant mortality rate (1/3rd weight). In this case, child wasting means low weight in relation to child's height, and stunting means deficiency of height in relation to age. Calculated each year by the International Food Policy Research Institute (IFPRI), the GHI highlights successes and failures in hunger reduction and provides insights into the drivers of hunger.

India ranked 94 among 107 countries in Global Hunger Index (GHI) 2020 and continues to be in the serious hunger category, though it has made some progress, particularly since the enactment of National Food Security Act. The neighbouring countries of Bangladesh, Myanmar and Pakistan too were in serious category but ranked higher than India. These three countries were ranked 75th, 78th and 88th respectively in 2020. GHI report. Nepal and Sri Lanka were 73rd and 64th position and were in the "moderate" hunger category.

The report said that 14% of India's population was undernourished, a 37.4% stunting rate among children under five and a wasting rate of 17.3 per cent. Under five mortality rate stood at 3.7 per cent. According to WHO report 35.8% of Indian children under the age of five have lower weight in relation to their height and about 33% are too short in relation to their age. About 35 crore people of India are suffering from hunger which is more than the combined population of Canada, Australia, Japan and Russian Federation.

The UN has set 2030 as the deadline to end global hunger and India with its massive population, has a big role to play. India has performed poorly in its efforts to eliminate hunger. However, India has set some targets to reduce hunger and malnutrition as is illustrated in **table 10.13**.

TABLE 10.13. India's targets for reducing hunger and malnutrition

Item	Current	Target for 2030
1. Ratio of rural households covered under public distribution system (PDS) to rural households in which the highest earner makes less than ₹ 5,000	1.01	1.29
2. Percentage of children under 5 years who are stunted	38.4	21.03
3. Percentage of pregnant women aged 15-49 who are anaemic	50.3	23.57
4. Production of food, kg of rice, wheat, coarse cereals produced annually per hectare	2509	5018
5. Percentage of underweight below 5 years	35.7	20.7
6. Reduction in prevalence of anemia in children (6-59 months)	58,4	19.3

Source : NITI Aayog

Keeping in view the picture depicted in **table 10.13**, the NITI Aayog has come out with a national strategy to fight hunger and malnutrition. The focus is on 100 poor performing districts in terms of stunting of which 29 are in Uttar Pradesh. Recognising that India pays an income penalty of 9-10% due to workforce stunted during childhood, the national nutrition strategy aims to bring nutrition to the centrestate of the National Development Agenda and has outlined the vision *Kuposhan Mukt Bharat* (Malnutrition Free India) reducing all forms of malnutrition by 2030.

According to the NITI Aayog, five worst states with respect to percentage of children under 5 years

Some glaring facts about hunger in India (as per International Food Policy Research Institute)

- 20 per cent is the prevalence of caloric under-nourishment in India.
- 25 is the number of Sub-saharan countries better off than India in terms of food security. We are worse off than even Pakistan.
- 4 African nations—Nigeria, Cameroon, Kenya and Sudan—have far lower per capita income than India but manage hunger better.
- 410 million is the number of people who are poor and food insecure in just 8 Indian states—more than many sub-Sarahan countries (according to an Oxford University report).
- There is not even a single state in India with low or even moderate levels of hunger.
- **Punjab** the food bowl of India, India's best performing state, falls in serious category and ranks behind Vietnam and Saudi Arabia.
- Madhya Pradesh, India's worst state, besides Gujarat, Chhattisgarh and Haryana are worse off than Sudan.
- Bihar and Jharkhand rank lower than Zimbabwe and Haiti.

who are stunted *i.e.* their height-for-age is below the WHO standards are Bihar (48.3%), Uttar Pradesh (46.3%), Jharkhand (45.3%), Meghalaya (43.8%) and Madhya Pradesh (42.0%). Five worst states with percentage of pregnant women aged 15-49 years who are anaemic are Jharkhand (62.6%), Bihar (58.3%), Haryana (55.0%), Tripura (54.4%) and Meghalaya (53.1%).

5. Poverty. Poverty is another serious population problem, but million dollar question is "who is a poor Indian". Different criteria have been followed by different committees and the number of families living below poverty line have varied widely from each other. For example in 2010 Wadhwa Committee put BPL (Below Poverty Line) families at 20 crores, while State Governments estimated 10.5 crore. Tendulkar Committee at 9.25 crores and World Bank at 7.5 crore BPL families in India. However, general parameters for determining poverty status are agreed upon by majority of investigators (see box).

STUDY PARAMETERS

Who were automatically counted as poor :

- Households without shelter
- Destitute or beggars
- Manual scavengers
- Primitive tribal groups
- Legally released bonded labour

Depreciation indicators

- Households with one room having kucha walls and roof.
- No adult member between 16 and 59 years of age.
- Female-headed household with no adult male between 16 and 59 years of age.
- No able-bodied adult member and no literate member above 25 years of age.
- Scheduled caste, scheduled tribe and landless households.

The Planning Commission estimates poverty using data collected by National Sample Survey Office (NSSO) every five years and defines poverty line on the basis of monthly per capita consumption expenditure (MPCE). The expert group headed by Prof. Suresh D. Tendulkar submitted its report in December 2009. This report computed the poverty line at all India level as MPCE of ₹ 447 for rural and ₹ 579 for urban areas in 2004-05. This survey was again conducted in 2009-10 and accordingly the poverty lines were decided at all India level as an MPCE of ₹ 673 for rural and ₹ 860 for urban areas in 2009-10. Based on these cut off figures, the percentage of people living below poverty line in the country has declined from 37.2 per cent in 2004-05 to 21.9 per cent in 2011-12. Even in absolute terms, the number of poor people had fallen by 137.8 million during this period. Of this 109.8 million were rural poor and 28.0 million were urban poor.

Table 10.14 shows that in all the major states poverty ratio is higher in rural areas as compared to urban areas. This is due to lack of employment opportunities in agriculture and allied fields. In contrast urban areas offer better employment opportunities in industries, trade, transport etc. It also shows that there are large variations in poverty ratio at the state level. In states of Assam, Bihar, Jharkhand and Odisha, more than 30% people are living in poverty conditions. The least percentage of poverty is found in Kerala where only 7.1% people are living in poverty conditions.

The Government, in July 2014, declared that those spending daily ₹ 32 in rural areas and ₹ 47 in urban areas should not be considered poor. This means that 29.5% of the Indian population lives below poverty according to a report by Rangarajan

TABLE 10.14. Poverty Headcount Ratio in major states of India (2011-12)

States	Rural	Urban	Total
1. Andhra Pradesh	10.9	3.8	9.2
2. Assam	33.9	20.5	31.9
3. Bihar	34.1	31.2	33.7
4. Chhattisgarh	44.6	24.8	39.9
5. Gujarat	21.5	10.1	16.6
6. Haryana	11.6	10.3	11.2
7. Himachal Pradesh	8.5	4.3	8.1
8. Jammu & Kashmir	11.5	7.2	10.4
9. Jharkhand	40.8	24.8	36.9
10. Karnataka	24.5	15.3	20.9
11. Kerala	9.1	4.9	7.1
12. Madhya Pradesh	35.7	21.1	31.7
13. Maharashtra	24.2	9.1	17.4
14. Odisha	35.7	17.3	32.6
15. Punjab	7.7	9.2	8.3
16. Rajasthan	16.1	10.7	14.7
17. Tamil Nadu	15.8	6.5	11.3
18. Uttar Pradesh	30.4	26.1	29.4
19. Uttarakhand	11.6	10.5	11.3
20. West Bengal	22.5	14.7	19.9
All India	25.7	13.7	21.9

Source: Economic Survey 2021-22, Statistical Appendix pp. 178-79

Note: Heal Count Ratio (HCR) is the percentage of a population that falls below Poverty Line.

Committee. According to the Millennium Development Goals (MDG) report released in July 2014, about 32.9 per cent of the world's poorest surviving on less than ₹ 80 ($1.25) a day live in India.

Figure 10.11 shows the poverty trends in India from 1951-52 to 2030. It is clear from the figure that poverty was widespread at the time of first census in 1951 after Independence. As much as 45 per cent of India's population was living in extreme poverty. Poverty level peaked at 62% in 1967-68 after which poverty had been decreasing consistently for two decades upto 1987-88. After that, it increased slightly and started a steep downward trend which is continuing till now. India is expected to be poverty free by 2030.

About 44 Indians are coming out of extreme poverty every minute. This is one of the fastest rates of poverty reduction in the world. As a result, India has finally shed the dubious distinction of being home to the largest number of poor, with Nigeria taking that unwanted position in May 2018. In a recent study published in 'Future Development', Brookings defined extreme poverty as living on less than $ 1.9 a day, and estimated that in India, extreme poverty could be eliminated altogether by 2030.

The Multidimensional Poverty Index (MPI) developed in 2010 by Oxford Poverty and Human Development Initiative (OPHI) is based on 10 indicators related to health, education and standard of living. A person is MPI poor of they are deprived in at least one-third of the weighted MPI indicators.

Using the Multidimensional Poverty Index (MPI), the UN released a report in September, 2018 according to which 271 million people moved out of poverty between 2005-06 and 2015-16, with poverty nearly halving-falling from 55% to 28% per cent. But there is still a huge discrepancy between states. The four poorest states—Bihar, Jharkhand, Uttar Pradesh and Madhya Pradesh—are still home to 196 million poor people.

The largest number of people who have come out of extreme poverty between 2005-06 and 2015-16 are 5 crores from Uttar Pradesh. This is followed by West Bengal (2.7 crore), Andhra Pradesh (2.6 crore), Maharashtra (2.1 crore) and Karnataka (2.0 crore).

6. Low Agricultural Productivity. In most parts of India, agriculture is tradition bound and is of subsistence type in which the entire agriculture produce is consumed by the family members of the farmer. A large percentage of farmers are poor and cannot afford costly inputs like agricultural machines, high yielding seeds, and chemical fertilizers etc. Moreover land tenancy, small and fragmented fields also create hindrances in agricultural growth. For the last so many years the annual rate of agricultural growth has been very slow and is not able to meet the growing demand of teeming millions. Further agricultural sector has failed to provide suitable employment opportunities to growing number of rural youth.

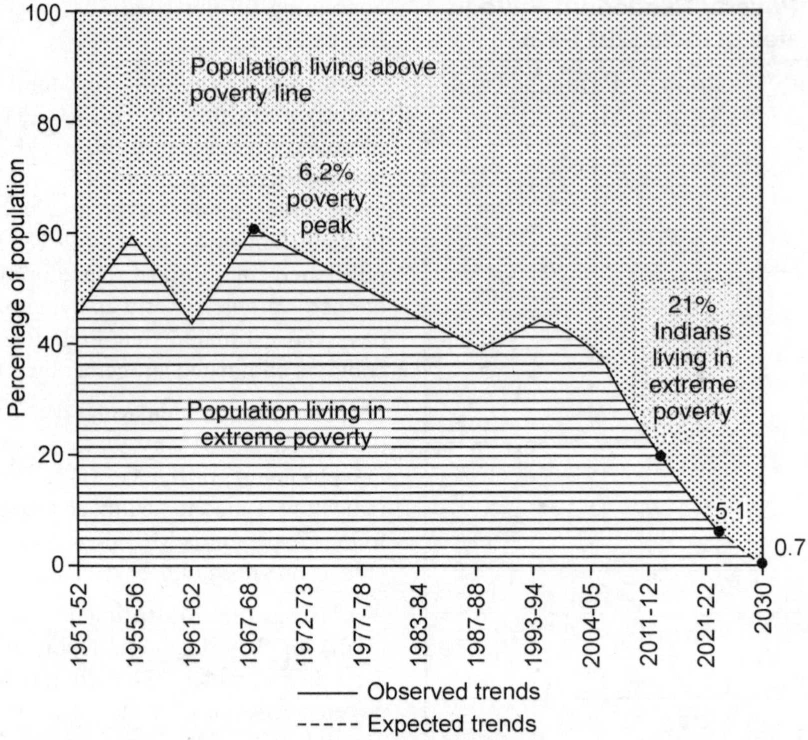

FIG. 10.11. Poverty trends in India

Note : Since the definition of extreme poverty has been changing over time, a strict comparison between different years is not valid. But long term trend is clear.

7. Slow Growth of Industries. Although industrial sector has grown faster than agricultural sector, yet this sector is unable to provide sufficient employment to our young population, particularly to those who migrate from rural to urban areas. The basic problem is that the labour which migrates from rural to urban areas largely consists of unskilled workers for whom there is little scope in the industrial sector. Financial constraints and lack of proper infrastructure are also great hindrances in industrial growth. A high percentage of Indian population consists of poor people and there is lack of proper market for industrial products.

8. Frequent Strikes and Bandhs. The fast growing population has resulted in underemployment, unemployment, hunger, malnutrition, poverty, deprivation and lack of basic necessities of life and has led to overall frustration, particularly among the youngsters and they become indisciplined. Consequently, there are frequent strikes in different spheres of life and industrial sector is the worst sufferer. This reduces the overall productivity of the people.

9. Religion, Tradition and Orthodoxy. Indian population is primarily religious minded society and followers of almost all religious can be found in this country. Some of the religions do not believe in family planning practice. This leads to unchecked growth of population which has its own implications. Further, Indian society is tradition-bound and this is more true of rural society. Orthodoxy and ignorance are the major social aspects of Indian population. All these traits are major obstacles in the way to adopt new technology and innovative ideas. Hence masses are still in a backward state and lead miserable life. In order to overcome these obstacles, large scale literacy and education campaign is necessary.

10. Terrorism and insurgency. India's socio-economic system has undergone drastic changes during the last few decades and economic disparities have increased tremendously. In the existing socio-economic scenario, a few people have become very rich while a large proportion of population is living a life of misery and deprivation. The youthful section of society becomes frustration in the absence of opportunities to earn a livelihood and the youngsters resort to anti-social activities like thefts, robberies, kidnapping, extortion, murder, etc. Some of them go to the extent of resorting to terrorism and insurgency. There are many terrorist groups operating in different parts of the country and the frequency as well intensity of terrorist attacks is increasing with the passage of time. A lot of insurgency is taking place in the peripheral states, particularly in the north-eastern states.

11. Environmental Degradation. Growing pressure of population on natural resources is leading to depletion of these resources and at the same time it has several problems regarding environment and ecology. Large scale destruction of forests for forest products and for making land available for other uses has resulted in drastic ecological imbalance. Shifting agriculture, known as Jhum culture, in the north-eastern states is the main cause of forest destruction there, because the duration of completing the *Jhum* cycle has been reduced due to increase in demand for agricultural crops caused by growth in population. Over irrigation in Punjab, Haryana and western parts of Uttar Pradesh has caused large scale depletion of ground water and increase in salinity and alkalinity in soil over vast areas. Air and water in (both surface and ground) have been polluted over large tracts in different parts of the country. Noise pollution has become a serious problem in large cities of India. Environmental pollution leads to ill health and cause several diseases which put undue pressure on health services.

12. Pressure on infrastructure and low standard of living. Rapidly growing population puts undue pressure on infrastructure, Educational institutes, hospitals, transport system always remain over crowded and there is acute shortage of housing amenities. Supply of drinking water and electricity is almost invariably erratic and condition of roads and streets is pathetic. Overall deprivation leads to social tensions, and the number of people involved in anti-social activities is increasing with each passing day. Thus the overall standard of living of the masses is very low and people are deprived of even the basic amenities.

Measures to Solve Population Problems

1. Rapid growth of population in India is the main problem concerning population of the country and all other problems are the off-shoots of this basic problem. If population growth is checked, all other problems will be automatically solved. Following few steps are suggested to solve this problem.

(*i*) People should be persuaded to adopt family planning methods and opt for small families.

(*ii*) Incentives should be given to those couples who opt for small families.

(*iii*) There should be a complete ban on child marriage and minimum age for marriage (18 years for girls and 21 years for boys) should be strictly followed and violators should be given severe punishment. System of child marriage is more prevalent in the rural areas and rural masses should be educated about the ill effects of child marriage.

(*iv*) There is close correlation between infant mortality and birth rate. Unfortunately both are high in India which adversely affect the family planning programmes. There is urgent need to bring these rates down considerably.

(*v*) Illiteracy and high birth rate go hand in hand and high birth rate leads to high growth rate of population. Although literacy rate has increased considerably during the last few decades still more than 25 per cent of India's population is illiterate. This section of society generally fails to understand in significance of population control and the population increases unchecked. It is often said that "*education is the best contraceptive*." Thus spread of education can be a great instrument in controlling the population growth.

(*vi*) As mentioned earlier, a large section of our population is suffering from hunger and malnutrition. This makes us a nation of unhealthy people who are unable to make much contribution to economic growth. Proper arrangements should be made to feed these teeming millions so that the nation moves faster on the path of economic growth. It is hoped that the food security bill passed by the Parliament in 2013 will go a long way to eradicate hunger and malnutrition from our country.

(*vii*) A section of our population, particularly in the rural areas, sufferers from the sense of social insecurity and a very high percentage of Indian parents opt for larger number of children who could look after them in their old age. Such a thinking has to be washed away by providing some measure of social security to this section of society.

The other methods to solve our population problems are briefly discussed below.

2. Special attention should be paid to agriculture for solving the food problem. Incentives should be given for larger use of agricultural inputs like high yielding varieties of seeds, fertilizers and farm mechanisation. From 1960s onwards, Green Revolution has been helping India to solve her food problem. But in the post Green Revolution era, growth of population outpaced the growth of food grains and India is again finding it difficult to provide proper food to her masses. We are looking forward to see the positive results of the Second Green Revolution.

3. Unproductive lands such as waste land, barren lands, etc. should be improved by providing proper inputs and should be brought under plough.

4. Different types of natural resources such as minerals, soils, water, natural vegetation, etc. should be judiciously used and wisely conserved so that they are easily available in sufficient quantities for the present and the future generations.

5. Incentives should be given to industrialization and urbanization so that employment and housing facilities could be provided to larger number of people.

6. Migration of the people from densely populated areas to sparsely populated areas should be encouraged so that regional imbalances in population distribution are removed or at least reduced. This will help in reducing pressure of population in crowded areas and using the resources of sparsely populated areas.

Population Policies

Rapidly growing population of India has forced our planners to frame a solid population policy to keep the rate of population growth at a reasonable level and the importance of such a policy was raised immediately after independence.

When the population policy was designed in the *First Five Year Plan (1951-56)* the population of India had already crossed 361 million mark. It was realised that this base of population was very large and it was difficult to check the trend of rapidly growing population within a limited time period. According to provision of the First Five Year Plan, the programme for family limitation and population control was design to :

(*i*) present an accurate picture of the factors responsible for rapid increase in population.

(*ii*) discover suitable techniques of family planning and devise suitable methods by which knowledge of these techniques could be widely disseminated, and

(*iii*) give advice on family planning as an integrated part of the service of government hospitals and public agencies.

However, the financial allocation for family planning programme was only ₹ 0.65 million which was so meager that it could not yield any tangible results.

A voluntary sterilization programme was introduced in the *Second Five Year Plan* (1956-57 to 1960-61) for which provision of ₹ 50 million was made. This account was mainly used to set-up 1650 family planning centres in different parts of the country.

In the *Third Five Year Plan* (1961-62 to 1965-66), logistics were provided for family planning, which motivated about one million people to accept sterilization.

In the *Fourth Five Year Plan* (1969-74), the main emphasis was on family planning programme in which a time bound programme of reducing the birth rate from 3.9 per cent to 2.3 per cent by 1979 was fixed. The outlay for family planning programme in this plan was raised to ₹ 2,860 million. This amount was spent to cover nearly 9 million couples under sterilization and about 6 million couples under other family planning methods. It is estimated that about 7 million births were averted during this period.

Strict population control policy followed by the central government led by Mrs. Indira Gandhi particularly during the emergency period was vigorously opposed by the masses and this led to an unprecedented defeat of the government in 1977 elections. Consequently a more pragmatic population policy was adopted in the Seventh (1985-1990) and Eighth Plan (1992-97) periods in which more emphasis was laid on persuation, publicity as well family and individual well-being.

Unfortunately the above mentioned population policies could not give the desired results and India's population kept on growing at a fast rate and almost took the shape of a *population explosion*. Compelled by the prevailing circumstances, the government came out with a solid population policy in 2000, a brief summary of which is given below :

National Population Policy, 2000. It is a very comprehensive policy which spells out 14 demographic goals and 12 strategic themes. The immediate objective of this policy was to meet all the unmet needs for contraception and health care for women and children. The medium-term objective was to bring the Total Fertility Rate (TFR) to replacement level (TFR of 2.1) by 2010, and the long-term object is to achieve population stabilisation by 2045. However, the subsequent events reveal that the goals set under the policy were rather over-ambitious. For example, National Population Policy (NPP) wanted Infant Mortality Rate (IMR) to be reduced to below 30 per thousand live births. But figures released by Office of Registrar General India in 2008 showed that IMR stood at a high of 53 per thousand live births which was nowhere near the desired figures of 30 per thousand. The situation in regard to other goals is also depressing.

Most scholars agree with the concluding paragraph of NPP, 2000 which is reproduced below.

"The vast numbers of the people of India can be its greatest asset if they are provided with the means to lead healthy and economically productive lives. Population stabilisation is a multi-sectoral endeavour requiring constant and effective dialogue among a diversity of stakeholders, and coordination at all levels of the government and society. Spread of literacy and education, increasing availability of affordable reproductive and child health services, convergence of service delivery at village levels, participation of women in the paid workforce, together with a steady, equitable improvement in family incomes, will facilitate early achievement of the socio-demographic goals. Success will be achieved if the Action Plan contained in the NPP 2000 is pursued as a national movement."

EXCERPTS

National Socio-Demographic Goals for 2010

1. Address the unmet needs for basic reproductive and child health services, supplies and infrastructure.
2. Make school education upto age 14 free and compulsory, and reduce dropouts at primary and secondary school levels to below 20 per cent for both boys and girls.
3. Reduce infant mortality rate to below 30 per 1000 live births.
4. Reduce maternal mortality rate to below 100 per 100,000 live births.
5. Achieve universal immunisation of children against all vaccine preventable diseases.
6. Promote delayed marriage for girls, not earlier than age 18 and preferably after 20 years of age.
7. Achieve 80 per cent institutional deliveries and 100 per cent deliveries by trained persons.
8. Achieve universal access to information/counselling, and services for fertility regulation and contraception with a wide basket of choices.
9. Achieve 100 per cent registration of births, deaths, marriagies and pregnancies.
10. Contain the spread of Acquired Immuno Deficiency Syndrome (AIDS), and promote greater integration between the management of reproductive tract infections (RTI) and sexually transmitted infections (STI) and the National AIDS Control Organisation.

11. Prevent and control communicable diseases.
12. Integrate Indian Systems of Medicine (ISM) in the provision of reproductive and child health services, and reaching out to households.
13. Promote vigorously the small family norm to achieve replacement levels of TFR.
14. Bring out convergence in implementation of related social sector programmes so that family welfare becomes a people centred programme.

Strategic Themes

Following 12 strategic themes must be followed to achieve the national socio-demographic goals for 2010.

I. Decentralised Planning and Programme Implementation. The 73rd and 74th Constitutional Amendment Act, 1992 made health, family welfare, and education a responsibility of village panchayats in the context of NPP 2000. As may as 33 per cent of elected panchayat seats were reserved for women to promote a gender sensitive, multi-sectoral agenda for population stabilisation that will "think, plan and act locally and support nationally". Panchayats demonstrating exemplary performance would be nationally recognised and honoured.

II. Convergence of Service Delivery at Village Level. Efforts at population stabilisation will be effective only if we direct an integrated package of essential services at the village and household levels. A flexible approach has to be promoted by extending basic reproductive and child health care through mobile clinics and counselling services particularly to remote, inaccessible or sparsely populated regions in the country like hilly and forested areas, desert regions and tribal areas.

III. Empowering Women for Improved Health and Nutrition. The complex socio-cultural determinants of women's health and nutrition have cumulative effects over a life time. Discrimination against girl child leads to her malnutrition and impaired physical development. This situation is compounded by early childbearing and consequent risk of serious pregnancy related complications. Women's risk of premature death and disability is the highest during their reproductive years. Shockingly maternal mortality ratio (MMR) in India is among the highest in the world. Programmes for Safe Motherhood, Universal Immunisation, Child Survival and Oral Rehydration have been combined into Integrated Reproductive and Child Health Programme.

IV. Child Health and Survival. India suffers from high rate of mortality and morbidity among infants and children below 5 years of age. For solving this problem, a National Technical Committee consisting of consultants in obstetrics, pediatrics, family health, medical research and statistics from academic, public health professionals, clinical practitioners etc. has to be set-up. The baby friendly hospital initiative (BFHI) should be extended to all hospitals, upto to sub-centre levels. Breast-feeding should be encouraged.

V. Meeting the Unmet Needs for Family Welfare Services. There are unmet needs in both rural and urban areas for contraceptives, supplies and equipment for integrated service delivery, mobility of health providers and patients, and comprehensive information. It is necessary to meet these requirements and strengthen health infrastructure at the village, sub-centre and primary health centre levels.

VI. Under-Served Population Groups. Following groups of population are under-served which need due attention of the government and the concerned organisations :

(*a*) **Urban slums** with little or no access to potable water, sanitation facilities and health care services.

(*b*) **Tribal Communities, Hill Area Populations and Displaced and Migrant Populations** in remote and low density areas.

(*c*) **Adolescents** representing about a fifth of India's population need protection from unwanted pregnancies and sexually transmitted diseases, particularly in rural areas.

(*d*) **Increased Participation of Men in Planned Parenthood.** In the past, population programmes have tended to exclude menfolk but now active participation of men is recognised in all efforts in planning families. Nearly 97 per cent of sterilisations are tubectomics and this gender imbalance needs to be corrected.

VII. Diverse Health Care Providers. Considering the large unmet goals, it is imperative to increase the numbers and diversify of the categories of healthcare providers.

VIII. Collaboration with and commitments from Non-Government Organisations and Private Sector. Government alone cannot reach each and every household and partnership of non-government voluntary organisations and private corporate sector with government must be encouraged.

IX. Mainstreaming Indian Systems of Medicine and Homeopathy. Indian system of medicine has sustained life in the country for centuries with minimal side effects and this branch of medicine will expand the pool of effective health care providers at low cost.

X. Contraceptive Technology and Research on Reproductive and Child Health. The government has to take steps to advance, encourage, and support medical, social science, demographic and behavioural science, research on maternal, child and reproductive health care issues. This will improve medical techniques relevant to country's needs.

XI. Providing for the Older Population. Keeping in view the increasing life expectancy it is estimated that proportion of population 60 years and above will increase considerably. Considering the weakening traditional support system, the elderly are becoming increasingly vulnerable, needing protection and care. Promotion of old age health care and support will also serve to reduce incentive for large families. National Policy for Older Persons was adopted by the Ministry of Social Justice and Empowerment in January, 1999.

XII. Information, Education and Communication. Information, education and communication (IEC) of family welfare messages must be clear, focussed and disseminated everywhere, including the remote corners of the country and in local dialects.

National Commission on Population

The Government of India constituted the National Commission on Population on May 11, 2000 with the Prime Minister as the Chairman and the Deputy Chairman of Planning Commission as its vice chairman. The Commission has the mandate to :

- review, monitor and give direction for the implementation of the National Population Policy with the view of achieving the goals it has set;
- promote synergy between health, educational, environmental and developmental programmes so as to hasten population stabilisation;
- promote inter-sectoral coordination in planning and implementation of the programmes through different agencies at the Centre and in the states; and
- develop a vigorous people's programme to support this national effort.

TABLE 10.15. Main health indicators in India

Particulars	NFHS-1 (1992-93)	NFHS-2 (1998-99)	NFHS-3 (2005-06)	NFHS-4 (2015-16)	NFHS-5 (2019-21)
Total Fertility Rate (Children per women)	3.4	2.9	2.7	2.2	2
Sex ratio at birth for children born in the last five years (females per 1,000 males)	—	—	914	919	929
Infant Mortality Rate (per 1000 livebirths)	78.5	67.6	57	40.7	35.2
Under-five Mortality Rate (per 1000 livebirths)	109.3	94.9	74.3	49.7	41.9
Institutional Birth (%)	26.1	33.6	40.8	78.9	88.6
Pregnant women age 15-49 who are anaemic (%)	—	51.8	57.9	50.4	52.2

NFHS = National Family Health Survey

Source: Economic Survey 2021-22, p. 378.

A Strategic Support Group consisting of secretaries of concerned sectoral ministries has been constituted as a standing advisory group to the Commission. Nine working groups were constituted to look into specific aspects of implementation of the programmes aimed at achieving the targets set in the National Population Policy.

Health Indicators

Health is one of the most important constituent of human development. Life free from illness and having reasonably long span is termed as healthy life.

Following are important health indicators :

1. Life expectancy at birth. This is an important health indictor which indicates the number of years a person (male and female) would live after birth. Due to improvement in health and sanitation services, life expectancy at birth in India has improved considerably during the last over four decades (**Table 10.14**).

Table 10.16 shows that the difference between female-male life expectancy has been in favour of males in 1970-75 whereafter it became in favour of females. It is also worth noting that the gap between female and male life expectancy has gradually increased between 1981-85 and 2001-06. It remained constant between 2001-06 and 2016-21. However, it is likely to increase in future.

TABLE 10.16. Life expectancy at birth by sex, India

Year	Males	Females	Difference Female–Male
1970-75	50.5	49.0	–1.5
1981-85	55.4	55.7	+0.3
1991-95	59.7	60.9	+1.2
2001-06	63.8	66.1	+2.3
2006-11	65.8	68.1	+2.3
2011-16	67.3	69.6	+2.3
2016-21	68.8	71.1	+2.3
2021-26*	69.8	72.3	+2.5

Source : (*i*) Premi Mahendra K. and Das Dependra Nath (2012) Population of India 2011, p. 47.

(*ii*) Bose Ashish (2010), India's Quest for Population Stabilisation, p. 89.

*Projected

2. Crude Birth Rate (CBR). Crude birth rate is expressed in terms of number of births in a year per thousand of mid year population. It is worth noting that only live births during the year are to be taken into account. It is calculated as under :

$$CBR = \frac{Bt}{P} \times 1000$$

where Bt = live births during the year

and P = estimated mid-year population

This is the simplest and most widely used measure of human fertility. Birth rate declined from 22.5 per thousand in 2009 to 19.7 per thousand in 2019. This is an indication of our efforts to lower birth rate and to avoid population explosion. According to 2019 figures, a low birth rate of less than 13 per thousand has been recorded in states of Goa, Nagaland and Tripura. In contrast, high birth rate of over 20 per thousand has been reported from states of Assam, Bihar, Chhattisgarh, Haryana, Jharkhand, Madhya Pradesh, Rajasthan, Uttar Pradesh and Meghalaya. Among the Union Territories, Andaman recorded the lowest (11 per thousand) and Daman and Diu recorded the highest (21.9 per thousand) birth rate (**Table 10.17**).

3. Total Fertility Rate (TFR). One of the major factors which affects population growth, is the rate of birth which is closely related to fertility rate (the other major factors are mortality rate, migration and longevity). This is expressed in terms of number of children below five years of age per thousand females of reproductive age group. It is calculated as under

$$FR = \frac{P_{0\text{-}4}}{P_{f\,15\text{-}49}}$$

where $P_{0\text{-}4}$ stands for the number of children a under 5 years of age and $P_{f\ 15\text{-}49}$ stands for women of childbearing age.

As mentioned earlier India has succeeded in reducing the fertility rate to a great extent. TFR was at a very high level of 5.9 immediately after Independence which reduced a low of 2.0 in 2019-21. This is below the replacement rate of which means that India's population growth will come down in the near future. Large scale variations in TFR are found at the state level. In 2019-21, TFR was a minimum of 1.1 in Sikkim and maximum of 3.0 in Bihar (**Table 10.18**)

TABLE 10.17. Statewise Birth rate and Death rate

States/UTs	Birth Rate		Death Rate	
	2009	2019	2009	2019
(1)	(2)	(3)	(4)	(5)
Andhra Pradesh	18.3	15.9	7.6	6.4
Assam	23.6	21.0	8.4	6.3
Bihar	28.5	25.8	7	5.5
Chhattisgarh	25.7	22.2	8.1	7.3
NCT of Delhi	18.1	14.4	4.4	3.2
Gujarat	22.3	19.5	6.9	5.6
Haryana	22.7	20.1	6.6	5.9
Jammu & Kashmir	18.6	14.9	5.7	4.6
Jharkhand	25.6	22.3	7	5.3
Karnataka	19.5	16.9	7.2	6.2
Kerala	14.7	13.5	6.8	7.1
Madhya Pradesh	27.7	24.5	8.5	6.6
Maharashtra	17.6	15.3	6.7	5.4
Odisha	21	18.0	8.8	7.1
Punjab	17	14.5	7	6.6
Rajasthan	27.2	23.7	6.6	5.7
Tamil Nadu	16.3	14.2	7.6	6.1
Telangana		16.7		6.1
Uttar Pradesh	28.7	25.4	8.2	6.5
Uttarakhand	19.7	17.1	6.5	6.0
West Bengal	17.2	14.9	6.2	5.3
Arunachal Pradesh	21.1	17.6	6.1	5.8
Goa	13.5	12.3	6.7	5.9
Himachal Pradesh	17.2	15.4	7.2	6.9
Manipur	15.4	13.6	4.7	4.3
Meghalaya	24.4	23.2	8.1	5.6
Mizoram	17.6	14.5	4.5	4.0
Nagaland	17.2	12.7	3.6	3.5
Sikkim	18.1	16.5	5.7	4.2
Tripura	14.8	12.8	5.1	5.5
Andaman & Nicobar Islands	16.3	11.0	4.1	5.3
Chandigarh	15.9	13.0	3.9	4.0
Dadra & Nagar Haveli	27	21.9	4.8	3.7
Daman & Diu	19.2	18.6	5.1	4.1
Lakshadweep	15	14.8	5.8	5.6
Puducherry	16.5	13.3	7	6.8
India	22.5	19.7	7.3	6.0

Note : 1. Andhra Pradesh includes Telangana for the year 2009 and Jammu & Kashmir includes Ladakh for the year 2019.
2. Birth rate and Death rate per 1000 population.

Source: Economic Survey 2021-22, Statistical Appendix p. 167.

4. Crude Death Rate (CDR). It is the simplest measure of mortality indicating the number of deaths in a particular year per thousand of population. It is expressed as under :

$$CDR = \frac{D}{P} \times 1000$$

where D = Number of deaths in a year
P = Estimated mid-year population

CDR in India declined from 42.6 per thousand in 1901-11 to a mere 6.0 per thousand in 2019. This is an indication of improvement in our health services and increased availability of health facilities.

Large scale variations are found at the state level. In 2019, it was lowest of 3.2 per thousand in NCT of Delhi and highest of 7.3 per thousand in Chhattisgarh (**Table 10.17**).

5. Infant Mortality Rate (IMR). Infant Mortality Rate refers to the death rate among infants and is calculated for connoting mortality among children of less than one year of age. It is expressed as under :

$$IMR = \frac{D_o}{B_l} \times 1000$$

where D_o = number of deaths of children under one year of age

and B_l = number of live births.

Thus IMR is the ratio between number of deaths among children under one year of age to the number of live births.

In the year 2015-16, the infant mortality rate was 40.7 which came down to 35.2 in 2019-21. This is a clear indication of improvement of child care in the country. This is of utmost importance because this section of population will determine the future course of population growth.

Large variations are found at the state level with respect to IMR. In 2019-21, the lowest IMR of 2.9 was reported from the Union Territory of Puducherry and Lakshadweep do not report any infant mortality. Among the states Kerala reported the lowest IMR of 4.4 per thousand. As expected Uttar Pradesh reported the highest IMR of 50.4 per thousand (**Table 10.19**).

TABLE 10.18. Statewise distribution of Total Fertility Rate in India

States/UTs	Total Fertility Rate (Children per women)	
	NFHS-4 (2015-16)	NFHS-5 (2019-21)
All India	2.2	2.0
Andhra Pradesh	1.8	1.7
Arunachal Pradesh	2.1	1.8
Assam	2.2	1.9
Bihar	3.4	3.0
Delhi	1.8	1.6
Goa	1.7	1.3
Gujarat	2.0	1.9
Haryana	2.1	1.9
Himachal Pradesh	1.9	1.7
Jammu and Kashmir	2.0	1.4
Karnataka	1.8	1.7
Kerala	1.6	1.8
Madhya Pradesh	2.3	2.0
Maharashtra	1.9	1.7
Manipur	2.6	2.2
Meghalaya	3.0	2.9
Mizoram	2.3	1.9
Nagaland	2.7	1.7
Odisha	2.1	1.8
Punjab	1.6	1.6
Rajasthan	2.4	2.0
Sikkim	1.2	1.1
Tamil Nadu	1.7	1.8
Tripura	1.7	1.7
Uttar Pradesh	2.7	2.4
West Bengal	1.8	1.6
Chhattisgarh	2.2	1.8
Jharkhand	2.6	2.3
Uttarakhand	2.1	1.9
Telangana	1.8	1.8
Andaman and Nicobar (UT)	1.4	1.3
Chandigarh (UT)	1.6	1.4
Dadra and Nagar Haveli & Daman Diu (UT)	2.1	1.8
Lakshadweep (UT)	1.8	1.4
Puducherry (UT)	1.7	1.5
Ladakh (UT)		1.3

Source: Economic Survey 2021-22, pp. 381-82.

TABLE 10.19. Infant Mortality Rates (per 1000 live births) across states and UTs of India

States/UTs	NFHS-4 (2015-16)	NFHS-5 (2019-21)
Andhra Pradesh	34.9	30.3
Arunachal Pradesh	22.9	12.9
Assam	47.6	31.9
Bihar	48.1	46.8
Delhi	31.2	24.5
Goa	12.9	5.6
Gujarat	34.2	31.2
Haryana	32.8	33.3
Himachal Pradesh	34.3	25.6
Jammu and Kashmir	32.4	16.3
Karnataka	26.9	25.4
Kerala	5.6	4.4
Madhya Pradesh	51.2	41.3
Maharashtra	23.7	23.2
Manipur	21.7	25.0
Meghalaya	29.9	32.3
Mizoram	40.1	21.3
Nagaland	29.5	23.4
Odisha	39.6	36.3
Punjab	29.2	28.0
Rajasthan	41.3	30.3
Sikkim	29.5	11.2
Tamil Nadu	20.2	18.6
Tripura	26.7	37.6
Uttar Pradesh	63.5	50.4
West Bengal	27.5	22.0
Chhattisgarh	54.0	44.3
Jharkhand	43.8	37.9
Uttarakhand	39.7	39.1
Telangana	27.7	26.4
Andaman and Nicobar (UT)	9.8	20.6
Chandigarh (UT)		31.8
Dadra and Nagar Haveli & Daman Diu (UT)	33.4	31.8
Lakshadweep (UT)	27.0	0.0
Puducherry (UT)	15.7	2.9
Ladakh (UT)		20.0
India	40.7	35.2

NFHS reports, MoH&FW; Economic Survey 2020-21 pp. 379-80.

Note: NFHS = National Family Health Survey

In NFHS-5, Jammu & Kashmir is Union Territory excluding Ladakh (UT)

Source: Economic Survey, 2021-22, pp 379-80.

6. Under Five Mortality Rate. This is another health indictor which is concerned with children below the age of 5 years. Drive to care for children especially in the rural areas, has paid rich dividends. It is heartening to note that under five mortality rate decreased considerably in the Post-Independent era, from 326 per thousand in 1951 in 49.7 in 2015-16 and 41.9 in 2019-21.

Large scale variations are found at the state level with respect to under five mortality rate. This is zero and 3.9 in the Union Territories of Lakshadweep and Puducherry respectively and 5.2 in the state of Kerala. As against this Madhya Pradesh, Chhattisgarh, Bihar and Uttar Pradesh have report high under five mortality rate of 49.2, 50.4, 56.4 and 59.8 per thousand respectively (**Table 10.20**).

7. Maternal Mortality Rate (MMR) is the indicator of health of women during pregnancy and delivery. According to a report issued by the Registrar General of India on 14 March 2022, MMR was 113 per 1,00,000 live births in 2016-18 which came down 103 per 1,00,000 live births in 2017-19. This means that there has been a considerable improvement in the maternal health. In all seven states reported high MMR in 2017-19. The states are Rajasthan, Uttar Pradesh, Madhya Pradesh, Chhattisgarh, Bihar, Odisha and Assam. These states reported 130 or more than this maternal death rate per 1,00,000 live births. Punjab, Uttarakhand and West Bengal reported 100-130 MMR. Karnataka and Haryana reported low MMR varying from 70 to 100. Spatial variations in MMR are due to variations in health services to pregnancy women.

HUMAN DEVELOPMENT

Growth and Development. Although both growth and development refer to changes over a period of time, yet these terms differ from each other. **Growth** is *quantitative* and value neutral which may have a positive or a negative sign. This means that the change may be either positive (increase) or negative (decrease). Development on the other hand is *qualitative* change which is always value positive. This means that development cannot take place unless there is an increment or addition to the existing conditions. Development occurs when positive growth takes place. Yet, positive growth does not

TABLE 10.20. Under Five Mortality Rates (per 1000 live births) across states and UTs of India

States/UTs	NFHS-4 (2015-16)	NFHS-5 (2019-21)*
Andhra Pradesh	40.8	35.2
Arunachal Pradesh	32.9	18.8
Assam	56.5	39.1
Bihar	58.1	56.4
Delhi	42.2	30.6
Goa	12.9	10.6
Gujarat	43.5	37.6
Haryana	41.1	38.7
Himachal Pradesh	37.6	28.9
Jammu and Kashmir	37.6	18.5
Karnataka	31.5	29.5
Kerala	7.1	5.2
Madhya Pradesh	64.6	49.2
Maharashtra	28.7	28.0
Manipur	25.9	30.0
Meghalaya	39.6	40.0
Mizoram	46.0	24.0
Nagaland	37.5	33.0
Odisha	48.1	41.1
Punjab	33.2	32.7
Rajasthan	50.7	37.6
Sikkim	32.2	11.2
Tamil Nadu	26.8	22.3
Tripura	32.7	43.3
Uttar Pradesh	78.1	59.8
West Bengal	31.8	25.4
Chhattisgarh	64.3	50.4
Jharkhand	54.3	45.4
Uttarakhand	46.5	45.6
Telangana	31.7	29.4
Andaman and Nicobar (UT)	13.0	24.5
Chandigarh (UT)		
Dadra and Nagar Haveli & Daman Diu (UT)	39.9	37.0
Lakshadweep (UT)	30.2	0.0
Puducherry (UT)	16.2	3.9
Ladakh (UT)		29.5
India	49.7	41.9

***Source*:** Economic Survey 2021-22. Statistical Appendix, p. 187.

always lead to development. Development occurs when there is a positive change in quality. For example, if the population of a city grows from five lakhs to ten lakhs over a period of time, we say the city has grown. But if the basic facilities of life such as housing, water, power, transport, severage, etc. remain the same or do not grow in proportion to population growth, then the growth has not been accompanied by development.

Before 1990s, a country's development was gauged as a measure of its economic progress. This implied that countries with better economic conditions were more developed while poor countries were less developed. This however, did not reflect the true nature of development because in many cases the benefit of economic growth did not reach the common man who deserved it the most. Quality of life of the people, the opportunities they have and the freedoms they enjoy are important aspects of development.

The above mentioned ideas were clearly spelt out for the first time in the late 1980s and early 1990s by two great economists of Asia. One was **Dr. Mahbub-ul-Haq** of Pakistan and the other was the **Nobel Laureate Dr. Amartya Sen** of India. **Dr. Mahbub-ul-Haq** created the Human Development Index (HDI) in 1990. According to him, *development is all about enlarging people's choices in order to lead long, healthy lives with dignity*. Dr. Haq believed that people are central to all development. The choices of the people are not fixed but keep on changing. The chief objective of development is to create conditions in which people can live meaningful lives.

A meaningful life is not just a long one. It must be a life with some purpose. This means that people must be healthy, be able to develop their talents, participate in society and be free to achieve their goals.

Dr. Amartya Sen expressed the opinion that an increase in freedom (or decrease in unfreedom) is the main objective of development. Increasing freedom is one of the most effective ways of bringing about development. He laid much emphasis on the role of social and political institutions and processes of increasing freedom.

Following three are the most important aspects of human development:

(*i*) Leading long and healthy life.

(*ii*) Attaining ability to gain knowledge.

(*iii*) Having enough means to be able to live a decent life.

In view of the above mentioned aspects, it is essential to have access to resources, health and education. But very often, people fail to attain capability and freedom to make even basic choices. This may be due to various reasons of which lack of knowledge, poverty, social discrimination etc. are important. Such a situation prevents people from leading healthy lives, being able to get educated and to lead a decent life. Therefore, in order to enlarge the choices of the people, it is essential to build their capabilities in the areas of health, education and access to resources. The choices become limited in the absence of these capabilities.

Generally speaking "Development is freedom". Development and freedom are often associated with modernisation, leisure, comfort and affluence. In the present day world computerisation, industrialisation, efficient transport and communication network, large education system, advanced and modern medical facilities, safety and security of individuals, etc. are considered as the symbols of development. Level of development is measured with respect to the availability and access to these modern things. But this is only one sided view of development which is often called the ***'western or euro-centric'*** view of development. For a postcolonial country like India, colonisation, marginalisation, social discrimination and regional disparity, etc. show the other face of development.

In India, development is a mixture of opportunities as well as neglect and deprivation. A few people in urban areas are enjoying all the facilities of modern life while on the other hand vast humanity living in rural areas and urban slums do not have even the basic amenities of life. People belonging to scheduled castes, scheduled tribes, landless agricultural labourers, poor farmers, slum dwellers, etc. are the worst sufferers. The condition of female population is even more pathetic.

Lack of development leads to deteriorating human conditions and results in environmental pollution. Air, soil, water and noise pollution are posing a great threat to the very existence of our society. Consequently, the poor are being subjected to three inter-related processes of declining capabilities; *i.e.* (1) social capabilities — due to displacement and weakening social ties (social capital), (2) environmental capabilities—due to pollution and, (3) personal capabilities—due to increasing incidence of diseases and accidents. This, in turn, has adverse effects on their quality of life and human development.

Why Human Development ?

According to **Paul Streeten**, human development is necessary due to following reasons :

1. The main purpose of human development is to improve the human conditions and to enlarge people's choices.
2. It is a major tool of achieving higher level of productivity. A well-nourished, healthy, educated, skilled, alert labour force is the most productive asset. Therefore, investments on these sectors are justified on grounds of productivity too.
3. It helps in reducing the rate of growth of population.
4. Human development is friendly to the physical environment also. Deforestation, desertification and soil erosion decline when poverty declines.
5. Improved living conditions and reduced poverty contribute to a healthy civil society, enhanced democracy and greater social stability.
6. Human development also helps in reducing civil disturbances in the society and in increasing political stability.

The Four Pillars of Human Development

Just as any building is supported by pillars, the idea of human development is supported by the concepts of equity, sustainability, productivity and empowerment.

1. Equity. Equity means making equal access to opportunities available to everybody. The opportunities available to people must be equal irrespective of their gender, race, income and in the Indian case, caste. Yet this is very often not the case and happens in almost every society. Normally, the poor and persons belonging to socially and economically backward groups fail to have access to equity.

2. Sustainability. Sustainability refers to continuity in the availability of opportunities. This means that each generation must have the same opportunities. Therefore, we must use our environmental, financial and human resources in such a way that our future generations are not deprived of these opportunities.

3. Productivity. Productivity means human labour productivity or productivity in terms of human work. It must be enriched by building capabilities in people. In fact, people of a country are its real wealth. As such sincere effort should be made to increase their knowledge and provide better facilities of health and education so that they can attain better work efficiency.

4. Empowerment. It refers to have power to make choices. Such power comes from increasing freedom and capability. Good governance and people-oriented policies are required to empower people. The empowerment of socially and economically disadvantaged groups is of special importance.

Indicators of Human Development

Although it is not possible to have a flawless quantitative measure of human development, the United Nations Development Programme (UNDP) has developed a composite index, now known as the Human Development Index (HDI). It includes (*i*) longevity of life, (*ii*) knowledge base, and (*iii*) a decent material standard of living. To keep the index simple, only a limited number of variables are included. Initially, life expectancy was chosen as an index of longevity, adult literacy as an index of knowledge and per capita Gross National Product adjusted for **Purchasing Power Parity (PPP)** as an index of decent life. These variables are expressed in different units. Therefore, a methodology was evolved to construct a composite index rather than several indices.

In India, three sets of indicators have been selected for preparing the Human Development Report. Among them, a core set of composite indices presents the state of human development for the society as a whole. Besides, Gender Equality Index has been estimated to reflect the relative attainments of women, and the Human Poverty Index to evaluate the state of deprivation in the society.

Several other variables have gradually been added to the above sets of indicators. Among them, *health indicators* related to longevity at birth rate, death rate with special reference to infant mortality and nutrition. *Social indicators* include literacy particularly female literacy, enrollment of school-going children, drop out ratio, and pupil-teacher ratio. *Economic indicators* are related to wages, income, and employment. Per Capita Gross Domestic Product, incidences of poverty and employment opportunity are also favoured indicators in this group. They are converted into a composite index to present the holistic picture of the Human Development.

WHAT IS HUMAN DEVELOPMENT?

"Human development is a process of enlarging the range of people's choices, increasing their opportunities for education, health care, income and empowerment and covering the full range of human choices from a sound physical environment to economic, social and political freedom."

Thus, enlarging the range of people's choices is the most significant aspect of human development. People's choices may involve a host of other issues, but, living a long and healthy life, to be educated and have access to resources needed for a decent standard of living including political freedom, guaranteed human rights and personal self-respect, etc. are considered some of the non-negotiable aspects of the human development.

Approaches to human development

Four important approaches to human development are :

1. The income approach;
2. The welfare approach;
3. Minimum needs approach;
4. Capabilities approach.

TABLE 10.21. Approaches to Human Development

1. Income Approach	This is one of the oldest approaches to human development. Human development is seen as being linked to income. The idea is that the level of income reflects the level of freedom an individual enjoys. Higher the level of income, the higher is the level of human development.
2. Welfare Approach	This approach looks at human beings as beneficiaries or targets of all development activities. The approach argues for higher government expenditure on education, health, social secondary and amenities. People are not participants in development but only passive recipients. The government is responsible for increasing levels of human development by maximising expenditure on welfare.
3. Basic Needs Approach	This approach was initially proposed by the International Labour Organisation (ILO). Six basic needs i.e., health, education, food, water supply, sanitation and housing were identified. The question of human choices is ignored and the emphasis is on the provision of basic needs of defined sections.
4. Capability Approach	This approach is associated with Prof. Amartya Sen. Building human capabilities in the areas of health, education and access to resources is the key to increasing human development.

Measuring Human Development

Human Development is measured in terms of Human Development Index (HDI). It measures average achievements in basic human development in one simple composite index and produces a ranking of countries on the basis of their performance in key areas of health, education and access to resources. These rankings are based on a score between 0 & 1 that a country earn from its record in key areas of health, education and access to resources.

1. Health. The indicator chosen to assess health is *life expectancy at birth*. It means that people have a better choice to lead longer and healthier life. Higher the life expectancy at birth, higher is the human development index.

2. Education. Education here involves adult literacy rate and gross enrolment ratio. It means that a country should have larger number of adults who are able to read and write and larger number of children enrolled in schools to be placed higher in human development index.

3. Access to resources. Access to resources is measured in terms of purchasing power (in U.S. dollars).

> The Human Development Report was first published in 1990 by the United Nations Development Programme (UNDP). Since then it is published every year. This report gives an interesting rank wise list of the member countries according to the level of human development. The Human Development index and the Human Poverty index are two important indices to measure human development used by the UNDP.

International Comparisons

Human Development is not necessarily related to size or per capita income of a country. It has been found that smaller and economically poorer countries are often ranked higher in terms of human development than the bigger and richer countries. For example Switzerland, Norway and Iceland are much smaller and poorer countries in terms of GDP than the U.S.A. But these small countries rank much higher than the U.S.A. with respect to HDI value.

TABLE 10.22. India's position in Global HDI 2021

Country	HDI in 2021		Life expectancy at birth (years)	Mean years of schooling	Gross national income per capita
	Rank	Value			
Switzerland	1	0.962	84	13.9	66,933
Norway	2	0.961	83.2	13.0	64,660
U.K.	18	0.929	80.7	13.4	45,225
Japan	19	0.925	84.8	13.4	42,274
U.S.A.	21	0.921	77.2	13.7	64,765
China	79	0.768	78.2	7.6	17,504
Brazil	87	0.754	72.8	8.1	14,370
South Africa	109	0.713	62.3	11.4	12,948
Indonesia	114	0.705	67.6	8.6	11,466
India	132	0.633	67.2	6.7	6,590
South Asian Region	—	0.632	67.9	6.7	6,481
World Average	—	0.732	71.4	8.6	16,752

Source: 2021/2022 Human Development Report, UNDP in Economic Survey, 2022-23, p. 149.

Similarly Sri Lanka, Trinidad and Tobago have smaller economies than that of India but have higher rank than India in terms of human development index (HDI). In India, Kerala has higher rank than Panjab and Gujarat despite having lower per capita income than these states.

Member countries have been divided into three broad groups on the basics of the human development scores earned by them. Countries with a score of above 0.8 are termed as having high index value. Countries with score of between 0.5 and 0.799 are placed in medium level of development. Countries having development index below 0.5 are said to have low level of human development.

Computing the HDI

To construct the Index, fixed minimum and maximum values have been established for each of the indicators :

- Life expectancy at birth : 25 years and 85 years;
- General literacy rate : 0 per cent and 100 per cent;
- Real GDP per capita (PPP$) ; PPP$ 100 and PPP$ 40,000.
- Purchasing power party.

Individual Indices are computed first on the basis of a given formula. HDI is a simple average of these three indices and is derived by dividing the sum of these three indices by 3.

With normalisation of the values of the variables that make up the HDI, its value ranges from 0 to 1. The HDI value for a country or a region shows the distance that it has to travel to reach the maximum possible value of 1 and also allows intercountry comparisons.

The closer a score is to one, the greater is the level of human development. Therefore, a score of

0.983 would be considered very high while 0.268 would mean a very low level of human development.

HDI of India. India has consistently improved with respect to human development, ever since the beginning of HDI reports by the UNDP. The country's HDI values increased from 0.427 in 1990 to 0.633 in 2021 an increase of nearly 50 per cent. The HDI report released on September 8, 2022 ranked India at the 132th position, out of a total of 191 countries. Movements in HDI are driven by changes in health, education and income. At the bottom are Niger (189), Central African Republic (188), South Sudan (187), Chad (186) and Burandi (185).

However, there are inter-state disparities which are reflected by the State level HDIs. The subnational Human Development Index (SHDI) for different States for period 1990 and 2017 released by UNDP show that all states have shown significant improvement in levels of human development. The minimum SHDI for the year 2017 is more than maximum SHDI for the year 1990 across all states. The 2017 HDI scores indicate that the States like Kerala, Goa, Himachal Pradesh and Punjab occupy the top four positions while the States like Bihar, Uttar Pradesh and Madhya Pradesh are at the bottom of the rankings.

REFERENCES

Agarwala, S.N. (1977) : *India's Population Problems,* Tata McGraw Hill, New Delhi.

Bhutani, Smita (1995) : *Demographic Dynamism in India*, Discovery Publishing House, New Delhi.

Bose, A. (1988) : *From Population to People, Vol. I,* B.R. Publishing Corporation, New Delhi.

Bose, Ashish (2010) : *India's Quest for Population Stabilisation,* National Book Trust, New Delhi.

Census of India (2011) : Provisional Population Totals, Paper 1 of 2011 India Series 1, Registrar General and Census Commissioner India, New Delhi.

Chandna, R.C. (2017) : *A Geography of Population,* Kalyani Publishers, Ludhiana.

Chatterjee, S.P. (1962) : "Regional Pattern of Density and Distribution in India." *Geographical Review of India*, 24, 1-28.

Clarke, John I (1972) : *Population Geography,* Pergamon Press, Oxford.

Davis, Kingslley (1951) : *The Population of India and Pakistan,* Princeton University Press, Princeton.

Economic Survey, 2020-21 and 2021-22, Govt. of India, Ministry of Finance.

Ghosh, S. (1970) : "Physical and Economic Factors in Population Distribution of Bihar", *The National Geographical Journal of India,* 16, 61-70.

Gosal, G.S. (1982) : "Recent Population Growth in India", *Population Geography,* Vol. 4, pp. 33-53.

India, 2022 : Publication Division, Ministry of Information and Broadcasting, Government of India, New Delhi.

Krishan, Gopal (1968) : "Distribution and Density of Population in Orissa", *National Geographical Journal of India,* 14, 250-257.

Krishan, Gopal (1984) : Spurts in India's Population Growth during the Twentieth Century–A spatial view, *Population Geography,* Vol. 6, Nos. 1 & 2, June-Dec.

Mehta, B.C. (1973) : "Spatial Distribution of Population in Rajasthan" *National Geographical Journal of India,* 19, 149-157.

Ministry of Health and Family Welfare, Department of Family Welfare, New Delhi; National Population Policy, 1976.

Ministry of Health and Family Welfare, Department of Family Welfare (2000), New Delhi; National Population Policy, 2000.

Mitra, Ashok (1978) : *India's Population : Aspects of Quality and Control,* Abhinav Publications, New Delhi.

Morris, A. (1998) : *Geography and Development,* University College of London Press, London.

Prakash, O. (1970) : "Pattern of Population in Uttar Pradesh" *National Geographical Journal of India,* 16, 150-160.

Premi, Mahendra K (2006) : Population of India, In the New Millennium : Census 2001, National Book Trust, New Delhi.

Premi, Mahendra, K (2009) : *India's Changing Population Profile,* National Book Trust, New Delhi.

Premi, Mahendra and Das Dipendra Nath (2012) : *Population of India 2011,* B.R. Publishing Corporation, Delhi.

Sinha, R.K. (2007) : *India 2005*, Shipra Publications, New Delhi.

Swami, S. (1974) : *Population Growth and Economic Development in Ashis Bose et. al (eds.), Population in India's Development 1947-2000*, Vikas Publishing House, New Delhi.

UNFPA : *The State of World Population,* United Nations Population Division.

United Nations Development Programme (UNDP) : *Human Development Report, 2019,* Oxford University Press.

Verma, S.D. (1956) : "Density and Pattern of Population in Punjab" *National Geographical Journal of India.*

Zlinsky, W. (1966) : *A Prologue to Population Geography,* Prentice Hall, Inc. N.J.

◆◇◆

CHAPTER

11

Population Composition

INTRODUCTION

Population composition is a vast field which refers to the physical, socio-cultural and economic attributes of population, such as ethnicity, tribes, language, religion, literacy and education, age, sex, economically active population and many more traits. The study of composition of population helps us in understanding the social, economic and demographic structure of population.

ETHNIC COMPOSITION

Our present day population is a conglomeration of people belonging to different racial groups with different ethnic backgrounds. These people entered India from different parts of the world at different points of time adopting different land and water routes. In fact India has been a melting pot of various races and tribes from times immemorial. Almost all the major races of the world are visible in India as a result of which the country is said to have a varied and diverse ethnic composition. The present day population of the country has been derived mainly from the following racial groups :

1. The Negritos

According to Hutton, the earliest occupants of India were the people of Negrito race. S.K. Chatterjee and S.M. Katre have expressed the view that Negroid people migrated to India from Africa and established their language on the soil of India. **A.C. Haddon** opines that Negrito features are met with particularly amongst the Andaman islanders and most probably the Uralis of Nilgiri hills, Kadors of Kochi, Pullayans of Palni hills, etc. Besides some tribes like the Angami Nagas in the north-east and the Badgis in the Rajmahal hills in Jharkhand, etc. possessing some physical traits reminiscent of the Negrito are seen. The Negrito race is characterised by short stature, dark chocolate brown skin, woolly hair, bulbous forehead, broad flat nose and slightly protruding jaws.

2. The Proto-Australoids

Hutton is of the opinion that the Proto-Australoids came to India from the East Mediterranean area (Palestine). They came soon after the Negritos. Today they constitute the bulk of population in many isolated and semi-isolated parts of central and southern India. The Veddahs, Malavedahs, Irulas and Sholagas are the true representatives of Proto-Australoids. The Bhils, Kols, Badagas, Korwas, Kharwars, Mundas, Bhumjis and Malpaharis of the highlands of Central India and the Chenchus, Kurumbas, Malayans and Yeruvas of South India may all be treated as Proto-Australoids.

Some anthropologists believe that the Proto-Australoids supported the Mediterraneans in building the Indus Valley Civilization. On their arrival in India, the Proto-Australoids pushed, pressed, displaced and supplanted the Negritos and forced them to shift to more inaccessible, remote and less hospitable areas, where they are found even today. In the process, there was some admixture of the two races, perhaps more so in the south than in the north. In physical appearance the Proto-Australoids more or less resemble the Negritos with the main exception that they do not have wooly hair like the Negritos. Their other physical characteristics are bulbous forehead, broad flat nose and slightly protruding jaws.

3. The Mongoloid

According to **Risley**, "On its northern and eastern frontier, India marches with the great Mongoloid region of the earth". Most of the anthropologists believe China to be the homeland of the Mongoloid race from where they were pushed southward into the Malaya peninsula and Indonesia. They entered India through the passes in the northern or eastern mountains. Hutton is of the opinion that the bulk of Burma (Myanmar) in any case is primarily Mongoloid, and any non-Mongoloid streams of migration that may have reached India through Myanmar have absorbed a vast quantity of Mongolian blood. There is also some evidence of a Mongoloid Melanesian intrusion from Oceania to Tamil Nadu and Kerala and probably that accounts for the occasional Mongoloid element noticed among the people of these states. Presently, they occupy large areas in Ladakh, Sikkim, Arunachal Pradesh and some other parts of east India. Some of the basic physical characteristics of the Mongoloid race include a round and broad head, face with very high cheek bones and a long flat nose, with little or no hair on the face and the body. The tribes of Garo, Khasi, Jaintia, Lipchas, Chakmas, Murmis, Naga and Daffla belong to the Mongoloid race.

The Mongoloid racial stock of India can be divided into two sub-groups as follows :

(*i*) Palaeo-Mongoloids

(*ii*) Tibeto-Mongoloids.

(i) Palaeo-Mongoloids are further divided into ***broad headed*** and ***long headed*** sub-types. They settled mainly along the fringes of the Himalayas in Assam and the Myanmar border.

(ii) Tibeto-Mongoloids have come from Tibet as their name suggests. They are mostly living in Bhutan and Sikkim, as well as in the north-western Himalayas and Trans Himalayan regions.

4. The Mediterraneans

The Mediterranean racial stock came to India from eastern Mediterranean region or South West Asia. They are believed to have migrated during the third and the second millennium B.C. This race has contributed much to the physical composition of peoples of India and also to its culture. They brought earlier forms of Austro-Asiatic languages and are believed to be the bearers of the earliest form of Hinduism in India. *Palaeo-Mediterrraneans* are considered to be the first and the most ancient of all the Mediterranean races to enter India. Their physical characteristics include medium stature, dark skin and long head. In all probability, they first settled in north-west India and started practising agriculture there. However, they were pushed into central and southern India by subsequent immigrants. Today the Palaeo-Mediterranean stock forms the bulk of population of south India and a considerable proportion of population in northern India. The Mediterraneans were the chief architects of the Indus Valley Civilization as is evident from the excavations of Mohanjo Daro and Harappa.

The Dravidians. To quote **Risley**, "The Dravidian race, the most primitive of the Indian types, occupies the oldest geological formation in India, the medley of forestclad ranges, terraced plateaus, and undulating plains which stretches, roughly speaking, from the Vindhyas to Cape Comorin (Kanniyakumari). On the east and west of the peninsular area the domain of the Dravidian is coterminous with the Ghats; while further north it reaches on one side to the Aravalis and on the other to the Rajmahal hills. Where the original characteristics have been unchanged by contact with Indo-Aryan or Mongoloid people, the type is remarkably uniform and distinctive."

5. Brachycephals

Brachycephal groups of races of India are characterised by broad heads. Coorgis and Parsis are representatives of the Brachycephals in India. These races are sub-divided into three major groups. They followed three different routes to enter India.

(*i*) Alpinoids,

(*ii*) Dinarics and

(*iii*) Armenoids.

(*i*) ***Alpinoids.*** Alpinoids came to India along the route passing through Baluchistan, Sind, Kathiawar, Gujarat, Maharashtra, Karnataka and Tamil Nadu.

(*ii*) ***Dinarics.*** Dinarics followed the Ganga valley and its delta as their route to enter India.

(*iii*) ***Armenoids.*** Chitral, Gilgit, Kashmir and Nepal formed the third route for the Armenoids of the Brachycephal group of races to enter India.

6. The Nordics

The Nordics constitute the last wave of migration into India. They spoke Aryan language and migrated to India sometime during the second millennium B.C. The main concentration of these people is in the north-western part of the country. They are a predominant type in Punjab, Haryana and Rajasthan. They are mostly represented among the upper castes of North India particularly in Punjab. The main characteristics of this race are long head, fair complexion, well developed nose and a well built, strong body.

CASTE GROUPS

India's present caste system owes its origin to the ***chatur varna*** which divided the population into four classes, viz. Brahmins, Kshatriyas, Vaishyas, and Shudras. This division was based on the occupation of the people and the complexion of the skin. In due course of time, the caste system in India became extremely hierarchical and rigid encouraging high caste people to exploit the low caste people. Unfortunately, even today, the Indian caste system is intensely hierarchical resulting in several social and economic problems. Today India has more than 3,000 castes.

For the sake of convenience, the castes of the country are divided into three groups. **Table 11.1** indicates that various castes are concentrated in various regions. In other words, their distribution is mainly regional. The regional dimensions of the caste system has led to regional differentiation in social geography of India.

Four major castes according to *varna* (literally colour) are briefly described as under :

1. Brahmans. Brahmans are at the top in *varna* hierarchy. Main castes of this *varna* are those of priests, teachers, custodians of social ritual practices and arbiter of correct social and moral behaviour. Geographically, they are the most ubiquitous because they officiate in a variety of rituals. Although ritually barred from cultivation, they have gathered large stretches of land through grants by local rulers and patrons by virtue of their traditional prestige. Thus they constitute a prominent land owning and money lending class. The services of Brahmans are constantly in demand by other castes for major ritual functions such as births, marriages, deaths, etc. Although Brahman households are a few in a typical village, yet they command a wide variety of services from other castes because they own a large part of agricultural land in the village. By and large Brahmans are better educated than other caste groups.

TABLE 11.1. Caste Groups

Caste

High (Brahmans, Kshatriyas and Vaishyas)	Middle (Cultivating Castes)	Low (Harijans or Scheduled Castes)
1. Saryupanies of Awadh	1. Jats of Haryana and Punjab	1. Chamars of U. P.
2. Namboodris of Kerala	2. Bhumihars of Bihar	2. Balais of W. Bengal
3. Chitpowans of Maharashtra	3. Reddys of Andhra Pradesh	3. Meghs of Gujarat
4. Chattopadhyayas of West Bengal	4. Vellalas of Karnataka	4. Mahars of Maharashtra
5. Iyengars of Tamil Nadu		5. Malls of Andhra Pradesh
		6. Adi-Dravidas of Tamil Nadu

2. Kshatriyas. Next to Brahmans are the Kshatriyas in *varna* ranking. They comprise very powerful castes as they are traditionally warriors and play a major role in defence. However, the role of defence is now largely submerged under the rule of land ownership. When castes form large majorities in a village, the village is generally known by the *jati* name of the Kshatriyas. There are several examples of *Rajput* villages (Rajasthan), *Jat* villages (Haryana, Uttar Pradesh), *Thakur* villages (Himalayan region) and *Nair villages* (Kerala).

3. Vaishyas. Vaishyas rank below Kshatriyas, but fall within the ambit of ritually high *varna*. Vaishyas are primarily engaged in farming and retail trade. Several prominent vaishya groups have established successful monolithic business all over the country.

4. Shudras. These belong to the lowermost class in *varna* ranking. They are mostly engaged in cultivation, and in a wide variety of artisan services such as carpentry, metal work, and basket weaving. However, they are debarred from several ritual privileges. Currently they form the bulk of the country's population.

SCHEDULED CASTES

At the bottom of the social ladder are the *niravasita* meaning "excluded" or the "exterior" castes, so called "*casteless*", officially "scheduled castes". Their occupations are unclean (with respect to upper castes) which include disposal of the dead, flaying of dead animals, menial work, cleaning, etc. Many of them are agricultural labourers, cultivators of small land holdings, petty labourers and industrial workers. To the upper castes, they are "untouchables". The practice of untouchability started with their original "unclean" occupations, and physical contact with upper castes was prohibited. They are forced to live in segregation particularly in rural India.

The term 'Scheduled Caste' appeared for the first time in Government of India Act, 1935 and the Government of India Scheduled Caste Order, 1936. Dr. B.R. Ambedkar preferred to call them '*Depressed classes*' while Mahatma Gandhi called them '*Harijans*'. However, the constitution of India has retained the term 'Scheduled Castes'.

It is clear from the above description that Scheduled castes have been a deprived, neglected and exploited lot in Indian society since the introduction of caste system in the country. It is, therefore, necessary to make a thorough study of these castes. Scheduled Caste is a heterogeneous group consisting of 542 castes. As regards their social and economic status *vis-à-vis* the rest of the population, they are homogeneous to some extent. Majority of them are poor, live in rural areas and are engaged in agriculture. They provide substantial support to agriculture which is the backbone of Indian economy.

Scheduled castes constitute an important segment of Indian population. The Scheduled Caste population has been increasing steadily since 1951, not only in absolute numbers but also in proportion to the total population. Scheduled Castes population was 64 million in 1961 which rose to 80 million in 1971, 105 million in 1981, 138.2 in 1991 (excluding Jammu and Kashmir where 1991 census could not be held), 166.6 million in 2001 and 201.4 million in 2011. The share of Scheduled Castes population has been growing almost consistently since 1951 at the rate of 1.1 per cent to reach 16.63 per cent of the total population in India in 2011.

Distributional Pattern of Scheduled Castes

A reference to **Table 11.2** shows that Scheduled Castes are not confined to specific areas; rather they are distributed all over the country. Uttar Pradesh had the largest number of 41.35 million Scheduled Caste people in 2011. This was followed by West Bengal where 21.46 million Scheduled Caste population. These two states account for nearly one-third (31.3 per cent) of the Schedules Caste population of the country. Other states with large size of the Scheduled Caste population are Bihar (16.5 million), Tamil Nadu (14.4 million), Andhra Pradesh and Telangana (13.8 million), Maharashtra (13.2 million), Rajasthan (12.2 million), Madhya Pradesh (11.3 million), Karnataka (10.4 million), Punjab (8.8 million) and Odisha (7.2 million). States with low concentration of scheduled castes population are Sikkim, Manipur, Meghalaya and Goa. Union territories of Daman and Diu and Dadra and Nagar Haveli also have low concentration of Scheduled Castes. No Scheduled Castes have been reported in the states of Arunachal Pradesh, Nagaland and Union Territories of Lakshadweep and Andaman and Nicobar Islands.

TABLE 11.2. Total Population and Percentage of Scheduled Caste and Scheduled Tribe : 2011 Census

Sl. No.	India / State/ Union Territory	Total Population	Scheduled Caste		Scheduled Tribe	
			Total Population	Percentage of total population	Total Population	Percentage of total population
	INDIA	**1,21,05,69,573**	**20,13,78,086**	**16.63**	**10,42,81,034**	**8.61**
1.	Jammu and Kashmir	1,25,41,302	9,24,991	7.37	14,93,299	11.91
2.	Himachal Pradesh	68,64,602	17,29,252	25.19	3,92,126	5.71
3.	Punjab	2,77,43,338	88,60,179	31.94	0	0.00
4.	Chandigarh*	10,55,450	1,99,086	18.86	0	0.00
5.	Uttarakhand	1,00,86,292	18,92,516	18.76	2,91,903	2.89
6.	Haryana	2,53,51,462	51,13,615	20.17	0	0.00
7.	NCT of Delhi*	1,67,87,941	28,12,309	16.75	0	0.00
8.	Rajasthan	6,85,48,437	1,22,21,593	17.83	92,38,534	13.48
9.	Uttar Pradesh	19,98,12,341	4,13,57,608	20.69	11,34,273	0.57
10.	Bihar	10,40,99,452	1,65,67,325	15.91	13,36,573	1.28
11.	Sikkim	6,10,577	28,275	4.63	2,06,360	33.79
12.	Arunachal Pradesh	13,83,727	0	0.00	9,51,821	68.79
13.	Nagaland	19,78,502	0	0.00	17,10,973	86.48
14.	Manipur	25,70,390	97,042	3.78	9,02,740	35.12
15.	Mizoram	10,97,206	1,218	0.11	10,36,115	94.43
16.	Tripura	36,73,917	6,54,918	17.83	11,66,813	31.76
17.	Meghalaya	29,66,889	17,355	0.58	25,55,861	86.15
18.	Assam	3,12,05,576	22,31,321	7.15	38,84,371	12.45
19.	West Bengal	9,12,76,115	2,14,63,270	23.51	52,96,953	5.80
20.	Jharkhand	3,29,88,134	39,85,644	12.08	86,45,042	26.21
21.	Odisha	4,19,74,218	71,88,463	17.13	95,90,756	23.85
22.	Chhattisgarh	2,55,45,198	32,74,269	12.18	78,22,902	30.62
23.	Madhya Pradesh	7,26,26,809	1,13,42,320	15.62	1,53,16,784	21.09
24.	Gujarat	6,04,39,692	40,74,447	6.74	89,17,174	14.75
25.	Daman and Diu*	2,43,247	6,124	2.52	15,365	6.32
26.	Dadra and Nagar Haveli*	34,33,309	6,186	1.79	1,78,564	51.95
27.	Maharashtra	11,23,74,333	1,32,75,898	11.81	1,05,10,213	9.35
28.	Andhra Pradesh and Telangana	8,45,80,777	1,38,78,078	16.41	59,18,073	6.99
29.	Karnataka	6,10,95,297	1,04,74,992	17.15	42,48,987	6.95
30.	Goa	14,58,545	25,449	1.74	1,49,274	10.23
31.	Lakshadweep*	64,473	0	0.00	61,120	94.79
32.	Kerala	3,34,06,061	30,39,573	9.09	4,84,839	1.45
33.	Tamil Nadu	7,21,47,030	1,44,38,445	20.01	7,94,697	1.10
34.	Puducherry*	12,47,953	1,96,325	15.73	0	0.00
35.	Andaman and Nicobar Islands*	3,80,581	0	0.00	28,530	7.49

*Union territory

Source : Computed from Census of India 2011 Report on revised data issue on C.D. in 2014.

TABLE 11.3. Growth of Scheduled Tribes Population in India

Census Year	Total Population of Scheduled Tribes in lakhs	Percentage of Scheduled Tribes to Total Population
1951	225	6.23
1961	302	6.87
1971	380	6.94
1981*	538	7.58
1991**	678	8.08
2001	843	8.20
2011	1,043	8.61

*Excluding Assam

**Excluding Jammu and Kashmir

Source : (*i*) Census of India 2001

(*ii*) Census of India 2011 Report released on C.D. in 2014.

It is clear from the above description that the highest concentration of Scheduled Castes is found in the alluvial plains of north India because they find ready employment as agricultural labourers. They are also found in large numbers in the delta plains of South India. In contrast, most of the north-eastern states and large parts of Jammu and Kashmir have very low concentration of Scheduled Castes.

In terms of the percentage of Scheduled Caste population to the total population, Punjab with 31.94 per cent tops the list. Other states with more than one-fifth of population being Scheduled Caste are Himachal Pradesh (25.19%), West Bengal (23.51%) and Uttar Pradesh (20.69%). As many as 10 states and union territories have 15 to 20 per cent of their population termed as Scheduled Castes. These are in descending order, Uttarakhand, Chandigarh, Rajasthan, Tripura, Karnataka, Odisha, NCT of Delhi, Andhra Pradesh (including Telangana), Tamil Nadu and Madhya Pradesh. In Jharkhand, Chhattisgarh and Maharashtra, the proportion of Scheduled Caste population to total population varies from 10 to 15 per cent. Kerala, Jammu and Kashmir, Gujarat, Assam and Sikkim have 5 to 10 per cent of the population as Scheduled Castes. Daman and Diu, Manipur, Dadra and Nagar Haveli, Goa, Arunachal Pradesh, and Meghalaya have less than five per cent Scheduled Caste population. Scheduled Castes are negligibly small in Mizoram, no castes have been scheduled for Arunachal Pradesh, Nagaland, Lakshadweep and Andaman and Nicobar Islands (**Fig. 11.1**).

Keeping in view the above description regarding the distribution of Scheduled Caste population, following two zones of high concentration are recognised :

(*a*) The Indo-Gangetic Plains. The Indo-Gangetic Plains are composed of rich alluvial soils. Water supply, suitable climate and weather provide opportunities for agricultural labourers to settle in the alluvial plains. Consequently, the Indo-Gangetic Plains are the zones of high concentration of scheduled castes who are primarily agricultural labourers.

(*b*) The East Coastal Plains. The East Coastal plains provide identical opportunities for the settlement of agricultural communities as they are available in the Indo-Gangetic Plains of the north. Hence, the scheduled castes are found concentrated on eastern coasts particularly in coastal areas of Tamil Nadu, Andhra Pradesh, and Odisha.

TRIBAL POPULATION

The tribes are the autochthonous or native people of the land who are believed to be the earliest settlers in the Indian Peninsula. They are generally called *adivasis,* implying original inhabitants. The ancient and medieval literature mention a large number of tribes living in India. Before the introduction of the caste system during the Brahminic Age, people were divided into various tribes. A tribe was a homogeneous and self contained unit without any hierarchical discrimination.

Growth of Tribal Population

The demographic study of tribal population has suffered seriously due to the adoption of arbitrary criteria for 'scheduling' the tribes. After the partition of the country in 1947, the tribal population, as determined by the Constitutional (Determination of Population) Order 1950 was reduced from 24.7 million in 1941 to about 17.9 million. The 1951 census, which was the first census after Independence, recorded a total tribal population of 1,91,19,054. This number of tribal population was according to the terms of the *Scheduled Tribes Order* of 1950. This led to strong resentment among certain tribes resulting in stresses and strains in the Indian polity. In pursuance of the 'Modification Order' of 1956, another 34,00,000 persons, belonging to tribal groups and left out earlier, were added and the number rose to 2,25,19,054. This accounted for 6.23 per cent of the total population of the country at that

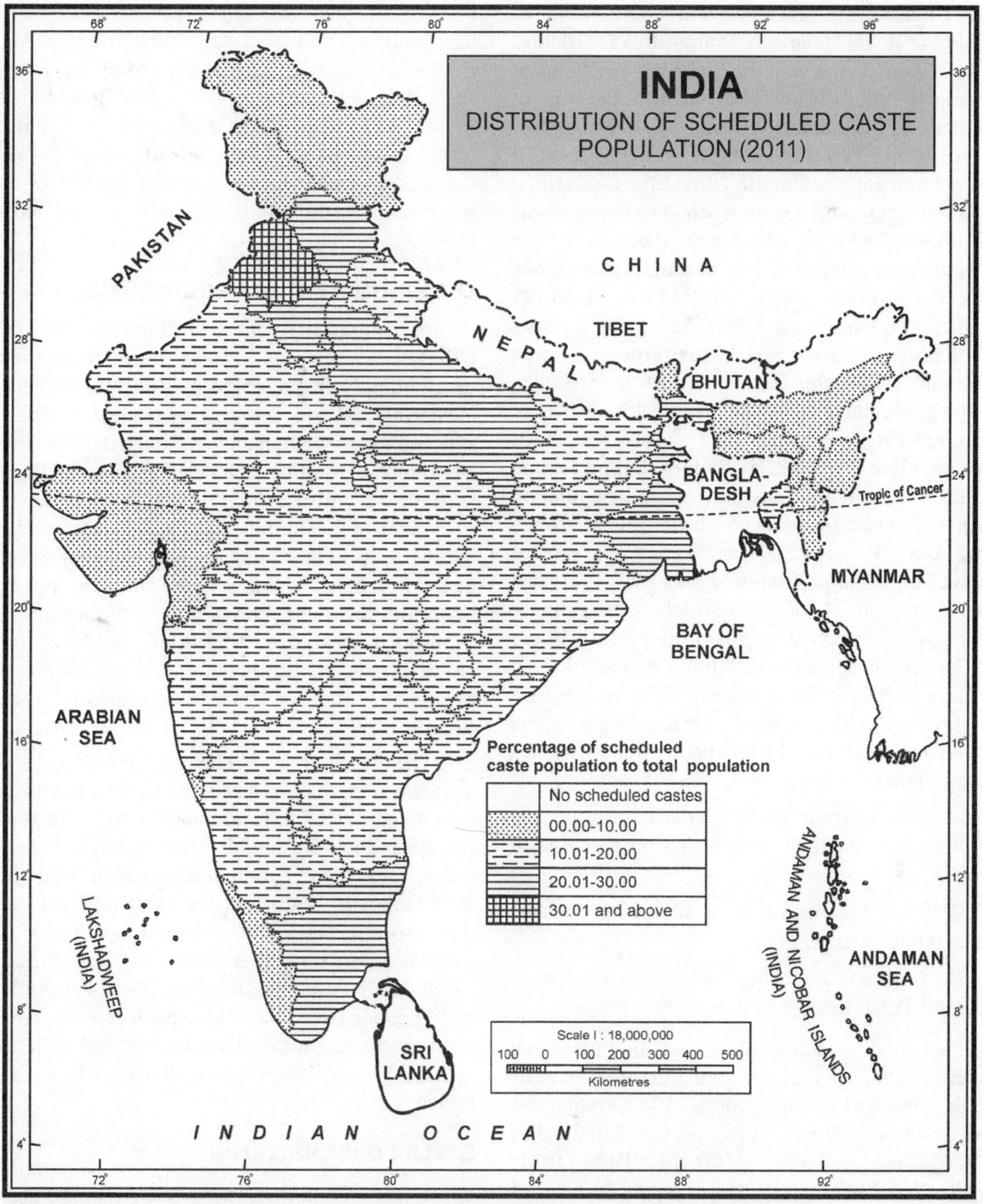

FIG. 11.1. India : Distribution of Scheduled Caste Population (2011)

time. According to 1961 census, there were 3,01,73,998 persons belonging to Scheduled Tribes accounting for 6.87 per cent of the total population of the country. The situation further changed during the 1961-71 inter censal period. Several communities of Uttar Pradesh, *viz*. Jaunsari, Tharus, Buxas, Bhutias and Rajis were included in the 'Schedule' in terms of Presidential Order of June 1967. Thus at the time of 1971 census, the number of Scheduled Tribes had risen to 3,80,15,162 accounting for 6.94 per cent of India's total population. In 1981, the total tribal population (excluding Assam where 1981 census was not held) was 5.38 crore which worked out to 7.58 per cent of the total population. The decennial growth rate during 1971-81 was abnormally high at 41.56 per cent. This was caused by the amendment in the lists of Scheduled Tribes in 1976 removing area restrictions. During this decade, the list of Scheduled Tribes was promulgated by the President in respect of Sikkim (22-6-1978). As per 1991 census, Scheduled Tribes population was 6,77,58,380 (excluding Jammu and Kashmir) and its share in the total population worked out to be 8.08 per cent. The decennial growth during 1981-91 was 25.91 per cent. The rate of growth of the tribal population during 1991-2001 had been recorded at 24.34 per cent, which was higher than the general population growth rate of 21.54 per cent. The tribal population of India increased from 8.43 crore in 2001 to 10.43 crore in 2011 registering a growth rate of 23.72 per cent during the decade. Table 11.3 shows the growth of tribal population as well as its share in the total population.

It is obvious from the above discussion that the growth of Scheduled Tribes population was due to following two reasons :

(*i*) There has been a rapid natural growth of tribal population and

(*ii*) Additions have been made to the list of Scheduled Tribes time and again.

There are divergent opinions regarding the actual number of tribal communities living in different parts of India. The 1961 census recorded 354 communities based on the scheduled castes and scheduled tribes lists (Modification Order, 1956), **N.K. Bose** (1971) and **B.K. Roy Burman** (1972) had put the number of Scheduled Tribes at 427 and 450 respectively. The Census of India had given the number of scheduled tribes as 573 in its paper-2 of 1992 of series 1. But a caution has been given that it should not be taken as the total number of tribes, as a tribe might have been notified in different states and union territories and counted as tribe more than once. For example, Bhil has been notified as a scheduled tribe in Andhra Pradesh, Gujarat, Karnataka, Madhya Pradesh, Maharashtra, Rajasthan and Tripura. Thus in the total of 573, Bhil has been added seven times, once in each of the seven states. Hence, this total does not reflect the actual number of tribes in the country. A fairly accurate estimate puts the number of tribes in India at 365.

Distribution of Scheduled Tribes

An appraisal of the distribution pattern of the tribal communities shows that their spatial distribution is characterised by a striking tendency of clustering and concentrating in pockets, which have suffered from isolation and are situated in areas where environmental setting is, by and large, not suitable for settled agriculture. Thus, most of the tribal communities live in hilly and forested tracts and other remote areas of the country. Constrained by the rigors of environment, which fostered physical and social isolation for ages, the tribal communities have developed their own traditional mode of living. However, the interaction between tribal and non-tribal people after Independence has changed the scenario to some extent.

The distribution of Scheduled Tribes is markedly different from that of Scheduled Castes. For example, Scheduled Tribes have a tendency to concentrate in remote and less hospitable areas, whereas Scheduled Castes show a very high concentration in the fertile Indo-Gangetic Plain and the Coastal Plains where they work as agricultural labourers. Further, it is interesting to note that while no caste has been scheduled in Arunachal Pradesh, Nagaland, Andaman and Nicobar Islands and Lakshadweep, no tribe has been scheduled in the states of Haryana and Punjab and the Union Territories of Chandigarh, Delhi and Puducherry.

State Level Patterns

There are wide variations in the state level distribution of tribal population. While, on one hand, no tribes have been scheduled in Punjab and Haryana, and the union territories of Delhi, Chandigarh and Puducherry, as much as 94.43 per cent of the total population in Mizoram and 94.79 per cent in Lakshadweep belongs to Scheduled Tribes. The other States/Union Territories with predominantly Scheduled Tribes population are : Nagaland (86.48), Meghalaya, (86.15), Arunachal Pradesh (68.79). Manipur, Chhattisgarh, Tripura and Sikkim also have substantial proportion of population as Scheduled Tribes where more than 30% of the population

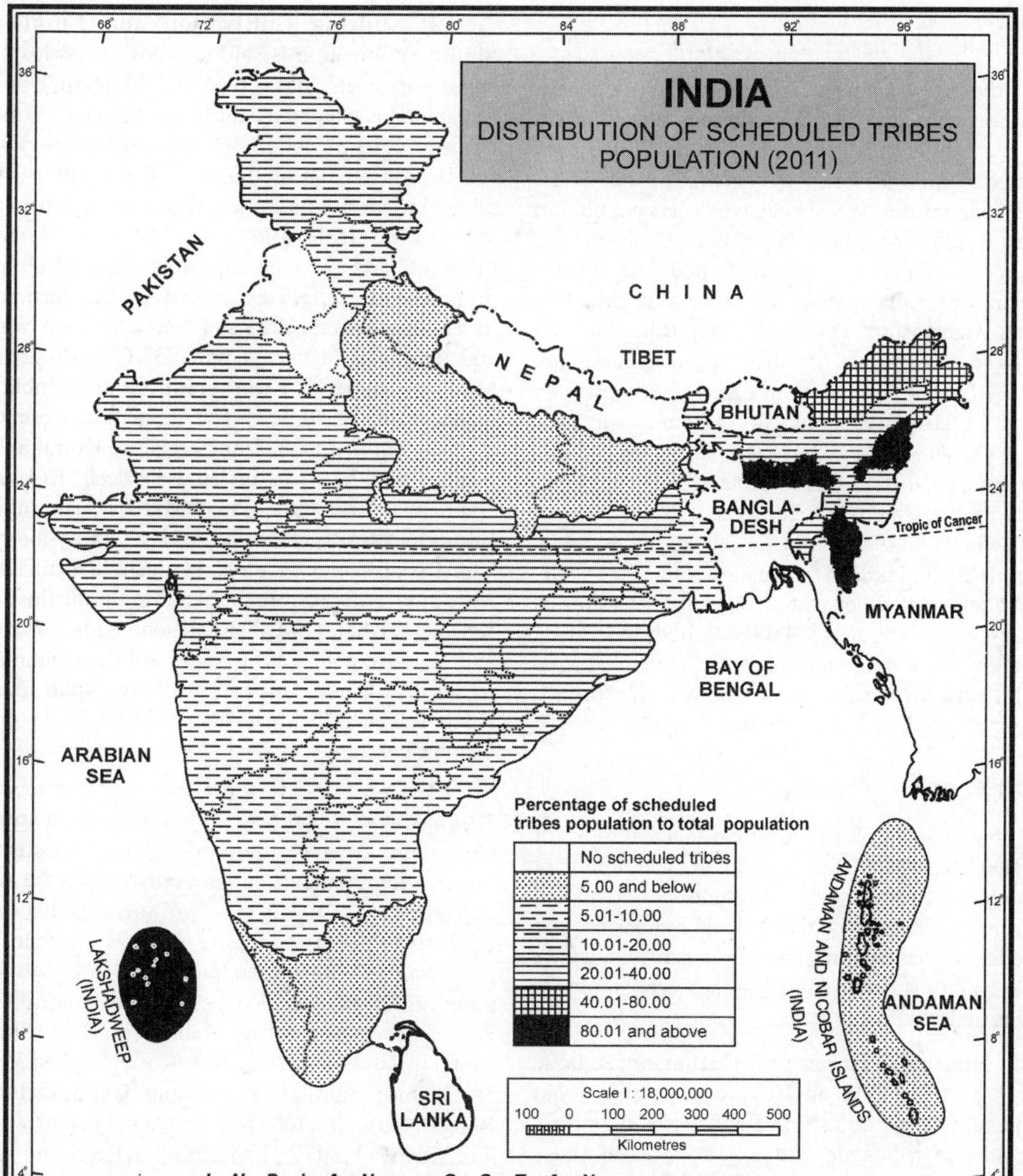

FIG. 11.2. India : Distribution of Scheduled Tribes Population

consists of Scheduled Tribes. Among the 15 major states, Chhattisgarh has the largest proportion of Scheduled Tribe population of 30.62 percent followed by Jharkhand 26.21 per cent Odisha (23.85%) and Madhya Pradesh (21.09%) Gujarat, Assam, Rajasthan, Jammu and Kashmir and Goa are the other five major States in which more than 10 per cent of the population belongs to Scheduled Tribes.

Of the total Scheduled Tribes population of about 10.43 crores counted in different States/UTs in India, 14.68 per cent is in Madhya Pradesh, 10.08 per cent in Maharashtra and 9.12 per cent in Odisha. Thus, roughly one-third of the Scheduled Tribes population of India lives in these three states only. Going a little further, 71 per cent of the Scheduled Tribes population of India lives in six states, viz. Madhya Pradesh, Maharashtra, Odisha, Gujarat, Rajasthan and Jharkhand. In terms of absolute numbers, the Scheduled Tribes population was 153.17 lakhs in Madhya Pradesh, 105.1 lakhs in Maharashtra, 95.9 lakhs in Odisha, 92.38 lakhs in Rajasthan, 89.17 lakhs in Gujarat, 86.46 lakhs in Jharkhand 54.18 lakhs in Andhra Pradesh (including Telangana), 52.96 lakhs in West Bengal, 42.49 lakhs in Karnataka, 38.84 lakhs in Assam, 25.56 lakhs in Meghalaya and 17.11 lakhs in Nagaland. All other states and Union territories together have about 45 lakhs Scheduled Tribe population; there being absolutely no scheduled tribe population recognised in Haryana, Punjab, Chandigarh, Delhi and Puducherry.

Tribal Economy

Tribal economy forms an important criterion for classifying Scheduled Tribes in India. The dominant economies of the tribes are : (1) Hunting, fishing and gathering, (2) Shifting cultivation and lumbering, and (3) Sedentary cultivation and animal husbandry. A brief account of tribes practising these economies is given as under :

(1) Hunting, Fishing and Gathering. A large number of tribes live in isolation in forests and depend upon hunting, fishing and food gathering. Some of the tribes depend exclusively on these occupations. The main tribes which practise these professions are the Raji in Uttar Pradesh; Kharia, Birhor, Korwa, Pariha and Birgias in Jharkhand; Kuki in West Bengal; Hill-Maria in Chhattisgarh, Juang in Odisha, Chenchu and Yanadi in Andhra Pradesh and Telangana; Koya, Reddi, Kadar and Paliyan in Tamil Nadu, Bhil, Garasia in Maharashtra and Gujarat; Bhil, Garasia and Sahariya in Rajasthan and Kuki, Konyak and Naga in Assam, Meghalaya, Nagaland and Arunachal Pradesh.

(2) Shifting Cultivation and Lumbering. Shifting cultivation is said to be as old as history of agriculture itself. In this type of cultivation, a piece of forest land is cleared by slash-and-burn technique and crops are grown. After 2-3 years, the fertility of the soil is reduced and the farmer shifts to another piece of land. This is the reason that it is called shifting agriculture. It is known by different names in different parts of the country. It is called *Jhum* or *jum* in north-east India, *kumari* in Western Ghats, *watra* in south-east Rajasthan and *penda, bewar* or *dahia* and *deppa* in different parts of Chhattisgarh and Madhya Pradesh. **Lumbering** involves obtaining wood from the forests. The main tribes practising shifting cultivation and lumbering are Koria, Saharia, Bhutias and Kharwar in Uttar Pradesh, Korwa and Asur in Jharkhand; Garo, Mal-Paharia in West Bengal; Maria, Gonda, Baiga and Dhora in Chhattisgarh and Madhya Pradesh; Saora, Khond, Kurumba and Bagola in Andhra Pradesh; Saora, Keria and Khond in Odisha; Khond, Gond, Kurumba and Muduwan in Tamil Nadu; Bhil and Garasia in Maharashtra and Gujarat; Kathodia in Rajasthan, Naga, Lakher, Chakmas, Garo, Riang, Notia, etc., in the north eastern states.

(3) Sedentary Cultivation and Animal Husbandry. Sedentary cultivation is a type of agriculture in which the farmer grows crops to meet his own requirements and not much is left for sale in the market. This is not an advanced type of cultivation and is generally practised by the tribal people along with animal husbandry. The main tribes adopting these professions are the Tharu, Maghi Khasa, Bhoksa, Kol and Bhotias in Uttar Pradesh; Munda, Ho, Oraon, Tamaria, Korwa and Santhal in Jharkhand; Santhal, Polia, and Bhumji in West Bengal; Parja, Bhatra, Baija and Gond in Chhattisgarh and Madhya Pradesh, Badaga, Irula, Parga and Malydi in Tamil Nadu, Badaga, Koya, Irula and Kota in Andhra Pradesh and Telangana; Bhil, Dubla, Raiwari, Barali, Koli, Dhamalia, etc. in Maharashtra and Gujarat, and Bhil, Garasiya and Meena in Rajasthan.

LANGUAGE AND DIALECT GROUPS

India is a land of vastness and continuity. It is now certain that the inhabitants of the country are not her original people. They entered India in different spans of time and got settled here. Most of them belong to the Asian parts—Central, Eastern and Western. It is natural that differences and variations exist in their languages and dialects owing to their coming into India from different parts of Asia. After coming into India, cultural mixing has taken place among various races and it led to the mixing of their languages and dialects to a great extent. Despite all this, people of different races and classes live in different parts of the country and they speak different languages and dialects.

Classification of Indian Languages

People of India speak a large number of languages which are broadly divided into the following four families :

1. Indo-European Family (Arya),
2. Dravidian Family (Dravida),
3. Austric Family (Nishada), and
4. Sino Tibetan Family (Kirata).

The above classification is based upon the number of people speaking each family of languages. The first is the Aryan family which is numerically and also culturally, the most important in India. About 73% of the Indian population speaks different languages of the Aryan family. Next comes the Dravidian family which is spoken by about 20 per cent of the Indian population. The Austric and the Sino-Tibetan languages are spoken by small percentage of people.

1. The Aryan Languages

This is the most important of all the families of languages and spoken by a little less than three fourths of the Indian population. The Aryan languages are divided into following two main branches :

(*i*) The Dardic Aryan Languages and
(*ii*) The Indo Aryan Languages

(*i*) The Dardic Aryan Languages. This group comprises a number of languages which are current among very small mountain communities in Kashmir. Out of India, it is spoken by small communities living on the frontier between Pakistan and Afghanistan. The Dardic languages fall into three branches : (*a*) Shina including Kashmiri, Shina proper and Kohistani; (*b*) Khowar or Chatrari or Chitrali and (*c*) Kafiristan (or Nuristani) dialects.

In Kashmir, there is Shina and Kashmiri, and some dialects allied to Kashmiri. Kashmiri appears to be based on a Dardic Aryan dialect. But it has been profoundly influenced by Sanskrit and the Prakrits from the very early times. Many scholars are of the opinion that Kashmiri is Indo-Aryan rather than Dardic. Most scholars consider Dardic to be just a branch of Indo-Aryan. These Dardic dialects are largely on the way to extinction. Kashmiri, however, is one of the recognised national languages of the Indian Union. Except Kashmiri, which is spoken by more than 20 lakh people, no other language of the Dardic Aryan languages is spoken by more than 7 thousand people.

(*ii*) The Indo-Aryan Languages. This is the second sub-group of the Aryan languages in which Hindi, Bengali, Punjabi, Rajasthani, Gujarati, Sindhi, Kachchi, Marathi, Oriya, Sanskrit, Assamese and Urdu are included. Based upon the regional distribution of the people speaking these languages, they are further grouped as under :

(*a*) *Northern Aryan Languages.* Languages of this group belong to the dialects spoken by the hill people in North India. They include Nepali, Central Pahari and Western Pahari Aryan languages.

(*b*) *North-Western Aryan Languages.* Khanda, Kachchi and Sindhi are the well-known Aryan languages which are spoken by the people living in the north-western part of the country.

(*c*) *Southern Aryan Languages.* Marathi and Konkani are the languages included in the Southern group of Aryan languages.

(*d*) *Eastern Aryan Languages.* The region of these languages lies in the eastern parts of the country. Bihari, Oriya, Bengali and Assamese languages constitute this group of Aryan languages.

(*e*) *East Central Aryan Languages.* Avadh, Bundelkhand and Chhattisgarh regions include these languages as the languages of the people living there. Avadhi, Bugheli and Chhattisgarhi are their languages.

(f) Central Aryan Languages. The central region of India is the region of Central Aryan languages. The major languages of this region are Hindi, Punjabi, Rajasthani and Alawari.

Among the Aryan languages, Hindi is the most important language, spoken by a large percentage people of the country. In every Indian state, Hindi-speaking people are commonly found. Even the illiterate people can speak and understand Hindi. This is the reason that Hindi has been given the status of the National Language the status of a national language.

2. Dravidian Languages

Dravidian languages are older than the Aryan languages. According to an estimate, Dravidians entered India much before the Aryans. Other estimates indicate that they are the original inhabitants of the country, who were driven away towards south by the Aryans at a later stage. Today, the Dravidian languages form a well knit family by themselves and unlike the Aryan, the Austric and the Sino-Tibetan languages, they have no relations outside the Indian subcontinent. The Dravidian languages fall into several groups. Two major groups are as under :

(*i*) The North Dravidian Languages. Telugu and a number of other languages such as various Gondi dialects, Kuruth or Oraon, Maler or Malpahariya, Kui or Kandh, Parji, Kolami and a few others are included in this group. Telugu is numerically the most important of all the Dravidian languages and has a very rich literature. This language has spread outside India also—in Myanmar, Indo-China and South Africa. It is usually called the *Italian of the East* by its admirers. Its vocabulary is much influenced by Sanskrit.

(*ii*) South Dravidian Languages. This group of languages includes Tamil, Kannada and Malayalam. A number of languages like Tulu, Kota, Kurgi (or Kedagu) and Toda are also included in this group. Tamil is spoken in large parts of Tamil Nadu. Outside India, it is spoken by a large number of people in Sri Lanka. This language has preserved the old Dravidian spirit in its original form to a great extent. Tamil literature goes back to many centuries before Christ. Tamil presents certain new literary types which are not found in Sanskrit and other Aryan languages. While it includes extensive Sanskrit element, this language has retained the purity of its Dravidian vocabulary to a much greater extent than any other cultivated Dravidian language.

Malyalam is currently the language of Kerala and Lakshadweep. It had its origin in the old Tamil about 1,500 years ago. The Old Tamil speech, started showing simplifications as early as 10th century A.D. Then it followed its own path away from its sister dialects. The speech of Kerala developed independently and became transformed into Malyalam. The first Malayalam writings are said to go back into the period from 13th century to 15th century, when it was established as an independent language. Malayalam has been influenced by Sanskrit more than any other language of India.

Kannada is the main language of the present Karnataka state. The literary cultivation of this language began from the middle of the first millennium A.D. Kannada has passed through three stages : (*a*) Old Kannada upto 13th century (*b*) Medieval Kannada upto 16th century and (*c*) Hosa Kannada which is the language of the present day.

3. Austric Languages

The Austric languages of India belong to the Austro-Asiatic sub-family. This category is further sub-divided into Munda and Mon-Khmer.

(*i*) Munda or Kol Languages. Munda languages are the largest of the Austric group of languages. They consist of fourteen tribal languages. The Kherwari is the major group, which is current in Eastern India (Chota Nagpur, Odisha, Chhattisgarh and West Bengal) and includes Santhali, Mundari, Ho, Birhor, Bhumiej, Korwa and Korku (or Kurku). Santhali, Mundari, and Ho languages have a noteworthy literature preserved orally, consisting of songs and mythological romantic stories.

(*ii*) Mon-Khmer Languages. Mon-Khmer group of Austric languages has two sub-groups—Khasi and Nicobari. Khasi languages are spoken by Khasi tribal people of Meghalaya, while Nicobari languages are the languages of the tribal people of the Nicobar islands. Khasi used to be written in Bengali-Assamese script about a century ago. Through the influence of Welsh Methodist missionaries, the Roman alphabet has been adopted for Khasi and some literature has been produced.

4. Sino-Tibetan Languages

The Sino-Tibetan languages are spoken by a variety of people. Depending upon the region of settlement, these languages are put into several groups and sub-groups. Sino-Tibetan languages have three major sub-divisions :

(*i*) The Tibeto-Himalayan.

(*ii*) The North-Assamese

(*iii*) The Assam-Myanmari (Burmese)

(*i*) The Tibeto-Himalayan Languages. This sub-division of the Sino-Tibetan group of languages is further sub-divided as the Himalayan group and the Bhutia group.

(*a*) *The Himalayan Group.* The Himalayan group consists of 4 languages. They are Chamba, Lahauli, Kinnauri and Lepcha. Kinnauri is the most widely spoken language of the Himalayan group.

(*b*) *The Bhutia Group.* Tibetan, Balti, Ladakhi, Lahauli, Sherpa and Sikkim Bhutia are included in the Bhutia group of Sino-Tibetan languages. Ladakhi has largest number of Bhutia speakers. It is followed by Sikkim Bhutia and the Tibetan languages in that order.

(*ii*) North Assam Languages. The North Assam branch of languages of the Sino-Tibetan group is also called the *Arunachal* branch. It consists of six languages, such as Aka, Daflta, Abor, Miri, Mishnil and Mishing. Largest number of people speak Miri language.

(*iii*) The Assam Myanmari Languages. This group of languages includes Boro or Bodo, Naga, Cochin, Kukichin and Myanmar groups. Naga is the largest spoken language of this group.

Besides these, the Sino-Tibetan group of languages has some other important languages. They are Manipuri, Garo, Tripuri, Mikir and Lusai. Lusai is also termed as Mizo.

Linguistic Regions. It is said that India is a (veritable) forest of languages. In the Linguistic Survey conducted during the British period it was concluded that there were 179 languages and 544 dialects in this region (Linguistic Survey of India, 1903-1928). The principal credit for this significant piece of work is given to its editor-in-chief Sir George A. Grierson. In this survey of modern Indian languages, he classified them into language families along historical (comparative) lines. Out of a total of 179 languages mentioned in the Linguistic Survey, 116 are small tribal speeches. They are spoken only on the northern and north-eastern fringes of India and are current among less than one per cent of the total population of the country. Nearly two dozen more are, likewise, insignificant speeches of other language groups; or they are languages not truly belonging to India.

The most comprehensive data on languages was collected at the time of 1961 Census. According to these census figures, there were 187 languages spoken in India. Out of these, as many as 94 languages are spoken by less than 10,000 persons each and 23 languages together account for 77 per cent of the total population of the country. Of these 23 languages, 15 languages in addition to English have been specified in the Eighth Schedule of the Constitution of India. Three more languages have been added to Eighth Schedule by a Parliamentary Act on 20th August, 1992. These languages are Nepali, Konkani and Manipuri. Later on, Maithili, Santhali, Dogri and Bodo were also included in the Schedule, making a total of 22 languages. The percentage of people speaking these languages is given in **table 11.4.**

It is worth noting that among all official languages, Hindi grew at the fastest rate. In 1971 about 36.9% of India's total population identified Hindi as their mother tongue. This percentage grew to 38.7% in 1981, 39.3% in 1991, 41.03% in 2001 and 43.6% in 2011. Further it may be noted that Hindi has about 60 regional dialects. Official Hindi which is based on *Khariboli* dialect is used by 61.0% of the Hindi speaking people. The percentage of Hindi speaking people using different dialects is Bhojpuri (9.6%), Rajasthani (4.9%), Chhattisgarhi (3.1%), Magadhi/Magahi (2.4%), Haryanvi (1.9%), Khortha/Khotta (1.5%), Marwari (1.5%), Bundeli/Bundel Khandi (1.1%), and Malvi (1.0%). The remaining 12% of Hindi speaking people speak other dialects.

Speakers of certain other languages left out of the Eighth schedule are demanding the inclusion of their languages in the list of the languages. Some of the important languages of this category are Sindhi, Rajasthani, Khasi and Gondi.

Language became a very important basis for the formation of states in India after Independence,

TABLE 11.4. Schedule Languages in Descending Order of Speakers' Strength-2011

Language	Percentage to total population 2011
1. Hindi	43.63
2. Bengali	8.03
3. Marathi	6.86
4. Telugu	6.70
5. Tamil	5.70
6. Gujarati	4.58
7. Urdu	4.19
8. Kannada	3.61
9. Odia	3.10
10. Malayalam	2.88
11. Punjabi	2.74
12. Assamese	1.26
13. Maithili	1.12
14. Santhali	0.61
15. Kashmiri	0.56
16. Nepali	0.24
17. Sindhi	0.23
18. Dogri	0.21
19. Konkani	0.19
20. Manipuri	0.15
21. Bodo	0.12
22. Sanskrit	N

thereby adding a new political meaning to the geographical distribution of languages. Today, most of the states of the Indian Union have been delimited on the basis of linguistic pattern of languages. However, the languages and dialects of the tribal people living in the north-eastern, eastern and central parts of the country do not fit in any linguistic pattern, as they are varying in number and their speakers live in unspecified areas or regions. In all, there are 12 languages in India, which are widely spoken. Based on these languages, there are twelve linguistic states, of which states delimited from the political point of view are depicted in **Table 11.5** and **Fig. 11.3**.

The relative importance of different languages had undergone drastic change in the historical times. Until the ninth century, Sanskrit was the language for administration and of the cultural elite. Today, Sanskrit is no longer a spoken language, although it is mother of many Indian language. Persian and later Urdu became the court and the administrative language of the Mughal and other Muslim rulers. In the eighteenth century when the Britishers established their supremacy and became the rulers of India, English replaced Persian as the language of courts, administration and higher learning. In the post-independence period, Hindi has emerged as the most popular language and is spoken by majority of people living particularly in north India. This is the official language of several Hindi speaking states like Uttar Pradesh, Uttarakhand, Bihar, Jharkhand, Madhya Pradesh, Chhattisgarh, Haryana, Rajasthan, and Himachal Pradesh. Including its variants like the Bihari, Rajasthani, Marwari, Magadhi, Chhattisgarhi, Maithili, Pahari, Bhojpuri, etc. Hindi ranks numerically third most widely spoken language of the world after Mandarin Chinese and English. As per 2011 Census figures, Hindi along with its recognised regional variants is spoken by 528 million Indians which is about 43.63% of the total population. The other major languages are Bengali (8.03%), Marathi (6.86%), Telugu (6.7%) and Tamil (5.7%). Gujarati, Urdu, Kannada, Malayalam, Oriya, Punjabi and Assamese are spoken by less than 5 per cent but more than one per cent of total population. Kashmiri,

TABLE 11.5. Linguistic Regions and Languages

Linguistic Region	State/U.T.
1. Kashmiri	Valley of Kashmir
2. Punjabi	Punjab and adjoining parts of Haryana
3. Hindi	U.P., Haryana, H.P., M.P., Chhattisgarh, Jharkhand, Uttarakhand, Delhi, Bihar and Rajasthan
4. Bengali	W. Bengal and Parts of Tripura
5. Assamese	Assam and other northeastern states
6. Oriya	Odisha
7. Gujarati	Gujarat
8. Marathi	Maharashtra, Goa
9. Kannada	Karnataka
10. Telugu	Andhra Pradesh and Telangana
11. Tamil	Tamil Nadu, Puducherry
12. Malayalam	Kerala, Lakshadweep

Nepali, Sindhi, Konkani, Dogri, Manipuri, Bodo and Sanskrit are spoken by less than one per cent population each.

It is worth mentioning here that although, the present political map of India represents more or less the language map of the country, the state boundaries

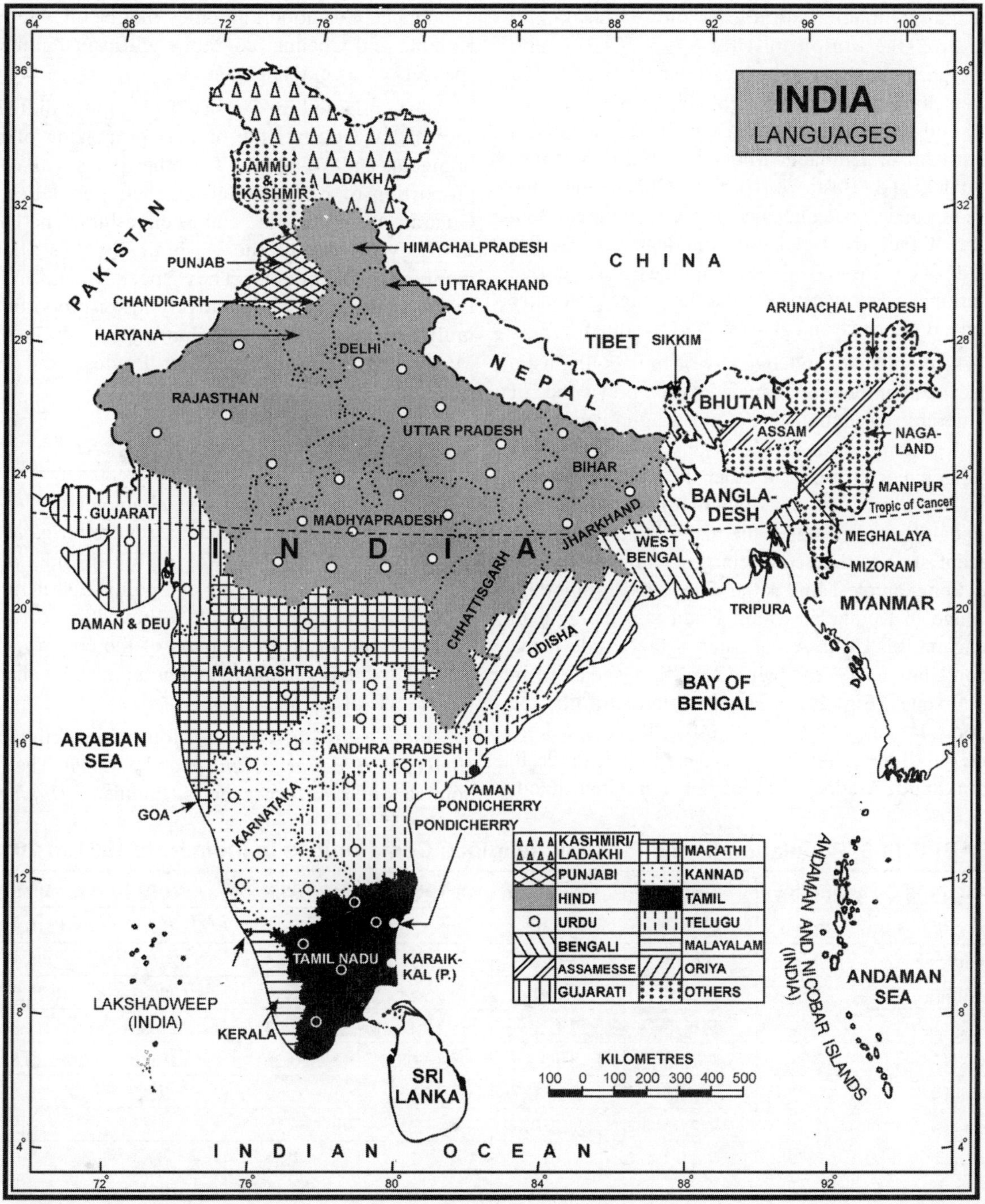

FIG. 11.3. Linguistic Regions

do not always correspond to linguistic boundaries. The example of Hindi belt may be cited here. This belt covers major states of Uttar Pradesh, Bihar, Jharkhand, Rajasthan, Madhya Pradesh, Chhattisgarh, Haryana, Himachal Pradesh and Uttarakhand including the union territories of Delhi and Chandigarh. In these areas, various languages and dialects have all been brought under the aegis of Hindi and without considering the spoken languages in Garhwal or Kumaon, Magadha or South Bihar, Rajasthan, and Bastar district of Chhattisgarh the entire region has been labelled as the Hindi area. As a matter of fact, the linguistic boundary itself is not well defined. Instead of being a sharp line, it is a transitional zone over which one language gradually loses its dominance and gives way to the other.

The complexity and diversity of India's linguistic landscape is further compounded by the prevalence of considerable bilingualism or multilingualism along the inter-state boundaries where two or three languages merge into each other. It is estimated that about 10 per cent of India's population is bilingual or trilingual. In several states, the major language of one adjacent state is the second largest language of the state. For example Tamil is the second most important language in Kerala; in Tamil Nadu it is Telugu; in Telangana it is Urdu. As a matter of fact, Urdu is the "second language" of over 43 million speakers in several states (**Fig. 11.3**). Hindi is claimed to be the "second language" by about 30 million speakers living out of the "Hindi belt". States of Uttar Pradesh, Uttarakhand, Madhya Pradesh, Bihar, Jharkhand, Chhattisgarh, Haryana, Himachal Pradesh and Rajasthan are included in the "Hindi belt". Further, nearly 10 per cent of the Telugu-speakers live outside the Telugu-speaking states of Andhra Pradesh and Telangana. Cosmopolitan cities like Delhi, Mumbai, Kolkata and Chennai contain significant number of speakers of several languages.

Data released by Census 2011 on multiliqualism shows that Goa has the highest proportion of both bilinguals and trilinguals. The other states with a high proportion of multilinguals in their population are Gujarat, Assam, Punjab and Maharashtra. The Union Territory of Jammu and Kashmir also has a larger number of bilinguals. They speak Kashmiri and Dogri. States with comparatively lower proportion of multilinguals include Chhattisgarh, Rajasthan, Madhya Pradesh, Bihar, Uttar Pradesh and West Bengal.

RELIGIOUS COMPOSITION

Religion is a very important characteristic of the Indian population and Indian masses are *religious par excellence*. Religion is a way of life in India and it affects the social, economic and political structure of society. Different religions followed by different people living in different parts of the country, have created diversity of culture and brought about changes in life style of the masses.

India is the birth place of four major religions—*Hinduism*, *Buddhism*, *Jainism* and *Sikhism*. The most dominant religion, however, is Hinduism. *Hindustan*,

TABLE 11.6. Population (in millions) by Religious Communities in India from 1961 to 2011

Religious Communities	1961	1971	1981	1991	2001	2011
All religious communities	439.23	548.16	683.33	846.42	1028.74	1210.19
Hindus	366.53	453.29	549.78	687.65	827.58	996.3
Muslims	46.94	61.42	75.51	101.59	138.19	172.2
Christians	10.73	14.22	16.16	19.64	24.08	27.8
Sikhs	7.85	10.38	13.08	16.26	19.21	20.8
Buddhists	3.26	3.81	4.72	6.39	7.95	8.4
Jains	2.03	2.60	3.21	3.35	4.23	4.5
Others	1.49	2.18	2.71	3.27	6.64	7.9
Religion not stated	0.11	0.36	0.60	0.42	0.73	2.9

Source : Census of India, 2011.

Note : Totals may not tally due to rounding off

the land of the Hindus, is one of the names by which India was known. Hinduism, a religion thousands of years old and whose origin is difficult to trace, evovled out of the varied Indian ways of life, so different and yet in some intangible way unified. The early pre-Vedic Hindu religion got modified in the Vedic period after the middle of the second millennium B.C.

In addition to being the birth place of four major religions of the world, India has embraced other world religions also. For example, Christianity and Islam came to India from other lands. The entry of these two religions in India goes back to almost the first days of their prophets. Syrian Christians appeared on the West Coast of India in the very first century of the Christian Era. The Arab traders brought Islam to the West Coast of India much before the Muslim conquest of this country. The prosecuted Jews and Zoroastrians (Parsis) found sanctuary on the Indian soil. Sikhism appeared on the religious scene of India only about five centuries ago.

There have been large scale changes in the religious composition of population due to conversions from one faith to another. Spatial pattern of distribution of different religious groups had undergone drastic changes due to large scale migrations as a result of partition of India in 1947. Before partition, Hindus accounted for 66.5 per cent of the population of the sub-continent and Muslims 23.7 per cent (census 1941). With partition, large number of Muslims migrated from India to Pakistan and Bangladesh and Hindus migrated to India from these countries. Consequently, the proportion of these two religious communities in the total population changed. The percentage of Hindus rose to 84.1 per cent and that of Muslims fell to 9.8 per cent (census 1951). Since then, the percentage of Hindu population has fallen marginally while that of Muslims has increased considerably. The percentage of Hindus had fallen from 84.4 per cent in 1961 to 79.8 per cent in 2011 while that of Muslims had increased from 9.9 per cent in 1961 to 14.2 in 2011. (*see* **Table 11.7**)

One heartening fact about religious composition of India is that different religious groups have coexisted for the last several centuries even during the troubled times when there have been clashes based on religion. India is the only country in the world where people belonging to different relgious faiths are co-existing in peace and harmony.

1. The Hindus

As mentioned earlier, the Hindus are the preponderant majority in most of the states and union territories of India. According to 2011 census figures there were 966.3 million Hindus which accounted for 79.8 per cent of the total population of the country.

TABLE 11.7. Proportion of Population by Religious Communities, India, 1961-2011

Religious communities	1961	1971	1981	1991	2001	2011
All religious communities	**100.0**	**100.0**	**100.0**	**100.0**	**100.0**	**100.0**
Hindus	84.4	83.5	83.1	82.4	81.4	79.8
Muslims	9.9	10.4	10.9	11.7	12.4	14.2
Christians	2.4	2.6	2.5	2.3	2.3	2.3
Sikhs	1.8	1.9	2.0	2.0	1.9	1.7
Buddhists	0.7	0.7	0.7	0.8	0.8	0.7
Jains	0.5	0.5	0.5	0.4	0.4	0.4
Others	0.3	0.4	0.4	0.4	0.7	0.7
Religion not stated	Neg.	Neg.	Neg.	Neg.	Neg.	0.2

*Excludes Jammu & Kashmir and Assam for all decades from 1961 to 2001.

Notes : 1. The Census 2001 population figures for India and Manipur exclude those of Mao Maram, Paomata and Purul sub-divisions of Senapati district of Manipur.

2. No Census conducted in Assam in 1981 and in Jammu & Kashmir in 1991.

3. Neg.—Negligible.

***Source* :** Census of India, 2011, The First Report on Religion Data Abstract.

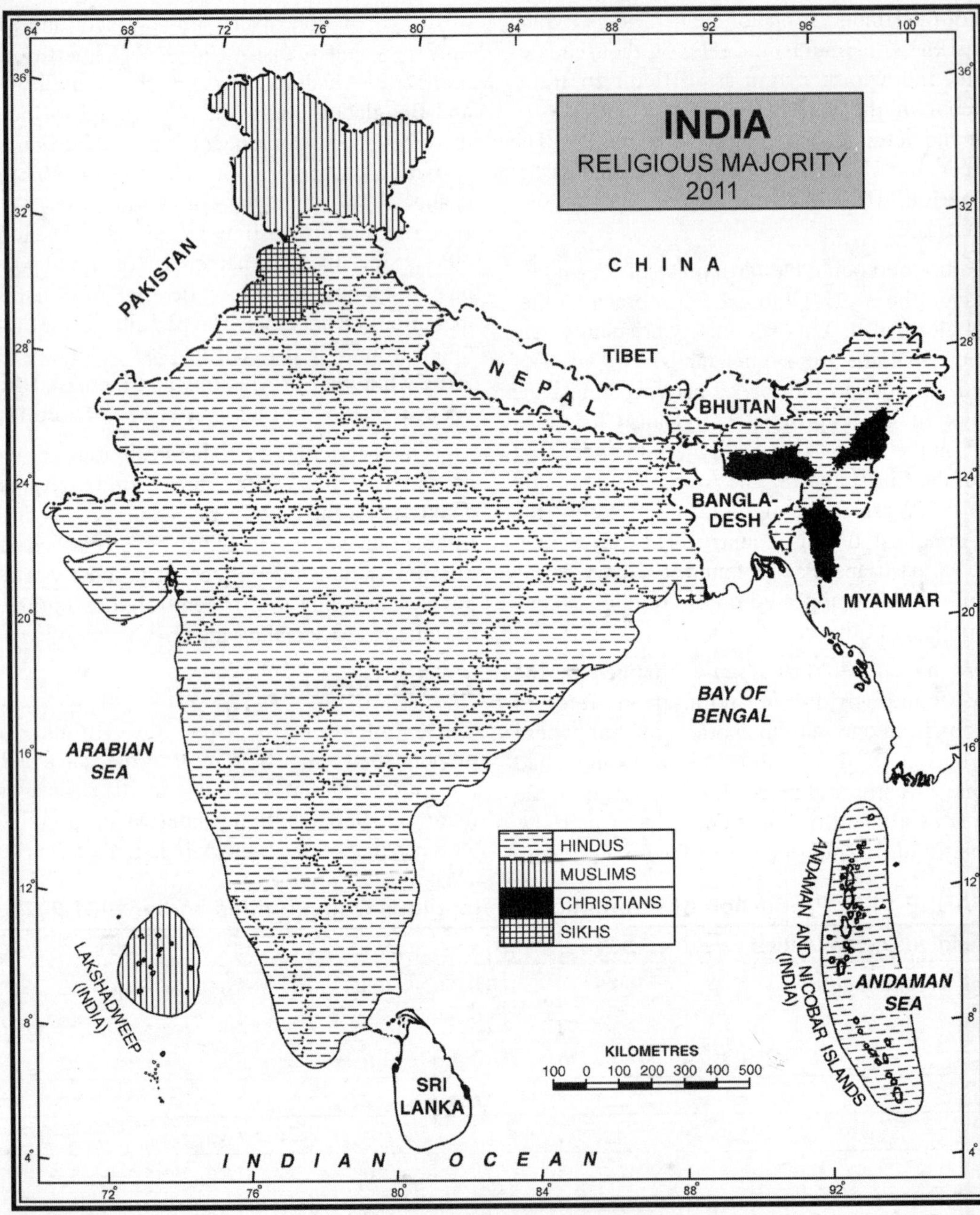

FIG. 11.4. India : Religious Majority

The Hindus of India account for about 14 per cent of the world population ranking below those professing Christianity and about equal with the followers of Islam, but much above other religions. However, the percentage of Hindus to total population of India is decreasing gradually. It came down from 83.5 per cent in 1971 to 79.8 per cent in 2011. This is largely because of comparatively lower growth rate of the Hindus and partly due to conversion of Hindus into other religious faiths.

The growth rate of the Hindus had been declining consistently since 1971. It was 24.2 per cent in

1971-81, 22.8 per cent in 1981-91 and fell down to 17.6 per cent in 2001-11.

The Hindus constitute the majority community in most states and union territories of India. While the Hindus comprised 79.8 per cent of the total population of India in 2011 they were 95.17 per cent in Himachal Pradesh, 93.63% in Odisha, 93.25% in Chhattisgarh and 90.89% in Madhya Pradesh. They comprise of 93.933% in the Union Territory of Dadra and Nagar Haveli (**Table 11.8**). However, the Hindus are in minority in certain states and union territory of Lakshadweep. They are outnumbered by

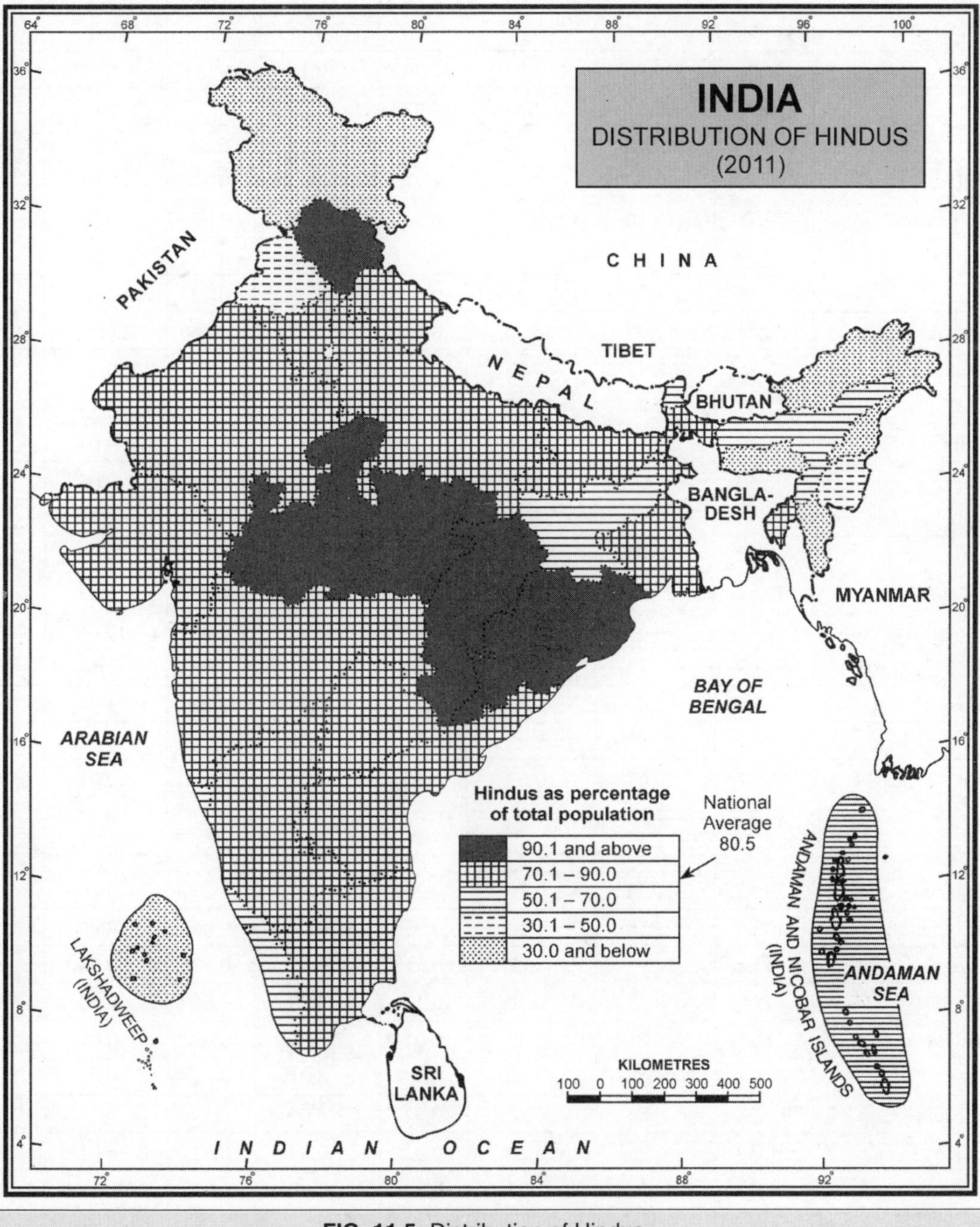

FIG. 11.5. Distribution of Hindus

TABLE 11.8. India : Percentage of Population of Religious Groups

States/Union Territories	Hindus	Muslims	Christians	Sikhs	Buddhists	Jains	Other Religions	Religions not stated
Jammu & Kashmir	28.44	68.31	0.28	1.87	0.90	0.02	0.01	0.16
Himachal Pradesh	95.17	2.18	0.18	1.16	1.15	0.03	0.01	0.12
Punjab	38.49	1.93	1.26	57.69	0.12	1.16	0.04	0.32
Chandigarh	80.78	4.87	0.83	13.11	0.11	0.19	0.02	0.10
Uttarakhand	82.97	13.95	0.37	2.34	0.15	0.09	0.01	0.12
Haryana	87.46	7.03	0.20	4.91	0.03	0.21	0.01	0.17
Delhi	81.68	12.86	0.87	3.40	0.11	0.99	0.01	0.08
Rajasthan	88.49	9.07	0.14	1.27	0.02	0.91	0.01	0.10
Uttar Pradesh	79.73	19.26	0.18	0.32	0.10	0.11	0.01	0.29
Bihar	82.69	16.87	0.12	0.02	0.02	0.02	0.0	0.24
Sikkim	57.76	1.62	9.91	0.31	27.39	0.05	2.67	0.30
Arunachal Pradesh	29.04	1.95	30.26	0.24	11.77	0.06	26.20	0.48
Nagaland	8.75	2.47	87.93	0.10	0.34	0.13	0.16	0.12
Manipur	41.39	8.40	41.29	0.05	0.25	0.06	8.19	0.38
Mizoram	2.75	1.35	87.16	0.03	8.51	0.03	0.07	0.09
Tripura	83.40	8.60	4.32	0.03	3.41	0.02	0.04	0.14
Meghalaya	11.53	4.40	74.59	0.10	0.33	0.02	8.71	0.32
Assam	61.47	34.22	3.74	0.07	0.18	0.08	0.09	0.16
West Bengal	70.54	27.01	0.72	0.07	0.31	0.07	1.03	0.25
Jharkhand	67.83	14.53	4.30	0.22	0.03	0.05	12.84	0.21
Odisha	93.63	2.17	2.77	0.05	0.03	0.02	1.14	0.18
Chhattisgarh	93.25	2.02	1.92	0.27	0.28	0.24	1.94	0.09
Madhya Pradesh	90.89	6.57	0.29	0.21	0.30	0.78	0.83	0.13
Gujarat	88.57	9.67	0.52	0.10	0.05	0.96	0.03	0.10
Daman & Diu	90.50	7.92	1.16	0.07	0.09	0.21	0.03	0.10
Dadra & Nagar Haveli	93.93	3.76	1.49	0.06	0.18	0.35	0.09	0.14
Maharashtra	79.83	11.54	0.96	0.20	5.81	1.25	0.16	0.25
Andhra Pradesh	88.46	9.56	1.34	0.05	0.04	0.06	0.01	0.48
Karnataka	84.00	12.92	1.87	0.05	0.16	0.72	0.20	0.27
Goa	66.08	8.33	25.10	0.10	0.08	0.08	0.02	0.21
Lakshadweep	2.77	96.58	0.49	0.01	0.02	0.02	0.01	0.10
Kerala	54.73	26.56	18.38	0.01	0.01	0.01	0.02	0.26
Tamil Nadu	87.58	5.86	6.12	0.02	0.04	0.11	0.01	0.26
Puducherry	87.30	6.05	6.29	0.02	0.04	0.11	0.01	0.17
Andaman & Nicobar	69.45	8.52	21.28	0.34	0.09	0.01	0.15	0.18

Source : Census of India, 2011.

the Muslims in Jammu and Kashmir and Lakshadweep, by Sikhs in Punjab, by Christians in Arunachal Pradesh, Meghalaya, Mizoram and Nagaland (**Fig. 11.4**). States and Union territories having Hindus less than 10 per cent of their total population are Nagaland (8.75%), Mizoram (2.75%) and Lakshadweep (2.77%).

2. The Muslims

The Muslims constitute the second largest religious community and the largest minority community of India. According to 2011 census figures, the Muslim population numbered 172.2 million which worked out to be 14.2 per cent of the total population of India. The special feature of the Muslim population is that it has shown an overall growth rate of 24.62 per cent during 2001-11. This percentage of growth rate was arrived at by including the Muslim dominated population of Jammu and Kashmir where census could not be conducted in 1991. This growth rate is much higher when compared to growth rate of Hindus. Muslims have rejected family planning vociferously on religious grounds and higher growth rate of the Muslims is the natural sequel.

There was no census in Assam in 1981 and in Jammu and Kashmir in 1991, but heads were counted in the whole country in 1961, 1971, 2001 and 2011. The comparison of the growth rates of Hindus and Muslims during this period makes an interesting study. The Hindu population that was at 453.3 million in 1971 which grew to 966.3 million in 2011 registering a growth rate of 113.17 per cent. The Muslim population grew from 61.4 million in 1971 to 172.2 million in 2011, thus registering an increase of 192.18 per cent. Even after making adjustment for population of Jammu and Kashmir, the growth of the Muslims is the highest as per 2011 census figures. This is a dangerous trend and is likely to disturb the demographic set up of the country. The north-eastern states have registered an exceptionally high growth rate of the Muslim population primarily due to exodus of Bangladesh Muslims into these states.

Regarding percentage of Muslim population to total population, Lakshadweep had the highest percentage of 96.58 per cent Muslims. This is followed by Jammu and Kashmir where 68.31 per cent of the total population are Muslims. The highest concentration of Muslims is found in the Kashmir Valley, where in certain areas the Muslims constitute well over 95 per cent of the total population. Assam has about 34.22 per cent of its population consisting of Muslims. In West Bengal, the Muslims account for 27.01 per cent of the total population. Murshidabad is the main area of concentration, where Muslims comprise over 55 per cent of the total population. Kerala has 26.56 per cent of its population comprised by the Muslims, where Malappuram is the main area of concentration. In Uttar Pradesh 19.26 per cent of the total population is Muslim. Most of the Muslims of Uttar Pradesh are concentrated in Rohilkhand and Upper Ganga plain, where the districts of Rampur, Bijnor, Moradabad, Bareilly, Pilibhit, Saharanpur, Muzaffarnagar, Meerut and Ghaziabad have fairly high proportion of Muslims. Muslims constitute less than five per cent of the total population in Punjab, Chandigarh, Himachal Pradesh, Sikkim, Arunachal Pradesh, Nagaland, Mizoram, Meghalaya, Odisha, Chhattisgarh, and Dadra and Nagar Haveli.

3. The Christians

The Christians form the third largest religious community in India. According to 2011 census figures there are about 27.8 million Christians living in India. Christians formed just 1.6 per cent of Indian population in 1941 but it rose to 2.3 per cent in 1951, perhaps due to partition of the sub-continent. In 1971, there were 2.6 per cent Christians in India which fell slightly to 2.5 per cent in 1981 and to 2.3 per cent in 2011. The north-eastern states are predominantly inhabited by Christians. For example, Christians constitute 87.93 per cent of the total population in Nagaland, 87.16 per cent in Mizoram, 74.79 per cent in Meghalaya and 41.29 per cent in Manipur. Christians are in large proportion in Goa and Kerala also, where they form 25.10 and 18.38 per cent of the total population respectively. But speaking in absolute figures, Kerala has the largest number of Christians, amounting to about one-fourth of the total Christian population of India. In Kottayam and Ernakulam districts of Kerala, Christians account for over 40 per cent of the total population. There are several other districts in this state where the Christian population ranges between 15 to 25 per cent. After Kerala the important states with Christian population are Tamil

Nadu, Arunachal Pradesh, Meghalaya, Andhra Pradesh (1.18 million), Jharkhand and Maharashtra.

4. The Sikhs

There were 20.8 million Sikhs in 2011. Sikhs constituted just 1.7 per cent of the total population of India in 1941. Their percentage increased to 1.9 in 1951 as a result of large scale migration of the Sikhs from Pakistan to India following partition of the country in 1947. Thereafter, the percentage of the Sikhs to total population of India has been increasing steadily. They constituted 1.9 per cent of the total population in 2001 but again fell to 1.7 per cent in 2011, perhaps due to increase in other religious communities. Needless to say that the Sikhs are an enterprising community and are scattered in almost all parts of the country. However, their largest concentration is in Punjab. According to 2011 census figures, Sikhs constitute about 57.7 per cent of the total population of the state. It is estimated that 75.9 per cent of the total Sikhs of the country live in Punjab alone. This is quite obvious because Sikhism took its roots in the soil of Punjab, where Guru Nanak, the founder of the Sikhism preached his teachings. The Sikhs have an absolute majority in the districts of Amritsar, Kapurthala, Ferozepur, Gurdaspur, Bhatinda, Patiala, Ludhiana, Faridkot, Fatehgarh Sahib and Mansa. In the neighbouring state of Haryana 1.2 million Sikhs are living according to 2011 census data. This amounts to 4.9 per cent of the total population of the state. Ambala, Kurukshetra, Karnal, Fatehabad and Sirsa are the main districts of Sikh population. Minor pockets of the Sikh concentration are found in the Tarai region of Uttarakhand and Uttar Pradesh as well as in Ganganagar, Alwar and Bharatpur districts of Rajasthan. In the union territory of Delhi, Sikhs account for 3.4 per cent of its total population.

The most important factor about the Sikhs is that their rate of growth has declined considerably. It was 32.0 per cent in 1961-71 which came down to 26.2 per cent in 1971-81 and 25.5 per cent in 1981-91. A steep decline was observed between 1981-91, 1991-2001 and 2001-11.

5. Buddhists

India's 8.4 million Buddhists constitute only 0.7 per cent of the total population of the country. Their growth rate had been changing in each censual decade. In 1961-71 it was only 17.0 per cent and shot up rather out of proportion to 36.0 per cent in 1981-1991. This was perhaps due to large scale conversion of people from other religions to Buddhism. However, it fell to 18.9 per cent in 2001-11.

The Buddhists are largely found in Maharashtra, Karnataka, Uttar Pradesh, West Bengal, Madhya Pradesh, Arunachal Pradesh and the Union Territory of Ladakh. They are also found to a lesser extent in Mizoram, Tripura and Himachal Pradesh. The concentration of Buddhists in the northern part of India is mainly because it is near here that the Buddhism originated and spread over the Himalayas. Maharashtra has the largest number of Buddhists. The large population of Buddhists in Maharashtra is mostly due to the wholesale conversion of a community of Harijans, following the advice of their leader, Dr. B.R. Ambedkar.

6. The Jains

4.5 million Jains of India are widely spread in the western parts of the country. Maharashtra, Rajasthan, Madhya Pradesh, Gujarat and Uttar Pradesh are important states. Nowhere do they account for more than 1.3 per cent of the total population. The Jains have maintained their growth rate above the national growth rate.

7. The Parsis

As per 2011 census, the Parsi population in the country is a little over 69 thousands. This is clearly a visible but extremely unfortunate decline of a rich civilization of Zoroastrians and its people. It is apparent from 2011 census results that urgent and drastic interventions are required by all concerned including the government and definitely the Parsi community leaders to ensure survival of Parsi population in India. Fertility improvement innovative initiatives rather than fertility control measures adopted by the community so far are possibly the need of the hour before it reaches a point of no return. It is expected that this loud and clear message from 2011 census results awaken the country and the Parsi community from the deep slumber it is possibly in and have a beneficial effect for them. About 90 per cent of this community is concentrated in the city of

Mumbai and the southern coastal Gujarat, around Surat.

SEX COMPOSITION

Sex composition of the human population is one of the basic demographic characteristics, which is extremely vital for any meaningful demographic analysis. Changes in sex composition largely reflect the underlying socio-economic and cultural pattern of a society in different ways. It is an important social indicator to measure the extent of prevailing equity between males and females at a given point of time. "The separate data for males and females are important for various types of planning and for the analysis of other demographic characteristics such as natality, mortality, migration, marital status, economic characteristics, etc. The balance of sexes affects the social and economic relationship within a community". (Chanana : 2012)

Sex composition is expressed with the help of a ratio known as sex ratio. Sex ratio in India is defined as "*number of females per 1,000 males in the population.*" It is expressed in the following form.

$$\text{Sex Ratio} = \frac{\text{Number of females}}{\text{Number of males}} \times 1000$$

Thus, a sex ratio of 1,000 implies complete parity between the two sexes. Ratios above 1,000 indicate excess of females over males; those below 1,000 indicate a deficit of females.

Sex ratio of population of a country is mainly the outcome of the interplay of sex differentials in mortality, sex selective migration, sex ratio at birth and at times the sex differential in population enumeration. According to figures of 2011 census, out of total population of 1210.1 millions, 623.7 millions are males and 586.4 millions are females. Thus, the overall sex ratio for Indian population is 940. This suggests that the number of females is quite less as compared to males. In other words the sex ratio in the country had always remained unfavourable to females. Moreover, barring some hiccups, it has shown a long term declining trend. The sex ratio at the beginning of the twentieth century was 972 and thereafter showed continuous decline until 1941. In 1951 there was a marginal increase of one point, but the country saw the sharpest decline of 11 points in sex ratio from 941 in 1961 to 930 in 1971. A slight improvement of 4 points in 1981 could not be maintained and there was a fall of 7 points from 934 in 1981 to 927 in 1991. Increase of six points from 927 in 1991 to 933 in 2001 and seven points in 2011 is a welcome improvement and it is hoped that the same trend will continue in the coming decades. However, it may be mentioned that the net deficit of females which was 3.2 million in 1901 has now widened to over 37.3 million at the time of 2011 census. **Table 11.9** and **Fig. 11.6** show the trends in sex ratio in India from 1901 to 2011.

Sex ratio in India is substantially low as compared to Russian Federation (1,167), Japan (1,055), Brazil (1,042), the USA (1,025), Nigeria (1,016) and Indonesia (988). The world sex ratio declined from 986 in 2001 to 984 in 2011.

Following are some important factors responsible for low and declining sex ratio.

1. More males are born than females. This is almost a worldwide phenomena and India is no exception. In a study of about two million births that took place in hospitals and health centres throughout India during 1949-58, the sex ratio at birth was found to be 942 for the country as a whole. The data collected by the Registrar General, India regarding six million live births that took place during 1981-91 in hospitals, health centres and institutions located mostly in urban areas in various parts of the country, gave the sex ratio at birth at 891. Thus there has been a drastic decline in the sex ratio at birth from the 1950s to 1980s.

According to the findings of the census of India, the imbalance in the number of males and females starts in the beginning. It is now a well established law of nature that the males exceed females at the time of birth. It is believed that 943-952 female births

TABLE 11.9. Sex Ratio-India : 1901–2011

Census Year	Sex Ratio	Census Year	Sex Ratio
1901	972	1961	941
1911	964	1971	930
1921	955	1981	934
1931	950	1991	927
1941	945	2001	933
1951	946	2011	940

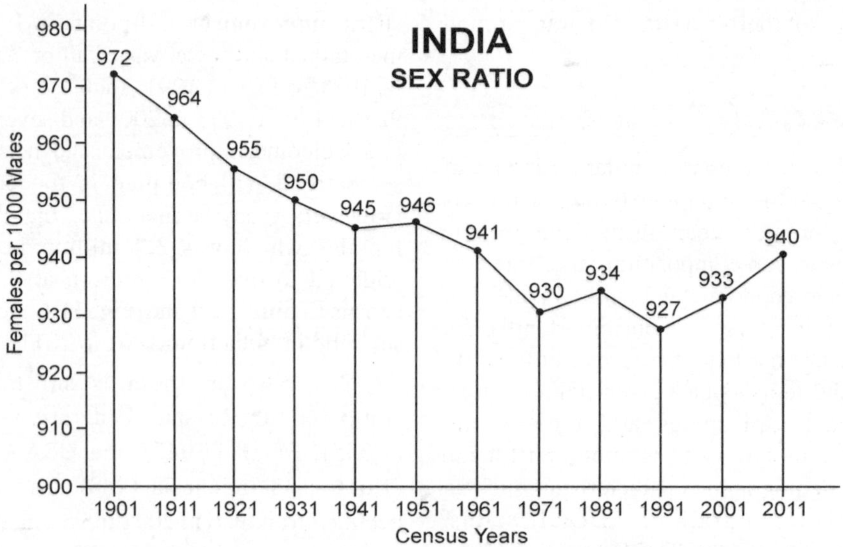

FIG. 11.6. Sex ratio in India from 1901 to 2011.

take place for every 1,000 male births, which in effect would mean that there is a deficiency of about 50 females per 1,000 males in every birth cohort. Many demographers believe that left to its own, this is an unalterable constant.

2. Males get preferential treatment while females are neglected. This results in higher female mortality. More females die in India at infancy, as well as during the reproductive period. In old age too the females suffer greater neglect than their male counterparts.

3. The practice of female infanticide in the past and the cognizant foeticide at present have also resulted in low sex ratio. Although it is difficult to assess the impact of these two practices in the absence of the relevant data, the craze for the male child is reflected in increasing number of sex determination tests and the resultant termination of pregnancy in case foetus happens to be a female. Thus the number of girl children continues to fall, rather drastically, as compared to boys. Such a skewed ratio of male-female at the time of birth is largely due to female foeticide.

Despite the Pre-Conception and Pre-Natal Diagnostic Technique (Prohibition of Sex Selection) Act being in force since 1996, these tests are being conducted particularly in Punjab, Haryana and Delhi, in clear violation of the act and the girl child is killed in the womb itself before she takes her birth. Obviously the law has failed to yield the desired result because of its inherent loopholes and faulty implementation.

4. With small family norms, many young couples do not go for a second child if the first child happens to be a male.

5. There has been a steady rise in dowry deaths in the recent past, although its exact impact on sex ratio cannot be specified in the absence of relevant data.

State Level Patterns of Sex Ratio

There are large spatial and temporal variations in sex ratio at state level. **Table 11.10** gives the trends in sex ratio for 110 years in between 1901 and 2011 in respect of all the States and Union territories, except Arunachal Pradesh, for which data are not available from 1901 to 1951. In 1901, there were as many as eleven States and Union territories that had sex ratio of more than unity. Among these, except Kerala all other States and Union territories have shown a downward slide. The major States that are largely responsible for the decline in the overall sex ratio in India are Uttar Pradesh, Bihar, Jharkhand, Odisha, Chhattisgarh, Madhya Pradesh, Gujarat, Maharashtra and Tamil Nadu. However, some of these states have shown some improvement with regard to sex ratio. Although the sex ratio in Punjab has been consistently low, it has shown a long term upward trend and has not contributed to the overall

TABLE 11.10. Sex Ratio (females per 1,000 males), 1901-2011

State/ UT Code	India / State / Union territory*	Census year											
		1901	1911	1921	1931	1941	1951	1961	1971	1981	1991	2001	2011
	INDIA	**972**	**964**	**955**	**950**	**945**	**946**	**941**	**930**	**934**	**927**	**933**	**940**
1	Jammu & Kashmir	882	876	870	865	869	873	878	878	892	896	892	883
2	Himachal Pradesh	884	889	890	897	890	912	938	958	973	976	968	974
3	Punjab	832	780	799	815	836	844	854	865	879	882	876	893
4	Chandigarh*	771	720	743	751	763	781	652	749	769	790	777	818
5	Uttarakhand	918	907	916	913	907	940	947	940	936	936	962	963
6	Haryana	867	835	844	844	869	871	868	867	870	865	861	877
7	Delhi*	862	793	733	722	715	768	785	801	808	827	821	866
8	Rajasthan	905	908	896	907	906	921	908	911	919	910	921	926
9	Uttar Pradesh	938	916	908	903	907	908	907	876	882	876	898	908
10	Bihar	1,061	1,051	1,020	995	1,002	1,000	1,005	957	948	907	919	916
11	Sikkim	916	951	970	967	920	907	904	863	835	878	875	889
12	Arunachal Pradesh	NA	NA	NA	NA	NA	NA	894	861	862	859	893	920
13	Nagaland	973	993	992	997	1,021	999	933	871	863	886	900	931
14	Manipur	1,037	1,029	1,041	1,065	1,055	1,036	1,015	980	971	958	978	987
15	Mizoram	1,113	1,120	1,109	1,102	1,069	1,041	1,009	946	919	921	935	975
16	Tripura	874	885	885	885	886	904	932	943	946	945	948	961
17	Meghalaya	1,036	1,013	1,000	971	966	949	937	942	954	955	972	986
18	Assam	919	915	896	874	875	868	869	896	910	923	935	954
19	West Bengal	945	925	905	890	852	865	878	891	911	917	934	947
20	Jharkhand	1,032	1,021	1,002	989	978	961	960	945	940	922	941	947
21	Odisha	1,037	1,056	1,086	1,067	1,053	1,022	1,001	988	981	971	972	978
22	Chhattisgarh	1,046	1,039	1,041	1,043	1,032	1,024	1,008	998	996	985	989	991
23	Madhya Pradesh	972	967	949	947	946	945	932	920	921	912	919	930
24	Gujarat	954	946	944	945	941	952	940	934	942	934	920	918
25	Daman & Diu*	995	1,040	1,143	1,088	1,080	1,125	1,169	1,099	1,062	969	710	618
26	Dadra & Nagar Haveli*	960	967	940	911	925	946	963	1,007	974	952	812	775
27	Maharashtra	978	966	950	947	949	941	936	930	937	934	922	925
28	Andhra Pradesh (including Telangana)	985	992	993	987	980	986	981	977	975	972	978	992
29	Karnataka	983	981	969	965	960	966	959	957	963	960	965	968
30	Goa	1,091	1,108	1,120	1,088	1,084	1,128	1,066	981	975	967	961	968
31	Lakshadweep*	1,063	987	1,027	994	1,018	1,043	1,020	978	975	943	948	946
32	Kerala	1,004	1,008	1,011	1,022	1,027	1,028	1,022	1,016	1,032	1,036	1,058	1084
33	Tamil Nadu	1,044	1,042	1,029	1,027	1,012	1,007	992	978	977	974	987	995
34	Puducherry*	NA	1,058	1,053	NA	NA	1,030	1,013	989	985	979	1,001	1038
35	Andaman & Nicobar Islands*	318	352	303	495	574	625	617	644	760	818	846	878

* Union Territory

Source : Census of India 2011, Provisional Populatin Totals Paper 1 of 2011, Series 1, pp. 1974-75.

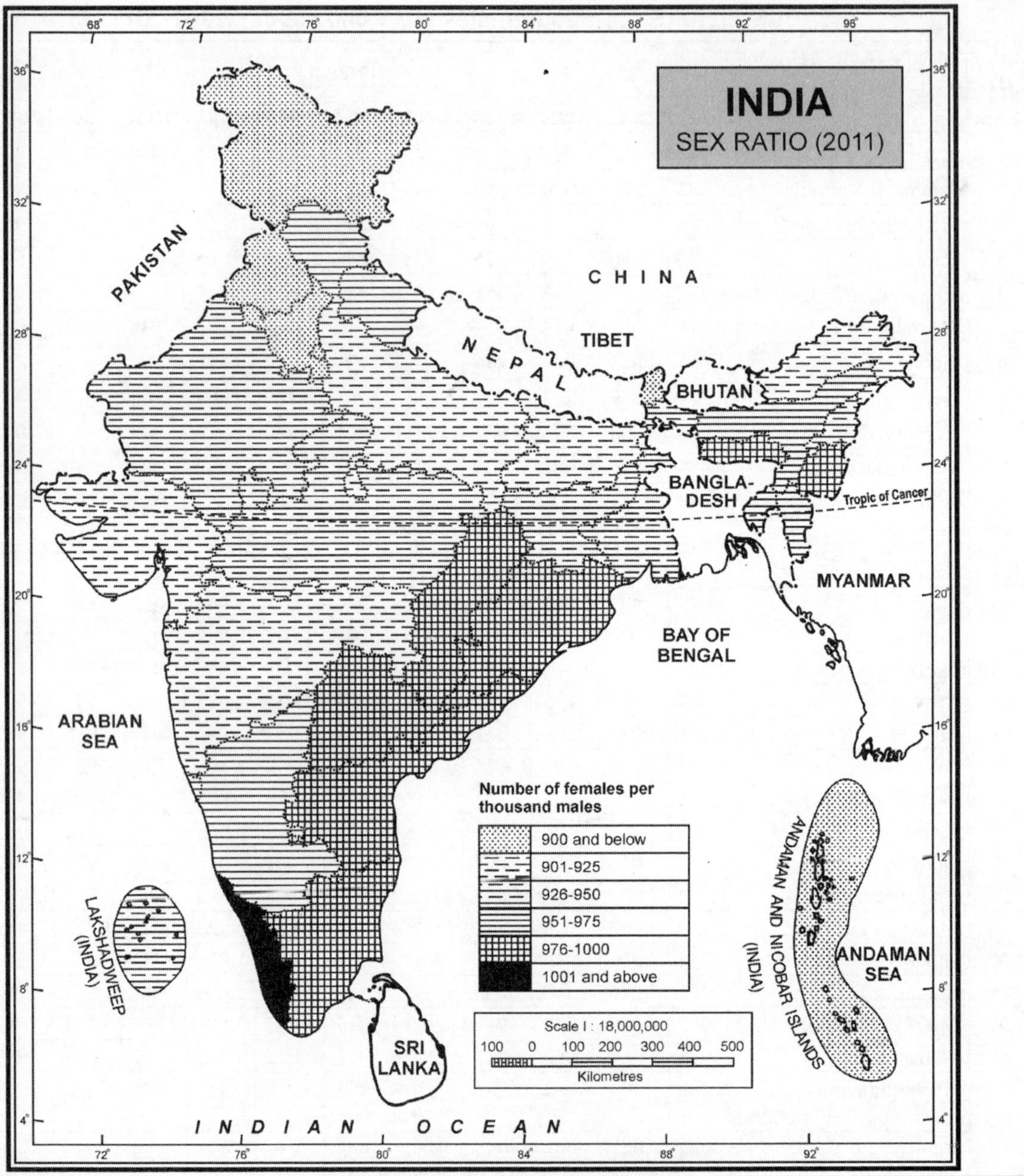

FIG. 11.7. India : Sex Ratio, 2011

deterioration in sex ratio of the country. In Rajasthan, the sex ratio kept fluctuating in a narrow band and always remained at a low level. Haryana, Andhra Pradesh, Telangana and Karnataka are the States where the sex ratio has remained more or less stagnant. In West Bengal the sex ratio declined sharply from 1901 to 1941 and then made a gradual turnaround on an upward path to reach 947 in 2011. Goa, Manipur, Mizoram, Odisha and Lakshadweep (except in 1911 and 1931) have invariably shown sex ratio of over one thousand upto 1961, although for different reasons. In spite of the fact that data for 1901, 1931 and 1941 in respect of Puducherry are not available, this Union Territory is supposed to have sex ratio in favour of females upto 1961. This union territory experienced a steep fall in sex ratio from 1,013 in 1961 to 979 in 1991. But in decades 1991-2001 and 2001-11 Puducherry made a remarkable recovery to reach a sex ratio of 1,038 in 2011 from 1001 in 2001. Thus in 2011, Kerala and Puducherry are the only two areas to record a sex ratio favourable to women. Sex ratio favourable to females has been shown by Bihar and Meghalaya upto 1921 and by Tamil Nadu upto 1951. Kerala is the only state which has shown more females than males throughout the census history of India. In 2011, Kerala with a sex ratio of 1,084 was at the top among all the states. This is a demographic trait which is more characteristic of the developed countries.

The other states and union territories with sex ratio higher than the national average in 2011 are Himachal Pradesh, Uttarakhand, Manipur, Mizoram, Tripura, Meghalaya, Assam, West Bengal, Odisha, Chhattisgarh, Andhra Pradesh (including Telangana), Karnataka, Goa, Lakshadweep, Kerala, Tamil Nadu and Puducherry. Thus 18 states and union territories have sex ratio above the national average. The remaining 17 states and union territories have sex ratio below the national average. Sex ratio of 950 is considered tolerable in the Indian context. Jammu and Kashmir, Punjab, Chandigarh, Haryana, Delhi, Sikkim, Daman and Diu, Dadra and Nagar Haveli, and Andaman and Nicobar Islands have a very low sex ratio of 900 or below. Five union territories of Andaman and Nicobar Islands, Delhi, Dadra and Nagar Haveli, Chandigarh and Daman and Diu have critically low sex ratio. None of them has sex ratio exceeding 878. In the states of Punjab and Haryana and the Union territories of Delhi and Chandigarh, extremely low sex ratio prevails due to large scale sex determination and female foeticide. Chandigarh and Delhi are highly urbanized union territories where people (mostly males) migrate in large numbers in search of livelihood, leaving their families (mostly females) in their native towns and villages.

One conspicuous improvement has been with respect to Andaman and Nicobar Islands, where the sex ratio was desperately low at 318 in 1901 and remained below 600 till 1941. This was perhaps due to the repressive policy of the British regime, when freedom fighters from the main land of India were sent as prisoners there most of whom were males. The

TABLE 11.11. India: Sex Ratio at Birth for children in the last five years (females per thousand males)

States/UTs	Sex Ratio at Birth for children born in the last five years (females per 1000 males)	
	NFHS-4 (2015-16)	NFHS-5 (2019-21)
All India	**919**	**929**
Andhra Pradesh	914	934
Arunachal Pradesh	926	979
Assam	929	964
Bihar	934	908
Delhi	812	923
Goa	966	838
Gujarat	906	955
Haryana	836	893
Himachal Pradesh	937	875
Jammu and Kashmir	923	976
Karnataka	910	978
Kerala	1,047	951
Madhya Pradesh	927	956
Maharashtra	924	913
Manipur	962	967
Meghalaya	1,009	989
Mizoram	949	969
Nagaland	953	945
Odisha	932	894
Punjab	860	904
Rajasthan	887	891
Sikkim	809	969
Tamil Nadu	954	878
Tripura	969	1,028
Uttar Pradesh	903	941
West Bengal	960	973
Chhattisgarh	977	960
Jharkhand	919	899
Uttarakhand	888	984
Telangana	872	894
Andaman and Nicobar (UT)	891	941
Chandigarh (UT)	981	838
Dadra and Nagar Haveli & Daman Diu (UT)	983	817
Lakshadweep (UT)	905	1,051
Puducherry (UT)	843	959
Ladakh (UT)		1,125

NFHS = National Family Health Survey

***Source*:** Economic Survey 2021-22 pp. 381-82.

situation, however, improved after Independence and this Union Territory showed a record sex ratio of 878 in 2011 throughout its census history.

Recent Trends

Sex ratio in India has shown remarkable improvement during the last few years. Sex ratio in total population has risen from 991 in 2015-16 to 1020 in 2019-21. More importantly sex ratio at birth (SRB), female children per thousand male children, born during the last five years has grown from 919 in 2015-16 to 929 in 2019-21. To prevent gender biased sex selective elimination, to ensure survival and protection of the girl child and to ensure education and participation of the girl child, the government has made specific interventions through ***Beti Bachao Beti Padhao Scheme***. Sex ratio at birth for children born in the last five years has improved in 2019-21 from 2015-16 in all states, except for Himachal Pradesh, Bihar, Jharkhand, Chhattisgarh, Odisha, Maharashtra, Tamil Nadu, Kerala, Meghalaya, Goa and Nagaland (**Table 11.11**).

LITERACY

It is necessary for a person to be literate before he becomes educated. Higher level of education provides dynamism to society and helps in social upliftment.

The Population Commission of United Nations considers the *ability, to both read and write a simple message with understanding in any language, a sufficient basis for classifying a person as literate.* The Census of India has adopted this definition with a little bit of modification. According to Census of India, "*person aged seven and above, who can both read and write with understanding in any language, is treated as literate*. A person, who can only read but cannot write, is not literate." In the Censuses prior to 1991, children below five years of age were necessarily treated as illiterates.

The age limit was raised to 7 years based on the advice of experts that the ability to read and write with understanding is not ordinarily achieved until that age. It was, therefore decided at the 1991 Census that all children in the age group 0-6, would be treated as illiterate by definition and the population aged seven years and above only would be classified as literate or illiterate. The same criterion has been retained in the Censuses of 2001 and 2011. It should be clearly understood that it is not necessary that to be treated as literate, a person should have received any formal education or acquired any minimum educational standard.

Literacy Rate—Definition

In earlier Censuses up to 1981, it was customary to work out the literacy rate taking into account the total population. Since literacy rate is more meaningful if the sub-population in the age group 0-6 is excluded from the total population. it was decided in 1991 to calculate literacy rate for the population seven years and above. The same concept has been retained in all Censuses since 1991.

The literacy rate taking into account the total population in the denominator has now been termed as ***'crude literacy rate'***, while the literacy rate calculated taking into account the 7 and above population in the denominator is called the ***effective literacy*** rate. The formula for computing crude **literacy rate** and effective literacy rate are as follows :

Crude Literacy Rate =

$$\frac{\text{Number of Literate persons}}{\text{Total Population}} \times 100$$

Effective Literacy Rate =

$$\frac{\text{Number of Literate persons aged 7 and above}}{\text{Population aged 7 and above}} \times 100$$

Effective literacy rate and literacy rate have been used interchangeably in this chapter.

Literates and Illiterates by Gender

As per the provisional population totals of Census 2011, out of the provisional total population of 1,210,193,422, the number of persons aged seven years and above is 1,051,404,135. Out of this, 778,454,120 are literate and 272,950,015 are illiterates There has been an increase of 186,504,094 persons in the age group seven years and above during 2001-2011, while 217,700,941 additional persons have become literate during the decade.

A significant milestone reached in Census 2011, is that the total number of illiterates has come down from 304,146,862 in 2001 to 272,950,015 in 2011, showing a decline of 31,196,847 persons.

One of the interesting features of Census 2011 is that out of total of 217,700,941 literates added during the decade, females (110,069,001) out number males (107,631,940). A reverse trend was noticed during 1991-2001. The decadal increase in number of literates among males is of 31.98 percentage points while the corresponding increase in case of females is of 49.10 percentage points.

A notable feature is that out of the total decrease of 31,196,847 in the number of illiterates, the females (17,122,197) out number males (14,074,650). The above two changes are a clear indication of the fact that the gender gap in literacy is shrinking in the country. This trend of rising female literates will has far reaching consequences on the development of society.

Table 11.12 gives the number of literates and illiterates among the population aged seven years and above in absolute figures for India at the 2001 and 2011 censuses.

Literacy Rate—Trends

The effective literacy rate for India in Census 2011, works out to 74.04 per cent. The corresponding figures for males and females are 82.14 and 65.46 per cent respectively. Thus three-fourths of the population of aged 7 years and above is literate in the country. Four out of every five males and two out of every three females in the country are literate. The country has continued its march in improving literacy rate by recording a jump of 9.21 percentage points during 2001-11. The increase in literacy rates in males and females is in the order of 6.88 and 11.79 percentage points respectively.

An extremely positive development in the decade 2001-11 is that the gap of 21.59 percentage points recorded between male and female literacy rates in 2001 census has reduced to 16.68 percentage points in 2011. Though the target set for the year 2011-2012 by the Planning Commission of reducing the gap to 10 percentage points has not been achieved, it is heartening that the reduction has been to the order of almost 5 percentage points.

Table 11.13 presents the effective literacy rate for the country by persons, males, females and the male-female gap since 1951.

TABLE 11.12. Number of literates and illiterates among population aged 7 years and above and their change—India : 2001 and 2011

Literates/illiterates		Persons	Males	Females
Population (aged 7 and above	2001	86,49,00,041	44,72,14,823	41,76,85,218
	2011	1,05,14,04,135	54,07,72,113	51,06,32,022
	Increase in 2011 over 2001	18,65,04,094	9,35,57,290	9,29,46,804
Literates	2001	56,07,53,179	33,65,71,822	22,41,81,357
	2011	77,84,54,120	44,42,03,762	33,42,50,358
	Increase in 2011 over 2001	21,77,00,941	10,76,31,940	11,00,69,001
Illiterates	2001	30,41,46,862	11,06,43,001	19,35,03,861
	2011	27,29,50,015	9,65,68,351	17,63,81,664
	Increase in 2011 over 2001	–3,11,96,847	–1,40,74,650	–1,71,22,197

Source : Census of India 2011, Provisional Population Totals, Paper 1 of Series 1, p. 100.

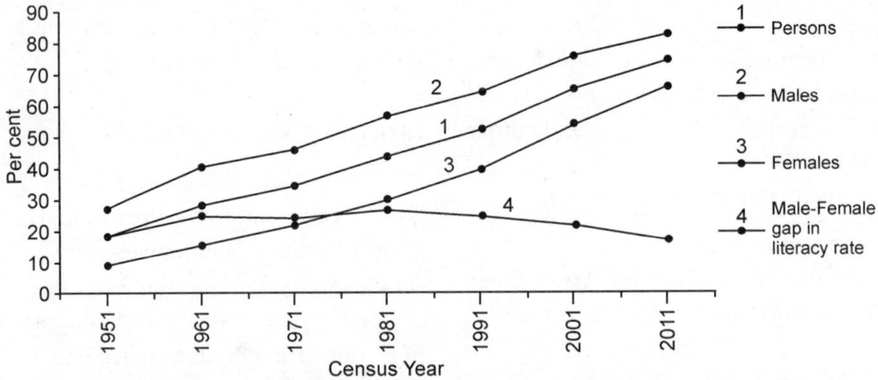

FIG. 11.8. Literacy rates in India.

TABLE 11.13. Literacy Rate in India 1951-2011

Census year	Persons	Males	Females	Male-Female gap in literacy rate
1951	18.33	27.16	8.86	18.30
1961	28.30	40.4	15.35	25.05
1971	34.45	45.96	21.97	23.98
1981	43.57	56.38	29.76	26.62
1991	52.21	64.13	39.29	24.84
2001	64.83	75.26	53.67	21.59
2011	74.04	82.14	65.46	16.68

Source : Census of India, Provisional Population Totals Paper 1 of Series 1 p. 102.

The above trends have been shown in **Fig. 11.8**. **Table 11.14** provides the crude literacy rate for India by sex during 1901-2011. The literacy rate designated as crude literacy rate in this statement has been computed with total population as base without removing the mandatory illiterate population aged 0-4 or 0-6 from the denominator. The crude literacy rate from 1901 onwards shows a consistent increase both for males and females.

The improvement in crude literacy rate has been phenomenal (48.22 percentage points) in post independent India. The corresponding increase in case

TABLE 11.14. Crude Literacy Rate in India by Sex : 1901 to 2011

Census Year	Crude literacy rate			Change in Percent points		
	Persons	Males	Females	Persons	Males	Females
1901	5.35	9.83	0.60	—	—	—
1911	5.92	10.56	1.05	0.57	0.73	0.45
1921	7.16	12.21	1.81	1.24	1.65	0.76
1931	9.50	15.59	2.93	2.34	3.38	1.12
1941	16.10	24.90	7.30	6.60	9.31	4.37
1951	16.67	24.95	7.93	0.57	0.05	0.63
1961	24.02	34.44	12.95	7.35	9.49	5.02
1971	29.45	39.45	18.69	5.43	5.01	5.74
1981	36.23	46.89	24.82	6.78	7.44	6.13
1991	42.84	52.74	32.17	6.61	5.85	7.35
2001	54.21	63.24	45.15	11.67	10.50	12.98
2011	64.32	71.22	56.99	9.81	7.98	11.84

Notes :

1. Figures upto 1941 are for undivided India.
2. Figures for 1981 exclude Assam as 1981 census could not be conducted in this state due to disturbed conditions.
3. Figures for 1991 census do not include Jammu and Kashmir, as no census was held in the state.

Source : Census of India 2011, Provisional Population Totals, Paper 1 of 2011, Series 1, p. 103.

TABLE 11.15. Ranking of States and Union Territories by literacy rate : 2011

Rank	Persons		Rank	Males		Rank	Females	
	State/ Union Territory	Literacy rate		State/ Union Territory	Literacy rate		State/Union Territory	Literacy rate
1	Kerala	93.91	1	Lakshadweep#	96.11	1	Kerala	91.98
2	Lakshadweep#	92.28	2	Kerala	96.02	2	Mizoram	89.40
3	Mizoram	91.58	3	Mizoram	93.72	3	Lakshadweep#	88.25
4	Tripura	87.75	4	Goa	92.81	4	Tripura	83.15
5	Goa	87.40	5	Tripura	92.18	5	Goa	81.84
6	Daman & Diu#	87.07	6	Puducherry #	92.12	6	Andaman &	
7	Puducherry #	86.55	7	Daman & Diu#	91.48		Nicobar Islands#	81.84
8	Chandigarh #	86.43	8	NCT of Delhi#	91.03	7	Chandigarh#	81.38
9	NCT of Delhi #	86.34	9	Himachal Pradesh	90.83	8	Puducherry#	81.22
10	Andaman &		10	Chandigarh#	90.54	9	NCT of Delhi#	80.93
	Nicobar Islands #	86.27	11	Andaman &		10	Daman & Diu#	79.59
11	Himachal Pradesh	83.78		Nicobar Islands#	90.11	11	Nagaland	76.69
12	Maharashtra	82.91	12	Maharashtra	89.82	12	Himachal Pradesh	76.60
13	Sikkim	82.20	13	Uttarakhand	88.33	13	Sikkim	76.43
14	Tamil Nadu	80.33	14	Sikkim	87.29	14	Maharashtra	75.48
15	Nagaland	80.11	15	Gujarat	87.23	15	Tamil Nadu	73.86
16	Manipur	79.85	16	Tamil Nadu	86.81	16	Meghalaya	73.78
17	Uttarakhand	79.63	17	Manipur	86.49	17	Manipur	73.17
18	Gujarat	79.31	18	Dadra & Nagar Haveli #	86.46	18	Punjab	71.34
19	Dadra & Nagar Haveli #	77.65	19	Haryana	85.38	19	West Bengal	71.16
20	West Bengal	77.08	20	Nagaland	83.29	20	Gujarat	70.73
21	Punjab	76.68	21	Karnataka	82.85	21	Uttarakhand	70.70
22	Haryana	76.64	22	West Bengal	82.67	22	Karnataka	68.13
23	Karnataka	75.60	23	Odisha	82.40	23	Assam	67.27
24	Meghalaya	75.48	24	Punjab	81.48	24	Haryana	66.77
25	Odisha	73.45	25	Chhattisgarh	81.45	25	Dadra & Nagar Haveli#	65.93
26	Assam	73.18	26	Madhya Pradesh	80.53	26	Odisha	64.36
27	Chhattisgarh	71.04	27	Rajashthan	80.51	27	Chhattisgarh	60.59
28	Madhya Pradesh	70.63	28	Uttar Pradesh	79.24	28	Madhya Pradesh	60.02
29	Uttar Pradesh	69.72	29	Assam	78.81	29	Andhra Pradesh	59.74
30	Jammu & Kashmir	68.74	30	Jharkhand	78.45		(including Telangana)	
31	Andhra Pradesh	67.66	31	Jammu & Kashmir	78.26	30	Arunachal Pradesh	59.57
	(including Telangana)		32	Meghalaya	77.17	31	Uttar Pradesh	59.26
32	Jharkhand	67.63	33	Andhra Pradesh	75.56	32	Jammu & Kashmir	58.01
33	Rajasthan	67.06		(including Telangana)		33	Jharkhand	56.21
34	Arunachal Pradesh	66.95	34	Arunachal Pradesh	73.69	34	Bihar	52.33
35	Bihar	63.82	35	Bihar	73.39	35	Rajasthan	52.66

#Union Territory.

Source : Census of India 2011, Provisional Population Totals, Paper 1 of 2011, Series 1 pp. 110-11.

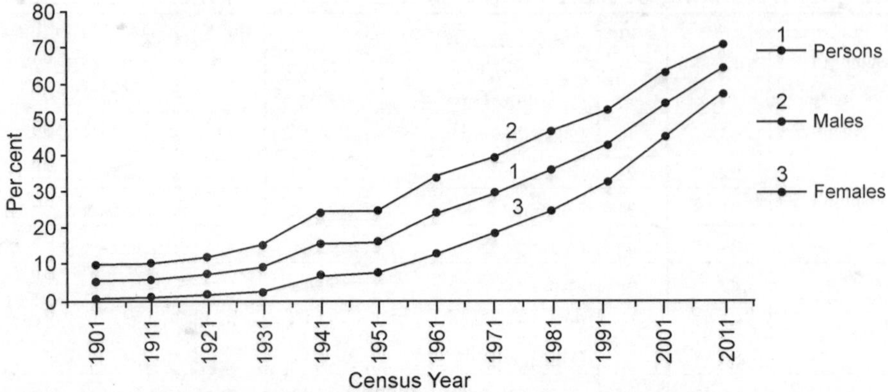

FIG. 11.9. India : Crude Literacy Rate by sex (1901-2011)

of males has been of 46.32 percentage points and among females it is of 49.69 percentage points. The crude literacy rate has increased by almost 10 percentage points during 2001-2011 decade. It surged forward by 12 percentage points in case of females while there was an increase of 8 percentage points in male crude literacy rate during the last decade. The gap in crude literacy rates of males and females has decreased from 18.09 percentage points in 2001 to 14.23 percentage points in 2011.

Regional Variations in Literacy Rates

States and Union Territories of India have been arranged in descending order with respect to overall effective literacy rates along with effective male and female literacy rates. Regional variations of overall, male and female literacy rates have been shown in **Fig. 11.10, 11.11** and **11.12** respectively. A brief description of overall male and female literacy rates is given below.

Overall Effective Literacy Rate

Kerala ranks first in the country with a literacy rate of 93.91 per cent, closely followed by Lakshadweep (92.28 per cent) and Mizoram (91.58 per cent). Bihar with a literacy rate of 63.82 per cent ranks last in the country preceded by Arunachal Pradesh (66.95 per cent) and Rajasthan (67.06 per cent). Among the major States, Maharashtra (82.91 per cent) comes after Kerala, followed by Tamil Nadu (80.33 per cent). The States and Union Territories with literacy rates below the National average (74.04 per cent) are Jammu and Kashmir in the North, Rajasthan in the West, Andhra Pradesh and Telangana in the South, Madhya Pradesh, Chhattisgarh, Uttar Pradesh in the Central, Bihar, Jharkhand, Odisha in the East and Arunachal Pradesh and Assam in the North-East of the country. Ten States and Union Territories viz., Kerala, Lakshadweep, Mizoram, Tripura, Goa, Daman & Diu, Puducherry, Chandigarh, NCT of Delhi and Andaman & Nicobar Islands have achieved literacy rate of above 85 per cent, the target set by Planning Commission for the year 2011-2012. The States and Union Territories, which have literacy rate below the National average in respect of all the three categories i.e., persons, males and females are Arunachal Pradesh, Chhattisgarh, Madhya Pradesh, Andhra Pradesh (including Telangana), Bihar, Rajasthan, Jammu and Kashmir, Jharkhand and Uttar Pradesh.

Male Effective Literacy Rate

Lakshadweep (96.11 per cent) holds the first rank in the country with respect to male literacy rate. Kerala (96.02 per cent) ranks second. Bihar (73.39 per cent) has recorded the lowest literacy rate in case of males preceded again by Arunachal Pradesh (73.69 per cent). The States and Union Territories with literacy rates below the National average for males (82.14 per cent) are Madhya Pradesh, Assam, Meghalaya, Arunachal Pradesh, Andhra Pradesh

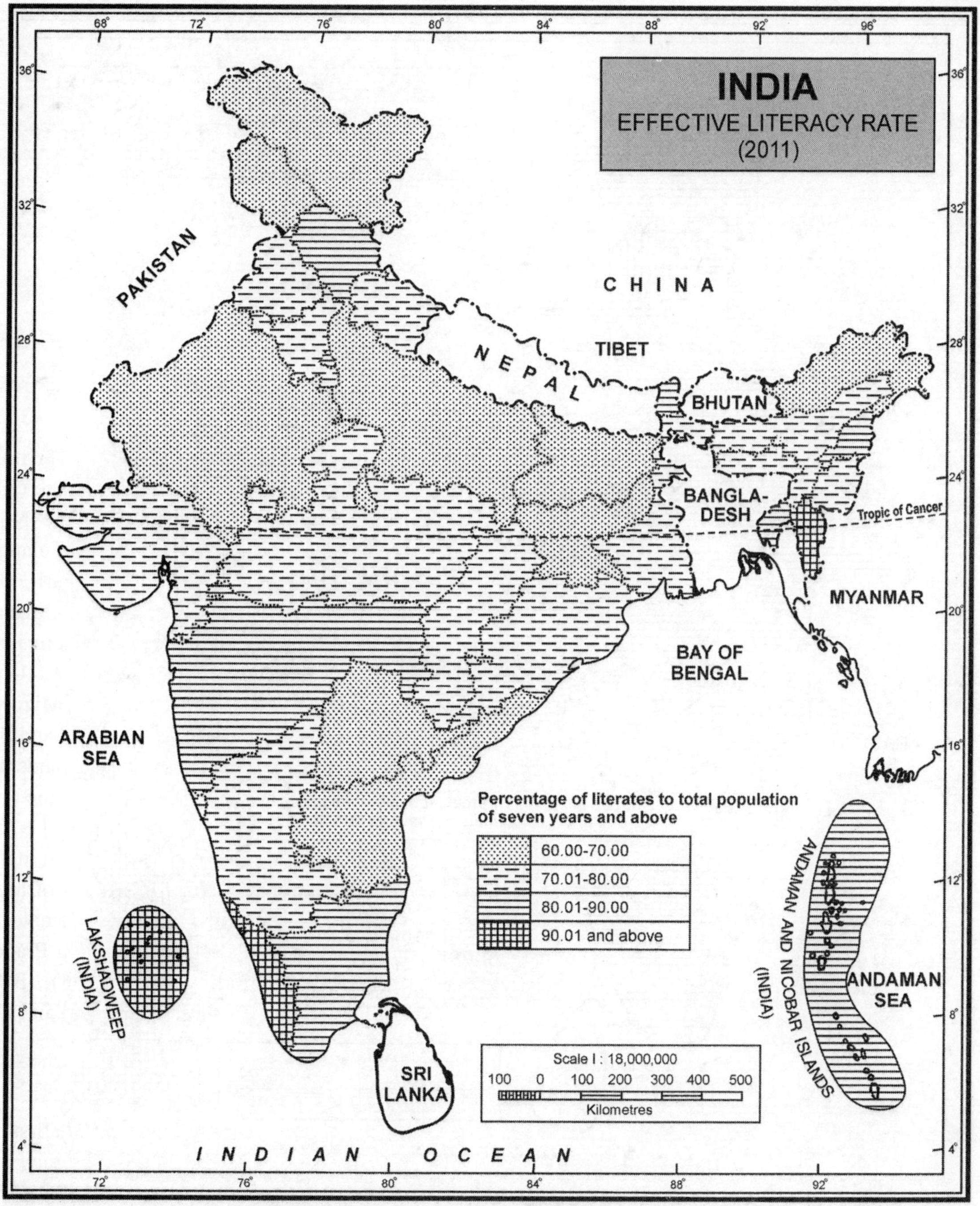

FIG. 11.10. India : Overall effective literacy rate (2011)

(including Telangana), Bihar, Rajasthan, Jammu and Kashmir, Chhattisgarh, Punjab, Jharkhand and Uttar Pradesh.

Female Effective Literacy Rate

Kerala holds the first rank in the country in female literacy with 91.98 per cent. Rajasthan (52.66

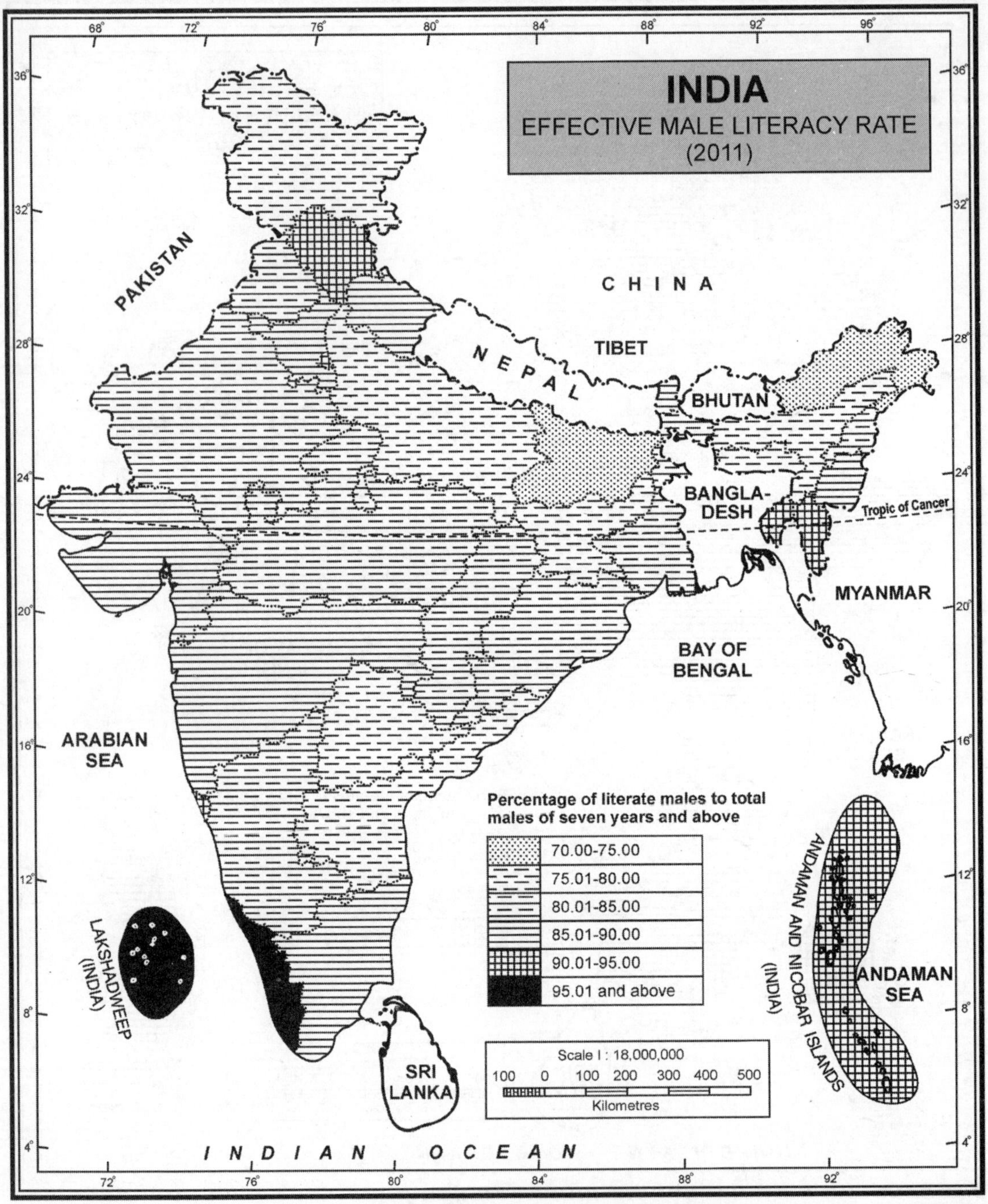

FIG. 11.11. India : Effective male literacy rate (2011).

per cent) has recorded the lowest female literacy rate followed by Bihar (53.33 per cent). Similarly, the States and Union Territories with female literacy rate below the National average (65.46 per cent) are Odisha, Chhattisgarh, Madhya Pradesh, Andhra Pradesh (including Telangana), Arunachal Pradesh, Uttar Pradesh, Jammu and Kashmir, Jharkhand, Bihar and Rajasthan.

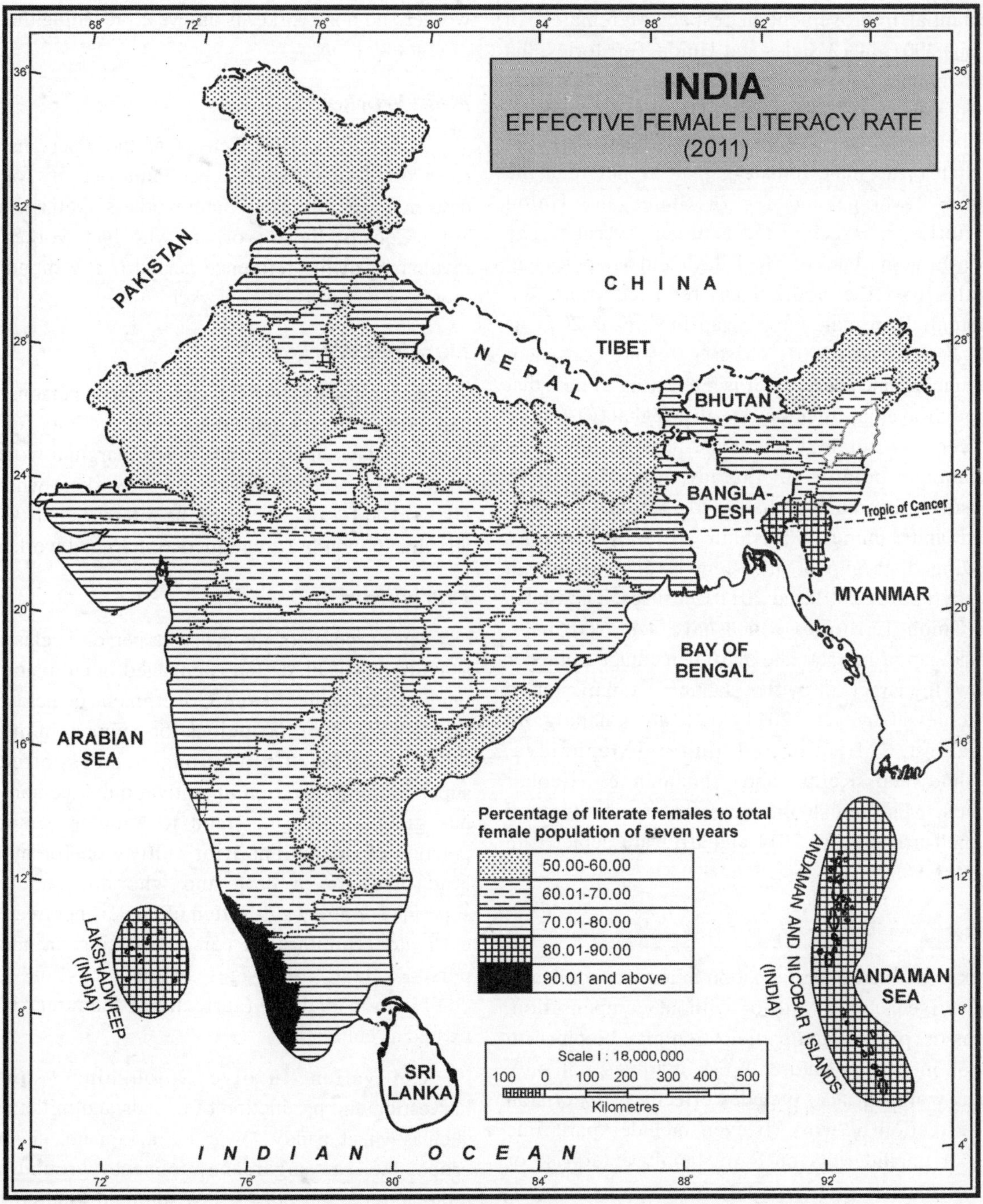

FIG. 11.12. India : Effective female literacy rate (2011)

Effective Literacy Rate-Gender Gap

At the National level, the male-female gap for Census 2001 stood at 21.59 whereas for Census 2011 it is only 16.68. The decadal difference in literacy rates for males and females stands at 6.88 and 11.79 percentage points respectively, indicating a substantial improvement in respect of females. In Census 2001, in 12 States and Union Territories, the male-female gap was higher than the National average and for 23 States and Union Territories, it was below the National average. In Census 2011, in 11 States, the male female gap is higher than the National average and for 24 States and Union Territories, it is below the National average. The North-Eastern States of Meghalaya and Mizoram and Kerala from the South have reported minimum differentials in male-female gap in Census 2011 as well as in Census 2001. These States are consistent forerunners since decades. It is interesting to note that although average literacy rate in Meghalaya (75.48 per cent) is not high the male-female gap (3.39 percentage points) in literacy rate is lowest. Rajasthan, Jharkhand, Chhattisgarh, Madhya Pradesh, Bihar and Jammu & Kashmir are at the bottom, showing huge gap in male-female differentials in literacy both at 2001 and 2011 Censuses. The States and Union Territories which have achieved male-female gap of literacy rate of 10 percentage points or below, the target set by the Planning Commission to be achieved by year 2011-2012, are Chandigarh, Nagaland, Mizoram, Tripura, Meghalaya, Lakshadweep, Kerala and Andaman & Nicobar Islands. Male-female literacy gaps for States and Union Territories for 2011 and 2011 are depicted in **table 11.16**.

WORK

Work is defined as participation in any economically productive activity with or without compensation, wages or, profit. Such participation may be physical and/or mental in nature. Work involves not only actual work but also includes effective supervision and direction of work. It even includes part time help or unpaid work on farm, family enterprise or in any other economic activity. All persons engaged in 'work' as defined above are workers. Persons who are engaged in cultivation or milk production even solely for domestic consumption are also treated as workers.

Reference period for determining a person as worker and non-worker is one year preceding the date of renumeration.

Main Workers

According to 1981 Census, all those persons who had worked in any gainful economic activity for 183 days in a year are called main workers. At the time of 2001 Census, those workers who had worked for major part of the reference period (*i.e.* 6 months or more) are terms as main workers.

Marginal Workers

According to 1981 Census, all those persons who had worked in any gainful economic activity for less than 183 days in a year are called marginal workers. At the time of 2001 census, those workers who had not worked for major part of the reference period (*i.e.* less than 6 months) are termed as marginal workers.

Cultivators

For purposes of the census a person is classified as cultivator if he or she is engaged in cultivation of land owned or held from Government or held from private persons or institutions for payment in money, kind or share. Cultivation includes effective supervision or direction in cultivation. A person who has given out her/his land to another person or persons or institution(s) for cultivation for money, kind or share of crop and who does not even supervise or direct cultivation of land, is not treated as cultivator. Similarly, a person working on another person's land for wages in cash or kind or a combination of both (agricultural labourer) is not treated as cultivator.

Cultivation involves ploughing, sowing, harvesting and production of cereals and millet crops such as wheat, paddy, jowar, bajra, ragi etc., and other crops such as sugarcane, tobacco, ground-nuts, tapioca, etc., and pulses, raw jute and kindred fiber crop, cotton, cinchona and other medicinal plants, fruit growing, vegetable growing of keeping orchards or groves, etc. Cultivation does not include the following plantation crops—tea, coffee, rubber, coconut and betel-nuts (areca).

TABLE 11.16. Literacy Rates and Male-Female Gap in Literacy (in percentage)

State/ UT code	India/State/ Union Territory	2001		Gap in literacy rate	2011		Gap in literacy rate
		Males	Females		Males	Females	
	India	**75.26**	**53.67**	**21.59**	**82.14**	**65.46**	**16.68**
01	Jammu & Kashmir	66.60	43.00	23.60	78.26	58.01	20.25
02	Himachal Pradesh	85.35	67.42	17.93	90.83	76.60	14.23
03	Punjab	75.23	63.36	11.87	81.48	71.34	10.14
04	Chandigarh#	86.14	76.47	9.67	90.54	81.38	9.16
05	Uttarakhand	83.28	59.63	23.65	88.33	70.70	17.63
06	Haryana	78.49	55.73	22.76	85.38	66.77	18.61
07	NCT of Delhi#	87.33	74.71	12.62	91.03	80.93	10.10
08	Rajashthan	75.70	43.85	31.85	80.51	52.66	27.85
09	Uttar Pradesh	68.82	42.22	26.60	79.24	59.26	19.98
10	Bihar	59.68	33.12	26.56	73.39	53.33	20.06
11	Sikkim	76.04	60.40	15.64	87.29	76.43	10.86
12	Arunachal Pradesh	63.83	43.53	20.30	73.69	59.57	14.12
13	Nagaland	71.16	61.46	9.70	83.29	76.69	6.60
14	Manipur	79.54	60.10	19.44	86.49	73.17	13.32
15	Mizoram	90.72	86.75	3.97	93.72	89.40	4.32
16	Tripura	81.02	64.91	16.11	92.18	83.15	9.03
17	Maghalaya	65.43	59.61	5.82	77.17	73.78	3.39
18	Assam	71.28	54.61	16.67	78.81	67.27	11.54
19	West Bengal	77.02	59.61	17.41	82.67	71.16	11.51
20	Jharkhand	67.30	38.87	28.43	78.45	56.21	22.24
21	Odisha	75.35	50.51	24.84	82.40	64.36	18.04
22	Chhattisgarh	77.38	51.85	25.53	81.45	60.59	20.86
23	Madhya Pradesh	76.06	50.29	25.77	80.53	60.02	20.51
24	Gujarat	79.66	57.80	21.86	87.23	70.73	16.50
25	Daman & Diu#	86.76	65.61	21.15	91.48	79.59	11.89
26	Dadra & Nagar Haveli#	71.18	40.23	30.95	86.46	65.93	20.53
27	Maharashtra	85.97	67.03	18.94	89.82	75.48	14.34
28	Andhra Pradesh (including Telangana)	70.32	50.43	19.89	75.56	59.74	15.82
29	Karnataka	76.10	56.87	19.23	82.85	68.13	14.72
30	Goa	88.42	75..37	13.05	92.81	81.84	10.97
31	Lakshadweep#	92.53	80.47	12.06	96.11	88.25	7.86
32	Kerala	94.24	87.72	6.52	96.02	91.98	4.04
33	Tamil Nadu	82.42	64.43	17.99	86.81	73.86	12..95
34	Puducherry#	88.62	73.90	14.72	92.12	81.22	10.90
35	Andaman & Nicobar Islands#	86.33	75.24	11.09	90.11	81.84	8.27

Union Territory

Source : Census of India, Provisional Population Totals, Paper 1 of 2011, Series 1, Pp. 116-17.

Agricultural Labourers

A person who works on another person's land for wages in money or kind or share is regarded as an agricultural labourer. She or he has no risk in the cultivation, but merely works on another person's land for wages. An agricultural labourer has no right of lease or contract on land on which she/he works.

Household Industry Workers

Household Industry is defined as an industry conducted by one or more members of the household at home or within the village in rural areas and only within the precincts of the house where the household lives in urban areas. The larger proportion of workers in the household industry consists of members of the household. The industry is not run on the scale of a registered factory which would qualify or has to be registered under the Indian Factories Act.

The main criterion of a Household industry even in urban areas is the participation of one or more members of a household. Even if the industry is not actually located at home in rural areas there is greater possibility of the members of the household participating even if it is located anywhere within the village limits. In the urban areas, where oraganised industry takes greater prominence, the Household Industry is confined to the precincts of the house where the participants live. In urban areas, even if the members of the household run an industry by themselves but at a place away from the precincts of their home it is not considered as a Household Industry. It should be located within the precincts of the house where the members live in the case of urban areas.

Household Industry relates to production, processing, servicing, repairing or making and selling (but not merely selling) of goods. It does not include professions such as pleader, doctor, musician, dancer, astrologer, *dhobi*, barber, etc., or merely trade or business, even if such professions, trade or services are run at home by members of the household. Some of the typical industries that can be conducted on a household industry basis are : Foodstuffs : such as production of flour, milking or dehusking of paddy, grinding of herbs, production of pickles, preservation of meat, etc. Beverages : such as manufacture of country liquor, ice cream, soda water etc., Tobacco Products : such as *bidis*, cigars, Textiles cotton, Jute, Wool or Silk, Manufacture of Wood and Wood Products, Paper and Paper Products, Leather and Leather Products, Petroleum and Coal Products : such as making foot wear from torn tyres and other rubber footwear, Chemical and Chemical Products : such as manufacture of toys, paints, colours, matches, fireworks, perfumes, ink, etc., Service and Repairing of Transport Equipments : such as cycle, rickshaw, boat or animal driven carts, etc.

Other Workers

All workers, *i.e.*, those who have been engaged in some economic activity during the last one year, but are not cultivators or agricultural labourers or in Household Industry are 'Other Workers (OW)'. The type of workers that come under this category of 'OW' include all government servants, municipal employees, teachers, factory workers, plantation workers, those engaged in trade, commerce, business, transport, banking, mining, construction, political or social work, priests, entertainment artists, etc. In effect, all those workers other than cultivators or agricultural labourers or household industry workers, are 'Other Workers'.

Non-Workers

A person who did not at all work during the reference period was treated as a non-worker. The non-workers broadly constitute students who did not participate in any economic activity paid or unpaid, household duties who were attending to daily household chores like cooking, cleaning utensils, looking after children, fetching water, etc. and are not even helping in the unpaid work in the family farm or cultivation or milching, dependent such as infants or very elderly people not included in the category of worker, pensioners those who are drawing pension after retirement and are not engaged in any economic activity. Beggars, vagrants, prostitutes and persons having unidentified source of income and with unspecified sources of subsistence and not engaged in any economically productive work during the reference period. Other category includes all Non-workers who may not come under the above categories such as rentiers, persons living on remittances, agricultural or non-agricultural royalty, unpaid work and persons who are seeking/available for work.

Work Participation Rate

Work participation rate is defined as the percentage of total workers (main and marginal workers) to total population.

$$\text{Work participation rate} = \frac{\text{Total Workers (Main + Marginal)}}{\text{Total Population}} \times 100$$

According to the 2011 census, the participation rate in India is 39.79 which is a slight improvement over 39.1 in 2001. This leaves a huge 60.21 per cent population as non-workers. Thus a large percentage of non-working population depends upon a low percentage of workers which is less than 40 per cent of the total population. This is not congenial to economic growth of the country. India's low percentage of working force is the outcome of her demographic, socio-cultural and economic structure. Among the major factors responsible for low percentage of work force are high birth rate and consequent large proportion of children below the age of 15, prejudices against females' education, mobility and employment, and incapability of the Indian economic structure to generate enough employment opportunities.

The composition of working force varies with sex, residence and age. The disparity in participation rates of males and females are quite glaring. According to 2011 census figures, while 53.26 per cent of males are recorded as main workers, the corresponding figure for females is only 25.51 per cent. Such a situation is the result of prejudices against females as discussed earlier. In a male dominated society as that of India, earning of bread still continued to rest on masculine shoulders and feminity is given a secondary place in social and economic activities. Similar variations are observed in the participation rates of rural and urban areas. The relatively low participation rates in urban areas is attributed to the differences in the nature of economy and society in urban and rural areas. The nature of jobs in urban areas is such for which education and skill is a pre-requisite. The acquisition of education and skill takes its own time and delays entry into work-force. Moreover, the urban society is comparatively more awakened and the incidence of child participation in the work force is low as compared to that of the countryside. In contrast to this, education and skill have little relevance in the countryside and the incidence of child and female participation in the workforce is more than that prevalent in the urban areas. This factor contributed a great deal to the relatively low participation rates in urban areas. The participation rates in urban areas would have been still lower, but for large scale in migration of workers from rural areas to urban areas.

The 1971 census divided the entire work force into nine industrial categories which had been adopted till the 1991 census. However, the main categories of work force at the time of 2011 census have been recognised as cultivators, agricultural labourers and household industry workers (*see* **Table 11.17**).

TABLE 11.17. Working Force in India, 2011

Sl. No.	Category	Percentage
1.	Participation rate	39.79
2.	Male participation	53.26
3.	Female participation	25.57
4.	Main workers to total workers	75.24
5.	Marginal workers to total workers	24.76
6.	Cultivators to total workers	47.43
7.	Agricultural labourers to total workers	17.89
8.	Household workers to total workers	2.56
9.	Other workers to total workers	38.7

Source : Data computed from Census of India 2011,. C.D. released in 2014.

Table 11.18 and **Fig. 11.13** depict the areal distribution of participation rate in India in 2011. It is observed that more than half the population is working in Himachal Pradesh and Sikkim. As many as 19 states and union territories have above 40 per cent participation rate which is higher than the

TABLE 11.18. India : Work Participation Rate and Occupation Structure, 2011

States/UTs	Work Partici-pation Rate (%)	Cultivators	% of Total Wor-kers	Agricultural Lab.	% of Total Wor-kers	Household Industries	% to Total Wor-kers	Other Workers	% of Total Wor-kers
INDIA	**39.8**	**11,86,92,640**	**24.6**	**14,43,29,833**	**30**	**1,83,36,307**	**3.8**	**20,03,84,531**	**41.6**
Jammu & Kashmir	34.5	12,45,316	28.8	5,47,705	12.7	1,72,586	4	23,57,106	54.5
Himachal Pradesh	51.9	20,62,062	57.9	1,75,038	4.9	58,719	1.6	12,63,603	35.5
Punjab	35.7	19,34,511	19.5	15,88,455	16	3,85,960	3.9	59,88,436	60.5
Chandigarh	38.3	2,578	0.6	1,687	0.4	4,799	1.2	3,95,072	97.8
Uttarakhand	38.4	15,80,423	40.8	4,03,301	10.4	1,14,312	3	17,74,239	45.8
Haryana	35.2	24,80,801	27.8	15,28,133	17.1	2,62,280	2.9	46,45,294	52.1
NCT of Delhi	33.3	33,398	0.6	39,475	0.7	1,81,852	3.3	53,32,324	95.4
Rajasthan	43.6	1,36,18,870	45.6	49,39,664	16.5	7,20,573	2.4	1,06,07,148	35.5
Uttar Pradesh	32.9	1,90,57,888	29.0	1,99,39,223	30.3	38,98,590	5.9	2,29,19,014	34.8
Bihar	33.4	71,96,226	20.7	1,83,45,649	52.8	14,11,208	4.1	77,71,904	22.4
Sikkim	50.5	1,17,401	38.1	25,986	8.4	5,143	1.7	1,59,608	51.8
Arunachal Pradesh	42.5	3,02,723	51.5	36,171	6.2	8,365	1.4	2,40,398	40.9
Nagaland	49.2	5,37,702	55.2	62,962	6.5	22,838	2.3	3,50,620	36.0
Manipur	45.1	4,57,891	39.5	1,11,061	9.6	89,495	7.7	5,00,606	43.2
Mizoram	44.4	2,29,603	47.2	41,787	8.6	7,852	1.6	2,07,463	42.6
Tripura	40.0	2,95,947	20.1	3,53,618	24.1	41,496	2.8	7,78,460	53.0
Meghalaya	40.0	4,94,675	41.7	1,98,364	16.7	20,488	1.7	4,72,092	39.8
Assam	38.4	40,61,627	33.9	18,45,346	15.4	4,91,321	4.1	55,71,396	46.5
West Bengal	38.1	51,16,668	14.7	1,01,88,842	29.3	24,64,124	7.1	1,69,86,701	48.9
Jharkhand	39.7	38,14,832	29.1	44,36,052	33.9	4,55,162	3.5	43,92,228	33.5
Odisha	41.8	41,03,989	23.4	67,39,993	38.4	7,83,080	4.5	59,14,527	33.7
Chhattisgarh	47.7	40,04,796	32.9	50,91,882	41.8	1,87,631	1.5	28,95,916	23.8
Madhya Pradesh	43.5	89,44,430	31.2	1,21,92,267	38.6	9,59,259	3.0	85,78,168	27.2
Gujarat	41.0	54,47,500	22.0	68,39,415	27.6	3,43,999	1.4	1,21,36,833	49.0
Daman & Diu	49.9	2,316	1.9	772	0.6	684	0.6	1,17,499	96.9
D & N Haveli	45.7	28,164	17.9	17,799	11.3	2,195	1.4	1,09,003	69.4
Maharashtra	44.0	1,25,69,373	25.4	1,34,86,140	27.3	12,25,426	2.5	2,21,46,939	44.8
Andhra Pradesh	46.6	64,91,522	16.5	1,69,67,754	43.0	14,39,137	3.7	1,45,24,493	36.8
Karnataka	45.6	65,80,649	23.6	71,55,963	25.7	9,13,227	3.3	1,32,22,758	47.4
Goa	39.6	31,354	5.4	26,760	4.6	14,708	2.5	5,04,426	87.4
Lakshadweep	29.1	0	0	0	0	264	1.4	18,489	98.6
Kerala	34.8	6,70,253	5.8	13,22,850	11.4	2,73,022	2.3	93,52,938	80.5
Tamil Nadu	45.6	42,48,457	12.9	96,06,547	29.2	13,64,893	4.2	1,76,64,784	53.7
Puducherry	35.7	12,099	2.7	68,391	15.4	7,892	1.8	3,56,586	80.1
A & N Islands	40.1	16,567	10.9	4,781	3.1	3,727	2.4	1,27,460	83.6

Source : Census of India, 2011.

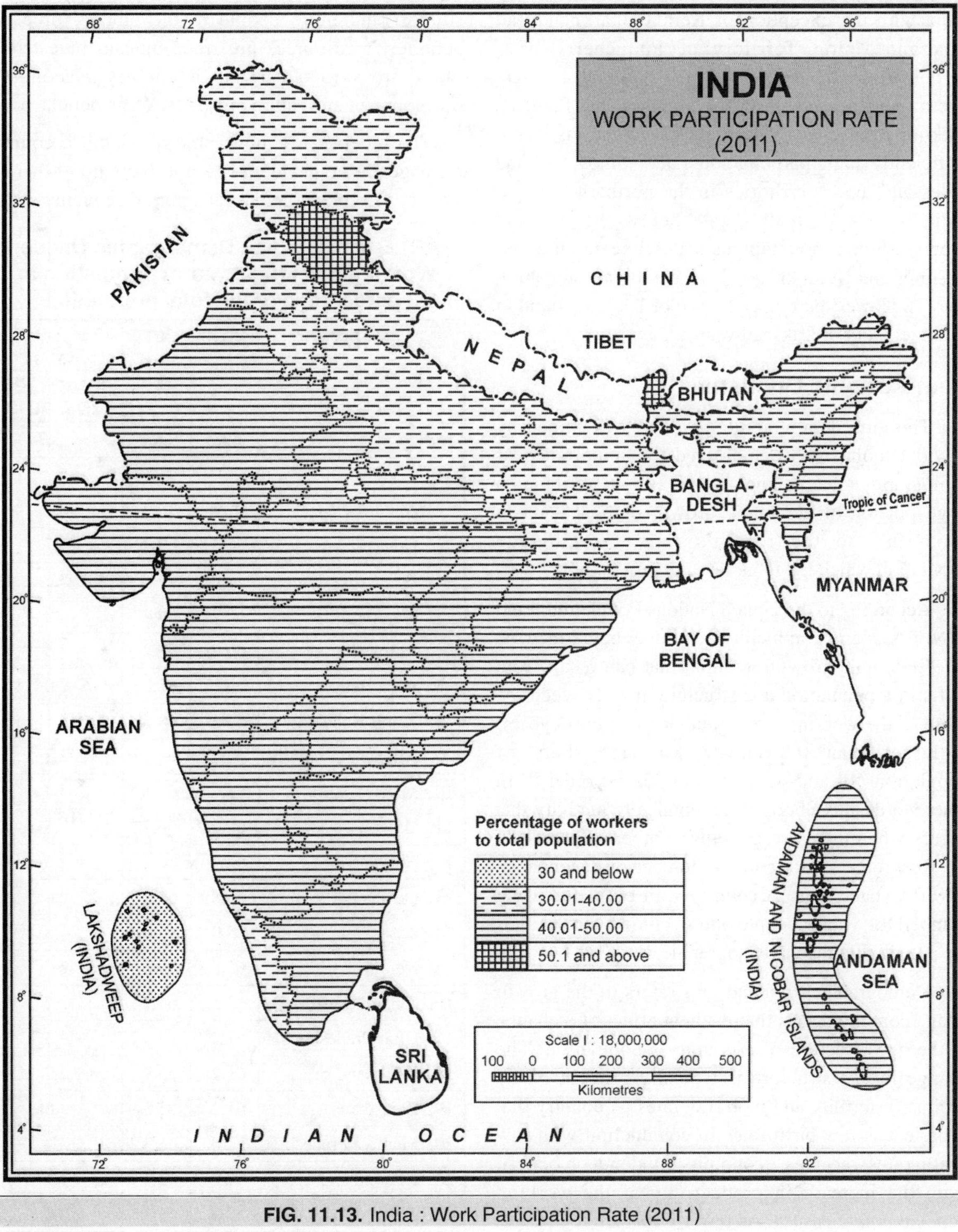

FIG. 11.13. India : Work Participation Rate (2011)

national average of 39.79. All states of the peninsular India with the exception of Kerala, Jharkhand and Goa and Union Territory of Puducherry have participation rate higher than the national average. The mountain areas of Mizoram, Himachal Pradesh, Sikkim, Arunachal Pradesh and Nagaland also have sufficiently high participation rate. Almost all the states and union territories in the northern plain of India have participation rate below the national average. Some mountainous states like Jammu and Kashmir and Uttarakhand also have low participation rate. The lowest participation rate of 29.09 is found in union territory of Lakshadweep.

Demographic Dividend

The term demographic dividend was coined by Devid Bloom, who emphasized the importance of demography to economic growth. Bloom attributed a large portion of economic growth of East Asia in 1965-1990 to the region's large working-age population which led to increase in productivity.

According to the United Nations Population Fund (UNFPA), demographic dividend may be defined as, "the economic growth potential that can result from shifts in a population age structure, mainly when the share of the working-age population (15 to 64 years) is larger than the non-working-age share of population (14 and younger and 65 and older)". In other words, it is a boost in economic productivity that occurs when there are growing number of people in the workforce, relative to the number of dependents. UNFPA stated that "A country with both increasing numbers of young people and declining fertility has the potential to reap a demographic dividend."

Demographic dividend thus refers to the growth in an economy that is the resultant effect of a change in the age structure of a country's population. The change in age structure is typically brought on by decline in fertility and mortality rates. A country that experiences low birth rates in conjunction with low death rates receives economic dividend or benefit from the increase in productivity of the working population that ensues. As fewer births are registered, the number of young dependents grows smaller relative to the working force. With fewer people to support and more people in the labour force the economy's resources are freed up and invested in other areas to accelerate a country's economic development and future prosperity of its populace.

To receive a demographic dividend, a country must go through a demographic transition where it switches from a largely rural agrarian economy with

TABLE 11.19. India : Demographic Dividend Working age (19-59 years) population as percentage of the total population

State/Year	2013	2018	Percentage points change 2018 over 2013
India	**63.3**	**66.0**	**2.7**
1. Telangana	N.A.	71.1	2.2
2. Andhra Pradesh	N.A.	70.9	2.1
3. Delhi	67.2	69.9	2.7
4. Jammu-Kashmir	66.0	69.6	3.6
5. Punjab	66.2	69.4	3.2
6. Karnataka	66.2	68.8	2.6
7. West Bengal	66.8	68.8	2.0
8. Tamil Nadu	66.1	68.6	2.5
9. Himachal Pradesh	64.4	67.2	2.8
10. Maharashtra	63.7	67.1	3.4
11. Gujarat	64.2	66.7	2.5
12. Haryana	64.4	66.7	2.3
13. Odisha	63.8	66.2	2.4
14. Assam	64.0	65.8	1.8
15. Kerala	63.8	65.8	2.0
16. Chhattisgarh	62.5	64.7	2.2
17. Uttar Pradesh	61.2	64.5	3.3
18. Jharkhand	61.2	63.9	2.7
19. Rajasthan	61.2	63.8	3.6
20. Madhya Pradesh	61.6	63.3	2.7
21. Uttarakhand	N.A.	62.4	1.7
22. Bihar	57.2	59.7	2.5

***Source*:** Sample Registration System's 2018 report.

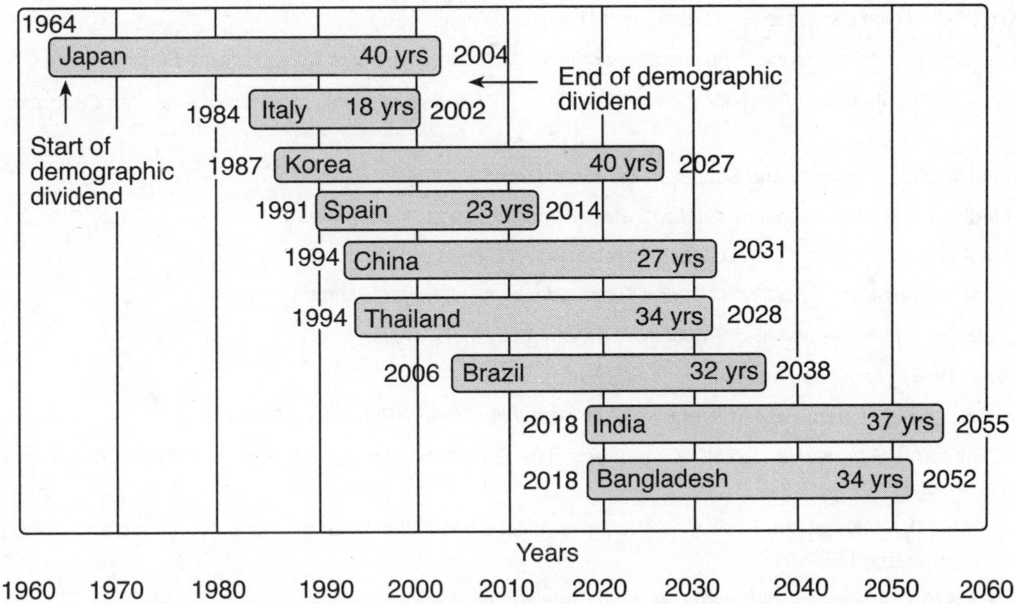

FIG. 11.14. Period of demographic dividend in some selected economics

high fertility and mortality rates to an urban industrial society characterized by low fertility and mortality rates. In the initial stages of this transition, fertility rates fall, leading to labour force that is temporarily growing faster than population dependent on it. All else being equal, per capita income grows more rapidly during this time too. This economic benefit is the first dividend received by a country that has gone through the demographic transition.

According to Sample Registration Systems 2018 report released in July 2020, "India's demographic dividend continues to swell with share of working age population in the country's total population increasing gradually." The report estimates that in 2018, nearly two-thirds of the population or 66% was between 15 and 59 years age group. An analysis of reports from 2013 to 2018 shows that this proportion has inched upeach year. Those of working age group constituted over two thirds of population in 12 of 22 major states (with population over one crore) in 2018 while in 2013 this was true in only four of the current states. Two states—Andhra Pradesh and Telangana had over 70 per cent of their population in the working age group, and Bihar was the lone state where less than 60% of the total population was in the working age group.

Generally speaking, states widely considered less developed had lower proportion of their population in the working age group. This is because most of these states have also had relatively high rates of fertility and larger proportion of dependent population.

According to United Nation Population Fund (UNFP), India's demographic dividend started in 2018 and will continue till 2055 (**Fig. 11.14**). This is fairly good and is in tune with other large economics.

REFERENCES

Bhardwaj, S.M. (1989) : *Hindu Places of Pilgrimage*, Berkelay.

Bhutani, Smita (1995) : *Demographic Dynamism in India*, Discovery Publishing House, New Delhi, Chapt. 5-7.

Bose Ashish (2010) India's Q: əst for Population Stabilisation, National Book Trust, New Delhi.

Bose, N.K. (2002) : *Tribal Life in India*, National Book Trust, New Delhi, India.

Bose, N.K. *et al.* (1973) : *The People*, The Gazetteer of India, Vol. one, Ministry of Education and Social Welfare Chapt. VI.

Brush, John E. : "Distribution of Religious Communities in India", *Annals of the Association of American Geographers,* 1948, pp. 81-96.

Census of India (1951) : *Literacy and Education Standard*, Paper 5, 1954.

Census of India (2011) : *Provisional Population Totals*, Paper 1 of 2011, India Series 1

Census of India (2011) : *Primary Census Abstract* released on C.D. in 2014.

Chakraborti, N. and Mukherji, F. (1971) : *Indian Tribes*, Saraswati Library, Calcutta.

Chandna, R.C. (2016) : *A Geography of Population*, Kalyani Publishers, Delhi.

Chatterji, S.K. (1945) : *Language and Linguistic Problem,* London.

Chatterjee, S.K. (1963) : *Language and Literature of Modern India*, Bengal Publishers, Calcutta.

Chatterjee, S.K. and S.M. Katre (1973) : *Languages*. The Gazetteer of India vol. one, Ministry of Education and Social Welfare, Chapter VII.

Dandekar, K. (1975) : Why has the Position of Women in India's population been declining, *Economic and Political Weekly,* No. 107 October, pp. 1663-67.

Dasgupta, J. (1970) : *Language Conflict and National Development,* Berkeley.

Desai, P.B. (1975) : *A Survey of Research in Demography*, Indian Council of Social Science Research, Popular Prakashan, Mumbai.

Dubey, R.N. and B.S. Negi (1974) : *Economic Geography of India*, Kitab Mahal, Allahabad, pp. 610-622.

Dubey, S.C. (1977) (ed.) : *Tribal Heritage in India : Ethnicity, Identity and Interaction*, Vikas Publishing House Pvt. Ltd., New Delhi.

Dutt, A.K. : "Religious Pattern of India", *Geo-Journal,* Vol. 3 (2), 1979, pp. 201-204.

Economic Survey, 2018-19, Vol. I, Ch. 7, pp. 136-40

Ghurye, G.S. (1959) : *The Scheduled Tribes*, Popular Prakashan, Mumbai.

Gopal, R. (1966) : *Linguistic Affairs of India*, Asia Publishing House, Mumbai.

Gosal, G.S. (1964) : *Literacy in India : An Interpretative Study*, Rural Sociology, Vol. 29, No. 3, pp. 261-277.

Govt. of India (1973) : *The Tribal People of India*, Ministry of Information and Broadcasting.

Haldar, Gopal (2000) : *Languages of India* (Translated by Tista Bagchi), National Book Trust, India, New Delhi.

Hassan, Mohammad Izar, (2005), Population Geography, Rawat Publications, Jaipur.

Husain, Majid (2011) : *Human Geography*, Rawat Publications, Jaipur.

India (2022) : A Reference Annual, Publication Division, Ministry of Information & Broadcasting, GOI, New Delhi.

Muthaiah, S. *et al.* (eds.) (1987) : *A Social and Economic Atlas of India,* Oxford University Press, New Delhi.

Krishan, G. (1980) : *Non Agricultural Workers in Rural India*, Asian Profile vol. 8, No. 2, April, pp.150-169.

Krishan, G. and Shyam, M. (1977) : *Literacy in India*, Geographical Review of India, Vol. 39, pp. 117-125.

Majumdar, D.N. (1958) : *Races and Cultures of India*, Asia Publishing House, Bombay.

Maurya, S.D. (2012) : *Human Geography,* Prayag Pustak Bhandar, Allahabad.

Premi Mahendra K. (2009), India's Changing Population, Profile, National Book Trust (NBT), New Delhi.

— — (2006) : Population of India (In the New Millennium : Census 2001), NBT, New Delhi.

Premi Mahendra K. and Dipendra Nath Das (2012) : Population of India, B.R. Publishing Corporation, Delhi.

Ramaswami Aiyar, C.P. *et al.* (1973) : *Religion*, The Gazetteer of India, Vol. 1, Ministry of Education & Social Welfare, Chapt. VII.

Raza Moonis and Aggarwal, Y.P. (1986) : *Inequalities in Levels of Literacy in India*, in Shafi M. and Raza M. (eds) *Spectrum of Modern Geography,* Concept Publishing Co., New Delhi, pp. 193-225.

Raza, Moonis and Ahmad Aijazuddin (1990) : *An Atlas of Tribal India*, Concept Publishing Company, New Delhi.

Rubenstein James, M. (2012) : *Contemporary Human Geography,* PHP Learning Pvt. Ltd., New Delhi.

Schermerharn, R.A. (1978) : *Ethnic Plurality in India,* Tuscon.

Schawartzherg, J.E. : The Distribution of Selected Castes in the North Indian Plains, "*Geographical Review,* Vol. 55, 1965, pp. 477-495.

Sharma, O.P. and D. Retherford (1993) : *Literacy Trends in the 1980s in India*, Occasional Paper No. 4 of 1993, Census of India, New Delhi.

Singh, K.S. (1972) (ed.) : *Tribal Situation in India*, Indian Institute of Advanced Study, Shimla.

Sopher David, E. (1967) : *The Geography of Religions,* Prentice Hall Inc., New Jersey.

Verma, R.C. (1995) : *Indian Tribes through the Ages*, Publication Division, Ministry of I & B, Govt. of India.

◆◇◆

CHAPTER 12

Migration Patterns

INTRODUCTION AND DEFINITION

Migration is a form of spatial mobility of population between one geographical unit and another involving a permanent change of residence (UN : 1958). The Census of India determines migration by place of birth or residence. *If a person was born at a place other than the place of enumeration, then he is treated as a migrant*. Of the three components of population change, migration holds a place of prominence—the other two components being *fertility* and *mortality*. Migration cannot be considered as a mere shift of people from one place of residence to another, as it is most fundamental to the understanding of continuously changing space-content and space-relationships of area (Gosal, 1961). Gill (1981) is of the opinion that movement over territories is a characteristic feature of all human populations irrespective of their stage of development. Bogue (1959) considers it an instrument of cultural diffusion and social integration, which yields more meaningful redistribution of population. Smith (1960) has talked about three-fold impact of migration on (*i*) the area of out-migration (*ii*) the area of in-migration and (*iii*) the migrants themselves. He has rightly remarked that areas of out-migration, in-migration and the migrants themselves never remain the same. The population of the area of out-migration decreases whereas the population of the area of in-migration increases. Migrants are also affected because there is a change in the residence of the migrants.

The studies regarding migration are seriously hampered due to lack of methodology and data constraints. Most scholars who write about migration theories and models recognise the very imperfect state of present-day theoretical and empirical knowledge of migration phenomenon (Germani, 1964). There is considerable agreement, that the study of migration has been hampered by the grave deficiencies in migration theories, which tend to be *time bound, culture bound and description bound* (Manglam and Schwarzweller: 1968). To some extent, this situation may be attributed to the greater complexity of migration as compared with the other two components of population change—fertility and mortality. According to **Jones** (1981), "Of the three components of population change, migration is the most difficult to conceptualize and measure". Because, it involves a change from the place of origin to the place of destination, migration has both a separative and an additive effect and both aspects are relevant to an understanding of why people move.

The data constraints are no less pronounced than the lack of methodology. The census of India does not provide any direct data on migration. It is only with the help of place of birth data that some idea of magnitude and direction of patterns of migration can be obtained. Although, post-independence censuses have tried to improve the quality and quantity of migration data, yet the census data are far from satisfactory for a reasonable analysis of migration.

Migration and Census of India. Along with other things, census data contain information about migration also. In fact, migration data were recorded at the time of first Indian census in 1881. It was based on the place of birth. In 1961, modification was made to include place of birth and duration of residence (if born elsewhere). In 1971, additional information on place of last residence and duration of stay at the place of enumeration were incorporated. Information on reasons for migration were incorporated in 1981 census and modified in consecutive censuses.

Following questions are asked about migration at the time of census enumeration:

- Is the person born in this village or town? If no, then further information is taken on rural/urban status of the place of birth, name of district and state and if outside India then name of the country of birth.
- Has the person come to this village or town from elsewhere? If yes, then further questions are asked about the status (rural/urban) of previous place of residence, name of district and state and if outside India then name of the country.

In addition, reasons for migration from the place of last residence and duration of residence in place of enumeration are also asked.

In the Census of India, migration is enumerated on the following two bases.

(*i*) Place of birth, if the place of birth is different from the place of enumeration (known as life-time migrant);

(*ii*) Place of residence, if the place of last residence is different from the place of enumeration (known as migrant by place of last residence).

Definition of Certain Terms Associate with Migration. *Place of origin* is the place which the migrant leaves, and, for that place, the person is an *out-migrant*. *Place of destination* is the place where the migrant arrives and for that place, the person is called *in-migrant*. Thus the same person is an out-migrant for the place of origin and in-migrant in the place of destination. When the migration takes place across international boundaries, the terms used are *emigration* and *immigration* to describe migration phenomenon, and migrant and immigrant to define a person.

Classification of migration. Migration can be classified on the basis of cause, distance and duration of time as illustrated below :

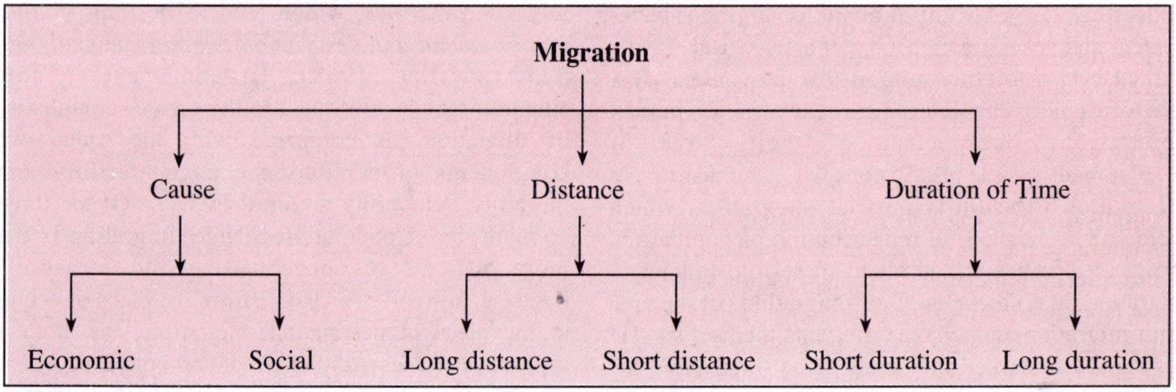

TYPES OF MIGRATION

Migration may be broadly classified as *international* and *internal*. India has experienced both, though at a much lesser scale as compared to some other countries of the world.

International Migration

Movement of population from one country to another, across the international borders, is called ***international migration***. International migration is of two types *viz.* out-migration of Indians to other countries and immigration into India from other countries.

OUT-MIGRATION

Instances of Indians migrating to other countries have been noticed in the historical times. In fact migration by Indians to other countries has a long history. During the reign of Ashoka the Great, several preachers travelled to south-east Asian countries to spread the Buddhist religion and popularize Indian art and culture. In the 19th century many Indians migrated to Mauritius, Burma (Myanmar), Sri Lanka, Malaysia, Fiji, Guyana, Surinam, Trinidad, South Africa and East Africa as labourers and settled permanently there. The first group of Indian labourers consisting of convicts and criminals was transported from Kolkata (Calcutta) in 1815 to work in sugarcane fields. After 1934, a large number of free workers were attracted by the mild oceanic climate of Mauritius. At present about three-fourths of the total population of Mauritius consists of people of Indian origin. There is a big colony of about 8000 Indians in the Reunion Islands near Mauritius. A large number of Indians also migrated to West Indies.

Indians also migrated to different countries of Africa during 18th and 19th centuries. Several merchants of Indian origin established their business with the help of Arab traders. A large number of Indians were encouraged by the European rulers to migrate to African countries and work there as labourers to clear the forests and provide land for agricultural purposes, to work in their plantations and to provide labour force for constructing railways. These labourers were followed by other professionals such as traders, shop-keepers and money lenders. Thus large Indian colonies came up in central and southern parts of Africa. The countries where Indian colonies were developed include Kenya, Uganda, Tanganika, Mozambique, Tanzania, Zimbabwe, Somalia, Ethiopia, Zambia, Natal, South Africa, etc. Most migrations to African countries originated from the states of Punjab, Gujarat, Maharashtra, Tamil Nadu and Kerala.

Indians started migrating to West Indies after 1840 and their main destinations were Guinea and Trinidad. Each has over two lakh Indians. Some Indians had migrated to Jamaica, Martinique and Guadeloupe also. Most of the Indians migrated to these islands as labourers to work in the agricultural fields. Majority of such migrants were from Eastern Uttar Pradesh and Bihar. At present, Indians constitute over 50% of the population of West Indies.

THREE WAVES OF INDIAN DIASPORA

There have been three waves of Indian Diaspora at three different times in the history of India. According to Migration Policy Institute diaspora is defined as "people of Indian origin who live outside the country but continue to exhibit some of India's ethnocultural characteristics."

1. The first wave took place during the British period when a large number of labourers were sent to Mauritius, Caribbean islands (Trinidad, Tobago and Guyana), Fiji and South Africa by British from Uttar Pradesh and Bihar; to Reunion Island, Guadeloupe, Martinique and Surinam by French and Dutch and by Portuguese from Goa, Daman and Diu to Angola, Mozambique to work as plantation workers. All such migrations were covered under the time-bound contract known as ***Girmit Act*** (Indian Emigration Act). These labourers were living in inhuman conditions and their living conditions were as bad as those of slaves.

2. The second wave of migrants went to settle in the neighbouring countries like Thailand, Malaysia, Singapore, Indonesia, Brunei and some African countries. This is a recent development under which professionals, artisans, traders, factory workers went out of the country in search of better quality of life. This trend is still continuing. In the 1970s, there was oil boom in West Asia and a large number of skilled and semi-skilled workers went there to avail of the opportunity Some entrepreneurs, store owners, professionals, businessmen, etc. went to western countries also.

3. The third wave started in 1960s and still continues to operate. In this period high profile professionals like doctors, engineers, etc. migrated out of India. In 1980s, software engineers, management consultants, financial experts, media persons, etc. moved out of the country to work in countries like the U.S.A., Canada, U.K., Australia, New Zealand, Germany, etc. After liberalisation in 1991, education and knowledge based Indian migration has made Indian Diaspora most powerful in the world.

By the end of the 19th country, a large number of Indians migrated to Sri Lanka and Malaysia to work as labourers in tea and rubber plantations respectively. At present there are about five million Indians in Sri Lanka which account for about one-third of the total population of that country. Similarly about 10 per cent population of Malaysia is of Indian origin. Most of the migrants to Sri Lanka and Malaysia are from Tamil Nadu and they had been fighting a long war for their rights. Some labourers from Bihar, Uttar Pradesh, West Bengal and Odisha also migrated to these countries.

Since 1970s, highly educated and skilled professionals have migrated from India to some of the most advanced countries such as the U.S.A., the U.K., Canada, Australia, New Zealand, Japan and some countries of Western Europe, leading to a serious problem of *brain drain*. A less educated work force had earlier migrated to these countries since Independence as labourers. During the last five decades or so, many Indians have migrated to the oil-rich countries of the Middle East. Some migration from India has taken place to South-east Asian and south American countries.

But global recession in 2007-08 dealt a big blow to migration and many Indians had to come back to India due to heavy reduction in job opportunities in North American and European countries. Out migration from India again picked up after normalization of world economy. But Indian migration to Australia has suffered a set back due to racial discrimination and violence against Indians. Again, a large number of Indian migrants had to come back to India due to Covid-19. Migration of people from India during historic times is shown in **Fig. 12.1**.

Partition of Indian subcontinent and International Migration

Partition of the Indian subcontinent in 1947 resulted in large scale migration of population from India to Pakistan and Bangladesh and *vice versa*. This has been termed as one of the greatest movements of human beings in the history of the world. In spite of its short duration, it had long term consequences for

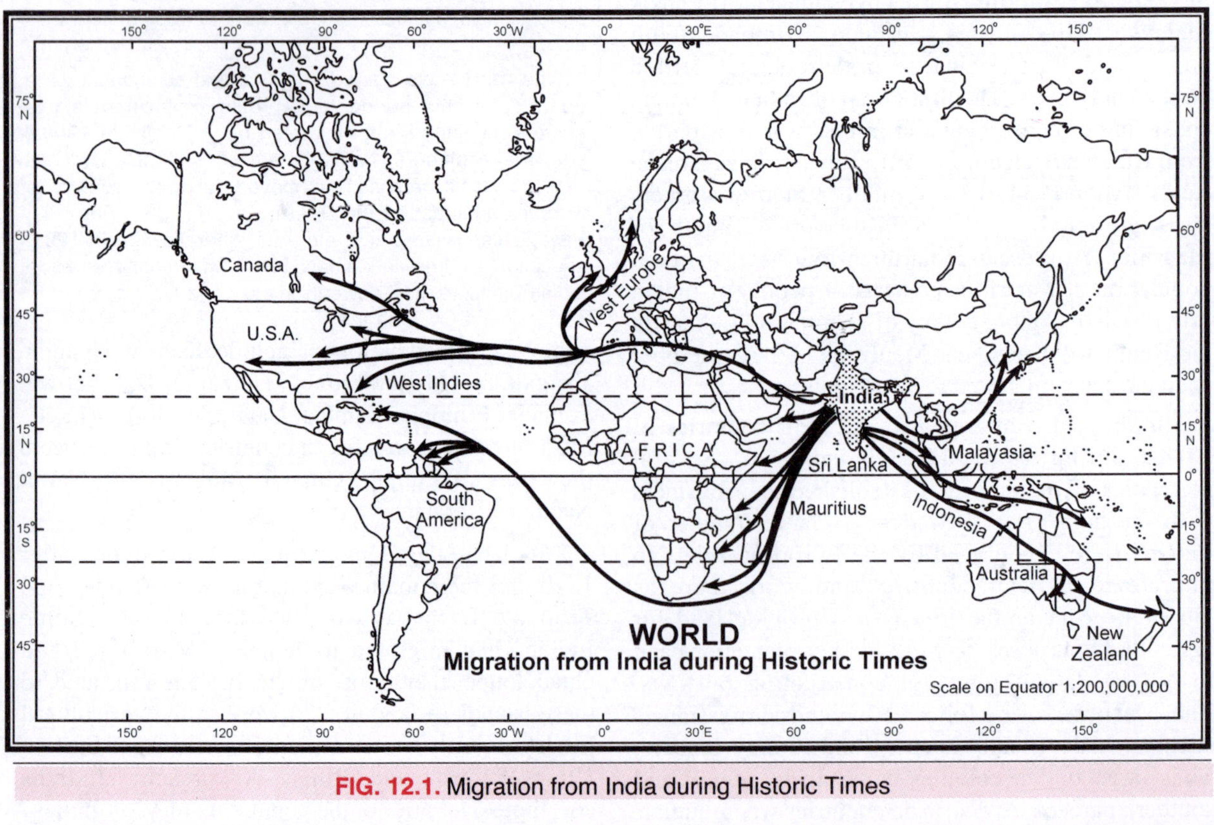

FIG. 12.1. Migration from India during Historic Times

the economy and society of the three concerned countries. About 14.5 million people moved across the borders. Hindus and Sikhs came to India from Pakistan and Bangladesh and Muslims from India went to these countries. In 1951, 7.3 million refugees were enumerated in India (4.7 million from Pakistan and 2.6 million from Bangladesh). The 1951 census of Pakistan enumerated 7.2 million *muhajirs* (refugees) from India. The patterns of migration due to partition of subcontinent are shown in **Fig. 12.2**.

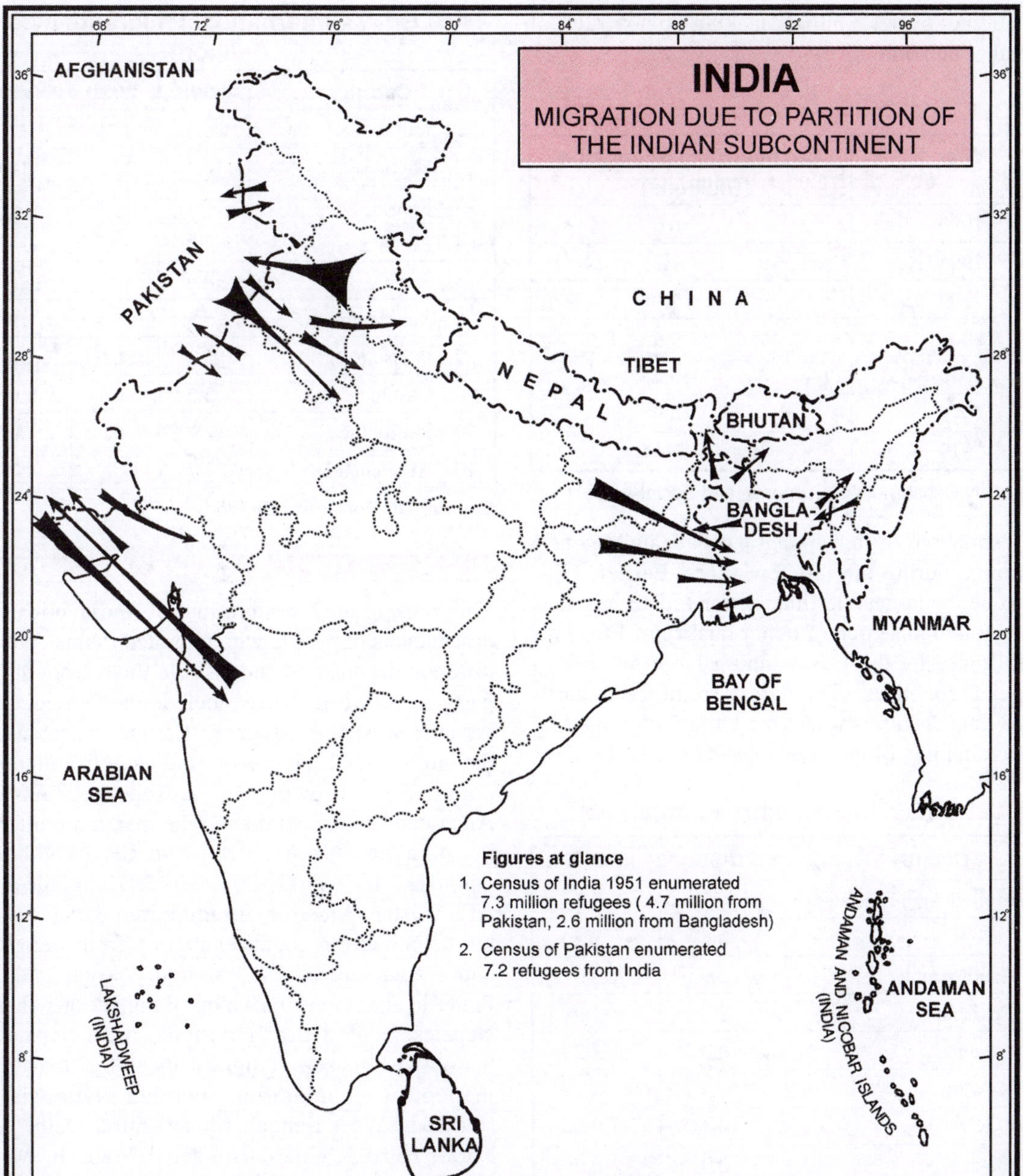

FIG. 12.2. Partition of Indian Subcontinent and International Migration

Recent Trends. Number of people migrating out of India has been rising steadily since 1990. The number of those who migrated from India to other countries of the world was a meagre 6.6 millions in 1990 which increased to 17.5 millions in 2019 (**Table 12.1**). But it is miniscule group compared to huge population of over 1.3 billion, making up just 1.2% of the total population.

TABLE 12.1. Indian diaspora (1990-2019)

Year	Out migrating population (millions)
1990	6.6
1995	7.2
2000	7.9
2005	9.6
2010	13.2
2015	15.9
2019	17.5

Source: UN Department of Economic and Social Affairs

Destinations of Indian migrants have undergone a sea change during the last few years. Earlier, U.K. used to be the largest destination for Indian migrants who migrated in search of better quality of life. This was followed by the U.S.A. and a large number of European countries. The latest trend is towards Middle East countries where the Indian migrants find ready availability of lucrative jobs (**Table 12.2**).

TABLE 12.2. Indians abroad (millions)

Country	1990	2017
1. United Arab Emirates	0.5	3.3
2. U.S.A.	0.5	2.3
3. Saudi Arabia	0.9	2.3
4. Pakistan	2.9	1.9
5. Oman	0.2	1.2
6. Kuwait	0.4	1.2
7. U.K.	0.4	0.8
8. Qatar	0.0	0.7
9. Canada	0.2	0.6
10. Nepal	0.4	0.4

Source: UN Population Division

India sends most migrants abroad

In 2020, about 18 million Indians were living in countries other than their birthplace. This is world's largest international migrant population and is followed by Mexico, Russia and China (**Table 12.3**).

TABLE 12.3. International migrants by origin (million)

Country	1990	2020	% change
1. India	6.6	17.9	170.0
2. Mexico	4.4	11.2	154.5
3. Russia	12.7	10.8	–15.0
4. China	4.2	10.5	147.2
5. Syria	0.6	8.5	1,263.4
6. Bangladesh	5.5	7.4	35.8
7. Pakistan	3.3	6.3	89.4
8. Ukraine	5.5	6.1	10.7
9. Philippines	2.0	6.1	202.8
10. Afghanistan	7.7	5.9	–23.8

Source: UN Population Division.

IN-MIGRATION

The present day population of India consists of descendants of people who migrated to this country at different pre-historic and historic times from different parts of the world. These include the Dravidians, the Aryans, Muslisms, Mughals, Europeans, etc. Most of the international migrants in India are from the Asian countries, followed by Europeans, Africans, Americans and Australians. The maximum migration in India has taken place from the neighbouring countries like Pakistan, Nepal, Bangladesh and Afghanistan. Most of the immigrants in India from the neighbouring countries mainly from Bangladesh and Nepal, are concentrated in Assam and Uttar Pradesh. Assam receives more Bangladeshis than the Nepalese while Uttar Pradesh has more Nepalese for the obvious reason of significance of distance factor in population migration. Foreign Nationals have settled in West Bengal, Maharashtra, Delhi, Tamil Nadu, Bihar, Punjab, Himachal Pradesh, Madhya Pradesh, Arunachal Pradesh, Odisha, Gujarat, Goa and Kerala.

Refugee Influx into India. Although India has not signed the 1951 Refugee convention, yet, India

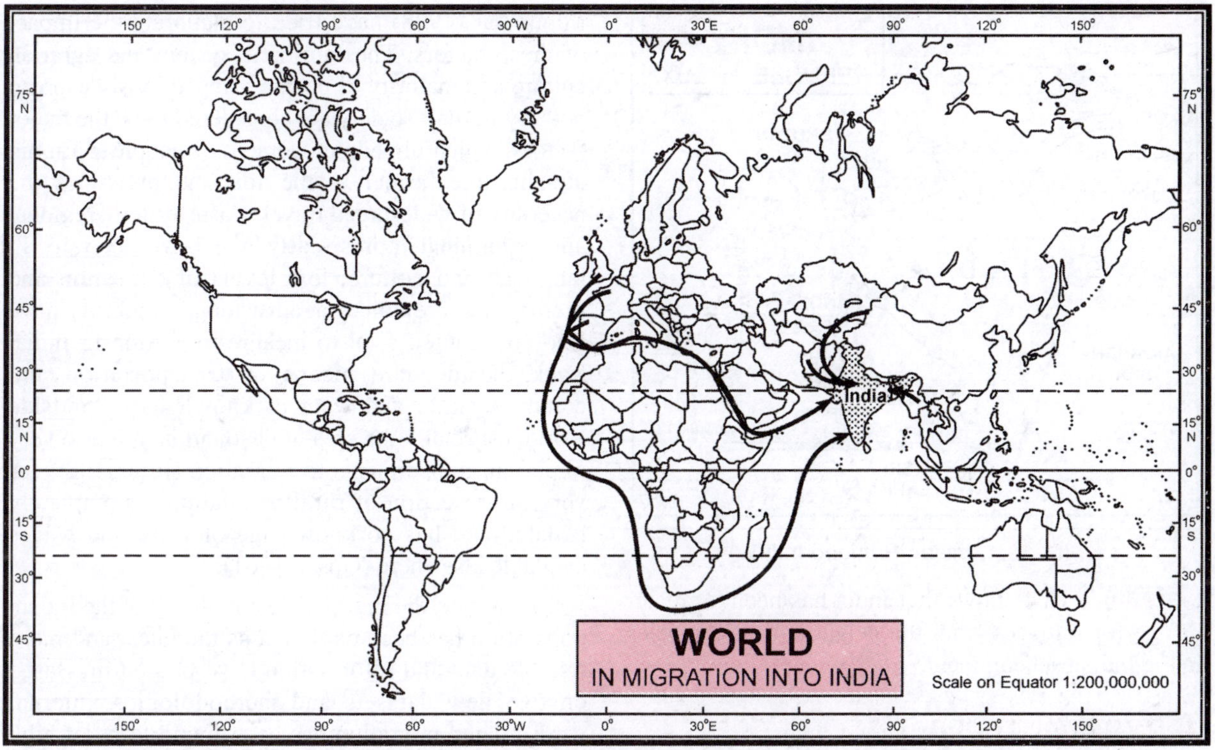

FIG. 12.3. In-migration into India

has been the main destination for refugees during the last about six decades and there has been a surge in refugees from the neighbouring countries like Tibet (China), Afghanistan, Sri Lanka, Myanmar and Bangladesh. The estimated figures say that there were more than 3 lakh refugees in India in 2011 making India one of the top 25 refugee recipients in the world. The following brief description gives an idea of refugee influx into India.

1. Tibetans. Dalai Lama was allowed to set up a government in-exile at Dharamsala (Himachal Pradesh) in 1959. This brought in a wave of Tibetan refugees and currently there are about 1,10,000 Tibetans in India. Of them 80,000 have resident permits and are offered low paying public works jobs by the Indian government.

2. Afghans. In 1979 about 60,000 Afghans fled to India when Soviet Union invaded that country. In 1992 and afterwards, thousands of Afghanistanis took refuge in India when Taliban captured power in that country. In 2007, UNHCR estimated 9,200 Afghans in India and 4,000 asylum seekers in the process of refugee certification. In 2021, Taliban again captured Afghanistan and thousands of Afghanis are seeking refuge in India.

3. Sri Lanka. The civil war in 1983 triggered first wave of Sri Lanka Tamil refugees to India. By 1995 India and UNHCR repatriated 100,000 Tamil Sri Lankans as the war ended. In 2008, 73,000 Sri Lankan refugees were living in 117 camps mostly in Tamil Nadu.

4. Myanmar. 50,000 people fled from Myanmar into Mizoram in 2004 in the wake of military oppression. In 2017, nearly 40,000 Rohingyas muslims, illegally entered into India from Myanmar.

5. Bangladeshis. About 50,000 Chakma refugees fled Bangladesh into Tripura in 1988.

Until December 2008, the Indian policy did not charge refugees a "visa or penalty fee" for a residence permit. This was one of the main reasons for surge of refugees into India. However, since then the numbers are dwindling. But thousands of Bangladeshis are sneaking illegally into India, making use of porous border between India and Bangladesh.

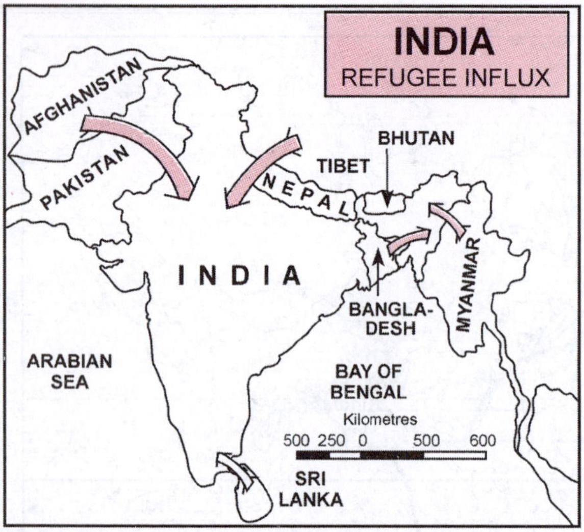

FIG. 12.4. India : Refugee Influx

Figure 12.4 shows that India has been a greater refuge for refugees from the neighbouring countries in the Indian subcontinent.

Internal Migration

The Indian population has been considered as one of the least mobile populations of the world (Davis 1951, Gosal 1961, Skeldon 1986). Migration rates in India are desperately low in comparison to most of the European and North American countries. Davis (1951) found from an analysis of place-of-birth statistics from the 1931 census that while 22.5 per cent of the native population in the United States in 1940 lived outside the state in which they were born, only 3.6 per cent of the Indian population in 1931 lived outside the state or province of birth. Blunt (1938) commented that nowhere in the world is the population so immobilised as it is in India. At every census, some 90 per cent of the people are enumerated in the district in which they were born; and of the rest, some seven per cent were born in neighbouring districts. This has been supported by Zachariah (1964) during his study of inter-state migration in India between 1902 and 1931. Several factors have been held responsible for the stubborn immobility of Indian population. These include factors of economic, social, cultural and demographic character. Several scholars like Kamath (1914), Wattal (1934), Chandershekhar (1951), Davis (1951), Gosal (1961), D'Souza (1964) and Majumdar and Majumdar (1978) have tried to elaborate the impact of these factors. The familistic system and agrarian culture and majority of people gripped by widespread poverty in rural society are considered to be the major factors, which impede movement of people. Farming attaches the farmer to the soil and precludes the necessity of widespread travel. Caste system, regional and communal mores, early marriage, diversity in language and culture, low levels of education and literacy and prejudices against female mobility have also contributed a lot to meagre migration in India. Lack of adequate means of transportation and communication and want of knowledge of outside world, especially the area of destination, are also very important impediments to migration (Raju, 1989). A very large section of rural population is completely isolated and has no knowledge of work and wages available elsewhere (Gosal, 1961).

However, the myth of least mobility of the Indian population has been exploded as the latest and more reliable data and information have poured in. Based on two new datasets and methodologies, internal work-related migration has been found to be of high level. (*i*) Analysing the changes in the same age cohort using Census data yields an annual inter-state migration of about 5 to 6.5 million between 2001 and 2011. (*ii*) Railway passenger data analysis suggests an annual inter-state migration flow of close to 9 million between 2011 and 2017. Clearly, rising growth after 1980s has led to an acceleration of labour migration flows as rewards of better economic opportunities have overcome the cost of moving. Historically, migration of people for work and education has been a phenomenon that accompanies the structural transformation of economies and has paved the way for release of "surplus labour" from relatively low-productive agricultural activities to sectors enjoying higher productivity (like industry, trade, transport etc.). The resulting remittance flows increase household spending in the receiving regions and further the economic development of less-developed regions.

The patterns of flow of people are broadly consistent with popular conception—"*less affluent states see more people migrating out while the most affluent states are the largest recipients of migrants*." Following three findings emerge on the basis of data analysis :

1. India is increasingly on the move—and so are Indians. The study shows that annually inter-state labour mobility averaged 5-6 million people between 2001 and 2011, yielding an inter-state migrant population of about 60 million and inter-district migration as high as 80 million. The first-ever estimates of internal work-related migration using railways data for the period 2011-2016 indicate an annual average flow of close to 9 million people between the states. Both these estimates are significantly grater than in annual average number of about 3.3 million suggested by successive censuses and higher than previously estimated by any study.

2. Migration is accelerating. In the period 2001-11, according to Census estimates, the annual rate of growth of labour migrants nearly doubled relative to previous decade, rising to 4.5 per cent per annum in 2001-11 from 2.4 per cent per annum in 1991-2001.

3. While internal political borders impede the flow of people, language does not seem to be a demonstrable barrier to the flow of people.

Magnitude and Patterns of Migration

Figure 12.5 shows the all India net annual passenger flows for India from 2011-12 to 2015-16. Net flows at the All-India level have averaged close to 9 million, peaking around 2013-14, considerably above level suggested by the Census.

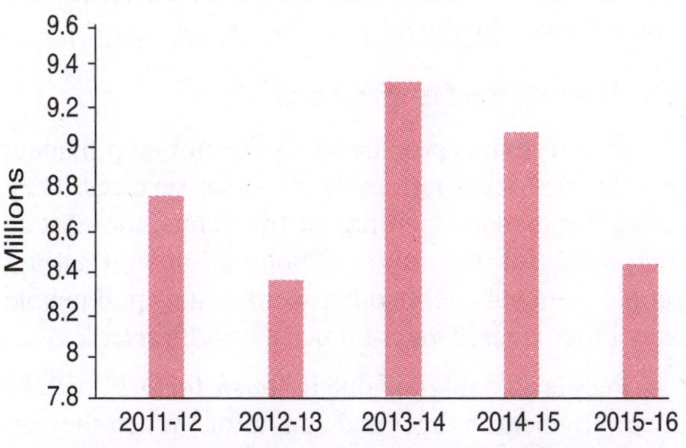

FIG. 12.5. Total Net Flows at All India Level
Source: Survey Calculations

According to the survey, the largest recipient was the Delhi region, which accounted for more than half the migration in 2015-16, while Uttar Pradesh and Bihar taken together account for half of total out-migrants. Maharashtra, Goa and Tamil Nadu had major net in-migration, while Jharkhand and Madhya Pradesh had net out-migration.

Figure 12.6 shows the largest inter-state migration routes. States like Delhi, Maharashtra, Tamil Nadu, and Gujarat attract large swathes of migrants from the Hindi heartland of Uttar Pradesh, Bihar, and Madhya Pradesh. Kolkata in West Bengal attracts migrants from nearby states of Jharkhand, Uttar Pradesh and Odisha making evident one of the laws of migration propounded by Ravenstein (1885)—"*There is a process of absorption, whereby people immediately surrounding a rapidly growing town move into it and the gaps they leave are filled by migrants from more distant areas, and so on until the attractive force is spent.*" There is an interesting dynamic between Gujarat and Maharashtra where Surat has started acting as a counter magnet region to Mumbai and attracts migrants from the neighbouring districts of Maharashtra. Other counter-magnet region dynamics are observed in Jaipur and Chandigarh (to Delhi).

Reasons for Migration

Data on reasons for migration were collected for the first time in the 1981 census. These reasons were classified into five categories *viz*. (1) employment, (2) education, (3) family moved (associational), (4) marriage and (5) others. Two new categories of reasons were added at the time of 1991 census. They were (1) business and (2) natural calamities like drought, floods, etc. Reasons for migrations were again revised on the eve of 2001 census. According to this scheme reasons for migration were classified into work/employment, business, education, marriage, moved after birth, moved with household and others.

At the time of 2011 census reasons for migration we cited as (*i*) marriage, (*ii*) moved with household, (*iii*) moved after birth, (*iv*) work (employment), (*v*) education, (*vi*) business, (*vii*) others. According to census 2011 figures there were 453.6 million migrants out of which 312.7 million were women and 140.9 million were men. About 38% of all Indians were migrants, up from 31% in 2001. It is also estimated that about 23% of men and 53% of women are migrants, compared with 18% and 45% in 2001.

1. Work/Employment. Among the reasons cited for migration by those surveyed, about 10% moved in search of employment during the decade ending 2011, down from 15% in 2001. Males migrate more in numbers than their female counterparts in search of a livelihood.

People migrate in large numbers from rural to urban areas in search of employment. The agricultural base of rural areas does not provide employment to all the people living there. Even the small-scale and cottage industries of the villages fail to provide employment to the entire rural folk. Contrary to this, urban areas provide vast scope for employment in industries, trade, transport and services.

2. Marriage. Marriage remains the biggest cause for migration. Every girl has to migrate to her in-law's place of residence after marriage. Thus, the entire female population of India has to migrate over short or long distance.

In 2011 marriage accounted for 49% migration, up from 44% in 2001.

3. Education is another important reason for migration. Rural areas, by and large, lack educational facilities, especially those of higher education and rural people have to migrate to the urban centres for this purpose. Many of them settle down in the cities for earning a livelihood after completing their education.

The share of those moving to pursue education in 2011, makes up only 2% of migrants, down from 3% recorded in 2001.

4. Moved after Birth. The census questionnaire includes a reason designated as "moved after birth" which is mainly children being born in a place—mother's village, or nearby town with hospital—and then moving back to the parental house. This share has increased from 7% in 2001 to 11% in 2011.

5. Moved with Household. About 21% of the migrants had said in 2001 that the reason for migrating was that their family was moving. They would be dependents, both young and old. This share has declined to 15% in 2011.

6. Business. Business seems to be a very insignificant factor responsible for migration, although it has great financial value. Only one per cent of the total migrants were reported to have migrated to pursue business in 2001 and this figured remained the same in 2011 also.

TABLE 12.4. Reasons for Migration (in per cent)

	Reason	2001	2011
1.	Marriage	44	49
2.	Moved with household	21	15
3.	Moved after birth	7	11
4.	Work/Employment	15	10
5.	Education	3	2
6.	Business	1	1
7.	Others	9	12

Source : Census of India, 2011.

7. Lack of security also causes migration. Political disturbances and interethnic conflicts drive people away from their homes. Large number of people had migrated out of Jammu & Kashmir and Assam during the last few decades due to disturbed conditions there. A large number of Kashmiri Pandits had to leave Kashmir in wake of prolonged ethnic conflict.

Similarly, thousands of people had to migrate out of Punjab during the period of terrorism by religious fundamentalists.

People also migrate on a short-term basis in search of better opportunities for recreation, health care facilities, legal advice or for availing service which the nearby towns provide.

'Pull' and 'Push' Factors

Urban centres provide vast scope for employment in industries, transport, trade and other services. They also offer modern facilities of life. Thus, they act as 'magnets' for the migrant population and attract people from outside. In other words, cities pull people from other areas. This is known as ***'pull factor'***.

People also migrate due to ***'push factors'*** such as unemployment, hunger and starvation. When they do not find means of livelihood in their home villages, they are 'pushed' out to the nearby or distant towns. Millions of people who migrated from their far-off villages to the big cities of Kolkata, Mumbai or Delhi did so because these cities offered them some promise for a better living. Their home villages had virtually rejected them as surplus population which the rural resources of land were not able to sustain any longer.

CONSEQUENCES OF MIGRATION

Consequences of migration are as varied and diverse as its causes. Migration affects both the area of origin of migration and the area of destination of the migrant population. Over and above, the people are affected in a number of ways. The consequences of migration can be broadly grouped as demographic, social and economic.

(*i*) Demographic Consequences

Migration brings about changes in the characteristics of the population in both the regions, *i.e.*, region of out migration and region of in-migration. It changes not only the age and sex composition of population but also affects the rate of growth of population. Generally, the proportion of the old, children and females increases in population of source areas due to out-migration. On the other hand, the proportion of these persons in the population of in-migration areas gets generally lowered. Migration is one of the major causes of high sex ratio in source areas and low sex ratio in the receiving areas. This happens because it is mostly the youthful male population which is involved in migration. Thus not only the number of people but also the structure of population in both the regions involved in migration is changed. This leads to change in rates of fertility, mortality and consequently in the growth of population. The source regions are depleted of the youthful population and this results in lowered rates of births and comparatively lower rates of growth. An inverse impact is observed in case of the population structure of the receiving areas.

(*ii*) Social Consequences

Migrants are very good agents of social change as they bring new ideas related to technologies, family planning, girl education, etc.

Migration results in intermixing of diverse cultures and leads to the evolution of composite culture. It breaks the narrow considerations and widens the mental horizon of the people. In the historic times, India received migrants belonging to different cultural groups which led to inter-mixing of one culture with the other. But at the same time, migration has serious negative consequences also. These include anonymity, which creates social vacuum and sense of dejection among individuals. Continued feeling of dejection may motivate people to fall in the trap of anti-social activities like crime and drug abuse.

(*iii*) Economic Consequences

The effects on the resource-population ratio is one of the major economic consequences of migration. This ratio changes in both the source regions and the receiving regions. The resource-population ratio may be such in an area which might be called either ***under populated*** or ***over populated or optimally populated***. The condition of under population means that population is too low to allow development and utilisation of its resources. On the other hand, over population leads to high pressure of population on resources and this condition generally results in low standards of living. A country having enough number of people to enable development and utilisation of its resources without lowering the quality of life is called ***optimally populated***. If the people are moving from an over populated area to an area of under population, the result is in the balancing of the resource-population ratio. On the other hand, if the migration is from an area of under population to over populated or optimally populated, the consequences may be harmful to both the areas.

Migration affects the occupational structure of the population in both the regions. Generally the proportion of working population in source areas is lowered and the same proportion in the receiving areas is increased. Thus the population of the receiving areas tends to become more productive and in the source areas it results in increasing the dependency ratio by reducing the proportion of the working people in the population. One of the serious consequences of migration is **'brain drain'.** This refers to the migration of the skilled persons from the poorer countries to the developed countries in search of better economic opportunities. An example can be of the migration of doctors and engineers etc. from India to the U.S.A. the U.K. and Canada. This type of migration does not alter the resource-population ratio significantly as the number of people involved in migration is not very large. However the quality of

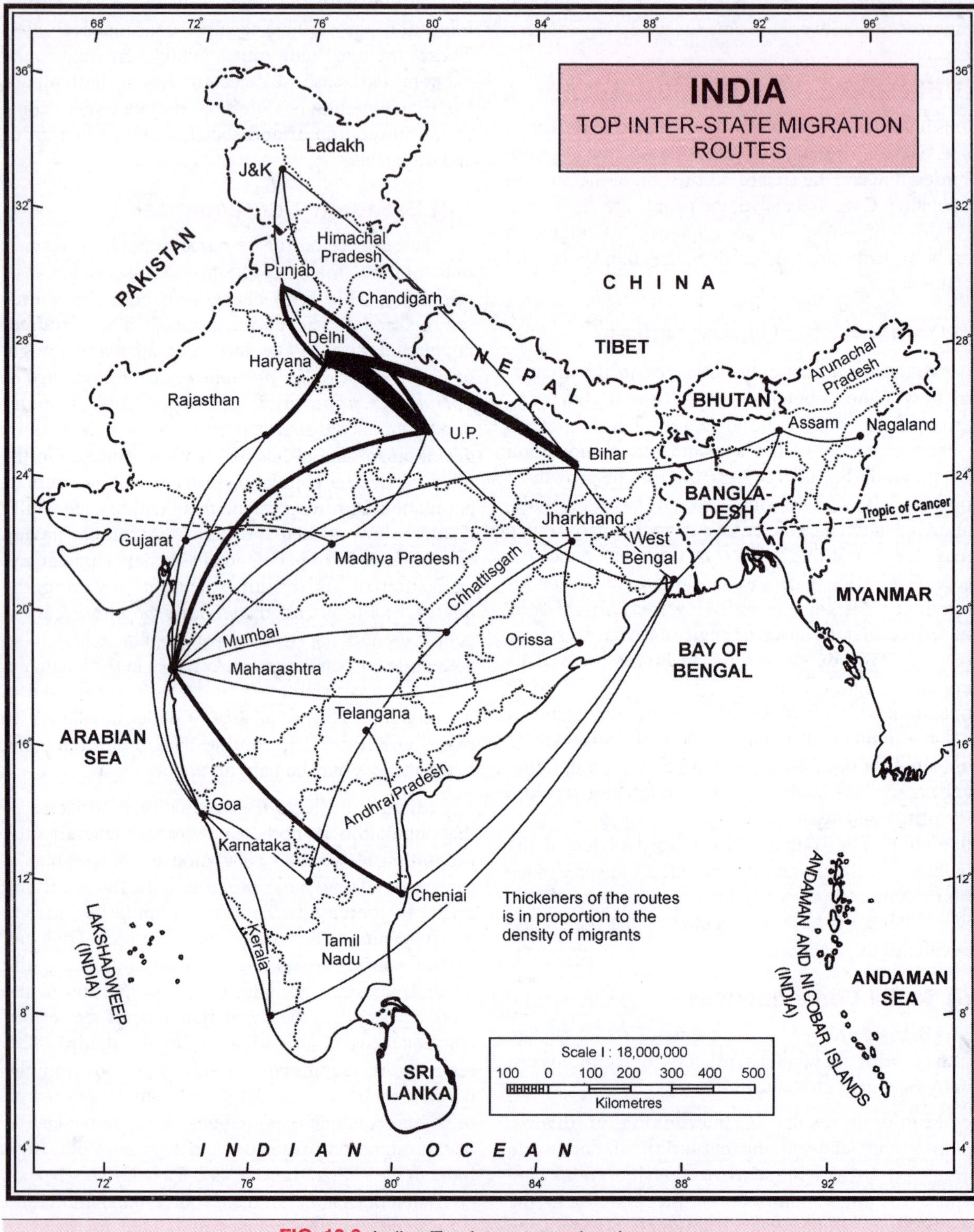

FIG. 12.6. India : Top inter-state migration routes

human resources in the source region suffers a lot. The resources of the source regions which are generally poorer countries can not be developed fully because of the huge size of the population.

Most people migrate for economic gain and economic benefit is the most important consequence of migration. People migrating out send remittance to their families at home and add to economic prosperity. Remittances from the international migrants are one of the major sources of foreign exchange. According to World Bank's Migration and Remittances Factbook 2008, India is the top receiver of remittances from abroad and received US $27.0 billion in 2007. This was followed by China (US $25.7 billion), Mexico (US $25.0 billion), Philippines (US $17.0 billion), France (US $12.5 billion), Spain (US $8.9 billion), Beligium (US $7.2 billion), Germany and UK (US $7.0 billion each), and Romania (US $6.8 billion). Punjab, Kerala and Tamil Nadu receive very significant amount from their international migrants. Remittances by internal migrants also plays a significant role in the economic growth of the source regions. These remittances are used for food, repayment of debts, treatment, marriages, children's education, agricultural inputs, construction of houses, etc. For thousands of the poor villages of Bihar, U.P., Odisha, Andhra Pradesh, Himachal Pradesh, etc. remittance works as life blood for their economy.

With the ushering in of the Green Revolution particularly in Punjab, Haryana and western part of Uttar Pradesh, large number of poor people migrated to these states from eastern part of Uttar Pradesh, Bihar, Madhya Pradesh and Odisha. These agricultural labourers had been a great source of remittance to their home villages. But a large number of Bihari labourers working in Punjab have gone back to their places as prospects of employment opportunities have increased there.

(*iv*) Environmental Consequences

Large scale rural-urban migration leads to overcrowding in the cities and puts tremendous pressure on the infrastructure. It also results in unplanned and haphazard growth of cities in which slums and shanty colonies are very common.

Over-crowding is also related to over-exploitation of natural resources and cities are facing serious problems of water shortage, air and water pollution, problem of sewage disposal and management of solid wastes.

(*v*) Other Consequences

Migration has a deep impact on the status of women. Generally, male members of the family migrate from rural to urban areas and leave their wives behind at home. This puts tremendous physical and mental pressure on the women. Although migration of women enhances their autonomy and role in economy yet it increases their vulnerability.

Migration enhances remittances to the source region but there is heavy loss of human resources, particularly those of skilled people. Market for advanced skills has increased considerably at the international level. As such most dynamic industrial economies are admitting and recruiting significant proportions of the highly trained professionals from poor regions. Consequently, the existing under-development in the source region gets reinforced.

MIGRATION STREAMS

Depending on place of birth (or last residence) and place of enumeration, migrants can be classified into following four migration streams, which are roughly indicative of migration distance.

1. **Intradistrict migrants** are the persons born (or with last residence) outside the place of enumeration but within the same district.
2. **Interdistrict migrants** are the persons born (or with last residence) outside the district of enumeration but within the same state.
3. **Interstate migrants** are the persons born (or with last residence) outside the state of enumeration but within India.
4. **Immigrants** are the persons born (or with last residence) outside the country.

Based on the rural or urban nature of the place of birth (or of last residence) and the place of enumeration, internal migrants can be further classified into following four migration streams :

(*a*) rural to rural (R – R)
(*b*) rural to urban (R – U)
(*c*) urban to urban (U – U)
(*d*) urban to rural (U – R)

(*a*) Rural-to-Rural (Rural Turn Over)

This stream of migration dominates over all other streams in terms of volume of migration. An outstanding feature of rural-to-rural stream of migration is the preponderance of female migrants. Studies made by Zachariah (1964), Bose (1965), Agarwal (1968), Narain (1975), Bhande *et al*. (1976), Premi (1976) and Kumar and Sharma (1979) have shown that this preponderance of female migrants is primarily due to the prevalence of patriarchal residence after marriage (marriage migration). According to Indian tradition, the girl has to move from her parents' residence to the residence of her in-laws and live with her husband.

Apart from marriage migration, there are several other factors which contribute to large scale rural-to-rural migration. Migration of agricultural labourers and movement of people to the newly reclaimed areas for agricultural purposes constitute the most important component of such migrations. In slack agricultural season, a large number of villagers move out to seek casual employment in irrigation projects, construction of roads, rail-roads and buildings and other miscellaneous menial jobs.

Normally, rural-to-rural migration originates from crowded areas of low productivity and is destined towards sparsely populated areas experiencing large scale developmental activities. Such a migration may take place even for longer distances and may result in permanent redistribution of population. The introduction of Green Revolution in Punjab, Haryana and western Uttar Pradesh in 1960s generated considerable migration of labour force from economically depressed areas of eastern Uttar Pradesh and Bihar to these areas. Neither the innate love of home nor the caste system, that figured in earlier explanations, could hold back the poor labourers from such migration. Similarly, thousands of villagers have moved to plantations in West Bengal and Assam as labourers. Sometimes decisions taken by the government also affect migration pattern. The resettlement of sikh immigrants from Pakistan in the *tarai region* of Uttar Pradesh is an outstanding example of such a migration. The Dandakarnya Project is another such example. A vast area of several thousand square kilometres in the districts of Koraput and Kalahandi (Odisha) and Bastar (Chhattisgarh) was carved out for agricultural development and to resettle thousands of displaced families. Resettlement projects in the northern parts of Rajasthan and in Andaman and Nicobar Islands have also resulted in movement of people. Peasants (owner-cultivators) move to large river valley project areas either on their own or under sponsorship of the government. Peasants who have experience of flow irrigation, voluntarily move to new river valley project areas, where they legally buy land from the locals and establish their settlements called ***camps***. Labourers also move as it becomes labour demand area (Vaidyanathan, 1972).

TABLE 12.5. Streams of Migration in India (2011)

Stream	Percentage of migrants
Rural to rural	50
Urban to rural	5
Urban to urban	17
Rural to urban	18

Note : The remaining 10% could not be classified as 'rural' or urban.

Source : Census of India, 2011.

(*b*) Rural-to-Urban

Rural-to-urban migration is next only to rural-to-rural migration in terms of volume of migration. Rural-urban migration is caused by both *push* of the rural areas as well as *pull* of the urban areas. In rural areas appalling poverty, unbearable unemployment, low and uncertain wages, uneconomic land holdings and poor facilities for health education, recreation and other services work as *push factors*. By contrast, the *pull* of urban areas may include better employment opportunities, regular and higher wages, fixed working hours, better amenities of living, facilities for education and socio-cultural activities (Chandna, 1992). The glamour of urban life and rigid caste system in the countryside have given further strength to push and pull factors. Both poor and rich from the countryside migrate to the urban areas under the influence of push and pull factors. While the poor migrate out of economic compulsion to eke out their living, the rich migrate due to their desire for better and greater comforts of life. In brief, rural-to-urban migration is an outcome of interplay of forces hostile to comfortable living in the villages and of availability of lucrative opportunities in urban areas (Raju, 1987).

Both males and females migrate from rural to urban areas but males have always outnumbered females in this stream of migration. *If rural-to-rural migration is woman-migration, rural to urban migration is man-migration*. The rural-to-urban migration has substantially affected the areal distribution of population and has contributed a great deal to urban growth in India. It has been estimated that about one-third of the total urban growth in India has been due to migration (United Nations 1984). This has led to over-crowding in urban centres resulting in great population pressure on the urban infrastructure. The growth and spread of slum areas, as a result of population pressure, is quite glaring and painful.

(*c*) Urban-to-Urban (Urban Turn Over)

This is a stream of migration which is believed to be dominated by the middle class people (Singh, 1980). Generally, people migrate from small towns with less facilities to large cities with more facilities. This is the reason that class I cities have grown at a much faster rate as compared to other towns. In fact, *small towns are constantly loosing to big cities*. The vacuum thus caused in small towns is filled by the subsequent in-migration from the surrounding rural areas. This migration forms a part of what is known as ***step-migration.***

(*d*) Urban-to-Rural (Pushback or Reverse Migration)

This stream has the lowest volume of migrants accounting for about five per cent only in 2011. However, there has been slight increase in this stream of migration in the recent past. Such a movement takes place at the advanced stage of urbanization when urban centres are characterised by over-congestion, haphazard growth, high cost of living, heavy pressure on public utility services, unemployment, etc. Usually females outnumber their male counterparts in this stream of migration. This is largely due to matrimonial alliance (Premi, 1978).

CONCLUSIONS

Following conclusions can be drawn from the study of migration patterns :

1. Nearly 85 per cent of the intra state migrants were born in rural areas, while 63 per cent were born in rural areas in inter-state migration category.
2. About three fourths of all migrants in the intra-state category were women; mainly because of marriage.
3. Over half of the intra-state migrants moved from rural to rural areas. Such people migrated in search of employment on farms or other establishments located in the rural areas.
4. Rajasthan, Uttar Pradesh, Bihar, Andhra Pradesh and Kerala are the major areas from where out-migration takes place.
5. West Bengal, Maharashtra, Assam, Punjab, Delhi, Chandigarh and Andaman and Nicobar Islands are mostly in migrating states or union territories.
6. Disparities in economic development seems to be the main cause of migration in case of males and marriage in case of females.
7. Densely populated rural areas with increasing pressure of population on land register out migration. Big cities, mining and industrial centres, plantation agriculture etc. are the main "Pull" factors behind in-migration.

Comparatively small states had larger share of inter-state migrants in their populations. By contrast, larger states such as Andhra Pradesh, Odisha, Tamil Nadu, Bihar, Uttar Pradesh displayed relatively smaller proportion of inter-state migrants.

The urbanized and small sized union territories of Chandigarh and Delhi had high percentage of inter-state migration.

REFERENCES

Agarwal, S.N. (1968) : *Socio-Economic and Demographic Characteristics of Rural Migrants*, Journal of the Institute of Economic Research 3(2), July.

Ali, A. (1980) : *Plantation to Politics, Study on Fiji Indians*, Suva.

Bhutani, Smita (1995) : *Demographic Dynamism in India*, Discovery Publishing House, New Delhi, Chapter 4.

Bogue, D.J. (1959) : *International Megration*, in Hauser, P.M. and Dumean, O.D. (Eds). The *Study of Population,* The University of Chicago Press, Chicago.

Bose, Ashish (1965) : *Why do People Migrate to Cities*, Yojana, January 26.

—— (1978) : *Rural to Rural Migration, A New Feature*, in The Hindu, December 13, Chennai.

Champion, A. and Fielding, A. eds. 1992, *Migration Process and Patterns* Vol. 1. Research Progress and Prospects, London.

Chand, Gyan (1972) : *Population Perspective*, Orient Longman Ltd., New Delhi.

Chandna, R.C. (2014) : *A Geography of Population*, Kalyani Publishers, 2014, New Delhi.

Chakaravarti, N.C. (1971) : *The Indian Minority in Burma : The Rise and Decline of an Immigrant Community*, Oxford Uni. Press.

Clark, H.A.V. 1975). *Migration and Development*, the Hague, Moutn.

Connel *et al.* (1976) : *Migration from Rural Areas : The Evidence from Village Studies*, Oxford University Press, New Delhi.

Davis, Kingsley (1951) : *Population of India and Pakistan*, Princeton University Press, Princeton.

D'Souza, V.S. (1964) : *Some Socio-cultural Aspects of Internal Migration in India in* Ashok Mitra (ed.) Indian Population Bulletin (Vol. 3), Office of the Registrar General of India, New Delhi.

Economic Survey 2016-17, Chap. 12, pp. 264-84.

Gill, M.S. (1981) : *Patterns of Migration in an under-developed Tract : Hissar District of Haryana, 1951-71.* Geographical Review of India, Vol. 43, No. 2 June.

Gore, M.S. (1970) : *Immigrants and Neighbourhoods,* Tata Institute of Social Sciences, Mumbai.

Gosal, G.S. (1961) : *Internal Migration in India—A Regional Analysis.* The Indian Geographical Journal, Vol. 36, No. 3, July-Sept.

Gosal, G.S. and Krishan, G. (1975) : *Patterns of Internal Migration in India* in L.A. Kosinki and R.M. Prothers *(ed.) People on the Move*, Methuen and Co. Ltd., London.

Jackson, J.A. (1969) : *Migration*, University Press, Cambridge.

Kumar, A. and Sharma, N. (1979) : *Bihar's Population on Move : Issues on Interdistrict Migration,* in Sinha, V.N.P. and Mandal R.B. (eds.) *Dimensions in Geography,* Associated Book Agency, Patna.

Lewandouski, Susan (1980) : *Migration and Ethnicity in Urban India*, Manohar, New Delhi.

Lewis, C.J. (1982) : *Human Migration,* London, Groom Helm/New York, St. Martin's Press.

Majumdar, P.S. and Majumdar, Illa (1978) : *Rural Migrants in an Urban Setting*, Hindustan Publishing Co., New Delhi.

Mehta, S. (1990) : *Spatial Mobility in India : Evolving Patterns, Emerging Issues and Implications,* Population Geography, Vol. 12 Nos. 1 & 2, June and December.

Mitra, Ashok (1967) : *Internal Migration and Urbanization in India*, Office of the Registrar General of India, New Delhi.

Muthiah, S. *et al.* (ed.) (1987) : *A Social and Economic Atlas of India*, Oxford University Press, New Delhi.

Oberoi, A. (1983) (ed) : *Policies and Internal Migration*, New York.

Ogden, P.E. (1984) : *Migration and Geographical Change,* Cambridge Press, Cambridge.

Prasada Rao, C.R. (1970) : *Rural-Urban Migration, A Clue to Rural-Urban Relations in India*, The Indian Journal of Social Work 30(4), January.

Premi, M.K. (1978) : *Out-migrating Towns : A study into the Nature, Causes and Consequences of Out-migration.* ICSSR, New Delhi.

Premi, M.K. (2009) : *India's Changing Population Profile,* National Book Trust, New Deli.

Raju, B.R.K. (1989) : *Developmental Migration* : *A Proecessual Analysis of Inter-State Rural-Rural Migration*, Concept Publishing Co., New Delhi.

Rao, Bhaskar N. (1967) : *Estimate of Migration on Four Major Cities of India, 1941-51 to 1951-61,* The Asian Economic Review 10(1).

Robinson Francis (1989) (ed) : *The Cambridge Encyclopaedia of India, Pakisatn, Bangladesh & Sri Lanka,* Cambridge University Press, Cambridge.

Sandesara, J.C. (1964) : *Migration and Metropolitan Living : A Study of Indian Cities,* Economic Weekly, 16(9) May.

Saxena, D.P. (1978) : *Rural-Urban Migration in India : Causes and Consequences,* Popular Prakashan, Mumbai.

Singh, S.N. and Yadav, K.N.S. (1974) : *A Study of Patterns of Rural-Urban Migration.* Demography of India 3(2).

Skeldon, R. (1986) : *On Migration Patterns in India During the 1970s,* Population and Development Review, Vol. 12 No. 4, Dec.

Zachariah, K.C. (1964) : *A Historical Study of Internal Migration in the Indian Sub-Continent, 1901-31,* Asia Publishing House, Mumbai.

Zelinsky, W. (1971) : *The Hypothesis of the Mobility Transition*, Geographical Review 61, No. 2.

◆◇◆

CHAPTER 13

Regional Development and Planning

DEVELOPMENT

Development implies overall improvement in economic, social and political conditions of a society. Recently one more dimension of conservation and improvement of environment has been added to the overall gamut of development. The quest for development and strategies for development could be related to :

(*i*) Rapidly increasing population and pressure on physical resources.

(*ii*) Growing demand for food and other necessities of life.

(*iii*) Quest for improving the general standards of living of the people specifically in the less developed parts of the world.

(*iv*) Decolonisation of large number of countries and their consequent emergence as independent countries. For example, India became an independent country in 1947, after which a lot of development has taken place.

(*v*) Reconstruction of former colonies that have become now independent sovereign states.

(*vi*) Regional disparities at the national and international levels.

The concept of development is not only area specific but also time specific. It means that concept of development changes from place to place and from time to time. Thus development process belies any permanent conceptualization. Different disciplines like economics, sociology, political science, history, geography, etc. often deal with development from their own point of view.

The largest contribution to the study of development has been made by economists. An economist looks at development in terms of economic progress. Economic progress is expressed in terms of enhancement in general productivity level, per capita income, efficiency of a worker, overall improvement in quality of life in general and elimination of poverty in particular. As such the economists have met with greater success in the formulation of theories and models and have earned greater acceptability in policy formulation.

Sociologists, on the other hand, believe that economic development was of little relevance in the absence of social development and social development must precede economic development. The main focus of social development is the development of people in terms of their mental relations and the institutional and structural changes in the society. The two most significant parameters of measuring social development are quality of life (*i.e.* social harmony and social cohesion) and social justice.

Political scientist's, conceptualization of development has its focus on political context. This is justified in view of the increasing role of government not only as promoter but also as the principal planner. Thus, power structure holds the pivotal position in political scientist's conceptualization of development.

A historian's concept of development lies in the formulation of theories and models that may explain economic, social and political history. Thus, a historian's focus is upon temporal changes in economic, social and political institutes and models, if any, would be temporal models.

Geography is an integrating discipline and offers a unique synthesis of development of natural and human resources. As such a geographer's conceptualization of development is much more comprehensive. It considers economic progress, social advancement, political development and environmental preservation.

Figure 13.1 gives an idea of geographer's conceptualization of the process of development. The figure shows that each social science such as sociology, economics, political science, geography, etc. has its own perception of the process of development. However, geographer's concept offers a unique synthesis of physical, social and economic development. The figure also shows four basic steps

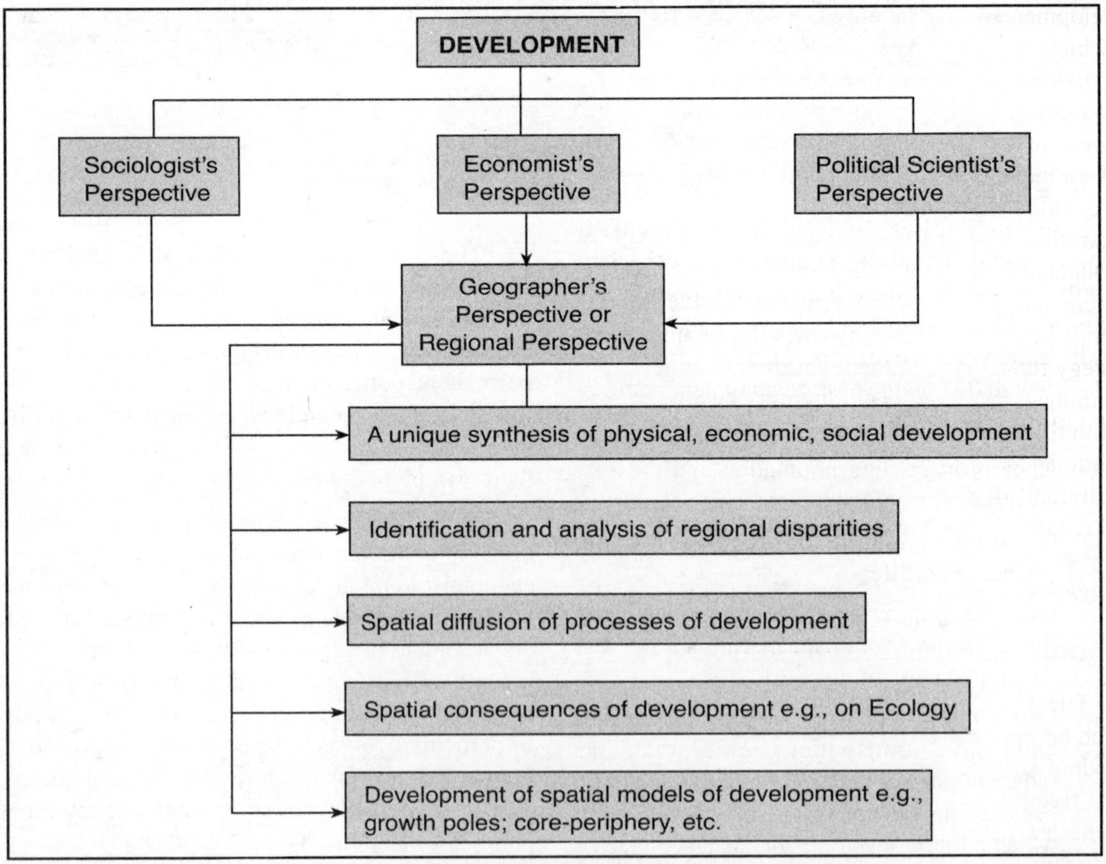

FIG. 13.1. Development

in geographic analysis of regional development. These are (*i*) identification and analysis of regional disparities, (*ii*) spatial diffusion of process of development, (*iii*) spatial consequences of development *e.g*. on ecology, and (*iv*) development of spatial models of development *e.g*. growth poles, core-periphery, etc.

In the last quarter of 20th century, a growing focus on the behavioural aspect of development has been observed. This is in tune with the behavioural revolution in geography. This thrust was the outcome of geographer's quest for role of human factor in development, more specifically human behaviour.

What is Planning ?

Planning has been defined as "the process of thinking through and implementing a set of appropriate actions to achieve some goals." Generally the goals are economic prosperity and social development. Thus planning is a device with the help of which we can achieve socio-economic prosperity by making optimum use of natural and human resources available to us. In other words, it is a process of socio-economic development which can make a place for good living and can help in keeping the ecology and environment in a healthy condition. According to the Planning Commission of India, "Planning involves the acceptance of a clearly defined system of objectives in terms of which to frame overall policies. It also involves the formation of a strategy for promoting the realisation of ends defined. Planning is essentially an attempt at working out a rational solution of problems, an attempt to coordinate means and ends; it is then different from traditional hit and miss methods by which reforms and reconstruction are often undertaken." (First Year Plan, p. 7). In fact, planning is a *process of human thought and action based upon that thought*.

Planning India

The foundation of planning in India was laid much before Independence during the heat of India's freedom struggle by the stalwarts of Indian thought and freedom movement like Mahatma Gandhi, Subhash Chandra Bose and Jawaharlal Nehru. The setting up of National Planning Committee in 1938 by the then Indian National Congress marks the beginning of thought for planning in pre-Independence era. As a consequences of this initiative, regular planning was started immediately after Independence.

The Planning Commission was set up by a Resolution of the Government of India in March 1951 in pursuance of declared objectives of the Government to promote a rapid rise in the standard of living of the people by efficient exploitation of the resources of the country, increasing production and offering opportunities to all for employment in the services of the community.

NITI Aayog

The National Institution for Transforming India (NITI Aayog) came into existence in 2015 replacing the existing Planning Commission which was established in 1950. The NITI Aayog is the successor to the Planning Commission. The new institution is envisaged to be a catalyst to the developmental process; nurturing an overall enabling environment, through a holistic approach to development going beyond the limited sphere of the public sector and Government of India. This is to be built on the foundation of an empowered role of states as equal partners in national development; underlying the principle of cooperative federalism. A knowledge hub of internal as well as external resources; serving as repository of good governance best practices, and a think tank offering domain knowledge as well as strategic expertise to all levels of government. A collaborative platform facilitating implementation; by monitoring progress, plugging gaps and bringing together the various ministries at the Centre and in states, in the joint pursuit of developmental goals.

Objectives

The Resolution setting up the NITI Aayog includes the following objectives: To evolve a shared vision of national development priorities, sectors and strategies with the active involvement of States in the light of national objectives. The vision of the NITI Aayog is to provide a framework 'national agenda' for the Prime Minister and the chief ministers to provide impetus to. To foster cooperative federalism through structured support initiatives and mechanisms with the states on a continuous basis, recognizing that strong states make a strong nation. To ensure, on areas that are specially referred to it, that the interests of national security are incorporated in economic strategy and policy. To pay special attention to the sections of our society that may be at risk of not benefitting adequately from economic progress. To design strategic and long term policy and programme frameworks and initiatives, and monitor their progress and their efficacy. The lessons learnt through monitoring and feedback will be used for making innovative improvements, including necessary mid course corrections. To create a knowledge, innovation and entrepreneurial support system through a collaborative community of national and international experts, practitioners and other partners. To offer a platform for resolution of inter-sectoral and inter-departmental issues in order to accelerate that implementation of the development agenda. To undertake other activities as may be necessary in order to further the execution of the national development agenda, and the objectives mentioned above.

First Five Year Plan (1951 to 1956). Regional and national problems became glaring during the initial phase of planning in India. This plan set out following five objectives for accomplishment :

(*a*) increasing the national income and improving the average standard of living of the people;

(*b*) increasing the pace of industrialisation of both the basic and heavy industry sector;

(*c*) increasing employment potential of various sectors;

(*d*) narrowing down both vertical and horizontal inequalities; and

(*e*) achieving self-reliance so as minimise dependence on foreign aid.

Since the First Five Year Plan was initiated immediately after Independence and partition of the Indian subcontinent, India faced the following serious problems in the beginning of this plan :

(*i*) resettlement of immigrants from Pakistan due to partition of the country in 1947.

(*ii*) acute shortage of food resulting in large scale import of foodgrains.

(*iii*) Checking the increasing rate of inflation.

Also there was immediate need to reduce the regional imbalances in economy caused by World War II and partition of the country.

Keeping in view above scenario, new towns were planned and old towns were expanded to rehabilitate immigrants from Pakistan. Top priority was given to irrigation and agriculture and establishment of power projects in the form of multi-purpose projects. The Plan aimed at increasing the rate of investment from 5 to about 7 per cent of the national income.

The national level problems were very serious and not much attention was paid to regional planning. But all the states were divided into five broad divisions which is considered to be the first step towards regional planning. Although regions were ignored in broad outlines of planning, yet problem areas like Damodar Valley were identified and importance of resource planning at the regional level was recognised.

Second Five Year Plan (1956-57 to 1960-61). The efforts made during the First Five Year Plan helped the country to tide over the economic crisis and increase agricultural production to some extent. Inflation rate was also controlled partially. Therefore, it was felt that industries should be given priority over agriculture. The plan aimed at promoting a pattern of development which would lead to the establishment of a socialistic pattern of society. It aimed at :

(*i*) an increase of 25 per cent in the national income;

(*ii*) rapid industrialization with particular emphasis on the development of basic and heavy industries;

(*iii*) large expansion of employment opportunities;

(*iv*) reduction of inequalities in income and wealth and a more even distribution of economic power.

The Plan aimed at increasing the rate of investment from about 7 per cent of the national income to 11 per cent by 1960-61. It laid emphasis on increasing the production of iron and steel, heavy chemicals including nitrogenous fertilizers and development of heavy engineering and machine building industry.

Third Five Year Plan (1960-61 to 1965-66). The first two Five Year Plans helped in generating the required infrastructure and Indian economy entered the "*take off stage*" at the beginning of Third Five Year Plan. Therefore this Plan aimed at securing self-reliance, self-generating economy and self-sustained growth. Its immediate objectives were to :

(*i*) secure over five per cent per annum increase in national income and at a same time ensure a pattern of investment that could sustain this rate of growth in the subsequent plan periods;

(*ii*) achieve sell sufficiency in food grains and also increase production of other crops to meet requirements of industry and export;

(*iii*) expand basic industries like steel, chemicals, fuel and power and establish machine building capacity so that requirements of further industrialization could be met within a period of about 10 years mainly from the country's own resources;

(*iv*) fully utilize the manpower resources and increase substantially the employment opportunities;

(*v*) establish progressively greater quality of opportunity and bring about reduction in disparities of income and wealth and more even distribution of economic power.

The Third Five Year Plan was *shifted from development to defence at a later stage due to invasion of India by China in 1962 and India's war with Pakistan in 1965*.

Annual Plans (1966-67 to 1968-69). Many problems cropped up by the end of the Third Five Year Plan which delayed the finalization of the Fourth Five Year Plan. These problems were : (*i*) Indo-Pakistan conflict in 1965; (*ii*) two successive years of severe drought (1965-66 and 1966-67); (*iii*) devaluation of the currency; (*iv*) general rise in prices; and (*v*) erosion of resources available for plan purposes. Consequently, three annual plans from 1966-67 to 1968-69 were prepared. Although annual plans were prepared, there was no overall five-year framework and the period of 1966-67 to 1968-69 between the Third and the Fourth Five Year Plans is termed as ***"Plan Holiday."***

Fourth Five Year Plan (1969-74). The primary aim of this plan was to accelerate the tempo of development and reduce the fluctuations in agricultural production as well as the impact of uncertainties of foreign aid. It sought to raise the standards of living through programmes designed to promote equality and social justice. The Plan laid particular emphasis on improving the conditions of the less privileged and weaker sections by providing facilities of employment and education. Provision was also made for reducing concentration of wealth, income and economic power to promote equity. The average annual growth rate was 3.4 per cent against the target of 5.7 per cent.

Fifth Five Year Plan (1974-79). This plan was prepared against the backdrop of severe inflationary pressure. Its two primary aims were (*i*) removal of poverty and (*ii*) attainment of self-reliance through promotion of higher rate of growth, better distribution of income and very significant set up in domestic rate of saving. Efforts were made to develop backward areas through the cooperation of the centre and state governments. The backward areas were divided into two classes *viz*. (*a*) areas where conditions of physical geography such as relief, climate, etc. were not much suited for human habitation and where people of typical culture were living and (*b*) economically backward areas where land man ratio was low, infrastructure was poor or resources for development ever scarce. Separate plans were prepared for both types of areas. This led to the development of concept of area specific planning and special programmes were chalked out for drought affected areas, hilly areas and for areas inhabited by the tribal people. Medium and minor irrigation projects were launched to meet the requirements of drought affected areas. Special plan was also prepared for afforestation, soil conservation, orchards, livestock dairy farming, road construction and drinking water. The plan targeted an annual growth rate of 5.5 per cent but actual growth rate was 5.0 per cent.

Four Annual Plans pertaining to this plan were completed. Then there was change of guards and Government headed by Janata Party replaced the Congress led Government at the centre. The new Government at the centre decided to end the Fifth Plan period with the close of the Annual Plan 1978-79.

Sixth Five Year Plan. There are two different types of Sixth Plan in the planning history of India. One of the Sixth Five Year Plans (1978-83) was prepared by the Government led by Janata Party which sought to reconcile the objectives of higher production with greater opportunity for employment. Provision was made for increasing employment opportunities in agriculture and allied activities, incentives for small scale and cottage industries and increasing income of low income groups.

As a result of 1980 general election, Janata Party was defeated and Congress (I) came back to power. The new government led by Congress (I) presented its own Sixth Five Year Plan (1980-85). The thrust of this plan was to expand the economic base and to reduce poverty. This was to be achieved by strengthening the agricultural and industrial base. Stress was laid on tackling inter-related problems through a systematic approach with greater management, efficiency and intensive monitoring in all sectors and active involvement of people in formulating specific schemes of development at local level and securing their speedy and effective implementation.

The achieved annual growth rate of 5.4 per cent was higher than the targeted annual growth rate of 5.2 per cent.

Seventh Five Year Plan (1985-90). This plan emphasized on policies and programmes, which aimed at rapid growth in food grains production, increase in employment opportunities and productivity within the framework of basic tenets of planning, namely growth, modernization, self-reliance and social justice. Production of food grains increased by 3.23 per cent compared to a long-term growth rate of 2.68 per cent between 1967-68 and 1988-89. To reduce unemployment and poverty, special programmes like ***Jawahar Rozgar Yojna*** were launched in addition to the existing programmes. Special attention was paid to small scale and food processing industries. The Gross Domestic Product (GDP) grew at an average rate of 5.8 per cent exceeding the targetted growth rate of by 0.8 per cent.

Annual Plans. The Eighth Five Year Plan could not take-off due to fast changing political situation at the centre. The new Government at the centre decided that the Eighth Five Year Plan would commence on 1st April, 1992 and that 1990-91 and 1991-92 would be treated as separate Annual Plans. The basic of the Annual Plans was on maximization of employment and social transformation.

Eighth Five Year Plan (1992-97). This plan was prepared in the backdrop of worsening inflation and Balance of Payments position. The Plan aimed at an average annual growth rate of 5.6 per cent and an average industrial growth rate of about 7.5 per cent. The salient features of this plan were (*a*) a faster economic growth, (*b*) a faster growth of manufacturing, agriculture and allied sectors, and (*c*) significant growth in exports and imports. Following programmes were initiated to achieve these goals :

(*i*) increase employment opportunities and achieve full employment by the end of 20th century;

(*ii*) control population growth with the cooperation of the people;

(*iii*) universalization of primary education and spread literacy among the people in age group 15-35 years;

(*iv*) providing drinking water and primary health services;

(*v*) development and diversification of agriculture for self-reliance in food grains and for export of agricultural commodities;

(*vi*) strengthen the infrastructure (power, transport, communication, irrigation), to sustain the tempo of growth.

The actual average annual growth rate was 6.8 per cent against the envisaged growth rate of 5.6 per cent.

Ninth Five Year Plan (1997-2002). The aim of this plan was to achieve the targeted GDP growth rate of seven per cent per annum and there was emphasis on seven identified **Basic Minimum Services (BMS).** These include (*i*) provision of safe drinking water; (*ii*) primary health services, (*iii*) universalization of primary education, (*iv*) public housing assistance to shelterless poor families, (*v*) nutritional support to children, (*vi*) connectivity of all villages and habitations, and (*vii*) streamlining the public distribution system with a focus on the poor. The plan also aimed at reduction in revenue deficit through a combination of improved revenue collections and control of non-essential expenditures. Following steps were proposed to achieve the targets set in this plan :

(*i*) priority to agriculture and rural development for eradication of poverty and generating adequate employment;

(*ii*) accelerating the growth rate of economy with stable prices;

(*iii*) ensuring food and nutrition security for all; particularly the vulnerable sections of society;

(*iv*) providing basic minimum services of safe drinking water, primary health care, universal primary education, shelter and road connectivity to all in a time bound manner;

(*v*) containing growth rate of population;

(*vi*) ensuring environmental stability;

(*vii*) empowering women, backward classes (scheduled castes, scheduled tribes, etc.) people belonging to minority communities as agents of economic and social change;

(*viii*) promoting and developing people's institutions like Panchayati Raj Institution, Cooperatives and self-help groups;

(*ix*) strengthening efforts to build self-reliance;

(*x*) strengthening foreign exchange.

Tenth Five Year Plan (2002-2007). This plan was approved by the National Development Council (NDC) which envisaged to double the per capita income in ten years and achieve a growth rate of eight per cent of GDP per annum. The plan also aimed at harnessing the benefits of growth to improve the quality of life by setting the following targets :

(*i*) Reduction in poverty rates from 26 per cent to 21 per cent by 2007.

(*ii*) Reduction in decadal population growth from 21.3 per cent in 1991-2001 to 16.2 per cent in 2001-11.

(*iii*) Growth in gainful employment, at least, to keep pace with addition to the labour force.

(*iv*) All children to be in school by 2003 and all children to complete five years of schooling by 2007.

(*v*) Reduce gender gaps in literacy and wage rates by 50 per cent.

(*vi*) Increase literacy rate from 65 per cent in 1999-2000, to 75 per cent in 2007.

(*vii*) Provide potable drinking water to all villages.

(*viii*) Reduction in infant mortality rate from 72 per thousand in 1999 to 45 per thousand in 2007.

(*ix*) Reduction in mortality ratio from 4 in 1979-2000 to 2 in 2007.

(*x*) Increase in Forest/Tree cover from 19 per cent in 1999-2000 to 25 per cent in 2007.

(*xi*) Cleaning the major polluted rivers.

The growth rate during this plan was 7.7 per cent against the target of 8 per cent.

Eleventh Five Year Plan (2007-2012). This plan provided a comprehensive strategy for inclusive development, building on growing strength of economy, while also addressing the weaknesses that have surfaced. It set a target for 9 per cent growth in five year period with acceleration during the period to reach 10 per cent by the end of the plan. It also covered 26 major indices of performance relating to poverty, health, education, women and children, infrastructure and environment.

The new priorities outlined in the plan relate to reviving dynamism in agriculture and building the necessary supportive infrastructure in rural areas, expanding access to health and education, especially in rural areas, undertaking programs for improving living conditions for the weaker sections and for improving their access to economic opportunity. Major thrust for infrastructure development was also included.

The plan provided a major expansion in irrigation and water management for increasing agricultural production. The National Food Security Mission aimed at increasing cereal and pulses production by 20 million tons over a period of five years.

In education, it was proposed to spend more than double the amount what was spent in the tenth plan. The plan aimed at providing improved broad based health care in rural areas through the National Rural Health Mission. The *Rashtriya Swasthya Bima Yojna* provided insurance cover against illness to the population below poverty line.

The plan emphasised on the need for energy conservation, increasing energy efficiency, and development of renewable sources of energy.

Twelfth Plan (2012-17). This plan fully recognised that the objective of development was broad-based improvement in the economic and social conditions of our people. However, rapid growth of GDP is an essential requirement for achieving this objective.

The Approach Paper to this plan, had set a target of 9 per cent average growth of GDP over the plan period. The Plan envisaged that the current slowdown in GDP growth could be reversed through strong corrective measures, including especially an expansion in investment with corresponding increase in savings to keep inflation pressure under control.

Regional Planning

As national planning is concerned with the development of the entire country, regional planning is concerned with the development of a particular region. It is neither economic planning alone, nor physical planning alone. Instead, it is such a planning whose core area of interest lies in the synthesis of physical, social, economic, political *et al.* interests of the concerned region. Thus regional planning is a specific type of planing, based on a specific planning structure (regional system), for inducting public action aimed at social well being. It implies that regional planing is concerned fundamentally with the society in the context of space (Chandna, 2008 : 33). Regional planning is concerned both with space and society as both are intimately interwoven with each other and form the basis of regional planning. Regional planning is, thus, a specific type of planning, based on a specific planning structure, the primary aim of which is the well being of the society and improvement in living standard. A regional planner is supposed to reorganise or reshape the regional system in such a way that the regional plan is able to serve the society to the best. Regional planning while confining to the planning of regional space can hardly afford to overlook the economic, socio-cultural and political dimensions of the ever evolving regional organism. Right location of each of the economic activities is the main concern of a regional planner.

Principles of Regional Planning

Regional planning is guided by the following seven principles :

1. The principle of vertical unity of phenomena
2. The principle of horizontal spatial unity
3. The principle of space-time continuum
4. The principle of comprehensive development
5. The principle of community development
6. The principle of equilibrium between social desirability and economic viability
7. The principle of ecological equilibrium.

1. The Principle of Vertical Unity of Phenomena. The principle of vertical unity of phenomena means that all phenomena in a region are inter-related and no single phenomena can change independently without influencing the other phenomena. **Figure 13.2** shows that vertical phenomena consists of four different spaces. On the base is the *real physical space* relating to lithosphere, hydrosphere and atmosphere (both known and unknown). On the real physical space, lies perceived physical space, as perceived by the regional planner. It may consist of all known physical phenomena concerning lithosphere, hydrosphere and atmosphere. Above the perceived physical space lies the human space consisting of all sorts of human phenomena amanating from the demographic structure. The interaction between the physical space and the human

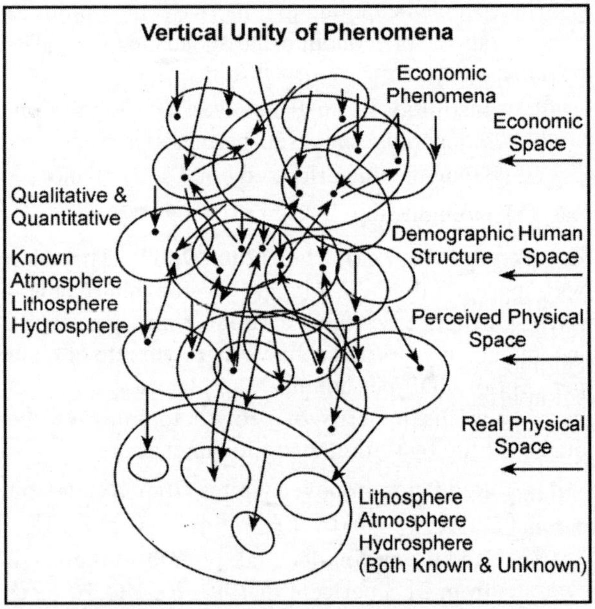

FIG. 13.2. Vertical unity of Phenomena

space gives rise to economic space. **Figure 13.2** shows the economic space, the human space and the physical space get integrated with each other. It implies that we have to realize the vertical unity of all phenomena and planning for any single phenomenon independent of other phenomena will not be in the real spirit of regional planning.

2. The Principle of Horizontal Spatial Unity. According to this principle, each region is a sub-system of the regional system whole. This means that different regions constituting the regional space do not exist in isolation, rather they co-exist in integration with each other as a part of the regional system whole.

The human body provides the best illustration of such an integration between sub-systems and the system. The entire human body as a whole functions as a system, which consists of a number of sub-systems like digestive system, renal system, respiratory system, nervous system etc. Each one of them constitutes a complete system in itself and yet is a sub-system of the body whole. If something goes wrong with any of these sub-systems say the digestive system, it will affect the functioning of the entire human body, as the person may fall sick and may not be able to perform even day to day functions. (Chandna, 2008 : 50)

Similarly a region may be a complete system in itself and yet may constitute a sub-system in the regional system whole. One thing happening in a region affects the regional system whole because all the regions are integrated with the whole just as different parts of the human body are integrated with the whole body. **Figure 13.3** shows how a single regional unit gets integrated with other regional units of a nation—state and the nation-states getting integrated with the international space.

Thus the principle of horizontal spatial unity implies that planning for any region cannot be done in isolation from the adjoining regions within the same nation. Similarly, the planning within the national space cannot be done in isolation from planning in international space. Therefore regional planning amounts to planning in universalism.

3. The Principle of Space-time Continuum. According to this principle, spatial reality has four dimensions of which three dimensions (length,

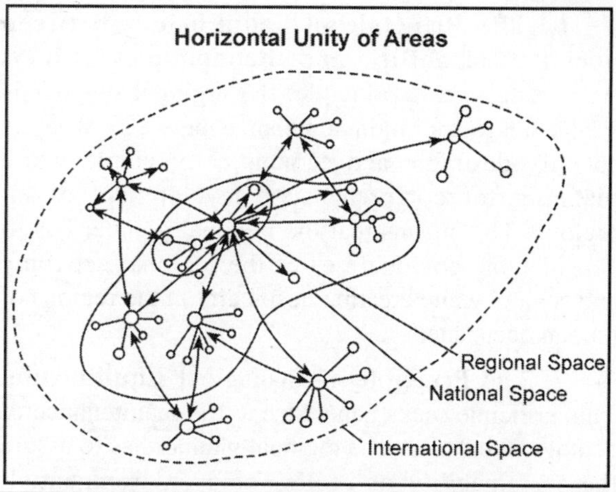

FIG. 13.3. Horizontal Unity of Areas

breadth and thickness) are of space and the fourth dimension is that of time. Also space and time cannot be separated from each other and time is considered as an integral part of space. A region is a living dynamic identity which operates simultaneously in the past, present and future like the human body's DNA. It uses the past experience for the present and preserves energy for the future. Therefore, the regional planner has to realise the fact that regional space is a continuously growing organic whole and the region changes through time not only in terms of space-content but also in terms of space relationships. Thus, the regional planner should be conscious of the fact that while he is planning in space, be is planning in time also.

4. The Principle of Comprehensive Development. This principle means that regional planning seeks to achieve the comprehensive development of the entire region, the regional system in its entirely. In other words, the regional planner seeks comprehensive development of all sectors of economy along with all segments of society.

5. The Principle of Community Development. This is the principle of equal opportunities to all for self development. The entire community is considered as an organic whole in which each individual is a sub-system of this whole. It is only through equal opportunities (like education, health, employment, etc.) to each individual that the whole society can be developed into community with the sense of belonging to each other.

6. The Principle of Equilibrium between Social Desirability and Economic Viability. According to this principle, the regional planner is supposed to maintain a balance between what is socially desirable and economically viable while planning for comprehensive development of all regions. This means that the regional planner has to consider the potentialities of the regional economy and society while creating hospitality in the regions of human occupance.

7. The Principle of Ecological Equilibrium. This principle seeks adherence to the maintenance of ecological balance. The regional planner has to assure the sustainability of ecology and environment. It means that he has to develop the regional space at his disposal only within the framework of ecological equilibrium. In simple words all the development plans in the region should be carried out in harmony with nature and without disturbing the ecological balance.

REGIONAL PLANNING IN INDIA

As mentioned earlier, the basic postulates of planning in India were conceived and formulated during the heat of India's freedom struggle and the First Five Year Plan was launched in 1951 *i.e.* within four years of India's Independence. The primary objectives of the plan were accelerating the pace of economic and social development, alleviating poverty and unemployment and improving the standard of living.

Unfortunately no recognition worth the name was given to regional planning in the first two Five Year Plans (from 1950-51 to 1960-61). Following could be the probable reasons for this complete negligence of Regional Planning in the early phase of planning period :

(*i*) There was no serious attempt to identify a regional scheme or a set of regions, which could form the basis of the planning activity in the country.

(*ii*) The leaders who took the reigns of the country immediately after Independence were perhaps too enthusiastic and too keen to initiate the process of planned development that could not wait for discovering the existing regional system/real region nature.

(*iii*) Perhaps the leaders of that time could not appreciate the utility of planning process based on regional system.

(*iv*) The leadership of the early plan period probably believed that as the developmental process starts, the new regional system may emerge in the country which could form the basis of future regional planning process.

The regional or spatial planning process in the country was most probably incorporated in the Third Five Year Plan when the philosophy of balanced regional development got appended to the era of Five Year Plans. Even during and after the Third Five Year Plan, the concept of Regional Planning could not much influence our planners because our plans are still conceived and termed as Five Year or Annual Plans rather than Regional Plans devised for different regions. Our planners have probably failed to realise the fact that identification of a regional scheme is fundamental to any regional planning. After the identification of a regional scheme, preparation of plans for each region has to be made. Care has to be taken that plan for any region is not to be prepared in isolation, rather in integration with other regions so as to achieve a comprehensive plan for the entire country.

NITI Aayog should supervise the process of regional planning at the regional level. Establishment of State Planning Boards at the state level is a step towards decentralisation of the country's planning and also to incorporate the spatial dimension to the planning process. In some of the states like Punjab, the State Planning Boards have taken a step forward to formulate the district level plans and even the block level plans.

Most of the Regional Plans are prepared according to the political boundaries of the states because official data are available according to the political divisions only. But political boundaries do not always coincide with the geographical boundaries and a planning process based on such political boundaries rarely succeeds. The Damodar Valley Corporation (DVC) and the National Capital Region (NCR) are two such examples which do not follow political boundaries. The Damodar Valley spreads over Jharkhand and West Bengal whereas the National Capital Region covers large areas of Union Territory of Delhi, Haryana, Uttar Pradesh and Rajasthan. Development plan for DVC has met with partial success while NCR plan has almost completely failed to achieve the target and serve the planned purpose.

The success of regional planning primarily depends upon :

(*i*) quality of planning education

(*ii*) degree of public awareness about the rationale for a planned effort

(*iii*) efficiency of planning institutions, and

(*iv*) proper comprehension and appreciation of the basic issues involved in regional planning by society itself.

In India, regional planning is still at its nascent stage and is not very popular with our planners. It has a long way to go in terms of its content and implementation before it is able to show its utility and be of some use to people at the regional level.

Integrated Rural Development Programmes

The integrated rural development programme was launched in 1977 when Janta Party formed the government at the centre after defeat of the Congress Party. This programme was initially implemented in 2000 blocks out of a total of 5004 blocks at that time. Its broader version was presented in 1978-79 through which 2300 blocks were benefitted. Another 300 blocks were brought under this programme in 1979-80.

The main objective of this programme was to provide employment opportunities to the poor rural people and to increase their assets for which provision for financial assistance was made. This programme covered poor people living below poverty line, which included small and marginal farmers, agricultural labourers, scheduled castes and scheduled tribes and economically backward people having an annual income of less than ₹ 11,000 at the time of the Eighth Five Year Plan. With a view to give maximum benefit to the most vulnerable families, it was decided that at least 50 per cent of the families should be from the scheduled castes and scheduled tribes. Besides, reserved benefit was 40 per cent for women and 3 per cent for handicapped people. Families rather than persons were selected and among families too, women were given priority.

The Integrated Rural Development Programme was implemented at the grassroot level through District Rural Development Agencies (DRDA) and Block Level Agencies (BLAs). DRDA was established in April 1, 1999 with the objective of strengthening the programme. The funding pattern of the DRDA administration is in the ratio of 75 : 25 between the Centre and the States for non-NE States and 90 : 10 for NE-States. In the case of Union Territories, it is hundred per cent under the scheme. The governing bodies *i.e.* DRDAs include local MPs, MLAs, Chairman of Zila Parishad, Heads of District Development Departments, representatives of Scheduled Castes and Schedule Tribes, and women. Staff of blocks is responsible for implementing the programme on the grassroot level. At the state level, this programme is implemented by the State Level Coordination Committee. Ministry of Rural Areas and Employment is responsible for providing financial assistance, framing policies, guiding the implementation of policies, and evaluating the programme at the central level. Some of the important programmes are National Rural Employment Programme (NREP), Training of Rural Youth for Self-Employment (TRYSEM), Development of Women and Children in Rural Areas (DWCRA), *Indira Awas Yojna* (IAY), Council for Advancement of People Action and Rural Technology (CAPART), *Jawahar Rojgar Yojna,* the *Ganga Kalyan Yojna,* etc.

Some of the important projects initiated under Integrated Rural Development Programmes are briefly describe as under :

1. Training of Rural Youth for Self-Employment (TRYSEM). Initiated on 15th August, 1979, this is an integral part of IRDP. Young people in the age group of 18-35 years are selected to become skilled workers and they are imparted technical training also. They are trained in such a way that they become capable of starting their own work for self employment. Young people are selected from families having income less than ₹ 3,500 per annum and poorest of the poor are given priority. Only one member is selected from each family. Preferential treatment is given to people belonging to Scheduled Castes, Scheduled Tribes, Ex-Army men and also those people who have participated in the National Adult Education for nine months.

2. Supply of Tools Kits in Rural Areas (STKRA). This scheme aims at providing quality implements to rural artisans so that they are able to improve quality and increase quantity of their products. As many as 7.46 lakh kits of implements costing ₹ 14.8 million were distributed among the rural artisans during 1992-93 to 1997-98.

3. Development of Women and Children in Rural Areas (DWCRA). This programme aimed at improving the income levels of women belonging to poor families. Groups of 5 to 10 women are constituted so that they can help each other in work and increase their income. The programme is based on equal contribution by the central and state governments as well as by UNICEF. Basic facilities like womens' health, education, child care and nutritious food are also provided.

4. Council for Advancement/Peoples Action and Rural Technology (CAPART). Set-up on 1st September, 1986, it is an autonomous organisation under the Ministry of Rural Development and acts as a funding agency for voluntary organisations. Its primary objective is to promote voluntary action through community participation and to propagate rural technologies for the benefit of downtrodden people (SC/ST/OBC and others) living in rural areas.

5. Jawahar Rojgar Yojna. This scheme came into being by the merger of National Rural Employment Programme (NREP) and Rural Landless Employment Guarantee Programme (RLEG) in April, 1989. The primary object of this scheme is to provide additional employment opportunities to unemployed and under-employed men and women and to increase their assets. Besides, this scheme aims at improving the living conditions of the villagers and to provide sustainable employment. The funding by the centre and the state governments for this scheme are in the ratio of 80 : 20. For schedule castes and schedule tribes, 22.5 per cent of the allocation is reserved. The other components of this scheme are assistance to wards of child labourers living below poverty line, disabled and food for work. This scheme was further strengthened in the Ninth Five Year Plan so that it could work more effectively.

6. Indira Aawas Yojna (IAY). This scheme was started as a sub-scheme of Jawahar Rojgar Yojna in May, 1985. It aims at providing financial assistance to extremely poor people belonging to SCs/STs, for constructing their houses and for repairing the dilapidated *kutcha* houses. Since 1995-96, this benefit has been extended to war widows and retired workers of military and para-military forces, subject to their qualifying the laid conditions. About 3 per cent of the allocated funds have been reserved for disabled people living below poverty line. People living below poverty line and belonging to minority communities have also been included among the beneficiaries from this scheme.

IAY is being implemented across the country (except Delhi and Chandigarh) and funding for this scheme is shared between the centre and the states in the ratio of 75 : 25. In case of union territories, the entire fund is provided by the centre. In case of north-eastern states the funding ratio between the centre and states is 90 : 10. With effect from April 1, 2010, the ceiling on construction assistance under IAY scheme is ₹ 45,000 per unit in the plain areas and ₹ 48,500 in the hilly/difficult areas. For upgrading *kutcha* house, the financial assistance upto ₹ 15,000 per unit.

During 2010-11, as against the target of constructing 29.69 lakh houses, only 14.57 lakh houses were constructed. Since its inception, about 286.88 lakh houses have been constructed with an expenditure of ₹ 85141.13 crore.

7. Swarnajayanti Gram Swarojgar Yojna/ National Rural Livelihood Mission (NRLM). This scheme was launched on 1st April, 1999 and is designed as a holistic self employment scheme aimed at providing sustainable income to rural BPL (below poverty line) families through income generating assets/economic activities so as to bring them above the poverty line.

Assistance under Swarnajayanti Gram Swarojgar Yojna to individuals or Self Help Groups (SHGs) is given in the form of Revolving Funds Assistance and Capital Subsidy by the government linked credit by the banks. Since its inception upto March, 2012, 43.34 lakh SHGs have been formed and 14.46 lakh SHGs have taken up economic activities.

8. Employment/Assurance Scheme. Poor labourers do not get any employment during the lean season in agricultural operations. The Government of India launched scheme for employment assurance on

2nd October, 1993 in some selected blocks. National Rural Employment Guarantee Act was passed which was implemented for 200 most backward districts of the country in its first phase. In the second phase, it was implemented for 130 districts in 2007-08 and for 274 districts in the third phase after 1st April, 2008.

This is the first law which guarantees employment opportunities to the poor people on such a large scale. Its other aims are to provide safety against droughts, deforestation, soil erosion etc. for sustainable growth. These are some of the factors responsible for poverty.

9. Million Wells Scheme. This programme was started in 1988-89 as a sub-programme of National Rural Employment Programme (NREP) and Rural Landless Employment Guarantee (RLGP). It continued to work as a sub-programme of Jawahar Rojgar Yojna from 1st April 1989 to 31st December 1995 and became a full-fledged independent programme on 1st January, 1996. The main purpose of this scheme is to provide irrigation facilities to small and marginal farmers living below poverty line and to bonded labourers who have been freed from the clutches of their masters. This scheme was basically meant for people belonging to SCs and BCs but poor people belonging to other castes were also included in 1993-94. There is provision for 3% reservation for physically handicapped persons. Panchayati Raj organisations are doing a wonderful job preparing the list of beneficiaries from this scheme.

10. Ganga Kalyan Yojna (GKY). This programme was initiated by the Central Government on 1st February 1997 to provide irrigation facilities to small and marginal farmers or groups of such farmers living below poverty line by utilising ground water through wells and tube-wells. The finances for this programme are provided by government subsidy or financial organisations. The funding ratio between the Centre and the States is 80 : 20. Priority is given to SCs and BCs.

11. National Social Assistance Programme (NSAP). Started on 15th August, 1995, this programme has the following three components :

(*i*) old age pension of ₹ 75 per month to people of age 65 years and above.

(*ii*) one time financial assistance to those families which have lost their only earning member; ₹ 5,000 in case of natural death and ₹ 10,000 in case of accidental death.

(*iii*) financial assistance to pregnant women living below poverty line for the birth of first two children under National Maternity Benefit Scheme (NMBS), provided the women is above 19 years of age.

12. Total Sanitation Campaign (TSC). Rural sanitation came into focus of the Government of India during the World Water Decade of 1980s. The Central Rural Sanitation Programme (CRSP) was started in 1986 for providing sanitation facilities in rural areas. Following are the main objectives of this campaign :

(*i*) Bring about an improvement in the general quality of life in the rural areas.

(*ii*) Accelerate sanitation coverage in rural areas.

(*iii*) Generate demand for sanitation facilities through awareness creation and health education.

(*iv*) Cover schools/anganwadis in rural areas with sanitation facilities and promote hygiene education and sanitary habits among students.

(*v*) Encourage cost effective and appropriate technologies in sanitation.

(*vi*) Eliminate open defecation to minimise risk of contamination of drinking water sources and food.

(*vii*) Convert dry latrines to pour flush latrines, and eliminate manual scavenging practice, wherever it is in existence in rural areas.

Panchayati Raj and Decentralised Planning

It is a strongly believed that Panchayati Raj is a very powerful weapon for rural economic, social and cultural development and also for decentralisation of rights and taking the process of democracy to the common and most humble man of the rural community. This enables people to become a part of the development process and increases the possibility of decentralised planning. Panchayati Raj is an important medium of politics, administration and planning.

India has a very old tradition of forming panchayats. Panchayats were recognised as the local unit of administration right from the beginning of the British rule in India. Later on, the Britishers tried to strengthen the institution of Panchayati Raj to divert the attention of our national leaders from the demand of complete Independence.

Mahatma Gandhi often used to say that real India lives in villages and the country cannot grow as a nation until the life style of the villagers is not improved. About one-third of our villages had Panchayats when India got freedom in 1947. The Congress Government at the centre took immediate steps to strengthen village Panchayats after Independence. The 40th amendment in the Constitution in 1950 stated that "The state shall take steps to organise village panchayats and to endow them with such powers and authority as may be necessary to enable them to function as units of self-government."

Programme of Community Development was started in the year 1952 which aimed at infusing awakening among the rural folks about economic planning and social reconstruction. But this programme could not yield the desired results. Consequently Block Advisory Committees were constituted which were renamed as Block Development Committees. This step also did not get much success. Then it was realised that the rural masses will not take interest in any scheme unless they are directly involved in framing and implementing any development programmes. Balwant Rai Mehta Committee was constituted to remove the shortcomings in the Panchayati Raj (See box).

The National Development Council accepted the recommendations of the Mehta Committee report in 1958 and different states started implementing the Panchayati Raj scheme.

Various Patterns of Panchayati Raj Institutes

Different states have different patterns of Panchayati Raj institutions. Some states have three tier panchayats at the village, block and district level while some states have two tier panchayat at the village and block levels and still some other states have only one tier Panchayati Raj.

A Committee to study the Panchayati Raj was constituted in 1957 under the chairmanship of Sh. Balwant Rai Mehta and it is known as Balwant Rai Mehta Committee. This committee submitted the report by the end of 1957. The report said that such programmes can succeed only if the local people at the village level are involved in it. It was recommended that Panchayati Raj institutions should be constituted immediately for decentralisation of democracy and development of the community.

Following were the main recommendations of this committee :

1. There should be three tier Panchayati Raj from village to district level. Gram Panchayat at the village level, Block Samiti at the Block level and Zila Parishad at the district level, they should be inter-related to each other.
2. Their rights and duties should be well defined.
3. They should have sufficient resources to discharge their duties in the desired manner.
4. Programmes regarding social and economic planning should be executed through these institutions.
5. In future all the facilities, rights, responsibilities and resources should be transferred under the Panchayati Raj system.

Panchayati Raj was formally inaugurated by Pandit Jawaharlal Nehru on 2nd October, 1959 in Nagaur district of Rajasthan. It had three constitutional parts *viz.* (*i*) Gram Panchayat, (*ii*) Panchayati Samiti and (*iii*) Zila Parishad. This system worked for about three decades after which Maharashtra and Gujarat also constituted three tier Panchayati Raj.

Karnataka introduced a system of Panchayati Raj under 1985 Ordinance which was different from the recommendations of Mehta Committee. The three tier system accepted by this state consisted of Zila Parishad at the district level, Taluka Panchayat Samiti at block/taluka level and Mandal Panchayat at mandal level.

Achievements of Panchayati Raj Institutions. Spatial and temporal changes keep on occurring in the structure of Panchayati Raj. For example, there are changes from time to time in tiers and duties of Panchayati Raj institutions. In Andhra Pradesh, Panchayati Raj institutions have done a wonderful job in the basic fields like education and health. In Rajasthan, these institutions played a major role with respect to development at the initial stage. Panchayati Raj institutions in Tamil Nadu did a commendable job in the fields of education, water supply, roads and nourishment. In Maharashtra and Gujarat, these organisations worked very hard for decentralisation of planning.

The greatest achievement of Panchayati Raj institutions is that villages have got rid of fear psyche and are no longer afraid of government officers. They can easily and directly talk to the concerned Block Development Officer (BDO) and get solutions to their problems. The main purpose of constituting panchayat was to develop self confidence among the villagers, provide them with basic facilities of life and persuade them to come forward for planning and development. Panchayati Raj has infused awareness about development, planning and democracy and now they can take their own decisions and prepare programmes which are beneficial for them.

The Failures of Panchayati Raj

Many weaknesses have been observed in Panchayati Raj during the last few decades as a result of which many failures of this system have come to light. Some of the major failures are listed below.

1. Lack of Conceptual Clarity. The greatest drawback of Panchayati Raj system is the lack of conceptual clarity. Some scholars recognise it just an administrative agency while others treat it as spread of democracy at the grass root level. Still some others see it as government at the rural level. All these concepts work simultaneously which creates hinderance in the developing of a clear and universally accepted concept.

2. Lack of Political Will. Critical analysis of the working of Panchayati Raj has revealed that these organisations do not have sufficient resources, their activities are insufficient and not upto the mark and proper attention is not paid to them. Because of these factors the achievements of Panchayati Raj institutions are not upto the desired level. Such a situation is largely attributed to lack of will power. In certain states, the Panchayat elections are not held at the scheduled time and often the elected panches are ignored. Normally, MPs and MLAs work against the elected members of the gram panchayats and the grassroot problems are often ignored. Generally the political leaders cultivate a feeling that the panchayat members are working against their interests. In such a scenario, the panchayat members are not able to receive political support and proper action is not taken on their plans.

3. Role of Bureaucracy. Bureaucracy in India enjoys immense powers and bureaucrats often deprive panchayat members of their role in the development process. Most of the government officers do not like to work under panchayat members and consider themselves as representatives of the government, rather than as part of the development mechanism.

4. Disillusionment about structural functioning. In most panchayats, only the wealthy and influential people get elected and those people do not bother much about the interests of the poor and downtrodden. Due to panchayat elections, the villagers are often divided into political and socials groups which leads to frequent and sometimes serious disputes. Such a situation creates great obstacle in the path to progress and the primary aim of Panchayati Raj is ignored.

In addition to the above mentioned factors it is felt that planning is primarily the job the planner and interference by panchayats causes shortcomings in planning and retards the pace of progress.

Ashok Mehta Committee

As a result of 1977 general elections, Congress Party was defeated and Janata Party came to power at the centre. In December 1977, the Janata Party Government at the centre constituted a 14 member Committee with Ashok Mehta its chairman. The report submitted by this Committee in August, 1978 said that it was wrong to think that Panchayati Raj does not cooperate in the process of development. Panchayati Raj has achieved many successes but they are not given much opportunities to be a great instrument of development. Panchayati Raj institutions should be constituted in such a way that it becomes a strong medium of economic, social and political development through peoples' participation.

Command Area Development

An area served by canals, wells, tube wells, tanks etc. for irrigation is known as command area. The irrigation potential created is not used to its optimum due to the following reasons :

1. Proper basic structure for carrying water from its source so the agricultural fields—like channel, drainage, etc. is lacking.

RECOMMENDATIONS OF ASHOK MEHTA COMMITTEE

Report submitted by Ashok Mehta Committee made about 132 recommendations. The main recommendations pertained to (*a*) functional necessity for decentralization of administration through under democratic supervision; (*b*) two-tier system of Panchayati Raj *viz.* the revenue District assuring the technical expertise of high order required for rural development and the Mandal Panchayat to be constituted by grouping a number of villages; (*c*) the Panchayati Raj Institutions to have compulsory powers of taxation and mobilize their own resources, thus reducing their dependence on diversion of funds from the state government. Certain taxes collected from the area such as provision tax, entertainment tax and special tax on land and building be transferred to Panchayati Raj Institutions; (*d*) open participation of political parties in Panchayati Raj affairs; (*e*) the creation of certain economic groups in the villages; (*f*) a regular social audit by a district level agency as well as a committee of legislators to check whether funds earmarked for these social and economic groups are actually spent on them; (*g*) the State legislators will have a committee on Panchayati Raj with adequate representation for schedule castes and tribes to cater to their needs and mitigate grievances of the weaker sections; (*h*) the State Government must not supersede the Panchayati Raj Institutions on partisan grounds. In case of an imperative supersession, election must take place within six months; (*i*) provision of urban amenities such as roads, potable water, medical care, employment and education in rural areas to neutralise pull for immigration to urban centres.

50 PER CENT RESERVATION FOR WOMEN

The central cabinet passed a resolution on 27th August, 2009 according to which 50 per cent seats in panchayats are reserved for women. According to data released in 2009, there were 28.18 lakh elected representatives in Indian panchayats out of which 36.87 per cent were women. After this resolution, more than 14 lakh women are supposed to the members of panchayats.

2. There is lack of proper agricultural system according to the ecological conditions.
3. There is lack of awareness on the part of farmers.
4. Lack of proper maintenance of canals results in loss of water.
5. There is over irrigation in areas near the source of canals which leads to water logging, salinity and alkalinity of vast area on one hand and shortage of water at the tail of the canals where agricultural fields do not get sufficient water.

Keeping in view the above problems, the Central Government started Command Area Development (CAD) programme in 1974-75 with the objective to bridge the gap between irrigation potential created and utilized through micro level infrastructure development for efficient water management and enhancement of agricultural production and productivity so as to improve socio-economic conditions of the farmers. This programme is implemented by Area Development Authority. Specialists from different departments work under this authority which makes it easy to achieve the fixed targets.

This programme was initially applied to Indira Gandhi Canal Command Area in 1974. In the Fifth Five Year Plan, 38 Area Development Authorities were constituted for 50 irrigation projects. These authorities provided irrigation facilities to 12.4 million hectares spread over 108 districts in 13 states. So far it has covered 332 irrigation projects with a Cultivable Command Area of nearly 29 million hectares.

Following are the main objectives of this programme :

1. Influential farmers with large expanses of land at their disposal get the maximum benefit of irrigation facilities and a small and marginal farmers are often deprived of irrigation facilities which they rightly deserve. ***Warabandi*** was introduced to solve this problem. It is a rotational system which assures equitable and timely supply of water to all farmers irrespective of the size of their holdings. This programme is implemented with the help of Panchayat Samities.
2. Construction of field channels and field-drains to reduce the pilferages and mis-use of water and making sure of effective use of water for irrigation. Central assistance up to 50 per cent (limited to the prescribed cost norms) is provided for irrigation and field-development activities.
3. Land has to be levelled so that water is spread uniformly in the agricultural fields.
4. Giving proper shape to fields and their boundaries through consolidation of holdings

so that the objective of optimum utilisation of water is achieved. Simultaneously provision of **chak roads** is also essential.

5. Educating farmers through demonstrations and imparting technical training for adopting new agricultural innovations and adopting cropping patterns according to natural environment as well as improving farming practices and maintaining fertility of soil.
6. Preparing plan for supply of inputs, credit, seeds, fertilisers, insecticides and pesticides to the farmers.
7. Promoting ancillary activities such as animal husbandry, forestry, poultry, marketing and processing facilities.
8. Updating land records.
9. Diversifying agriculture to make it more profitable.
10. To lay more emphasis on the cultivation of more remunerative crops like oil-seeds, pulses, green-manure crops.
11. To use ground water for compensating the shortage of surface water.
12. Reclamation of the water-logged, saline and alkaline areas rendered uncultivable due to over irrigation by canals.
13. To introduce participatory management of irrigation.

The Command Area Development programme was restructured and renewed as Command Area Development and Water Management (CAD&WM) with effect from 1st April, 2004. During the mid-term appraisal, the Planing Commission has emphasised the need for implementing CAD and WM programme *paripassu* with creation of infrastructure. Under this programme, financial assistance is being provided to the State governments on 50 : 50 basis for construction activities and on 75 : 25 basis to carry out training programmes for field functionaries/ farmers, monitoring, evaluating, adaptive trials and demonstration.

At the initial stage, 60 major and medium irrigation projects were taken up under the CAD programme, covering a Culturable Command Area (CCA) of about 15.0 million hectares. At present there are 145 projects covering a CCA of 16.02 million hectares. The scheme had been implemented as a State Sector Scheme during the Eleventh Five Year Plan (2008-09 to 2011-12).

An area of about 18.06 million hectares was covered under the programme, upto the end of March, 2007 and another 1.62 million hectares had been covered between March 2007 and March 2011.

The National Water Policy, 2002, stresses on participatory approach in water resources management. There is a great realisation of the fact that participation of beneficiaries greatly enhances the optimal use of the irrigation facilities and proper use of irrigation water. The participation of farmers in the management of irrigation would give responsibility for operation and maintenance, collection of water charges from the areas under the jurisdiction of Water Users Associations (WUAs) and redressal of petty grievances.

Reports from various states reveal that the CAD programme has made positive impact on some important indicators, like increase in irrigated area, productivity and production, irrigation efficiency etc. In spite of all these efforts, the problem of water logging has been reported in many irrigated commands. For reclaiming the water logged areas, 579 schemes in 9 states *viz.* Bihar, Gujarat, Madhya Pradesh, Jammu and Kashmir, Karnataka, Kerala, Odisha, Maharashtra and Uttar Pradesh have been approved for reclaiming 78.81 thousand hectares of water logged area; of which about 52.11 thousand hectares of water logged area has already been reclaimed.

The programme is being implemented under Pradhan Mantri Krishi Sinchai Yojna (PMKSY)—***Har Khet Ko Pani***—from 2015-16.

Indira Gandhi Canal Command Area Development Programme

Indira Gandhi Canal Project is a living example of man's effort to transform a desert land into a green land of prosperity and plenty. It is one of the largest canal systems of the world. The command area of this canal is located in dry desert lands of Ganganagar, Bikaner, Jaisalmer, Barmer, Jodhpur and Churu districts in Rajasthan. It stretches over 23,725 sq km having 525 kms length and 45 kms breadth along the international boundary between India and Pakistan.

The Indira Gandhi Main Canal runs parallel to the Pakistan border for an approximate distance of 38 kms from north-east to south-west.

The Canal originates from Harike Barrage near the confluence of Satluj and Beas rivers in Ferozepur district of Punjab. The plan for this canal was prepared in 1957-58 and the work on this project started on 31st March 1958. The canal does not do any irrigation in Punjab and is known as Rajasthan Feeder. The total length of Rajasthan Feeder is 204 km. The main canal is 40 metres wide at bottom and 6.4 metres deep. The carrying capacity of canal is 18,500 cusecs of water at its head. According to a proposal in 1981, Rajasthan was allocated 8.6 million acre feet of Ravi-Beas surplus water. The Indira Gandhi Canal envisages the utilisation of 7.6 million acre feet of water allocated to Rajasthan.

The head of the main canal is located near Masitanwali in Hanumangarh district. The tail of the 445 km long main canal is located near Mohangarh in Jaisalmer district. The command area of the canal is further extended till Gadra Road in Barmer district, through Sagarmal Gopa branch.

Construction work of the project has been carried out in two stages. Water was released in the main canal on 11 October 1967 and reached its tail on 1 January 1987. The main canal has two kinds of branches and distributaries. The right bank branches of the canal are flow channels as the land west of the main channels slopes down gently towards Pakistan

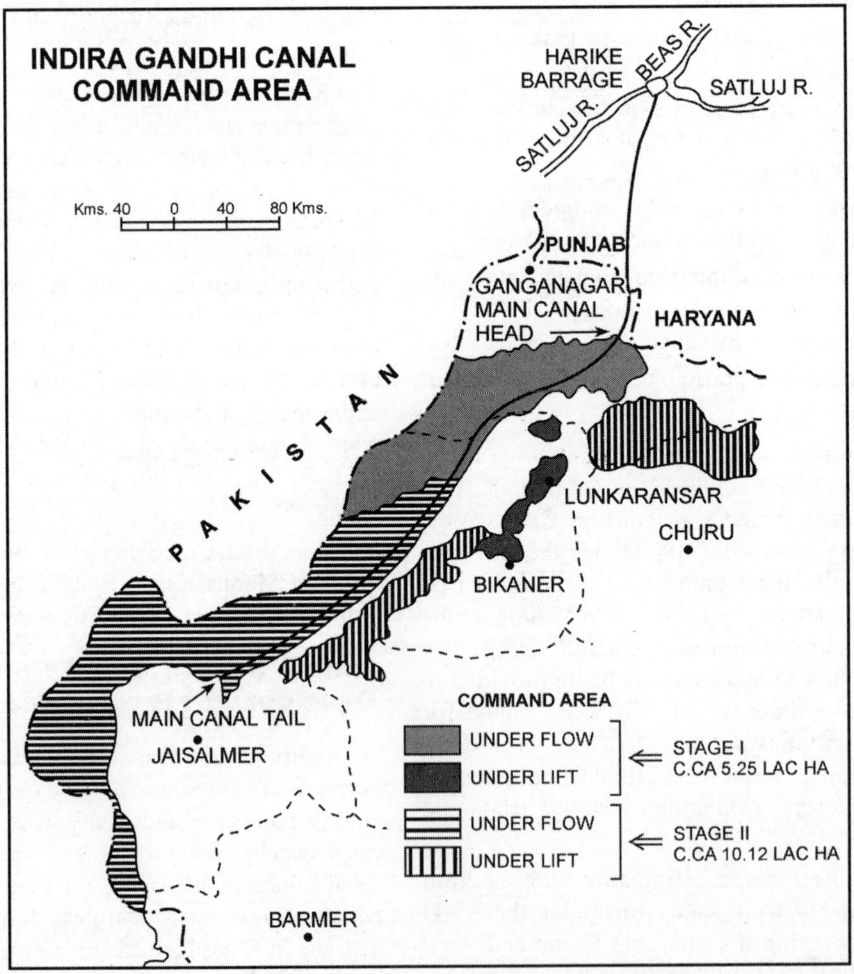

FIG. 13.4. Indira Gandhi Canal Command Area

border. The left bank branches except Rawatsar Branch which takes off from the head of the main canal, are lift channels. This is because of the fact that area towards the south-east of the main canal slopes towards the canal and water has to be lifted against the slope of the land.

Stage I. Construction work of Stage I has been completed at a cost of ₹ 246 crore. It included the construction of 204 kms long feeder, 189 kms of the main canal and 2,960 kms long distribution system. Stage I has five flow branches and one lift canal covering south and south-western part of Ganganagar district and north and northwestern parts of Bikaner district. Nearly 4.79 lakh hectares of land is provided with flow irrigation and 0.46 lakh hectares get lift irrigation. The irrigation potential on full development has been estimated to be 5.78 lakh hectares. The project plan of Stage I envisages intensive irrigation with an irrigation intensity of 110%. *Irrigation intensity is expressed as percentage ratio between gross irrigated area and culturable command area of the project.*

The above facilities result in annual food production of 14.50 lakh tonnes [*See* **Table 13.1**].

Stage II. Stage II of the project includes the construction of 256 kms long main canal and 4,800 kms long distributaries. According to the revised plan, this stage provides extensive irrigation. Extensive irrigation means reducing per acre allowance of water and providing irrigation to maximum cultivated area. Irrigation intensity of this region is *80%* that amounts to provide irrigation to 80% of culturable command area. This will prevent water-logging and soil-salinity and help in growing light irrigated crops. Stage II proposes to develop irrigated pastures on an area of about 3.66 lakh hectares. This would help in providing benefits of irrigation to aboriginal nomadic communities, develop animal husbandry and arrest desertification. This stage of the project envisages to provide flow irrigation to 7 lakh hectares and lift irrigation to 3.12 lakh hectares of land. This will lead to annual food production of 22.50 lakh tonnes (*See* **Table 13.1**).

Command Area Development Programme. The Command Area Development Programme is an integrated area development approach towards the command area of major and medium irrigation projects in the country. This programme is aimed at bridging the gap between created irrigation potential and its utilisation in the command areas of major and medium irrigation projects. The Fifth Five-Year Plan document emphasized the need of implementing this programme in all the command areas of major and medium projects in the country. The importance of Command Area Development (CAD) is all the more important in a desert areas because it requires proper use of water and ecological development for the following reasons :

(*i*) Conveyance loss of water is 30 to 50% below outlet level in sandy soils. This results in under-utilisation of water and leads to water-logging and soil-salinity. Conse-quently, it adversely affects agricultural production.

(*ii*) This is a newly settled area and requires civic amenities as well as modern agricultural inputs.

(*iii*) There is large-scale siltation in the main canal and its distributaries by wind. In addition, cultivable land is also eroded.

The Command Area Development Programme was introduced in the Indira Gandhi Canal Command Area in 1974 and it was entrusted with following tasks :

(*a*) On-farm development which includes surveying and planning water course lining, land levelling, shaping and reclamation of degraded lands.

(*b*) Afforestation and pasture development which includes canal side and roadside plantation, block plantation near new settlements, sand dune stabilisation and pasture development on culturable waste land.

(*c*) Providing communication and civic amenities which includes construction of roads, connecting the settlements with markets, construction of new markets and supplying drinking water.

(*d*) Availability of modern agricultural inputs including ensuring supply of HYV seeds, chemical fertilizers, insecticides and pesticides and providing agricultural extension and training facilities to the farmers.

TABLE 13.1. Salient Features of Indira Gandhi Canal Project

Particulars	Unit	Stage I	Stage II	Total
1. Length of the Main Canal :				
(*i*) Indira Gandhi Feeder	kilometres	204	—	204
(*ii*) Indira Gandhi Main Canal	kilometres	189	256	445
Total		**393**	**256**	**649**
2. Length of Distribution System	kilometres	2,960	4,800	7,760
3. Culturable Command Area :				
(i) Under flow irrigation	Lakh hectares	4.79	7.00	11.79
(*ii*) Under lift irrigation	Lakh hectares	0.46	3.12	3.58
Total		**5.25**	**10.12**	**15.37**
4. Irrigation Potential on Full Development	Lakh hectares	5.78	8.10	13.88
5. Irrigation Intensity	Per cent	110	80	90
6. Water Requirement	Million acre feet	3.59	4	7.59
7. Drinking and Industrial Use of Water	Cusecs	300	900	1200
8. Cost	Crore ₹	246	1,420	1,666
9. Annual Food Production	Lakh tonnes	14.50	22.50	37.00

Agricultural Development. Crops could be grown only in the *kharif* season in this dry area and a large tract of cultivable land was left as culturable waste and fallow land. Introduction of irrigation has dramatically increased the net sown area (NSA) and area sown more than once, *i.e.*, double cropped area. Drought-resistant crops like *bajra, guar, moong, moth* and *gram* used to occupy about 95% of the gross cropped area of this region before the introduction of irrigation. Cropping pattern has undergone drastic changes as a result of irrigation. Commercial crops, such as cotton, groundnut, wheat and mustard have become very important occupying nearly two-thirds of the gross cropped area in Stage I. Wheat alone occupies about one-fourth of the gross cropped area. The yield of cotton, groundnut, paddy and wheat is increasing consistently, while yield of *guar*, gram and mustard is fluctuating. The main reason for the fluctuations of yield level of these crops is that they are not fully irrigated.

The production of cotton, groundnut, wheat and mustard has increased considerably. This is due to increase in area under these crops as well as their increase in yield per hectare as a result of irrigation and use of better seeds.

Impact of Irrigation on Environment. Whereas irrigation has increased the agricultural production tremendously, it has resulted in environment degradation in the form of water logging and soil-salinity. The ground-water table is rising at an alarming rate of 0.8 metre per year in most parts of Stage I. About 25% of land under the command area near Ghaggar basin is critical area as the ground water level in this area is less than 6 metres below surface level. In a large part of the command area in Stage I, soil-salinity has arisen because of water-logging and the presence of salt in the soils. This has adversely affected the soil-fertility and agricultural productivity. This problem is more serious in the command area of Stage II where irrigation was introduced in mid-eighties. This part of the command area is underlain by hard pan of calcium carbonate and clay at a depth of few metres which causes parched water table and water-logging.

Watershed Management

Watershed management is a very important component of water policy and planning. *Watershed is the line separating headstreams which flow in different river systems. It may be as sharply defined as crest of a ridge or it may be indetermined as is the case in a low undulating area.* The water divide determines the boundary of the watershed. Watershed becomes a practical agricultural unit in two different situations (*i*) when it is used as drainage line or flood control and (*ii*) when it is accepted as a unit planning for hydro-electric production, irrigation, drinking water, etc.

The All India Soil and Land use Survey Organisation of the Department of Agriculture, Government of India, has finalised a nationwide system of delineation and codification of watersheds. According to this system, watersheds have been divided into the following four categories depending upon their size :

(*i*) basins, (*ii*) catchments, (*iii*) subcatchments, and (*iv*) watersheds.

As per computation made by this organisation, India has 35 river basins, 12 catchments, 500 sub-catchments, and 320 watersheds. Depending upon their area, watersheds are divided into following four categories :

(*i*) Sub-watersheds (10,000 to 20,000 hectares)
(*ii*) Milli-watersheds (1,000 to 10,000 hectares)
(*iii*) Micro-watersheds (100 to 1,000 hectares)
(*iv*) Mini-watersheds (1 to 100 hectares)

The watershed management is a multi-departmental and comprehensive programme which aims at the optimum utilisation of the available water and other resources to achieve the following goals :

(*i*) increasing agriculture productivity
(*ii*) promoting allied activities
(*iii*) generating additional income from social forestry
(*iv*) checking soil erosion
(*v*) developing cottage industries
(*vi*) maintaining ecological balance.

Integrated Watershed Management Programmes (IWMP). It has been estimated that about 52 per cent of cultivated area in India is rainfed which suffers from cronic problem of water shortage, low agricultural productivity, land degeneration, poverty, malnutrition, etc. Watershed development programme is considered to be an effective tool to solve these problems.

Under the aegies of the Planning Commission, National Rainfed Area Authority (NRAA), framed Common Guidelines for watershed programmes based on Parthasarathy Committee Report. Accordingly, a single modified programme called Integrated Watershed Management Programme (IWMP) was launched in 2009-10 which is being implemented as per Common Guidelines for Watershed Development Projects, 2008.

Impact Assessment Studies of Watershed Projects. A study entitled "Comprehensive Assessment of Watershed Programmes in India" has been assigned to International Crops Research Institute for the Semi-Arid Tropics (ICRISAT), Hyderabad to assess the impact of various watershed development programmes in India, which among other things, gave the following report :

- Soil loss of 1.1 tonne/hectare/year was prevented
- Additional water storage capacity of about 38 hectares metre was created in 9500 hectares watershed.
- The area under irrigation increased by 52 per cent and the cropping intensity increased by 35.5 per cent.
- Low income regions were more benefitted than high income regions.
- Benefits were more in areas receiving between 700 mm and 1000 mm of annual rainfall with the available technologies.
- Peoples' participation produced better results.
- The macro, watersheds (area more than 1000 hectares) perform better than micro watersheds (area below 500 hectares).

PLANNING FOR BACKWARD AREA

It is difficult to define and identify backward areas because several parameters are used for this purpose. Generally areas with low per capita income are treated as backward areas. But this is not a very reliable measure. Various other measures have to be applied for proper identification of backward areas. Some such measures can be (*i*) the percentage of the scheduled caste and schedule tribe population to the total population of the area, (*ii*) the ratio of population to the cultivated land, (*iii*) the percentage of the working force engaged in agriculture, (*iv*) the ratio of urban to rural population, (*v*) availability of transport, communications, and other services, (*vi*) availability of water, electricity, and other facilities, (*vii*) level of literacy, etc. During the first decade of planning no attempt was made to specify the backward areas and pockets of underdevelopment taking into account these various indicators. Perhaps the Committee on the Dispersal of Industries, set up by the Small Scale Industries Board in 1960, was the first to identify the backward areas for purposes of rural industrialisation. The criteria for backwardness suggested by this Committed were as follows :

1. Poverty of the people as indicated by
   (*a*) Low per capita income; and
   (*b*) Low per capita consumption.
2. High density of population in relation to development of productive resources and employment opportunities as indicated by the following factors :
   (*a*) High ratio of population to cultivable land (50% above the national average should be considered as backward);
   (*b*) Low percentage of population engaged in output (50% or more below the national average should be considered as backward);
   (*c*) Absence or under-exploitation of other natural resources, *viz*., minerals, forest and animals;
   (*d*) Low percentage of population engaged in secondary and tertiary activities (25% below the national average should be considered as backward);
   (*e*) Low ratio of urban to rural population (districts where the ratio was less than 50% of the national average might be considered as backward);
   (*f*) Low percentage of factory employment (50% below the national average might be considered as backward).
3. Poverty of communications as indicated by small lengths of railways and metalled roads per hundred square mile (districts where the railway and road mileage fall below 50% of the national average might be considered as backward).
4. High incidence of unemployment or gross underemployment.
5. Consumption of electric power.

At the time of formulating the Fourth Plan (1966-1971) the Planning Commission appointed a Study Group to suggest the criteria for identifying the backward areas. The Group suggested the following indicators :

1. Total population and density of population;
2. Number of workers engaged in agriculture including agricultural labourers as percentage of total workers;
3. Total area per agricultural worker;
4. Cultivable area per agricultural worker;
5. Net area sown per agricultural worker;
6. Percentage of gross irrigated area to net sown area;
7. Percentage of area sown more than once to net sown area;
8. Per capita (rural population) gross value of agricultural output;
9. Establishments (manufacturing and repair) using electricity :
   (*a*) Total
   (*b*) Household
   (*c*) Non-household
10. Number of workers per lakh of population employed in registered factories;
11. Mileage of surfaced roads :
    (*a*) Per 100 sq milcs
    (*b*) Per lakh of population
12. Number of commercial vehicles registered in a district;
13. Percentage of literate population :
    (*a*) Men
    (*b*) Women
14. Percentage of school going children :
    (*a*) Boys
    (*b*) Girls in age groups of
    (*i*) 6-11 years, and (*ii*) 11-14 years.
15. Number of seats per million population for technical training :
    (*a*) Craftsmen
    (*b*) Diploma level
16. Hospital beds per lakh of population.
17. Per capita income.

In the year 1968, the Planning Commission set up two Working Groups for studying the regional imbalances. One group was for recommending the criteria for the identification of backward area and the other for recommending the fiscal and financial incentives for starting industries in the backward areas. The committee for specifying the criteria for identification of backward areas was set up under the chairmanship of B.D. Pande (and is popularly known as the Pande Committee) and the committee for recommending the fiscal and financial incentives in the backward areas was set up under the chairmanship of N.N. Wanchoo (and is popularly known as the Wanchoo Committee).

The criteria for the identification of industrially backward States and Union territories adopted by the Pande Committee were as follows :

1. Total per capita income.
2. Per capita income from industry and mining.
3. Number of workers in registered factories.
4. Per capita annual consumption of electricity.
5. Length of surfaced roads in relation to :
   (*a*) the population, and
   (*b*) the area of the State.
6. Railway mileage in relation to :
   (*a*) the population, and
   (*b*) the area of the State.

Based on the above described criteria, the Pande Committee classified ten backward states. These states are Andhra Pradesh, Assam, Bihar, Himachal Pradesh, Jammu & Kashmir, Madhya Pradesh, Nagaland, Odisha, Rajasthan and Uttar Pradesh. All Union territories except Chandigarh, Delhi and Puducherry were also classified as industrially backward.

This committee identified 238 districts all over India comprising 60% of the country's area and an equal proportion of its population as industrially backward.

The Wanchoo Committee recommended following incentives to develop industries in backward areas :

(*a*) Grant of higher development rebate to industries located in backward areas.

(*b*) Grant of exemption from income-tax, including corporate tax, for 5 years after providing for the development rebate.

(*c*) Exemption from the payment of import duties on plant and machinery, components, etc., imported by units set up in backward areas.

(*d*) Exemption from excise duties for a period of 5 years.

(*e*) Exemption from sales tax, both on raw materials and finished products, to units set up in specified backward areas for a period of 5 years, from the date of their going into production.

(*f*) Transport subsidy for taking out the finished products for a period of 5 years. Upto 400 miles (650 km) the distance should be considered as normal and beyond that the transportation cost for finished products should be subsidized for such backward areas as may be selected in the States of Assam, Nagaland, Manipur, Tripura, NEFA (present Arunachal Pradesh) and Andamans. The transport subsidy should be equivalent to 50% of the cost of transportation in case of the backward areas specified in J&K State.

The National Development Council (NDC), decided in September, 1969 that the recommendations of the Pande Committee should be applied not only to backward districts in backward states only, rather such concessions should be made available to selected backward areas in all the States and Union Territories. The criteria for the selection of the industrially backward districts in the States and Union territories were to be decided by the Planning Commission in consultation with the financial institutions and the State governments, in the light of the two sets of criteria recommended by the Pande Committee. The following set of criteria was evolved by the Planning Commission for the purpose of identification of the industrially backward districts to qualify for concessional finance :

1. Per capita foodgrains/commercial crops production depending on whether the district is predominantly a producer of foodgrains/cash crops. (For inter-district comparisons, conversion rates between

foodgrains and commercial crops may be determined by the State government on a pre-determined basis where necessary).

2. Ratio of agricultural workers to population.
3. Per capita industrial output (gross).
4. Number of factory employees per lakh of population or alternatively number of persons engaged in the secondary and tertiary activities per lakh of population.
5. Length of surfaced roads in relation to population or railway mileage in relation to population.

On the basis of these criteria, 241 districts (later raised to 247) all over the country were classified as industrially backward.

Location of Large Public Sector Projects. Several public sector projects were located in the backward areas for their economic development. Such projects included steel plants, fertilizer plants, heavy electrical plants in states like Odisha, Chhattisgarh, Jharkhand and Madhya Pradesh.

Industrial Estates Programme. This programme was initiated in 1955 when 12 estates were sanctioned. The construction, organization, and management of ten of these estates was entrusted to the State governments while the development of two estates (one situated in Naini at Allahabad and the other in Okhla at Delhi) was the responsibility of the National Small Scale Industries Corporation. The ten estates entrusted to the respective State governments were situated at Rajkot in Gujarat, Guindy and Virudhunagar in Tamil Nadu, Kanpur and Agra in Uttar Pradesh and Palghat, Trivandrum (Thiruvananthapuram), Kottayam, Quilon and Trichur in Kerala. The State governments were sanctioned a loan of ₹ 57.89 lakhs and a grant of ₹ 0.495 lakhs each for the purpose.

The principal objectives of the programme, varying to some degree with time and place, have been :

(*i*) to encourage the growth of small scale industries;

(*ii*) to shift small scale industries from congested areas to estate premises with a view to increasing their productivity;

(*iii*) to achieve decentralised industrial development in small towns and large villages; and

(*iv*) to encourage the growth of ancillary industries in the townships, surrounding major industrial undertakings, both in public and private sectors.

Desert Development Programme

The Desert Development Programme (DDP) was launched in 1977-78 to reduce the adverse effects of the natural conditions prevailing in the desert areas. It is applied to 235 blocks of 40 districts in seven states. As many as 15,746 plans had been sanctioned between 1995-96 and 2007-08, which benefited 78.73 lakh hectares of land. A sum of ₹ 2103.29 crore was spent on these plans. The seven states which were benefitted from this programe are Rajasthan, Haryana, Gujarat, Himachal Pradesh, Jammu and Kashmir, Andhra Pradesh and Karnataka. These states have hot sandy or cold deserts. The funding programme between centre and states is in the ratio of 75 : 25 for hot sandy deserts but the Central Government bears the entire cost in case of cold deserts. The primary aim of this project is to encourage proper use of natural resources, increase employment opportunities, remove poverty and illiteracy and raise the living standard of the common man.

Problems of Desert Areas

1. Most of the desert areas receive scanty rainfall and enough water is not available for crops. Vast areas of Rajasthan and the adjoining parts of Haryana and Gujarat receive less than 40 cm annual rainfall. The amount of annual rainfall is less than 20 cm in Barmer and Jaisalmer districts in the western part of Rajasthan. Union Territory of Ladakh also receives less than 20 cm annual rainfall. Most parts of Andhra Pradesh, Telangana and Karnataka are located in the rain shadow area of the Western Ghats where the amount of annual rainfall is less than 60 cm. Lesser amount of rainfall is not the only problem, rainfall variability is also very high. Rainfall variability is as high as 50 per cent in Rajasthan and its neighbouring

areas and 40 per cent in the Leh-Ladakh region. Under such conditions, it is not an easy task to depend on natural amount of rainfall and people have to face famine conditions whenever the actual rainfall is below average.

2. In most parts of Rajasthan and its neighbouring areas, the soils are sandy which are easily blown away by winds. This leads to soil erosion and low agricultural productivity.
3. Only those crops can be grown which need lesser amount of water for their proper growth. These include coarse grains such as *jowar, bajra,* barley and pulses. They have lesser market value and their yields are also very low. With the construction of the Indira Gandhi Canal in the western part of Rajasthan, the cropping pattern has changed and wheat and at certain places rice cultivation has become possible.
4. Most of the farmers in the desert areas are poor and cannot afford modern inputs like high yielding varieties of seeds, fertilizers, pesticides and new implements.
5. In the event of a drought condition, there is acute shortage of food and fodder and people are force to migrate to other places in search of food, fodder and livelihood.
6. The infrastructural facilities like market, storage, transport, etc. are badly lacking.

Main Objectives

1. To check the spread of desert and increase productivity.
2. To promote dry farming and increase productivity.
3. To make optimum use of natural resources of the desert areas and raise the living standard of the common man.
4. To generate more employment and enhance per capita income so that people can lead a comfortable life.

Strategies for Development

1. The most important strategy is to check the spatial growth of desert through shelter belt and to stabilize sand dunes by intensive plantation.
2. The Desert Development Programme has been made an integral part of the 20-point Programme of the government so that maximum benefit reaches the maximum number of people.
3. Under the dry farming programme, crops needing lesser amount of water have been promoted. Course grains (*jowar, bajra,* barley), pulses, oil-seeds, etc. are such crops.
4. New technologies are promoted to increase the production of fodder.
5. Proper use of resources has been encouraged to maintain ecological balance.
6. Much emphasis has been laid on the use of eco-friendly pesticides and wedicides.
7. Irrigated agriculture and animal husbandry have been given priority in the cold deserts of Ladakh and Spiti.
8. Various devices for water harvesting have been used to meet the water shortages.
9. In addition to growing crops, allied activities like milk production, poultry, horticulture, selviculture, sheep rearing, bee-keeping etc. have been suggested to help farmers, especially small and marginal farmers.
10. Emphasis has been laid on agro-forestry, social-forestry etc. to check soil erosion and to maintain ecological balance.
11. Plan has been prepared to reclaim water logged areas in the Indira Gandhi Canal Command Area.
12. Programme to involve Non-Government Organisations (NGOs) in desert development programmes.

Drought Prone Area Development Programme

On an average, about 12 per cent of population and 16 per cent of the total area of India is affected by droughts and a severe drought occurs after every 4-5 years. At the micro level, one part of the country or another is affected by a drought almost every year. Drought causes shortage of food and fodder and lakhs of people and animals become victims of hunger and malnutrition. Hence solid steps are required to be taken to solve the problem of droughts. Drought Prone Area Programme ((DPAP) was initiated in 1973-74. Currently 972 blocks in 193 districts of 16 states are being benefited by this programme. Between 1995-96 and 2007-08, as many as 27439 schemes covering a vast area were prepared at the cost of ₹ 2837.81 crores. The primary aim of this programme is to make provisions for optimum use of land, water and animals resources, maintain ecological balance and to raise the income level of the poor section of society. Following are the salient features of this programme :

1. Management and conservation of water resources.
2. Land reforms.
3. Conservation of soil and soil moisture.
4. Special emphasis on forestation and social forestry.
5. Development of pasture lands and sheep rearing.
6. Development of livestock and dairy.
7. Restructuring the cropping pattern and changes in the farming process.
8. Cleaning of ponds, canals, rivers and other water reservoirs.
9. Special assistance to small and marginal farmers as well as to agricultural labourers.
10. Availability of infrastructural fund and development of citizens' conveniences.
11. Development of allied activities, in addition to the main agricultural activity.

Since 1995-96, this programme is being carried on for watershed administration based on new guidelines in which local people are also involved. The primary object of this programme is to minimise the problems of areas prone to low rainfall and drought conditions for which coordinated programme involving local people and based on new techniques is adopted. The main emphasis is laid on irrigation, forestation, development of pasture lands, rural electrification, rural road connectivity, markets for agricultural products, loans to farmers and services. These factors help in strengthening infrastructure and reducing the impact of drought.

Different systems for implementing this programme prevail in different states. For example, this work is done by the District Planning Board in Maharashtra. In Karnataka this work is done at the district level under the chairmanship of the Divisional Commissioner while in Tamil Nadu, it is implemented by the District Development Corporation.

Shortcomings of the Drought Prone Area Development Programme

1. The process of fixing the targets is faulty. In many areas small and marginal farmers as well as agricultural labourers do not get benefit from the programme; and this section of society is in dire need of assistance in such a situation.
2. Programmes are often delayed due to shortage of working force with the requisite skill.
3. There is a general lack of experience to run such a programme.
4. The expected results are not obtained due to lack of information.
5. Dualism and dichotomy in efforts result in lower capabilities.

Hill Area Development Programme

According to 1951 Census data, about 18.6 per cent of India's total area is termed as hilly area (305-2135 metres) and as per 2001 census, these hilly areas are divided into following two broad categories :

(i)* The Hilly Areas that are along the boundaries of the States and Union Territories.** This category includes Ladakh, Jammu and Kashmir, Himachal Pradesh, Uttarakhand, Darjeeling hills areas of West Bengal, and north-eastern states (Assam, Arunachal Pradesh, Nagaland, Mizoram, Manipur, Tripura, Meghalaya and Sikkim). These are termed as ***Special Category States and most part of the outlay for development of these areas is provided in the shape of Central Assistance.

***(ii)* Hilly Areas which are parts of States.** These hilly areas form part of one or more than one states. The main examples of this category of hilly areas are the Aravalli, the Vindhyas, the Satpura, the Mahadev, the Maikala, the Ajanta, the Western Ghats and the Eastern Ghats. All these hilly areas are situated in the peninsular plateau area and they spread in different directions. The development plans for these areas are prepared and executed by the concerned state governments.

For the integrated development of the North-eastern region, the Central Government set up the North-Eastern Council by an Act of Parliament. This council prepares development plans for one or two states or for the whole of North-eastern region. The North-eastern council has played a significant role in the development of inter-regional programmes of power generation and its transmission, construction of roads, agriculture, animal husbandry, fisheries, forestry, etc. This council is also engaged in research, experiment and human development as well as in educating the people.

Some of the typical hilly areas from the development planning point of view are Karbi Anglong and Cachar in Assam (area 15,200 sq km) Darjeeling in West Bengal (area 2,400 sq km), Dehra Dun, Pauri Garhwal, Chamoli, Uttarkashi, Almora, Pithoragarh, Tehri Garhwal, and Nainital (area 51,100 sq km) in Uttarakhand. Although the concerned state governments are responsible for the development of these areas, yet need for central assistance was felt during the Fifth Five Year Plan, which was granted keeping in view the area and population of the concerned regions. The concept of sub-plans was also encouraged so that they are made complementary to each other. Another major hilly area is the Western Ghats which spreads over 132 talukas (area 1,37,000 sq km) in Maharashtra, Goa, Karnataka, Kerala and Tamil Nadu. The Central Government provides financial assistance for the development of the Western Ghats.

The problems of hilly areas are quite different from those of the plain areas. The rugged topography and different climate of hilly areas give rise to a different set of economic, social and cultural conditions. These conditions require special planning for the hilly areas for which detailed information regarding economic, social, cultural, and political conditions as well as that of natural and human resources is very essential. Special attention is paid to the optimum utilisation of land, mineral, water, and biotic resources.

Active participation of the people, particularly of the women, is essential for the successful implementation of the plan. Peoples' involvement can be ensured by encouraging the concept of ***social fencing***, which implies a voluntary and self imposed discipline in managing resources of society at a local level. Under the development plan subsistence agriculture, plantation agriculture, food grains culture, animal husbandry, poultry, bee-keeping, pig farming, forestry, etc. can be developed by using the local resources. Cooperative societies and Farmers' Service Societies are strengthened. Plantation Agriculture (tea, coffee, spices, etc.), agricultural forestry and social forestry are given priority under the forestry programme. Linkages are established between fruit producing and fruit consuming areas.

In some of the areas of the north-east region, the tribal people are practicing shifting agriculture locally known as ***Jhuming.*** This type of agriculture causes heavy damage to forest and soil resources and disturb the ecological balance of the concerned area. Under the Hill Area Development Programme, the practice of shifting agriculture is discouraged and in its place, plantation crops such as tea, coffee, rubber, etc. are encouraged. This change from shifting to plantation agriculture saves soil and forest resources and helps the tribal people in leading a settled life. These tribal people become owners of plantations in due course of time.

Planning is also done for animal husbandry depending upon the number of animals and extent of pasture lands. Provisions are made for improving the breeds, health and productivity of the animals and plans are also prepared for processing and marketing animal products.

Major thrust of the Hill Area Development Programme is to promote the following :

- Agriculture—sedentary, plantation (tea, coffee, species, rubber, coconut, etc.)
- Horticulture, especially banana, apple, grapes, and citrus fruits.
- Animal husbandry, poultry, bee-keeping and piggery.
- Conservation of soil and forest resources.
- Cottage and village industries.
- Employment to the rural youth.
- Strengthen co-operatives and farmers' service societies.
- Discourage Jhuming.

The hilly areas are more suitable for industries that require pollution free atmosphere, relatively cool and comfortable climate, are based on high skill and are capable of high value addition. Electronics, watch-making, optical glasses, medicines, folding furniture are some of such industries which flourish in these areas. Cottage industries like handloom and carpet-making are also carried on. Besides, tourist industry is a very important economic activity. The natural beauty of the hilly areas attracts tourists in large numbers from far and wide places and tourism has great potential in these areas. Solid planning is required to harness this potential.

Vast hilly areas, particularly in the Himalayan region, provide ideal conditions for the growth of pharmaceutical herbs, fruits, flowers and other useful plants. Thus, pharmaceutical industry has a huge potential in these areas. In fact many such industries have already been developed in the Shiwalik and the Lesser Himalayan region. The wild life of this area is an added attraction as it adds to the bio-diversity of the hilly areas. Therefore, integrated planning for conserving botanical and zoological wealth of these areas is very essential, under which there should be provision for national parks, wild life sanctuaries and biosphere reserves.

For any successful planning for the hilly areas, extensive field survey is of utmost importance because we get a wealth of information from such surveys. These surveys need to be updated periodically so that latest ground realities are made available to the planner. For such survey, modern techniques like remote sensing, air photographs, Global Positioning System (GPS) etc. are very useful. Planning can be at the regional, sub-regional, taluka, block or even at settlement level. Peoples' participation is very essential for the success of such plans. The plan should be eco-friendly and peoples' awareness is also important. There should be provision for compensating the loss of natural resources. For example reforestation should be made mandatory for compensating the loss of forests if a paper mill is set up in any one of these areas. Wastage of water can be avoided by constructing small dams and storing water in tanks, lakes and other reservoirs. Hilly areas are basically more suitable for perennial commercial crops like tea, coffee and fruits and are deficient in the production of food grains. Thus, there should be sufficient arrangement for storing food grains so that common man, along with those who are willing to stop the practice of shifting agriculture may get the right quantity of food at the right time.

There are regional disparities of development between the hilly areas and plain areas which has led to separatist sentiments in the peripheral hilly areas and there is lot of insurgency particularly in the north-eastern region. Most of these areas are inhabited by the scheduled tribes and their development is necessary. The only way to bring them in the main stream of the century is to improve their economic social and cultural conditions. Thus there is dire necessity to prepare and implement solid development plans to bring about positive change in their life style.

Tribal Area Development

Defining a tribe is a difficult task, and no definition in India has proved satisfactory and precise. In general, the tribes are expected to possess some, if not all, of the following characteristics :

(*i*) their roots in the soil date back to a very early period.

(*ii*) they live in relative isolation of hills and forests.

(*iii*) their sense of history is shallow in the sense that after some generations, the remembered history tends to shade off into mythology because they usually do not keep a written record of their history.

(*iv*) they have a low level techno-economic development.

(*v*) in terms of their cultural ethos (language, institutions, belief and customs), they stand out from other sections of society.

(*vi*) even if they are not egalitarian, they are at least non-hierarchic and undifferentiated.

On the basis of above characteristic features, the Government of India has included 427 communities in the Scheduled Tribes, given in he Eighth Schedule of the Constitution of India. Numerically, the Bhils and Santhals are the most important tribal groups of India, each with a population of more than 35 lakhs. Next to them are Minas, the Mundas and the Oraons, each having a population of over ten lakhs. The Hos, the Khonds and the Kols are next to them, each with a population of over five lakhs. After that, there are 42 tribes, each having a population varying from one to five lakhs.

Economy. Economy is the basic parameter for identifying and classifying the Tribal people. They are mainly involved in hunting, fishing, gathering, animal husbandry and agriculture (mainly shifting and sedentary cultivation). Tribes of Andaman and Nicobar Islands depend on gathering and hunting. On the mainland of India, almost all the tribal people living in forests also depend on gathering and hunting. Birhor, Kharia, Kadar, Chenchu, etc. are some of the tribes living in the forests of Central India and mainly depend on gathering various roots and tubers, fruits, honey, etc. They collect these items for their personal use or for exchange with agricultural products. The Todas of the Nilgiri Hills are pastoral tribes. They rear buffaloes and live on buffalo products. However most of the tribal people are engaged in agriculture although their method of cultivation is primitive and their agricultural productivity is low. It may be mentioned that about 90 per cent Bhils, Gonds, Santhals, Mundas, Oraons, Nagas, Khasi, Mizo, etc. are engaged in agriculture.

Most of the economic activities of the tribal people are low grade primary activities which do not provide enough means for their survival. Thus majority of the tribal people are not able to enjoy the basic facilities of good health, education and are still at lower level of economic development.

Tribal Development through Planning. The policy of the Central Government immediately after Independence was to focus on improving the economic condition of the Schedule Tribes and provide security to their interests. For the first time in 1954, about 43 special multi-purpose development schemes were implemented. In the year 1957 the tribal block development plan was initiated, which aimed at improving the economy, education and communication for the tribal people. One development block consisted of 150-250 sq miles of area and 25,000 population. The Shibu Ao Committee constituted in 1969 recommended improvements in land tenure, loans, and economic conditions with respect to Scheduled Tribes.

Programmes initiated by the Central and State Governments

The Central Government and the State Governments have initiated a large number of programmes for improving the lot of the tribal people and for bringing them in the mainstream of the country. Some of the important programmes are named as under :

Centrally Sponsored Programmes

1. Scholarship to all the students (both males and females) after matric.
2. Special provision of hostels for girl students.
3. Pre-examination training.
4. Establishment of Tribal Development Blocks (TDBs).
5. Encouragement to cooperation.
6. Research, Training and Special Projects.
7. Improvement of working and living condition of those tribals who are working in unhygeinic occupations.
8. Setting up centres of Coaching-cum-Guidance.
9. Grants for All-India non-official organisations doing welfare work among the Scheduled Tribes and Schedules Castes.

Programmes Sponsored by the State Governments

1. Grant of pre-matric scholarships and stipends
2. Exemption from tuition and examination fees
3. Mid-day-meal
4. Provision for educational equipments
5. Setting up of Ashram Schools
6. Grants for the construction and maintenance of hostel and school buildings
7. Providing land and irrigation facilities
8. Provision for better seeds, fertilizers and agricultural implements
9. Bullock carts for transportation
10. Development of cottage industries in addition to agriculture for supplementing the income
11. Encouragement to cooperation

12. Development and improvement of communication system
13. Control on spread of shifting agriculture
14. Encouragement to animal husbandry and supply of poultry, sheep, pigs, goats, etc.
15. Health services at the village level.
16. Programme for providing drinking water.
17. Provision for house sites and houses.
18. Legal aid to resolve disputes.
19. Grants-in-aid to non-official organisations working at the State level.

Areas of Development

1. Agriculture. About 90 per cent of the tribal people are cultivators but majority of them are engaged shifting and/or sedentary or rudimentary cultivation. Their fields are of small size and their implements are old traditional and inefficient. The irrigation facilities are badly lacking. Consequently, their agricultural productivity is very low and many of them are not able to produce which may be considered sufficient even for their own family members. Economic gains from agriculture are simply out of question in such circumstances. Keeping these facts in view, the Central and the State Governments have prepared many plans for providing better seeds, fertilizers, implements, credit and irrigation facilities and making arrangements for land reforms, soil conservation, land reclamation, etc.

2. Land Distribution and Land Alienation. Land distribution is very uneven and irrational among the Scheduled Tribes and whatever little land they own, that is grabbed by the moneylenders, *Sahukars* and other non-tribal people. They lend money to the tribal people at very high rates of interest and when they are not able to pay back the loan, their land is transferred to the moneylenders. This transfer of land from the tribal to non-tribal people on the basis of non-payment of loan is known as ***land alienation***. This process has been operating in the tribal belts of India for a pretty long time and many tribals have little assets left that they had inherited from their fore-fathers. To safeguard against this trend many State governments have adopted several measures. Some of the remedial measures are as follows :

(*i*) In Gujarat, Odisha and Rajasthan, there is restriction on transfer of land from tribals to non-tribals by sale.

(*ii*) In West Bengal, there is restriction on transfer of tribal land to any person (whether tribal or non-tribal).

(*iii*) Tribal land cannot be transferred to any one by any means.

In spite of all the above mentioned legal safeguards, the government agencies have failed to check land alienation which is going on rampantly.

3. Industrialisation. Tribal people are generally engaged in cottage industries and subsidiary occupations. The activities of their cottage industry include making conventional and crude implements, weaving, bee-keeping, oil extracting, sericulture, *palm-gurs,* etc.

After Independence, some integrated iron and steel plants were set up in tribal areas. These include iron and steel plants at Durgapur, Rourkela, Bhilai, Bokaro, etc. Many illiterate, unskilled and semi-skilled workers find jobs in these iron and steel plants. Technical training is also being provided to tribals to enable them to work as skilled labour in these and other projects.

4. Education. The Central and State Governments are spending huge amounts of money to educate the tribal people and to bridge the gap between the tribals and the non-tribals. There is provision for scholarships, stipends, hostels, free stationery and books, free boarding, mid-day meal etc. Coaching and guidance facilities are available even to those candidates who are preparing for competitive examinations. Scholarships are also granted for higher studies abroad.

5. Employment. There is 7.5 per cent reservation for candidates belonging to Scheduled Tribes in all the vacancies created by the Central and the State Governments. The reservation in direct recruitment to Class III and Class IV categories of posts which normally attract candidates belonging to a locality or a fixed region are fixed in proportion to the population of the Scheduled Tribes in the respective State and Union Territory.

6. Cooperation. The tribal people are almost invariably exploited by the money lenders, forest contractors and other non-tribals and there is urgent need to save the tribals from the clutches of these exploiters.

7. Transport and Communication. Most of the tribal people are living in hilly, mountainous or forested areas which are inaccessible and isolated. Such a situation creates great hindrance in the development of their life style. Therefore, it is necessary that areas inhabited by the tribes should be provided with transport (roads and railways) and proper means of communication so that they are absorbed and assimilated in the national stream.

8. Marketing. The tribal people are known for their handicrafts and produce a large variety of goods which do not need modern sophisticated technology. They are also well trained in collecting a variety of goods from the forests, some of which are very rare. They are exploited by the purchasers because the purchasers do not offer them a remunerative price for their goods. In order to overcome this difficulty, proper marketing facilities should be provided to them.

9. Tribal Development Agency Projects. The Tribal Development Agency Projects were started in 1971-72 in some areas on an experimental basis. The work of these agencies is implemented in some selected districts under the chairmanship of the district collector. Such agencies have full-time project officers and other district officers as their members.

10. Other Schemes. In addition to the above mentioned schemes, several other schemes are also undertaken to improve the living standard of the tribals by providing them proper facilities of education, health, sanitation, etc.

Critical Appraisal of the Tribal Area Development Plan

The Tribal Area Development Plan has achieved great success in serving the tribals, improving their lot and bringing them in the main stream of the Indian society, yet some of the glaring shortcomings have been noticed, as is clear from the following points :

- **Achievements of the plans**
  1. Several policies adopted by the Central and the State Governments have helped in uplifting the tribal people and changing the mode of thinking and attitude of many tribals.
  2. The educational policies, reservation for tribals in government jobs, help in the field of agriculture, co-operation, communications, industrialisation, social service, etc. have brought tribals closer to the main stream of national life.
  3. Tribal markets have been linked with the main stream markets which has given greater opportunities for economic progress to tribals and creating commercial aptitude among them.
- **Failures of the Plan**
  1. Benefits of the government policies could not percolate to the lower strata of the tribal community and remained concentrated in the upper crust of the tribals.
  2. The Tribal Development Blocks (TDBs) are usually criticised because they have not bothered about the participation of the tribal people in the preparation and implementation of the plan.
  3. The rigid, self-centred and unchanging bureaucracy has failed to appreciate the problems of the tribals and to fulfill the role of development agency.
  4. Local politicians also do not work much to safeguard the interests of the tribals and always work in their own interest.
  5. Large industrial projects in the tribal areas have resulted in large scale displacement of the tribals.

Area Development Programme

This is an integrated programme which has been prepared for areas with more than 50 per cent or above of its population consisting of tribals. Emphasis has been laid in the sub-plan with the following aims in mind :

1. Narrowing the gap in the level of development between the tribals and non-tribals.
2. Improving the standard of living of the tribals.

Tribal Sub-plan

Different tribal areas have different environment and bio-diversity and their problems are also different from one another. Therefore different methods have to used for solving the tribal problems in different parts of the country. Taking these facts into consideration, Tribal Sub-plan was proposed during the Fifth Five Year Plan. Tribal areas of 19 States and 2 Union Territories were identified under this sub-plan. Financial assistance for this programme was provided by the Central Government and the State Governments.

- **Aims of the Tribal Sub-Plan**

1. To relieve the tribal people from their exploitation by the money lenders and contractors.
2. To improve the economic, social and cultural position of the tribal people.
3. To bridge the gap between the development of tribals and non-tribals.
4. To evaluate the resources from different sources.
5. To prepare programmes for the Tribal Sub-plan.
6. To develop proper structure of administration.

- **Strategies to Achieve the Goals**

1. Checking the exploitation of the tribals at all levels.
2. Developing the tribal economy by reforms in agriculture, land reforms; soil conservation, forest conservation and afforestation, irrigation, etc.
3. Improving the infrastructure by providing facilities of education, health, training, electricity, etc.
4. Understanding the problems of displaced, migrated, liberated bonded labourers, forest dwellers, Jhum cultivators and other backward groups and solving these problems.
5. Arranging for remunerative prices of tribal products.
6. Improving the environmental conditions.

Three Tier Structure of the Tribal Sub-plan

The Tribal Sub-plan has three tier structure involving micro, meso and macro levels of planning. The ***micro-region*** is coterminous with development block, ***meso region*** is contiguous to the development block having a population of 3 to 5 lakhs and its boundaries are generally determined by the boundaries of talukas or tehsils. The ***macro region*** is a bigger tribal belt which may contain the entire tribal area of a state. The functions at the micro, meso and macro levels have been visualised as follows :

1. Micro Level. Functions at the micro level include :

(*i*) education upto higher secondary level
(*ii*) elementary and complementary health services
(*iii*) agricultural research extension
(*iv*) supply of agricultural inputs
(*v*) minor irrigation schemes
(*vi*) elementary veterinary services
(*vii*) multipurpose co-operatives
(*viii*) local panchayats
(*ix*) household industry
(*x*) village approach roads.

2. Meso Level. Functions included at the meso level are :

(*i*) higher general education
(*ii*) technical and vocational training
(*iii*) manpower planning and employment services
(*iv*) advanced health services with referral facilities
(*v*) agricultural research extension
(*vi*) seed multiplication forms
(*vii*) soil conservation and land management
(*viii*) apex interpreted credit marketing structure with adequate storage and buffer stock facilities.
(*ix*) development of road and communication infrastructure connecting markets with state/district highways
(*x*) distribution network of power, rural electrification, etc.
(*xi*) local resource based industries with adequate market linkages
(*xii*) forest management

(*xiii*) horticulture development

(*xiv*) complementary development programme in hinterland and the bigger industries

(*xv*) medium irrigation projects

(*xvi*) research strategies and evaluation.

3. Macro Level. The macro level functions include :

(*i*) coordination of activities in tribal development projects

(*ii*) agricultural research on regional basis.

(*iii*) direction of various sectoral programmes in the project

(*iv*) major irrigation projects

(*v*) river valley development

(*vi*) industrial and mineral development

(*vii*) marketing support projects

(*viii*) evaluation

Some of the problems are still, un-attended and need immediate attention. These problems are : (*i*) land alienation, (*ii*) indebtedness and exploitation, (*iii*) bonded labour, (*iv*) low standard of education, (*v*) low agricultural production and productivity, (*vi*) inadequate supply of essential consumer goods, (*vii*) land reclamation, (*viii*) inadequate irrigation facilities, (*ix*) shortage of drinking water, (*x*) inadequate facilities of communication, (*xi*) shortage of houses, (*xii*) inadequate marketing facilities, (*xiii*) lack of employment opportunities, (*xiv*) under-developed state of health services.

Critical Evaluation of Tribal Sub-Plan

Although this plan has many merits and has served the intended purpose to a great extent, yet it has some of the glaring drawbacks, as is clear from the following points :

1. This sub-plan is just a part of the departmental plan of the state plan and does not enjoy the privilege of independent functioning.
2. Since it has been integrated with the state planning, it has failed to address the local problems.
3. The basic structure of the ground level information regarding development is overlooked.
4. The pattern of administrative unity within the sub-plan is lacking.

Suggestions for Improvement

1. The Tribal Sub-plan should be separated from the main State Plan and a separate plan for the tribal people should be prepared.
2. The Planning Commission should make separate financial provision for this plan so that it is not financially dependent on the State plan.
3. The department for Tribal Development should act as a coordinating unit in each state.
4. There should be separate financial provision for each department so that local problems are solved.
5. Participation of the tribal people is very essential for the proper preparation and implementation of the Sub-Plan. This will help in understanding the real problems of the tribal people and finding solution to such problems.

Strategies for Development

1. Problems of Agricultural Development. Developmental programme for increasing agricultural production and productivity should include : (*i*) increase in irrigation facilities; (*ii*) soil conservation; (*iii*) provisions of modern agricultural inputs at affordable prices; (*iv*) arranging for economic size of the holdings; (*v*) improving agricultural marketing facilities and (*vi*) making tribal farmers aware of the latest techniques used in agriculture.

2. Problem of Land Alienation. A large part of land belonging to the tribal people has been grabbed by the money lenders and forest contractors. Some strong steps are to be taken to check this trend which include : (*i*) legal protection to the tribals; (*ii*) making tribals conscious of their rights; (*iii*) improving and up dating the land records. Unfortunately there are no land records in some of the tribal areas.

3. Development of Transport and Communication. Roads and different modes of communication need to be improved to provide better accessibility, social inter action and opportunities of economic well being to the tribals.

4. Forest Policy. Most activities of the tribal people are based on forests. Earlier forest policies could not be of much help to these people. However, new forest policy of 2018 holds some hope for these people.

5. Development of Allied Agricultural Activities. Allied agricultural activities such as dairying, poultry, bee-keeping, animal husbandry, piggery, etc. should be developed to supplement the income of the tribals.

6. Development of Public Distribution System. Development of public distribution system is needed for proper supply of essential consumer goods to the tribals.

7. Development of Cottage and Small Scale Industries. Cottage and small scale industries can provide additional employment and these industries should be developed at least in some selected growth centres. These industries can be useful especially to those who are displaced from their original habitats by development projects like dams, large scale industries; transport routes, etc.

8. Provisions of Social Services. Social amenities like medical and health facilities, nutrition, drinking water, housing, etc. are to be provided to the tribals so that they are able to lead a reasonably good life.

9. Co-operation. Co-operative movement and Panchayati Raj institutions can go a long way in improving the lot of the tribals because they have a strong community feeling and long tradition of democratic institutions.

10. Educational Facilities. Provision of education facilities is of utmost importance to have a sustainable economic development among the tribal people.

11. Prevention of Exploitation. Poor, ignorant, illiterate and helpless tribals have been exploited since a long-time by the money lenders, landlords, forest contractors, government officials, panchayat members, etc. It is very difficult to put a check on such a large variety of powerful people but this is necessary if we want to see our tribals leading a peaceful and respectful life.

12. Change in Administrative Set-up. Administrative set-up should be changed in such a way that it becomes more responsive to the needs of the tribal people.

13. Periodic Evaluation. All the plans and programmes meant for the tribals need periodic evaluation so that old programmes may give way to new programmes and 'timely correction' is assured.

Multi-Level Planning

Planning is the process which may be conceived in the form of either single level or multi-level. In India, single level plan is formulated at the central level while in multi-level plan, states, districts, blocks and villages play their respective roles. Under the multi-level planning, formulation of the plans takes place at different levels and the combination of plans at different levels gives a macro system of planning. Multi-level planning may be defined as "*planning for a variety of regions which together form a system and subordinate systems*." Multi-level planning is more effective than the single level planning because every region (state, district, block or village) is involved in the planning process and there is direct participation of the people at the grass roots level.

History of Multi-level Planning in India

During the first three Five Year Plans, planning was mainly formulated by the Central Government and implementation of the plans was also the responsibility of the Central Government. These plans were prepared by the Planning Commission of India which was governed by the Central Government. Some State Governments prepared their own plans during the Fourth Five Year Plan. B.R. Mehta Committee (1957) and Administration Reforms Committee (1967) recommended planning at the district level. The Government of India issued guidelines for district level planning in 1969 and some states started preparing their own state level plans following these guidelines. For example, Maharashtra and Gujarat prepared state level plans in 1972 and 1979 respectively. Emphasis was laid on decentralization while preparing state level plans in 1972. During 1978 and 1983, block level planning was encouraged. This enabled the concerned agencies to gain experience in multi-level planning and helped in concentrating on the proper utilisation of the local resources.

In 1977-78. Dantewala Committee and Ashok Mehta Committee were constituted to suggest ways and means for implementing plans in the right perspective. Dantewala Committee suggested that the plans should be implemented at the block level. The activities involved in the block level planning should include minor irrigation projects, poultry, soil conservation, water management, social facilities, etc. According to Dantewala Committee, block should be basic unit for planning because it has small size having limited population. These factors are quite convenient for establishing a close contact between the planners and the people. Ashok Mehta Committee emphasized upon the role of Panchayati Raj and stressed upon planning at the village level. On the recommendations of these two committees more emphasis was laid on rural development and rural planning during the Sixth Five Year Plan (1978-83).

The Present Day Multi-Level Planning in India

At present, planing is done at five different levels *viz*. (1) Centre, (2) State, (3) Districts, (4) Blocks, and (5) Villages. Before the 73rd and 74th Constitutional amendments in 1992, the task of plan formulation was primarily carried out by the centre and the state governments and the other levels (namely district, block and village) of planning process came into picture only at the implementation stage. But after these Constitutional Amendments, the lower levels of planning have been duly recognised and financial assistance has been provided to multi-level planning. Different levels of planning are briefly described as under :

1. The Central Level Planning. Also called First Level Planning, this is the first and the foremost level of planning which is formulated by the Central Government,. The central plans are prepared by the Planning Commission of India headed by the Prime Minister. Its members are some central ministers and 3-4 planning experts. The final assent to the plan is granted by the President of India who acts on the advice of the Council of Ministers which is binding on the President.

The business of the government is transacted through a three tier set-up. The ***first layer*** is constituted by the Prime Minister, Cabinet Ministers, Minister of State, Deputy Ministers and Parliamentary Secretaries. The ***second layer*** is constituted by the secretariat organisation of the ministry with secretary as its head. The ***third layer*** consists of executive organisation of the department.

2. The State Level Planning. This is also known as Second Level Planning under which all the states are entitled to frame and execute their own plans aimed at economic growth and in the interest of the general public. The plan is prepared by the State Planning Commission headed by the Chief Minister and is sent to the State Governor for his assent on the advice of the Council of Ministers which is binding on the Governor. The Council of Ministers works through the secretariat which is headed by a secretary. The State Planing Commission, in addition to the Chief Minister, has some ministers and a few professional experts in planning which assist the Council of Ministers in the planning process. The Council of Ministers has various departments functioning under it. These departments can be classified into the following three categories :

1. Development departments (agriculture, animal husbandry, rural development, public works and industries).
2. Social welfare department (health, education and social welfare).
3. Coordinating departments (home, revenue, finance, and planning).

Like the Central Planning, the State Level Planning is also of five years. However, the finances for the State Level Planning are to be arranged by the concerned state although there is a provision for financial assistance and loans from the centre.

The Constitution of India has provision for division of powers between the Centre and the States with reference to planning so that the chances of confrontation between the two may be minimised. The Seventh Schedule and Article 246 of the Constitution of India gives three lists of subjects (*i*) ***Union List*** (containing 97 items), ***State List*** (containing 66 items), and ***Concurrent List*** (containing 49 items).

The Central Government enjoys the power to legislate on the subject given in the Union List whereas the State Governments have power to legislate on the subjects contained in the State list. But both the Central and State Governments have powers to legislate on the subjects contained in the Concurrent List. However, the Central Law prevails whenever there is a conflict between the Centre and the States. According to the Prevailing law, organised activities like minerals, industries, railways, and telecommunications come under the Centre's purview while agriculture, collection of land revenue, irrigation, power, public health, education, local self government and several other important subjects are the responsibilities of the State Governments.

3. The District Level Planning. This is the Third Level of Planning where local planning starts. The District Level Planning is the responsibility of the District Planning Committee which is directly under the control of the district collector (also known as deputy commissioner in some states). He is responsible for maintaining law and order in the district and also acts as a coordinator among various departments. The main departments are those of agriculture, irrigation, animal husbandry, fishing, dairying, forestry, education, health, public works, industries, cooperation, social welfare, transport, panchayati raj, etc. Besides there are offices of the autonomous bodies, like those of Electricity Board, State Road Transport etc. Municipality for Urban Planning and Gram Panchayat for Rural Planning play an important role.

4. Block Level Planning. The Block Level Planning is lower than the District Level Planning but higher than the Local or Panchayat Level Planning. This level of planning was started right in the First Five Year Plan. For planning purposes, each district was divided into a number of blocks. Each block consisted of about 100 villages and a population of about 60,000. Participation of the local people is very important and decisions by the local people are intended to use the local resources in a better way. This level of planning is executed by the ***samiti*** of the villagers headed by the Block Development Officer (BDO). The general supervision of blocks is done by ***Block Samitis*** under the chairmanship of the ***Block Pramukh*** and elected representatives.

The institutional set up consists of a Block Development Officer (BDO) and five extension officers each of whom assists the BDO in the field of agriculture, animal husbandry, co-operation panchayats and rural industries. Apart from this, there is an oversear, a social education organiser, a progress assistant and village level workers on the staff. Also there are veterinary stockman, a medical officer, a sanitary inspector and a lady health visitor. Some states have an extension officer for programmes related to women and children.

The primary aim of the Block Level Planning is to understand the problems of the local people, to find ways and means to solve those problems and formulate as well as execute the development plans. Proper understanding of the local problems helps in optimum utilisation of the local resources. It generates employment opportunities and the way for planning. In fact the entire strategy of the Block Level Planning is based on employment planning, growth-centre planning and credit planning. The main activities of this type of planning include (*i*) agriculture and allied activities, (*ii*) soil conservation and water management, (*iii*) animal husbandry and poultry, (*iv*) fisheries, (*v*) forestry, (*vi*) processing of agricultural produce, (*vii*) organising input supply, credit and marketing, (*viii*) cottage and small scale industries, (*ix*) local infrastructure, (*x*) social services (drinking water supply, health and nutrition, education, housing, sanitation, local transport and welfare programme) and (*xi*) training of local youth and updating of skills of local population.

5. Local or Panchayat Level Planning. This is the lowest level of planning for which the Village Development Officer (VDO) and secretary are responsible. It is supervised by the *Gram Sabha* which is headed by the *Gram Pradhan*. It formulates plans for the welfare of the villagers. Funds are directly allocated from the centre for executing rural development programmes such as Integrated Rural Development Programme (IRDP) and Jawahar Rojgar Yojna (JRY). In addition, the *Panchayat* is also responsible for (*i*) promotion of agriculture; (*ii*) rural industries, (*iii*) health, (*iv*) maternity, women and child welfare, (*v*) common grazing grounds, (*vi*) village roads, tanks, wells, (*vii*) sanitation, and (*viii*) socio-economic development programmes.

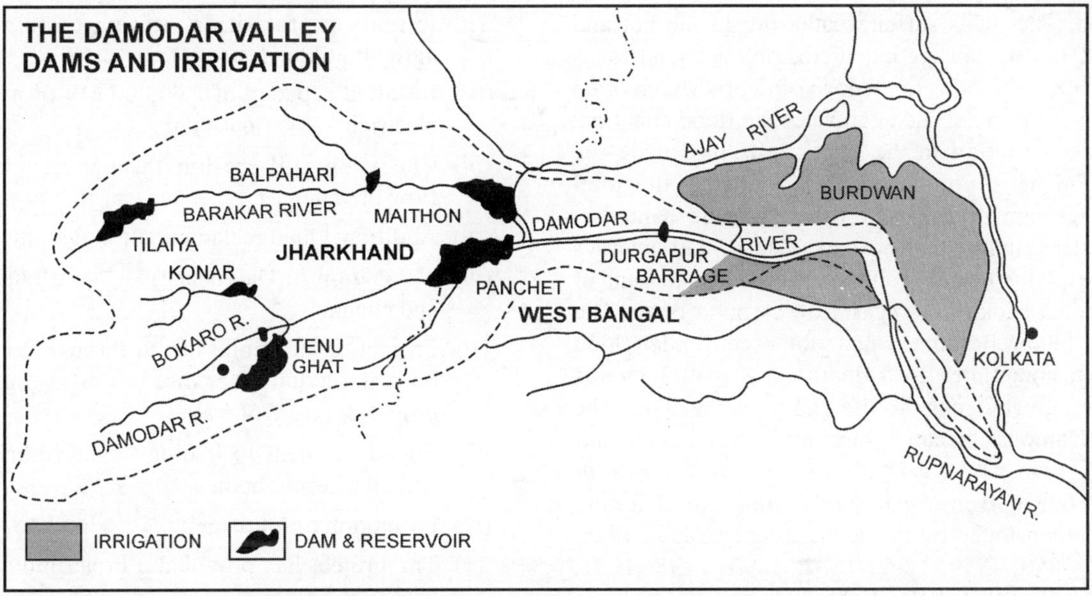

FIG. 13.5. The Damodar Valley : Dams and Irrigated Area

DAMODAR VALLEY CORPORATION (DVC)

The Damodar river is a tributary of the Hugli river. It flows more or less in the west to east direction through Jharkhand and West Bengal. Its total length from its source in the hills of Chotanagpur plateau in Jharkhand to its confluence with Hugli in West Bengal is about 541 kms, 270 kms of which is in Jharkhand and the remaining 271 kms lies in West Bengal. In its upper reaches, it is known as Deonadi. It drains Ranchi, Hazaribagh, Dhanbad and Santhal Parganas districts of Jharkhand and Bankura and Bardhman districts of West Bengal. It takes a southerly turn from Bardhman town and joins river Hugli about 50 kms upstream from Kolkata. Damodar valley covers an area of 24,235 sq km in Jharkhand and West Bengal. Bokaro, Barakar, Konar are its important tributaries.

Physical Setting. Damodar valley runs through the middle of Chotanagpur Plateau in Jharkhand. It lies between the plateaus of Chotanagpur and Hazaribagh in the north and of Ranchi in the south. The average elevation of this plateau complex varies between 300 and 600 metres. The river originates at an altitude of 510 metres above mean sea-level and joins the Hugli at an altitude of less than 30 metres above the mean sea-level. It leaves plateau at a height of about 150 metres and reaches the plains in Bardhman district **(Fig. 13.5)**.

The Damodar valley is a rift valley or a sunken trough, bounded by broken and tilted edges of the plateaus. As there are sudden descents from one level of plateau surface to the other, the whole topography of the surrounding area has an undulating nature. The break points, where tributary streams join the main Damodar river, have provided suitable sites for power houses. The residual hills of hard rocks left here and there in the basin have been joined to construct dams across the streams. The undulating nature of topography and faulted nature of strata have also favoured the extraction of coal.

Climate. The region receives annual rainfall of about 125 cms. This is monsoon rainfall which is concentrated between June and September. As the area lies in the path of tropical cyclones from the Bay of Bengal, there are chances of stormy rainfall many a times. Generally, the amount of rainfall is more on upper slopes than at the valley floor and more in the west than in the east.

Hydrology and Water-Power Development. The erratic monsoonal rainfall and heavy downpour due to tropical cyclones poses serious problem of floods. The channel of the river is unable to accommodate the flow of water during the rainy season. The rocks suffer weathering in the hot and humid climate causing much erosion over sandstones and slates. This adds to the load of debris which often chokes the channel and aggravates the flood situation. The deforestation in the upper hills has made the problem more complicated. As soon as the total drainage converges and crosses through a narrow bottleneck amidst the rocks at Asansol and emerges on the plains near Bardhman, the heavy amount of sediments choking its peak flow is burst open. The plains below Bardhman get submerged under flood waters, sometimes to a depth of 2 to 2.5 metres causing great damage to life and property. Thus, the river Damodar became notorious for its floods and was known as ***'River of Sorrow'*** or ***'Sorrow of Bengal'*** and even ***'Sorrow of region'***. Its notoriety was demonstrated by the devastating floods in 1823, 1848, 1856, 1859, 1863, 1882, 1890, 1898, 1901, 1905, 1907, 1913, 1916, 1923, 1935 and 1943. While major floods occur at intervals, minor floods are experienced almost every year. The sediments brought by the Damodar create the problem of sedimentation in the Hugli which, in turn, endangers the Kolkata port.

In order to control the floods and other related problems, the Central Government, in consultation with the state governments of the then Bihar and West Bengal, worked out a unified development project for Damodar basin. The Damodar Flood Enquiry Committee suggested a comprehensive plan. This plan was based on the memorandum submitted by W.L. Voorduin, an engineer with the Tennessee Valley Authority (TVA) in the USA. The Damodar Valley Corporation (DVC) was established on 18th February 1948 to execute the Damodar Valley Project. The original plan was to construct seven major dams. These dams were Aiyar and Panchet Hill on the Damodar river; Maithon, Belpahari and Tilaiya on the Barakar river; Konar on the Konar river and Bokaro on the Bokaro river. But the DVC has constructed only four dams (Tilaiya, Maithon, Konar and Panchet). **[Fig. 13.5]**

Benefits from the Project. Damodar Valley Project is a big landmark in the economic development of this region. Following are the main benefits drawn from the project :

(*i*) Flood control in the flood prone areas of Jharkhand and West Bengal.

(*ii*) Irrigation facilities to vast area of agricultural land.

(*iii*) Installed capacity of 2,60,000 kW of hydro-electricity at various dam sites.

(*iv*) Check on soil-erosion through regulated flow of water.

(*v*) Additional land reclamation for agriculture.

(*vi*) Navigation in Damodar river, its tributaries and channels.

(*vii*) Promotion of public health through control on malaria and other diseases as a result of proper drainage of water.

(*viii*) Encouragement to fishing in the reservoirs and other water bodies.

(*ix*) Promotion of tourism.

(*x*) The project has provided a broad industrial base to the area.

Soils. Soils in the Damodar Valley are heavy clays and loams. They are deep and heavy on flat coal layers or low-lying tracts and light, coarse and thin in the uplands. Most of the land in the upper part of the basin in Chotanagpur region of Jharkhand is of relatively low fertility. The terrain is undulating and lacks irrigation facilities.

Forests*:* Over 40% of the region in Chotanagpur is under forests. There are valuable forest products. Besides *sal* wood, minor produce is collected from a number of other trees. Forest cover is poor on high hill slopes, having only palm, *ber, sabai* grass, *dhak,* bamboo and thorny plants.

Soil-Conservation and Afforestation. The Damodar Valley Corporation has been assigned the task of soil-conservation and afforestation. The main objective of soil conservation and afforestation in the catchment area is to reduce the soil erosion and save the reservoirs from heavy siltation. The problem of soil-erosion is being tackled on watershed basis. The measures adopted for soil-conservation are survey of soils taking into account the various 'physio-chemical properties of the soils, degree of slope, extent of erosion, present land use and suitability of irrigation', demonstration of better methods of land management and assessing the soil-fertility.

Afforestation within the catchment of Damodar valley is being carried out by the Forest Division of the Corporation as well as the forest departments of the state governments of Jharkhand and West Bengal.

Agriculture. The agricultural pattern is similar to that prevailing in other parts of Chotanagpur plateau in Jharkhand. For most parts of the Damodar basin, it is one-crop culture, *i.e.,* only one crop is grown in one agricultural year. Rice is the main crop of lowlands while uplands are more suitable for maize and pulses. These crops are just sufficient to meet the local requirements only.

The agricultural productivity of the area is low. The population density is not very high but limited nature of arable land available to the peasants has brought down the land-man ratio. There is only 0.1 to 0.4 hectares of land of poor type of net sown area for each individual peasant. In Jharkhand portion, besides some fairly big stretches of cultivated blocks in the lower portion of upper valley floor, Dhanbad district in its lower portion has the most extensive cultivation, amounting to 23% of its total area. The progress of coal mining, its heaps, mining settlements, a dense network of communication lines, thick forests and disturbance of local drainage because of huge mining operation, have all damaged the basin's cultivation.

Mineral Resource Base. The Damodar valley region is the most mineralised region of our country. As a matter of fact, the Chotanagpur plateau, in which this valley is located, is known as the ***'mineral heartland of India'*** **[Fig. 13.6].** Some important minerals are briefly described as under :

Coal. The Gondwana sedimentary rocks in the valley contain rich coal deposits. The richest, largest and most productive coal-fields of the country lie in this region. The Damodar valley has about 36% of the country's reserves of Gondwana coal and produces 19 per cent the output of coal in India. Most of the coal-fields are situated in the main valley of Damodar to the north of the river.

The main coal-field are those of Raniganj, Jharia, Bokaro, Ramgarh, Karanpur and Chandrapur. Some of the important coal-fields of this region are shown in **Fig. 13.6**.

Other Minerals. Apart from coal, this region produces a large variety of other minerals, such as iron-ore, bauxite, copper, lead, mica and manganese. Fire clay is an important mineral, capable of resisting high temperature and a raw material for the manufacture of fire bricks and other refractories. This region produces about 10% of India's total production. Graphite in Dhanbad, sands in Damodar, crystalline

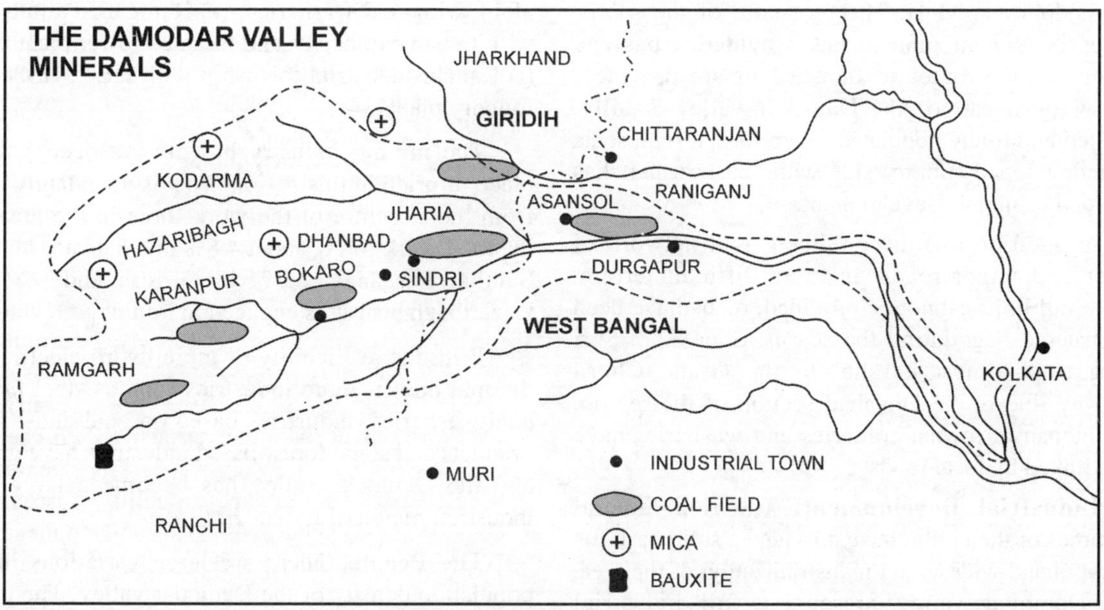

FIG. 13.6. Damodar Valley : Minerals

limestone in the western parts, iron-ore in Dhanbad and scattered quartz elsewhere are also important.

Thermal Power. Thermal power stations have been set-up at Bokaro, Chandrapura, Durgapur, Sindri, Jamshedpur, Burnpur and Sitapur based on the locally available coal. They meet the power requirements of most parts of the Damodar river drainage basin.

Development of Transport Infrastructure. The region has a dense network of railways and roadways. Towards the lower part of Damodar basin, there is a clear convergence of railways from north, north-west, west and south-west. It has the locational advantage of being near to Kolkata. Dhanbad has become the focus of convergence of rails and roads. The grand chord line joining Delhi and Kolkata passes through the northern part of the Damodar valley. This line is now fully electrified. Besides this, many short railway lines criss-cross this area. A number of routes diverge towards west of Asansol. The watershed between the Ajai and the Damodar rivers is not very prominent, hence the railway link to Patna follows it and a branch line links Giridih. The railway line linking Kolkata-Asansol to Gaya-Patna passes through the upper Barakar river valley.

In the upper portion of the valley, the railway line from Dehri-On-Son via Daltonganj runs along the coal-fields belt. A branch line extends southwards to Jamshedpur via Muri. The recession of the scarp-slopes of Ranchi plateau has provided a passage through which routes to Jamshedpur are provided. The western part of the Damodar valley is hilly, forested and thinly populated. Therefore, it represents an undeveloped country-side while eastern part has recorded economic development.

In addition to railways, there is a network of major and minor roads. In more difficult terrain, where rail links cannot be provided, roads have been constructed negotiating the scarps. Sher Shah Suri Marg runs almost parallel to the Grand Chord Railway line in the Dhanbad section of the region. The industrial towns, collieries and washeries have been linked by local roads.

Industrial Development. The rich mineral resources of the region have provided a solid base for the rapid and widespread industrialization of the area. The Damodar valley presents a full industrial landscape which is comparable to any advanced country of the world. The valley is full of large and medium-sized factories. The Sindri fertilizer factory and cement factories at Sindri and Khalari, coal-washeries in Dhanbad, engineering works at Kumardhubi, refractory works at Dhanbad, glass and lead smelting factory works at Tundoo, aluminium factories near Asansol and at Muri, copper smelting at Ghatshila, locomotives at Chittaranjan and steel plants at Durgapur, Bokaro and Jamshedpur combine to make it a great industrial region. Government has contributed a lot towards the industrial growth of the region by setting up a number of public sector undertakings. Durgapur and Bokaro steel complexes, Heavy Engineering Corporation Ltd. at Ranchi, Sindri Fertilizer Plant and Indian Explosives near Hazaribagh provide some such examples. Six industrial complexes have emerged over time in the Damodar valley region. These are : (*a*) Durgapur industrial complex having thermal power plant, coke oven, iron and steel, mining machinery and cement machinery and electrical industries; (*b*) Asansol-Kulti-Burnpur industrial complex with steel factory at Burnpur, Chittaranjan Locomotives, Kulti Foundry Works, and Kumardhubi Engineering Works; (*c*) Dhanbad-Sindri-Jharia industrial area having Sindri Fertilizer Industry, chemical and coal-based industries; (*d*) Bokaro industrial area consisting of steel plant, thermal plants at Bokaro and Chandrapura along with coal washeries; (*e*) Ramgarh-Patratu area with Indian explosives and thermal power plants; and (*f*) Ranchi industrial township with HEC, HMT and foundry plants.

The mining industry has always occupied the place of pride in this region. This is abundantly clear from the structure of the work force in the area. Of the total work force, about 4.8% in Dhanbad, 6.3% in Santhal Parganas, 8.7% in Giridih and 9.8% in Hazaribagh district is engaged in mining.

With the availability of more hydro-electric and thermal power, more industries such as steel plants, coal-washeries, industries based on coal and its by-products and some forest-based industries have grown and the Damodar valley has become a large size industrial region of India.

The People. There are large variations in the population density of the Damodar valley. The upper catchment area is hilly and forested where low population density is found. In contrast, the lower part of the valley supports high population density and has experienced more urbanisation. The process of urbanisation is closely related to the rate of industrialisation. As such, the areas having more industries are more urbanised. A major part of the population especially of the Chotanagpur plateau is tribal which has its own problems of development.

The Problems. The Damodar Valley Project was primarily conceived for flood control and this target has not been achieved fully. Heavy floods did occur in 1959, 1970-71 and 1978, though their ferocity has been much reduced. Only four dams have been constructed in place of the original suggestion of seven dams. The over-utilisation of forests and lack of afforestation have aggravated the siltation in the dams. Thus, the first line of defence in the form of forests and the second line of defence in the form of dams have failed to eliminate the problem of floods altogether. Hydroelectricity has to be supplemented by thermal electricity because the flood-control dams cannot be taken as power-generating dams in view of the creation of flood cushion before monsoon every year. However, the DVC has been an important example of integrated regional development. Its efficiency can be improved with better management and foresight.

Comparison of Damodar Valley with Ruhr Valley. There are so many similarities between Damodar Valley of India and Ruhr Valley of Germany. Damodar Valley is an important coal producing region of India and Ruhr Valley contributes major part of coal in Germany. Both the areas have high level growth of coal-based industries. The other industries are also fairly developed. Transport facilities in the shape of railways, roadways and waterways are abundantly available to both the areas. Looking at these similarities, the Damodar river basin is often called the *'Ruhr of India'*.

NATIONAL CAPITAL REGION

Delhi, the national capital of India, has registered a phenomenal growth in population in the twentieth century, especially after independence. **Table 13.2** gives an idea of population growth of Delhi.

According to the figures released by the Election Commission of India, Delhi's population stood at 2.01 crore in January 2020, which is more than the population of 140 countries/territories. In other words 140 countries/territories of the world have less population than Delhi.

The vast economic opportunities available in Delhi have attracted large number of migrants not only from its immediate neighbourhood but also from the far-off places in the country. This large-scale influx of people has put heavy strain on infrastructural facilities of the city which include among other things, housing, employment, transport, electricity, water, sewerage, education, medical treatment, etc. The concept of National Capital Region was conceived to reduce pressure of increasing population on these infrastructural facilities. Any strategy of containing the growth of Delhi within limits will have to be taken within the regional frame in which Delhi exists. The region in the immediate hinterland of Delhi, within which the development had to be planned in order to release pressure from Delhi, is known as *'National Capital Region'*. The Master Plan of Delhi was prepared in 1959 and was finally approved by the Government of India in 1962. The Master Plan contained amongst others a recommendation of setting a statutory NCR Planning Board. Initially, this board was constituted as an advisory body which was reconstructed in 1973. Its task was of coordinating the development of urban and rural areas in the National Capital Region under a

TABLE 13.2. Population of Delhi Urban Area

Census Year	Population
1901	2,14,115
1911	2,37,944
1921	3,04,420
1931	4,47,442
1941	6,95,684
1951	14,37,134
1961	23,89,408
1971	36,47,023
1981	57,29,283
1991	83,75,188
2001	1,38,50,507
2011	1,67,53,235

comprehensive regional plan and to secure the collaboration of the concerned state governments in implementing the plan.

However, it was only in 1985 that a statutory organisation, named as National Capital Region Planning Board could be instituted through the enactment of NCRPB Act with a view to plan, implement and supervise regional development planning in the NCR. The plan was prepared on the basis of expert studies and deliberations between central and the agencies of the concerned state governments and was finally approved by the NCRPB on November 3, 1988 for its immediate implementation.

The Physical Extent of NCR. The National Capital Region extends over the Union Territory Delhi and adjoining parts of Haryana, Uttar Pradesh and Rajasthan. The administrative units constituting the NCR are as under :

(*i*) Union Territory of Delhi (1,483 sq km)

(*ii*) Haryana sub-region comprising 14 districts of Faridabad, Gurugram, Nuh, Rohtak, Sonepat, Rewari, Jhajjar, Charkhi Dadri, Panipat, Palwal, Mahendragarh, Bhiwani, Karnal and Jind.

(*iii*) The Uttar Pradesh sub-region comprising eight districts of Meerut, Ghaziabad, Gautam Budh Nagar (NOIDA), Bulandshahr, Baghpat, Muzaffarnagar, Hapur and Shamli

FIG. 13.7. National Capital Region

(*iv*) The Rajasthan sub-region comprising two districts Alwar and Bhratpur. 'With an area of 8,380 sq km. Alwar is the largest district in NCR.

TABLE 13.3. Physical extent of NCR

Political extent	Number of districts	Area (sq. km.)	Population (2011) in crores
Delhi	11	1,483	1.7
Haryana	14	25,327	1.1
Uttar Pradesh	8	14,826	1.5
Rajasthan	2	13,447	0.4
Total	35	55,083	4.7

Note: There is proposal to include Kotputli, Shahpur, Manoharpura and Chandwaji towns of Jaipur district of Rajasthan. This will lead to extension of physical extent of NCR.

Aims and Objectives

The aims and objectives of the NCRPB can be summed up in the following points :

(*i*) to contain the growing population of Delhi within the manageable limits through integrated development of all the components of the NCR.

(*ii*) to divert the population to ring towns like Faridabad, Ghaziabad, Ballabhgarh, Gurgaon, Bahadurgarh, Sonepat, Narela, Loni, Modinagar.

(*iii*) to re-model the pattern and functional character of the settlements within the NCR through guided growth so as to enable them to play a constructive and co-operative role in the planned development of the NCR.

(*iv*) to decentralise economic activities through proper development of ring towns and also through certain checks in the Delhi metropolitan area.

(*v*) to make arrangements for dispersal of industries and decentralisation of public undertakings including government offices from Delhi to the ring towns.

(*vi*) to provide cheap and efficient transportation system for dispersal of population and economic activities, and

(*vii*) to accomplish a balanced and harmonious growth of urban and rural component of the NCR.

Main Problems of NCR

(*i*) **In-migration.** There is large scale in-migration both from the neighbouring as well as from far off places. Approximately 2 lakh in-migrants move into the NCR every year from outside.

(*ii*) **Widening Gap between Demand and Supply of Essential Services.** The gap between demand and supply of essential services like water, power, transport, waste disposal, waste treatment and management of solid waste is widening at rapid rate because demand for these services is always outpacing the supply.

(*iii*) **Congestion.** Delhi is suffering from the acute problem of congestion and rapidly growing slum areas.

(*iv*) **Increasing Employment Opportunities.** Delhi is acting as a '*great magnet*' to attract job seekers of all the categories which is great '*pull factor*' for in-migrants. It leads to over population and allied problems.

(*v*) **Collapse of Infrastructure.** The perennial and heavy inflow of population has put heavy burden on the infrastructure which is collapsing under its weight.

(*vi*) **Deteriorating Quality of Life.** Due to ever growing demand by the fast growing population and limited supply of almost all essential commodities, the quality of life is fast deteriorating.

(*vii*) **Increase in Crime Rate.** Most of the in-migrants are unskilled or semi-skilled workers. They do not get proper jobs and resort to all sorts of crimes.

Planning Strategies and Proposals

The regional plan for NCR listed the following strategies and proposals for meeting the challenges in this area :

(*i*) To decelerate the growth of Delhi city.

(*ii*) To control the growth rate of Delhi Metropolitan area.

(*iii*) To develop regional centres with infrastructure and adequate network of services.

(*iv*) To bring down the population growth rate of NCR to 1.2 per cent per annum from the existing 3 to 4 per cent per annum.

(*v*) To bring down the volume of in-migration to 84,000 per annum as against the existing inflow of 2 lakh per annum.

(*vi*) To diversify the functional character of other towns of the NCR.

(*vii*) To revitalise the economy of stagnating urban centres of the NCR.

(*viii*) To select priority towns and develop them on priority basis for a balanced development of the NCR.

(*ix*) To integrate rural-urban development process within the NCR.

(*x*) To develop selected regional centres namely Meerut, Bulandshahr, Panipat, Rewari, Alwar, Bhiwadi Complex, Khurja Complex, Dharuhera, Rohtak, Palwal, Karnal, Jind, etc. to accommodate Delhi-bound potential migrants by creating employment opportunities and the secondary and tertiary sectors in these regional centres.

(*xi*) To develop a number of sub-regional centres at focal points such as divisional head-quarters, agricultural markets, service centres, and industrial centres.

The Tasks Ahead

(*i*) Distribution of Population. The development of some nodal points in NCR is necessary to regulate the population. Some nodal points have already been identified. These are Meerut, Hapur, Bulandshahar, Khurja, Palwal, Rewari-Bhiwadi-Dharuhera Complex, Rohtak, Panipat, Karnal, Jind and Alwar. These centres have the potential to attract migrants.

(*ii*) Dispersal of Economic Activities. The dispersal of economic activities is the basic necessity to reduce the population pressure on Delhi. This requires the generation of employment opportunities in different towns of the NCR. The economic activities for which the dispersal plan is envisaged are industries, Central Government and public sector offices, wholesale and distributive trade and commerce. Medium and large-scale industries have to be restricted in Delhi and incentive for their dispersal have to be given in the form of industrial estates in the selected towns. The government offices and public enterprises which do not perform ministerial functions, protocol functions and liaison functions should be shifted outside Delhi. The wholesale trade which does not serve the consumers in Delhi should also be shifted elsewhere in the NCR.

(*iii*) Land Use in the NCR. Agriculture is the main economic activity in the NCR. Growing urbanisation is fast encroaching upon the agricultural land, thus adversely affecting the agricultural output. Urban settlements should be allowed only on less fertile, barren or wastelands. With the reduction in agricultural land only intensification of agriculture will maintain our food production. The development and conservation of forest area as well as social forestry will have to be encouraged. Care should be taken that development does not harm the environment. The sites of scenic beauty, archaeological sites, parks, national parks, bird sanctuaries within the region should be developed and strengthened in order to provide recreational avenues. A buffer zone of green belt between Delhi Urban Area and Delhi Metropolitan Area has been proposed to be created. The Aravalli hills extending from Delhi to Alwar provides ideal extent for developing a natural forest through conservation.

(*iv*) Regional Transport System. Cheap and efficient transportation network increases the flow of passengers and goods and helps in their dispersal. There is proposal to have one inner grid of roads in the NCR connecting Sonipat-Bagpat-Meerut-Hapur-Bulandshahar - Sikandarabad - Faridabad - Rohtak - Gurgaon-Jhajjar-Gohana-Sonipat. The outer grid will link Panipat-Muzaffarnagar-Meerut-Hapur-Bulandshahar-Khurja-Palwal-Rewari-Jhajjar-Gohana-Rohtak- Panipat.

There is proposal also to develop regional rail bypass connecting Meerut-Hapur-Bulandshahar-Khurja-Palwal-Sohna-Rewari-Jhajjar and Rohtak. The metre gauge railway line in Rajasthan and Haryana sectors, has been converted into high capacity broad gauge system. The EMU services on the ring system of rail tracks may increase the flow of commuters and materials.

To improve the transportation system and to reduce the pressure of passengers on it, Delhi introduced Metro Rail System way back in 2002 and the first 8.3 km long Metro Rail between Shahdara and Tis Hazari was started on December 29, 2002. Today, it has a vast network 389 km and its services are available for almost all parts of Delhi. Metro Rail services to some of the satellite towns like Gurgaon, NOIDA, Faridabad and Bahadurgarh has been started and some other towns are also expected to be connected by Metro Rail.

A Highway Corridor Zone is proposed with a minimum width of 500 metres inclusive of green belt on either side of the right-of-way (ROW) along the National Highway (NH) 1. 2, 8, 10, 24, 58 and 91 converging at Delhi to enable a planned and regulated development along the stretches of these highways that are outside the controlled/development/ regulated areas. In addition to the above, the Highway Corridor Zones along the NH 71, 71 A, 7IB, 119, 93, 235, 11A and Expressways have been proposed.

The Yamuna Express Highway from Delhi to Agra has offered fast movement of road traffic and greater connectivity between the two cities. The Western Peripheral Expressway or Kundli-Manesar-Palwal Expressway in 6 lane, 135.6 km long and Eastern Peripheral Expressway or Kundi-Ghaziabad-Palwal Expressway (135 km) have become operative and have given much need relief to Delhi.

Planning the Physical Infrastructure. The infrastructure in the NCR will have to be properly developed if it is to function as one integrated region. Adequate uninterrupted power supply is essential in all the urban and rural areas for industries, transport and for domestic use. The Narora Atomic Power Plant may help in meeting the power shortage to some extent. Power tariff will have to be adjusted in such a way that Delhi becomes less attractive for industrial and commercial power. Another important element of infrastructure is the availability of drinking water to urban and rural areas. Efforts will have to be made to provide drinking water in sufficient quantities. The strict enforcement of Water Pollution Control Act, 1974 and Environmental Protection Act, 1986 may go a long way to restrict the wastage of this cyclic resource. The problem of sanitation, sewerage disposal and storm water drainage will have to be tackled in the urban areas as a first priority. Low-cost sanitation systems may be introduced in urban areas where such a facility does not already exist. The NCR Planning Board is considering to introduce the model of Sulabh International for the disposal of sewerage.

Counter Magnets

Counter-magnet towns are identified as those that can be developed as alternative centres of growth and attract migrants to them rather than Delhi. Promoting growth of counter magnet towns are the principal components of the strategy to reduce both migration and population explosion in the Delhi metropolitan area.

These towns are located in six states—Uttarakhand, Madhya Pradesh and Punjab, in addition to Uttar Pradesh, Haryana and Rajasthan, which are already part of the national capital region.

The criteria for selecting counter magnet towns are : that they should not be within approximately 250 kilometres from Delhi, should have their own established roots and potential for growth and should not be centres of either religious, strategic or environmental importance.

The major counter magnets are Hisar, Ambala, Yamuna Nagar, Karnal in Haryana, Bareilly, Moradabad, Saharanpur, Allahabad, Kanpur in Uttar Pradesh, Patiala, Ludhiana, Bathinda in Punjab, Jaipur, Kota, Sikar, Ajmer in Rajasthan, Gwalior in Madhya Pradesh and Dehradun in Uttarakhand.

Regional Planning and Development of Island Territories

India has two groups of islands which are treated as Union Territories. In the Bay of Bengal are the Andaman and Nicobar Islands while the Lakshadweep Islands are located in the Arabian Sea. The island territories and the planning strategies for their socio-economic development are briefly described as under :

The Andaman and Nicobar Islands Union Territory. The Andaman and Nicobar archipelago comprising of 572 islands, islets and rocks is situated about 1200 km off the south-eastern coast of India in the Bay of Bengal. Together they constitute one of the Union Territories of India, and are divided into two districts. Andaman group of islands is in the north of 10° channel while Nicobar group of islands is in the south of this channel. The two are separated by about 160 km of sea. Of the 550 islands in the Andaman group, only 24 are inhabited. The Nicobar group has 22 islands out of which 12 islands are inhabited. The total area of this Union Territory is 8249 sq km out of which Andaman Islands account for 6596 sq km and Nicobar Islands spread over 1653 sq km.

Being close to the equator and surrounded by the sea on all sides, these islands have equatorial type of climate. The precipitation is heavy and often exceeds 200 cm annually. Both north-east and south-west monsoons bring rainfall to these islands. It rains for about eight months in a year. The temperature is moderate and relative humidity is high. Heavy rainfall is also caused by tropical cyclones.

These islands are believed to be a continuation of Arakan Yoma mountain range of Myanmar and have undulating terrain with ridges in north-south direction although a few hills run in east-west direction also. Deep inlets and creeks exist between the main ridges. There are a few flatlands and perennial streams. Ground water reserve is limited and soil is mostly acidic and poor in nutrients. The soil types vary from clay to loamy sand.

The cumulative land area is 8,24,900 hectares out of which 7,49,400 hectares, including 1,097,046 hectares of tribal reserve area is under forest cover. These islands are blessed with a unique luxuriant evergreen tropical rainforest canopy comprising of Indian, Myanmarese, Malaysian and endemic floral strain. So far, about 2,200 varieties of plants have been recorded out of which 200 are endemic and 1,300 do not occur in the mainland India. The North Andaman is characterised by the wet evergreen type, with plenty of woody climbers. The Middle Andamans harbour mostly moist deciduous forests while the South Andaman forests have a perfused growth of epiphytic vegetation, mostly ferns and orchids. The Nicobar Islands also have a rich variety of evergreen tropical rainforests.

Andaman Forests abound in plethora of timber species numbering 200 or more, out of which about 30 varieties are considered to be commercial. The major commercial timber species are Gurjan and Padank. The holy Rudraksha and aromatic Dhoop/Resin trees also occur here. These islands have about 50 varieties of mammals, 270 varieties of birds and 225 species of butterflies.

The original inhabitants of Andaman and Nicobar Islands, account for only 12 per cent of the total population. The main aborigines of Andaman group of islands are Sentinelese, Jarawas, Onges, and Great Andamanese of Negrito origin; whereas the Nicobarese and Shompens are off-shoots of Mongoloid stock and live in the Nocobar group of Islands. The Sentinelese are the sole inhabitants of the North Sentinel Island, about 34 km west of South Andaman. They live in complete isolation, are hostile and semi-nomadic. The Jawaras live in the reserve forest belt of about 639 sq km in the western coast of South and Middle Andaman. The Ongnes live in Little Andaman, an island 130 km south of Port Blair. The vanishing Great Andamanese are settled at Strait Island which is about 46 km from Port Blair. The Shompens inhabit 119 sq km in the interior forest area of Great Nicobar. As per 2011 census figures, the total population of the Union Territory of Andaman and Nicobar Islands is 3,79,944 with a population density of 46 persons per sq km.

Out of a total area of 8,24,900 hectares, only 48,700 *i.e.* 5.9 per cent is used for agriculture. It shows that aborigines are mainly living in forests and little importance is given to agriculture. Paddy is the main food crop and major part of its cultivation is in the Andaman group of Islands. Pulses, oilseeds and vegetables are also grown. Coconut and arecanut are the major cash crops.

There is no major industry although 1375 small scale, village and handicraft units are working. Two fish processing units are export oriented. Three shell and wood based handicraft units are also functioning. The small scale industrial units are manufacturing polythene bags, PV conduit pipes and fittings, paints and varnishes, fibre glass, flour, soft drinks and beverages, shell crafts, bakery, rice milling, furniture, etc. There are only four medium-sized industrial units. The Andaman and Nicobar Islands Integrated Development Corporation is doing a lot for the development of these islands and is taking active part in promoting tourism, fisheries and industrial financing. It also functions as the authorised agency for Alliance Air/Jet Airways.

The development of the Andaman and Nicobar islands is closely related to their strategic location in the Bay of Bengal. These island are extensive enough to be developed as a strong naval base and associated industries. Deep sea fishing can be launched from these locations and related industries can be set up here to diversify and enrich the economy of the people living there. Strong linkage with the mainland is necessary in any programme of development. The efficiency of distribution channels of goods and services from Port Blair to different islands also needs to be improved.

Lakshadweep. Lakshadweep is a group of islands of coral origin in the Arabian Sea. The total area of the Union Territory is 32 sq km. The main islands are Minicoy (4.53 sq km), Androth (4.32 sq km), Karavathi (3.49 sq km), Kadamath (3.03 sq km) Agathi (2.73 sq km) and Amini (2.51 sq km). Minicoy by virtue of its strategic location, is important for the naval defence of the country. All the islands have low level and rise no more than 5 metres above sea level. There are not much variations in the climate and weather remains muggy throughout the year. Beneath a thin layer of vegetal humus there is fine coral sand extending at the surface of all the islands. The average distance from the Kerala coast varies from 200 to 300 km and the shortest distance from Calicut is 108.78 km.

According to census of India 2011, the total population of this Union Territory is 64,429 and only one-fourth of the area is inhabited. However these islands have the fifth highest density of population which was 2013 persons per sq km in 2011. The inhabitants of these islands are ethnically similar to the Kerala's Malayali people, and were influenced by the Arab Traders. However, the people of Minicoy are different from those of other islands. They are descendants of early settlers from Sri Lanka who were originally Buddhists and were later converted to Islam. More than 95 per cent of the people follow Islam.

Cultivation of coconut is the main agricultural activity and coconut is the main wealth of the people. The total area under coconut cultivation is about 2,600 hectares. Another perennial crop is bread fruit, citrus fruit, drumsticks, coarse grains, pulses, vegetables, etc. are grown in kitchen gardens. Agriculture Department introduced the use of artificial manures, vegetable seeds and agricultural implements during the Third Five Year Plan (1961-66). Paddy cultivation was started on experimental basis in 1967.

There are no large scale industries but household industries like coir-spinning and copra making form the backbone of the economy. Coir Training-cum-Production Centres have been established where improved methods of coir twisting have been introduced. Fishing is also an important activity. Some hosiery and handloom factories have also been set up. The scenic beauty of these islands attracts tourists, both domestic and international. The only airport is the Agathi Aerodrome in Agathi Island which connects these islands to the mainland by air. The inter island transport is by boats or helicopters.

Development Strategies

The inhabitants of islands in the Bay of Bengal and Arabian Sea are suffering from the perpetual problem of poverty, hunger, starvation, malnutrition, illiteracy, unemployment, housing and are leading a low standard of living. The tribals of these islands have been an extremely exploited lot for the last so many years by the non-tribals. Considering the prevailing condition, the Government of India initiated special steps to improve the living conditions of these people. Some major steps are described as under :

1. Most of the people living here especially in the Andaman and Nicobar Island are tribals who heavily depend on forests. As such it is necessary to conserve the forest wealth of these islands. Also it is necessary to develop more national parks and biosphere reserves.
2. There are mangrove swamps and wetlands in the creeks and inland waters which are facing great threat due to the increasing interference of man. These ecosystems are in a delicate balance and this balance has to be maintained at any cost.
3. Most of the islands both in the Bay of Bengal and in the Arabian Sea are of coral origin which are at risk of destruction. These coral ecosystems need to be conserved.
4. Soils are at great risk of erosion and degradation due to heavy rainfall in these islands. This problem can be solved to a great extent by afforestation.

5. Forest based cottage industries and handicrafts should be encouraged because these industries have great potential of providing employment to the inhabitants of these islands.
6. The natural beauty of these islands has great potential to attract tourists from far and wide places, both domestic and foreign. This potential should be harnessed to its maximum so that the living standard of the local people is improved.
7. The buildings should be made earthquake proof because the Andaman and Nicobar group of Islands lies in the highest seismic risk zone V according to Indian Standards.
8. Coastal creeks, attols and bays in the coasts of these islands offer great opportunities to develop fisheries. Thus fishing can provide large scale employment and all possibilities of developing fisheries should be explored.
9. Although a large number of tribes are living in these islands, their population is fast declining. Jarawa, Janget, Orge, Sentinelese etc. are some such tribes. **Table 13.4.** gives the desperate picture of population of some of the tribes which are struggling for their survival.

TABLE 13.4. Population of Tribals in Andaman and Nicobar Islands

Name of the Tribe	Estimated Population
1. Jarawa	350
2. Shompen	250-300
3. Sentinelese	250
4. Orge	98
5. Great Andamanese	39

Source : Samudrika (2001) pp. 14-15.

10. Plantation of coconut, arecanut, cincona, tea, pineapple etc. should be encouraged as they can enhance the income and living standard of these people.

If the above mentioned steps are taken together, they can complement one another and help in improving the living conditions of the inhabitants of these island groups.

REFERENCES

Agarwal, A.N. (1995) : *Indian Economy, Problems & Development and Planning,* Vishwa Prakashan, New Delhi.

Ahmad, E. (1965) : *Bihar : A Physical, Economic and Regional Geography,* Ranchi University, Ranchi.

Ahuja, Ram (1992) : *Social Problems in India,* Rawat Publications, Jaipur.

Alagh Yoginder, "*Regional Aspects of Indian Industrialization,* University of Bombay, Bombay, Economic Series, No. 25, 1972.

Bagchi, Kanangopal (1977) : "*The Damodar Valley Development and its Impact on the Region*" In Allen G. Noble and Ashok K. Datt (eds.) : *Indian Urbanization and Planning : Vehicles of Modernization,* Tata McGraw Hill Publishing Co. Ltd., New Delhi.

Balakrishna (1948) : *Regional Planning in India,* Bangalore Printing and Publishing Co. Ltd. Bangalore.

Basu, N.G. (1987) : *Forests and Tribals,* Manisha Granthalaya Pvt. Ltd. Calcutta.

Bhat, L.S. (1971) : *Regional Planning in India,* S.P.S. Calcutta.

Bhattacharya, S.N. (1970) : *Community Development* : An Analysis of the Programme in India, Academic Publishers, Calcutta.

Bicamic, R. (1967) : *Problems of Planning,* East and West, Mouton, London.

Blowers A. (ed.) (1993) : *Planning for Sustainable Environment,* Town and Country Planning Association, London.

Bose, Nirmal Kumar (2002), *Tribal Life India,* National Book Trust of India.

Branch, M.C. (1988) : *Regional Planning; Introduction and Explanation,* Pareger, New York.

Census of India, 2011 : *Provisional Population Totals,* Paper 2 of 2011.

Census of India 2011 : *Primary Census Abstract,* released on C.D. in 2013.

Chakraborti, M and Mukherjee D : *Indian Tribes,* Concept Publishing Company, New Delhi.

Chand Mahesh and V.K. Puri (1983) : *Regional Planning in India,* Allied Publishes, Pvt. Ltd., New Delhi.

Chandna, R.C. (2008) : *Regional Planning and Development,* Kalyani Publishers, New Delhi.

Chandrasekhara, C.S. and Sundaram, K.V. (1968) : *Planning Regions of India,* Town and Country Planning Organisation.

Chadrasekhara C.S. (1972) : "*Balanced Regional Development and Planning Regions*", *in Economic and Socio-cultural Dimensions of Regionalisation,* edited by A. Chandrasekhara, Census of India, New Delhi.

Chatterjee, S.P. (1967) : "*Region, Regionalism and Economic Regionalisation*", The Geographer, Vol. XIV.

Chaudhury, Jayasri Rai (2009) : *An Introduction to Development and Regional Planning, with special reference to India,* Orient Black Swan, Hyderabad.

Chaudhury, M.D. (1977) : *The Problems of Multi-Level Planning and Decision-making in India in Antoni Kuklinski* (ed), *Social Issue in Regional Policy and Regional Planning,* Mouton & Co. Hague.

Chaudhury, P. (1979) : *The Indian Economy, Poverty and Development,* Vikas Publishing House, Delhi.

Damodar Valley Corporation, *Eight Years of DVC* (Calcutta, 1956)

Damodar Valley Corporation, *Annual Report,* 1977-78.

Dev and Ranade (1997), Poverty and Public Policy : A Mixed Record in *India Development Report* 1997 Ed. Kirit S. Parikh, Oxford University Press, Delhi.

Dreze, Jean and Sen Amartya (1996) : *Economic Development and Social Opportunity,* Oxford University Press, Delhi.

Dutta-Chaudhary (1971) : "*Regional Planing in India*" in *Issues in Regional Planning* edited by Dunham and Hillhorst, Monton, London.

Dubey, K.N. (ed.) (1990), *Planning and Development in India,* Asia Publishing House, N. Delhi.

Economic Survey 2022-23, Govt. of India.

Friedmann, J. and Alonso, W (eds) (1964) : *Regional Development and Planning : A Reader,* MIT Press, Boston.

Geddes, P. (1968) : *Cities in Evolution* (New Revised Edition), Benn

George K.K. : *Development of Backward Regions : Strategy Destorted,* The Economic Times, 1 & 2 October, 1976.

Gilkson, P. (1955) : *Regional Planning and Development,* Leiden, Netherlands.

Gili, Richard J. (1975) : *Economic Development : Past and Present,* Prentice Hall of India Pvt. Ltd., New Delhi.

Glasson, John (1975) : *An Introduction to Regional Planning,* Hutchinson and Co. (Publishers) Ltd., London.

Govt. of India (1985) : *Report of Committee on Rehabilitation of Displaced Tribals due to Development Projects,* Ministry of Home Affairs, New Delhi.

Govt. of India (1988) : *Regional Plan 2001*—National Capital Region, NCRIB, Ministry of Urban Development, New Delhi.

Govt. of India (1974) : *Report of Commissioner for Scheduled Castes and Schedule Tribes,* 1971-72 and 1972-73.

Griffin, K. (1976) : *Land Concentration and Rural Poverty.*

Hilhorst, J. (1968) : *Multi-disciplinary Aspect of Regional Development,* OECD, Paris.

Hufsahmidt, M.M. (1969) : *Regional Planning, Challenge and Prospects*, Preager and Co. New York.

India 2022, *A Reference Annual,* Publication Division, Ministry of Information and Broadcasting, Govt. of India.

Jagdish N. Bhagwati and Padma Desai (1970) : *India : Planning for Industrialisation,* Oxford University Press, London.

Lokhoria, R.N. "*Tax Incentives for Backward Areas*", The Economic Times, 17 December, 1974.

Malgankar, P.D. and Chiara, B.M. "*Regional Development* : *Where and How,* in L.K. Sen (ed), *Reading on Micro-Level and Rural Growth Centres,* National Institute of Community Development, Hyderabad.

Menon, K.S.V. : "*Backward Areas Development,* "the *Economic Times,* 27 May, 1979.

Misra, R.P., Sundaram, K.V. Prakasa Rao, V.L.S. (1974) : *Regional Development Planning in India, A New Strategy,* Vikas Publishing House, New Delhi.

Misra, R.P. (ed) (1992) : *Regional Planning, Concepts, Techniques, Policies and Case Studies,* Concept, New Delhi.

Misra, S.K. and Puri, V.K. (1997) : *Indian Economy,* Himalaya Publishing House, Mumbai.

Mitra, Ashok (1965) : *Levels of Regional Development in India*, Census of India, New Delhi.

Mustafa, Arif M. and James Paul (ed) (2001) : *Understanding Andaman and Nicobar Islands through Samudrika,* Naval Marine Museum.

Nair, M.N.V., Srinivasan, V. and Kumar, S.P. (1977) : *Need for New Policy Perspectives and Regional Development in Backward Area Development : Strategies and Policies,* Management Development Institute, New Delhi.

Nath, V. (1965) : *Resource Development Regions and Divisions of India,* Planning Commission, New Delhi.

Nath, V. : *Regional Development in Indian Planning,* IN *Economic and Political Weekly,* Jan., 1970 *Annual Volume,* pp. 255-56.

Prakasa Rao, V.L.S. (1963) : *Regional Planning,* Asia Publishing House, London.

Patel, V.G. (1969) : *An Analysis of Plan Implementation in India,* Balgovind Prakashan, Ahmedabad.

Pathak, C.R. (1969) : "*Damodar Valley Region*—A Case Study in Regional Development in R.P. Misra, *Regional Planning,* Mysore University, Mysore.

Pathak, C.R. and Kundu, Amitabh (1973) : Regionalisation and Regional Planning—An Axionnate Approach, in *Indian Journal of Regional Science.*

Pathak, Mahesh, T. "*Development of Backward Areas : Problems and Prospects,* Artha Vikas, Vol. X, No. 1-2, January-July, 1973.

Peter Hall (1970) : *Theory and Practice of Regional Planning,* Pamberton Publishing Co. Ltd., London.

Planning Commission, Government of India, *Identification of Backward Areas,* 1969.

Raj, K.N. : "*Growth and Stagnation in Indian Industrial Development,* Mainstream, 21 January, 1976.

Rao Hemlata (1977, June) : "*Identification of Backward Regions and Trends in Regional Disparities*", In *India Artha Vijnana, Vol. 19, No. 2.*

Reddy, Venugopal, Y. (1979) : *Multilevel Planning in India,* Vikas Publishing House, New Delhi.

Sandhu Urmila : "*Identification of Backward Regions—A Brief Critical Review, Margin Vol. 9, No. 2, January, 1977.*

Sen, L.K. (1972) : *Reading on Micro Level Planning and Rural Growth Centres,* National Institute of Community Development, Hyderabad.

Sengupta, P. (1967) : *Principles and Techniques of Regional Planning.* In *The Geographer Vol. 14.*

Sengupta P. and Sdasyuk G. (1968) : *Economic Regionalisation of India : Problems and Approaches,* Census of India, Monograph edited by A Mitra, New Delhi.

Singh, R.L. (ed) (1971) : *India : A Regional Geography,* Ch. 29, National Geographical Society of India, Varanasi.

Subramanyam, K.V. : "*D.V.C. and Flood Control*" Economic and Political Weekly, 11 November, 1978.

Sundaram, K.V. (1978) : "*Delhi : The National Capital*" In: *Million Cities of India* Ed. R.P. Misra, Vikas Publishing House Pvt. Ltd., New Delhi.

Sundaram, K.V. and Misra R.P. (1976) : *Micro Level Planning and Development Process,* Vol. I, University of Mysore.

Vishnu Bhagwan and Vandana Mohla (2012) : *Indian Government and Policies,* Kalyani Publishers, New Delhi.

Walter Isard (1960) : *Method of Regional Analysis,* M.I.T. Press, Cambridge.

◆◇◆

CHAPTER 14

Settlements

INTRODUCTION

Shelter is the third most essential and basic need of man after food and clothing. Man builds houses to lead a settled life and the place where he builds houses is known as ***settlement***. Thus settlement is a permanently occupied human dwelling place which indicates a community of dwellings and associated buildings ranging from a small hamlet to metropolitan city or megalopolis. In the opinion of **A.N. Chark** (1989), even a single dwelling may be termed as a settlement although the term is normally applied to a group of dwellings. By constructing settlements, man tries to adopt himself to his physical and cultural environment. According to **Perpillon** (1986 : 433), "Settlement is man's first step towards adopting himself to his environment. A settlement is essentially an occurrence unit consisting of houses, served by roads and streets." According to **R.L. Singh** (1961), "Settlement as an occupance unit represents an organised colony of human beings including the buildings in which they live or work or store or use them otherwise and the tracts or streets over which their movements take place.

CLASSIFICATION OF SETTLEMENTS

Settlements may be classified on the basis of their functions, size, morphology, number and density of houses, etc. but the most widely used criteria is the function carried out by the inhabitants of a settlement. On this basis two types of settlements are recognised *viz.* (1) rural settlements, and (2) urban settlements.

Rural Settlements

Rural settlements refer to clusters of dwellings called 'villages' together with the surrounding land from which the inhabitants derive their sustenance. Village has been defined in different ways by various authors. According to **Richthofen**, "They (Villages) are groups of families, united by common descent, or at least having rites in common, who cleave to one another because of the necessity for cooperating in the cultivation of the same crops." **Blache** has termed village as the expression of a type of community larger than family or clan. To quote **Bruhnes**, "Village is a term applied to a geographical fact—the collection of houses and residents of the most numerous aggregations". **Zelinskey** pointed out that an individual village is not simply a collection of farms but a definite social entity. The village has a number of characteristics—territorial, ethnic, cultural—which defines it as an entity. **M. deClerck** describes "a village is above all, a socio-psychological environment, where everybody knows everybody else, where everyone's attitude is strictly controlled by the group; the cultural system defines the patterns of behaviour and the habits of the individuals coincide with village custom".

In India, however, officially a village stands for the area demarcated as a *mauza* meaning "parcel of ground with definite boundaries for revenue purposes without clear and consistent regard for its population". A revenue village thus defined, is a definite administrative unit and includes one or more clusters of dwellings together with the land territory under its possession. Each village has a distinctive place name which helps in distinguishing one village from the other. The hamlet, locally named as *faliya*, *para*, *dhana*, *dhani* etc. means a separate aggregate of houses within the village boundary, sometimes having a name and always forming a part of the whole.

Classification of Villages

Although there are several ways of classifying villages, classification based on size, population and area of land is the most popular and accepted classification of villages. On the basis of this criteria, following classes of villages are recognised.

1. Pura. The place where a habitat had been in ancient days and where chief habitats are all around or nearby areas is called pura. This habitat can thus be called the nucleus of the area. In agricultural areas solitary habitats generally become "puravas" which under favourable circumstances grow into villages.

2. Khas. The word khas is used for the main village or sadar. Sometimes people of the main village settle at some distance and call this new village after their main village. When the population increases, the word khas is used for the village from where people spread around.

3. Kalan. The word kalan is used for large villages and used at the end of the name of the village *i.e. Bound Kalan*. People of several classes and castes live in such villages.

4. Khurd. The word khurd is a degraded form of the urdu word *barkhurdar* (meaning son or small). Therefore this word is used for small villages *i.e.* Dumarkha Khurd.

5. Khera. This word is used for small colonies and also for the higher land of the village. Khera has great social value for the village community because almost all the shows such as *Ramlila*, *Nautanki*, etc. take place here. In some areas the word khera is used for those places where the ruins of an ancient fort are found.

6. Nanglay. It comprises a group of small villages where one village is surrounded by several satellite villages.

Types of Rural Settlements

Before we proceed to discuss settlement types, it is worth making a distinction between settlement types and settlement patterns. In geographical literature, these terms convey various meanings, sometimes synonymous, at others interchangeable and in yet others as one being element/part of the other. But, actually, these are neither of them. In simple language, *types of rural settlements imply the degree of dispersion or nucleation of the dwellings whereas the patterns refer to geometrical shapes formed by the arrangement of dwellings.*

Various authors have suggested different schemes of discussing settlement types. **Finch, Trewartha** *et al.* refer to two primary types of settlements. (*i*) the isolated or dispersed and (*ii*) the nucleated. These are two extreme types of groupings, wherein isolated settlements refer to a single family residence and the nucleated settlements refer to a group of dwellings clustered almost in the centre of the village lands.

D.C. Money has given three broad classes of settlements *viz.* (*i*) the single large nucleated village, (*ii*) hamlets scattered throughout the countryside and (*iii*) single homesteads. **Enayat Ahmed** gives four types : (*i*) compact, (*ii*) cluster and hamlet type, (*iii*) fragmented or hamleted and (*iv*) dispersed settlement. **R.L. Singh** discerns four main types : (*i*) compact settlements, (*ii*) semi-compact or hamleted cluster, (*iii*) semi-sprinkled or fragmented or hamleted settlements and (*iv*) sprinkled or dispersed type. On the basis of number of villages, hamlets and number of occupance units, **R.B. Singh** identified four types of settlements. They are (*i*) compact, (*ii*) semi-compact, (*iii*) hamleted and (*iv*) dispersed or scattered type.

(*i*) Compact settlements. If the number of villages equals the number of hamlets in an area unit, the settlement is designated as compact. Such settlements are found throughout the plateau region of Malwa, in the Narmada Valley, Nimar upland, large parts of Rajasthan, paddy lands in Bihar, Uttar Pradesh, Vindhyan Plateau and several other cultivated parts of India. In such villages all the dwellings are concentrated in one central site. The inhabitants of the village live together and enjoy the benefits of community life. Such settlements range from a cluster of about thirty to hundreds of dwellings of different forms, sizes and functions. Their size varies from 500 to 2,500 persons in sparsely populated parts like Rajasthan to over 10,000 in the Ganga plain.

Different geographical factors contribute to the growth and development of compact rural settlements. In the fertile plains of north India they develop in areas of intensive subsistence agriculture. The site of such settlements in this vast plain is generally *bhangar* area free from the annual floods, water points (oisis) in the Thar desert region and Ox-bow lakes in the Middle Ganga Plain. Availability of water is a very important factor in the siting of such settlements. In vast tracts of Punjab, Haryana, western part of Uttar Pradesh, Rajasthan and the Deccan Trap, peasants flock to the perennial sources of water like wells, canals, ponds, etc. In the tribal areas, these settlements are constructed on hill tops, ridges and spurs, etc. Such sites are selected from the defence point of view.

***(ii)* Semi-compact or Hamletted settlements.** If the number of villages equals more than half of the hamlets, it is semi-compact settlement. The dwellings in such settlements are not very closely knitted and are huddled together at one common site. It covers more area than the compact settlements; the hamlets occupy new sites near the periphery of the village boundary.

The chief characteristic of these settlements is an early recognised site and one or more than one hamlets which are closely linked with the main site, foot-paths, cart-tracks or roads,. The Purvas and Mazras (hamlets) in the near neighbourhood of the main village grow due to increase in population of the main settlement. The pressure of population on the main site forces many families to shift and built their houses outside the main village. The families which shift outside the main village are generally those of agricultural labourers, artisan castes and other poor people. Care is taken that the *Purvas* and *Mazras* built by such people are located close to the main village considering the close social links among the village communities and the prevailing economic interdependence.

Semi-compact or halmetted rural settlements are found both in plains and plateaus depending upon the environmental conditions prevailing there. These are very common in the *khadar* areas of the Ganga, the Yamuna, the Brahmaputra and several of their tributaries, the *Bet* banks of Punjab, the *Tarai* region of Uttar Pradesh and Uttarakhand and in the deltas of the peninsular rivers of the Mahanadi, the Godavari, the Krishna and the Cauvery.

***(iii)* Hamleted settlements.** If the number of villages is equal to half of hamlet number, it is a hamlet settlement. The hamlets are spread over the area with intervening fields and the main or central settlement is either absent or has feeble influence upon others. Often the original site is not easily distinguishable and the morphological diversity is rarely noticed. Such settlements are found in West Bengal, eastern Uttar Pradesh, Madhya Pradesh and coastal plains.

***(iv)* Dispersed settlements.** If the number of villages is less than half the number of hamlets, the settlement is regarded as dispersed. The inhabitants of dispersed settlements live in isolated dwellings scattered in the cultivated fields. Individualism, sentiments of living freely, custom of marriage relations are conducive to such settlements. However, these dwellings are deprived of neighbourhood, communal interdependence and social interaction. Dispersed settlements are found in tribal areas covering central part of India, eastern and southern Rajasthan, Himalayan slopes and land with dissected and uneven topography. Homesteads or farmsteads of wheat producing areas in Punjab, Haryana and western Uttar Pradesh also belong to this category.

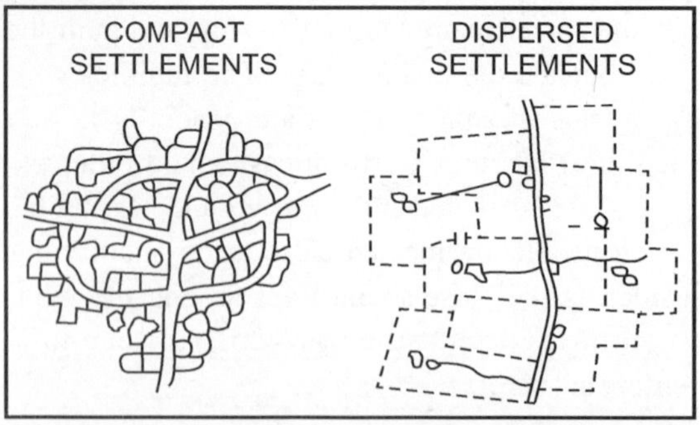

FIG. 14.1. Compact and Scattered Settlements

Difference between Compact Settlements and Dispersed Settlements

Compact Settlements	Dispersed Settlements
1. Compact settlements are mainly found in fertile plains and river valleys.	Scattered settlements are mainly found in hills, plateaus, highlands, forests and arid and semi-arid lands.
2. The main occupation is agriculture.	Animal grazing and lumbering are the main occupations.
3. Houses are built in close vicinity to each other and have lesser living space.	Houses are isolated and scattered over the land. They provide more living space.
4. Size of the fields is small.	Fields are large.
5. Streets are dirty due to lack of proper drainage.	These settlements are quite neat and clean.
6. The inhabitants of compact settlements work and defend themselves collectively.	People of dispersed settlements lead isolated life.

RURAL SETTLEMENT PATTERNS

Pattern refers to geometrical form and shape of the settlement and different settlements have different types of pattern depending on the site and historical background of the settlement. **Embryos Jones** defined pattern of settlement as the relationship between one house or building and another. The pattern of a settlement provides a picture of its shape and a distinct pattern gets its name *i.e.* linear, elongated, square, etc. Pattern also refers to two dimensional geometrical arrangement of rural settlements in an area. Sometimes, a rural settlement may not depict any geometrical shape and the pattern may be termed as ***non-geometrical***. Thus two patterns—geometrical and non-geometrical—are easily discernible on a large scale map or an airphoto of the concerned area.

Following are some of the most common patterns of rural settlements found in India.

1. Linear pattern. This is also known as ***ribbon*** or ***string*** pattern. The main street of the village runs parallel to a road, railway line or water front and most of the village shops are located on this road. Some villages of fishermen along the coast have their houses where water level at high tide determines their location (**Fig. 14.2**). Such pattern is found largely in

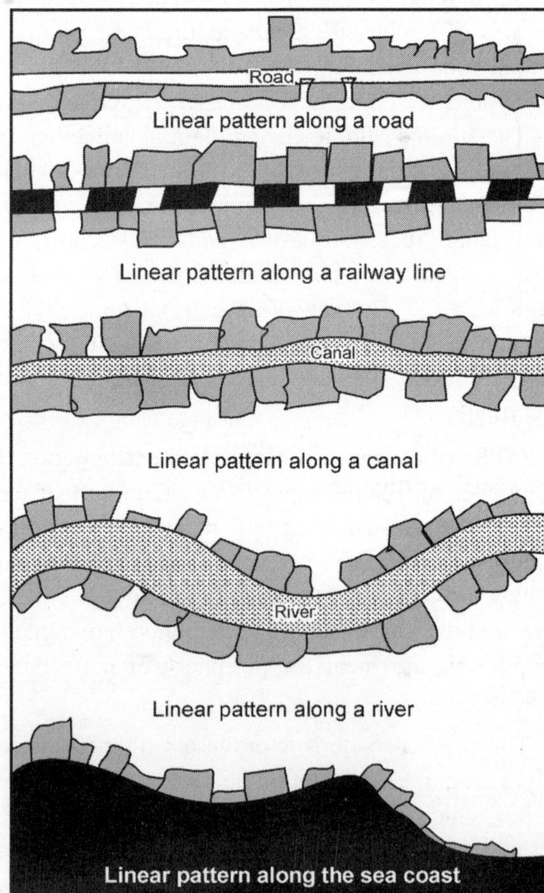

FIG. 14.2. Linear Patterns

middle and lower Ganga plain, parts of the Himalayan region, along the coasts, especially in Malabar and Konkan and north-east Indian states.

2. Checkerboard Pattern. This pattern develops at a place where two roads or some other mode of transport meet almost at right angles. The point of intersection of two main roads is the focus of growth of the settlement. The other roads or streets are parallel to the main roads and also parallel to each other. They meet each other at right angles (**Fig. 14.3**). Such villages are found in large numbers in the fertile plain of north India. They are highly concentrated in the Ganga-Yamuna Doab. In south India, the checkerboard pattern is a quite common in Tamil Nadu, Karnataka, Andhra Pradesh and Telangana.

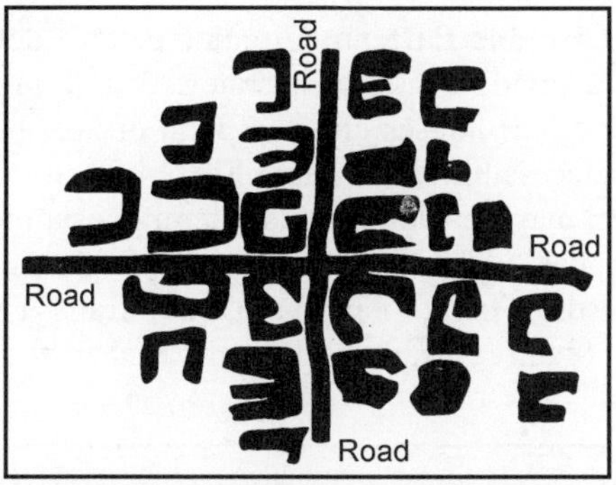

FIG. 14.3. Checkerboard Pattern

3. Rectangular Pattern. The geometry of these settlements is largely rectangular and sometimes square. Such settlements normally develop in the fertile plains of north India where intensive cultivation is practised. These settlements have straight streets which meet each other at right angles. They are connected to each other by roads or foot-paths (**Fig. 14.4**). In India, the rectangular pattern is the heritage of our old system of land measurement in '*Bighas*' which acted as a major factor because the entire cultivated plots and orchards were designed in rectangular shape.

According to **J.P. Misra** (1989 : 126) "The rectangular shape of the fields facilitates ploughing process and making cart-tracts and foot-paths and hence it is maintained by peasants at all costs. Furthermore, the rectangular shape of plots adjusts fairly several rows of houses and render them more spacious." The village streets also confirm to the field patterns and give strength to the rectangular pattern. It may be mentioned that the village is not able to maintain its square shape for a long time because it grows in a particular direction, especially along the transport routes as the time passes.

In certain cases there is a vacant space within the village and the pattern is said to be 'hollow rectangle'. This vacant space may be ascribed to the site of old places, religious places or some water body. This space is the common land and is used by the entire village community.

They are abundantly found in Punjab, Haryana and western part of Uttar Pradesh. In Rajasthan, the Indira Gandhi Canal Command Region has a large number of rural settlements planned as rectangles. Their highest concentration is found in Suratgarh, Hanumangarh and Ganganagar districts. In south India, the deltas of the peninsular rivers have a large number of villages of rectangular shape.

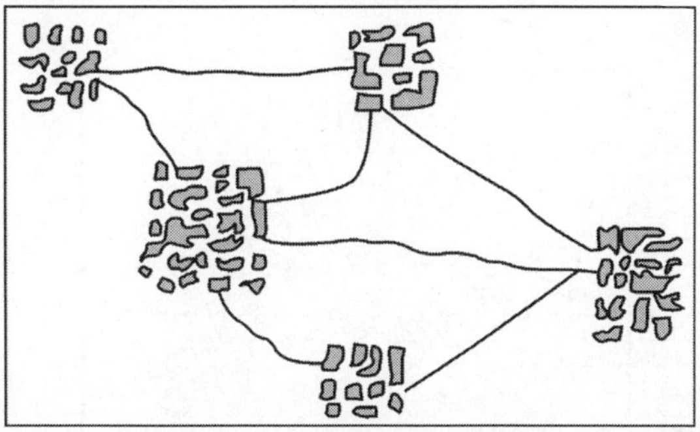

FIG. 14.4. Rectangular Pattern

4. Radial Pattern. Radial pattern develops at a nodal place where number of roads and streets or foot-paths coming from different directions converge. Conversely the roads, streets and foot-paths radiate from a nodal place to different directions and give rise to radial pattern. Houses are built along the transport routes and a radial pattern comes into being. Obviously the central or the nodal part is a vital point in the evolution of the settlement and is occupied by the landlord or village head or some religious place.

Some socio-cultural buildings may also come at this place.

About one-third of Indian villages have radial pattern. Most of them are found in the northern plain of India.

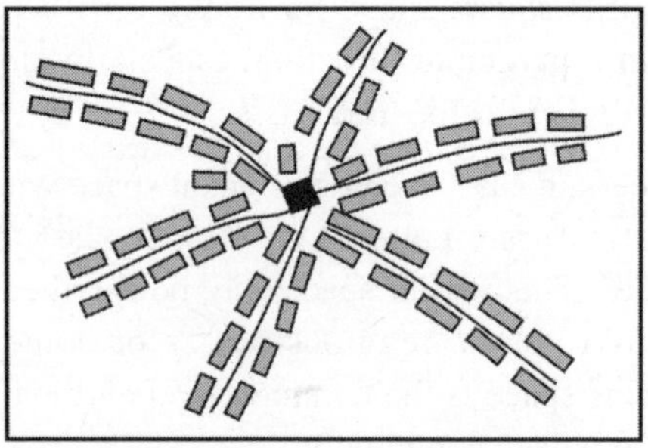

FIG. 14.5. Radial Pattern

5. Star-like Pattern. It is a refined form of radial pattern. It develops when the space between the transport routes like roads, streets and foot-paths is occupied by houses and other buildings. The building process of the houses starts from the nodal place and spreads in all directions. However, the maximum building process takes place along the transport routes and the entire village takes the shape of a star (**Fig. 14.6**).

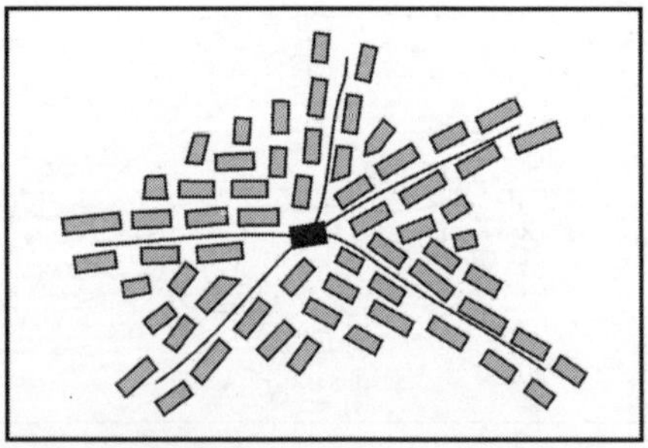

FIG. 14.6. Star-like Pattern

This pattern is mainly found in the fertile areas of Punjab, Haryana and western part of Uttar Pradesh where there is a dense network of roads.

6. Triangular Pattern. This pattern is the result of hindrance occurring on three sides. This hindrance may be physical or cultural or both. The growth of settlement is restricted on three sides by such hindrances and the settlement assumes a triangular pattern. Such a pattern usually develops at the confluence of two rivers or two roads. The lateral expansion of dwellings at the confluence is restricted by the rivers and triangular shape of the dwelling emerges (**Fig. 14.7**).

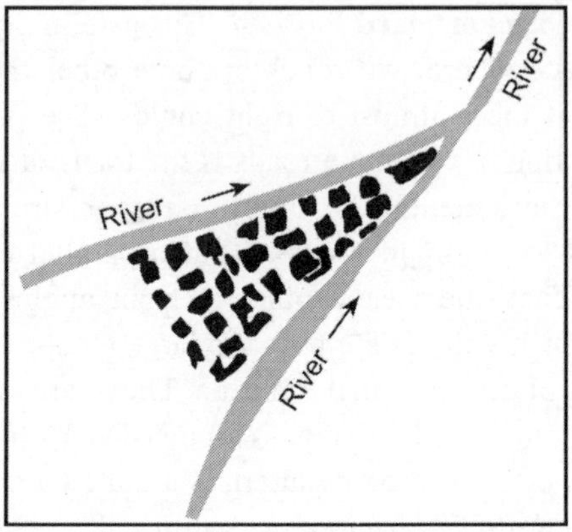

FIG. 14.7. Triangular pattern

7. Circular Pattern. Circular pattern develops around a pond, a lake or a crater. People prefer to construct their houses close to pond or lake because of easy accessibility of water (**Fig. 14.8**). Sometimes a pattern may develop around a temple or a mosque. Villages having circular pattern are found in the upper part of the Ganga-Yamuna Doab, trans-Yamuna region, Madhya Pradesh, Punjab, Maharashtra and Gujarat.

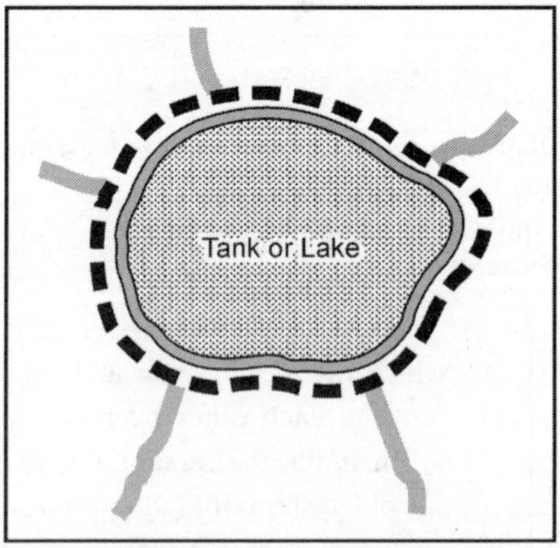

FIG. 14.8. Circular Pattern

8. Semi-circular Pattern. Villages which grow along the river meanders, oxbow lakes or a lake located at the foothill assume semi-circular shape. The Ganga and its several tributaries have semi-circular villages (**Fig. 14.9**).

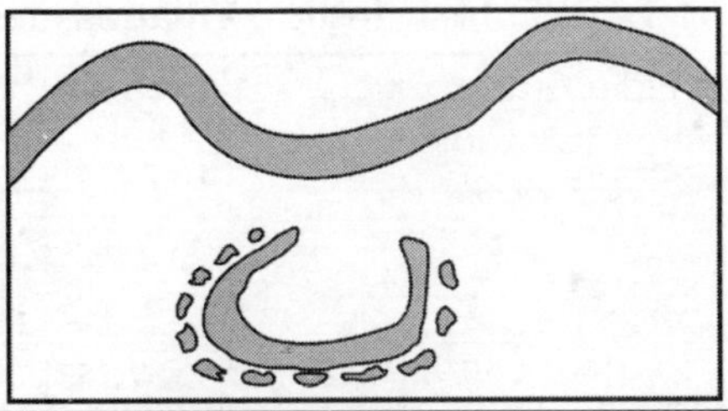

FIG. 14.9. Semi-circular Pattern

9. Arrow Pattern. The villages which grow at the end of a cape, or at a sharp bend of a meandering river or a lake often assume arrow shape. The main concentration of houses is along the roads and the number of houses increases away from the arrow point (**Fig. 14.10**). In India, such settlements are found at Kanniyakumari, Chilka lake, Gulf of Khambhat and along the banks of Sonar river in Madhya Pradesh and Burhi Gandak in Bihar.

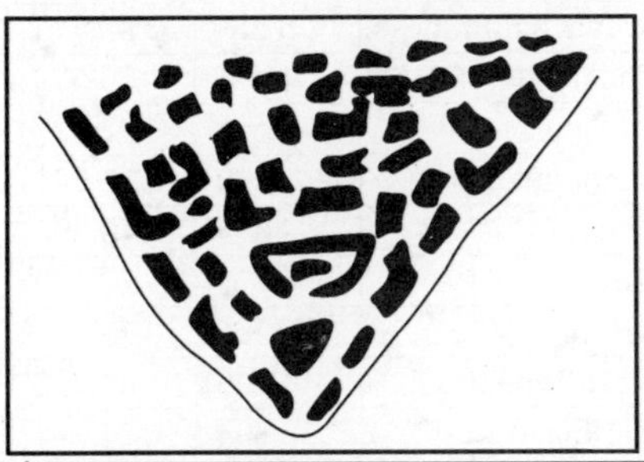

FIG. 14.10. Arrow Pattern

10. Nebular Pattern. Settlements resembling the shape of a nebula are said to have nebular pattern. The roads in such settlements are generally circular which end at the centre or nucleus of the village. These settlements are usually of small size and are found in hilly and undulating areas of Himachal Pradesh, Uttarakhand, Vindhyachal, Kandi lands of Jammu and Kashmir and to a lesser extent in the Ganga-Yamuna Doab.

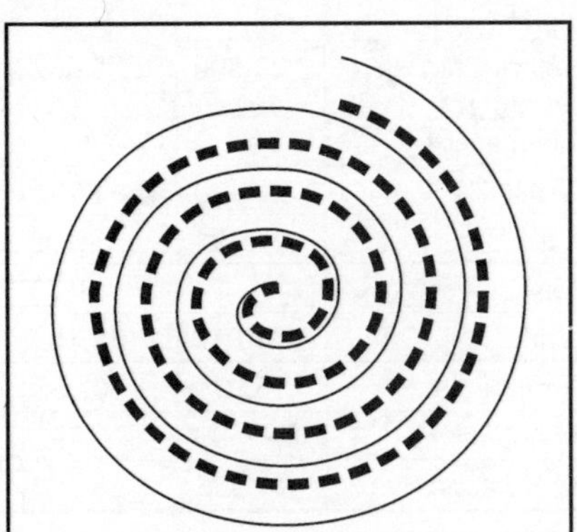

FIG. 14.11. Nebular Pattern

11. Terraced Pattern. Villages having terraced pattern are mostly found on the hill slopes. These slopes are cut and converted into terraces for cultivation. Farmers construct their houses along the terraces and terraced pattern comes into being. Houses in these settlements can be close to each other or they be distant apart. They are constructed at different heights depending on the slope of land. They are usually constructed near a spring or some stream. Such villages may be seen along the hill slopes in Jammu and Kashmir, Himachal Pradesh, Uttarakhand, Sikkim, Arunachal Pradesh, Mizoram and Western Ghats.

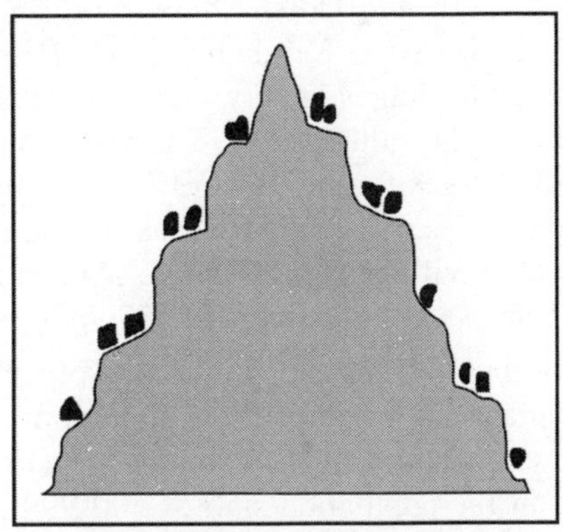

FIG. 14.12. Terraced Pattern

12. T-Shape Pattern. Sometimes the routes meet in such a way that they make a 'T' junction. People

FIG. 14.13. T-shape Pattern

start making their houses along the routes extending in all the three directions and T-shape pattern develops.

Some villages do not have any particular shape and are called shapeless or amorphous villages. Such villages are found in south-west of Uttar Pradesh, Punjab, Rajasthan, Malwa Plateau, Chota Nagpur Plateau, Madhya Pradesh, Andhra Pradesh, etc.

Size, Spacing and Distribution of Rural Settlements

According to census of India 2011, there are 640,867 villages in the country including un-inhabited villages (**Table 14.1**). The largest number of 106,704 (*i.e.* more than 16.6 per cent) are in Uttar Pradesh alone. In fact the whole of Ganga plain is dotted with villages of varied sizes. In the peninsular plateau, Madhya Pradesh and Odisha have more than fifty thousand villages each. Kerala has only 1018 villages but the size of villages is very large. Goa is a small state and has 334 villages only. This is the minimum number of villages for any state of India. Among the union territories, Chandigarh has the minimum number of 5 villages only, while Andaman and Nicobar Islands have maximum of 555 villages.

According to Census of India, size of villages is determined on the basis of their population. **Table 14.2** shows that the highest number of villages in India are in the population slab of 500-999 persons per village. This category of villages accounts for about one-fourth of the total villages in India. There is a gradual decrease in number of villages both in higher and lower population slabs. However, this decrease is more conspicuous in the higher population slabs. For example, only 0.78 per cent of the total villages in India are having population above 10,000 persons. There are a few exceptions to this general observation. For example, Kerala has only three villages with population 500-900 but 797 *i.e.* 78.4% of total villages of the state have population more than 10,000 persons. Even some of the larger states like Uttar Pradesh, Madhya Pradesh, Rajasthan, Maharashtra, Andhra Pradesh, Karnataka, etc. having pre-dominantly rural population do not have so many villages with population exceeding ten thousand.

The above mentioned aggregate pattern of the size of the villages varies greatly from one region to another owing to diversified physiography, climate

TABLE 14.1. India : Villages

State Code	India/State/ Union Territory#	Villages*	Area sq. km	Village Density (No. of villages per 100 sq. km)
	INDIA	**640,867**	**32,87,263**	**19**
01	Jammu & Kashmir	6,551	2,22,236	3
02	Himachal Pradesh	20,690	55,673	37
03	Punjab	12,581	50,362	25
04	Chandigarh#	5	114	4
05	Uttarakhand	16,793	53,484	31
06	Haryana	6,841	44,212	15
07	NCT of Delhi#	112	1,483	8
08	Rajasthan	44,672	3,42,239	13
09	Uttar Pradesh	106,704	2,38,566	45
10	Bihar	44,874	94,163	48
11	Sikkim	452	7,096	6
12	Arunachal Pradesh	5,589	83,743	7
13	Nagaland	1,428	16,579	9
14	Manipur	2,588	22,327	12
15	Mizoram	830	21,081	12
16	Tripura	875	10,492	8
17	Meghalaya	6,839	22,429	30
18	Assam	26,395	78,438	34
19	West Bengal	40,203	88,752	45
20	Jharkhand	32,394	79,714	41
21	Odisha	51,313	1,55,707	33
22	Chhattisgarh	20,126	1,36,034	15
23	Madhya Pradesh	54,903	3,08,000	18
24	Gujarat	18,225	1,96,024	9
25	Daman & Diu#	19	112	17
26	Dadra & Nagar Haveli#	65	491	13
27	Maharashtra	43,663	3,07,713	14
28	Andhra Pradesh including Telangana	27,800	2,75,069	10
29	Karnataka	29,340	1,91,791	15
30	Goa	334	3,702	9
31	Lakshadweep#	21	32	65
32	Kerala	1,018	38,863	3
33	Tamil Nadu	15,979	1,30,058	12
34	Puducherry#	90	492	18
35	A&N Islands#	555	8,249	7

*includes un-inhabited villages.

Source : Data computed from Census of India, *Provisional Population Totals* Paper 2 of 2011, Series 1, p. 40.

TABLE 14.2. Distribution of Villages according to Population 2011 Census and Total Number of Inhabited Villages

States/UTs No.	10,000 and above	5,000-9,999	2,000-4,999	1,000-1,999	500-999	200-499	Less than 200	Total No. of inhabited villages.
1. Jammu and Kashmir[@]	23	212	1152	1641	1521	1275	513	6337
2. Himachal Pradesh	1	12	219	832	2459	5898	8461	17882
3. Punjab	41	330	2227	3471	3237	1903	959	12168
4. Chandigarh	0	3	2	0	0	0	0	5
5. Uttarakhand	21	96	471	824	1826	4684	7823	15745
6. Haryana	121	594	2256	1967	1035	437	232	6642
7. NCT of Delhi	6	9	55	21	6	2	4	103
8. Rajasthan	120	832	5802	10530	12421	8869	4690	43264
9. Uttar Pradesh	545	3432	21013	28020	23381	13591	7832	97814
10. Bihar	1129	3216	10128	10076	7536	4584	2404	39073
11. Sikkim	0	4	42	127	150	75	27	425
12. Arunachal Pradesh	0	1	21	112	325	1026	3773	5258
13. Nagaland	1	30	158	240	337	423	211	1400
14. Manipur	7	28	175	260	417	768	794	2379
15. Mizoram	0	2	43	105	233	236	85	704
16. Tripura	18	131	375	196	80	52	11	863
17. Meghalaya	0	7	82	254	953	2515	2648	6459
18. Assam	21	309	3304	5718	6076	5649	4295	25372
19. West Bengal	417	1716	7552	8731	8574	6848	3640	37478
20. Jharkhand	35	224	2365	5215	7948	8219	5456	29492
21. Odisha	10	195	3058	7902	11978	13271	11263	47677
22. Chhattisgarh	2	88	1913	5392	6142	4252	1778	19567
23. Madhya Pradesh	31	583	5084	12515	16339	11943	5434	51929
24. Gujarat	106	963	4781	5566	3891	1900	576	17843
25. Daman and Diu	0	5	4	5	3	2	0	19
26. Dadra & Nagar Haveli	1	6	28	18	6	5	1	65
27. Maharashtra	310	1175	7170	12154	11144	6426	2582	40961
28. Andhra Pradesh	482	1923	7158	6397	4245	3243	2838	26286
29. Karnataka	164	836	4433	6492	7039	5296	3137	27397
30. Goa	1	14	87	75	57	57	29	320
31. Lakshadweep	0	1	2	0	0	1	2	6
32. Kerala	797	146	58	11	3	1	1	1017
33. Tamilnadu	209	1490	5100	4231	2376	1171	472	15049
34. Puducherry	3	26	46	13	2	0	0	90
35. Andaman and Nicobar Islands	0	2	24	53	61	78	178	396
INDIA[@*]	**4682**	**18641**	**96388**	**139134**	**141761**	**114730**	**82149**	**597485**

Note :

@ India and Jammu & Kashmir excludes the villages of the areas under unlawful occupation of Pakistan and China where Census could not be taken.

* India and Manipur figures excludes those of the three sub-divisions viz. Mao Maram, Paomata and Purul of Senapati district of Manipur as census results of 2011 census in these sub-divisions were not included due to technical and administrative reasons.

Source : India 2017, Reference Annual, p. 26

and cultural set up. The hilly areas are not suitable for large villages due to difficult terrain and have small villages, most of them having population less than 1000 persons. States and Union Territories of Sikkim, Arunachal Pradesh, Mizoram, Meghalaya, Chandigarh, Daman and Diu, Lakshadweep and Andaman and Nicobar Islands did not have even a single village with population exceeding ten thousand while Himachal Pradesh, Nagaland, Dadra and Nagar Haveli and Goa had one village each belonging to this category. On the contrary, the Great Plain of India has large number of medium and large sized villages.

The spacing of the villages is also affected by physiography and other environmental factors. In the hilly areas of Jammu and Kashmir, Himachal Pradesh, Sikkim, Arunachal Pradesh, Mizoram etc. the spacing of villages is very high and sometimes it reaches upto 5 km. In the Ganga Plain, on the other hand, the spacing of villages is low and is often less than 2 km. To conclude, we can say that *high spacing with low density is the character of mountainous areas while low spacing with high density is the character of plain areas*.

According to 2011 census figures 68.84 per cent or more than two-thirds of India's population lives in more than 6.4 lakh villages and still larger number of purvas, nanglas and isolated houses. For the sake of convenience and simplicity, Indian rural settlements are divided into three categories *viz.* (1) dispersed settlements, (2) nucleated settlements and (3) linear settlements. A brief description of these three types of settlements is given below :

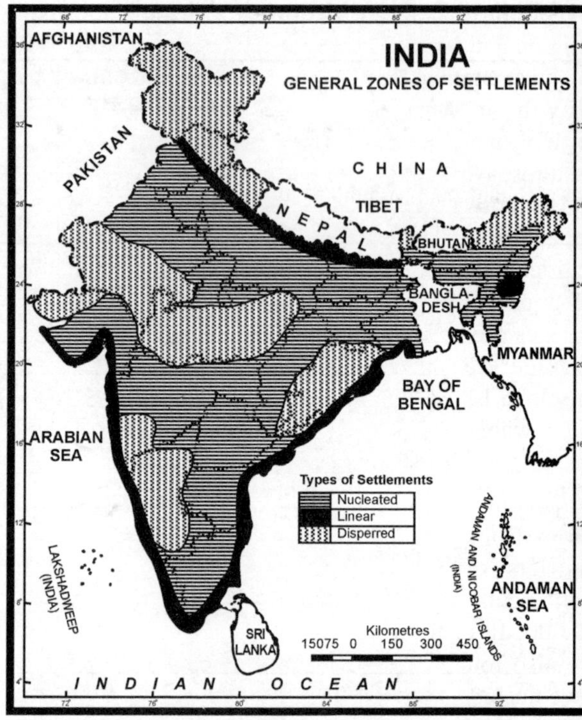

FIG. 14.14. India : General Zones of Rural Settlements

1. Dispersed Settlements

Although India is primarily a country of nucleated settlements, yet dispersed settlements can be found in different parts of the country. Following areas are specifically identified with dispersed settlements :

(*i*) Ridges and spurs in the hilly and mountain areas of Jammu & Kashmir, Ladakh, Uttarakhand, Sikkim and Arunachal Pradesh.

(*ii*) Forest areas of Assam, Arunachal Pradesh, Nagaland, Manipur, Mizoram and Tripura.

(*iii*) Higher areas of Western Ghats from Satara in Maharashtra to Kerala.

(*iv*) Flood prone areas in Uttar Pradesh and Bihar.

(*v*) Wheat and sugarcane producing areas of Punjab and Haryana.

(*vi*) Western Malwa plateau.

(*vii*) Along small water bodies in West Bengal.

(*viii*) Thar desert, Bikaner, Jaiselmar, Barmer of Rajasthan and adjoining parts of Gujarat.

(*ix*) Hilly areas in Odisha.

It is worth noting that the isolated settlements consisting of one house or farmstead are rarely found in India. Most of dispersed settlements consist of 2 to 5 or 10 huts.

2. Nucleated Settlements

Nucleated settlements are found in most parts of India. The largest concentration of such settlements is found in the northern plain of India consisting of Punjab, Haryana, Uttar Pradesh, Bihar, West Bengal and the Brahmaputra valley of Assam. Level land, fertile soil, water resources, cooperation and interdependence in agricultural operations, small and fragmented agricultural fields, caste and clan considerations, social and economic cohesion, religion and orthodoxy and security considerations in the past are the main factors contributing to the growth of nucleated settlements. Nucleated settlements of Rajasthan are due to agricultural land, pastures, availability of water at a few selected places and security requirements. Nucleated settlements here appear to be in response to peoples' reaction to political instability and social insecurity which had been prevailing for so many centuries. Here the villages are of larger size and are distant apart. The average distance between these settlements in Jaiselmar is about 10 km which is the largest distance in the whole of Rajasthan.

In south India, the Malwa plateau has nucleated settlements of larger size whose average population varies from 500 to 5000. Sehore, Indore and Betul districts of Madhya Pradesh have still larger settlements with populations varying from 5,000 to 10,000. In certain areas semi-nucleated settlements are also found. Bundelkhand and Baghelkhand also have nucleated settlements of larger size. Settlements of Bundelkhand are comparatively larger and are mostly located near towns. Chhattisgarh plain has comparatively smaller settlements where about two-thirds of the total settlements have population less than 500. In Mayurbhanj, Kendujhar and Sundergarh districts of Odisha, nucleated settlements are found on the valley slopes. Such settlements are found in the Mahanadi basin also. The black cotton soil of Maharashtra also support nucleated settlements. In the Karnataka plateau, the Malnad hills have infertile soil but they are covered with teak, sal and sandal forests. Consequently the nucleated settlements of this area are of small size. But the northern plain of Karnataka has larger settlements having population varying from 2,000 to 10,000. Andhra Pradesh, Telangana and Tamil Nadu also have nucleated settlements of varying sizes. Kerala has the unique distinction of having settlements of large sizes where more than two-thirds of the settlement have population over ten thousand.

3. Liner Settlements

These are elongated settlements with larger length and minimum breadth. Such settlements usually develop along a road, a canal, a river, embankments made by the old course of a river or along the sea coast. In Assam, Tripura, West Bengal and Odisha, linear settlements can be seen stretching for miles together along both the banks of rivers and canals. Houses in these settlements face the river water front and face each other across the water course. In West Bengal, linear settlements along the courses of the dead rivers in Murshidabad, Nadia, as well as in North and South Twenty Four Parganas can still be seen. Odisha (Balasore), Tamil Nadu (Thanjavur, Ramnathpuram, Tirunelveli) and some parts of Maharashtra also have linear settlements. Over six thousand kilometre long coastline of India is dotted with thousands of fishermen's villages most of which present linear pattern. Deltas and estuaries of big rivers have long queues of linear settlements. In north India, there is continuous belt of linear settlements along the foothills of the Himalayas stretching over the Union Territory of Jammu & Kashmir and states of Himachal Pradesh, Uttarakhand, Uttar Pradesh, Bihar and Sikkim.

Regional Distribution of Rural Settlements

From the physiographic point of view, India can be divided into four major divisions *viz*. The Himalayas, the Northern plains, the Paninsular plateau and the coastal areas and the distribution of rural settlements can be described on the basis of these physiographic divisions.

1. Rural Settlements in the Himalayan Region. The type of rural settlements in the Himalayan region varies according to altitude, slope and general relief of the land as well as on the climatic conditions. In the inter-mountain valleys such as those of Kashmir, Kullu-Manali and Dun valley most of the settlements are either compact or semi-compact. At higher altitudes, the undulating topography, cold climate and shortage of agricultural land favour dispersed or scattered settlements. Some of the migratory tribal people like Gujjars and Bakarwals of Kashmir, Bhutias of Himachal Pradesh and Uttarakhand and Lepehas of Sikkim practise *transhumance i.e.* they move to higher areas in summer to graze their animals and come back to lower areas to avoid severe winter conditions of the higher areas. Thus they are forced to have houses in both the areas. They build permanent houses in lower areas where they spend their winter season and are engaged in agriculture and temporary houses in the alpine pastures at higher altitudes where they graze their animals in summer.

In the Eastern Himalayas, a number of states like Arunachal Pradesh, Nagaland, Mizoram and Tripura have small and widely dispersed settlements. This is due to typical physical and social environment of this region which is reflected by rugged topography, heavy rainfall, dense forests and multiplicity of tribes with different traditions and dialects. In Nagaland, most of the villages are found on the flat tops of hills, spurs and ridges located at altitudes varying from 1,000 to 2,000 metres above sea level. Such sites are selected from the point of view of security against invaders. In Manipur, Kukis tribal people are engaged in shifting agriculture locally known as ***Jhuming*** and construct their temporary huts in Jhum fields. However, some permanent houses may also be found. In Mizoram, linear settlements are found in valleys, on hill tops and along the roads. These settlements are built by the Lushai tribes. The Chakma tribes of Tripura build settlements which are similar to those build by the Lushais of Mizoram. In the Khasi and Jaintia hills region of Meghalaya, the settlements range from isolated homesteads to dispersed and semi-compact settlements. These settlements are generally located along the hill slopes near springs and other water bodies.

2. Rural Settlements in the Northern Plain of India. The Northern Plain of India has been formed by the depositional work of mighty rivers like the Indus, the Ganga, the Brahmaputra and hundreds of their tributaries. As such, this is one of the most fertile plain of the world which has been intensively cultivated for the last several centuries. People lead settled life and have built permanent settlements. The plain of Punjab, Haryana and western part of Uttar Pradesh has large sized compact settlements which are uniformly spaced. The *Bhangar* tracts are free from floods and have compact and semi-compact settlements while the *Khadar* and *Bet* lands are flooded almost every year during the rainy season and most of the settlements in these areas are semi-compact or scattered. The eastern part of Rajasthan, adjoining Punjab and Haryana also has compact settlements of large size. But the western parts of Rajasthan, to the west of the Aravali range, particularly Jaiselmer and Barmer, are characterised by aeolian topography, especially the sand dunes. Therefore most of the settlements here are of semi-compact type.

In Uttar Pradesh and Uttarakhand, there is tarai tract to the south of the Bhabar belt extending in the east-west direction. This tract is marked by marshy lands, seasonal floods and thick forests (many of them have been cleared for agriculture). Under these circumstances, the rural settlements in this tract are largely semi-compact and are of medium size.

Most of the rural settlements in the Middle Ganga Plain are semi-compact and are of small size. But since the land is fertile and can support higher density, the settlements in this area are closely spaced. In the Lower Ganga Plain, the types of rural settlements are controlled by the hydrological characteristic and a variety of rural settlement patterns may be noticed. However, scattered settlements are very common in the *Duars* and *Sundarbans Delta*.

In the Brahmaputra valley, the villages are of small size and are generally oriented along the river levees and transport routes. The houses are separated by bamboo fences. The low lying areas are prone to annual floods where boat is the only mode of transport during the rainy season. Therefore the *Machan* type of houses are constructed on wooden pillars.

3. Rural Settlements in Peninsular India. Peninsular India is one of the oldest land masses of the world and is characterised by a greater diversity of topography. Therefore, this part of the country has a large variety of rural settlements. The Aravalli Hills receive scanty rainfall and the rocky terrain does not support high density of population and human settlements. Thus the settlements are small and are widely dispersed. Mewar and Alwar have dissected hills where isolated farmsteads are dotted along the narrow valleys. These are called ***Dhanis*** in the local language. Jaipur and its neighbouring areas have compact and semi-compact settlements. In the Udaipur division of Rajasthan, there is preponderance of isolated and dispersed settlements. Settlements of Bundelkhand are by and large scattered.

In Chotanagpur plateau of Jharkhand, there are clustered and semi-clustered rural settlements whereas the settlements in Baghelkhand are largely compact. Compact settlements are also found in the lava plateau of black soils in Maharashtra where most of the settlements are located near the source of water. But the tract between Pune and Solapur is dotted with semi-compact and dispersed settlements.

In Karnataka, the Maidan region has widely spaced compact settlements whereas hilly areas of Malnad has semi-compact settlements. In Andhra Pradesh, the coastal and Rayalseema regions have compact settlements while Telangana has compact and semi-compact settlements. The higher parts of whole of Sahyadris stretching from the Tapi valley in the north to Kanniyakumari in the south is dotted with isolated settlements.

4. Rural Settlements in the Coastal Regions. The entire length of Indian coast-line is dotted with villages of fishermen of varied shapes and sizes. However, there is some difference between the settlements of the east coast and those of west coast. In the east coastal region compact settlements have come up on the elevated basin of deltas of large rivers like the Mahanadi, the Godavari, the Krishna and the Kaveri. These compact settlements are the result of intense cultivation on the fertile lands of these deltas. Since these settlements are located on the elevated levees, they do not face the danger of flood.

Along the western coast there is Rann of Kachchh in the north where one finds semi-compact and scattered settlements of small sizes. The Konkan coastal region is marked by compact and semi-compact settlements. The Malabar coast of Kerala is unique in the sense that it has compact settlements of very large size. However scattered and dispersed settlements are also found in plantations of coconut and cashew nut.

Density of Rural Settlements

Density of rural settlements is defined as the number of villages per 100 square kilometres of land area. A look at **table 14.1** reveals that average density of rural settlements in India is 19 villages per 100 sq. km. Among the larger states Bihar has the highest density of 48 villages per 100 sq km. This is followed by Uttar Pradesh and West Bengal (45 each) and Jharkhand (41). These states are said to have some of the highest densities in India. Next in order are Himachal Pradesh (37), Assam (34), Odisha (33) and Uttarakhand (31). The lowest densities of 10 or less then ten villages per hundred sq km are found in Jammu and Kashmir (3), Kerala (3), Sikkim (6), Arunachal Pradesh (7), Tripura (8), Mizoram (8), Gujarat (9) and Andhra Pradesh including Telangana (10). The rest of the states have moderate densities varying from 11 to 30 villages per 100 sq km. Among the Union Territories, Lakshadweep has the highest density of 65 villages per 100 sq km which is the highest in the whole of India. The lowest is found in the Union Territory of Chandigarh where there are only 4 villages per 100 sq km.

Morphology of Rural Settlements

Morphology of rural settlements is concerned with their internal and external structure and identification, classification, regionalisation and analysation of its components. Thus rural morphology tries to find answers to meaningful basic questions like what is, where and why ? **Carl Sauer** opines that rural morphology is based on three postulates *viz.* (*i*) organic or quasi-organic, (*ii*) functional equivalent and (*iii*) agglomeration and placement of structural elements. Thus rural morphology deals with the study of built up area consisting of houses and streets etc. and the open land surrounding the built up area where agriculture or some other primary activity is carried on. The famous settlement geographer **C.A. Doxiadis** has suggested four main parts of a rural settlements (**Fig. 14.16**) :

(*i*) Homogenous part, agricultural fields etc.
(*ii*) Central part, built up area or *Basti*.
(*iii*) Circulatory part, roads, streets, footpaths, etc.
(*iv*) Special part, school, temple, mosque, church, *panchayat-ghar,* cultural place, etc.

The built-up area consists of houses, bazars (shops), streets, roads, schools, religious places, cultural places, *panchayat-ghar,* etc.

Almost all the rural settlements have built-up area at the centre which is surrounded by agricultural land or grazing land. These two types of areas are connected to each other by *Chak-roads*, footpaths, etc. Normally the nucleus of the settlement is located on a higher site which is occupied by the first settlers (normally people belonging to upper castes or landlords locally known as ***Zamindars***). Once the nucleus is properly inhabited, the village starts expanding around this nucleus with the passage of time. Some hamlets may develop near the main

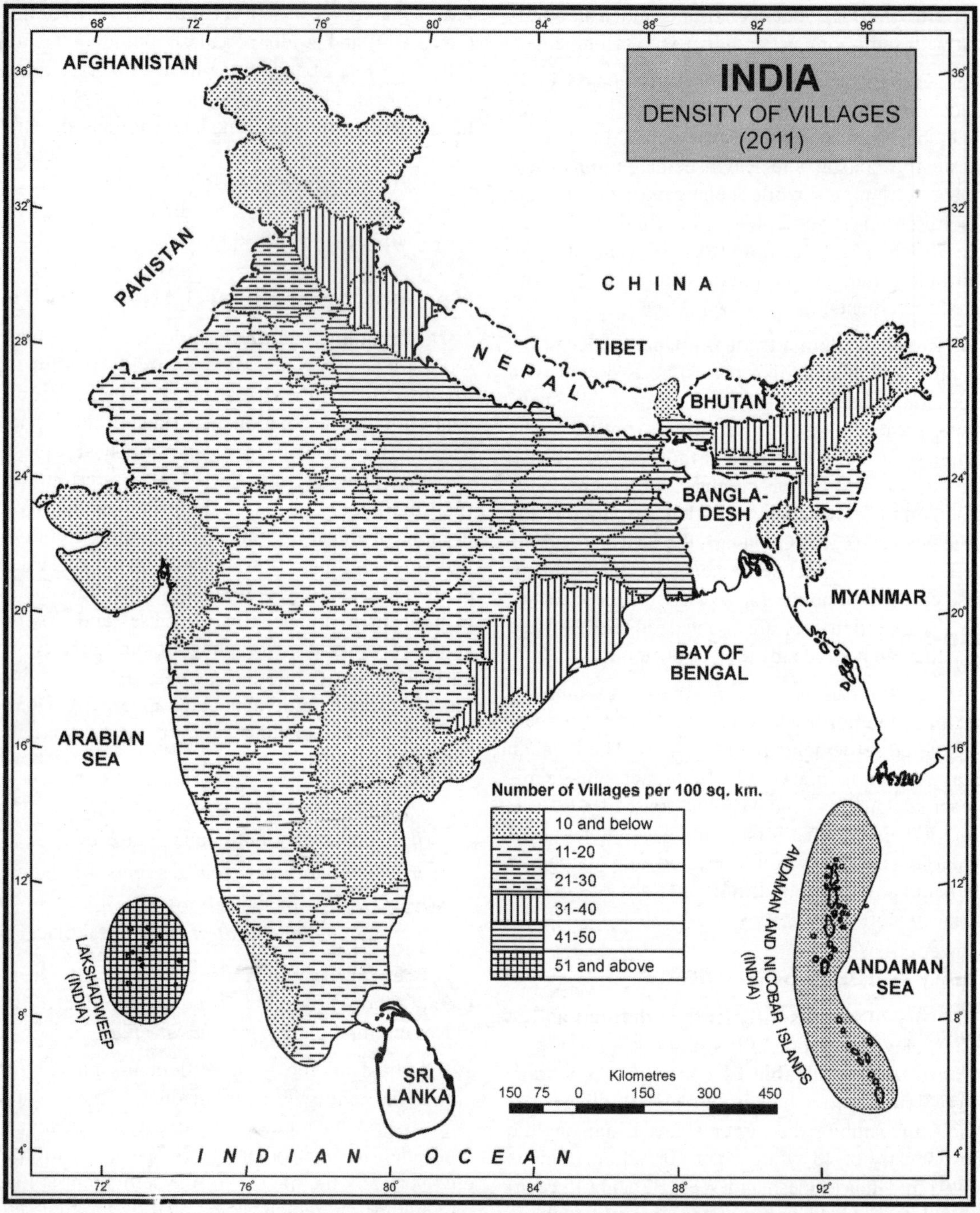

FIG. 14.15. India : Density of Villages (2011)

settlement. Such hamlets are generally inhabited by the people belonging to lower castes and landless agricultural labourers. The socio-economic activities are maximum at the nucleus and their intensity decreases with the increasing distances from the nucleus.

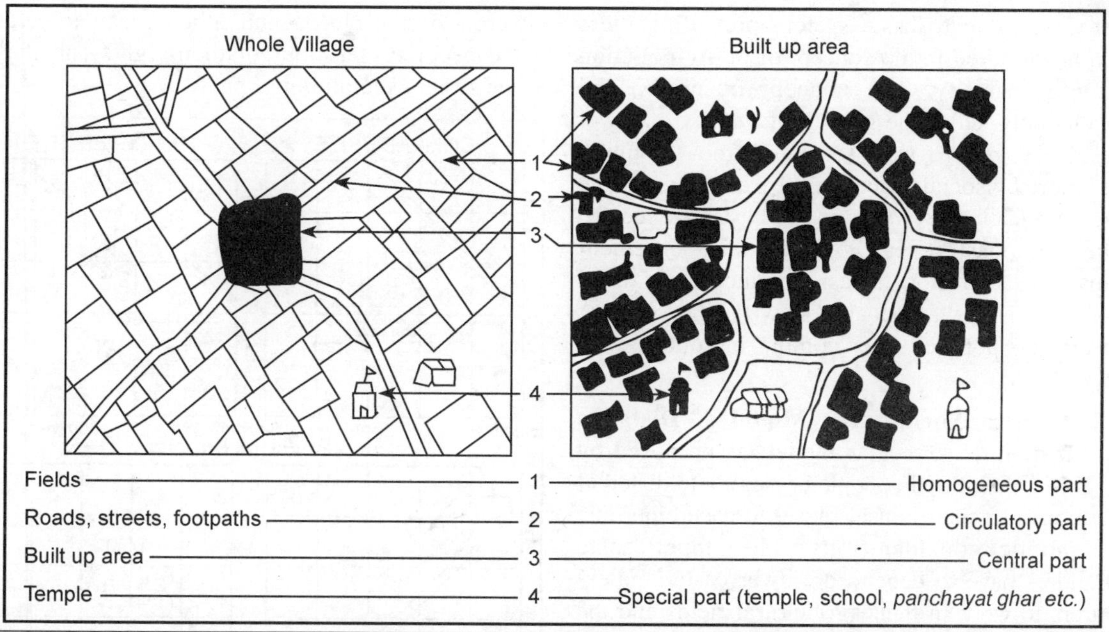

FIG. 14.16. Morphology of an imaginary rural settlement after Doxiadis

Socio-Spatial Structure of Rural Settlements

The socio-spatial morphology of any rural settlement can be explained on the basis of following two concepts :

1. Concept of functional space
2. Concept of social space

According to the views expressed by **Prof. Kashi Nath Singh**, there is almost complete segregation between upper castes (Brahmins, Rajputs, Kayathas) and lower castes (scheduled castes). This segregation is caused due to religious considerations and the distance between two types of castes is maximum. On the contrary, such distances are minimum in a secular set up.

Based on caste considerations, people belonging to upper castes occupy the best site at the central place. Their houses are very large in which each married woman is provided with a separate apartment. Such houses have spreading courtyards also. On the contrary, the lower castes, especially the scheduled castes, are generally poor and have a single room hut/house which is share by all the family members. Sometimes the family members have to share these small huts with domestic animals also. These huts are invariably made of locally available cheap materials. Normally houses of low castes are located on the peripheries of the village at a certain distance from the houses of the upper castes. Hamlets of lower castes are called ***Purvas*** which are known by different names in different parts of the country. *Ahyiron, Chamarkheda, Jatvara, Ladhian, Posiwada, Machpara,* etc. are some of the common names for ***Purvas***. The hamlets and their inhabitants are closely linked with the main site of the village under *Jajmani* system and act like a single functional unit. **Prof. Kashi Nath Singh** found both concepts of social discrimination and functional unity in his Socio-Spatial Structure of Rural Settlements and propagated two models based on religious-ritual dominance and secular norms which are briefly described as below :

1. The religious-ritual dominance model of distance maximization. The rural Hindu society is dominated by powerful social structure and is divided into a number of castes. Each caste performs a specific socio-economic function. For example, Brahmins perform religious rituals, Kshatriyas are responsible for security, Vaishyas are engaged in trade, commerce and agriculture, and Shudras are supposed to provide low grade services to the other castes. They are debarred from several ritual privileges which are enjoyed by the upper castes. The entire rural society is divided into upper castes and lower castes due to caste system prevailing there which has resulted in the concept of purity-pollution and untouchability. The concept of purity and untouchability led to the growth of *purvas* of lower castes. These *purvas* were separated from the upper castes more by social space them by any appreciable barrier. Normally the *purvas* of low castes are located in the south-east or east of the main settlement because this direction is less conducive for wind movement. This was because it was believed that even air gets polluted after coming in contact with *Shudras* body.

2. Secular Norms as a Model of Distance Minimization. In a secular model, as proposed by Prof. Kashi Nath Singh, people belonging to different castes construct their houses closer to each other due to economic considerations. The upper caste landholders have to depend on lower caste landless labourers to work in their agricultural fields and the lower caste labourers heavily depend on the upper caste landlords for their livelihood. Thus there is inter-dependence between upper and lower castes and the distance between two sections of society is minimized. This leads to the emergence of a compact, nucleated and unified settlement. **Fig. 14.17** shows the religious-ritual model and secular dominance model as proposed by Prof. Kashi Nath Singh. However, it is observed that the power enjoyed by a person in the rural society is not always by the upper castes *i.e.* Brahmins, but by the person who owns land. According to **Oscar Lewis** (1965; 81) "While the land owners are generally of higher caste in Indian villages, it is their position as land owners, rather than caste membership what gives them power and status." For example if *Brahmin* is a land owner in a village, then *Brahmins* will get the place of honour, but if a *Thakur* is the land owner, then *Thakurs* will get the place of honour and they will provide livelihood to *Brahmins* along with other castes.

Both the above mentioned models leave their own impact on the morphology of Indian village independently. But it is the joint impact of these models which gives a distinct morphology to a village. The old socio-economic system is gradually giving way to the new one and *Jajmani* system is becoming a thing of the past. The rigid caste system is losing its grip and new socio-economic structure is coming up. With the spread of education and general awareness, the *Dalits* and other backward classes have become conscious of their rights and the old caste system is crumbling.

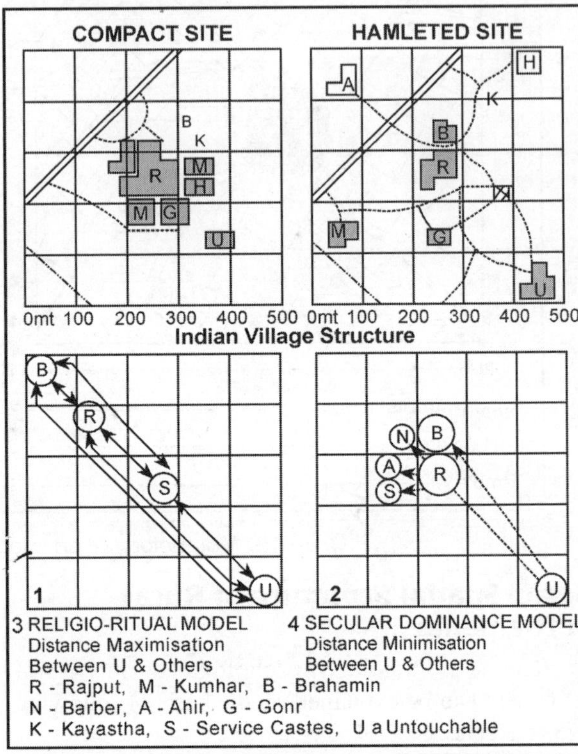

FIG. 14.17. Hypothetical Socio-Spatial Structure of Rural Settlement (After Kashi Nath Singh)

Rural Houses and House Types

House is a shelter built by man to protect himself from the vagaries of climate and to meet the basic physiological requirements of the body. Thus the house is a universal feature of the inhabited world.

Structurally, a house consists of a roof, supported by walls with a door. The census of India defines a 'census house' as a building or a part of a building having a separate main entrance from the road, common courtyard or staircase, etc. used or recognised as a separate unit.

Almost everywhere in rural India, the houses are made of locally available building materials such as stones, mud, unburnt bricks, bamboos, wood reeds, leaves, grasses, etc. This is more true of the poor people who cannot afford the luxury of building materials other than the locally available ones. These are invariably ***kutcha houses.*** However, some rich people in the rural areas can afford to build houses using burnt bricks and cement. These are called ***pucca houses.***

According to Census of India, houses have been classified as *Pucca, Semi-pucca*, and *Kutcha* according to the types of materials used in the construction of walls and roof of the house. For the purpose of this classification, the criteria adopted by National Building Organisation has been made use of. The basis of the classification is as below :

Those houses which have both walls and roof made of pucca materials are classified as *pucca*. When both walls and roof are made of kutcha materials, the house is classified as *kutcha*. If either walls or roof is made of pucca material and the other of kutcha material, then the house is classified as *semi-pucca*.

TABLE 14.3. House types and building materials

Category	Materials used for construction
	Wall
Pucca	Burnt bricks, G.I. sheets or other metal sheets, stone, cement, concrete.
Kutcha	Grass, leaves, reeds, bamboo, mud, wood, un-burnt bricks.
	Roof
Pucca	Tiles, slate, shingle, corrugated iron, zinc or other metal sheets, asbestos, cement sheets, bricks, lime and stone, stone and RBC/RCC, concrete.
Kutcha	Grass, leaves, reeds, bamboo, thatch, mud, un-burnt bricks, wood.

URBAN SETTLEMENTS AND TRENDS IN URBANISATION

What is an Urban Place ?

It will be pertinent to know and understand an urban place before one proceeds to discuss trends in urbanization for which defining an urban place is most essential.

According to Census of India, 2011, an urban area comprises of two types of administrative units—Statutory Towns and Census Towns.

(*a*) **Statutory Towns.** All administrative units that have been defined by statute as urban like Municipal Corporation, Municipality, Cantonment Board, Notified Town Area Committee, Town Panchayat, Nagar Palika etc., are known as Statutory Towns.

(*b*) **Census Towns.** Administrative units satisfying the following three criteria simultaneously are treated as Census Towns:

(*i*) It should have a minimum population of 5,000 persons*;

(*ii*) At least 75 per cent of the male main working population should have been engaged in non-agricultural pursuits; and

(*iii*) It should have a density of population of at least 400 persons per sq km (1,000 per sq mile).

*For the purpose of identification of places that qualify to be classified as 'Census Towns', all villages with a population of 4000 and above as per the Census 2001, a population density of 400 persons per sq km and having at least 75 per cent of male main working population engaged *in non*-agricultural activity were considered.

(*c*) **City.** Towns with population of 1,00,000 and above are categorised as cities.

Source : Census of India, 2011, Tables on Houses, Household Amenities and Assets, Series 1, p. lix.

Out Growth. An Out Growth (OG) is a viable unit such as a village or a hamlet or an enumeration block made up of such village or hamlet and clearly identifiable in terms of its boundaries and location. Some of the examples are railway colony, university campus, port area, military camp, etc., which have come up near a statutory town outside its statutory limits but within the revenue limits of a village or villages contiguous to the town. While determining the outgrowth of a town, it has been ensured that it possesses the urban features in terms of infrastructure and amenities such as pucca roads, electricity, taps, drainage system for disposal of waste water, etc. educational institutions, post offices, medical facilities, banks, etc. and physically contiguous with the core town of the UA. Each such town together with its outgrowth(s) is treated as an integrated urban area and is designated as an 'urban agglomeration'. In Census 2011, 474 UAs with 981 OGs were identified as against 384 UAs with 962 OGs in Census 2001.

Urban Agglomeration

Each town together with its outgrowth(s) is treated as an *urban agglomeration*. Thus the concept of *Town Group*, which was adopted in 1961 to obtain a broad picture relating to urban spread, was refined and replaced by the concept of *urban agglomeration in 1971* to obtain better feed back in regard to urban contiguity, process and trends of urbanization and other related matters. An "urban agglomeration" denotes *a continuous urban spread and normally consists of a town and its adjoining urban outgrowths (OGs), or two or more physically contiguous towns together with contiguous well recognised outgrowths, if any, of such towns*. This concept has remained operative in the later censuses without any change or modification.

For the purpose of delineation of Urban Agglomerations, following criteria are taken as pre-requisites :

(*a*) The core town or at least one of the constituent towns of an urban agglomeration should necessarily be a statutory town.

(*b*) The total population of all the constituents (*i.e.* towns and outgrowths) of an urban agglomeration should not be less than 20,000 (as per 2001 Census). In varying local conditions, there were similar other combinations which have been treated as urban agglomerations satisfying the basic conditions of contiguity. Examples are Greater Mumbai UA, Delhi UA, etc.

With the above mentioned basis two criteria having been met the following are the possible different situations in which Urban Agglomerations should be constituted :

(*i*) a city or town with one or more contiguous outgrowths

(*ii*) two or more adjoining towns with their outgrowth

(*iii*) a city and one or more adjoining towns with their outgrowths all of which form a continuous spread.

Number of statutory towns, census towns, urban agglomerations and outgrowths in 2011 and 2001 is given in **Table 14.4.**

TABLE 14.4. Number of UAs/Towns and Out Growth (OGs)

Type of Towns/UAs/OGs	Number of Towns	
	2011 Census	2001 Census
1. Statutory Towns	4,041	3,799
2. Census Towns	3,892	1,362
3. Urban Agglomerations	474	384
4. Out Growths	981	962

Source : *Census of India 2011*, Tables on Houses, Household Amenities and Assets, Series 1, p. lx.

Metropolitan Cities. Cities with population of one million and over are known as ***metropolitan cities***. The main activities in these cities are industries, trade, commerce, transport, cultural and political. The number of metropolitan cities increased from 12 in 1981 to 53 in 2011.

Mega Cities. Cities with population of 5 million and above are known as mega cities according to Census of India. But United Nations considers mega cities as those that have a population of 10 million and above. In India, Greater Mumbai, Kolkata and Delhi are examples of mega cities.

Conurbation. This term was coined by **Patrick Geddes**. It comes into being by coalescence of urban settlements which were separated by open space in the past. The coalescence usually occurs through Ribbon Development along the main inter-urban transport routes. Mumbai, Delhi and Kolkata are good examples of conurbations in India.

Megalopolis. It is a Greek word which is derived by combining two terms 'great' and 'city'. It is just like conurbation and is formed when a large city sprawls and brings into its fold, the smaller adjacent towns and cities. This term was first used by **Gottman** in 1964 to describe the urban scene of the north-eastern board of the U.S.A. He identified a large conurbation like mass of linked built-up areas (and yet containing much more open land) extending over 960 km (600 miles) from north of Boston to Norfolk in Virginia. Tokyo and New York are good examples of megalopolis.

Size Class of Cities. The Census Department of Government of India has classified the urban places under the following six categories :

Size class	Population
I	100,000 & above
II	50,000-99,999
III	20,000-49,999
IV	10,000-19,999
V	5,000-9,999
VI	Less than 5,000

Urban Growth and Development in India

1. Cities of Ancient and Medieval Periods. India has a long history of urbanization. The first phase of urbanization is traced in the Indus valley which is associated with the Harappan urbanisation. The cities of Harappan civilization flourished for about 600 years, between 2350 B.C. and 1750 B.C. Two major towns of the Indus civilization *viz.* Harappa and Mohenjodaro are in Pakistan now. Important towns of the Harappan culture located in India are Lothal, Surkotada, Rangpur and Rojdi (Gujarat), Kalibangan (Rajasthan), Banwali (Haryana) and Rupar (Punjab). This period was followed by a prolonged period of about one thousand years during which there is no evidence of urbanization. From around 600 B.C. onwards, towns and cities grew in association with two cultural streams *viz.* the Aryan civilization in the North and the Dravidian civilization in the South. The major cities which grew during this period were Hastinapur, Mathura, Ayodhya, Kapilavastu, Kusinagar, Vaishali, Patliputra, Varanasi, Rajgir and Champa in north India and Ujjain, Mahishamati, Nagarjunakonda, Kancheepuram, Puhar, Uraiyur, Madurai, Korkai and Vanji in south India. Cities grew both in number and size during the Mauryan and post Mauryan periods (from 300 B.C. to A.D. 600) in north, as well as in south India. Northern part of the country saw a decline in urban growth during the post-Gupta period from A.D. 600 to about A.D. 1000. Later on, the invasion of Turks followed by the Sultanate rule helped in revival of urban development in India. This phase started around 11th century A.D. and continued with some changes till the fall of the Mughal Empire *i.e.* till the 17th century A.D. The cities which grew and flourished during this period are Ludhiana, Hisar, Bikaner, Jaipur, Jodhpur, Udaipur, Kota, Chittaurgarh, Moradabad, Agra, Jaunpur, Ahmedabad, Indore, Raipur, Aurangabad, Ahmadnagar, Pune, Gulbarga, Bijapur, Vijaynagar, Hyderabad (Golcunda) and Mysore.

2. Cities of British Period. With the arrival of the British East India Company, the nature of urbanization process changed remarkably. The major contributions of the British to the Indian urban scene were : (1) the creation of three metropolitan port cities of Mumbai (Bombay), Kolkata (Calcutta) and Chennai (Madras), (2) creation of a chain of hill stations in the Himalayan region and in South India, (3) the modification of the urban landscape of the existing cities with the introduction of civil lines and cantonments, (4) the introduction of railways and modern industry, and (5) the improvements in urban amenities and administration. Some of the major towns that came up during the British period are :

Hill stations. Shimla, Mussoorie, Dehradun, Almora, Nainital, Darjeeling and Shillong in the north and Ooty, and Kodaikanal in the south.

Port cities. Mumbai, Panaji, Marmagao, Cannanore, Alleppey, Karaikal, Puducherry, Chennai and Kolkata.

Industrial cities. Kanpur, Dhanbad, Jamshedpur, Asansol, Bhadravati.

Transportation cities. Bhusawal, Jalgaon, Siliguri.

Other cities. Sriganganagar, Bhilwara, Ranchi.

It may be mentioned here that all the above cities were not built by the Britishers alone but were the result of efforts made by the Britishers, the Portuguese, the Dutch, the French and the Indians.

It is clear from the above discussion that urbanization in India passed through different phases of history and around the year 1800, India had 16 cities with a population of one lakh or more and about 1,500 towns. About 11 per cent of the total population of the country lived in urban areas at that time. At the time of first Census in 1872, the urban population declined to 8.7 per cent. The number of cities with a population of one lakh or more still remained at 16 and only 43 places had a population of 50,000 or more. Thus major part of the 19th century was marked by stagnation or decline in urbanization in India. This was primarily due to the lack of interest on the part of Britishers and ushering of the *Industrial*

Revolution in England in the latter half of the 18th century. As a matter of fact, the cities of Kolkata, Mumbai and Chennai built by the Britishers showed remarkable growth while most of the pre-British cities desperately declined.

3. Cities of Post-Independence Period. After India got Independence in 1947, a greater need for economic development was felt and efforts were made for the overall growth of the country. This naturally called for establishing new towns and cities and growth of the pre-existing cities. The central and state governments made strenuous efforts to set-up new towns. Many new states were created and such states preferred to have new planned cities as their capitals. Chandigarh is the best example of such cities. Gandhinagar, Itanagar, Dispur, and Bhubaneshwar are other examples of capital cities. The industrial sector also progressed rapidly after Independence and many industrial towns come up after Independence. Iron and Steel Industry gave birth to new towns of Durgapur, Rourkela, Bhilai and Bokaro. Vishakhapatnam and Vijaynagar (Hospet) also grow due to growth of iron and steel industry.

Noonmati, Barauni, Haldia, Digboi, Koyali, etc. are some of the towns which originated or grew because of oil refineries. Pinjore in Haryana owes its origin and growth to Hindustan Machine Tools. Some of the towns and cities have a large number of industries working simultaneously. The list of such towns and cities is very large. However, names of Sindri, Chittaranjan, Nangal, Neyveli, Modinagar, Ghaziabad, Faridabad, Yamunagar, Dalmiapuram, Pimpri, Namrup etc. are worth mentioning. Some of the old towns have grown in due course of time around metropolitan cities and are known as satellite towns. Ghaziabad, Faridabad, Gurgaon, Noida, Bahadurgah, Sonipat around Delhi are some of such satellite towns. Towns of small and medium size have developed in almost all parts of the country.

Urbanisation

The process of society's transformation from a pre-dominantly rural to a predominantly urban population is known as '***urbanisation***'. It includes two things—an increase in the number of people living in urban settlements, and an increase in the percentage of the population engaged in non-agricultural activities, living in such places.

Trends in Urbanisation. Trends in urbanistion in India from 1901 to 2011 are shown in **Fig. 14.18** and are further elaborated in **table 14.5**. This table shows that the number of towns/urban agglomerations increased by more than four times from 1,915 in 1901 to 7,935 in 2011. There was steady increase in number of towns till 1951, but due to more rigorous tests applied in 1961 to determine whether a place qualified to be treated as a town or not, many urban places were declassified and hence, the number of towns declined from 3,035 in 1951 to 2,657 in 1961.

Table 14.5 also shows that the total population living in urban areas as well as the percentage of urban population to total population of India had been gradually increasing since 1901. For example, only 25.94 million people lived in towns till 1911 and by 2011 the urban population of India increased by more than fourteen times to 377.11 million. At the time of the first census after Independence *i.e.* 1951,

TABLE 14.5. Urbanization in India, 1901-2011

Year	Population (in million)		Urbanisation rate (per cent)	No. of cities/ towns	Decadal growth
	Total	Urban			
1901	238.40	25.85	10.84	1915	—
1911	252.09	25.94	10.29	1864	0.35
1921	251.32	28.09	11.18	2018	8.27
1931	278.98	33.46	11.99	2188	19.12
1941	318.66	44.15	13.86	2392	31.97
1951	361.10	62.44	17.29	3035	41.42
1961	439.09	78.94	17.97	2657	26.41
1971	548.23	109.11	19.91	3081	28.23
1981	683.33	159.46	23.34	3981	46.14
1991	846.39	217.55	25.70	4615	37.52
2001	1028.61	286.12	27.82	5161	31.52
2011	1210.19	377.11	31.16	7935	31.80

Notes 1. As the 1981 census was not conducted in Assam, the 1981 total and urban population figures for India include interpolated figures for that state.

2. The 1991 census was not held in Jammu and Kashmir. The 1991 total and urban population figures for India include the interpolated figures for that state.

Source : (*i*) Premi (2012 : 69); (*ii*) Census of India 2011, Series 1, India, *Provisional Population Totals,* Paper 2 of 2011.

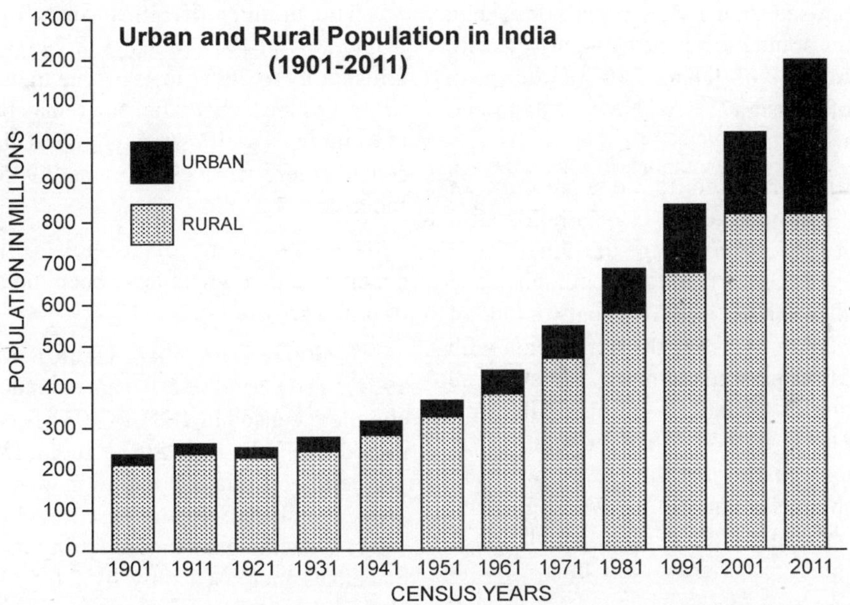

FIG. 14.18. Urban and Rural Population in India (1901-2011)

the population living in towns was 62.44 million. Since the 1961 census, however, the urban population and number of towns/urban agglomerations had increased steadily. There has been more than three times growth in urban population in four decades which increased from 109.11 million in 1971 to 377.11 million in 2011.

The percentage of urban population to total population has also recorded a gradual increase from one decade to another with the only exception of a decline from 10.84 per cent in 1901 to 10.29 per cent in 1911. This decline is largely attributed to the devastating plague epidemic of 1911 which spread mainly in urban areas and brought an exodus of urban population to rural areas. On the other hand, the World War II (1939-45) and partition of the country in 1947 were mainly responsible for a sudden spurt in urban growth during 1931-41 and 1941-51. Partition of the country resulted in a massive influx of refugees into India from West Pakistan (now Pakistan) and East Pakistan (now Bangladesh). Refugees from West Pakistan settled down mainly in Delhi and Punjab (including present Haryana) and to a lesser extent in western Uttar Pradesh, while those from East Pakistan found their way in Kolkata and its suburbs as well as in Assam and Tripura. In all, 14 new towns were built to accommodate refugees migrating from Pakistan. In addition, refugee colonies (new townships) were established near existing cities. As many as 19 places in Punjab, Haryana and Delhi were selected for locating these townships (generally called Model Towns). However, the pace of urbanization slowed down during 1951-61 as a result of declassification of large number of towns. The percentage of urban population to total population increased only by 0.68 per cent from 17.29 percent in 1951 to 17.97 per cent in 1961. As mentioned earlier, the number of towns/urban agglomerations actually declined during this period. Further, it is worth mentioning that the pace of urbanization slowed down during 1981-91. The decennial growth rate had declined from 46.14 per cent in 1981 to 37.52 per cent in 1991. According to 2011 census, 377.11 million persons or 31.16 per cent of the total population of India is termed as ***urban***.

In spite of the above mentioned developments, India is still one of the least urbanised countries of the world. Only 31.16% of India's population (census 2011) is urban whereas at present world's 55% population lives in urban areas. India was previously more urbanised than China where 21% population was urban. But that country has left India far behind with 41% of its population as urban. Even Pakistan has 34% of its total population living in urban areas and is more urbanised than India. India's less than one-third of urban population is no comparison with 79% of the USA, 78% of Japan, 74% of European countries, 73% of Russia, 78% of New Zealand and 91% of Australia.

India is still considered to be a country of villages. But **R. Ramachandran** (1995) holds altogether a different view. According to him, "India is often portrayed as a land of villages and hamlets, nevertheless, in reality, it is equally a land of towns and cities. With over 12,000 settlements with populations of 5,000 persons or more, India has an urban infrastructure of gigantic magnitude". In terms of absolute numbers, India's urban population far exceeds the total population of the U.S.A. which is the third most populous country of the world. Further, it may be mentioned that India's urban population is the world's second largest after that of China.

Table 14.5 shows that there had been gradual growth of urban population before independence. The decennial growth rate was at a very low level of 0.35 per cent in 1901 which shot up to 41.42 per cent in 1951. Thereafter, varying trends have been observed with respect to growth rates. The maximum 46.14 per cent was in 1981 which declined to 31.80 per cent in 2011. However, such percentages can be misleading because the total urban population was much less in the pre-independence period as compared to the post-independent period.

The number of cities/towns increased rapidly from 1,864 in 1911 to 3,035 in 1951. But this number fell sharply to 2,657 in 1961 due to declassification of several towns. Thereafter, there has been steep rise in the number of cities/towns and at the time of 2011 census, there were as many as 7,935 cities/towns in India.

The forgoing discussion brings us to the conclusion that where have been three distinct phases of urban growth in India.

1. Period of Slow Urban Growth (before 1931). The period of 50 years extending from the first complete census in 1881 to 1931 is considered as the period low urban growth in India. In 1881, only 9.3 per cent of India's population was living in urban areas which slowly increased to 11.99 per cent in 1931. Thus the growth rate during the first fifty years of census increased only by 2.69 per cent which is negligibly small as compared to the later increase in growth rates. This slow rate of urban growth is attributed to a large number of factors, but natural disasters like droughts, floods, famines, epidemics had been the major causes. These factors led to high mortality rate and retardation of urban growth.

2. Period of Medium Growth Rate (1931-61). Period of thirty years from 1931 to 1961 is termed as period of medium growth. There was more than two-fold increase in urban population from 33.46 million in 1931 to 78.94 million in 1961 and the percentage

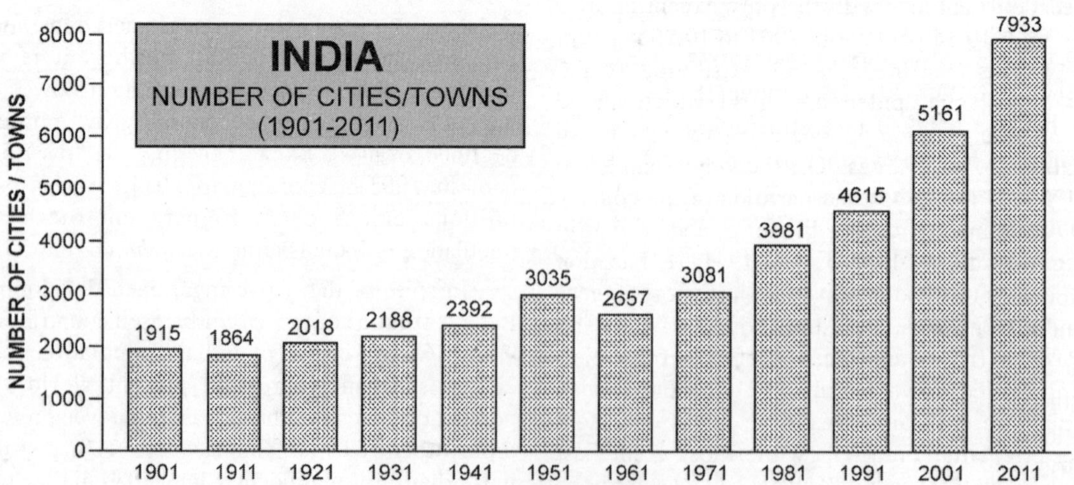

FIG. 14.19. India : Number of Cities/Towns

of urban population to total population also increased from 11.99 to 17.97 during the same period. This was because of major thrust given by the Government of India and many industrial towns and state capital towns came up immediately after Independence. World War II (1939-45) and partition of the Indian subcontinent in 1947 gave a major thrust to urbanisation in India. Several new towns were set-up to accommodate displaced persons from Pakistan. As a result of declassification of several towns in 1961, the number of towns fell from 3035 in 1951 to 2657 in 1961. Therefore *1951-61 is termed as inactive decade from urbanisation point of view.*

3. Period of Rapid Growth (1961-2011). During the period of 50 years from 1961 to 2011, India witnessed rapid growth in urbanisation and urban centres. The urban population saw more than four- fold increase from 78.94 million in 1961 to 377.11 million in 2011 and the percentage of urban population also shot up from 17.67 in 1961 to 31.16 in 2011. The number of towns which fell from 3,035 in 1951 to 2,657 in 1961 sharply increased in to 7,933 in 2011. This is a reflection of India's economic growth history and a major change in the demographic set up of the country as a large number of people are migrating from rural to urban areas in search of livelihood and better quality of life.

Recent Trends and Future Projections

The momentum of rapid urban growth in India observed during 1961-2011 is gaining strength and India is being urbanized at a still rapid pace after 2011. **Figure 14.20** shows that percentage of urban population is growing very fast while percentage of rural population is decreasing in the same ratio. It is estimated that more than fifty per cent of Indian population will live in urban areas by 2050. India will add about 416 million persons to its urban population between 2018 and 2050. This will be the fastest growth in urban population in the world. At 17 per cent of global urban growth, India will be far ahead of China (10%) and Nigeria (8%). Further it is estimated that India will have 7 megacities with 10 million or more population each by 2080.

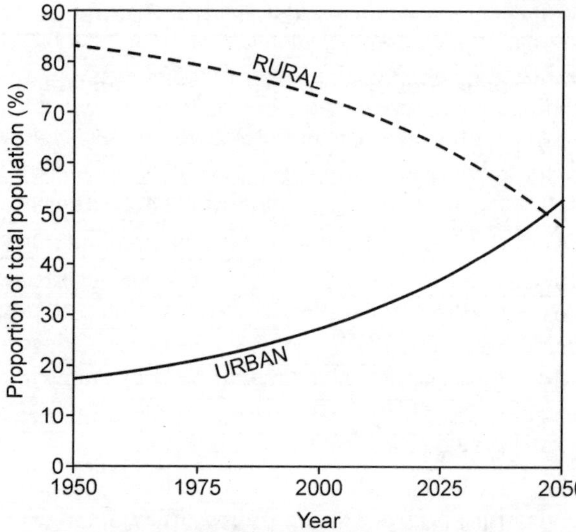

FIG. 14.20. Percentage of urban and rural population in India

Spatial Patterns of Level of Urbanisation

Table 14.6 and **Fig. 14.21** show that the level of urbanisation varies widely among the states. Goa is the most urbanised state where 62.17 per cent of the population lives in urban areas. This is followed by Mizoram where 51.51% of total population of the state lives in towns/cities. Among the larger states Tamil Nadu with 48.45 of its urban population is the most urbanised state. This state is followed by Kerala (47.7%), Maharashtra (45.23%), Gujarat (42.58%), Karnataka (38.57%) and Punjab (37.49%). The other states with percentage of urban population higher than the national average of 31.16 are Haryana, West Bengal and Andhra Pradesh. On the other end of the scale, Himachal is the least urbanised state where only 10.04 per cent of the population lives in urban areas. The other states where level of urbanisation is below 20 per cent are Bihar (11.30%), Assam (14.08%), and Odisha (16.68%). The other states with level of urbanisation below the national average are Jammu and Kashmir, Uttar Pradesh, Sikkim, Arunachal Pradesh, Nagaland, Tripura, Meghalaya, Chhatisgarh and Madhya Pradesh.

Among the Union Territories, Delhi is the most urbanized where 97.50 per cent of the population is classified as urban. Next in order of level of

TABLE 14.6. Spatial patterns of urbanisation in India

State/ UT Code	India/State/Union Territory#	Population 2011			Percentage share of Total Population 2011		Percentage Decadal Growth 2001-2011		
		Total	Rural	Urban	Rural	Urban	Total	Rural	Urban
	INDIA	**1,210,193,422**	**833,087,662**	**377,105,760**	**68.84**	**31.16**	**17.64**	**12.18**	**31.80**
01	Jammu & Kashmir	12,548,926	9,134,820	3,414,106	72.79	27.21	23.71	19.77	35,66
02	Himachal Pradesh	6,856,509	6,167,805	688,704	89.96	10.04	12.81	12.50	15.64
03	Punjab	27,704,236	17,316,800	10,387,436	62.51	37.49	13.73	7.58	25.72
04	Chandigarh#	1,054,686	29,004	1,025,682	2.75	97.25	17.10	–68.51	26.86
05	Uttarakhand	10,116,752	7,025,583	3,091,169	69.45	30.55	19.17	11.34	41.86
06	Haryana	25,353,081	16,531,493	8,821,588	65.21	34.79	19.90	10.00	44.25
07	NCT of Delhi#	16,753,235	419,319	16,333,916	2.50	97.50	20.96	–55.61	26.56
08	Rajasthan	68,621,012	51,540,236	17,080,776	75.11	24.89	21.44	19.05	29.26
09	Uttar Pradesh	199,581,477	155,111,022	44,470,455	77.72	22.28	20.09	17.81	28.75
10	Bihar	103,804,637	92,075,028	11,729,609	88.70	11.30	25.07	23.90	35.11
11	Sikkim	607,688	455,962	151,726	75.03	24.97	12.36	–5.20	153.43
12	Arunachal Pradesh	1,382,611	1,069,165	313,446	77.33	22.67	25.92	22.88	37.55
13	Nagaland	1,980,602	1,406,861	573,741	71.03	28.97	–0.47	–14.59	67.38
14	Manipur	2,721,756	1,899,624	822,132	69.79	30.21	18.65	10,58	42.74
15	Mizoram	1,091,014	529,037	561,977	48.49	51.51	22.78	18.20	27.43
16	Tripura	3,671,032	2,710,051	960,981	73.82	26.18	14.75	2.13	76.08
17	Meghalaya	2,964,007	2,368,971	595,036	79.92	20.08	27.82	27,04	31.03
18	Assam	31,169,272	26,780,516	4,388,756	85.92	14.08	16.93	15.35	27.61
19	West Bengal	91,347,736	62,213,676	29,134,060	68.11	31.89	13.93	7,73	29.90
20	Jharkhand	32,966,238	25,036,946	7,929,292	75.95	24.05	22.34	19.50	32.29
21	Odisha	41,947,358	34,951,234	6,996,124	83,32	16.68	13.97	11.71	26.80
22	Chhattisgarh	25,540,196	19,603,658	5,936,538	76.76	23.24	22.59	17.75	41.83
23	Madhya Pradesh	72,597,565	52,537,899	20,059,666	72.37	27.63	20.30	18.38	25.63
24	Gujarat	60,383,628	34,670,817	25,712,811	57.42	42.58	19.17	9.23	35.83
25	Daman & Diu #	242,911	60,331	182,580	24.84	75.16	53.54	-40.18	218.37
26	Dadra & Nagar Haveli#	342,853	183,024	159,829	53.38	46.62	55.50	7.64	216.73
27	Maharashtra	112,372,972	61,545,441	50,827,531	54.77	45.23	15.99	10.34	23.67
28	Andhra Pradesh (including Telangana)	84,665,533	56,311,788	28,353,745	66.51	33.49	11.10	1.64	36.26
29	Karnataka	61,130,704	37,552,529	23,578,175	61.43	38.57	15.67	7.63	31.27
30	Goa	1,457,723	551,414	906,309	37.83	62.17	8.17	-18.56	35,15
31	Lakshadweep#	64,429	14,121	50,308	21.92	78.08	6,23	-58.08	86.55
32	Kerala	33,387,677	17,455,506	15,932,171	52.28	47.72	4.86	-25.96	92.72
33	Tamil Nadu	72,138,958	37,189,229	34,949,729	51.55	48.45	15,60	6.49	27.16
34	Puducherry#	1,244,464	394,341	850,123	31.69	68.31	27.72	21,07	31.07
35	Andaman & Nicobar Islands#	379,944	244,411	135,533	64.33	35.67	6.68	1.86	16.64

Source : Census of India 2011, Paper 2, Volume 1 of 2011, India Series 1, pp. 2, 8.

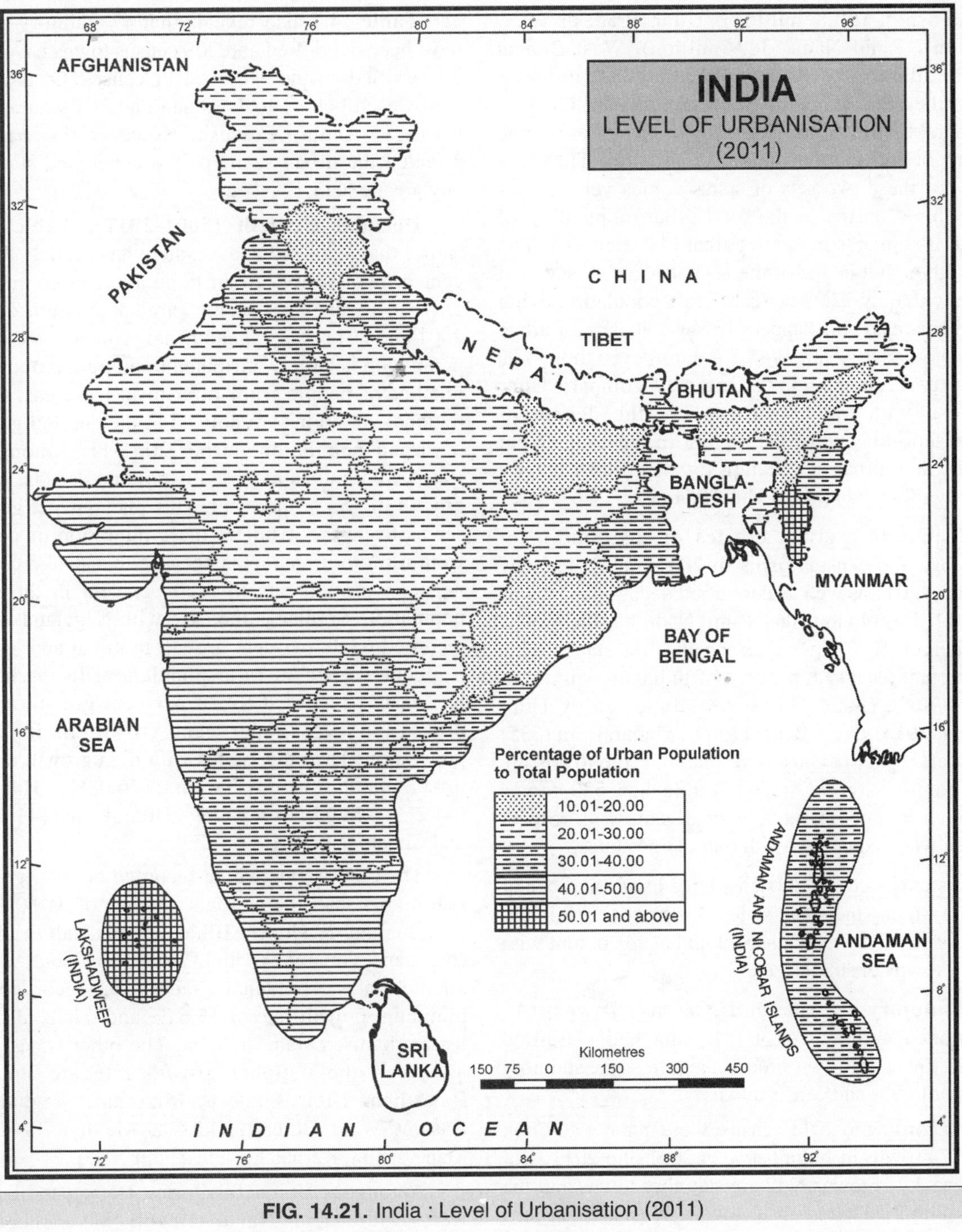

FIG. 14.21. India : Level of Urbanisation (2011)

urbanisation are Chandigarh (97.25%), Lakshadweep (78.08%), Daman and Diu (75.16%), Puducherry (68.31%) and Dadra and Nagar Haveli (46.62%). The Union Territories of Andaman and Nicobar Islands is the least urbanised where only 35.67% of the population lives in urban areas.

Of the total urban population of India, more than one-half lives in just five states. These states are

Maharashtra (50.8 million), Uttar Pradesh (44.5 million), Tamil Nadu (34.9 million), West Bengal (29.1 million) and Andhra Pradesh (28.4 million). The other five states of Madhya Pradesh, Gujarat, Karnataka, Bihar and Rajasthan, account for one fourth of India's total urban population. Thus, the total for these two sets of states comes very high—over three-fourths of the total urban population of India. It is interesting to note that U.P. figures in this list, although it is one of the least urbanised states of India with only 22.28 per cent of its population living in urban areas. This happens because its size of urban population is very large (44.5 million) even though its urban percentage is low. One may compare Uttar Pradesh with Maharashtra where the degree of urbanisation is one of the highest in the country but its absolute urban population (50.8 million) is only slightly more than that of Uttar Pradesh.

Table 14.7 gives an idea of distribution of statutory and census towns in 2001 and 2011, new towns in 2011 as well as denotified/declassified towns in 2011. It is obvious that Tamil Nadu has the largest number of 1097 towns among the big states. This means that over 13.8 per cent of India's towns are in Tamil Nadu alone. This state is followed by Uttar Pradesh (915), West Bengal (909), Maharashtra (535) and Kerala (520). Thus more than half the towns are located in just five states. Kerala has 520 towns although it is comparatively small state with respect to area when compared with other large states.

As expected, small states have lesser number of towns. Among the small states Goa has the maximum number of 70 towns while Sikkim has to content with only 9 towns of small sizes.

Statutory Towns and Census Towns. As mentioned earlier, Census of India had classified Indian towns into two broad categories *viz.* statutory cities or towns and census towns.

According to 2011 census, there is a net addition of 2,774 towns in India between 2001 and 2011. Out of these 2,774 towns, 242 are statutory towns and the remaining 2,532 are census towns. West Bengal tops the list with 537 towns followed by Kerala (362), Tamil Nadu (272) and Uttar Pradesh (208). The recognition of these new towns has resulted in the accelerated rate of urbanisation and have changed the urban map of some states substantially.

Table 14.7 also reveals that 47 statutory towns have been denotified and 208 census towns have been declassified on the eve of 2011 census. Of these 27 towns each belong to Jharkhand and Gujarat and 24 towns to Andhra Pradesh. However this has not affected the urban scene of the concerned states to any great extent.

Decadal Growth (2001-2011). Table 14.6 shows that at all India level, there has been 31.80 per cent growth in the urban population in contrast to only 12.18 per cent in the rural population during 2001 and 2011. This phenomenal growth in the urban population is largely attributed to migration of a large number of people from rural to urban areas. Great regional variations are noticed in the decadal growth of urban population during 2001-11. Among the states, Sikkim has recorded maximum decadal urban growth of 153.43 per cent. This phenomenal growth seems to be the result of mass migration of people from rural to urban areas because this state recorded negative growth rate of 5.20 per cent in the rural population. Similar is the case with Nagaland which recorded 67.38 per cent growth in urban and –14.59 per cent growth in rural population. In the south, Kerala has recorded 92.72 per cent growth in the urban and –25.46 per cent growth in its rural population. The other states with high growth rate in urban population are Tripura (76.08%), Haryana (44.25%), Manipur (42.74%), Uttarakhand (41.86%) and Chhattisgarh (41.83%).

The other states which recorded decadal growth rate higher than the national average of 31.80% are Jammu and Kashmir, Bihar, Arunachal Pradesh, Jharkhand, Gujarat, Andhra Pradesh and Goa. Among all the states, Himachal Pradesh has recorded the minimum growth rate of 15.64% and this is also the least urbanised state in India. The other states with less than the national growth rate are Punjab, Rajasthan, Uttar Pradesh, Mizoram, Meghalaya, Assam, West Bengal, Odisha, Madhya Pradesh, Maharashtra, Karnataka, and Tamil Nadu.

Among the Union Territories, Daman and Diu as well Dadra and Nagar Haveli have recorded exceptionally high growth of 218.37 and 216.73% respectively. This is largely due to large scale migration of the people from rural to urban areas both from within these territories and from the surrounding areas. Lakshadweep has also recorded very high

TABLE 14.7. Statewise distribution of 2001 and 2011 censuses, new towns of 2011 census and denotifies/declassified towns in 2011

State/ Union Territories	2001 Number of Towns			2011 Number of Towns			New Towns in 2011 census		2011 towns minus new towns		Denotified/ declassified towns		
	Total	S	C	Total	S	C	S	C	S	C	Total	S	C
India	**5,161**	**3,799**	**1,362**	**7,935**	**4,041**	**3,894**	**289**	**2,740**	**3,752**	**1,154**	**255**	**47**	**208**
Big States													
Andhra Pradesh (including Telangana)	210	117	93	353	125	228	19	159	106	69	35	11	24
Assam	125	80	45	214	88	126	8	86	80	40	5	0	5
Bihar	130	125	5	199	139	60	17	56	122	4	4	3	1
Chhattisgarh	97	75	22	182	168	14	93	10	75	4	18	0	18
Gujarat	242	168	74	348	195	153	41	106	154	47	41	14	27
Haryana	106	84	22	154	80	74	11	52	69	22	15	15	0
Himachal Pradesh	57	56	1	59	56	3	0	3	56	0	1	0	1
Jammu & Kashmir	75	72	3	122	86	36	14	33	72	3	0	0	0
Jharkhand	152	44	108	228	40	188	2	107	38	81	33	6	27
Karnataka	270	226	44	347	220	127	4	94	216	33	21	10	11
Kerala	159	60	99	520	59	461	0	362	59	0	1	0	1
Madhya Pradesh	394	339	55	476	364	112	26	64	338	48	8	1	7
Maharashtra	378	251	127	535	256	279	7	171	249	108	21	2	19
Odisha	138	107	31	223	107	116	0	86	107	30	1	0	1
Punjab	157	139	18	217	143	74	7	60	136	14	7	3	4
Rajasthan	222	184	38	297	185	112	0	80	184	32	6	0	6
Tamil Nadu	832	721	111	1097	721	376	0	272	721	0	1	–16	17
Uttar Pradesh	704	638	66	915	648	267	9	208	639	59	6	–1	7
Uttarakhand	86	74	12	116	74	42	0	31	74	11	1	0	1
West Bengal	375	123	252	909	129	780	5	537	124	243	8	–1	9
Small States													
Arunachal Pradesh	17	0	17	27	26	1	10	1	16	0	1	–16	17
Goa	44	14	30	70	14	56	0	26	14	30	0	0	0
Manipur	33	28	5	51	28	23	0	19	28	4	1	0	1
Meghalaya	16	10	6	22	10	12	0	6	10	6	0	0	0
Mizoram	22	22	0	23	23	0	1		22	0	0	0	0
Nagaland	9	8	1	26	19	7	10	7	9	0	0	–1	1
Sikkim	9	8	1	9	8	1	0	1	8	0	1	0	1
Tripura	23	13	10	42	16	26	3	23	13	3	7	0	7
Union Territories													
A & N Islands	3	1	2	5	1	4	1	2	0	2	1	1	0
Chandigarh	1	1	0	6	1	5	0	5	1	0	0	0	0
D. & N. Haveli	2	0	2	6	1	5	0	5	1	0	1	–1	2
Daman and Diu	2	2	0	8	2	6	0	6	2	0	0	0	0
Lakshadweep	3	0	3	6	0	6	0	3	0	3	0	0	0
NCT of Delhi	62	3	59	113	3	110	0	55	3	55	4	0	4
Puducherry	6	6	0	10	6	4	0	4	6	0	0	0	0

Note: S—Statutory. C—Census.

Source : Mahendra Premi and Dipendra Nath Dass. Population of India 2011, B.R. Publishing Corporation, Delhi (2012) pp. 93-95.

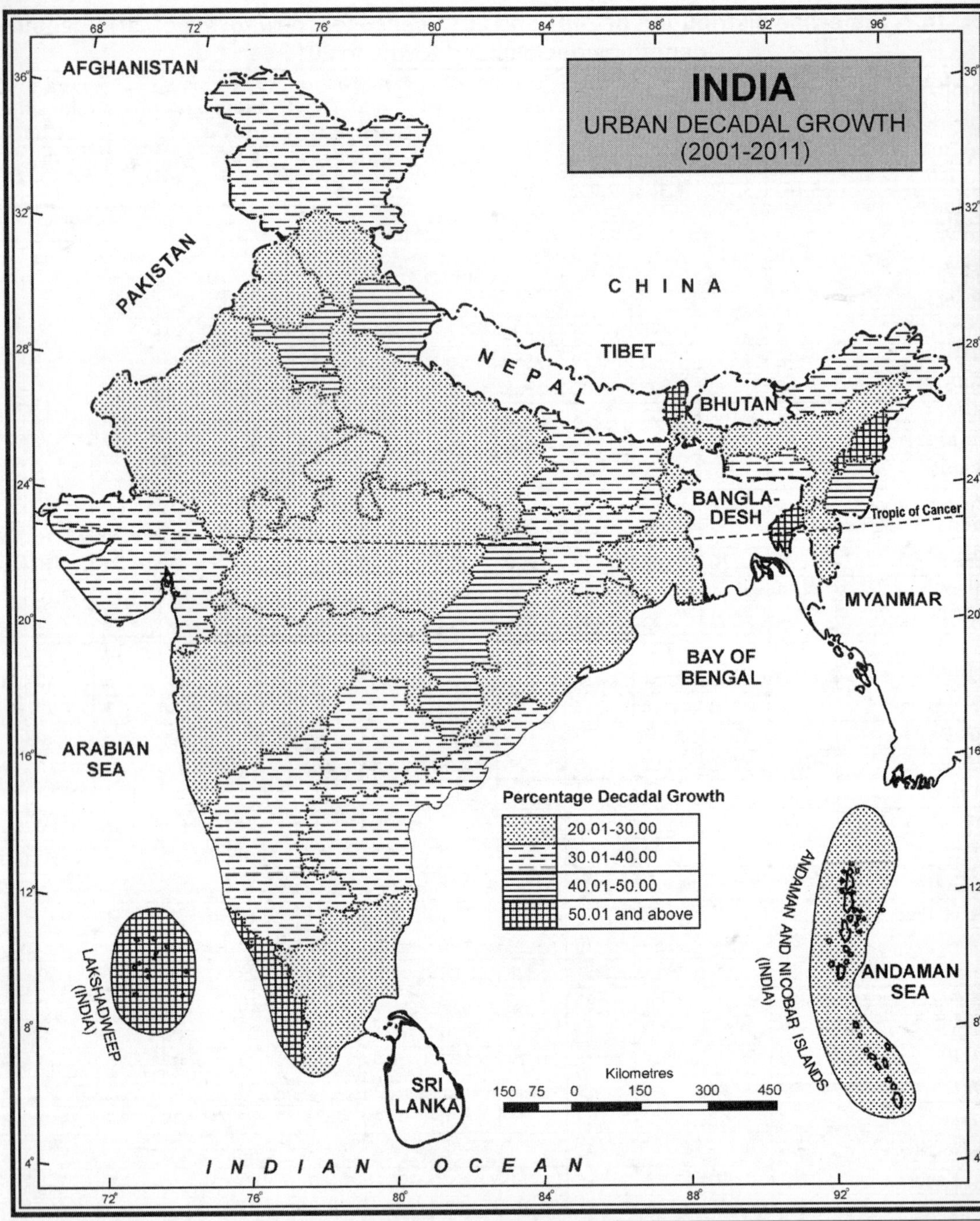

FIG. 14.22. Urban Decadal Growth

growth rate of 86.55 per cent. Puducherry had recorded below national average growth rate. In the north, Chandigarh and Delhi have recorded low percentage of growth rates of 26.86% and 26.56% respectively which may be due to larger number of people already living in these union territories. Andaman and Nicobar Islands have recorded the lowest urban growth rate of 16.64 per cent. **Fig. 14.22** shows the areal variations in the urban decadal growth rate.

METROPOLIZATION (OR METROPOLITANIZATION)

Metropolization refers to the growth of metropolitan centres rooted in industrial and tertiary economic base. A metropolis is a distinct form of settlement, charateristically with sprawling of its built-up area and includes its inter-dependent nearby villages and even towns. The metropolitan centres are a class by themselves, characterized by large scale consumption and a large quantum of flow of people, goods, services and information (Prakasa Rao, 1983). According to **R. Ramchandran** (1995), "Metropolization is essentially a product of the centralization of administrative, political and economic forces in the country at the national and state capitals. Metropolization is also a product of intense interaction between cities and the integration of national economy and urban centres into a viable inter-dependent system". The pace of metropolization depends upon the rate of direct migration of rural folk to metropolitan cities as well as from the smaller towns. In fact migration constitutes the very foundation of the process of urbanization and is recognised as the chief mechanism by which urban centres continue to grow. One view is that urbanization stops when migration to urban centres stops (Prakasa Rao, 1983). However, natural growth cannot be overlooked. Sixty per cent of the urban growth during 1981-91 is attributed to the factor of natural increase (Premi, 1991).

The Census of India has defined metropolitan *as an urban agglomeration/city having a population of one million and above*. These are also called the *million plus cities*. Metropolization in India is primarily a phenomenon of the post-independence era. In 1901 Kolkata (Calcutta) was the only metropolitan city in the whole of India. In 1911, Mumbai (Bombay) joined Kolkata. The number of such cities remained at 2 for another three decades till 1941. But at the time of the first census after independence in 1951 their number increased to 5 with population of Delhi, Chennai (Madras) and Hyderabad crossing one million mark. In the next 30 years from 1951 to 1981, the cities which joined the million plus group of cities were Ahmedabad and Bangalore in 1961, Kanpur and Pune (Poona) in 1971 and Lucknow, Nagpur and Jaipur in 1981. Thus the number of million plus cities rose to 7 in 1961, 9 in 1971 and 12 in 1981. The decade 1981-91 recorded a phenomenal growth and the number of such cities rose to 23 in 1991. The growth of metropolitan cities was further accelerated and the number of metropolitan cities stood at 35 in 2001 and 53 in 2011 (**Table 14.8**). Obviously these cities are characterised

TABLE 14.8. Urban Agglomerations/Cities having Population of more than a Million, 1901-2011

Census Year	Number of metropolitan cities	Population (in million)	Average Population per UA/city (in million)	Population of million plus urban agglomerations/cities as percentages of India's urban population	Decadal variation of population (per cent)
1901	1	1.51	1..51	5.84	—
1911	2	2.76	1.38	10.65	83.02
1921	2	3.13	1.56	11.14	13.24
1931	2	3.41	1.70	10.18	8.86
1941	2	5.31	2.65	12.02	55.79
1951	5	11.75	2.35	18.81	121.32
1961	7	18.10	2.59	22.93	54.10
1971	9	27.83	3.09	25.51	53.75
1981	12	42.14	3.51	26.41	51.35
1991	23	70.66	3.07	32.54	67.76
2001	35	107.88	3.08	37.81	58.83
2011	53	—	—	—	—

by high concentration of population. Together, these cities provided home to 107.88 million people in 2001 which accounted for about 37.81 per cent of the total urban population of India. The average population per urban agglomeration/city of million plus category has almost doubled from 1.51 million in 1901 to 3.08 million in 2001. The decadal variation had also changed largely from 8.86 per cent in 1931 to 121.32 per cent in 1951. The exceptional increase in 1951 is attributed to influx of refugees from West Pakistan (now Pakistan) and East Pakistan (now Bangladesh) to metropolitan cities of Mumbai, Kolkata and Delhi.

According to 2011 Census, India has 53 metropolitan cities. Details of these 53 metropolitan cities are given in **table 14.9**.

According to projections made by the Census Department, India will have 87 metropolitan cities in 2039 in which 255 million people will be living. In 2011, India's top ten metropolitan cities had 93 million people and this figure is likely to be 149 million in 2039.

MORPHOLOGY OF INDIAN CITIES

The term 'morphology' has been derived from two Greek words which literally means, "to describe the shape or layout." According to **L.D. Stamp** (1961), "Morphology is the science of form and structure and development which influences the form. **R.E. Murphy** (1966) suggests, "In morphological studies the form elements of the urban area are of chief concern; the arrangement of streets and rail-roads, the form of building, infact, the whole urban landscape." These introductory definitions of morphology help us to conclude that morphology includes three major constituents : (1) internal layout and external shape, (2) internal functional structure or land use and (3) demographic structure. Thus there are three forms of urban morphology. Prof. Enayat Ahmed (1957) has rightly described the urban morphology when he says "from the view point of morphology we are mainly concerned with the ground plan built and aspect of town...... Plan may be viewed from two angles, many concentrate on the internal plan pattern of the town by which is meant the relative arrangements of streets and built-ground. We may also concentrate on the external outline of the plan." In short urban morphology comprises the

TABLE 14.9. India : Population of Million plus Cities/Urban Agglomeration, 2011

Sl. No.	Name of Urban Agglomeration	Population
1.	Srinagar UA	1,273,312
2.	Ludhiana (M Corpl)	1,613,878
3.	Amritsar UA	1,183,705
4.	Chandigarh UA	1,025,682
5.	Faridabad (M Corp.)	1,404,653
6.	Delhi UA	16,314,838
7.	Jaipur (M Corp.)	3,073,350
8.	Jodhpur UA	1,137,815
9.	Kota (M Corp.)	1,001,365
10.	Kanpur UA	2,920,067
11.	Lucknow UA	2,901,474
12.	Ghaziabad UA	2,358,525
13.	Agra UA	1,746,467
14.	Varanasi UA	1,435,113
15.	Meerut UA	1,424,908
16.	Allahabad UA	1,216,719
17.	Patna UA	2,046,652
18.	Kolkata UA	14,112,536
19.	Asansol UA	1,243,008
20.	Jamshedpur UA	1,337,131
21.	Dhanbad UA	1,195,298
22.	Ranchi UA	1,126,741
23.	Raipur UA	1,122,555
24.	Durg-Bhillainagar UA	1,064,007
25.	Indore UA	2,167,447
26.	Bhopal UA	1,883,381
27.	Jabalpur UA	1,267,564
28.	Gwalior UA	1,101,981
29.	Ahmadabad UA	6,352,254
30.	Surat UA	4,585,367
31.	Vadodara UA	1,817,191
32.	Rajkot UA	1,390,933
33.	Greater Mumbai UA	18,414,288
34.	Pune UA	5,049,968
35.	Nagpur UA	2,497,777
36.	Nashik UA	1,562,769
37.	Vasal Virar City (M Corp.)	1,221,233
38.	Aurangabad UA	1,189,376

Table Contd.

Sl. No.	Name of Urban Agglomeration	Population
39.	Hyderabad UA	7,749,334
40.	GVMC (MC)	1,730,320
41.	Vijayawada UA	1,491,202
42.	Bangalore UA	8,499,399
43.	Kochi UA	2,117,990
44.	Kozhikode UA	2,030,519
45.	Thrissur UA	1,854,783
46.	Malappuram UA	1,698,645
47.	Thiruvananthapuram UA	1,687,406
48.	Kannur UA	1,642,892
49.	Kollam UA	1,110,005
50.	Chennai UA	8,696,010
51.	Coimbatore UA	2,151,466
52.	Madurai UA	1,462,420
53.	Tiruchirappalli UA	1,021,717

Source : Census of India, 2011, Provisional (Website http://www.censusindia.gov.in)

structure of the city, residential areas, industrial estates, central business district (main market), open space, down town (old city), railways, roads and streets, parks and play grounds, cultural centres, water bodies, etc. Prof. **Enayat Ahmed** has attached great importance to morphology of towns. According to him "The morphological aspect is highly important and diagnostic. As geomorphic landscape is function of structure, process and stage, the morphology of a town is largely the function of site factors, historical viscissitutes, impact of successive cultures and economic settings and development".

Morphology of Indian towns and cities has changed considerably with reference to time and space. In fact no two towns have same morphology and morphology of the same town changes with the passage of time. For sake of simplicity, the morphological study of Indian towns can historically be described in the following three phases :

1. Morphology of Ancient Towns. Most of the ancient towns of India are associated with the Indus Valley Civilization, ample proof of which is available from the excavations of Harappa and Mohenjodaro. Towns of Indus valley civilization has gridiron type of morphology which is also known as chequer board morphology. In this type of morphology, the main streets are parallel to one another and meet at right angles. They are generally oriented north-south and east-west direction and the town is generally of rectangular shape. These roads divide the town into different rectangular blocks of almost equal sizes and buildings are constructed on both sides of the roads. The rectangular plan of the towns was designed to meet the needs of some sort of ceremony—religious or secular or both in which the terrace or terraces played a significant role, and to which the professional access was required. This type of morphology can be seen in towns like Ayodhya, Kanchipuram, Lumbini, Madurai, Mathura, Rajgir, Varanasi and Ujjain.

This type of morphology has its merits and demerits. The greater merit is that all the roads receive almost equal amount of air and sun-shine. However, since the roads are parallel to and intersect each other at right angles, it requires lots of time and labour to reach from one place to another. Visibility at the road crossings is limited and accidents occur frequently. Sometimes diagonals are used across the rectangles. This provision increases visibility but the shape of buildings is distorted and they often become triangular in shape.

2. Morphology of the Medieval Towns. In majority of the medieval towns, the grid-pattern of the ancient towns was adopted in which forts, city walls, markets, religious places to water bodies and *chowks* were added. This morphology is found in the cities of Agra, Ahmedabad, Ahmadnagar, Aligarh, Aurangabad, Bijapur, Bulandshahr, Delhi, Chittaurgarh, etc.

3. Morphology of Modern Towns. Rapid pace of industrialisation and urbanisation in India after Independence gave birth to a large number of new towns and changed the landscape of several pre-existing towns. Large scale migration from rural to urban areas in search of livelihood drastically changed the morphology of the Indian towns. Many old cities became over-crowded and rich people from old cities constructed their houses in the peripheral areas to escape over-crowding in the centre and to enjoy the open environment which the peripheries offered them. The negative impact of rural-urban migration was the growth of slum and squatter colonies in different parts of the cities and disparity between the rich and the poor became quite conspicuous. The British rulers built many towns reflecting British architect and after Independence, these towns grew according to Indian architecture. Thus, these towns reflect a mixture of British and Indian architecture. Unplanned and unchecked growth of some cities has led to degradation of urban life and different parts of such cities are marked by narrow tortuous lanes often with blind curves.

The most outstanding feature of cities built by the Britishers is the Central Business District (CBD). CBD is also known as ***Central Traffic District***, ***Down Town Commercial Core***, ***City Centre***, etc. It lies at the heart of the city where one finds the greatest concentration of offices and retail stores reflected in the city's highest land values and tallest buildings. This is also the chief focus of transport routes. Different floors of the multi-storeyed buildings are used for different purposes. The ground floor is invariably used by the retailers while other floors may be used by wholesalers, offices and even for residential purposes. There is almost a complete absence of independent residential buildings and permanent population. This *area throbs with activity during the day time and is a dead heart at night.* Usually the CBD is at the centre of the city but growth of city in due course of time may result in location of CBD at a place other than its centre. Thus CBD should not necessarily to be located at the centre of the city but should be the business centre of the city. Most of the million cities do have more than one CBD. For example, Connaught Place, Karol Bagh and Chandni Chowk can be treated as three different CBDs of Delhi.

CHIEF CHARACTERISTICS OF CBD

- Dominance of trade
- Concentration of transportation routes
- Concentration of offices
- Multi-storey buildings
- Different use of each floor of multi-storey buildings
- High land values
- High intensity of land use
- Absence of manufacturing industries, residential buildings and permanent population

The British rulers built in or near the pre-existing towns several features such as cantonments, civil lines, educational institution, hospitals, railway colonies, barracks, big bungalows, residential quarters, etc. These were well planed and equipped with modern facilities available at that time. Such structures clearly left the impact of the British architecture. Since these towns have both the Indian and the British architecture, a greater variety is noticed in their morphology.

Morphology of Pryagraj (Allahabad). Located at the confluence of the holy rivers of the Ganga and the Yamuna in Uttar Pradesh, this cultural, educational and administrative urban centre was known as ***Prayag*** in the ancient times and this name is still popular with many social and religious organisations. The Government of Uttar Pradesh has official restored its old name *i.e.* Pryagraj in 2018. Prayag became an important city in the third century B.C. The construction of the present city was initiated by Akbar in the last quarter of the 16th century. However, the actual development of the present city began in the 19th century during the British rule. This city progressed very fast when it became capital of United Provinces of Agra and Awadh in 1858. Establishment of High Court in 1868 further boosted its growth. New colonies came up after Independence and the city expanded considerably. It now covers an area of more than 85 sq km and is the home of more than one million people.

Following are some of the important functional zones of this city (**Fig. 14.23**).

1. Commercial Area. Locally known as ***Chowk Area,*** the commercial area of Allahabad is located around clock tower (Ghantaghar) along the Grand Trunk (G.T.) Road. This area may be called the Central Business District (CBD) of Allahabad. Roads radiate in all directions from the Chowk area. Shops and business establishments are located along the roads and in the area between the roads. Civil Lines business area is about 3 km north-west of Chowk area. Another business area, known as ***Katra***, lies at a distance of about 3 km north-east of Civil Lines. The expansion of city has given birth to additional business areas like Phaphaman grain market and Transport Nagar. Business areas on a small scale have come up in the residential areas such as Allahpur, Teliarganj, Rajapur, etc.

2. Residential Area. The main residential area of Allahabad is in the old city. This is an over-crowded area with very high density of population. On an average, more than 10 inhabitants live in one house

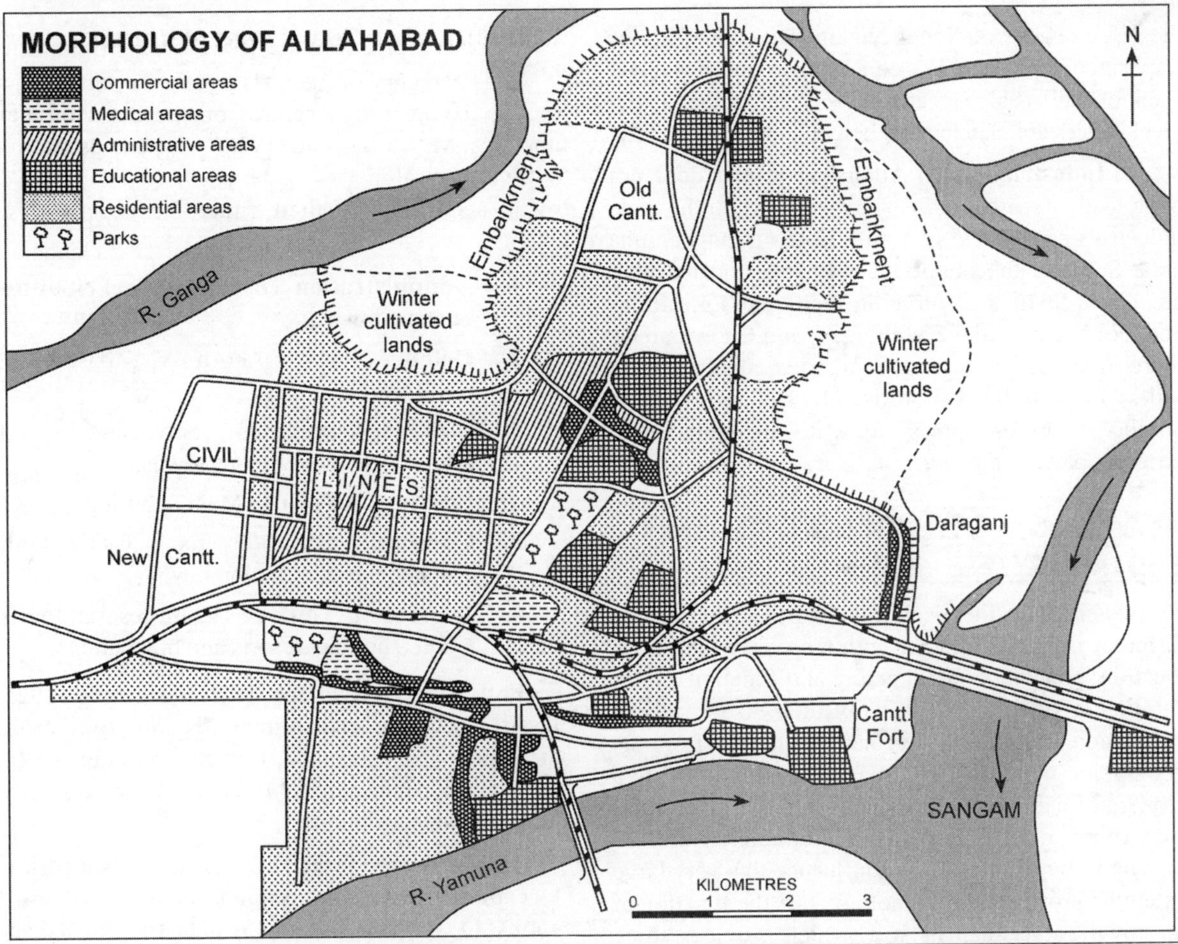

FIG. 14.23. Morphology of Allahabad

and more than 40 per cent households live in one room only. Some new colonies have been built, particularly between the Ganga and the Yamuna rivers. Of these, Civil Lines is a planned area which provides most of the facilities required for a reasonably good living. However, most of the new colonies have been built in an unplanned and haphazard manner and lack the basic facilities of markets, schools, hospitals, play grounds, parks, parking areas for vehicles, wide roads and open space, community centres, etc.

3. Educational Area. Allahabad holds a position of pride in the field of education, and educational institutes cover a substantial part of urban land-use of the city. Almost all important educational institutes are located in the eastern part of the city extending from Phamphamau in the north to Naini in the south. The major institutions located in this area University of Allahabad, Engineering College, I.E.R.T. Medical College, Purushotam Das Tandon Open University, Agricultural Institute, Indian Institute of Information Technology and a number of degree colleges, intermediate colleges, training institutes and high schools.

4. Administrative Area. Allahabad is an important centre of administration where several offices of the state and central government are located. The main administrative area lies in the western part of the Civil Lines. This area hosts the Allahabad High Court, Board of Revenue, Offices of the Accountant General, Director of Education, Intermediate Board, Higher Education Commission, etc. Some of the administrative areas are in the northern and eastern sides of old city also. Besides, these are offices of Nagar Nigam, Divisional Police Superintendent, P.W.D., etc. Main branch of State Bank of India is also located here. Some of the district level offices are also located here.

5. Industrial Area. Allahabad proper does not have well defined industrial area although some industries are located in Naini, in the trans-Yamuna area. Some of the important industries located here are glass, textiles. Ayurvedic medicines, machines and tool making. In the Multhiganj and Chowk areas, there are food processing industries such as pulse mills, flour mills, oil mills, etc. A considerable number of printing press industries are scattered in different parts of the city.

FUNCTIONAL CLASSIFICATION OF INDIAN CITIES

No two cities are identical in the world. All cities are different from one another with respect to their size, location, population, architecture and function. Cities can be classified on the basis of their size, location, population, morphology, formation, etc. But from the geographic point of view, functions performed by a city are of paramount important for classifying cities. According to **R.E. Dickinson**, "Functions are the driving force of city life, and influence to a very large extent its growth and morphology. It is the function of the town, rather than its size, which provides some basis for comparison.

Classifying cities on the basis of their functions is a very difficult task because people performing different functions live in almost all the cities. Cities are often classified on the basis of the most important function performed by them. For example, Chandigarh is an administrative city, while Gurgaon is an industrial city and Kurukshetra is a city of religion, culture and education. But this type of classification is not based on some scientific method, rather it is based on general observation. There is need to apply some scientific methods so that a clear picture of functional classification of cities emerges. Methods used to classify towns on the basis of functions performed by them are categorised into following three classes :

1. Empirical or Qualitative Methods
2. Empirical-cum-Statistical Methods
3. Pure Statistical or Quantitative Methods

1. Empirical or Qualitative Methods

This category of methods denote the initial stage for classifying urban centres on the basis of their functions. M. Aurousseau (1921) was perhaps the first to use empirical method and recognised six classes of towns according to their functions described as under :

1. **Administration Towns :** Capital cities and revenue towns.
2. **Defence Towns :** Fortress, garrison, and naval towns.
3. **Cultural Towns :** University, cathedral, art, pilgrimage and religious towns.
4. **Production Towns** : Manufacturing towns.
5. **Communication Towns :** Collection, transfer and distribution towns.
6. **Recreation Towns :** Health resorts, tourist resorts, holiday resorts and hill stations.

Following Aurousseau, R.D. Mackenzie (1925) classified American communities into four broad categories : (*i*) primary service community, (*ii*) commercial community, (*iii*) industrial community, and (*iv*) other communities.

James divided Indian towns into six categories : (1) capital towns, (2) religious centres, (3) army centres, (4) market centres, (5) industrial towns, and (6) port towns. **V. Nath** divided Indian towns into eight categories in his essay on "Urbanization in India with special reference to Growth of Cities." But this division was based on the economic aspect of towns.

2. Empirical-cum-Statistical Methods

Also known as Semi-quantitative methods, these methods are mixture of empirical or qualitative methods on one hand and pure statistical or quantitative methods on the other hand. As such, these methods symbolize a transition phase between the earlier empirical and later quantitative methods.

Ogburn (1937), presented a functional classification of towns on the basis of percentage of population engaged in trade, manufacturing and transportation. **Chancy D. Harris** (1943) presented "A Functional Classification of Cities of the United States" in which he recognised nine functions performed by towns. He argued that all cities were more or less multi-functional and based his classification of the cities on the function of the greatest importance. Each of the nine principal types was designated by a letter : (*i*) Manufacturing cities (M′ and M), (*ii*) Retail centres (R), (*iii*) Diversified cities (D), (*iv*) Wholesale centres (W), (*v*) Transport centres (T), (*vi*) Mining towns (S), (*vii*) University towns (E), (*viii*) Resort and retirement towns (X), (*ix*) Other towns including political (P). He fixed varied minimum percentages of workers for different functional groups. For example, a city was considered ***manufacturing*** if atleast 60 per cent of the total employees are engaged in manufacturing. For retail centre at least 50 per cent of the total employment should be in retail business. If the employment is less than 60% in manufacturing, 20% in wholesale and 50% in retail, the town is considered to be multi-functional.

Harris method was followed with some modifications by **Kneedler Ohlson** (1945), Victor Jones (1954) and Hart (1955).

The greatest disadvantage of these methods is that the parameters adopted are arbitrary and are not applicable in all the circumstances.

3. Pure Statistical or Quantitative Methods

As their name indicates, these methods are dependent purely on statistical analysis and are used for all types of cities. Statistical methods are fixed, reliable and easy to understand and have become very popular by virtue of these qualities. Although attempts to use these methods were made earlier also, the first popular attempt was made by **L.L. Pownall** in 1953 for studying the functions of New Zealand towns. He classified towns into six functional groups : (*i*) manufacturing, (*ii*) building and construction, (*iii*) primary industry, (*iv*) transport and communications, (*v*) distribution and financial, and (*vi*) personal services.

The most popular scheme was put by **Howard J. Nelson** when he presented his scheme, "A Service Classification of American Cities" in 1955. He based his classification on the data derived from the 1950 census and selected 897 towns each with population over 10,000. He recognised 24 different activities and divided them into 9 major functional groups. The percentage of the total labour force in each of the 897 cities was calculated and thus, arithmetic mean was determined for each functional group for all the cities together. The average employment in different groups ranged from 1.62 per cent for mining to 27.07 per cent for manufacturing. He selected **standard deviation (S.D.)** and calculated S.D. from the mean for each of the nine functional groups. In this way cities were grouped in their appropriate categories : Nine functions recognised were (*i*) mining, (*ii*) manufacturing, (*iii*) transport and communication, (*iv*) wholesale trade, (*v*) retail trade, (*vi*) finance, (*vii*) insurance and real estate, (*viii*) personal services and (*ix*) professional services and public administration.

Contribution of Indian Geographers

Indian geographers have made their own contribution in classifying Indian cities. Some of the outstanding contributions are briefly described as under :

V.A. Janki (1954) in essay on "Functional Classification of Urban Settlements in Kerala" divided urban settlements into five categories : (*i*) Administration Towns, (*ii*) Commercial and Industrial Towns, (*iii*) Trade Towns (centre for collection and distribution of agricultural products), (*iv*) Religious Towns and (*v*) Plantation Towns. Following Nelson **Kashi Nath Singh** (1959) presented "Functional Classification of Towns of U.P. in which many towns remained unclassified due to shortcomings in the methodology. **Amrit Lal** (1959) adopted a new technique while presenting his essay on "Some Aspects of Functional Classification of Cities and Proposed Scheme of Classifying Indian Cities." He used location quotient (LQ) method for functional classification of Class I cities of India and concluded that most of Class I cities of India are multifunctional. **V.L.S. Prakasa Rao** (1965) used "Least Square Regression" method for classifying towns. **A. Ramesh** (1965) adopted Nelson's method for classifying towns of Tamil Nadu. **Quazi Ahmed** (1965) used 62 variables for classifying 102 Indian cities according to functions performed by them. **S.M. Raffullah** (1965) adopted a new technique and methodology and suggested that cities perform not one function but many functions simultaneously. Therefore, cities should be classified not on the basis of just one function, rather the classification should be based on a number of functions performed by the cities. He suggested that functions can be categorised into a number of classes : (*i*) Monominal function, (*ii*) Binominal function, (*iii*) Trinominal function, (*iv*) Quadrinominal function, and (*v*) Polynominal function. **M.P. Saxena** (1966) presented a new formula for classifying towns of the Yamuna-Hindon region. He based his classification on Census 1961 data and divided the functions into nine classes and presented four combinations of those nine classes. They are: (*i*) primary, (*ii*) trade, (*iii*) industry and (*iv*) miscellaneous. **Onkar Singh** (1967) also used 1961 Census data and expressed the view that urban functions are : (*i*) isolated and (*ii*) integrated. He used the functional index and specialisation index as explained below :

$$\text{Functional Index} = \frac{P \times P}{MP}$$

$$\text{Specialisation Index } \sum \left[\frac{P \times P}{MP}\right] \div 100$$

where P = No. of workers in a particular function in the city

MP = No. of workers in a particular function in all the cities.

Om Prakash Singh (1968) followed Nelson's method for classification of towns of Uttar Pradesh. He introduced following two formulae :

(*i*) Functional Specialization Index or F.S.I.

$$= \frac{Cs \times 100}{Rs} \times Mf$$

(*ii*) Functional Centrality Index or F.C.I.

$$= \frac{Cf \times 100}{Rf}$$

where Cs = Total population of the town

Mf = Mean functional percentage of the function of all towns in the region

Rs = Total population of the region

Rf = Population of the region engaged in the same function.

Ashok Mitra (1971) also used 1961 Census data for classifying towns of India. Census of India gives details of population engaged in nine different types of functions. Ashok Mitra excluded first two functions in his classification of towns because they are concerned with agriculture and agricultural labourers which are purely rural functions. He considered the remaining seven functions *viz.* (3) lumbering, fishing, mining, plantation, (4) house-hold industry, (5) manufacturing industry, (6) construction, (7) trade and commerce, (8) transport and communication, and (9) service. Considering the number of people engaged in these seven functions to be 100, calculation for people engage in each function was made. Based on these calculations he classified 2462 towns/cities of India. According to his findings 83 were mining and plantation towns, 496 were household industry towns, 374 were manufacturing towns, 24 were construction towns, 244 were trade and commerce towns, 78 were transport towns and 1163 were service towns.

V.P.P. Sinha (1976) recognised the significance of regional elements while classifying the towns of the Chotanagpur Plateau. **R.N. Singh** and **Saheb Deen** (1976) successfully removed the shortcomings of Nelson's method in their essay entitled "A Functional Typology of Urban Centres of Eastern Uttar Pradesh." **Ram Pyare** (1980) followed more or less similar methodology in his article entitled "Functional Classification of Towns of Bundelkhand." Like Ashok Mitra he based his classification on nine functions suggested by Census of India and used the Functional Specialization Index (Si) = 100 *e*/*w* where *e* is the number of people engaged in one function and *w* is the number of total workers in the town. Based on his calculations, he classified 37 towns of Bundelkhand. He used indices for different classes of towns explained as follows : A = Agriculture, M = Mining, H = House-hold industry, Mf = Manufacturing, C = Construction, Tc = Transport and communication, T = Trade.

On the basis of different methodologies suggested by different scholars, following simple functional classification of Indian cities is suggested :

1. Administrative Cities. The main function of an administrative city/town is to administer a given area which may be a district, a state or the whole country. From this point of view all the state capitals in India are administrative cities. Some of the district headquarters may also fall in the category of administrative city. In India, New Delhi is the largest administrative city because it is the capital of the whole country. At the state level, Chandigarh,

Bengaluru, Chennai, Hyderabad, Jaipur, Lucknow, Mumbai, Patna, Bhopal, Gandhinagar, Itanagar, Dispur, Jaipur, Aizawl, Kohima, Raipur, Panaji, Ranchi, Thiruvananthapuram, Imphal, Shillong, Shimla, Bhubaneshwar, Gangtok, Agartala, Kolkata are important administrative centres.

2. Production Centres. Cities/towns associated with industrial production are often referred to as production centres. Since they are concerned with industrial production they are popularly known as ***industrial cities/towns***. These towns are well connected with their surrounding areas because industries depend on raw materials and transport facilities. Indian industries have progressed a lot after Independence and a large number of industrial towns have come up in the post-Independence era. Ahmedabad, Jamshedpur, Durgapur, Rourkela, Bhilai, Bokaro, Bhadravati, Sindri, Coimbatore etc. are important industrial towns of India.

3. Transport Centres. These centres develop at the following locations :

(*i*) Where rail transport ends and road transport starts. Most of such towns are located up the foothills of the Himalayas. Kalka, Rishikesh, Kotdwar, Kathgodam, Tanakpur are examples of such towns.

(*ii*) Where sea route ends and inland water or land route starts. These towns have facilities of loading and unloading and are equipped with godowns. Mumbai, Kolkata, Chennai, Kandla, Calicut, Vishakhapatnam, etc. are cities of this type.

(*iii*) Where inland transport routes converge for example Agra, Dhulia, Mughalsarai (Deen Dayal Upadhyay), Itarsi, Katni, etc.

4. Trade Centres. Trade centres primarily depend upon trade for their growth. India has hundreds of big and small towns/cities which can be termed as ***trade centres***. Small towns depend on agricultural commodities and develop as grain markets. Big cities carry on various types of trade. Mumbai is the largest trade centre and is known as the ***'Commercial Capital'*** of the country. Kolkata, Chennai, Bengaluru, and a large number of other cities are popular as trade centres.

5. Mining Towns. Many towns have developed due to mining and are known as ***mining towns***. For example, Jharia, Raniganj, Singrauli, etc. have developed due to coal mining. Digboi, and Ankleshwar due to oil drilling and Khetri due to copper mining.

6. Defence Towns. The main function of defence towns is to provide security and arrange for defence of the country. In the mediaeval period Mughals and Rajputs had built castles for defence purposes, and those castles developed into towns at a later stage. Chittaurgarh, Ajmer, Jodhpur, Ranthambhor, Gwalior, Chanderi, etc. are example of fort towns. If a town has '*garh*' attached to its name, then if must be fort town because 'garh' in Hindi is used for a fort.

Latter on British rulers had set-up a large number of contonments for defence purposes. Then came the age of air battles and air-fields in various parts of the country were built. In addition harbours and navel headquarters were also built to strengthen the defence preparedness of the country. Ambala, Jalandhar, Ferozepur, Meerut, MOHO, Pathankot, Udhampur, Adampur, Halwara, Jamnagar, Vishakhapatnam, etc. are some of the examples of defence towns.

7. Education Centres. Some towns grow due to educational facilities and become popular because they impart quality education. Roorkee, Varanasi, Aligarh, Pilani, etc. are famous as educational centres.

8. Religious and Cultural Centres. Some towns/cities are famous for their religious, cultural and spiritual endowments and attract devotees even from far-off places. Even during the ancient period, towns have been built around religious places like temples, mosques and churches. Some towns have come-up near *gurudwaras* also. Some of the major places of religious importance are Haridwar, Rishikesh, Mathura, Vrindavan, Allahabad, Varanasi, Dwarka, Rameshwaram, Ayodhya, Badrinath, Kedarnath, Kurukshetra, Ajmer, Pushkar, Nashik, Amritsar, Bodh-Gaya, etc.

9. Tourist Resorts. Tourist resorts are generally located in areas of healthy climate, natural beauty or some man-made features which provide facilities for enjoyment, fun and sports. Such places are usually found in hilly areas or on the bank of a water body. Among the hill stations, Srinagar, Pahalgam, Gulmarg, Dalhousie, Dharamsala, Leh-Ladakh, Kullu, Manali, Nainital, Darjeeling, Gangok, Shimla, Ooty, Mt. Abu, Pachmadhi, Ranikhet, Khandala,

Matheran, Mahabaleshwar, etc. are important. On the sea coast of India there are beautiful beaches at Gopalpur, Chennai, Goa, and Mumbai.

10. Residential Towns. Some of the town are developed with the primary purpose of providing residential accommodation to urban people. A large number of colonies and satellitic towns have been built to accommodate the growing population of big cities like Delhi, Mumbai, Kolkata, Chennai, Bengaluru, Hyderabad, etc. Panchkula and Mohali near Chandigarh and Partapur near Meerut are primarily residential towns.

11. Seaports. Seaports are developed to facilitate imports and exports. Kandla, New Mangalore, Kochi, New Tuticorin, Paradweep etc. are some of the examples of seaports.

12. Multifunctional Cities. Most cities of India are performing more than one or two functions simultaneously and are termed as multifunctional cities. For example a capital city may also be a centre for trade and commerce, industry, transport, tourism, Big cities like Delhi, Mumbai, Kolkata, Chennai, Bengaluru, Hyderabad, etc. are performing highly diversified functions and are known as multifunctional cities.

CONURBATIONS

As mentioned earlier in this chapter, the term conurbation was coined by Patrick Geddes in 1915 for an extensive urban area, usually resulting from the coalescence of several originally separate expanding towns or urban nuclei. This coalescence is usually along the main inter-urban transport routes. The term conurbation perhaps has been formed by compounding two words 'continuous' and 'urbation'. Thus a conurbation means a continuous stretch of urban development over a considerable area. According to **Mayer and Kohn** (1967), "Tentacles of urbanization, in many areas, have reached towards one another, and numerous cities have converged into larger masses, conurbation or megalopolitan concentrations. **C.B. Fawcett** devoted much attention to the studies of conurbation during his studies of population of Britain on the basis of 1921 and 1931 census figures. According to his definition "A conurbation is an area occupied by continuous series of dwellings, factories and other buildings, harbours and docks, urban parks and playing fields, etc., which are not separated from each other by rural land; though in many cases in this country such an area includes exclosures of rural land which is still in agricultural occupation."

Currently different expressions are used to express large urban area in view of the rapidly increasing urban sprawl. These terms are ***'urban agglomeration', 'urban aggregate', 'urban complex', 'metropolitan area', 'megalopolis', 'metropolitan labour area', 'conglomeration', 'urban region', 'town aggregate', 'urban tract'***, etc. However, the term conurbation is commonly used only for that group of contiguous towns or cities which are under separate civic administrations.

Process of formation of conurbation. Conurbation is an indication of climax in the growth of an urban area. There is rapid growth of industries, trade, transportation and economic activities with the development of science and technology and growth as well spread of urban areas is a natural sequel. Interaction between two neighbouring cities increases and distance between them is reduced. There comes a time when continuity between the neighbouring cities is complete and there is practically no agricultural rural land in-between them. Normally a conurbation is formed under the following three situations :

1. When only one city expands and encompasses the neighbouring rural and urban areas, for example London.
2. When two neighbouring cities expand and meet each other leaving practically no open space between the two, for example Hyderabad-Secunderabad and Kolkata-Haora.
3. When more than two cities expand and meet one another, for example Delhi, Faridabad, Ballabhgarh, Palwal, Gurgaon, NOIDA.

A.E. Smailes has termed these conurbations as uninuclear, binuclear and polynuclear (**Fig. 14.24**).

The world's first and the largest conurbation is in the eastern sea-board of the USA which represents the coalescence of million cities of Portland, Boston, New Haven, New York, Philadelphia, Baltimore, Washington D.C., New Port, Columbia, Richmond and Columbus.

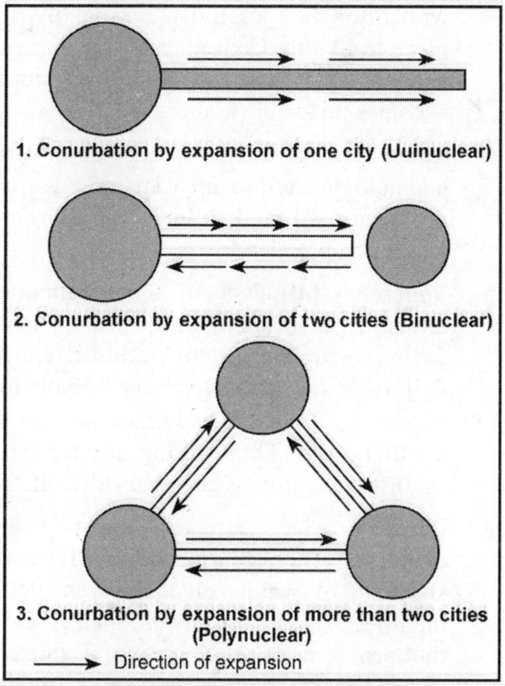

FIG. 14.24. Origin and Growth of Conurbation

Conurbations in India

With only 31.16 per cent of her total population as urban as per 2011 census figures, India is considered to be one of the least urbanised countries of the world. But her massive urban population of over 377 millions is more than the total population of any country of the world except that of China and India boasts of her 53 million cities. As such, it is but natural to expect the growth of conurbations in different parts of the country. One such conurbation has developed in West Bengal along the banks of the Hugli river where Kolkata is the main city. Most of the cities of this conurbation are based on jute industry. Most of the jute mills of West Bengal are located in a narrow belt about 100 km long and 3 km wide. Besides Kolkata, Haora, Titanagar, Jagatdol, Budge Budge, Bhadreshwar, Rishra, Bansbaria, Shamnagar, Naihati, etc. are other important cities of this conurbation (**Fig. 14.25**). Group of experts making geographical study of this conurbation has taken into account the administrative units while demarcating the boundaries of the conurbation. This group has also considered the villages located in the non-municipal corporation for this purpose. Accordingly Kolkata conurbation embraces 30 municipal corporations, 2 corporations, 2 cantonment areas and 2 areas without cantonment spread in four districts.

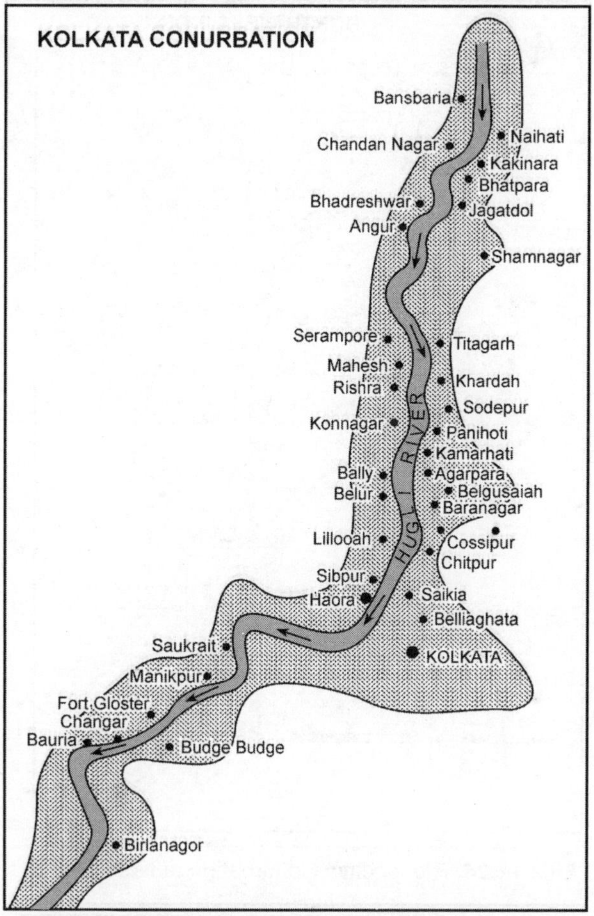

FIG. 14.25. Kolkata Conurbation

A similar conurbation is coming up in the north-western part of India. It extends from Agra in the south to Delhi in the north. At Delhi it is bifurcated into two branches. One branch goes via Meerut and Muzaffarnagar to Saharanpur in U.P. and to Dehra Dun in Uttarakhand while the other branch goes to Ambala in Haryana. At Ambala it is further bifurcated in two sub-branches, one sub-branch goes to Amritsar through Ludhiana and Jalandhar and the second sub-branch goes to Chandigarh and Kalka. Currently these cities/towns appear to be at certain distances and some agricultural land is also seen between them. But these towns/cities are fast spreading and are likely to form a very big conurbation in the near future (**Fig. 14.26**)

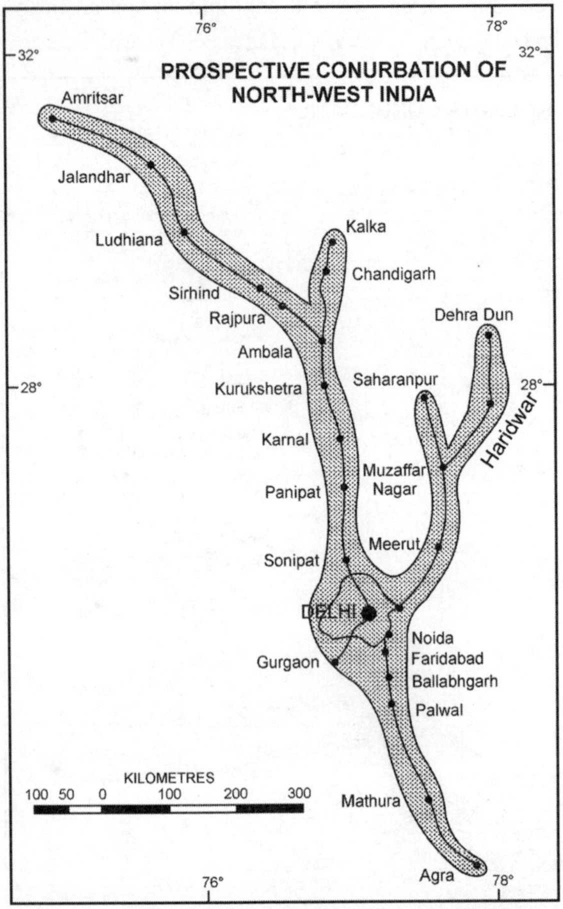

FIG. 14.26. Prospective Conurbation of North-west India

Problems of Conurbations

1. Most of the conurbations are growing at an alarming rate and the growing population puts great pressure on the urban infrastructure. The basic facilities are not available to the people and life in conurbation is less enjoyable and more tense.
2. There is unplanned growth of urban areas due to large scale rural to urban migration of the people. This process creates lots of problems for the urban people.
3. Most of the people who migrate from rural to urban areas are unskilled and semi-skilled who often fail to find a reasonably good place to live in. Consequently there is rapid growth of slums and squatters and life becomes miserable in these areas.
4. There are growing trends of poverty, unemployment, insecurity and crimes among the conurbation dwellers due to rapidly increasing population.
5. Sphere of influence of the neighbouring cities/towns often overlap which creates serious problems of administration. Criminals cross the territory of sphere of influence of one city and enter the sphere of another city. Thus taking proper action against the criminals becomes difficult.
6. Because of large size and high population density of conurbation, there is always shortage of water, electricity as well as facilities of transport, sewer, open space and the problem of environmental pollution becomes acute.

Metropolitan Regions of India

Metropolitan region is a vast urbanised region which consists of a large city and its surrounding or accompanying towns or suburbs. It is till now an obscure, unclear and loosely used term because no precise parameters have been fixed to define and delineate such regions in India. However, this term is used in the U.S.A. which may be treated as equivalent to a conurbation.

Three metropolitan cities of India, namely Mumbai, Kolkata and Chennai were developed on the Indian sea coast by the British rulers for their own benefit. They used these three cities as ports for international trade. Britain used to carry raw materials from India to Britain and send back the manufactured goods from Britain to India. As a result of the trade benefits, these three port cities become the economic hubs of the country in the beginning of 19th century and ruled over the economic scene of the country till the beginning of 20th century. In the year 1911, the British rulers shifted British India's capital from Kolkata to New Delhi and this city soon became the largest centre of political, economic and social activities in India. In fact Delhi recorded the growth rate which was much faster than any other large city of India and today it overshadows the remaining three coastal metropolitan cities of the country. Thus, at present there are four metropolitan cities in India which together, account for over 85 per cent of the richest people of the country. They cover all the four directions of the country. In the north is Delhi, in the south there is Chennai, the west is dominated by Mumbai and east is served by Kolkata. These four cities are indicators of India's accelerated rate of urbanisation and economic growth. The areas of influence of these four major cities are determined on the basis of the following parameters and are shown in **Fig. 14.27.**

1. Air travel.
2. Railway passengers and commodity flow.
3. Long distance telephone calls.

However, the landline phones have lost much of their relevance as the mobile phones are becoming more popular.

1. Delhi. Being the capital of India, Delhi is the largest centre of politics and administration. Its economic influence is also felt over large areas of north-western and in some parts of central India. It covers the Union Territories of Jammu and Kashmir and Ladakh and states of, Punjab, Haryana, Himachal Pradesh, Uttarakhand and large parts of Rajasthan, Madhya Pradesh and Uttar Pradesh.

2. Mumbai. This is the largest city of India and is popular as the commercial capital of the country. This metropolitan region covers the whole of Maharashtra, Gujarat and Goa, southern part of Madhya Pradesh and contiguous parts of Chhattisgarh, Karnataka and Rajasthan.

3. Kolkata. This is the first city which was developed by the East India Company and remained the capital of British India upto 1911. This was the first city of India to cross one million mark at the time of 1901 census and remained the largest city of India till 1981 after which it was superseded by Mumbai. The Kolkata metropolitan region covers most parts of East India which encompasses all the north-eastern states (Sikkim, Assam, Arunachal Pradesh, Nagaland, Manipur, Mizoram, Tripura, Bihar, Jharkhand, West Bengal, Odisha and eastern parts of Uttar Pradesh and Chhattisgarh.

4. Chennai. Chennai is the fourth largest city of India which is on the eastern coast of the peninsular India. Its metropolitan region covers five southern states of Tamil Nadu, Andhra Pradesh, Telangana Karnataka and Kerala as well as southern tips of Chhatisgarh and Odisha.

Within the above mentioned four primary metropolitan areas, there are several secondary metropolitan areas also. For example Delhi primary metropolitan area has Lucknow and Jaipur secondary metropolitan areas. Similarly Mumbai has Nagpur, Pune and Ahmedabad, and Chennai has Bengaluru and Hyderabad secondary metropolitan areas (**Fig. 14.27**)

Urban Sprawl

Growth in population and areal expansion of a city in the countryside is known as ***urban sprawl***. Almost all the cities of the world grow, both with respect to population and area with the passage of time making urban sprawl more or less a continuous process. Following three factors are responsible for urban sprawl.

1. Natural growth of population (difference between birth rates and death rates).
2. Large scale migration of the people from rural to urban areas.
3. Migration of the people from small towns to big cities.

The above mentioned three factors cause rapid increase in population and area of cities and often these factors work in combination. However, there are temporal and spatial variations in the relative importance of these factors. In certain cities, natural growth of population may be the dominating factor but in most of the cities it is the migration of the people from rural to urban areas which plays a significant role in the process of urban sprawl.

The first large flow of migration from rural to urban areas was during the "depression" of late 1930s when people migrated in search of jobs. Later, during the decade 1941-51, another one million persons moved to urban places in response to wartime industrialisation and partition of the country in 1947. During 1991-2001, well over 20 million people migrated to cities. The greatest pressure of the immigrating population has been felt in the central

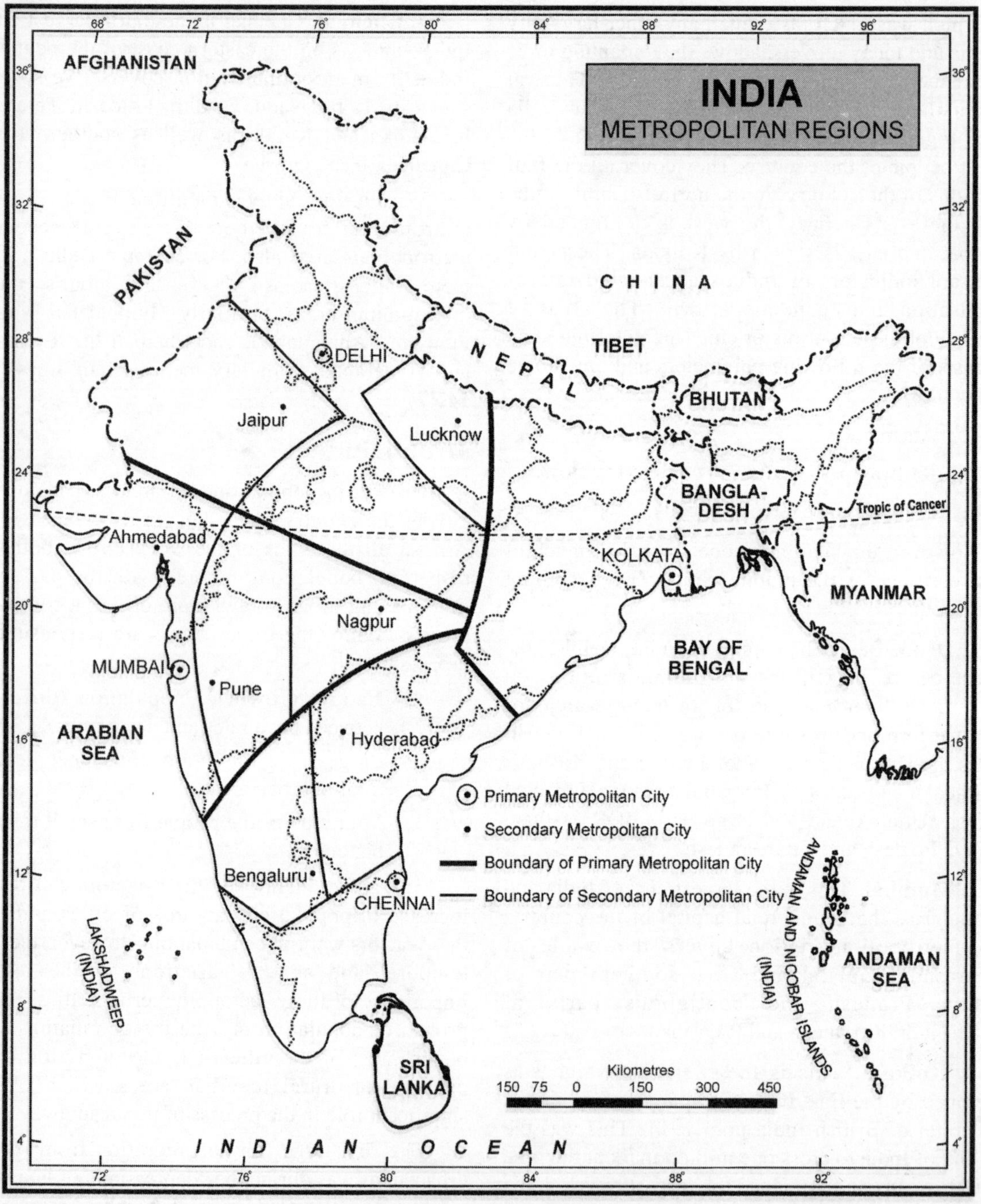

FIG. 14.27. India : Metropolitan Regions

place of the city (the "old city") where the immigrants flock to their relatives and friends before they search for housing. Population densities beyond the "old city" decline sharply. **Brush** (1968) has referred to this situation in the central parts of the cities as ***"urban impulsion"*** which results from concentration of people in the centre of the city close to their work and shopping. Incidentally many of the fastest growing urban centres are large cities. This is due to the fact that such large cities act as magnets and attract large number of immigrants by dint of their employment opportunities and modern way of life. Such ***hyperurbanisation*** leads to projected cities, sizes of which defies imagination. Delhi, Mumbai, Kolkata, Chennai, Bengaluru, etc. are examples of urban sprawl due to large scale migration of people from the surrounding areas.

In several big cities wealthy people are constantly moving from the crowded centres of the cities to the more pleasant ***suburbs*** where they can build larger houses and enjoy the space and privacy of a garden around the house. In some cities, the outskirts are also occupied by squatters who build makeshift shacks on unused land although they have no legal right to the land. The difficulty of restricting town growth in either case is immense and most towns and cities are surrounded by wide rings of suburbs.

Historically suburbs have grown first along the major roads leading into the town. This type of growth is known as ***ribbon settlement.*** Such sites are first to be developed because of their location near the road gives them greater accessibility. But soon the demand for suburban homes causes the land between ribbon settlements to be built and made accessible by constructing new roads. This type of development is known as ***'infil'.*** Simultaneously small towns and villages within the commuting distance of major cities are also developed for residential purposes. In this way towns are continuously growing and in some areas the suburbs of a number of neighbouring towns may be so close together as to form an almost continuous urban belt which is called ***conurbation.*** Urban sprawl is taking place at the cost of valuable agricultural land.

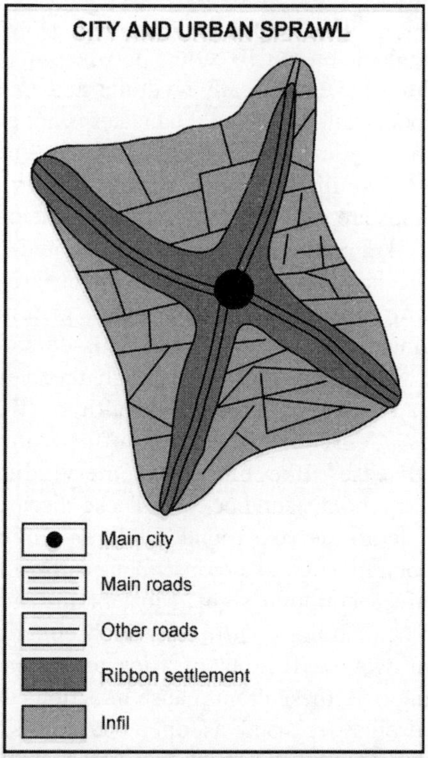

FIG. 14.28. Urban Sprawl

Urban sprawl has its own advantages and disadvantages. Whereas urban sprawl provides livelihood and shelter to lakhs of people, the problems created by urban sprawl are complicated and solving such problems is an uphill task. Some of the problems created by urban sprawl are briefly described below :

1. Urban sprawl is generally caused by large scale immigration which leads to steep rise in land prices. People with moderate incomes are unable to afford high land prices and their life becomes miserable.
2. There is heavy pressure of population on the infrastructure facilities such as water, electricity, transport, sewerage, education and health services, etc. and people have to face multiple problems in their daily life.
3. Most immigrants come to big cities in search of jobs. They often fail to find suitable jobs and add to the problem of unemployment.
4. Slums and squatter settlements come up in different parts of cities. This leads to environmental degradation and lower standard of living.
5. Unemployment, particularly among the youngsters, leads to several social evils like thefts, decoities, kidnappings, murders, rapes, etc. which creates many problems for the common man.
6. Valuable fertile agricultural land is encroached upon by urban sprawl.
7. Increasing urbanisation is also associated with industrialisation and both of them lead to environmental degradation and environmental pollution.

Slums and Squatter Settlements

The natural sequel of unchecked, unplanned and haphazard growth of urban areas is the growth and spread of slums and squatter settlements which present a striking feature in the ecological structure of Indian cities, especially of metropolitan centres.

The rapid urbanisation in conjunction with industrialisation has resulted in the growth of slums. The proliferation of slums occurs due to many factors, such as, the shortage of developed land for housing, the high prices of land beyond the reach of urban poor, a large influx of rural migrants to the cities in search of jobs, etc. In spite of several efforts by the Central and State Governments to contain the number of slum dwellers, their growth has been increasing sharply exerting tremendous pressure on the existing civic amenities and social infrastructure.

In India slums have been defined under section 3 of Slum Areas (Improvement and Clearance) Act 1956 as

(*i*) Area in which buildings are of poor quality.

(*ii*) Area in any respect unfit for human habitation.

(*iii*) Area by reason of dilapidation, overcrowding, faulty arrangement and design of such buildings, narrowness or faulty arrangement of streets, lack of ventilation, light, sanitation facilities or any combination of these factors, which are detrimental to safety, health and morals.

The following criteria characterise an area as slum :

(*i*) All areas notified "Slum" by state govt. under any Act.

(*ii*) All areas recognised as slum by state govt. which have not been formally notified as slum under any Act.

(*iii*) A compact area of at least 300 population or about 60-70 households of poorly built congested tenements in unhygienic environment usually with inadequate infrastructure and lacking in proper sanitary and drinking water facilities.

Census of India defines slums as "residential areas where dwellings are unfit for human habitation" because they are dilapidated, cramped, poorly ventilated unclean or "any combination of these factors which are detrimental to safety and health."

Socially, slums tend to be isolated from the rest of the urban society and exhibit pathological social symptoms (drug abuse, alcoholism, crime, vandalism and other deviant behaviour). The lack of integration of slum inhabitants into urban life reflects both, the lack of ability and cultural barriers. Thus the slums are not just huts and dilapidated buildings but are occupied by people with complexities of social-networks, sharp socio-economic stratification, dualistic group and segregated spatial structures.

In India, slums are one or two-room hutments mostly occupying government and public lands. The houses in slums are built in mud or brick walls, low roofs mostly covered with corrugated sheets, tins, bamboo mats, polythenes, gunny bags and thatches, devoid of windows and ventilators and public utility services. Slums have invariably extreme unhygienic conditions. They have impoverished lavatories made by digging shallow pits in between three or four huts and with sackcloth as a curtain, hanging in front. When the pit overflows excreta gets spread over the surrounding area and is rarely cleaned. The children cultivate the habit of defecating anywhere in the slum area. Slums have practically no drains and are marked by cesspools and puddles. Piped water is not available to slum dwellers and they mainly depend upon shallow handpumps for water supply. Such handpumps are generally dug in the middle of a stale dirty pool. People wash their clothes and utensils under the handpumps. The entire muck around the handpump percolates into the ground and contaminates the ground water. This contaminated ground water is taken out through the handpump which adversely affects the health of the slum dwellers. Consequently people suffer from water-borne diseases like blood dysentery, diarrhoea, malaria, typhoid, jaundice, etc. These diseases stalk the people all the year round. Children with bloated bellies or famished skeletons, many suffering from polio, are a common sight. Most of the slums are located near drains (*Nullahs*) which contain filthy stagnant water. Billions of flies and mosquitoes swarming over these drains cause infectious diseases. These drains are used as open lavatories by the inhabitants and are always choked. Such drains (*Nullahs*) pose serious threat to health of the people.

TABLE 14.10. Number of Town and Households—2011

State Code	India/State/ Union territory #	Number of towns			Urban households	Slum households	
		Total	Statutory	Slum reporting		Absolute	Percentage
1	2	3	4	5	6	7	8
	INDIA	**7,933**	**4,041**	**2,543**	**78,865,937**	**13,749,424**	**17.4**
01	Jammu & Kashmir	122	86	40	517,168	96,990	18.8
02	Himachal Pradesh	59	56	22	166,043	14,240	8.6
03	Punjab	217	143	71	2,094,067	296,482	14.2
04	Chandigarh #	6	1	1	228,276	22,080	9.7
05	Uttarakhand	115	74	31	592,223	89,398	15.1
06	Haryana	154	80	75	1,751,901	325,997	18.6
07	NCT of Delhi #*	113	3	22	3,261,423	383,609	11.8
08	Rajasthan	297	185	107	3,090,940	383,134	12.4
09	Uttar Pradesh*	915	648	260	7,449,195	992,728	13.3
10	Bihar	199	139	71	2,013,671	194,065	9.6
11	Sikkim	9	8	7	35,761	8,612	24.1
12	Arunachal Pradesh	27	26	5	65,891	4,005	6.1
13	Nagaland	26	19	7	115,054	15,268	13.3
14	Manipur	51	28	NS	171,400	NS	NS
15	Mizoram	23	23	1	116,203	16,240	14
16	Tripura	42	16	15	235,002	33,830	14.4
17	Meghalaya	22	10	6	116,102	10,936	9.4
18	Assam	214	88	31	992,742	48,122	4.8
19	West Bengal	909	129	122	6,350,113	1,393,319	21.9
20	Jharkhand	228	40	31	1,495,642	79,200	5.3
21	Odisha	223	107	76	1,517,073	350,306	23.1
22	Chhattisgarh	182	168	94	1,238,738	395,297	31.9
23	Madhya Pradesh	476	364	302	3,845,232	1,086,692	28.3
24	Gujarat	348	195	96	5,416,315	360,291	6.7
25	Daman & Diu #	8	2	NS	47,631	NS	NS
26	D & N Haveli #	6	1	NS	37,655	NS	NS
27	Maharashtra	534	256	187	10,813,928	2,449,530	22.7
28	Andhra Pradesh and Telangana	353	125	124	6,778,225	2,421,268	35.7
29	Karnataka	347	220	206	5,315,715	728,277	13.7
30	Goa	70	14	3	198,139	4,846	2.4
31	Lakshadweep #	6	0	NS	8,180	NS	NS
32	Kerala	520	59	19	3,620,696	54,849	1.5
33	Tamil Nadu	1,097	721	504	8,929,104	1,451,690	16.3
34	Puducherry #	10	6	6	206,143	35,070	17
35	A & N Islands #	5	1	1	34,346	3,053	8.9

Note : 'NS' indicates slum not reported.

***Delhi includes 19 Census towns and Uttar Pradesh includes 1 Census Town.**

Source : Census of India 2011, Housing Stock, Amenities and Assets in Slums, Series 1, p. ci.

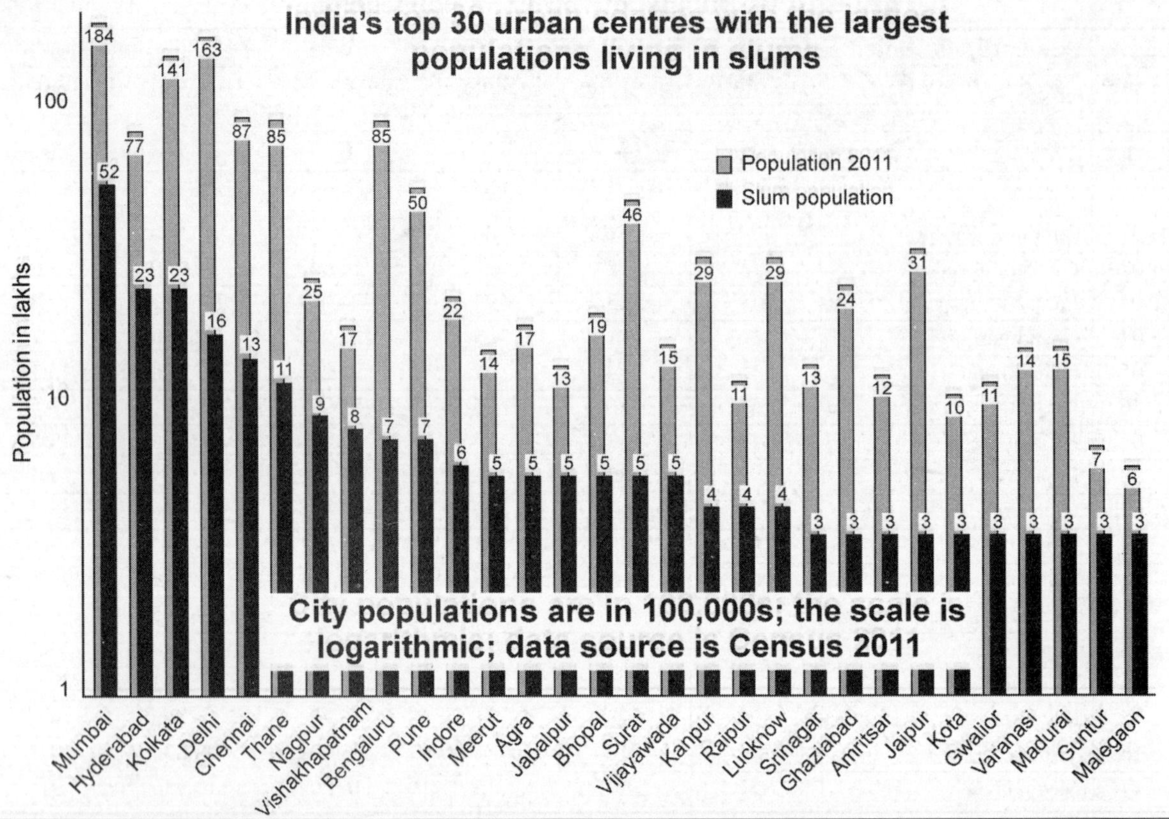

FIG. 14.29. India's top 30 urban centres with the largest populations living in slums

Slums are known by different names in different cities. They are called *bustees* in Kolkata, *jhuggi-jhoparies* in Delhi, *Jhoparpattis* or *Chawl* in Mumbai and *Cheri* in Chennai.

Squatter Settlements

No clear-cut distinction can be drawn between slums and squatter settlements in practice except that slums are relatively more stable and are located in older, inner parts of cities compared to squatter settlements which are relatively temporary and are often scattered in all parts of the city, especially outer zones where urban areas merge with their rural hinterland. Normally, squatter settlements contain makeshift dwellings constructed without official permission (*i.e.,* on unauthorised land). Such settlements are constructed by using any available material such as cardboards, tin, straw mats or sacks. Squatter settlements are constructed in an uncontrolled manner and badly lack essential public services such as water, light, sewage. Such an environment leads to several health problems. Determining size of squatter settlement is a difficult job. Some may occur singly or in small groups of 10-20 dwellings while others occur in huge agglomerations of thousands of houses. They can occur through organised rapid (almost overnight) invasions of an area by large number of people or by gradual accretion, family by family.

Squatter settlements have following three characteristics in common.

Physical Characteristics : Due to inherent 'non-legal' status, a squatter settlement has services and infrastructure below the adequate minimum levels. As such water supply, sanitation, electricity, roads, drainage, schools, health centres, and market places are either absent or arranged informally.

Social Characteristics : Most of the squatter households belong to lower income group. They are predominantly migrants, but many are also second or third generation squatters.

Legal Characteristics **:** Such settlements lack land ownership.

From the above discussion it is clear that *squatter* refers to legal position of the settlement and *slum* refers to the condition of a settlement.

A distinction has to be drawn between squatter settlements and *shanty towns*. Illegality of tenure is the hallmark of the squatter settlement but shanty huts or mean dwellings are defined by their fabric. *Shanty* towns result mainly from massive rural-urban migration and from the inability of city authorities to provide sufficient housing facilities and employment for the vast influx of people from rural to urban areas.

Indian cities abound with slums which have been termed as ***'eyesores'***, a ***'rash'*** on city landscape, 'a blot on civilization' etc. But actually they are much more health hazards for its unfortunate poverty stricken inhabitants and also for the city as a whole. The most shocking aspect is that slums are growing at an accelerated rate.

CAUSES OF SLUM GROWTH

- Rapid growth of urban population.
- Unemployment in rural areas.
- Mass migration of unskilled and semi-skilled workers from rural to urban areas.
- Limited employment opportunities to immigrant rural population in urban areas.
- Limited land and high land values in urban areas.
- Shortage of cheap residential accommodation in urban areas.

CHIEF CHARACTERISTIC FEATURES OF SLUMS

- Slums grow illegally on government land.
- Slums are almost invariably occupied by poor people belonging to lower strata of society.
- Houses in slum areas are very small and most of them consist of one room without separate toilet and kitchen.
- Houses are constructed with poor materials consisting of mud or bricks with thatched roofs or roofs covered with corrugated sheets, tins, bamboo mats, polythenes, gunny bags etc.
- Houses badly lack in ventilation and sanitation.
- Public latrines are practically absent and people, especially children defecate anywhere in the slum area.
- Infrastructural facilities such as electricity, piped (running water), sewage, garbage removal and roads are almost completely absent.
- Hand-pumps are used for water supply.
- People take bath and wash their clothes and utensils with water obtained from the hand-pumps.
- Muck around hand pumps percolates into the ground and contaminates ground water. People fall ill after consuming this water.
- There is shortage of education and health facilities and most people are illiterate and have poor health.
- Sex ratio is low in slum areas.
- There is obnoxious smell in the surroundings.

Census of India, for the first time in 2001, came out with detailed data on slum population in India. According to data released by Census of India 2011. types of households have been considered as an important criterion for evaluating the position of slum dwellers in 2011 census. Out of a total of 78.87 million households, as many as 13.75 million households or 17.4% of the total have been reported as slum households. Maharashtra had reported maximum of 2.45 million slum households followed by Andhra Pradesh and Telangana (2.42 million), Tamil Nadu (1.45 million) and West Bengal (1.39 million). State of Manipur and union territories of Daman and Diu, Dadra and Nagar Haveli and Lakshadweep have not reported any slum households (**Table 14.10 and Fig. 14.30**). Entirely different picture emerges if we consider the percentage of slum households to total households at the state/union territory level. The highest percentage (35.7%) of slum households are in Andhra Pradesh including Telangana. This is followed by Chhattisgarh (31.9%) and Madhya Pradesh (28.3%). The lowest percentage of 1.5% has been reported by Kerala.

Figure 14.29 shows that Mumbai has the largest number of 52 lakh slum dwellers which is 28.26 per cent of the total population of city's Municipal Corporation population. This is followed by Hyderabad (23 lakh), Kolkata (23 lakh), Delhi (16 lakh) and Chennai (13 lakh). If we consider the percentage of population living in slums to the total population of the concerned cities, then Hyderabad is at the top where 29.87 per cent population is living in slum areas. This is followed by Mumbai (28.26%), Kolkata (16.31%), Chennai (14.94%) and Delhi (9.8%). Slum population of the remaining big cities can be seen from **Fig. 14.29**. **Dharavi** slum in Central Mumbai is the second largest slum of Asia. Earlier

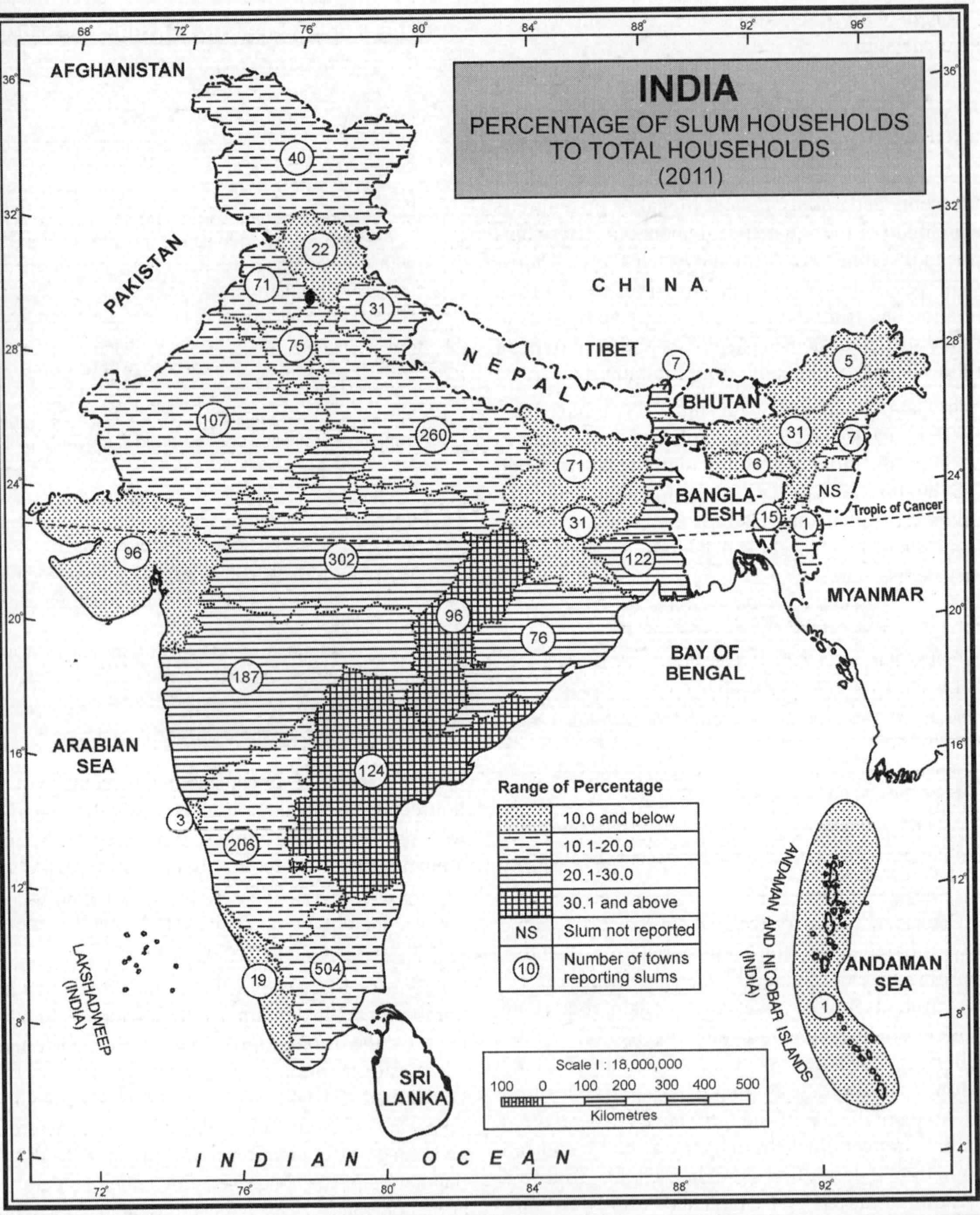

FIG. 14.30. Slum Population, 2011

Dharavi of Mumbai was the largest slum of Asia, but according to Human Development report released by the United Nations Development Programme (UNDP) in 2009, 'Orangi in Karachi' is the largest slum of Asia. Here some of the side allays and lanes are so narrow that not even a bicycle can pass. The whole

TABLE 14.11. Comparison of slum and non-slum households with respect to basic amenities in some selected cities of India (2011)

City (Figures in %)	Tap Water		Toilet		Electricity		Cell Phone	
	S	NS	S	NS	S	NS	S	NS
Delhi	86.7	89.9	50.6	95.8	97.8	99.6	66.1	63.4
Kolkata	90.8	87.0	92.0	96.2	95.5	96.5	70.1	63.5
Greater Mumbai	96.1	97.3	32.8	75.1	95.6	98.3	76.6	53.5
Bengaluru	84.3	80.2	86.8	97.8	96.6	98.4	71.0	66.5
Ahmedabad	88.2	86.6	61.3	94.6	91.5	98.9	48.8	63.1
Pune	98.8	98.4	35.8	90.6	96.2	98.9	73.4	62.0
Chennai	78.2	83.7	91.0	97.4	98.4	99.3	70.0	59.2

S = Slum house, NS = Non-slum household.
Source : Census of India 2011.

neighbourhood consists of tenement buildings, two or three storey high with rusty iron stairways to the upper part, where a single room is rented by a whole family, sometimes twelve or more people. In this place of shadowless, treeless sunlight, uncontrolled garbage, stagnant pools of foul water, the only non-human creatures are the shining black crows and long gray rats. Dharavi was an arm of the sea, that was filled by waste, largely produced by the people who have come to live there.

SOME GLARING FACTS ABOUT SLUMS IN INDIA

- 17 per cent of India's urban population lives in slums.
- Maharashtra accounts for 23 per cent of total slum population followed by Andhra Pradesh and West Bengal.
- Half of slum population lives in 53 million plus cities.
- Maharashtra, Karnataka and Andhra Pradesh have largest number of slums.
- Slum population has risen from 52 million in 2001 to 65 million in 2011.

However the UN-HABITAT report entitled "*State of the World Cities 2010/2011, Bridging the Urban Divide*" released at the World Urban Forum in Rio de Janeiro in March, 2010 has shown a silver lining. According to this report, China and India have lifted at least 125 million people out of slum conditions. Out of this, 59.7 million people are in India and the rest are in China.

The Census of India 2011 report released on March 21, 2013, had pointed a very bright picture of living conditions in slum areas of some major cities. The report says that living conditions in slums are no different from that of non-slums. In fact there are great similarities as far as access to basic amenities like drinking water and electricity is concerned. India's first-ever census of household amenities and assets in slums has revealed that slum dwellers are also spending more on TV sets, computers and mobile phones rather than sanitation.

In 2011, of the 13.7 million slum households, 74% have access to tap water as against 70% in non-slums. As many as 90% have access to electricity as against 92% in non-slums. Also more slum dwellers—77% have access to permanent houses. The figures indicate the narrowing gaps between slums and non-slums.

Highlights about slums according to provisional figures released by Census of India 2011 in 2013

- 13.71 million urban households live in slums which is 17.4 per cent of total households in the country.
- There are 19 million plus cities where more than 25 per cent households live in slums.
- 35 per cent of India's slum households are in Maharashtra and Andhra Pradesh.
- Maharashtra has 2.45 million slum households which is two times more than 1.19 million slum households in Uttar Pradesh and Bihar put together.
- 64 per cent of India's slum households live in five states *viz.* Maharashtra, Andhra Pradesh, Tamil Nadu, West Bengal and Madhya Pradesh.

In an indication of an increase in purchasing power, the report also reveals that 70.2% of slum families own their houses at compared to 69% in non-slum households. Further 70% of slum families have TV sets, 63% have mobile phones and 10.4% computers.

Overall there are 19 million plus cities where more than 25% households live in slums.

TOWN PLANNING

Introduction and Necessity of Town Planning

Urbanisation has increased at rapid pace during the last few decades and size as well as number of cities have increased at an unprecedented rate. This phenomenon has resulted into a large number of complex problems in urban areas. Uncontrolled, unplanned and haphazard growth of cities has led to tremendous pressure of population on the basic infrastructural facilities such as housing, transport, sewer, education, health, water, electricity, etc. and the city dwellers have to face a lot of problems, some of which have really become very complicated. With the increasing number of vehicles, slums and squatter settlements, industries, urban environment has become polluted beyond the permissible limits which has made life of the urban people miserable. Keeping in view these problems, town planning has become of paramount importance. Its implies that new towns should be dwelled in a planned way and proper plans should be made to improve the condition of old cities. By town planning, we will not only make life of the present generation comfortable but will secure the life of the future generations also. From the historic point of view, it is very aptly said that town planning is as old at towns themselves. Romans planned city of Rome more than two thousand years ago and this city has the pride privilege of being the first planned the city of the world. The present day problems of towns have made town planning as the foundation of towns on which the infrastructure rests and the future prospects of towns depend.

Definition of Town Planning

Town planning is a multi-disciplinary subject and specialists from different fields make their respective contribution to the process of town planning. For example sociologists, economists, geographers, artists, politicians, engineers, demographers, etc. view town planning from their respective angles. The role of geographers in town planning has become very important during the last few decades and many geographers have become professional town planners in India. In fact, *town planning is an art of which geography is a science.* Several geographers have defined town planning in their own way (see box).

1. "The underlying idea of Town Planning is the welfare of the citizens and raise the standard of living of the people." —**L.D. Stamp**
2. "Town and country planing is concerned with the use and development of land." —**Jackson**
3. "The touchstone of planning is the accommodation of several units to make a complete and harmonious whole." —**H.M. Mayer**
4. "City planing is simply the exercise of such foresight as will promote the orderly and sightly development of a city and its environs along rational lines with due regards for health, amenity and convenience and commercial and industrial advancement." —**Nelson, P. Lewis**

Aims and Objectives of Town Planning

There can be several aims and objectives of town planning but for the sake of convenience, three objectives are considered important : (1) beauty, (2) health, and (3) convenience.

1. ***Beauty*** of a town implies that the town should look beautiful and attractive and its beauty should not fade away with the passage of time. To maintain the beauty of a town, there should be proper provision for cleanliness, drainage, light, roads, buildings, etc. The initial beauty of a town quickly fades away if proper arrangements for maintaining this beauty are not made.

2. ***Health*** of the inhabitants of a town is the main pillar on which the town's progress depends. People need pollution free environment and nutritious food to maintain good health. Industries and other polluting establishments should be placed in such a corner of the city that they cause least harm to the people living the town. Wind rose diagrams can be great help in this regard.

3. ***Convenience*** is an important need of man and all activities are performed according to man's convenience. If a town looks beautiful and proper provision for the people's health is also taken care of the town cannot be termed as planned town if convenience of the people is ignored. For example, if an industrial worker has to spend 1-2 hours to reach his place of work from his place of residence, it will be very inconvenient for him. Things can become very convenient for him, if arrangement for his residence is made near the factory or if proper arrangements for cheap and efficient transport are made.

Thus beauty, health and convenience are three basic components of a planned town. Chief aims and objectives of town planning can be summed up as follows :

1. Proper plan should be made for the chronological growth of the town so that there is no obstacle in the future urban sprawl.
2. Land surrounding the town should be included in the plan of the town so that town can spread depending upon the nature of land.
3. Buildings should be planned according to convenience of the people and there should be no scope for unplanned buildings.
4. Growth of town should be planned according to civilization, culture, rituals or social structure of its inhabitants.
5. There should be proper provision for trade, commerce and other economic activities so that town could be provided with a socio-economic base.
6. Requirements for the coming at least fifty years should be considered while preparing a plan for a town.
7. Roads and buildings should be specious and there should be open space at the corners.
8. The buildings should be designed in such a way that fresh air and sufficient light can enter in the building. These things are essential for maintaining good health.
9. Due consideration must be given to convenience, comfort and health of the people while preparing plan for a town.

Subject and Scope of Town Planning

Number of size of towns is fast increasing all over the world including India which has increased the importance of town planning considerably. Therefore, it is necessary to know the subject matter of town planning. The subject and scope of town planning has become very vast with the increasing trend in urbanisation. **A. Augiston** of France has mentioned four main elements of town planning. (1) trade, (2) industry, (3) administration and (4) residence. Thus, according to him, a town should be planned in such a way that it is capable of providing scope for growth in trade and industries, its administration base is solid and it provides sufficient scope for residential areas. This plan has not mentioned anything about facilities of health and education which are essential parts of urban life. **Arthur B. Galion** has divided the subject matter of town planning into two broad classes. (1) *Planning for land-use* which is concerned with the use of land for residence, trade, industry open spaces, etc. Density of land use is determined according to houses and population. (2) *Planning for transport and communication* which requires proper plans for highways and other roads, railways, airports, water-ways and communication network.

Special attention is to be paid to the convenience of the inhabitants and possibilities of future growth of the town. Department of Town and Country Planning has prepared master plans for several towns in India. According to the views expressed by Indian town planners, land use map must be prepared to begin with and then the town plan should be proposed according to the problems of the concerned town. Following are the special features of town planning :

1. Special attention is to be paid to the physical growth of the town.
2. Town plan should have proper consideration for future growth.
3. It should have a broad base.
4. The present population should find it convenient and comfortable to live in the town.

Principles of Urban Planning

Town planning is always bound by certain principles, some of these are briefly described below :

1. Provision for separate area for industries. There is a close relationship between industrialisation and urbanisation and both go hand-in-hand with each other. In some of the towns, industries are set-up in residential areas which is against the principle of town planning. Industries should be located in such an area which is accessible from other parts of the town and at the same time it has least effect on air, water, land and sound pollution.

2. Division of town into functional zones. The whole town is divided into functional zones and each zone is meant for a particular function. Thus there are residential zones, industrial zones, trade and commerce zones, medical zones, educational zones, etc. They are located in such a way that they are complimentary to rather than competitive against each other.

3. Planning for transport. Transport is the main artery and forms the life-line of a town which facilitates the movement of people and goods from one part of the town to another. Therefore, meticulous planning of transport system of a town is of paramount importance. Roads should be broad and clean and there should be separate lanes for separate modes of transportation. Pedestrians need special attention because they are the first victims in any road accident. The road network should be planned in such a way that people do not find much difficulty in travelling from one part of the town to another. Roads coming in and going out of the town are to be properly planned so that greatest accessibility is achieved. Removal of unwanted speed breakers, provision of traffic signals, fly-overs, under-passes, etc. can go a long way in the smooth flow of traffic in big cities.

Local railways also play a significant role in big cities like Mumbai, Delhi, Kolkata, Chennai, etc. Mumbai has the best network of local railways in India, but even there, one finds that local trains are unable to meet the increased pressure of traffic and are almost invariably over-crowded. Delhi metro rail has brought much needed relief to commuters and the success of this railway has encouraged many of the million cities to develop their own metro rail systems.

4. Slums and Squatter Settlements. Slums and squatter settlements grow in towns and cities when unskilled and semi-skilled workers migrate from rural to urban areas and do not find a suitable job there. Under the plan, the slum dwellers are provided with free or heavily subsidised alternate residential accommodation. In certain cases where it is not possible to remove slums and squatter settlements, basic facilities like water, electricity, roads, sewer, garbage disposal, etc. should be proved so as to enable the slum dwellers to lead a reasonably good life.

5. Provision for Public Places. Any good town plan is expected to make provision for public places like parks, play grounds, clubs, community centres, religious places so that people, particularly children and senior citizens can walk, play and spend their free time.

6. Stratified Residential Areas. Normally three types of people live in towns which are labelled as upper class, middle class and lower class according to their financial status. Each class should be allotted separate residential area so that people belonging to one class are able to mix with one another and enjoy community life.

7. Green Belt. Green belt is usually provided to check the uncontrolled and irregular growth of a town. Usually it is circular area which surrounds the town from all sides and acts as a buffer area between urban and rural land use. In addition to check the uncontrolled, unplanned and irregular growth of the town, green belt also helps in providing pleasant and healthy environment to the people. Scope for future planned growth of the town is also there in the green belt (**Fig. 14.31**).

8. Land Use. In a planned town, specific piece of land is earmarked for a specific purpose and it is assured that the land is actually used for the purpose for which it is allotted. Height, spacing and design of buildings is also fixed and is strictly adhered to.

9. Open Spaces. Open spaces are needed by all sections of society. Therefore, distribution of parks and playgrounds is planned in such a way that everyone is able to lead a life of virtue.

10. Urban Decentralisation. Satellite towns are developed around big cities to decentralise urban population so that high concentration and overcrowding in the city centre is reduced. Satellite towns are new, open and cheaper and act as great mechanism for decentralisation.

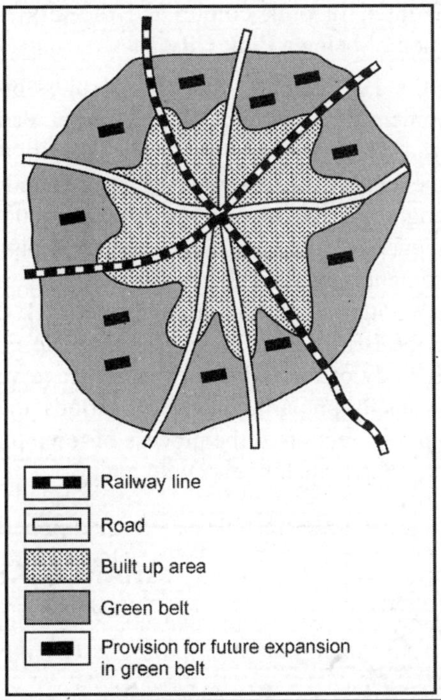

FIG. 14.31. Green Belt

11. Renewal of Cities. This is a very important aspect of town planning according to which master plan for 10-20 years is prepared and old land use is improved upon.

12. Rehabilitation and re-development. This principle is used for improving old towns. Under this principle, old buildings are improved. Critically old buildings, which are dangerous, are demolished and new buildings are constructed in their place.

13. Development of City Centre. City centre which is usually located at the centre of the city is called the Central Business District (CBD). This area is characterised by tall buildings, high density of land use, convergence of transport routes, high concentration trade and overcrowding. Such a situation leads to a large number of problems. Under the planned development, ways and means are devised to solve these problems.

14. Provision for future growth of towns. Plan for future growth and sprawl of urban areas is prepared under this principle. While preparing the plan for future growth of the city, direction and limit of growth are fixed for residential, industrial and commercial areas and priorities for different varieties of land use are also fixed. For metropolitan cities like Delhi, Mumbai, Kolkata, Chennai, etc. plans for future development of satellite towns, linear towns, dispersed towns and radial towns are prepared.

TOWN PLANNING IN INDIA

India has a long tradition of town planning. Our old scriptures have many references of planned towns wherein markets, roads, streets, temples, public buildings, residential places, secretarial centres, parks etc. have been mentioned as parts of planned towns. Excavations of Harapa and Mahenjodaro in the Indus Valley are solid proofs of town planning of higher level in India. Four main roads and several other roads and streets of Mohenjodaro meet at right angles giving a typical rectangular shape to the town. The width of main road and the other roads was 10 and 7 metres respectively.

Several towns were developed during the *Vedic* period. Some of the important town built during this period are Kannauj, Ayodhya, Mathura, Janakpuri, Hastinapur, Vaishali, Patliputra, Madurai, Sanchi, Kanchipuram, etc. Majority of these towns were the capitals of their respective kingdoms and were built very meticulously in a planned way. Different parts of the towns were used for different purposes and division of land use pattern was planned in a scientific way. The length and breadth of Ayodhya at that time was 600 and 1500 yards respectively. The town contained beautiful roads and houses and provision for drinking water was made throughout the town.

Patliputra (present Patna) is another example of town planning in ancient time. It was 12 miles (20 km) long and 2 miles (3.2 km) wide and had one fort with 60 doors. The King's palace was in the centre of the city and this palace was surrounded by parks, fountains, ponds, etc. which gave an aesthetic look to the place. The whole city was encircled by a water channel from security point of view.

Delhi had been the target of invaders throughout its history and had the privilege of being the capital of the country a number of times. Its beginning was made at Indra Prastha which had several palaces, parks, gardens, tanks and beautiful buildings. The Mughal emperor Shahjahan built the city of Shahjahanabad on the bank of the river Yamuna which is now known as Old Delhi. Chandni Chowk was at the heart of the city. The town was spread over 1240 acres of land meant for about 60,000 inhabitants. It had a protection wall on all the sides which had four gates. Currently they are known as Delhi Gate, Kashmiri Gate, Ajmiri Gate and Lahori Gate. It had broad roads which were meeting each other at right angles. A large number of mosques, parks, gardens and fountains enhanced the beauty of the city. Other towns built by the Mughal emperors include Srinagar, Agra and Fatehpur Sikri.

Jaipur, the present capital of Rajasthan, built by the Rajput king Maharaja Jai Singh in 1727 is an outstanding example of planned city. Its broad and beautiful roads and beautiful buildings add to the beauty of the city. All buildings of the original city were painted in pink colour and the city has the distinction of being a **Pink City.**

In the 19th century, the British rulers prepared development plans for cities like Mysore, Vadodara, Kolkata, Delhi, etc. for their administrative convenience. Most of the plans were prepared by the army engineers and special attention was paid to the convenience of army personnels while preparing the development plans. In 1915 the then governor of Madras invited a great town planner Patrick Geddes to help the Britishers in town planning. He made an indepth study of the Indian towns and gave valuable suggestions for preparing plans for various towns. It gave a new direction to the process of town planning in India. The British rulers could not afford to ignore

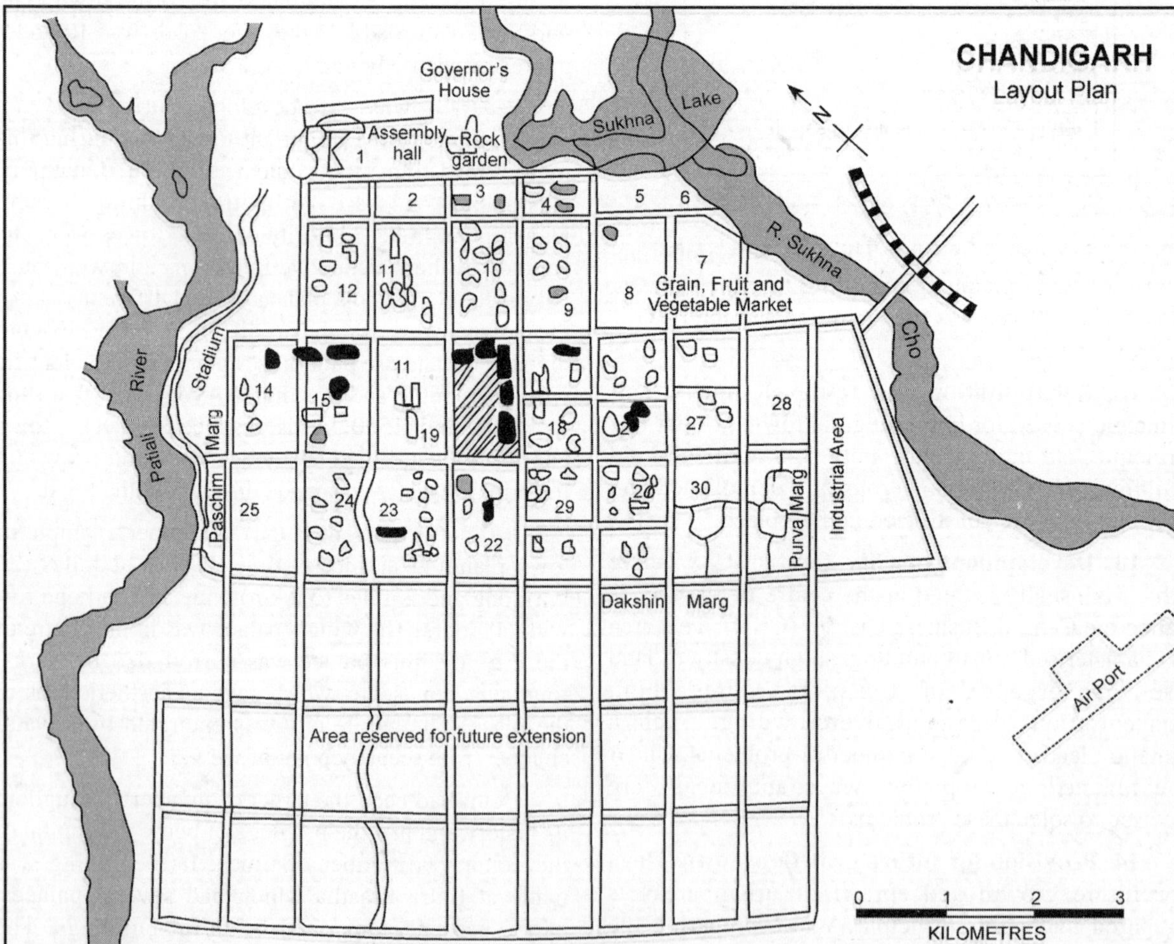

FIG. 14.32. Layout Plan of Chandigarh

the strategic importance of Delhi and decided to shift the capital of India from Kolkata (Calcutta) to Delhi in 1911. Plan for developing New Delhi on a piece of land about 5 miles long and 4 miles broad between the Delhi Ridge and the Yamuna river was prepared by Delhi Town Planning Committee in 1912. Lutyans was the chief architect of this plan and New Delhi is still known as Lutyen's Delhi. Delhi has been one of the fastest growing urban areas in the country. From a little over 2 lakhs in 1901 and 4.5 lakh in 1931 the city took a leep to 7 lakh in 1941. In the first four years after partition, it is estimated that about 1,00,000 left for Pakistan but thrice that number came in and by 1951, Delhi's population had doubled. By 1961, the urban area recorded a population of 2.3 million which rose to 16.75 million in 2011. At the initial stage, New Delhi and Old Delhi were separate urban identities and there was a vast open area between the two. Both the areas have widened with the passage of time and at present there is no open space between the two.

In the year 1907 Jamshedji Tata established Tata Iron and Steel Company (TISCO) at Saktchi in Singhbum district of Jharkhand and prepared development plan for the industrial town of Jamshedpur in 1911 which was revised in 1920, 1936 and 1943. According to the plan, the whole town is divided into two parts and 12 zones. One part is industrial area and the other is residential area. According to the plan, houses are not close to each other, rather they are wide apart, separated by sufficient open space. Due provision for entertainment, market, education and medical facilities has been made.

Town Planning in the Post-Independence Period

India witnessed rapid growth of urbanisation after Independence in 1947. Several new towns came up and many old towns were remodelled. New towns were direly needed for immediate settlement of migrants from Pakistan as a result of partition of the country. New colonies named as '*model towns*' were establishes in old towns to provide shelter to the immigrants. A number of industrial towns was also set up to boost the industrial growth. Durgapur, Rourkela, Bhilai, Bokaro, Chittaranjan, etc. are examples of such towns. Many new capital towns like Chandigarh, Bhubaneshwar, Gandhinagar, Dispur, Itanagar were also established for administrative purposes. All these towns have been constructed according to pre-determined plans and they are examples of planned towns in India.

Chandigarh. Chandigarh is the proud city of Independent India and is considered to be one of the most beautiful cities of the world. Need for the new capital of Punjab was felt immediately after partition of the country and its plan was prepared at the initiative of the then chief minister of Punjab, Sardar Partap Singh Kairon. Its construction started in 1952 and it became a union territory and capital of Punjab and Haryana on 1st November 1966 when the erstwhile Punjab was divided into present Punjab and Haryana. It was designed to accommodate 1,50,000 inhabitants but is remained a great attraction for people of the neaghouring states and it is one of the million plus cities of India according to 2011 census.

The city has a rectangular plan in which the roads are designed according to gridiron or chequerboard pattern and they meet each other at right angles (**Fig. 14.32**). According to the original plan, the city is spread on an area of 3642 hectares between Patiali river in the west and Sukhna Cho in the north-east and has 30 sectors, out of which 24 are residential sectors and the remaining six are meant for other purposes. Sector 17 is meant for trade and commerce and may be termed as Central Business District (CBD) of the city. Sector 14 has the campus of Punjab University and sector 11 and 12 host other educational institutes. Important buildings like assembly hall, secretariat, and governor's house are in the northern part of the city. Adjacent to the Sukhna Cho are grain, fruit and vegetable markets while industrial area is in the extreme east corner of the city. Sukhna Lake, Rock Garden, Rose Garden etc. are the main tourist attractions of the city. The southern part of the city has been kept vacant for future expansion of the city.

PROBLEMS OF URBANISATION IN INDIA

Although India is one of the less urbanized countries of the world with only 31.16 per cent of her population living in urban agglomerations/towns, this country is facing a serious crisis of urban growth at the present time. Whereas urbanisation has been an instrument of economic, social and political progress, it has led to serious socio-economic problems. The sheer magnitude of the urban population, haphazard and unplanned growth of urban areas, and a desperate lack of infrastructure are the main causes of such a situation. The rapid growth of urban population both natural and through migration, has put heavy pressure on public utilities like housing, sanitation, transport, water, electricity, health, education and so on. Poverty, unemployment and under employment among the rural immigrants, beggary, thefts, dacoities, burglaries and other social evils are on rampage. Urban sprawl is rapidly encroaching upon the precious agricultural land. The urban population of India had already crossed the 377 million mark by 2011 which is more than the total population of the U.S.A., third most populous country of the world after China and India. By 2050, more than 50 per cent of India's population is expected to live in urban areas. Following problems need to be highlighted.

1. Urban Sprawl

Most cities of India are growing at a rapid pace, both with respect to population and area. This is known as urban sprawl which creates a large number of problems. These problems have already been discussed in this chapter and need not be repeated.

2. Overcrowding

Overcrowding is a situation in which too many people live in too little space. Overcrowding is a logical consequence of over-population in urban areas. It is naturally expected that cities having a large size of population squeezed in a small space must suffer from overcrowding. This is well exhibited by almost all the big cities of India. For example, Mumbai has one-sixth of an acre open space per thousand population though four acres is the suggested standard by the Master Plan of Greater Mumbai. Metropolitan cities of India are overcrowded both in 'absolute' and 'relative' terms. Absolute in the sense that these cities have a real high density of population; relative in the sense that even if the densities are not very high the problem of providing services and other facilities to the city dwellers makes it so. Delhi has a population density of 11,297 persons per sq km (Census 2011) which is the highest in India. This is the overall population density for the National Capital Territory of Delhi. Population density in central part of Delhi could be much higher. This leads to tremendous pressure on infrastructural facilities like housing, electricity, water, transport, employment, etc. Efforts to decongest Delhi by developing ring towns has not met with the required success.

3. Housing

Overcrowding leads to a chronic problem of shortage of houses in urban areas. This problem is specifically more acute in those urban areas where there is large influx of unemployed or underemployed immigrants who have no place to live in when they enter cities/towns from the surrounding areas. An Indian Sample Survey in 1959 indicated that 44 per cent of urban households (as compared to 34 per cent of rural families) occupied one room or less. In larger cities the proportion of families occupying one room or less was as high as 67 per cent. (Roy Turner, 1962). Moreover, the current rate of housing construction is very slow which makes the problem further complicated. Indian cities require annually about 2.5 million new devellings but less than 15 per cent of the requirement is being constructed.

As per 2011 Census figures, there are 110.1 million houses for a total urban population of over 377 million which means that nearly four persons live in one house. Only 68.5 per cent of the total houses are in good condition and remaining are just livable or dilapidated. **Table 14.12** shows that about two-thirds of the houses are just one room or two room dwellings, many of them without a kitchen or a

TABLE 14.12. Households by Number of Dwelling Rooms

Size of houses	Percentage of the total houses
1. No exclusive room	3.1
2. One room	32.1
3. Two rooms	30.6
4. Three rooms	18.4
5. Four rooms	9.3
6. Five room	3.2
7. Six rooms and above	3.3
Total	100.0

Source : Census of India 2011. Tables on Houses, Household Amenities and Assets, Series 1, p. XIV.

bathroom. The table also shows that the percentage of houses decreases with the increase in number of rooms.

Several factors are responsible for the above mentioned sad state of affairs with respect to housing problems faced by the urban people. The major factors are shortage of land, building materials and financial resources, inadequate expansion of pubic utilities into sub-urban areas, poverty and unemployment of urban immigrants, strong caste and family ties and lack of adequate transportation to sub-urban areas where most of the vacant land for new construction is located.

4. Unemployment

The problem of unemployment is no less serious than the problem of housing mentioned above. Urban unemployment in India is estimated at 15 to 25 per cent of the labour force. This percentage is even higher among the educated people. It is estimated that about half of all educated urban unemployed are concentrated in four metropolitan cities (Delhi, Mumbai, Kolkata, Chennai). Furthermore, although urban incomes are higher than the rural incomes, they are appallingly low in view of high cost of living in urban areas.

One of the major causes of urban unemployment is the large scale influx of people from rural to urban areas. Rural-urban migration has been continuing for a pretty long time but it has not always been as great a problem as it is today. The general poverty among the rural people pushes them out to urban areas to migrate in search of livelihood and in the hope of a better living. But the growth of economic opportunities fail to keep pace with the quantum of immigration. The limited capacity of urban areas could not create enough employment opportunities and absorb the rapid growth of the urban labour force. Efforts made by the central and the state governments to create employment opportunities in rural areas and to check the large scale rural-urban migration have not met with much success.

5. Slums and Squatters

Slums and squatter settlements grow due to unchecked, unplanned and haphazard growth of city and also due to low income levels of the city dwellers. This creates many problems which have already discussed in this chapter.

6. Transport

With traffic bottlenecks and traffic congestion, almost all cities and towns of India are suffering from acute form of transport problems. Transport problems increase and become more complex as the town grows in size. With its growth, the town performs varied and complex functions and more people travel to work or shop. As the town becomes larger, even people living within the built-up area have to travel by car or bus to cross the town and outsiders naturally bring their cars or travel by public transport. Wherever trade is important, commercial vehicles such as vans and trucks will make problem of traffic more complicated.

Since most of the commercial activities of the towns are concentrated in the Central Business District (C.B.D.), the centres are areas of greatest congestion. However, other parts of the town are not free from traffic congestion. Such areas include the roads leading to factories, offices, schools, etc., which will be thronged with people in morning and evening; minor shopping centres which grow up in the suburbs; sporting arenas, entertainment districts which will be busy at night, roads leading to residential and dormitory towns which will be busy when commuters flock to the cities in the morning to work and return home in the evenings. Such congestion becomes greater when the centre is built up in tall skyscraper blocks whose offices sometimes employ thousands of workers, because at the end of the office hours everyone leaves the building within a short span of time to make his way home. This puts tremendous pressure on public transport and causes journeys to take much longer period than they normally would. In most cities the rush hour or peak traffic hour lasts for about two hours and during that period buses and trains are crammed to capacity, roads are overcrowded with vehicles and the movement of traffic becomes very slow.

In other towns, the narrowness of the streets, which were built long before the motorised transport and lack of parking facilities are the main cause of congestion. Cars may be parked along the edges of the roads restricting movement to a narrow lane and the multiplicity of narrow streets, sharp corners and long waiting period to turn into lanes of traffic may slow down the movement and thus create even greater congestion.

The traffic scenario in almost all the Indian cities presents a pathetic picture with Mumbai still having the best city transport system and Chennai, Ahmedabad and Pune being reasonably well served by local transport system. In all other cities, if one does not own a personal vehicle, great hardship is experienced in moving about in the city. Apart from that, the level of incomes and affordability of Indian masses is very low and the citizens are not able to pay an economic fare for use of public transport system. Therefore, all city bus services sustain such heavy losses that they cannot really expand or even maintain a fleet adequately to meet the city needs. Moreover, mixture of vehicles causes uncontrollable chaos on the roads. Free movement of stray cattle and domestic animals on the roads adds to traffic problems and often cause accidents. Heavy traffic and congestion leads to slow movement of traffic, fuel wastage environmental pollution and loss of precious time.

A study of traffic problem in Delhi will aquant us to traffic scenario in the rest of urban India. Already there are more than one crore vehicles on Delhi road (2020). The road length, however has not increased proportionately. Introduction of metro rail eased traffic congestion to begin with but the number of metro rail users is far out pacing the facilities which this mode of transportation is able to provide.

Similar conditions prevail in most of the Indian cities. In Kolkata, metro rail and Vivekanand Setu were constructed to ease traffic flow. But traffic congestion in several old localities and near Haora bridge is almost a daily routine. In Ahmedabad, the speed of vehicles comes down to 5 km/hr on Gandhi Marg and several other roads due to congestion and overcrowding.

7. Water

Water is one of the most essential elements of nature to sustain life and right from the beginning of urban civilisation, sites for settlements have always been chosen keeping in view the availability of water to the inhabitants of the settlement. However, supply of water started falling short of demand as the cities grew in size and number. Today we have reached a stage where practically no city in India gets sufficient water to meet the needs of city dwellers. In many cities people get water from the municipal sources for less than half an hour every alternate day. In dry summer season, taps remain dry for days together and people are denied water supply at a time when they need it the most. The individual towns require water in larger quantities. Many small towns have no main water supply at all and depend on such sources as individual tubewells, household open wells or even rivers. Accelerated Urban Water Supply Programme (AUWSP) was launched to provide water to towns with population of less than 20,000. Keeping in view the increased demands for water by the urban population, Central Public Health and Environmental Engineering Organisation (CPHEEO) fixed 125-200 litres of water per head per day for cities with a population of more than 50,000, 100-125 litres for population between 10,000 and 50,000 and 70-100 litres for towns with a population below 10,000. The **Zakaria Committee** recommended the water requirement per head per day 204 litres for cities with population between 5 lakh and 2 million and 272 litres for cities with population more than 2 million. This amount of water is supposed to be used for drinking, kitchen, bathing, cloth washing, floor and vehicle washing and gardening. Sadly majority of the cities and towns do not get the recommended quantity of water. Gap in demand and supply of water in four metro cities, *viz.*, Mumbai, Kolkata, Delhi and Chennai varies from 10 to 20 per cent. The condition is still worse in small cities and towns. To meet the growing demand for water, many cities are trying to tap external sources of water supply. Mumbai draws water from neighbouring areas and from sources located as far as 125 km in the Western Ghats. Chennai uses water express trains to meet its growing demand for water. Bengaluru is located on the plateau and draws water from Cauvery river at a distance of 100 km. Water for Bengaluru has to be lifted about 700 metres with help of lifting pumps. Hyderabad depends on Nagarjuna Sagar located 137 km away. Delhi meets large part of its water requirements from Tajewala in Haryana. Water is also drawn from Ramganga as far as 180 km. Under the proposed scheme it will meet its growing requirements of water from Tehri, Renuka, and Kishau barrages.

8. Sewerage Problems

Urban areas in India are almost invariably plagued with insufficient and inefficient sewage facilities. **Not a single city in India is fully sewered.** Resource crunch faced by the municipalities and unauthorised growth of the cities are two major causes of this pathetic state of affairs. According to latest estimates, only 35-40 per cent of the urban population has the privilege of sewage system. Most of the cities have old sewerage lines which are not looked after properly. Often sewerage lines break down or they are overflowing. Most cities do not have proper arrangements for treating the sewerage waste and it is drained into a nearby river (as in Delhi) or in sea (as in Mumbai, Kolkata and Chennai), thereby polluting the water bodies.

In most Indian cities, water pipes run in close proximity to sewer lines. Any leakage leads to contamination of water which results in the spread of several water borne diseases.

9. Trash Disposal

As Indian cities grow in number and size the problem of trash disposal is assuming alarming proportions. Huge quantities of garbage produced by our cities pose a serious health hazard. Most cites do not have proper arrangements for garbage disposal and the existing landfills are full to the brim. These landfills are hotbeds of diseases and innumerable poisons leaking into their surroundings. Wastes putrefy in the open inviting disease carrying flies and rats and a filthy, poisonous liquid, called leachate, which leaks out from below and contaminates ground water. People who live near the rotting garbage and raw sewage fall easy victims to several diseases like dysentery, malaria, plague, jaundice, diarrhoea, typhoid, etc.

10. Urban Crimes

Modern cities present a meeting point of people from different walks of life having no affinity with one another. Like other problems, the problem of crimes increases with the increase in urbanisation. In fact the increasing trend in urban crimes tends to disturb peace and tranquility of the cities and make them unsafe to live in, particularly for the women. Growing materialism, consumerism, competition in everyday life, selfishness, lavishness, appalling socio-economic disparities and rising unemployment and feeling of loneliness in the crowd are some of the primary causes responsible for alarming trends in urban crime. Not only the poor, deprived and slum dwellers take to crime; youngsters from well-to-do families also resort to crime in order to make a fast buck and for meeting requirements of a lavish life style. Occasional failures in life also drag youngsters to crime. The problem of urban crime is becoming more complicated in the present day world because criminals often get protection from politicians, bureaucrats and elite class of the urban society. Some of the criminals reach high political positions by using their money and muscle power.

According to study made by Dutt and Venugopal (1983), violent urban crimes like rape, murder, kidnapping, dacoity, robbery, etc. are more pronounced in the northern-central parts of the country. Even economic crimes (like theft, cheating, breach of trust, etc.) are concentrated in the north-central region. Poverty related crimes are widespread with main concentration in the cities of Patna, Darbhanga, Gaya and Munger.

However, the latest surveys show that Mumbai and Delhi figure in 35 cities that have high crime rate. As much as 31.8 per cent of citizens in Mumbai and 30.5 per cent in Delhi have been victims of crime. Sexual assault was higher in Mumbai (3.5 per cent) as compared to Delhi (1.7 per cent). Both cities score poorly on corruption, with 22.9% in Mumbai being exposed to bribery as compared to 21% in Delhi.

11. Problem of Urban Pollution

With rapid pace of urbanisation, industries and transport systems grow rather out of proportion. These developments are primarily responsible for pollution of environment, particularly the urban environment. The problem of environmental pollution has already been discussed in details in Chapter 8 and need not be repeated here.

REFERENCES

Adams R, Mc. C., (1968), *The Evolution of Urban Society,* Chicago University, Chicago.

Ahmed, E. (1952) : "Rural Settlement Types in Uttar Pradesh", Annals. Assoc. of American Geographers, Vol. 42, pp. 223-246.

Ahmed E. (1954), *Geographical Essays in India,* Patna.

Alam, S.M. (1966), *Hyderabad-Secundarabad (Twin Cities)* : A Study in Urban Geography, Allied Publishers, Bombay.

Bala, Raj (1986) : *Trends in Urbanization in India*, Rawat Publications, Jaipur.

Berry, B.J.L. (1970) : *Geographical Perspective on Urban Systems,* U.S.A.

Berry, B.J.L. and Spodek, H. : "Comparative Ecologies of Large Indian cities", *Economic Geography,* Vol. 47 (suppl.), pp. 266-285.

Bose, A. (1973) : *Studies in India's Urbanization,* Tata McGraw Hill Publishing Co. Bombay.

Bose, Ashish (1980) : *India's Urbanisation 1901-2001*, Tata McGraw Hill, New Delhi.

Bose, N.K. (1965) : Calcutta : A premature Metropolis, *Scientific American,* Sept., 1965.

Breese, G. (1966) : *Urbanization in Newly Developing Countries,* Prentice Hall.

Brush, J.E. : "The Spatial Patterns of Population in Indian Cities "*Geographical Review*", Vol. 587, 1968, pp. 362-391.

Carter, Harold (1975) : *The Study of Urban Geography,* Edward Arnold, London.

Census of India 2011, *Housing Stock, Amenities and Assets in Slums,* Tables Based on Housing and Housing Census Series 1.

Census of India 2011, *Provisional Population Totals,* Paper 1, of 2011 India, Series 1.

Census of India 2011, *Provisional Population Totals,* Paper 2, Volume 1 of 2011, Rural-Urban Distribution, India Series 1.

Census of India 2011, *Tables on Houses, Household Amenities and Assets,* India Series 2.

Doxiadis, C.A. (1968) : *Ekistics, An Introduction to Science of Human Settlements,* Oxford University Press, New York.

Dube, K.K. and A.K. Singh (ed.) (1988) : *Urban Environment in India*, Inter India Publications, New Delhi.

Economic Survey 2017-18, Govt. of India, Ministry of Finance.

Faweett, C.B. (1932) : "*Distribution of Urban Population in Great Britain, 1931*", Geographical Journal, 79,100-116.

Finch, V.C. *et al.* (1977) : *Elements of Geography, Physical and Cultural*, McGraw Hill, New Delhi.

Fukutake, T. (1967) : *Asian Rural Society,* Seattle.

Geddes, P. (1957), *City in Evolution,* London.

Gibson, M.S. and Langstaff, M.J. (1982) : *An Introduction to Urban Renewal,* Hutchinson, London.

Gosal, Gurdev Singh (1999) (ed) : *4th Survey of Research in Geography,* ICSSR, New Delhi.

Gottmann, J. (1964) : *Megalopolis* : *The Urbanized North-eastern Seaboard of United States,* The MIT Press, Cambridge.

Harris, C.D. (1943) : "*A Functional Classification of Cities in the United States*", Geographical Review, 33, 86-99.

India 2021 : *A Reference Annual,* Publications Division, Ministry of Information and Broadcasting, Govt. of India, New Delhi.

India : Town and Country Planning Organisation, "Land Use Pattern of India's Cities and Towns", *Urban and Rural Thought,* Vol. II, 1965, pp. 188-190.

James, H.E. (1930) : *Urban Geography of India,* Bulletin of Geographical Society of Philadelphia, 28, 101-122.

Janki, V.A., Sayed, Z.A. (1962) : *The Geography of Padra Town,* Baroda University.

Jonston, R.J. (1985) : *Urban Geography, City Structure in Progress in Human Geography.*

Kaushik, S.D. (1959) : "*Types of Human Settlements in Jaunsar Himalaya*", Geographical Review of India, 21, 1-18.

Kohn, C.F. (1954) : "Settlement Geography" In: Jones, P.E. and Jones, C.F. (eds.) : *American Geography—Inventory and Prospectus,* Syracuse University Press.

Krishan, Gopal and Singh, Nina (1996) : *Urbanization—Emerging Scenario and New Challenges*, In: Kulwant Singh and Florian Steinberg (ed.), *Urban India in Crisis,* New Age International Ltd., New Delhi.

Kundu, A. (1992) : *Urban Development and Urban Research in India*, Khanna Publishers, New Delhi.

Lakshman, T.K. and Narayan, B.K. (1984) (eds) : *Rural Development in India*, Himalayan Publishing House, Mumbai.

Lewis, J.P. (1969) : *Quiet Crisis in India,* Washington, D.C.

Manzoor Alam, S. and Fatima Ali (ed) (1987) : *Poverty in Metropolitan Cities,* Concept Publishing Co., New Delhi.

Maurya S.D. (1989) : *Urbanization and Environmental Problems,* Chugh Publications, Allahabad.

Maurya, S.D. (2012) : *Human Geography,* Prayag Pustak Bhawan, Allahabad.

Mayer, Harold M and Kohn Clyde, F. (1967) : *Readings in Urban Geography,* University of Chicago, Chicago.

Mill, E.S. (1980) : *Three Essays on Urbanization in India*, Princeton University (Mimeo).

Misra, R.P. (1978) : *Million Cities of India,* Vikas Publishing House, Pvt. Ltd. New Delhi.

Mitra, A. (1973) : *Functional Classification of Towns,* Institute of Economic Growth.

Mookherjee, D. *et al.* (1973) : *Urbanization in Developing Economy, Indian Perspectives and Patterns,* Berkelay.

Mukherji, Shekhar, (2000) : Distress Migration and Urban Decay in India : Phenomena, Process and Planning Strategies, paper presented in International conference on Land-use change, Environment and Sustainable Development, Dept. of Geography, Banaras Hindu University, Varanasi.

Muttiah, S. *et al.* (eds.) (1987) : *A Social and Economic Atlas of India,* Oxford University Press, New Delhi.

Nagpaul, Hans (1996) : *Modernization and Urbanization in India—Problems and Issues*, Rawat Publications, Jaipur.

Nangia, S. (1976) : *Delhi Metropolitan Regions*, K.B. Publications, New Delhi.

National Institute of Urban Affairs (1988) : *State of India's Urbanization*, New Delhi.

Neale, C. (1986) : *Developing Rural India,* Riverdale, AM.

Negi, B.S. : *Rural Geography*, Kedar Nath Ram Nath, Meerut.

Nelson, H.J. (1955) : "*A Service Classification of American Cities*", Economic Geography, 3, 189-210.

Piggo, S. (1950) : *Pre-Historic India,* Penguin Books.

Pownall, L.L. (1953) : "*Functions of New Zealand Towns*", A.A.A.G. 43/4, 332-350.

Prakasa Rao, V.L.S. (1964) : *Towns of Mysore State,* Asia Publishing House, Delhi.

Prakasa Rao, V.L.S. (1983) : *Urbanization in India : Spatial Dimensions*, Concept Publishing Co., New Delhi.

Premi, M.K. (1986) : *India's Urban Scene and its Future Implications*. Demography India, 20.

Rafiullah, S.M. (1965) : '*A New Approach to the Functional Classification of Towns*'. The Geographer, Vol. XII.

Raza, M,. *et al.* (1979) : *Spatial Organization and Land Utilization.* In: India in R.P. Misra and K.V. Sundaram (eds.) Rural Area Development, Perspective and Approaches, Sterling New Delhi.

Rees, H. (1942-44) : *A Functional Classification of Towns,* Journal of Manchester Geographical Society, 52, 26-32.

Ramachandran, R. (1995) : *Urbanization and Urban Systems in India*, Oxford University Press, New Delhi.

Sharma, R.C. (1975) : *Settlement Geography of Indian Desert,* Kumar Brothers, New Delhi.

Siddhartha, K. and Mukherjee, S. (2001) : *Cities, Urbanization and Urban Systems,* Kisalaya Publications, New Delhi.

Singer, M. (ed.) (1969) : *Traditional India : Structure and Change,* Philadelphia.

Singh, K.N. (1972) : *An Approach to the Study of the Morphology of the Indian Village in Rural Settlements in Monsoon Asia,* NGSI, Varanasi, pp. 204-14.

Singh, Kulwant and Steinberg Florian (ed) (1996) : *Urban India in Crisis*, New Age International Limited Publishers, New Delhi.

Singh L.R. (2003) : *Fundamentals of Human Geography,* Sharda Pustak Bhawan, Allahabad.

Singh, O.P. (1969) : *A Study of Central Places in Uttar Pradesh,* Varanasi.

Singh, R.B. (1975) : *Rajput Clan Settlements in Varanasi Distt.* N.G.S.I. Varanasi Pub. No. 12.

Singh R.L. (1955) : *Banaras* : A Study in Urban Geography, Nand Kishore Brothers, Banaras.

Singh R.L. (1961) : *Meaning, Objectives and Scope of Settlement Geography,* National Geographical Journal of India, 7, 12.

Singh, R.L. *et al.* (1975) : *Readings in Rural Settlement Geography*, National Geog. Society of India, Varanasi.

Singh, R.L. (1985) : Slums of Allahabad : A Socio-Economic Profile, *National Geographer,* Vol. 20, No. 2, pp. 97-98.

Singh, R.Y. (1994) : *Geography of Settlements*, Rawat Publications, Jaipur.

Sinha, V.N.P. (1976) : *Chotanagpur Plateau—A Study in Settlement Geography.*

Sivaramakrishnan, K.C. (1978) : '*Indian Urban Scene*', Indian Institute of Advanced Study, Shimla.

Smailies, A.E. (1962) : *The Geography of Towns,* Hutchinson University Library, London.

Smailies, A.E. (1973) : *Indian City Structure, Urban Geography Proceedings,* Varanasi Symposium, pp. 50-59.

Smith, A.E. (1965) : *Methods and Purpose in Functional Town Classification,* A.A.A.G. 55/3, 39-48.

Sovani, N.V. (1966) : *Urbanization and Urban India*, Asia Publishing House, Mumbai.

Spate, O.H.K and Ahmad, E., "Five Cities of the Gangetic Plain" *Geographical Review,* Vol. 40, 1950, pp. 260-278.

Srinivas, M.N. (ed.) (1955) : *India's Villages*, Bombay.

Sunderam, V.K. (1977) : *Urban and Regional Planning in India*, Vikas Publication, New Delhi.

Taylor, G. (1949) : *Urban Geography,* Methuen, London

Tewari, Vinod K., Weinstein Jay A. and Prakasa Rao V.L.S. (1986) : *Indian Cities : Ecological Perspective*, Concept Publishing Co., New Delhi.

Thakur, V.K. (1981) : *Urbanization in Ancient India*, Abhinav Prakashan, New Delhi.

Tirtha Ranjit and Krishan Gopal (2001) : *Geography of India,* Rawat Publications, Jaipur, Ch. 7.

Turner, Roy (ed.) (1962) : *India's Urban Future*, University of California Press, Berkeley/Los Angles.

Verma, L.N. (2006) : *Urban Geography,* Rawat Publications, Jaipur.

◆◇◆

CHAPTER 15

Multipurpose Projects

A multipurpose project is that which simultaneously serves several purposes. A dam built across a river often serves more than one purpose at a time and is termed as a multipurpose project. Flood control, irrigation, hydroelectric generation, navigation, fishing and tourism are some of the chief aims of a multipurpose project. The development of multi-purpose projects in India since the beginning of planning era in 1951 has been the salient feature of the eoncomic growth of the country. Some of the important multipurpose projects are shown in **Fig. 15.1.**

1. BHAKRA NANGAL PROJECT

It is the largest and the most important multipurpose project named after the two dams built at Bhakra and Nangal on the Satluj river. It is a joint venture of the Punjab, Haryana and Rajasthan states designed to harness the precious water of the Satluj for the benefit of the concerned states. The project comprises of (*i*) Two dams at Bhakra and Nangal, (*ii*) Nangal Hydel Channel, (*iii*) Power houses with a combined installed capacity of 1,204 megawatt (M.W.) (*iv*) Electric transmission lines and (*v*) Bhakra canal system for irrigation.

The Bhakra Dam is one of the highest straight gravity dams in the world. It has been constructed on the Satluj river at the site of Bhakra gorge near Rupnagar (Ropar). The dam is 226 metres high and 518 metres long with its maximum width at the base as 362 metres. The dam has created a huge reservoir of water which is 88 km long and 8 km wide with a storage capacity of 986.8 crore cubic metres. This reservoir is named as Gobindsagar Lake after Guru Gobind Singh, the tenth guru of the Sikh community.

The Nangal Dam has been constructed at Nangal about 13 km downstream of the Bhakra dam. This 29 metre high, 305 metre long and 121 metre wide dam is an auxiliary dam which serves as a balancing reservoir for taking up daily fluctuations from the Bhakra Dam.

The Nangal Hydel Channel is 64.4 km long, 42.65 m wide and 6.28 m deep canal. It has been cemented throughout its length to avoid seepage. It is one of the longest cemented canals of the world. It takes off from the left bank of the river and flows through rugged topography of steep slope. There is a fall of 70 metres in elevation within a distance of 64 km. Its main function is to turn the turbines of power houses located below the Nangal Dam but it also supplies water to the Bhakra irrigation canals.

Power houses have been built to generate hydroelectricity from water of the Satluj river. One power house has been built at Ganguwal about 19 km downstream from Nangal. Another power house has been constructed at a distance of 29 km from the Nangal dam. This is known as the Kotla power house. The Ganguwal and the Kotla power houses have 2 units of 24 MW each and one unit of 29 MW. The third power house is at Rupnagar. Two power houses have been constructed at Bhakra dam, one of

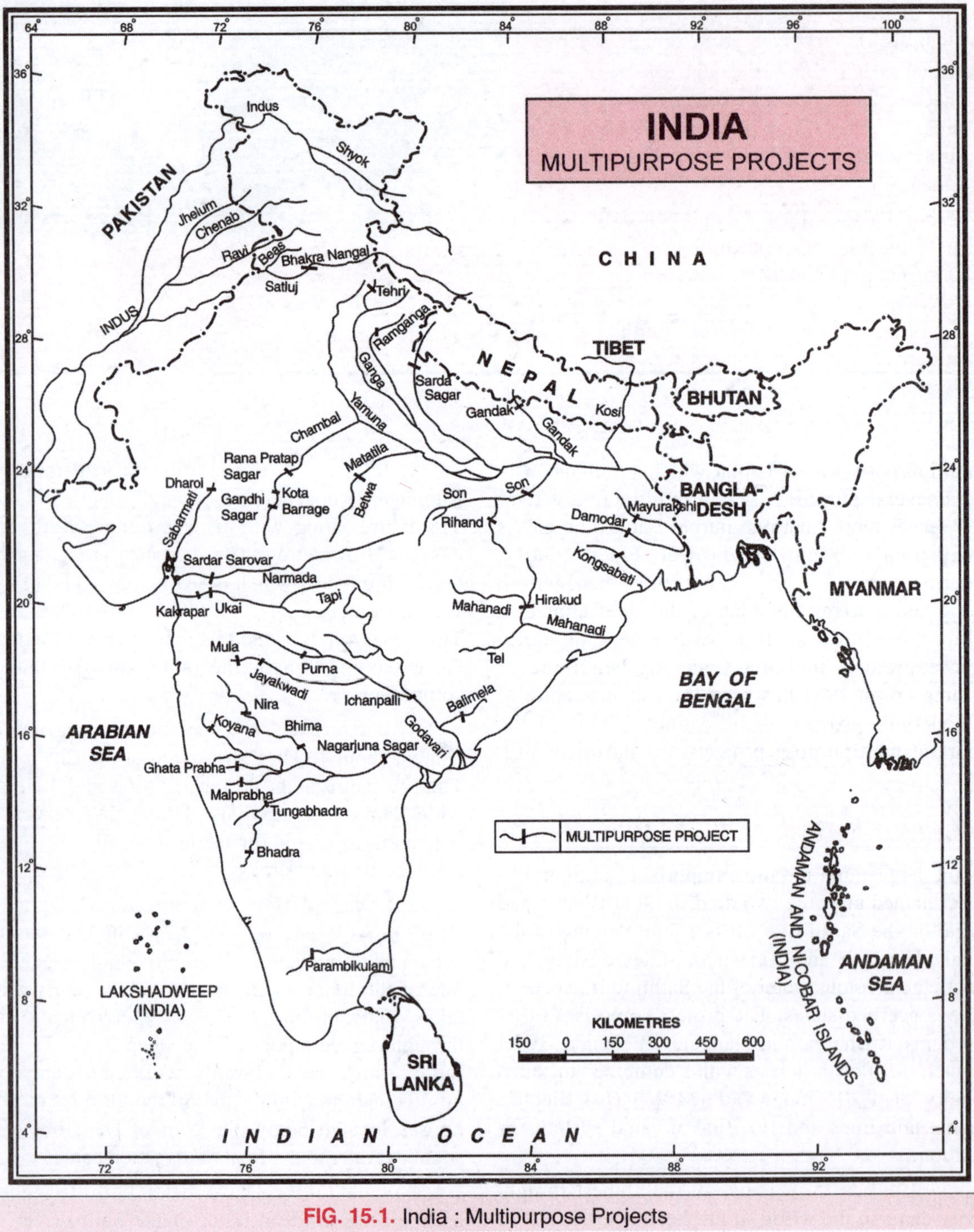

FIG. 15.1. India : Multipurpose Projects

which is on the left and the other is on the right bank of the river. The installed capacity of these two power houses is 450 MW and 600 MW respectively.

The main Bhakra canal is 174 km long. The length of the canal system and that of the distributaries is 1,104 km and 3,360 km respectively. This canal system commands a gross area of about 27 lakh hectares and provides irrigation to about 15 lakh hectares. Of this 37.7 per cent is in Punjab, 46.7 per cent in Haryana and the remaining 15.6 per cent is in Rajasthan.

Transmission Lines have been laid to carry hydroelectricity to the consuming centres. A total of 3,680 km long transmission lines are used to supply power to Rupnagar, Ludhiana, Patiala, Rajpura, Nabha, Moga, Ferozepur, Fazilka, Faridkot, Muktsar, Jalandhar, Hoshiarpur, Kapurthala and Pathankot districts in Punjab; Ambala, Panipat, Hisar, Bhiwani, Rewari, Rohtak, Panchkula, Kalka, Gurgaon, Faridabad and Palwal districts in Haryana; Delhi; Kasauli, Kangra and Shimla in Himachal Pradesh and Rajgarh and Rattangarh in Rajasthan.

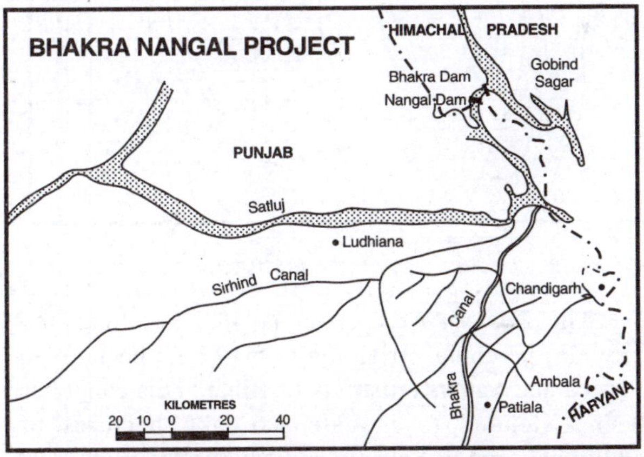

FIG. 15.2. Bhakra Nangal Project

> ***"Bhakra Nangal Project is something tremendous, something stupendous, something which shakes you up when you see it. Bhakra today is the symbol of India's progress."*** **—Nehru**

Some Interesting facts about Bhakra-Nangal Project

1. About 14 million cubic metres of concrete and other building materials have been used for the completion of this project. This material is more than double the material used to build the *grand pyramids* of Egypt.

2. The number of tiles used in the main Bhakra canal is so large that if they are arranged in a straight line, the length of the line so formed will be seven times the length of the equator.

3. Seventy million man-days have been used to construct the canals of the Bhakra-Nangal project.

4. About 95.2 million cubic metres of earth has been excavated for constructing the canals of this project. This amount of soil is enough to construct a 6 metre wide road at 1 metre elevation from New Delhi to New York.

5. This project has helped in obtaining additional 1.3 million tonnes of foodgrains a year, 0.8 million tonnes of cotton, 0.5 million tonnes of sugarcane and 0.1 million tonnes of oilseeds. No other river valley project in the world has so much potential as this project.

However, Bhakra Nangal project like all other river valley projects, suffers from the problem of silting. Water coming from the higher reaches deposits its silt at the bed of the reservoir, thereby reducing its capacity to store water. The capacity of the reservoir was 6.03 million acre feet in 1963 which was reduced to 5.5 million acre feet in 1988, thereby causing a reduction of over half a million acre feet in a short span of only 25 years.

2. THE DAMODAR VALLEY PROJECT

(See Chapter 13)

3. THE HIRAKUD DAM PROJECT

It is an ambitious project of Odisha under the auspices of which a 61 metre high and 4,801 metre long dam has been built on the river Mahanadi at Hirakud about 14 km upstream off the city of Sambalpur. This is one of the longest dams in the world with a gross storage capacity of 8,100 million cubic metres over an area of 630 sq km. Two other dams have been constructed on the Mahanadi—one at Tikrapara and the other at Naraj, a few kilometres west of Cuttack. These three dams provide irrigation to 1 million hectares of land in Sambalpur, Bolangir, Puri and Cuttack districts, generate 3.5 lakh kW of electricity and also offer navigation facilities. The whole of the Mahanadi valley particularly Sambalpur district, Sonapur and the delta region are benefited by this project. The areas served by the Hirakud project are very rich in minerals like iron ore, bauxite, manganese, graphite, chromite, mica and several other useful minerals which require large supply of hydroelectricity for their exploitation. Hirakud project provides the required power to these areas and helps their economic growth. Besides, this project supplies power to a large number of industries and urban centres.

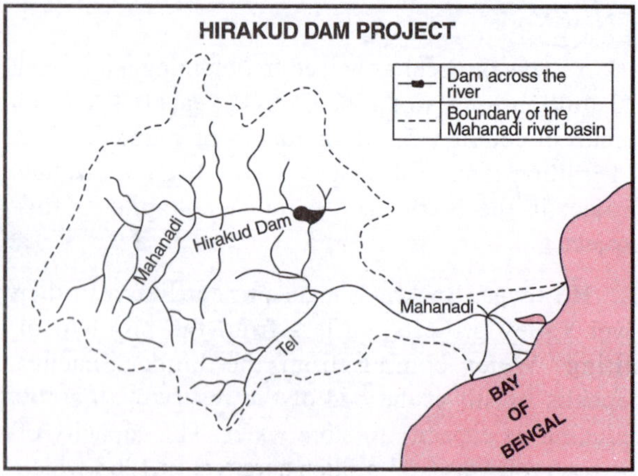

FIG. 15.3. Hirakud Dam Project

4. THE KOSI PROJECT

The Kosi river had earned the dubious name of *sorrow of Bihar* due to its devastating floods and frequent changes in its course. In order to tame the river and save precious lives and property from its annual fury, an agreement was signed with Nepal in 1954 (revised in 1961) and work on the project was started in 1955. Thus it is an international project. The entire plan has been executed by India but the benefits are being shared both by India and Nepal. The chief objectives of this project are irrigation, flood control, power generation, land reclamation, fishing and navigation.

The entire work on this project comprises three units *viz.*, a barrage near Hanumannagar in Nepal, flood embankments on both sides of the river and construction of irrigation canals.

The 1149 metre long and 72 metre high concrete barrage at Hanumannagar in Nepal was constructed in 1965. About 270 km long embankments on either side of the river were completed much earlier in 1959. The eastern and the western flood embankments were later extended by 25.76 km and 4.0 km respectively. This extension has provided protection to an area of 15,190 hectares of land in the lower reaches of the river from recurring submergence by floods.

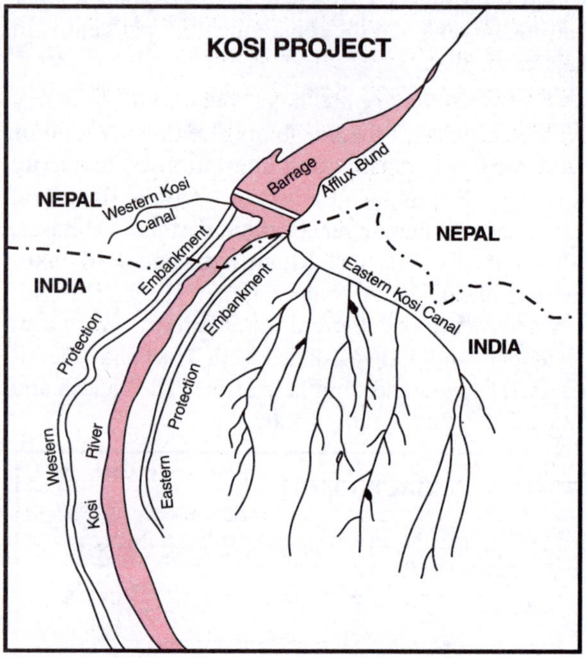

FIG. 15.4. Kosi Project

The *Eastern Kosi* canal is 43.5 km long and provides perennial irrigation to 5.19 lakh hectares in Purnea and Saharsa districts of Bihar. This canal has been extended to irrigate 1.6 lakh hectares of additional land in Saharsa and Munger districts. The 113 km *Western Kosi* Canal takes off from the right bank of the *Kosi barrage* and irrigates about 3.25 lakh hectares in Darbhanga district of Bihar and 12,120 hectares in Satpari district of Nepal. The length of the Rajpur Canal is 9.6 km only but it irrigates 1.6 lakh hectares in Saharsa and Munger districts. It takes off from the Eastern Kosi main canal. The power house at Eastern Kosi canal has an installed capacity of 20 M.W. Electricity generated by this power house is shared on 50 per cent basis by India and Nepal. There is a plan to connect the power house of Kosi to the power houses of the Damodar Valley Project and set up a grid to supply electricity to larger areas.

5. THE RIHAND VALLEY PROJECT

This is by far the most important multipurpose project of Uttar Pradesh. It consists of 934 m long and 92 m high straight gravity concrete dam across the Rihand river (a tributary of the Son) near Pipri in Mirzapur district. The reservoir created by this dam has been named as Gobind Ballabh Pant Sagar. It spreads over an area of 466 sq km and is the largest man made reservoir in India. It has the capacity to hold 11.4 lakh hectare metres of water. Another dam about 25 km north of the Rihand Dam has been constructed at Obra. The power house near the Rihand dam has 6 units with a 5 MW capacity each. Besides, power house at Obra also has 6 units, each having a capacity of 50 M.W. This power is supplied to the eastern parts of Uttar Pradesh, western parts of Bihar and northern parts of Madhya Pradesh. The Son river valley, within a radius of 200 km from Pipri, is blessed with vast deposits of coal, iron ore, limestone, mica, bauxite, feldspar, etc. The hydroelectric power generated by the Rihand Valley Project goes a long way to help in exploiting these mineral resources and usher into a new era of economic prosperity.

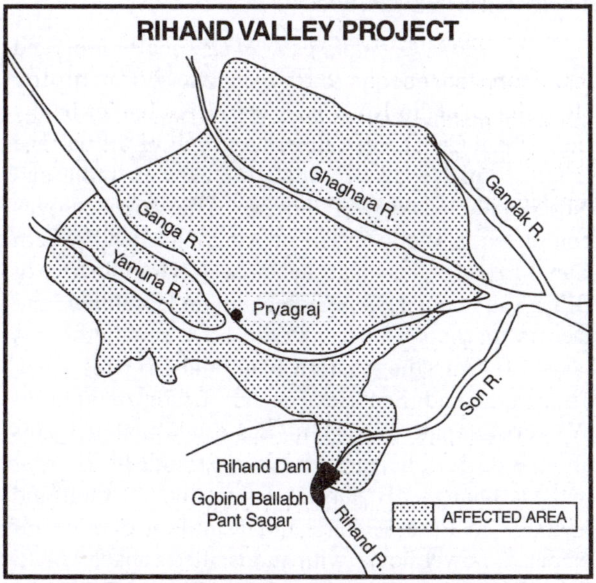

FIG. 15.5. The Rihand Valley Project

About 5,000 wells in Uttar Pradesh have been energised. Another 2.25 lakh hectares of land has been provided with tube-well irrigation in Bihar.

Flood control in Son valley, fishing in Gobind Ballabh Pant Sagar, tourism, prevention and control of soil erosion in Baghelkhand region are some of the other benefits from this project.

6. THE CHAMBAL VALLEY PROJECT

This is a joint venture of Rajasthan and Madhya Pradesh initiated in 1954 on the Chambal river (the main tributary of the Yamuna). The project aims at harnessing the Chambal river for irrigation, power generation and for prevention and control of soil erosion in the valley. The project has been executed in three successive stages.

The first stage consists of construction of the 64 m high and 514 m long Gandhisagar dam about 8 km downstream of the Chaurasigarh fort in Bhanupura tehsil at the border of M.P. and Rajasthan. Constructed in 1960, the dam has created the Gandhisagar reservoir which spreads over an area of

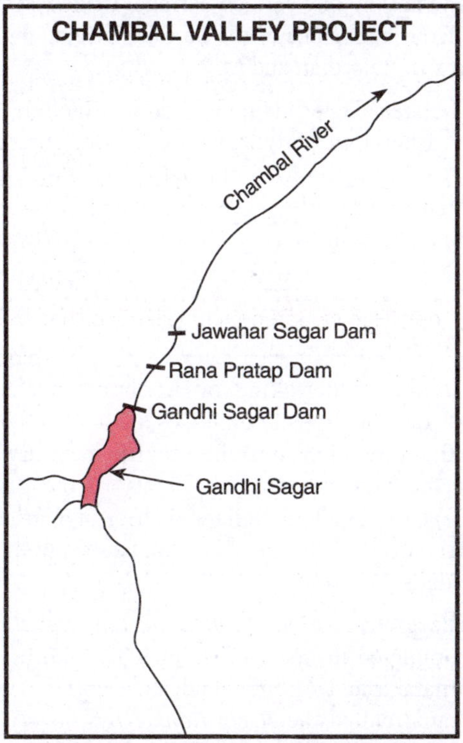

FIG. 15.6. The Chambal Valley Project

688 sq km. It has a capacity to hold 692 crore cubic metres of water which provides irrigation to 4.44 lakh hectares. Five units of 23,000 K.W. capacity each have been set up at the dam site.

The second stage includes the construction of the 54 m high and 1,143 m long *Rana Pratap masonry dam* at Rawatbhata about 56 km downstream of the Gandhisagar dam. It has a gross storage capacity of 290 crore cubic metres spreading over an area of 198 sq km. It provides irrigation to 1.2 lakh hectares of land. The Rana Pratap power station is located on the left bank just at the toe of the dam. Four generating units of 43,000 KW. each are installed here.

The third stage consists of the construction of a 45 m high and 548 m long gravity dam, known as the *Jawahar Sagar dam or Kota dam*, about 29 km upstream of Kota city. It was completed in 1971-72. The reservoir created by this dam has a potential of 68 million cubic metres. Three generating units of 33,000 K.W. each have been installed here.

Kota barrage was constructed in 1960, at a distance of less than 1 km from Kota. This is 36 m high and 600 m long earthen barrage. Irrigation canals taken from both sides of the barrage irrigate 4.4 lakh hectares in Rajasthan and M.P.

The areas benefited by this project include Kota, Bundi, Bharatpur, Jaipur, Sawai Madhopur, Tonk, Ajmer, Pali, Bhilwara, Sirohi and Udaipur districts of Rajasthan and Mandsaur, Indore, Ujjain, Gwalior and Ratlam districts of M.P.

7. TUNGBHADRA MULTIPURPOSE PROJECT

It is a joint undertaking of Andhra Pradesh and Karnataka. The project comprises a 2,441 metre long and 50 metre high straight gravity masonry dam across the Tungabhadra (a tributary of the Krishna River) at Mallapur in Bellary district of Karnataka, two irrigation canals and power houses on both sides of the dam.

The 349 km long *Right Bank Low Level* canal was completed in 1963. It irrigates 3.5 lakh hectares in Karnataka and Andhra Pradesh. Another 196 km long canal called the *Right Bank High Level Canal* has also been completed and irrigates 36 thousand hectares. The 227 km long *Left Bank Low Level Canal* was also completed in 1963. It irrigates 1.82 lakh hectares of land.

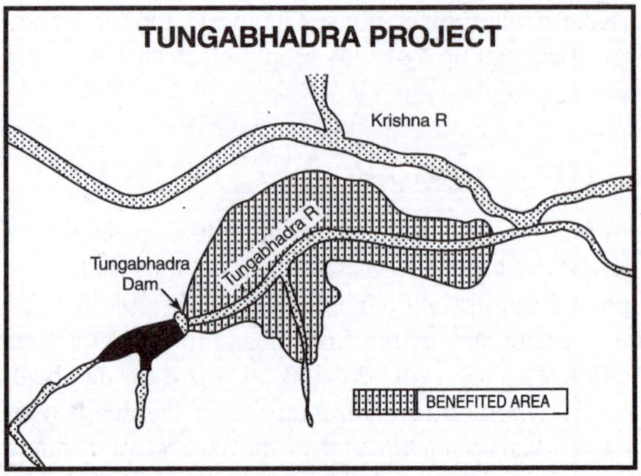

FIG. 15.7. Tungabhadra Project

Of the three power stations, two are situated at the foot of the dam (one on either side of the dam) and the third is at 15th km of the Right Bank Low Level Canal. All the three power stations have a potential of 126 M.W.

8. THE GANDAK PROJECT

This project is jointly executed by Bihar and Uttar Pradesh. Nepal also derives some benefit according to an agreement signed in 1959.

This project comprises a 747.37 metre long and 9.81 metre high barrage at Bhansoloten (Balmikinagar) in Bihar near the Nepal border below the Tribeni Canal Head Regulator (half of the barrage length is in Nepal), 4 canals, (2 each in India and Nepal) and a power house. The Project was completed in 1966-67. The 66 km long Main Western Canal irrigates 4.84 lakh hectares in Saran district of Bihar and about 1.88 lakh hectares in Gorakhpur and Deoria districts of Uttar Pradesh. A separate canal takes off from the western bank and irrigates about 16.4 thousand hectares in the Bhairwa area of Western Nepal. The Main Eastern Canal irrigates about 6.03 lakh hectares in Champaran, Muzaffarpur and Darbhanga districts of Bihar and 42 thousand hectares in Parasa, Bara and Rautuhat districts of Nepal. A power house with an installed capacity of 15 M.W. on the Main Western Canal has been commissioned and has been gifted to Nepal.

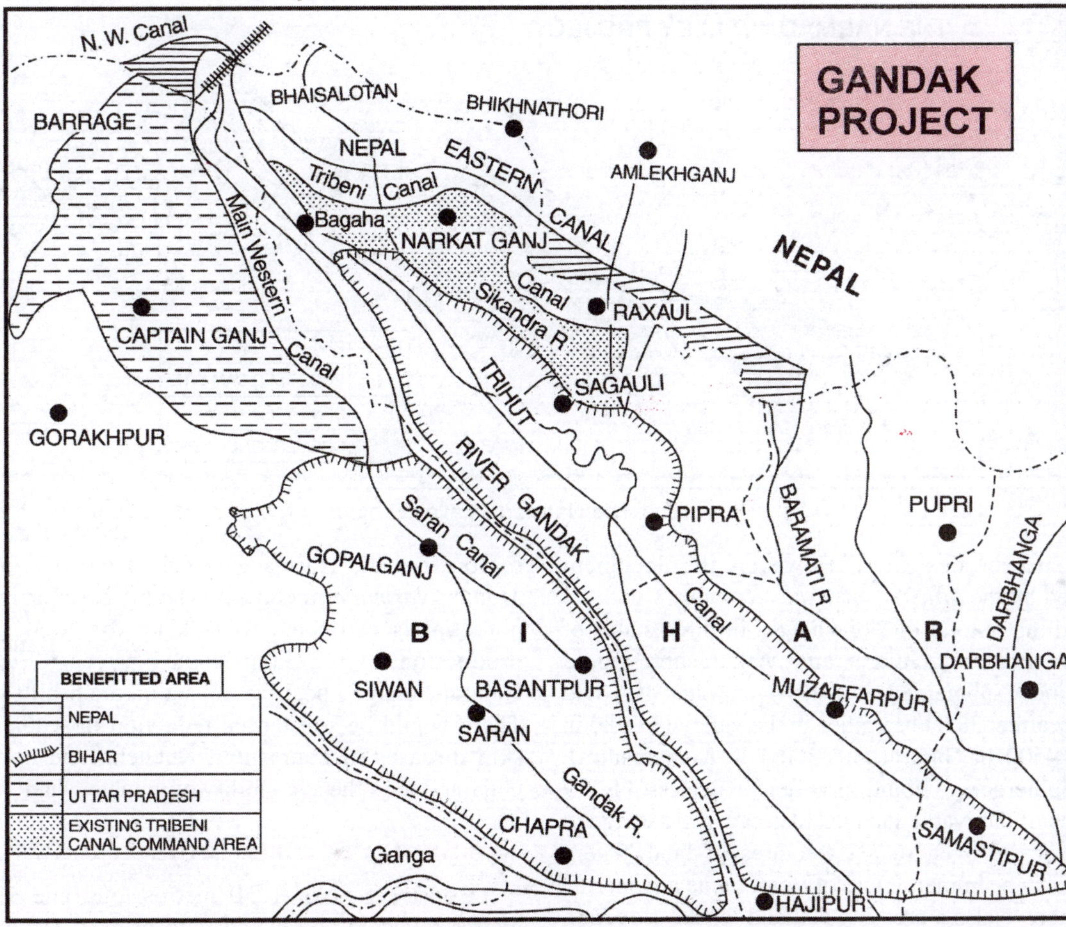

FIG. 15.8. The Gandak Project

9. THE NARMADA VALLEY PROJECT

Originating in the Amarkantak plateau of Madhya Pradesh, the Narmada is the fifth largest river of India and the largest among the west flowing rivers of the Peninsula. The volume of average annual flow of water is 40,700 million cubic metres, 90 per cent of which flows during the monsoon season. Only 5 per cent of this flow is utilized and 95 per cent flows unutilized to the Gulf of Khambhat. The Narmada Valley Project aims at harnessing this flow for the economic prosperity of the concerned areas. This is going to be one of the largest river valley projects of the world because the entire project includes the construction of 30 major, 135 medium and 3,000 minor dams on the river. Conceived in 1945-46 the project has been surrounded by controversies which have become very sharp after 1987 when the construction started and in the present day context this project is more controversial than any other river valley project of the world. There seems to be a tug of war between the environmentalists on one hand, and the protagonists of the project on the other. The environmetalists have predicted doom as, according to them, the project would lead to waterlogging, soil erosion, destruction of forests, cultivable land and wild life and also trigger health problems. They plead that the reservoirs created by two main dams *viz*. the *Sardar Sarovar dam and the Narmada Sagar Dam* will submerge 480 villages and displace 1.5 lakh persons and 8 lakh cattle heads. The Bargi Dam, a part of the giant *Narmada Sagar-Sardar Sarovar* scheme, has already dislocated about one lakh people in 162 villages in M.P. This dam was commissioned in 1968. Besides 50,000 hectares of reserve forests

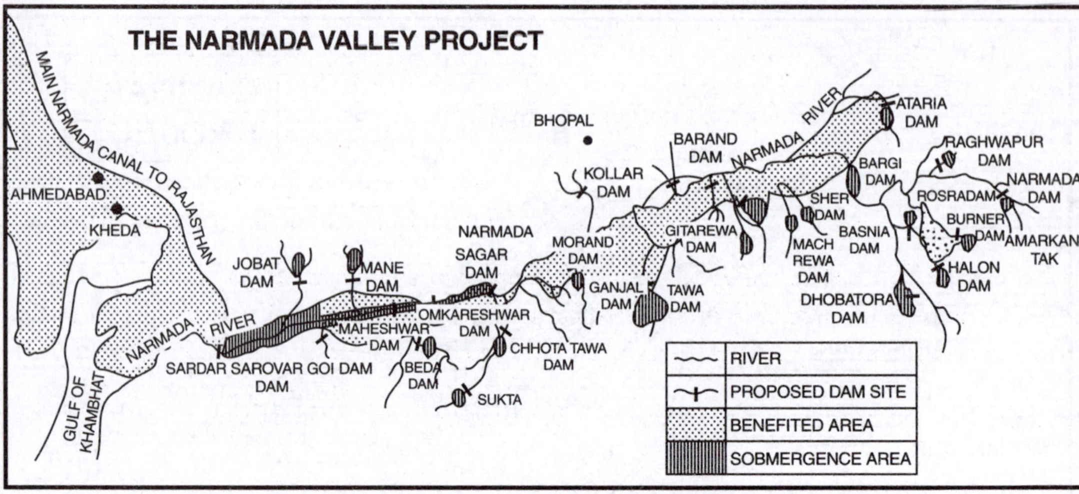

FIG. 15.9. The Narmada Valley Project

would also be destroyed. However, the document prepared by the *Ministry of Water Resources* in 1993 showed that a total of 245 villages in three states of Madhya Pradesh, Gujarat and Maharashtra would come under submergence of the river water. It would directly affect 38,044 families in the valley (31,080 in M.P., 4,500 in Gujarat and 2,464 in Maharashtra). The submergence would also render useless 11,279 hectares of cultivable land besides causing damage to forest spread over 10,719 hectares of land. There would also be immense loss of flora, fauna and marine life. There is also a pressing demand for the reduction of height of the *Sardar Sarovar Dam* from 455 ft to 436 ft. to save over 38,000 people from rehabilitation and 25,000 acres of land from getting submerged.

On the other hand, the protagonists of the project plead that this project would augur an impetus for a commendable economic development of the region. There is no denying the fact that the Narmada scheme is designed to benefit the people in not only Gujarat but also those in Madhya Pradesh and Maharashtra, with Rajasthan tagged as a marginal beneficiary to get more drinking water. The dams on the river are designed to produce about 3,000 M.W. of hydroelectricity. The canal system to be built in the integrated scheme is expected to provide irrigation facility to 17.92 lakh hectares of land in Gujarat, 1.40 lakh hectares in Madhya Pradesh and 73,000 hectares in Rajasthan. Although 87% of the Narmada flow is in M.P., 11.5% in Gujarat and 1.5% in Maharashtra, Gujarat is the main beneficiary state. This is the reason that the Narmada is called the *life line or Mother Narmada* in Gujarat. The increased irrigation potential is estimated to jack up the food grains production by 43 lakh tonnes a year. Further, Gujaratis claim a positive impact on the health scene. There would be substantial reduction in scabies and skin diseases in Saurashtra, Kuchchh and northern Gujarat due to the availability of potable water.

The Sardar Sarovar Project (SSP)

Construction of 1250 metres long and 121.92 metres tall dam completed upto brim in December, 2006 in the Lower Narmada Valley in Gujarat (Fig. 15.9). It has create a reservoir having a potential of 77 lakh acre feet. It has provide irrigation to 18 lakh hectares in Gujarat which has helped in increasing the farm production by 82 lakh tonnes. About 10.6 lakh hectare metre water has been made available for domestic and industrial use to 131 cities and towns and to 8,720 villages. Two power stations produce 1,450 M.W. of hydroelectricity. An area of 4,650 hectares has been brought under forest to partly compensate for 10,713 hectares which has been submerged under the reservoir. It will provided employment to 4 lakh persons during the course of construction and to about 6 lakh persons after the completion due to increase in agriculture, animal husbandry, dairy, fertilizer industry and other allied occupations. The initial cost of the project was put at ₹ 5,793 crore which has now escalated to about ₹ 13,400 crore.

HIGHLIGHTS OF SARDAR SAROWAR PROJECT

- Generation of 1450 MW Hydro Power achieved.
- Beneficial to Gujarat, Rajasthan, Madhya Pradesh and Maharashtra.
- Irrigation facility in 18 lakh hectares in Gujarat and 2.50 lakh hectares in Rajasthan.
- Drinking water facility to 2 crores population of Gujarat and 15 lacs population of Rajasthan.
- Main Canal of 458 kms has been constructed and Branch Canals of 2,585 kms.
- 5,112 kms distributaries.
- 18,453 kms minor and 48,058 kms sub minor canals.
- Narmada Dam site is emerging as a favourite and pleasing Tourist Place, more than 12,00,000 Tourists visit the Narmada Dam site every year.
- Share of water after completion of the project will be by Madhya Pradesh (65.18%), Gujarat (32.14%), Rajasthan (1.79%) and Maharashtra (0.89%).
- States' share of 1,450 MW hydro power will be Madhya Pradesh (57%), Maharashtra (27%) and Gujarat (16%).

The Narmada Sagar Dam Project

The foundation stone of the Narmada Sagar Dam on the Narmada in Madhya Pradesh was laid on 3rd October, 1984. When completed, it will completely submerge 89 villages and 91 thousand hectares of land in East Nimar (Khandwa), Dewas and Hoshangabad districts. Besides, 105 villages will be partially submerged (**Fig. 15.9**). About 900 sq km of black soil fertile land and 40,000 hectares of forest land would also be submerged in water. On the positive side this project will usher into a new era in the form of added irrigation, water power generation, potable water and water for industrial purposes, fishing, tourism, etc.

On June 12, 2014, the Narmada Control Authority (NCA) permitted the Gujarat government to raise the height of the Narmada dam by about 17 metres from 121.92 metres to 138.68 metres which will also be the height of the reservoir in closed gates condition. With this increase in height the storage capacity of the reservoir will go up from the existing 1.27 million acre feet to 4.75 million acre feet.

10. THE NAGARJUNASAGAR PROJECT

Started in 1955-56, the Nagarjuna Sagar Project comprises the construction of a 124.7 metre high and 1450 metre long concrete dam on the Krishna river in Nalgonda district of Telangana about 114 km to the south-east of Hyderabad. The reservoir behind the dam spreads over an area of 118 sq km with a storage capacity of 1,156 crore cubic metre of water. Two canals have been taken, one each from either side of the dam. The 349 km long *Jawahar canal* on the right side of the dam has a capacity to irrigate about 4 lakh hectares in Guntur, Prakasam (Ongole) and Nellore districts. The 357 km long *Lal Bahadur canal* irrigates about 3 lakh hectares in Khammam, West Godavari and Krishna districts. A power house with two units of 50 M.W. each was set up at the toe of the dam in 1970. It supplies electricity to Nalgonda, Mahaboobnagar and Hyderabad districts for various purposes.

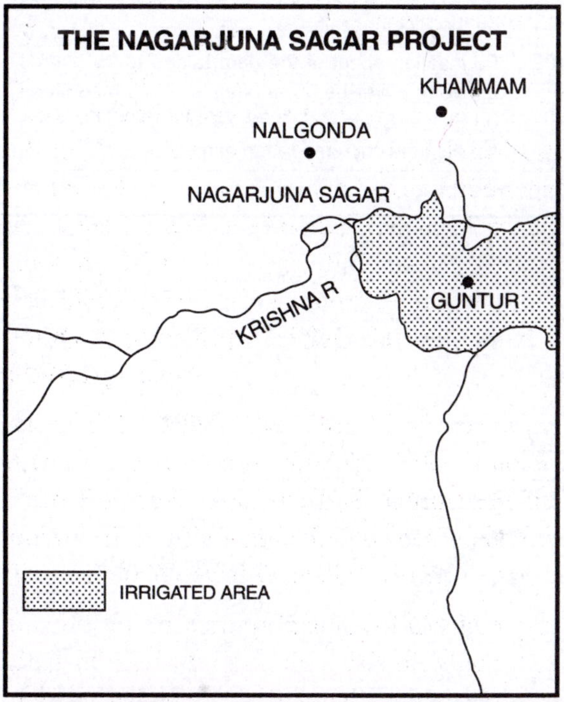

FIG. 15.10. Nagarjunasagar Dam Project

11. TEHRI DAM PROJECT

This dam in Tehri Garhwal was conceived in 1949. Situated at the confluence of the Bhagirathi and the Bhilganga rivers in Garhwal district of Uttarakhand, this 2160.5 m long Tehri dam, said to be the highest in Asia, was commissioned in 1972, but work on it began only in 1978. And ever since,

MAJOR EVENTS CONCERNING TEHRI DAM

1949 Tehri dam on the Bhagirathi river conceived.

1961 Survey started for site selection.

1978 Construction starts for 600 M.W. power project.

1979 Installed capacity raised to 1,000 M.W.. Antidam protests starts and work on the project suspended for seven years.

1986 Union Cabinet approves Tehri as a joint project of Centre and Uttar Pradesh government. Assistance from erstwhile USSR sought. Writ petition filed in the Supreme Court against the project on grounds of safety and environmental impact. Installed capacity raised to 2400 MW.

1988 Tehri Hydro Development Corporation formed.

1990 Experts conclude that Tehri dam is safe even in the event of a severe earthquake of 8 magnitude on the Richter scale.

1991 Foundation sheet of the dam raised to 15 metres above the river bed. Earthquake of 6.6 magnitude on Richter Scale rocks Uttarkashi on 20th October. No damage occurred to the dam.

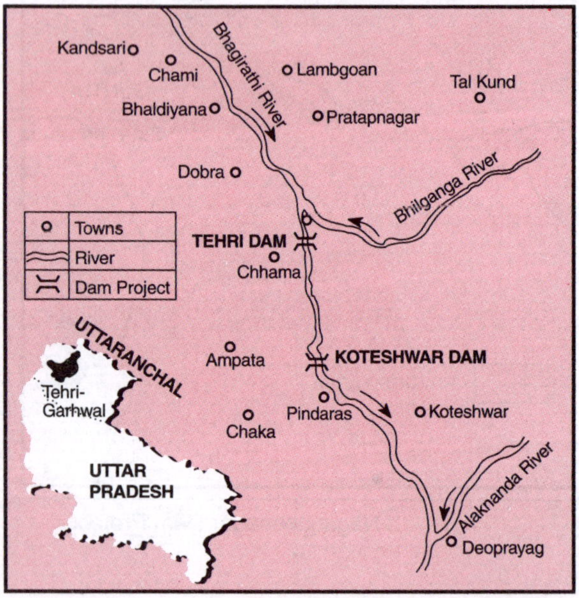

FIG. 15.11. Tehri Dam

environmental and safety aspects threatened the project. This rock-fill dam, generates 3,500 MW of electricity, irrigates 270,000 ha of arid land between the Ganga and Yamuna. A gigantic reservoir has been created over an area of 4,200 ha which feeds an underground power house with a racing torrent generating enough electricity. The initial project cost of ₹ 192 crore (in 1978), reached ₹ 6,000 crore in 1998.

The 2,400-MW project has submerged town of Tehri, which was well known as a seat of Garhwali culture, and 23 villages in its vicinity. About 72 other villages along the river have also been partly submerged. People from 21 villages have already been displaced for constructing the New Tehri township, thus displacing about 70,000 people from their ancestral lands. The dam would also flood 1,000 ha of cultivated land, 1,000 ha of forest land and 2,000 ha of pasture land.

12. THE BEAS PROJECT

A joint venture of Punjab, Haryana and Rajasthan, the project consists of (*i*) Beas-Satluj link and (*ii*) Pong dam on the Beas. The Beas-Satluj link involves the construction of a 61 metre high diversion dam at Pandoh in Himachal Pradesh, about 27 km long water conductor system comprising two tunnels 12-13 km long each and an open hydel channel and a power plant at Dehar with an installed capacity of four units of 165 M.W. each. It has the capacity to provide irrigation to about 5.25 lakh hectares in Punjab and Haryana.

The second unit includes a 116 metre high dam at Pong in the Dhaoladhar range near Pong village. It is mainly an irrigation scheme which is intended to ensure extension of perennial irrigation to about 21 lakh hectares in Punjab, Haryana and Rajasthan. It also provides for a power plant of 240 M.W. capacity with provision of 2 future units of 60 M.W. each. The total installed capacity of the Beas complex is 1020 M.W. including one unit of 120 M.W. at Bhakra right bank power house.

Apart from the above major multipurpose projects, there are several other projects including the Mahi project, Mayurakshi, Rampadsagar, Ramganga, Machkund, and many more which have been taken up to boost the economy of the country.

REFERENCES

Agarwal, Anil; Narain, Sunita and Sen, Srabane (eds.) (1999) : Citizens' Fifth Report, State of India's Environment, Centre for Science and Environment, New Delhi, Chap. 4.

A Social and Economic Atlas of India (1987) : Oxford University Press, New Delhi.

Irrigation and Power Projects (1970) : Five Year Plans, Publications Division, New Delhi.

Johnson, B.L.C. (1979) : *India : Resources and Development*, New York.

Malik, Ashok : *Ma Narmada's Children*, The Hindustan Times, July 2, 1994.

Our River Valley Projects (1961) : Publication Division, Govt. of India, New Delhi.

Qureshy Ahtesham, *Human Aspect of Narmada Project.* The Hindustan Times, August 8, 1993.

Report of Irrigation Commission (1972) : Vol. III (Part I), New Delhi.

◆◇◆

CHAPTER 16

Irrigation

Water is an important input for successful agriculture. Water may be available to crops in the natural course by rainfall or it may be supplied to the agricultural fields artificially by human efforts. *The process of supplying water to crops by artificial means such as canals, wells, tube-wells, tanks, etc. from the sources of water such as rivers, tanks, ponds or underground water is called irrigation.*

NEED FOR IRRIGATION

The geographical conditions, especially the nature of monsoon rainfall, in India make irrigation indispensable for sustainable agricultural development. Unfortunately, rainfall in India is **uncertain, unreliable, irregular, variable, seasonal and unevenly distributed.** The main rain bringing south-west monsoon often fails to keep its date. It may come either before or after the scheduled date of arrival. Normally speaking, the rainfall keeps its date of arrival and withdrawal only in one out of five years. The amount of rainfall may also vary greatly from the normal. Excess rainfall may cause floods but deficient rainfall forces the farmers to resort to irrigation. Ironically, the variability of rainfall is very high in areas of low rainfall. The north-western parts of the country, especially Punjab, Haryana, Rajasthan and western parts of Uttar Pradesh often suffer from high variability of rainfall. There are large variations in the spatial distribution of rainfall. On the one end of the scale, there are areas in Meghalaya which receive more than 1000 cm of annual rainfall while on the other end there are parts of the Thar Desert which receive less than 10 cm of rain in a year. Only 30.2 per cent of the cultivated area in India receives sufficient rainfall where the annual rainfall exceeds 100 cm. About 35.7 per cent of the cultivated area receives 75 to 100 cm of annual rainfall and 34.1 per cent of the cultivated area receives less than 75 cm of annual rainfall. Therefore, it is clear that about two thirds of the total cropped area needs irrigation facilities. Even in areas of high rainfall, irrigation is necessary to further increase farm productivity. Indian rainfall is characterised by *monsoon gaps*. Consequently it may not rain for two or more weeks during the rainy season and the crops may be badly damaged in the absence of irrigation facilities. The chief characteristic of the Indian monsoon is that it is seasonal. About 75 per cent of the rainfall in India is caused by the south-west monsoons which are active only for 3-4 months in a year. The remaining 8-9 months are marked by dry season when irrigation is badly needed for successful growing of the crops. The duration of dry season varies from 5 months in Kerala to over 9 months in north-west India. Moreover, rainfall in most parts of India is torrential. As the popular saying goes, **it pours, it never rains in India.** This leaves little opportunity for soil to absorb water and much of the surface water goes waste. This loss of water due to wasteful flow has to be compensated by irrigation.

Apart from the vagaries of monsoon rain as described above, there are certain crops such as rice, sugarcane, jute, cotton, chillies, etc. which require more water and have to be provided with irrigation even in areas of heavy rainfall. It is estimated that yields of irrigated crops are 50 to 100 per cent higher than those of the unirrigated crops under similar geographical conditions. With the introduction of high yielding varieties (HYV) of seeds and heavy doses of chemical fertilizers since the second half of 1960s, irrigation has become an indispensable ingradient of Indian agriculture. The ever increasing population leads to more intensive agriculture which needs more irrigation facilities, along with other inputs.

To raise the productivity of agriculture in India there is need to expand the acreage under irrigation along with adoption of appropriate technologies for efficient utilisation of water through suitable pricing. First, adoption of irrigation technologies which improve efficiency in the use of water is imperative in a scenario where flood irrigation has resulted in wastage of water. Second, focus on efficient irrigation technologies is important with increasing water shortages owing to climate change and indiscriminate wastage of water in agriculture and other uses. Having '*more crop per drop*' through efficient irrigation technologies should be the motto to improve productivity in agriculture which can ensure food and water security in the future. The significance of irrigation can be seen in **Figure 16.1** which compares the yield per crop under rain fed and irrigated agriculture.

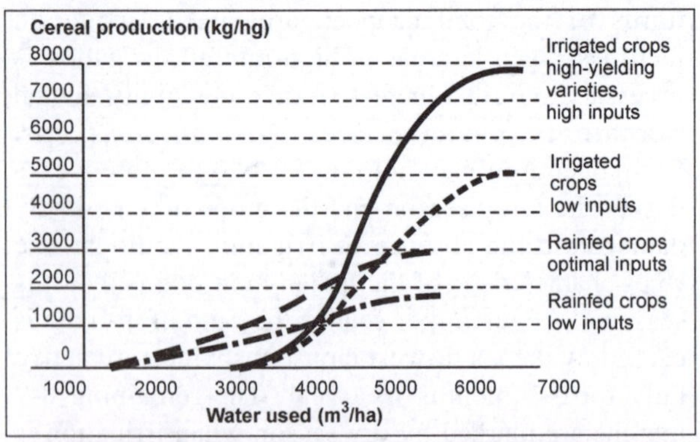

FIG. 16.1. Comparative Yields of Cereals under Rainfed and Irrigated Farming

Source : FAO, 2002.

GEOGRAPHICAL FACTORS FAVOURING IRRIGATION

Certain geographical factors have helped in developing irrigation in different parts of India. The northern plain of India has extremely rich fertile soils deposited by the mighty rivers flowing from the Himalayan ranges. In fact, the Indo-Gangetic plain is considered to be one of the most fertile plains of the world. The slope of the land is so gentle that canals can carry the irrigation water to far off places. The soft and friable nature of the soil makes it easy to dig canals and to sink wells. The deep clay in the sub-soil acts as reservoir for rain water which percolates through the porous alluvium. There is thus, a large quantity of ground water which is taken out for irrigation through wells and tube wells. The area is blessed with a large number of perennial rivers which provide water for irrigation throughout the year. However, the rate of flow of water fluctuates with the change of season. In the Peninsular plateau area, the rocks are hard and the surface relief is uneven. Therefore, it is difficult to have canals and wells and tanks are used for irrigation.

Types of Irrigation Projects

Irrigation Projects in India are classified into three categories *viz*. Major, Medium and Minor Irrigation. Projects which have a Cultivable Command Area (CCA) of more than 10,000 hectare are termed as ***Major Projects***, those Irrigation Projects which have a CCA of less than 10,000 hectare but more than 2,000 hector are termed as ***Medium projects*** and those Irrigation Projects which have a CCA of 2,000 hectare or less are known as ***Minor projects***.

TYPES OF IRRIGATION TECHNIQUES

Various types of irrigation techniques differ in how the water obtained from the source is distributed within the field. In general, the goal is to supply the entire field uniformly with water, so that each plant has the amount of water it needs, neither too much nor too little. The various irrigation techniques are as under :

1. Surface Irrigation. In surface irrigation systems, water moves over and across the land by simple gravity flow in order to wet it and to infiltrate into the soil. Surface irrigation can be subdivided into furrow, border strip or basin irrigation. It is often called flood irrigation when the irrigation results in flooding or near flooding of the cultivated land.

2. Localized Irrigation. Localized irrigation is a system where water is distributed under low pressure through a piped network, in a pre-determined pattern, and applied as a small discharge to each plant or adjacent to it. Drip irrigation, spray or micro-sprinkler irrigation and bubbler irrigation belong to this category of irrigation methods.

3. Drip Irrigation. Drip irrigation, also known as ***trickle irrigation*****,** functions as its name suggests. Water is delivered at or near the root zone of plants, drop by drop. This method can be the most water-efficient method of irrigation, if managed properly, since evaporation and runoff are minimized. In modern agriculture, drip irrigation is often combined with plastic mulch, further reducing evaporation, and is also the means of delivery of fertilizer.

4. Sprinkler Irrigation. In sprinkler or overhead irrigation, water is piped to one or more central locations within the field and distributed by overhead

TABLE 16.1. Net area irrigated from different sources and gross irrigated area—All India

(Thousand Hectares)

Year	Source of Irrigation							Net Irrigated Area (Col. 4 to 8)	Gross Irrigated Area	Area Irrigated more than once (Col. 10-9)
	Canals			Tanks	Tube-Wells	Other Wells	Other source			
	Govt.	Private	Total							
1	2	3	4	5	6	7	8	9	10	11
1996-97	16,889	220	17,109	2,821	19338	12,457	3,388	55,113	76,025	20,912
1997-98	17,186	211	17,397	2,597	19,680	12,431	3,106	55,211	75,670	20,459
1998-99	17,099	212	17,311	2,795	21,394	12,606	3,329	57,435	78,670	21,235
1999-00	17,247	194	17,441	2,539	22,042	12,597	2,912	57,531	79,216	21,685
2000-01	15,809	203	16,012	2,466	22,566	11,252	2,909	55,205	76,187	20,982
2001-02	14,993	209	15,202	2,196	23,245	11,952	4,342	56,937	78,371	21,434
2002-03	13,867	206	14,073	1,811	25,627	8,728	3,659	53,898	73,056	19,158
2003-04	14,251	206	14,458	1,916	26,691	9,693	4,299	57,057	78,042	20,985
2004-05	14,553	214	14,766	1,734	25,235	9,956	7,538	59,229	81,078	21,849
2005-06	16,490	227	16,718	2,083	26,025	10,044	5,966	60,837	84,279	23,442
2006-07	16,802	224	17,027	2,078	26,942	10,698	5,999	62,743	86,752	24,008
2007-08	16,531	217	16,748	1,973	28,496	9,864	6,107	63,189	88,057	24,869
2008-09	16,686	195	16,881	1,981	28,366	10,389	6,020	63,637	88,895	25,258
2009-10	14,786	188	14,975	1,585	28,370	9,991	7,024	61,945	85,087	23,142
2010-11	15,475	171	15,646	1,979	28,543	10,629	6,869	63,665	88,940	25,275
2011-12	15,837	172	16,008	1,917	29,943	10,594	7,245	65,707	91,786	26,079
2012-13	15,512	165	15,667	1,752	`30,543	10,762	7,552	66,285	92,247	25,961
2013-14(P)	16,115	163	16,278	1,841	31,128	11,311	7,557	68,116	95,771	27,656
2014-15(P)	16,020	163	16,812	1,723	31,606	11,354	7,519	68,383	96,457	28,074

Source **:** EnviStats India, 2018, p. 185.

(P) : Provisional

high-pressure sprinklers or guns. A system utilizing sprinklers, sprays, or guns mounted overhead on permanently installed risers is often referred to as a solid-set irrigation system. Higher pressure sprinklers that rotate are called rotors and are driven by a ball drive, gear drive, or impact mechanism. Guns are used not only for irrigation, but also for industrial applications such as dust suppression and logging. Sprinklers can also be mounted on moving platforms connected to the water source by a hose. Automatically moving wheeled systems known as traveling sprinklers may irrigate areas such as small farms, sports fields, parks, pastures, and cemeteries unattended.

5. Sub-irrigation. Sub-irrigation also sometimes called seepage irrigation has been used for many years in field crops in areas with high water tables. It is a method of artificially raising the water table to allow the soil to be moistened from below the plant root zone. Often these systems are located on permanent grasslands in lowlands or river valleys and combined with drainage infrastructure. A system of pumping stations, canals, weirs and gates allows it to increase or decrease the water level in a network of ditches and thereby control the water table. Sub-irrigation is also used in commercial greenhouse production, usually for potted plants. Water is delivered from below, absorbed upwards, and the excess collected for recycling.

GROWTH AND UTILIZATION OF IRRIGATION

Irrigation has been practised in India since ancient times. The Hindu monarchs, the Mughal emperors and the British rulers exhibited great engineering feats to develop irrigation at different times in the history of India. In the pre-Independence period, the undivided India had some of the best irrigation systems in the world. Much of the canal irrigated area of the Satluj and the Indus system went to Pakistan after partition.

Sustained and systematic programme for development of irrigation facilities was taken up with the advent of Planned Development in 1951. A broad-assessment of the area that can be ultimately brought under irrigation, both by surface and ground water, made by the various States in sixties has indicated that ultimate irrigation potential of the country would be of the order of 113 m.ha (million hectare). However, the ultimate potential is 139 m.ha, the increase being primarily due to upward revision in assessed potential of minor ground water schemes and minor surface water schemes. Minor irrigation projects have both surface and ground water as their source, while Major and Medium projects mostly exploit surface water resources.

SOURCES OF IRRIGATION

Different sources of irrigation are used depending upon the topography, soils, rainfall, availability of surface or ground water, nature of rivers (whether perennial or non-perennial), requirements of crops, etc. The main sources of irrigation used in different parts of the country are (*i*) Canals; (*ii*) Wells and tube wells (*iii*) Tanks and (*iv*) Others (Dongs, Kuhls, springs, etc.)

TANK IRRIGATION

A tank consists of water storage which has been developed by constructing a small bund of earth or stones built across a stream. The water impounded by the bund is used for irrigation and for other purposes. Some tanks are built partly as dugouts and partly by enclosing bunds. Tanks are of varying size but most of the tanks are of small size and are built by individual farmers or groups of farmers. There are about 5 lakh big and 50 lakh small tanks irrigating nearly 1723 thousand hectares of agricultural land. The ratio of tank irrigated land to the total irrigated area of the country has reduced from 14 per cent in the 1960-61 to about 2.52 per cent in 2014-15, primarily due to increase in well and tubewell irrigation and partly due to fall in the tank irrigation. For example, tanks irrigated 2821 thousand hectares of agricultural land in 1996-97 which was reduced to 1723 thousand hectares in 2014-15. In contrast tube wells irrigated 19,338 thousand hectares of agricultural land in 1996-97 which increased to 31,606 thousand hectares in 2014-15.

Tank irrigation is popular in the peninsular plateau area where Andhra Pradesh, Telangana and Tamil Nadu are the leading states. Andhra Pradesh and Telangana have 727 thousand hectares irrigated by tanks. These two states have about 42.2 per cent of tank irrigated area of India. About 16 per cent of the total irrigated area of these states is irrigated by tanks. The drainage areas of the Godavari and its tributaries have large number of tanks. Nellore and Warangal are the main districts of tank irrigation. Tamil Nadu has an area of 589 thousand hectares under tank irrigation. This is over 34.2 per cent of tank irrigated area of India and about one-fifth of the total irrigated

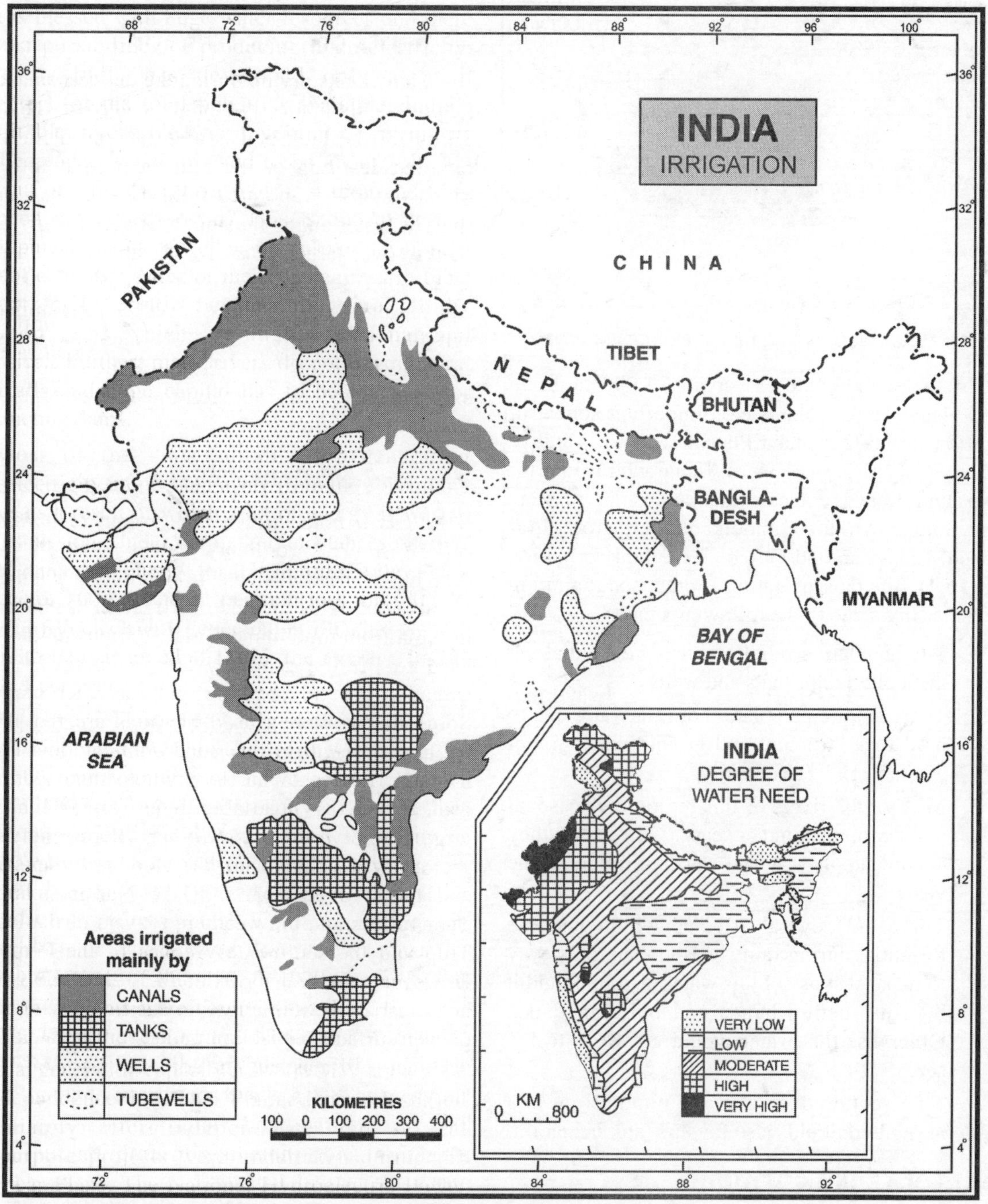

FIG. 16.2. India : Irrigation

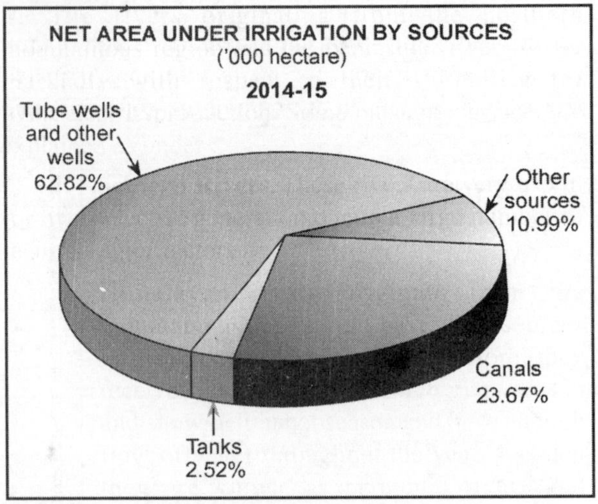

FIG. 16.3. Net area irrigated in India by source

area of the state. There are about 24,000 tanks in Tamil Nadu. Tanks comprise an important source of irrigation in the Karnataka Plateau, eastern Madhya Pradesh, Chhattisgarh, eastern Maharashtra, interior Odisha and Kerala. Outside the Peninsular plateau, West Bengal, Bihar, Bundelkhand area of Uttar Pradesh, Rajasthan and Gujarat have tank irrigation (**Fig. 16.4**). The tank irrigation is practised mainly in peninsular India due to the following reasons.

1. The undulating relief and hard rocks make it difficult to dig canals and wells.
2. There is little percolation of rain water due to hard rock structure and ground water is not available in large quantity.
3. Most of the rivers of this region are seasonal and dry up in summer season. Therefore, they cannot supply water to canals throughout the year.
4. There are several streams which become torrential during rainy season. The only way to make best use of this water is to impound it by constructing bunds and building tanks. Otherwise this water would go waste to the sea.
5. The scattered nature of population and agricultural fields also favours tank irrigation.

Merits of Tank Irrigation

Most of the tanks are natural and do not involve heavy cost for their construction. Even an individual farmer can have his own tank. Tanks are generally constructed on rocky bed and have long life span. In many tanks, fishing is also carried on. This supplements both the food resources and income of the farmer.

Demerits of Tank Irrigation

Many tanks dry up during the dry season and fail to provide irrigation when it is needed the most. Silting of the tank bed is a serious problem and it requires desilting of the tank at regular intervals. Much water is evaporated from the large expanse of shallow water and is thus not available for irrigation. Tanks cover large areas of cultivable land. In many areas, other sources of irrigation have been adopted and the dry beds of tanks have been reclaimed for agriculture. Moreover, lifting of water from tanks and carrying it to the fields is a strenuous and costly exercise which discourages the use of tanks as a source of irrigation.

WELLS AND TUBEWELLS

A well is a hole dug in the ground to obtain the subsoil water. An ordinary well is about 3-5 metres deep but deeper wells upto 15 metres and even more are also dug. This method of irrigation has been used in India from time immemorial. Various methods are used to lift the ground water from the well for irrigation, drinking, bathing and for other purposes. Some of the widely used methods are the *persian wheel*, *reht*, *charas* or *mot*, and *dhinghly* (lever). Well irrigation is popular in areas where sufficient sweet ground water is available. These areas include a large part of the Great Northern Plain, the deltaic regions of the Mahanadi, the Godavari, the Krishna and the Cauvery, parts of the Narmada and the Tapi valleys and the weathered layers of the Deccan Trap and the sedimentary zones of the Peninsula. However, the greater part of the Penisnular India is not suitable for well irrigation due to rocky structure, uneven surface and lack of underground water. Large dry tracts of Rajasthan, the adjoining parts of Punjab, Haryana, and Gujarat and some parts of Uttar Pradesh have brackish ground water which is not fit for irrigation and human consumption and hence unsuitable for well irrigation.

There were about 5 million wells in 1950-51 and their number increased to about 17.9 million in

2013-14. Well irrigation accounts for about 16.6 per cent of the net irrigated area in the country. It accounted 59.78 lakh hectares in 1950-51 which rose to about 113.54 lakh hectares in 2014-15 thereby registering about two fold increase in well irrigation. Uttar Pradesh has the largest area under well irrigation

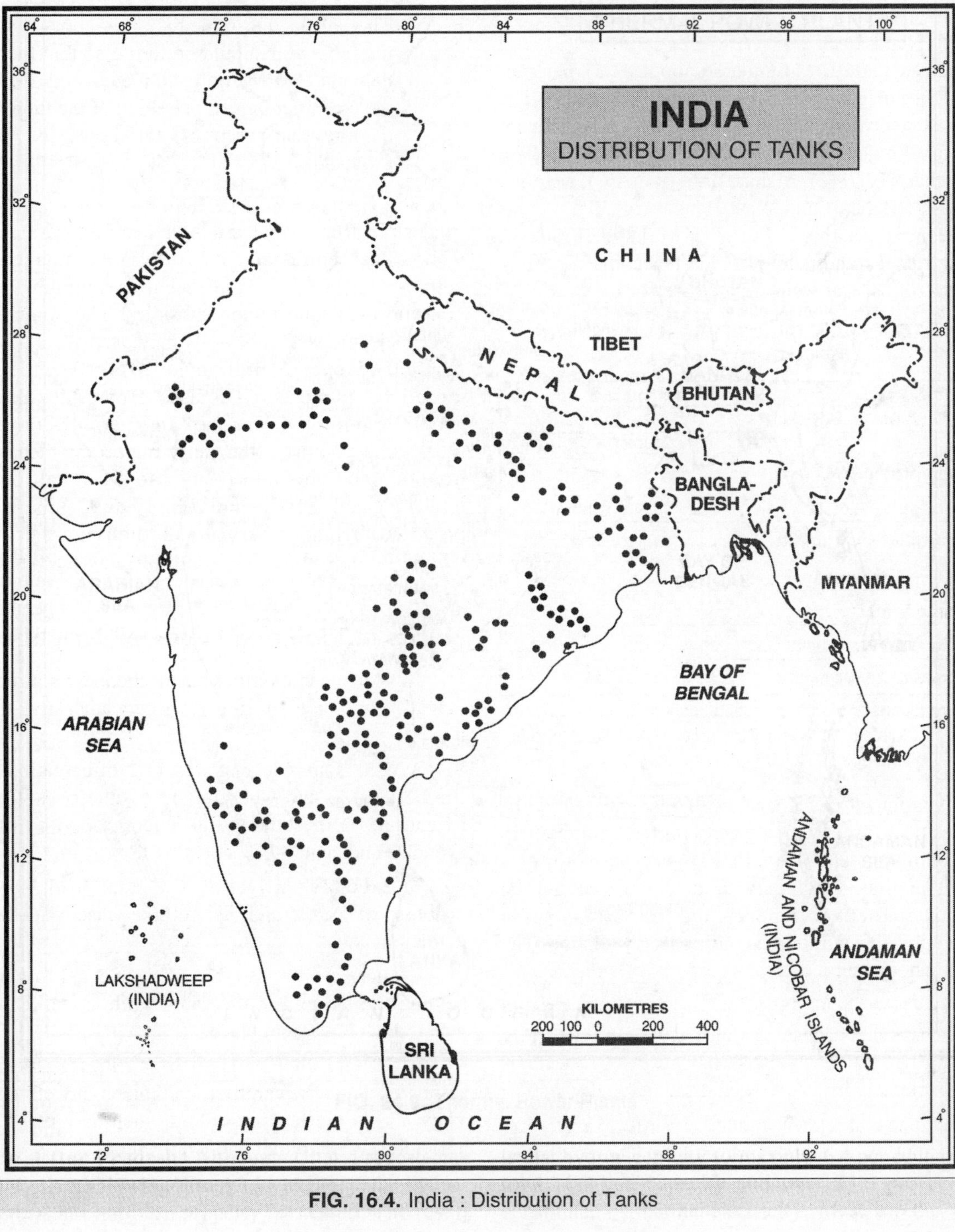

FIG. 16.4. India : Distribution of Tanks

and accounts for about 28.19 per cent of the well irrigated area of India. This is followed by Rajasthan (10.44%), Punjab (8.65%), Madhya Pradesh (7.97%), Gujarat (7.34%) Bihar (6.29%), Andhra Pradesh (5.87%), Maharashtra (5.75%), Haryana (4.41%), Tamil Nadu (4.35%), West Bengal (4.19%) and Karnataka (3.06%). In Gujarat, about 82.31 per cent of the net irrigated area is under well irrigation. The other states where well irrigation plays a significant role are Punjab (79.96%), Uttar Pradesh (73.22%), Rajasthan (70.78%), Maharashtra (64.62%), Madhya Pradesh (64.11%) and West Bengal (59.35%). Uttar Pradesh, Rajasthan, Punjab, Madhya Pradesh, Gujarat, Bihar and Andhra Pradesh account for three-fourths of the total well irrigated area of India.

TABLE 16.2. India : Number of Minor Irrigation Schemes

Name of Scheme	Number (Millions)	
	2006-07	2013-14
Groundwater		
1. Digwell	8.8	9.2
2. Shallow + Medium	9.1	9.1
3. Deep Tubewell	1.5	2.6
Surface Water		
1. Surface flow schemes	0.6	0.6
2. Surface Lift Schemes	0.7	0.6

Source : Census of minor irrigation schemes report 2017 by Ministry of Water Resources, River Development and Ganga Rejuvenation.

A *tubewell* is a deeper well (generally over 15 metres deep) from which water is lifted with the help of a pumping set operated by an electric motor or a diesel engine. Obviously, a tube well cannot be constructed everywhere and requires some geographical conditions favouring its installation. The main factors are :

(*i*) There should be sufficient quantity of ground water because a tube well can generally irrigate 2 hectares per day against 0.2 hectares per day irrigated by an ordinary well.

(*ii*) The water level should be nearly 15 metres. If the water table is more than 50 metres deep the cost of pumping out water from the tube well becomes uneconomic.

(*iii*) There should be regular supply of cheap electricity or diesel so that water from the tube well can be taken out at the hour of need.

(*iv*) The soil in the immediate neighbourhood of the tube-well should be fertile so that there is demand for irrigation and the cost involved in the construction and operation of the tube well can be recovered by the increased farm production.

The first tubewell of India was sunk in Uttar Pradesh in 1930. Till 1951 India had just 2,500 tube-wells. The central and the state governments are helping the farmers by distributing pumping sets, granting loans and giving subsidies. The number of electrical pumpsets/tubewells increased from 2 lakh in 1960 to over 5 million in 2011-12 while the dieselized pumpsets increased from 2.3 lakh to about 3.6 million during the same period. Currently tube wells comprise the most important means of irrigation and provide irrigation to about half of the total irrigated area in India. In 1996-97 tube wells provided irrigation to 19,338 thousand hectares which increased to 31,606 thousand hectares in 2014-15.

Merits of Well and Tubewell Irrigation

1. Well is the simplest and cheapest source of irrigation and the poor Indian farmer can easily afford it.

2. Well is an independent source of irrigation and can be used as and when the necessity arises. Canal irrigation, on the other hand, is controlled by other agencies and cannot be used at will.

3. Excessive irrigation by canal leads to the problem of *reh* which is not the case with well irrigation.

4. There is a limit to the extent of canal irrigation beyond the tail end of the canal while a well can be dug at any convenient place.

5. Several chemicals such as nitrate, chloride, sulphate, etc. are generally found mixed in well water. They add to the fertility of soil when they reach the agricultural field along with well water.

6. The farmer has to pay regularly for canal irrigation which is not the case with well irrigation.

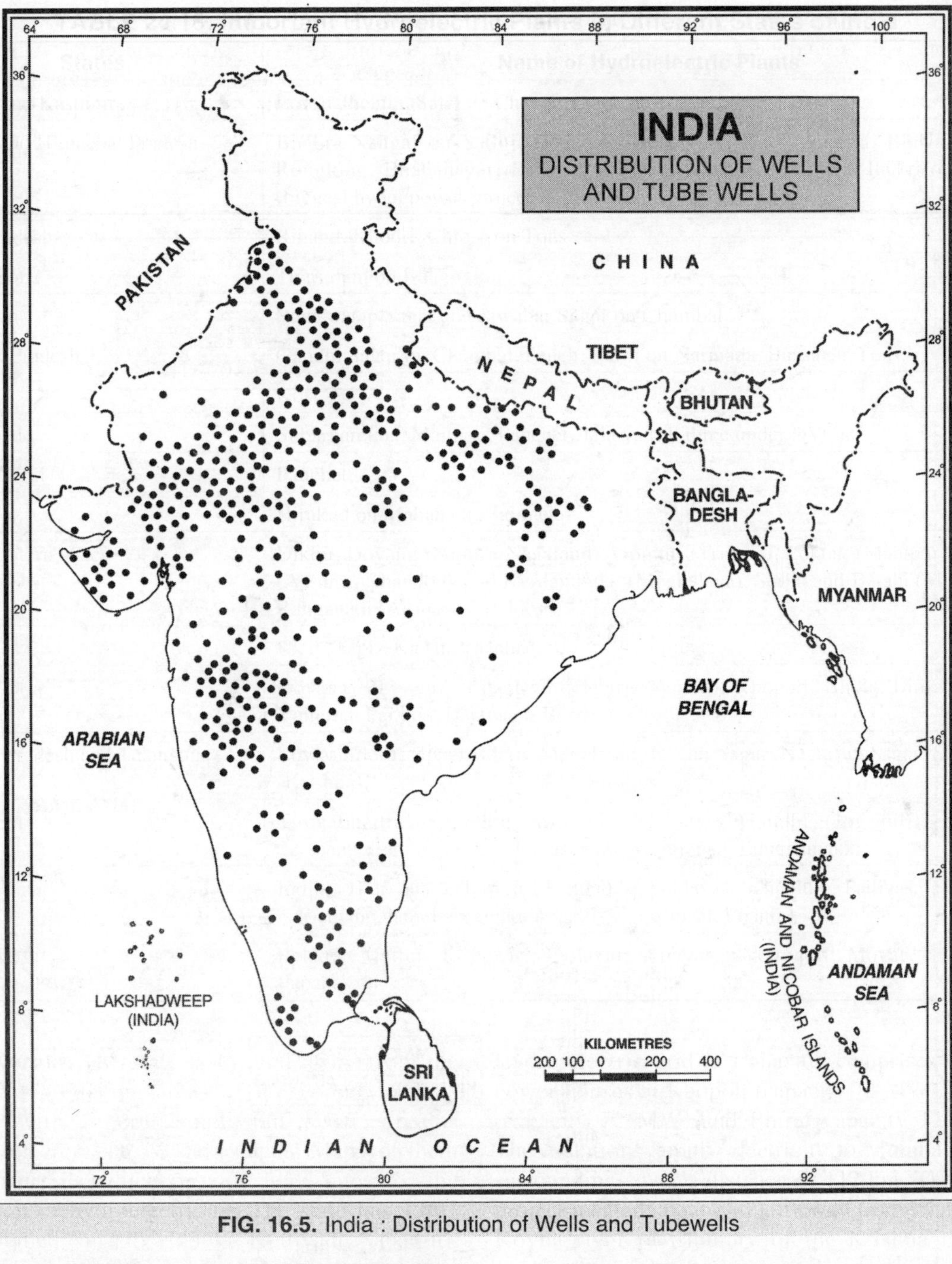

FIG. 16.5. India : Distribution of Wells and Tubewells

Demerits of Well and Tubewell Irrigation

1. Only limited area can be irrigated. Normally, a well can irrigate 1 to 8 hectares of land.

2. The well may dry up and may be rendered useless for irrigation if excessive water is taken out of it.

3. In the event of a drought, the ground water level falls and enough water is not available in the well when it is needed the most.

4. Tubewells can draw a lot of groundwater from its neighbouring areas and make the ground dry and unfit for agriculture.

5. Well and tube well irrigation is not possible in areas of brackish groundwater.

CANALS

Canals used to be the most important source of irrigation upto 1960s, but in the 1970s they yielded first place to wells and tube wells and now constitute the second most important source of irrigation in India. The percentage of canal irrigation area to total irrigated area in the country has fallen from about 39.77 per cent in 1950-51 to 23.67 per cent in 2014-15.

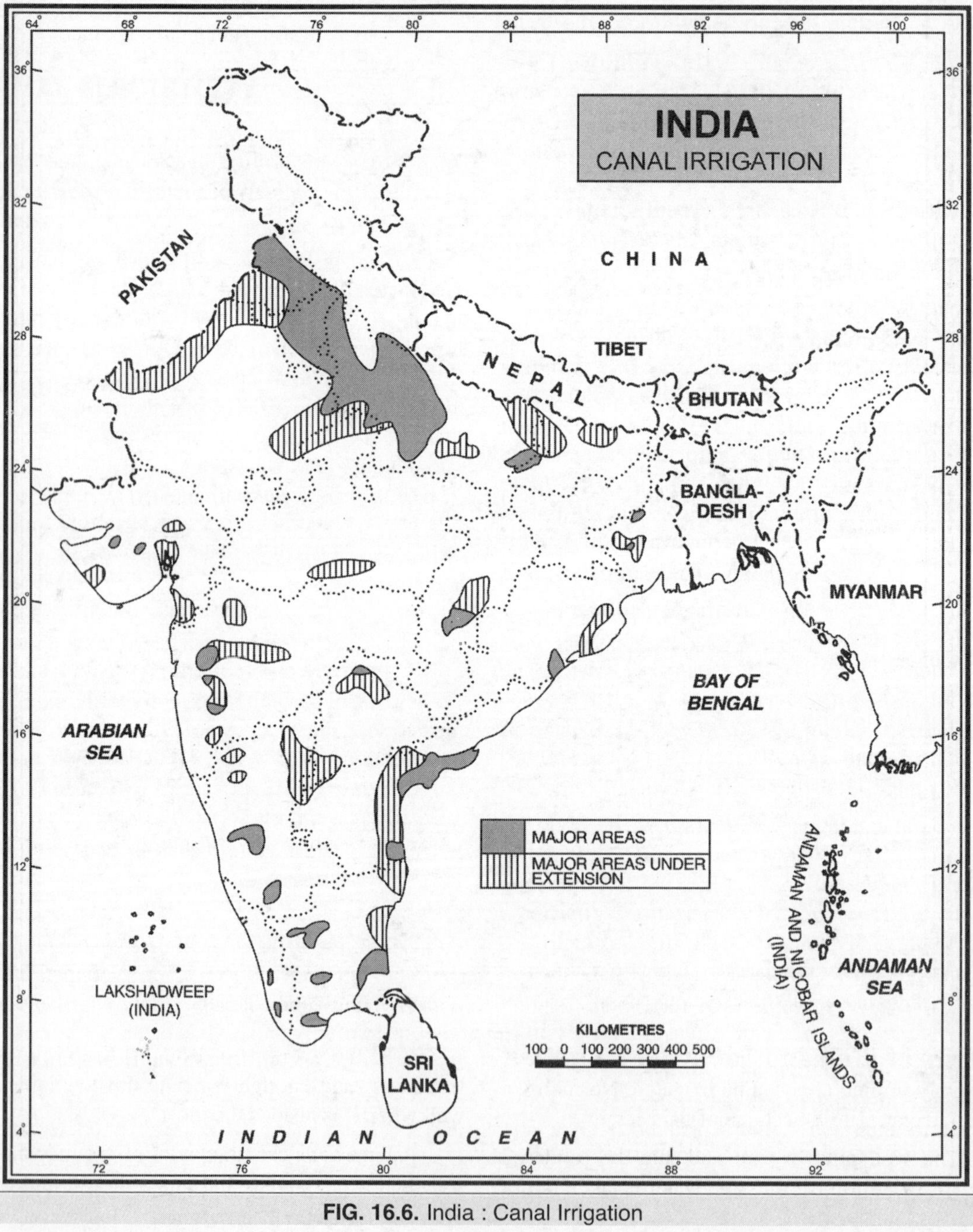

FIG. 16.6. India : Canal Irrigation

Canals can be an effective source of irrigation in areas of low level relief, deep fertile soils, perrenial source of water and extensive command area. Therefore, the main concentration of canal irrigation is in the northern plain of India, especially the areas comprising Uttar Pradesh, Haryana and Punjab. The digging of canals in rocky and uneven areas is difficult and uneconomic. Thus the canals are practically absent from the Peninsular plateau area. However, the coastal and the delta regions in South India do have some canals for irrigation.

Broadly speaking, canals in India are of two types, *viz.* (*i*) *Inundation canals*, which are taken out from the rivers without any regulating system like weirs etc. at their head. Such canals provide irrigation mainly in the rainy season when the river is in flood and there is excess water. When the rainy season is over, the flood in the river subsides, the level of water falls below the level of the canal head and the canal dries up. Some canals taken off from the Satluj in Punjab were of this type. Since irrigation from this type of canals is uncertain, they have been converted in perennial canals. (*ii*) *Perennial canals* are those which are taken off from perennial rivers by constructing a barrage across the river. Most of the canals in India today are perennial.

The net area under government canal irrigation is about 16.02 million hectares. The main canal irrigated areas are in the northern plains of India where Uttar Pradesh, Punjab, Haryana, Rajasthan and Bihar account for about 60 per cent of the canal irrigated area of the country. In south and central India, Andhra Pradesh, Maharasthra, Karnataka, Madhya Pradesh, Chhattisgarh, Odisha and Tamil Nadu are important states of canal irrigation.

Uttar Pradesh

Canals constitute an important source of irrigation in Uttar Pradesh. The state is drained by perennial rivers originating in the snow covered Himalayan ranges and is blessed with fertile soils. But the amount of rainfall, especially in western parts of the state, is not sufficient for sustained agricultural growth. Therefore, a large number of canals have been constructed to provide regular supply of sufficient water to the crops. Uttar Pradesh has about one-third of the total canal irrigated area of the country. Over one-fourth of the net irrigated area of the state is irrigated by canals. Following are the main canals.

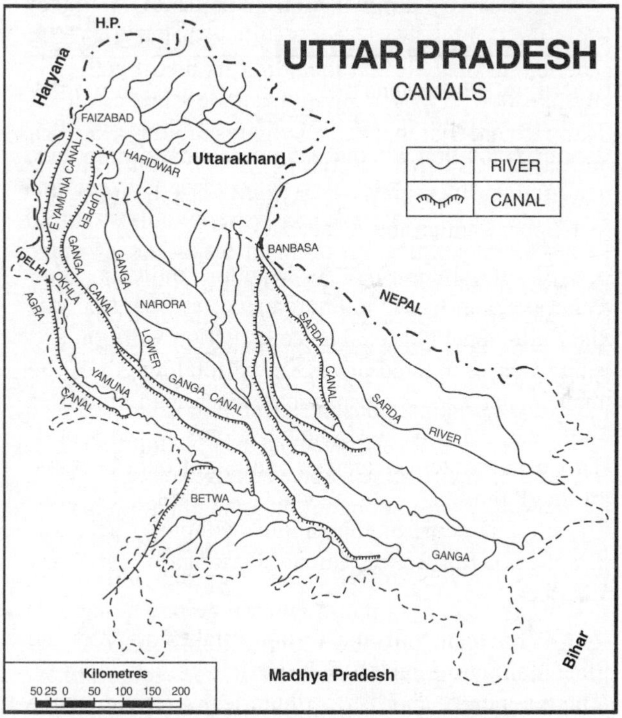

FIG. 16.7. Canals of Uttar Pradesh

1. Upper Ganga Canal. This canal takes off from the Ganga at Kankhal (Hardwar). The construction of this canal commenced in 1842 and it was completed in 1854. The main canal is 342 km long while the length of its distributaries is about 6,200 km. During the first 32 km of its course, between Haridwar and Roorkee, it passes through a broken country so that at some places it is taken over the rivers and at others below the rivers. It irrigates about 7 lakh hectares of land in the upper part of the Ganga-Yamuna Doab. Districts of Saharanpur, Meerut, Ghaziabad, Bulandshahar, Aligarh, Mathura, Etah, Kanpur, Mainpuri, Farrukhabad and Fatehpur get benefit from this canal. Its main branches are Devbandh, Anupshahar, Motta and Hathras. It joins with the Lower Ganga Canal at Mainpuri and the water in this canal is considerably increased. Further beyond, these two canals flow separately.

2. Lower Ganga Canal. This canal was taken from the Ganga near Narora (Bulandshahar) in 1878. The length of the canal including its distributaries is about 6000 km. Its main branches are Etawah, Kanpur and Fatehpur. It irrigates about 4.6 lakh hectares in the districts of Bulandshahar, Farrukhabad, Mainpuri, Aligarh, Etah, Etawah, Fatehpur, Kanpur and Allahabad.

3. Sharda Canal. As its name indicates, this canal is taken from the Sharda river at Banbasa near the Indo-Nepal border. The construction work on this canal was completed in 1928. The total length of the main canal and its distributaries is 13,624 km. It is thus one of the longest canal systems of the world. This canal system irrigates about eight lakh hectares of land mainly in Allahabad, Sultanpur, Pilibhit, Bareilly, Hardoi, Shahjahanpur, Sitapur, Lucknow, Barabanki, Rai Bareli, Unnao, Partapgarh and Kheri districts.

4. Eastern Yamuna Canal. It takes off from the river Yamuna at Faizabad. It was constructed in 1831. The main canal and its distributaries cover a distance of 1,450 km and irrigate about 2 lakh hectares of land in the districts of Saharanpur, Muzaffarnagar, Meerut and Ghaziabad. It again joins the Yamuna river at Delhi and irrigates a part of the union territory also.

5. Agra Canal. This canal is taken from the right bank of the Yamuna at Okhla (Delhi). It was built in 1874 and irrigates about 1.5 lakh hectares in Agra and Mathura districts in U.P., Faridabad and Palwal in Haryana, Bharatpur in Rajasthan and also parts of union territory of Delhi.

6. Betwa Canal. Built in the Third Five Year Plan, this canal takes off from the Betwa river about 56 km south-west of Jhansi. It irrigates about 1.2 lakh hectares in Jhansi, Jalaun and Hamirpur districts.

Apart from the above mentioned major canals, some other canals such as Ken, Chambal, Dhasan and Son canal also irrigate some areas in the southern part of Uttar Pradesh.

Punjab

In early days, canal irrigation accounted for about 39.14 per cent of the total irrigated area in Punjab but the share of canal irrigation fell down to less than 19 per cent in 2009-10. Following are some of the important canals of Punjab.

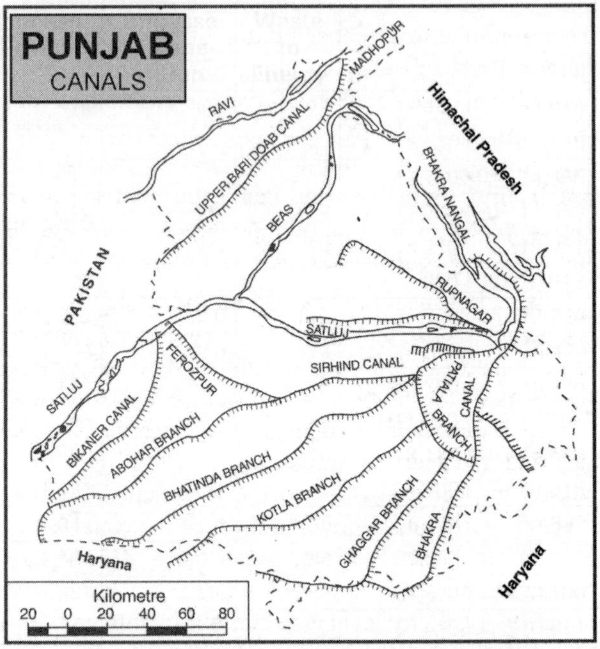

FIG. 16.8. Punjab : Canals

1. Upper Bari Doab Canal. This canal is taken from the Ravi river at Madhopur near Pathankot. Construction of this canal started in 1849 and completed in 1859. It irrigates about 3 lakh hectares in Gurdaspur and Amritsar districts.

2. Sirhind Canal. This canal was taken from the Satluj river at Rupnagar (Ropar) in the year 1886-87. The total length of the canal along with its distributaries is 6,115 km. Its main branches are the Patiala, Abohar, Bathinda, Kotla and Ghaggar. It irrigates about 7 lakh hectares in Patiala, Sangrur, Bathinda, Ludhiana and Ferozepur districts. The Kotla and the Ghaggar branches provide irrigation to Hissar, Sirsa and Fatehabad districts of Haryana also. In order to augment the supply of water, the *Sirhind Feeder Canal* was completed in 1960. It takes off from the Ferozepur Feeder at its 18th km at Malanwala. It is 142 km long and supplies water to Abohar branch of the Sirhind Canal. This water is drawn from the Satluj and Beas rivers which used to go unused to Pakistan. It also provides irrigation to Ferozepur, Faridkot and Muktsar districts in Punjab and to some parts of Rajasthan.

3. Bhakra Canal. It draws water from the Bhakra dam built across the Satluj river. It was completed in 1954. It irrigates a vast area of about 15 lakh hectares in Punjab, Haryana and Rajasthan. In Punjab, Ludhiana, Patiala, Sangrur, Jalandhar, and Ferozepur districts are benefited by this canal.

4. Bist Doab Canal. It is a part of the Bhakra-Nangal Project. The total length of this canal along with its distributaries is 1,090 km. It irrigates about 4 lakh hectares in Jalandhar and Hoshiarpur districts.

Bikaner and Beas are other projects which provide irrigation to parts of Punjab and the neighbouring states of Haryana and Rajasthan.

Haryana

Haryana depends upon canal irrigation for its agricultural prosperity to a great extent. About 49.89 per cent of the irrigated area in Haryana is irrigated by canals. Following are the main canals.

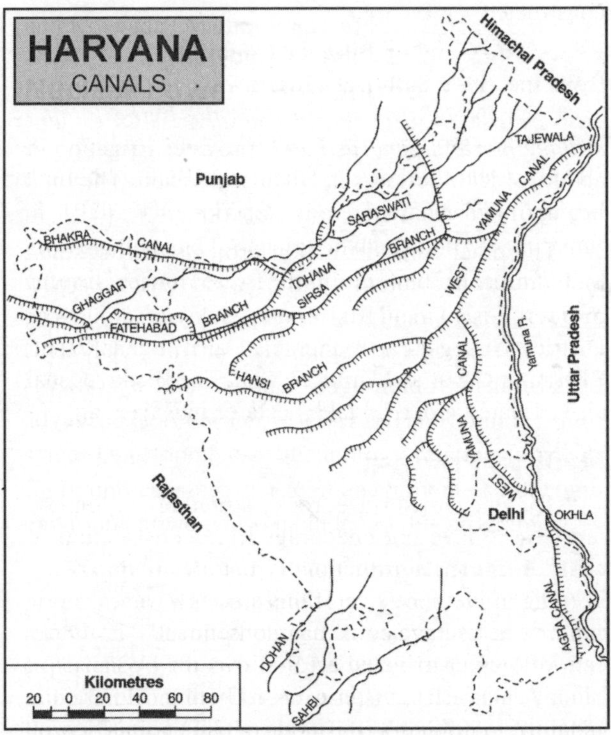

FIG. 16.9. Haryana : Canals

1. The Western Yamuna Canal. It takes off from the right bank of the Yamuna at Tajewala. It was built by Feroze Shah Tughlak. The total length of the canal along with its distributaries is 3,200 km and it irrigates about 4 lakh hectares in Ambala, Kurukshetra, Karnal, Panipat, Rohtak, Hissar, Sirsa, Fatehabad and Jind districts. Its important branches are, the Delhi, the Hansi and the Sirsa branch.

2. Bhakra Canal. After irrigating Punjab areas, the Bhakra canal enters Haryana near Tohana and irrigates large parts of Hisar, Fatehabad and Sirsa districts. Its main branches are the Fatehabad, the Ratia, the Rori, the Barwala and the Tohana branch.

3. Jui Canal. This is a lift irrigation scheme designed to irrigate the semi-desert region of Bhiwani and adjoining areas. This 169 km long canal irrigates about 32 thousand hectares.

4. Gurgaon Canal. It takes off from the Yamuna at Okhla in Delhi. Started in 1970, this canal irrigates 1.2 lakh hectares in Gurgaon, Rewari, Faridabad and Palwal districts.

Andhra Pradesh

With 10.31% of the total canal irrigated area of India, Andhra Pradesh is next only to Uttar Pradesh and tops in South India in so far as area under canal irrigation is concerned. Canal irrigation accounts for about 36.42 per cent of the net irrigated area of the state. The major canals of Andhra Pradesh are taken off from the Krishna, the Godavari and the Tungabhadra rivers and the major canal irrigated areas are in the deltas and the coastal regions.

The Godavari delta project comprises of two weirs — the Dowlaiswaram and the Ralli, which were completed in 1846. From these, right bank canal and the delta canal have been taken to irrigate about 4.5 lakh hectares. The *Krishna delta project* comprise Vijayawada anicut. Sunkesula anicut across the Krishna and the Tungabhadra irrigates about 4.5 lakh hectares. The *Kurnool-Cuddapah* canal was taken off from the Tungabhadra in 1816. It irrigates about 1.2 lakh hectares in Kurnool and Cuddappah districts. The *Nagarjunasagar Project* consists of a dam across the Krishna River in Nalgonda district from where two canals have been taken off. The right bank canal is 204 km long and irrigates 6.7 lakh hectares in Guntur, Nalgonda (in Telangana), Kurnool and Krishna districts. The left bank canal is 171 km long and irrigates 3.56 lakh hectares in Khammam (in Telangana), Krishna and West Godavari districts.

Bihar

Rainfall in several parts of Bihar is inadequate, unreliable and uncertain leading to severe and frequent droughts. This has made canal irrigation an important part of agricultural practice in Bihar. Canal irrigation accounts for over 31 per cent of the total area of the state.

1. Son Canals. The Eastern Son Canal was taken from the Son river at Varun in 1857. This 130 km long canal irrigates 2.5 lakh hectares in Patna and Gaya districts. The Western Son Canal has been taken from this river at Dehri. It provides irrigation to Shahabad district.

2. Kosi Canals. Two canals known as the Eastern and the Western Kosi Canal have been taken from the eastern and the western banks of the river respectively by constructing a multipurpose dam near the Indo-Nepal border. These two canals have the irrigation potential of 4.5 and 3.5 lakh hectares respectively. Purnea, Muzaffarpur, Darbhanga, Champaran and Saran districts are benefited by this project. Besides, 4 lakh hectares are irrigated in Nepal.

3. Gandak Canals. A 743 metre long dam has been built across the Gandak at Balmiki Nagar from where two canals have been taken off. The eastern canal is called Tirhut. It is 250 km long and irrigates about 7 lakh hectares in Champaran, Darbhanga, and Muzaffarpur districts. The western canal is known as Saran Canal and irrigates 7.6 lakh hectares of Saran district of Bihar and Deoria and Gorakhpur districts of Uttar Pradesh. Another Canal *Triveni Canal* has been taken from Triveni to irrigate about one lakh hectares of Champaran district.

West Bengal

Although most parts of West Bengal receive sufficient rainfall and do not require irrigation, still some parts of the state do feel the necessity of irrigation. More than one third of the net irrigated area is irrigated by canals. The *Mayurakshi* project was completed in 1951 at Tilpara. Two canals have been taken from this project to irrigate large areas in Bir Bhum, Murshidabad and Bardhaman districts. The main canal from the Kaugsabati completed in 1965 irrigates 1.2 lakh hectares in Bankura and Medinipur districts. The canals from the Damodar river built by Damdar Valley Corporation irrigate over five lakh hectares in West Bengal. The 520 km long *Medinipur Canal* has been taken off from the *Kasai* at Medinipur and irrigates about ten thousand hectares.

Rajasthan

Most of Rajasthan is a desert area and largely depends upon irrigation for successful growth of crops. Canals form an important source of irrigation and account for about 27.6 per cent of the net irrigated area of the state.

1. Indira Gandhi Canal. (see Chapter 13).

2. Chambal Project. This is a joint venture of Rajasthan and Madhya Pradesh. Under this project, *Gandhi Sagar Dam* has been constructed. Canals taken off from this dam irrigate about 5.15 lakh hectares in Rajasthan and Madhya Pradesh. In the second stage, *Rana Partap Dam* has been constructed which provides irrigation to 1.2 lakh hectares. In the third stage *Jawahar Sagar Dam* has been constructed.

3. The Gang or Bikaner Canal. It was taken off from the river Satluj at Husainwala in 1928. This canal is *a blood transfusion from the living Punjab into the moribund marusthal*. It provides irrigation to about 3.4 lakh hectares in Bikaner division. The total length of this canal system is 1,280 km.

The other canal irrigation projects of Rajasthan include Jawai project on the Jawai river, Parbati project near Dhaulpur, Gudha project in Bundi District, Ghaggar canal, Pichuna Canal, Banas Canal, Bharatpur Canal and Urmila Canal. The Bhakra Canal also irrigates about 2.3 lakh hectares in Rajasthan.

Madhya Pradesh

The unevern surface, rocky structure and limited fertile soil have not encouraged the construction of canals and canals irrigation is not much important, keeping in view the size of this state. However, a few canal systems have come up. Consequently 19.49 per cent of the net irrigated area is irrigated by canals in Madhya Pradesh is under canal irrigation. Most of the canal irrigated area is in Gwalior, Bhind and Morena districts. The *Chambal project* provides irrigation to about 2.83 lakh hectares in these districts. The *Weinganga* canal takes off from the Weinganga river and irrigates 4,000 hectares of land in Balaghat and Seoni districts. Besides it provides irrigation to

Bhandara district of Maharashtra. The *Barna* project involves the construction of a barrage across the Barna river (a tributary of the Narmada). The canal taken out from this barrage irrigates about 64,400 hectares of land in Raisen district. The *Tawa* project canals originate from the barrage constructed across the Tawa river, and irrigate about 3 lakh hectares of land in Hoshangabad district.

Chhattisgarh

The geographical conditions of Chhattisgarh are more or less similar to those prevailing in Madhya Pradesh which are not favourable for canal irrigation. Even then this state has some areas under canal irrigation. The *Mahanadi canal* takes off from Mahanadi river at Rudri in Raipur district. A subsidiary reservoir has been built at Maramsilli. The length of the main canal and distributaries is 215 km and 1,175 km respectively. It irrigates about 3 lakh hectares. The *Tandula canal* takes off from the confluence of the Tandula and Sukha rivers by building two separate earthen dams. It irrigates about 1.1 lakh hectares in Raipur and Durg districts.

Odisha

Canals irrigate about 878 thousand hectares of land which is nearly 25% of the net irrigated area of the state. Canals taken off from the *Hirakud dam* on the *Mahanadi*, form a major irrigation network and provide irrigation to about 2.4 lakh hectares in Bolangir and Sambalpur districts. The 3,650 km long canal system in the Mahanadi delta region provides irrigation to about 4 lakh hectares in Cuttack and Puri districts. Taldanda Canal irrigates 62 thousand hectares in Mahanadi catchment area.

Karnataka

Over 36.5 per cent of the net irrigated area in Karnataka is irrigated by canals. The *Ghatprabha* valley scheme developed on the Ghatprabha river is the most important irrigation project and irrigates about 3.2 lakh hectares in Belgaum and Bijapur districts. Canals of the *Tungbhadra project* irrigate about 2.7 lakh hectares in Bellary and Raichur districts. The canals of *Malprabha* project irrigate about one lakh hectares in Belgaum, Dharwar and Bijapur districts and those of *Bhadra Project* also irrigate one lakh hectares in Shimoga district. The *Visveswaraya Canal* taken off from the *Krishnaraja Sagar Dam* on the Cauvery irrigates about 50 thousand hectares in Mysore and Mandya districts.

Tamil Nadu

Tamil Nadu has about 29 per cent of its net irrigated area under canal irrigation. The most important canal system lies in the Cauvery delta where 6,400 km long canals irrigate about 4 lakh hectares in Thanjavur and Tiruchchirapalli districts. The *Mettur Canal system* of the Mettur dam on the Cauvery river irrigates about 1.2 lakh hectares in Salem and Tiruchchirapalli districts. The *Lower Bhawani Project* canal system irrigates about 79 thousand hectares in Coimbatore district. *Parambikulam Aliyar and Manimuthar* are other projects.

Maharashtra

Over 35 per cent of the net irrigated area in Maharashtra is irrigated by canals. Over 3,000 km long canals irrigate over 4 lakh hectares. The right bank and the left bank canals of the *Mutha Project* across Mutha river (a tributary of the Bhima river) at Khadakvasla irrigate about 45 thousand hectares in Pune district besides providing potable water to Pune and Kirkee. The canals of the Nira project irrigate over 66 thousand hectares in Pune, Satara and Solapur districts. Two canals taken out of the Godavari at *Darana dam* irrigate about 27 thousand hectares in Nashik and Ahmednagar districts. Canals taken out from *Gangapur dam* on the Godavari irrigate 33 thousand hectares in Nashik district. The *Pravara* canals take off from *Bhandardara masonary* dam at Pravara river. Both the left bank and the right bank canals irrigate about 34 thousand hectares in Ahmednagar. The Tapi canal system is a joint venture of Maharashtra and Gujarat and irrigates about 3 lakh hectares. The other irrigation canals are those of Mula, Vir, Purna, Girna, Jayakawadi, Warna, etc. Some other projects are at different stages of their completion.

Gujarat

Mahi project stage I and II is designed to irrigate over 2 lakh hectares in Kheda and Panchmahals districts. Canals of *Ukai* project on the *Tapi* river irrigate over one lakh hectares. The other important irrigation projects of Gujarat are *Rudramala*, *Ozat*, *Dantiwada*, *Panam* and *Kaprapara*.

Merits of Canal Irrigation

1. Most of the canals provide perennial irrigation. This saves the crops from drought conditions and helps in increasing the farm production.

2. Canals carry a lot of sediment brought down by the rivers. This sediment is deposited in the agricultural fields which adds to the fertility of soil.

3. Some of the canals are parts of multipurpose projects and, therefore, provide cheap source of irrigation.

4. Although the initial cost involved in canal irrigation is much higher, it is quite cheap in the long run.

Demerits of Canal Irrigation

1. The canal water soaks into the ground and leads to the problem of *water-logging* along the canal route.

2. Excessive flow of water in the fields raises the ground water level. Capillary action brings alkaline salts to the surface and makes large areas unfit for agriculture. Vast areas in Punjab, Haryana and Uttar Pradesh suffer from the problem of '*reh*' caused by canal irrigation. Similarly about 36,000 hectares have been rendered useless in Nira Valley of Maharasthra due to high concentration of salts in the soil resulting from canal irrigation.

3. The marshy areas near the canals act as breeding grounds of mosquitoes which result in widespread malaria.

4. Many canals overflow during rainy season and flood the surrounding areas.

5. Canal irrigation is suitable in plain areas only.

INTENSITY OF IRRIGATION

Intensity of irrigation is defined as *the percentage of net irrigated area to the net sown area*. Map in Fig. 16.10 shows that the regional variations in the intensity of irrigation are great and at once impressive. These variations are due to varied geographical conditions in different parts of the country. Rugged mountains, sandy deserts and rocky terrains without aquifers have very poor facilities of irrigation. On the other hand, fertile alluvial plains with perennial rivers and potable groundwater as well as areas of less than 125 cm of annual rainfall are by far, the areas of high percentage of irrigation. The highest intensity of irrigation exists in the Kashmir Valley, large parts of Punjab and Haryana, the Ganga-Yamuna Doab of Uttar Pradesh, Western part of the south Bihar Plain, Birbhum (West Bengal), Lakhimpur (Assam), the Godavari Krishna Deltas and Chengalpattu district (Tamil Nadu). The intensity of irrigation in these areas is invariably above 60 per cent and in some parts of Punjab it exceeds 75 per cent. Dry areas of Union Territory of Ladakh and Lahul and Spiti district in Himachal Pradesh cannot raise crops without irrigation. Although exact data are not available, these areas are believed to have 100 per cent intensity of irrigation. Large parts of the northern plain and east coastal plain have average intensity varying from 30-60 per cent. Parts of Brahmaputra Plain, the Chambal Valley and those of the Peninsular plateau have low intensity of irrigation varying from 15 to 30 per cent. The areas of low intensity are those which either do not need irrigation by virtue of high and dependable rainfall or they have not been able to develop irrigation facilities due to unfavourable geographical conditions such as rugged topography, hard rocks, infertile soils, lack of surface and ground water, etc. In this category are included large parts of Rajasthan to the west of the Aravali Range, parts of Bihar plain, central part of Peninsular plateau, the Maharasthra and Kerala coasts, Manipur, Mizoram and Tripura. Andaman and Nicobar Islands have zero per cent intensity of irrigation due to adequate rainfall throughout the year.

Accelerated Irrigation Benefits Programme (AIBP)

This programme was launched in 1996-97 to provide central assistance to major/medium irrigation projects in the country, with the objective to accelerate implementation of such projects which were beyond resource capability of states or were in advanced stage of completion. Priority was given to those projects which were started in pre-Fifth and Fifth Plan period and also to those which were benefiting tribal and drought prone areas. From the year 1999-2000 onwards, Central Loan Assistance under this programme was also extended to special category states (North-Eastern States and Hilly States of

Himachal Pradesh, Sikkim, Uttarakhand, Union Territory of Jammu & Kashmir, and some districts of Odisha). Since its inception, 297 irrigation/multi-purpose projects have been included for funding under AIBP. **Table 16.3** gives details of irrigation potential created under AIBP.

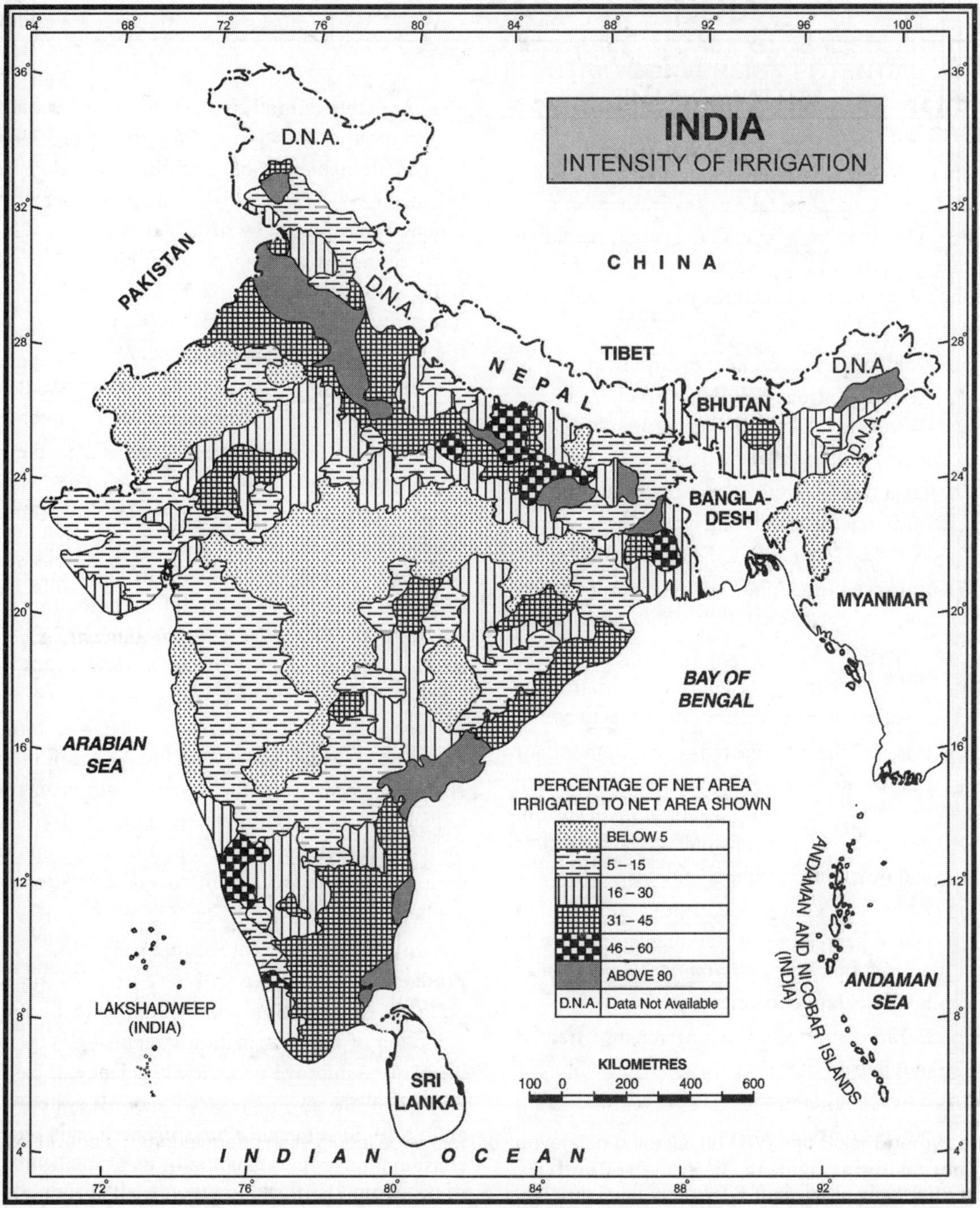

FIG. 16.10. Intensity of Irrigation

TABLE 16.3. Irrigation Potential created under Accelerated Irrigation Benefit Programme (AIBP)—All major and medium projects ('000 Hectare)

Year	1997-98	2006-07	2016-17	2017-18	2018-19	2019-20	2020-21
Potential	199.55	550.01	514.74	702.73	610.40	317.16	129.94

Source : Agricultural Statistics at a Glance, 2021, pp. 346-47.

Pradhan Mantri Krishi Sinchayee Yojana

This scheme was launched in 2015-16 which aimed at enhancing physical access of water on farm and expand cultivable area under assured irrigation, improve on farm water use efficiency, introduce sustainable water conservation practices, etc. The main objective of the scheme are :

(*i*) Achieve convergence of investments in irrigation at the field level (preparation of district level and, if required, sub-district level water use plans).

(*ii*) Enhance the physical access of water on the farm and expand cultivable area under assured irrigation (Har Khet ko Pani).

(*iii*) Enhance the adoption of precision-irrigation and other water saving technologies (more crop per drop).

(*iv*) Enhance recharge of aquifers and introduce sustainable water conservation practices.

(*v*) Ensure the integrated development of rainfed areas using the watershed approach towards soil and water conservation, generation of ground water, arresting runoff and providing livelihood options.

(*vi*) Promote extension activities relating to water harvesting, water management and crop alignment for farmers and grass root level field functionaries.

(*vii*) Explore the feasibility of reusing treated municipal water for peri-urban agriculture.

(*viii*) Attract greater private investments in precision irrigation.

Increasing Irrigation Water Productivity (IWP) in Agriculture. India is basically an agricultural country, and agriculture is primarily dependent on irrigation in the background of erratic and unreliable rainfall by the monsoons. About 89% of the groundwater is used for irrigation. There is large scale depletion of groundwater resources and the major concern is whether the present practice of groundwater use can be sustained as the groundwater level continues to drop at alarming rate. By 2050, India will be in the global hot spot for 'water insecurity'. So appropriate mechanism needs to be framed for economical use of water.

The cropping pattern in India is highly skewed towards water intensive crops which puts heavy strain or our water resources. There is a good deal of divergence between land productivity and irrigation water productivity in major sugarcane producing states of India like Tamil Nadu, Karnataka, Maharashtra and Andhra Pradesh. These states have high land productivity and low irrigation water productivity. *Irrigation water productivity is defined as the ratio of crop output to the irrigation water applied by farmer/irrigation system either through surface canals, tank, pond or well and tubewell during the crop growth.* Thus irrigation is an economic activity and the farmer has to incur certain expenditure to apply water. This is a practical indicator which helps in estimating the crop output obtained with respect to the actual irrigation water applied by the farmer.

In order to improve Irrigation Water Productivity (IWP), we will have to apply improved methods of irrigation and irrigation technologies. Adoption of micro-irrigation systems (MI) is one of the ways to improve water use efficiency. It has been found that states with penetration of MI systems have almost 40-50 per cent savings in energy and fertilizer consumption. A combination of measures which suit the local agro-economic context need to be applied to improve irrigation productivity in agriculture which will reflect sustainable water use in agriculture.

REFERENCES

Dhawan, B.D. (1990) : *Studies in Minor Irrigation with Special Reference to Ground Water.* Commonwealth Publishers, New Delhi.

Dikshit, G.S., G.R. Kupuswamy and S.K. Mohan (1993) : *Tank Irrigation in Karnataka : A Historical Study,* Gandhi Sahitya Sangh, Bangalore.

Economic Survey 2018-19, pp. 177-78.

Govt. of India (2021) : *Agricultural Statistics at a Glance,* Ministry of Agriculture and Farmers Welfare.

Govt. of India (1965) : *Development of Irrigation in India,* Central Board of Irrigation and Power.

Govt. of India (2018) : **EnviStats.**

Govt. of India (1970) : *Irrigation and Power Projects*, Ministry of Irrigation and Power, New Delhi.

Govt. of India (1971) : *Irrigation Atlas of India*, Ministry of Education and Scientific Research, New Delhi.

India (2024) : *A Reference Annual*, Ministry of Information and Broadcasting, New Delhi.

Jain, Bharatial H. (2004) : Irrigation—Tapping nature's bounty, The Hindu Survey of Indian Agriculture, 2004, Kasturi & Sons Ltd., Chennai, pp. 138-139.

NATMO (1990) : *Irrigation Atlas of India*, 2nd Ed. National Atlas & Thematic Mapping Organisation, Kolkata.

Report of the Irrigation Commission (1972) : Vol. I, Ministry of Irrigation and Power, New Delhi.

Sharma, T.C. (1987) : "Progress of Irrigation in Uttar Pradesh Plains : Inter-regional contrasts" *Agricultural Situation in India,* Vol. 42, No. 11, Feb. in Noor Mohammad (ed.), (1992), *New Dimensions in Agricultural Geography,* Vol. 5, Socio-Economic Dimensions of Agriculture, Concept, New Delhi.

Sivanappan, R.K. (2004) : Water Management Efficient Storage Method, The Hindu Survey of Indian Agriculture—2004; Kasturi & Sons Ltd., Chennai, pp. 135-137.

CHAPTER

17

Water Resources

Water is one of the most precious natural resources and a key element in the socio-economic development of a country. A person can live without food for a month, but only for a week without water. Nothing will quench thirst the way water can. Water is an essential part of the modern day life. It is used for drinking, bathing, washing, irrigation, industries and a host of other purposes.

About 71 per cent of the earth's surface is covered by water and that is why our earth is called the ***'watery planet'.*** In fact earth is the only planet in the entire solar system which contains water and sustains life. No other planet in the solar system has, so far, shown any trace of water and all the planets, except the earth, are lifeless. But water on the earth surface is distributed in such a way that only a small fraction of total water available on the earth is useful for human consumption.

INDIA'S WATER RESOURCES

Rainfall is the main source of fresh water in India. From precipitation alone (including snowfall), India receives 4,000 km^3 water per annum. Of this, monsoon rainfall from June to September alone accounts for about 3,000 km^3. A good part of it is lost through the process of evaporation and plant transpiration. Large part of water percolates into the ground and is available to us in the form of ground water.

As per National Commission on Integrated Water Resources Development (NCIWRD) report, the total water availability in the country received through precipitation is about 4000 Billion Cubic Metre (BCM) per annum. After evaporation, 1869 BCM water is available as natural runoff. Due to geological and other factors, the utilizable water availability is limited to 1122 BCM per annum comprising 690 BCM of surface and 432 BCM of replenishable ground water. Obviously water is available in two different forms, *viz.,* (1) surface water, and (2) ground water.

Surface Water

Surface water is available on the surface of the earth in the form of rivers, lakes, ponds, canals, etc. However, rivers comprise the most important source of surface water. India is blessed with a large number of major, medium and small sized rivers. As many as 13 of them are classified as major rivers whose total catchment area is 252.8 million hectares (m. ha). This is about 83 per cent of the total area of all drainage

TABLE 17.1. Surface Water Resources in Major River Basins of India

Basin	Average annual Run-off km^3	Utilisable Flow km^3	Storage completed km^3
1. Indus	73	46	14.52
2. Ganga	501	250	37.4
3. Brahmaputra	537	24	1.09
4. Barak Sub-basin	60	—	—
5. Mahanadi	67	50	8.93
6. Brahmani	36	18.1	4.29
7. Godavari	119	76	17.27
8. Krishna	68	58	32.23
9. Cauvery	21	19	7.25
10. Penner	6.81	6.81	2.37
11. Narmada	41	34.5	3.02
12. Tapi	18	14.5	8.68
13. Sabarmati	3.8	1.91	1.3
14. Mahi	41	34.5	3.02

Source : India 1992 : A Reference Annual, p. 418.

basins. Of the major rivers, the Ganga-Brahmaputra-Meghna system is the biggest with catchment area of about 110 million hectares (m. ha) which is more than 43 per cent of the catchment area of all major rivers in the country. The other major rivers with catchment area more than 10 m. ha are those of the Indus (32.1 m. ha), Godavari (31.3 m. ha), Krishna (25.9 m. ha), and Mahanadi (14.2 m. ha). The catchment area of medium rivers is about 25 m. ha. It is worth mentioning that about 40 per cent of utilisable surface water resources are in the Ganga-Brahmaputra-Meghna system. An idea of magnitude of our water resources can be had from **Table 17.4**.

Interlinking the Rivers

Although India has vast surface water resources, the same are very unevenly distributed over time and space. While some river basins have vast catchment areas and carry enormous quantities of water, others are small and have comparatively small quantities of water. Most of the Himalayan rivers are large and originate in the snow covered high altitude areas of the Himalayan ranges. As such they carry sufficient water throughout the year and are called ***perennial rivers.*** In contrast, the rivers of the peninsular India are seasonal. They carry no or very little water in the dry summer season. During the rainy season, most of the rivers are flooded and large part of water flows down the slope to the sea. Thus much of precious water is wasted and is not available for use. As against this most rivers have insufficient flow of water during the dry season and acute scarcity of water is felt in almost all parts of the country. Besides there is a chronic problem of floods in one part and droughts in other part of the country. Even dry areas like Rajasthan and Gujarat may have floods and wet areas like West Bengal may confront a situation of drought. Keeping these and many more problems in mind, the idea of interlinking rivers through inter-basin linkages or through national grid has been mooted. Such projects aim at reducing disparities in different river basins by transferring water from 'surplus' basins to 'deficit' basins. Some of the important projects are briefly described as under :

The Ganga-Cauvery Proposed Link Canal

This project was prepared by the United Nations team at the request of the Government of India. It was proposed by the then irrigation minister **Dr. K.L Rao** in 1950s. The project aimed at reducing the impact of floods in the Ganga basin and supply water to central and eastern parts of the country which suffer from chronic problem of water shortage. This link canal was to take off from near Patna, pass through the basins of the Son, the Narmada, the Tapi, the Godavari, the Krishna and the Pennar rivers, and join the Cauvery upstream of the Grand Anicut. Water from Patna barrage will be lifted by large pumps to a point near the boundary of the basins of the Ganga and the Narmada from where it will be distributed by gravity via dug up canals or through existing rivers to the west or south. Flood waters of the Narmada and the Godavari could also be used. However, it was pleaded that water from the Ganga for the inter-basin transfer could be drawn only during four months of rainy season from July to October when the flow of water in the Ganga river exceeds 2,850 cumecs (1,00,000 cusecs).

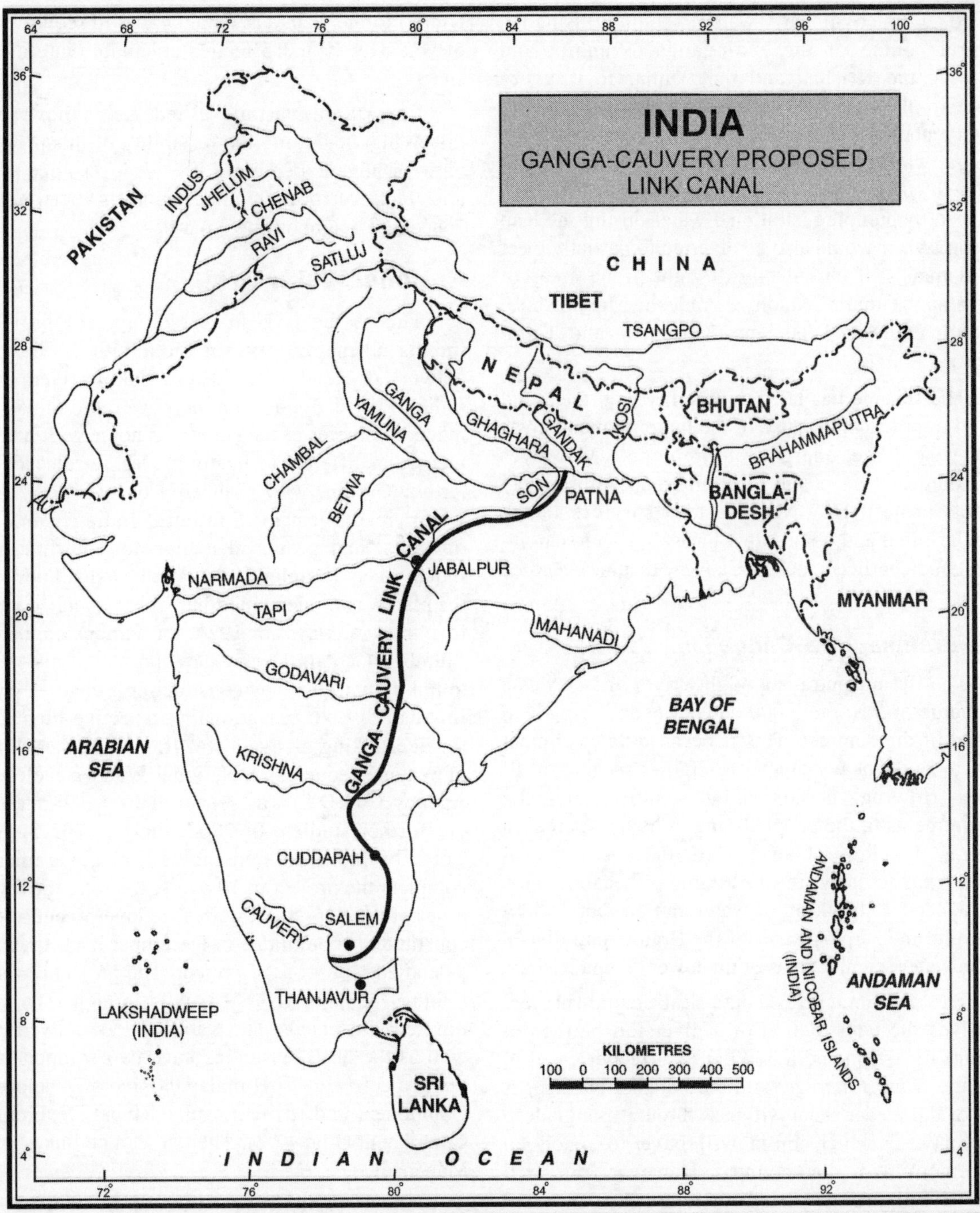

FIG. 17.1. Ganga-Cauveri Link Canal

The proposed Ganga-Cauvery Link Canal was to have been 2,640 km long, withdrawing 60,000 cusecs. from the flood flows of the Ganga for about 150 days in the year, and would have involved a lift of a substantial part of water over 450 metres. From Patna, water would be pumped into a series of reservoirs between water sheds of the Narmada and the Son. From Bargi reservoir on the Narmada (423 m), a linked aqueduct will carry water to south utilising the natural course of the Wainganga, Pranhita and crossing the Krishna and the Pennar to Cauvery upstream the Upper Anicut. It was also proposed to supply about 300 cumecs (10,000 cusecs) of the Ganga water to different parts of Bihar, Uttar Pradesh, Jharkhand, Chhattisagarh and Madhya Pradesh by pumping additional water during the lean season. Water would also be diverted to partially meet the demands of chronically drought prone areas of Rajasthan, Gujarat, Madhya Pradesh, Maharashtra, Andhra Pradesh, Telangana, Karnataka and Tamil Nadu.

The scheme has been thoroughly examined and found impractical because of the huge financial costs and very large energy requirements. Moreover environmental issues put great hindrances in the way of this project. However, the idea survives in the popular mind and comes up whenever water scarcity is felt and conflicts (such as Cauvery dispute) become acute.

The Brahmaputra-Ganga Link Canal

The Brahmaputra is a mighty river and carries a discharge of 3,500 to 5,000 cumecs even during lean period of dry summer. This is because its catchment area receives heavy annual rainfall. The amount of water flowing in this river is more than the requirements of the people living in its basin area. In contrast the lower Ganga basin faces scarcity of water, particularly in the summer season. This unbalanced distribution of water can be rectified by diverting the surplus water of the Brahmaputra basin to the water scarcity areas of the lower Ganga basin.

The Brahmaputra-Ganga link canal project involves the construction of a diversion barrage at Dhubri in Assam and a 320 km long feeder canal to link the Dhubri barrage with the Farakka barrage. A part of the feeder canal will pass through Bangladesh territory for which India will have to reach an agreement with that country. However, this will benefit Bangladesh also because a part of water available in the feeder canal can be used for irrigation or for augmenting the water flow in the Padma river. It is also expected to boost inland navigation in both the neighbouring countries. It is estimated that when completed, this project will be able to divert about 1,150 cumecs of water from the Brahmaputra at Dhubri to the Ganga at Farakka. It will require lifting of water by 10 to 15 metres at suitable intermediate sites.

Experts in various allied fields have grave misgivings regarding the feasibility of this project in view of huge expenditure involved, resource crunch and lack of proper understanding between the concerned neighbouring countries.

National Water Grid

This is the largest ever thought of project of linking all major rivers of India with the help of a network of canals. The idea of linking rivers across India to solve flood, drought, power and other water related problems of the country is not new. It has been mooted in different forms for the last few decades. Sir Arthur Cotton, who pioneered the development of water resources in Southern India from 1839 onwards, had proposed a plan for interlinking of Indian rivers for inland navigation. A small portion of the plan was implemented but was abandoned later in favour of railways. In 1974, the famous engineer of Mumbai, Captain Deen Shaw, presented his scheme to link Indian rivers by constructing a chain of canals. In August, 1980, a National Perspective Plan (NPP) for interlinking of rivers was framed by the Ministry of Water Resources. A National Water Development Agency (NWDA) was set up in July, 1982 to carry out further studies. In 2002, the Hon'ble Supreme Court of India ordered the central government to complete the project in 10 years. Keeping in view the order of the Supreme Court, the Government of India constituted a committee in December 2002 to prepare a detailed report of the project. The project envisages linking 26 major rivers of India by constructing 30 different link canals. The National Perspective Plan as well as the NWDA studies have two components of the project *viz*., (*a*) Himalayan Rivers Development Component and (*b*) Peninsular Rivers Development Component (**Fig. 17.2**). The two can be linked on the Mahanadi.

(a) The Himalayan Rivers Development Component

The Himalayan Rivers Development Component envisages construction of storage reservoirs on the principal tributaries of Ganga and Brahmaputra in India, Nepal and Bhutan alongwith interlinking canal

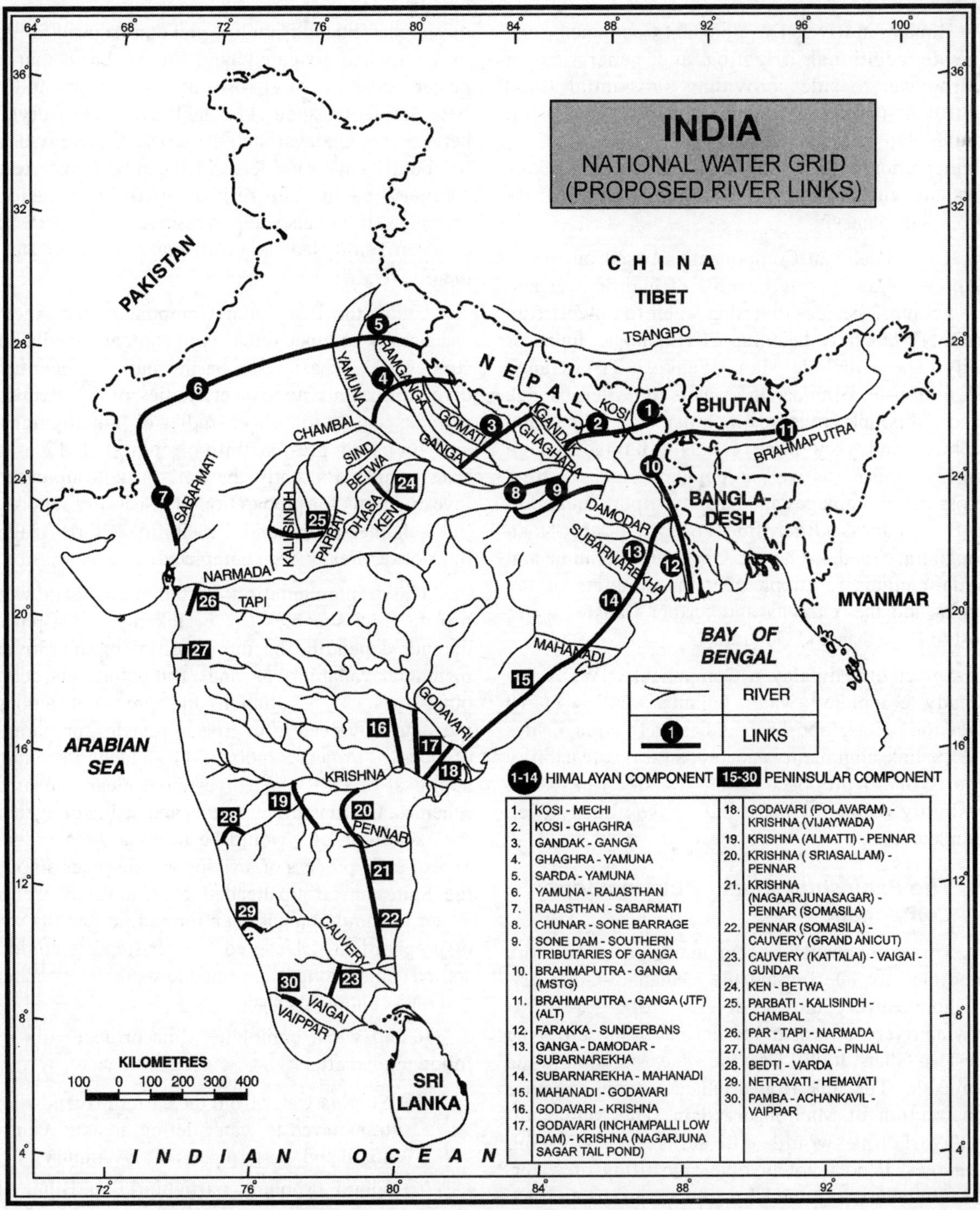

FIG. 17.2. India : Proposed River Links (The Garland Canal System)

systems to transfer surplus flows of the eastern tributaries of the Ganga to the west, apart from linking of the main Brahmaputra and its tributaries with the Ganga and the Ganga with the Mahanadi.

The Himalayan Rivers Development Component shall provide additional irrigation and generation of hydropower, besides providing substantial flood control in the Ganga and Brahmaputra basins. It would also provide the necessary discharge for augmentation of flows at Farakka to flush Kolkata port and enhance the inland navigation facilities across the country.

The Himalayan Component is based on multi-purpose storage giving benefits of hydropower and flood control, besides diverting water to downstream links. NWDA have taken up 11 Himalayan links for study. These include Manas-Sankosh-Tista-Ganga link to transfer Brahmaputra waters to Eastern Ganga Basin. This link envisages high dams on Manas and Sankosh and very large canals running through densely forested as well as populated areas cutting across major drainages and narrow strip of land north of Bangladesh. Other lines proposed are Kosi-Ghaghara, Gandak-Ganga, Ghaghara-Yamuna and Sarda-Yamuna to supplement the supplies of the Ganga and the Yamuna and further transfer water west to Rajasthan and Gujarat.

Under the Himalayan Component, NWDA has already completed water balance studies at 19 diversion points, toposheet studies of 16 storage sites and 19 link alignments and prepared pre-feasibility reports of 14 proposed water transfer links. The feasibility report for these links have already been completed.

(b) The Peninsular Rivers Development Component

This component has four major parts but more important are interlinking the Mahanadi-Godavari-Krishna-Cauvery and the diversion of a few west-flowing rivers towards the peninsular basin across the Western Ghats. It will ease the water situation in the peninsula. The link from Mahanadi would require construction of Manibhadra dam. The Mahanadi-Godavari link would cut across east-flowing drainages. It does not involve any lifting of water. The Godavari is joined by two major tributaries, Indravati and Penganga downstream of the major storage at Shri Ramsagar. Even after meeting downstream basin requirements, the Godavari has surplus water in this reach which can be stored in the available dam site of Inchampalli. The benefits to those rehabilitated from low yield unirrigated farming to high yield irrigated land are so large that the project deserves high priority in implementation. After irrigating about 11.5 lakh hectares of dryland between the Godavari and the Krishna, there will still be about 3 km^3 water left which can be transferred to Cauvery basin and relieve it from occasional shortages. It will also help to resolve the bitter dispute between Tamil Nadu and Karnataka over the water of the Cauvery river.

Under the Peninsular Component, the Agency has already completed data collection and the balance studies of 137 basins/sub-basins and at 77 identified diversion points, toposheet studies of 72 identified storages and 30 toposheet studies of link alignments, and prepared pre-feasibility reports of 17 water transfer links. Presently, the work of field surveys and investigations for preparation of feasibility reports of link schemes is on hand. Feasibility reports of these links have already been completed.

The implementation of the inter-basin water transfer link schemes can be taken up in a phased manner depending on the priorities of the Government and availability of funds. But before this, certain other steps, *viz.*, negotiations and agreements amongst the states involved in interbasin transfer, preparation of Detailed Project Reports (DPRs), techno-economic appraisal of DPRs and investment clearance of the schemes, funding arrangements and fixing of agencies for execution, etc. would be necessary. In order to expedite the process of arriving at consensus amongst the States, a Group headed by Chairman, Central Water Commission has been formed, so that the work of preparation of Detailed Project Reports could be started. This Group is expected to work independently and on a continuous basis.

If and when completed, this project will give following benefits :

(*i*) Surplus water from the eastern rivers will be transferred to water deficit areas of central, south and western parts of the country.

(*ii*) Flood problem, particularly in Bihar and Assam will be solved to great extent.

(*iii*) 34,000 megawatt (34 million kW) hydroelectricity would be generated. This electricity will be used for irrigation and other purposes.

(*iv*) The implementation of the programme of interbasin water transfer is expected to provide additional irrigation benefit of 35 Mha which will be over and above the ultimate irrigation potential of about 140 mha envisaged from major, medium and minor irrigation projects.

(*v*) Chronically drought affected areas will get sufficient water for irrigation, drinking and other purposes.

(*vi*) As much as 1275.74 billion cubic metre (BCM) water wastefully flowing in rivers originating in the glaciers of the Himalayas will be properly utilised.

(*vii*) The project will provide ample opportunities for inland navigation and thus reduce pressure on rail and road transport.

(*viii*) It will be of great help in resolving inter-state water disputes.

(*ix*) A minimum flow of water will be ensured in water deficit rivers.

However, the project has been criticized on several grounds and several experts have questioned even the feasibility of such a gigantic project. Following are some of the objections raised against this project.

(*i*) The project will involve a huge cost of ₹ 5,60,000 crore or $ 120 billion which amounts to a quarter of the country's current GDP. A developing country like India can hardly afford such a huge investment in just one project.

(*ii*) The project will entail the construction of several major dams and lengthy canals cutting across various river basins. This is not an easy task and will require engineering skills of high calibre.

(*iii*) The project aims at transferring water from water surplus rivers to water deficit rivers. But hydrologists believe that there are hardly any surplus waters anywhere in the country.

(*iv*) The construction of big dams and long canals will destroy forested areas, fertile soils and agricultural lands and disturb the ecological balance.

(*v*) No provision worth the name, has been made to resettle the displaced people. An estimated over 30 million people have been displaced by development projects since Independence.

(*vi*) Some experts have expressed doubt about the capability of the project. For example, if water available in the deficit Sabarmati basin is only 300 cubic metres per capita (CMPC), then it is difficult to comprehend how a further 1,400 CMPC can be brought from a surplus basin to solve the problem.

(*vii*) Alternative measures of managing and conserving water resources can prove to be more useful than the proposed project of interlinking the rivers. According to the Central Ground Water Board (CGWB), 37 BCM of ground water can be recharged locally at only ₹ 24,500 crore. Then what is the validity of investing ₹ 5,60,000 crore in obtaining a mere 175 BCM from interlinking of rivers.

(*viii*) Jayanta Bandhopadhyay of Centre for Development and Environment Policy at the Indian Institute of Management (Kolkata) questions the validity of such a project involving high cost and comparatively little returns. According to him, arresting the natural flow of rivers on this gigantic scale could spell "the death knell" of mangroves in the Delta region of West Bengal and Bangladesh and some of the richest fisheries could be lost forever. Salinity would also make inroads into the region, affecting thousands of hectares of arable land. Also the link scheme did not envisage carrying water to some of the most deprived sections of population, including tribals and those living in the highlands.

(*ix*) Neighbouring countries like Bangladesh and Nepal have not been consulted although these countries, along with China, would be affected by the project.

(*x*) Dr. Bharat Singh, Professor Emeritus, Water Resources Development IIT, Roorkee has simply termed it "*A big dream of little logic*". According to him, "there is no convincing argument which can justify taking up the river linking project in its entirety."

In spite of the above mentinoed difficulties, efforts are being made for interlinking some of the rivers where geographical conditions permit the feasibility of such projects and where cost factors are also within the means. The links namely (*i*) Ken-Betwa link, (*ii*) Parbati-Kalisindh-Chambal link, (*iii*) Godavari (Polavaram)—Krishna (Vijayawada) link, (*iv*) Dhamanganga-Pinjal link and Par-Tapi-Narmada link have been identfiied as priority links.

1. Ken Betwa Link. A tripartite Memorandum of Understanding (MoU) was signed between the Union Minister of Water Resources and Chief Ministers of Uttar Pradesh and Madhya Pradesh on 25th August, 2005 for preparation of detailed project report of Ken-Betwa link and the report was completed by the National Water Development Agency (NWDA) in December 2008. The project aims to transfer surplus water from Ken to Betwa. The region spreads across Jhansi, Banda, Lalitpur and Mahoba districts of Uttar Pradesh and Tikamgarh, Panna and Chhatarpur districts of Madhya Pradesh. The project involves building of 77 metre tall Daudhan dam and a 230 km long Canal.

The project will benefit the perennially water-short Bundelkhand region and other associated regions of Bina and upper Betwa basins. Besides it will provide drinking water to 62 lakh people and irrigation facilities to the drought prone areas. It will also generate 100 MW hydro power and 27 MW solar power. The project will provide year round employment to Bundelkhand region and control forced migration to far flung areas for livelihood. The assured drinking water will also uplift health standard of the local people.

However, the project will have some environmental and resettlement and rehabilitation issues. Daudhar reservoir will cause submergence of about 9,000 hectares of land of which 5,800 hectares will be forest land.

2. Parbati-Kalisindh-Chambal Link. It is joint venture of Madhya Pradesh and Rajasthan for which both the states had agreed on 25th August, 2005. Madhya Pradesh agreed to transfer water of the Chambal river to the Banas river. Both the states agreed to make full use of 12 million acre foot water of the Chambal basin. Currently both the states are using only 3.5 million acre foot water. There is a proposal to use the remaining 8.5 million acre foot of water also. On completion, this project will provide irrigation facilities to 2.21 lakh hectares of agricultural land in Madhya Pradesh and Rajasthan. In addition, drinking water would also be provided to neighbouring areas of Madhya Pradesh and Rajasthan. The main right canal starting from Kota barrage would be repaired so that farmers at the tail end are not deprived of water for irrigation.

3. Par-Tapi-Narmada Link. This project aims at transferring surplus water of the Western Ghats to the water deficit region of Saurashtra and Kachchh. It includes seven reservoirs proposed in north Maharashtra and South Gujarat. The water from these reservoirs would be taken through a 395 km long canal including 33 km length of feeder canals. Besides providing irrigation to enroute command and Narmada command, the link will generate 93 MKWh power.

4. Godavari-Krishna Link. This project envisage the diversion of 247 thousand million cubic feet (tmcf) of unutilised water in Indravati sub-basin of the Godavari basin. Water will be lifted from Godavari river to Nagarjuna Sagar dam and further south to meet the demands of Krishna, Pennar and Cauvery basins. The project will provide irrigation to 3.45-5.04 lakh hectares of land in Prakasam, Nellore, Krishna, Guntur and Chittoor districts of Andhra Pradesh.

5. Mahanadi - Godavari - Krishna - Pennar-Cauvery-Vaigai-Gundar Link System. The consesus building for eight more links under Mahanadi-Godavari-Krishna-Pennar-Cauvery-Vaigai-Gundur Linkage System has been initiated with the concerned states for preparing detailed project report. The National Water Development Agency (NWDA) has received 36 proposals of Inter-State links from 7 States out of which Pre-feasibility Reports (PFRs) of 15 intra-state links have been completed.

Ground Water

A part of the rain water percolates in the rocks and soils and is available to us as ground water. The assessment of water resources in India dates back to 1949. **Dr. A.N. Khosla** (1949) estimated the total average annual run-off of all river systems in India at 167.4 m. ha m (million hectare metre) based on empirical formula which included both surface and ground waters. Since then, several attempts have been made to assess the ground water resources in the country. The National Commission on Agriculture (1976), assessed the total ground water of the country as 67 m. ha m, excluding soil mixture. The usable ground water resource was assessed as 35 m. ha m of which 26 m. ha m was considered as available for irrigation. The first attempt to estimate the ground water resources on scientific basis was made in 1979 when a High Level Committee, known as Ground Water Over Exploitation Committee was constituted by Agriculture Refinance and Development Corporation (ARDC). Based on the norms for ground water resources computations recommended by this committee, the State Governments and the Central Ground Water Board computed the gross ground water recharge as 46.79 m. ha m and the net recharge (70% of the gross) as 32.49 m. ha m. The norms recommended by the Ground Water Estimation Committee (1984) are currently utilized by the Central Ground Water Board and the State Ground Water Departments to compute the ground water resources. Based on the recommendations of this committee, the annual replenishable ground water resources in the country work out to be 45.33 m. ha m. Keeping a provision of 15% (6.99 m. ha m) for drinking, industrial and other uses, the utilisable ground water resource for irrigation was computed at 38.34 m. ha m per year.

The methodologies adopted for computing ground water resource are generally based on the hydrological techniques. The main items of supply and disposal of ground water are listed below :

TABLE 17.2. Ground Water Resource Potential as per Basin

Sl. No.	Basin	Total Replenishable Ground Water Resource (M.C.M./Yr)	Provision of Domestic Industrial and Other Uses (M.C.M./Yr)	Available for irrigation (M.C.M./Yr)	Net Draft (M.C.M./Yr)	Balance for Future Use (M.C.M./Yr)	% Level of Ground Water Development
1.	Brahmaputra	26545.69	3981.35	22564.34	760.06	21804.29	3.37
2.	Brahmani with Baitarni	4054.23	608.13	3446.09	291.22	3154.88	8.45
3.	Cambal Composite	7187.25	1078.09	6109.16	2449.06	3660.10	40.09
4.	Caveri	12295.71	1844.35	10451.35	5782.85	4668.50	55.33
5.	Ganga	170994.74	26030.47	144964.26	48593.67	96370.56	33.52
6.	Godavari	40649.82	9657.69	30992.12	6054.23	24937.90	19.53
7.	Indus	26485.42	3053.95	23431.47	18209.30	5222.17	77.71
8.	Krishna	26406.97	5578.34	20828.63	6330.45	14498.19	30.39
9.	Kutch & Saurashtra	11225.09	1738.10	9486.99	4851.87	4791.02	51.14
10.	Madras & Southern	18219.72	2732.95	15486.77	8933.25	6553.52	57.68
11.	Mahanadi	16460.55	2471.10	13989.45	972.63	13016.81	6.95
12.	Meghna	8516.69	1277.48	7239.21	285.34	6953.87	3.94
13.	Narmada	10826.54	1653.75	9172.79	1994.18	7178.61	21.74
14.	Northeast Composite	18842.61	2826.39	16016.22	2754.93	13261.29	17.20
15.	Pennar	4929.29	739.39	4189.89	1533.38	2656.51	36.60
16.	Subarnarekha	1819.41	272.91	1546.50	148.06	1398.43	9.57
17.	Tapi	8269.50	2335.79	5933.0	1961.33	3972.38	33.05
18.	Western Ghat	17693.72	3194.78	14499.18	3318.12	11181.06	22.88
	Total	**431422.93**	**71075.02**	**360348.15**	**115223.93**	**245280.08**	**31.92**

Source : Central Ground Water Board in EnviStats, India, 2018, p. 224. MCM/yr : Million Cubic Metre/year

Note: Totals may not tally due to rounding off.

1. Items of supply to ground water reservoir

(*i*) Precipitation infiltration to the water table.

(*ii*) Natural recharge from streams, lakes and ponds.

(*iii*) Ground water inflow into the area under consideration.

(*iv*) Recharge from irrigation, reservoirs and other schemes especially designed for artificial recharge.

2. Items of disposal from ground water reservoir

(*i*) Evaporation from capillary fringe in areas of shallow water table, and transpiration by vegetation.

(*ii*) Natural discharge by seepage and spring flow to streams, lakes and ponds.

(*iii*) Ground water outflow.

(*iv*) Artificial discharge by pumping or flowing wells or drains.

Details of basin-wise ground water resources potential are given in **table 17.2.**

Groundwater Depletion in India. India is the world's largest groundwater extractor, pumping out nearly 25% of the global groundwater usage annually. Extraction of groundwater is far in access of its replenishment by rain water and the rate of extraction is increasing with each passing day. According to the Ministry of Jal Shakti, almost 22% of the groundwater (assessed units) in the country has either dried up or is in 'critical' and 'over-exploited' categories. The 'Dynamic Groundwater Resources of India' report of the Central Groundwater Board (CGWB) showed that 1499 out of 6881 assessed units (blocks/mandals/taluks) in 2017 came under 'over-exploited' (1186 units) and critical (313 units) categories. The highest number of water stressed blocks were in Tamil Nadu (541), followed by Rajasthan (218), Uttar Pradesh (139), Telangana (137), Punjab (111) and Haryana (81).

The CGWB report showed that 'over-exploited' areas were mostly concentrated in parts of Punjab, Haryana, Delhi and western Uttar Pradesh (where even though replenishable resources are abundant, there has been indiscriminate withdrawal of groundwater leading to over-exploitation); parts of Rajasthan and Gujarat (where due to arid climate, groundwater recharge is limited); and parts of Karnataka, Andhra Pradesh, Telangana and Tamil Nadu where due to inherent properties of crystalline aquifers, the groundwater availability is low.

Over exploitation of groundwater has led to fall in the water table to a dangerous level. In some parts of Punjab, Haryana and Western Uttar Pradesh, the groundwater level is falling at the rate of one metre per annum and many aquifers may dry up sooner than thought. **Figure 17.3** shows that there has been a considerable fall in the groundwater level during forty years from 1980 to 2020.

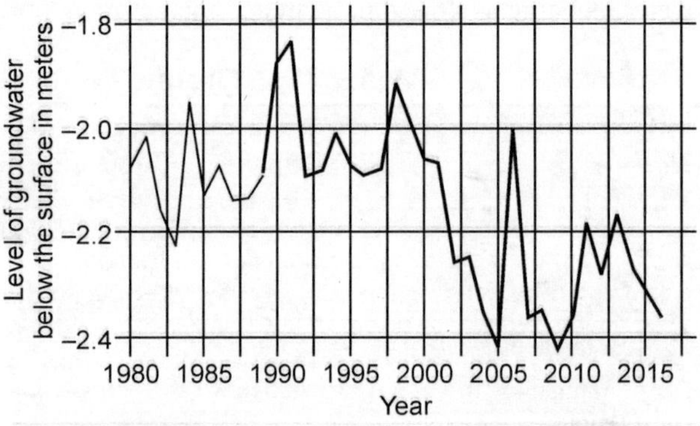

FIG. 17.3. Groundwater depletion in India

The top ten water stressed states/union territories in India are Haryana, Rajasthan, Uttar Pradesh, Punjab, Gujarat, Uttarakhand, Madhya Pradesh, and Union Territories of Jammu and Kashmir, Puducherry and Chandigarh.

According to Asian Water Development Outlook, 2016, almost 89% of the groundwater is used for irrigation. Rice and sugarcane are two water guzzing crops and depend heavily on groundwater. These two crops consume more than 60% of irrigation water available in the country, thereby reducing water availability for other crops.

Wheat is the second most important food crop of India and this crop also uses large quantities of ground water. On an average it requires 2800 litres of water to produce one kilogram of rice and 1654 litres to produce one kilogram of wheat. These quantities can vary depending on climate and soil conditions. For example, Punjab is using three times more water to produce a kg of rice compared to Bihar and more than twice compared to West Bengal. At present India is exporting more than 10 trillion litre of virtual water through export of basmati rice. To make things still worse, these crops have expanded in water scarce areas consuming larger amounts of irrigation water per hectare. The best solution for this complex problem is to bring about a change in the cropping pattern. For example, replacing rice and wheat with crops like maize, millets, sorghum, etc. can reduce irrigation water demand by one-third.

Hydrological Situation

India is a vast country having diversified geological, climatological and topographic set up, giving rise to divergent ground water situations in different parts of the country. The prevalent rock formations, ranging in age from the Archaean to the Recent, which control occurrence and movement of ground water, are widely varied in composition and structure. Variations of land forms varying from the rugged mountainous terrain of the Himalayas to the flat and featureless alluvial plains of the north, river valleys and coastal tracts, and aeolian deserts of Rajasthan are no less important. The topography and rainfall virtually control runoff and ground water recharge.

The high relief areas of the northern and north-eastern regions, the Aravali range of Rajasthan, and peninsular regions with steep topographic slope and characteristic geological set-up offer high run-off and little scope for rain water infiltration. The ground water potential in these terrains are limited to intermontane valleys.

The large alluvial tract in the Indus-Ganga-Brahmaputra plains, extending from Punjab in the west to Assam in the east, constitutes one of the largest and the most potential ground water reservoir in the world. The acquifer systems are extensive, thick, hydraulically interconnected and moderate to high yielding. To the north of this tract, all along the Himalayan foot hills, occur the linear belt of Bhabar piedmont deposits, and the Tarai belt down the slope with characteristic autoflowing conditions.

Almost the entire peninsular India is occupied by a variety of hard and fissured formations with patches of semi-consolidated sediments in narrow intracratonic basins. Rugged topography, compact and fissured nature of the rock formations, combine to give rise to discontinuous acquifers with limited to moderate yield potentials. The near surface, weathered mantle forms the all important ground water reservoir, and the source for circulation of ground water through underlying fracture systems. In the hard rock terrain, deep weathered pediments, low-lying valleys and abandoned river channels generally contain adequate thickness of porous material, to sustain ground water development under favourable hydrometeorological conditions. Generally, the potential water saturated systems occur down to 100 metre depth. The friable semi-consolidated sandstones also form moderate yielding aquifers and auto flowing zones in these formation are not uncommon.

The coastal and deltaic tracts in the country form a narrow linear strip around the peninsular plateau. The eastern coastal and deltaic tract and the estuarine areas of Gujarat are receptacles of thick alluvial sediments. Though highly productive aquifers occur in these tracts, salinity hazards impose quality constraints for ground water development.

The above description leads as to the conclusion that the ground water resources are influenced by a number of natural conditions of which climate (particularly rainfall and temperature), relief features (topography), geological structure and hydrological setup are of outstanding significance. Accordingly Dr. R.L. Singh (1971) has divided India into 8 ground water provinces (**Fig. 17.4**), described as under :

1. **Pre-Cambrian Crystalline Province.** It extends over half of the country's geographical area covering Tamil Nadu, Andhra Pradesh, Telangana, Karnataka, Maharashtra, Dandakaranya, Bundelkhand and Aravali range. This province is deficient in ground water resources.
2. **Pre-Cambrian Sedimentary Province.** It extends over Cuddapah and Vindhyan basins where the rocks belong to Cuddapah and Vindhyan systems. This province is also not much suitable for ground water development and contains inadequate amount of ground water.

3. **Gondwana Sedimentary Province.** The Gondwana sedimentary rocks of the Barakar and Godavari river basins contain good aquifers of ground water.

4. **Deccan Trap Province.** These are 1,200 metre thick covering of impermeable basalt over the surface which obstructs percolation of water. As such, the whole province is

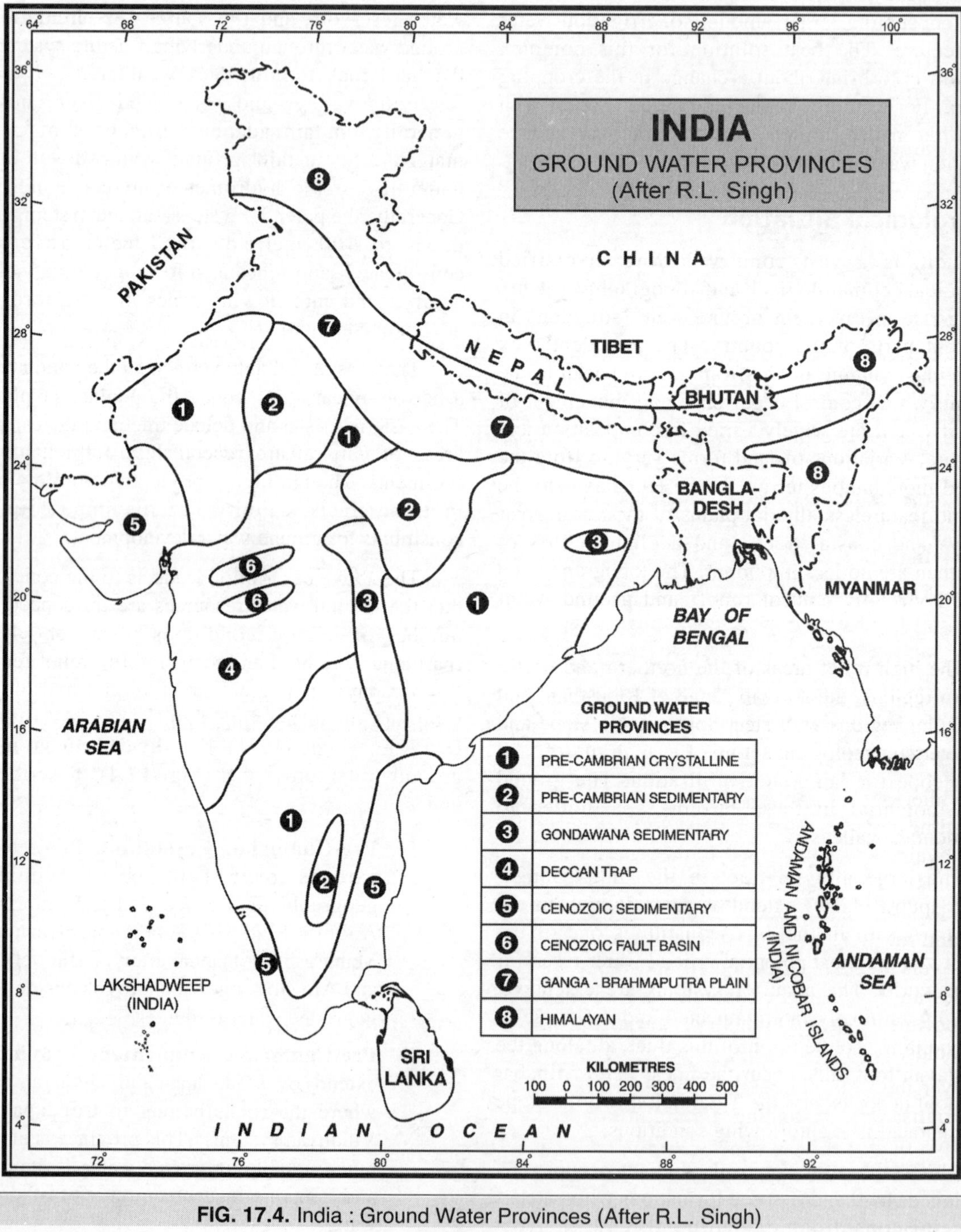

FIG. 17.4. India : Ground Water Provinces (After R.L. Singh)

deficient in ground water resources. The only aquifers preserved are in the fractures where secondary porosity develops in the weathered *moorums* at times, in the intertrappean beds sandwiched between two impermeable strata as also in the vesicles and amygdales.

5. **Cenozoic Sedimentary Province.** This province includes the Andhra Pradesh, Tamil Nadu, Kerala and Gujarat coasts. These areas have tertiary sandstones and the province as a whole has good acquifers.
6. **Cenozoic Fault Basin.** The rift zone of the Narmada, the Purna and the Tapi provides good resource of ground water in its 80-160 metre thick alluvial cover of sand, silt and clay.
7. **Ganga-Brahmaputra Alluvial Province.** This is the richest ground water province of the country. The bhabar, tarai and the axial belts are well defined. The streams disappearing in the unassorted materials of the bhabar zone seep out in the tarai belt. Moreover, the ground water table is also high.
8. **Himalayan Province.** This complex structural and geographic unit is not very significant with respect to ground water resources. Local springs are common but wells are a rare feature.

Water Scarcity

While water is a renewable resource, it is at the same time a finite resource. The total quantity of water available on the globe is the same as it was thousands of years ago. It is important to appreciate the fact that only 3 per cent of the world's water is fresh and roughly one-third of it is inaccessible. The rest is very unevenly distributed and the available supplies are increasingly contaminated with wastes and pollution from industry, agriculture and households.

Over the years, increasing population, growing industrialisation, expanding agriculture and rising standards of living have pushed up the demand for water. Efforts have been made to collect water by building dams and reservoirs and creating ground water structures such as wells. Recycling and desalination of water are other options but cost involved is very high. However, there is a growing realisation that there are limits to '*finding more water*' and in the long run, we need to know the amount of water we can reasonably expect to tap and also learn to use it more efficiently.

It is human nature that we value things only when they are scarce or are in short supply. As such we appreciate the value of water once the rivers, reservoirs, ponds, wells, etc. run dry. Our water resources have now entered an era of scarcity. It is estimated that thirty years from now, approximately once-third of our population will suffer from chronic water shortages. The increasing demands on fresh water resources by our burgeoning population and diminishing quality of existing water resources because of pollution and the additional requirements of serving our spiralling industrial and agricultural growth have led to a situation where the consumption of water is rapidly increasing and the supply of fresh water remains more or less constant. It may be maintained that the water available to us is the same as it was before but the population and the consequent demand for water has increased manifold. The consequences of scarcity will be more drastic in arid and semi-arid regions. Water shortage will also be felt in rapidly growing coastal regions and in big cities. Several cities are already, or will be, unable to cope with the demand of providing safe water and sanitation facilities to their inhabitants.

Indicators of water stress and scarcity are generally used to reflect the overall water availability in a country or a region. When the annual per capita of renewable fresh water in a country or a region falls below 1,700 cubic metres, it is held to be situation of *water stress*. If the availability is below 1,000 cubic metres, the situation is labelled as that of *water scarcity*. And when the per capita availability falls below 500 cubic metres, it is said to be a situation of *absolute scarcity* (Engelman and Roy, 1993). These are also the findings of a study conducted by the Tata Energy Research Institute (TERI). This concept has been propounded by Malin Falkenmark on the premise that 100 litres a day (36.5 cubic metres a year) is roughly the minimum per capita requirement for basic household needs and to maintain good health, roughly 5 to 20 times that amount is needed to satisfy the requirement of agriculture, industry and energy.

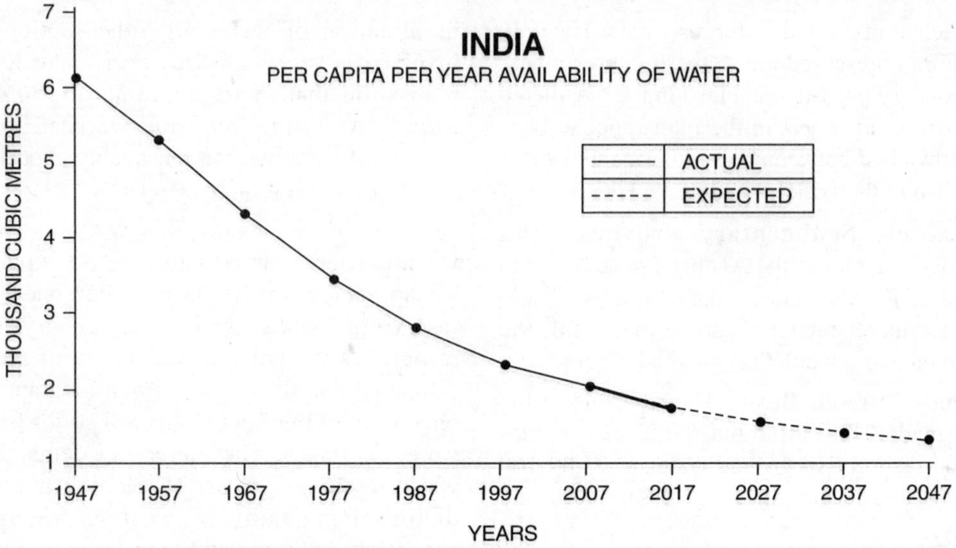

FIG. 17.5. India : Declining availability of water

TABLE 17.3. Trends of water availability in India

Year	Population (in billions)	Per capita water availability (in cubic metres per year)
1947	—	6042
1951	0.36	5177
2001	1.03	1820
2011	1.21	1544
2015	1.33	1441
2021	1.35	1421
2031	1.46	1306
2041	1.56	1225
2051	1.63	1174

Source: CWC, "Reassessment of Water Availability in India, using Space Inputs.

At the time of Independence, *i.e.*, in 1,947, the per capita availability of water in India was 6,042 cubic metres a year. It came down to 5,177 cubic metres a year in 1951, to 1,820 cubic metres a year in 2001 and 1,544 in 2011. As per figures released by Ministry of Jal Shakti, India's per capita availability of water has been pegged at 1,541.5 cubic metres a year, including non-personal consumption, such as irrigation—notches below the international threshold of 1700 cubic metres a year. India has only 4 per cent of the world's water resources, and she has to support over 17 per cent of the world's population and over 20 per cent of livestock. According to mid term appraisal (MTA) of the 10th Plan, per capita availability of water is likely to fall down to 1,340 cubic metre in 2025 and 1,174 cubic metre in 2050. **Figure 17.4** shows the probable trends in per capita water availability for hundred years from 1947 to 2047. The problem of water shortage is further complicated when we look at the areal distribution of water resources with reference to population. From as high as 18,417 cubic metres in the Brahmaputra valley, per capita water availability comes down to a low of 411 cubic metres in the east-flowing rivers between Pennar and Kanniyakumari. Even within the Ganga basin, the availability varies from 740 cubic metres in the Yamuna to 3,379 cubic metres in the Gandak basin (Chitale, 1992).

According to 2011 Census figures 85.5 per cent of India's households have access to safe drinking water (tap/hand pump/tubewell). At 91.4 per cent, urban people are better placed than 82.7 per cent rural households. Punjab has the highest percentage of 97.6 households enjoying the facility of safe drinking water. As much as 96.7 per cent rural and 98.9 per cent urban households in Punjab have access to safe drinking water. Meghalaya is the worst placed where only 44.7 per cent of the total households have access to safe drinking water. The figures for rural and urban

TABLE 17.4. Householders with access to Safe Drinking Water in India in 2011 (per cent)

States/Union Territories	Total	Rural	Urban
Andhra Pradesh	90.5	88.6	94.5
Arunachal Pradesh	78.6	74.3	91.3
Assam	69.9	68.3	78.2
Bihar	94.0	93.9	94.7
Chhattisgarh	86.3	84.1	93.9
Gujarat	90.3	84.9	97.0
Haryana	93.8	92.0	96.7
Himachal Pradesh	93.7	93.2	97.8
Jammu & Kashmir	76.8	70.1	96.1
Jharkhand	60.1	54.3	78.4
Karnataka	87.5	84.4	92.3
Kerala	33.5	28.3	39.4
Madhya Pradesh	78.0	73.1	92.1
Maharashtra	83.4	73.2	95.7
Manipur	45.4	38.1	60.8
Meghalaya	44.7	35.1	79.5
Mizoram	60.4	43.4	75.8
Nagaland	53.8	54.6	51.8
Odisha	75.3	74.4	79.8
Punjab	97.6	96.7	98.9
Rajasthan	78.1	72.8	94.3
Sikkim	85.3	82.7	92.2
Tamil Nadu	92.5	92.2	92.9
Tripura	67.5	58.1	91.9
Uttar Pradesh	95.1	94.3	97.9
Uttarakhand	92.2	89.5	98.7
West Bengal	92.2	91.4	93.9
Andaman & Nicobar Islands	85.5	78.2	98.1
Chandigarh	99.3	98.7	99.4
Dadra & Nagar Haveli	91.6	84.3	98.4
Daman & Diu	98.7	97.8	99.0
Delhi	95.0	87.9	95.2
Goa	85.7	78.4	90.4
Lakshadweep	22.8	31.2	20.2
Puducherry	97.8	99.6	97.0
All India	**85.5**	**82.7**	**91.4**

Source : Economic Survey 2022-23 Statistical Appendix, p. 173.

areas are 35.1 and 79.5 per cent respectively. Among the Union Territories, Chandigarh is the best placed where 99.3 per cent of the households have access to safe drinking water, while the people of Lakshadweep are very unlucky where only 22.8 per cent of the total households have access to safe drinking water (**Table 17.4**).

However, the above description regarding percentage of households having access to safe drinking water could be misleading and the real picture emerges only when we look at the individual cities. A survey conducted by Tata Institute of Social Science (TISS) showed 50 lakh households in Mumbai, Delhi, Kolkata, Ahmedabad, Hyderabad, Kanpur and Madurai are water deficient (see **Table 17.5**).

TABLE 17.5. Number of Water Deficient Households

City	Number of water deficient households
Mumbai	17,35,756
Delhi	14,73,114
Kolkata	4,36,905
Ahmedabad	4,29,199
Hyderabad	4,13,881
Kanpur	3,29,714
Madurai	1,48,533

Source : Tata Institute of Social Sciences (TISS) Survey, 2005.

Table 17.6 shows that several cities particularly the southern cities are most water deficient. Chennai and Bengaluru suffer from 53.8 and 39.5 per cent deficiency respectively. Hyderabad in Telangana has 24.2% deficiency while Vishakhapatnam in Andhra Pradesh suffers from extreme deficiency of 91.8 per cent. In the north, Delhi records 29.8 per cent water deficiency and Lucknow, 27.3 per cent. Central India is more water- deficient than the north with wide regional variations. For example, Bhopal is 26.4 per cent water-deficient while Indore and Jabalpur record rates of 72.8 and 65.4 per cent deficiency respectively. Mumbai in the west, with deficiency rate of 43.3 per cent, is similarly situated as Kolkata which clocks at 44 per cent.

Nearly 40 per cent of water demand in urban India is met by ground water. So ground water tables in most cities are falling at alarming rate of 2-3 metres per year. Another factor is water leakage. Delhi loses at least 35 per cent of its water due to leakages in its 83.0 km long pipeline network. Mumbai loses about 30 per cent of its water due to leakage.

TABLE 17.6. Demand, Supply and Deficiency of Water in Selected Cities of India in Million Litres per Day (MLD)

City	Demand	Supply	Deficiency (%)
Mumbai	4,300	3,600	43.3
Delhi	3,830	2,950	29.8
Kolkata	2,258	1,568	44.0
Chennai	3,000	1,950	53.8
Hyderabad	956	770	24.2*
Indore	318	184	72.8
Bengaluru	1,200	860	39.5
Lucknow	560	440	27.3
Jabalpur	239	144.5	65.4
Vishakhapatnam	305	159	91.8**

*least deficient **most deficient

***Source*:** Abdul Shaban (2005), "Water Consumption Patterns in Domestic Households in Major Cities in India. Tata Institute of Social Sciences (TISS), Mumbai.

According to a study released by the World Wide Fund for Nature (WWF) on November 2, 2020. India has 31 cities exposed to serious water security risks. Thus India faces a greater challenge than the continents of Africa or Europe. Consequently India faces the second highest basin risk in the world, only better than Palestine.

The analysis, 'Water Risk Filter', focuses on two things—water stress cities face now and the projected increase by 2050. The study says, "If you go just by water risk now and that projected for 2050, Indian cities dominate the list of cities facing very high water risk."

On a scale of 1 to 5, anything over 3 is 'high risk' and over 4 is 'very high risk'. Of the 31 Indian cities that are under water stress, 26 scored over 4, meaning that they face 'very high risk' right now.

WATER WOES IN INDIA

- The number of households has increased from 24.1 crore in 2001 to 33.1 crore in 2011 and per capita availability of water has reduced from 1820 cubic metres/yr in 2001 to 1545 cubic metre/yr—notches below the international threshold of 1700 cubic metres/yr.
- 30% of rural Indians lack drinking water supply (World Bank, UNICEF).
- According to National Sample Survey Office (NSSO), 57% of the rural women in India have to walk up to 5 km every day to fetch potable water as compared to just 21% in Urban areas.
- Following findings of the report released by National Sample Survey Office (NSSO) in August, 2014 are worth noting :
  (*a*) 54% of rural woman in India had to travel between 200 metres and 5 kilometres daily to fetch drinking water in 2012.
  (*b*) They walked 20 minutes a day, on an average, and spend another 15 minutes at the source of water.
  (*c*) Every second woman in rural India walked average 173 km to fetch potable water in 2012 making her track 25 km longer than what it was in 2008-09.
  (*d*) Every second woman has to spend 210 hours in a year for fetching water which means a loss of 27 days' wages for these households. Collectively, these women cover 64,000 times the distance between the earth and the moon.
  (*e*) Water crisis is imminent in villages because of over harvesting of ground-water resources. About 80% of the country's drinking water needs are met by groundwater.
  (*f*) 75% of the women in states like Chhattisgarh, Manipur, Odisha and Jharkhand have to travel long distances for drinking water. Average time taken for this exercise is highest in Jharkhand (40 minutes) followed by Bihar (33 minutes) and Rajasthan (30 minutes).
- India records worlds highest per capita water-borne diseases even more than some of the least developed nations.
- In most of the large cities about one-third of water never reaches the consumer because of leaks and poor maintenance. According to Centre for Science and Environment, over 35% of water in Delhi and about 30% in Mumbai is lost because of leakage.
- India is home to 17.5% of world's population but has only 4% of water.

Water Disputes and Conflicts

Any commodity which is in short supply is likely to cause disputes and conflicts and water shortage is no exception. Water disputes and conflicts are taking place at the national (inter-state) and international levels in the present day world. Experts believe that the biggest potential destabiliser in the world is water scarcity. Former UN Secretary General, Boutros Bautros Ghali had warned in 1980s that future wars could be fought over water. His successor Kofi Annan was also worried about the fierce competition over water resources that contained the seeds of violent conflicts. Ismael Serageldin, Vice President, World Bank, had predicted in 1995 that "if wars of this century were fought over oil, the wars of the next century will be fought over water." Whereas 'oil' had been the bone of contention in many wars of the yester years, the prediction of Mrs. Elizabeth Dowdeswell the "*Major clashes over dwindling supplies of water may well constitute the source of future conflicts between nations*", seems to be coming true, with the waging national and international disputes over water sharing, blowing up to the proportion of an imminent Third World War. It is now feared that "*Third World War if fought, will not be fought for territorial gains or political supremacy but for water.*"

Potential conflicts are likely where rivers and lakes are shared by more than one country. The Nile, the Jordan, the Indus, the Ganga, the Brahmaputra and the Mekong are some of these. In times of water stress and shortages, regions will face water refugees from one region to the other within the country or between two countries.

There could be wars for control of water supplies; or water resources or systems used as a weapon during military conflict; or used as political goal; terrorists could threaten using water resources as a weapon of coercion. Water systems themselves could be targets of military action. Then, with multinational giants having entered the business of supplying water privately for profit, there could be wars for entrepreneurial control. With mismanagement of water resources, mighty rivers can become mere rivulets, unable to reach the sea. With taps run dry and crops wither away, there would be upheavals—mixed as they would be with regional, caste, sectarian and communal colour. All this may be difficult to imagine, but this is calamity about to happen. It is nightmare about to come true.

Inter-state River Water Disputes in India

Most rivers of India are plagued with interstate disputes. Almost all the major rivers of the country are inter-state rivers and their waters are shared by two or more than two states. After independence, demand for water had been increasing at an accelerated rate due to rapid growth of population, agricultural development, urbanisation, industrialisation, etc. These developments have led to several inter-state disputes about sharing of water of these rivers. Following interstate river water disputes are worth mentioning.

(*i*) Cauvery water dispute between Tamil Nadu, Karnataka and Kerala.

(*ii*) The Krishna water dispute between Maharashtra, Karnataka and Andhra Pradesh.

(*iii*) The Tungabhadra water dispute between Andhra Pradesh and Karnataka.

(*iv*) The Aliyar and Bhivani river water dispute between Tamil Nadu and Kerala.

(*v*) The Godavari river water dispute between Andhra Pradesh, Madhya Pradesh, Chhattisgarh, Odisha and Karnataka.

(*vi*) The Narmada water dispute between Gujarat, Maharashtra, Madhya Pradesh and Rajasthan.

(*vii*) The Mahi river dispute between Gujarat, Rajasthan and Madhya Pradesh.

(*viii*) The Ravi and Beas river water dispute between Punjab, Haryana, Himachal Pradesh, Rajasthan, Jammu and Kashmir and Delhi.

(*ix*) The Satluj-Yamuna Link canal dispute between Punjab, Haryana and Rajasthan.

(*x*) The Yamuna river water dispute between Uttar Pradesh, Haryana, Himachal Pradesh, Punjab, Rajasthan, Madhya Pradesh and Delhi.

(*xi*) The Karmanasa river water dispute between Uttar Pradesh and Bihar.

(*xii*) The Barak river water dispute between Assam and Manipur.

Efforts are being made to resolve disputes through negotiations amongst the basin states with the assistance of the Central Government. Many of these interstate river water disputes have been settled on the basis of equitable apportionment which is the universally accepted principle. Adjudication through appointment of water disputes tribunals is also resorted to as and when required. So far, the following tribunals have been appointed to resolve inter-state water disputes :

(*i*) The Godavari Water Disputes Tribunal
(*ii*) The Krishna Water Disputes Tribunal
(*iii*) The Narmada Water Disputes Tribunal
(*iv*) The Ravi and Beas Water Disputes Tribunal
(*v*) The Cauvery Water Disputes Tribunal
(*vi*) New Krishna Water Disputes Tribunal

In a developing country like India, the inter-state river water disputes have to be resolved quickly and amicably. This is most urgent for the proper utilisation of water resources and economic growth.

International Cooperation

Several big rivers of India like the Indus, the Ganga and the Brahmaputra flow through the territory of some neighbouring countries also in addition to their flow through the Indian territory. These rivers call for cooperation between India and her neighbouring countries. India has taken several initiatives to create an atmosphere of cooperation and mutual understanding with the neighbouring countries. A few examples are described below :

1. The Indus Water Treaty. This treaty is concerned with sharing of the waters of the Indus and its tributaries between India and Pakistan. Signed by the two concerned countries on September 19, 1960, this treaty was reached through the arbitration of International Bank of Reconstruction and Development. According to the provisions of this treaty, India has the exclusive rights to use the waters of three eastern rivers (Sutluj, Beas and Ravi), leaving out remaining three rivers (Chenab, Jhelum and Indus) to Pakistan which will also take care of water requirements of the Indian Union Territory of Jammu and Kashmir. However, India can utilise only 20 per cent of the total discharge of the Indus river system under the regulations of the Indus Water Treaty.

2. India-Bangladesh Cooperation. An Indo-Bangladesh Joint Rivers Commission (JRC) is functioning since 1972, the year that country came into being. This Commission was set up to ensure the most effective joint efforts in maximising the benefits from common river systems.

The signing of the Treaty between India and Bangladesh on the Ganga water sharing at Farakka on 12th December, 1996, ushered a new era of cooperation in water resources sector between the two countries. Under the agreement India and Bangladesh would share alternately for 10 days each 35,000 cusecs of waters during the lean season (1 March to 15 May) to fulfill their water needs.

Discussions are also continuing for sharing of waters of the Tista river.

3. India Nepal Cooperation. A Nepal-India Joint Committee on Water Resources (JCWR) is functioning with the mandate to act as Umbrella Committee of all committees—groups. A Treaty on Integrated Development of Mahakali was signed between India and Nepal in February, 1996 which came into force in June 1997 (*Mahakali Treaty*). Pancheshwar Multipurpose Project on river Mahakali (known as Sarda in India) is the centrepiece of Mahakali Treaty. Agreement has also been reached with Nepal to take up joint field investigations, studies and preparation of Detailed Project Report of Sapta Kosi High Dam Multipurpose Project and Sun Kosi Storage Dam-cum-Diversion Scheme.

For dealing with the problems of inadvertent inundation caused by the construction of various works on the border rivers between India and Nepal, a Standing Committee on Incubation Problems (SCIP) between India and Nepal is also functioning since 1986. This committee has been set up for identifying the problem areas and suggest possible solutions on a continuing basis.

4. India-China Cooperation. A Memorandum of Understanding (MoU) between India and China was signed in the year 2002. It makes provision of hydrological information, namely rainfall, water level, discharge and other relevant information on Yaluzangbu/Brahmaputra river in flood season by China to India from 1 June to 15 October every year. The Chinese side had started transmitting the relevant data in June, 2002. The information received from China is very useful for making advance forecasts for floods in the Brahmaputra river in India. The

Government of India has also taken up the matter with China for establishing additional hydrological stations on Langquinzagbu (Satluj) and Palongzangbu tributary of Brahmaputra.

5. India-Bhutan Cooperation. A Joint Expert Team (JET) consisting of officials from India and Bhutan is functioning since 1979 to review the progress and other requirements of the "Comprehensive Scheme for Establishment to Hydro-meteorological and Flood Forecasting Network on rivers common to India and Bhutan." A network of 35 hydro-meteorological/meteorological stations has been established in Bhutan. Data received from these stations is used by India for flood forecasting. Bhutan has agreed to join a sub-regional plan for sharing river waters and power with India and Bangladesh. That country is willing to divert 12,000 cusecs of water from the Sankosh river to the Tista river and from the Tista to Farakka barrage to be shared by India and Bangladesh. India has also agreed to purchase 4,000 megawatt of hydroelectricity from Bhutan. This power strengthens our National Power Grid and meets the power needs of the north-eastern region.

Conservation of Water Resourses

Water is an important natural resource and is the very basis of our life. We use water for drinking, irrigation, industry, transport and for the production of hydro-electricity. Water is a *cyclic resource* which can be used again and again after cleaning. The best way to conserve water is its judicious use. A large quantity of water is used for irrigation and there is an urgent need for proper water management in irrigation sector. Over-irrigation through canals has led to waterlogging in western Uttar Pradesh, Punjab, Haryana and Hirakud command area. Seepage along the canals can be checked by lining them. The overdraft by tube-wells has resulted in lowering of water table in a number of villages in Haryana, Punjab and western Uttar Pradesh. In arid areas, wherever water has been brought for irrigation, saline and alkaline tracts have emerged, rendering the soil infertile. Wasteful use of water should be checked. Sprinkler irrigation and drip irrigation can play a crucial role in conserving scarce water resources in dry areas. Drip irrigation and sprinklers can save anywhere between 30 to 60 per cent of water. Only 0.5 per cent—nearly half of this in Maharashtra—is under drip irrigation and 0.7 per cent under sprinklers. There is large-scale pollution of water as a result of industrialisation and urbanisation. This trend has got to be checked. Although one-eighth of India is declared as flood prone, there are several thousand villages in India which do not have potable drinking water. The basins should be treated as one unit for planning water utilization. Dry farming should be practiced in dry areas. The experimentation under the National Watershed Development Programme for Rainfed Agriculture is being carried on since 1986-87.

There is a great demand for water in industries and the industrial sector offers great opportunities to conserve water. The economy in water-use in this sector will have two benefits. Firstly, the saved water may be used to meet the demand in other sectors. Secondly, the affluents thrown in the water bodies will be less. Water in most industries is used for cooling purposes, thus, it is not necessary to use fresh potable water. Instead, the recycled water may be used for this purpose. By using the recycled water over and over again, fresh water can be conserved.

Demand of water for domestic use can also be reduced. For example, in most urban areas about 12.5 litres of water is used in one flushing. In some cities cisterns requiring only 5 to 7 litres of water in one flushing are now used. Thus if each urban individual adopts smaller cisterns, the amount of water consumption for flushing can be reduced to half. Similarly, if raw water is used for cleaning, gardening, etc., a lot of fresh potable water can be saved. Water used in kitchen sink, wash basin and in bathroom can be collected into a tank and reused for flushing toilet and for gardening also.

Rain Water Harvesting

Rain water harvesting is one of the most effective methods of water management and water conservation. It is the term used to indicate the collection and storage of rain water used for human, animals and plant needs. It involves collection and storage of rain water at surface or in sub-surface aquifer, before it is lost as surface run off. The augmented resource can be harvested in the time of need. Artificial recharge to ground water is a process by which the ground water reservoir is augmented at a rate exceeding that under natural conditions of replenishment. The collected water is stored and pumped in a separate pipe distribution. This is a very useful method for a developing country like India in reducing the cost and the demand of treated water and also economising the treatment plants operation, maintenance and distribution costs.

22nd March is observed as "World Water Day" out of 1992 UN Resolution.

Need

- To overcome the inadequacy of surface water to meet our demands.
- To arrest decline in ground water levels.
- To enhance availability of ground water at specific place and time and utilize rain water for sustainable development.
- To increase infiltration of rain water in the subsoil which has decreased drastically in urban areas due to paving of open area.
- To improve ground water quality by dilution.
- To increase agricultural production.
- To improve ecology of the area by increase in vegetation cover etc.

Advantages

- The cost of recharge to sub-surface reservoir is lower than surface reservoirs.
- The aquifer serves as a distribution system also.
- No land is wasted for storage purpose and no population displacement is involved.
- Ground water is not directly exposed to evaporation and pollution.
- Storing water under ground is environment friendly.
- It increases the productivity of aquifers.
- It reduces flood hazards.
- Effects rise in ground water levels.
- Mitigates effects of drought.
- Reduces soil erosion.

The New Delhi-based Centre for Science and Environment estimates that merely capturing the rain water and run off on 2 per cent of India's land area could supply 26 gallons of water per person per day. As such much effort is being made to popularize the concept of rain water harvesting at the grass roots level.

Potential Areas

- Where ground water levels are declining on regular basis.
- Where substantial amount of aquifer has been de-saturated.
- Where availability of ground water is inadequate in lean months.
- Where due to rapid urbanization, infiltration of rain water into subsoil has decreased drastically and recharging of ground water has diminished.

Methods and Techniques

The methods of ground water recharge are mainly :

Urban Areas

Roof top rain water/storm run off harvesting through

(*i*) Recharge Pit
(*ii*) Recharge Trench
(*iii*) Tubewell
(*iv*) Recharge Well

Rural Areas

Rain water harvesting through

(*i*) Gully Plug
(*ii*) Contour Bund
(*iii*) Gabion Structure
(*iv*) Percolation Tank
(*v*) Check Dam/Cement Plug/Nala Bund
(*vi*) Recharge Shaft
(*vii*) Dugwell Recharge
(*viii*) Ground Water Dams/Subsurface Dykes

Urban Areas

In urban areas, rain water available from roof tops of buildings, paved and unpaved areas goes waste. This water can be recharged to aquifer and can be utilized gainfully at the time of need. The rain water harvesting system needs to be designed in a way that it does not occupy large space for collection and recharge system. Roof top rain water harvesting can be a very effective tool to fight the problem of water shortage particularly in urban areas. Roof top rain water harvesting depends upon the amount of rainfall and the roof top area. More the amount of rainfall, more is the harvested water from roof top. Similarly, larger amount of roof top rain water is harvested from roofs with large area. **Table 17.7** gives the amount of harvested water from roof top in cubic metres in relation to the amount of rainfall in millimetres and the roof top area in square metres.

A few techniques of roof top rain water harvesting in urban areas are described as under :

(i) Roof Top Rainwater Harvesting through Recharge Pit

- In alluvial areas where permeable rocks are exposed on the land surface or at very shallow depth, roof top rain water harvesting can be done through recharge pits.
- The technique is suitable for buildings having a roof area of 100 sq m and are constructed for recharging the shallow aquifers.
- Recharge pits may be of any shape and size and are generally constructed 3 to 5 m long 1 to 2 m wide and 2 to 3 m deep which are back filled with boulders (5-20 cm), gravels (5-10 mm) and coarse sand (1.5-2 mm) in graded

TABLE 17.7. Availability of Rain Water through Roof Top Rain Water Harvesting

Rainfall (mm)	100	200	300	400	500	600	800	1000	1200	1400	1600	1800	2000
Roof top area (sqm)	Harvested Water from Roof top (cum)												
20	1.6	3.2	4.8	6.4	8	9.6	12.8	16	19.2	22.4	25.6	28.8	32
30	2.4	4.8	7.2	9.6	12	14.4	19.2	24	28.8	33.6	38.4	43.2	48
40	3.2	6.4	9.6	12.8	16	19.2	25.6	32	38.4	44.8	51.2	57.6	64
50	4	8	12	16	20	24	32	40	48	56	64	72	80
60	4.8	9.6	14.4	19.2	24	28.8	38.4	48	57.6	67.2	76.8	86.4	96
70	5.6	11.2	16.8	22.4	28	33.6	44.8	56	67.2	78.4	89.6	100.8	112
80	6.4	12.8	19.2	25.6	32	38.4	51.2	64	76.8	89.6	102.4	115.2	128
90	7.2	14.4	21.6	28.8	36	43.2	57.6	72	86.4	100.8	115.2	129.6	144
100	8	16	24	32	40	48	64	80	96	112	128	144	160
150	12	24	36	48	60	72	96	120	144	168	192	216	240
200	16	32	48	64	80	96	128	160	192	224	256	288	320
250	20	40	60	80	100	120	160	200	240	280	320	360	400
300	24	48	72	96	120	144	192	240	288	336	384	432	480
400	32	64	96	128	160	192	256	320	384	448	512	576	640
500	40	80	120	160	200	240	320	400	480	560	640	720	800
1000	80	160	240	320	400	480	640	800	960	1120	1280	1440	1600
2000	160	320	480	640	800	960	1280	1600	1920	2240	2560	2880	3200
3000	240	480	720	960	1200	1440	1920	2400	2880	3360	3480	4320	4800

Source : Rainwater Harvesting Techniques to Augment Ground Water (2003), Central Ground Water Board, Ministry of Water Resources.

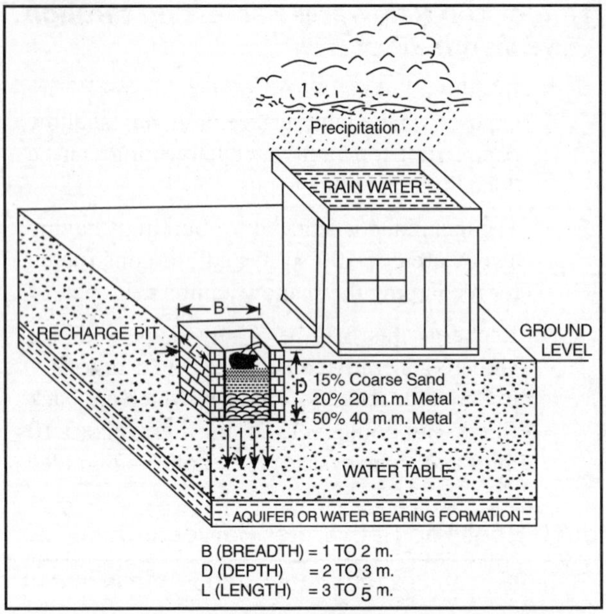

FIG. 17.6. Roof top rain water harvesting through recharge pit

form—boulders at the bottom, gravels in between the coarse sand at the top so that the silt content that will come with runoff will be deposited on the top of the coarse sand layer and can easily be removed. For smaller roof area, pit may be filled with broken bricks/cobbles.

- A mesh should be provided at the roof so that leaves or any other solid waste/debris is prevented from entering the pit and a desilting/collection chamber may also be provided at the ground to arrest the flow of finer particles to the recharge pit.
- The top layer of sand should be cleaned periodically to maintain the recharge rate.
- By-pass arrangement be provided before the collection chamber to reject the first showers.

(ii) Roof Top Rain Water Harvesting through Recharge Trench

- Recharge trenches are suitable for buildings having roof area of 200-300 sq m and where permeable strata is available at shallow depths.
- Trench may be 0.5 to 1 m wide, 1 to 1.5 m deep and 10 to 20 m long depending upon availability of water to be recharged.

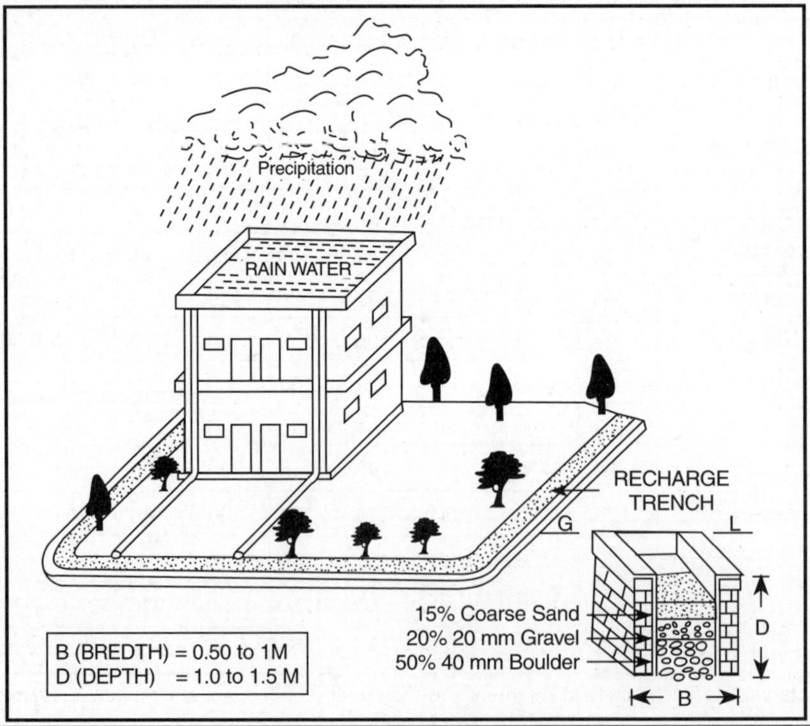

FIG. 17.7. Roof top rain water harvesting through recharge trench

- These are back filled with boulders (5-20 mm), gravels (5-10 mm) and coarse sand (1.5-2 mm) in graded form—boulders at the bottom, gravel in between and coarse sand at the top so that the silt content that will come with runoff will be deposited on the top of the sand layer and can easily be removed.
- A mesh should be provided at the roof so that leaves or any other solid waste/debris is prevented from entering the trench and a desilting/collection chamber may also be provided on ground to arrest the flow of finer particles to the trench.
- By-pass arrangement be provided before the collection chamber to reject the first showers.
- The top layer of sand should be cleaned periodically to maintain the recharge rate.

(iii) Roof Top Rain Water Harvesting through Existing Tubewells

- In areas where the shallow acquifers have dried up and existing tubewells are tapping deeper acquifer, roof top rain water harvesting through existing tubewell can be adopted to recharge the deeper aquifers.
- PVC pipes of 10 cm diameter are connected to roof drains to collect rain water. The first roof runoff is let off through the bottom of drain pipe. After closing the bottom pipe, the rain water of subsequent rain showers is taken through a T to an online PVC filter. The filter may be provided before water enters the tubewell. The filter is 1-1.2 m in length and is made up of PVC pipe. It's diameter should vary depending on the area of roof, 15 cm if roof area is less than 150 sq m and 20 cm if the roof area is more. The filter is provided with a reducer of 6.25 cm on both the sides. Filter is divided into three chambers by PVC screens so that filter material is not mixed up. The first chamber is filled up with gravel (6-10 mm), middle chamber with pebbles (12-20 mm) and last chamber with bigger pebbles (20-40 mm).
- If the roof area is more, a filter pit may be provided. Rain water from roofs is taken to collection/desilting chambers located on ground. These collection chambers are interconnected as well as connected to the filter pit through pipes having a slope of 1 : 15. The filter pit may vary in shape and size depending upon available run off and are back-filled with graded material, boulder at the bottom, gravel in the middle and sand at the top with varying thickness (0.30-0.50 m)

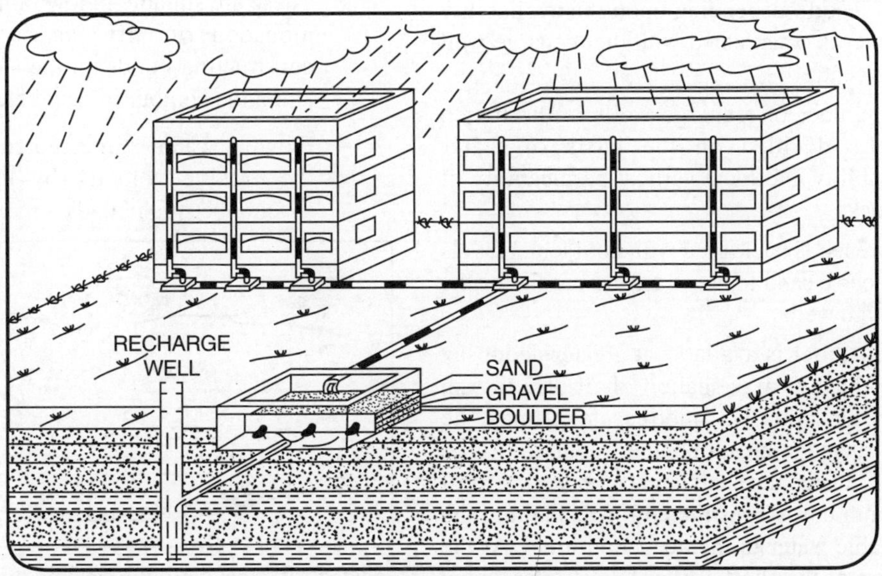

FIG. 17.8. Roof top rain water harvesting through existing tubewells

and may be separated by a screen. The pit is divided into two chambers, filter material in one chamber and other chamber is kept empty to accommodate excess filtered water and to monitor the quality of filtered water. A connecting pipe with recharge well is provided at the bottom of the pit for recharging of filtered water through well.

(iv) Roof Top Rain Water Harvesting through Trench with Recharge Well

- In areas where the surface soil is impervious and large quantities of roof water or surface runoff is available within a very short period of heavy rainfall, the use of trench/pits is made to store the water in a filter media and subsequently recharge to ground water through specially constructed recharge wells.
- This technique is ideally suited for area where permeable horizon is within 3 m below ground level.
- Recharge well of 100-300 m diameter is constructed to a depth of at least 3 to 5 m below the water level. Based on the lithology of the area well assembly is designed with slotted pipe against the shallow and deeper aquifer.
- A lateral trench of 1.5 to 3 m width and 10 to 30 m length, depending upon the availability of water is constructed with the recharge well in the centre.
- The number of recharge wells in the trench can be decided on the basis of water availability and local vertical permeability of the rocks.
- The trench is backfilled with boulders, gravels and coarse sand to act as a filter media for the recharge wells.
- If the aquifer is available at greater depth say more than 20 m, a shallow shaft of 2 to 5 m diameter and 3-5 metres deep may be constructed depending upon availability of runoff. Inside the shaft a recharge well of 100-300 mm dia is constructed for recharging the available water to the deeper aquifers. At the bottom of the shaft a filter media is provided to avoid choking of recharge well.

Rural Areas

In rural areas, rain water harvesting is taken up considering watershed as a unit. Surface spreading techniques are common since space for such systems is available in plenty and quantity of recharged water is also large. Following techniques may be adopted to save water going waste through slopes, rivers, rivulets and nalas.

(i) Rain Water Harvesting through Gully Plug

- Gully plugs are built using local stones, clay and bushes across small gullies and streams running down the hill slopes carrying drainage to tiny catchments during rainy season.
- Gully plugs help in conservation of soil and moisture.
- The sites for gully plugs may be chosen whenever there is a local break in slope to permit accumulation of adequate water behind the bunds.

(ii) Rain Water Harvesting through Contour Bund

- Contour bunds are effective method to conserve soil moisture in watershed for long duration.
- These are suitable in low rain fall areas where monsoon run off can be impounded by constructing bunds on the sloping ground all along the contour of equal elevation.
- Flowing water is intercepted before it attains the erosive velocity by keeping suitable spacing between bunds.

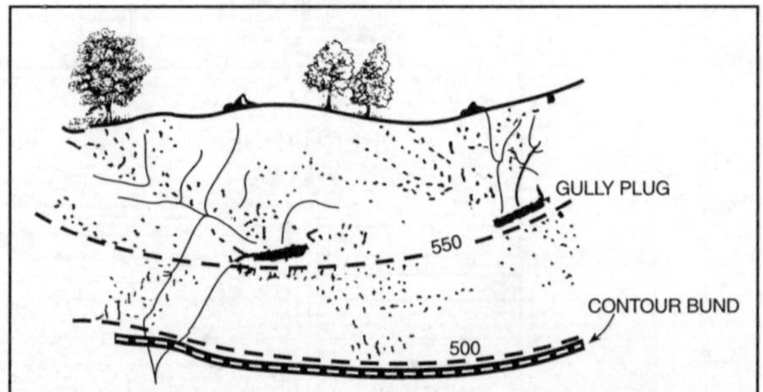

FIG. 17.9. Rain water harvesting through gully plug and contour bund

- Spacing between two contour bunds depends on the slope of the area and the permeability of the soil. Lesser the permeability of soil, the close should be spacing of bunds.
- Contour bunding is suitable on lands with moderate slopes without involving terracing.

(iii) Rain Water Harvesting through Gabion Structure

- This is a kind of check dam commonly constructed across small streams to conserve stream flows with practically no submergence beyond stream course.

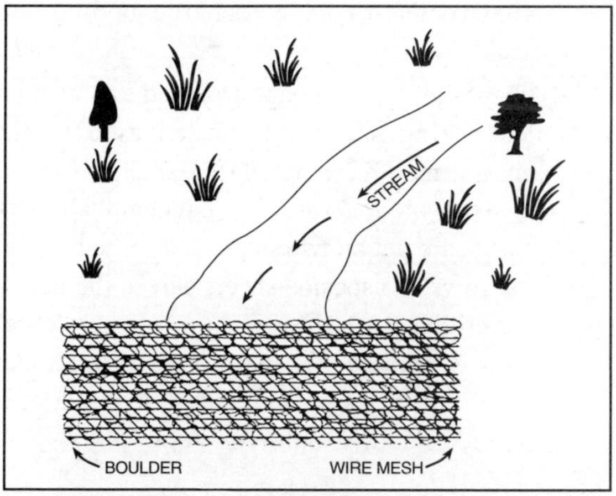

FIG. 17.10. Rain water harvesting through Gabion Structure

- A small bund across the sream is made by putting locally available boulders in a mesh of steel wires and anchored to the stream banks.
- The height of such structures is around 0.5 m and is normally used in the streams with width of less than 10 m.
- The excess water over flows this structure storing some water to serve as source of recharge. The silt content of stream water in due course is deposited in the interstices of the boulders in due course and with growth of vegetation, the bund becomes quite impermeable and helps in retaining surface water run off for sufficient time after rains to recharge the ground water body.

(iv) Rain Water Harvesting through Percolation Tank

- Percolation tank is an artificially created surface water body, submerging in its reservoir a highly permeable land so that surface runoff is made to percolate and recharge the ground water storage.
- Percolation tank should be constructed preferably on second to third order streams, located on highly fractured and weathered rocks which have lateral continuity down stream.

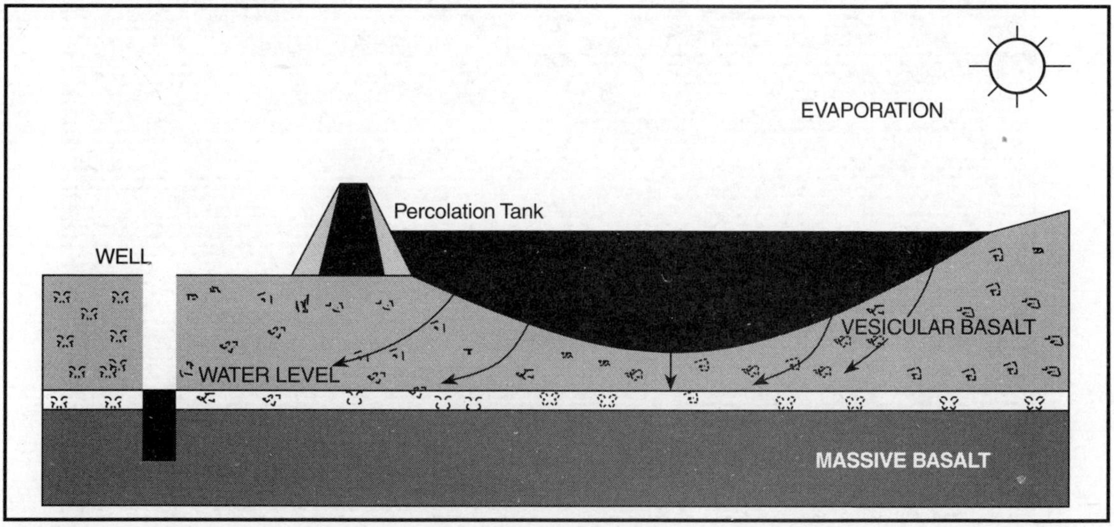

FIG. 17.11. Rain water harvesting through Percolation Tank

- The recharged area down stream should have sufficient number of wells and cultivable land to benefit from the augmented ground water.
- The size of percolation tank should be governed by percolation capacity of strata in the tank bed. Normally percolation tanks are designed for storage capacity of 0.1 to 0.5 MCM. It is necessary to design the tank to provide a ponded water column generally between 3 and 4.5 m.
- The percolation tanks are mostly earthen dams with masonry structure only for spillway. The purpose of the percolation tanks is to recharge the ground water storage and hence seepage below the seat of the bed is permissible. For dams upto 4.5 m height, cut off trenches are not necessary and keying and benching between the dam seat and the natural ground is sufficient.

(v) Rain water Harvesting through Check Dams/Cement Plugs/Nala Bunds

- Check dams are constructed across small streams having gentle slope. The site selected should have sufficient thickness of permeable bed or wheathered formation to facilitate recharge of stored water within short span of time.
- The water stored in these structures is mostly confined to stream course and the height is normally less than 2 m and excess water is allowed to flow over the wall. In order to avoid scouring from excess run off, water cushions are provided at downstream side.
- To harness the maximum run off in the stream, series of such check dams can be constructed to have recharge on regional scale.
- Clay filled cement bags arranged as a wall are also being successfully used as a barrier across small nalas. At places, shallow trench is excavated across the nala and absestos sheets are put on two sides. The space between the rows of asbestos sheets across the nala is backfilled with clay. Thus a low cost check dam is created. On the upstream side clay filled cement bags can be stacked in a slope to provide stability to the structure.

FIG. 17.12. Rain water harvesting through check dams/cement plugs/nala bunds

(vi) Rain Water Harvesting through Recharge Shaft

- This is the most efficient and cost effective technique to recharge unconfined acquifer overlain by poorly permeable strata.
- Recharge shaft may be dug manually if the strata is of non-caving nature. The diameter of shaft is normally more than 2 m.
- The shaft should end in more permeable strata below the top impermeable strata. It may not touch water table.
- The unlined shaft should be backfilled, initially with boulders/cobbles followed by gravel and coarse sand.
- In case of lined shaft the recharge water may be fed through a smaller conductor pipe reaching up to the filter pack.
- These recharge structures are very useful for village ponds where shallow clay layer impedes the infiltration of water to the aquifer.
- It is seen that in rainy season village tanks are fully filled up but water from these tanks does not percolate down due to siltation and tubewells and dugwells located nearby remain dried up. The water from village tanks gets evaporated and is not available for beneficial use.
- By constructing recharge shaft in tanks, surplus water can be recharged to ground water. Recharge shafts of 0.5 to 3 m diameter and 10 to 15 m deep are constructed depending upon availability of quantum of water. The top of shaft is kept above the tank bed level preferably at half of full supply level. These are back filled with boulders, gravels and coarse sand.
- In upper portion of 1 or 2 m depth, the brick masonry work is carried out for the stability of the structure.
- Through this technique all the accumulated water in village tank above 50% full supply level would be recharged to ground water. Sufficient water will continue to remain in tank for domestic use after recharge.

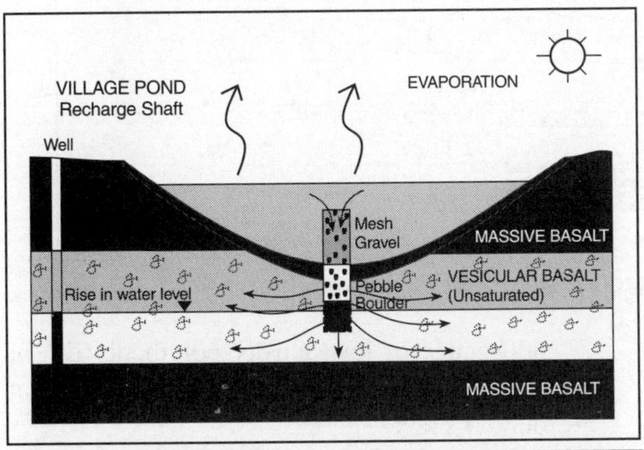

FIG. 17.13. Rain water harvesting through recharge shaft

(vii) Rain Water Harvesting through Dugwell Recharge

- Existing and abandoned dug wells may be utilized as recharge structure after cleaning and desilting the same.
- The recharge water is guided through a pipe from desilting chamber to the bottom of well or below the water level to avoid scouring of bottom and entrapment of air bubbles in the acquifer.
- Recharge water should be silt free and for removing the silt contents, the runoff water should pass either through a desilting chamber or filter chamber.
- Periodic chlorination should be done for controlling the bacteriological contaminations.

(viii) Ground Water Dams or Sub-surface Dykes

- Sub surface dyke or under-ground dam is a subsurface barrier across stream which retards the base flow and stores water upstream below ground surface. By doing so, the water level in upstream part of ground water dam rises saturating otherwise dry part of aquifer.
- The site where sub-surface dyke is proposed should have shallow impervious layer with wide valley and narrow outlet.
- After selection of suitable site, a trench of 1-2 m wide is dug across the breadth of stream

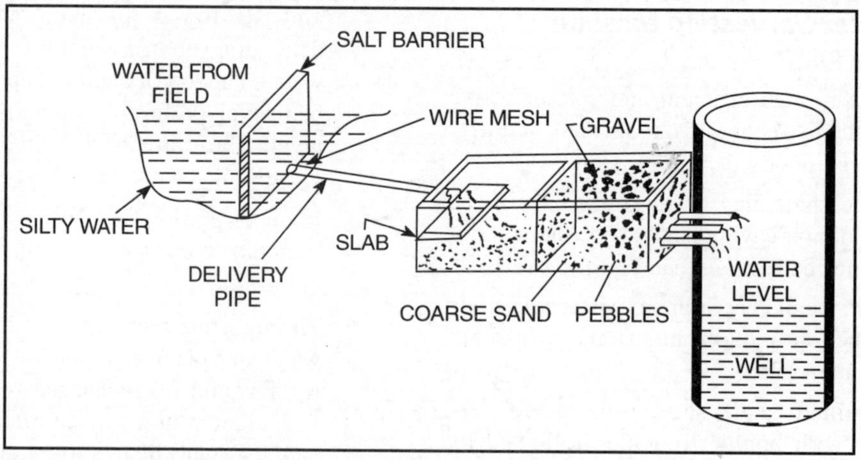

FIG. 17.14. Rain water harvesting through dugwell recharge

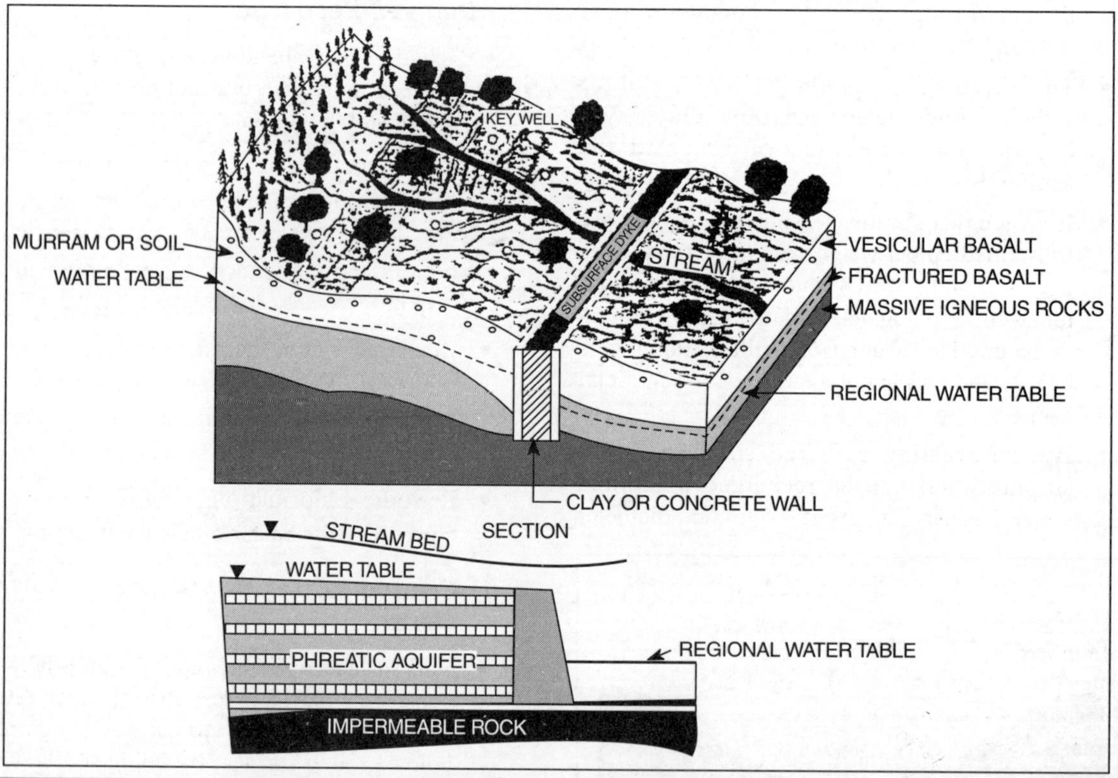

FIG. 17.15. Rain water harvesting through ground water dams or sub-surface dykes

down to impermeable bed. The trench may be filled with clay or brick/concrete wall upto 0.5 m below the ground level.

- For ensuring total imperviousness, PVC sheets of 3,000 PSI tearing strength at 400 to 600 gauge or low density polythene film of 200 gauge can also be used to cover the cut out dyke faces.
- Since the water is stored within the aquifer, submergence of land can be avoided and land

above the reservoir can be utilized even after the construction of the dam. No evaporation loss from the reservoir and no siltation in the reservoir takes place. The potential disaster like collapse of the dams can also be avoided.

National Water Policy (2012)

The first National Water Policy was adopted in September 1987. It was reviewed and updated in 2002 and later in 2012. The National Water Policy 2012 was adopted to conserve, develop and better manage water resources in the country. Some of the important recommenda-tions of this policy are :

1. Emphasis on the need for national framework law.
2. Comprehensive legislation for optimum development of inter-state rivers and river valleys.
3. Evolving a system of benchmarks for water uses for different purposes to ensure efficient use of water.
4. Setting up of Water Regulatory Authority in each state.
5. Incentivizing recycle and reuse of water.

The resources projects and services; incentivizing water saving in irrigation through methods like aligning cropping pattern with natural resources endowments, micro irrigation, automated irrigation operation, evaporation—transpiration reduction, etc., undertaking conservation of rivers, river corridors, water bodies and infrastructure in a scientifically planned manner through community participation; and arresting ground water levels in over-exploited areas.

Jal Kranti Abhiyan

Ministry of Water Resources, River Development and Ganga Rejuvenation celebrated Jal Kranti Abhiyan during 2015-16 to consolidate water conservation and management in the country through a holistic approach involving all stakeholders, making it a mass movement. The main objectives of programme are :

(*a*) strengthening gross root involvement of all stakeholders including Panchayati Raj institutions and local bodies in water security and development schemes.

(*b*) encouraging the adoption/utilization of traditional knowledge in water resources conservation and its management.

(*c*) changing livelihood security through water security in rural areas.

REFERENCES

Census of India 2011 : Tables on Houses, Household Amenities and Assets, India Series 1.

Central Ground Water Board, Ministry of Water Resources, GOI : *Ground Water Resources of India* (1995).

Central Ground Water Board, Ministry of Water Resources, GOI : *Ground Water Statistics* (2013).

Central Ground Water Board, Ministry of Water Resources, GOI : *Rain Water Harvesting Techniques to Augment Ground Water* (2013).

Central Ground Water Board, Ministry of Water Resources, GOI : *Save Water for Every One* (1997).

Centre for Science and Environment (2012) : Report on Excreta Matter.

Chellancy Brahma (2014) : Water, Peace and War, Oxford University Press, New Delhi.

Dakshinamurti, C. *et al.* (1973) : *Water Resources of India and Their Utilization in Agriculture*, New Delhi.

Dasgupta, S.P. (1989) : India's Water Resource Potential, in S.P. Dasgupta (ed.) *Man and Ecology,* School of Fundamental Research, Calcutta.

D'sMonte, Dorryl : *Current of Dissent,* Hindustan Times, August 19, 2003, New Delhi.

Economic Survey 2016-17 and 2017-18, Government of India, New Delhi.

EnviStats, India, 2018.

Gurjar, R.K. and Jat B.C. (2008) : Geography of Water Resources, Rawat Publications, Jaipur.

India 2022 : *A Reference Annual,* Ministry of Information and Broadcasting, New Delhi.

Khan, K.U., Save the Elixir of Life, *Science Reporter,* Vol. 41, No. 6, June 2004.

Kothari Ashish and Kohli Kanchi : *When Logic doesn't Wash,* Hindustan Times, June 13, 2003.

Krishna, S. Ananda. Is the Water Bank Tricking for India ? *Science Reporter,* Vol. 41, No. 6, June, 2004.

Mathur, P.C. : Performance Improvement of Irrigation Sector : Issues and Options in *Water Asia 1999,* International. Conference and Exhibition, New Delhi.

National Sample Survey Office (NSSO) (2012) : Survey of Water Resources.

NATMO (1990) : *Water Resources Development Atlas of India,* National Atlas of Thematic Mapping Organisation, Kolkata.

Neogi, Saikat, *Lifeline for a pittance,* Hindustan Times, June 27, 2005, New Delhi.

Pathak, B.D. (1984) : *Exploration and Exploitation of Ground Water with Special reference to Recharge* in A. Ramesh (ed.) Resource Geography (Contribution to Indian Geography), Heritage Publishers, New Delhi.

Prasad, R.S., Water Resource Management—Integrated Development of river Basin in *Water, Asia 2000,* International Conference and Exhibition on Asian Water Industry, New Delhi.

Prasad, R.S., Singh Ravinder and Sharma, R.P. Water Resources Assessment for Medium and Long Term Availability and Demand for its Diverse Uses, *in Water Area, 1999,* International Conference and Exhibition, New Delhi.

Rao, K.L. (1975) : *India's Water Wealth,* Orient Longman, New Delhi.

Roy, Dunu : *Garland of Hype (River-Linking A Misplaced Technological Fantasy)* : Times of India, August 27, 2004, New Delhi.

Sharma, S.K. : *Ground Water Management,* Paper presented at the Indo-French Water Week on 19-20th November 1999 at New Delhi.

Sharma, S.K. : *Rain Water Harvesting—An Alternative Technology for Fresh Water Augmentation,* Published in 'National Seminar on Rain Water Harvesting', May 22nd and 23rd, 2000, New Delhi.

Singh, Bharat : *A high dream of little logic,* Hindustan Times, March 9, 2003.

Singh Karan and Sameed Cooshalli : Dwindling Table : *Cities dig deeper,* Hindustan Times, June 27, 2005, New Delhi.

Singh, R.L. (ed.) (1971) : *India : A Regional Geography,* National Geographical Society of India, Varanasi.

Singh, Shivani : *A Balance Gone Awry,* Hindustan Times, June 27, 2005, New Delhi.

Sood, Vikram : *Drying and Cracking (The prospect of wars breaking out over water in near future is very real),* The Hindustan Times, July 30, 2004.

Status of Water Quality in India, 2011, Central Pollution Control Board, Delhi.

Suresh, V. : Issues and Initiatives for Promotion of Sustainable Drinking Water Supply Systems in *Water Asia 2000,* International Conference and Exhibition on Asian Water Industry, New Delhi.

Tikeeu, Alok : *Scope for Optimum Harvest,* Hindustan Times, June 27, 2005, New Delhi.

Vasani, P. Rupesh : Domestic Roof Water Harvesting—Need of the Hour in *Water Asia 2000,* International Conference and Exhibition on Asian Water Industry, New Delhi.

◆◇◆

CHAPTER 18

Biotic and Marine Resources

INTRODUCTION

Biotic resources consist of living creatures which include plants, animals (both wild and domestic), birds, animals living in water, and above all man himself. Considering the space constraint we will confine our attention to the description of domestic animals, poultry birds and animals living in water only.

LIVESTOCK

Livestock includes domestic animals such as cattle, buffaloes, sheep, goats, horses, ponies, donkeys, camels, pigs, etc. India's animal wealth is both large and varied. In most parts of India, killing of animals is taboo from religious point of view, as a result of which animals of various descriptions have outgrown in numbers.

According to 20th livestock census conducted in 2019, India had 192.49 million cattle, 109.85 million buffaloes, 74.26 million sheep, 148.88 million goats, and 851.81 million poultry birds. Thus India far exceeds any other country of the world in so far as the number of animals is concerned. India boasts of about half of the buffaloes, one fifth of the cattle and one sixth of the total goat population of the world. But this quantitative supremacy in no way speaks of the quality of our animals. Our animals are of poor quality in terms of stamina for draught and milk yield. Several factors are responsible for this sad state of affairs. Livestock rearing is a highly neglected industry in India and is always treated as an occupation subsidiary to crop production. Indian agricultural scene is primarily dominated by subsistence agriculture and commercial grazing is conspicuous by its absence. Every farmer has a few animals which are fed on straw and other farm by-products. Only 3.33 per cent of our total area is termed as grassland which is too small to feed such a vast population of livestock. Animal husbandry accounts for nearly one-fourth of the total income from the agriculture sector and its contribution to the national income does not exceed 5 per cent.

However animal husbandry sector provides large scale self-employment opportunities to the rural people and plays an important role in supplementing family income and generating gainful employment, particularly among the landless labourers, small and marginal farmers and women. It also provides cheap nutritional food to millions of Indians by producing meat, milk, eggs, etc. Livestock is a great source of raw material byproducts such as hides and skins, bones, blood, fat, etc. Livestock is the best insurance against the vagaries of nature like drought, famine and other natural calamities.

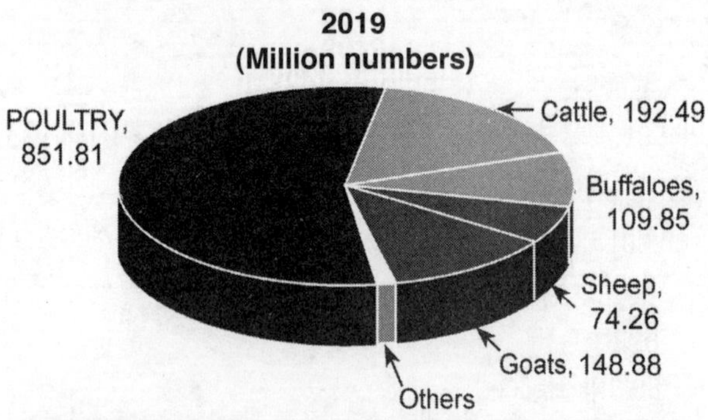

FIG. 18.1. India : Composition of Livestock

CATTLE

India has the largest number of cattle in the world. In 2019, India had 192.49 million cattle which account for about one fifth of the world's total number of cattle and about 35.9 per cent of the total livestock population of India. Thus the number of cattle possessed by India is simply amazing. Further the cattle have registered an increase of 24 per cent from 155.3 million in 1951 to 192.49 million in 2019. Since 2012 Census, West Bengal has surpassed Uttar Pradesh and Madhya Pradesh in cattle population. According to 2019 Census, West Bengal has 19 million cattle, followed by Uttar Pradesh (18.8 million), Madhya Pradesh (18.7 million), Bihar (16.3 million) and Maharashtra and Rajasthan (13.9 million each). The other states with considerable number of cattle are Jharkhand, Assam, Chhattisgarh, Odisha, Gujarat and Tamil Nadu (**Fig. 18.2**).

It has already been mentioned that Indian cattle are, by and large, of poor quality. The average yield of milk per cow is just one litre per day whereas this yield is 30-40 litres in some of the advanced countries like New Zealand, Denmark and Holland. This is the reason that Indian cow is often called *tea-cup-cow*.

Although majority of Indian cattle are of poor quality, there are 14 well defined excellent breeds of milch cattle and 12 well known draught and disease resistant breeds of cattle. Most of the good milch, draught or dual purpose breeds are found in dry northern, north-western and southern parts of the country. The humid regions generally do not support good breeds of cattle.

Milch Breeds. *Gir*, *Sindhi*, *Red Sindhi*, *Sahiwal*, *Tharparkar* and *Deoni* are some of the outstanding breeds of milch cattle. The ***Gir*** is a native of Saurashtra and is now found in several parts of Gujarat and adjoining Rajasthan. ***The Sindhi*** breed is mainly raised in Gujarat, Rajasthan and Maharashtra although it can be raised in several other parts of the country due to its disease resistant quality. The ***Red Sindhi*** breed has a distinct red colour and hails from Sind in Pakistan. The ***Sahiwal*** breed has its origin in the Montogomery district of Pakistan and is widely raised in Punjab, Haryana, Rajasthan, Uttar Pradesh and Delhi. ***The Deoni*** breed is widely raised in Telangana and Andhra Pradesh.

Draught Breeds. Among the important draught breeds are included the *Nagori*, *Bauchaur*, *Kenkatha*, *Malvi*, *Kherigarh*, *Hallikar*, *Khillari*, *Amritmahal*, *Kangayam*, *Ponwar*, *Bargur*, and *Siri*. The ***Nagori*** breed is a native of Jodhpur and is found in large parts of Rajasthan, Haryana, U.P. and M.P. ***The Bauchaur*** breed is mainly found in Bihar. The ***Malvi*** is largely concentrated in the dry western parts of Madhya Pradesh. ***The Kenkatha or Kenwariya*** breed hails from Banda district of Uttar Pradesh and neighbouring areas of Madhya Pradesh. Kheri district of Uttar Pradcsh is the habitat of the ***Kherigarh*** *breed*. The ***Halikar*** and ***Amritmahal*** breeds are indigenous to Tumkur, Hassan and Mysore districts of Karnataka but are spread all over the Peninsular India. Solapur and Satara districts comprise the home of the ***Khillari*** breed. The ***Bargur*** and the ***Kangayam*** breeds are the natives of Coimbatore district of Tamil Nadu. The ***Siri*** breed flourishes well in the hilly areas of Darjeeling and Sikkim.

Dual Purpose Breeds. Cattle of these breeds are used both for milk and for draught purposes. The cows are fairly good yielders of milk while bullocks are good for draught. *Tharparkar*, *Haryana*, *Mewati*, *Kankrej*, *Rath*, *Nimari*, *Dangi*, *Gaolao*, *Krishna Valley* and *Ongole* are important breeds of this category. ***The Tharparkar*** breed hails from the Sind province of Pakistan and is found in large parts of Gujarat and Rajasthan. As its name indicates, the ***Haryana breed*** is the much prized cattle of Haryana and is also found in the neighbouring parts of Delhi and Western Uttar Pradesh. ***The Mewati***

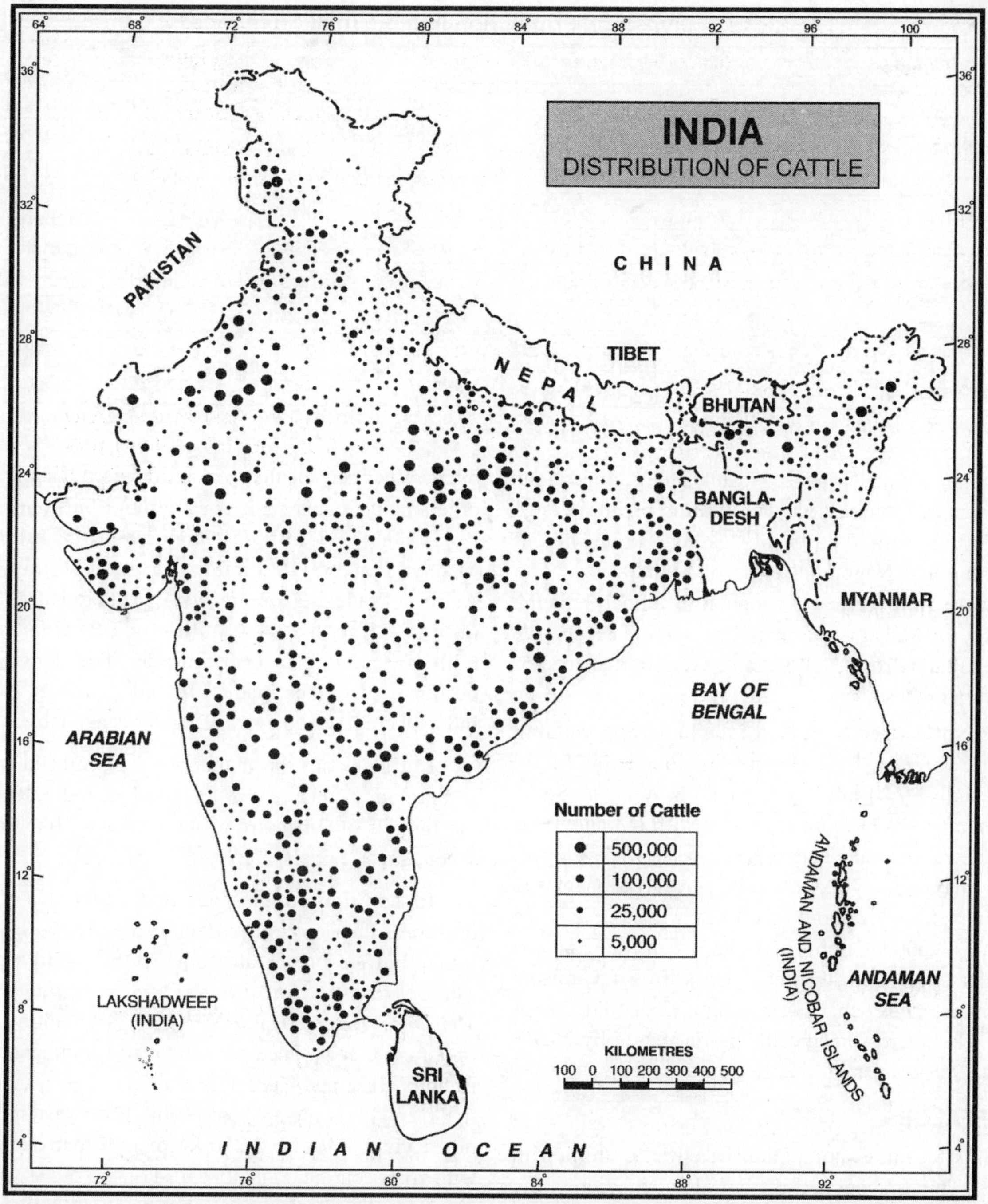

FIG. 18.2. India : Distribution of Cattle

breed is found in Mathura district of Uttar Pradesh and Bharatpur and Alwar districts of Rajasthan. ***The Kankrej*** breed is indigenous to Gujarat plains. ***The Rath*** breed is a mixture of *Haryana*, *Nagori* and *Mewati* breeds and is found in the areas dominated by these breeds. The ***Nimari*** breed is largely found in the Narmada Valley in East and West Nimar districts of Madhya Pradesh. The ***Dangi*** breed is normally

TABLE 18.1. Livestock Population in India (million numbers)

Species	1951	1972	2003	2007	2012	2019
Cattle	155.30	178.30	185.18	199.08	190.90	192.49
Buffalo	43.40	57.40	97.92	105.34	108.70	109.85
Sheep	39.10	40.00	61.47	71.56	65.07	74.26
Goats	47.20	67.50	124.36	140.54	135.17	148.88
Horses and Ponies	1.50	0.90	0.75	0.61	0.63	0.34
Camels	0.60	1.10	0.63	0.52	0.40	0.25
Pig	4.40	6.90	13.52	11.13	10.29	9.06
Total Livestock	292.80	353.60	485.00	529.70	512.21	535.78
Poultry	73.50	138.50	489.1	648.88	729.21	851.81

Source : *Department of Animal Husbandry and Dairying,* 20th Livestock Census (2019).

found in Kolaba, Thane, Nashik and Ahmednagar districts of Maharashtra. The ***Gaolao*** breed belongs to Chhindwara district of Madhya Pradesh and Wardha and Nagpur districts of Maharashtra. The ***Krishna valley*** breed is popular in upper Krishna Valley in Maharashtra and Karnataka. The ***Ongole*** breed hails from Nellore and Guntur districts of Andhra Pradesh.

Exotic Breeds. Some of the high milk yielding exotic breeds have been developed in India, especially in 20 military farms in hilly areas. Some foreign breeds have been crossed with Indian breeds and new breed called ***cross breed*** has been developed. The maximum yield of milk per lactation at the military farms is about 6,000 kgs while the average yield is 2,600 kg. Some of the important exotic breeds are Jersey, Holstein-Friesian, Swiss-Brown, Gurnsey, German Feleckvich, and Ayreshire. The population of total exotic/crossbreed cattle has increased by 26.9% in 2019 as compared to 2012 Census.

BUFFALOES

India's buffaloes population was 109.85 million in 2019. This is about half the buffalo population of the world and about 20.5 per cent of the total livestock of India. Buffaloes thrive best in areas of warm and humid climate. Buffaloes are reared mainly for milk but some buffaloes are used as draught animals in certain parts of the country. Uttar Pradesh has the largest number of about 33 million buffaloes which is over 30 per cent of the total buffaloes of India. The other major states with considerable buffaloes are Rajasthan (13.7 million), Gujarat (10.5 million), Madhya Pradesh (10.3 million), Bihar (7.7 million), Andhra Pradesh (6.2 million), Maharashtra (5.6 million), Haryana (4.4 million), Telangana (4.2 million) and Punjab (4.0 million). The density of buffaloes is higher in the alluvial plains of North India where large quantity of fodder is available.

Buffaloes are the main source of milk in India and provide 46 per cent of total milk of India. Rest of the milk is obtained from other animals like cattle sheep, goats, camels, etc.

In addition to the fact that India has more buffaloes than any other country of the world; the Indian buffalo-breeds are some of the world's best. *The Murrah*, *Bhadawari*, *Jaffarabadi*, *Surti*, *Mehsana*, *Nagpuri* and *Nili Ravi* are among the important breeds. The **Murrah** breed is indigenous to Rohtak, Hisar and Gurugram districts of Haryana and to the neighbouring areas of Delhi. Buffaloes of this breed yield 1,400 to 2,270 kg of milk per lactation with 7 per cent fat content against only 4.5 per cent in cow's milk. The Murrah male buffaloes are good draught animals. The **Bhadawari** breed belongs to Agra and Etawah districts of Uttar Pradesh and the neighbouring parts of Madhya Pradesh and Rajasthan. The **Jaffarabadi** breed hails from the Gir forest of Gujarat. The buffaloes of this breed are quite massive and yield about 2,500 kg of milk per lactation. The **Surti** breed comes from the Gujarat plains and gives about 1,655 kg of milk per lactation. The **Nagpuri** originates from Nagpur. The **Nili Ravi** breed belongs to Ferozepur district of Punjab and yields about 1,600 kg of milk in one lactation.

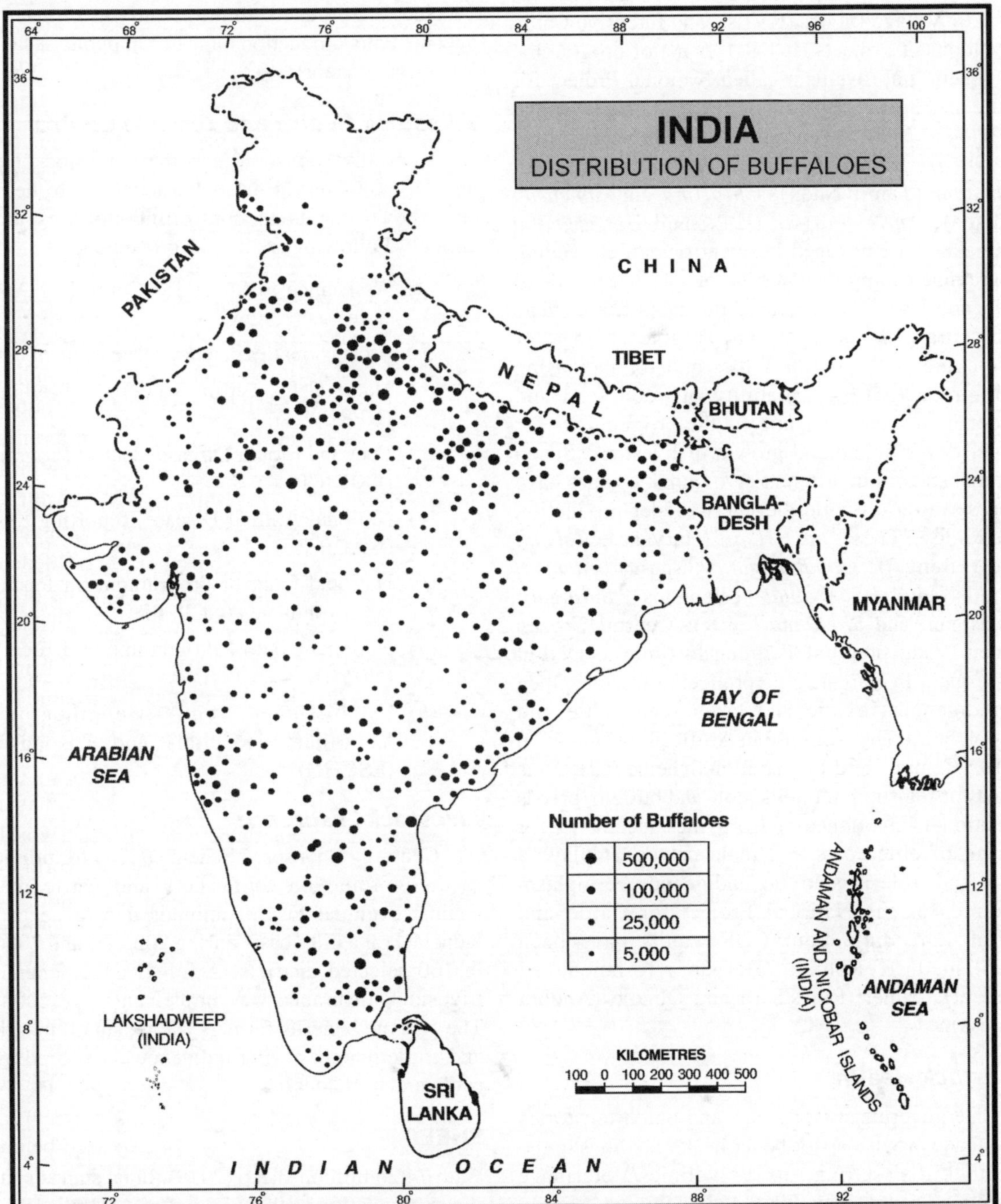

FIG. 18.3. India : Distribution of Buffaloes

Cattle and Buffalo Development Programmes

The important schemes initiated by the government for the development of cattle and buffaloes are Key Village Blocks (KVBs) and Intensive Cattle Development Projects (ICDPs). A major programme for genetic improvement called National Project for Cattle and Buffalo Breeding (NPCBB) was launched in October, 2000. Seven central cattle breeding farms at *Suratgarh* (Rajasthan), *Dhamrod* (Gujarat), *Alamadhi* (Tamil Nadu), *Chiplima* and *Suncheda* (Odisha), *Andeshnagar* (U.P.) and *Hessarghatta* (Karnataka) are engaged in scientific cattle breeding programme and progeny testing of selected breeds of indigenous as well as exotic cattle and buffaloes. The embryo transfer technique is the latest technology for developing cattle and buffaloes and for increasing their productivity. Availability of good feed and fodder in sufficient quantity is the pre-requisite for proper development and growth of livestock. Seven regional stations in different agro-climatic zones have been set up to provide the latest fodder production technology. These are *Hisar* (Haryana), *Kalyani* (West Bengal), *Gandhinagar* (Gujarat), *Alamadi* (Tamil Nadu), *Hyderabad* (Telangana), *Suratgarh* (Rajasthan) and *Shehoma* (J&K). Central Frozen Semen Production and Training Institute located at Hessergatta in Karnataka is producing frozen semen of indigenous, exotic and crossbreed cattle and Murrah buffalo bulls for use in Artificial Insemination (A.I.) : Central Herd Registration Scheme (CHRS) is meant for registration of elite cow and buffalo, breeds of national importance and provides incentives for rearing of elite cows and male calves. It plays a significant role in sourcing indigenous germplasm required for the National Project for Cattle and Buffalo Breeding. Four CHRS units have been established at Rohtak (Haryana), Ahmedabad (Gujarat), Ajmer (Rajasthan) and Ongole (Andhra Pradesh).

Livestock Health

The governments of states and union territories have made provisions for better health facilities to the livestock by setting up Polyclinics/Veterinary Hospitals/Dispensaries/First Aid including Mobile Veterinary Dispensaries so that morbidity and mortality rate among the animals is minimised. One Central and five Regional Disease Diagnostic Laboratories are also functioning. For controlling major livestock and poultry diseases by way of prophylactic vaccination, the required quantities of vaccines are produced in the country at 27 veterinary vaccine units production units (20 in public sector and 7 in private sector).

Livestock Health and Disease Control

Centrally sponsored scheme 'Livestock Health and Disease Control' has been launched to help the state governments in controlling diseases among animals. Following are its major components :

(*a*) Assistance to States for Control of Animal Diseases (ASCAD)

(*b*) Professional Efficiency Development (PED)

(*c*) National Project on Rinderpest Eradication (NPRE)

(*d*) Foot and Mouth Disease Control Programme (FMD-CP)

(*e*) National Animal Disease Reporting System (NADRS)

(*f*) National Control Programme on Peste des petits ruminants (NCPPPR)

(*g*) National Control Programme on Brucellosis (NCPB)

(*h*) Establishment and strengthening of Veterinary Hospitals and Dispensaries (ESVHD)

Livestock Insurance

Centrally Sponsored Scheme (CSS) for providing assured protection to farmers and cattle rearers against eventual loss of animals due to death was launched on a pilot basis during 2005-06 and 2006-07 in 100 selected districts. A full fledged scheme on Livestock Insurance was implemented in 2008-09. The scheme benefits farmers (large, small, marginal) and cattle rearers having indigenous/crossbred milch cattle and buffaloes.

SHEEP

With 74.26 million sheep (2019), India stands sixth in sheep population in the world after Australia, Russia, China, Argentina and New Zealand. More than 4 per cent of the world's sheep are reared in India. *Most of the sheep are raised in regions which are too dry, too stony or too mountainous to be too good for agriculture or for cattle rearing*. Most of the Indian sheep are of poor quality yielding inferior wool in less quantity. Their yield of mutton is also very low. However, some of the good breeds are found in the northern temperate region. By virtue of their sheer numbers, sheep occupy an important place in our economy because they provide us with wool, mutton and skins. About five million households are engaged in the rearing of sheep and other allied activities. The largest number of 19.1 million sheep are found in Telangana which account for over 25% of the total sheep population of India. This is followed by Andhra Pradesh, Karnataka, Rajasthan, Tamil Nadu, Maharashtra, Jammu and Kashmir, Uttarakhand, Himachal Pradesh, Gujarat, Bihar and Uttar Pradesh. The distribution of sheep may be properly studied by dividing the sheep areas into following regions :

1. The Temperate Himalayan Region. It comprises of Jammu and Kashmir, Ladakh, Himachal Pradesh and Uttarakhand. The entire region has temperate climate which is quite suitable for good quality sheep. Excellent pastures exist on the hill slopes. India's best quality sheep are reared in the Kulu, Kangra and Chamba districts of Himachal Pradesh and Kashmir Valley at altitudes varying from 2,000 to 3,000 metres. The shephards practise *seasonal transhumance*, going up in summer and moving down in winter. There are about 55 lakh sheep producing over 5,000 tonnes of superior quality wool in this region.

2. The Dry North-Western Region. This region includes Rajasthan and neighbouring parts of Punjab, Haryana, western Uttar Pradesh, Gujarat and Madhya Pradesh. There are more than 13 million sheep contributing to about half of the total wool production of India. However, the wool is of comparatively inferior quality and the yield of wool per sheep is lower than that of the Himalayan region.

3. The Semi-arid Southern Region. This region comprising Maharashtra, Karnataka, Andhra Pradesh, Telangana, Tamil Nadu and parts of Madhya Pradesh supports about 23.5 million sheep which is about one-third of the total number of sheep found in India. In spite of the largest number of sheep, this region produces only 11,000 tonnes of inferior quality rough wool. About 50 per cent of the sheep of this region are raised for mutton and produce no wool.

4. The Humid Eastern Region. This region, comprising Bihar, West Bengal, Assam and Odisha, has humid climate which is not favourable for sheep rearing. There are about 30 lakh sheep which are mainly reared for producing mutton. The per sheep wool and the total production of wool are lower than that of any other region.

Development of Sheep

Development of sheep is necessary to meet the growing demand for wool and mutton in the country and for a possible export of these commodities. This can be done by scientific breeding of the sheep. The breeding policy envisages selective breeding of important carpet wool breed and cross-breeding involving suitable exotic breeds with coarse carpet type. India is importing large number of exotic fine wool breed sheep in a phased manner. So far nearly 10 thousand fine wool sheep have been imported from the USA, Australia and Russia for increasing the production of quality wool. A central Sheep Breeding Farm with exotic breeds of sheep has been established at Hisar in Haryana. At present, it is having a breeding programme with pure exotic breed as well as cross breeding for production of superior rams. It has distributed over 12,000 exotic/cross breed rams to different states.

GOATS

Goat is called the *poor man's cow* because it can be cheaply reared on meagre grass of poor quality. It is the major supplier of mutton along with milk, hair and skins. The number of goats increased sharply from 47.2 million in 1951 to 148.88 million in 2019. Goats are found in larger number as compared to sheep and are next only to cattle. About one-sixth of the world's goats are reared in India. Although goats are found in almost all parts of the country, their major concentration is in Rajasthan (20.8 million), West Bengal (16.3 million), Uttar Pradesh (14.5 million), Bihar (12.8 million) and Madhya Pradesh (11.1 million). These states account for more than half the goats of India.

About 90 per cent of goats in the country are *desi* or non-descript, mostly found in the Deccan Plateau. But there are some outstanding breeds which are found in some specific areas. The Himalaya or *Angora* goat which is also known as the *Chamba*, *Gaddi*, *Chegu* or Kashmiri breed is reared in Kashmir and Himachal Pradesh. It produces soft warm hair. The *Pashmina* reared in Kashmir and Kulu valley is world renowned for its pashmina hair known as ***Mohair.*** The yield of hair per goat varies from 21 to 56 grams per year. The *Jamunapuri* is the breed found between the rivers Yamuna and the Chambal. It is a dual purpose breed providing meat and milk. The *Barabari* breed of western Uttar Pradesh and Haryana can yield upto 2.5 kg of milk per day. Among the other breeds are the *Beetal* of Punjab, the *Marwari*, *Mehsana*, *Kathiawar*, and *Zalwadi* of Rajasthan, Gujarat and Madhya Pradesh and the *Barari*, *Surti* and *Deccani* of the Peninsular India. Several important foreign breeds such as *Alpine*, *Nubian*, *Saanen*, *Toggenberg* and *Angora* have been used for cross breeding with the local breeds.

PIGS

There are about 9 million pigs providing about 5 per cent of India's meat production in the form of pork. In a poor and thickly populated country like India, pig rearing is an important activity because pig provides rich meat at low cost. Pig farming plays an important role in improving the socio-economic status of sizeable section of weaker rural communities especially in north-eastern states where every rural family rears pigs for meat. Pig farms have been set up for improving the quality of pigs. At present, there are about 100 pig farm units in the country run by state governments maintaining about 29,000 pigs.

The Central Government has prepared a scheme to assist farmers/landless labourers/cooperatives and tribals, particularly in the North-eastern States by rearing pigs under stall fed conditions for quality pork production and organised pork marketing in rural and semi-urban areas. The main objectives of the scheme are :

(*a*) Encourage commercial rearing of pigs by adopting scientific methods and creation of infrastructure.

(*b*) Production and supply of improved germ plasm.

(*c*) Organising stakeholders to popularize scientific practices.

(*d*) Create supply chain for meat industry.

(*e*) Encourage the value addition for better income.

HORSES AND PONIES

There were 34 lakh horses and ponies in the country in 2019. They have lost much of their importance with the increasing use of automobiles for transport. But in remote hilly areas, horses and ponies are the only means of transport. About one-fourth of the total horses and ponies are found in Uttarakhand, Himachal Pradesh, Bihar, and Jammu and Kashmir. Some of the important indigenous breeds include *Marwari*, *Kathiawari*, *Manipuri*, *Bhutani*, *Spiti* and *Chummarti*. Some other breeds have been developed by cross-breeding the indigenous breeds with the Arabian and English breeds.

DONKEYS AND MULES

Donkeys and mules are used as beasts of burden, especially in those areas where modern modes of transportation cannot be used. Most of the donkeys are found in Rajasthan, Uttar Pradesh, Punjab, Gujarat and Tamil Nadu. Mules are derived from the cross-breeding of mares and donkeys. The largest concentration of mules is found in Uttar Pradesh and West Bengal.

CAMELS

Camel is an extremely useful animal for draught and transport purposes in the arid lands and is called ***ship of the desert.*** There were about 25 lakh camels in the country in 2019, about two-thirds of which were concentrated in Rajasthan alone. The rest of the camels are found in the arid and semi-arid areas of Punjab, Haryana and Gujarat.

LIVESTOCK PRODUCTS

Livestock provide us with a large variety of products which are very useful in our everyday life. Animal products help in increasing the national income and in the upliftment of the rural masses. Livestock products play a significant role in the national economy of India. The gross value of output from livestock sector (at current prices) is estimated to account for 25 per cent of the total value of output from agriculture sector. This excludes the contribution of animal draught power. Animal husbandry plays a significant role in rural economy by providing gainful employment particularly to small/marginal farmers and agricultural labourers and more so for people living in drought prone, hilly, tribal and other poorly developed areas, where crop production on its own may not sustain them fully. The major animal products include milk, meat, wool, hides and skins, dung, etc. Several industries mainly depend on the animal products. Over and above, animals render extremely useful service in different agricultural operations. Draught animals plough 100 million hectares of farmland forming over 70 per cent of the cultivated area. They also transport 25,000 million tonne kilometres of freight. Without animals, no cultivation would be possible, fields would remain unploughed, stores and bins stand empty and food and drink loose half their savour, for in a vegetarian country like India what can be worse than to have no milk, butter or *ghee*.

Milk

Milk is primarily obtained from buffaloes, cows and to some extent from goats and sheep. About two-thirds of buffalo milk is produced in Uttar Pradesh, Punjab, Haryana, Andhra Pradesh, Telangana and Madhya Pradesh. Similarly, about two-thirds of cow milk is produced in Uttar Pradesh, Tamil Nadu, Madhya Pradesh, Maharashtra, Punjab, Kerala, Karnataka and Bihar.

White Revolution

The production of milk immediately after Independence was at a very low level and remained almost stagnant between 1947 and 1970 with animal growth rate of merely one per cent. As a result of concerted efforts by centre and state governments, the milk production increased rapidly. Milk production increased from a meager 17 million tonnes in 1951 to a staggering 198.4 million tonnes in 2019-20 thus registering over 11 times growth within a span of about seven decades (**Table 18.2**). This phenomenal increase in milk production after 1970 has been termed as ***White Revolution,*** similar to Green Revolution in agriculture. India is now the world's largest producer of milk. The dairy sector is the largest contributor in the agricultural sector to the nation's GDP.

Operation Flood

The phenomenal increase in milk production (**Table 18.2**) has also been termed as ***Operation Flood***. It started with the establishment of the National Dairy Development Board (NDDB) in 1965. This board was set up to promote, plan and organise dairy development through cooperatives. These cooperatives were envisaged as democratic institutions, owned and managed by rural producers and were sensitive to the producers' demands. The dairy development programme through cooperative societies was first established in the Anand District of Gujarat which was most successful. Consequently Anand model of dairy development was replicated in other parts of the country. The cooperative societies helped the milk producers by providing financial and technical assistance and in the sale of milk at remunerative prices.

The NDDB launched 'Operation Flood' in 1970 with commodity gifts from the European Economic Community. Those gifts included skimmed milk, milk powder and butter oil. The operational system of 'Operation Flood' consists of multi-tiered cooperative structure with Primary Village Cooperative Societies at the base. District Units at the district level, State Federations at the state level and the National Cooperative Dairy Federation of India as the apex body for milk cooperative societies. The cooperative societies have a well organised system of carrying milk from the producers to the consumers. After being collected at the village collecting centre, milk is

TABLE 18.2. Phenomenal Increase in Milk Production in India (1950-51 to 2019-20)

Year	1950-51	1960-61	1970-71	1980-81	1990-91	2000-01	2010-11	2017-18	2018-19	2019-20	2020-21	2021-22
Production (million tonnes)	17.0	20.0	22.0	31.6	53.9	80.6	121.8	176.3	187.7	198.4	209.96	221.06

Sources : *(i) Economic Survey 2022-23, Statistical Appendix, p. 45.*

promptly transported to the dairy plant at the milk chilling centre. The timings of collecting milk are rigidly maintained by the village society, truck operators and milk is quickly transported to dairy plants. The chilling plants are managed by producers' cooperative unions which purchase milk directly from the producers, thus eliminating the middlemen. Before setting up of these societies, middlemen were indulging in maximum exploitation of the milk producers.

Today, 'Operation Flood' is considered to be the largest dairy development programme in the world.

Objectives

Following are the main objectives of 'Operation Flood' :

1. Forming cooperative societies.
2. Procurement, transportation and storage of milk at the chilling plants.
3. Production of milk products and management of their marketing.
4. Provision of cattle feed.
5. Facilities for superior breeds of cows and buffaloes, health services, veterinary treatment and artificial insemination.
6. Provision of extension services.

Phases of the White Revolution

The White Revolution is usually examined under the following three phases :

Phase-I (1970-81). The objective of this phase was to set up dairy cooperatives in 18 milk sheds in 10 states to provide milk to four metropolitan cities *i.e.,* Mumbai, Kolkata, Delhi and Chennai. The important step in this phase was the setting up of 4 Mother Dairies in Mumbai, Kolkata, Delhi and Chennai. By the end of this phase, there were 13,000 village dairy cooperatives covering 15 lakh farmer families.

Phase II (1981-85). This phase coincided with the Sixth Five Year Plan and was designed to be built on the foundation of Phase-I. Dairy development programme was extended to three states of Karnataka, Rajasthan and Madhya Pradesh. The programme aimed at organising in 144 more cities, proper provision for fodder to feed the milch animals, control on animal diseases, improving the breeds and providing facilities to the milkmen. A new vaccine called 'Raksha' was developed by the Research Institute at Hyderabad to control cattle diseases. By the end of Phase II, there were 136 milk sheds and 34,500 village dairy cooperatives with a membership exceeding 36 lakhs.

Phase III. Started in 1985, this phase aimed at setting up 170 milk centres to benefit 250 districts in 22 states. The emphasis was on consolidating the gains of earlier two phases by improving productivity and efficiency of the cooperative dairy sector and its institutional base for sustainable development. By September 1996, about 73,300 dairy cooperative societies had been organised in 170 milksheds and the membership rose to ten million milkmen.

Achievements of White Revolution

1. India has become the largest producer of milk in the world.
2. There has been significant increase in the total production of milk and per capita availability of milk in the country.
3. The increase in milk and milk products has reduced the import of these products substantially and helped in saving the precious foreign exchange. In fact, India is now in a position to export some of the milk products to a number of countries.
4. The small and marginal farmers and landless labourers have been especially benefited from the White Revolution. About two-thirds of the milk supplies under "Operation Flood" comes from the small and marginal farmers and landless labourers.
5. 'Operation Flood' has helped in increasing the income levels of small and marginal farmers as well as of the landless labourers. About 9 million small farmers in 70,000 villages are earning jointly an incremental income of about ₹ 2,500 crore annually.
6. Dairy industry and infrastructure have been expanded and modernised. A Milk Grid has been activated to offset regional and seasonal imbalance in milk production. A stable structure has now been developed to safeguard against political instability.

7. To ensure success of 'Operation Flood' programme, research centres have been set up at Anand, Mehsana, Palampur (Banaskantha) all in Gujarat. Besides, three regional centres are working at Siliguri, Jalandhar and Erode.
8. A centrally sponsored scheme for livestock insurance was initiated in 2006-07 and is being implemented in all the states with twin objectives of providing protection mechanism to farmers and cattle rearers against any eventual loss of their animals due to death and to demonstrate the benefit of the insurance of livestock and popularize it with ultimate goal of attaining qualitative improvement in livestock and its products. The scheme benefits farmers and cattle rearers with indigenous/cross-breed milk cattle and buffaloes in 300 selected districts.
9. Genetic improvement of milch cattle has been made possible by cross-breeding and this process has become very common in India.

As a result of the above mentioned developments, the per capita availability of milk has increased from 266 gram/day in 2008-09 to 394 gram/day in 2018-19, recording over 48% increase within a short span of one decade (Table 18.3). This has happened in spite of the rapid growth in population.

Problems

White Revolution has several achievements to its credit and has increased the milk production to a great extent and has infused new life in producers and consumers alike. In spite of all its achievements, it has to face some problems as described below :

1. Collection of milk from remote and inaccessible areas is difficult, expensive, time consuming and economically unviable. Under these circumstances, the producers do not get remunerative price for their milk but the consumers have to buy milk at much higher cost.
2. In most villages, the cattle are kept under unhygienic conditions which results in lesser quantity of lower quality milk.
3. The existing marketing facilities are inadequate. In most villages, milk is converted into ***ghee*** which is not much remunerative. Therefore, there is need to improve the marketing infrastructure.
4. Although India possesses some of the best breeds of cattle in the world, yet most of the breeds are inferior which yield low production. There is great scope for improving the cattle breeds.
5. The extension service programme is not effective and needs improvement.

Prospects

Keeping in view the number of milch cattle and increasing demand for milk and milk products, it can be safely said that dairy development in India has great future. There are immense possibilities of capturing the international market with reference to liberalisation and global trade. Several corporate sector firms are taking advantage of the existing situation of liberalisation and globalisation. Such a situation can be encashed by strengthening the infrastructure and by increasing the production. India is capable of exporting milk products after meeting her domestic requirements by increasing quantity and improving quality of its products. The government has constituted Technology Mission for Dairy Development and Anand Model Cooperatives are being promoted to cover about 60 per cent of the total area of the country.

Intensive Dairy Development Programme (IDDP) was launched in 1993-94 for helping Non-

TABLE 18.3. India : Per Capita availability of Milk (Gram/day)

Year	2008-09	2009-10	2010-11	2011-12	2012-13	2013-14	2014-15	2015-16	2016-17	2017-18	2018-19
Availability of milk	266	273	281	290	299	307	322	337	355	375	394

Sources : *Economic Survey 2019-20*, p. 205.

Operation Flood, hilly and backward areas. Following are the main objectives of the IDDP :

(*i*) development of milch cattle

(*ii*) increasing milk production by providing technical input services

(*iii*) procurement, processing and marketing of milk in a cost effective manner

(*iv*) ensure remunerative prices to the milk producers

(*v*) generate additional employment opportunities

(*vi*) improve social, nutritional and economic status of residents of comparatively more disadvantaged areas.

The scheme was modified in March 2005 and also extended to some districts of Operation Flood Programme.

A comprehensive new scheme National Programme on Bovine Breeding and Dairy Development was launched with the objective of enhancing milk production and productivity in a sustainable manner. The National Dairy Plan Phase-I was launched in March 2012 with the objective of improving productivity of milch animals, strengthening and expanding village-level infrastructure for milk procurement and providing producers greater access to market in dairy sector continues.

Production of Milk at the State Level. Uttar Pradesh is the largest producer accounting for over 16 per cent of the total milk production of India. This is followed by Rajasthan (12.7%) and Madhya Pradesh (8.3%). Although Gujarat has the distinction of being the birth place of the "White Revolution" in India, this state produces only 7.7% of India's milk and has to content with fourth position among the major milk producing states of the country. The other important milk producing states of India are Andhra Pradesh, Punjab, Maharashtra, Haryana, Bihar, Karnataka and Tamil Nadu (**Fig. 18.4**).

Meat

The annual meat production in the country is 7.7 million tonnes of which 54 per cent is obtained from goats and sheep, 26 per cent from cattle, buffalo, and 7 per cent from pigs. The remaining 13 per cent comes from poultry birds. As about two-thirds of the Indian population is vegetarian and a large number of non-vegetarian people do not eat all types of meat due to religious sentiments, the production of meat is not much in spite of the fact that India has a very large livestock population. The annual per capita consumption of meat in India is less than 2 kg which

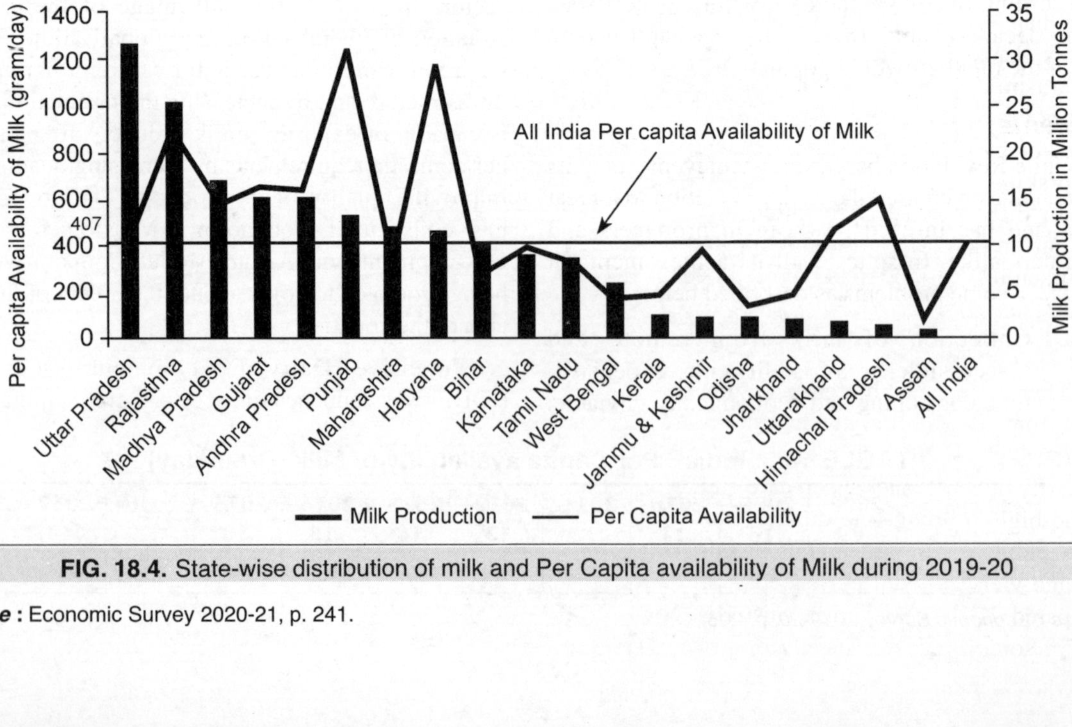

FIG. 18.4. State-wise distribution of milk and Per Capita availability of Milk during 2019-20

Source : Economic Survey 2020-21, p. 241.

is too small compared to over 100 kg in Australia, Argentina and New Zealand.

There are about 3,600 recognised slaughter houses in the country in which over five crore animals are slaughtered annually. Besides there are 32,000 illegal slaughter houses. Asia's biggest modern slaughter house is operating in Deonar, Mumbai while another abattoir of the Goa meat complex is functioning at Usgaon. There are 21 export-oriented modern abattoirs and 57 meat processing plans registered with Agriculture and Processed Food Products Export Development Authority exporting raw meat (chilled and frozen) to about 63 countries. In 1990-91 the country had exported meat worth ₹ 140 crore. The figure rose to ₹ 23,347 crore in 2019-20.

Among the major states Uttar Pradesh is the largest meat producing state of India and produces over 15 per cent of the total meat production of the country. This is followed by Maharashtra, West Bengal, Andhra Pradesh, Telangana, Tamil Nadu, Haryana, and Kerala.

Hides & Skins

India produces about 2 crore pieces of cattle hides and about 65 lakh pieces of buffalo hides accounting for about 15 per cent of the world's hides. More than 50 per cent of Indian hides are produced in Uttar Pradesh, Madhya Pradesh, Gujarat, Maharashtra, Bihar, West Bengal, Telangana and Andhra Pradesh. India also produces about one crore pieces of goat skins and about two crore pieces of sheep skins every year. Indian skins are one of the best in the world and are much in demand in the international market. India exported leather and manufactures worth ₹ 16,960 crore in 2019-20.

Wool

There has been increase in the production of wool from 27.5 million kgs in 1950-51 to about 41.5 million kg in 2017-18. This has been possible primarily due to proper sheep care and cross breeding of the high yielding fine quality exotic breeds with low yielding coarse indigenous breeds. However, the average yield of wool per sheep is still very low as compared to that obtained in Australia, New Zealand and in some other advanced countries. A Bikaneri sheep gives an average of 1 kg per clip while a fine Merino sheep gives 3 to 15 kg per clip. About 34 per cent of Indian wool is produced in Rajasthan. The other important producers are Jammu and Kashmir (18%), Telangana (11%), Karnataka (10%), Gujarat, Himachal Pradesh, Maharashtra and Uttar Pradesh.

The current wool production in India is 41.5 million kg against a total requirement of about 80 million kg per annum. Therefore, India has to import a large quantity of wool, especially the superior quality wool. India also exports raw wool in small quantity to the USA and UK.

Sericulture

Sericulture includes cultivation of mulberry tree and rearing of silk worms. Most of silk in India is obtained from silk worm which feeds on the mulberry leaves. On an average 1 kg of silk needs about 150 kg of mulberry leaves. As per current figures, about 4.5 lakh hectares of area is under mulberry cultivation. Silk production is mainly confined to areas between 15° and 34° N latitudes. The climate of the areas encompassed within this latitudinal zone is congenial to the growth of mulberry trees and includes Karnataka, Andhra Pradesh, Telangana, Tamil Nadu, West Bengal and Jammu and Kashmir. These areas collectively account for 97 per cent of the total area under mulberry cultivation and 95 per cent of raw silk production in the country. Of late, sericulture has gained ground in some non-traditional areas also.

Sericulture is a labour-intensive agro-based cottage and industry which provides gainful employment to about 7.25 million people in rural and semi-urban areas in India. Of these, a sizeable number of workers belong to schedule castes/scheduled tribes and economically weaker sections of society. There is substantial involvement of women in this industry.

The major fields of employment cover silk worm seed producers, farmers-cum-rearers, twisters, hand spinners of silk waste and traders. Mulberry plantation also requires high input of human labour. One hectare of mulberry plantation generates employment for a minimum of 13 persons annually.

Two types of silk are produced in India, *viz.* mulberry and non-mulberry. The distinction arises from the rearing of silk worms either upon mulberry leaves or on leaves of other plants. Karnataka and Andhra Pradesh are the leading producers which account 43% and 32% of the total production respectively. The other major producers are West Bengal, Jammu and Kashmir, Tamil Nadu, Telangana and Assam.

Vanya (non-mulberry) silk comprising Tasar, Eri and Muga are produced in Jharkhand, Chhattisgarh, Madhya Pradesh, Odisha, Bihar, West Bengal, Assam, Meghalaya, Manipur and Nagaland. Different types of ***Vanya*** silk and spun silk and noil yarn are further defined as follows :

1. **Tasar Silk.** This silk is released from cocoons of silkworms belonging to saturniidae family which are fed on leaves of oak, Asan and Arjuna trees. It is mainly produced in Jharkhand, Bihar, Chhattisgarh, Madhya Pradesh, Odisha, West Bengal and Andhra Pradesh.
2. **Eri Silk.** It is spun from cocoons of silkworms belonging to saturaiidae family which are fed on castor leaves. Eri silk is mainly produced in Assam, Bihar, Manipur, Meghalaya, Nagaland and West Bengal.
3. **Muga Silk.** This type of silk is produced only in Assam from cocoons of silkworms belonging to saturniidae family which are fed on Som and Soalu leaves.
4. **Spun Silk Yarn.** As the name indicates, spun yarn is composed of silk filaments of lengths varying from 1 to 20 cm produced by breeding, dressing and spinning the silk waste which is the by-product of the raw silk reeling industry.
5. **Noil Yarn.** It is a short, staple residue obtained during dressing operations in silk spinning from silk waste. It is a by-product of spun silk industry. This can be spun into Noil yarn of coarse counts.

Poultry Farming

The term 'poultry' refers to domestic fowls which are reared for their flesh, eggs and feathers and includes chickens, ducks, geese, turkeys, etc. Poultry farming has assumed much importance due to the growing demand of poultry products, especially in urban areas, because of their high food value. There has been a rapid increase in the number of poultry birds. Andhra Pradesh has the largest number of poultry birds accounting for about 18.2 per cent of the total poultry birds of India. This is followed by Tamil Nadu, Telangana, Maharashtra, West Bengal and Karnataka.

Poultry sector has emerged from entirely unorganised farming practice to commercial production system with state-of-the-art technological innovations. It involves small capital investment; provides direct and indirect employment to the people and is a potent tool for subsidiary income generation for landless and marginal farmers.

At present, India is among the top five chicken meat producing countries of the world. In the year 2020-21 India produced more than two million metric tonnes of poultry meat. The production of eggs increased from 1832 million in 1950-51 to 1,14,419 million in 2019-20 (**Table 18.4**).

TABLE 18.4. Production of Eggs in India

Year	1950-51	1960-61	1970-71	1980-81	1990-91	2000-01	2010-11	2017-18	2018-19	2019-20	2020-21	2021-22
No. of eggs in millions	1,832	2,881	6,172	10,060	21,101	36,632	63,024	95,217	1,03,804	1,14,383	1,22,050	1,29,600

Source : *Economic Survey,* 2022-23, Statistical Appendix, p. 45.

TABLE 18.5. India : Per capita availability of Eggs (number/annum)

Year	2008-09	2009-10	2010-11	2011-12	2012-13	2013-14	2014-15	2015-16	2016-17	2017-18	2018-19	2019-20
Availability of eggs	48	51	53	55	58	61	63	66	69	74	79	86

Source : *Economic Survey,* 2019-20, p. 205, Economic Survey 2020-21, p. 241.

Per capita availability of eggs has also increased from 48 per annum in 2008-09 to 86 per annum in 2019-20, recording a 80% increase in a short span of about one decade (**Table 18.5**). This has happened in spite of rapid increase in population.

After the initial boost given by the Central Poultry Breeding Farms to promote poultry farming on commercial lines in the country, the poultry sector has flourished and now about 70 per cent of the production management and marketing is under highly organised sector. Four Central Poultry Development Organizations (CPDOs) are at Bengaluru (Southern Region), Bhubaneshwar (Eastern Region), Chandigarh (Northern Region) and Mumbai (Western Region) are functioning to provide necessary services including training the farmers to upgrade their skills. They are playing a pivotal role for implementing the policies of the Government with respect to poultry. The mandate of these organisations has specially been re-oriented to focus on improved indigenous birds, which lay on an average 180-200 eggs per annum and have vastly improved Feed Conversion ratio in terms of the feed consumption and weight gained. CPDO Bengaluru is also imparting training to in-service personnel from within the country as well as from overseas. Diversification with species other than chicken like ducks, turkey, guinea, fowl, Japanese squail and emu is also undertaken. The Central Poultry Performance Testing Centre (CPPTC) at Gurugram is entrusted with responsibility of testing the performance of layer and broiler varieties. The "Poultry Development" scheme comprising three components, namely Assistance to State Poultry Farms, Rural Backyard Poultry Development, and Poultry Estates, is being implemented. Further, to encourage entrepreneurship skills of individuals, a central-sector 'Poultry Venture Capital Fund' scheme is also being implemented on capital subsidy mode since 1 April, 2011, covering various poultry activities.

Andhra Pradesh is the largest producer of eggs producing about 18.7 per cent of total eggs of India. This is followed by Tamil Nadu (18.3%), Telangana (13.3%), West Bengal (8%), Maharashtra (5.9%) and Haryana (5.87%). The other major producers are Punjab, Karnataka, Odisha and Kerala.

FISHERIES

Introduction

Fishing is one of the oldest occupations of man. Man learnt fishing much before he could learn something about agriculture. Fishing has assumed much importance in view of the rapidly increasing population and depleting land resources. Fish provides protein rich food and is also a big source of vitamins A, B and D. There are about 30,000 species of fish in the world out of which about 18,000 are found in India. Fish forms an important part of diet of the people living in the coastal areas of Kerala, West Bengal, Odisha, Andhra Pradesh, Tamil Nadu, Maharashtra, Karnataka, Goa and Gujarat. Inedible fish is also a rich source of animal protein for livestock feeding. Fish scales and fishery wastes are also a source of organic manure. The fish catch in India is of two types :

1. Sea or Marine Fisheries. It includes coastal, off-shore and deep sea fisheries. The marine fisheries resources are spread along the country's vast coastline and 2.02 million sq km Exclusive Economic Zone (EEZ) and 0.53 million sq km continental shelf area.

2. Inland or Fresh Water Fisheries. Rivers, lakes, canals, reservoirs, ponds, tanks, etc. contain fresh water and provide fresh water fisheries. Inland fisheries also include those obtained from estuaries, delta channels, back-waters, lagoons and coastal lakes.

Fish Production in India

India is the second largest fish producing country in the world after China and accounts for 7.58 per cent of the global production. Fish production in India reached all-time high of 14164 thousand tonnes in 2019-20. The fisheries sector contributes 1.24 per cent to the Gross Value Added and 7.28 per cent to the agricultural Gross Value Added. The livelihood opportunity provided by this sector has been instrumental in sustaining incomes of 28 million people in India, especially the marginalized and vulnerable communities, and has promoted meaningful socio-economic development.

Recognising the importance of the sector, an independent Department of Fisheries has been created in 2019 to provide sustained and focused attention towards the development of fisheries sector.

Realizing the potential, scope and importance of the fisheries sector, new flagship scheme Pradhan Mantri Matsya Sampada Yojna (PMMSY) was launched in May, 2020 as a part of Atma Nirbhar Bharat Package by the Government of India with an estimated investment of ₹ 20,050 crores comprising central share of ₹ 9,407 crore, state share of ₹ 4,880 crore and beneficiaries contribution of ₹ 5,763 for a period of five years from 2020-21 to 2024-25. PMMSY aims to enhance fish production to 220 lakh tonnes by 2024-25 at an annual growth rate of about 9 per cent. The ambitious scheme will result in doubling the export earnings to ₹ 1,00,000 crores and generate direct and indirect employment opportunities of about 55 lakhs in the fisheries sector over a period of five years. The production of fish in selected years is shown in **Table 18.6**.

TABLE 18.6. Production of Fish in India ('000 tonnes)

Year	Marine	Inland	Total
1950-51	534	218	752
1960-61	880	280	1,160
1970-71	1,086	670	1,756
1980-81	1,555	887	2,442
1990-91	2,300	1,536	3,836
2000-01	2,811	2,845	5,656
2010-11	3,250	4,981	8,231
2015-16	3,600	7,162	10,762
2016-17	3,625	7,806	11,431
2017-18	3,688	8,902	12,590
2018-19	3,853	9,720	13,573
2019-20	3,727	10,437	14,164
2020-21	—	—	14,725
2021-22	—	—	16,253

Source : (*i*) Agricultural Statistics at a Glance, 2018, pp. 400-01.
(*ii*) Agricultural Statistics at a Glance, 2021, p. 181.
(*iii*) Economic Survey 2022-23, Statistical Abstract, p. 45

It is clear from **table 18.6** that before 2000-01, marine fish production was much higher than the inland fish production. But after that, inland fish production has been much higher than marine fish production. This means that inland fish production has increased at a much faster pace as compared to marine fish. Thus 2000-01 is a watershed year so far as marine fish production and inland fish production are concerned.

Marine Fisheries

India has a coastline of over 7,517 km including that of Andaman and Nicobar Islands and Lakshadweep Islands and its continental shelf spreads over 3,11,680 sq km. This entire area is suitable for marine fisheries. It is estimated that about 75 per cent of the marine fish landings are on the West coast and only 25 per cent is contributed by East coast.

There is a vast scope of increasing the marine fish catch in India. It is worth mentioning here that Indian Ocean is the least exploited of all the oceans of the world so far as fishing is concerned. The important fish caught along the coast are *shark*, *sardine*, *herring*, *anchovies*, *Mumbai duck*, *fly fish*, *ribbon fish*, *mackerel* and *Indian salmon*. Mackerel accounts for about one-third of the total catch while herring and prawn account for 15 per cent and 9 per cent respectively. Kerala, Maharashtra, Tamil Nadu, Karnataka, Andhra Pradesh, Odisha, West Bengal and Gujarat are the main producers of marine fish.

India's offshore and deep sea fish catch is very poor considering the marine potential of 20-25 million tonnes annually. Only 10-12 per cent is caught at present. Only 11 per cent of the potential fishing grounds are more than 200 metres deep, even then we are not able to make proper use of the natural advantage. Following are the main reasons for this sad state of affairs.

(*i*) India has tropical climate in which fish cannot be preserved for a long time and spoils very quickly. Heavy expenditure on refrigeration and deep freezing increases the market price of the fish.

(*ii*) Indian coast does not have many gulfs, bays, estuaries and backwaters as is the case with Norway. As such, it lacks good fishing grounds.

(*iii*) Marine fishing in India is a seasonal phenomena. Strong winds during the monsoon season accompanied by tropical cyclones often hinder fishing operations.

(*iv*) Majority of Indians are vegetarians and do not eat fish.

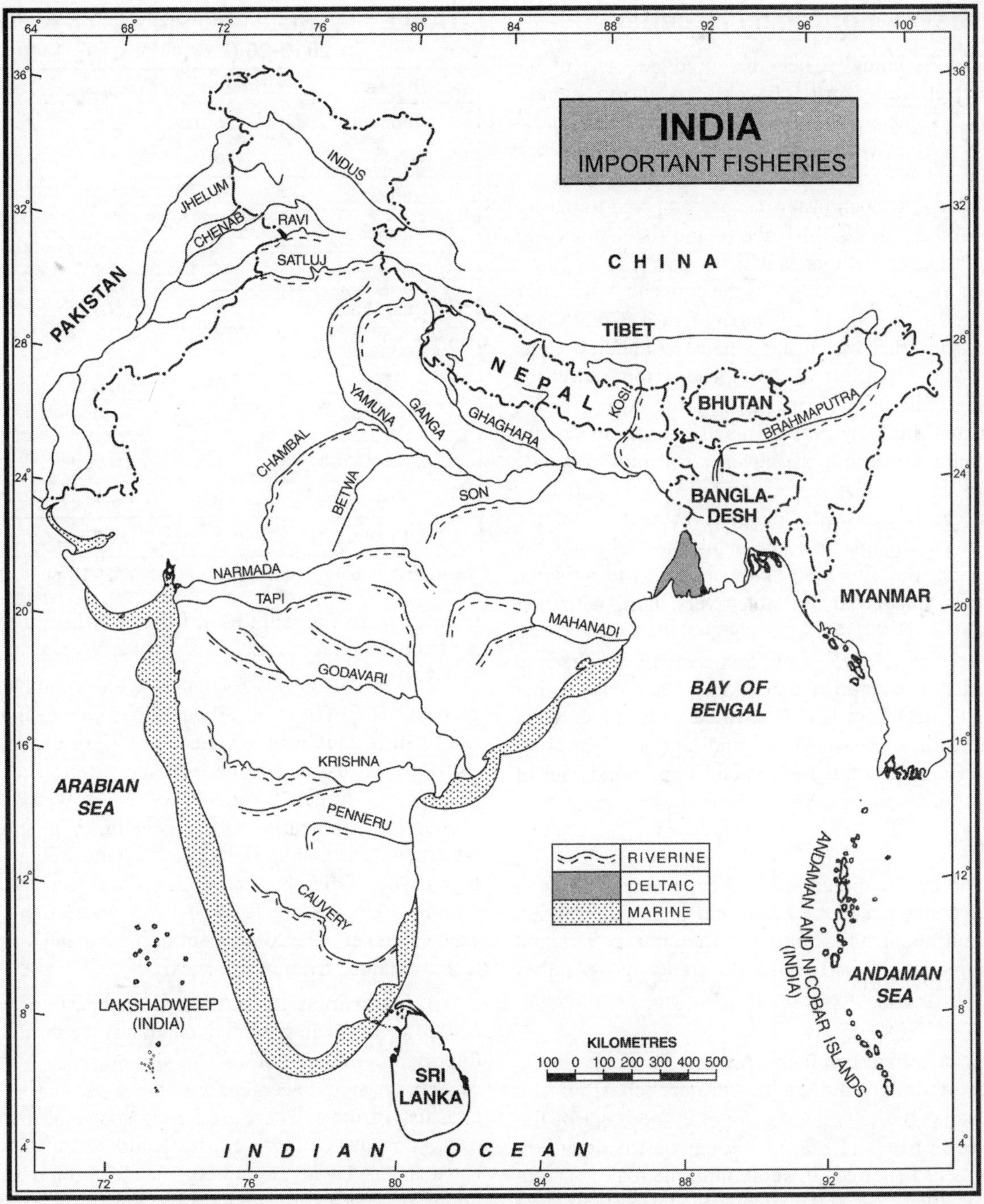

FIG. 18.5. India : Major Fisheries

(*v*) About 60 per cent of the fishermen still use small non-mechanised boats. They normally do not venture beyond 10 km from the coast and in waters more than 18 metres deep. They have to come back to the coast at night.

(*vi*) Lack of landing, freezing, canning, transport and organised markets are other handicaps faced by fishing in India.

Fresh Water or Inland Fisheries

India's inland fishery resources are one of the richest in the world. Rivers, irrigation canals, reservoirs, lakes, tanks, ponds, delta channels, back-waters, lagoons, estuaries, etc. provide the foundation for inland fisheries. The inland fish production has increased at a much faster rate as compared to marine fish. In the year 1950-51, the inland fish caught was only 218 thousand tonnes which was less than 29 per cent of the total fish caught in the country. After half a century *i.e.* in 2000-01 the inland fish was more than the marine fish which happened for the first time. After that, the inland fish production increased rapidly whereas marine fish production almost stagnated. In 2018-19, the inland fish accounted for 9.71 million metric tonnes against 3.71 million metric tonnes of marine fisheries. India's riverine fishery resources comprise the major river systems of the Ganga, the Brahmaputra, the Indus, the Mahanadi, the Narmada, the Tapi, the Godavari and the Krishna. The total length of the fishable rivers, along with their tributaries is 27,359 km. The length of irrigation channels is 1,12,654 km. The dam reservoirs and small lakes cover an area of 29 lakh hectares. Another 26 lakh hectares of land is covered by brackish water in the form of coastal lakes and lagoons. About 16 lakh hectares are covered by tanks and ponds out of which 6 lakh hectares are used for pisciculture.

Distribution

Although fishing is carried on in almost all parts of the country, about 97 per cent of India's marine fish and about three fourth of the country's inland fish are caught in the littoral states like Andhra Pradesh, West Bengal, Gujarat, Kerala, Tamil Nadu, Maharashtra .

1. Andhra Pradesh. Andhra Pradesh is the largest fishing state on the eastern coast of the Peninsular India producing about 29 per cent of the total fish of India. Its 960 km long coast is dotted by villages of fishermen in addition to the major fishing centres like, Vishakhapatnam, Machilipatnam and Kakinada. About 84 thousand fishermen use 580 mechanized and over 36,000 non-mechanized boats for fishing. There are 1,169 primary cooperatives to handle the processing of fish. *Oil sardine*, *mackerel*, *silver bellies*, *ribbon fish*, *catfish* and *soles* are important varieties of fish caught here. A large quantity of fish is sent to West Bengal where there is a great demand for fish.

TABLE 18.7. State-wise Production of Fish in 2019-20 (Lakh Tonnes)

States	Inland	Marine	Total
1. Andhra Pradesh	36.10	5.64	41.74
2. West Bengal	16.19	1.63	17.82
3. Gujarat	1.58	7.01	8.59
4. Odisha	6.60	1.58	8.18
5. Tamil Nadu	1.74	5.83	7.57
6. Uttar Pradesh	6.99	0.00	6.99
7. Kerala	2.05	4.75	6.80
8. Bihar	6.41	0.00	6.41
9. Karnataka	2.29	4.03	6.32
10. Chhattisgarh	5.72	0.00	5.72
11. Maharashtra	1.18	4.43	5.61
All India	**104.37**	**37.27**	**141.64**

Source: Agricultural Statistics at a Glance, 2021, pp. 182-83.

2. West Bengal. West Bengal is the second larget producer of fish in India. About 12 per cent of India's total fish is caught in West Bengal. Here the inland fish far exceeds the marine fish. West Bengal also is the second largest producer of fresh water fish accounting for nearly 15 per cent of India's fresh water fish. The Ganga Delta is inhabited by about one lakh fishermen who are engaged in fresh water fishing. Bengalis are fond of fresh water fish. The state can meet only 20 per cent of its demand and the rest is obtained from other states.

3. Gujarat. With 1,000 km long coast and over 65,000 sq km fishing ground, Gujarat is the third most important producer of fish in the northern part of western coast and accounts for over 6 per cent of the total fish of India. There are 52 big and small fishing ports on the Gujarat coast. Kandla, Porbandar, Navabandar, Dwarka, Umbergaon and Jaffarabad are important centres. Besides several small towns and a large number of villages also contribute to fishing. About 1.5 lakh fishermen are engaged in catching about 7 lakh tonnes of fish every year. *Mumbai duck*, *pomfret*, *jew fish*, *Indian salmon*, *tunnies*, *grey mullet*, *mackerel*, *eel*, *shark*, etc. are important fish varieties caught here. Local consumption being very low,

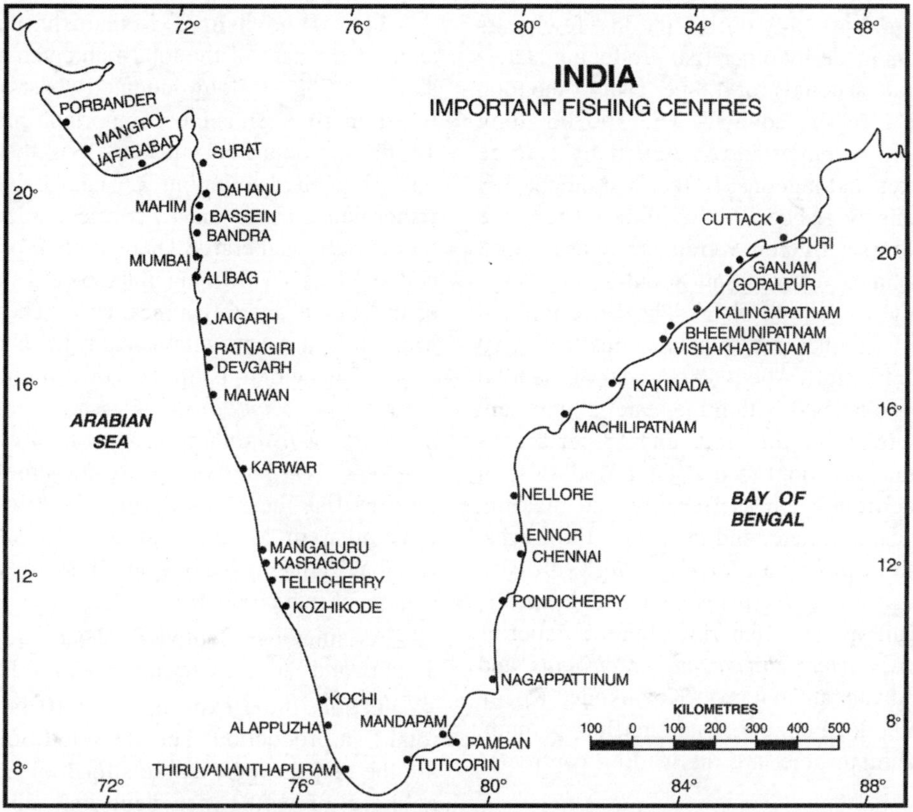

FIG. 18.6. India : Important Fishing Centres

about 97 per cent of the fish is transported to big markets like Mumbai, Kolkata and Delhi. Some fish is exported to Sri Lanka, Myanmar, Mauritius and Singapore.

4. Odisha. Odisha's 720 km long coast provides about 8.18 lakh tonnes of fish every year which is about 5.8 per cent of the total fish catch of India. Cuttack, Puri, Sambalpur and Balasore are important fishing centres. Chilika lake is the most intensively fished area for inland fisheries. The fish from Chilika lake is of superior quality and is in great demand.

5. Tamil Nadu. Tamil Nadu is the fifth largest producer and produces about 5.3 per cent of India's fish. The 1,000 km long coast of Tamil Nadu is not much indented which hinders fishing to some extent and encourages the use of mechanized boats. Presently, 96,500 persons are engaged in fishing along the Tamil Nadu coast using 2,757 mechanized boats and 43,343 country craft. *Mackerel*, *ribbon fish*, *catfish* and *soles* are the main varieties of marine fish caught here. Chennai is the largest centre. Other important centres are Tuticorin, Ennore, Cuddalore, Mandapam and Nagpattinam. The Tamil Nadu coast is dotted by about 300 villages of fishermen who solely depend upon fishing for their livelihood. Fish processing is done in 46 freezing, 60 cold storage, 3 canning and 6 fish meal units.

6. Uttar Pradesh. Uttar Pradesh is totally landlocked state and does not have any access to sea. Therefore, there is no marine fish production in the state. However, this state is blessed with the largest river of India, *i.e.*, the Ganga and its large number of its tributaries. Thus the entire fish production in this state consists of inland fisheries. Apart from the rivers, this state has a close network of canals which offer ample opportunities for inland fishing. From an insignificant producer, Uttar Pradesh has become one of the leading fish producing states of India.

7. Kerala. Kerala used to be an important fishing state of India but the relative importance of this state has come down drastically during the last few years due to progress made by other fish producing states. Currently Kerala accounts for 4.8 per cent of the total fish production of the country. The 590 km long Kerala coast is indented and is dotted by a large number of lakes and lagoons. In fact Kerala coast is the most intensely fished area of India. There are about 1.35 lakh fishermen in Kerala. They use over 3 thousand mechanised boats and about 27 thousand country craft to catch fish. The mechanised production is 32 tonnes per boat as against about 10 tonnes by country craft. About 60 per cent of the total fish caught is consumed within the state, 22 per cent sold to other states and the remaining 18 per cent is exported. Kerala has about 85 per cent of India's total processing facilities such as freezing, ice making, cold storage, canning, etc. and processes the largest amount of fish in the country. *Oil sardines, prawns, horse mackerel, soles, lactarius* and *shark* are the most important marine fisheries. Inland fisheries include *prawns, barbus, clarius, mallet, etroplus,* and *murrels*. Thiruvananthapuram, Kozhikode, Kochi, Kollam, Azhikode, Ponnani, Kannur, Baliapatnam, Baypore, Vizhinjam, etc. are the leading centres of fish production.

8. Bihar. Like U.P., Bihar is also a landlocked state with a large number of rivers and their tributaries. Besides Bihar has a large number of canals also. Thus the entire production is that of inland fisheries.

9. Karnataka. Karnataka produces nearly 4.4 per cent of India's fish. Mangaluru, Karwar, Ankola, Kumta, Honawar, Bhatkal, Majali, Bingi, Chendia, Gangolli, Malpe, Udiayawar and Bokapatnam are the main centres. About 20,000 fishermen using 6,500 boats are engaged in fishing. *Sardine*, *mackerel*, *shark*, and *seer*, are the principal varieties caught here. The creeks of Netravati, Sharavati and Kali rivers are the most intensively used areas for inland fishing. Besides there are 30,000 minor tanks, 2,700 major tanks, 17 reservoirs and 6,885 km long stretch of rivers used for fishing.

10. Chhattisgarh. Like Uttar Pradesh and Bihar, Chhattisgarh is a land looked state and does not produce any marine fish. However, this state is drained by Mahanadi and its tributaries and has a large number of tanks. In 2019-20, this state produced about 4 per cent of the total fish of India.

11. Maharashtra. Maharashtra accounts for about 4 per cent of the total fish catch of India. This state has 750 km long indented coast, as a result of which marine fish is more important than inland fish. There are about 250 villages along the Maharashtra coast inhabited by about 2.6 lakh fishermen. These fishermen operate on 4,718 mechanized boats and 5,662 traditional crafts. There are 530 primary fishing cooperatives, 41 freezing units, 6 cold storage units, 3 canning units and 6 fish meal units. The sea along the Maharashtra coast remains calm for about 7 months in a year which is quite convenient for fishing operations. *Mumbai duck, white pomfret, black pomfret, jew fish, Indian salmon, tunnies, grey mullet, mackerel, eel,* and *shark* are the chief varieties of marine fish here. Estuarine fisheries have been developed in Mahim and other creeks. Mumbai, Kolaba, Ratnagiri, Alibag and Bassein are the leading centres of fish production.

Assam, Goa, Madhya Pradesh, Punjab, Haryana, Tripura and Punducherry are other producers.

Trade. India exports about 10 per cent of the total fish production. The USA and South-East Asia are the major export markets for Indian seafood with a share of 34.81 per cent and 23.67 per cent respectively. There has been a spectacular increase in export of marine products in thc recent years (**Table 18.8**). Marine products have been identified as a major thrust area for exports during the Eighth Plan and it continues till now. India exported 1168 thousand tonnes of fish and fish preparations worth ₹ 44,176 crore in 2020-21.

TABLE 18.8. India: Exports of fish and fish preparations

Year	Quantity (Thousand tonnes)	Value (₹ crores)
1960-61	19.9	5
1970-71	32.6	31
1980-81	69.4	217
1990-91	158.9	960
2000-01	502.6	6,367
2010-11	825.3	11,917
2019-20	9511.0	45427
2020-21	1168	44,176
2021-22	1398	57,910

Source: Economic Survey 2022-23, Statistical Appendix, pp. 116-18.

Marine Fishing Policy 2004

A comprehensive Marine Policy was launched in November 2004 to facilitate sustainable deep sea fishing. In the inland sector, the potential for fishery development in East and North-Eastern States is immense. Development of fisheries can go a long way in tackling the problem of food as well as unemployment in these states.

Following are the main features of Marine Fishing Policy 2004 :

The policy objectives are : (1) to augment marine fish production of the country up to the sustainable level in a responsible manner so as to boost export of sea food from the country and also to increase per capita fish protein intake of the masses, (2) to ensure socio-economic security of the artisan fishermen whose livelihood solely depends on this vocation, (3) to ensure sustainable development of marine fisheries with due concern for ecological integrity and bio-diversity.

National Policy on Marine Fisheries, 2017. It has replaced the Marine Fishing Policy, 2004 and aims at fully utilising the fishing potential of marine fish resources and lays emphasis on bringing sustainable utilisation of fisheries wealth from marine and other aquatic resources.

Some of the important features of this policy include the use of Information Technology (IT) and space technology for improving the capacities of fishing community, strengthening the monitoring, control and surveillance system by introducing chip-based smart registration cards for fishermen and their fishing vessels, and training of fisherman to avoid crossing the International Marine Boundary Line (IMBL).

The new policy also focuses on maximising fleet-size, mainstreaming biodiversity conservation in production processes, species-specific and area-specific management plans and spatial and temporal measures for sustainable utilisation of resources. It aims at encouraging an integrated approach on fisheries management, besides blending traditional knowledge and scientific business principles.

Blue Revolution. The term 'blue revolution' is used to describe the phenomenal increase in fish production from a meager 752 thousand tonnes in 1950-51 to a staggering 9,511 thousand tonnes in 2019-20 by adopting a package of methods. It was coined and started in 1970 after the success of the Green Revolution in agriculture when the Central Government sponsored the Fish Farmers Development Agency (FFDA).

The fisheries industry has witnessed a massive transformation form the traditional subsistence type of enterprise to a multi-crore business industry with improved modern infrastructure. Motorised boats have replaced the traditional hand driven inefficient boats and big trawlers are a common site in the Indian seas. In fact Blue Revolution has brought a conspicuous improvement in fishing industry by adopting new techniques of fish breeding, fish rearing, fish marketing and fish export. Refrigeration and fast transport system have opened up new vistas for exploring the untapped markets and have made fish a profitably marketable commodity.

AQUACULTURE

Also known as **aquafarming**, aquaculture refers to the production of aquatic animals and plans under controlled conditions. According to Food and Agriculture Organisation (FAO) "acquaculture is understood to mean the farming of aquatic organisms including fish, crustaceans, molluscs and aquatic plants." Although aquaculture has a long history and has been practised in China since, 2500 BC, it is a recent phenomenon in India. Aquaculture can be classified into two major categories *viz.* ***freshwater aquaculture*** and ***coastal aquaculture.*** Coastal aquaculture can be further subdivided into two categories; namely ***sea farming*** and ***brackish water aquaculture.*** Sea farming is concerned with the culture of organisms in open coastal waters and bays. Brackish water aquaculture, on the other hand, refers to land based farming systems using salt water from creeks, estuaries as well as from the coastal seas.

To encourage research and development in aquaculture, the Indian Council of Agricultural Research (ICAR), New Delhi, reorganised the fisheries research institute in 1987 and three different institutes were established. These are : (*i*) Central Institute of Freshwater Aquaculture (CIFA) at Bhubaneshwar (Odisha), (*ii*) the Central Institute of

Brackish water Aquaculture (CIBA) at Chennai (Tamil Nadu) and (*iii*) the National Research Centre for Cold Water Fisheries (NRCCWF) at Bhimtal (Nainital, Uttarakhand).

Types of Aquaculture

1. Freshwater Aquaculture. The Central Institute of Freshwater Aquaculture (CIFA) has contributed a lot for breeding and rearing the air-breaching cat-fish (popularly known as magur) in vast areas of the country. This institute has adopted biotechnological approach and the fish are implanted with hormonal pellets to advance maturity. Freshwater shark has also been bred and reared by this institution. This is a very popular fish in the north-eastern part of India particularly in Manipur. Freshwater prawn or shrimp is an important form of aquaculture and prawn is raised for human consumption.

The initial step for freshwater aquaculture in India was taken with the setting up of Pond Culture Division at Cuttack in 1949 under the name of the Centre of Central Inland Fishers Research Institute (CIFRI). This type of aquaculture is popular particularly in the eastern states of West Bengal, Odisha and Andhra Pradesh. However, Punjab, Haryana, Assam and Tripura are also taking up this culture in a big way.

2. Brackish water Farming. This type of aquaculture has a long history. It is confined mainly to coastal wetlands impounded by human, locally known as ***bheries*** in West Bengal. It is also practised in salt resistant deepwater paddy fields in the coastal areas of Kerala, locally known as ***pokkali.*** These systems have sustained high production level although there has not been much additional input. This type of aquaculture got its due recognition after the initiation of an All India Coordinated Research Project (AICRP) in 'Brackish water Fish Farming' by the Indian Council of Agricultural Research (ICAR) in 1973. This project is credited for developing several technologies concerning fish and shrimp farming.

3. Shrimp Farming. Shrimp farming is a type of aquaculture in which marine shrimps are produced for human consumption. There has been phenomenal increase in area under shrimp farming due to development of more commercial hatcheries. The formation of Brackishwater, Fish Farmers' Development Authority (BFDA) in marine states supplemented by various government programmes proved to be a great help to shrimp farming. Subsidies from the Marine Products Export Development Authority (MPEDA), loans from commercial banks gave financial support to this farming. Demonstration of semi-intensive farming technology provided the required technical support. The Central Marine Fisheries Research Institute (CMFRI) initiated studies on maturation in early 1970s. In late 1980s Marine Products Export Development Authority (MPEDA) established the Andhra Pradesh Shrimp Seed Production and Research Centre and the Odisha Shrimp Seed Production and Research Centre to facilitate shrimp farming in these two states and to benefit other states also. Obviously Andhra Pradesh, Odisha and Tamil Nadu have developed shrimp farming in a big way. Nellore district of Andhra Pradesh is such a prolific producer of shrimp that it has earned the distinct name of ***Shrimp Capital of India.*** A number of farm holdings are located in Kerala and West Bengal.

4. Mariculture. Mariculture is that branch of aquaculture which is concerned with the cultivation of marine organisms for food and other products in the open ocean, enclosed seas or a section of ocean, tanks, ponds or raceways filled with seawater. Fish, prawns and oysters are the main food products. Non-food products include fish-meat, nutrient agar, jewelleries (cultured pearls) and cosmetics. Mandapam (in Tamil Nadu) centre of Central Marine Fishers Research Institute made the earliest attempt at mariculture in 1958-59 with the culture of milkfish. This institute has developed different types of technologies for a number of species including oysters, mussels and clams among sedentary species, as well as for shrimps and finfish.

5. Algaculture. This type of aquaculture involves farming of varied species of algae. Majority of algae that are cultivated belong to the category of micro algae also referred to as phytoplankton, microphytes, or planktonoic algae.

6. Integrated Multi-Trophic Aquaculture (IMTA). In this type of aquaculture, by-products (wastes) from one species are recycled as inputs for another. Fed aquaculture (*e.g.* fish, shrimp) is combined with inorganic extractive (*e.g.* seaweed) and organic extractive (*e.g.* shellfish) aquaculture to create balanced systems for environmental and economic sustainability as well as for social acceptability.

7. Fish Farming. Fish farming involves raising of fish in tanks or other water enclosures on a commercial scale. Products of fish farming are primarily used as food items. Salmon, catfish, cod, carp, trout, tilapia etc. are the chief species of fish reared under fish farming.

8. Seaweed culture. Marine algae are better known as seaweeds. Like other photosynthetic plants, seaweeds need sunlight. Little sunlight penetrates depths greater than 15 metres, so most of the seaweeds grow in shallow waters around shores or reefs. They are divided into three groups according to colour, brown, green and red. They provide food to tiny creatures, most of which filter dead particles from water. At present seaweed accounts for about 30 per cent of world aquaculture production. The Indian Ocean abounds in seaweeds, one-third of which occurs along the Indian coast. Seaweed has a wide range of application in the fields of food, textile, cosmetic, pharmaceutical, fodder, fertilisers, etc. due to which its demand has increased tremendously in the recent past. Seaweed are rich in vitamins, minerals, trace elements and bioactive substances and have became an important ingredient of human food. There is vast scope for developing seaweed food in India.

9. Sewage-fed fish culture. Sewage-fed fish culture is an old practice in ***bheries*** in West Bengal. This is a culture in which fish are reared in sewage water. It involves multiple stocking and multiple harvesting approaches.

10. Paddy-cum-fish culture. In many parts of the country, fish are reared in paddy fields which are flooded with water for paddy sowing. It is practised in medium to semi-deep water paddy fields in low land areas. Strong dykes or field boundaries are constructed to prevent the escape of cultivated fish during floods. Although the system largely depends upon natural stocking, modern farming techniques are also practised. These techniques involve major and minor carps stocked at the densities of 5,000 to 10,000 per hectare are practised in several parts of the country. Fresh water prawns are also cultivated along with paddy-cum-fish.

Apiculture (Beekeeping)

The term apiculture has been derived from Latin *apis* which means a bee. Thus apiculture is the practice of developing and maintaining bee colonies by man. Beekeeping is practised mainly for honey, beeswax and for pollinating crops. Honey bees have tendency to live in colonies and as such they are kept in hives. Various types of hives are used depending on the local conditions and choice of the beekeepers. Fixed frame hives, movable frame hives and top bar hives are some of the popular types of hives used in the present day world. Usually there are three classes of bees in a colony : (*i*) a queen which is normally the only breeding female, (*ii*) about 30 to 50 thousand female workers which serve the queen and build hives and (*iii*) large number of male drones which may run into thousands in spring to a very few due to death in cold season. Beekeepers are generally categorised in following three classes :

(*i*) Commercial beekeepers, *i.e.*, beekeeping as primary source of income as in Himachal Pradesh.

(*ii*) Sideliners who keep bee keeping as a secondary source of income; in different parts of the country.

(*iii*) Migratory beekeepers as of Kanniyakumari.

Beekeeping has become a reasonably important occupation which has great capacity to supplement the income of the farmers. It provides employment to rural workers and helps in improving the quality and quantity of crops. Honey bees are used to increase the production of plantation crops. Cross-pollination by bees is very useful for crops. In addition to the significant role played by bees in crop production, they are a greater source of honey and wax. Currently beekeeping is carried on over a cultivated area of about 50 million hectares for growing vegetables, fruits, oilseeds, pulses, etc. and 3-4 colonies of honey bees are required per hectare of cultivated land.

Honey is the most important product of beekeeping. It contains vitamins and a large number of nutritions elements. Each 100 grams contains 138 mg potassium, 17 mg sodium, 13 mg calcium, 5 mg phosphorous and 1.5 mg iron. Further, it is estimated that 100 gm of honey gives 320 calories of energy. Traditionally honey is known to increase body resistance to diseases and help in weight management when taken regularly.

India's honey production in 2017-18 was 1.05 lakh metric tonnes as compared to 35,000 metric tonnes in 2005-06. In 2017-18, India had 35 lakh bee colonies as compared to 8 lakh colonies in 2005-06. On an average, one hive yields about 5 kg of honey which varies from 1.8 to 2.3 kg in south India to 5.5 to 6.8 kg in north India. It has been estimated that as much as 50 kg of honey can be obtained from one hive in one year by using modern technology. Only 30 per cent honey is collected by Khadi Gramodyog and commercial beekeepers and the remaining 70 per cent is collected by the tribal people. Most of the honey is collected in Himachal Pradesh, Maharashtra and Uttar Pradesh. Tamil Nadu, Karnataka, Kerala, Andhra Pradesh, Telangana, West Bengal, Chhattisgarh, Madhya Pradesh etc. are some other honey producing states.

There are vast potentialities for increasing the production of honey in India and for exporting it to European countries, USA and Japan. Indian honey is in great demand in the international market due to its high quality and low cost.

Species of Honey Bees. A large number of species of honey bees are found in India because of diversity in topography, climate and flora in this vast country. Some of the important species are briefly described as under :

Apis cerana are found in large areas especially in high Himalayan region where beekeepers log hives in their houses. In the South India, *Apis cerana* beekeeping is undertaken under the supervision of the Bee Research and Training Institute of Pune and Khadi and Village Industries Commission at Mumbai. This type of beekeeping is also practised in Mahabaleshwar hills of Maharashtra, parts of West Bengal, north-eastern states of Arunachal Pradesh and Sikkim, as well as in large parts of Karnataka, Kerala and Tamil Nadu.

Apis dorsata is found in higher reaches of the Himalayas, in the Terai region of Uttar Pradesh and Himachal Pradesh in North India. In the central part of India, thick forests provide congenial conditions for *Apis dorsata* where this breed of honey bees is kept mainly by the tribals. Mangrove forests of Sundarbans in West Bengal and Eastern Ghats in Andhra Pradesh provide ideal conditions for *Apis dorsata*. In Himachal Pradesh, farmers keeping *Apis dorsata* practise **transhumance**. They move to higher reaches of the mountains in summer with thick loads of bee colonies for pollination of apple orchards and come down to plains in winter for placing the bee hives amongst eucalyptus trees and sunflower plants.

Apis mellifera are steadily increasing in Himachal Pradesh, Punjab, Bihar and Madhya Pradesh.

REFERENCES

Agricultural Statistics at a glance, 2021, Govt. of India, Ministry of Agriculture.

Ayyappan, S and Krishna, M (2007) : Inland Fishers—Changing Consumption Patterns, The Hindu Survey of Indian Agriculture, 2007, Kasturi and Sons Ltd. Chennai, pp. 101-102.

Ayyappan, S. and Biradar, R.S. (2004) : Inland Fisheries—Enhancing Global Competition, *The Hindu Survey of Indian Agriculture,* 2004, Kasturi and Sons Ltd., Chennai, pp. 91-95.

Balram, N. (2007) : *Livestock—Need to boost export share,* The Hindu Survey of Indian Agriculture, Kasturi and Sons Ltd., Chennai, pp. 94-97.

Dayal, P. (1984) : *Distribution of Cattle and Problem of Fodder Supply in Bihar,* in A. Ramesh (ed.), *Resource Geography,* Heritage Publishers, New Delhi.

Desai, Anuradha (2004) : Poultry—Problems and Prospects, *The Hindu Survey of Indian Agriculture*, 2004, Kasturi & Sons Ltd., Chennai, pp. 94-95.

Economic Survey, 2018-19 and 2017-18, Oxford University Press, GOI, New Delhi.

Govt. of India, *Economic Survey 2022-23.*

India (2024) : *A Reference Annual,* Publication Division, Ministry of I&B, Govt. of India.

Jana, R.K. and Sena R.K. (2004) : Aquaculture—Overwhelming Growth, *The Hindu Survey of Indian Agriculture,* 2004, Kasturi and Sons Ltd., Chennai, pp. 101-103.

Kadirvel, R. (2004) : Livestock—Towards Open Economy, *The Hindu Survey of Indian Agriculture,* 2004, Kasturi & Sons Ltd., Chennai, pp. 110-113.

Kaura, R.L. (1952) : *Indian Breeds of Livestock Including Pakistan*, Prem Publishers, Lucknow.

Muthiah, S. *et al.* (eds.) : *A Social and Economic Atlas of India*, Oxford University Press, New Delhi.

Narhari, D. (1996) : *Commercial Broiler Production*, Emkay Publications, Delhi.

Panikkar, N. Kesava (1984) : *Development of Indian Marine Fisheries, in* A. Ramesh *(ed.)*, *Resource Geography*, Heritage Publishers, New Delhi.

Patel, Amrita (2004) : Dairying—Improving Milk Quality, *The Hindu Survey of Indian Agriculture,* 2004, Kasturi & Sons Ltd., Chennai, pp. 107-109.

Raghava, Kurup C.G. (1967) : (ed.) *Handbook of Animal Husbandry*, I.C.A.R., New Delhi.

Randhawa, M.S. (1958) : *Agriculture and Animal Husbandry*, I.C.A.R., New Delhi.

Sharma, Nagendra and Kumar, Shalender (2004) : Sheep and Goat—Yawning Productivity Gap, *The Hindu Survey of Indian Agriculture,* 2004, Kasturi & Sons Ltd., Chennai, pp. 115-117.

Singh, Harbans. (1966) : *Domestic Animals— India*, National Book Trust, New Delhi.

Singhal, N.C. (1992) : Marine Fisheries : Joint Efforts for Big Catch, *The Hindu Survey of Indian Agriculture,* 1992, Kasturi and Sons Ltd., Chennai, pp. 171-176.

Srivastava, U.K. and Dharma, M. (ed.) (1985) : *Fisheries Development in India: Some Aspects of Policy Management*, Concept Publishing Co., New Delhi.

Sundararajan, B. (2007) : *Poultry—Logistics Plays Vital Role*, The Hindu Survey of Indian Agriculture, 2007, Kasturi and Sons Ltd. Chennai, pp. 92-93.

Yadava, Y.S. (2004) : Marine Fisheries—Exploiting deep sea-avenue, *The Hindu Survey of Indian Agriculture,* 2004, Kasturi and Sons Ltd., Chennai, pp. 91-95.

Yadava, Y.S. (2007) : *Marine Fishers—Signs of Growth Visible*, The Hindu Survey of India Agriculture (2007), Kasturi and Sons Ltd., Chennai, pp. 98-100.

◆◇◆

CHAPTER 19

Land Utilization

INTRODUCTION

As in all other countries, land in India is put to various uses. The utilization of land depends upon physical factors like topography, soil and climate as well as upon human factors such as the density of population, duration of occupation of the area, land tenure and technical levels of the people. There are spatial and temporal differences in land utilization due to the continued interplay of physical and human factors. India has total geographical area of about 328.73 million hectares but statistics pertaining to land utilization were available for about 307.79 million hectares in 2018-19.

1. Net sown area

Cropped area in the year under consideration is called ***net sown area***. This area has a special significance in an agricultural country like India because agricultural production largely depends upon this type of land. There is an urgent need to increase the net area sown for meeting the food and other requirements of rapidly increasing population in India; although there is not much scope for increasing area under this category due to natural limitations such as topography, soils, climate, etc. The net sown area increased from 118.75 million hectares in 1950-51 to all time high of 142.87 million hectares in 1990-91, after which varying trends have been found. It stood at 139.35 million hectares in 2018-19 (**Table 19.1**)

Net sown area accounted for about 45.27 per cent of the total reporting area of India (in 2018-19) against the world average of about 32 per cent. This is much higher than 40 per cent in the USA, 25 per cent in Russia, 16 per cent in Brazil and only 6 per cent in Canada. But the per capita cultivated land has gone down drastically from 0.53 hectares in 1951 to 0.10 hectares in 2018-19. This is a serious trend and can be checked only by population control.

Rajasthan has the largest net sown area of 17.5 million hectares which is about 12.5 per cent of the total reporting net sown area of India. This is followed by Maharashtra (17.35 million hectares).

It may be noted that agricultural prosperity does not depend as much as on the total net sown area as it does on the percentage of net sown area to the total reporting area. There are large variations in the proportion of net sown area to total reporting area from one state to another. Punjab and Haryana had some of the highest proportions of 81.8 and 79.1 per cent respectively while Arunachal Pradesh had 3 per

TABLE 19.1. Pattern of Land Utilization in India

(million hectares)

Year	Total reported area (according to village papers)	Net area sown	Area sown more than once	Total cropped area	Forests	Not available for cultivation	Permanent pastures and other grazing lands	Land under Miscellaneous tree crops and groves (not included in net sown area)	Culturable waste	Fallow lands other than current fallows	Current fallows
1950-51	284.32	118.75	13.15	131.89	40.48	47.52	6.68	19.83	22.94	17.45	10.68
1960-61	298.46	133.20	19.57	152.77	54.05	50.75	13.97	4.46	19.21	11.18	11.64
1970-71	303.75	140.86	24.93	165.79	63.83	44.61	13.26	4.37	17.50	8.73	10.60
1980-81	304-16	140.29	34.63	172.63	67.46	39.55	11.99	3.58	16.74	9.72	14.83
1990-91	304.86	142.87	42.87	185.75	67.81	40.48	11.40	3.82	15.00	9.66	13.70
2000-01	305.19	141.34	44.00	185.74	69.84	41.23	10.66	3.44	13.63	10.27	14.76
2010-11	307.48	141.56	56.00	197.56	71.59	43.58	10.30	3.20	12.65	10.32	14.28
2015-16	307.75	139.51	57.55	197.05	71.87	44.02	10.26	3.09	12.29	11.31	15.41
2016-17	308.32	139.42	60.79	200.20	72.02	44.82	10.34	3.12	12.24	11.27	15.09
2017-18	307.77	139.18	60.81	199.99	72.05	44.32	10.34	3.17	12.29	10.62	14.81
2018-19	307.79	139.35	57.97	197.32	72.01	44.51	10.38	3.15	12.22	11.63	14.53

Source : Agricultural Statistics at a Glance, 2021, pp. 330-33.

cent only in 2018-19. Large parts of the Satluj, Ganga plains, Gujarat plains, Kathiawar plateau, Maharashtra plateau and West Bengal basin have high proportion of cultivated area. This is largely due to gentle slope of the land, fertile alluvial and black soils, favourable climate, excellent irrigation facilities and high density of population. In contrast, mountainous and hilly areas in the Himalayan region and some of the drier tracts are not much suited to farming because of rugged topography, unfavourable climate and infertile soils.

2. Area sown more than once

As the name indicates, this area is used to grow more than one crop in a year. According to **Table 19.1,** the total cropped area has increased from 185.74 million hectares in 2000-01 to 197.32 million hectares in 2018-19. This means that the area sown more than once has increased from 44.00 million hectares in 2000-01 to 57.97 million hectares in 2018-19. Thus there is a net increase of about 14 million hectares in 'area sown more than once' in a short span of 18 years. This is a healthy trend and this pace of increasing area sown more than once should be kept if we really want to be self-sufficient in food grains and succeed in our mission of food security in view of the rapidly increasing population. This type of area comprises of land with rich fertile soils and regular water supply. It is clear that the percentage of area sown more than once is rather low in India as a whole. This is attributed to infertile soils, deficiency of moisture and insufficient use of manures and fertilizers. This type of land is of special significance. Since almost all the arable land has already been brought under plough, the only course left to increase the agricultural production is to increase the intensity of cropping which can be done by increasing the area sown more than once. Large tracts of the Indo-Ganga plain in Punjab, Haryana, Uttar Pradesh and Bihar and in coastal regions have large percentage of area sown more than once.

3. Forest Area

This area includes all land classified either as forest under any legal enactment, or administered as forest whether state owned or private and whether wooded or maintained as potential forest land. The area of crops grown in the forest and grazing lands or areas open for grazing within the forests remain included under the forest area. Forests cover about 23.45 per cent of the reported area which is a definite improvement against 14 per cent in 1950-51. However, 23.45 per cent of forest land to the total reporting area is not sufficient for a tropical country like India where about 33 per cent of the total land should be under forests. This will require massive tree plantations and vigorous restrictions on the reckless felling of trees. According to the expert committee recommendations, much of the area reclaimed from the forest for agriculture should be retired from cultivation and brought back under forests to save the land from the adverse effects of deforestation.

4. Land not available for cultivation

This class consists of two types of land *viz*. (*i*) land put to non agricultural uses and (*ii*) barren and unculturable waste. The area put to non-agricultural uses includes land occupied by villages, towns, roads, railways or under water *i.e.* rivers, lakes, canals, tanks, ponds, etc. The barren land covers all barren and uncultivated lands in mountains and hill slopes, deserts and rocky areas. These areas cannot be brought under plough except at high input cost with possible low returns. The amount of this land has been variable right from 1950-51 to 2014-15, the data for which are available. Land not available for cultivation increased from 41.48 million hectares in 2000-01 to 44.51 million hectares in 2018-19 and accounted for 14 per cent of the total reported area in 2018-19. The largest amount of land in this category is in Rajasthan, followed by Gujarat, Madhya Pradesh, Uttar Pradesh, Andhra Pradesh and Maharashtra.

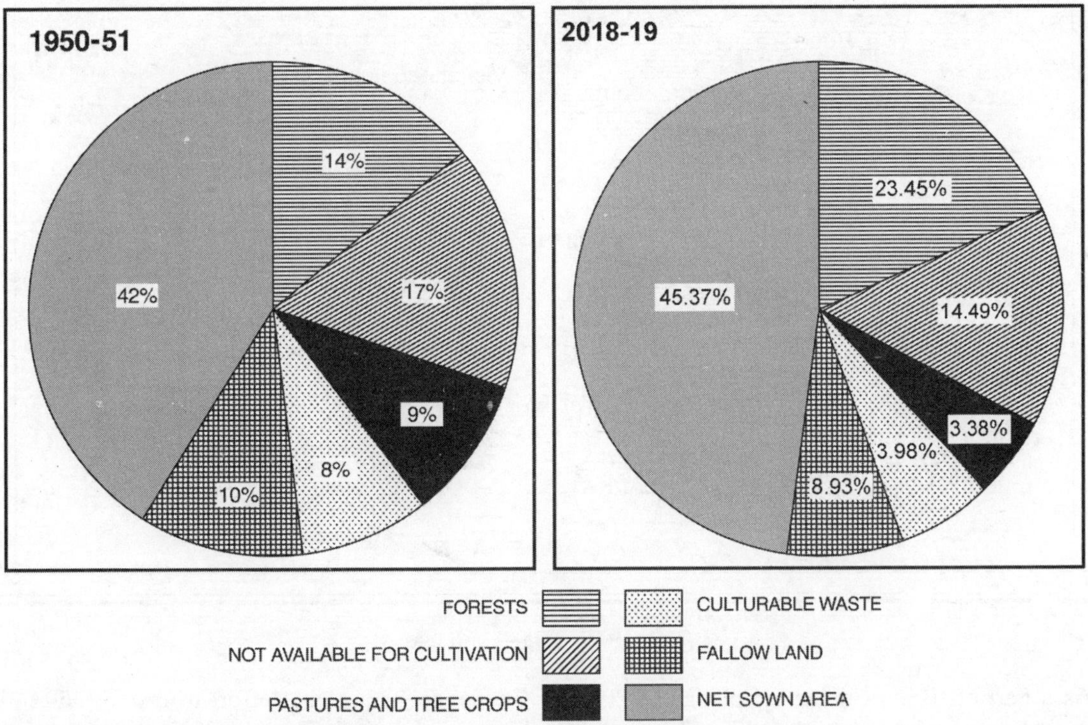

FIG. 19.1. India : Land use 1950-51 and 2018-19

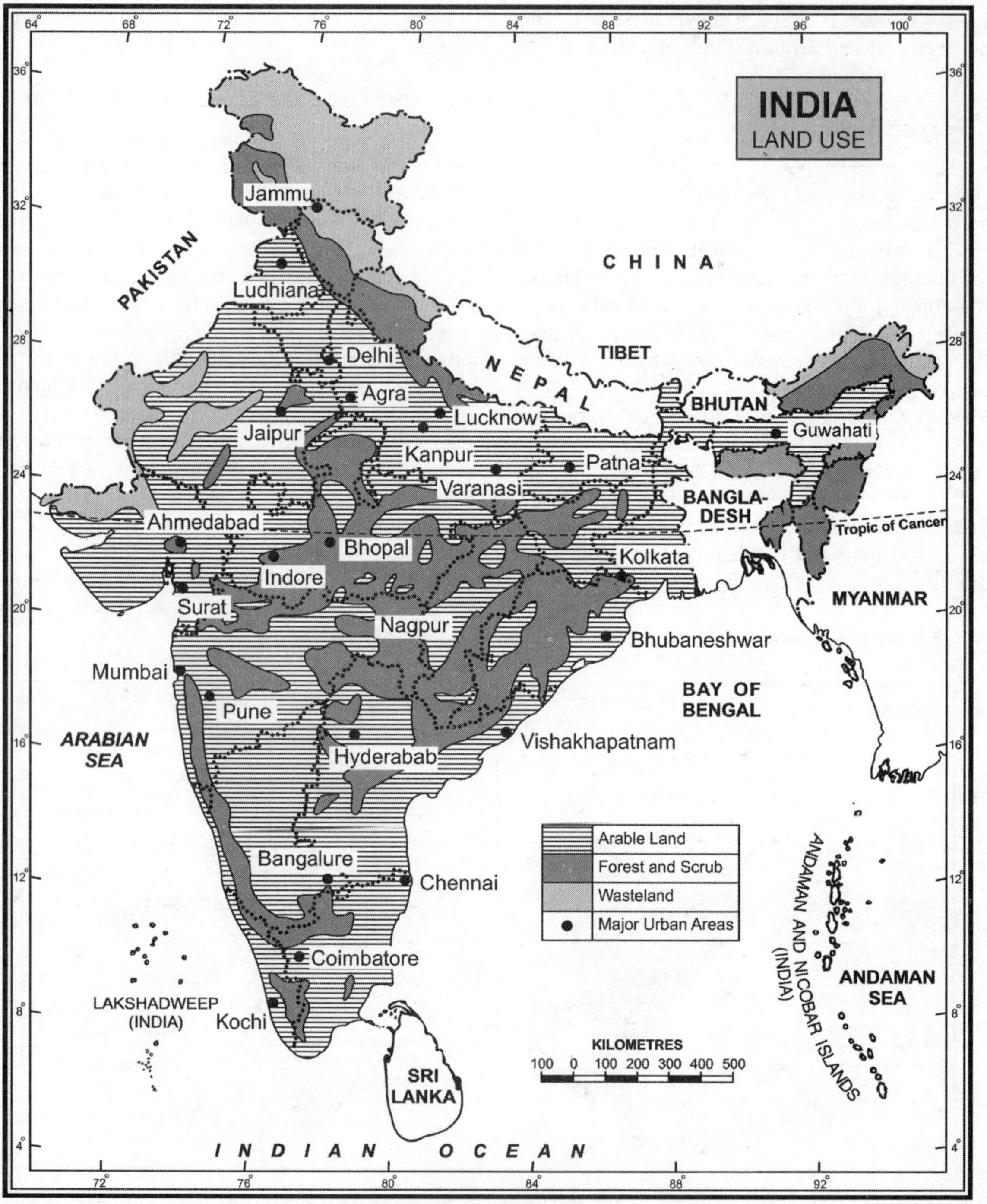

FIG. 19.2. India : Land use

As a part of 'Bonn Challenge' pledge of 2015, India is planning to restore 21 million hectares of degraded and deforested land by 2030. The 'Bonn Challenge' is a global effort to restore 350 million hectares of degraded land by 2032.

TABLE 19.2. Pattern of Land Utilisation in India (2018-19) *(thousand hectares)*

	State/Union Territory	Reporting area	Forest	Land not available for cultivation	Permanent pastures and other grazing lands	Land under Miscellaneous tree crops and groves (not included in net sown area)	Cultureable waste land	Fallow lands other than current fallows	Current fallows	Net area sown	Total cropped area
1.	Andhra Pradesh	16,297	3,688	3,402	208	155	412	936	1,446	6,049	7,297
2.	Arunachal Pradesh	7,172	6,639	62	18	34	61	61	33	234	320
3.	Assam	7,844	1,853	2,515	172	222	160	92	108	2,723	4,004
4.	Bihar	9,360	622	2,150	15	248	44	118	995	5,167	7,406
5.	Chhattisgarh	13,788	6,304	1,027	888	2	366	256	266	4,679	5,614
6.	Goa	361	125	37	1	1	53	—	16	128	150
7.	Gujarat	19,069	1,834	3,723	851	4	1,960	16	379	10,302	11,429
8.	Haryana	4,371	35	445	74	21	54	35	105	3,601	6,605
9.	Himachal Pradesh	4,577	1,125	1,133	1,502	71	120	19	65	542	914
10.	Jammu & Kashmir	4,130	2,361	549	108	63	158	36	121	713	1,124
11.	Jharkhand	7,970	2,239	1,285	127	146	380	1,086	1,427	1,281	1,824
12.	Karnataka	19,050	3,073	2,274	872	251	403	571	940	10,664	13,551
13.	Kerala	3,886	1,082	570	—	2	96	46	57	2,034	2,571
14.	Madhya Pradesh	30,756	8,700	3,619	1,316	21	964	467	493	15,205	20,115
15.	Maharashtra	30,758	5,174	3,542	1,323	272	924	1,260	1,448	16,815	19,139
16.	Manipur	2,161	1,685	27	1	6	1	0	0	441	441
17.	Meghalaya	2,196	910	275	—	165	385	153	57	255	312
18.	Mizoram	2,039	1,585	75	11	41	7	127	47	145	187
19.	Nagaland	1,653	863	112	—	73	62	111	49	384	529
20.	Odisha	15,430	5,814	2,419	523	251	591	738	1,089	4,006	4,527
21.	Punjab	5,033	250	545	4	11	12	6	86	4,119	7,851
22.	Rajasthan	34,287	2,760	4,375	1,168	26	3,784	2,106	1,789	17,778	25,313
23.	Sikkim	441	334	10	—	4	4	5	7	77	162
24.	Tamil Nadu	13,033	2,157	2,660	108	226	323	1,931	1,047	4,582	5,672
25.	Telangana	11,208	2,698	1,443	299	112	179	751	1,064	4,660	5,774
26.	Tripura	1,049	629	148	1	10	3	1	1	256	487
27.	Uttar Pradesh	24,170	1,715	3,611	70	269	388	594	987	16,538	26,859
28.	Uttarakhand	6,002	3,812	435	208	395	328	89	88	648	1,029
29.	West Bengal	8,684	1,175	1,892	2	47	13	6	301	5,248	9,960
30.	Andaman & Nicobar	757	717	9	4	4	3	3	3	15	41
31.	Chandigarh	7	0	5	—	0	0	0	0	1	1
32.	Dadra & Nagar Haveli	44	20	4	0	—	0	0	0	20	20
33.	Daman & Diu	3	0	—	0	0	0	0	0	3	3
34.	Delhi	147	1	93	0	1	10	8	12	22	59
35.	Laskshadweep	3	—	0	—	—	—	—	—	3	3
36.	Puducherry	49	0	20	—	1	5	3	4	15	27
	All India	**3,07,787**	**72,011**	**44,512**	**10,376**	**3,154**	**12,219**	**11,633**	**14,531**	**1,39,351**	**1,97,320**

Source : Agricultural Statistics at a Glance, 2021, pp. 339-45.

5. Permanent pastures and other grazing lands

A total area of 10.37 million hectares is devoted to permanent pastures and other grazing lands. This amounts to about 3.37 per cent of the total reporting area of the country. Grazing takes place mostly in forests and other uncultivated land wherever pasturage is available. The area presently under pastures and other grazing lands is not sufficient keeping in view the large population of livestock in the country. About one-third of the reporting area in Himachal Pradesh is under pastures. The proportion varies from 4 to 10 per cent in Madhya Pradesh, Karnataka, Gujarat, Rajasthan, Maharashtra and Odisha. It is less than 3 per cent in the remaining parts of the country.

6. Land under miscellaneous tree crops and groves

Land under miscellaneous tree crops and groves includes all cultivable land which is not included under net area sown, but is put to some agricultural use. Land under casuarina trees, thatching grass, bamboo, bushes, other groves for fuel, etc. which are not included under orchard are classed under this category. Land under this category declined sharply from 19.8 million hectares in 1950-51 to only 4.46 million hectares in 1960-61 and further to 4.29 million hectares in 1970-71. Thus the percentage of this land fell from 6.97 per cent in 1950-51 to a mere 1.49 per cent in 1960-61 and further to 1.41 per cent in 1970-71. After that juncture the area under miscellaneous tree crops and groves has shown varying trends and stood at 3.1 million hectares or 1 per cent of the total reporting area in 2018-19. Uttarakhand has the largest area under this categories, followed by Uttar Pradesh, Jharkhand, Bihar and Assam.

7. Culturable Waste Land

The "wasteland survey and reclamation committee" defines "culturable waste" as *the land available for cultivation but not used for cultivation for one reason or the other*. This land was used in the past but has been abandoned for some reason. It is not being used at present due to such constraints as lack of water, salinity or alkalinity of soil, soil erosion, waterlogging, an unfavourable physiographic position, or human neglect. Reh, bhur, usar, and khola tracts of Uttar Pradesh, Punjab and Haryana as well as in several other parts of the country were used for agriculture in the past but had to be abandoned due to some deficiencies in the soil resulting from faulty agricultural practices. The land under this category has declined considerably from about 22.9 million hectare in 1950-51 to 13.8 million hectare in 1999-2000. In 2018-19, the culturable waste land was 12.22 million hectares which was estimated to be about 4 per cent of the total area about which statistics were available. This decline in the wasteland is due to some land reclamation schemes launched in India after Independence. About one-sixth of the total reporting area in Goa is termed as culturable waste. Rajasthan has 3.8 million hectares of cultivable waste land which is about 37 per cent of the total waste land of India. The other states with considerable culturable waste land are Gujarat, Madhya Pradesh, Uttar Pradesh and Maharashtra. The cultivable waste, if brought under cultivation can be an important factor in augmenting the country's agricultural production. However, in the interest of long term conservation and maintenance of ecobalance, this land should be put under afforestation and not under crop farming. National Remote Sensing Agency (NRSA), Hyderabad is making valuable contribution in mapping the wastelands in India through satellite imageries.

8. Fallow lands

This category includes all that land which was used for cultivation but is temporarily out of cultivation. Fallow land is of two types *viz*., *current fallow* and *fallow other than current fallow*. Fallow of one year is called ***'current fallow'*** while that of 2 to 5 years is classified as 'fallow other than current fallow'. Fallow land is left uncultivated from 1 to 5 years to help soil recoup its fertility in the natural way depending upon the nature of soil and the nature of farming. There have been varying trends in the extent of current fallow and it amounted to 4.9 per cent of the reported area in 2014-15. But there had been a sharp decline in fallow lands other than current fallow. In the year 2018-19 about 3.8 per cent of the reported areas was found to be 'fallow land other than current fallow'. The largest area of over 12 million hectare of 'fallow land other than current fallow' is in Rajasthan. This state also has the largest area of 1.8 million hectares termed as current fallow.

There is need to reduce the extent and frequency of fallow land in order to increase agricultural production. This can be done by proper dose of fertilizers, providing irrigation facilities, crop rotation and combination and several other similar farm techniques.

REFERENCES

Agricultural Statistics at a glance, 2021. Govt. of India, Ministry of Agriculture.

Banerjee, Bireswar (1992) : *Strategies for Restoration of Wastelands in India,* in Mehdi Raza (ed.) *Development and Ecology*, Rawat Publications, Jaipur.

Bhumbla, D.R. and Khare, A. (1994) : *Estimate of Wastelands in India*, Society for promotion of Wastelands Development, New Delhi.

Classification of area and irrigated area (1966-67) : Directorate of Economics and Statistics, Ministry of Food and Agriculture, New Delhi.

Gautam, N.C. and Narayan, L.R.A. (1988) : *Wastelands in India*, Pink, Mathura.

Indian Agriculture in Brief (1986) : Directorate of Economics and Statistics, Ministry of Agriculture, New Delhi.

Negi, B.S. : *Rural Geography,* Kedar Nath Ram Nath, Meerut.

Singh, Abha Laxmi. (1985) : *The Problem of Wastelands in India*, B.R. Publishing Corporation, Delhi.

Singh, H.S. (1988) : *Land use in relation to Agro-climatic Zones in India in Proc. National Symp.* Remote Sensing in Rural Development, H.A.U., Hisar.

Singh, Jagdish (1991) : *Optimum Land use in Eco-Development Perspective* in Jaymala Diddee, *Emerging Trends in Indian Geography*, Rawat Publications, Jaipur.

Study and Survey and Reclamation of Ravines in India (1955) : Planning Commission, New Delhi.

Study of Wastelands including Saline, Alkaline and Water-logged Lands and their Reclamation Measures (1962) : Planning Commission, New Delhi.

Yadav, H.R. (1990) : *Reclaiming Wastelands*, Concept Publishing Co., New Delhi.

◆◇◆

CHAPTER 20

Agriculture

INTRODUCTION

Agriculture includes raising of crops from the land, animal husbandry, agroforestry and pisciculture. India is pre-eminently an agricultural country. Agriculture has been practised in India since times immemorial. It plays a vital role in the economy of India. Till 1971, about 80 per cent of India's population lived in rural areas and depended directly or indirectly on agriculture. It contributed about 45 per cent of Gross Domestic Production (GDP) at that time. The relative importance of agriculture has reduced considerably since then due to rapid development of other occupations such as mining, manufacturing, transport, trade and services. Today, agriculture and allied sectors contribute nearly 17.4 per cent of Gross Value Added, while about 54.6 per cent of the population is dependent on agriculture for their livelihood, and it still forms the hub of India's economy. In addition to providing food and fodderv to large population of human beings and livestock respectively, agriculture is the main source of raw materials for several key industries. Sugarcane, cotton, jute and oil seeds are some of the outstanding agricultural raw materials used in industries.

India is a unique country from agricultural point of view. Its enormous expanse of level plains, rich soils, high percentage of culturable land, wide climatic variety with adequate aggregate rainfall combined with sufficient temperature, ample sunshine and long growing season provide solid base to agriculture.

A healthy and advanced agriculture creates demand for several industrial products like tractors, harvestors, threshers, chemical fertilisers, pesticides, etc. Moreover, income generated in the agricultural sector creates ready market for various manufactured goods. Thus agriculture has double relation with industry. It acts as a supplier of raw materials to the agro-based industries and as consumer of industrial products. It goes without saying that the prosperity of industrial sector largely depends upon the agricultural prosperity. In fact, prosperity of the entire nation depends upon the prosperity of agriculture.

Agricultural sector also contributes a lot to the export trade of India. Bulk of India's export trade consists of agricultural products and agro-processed products. The major agricultural commodities of export are tea, coffee, cashew kernels, raw cotton, oil cakes, tobacco, spices, fruits and vegetables. There is great need to increase agricultural production so that sufficient exportable surplus commodities are available after meeting our domestic requirements.

From the above discussion, it can be concluded that agriculture furnishes the central sinew of Indian economy. A prosperous farmer means a prosperous nation.

SALIENT FEATURES OF INDIAN AGRICULTURE

Indian agriculture has its own peculiarities. Some of the outstanding features of Indian agriculture are mentioned as follows.

1. Subsistence agriculture. Most parts of India have subsistence agriculture. The farmer owns a small piece of land, grows crops with the help of his family members and consumes almost the entire farm produce with little surplus to sell in the market. This type of agriculture has been practised in India for the last several hundreds of years and still prevails in spite of the large scale changes in agricultural practices after Independence.

2. Pressure of population on agriculture. The population in India is increasing at a rapid pace and exerts heavy pressure on agriculture. Agriculture has to provide employment to a large section of work force and has to feed the teeming millions. While looking into the present need of food grains, we require an additional 12-15 million hectares of land to cope with the increasing demands. Moreover, there is rising trend in urbanization. As much as 31.16 per cent of the Indian population lived in urban areas in 2011 and it is estimated that over half of the total population of India would be living in urban areas by 2050 A.D. This requires more land for urban settlements which will ultimately encroach upon agricultural land. It is now estimated that about 4 lakh hectares of farm land is now being diverted to non-agricultural uses each year. Due to rapidly increasing population and fast moving urbanisation, per capita availability of cultivable land has decreased from 0.53 hectares in 1951 to 0.10 hectares in 2018-19. It is feared that this will further decreased in times to come.

3. Importance of animals. Animal force has always played a significant role in agricultural operations such as ploughing, irrigation, threshing and transporting the agricultural produce. Complete mechanisation of Indian agriculture is still a distant goal and animals will continue to dominate the agricultural scene in India for several years to come.

4. Dependent upon monsoon. Indian agricluture is mainly dependent upon monsoon which is uncertain, unreliable and irregular. In spite of the large scale expansion of irrigation facilities since Independence, less than 49 per cent of the net area sown is provided by irrigation and the remaining 51% of the cropped area has to bear the brunt of the vagaries of the monsoons.

5. Variety of crops. India is a vast country with varied types of relief, climate and soil conditions. Therefore, there is a large variety of crops grown in India. Both the tropical and temperate crops are successfully grown in India. Very few countries in the world have a variety of crops comparable to that produced in India.

6. Predominance of food crops. Since Indian agriculture has to feed a large population, production of food crops is the first priority of the farmers almost everywhere in the country. More than two-thirds of the total cropped area is devoted to the cultivation of food crops. Area under foodgrains increased from 121.05 million hectares in 2000-01 to 129.3 million hectares in 2020-21 and there is not much scope for further increase in area under foodgrains because more than 90 per cent of the net sown area is already under foodgrains.

7. Insignificant place given to fodder crops. Although India has the largest population of livestock in the world, fodder crops are given a very insignificant place in our cropping pattern. Less than four per cent of the reporting area is devoted to permanent pastures and other grazing lands. This is due to pressing demand of land for food crops. The result is that the domestic animals are not properly fed and their productivity is very low compared to international standards.

8. Seasonal pattern. India has three major crop seasons.

(*i*) ***Kharif*** season starts with the onset of monsoons and continues till the beginning of winter. Major crops of this season are rice, maize, jowar, bajra, cotton, sesamum, groundnut and pulses such as *moong*, *urad*, etc.

(*ii*) ***Rabi*** season starts at the beginning of winter and continues till the end of winter or beginning of summer. Major crops of this season are wheat, barley, jowar, gram and oil seeds such as linseed, rape and mustard.

(*iii*) ***Zaid*** is summer cropping season in which crops like rice, maize, groundnut, vegetables and fruits are grown. Now some varieties of pulses have been evolved which can be successfully grown in summer.

9. Mixed Cropping. Mixed cropping is one of the chief characteristics of Indian agriculture particularly in the rain-fed areas. Sometimes four to five crops are grown simultaneously in the same field and in areas of Jhuming (shifting agriculture) ten to fifteen crops are mixed in one field. The popular crops are millets, maize and pulses in the *kharif* season and wheat, gram and barley in the *rabi* season. This is done to ensure good agricultural production keeping in view the vagaries of the monsoon rainfall and uncertain weather conditions. If the amount of rainfall is good, rice crop will give better output and if there is failure of the moonsoon rains, then less water requiring crops such as maize, millets and pulses will give better yields. Mixed cropping is a characteristic of subsistent agriculture and this practice reduces the overall agricultural output and per hectare yield.

10. High percentage of reporting area under cultivation. In the year 2018-19, 139.35 million hectares was the net sown area out of total reporting area of 307.8 million hectares. Thus nearly 45.3 per cent of the total reporting area is under cultivation. This is a very high percentage when compared to some of the advanced countries like U.S.A., Japan, and Canada.

11. Labour intensive. In large parts of India, agriculture is labour intensive as most of agricultural operations like ploughing, levelling, sowing, weeding, pruning, sprinkling, spraying, harvesting, threshing, etc. are done by the farmers and their animals. Mechanisation of farming is prevalent in Punjab, Haryana, and western part of Uttar Pradesh and in these area too, it is the privilege of the rich farmers only. Farm mechanism is picking up in Uttarakhand, Gujarat and Maharashtra also where limited areas use farm machinery.

PROBLEMS OF INDIAN AGRICULTURE AND THEIR SOLUTIONS

Indian agriculture is plagued by several problems; some of them are natural and some others are man-made. Some of the major problems and their possible solutions have been discussed as follows.

1. Small and Fragmented Land-holdings. The seemingly abundance of net sown area of 139.35 million hectares and total cropped area of 197.3 million hectares (2018-19) pales into insignificance when we see that it is divided into economically unviable small and scattered holdings. The average size of holdings was 2.28 hectares in 1970-71 which was reduced 1.08 hectares in 2015-16. The size of the holdings will further decrease with the infinite sub-division of the land holdings. The problem of small and fragmented holdings is more serious in densely populated and intensively cultivated states like Kerala, West Bengal, Bihar and eastern part of Uttar Pradesh where the average size of land holdings is less than one hectare and in certain parts it is less than even 0.5 hectare. It has been estimated that over two-thirds of the landholdings are less than one hectare and only 0.57 per cent are over 10.0 hectares in size.

The main reason for this sad state of affairs is our inheritance laws. The land belonging to the father is equally divided among his sons. According to new inheritance law, even daughters are entitled to share the father's property. This distribution of land does not entail a collection or consolidated one, but its nature is fragmented. Different tracts have different levels of fertility and are to be distributed accordingly. If there are four tracts which are to be distributed between two sons, both the sons will get smaller plots of each land tract. In this way the holdings become smaller and more fragmented with each passing generation.

Sub-division and fragmentation of the holdings is one of the main causes of our low agricultural productivity and backward state of our agriculture. A lot of time and labour is wasted in moving seeds, manure, implements and cattle from one piece of land to another. Irrigation becomes difficult on such small and fragmented fields. Further, a lot of fertile agricultural land is wasted in providing boundaries. Under such circumstances, the farmer cannot concentrate on improvement.

The only answer to this ticklish problem is the *consolidation of holdings* which means the reallocation of holdings which are fragmented, the creation of farms which comprise only one or a few parcels in place of multitude of patches formerly in the possession of each peasant. But unfortuantely, this plan has not succeeded much. Although legislation for consolidation of holdings has been enacted by almost all the states, it has been implemented only in Punjab, Haryana and in some parts of Uttar Pradesh.

2. Seeds. Seed is a critical and basic input for attaining higher crop yields and sustained growth in agricultural production. Distribution of assured quality seed is as critical as the production of such seeds. Unfortunately, good quality seeds are out of reach of the majority of farmers, especially small and marginal farmers mainly because of exorbitant prices of better seeds. In order to solve this problem, the Government of India has taken several steps so that quality seeds are made available to farmers in sufficient quantity at reasonable prices. But the benefit of schemes launched by the government still remain out of reach of the small and marginal farmers.

3. Manures, Fertilisers and Biocides. Indian soils have been used for growing crops over thousands of years without caring much for replenishing. This has led to depletion and exhaustion of soils resulting in their low productivity. The average yields of almost all the crops are among the lowest in the world. This is a serious problem which can be solved by using more manures and fertilisers. Manures and fertilisers play the same role in relation to soils as good food in relation to body. Just as a well-nourished body is capable of doing any good job, a well-nourished soil is capable of giving good yields. It has been estimated that about 70 per cent of growth in agricultural production can be attributed to increased fertiliser application. Thus increase in the consumption of fertilisers is a barometer of agricultural prosperity. However, there are practical difficulties in providing sufficient manures and fertilisers in all parts of a country of India's dimensions inhabited by poor peasants. Cow dung provides the best manure to the soils. But its use as such is limited because much of cow dung is used as kitchen fuel in the shape of dung cakes. Reduction in the supply of fire wood and increasing demand for fuel in the rural areas due to increase in population has further complicated the problem. Chemical fertilisers are costly and are often beyond the reach of the poor farmers. The fertiliser problem is, therefore, both acute and complex.

It has been felt that organic manures are essential for keeping the soil in good health. The country has a potential of 650 million tonnes of rural and 16 million tonnes of urban compost which is not fully utilized at present. The utilization of this potential will solve the twin problem of disposal of waste and providing manure to the soil.

The government has given high incentive especially in the form of heavy subsidy for using chemical fertilisers. There was practically no use of chemical fertilisers at the time of Independence. As a result of initiative by the government and due to change in the attitude of some progressive farmers, the consumption of fertilisers increased tremendously.

Pests, germs and weeds cause heavy loss to crops which amounted to about one-third of the total field produce at the time of Independence. Biocides (pesticides, herbicides and weedicides) are used to save the crops and to avoid losses. The increased use of these inputs has saved a lot of crops, especially the food crops from unnecessary wastage. But indiscriminate use of biocides has resulted in wide spread environmental pollution which takes its own toll.

4. Irrigation. Although India is the second largest irrigated country of the world after China, only 49 per cent of the total cropped area is under irrigation. Irrigation is the most important agricultural input in a tropical monsoon country like India where rainfall is uncertain, unreliable and erratic. India cannot achieve sustained progress in agriculture unless and until more than half of the cropped area is brought under assured irrigation. This is testified by the success story of agricultural progress in Punjab, Haryana and western part of Uttar Pradesh where over half of the cropped area is under irrigation. Large tracts still await irrigation to boost the agricultural output. However, care must be taken to safeguard against ill effects of over irrigation especially in areas irrigated by canals. Large tracts in Punjab and Haryana have been rendered useless (areas affected by salinity, alkalinity and waterlogging), due to excessive irrigation. In the Indira Gandhi Canal command area also intensive irrigation has led to sharp rise in sub-soil water level, leading to waterlogging, soil salinity and alkalinity.

5. Lack of Mechanisation. In spite of the large scale mechanisation of agriculture in some parts of the country, most of the agricultural operations in larger parts are carried on by human hand using simple and conventional tools and implements like wooden plough, sickle, etc. Little or no use of machines is made in ploughing, sowing, irrigating, thinning and pruning, weeding, harvesting, threshing and transporting the crops. This is specially the case with small and marginal farmers. It results in huge wastage of human labour and in low yields per capita labour force.

The overall farm mechanization in India is only 40 per cent which is very low compared to 60 per cent in China, 75 per cent in Brazil and 95 per cent in the U.S.A. A study by NABARD (National Bank of Agriculture and Rural Development) in 2018 on farm mechanization has identified that economies of operation due to small holdings, access to power, credit cost and procedures, uninsured markets and low awareness being some of the important reasons for low rate of agricultural mechanization. The mechanization levels for major crops in India are shown in **table 20.1**.

TABLE 20.1. Mechanization levels for major crops (field operations) in India 2018-19 (in per cent)

Crop	Seed bed preparation	Sowing/ planting/ transporting	Weeding and interculture and plant protection	Harvesting and threshing
1. Rice	70	20	30	60
2. Wheat	70	60	50	70
3. Maize	60	40	30	30
4. Sorghum and millets	50	30	15	10
5. Pulses	50	40	20	25
6. Oilseeds	50	40	20	25
7. Cotton	50	30	25	0
8. Sugarcane	55	10	20	10

Source : Department of Agricultural Research and Education (DARE) in Economic Survey, 2019-20, p. 198.

There is urgent need to mechanise the agricultural operations so that wastage of labour force is avoided and farming is made convenient and efficient. Agricultural implements and machinery are a crucial input for efficient and timely agricultural operations, facilitating multiple cropping and thereby increasing production.

Some progress has been made for mechanising agriculture in India after Independence. Need for mechanisation was specially felt with the advent of Green Revolution in 1960s. Strategies and programmes have been directed towards replacement of traditional and inefficient implements by improved ones, enabling the farmer to own tractors, power tillers, harvesters and other machines. Strenuous efforts are being made to encourage the farmers to adopt technically advanced agricultural equipments in order to carry farm operations timely and precisely, and to economise the agricultural production process.

6. Agricultural Marketing. Agricultural marketing still continues to be in a bad shape in rural India. In the absence of sound marketing facilities, the farmers have to depend upon local traders and middlemen for the disposal of their farm produce which is sold at throw-away price. In most cases, these farmers are forced, under socio-economic conditions, to carry on distress sale of their produce. In most of small villages, the farmers sell their produce to the money lender from whom they usually borrow money. According to an estimate 85 per cent of wheat and 75 per cent of oil seeds in Uttar Pradesh, 90 per cent of jute in West Bengal, 70 per cent of oilseeds and 35 per cent of cotton in Punjab is sold by farmers in the village itself. Such a situation arises due to the inability of the poor farmers to wait for long after harvesting their crops. In order to meet his commitments and pay his debt, the poor farmer is forced to sell the produce at whatever price is offered to him. The Rural Credit Survey Report rightly remarked that the producers in general sell their produce at an unfavourable place and at an unfavourable time and usually they get unfavourable terms. In the absence of an organised marketing structure, private traders and middlemen dominate the marketing and trading of agricultural produce. The remuneration of the services provided by the middlemen increases the load on the consumer, although the producer does not derive similar benefit.

Many market surveys have revealed that middlemen take away about 48 per cent of the price of rice, 52 per cent of the price of groundnuts and 60 per cent of the price of potatoes offered by consumers.

In order to save the farmer from the clutches of the money lenders and the middle men, the government has come out with regulated markets. These markets generally introduce a system of competitive buying, help in eradicating malpractices, ensure the use of standardised weights and measures and evolve a suitable machinery for settlement of disputes thereby ensuring that the producers are not subjected to exploitation and receive remunerative prices.

7. Inadequate Storage Facilities. Storage facilities in the rural areas are either totally absent or grossly inadequate. Under such conditions the farmers are compelled to sell their produce immediately after the harvest at the prevailing market prices which are bound to be low. Such distress sale deprives the farmers of their legitimate income.

The Parse Committee estimated the post-harvest losses at 9.3 per cent of which nearly 6.6 per cent occurred due to poor storage conditions alone. Scientific storage is, therefore, very essential to avoid losses and to benefit the farmers and the consumers alike. At present there are number of agencies engaged in warehousing and storage activities. *The Food Corporation of India* (F.C.I.), the *Central Warehousing Corporation* (C.W.C.) and *State Warehousing Corporations* are among the principal agencies engaged in this task. These agencies help in building up buffer stock, which can be used in the hour of need. The Central Government is also implementing the scheme for establishment of national *Grid of Rural Godowns* since 1979-80. This scheme provides storage facilities to the farmers near their fields and in particular to the small and marginal farmers. The Working Group on additional storage facilities in rural areas has recommended a scheme of establishing a network of *Rural Storage Centres* to serve the economic interests of the farming community.

8. Inadequate Transport. One of the main handicaps with Indian agriculture is the lack of cheap and efficient means of transportation. Even at present there are lakhs of villages which are not well connected with main roads or with market centres. Most roads in the rural areas are *Kutcha* (bullock-cart roads) and become useless in the rainy season. Under these circumstances the farmers cannot carry their produce to the main market and are forced to sell it in the local market at low price. Linking each village by metalled road is a gigantic task and it needs huge sums of money to complete this task.

9. Scarcity of Capital. Agriculture is an important industry and like all other industries it also requires capital. The role of capital input is becoming more and more important with the advancement of farm technology. *Since the agriculturists' capital is locked up in his lands and stocks*, he is obliged to borrow money for stimulating the tempo of agricultural production. The main suppliers of money to the farmer are the money-lenders, traders and commission agents who charge high rate of interest and purchase the agricultural produce at very low price. *All India Rural Credit Survey Committee* showed that in 1950-51 the share of money lenders stood at as high as 68.6 per cent of the total rural credit and in 1975-76 their share declined to 43 per cent of the credit needs of the farmers. This shows that the money lender is losing ground but is still the single largest contributor of agricultural credit. Rural credit scenario has undergone a significant change and institutional agencies such as *Central Cooperative Banks, State Cooperative Banks, Commercial Banks, Cooperative Credit Agencies* and *some Government Agencies* are extending loans to farmers on easy terms.

10. Tradition Bound. In spite of some breakthroughs, Indian agriculture remains tradition bound even at the dawn of 21st century. Established centuries ago, the structures of a self-contained rural economy, founded in caste-based occupational land tenure, made complex by absentee and parasitic landlords still continues. The tradition bound institutions have been the greatest hindrance in the way of modernisation and Indian agriculture has been rather slow in responding to new innovative ideas.

11. Primitive Technology. A large proportion of Indian farmers use primitive technology which hinders the requisite progress in agricultural production. They are hand tools like sickle, hoe, etc. and drought animals like bullocks, male buffaloes, camels, etc. as source of motive power in agricultural operations.

Although agricultural machinery is replacing the animal and human power, yet the pace of progress is very slow and use of agricultural machinery is the privilege of a few rich farmers in selected states like Panjab, Haryana, Uttar Pradesh only.

12. Dependent on Monsoon Rainfall. In large parts of India irrigation facilities are either totally absent or are only partially available and agriculture depends on monsoon rainfall. Unfortunately Indian monsoon rainfall is highly erratic and least dependable. It varies in time and space and variability of rainfall is the highest in areas of least rainfall. Whenever rain fails or there is deficiency of rainfall, the agricultural production drops to a miserably low level. There is overall scarcity of agricultural products in the market and the prices of agricultural products reach sky high. In extreme cases famine conditions prevail and humans and livestock die of hunger and starvation.

13. Lack of Crop Diversification. Crop diversification means growing a large number of crops and reducing dependency on a single crop. Unfortunately in India more emphasis is laid on food crops and other crops are given a secondary status. Although top priority to food crops is necessary in the back drop of fast growing population, neglecting other crops is detrimental to balanced growth of agriculture. Only 3.3 per cent of the reported area is under fodder crops which is very insignificant in view of the fact that India has the largest number of livestock in the world. Further, nitrogen fixing leguminous crops are ignored which leads to imbalance in the composition of soil and reduction in the soil fertility.

A diversified cropping pattern helps in mitigating risks faced by farmers in terms of price shocks and production/harvest losses. With 9.6% of the global net cropland area, India has tremendous potential for crop diversification and to make farming a sustainable profitable economic activity.

Index of Crop Diversification for major states and All India has be computed on the basis of the following formula used by Gibbs and Martin (Economic Survey 2017-18 Vol. 2 p. 104)

$$\text{Index of Crop Diversification} = 1 - [\sum x^2/(\sum x)^2]$$

where x is the percentage of total cropped area under an individual crop.

The index value ranges between 0 and 1 and higher the value, greater the diversification. It has been observed that there has been a declining inter-temporal behaviour crop diversification in Chhattisgarh, Haryana, Madhya Pradesh, Odisha, Punjab, and Uttar Pradesh. Among these states Odisha has shown a sharp decline in index of cropping pattern from 0.740 in 1994-95 to 0.340 in 2014-15. This state has about 80% of the cropped area under rice, around 10% under pulses and nearly 4% under other food crops. In Punjab, wheat and paddy cover about 83% of the cultivable area. The issues related to monoculture as witnessed in Odisha are declining productivity, lower fertiliser response ratio, degradation of soil health and declining profitability of cultivation. The crop diversification scenario for India as a whole appears to be almost stable from 1994-95 to 2020-21, data for which have been processed.

There is an urgent need to diversify into high value crops and horticulture crops for which Govt. has taken several measures. Crops Diversification Programme is being implemented by the Government in original green revolution states *viz.*, Punjab, Haryana and Western Uttar Pradesh to diversify paddy area towards less water requiring crops like oilseeds, pulses, coarse cereals, agro-forestry and shifting of tobacco farmers to alternate crops/cropping system in tobacco growing states *viz.*, Andhra Pradesh, Bihar, Gujarat, Karnataka, Maharashtra, Odisha, Tamil Nadu, Telangana, Uttar Pradesh and West Bengal.

14. Low Productivity. In spite of the rapid strides made by India in agricultural field, particularly after the advent of the Green Revolution in 1960s, agricultural productivity in India still remains at a low level. Yield per hectare of almost all the crops is much lower as compared to international standards. This is due to low fertility of soil and little care to replenish is through fertilisers, green manure, fallowing, crop rotation, etc. Other inputs like machinery, irrigation, better seeds etc. are also limited to a few selected areas and to a few rich farmers.

15. Government Apathy. Indian agriculture has been the victim of negligence and step motherly treatment by the government. Although agriculture got highest priority in the First Five Year Plan in view of the acute shortage of foodgrains immediately after partition of the country in 1947, agriculture has not been given its due importance and more emphasis was laid on industrial growth in the subsequent plans. Farmers do not get remunerative price of their products and most of them permanently remain under debt. Even in some of the so-called rich areas from agriculture point of view like Punjab and Maharashtra, a large number of farmers have committed suicide out of distress and depression. However, the government has become slightly more sensitive to the problems of the farmers and some recognition has been given to agriculture in Tenth and Eleventh Five Year Plans. The present government is more sensitive to the agricultural problems and several steps have been taken to address the problems faced by the farmers.

16. Lack of Definite Agricultural Land use Policy. There is no definite policy concerning agriculture and land use at the national or regional level and the farmers grow one or the other crop at their own sweet will. It often leads to excess or scarcity of particular crops. In the event of excess crop the farmers are forced to sell their produce at throw away prices. On the contrary consumers are the main sufferers when there is shortage of a particular crop.

17. Low Fertility of Soils. Indian soils have been used for cultivation for the last hundreds of years without much care to restore their fertility. Most of the Indian soils are exhausted and are not capable of giving high yields. They lack in various chemicals and humus which are necessary for high rate of productivity in the agricultural field.

18. Soil Erosion and Soil Degradation. Wrong agricultural practices coupled with reckless felling of trees has led to large scale soil erosion and soil degradation both by water and by wind. Rain water washes away huge amounts of fertile top soil in areas of heavy rainfall during the rainy season and in areas of scanty rainfall, strong winds blow away the fertile top soil.

19. Low Status of Agriculture in Society. In large parts of India agriculture is not given its due place of honour and is considered to be a profession of low status. This leads to disappointment and lack of enthusiasm amongst farmers. Younger generation belonging to families of farmers are no more interested in agricultural profession and tend to opt for petty jobs in government offices. Besides, rich farmers invest their agricultural profits in more lucerative non-agricultural sectors. Rural youth migrate to urban areas in search of non-agricultural or white colored jobs and many of them end up in slums, ghettoes and shanty colonies.

20. Land Tenancy. In many parts of India the actual tillers are not the owners of land and they are forced to till the land of absentee landlords. There are big landlords who own vast stretches of land but do not till the land themselves. The poor landless tanent cultivators do not take much interest in the development of agriculture as a result of which the yields of almost all the crops are at a miserably low level.

21. Lack of Agricultural Research, Education, Training and Extension Services. Although a number of research institutions were established immediately after Independence, and many advancements in agricultural research have been made since then, yet agricultural research hardly matches international standards. Further there is lack of coordination between research laboratories and farms and there is a big vacuum between the two. Farmers, especially small and marginal farmers, are deprived of the benefit of the new findings of research laboratories. In a similar way hardly any attention is paid to educate the farmers about the new techniques of agriculture for increasing the farm production. There is need to raise an army of trained and dedicated workers which can help the farmers in coming out of the age old tradition bound agriculture and adopt new innovative ideas and enhance their production and income.

Farm Suicides. Due to the above mentioned drawbacks in Indian agriculture, farmers, especially small and marginal farmers, remain under debt as they do not find agriculture, as it is, a profitable business. Once the farmer takes loan, especially from the moneylender in the village at high rate of interest, he is unable to pay back the debt and is forced to commit suicide. The number of farmers committing suicide is increasing with the passage of time. As many as 20 thousand farmers committed suicide in 2018. Maharashtra reported the largest number of 8,148 farmers who committed suicide in 2018. This was followed by Telangana (2,888), Karnataka (2,730), Madhya Pradesh (2,006) and Chhattisgarth (1,882).

DETERMINANTS OF AGRICULTURE

Agriculture in India is determined by a set of factors. Some of the important factors are listed below.

1. Physical factors. Relief, climate, soil.

2. Institutional factors. Size of farm holdings, land tenure, land reforms.

3. Infrastructural factors. Irrigation, power, roads, credit, marketing, insurance and storage facilities.

4. Technological factors. High yielding varieties of seeds, chemical fertilisers, insecticides and farm machinery.

The above mentioned factors affect the level of agricultural development, cropping patterns and agricultural productivity in a region. A brief review of these factors is given below.

1. PHYSICAL FACTORS

Man's agricultural activities depend to a great extent on the physical environment in which he lives although he often tries to minimize the restrictions imposed by various factors of physical environment. For example, he has been able to extend agriculture to arid and semi-arid lands with the help of irrigation, restore soil fertility by using manures and chemical fertilisers and save crops from pests and insects by applying pesticides and insecticides. But these ventures have a limited effect and physical environment has a much deeper impact on agricultural development and agricultural productivity. This is amply clear from the following description regarding the impact of relief, climate, and soils on agriculture.

1. Relief

Relief is the difference of height between the lowest and the highest point in the region and is expressed in terms of height above mean sea level and gradient of slope.

Effect of Altitude on Agriculture. There is gradual decrease in temperature with increasing atitude which affects agricultural productivity, types of crops and agricultural operations. Depending on altitude following land features and associated crops can be identified.

(*a*) Plains. Plains are the most preferred areas for agriculture due to their low altitude and gentle slope. Agricultural operations like ploughing, sowing, weeding and thinning, harvesting, etc. are much easier in plain areas than in plateaus or mountains. Plains are usually blessed with fertile soils which offer greater opportunities for agricultural growth. This is the reason the Great Plain of North India is an area of intensive agriculture where rice and wheat are the major food crops. Besides, a large number of commercial crops like cotton, jute, sugarcane, pulses, coarse grains etc. are also grown. In the coastal plains rice is the main crop and orchards of coconut are found all along in coast.

(*b*) Plateaus. Plateaus are also useful for agriculture but their suitability for agriculture is less as compared to that of plains. The peninsular plateau of India provides suitable conditions for cultivation of a number of crops including rice, wheat, cotton, sugarcane, tobacco and some plantation crops. However, only selected parts of this vast plateau are suitable for cultivation and several parts of this plateau are characterized by rugged topography, dissected surface, steep scarps, shallow soils, bare rocks and intervening hill slopes.

(*c*) Mountains. Mountains are least preferred from agricultural point of view because of high altitude, steep slope, immature soils, low temperature, low pressure, rarified air and shortage of oxygen. Temperature decreases regularly with altitude which determines the types of crops grown on different altitudes and practically no agriculture is possible above 3500 m in the Himalayas. Sher Singh Dhillon (1973 : 75) has identified six fold altitudinal agro-climatic zones based on seasonal rhythm and temperature involving different agricultural practices in the Western Himalayas (**Fig. 20.1**). On the slopes of the Eastern Himalayas and those of the Nilgiri Hills of South India, the plantations dominate the agricultural scene. In South India, coffee plantations are also seen along with tea plantations. In several hilly areas, hill slopes are used for terraced cultivation wherein rice and other crops are grown.

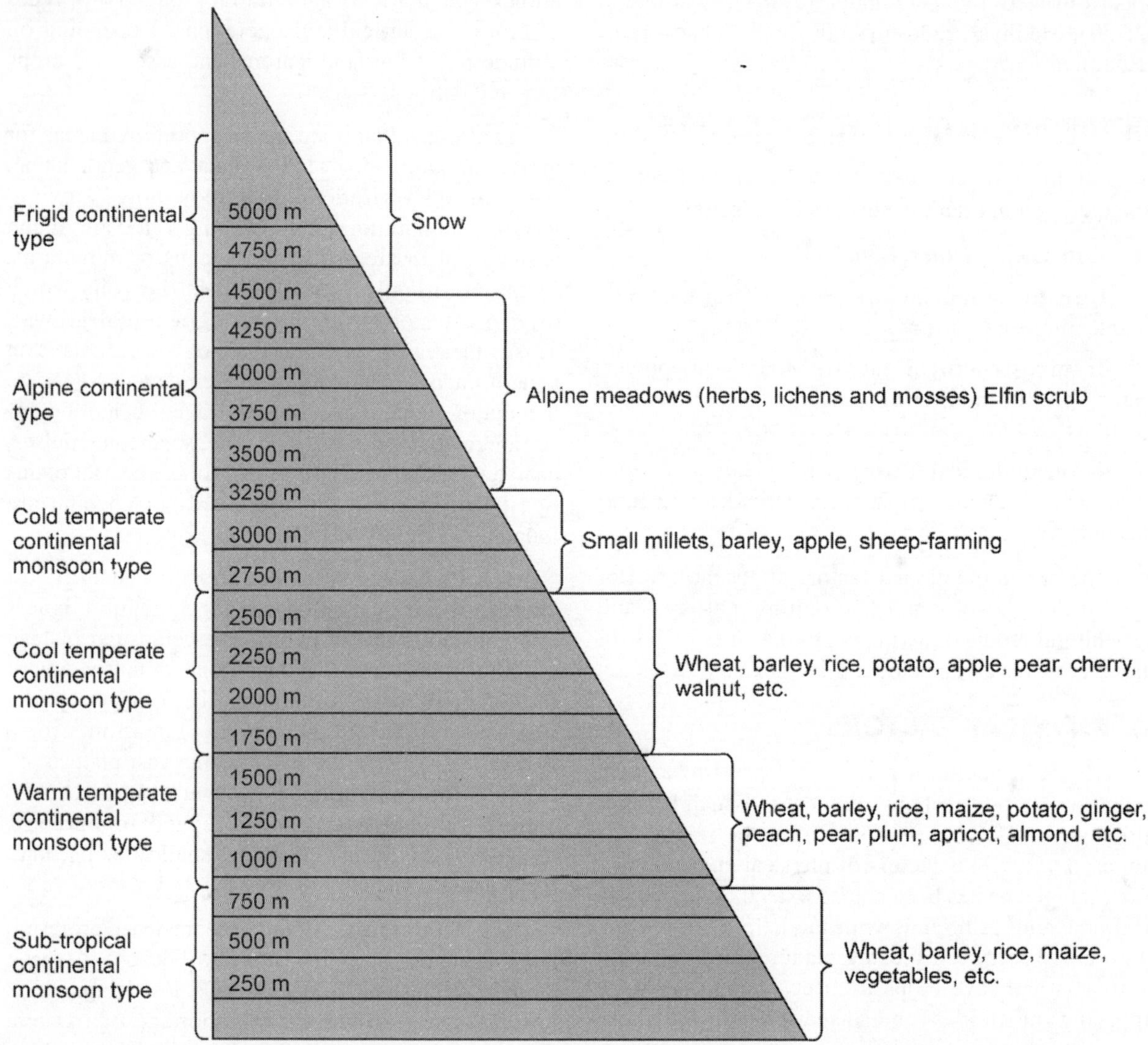

FIG. 20.1. Altitudinal agroclimatic zones in the Western Himalayas (after Dhillon)

2. Climate

Of all the physical factors, climate is the most potent determinant of agricultural land use and cropping pattern. Various elements of climate like temperature, humidity, sun shine and winds influence the agricultural activities in their own respective ways.

I. Temperature. Temperature is a dominant factor which determines the growth of plants. In the absence of suitable temperature conditions germination of seeds and growth of plants are retarded. Temperature regulates all the chemical and physical processes of plant metabolism. The metabolic processes begin at certain minimum temperature and increase with rise in temperature until they reach a maximum at a temperature. which is known as ***optimum temperature.*** Further rise in temperature above the optimum level leads to slow down in metabolic activity until it ceases at a temperature called the ***maximum.*** Each species has its own minimum and maximum temperature beyond which its life activity ceases (Kochhar 1967). Each species has also the optimum temperature at which the plant growth is maximum. The minimum, the maximum and the optimum temperature required by plant species are known as ***cardinal temperature points***. **Wilsie** (1962) has suggested cardinal temperature points for some selected crops which are reproduced in **table 20.2**.

Each and every crop needs a certain number of a effective heat units for germination, growth, stalking, maturing and ripening. This is known as ***thermal constant*** which varies from crop to crop. Therefore, temperature above the minimum is effective in furthering the growth of a plant towards maturing and ripening. According to **Schimper** the crucial air temperature is 6°C above which the plants usually start growing. The ideal temperature conditions for crop production are from 18.3°C to 23.9°C.

The above mentioned temperatures conditions affecting different crops are reflected in the areal distribution of crops in India. Low winter temperatures in the N.W. India encourage in cultivation of wheat crop while major rice producing areas are confined to southern and eastern parts of India. However, rice cultivation has become popular in Punjab, Haryana and Uttar Pradesh also due to availability of irrigation facilities. Sugarcane is grown both in North and South India although temperature conditions are more congenial in South India. This is the reason that sugarcane cultivation is slowly shifting from North India to South India. It is because of temperature difference that wheat is grown as a winter crop in Punjab and Haryana and as a summer crop in Ladakh because winter temperature in Ladakh is too low to permit the growth of any crop.

TABLE 20.2. Cardinal temperature points for selected crops (°C)

Crops	Minimum Temperature	Optimum Temperature	Maximum Temperature
Rice	10-12	30-32	36-38
Wheat	4-5	25	30-32
Sugarbeet	4-5	25	28-30
Maize	8-10	31.5-35	38-40
Lentil	4-5	30	36
Tobacco	12.7-14	28	35
Sugarcane	21.1	32.2-37.7	38.5

Source : Wilsie (1962).

II. Frost. Sudden frost is enemy number one for a number of crops such as cotton and sugarcane. Such crops need frost free season which falls between the end of spring and the beginning of autumn. This period is generally recognized as ***growing season*** which is different for different crops. Statistically, frost-free season refers to all those days when the temperature continuously remains above the freezing point. The correct information of such a period helps farmers to plan their crops and also to have an idea of the extent of damage that could occur. In South India, there is no danger to crops from frost but in North India, the winter temperatures are very low and frost is common phenomenon.

III. Winds. Winds affect the growth of plants both directly and indirectly. The direct effect of strong winds is entirely of mechanical nature. In the months of May and June, strong winds raise a lot of dust and trees are uprooted, or leaves are stripped off and branches and stems of plants twist and break. A lot of raw fruits, such as mangoes, fall down and are bruished. Banana plantations need extra care, particularly when the plantations have borne fruit. Fruit trees are much damaged in the coastal regions whenever tropical cyclones originating in the Bay of Bengal and the Arabian Sea hit these areas.

The indirect effect of winds is apparent on plants physiology. Transpiration from plants adds moisture to the surrounding air. Crops gradually dry up and the moisture diffuses into the surrounding atmosphere. The locally saturated atmosphere retards further transpiration and promotes the plant growth. However, if the wind is dry and strong, it will deprive the plants of its moisture and may cause heavy damage to plants.

Winds are also great agent of erosion, transportation and deposition of soil. In dry areas winds carry soil to such an extent that, sometimes, basal rock is visible and the land becomes unfit for cultivation. Such a phenomenon can often be observed in Rajasthan and the adjoining areas of Haryana and Gujarat.

IV. Snow. Snow has a deep impact on livestock and cropping in the higher reaches of the Himalayan region. Snowy winters in these areas cause a severe loss to pastures due to thick snow cover on the grains. Hill sheep farming is adopted to snow conditions and flocks are moved to safer places. **Transhumance** is a common phenomenon in Jammu and Kashmir and in Himachal Pradesh. In these areas, shepherds move down slopes along with their flocks in winter and up slopes in summer to make use of pasture lands. Winter snow blocks transport routes of all kinds and agricultural fields as well markets become inaccessible. All agricultural activities remain suspended and are resumed only with the onset of summer.

Snow has a number of benefits also for agriculture. It provides cool climate to orchards, especially those of apples at sufficiently lower altitudes. Snow melt in summer provides large quantities of water for irrigation and for other purposes in daily life. Some of the snow melt water is available in the form of springs which is very rich in mineral contents and acts as fertiliser to most of the cereal crops. The usual snow cover over standing potato fields in Himachal Pradesh makes farmers happy. The local adage "A good quantum of snow means a bumper potato crop in hand'' still holds goods.

V. Humidity or Moisture. Humidity is of paramount importance as all crops need moisture for their proper growth. It has already been mentioned that for every crop, there is optimum temperature. Similarly for each crop there is optimum moisture. Maintenance of optimum moisture in the soil is very important because plants draw moisture from the soil through root system. Excessive moisture reduces the quantity of oxygen in the soil and increases the formation of compounds that are toxic to plant roots. Such a situation leads to stunted growth of plants. If the surface water goes to the subsoil in large quantities, it will deplete plant nutrients and hinder the development of plants. This problem can be solved by proper drainage of the ill-drained areas. On the other extreme, there can be shortage of moisture in the soil due to scarcity of rainfall which can be compensated by irrigation.

VI. Rainfall. Rainfall is the single most dominant climatic element influencing the intensity and location of farming systems and farmer's choice of enterprises. It also becomes a climatic hazard to farming if it is characterized with scantiness, concentration, intensity, variability and unreliability.

The amount of annual rainfall has a great impact on the cropping pattern in different parts of the country. For example rice is the main crop in areas of heavy rainfall whereas wheat is an important crop in areas of lesser rainfall. In fact, isohyet (line joining places of equal rainfall) of 100 cm roughly divides the rice and wheat producing areas in India. Areas with more than 100 cm annual rainfall are predominantly rice producing areas and areas with less than 100 cm annual rainfall are mainly wheat producing regions. In areas of scanty rainfall of less than 50 cm annual rainfall, crops requiring lesser amount of water are grown. Jowar, bajra, maize, etc. are such crops.

In addition to total amount of annual rainfall, seasonal distribution of rainfall is also very important. Most of the crops need sufficient amount of rainfall during the period of growth and practically dry season at the time of ripening. For example, rainfall in the month of September in northern India is extremely important for the cultivation of rice crop. Winter rainfall in north-western part of India is very useful for wheat crop. Ragi crop is adversely affected in south India if there is no rainfall in the months of October and November. Drought conditions prevail in the event of less than normal rainfall. Drought leads to crop failure and famine. States of Rajasthan, Gujarat, Madhya Pradesh, Chhattisgarh, Jharkhand, Maharashtra, Andhra Pradesh, Telangana, Karnataka and parts of Punjab, Haryana, Uttar Pradesh, Bihar, Tamil Nadu etc. are often affected by drought conditions.

Rainfall effectiveness is as important as amount of total rainfall received in a particular area. It is usually expressed as the actual total rainfall minus the total possible evaporation. Thus the entire amount of water received from rainfall is not available to plant and it is in this context that rainfall effectiveness becomes import. Rainfall effectiveness depends on the total amount of rainfall, as well as temperature and moisture in the air and the wind velocity. It increases with the amount of rainfall and moisture present in air and decreases with increase in air temperature and wind velocity. Some water evaporates from the leaves of the plants. This is known as ***potential evapo-transpiration.*** Therefore, while assessing the rainfall effectiveness, total amount of rainfall, evaporation and potential evapotrans-piration have to be taken into account.

VII. Sunlight and Sunshine. Sunlight and sunshine are important to plants in two different ways. (*i*) they are useful in photosynthesis and (*ii*) they help in production of chlorophyll. Photosynthesis involves production of special sugar based substances which keep plants alive. It is generated by the action of sunlight on green matter (chlorophyll) in leaves. Thus all plants depend on sunlight for food and growth. Better the sunlight, more stout and more compact the individual plants will be. Some plants like sunflower, are so sensitive that its floral part moves from east to west with the movement of the sun.

3. Soils

Soil is of utmost importance to agriculture as each and every plant has its roots in the soil and growth in the soils itself. Soils constitute the physical base for any agricultural enterprise. Great civilizations have almost invariably flourished in areas of fertile soils. **In fact, history of civilization is the history of the soil.** The agricultural productivity and hence our standard of living largely depends on a combination of physical and chemical characteristics of soils. Physical characteristics of soils include their texture, structure and porosity as well as their colour and temperature. Chemical composition shows ingradients present in the soil which exist in shape of some compounds. These and some of the biological characteristics determine the fertility of the soils which is basic to the proper growth of agricultural crops.

Each crop has some specific requirements of soil and its fertility and each soil is suitable to selected group of crops (**Table 20.3**).

TABLE 20.3. Suitability of different soils for different crops

Soils	Crops
1. Alluvial soils	Wheat, maize, barley, gram, oilseeds, pulses, sugarcane.
2. Clayey loams, fine and heavy soils	Rice, jute
3. Volcanic black soils or *regur*	Cotton, wheat, oilseeds
4. Sandy loams and sandy soils	Jawar, bajra, groundnut, guar, pulses (green gram, red gram, black gram etc.)
5. Red and yellow soils	Jawar, groundnut, sugarcane.

Saline and alkaline soils are the result of wrong irrigation practices (over irrigation) and are not fit for cultivation unless they are reclaimed by using chemical fertilisers and biological manure. Use of gypsum can be very effective. These are cumbersome processes and involve high cost.

2. INSTITUTIONAL FACTORS

Institutional factors is the second important groups of determinants of agriculture in India. These include land tenure and land tenancy, size of holding and land reforms.

1. Land Tenure and Tenancy. Land tenure and tenancy is a system under which type of land ownership is determined. In the ancient days and in primitive societies like those of shifting cultivators, system of collective ownership was prevalent and land belonged to the entire community. At a later stage land ownership rights were vested in king, government, zamindars or talukdars. The Mughals established a comprehensive system of land revenue administration wherein Jogirdass or Zamindars were responsible for collection of land revenue from the cultivators. The Britishers introduced the system of permanent settlement or Zamindari, Ryotwari and Mahalwari with a hope to make the system of revenue collection more effective and to improve agricultural production. After Independence Zamindari was abolished and a number of legislations were passed to restore the right of land to the tillers which could meet only partial success. Also not much success has been achieved to prevent division and fragmentation of holdings.

I. Land Holdings

As mentioned earlier in this chapter, the average size of land holding is very small. **Fig. 20.2** shows that it got reduced from 2.28 hectares in 1970-71 to 1.08 hectare in 2015-16. **Table 20.4** brings to light the glaring fact that over 68 per cent of the holdings are less than one hectare with average size as low as 0.38 hectare. This size is too small to be economically viable for any effective method of agriculture. The owners of small holdings are not economically strong enough to afford irrigation, fertilisers, insecticides, pesticides, better seeds and farm machinery. Such a situation leads to low yields which India, with fast growing population, can ill afford.

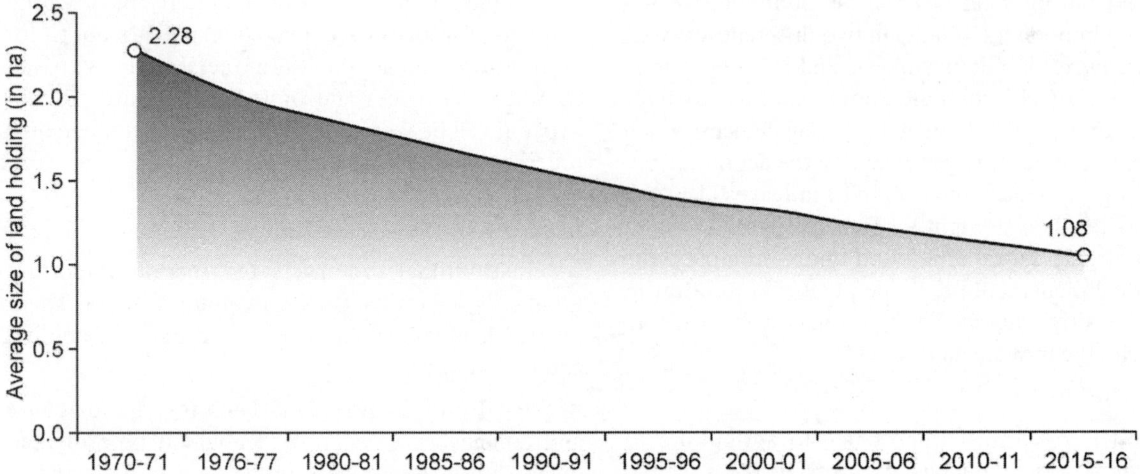

FIG. 20.2. Shrinking farms

***Source*:** Agriculture Census 2015-16 released in 2018 by Ministry of Agriculture and Farmers Welfare.

TABLE 20.4. Number and Area of Operational Holdings by Size Group

No. of Holdings : ('000 Number)
Area Operated : ('000 Hectares)
Average size: (Hectares)

Category of Holdings	Number of Holdings			Area			Average Size of Holdings		
	2005-06	2010-11	2015-16	2005-06	2010-11	2015-16	2005-06	2010-11	2015-16
1	2	3	4	5	6	7	8	9	10
Marginal (Less than 1 hectare)	83,964 (64.8)	92,826 (67.1)	1,00,251 (68.45)	30,026 (20.2)	35,908 (22.5)	37,923 (24.03)	0.38	0.39	0.38
Small (1.0 to 2.0 hectares)	23,930 (18.5)	24,779 (17.9)	25,809 (17.62)-	33,101 (20.9)	35,244 (22.1)	36,151 (22.91)	1.38	1.42	1.40
Semi-Medium (2.0 to 4.0 hectares)	14,127 (10.9)	13,896 (10.0)	13,993 (9.55)	37,898 (23.9)	37,705 (23.6)	37,619 (23.84)	2.68	2.71	2.69
Medium (4.0 to 10.0 hectares)	6,375 (4.9)	5,875 (4.2)	5,561 (3.80)	36,583 (23.1)	33,828 (21.2)	31,810 (20.16)	5.74	5.76	5.72
Large (10.0 hectares and above)	1,096 (0.8)	973 (0.7)	838 (0.57)	18,715 (11.8)	16,907 (10.6)	14,314 (9.07)	17.08	17.38	17.07
All Holdings	**1,29,222 (100.0)**	**1,38,348 (100.0)**	**1,46,454 (100.0)**	**1,58,323 (100.0)**	**1,59,592 (100.0)**	**1,57,817 (100.0)**	**1.23**	**1.15**	**1.08**

Figures in parenthesis show percentages

***Source*:** Agricultural Statistics at Glance 2021 p. 356.

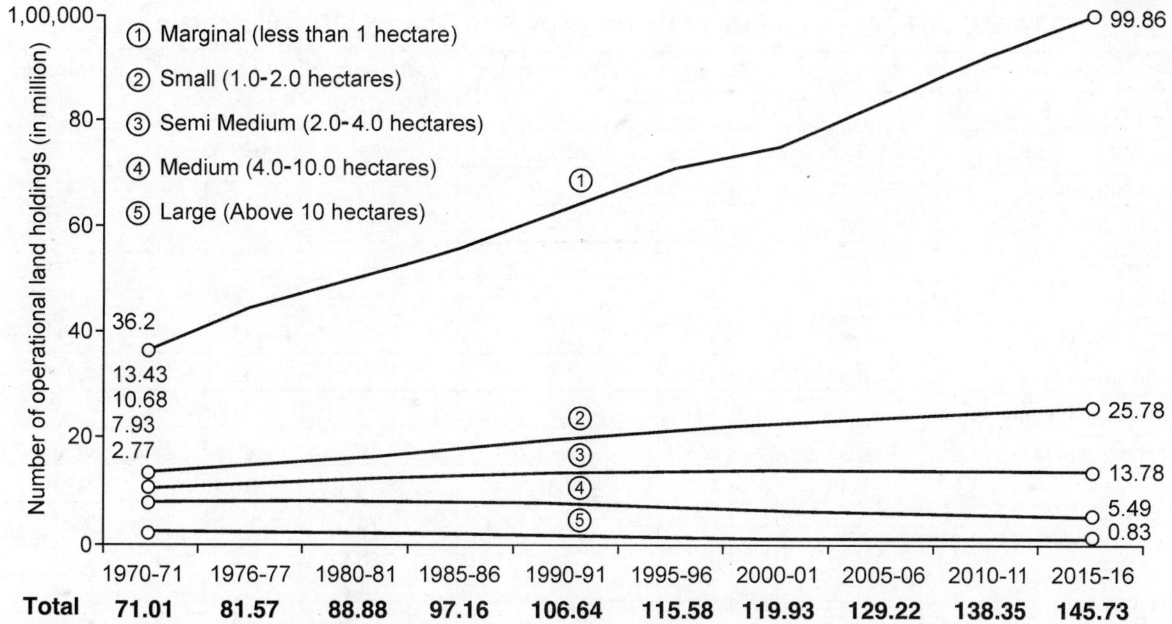

FIG. 20.3. Number of operational land holdings in India

Economic Holding

The Congress Agarian Reforms Committee (1949) defined "an economic holding as one which could provide a reasonable standard of living to the cultivator and give full employment for a family of normal size". The concept of economic holding has been used to fix the upper limit for size of holdings because no zamindar can possess more land than three times the economic holding.

Factors affecting the Size of Economic Land Holding. Size of economic land holding depends on the number of factors which keep on changing from one place to another. Some of the normal factors are briefly described below:

1. Fertility of Soil. Even a small piece of land with fertile soil can feed all the members of the farmer's family. But even a large piece of land is unable to feed the farmer's family members if the soil is not fertile and the farmer needs a larger piece of land for his sustenance. This is the main reason of smaller holdings in more fertile lands and larger holdings in less fertile lands. Simply speaking, size of lands holding is inversely proportioned to the fertility of the soil.

2. Irrigation. Crops get the required supply of water at the time of need in irrigated areas. The yields are higher and there is no fear of crop failure. Thus a small piece of land can enable the farmer to produce enough for his family members. On the contrary crops have to depend on rainfall for their growth in area lacking irrigation facilities and rainfall in India is very erratic. The yields are low and the danger of crop failure looms large. Under such a condition, the farmer needs larger piece of land to feed his family members. Therefore, holdings are usually of large size in areas lacking in irrigation.

3. Type of Agriculture. In area of intensive cultivation even a small piece of land can give high production and holdings are generally of small size. In contrast, in areas of extensive agriculture the yields are low and a large piece of land is required to sustain the farmer and his family members. Thus the size of holdings is quite large in such areas.

4. Farm Mechanisation. Use of farm machinery is not useful practically and economically on small holdings. Hence the size of holdings is large in areas where modernization of farming activities has taken place. Conversely, it can be said that mechanization of farming is possible only in areas of large holdings. One man can manage more than 50 hectares of agricultural land with the help of modern machines like tractors, harvesters, threshers etc., but a

TABLE 20.5. Average Size of Holdings by Size Group, 2015-16 (Hectares)

State/UT	Marginal	Small	Semi-Medium	Medium	Large	All Holdings
1	2	3	4	5	6	7
A and N islands	0.46	1.48	2.74	4.49	39.53	1.78
Andhra Pradesh	0.40	1.42	2.67	5.49	18.71	0.94
Arunachal Pradesh	0.53	1.27	2.67	5.82	15.59	3.35
Assam	0.42	1.41	2.73	5.17	72.80	1.09
Bihar	0.25	1.25	2.60	5.29	14.48	0.39
Chandigarh	0.43	1.37	2.91	5.40	15.92	1.22
Chhattisgarh	0.43	1.41	2.67	5.67	16.10	1.24
Dadar and Nagar Haveli	0.53	1.34	2.75	5.79	15.68	1.38
Daman and Diu	0.22	1.34	2.71	5.89	13.11	0.36
Delhi	0.45	1.40	2.81	5.72	16.13	1.39
Goa	0.29	1.66	3.25	6.75	28.08	1.10
Gujarat	0.53	1.45	2.76	5.66	14.80	1.88
Haryana	0.49	1.46	2.89	6.07	19.04	2.22
Himachal Pradesh	0.40	1.40	2.71	5.64	15.85	0.95
Jammu and Kashmir	0.33	1.38	2.65	5.41	20.40	0.59
Jharkhand	0.41	1.36	2.72	5.64	15.33	1.10
Karnataka	0.43	1.40	2.67	5.69	15.35	1.36
Kerala	0.12	1.34	2.54	5.32	51.04	0.18
Lakshadweep	0.17	1.21	2.43	5.45	17.53	0.27
Madhya Pradesh	0.49	1.41	2.70	5.67	14.83	1.57
Maharashtra	0.46	1.33	2.59	5.59	16.66	1.34
Manipur	0.53	1.29	2.48	4.89	11.09	1.14
Meghalaya	0.46	1.32	2.73	5.55	16.17	1.29
Mizoram	0.60	1.28	2.29	4.65	12.82	1.25
Nagaland	0.56	1.24	2.68	5.84	14.67	4.87
Odisha	0.57	1.58	2.75	5.55	21.70	0.95
Puducherry	0.29	1.45	2.73	5.60	17.38	0.62
Punjab	0.60	1.40	2.67	5.67	14.85	3.62
Rajasthan	0.48	1.42	2.82	6.10	17.04	2.73
Sikkim	0.41	1.39	2.72	5.68	17.20	1.27
Tamil Nadu	0.35	1.39	2.69	5.59	21.84	0.75
Telangana	0.44	1.40	2.60	5.48	14.22	1.00
Tripura	0.30	1`.46	2.65	5.07	14.82	0.49
Uttar Pradesh	0.38	1.39	2.71	5.51	14.98	0.73
Uttrakhand	0.43	1.39	2.68	5.44	26.22	0.85
West Bengal	0.49	1.60	2.74	4.81	361.08	0.76
Total	**0.38**	**1.40**	**2.69**	**5.72**	**17.07**	**1.08**

Source: Agricultural Statistics at a Glance, 2021, p. 359.

person cannot manage more than 6 to 8 hectares of agricultural land if he has to work with his hands or with primitive tools.

5. Type of Crops. Type of crops grown in a particular holding also determines its size to a great extent. For instant, if the farmer grows vegetables, even a small piece of land measuring two hectares is enough to give a reasonably good employment to the farmer and his family members throughout the year. In contrast, the same family requires at least eight hectares of land if the farmer grows wheat in his agricultural land.

Areal Distribution of Size of Holdings. Table 20.5 shows the average size of holdings by size group in different states and union territories of India in 2015-16. The last column of the table shows average size of all holdings. This column reveals that the size of holdings varies widely from as small as 0.18 hectares in Kerala to as large as 4.87 hectares in Nagaland although the average size for the country as a whole is 1.08 hectares. A large number of states have average size of holding much smaller than the national average. In states and union territories like Andhra Pradesh, Bihar, Daman and Diu, Himachal Pradesh, Jammu and Kashmir, Kerala, Lakshadweep, Odisha, Puducherry, Tamil Nadu, Tripura, Uttar Pradesh, Uttarakhand and West Bengal, the average size of holdings is less than one hectare. This is really a depressing situation because no efficient cultivation, worth the name, is possible on such small holdings. The remaining states and union territories have holdings larger than the national average but the condition is not very satisfactory in these states also because in a large number of states; the holdings are less than two hectares in size. Gains of consolidation of holdings in Punjab, Haryana, Uttar Pradesh, Madhya Pradesh, etc. have been nullified due to redistribution of land as a result of inheritance law. Large size of holdings in states like Nagaland and Arunachal Pradesh does not mean much because of rough terrain, thick forests and shifting agriculture. Moreover, large section of society lives in the form of clans which do not lead to division and fragmentation of holdings. Similarly large size of holdings in Rajasthan, is not of much use due to vast areas under desert conditions.

Fig. 20.4 shows the areal distribution of holdings in India characterized by lower level of productivity. This makes it necessary for the farmers to maintain large size of holdings.

II. Land Reforms

Two primary objectives of land reforms are (*i*) to make optimum use of limited land resources so that maximum benefit is drawn from labour and capital inputs and (*ii*) fixing the size of land holdings and redistributing the surplus land to landless and small farmers so that the actual tiller of the land feels secure and works hard to increase the agricultural production. Therefore, the scope of land reforms includes the following :

1. Abolition of intermediaries
2. Tenancy reforms
3. Ceiling on land holding and redistributing the surplus land to landless agricultural labourers and small farmers.
4. Agrarian reorganisation including consolidation of holdings and prevention of their sub-division and fragmentation.
5. Organisation of co-operative farms.
6. Improvement in the system of land record keeping.

1. Abolution of Intermediaries

Before Independence, the British rulers had introduced three categories of land tenure systems, namely Zanindari, Mahalwari and Ryotwari.

(*a*) Zamindari or Permanent System. Although Zamindari system existed during the period of Mughal Empire, the Zamindars were mere functionaries for collecting revenue on behalf of the Mughal Emperor and his representative or Diwan. The Diwan in turn would supervise the activities of he Zamindar to make sure of the rightful collection of revenue. This system underwent significant change after East India Company of the Britishers was awarded the diwani (overlordship) of Bengal by the Mughal Emperor following the Battle of Plassey in 1765. In the year 1793, **Lord Cornwallis** introduced 'Permanent Settlement' according to which feudal lords (jamindars, jagirdars, etc.) were declared proprietors of land on the condition of fixed revenue payments to the British regime. Under this arrangement landlords were expected to pay 90 per cent of rent as revenue to the government in perpetuity and 10 per cent was left with them as collection expenditure. This system prevailled over most of North India including the present day Uttar Pradesh, (except Awadh and Agra), Bihar, West-Bengal most of the Odisha and Rajasthan (except Jaipur and Jodhpur), and covered nearly 58 per cent of the total area cultivated. The landlords, in turn, created a hierarchy of intermediaries who were responsible for collecting rent from the cultivators. These landlords were living a luxurious life in the towns while the poor cultivators had to bear the burden of a large number of intermediaries.

In addition to permanent Zamindari with fixed land revenue in perpetuity there was temporary settlement of land revenue which used to be assessed for period ranging from 20 to 40 years and was subjected to revision. The East India Company introduced a system which could yield a regular flow of rent.

The British rulers argued that the Zamindars represent higher strata of rural society and assigning the job of rent collection will enhance the revenue and increase agricultural production. But this argument proved to be a myth. Demand for land increased due to increase in population and fall in rural industries and the Zamindars started collecting rent at higher rates. The system which was introduced to improve agriculture gave birth to ***absentee landlords*** and their number increased regularly. Thus there grew an intermediary between the state and actual tiller who was interested in land only to the extent of extracting exorbitant rent. So the Zamindari system represented large scale exploitation of the tillers by the absentee landlords and Indian agriculture was degraded to subsistence farming with low productivity.

(*b*) **Ryotwari.** The British rulers did not improve the above mentioned system of settlement in the rest of India after it came under their rule. Captain Munro introduced the second system of revenue collection known as the Ryotwari system of land settlement. It was first introduced in Madras in 1792 and in Bombay in 1817-18. Under this system, the individual cultivators (ryots or raiyats) were reorganised as proprietors of their land with rights to sub-let, mortgage and transfer their land by gift or sale. The individual land holder was directly responsible to the state for payment of land revenue. The settlement was concluded for a short period and the government reserved to itself the right to enhance the assessment for which no specific guidelines were laid down by law. This system of settlement was a complete contradiction to the prevailing old system where the land revenue was permanently fixed by custom and land was not a transferrable property. However, it reflected a kind of Zamindari system because the ryot was at liberty to sublet his land to the tiller. This system was known at ***batai***.

The Ryotwari system accounted for about 38 per cent of the total cultivated area of the country and was prevalent in most of South India, including the present day Maharashtra, Karnataka, Tamil Nadu, Kerala, Andhra Pradesh, Telangana, and most of Madhya Pradesh and Assam. The princely states of Jaipur, and Jodhpur in Rajasthan also fell under Ryotwari type systems. Pockets of ***zamindari***-type tenure existed within these ryotwari areas, particularly where administered by local rajas or nawabs.

(*c*) **Mahalwari.** Mahalwari was the third system of land settlement under which the village lands were held jointly by the village communities, the members of which were jointly and severally responsible for payment of land revenue. Peasant farmers contributed shares of total revenue demand for the village (mahal) in proporosition to their respective holdings. The state was initially entitled to as much as 83 per cent of gross produce in revenue, although this was reduced to 66 per cent at a later stage. The Mahalwari system was introduced between 1820 and 1840 in Punjab (including both present day Punjab in Pakistan and India, and Haryana), parts of the present day states of Madhya Pradesh, and Odisha and Princely states of Awadh and Agra in Uttar Pradesh. It covered about 5 per cent of cultivated area. Usuallly the village *lumbardar* was entitled to collect the revenue for which he received *panchortra i.e.,* 5 per cent as commission.

Zamindari, Ryotwari and Mahalwari had ill-effects on land tenure and led to low productivity in Indian agriculture. For example, legislation introduction in Ryotwari and Mahalwari areas during 1850s enabled money-lenders to recover debts on loans secured on land holdings. Indebtness grew out of proportion due to exorbitantly high rates of revenue and dispossession of land led to rapidly rising tenacy.

Thus, the rural society was polarized into landlords and rich peasants versus tenants and agricultural labourers and the distribution of land became highly unequal. According to Royal Commission on agriculture, 1924-25, as much as 86 per cent of the cultuivated area was held by a minority of 12 per cent of rich cultivators. In Punjab, by 1939, 2 per cent of land owners held 38 per cent of cultivated area. Since the actual cultivator had the right to land as property and the land became transferable under Ryotwari, there was larger scale alienation of land to non-agriculturist money-lenders due to excessive increase in revenue demand and increasing indebtness. In this way land gradually began to pass from the cultivators to money-lenders who were not much interested in improving the agricultural production.

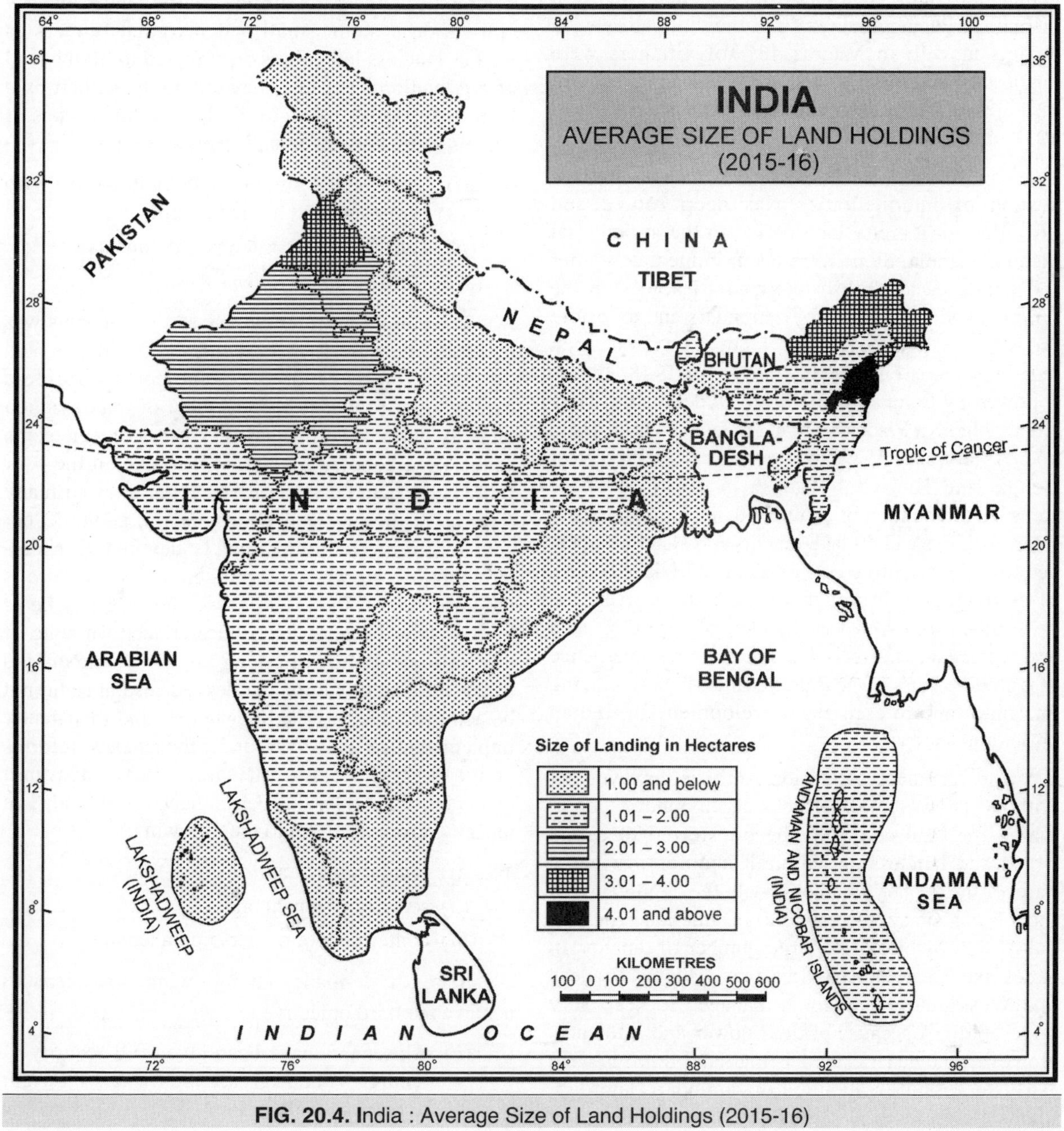

FIG. 20.4. India : Average Size of Land Holdings (2015-16)

In **Zamindari** areas, rural society was even more hierarchically divided between landlords, tenants with heredity rights (raiyats), sub-tenants, share croppers and agricultural labourers, and land distribution was even more unequal than in ryotwari areas (Bhalla 2007 : 21). In spite of a large number of legislations passed between 1859 and 1937 to help the cultivators very little benefits of reduction of rent reached the actual cultivator. The state of peasantry was quit dismal. In Punjab about 40% of the cultivators were tenants at will. In Madras 48% of villagers were landless labourers. In Bengal, there was an increase of 62% in number of rent recoveries between 1921 and 1931.

The above description makes it clear that the system of intermediaries was deep rooted and prevailed for a pretty long period in the agricultural history of India. Steps were taken immediately after Independence to abolish intermediaries, to stop the tyranmy and exploitation by zamindars and to restore the tillers' right over the land. The first such legislation was enacted in Madras in 1948. It was followed by Bombay in 1949-50, Hyderabad in 1951 Bihar, Madhya Pradesh, Uttar Pradesh, Central India and Assam in 1952 and Karnataka, Delhi, West Bengal and Himachal Pradesh in 1954-55. These steps led to abolition of about 2,60,000 zamindars and intermediaries and 30 lakh tenants and share croppers acquired ownership rights of about 25 lakh hectares of cultivated land throughout the country. Also the government acquired vast starches of forests, barren and waste land. Further, it resulted in the emergence of a middle class of peasantry which is now playing an important role in the development of Indian agriculture.

The Zamindari Abolition Act was however not without its inherent drawbacks and loopholes due to which the land could not be transferred to actual tillers and landless agricultural labourers. In many parts of the country, landlords with the connivance of local bureaucracy, were able to resume land for self-cultivation by ejecting a large number of tenants. In states like Bihar, Odisha, Rajasthan, Madhya Pradesh and West Bengal, landlords managed to keep very large holdings because of their power and influence. However, the general view is that land reforms were fairly successful in abolition of zamindari, jagirdari, inams etc. in more parts of India. Extent of tenancy declined considerably and self-cultivation became the dominant mode of production in most parts of the country.

2. Tenancy Reforms

Before Independence, the British rulers had introduced Zamindari, Ryotwari and Mahalwari system under which tenancy cultivation was very common in large parts of the country. Under tenancy cultivation system, small and marginal farmers as well as landless laboureres were forced to till the land of big landlords because they did not have their own land in sufficient measure. Following three types of tenants were recognized under this system:

(*i*) Occupancy or permanent tenants with permanent and heritable rights.

(*ii*) Tenants at will or temporary tanents.

(*iii*) Sub-tenants *or Shikmi-Kisan*.

The condition of temporary and sub-tenants was miserable because they were allowed to till the land on adhoc basis and were subject to ruthless exploitation. Frequent enhancement of rent, eviction on petty grounds and begari (free service) were some of the ways adopted by landlords to exploit the poor farmers and landless labourers. Even after so many reforms in the post Independence era, about 20 per cent of the agricultural land is devoted to share-cropping (*batai*) where 50 per cent of the produce is the normal rent. In certain cases, the rent may be as high as two-thirds of the total produced. In spite of the **National Policy** of giving full ownership of land to the actual cultivators, the desired results could not be achieved due to weak legislations and their faulty implementation. However, under the tenancy reforms about 112.92 lakh cultivators have acquired ownership right over 62.13 lakh hectares of land. The tenancy reforms pertain to the following:

(*i*) regulation of rent

(*ii*) security of tenant and

(*iii*) conferment of ownership on tenants

States have made the following provisions to achieve the third objective

(*a*) All tenants have been given full security of tenure, without giving the owners the right of personal cultivation.

(*b*) Owners have been given the right to resume a limited area (not more than a family holding in any case) subject, however, to conditions that a minimum area is left with the tenant.

(*c*) A limit has been placed on the extent of land with a land-owner may resume, but the tenant is not entitled to retain minimum area of cultivation in all cases.

Regulation of Rents. As mentioned earlier, the tenants were supposed to pay very heavy rent ranging from 1/2 to 2/3rd of the total produce which left the poor farmers with little produce and led to overall extreme poverty among the majority of the farmers. As a result of legislations passed by a large number of states, the maximum rates of rent have been fixed at levels not exceeding 1/4th to 1/5th of the gross produce in all the States except in Andhra Pradesh, Haryana and Punjab. In some of the States, as in Gujarat, Maharashtra and Rajasthan, the maximum rent now stands at one-sixth of the produce. In Assam, Kerala, Odisha and Union Territories, the rent payable is about one-fourth of the produce or less. In several States, the normal level of rent is still about a third of the produce.

Security of Tenure. According to **Sir Arther Young,** "Give a man the secure possession of a bleak rock and he will turn it in to a garden, give him nine years lease of a garden and he converts it into a desert." It clearly implies that the tenant, who is the actual cultivator of land should be provided security against eviction and exploitation. In the absences of proper security he tries to get maximum output with minimum input. He tends to spend minimum on irrigation, fertilisers, better seeds and field boundaries and the lack of these inputs adversely affect the agricultural production. Therefore, the tenant should be provided with sufficient security for which strong legislation should be enacted. Unfortunately, the legislation and its implementation had been very weak and could not improve the conditions of tenancy to the desired level. This is because the social and economic position of tenants in the village is weak and it becomes difficult for them to seek the protection of law. Moreover, resorting to legal process is costly and generally beyond the means of tenants. Thus, despite the legislation, the scales are weighted in favour of the continuance of existing terms and conditions. However, some progress could be recorded in some selected states like West Bengal, Kerala and Karnataka. In West Bengal 14 lakh share-croppers benefited from ***Operation Barga***. In Kerala 24 lakh tenant farmers achieved ownership of land as a result of efforts made by the tenants associations. Land tribunals set up in Karnataka to settle tenancy issues decided in favour three lakh tenants involving 11 lakh acres of land.

More than 20 million ***adivasis*** living in forests have been clamouring for rights over forest land and forest products for a very long period. The Scheduled Tribes and other Forest Dwellers (Recognition of Forest Rights) Act 2006 has granted heritable, inalienable right to use 2.5 hectares of forest land per family to all legitmate dwellers in forests. The Act came into force in January, 2008.

A new form of tenancy known as ***commercial tenancy*** had become popular in the agrarian economy of India. In areas where agriculture is technologically more advanced, small farmers lease their land to bigger farmers because they cannot afford high technology inputs. For example, in Punjab this form of tenancy stands in stark contrast to the semi-feudal subsistency tenancy of the yore.

3. Ceiling on Landholdings

Imposition of ceiling on landholdings and distribution of surplus land among the landless labourers was the third most important objective of land reforms legislation in India. It envisaged that all land more than a certain specified limit belonging to landlords would be taken over by the state and allotted to small farmers to make their holdings economic or to landless labourers to meet their demand for land. The idea of land redistribution through fixation of land ceiling has gained wide acceptance in India. The strategy basically is to ration land in such a way that poor farmers are not deprived of the land (the basic asset) while big landlords enjoy the privilege of having vast stretches of land. Ceiling laws were enacted and enforced in two phases. The first phase was from 1960 to 1972 during which ceiling legislation largely treated land holder as the unit of application. In the second phase after 1972 it was decided to have family as the basis of holding. **Table 20.6** depicts the ceiling limit on different categories of land in different states.

Till 1972 *i.e.*, under the old ceiling laws only 9.30 lakh hectares of land was declared surplus out of which only 5.26 lakh hectares were redistributed among small and marginal farmers as well as landless labourers.

The Conference of Chief Ministers held in July, 1972 approved the following guidelines for implementation of land ceiling.

(*i*) For irrigated land with two crops per year, the ceiling was fixed from 4.05 to 7.28 hectares.

(*ii*) For irrigated land with one crop a year, the ceiling was fixed at 10.93 hectares.

(*iii*) In case of inferior dry land with practically no irrigation, the ceiling was put at a higher level of 21.85 hectares.

(*iv*) The unit of application should be a family of five members. Additional land may be allotted for each member in excess of five members subject to a maximum of twice the limit of the celings.

(*v*) The celing process was not to be applied to land under plantation crops (tea, coffee, rubber, spices, coco, etc.) as well as under industrial and commercial establishments.

(*vi*) State governments were allowed to exempt religious, charitable and educational trusts of public nature from the purview of ceiling.

(*vii*) While distributing the surplus land, priority should be given to landless agricultural labourers especially to those belonging to scheduled castes and scheduled tribes.

(*viii*) Compensation payable for surplus land as a result of ceiling should be fixed well below the market value so that it is within the capacity of the new allottees.

(*ix*) The compensation may be fixed in graded slabs and preferably in multiples of land revenue payable for the land.

The Conference of Chief Ministers (1972) was followed by Regional Ministers Conference in 1986 which suggested some changes in the ceiling on land holdings (**Table 20.6**).

In spite of ceiling laws the progress of acquiring land from big landlords and redistributing it among the poor farmers has not been satisfactory.

Failure of land ceiling laws is attributed to loopholes caused by the following :

(*i*) Provision for holding land upto twice the ceiling limits by families with more than 5 members.

(*ii*) Provision to give separate ceiling limited to major sons of the family.

(*iii*) Provision to treat every share-holder of a joint family as a separate unit for ceiling limits.

(*iv*) Exemption of plantation gardens, religious and charitable institutions from the provisions of ceiling.

(*v*) Individual states allowed landlords to retain control over land holdings, mostly through *benami* transactions namely transactions that transferred holdings in the names of deceased or ficticious persons registered with the connivance of the village *patwari*.

(*vi*) Misuse of exemption and misclassification of land.

(*vii*) Non-application of appropriate ceiling to lands newly brought under irrigation by public investment.

Furthermore, it is opined that lack of accurate updated records of rights on land was a major constraint on effective implementation of land ceilings as also tenancy reforms.

Conclusions. Following three conclusions are drawn :

(*i*) The share of small and marginal farmers both in ownership holdings and total area owned is increasing at a rapid pace. For instance, during 1953-54 the small and marginal farmers owning less than 2 hectares of land constituted 51.64% of the total holdings but accounted for only 16.31% of the total owned area. By 2003, the share of these farmers in the number of ownership holdings had increased to 80.4% and their share of area had increased to 43.43%.

(*ii*) The proportionate share of large farmers both in total holdings and land owned has declined rapidly over the years.

TABLE 20.6. Ceiling on Land Holdings

(In Acres)

State	Irrigated Land with two crops	Irrigated Land with one crop	Dry land
1	2	3	4
As recommended in 1972 National Guidelines	10-18	27	54
Proposed in Agenda Notes 1985 of Regional Ministers Conference	12	18	30
Andhra Pradesh	10 to 18	15 to 27	35 to 54
Assam	17	17	17
Bihar	15 to 18	25	30 to 45
Gujarat	10 to 18	15 to 27	20 to 54
Haryana	18	27	54
Himachal Pradesh	10	15	30 to 70
Jammu and Kashmir	9 to 12.5	9 to 12.5	15 to 23 (In Ladakh 19)
Karnataka	10 to 20	25 to 30	54
Kerala	12 to 15	12 to 15	12 to 15
Madhya Pradesh	18	27	54
Maharashtra	18	27	54
Manipur	12	12	15
Mizoram	—	—	—
Odisha	10	15	30 to 45
Punjab	17	27	51
Rajasthan	18	27	54 to 175
Tamil Nadu	12	30	60
Sikkim	12.5	12.5	50
Tripura	10	10	30
Uttarakhand	18	27	45
Uttar Pradesh	18	27	45
West Bengal	12	12	17
Andaman and Nicobar Islands	—	—	—

Notes : 1. The actual limits for lands in Karnataka and Uttar Pradesh are higher due to classification of land.
2. The actual ceiling limits in Himachal Pradesh and Rajasthan are higher due to hilly terrain and desert lands.
3. 1 Acre = 0.404686 Hectare.

Source : Agricultural Statistics at a Glance 2021 p. 349.

(*iii*) Semi-medium and medium farmers are the main gainers in terms of share of owned land although their share in total ownership holding has declined considerably.

4. Consolidation of Holdings

As mentioned earlier, most of the land-holdings in India are small and fragmented. This results in large scale wastage of time and energy of the farmer because he has to operate his several small holdings scattered in different parts of the village agricultural land. The only way to take the farmer out of such an unwanted situation is consolidation of holdings which involves bringing together in one compact block all plots of land of a farmer which are scattered in different parts of the village. Under this scheme all land of the village is first pooled into one compact block and then divided into smaller blocks called *chaks* and then allotted to individual farmer. Laws for this purpose have been passed in 15 states.

Historically, 4.5 million hectares of land was brought under consolidation upto 31st January, 1956. This figure rose to 33 million hectares in 1972 and by the end of the Sixth Five Year Plan 45 million hectares were consolidated. Thus about one-fourth of the consolidable land was brought under consolidation. But consolidation of holdings has shown its impact in a few selected areas only. Punjab, Haryana and Western part of Uttar Pradesh are some such areas. This process has not even started in Rajasthan and some south Indian states. Among the eastern states only Bihar and Odisha have started this process. (Planning Commission, Seventh Five Year Plan-1980-85, p. 115). According to the data collected by the Ministry of Rural Development, consolidation was completed for 1663 lakh acres till 31st March, 2004. Out of this Maharashtra, 527 lakh acres or 32.2% and Uttar Pradesh 482 lakh acres or 29.4% were the main achievers. The programme of consolidation of holding has succeeded to a great extent in Punjab, Haryana and Madhya Pradesh. Efforts are being made to implement this programme in Bihar and Jammu and Kashmir. It has not been implemented in West Bengal and Assam. The states of Andhra Pradesh, Tamil Nadu, Kerala, Puducherry (U.T.) and all north-eastern states have not even passed the laws necessary for this purpose. Till now only 49% of the agricultural land of India has been consolidated and the remaining 51% is yet to be consolidated. According to 2004-05 report of the Ministry of Rural Development no consolidation has been done in 9000-10,000 villages in Uttar Pradesh. These facts and figures indicate that consolidation of holdings has not yielded the desired result for which following factors may be responsible.

1. Farmers are emotionally attached to their ancestral land and are generally not willing to take advantage of this scheme.
2. Farmers possessing more fertile land are not in favour of this scheme for fear of being allotted infertile land after consolidation.
3. Consolidation of holding is a lengthy and cumbersome process. The government officials who are responsible for implementing the scheme of consolidation of holdings are usually slow, careless and corrupt.
4. Cooperation of the farmers was badly lacking with respect to implementation of this scheme.
5. A large number of farmers were unhappy with the redistribution of land after consolidation. This led to large scale litigation and many cases concerning land disputes are still pending in different courts. Such a situation has vitiated the serene social atmosphere of the rural areas.
6. Under the existing laws of inheritance, the land belonging to father is divided among his heirs and land holdings continue to be divided and fragmented. Thus the initial gain of consolidation of holdings is nullified and need for a new scheme of consolidation of holdings is badly felt.
7. In every consolidation about 5 to 10 per cent of the village land is taken out for providing house sites to weaker sections of society, approach (*chak*) roads and village utility services. Thus if the process of consolidation of holdings is repeated 3 to 4 times, a sizeable portion of agricultural land is lost to non-agricultural uses.
8. The cost of consolidation of holdings is realised from the farmers who are generally unwilling to bear this cost and are usually against this scheme. Further, it dries up their financial resources and has adverse effect on their economy.
9. It has been observed that small farmers usually get inferior quality of land with low fertility and low productivity. Since they do not possess enough money power, they are unable to bribe the corrupt officials and also fail to get proper justice in the court.

5. Cooperative Farming

Cooperative farming is a type of farming in which the farmers pool their resources and distribute the agricultural products in proportion to their resources. It helps in efficient farming and small farmers get the benefits which are usually available to big farmers. Problem of small and fragmented holdings is solved and economic position of poor farmers is improved. According to Planning Commission "Cooperation represents institutionalization of the principle and impulse of mutual aid. It has the merit of combining freedom and opportunity for the small man with the benefit of large scale management and organisation.''(Draft Fifth Five Year Plan Vol. II, p. 76.) Although cooperative movement started way back in 1882, more emphasis on this type of farming was laid after Independence only. But like other institutional factors, cooperative farming also could not achieve its target and could not yield the desired results. About 60% of the cooperative societies are not functioning satisfactorily.

6. Land Records

Updating of land records is most essential to implement the land reforms in their right earnestness. In India land records maintenance is highly unsatisfactory and is responsible for lots of land disputes leading to large scale litigation in the rural areas. Centrally sponsored scheme for updating land records and strengthening the revenue administration was introduced during the Seventh Plan. The Central Government provides financial assistance to states and union territories for using modern technology such as photo maps, areal photographs, remote sensing, computers, etc. A committee on Revitalization of Land Revenue Administration submitted its report on March 2, 1995 on which necessary follow-up action was initiated by the Government of India and the States/Union Territories. Several legislations have been passed regarding land reforms and ceiling on land holdings. Till now, more than 250 land laws have been included in the Ninth Schedule of the Constitution.

3. INFRASTRUCTURAL FACTORS

Infrastructural factors include irrigation, power, roads, credit, insurance, marketing, storage facilities etc.

1. Irrigation. Need, feasibility and sources of irrigation in India have been described in details in Chapter 16. Expansion of irrigation facilities along with other infrastructural factors is the main strategy for increasing agricultural production. It is one of the most important inputs required at different critical stages of plant growth of various crops for optimum growth. Although India has made considerable progress with respect to irrigation during the planned period, more than half of the cultivated area is still without irrigation which entirely depends on monsoon rainfall for agricultural production. Since the monsoon rainfall is very erratic, there are large scale spatial and temporal variations in agricultural outputs. The adage that *Indian agriculture is a gamble in the monsoons still holds good for large part of India*. However, irrigation has made a major contribution to solve the food problem in India. It is estimated that about 60% of the food grains production comes from the irrigated areas and the remaining 40% from rainfed areas. Yields of almost all the crops are much higher in irrigated areas as compared to rainfed areas.

The growth of irrigation facilities has failed to keep pace with the increasing demand for irrigation. This has been particularly the case in major/medium irrigation potential. The chief reasons for low progress in major/medium irrigation were (*a*) long gestation period required for these projects (*b*) financial constraints as these projects involve heavy cost (*c*) inter-state disputes about water allocation of almost all major rivers (*d*) problems of acquisition of vast stretches of land and (*e*) resettlement and rehabilitation of the oustees.

The other major problem is the widening gap between potential created and potential utilized. The main gap occurs in major/medium projects. This happens due to lack of command area development, cropping pattern changes (shift to higher water consuming crops like rice cultivation in Punjab and Haryana) and lack of field channels. Regarding minor irrigation, although over 70% of the groundwater potential has been utilized, there are very serious problems of over-exploitation of ground water and its rapid depletion, particularly in the north-western part of the country.

There are large variations in percentage of irrigated area under different crops. (**Table 20.7**). For example 54.3% of the area under foodgrains is irrigated and this percentage varies from a meager 9.4% in case of bajra to 95.3% in case of wheat. Rice, is the most important food crop of India and has 62.7% area under irrigation. Among the other crops sugarcane gets the maximum irrigation benefit because 96.6% of the sugarcane area gets irrigation facilities. About 45% of the cotton area is irrigated. Oil seeds are usually grown in the rainfed areas and less than one-third of the area under oil seeds is provided with irrigation.

Another major problem regarding irrigation is low efficiency of water use which ranges from 38 to 40% for canal irrigation against an ideal value of 60%. Efficiency of water use in case of ground water irrigation schemes is about 60 percent. Low water use efficiency results in low agricultural productivity, iniquity in water supply to tail enders and water logging and salinity in water surplus areas. Among the several reasons for low efficiency of water use are dilapidated condition of irrigation canal system due to silting, weed growth and breakage of regulating structures leading to over-use of water. In several old delta systems like those of the Mahanadi, the Godavari and the Cauvery, irrigation is practiced by field-to-field flooding. Concessional water rates also lead to over use of canal water and results in low efficiency of water use. Dilapidated system of irrigation, operation and maintenance charges, low water rates and participatory irrigation system are all inter-related. Therefore the entire problem of low efficiency of water use should be tackled as a package of measures so that enough water is made available to crops and agricultural productivity is increased. The package should include modernization, conjunctive use through shallow augmentation tube wells, provision of temper-proof outlets, replacement of old canal road bridges, development of canal banks as roads for maintenance of canals and improving the rural transport system, promotion of water saving devices like sprinkler and drip irrigation systems through tax concession and subsidy-cum-loan schemes (Bhalla 2007: 130-31).

TABLE 20.7. Percentage Coverage of Irrigated Area under Principal Crops during 2018-19

Crop	Percentage of area under irrigation
Rice	62.7
Jowar	10.1
Bajra	9.4
Maize	26.3
Wheat	95.3
Barley	74.1
Total cereals	**63.0**
Gram	49.6
Tur or Arhar	5.2
Total Pulses	**23.2**
Total Foodgrains	**54.3**
Sugarcane	96.6
Groundnut	30.9
Rape-seed and Mustard	83.1
Soya-bean	0.5
Sun-flower	30.8
Total oilseeds	**31.4**
Cotton	45.1
Tobacco	67.9
All crops	**52.0**

Source : Agricultural Statistics at a Glance, 2021, pp. 337-38.

2. Power. Indian agriculture is going the industries way as more and more agricultural operations are depending on regular supply of cheap power in the form of hydro electricity and diesel. There is a linear relationship between availability of farm power and farm yield and Government has decided to enhance farm power availability from 2.02 kW/ha in 2016-17 to 4.0 kW/ha by the end of 2030 to cope will increasing demand for foodgrains. However, the farm power availability is much lower than Korea (7 + kW/ha), Japan (14 + kW/ha) and the U.S.A. (6 + kW/ha). Thus it is concluded that more farm power should be made available to the farmers to upscale farm productivity so as to grow more foodgrains in view of stagnation in increase in net area sown.

Table 20.8 shows that there has been increasing trend with respect to consumption of electricity in agriculture. It increased from 23,422 GWh in 1985-86 to 2,11,295 GWh in 2019-20, thus registering a more than nine times increase in a span of 34 years. Decline in percentage of electricity consumed in agriculture is not due to fall in total electricity consumption in agriculture but because of increase in electricity consumption in other fields like domestic, industrial and commercial.

There are large spatial variations in the total consumption of electricity and percentage of share of

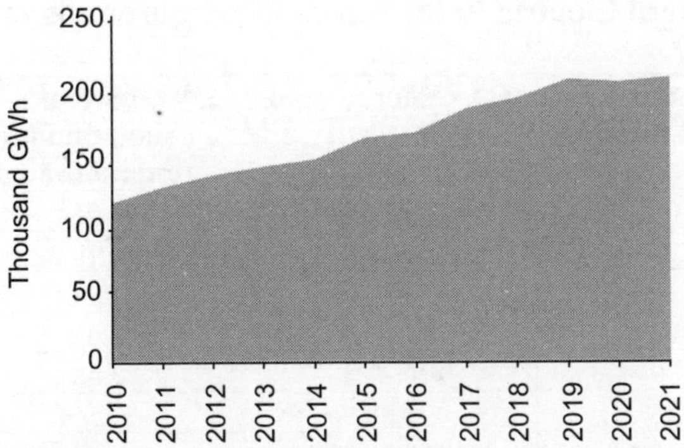

FIG. 20.5. Electricity consumption in agriculture
Source: Economic Survey, 2022-23, p. 245.

TABLE 20.8. Consumption of Electricity for Agricultural Purposes

Year	Consumption for Agricultural purposes (GWh)	Total Consu-mption (GWh)	% Share of Agriculture consumption to total consumption
1985-86	23,422	1,22,999	19.04
1990-91	50,326	1,90,357	26.44
1995-96	85,732	2,77,029	30.95
2000-01	84,729	3,16,600	26.76
2011-12	1,40,960	6,72,933	20.95
2015-16	1,73,185	8,63,364	20.06
2016-17	1,91,151	9,14,093	20.91
2017-18	1,99,247	9,73,131	20.47
2018-19	2,13,409	10,37,518	20.57
2019-20	2,11,295	10,52,346	20.08

GWh = Giga Watt Hour
Source : Agricultural Statistics, at a Glance 2021, p. 298.

consumption of electricity in the agricultural field. So far as total consumption of electricity for agriculture is concerned. Maharashtra is at the top with more than 29.3 thousand GWh. The other major consuming states are Karnataka (21.8 thousand GWh), Rajasthan (25.7 thousand GWh), Madhya Pradesh (23.0 thousand GWh), Gujarat (11.99 thousand GWh), Telangana (20.8 thousand GWh), Uttar Pradesh (18.2 thousand GWh) and Andhra Pradesh (15.3 thousand GWh).

Most of the north-eastern states are least developed with respect to total consumption of electricity for agricultural purposes.

Regarding percentage share of agricultural consumption to total consumption of electrically, Rajasthan with over 41.43% is at the top. This is followed by Madhya Pradesh (39.26%), Karnataka (35.70%), Telangana (35.07%), Andhra Pradesh (26.85%), Haryana (23.91%), Punjab (23.56%) and Maharashtra (23.31%).

In order to increase the efficiency and productivity of agricultural operations, cheap and regular supply of electricity to farmers should be assured. Already some state electrically boards are facing heavy financial burden due to highly subsidized or almost free electricity to farmers. However, huge power shortage and erratic power supply with long declared and undeclared cuts in several parts of the country is the major cause of farmers' agitation in some of the agriculturally advanced states like Punjab, Haryana and western part of Uttar Pradesh where tubewells are the major source of irrigation. The problem of power shortage is likely to be further complicated with sharp increase in demand for power in agriculture and other fields of economic activities. Therefore, there is an urgent need to increase power supply and make it more regular.

3. Roads. Road connectivity to villages is very essential for economic and social development of rural areas. Farmers have to frequently visit the nearby urban markets to purchase seeds, fertilisers, insecticides, pesticides, farm implements etc. and also articles of daily use. They also go the urban markets to sell their agricultural products. Village roads are lifelines for the rural folks particularly when sale of perishable commodities like vegetables, fruits, milk etc. is involved.

Although the total length of village roads is about 41.67 lakh kilometres, it is not sufficient keeping in view the size of the country and the percentage of people living in the rural areas. Further only about 15% of the village roads are surfaced. Considering the important role played by the village roads for the overall development of the rural areas, the Government of India launched hundred per cent centrally sponsored scheme of Pradhan Mantri Gram Sadak Yojna (PGSY) in December 2000 with the

TABLE 20.9. Region-wise/State-wise consumption of Electricity for Agricultural purposes in 2019-20

Region	States/UTs	Consumption for Agricultural Purpose (GWh)	Total Energy Sold (GWh)	% Share of Consumption for Agriculture
Northern	Chandigarh	1.02	1,603.81	0.06
	Delhi	37.36	29,812.02	0.13
	Haryana	10,371.16	43,382.05	23.91
	Himachal Pradesh	56.73	9,123.99	0.62
	Jammu & Kashmir	368.00	10,622.85	3.46
	Punjab	11,581.56	49,168.06	23.56
	Rajasthan	25,664.75	61,941.93	41.43
	Uttar Pradesh	18,200.05	94,932.13	19.17
	Uttarakhand	191.35	12,021.35	1.59
	Sub-Total (NR)	**66,471.96**	**3,12,608.20**	**21.26**
Western	Chhattisgarh	5,082.35	22,793.08	22.30
	Gujarat	11,989.83	92,245.12	13.00
	Madhya Pradesh	23,016.94	58,628.43	39.26
	Maharashtra	29,291.23	1,25,644.95	23.31
	Daman & Diu	4.34	2,495.67	0.17
	Dadra & Nagar Haveli	5.25	6,287.98	0.08
	Goa	27.29	4,104.19	0.66
	Sub-Total (WR)	**69,417.23**	**3,12,199.42**	**22.23**
Southern	Andhra Pradesh	15,338.12	57,118.77	26.85
	Telangana	20,820.54	59,364.54	35.07
	Karnataka	21,823.07	61,130.64	35.70
	Kerala	360.47	23,148.94	1.56
	Tamil Nadu	13,828.26	95,919.07	14.42
	Puducherry	55.32	2,713.78	2.04
	Lakshadweep	0.00	48.42	0.00
	Sub-Total (SR)	**72,225.77**	**2,99,444.17**	**24.12**
Eastern	Bihar	932.94	22,708.58	4.11
	Jharkhand (estimated)	193.91	21,997.03	0.88
	Odisha	579.15	20,715.89	2.80
	West Bengal	1,398.75	50,481.54	2.77
	Sikkim	0.00	411.35	0.00
	A. & N. Islands	1.12	268.50	0.42
	Sub-Total (ER)	**3,105.87**	**1,16,582.90**	**2.66**
North Eastern	Arunachal Pradesh	0.03	418.69	0.01
	Assam	37.00	7,109.00	0.52
	Manipur	2.18	631.16	0.35
	Meghalaya	0.31	1,259.48	0.02
	Mizoram	0.14	486.40	0.03
	Nagaland	0.00	692.72	0.00
	Tripura	34.38	914.22	3.76
	Sub-Total (NER)	**74.05**	**11,511.67**	**0.64**
	Total (All India)	**2,11,294.89**	**10,52,346.36**	**20.08**

GWh : Giga Watt hour

Source : Agricultural Statistics at a Glance, 2021, pp. 299-301.

objective of providing connectivity by all weather roads to about 1.70 lakh unconnected habitations with population of 500 persons and above in plain areas and 250 persons and above in hill states, tribal areas and desert and left wing extremist (LWE) affected areas. All the villages under this scheme have been provided with road connectivity.

Government of India has identified 'Rural Roads' as one of the six components of '*Bharat Nirman*' with a goal to provide connectivity to all eligible unconnected habitations with a population of 1000 persons and above in plain areas and 500 persons and above in hilly or tribal areas with all weather roads.

4. Agricultural Credit. Agricultural credit plays a significant role in improving agricultural production and productivity and mitigating the distress of farmers. Indian farmers need credit both for working capital and for investment. Demand for both short-term and long term credit started rising rapidly with the onset of Green Revolution in mid 1960s. It happened due to the fact that farmers had to purchase costly inputs like fertilisers, better seeds, pesticides, etc. Money was also required for land improvements, purchase of farm machinery, to build farm structures and dig wells.

The sources from which the farmers can borrow money can be divided into two categories viz. (*i*) non-institutional and (*ii*) institutional. The main non-institutional sources are the village money-lenders who have been operating for several centuries. Borrowing money from the money-lenders is very easy but they charge high rate of interest from the farmers. Often, the farmers are unable to repay loan and interest from their routine earning and have to mortgage their land to do so. Once mortgaged, the farmer finds it extremely difficult to get back his land.

In order to free the farmers from the clutches of the money-lenders, the Government of India initiated institutional sources which include cooperative banks, commercial banks and Regional Rural Banks (RRBs). These institutions provide loans for productive purposes at much lower rates of interest. The state governments also provide *taccavi* loans to farmers and extend financial support to State Cooperative Banks and Land Development Banks.

Although institutional credit system was initiated in the beginning of 20th century, much headway was not made till Independence. In the post-Independence era, credit institutions were developed in the following three phases.

(*i*) The ***first phase*** ranged from 1947 to 1969 during which cooperative agencies were the main institutions for providing loans to the farmers.

(*ii*) The ***second phase*** was between 1969 and 1975 during which the commercial banks were to provide agricultural credit and supplement credit by cooperatives.

(*iii*) The ***third phase*** started in 1975. During this phase, the Regional Rural Banks (RRBs) were established to provide credit to small and marginal farmers and also the weaker sections of society.

The cooperative credit services and credit structures. It consists of two wings viz., (*i*) short term credit structure and (*ii*) long term credit structure. The ***short term credit*** structure is designed to provide credit for production purposes and is based on three-tier pattern consisting of (*a*) State Cooperative Banks (SCBs) (*b*) District Central Cooperative Banks (DCCBs) and (*c*) Primary Agricultural Credit Societies (PACs). The ***long term credit*** is meant for land development and other capital expenditure. It is provided by the Primary Cooperative Agricultural and Rural Development Banks (PCARD) and State Cooperative Agricultural and Rural Development Banks (SCARD).

Finance to cooperatives is provided by National Bank for Agricultural and Rural Developmnent (NABARD) which was established in 1982. Three main functions of NABARD are development, credit and supervision.

Commercial Banks. Commercial banks started extending agricultural credit after the nationalization of 14 banks in 1969. These banks were to lend 18% of their total credit to agriculture as a priority sector and this mode of credit increased rapidly. In 1969, out of a total of 8,262 bank branches, only 1860 were located in rural or semi-urban areas and there were only 2 lakh agricultural borrowing accounts. By 2004-2005, the number of agricultural borrowing accounts increased to over 16 million and the loans issued amounted to ₹ 72,886 crores. By 2005-06,

TABLE 20.10. Flow of Institutional Credit to Agriculture Sector (₹ in crore)

Particulars/Agency	2014-15	2015-16	2016-17	2017-18	2018-19	2019-20	2020-21	2021-22
Co-operative Banks	1,38,469	1,53,295	1,42,578	1,50,321	1,52,340	1,57,367	1,94,704	8,98,869
Regional Rural Banks (RRBs)	1,02,483	1,19,260	1,23,216	1,41,216	1,49,667	1,65,326	1,90,682	1,47,064
Commercial Banks	6,04,376	6,92,954	7,99,781	8,71,080	9,54,823	10,70,036	1,90,012	1,51,173
Total	**8,45,328**	**9,15,509**	**10,65,755**	**11,62,617**	**12,56,830**	**13,92,729**	**15,75,398**	**11,97,105**

Source: Agricultural Statistics at a Glance, 2021, p. 302.

commercial banks accounted for 69.52% of the total agricultural credit.

Regional Rural Banks (RRBs). These were set up in 1975 with the main objective of taking banking to the doorsteps of the rural masses, particularly in those areas where banking facilities were not existing. The idea was to serve the weaker sections of society. Initially only 5 RRBs were set up one each at Moradabad, Gorakhpur, Bhiwani, Jaipur and Malda. The number of RRBs increased to 1996 by 2004 with 14,000 branches spreading over 516 districts.

Rural Infrastructural Development Fund (RIDF). Established in 1995-96, RIDF aims at providing loans to State Governments and State owned corporations to enable them to complete ongoing rural infrastructural projects.

Kisan Credit Card (KCC) Scheme. This scheme was introduced in 1998-99 to provide adequate and timely credit to the farmers from the banking system for their cultivation needs in a flexible, hassle-free and cost effective manner. The farmers can use their cards for purchasing agricultural inputs like fertilisers, pesticides, seeds, etc. and also draw cash for their production needs. Credit limits are fixed on the basis of size of operational land holding, cropping pattern, scale of finance, etc. It has been adopted by 27 commercial banks, 378 District Central Cooperative Banks/State Cooperative Banks and 196 regional rural banks. The number of credit cards issued had risen to over 17 crore upto 31st March. 2018. Table 20.11 gaves bankwise distribution of Kisan Credit as on 31st March, 2018.

TABLE 20.11. Distribution of Kisan Credit Cards according to categories of banks upto 31st March, 2018

Category of Banks	Number of Kisan Credit Cards (lakhs)
1. Cooperative Banks	550.20
2. Regional Rural Banks	289.49
3. Commercial Banks	926.99
Total	1,766.68

Source : Agricultural Statistics at a glance, 2018, pp. 340.

A revised Kisan Credit Card Scheme was introduced in March, 2012 in which KCC pass book has been replaced by an ATM-cum-debit card to all eligible and willing farmers in a time bound manner.

Kisan Credit Card Scheme 2021. Farmers have been issued Kisan Credit Cards which enables them to draw credit of ₹ 1,60,000. This helps the farmers to look after their agricultural processes in a better way. Recently animal rearers and fisherman have also been included in this scheme. The farmers can get loan at a low 4% rate of interest.

Self Help Groups (SHG) Bank Linkage. This programme was initiated in 1992 for improving the flow of credit to poor sections of the society and it has emerged as the largest and the fastest growing micro-finance programme in the country. As many as 560 banks are involved in this programme. These include 48 commercial banks, 196 RRBs and 316 cooperative banks.

Interest Subvention Scheme (ISS). This scheme is designed to help small and marginal farmers who are the major borrowers. Operational since 2006-07, this institutional credit helps in delinking the farmers from non-institutional sources of credit, where they are compelled to borrow at usurious rates of interest. Under this scheme, the farmers can avail concessional crop loans of upto ₹ 3 lakh at 7 per cent rate of interest. It also provides for an additional subvention of 3 per cent for prompt repayment within a period of one year from the date of advance.

As a measure to check distress sale, post harvest loans for storage in accredited warehouses against Negotiable Warehouse Receipts (NWRs) are available for upto 6 months for Kisan Credit Card (KCC) holding small and marginal farmers. In order to give relief to small and marginal farmers who would have to borrow at 9 per cent for post harvest storage of their produce, the central Govt. has approved an interest subvention of 2 per cent *i.e.* an effective interest rate of 7 per cent for loans upto 6 months.

To provide relief to be the farmers affected by natural calamities, the interest subvention of 2 per cent will provided to Banks for the first year on the restructured amount.

5. Agricultural Insurance. Agricultural insurance is necessary to provide financial compensation to farmers for losses in crop yields due to non-preventival risks including natural calamities like floods, droughts, etc., pests and diseases and adverse weather conditions. It is a useful means of encouraging farmers to diversity to high value crops like fruits and vegetables which are susceptible to large fluctuations in output because of changes in weather (Bhalla 2007: 170). The government of India is currently implementing the following Central Crop Insurance Schemes :

(i) National Agricultural Insurance Scheme (NAIS). This scheme was introduced in the rabi season of 1999-2000 with a view to providing insurance cover and financial support to farmers in the event of failure of crop due to natural calamities such as drought, flood, fire, pests and diseases. The scheme covers a wide range of crops including all food crops (cereals, millets and pulses), oilseeds and annual commercial/horticultural crops. At present the scheme is being implemented in a large number of states and union territories. In 2015-16, over 27 crore farmers, and 39 crore hectares of land were covered under the scheme and a sum of ₹ 4,67,16,920.37 lakh was insured.

(ii) Modified National Agricultural Insurance Scheme (MNAIS). With the aim of further improving crops insurance schemes the MNAIS is under implementation on pilot basis in 50 districts of 16 states from rabi 2010-11 season. The major changes introduced in this scheme *inter-alias* include village panchayat as a unit area of insurance for major crops, rationalization of calculation of threshold/guaranteed yield actuarial premium with adequate subsidy etc. Since the inception of the scheme from rabi 2010-11 to rabi 2016, 276.4 lakh farmers have been covered over an area of 298 lakh hectare insuring a sum of ₹ 61,661.45 crore. Claims of ₹ 5,191.17 crore have been paid against premium of ₹ 4,970.2 crore benefitting about 95 lakh farmers.

(iii) Pilot Weather Based Crop Insurance Scheme (WBCIS). This scheme was implemented in Kharif 2007 season with the intention to provide insurance protection to farmers against adverse weather incidence such as deficit or excess rainfall, high or low temperature, and humidity that are deemed to adversely affect crop production. The scheme is based on actuarial rates of premium but to make it more attractive, premium actually charged from the farmers has been restricted to at par with NAIS. From kharif 2007-08 to rabi 2015-16, 724.25 lakh farmers cultivating an area of about 895.8 lakh hectares with a sum insured of about ₹ 12,413 crore have been covered under the scheme.

(iv) Coconut Palm Insurance Scheme (CPIS). This scheme was approved on profit basis during the year 2009-10 and 2010-11 in selected areas of Andhra Pradesh, Goa, Karnataka, Kerala, Tamil Nadu, Maharashtra, Odisha and West Bengal. The sum insured is based on the input cost of the plantation and the age of the specific plant and it varies from ₹ 60 per palm in the age groups of 4-15 years to ₹ 1150 per palm in the age group of 16 to 40 years. The premium rate varies from ₹ 4.2 per palm in the age group of 4 to 15 years to ₹ 5.75 per palm in the age group of 16 to 40 years. Fifty per cent is contributed by the Central Government, 25 per cent by the concerned State Government and the remaining 25 per cent by the farmers. From its inception in 2009-10 to 2017-18 1,24,459 farmers with 57,41,623 palms were insured and the sum insured was ₹ 50,636.70 lakh.

(v) Pradhan Mantri Fasal Bima Yojana (PMFBY). This scheme was initiated in the kharif season of 2016. It is a very important initiative by the Central Government to provide a comprehensive risk solution at the lowest uniform premium across the country for farmers. As an end to end risk mitigation for farmers, the scheme extends coverage for the entire cropping cycle from pre-sowing to post-harvest including coverage for losses out of prevented sowing and mid-season adversities. Individual farm level losses arising out of localized calamities and post harvest losses are also covered due to perils such as inundation, cloudburst and natural fire. The average sum insured per hectare has increased from ₹ 15,100 during the pre-PMFBY Schemes to ₹ 40,700 under PMFBY.

The scheme was revamped in February 2020 and was made voluntary for all farmers. Further, the states have also been provided flexibility to rationalize the sum assured so that adequate benefits can be availed by farmers. The scheme covers over 5.5 crore farmer applications year on year.

6. Marketing. Although agricultural markets have existed in India for the past hundreds of years, they were largely unorganized and informal and could not play important role to help the farmers and the consumers. The farmers had limited choice because they had little surplus to sell and sold their crops to the village trader-cum-money lenders to pay off their rents, debts and to meet other cost requirements. Sometimes, farmers could venture to take their produce to a market in a nearby town where they could hardly get remunerative price for their produce due to underdeveloped nature of the markets, monopoly of a few traders, illiteracy of the farmers, malpractices like under weigthment, illicit deductions etc. and interlocking of credit and marketing. Simultaneously, price paid by the consumers used to be very high due to high profit of the traders and other middle men. It is estimated that producers received only 53% of the price of rice, 39% of the price of vegetables and 35% in the case fruit.

Although Agricultural Produce (Grading and Marketing) Act was passed way back in 1937, the rural development of agricultural marketing took place in the mid-1960s when the marketed surplus become much larger due to increased production as a result of effect of the Green Revolution. According to the Sub-Group an Estimation of Marketed Surplus Ratio Government of India (2004), 43% production of rice and other cereals and 51.5% of that of wheat were marketed during 2002-03. For pulses and oilseeds the share was 72.4% and 79.6% respectively. Commercial crops recorded a much higher marketed surplus. For example marketed surplus was about 93% in case of sugarcane and cotton. These developments necessitated the growth of agricultural markets and all the states of India enacted legislation known as the State Agricultural Produce (Markets) Act with a view to having regulated markets. The number of regulated markets was only 286 in 1950 and their number stood at 7,157 as on 31st March, 2010. The advent of regulated markets has helped in mitigating the market handicaps of producers/sellers at wholesale assembling level, but rural periodic markets in general, and tribal markets in particular remain out of its development ambit.

Organised marketing of agricultural products has been promoted through a network of regulated markets to ensure reasonable gains to farmers and consumers by creating a market environment conductive to fair play of supply and demand. A modal Agricultural Produce Market Committee (APMC) Act was prepared in 2003 for bringing about reforms in agricultural marketing. Many state governments have brought about amendments in APMC Act to further benefit the farmers and the consumers alike. But some states are yet to implement the provisions of the APMC Act.

Recently a major initiative has been taken to promote modern terminal markets for fruits, vegetables and other perishables in important urban centres of the country. These markets would be provided with state-of-the-art infrastructure facilities for electronic auction, cold storage chains and logistics and operate through primary collection centres conveniently located in producing areas to allow easy access to farmers. They are envisaged to operate on "Hub and Spoke'' format wherein the Terminal market (the Hub) would be linked to a number of collection centres (the Spokes) conveniently locate in key production centres to allow easy access to farmer for marketing their produce. These markets would be set up under the scheme of National Horticulture Mission (NHM).

Agriculture sector requires competitive and well functioning markets with alternate choices to the farmers. Reforms in agricultural markets were initiated to ease out restrictive and monopolistic approach of state and 2-3 grades were prescribed for each commodity. Grades help farmers/traders to get prices of agricultural commodities as per their quality and the consumers get the desired quality. As many as

TABLE 20.12. Procurement of food crops in India according to Marketing year (thousand tonnes)

Crop/ year	2013-14	2014-15	2015-16	2016-17	2017-18	2018-19	2019-20	2020-21	2021-22
Rice (Oct.-Sept.)	31,845	32,040	34,218	38,106	38,174	44,399	51,826	60,078	10,236
Wheat (April-March)	25,092	28,131	28,088	22,961	30,824	35,795	34,132	38,992	43,344

Source : Agricultural Statistics at a Glance, 2021, p. 244.

105 Grading and Marketing Rules (GMR) covering 213 commodities have been notified under the provisions of Agricultural Produce (Grading and Marketing) Act, 1937. These include fruits and vegetables, cereals, pulses, oilseeds, vegetable oils, ghee, spices, honey, etc.

The Department of Agriculture and Cooperation (DAC) has the following three organizations dealing with marketing under its administrative control.

(i) Directorate of Marketing and Inspection (DMI) has its head offices at Faridabad (Haryana). Branch Head Office at Nagpur (Maharashtra), 11 Regional Offices and Central Agmark Laboratory at Nagpur.

(ii) Ch. Charan Singh National Institute of Agricultural Marketing (NIAM) set up Japiur on 8th August, 1988 with a mandate for Training Research Consultancy and Education in the field of Agriculture Marketing.

(iii) Small farmers Agri-Business Consortium (SFAC) registered in January, 1994 for innovative ideas, generating income and employment in rural areas by promoting private investment in agricultural project.

7. Storage and Warehousing. Farmers often find it difficult to retain their production due to lack of storage facilities and are forced to opt for distress sale in the market. Storage facilities are, therefore, of utmost importance to help the farmers for getting remunerative prices and to facilitiate regular flow of agricultural products to the consumers. Considerable storage capacity has been created in the country after Independent and particularly in the post Green Revolution era, both at the national and the state levels. Whereas the Food Corporation of India (FCI) and the Central Warehousing Corporation (CWC) have created large storage facilities at the centres of all India importance, the State Governments and the State Warehousing Corporations (SWCs) have built storage at centres of State/district level importance. FCI is the main agency which provides storage facilities for agricultural products. In addition to its own godowns, its hires storage capacity from other sources such as CWCs, SWCs and private parties.

Although there had been considerable increase in storage capacity, the storage system in India came under severe stress after 2001-02 when marketable surplus in agriculture increased rapidly and the total foodgrains status with the FCI reached levels ranging from 50 to 60 million tonnes. Shortage of storage facilities results in damage of huge quantities of foodgrains. Large quantities of foodgrains are stored in open a large portion of wheat is exposed to rain and rodents.

Table 20.13 shows region-wise/state-wise storage capacity in India. In order to meet the requirement of all time high stock levels the FCI resorted to short term hiring to efficiently manage the stocks. Recognizing the problem of acute shortage of storage capacity, the government has set up a High Level Committee to look into the storage issues. Additional storage capacity has to be created at suitable locations in order to meet the challenge of achieving broad objectives of food security. In this regard, the Government has formulated a scheme for construction of godowns under Private Entrepreneurs Guarantee (PEG) Scheme. Under the PEG scheme, the FCI guarantees 10 year usage of storage capacities to private investors and nine years to Central Warehousing Corporation and State Warehousing Corporations.

Gramin Bhandaran Yojna (GBY) or Rural Godowns Scheme (RGS) was launched as on April 1, 2001 to create a network of rural godowns and to enable the farmers to retain the produce till the market prices are favourable and also to meet their credit requirements. The project for construction of rural godowns can be taken up by individuals, farmers,

TABLE 20.13. Region-wise storage capacity in India (Lakh Metric Tonnes)

Region/States	March 2016	March 2017	March 2018	March 2019	March 2020	March 2021
East Zone : Bihar, Odisha, West Bengal, Sikkim, Jharkhand	45.98	50.44	60.26	60.20	54.55	52.32
North-east Zone : Assam, Arunachal Pradesh, Tripura, Manipur, Mizoram, Meghalaya, Nagaland	9.02	6.04	7.34	7.14	6.37	6.42
North Zone : Delhi, Haryana, Himachal Pradesh, Jammu & Kashmir, Punjab, Chandigarh, Rajasthan, Uttar Pradesh, Uttarakhand	466.79	439.33	390.12	446.31	448.66	441.47
South Zone : Andhra Pradesh, Telangana, Kerala, Tamil Nadu, Karnataka, Puducherry, Andaman & Nicobar Islands, Lakshadweep	97.40	93.16	116.07	110.32	98.59	97.34
West Zone : Gujarat, Maharashtra, Goa, Madhya Pradesh, Chhattisgarh, Daman and Diu, Dadra and Nagar Haveli	195.65	183.96	269.24	231.71	147.78	220.41
Grand Total	814.84	772.93	843.03	855.68	755.94	817.96

Source: Agricultural Statistics at a Glance, 2021, pp. 270-71.

groups of farmers/growers, partnership/proprietory firms, Non-Government Organisations, Self Help Groups (SHGs), Companies, Corporations, Cooperatives, Local bodies other than Municipal Corporations, Federations, Agricultural Produce Market Committees, Marketing Boards and Agro Processing Corporations in the entire country. The Government provides subsidy for constructing these godowns. The scheme has helped the farmers to store their agricultural products near their fields and obtain pledge loans and marketing credit from banks, thereby avoiding distress sale immediately after the harvest.

4. TECHNOLOGICAL FACTORS

Technological factors pertain to high yielding varieties of seeds, chemical fertilisers, insecticides, pesticides and farm machinery, etc.

Seeds. High Yielding Varieties (HYV) of seeds comprise one of the most important inputs for enhancing agricultural productivity and production. Efficiency of other agricultural inputs such as fertilisers, pesticides and irrigation is largely determined by it. Seed quality is estimated to account for 20-25 per cent of productivity. In addition to their genetic potential for high grain yield, the other important attributes of HYV seeds are (*a*) their high level of responsiveness to fertilisers and other inputs. (*b*) their high grain stalk ratios, and (*c*) photo-period insensitivity and short period maturity in most cases. The new seed varieties being short maturing also enable double cropping, have dwarf stems which are tough enough to carry heavy load, are resistant to wind damage, and have large leaf surface for faciliting photosynthesis. (Bhalla 2007: 101).

The High Yielding Variety Programme (HYVP) was launched in 1966-67 and covered 18.9 lakh hectares or 2.2% of the total cropped area of the six involved crops (rice, wheat, jowar, bajra, maize and ragi) in that year. Since then, the production and consumption of HYV seeds has been increasing steadily. At present more than four-fifth of the total cropped areas under these crops enjoys the facility of HYV seeds. The Central Government has been addressing this issue through various programmes/ schemes. This includes the Indian Seed Programme involving the participation of the Central and State Governments. Indian Council of Agricultural Research (ICAR), State agricultural universities, cooperatives and the private sector and farmers and plant breeders. Currently the seed production system consists of production of breeder, foundation and certified seeds and their distribution. ***Breeder seeds*** represent the first stage of quality seed development and are prepared by the ICAR and some selected agricultural universities. ***Foundation seeds*** represent the second stage of quality seed development and are produced from the breeder seeds. These seeds are produced by the National Seeds Corporation (founded in 1963) and some selected agricultural universities. The seed multiplication and supply is done through a large number of agencies such as Central Government Seed Farms, State Seed Farms, State Departments of Agriculture, private seed producers and Seed Producers' Corporatives. Private producers produce seeds under strict supervision of scientists in the universities. ***Certified seed*** is the ultimate stage in seed production chain and is the progeny of foundation seed. The seeds have to be certified by the competent authorities before they are released to farmers for use in their fields. **Table 20.14** shows the progress in production and distribution of HYV seeds.

The Seed Act 1966 provides for legislative framework for regulation of quality seeds sold in the country. Considering the vital importance of the seed industry in promoting agricultural growth, the Ministry of Agriculture has been proposing replacement of the existing Seeds Act, 1966 by a suitable legislation. The Seeds Bill 2004 aims to regulate the quality of seeds and replaces the Seeds Act, 1966. Seed Bill 2019 aims to regulate the quality of seeds sold and facilitate the production and supply of these seeds to farmers.

A Central Sector Scheme for Development Strengthening of Infrastructure Facilities for production and distribution of quality seeds with the aim of making quality seeds of various crops available to farmers at affordable price is under implementation since 2005-06. As a result of this initiative, availability of certified quality seeds has increased from 126.75 lakh quintals in 2005-06 to 352.01 lakh quintals in 2017-18. For achieving timely availability of seeds at affordable price to farmers in hilly/remote areas of north-eastern states, a transport Subsidy on Movement of Seeds scheme is in operation.

In general, food crops have been given top priority so that sufficient foodgrains are made available to the fast growing population. Oilseeds have also received considerable attention because India heavily depends on imports of oilseeds to meet her growing demand.

Merits of HYV Seeds. A brief description of merits of HYV seeds has been given in the beginning of section on HYV seeds. However, a detailed description is required to appreciate the merit of these seeds and their contribution to enhancement in agricultural productivity will be very useful at this stage :

1. Suitable for use of Fertilisers. HYV seeds are suitable for use of fertilisers. It has been observed that new seeds give much higher yields as compared to old seeds for the same amount of fertilisers input. This is the reason that areas using higher amounts of new seeds are also using fertilisers in larger quantities.

2. Shorter Life Cycle. HYV seeds have shorter life cycle. These seeds give early maturing crops and give greater opportunities to a farmer to venture for multiple cropping. For example, new seeds of rice and wheat complete life cycles in 100 and 110 days respectively in contrast to 130 and 150 days respectively in case of traditional varieties of seeds. Thus new seeds help in increasing the farm production and economise on land.

3. Economical on Irrigation. Although new seeds require large amount of irrigation for proper growth of crops, yet the per quintal requirement of irrigation is much lower as compared to old seeds. This is due to the fact that per hectare yields in quintals is much higher in case of new seeds when compared with those of old seeds.

4. Employment. Agriculture based on HYV seeds requires more labour per unit area and generates more employment. Prior to the introduction of new seeds in the Indian agriculture, farmers were largely dependent on monsoon rainfall. This was particularly the case in the Northern-Plains of India. The farmers could not find any employment during the period between the harvesting of rabi crops and sowing of kharif crops and were forced to sit idle from April to June. With the introduction of new seeds, the farmers can grow second crop immediately after harvesting the first and are thus able to find employment throughout the year.

5. Easy to Adopt. It does not require any special skill to adopt HYV seeds. They can be easily adopted by any class of farmers, whether big landlords or small farmers after elementary training. However, a little adjustment of dates of sowing the wheat crop is required because this crops requires relatively cool temperature at the time of

TABLE 20.14. Production of Breeder and Foundation Seeds and Distribution of Certified/ Quality Seeds

Year	2015-16	2016-17	2017-18	2018-19	2019-20	2020-21
Production of Breeder seeds (Thousand Quintals)	90.37	110.71	105.08	104.26	92.7	91.16
Production of Foundation Seeds (Lakh Quintals)	14.95	22.09	19.54	18.00	22.25	24.12
Distribution of Certified/Quality Seeds (Lakh Quintals)	304.04	348.58	352.01	320.41	383.72	421.09

Source : Agricultural Statistics at Glance, 2021, pp. 274-75.

TABLE 20.15. Crop-wise Distribution of Certified/Quality Seeds (lakh quintals)

Crops/Year	2015-16	2016-17	2017-18	2018-19	2019-20	2020-21
Cereals (wheat, paddy, maize, jawar, bajra, ragi, barley)	194.95	229.11	238.00	206.87	229.98	246.00
Pulses (gram, lentil, peas, urad, moong, arhar, cowpea, others)	22.71	29.47	23.54	31.80	34.16	32.70
Oilseeds (groundnut, rapeseed and mustard, til, sunflower, linseed, castor seeds, safflower, others)	47.44	49.97	57.23	48.26	53.09	52.26
Fibre crops (cotton, jute, mesta, others)	2.49	2.17	2.46	2.46	3.21	3.04
Other miscellaneous crops	36.45	0.71	30.77	31.02	63.27	87.37
Total	**304.04**	**348.04**	**352.01**	**320.41**	**383.41**	**421.09**

Source : Agricultural Statistics at a Glance, 2021, pp. 295-96.

sowing and during early period of its growth. This is due to the fact that wheat based on new seeds is of early maturing variety and has to be sown a little later. For example, in Punjab, Haryana and western part of Uttar Pradesh wheat based on old seeds was sown in the beginning of October, but wheat based on new seeds is now sown in November-December in these areas. In contrast, rice is a kharif crop and does not require much adjustment in time of sowing. However, introduction of new seeds has enabled the farmers to take two crops of rice in place of only one crop obtained earlier on the basis of old seeds.

Constraints in Adoption of New Seeds

Although new HYV seeds have entirely transformed agricultural scenario in India and have made India from a food deficit to a food surplus country even in the wake of rapidly increasing population, yet there are some constraints in the use of these seeds. Some of the major difficulties in adopting these seeds are briefly described as under :

1. Irrigation. New seeds do not tolerate dry weather and water shortage and need copious irrigation for the successful growth of crops based on these seeds. The monsoon rainfall occurs only for 3-4 months in the whole duration of the year and remaining 8-9 months are almost completely dry. Thus Indian agriculture is heavily dependent on irrigation and this dependency has further increased with the introduction of new seeds. New seeds have led to high crop intensity which requires much higher level of irrigation. The new seeds need right amount of irrigation at right time during the crop's growth. Under irrigation or over irrigation or irrigation at inappropriate time can hinder the proper growth of crops. It has been estimated that in areas like Punjab, Haryana and western part of Uttar Pradesh, 50 per cent increase in the wheat yields can be obtained by irrigation without other inputs like fertilisers etc. The first irrigation of wheat around the third week of sowing alone can increase the yields by as much as 30 per cent. Other inputs like fertilisers, pesticides, insecticides etc. are of no use if proper irrigation at the proper time is not provided to crops.

2. Fertilisers. Indian soils, particularly those of the Great Plains of North India, have been tilled for the last thousands of years. As such they are exhausted and they have lost much of their fertility. This loss of fertility is to be compensated by using chemical fertilisers because conventional manures like cowdung, compost and green manure are not able to cope with the increasing demand. The introduction of new seeds has further increased the demand for chemical fertilisers because these seeds require higher inputs of fertilisers for giving higher crop yields. As mentioned earlier, new seeds have dwarf stems which are strong enough to carry heavy load, are resistant to wind damage, and have large leaf surface to facilitate photosynthesis. All these characteristics require heavy dose of chemical fertilisers. This is the reason that demand for chemical fertilisers has increased considerably since mid-1960s.

3. Insecticides and Pesticides. Crops based on new seeds are very delicate and are highly susceptible insects, pests and diseases. This is due to the fact that new seeds require heavy dose of irrigation, fertilisers and air as well as soil contain a lot of moisture and chemicals. These conditions lead to fast growth and multiplication of insects and pests in hot and humid climate of India. These insects and pests cause heavy damage to crops and reduce their yields substantially. This problem can be solved by spraying the crops with insecticides and pesticides at required intervals. For this purpose, the farmers are supposed to have adequate knowledge of the crop diseases and the types of chemicals to finish such diseases. The diseases may spread from one field to another as from one village to another village. Therefore quick and sufficient steps must be taken to spray the crops with the required insecticides and pesticides. Otherwise the insects and pests may damage the crop partially or wholly.

4. Capital Constraint. Use of new seeds is capital intensive because their use needs heavy inputs of irrigation (canal, tube wells), chemical fertilisers, insecticides and pesticides and agricultural machinery (tractors, harvesters, threshers, sprayers etc.). All these inputs are complimentary to one another and one cannot be of any use in the absence of the others. Their purchase requires sufficient capital which majority of the farmers, especially small and marginal farmers, do not possess. They should be extended loans at easy terms by different agencies like State Cooperative Banks, District Central Cooperative Banks, Primary Agriculture Credit Societies, etc. Unfortunately, it is the rich farmers who get the maximum advantage of loan facilities due to their strong influence and corrupt officials and poor farmers are usually deprived of such facilities. Therefore, there is urgent need to strengthen the agencies extending loans to farmers so that the most needy farmers get their due.

5. Mechanization. Use of modern agricultural machines is also essential for successful cultivation of HYV seeds. Raising of 2-3 crops is possible only by applying modern technology. The traditional farm implements like sickel, hoe, wooden plough, bullock cart, etc. are more labour intensive and less efficient. Modern machines like tractors, harvesters, threshers, tillers, sprayers, pumping sets, etc. are less labour intensive and more efficient. Since the farm machines are more efficient, they have become big time savers and the farmer gets sufficient time to prepare for the second crop after harvesting the first crop. Thus it is possible for the farmer to take more than one crop during the course of year from the same field. Farm mechanization also helps in optimum utilization of the complementary inputs like irrigation, fertilisers, insecticides, pesticides, etc. Availability of power in the form of hydroelectricity or diesel at cheap rates is of paramount importance for mechanisation of Indian agriculture. Power gives motion to farm machines in the absence of which the entire farm machinery comes to a stand still.

Farm mechanization and crop productivity have direct correlation as farm mechanization saves time and labour, reduces drudgery, cut down production cost in the long run, reduces post harvest losses and boosts crop output and farm income. Use of improved implements has potential to increase productivity up to 30 per cent and reduce the cost of cultivation up to 20 per cent. At present, Indian farmers are adopting farm mechanization at a faster rate in comparison to the recent past. Although, the sale of tractors is not the only measure of farm mechanization but to a great extent it reflects the level of mechanization. Indian tractor industry have emerged as the largest in the world and account for about one-third of total global tractor production.

According to World Bank estimates, half of the Indian population would be urban by the year 2050. It is estimated that percentage of agricultural workers of total work force would drop from 58.2 per cent in 2001 to 25.7 per cent in 2050. Thus, there is a need to enhance the level of farm mechanization in the country.

6. Marketing and Storage Facilities. The use of HYV seeds leads to high agricultural productivity which requires adequate facilities of marketing and storage. Food Corporation of India (FCI) was established by the Government of India for storing foodgrains which has helped in storing foodgrains after harvesting and releasing these foodgrains at the time of need. Usually the farmers take their agricultural products to the nearby urban market and sell them there. They also purchase seeds, fertilisers, pesticides, insecticides, farm machinery and things of daily use from these markets. For successful marketing of the agricultural products, it is essential that villages should be connected to the nearby urban markets by all weather metalled roads. A well developed transport system acts as blood veins for the producers and the consumers alike. Sale of perishable commodities like milk, vegetables, fruits, flowers, etc. in the urban markets from the surrounding rural areas is simply impossible in the absence of a cheap and efficient transport network.

7. Extension Services. HYV seeds are very sensitive to irrigation, fertilisers, pesticides, insecticides, sowing, weeding and thinning, etc. and mismanagement of any one of these inputs can adversely affect the farm productivity. For all these things to function properly, the farmers need to be adequately educated. This requires extension services by well trained, efficient and dedicated agents. Full utilization of extention services requires perfect understanding and coordination between farmers extension agents, farm supervisors, researchers and agricultural scientists. Any slackness on any front will lead to low agricultural productivity and overall poverty of farmers and aggravate the food problem.

8. Human Factor. Like all other major inputs, adoption of HYV seeds depends very much on human factor. Utilization of facilities like fertilisers, pesticides, insecticides and farm machinery for proper use of HYV seeds much depends on attitude of the farmers. There are certain farmers who are open to new ideas and technologies and adopt them easily. On the other hand there are other farmers who are conservative and orthodox in their approach towards life and do not accept new ideas in agriculture. In almost all the villages of India there are some progressive farmers who have improved their farm production and their standard of living while their conservative and orthodox counterparts are still living a life of poverty, deprevation, hunger and starvation. Unfortunately, farmers in large parts of India are still uneducated and do not adopt new technologies easily. This is the reason that in spite of large scale progress in Indian agriculture after 1960s, Indian agricultural productivity is still far behind the world averages. The effect of Green Revolution has been maximum in Punjab and Haryana because farmers in these states are progressive and hardworking and adopt new technologies with ease and comfort. These two states have recorded unprecedented increase in the production of wheat after the advent of the Green Revolution. A similar progress has been noted in the coastal areas of Andhra Pradesh and Tamil Nadu with respect to increase in rice production because the farmers in these areas have easily adopted new techniques of rice cultivation.

Deficiencies of HYV Seed Programme

Although HYV seed programme has contributed a great deal in increasing the agricultural productivity. some glaring deficiencies in this programme have been noticed. Lack of appropriate scientific tests by the concerned government agencies before buying the seeds from seed farmers and distribution of seeds by some unscrupulous private dealers have resulted in reduced yields, variability in performance and increased growth of pests and insects. All these factors have exposed the farmers to bigger risks than ever before. Following concerns have become more serious at present :

(*i*) Inadequate inspection and certification of seeds has resulted in mushrooming seed production under lax scientific conditions. There is flood of non-certified seeds marked as "Truthfully Marked Seeds" in the market and the farmers have to bear the brunt of this unhealthy practice.

(*ii*) Non-availability of good quality seeds has led to increasing tendency for farmers to produce their own seeds. The seed replacement rate continued to be much below 20% for most crops. As a result there is deterioration of quality seeds and the yields are adversely affected.

(*iii*) The major emphasis of seed programme has been on the development of better seeds for wheat and rice and other crops have received much less attention.

(*iv*) Throwing seed trade open to multinationals has led to entry of genetically modified seeds in some high value crops like cotton, vegetables and fruits. Most important was Bt. cotton. Introduction of this seed led to phenomenal increase in yields of cotton in the beginning to but a stage of saturation reached soon. The yield of cotton increased from 106 kg/ha in 1970-71 to 499 kg/ha in 2010-11 and fell to 477 kg/ha in 2017-18. Moreover due to lack of affective legislation, some unscrupulous traders started selling spurious seeds to the farmers. This combined with bad weather led to ruination of many cotton farmers driving some to commit suicide. (Bhalla 2007:104). Hundreds of farmers growing cotton have committed suicide in Maharashtra and other states due to crop failure.

Fertilisers

Fertilisers are a crucial input for increasing the farm productivity and new HYV seeds are of no use without fertilisers. High level of plant nutrients is essential to achieve the objective of increasing farm production through high cropping intensity under multiple cropping programme. Green Revolution would have remained a dream for India but for the introduction of fertilisers, in spite of large scale use of HYV seeds and irrigation. In fact HYV seeds fertilisers and irrigation is that ***trilogy*** which has led to unprecendented increase in agricultural production.

Production, Import and Consumption of Fertilisers. India is the third largest producer of fertilisers after China and USA and second largest consumer after China in the world. There are three main types of chemical fertilisers which are known as Nitrogenous (N), Phosphatic (P) and Potassic (K) fertilisers. Nitrogeneous fertilisers are the most important and account for about three-fourth of the total fertilisers production in the country. The remaining one-fourth are phosphatic and potassic fertilisers. The production of all fertilisers (NPK) increased from 90.45 lakh tonnes in 1990-91 to 181.09 lakh tonnes in 2017-18. Thus the production more than doubled within a short span of less than two decades. Similarly imports and consumption also have recorded phenomenal increase (**Table 20.16**). Uttar Pradesh is the largest consumer with a consumption of 4,655 thousand tonnes in 2017-18. This is followed by Maharashtra consuming more than 2,943 thousand tonnes. Most of the north-eastern states are least dependent on chemical fertilisers. Entirely different picture emerges when we look at

TABLE 20.16. Consumption, Production and Import of Fertilisers *(lakh tonnes)*

Year	Consumption				Production				Imports			
	N	P	K	Total	N	P	K	Total	N	P	K	Total
1990-91	79.97	32.21	13.28	125.46	69.93	20.52	0.00	90.45	4.14	10.16	13.28	27.58
1995-96	98.23	28.98	11.56	138.77	87.77	25.58	0.00	113.35	19.93	6.47	13.15	39.55
2005-06	127.23	52.04	24.14	203.41	113.55	42.21	0.00	155.76	13.85	11.21	27.47	52.53
2010-11	165.58	80.50	35.14	281.22	121.57	42.23	0.00	163.80	44.92	38.02	40.69	123.63
2015-16	173.72	69.79	24.02	267.53	134.16	43.94	0.00	178.10	50.68	28.88	20.53	100.09
2016-17	167.35	67.05	25.08	259.49	133.54	45.95	0.00	179.49	33.88	21.29	23.18	78.35
2017-18	169.58	68.54	27.79	265.91	133.86	47.23	0.00	181.09	35.88	20.47	28.95	85.30
2018-19	179.02	68.57	25.29	272.88	132.98	45.94	0.00	178.92	47.01	31.67	26.29	104.97
2019-20	191.00	76.62	26.07	293.69	136.85	47.91	0.00	184.76	51.91	24.13	22.6	98.84
2020-21	204.04	89.78	31.54	325.36	137.15	47.39	0.00	184.54	56.33	25.43	26.7	108.46

Source : Agricultural Statistics at a Glance, 2021, pp. 277-78.

TABLE 20.17. State-wise Consumption of Fertilisers in terms of Nutrients (N, P and K) 2020-21

Sl. No.	State/Zone	Total Consumption (Thousand tonnes)				Consumption in kg/hectare			
		N	P	K	Total	N	P	K	Total
South Zone									
1.	Andhra Pradesh	1,139.63	618.59	267.72	2,025.94	116.80	63.40	27.44	207.64
2.	Telangana	1,169.73	474.60	173.22	1,817.55	129.06	52.36	19.11	200.53
3.	Karnataka	1,179.45	688.20	355.19	2,222.84	71.91	41.96	21.65	135.52
4.	Kerala	88.06	37.96	75.09	201.11	17.35	7.48	14.79	39.62
5.	Tamil Nadu	644.86	269.32	206.72	1,120.90	90.83	37.93	29.12	157.88
6.	Puducherry	8.48	2.16	1.55	12.19	217.07	55.29	39.68	312.04
7.	A & N Islands	0.13	0.00	0.00	0.13	3.24	0.00	0.00	3.24
8.	Lakshadweep	0.00	0.00	0.00	0.00	—	—	—	—
	SZ Total	**4,230.34**	**2,090.83**	**1,079.49**	**7,400.66**	**89.10**	**44.04**	**22.74**	**155.88**
West Zone									
9.	Gujarat	1,349.67	470.81	136.93	1,957.41	99.47	34.70	10.09	144.26
10.	Madhya Pradesh	1,732.70	1,020.62	140.19	2,893.51	57.73	34.00	4.67	96.40
11.	Chhattisgarh	525.09	271.99	74.53	871.61	78.14	40.47	11.09	129.70
12.	Maharashtra	1,699.85	1,127.93	585.82	3,413.60	64.33	42.69	22.17	129.19
13.	Rajasthan	1,242.56	509.77	24.66	1,776.99	47.17	19.35	0.94	67.46
14.	Goa	1.77	0.93	0.96	3.66	18.12	9.52	9.83	37.47
15.	Daman & Diu	0.00	0.00	0.00	0.00	0.00	0.00	0.00	0.00
16.	D & N Haveli	0.52	0.36	0.00	0.88	28.73	19.89	0.00	48.62
	WZ Total	**6,552.16**	**3,402.41**	**963.09**	**10,917.66**	**63.50**	**32.97**	**9.33**	**105.80**
North Zone									
17.	Haryana	1,107.03	318.45	39.19	1,464.67	167.05	48.05	5.91	221.02
18.	Punjab	1,495.18	382.65	55.42	1,933.25	190.81	48.83	7.07	246.71
19.	Uttar Pradesh	3,963.73	1,426.83	238.43	5,628.99	133.06	47.90	8.00	188.96
20.	Uttarakhand	120.68	28.99	8.37	158.04	99.39	23.87	6.89	130.15
21.	Himachal Pradesh	37.75	11.25	10.26	59.26	41.20	12.28	11.20	64.68
22.	Jummu & Kashmir	114.10	37.56	28.84	180.50	59.73	19.66	15.10	94.49
23.	Delhi	10.71	1.70	0.29	12.70	110.46	17.53	2.99	130.98
24.	Chandigarh	0.00	0.00	0.00	0.00	—	—	—	—
	NZ Total	**6,849.18**	**2,207.43**	**380.80**	**9,437.41**	**141.54**	**45.62**	**7.87**	**195.03**
East Zone									
25.	Bihar	1,285.28	457.97	167.71	1,910.96	136.42	48.61	17.80	202.83
26.	Jharkhand	144.19	58.24	5.72	208.15	34.62	13.98	1.37	49.98
27.	Odisha	347.24	178.88	85.07	611.19	38.67	19.92	9.47	68.06
28.	West Bengal	800.67	518.55	418.39	1,737.61	83.93	54.36	43.86	182.15
	EZ Total	**2,577.38**	**1,213.64**	**676.89**	**4,467.91**	**80.28**	**37.80**	**21.08**	**139.16**
North East Zone									
29.	Assam	175.63	54.76	49.23	279.62	35.68	11.13	10.00	56.81
30.	Tripura	7.27	6.18	2.59	16.04	13.82	11.75	4.92	30.50
31.	Manipur	10.73	2.54	1.61	14.88	29.13	6.89	4.37	40.39
32.	Meghalaya	0.00	0.00	0.00	0.00	—	—	—	—
33.	Nagaland	0.30	0.04	0.00	0.34	—	—	—	—
34.	Arunachal Pradesh	0.00	0.00	0.00	0.00	—	—	—	—
35.	Mizoram	0.99	0.13	0.00	1.12	7.76	1.02	0.00	8.78
36.	Sikkim	0.00	0.00	0.00	0.00	—	—	—	—
	NE Total	**194.92**	**63.65**	**53.43**	**312.00**	**32.13**	**10.49**	**8.81**	**51.44**
	ALL INDIA	**20,403.98**	**8,977.96**	**3,153.7**	**32,535.64**	**86.01**	**37.85**	**13.29**	**137.15**

Source : Agricultural Statistics at a Glance, 2021, pp. 279-84.

per hectare consumption of fertilisers. From this point of view, Telangana is at the top consuming 264.46 kg/ha. Again, north-eastern states are the least consumers per hectare. Among the union territories. Daman and Diu uses over 35 kg/hectare and Lakshadweep does not use chemical fertilisers at all. (**Table 20.17**).

India meets 80 per cent of its urea requirement through indigenous production but is largely import dependent for meeting her requirements of potassic (K) and phosphate (P) fertilisers. The government has notified the New Investment Policy 2012 (NIP-2012) in the urea sector which will encourage investments leading to increase in indigenous capacities, reduction in import dependence and savings in subsidy due to import substitution at prices below import parity price (IPP). The New Urea Policy-2015 was notified with the objectives of maximising indigenous urea production, promoting energy efficiency in urea production; and rationalising subsidy burden on the government.

Although India has progressed a lot with regard to production and consumption of fertilisers, we still lag far behind several countries of the world so far as consumption of fertilisers per hectare is concerned (**Table 20.18**). Even some of the neighbouring countries like China and Bangladesh consume much more fertilisers than India. Most of the European countries, Egypt in Africa, Chile in South America, Japan in Asia and New Zealand in Oceanic use much larger quantities of fertilisers than India. This brings us to the conclusion that there is much scope for increasing the consumption of fertilisers so that our agricultural productivity increases and we are able to meet the growing needs of rapidly increasing population. However, proper consideration should be given to environmental problems arising out of thoughtless use of fertilisers. Already we are facing so many environmental problems resulting from unscientific use of fertilisers.

Subsidy on Fertilisers. Under the Nutrient Based Subsidy (NBS) scheme for phosphatic and potassic (P&K) fertilisers implemented in 2010, a fixed amount of subsidy, decided on annual basis, is provided to each grade of P&K fertilisers, depending upon its nutrient content. An additional subsidy is also provided to secondary and micro-nutrients.

TABLE 20.18. Fertilisers Consumption per hectare of Arable Land in selected Countries in 2016

Country	Consumption/Fertilisers (kg/hectare)
1. Egypt	427.0
2. Chile	243.7
3. Bangladesh	268.2
4. China	354.1
5. India	153.1
6. Japan	230.9
7. Republic of Korea	291.2
8. Malaysia	232.7
9. Vietnam	242.7
10. Belarus	169.1
11. France	154.5
12. Germany	193.9
13. Netherlands	231.1
14. U.K.	252.4
15. New Zealand	1,400.9

Source : Agricultural Statistics at a Glance 2018, pp. 319-20.

Under this scheme, manufacturers/marketeers are allowed to fix the maximum retail price (MRP).

Economizing the use of Fertilisers and Pesticides. For the small and marginal farmers, the costs of fertilisers are key determinants of profitability of farming. Since 2002-03, the fertiliser consumption had continually increased till 2010-11. After that varying trends have been found but the general trend has been towards declining use. Further, the **fertiliser response ratio** is showing a declining trend. Declining fertiliser response ratio is an indicator of declining responsiveness of soil fertility to fertiliser application. According to Department of Fertilisers, the declining fertiliser response ratio in Indian agriculture is due to inadequacy and imbalance in fertiliser use, increasing multinutrient deficiency, lack of farmer's awareness about balanced plant nutrition and poor crop management. The improvement in fertiliser use efficiency (*i.e.* yield per unit of fertiliser input) requires farmers' knowledge regarding the right product, dosage, time and method of application. Some of the suggested measures are the use of optimal dose based on soil health status,

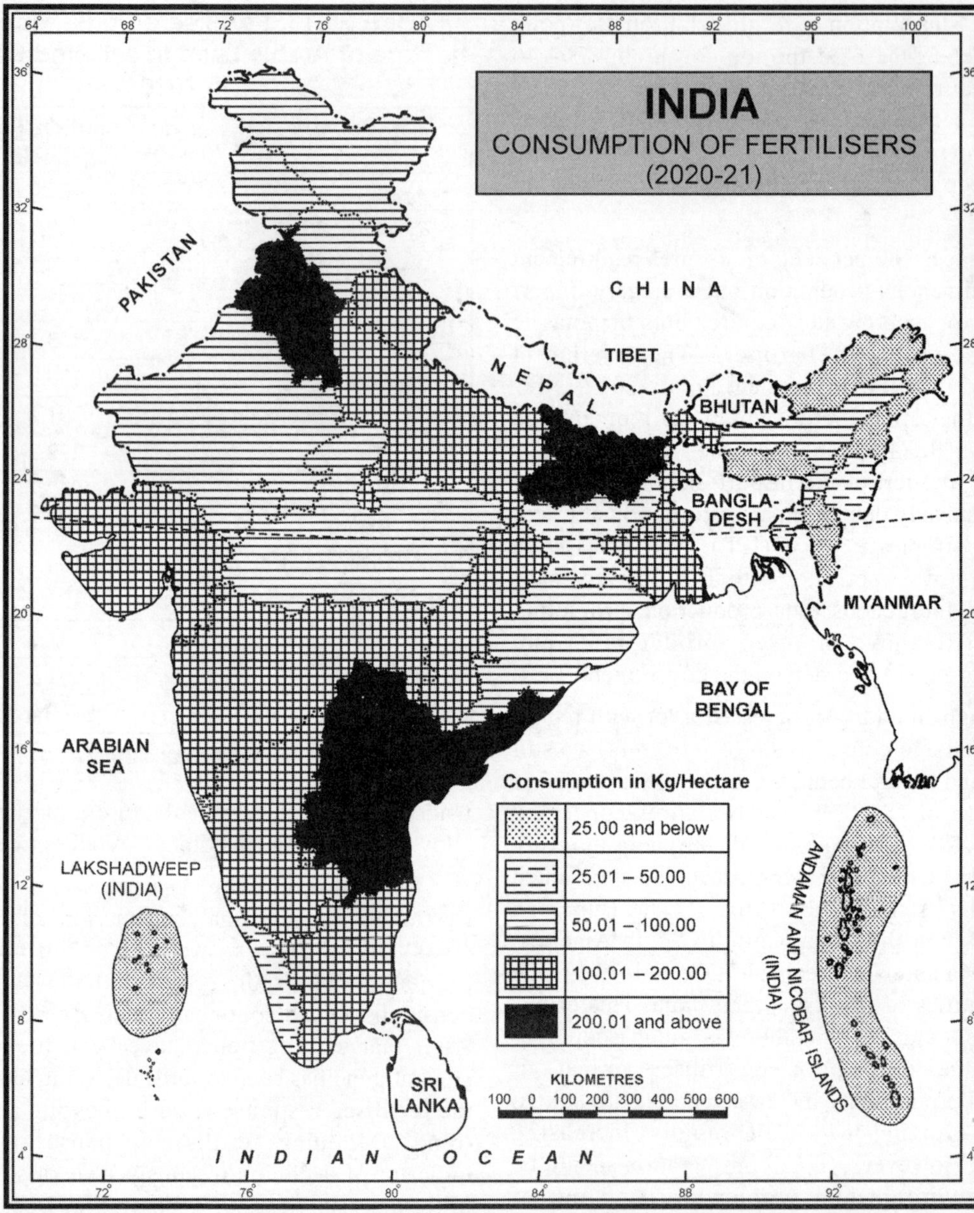

FIG. 20.6. India : Consumption of fertilisers in kilograms per hectare (2020-21)

promotion of neem-coated urea, promotion of micronutrients, promotion of organic fertilisers, and promotion of water soluble fertilisers.

Unbalanced Use of Fertilisers. One of the major problems of fertilisers in India is their unbalanced use. According to recommendation of the agricultural scientists, the ideal ratios in which nitrogeneous, phosphatic and potassium (NPK) fertilisers should be used is 4:2:1. This ratio was 6.0:2.4:1 in 1990-91. The prices of different fertilisers were changed after the Government of India adopted the policy of liberalization in 1991 which resulted in highly unbalanced use of fertilisers and it rose to 9.7:2.9:1 in 1993-94. Some of the agriculturally advanced states like Punjab, Haryana, Uttar Pradesh, Andhra Pradesh and Tamil Nadu showed still greater unbalanced use of NPK fertilisers. **Table 20.19** gives an idea of unbalanced use of fertilisers in Bihar, Haryana and Punjab. This led to a serious degradation of soil nutrients. In order to remove this unbalanced use of fertilisers, the Government of India come out with subsidies of varying extent to different types of fertilisers. In spite of steps taken to correct the unbalanced use, the NPK ratio stood at 8.2 : 3.2 : 1 in 2012-13 which still remained skewed in favour of N and P against K. Use of urea in higher proportion adversely affects the soil profile. Proper growth of plants occurs only when the soil contains all sixteen nutrients. The soil should have sufficient quantity of sulphur, zinc and calcium in addition to NPK.

TABLE 20.19. Unbalanced use of NPK fertilisers

	2009-10	2010-11	2012-13
Bihar	5.3 : 1.5 : 1	5.8 : 1.9 : 1	12.3 : 3.6 : 1
Haryana	15.9 : 5.5 : 1	20.4 : 6 : 1	61.4 : 18.7 : 1
Punjab	18.4 : 5.9 : 1	19.1 : 5.4 : 1	61.9 : 19.3 : 1
All India	4.3 : 2 : 1	5.0 : 2.4 : 1	8.2 : 3.2 : 1

Source : Economic Survey 2013-14, p. 144.

Fatigue of the Green Revolution and stagnation in agricultural growth in 1990s was largely due to unbalanced used of fertilisers. Crop productivity has not increase in proportion to the use of fertilisers in areas of higher consumption of fertilisers. It seemed that crop production in India has reached its zenith with reference to the present day level of technological advancement but is still at a very low level when compared with some other advanced countries.

Fertiliser Quality Control. The Central Government ensures the quality of fertilisers through Fertilisers (Control) Order (FCO) issued under Essential Commodities Act 1953 to regulate the trade, price, quality and distribution of fertilisers in the country. The State Governments are the enforcement agencies for implementation of provisions of FCO, 1985. The order prohibits the manufacturing, import and sale of any fertiliser, which does not meet prescribed standards.

Farm Mechanisation

Like other inputs such as HYV seeds, fertilisers, irrigation, etc., farm mechanization is very important and has immense potential for improving farm productivity. It has been estimated that adoption of appropriate mechanization of farm operations can increase farm productivity by 30 per cent, cropping intensity by 5-20 per cent, fertilisers and chemicals upto 15-20 per cent and time labour upto 20-30 per cent. Unfortunately, a vast majority of Indian farmers still use primitive and inefficient farm implements thereby causing great hinderance in the path to agricultural progress. The wooden plough and the bullock cart are still most commonly used implements in large parts of the country. However, some progress has been made towards farm mechanization and impressive achievement is visible in the field of power-operated irrigation pumps. At present about 84 lakh electric pumps/tube wells are operating in different parts of the country. Tractors are the main power source for various farm operations and India is the world leader in tractor promotion. The number of tractors sold in the country increased from a modest 248 thousand in 2004-05 to 797 thousand in 2017-18. Uttar Pradesh has the largest number of tractors and accounts for over 16% of all tractors in India. This is followed by Maharashtra (10.3%), Madhya Pradesh (10.2%), Rajasthan (9.1%) and Gujarat (6.9%). Bihar, Telangana, Haryana, Karnataka, Punjab and Andhra Pradesh also have sufficient number of tractors. Tillers, harvesters, threshers etc. are other important machines being used in agricultural operations.

Small, fragmented and shrinking size of the holdings and low purchasing power of small and marginal farmers are making individual ownership of agricultural machinery progressively uneconomical. This requires steps for setting up of custom-hiring centres/high-tech machinery banks so that small and marginal farmers can reap the benefits of farm mechanization. So far, advantages of farm mechanization are available to rich landlords only who own vast stretches of land and have sufficient financial resources to purchase costly farm machinery. To help the poor farmers, the government initiated a Sub-Mission on Agriculture Mechanisation in the Twelfth-Five Year Plan. Introduction of technologically advanced equipments through extension and demonstration besides institutional credit has also been taken up. A huge industrial base for manufacturing agricultural machines has also been developed.

TABLE 20.20. Year-wise sale of Tractors and Power tillers in India (Thousands)

Year	2004-05	2010-11	2015-16	2016-17	2017-18	2018-19	2019-20	2020-21	2021-22
Tractors	248	545	571	662	797	890	880	899	842
Power Tillers	17	55	46	45	52	52	45	54	—

Source: Agricultural Statistics at a Glance 2021, p. 327.

CROPPING PATTERNS

Cropping pattern refers to proportion of area under different crops at a given point of time. It gives an idea of relative importance of different crops in a region or a country at specific time. Cropping pattern is not static; it changes with reference to time and space. Any change in cropping pattern reflects the change in land used to grow different types of crops. At the time of Independence, about three-fourths of the total cropped area was under food crops. There was very little diversification in the pattern of cropping and almost entire Great Plain of North India was dominated by food crops including sugarcane. Cotton in Maharashtra, tea in Assam and jute in West Bengal were the other major crops.

There was no major change in cropping pattern in the country during the first three Five Year Plans from 1951 to 1966. However, food crops gained further importance with the advent of the Green Revolution and due to rapid increase in population. As much as 82% of the cropped area was devoted to food crops. The major increase in area, yield and production was in case of wheat and rice, two main food crops of India. In due course of time diversification in the cropping pattern started showing its impact and in 1990-91, the pattern became rather complex as non-food crops became more prominent. Rice and wheat covered about 40% of the cropped area while oilseeds became more prominent occupying 13% and pulses occupied 14 per cent of the cropped area.

Identification of cropping pattern becomes easy when we work out the cropping pattern of area by considering the dominant (primary), major, secondary and minor crops. Normally dominant crops are those that are first rank crops; major crops occupy over 15 per cent of the cropped areas, secondary crops 5 to 15 per cent and minor crops less than 5 per cent. Thus the number of crops and their relative strength in land occupance makes for emergence of any number of cropping pattern regions. For the sake of simplicity minor crops are usually eliminated and are not considered while reckoning with cropping pattern.

In any cropping pattern, each crop is given its position in terms of percentage in relation to the total cropped area of an areal unit. This is expressed as

$$C_p = \left(\frac{C_a}{N}, \frac{C_b}{N}, \frac{C_c}{N} \cdots\cdots \frac{C_z}{N}\right) \times 100$$

where, C_p represents the cropping pattern : C_a, C_b, C_c etc. the cropped area under crops a, b, c ... in an enumeration unit and N is the total cropped area in the same unit.

A review of the cropping pattern of India highlights the fact that cropping pattern in the country is still tradition bound in which food crops occupy prominent place and little attention is paid to the commercial crops. Even among the food crops, rice is the most important crop which is the staple food of people living in the southern part of the country. Wheat is the second most important food crop which is primarily grown in the north-western part of the country. Coarse grains like jowar, bajra, maize, barley, ragi, etc. are given comparatively less importance. The cropping pattern in a large number of regions in India is typical of an underdeveloped agricultural economy in which most of the cultivated area is devoted to subsistence foodgrains. These are mainly produced for domestic use and there is hardly any surplus to be sold in the market. In spite of the unprecedented increase in the production of foodgrains due to positive effects of the Green Revolution, food surplus is still a far cry in view of the fast increasing population. Cash crops constitute only a negligible percentage of the cropped area.

Factors affecting cropping pattern

It is but natural to find regional variations in cropping patterns in a vast country like India. These variations are due to geographical, economic and political factors.

I. Geographical Factors

Geographical factors like relief, soil, temperature

and rainfall bring about regional variations in cropping patterns in their own way.

1. Relief. Most of the crops prefer plain areas where vast stretches of land are available for cultivation and agricultural operations are much easier as compared to hilly areas. But there are some crops like tea and coffee, which require large amount of water for their growth but stagnent water is harmful for their roots. Therefore, these crops are grown on hill slopes where rain water can easily drain down the slope. Some of the hill slopes are cut into terraces and terraced cultivation is practised there. Rice is the main crop on these hill terraces.

2. Soil. Rice is mainly grown in clayey soils while loamy soils are best soils for wheat. The regur soil of the Deccan Plateau is ideal for cultivation of cotton. Coarse grains such as jowar, bajra, maize, ragi, barley etc. are grown in inferior soils which include light sandy soils, light black soils, red and laterite soils etc. Fertility of soil is a major determinant of crop productivity and hence of cropping pattern in any region. Delta soils of West Bengal are renewed by floods every year and are very fertile. The Ganga-Brahmaputra delta is world renowned for jute cultivation. These fertile delta soils enable the farmers to take 2-3 and sometimes even four crops in a year. In fact the whole of Great Plain of North India is drained by mighty rivers like the Ganga, the Indus and the Brahmaputra and is blessed with some of the most fertile soils of the world. It is because of its fertile soils that this region is agriculturally the most productive part of the country. Soils of the Darjiling hills in West Bengal contain sufficient quantities of humus, iron, potash and phosphorus which are necessary for tea bush to grow. This is the reason that Darjiling tea is famous all over the world for its high quality.

3. Temperature. Temperature has a great bearing on the cropping pattern of an area because each crop requires specific temperature for its growth and ripening. Most crops require lower temperature at the time of sowing and higher temperature at the time of ripening. Some crops require higher temperature and are sown in the summer season. These are known as ***kharif*** crops. There are other crops which require lower temperature and are sown in the winter season. These are known as ***rabi*** crops. India's two most important food crops viz. rice and wheat are kharif and rabi crops respectively.

4. Rainfall. Rainfall is one of the main factors that affects the choice of crops and the cropping pattern of any place largely depends on the amount and distribution of rainfall. In areas where land use depends exclusively on rainfall, the main determining factor is the duration and amount of rainfall. Depending on the amount of annual rainfall, following three types of cropping patterns can be recognized.

(*a*) Cropping pattern in Areas of Heavy Rainfall. Areas receiving more than 150 cm annual rainfall are termed as areas of heavy rainfall. They include most parts of East India and the west coastal plains. Rice is the most important crop of these areas because it requires plenty of water for its successful growth. Tuber crops, plantation crops and some cereal crops like maize and ragi are also grown. Vast areas are marked by monoculture of rice though there is considerable scope for diversifying the cropping pattern. Animal population is fairly high due to availability of grazing area.

(*b*) Cropping Pattern in Areas of Medium Rainfall. Areas receiving 75 to 150 cm annual rainfall are termed as areas of medium rainfall. Areas in the vicinity of 150 cm annual rainfall isohyets are suitable for the cultivation of rice while those near 75 cm annual rainfall isohyets usually grow maize, cotton and soyabeans. These areas are rich in natural resource but people living here are poverty sticken due to poor management of resources. Eastern part of Uttar Pradesh, Bihar, Odisha, eastern parts of Madhya Pradesh and Vidarbha region of Maharashtra receive 30 cm or more rainfall in July-August and 20-30 cm in June and September. Vast areas are to be left as fallow land in kharif season to enable the soil to recoup its lost fertility. Wheat is the principal rabi crop. Different crops require different amount of water and mixed cultivation is often practiced. Rice is grown if sufficient rainfall is received in September, otherwise millets are the natural priority. Jawar is the main crop in areas of lesser rainfall. In addition cotton, soyabean and pulses are also grown.

Areas of medium rainfall have vast potential for improving agriculture and changing the cropping patterns. The main objective should be to fit the crops into a climate rhythm prevailing in those areas. The risk in forming can be reduced considerably by growing less moisture demanding crops such as maize, soyabean and jowar.

(*c*) Cropping Pattern in Areas of Low Rainfall. These areas receive 25 to 75 cm annual rainfall and stretch in a long belt running from Kashmir in the north to Kanniyakumari in the south. Beside scarcity of rainfall, variability of rainfall is high and only those crops are grown which can survive in dry conditions and tolerate large variations in the amount of rainfall. Therefore, the major crops in this belt are millets, jowar, and bajra in the northern, jowar in central and ragi in the southern part. Wheat is the main rabi crop which is grown in irrigated areas. Mixed cropping is very common in which pulses are mixed with cereals. Groundnut is an important commercial crop grown in dry areas. Sunflower, rapeseed and mustard are other oilseeds. Cropping has been developed in such a way that no one crop dominates as is the case in areas of heavy rainfall.

II. Economic Factors

Irrigation is the most dominating of all the economic factors. The other major economic factors are those of size of land holdings, sale price of crops and income of farmers, insurance and investment.

1. Irrigation. Irrigation is an important input and assumes greater significant in arid and semi-arid areas. Rainfall is scanty and erratic in these areas and crops cannot sustain without irrigation. Irrigation provides the right amount of water at the right time and thus saves crops from the vagaries of weather. Now a days, irrigation is widely used even in areas of higher rainfall so that higher yields could be obtained. In the semi-arid areas of Punjab, Haryana and western part of Uttar Pradesh, wheat is the traditional crop which has been grown in these areas for hundreds of years. With the increase in irrigation facilities after the mid-1960s these areas have become major rice producing areas and the cropping pattern has undergone major changes.

2. Size of Land Holdings. Size of land holdings has a direct bearing on the cropping pattern. In case of small holdings, the priority of the farmers would be to grow foodgrains for his family members and opt for commercial crops only if he has some surplus land over and above his personal requirements. Obviously farmers with large holdings can opt for cash crops and help in crop diversification, leading to changes in the cropping pattern.

3. Sale price of crops and Income of Farmers. The farmers want to sell their crops at highest rate and get highest amount for their products so that their income level improves. Farmers in Punjab, Haryana and western part of Uttar Pradesh are traditionally wheat growers but they started rice cultivation with effect from second half of 1960s because rice could fetch higher prices in the market. A large number of tube wells have been sunk to irrigate rice crop in these areas and the ground water is depleting at an alarming rate. Unmindful of this dangerous trend in depletion of the ground water resources, farmers continue grow rice crop, rather vigorously, to take maximum financial advantage of irrigation facilities. In certain areas where the farmers are unable to get remunerative prices for their products, they opt for cash crops.

4. Insurance. Indian monsoon climate is characterized by wide variations from the normal weather conditions and the farmers run heavy losses in the event of bad weather when the crops are damage partially or wholly. The farmers become bankrupt and are unable to repay their loans. In extreme cases they commit suicide out of frustration. Hundreds of cotton growers have committed suicide in the otherwise agriculturally rich areas of Maharashtra, Andhra Pradesh, Gujarat and Punjab. Under such circumstances it is of utmost importance that the farmers are provided with crop insurance at easy terms so that they can pursue agriculture without any fear and most appropriate cropping pattern is adopted. The Government has taken several steps to provide crops insurance coverage to farmers in the recent past.

5. Investment. Changing the cropping pattern requires huge investment because large sums of money are required for irrigation, seeds, fertilisers, farm machinery etc. The main reason of groundnut replacing cotton in Madhya Pradesh is easy availability of better quality groundnut seeds to the farmers. Even otherwise, groundnut has short maturing period than cotton and farmers get good return after selling groundnut in the market.

III. Political Factors/Government Policies

Government policies can influence the cropping pattern in any region to a great extent. Government can encourage any crop by providing subsidy on seeds, fertilisers, electricity etc. and discourage any other crop by putting restrictions on it. Increase in the production of foodgrains and decrease in the cultivation of tobacco, indigo, poppy are largely the result of government policies. Moreover, government can change the cropping pattern by strengthening rural road transportation. Farmers tend to opt for cash crops like vegetables and fruits when the villages have direct road link to the neighbouring urban market.

AGRICULTURAL PRODUCTIVITY

Although usually treated as synonymous with agricultural efficiency, by several authors, agricultural productivity has a special meaning in agricultural geography and is defined as total agricultural output per unit of cultivated area per agricultural worker or per unit of input in monetary values. These may be referred to as 'land productivity', 'labour productivity' and 'capital productivity', respectively.

Improvement in agricultural productivity is generally the result of a more efficient use of the factors of production, viz., environment, arable land, labour, capital and the like. Productivity, which may be industrial or agricultural, is a difficult theme, both in concept and in terms of measurement of its level. Therefore, any definition that is adopted is bound to suffer from certain weaknesses. It is important to remember, however, that productivity is a physical rather than a value concept, which describes the relationship between the output and the major inputs like land, labour and capital. It is essentially a measure of the efficiency with which the inputs are utilized in production. The regional differences in agricultural productivity are the result partly of natural advantages of abiotic environment (soils and climate) and partly of farming efficiency as controlled by cultural ecology. Farming efficiency refers to the properties and qualities of various inputs and the manner in which they are combined and put to use for production. Increase in agricultural productivity is largely related to the choice of inputs, and their relative quantities, and the techniques and the skill with which they are used in the production processes. (Singh and Dhillon, 1984: 226)

The level of agricultural productivity, as a concept, means the degree to which the economic, cultural, technical and organizational variables (*i.e.*, the man-made frame) are able to exploit the abiotic resources of the area for agricultural production (Singh, 1979).

Computation and delineation of agricultural productively is of great significance because of its following advantage :

(*i*) It helps in ascertaining the relative productivity of the component areal units of a region.

(*ii*) Agriculturally weakers areas with lower productivity can be identified.

(*iii*) The present agricultural productivity helps in assisting the past development.

(*iv*) A study of present productivity provides a sound base for future agricultural planning.

Several techniques adopted for computing efficiency in level of agricultural productivity per unit area per unit of time, or per unit of farm work force, etc. are detailed below :

1. Assessing the value of agricultural production per unit area.
2. Measuring production per unit of farm labour or man-hour.
3. Determining output in relation to input, or output-input ratio and profitability of farming measured in terms of the return for the sum total of human efforts or paid-out-cost in relation to the output (Khusro, 1964)
4. Expressing production of agriculture in terms of grain equivalents per head of population (Buck, 1967; E. de Vries, 1967; Clark and Haswell, 1967).
5. Considering output per unit area or yields per hectare after grading them in ranking order, thereby deriving the ranking coefficient (Kendall, 1939; Stamp, 1960; Shafi, 1960).
6. Giving weightage to the ranking order of the output per unit area with the percentage share under each crop (Sapre and Deshpande, 1964; Bhatia, 1967).
7. Using the carrying capacity of land in terms of population (Stamp, 1958, 1967).
8. Determining an index of productivity (Enyedi, 1964; Shafi, 1972, 1974).

9. Calculating the index number of agricultural efficiency by expressing the per unit area carrying capacity (in terms of population) of the component enumeration unit as a percentage of the per unit area carrying capacity for the entire region (Singh 1972, 1974).
10. Computing the crop yield and concentration indices ranking coefficient (Singh, 1976).
11. Involving the area, production and price of each cultivated crop in each of the constituent areal units of the region, and then relating the out turn in terms of money of the unit to the corresponding productivity of the region (Hussain, 1976).
12. Delimiting agricultural productivity by computing the intensity and spread indices of three variables, i.e. (*i*) yield, (*ii*) grain equivalents, and (*iii*) cropping system (Singh, V.R., 1979).
13. Assessing net income (farm business income) in rupees per hectare of cropped area or per adult-male unit of farm family work force (Singh, Jasbir *et al.*, 1982).

The above mentioned techniques differ widely from one another and none of them gives a flowless assessment of agricultural productivity. The first three techniques seem to require such statistics as are not readily available and even easily accessible in most of the underdeveloped and developing countries of the world. Statistics, though available at the farm level in some states of India, do not seem to be adequate for area-analysis of agricultural efficiency. Technique (3), however, has little validity in a subsistence farm economy where (*i*) foodgrains dominate and constitute 75 to 85 per cent of all agricultural production, (*ii*) the major output is retained for domestic consumption, and (*iii*) most of the inputs are provided by the farmer himself. However, Khusro (1964) has expressed himself in favour of paid-out-cost in relation to output as a measure of farm efficiency. The efficiency and profitability of agriculture measured by the surplus or deficit of output over paid-out-cost excluding the *imputed value* of farm family labour is an ideal method of determining farm efficiency at the level of different size-class of operational holdings.

Technique 4 of expressing production of agriculture in teutrs of grain equivalents per head of population was used for the first time by Buck in 1967. He realized that in a subsistence agricultural economy as in China, or for that matter in India productivity expressed in terms of monetary value has no meaning because the crops may be grown for local domestic consumption, and only a small portion of it may be sold for cash. To him, therefore the natural unit for measuring the level of agricultural productivity in such a community appeared to be the conversion of kilograms of agricultural products into grain equivalents. He considered all grains to be equal in food value. He converted agricultural products other than foodgrains into grain equivalents according to the local market price at which they are exchanged against foodgrains which are grown predominantly in a locality. Buck's grain equivalent device for measuring agricultural progress was modified and also used by **E. de Vries**. He expressed all output of grains in Asian countries in terms of *milled-rice equivalents* per head of total population. This is a better method because different types of grains are converted into rice equivalents depending on the local market price of each grain. In contrast **Buck** counted all grain equally. The main drawback in **E. de Vries** method is that he worked out rice equivalents per head of total population without caring for the fact that in many instances the effective population which contributes towards agricultural efficiency is the rural population widely engaged in agriculture.

Clark and **Haswell** (1967) used a system in terms of wheat equivalents per person. In this scale total agricultural production is expressed in terms of economic wheat equivalents/person/annum.

Technique 5 of considering output per unit area or yield per hectare was developed by Kendall in 1939 in which he devised a system of ranking coefficient. He explained by taking into account the hectare yield of ten leading crops for 48 administrative counties of England. Stamp (1960) applied Kendall's *ranking coefficient* technique for international comparisons, by selecting twenty countries and nine major crops. In India Shafi (1960) too used this technique, for determining the efficiency of the districts comprising the state of Uttar Pradesh by considering the hectare yield of eight foodgrain crops.

Kendall followed the following procedure for determining agricultural productivity :

(*i*) the enumeration units are ranked in order of output per heactare for each of the selected crops.

(*ii*) the ranks occupied by each unit in respect of the selected crops added.

(*iii*) sum of ranks of each unit is divided by the number of selected crops.

Technique 6 of giving weightage to ranking order of the output per unit area with the percentage share under each crop was devised by Sapre and Deshpande in 1964. It was a modification of Kendall's technique in the sense that this technique used *weighted average ranks* instead of *simple average ranks*. The weighted ranks for various crops is proportionate to the percentage of cropland under each crop.

Bhatia in his study in 1967 made the following two assumptions :

(*i*) hectare yields express all the physical and human factors connected with the production of crops.

(*ii*) sharing of cropland among the various crops reflects various factors involved in land utilization.

A weighted average of the yield efficiency of all crops in component regional unit, where the weights are proportionate to the share of cropland devoted to each crop, would give a measure of overall agricultural efficiency of the component regional unit relative to the entire region. This may be expressed as:

$$I_{ya} = \frac{Y_c}{Y_r} \times 100$$

where I_{ya} is the yield index of crop *a*; Y_c is the hectare yield of crop '*a*' in the component areal unit, and Y_r is the hectare yield of crop '*a*' in the entire reigion; and

$$E_i = \frac{I_{ya} \times C_a + I_{yb} \times C_b + \ldots + I_{yn} \times C_n}{C_a + C_b + \ldots + C_n}$$

where 'E_i' is the agricultural efficiency index, I_{ya}, I_{yb}, , ..., I_{yn} are the yield indices of various crops, and C_a, C_b, ... C_n are the percentages of cropland under the different crops.

With this technique a picture of overall performance of an area can be had, but this does not give any weightage to the harvested area under different crops which actually contribute to establish agricultural efficiency.

Technique 7 using the carrying capacity of land in terms of population was devised by L.D. Stamp in 1958 and modified in 1967. The idea of using carrying capacity struck him in the backdrop of fast increasing population pressure on land resources. He had taken into account, first a *standard nutrition unit*, that is, to what extent food and land are required to support one average human being and to produce that much amount of food respectively; second the *caloric value* of some leading food crops since 90 per cent of the world population depends upon foodgrains. In this approach the production of crops is converted into calories, which then can be used to measure the *optimum carrying capacity* of land in terms of population. In a sense carrying capacity is a measure of farming efficiency.

Technique 8 of determining an index of productivity was suggested by **Enyedi** in 1964. **Shafi** (1972 and 1974) also adopted this approach to determine the productivity indices in respect of twelve food crops of India. Enyedi's formula of productivity index is :

$$\text{Productivity Index} = \frac{Y}{Y_n} \div \frac{T}{T_n}$$

where, Y is the total production of the selected crop in unit area.

Y_n is the total production of the same crop on national scale,

T is the total cropped area of the unit area. and

T_n is the total cropped area on national scale.

Technique 9 is a modification of technique 7 in which Jasbir Singh (1972, 74) calculated the index number of agricultural efficiency by expressing per unit area carrying capacity of the component enumeration unit as a percentage of the per unit area carrying capacity for the entire region. He opined that actual production of foodgrains is more important particularly in those countries which suffer from food shortages.

After making some allowances for the quality of production (*i.e.*, to change the crop production into caloric values) and the usages and wastages in the form of seed, cattle feed, storage losses, etc. it can be inferred that higher the caloric output per unit area greater is the carrying capacity of land in terms of population and hence the efficiency of farming. The carrying capacity of agricultural areas, *i.e.*, their maximum caloric return from the soils, can be considered as their index of efficiency. The variations in caloric productivity per unit area are the result of differential interaction between the combination of environmental conditions and the combination of human activities.

In 1974 **Jasbir Singh**, determined caloric output (C_o) available for ingestion per unit area under food crops and oil seeds after making a careful total deduction for usages and wastages which add up to 16.80 per cent of the total production. He then proceeded to compare the calorific output with the weighted average standards nutrition (S_n) for ingestion in calories/person/annum. Finally, the carrying capacity (C_p) was expressed as :

$$C_p = \frac{C_o}{S_n}$$

For an objective measurement of agricultural efficieny the carrying capacity per unit area in the component enumeration unit may be expressed as a percentage of the carrying capacity in the entire region to obtain index numbers, which would give a measure of the agricultural efficiency of the component enumeration unit relative to the entire region. The above may also be read as :

$$I_{ae} = \frac{C_{p_c}}{C_{p_r}} \times 100$$

where, I_{a_e} is the index number of agricultural efficiency of an enumeration unit,

C_{p_c} is the carrying capacity in terms of population in the component enumeration unit, and

C_{pr} is the carrying capacity in the entire region.

Technique 10 was also introduced by Jasbir Singh in 1976 in which he computed the crop yield and concentration indices ranking coefficient in order to assess the regional differences in levels of food produced and to determine the weaker areas form the point of view of agricultural productions. He used average food crop yields and proportions of these crops in the total harvested area as twin-elements for measuring the index of the level of food production. For an objective measurement of the level of agricultural productivity, the relative crop yield and concentration indices arranged in ranking order and computed into average coefficient, would give a measure which one may call *the crop yield and concentration indices ranking coefficient.* The procedure may be explained as follows :

$$Y_i = \frac{Y_{a_e}}{Y_{a_r}} \times 100$$

where, Y_i is the crop yield index,

Y_{a_c} is the average yield per hectare of crop *a* in the component enumeration unit, and

Y_{a_r} is the average yield of the crop *a* in the entire region or country.

$$C_i = \frac{P_{a_e}}{P_{a_r}} \times 100$$

where C_i is the crop concentration index,

P_{a_e} is the percentage strength of crop '*a*' in the total harvested area in the component enumeration unit, and

P_{a_r} is the percentage strength of crop '*a*' in the total harvested area in the entire region or country.

The crop yield and concentration indices thus derived for all the regional units and the crops are ranked separately. Yields and concentration ranks for individual crops are added and thereafter divided by 2, thus giving the *crop yield and concentration indices ranking coefficient*. The equation is :

$$\text{Crop yield and concentration indices ranking coefficient for crop 'a'} = \frac{\text{Crop yield index ranking of crop 'a'} + \text{Crop concentration index ranking of crop 'a'}}{2}$$

The result thus derived will give us an idea of the level of agricultural productivity, the lower the ranking coefficient, the higher the level of agricultural productivity and *vice versa*.

Technique 11 was devised by Hussain in 1976 while establishing agricultural productivity of the Satluj-Ganga plains. He converted the agricultural production into money value of the regional unit in proportion to the whole region. He argued that other things remaining the same, higher the return in terms of value in money, greater is the productivity of land.

V.R. Singh (1979) used 12th technique in technique 13 which is concerned with delimiting agricultural productivity by computing the intensity and spread indices of three variable, namely (*i*) yield, (*ii*) grain equivalents and (*iii*) cropping system.

Jasbir Singh *et al.* (1985) followed technique 13 while studying the agricultural productivity in Haryana. The technique involves assessing net income (farm business income) in rupees per hectare of cropped area or per adult male unit of farm family work force. Adult male unit (adult male equivalent) for standardization 2 children (7-15 age group), 1.5 female (15-59 age group and 1 male (15-59 age group) may be considered equal to one adult male unit. It was found that agriculturally productive regions registered highest net income per hectare of the cropped area. Besides net income increases with an increase in the size of the farm with a few exceptions here and there.

AGRICULTURAL INTENSITY OR INTENSITY OF CROPPING

Agricultural production and productivity, can be increased in two ways; by expanding the cropped area and by increasing the intensity of cropping. Since much of the physically suitable land for cultivation is already under plough, there is little scope for expansion of net sown area. The only alternative left is the intensification of cropping. *The intensity of cropping refers to the number of crops raised on a field during an agricultural year. The total cropped area as percentage of the net sown area gives a measure of cropping intensity.*

$$\text{Thus, cropping intensity} = \frac{\text{Total cropped area}}{\text{Net sown area}} \times 100$$

The above formula can be illustrated by taking example of a village. Suppose the total cropped area (net sown area + area sown more than once) in the village is 1,200 hectares (700 hectares in the *rabi* season and 500 hectares in the *kharif* season) and the net sown area in the entire village is 1,000 hectares, then the agricultural intensity in this village will be :

$$\frac{700 + 500}{1{,}000} \times 100 = 120\%$$

Agricultural intensity in India has undergone significant change after Independence.

Table 20.21 shows that there has been gradual increase in the cropping intensity between 1950-51 and 2018-19. This is an indication of our success in increasing agricultural production by bringing more area under **'area sown more than once'** category. This has been made possible due to a variety of inputs like irrigation, fertilizers, early maturing varieties of seeds and farm modernization.

In 2018-19, cropping intensity for India as whole was calculated to be 141.60% which is just an average figure. There are large variations in intensity of cropping at the state level as indicated in **Table 20.22** and **Figure 20.7**. It varies from 100 per cent in Manipur to 183.4 per cent in Haryana, 189.8 per cent in West Bengal and 190.6 per cent in Punjab. Union Territories of Delhi and Andaman and Nicobar have reported exceptionally high cropping intensity of over 260 per cent. It is easy to understand higher rate of cropping intensity in agriculturally advanced states

TABLE 20.21. Temporal change in cropping intensity in India (million hectares)

Year	1950-51	1960-61	1970-71	1980-81	1990-91	2000-01	2010-11	2014-15	2015-16	2016-17	2017-18	2018-19
Net area sown	118.75	133.20	140.86	140.29	142.87	141.34	141.56	140.13	139.51	139.42	139.18	139.35
Total cropped area	131.89	152.77	165.79	172.63	185.74	185.34	197.68	198.38	197.05	200.20	199.99	197.32
Cropping Intensity (%)	111.07	114.69	117.69	123.05	130.01	131.13	139.64	141.57	141.25	143.60	143.69	141.60

Source : Agricultural Statistics at a Glance, 2021, p. 333.

TABLE 20.22. State level variations in cropping intensity in India (2018-19)

State/Union Territory	Net area sown (Thousand hectares)	Total cropped area (Thousand hectares)	Cropping Intensity (%)
1. Andhra Pradesh	6,049	7,418	122.1
2. Arunachal Pradesh	234	320	136.7
3. Assam	2,723	4,004	147.0
4. Bihar	5,167	7,406	143.3
5. Chhattisgarh	4,679	5,614	120.0
6. Goa	128	150	117.4
7. Gujarat	10,302	11,429	110.9
8. Haryana	3,601	6,605	183.4
9. Himachal Pradesh	542	914	168.5
10. Jammu & Kashmir	713	1,124	157.6
11. Jharkhand	1,281	1,824	142.5
12. Karnataka	10,664	13,551	127.1
13. Odisha	4,006	4,527	113.0
14. Punjab	4,119	7,851	190.6
15. Rajasthan	17,778	25,313	142.8
16. Sikkim	77	162	210.2
17. Tamil Nadu	4,582	5,672	123.8
18. Tripura	256	487	190.6
19. Uttarakhand	648	1,029	158.9
20. Uttar Pradesh	16,538	26,859	162.4
21. Kerala	2,034	2,571	126.4
22. Madhya Pradesh	152.5	26,115	171.8
23. Maharashtra	16,815	19,139	113.8
24. Manipur	441	441	100.0
25. Meghalaya	255	312	122.5
26. Mizoram	145	187	129.2
27. Nagaland	384	529	137.9
28. West Bengal	5,248	9,960	189.8
29. Telangana	4,660	5,774	123.9
30. Andaman & Nicobar Islands	15	41	277.2
31. Chandigarh	1	1	114.0
32. Dadra and Nagar Haveli	20	20	100.2
33. Daman & Diu	3	3	100.4
34. Delhi	22	59	269.0
35. Lakshadweep	3	3	100.9
36. Puducherry	15	27	174.5
All India	**1,39,351**	**1,97,320**	**141.6**

Source : Agricultural Statistics at a Glance 2021, pp. 339-345.

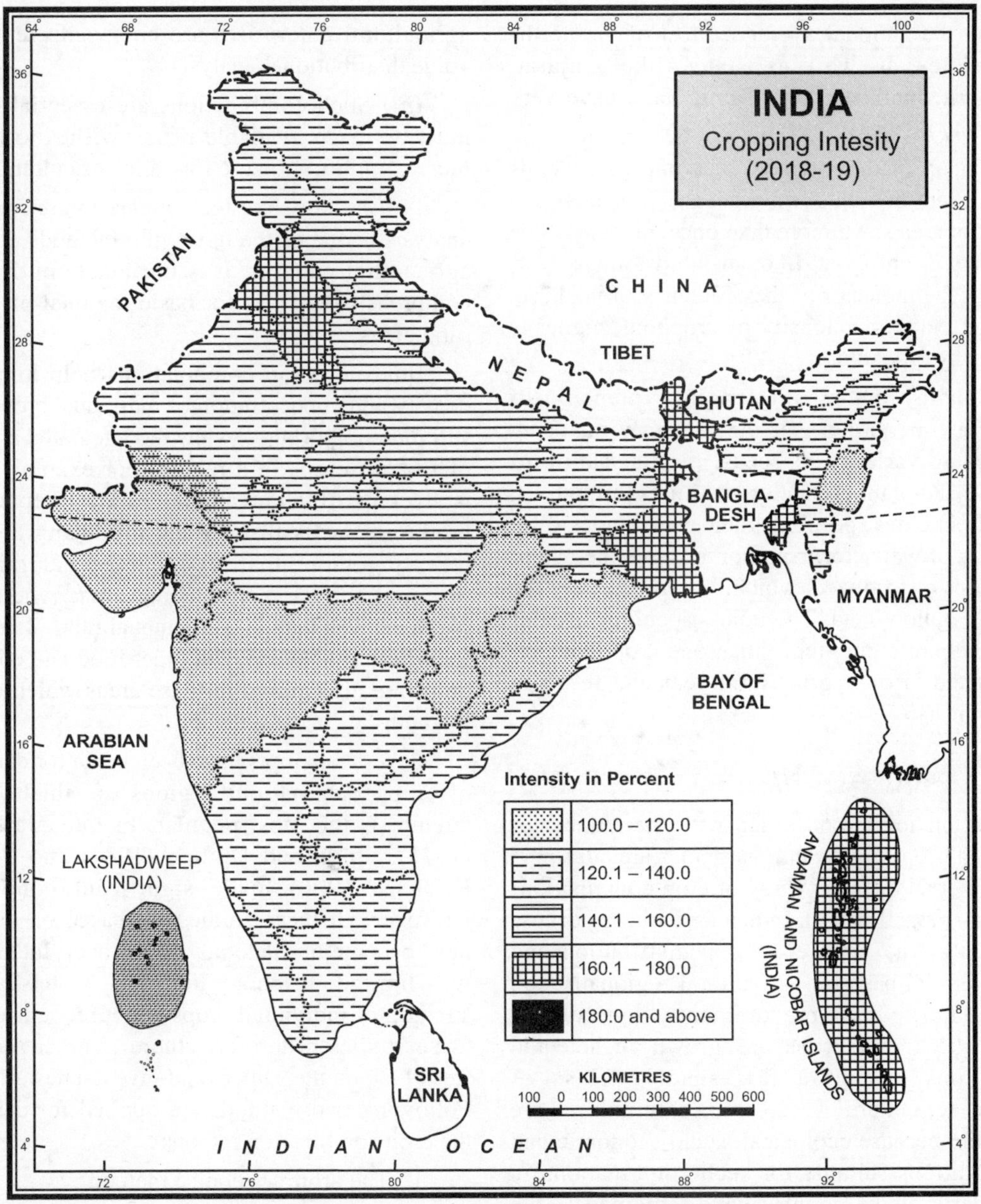

FIG. 20.7. India : Cropping Intensity (2014-15)

like Punjab and Haryana where Green Revolution has shown its maximum impact with respect to two basic foodgrains like wheat and rice. Similarly, some parts of West Bengal get three crops of rice in a year and the cropping intensity is high there. Most parts of Delhi are engaged in cultivation of vegetables which have ready local market in one of the most urbanized areas of the country. Both rabi and kharif vegetables are grown which keep the farmers busy throughout the year and high cropping intensity is recorded. But high cropping intensity in the hilly state of Sikkim is a real surprise which needs further investigation. On the whole 16 states and union territories have cropping intensity higher than the national average of 141.56 per cent. The remaining 20 states and union territories have cropping intensity lower than the national average. Most of the north-eastern states as well as states like Gujarat, Chhattisgarh, Jharkhand and Tamil Nadu have very low cropping intensity.

The index of the intensity of cropping depends upon the extent of area sown more than once. Higher the extent of area sown more than once, higher will be the intensity of cropping. In other words, intensity of cropping is the indicator of the efficiency of land use. Higher the index of intensity of cropping, higher is the efficiency of land use.

The main factors influencing intensity of cropping are irrigation, fertilizers, early-maturing high yielding varieties of seeds, mechanisation of agriculture and plant protection measures through the use of insecticides, pesticides and weedicides. The availability of water for irrigation ensures the use of higher doses of fertilizers which, in turn, reduces the extent of fallow land. The quick-ripening varieties of seeds help in taking more than one crop from the same field in one agricultural year and result in higher intensity.

CROP COMBINATION

Crop combination is concerned with the number or diversity of crops grown in a particular area during a specific interval of time. Study of crop combinations has a greater significant in Indian agriculture because it presents a true picture of crop distribution and provides a solid base for agricultural regionalisation of the country. In a predominantly agricultural country like India, crops are grown in selected combinations, except a few small pockets of plantation agriculture. These crops assemblages are fairly stable because ecological, socio-economic and other conditions suitable for such combinations in various parts of the country have remained stable since long. The relative positions and strengths of different crops or livestock in different enumeration units can better be comprehended with combination analysis. **Weaver** (1954) listed the following advantages of studying crop combinations:

(*i*) The establishment of crop, livestock or ranking combinations is essential for an adequate understanding of geography of crops, livestock, etc. that hold variable ranks in the combinations.

(*ii*) The combinations of crops, livestock or agricultural enterprises are composite realities that guide distributional analysis.

(*iii*) Such combinations are essential, and they must be made available if one wishes to build still more complex structure for valid agricultural regions.

In fact, crop, livestock or agriculture combination analysis is one of the most vital methods of studying agricultural patterns. It is invaluable in providing a comprehensive basis for basic regional planning for rural areas.

Study of crop combinations help knowing the density and concentration of individual crops and it is still more important to view the integrated assemblage of various crops in a region. For example in India, isohyet of 100 cm annual rainfall separates rice producing areas from wheat producing areas. Area receiving more than 100 cm annual rainfall are predominantly rice producing areas whereas those receiving less than 100 cm annual rainfall are popular for wheat cultivation. But other food and commercial crops are also grown in these areas which cannot be ignored.

Several attempts have been made for demarcation of crop combinations regions of which Weaver's attempt is the most popular. In India Jasbir Singh (1971), Singh and Dhillon (1984) and Tirath and Krishan (1996) made significant contributions. However, all these studies are based on data which have become obsolete now. Moreover, their coverage was limited, often to the crop groups and areal strength of individual crops is not recognised well in these studies. Therefore, there is great scope for a fresh look on the subject and give it a new orientation. Following two methods are applied for determining the crop combination regions :

1. The arbitrary choice method
2. The statistical method.

1. Arbitrary Choice method

In this method, first two or three crops grown in an area are included and the remaining are excluded from the combination. When only one crop covers maximum percentage of area, then only the first crop is shown on the map. Similarly, first two or three crops are shown in two crop or three crop combination on the crop combination map of an area. This fact is universally known that rice growing areas are different from wheat growing area in India. It implies that rice growing areas do not grow wheat and wheat growing area do not grow rice. (In this context, it should be noted that traditionally wheat growing areas of Punjab, Haryana and western part of Uttar Pradesh have started growing rice by taking advantage of irrigation facilities available in these areas). Therefore macro level production zones are demarcated on the basis of these two major crops only. Micro regions are delineated on the basis of other crops which are important at the micro level. This method suffers from the following limitations :

1. Only first two or three crops are taken for consideration and other crops are excluded from the combination irrespective of the percentage of area covered by them and also their monetary value. Therefore this is an unscientific method of determining crop combination.
2. There is always a fear of repetition while presenting and analysing the crop combinations.
3. This is not a very sensitive method because subjectivity often overpowers the objective analysis.
4. The number of crop combinations increases with the number of crops which makes the process very complicated.
5. Sometimes the cumulative area covered by the crops which are included in the combination is less than 50% and a large number of crops of lesser importance are not included. This leads to erroneous results.

In spite of its above mentioned limitations, this is a widely used method because of its simplicity.

Statistical Method

Since this method is based on statistical analysis, it is more scientific, accurate and hence more popular than the arbitrary choice method. It has greater capacity to handle the strongly developed cropping diversity or even agricultural diversification in an area. It was proposed by **J.C. Weaver** in 1954 for studying the complex structure of crop-combination regions and the Middle West (U.S.A.). He computed the percentage of the total cultivated area in each of the 1081 counties covered in his research work. He adopted such an approach that could provide an objective, constant and precisely repeatable procedure and would yield comparative results for different years and localities. First of all, he determined the percentage of each of the selected crop to the total cropped area. Then he considered each percentage against a standard norm and determined the right crop combinations with the help of theoretical standard deviations. The theoretical curve for the standard measurement was employed as follows :

Monoculture = one crop accounts for 100.00 per cent of total harvested or cropped area.

2 crop combination = each crop 50 per cent
3 crop combination = each crop 33.33 per cent
4 crop combination = each crop 25.00 per cent
5 crop combination = each crop 20.00 per cent
6 crop combination = each crop 16.67 per cent
7 crop combination = each crop 14.29 per cent
8 crop combination = each crop 12.50 per cent
9 crop combination = each crop 11.11 per cent
10 crop combination = each crop 10.00 per cent

For accurately comparing the actual percentages within the individual regional units with the theoretical distribution, following formulae concerning variances and standard deviation were used :

1. Variance = $\frac{\sum d^2}{n}$

2. Standard deviation (SD) = $\sqrt{\frac{\sum d^2}{n}}$

where d = difference between actual crop percentage in a given county (regional unit) and the percentage in the theoretical distributions

and n = number of crops in a given combination.

Weaver's technique can be illustrated by following example :

Suppose in a region the area occupied by crops in percentages is as given in the following table :

Crop	Percentage of area
Wheat	40
Jowar	30
Bajra	15
Pulses	10
Maize	5

If it is one crop region, then $\frac{(100-40)^2}{1} = 3{,}600$

If it is two crop region, then

$$\frac{(50-40)^2+(50-30)^2}{2} = \frac{100+400}{2} = 250$$

If it is three crop region, then

$$\frac{(33.33-40)^2+(33.33-30)^2+(33.33-15)^2}{3}$$

$$= \frac{44.89+11.08+335.99}{3} = \frac{391.96}{3} = 130.65$$

If it is four crop region, then

$$\frac{(25-40)^2+(25-30)^2+(25-15)^2-(25-10)^2}{4}$$

$$= \frac{225+25+100+225}{4} = \frac{575}{4} = 143.75$$

If it is five crop region, then

$$\frac{(20-40)^2+(20-30)^2+(20-15)^2+(20-10)^2+(20-5)^2}{5}$$

$$= \frac{400+100+25+100+225}{5} = \frac{850}{5} = 170$$

It is clear from the above computation that the minimum deviation from the normal curve is in three crop-region. Therefore the concerned area is three crop region (wheat, jowar and bajra).

Weaver's technique suffers from the following drawbacks :

(*i*) Sometimes, there may be crops covering insignificant area but they might be of great importance for that particular region. However, weaver tried to solve this problem by imprinting the name of such crop(s) on the map after the crop combination has been prepared.

(*ii*) Weaver himself admitted that this technique occasionally tends to show lowest deviation for a crop combination which includes even a crop occupying as less as one per cent of the total harvested cropland.

(*iii*) This method gives most unwieldy combinations for areas of high crop diversification.

(*iv*) It involves laborious calculations.

Doi (1959), **Peter Scot**, **Thomas** (1963), **Coppock** (1964), **Singh** (1974) and several other scholars have tried to modify Weaver's technique with the help of standard statistical algorithm, namely the *least squares*. Doi presented a modification of Weaver's technique by substituting the variance $\left(\frac{\sum d^2}{n}\right)$ or least standard deviation $\left(\sqrt{\frac{d^2}{n}}\right)$ of Weaver with the sum of square deviations ($\sum d^2$) *i.e.*, of actual percentages from the theoretical distributions. **Peter Scot** made some modifications in Weaver's technique by suggesting that animals are as important as crops while determining the crop combination regions of Tasmania. **Thomas** modified the version of minimum variance or least standard deviations approach of Weaver. His technique was designed to make use of data available for all crops in an enumeration unit for the calculation of variances for crop combinations. **J.T. Coppock** modified Scot's method by suggesting conversion of all animals on the farm to a common unit according to the quantity of food taken by the animals and then each type to be expressed as a percentage of the total animals. He even prepared the following scale of conversion for the type of animals and the common unit on the basis of the quantity of food taken by them.

Types of animals	Number of animals on the basis of food units
Horses	1
Cows, oxen, buffaloes	1
Other animals (more than 1 year but less than two years)	2/3
Other animals (less than 1 year)	1/3
Sheep	1/15
Chicken (more than 6 months)	1/50
Chicken (less than 6 months)	1/200

Coppock demarcated 11 crop regions (of first order) and 38 crop combinations by using the least square method. These 11 crop regions were based on first rate crops. These crops included rice, jowar, wheat, maize, bajra, pulses, ragi, barley, cotton, groundnut and tea. These crops are grown over larger areas while other crops are grown over limited areas.

Generalised Crop Combinations in India. Although crop combinations keep on changing with time and space, yet generalised crop combinations can be visualised as under :

1. Rice. Rice is the staple food of a vast majority of India's population and it is grown as the first crop in the following seven regions.

(i) Monoculture of rice. The areas exclusively growing rice are called areas of rice monoculture. These areas include large parts of Chhattisgarh and adjoining areas of east Madhya Pradesh, Chotanagpur Plateau, West Bengal, delta regions of the Krishna, the Godavari and the Cauvery rivers, coastal areas of Odisha, the Brahmaputra valley, Tripura, Manipur and Nagaland in the north-east region and Andaman and Nicobar Islands in the Bay of Bengal.

(ii) The West Coastal Plains. It includes the Malabar and the Konkan coastal regions. Besides rice, these areas grow rubber, coconut, ragi, vegetables and fodder crops.

(iii) The East Coastal Plains. It includes the non-deltaic coastal regions of Tamil Nadu and Andhra Pradesh. Crops other than rice include groundnut, bajra, jowar and maize.

(iv) Middle Ganga Valley in eastern Uttar Pradesh and Bihar. Rice is the first ranking crop which is followed by wheat, pulses, sugarcane, barley and maize.

(v) South Karnataka Plateau. In this region, rice is followed by coffee, ragi, pulses, cardamom, coconut and citrus fruits.

(vi) Hilly areas of northern part of West Bengal. This area spreads in the Jalpaiguri district where rice is the first ranking crop and maize is the second ranking crop.

(vii) Meghalaya Plateau. Potato, maize and cotton are the crops grown in this area in addition to the first ranking crop of rice.

2. Wheat. Wheat is the second most important food crop of India next only to rice. Nowhere in India, one would find monoculture of wheat. Hence areas having more than 40 per cent the cropped area are called wheat regions. Following four wheat regions may be recognised :

(i) The Ganga-Yamuna Doab. This is the most prolific wheat growing region of the country. Apart from wheat, rice, maize, sugarcane, bajra, pulses and fodder are the main crops.

(ii) Eastern Haryana. Pulses, bajra, jowar sugarcane, and fodder crops are grown in combination with wheat.

(iii) Himachal Pradesh and the neighbouring areas of Punjab. It is a plain tract adjoining the hilly area where pulses, jowar, bajra, sugarcane, and fodder are the main crops in addition to wheat.

(iv) Rest of Punjab. The rest of Punjab, excluding the foothill areas is famous for maize, rice, and pulses in combination with wheat.

3. Jowar. This is the first ranking crop in the following five regions.

(i) Tamil Nadu highland (Salem-Coimbatore). The other crops grown in this region are groundnut, bajra, rice, ragi, barley, pulses and cotton.

(ii) North Karnataka Plateau and western Maharashtra. Here jowar is followed by bajra, ragi, barley, rice, groundnut and pulses.

(iii) North Maharashtra and Madhya Pradesh. Crop combination found here is that of jowar, pulses, wheat, cotton and rice.

(iv) Telangana and Chandrapur. Rice and pulses are the main crops in addition to jowar.

4. Maize. Maize is the first ranking crop in the following two regions :

(i) South-east Rajasthan and contiguous areas of Madhya Pradesh and Gujarat. This area is known for wheat, rice, groundnut, gram, fodder and pulses in combination with maize.

(ii) Himachal Pradesh and hills of Kashmir. Apart from maize, this region grows plantation crops on the hill slopes and rice in the valleys.

5. Bajra. This is the first ranking crop of large parts of Rajasthan, Kachchh region of Gujarat, and some parts of Uttar Pradesh and Haryana. Wheat fodder and pulses are grown in combination with bajra.

6. Ragi. This is the first ranking crop of South Karnataka and Dharmapuram district of Tamil Nadu. The other crops grown in combination with ragi are rice, pulses, groundnut, jowar and coconut.

7. Barley. This is the first ranking crop of Lahul-Spiti and Kinnaur districts of Himachal Pradesh. *Jai* and wheat are grown in combination with barley in these districts of higher mountain ranges.

8. Cotton. Cotton is grown as the first ranking crop in eastern and northern parts of Maharashtra, eastern parts of Gujarat and some parts of Madhya Pradesh. This is primarily based on the black cotton soil (regur of these areas). The other crops grown in combination with cotton are jowar, bajra, groundnut and fodder.

9. Groundnut. This is the first ranking crop of Anantpur-Cuddapah and Chittoor districts of Andhra Pradesh and Kathiwar region of Gujarat. Jowar, bajra and some other dry crops are grown in combination with groundnut.

10. Tea. Tea is the first ranking crop on the hill slopes of Darjeeling district of West Bengal and Nilgiri district of Tamil Nadu. Tea represents more or less monoculture in Darjeeling but it is associated with coffee and vegetables in Nilgiri.

11. Pulses. Pulses comprise the first ranking group of crops in comparatively elevated areas south of the Ganga river and in the north-eastern parts of Rajasthan. These are also the first ranking crops in some parts of Odisha, Telangana and Andhra Pradesh.

Factors affecting Crop Combinations

Most parts of India are characterised by 3-4 crop combination although 7 crop combinations are also very common. The factors which affect crop combinations in different parts of the country include physiography, soil fertility, amount of rainfall, weather conditions, irrigation and priorities of the farmers. Increase in irrigation facilities in arid and semi-arid areas has considerably changed the crop combinations in large parts of the country. The number of crops included in crop combinations is limited in areas of extremely dry, extremely wet, extremely cold and extremely hot climate. Other factors remaining the same, areas receiving 30-100 cm annual rainfall have higher crop diversity as compared to the other areas. Crop diversity is also large in hilly areas of the Himalayas.

LAND CAPABILITY

Land capability takes into account the existing fertility and productivity at the current level of technology and the potentiality of land for production of crops on the basis of physical and chemical characteristics of the soil. (Shafi 2006 : 485). In order to understand land capability, it is essential to conduct *land capability survey* which gives direct information in connection with the soil potentiality of different areas. The land *capability classification map* is one of the basic documents for preparing any development plan for agriculture. Land capability classification investigates into the physical characteristics of the land, its soil qualities and farm management qualities. In view of food shortages, it is very essential to make an inventory of the soil resources to ascertain which uncultivated land can be brought into production and which cultivated soils can be made to produce more than what they do at present. Land capability has to be surveyed in view of the fact that some areas are more fertile than others. **L.D. Stamp** expressed the view, during his Land use Survey of Britain in 1930-31, that it is more important to know about the land capability rather than present status of land use.

There is basic difference between land classification and land capability. Under 'land classification', land is divided into different categories. Land capability, on the other hand involves a scientific appraisal of several factors, of which physiography, climate and physical as well chemical characteristics of soils are important. In addition, some social and economic factors are also considered. Maps based on land capability surveys are helpful in identifying the problem and the potential areas. It gives information about those areas which can provide higher level of production by using modern technology and by arranging for better administration.

In a developing and over-populated country like India, it is of utmost importance to know how much increase in farm production can be achieved by improving technological inputs. Land capability classification is of great help in assessing the capability of different soils and to take proper steps to remove the deficiencies. Land capability survey was devised by the Commonwealth Scientific and Industrial Research Organisation (CSIRO) for delimiting land capability regions of Australia. Subsequently such surveys were conducted in European and Aglo-American countries and in some of the Third World countries. The United States Department of Agriculture (1951) has recognised eight classes, with a primary division into land suited to cultivation and land not suited to cultivation. *Hill* and *Partilance* presented land capability classification of Ontario State of Canada in 1960 which aimed at preparing a social plan for land use.

In India soil survey was conducted with the primary aim to achieve land classification on the lines adopted by the U.S. Department of Agriculture. But this system was more suited to the American conditions and did not yield the desired results in Indian conditions. Consequently, various workers and organizations tried to evolve a land capability or suitability classification that is suited to the conditions and planning purposes in a developing country like India. One such example was an attempt made in India by the All India Land-Use Survey Organisation in 1960. Many inadequacies in the Soil Survey were removed in the subsequent years and revised Soil Survey Manual was published in 1970. The All India Soil and Land Use Survey Organization (1970) has identified eight different land-use capability classes with a broad classification into :

(*i*) land suitable for cultivation, and (*ii*) land not suitable for cultivation; these are presented here in summary form :

***Land suitable for cultivation* :**

Class I. Very good cultivable land with no special difficulty in farming.

Class II. Good cultivable land which needs protection from erosion or floods, drainage improvement, and conservation of irrigation water.

Class III. Moderately good cultivable land where special attention has to be paid to erosion control, conservation of irrigation water, intensive drainage, and protection from floods.

Class IV. Fairly good land suited for occasional or limited cultivation needs intensive erosion control, intensive drainage, and very intensive treatment to overcome soil limitations.

Land not suitable for cultivation :

Class V. Very well suited for grazing but not for arable farming, needs protection from gullying.

Class VI. Well suited for grazing or forestry but not for arable farming.

Class VII. Fairly well suited for grazing or forestry, but not for arable farming.

Class VIII. Suited only for wild life, recreational facilities and protection of water supplies.

Evidently, the suitability classification given above is valuable for the assessment of land productivity. It is a classification in terms of the limitations of soils for agricultural land use. It is also an interpretive grouping according to capability. Its utility for development plan can be enhanced if due consideration is given to the physical factors.

DRY FARMING

As its name indicates, dry farming is practised in the dry parts of India. The western half of India is characterised by cry climate where arid and semi-arid conditions prevail. The average annual rainfall in these areas is less than 75 cm and the isohyet of 75 cm annual rainfall marks its eastern boundary. The area of dry farming extends from Ladakh in the north to Kanniyakumari in the south with a small gap in Himachal Pradesh and Jammu region of Jammu and Kashmir. The Bay of Bengal monsoons are almost exhausted when they reach western part of Uttar Pradesh, Haryana, Punjab and Rajasthan and cause little rainfall. The northern parts of Jammu and Kashmir are almost out of reach of monsoon winds and Ladakh region is known as cold desert. South of the Tapi river upto Kanniyakumari, the whole region is in the rain shadow area of the Western Ghats and receives less than 75 cm annual rainfall from the Arabian sea monsoons. Map in **figure 20.8** shows that dry farming area comprises a continuos belt-encompassing whole of Rajasthan, Punjab and Haryana, and large parts of Gujarat, western parts of Madhya Pradesh, and Uttar Pradesh, Vidharbha and Marathwada regions of Maharashtra, plateau region of eastern Karnataka, Rayalseema and Telangana and large parts of Tamil Nadu except coastal areas north of Ramnathpuram. Apart from the main belt the north eastern parts of Jammu and Kashmir are also known for dry farming. It is estimated that dry farming is practised on 31.7 million hectares which accounts for about 22% of the total agricultural land of the country.

Salient Features of Dry Farming

(*i*) **Dry Climate.** As mentioned earlier, dry farming is practised in dry areas of the country where annual rainfall is less than 75 cm. The eastern parts of this belt receiving nearly 70 cm of annual rain are areas of sub-humid climate which are deficient in humidity for about 8 months in the year. Areas with 25-50 cm of annual rainfall have semi-arid climate which suffer from deficiency in humidity throughout the year. Jaisalmer and Barmer districts of Rajasthan receive less than 25 cm of annual rainfall and are completely dry areas.

It is now abundantly clear that these areas suffer from the problem of scarcity of rainfall and shortage of water is a perpetual problem. In addition, variability of rainfall is very high which varies from 20 to 60 per cent. Rainfall is seasonal and occurs only in 3-4 months, leaving 8-9 months almost completely dry. Moreover, rainfall is erratic and unreliable. Late arrival or early retreat of the monsoons increase unreliability of rainfall. Break in the monsoon is also harmful for the crops. There is almost complete crop failure in the event of a drought.

(*ii*) **Lack of Irrigation.** Most areas of dry irrigation lack in irrigation and have to depend on rainfall. Punjab, Haryana and western parts of Uttar Pradesh have canal and tubewell irrigation facilities. However, keeping in view the change in the cropping pattern (introduction of rice cultivation) and high level of agricultural intensity, irrigation facilities in these areas are found to be inadequate. Rajasthan does not have any major river flowing through its territory and ground water is brackish. There is, thus, not much scope of canal and tube well irrigation. The Indira Gandhi Canal, taken from confluence of Satluj and Beas in Punjab, has provided irrigation facilities to western parts of Rajasthan. This has led to increase in agricultural productivity and change cropping pattern. Most parts of South India are made up of hard rocks and lack in ground water resources. Sinking of tube wells is also difficult in hard rocks. As a result of these geographical conditions, tanks are the major source of irrigation. These tanks dry up during the dry season and fail to provide water when the crops need it the most.

(*iii*) **Subsistence Agriculture.** Dry farming is mainly practised by small and marginal farmers who work in the field along with their family members. The agricultural production is just sufficient to feed the family members of the farmer and there is not much surplus for sale in the market. These farmers lack financial resources and are not able to buy modern inputs like better seeds, fertilizers and farm machinery. They have to face hunger, starvation and unemployment whenever there is any drought.

Methods of Dry Farming

(*i*) Field is ploughed repeatedly in order to conserve moisture in the soil. This process is necessary especially during the raining season so that rain water is properly utilized.

(*ii*) Soil fertility is reduced when cultivation is continued for a number of years. This problem becomes serious particularly in the background of farmer's inability to make use of fertilizers and manures. In order to solve this problem, land is left as fallow land. It helps in recuperation of soil fertility.

(*iii*) Pulverisation of soil is done before sowing. This process converts soil into small particles which permits easy flow of water in the soil and plant roots find ample opportunity to grow in strength and support the plants.

(*iv*) Hoeing, weeding and pruning are done at regular intervals. Hoeing allows air to enter the soil space and helps in plant growth. It is generally done before sun-rise so that night dew can enter soil and provide moisture to plants. Through weeding, unwanted plants in the fields are removed and pressure on the soil to feed the plants is reduced. Pruning removes unwanted parts of the plants and wanted plants find ample opportunity to grow.

(*v*) Land is usually covered with straw to reduce evaporation of the soil moisture and to control soil erosion by wind.

(*vi*) In order to improve their income level, the farmers keep themselves engaged in allied agricultural activities such as livestock keeping and dairying.

Significance of Dry Farming

Dry land is an agriculturally dominant region wherein more than two-thirds of the total population (against national average of 54.6%) depends on agriculture. Here agriculture occupies 65% of the

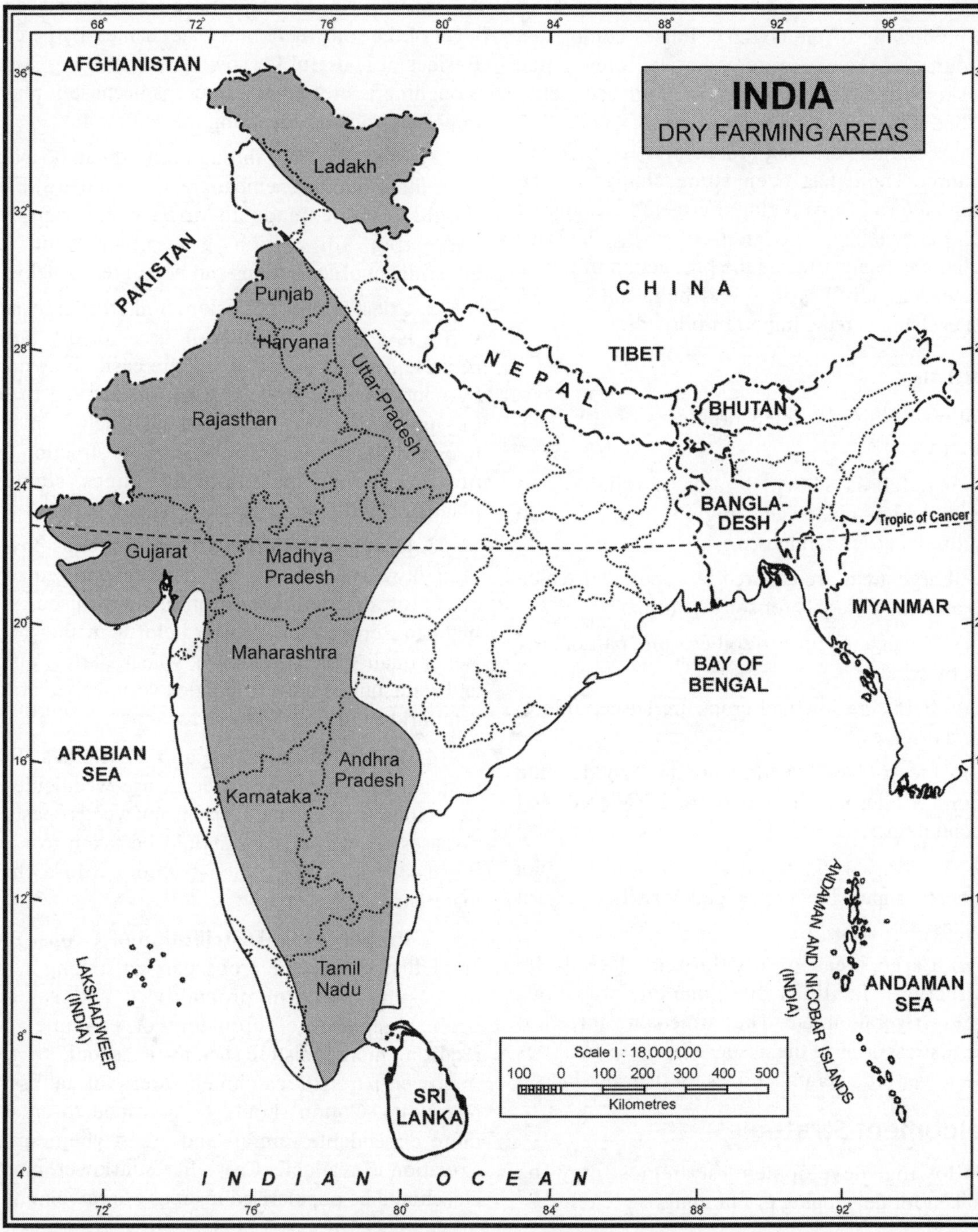

FIG. 20.8. India : Dry Farming Areas

total area while 2.9% of the total geographical area is occupied by permanent pastures. Although this region is plagued with a large number of problems like scanty rainfall, inadequate irrigation, lack of basic inputs like better seeds, fertilizers, and farming machinery, etc., as well as widespread poverty among the farmers, yet this region accounts for 85% of pulses production of the country. Pulses comprise a great source of protein in a vegetarian country like India. This region also produces 75% oilseeds, 80% maize, and 95% jowar and bajra of India. About 40% foodgrains of India are also produced by this type of agriculture. There has been some change in the cropping pattern of this region. Farmers have started to give priority to main foodgrains like wheat and rice which has adversely affected the production of pulses and oilseeds. India has to spend huge amounts of foreign exchange on the import of oilseeds.

Problems

Following are the main problems of dry zone agriculture :

(*i*) Rainfall is scarce, erratic, unreliable and uncertain which makes this region susceptible to floods, droughts and famines.

(*ii*) Large areas are covered by sandy soil which lack in nutrient materials for soil fertility.

(*iii*) The area is prone to soil erosion, particularly erosion by wind.

(*iv*) Yields are low and crops are susceptible to pests and diseases.

(*v*) Due to lack of moisture and inadequate irrigation facilities it is difficult to use HYV seeds and new technologies.

(*vi*) Majority of the farmers are poor and cannot afford costly inputs like better seeds, fertilizers, farm mechanies, etc.

(*vii*) Large areas of dry farming lack basic infrastructural facilities like market, transport, storage, refrigeration, etc. The farmers are forced to opt for distress sale in the absence of these facilities and fail to get remunerative prices for their products.

Development Strategies

Following development strategies may be undertaken for developing dry farming.

1. Conservation and Optimum use of Water Resources. The greatest problem faced by dry farming is scarcity of water for which water conservation and optimum use of water resources is of paramount importance. Water present in the soil can be saved by deep and surface ploughing. Deep ploughing is suitable for kharif crops because it breaks deeper layer of the soil where rain water can easily percolate. Besides, it is useful for sowing and weeding. Surface ploughing is useful for rabi crops because it plays an important role in conserving the soil moiture.

Wastage of water through run off can be avoided by contour bunding, embankments and dams. Efforts should also be made to arrange for rain water harvesting for which government and non-government organisations can be of greater help.

2. Checking Soil Erosion. Soil erosion by strong winds is a serious problem in areas of dry farming because loose particles of soil are easily blown away by wind. Several strategies can be adopted to solve this problem. Important strategies include planting trees in rows so that they can act as obstruction to the blowing wind, making sand dunes stable by plantation and ploughing along he contours.

3. Drought Resistant Crops. As far as possible, drought resistant crops should be encouraged so that effect of drought condition is minimised. Early maturing crops can also be helpful in this regard. Better quality seeds of maize, wheat, barley, mustard and some pulses have responded very well to arid and semi-arid conditions in these areas.

4. Use of Weedicides and Pesticides. Proper arrangement should be made to use weedicides and pesticides to save the crops from weeds, pests and diseases. However, care should be taken to replace these chemicals by eco-friendly bio-technical materials.

5. Proper Areal Distribution of Crops. Proper areal distribution of crops can go a long way in increasing the farm productivity and safeguard against vagaries of climate. For example, crops requiring more moisture such as rice should be grown in lower parts of catchment areas of tanks and reservoirs. Cotton should be confined to areas of more dependable rainfall and areas where sprinkle irrigation is available. Crops like sunflower, mustard, etc. should be popularised in areas of scarce rainfall.

6. Diversification of Crops. It is always better to opt for crop diversification rather than choosing a few selected crops. This can solve the twin problem of uncertainty of weather and soil degradation. If one crop is destroyed due to unsuitable weather conditions, the other crop may compensate for the loss. For example, legumes can be grown along with foodgrains. Legumes add nitrate to the soil.

7. Use of Manures. Majority of the farmers in dry farming areas are poor and cannot afford to purchase costly chemical fertilizers. Under such conditions, the farmers should be encouraged to use cow dung and other biological manures. This will economise agricultural inputs and also save environment from pollution.

8. Alternative land use planning. Following two alternative land use planning strategies are suggested to avoid crop failure and to increase income of the farmers :

(*i*) Encourage agro-forestry and devote large areas to legumes.

(*ii*) Develop grazing lands and social forestry in marginal lands and in cultivable wastes.

Government Schemes. The Government has initiated the following schemes to develop dry farming.

1. It forms an important part of the 20 point programme.

2. Identification of 4,609 micro-watersheds has been done for comprehensive and systematic development covering an area of about 35,45,000 hectares. Each micro-watershed covers an area of about 1,000 hectares. The programme includes scientific management of rain water, land development, use of drought resistant seeds, plantation of trees, and development of animal husbandry and allied activities. As many as 46 model watersheds have been identified for development as focal points.

3. A pilot project, taken up with the assistance of the World Bank, envisages the development of identified watersheds of about 25 to 30 thousand hectares in Madhya Pradesh, Maharashtra, Andhra Pradesh, Telangana and Karnataka.

4. Crops requiring less moisture such as maize, sunflower, soyabean, mustard and coarse grains, are being promoted in rain shadow areas of Karnataka.

5. A national waterlogging area development plan was initiated during the Seventh Five Year Plan in 99 selected districts of 16 states.

6. Agro-forestry, agro-horticulture and silvipasture, etc. have been initiated by the government for alternative land use systems.

AGRICULTURAL REGIONALISATION

India is a vast country with varied physical and socio-economic features which give rise to different types of agriculture in different parts of the country. Therefore, it is desirable that an agricultural regional study should be undertaken to understand the whole gamut of agriculture in India. *Agricultural region is defined as an area which depicts homogeneity with respect to agricultural land use and cropping pattern.* It may be identified on the basis of crops and livestock and the ecological factors which affect these important elements of agricultural. An agricultural region depicts broad similarities in the nature of crops, crop combination patterns, methods of cultivation, quality and quantity of inputs and orientation of the farming activities. Such a region is generally used as a device for selecting and investigating regional groupings of the complex agricultural phenomena found in different parts of the country. Agricultural region is not static but a dynamic concept which changes in space and time. The introduction of rice cultivation in Punjab and Haryana and that of wheat cultivation in the lower Ganga plain as a consequence of the Green Revolution are examples of such changes.

Agricultural regions have the following four characteristic features :

(*i*) they have location.

(*ii*) their boundaries are not well defined and they have transitional belts.

(*iii*) they may be either formal or functional.

(*iv*) they may be hierarchically arranged.

As mentioned above, agricultural regions do not have sharply defined boundaries. Rather two neighbouring agricultural regions have a transitional zone between them because characteristics of one region give way to those of the other region slowly and gradually. Thus precise delineation of the boundary of agricultural regions is a difficult task for which following techniques may be applied.

(*i*) Empirical Technique

(*ii*) Single-element Technique

(*iii*) Multi-element (Statisical) Technique

(*iv*) Quantitative-cum-Qualitative Technique

(*i*) **Empirical Technique.** This technique was first of all used by **Oliver E. Baker** for demarcating agricultural belts of the U.S.A. He was able to demarcate the Cotton Belt, the Corn Belt, and the winter as well as the spring wheat belts of the U.S.A. on the basis of the observed data. Since this technique is largely based on the experience of the farmers, it just gives a generalised picture of the cropping patterns and agricultural regions. This technique was used by a number of scholars like **C.F. Jones**, **Taylor**, **Van Valkenberg**, and **G.B. Cressey** for demarcating agricultural regions in different parts of the world. However, the technique has been criticised by a large number of scholars because it is not objective and lacks scientific approach.

(*ii*) **Single Element Technique.** This is an arbitrary technique in which only the first ranking crops are plotted for demarcating agricultural regions. Its main drawback is that it conceals the position and importance of other crops grown in that particular region. It is not of much use in India because most parts of the country grow crops in combination and not in isolation.

(*iii*) **Multi-Element (or Statistical) Technique.** This technique is more objective and scientific because it involves more than one crop. The agricultural regions may be demarcated with the help of the following :

(*i*) Cropping patterns, crop concentration, and crop diversification

(*ii*) Crop combination

(*iii*) Regional patterns of agricultural productivity.

Several scholars like Shafi, Jasbir Singh, Aujer Hussain, Sapre, Deshpande, Tiwari, etc. have used this technique for demarcating agricultural regions of India with fairly good results. However, non-availability of reliable data puts constraints for applying this technique.

(*iv*) **Quantitative-cum-Qualitative Technique.** This technique involves consideration of physical, socio-economic, cultural and political factors. The physical traits include (*i*) relief, (*ii*) climate, (*iii*) surface and sub-soil water, (*iv*) soil, (*v*) sub-soil and (*vi*) vegetation. The six functional traits are (*i*) rural population, (*ii*) cultural and religious values, (*iii*) technology, (*iv*) farming operations, (*v*) dependent rural population and (*vi*) degree of commercialisation. In this method also non-availability of reliable data is a big hindrance.

Several scholars have attempted to divide India into agricultural regions of which mention may be made of **E. Simkins** (1926), **D. Thorner** (1956), **L.D. Stamp** (1958), **M.S.A. Randhawa** (1958), **Chen Han Sen** (1959), **O.H.K. Spate** and **A.T.A. Learmonth** (1960), **Ramchandran** (1963), **E. Siddiqui** (1967), **P. Sengupta** and **G. Shasyuk** (1968), **O. Slampa** (1968), **B.L.C. Johnson** (1969 and 1979), **R.L. Singh** (1971) and **Jasbir Singh** (1975). Planning Commission of India has made its own contribution for delineating agro-climatic regions of India. The following section of this chapter gives a brief description of some of the important agricultural regions of India.

AGRO-CLIMATIC REGIONS

Climate is one of the most potent factors which influence the agricultural scenario of a region. It plays an important role in involving crop ecology of a region and is responsible for regional variations in agriculture. Such variations are more prominent in a large country like India where there are large variations in climatic elements. Effects of climatic elements are reflected in crop calendars, crop productivity and cropping patterns in different parts of the country. Several scholars have used climate as the basis of dividing India in agro-climaic regions. Some of the outstanding attempts have been those of **Randhawa** (1958), **Slampa** (1968) and **Sengupta** and **Shasyuk** (1967). The latest attempt has been made by the Planning Commission of India in 1989 in association with the National Remote Sensing Agency (NRSA) which divides into agro-climatic regions on the basis of commodity of agro-climatic factors like soil type, rainfall, temperature, water resources, etc. The development profile for each region is formulated through an optimal mix of land stock management, crop production, animal husbandry, horticulture, forestry and agro-processing activities.

Objectives. The primary objectives of agro-climatic regional planning are to optimise agricultural production, increase in farm income and create

FIG. 20.9. India : Agro-climatic Regions

more employment opportunities through scientific utilisation of agricultural and allied resources. Thus the Planning Commission of India laid down four primary objects for agro-climatic regional planning.

These objectives were : (*a*) to attempt a broad demand-supply balance of major commodities at the national level but based on a careful analysis of the potential and prospects of the various zones; (*b*) to maximize the net income of the producers; (*c*) to generate additional employment, particularly for landless labourers, (*d*) to provide a frame-work for scientific and sustainable use of natural resources, particularly land, water and forests in the long run.

Keeping the above mentioned objectives in mind, the Planning Commission divided India into 15 major agro-climatic regions in 1989. These regions are :

1. The Western Himalayas
2. The Eastern Himalayas
3. The Lower Gangetic Plains
4. The Middle Gangetic Plains
5. The Upper Gangetic Plains
6. The Trans Gangetic Plain
7. The Eastern Plateaus and Hills
8. The Central Plateaus and Hills
9. The Western Plateaus and Hills
10. The Southern Plateaus and Hills
11. The East Coastal Plains and Hills
12. The West Coastal Plains and Western Ghats
13. The Gujarat Plains and Hills
14. The Western Dry Region
15. The Islands Region

In addition to resource considerations and land productivity level, relative pressure on land and environmental factors have been considered in framing this typology necessary to identify the zones and providing general guidelines for overall agricultural development.

It may be noticed that different agro-climatic regions have different cropping patterns and crop specialisation varies from one set of regions to another. For example rice has main concentration in zone 3 (Lower Ganga Plain), 4 (Middle Ganga Plain), 7 (Eastern Plateau) and 11 (Eastern Coast), while concentration of wheat is found in 4 (Middle Ganga Plain), 5 (Upper Ganga Plain), 6 (Trans Ganga Plain) and 8 (Central Plateau). Similarly other crops like jowar, pulses, oilseeds, cotton, sugarcane and fruits and vegetables have their own respective zones of concentration. It may also be seen that crop specialisation is more in certain crops like sugarcane, cotton, jowar, and wheat. The distribution is most widespread in case of rice, pulses, fruits, and vegetables.

A brief description of 15 Agro-climatic regions is given below :

1. Western Himalaya. The Western Himalayan Region stretches over Ladakh, Jammu and Kashmir, Himachal Pradesh and Uttarakhand. This region has great variations in topography which consists of lofty mountains, deep valleys, steep slopes and a large number of topographical features of local importance. The climate of this region is characterised by mild summer with July average temperature varying from 5°C to 30°C and severe winter with January temperature varying from 0°C to –4°C. The mean annual rainfall varies from 75 cm to 150 cm, except in Ladakh where it is below 30 cm. The valleys like those of Kashmir and Kullu as well as duns like those of Dehra Dun are covered with thick layers of alluvium while hill slopes have thin brown soils. The region is drained by a number of perennial rivers which originate in the snow covered mountain ranges. These include the Ganga and its major tributary the Yamuna and the Indus and its major tributaries like the Jhelum, the Chenab, the Beas and the Satluj. Some of the rivers are utilised as sources of canal irrigation and hydro-electric power. Unfortunately, the increased human interference in the natural set-up of the region has led to overall deterioration of ecological system. Rapid and unplanned construction activities, unscrupulous mining and reckless felling of trees have made the region more prone to floods, droughts, landslides receding glaciers and environmental pollution. The Uttarakhand tragedy of 15-16 June, 2013 is on eye opener in this regard.

Rice, maize, wheat, barley and vegetables are grown in the terraced fields on the hill slopes. But the plantations do not need terracing and are set up along the hill slopes. The region is well known for temperate fruits like apples, peach, pears, almond, walnut, etc. The environmental conditions of this region are quite congenial to development of garden and plantation crops and there is much scope for improvement in this regard. Alpine pastures above 2000 m are known as *dhoks* or *margs* which are used by Gujjars and Gaddis for rearing sheep and other animals.

This region requires a more judicious land use planning for increasing farm productivity. There should be rational allotment of land for different purposes like cultivation of crops, horticulture, pastures and forestry. The most suitable recommendations are (*a*) agriculture on land with upto 30 per cent slope, (*b*) horticulture/fodder development on land having slope 30-50 per cent and (*c*) all land over 50 per cent slopes under tree cover.

2. The Eastern Himalaya. This region encompassed the eastern part of the Himalayas consisting of Sikkim, Darjeeling hill area of West Bengal, Assam Hills, Arunachal Pradesh, Nagaland, Manipur, Mizoram, Tripura and Meghalaya. It is characterised by rugged topography, steep slopes, thick forests and swift flowing rivers. The climate is sub-humid with annual rainfall over 200 cm. The July and January temperatures vary from 25°C to 30°C and 10°C to 20°C respectively. The soil is red-brown and less fertile. About one-third of the total cultivated area is under shifting agriculture locally known as *Jhuming*. Rice, maize, potato and fruits (orange, pine apple, lime, lichi, etc.) are the main crops. Tea plantations are found on the hill slopes of Assam and Darjeeling.

Shifting agriculture is causing heavy damage to soil and forest resources and there is urgent need to check this type of agriculture and replace it by other forms of agriculture like subsistence type of agriculture. Soil erosion can also be checked by developing terraced cultivation and planting more trees, especially on the hill slopes. Also, there is vast scope for improving infrastructural facilities. There is too much wastage of water by surface run-off. Steps should be taken to check it and water thus conserved should be used for minor irrigation. A programme to grow fruits above 30% slopes in the hills needs to be framed. Supportive activities like sericulture, handicrafts, poultry, piggery, etc. should be encouraged to supplement the income of the farmers. A long-term quality seed production, suiting the ecological conditions of the region, needs to be implemented on priority basis.

3. The Lower Gangetic Plains. This region spreads over eastern part of Bihar, West Bengal (excluding Darjeeling hilly area) and Brahmaputra valley in the Assam. It is made up of rich alluvial soil deposited by big rivers. The Ganga delta is the largest in the world. Most of the area is characterised by extremely gentle slope and oxbow lakes. The area has hot and humid climate where the annual rainfall varies from 100 to 200 cm and January and July temperatures range between 12°C to 18°C and 25° to 30°C respectively. The region has enough storage of ground water and the water table is very high. The region suffers from waterlogging and marshy areas are found over vast stretches of land. Wells and canals are the main sources of irrigation. Hot and humid climate and rich alluvial soils provide ideal conditions for the cultivation of rice and jute crops. In some parts of West Bengal, three successive crops of rice (Aman, Aus, and Boro) are grown in a year. The Hugli basin is world nonowned for jute cultivation. Wheat has become popular as a winter crop as a consequence of the Green Revolution. Maize, pulses, potato, etc. are other important crops. The development strategy includes improvement in rice farming, horticulture (banana, mango and citrus fruits), pisciculture, poultry, livestock, forage production and seed supply. A large number of farmers engage themselves in fishing in ponds and reservoirs.

4. The Middle Gangetic Plains. It spreads over eastern part of Uttar Pradesh and large part of Bihar. It is a gently sloping plain which is made up of fertile alluvial soil that has been deposited by the mighty Ganga and its tributaries. This is an area of hot and humid climate where the annual rainfall is 100-150 cm. The temperature ranges from 25°C to 40°C in July and 10° to 25°C in January. The region has vast potential of ground and run-off water from the perennial rivers which is used for irrigation through tube wells and canals. Rice, maize, millets etc. are the main *kharif* crops while wheat, gram, barley, peas, mustard and potato are important *rabi* crops. Mango, guava, lichi, banana etc. are the main fruit crops.

There is vast scope for improvement of agriculture in this region. Efforts should be made to improve the yield of kharif paddy crop which occupies over 40 per cent of the gross cropped area of the region. Diversification of crops by giving more area to vegetables, fruits, and floriculture can yield better results. Dairying, silviculture, agro forestry, and pisciculture can also help a lot in supplementing the income of the farmers.

5. The Upper Gangetic Plain. This plain stretches over the Ganga-Yamuna doab, Lucknow division and Rohilkhand of western Uttar Pradesh and Hardwar and Udham Singh Nagar districts of Uttarakhand. This is more or less a plain area with gentle slope which has been built by fertile soils deposited by the Ganga and its tributaries. This is a region of sub-humid continental climate where the annual rainfall varies from 75 to 150 cm. The temperatures vary between 10°C and 25°C in January and between 25°C and 40°C in July. The soil varies from sandy loam to clayey loam. The area has developed adequate facilities of canal and tube well irrigation. This is an intensive agricultural region where wheat, rice, sugarcane, millets, maize, pulses, gram, barley, oilseeds and cotton are widely grown. Like Punjab and Haryana, this area has taken maximum advantage of the Green Revolution.

This region has great potential for improvement in agriculture and efforts should be made to increase the agricultural productivity of this region. The strategy should include greater emphasis on mixed cropping, horticulture, floriculture, judicious use of irrigation facilities, lining of canals to check seepage, reclaiming saline/alkaline soils, improving cattle breeds and bringing more areas under fodder crops.

6. The Trans-Ganga Plain. This plain consists of Punjab, Haryana, Delhi, Chandigarh and Ganganagar district of Rajathan. The plain is made up of highly productive alluvial soils which have been deposited by the rivers draining this region. The climate has semi-arid characteristics where the annual rainfall varies from 40 to 100 cm. Most of the rainfall is received form the south-west monsoons during the summer season but some rainfall is also brought by the western disturbances during the winter season. Being an area of continental climate, the region experiences extremes of temperature which may soar to 45°C during day in May/June and dip to almost freezing point during night in December/January. However, the average temperatures for January and July vary from 10°C to 20°C and 25°C to 40°C respectively. The perennial rivers draining this region provide ample opportunities for canal irrigation. Besides, the region has lakhs of tube wells. In fact, tube wells irrigate much larger area as compared to canals. Both Punjab and Haryana have some of the highest intensities of agriculture. Wheat, rice, sugarcane, cotton, gram, maize, millets, pulses and oilseeds are the main crops. The area has the distinction of introducing the Green Revolution in India when HYV seeds of wheat and rice were introduced in the mid-1960s. Along with seeds, this area was the first to adopt other modern methods of farming like fertilizers, irrigation and farm mechanisation. However, the region is facing serious problems of waterlogging, salinity, alkalinity and soil degradation due to over irrigation by canals. The region is also facing a serious crisis of falling water table due to over exploitation of ground water through tube-wells.

Although this is a highly productive part of the country with respect to agriculture, yet it can be made even more productive if some corrective measures are taken. Wadia (1996 : 100) has suggested the following measures :

(*i*) diversion of 5 per cent of rice-wheat area to other crops like maize, pulses, oilseeds, and fodder.

(*ii*) development of genotypes of rice, maize and wheat with inbuilt resistance to pests and diseases.

(*iii*) promotion of horticulture besides pulses like tur and peas in upland conditions.

(*iv*) cultivation of vegetables in the vicinity of industrial clusters.

(*v*) supply of quality seeds of vegetables and planting materials for horticulture crops.

(*vi*) development of infrastructure for transit godowns and processing to handle additional fruit and vegetable production.

(*vii*) implementation of policy and programmes to increase productivity of milk and wool.

(*viii*) development of high quality fodder crops and animal feed by stepping up area under fodder production by 10 per cent.

7. The Eastern Plateaus and Hills. The region includes the Chotanagpur plateau, Rajmahal Hills and Chhattisgarh plains. It consists of red and yellow soils with occasional patches of laterites and alluviums. Most of the soils are not very productive. Fertile soils are found only in the Mahanadi basin. The region receives annual rainfall varying from 75 cm to 150 cm. The mean January and July temperatures range from 10°C to 25°C and 25°C to 40°C respectively. The region is deficient in surface water due to non-perennial streams and ground water due to hard and impermeable rocks. The agriculture is mainly rain-fed in which rice, maize, millets, *ragi,* gram, oilseeds, tobacco, potato, etc. are the main crops. Fallowing is the common practice in areas lacking irrigation.

This area needs improvement in agriculture. The strategy for this may include larger use of HYV seeds, cultivation of high value crops like oilseeds, pulses, vegetables, improvement of indigenous breeds of cattle and buffaloes, water-harvesting and water-shed development, soil conservation, renovation of old tanks and excavation of new tanks and rehabilitation of degraded forest land.

8. The Central Plateaus and Hills. Most of this region spreads over eastern part of Madhya Pradesh and adjoining parts of Rajasthan and includes Bundelkhand, Baghelkhand, Bhander plateau, Malwa plateau and Vindhyanchal hills. The region is characterised by semi-arid climatic conditions where the annual rainfall varies from 50 to 100 cm. The mean January and July temperatures are 10°C to 25°C and 25°C to 40°C respectively. The soils are mixed red, yellow and black in which crops like millets, wheat, gram, pulses, oilseeds, cotton, sunflower etc. are grown. Water scarcity is the main problem of this region. Among the main strategies for improving agriculture are water conservation through devices like sprinklers and drip system, opting for dry farming, crop diversification, dairy development and poultry farming.

9. The Western Plateaus and Hills. The region spreads over the southern part of Malwa plateau and the Deccan plateau of Maharashtra. This is a region of black soil known as *regur*. The region is characterised by semi-arid climate with average annual rainfall varying from 25 cm to 75 cm. The January and July temperatures range from 5°C to 25°C and 25°C to 40°C respectively. Irrigation facilities lack badly and only a little over 12% of the cropped area enjoys irrigation facilities. Therefore, most of the crops have to depend on rainfall and mostly drought resistant crops are grown. Jowar, cotton, sugarcane, rice, bajra, wheat, gram, pulses, potato, groundnut, oilseeds are the main crops. Agricultural production can be increased by adopting water saving devices like sprinkle irrigation, drip system, water harvesting, replacing low value crops by high value crop, providing market and storage facilities, developing dairy and poultry farming.

10. The Southern Plateaus and Hills. It includes southern Maharashtra, Karnataka, Telangana, Andhra Pradesh and northern Tamil Nadu. The area has semi-arid climate with annual average rainfall of 50-100 cm. Temperatures for January and July are 13°C to 21°C and 25°C to 40°C respectively. Since the rainfall is less and the temperatures remain high throughout the year, this is essentially an area of dry farming where millets, pulses, oilseeds, coffee, tea, cardamom are the main crops. Development of poultry, dairy farming, horticulture, restoration of tank irrigation, use of water saving devices and diversion of jowar/bajra/ragi to groundnut/sesamum/sunflower can improve the situation.

11. The East Coastal Plains and Hills. This region extends all along the eastern coast from Baleshwar in Odisha to Kanniyakumari in Tamil Nadu. Its northern part is known as Northern Circars and southern part is called Coromandal coast. It has been formed by the depositonal work of rivers like Mahanadi, Godawari, Krishna and Cauvery and the deltas of these rivers are the chief characteristic features of this region. The region has sub-humid marine climate where the annual rainfall varies between 75 cm and 150 cm. Since the climate of this coastal area is influenced by the Bay of Bengal, there is not much difference between summer and winter temperatures. Here, the January and July temperatures vary from 20°C to 30° and 25°C to 35°C respectively. The soils are alluvial, loam and clay. These soils suffer from the serious problem of alkalinity. The main crops are rice, jute, tobacco, sugarcane, maize, millets, pulses, groundnut, and oilseeds.

The main strategies for agricultural development in this region are discouraging mono-culture of rice and encouraging crop diversification, replacing coarse grains by high value crops like pulses and tobacco, increasing cropping intensity, cultivation of spices like pepper and cardamom, proper use of water resources and improving tank irrigation system, treating areas suffering from alkalinity and water logging and increasing potential of animals and fisheries.

12. The West Coastal Plains and Western Ghats. This region extends from the Tapi estuary in the north to Kanniyakumari in the south and covers coastal areas of Maharashtra, Karnataka and Kerala. Its northern part is known as Konkan and southern part is called Malabar. The Sahyadri range running in north-south direction is also included in this region. South-west monsoons originating in the Arabian Sea strike against the Western Ghats and cause heavy rain. The average annual rainfall exceeds 200 cm. The mean January and July temperatures are 18°C to 30°C and 26°C to 32°C respectively. Therefore the climate of this region may be termed as hot and humid. Rice, coconut, oilseeds, sugarcane, millets, pulses and cotton are the main crops. Slopes of the Western Ghats are famous for plantation corps.

The strategies for development include devoting more area to high value corps like spices, pulses, fruits and coconut, improvement in drainage, sinking of dug wells, improvement in infrastructure, and promotion of prawn culture in brackish waters.

13. The Gujarat Plains and Hills. This region encompasses the plains and hills of Kathiawar and the fertile valleys of Mahi and Sabarmati rivers. This is an arid and semi-arid region where the average annual rainfall varies from 50 cm to 100 cm. The amount of rainfall increases from west to east. January and July temperatures are 15°C to 30°C and 25°C to 42°C respectively. Different types of soils are found in this region. For example, soils are regur in the plateau region, alluvial in the coastal plains and yellow to red in the Jamnagar area. Groundnut, cotton, rice, millets, oilseeds, wheat and tobacco are the main crops. Wheat is the main rabi crop in irrigated areas. Kachchh is a dry area and is not much suitable for agriculture. The whole region is famous for production of oilseeds.

The development strategies should include surface and ground water management, rain-water harvesting, dry land farming, agro-forestry development, wasteland development and development of fishing and acquaculture in coastal zone and river deltas.

14. The Western Dry Region. This region stretches over the western part of Rajasthan (west of the Aravali range) which spreads over the waste desert of Marwar and the Mewar hills. This is totally an arid region where the annual rainfall does not exceed 25 cm. The January temperature ranges between 5°C and 20°C (and sometimes even touches the freezing point) which soars to over 40°C in July. Most of the region is sandy desert and is characterised by sand dunes. Luni river flows through the Jodhpur basin. Rainfall is scarce and erratic and droughts and famines are very common. Bajra, jowar and moth are the chief *kharif* crops while wheat and gram are grown in the *rabi* reason. However, livestock rearing is the main occupation of the people. Camel, goat and sheep are the main animals. Irrigation by the Indira Gandhi canal has changed the cropping pattern and raised the income levels of the farmers.

The development strategies should include enhancement in irrigation, promoting use of fertilizers and better seeds, increasing yields of fruits like date, palm, watermelon and guava and improving pastoral system.

15. The Islands Region. This region includes Andaman and Nicobar Islands in the Bay of Bengal and Lakshadweep Islands in the Arabian Sea. Being near the equator and surrounded by see on all sides, these islands have typically uniform equatorial climate where the average annual rainfall exceeds 200 cm and January and July temperatures are 25°C and 30°C respectively. The soils vary from sandy along the coast to clayey loam in valleys and lower slopes. Rice is the main crop which is followed by maize, jowar, bajra, pulses and plantation crops like arecanut, cassava, turmeric, etc. Nearly half of the cropped area is under coconut. A large part of area is covered by forests and agriculture is in a backward state.

The main development strategies should lay more emphasis on crop improvement, water management and fisheries. The farmers should be encouraged to take two crops of rice in a year by adopting modern methods of agriculture. Old and less productive coconut trees should be replaced by young and more productive trees.

Agro-Ecological Regions of India

The concept of 'agro-ecological regions' is a modified and improved version of 'agro-climatic regions'. Several scholars have tended to use these two terms in the same way but there is basic and distinct difference between the two. Food and Agriculture Organisation (FAO) classified the difference between these two terms in 1983 and stated that agro-climatic region is a land unit in terms of major bio-climatic and length of growing period and which is climatically suitable for certain range of crop cultivation. On the other hand, agro-ecological region is a land unit carved out of agro-climatic region when superimposed on land form and soil condition that acts as modifier of length of growing period. Therefore, there may be a few agro-ecological regions within an agro-climatic region, depending on soil conditions. This approach has been used for the purpose of delineating agro-ecological regions of India.

Methodology. For preparing agro-ecological region map, the soil-scap (soil map) is superimposed on bio-climatic map. On the resultant map, the growing period map is incorporated by using Geographical Information System (GIS) technology.

For demarcating agro-ecological regions of India, the agro-climatic regions of the country have been sub-divided on the basis of soil types. Thus the maso regions are obtained and the length of growing period (LGP) has been superimposed on them. By using this method, India has been divided into 20 agro-ecological regions (**Fig. 20.10**) and 60 agro-ecological sub-regions. A brief description of 20 agro-ecological regions is given below :

1. The Western Himalayan Cold and Dry Region. Located in the Union Territory of Ladakh, this region is out of reach of the south-west monsoon and receives less than 20 cm annual rainfall. The region has very severe winters due to its high altitude. Thus this region is known as **cold desert.** This region is not very useful for cultivation because the growing period here is less than 90 days.

2. The Western Plain Kachchh and part of Kathiawar Peninsula Region. This is hot and dry region where the annual rainfall is less than 40 cm. In the vast area of Jaisalmer, Barmer and Bikaner, the annual rainfall is less than even 20 cm. Similarly Kachchh and Kathiawar areas also receive scanty rainfall and have dry climate. The isohyet of 25 cm annual rainfall divides the entire region into two parts *viz.* (*i*) dry desert (Marusthali) and (*ii*) semi-arid transitional zone. The summer temperature soars to over 40°C as a result of which it is known as hot and dry region. The area has vast stretches of sandy and saline soils which are not much suited to agriculture. Land use data for 12 western districts of Rajasthan from 1951-52 to 1971-74 had shown that most of the area is unfit for cultivation and about 50 per cent of the land is either fallow or waste land. Limitation for cultivation can be appreciated by the fact that growing period here is less than 90 days. The Indira Gandhi canal has created problems of waterlogging, salinity and alkalinity.

3. The Deccan Plateau Hot Arid Region. This region has red and black soils. This region receives scanty rainfall because it is located in the rain-shadow area of the Western Ghats. However, black soil is quite fertile and supports agriculture wherever tank irrigation is available. The growing period is less than 90 days due to scanty rainfall.

4. The Northern Plains and Central Highlands including Aravalis. This is a hot semi-arid ecoregion where annual rainfall varies from 40 to 70 cm. The Northern Plain area has alluvial soil but the soils of the Central Highlands and the Aravali region have been derived from the decay and decomposition of the parent rocks. Here the growing period varies from 90 to 150 days.

5. The Central Mawa Highlands, Gujarat Plains and Kathiawar Peninsula. This is also a hot and semi-arid region where the annual rainfall is 40 to 60 cm. The summer temperatures are very high due to which this region has been termed as hot and semi-arid area. The region has medium to deep black soils which are capable of giving high yields if sufficient water is available. Here also the growing period varies between 90 and 150 days.

6. The Deccan Plateau. It lies in the rain shadow area of the Western Ghats in the states of Maharashtra and Karnataka and receives about 60 cm annual rainfall. It is termed as hot and semi-arid region due to high temperature during the summer season. This region is blessed with rich fertile black soil which is ideal for cotton cultivation. The growing period varies from 90 to 150 days.

7. The Deccan (Telangana) Plateau and the Eastern Ghats. This is also located in the rain shadow area of the Western Ghats and receives less rainfall. Most of it lies in Telangana and Andhra Pradesh. This hot and semi-arid region has red and black soils which are used for agriculture with the help of irrigation. The region has a growing period varying from 90 to 150 days.

8. The Eastern Ghats, Tamil Nadu Plateau and Deccan (Karnataka). This is also a hot and semi-arid region where red and loamy soils are found Agriculture is based on tank irrigation. Growing period is 90-150 days.

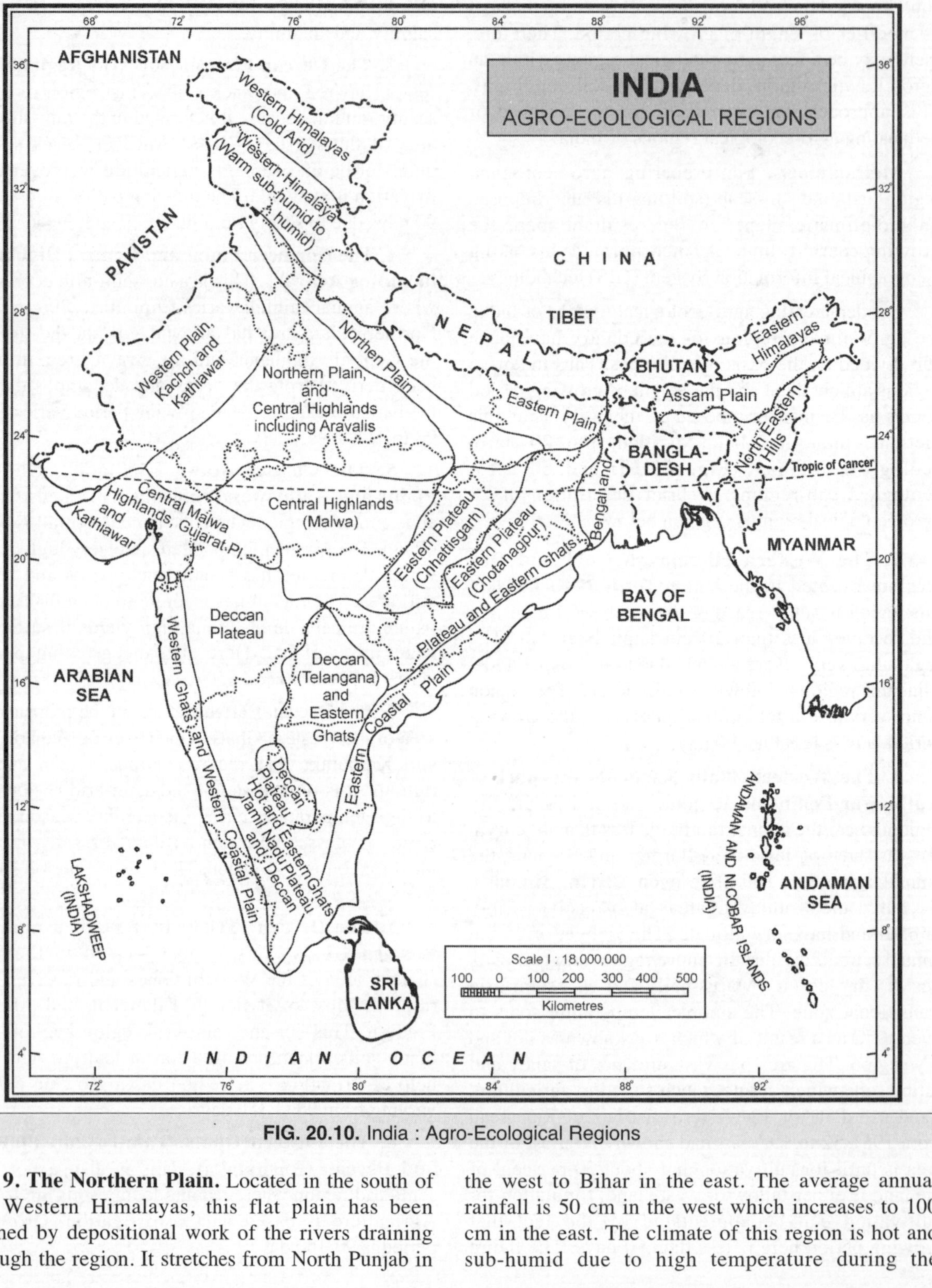

FIG. 20.10. India : Agro-Ecological Regions

9. The Northern Plain. Located in the south of the Western Himalayas, this flat plain has been formed by depositional work of the rivers draining through the region. It stretches from North Punjab in the west to Bihar in the east. The average annual rainfall is 50 cm in the west which increases to 100 cm in the east. The climate of this region is hot and sub-humid due to high temperature during the summer season. The region is blessed with rich alluvial soils. Wheat in the west and rice in east are the main crops. Several other crops are also grown depending on the local conditions. The growing period varies from 150 to 180 days.

10. The Central Highlands (Malwa, Bundelkhand and Eastern Satpura). This is a hot and sub-humid region with black and red soils. Most of this region is in Madhya Pradesh and Vidarbha region of Maharashtra. Irrigation is required for successful growth of crops. The growing period is between 150 and 189 days.

11. Eastern Plateau (Chhattisgarh). Major part of this region lies in Chhattisgarh and has hot and sub-humid climate. The region has red and yellow soils and is well known for rice cultivation. In fact this region has earned the destination of *rice bowl of India* due to its intensive rice cultivation. The growing period in this region varies from 150 to 180 days.

12. The Eastern (Chotanagpur) Plateau and Eastern Ghats. This agro-ecological region includes major part of the Chotanagpur plateau of Jharkhand and the Eastern Ghats area of Odisha. With the annual rainfall of about 150 cm, this region has hot and sub-humid climate. The red and laterite soils of this region are not much suitable for agriculture although it has a long growing season from 150 to 180 days.

13. The Eastern Plain. This agro-ecological region is in the north-east of the Northern Plain and stretches over northern parts of Uttar Pradesh and Bihar. The northern part of this region consists of Bhabar and Tarai. The whole region has been formed by the depositional work of the perennial rivers originating in the snow covered Himalayas. There is plenty of fertile alluvial soil which has provide a solid base to agriculture in this region. The region has a long growing period of 180 to 210 days which enables farmers to take two crops per year.

14. The Western Himalayas (Warm sub-humid to humid). This agro-ecological region lies between the Western Himalayan cold arid region in the north and the North Plain region in the south and stretches over Uttarakhand, Himachal Pradesh and Jammu and Kashmir. It is different from the first region because that region is cold and arid whereas this region has warm sub-humid to humid climate. It also includes perhumid ecoregion. Agriculture in this region is not much developed in spite of long growing season of 180-210 days. This is due to rugged topography of the area. However, the ecological conditions of this region are very congenial to orchards and plantation crops. Consequently, a large variety of temperate fruits are grown here. Brown and podzole soils give natural support to fruit cultivation in this region.

15. The Bengal and Assam Plain. This region includes the entire plain area of West Bengal and the Brahmaputra valley of Assam. The area receives about 150 cm annual rainfall from the south-west monsoons originating in the Bay of Bengal. The climate of this region is termed as hot sub-humid to humid (inclusion of perhumid). These climatic conditions coupled with rich fertile alluvial soils deposited by the rivers provide solid base to rice cultivation in this region. The Ganga-Brahmaputra delta is world famous for jute cultivation. Long growing period of 180-210 days provide ample scope for more than one crop.

16. The Eastern Himalayas. It includes Sikkim, Arunachal Pradesh and hilly areas of Assam. It is warm perhumid ecoregion with brown and red soils. The growing period over 210 days.

17. The North-Eastern Hills (Purvanchal). It spreads over the hills of Nagaland, Manipur, Mizoram, Tripura and Meghalaya. With average annual rainfall of 150-250 cm, this is the region of warm perhumid climate. Although the region has a long growing period of over 210, the red and laterite soils do not support much agriculture. Most of the area is covered by forests.

18. The Eastern Coastal Plain. It spreads over vast area along the east coast of the country from Subarnarekha river in the north to Kanniyakumari in the south. This coastal plain runs parallel to the direction of the south west monsoon originating in the Bay of Bengal and does not receive much rainfall. Therefore, this area has hot sub-humid to semi-arid climate. Big rivers like Mahanadi, Godavari, Krishna and Cauvery have formed large deltas of fertile alluvial soils. These deltas are famous for rice cultivation. There are large variations in agricultural calendar because the growing period varies from 90 to 210 days.

19. The Western Ghats and Western Coastal Plain. The Western Ghats run parallel to the west coast of India in a north-south direction and offers an effective obstruction in the way of the south-west monsoons originating in the Arabian sea. This results in copious rainfall of over 200 cm in the western coastal plains and on the western slopes of the Western Ghats. Thus this is an area of hot humid perhumid climate. It has laterile and alluvium soils. Heavy rains and alluvial soils coupled with long growing season of 90 to 210 days provide suitable conditions for rice cultivation.

20. The Islands of Andaman and Nicobar and Lakshadweep. The islands of Andaman and Nicobar are located in the Bay of Bengal and those of Lakshadweep are in the Arabian sea where hot humid to perhumid climate is found. These islands have red loaming and sandy soils and the growing period is over 210 days.

REFERENCES

Acharya, S.S. and N.L. Agarwal (1987) : *Agricultural Marketing in India,* Oxford and IBM Publishing Co., New Delhi.

Agricultural Statistics at a Glance, 2016, Govt. of India, Ministry of Agriculture and Farmers Welfare.

Ahmed, A. and M.F. Siddiqui (1967) : *Crop Association Patterns in the Luni Basin,* Geographer, 14, pp. 69-80.

Andeson, J.R. (1970) : *A Geography of Agriculture,* WMC Brown Co., Dubuqu, Iowa.

Appu, P.S. (1996) : *Land Reforms in India : A Survey of Policy Legislation and Implementation,* Vikas, New Delhi.

Appu, P.S. (1975) : "Tenancy Reforms in India", *Economic and Political Weekly,* Special Number, August, 1975, pp. 1339-75.

Awasthi, U.S. (2004) : *Fertilizer—Pragmatic Views, Concerns,* The Hindu Survey of Indian Agriculture, 2004 Kasturi and Sons Ltd., Chennai, pp. 145-146.

Banerjee, Abhijit V. and Maitreesh Ghatak (1996) : Empowerment and Efficiency : *The Economics of Tenancy Reform,* Minco MIT and Harvard University.

Banerjee, Bireswar (1991) : *Models of Agricultural Development : The Indian Experience* in Jayamala Diddee (*ed.*), *Emerging Trends in Indian Geography,* Rawat Publication, Jaipur.

Bansil, P.C. (1977) : *Agricultural Problems in India,* Vikas Pub., New Delhi.

Basu, D.N. and G.S. Guha (1996) : *Agro-Climatic Regional Planning in India,* Vol. 1, Concepts and Application, New Delhi.

Bergmann, T. (1983) : Co-operation in India : Selected aspects in A.M. Khusro (ed.) *Readings in Agricultural Development,* Allied Publishers, Bombay.

Bhalla, G.S. (2004) : *Globalisation and Indian Agriculture, State of the Indian Farmer : A Millennium Study*, Vol. 19, Academic Foundation, N. Delhi.

Bhalla, G.S. (2006) : *Condition of Indian Peasantry,* National Book Trust, New Delhi.

Bhalla, G.S. (2007) : *Indian Agriculture Since Independence,* National Book Trust, New Delhi.

Bhalla, G. S., Alag, Y.K. (1979) : *Performance of Indian Agriculture,* Sterling, New Delhi.

Bhalla, G.S. and Gurmail Singh (2001) : *Indian Agriculture : Four Decades of Development,* Sage, New Delhi.

Bhandari, Suparas (2004) : *Insurance—Assuring Livelihood Security,* The Hindu Survey of Indian Agriculture, 2004, Kasturi and Sons Ltd., Chennai, pp. 170-171.

Bhatia, Kumar (2004) : *Seed Industry—Achieving Varietal diversity,* The Hindu Survey of Indian Agriculture, 2004, Kasturi and Sons Ltd., Chennai, pp. 161-163.

Bhatia, S.S. (1965) : "Patterns of Crop Concentration and Diversification in India", *Economic Geography,* Vol. 41, pp. 40-56.

Bhatia, S.S. (1967) : "A New Method of Agricultural Efficiency in Uttar Pradesh, *Economic Geography,* Vol. 43, pp. 244-260.

Blyn, G. (1966) : *Agricultural Trends in India,* Philadelphia.

Brezeale, E.L. and W.T. McGeorge, "Influence of Atmospheric Humidity on Root Growth," *Soil Science,* 76 (1953), pp. 361-365.

Buck, J.L. (1967) : *Land Utilization in China* (Volume I), Nankin, University of Nankin.

Centre for Science and Environment, State of India's Environment (2019) in Figures pp. 28-29.

Chang, Jen-Hu (1968) : *Climate and Agriculture : An Ecological Survey,* Aldine Publishing Co. Chicago.

Coppock, J.T. (1964) : "Crop Livestock and Enterprise Combinations in England and Wales", *Economic Geography,* Vol. 40, pp. 65-81.

Coppock, J.T. (1968) : Changes in land use in Great Britain, in *Land Use and Resources : Studies in Applied Geography* (London Institute of British Geographers), Special Publication, No. 1.

Coppock, J.T. (1971) : *An Agricultural Atlas of England and Wales,* G. Bell and Sons, London.

Dandekar, V.M. (1994) : The Indian Economy, 1947-92, Vol. I on *Agriculture,* Sage, New Delhi.

Dantwala, M.L. (1959) : "Small Farmers not Small Farmers", *Indian Journal of Agricultural Economics,* 15, pp. 57-58.

Dantwala, *et al.* (1973) : *Agricultural Development since Independence,* Oxford University Press, New Delhi.

Dantwala, M.L. and L.B. Dorde (1965) : "The Economic Cultivator, *Indian Journal of Agricultural Economics* (1945-65), Selected Readings.

Dantwala, M.L. ***et al.*** (eds.) (1986) : *Agricultural Development since Independence,* Oxford University Press, New Delhi.

Dantwala, M.L. (1996) : *Dilemmas of Growth* in Pravin Visaria *et al.* (eds.) Sage, New Delhi.

Dayal, E. (1967) : "Crop Combination Regions : A Case Study of the Punjab Plains", *Tijdschrift Voor Econ.* En Soc. Geografic. Vol. 58, pp. 39-47.

Dhawan, B.D. (1990) : *Studies in Minor Irrigation with Special Reference to Ground Water,* Commonwealth Publishers, New Delhi.

Dikshit K.R. (1973) : "Agricultural Regions of Maharashtra" *Geographical Review of India,* Vol. 35, pp. 387-391.

Edwards, G.A.B. (2004) : *Farm Mechanisation—Enhancing ploughing efficiency,* The Hindu Survey of Indian Agriculture, 2004, Kasturi and Sons Ltd., Chennai, pp. 149-151.

Eitenne, G. (1968) : *Studies in Indian Agriculture,* Bombay.

Enyedi, G.Y. (1964) : *Geographical Types of Agriculture,* Budepest, Applied Geography in Hungary.

Framji, K.K. (1974) : "Review of Waterlogging and drainage problems and their solutions in India; *International Commission on Irrigation and Drainage Bulletin of January,* 1974, pp. 36-40.

Gasson, R. "Occupation Chosen by sons of Farmers", *Journal of Agricultural Economics* 19 (1968), pp. 317-326.

Government of India (1985-86) : *All India Report on Agricultural Census, 1985-86,* Department of Agriculture and Cooperation, New Delhi.

Government of India (2021) : Agricultural Statistics at a Glace, Ministry of Agriculture.

Government of India : *Economic Survey* 2018-19, 2019-20, 2020-21, 2021-22 and 2022-23 issues.

Government of India : *A Reference Annual,* 2019, 2020, 2021 and 2022 issues.

Government of India (1996) *Implementation of Land Reforms,* Planning Commission, New Delhi.

Government of India (1998) : *Agro-Climatic Regional Planning* : *An Overview,* Planning Commission, New Delhi.

Government of India : Survey of Land Holdings in India, Various Rounds NSSO.

Gragor, Howard F. (1970) : *Geography of Agriculture : Themes in Research,* Prentice Hall, London.

Gulati, Ashok and Sudha Narayan, (2003) : *The Subsidy Syndrom in Indian Agriculture,* Oxford University Press, New Delhi.

Gupta, P. Sen and Sadasyuth, G. : *Economic Regionalisation of India : Problems and Approaches,* Census of India 1961, Monograph Series Vol. I, No. 8 (1968), New Delhi.

Hassis, B. : "*Innovation adoption and Indian Agriculture*"*—the High Yielding Varieties.*

Hildreth, A.C. *et al.* : "Effects of Climatic factors on sowing plants." Water : *The United States Department of Agriculture, Yearbook* 1941, Washington.

Hussain, M. (1972) : "Crop Combination Regions of Uttar Pradesh : A study in Methodology," *Geographical Review of India,* 34, pp. 134-156.

Hussain, M. (1976) : "A new approach to the agricultural productivity regions in the Satluj-Ganga Plains of India." *Geographical Review of India,* 36, pp. 230-236.

Indian Council of Agricultural Research (1970) : Handbook of Agriculture, New Delhi.

Kendall, M.G. : "The Geographical distribution of crop productivity in England", *Journal of Royal Statistical Society,* 169 (1939) pp. 24-28.

Khan, W. and R.N. Tripathy, *Intensive Agriculture and Modern Inputs, Prospects of Small Farmers,* A Study in West Godavari District, Hyderabad Institute of Community Development, 1972.

Khusro, A.M. : "Returns to Scale in Indian Agriculture", *Indian Journal of Agricultural Economics,* 19 (1964), pp. 52-53.

Khusro, A.M. (1973) : (*ed.*) *Economics of Land Reforms and Farm Size in India,* Macmillan, New Delhi.

Klages, K.H.W. (1958) : *Ecological Crop Geography,* Macmillan, New York.

Kochhar, P.L. (1967) : *Plant Ecology, Genetics and Evolution,* Atma Ram and Sons, New Delhi.

Kahlon, A.S. and D.S. Tyagi (1983) : *Agricultural Price Policy in India,* Sage, New Delhi.

Kohl, R.L. (1958) : *Marketing of Agricultural Products,* Macmillan, New York.

Land Reforms : *Defects in Agrarian Structure as Obstacles to Economic Development* (1951), United Nations Publications.

Millor, J.W.&R. Ahmed (1988), *Agricultural Price Policy for Developing Countries,* Johns Hopkins University Press, Ballimore and London.

Mamoria, C. B. (1970) : *Agricultural Problems in India,* Kitab Mahal, Allahabad.

Mohammed, A. (1978) : *Studies in Agricultural Geography,* Rajesh Publication, New Delhi.

Morgan, W.B. and R.J.C. Munton (1971) : *Agricultural Geography,* Methuen, London.

Munton, R.J.C. (1969) : *The Economic Geography of Agriculture in Trends in Geography,* U. Cooke and James H. Johnson (ed.), Pergamon, London.

NABARD (2001) : *Report on High Power Expert Committee on Rural Credit.*

NABARD (2001) : *Handbook on Kisan Credit Card Scheme,* Bombay.

Nadkarni, M.V. (1991) : *The Crisis of Increasing Costs in Agriculture. Is there any way out* ? in Nadkarni *et al.* (*ed.*), *India, the Emerging Trends.*

Nanavati, M.B. and Anjana, J.J. (1970) : *The Indian Rural Problem*, ISAE, Bombay.

Nayak, G. : *Fertilizer Consumption,* Economic Times, New Delhi, May 6, 1993.

Oldenburg, Phillip (1990) : Land Consolidation as Land Reforms in India, *World Development,* 18(2).

Paroda, R.S. (2008) : Rainfed Farming : *Strategy for Productivity Growth. THSOIA* pp. 47-49.

Randhawa, M.S. (1958) : *Agriculture and Animal Husbandry in India* ICAR, New Delhi.

Rao, A.P. (1967) : *Size of Holding and Productivity,* Economic and Political Weekly, 11 Nov., 1967.

Rao Hanumantha, C.H. (1975) : *Technological Change and Distribution of Gains in Indian Agriculture,* Macmillan, New Delhi.

Rao Hanumantha, C.H. (1979) : *Farm Mechanisation* in M. Shafi (ed.) *Agricultural Development of India*, Bombay.

RBI (1990) : *A Review of the Agricultural Credit System in India :* Report of the Agricultural Credit Review Committee, Bombay.

Sapre, S.G. and V.D. Deshpande (1964) : "Inter-district Variations in Agricultural Efficiency in Maharashtra State, *Indian Journal of Agricultural Economics* 19, pp. 242-252.

Sen, Abhishwar (1958) : "Influence of Climatic Factors on Salinity of Indian Soils", *Indian Journal of Agricultural Science* 28, pp. 67-71.

Sharma, B.L. (1996) : *Sustainable Agricultural Development* in Majid Husain *et al.* (ed.), *Environmental Management in India,* Manak Publications, New Delhi.

Sharma, D.P. and Desai, U.V. (1980) : *Rural Economy of India*, Vikas, New Delhi.

Shafi, M. "Measurement of Agricultural Efficiency in Uttar Pradesh", *Economic Geography* 366 (1960) pp. 296-30.

Shafi, M. "Measurement of agricultural productivity of the Great Plains" *The Geographer* 19 (1972), pp. 4-13.

Shafi, M. "Perspective of Measurement of Agricultural Productivity" *The Geographer* 21 (1974) pp. 6-8.

Shafi, M. (2006) : *Agricultural Geography,* Pearson Education, Delhi.

Sharma T.C. (1972) : "Patterns of Crop Land use in Uttar Pradesh", *The Deccan Geographer,* Vol. 10, pp. 1-7.

Sharma T.C. Trends in Cropping Intensity in Karnataka : A Regional Profile, *Transactions Journal of the IIG,* Vol. 19, No. January-June, pp. 17-30.

Sharma T.C. (1999) : *Technological Change in Indian Agriculture,* Rawat Publications, Jaipur.

Sharma, V.K. "Water balance in the Haryana, Punjab Plain—A Climatological Approach", *National Geographical Journal of India,* 20 (1974), pp. 145-159.

Singh Amarjit and A.N. Sadhu (1986) : *Agricultural Problems in India,* Himalaya Publishing House, Bombay.

Singh G.B. (1979) : *Transformation of Agriculture,* Vishal Publications, Kurukshetra.

Singh Jasbir. "A new technique for measuring agricultural productivity in Haryana (India)", The *Geographer* 19 (1972) pp. 15-33.

Singh Jasbir (1974) : *An Agricultural Atlas of India,* Vishal Publication, Kurukshetra.

Singh Jasbir and S.S. Dhillon (1984) : *Agricultural Geography*, Tata McGraw Hill, New Delhi.

Singh, Surendra (1994) : *Agricultural Development in India : A Regional Analysis,* Kaushal Publications, Shillong.

Singh Surjit and Vidhya Sagar (2004) : *Agricultural Credit in India,* State of the Indian Farmer, Ministry of Agriculture, Academic Evaluation, New Delhi.

Singh V.R. "A method for analysing agricultural productivity", *Transactions* 1 (1979) pp. 39-46.

Subramaniam, A.R. (1991) : *Agroclimatic Studies in India* in Jaymala Diddee *(ed.), Emerging Trends in Indian Geography,* Rawat Publications, Jaipur.

Seymons, L. (1967) : *Agricultural Geography,* B. Bell, London.

Tata S.N. (ed) (1980) : *Handbook of Agriculture,* ICAR, New Delhi.

The Hindu (2012) : Survey of Indian Agriculture.

Thorner, D. (1956) : Rationale of Regional Variations of Agrarian Structure in India, *Indian Journal of Agricultural Economics,* pp. 46-52.

Weaver, J.C. "Crop Combination regions in the middle West" Geographical Review XLIV (1954), pp. 176 and 181.

◆◇◆

CHAPTER 21

Green Revolution

INTRODUCTION

Green Revolution owes its origin in the finding of new dwarf variety of wheat seed by **Dr. Norman Earnest Borlaug**. He was incharge of Wheat Development Programme in Mexico in the 1950s and was the genetic architect of the dwarf wheat. Earlier, he and Dr. Hassar had conducted, in the late 1940s, most relentless breeding programme choosing the best of wheat genes in the world. His efforts at breeding a suitable dwarf variety were crowned with success by 1951 in Mexico and that country became self-sufficient in food by 1956. Later on, the Japanese wheat variety NORIN-10 was crossed with the Mexican improved varieties and the first break through came in 1961 when the Mexican farmers could obtain yields as high as 7,000 kg per hectare which was about $2^1/_2$ times the previous varieties. Green Revolution in rice was triggered off by intense upsurge in rice research resulting from the establishment of International Rice Research Institute (IRRI) at Manila. Some work on rice had been done in Taiwan also.

Although seeds of the Green Revolution were sown in early 1950s in Mexico, the term Green Revolution was first used by the then Administrator of the U.S. AID, **William S. Gadd** on 8 March, 1968 in Washington D.C. when he addressed the Society for International Development on the subject *Green Revolution—Accomplishments and Apprehensions*.

In India, the seeds of Green Revolution were first field tested in the drought year of 1964-65. They were introduced to the Indian scientists by **Dr. Borlaug** in 1963. He had predicted in 1961 that India could double her wheat production in one decade. India received 100 kg seeds each of four dwarf and semi-dwarf varieties. These seeds were planted in different soils in Delhi, Ludhiana, Pusa and Kanpur. The yield was over 4,000 kg per hectare which was about four times the yield of local varieties. These varieties were released for general cultivation after experimentation, multiplication and demonstrations by Indian scientists in about 100 different farmers' fields. In 1966, about 16,000 tonnes of seeds were imported for cultivating about 4 lakh hectares of land. High Yielding Varieties Programme (HYVP) was introduced in the kharif season of 1966. The production of foodgrains in 1967-68 was 25 per cent higher than that of 1966-67. This increase was more than the increase recorded in the preceding 16 years of plan period. This unprecedented increase in production was nothing less than a revolution and it was termed as *Green Revolution*. In the words of **Dr. Hassar**, *The Green Revolution is the phrase generally used to describe* *the spectacular increase that took place during 1968 and is continuing in the production of foodgrains in India*. Unfortunately, Green Revolution left its impact only in Punjab, Haryana and Western U.P. in respect of wheat production and Andhra Pradesh and Tamil Nadu in respect of rice production. There seems to be no valid reason why other states cannot follow suit and get the benefit of Green Revolution. Sudhir Sen was quite justified when he said, *The Green Revolution is not a misnomer nor is it a fancy phrase; it is already much a reality. It has not only solved the food problem of India and other developing countries but it has brought the solid assurance that the problem can be solved. It has given them a breathing space in a period of spiralling population, to come to grips with the problem and set their economic house in order*.

INTENSIVE AGRICULTURE DEVELOPMENT PROGRAMME (IADP)

Popularly known as Package Programme, the IADP was taken up in 1961 and it paved the way to an *Adoption Breakthrough* for *Green Revolution* in India. It aimed at diffusing technical know-how, credit and agricultural technology to step up agricultural production in selected districts so that it may have demonstration effect. In the words of **R.N. Chopra**, "It was a historic decision which yielded solid results. The relevance of IADP experiment (1961-64) arises from the fact that it provided a crucial and much needed 'adoption breakthrough'—adoption of modern techniques of agricultural growth which, when coupled with the epoch-making '*biological breakthrough*', gave birth to the Green Revolution in India". The pilot project was launched first in 7 districts of 7 different states. Out of these, four were rice producing, two wheat producing and one millets producing. The districts were Shahabad in Bihar, Thanjavur in Tamil Nadu, Aligarh in U.P., West Godavari in Andhra Pradesh, Ludhiana in Punjab, Raipur in Chhattisgarh and Pali in Rajasthan. Later 8 more districts were selected for this programme. The programme was a grand success and it created new dynamism in the stagnant countryside. It was extended to a total of 114 districts in 1964-65. According to **C. Subramaniam**, "It is one of the miracle stories of modern development that the allegedly backward tradition-bound Indian farmer has been so responsive to new technology. This has been in a large measure due to the poineering efforts of the IADP through which the package approach to agriculture was introduced." It is worth mentioning that 64 per cent of the growth in the output of foodgrains in the IADP districts was due to capital and knowledge (HYV seeds, fertilizers and pesticides, etc.) — *viz*., factors other than land and labour inputs.

COMPONENTS OF THE GREEN REVOLUTION

Following are the 12 components of the Green Revolution :

1. High Yielding Varieties (HYV) of seeds.
2. Irrigation (*a*) surface and (*b*) ground.
3. Use of fertilizers (chemical).
4. Use of insecticides and pesticides.
5. Command Area Development (CAD).
6. Consolidation of holdings.
7. Land reforms.
8. Supply of agricultural credit.
9. Rural electrification.
10. Rural roads and marketing.
11. Farm mechanisation.
12. Agricultural universities.

Some of the above mentioned components have already been discussed in the earlier chapters and will need little description, while some other components will be discussed in detail here.

It must be noted that majority of the components do not act in isolation, rather they are closely inter-related and heavily dependent upon one another. For example, HYV seeds are highly responsive to use of fertilizers and are equally vulnerable to pest attacks and growth of useless weeds. Their full potential cannot be developed without the requisite supply of water. The shorter maturing period enables the farmers to obtain more than one crop in a year from the same piece of land. This would require hastening of the harvesting operations so that land is quickly prepared for the second crop. This will require mechanisation of farming. In order to make optimum use of the farm inputs, the farmer must know the *why, where, what, when and how much* of each for which there is ample scope.

1. High Yielding Varieties (HYV) of seeds. According to **R.N. Chopra**, "The high yielding variety seeds are major input of agricultural production under the Green Revolution technology. Their main characteristic is increased responsiveness to chemical fertilizers, their period of maturing is short, it helps double cropping; their short stems can easily carry fertilizer load, resist wind damage, their large leaf surface helps the process of photosynthesis." According to **Sunil Kumar Munsi**, "The HYV seeds were perhaps the single most important input in the Indian Green Revolution. All other inputs were linked with HYV." **M.S. Swaminathan** has remarked that apart from erasing the 'begging bowl' image of our country, the most important gain has been the saving of forests and land, thanks to the productivity improvement associated with high yielding varieties. The development of HYV seeds of wheat in 1960s and those of rice in 1969-70 laid the foundation for Green Revolution in India. **Bandhu Das Sen** has rightly remarked that they play the role of modernisers of agriculture like engines of change, capable of transforming a traditional farmer into a commercial producer. They act as part of steam engine (for industrial revolution) to ignite an agrarian revolution in poor countries. Thus the HYV programme brought about a major change—a transformation affecting almost every aspect of Indian agriculture. In words of **Dantwala**, "widespread adoption of HYVs has helped to step up cereal production, stimulated investment and substantially increased the use of modern inputs." **The Pearson Commission Report** hailed it as *one of the authentic marvels of our time*. Its most important effect was to be seen in the attainment of self-sufficiency in cereals, which enabled us to have a break from the *ship-to-mouth* situation and move forward ahead of population.

Details of seed production, distribution and use have been given in Chapter 20 and need not be repeated here.

2. Irrigation. Irrigation is the second most important component of Green Revolution technology after HYV seeds. Assured and regular supply of sufficient water to crops not only adds to production, it also assures stability in production. Indian rainfall being unreliable, irregular and seasonal, there is urgent need to expand irrigation potential to meet the requirements of the Green Revolution strategy. Irrigation is a pre-condition for successful introduction of HYV seeds even in areas of heavy rainfall. The success in use of HYV seeds lies in availability of water at the right time and in the right quantity for which **B.B. Vohra** had laid more emphasis on ground water rather than on surface water. The ground water gives the advantage of *push-button* irrigation, made possible by a pumpset or a tube well and is completely under farmer's own control. Appreciating the role of ground water in the success of Green Revolution, **Vohra** has preferred to call it the *Ground Water Revolution*. However, there is serious threat of depletion of ground water due to over-exploitation when the rate of drawl of ground water is higher than the rate at which it is replenished. In many districts of Haryana and Punjab, the ground water exploitation is very high due to tubewell irrigation.

The available potential, the growth and utilization of irrigation in India has been discussed at length in earlier chapters and need not be repeated here.

3. Use of Fertilizers (Chemical). The use of chemical fertilizers has been the third most important input of Green Revolution after HYV seeds and irrigation, *rather the three are tied together*. In fact use of HYV seeds needs heavy dose of irrigation and fertilizers to give high yields. Since the entire culturable land has already been brought under plough and there is practically no scope for bringing any new areas under cultivation, further increase in foodgrains production can be achieved only by multiple-cropping which heavily leans on the *trio* of the basic inputs, *viz.* HYV seeds, irrigation and chemical fertilizers.

To encourage balanced fertilizer use and make fertilizers available to farmers at affordable prices, the Central Government determines and notifies the selling price of urea as well as decontrolled P&K fertilizers. The current selling prices of urea and P&K fertilizers are less than the cost of production, the difference between the selling price and the cost of production as assessed by the government is borne as subsidy.

Although the use of fertilizers has considerably increased over the years, this increase is more prominent in areas where Green Revolution has shown its impact. In 1970, southern India was leading in consumption of fertilizers, but later on northern

India, particularly, Punjab, Haryana and Uttar Pradesh, became the main consumers.

In spite of the fact that India is the third largest consumer of chemical fertilizers in the world, after the China and U.S.A., per hectare consumption still remains low compared to the world averages. This means that there is still large scope for using chemical fertilizers, increasing the yields and converting the dreams of Green Revolution into reality.

Details of production, distribution, import and utilization have been given in Chapter 20.

4. Use of Insecticides and Pesticides. Though intensive use of irrigation and fertilizers under the Green Revolution technology has increased the farm production, it has also given birth to the problem of pests, insects, weeds, rodents, etc. The monoculture promoted by the Green Revolution technology is more vulnerable to the insects and pests. These pests, weeds and diseases are to be checked by proper doses of insecticides, pesticides and weedicides. Pest surveillance should be an integral part of crop production. The first disease surveillance in this country related to wheat diseases in 1966-67 followed by adhoc rice survey and surveillance of pests and diseases in 1970, 71 and 72. According to annual report of Ministry of Agriculture (1983-84), over 30 million hectares of cropped area in the country is affected by various pests and diseases, taking an annual toll of 5 to 25 per cent of the agricultural production. There has been a tremendous increase in the use of different types of biocides in the areas under plant protection. The regional distribution makes it clear that areas with Green Revolution technology are the main consumers of pesticides. For example, Punjab, Haryana, Andhra Pradesh and Tamil Nadu consumed over 60 per cent of the country's pesticides in 2018-19.

5. Command Area Development (CAD). Command Area Development Programme is a centrally sponsored scheme which was launched in January 1975. Its aim was to bridge the gap between potential created and utilized in selected major/medium irrigation projects of the country for optimising agricultural production from the irrigated land. The programme covers the following components :

(*i*) *On-farm development (OFD)* works which include soil surveys, land shaping, construction of field channels, field drains, farm roads, re-alignment of field boundaries (where possible consolidation of holdings should also be combined), introduction of *warabandi* to ensure equitable and assured supply of water to each and every farm holding, supply of all inputs and services including credit and strengthening of extension services.

(*ii*) Selection and introduction of suitable cropping pattern.

(*iii*) Development of groundwater to supplement surface water.

(*iv*) Development and maintenance of the main and intermediate drainage system.

(*v*) Modernisation, maintenance and efficient operation of the irrigation system upon the outlet of one cusec capacity.

Initially, 60 major and medium irrigation projects were taken up under CAD programme, covering a Culturable Command Area (CCA) of about 15 million hectares. After inclusion of new projects, deletion of completed projects and clubbing of some projects, there are now 149 projects under implementation. The programme was restructured and renamed as Command Area Development and Water Management (CADWM) programme with effect from 1st April, 2004. The scheme had been implemented as State-Sector Scheme during XI Five Year Plan (2008-09 to 2011-12).

6. Consolidation of Holdings. Small and fragmented land holdings have been one of the main obstacles in the progress of agriculture in India. Consolidation of holdings has been introduced to solve this problem. A detailed discussion on the reducing size of holdings, their fragmentation and consolidation thereof, has been given in Chapter 20.

7. Land Reforms. Immediately after Independence, it was felt that land reforms must be brought in to improve the agricultural situation in the country. Absentee landlordism, tenancy-at-will and share cropping could not help in inculcating interest among the farmers to make investments in farm inputs and adopt new farm technology. In 1947 half of India was under *Zamindari System* in which 80 per cent of the land was in the hands of the absentee landlords. The *Zamindar* used to exploit the farmers who used to till the land. Soon after Independence, the slogan of *land to the tiller* was raised and steps were taken for the abolition of the Zamindari. Consequently, tenants became owners of land. They started taking interest and pains to increase the farm production. *Raitwari* system prevailed in Madras, Bombay, Assam and Punjab. Under this system the peasant was the owner of land and paid rent directly to the Government. The rent was usually half of the net produce. A fixed amount of rent was to be paid irrespective of the condition of the crops. In the event of crop failure the peasant was obliged to pay rent by incurring debt against mortgage. Ultimately the land passed into the hands of the money lender who had no real interest in cultivation. *Mahalwari* was another system in which a chosen peasant (*Lambardar*) was responsible for depositing the rent varying from 40 to 70 per cent of the produce. These systems were to be abolished in the interest of better agricultural performance. Another measure taken by the government was the enforcement of land ceiling act. Under this act a farmer cannot own more land than the ceiling limits. This resulted in the re-distribution of surplus land which proved beneficial to lakhs of landless farmers. After obtaining the ownership rights, farmers worked whole-heartedly on their farms and this led to a tremendous increase in agricultural production. **Dr. Randhawa** has beautifully summed up the benefits of land reforms. According to him, "A stable and restructured rural base with an equitable tenurial system paved the way to Green Revolution and can be accredited with its blooming to the present stage."

8. Supply of Agricultural Credit. In the words of **R.N. Chopra** "Credit is the most crucial input in all agricultural developmental programmes. The other inputs *viz.*, technology, HYV seeds, fertilizers, pesticides, irrigation water and machinery—all depend on the availability of credit". A large percentage of Indian farming community consists of small and marginal farmers who do not have their own resources to invest in agriculture. They depend upon agricultural credit to carry on most of their agricultural operations. Earlier they used to get loan from the money-lender who used to charge very high rate of interest. Now Cooperatives, Commercial Banks and Regional Rural Banks extend loans to farmers on easy terms. The role played by different agencies in extending loan to farmers has been discussed in Chapter 20.

9. Rural Electrification. Rural electrification is one of the essential inputs in modern agricultural system. Electricity makes a significant contribution to development of agriculture. It is a cheap source of energy which can be used for lifting water by tubewells/pumpsets, processing and preserving agricultural produce, sprinkler irrigation and so many other farm operations. The development of ground water, so vital for Green Revolution, requires uninterruped supply of elctricity at cheaper rates.

Realising the importance of electricity for the proper growth and development of agriculture, a massive programme of rural electrification was taken up immediately after Independence. At the time of Independence only 1,300 villages had been electrified and only 6,400 energised pump sets were working in the entire country. Currently almost all the villages are electrified.

Haryana was the first state to electrify all its 6,759 villages in 1970. On 29th April, 2018, the Central government claimed that all the 5,97,464 inhabited villages in the country have been electrified Manipur's Leisang was the last village to be electrified. But it still leaves about 17.4% rural households without electricity because a village is considered electrified if 10% of its households and public places such as schools, panchayat office and health centre have access to electricity.

10. Rural Roads and Marketing. Rural roads and marketing have already been discussed in Chapter 20. They constitute an important segment of infrastructure to improve agricultural productivity under the Green Revolution programme. Rural roads are very essential for connecting the villages to the neighbouring markets and villages. Unfortunately, there is still a big gap between the requirement and availability of village roads. Road network upto town level is fairly satisfactory. The weakest point is that of rural roads.

Marketing is essential for progressive agriculture. Regulated markets enable the farmer to sell his agricultural produce and to purchase farm implements and tools, fertilizers, pesticides and other agricultural inputs as well as goods of every day use. The farmer can go to the market with his produce, sell it and on his return journey he can bring the goods required for agriculture or in every day life. In this way the farmer can save on his return transport and avoid unnecessary waste of time, energy and money. Ideally speaking the market place should be within a distance of 5 km from the village. In the words of **Leo E. Holman**, "Marketing is the part and parcel of a modern productive process, the part at the end that gives point and purpose to all that had gone before. Benefits from mechanisation can be minimised if corresponding improvements are not made in the marketing system."

11. Farm Mechanisation. Much success of the Green Revolution depends upon farm mechanisation. Mechanisation saves a lot of human labour and quickens the farm operations, thereby adding to the farm efficiency and productivity. For a detailed account of farm mechanisation see Chapter 20.

12. Agricultural Universities. Agricultural universities and other agricultural institutes are primarily engaged in agricultural research and passing on the research findings to the farmers. A good deal of research and extension work done by these universities has paid rich dividends in the agricultural field. Success of Green Revolution largely depends upon the work done by these universities. Punjab, Haryana and Uttar Pradesh, are the best examples of such a progress.

IMPACT OF GREEN REVOLUTION

Like other developing countries, Green Revolution has influenced the economy and way of life in India to a great extent as is evident from the following points :

1. Increase in Agricultural Production. The introduction of Green Revolution in 1967-68 has resulted in phenomenal increase in the production of agricultural crops especially in foodgrains. From 1967 onwards, the *Green Revolution* aimed at bringing about a *Grain Revolution*. Among the food grains too, it is the wheat crop which drew maximum benefit from Green Revolution. The production of wheat increased from 23.83 million tonnes in 1970-71 to 107.6 million tonnes in 2019-20 thereby recording about five fold increase. while the overall increase in the production of foodgrains was about 2.5 times in the corresponding period. On account of this reason, it is said that the Green Revolution in India is largely the *Wheat Revolution*. The area, production and yield of foodgrains were 12.57 million hectares, 10.4 million tonnes and 827 kg/hectare in 1965-66. The corresponding figures rose to 31.6 million hectares, 109.5 million tonnes and 3464 kg/hectare in 2020-21.

2. Diffusion of Rice and Wheat Cultivation to non-traditional areas. Since the success of the Green Revolution depends on the basic inputs like better seeds, fertilizers and irrigation, it has led to diffusion of crops, particularly two major food-crops *viz.* rice and wheat, to the areas hitherto unknown for their cultivation. For instance, West Bengal and Bihar had been traditional producers of rice and had the distinction of being called as **'rice heartland'**. But the virtues of the Green Revolution have helped in spreading rice cultivation to semi-arid areas of Punjab, Haryana and western part of Uttar Pradesh, thereby changing the cropping pattern of these areas. This feat has been achieved primarily due to increase in irrigation facilities supplemented by availability of better seeds and fertilizers. However, contribution of enterprising farmers cannot be under-estimated. Besides, rice cultivation has become popular in vast areas in South India which were producing very little or no rice before 1965-66 (**Fig. 21.1**). The area, production and yield of rice were 35.47 million hectares, 30.59 million tonnes and 862 kg/ha respectively in 1965-66 and the corresponding figures increases to 43.8 million hectares, 116.4 million tonnes and 2,659 kg/ha in 2018-19.

Similarly wheat cultivation has spread to vast areas in eastern Uttar Pradesh, Madhya Pradesh, Rajasthan, and some parts of Maharashtra, Gujarat and West Bengal. Although diffusion of wheat cultivation started from its heartland right with the beginning of the Green Revolution in the mid-1960s, yet the major part of diffusion in Uttar Pradesh, Madhya Pradesh, Rajasthan, Gujarat, Bihar and West Bengal was observed in 1970s (**Fig. 21.2**).

The area, production and yield of wheat were 12.57 million hectares, 10.40 million tonnes and 827 kg/ha respectively in 1965-66 which rose to 31.5 million hectares, 107.6 million tonnes and 3,421 kg/ha in 2019-20 respectively. This is a great

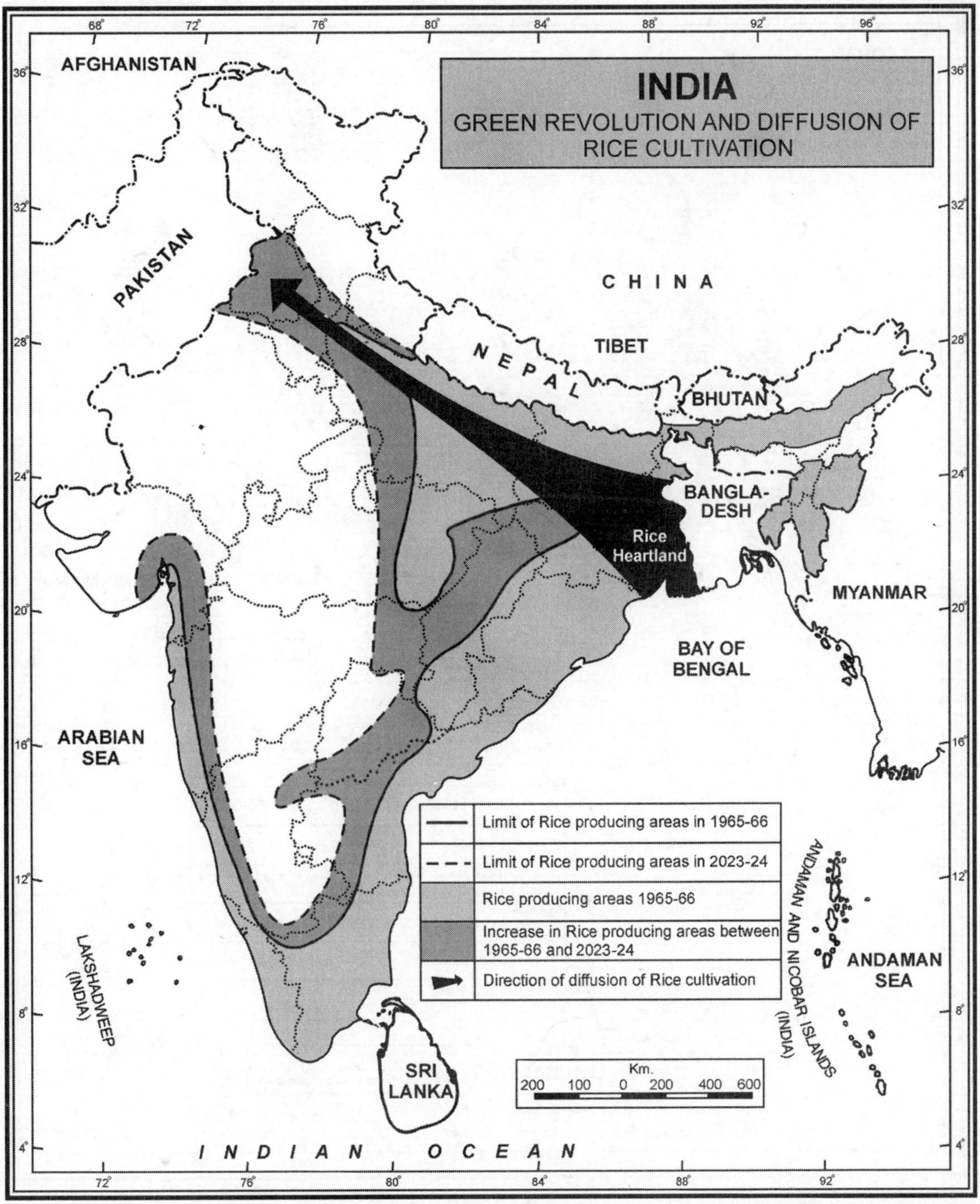

FIG. 21.1. India : Green Revolution and diffusion of rice cultivation.

achievement indeed. However, much more has to be done to meet our growing demand for foodgrains.

3. Prosperity of Farmers. With the increase in farm production the earnings of the farmers also increased and they became prosperous. This has, especially, been the case with big farmers having more than 10 hectares of land.

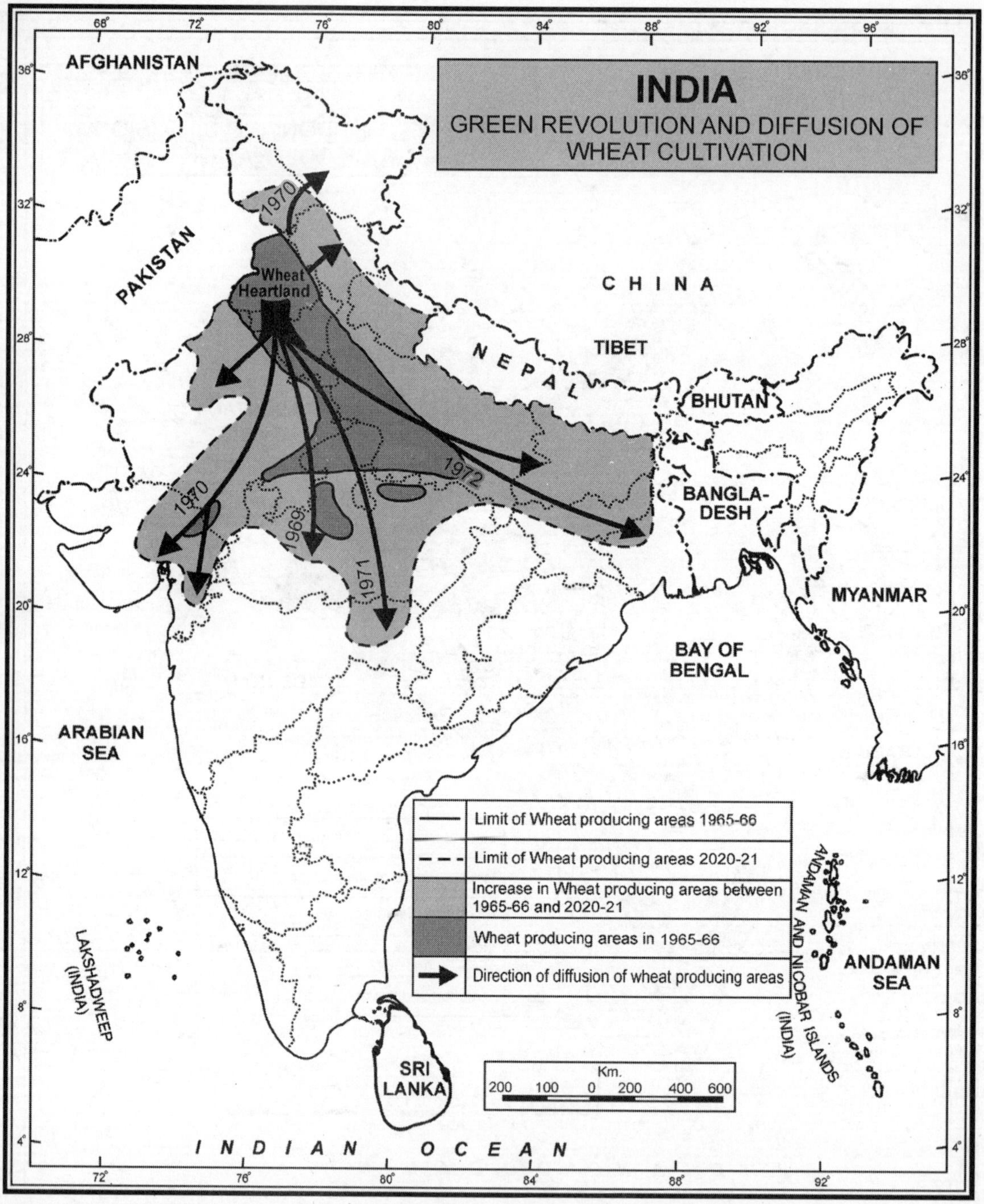

FIG. 21.2. India : Green Revolution and diffusion of wheat cultivation

4. Reduction in import of foodgrains. The main benefit of Green Revolution was the increase in the production of foodgrains, as a result of which there was a drastic reduction in their imports. We are now self-sufficient in foodgrains and have sufficient stock in the central pool. Sometimes we are in a position to export foodgrains also. The per capita net availability of foodgrains has also increased from 408.1 grams per day in early 1966 to the level of 494.1 grams in 2018, this inspite of the rapid increase in population. In the words of **Dantwala**, *Green Revolution has given a breathing time. As a result, there will be relief from anxiety of food shortage and the planners will concentrate more on Indian planning*.

5. Capitalistic Farming. Big farmers having more than 10 hectares of land have tended to get the maximum benefit from Green Revolution technology by investing large amounts of money in various inputs like HYV seeds, fertilizers, irrigation machines, etc. This has encouraged capitalistic farming.

6. Ploughing back of profit. The introduction of Green Revolution helped the farmers in raising their level of income. Wiser farmers ploughed back their surplus income for improving agricultural productivity. This led to further improvement in agriculture. According to a study conducted by Punjab Agriculture University, Ludhiana farmers plough back about 55 per cent of their income for agricultural progress.

7. Industrial Growth. Green Revolution brought about large scale farm mechanisation which created demand for different types of machines like tractors, harvestors, threshers, combines, diesel engines, electric motors, pumping sets, etc. Besides, demand for chemical fertilizers, pesticides, insecticides, weedicides, etc. also increased considerably. Consequently, industries producing these items progressed by leaps and bounds. Moreover, several agricultural products are used as raw materials in various industries. These industries are known as *agro based industries*. Textile, sugar, vanaspati, etc. are some outstanding examples of agro based industries.

8. Rural Employment. While on one hand, large scale unemployment was feared due to mechanization of farming with the introduction of Green Revolution technology in India, there was an appreciable increase in the demand for labour force due to multiple cropping and use of fertilizers. According to **Gobind Thukral**, "Green Revolution has generated lakhs of new jobs in Punjab. Almost 15 lakh poor people from the impoverished regions of Bihar, eastern Uttar Pradesh and Odisha work here. They not only earn their bread and butter, but take back home new ideas and technology". As per findings of **Bhalla** and **Chadha** in respect of Punjab, "The drive towards mechanization was caused mainly by the scarcity of labour and relatively high wage rates especially during peak agricultural operations." During the early years of the Green Revolution, a large number of farm labourers had migrated from Bihar and eastern Uttar Pradesh to Punjab where they found better opportunities of earning a livelihood. However, reverse trend has been observed during 2010s as a large number of labourers from these areas are going back to their home state because they are now finding employment opportunities there caused by overall development including impact of Green Revolution. Besides a large number of labourers rushed back to their homes in the wake of Covid-19 and never returned back to work in Punjab.

9. Change in the Attitude of Farmers. The Indian farmer had remained illiterate, backward and traditional and had been using conventional methods of cultivation since the earliest times. But Green Revolution has brought about a basic change in his attitude towards farming. The way he has readily adopted the Green Revolution technology has exploded the myth that the Indian farmer is basically tradition bound and does not use new methods and techniques. **Wolf Ladejinsky** has rightly concluded that, "Where the ingredients for new technology are available, no farmer denies their effectiveness. The desire for better farming methods and a better standard of living is growing not only among the relatively small number of affluent farmers using the new technology, but also among countless farmers still from outside looking in."

DEMERITS OR PROBLEMS OF GREEN REVOLUTION

Green Revolution is a unique event in the agricultural history of Independent India. This has saved us from the disasters of hunger and starvation and made our peasants more confident than ever before. But it has its own inherent deficiency segments. Ever since its inception, the income gap between large, marginal and small farmers has increased, gap between irrigated and rainfed areas has widened and some crops have benefitted more than the others, sometimes even at the cost of other crops. *It is neither product-neutral nor region-neutral and leaves uneven effects of growth on products, regions and classes of people*. This has given birth to a plethora of socio-economic problems. According to **Radha Krishna Rao**, "The spiraling prices of fertilizers, the tendency to use them frequently and the stagnant wheat and rice yields in Punjab and Haryana have combined to confirm that Green Revolution has reached ripened old age". The fatigue of the Green Revolution is already visible. Still the main lacuna in the Green Revolution is that up till now it is an unfinished task. Some of the demerits or problems of Green Revolution are briefly discussed as under :

1. Inter-Crop Imbalances. The effect of Green Revolution is primarily felt on foodgrains. Although all foodgrains including wheat, rice, jowar, bajra and maize have gained from the Green Revolution, it is wheat which has benefited the most. It has wrested areas from coarse cereals, pulses and oilseeds. The HYV seeds in latter crops have either not been developed so far at all, or they are not good enough for farmers to risk their adoption. Consequently, their cultivation is fast becoming uneconomic and they are often given up in favour of wheat or even rice. The result is that an excess of production in two main foodgrains (wheat and rice) and shortages in most others today prevail side by side. Major commercial crops like cotton, jute, tea and sugarcane have also remained almost untouched by the Green Revolution. This is not good for a balanced growth of Indian agriculture. Central Government has taken some steps to remove these imbalances.

2. Regional Disparities. Green Revolution technology has given birth to growing disparities in economic development at inter and intra regional levels. It has so far benefited only 40 per cent of the total cropped area and 60 per cent is still untouched by it. The most benefited areas are Punjab, Haryana and western Uttar Pradesh in the north and Andhra Pradesh and Tamil Nadu in the south. The major benefit in these two regions has been with respect to increase in wheat and rice cultivation respectively (**Fig. 21.3**). It has hardly touched the Eastern region, including Assam, Bihar, West Bengal and Odisha and arid and semi-arid areas of Western and Southern India. In short, it failed to take care of areas like rainfed, hilly, coastal, dry and arid zones of the country which could be developed for production of exportable items like fruits, honey, mushroom, milk, meat, etc. In short, Green Revolution affected only those areas which were already better placed from agricultural point of view. Thus the problem of regional disparities has further aggravated as a result of Green Revolution. The ratio between the lowest and highest yield-rates among the states for the 1975-78 period amounted to 1 : 3.2 in paddy, 1 : 3.7 in wheat, 1 : 3.4 in cereals, 1 : 3.2 in pulses, 1 : 3.2 in food grains, 1 : 3.0 in oilseeds, 1 : 3.2 in sugarcane, 1 : 4.9 in cotton and 1 : 1.6 in jute. Study of some sample surveys recently conducted by the Indian Agricultural Statistics Research Institute (IASRI) revealed that the single most important factor is the '*input differential*' which alone can explain extreme yield variations even under similar physical and cultural conditions. According to a study by **Bhalla** and **Alagh**, 69 districts with a relatively high productivity levels account for 20 per cent of the cultivated area and 36 per cent of output, consume 44 per cent of fertilizers, employ 50 per cent of tractors and 45 per cent of irrigation pumps and have 38 per cent of India's gross irrigated area.

Regional disparities in crop yields can be reduced by evolving suitable disease resistant high-yield strains of paddy for most eastern parts and by developing irrigation facilities and a suitable dry farming technology for the arid and semi-arid western and southern regions.

3. Increase in Inter-Personal Inequalities. It has been observed that it is the big farmer having 10 hectares or more land, who is benefited the most from Green Revolution because he has the financial resources to purchase farm implements, better seeds, fertilizers and can arrange for regular supply of irrigation water to the crops. As against this, the small and marginal farmers do not have the financial resources to purchase these farm inputs and are deprived of the benefits of Green Revolution Technology. There were about 1,45,727 thousand holdings in India in 2015-16 out of which only 0.57 per cent exceeded 10 hectares in size. **Francine R. Rankel** has concluded from his study of Ludhiana (Punjab), West Godavari (Andhra Pradesh), Thanjavur (Tamil Nadu), Palghat (Kerala) and Bardhaman (West Bengal) that the greater beneficiaries are those farmers who own 10 to 12 hectares of land. Similar conclusion was drawn by **G.R. Saini** from his study of Ferozepur (Punjab) and Muzaffarnagar (U.P.). **G.S. Bhalla** and **G.K. Chadha** have found out that Green

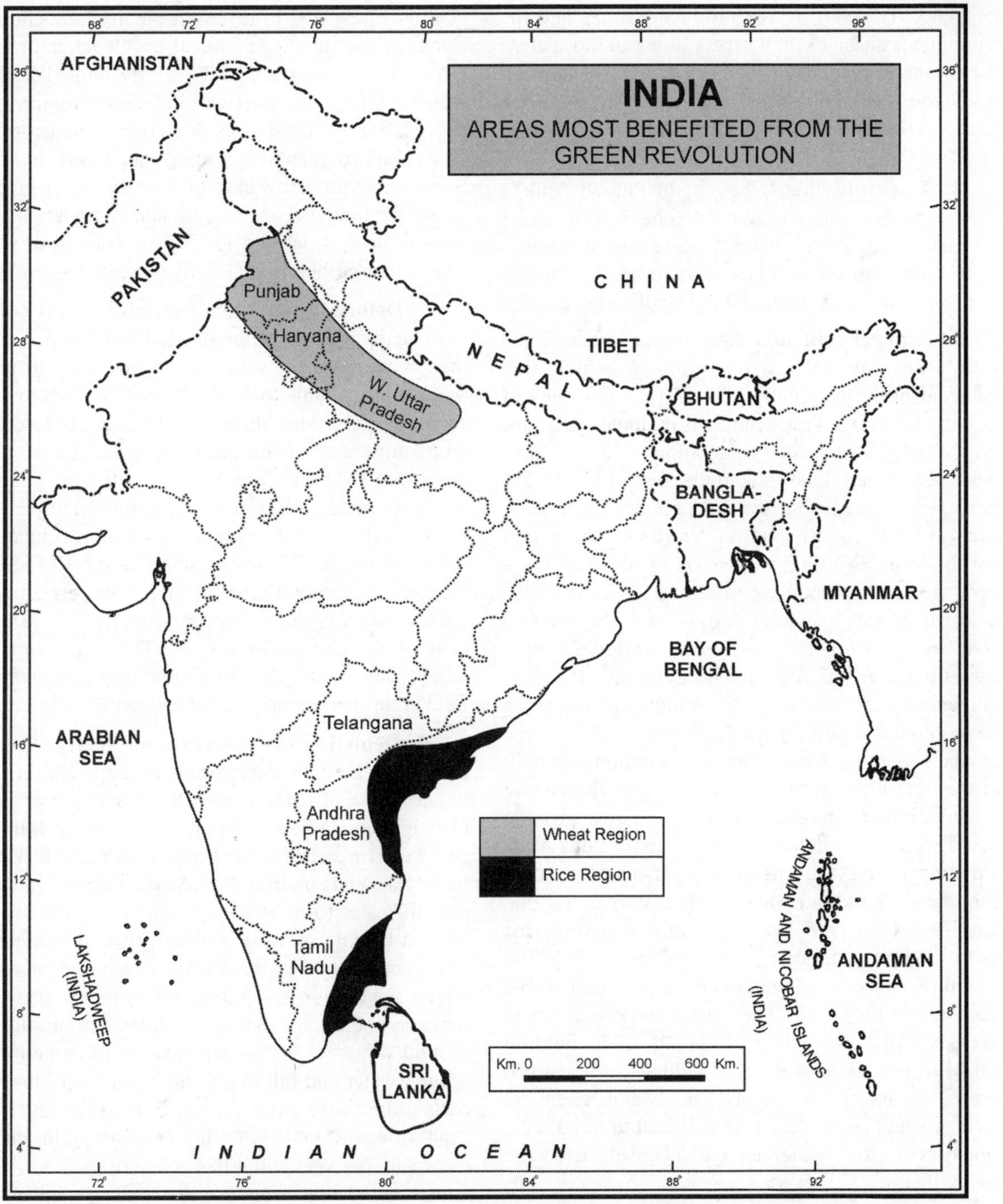

FIG. 21.3. Areas most benefitted from the Green Revolution

Revolution has benefited the farmers in general but one-third of them are small farmers with 2.5 acres (about one hectare) of land and are living below poverty line. Another 24.0 per cent of the farmers own 2.5 to 5.0 acres (1-2 hectares) of land and they are also living below poverty line. The land holdings are generally small in rice producing areas and the economic position of the farmers living in those areas is extremely miserable. In short, Green Revolution has *made the rich richer and rendered the poor poorer resulting in wide-spread social and economic tensions*.

4. Unemployment. Except in Punjab, and to some extent in Haryana, farm mechanization under Green Revolution has created widespread unemployment among agricultural labourers in the rural areas. The worst hit are the poor and the landless people.

5. Other Problems. Agriculture under Green Revolution has not grown at a rate which was expected in the beginning. The differential rates of growth of different crops and their regional variations have already been discussed. Some scholars have expressed serious doubts about the capability of HYV seeds itself. Analysing the role played by *miracle seeds* in the Green Revolution, **Vandana Shiva** says that the term HYV is a *misnomer*. In actuality, these seeds are highly responsive to certain key inputs such as fertilizer and irrigation and as such they should have been called highly *responsive* varieties. Shiva says that there is increasing evidence that the indigenous varieties could also be high yielding given the required doses of inputs. According to Shiva, "the inevitability of the Green Revolution option was built on neglecting the other avenues for increasing production that is more ecological such as improving mixed cropping systems, improving indigenous seeds and improving the efficiency of use of local resources." **Vandana Shiva** further comments that "having destroyed nature's mechanisms for controlling pests through the destruction of diversity, the 'miracle seeds' of the Green Revolution became mechanisms for breeding new pests and creating new diseases". In a case study of Punjab, **M.K. Sekhon** and **Manjeet Kaur** of P.A.U. Ludhiana have warned against the excessive use of groundwater, chemical fertilizers and pesticides. This will lead to large scale depletion of groundwater and will adversely affect the health of soil.

ECOLOGICAL IMPLICATIONS OF GREEN REVOLUTION

Apart from its socio-economic implications, Green Revolution has left a deep impact on various ecological aspects in different parts of the country. Some of the major ecological problems created by Green Revolution are soil salinity and alkalinity, waterlogging, desertification, soil erosion and degradation, deforestation, depletion of underground water, environmental pollution and health hazards. Some of the problems like soil salinity, water-logging desertification and soil erosion and degradation have already been described in Chapter 6 on soil, while some other problems are briefly discussed as under :

1. Deforestation. More cultivable land is needed to increase agricultural production and forests are the main victims as trees are felled recklessly to make more land available for agriculture. This leads to large scale deforestation which has its own ill-effects on environment and ecology. Forests have been destroyed over vast areas in the northern plain of India to make room for agriculture. Punjab and Haryana, the 'Granary of India' are almost completely devoid of forests. These two states have 5.19 and 3.53 per cent of the total area under forests respectively. Uttar Pradesh is slightly better with just over 7.02 per cent of its area under forests. This has happened largely due to our quest for a more land to grow more food grains particularly wheat and rice.

2. Depletion of Underground Water. Success of Green Revolution primarily depends on the intensive use of HYV seeds which require high input of irrigation. In semi-arid areas of Punjab, Haryana and western part of Uttar Pradesh, wheat cultivation largely depends on irrigation. Since canal irrigation is not sufficient to meet the growing demand, farmers depend heavily on tube well irrigation. Introduction of rice cultivation in these areas as a consequence of Green Revolution has put heavy strain on tube well irrigation. Farmers indulge in over exploitation of ground water resources which leads to depletion of ground water and fall in ground water-table. In some parts of north-west India (Punjab, Haryana and Uttar Pradesh) water table is falling at an alarming rate of one meter per year and it has gone critically low. It is apprehended that ground-water will dry up in due course of time if the present trend is not checked immediately.

3. Environmental Pollution. Success of Green Revolution is primarily based on use of HYV seeds which perform well only if other inputs like chemical fertilizers (NPK), insecticides and pesticides are also used. Fertilizers help in getting higher yields per hectare while insecticides and pesticides are used to save crops from insects, pests and diseases. Care has to be taken to use fertilizers, insecticides and pesticides upto a permitted limit. Indiscriminate use of these chemicals leads to environmental pollution. When insecticides are sprayed on the crops, some part of it enters the surrounding air and pollutes it. Some of the chemicals enter the soil and destroy the micro-organisms. These micro-organisms are vital to maintain the fertility of the soil. Some of the chemicals enter water, are dissolved in it, and pollute both surface as well as ground water. Thus, it can be easily said that whereas chemical fertilizers, insecticides and pesticides have saved us from hunger and starvation, they have given birth to serious problems of environmental pollution.

4. Noise Pollution. Farm mechanisation is one of the important inputs for success of Green Revolution. Increasing use of tractors, harvesters, threshers, crushers, etc. makes lots of noise and disturb rural peace and tranquillity.

5. Health Hazards. Indiscriminate use of chemical fertilizers, insecticides and pesticides has resulted in a large number of health hazards. These chemicals pollute air, water and land and cause many ailments. Some of the poisonous chemicals enter our food chain through fruits and vegetables and result in ill-health. According to Indian Council of Medical Research, traces of lead, zinc and copper are found in milk and vegetables due to use of these chemicals on the crops. In areas of high irrigation intensity by canals and tube wells, in Punjab, Haryana, Rajasthan and western Uttar Pradesh, there are vast tracts of waterlogged areas which provide ideal conditions for mosquito breeding. These mosquitoes cause a large number of ailments including malaria, dengue and chikungunya.

Suggestions for Stability of Green Revolution

Following suggestions are worth noting for stability of Green Revolution :

1. Larger Area. Till now the impact of Green Revolution is seen mainly in Punjab, Haryana and Western part of Uttar Pradesh in the north and Andhra Pradesh and Tamil Nadu in the south. The rest of the country is deprived of the benefits of Green Revolution. This is not a satisfactory condition. More areas should be brought under Green Revolution to draw the maximum benefit. There is urgent need to increase agricultural production in eastern and southern parts of the country.

2. More Crops. So far Green Revolution has mainly been 'Grain Revolution' and amongst grains too, it is the wheat crop which has drawn maximum benefit. This is the reason that sometimes Green Revolution is also called 'Wheat Revolution'. This is a lopsided development and the progress should not be allowed to confine only to wheat culture. Other crop like rice and other food crops as well as some commercial crops should also be covered by Green Revolution. There is urgent need to increase the production of rice, pulses, cotton, tea, sugarcane, jute, etc.

3. Irrigation. Much success of Green Revolution depends on the availability of irrigation. HYV seeds need heavy doze of irrigation. In this context, micro irrigation projects like wells, tube wells, tanks, etc. are very important. More than half of India's irrigation is provided by micro irrigation projects.

4. Green Revolution for Small Farmers. About 85% of India's farmers are small or marginal farmers and they are least benefited by Green Revolution. On the contrary, farm mechanisation has rendered many agricultural labourers jobless. Following suggestions are made to improve the lot of small farmers.

(*i*) Land reforms should be implemented immediately and surplus land should be distributed among small farmers.

(*ii*) Credit at low rate of interest should be provided to the small farmers so that they can purchase HYV seeds, fertilisers and farm implements.

(*iii*) Large farm machines should be provide on rent.

(*iv*) Cooperative farming should be encourage to the benefit of small landholders.

5. Multicrop System. Multicrop system should be encouraged to draw maximum benefit from inputs. This will increase the farm production and reduce unemployment among farmers.

6. Integrated Farm Policy. Integrated farm policy is a must for success of Green Revolution. Under this policy, farmers are provided with HYV seeds, fertilisers, insecticides, pesticides, farm machinery, etc. at reasonable rates. This is quite beneficial to small farmers.

7. Modification of New Agricultural Strategy. Two modifications are necessary in the New Agricultural Policy—(*i*) Fertilisers in sufficient quantities should be supplied to areas of deficient fertilisers so that farm production may be improved. (*ii*) For production of fertilisers, coal should be used in place of oil. This is because India has large deposits of coal whereas oil has to imported.

8. Institutional Reforms. Institutional reforms play a vital role for the success of Green Revolution. Land records need to be updated. Surplus land resulting from ceiling of land holdings should be distributed to the small farmers. Interests of the cultivators should be protected. Further, division and disintegration of land holdings should be checked and consolidation of holdings should be completed. These steps will benefit the poor farmers.

SECOND GREEN REVOLUTION

As mentioned earlier, the fatigue of the First Green Revolution has become conspicuous and the production as well yields of almost all the crops has reached the plateau stage. Declining productivity of a number of crops due to unsustainable agricultural practices over the years and a galloping rate of population growth have once again put a severe strain on the food supply situation in the country. The food safety net for our rapidly growing population requires enhanced production and productivity in the form of a Second Green Revolution. In fact, a Second Green Revolution has become imperative keeping in view the tardy growth of food grains in proportion to the population growth. Further, special attention is required for achieving higher production and productivity levels in pulses, oilseeds, fruits and vegetables which remained untouched in the First Green Revolution but are essential for our nutritional security. In this regard, achieving high production of poultry, meat and fisheries is also essential. The First Green Revolution also ignored certain areas like rainfed, hilly, coastal and arid zones which could be developed for producing fruits, milk and meat. Thus, a number of shortcomings have been noticed in the First Green Revolution and it is right time to strive for the Second Green Revolution. Even Norman Barlaug, the chief architect of the First Green Revolution noted upon receiving the 1970 Nobel Peace Prize, that the Green Revolution represented only a **"temporary success."**

The Second Green Revolution differs from the First Green Revolution because in the First Green Revolution the main emphasis was on increasing the production of food grains, often without much caring about environment and ecology. The Second Green Revolution on the other hand, refers to practising sustained agriculture. In order for development to be sustainable it must meet the needs of the present without compromising the ability of future generations to meet their own needs." (Brundtland Commission on Environment and Development, 1987). Thus sustainable agriculture involves protecting natural resources from becoming increasingly degraded and polluted, and using production technologies that conserve and enhance the natural resource base of crops, forests, inland and marine fisheries.

As of now, the outlook for a Second Green Revolution seems uncertain because most of the increase in food supplies has to come from currently cultivated land as all the land fit for cultivation has already been brought under plough using the current level of technology. Therefore raising the level of productivity will require new technologies and better farming practices. Besides green technologies will have to be specially focussed on dry land agriculture and to benefit small and marginal farmers. Improving soil health by taking care of physical, chemical and biological characteristics of soil is equally important. Also of vital concern are water harvesting, water conservation and sustainable and equitable use of water. Besides, there is need to pay more attention to issues such as access to affordable credit and to life and crop insurances reform. Equally important are development and dissemination of appropriate technologies and improved opportunities, infrastructure and regulations for marketing of agricultural products. Thus India is at a juncture where further reforms are urgently required to achieve greater efficiency and productivity in agriculture for sustaining growth.

Strategies for Second Green Revolution

Following strategies have been suggested for success of Second Green Revolution :

(*i*) Micro-irrigation System. All ground water and surface schemes having Culturable Command Area (CCA) upto 2,000 hectares individually are known as minor irrigation schemes. These include dug-wells, private shallow tube-wells, deep public tube-wells, boring and deepening of dug-wells and small surface water development works through diversion schemes, storage schemes and lift irrigation. The government has accepted the recommendations of the task force on micro-irrigation and has planned to make heavy investment in this field.

Micro-irrigation technology enables optimal synergies of three components of Green Revolution—improved seeds, water and fertilizers. It helps in direct and concentrated application of water to root zones of crops, through specially designed emitters. Some countries have already started using this technology for optimum use of water resources thereby improve their productivity. In India, the task force on micro-irrigation has recommended an increase of area under micro-irrigation from the current 1.3 million hectares to 69 million hectares.

(*ii*) Organic Farming. Ill-effects of progress based on synthetic agrochemicals under Green Revolution have been documented several times by a large number of competent authorities. Their adverse effects on environment and public health are almost universally known now. The cost involved in minimising these effects is supposed to be enormous. Increasing consciousness about conservation of environment as well as health hazards caused by agrochemicals has brought major shift in consumer preference towards food quality, particularly in higher strata of society. Consumers are increasingly looking forward to safe and hazard free organic food. Globally, demand for organic food is increasing at the rate of 20-25 per cent per annum. Worldwide, over 130 countries produce certified organic products in commercial quantities. It is high time that India discards synthetic agrochemicals and opt for organic farming.

(*iii*) Precision Farming. As mentioned earlier, agriculture is the backbone of Indian economy and provides food to teeming millions. But the million dollar question is, how long our agriculture will be able to feed the fast growing population at the current level of technology. Agricultural technology available to us at the time of Independence could not have met our food requirements without First Green Revolution. Similarly, the present technology will not be able to produce sufficient food and other agricultural products in the coming future. Thus we will have to develop new technologies which can revolutionize our agricultural productivity and meet the growing demand for agricultural products.

The term "Precision Farming" and "Precision Agriculture" is capturing the imagination of the people concerned with the production of food, feed and fiber. It promises the scenario of increasing productivity, decreasing production costs and minimizing the adverse effects of farming on environment.

(*iv*) Green Agriculture. It is a system of cultivation which is based on integrated pest management, integrated nutrient supply, and integrated natural resource management systems. It does not exclude the use of minimum essential quantities of mineral fertilizers and chemical pesticides. It is widely practised and promoted in China and is likely to pick up in India in the near future.

(*v*) Eco-agriculture. Eco-agriculture has been defined as an approach that brings together agricultural development and conservation of biodiversity as explicit objectives in the same landscape. Its primary aim is to reinforce the mutual relationship between agricultural productivity and conservation of nature. It is capable of bringing together the most productive elements of modern agriculture, new ecological insights and knowledge of local people who have lived in harmony with nature for thousands of years.

(*vi*) White Agriculture. This system of agriculture is based on a substantial use of microorganisms, particularly fungi. The concept of white agriculture originated in China in 1986 and is picking up in India. In this context, white refers to the white-coated scientists and technicians who perform high tech processes to produce food directly from micro-organisms or use them to improve green agriculture.

IMPORTANT CONSIDERATIONS FOR SECOND GREEN REVOLUTION

- More attention of the government and non-government organisations is needed towards agriculture.
- Benefits of research and development must reach the farmers at the ground level for increasing yields and agricultural production.
- Improvement of marketing by strengthening infrastructure.
- Soil health enhancement.
- Better irrigation and water conservation strategies.
- Access to affordable credit.
- Private public partnership and decentralization of partnership by the poor farmers in the development programmes.
- Food security to people living below poverty line (BPL), small and marginal farmers and landless labourers through the following three pronged strategy.
  (*a*) Increase in income and agricultural wage by increasing farm productivity.
  (*b*) Provision of additional off-farm and non-farm employment.
  (*c*) Effective and strong public distribution system (PDS).
- Assured and remunerative marketing for farm products.

The future of the Indian agriculture largely depends upon our ability to increase the productivity, particularly of small holdings without much damage to environment and ecology so that the object of sustainable development is achieved. *Transforming green revolution into evergreen revolution using one or more than one method described above will usher in a win-win situation both for agriculture and ecosystems*. Extending the benefit of science and technology to areas and farmers that were ignored during the First Green Revolution combined with other facilities can enable us to achieve sustainable food security to all Indians.

Realising the gravity of situation with respect to agricultural production, population growth and food security, and appreciating the limitations of the First Green Revolution, the Government of India has initiated steps to usher in the Second Green Revolution. The Approach Paper to the Eleventh Five Year Plan (2007-12) has highlighted a holistic framework and suggested the following strategies to raise agricultural productivity :

1. Doubling the rate of growth of irrigated area.
2. Improving water management, rainwater harvesting, and watershed development.
3. Making efforts to reclaim degraded land and focusing on soil quality.
4. Bridging the knowledge gap through effective extension of information technology.
5. Laying more emphasis on crop diversification and giving importance to high value outputs, *e.g.* fruits, vegetables, flowers, herbs and spices, medicinal plants, bamboo, bio-diesel, but with adequate measures to ensure food security.
6. Providing easy access to credit at affordable rate of interest.
7. Improving the infrastructure and functioning of markets.
8. Refocusing on land reforms issues.
9. Laying emphasis on the cultivation of pulses including hybrid varieties.
10. Focusing on the development of area specific seeds and their application.
11. Paying more attention to rainfed areas, drought-prone areas, and drought resistant, and those amenable to biotechnological application.

Bringing Green Revolution to Eastern India (BGREI). This is a sub-scheme of Rashtriya Krishi Vikas Yojna (RKVY) which was started in 2010-11. Its primary aim is to improve the rice based cropping system in the eastern states with the objective of increasing crop productivity through promotion of recommended agricultural technologies and package of practices. Presently this sub-scheme covers seven states of east India. These states are Assam, Bihar, Chhattisgarh, Jharkhand, Odisha, Eastern Uttar Pradesh and West Bengal. Focussed efforts with scientific back-up approach led to record estimated production of 58.5 million tonnes of rice in implementing states during 2012-13 against average of 48.8 million for 2006-07 to 2010-11 (**Table 21.1**).

TABLE 21.1. Production of rice in seven eastern states under BGREI (Lakh tonnes)

State	Normal (2006-07 to 2010-11)	2010-11	2011-12	2012-13
1. Assam	38.63	47.37	45.16	45.81
2. Bihar	43.40	31.20	71.63	73.43
3. Chhattisgarh	56.26	61.59	60.28	66.68
4. Jharkhand	24.27	11.10	31.31	33.33
5. Odisha	69.89	68.28	58.67	74.82
6. Uttar Pradesh	117.60	119.92	140.22	141.26
7. West Bengal	143.29	130.46	146.06	149.72
Total for seven States	488.29	469.74	552.73	584.45
All India	948.61	959.70	1,053.01	1,042.16

Sources **:** India : A Reference Annual, 2014, p. 72.

REFERENCES

Bardhan, K. Pranab. "Green Revolution and Agricultural Labourers", *Economic and Political Weekly,* Vol. 5, Nos. 29-31, July, 1970.

Bayliss Smith, Tim P. and Sudhir Wanmali (*eds.*) (1984) : *Understanding Green Revolution*, Cambridge University Press, Cambridge.

Bhalla, G.S. (2007) : *Indian Agriculture Since Independence,* National Book Trust, New Delhi.

Bhalla G.S. and Alagh, Y.K. (1979) : *Performance of Indian Agriculture : A Districtwise Study*, Sterling Publishers, New Delhi.

Bhalla, G.S. and Chadha, G.K. (1983) : *Green Revolution and the Small Peasant : A Study of Income Distribution among Punjab Cultivators*. Concept Publishing Company, New Delhi.

Chakravarti, A.K. : "Green Revolution in India." Annals of the A.A.G., Vol. 63, 1973, pp. 319-30.

Chopre, R.N. (1986) : *Green Revolution in India* : *The Relevance of Administrative Support for its Success* : *A Study of Punjab, Haryana, U.P. & Bihar*, Intellectual Publishing House, New Delhi.

Dantwala *et al. (eds)* (1986) : *Agricultural Development Since Independence*, Oxford & IBH, New Delhi.

Dasgupta, Biplab. (1980) : *The New Agrarian Technology and India*. The Macmillan Co. of India Ltd., Chennai.

Farmer, B.H. (*ed.*) (1974) : *Green Revolution?* Macmillan Press, London.

— —, (*ed.*) (1980) : *Green Revolution in India*, Macmillan Press, London.

Frankel, Francine, R. (1971) : *India's Green Revolution : Economic Gains and Political Costs*, Princeton University Press, New Jersey.

Gilbert, Eleine : *Green Pastures*, India Today, 30 April, 1984, New Delhi.

Government of India (2021) : Agricultural Statistics at a Glance.

Government of India (2022-23) : Economic Survey.

Government of India (2007-12) : Eleventh Five Year Plan.

Government of India (2014 and 2021) : India : A Reference Annual.

Jain, N.C. : *Towards a Greener Revolution*, The Hindustan Times, 14-10-96.

Kumar Sharan (2008 : *Towards a Second Green Revolution* Science Reporter, National Institute of Science Communication and Information Resources (NISCAIR), CSIR, New Delhi.

Ladejinsky, W. : "Ironies of India's Green Revolution", *Foreign Affairs,* Vol. 48, 1970, pp. 758-768.

Ladejinsky, Wolf : *How Green is the Green Revolution?* Economic and Political Weekly, Vol. 8, No. 52, December 29, 1973.

Leo, E. Holman. (1969-70) : *Improving the Handling of Grain in Modern Wholesale Market*, The Ford Foundation, New Delhi.

Munsi, Sunil Kumar (1994) : *Sustainability in Indian Agriculture and the Dunkel Proposals* in Mohammad Shafi and Mehdi Raja (*ed*). *Geography of Environment*, Rawat Publications, Jaipur.

Ramakrishnan, P.S. (2004) : *Ecology and Sustainable Development,* National Book Trust, New Delhi.

Randhawa, M.S. (1974) : *Green Revolution—A Case Study of Punjab,* Vikas Publications, New Delhi.

Rao, Hanumantha C.H. (1980) : *Technical changes and Distribution of Grains in Indian Agriculture,* Macmillan, New Delhi.

Rao, Radha Krishan : *Green Revolution Goes Grey*, The Hindustan Times, 23-03-95, New Delhi.

Saini, G.R. : *Green Revolution and Distribution of Farm Incomes*, Economic and Political Weekly, 27 March, 1976.

Sekhon, M.K. and Kaur, Manjeet (1996) : *Green Revolution in Punjab : Other side of the Coin in Agricultural Situation in India.* Vol. I, Feb. 1996, Directorate of Economics & Statistics, Deptt. of Agriculture & Cooperation, Ministry of Agriculture.

Sen, Bandhu Das : (1974) : *Green Revolution in India — A Perspective*, Wiley Eastern Pvt. Ltd., New Delhi.

Sen, Sudhir (1974) : *A Richer Harvest*, Tata McGraw Hill, New Delhi.

— —, (1975) : *Reaping the Green Revolution*, Tata McGraw Hill, New Delhi.

Shafi, Mohammad (2006) : Agricultural Geography, Pearson Education, New Delhi, Ch. 12.

Sharma, H.S. and Srikumar Chattopadhyay (1998) (eds.) : *Sustainable Development, Issues and Case Studies,* Concept Publishing Company, New Delhi.

Sharma, T.C. (1997) : "How Green is the Green Revolution in Karnataka" ? A Regional Study, *Transactions* Journal of the IIG Vol. 19, No. 1, January, Pune.

Sharma, T.C. and Coutinho, O. (1989) : *Green Revolution Gaps : A Geographical Analysis*, Rawat Publications, Jaipur.

Shiva, Vandana (1991) : *The Violence of Green Revolution*, Third World Network, Malaysia.

Singh, D.P. : *The Impact of Green Revolution on the Poor*, Agricultural Situation India, Vol. 75, No. 5, August 1980.

Singh, Jasbir (1974) : *Green Revolution in India : How Green is it* ? Vishal Publishing, University Campus, Kurukshetra.

Subramaniam, C. (1979) : *A New Strategy in Agriculture*, Vikas, New Delhi.

Swaminathan, M.S. : *Keeping the Green Revolution green.* The Hindustan Times, 21-7-96.

Thukral, Govind : *Fruits of the Green Revolution*, The Hindustan Times, 1-6-95.

Venkateswaran (1983) : *Dynamics of Green Revolution in India* : Agricole Publishing Academy, New Delhi.

Vohra, B.B. (1975) : *Land and Water Management Problems in India*, Department of Personnel and Administrative Reforms, New Delhi.

Vyas, V.S. "Green Revolution—A Promise and Problems", Industrial India, Annual Number, 1969, p. 35.

Wolf Ladejinsky, "How Green is India's Green Revolution ?" Economic and Political Weekly, 29 December, 1993, p. A-134.

◆◇◆

CHAPTER 22

Major Crops

With varied types of relief, soils, climate and with plenty of sun-shine and long growing season, India is capable of growing almost each and every crop. Crops requiring tropical, sub-tropical and temperate climate can easily be grown in one or the other part of India. Indian crops can be divided into following categories.

1. Food Crops. Rice, Wheat, Maize, Millets—Jowar, Bajra, Ragi; Pulses—Gram, Tur (Arhar).

2. Cash Crops. Cotton, Jute, Sugarcane, Tobacco, Oilseeds, Groundnut, Linseed, Sesamum, Castorseed, Rapeseed, Mustard, etc.

3. Plantation Crops. Tea, Coffee, Spices—Cardamom, Chillies, Ginger, Turmeric; Coconut, Arecanut and Rubber.

4. Horticulture. Fruits—Apple, Peach, Pear, Apricot, Almond, Strawberry, Walnut, Mango, Banana, Citrus Fruits, Vegetables.

FOOD CROPS

Agriculture forms the back-bone of Indian economy and food crops form the back-bone of Indian agriculture. Food crops cover about three-fourths of the total cropped area in the country and contribute to about half of the total value of agricultural production. No part of the country is without food crops. They are grown in all parts of the country either as a sole crop in or combination with other crops. The importance of food crops grows with the growth of population because these crops provide the basic and essential food for man.

RICE

Rice is the most important food crop of India covering about 30 per cent of the total cropped area and providing food to about half of the Indian population. This is the staple food of the people living in the eastern and the southern parts of the country, particularly in the areas having over 150 cm annual rainfall. There are about 10,000 varieties of rice in the world out of which about 4,000 are grown in India.

Rice is life for thousands of millions of people. In Asia alone, more than 2,000 million people obtain 60 to 70 per cent of their calories from rice and its products. Recognising the importance of this crop, the United Nations General Assembly declared 2004 as the "International Year of Rice" (IYR). The theme of IYR—"*Rice is life*" reflects the importance of rice as a primary food source, and is drawn from an understanding that rice-based systems are essential for food security, poverty alleviation and improved livelihood.

Conditions of Growth

Temperature, rainfall, relief, soil and labour are five major geographical factors which determine the growth of any crop in any area. Following geographical conditions favour the cultivation of rice.

(*i*) Temperature. Rice is a tropical plant and requires high heat for its successful growth. The temperature should be fairly high at mean monthly of 24°C. It should be 20°-22°C at the time of sowing, 23°-25°C during growth and 25°-30°C at the harvesting time.

(*ii*) Rainfall. Along with high heat, rice requires high level of humidity for its successful growth. The average annual rainfall required for rice is 150 cm. It is the dominant crop in areas of over 200 cm annual rainfall and is still an important crop in areas of 100-200 cm rainfall. The 100 cm *isohyet* forms the limit of rice in rainfed areas. In areas receiving less than 100 cm annual rainfall, rice can be grown with the help of irrigation, as is done in Punjab, Haryana and western U.P. About 40 per cent of rice crop in India is raised under irrigation. However, it is the temporal distribution of rainfall, rather than the total amount of annual rainfall which is more decisive. The rainfall should be fairly well distributed throughout the year and no month should have less than 12 cm of rainfall. Lesser amount of rainfall is required as the harvesting time approaches.

(*iii*) Relief. The fields must be flooded under 10-12 cm deep water at the time of sowing and during early stages of growth. Therefore, the fields must be level and have low mud walls to retain water. This peculiar requirement of rice makes it primarily a crop of plain areas. Rice grown in well watered lowland plain areas is called ***wet*** or ***lowland rice***. In hilly areas, the hill slopes are cut into terraces for the cultivation of rice. Such a cultivation in which the hill slopes are cut into terraces is called ***terraced cultivation***. The supply of water to the hill terraces is not as much as in the plain areas and the rice grown in hilly areas is called ***dry*** or ***upland rice***.

(*iv*) Soil. Rice can be grown on a variety of soils including silts, loams and gravels and can tolerate acidic as well as alkaline soils. However, deep fertile clayey or loamy soils which can be easily puddled into mud and develop cracks on drying are considered ideal for raising this crop. Such soil requirements make *it dominantly a crop of river valleys, flood plains, deltas and coastal plains and a dominant crop there*. High-level loams and lighter soils can be used for quick maturing varieties of rice. Black lava soil is also useful for rice cultivation.

(*v*) Labour. Rice culture is not much suited to mechanisation and is called **'*hoe-culture*'**. Most of the work in preparing the seed-bed, in broadcasting seeds, or in transplantation of plants from nurseries to the fields, in harvesting and in winnowing operations is done by human hand. Thus it is a *labour intensive cultivation* and requires large supply of cheap labour for its successful cultivation. It is, therefore, primarily grown in areas of high population density which provide abundant labour and at the same time, offer ready market for its consumption. In most rice producing states, labour is locally available but in Punjab and Haryana, rice cultivation mainly depends upon the migrant labourers from Bihar and eastern U.P.

Considering the above mentioned conditions of growth of rice, **G.B. Cressey** has rightly remarked that *rice needs plenty of heat, plenty of rain, plenty of alluvium and plenty of labour to provide plenty of food for plenty of people. There is no other food crop which is so plentiful as rice in India.*

Methods of Rice Cultivation

Following methods of rice cultivation are practised in India.

1. Broadcasting method. Seeds are sown broadcast by hand. This method is practised in those areas which are comparatively dry and less fertile and do not have much labour to work in the fields. It is the easiest method requiring minimum input but its yields are also minimum.

2. Drilling method. Ploughing of land and sowing of seeds is done by two persons. This method is mostly confined to peninsular India.

3. Transplantation method. This method is practised in areas of fertile soil, abundant rainfall and plentiful supply of labour. To begin with, seeds are sown in nursery and seedlings are prepared. After 4-5 weeks the seedlings are uprooted and planted in the

TABLE 22.1. Rice Cropping Seasons in India

Crop	Local name	Sowing	Harvesting	Percentage of area	Percentage of Production
Autumn (Kharif)	Aus or Kar	May-June	Sept-Oct.	39.4	43.97
Winter (Rabi)	Aman, Sali or Karthika	June-July	Nov.-Dec.	54.2	48.79
Summer (Spring)	Boro or Dalua	Nov.- Dec.	March-April	6·4	7·24

field which has already been prepared for the purpose. The entire process is done by hand. It is, therefore, a very difficult method and requires heavy inputs. But at the same time it gives some of the highest yields.

4. Japanese method. This method includes the use of high yielding varieties of seeds, sowing the seeds in a raised nursery-bed and transplanting the seedlings in rows so as to make weeding and application of fertilizers easy. It also involves the use of a heavy dose of fertilizers so that very high yields are obtained. The Japanese method of rice cultivation has been successfully adopted in the main rice producing regions of India.

Rice Cropping Seasons

Rice is grown almost throughout the year in hot and humid regions of eastern and southern parts of India where two to three crops in a year are not uncommon. But in the northern and hilly parts of the country, the winters are too cold for rice cultivation and only one crop is grown in those areas. **Table 22.1** gives the period of sowing and harvesting the rice crop.

Production

India is the second largest producer and consumer of rice in the world after China and accounts for 21.65 per cent of the world's total rice production. **Table 22.2** gives the trends in production of rice for selected years. A look at this table shows that there has been considerable increase in production, area and yield of rice in India.

In a span of about seven decades from 1950-51 to 2021-22, the area, production and yield have increased by about one and a half times, five and a half times and four times respectively. It is interesting to note that the rate of increase in production is much higher than the rate of increase in area under rice cultivation. This is due to the increase in yields as a result of better inputs and farm practices. Thus, there has been a modest gain in extent of cultivated area but a substantial gain in yield and production. Increased irrigation facilities in drier areas, reclamation of waterlogged soils and introduction of new high-yielding strain crops (particularly in Punjab, Haryana and Tamil Nadu) made this possible.

In spite of the spectacular progress, our yield of 2,809 kg per hectare (2021-22) is much lower compared to 6,866 kg in China, 7,537 kg in the U.S.A., 6,667 kg in Japan and 6,878 kg in Republic of Korea. This means that there is still a vast scope for increasing production. This will have to be done by increasing yields because scope for increasing area under rice crop is negligibly small.

Distribution

Rice is grown in varying degrees in almost all parts of the country excepting higher parts of the

TABLE 22.2. Area Production and Yield of Rice in India

Year	1950-51	1960-61	1970-71	1980-81	1990-91	2000-01	2010-11	2019-20	2020-21	2021-22
Area (Million hectares)	30.81	34.13	37.59	40.15	42.69	44.71	42.86	43.66	45.07	46.4
Production (Million tonnes)	20.58	34.58	42.22	53.63	74.29	84.98	95.98	118.7	122.27	130.3
Yield (kg/hectare)	668	1013	1123	1336	1740	1901	2239	2722	2713	2809

Source : (*I*) Agricultural Statistics at a Glance, 2021, pp. 48-49.
(*II*) Economic Survey 2022-23, Statistical Appendix, pp. 37-39.

Himalayan ranges exceeding 2,500 metres in altitude, Marusthali part of Rajasthan, Kachchh-Saurashtra, Malwa and Marathwada regions due to various geographical constraints. The premier rice producing areas include the lower and the middle Ganga Plains, the east and the west Coastal Plains, the Brahmaputra valley and parts of the Peninsular plateau. Punjab, Haryana and Uttar Pradesh have assumed considerable importance after the introduction of the Green Revolution. **Table 22.3** gives vital data about the statewise distribution of rice among the major producers.

It is clear that about half of rice production in India is contributed by four states namely W. Bengal, Punjab, Uttar Pradesh and Andhra Pradesh . The other major producers are Bihar, Tamil Nadu, Chhattisgarh, Odisha, Assam, Haryana, Telangana, Karnataka, Jharkhand, Madhya Pradesh, Maharashtra, Gujarat and Kerala.

1. West Bengal. West Bengal is the largest producer of rice in India contributing over 13 per cent of the total rice production from 12.38 per cent of the rice producing area of the country. Rice occupies three-fourths of the total cropped area in the lower Ganga Plains. Suitable climatic conditions and large scale alluvial deposits have helped in the cultivation of rice on such a scale. Over two-thirds of the production comes from Medinipur, Bardhaman, North and South Twenty Four Parganas, Bankura, Birbhum and West Dinajpur districts. The other producing districts are Haora, Hugli, Jalpaiguri, Coochbehar and Malda. As many as three crops called ***aman***, ***aus*** and ***boro*** are grown in a year. They account for about 78 per cent, 20 per cent and 2 per cent of the total production respectively.

2. Uttar Pradesh. Uttar Pradesh has recorded unprecedented progress in the production of rice during the last five decades. Earlier this state used to produce only 6 to 8 per cent of the country's rice but according to 2020-21 figures, Uttar Pradesh had 12.6 per cent of the country's rice producing area and accounted for 12.81 per cent of the total rice production of the country, thus occupying the second place among the top rice producing states. About one-fourth of the total cultivated area of the state is devoted to rice cultivation. The unusual interest shown by the farmers in the rice culture; supported by the easy availability of HYV seeds, fertilizers and

TABLE 22.3. State-wise Area, Production and Yield of Rice in India (2020-21)

Sl. No.	States	Area		Production		Yield (kg/hectare)
		Million hectares	%age of all India	Million tonnes	%age of all India	
1.	West Bengal	5.58	12.38	16.65	13.62	2,984
2.	Uttar Pradesh	5.68	12.60	15.66	12.81	2,759
3.	Punjab	2.79	6.19	12.18	9.16	4,366
4.	Odisha	4.03	8.95	8.77	7.17	2,173
5.	Andhra Pradesh	2.32	5.16	7.89	6.45	3,395
6.	Telangana	2.31	5.14	7.70	6.30	3,327
7.	Tamil Nadu	2.04	4.52	7.28	5.96	3,574
8.	Chhattisgarh	3.79	8.41	7.16	5.86	1,889
9.	Bihar	3.02	6.71	6.88	5.63	2,276
10.	Assam	2.36	5.24	5.26	4.30	2,224
	Others	11.13	24.70	26.84	21.95	2,411
	All India	**45.07**	**100.00**	**122.27**	**100.00**	**2,713**

Note: States have been arranged in descending order of production in 2020-21.

Source : Agricultural Statistics at a Glance, 2021, p. 50.

uninterrupted supply of irrigation are the chief factors responsible for this progress. As many as 50 districts of the state are producing rice out of which 7 are major producers. The main producing districts are Gorakhpur, Bareilly, Muzaffarnagar, Kheri, Faizabad, Barabanki, Banda, Varanasi, Pilibhit, etc.

3. Punjab. Punjab is traditionally a wheat producing state but the enterprising farmers of Punjab have made full use of package technology including perennial irrigation water by canals and tube wells, HYV seeds and fertilizers. This has resulted in widespread change in the cropping pattern of Punjab and the state has become the third largest producer of rice, in spite of its small size. Punjab now gives over 4,366 kg/hectare which is the highest yield for any state of India. It is far above the other traditional rice producing states. Since 1953-54, Punjab has recorded an annual growth rate of over 12 per cent as against only 2.8 per cent for the country as a whole. As many as 12 districts of Punjab are producing rice but major part of production comes from Patiala, Firozepur, Ludhiana, Sangrur, Amritsar, Faridkot and Jalandhar.

4. Odisha. Odisha produces over about 7.2 per cent rice of India. Although about two-thirds of the total cropped area of Odisha is devoted to rice, the total production is low due to low yield of only 2,173 kg/hectare. Over 90 per cent of the state's rice comes from Sambalpur, Koraput, Ganjam, Cuttack, Puri, Bolangir and Mayurbhanja.

5. Andhra Pradesh. With about 6.45 per cent of the rice production and about 5 per cent of the rice area of the country, Andhra Pradesh is the fifth largest producer of rice in India. About one-fifth of the total cropped area of the state is under rice cultivation. The Godavari-Krishna Delta and the adjoining coastal plains form one of the most outstanding rice producing tracts of the country. The yields and production have increased considerably with the introduction of *Package Technology* (Green Revolution). In 2020-21, Andhra Pradesh recorded yield of 3,395 kg/hectare against India's average of 2,713 kg/hectare. This is the third highest yield in India after that of Punjab (4,366 kg/ha) and Tamil Nadu (3,357 kg/ha). Almost all districts of Andhra Pradesh are producing rice out of which West Godavari, East Godavari, Krishna, Guntur, Srikakulam, Nellore, Prakasam, Anantapur and Chittoor are the major producers. In fact West Godavari, East Godavari and Krishna are three most important rice producing districts not only of Andhra Pradesh but of the whole of India.

6. Telangana. Although Telangana produces about 6.3 per cent rice of India. This state gives yields which are much higher than many other states. Mehbubnagar, Nalgonda, Khammam, Varangal, etc. are the main rice producing districts.

7. Tamil Nadu. Although Tamil Nadu produces only about 6 per cent rice of India, the state has the distinction of giving high yield of 3,574 kg/hectare which is second only to that of Punjab. This is mainly due to the effect of Green Revolution under which HYV seeds, fertilizers and irrigation facilities have been made available to the farmers. About 37 per cent of the cropped area of this state is under rice cultivation. Thanjavur in the Cauvery delta is the second largest rice producing district of the country which produces about 2.8 per cent rice of India and accounts for about 25 per cent rice produced by the state. Nagapattinum, Ramnathapuram, Thiruvarur, Virudhunagar, Pudukottai, etc. are the other important rice producers.

8. Chhattisgarh. Chhattisgarh basin drained by the Mahanadi and its tributaries is the main rice producing region in this state. Although Chhattisgarh accounts for about 8.41 per cent of the rice area of the country, this state produces only 5.86 per cent of the country's total rice. This is primarily due to low yields of rice in this state. In fact the average yield of rice in Chhattisgarh is 1,889 kg/hectare only which happens to be one of the lowest in the country. Bastar, Durg, Raigarh, Sarguja, Rajnandgaon, Kanker, Dantewara, Bilaspur, Janjgir, etc. are the main rice producing districts.

9. Bihar. Bihar has slipped from fourth position in 1990-91 to ninth position in 2020-21 among the rice producing states of India. This is partly due to increase in production in other states and partly due to carving of new state of Jharkhand out of Bihar. Although Bihar has 6.7 per cent of the rice area of the country and about two-thirds of the cultivated area of the state is under rice cultivation, the state produces only 5.6 per cent of the total rice of India. This is due to low average yield of only 2,276 kg per hectare against 2,713 kg per hectare for the country as a whole. About 25 districts of Bihar are producing rice but the main producing districts are Rohtas, Bhojpur, Purnea, Paschim Champaran, Purab Champaran, Aurangabad, Gaya, Bhagalpur, Patna and Gopalganj.

10. Assam. Rice is grown on three-fourths of the cropped area of Assam. Most of the rice is produced in the Brahmaputra, Surma, and Barak valleys. Some rice is produced on the hill slopes by terraced cultivation. Kamrup, Sibsagar, Goalpara, Darrang, Nowgong and Cachar are the main producing districts.

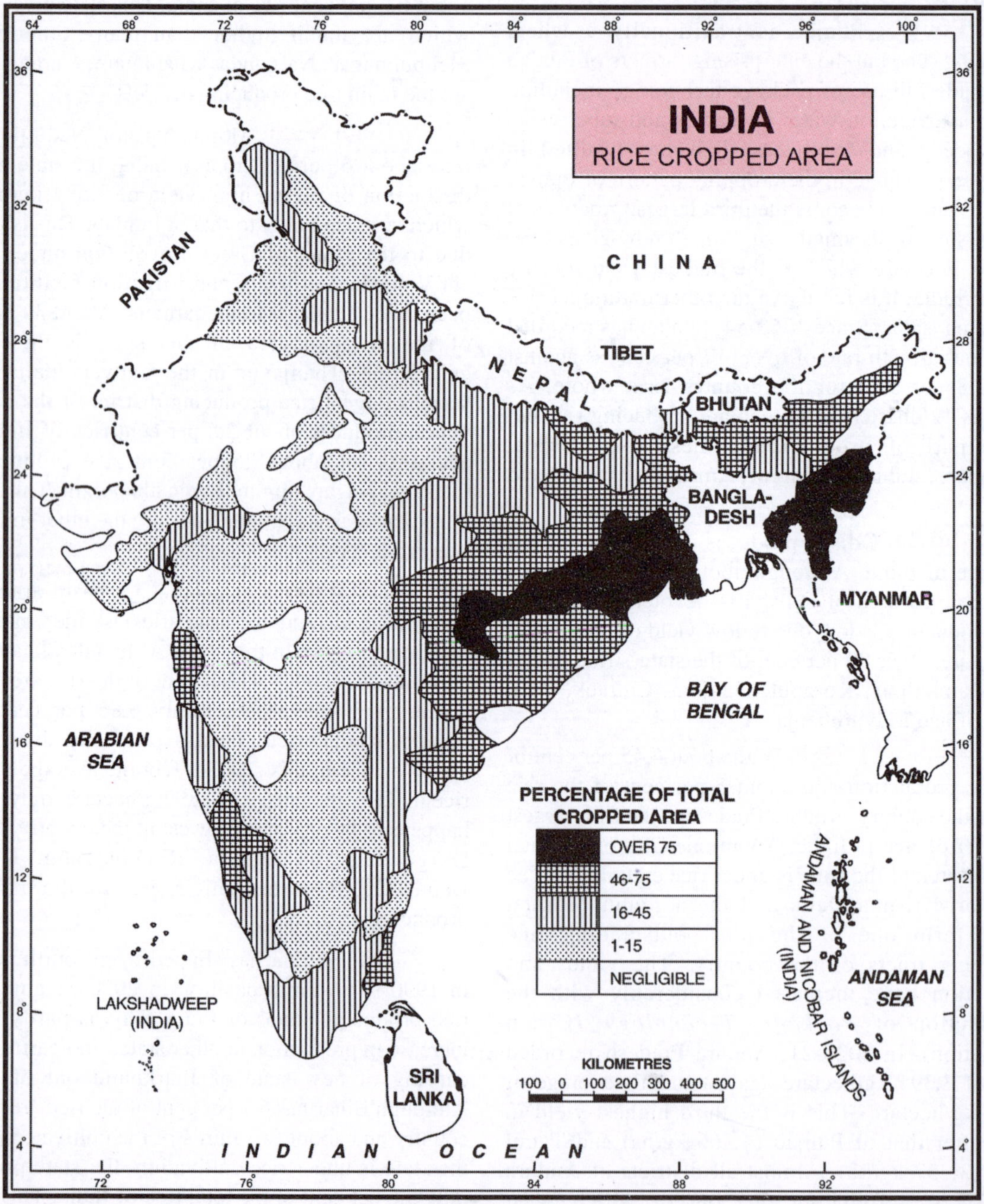

FIG. 22.1. India : Rice cropped area

11. Haryana. Like Punjab, Haryana is traditionally a wheat producing state. But the cropping pattern has undergone an unprecedented change due to large scale extension of irrigation and provision of HYV seeds and chemical fertilizers, coupled with the progressive outlook of the Haryana farmers. Only a few years ago Haryana was not on the rice map of India but rapid progress during the last five decades has enabled Haryana to occupy an important place among the rice producing states of India. Haryana has a high yield of 3,181 kg per hectare which is the fourth highest after that of Punjab, Tamil Nadu and Andhra Pradesh. Kurukshetra, Karnal, Ambala, Kaithal, Panipat and Yamunanagar are the important rice producing districts.

12. Madhya Pradesh. Madhya Pradesh produces 3.65 per cent rice of India and her yield of 2,026 kg/ha is very low as compared to the main rice producing states. Major part of production comes from river valleys where fertile soils are available.

13. Jharkhand. This state produces about three per cent rice of India. Ranchi, Paschim Singhbum, Purb Singhbum, Lohardaga and Gumla are the chief rice producing districts.

14. Karnataka. Karnataka has made rapid progress in rice cultivation during the last few years. In Karnataka, rice is mainly grown in the valleys of the Wainganga, the Tungabhadra and the Cauvery and in the northern red soil areas. Tumkur, Dakshina Kannada, Shimoga, Mandya, Uttar Kannada, Mysore, Raichur and Kodagu are the main producers.

15. Maharashtra. Rice is grown in the Konkan coastal area, on the Ghats and in some eastern parts. Chandrapur, Raigad, Thane, Bhandara, Kolhapur, etc. are important rice producing districts.

The other producers include Gujarat, Kerala, North Eastern hill states (Tripura, Manipur, Arunachal Pradesh, Mizoram, Sikkim), Himachal Pradesh, Uttarakhand, Jammu & Kashmir and Goa.

Trade

Large producers of rice are its large consumers also and there is little surplus for trade. However some inter-state trade is carried on and about ten per cent of the total production enters trade. Punjab, Haryana, Tamil Nadu, Andhra Pradesh and Uttar Pradesh are surplus states and supply rice to deficit states like West Bengal, Maharashtra, Gujarat, Kerala and Delhi.

Even in the face of huge home consumption of rice, it is amusing to note that rice exports from India have grown steadily during the last decade (**Table 22.4**). India now occupies second position in rice exports, next only to Thailand, among the rice trading countries of the world. However, the surplus production scenario has no room for complacency, keeping in view the rapid growth of population, per capita availability of 197 grams/day in 2020-21 and the vagaries of monsoon. It is estimated that the rice demand will be a whopping 140 million tonnes in 2025. This projected demand can only be met by maintaining steady increase in production over the years.

Table 22.4 shows that export of rice has rapidly increased both in terms of quantity and value of exports. Punjab, Haryana and western Uttar Pradesh produce some of the best qualities of Basmati rice for which there is great demand in the international market.

WHEAT

Next to rice, wheat is the most important foodgrain of India and is the staple food of millions of Indians, particularly in the northern and north-western parts of the country. It is rich in proteins, vitamins and carbohydrates and provides balanced food.

TABLE 22.4. Export of Rice from India

Year	1970-71	1980-81	1990-91	2000-01	2019-20	2020-21	2021-22
Quantity ('000 tonnes)	32.8	131.6	505.0	1534.4	9511.0	17779	21232
Value (₹ crore)	5	165	462	2943	45427	65405	72116

Source : Economic Survey 2022-23, Statistical Appendix pp. 116-18

Conditions of Growth

Conditions of growth for wheat are more flexible than those of rice. In contrast to rice, wheat is a rabi crop which is sown in the beginning of winter and is harvested in the beginning of summer. The time of sowing and harvesting differs in different regions due to climatic variations.

(*i*) Temperature. Wheat is primarily a crop of mid-latitude grasslands and requires a cool climate with moderate rainfall. The ideal wheat climate has winter temperature of 10° to 15°C and summer temperature varying from 21°C to 26°C. The temperature should be low at the time of sowing but as the harvesting time approaches higher temperatures are required for proper ripening of the crop. But sudden rise in temperature at the time of maturity is harmful.

(*ii*) Rainfall. Wheat thrives well in areas receiving an annual rainfall of about 75 cm. Annual rainfall of 100 cm is the highest limit of wheat cultivation. The isohyet of 100 cm marks the boundary between wheat growing areas on one hand and rice growing areas on the other. In areas of less than 50 cm annual rainfall, irrigation is necessary for its successful growth. In fact, wheat can be grown in areas with as little as 20-25 cm annual rainfall provided proper irrigation facilties are available. About 5 to 7 waterings are required in irrigated areas depending upon the amount of rainfall. While prolonged drought, especially in rainfed areas, at the time of maturity is harmfull, light drizzles and cloudiness at the time of ripening help in increasing the yield. Frost at flowering time and hail storm at the time of ripening can cause heavy damage to the wheat crop.

(*iii*) Relief. Wheat is grown in large levelled farms and requires vast plain areas for its successful cultivation. This is the reason that the Great Plain of North India is such a prolific producer of wheat in India.

(*iv*) Soil. Although wheat can be grown in a variety of soils, well drained fertile, friable loams and clay loams are the best suited soils for wheat cultivation. It also grows well in the black soil of the Deccan plateau.

(*v*) Labour. Wheat cultivation is an extensive type of farming which is highly mechanized and requires comparatively less labour.

To sum up wheat requires a combination of factors including *cool climate with moderate rainfall, flat and well drained plain areas, fertile friable loam and heavy inputs in the form of irrigation, HYV seeds, fertilizers and mechanization*.

Production

India is the second largest producer of wheat in the world next only to China, and accounts for 12.32 per cent of the total production of wheat in the world.

Wheat is grown on over 20 per cent of the cropped area of India. **Table 22.5** shows the production trends of wheat in India.

It is clear from the table that all the three aspects of the crop *i.e.* production, area and yield have recorded rapid growth particularly after the introduction of the Green Revolution strategy in 1967. The production had more than doubled from 11 million tonnes in 1960-61 to 23.83 million lakh tonnes in 1970-71. During the same period the area under wheat had increased by over 41 per cent and yield had increased by 53.6 per cent. The development of new varieties of seeds has brought about a real revolution in wheat production. The phenomena of overall development in wheat farming is still continuing although the pace of progress has

TABLE 22.5. Area, Production and Yield of Wheat in India

Year	1950-51	1960-61	1970-71	1980-81	1990-91	2000-01	2010-11	2018-19	2019-20	2020-21	2021-22
Area (Million hectares)	9.75	12.93	18.24	22.28	24.17	25.73	29.07	29.33	31.36	31.61	30.5
Production (Million tonnes)	6.46	11.00	23.83	36.31	55.14	69.68	86.87	103.60	107.86	109.52	106.8
Yield (kg/hectare)	663	851	1307	1630	2281	2708	2988	3534	3440	3464	3507

Source : (*i*) Agricultural Statistics at a Glance, 2021, pp. 51-52.

(*ii*) Economic Survey 2022-23, Statistical Appendix, pp. 37-39.

slowed down with Green Revolution reaching its mature stage. In spite phenomenal improvement in wheat culture in India, our yield of 3,507 kg/hectare (2021-22) is still very low as compared to that of some other wheat producing countries like China, France, Germany, Kazakhstan, Egypt and Italy. It is estimated that yield can be raised upto 4,000 kg/hectare in irrigated areas and upto 2,000 kg/hectare in unirrigated areas by using appropriate location specific technology including better quality seeds, proper fertilizers and control of weeds, pests and diseases. Further, there is vast scope for extending wheat cultivation to non-traditional areas like Assam valley and in Odisha. This can be done by timely harvest of the kharif crops and by reducing the extent of fallow land. West Bengal has already started growing wheat in sufficient quantity.

Distribution

Wheat production is mainly confined to North-Western parts of the country. **Table 22.6** gives the distribution pattern of wheat in India.

Uttar Pradesh, Madhya Pradesh, Punjab and Haryana are the four prominent wheat producing states. These states account for over three-fourths of the wheat area and produce about three-fourths of the total wheat production in India. In fact, Punjab, Haryana and the contiguous western parts of U.P. have earned the distinction of being called the '*Granary of India*'. The other major wheat producing states are Rajasthan, Bihar, Gujarat, Maharashtra, Uttarakhand and Himachal Pradesh.

1. Uttar Pradesh. Uttar Pradesh is the largest wheat producing state of India accounting for about one-third of area and production of wheat of the country. In 2020-21, this state produced 35.5 million tonnes of wheat. Fine alluvial soil deposited by the mighty Ganga and its several big and small tributaries and a close network of canals, supplemented by large number of tube wells have helped U.P. to occupy the top position. More than half of the wheat area lies in the Ganga-Ghagra doab. Next in importance is the Ganga-Yamuna doab. These two doabs account for about 75 per cent wheat of U.P. About 55 districts of Uttar Pradesh produce wheat out of which 43 are the important producers. Saharanpur, Muzaffarnagar, Meerut, Moradabad, Rampur, Budaun, Etawah, Hardoi, Bahraich, Kheri, Gonda, Basti, etc. are the main producing districts. However, wheat production to the east of Varanasi declines due to high rainfall and heavy soils.

TABLE 22.6. Area, Production and Yield of Wheat in India (2020-21)

Sl. No.	States	Area		Production		Yield (kg/hectare)
		Million hectares	%age of all India	Million tonnes	%age of all India	
1.	Uttar Pradesh	9.85	31.16	35.50	32.42	3,604
2.	Madhya Pradesh	6.39	20.20	17.62	16.08	2,758
3.	Punjab	3.53	11.15	17.14	15.65	4,862
4.	Haryana	2.56	8.08	12.36	11.28	4,836
5.	Rajasthan	3.00	9.50	11.04	10.08	3,676
6.	Bihar	2.22	7.02	6.34	5.79	2,855
7.	Gujarat	1.02	3.22	3.26	2.98	3,204
8.	Maharashtra	1.31	4.13	2.33	2.13	1,782
	Others	1.75	5.54	3.94	3.60	2,250
	All India	**31.61**	**100.00**	**109.52**	**100.00**	**3,464**

Note: States have been arranged in descending order of production in 2020-21.

Source : Agricultural Statistics at a Glance, 2021, p. 53.

2. Madhya Pradesh. Madhya Pradesh is the second largest wheat producing state of India. The state produces about 16% of India's wheat from about 17.6% of the wheat area of the country. This is an indication of low yield which is 2,758 kg/hectare. Steps have to be taken to increase the yield so that the

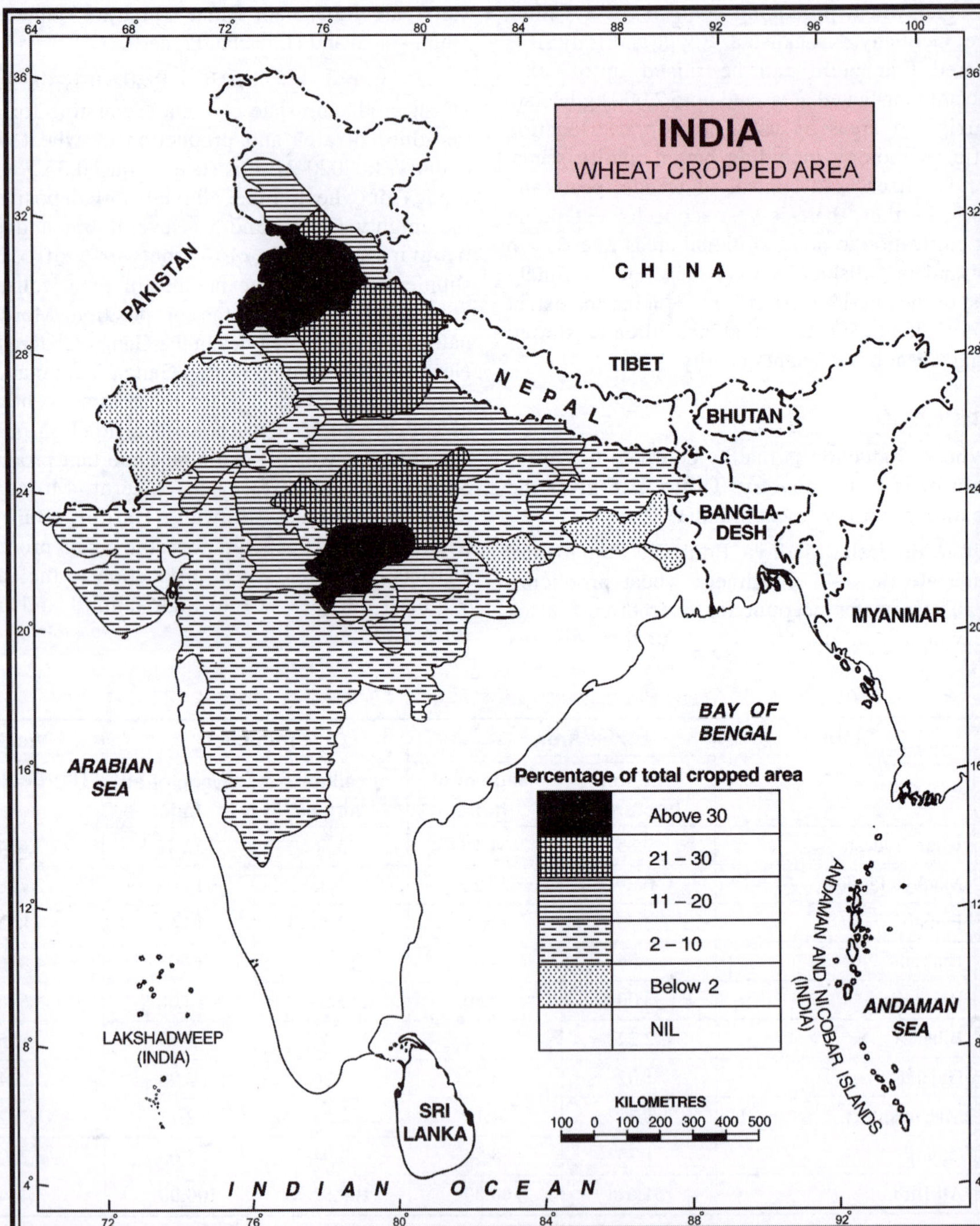

FIG. 22.2. India : Wheat cropped area

state occupies a prestigious position among the wheat producing states of India. Sagar, Vidisha, Tikamgarh, Morena, Sehore, Gwalior, Guna, Satna, Bhind and Chhatarpur are important wheat producing districts.

3. Punjab. Although very small state as compared to Uttar Pradesh and Madhya Pradesh, Punjab has emerged as a very important producer of wheat in India. The Green Revolution strategy has helped Punjab in making rapid strides in wheat production. In fact, *Punjab has drawn maximum benefit from Green Revolution and in Punjab too it is the wheat crop which has been benefited the most.* The excellent irrigation system provided by a close network of canals and the tube wells is supplemented by light rainfall associated with the *western disturbances*. The fertile alluvial soil brought by the rivers of the Indus system is ideal for wheat production. Over and above, the Punjab farmer is very enterprising and is always willing to adopt the new farm technologies. Punjab accounts for about 15.65 per cent of the wheat production and 11.15 per cent of wheat area of India.

This state gives the highest yield of 4,862 kg/ha. In 2020-21, Punjab produced 17.14 million tonnes of wheat, thus occupying third position among the major wheat producing states of India. Punjab has 12 leading wheat producing districts. Jalandhar, Ludhiana, Sangrur, Bathinda, Amritsar, Firozepur, Faridkot, Mansa, Kapurthala, Fatehgarh Sahib, Rupnagar and Patiala are the main producing districts. The state has a large surplus and contributes a lot of wheat to the central pool.

4. Haryana. The physical and human conditions for wheat cultivation in Haryana are the same as those prevailing in Punjab, although to a lesser degree. The impact of Green Revolution is clearly visible in Haryana also. Presently, Haryana accounts for about 8 per cent of the wheat area of India and produces over 11 per cent of the total wheat of the country. The yield at 4,836 kg/hectare is the second highest in the country after that of Punjab. Karnal, Kurukshetra, Ambala, Kaithal, Panipat, Sonipat, Rohtak, Jind, Hisar, Sirsa, Fetehabad and Gurgram are important producing districts.

5. Rajasthan. Vast stretches of sandy desert, scarcity of rainfall and paucity of irrigation facilities have been restricting wheat cultivation in Rajasthan since long. But some of the irrigation projects initiated after Independence, especially the Indira Gandhi Canal, have brought about considerable improvement in the cropping pattern of the state. Currently, Rajasthan accounts for about 10 per cent of the total wheat production and about 9.5 per cent of wheat area of India. Over 20 districts are producing wheat and 11 are major producers. Ganganagar, Hanumangarh, Bharatpur, Kota, Alwar, Jaipur, Chittaurgarh, Tonk, Sawai Madhopur, Udaipur and Pali are important wheat producing districts of Rajasthan.

6. Bihar. Bihar accounts for 5.79 per cent of wheat production and about 7 per cent of wheat area of India with a low yield of 2,855 kg/hectare. This means that there is need to improve the situation with respect to yield. Most of the wheat is produced in the *North Bihar Plain*. Rohtas, Bhojpur, Saran, Nalanda, Paschim and Purba Champaran, Siwan and Begusarai are the main districts.

Others. Gujarat, Maharashtra, Uttarakhand, Himachal Pradesh, West Bengal, Jammu and Kashmir, Jharkhand, Karnataka, Assam and Telangana are the other important producers of wheat in India. In *Gujarat*, wheat is mainly produced in Mahi and Sabarmati valleys. Mahesana, Junagadh, Bhavnagar, Amreli, Bharuch, Rajkot and Kheda are important wheat producing districts. In *Maharashtra*, wheat is produced in the valleys of the Wardha, Tapi, Godavari, Bhima, Purna and Krishna. *West Bengal* has made significant progress both in area and production of wheat with the introduction of new technology. Most of the production comes from Birbhum, Burdwan, Murshidabad and Nadia districts. In *Uttarakhand*, most of the wheat production comes from river valleys. In *Himachal Pradesh* wheat is produced mainly in Kangra, Mandi, Sirmaur and Una districts. Srinagar, Baramula, Doda, Anantnag, Jammu and Punch are the main producers in *Jammu* and *Kashmir*. Some wheat is also produced in Bijapur, Raichur, Belgaum and Dharwar districts of *Karnataka*.

Trade

About one-third of the total production of wheat enters trade. Punjab, Haryana, Uttar Pradesh, Rajasthan and Madhya Pradesh are surplus states and supply wheat to deficit states like Maharashtra, West Bengal, Bihar and the Union Territory of Delhi.

India is both an importer and an exporter of wheat depending on the domestic production and requirement. Imports become necessary in the event of deficit rainfall and fall in production. However, India exports some quantity of wheat whenever there is sufficient rainfall and we get bumper crop.

MAIZE

Maize is an inferior grain which is used both as food and fodder. Its grain provides food and is used for obtaining starch and glucose. Its stalk is fed to cattle.

Conditions of Growth

Maize can be grown under varied climatic and soil conditions. Maize is mainly a rainfed kharif crop which is sown just before the onset of monsoon and is harvested after retreat of the monsoon. In Tamil Nadu it is a rabi crop and is sown a few weeks before the onset of winter rainy season in Sept. and Oct.

Following conditions help in the growth of maize:

(*i*) Temperature. This crop usually grows well under temperatures varying from 21°C to 27°C, although it can tolerate temperatures as high as 35°C. Frost is injurious to maize and this crop is grown only in those areas where there are about four and a half frost free months in a year.

(*ii*) Rainfall. It requires 50-100 cm of rainfall and it cannot be grown in areas of more than 100 cm rainfall. In areas of lesser rainfall, the crop is irrigated. For example, more than half of the maize area in Punjab and Karnataka is irrigated. Long dry spell during the rainy season is harmful for maize. Sunshine after showers is very useful for maize. Cool and dry weather helps in ripening of the grain.

(*iii*) Relief. Well drained plains are best suited for its cultivation, although it grows in some hilly areas also.

(*iv*) Soil. Fertile well-drained alluvial or red loams free from coarse materials and rich in nitrogen are the best soils for its successful growth.

(*v*) Labour. Sufficient labour should be available at the time of sowing and harvesting.

Production

Maize is an important cereal of India and is grown over 6 per cent of the net area sown of the country. There have been large variations in the production of maize in India since Independence. It was only 1.7 million tonnes in 1950-51 which rose to 4.1 million tonnes in 1960-61 and 7.5 million tonnes in 1970-71. Thereafter, variable trends in maize production had been observed till 1995-96. However, production of maize increased rapidly from 9.53 million tonnes in 1995-96 to 33.6 million tonnes in 2021-22. Thus production of maize in India more than trippled within a short span of nearly two and a half decades. **Table 22.7** gives production figures for maize with respect to some selected years.

Distribution

Table 22.8 shows that about three-fourth of the maize is produced in states of Karnataka, Madhya Pradesh, Maharashtra, Tamil Nadu, West Bengal, Rajasthan, Bihar and Andhra Pradesh. **Karnataka** is the largest producer contributing about 16.45 per cent of the total maize production of India. Belgaum, Chitradurga, Bijapur, Kolar, Bengaluru and Mysuru are the important maize producing districts in Karnataka. In **Madhya Pradesh**, most of the production comes from Madhya Bharat Pathar, with

TABLE 22.7. Area, Production and Yield of Maize in India

Year	1950-51	1960-61	1970-71	1980-81	1990-91	1995-96	2000-01	2016-17	2017-18	2018-19	2019-20	2020-21	2021-22
Area (Million hectares)	3.16	4.41	5.85	6.01	5.90	5.98	6.61	9.63	9.38	9.03	9.57	9.86	10.0
Production (Million tonnes)	1.73	4.08	7.49	6.96	8.96	9.53	12.04	25.90	28.75	27.72	28.77	31.51	33.6
Yield (kg/hectare)	547	926	1279	1159	1518	1595	1822	2689	3065	3070	3006	3195	3349

Source: (*i*) Agricultural Statistics at a Glance, 2021, pp. 63-64.
(*ii*) Economic Survey 2022-23, Statistical Appendix, pp. 37, 39.

TABLE 22.8. Distribution of Area, Production and Yield of Maize in India (2020-21)

Sl. No.	States	Area		Production		Yield (kg/hectare)
		Million hectares	%age of all India	Million tonnes	%age of all India	
1.	Karnataka	1.68	17.00	5.18	16.45	3,092
2.	Madhya Pradesh	1.46	14.82	3.58	11.37	2,451
3.	Maharashtra	1.15	11.62	3.44	10.91	3,000
4.	Tamil Nadu	0.40	4.04	2.72	8.63	6,820
5.	West Bengal	0.37	3.70	2.45	7.76	6,702
6.	Rajasthan	1.00	10.09	2.27	7.20	2,281
7.	Bihar	0.65	6.58	2.22	7.06	3,427
8.	Andhra Pradesh	0.30	3.07	1.95	6.19	6,438
9.	Uttar Pradesh	0.77	7.85	1.80	5.72	2,331
10.	Telangana	0.26	2.64	1.75	5.55	6,721
	Others	1.83	18.59	4.14	13.15	2,259
	All India	**9.86**	**100.00**	**31.51**	**100.00**	**3,195**

Note: States have been arranged in descending order of production in 2020-21.

Source : Agricultural Statistics at a Glance, 2021, p. 65.

Indore, Ratlam, Dhar, Mandsaur, Dewas, Ujjain and Jhabua as the main producing districts. The geographical conditions are not much favourable for the cultivation of maize and the state gives low yield of 2,451 kg/ha. In **Maharashtra,** districts of Solapur, Dhule and Osmanabad are important producers. **Tamil Nadu** is the fourth largest producer of maize contributing about 8.6 per cent of total production of India. Areas receiving comparatively less rainfall in the western part of the state are the main producers. The plateau region of **Telangana** is well known for maize cultivation. Nizamabad, Medak and Karimnagar are important maize producing districts. More than one-third of the crop is raised irrigated and almost entirely with HYV seeds. At one stage, **Bihar** was the largest producer of maize but this state has lost much of its importance as a major maize producer in the country. Almost all the districts of the north Ganga plain produce maize but the major production comes from Samastipur, Begusarai, Bhagalpur, Purnea, Purbi Champaran and Siwan districts. **Andhra Pradesh** has recently emerged as an important maize producing state. Kurnool, Prakasham, East Godavari, West Godavari, Nelllore, etc. are important maize producing districts. The arid lands of **Rajasthan** are specially suited to maize cultivation where it is grown in Udaipur, Bhilwara, Dungurpur, Chittaurgarh and Banswara districts. This state gives one of the lowest yields among all the major maize producing states of India. At one stage, the Upper Ganga Plain of **Uttar Pradesh** was an important producer of maize in the state. In Uttar Pradesh, maize is grown in as many as 25 districts but Bulandshahar, Jaunpur, Ghaziabad, Bahraich, Farrukhabad and Gonda are the main producing districts. In **Gujarat**, Mahsana, Banaskantha, Rajkot and Kheda districts in the valleys of the Sabarmati and Mahi rivers are the main producers and together contribute over 55 per cent of the state's production. The hilly areas of **Himachal Pradesh** are also well suited to maize cultivation. Kangra, Mandi, Sirmaur and Chamba districts occupy an important position in the production of maize.

Among the other producers are Odisha, Punjab, Chhattisgarh, Jharkhand and Jammu and Kashmir. In Punjab, cultivation of maize has given place to other kharif crops and its production has drastically fallen. Still Jalandhar, Kapurthala, Rupnagar, Ludhiana, Amritsar, Faridkot and Patiala are important maize producing districts.

MILLETS

Millets are short duration (3-4 months) warm weather grasses grown in those inferior areas where main food crops like rice and wheat cannot be successfully grown. They provide food for the poor people and fodder for cattle. *Jowar, bajra, ragi, korra, kodon, kutki, sanwa, haraka, varagu, bauti,* and *rajgira* are some of the important millets grown in India.

JOWAR (SORGHUM)

Jowar plays a significant role in feeding the teeming poor millions in the rural areas of India. **Dr. Voelkar** has spoken very highly of nutritive value of jowar as a fodder.

Conditions of Growth

Jowar is grown both as *kharif* as well as a *rabi* crop. As a *kharif* crop, it grows well in areas having mean monthly temperature of 26°C to 33°C. However, the *rabi* crop can be grown in areas where the mean monthly temperature does not fall below 16°C. It requires more than 30 cm rainfall during the growing period and does not grow where the rainfall exceeds 100 cm. Jowar is *par excellence a rainfed crop of dry farming areas where irrigation is not used*. Both excessive moisture and prolonged droughts are harmful for its proper growth. Though it can be grown in a variety of soils including loamy and sandy soils, clayey deep regur and alluvium are the best suited soils for jowar. Most of the crop is grown in plain areas but it can also be raised on gentle slopes upto 1,200 metres height.

Production and Distribution

Jowar has suffered severely at the hands of other favoured crops. The area under jowar increased slightly from 15.57 million hectares in 1950-51 to 18.41 million hectare in 1960-61. Thereafter, it has been fluctuating but the general trend has been towards its reduction. Trends of production area and yields of jowar are shown in **Table 22.9**. India produced 4.8 million tonnes of jowar from 4.2 million hectares of land with an average yield of 1128 kg/hectare in 2020-21. Yield of 1128 kg/hectare in 2020-21 was a record till now.

It is clear from **Table 22.10** that **Maharashtra** far excels all other states and produces about 37 per cent of the total jowar production of India. As many as 22 districts of Mahrashtra produce jowar but Osmanabad, Nanded, Yavatmal, Buldhana, Parbhani, Kolhapur, Amravati, and Ahmednagar are important producing districts. In the Maharashtra plateau region, jowar is the staple food of the people and two crops in a year are raised here. First is sown just before the onset of the monsoon and the second is sown after the retreat of the monsoon. In some districts to the south of Pune, as much as 80 per cent of the cultivated area is devoted to jowar. The stalks are usually more than 2 metres long and are used as fodder for the cattle. **Karnataka** with about 18 per cent of India's jowar production is the second largest producer. Jowar is widely grown in the north-eastern parts of the Karnataka plateau. About 80 per cent of Karnataka's production comes from Dharwar, Bijapur, Raichur, Gulbarga, Belgaum, Chitradurga and Bidar districts. **Rajasthan** is the third largest producer of jowar in India contributing over 12 per cent of the total production of the country. Dry climate and sandy soils of the state provide favourable conditions for cultivation of jowar. Kota, Sawai Madhopur, Jaipur, Tonk, etc. are important jowar producing districts. In **Tamil Nadu** jowar is the second most important food crop after rice. Most of the production comes from Coimbatore, Tiruchirapalli, Madurai and Dharmapuri districts. **Andhra Pradesh** produces about 7.7%

TABLE 22.9. Production, Area and Yield of Jowar in India

Year	1950-51	1960-61	1970-71	1980-81	1990-91	2000-01	2017-18	2018-19	2019-20	2020-21	2021-22
Area (Million hectares)	15.57	18.41	17.37	15.81	14.36	9.86	5.02	4.09	4.82	4.24	30.5
Production (Million tonnes)	5.50	9.81	8.11	10.43	11.68	7.53	4.80	3.48	4.77	4.78	4.2
Yield (kg/hectare)	353	533	466	660	814	764	956	849	989	1,128	1110

Source : (*i*) Agricultural Statistics at a Glance, 2021, pp. 57-58.
(*ii*) Economic Survey, 2022-23, Statistical Appendix, pp. 37-39.

TABLE 22.10. Distribution of Jowar in India (2020-21)

Sl. No.	States	Area		Production		Yield (kg/hectare)
		Million hectares	%age of all India	Million tonnes	%age of all India	
1.	Maharashtra	1.94	45.63	1.76	36.85	911
2.	Karnataka	0.74	17.52	0.88	18.43	1,187
3.	Rajasthan	0.56	13.20	0.59	12.35	1,055
4.	Tamil Nadu	0.41	9.69	0.45	9.41	1,096
5.	Andhra Pradesh	0.12	2.83	0.37	7.70	3,070
6.	Uttar Pradesh	0.17	4.10	0.27	5.74	1,578
7.	Telangana	0.10	2.43	0.19	4.00	1,855
8.	Madhya Pradesh	0.11	2.55	0.18	3.69	1,636
	Others	0.09	2.06	0.09	1.82	998
	All India	**4.24**	**100.00**	**4.78**	**100.00**	**1,128**

Note: States have been arranged in descending order of production in 2020-21.

Source : Agricultural Statistics at a Glance, 2021, p. 59.

total jowar of India. Kurnool, Cuddapah, Anantpur, Guntur, Prakasam, etc. are the main jowar producing districts. In **Uttar Pradesh,** jowar is grown as a fodder crop in south-western part of the state. **Telangana** produces about 4.0 per cent jowar of India. Mahbubnagar, Karimnagar, Warangal and Adilabad are important jowar producing districts. **Madhya Pradesh** has lost much of its ground as a producer of jowar. However, Ujjain, Dewas, Sajapur, West and East Nimar and Mandsaur contribute a lot for jowar production in the state. In Gujarat it is grown as a fodder crop in the districts of Surat, Bharuch, Mehsana and Vadodara.

BAJRA (BULL RUSH MILLET)

Bajra is the second most important millet which is used as food in drier parts of the country. It is also widely used as fodder as its stalks are fed to cattle. In certain areas the stalk is used for thatching purposes.

Conditions of Growth

Bajra is a crop of dry and warm climate and is grown in areas of 40-50 cm of annual rainfall. It seldom grows in those areas where the annual rainfall exceeds 100 cm. The ideal temperature for its growth is 25°-30°C. Bright sunshine after light showers is very useful in early stages of its growth. Bajra can be grown on poor light sandy soils, black and red soils and on upland gravely soils. It is a kharif crop which is sown between May and September and harvested between October and Feb./March. It is sown either as a pure or mixed crop with cotton, jowar and ragi. It is a rainfed crop and is seldom irrigated.

Production and Distribution

There had been wide fluctuations in the production of bajra from a minimum of 2.6 million tonnes in 1950-51 to a maximum of 10.37 million tonnes in 2010-11. After that there was decline in the production till 2018-19. But the situation improved thereafter and a production of 9.6 million tonnes was recorded in 2021-22. The yields have also varied widely from a minimum of 288 kg/hectare in 1950-51 to a maximum of 1,436 kg/hectare in 2021-22. Large scale variations in area under bajra cultivation have also been observed (**Table 22.11**). The maximum area of 12.49 million hectares was recorded in 1970-71 which fell to 6.7 million hectares in 2021-22.

About 80 per cent of India's bajra comes from four states of Rajasthan, Uttar Pradesh, Haryana and Gujarat. These four states also account for over 80 per cent of area under bajra. **Rajasthan** is the largest producer which accounts for 41 per cent of the production and 57.1% of the area under bajra. The leading bajra producing districts are Barmer, Nagaur, Jalore, Jodhpur, Pali, Sikar, Churu, Ganganagar, Hanumangarh, Bikaner, Alwar, Bharatpur, Jaipur,

TABLE 22.11. Area, Production and Yield of Bajra in India

Year	1950-51	1960-61	1970-71	1980-81	1990-91	2000-01	2017-18	2018-19	2019-20	2020-21	2021-22
Area (Million hectares)	9.02	11.47	12.49	11.66	10.48	9.83	7.48	7.10	7.54	7.57	6.7
Production (Million tonnes)	2.60	3.28	5.33	5.34	6.89	6.76	9.21	8.66	10.36	10.86	9.6
Yield (kg/hectare)	288	286	426	458	658	688	1,231	1,219	1,374	1,436	1,436

Source : (*i*) Agricultural Statistics at a Glance, 2021, pp. 60-61.
(*ii*) Economic Survey 2022-23, Statistical Appendix, pp. 37-39.

TABLE 22.12. Distribution of Bajra in India (2020-21)

Sl. No.	States	Area		Production		Yield
		Million hectares	%age of all India	Million tonnes	%age of all India	(kg/hectare)
1.	Rajasthan	4.32	57.10	4.53	41.71	1,049
2.	Uttar Pradesh	0.91	11.99	2.01	18.54	2,221
3.	Haryana	0.57	7.52	1.35	12.43	2,372
4.	Gujarat	0.46	6.04	1.00	9.22	2,192
5.	Madhya Pradesh	0.33	4.32	0.74	6.79	2,256
6.	Maharashtra	0.64	8.42	0.64	5.88	1,003
7.	Karnataka	0.22	2.92	0.29	2.65	1,303
	Others	0.13	1.69	0.30	2.77	2,360
	All India	**7.57**	**100.00**	**10.86**	**100.00**	**1,436**

Note: States have been arranged in descending order of production in 2020-21.
Source : Agricultural Statistics at a Glance, 2021, p. 62.

Jaisalmer, Jhunjhunu and Sawai Madhopur. **Uttar Pradesh** produced 2 million tonnes (18.5 per cent of India) in 2020-21. The main production comes from Mathura, Agra, Aligarh, Badaun, Moradabad, Etah, Etawah, Bulandshahar, Shahjahanpur, Mainpuri, Pratapgarh, Ghazipur, Farrukhabad, Pryagraj and Kanpur. **Haryana** produced 1.35 million tonnes (12.43% of total for India) in 2020-21. Most of the production comes from the drier south-western part of the state contiguous to Rajasthan. Mahendragarh, Rewari, Gurgaon, Rohtak and Hissar are important bajra producing districts. **Gujarat** is the fourth important producer, where 1.0 million tonnes (9.2 per cent of India's total) of bajra was produced in 2020-21. Most of the crop is grown in sandy tracts with maximum concentration in semi-arid parts of the state. Kachchh, Mahesana, Kheda, Bhavanagar, Amreli, Banaskantha, Surendranagar, Sabarkantha, Jamnagar, Rajkot, Junagadh districts are important producers. **Madhya Pradesh** is the fifth largest producing state of bajra in India. Morena, Gwalior, Shivpuri, Jabalpur, Shahdol, etc. are important producing districts. **Maharashtra** has shifted from first position in 2002-03 to sixth position in 2020-21 partly due to decrease in production in this state and partly due to increase in production in other states. In Maharashtra, bajra is mainly grown in the central plateau having poor soils and dry climate. Nashik, Dhule, Satara, Pune, Sangli, Aurangabad, Solapur, Jalgaon and Ahmednagar are the main producing districts. Among the other producers are Tamil Nadu, Karnataka, Andhra Pradesh, Telangana and Jammu and Kashmir.

RAGI

Ragi is another important millet mainly grown in drier parts of south India with some parts of north India also contributing a small quantity. It requires 20°-30°C temperature and 50-100 cm rainfall. It is raised on red, light black and sandy loams as well as on well drained alluvial loams. It is a rainfed kharif crop which is sown between May and August and harvested between September and January.

There had been varying trends in area, production and yield of ragi in India. These variations are the result of varying weather conditions and choices of farmers regarding selection of crops.

The total production of Ragi was 23.6 lakh tonnes in 2019. **Karnataka** is the largest producer accounting for 17.24 lakh tonnes (73.23 per cent) from 9.38 lakh hectares (61 per cent). Bengaluru, Kolar, Hassan, Tumkur, Chitradurga and Mysuru districts account for over 80 per cent production of Karnataka. **Uttarakhand** is the second largest producer but lags far behind Karnataka with only 1.74 lakh tonnes (7.4 per cent). **Tamil Nadu** is the third largest producer with 1.33 lakh tonnes of ragi production. Dharmpuri, Salem, Coimbatore, Ramnathapuram, Madurai, and Nilgiri districts are the main producers. **Maharashtra** (1.32 lakh tonnes), **Andhra Pradesh** (0.79 lakh tonnes) are some other important producers.

BARLEY

Barley, like millets, is considered as an inferior grain. Besides food, it is used for manufacturing beer and whisky.

Conditions of Growth

It does not tolerate high heat and high humidity and thrives well in areas where the temperature is 10°-15°C for about three months in a year and the rainfall varies from 75 cm to 100 cm. Light clay and alluvial soils are best suited for its cultivation. It is grown as a rabi crop in the Great Plains and valleys of the western Himalayas. It can be grown upto an altitude of 1,300 metres as in Uttarakhand.

Production and Distribution

Over the years, fluctuating trends have been observed in the production of barley but the production has definitely declined by over 56 per cent in four and a half decades from 28.19 lakh tonnes in 1960-61 to 12.5 lakh tonnes in 2015-16. More conspicuous is the decline in area under barley which had come down from 32.05 lakh hectares in 1960-61 to a meagre 6.40 lakh hectares in 2015-16. However, the yield has recorded two and a half times increase during the same period. It was just 8.8 quintals/hectare in 1960-61 and rose to 19.7 quintals/hectare in 2015-16. The probable cause of decline in area and production is more importance being given to main food crops, particularly to wheat. This is not good for balanced growth of agriculture and must be checked.

Rajasthan is the largest producer accounting for over 40 per cent production of the country. The major part of the production comes from Pali, Tonk, Sawai Madhopur, Alwar, Bharatpur, Sikar, Jaipur, Udaipur, Ajmer, Bhilwara, Ganganagar and Hanumangarh districts. **Uttar Pradesh** is the second largest producer contributing 33.8 per cent of the total India. Aligarh, Agra, Bulandshahar, Mathura, Fatehpur, Pryagraj, Varanasi, Mirzapur, Jaunpur, Ghazipur, Gorakhpur, Ballia, Partapgarh, Azamgarh etc. are the main producing districts. Thus nearly three fourth barley comes from just two states. Some barley is also produced in **Madhya Pradesh** (100 thousand tonnes). **Haryana** (76 thousand tonnes), **Punjab** (63 thousand tonnes), **Chhattisgarh** (19 thousand tonnes), Uttarakhand (18 thousand tonnes), Himachal Pradesh (18 thousand tonnes), Bihar (18 thousand tonnes) and West Bengal.

PULSES

Pulses include a number of crops which are mostly leguminous and provide much needed vegetable proteins to a largely vegetarian population of India. They serve as an excellent forage and grain concentrates in the feed of cattle. Pulses have the capacity to fix atmospheric nitrogen in the soil and are normally rotated with other crops to maintain or restore soil fertility. Though gram and *tur* (arhar) are the more important pulses, several other pulses such as *urd* (black gram), *mung* (green gram), *masur* (lentil), *kulthi* (horse gram), *matar* (peas), *khesri* and *moth* are also grown.

GRAM

Gram is the most important of all the pulses and accounts for 37 per cent of the production and 28.28 per cent of the total area of pulses in India. It can be grown in a wide range of climatic conditions but it prefers mild cool and comparatively dry climate with

20°-25°C temperature and 40-50 cm rainfall. It grows well on loamy soils. It is a rabi crop, which is sown between September and November and is harvested between February and April. It is cultivated as pure or mixed with wheat, barley, linseed or mustard. Mixed cropping helps to check the gram blight to some extent.

Production

Like other food crops of second order, gram has also suffered a lot at the hands of wheat. The area under gram cultivation fell from a high of 10.33 million hectares in 1959-60 to an extremely low of 5.19 million hectares in 2000-01. Similarly production of gram fell from 7.62 million tonnes in 1958-59 to a very low of 3.63 million tonnes in 1987-88. However, some recovery has been made both with respect to area and production of gram during the last one decade although varying trends have been observed in area as well as production of gram. But the yield per hectare has recorded a significant increase from 482 kg in 1950-51 to 744 kg in 2000-01 and 1,260 kg in 2021-22 (**Table 22.13**).

Distribution

Although gram is cultivated in several parts of India, most of the gram comes from Madhya Pradesh, Maharashtra and Rajasthan. These three states account for about two-thirds of the total gram production of the country (**Table 22.14**).

Madhya Pradesh is the largest producer with 3.13 million tonnes (26%) to its credit. Vidisha, Bhind, Morena, Chhatarpur, Jabalpur, Narsingpur, Dhar, Hoshangabad, Raisen, Gwalior, Ratlam, Ujjain, Guna, Sagar, Shivpuri and Datia are significant

TABLE 22.13. Area, Production and Yield of Gram in India

Year	1950-51	1960-61	1970-71	1980-81	1987-88	2000-01	2017-18	2018-19	2019-20	2020-21	2021-22
Area (Million hectares)	7.57	9.28	7.84	6.58	5.77	5.19	10.56	9.55	9.70	9.85	10.9
Production (Million tonnes)	3.65	6.25	5.20	4.33	3.63	3.86	11.38	9.94	11.08	11.99	13.8
Yield (kg/hectare)	482	674	663	657	629	744	1078	1041	1142	1217	1260

Source : (*i*) Agricultural Statistics at a glance 2021, pp. 69-70.
(*ii*) Economic Survey, 2022-23, Statistical Appendix, pp. 37-39.

TABLE 22.14. Distribution of Gram in India (2020-21)

Sl. No.	States	Area		Production		Yield (kg/hectare)
		Million hectares	%age of all India	Million tonnes	%age of all India	
1.	Madhya Pradesh	2.10	21.35	3.13	26.09	1,488
2.	Maharashtra	2.15	21.78	2.37	19.77	1,105
3.	Rajasthan	2.11	21.45	2.32	19.37	1,099
4.	Gujarat	0.82	8.29	1.28	10.67	1,568
5.	Uttar Pradesh	0.61	6.20	0.84	7.01	1,376
6.	Andhra Pradesh	0.47	4.76	0.53	4.44	1,135
7.	Karnataka	0.72	7.26	0.50	4.17	700
8.	Jharkhand	0.27	2.70	0.33	2.79	1,257
9.	Chhattisgarh	0.30	3.06	0.27	2.23	887
	Others	0.31	3.16	0.41	3.46	1,332
	All India	**9.85**	**100.00**	**11.99**	**100.00**	**1,217**

Note: States have been arranged in descending order of production in 2020-21.
Source : Agricultural Statistics at a Glance, 2021, p. 71.

producers. **Maharashtra** is the second largest producer and produces about 2.4 million tonnes amounting to about 19.77 per cent of India's production. The bulk of gram comes from Osmanabad, Aurangabad, Solapur, Parbhani, Ahmednagar, Beed, Nanded and Nashik districts. **Rajasthan** produces 2.3 million tonnes and accounts for about 19.4 per cent of India's production. Gananagar, Alwar, Bharatpur, Jaipur, Sawai Madhopur, Udaipur, Banswara, Jaiselmer, Kota, Tonk, Churu, Jhunjhunu and Ajmer are the main producing districts. Gujarat has progressed a lot with respect to production of gram and now occupies fourth position among the major gram producing states of India. In 2020-21 this state produced about 1.3 million tonnes of gram which amounted to about 10.7 per cent of all India production. In **Uttar Pradesh,** the main producing districts are Banda, Hamirpur, Jalaun, Jhansi, Kanpur, Pryagraj, Fetehpur, Sitapur, Barabanki, Agra, Lalitpur and Sultanpur. In **Andhra Pradesh,** most of the production comes from the Ravalseema region. **Karnataka** has stepped down from second position in 2015-16 to seventh position in 2020-21 and now accounts for 4.2 per cent of the total production of India. In 2020-21, Karnataka recorded the lowest yield of 700 kg/a due to adverse weather conditions. Gulbarga, Bidur, Bijapur, Yadgir and Raichur are the main producing districts. In **Jharkhand,** Chotanagpur plateau is the main producer. The Central part of **Chhattisgarh** comprising of Rajnandgaon, Durg, Raipur, etc. are known for gram production. The other important producing states are Bihar and Telangana. Gram in small quantities is also produced in Haryana, Odisha and West Bengal.

TUR or ARHAR (Pigeon Pea or Red Gram)

Tur is the second most important pulse of India next to gram. It is chiefly grown as a kharif crop but in areas of mild winters it is grown as a rabi crop also. It is grown as a dry crop mixed with other kharif crops like jowar, bajra, ragi, maize, cotton, groundnut, etc. and is seldom grown as a single crop. Its conditions of growth are more or less similar to those of other pulses and millets.

Production

Table 22.15 shows that there have been large scale temporal variations with respect to area, production and yields of tur. In 2021-22, tur was grown on 5 million hectares of land and total production was 4.3 million tonnes. Yield was 869 kg/hectare.

Distribution

Maharashtra is the largest producer of tur in India with about 30 per cent production from over 25 per cent area under tur. Most of production comes from Amrawati, Wardha, Akola, Yavatmal, Beed, Nagpur, Usmanabad, and Parbani districts. **Karnataka** with about 29 per cent of the production is the second largest producer. Gulbarga, Bellary and Bidar districts of the North Maidan are the main producers. **Telangana** is the third largest producer of tur in India contributing about 7.8 per cent of the total production of the country. **Uttar Pradesh** produces about 6.8 per cent tur of India. Varanasi, Jhansi, Pryagraj and Lucknow are the main producers. **Jharkhand** has the climatic conditions favourable for

TABLE 22.15. Area, Production and Yield of Tur in India

Year	1950-51	1960-61	1970-71	1980-81	1990-91	2000-01	2017-18	2018-19	2019-20	2020-21	2021-22
Area (Million hectares)	2.18	2.43	2.66	2.84	3.59	3.63	4.44	4.55	4.53	4.80	5.0
Production (Million tonnes)	1.72	2.07	1.88	1.96	2.41	2.25	4.29	3.32	3.89	4.28	4.3
Yield (kg/hectare)	788	849	709	689	673	618	967	729	859	892	869

Source : (*i*) Agricultural Statistics at a Glance, 2021, pp. 72-73.
(*ii*) Economic Survey, Statistical Appendix, pp. 37-39.

TABLE 22.16. Distribution of Tur in India (2020-21)

Sl. No.	States	Area		Production		Yield (kg/hectare)
		Million hectares	%age of all India	Million tonnes	%age of all India	
1.	Maharashtra	1.24	25.74	1.28	29.87	1,035
2.	Karnataka	1.63	33.96	1.24	28.90	759
3.	Telangana	0.43	8.91	0.33	7.77	778
4.	Uttar Pradesh	0.30	6.16	0.29	6.83	988
5.	Jharkhand	0.25	5.21	0.28	6.64	1,138
6.	Madhya Pradesh	0.23	4.71	0.26	5.98	1,134
7.	Gujarat	0.23	4.78	0.25	5.91	1,104
8.	Odisha	0.13	2.70	0.12	2.89	953
	Others	0.38	7.85	0.22	5.20	591
	All India	**4.80**	**100.00**	**4.28**	**100.00**	**892**

Note: States have been arranged in descending order of production in 2020-21.
Source : Agricultural Statistics at a Glance, 2021, p. 74.

TABLE 22.17. Imports of Pulses

Year	2012-13	2013-14	2014-15	2015-16	2016-17	2017-18	2018-19	2019-20	2020-21
Quantity ('000 tonnes)	4013.24	3177.89	4584.85	5797.71	6609.49	5607.53	2,527.88	2,898.08	2,466.16
Value (₹ crore)	13344.43	11036.75	17062.94	25619.06	28523.18	18748.57	8,035.30	10,221.45	11,937.59

Source : Agricultural Statistics at a Glance, 2021, pp. 186-89.

tur cultivation and get the highest yield of 1,138 kg/hectare. In **Madhya Pradesh,** most of the production comes from Hoshangabad, Chhindwara, Betul, East Nimar, Dewas, Sidhi, Narsimhapuram, and Bhind districts. Madhya Pradesh has the second high yield of 1,134 kg/hectare. **Gujarat** has emerged as an important producer of tur in the recent past. In 2020-21, this state produced about 6 per cent of the total tur of India. The other producing states include Chhattisgarh, Odisha, Andhra Pradesh and Tamil Nadu.

Imports. Production of pulses always falls short of demand for pulses in India and the country has to spend huge amounts of money on their imports (**Table 22.17**).

CASH CROPS

Cash crops are those crops which are grown for sale in the market either in raw form or in semi-processed form. Thus cash crops have special characteristic of earning cash for the farmer. Prominent among the cash crops are cotton, jute, sugarcane, tobacco and oilseeds. These crops play a significant role in the economy of the country. This is evident from the fact that they occupy only 15 per cent of the total cropped area of the country but account for over 40 per cent of the agricultural production by value. Besides they provide raw materials to a large number of industries.

COTTON

Cotton is the most important fibre crop not only of India but of the entire world. It provides the basic raw material (*cotton fibre*) to cotton textile industry. Its seed (*binola*) is used in vanaspati industry and can also be used as part of fodder for milch cattle to get better milk.

Conditions of Growth

Following geographical conditions are congenial for proper growth of cotton.

(i) **Temperature.** Cotton is the crop of tropical and sub-tropical areas and requires uniformly high temperature varying between 21°C and 30°C. The growth of cotton is retarded when the temperature falls below 20°C. Frost is *enemy number one of the cotton plant* and it is grown in areas having at least 210 frost free days in a year.

(ii) **Rainfall.** The modest requirement of water can be met by an average annual rainfall of 50-100 cm. However, it is successfully grown in areas of lesser rainfall with the help of irrigation. About one-third of the total area under cotton cultivation is irrigated. About 80 per cent of the total irrigated area under cotton is in Punjab, Haryana, Gujarat and Rajasthan. Moist weather and heavy rainfall at the time of boll-opening and picking are detrimental to cotton as the plant becomes vulnerable to pests and diseases. High amount of rainfall in beginning and sunny and dry weather at ripening time are very useful for a good crop.

(iii) **Soil.** Cotton cultivation is closely related to deep black soils (*regur*) of the Deccan Plateau and the Malwa Plateau and those of Gujarat. It also grows well in alluvial soils of the Satluj-Ganga Plain and red and laterite soils of the peninsular region. Cotton quickly exhausts the fertility of soil. Therefore, regular application of manures and fertilizers to the soil is very necessary.

(iv) **Relief.** Flat and level plain areas are best suited for cotton cultivation.

(v) **Labour.** Picking is a crucial period from the labour point of view. Since picking of cotton is not yet mechanized in India, a lot of cheap and efficient labour is required at this time. Normally the picking season is spread over a period of about three months.

Types of Cotton

Three broad types of cotton are generally recognised on the basis of the length, strength and structure of its fibre.

1. Long staple cotton. It has the longest fibre whose length varies from 24 to 27 mm. The fibre is long, fine and lustrous. It is used for making fine and superior quality cloth. Obviously, it fetches the best price. There has been rapid progress in the production of long staple cotton since Independence. About half of the total cotton produced in India is long staple. It is largely grown in Punjab, Haryana, Maharashtra, Tamil Nadu, Madhya Pradesh, Gujarat and Andhra Pradesh.

2. Medium staple cotton. The length of its fibre is between 20 mm and 24 mm. About 44 per cent of the total cotton production in India is of medium staple. Rajasthan, Punjab, Tamil Nadu, Madhya Pradesh, Uttar Pradesh, Karnataka and Maharashtra are its main producers.

3. Short staple cotton. This is inferior cotton with fibre less than 20 mm long. It is used for manufacturing inferior cloth and fetches less price. About 6 per cent of the total production is of short staple cotton. U.P., Andhra Pradesh, Rajasthan, Haryana and Punjab are its main producers.

Production

India has the largest area under cotton cultivation in the world though she is the world's second largest producer of cotton after China. Currently it is grown over 8.6 per cent of the net sown area. **Table 22.18** shows the trends in the production of cotton in India.

TABLE 22.18. Area, Production and Yield of Cotton in India

Year	1950-51	1960-61	1970-71	1980-81	1990-91	2000-01	2017-18	2018-19	2019-20	2020-21	2021-22
Area (Million hectares)	5.88	7.61	7.61	7.82	7.44	8.53	12.59	12.61	13.48	13.01	11.9
Production (Million bales*)	3.04	5.60	4.56	7.01	9.84	9.52	32.81	28.04	36.07	35.38	31.2
Yield (kg/hectare)	88	125	122	152	225	190	443	378	455	462	445

*1 bale = 170 kg.

Source : (*i*) Agricultural Statistics at a glance, 2021, pp. 1993-94.
(*ii*) Economic Survey, Statistical Appendix, pp. 37-39.

Bt Cotton

Bt stands for *Bacillus thuringiensis,* the bacterium whose toxin is produced by Bt cotton after genetic alteration. The first Bt protected cotton crops were tested in the fields of U.S.A. and it entered in U.S. cultivation in 1995. China welcomed Bt cotton in 1997. Later 13 countries, including India, followed the cultivation of Bt cotton. In India, it was released for commercial cultivation in 2002. The initial field trials of Bt cotton showed extremely encouraging results as the yield of Bt cotton was *80 per cent more than that of non-Bt variety although it required three times less spraying and 70 per cent less insecticide*. The area under cotton crop increased sharply from 8.53 million hectares in 2000-01 to 13.0 million hectares in 2020-21 and the production also registered more than three times increase from 9.52 million bales (1 bale = 170 kg) in 2000-01 to over 35.4 million bales in 2020-21. A number of studies conducted on Bt cotton before and after commercialization have shown the following benefits :

(*i*) Higher cotton yield owing to effective control of bollworms.

(*ii*) Drastic reduction in application of chemical insecticides for bollworm control.

(*iii*) Higher profit to farmers.

(*iv*) Conservation of biological control agents and other beneficial organisms.

The area under Bt cotton increased from a minimal of 1 per cent in 2003-04 to a staggering 95 per cent in 2015-16 (**Table 22.19**). In spite of large area under Bt cotton, the yields have not registered any increase worth the name. This shows that the new Bt cotton is loosing its sheen. According to K.R. Kranthi, Director, Central Cotton Research Institute (CICR), "Although the GM variety, officially allowed in 2002, sharply raised yields initially, productivity is now projected to decline and signs of diminishing returns are visible now." In fact Bt toxin is a narrow spectrum bio-insecticide as it controls only bollworm. Cotton attracts over 130 different species of pests. The biodiversity of cotton pests varies from insects like ball weevil, grey weevil, cotton jassid, cotton aphids, cotton thrip to Nematodes like root-rust.

The above mentioned situation needs to be improved for which better technology (seeds, fertilizers, pesticides, insecticides and irrigation) is required. Maharashtra has the largest area of 20 lakh hectares under Bt cotton, followed by Andhra Pradesh (7.6 lakh hectares), Gujarat (3.3 lakh hectares) and Madhya Pradesh (3.1 lakh hectares). All these states are in South India : In North India Punjab and Haryana have 2.8 lakh hectares and 0.42 lakh hectares respectively under Bt cotton. The biggest concern about large scale adoption of Bt cotton in India is that the bollworm (a serious pest disease of cotton) may develop resistance to this transgenic crop as has happened in some parts of China. It is worth noting that almost 65 per cent of the area under cotton is rainfed with erratic and poorly distributed rains during the cropping season. It is subject to severe attack of pests and diseases.

Distribution

India has the sole distinction of growing all the three cultivated species of cotton and their intra- and inter-specific hybrids. In India, cotton is grown in three distinct agro-ecological zones, *viz*., Northern (Punjab, Haryana and Rajasthan), Central (Gujarat, Maharashtra and Madhya Pradesh) and Southern zone (Andhra Pradesh, Tamil Nadu and Karnataka).

TABLE 22.19. Area under Bt Cotton

Year	2002-03	2003-04	2004-05	2005-06	2006-07	2007-08	2008-09	2009-10	2010-11	2011-12	2015-16
Percentage of area under Bt	0	1	6	18	41	62	84	85	85	90	95

Source : Cotton Corporation of India.

Table 22.20 shows the statewise distribution of cotton in India.

1. Maharashtra. Maharashtra is the largest producer and produces over 27 per cent of the total cotton production of India. Maharashtra is a traditional producer of cotton. The lava soil of deccan plateau is world renowned for cotton production and is popularly known as the ***black cotton soil***. Over 80 per cent of the production comes from Khandesh, Vidarbha and Marathwada regions comprising the districts of Yavatmal, Nanded, Amravati, Parbhani, Wardha, Jalgaon, Akola, Buldhana, Nagpur, Dhule, etc. However, Maharashtra suffers from low productivity as this state gives the lowest yield of 380 bales/hectare among the cotton producing states of India.

2. Gujarat. Accounting for over 20% of the total production and nearly 17 per cent of the cotton area of the country, Gujarat is the second cotton producing state of India. The average yield is 544 kg/hectare. With 'black cotton soil' 1.5 metre deep in some parts and with 80-100 cm annual rainfall Gujarat provides favourable conditions for cotton cultivation. Two-thirds of the production comes from the Gujarat plains including Bharuch, Surendernagar, Vadodra and Ahmedabad districts. Mahesana, Kheda, Sabarkantha, Surat, Amreli and Panchmahals are other major producers.

3. Telangana. Soil and climatic conditions are very much congenial to cotton cultivation in Telangana plateau. At present Telangana is the third largest cotton producing state in India and accounts for about 17 per cent of the total cotton production of the country. Major part of the production comes from Nalgonda, Khamam and Mehbubnagar.

4. Rajasthan. Rajasthan accounts for about 9 per cent of the production and 6.2 per cent of the area of the country. Ganganagar is the most important cotton producing district of Rajasthan and accounts for over 80 per cent of the state's production. This district is contiguous to the cotton producing areas of Punjab and Haryana and enjoys the same advantages. The remaining cotton of Rajasthan comes from Bhilwara, Ajmer, Chittaurgarh, Jhalawar, Pali and Hanumangarh districts.

TABLE 22.20. State-wise Distribution of Cotton in India (2020-21)

Sl. No.	States	Area		Production		Yield (kg/hectare)
		Million hectares	%age of all India	Million bales*	%age of all India	
1.	Maharashtra	4.29	32.95	9.59	27.10	380
2.	Gujarat	2.27	17.46	7.27	20.55	544
3.	Telangana	2.36	18.14	5.99	16.94	432
4.	Rajasthan	0.81	6.21	3.21	9.06	675
5.	Karnataka	0.82	6.30	2.32	6.56	481
6.	Haryana	0.74	5.69	1.82	5.15	419
7.	Madhya Pradesh	0.57	4.40	1.78	5.04	530
8.	Andhra Pradesh	0.61	4.66	1.60	4.53	450
	Others	0.55	4.20	1.79	5.07	559
	All India	**13.01**	**100.00**	**35.38**	**100.00**	**462**

*1 bale = 170 kg.

Note: States have been arranged in descending order of production in 2020-21.

Source : Agricultural Statistics at a Glance, 2021, p. 95.

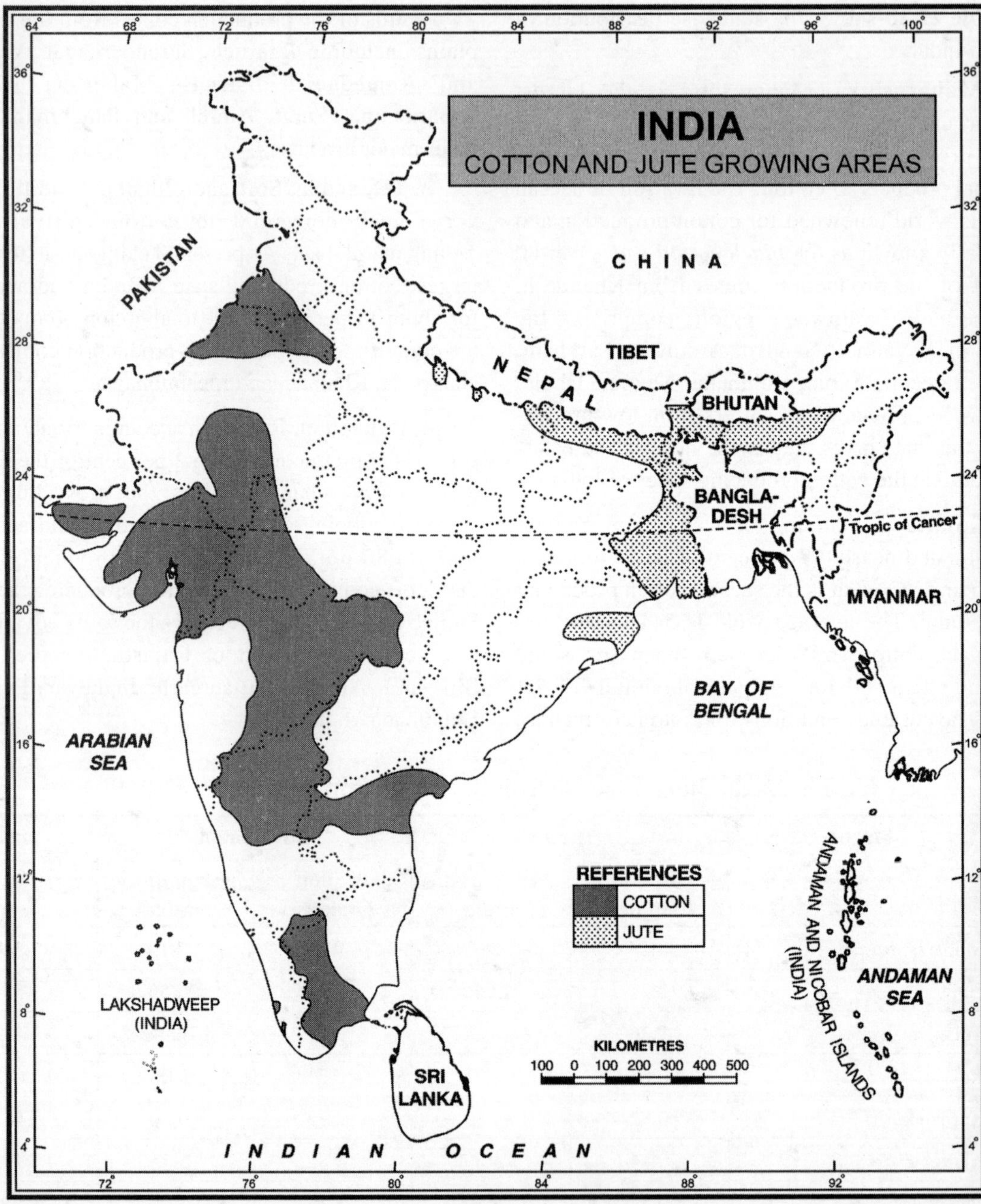

FIG. 22.3. India : Fibre Crops (Cotton & Jute)

5. Karnataka. This state produces 6.6 per cent cotton of India from 6.3 per cent of India's area under cotton cultivation. The North Karnataka plateau is the main area of cotton cultivation. Dharwad, Raichur, Bellary and Gulbarga are the main producing districts.

6. Haryana. Accounting for 5.15 per cent production and 5.7 per cent of hectarage, Haryana is a very important cotton producing state in north-west India. In the year 2020-21, Haryana produced 1.82 million bales. About 80 per cent of the production comes from Hisar, Sirsa and Fatehabad districts

TABLE 22.21. India : Imports and Exports of Cotton

Year	1960-61	1970-71	1980-81	1990-91	2000-01	2019-20	2020-21	2021-22
Imports								
Quantity ('000 tonnes)	237.1	139.1	N.A.	0.2	212.3	744	231	221
Value ₹ crore	82	99	N.A.	1	1,185	9,371	2,861	2,861
Exports								
Quantity ('000 tonnes)	32.6	32.1	131.6	374.4	30.2	657.8	1,214	1,259
Value ₹ crore	12	14	165	846	224	7,540	12,968	21,007

Source : Economic Survey 2021-22 and 2022-23 Statistical Appendix, pp. 112-118.

DESI BT COTTON

The Genetic Engineering Approval Committee (GEAC) has approved a new Bt cotton variety for commercial planting, making it India's first genetically modified (GM) to be developed by the public sector. The new cotton variety called *Bikaneri Narma* was developed by the Central Institute of Cotton Research (CICR), Nagpur and the University of Agricultural Sciences in Dharwar (Karnataka). The variety contains the gene for the Bt cry 1 AC protein and has been approved by GEAC for release in the North, Central and South Cotton Growing Zones of the country. Farmers have been using it since 2009. It is too early to draw any conclusion about the impact of desi Bt cotton.

which are contiguous to the major cotton producing districts of Punjab. Like Punjab, most of the production is in the form of the American long staple varieties. Bhiwani, Jind, Rohtak and Ambala are other producing districts.

7. Madhya Pradesh. More than 80 per cent of the production comes from Malwa where there are vast tracts of lava soil. East Nimar, West Nimar, Ujjain, Shajapur, Dewas, Dhar, Ratlam, Rajgarh, Indore and Bhopal are the main producers.

8. Andhra Pradesh. Andhra Pradesh accounts for about 4.53 per cent of production and over 4.6 per cent of hectarage of India. Two-thirds of the production of Andhra Pradesh comes from two districts, namely Guntur and Prakasam. Kurnool and Anantapur contributes the rest.

9. Punjab. Punjab has slipped from first position in 1990-91 to 9th position in 2020-21 as a producer of cotton in India. Punjab has the distinction of producing some of the best qualities of cotton in India. All this has been made possible due to fertile alluvial soils, a close network of irrigation facilities, heavy dose of fertilizers and pesticides and above all the enterprising spirit of the farmers. Most of the cotton production comes from the Malwa region of the state. This region contributes nearly 95 per cent of Punjab's cotton. Cotton is known as "white gold" in this region. Bathinda, Faridkot, Firozepur and Sangrur are the major producing districts and account for over three fourths of Punjab's total production of cotton. Ludhiana, Muktsar, Moga, Mansa and Fatehgarh Sahib are other cotton producing districts.

10. Tamil Nadu. Tamil Nadu contributes about 1.22 per cent of the total production with about 1.2 per cent of the total area of the country. Coimbatore, Salem, Madurai, Tiruchirapalli, Ramnathapuram, Vellore, Chengalpattu and Tirunelveli are the main producing districts.

Uttar Pradesh, Kerala, Odisha, Meghalaya and Mizoram also produce cotton in small quantities.

Trade

India is an exporter as well as importer of cotton. India exports inferior quality cotton mainly to U.K., where it is mixed with superior quality cotton. India has been a big importer of superior quality long staple cotton mainly from the USA, Russia, UAR, Sudan and Kenya (**Table 22.21**).

Considering the major provisions of World Trade Organisation (WTO) *vis-a-vis* India's position, the points that emerge to be of immediate concern in enhancing the productivity and quality of Indian cotton and making it competitive globally are : (*i*) bringing down the cost of cultivation and enhancing its productivity and quality, (*ii*) rendering our cotton globally attractive, (*iii*) keeping Indian cotton free of trash content.

JUTE

Next to cotton, jute is the second important fibre crop of India. Jute is in great demand because of the cheapness, softness, strength, length, lustre and uniformity of its fibre. It is used for manufacturing a large variety of articles such as gunny bags, hessian, ropes, strings, carpets, rugs and clothes, tarpaulins, upholstery and decoration pieces.

Conditions of Growth

Jute is the crop of hot and humid climate. It requires high temperature varying from 24°C to 35°C and heavy rainfall of 120 to 150 cm with 80 to 90 per cent relative humidity during the period of its growth. Small amount of pre-monsoon rainfall varying from 25 cm to 55 cm is very useful because it helps in the proper growth of the plant till the arrival of the proper monsoon. Incessant and untimely rainfall as well as prolonged droughts are detrimental to this crop. Rainfall between 2.5 to 7.5 cm in a month, during the sowing period, is considered to be sufficient. Occasional showers varying from 2 to 3 cm at intervals of a week's time during the growing period are very useful. Large quantity of water is required not only for growing the jute crop but also for processing the fibre after the crop is harvested. Light sandy or clayey loams are considered to be best suited soils for jute. Since jute rapidly exhausts the fertility of soil, it is necessary that the soil is replenished annually by the silt-laden flood water of the rivers. Large supply of cheap labour is also necessary for growing and processing the jute fibre.

Method of Cultivating and Processing of Jute

Jute is generally sown in February on lowlands and in March-May on uplands. The crop takes 8-10 months to mature but different varieties take different time to mature. The harvesting period generally starts in July and continues till October. The plants are cut to the ground and tied into bundles. Sheafs of jute stocks are then immersed in flood water or ponds or stagnant water for about 2 to 3 weeks for retting. High temperature of water quickens the process of retting. After retting is complete, the bark is peeled from the plant and fibre is removed. After this, stripping, rinsing, washing and cleaning is done and the fibre is dried in the sun and pressed into bales. All this process is to be done by human hand for which availability of plenty of labour at cheap rates is very essential. Luckily, this labour is readily available because jute is cultivated in areas of high population density.

Production

India suffered a great setback in the production of jute as a result of partition of the country in 1947 because about 75 per cent of the jute producing areas went to Bangla Desh (East Pakistan at that time). Fortunately, most of the jute mills remained in India. Strenuous efforts were made to increase production and area under jute, immediately after partition to feed our starving jute mills in the wake of short supply of raw jute. Currently India is the largest producer of jute in the world.

Table 22.22 shows that there had been rapid increase in area, production and yield of jute during three decades between 1950-51 and 1980-81 after which varying trends in almost all the three aspects (area, production and yield) have been observed. This is perhaps due to changes in weather conditions and pressure on land for other crops like rice. In fact there is tough competition for land particularly between jute and rice in the delta region.

The significant role played by jute in the country's economy can be assessed from the fact that more than 4 million farm families are involved in jute farming and majority of them belong to small and marginal categories. Cultivation of jute generates employment (seasonal) of more than 10 million mandays per season. Besides 0.5 million people are involved in raw jute and finished goods trading and ancillary activities. Currently India accounts for about 56 per cent of world jute production as compared to only 25 per cent produced by Bangladesh. Research and development work carried by the agricultural scientists during the last few years has not only resulted in increasing yield of the fibre but also in improvement of the fibre quality and shortening of cultivation period. Investigations reveal that the crop

TABLE 22.22. All India, Area, Production and Yield of Jute and Mesta

Year	1950-51	1960-61	1970-71	1980-81	1990-91	2000-01	2017-18	2018-19	2019-20	2020-21	2021-22
Area (Million hectares)	0.57	0.90	1.08	1.30	0.91	1.02	0.74	0.70	0.67	0.66	0.7
Production (Million bales*)	3.31	5.26	6.19	8.16	8.29	9.23	10.03	9.82	9.88	9.56	10.3
Yield (kg/hectare)	1,043	1,049	1,032	1,130	1,646	1,634	2,435	2,508	2,641	2,595	2,709

*1 bale = 180 kg.

Scurce : (*i*) Agricultural Statistics at a Glance, 2021, pp. 96-97.
(*ii*) Economic Survey 2022-23, Statistical Appendix, pp. 37-39.

pattern jute-paddy-potato is more profitable for the farmers than say paddy-potato-sesame.

Distribution

Table 22.23 shows that over 98 per cent of the total jute of India is produced in just three states of West Bengal, Bihar and Assam.

1. West Bengal. West Bengal is the undisputed king of jute production in India accounting for about three-fourth of the production and nearly two-thirds of the area under jute. Here hot and humid climate and alluvial, loamy soil coupled with cheap abundant labour provide conditions *par excellence* for the growth of jute. Almost all parts of the state are producing some jute but its cultivation decreases in the north sub-Himalayan region, towards the south in the Ganga delta where land is too low for jute and towards the west where the rocky ground of the Deccan plateau is more marked than the Ganga alluvium. However, major part of the production comes from Nadia, Murshidabad, 24 Parganas, Koch Bihar, Jalpaiguri, Hugli, Dakshin Dinajpur, Bardhaman, Maldah, Paschim Medinapur and Purba Medinapur districts. The entire jute production is consumed in the jute mills located in the Hugli basin. In 2020-21, West Bengal produced 7.61 million bales (79.68% of India) of jute from 0.52 million hectares (78.24%) of India giving a high yield of 2,643 kg/ha.

2. Bihar. Bihar is the second largest producer but lagging far behind West Bengal in the production of jute accounting only for about 10.2 per cent of the production and about 8.3 per cent of the area of the country under jute. Purnea is the largest producing district accounting for 60 per cent of Bihar's production. Katihar, Saharsa and Darbhanga are the other producing districts.

3. Assam. With about 8.33 per cent of the production and 10.26 per cent of the area of the country, Assam is the third largest jute producing

TABLE 22.23. Distribution of Jute and Mesta in India (2020-21)

Sl. No.	States	Area: Million hectares	Area: %age of all India	Production: Million bales*	Production: %age of all India	Yield (kg/hectare)
1.	West Bengal	0.52	78.24	7.61	79.68	2,643
2.	Bihar	0.06	8.35	0.98	10.20	3,172
3.	Assam	0.07	10.26	0.80	8.33	2,108
	Others	0.02	3.16	0.17	1.79	1,470
	All India	**0.66**	**100.00**	**9.56**	**100.00**	**2,595**

*1 bale = 180 kg.

Note: States have been arranged in descending order of production in 2020-21.

Source : Agricultural Statistics at a Glance, 2021, p. 98.

TABLE 22.24. India : Imports and Exports of Jute

Year	Imports of Raw Jute		Exports of Jute Products	
	Quantity ('000 tonnes)	Value (₹ crore)	Quantity ('000 tonnes)	Value (₹ crores)
1960-61	100.4	8	790.0	135
1970-71	0.7	—	560.0	190
1980-81	8	1	660.0	330
1990-91	32.1	20	220.0	298
2000-01	67.3	84	—	932
2019-20	77	350	—	2,428
2020-21	29	179	2,741	371
2021-22	29	179	790.0	135

Source : Economic Survey 2022-23, Statistical Appendix pp. 112-118.

state of India. The main concentration is in the Brahmaputra and Surma valleys. Goalpara, Kamrup, Nowgong, Darrang and Sibsagar are the main producing districts.

Others. Among the other producers, is **Andhra Pradesh** where delta area is the main producer. In **Odisha**, Cuttack, Puri and Bolangir are the main producers. In **Uttar Pradesh**, areas along the Himalayan foothills including Kheri, Bahraich and Sitapur districts are the main producers. Some jute is also produced in Maharashtra, Kerala, Madhya Pradesh, Tripura and Meghalaya.

Trade

India's production of jute always falls short of her requirements and it is imported to feed our jute mills. Bangladesh is the chief supplier of jute to India. India imports raw jute and exports jute hessian. There are large temporal variations in the imports of raw jute and exports of jute hessian (**Table 22.24**).

Being a natural fibre, jute is biodegradable and as such "environment friendly". The principal products can be reused and, as a result, many have a secondary value for other users. Despite such positive features, the world market for jute has remained depressed. The primary cause of such a situation is the development of substitutes like plastic products, synthetic fibres, paper package, etc.

SUGARCANE

Sugarcane belongs to bamboo family of plants and is *indigenous to India*. It is the main source of sugar, *gur* and *khandsari*. About two-thirds of the total sugarcane produced in India is consumed for making *gur* and *khandsari* and only one-third of it goes to sugar factories. It also provides raw material for manufacturing alcohol. Bagasse, the crushed cane residue, can be more beneficially used for manufacturing paper instead of using it as fuel in the mills. It is also an efficient substitute for petroleum products and a host of other chemical products. A part of it is also used as fodder. Sugarcane accounts for the largest value of production and holds an enviable position among all the commercial crops in India. Obviously, it is the first choice of the farmers, wherever geographical conditions favour its growth.

Conditions of Growth

It is a long duration crop and requires 10 to 15 and even 18 months to mature, depending upon the geographical conditions. The geographical conditions favouring the cultivation of sugarcane are described below.

***(i)* Temperature.** It requires hot and humid climate with average temperature of 21°-27°C. Temperature above 20°C combined with open sky helps in acquiring juice and its thickening. Frost is detrimental to sugarcane. Therefore, it must be harvested before frost season, if it is grown in northern parts of the country where winters are very cold and frost is a common phenomenon. On the other hand, hot dry winds are also inimical to sugarcane.

***(ii)* Rainfall.** Too heavy rainfall results in low sugar content and deficiency in rainfall produces fibrous crop. It usually grows well in areas receiving annual rainfall varying from 100 to 150 cm. Irrigation is required in areas receiving lesser rainfall than the prescribed limit. Short cool dry winter season during ripening and harvesting is ideal.

***(iii)* Soil.** It can grow on a variety of soils including loams, clayey loams, black cotton soils, brown or reddish loams and even laterites. In fact, sugarcane can tolerate any kind of soil that can retain moisture. But deep rich loamy soils are ideal for its growth. The soil should be rich in nitrogen, calcium and phosphorus but it should not be either too acidic or too alkaline. Sugarcane exhausts the fertility of the soil quickly and extensively and its cultivation requires heavy dose of manures and fertilizers.

***(iv)* Relief.** Flat plain or level plateau is an advantage for sugarcane cultivation because it facilitates irrigation and transportation of cane to the sugar mills.

***(v)* Labour.** It is a labour intensive crop requiring ample human hands at every stage *i.e.* sowing, hoeing, weeding, irrigating, cutting and carrying sugarcane to the factories. Therefore, cheap abundant labour is a prerequisite for its successful cultivation.

Production

India is the second largest country of the world with respect to area and production of sugarcane next only to Brazil. The cane production registered a dramatic increase of 93 per cent in the decade 1951-61 as a result of diversification of agriculture but this spurt slackened to 14.9 per cent growth between 1960-61 and 1970-71 mainly as a result of the farmers' withdrawal of land under cane owing to internal market fluctuations. However, production began looking up again with the establishment of a number sugar mills during the decade 1971-81 and growth rate was 22 per cent. The production of sugarcane reached a high level of 355.52 million tonnes in 2006-07 after which varying trends have been observed. In the year 2021-22, production of sugarcane in India stood at 431.8 million tonnes.

As in case of production, area under sugarcane cultivation registered a rapid increase from 1.7 million hectares in 1950-51 to 4.88 million hectares in 2010-11 after which the area under sugarcane cultivation increased at a slow rate. In fact, large variations in area under sugarcane have been observed since 2010-11 onwards. In 2021-22 sugarcane was grown on 5.1 million hectares.

The yield of sugarcane doubled in four decades increasing from just 33,422 kg/hectare in 1950-51 to 65,395 kg/hectare in 1990-91. In 2021-22 the yield was 84,000 kg/hectare. Our yields stand nowhere when compared with some of the best in the world. As against India's 84 tonnes/ha in 2021-22, yield of Colombia (86.4 tonnes/ha), Guatimala (129 tonnes/ha) are much higher. Three new breeds of sugarcane namely Co238, Co239 and Co118 have been developed by Coimbatore based Sugarcane Breeding Institute which are likely to give 10 tonnes more cane per hectare and bringing India at par with Brazil, a country which produces largest quantity of sugarcane. Lack of fertilizers, uncertain weather conditions, inadequate irrigation, poor varieties of cane, small and fragmented holdings and backward methods of cultivation are some of the major causes of low yields in India.

Sugarcane is a water-guzzling crop and is often blamed for worsening water crisis in parched areas of India. In order to solve the problem of water crisis, Indian Council of Agricultural Research (ICAR) has developed drought tolerant genetically modified

TABLE 22.25. Area, Production and Yield of Sugarcane in India

Year	1950-51	1960-61	1970-71	1980-81	1990-91	2000-01	2017-18	2018-19	2019-20	2020-21	2021-22
Area (Million hectares)	1.71	2.62	2.62	2.67	3.69	4.32	4.74	5.06	4.60	4.86	5.1
Production (Million tonnes)	57.05	110.00	126.37	154.25	241.05	295.96	379.90	405.42	370.50	399.25	431.8
Yield (kg/hectare)	33,422	45,549	48,322	57,844	65,395	68,578	80,198	80,105	80,497	82,205	84,000

***Source* :** (*i*) Agricultural Statistics at a Glance, 2021, pp. 99-100.
(*ii*) Economic Survey 2022-23, Statistical Appendix, pp. 37-39.

(GM) sugarcane that will not need huge quantities of water. If accepted by conservative farmers, this variety of sugarcane can solve the problem of water shortage and help in increasing the yields.

Sugarcane Research Institute set up at Coimbatore in Tamil Nadu started functioning in 1912. It evolved better varieties of cane, particularly for northern India. One of the important achievements of this institute was introduction of *ratooning* which became very popular in India. In this system the sugarcane is cut leaving the root intact in the soil. Ratoon crop is the second or any other successive crop obtained from the roots left over in the field from the first crop. This is widely practised in different parts of the country due to its low cost of production and relatively shorter maturation period, because cost inputs and time are saved as there is no need for fresh sowing and growing of roots. However, productivity decreases with each passing year and ratooning becomes uncommercial after one or two years.

Distribution

On the basis of study of conditions of growth for sugarcane as mentioned in the above paragraphs, following three distinct belts of sugarcane cultivation can be identified.

(*i*) The Satluj-Ganga plain from Punjab to Bihar containing 51 per cent of the total area and 60 per cent of the country's total production.

(*ii*) The black soil belt from Maharasthra to Tamil Nadu along the eastern slopes of the Western Ghats.

(*iii*) Coastal Andhra and the Krishna Valley.

Here, it is worth drawing a comparison between sugarcane cultivation in the northern and the southern parts of India. In northern plain of India, the summer temperatures ranging from 30° to 35°C and dry scorching winds called '*loo*' in May and June hamper the normal growth of the cane. In the winter months of December and January the sugarcane crop is likely to be damaged by excessively cold weather accompanied by frost. Consequently the yield/hectare is low. In south India, on the other hand, the absence of '*loo*' during the summer and reasonably high temperature during the frost free winter, coupled with the maritime winds in the coastal areas are some of the climatic factors which are extremely beneficial to this crop. The paradoxical character of sugarcane cultivation in India is that whereas south India offers more favourable climatic conditions for the growth of sugarcane, the most important sugarcane belt lies in north India. There are two reasons for such a contradictory situation. Before the World War I, this area was mainly used for growing indigo which was the most favourite cash crop with the farmers at that time. But with the introduction of cheap aniline dyes, indigo lost its market and its cultivation had to be discontinued after the First World War. Consequently, its place was taken by sugarcane cultivation. Another reason is that sugarcane has to face tough competition for land from a number of other cash crops such as cotton, tobacco, groundnut, coconut, etc. in the south.

1. Uttar Pradesh. Table 22.26 clearly shows that Uttar Pradesh far excels all other states with regard to production and area under sugarcane. The state accounts for about 44.5 per cent of the production and about 45 per cent of the area under sugarcane cultivation of India in 2020-21. In fact, all parts of the state except a few dry areas in the west and south-west produce sugarcane to some extent. Vast alluvial plains, congenial climate and large scale use of irrigation and fertilizers are some of the important factors which have helped U.P. to acquire this status. The largest concentration is in the upper Ganga-Yamuna Doab, Rohilkhand and trans-Saryu areas which together produce about 70 per cent of sugarcane produced in this state. Western part of the state forms the core of sugarcane production in the country. As many as 30 districts of U.P. produce sugarcane. However, Muzaffarnagar, Meerut, Bijnor, Moradabad, Saharanpur, Kheri, Deoria, Bulandshahr, Ghaziabad, Bareilly and Sitapur are the important sugarcane producing districts.

2. Maharashtra. Though Maharashtra is the second largest producer of sugarcane in India, this state lags far behind U.P. with respect to area and production accounting for about 25.5 per cent of the production and 23.5 per cent of the area of the country. But Maharashtra is in a superior position with respect to recovery of sugar and duration of crushing period. Further, Maharashtra gives high yield of 88,900 kg/ha. Most of the sugarcane is grown

TABLE 22.26. Distribution of Sugarcane in India (2020-21)

Sl. No.	States	Area		Production		Yield (kg/hectare)
		Million hectares	%age of all India	Million tonnes	%age of all India	
1.	Uttar Pradesh	2.18	44.89	177.67	44.50	81,500
2.	Maharashtra	1.14	23.53	101.59	25.45	88,900
3.	Karnataka	0.44	9.12	42.09	10.54	95,000
4.	Gujarat	0.22	4.43	15.85	3.97	73,687
5.	Tamil Nadu	0.12	2.57	12.80	3.21	1,02,730
6.	Bihar	0.22	4.51	10.71	2.68	48,918
7.	Haryana	0.10	2.04	8.53	2.14	86,179
8.	Punjab	0.09	1.84	7.49	1.87	83,820
9.	Uttarakhand	0.08	1.73	6.96	1.74	82,900
10.	Madhya Pradesh	0.11	2.26	5.88	1.47	53,450
11.	Andhra Pradesh	0.06	1.13	4.12	1.03	74,884
	Others	0.09	1.96	5.56	1.39	58,545
	All India	**4.86**	**100.00**	**399.25**	**100.00**	**82,205**

Note: States have been arranged in descending order of production in 2020-21.

Source : Agricultural Statistics at a Glance, 2021, p. 101.

on black lava soil with the help of irrigation. Most of the production comes from Ahmednagar, Kolhapur, Pune, Nashik, Solapur, Sangli, Satara, Osmanabad and Aurangabad districts.

3. Karnataka. With over 10 per cent of the production and over 9 per cent of the area, Karnataka is the third largest sugarcane producing state of India. Most of the sugarcane is grown in river valleys with the help of irrigation. Belgaum, Mandya, Mysuru, Bijapur, Shimoga and Chitradurga are important producing districts.

4. Gujarat. Gujarat produces about four per cent sugarcane from nearly 4.4 per cent area of India. Its recovery of 10.31 per cent of sugar is one of the highest among the major sugar cane producing states of India. Surat, Bhavnagar, Rajkot, Junagadh and Jamnagar are the important sugarcane producing districts.

5. Tamil Nadu. Though Tamil Nadu accounts only for about 3.2 per cent of the production and nearly 2.5 per cent of the area under sugarcane of the country, this state has unique distinction of giving the highest yield of 1,02,730 kg/hectare. Over 80 per cent of the production comes from Tirupur, Salem, Tiruchirapalli, Dharampuri and Coimbatore districts. The rest is contributed by Madurai, Thanjavur and Ramnathapuram districts.

6. Bihar. Bihar's main sugarcane producing areas comprise a continuation of the main sugarcane producing belt of U.P. In the recent years, area under sugarcane has decreased because much of the sugarcane area has been shifted to wheat cultivation. Champaran, Gaya, Saran, Muzaffarpur, Darbhanga and Patna are the main producing districts.

7. Haryana. Haryana has progressed a lot in sugarcane cultivation during the last few years mainly due to the extension of irrigation facilities. It is an important cash crop in the fertile areas of Karnal, Kaithal, Ambala, Kurukshetra, Rohtak, Hisar, Panipat, Sonepat, Gurugram, Faridabad and Palwal districts. However, it avoids western dry districts.

8. Punjab. Sugarcane cultivation in Punjab has suffered a lot on account of shift in favour of wheat after the introduction of Green Revolution strategy.

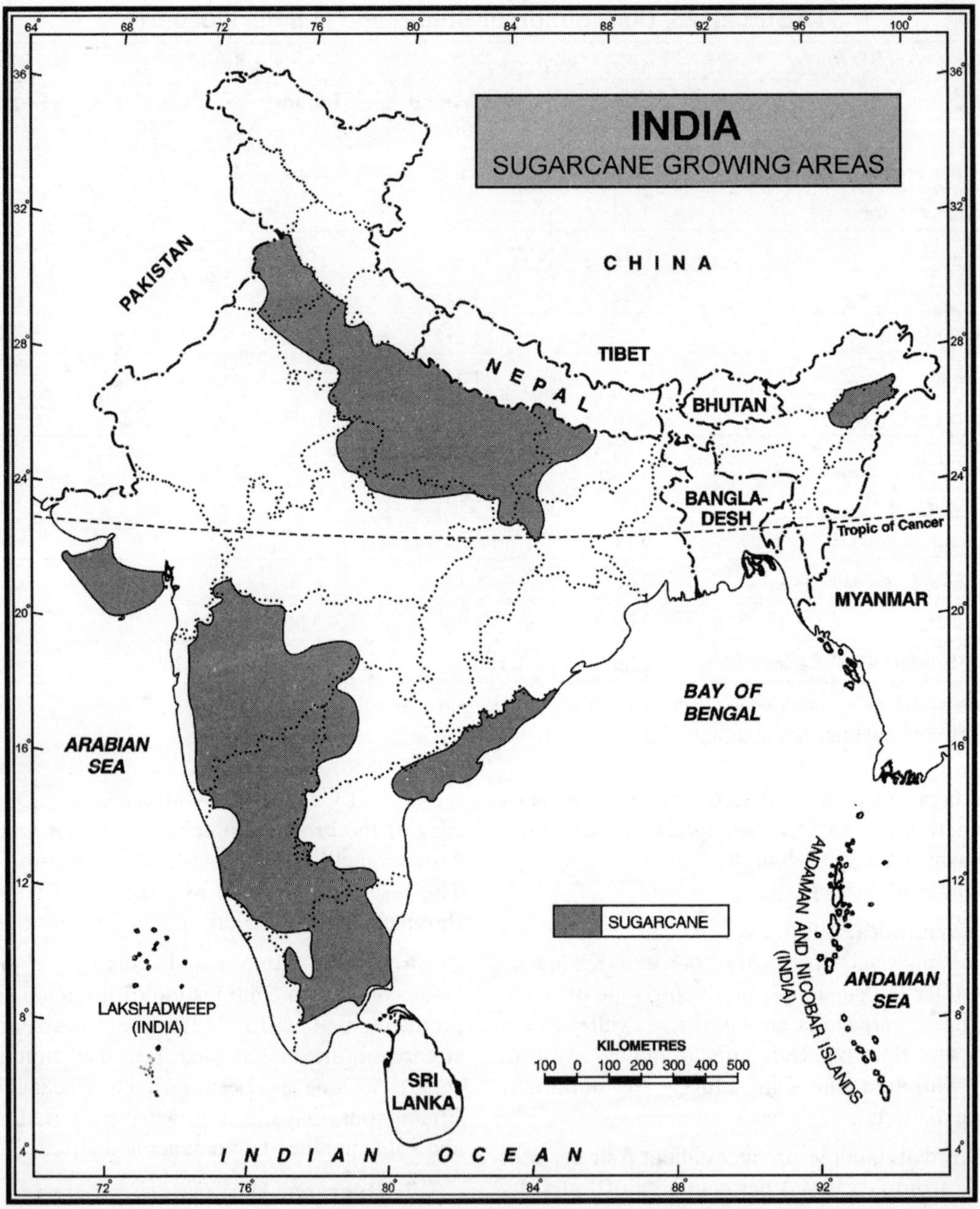

FIG. 22.4. India : Sugarcane Growing Areas

Still Gurdaspur, Jalandhar, Sangrur, Rupnagar, Patiala, Ludhiana, Firozepur, Amritsar districts are important producers of sugarcane in Punjab.

9. Uttarakhand. Most of Uttarakhand is a hilly and mountainous area and is not suitable for sugarcane cultivation. However, parts of Udham Singh Nagar, Haridwar, Nainital and Dehra Dun districts are plain areas or areas located at the foothills. As such these districts help the state to produce about 1.74 per cent sugarcane of India. Udham Singh Nagar is basically a tarai area which is very rich from agriculture point of view. This district contributes major part of sugarcane production of the state.

10. Madhya Pradesh. This state produces about 1.5% sugarcane of India. Gwalior, Morena and Ratlam are the main producing districts.

11. Andhra Pradesh. Most of the cultivation is done in the coastal areas having fertile soil and suitable climate. West Godavari, East Godavari, Vishakhapatnam, Krishna, and Srikakulam are important producers.

Others. Koraput, Cuttack, Puri and Sambalpur in *Odisha*; Brahamputra valley in *Assam*, Burdwan, Birbhum, Hooghly, Malda, North 24 Parganas, Nadia and Murshidabad in *West Bengal* and Ganganagar, Kota, Chittaurgarh, Bundi, Bhilwara and Udaipur districts in *Rajasthan* are other important producers of sugarcane in India. Western part of *Telangana* also produces sugarcane.

TOBACCO

Tobacco is another important cash crop of India. This was brought to India by the Portuguese in 1508. Since then its cultivation has spread to different parts of the country and at present India is an important producer of tobacco in the world. Tobacco is mainly used for smoking in the form of cigarette, *bidi*, cigar, cheroot and *hookah* and is chewed also. It is also used for manufacturing insecticides. Normally speaking, tobacco is to be grown in rotation with other crops but the cash returns from this crop are so high that it is grown year after year on the same land in many parts of India. Surprisingly, it is the common opinion in many parts of Andhra Pradesh and Maharashtra that the quality of tobacco is much improved if it is continuously grown for many years on the same piece of land.

Conditions of Growth

***(i)* Temperature.** It is a plant of tropical and sub-tropical climates and can withstand a wide range of temperature varying from 16° to 35°C. Frost is injurious to its growth.

***(ii)* Rainfall.** It normally requires 100 cm of annual rainfall but it can also be successfully grown in areas of 50 cm annual rainfall provided the rainfall is fairly distributed. Irrigation is required in areas of lower and erratic rainfall. Bright rainless weather is helpful at the curing stage.

***(iii)* Soil.** Well drained friable sandy loams, not too rich in organic matter but rich in mineral salts allowing full development of roots are best suited for tobacco. Soil is more important than climate. In fact, soil rather than climate is the determining factor for its geographical distribution.

***(iv)* Relief.** It can be grown from low lying flat plains up to a height of 1800 metres.

***(v)* Labour.** Cheap and abundant labour is required at all stages of its cultivation, starting from field preparation, transplantation, weeding, harvesting, processing and preparing it for the market.

Production

With a production of 801.4 thousand tonnes in 2019-20, India is the second largest tobacco producing country of the world after China (2,574 thousand tonnes) and accounts for 12.23 per cent of the world's total tobacco production. The production of tobacco had been slowly increasing and had more than doubled from 0.26 million tonnes in 1950-51 to 0.56 million tonnes in 1990-91 after which varying trends in production have been observed. Similarly, the area under tobacco cultivation had increased from 0.36 million hectares in 1950-51 to 0.41 million hectares in 1990-91. The area suddenly fell to 0.26 million hectares in 2000-01 but picked up very fast to reach a high level of 0.40 million hectares in 2019-20. The yield per hectare had also increased by about three times from 731 kg in 1950-51 to 2,317 kg in 2017-18. Trends of production, area and yield of tobacco in India from 1950-51 to 2019-20 are shown in **Table 22.27**.

Distribution

Although tobacco is grown in as many as 15 states of India, only two states *viz.* Gujarat and Andhra Pradesh account for over two-third of the production and 61 per cent of the hectarage under tobacco in the country.

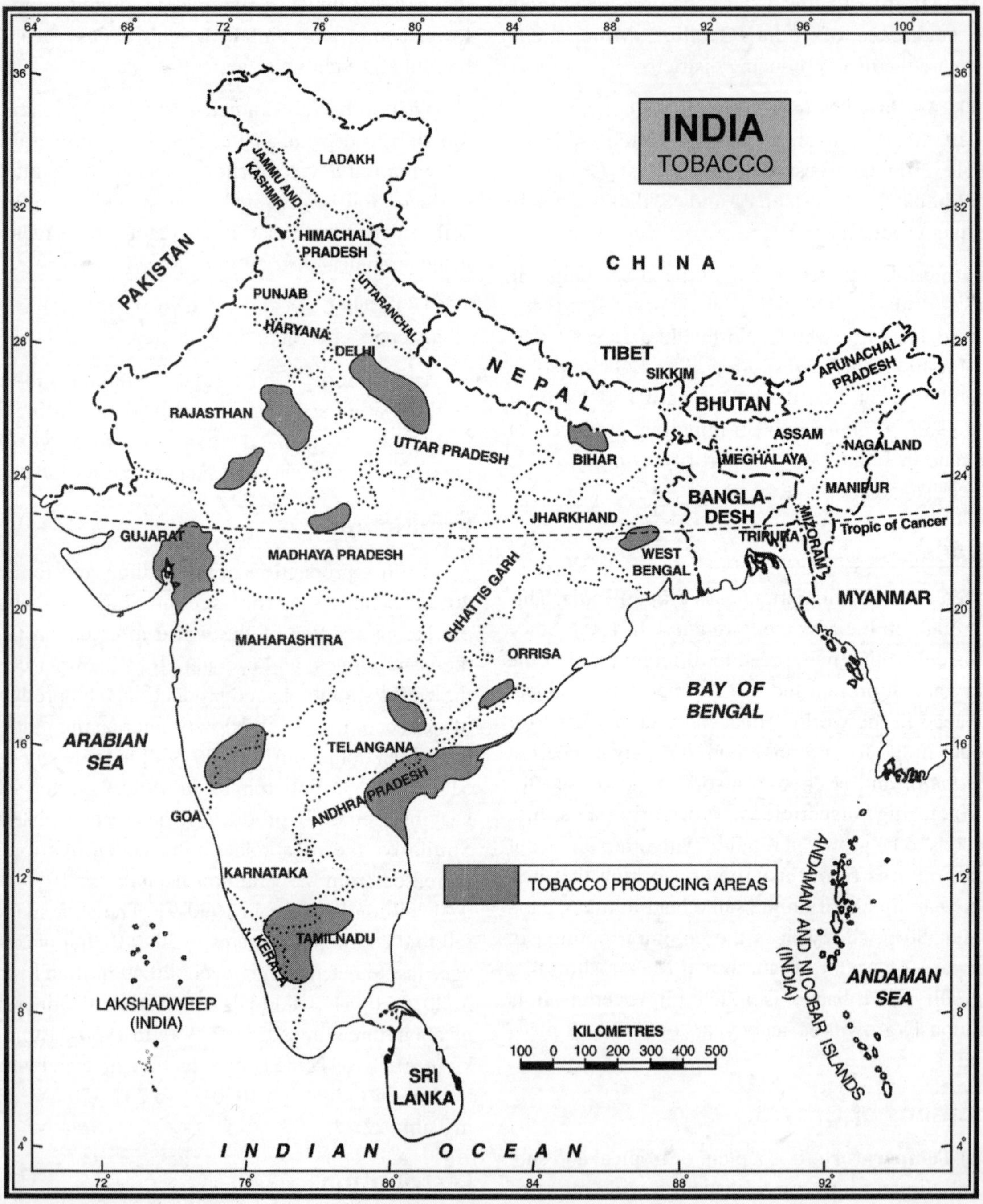

FIG. 22.5. India : Tobacco

1. Gujarat. According to 2019-20 figures, Gujarat produced 382.65 thousand tonnes (47.75% of India) of tobacco from 162 thousand hectares (40% of India) of land giving the yield of 2,358 kg/hectare. About 90 per cent of Gujarat's tobacco comes from Kheda and Vadodara districts.

TABLE 22.27. Area, Production and Yield of Tobacco in India

Year	1950-51	1960-61	1970-71	1980-81	1990-91	2000-01	2017-18	2018-19	2019-20
Area (Million hectares)	0.36	0.40	0.45	0.45	0.41	0.26	0.41	0.41	0.40
Production (Million tonnes)	0.26	0.31	0.36	0.48	0.56	0.34	0.95	0.66	0.80
Yield (kg/hectare)	731	766	810	1,065	1,353	1,318	2,317	1,610	1,982

Source : Agricultural Statistics at a Glance, 2021, pp. 102-03.

2. Andhra Pradesh. Andhra Pradesh has been the traditional producer of tobacco in India. In the year 2019-20, Andhra Pradesh accounted for 185 thousand tonnes (23% of all India) of tobacco from 85 thousand hectares (21% of all India) of land with an average yield of 2,176 kg/hectare. West Godawari, East Godavari, Krishna, Kurnool and Nellore are the main producing districts.

3. Uttar Pradesh. This is the third largest tobacco producing state of India. This state has rich fertile soils and an excellent irrigation system and has the dinstinction of giving the highest yield of 2,800 kg/ha in 2019-20. Farukhabad, Etah, Mainpuri, Meerut, Bulandshahr, Varanasi, Saharanpur, etc. are important tobacco producing districts.

4. Karnataka. Karnataka produces over 10 per cent tobacco of India. Belgaum and Mysuru are important tobacco producing districts.

Others. West Bengal, Telengana, Bihar, Punjab, Haryana, etc. are the other producing states of tobacco in India.

Trade

About 80 per cent of the total production of tobacco is used within the country and the remaining 20 per cent is exported. India is world's fourth largest exporter of tobacco. In 2020-21, the value of tobacco exports from India was ₹ 3,840 crore. Presently India exports tobacco to about 60 countries. Russia and U.K. purchase about two-third of our total tobacco exports. The other important buyers of Indian tobacco are Japan, Egypt, Sri Lanka, Nepal, Indonesia, Germany

TABLE 22.28. Production of Tobacco in India (2019-20)

Sl. No.	States	Area		Production		Yield (kg/hectare)
		Thousand hectares	%age of all India	Thousand tonnes	%age of all India	
1.	Gujarat	162.26	40.13	382.65	47.75	2,358
2.	Andhra Pradesh	85.00	21.02	185.00	23.08	2,176
3.	Uttar Pradesh	35.00	8.66	98.00	12.23	2,800
4.	Karnataka	90.00	22.26	83.00	10.36	922
5.	West Bengal	15.15	3.75	20.46	2.55	1,350
6.	Bihar	10.27	2.54	18.03	2.25	1,756
	Others	6.61	1.63	14.28	1.78	2,161
	All India	**404.29**	**100.00**	**801.42**	**100.00**	**1,982**

Note: States have been arranged in descending order of production in 2019-20.

Source : Agricultural Statistics at a Glance, 2021, p. 104.

TABLE 22.29. India : Exports of Tobacco

Year	1960-61	1970-71	1980-81	1990-91	2000-01	2019-20	2020-21	2021-22
Quantity (Thousand tonnes)	47.0	49.8	91.3	87.1	108.3	181.8	178	196
Value (₹ crore)	16	33	14	263	871	3,761	3,840	4,250

Source : Economic Survey 2022-23, Statistical Appendix, pp. 116-18.

and Singapore. About 90 per cent of the tobacco export trade is handled by Chennai alone, the rest being handled by Kolkata, Mumbai and Vishakhapatnam. To make Indian tobacco more competitive in the world market in terms of price and quality, varieties with enhanced productivity and desirable quality traits should be developed.

OILSEEDS

Oilseeds constitute a very important group of commercial crops in India. The oil extracted from oilseeds form an important item of our diet and are used as raw materials for manufacturing large number of items like paints, varnishes, hydrogenated oil, soaps, perfumery, lubricants, etc. Oil-cake which is the residue after the oil is extracted from the oilseeds, forms an important cattle-feed and manure.

India has the largest area and production of oilseeds in the world. Nine major oil seeds *viz*., groundnut, sesamum, rapeseed and mustard, linseed, safflower, castor seed, sunflower, and soyabean occupied 28.8 million hectares (2020-21). **Table 22.30** shows that there had been gradual increase in area, production and yield of oilseeds, with the passage of time, some short lived variations notwithstanding.

It must, however, be noted that the production of oilseeds has always fallen short of our demand and there has always been a need to import oilseeds or their products for meeting the demand of our evergrowing population.

With limited scope of bringing additional area under oilseeds, increase in oilseed production will have to come primarily from land saving technologies highlighting a combination of high yield plant type, standard crop management practices and balanced crop nutrition.

There have been large scale regional variations in area, production and productivity changes in oilseeds. Only a few states like Haryana, Madhya Pradesh, Rajasthan and West Bengal increased their oilseed production both through area expansion and productivity improvement. States like Maharashtra, Tamil Nadu and Himachal Pradesh increased their oilseed output mainly through productivity improvement. In some states like Odisha, area productivity and production declined sharply.

Table 22.31 shows that Rajasthan, Maharashtra, Madhya Pradesh and Gujarat are the main producers of nine oilseeds and account for over 70% of the area and more than three-fourths of the production of India. Tamil Nadu gives maximum yield of 2,073 kg/hectares. Other producers include Andhra Pradesh, Uttar Pradesh, Haryana, Karnataka, West Bengal, Odisha, Assam, Bihar and Punjab.

TABLE 22.30. Area, Production and Yield of Nine Oilseeds in India

Year	1950-51	1960-61	1970-71	1980-81	1990-91	2000-01	2017-18	2018-19	2019-20	2020-21	2021-22
Area (Million hectares)	10.73	13.77	16.64	17.60	24.15	22.77	24.51	24.79	27.14	28.79	29.2
Production (Million tonnes)	5.16	6.98	9.63	9.37	18.61	18.77	31.46	31.52	32.22	36.10	37.7
Yield (kg/hectare)	481	507	579	532	771	810	1,284	1,271	1,224	1,254	1,292

Source : (*i*) Agricultural Statistics at a Glance, 2021, pp. 78-79.
(*ii*) Economic Survey, 2022-23, Statistical Appendix, pp. 37-39.

TABLE 22.31. Distribution of Nine Oilseeds in India (2017-18)

Sl. No.	States	Area: Million hectares	Area: %age of all India	Production: Million tonnes	Production: %age of all India	Yield (kg/hectare)
1.	Rajasthan	5.19	18.04	7.94	22.00	1,529
2.	Maharashtra	4.81	16.70	6.69	18.52	1,391
3.	Madhya Pradesh	8.13	28.23	6.68	18.50	822
4.	Gujarat	3.35	11.65	6.18	17.11	1,842
5.	Haryana	0.65	2.27	1.31	3.63	2,005
6.	Uttar Pradesh	1.20	4.16	1.26	3.49	1,054
7.	Karnataka	1.19	4.13	1.20	3.33	1,012
8.	West Bengal	0.93	3.23	1.17	3.24	1,260
9.	Tamil nadu	0.48	1.65	0.99	2.73	2,073
	Others	2.87	9.95	2.69	7.44	938
	All India	**28.79**	**100.00**	**36.10**	**100.00**	**1,254**

Note: States have been arranged in descending order of production during 2020-21.
Source : Agricultural Statistics at a Glance, 2021, p. 80.

GROUNDNUT

Groundnut is the most important oilseed of India and accounts for a little less than half of the major oilseeds produced in the country. Groundnut kernels are rich in proteins and vitamins and have high calorific value. It contains 40-50 per cent oil which is mainly used as edible oil in its pure form or in hydrogenated vanaspati form. The groundnut oil is also used for manufacturing margarine, medical emulsions, soap and toilet requisites. Groundnut is eaten raw, roasted, sweetened or salted. Its oil cake is used as an important rich cattle feed. It serves as an important rotation crop because it syntehsizes atmospheric nitrogen and increases the fertility of soil.

Conditions of Growth

It thrives best in the tropical climate and requires 20°-30°C temperature and 50-75 cm rainfall. Isohyet of 100 cm marks the upper limit of groundnut cultivation. It is highly susceptible to frost, prolonged drought, continuous rain and stagnant water. Dry winter is needed at the time of ripening. It can be grown both as a *kharif* and as a *rabi* crop but 91 per cent of the total area under groundnut is devoted to *kharif* crop. Well drained light sandy loams, loams,

TABLE 22.32. Area, Production and Yield of Groundnut in India

Year	1950-51	1960-61	1970-71	1980-81	1990-91	2000-01	2017-18	2018-19	2019-20	2020-21	2021-22
Area (Million hectares)	4.49	6.46	7.33	6.80	8.31	6.56	4.89	4.73	4.83	6.09	5.7
Production (Million tonnes)	3.48	4.81	6.11	5.01	7.51	6.41	9.25	6.73	9.95	10.21	10.1
Yield (kg/hectare)	775	745	834	736	904	977	1,893	1,422	2,063	1,676	1,759

***Source*:** (*i*) Agricultural Statistics at a Glance, 2021, pp. 81-82.
(*ii*) Economic Survey, 2022-23, Statistical Appendix, pp. 37-39.

red, yellow and black cotton soils are well suited for its cultivation.

Production and Distribution

India is the second largest producer of groundnut contributing about 16.6% of the world's total production but is way behind China's 36.43 per cent. **Table 22.32** shows that there had been large scale temporal variations in area production and yields of groundnut depending on the farmer's choice of crops and weather conditions. It must be noted that groundnut is primarily a rainfed crop and there are bound to be fluctuations in its production, area and yield depending upon the amount of rainfall and its temporal distribution.

Table 22.33 shows that Gujarat, Rajasthan, Tamil Nadu and Andhra Pradesh are the four main producers. These four states together account for over three fourth of total production of India. ***Gujarat*** is the largest producer contributing about 40 per cent of India's total production. About 60 per cent of the state's total production comes from Junagadh, Jamnagar, Amreli, Bhavnagar, Rajkot, Sabarkantha, Panchmahals and Surat districts. ***Rajasthan*** produces about 19 per cent of India's groundnut. Chittaurgah, Bhilwara, Tonk, Jhalawar, and Jaipur are the main producing districts. ***Tamil Nadu*** is the third largest producer accounting for over 9 per cent of the total groundnut produced in India. Salem, Tiruchirapalli and Coimbatore districts produce half of the state's total output. ***Andhra Pradesh*** is the fourth important producer of groundnut in India and accounts for over about 7.6 per cent of India's total production. About 50 per cent of the state's production comes from Chittoor, Kurnool and Anantpur districts, though other districts also produce sufficient groundnut. ***Karnataka*** produces over 6 per cent of India's groundnut. Dharwar, Gulbarga, Belgaum, Bellary, Kolar, Tumkur, Raichur and Mysuru districts supply about three-fourths of the state's total production. ***Madhya Pradesh*** accounted for about 5.6 per cent of India's groundnut production in 2020-21. Mandsaur, West Nimad and Dhar are important groundnut producing districts. ***Maharashtra*** produces about 4.3 per cent of India's groundnut. Jalgaon, Dhule, Kolhapur, Satara, Osmanabad, Yavatmal, Nashik and Amravati districts produce about two-thirds of the state's production. West Bengal, Uttar Pradesh, Punjab, Odisha, Haryana, Bihar, Kerala also produce groundnut in small quantities.

TABLE 22.33. Distribution of Groundnut in India (2020-21)

Sl. No.	States	Area		Production		Yield (kg/hectare)
		Million hectares	%age of all India	Million tonnes	%age of all India	
1.	Gujarat	2.16	35.50	4.13	40.42	1,908
2.	Rajasthan	0.86	14.04	1.93	18.91	2,256
3.	Tamil Nadu	0.41	6.71	0.94	9.25	2,310
4.	Andhra Pradesh	0.87	14.28	0.78	7.62	894
5.	Karnataka	0.70	11.49	0.68	6.62	966
6.	Madhya Pradesh	0.37	6.02	0.58	5.64	1,569
7.	Maharashtra	0.34	5.57	0.44	4.30	1,294
	Others	0.39	6.38	0.74	7.23	1,897
	All India	**6.09**	**100.00**	**10.21**	**100.00**	**1,676**

Note: States have been arranged in descending order of production in 2020-21.

Source : Agricultural Statistics at a Glance, 2021, p. 83.

Trade

About 75 per cent of the total production enters the interestate trade—the main traders being Andhra Pradesh, Tamil Nadu, Gujarat, Maharashtra, Karnataka and Punjab. India's capacity to export groundnut and its products has drastically been reduced due to increased domestic consumption resulting from rapid population growth. However, groundnut cake is still exported to U.K. and other European countries and to Russia.

RAPESEED AND MUSTARD

Next to groundnut, rapeseed and mustard are two most important oilseeds of India. The oil content of these seeds is 25-45% which is used as a cooking medium, preservative for pickles, lubricants and toiletteries. Oil cake forms an important cattlefeed and is also used as manure.

Conditions of Growth

Like wheat and gram, they thrive only in cool climate of the Satluj-Ganga plain and very small quantity is grown in the peninsular India. They are mainly grown as rabi crop in pure or mixed form with wheat, gram and barley.

Production and Distribution

India and is world's third largest producer after Canada and China, accounting for about 10% of world's production. There has been nearly four-fold increase in their production in three decades from 1960-61 to 1991 after which varying trends of production have been noticed. **(Table 22.34)**

TABLE 22.34. Area, Production and Yield of Rapeseed and Mustard in India

Year	1950-51	1960-61	1970-71	1980-81	1990-91	2000-01	2017-18	2018-19	2019-20	2020-21	2021-22
Area (Million hectares)	2.07	2.88	3.32	4.11	5.78	4.48	5.98	6.12	6.86	6.69	8.1
Production (Million tonnes)	0.76	1.35	1.98	2.30	5.23	4.19	8.43	9.26	9.12	10.11	11.7
Yield (kg/hectare)	368	467	594	560	904	936	1,410	1,511	1,331	1,511	1,458

Source : (*i*) Agricultural Statistics at a Glance, 2021, pp. 84-85.
(*ii*) Economic Survey, 2022-23, Statistical Appendix, pp. 37-39.

TABLE 22.35. Distribution of Rapeseed and Mustard in India (2020-21)

Sl. No.	States	Area		Production		Yield (kg/hectare)
		Million hectares	%age of all India	Million tonnes	%age of all India	
1.	Rajasthan	2.72	40.60	4.51	44.57	1,659
2.	Madhya Pradesh	0.77	11.45	1.31	12.98	1,713
3.	Haryana	0.63	9.42	1.28	12.64	2,027
4.	Uttar Pradesh	0.70	10.48	0.99	9.79	1,412
5.	West Bengal	0.59	8.86	0.72	7.12	1,215
6.	Gujarat	0.21	3.18	0.42	4.16	1,976
7.	Jharkhand	0.43	6.44	0.35	3.51	823
8.	Assam	0.29	4.30	0.18	1.80	633
	Others	0.35	5.28	0.35	3.44	986
	All India	**6.69**	**100.00**	**10.11**	**100.00**	**1,511**

Note: States have been arranged in descending order production in 2020-21.
Source : Agricultural Statistics at a Glance, 2021, p. 86.

Rajasthan excells all other states with respect to production of mustard and rapeseed and produces nearly 44 per cent from over 40 per cent of area under these crops in India. Ganganagar, Bharatpur, Kota, Alwar, Jaipur, Chittaurgarh, Tonk, Sawai Medhopur, Udaipur, Pali districts contribute the major part of production and cover large area. **Madhya Pradesh** is the second largest producer. Bhind, Morena and Gwalior are the main producing districts in this state. **Haryana** is the third largest producer and has the distinction of giving the highest yield of 2,027 kg/hectare. Hisar, Sirsa, Fatehabad, Mahandergarh, Rewari, Gurugram, Mewat, Bhiwani are the chief producers. **Uttar Pradesh** is the traditional producer but has surrendered its top position during the last few years, partly due to fall in its own production and partly due to increase in production in other states. Agra, Faizabad, Meerut, Kanpur, Saharanpur, Etah and Etawah are the chief producing districts. **West Bengal** has also emerged as a major producer. West Dinajpur, Birbhum, Nadia and Murshidabad districts contribute the bulk of production. Gujarat, Jharkhand, Assam, Bihar, Punjab, Odisha, etc. are some of the other producing states.

PLANTATION CROPS

Plantation crops are those crops which are grown on plantations covering large estates. Unlike other crops, they are not annual crops and usually take 3-5 years to bear returns after they are sown. But once they start bearing fruit, they continue to do so for 35-40 years. They require heavy initial capital investment and high level technology for their growth and processing. They cover small area in India but are of high economic value. Tea, coffee and rubber are the principal plantation crops but spices are also included in this category. Tea and coffee crops are also famous as *beverage crops*. There are over 30,000 plantations in India giving full or part time employment to over 20 lakh persons. Most of the plantations are under tea, coffee and rubber.

TEA

Tea is the dried leaf of a bush. It contains theine and when added to boiling water along with sugar and milk, it gives a very low cost and stimulating drink. Thus it is the most important **beverage crop** of India.

Tea bush is supposed to be indigenous to China but it was reported by Major Robert Bruce in 1823 that indigenous tea bushes grew wild on the hill slopes of upper Assam. In the year 1840, tea seeds were imported from China and commercial tea plantations were set up in the Brahmaputra valley. To begin with, tea plantations were confined to Upper Assam only but later on, new areas such as lower Assam and Darjeeling were also opened up to tea plantations and by 1859, there were 30 tea plantations in Assam alone. Later on, tea plantations were also set up in Nilgiri Hills of South India, Tarai along the foothills of the Himalayas and in some places in Himachal Pradesh.

Conditions of Growth

Tea bush is a tropical and sub-tropical plant and thrives well in hot and humid climate. *There is a very close relation between climate, the yield and the quality of tea.*

(*i*) **Temperature.** The ideal temperature for its growth is 20°-30°C and temperatures above 35°C and below 10°C are harmful for the bush. Alternate waves of warm and cool winds are very helpful for tea leaves. Tea is a shade-loving plant and develops more vigorously when planted along with shady trees.

(*ii*) **Rainfall.** It requires 150-300 cm annual rainfall which should be well distributed throughout the year. While prolonged dry spell is harmful for tea, high humidity, heavy dew and morning fog favour rapid development of young leaves.

(*iii*) **Soil.** Tea bush grows well in well drained, deep, friable loams. However, virgin forest soils rich in humus and iron content are considered to be the best soils for tea plantations. Relatively large proportion of phosphorus and potash in the soil gives special flavour to tea as is the case in Darjeeling. In order to increase the yield, proper dose of nitrogenous fertilizers such as ammonium sulphate should be applied to soil.

(*iv*) **Relief.** Although tea requires heavy rainfall for its growth, stagnant water is injurious to its roots. It is, therefore, grown on hill slopes where water drains away easily and waterlogging does not take place. However, it grows equally well in the valleys if the drainage is good. Most of the tea plantations in

India are found at elevations ranging from 600 to 1,800 metres above mean sea level.

(v) Labour. Tea is a labour intensive crop and requires abundant supply of cheap and skilled labour, especially at the time of plucking the tea leaves. This is a tedious process which requires skilled manipulation of fingers for plucking two leaves and a bud at a time. For this purpose, women labourers are employed in large numbers. Currently, tea industry provides employment to one million workers. Through its forward and backward linkages another 10 million people derive their livelihood from tea. It is one of the largest employers of women among organised industries of India. Women constitute over 50 per cent of the total workforce.

Method of Cultivation

Tea gardens are set up on the cleared hill slopes where shade trees are planted in advance. Seeds are sown in the germination beds and the saplings transplanted to the garden. The garden is regularly hoed and weeded so that tea bush grows without any hindrance. Use of manures and fertilizers is a common practice in the gardens. Oil cakes and green manures are widely used. Pruning of the plant is an essential part of tea cultivation. It helps in maintaining the proper shape of tea bush to a height of about one metre with about the same diameter. The aim of pruning is to have new shoots bearing soft leaves in plenty and to facilitate the plucking of leaves by women labourers from the ground.

Production

Tea cultivation does not have long tradition in India as it started in the middle of nineteenth century only, when first tea plantations were established in Assam. However, tea cultivation has shown steady progress right from the beginning. The progress of tea cultivation in India from 1990-91 to 2020-21 is shown in **Table 22.36**. At present, India is the third largest producer of tea in the world, next to China and Turkey and contributes 17.1 per cent of the world's tea production.

Distribution

Tea cultivation in India is highly concentrated in a few selected pockets. Following three areas of tea cultivation are identified according to their importance as tea producers and their location.

(1) North-East India (2) South India (3) North-West India.

1. North-Eastern India

It is more or less a triangular area mainly in Assam and West Bengal. It extends from 23°N to 28°N latitudes and 88°E to 96°E longitudes. This is the most important tea producing region of India accounting for about three-fourths production and about the same percentage of area under tea production. Tea plantations are small in number (see **Table 22.37**) but fairly large in size, generally more than 200 hectares.

Assam. Assam is the largest producer of tea accounting for over 51 per cent of the production and over 55 per cent of area under tea cultivation in India. The average yield is 17.4 quintals/hectare. Two destinct areas of tea production can be identified.

(a) The Brahmaputra Valley extending from Sadiya to Goalpara comprises the main tea producing belt. It accounts for 44 per cent of India's tea from 40 per cent of tea area of the country. There are 676 tea estates mainly in the districts of Dibrugarh, Lakhimpur, Sibsagar, Darrang, Kamrup, Nagaon, and Goalpara. With summer temperature of 30°C and winter temperature never falling below 10°C, frost free weather throughout the year and 300-400 cm

TABLE 22.36. Area, Production and Yield of Tea in India

Year	1990-91	2000-01	2010-11	2016-17	2017-18	2018-19	2019-20	2020-21	2021-22
Area (Million hectares)	0.4	0.5	0.6	0.6	0.6	0.6	0.6	0.6	0.6
Production (Million tonnes)	0.7	0.8	1.0	1.3	1.3	1.4	1.4	1.3	1.3
Yield (kg/hectare)	1,794	1,673	1,712	2,165	2,210	2,251	2,138	2,016	2,113

Sources : (*i*) Economic Survey 2019-20, pp. A34-36; (*ii*) Economic Survey 2022-23; Statistical Appendix, pp. 37-39.

annual rainfall extended over 9 months; the area provides ideal climatic conditions for tea cultivation. Tea estates are located on the raised grounds (upto 450 metres) so that annual inundations and stagnant water during the rainy season do not harm the crop.

(*b*) ***Surma Valley*** is the second important tea producing area in Assam. This valley, lying in Cachar district, produces about 5 per cent of country's tea from 9 per cent of land under this crop. Here the tea gardens are scattered over small mounds called *teelas* or *bheels* or well drained flats along the river and its tributaries. Here rainfall is 300-400 cm and no month is completely dry.

West Bengal. West Bengal is the second largest producer contributing over 23 per cent of India's tea from about 24 per cent of the country's total area under tea cultivation. Entire tea of West Bengal is produced in three northern districts of Darjeeling, Jalpaiguri and Koch Bihar. These districts are contiguous to the main tea producing belt of Assam. Tea producing areas of West Bengal are divided into two geographical regions.

(*a*) ***The Duars*** in Koch Bihar and Jalpaiguri districts is a 16 km wide strip at the foot of the Himalayas. Here tea is raised on slightly elevated areas where suitable slope for proper drainage is available. Tea estates are found upto a height of 900-1,200 m.

(*b*) ***Darjeeling*** district is well known all over the world for its most exquisite aromatic tea. Annual rainfall of 300 cm; moderate temperature and fertile soils give special flavour to tea although yields are quite low, generally below 20 quintals/hectare. Tea estates are found within 900-1,800 m elevation beyond which the temperature is low and does not support tea cultivation. According to a study conducted by Tea Board in 2002, land under tea cultivation can be increased by more than 5 per cent. In some of the smaller gardens of Darjeeling, land under tea cultivation can be increased by about 22 per cent.

Some tea gardens are also found in Tripura, Arunachal Pradesh and Manipur in north-east India.

2. South India

In South India tea is produced in Nilgiri, Cardamom, Palni and Anaimalai hills in Tamil Nadu, Kerala and Karnataka states extending from 9°N to 14°N latitudes. This region accounts for 25 per cent production and about 18 per cent of area under tea in India. Tea gardens are mostly located on the hill slopes of the Western Ghats between 300 and 1,800 m altitude. The tea estates are quite large in number (*see* **Table 22.37**) but quite small in size. The temperatures are uniformally high and the annual rainfall exceeds 400 cm. There is no fear of frost in south India and weather conditions are quite congenial. The productivity is generally 15-25 quintals/hectare, although the quality of tea is inferior.

TABLE 22.37. Distribution of Tea in India (2020-21)

States	Area		Production		No. of tea gardens
	'000 hectares	%age of all India	'000 tonnes	%age of all India	
1. Assam	272	55.01	575	51.08	790
2. West Bengal	115	23.99	345	23.17	315
3. Tamil Nadu	76	12.53	205	16.70	6,795
4. Kerala	37	5.07	72	7.22	4,215
5. Tripura	8	1.53	8	0.86	15
6. Karnataka	2	0.38	5	0.54	25
Others	13	1.49	4	0.43	13,635
All India	**573**	**100.00**	**1,214**	**100.00**	

Source : Data collected from different sources.

TABLE 22.38. Exports of Tea and mate from India

Year	1960-61	1970-71	1980-81	1990-91	2000-01	2019-20	2020-21	2021-22
Quantity (thousand tonnes)	199.2	199.1	229.2	199.1	202.4	254.8	213.0	209
Value (₹ crore)	124	148	426	1,070	1,976	5,851	5,604	5,597

Source : Economic Survey 2022-23, Statistical Appendix pp. 116-18.

But some of the south Indian teas have a good combination of taste and flavour.

In South India, **Tamil Nadu** is the largest producer of tea accounting for over 16 per cent of total tea production of India from over 12 per cent of the land. Nilgiri and Anaimalai produce 46 per cent and 33 per cent of Tamil Nadu's tea respectively. **Kerala** is another important producer of tea in South India accounting for 7.22 per cent of the total production of India. Kottayam, Kollam and Thiruvananthapuram are the main tea producing districts. Some tea is produced in Hassan and Chikmagalur districts of **Karnataka.** This state has the distinction of giving the highest yield of over 25 quintals per hectare.

3. North West India

Some of tea is produced in Dehra Dun, Almora and Garhwal districts of Uttarakhand and in Kangra Valley and Mandi district of Hiamchal Pradesh. Green tea is produced in Kangra valley of Himachal Pradesh.

Tea in small quantity is also produced in Ranchi and Hazaribagh districts of Chota Nagpur plateau in Jharkhand.

Trade

Besides being the third largest producer, India is also one of the leading exporters of tea in the world. India had a long tradition of being the largest exporter of tea in the world, but her predominant position as an undisputed leader of tea exporting country has been severely shattered by fast increasing domestic consumption and by tough competition by some other tea exporting countries in the world market. Despite impressive increase in the production of tea from 0.7 million tonnes in 1990-91 to 1.3 million tonnes in 2021-22, our exports have been varying from 199 thousand tonnes to 209 thousand tonnes between 1960-61 and 2021-22.

Our present per capita consumption of about 0.6 kg is much lower than that of 0.8 kg in Australia, 0.9 kg in Japan, 1.0 kg in New Zealand, 1.5 kg in Ireland and 2.4 kg in United Kingdom. If our per capita consumption also increases along with the increase in population, for which there is a vast scope, we shall be left with very little surplus for export. Internal consumption of tea has already grown from a modest of 319 million kg in 1981 to a staggering figure of 1,010 million kg in 2017-18.

Any attempt to increase production and thereby increase export after meeting the increasing domestic demand will largely involve new seeds and techniques of tea cultivation because prospects of acquiring virgin land for expanding tea plantations are extremely bleak. A study by the Tea Board in 2007 has revealed that 38 per cent of all tea bushes are over 50 years and another 90 per cent are in the age group of 40-50 years, and are not capable of giving high yields. There is urgent need to replace the old bushes with new plants. TV-29, a high yielding clone of tea was introduced in 1991. This seed gives four times as large as the average yield given by the conventional seeds and holds out great possibilities for increasing tea production in India. Old bushes giving low yields are likely to be replanted gradually with high yielding tea plants. In order to achieve this goal of meeting the growing demand in the home market and to produce surplus for export, a plan for additional 5,000 hectares of new plantations in the non-traditional areas, replanting of 20,000 hectares of old bushes by new plants and rejuvenation of 40,000 hectares by irrigation, proper drainage and improved cultural practices has been chalked out.

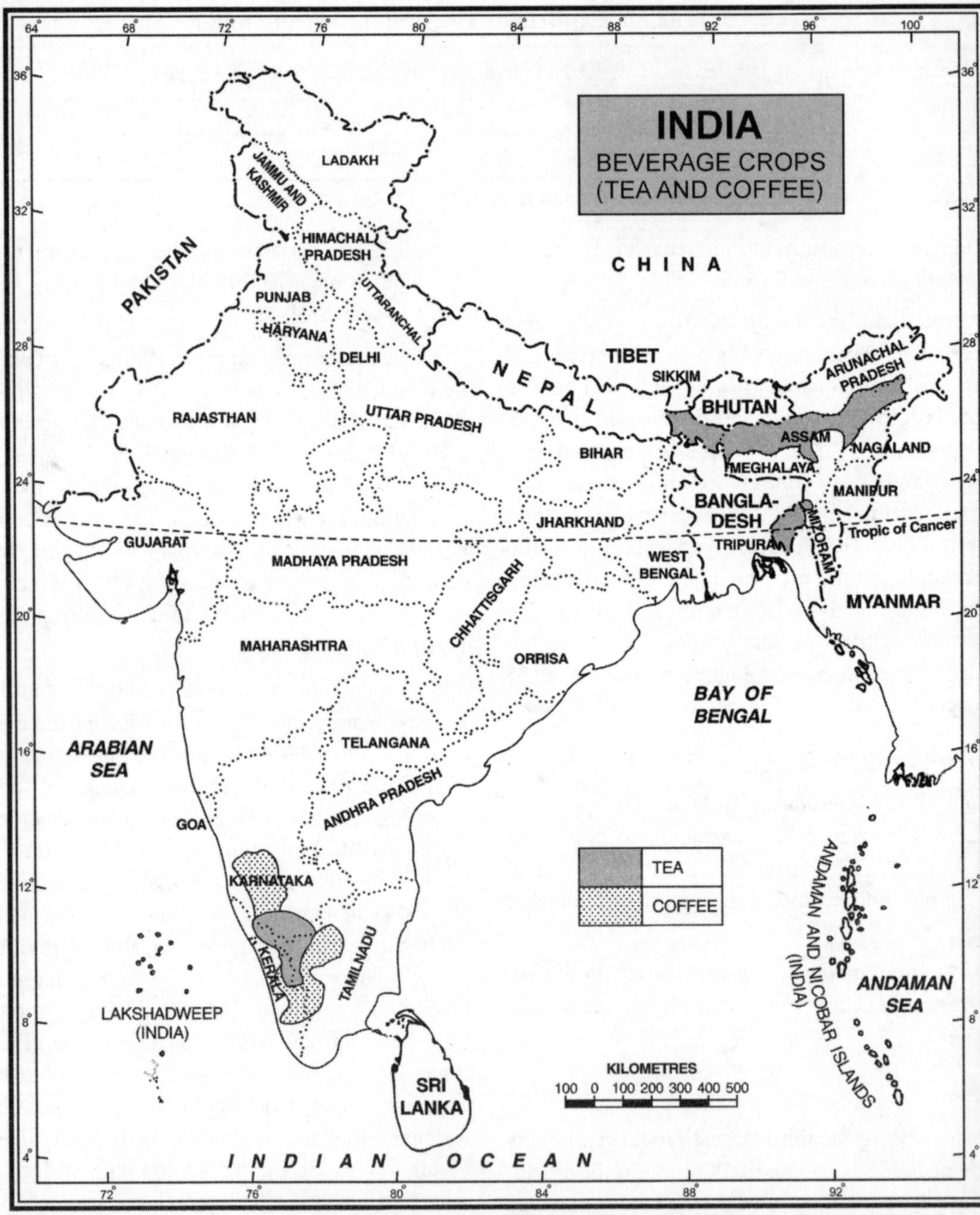

FIG. 22.6. India : Beverage Crops (Tea and Coffee)

In addition to the increasing domestic demand and dwindling exportable surpluses, India has to face tough competition from other tea exporting countries especially from Sri Lanka, China, Japan, Vietnam, Indonesia and some African countries like Kenya. India's share in the world market has come down from 45 per cent in 1950-51 to about 10 per cent in 2018-19. Exports from Sri Lanka occasionally surpass the exports from India. India exports tea to as many as 80 countries but U.K., Russia, the USA, Germany,

TABLE 22.39. Imports of Tea by India

Year	2009-10	2010-11	2011-12	2012-13	2013-14	2014-15	2015-16	2016-17	2017-18
Quantity (thousand tonnes)	34.46	20.82	22.43	22.30	22.74	28.39	23.72	24.89	24.94
Value (₹ crore)	276.52	202.00	218.91	274.52	291.68	388.66	377.47	338.35	356.99

Source : Agricultural Statistics at a Glance, 2018, pp, 267-69.

Australia, Afghanistan, Ireland, Sudan, Iran, Iraq, Egypt, etc. are our main customers. Kolkata is the chief port of tea export from India. The other major ports through which tea is exported are Chennai, Mangaluru and Kochi.

Of late, some quantity of tea is being imported for blending and re-exports. Large quantities of inferior quality tea are imported and re-exported, severely affecting India's quality image in the international market. **Table 22.39** shows the trends in tea imports by India.

Over 80 per cent of the imported tea was supplied by Vietnam, Indonesia and Sri Lanka. Vietnam was the largest source accounting for 55 per cent of the total imports.

COFFEE

Coffee is the second most important **beverage crop** of India next only to tea. It is indigenous to Abyssinia Plateau (Ethiopia) from where it was taken to Arabia in 11th century. From Arabia, its seeds were brought to India by Baba Budan in the 17th century and were raised in the Baba Budan Hills of Karnataka. British planters took keen interest in coffee plantations and large coffee estates were established near Chikmagalur (Karnataka) in 1826, in Manantody (Waynad) and Shevoroys in 1830 and Nilgiris in 1839. Currently, there are over 52,000 coffee gardens giving employment to 2.5 million persons.

Conditions of Growth

Coffee plant requires hot and humid climate.

***(i)* Temperature.** The most suitable temperature for proper growth of coffee varies from 15 to 28°C. It does not tolerate frost, snowfall, high temperature above 30°C and strong sun shine and is generally grown under shady trees.

***(ii)* Rainfall.** It requires annual rainfall varying from 150 to 250 cm. The rainfall should be fairly distributed throughout the year. Prolonged drought is injurious to coffee. Dry weather is necessary at the time of ripening of the berries.

***(iii)* Soil.** Well drained, rich friable loams containing good deal of humus and minerals like iron and calcium are ideal for coffee cultivation. The soil must be properly manured to retain and replenish fertility and to increase productivity.

***(iv)* Relief.** Stagnant water is harmful and this crop is grown on hill slopes at elevations from 600 to 1,600 metres above mean sea level. Northern and eastern aspects of slopes are preferred as they are less exposed to strong afternoon sun and the south-west monsoon winds.

***(v)* Labour.** Coffee cultivation requires plenty of cheap and skilled labour for various operations including sowing, transplanting, pruning, plucking, drying, grading and packing of coffee.

Production and Distribution

India produces about 3.7 per cent of world's coffee on almost the same percentage of coffee area. Thus India is an insignificant producer of coffee and stands nowhere when compared with Brazil (25%), Colombia (15%) and Indonesia (7%). Area, production and yield of coffee in India are shown in **Table 22.40**. Currently, India is the seventh largest producer of coffee in the world after Brazil, Vietnam, Colombia, Indonesia, Ethiopia, and Mexico.

Coffea Arabica and *Coffea Robusta* are the two main varieties of coffee grown in India accounting for 49 per cent and 51 per cent of area respectively under coffee.

The restricted agro-climatic conditions have forced the coffee plantations to confine themselves to small area in south India comprising hill areas around Nilgiris. Almost the entire production is shared by three states namely Karnataka, Kerala and Tamil Nadu.

TABLE 22.40. Area, production and yield of coffee in India

Year	2012-13	2013-14	2014-15	2015-16	2016-17	2017-18	2018-19	2019-20	2020-21	2021-22
Area (million hectares)	0.4	0.04	0.4	0.4	0.4	0.4	0.4	0.4	0.4	0.4
Production (million tonnes)	0.3	0.3	0.3	0.3	0.3	0.3	0.3	0.3	0.3	0.3
Yield (kg/hectare)	766	727	766	766	761	765	767	713	790	797

Source : Economic Survey, 2022-23, Statistical Appendix, pp. 37-39.

TABLE 22.41. Export of Coffee from India

Year	1960-61	1970-71	1980-81	1990-91	2000-01	2019-20	2020-21	2021-22
Quantity (thousand tonnes)	19.7	32.2	87.3	86.5	184.9	257.0	245.0	333
Value (₹ crore)	7	25	214	252	1,185	5,237	5,340	7,614

Source : Economic Survey 2022-23, Statistical Appendix, pp. 116-18.

Karnataka is the largest producer accounting for over 71 per cent of total coffee production and 59 per cent of the area under coffee in India. This state also gives the highest yield of 9.6 quintals/hectare. Most of the 4,650 plantations are at about 1,370 metres above sea level where annual rainfall is 125-150 cm. Kodagu and Chikmagalur districts account for over 80 per cent of the state's total output. The other important coffee producing districts are Shimoga, Hassan and Mysore. **Kerala** is the second largest producer of coffee but lags far behind, accounting only for about 22.27 per cent of the total production of the country. Its yield of 7.6 quintals/hectare is also low as compared to 9.6 quintals/hectare of Karnataka. Most of the coffee plantations are at an altitude of 1,200 m where annual rainfall is over 200 cm. Kozhikode, Waynad, Malappuram, Kollam, Kannur and Palakkad are the chief producing districts. **Tamil Nadu** is the third largest producer where India's 6.5 per cent coffee is produced. About half of Tamil Nadu's coffee is produced in Nilgiri district. The other districts are Madurai, Tirunelveli, Salem and Coimbatore.

Some coffee is also grown in Satara and Ratnagiri districts of Maharashtra. In line with the national policy of tribal development, coffee cultivation is being encouraged in such non-traditional areas as Andhra Pradesh, Odisha, Maharashtra, the north-eastern states and Andaman and Nicobar Islands.

Trade

Among the plantation crops, coffee has made significant contribution to the Indian economy during the last 60 years. Although India contributes only a small per cent of the world production, Indian coffee has created a niche for itself in the international market, particularly Indian Robustas, which are highly preferred for their good blending quality. Arabic coffee is also well received in the international market. In short Indian coffee is well known for its quality and is much in demand in the international market. Therefore, India exports coffee to a large number of countrics including U.K., the U.S.A., Russia, Australia, Iraq and a large number of countries of continental Europe. Chennai, Mangaluru and Kozhikode are the chief ports of export. Indian coffee exports have registered significant increase, both in terms of quantity and earnings during the last few years (*see* **Table 22.41**). In 2020-21 India exported 245 thousand tonnes of coffee which was over 80 per cent of the total production. The total earnings from the export of coffee in that year amounted to ₹ 5,340 crore.

RUBBER

Rubber is a coherent elastic solid obtained from latex of a number of tropical trees of which *Hevea brasiliensis* is the most important. Rubber is used for a variety of purposes from erasing pencil marks to manufacturing of tyres, tubes and a large number of industrial products. The first rubber plantations in India were set up in 1895 on the hill slopes of Kerala. However, rubber cultivation on a commercial scale was introduced in 1902.

Conditions of Growth

Rubber tree (*Hevea brasiliensis*) is a quick growing tall tree acquiring 20-30 metre height. It begins to yield latex in 5-7 years after planting. It requires hot and humid climate with temperature of 25°-35°C and annual rainfall of over 200 cm. The rainfall should be well distributed throughout the year. Dry spell, and low temperatures are harmful. Daily rainfall followed by strong sun is very useful. Deep well drained loamy soils on the hill slopes at elevation ranging from 300 to 450 metres above mean sea level provide best conditions for its growth. The yields decline at higher elevations and no rubber plantations are found above 700 m elevation.

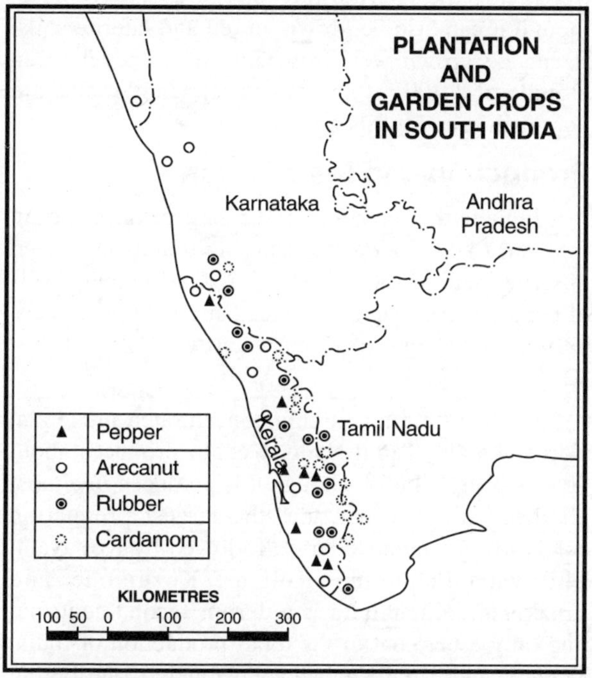

FIG. 22.7. Plantation and Garden Crops in South India.

Production and Distribution

India is the third largest natural rubber producing country of the world, next to Thailand and Indonesia, producing about 9 per cent of the global output. From about 200 hectares in 1902-03, the total area under rubber plantations increased to about 0.8 million hectares in 2020-21. Similarly, the production that was a mere 80 tonnes in 1910 increased to about 0.7 million tonnes in 2020-21. (**Table 22.42**).

Though the rubber plantation sector was dominated by large estates during the initial five decades, it has subsequently undergone important structural transformation leading to dominance of small holdings. Today small holdings account for 88 per cent of area and production of rubber in India. The average size of small holding is 0.49 hectares only. However, the average productivity realised by small holders is much higher than that produced by the estates. **Table 22.42** gives the production trends of rubber in India.

Almost entire rubber is produced in Kerala, Tamil Nadu and Karnataka.

Kerala is the largest producer of natural rubber producing 92 per cent of total rubber production of India. Kottayam, Kollam, Ernakulam, Kozhikode districts produce practically all the rubber of this state. **Tamil Nadu** is the second largest producer of rubber but lags far behind Kerala producing only 3 per cent of the total Indian production. Nilgiri, Madurai, Kanniyakumari, Coimbatore and Salem are the chief rubber producing districts of Tamil Nadu. **Karnataka** produces less than 2 per cent of total Indian production. Chikmagalur and Kodagu are the

TABLE 22.42. Area, production and yield of Rubber in India

Year	2011-12	2012-13	2013-14	2014-15	2015-16	2016-17	2017-18	2018-19	2019-20	2020-21	2021-22
Area (Million hectares)	0.7	0.8	0.8	0.8	0.8	0.8	0.8	0.8	0.8	0.8	0.8
Production (Million tonnes)	0.9	0.9	0.8	0.6	0.6	0.7	0.7	0.7	0.7	0.7	0.8
Yield (kg/hectare)	1,229	1,205	994	811	693	845	845	792	866	869	938

Sources : Economic Survey 2022-23, Statistical Appendix, pp. 37-39.

main producing districts. **Tripura** is the fourth largest producer but contributes less than 2% of India's rubber. Andaman & Nicobar Islands also produce small quantities of rubber.

Trade

In India, consumption of rubber is almost always higher than the production and the production-consumption gap is likely to widen as the rate of consumption is faster than that of production. The per capita consumption of rubber in India is less than one kg as against 14 kg in the developed countries. This calls for larger quantities of import of both natural and synthetic rubber.

Considering the long-term requirements of natural rubber in the country, the Rubber Board of India has conducted exploratory surveys and has identified North Eastern region, parts of West Bengal, Odisha, Maharashtra and Karnataka as potential regions of rubber cultivation. Rubber can be grown in an area of 1.2 million hectares with appropriate refinements in agro-management practices in these regions. In the North-Eastern regions about 4.5 lakh hectares of land can be brought under rubber plantations.

SPICES

India has a glorious past, pleasant present and a bright future with respect to production and export of spices. Pepper, cardamom, chillies, turmeric and ginger are some of the important spices produced in India. India is a great exporter of spices. During the past few years, there has been a steady increase in area and production of spices in India. The annual growth rate in area and production of spices is estimated to be 3.6 and 5.6 per cent respectively.

PEPPER

Among the spices, black pepper, "the king of spices" is the most important dollar earning crop which has a decisive role in our national and state economies. It is an important spice which is used for flavouring foodstuffs. It is a tropical plant which requires a minimum of 10°C and maximum of 30°C temperature. A well distributed rainfall of 200-300 cm helps its growth. The plant grows as a vine and needs support of other trees for its growth. It thrives well on deep, friable, well drained loamy soils, overlaying the lateritic hill tops of Western Ghats, though it can also be grown on red and laterite soils. It can be grown well from almost sea level to an altitude of 1,200 m but the coastal sandy plains are not much suited for its growth.

Production and Distribution

India is the second largest producer of pepper in the world after Indonesia. Varying trends have been observed with respect to area, production and yields of pepper depending on the weather conditions. These variations become obvious when we look at **table 22.43**.

Its distribution is highly concentrated in Kerala, Karantaka and Tamil Nadu. **Kerala** produces about 90% pepper of India. Though it is produced in almost all the districts of Kerala, the largest production comes from Kannur district, followed by Kottayam, Thiruvananthapuram, Kollam, Kozhikode and Ernakulam. **Karnataka** is a distant second contributing only 3 per cent of the total production of India. Kodagu and Uttar Kannad are the major contributing districts. **Tamil Nadu** also produces small quantity of pepper.

About one-third of total production of pepper finds its way to the foreign markets. India exports pepper to as many as 80 countries. The Russian Federation, the USA, Canada, East and South European countries, Sudan and Egypt are the main buyers. However, India's hitherto unquestioned status throughout history as the largest producer and exporter of black pepper in the world is now under threat with preliminary forecast on production and

TABLE 22.43. Production of Pepper in India

Year	1997-98	2000-01	2010-11	2015-16	2016-17	2017-18
Area (thousand hectares)	181.5	213.9	184.00	128.59	132.35	136.10
Production (thousand tonnes)	57.3	63.7	52.00	54.64	72.25	69.60
Yield (kg/hectare)	316	298	183	425	546	511

Source : Agricultural Statistics at a Glance, 2018, p. 203-04.

export data from Vietnam. India has to face this challenge to maintain her supremacy with respect to exports of pepper.

CARDAMOM

Cardamom is known as the 'queen of aromatic spices' and is mainly used for masticatory, flavouring and for medicines. It grows well in climate of high heat and humidity with temperature ranging between 15°C and 32°C and a fairly well distributed annual rainfall of 150-300 cm. Well drained forest loams, deep red and laterite soils with plenty of humus and leaf mould are some of the soils which are ideally suited to its successful growth. Tropical rain forests at an altitude of 800-1,600 metres above sea level provide the most congenial environment for its growth. It is a shade loving plant and is grown under shady trees.

Production and distribution

India produces a major part of the world's total cardamom. The production and productivity of cardamom in India have increased in spite of decrease in cardamom area from 94.3 thousand hectares in 1997-98 to 84 thousand hectares in 2017-18. The production of cardamom in India was 28.80 thousand tonnes in 2017-18. Demand for cardamom is increasing in the world market. To meet this demand the productivity has been increased from the level of 123 kg/ha in 1997-98 to 343 kg/ha in 2017-18.

The entire production comes from three states *viz*., Kerala, Karnataka and Tamil Nadu and these states contribute 53, 42 and 5 per cent respectively of the total production of India. (*see* **Fig. 22.7**). In **Kerala**, the crop is largely concentrated in the Cardamom hills. Idukki, Palakkad (Palghat), Kozhikode and Kannur are the leading producing districts. In **Karnataka** the main producing districts are Kodagu, Hassan, Chikmagalur, Uttara and Dakshin Kannad. Madurai is the most outstanding district of **Tamil Nadu.** This is followed by Salem, Coimbatore, Ramnathapuram, the Nilgiris and Tirunelveli.

About half of the total production is exported. More than 60 countries buy our cardamom but our major customers are Saudi Arabia, Russia, U.K., Germany, Sweden, Finland, Kuwait, Afghanistan and Bahrain. India enjoyed near monopoly in area, production and export of cardamom upto early 1980s. As Guatemala stepped up its production from mid-1980s, India was relegated to second position in productivity and export of cardamom. India needs to boost up its production to regain the lost eminence of cardamom trade. Presently Guatemala has emerged as the top producer and exporter of cardamom sharing 90 per cent of the world export of cardamom.

Chillies

Chilli is another important spice produced in India. It requires temperatures ranging from 10° to 30°C and moderate annual rainfall of 60 to 125 cm Too scarce or too heavy rainfall is harmful. It can be grown on a wide variety of soils including black cotton soil, and different types of loamy soils. It can be grown upto elevations of 1,700 metres.

The production of chillies has increased from 870 thousand tonnes in 1997-98 to 2,301 thousand tonnes in 2017-18. The yield per hectare had also increased by about three times from 1,035 kg in 1997-98 to 2,845 kg in 2017-18. Although all states of India produce some quantity of chillies, **Andhra Pradesh** with 50 per cent of the all India production was the largest producer in 2017-18. Guntur, East Godawari and West Godavari, Prakasam are the main chilli producing districts. Warangal, Khammam and Karimnagar are the main chilli producing districts in **Telangana**. **Maharashtra** and **Odisha** produce equal amount of chilli, although way behind Andhra Pradesh. The other major producers were Rajasthan, West Bengal, Karnataka, Tamil Nadu, Gujarat, Uttar Pradesh, Madhya Pradesh and Assam.

TABLE 22.44. Production of Cardamom in India

Year	1997-98	2000-01	2010-11	2015-16	2016-17	2017-18
Area (thousand hectares)	94.30	92.40	87.00	85.66	84.68	84.00
Production (thousand tonnes)	11.7	14.5	16.00	24.07	27.66	28.80
Yield (kg/hectare)	123	157	184	281	327	343

Source : Agricultural Statistics at a Glance, 2018, p. 203.

Most of the chillies produced in India are consumed within the country and only 5 to 7 per cent are exported, mainly to Sri Lanka, the USA and Russia.

Ginger

Ginger is used both as a spice and for making medicines. It is grown in tropical and sub-tropical climates and requires 10° to 25°C temperature and 125-250 cm rainfall. Well drained sandy, clayey or red loams and laterites are best suited soils for its cultivation. It can be grown from sea level to an altitude of 1,300 m above sea level.

India is the largest producer of ginger in the world producing about 80 per cent of the world production. The production of ginger has increased from 252.1 thousand tonnes in 1997-98 to 1,069 thousand tonnes in 2016-17. The area occupied by this crop also increased considerably from 75.6 thousand hectares in 1997 to 168 thousand hectares in 2016-17. A corresponding increase in yield per hectare has also been recorded from 3,335 kg in 1997-98 to 6,382 kg in 2016-17. This shows the overall progress made by this crop. (**Table 22.46**).

Meghalaya, Andhra Pradesh, Kerala, Sikkim, Odisha, Mizoram, Karnataka etc. are the main producers.

About 80-90 per cent of the total production of ginger is consumed within the country and still India is a major exporter of ginger in the international market and accounts for about half of the total world trade. About 80 per cent of our exports go to the West Asian countries.

Turmeric

Turmeric is the native of the tropical lands of South-East Asia. It is an important condiment and is used in dyes and medicines also. It requires tropical climate and well drained sandy and clayey loams, medium black, red or alluvial soils for its growth. India is an important producer of turmeric in the world. The production almost doubled from 549 thousand tonnes in 1997-98 to 1,076.6 tonnes in 2017-18. The area under turmeric also increased from 139.7 thousand hectares in 1997-98 to 222.7 thousand hectares in 2017-18. The yield per hectare did not lag behind and increased from 3,931 kg in 1997-98 to 4,862 kg in 2017-18.

Andhra Pradesh is the largest producer, producing about half of the total production of India. Guntur and Cuddapah districts account for two-thirds of the state's total production. Second place is occupied by **Tamil Nadu,** which produces more than 12 per cent turmeric of the country. Coimbatore accounts for 60 per cent of the state's production. **Karnataka** is the third largest producer accounting for 10.5% of the total production. Mysuru and Belgaum are the outstanding producers. **Odisha** produces 9.7% where Phulabani and Koraput are the main producing districts. The other producers are *Gujarat*, *West Bengal*, *Maharashtra*, *Bihar*, *Assam* and *Tripura*.

TABLE 22.45. Production of Chillies (dried) in India

Year	1997-98	2000-01	2010-11	2015-16	2016-17	2017-18
Area (thousand hectares)	840.60	836.50	792.00	811.14	840.11	808.90
Production (thousand tonnes)	870.10	983.70	1,223.00	1,520.39	2,096.10	2,301.10
Yield (kg/hectare)	1,035	1,176	1,544	1,874	2,495	2,845

Source : Agricultural Statistics at a Glance, 2018, pp. 207-02.

TABLE 22.46. Production of Ginger in India

Year	1997-98	2000-01	2010-11	2015-16	2016-17
Area (thousand hectares)	75.6	86.2	149.00	163.62	167.59
Production (thousand tonnes)	252.1	288.0	702.00	1,109.18	1,069.55
Yield (kg/hectare)	3,335	3,341	4,711	6,779	6,382

Source : Agricultural Statistics at a Glance, 2018, p. 199.

About 90 per cent of the total production is consumed within the country and only 10 per cent is exported. The leading buyers of Indian turmeric are the USA, Russia, Japan, Sri Lanka and Singapore.

Arecanut

It is a hard nut which is cut into small pieces and used for chewing with betel leaves, lime and catechu. It is also used in Hindu religious ceremonies. Its stem is used for construction purposes and leaves for thatching.

Areca is a tropical tree which, on maturity, attains a height of 20-25 metres. It flourishes well in areas having 15° to 35°C temperature and 200-300 cm annual rainfall. It grows on a variety of soils ranging from well-drained laterite, red loamy to alluvial soils. Its cultivation can be done from sea level to 1,000 metres.

India is the largest producer of arecanut in the world. In the year 2017-18 India produced over 809 thousand tonnes of arecanut from 494 thousand hectares of land. About 40 lakh people who are engaged in cultivating, curing, processing and trading of arecanut earn their livelihood from this nut. Kerala, Karnataka, Tripura, Assam and Meghalya are the major producing states accounting for about 90 per cent of the total output of India. **Kerala** is the largest producer accounting for 37 per cent of the Indian production. Kannur, Malappuram, Kollam, Kozhikode and Thrissur are the main producing districts of Kerala. In **Karantaka,** it is grown in Dakshina Kannada, Uttar Kannada, Chickamagalur, Shimoga and Tumkur districts. **Assam** produces about one-fourth of India's arecanut where Kamrup, Sibsagar and Darrang districts are important producers. Ratnagiri and Kolaba districts of **Maharashtra,** Coimbatore and Salem districts of **Tamil Nadu** are other important producers. West Bengal, Andhra Pradesh, Goa and Puducherry also produce some arecanut.

Most of the arecanut is consumed within the country especially in the south Indian states and only a small quantity is exported mainly to Nepal, UAR, Pakistan, Saudi Arabia, Aman, Kenya and Singapore.

Coconut

Coconut palm is a very useful tree which gives us several items of every day use. Coconut is used for the manufacturing of ***copra*** which gives us oil. This oil is used for cooking and for several other purposes. The tree trunk gives timber and the shells of the nut are used as fuel. Leaves are used for various purposes such as making of mats, baskets, screens, etc. and for roofing the huts. Juice of green nuts serves as a sweet health drink. *Gur*, sugar, *toddy* and vinegar are made from the juice collected from coconut spaths. Coconut oil-cake serves as food for poor people and is also fed to cattle. Being a perennial crop, coconut has distinct features such as long period of economic life span of more than 60 years and long gestation period of 5-7 years.

Coconut is a tropical crop and is grown where temperature is 25° to 30°C and a fairly well distributed annual rainfall of 125 to 130 cm. In a few

TABLE 22.47. Production of Turmeric in India

Year	1997-98	2000-01	2010-11	2015-16	2016-17	2017-18
Area (thousand hectares)	139.70	191.70	195.00	184.00	221.78	222.70
Production (thousand tonnes)	549.20	714.30	993.00	830.00	1,056.10	1,076.60
Yield (kg/hectare)	3,931	3,726	5,092	4,511	4,762	4,862

Source : Agricultural Statistics at a Glance, 2018, p. 201.

TABLE 22.48. Production of Arecanut in India

Year	1997-98	2000-01	2010-11	2015-16	2016-17	2017-18
Area (thousand hectares)	273.70	375.20	400.30	474.36	454.65	494.60
Production (thousand tonnes)	333.50	373.10	478.00	713.84	722.85	809.20
Yield (kg/hectare)	1,218	1,157	1,195	1,505	1,590	1,636

Source : Agricultural Statistics at a Glance, 2018, p. 198.

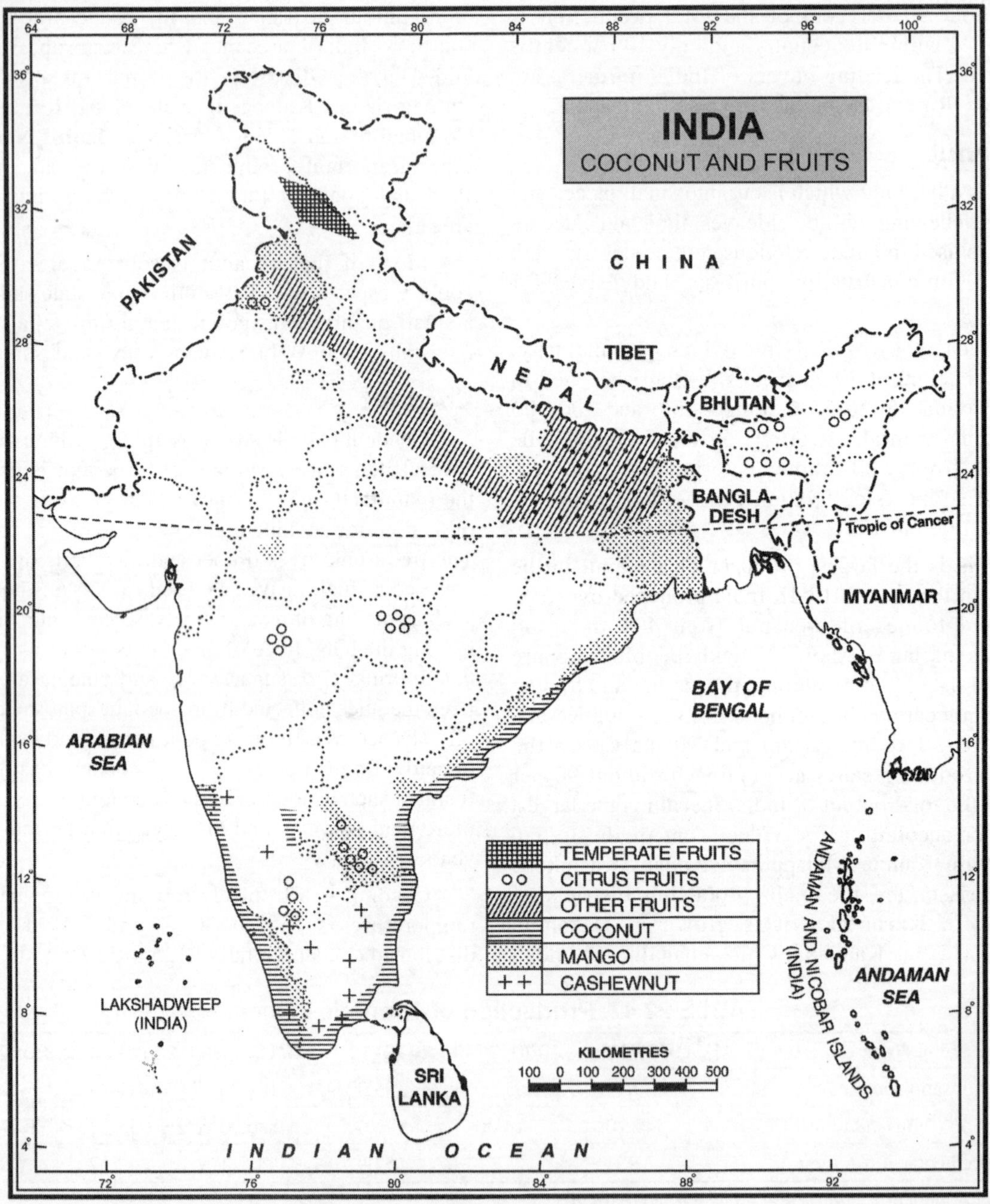

FIG. 22.8. India : Coconut and Fruits

places, especially in Odisha, coconut is grown with as little as 100 cm annual rainfall. Frost and drought are very harmful to coconut. It is predominantly grown under rainfed condition in Kerala and parts of coastal Karnataka and Tamil Nadu. In rest of the country it is mainly grown under irrigation. Well drained rich loamy soils are best suited for its cultivation. It grows well on sandy loams along sea-coasts and in adjoining river valleys.

India is the third largest coconut producing country of the world next to Philippines and Indonesia. It is an important plantation crop grown in the southern part of India. There has been considerable increase in area, production and yield of coconut in India over the years (**Table 22.49**).

Kerala is the largest producer accounting for about 35% of the total production of India in 2020-21. Though coconut is grown throughout the state, largest production comes from Thrissur, Ernakulam, Kollam, Thiruvananthapuram, Alappuzha, Kannur, Kottayam and Malappuram districts. These districts produce more than three-fourths of the total production of Kerala. **Tamil Nadu** is the second largest producer of coconut producing about 25 per cent of the total production of India in 2020-21. Though every district of Tamil Nadu produces some quantity of coconut, over half the production of the state comes from Thanjavur, Coimbatore and Salem districts. **Karnataka** is the third largest producer accounting for over 23 per cent of the total production of the country. About 80 per cent of Karnataka's coconut is produced by Uttara Kannada, Dakshina Kannada, Tumkur, Hassan, Chikmagalur and Chitradurg districts. **Andhra Pradesh's** production mainly comes from East Godavari, West Godavari and Srikakulam districts. The other producers are *West Bengal, Odisha, Gujarat, Maharashtra, Assam, Bihar, Andaman and Nicobar Islands, Lakshadweep* and *Puducherry*.

The export of coconut products indicate that India has done well in coir and coir products, while in the export of other coconut based products, its share is very negligible compared to other coconut growing countries of the world. However, the export of coconut products showed considerable increase from 1990-91 to 2020-21. The global demand for coconut and coconut products has increased considerably. It is estimated that 9 per cent increase per annum for fresh

TABLE 22.49. Production of Coconut in India

Year	1991-92	2001-02	2010-11	2015-16	2016-17	2017-18	2017-18	2018-19	2019-20	2020-21
Area (thousand hectares)	1,529	1,893	1,896	2,088	2,082	2,079	2,097	2,151	2,173	2,185
Production (thousand tonnes)	10,080	12,822	10,840	15,256	16,486	16,228	16,413	14,682	14,006	14,572
Yield (kg/hectare)	6,593	6,773	5,717	7,305	7,918	7,804	7,828	6,826	6,445	6,670

Source : Agricultural Statistics at a Glance, 2021, p. 171.

TABLE 22.50. Distribution of Coconut in India (2020-21)

Sl. No.	States	Area		Production		Yield
		'000 hectares	%ge of all India	'000 tonnes	%ge of all India	kg/hectares
1.	Kerala	760.72	34.82	5,114.00	35.09	6,723
2.	Tamil Nadu	437.57	20.03	3,710.63	25.46	8,480
3.	Karnataka	642.47	29.41	3,392.08	23.28	5,280
4.	Andhra Pradesh	114.31	5.23	880.24	6.04	7,701
5.	Maharashtra	29.95	1.37	361.15	2.48	12,059
6.	Odisha	52.35	2.40	274.61	1.88	5,246
7.	West Bengal	31.84	1.46	272.33	1.87	8,553
8.	Gujarat	49.31	2.26	203.97	1.40	4,136
9.	Assam	25.03	1.15	147.19	1,01	5,880
10.	Bihar	20.82	0.95	127.31	0.87	6,116
	Others	20.33	0.93	88.61	0.61	4,358
	All India	**2,184.70**	**100.00**	**14,572.12**	**100.00**	**6,670**

Note: States have been arranged in descending order of production in 2020-21.

Source : Agricultural Statistics at a Glance, 2021, p. 172.

coconuts, 45 per cent increase for coconut milk, 45 per cent for activated carbon, eight-fold increase for coir products and 100 per cent increase in coir dust will be recorded in the next 5 to 10 years.

HORTICULTURE

The importance of horticulture in improving the productivity of land, generating employment, improving economic condition of the farmers and entrepreneurs, enhancing exports and, above all providing nutritional security to the people is widely acknowledged. Horticulture sector, which includes fruits, vegetables, spices and floriculture, among others covered 27,586 thousand hectares of land in 2020-21. The horticulture sector contributed about 31 per cent of GDP from agriculture in 2020-21. In 2020-21, India accounted for about 10 per cent of the global production of fruits and was the second largest producer of fruits in the world. Immense agro-climatic diversity enables India to grow a large variety of horticulture crops. The country holds first position in global production of bananas, mangoes and cashew and is among the first ten in citrus fruits, pineapple and apple production. India holds first position in global production of cauliflower, second in onion, third in cabbage and is among the top ten in production of potato, tomato, and green peas.

Cashewnut

Cashew tree is grown for cashewnut and cashew apple. Cashew kernel is used as a dry fruit delicacy. It is extracted from the nut by roasting, shelling and peeling. The oil drawn from the shell is used for manufacturing paints and varnishes. Juice from the cashew apple is used for making syrup, jam, squash and wine.

Cashew requires average temperature between 16°C and 25°C and a wide range of rainfall from 50 to 350 cm. It can be easily grown on poor rocky soils. It grows well on laterite soils on the west coast and on sandy soils on the east coast. The average yield of cashew kernel varies from 600 to 700 kg/hectare.

Cashew plantations were introduced in India by the Portuguese on the western coast of the country during 16th century. At present, India holds first position in the world in the production of cashew and it is now widely grown on western and eastern coasts. In 2020-21, India produced 7.4 lakh tonnes of cashew from 11.6 lakh hectares of land. In 2020-21, Andhra Pradesh (14.3%), Odisha (12.1%), Karnataka (10.9%), Kerala (10.8%) and Tamil Nadu (8.7%) were the major producers of cashewnut. The other producers were Goa, West Bengal and Jharkhand.

India is the largest exporter of cashewnut kernel in the world. She meets about 90 per cent of the world demand of cashewnuts. In 2020-21, India exported 74 thousand tonnes of cashew kernel valued at ₹ 3,132 crore. India exports cashew kernel to about 50 countries, the chief among which are the U.S.A., Russia, the U.K., Germany, Canada, Australia, the Netherlands, Malaysia and Japan.

Mango

Mango is the native of monsoon lands and is grown in areas with temperature from 20°C to 30°C and rainfall 75 to 250 cm. It can grow in almost all soils of India but prefers rich clayey loams. It is largely grown in groves especially near towns and villages where it has a ready market.

India is the largest producer where about 1,000 varieties of mango are grown. The main varieties are *dashari*, *langda*, *chausa*, *sapheda*, *fajli*, *malda*, *mohanbhog*, *alfonso* and *shahpasand*. Mango occupied an area of 2,325 thousand hectares and production was about 20,822 thousand tonnes in 2020-21. India contributes over 54% of world production of mango. Uttar Pradesh, Bihar, Andhra Pradesh, West Bengal, Odisha, Kerala, Tamil Nadu, Maharashtra, Gujarat and Karnataka are its main producers.

Most of the mango produced is consumed within the country as a delicious fruit and for making pickles, *chutneys* and squashes. Even then, India is the largest exporter of mango in the world. Major importers of Indian mango are Bangladesh (33.76%), UAE (27.21%), Saudi Arabia (6.83%) and U.K. (4.95%).

Apple

Apple is a temperate fruit crop. It requires average temperature from 21° to 24°C during the active growing season. For optimum growth, 100-125 cm rainfall, well-distributed throughout the growing season is considered most favourable. The orchard localities should be free from hail storms and frost. Low temperature, rain, fog and cloudy weather hampers its proper growth at the time of maturity.

These conditions are found on hill slopes at altitudes ranging from 1,500-2,700 m above mean sea level.

Loamy soils, rich in organic matter and having good drainage are most suitable for apple cultivation. Soil should be free from hard substrata and waterlogging.

In most areas apple orchards have replaced millet crop which is a low value crop. This has raised the income levels and standard of living of the farmers. In addition it has generated employment in processing, packing and transporting apples.

Apple is a nutritive fruit and is in great demand throughout the country, particularly in urban markets. At the same time, it is a perishable commodity and has to be brought to the market as early as possible. This requires quick and efficient means of transport.

The main areas of apple production are Kullu, Kinnaur and Shimla districts in Himachal Pradesh, the Kashmir Valley and hilly areas of Uttarakhand. In 2020-21, 2,057 thousand tonnes of apples were produced from 312 thousand hectares of land.

Banana

Banana is primarily a tropical and sub-tropical crop, requiring average temperature of 20° to 30°C throughout in growth period. The rainfall should be fairly above 150 cm. The banana tree grows well in rich, well drained soil with ample moisture and humus.

India is the largest producer of banana in the world and accounts for 29.7% of the world production. Banana ranks third in area covering 12.46% of the total fruit area but first in total production accounting for over one-third of total fruit production. The production more than doubled from 12,642.5 thousand tonnes in 1997-98 to 33,379 thousand tonnes in 2020-21. Although banana growing is spread all over India, peninsular India provides ideal conditions for its cultivation. Tamil Nadu, Gujarat, and Andhra Pradesh are the main producers and account for 14.2%, 13.8% and 13.4% respectively of all India production. The other producers are Maharashtra, Karnataka, Kerala and Assam, Most of banana is produced on a small scale under different production systems. Some banana is exported. The main destination of banana export is the UAE.

Orange

Orange is the most important citrus fruit which is widely grown both in north and south India. Soil seems to be a more important factor for orange than climate. Most of orange orchards are rainfed and are located at heights from 600 to 1,500 m. Well-drained, even textured sandy loams which permit root penetration upto 2-4 metres are best for orange cultivation.

Although orange is grown in almost all the states, its cultivation is more prominently concentrated in the hilly region of Uttarakhand, Kangra valley of Himachal Pradesh, Darjeeling in W. Bengal, Khasi and Jaintia Hills in Meghalaya. Hyderabad district in Telangana, Kodagu district of Karnataka, Waynad district of Kerala, Nilgiri district of Tamil Nadu and Nagpur and Pune districts of Maharashtra. In

TABLE 22.51. Production of Important Fruits in India

Name of the fruit	2018-19		2019-20		2020-21	
	Area	Production	Area	Production	Area	Production
Apple	308	2,316	310	2,814	312	2,057
Banana	866	30,460	897	32,594	923	33,379
Mandarin (Orange)	469	6,243	454	6,136	462	6,026
Sweet Orange (Mosambi)	187	3,266	217	4,040	203	3,825
Grapes	140	3,041	150	3,181	152	3,229
Mango	2,296	21,378	2,294	20,317	2,325	20,822
Peach	19	118	19	119	18	118
Pear	42	300	43	290	42	280
Papaya	149	6,050	142	5,780	150	6,063

Area in '000 hectares. Production in '000 tonnes. ***Source :*** Agricultural Statistics at a Glance, 2021, p. 158.

2020-21, India produced 6,026 thousand tonnes of oranges from 469 thousand hectares of land.

Grape

Grape is a sub-tropical plant and requires long summer, short winter, a moderately fertile well drained soil, relatively low water supply during the growing period and a bright sunshine during the three months in which the fruit matures. In northern India, the plant gives only one crop during summer but in south India the plant grows throughout the year and yields two crops a year, one in March-April and the other in August-September.

Grapes can be grown any where in India but there are certain areas where this fruit is grown more intensively. Among the major producing states are Uttarakhand, Himachal Pradesh, Jammu and Kashmir and Punjab in the north and Maharashtra, Andhra Pradesh, Telangana, Tamil Nadu and Karnataka in the south. Some grape is exported. The main export destinations for grape are Bangladesh, the Netherlands, U.K. and U.A.E. In 2020-21, India produced 3,229 thousand tonnes of grapes from 152 thousand hectares of land.

Peach

Peach is more of a luxary than apple, due to its more perishable nature. It grows well in temperate climate. The main areas of peach cultivation are in Himachal Pradesh, Uttarakhand and Kashmir Valley. It cannot be transported over long distances due to its highly perishable nature and is primarily used in local market. However, some peach is carried over long distances at heavy transportation costs.

Pear

Pear is another temperate fruit, mainly grown in Kashmir, Kumaon region of Uttarakhand and Himachal Pradesh in the north and the Nilgiri hills in the south. These areas offer suitable conditions of cold winters, cool summers, moderate rainfall, high percentage of cloudiness and mist.

Apricot

Apricot is also a temperate fruit which requires 130 to 200 cm rainfall, moderate temperatures and humid nights. It requires cool weather with abundant moisture during early parts and moderately high temperature during the later part of its growth. It is mainly grown in Kashmir valley, Himachal Pradesh and Kumaon region of Uttarakhand.

Strawberry

Strawberry requires above 16°C temperature during its growing season and a lot of water because strawberry fields are submerged under 10 cm of fresh and slowly moving water for atleast three months. Any fertile soil which can retain water for a sufficiently long time is suitable for its cultivation but sandy loams are best suited for it. The hilly areas of J & K, H.P. and Uttarakhand are the main producers. Jeolikote in Nainital district is the largest producer of strawberry, producing nearly all the canned strawberries.

Almond and walnut are two other temperate fruits grown mainly in the Kashmir Valley, Himachal Pradesh and Kumaon-Garhwal region of Uttarakhand.

VEGETABLES

India also grows a large variety of vegetables. With growing population, health care and urbanization, demand for vegetables has increased considerably after Independence. With 100 million tonnes of annual production of vegetables, India ranks as the highest producer of vegetables in the world. Most of the vegetables are short duration crops as a result of which two to three crops are raised from the same piece of land in one year. Since, different vegetables are grown in different seasons, the process of vegetable cultivation continues throughout the year. Most of the vegetables are grown around urban areas where they find ready market for their sale. A strong vegetable sector in India will lead to economic growth throughout the country. Cultivating vegetables provides more jobs compared to cereal production. It will diversify and generate farm income larger than other products. India grows an amazing number of different vegetables—175 different types in all—but potato, tomato, onion, cabbage and cauliflower account for 60 per cent of the total production. The other important vegetables are cucumber, carrot, brinjal, mushroom, coloquintida, pumpkin, gourd, and many more. It is projected that the domestic vegetable requirements will increase from current level of 83-91 million tonnes in 2018-19 to 151-183 million tonnes in 2030. This demand will have to be fulfilled by increasing home production or by resorting to imports.

REFERENCES

Ali, Masood and Shivkumar (2004) : Pulses—Amazing nutrient composition, *The Hindu Survey of Indian Agriculture* (2004), Kasturi & Sons Ltd., Chennai, pp. 39-43.

Chada, K.L. (2004) : Horticulture—Diversifying strategies, *The Hindu Survey of Indian Agriculture* (2004), Kasturi & Sons Ltd., Chennai, pp. 119-121.

Desalphine, S.M. (2004) : Natural Rubber—Recent trends, *The Hindu Survey of Indian Agriculture* (2004), Kasturi & Sons Ltd., Chennai, pp. 60-62.

Dubey, R.N. (1974) : *Economic Geography of India*, Kitab Mahal, Allahabad.

Government of India (2021), Agricultural Statistics at a Glance.

Government of India (2024), A Reference Annual.

Government of India : Economic Survey 2022-23.

Hegde, D.M. (2004) : Oilseeds—Becoming self-reliant, *The Hindu Survey of Indian Agriculture* (2004), Kasturi & Sons Ltd., Chennai, pp. 45-47.

Johnson, B.L.C. (1969) : *South Asia*, Heinemann Educational Books Ltd., London.

Krishna Kumar, A.K. (2004) : Rubber—Mooting cost effective measures, *The Hindu Survey of Indian Agriculture* (2004), Kasturi & Sons Ltd., Chennai, pp. 63-65.

Mishra, B. (2004) : Rice—Exploring new opportunities, *The Hindu Survey of Indian Agriculture* (2004), Kasturi & Sons Ltd., Chennai, pp. 29-31.

Muthiah, S. *et al. (ed)* (1987) : *A Social and Economic Atlas of India*, Oxford University Press, Delhi.

Muthaiah, S. *et al.* (1990) : *An Atlas of India*, Oxford University Press, Delhi.

Peter, K.V. *et al.* (2004) : Spices—Future Prospects, *The Hindu Survey of Indian Agriculture* (2004), Kasturi & Sons, Chennai, pp. 56-59.

Rai, Mathura and Pandey, A.K. (2004) : Hybrid Vegetables—Meeting Global Trade, *The Hindu Survey of Indian Agriculture* (2004), Kasturi & Sons Ltd., Chennai, pp. 129-131.

Rajgopal, V. *et al.* (2004) : Coconut Industry—Improving genetic produce, *The Hindu Survey of Indian Agriculture* (2004), Kasturi & Sons Ltd., Chennai, pp. 67-69.

Rathinam, P. (2004) : Coconut—Making industry competitive, *The Hindu Survey of Indian Agriculture* (2004), Kasturi & Sons Ltd., Chennai, pp. 70-72.

Robinson, Francis *et al.* (ed.) (1989) : *The Cambridge Encyclopedia of India, Pakistan, Bangladesh and Sri Lanka*, Cambridge University Press, Cambridge.

Seetharama, N. & Rao, B. Dayakar (2004) : Cereals—Sustaining Nutritional security, *The Hindu Survey of Indian Agriculture* (2004), Kasturi & Sons Ltd., Chennai, pp. 37-38.

Sen, H.S. & Saha Depankar (2004) : Jute and Mesta—Insight into future trade; *The Hindu Survey of Indian Agriculture* (2004), Kasturi & Sons Ltd., Chennai, pp. 81-85.

Shanmugasundaram, S. (2004) : Vegetable—Surmountable challenges, *The Hindu Survey of Indian Agriculture* (2004), Kasturi & Sons Ltd., Chennai, pp. 126-127.

Shetty, Nagaraj (2004) : Cocoa and Arecanut—Offering rich biodversity, *The Hindu Survey of Indian Agriculture* (2004) : Kasturi & Sons Ltd., Chennai, pp. 86-87.

Shikamong, S.D. (2004) : Fruits—Lucrative export possibilities, *The Hindu Survey of Indian Agriculture* (2004), Kasturi & Sons Ltd., Chennai, pp. 122-125.

Shoran, Jag *et al.* (2004) : Wheat—New varieties and production, *The Hindu Survey of Indian Agriculture* (2004), Kasturi & Sons Ltd., Chennai, pp. 33-35.

Singh, Gopal (2003) : *A Geography of India*, Atma Ram & Sons, Delhi.

Singh, Jasbir (1974) : *An Agricultural Atlas of India : A Geographical Analysis*, Vishal Publications, Kurukshetra.

Singh, Kapil Deo (2004) : Tobacco—Changing scenario, *The Hindu Survey of Indian Agriculture* (2004), Kasturi & Sons Ltd., Chennai, pp. 88-89.

Singh, Phundan (2004) : Cotton—Emphasising on modernisation, *The Hindu Survey of Indian Agriculture* (2004), Kasturi & Sons Ltd., Chennai, pp. 77-80.

Tala, S.N. (ed) (1980), Handbook of Agriculture, ICAR, New Delhi.

The Hindu (2007, 2010 and 2014) : Survey of Indian Agriculture.

◆◇◆

CHAPTER 23

Mineral Resources

MINERALS AND MINING

Open cast mining and *underground mining* are two chief methods of mining practised in India. Open cast mining is more useful for minerals found just below the surface while underground mining is done to extract minerals found at greater depths. Drilling and pumping is practised for extracting oil and natural gas.

MINERAL WEALTH OF INDIA

India is endowed with a rich variety of minerals. Large size and diverse geological formations have favoured India in providing a wide variety of minerals. According to **Meher D.N. Wadia**, "The mineral wealth of India, though by no means inexhaustible, is varied enough to provide for sound economic and industrial development of the country but has at the same time, certain important deficiencies. It has been estimated that nearly 100 minerals are known to be produced or worked in India, of which nearly 30 may be considered more important including several which although comparatively unimportant in quantity today are capable of material development in future with expansion of industries. The country has fairly abundant reserves of coal, iron and mica, adequate supplies of manganese ore, titanium and aluminium, raw materials for refractories and limestone; but there is a deficiency in ores of copper, lead and zinc. There are workable deposits of tin and nickel." India earns a lot of foreign exchange by exporting a large variety of minerals such as iron ore, titanium, manganese, bauxite, granite and a host of other minerals. At the same time India has to depend upon imports to meet her requirements of some other minerals such as copper, silver, nickel, cobalt, zinc, lead, tin, mercury, limestone, platinum, graphite and so many other minerals.

DISTRIBUTION OF MINERALS AND MINERAL BELTS

The most striking feature of the Indian minerals is their uneven distribution. Some areas are very rich in minerals while some others are completely devoid of this valuable asset. The high rainfall areas of India lack in limestone, gypsum and salts which are soluble in water. The northern plains of India have thick layers of alluvium which has completely concealed the bedrocks. This region of the country is poor in mineral resources. The Himalayas have a variety of rocks but its geological structure is too complex. The exploitation of minerals in this mountainous terrain is not economically viable not only due to small quantity available at any one location but also due to difficult terrain, lack of transportation, sparse population and adverse climatic conditions. It is, thus, evident that our rich mineralised zone with relatively sizeable quantities is confined to the old, crystalline rock structures of plateaus and low hills of peninsular India. This has resulted in the emergence of well defined mineral belts which are briefly described as under :

1. The North-Eastern Peninsular Belt. This belt comprising of Chhota Nagpur plateau, and Odisha Plateau in the states of Jharkhand, West Bengal and Odisha is the richest mineral belt of India. It contains large quantities of coal, iron ore, manganese, mica, bauxite, copper, kyanite, chromite, beryl, apatite and many more minerals. In fact you ask for any major mineral of India and you will find it in this belt. Thus it is a mineral region *par excellence*. The Chota Nagpur plateau is known as the *mineral heart land* of India. According to **Wadia**, this region possesses India's 100 per cent Kyanite, 93 per cent iron ore, 84 per cent coal, 70 per cent chromite, 70 per cent mica, 50 per cent fire clay, 45 per cent asbestos, 45 per cent china clay, 20 per cent limestone and 10 per cent manganese. However, many changes have taken place in the recent years.

2. Central Belt. This belt encompassing parts of Chhattisgarh, Madhya Pradesh, Telangana, Andhra Pradesh and Maharashtra is the second largest mineral belt of India. Large deposits of manganese, bauxite, limestone, marble, coal, gems (panna), mica, iron ore, graphite, etc. are available here. A comprehensive geological survey is still needed to know about the mineral wealth of this belt.

3. The Southern Belt. It covers mostly the Karnataka plateau but extends over the contiguous Tamil Nadu upland. It is more or less similar to the north-eastern peninsular belt, as far as deposits of ferrous minerals and bauxite are concerned but it lacks coal deposits excepting lignite at Neyveli. It also does not have mica and copper deposits. Therefore, its mineral diversity is not as pronounced as that of the north-eastern peninsular belt.

4. The South-Western Belt. Western Karnataka and Goa are included in this belt. It has deposits of iron ore, garnet and clay.

5. The North-Western Belt. This belt extends along the Aravalis in Rajasthan and in adjoining parts of Gujarat. This belt has developed recently and is gradually becoming a productive region, holding great promise for the mining of non-ferrous metals (copper, lead, zinc), uranium, mica, stealite, beryllium and precious stones (aquamarine and emerald). Gujarat is fast becoming an important producer of petroleum, besides producing gypsum, manganese, salt, bauxite, etc.

Outside the main belts described above, minerals in some other parts of the country are scattered here and there. Assam has reserves of petroleum and lignite. The Himalayan region has some deposits of coal, bauxite, copper, slate, etc. Mumbai High and Godavari basin have reserves of oil and natural gas.

With the advancement of technology, even the sea bed is being exploited for minerals. There are great possibilities of obtaining large quantities of oil and natural gas, uranium, manganese, copper, zinc, lead and so many other minerals from the sea bed.

TYPES OF MINERALS

Normally two types of minerals are recognised :

(*i*) Metallic Minerals. These minerals contain metal. Iron ore, copper, manganese, nickel, etc. are important examples of metallic minerals.

Metallic minerals are further sub-divided into ferrous and non-ferrous minerals.

(a) Ferrous Minerals. These minerals have iron content. Iron-ore, manganese, chromite, pyrites, tungsten, nickel, cobalt, etc. are important examples of ferrous minerals.

(b) Non-ferrous Minerals. These minerals do not have iron content. Gold, silver, copper, lead, bauxite, tin, magnesium, etc. are important examples of non-ferrous minerals.

(*ii*) Non-metallic Minerals. These minerals do not contain metal. Limestone, nitrate, potash,

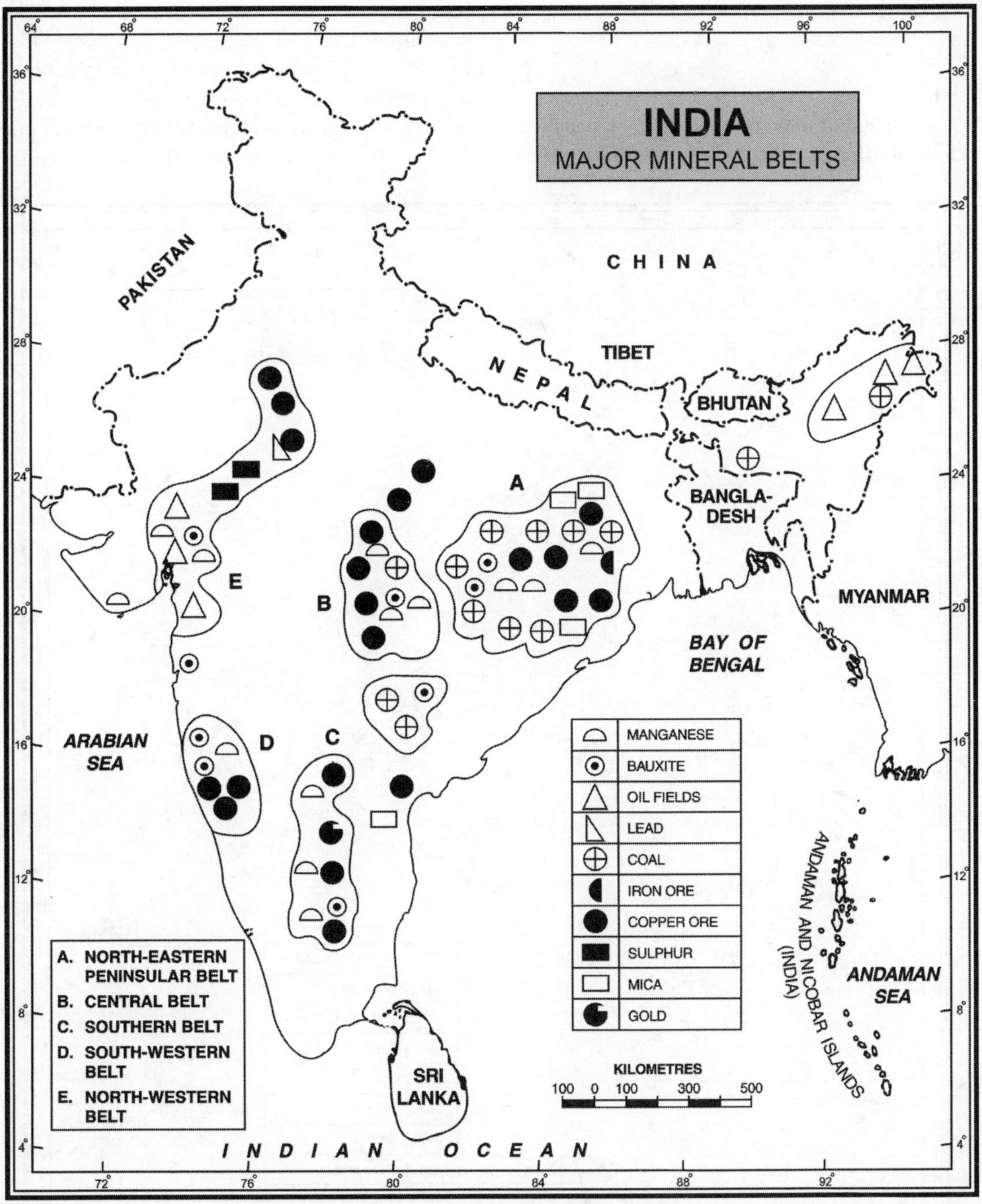

FIG. 23.1. India : Major Mineral Belts

dolomite, mica, gypsum, etc. are important examples of non-metallic minerals. Coal and petroleum are also non-metallic minerals. They are used as fuel and are also known as *mineral fuels*.

METALLIC MINERALS

Metallic minerals form an important section of mining activity in India and provide solid base to metallurgical industries in the country.

Fig. 23.2 shows the distribution of metallic minerals in India.

Iron Ore

Iron is a metal of universal use. It is the backbone of modern civilisation. It is the foundation of our basic industry and is used all over the world. The standard of living of the people of a country is judged by the consumption of iron. Iron is taken out from mines in the form of iron ore. Different types of iron ore contain varying percentage of pure iron. Following four varieties of iron ore are generally recognised.

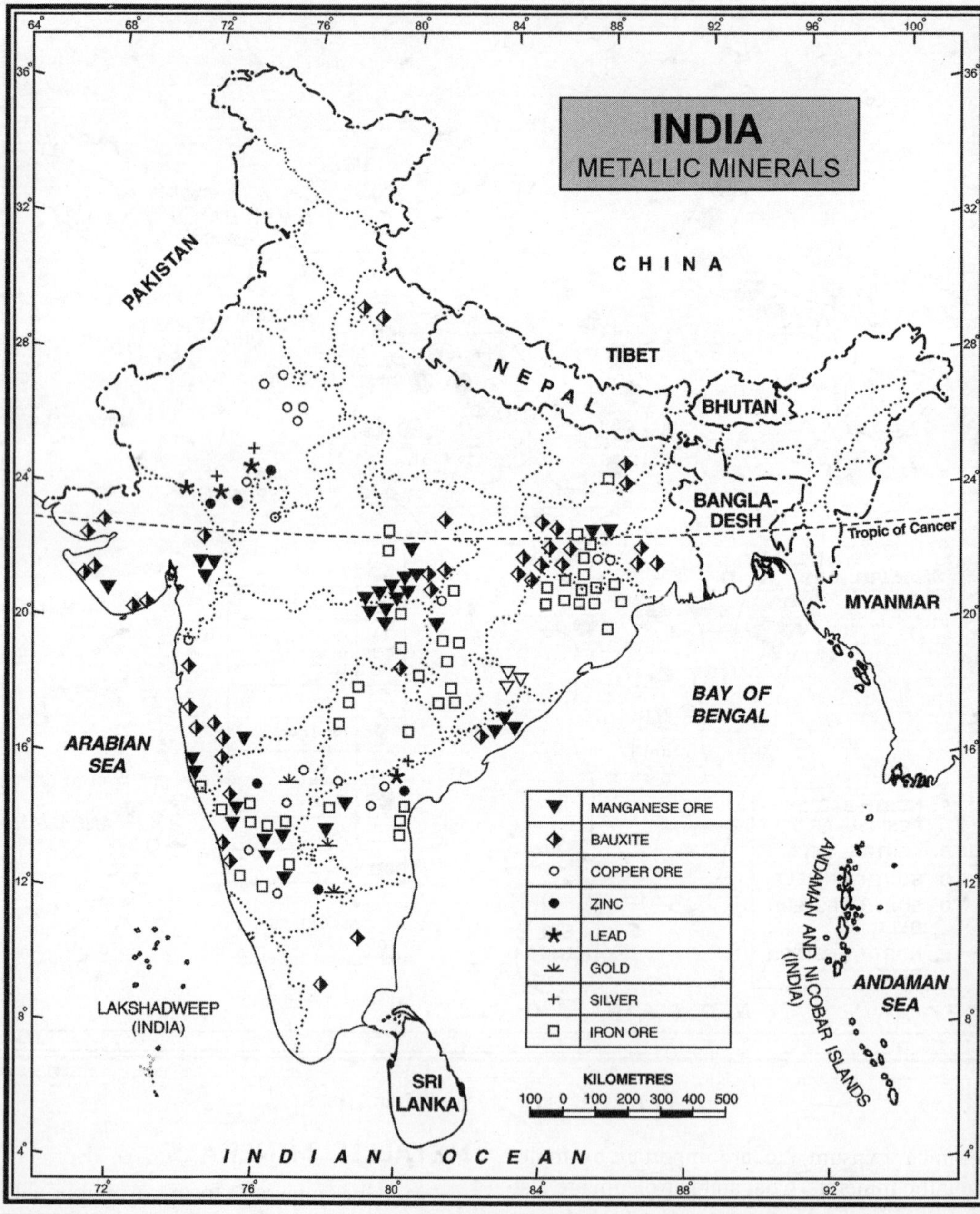

FIG. 23.2. India : Metallic Minerals

1. Haematite. This is the best quality of iron ore with about 70 per cent metallic content and occurs as massive, hard compact and bumpy ore having reddish or coral red colour. Most of the haematite ores are found in Dharwad and Cuddapah rock systems of the peninsular India. Over 80 per cent of the haematite ores are concentrated in eastern parts of India comprising of important iron ore producing states of Odisha, Jharkhand, Chhattisgarh and Andhra Pradesh. In the western section, the major concentration is in Karnataka, Maharashtra and Goa.

2. Magnetite. Also known as ***'black ore'***, due to blackish colour, this is the second best ore, next only to haematite with metallic content varying from 60 to 70 per cent. Like haematite, magnetite ores occur in the Dharward and Cuddapah systems of the peninsular India. Magnetite ores have magnetic quality as a result of which they are known as ***magnetite ores***. Most of the reserves are found in Karnataka, Andhra Pradesh, Rajasthan, Tamil Nadu and Kerala.

3. Limonite. Limonites are inferior ores, yellowish in colour, which contain 40 to 60 per cent iron metal. These are found in Damuda series in Raniganj coal field, Garhwal in Uttarakhand, Mirzapur in Uttar Pradesh and Kangra valley of Himachal Pradesh. Though poor in quality, these have the advantage of easy and cheap mining.

4. Siderite. Also known as '*iron carbonate*' this type of iron ore is of inferior quality and contains less than 40 per cent iron. It also contains many impurities and its mining is not economically viable. However, it is self fluxing due to presence of lime.

Reserves. Hematite and magnetite are the two most important iron ores in India. According to United Nations Framework Classification, the total reserves of haematite ore are estimated at 24,058 million tonnes. About 89 per cent of magnetic ore deposits occur in the southern sector. Major sources of haematite ore located in Odisha, 9,409 million tonnes (39%), Jharkhand 4,710 million tonnes (20%), Chhattisgarh 4,592 million tonnes (19%), Karnataka 2,836 million tonnes (12%), and Goa, 1198 million tonnes (5%). The balance resources are spread in Andhra Pradesh, Assam, Bihar, Maharashtra, Madhya Pradesh, Meghalaya, Rajasthan and Uttar Pradesh.

The total reserves of magnetite ore as per United Nations Framework Classification (UNFC) are estimated at 11,025 million tonnes. About 94 per cent of magnetite ore resources are located in just four states namely, Karnataka 7,802 million tonnes (69%), Andhra Pradesh 1,472 million tonnes (13%), Rajasthan 795 million tonnes (7%) and Tamil Nadu 529 million tonnes (5%). The remaining about 4 per cent are found in Assam, Bihar, Goa, Jharkhand, Kerala, Maharashtra, Meghalaya and Nagaland.

Production and Distribution. Production of iron ore keeps on changing from year to year as is clear from **table 23.1**.

TABLE 23.1. Production of iron ore in India

Year	Production ('000 tonnes)	Value (₹ '000)
2019-20	2,44,083	49,64,30,578
2020-21	2,05,041	5,27,242,469
	2,53,973	96,38,13,280

Source: Economic Survey, 2022-23, Statistics Appendix, p. 56.

It is worth mentioning that significant changes have taken place in the distribution pattern itself during the last few years. Goa occupied the first position among the major iron ore producing states for over a decade, but has been overtaken by Karnataka, Odisha and Chhattisgarh in due course of time. At present, over 96 per cent of India's iron ore is produced by just five states of Odisha, Chhattisgarh, Karnataka, Jharkhand and Goa. This fact speaks volumes of high concentration of iron ore reserves and their lopsided distribution in the country.

1. Odisha. Odisha produces over 50 per cent iron ore of India. The most important deposits occur in Sundargarh, Mayurbhanj, Cuttack, Sambalpur, Keonjhar and Koraput districts. India's richest haematite deposits are located in Barabil-Koira valley where 100 deposits are spread over 53 sq km. The ores are rich in haematites with 60 per cent iron content. Sizeable deposits occur near Gorumahisani, Sulaipat and Badampahar in Mayurbhanj district; Banspani, Tahkurani, Toda, Kodekola, Kurband, Phillora and Kiriburu in Keonjhar district; near Malangtoli, Kandadhar Pahar, Koira and Barsua in Sundargarh district, Tomka range between Patwali and Kassa in Sukind area of Cuttack district, Daitari hill along the boundary between Keonjhar and

Cuttack districts, Hirapur hills in Koraput district and Nalibassa hill in Sambalpur district.

TABLE 23.2. Production of Iron Ore in India (2016-17)

State	Production (thousand tonnes)	Percentage of all India
1. Odisha	99,614	51.86
2. Chhattisgarh	31,062	16.17
3. Karnataka	26,363	13.72
4. Jharkhand	21,335	11.11
5. Goa	8,933	4.65
6. Madhya Pradesh	1,730	0.90
7. Maharashtra	1,321	0.68
8. Rajasthan	1,228	0.65
Others	495	0.26
Total	**1,92,081**	**100.00**

Source : Data computed from Indian Minerals Yearbook, 2017

2. Chhattisgarh. Chhattisgarh, has about 18 per cent of the total iron ore reserves of India. This state produced about 16 per cent of the total iron ore production of the country in 2016-17. The iron ores are widely distributed, the prominent deposits being those of Bastar and Durg districts. The reserves in these districts are estimated to be of the order of 4,064 million tonnes. These reserves are of high grade ore, containing over 65 per cent iron. Bailadila, Dantewada, and Dalli Rajhara are important producers. In Bailadila, 14 deposits are located in 48 km long range running in north-south direction. With estimated reserves of about 1,422 million tonnes, the Bailadila mine is the largest mechanised mine in Asia. An additional ore benefication plant with a capacity of 7.8 million tonnes has been set up in Bailadila. A 270 km long slurry pipeline carries the ore from the Bailadila pithead to the Vizag plant. This has reduced the pressure on road route to a great extent. Bailadila produces high grade ore which is exported through Vishakhapatnam to Japan and other countries where it is in great demand. The Dalli-Rajhara range is 32 km long with iron ore reserves of about 120 million tonnes. The ferrous content in this ore is estimated to be 68-69 per cent. The deposits of this range are being worked by the Hindustan Steels Plant at Bhilai. Raigarh, Bilaspur, and Surguja are other iron ore producing districts.

3. Karnataka. Karnataka is the third largest producer and accounts for nearly 14 per cent of the total iron ore produced in India. In Karnataka production of iron ore has increased by about three times since 1980. Iron ores are widely distributed in the state, but high grade ore deposits are those of Kemmangundi in Bababudan hills of Chikmagalur district and Sandur and Hospet in Bellary district. Most of the ores are high grade haematite and magnetite. The other important producing districts are Chitradurga, Uttar Kannad, Shimoga, Dharwar and Tumkur.

4. Jharkhand. Jharkhand accounts for 25 per cent of reserves and over 11 per cent of the total iron ore production of the country. Iron ore mining first of all started in the Singhbhum district in 1904 (then a part of Bihar). Iron ore of Singhbhum district is of highest quality and will last for hundreds of years. The main iron bearing belt forms a range about 50 km long extending from near Gua to near Pantha in Bonai (Odisha). The other deposits in Singhbhum include those of Budhu Buru, Kotamati Buru and Rajori Buru. The well known Noamandi mines are situated at Kotamati Buru. Magnetite ores occur near Daltenganj in Palamu district. Less important magnetite deposits have been found in Santhal Parganas, Hazaribagh, Dhanbad and Ranchi districts.

5. Goa. Production of iron ore in Goa started quite late and it is a recent development. Starting from a non-entity, this state rose to the highest position and remained the largest producer of iron ore in India for several years. However, this state was surpassed by Odisha, Chhattisgarh, Karnataka and Jharkhand in succession and currently Goa is the fifth largest iron ore producing state of India. Goa now produces about 4.6 per cent of the total production of India. In 1975, the Geological Survey of India located 34 iron bearing reserves which estimated the total ore deposits of 390 million tonnes. There are nearly 315 mines in North Goa, Central Goa and South Goa. Important deposits occur in Pirna-Adolpale-Asnora, Sirigao-Bicholim-Daldal, Sanquelim-Onda, Kudnem-Pisurlem and Kudnem-Surla areas in North Goa; Tolsia-Dongarvado-Sanvordem and Quirapale-Santone-Costi in Central Goa; and Borgadongar, Netarlim, Rivona-Solomba and Barazan in South

Goa. The richest ore deposits are located in North Goa. These areas have the advantage of river transport or ropeways for local transport and that of Marmagao port for exporting the ore. Most of Goa's iron ore is exported to Japan. Most of the ore is of low grade limomite and siderite. Most of the mines are open-cast and mechanised which result in efficient exploitation of iron ore in spite of its inferior quality. About 34,000 people earn their livelihood from iron ore mining and allied activities in Goa.

Others. Apart from the above mentioned major iron ore producing states, iron ore in small quantities

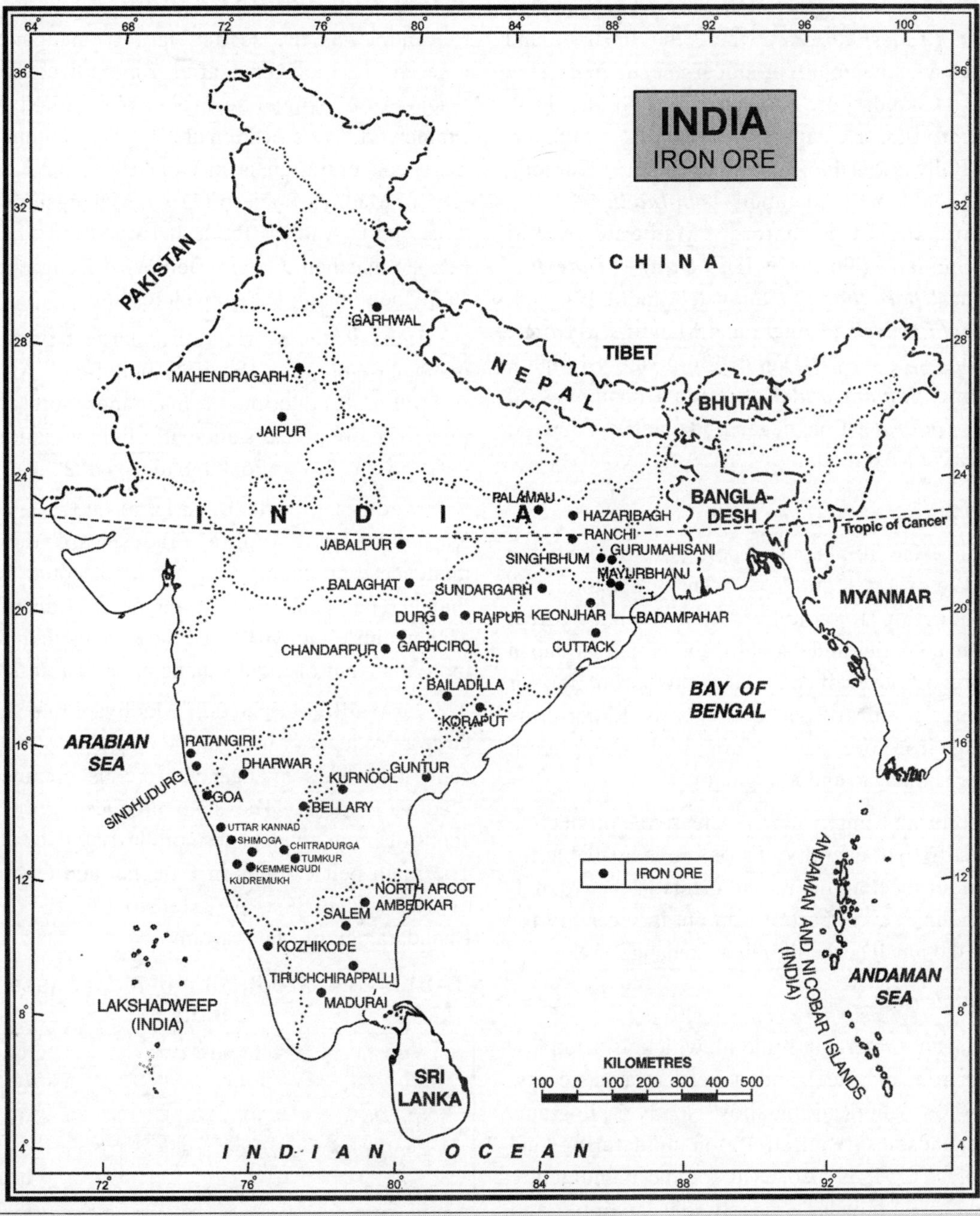

FIG. 23.3. India : Iron Ore.

TABLE 23.3. Exports of Iron Ore from India

Year	1960-61	1970-71	1980-81	1990-91	2000-01	2010-11	2020-21	2021-22
Quantity (Million tonnes)	3.2	21.2	22.4	32.5	20.2	46.9	58	26
Value (₹ crore)	17	117	303	1,049	1,634	21,416	36,254	24,145

Source : The Economic Survey 2022-23, Statistical Appendix, pp. 116-18.

is produced in some other states also. They include ***Madhya Pradesh*** (0.9%) where the Jabalpur and Balaghat are the main producers, ***Maharashtra*** (0.7%) : Chandrapur, Ratnagiri and Sindhudurg, ***Rajasthan*** (0.65%), Jaipur, Udaipur, Alwar, Sikar, Bundi, Bhilwara. ***Andhra Pradesh*** (0.25%) Kurnool, Guntur, Cuddapah, Anantpur. ***Tamilnadu*** : Salem, Tiruchirapalli, Coimbatore, Madurai, Nellai Kattabomman (Tirunelveli); ***Uttar Pradesh*** : Mirzapur, ***Uttarakhand*** : Garhwal, Almora, Nainital; ***Himachal Pradesh*** : Kangra and Mandi; ***Haryana*** : Mahendragarh; ***West Bengal*** : Burdwan, Birbhum, Darjeeling; ***Jammu and Kashmir*** : Udhampur and Jammu; ***Gujarat*** : Bhavnagar, Junagadh, Vadodara; and ***Kerala*** : Kozhikode.

Exports

India is the fifth largest exporter of iron ore in the world. We export about 25 per cent of our total iron ore production to countries like Japan, Korea, European countries and lately to Gulf countries. Japan is the biggest buyer of Indian iron ore accounting for about three-fourths of our total exports. Major ports handling iron ore export are Vishakhapatnam, Paradip, Marmagao and Mangaluru.

Efforts are being made to increase the production so that sufficient quantity of iron ore is available for export after meeting the requirements of the expanding home market. Export of iron ore is necessary for earning the much needed foreign exchange.

Manganese

It is an important mineral which is used for making iron and steel and it acts as a basic raw material for manufacturing alloy. Nearly 6 kilograms of manganese is required for manufacturing one tonne of steel. It is also used for the manufacture of bleaching powder, insecticides, paints, and batteries.

Production and Distribution

India has the second largest manganese ore reserves in the world after Zimbabwe. The total reserves of manganese ore as per United Nations Framework Classification are 504 million tonnes. The main concentration is in Odisha (34%), Karnataka (25%), Madhya Pradesh (12%), Maharashtra (12%), Goa (7%), Andhra Pradesh (6%), and Jharkhand (2%). Rajasthan, Gujarat and West Bengal together share the remaining 3 per cent resources.

India is the world's fifth largest producer of manganese ore after Brazil, Gabon, South Africa and Australia. Production of manganese ore in India remains more or less static with slight variations from year to year. It was 2688038 tonnes in 2020-21.

Madhya Pradesh, Maharashtra, Odisha, Karnataka and Andhra Pradesh are the major manganese producing states which account for more than 97 per cent of the total production of India. Maharashtra and Madhya Pradesh together produce more than half of India's manganese (**Table 23.5**).

1. Madhya Pradesh. Madhya Pradesh is the largest producer of manganese in India. About 27 per cent of India's manganese ore is obtained from Madhya Pradesh. The state produced only 11 per cent of India's manganese ore just two decades ago. The main belt extends in Balaghat and Chhindwara districts. It is just an extension of the Nagpur-Bhandara belt of Maharashtra.

TABLE 23.4. Production of Manganese ore in India

Year	Production in tonnes	Value (₹ '000)
2019-20	29,10,186	1,88,49,100
2020-21	27,03,313	1,74,15,460
2021-22	26,95,991	2,22,40,539

Source: Economic Survey 2022-23, Statistical Appendix, p. 56.

2. Maharashtra. It produces about 25 per cent of India's manganese ore. The main belt is in Nagpur and Bhandara districts. High grade ore is found in Ratnagiri district also.

3. Odisha. Odisha is the third largest producer and produces over 24 per cent manganese ore of India. It is obtained from Gondite deposits in Sundargarh district and Kodurite and Khondolite

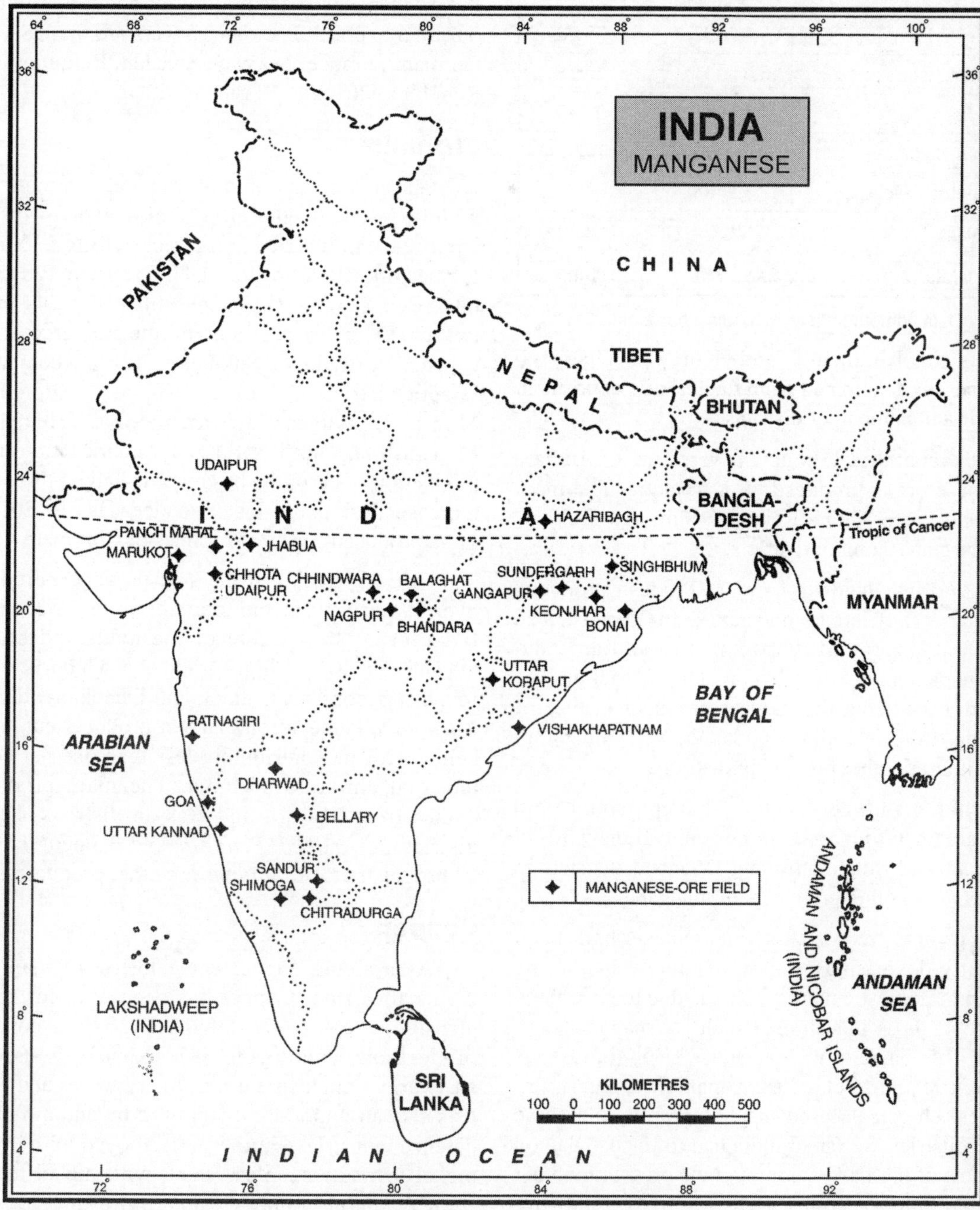

FIG. 23.4. India : Manganese

TABLE 23.5. Distribution of Manganese in India 2016-17

State	Production in tonnes	Percentage of all India Production
1. Madhya Pradesh	6,48,132	27.08
2. Maharashtra	6,04,301	25.25
3. Odisha	5,87,511	24.55
4. Karnataka	2,61,374	10.92
5. Andhra Pradesh	2,32,488	9.71
Others	59,376	2.49
All India	**23,93,182**	**100.00**

Source : Data computed from Minerals Yearbook, 2017.

deposits in Kalahandi and Koraput Districts. Manganese is also mined from the lateritic deposits in Bolangir and Sambalpur districts.

4. Karnataka. About 11 per cent of India's manganese ore is produced by Karnataka. The main deposits occur in Uttara Kannada, Shimoga, Bellary, Chitradurg and Tumkur districts.

5. Andhra Pradesh. Andhra Pradesh produced more than 9% of India's manganese in 2016-17. The main belt is found between Srikakulam and Vishakhapatnam districts. Srikakulam district has the distinction of being the earliest producer (1892) of manganese ore in India. Cuddapah, Vijayanagaram and Guntur are other producing districts.

Other producers. Goa, Panchmahals and Vadodara in *Gujarat*, Udaipur and Banswara in *Rajasthan* and Singhbhum and Dhanbad districts in *Jharkhand* are other producers of manganese.

Exports. Exports of manganese had been constantly decreasing because of rapidly increasing demand in domestic market. This is due to expansion of some of those industries, which use manganese as one of the basic raw materials. Such industries include iron and steel industry, manufacturing of dry batteries, chemicals used in photography and some other industries. So far so India had to impose ban on the export of high and medium grade ores with 35 per cent manganese content in 1971 to feed our industries. India is now exporting low grade ores with less than 35 per cent manganese content for which there is not much demand in the international market. It is worth mentioning here that the share of manganese ore exported to total production of the ore had fallen from 88.86 per cent in 1970-71 to less than 3 per cent in 2016-17. Japan is the largest buyer of Indian manganese accounting for about two-thirds of our total exports. The other buyers are the USA, UK, Germany, France, Norway, Sweden, Belgium, Czech Republic, Slovakia, Ukraine, etc.

Chromite

Chromite is an oxide of iron and chromium. It is widely used in metallurgical, refractories and chemical industries. As per United Nations Framework Classification (UNFC) system (2015), the total reserves of chromite are estimated at 344 million tonnes. Over 96 per cent of the resources are in Odisha, mostly in the Sukinda valley in Keonjhar and Jajapur districts. Minor deposits are scattered over Manipur, Nagaland, Karnataka, Jharkhand, Maharashtra, Tamil Nadu and Andhra Pradesh. The production of chromite has recorded about three times increase during the last two decades. In 2016-17, India produced 37,27,777 tonnes of Chromate.

Odisha is almost the sole producer, producing over 99 per cent of the total chromite production of India. Over 85 per cent of the total production of Odisha consists of high grade ore which is mainly found in Keonjhar, Cuttack and Dhenkanal districts. ***Karnataka*** is the second largest producer but lags far behind Odisha contributing less than one per cent of the total chromite of India. The main production comes from Mysuru and Hassan districts. Krishna district of *Andhra Pradesh* and Tamenglong and Ukhrul districts of *Manipur* are other producers.

Copper

As a metal, copper came in use of man much earlier than iron. Copper has been used for making utensils and coins since long. Being a good conductor of electricity and ductile, it is extensively used in a vast variety of electrical machinery, wires and cables. It is also an important metal used by automobile and defence industries. Further, it is alloyed with iron and nickel to make stainless steel, with nickel to make 'morel metal' and with aluminium to make 'duralumin'. When alloyed with zinc it is known as 'brass' and with tin as 'bronze'.

Copper ore is found in ancient as well as in younger rock formations and occurs as veins, as dissemination and as bedded deposits. Mining for copper is a costly and a tedious affair because most of the copper ores contain a small percentage of the metal. Against the international average of metal content (in the ore) of 2.5 per cent, Indian ore grade averages less than one per cent.

Production and Distribution

The total reserves of copper in the country are estimated at 1511.50 million tonnes with about 12.16 million tonnes of copper metal (2015). Rajasthan has 813.13 million tonnes ore (54%) containing 4.48 million tonnes of copper metal. This is followed by Madhya Pradesh 283.43 million tonnes (19%) containing 3.42 million tonnes of copper metal and Jharkhand 295.39 million tonnes ore (20%) containing 3.28 million tonnes of copper metal. The rest 7 per cent of reserves are accounted for by Andhra Pradesh, Gujarat, Haryana, Karnataka, Maharashtra, Meghalaya, Nagaland, Odisha, Sikkim, Tamil Nadu, Telangana, Uttarakhand and West Bengal.

TABLE 23.6. Production of Copper in India (in tonnes)

Year	Copper ore	Copper concentrates
2019-20	39,52,472	1,24,586
2020-21	32,72,915	1,08,718
2021-22	35,69,632	1,14,420

Source: Economic Survey 2022-23, Statistical Appendix, p. 56.

The entire copper of India is produced by three states namely Madhya Pradesh, Rajasthan and Jharkhand.

1. Madhya Pradesh. Madhya Pradesh has become the largest producer of copper in India surpassing Karnataka, Rajasthan and Jharkhand in succession. In the year 2016-17 the state produced 62 per cent of the total copper production of the country. The state is blessed with a fairly large belt in Taregaon area, in Malanjkhand belt of Balaghat district. This district has recoverable reserve of 84.83 million tonnes of copper ore having 1,006 thousand tonnes of metal. Reserves of moderate size are also found in Kherlibazar-Bargaon area of Betul district. Some other areas are also reported to have copper ore reserves.

2. Rajasthan. Rajasthan has also progressed a lot with respect to production of copper and is now the second largest producing state in India accounting for over 29 per cent of the total production of the country. Most of the copper reserves are found along the Aravali range, spread over the districts of Ajmer, Alwar, Bhilwara, Chittaurgarh, Dungarpur, Jaipur, Jhunjhunu, Pali, Sikar, Sirohi and Udaipur. The Khetri-Singhana belt in Jhunjhunu district is the most important copper producing area. This belt runs in north-east to south-west direction over a distance of 80 km from Singhana to Raghunathgarh with average width varying from 3 to 5 km. The annual output of copper ore at Khetri is 1.8 million tonnes yielding around 16,000 tonnes of metal. The Kho-Dariba area about 48 km to the south-west of the Alwar city and Delwara-Kirovli area about 30 km from Udaipur are other important producers. In Kishangarh area of Ajmer district, 2.5 million tonnes of copper ore, having 0.60 per cent copper, have tentatively been estimated.

3. Jharkhand. Jharkhand, used to be the largest producer of copper till early 1980s but it has lost much importance and has slipped to third position, partly due to fall in its own production and mainly due to increased production of other states. The state's share of copper ore production has fallen from 62 per cent of the nation's total production in 1977-78 to a desperate 8 per cent in 2016-17. The main copper belt extends over a distance of 130 km. Singhbhum is the most important copper producing district where Rakha, Kendadih, Surda, Dhobani, Mosabani and some other areas have proved reserves of 58.044 million tonnes from which 1,480.12 thousand tonnes of metal may be recovered. Hasatu, Baraganda, Jaradih, Parasnath, Barkanath, etc. in Hazaribagh district; Bairakhi in Santhal Parganas area and some parts of Palamu and Gaya districts are also reported to have some deposits of copper ore.

Imports. The production of copper ore in the country always falls short of our requirements and India has to import copper from other countries. The major part of supply comes from the USA, Canada, Zimbabwe, Japan and Mexico. The quantity of import varies from year to year depending upon demand and supply.

Nickel

Nickel does not occur free in nature and is found in association with copper, uranium and other metals. It is used as an important alloying material. When alloyed with iron, rust proof stainless steel of superior quality is obtained, from which utensils and sheets are made. Because of its greater hardness and tensile strength nickel steel is used for manufacturing armoured plates, motor cars, bullet jackets and in naval construction. When alloyed with copper or silver, it is used for making coins. Nickel-aluminium alloys are used for manufacturing aeroplanes and internal combustion engines. Metallic nickel is used for making storage batteries and as a catalyst for hydrogenation or hardening of fats and oils intended for use in soap and foodstuffs and in making *vanaspati*.

Important occurrences of nickeliferous limonite are found in the Sukinda valley of Jajapur district, Odisha, where it occurs as oxide. A suitable process has been developed for its utilization. Nickel also occurs in sulphide form along with copper mineralization in east Sighbhum district, Jharkhand. In addition, it is found associated with uranium deposits at Jaduguda, Jharkhand and process is being developed for its recovery. Other important occurrences of nickel are in Karnataka, Kerala and Rajasthan. Polymetallic sea nodules are another source of nickel. The total resources of nickel ore have been estimated to be over 189 million tonnes in the country. About 92 per cent resources; *i.e.*, 175 million tonnes are in Odisha. The remaining 8 per cent resources are distributed in Jharkhand (9 million tonnes) and Nagaland (5 million tonnes). Nominal resources are reported from Karnataka (0.23 million tonnes).

Lead and Zinc

Lead is a widely used metal due to its malleability, softness, heaviness and bad conductivity of heat. The most important industrial use of lead is as a constituent in alloys such as type metal, bronzes and anti-friction metal. Lead oxide is used in lead sheeting, cable covers, ammunition, paints, glass making and rubber industry. It is also made into sheets, tubes and pipes which are used in buildings, especially as sanitary fittings. It is now increasingly used in automobiles, aeroplanes, and calculating machines. Lead nitrate is used in dyeing and printing.

Lead does not occur free in nature, rather it occurs as a cubic sulphide known as galena. Galena is found in veins in limestones, calcareous slates and sandstones and occasionally in metamorphic rocks or in association with volcanic rocks.

Zinc is a mixed ore containing lead and zinc and is found in veins in association with galena, chalcopyrites, iron pyrites and other sulphide ores. It is mainly used for alloying and for manufacturing galvanized sheets. It is also used for dry batteries, white pigments, electrodes, textiles, die-casting, rubber industry and for making collapsible tubes containing drugs, pastes and the like.

The total resources of lead and zinc ores are estimated at 749.46 million tonnes. Of these, 106.12 million tonnes (14 per cent) fall under 'reserves' while balance 643.34 million tonnes (86 per cent) are classified as 'remaining resources'. The total metal content in resources is 11.55 million tonnes lead and 36.66 million tonnes zinc. Besides, 118.45 thousand tonnes lead and zinc metal resources are available. In terms of reserves, 224 million tonnes of lead metal and 12.45 million tonnes of zinc metal are estimated. Rajasthan is endowed with the largest resources of lead-zinc ore amounting to 670.34 million tonnes (89.44 per cent), followed by Andhra Pradesh 22.69 million tonnes (3.02 per cent), Madhya Pradesh 14.84 per million tonnes (1.98 per cent), Bihar 11.43 million tonnes (1.52 per cent) and Maharashtra 9.27 million

TABLE 23.7. Production of Lead and Zinc Concentrates in India (Tonnes)

Year	Lead concentrates		Zinc concentrates	
	Production (Tonnes)	Value (₹ '000)	Production (Tonnes)	Value (₹ '000)
2019-20	35,176	1,82,60,832	1,44,68,204	6,04,38,504
2020-21	3,76,923	1,88,10,483	15,13,996	6,31,27,109
2021-22	3,68,040	2,23,66,174	15,94,087	8,18,15,820

Source : Economic Survey 2022-23, Statistical Appendix, p. 56.

tonnes (1.24 per cent). Resources are also established in Gujarat, Meghalaya, Odisha, Sikkim, Tamil Nadu, Uttarakhand and West Bengal.

Almost the entire production comes from Rajasthan.

Tungsten

It is a valuable metal of which the chief ore is wolfram. This metal possesses some special properties as a result of which it is almost indispensable in several industries. Its most important property is that of *self-hardening* which it imparts to steel. Over 95 per cent of the worlfram is used by the steel industry. Steel containing the requisite proportion of tungsten is mainly used in manufacturing amunitions, armour plates, heavy guns, hard cutting tools, etc. Tungsten is easily alloyed with chromium, nickel, molybdenum, titanium, etc. to yield a number of hard facing, heat and corrosion resistant alloys. It is also used for various other purposes such as electric bulb filaments, paints, ceramics, textiles, etc.

The total resources of tungsten ore in the country, have been estimated at 87.4 million tonnes containing 142.094 tonnes tungsten (WO_3) content. These resources are mainly distributed in Karnataka (42 per cent), Rajasthan (27 per cent), Andhra Pradesh (17 per cent) and Maharashtra (9 per cent). Remaining 5 per cent resources are in Haryana, Tamil Nadu, Uttarakhand and West Bengal. In Kuhi-Khobana-Agargaon belt, GSI has identified seven mineralised zones in Sakoli basin in Bhandara and Nagpur districts of Maharashtra. The yearly production is 45 to 50 thousand kg of ore. This is not sufficient to meet our requirements and part of the domestic requirements are met by imports.

Bauxite

Bauxite is an important ore which is used for making aluminium. It is an oxide of aluminium (name derived after Le Beaux in France). It is not a specific mineral but a rock consisting mainly of hydrated aluminium oxides. It is a clay-like substance which is pinkish, whitish or reddish in colour depending on the amount of iron content.

Production and Distribution

Total resources of bauxite as per UNFC in the country are placed at 3,897 million tonnes in 2015. These resources include 657 million tonnes reserves and 3,240 million tonnes remaining resources. By grades, about 81 per cent resource are of metallurgical grade. The resources of refractory and chemical grades are limited and together account for about 5 per cent. Among states, Odisha alone accounts for 51 per cent of country's resources of bauxite followed by Andhra Pradesh 16 per cent, Gujarat 9 per cent, Jharkhand 6 per cent, Maharashtra 5 per cent each and Chhattisgarh and Madhya Pradesh 4 per cent. Major bauxite resources are concentrated in the east coast in Odisha and Andhra Pradesh.

Odisha, Gujarat, Jharkhand, Maharashtra, Chhattisgarh, Tamil Nadu and Madhya Pradesh are the main bauxite producing states in India.

1. Odisha. Odisha is the largest bauxite producing state accounting for more than 48 per cent of the total production of India. The main bauxite belt is in Kalahandi and Koraput districts and extends further into Andhra Pradesh. This 300 km long, 40 to 100 km wide and 950 to 1300 metre thick belt is the largest bauxite bearing region of the country. The main deposits occur in Kalahandi, Koraput, Sundargarh, Bolangir and Sambalpur districts. The important mining areas include Chandgiri, Baphalimoli Parbat, Kathakal, Manjimali, Pasenmali, Kunnumali, Kodingandi, Pottangi and Karalput in Kalahandi and Koraput districts. The new aluminium plant at Damanjodi provides ready market for bauxite of this area. The aluminium plant at Doraguda provides further impetus.

TABLE 23.8. Production of Bauxite in India

Year	Production in tonnes	Value (₹ '000)
2019-20	2,18,25,227	1,62,66,333
2020-21	2,03,80,548	1,68,93,448
2021-22	2,24,93,947	2,47,67,048

Source: Economic Survey 2022-23, Statistical Appendix, p. 56.

2. Gujarat. Gujarat is the second largest producer and produces over 23 per cent of the total bauxite of India, found in Jamnagar, Junagadh, Kheda, Kachchh, Sabarkantha, Amreli and Bhavnagar districts. The most important deposits occur in a belt which is 48 km long and 3 to 4.5 km wide lying between the Gulf of Kachchh and the Arabian sea through Bhavnagar, Junagadh and Amreli districts.

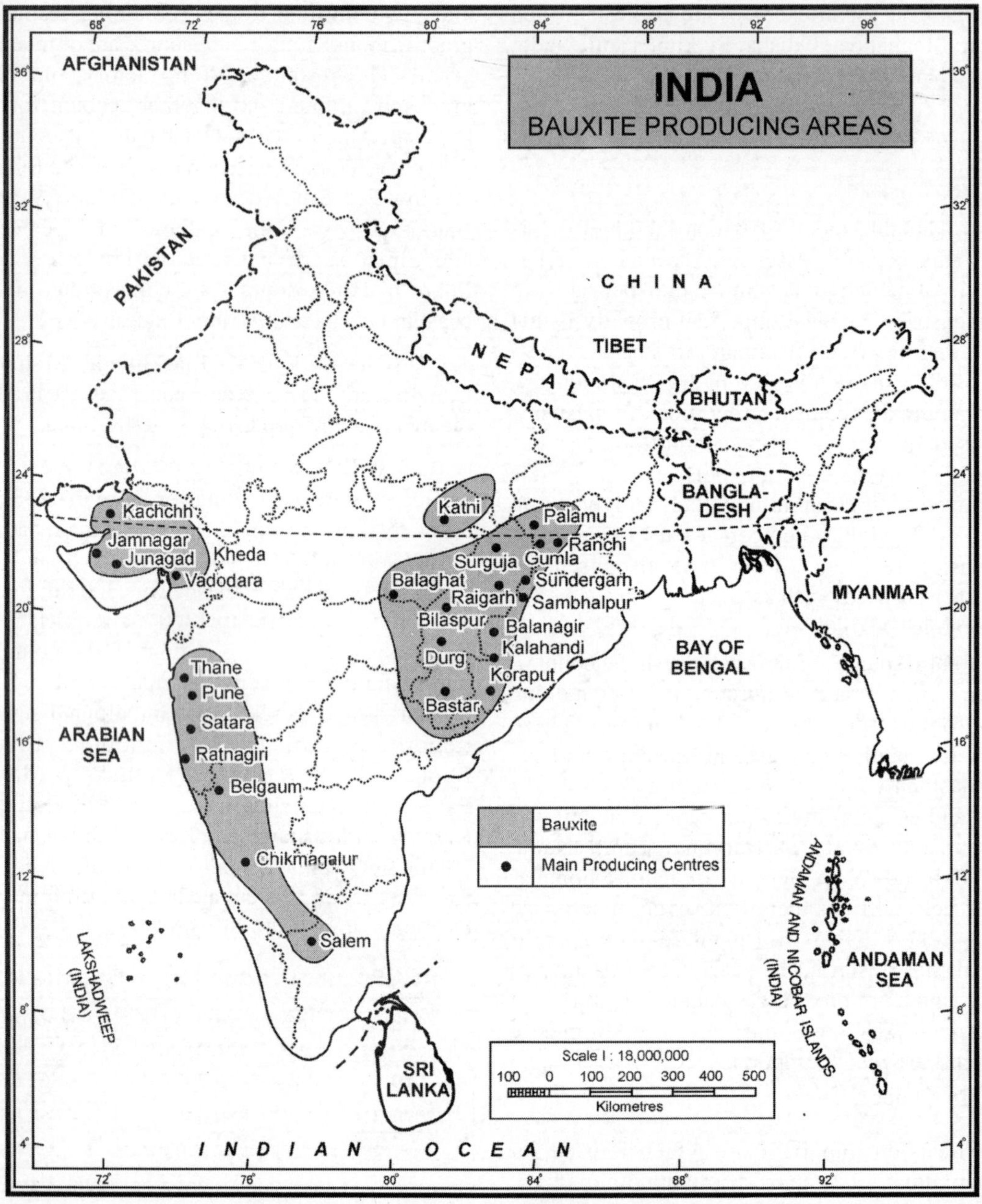

FIG. 23.5. India : Bauxite Producing Areas

3. Jharkhand. Jharkhand is an important bauxite producing state of India accounting for over 9 per cent of the total production. The reserves of this state are found in extensive areas of Ranchi, Lohardaga, Palamu and Gumla districts. Some bauxite is also found in Dumka and Munger districts. High grade ore occurs in Lohardaga and adjoining areas.

4. Chhattisgarh. Chhattisgarh produces about 8 per cent bauxite of India. The Maikala range in Bilaspur, Durg districts and the Amarkantak plateau region of Surguja, Raigarh and Bilaspur are some of the areas having rich deposits of bauxite.

5. Maharashtra. Maharashtra accounts for about 7.9 per cent of the total bauxite produced in

India. The total recoverable reserves in the state, have been estimated to be of the order of 87.7 million tonnes. The largest deposits occur in Kolhapur district capping the plateau basalts. Udgeri, Dhangarwadi, Radhanagari and Inderganj in Kolhapur district contain rich deposits with alumina content 52 to 89 per cent. The other districts with considerable deposits are Thane, Ratnagiri, Satara and Pune.

6. Madhya Pradesh. Amarkantak plateau area, the Maikala range in Shahdol, Mandla and Balaghat districts and the Katni as well as Jabalpur district are the main producers.

Goa and Karnataka also produce small quantities of bauxite.

TABLE 23.9. Distribution of bauxite in India 2016-17

State	Production in tonnes	Percentage of all India production
1. Odisha	1,19,90,035	48.61
2. Gujarat	58,18,467	23.59
3. Jharkhand	22,89,825	9.28
4. Chhattisgarh	19,54,233	7.92
5. Maharashtra	19,46,042	7.89
6. Madhya Pradesh	6,58,375	2.67
Others	7,655	0.04
All India	2,46,64,632	106.00

Source : Data computed from Indian Mineral Yearbook, 2017.

Trade. As much as 80 per cent of the bauxite is used for producing aluminium. India's exports of bauxite have been reduced considerably due to increasing demand in the home market. Still India manages to export small quantities of bauxite. The main buyers of Indian bauxite are Italy (60%), U.K. (25%), Germany (9%) and Japan (4%).

Pyrites

Pyrite is a sulphide of iron. Its economic value lies in its being the chief source of sulphur and not as an ore of iron because the high proportion of sulphur in it is injurious to iron. Sulphur is very useful for making sulphuric acid which in turn is used in several industries such as fertilizers, chemicals, rayon, petroleum, steel, etc. Elemental sulphur is useful for manufacturing explosives, matches, insecticides, fungicides and for vulcanising rubber.

Pyrite is widely distributed and occurs in many formations from the oldest crystalline rocks to the youngest sediments. The total recoverable reserves of pyrite are placed at 85.48 million tonnes of which 15.46 million tonnes are under proved category. The main deposits are found in the lower Son Valley at Amjor, Kasisiyakoh, and Kurriari in *Bihar*, in Chitradurga and Uttar Kannada districts of *Karnataka* and the pyritous coal and shale of Assam coalfields. The other deposits are in Rajasthan (Sikar), Tamil Nadu (Salem, Coimbatore, Nilgiri), Himachal Pradesh (Shimla), Meghalaya (Khasi and Jaintia Hills), Madhya Pradesh (Bilaspur, Chhindwara), Chhattisgarh (Durg and Surguja), Odisha (Mayurbhanj and Sundargarh), Uttarakhand (Garhwal, Tehri Garhwal and Almora), West Bengal (Jalpaiguri) and Arunachal Pradesh (Lower Subansiri). Production of pyrites had decreased from 1.4 lakh tonnes in 1996-97 to 9.5 thousand tonnes in 2011-12.

Gold

It is a valuable metal which occurs in auriferous lodes and some of it is found in sands of several rivers. It is used for making ornaments and is known as international currency due to its universal use.

Production and Distribution

The total resources of gold ore in the country were estimated at 527.96 million tonnes in 2015. Out of these, 17.23 million tones were placed under reserves category and the remaining 510.73 million tones under remaining resources category. Total resources of gold (primary), in terms of metal, stood at 654.74 tonnes. Out of these, 70.09 tonnes were placed under reserves category and 584.65 tonnes under remaining resources category. The resources include place-type gold ore in Kerala estimated at 26.12 million tonnes containing 5.86 tonnes gold metal. Among the states, the largest resources in terms of the metal ore (primary) are located in Bihar (44 per cent) followed by Rajasthan (25 per cent), Karnataka (21 per cent), West Bengal and Andhra Pradesh (3 per cent each), Telangana and Madhya Pradesh (2 per cent each). Remaining resources of ore are located in Chhattisgarh, Jharkhand, Kerala,

Maharashtra and Tamil Nadu. In terms of metal content, Karnataka has the highest followed by Rajasthan, Bihar, Andhra Pradesh, Jharkhand, etc.

In February 2020, Uttar Pradesh claimed to discover 3000 tonnes of gold reserves in Sonbhadra district. This could go upto 5,200 tonnes according to estimate made by the government agencies. However, this claim has been disputed by the Geological Survey of India (GSI). The matter is under investigation and final result is awaited.

There are three gold fields in the country, namely Kolar Gold Field, Kolar district, Hutti Gold Field in Raichur district (both in Karnataka) and Ramgiri Gold Field in Anantpur district (Andhra Pradesh).

Karnataka is the largest producer of gold in India. The main reserves are in Kolar, Dharwad, Hassan and Raichur districts. Some gold reserves are also reported from a number of scattered localities in Gulbarga, Belgaum, Bellary, Mysuru, Mandya, Chikamagalur and Shimoga districts. This state at one time used to produce about 88.7 per cent of India's gold.

Although every district of Karnataka has some reserves of gold, the most important reserves are found in Kolar Gold Fields, Kolar district which produces about 57.75 per cent of the total production of the state. The deposits in Kolar Gold Field occur in 80 km long (north-south) and 3-4 km wide belt in which gold bearing quartz veins are confined to a 6-7 km section near Marikuppam. The first mining operations in the Kolar Gold Fields started in 1871 and the area still continues to be a major supplier of gold in India. About 3,539 thousand tonnes of gold ore having 17,738 kg of gold content have been proved during the recent surveys conducted by the Geological Survey of India for locating new lodes. The main gold mine at Kolar is one of the deepest mines of the world and the production from this mine is decreasing day by day due to heavy cost of extraction. Moreover, most of the gold has already been taken out and very little gold is left in the mine. Now gold is available at depth of more than 3,000 metres and it is not economically viable to extract gold at this depth.

Next to Kolar, but far below in production, are the Hutti mines in Raichur district. In the Hutti gold field the gold mining belt is 3.7 km long and 1,220 m wide with six auriferous quartz reefs. It produced 593.3 kg of gold in 1915, but the production fell and the mine had to be closed down in 1920. The main problem with the Hutti mines is the low grade of ore. The mine reopened in 1948 and has been operating irregularly since then. The *in situ* gold reserves of gold fields of Raichur district are estimated at 4.5 million tonnes with a total gold content of about 45,000 kg.

Some gold is found in the Gadag field in Dharwad district. New fields have been discovered in Ballara (Tumkur district), Kempinkole (Hassan district), Honnoli (Shimoga district), Siddarhalli (Chikmagalur district), and Munghur (Gulbarga district) areas.

Andhra Pradesh. Though lagging far behind Karnataka, Andhra Pradesh is the second largest producer of gold in India. On the basis of the detailed mapping done by the Geological Survey of India in recent years, a total of 7.06 million tonnes of ore and 37,025 kg of gold metal have been assessed in the state. The main deposits are found in Ramagiri in Anantapur district. However, this field is almost exhausted. The other areas of gold deposits are Bisanattam and Palachchur in Chittoor district and Jonnagiri in Kurnool district.

TABLE 23.10. Production of Gold in India

Year	Production in kg.	Value (₹ '000)
2019-20	1,742	64,95,723
2020-21	1,127	54,75,469
2021-22	1,251	60,11,677

Source: Economic Survey 2022-23, Statistical Appendix. p. 56.

Alluvial Gold

Apart from the gold mines in the above mentioned areas, some gold is collected from the sands and gravels of several rivers. Gold is often liberated from the rocks by weathering and its particles get concentrated at certain places in the rivers. Such deposits are called placer deposits from which gold is recovered by *panning*. Although very small in quantity, this type of gold is widely spread in a large number of rivers.

Jharkhand. In addition to the above mentioned two states, Jharkhand is an important producer of gold in India. Jharkhand has both alluvial and native gold. Alluvial gold is obtained from the sands of the *Subarnarekha* (gold streak) river, as its name suggests, Sona nadi in Singhbhum district and the streams draining the Sonapat valley. Native gold is found near Lowa in Singhbhum district and in some other parts of Chota Nagpur plateau.

Kerala. The river terraces along the Punna Puzha and the Chaliyar Puzha have tracts of gold. Alluvial gold is found in the Ambankadava Puzha, Chaliyar Puzha and in the rivers near Mannarkkat.

Small quantities of gold are collected from rivers in Shimla and Bilaspur in Himachal Pradesh, Kargil area along the terraces of the Indus river and in alluvial and morainic deposits of Dras river in Ladakh, Balaghat, and Seoni districts in Madhya Pradesh, Bastar, Raipur and Raigarh in Chhattisgarh and parts of Purulia district of West Bengal.

Silver

Silver is another precious metal produced in India. It is valued next only to gold for making ornaments due to its softness and attractive white colour. It had been an important currency metal in several parts of the world. It is also used in the manufacture of chemicals, electroplating, photography and for colouring glass, etc.

The chief ore minerals of silver are *agentine*, *stephanite*, *pyrargyrite* and *proustite*. It is found mixed with several other metals such as copper, lead, gold, zinc, etc.

India is not a major producer of silver in the world. The main production comes from Zawar mines in Udaipur district of Rajasthan. Here, silver is obtained as a by-product during the concentration and smelting of galena ore in Hindustan Zinc Smelter. The silver content varies from 171.4 gm to 774.5 gm per tonne of zinc and lead concentrates respectively. The Tundoo Lead Smelter in Dhanbad district of Jharkhand is another important producer of silver as a by-product of lead. Some silver is produced by Kolar Gold Fields and Hutti gold mines in Karnataka during refining of gold. The Hindustan Copper Ltd. at Maubhandar smelter in Singhbhum district of Jhakhand obtains silver from copper slimes. Silver is also produced by Vizag Zinc smelter in Andhra Pradesh from the lead concentrates.

TABLE 23.11. Production of Silver in India

Year	Production in kg.	Value (₹ '000)
2019-20	6,09,340	2,56,16,104
2020-21	7,05,796	4,26,64,424
2021-22	6,47,140	4,21,23,586

Source: Economic Survey 2022-23, Statistical Appendix, p. 56.

Traces of silver occur in Hazaribag, Palamu, Ranchi and Singhbhum districts of Jharkhand; Cuddapah, Guntur and Kurnool districts of Andhra Pradesh; Vadodara in Gujarat, Bellary district of Karnataka, Baramula district of Jammu and Kashmir and Almora district of Uttarakhand.

NON-METALLIC MINERALS

India also produces a large number of non-metallic minerals although only a few of them have assumed as much industrial and economic importance as is done by the metallic minerals. However, they are used in a large variety of industries; the major industries being cement, fertilizers, electricals, etc.

Mica

Mica has been used in India since ancient times as a medicinal item in Ayurveda and is known as ***abhrak.*** With the development of electrical industry, mica found new vistas of use. Its insulating properties have made it a valuable mineral in electrical and electronics industry. It can withstand high voltage and has low power loss factor. The three major types of mica found in India are : *muscovite, phlogopite* and *biotite*.

Reserves

Most important mica-bearing pegmentities occur in Andhra Pradesh, Bihar, Jharkhand, Maharashtra and Rajasthan. Occurrences of mica pegmatites are also reported from Gujarat, Haryana, Karnataka, Kerala, Odisha, Tamil Nadu and West Bengal. The total resources of mica in the country in 2015 are estimated at 6,35,302 tonnes. Andhra Pradesh leads with 40 per cent share in country's total resources followed by Rajasthan (28 per cent), Odisha (16 per cent), Maharashtra (13 per cent), Bihar (2 per cent) and balance (less than 1 per cent) in Jharkhand and Telangana.

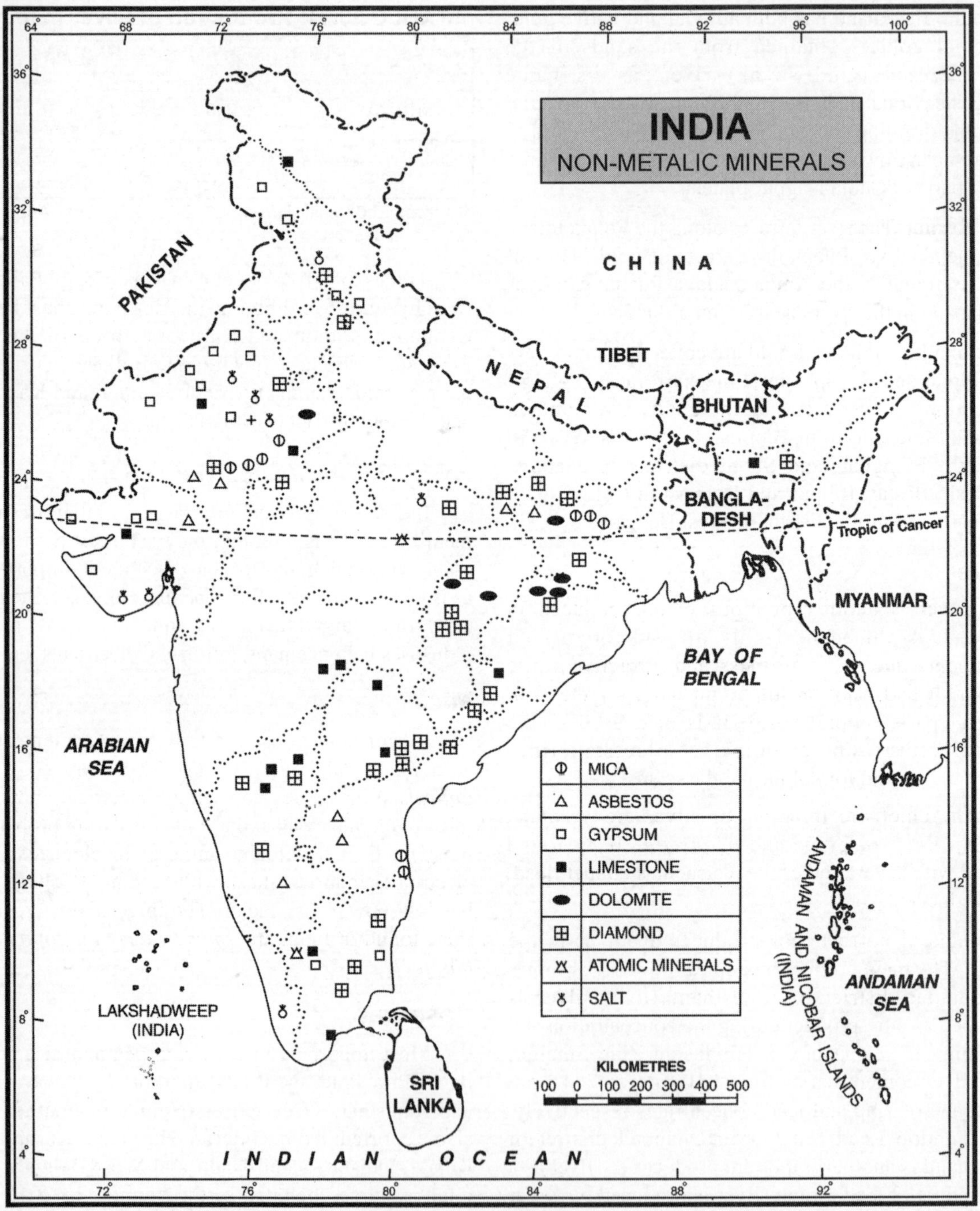

FIG. 23.6. India : Non-Metallic Minerals

Production and Distribution

India has a near monopoly in the production of mica, producing about 60 per cent of world's total production. Production was just 772 tonnes in 1947-48 which increased to about ten thousand tonnes within three years. The production increased at a rapid pace up to 1960-61 and there was a record production of 28,347 tonnes in that year. But afterwards it showed a declining trend and the production came down to 1,807 tonnes in 2011-12.

This decrease is the result of the fall in its demand in the international market. Earlier, there was no substitute for mica. Now materials like plastics and synthetics have been developed which can be used as substitutes for mica.

About 95 per cent of India's mica is found in just three states of Andhra Pradesh, Rajasthan and Jharkhand. Some mica is produced in Bihar also.

1. Andhra Pradesh. Andhra Pradesh is the largest mica producing state of India. In 2011-12, this state produced 1,694 tonnes of mica which was more than 93 per cent of all India production. In the recent years, share of Andhra Pradesh has progressed in respect to mica production. The mica belt lies in Nellore district and is 100 km long and 25 km wide. Nellore mica is generally light green in colour; it is generally stained and spotted. The other districts with workable mica deposits are Vishakhapatnam, West Godavari and Krishna (Tiruvur). Shah mine in Gudur taluka is the deepest with mining being done at 300 m depth.

2. Rajasthan. Although occupying the second position among the mica producing states of India, Rajasthan is not as important a producer as Andhra Pradesh is. This state produced 113 tonnes or 6.3 per cent of India in 2011-12. The main mica belt extends from Jaipur to Udaipur. This is 322 km long with an average width of 96 km. This belt broadens around Kumbhalgarh and Bhilwara. The main producing districts are Bhilwara, Jaipur, Tonk, Sikar, Dungarpur and Ajmer.

3. Jharkhand. Jharkhand is the third largest producer of mica in India. Mica in Jharkhand is found in a belt extending for about 150 km in length and 32 km in width from eastern part of Gaya district of Bihar across Hazaribagh, Girdih and Munger to Bhagalpur district. This belt contains the richest deposits of high quality ruby mica. The main centres of mica production in this belt are Kodarma, Dhorhakola, Domchanch, Dhab, Gawan, Tisri, Chakai and Chakapathal. Outside the main mica belt, mica occurs in Dhanbad, Palamu, Ranchi and Singhbum districts.

Other producers. The other areas with small deposits of mica are *Gujarat* (Banaskantha, Vadodara, Sabarkantha), *Kerala* (Alleppey and Kollam), *Tamil Nadu* (Nilgiri, Coimbatore, Salem and Tiruchirapalli), *Madhya Pradesh* (Balaghat and Chhindwara), *Chhattisgarh* (Bilaspur, Bastar and Surguja) and *Uttar Pradesh* (Mirzapur). Some deposits have also been reported from Odisha, Haryana, Himachal Pradesh and West Bengal.

Exports. India is not only the largest producer but also the largest exporter of mica in the world. In spite of the threat from synthetic mica, certain grades of Indian mica will remain vital to the world's electrical industries.

Most of the exports are routed through the ports of Kolkata and Vishakhapatnam. Japan (19%), the USA (17%), U.K. (7%), Norway (7%), Russia, Poland, Germany, Czech Republic, Slovakia, Hungary, France, the Netherlands, etc. are the main buyers purchasing about 80 per cent the total mica exported by India. Exports of Indian mica have increased considerably (*see* **Table 23.12**).

Limestone

Limestone is associated with rocks composed of either calcium carbonate, the double carbonate of calcium and magnesium, or mixture of these two

TABLE 23.12. Export of Mica from India

Year	1960-61	1970-71	1980-81	1990-91	2000-01	2019-20	2020-21	2021-22
Quantity (Thousand tonnes)	28.4	26.7	16.7	42.0	63.2	115.7	143	151
Value (₹ crore)	—	16	18	35	64	372	472	511

Source : Economic Survey 2022-23, Statistical Appendix, p. 116-18.

constituents. In addition to the main constituents of calcium and magnesium carbonates, limestone also contains small quantities of silica, alumina, iron oxides, phosphorus and sulphur. Limestone deposits are of sedimentary origin and exist in almost all the geological sequences from pre-cambrian to recent except in Gondwana.

The total resources of limestone of all categories and grades are estimated at 2,03,225 million tonnes in 2015. Karnataka is the leading state having 27 per cent of total resources followed by Andhra Pradesh and Rajasthan (12 per cent each), Gujarat (10 per cent), Meghalaya (9 per cent), Telangana (8 per cent) and Chhattisgarh (5 per cent). The total production of limestone was 3,13,196 thousand tonnes in 2016-17. As much as 87.22% of the total output in 2016-17 was contributed by eight principal states, *viz*., Rajasthan (21.02%), Madhya Pradesh (13.31%), Andhra Pradesh (12.10%), Gujarat (8.79%), Karnataka (8.21%), Telangana (8.18%), Chhattisgarh (8.03%) and Tamil Nadu (7.58%). The remaining 12.78% of the total production was shared by other limestone producing states.

Limestone is used for a large variety of purposes. Of the total consumption, 75 per cent is used in cement industry, 16 per cent in iron and steel industry and 4 per cent in the chemical industries. Rest of the limestone is used in paper, sugar, fertilizers, glass, rubber and ferromanganese industries.

Trends in production are shown in **Table 23.13**.

TABLE 23.13. Production of Limestone in India

Year	Production ('000 tonnes)	Value (₹ '000)
2019-20	3,59,464	8,88,90,081
2020-21	3,49,120	8,64,84,948
2021-22	3,92,760	9,73,49,554

Source: Economic Survey 2022-23. Statistical Appendix, p. 56

1. Rajasthan. Rajasthan has about 12 per cent of the reserves and produces over 21 per cent of the total limestone of India. Jhunjhunu, Banswara, Jodhpur, Sirohi, Bundi, Ajmer, Bikaner , Dungarpur, Kota, Tonk, Alwar, Sawai Madhopur, Chittaurgarh, Nagaur, Udaipur and Pali are the main producing districts.

2. Madhya Pradesh. Madhya Pradesh is the second largest producer of limestone and accounts for over 13 per cent of the total limestone production of India. Large deposits occur in the districts of Jabalpur, Satna, Betul, Sagar, Damoh and Rewa. The total reserves of all grades of limestone are estimated to be over 1,500 million tonnes.

3. Andhra Pradesh. Andhra Pradesh possesses about 12 per cent of the total reserves of the limestone in the country. This state produced about 12.1% per cent of India's limestone in 2016-17. Extensive deposits occur in Cuddapah, Kurnool, Guntur, Krishna.

4. Gujarat. Gujarat has about 10 per cent of the reserves and produces more than 8 per cent of the total limestone of India. High grade limestone deposits occur in Banaskantha district. The other important producing districts are Amreli, Kachchh, Surat, Junagadh, Kheda and Panchmahals.

5. Karnataka. Gulbarga, Bijapur and Shimoga districts of Karnataka possess about one-third of cement grade limestone of India. Currently, this state produces a little more than 8 per cent of the t otal limestone of India. The main producing districts of all grades of liemstone are Gulbarga, Chitradurg, Tumkur, Belgaum, Bijapur, Mysuru and Shimoga.

6. Telangana. This state produces over 8% limestone of India. Nalgonda, Adilabad, Warangal, Mehbubnagar and Karimnagar are the major producing districts.

7. Chhattisgarh. Chhattisgarh accounts for 8 per cent of total limestone of India. Deposits of limestone occur in Bastar, Bilaspur, Raigarh, Raipur and Durg districts.

8. Tamil Nadu. Large scale reserves in Ramnathapuram, Tirunelveli, Tiruchchirappalli, Salem, Coimbatore, Madurai, and Thanjavur districts enable Tamil Nadu to provide more than eight per cent limestone of the country. Most of the deposits, except those of Salem district, are of cement grade limestone.

Others. Apart from the above mentioned major producers, liemstone is also produced in *Sikkim*, *Maharashtra* (Yavatmal, Chandrapur, Nanded and Ahmadnagar), *Odisha* (Sundargarh, Sambalpur and

Kalahandi), *Jharkhand* (Palamu, Hazaribagh, Ranchi and Singhbhum), *Himachal Pradesh* (Bilaspur, Kangra and Chamba), *Haryana* (Mahendragarh and Ambala), *Assam* (Nagaon and Sibsagar), *Meghalaya* (Garo, Khasi and Jaintia hills), *Uttar Pradesh* (Mirzapur, Lucknow and Unnao), *Uttarakhand* (Dehra Dun and Mussoorie), *West Bengal* (Darjeeling and Jalpaiguri) and *Jammu and Kashmir* (Anantnag and Jammu).

Dolomite

Limestone with more than 10 per cent of magnesium is called dolomite; when the percentage rises to 45, it is *true dolomite*. The economic uses of dolomite are chiefly metallurgical; as refractories, as blast furnace flux as a source of magnesium salts and in fertilizer and glass industries. Iron and Steel industry is the chief consumer of dolomite accounting for over 90 per cent consumption followed by fertilizer (4%), ferro-alloys and glass (2% each), alloy steel (1%) and others (1%). Dolomite is widely distributed in the all parts of the country.

Total resources of dolomite are placed at 8,415 million tonnes, out of which 679 million tonnes are placed under reserves category and the balance 7,736 million tonnes under remaining resources category. Major share of about 88 per cent resources is distributed in eight states : namely, Madhya Pradesh (27 per cent), Andhra Pradesh (15 per cent), Chattisgarh (11 per cent), Odisha (10 per cent), Karnataka and Rajasthan (7 per cent each), Gujarat (6 per cent) and Maharashtra (5 per cent). The remaining (12 per cent) resources are distributed in Arunachal Pradesh, Jharkhand, Haryana, Sikkim, Tamil Nadu, Uttarakhand, Uttar Pradesh and West Bengal.

Chhattisgarh, Odisha, Andhra Pradesh, Karnataka, and Madhya Pradesh are the main dolomite producing states which account for more than 90 per cent of India'a total production. Chattisgarh and Odisha are the two outstanding states which produce more than half of India's dolomite.

1. Chhattisgarh. Chhattisgarh is the largest dolomate producing state of India which contributes more than 30 per cent of the total production of the country. The recoverable reserves of all grades of dolomite are estimated to be 1,638 million tonnes. The main deposits occur in Bastar, Bilaspur, Durg and Raigarh districts.

2. Odisha. Odisha is the second largest producer of dolomite and accounts for over 21 per cent of the total production of dolomite in India. The total reserves of recoverable dolomite of all grades are of the order of 562.6 million tonnes out of which 256 million tonnes are in Birmitrapur locality alone. The main deposits occur in Sundargarh, Sambalpur and Koraput districts. In Gangapur area, they occur near Sukra and extend for a total length of about 100 km.

3. Andhra Pradesh. This is the third largest dolomite producing state of India contributing more than 18 per cent of the total production. Anantapur, and Kurnool are the main producing districts. In the neighbouring **Telangana** state, Khammam is the leading dolomite producing district.

4. Karnataka. Karnataka produces slightly more than ten per cent dolomite of India. Belgaum, Bijapur, Chitradurga, Mysore, Uttar Kannada and Tumkur contribute major part of the state's production.

5. Madhya Pradesh. This state has vast reserves of dolomite and accounts for more than 6 per cent of India's production. Most of the deposits occur along the Vindhya Range.

6. Rajasthan. Rajasthan produces over 4 per cent of India's dolomite. Ajmer, Alwar, Bhilwara, Jaipur, Jaisalmer, Jhunjhunu, Jodhpur, Nagaur, Pali, Sawai Madhopur, Sikar and Udaipur are the main producing districts.

7. Jharkhand. Dolomite occurs in bands to the north of Chaibasa in Singhbhum district. Some dolomite also occurs in Palamu district.

Others. Among the other producers are Gujarat (Junagad, Kheda, Jamnagar and Amreli), West Bengal (Jalpaiguri), Uttar Pradesh (Banda and Mirzapur), Uttarakhand (Dehra Dun, Nainital and Tehri Garhwal), Arunachal Pradesh (Kamang), Haryana (Mahendragarh), Himachal Pradesh (Shimla, Solan and Mandi), Maharashtra (Nagpur, Chandrapur and Yavatmal) and Tamil Nadu (Salem and Triunelveli).

Asbestos

Two quite different minerals are included under this name; one, a variety of amphibole, resembling tremolite and the other, more important, a fibrous variety of serpentine (chrysotile). Chrysotile is more important variety and accounts for 80 per cent of the asbestos of commercial use.

Asbestos has found great commercial value due to its fibrous structure, its capability to be readily separated into filaments of high tensile strength and its great resistance to fire. It is widely used for making fire-proof cloth, rope, paper, millboard, sheeting, belt, paint, etc. and for manufacturing fire proof safes, insulators, felts, etc. It is also used in making aprons, gloves, curtains, brake-linings in automobiles and insulating mats. Asbestos cement products like sheets, slates, pipes and tiles are used for building purposes. When asbestos is brittle, it is made into filter pads for filtering acids, organic liquids and other chemicals. Mixed with magnesia, it is used for making 'magnesia bricks' used for heat insulation.

Reserves of all categories and varieties of asbestos are placed at about 1.046 million tonnes. The production was at low key till 1971 but picked up after that. From a record production of 42,699 tonnes in 1993-94, it fell drastically to 27,180 tonnes in 1996-97 and further to 387 tonnes in 2012-13.

Two states of Rajasthan and Andhra Pradesh produce almost the whole of asbestos of India. ***Rajasthan*** is the largest producer. This state produced over 94 per cent of the total production of India. Important occurrences are known in Udaipur, Dungarpur, Alwar, Ajmer and Pali districts. In **Andhra Pradesh,** asbestos of fine quality occurs in Pulivendla *taluk* of Cuddapah district. There are several occurrences between Chitravati and Papaghani rivers. The 15 km long zone between Lopatanuthula and Brahamnapalle is most promising. In **Karnataka,** the main deposits occur in Hassan, Mandya, Shimoga, Mysuru and Chikamaglur districts.

In addition to the above mentioned major producers, some asbestos is produced in Jharkhand, Madhya Pradesh, Chhattisgarh, Tamil Nadu, Gujarat, Uttarakhand and Nagaland also.

Magnesite

It is an alteration product of dunites (peridotite) and other basic magnesian rocks. It is primarily used for manufacturing refractory bricks. It is also used as a bond in abrasives and in the manufacture of special type of cement for artificial stone, tiles, fire-proof floorings and for extraction of the metal magnesium. Steel industry also uses magnesite.

The total reserves/resources of magnesite are about 394 million tonnes. Substantial quantities of resources are established in Uttarakhand (59 per cent), followed by Rajasthan (14 per cent) and Tamil Nadu (12 per cent). Resources are also located in Andhra Pradesh, Himachal Pradesh, Jammu and Kashmir, Karnataka and Kerala. Occurrences of magnesite in Tamil Nadu are low in lime and high in silica whereas those of Uttarakhand are high in lime and low in silica.

TABLE 23.14. Production of Magnesite in India

Year	Production in tonnes	Value (₹ '000)
2019-20	1,02,554	3,51,947
2020-21	74,661	3,14,676
2021-22	1,13,495	4,50,169

Source: Economic Survey 2022-23. Statistical Appendix, p. 56

Production of magnesite recorded more than four-fold increase in forty years from 1.19 lakh tonnes in 1951 to 5.28 lakh tonnes in 1990-91 after which production regularly declined and stood at 78,144 tonnes in 2020-21.

Tamil Nadu, Uttarakhand, Karnataka and Rajasthan produce almost the entire magnesite of India.

Tamil Nadu is the largest producer of magnesite in India. In 2011-12, this state produced more than two-thirds magnesite of India. Tamil Nadu has one of the largest deposits of magnesite in the world and the largest in India are found at Chalk Hills near Salem town. Some other deposits occur in Coimbatore, Dharmapuri, Nilgiri, Periyar and Tirunelveli. In **Uttarakhand,** Almora district with estimated reserves of 250 lakh tonnes has a 3.2 km long and 6 to 9 metre thick belt between Agar and Chahana. Chamoli district is also reported to have some deposits of magnesite. In **Karnataka**, magnesite veins are found in Hassan, Mysuru and Kodagu districts. Ajmer, Udaipur and Pali districts in **Rajasthan,** Chamba in **Himachal Pradesh** and Udhampur in **Jammu and Kashmir** also produce some magnesite.

Kyanite

Kyanite occurs in metamorphic aluminous rocks and is primarily used in metallurgical, ceramic, refractory, electrical, glass, cement and a number of other industries due to its ability to stand high temperatures. It is also used in making sparking plugs in automobiles.

India has the largest deposits of kyanite in the world. All the three grades of kyanite are found here. The first grade has an aluminium content from 62 to 64 per cent, the second grade from 58 to 62 per cent and the third grade from 54 to 58 per cent.

The total resources of kyanite in 2015 are placed at 105 million tonnes. Statewise, the share of Telangana is 46 per cent followed by Andhra Pradesh 30 per cent, Karnataka 13 per cent and Jharkhand 7 per cent. Remaining 4 per cent resources are in Maharashtra, Kerala, Rajasthan, Tamil Nadu and West Bengal.

The production of kyanite has been fluctuating since 1951 but there has been falling trend since 1990-91. The total production of Kyanite in 2020-21 was 4,925 tonnes. Jharkhand and Maharashtra produce practically the whole of kyanite of India.

TABLE 23.15. Production of Kyanite in India

Year	Production in tonnes	Value (₹ '000)
2019-20	3,498	12,728
2020-21	4,925	9,251
2021-22	9,432	17,578

Source: Economic Survey 2022-23. Statistical Appendix, p. 56.

1. Jharkhand is the largest producer of kyanite. This state produced 4,011 tonnes of kyanite in 2011-12 which was about 98 per cent of the total kyanite produced in India. Ores with high degree of purity with percentages of aluminium silicate reaching 95 to 97 are found in the Singhbhum district. Here kyanite occurs mainly as kyanite quartz rock and as massive kyanite-rock in beds of considerable size in the Archaean schists. This rock forms a bed nearly 130 km in length stretching from Lapsa Buru to Kharsawan in Saraikela. This has been the largest in the world and also the best in quality but now it is nearly exahusted. The other deposits in this district occur near Ghagidih, Badia, Bakra, Mohanpur, Jagnathpur, Bhakar, Hatiland, Singpura, Daontauri, Padampur and Shirbai. Small deposits are reported from Dhanbad and Ranchi districts also.

2. Maharashtra. Maharashtra produced only 45 tonnes or 1.4 per cent of the total kyanite produced in India in 2011-12 and occupied a distant second position among the major producing states of the country. Most of the reserves are in the districts of Bhandara and Nagpur where the total reserves of all grades are estimated at 70,652 tonnes.

Others. Some deposits of kyanite are also reported from **Odisha** (Dhenkanal, Sundargarh and Mayurbhanj), **Rajasthan** (Udaipur, Bhilwara and Dungarpur), **Tamil Nadu** (Kanniyakumari, Tiruchirapalli), **Andhra Pradesh** (Nellore). Telangana (Khammam), Karnataka (Chickmagalur, Chitradurga, Mandya, Mysuru, Dakshin Kannada and Shimoga)

Sillimanite

The occurrence and uses of sillimanite are almost the same as those of kyanite. The total resources of sillimanite are placed at 70.2 million tonnes (2015). The resources are located mainly in Tamil Nadu (25 per cent), Odisha (25 per cent), Uttar Pradesh (16 per cent), Andhra Pradesh (13 per cent), Kerala (10 per cent) and Assam (7 per cent). Remaining 4 per cent resources are in Jharkhand, Karnataka, Madhya Pradesh, Maharashtra, Meghalaya, Rajasthan and West Bengal.

Odisha, Kerala, Maharashtra and Rajasthan produce practically the whole of sillimanite of the country.

Odisha is the largest producer of sillimanite in India. This state produces as much as 56 per cent sillimanite of India. About 16.48 million tonnes of sillimanite reserves have been reported mainly from Ganjam district. **Kerala** is the second largest producing state which contributes about one-third of India's sillimanite. The beach sands of Kerala contain 5 to 6 per cent of sillimanite. Kozhikode, Palakkad, Ernakulam and Kottayam districts are major producers. In **Maharashtra** all grades of sillimanite have been reported from the Bhandara district, where the reserves are reported to be of the order of 2,32,055 tonnes. Small quantity of sillimanite is produced in **Rajasthan** also. Udaipur is the main producing district. **Karnataka** has about 85 thousand tonnes reserves of sillimanite mainly in Hassan, Mysuru and Dakshin Kannad. The recoverable reserves in **Meghalaya** are estimated to be 77,246 tonnes of all grades mainly confined to the Khasi hills. The other areas with some sillimanite reserves are **Assam** (Karbi-Anglong), **Madhya Pradesh** (Sidhi), **West Bengal** (Darjeeling, Bankura and Purulia) and **Tamil Nadu** (Kanniyakumari, Tirunelveli, Tiruchchirappalli).

TABLE 23.16. Production of Sillimanite in India

Year	Production in tonnes	Value (₹ '000)
2019-20	13,221	37,903
2020-21	11,110	13,987
2021-22	3,432	7,973

Source: Economic Survey 2022-23. Statistical Appendix, p. 56.

Gypsum

Gypsum is a hydrated sulphate of calcium which occurs as white opaque or transparent mineral in beds or bands in sedimentary formations such as limestones, sandstones and shales. In some cases it occurs as transparent crystals associated with clays. It is mainly used in making ammonia sulphate fertilizer and in cement industry. It is an essential constituent of cement, though its proportion is only 4-5 per cent. It is also used in making plaster of Paris, moulds in ceramic industry, nitrogen chalk, partition blocks, sheets, tiles, plastics, etc. It is conveniently applied as surface plaster in agriculture for conserving moisture in the soil and for aiding nitrogen absorption.

The total resources of mineral gypsum in the country are estimated at 1,330 million tonnes (2015) of which 37 million tonnes have been placed under 'reserve' and 1,293 million tonnes under 'remaining resources' category. Of the total resources, fertilizer/pottery grade accounts for about 80 per cent and cement/paint grade 13 per cent. The unclassified and not-known grades together account for 5 per cent resources. The remaining two per cent of resources is shared by surgical plaster and soil reclamation grades. Of the states, Rajasthan alone accounts for 81 per cent resources and Jammu and Kashmir 14 per cent resources. The remaining 5 per cent resources are in Tamil Nadu, Gujarat, Himachal Pradesh, Karnataka, Uttarakhand, Andhra Pradesh and Madhya Pradesh.

Rajasthan is by far the largest producer of gypsum in India. This state produced 3,159.7 thousand tonnes of gypsum in 2011-12 which was over 90 per cent of India's production. The main deposits occur in the Tertiary clays and shales of Jodhpur, Nagaur and Bikaner. Jaisalmer, Barmer, Churu, Pali and Ganganagar also have some gypsum bearing rocks. The remaining gypsum is produced by Tamil Nadu, Jammu and Kashmir, Gujarat and Uttar Pradesh in order of production. **Tamil Nadu** has most of its deposits in Tiruchirapalli district between Chittali in the north and Tappay and Periyakurukhai in the south. Some gypsum is also reported from Coimbatore district. Minor occurrences are found along the coast and in some salt pans in Terunelveli, Ramanathapuram, and Viluppuram districts. The recoverable deposits are estimated to be 16.35 million tonnes. **Jammu and Kashmir** has estimated reserves of 112.9 million tonnes. The main deposits occur in Baramula and Doda districts and in Uri. **Gujarat** has total reserves of 10.3 million tonnes mainly confined to Bhavnagar, Junagadh, Jamnagar and Kachchh districts. Minor deposits occur in Kheda and Surendranagar districts. In **Uttarakhand**, the main deposits of gypsum are found in Tehri Garhwal, Dehra Dun and Musoorie.

Some gypsum is also produced in **Andhra Pradesh,** (Nellore, Guntur, Prakasam), **Himachal Pradesh** (Spiti, Sirmur, Chamba), **Karnataka** (Gulbarga) and **Madhya Pradesh** (Shahdol).

In addition to mineral gypsum, water and phosphoric acid plants are important sources of by product gypsum. Marine gypsum is recovered from salt pans during the processing for common salt in coastal regions particularly of Gujarat and Tamil Nadu. Phospho-gypsum is obtained as a by-product while manufacturing phosphoric acid whereas fluro-gypsum is obtained while manufacturing aluminium flouride and hydro-fluoric acid. Similarly, boro-gypsum is the by-product of refining calcium borates. It is worth mentioning that the recovery of by-product phospho-gypsum, fluoro-gypsum, and marine gypsum together is higher than mineral gypsum.

Diamonds

Diamonds have been highly valued and cherished throughout the ages because of their brilliance, adamantine, lustre, transparency and hardness. They are widely used for ornaments and for polishing the surface metals, minerals and gem cutting. The most important industrial use of diamonds is in cutting-edges of drills used for exploration and mining of minerals.

As per the United Nations Framework Classification (UNFC) system as in 2015 all India resources of diamond are placed at around 31.84 million carats. Madhya Pradesh accounts for 90.18 per cent resources followed by Andhra Pradesh 5.72 per cent and Chhattisgarh 4.10 per cent.

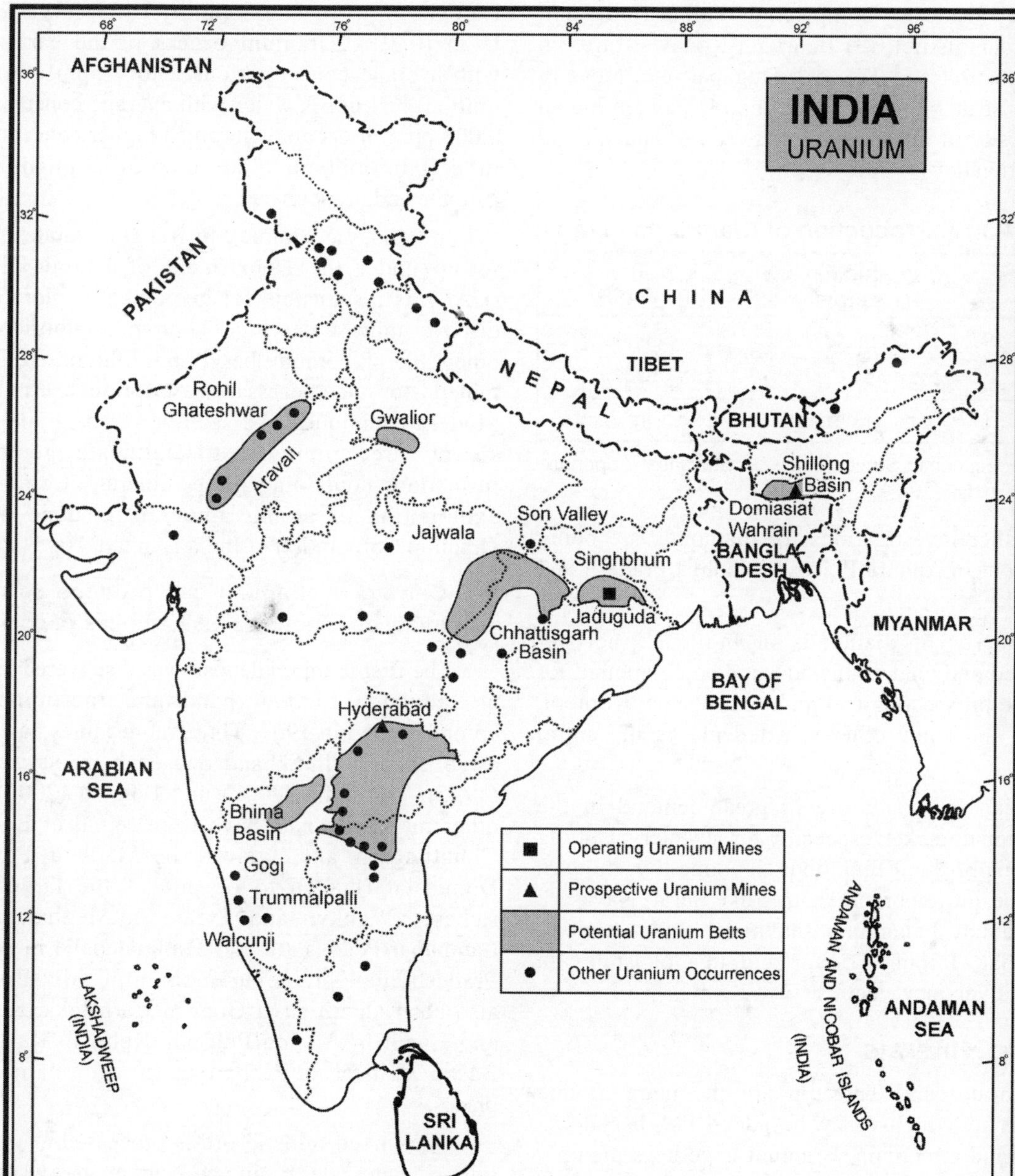

FIG. 23.7. India : Uranium

The diamond fields of India are grouped into four regions. (1) South Indian tract of Andhra Pradesh, comprising parts of Anantpur, Cuddapah, Guntur, Krishna and Kurnool districts and Mahaboobnagar district in Telangana; (2) Central Indian tract of Madhya Pradesh, comprising Panna belt and Chhatarpur district; (3) Behradin-Kodawali area in Raipur district and Tokapal, Dugapal, etc. areas in Bostar district of Chhattisgarh and (4) Eastern Indian tract mostly of Odisha lying between Mahanadi and Godavari valleys.

TABLE 23.17. Production of Diamonds in India

Year	Production in carats	Value (₹ '000)
2018-19	38,437	5,39,062
2019-20	28,816	3,98,070
2020-21	13,917	2,20,304

Source: Economic Survey 2021-22. Statistical Appendix, p. 56.

Further investigations for diamonds are being carried out in Andhra Pradesh, Madhya Pradesh and Karnataka.

Reserves of diamonds in India are not yet exhausted and modern methods are being applied for intensive prospecting and mining in Panna, Kurnool, Bellary and some other selected places in central India.

Indian diamonds are in great demand in the international market, especially for jewellery. Cutting and polishing of diamonds is done by modern techniques at important centres like Surat, Navasari, Ahmedabad, Palanpur, Bhavnagar and Mumbai. Khambhat, Jaipur, Trichur and in Goa state are comparatively new centres.

Atomic Minerals

Uranium and Thorium are the main atomic minerals to which many be added *i.e.*, beryllium, lithium and zirconium. Uranium is perhaps the only chemical element that has been dominating the world strategic scene since the mid-twentieth century. Discovered more than 200 years ago in mineral known as pitchblende by a German chemist Martin Heinrich Klaproth, it derives its name from the planet Uranus, which was discovered only a few years earlier in 1781. Henry Beaquerel, a French physicist discovered the new property of radioactivity in uranium which opened a new era in nuclear technology. Later it was discovered that uranium with atomic number 92 occurs in nature in two principal isotopic forms of mass number U-238 (99.3%) and U-35 (0.7%). Uranium occurs in the earth crust with average concentrations 2 to 4 ppm (parts per million) and in sea water with average concentration 0.003 ppm. It occurs with much higher concentration in certain minerals like uraninite, pitchblende, autunite and uranophane.

In India, the Atomic Mineral Directorate (AMD) set-up under the Department of Atomic Energy (DAE), is responsible for geological exploration to discover mineral deposits like uranium, thorium, and others like zirconium, beryllium, lithium, etc. These minerals provide necessary base for nuclear industry. Modern techniques like satellite pictures and aerial survey are employed to delineate geological formations containing these minerals. Commercial exploitation of atomic minerals is done by the Uranium Corporation of India Limited.

One kg of uranium can produce as much electricity as is done by 1,500 tonnes of coal.

The first uranium deposit was discovered in 1951 at Jaduguda in Jharkhand and the mine was commissioned in 1967. Three more mines have been established in Jharkhand one each in Narwapahar (1995), Turamdih (2002) and Bagjata (2007). The other major deposits have been found at Bodal in Chhattisgarh and Jajawal in Madhya Pradesh; Domiasiat (known to be one of the largest and richest), Wahkyn and Tyrani in Meghalaya and Lambapur-Peddagattu and Tummalapalle in Andhra Pradesh. Favourable uranium mineralization has also been identified at Gogi in Karnataka, Kuppam and Gandi in Andhra Pradesh, Rohil in Rajasthan, Bastar district in Chhattisgarh and many other places.

The mixed mineral ore is processed at uranium mills at Jaduguda, Batin and Narwapahar where it is converted into uranium oxide (U_3O_8), popularly known as "Yellow Cake".

The country has an estimated uranium oxide reserve 1,17,800 metric tonnes in six states (**Table 23.18**).

A study of samples of magnetic rocks from the Nubra-Shyok valley in Lakakh has found uranium (0.31-5.36 per cent) and thorium (0.76-1.43 per cent). The study made in the isotope laboratory of the University of Tuebingen in Germany in 2007 says that uranium found in these rocks is exceptionally high quality when compared with 0.1 per cent or even less in ores present elsewhere in the country.

TABLE 23.18. Distribution of Uranium reserves (2011-12)

	State	Uranium reserves (metric tonnes)
1.	Jharkhand	46,700
2.	Andhra Pradesh	22,000
3.	Karnataka	21,000
4.	Meghalaya	16,400
5.	Rajasthan	2,900
6.	Chhattisgarh	2,800
.	Total	1,17,800

Source : Department of Atomic Energy.

Timeline

However, the largest source of uranium comprise the monazite sands, both beach and alluvial. Although monazite sands occur on east and west coasts and in some places in Bihar, the largest concentration of monazite sand is on the Kerala coast. Over 15,200 tonnes of uranium is estimated to be contained in monazite. Some uranium is found in the copper mines of Udaipur in Rajasthan.

In spite of all the above mentioned developments, India's contribution to uranium production is negligibly small. About 65% of uranium resources are confined to only five countries, namely Australia, Kazakhistan, Russian Federation, Canada and Niger. The other important countries having recognisable resources are Namibia, South Africa, Brazil, U.S.A. and China.

Thorium is also derived from monozite which contains 10 per cent thoria and 0.3 per cent urania. The other mineral carrying thorium is thorianite. The known reserves of thorium in India are estimated to be between 4,57,000 and 5,08,000 tonnes. Kerala, Jharkhand, Bihar, Tamil Nadu and Rajasthan are the main producers.

Beryllium oxide is used as a 'moderator' in nuclear reactors for atomic power generation. India has sufficient reserves of beryllium to meet her requirement of atomic power generation.

Lithium is a light metal which is found in lepidolite and spodumene. Lepidolite is widely distributed in the mica belts of Jharkhand, Madhya Pradesh and Rajasthan as well as in Bastar region of Chhattisgarh, ***Zirconium*** is found along the Kerala coast and in alluvial rocks of Ranchi and Hazaribagh districts of Jharkhand.

Salt

It is an important mineral which is used in chemical industry. Sodium chloride, known as *common salt*, is used as a food item. Salt is obtained from sea water, brine springs, wells and salt pans in lakes and from rocks. Rock salt is taken out in Mandi district of Himachal Pradesh and in Gujarat. It is less than 1 per cent of the total salt produced in India. Sambhar lake in Rajasthan produces about 10 per cent of our annual production. Sea brine is the source of salt in Gujarat, Maharashtra and Tamil Nadu. Gujarat coast produces nearly half of our salt. Production of salt in India increased from 30 lakh tonnes in 1951 to 100 lakh tonnes in 2011-12. In 2005-06, 1,871 tonnes rock salt worth ₹ 29.54 lakh was produced.

CONSERVATION OF MINERAL RESOURCES

Conserving the mineral resources is the most serious problem because they are exhaustible resources. Once taken out of the mines, minerals are used for ever. This is the reason that mining is often called a *robber industry*.

The mineral resources are being exploited at an accelerated rate due to advancement in mining technology. India exports a large number of minerals to earn the much needed foreign exchange. But it will be much better if we export goods manufactured from the minerals rather than exporting minerals in their raw form. Minerals can be conserved by bringing in efficiency in mining technology as well as in the technology of benefication. There are many 'cyclic' minerals such as iron, aluminium, copper, brass, tin, etc. Recycling of these minerals can help in reducing the waste. Japan, Britain, Italy, etc. are some of the countries which are using scrap iron on a large scale for iron and steel industry. Scarce minerals may be saved by substituting them by those minerals which are cheaper and are found in abundance. The best example is that of aluminium which is now extensively used in electrical industry in place of copper.

REFERENCES

Coggin Brown, J. and Dey, A.K. (1955) : *India's Mineral Wealth*, Oxford University Press.

Economic Survey 2018-19.

Govt. of India : Envistats India, 2018

Govt. of India : India; A Reference Annual, 2024.

Govt. of India : Indian Minerals Quarterly : *April 1993*, Geological Survey of India.

Govt. of India : Minerals Yearbnook, 2017.

Govt. of India : Mineral Wealth of India, 2013

Murthy, M.S.S. : Uranium : The Strategic Mineral, *Science Reporter*, September 2008, pp. 28-32

Negi, Balbir Singh (1984) : *Mineral Resources of Kumaon and Garhwal in* A. Ramesh (ed.) *Resource Geography*, Heritage Publishers, New Delhi.

Wadia, D.N. (1975) : *Geology of India*, Chap. XXVI Economic Geology, Tata McGraw Hill, New Delhi.

Wadia, Mehar, D.N. (2015) : *Minerals of India*, National Book Trust of India, New Delhi.

◆◇◆

CHAPTER

24

Energy Resources

Energy is the primary input in the production of goods and services. *The wheels of progress move with the flow of energy*. One of the critical elements in raising the standard of living of a country's population is the provision of affordable and reliable energy services in sufficient quantities. More regular and ample is the availability of energy, more even will be the path to economic prosperity. The role of energy has significantly increased with the increase in industrialization and urbanization in the present day society. From its early role, which was confined to kitchen as a fuel for household cooking, energy is now a major input in sectors such as industry, commerce, transport and telecommunications, besides the wide range of services required in the household sector.

Depending upon its source and utilization, energy can be divided into two broad classes *viz*. (*i*) *traditional or non-commercial*, and (*ii*) *commercial energy*. The non-commercial energy includes fire-wood, charcoal, cow dung, agricultural wastes and also animal power. The commercial sources of energy comprise coal, oil, natural gas, hydro-electricity, nuclear power, as well as wind and solar power. Energy may also be classified as *conventional* and *non-conventional* depending upon its nature. Coal, petroleum, natural gas and electricity are the main sources of *conventional energy* while solar, wind, tidal, geothermal energy and biogas, etc. are some of the outstanding examples of *non-conventional energy*.

CONVENTIONAL SOURCES OF ENERGY

As mentioned in the preceding paragraph, coal, petroleum, natural gas and electricity are conventional sources of energy. A brief description of their production, distribution and consumption is given in the following pages.

COAL

Coal is an inflammable organic substance, composed mainly of hydrocarbons, found in the form of sedimentary rocks and capable of being used as fuel to supply heat or light or both. It also contains volatile matter, moisture and ash in varying proportions. Combustible matter in coal consists of carbon and hydrogen.

Coal was, is and will continue to be the main-stay of power generation in India for a longtime. It constitutes about 70 per cent of total commercial energy consumed in the country. The power sector and industries account for 94 per cent of total consumption. Manufacturing of iron and steel and a variety of chemicals largely depend upon the availability of coal. Due to its high utility as a source of energy and as a raw material for a large number of industries, it is often called ***black gold***. A recent study conducted by energy experts shows that the world coal reserves are six times the known reserves of oil and coal has been described as the *bridge into the future*.

Origin of Coal

Coal has originated from the organic matter wood. Large tracts of forest lands were buried under sediments in the geological past *i.e.* in the Carboniferous age. Wood was burnt and decomposed due to heat from below and pressure from above. During the process of decomposition of wood, hydrogen originates in the form of methane and water, oxygen in the form of carbon dioxide and water. During the process of change from wood to coal, the amount of oxygen and nitrogen decreases and the proportion of carbon increases. The capacity of coal to give energy depends upon the percentage of carbon contained in it. The percentage of carbon in coal depends upon the duration and intensity of heat and pressure on wood.

Varieties of Coal

Depending upon its grade from highest to lowest, following four varieties of coal are generally recognized.

1. Anthracite Coal. This is the best quality of coal and contains 80 to 95 per cent carbon. It has very little volatile matter and negligibly small proportion of moisture. It is very hard, compact, jet black coal having semi-metallic lustre. It ignites slowly and burns with a nice short blue flame. It has the highest heating value and is the most prized among all the varieties of coal. In India, it is found only in Jammu and Kashmir and that too in small quantity.

2. Bituminous Coal. This is the most widely used coal. It derives its name after a liquid called bitumen released after heating. It varies greatly in composition-in-carbon content (from 40 to 80 per cent)–and moisture and volatile content (15 to 40 per cent)–so that it is often sub-divided into several minor divisions such as sub-bituminous and bituminous coals. It is dense, compact, and is usually of black colour. A good bituminous coal is composed of alternate dull and bright bands. It does not have traces of original vegetable material from which it has been formed. Its calorific value is very high due to high proportion of carbon and low moisture content. By virtue of this quality, bituminous coal is used not only for steam raising and heating purposes but also for production of coke and gas. Most of the bituminous coal is found in Jharkhand, Odisha, West Bengal, Chhattisgarh and Madhya Pradesh.

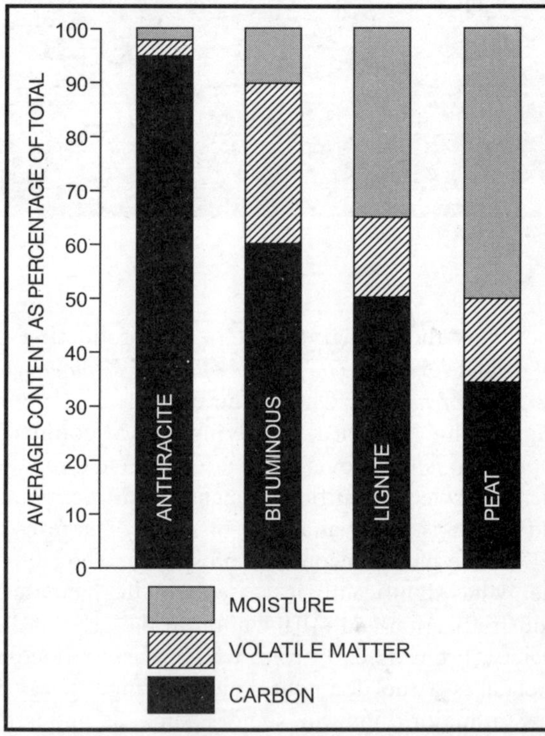

FIG. 24.1. Constituents of different types of coal

3. Lignite. Also known *as **brown coal***, lignite is a lower grade coal and contains about 40 to 55 per cent carbon. It represents the intermediate stage in the alteration of woody matter into coal. Its colour varies from dark to black brown. It is friable and pyritious. Its moisture content is high (over 35 per cent) so that it gives out much smoke but little heat. Its typical qualities make it liable to disintegrate on exposure and even to spontaneous combustion. It is found in Palna of Rajasthan, Neyveli of Tamil Nadu, Lakhimpur of Assam and Karewa of Jammu and Kashmir.

4. Peat. This is the first stage of transformation of wood into coal and contains less than 40 per cent carbon, sufficient volatile matter and lot of moisture. It is seldom sufficiently compact to make a good fuel without compressing into bricks. Left to itself, it burns like wood, gives less heat, emits more smoke and leaves a lot of ash after burning.

Occurrence of Coal in India

The coal bearing strata of India are geologically classified into two main categories, *viz.*, the Gondwana coal fields and the Tertiary coal fields.

(*a*) **Gondwana Coal.** Gondwana coal contributes overwhelmingly large proportion of both the reserves and production of coal, accounting for 98 per cent of the total reserves and 99 per cent of the production of coal in India. It is the store house of India's metallurgical as well as superior quality coal. Of the 113 major coal fields found all over India, 80 are located in the rock systems of the lower Gondwana Age. There are about 75 separate basins covering an area of 77,700 sq km mainly confined to the Peninsular India. The size of these basins varies from 1 sq km to 1,550 sq km. These basins occur down in the valleys of certain rivers *viz*., the Damodar (Jharkhand-West Bengal); the Mahanadi (Chhattisgarh-Odisha); the Son (Madhya Pradesh-Jharkhand); the Godavari and the Wardha (Maharashtra-Andhra Pradesh); the Indravati, the Narmada, the Koel, the Panch, the Kanhan and many more.

Gondwana coal is said to be about 250 million years old. It includes coking as well as non-coking and bituminous as well as sub-bituminous coal. Anthracite is generally not found in the Gondwana fields. The volatile compounds and ash (usually 13 to 20, rising to as much as 25 to 30 per cent) are present in too large a proportion to allow the carbon percentage to rise above 55 to 60; generally much less than that. The Gondwana coal is almost free from moisture, but it contains sulphur and phosphorus in small variable quantities.

It is possible that some coal bearing Gondwana rocks are hidden beneath the great pile of lava of the Deccan trap. At several places, chiefly in the Satpuras, denudation has exposed coal bearing Gondwana strata, from which it can be inferred that considerable quantities of valuable coal can be obtained from these areas. The Damuda series (*i.e.* Lower Gondwana) possesses the most valuable and best worked coalfields and accounts for about 80 per cent of the total coal production in India.

(*b*) **Tertiary Coal.** The tertiary rock system bears coals of younger age; from 15 to 60 million years and are mainly confined to the extra-Peninsula. This coal generally has low carbon and high percentage of moisture and sulphur.

Reserves. According to the Geological Survey of India, the coal reserves of India as on 1 April, 2017 have been estimated at 3,15,149 million tonnes. State-wise distribution of coal reserves is given in **Table 24.1**.

TABLE 24.1. Distribution of Coal Reserves in India as on 1-4-2017

State	Total reserves (million tonnes)	Percentage of all India reserves
Gondwana Coal Fields		
1. Jharkhand	82,440	26.29
2. Odisha	77,285	24.65
3. Chhattisgarh	56,661	18.07
4. West Bengal	31,667	10.09
5. Madhya Pradesh	27,673	8.83
6. Telangana	21,464	6.85
7. Maharashtra	12,259	3.91
Others	4,112	1.31
Total	**3,13,561**	**100.00**
Tertiary Coal Fields		
1. Meghalaya	576	36.27
2. Assam	511	32.18
3. Nagaland	410	25.82
4. Arunachal Pradesh	91	5.73
Total	**1,588**	**100.00**

Total (Gondwana + Tertiary) = 3,13,561 + 1,588 = 3,15,149
Source : EnviStats India 2018, pp. 133-34.

The **table 24.1** shows that the distribution of coal reserves is highly uneven. Over four-fifths of coal reserves are concentrated in just four states of Jharkhand, Odisha, Chhattisgarh and West Bengal. Jharkhand and Odisha have more than half of the coal reserves of India. Tertiary coal fields are mainly

TABLE 24.2. Reserves of coal as on 1st April (Million tonnes)

Year	Proved	Initiated	Inferred	Total
Gondwana Coal				
2010	10,9320	1,30,564	35,559	2,75,444
2011	1,13,408	1,37,372	33,590	2,84,370
2012	1,17,551	1,42,070	32,384	2,92,005
2013	1,22,588	1,42,532	32,301	2,97,421
2014	1,25,315	1,42,407	32,350	3,00,072
2015	1,31,020	1,43,142	30,941	3,05,103
2016	1,37,493	1,39,052	30,764	3,07,309
2017	1,42,464	1,30,212	31,885	3,13,561
Tertiary Coal				
2010	478	90	799	1367
2011	594	99	799	1,492
2012	594	99	799	1,493
2013	594	99	799	1,493
2014	594	99	799	1,493
2015	594	99	799	1,493
2016	594	99	799	1,493
2017	594	99	895	1,588

Source: EnviStats, India 2018, pp. 132-34.

confined to the north-eastern states of Meghalaya, Assam, Nagaland and Arunachal Pradesh.

Lignite. As on 1 April, 2017, the total reserves of lignite in India have been estimated at 44.59 billion tonnes by Geological Survey of India. Over 90 per cent of the reserves of lignite are concentrated in Tamil Nadu. Neyveli area of Cuddalore district in Tamil Nadu has 4,150 million tonnes of lignite reserves out of which 2,360 million tonnes have been proved. The other lignite reserves of considerable importance in Tamil Nadu are those of Jayamkondacholapuram of Trichy district, Mannargudi and East of Veeranam. Lignite reserves have been identified in Rajasthan, Gujarat, Jammu & Kashmir, Kerala, West Bengal and Puducherry also.

Although the reserves and production of coal have increase with the passage of time the production coal has not increased enough to meet the growing demand of expanding industries.

Production

Although the first coal mine was opened in 1774 at Raniganj in West Bengal, the real beginning was made in 1814 in the same area. Coal mining industry registered a much faster growth after Independence (see **Table 24.3**).

TABLE 24.3. Production of Coal (including lignite) in India in million tonnes

Year	Cocking		Non-cocking	Total	Lignite	Total (coal + lignite)
	Metallurgical	Non-metallurgical				
1950-51	NA	NA	NA	32.30	NA	NA
1960-61	17.0	NA	38.2	NA	NA	NA
1970-71	17.8	NA	55.1	NA	3.4	NA
1980-81	24.6	8.0	81.3	113.9	5.1	119.0
1990-91	24.1	21.2	166.4	211.7	13.8	225.5
2000-01	19.3	11.8	278.6	309.6	23.0	332.6
2010-11	17.7	31.9	483.2	532.7	37.7	570.4
2015-16	14.3	46.5	578.3	639.2	43.8	683.0
2016-17	15.2	46.4	596.2	657.8	45.3	703.1
2017-18	33.9	6.3	635.5	675.4	46.6	722.0
2018-19	35.1	6.0	687.6	728.7	44.3	773.0
2019-20	35.4	17.4	677.9	730.8	42.1	772.9

NA = Data Not Available

Source : Economic Survey 2022-23. Statistical Appendix, p. 46.

The production rose steadily from 119.02 million tonnes in 1980-81 to 772.9 million tonnes in of coal and lignate in 2019-20.

The nationalisation of coal industry in 1973-74 gave much needed impetus to coal and lignite production and India is now the third largest coal producer in the world after China and the USA. At present, the coal industry provides employment to nearly seven lakh persons.

Distribution of Coal in India

Majority of the coal-fields are found in the eastern part of India particularly to the east of 78° E longitude. Maximum concentration of coal fields is in the north-eastern part of the Peninsular plateau of India comprising parts of Jharkhand, Chhattisgarh, Odisha and eastern Madhya Pradesh and western part of West Bengal adjoining Jharkhand. Southern part of Madhya Pradesh, eastern part of Andhra Pradesh and eastern part of Maharashtra also have large deposits of coal (**Fig. 24.2**).

Table 24.4 makes it clear that about three-fourth of India's coal is produced by four states of Chhattisgarh, Jharkhand, Odisha and Madhya Pradesh. More than 40 per cent of India's total coal production comes from just two states of Chhattisgarh and Jharkhand.

TABLE 24.4. Distribution of Coal in India (2016-17)

State	Production in million tonnes	Percentage of all India production
1. Chhattisgarh	143.85	21.70
2. Odisha	139.36	21.03
3. Jharkhand	126.44	19.08
4. Madhya Pradesh	105.01	15.84
5. Telangana	59.53	8.98
6. Maharashtra	40.56	6.12
7. West Bengal	27.67	4.17
8. Uttar Pradesh	16.06	2.42
Odisha	4.31	0.66
Total	662.79	100.00

Source : Data computed from EnviStats India, 2018, p. 138.

Gondwana Coalfields

As mentioned earlier, the Gondwana Coal fields are exclusively found in the Peninsular plateau of India. State-wise major Gondwana coal producing areas are described as under :

1. Chhattisgarh. Chhattisgarh holds the third position with respect to coal reserves but occupies the first position, so far as production is concerned. This state has 18 per cent of the coal reserves and produces over 21 per cent coal of India. Most of the coal fields of Chhattisgarh are located in the northern part of the state. The *Korba* coalfield stretches over an area of 515 sq km in the valleys of Hasdo (a tributary of the Mahanadi) and its tributaries (Ahram and Kurang) in Korba district. Coal occurs in the Barakar measures with total thickness of 700 m. Most of the coal from the field is sent to thermal power plant at Korba. The *Bisrampur* coalfield lies in Surguja district. With total reserves of 542 million tones this field has coal seams of thickness varying from 30 cm to 1.8 m. The *Hasdo-Arand* coalfield extends from Rampur in Surguja district to Arand valley in Bilaspur district and covers an area of about 1004 sq km. The coal reserves in this field are estimated at 4,321 million tonnes. The coal seams have average thickness of 2.5 m to 7.0 m. *Chirmiri* coalfield in Surguja district spreads over an area of 128 sq km. The total reserves of this field are estimated at 362 million tonnes. There are four coal seams in this field out of which three seams contain good quality coal. *Lakhanpur* coalfield lies south of Bisrampur coalfield and spreads over Surguja, Koriya, Korba and Bilaspur districts. Here the coal seams are 1 to 3 m thick. *Jhilmili* coalfield occupies a total area of 106 sq km. Being an extension of Sahagpur coalfield of Shahdol district (in Madhya Pradesh), most of it lies in Koriya district of Chhattisgarh. It has five coal seams which belong to Talcher and Barakar measures. The coal is non-coking type and has high proportion of ash. *Johilla* coalfield, lying in the Johilla valley, covers an area of about 38 sq km. The reserves are estimated at 311 million tonnes. *Sonhat* coalfield in Surguja district has superior quality coal. Kutkona, Charch and Sardih coal bearing strata have high grade coal. *Tatapani-Ramkola* coalfields lie between Kanhar and Rehar rivers in the north-eastern part of Surguja district. Coal of Tatapani coalfields belongs to the Damuda series.

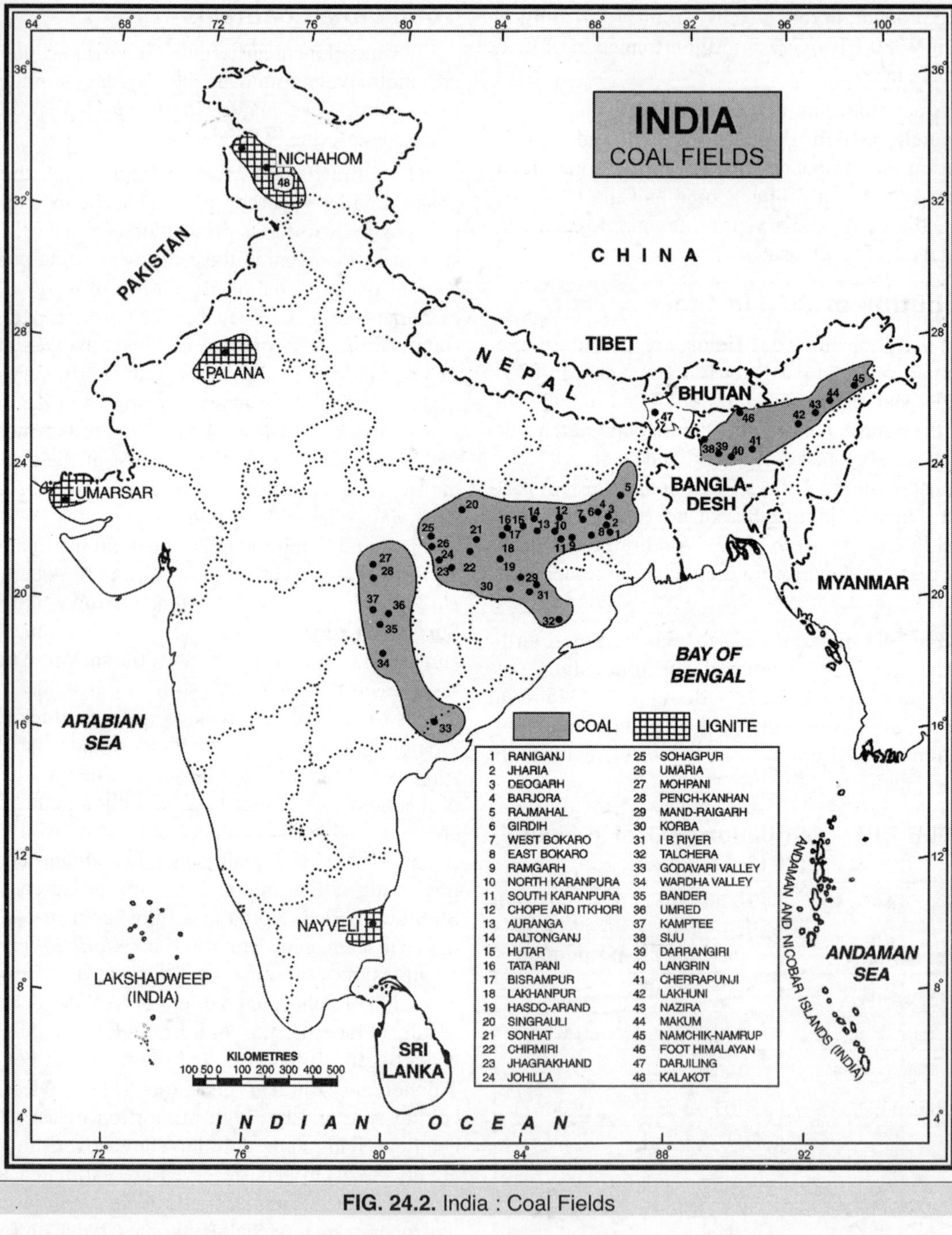

FIG. 24.2. India : Coal Fields

2. Odisha. Odisha is the second largest state with respect to reserves and production of coal in India contributing a little over 21 per cent of the total coal production of the country. Most of the deposits are found in Dhenkanal, Sambalpur and Sundargarh districts. The *Talcher* field stretching eastward from Talcher town to Rairkhol in Dhenkanal and Sambalpur districts ranks second in reserves (24,374 million tonnes) after Raniganj. This field covers an area of 578 sq km. Two workable seams of coal 3 to 4 metre thick had earlier been discovered in an area of 295 sq km near Talcher town. Coal from this field is most suitable for steam and gas production. Most of the coal is utilised in thermal power and fertilizer plants at Talcher. The *Rampur-Himgir* coalfields in the districts of Sambalpur and Sundargarh spread over an area of 520 sq km. Coal occurs here in middle and lower Barakar seams. Bulk of the coal is of inferior quality. The **Ib river** coalfield covers a total area of 512 sq km in Sambalpur and Gangpur districts. The inferred reserves in this area are estimated at 1,754 million tonnes. The coal belongs to the middle and lower Barakar systems in which Rampur, Lajkuria and Gamadera seams are important. Much of the coal is of inferior quality with about 50 per cent fixed carbon.

3. Jharkhand. Jharkhand is the richest state with respect to reserves but has conceded the first two places to Chhattisgarh and Odisha. This state was the largest producer of coal in India till recent past. Jharkhand has over 26 per cent of the coal reserves and produces more than 19 per cent coal of India. In the year 2016-17, Jharkhand produced 126.44 million tonnes of coal.

Most of the coal fields are located in a narrow belt running in east-west direction almost along the 24°N latitude (**Fig. 24.3**). There are 21 prominent coal fields in Jharkhand of which 8 are in Dumka (Santhal Parganas), 7 in Hazaribagh and 3 each in Dhanbad and Palamu. Amongst these, Jharia, Bokaro, Girdih and Karanpura are outstanding. The *Jharia coal-field* lies to the south-west of Dhanbad city and covers an area of 453 sq km. It is one of the oldest and the richest coal fields of India and has been recognised as the ***store house*** of the best metallurgical coal in the country. The total estimates of all grades of coal upto a depth of 900 metre are estimated to be 16,985.69 million tonnes. The *Bokaro coal-field* in Hazaribagh district lies within 32 km of western end of the Jharia coal field. It is a long but narrow strip in the catchment area of the Bokaro river spreading over an area of 674 sq km. The entire Bokaro coalfield is

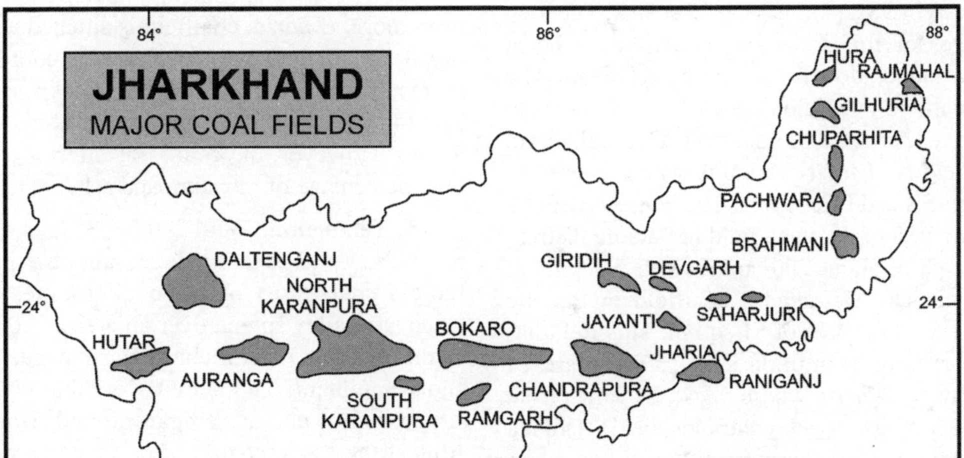

FIG. 24.3. Jharkhand : Major coal fields

divided into two parts *viz*. West Bokaro and East Bokaro. The reserves in West Bokaro upto a depth of 900 metre have been estimated at 4,473.73 million tonnes with seams ranging from 3.5 to 11 metre thick. East Bokaro has 4,246.32 million tonnes of reserves upto a depth of 600 metre. The *Girdih* (also known as ***Karharbari***) coalfield lies to the south-west of Girdih in Hazaribagh district. Spreading over an area of 28.5 sq km this field has three main seams of varying thicknesses: (*i*) the Lower Karharbari (*ii*) the Upper Karharbari and (*iii*) the Badhua seams. The Lower Karharbari seam is 3 to 7.5 metre thick and gives out the finest coking coal in India for metallurgical purposes. The *Karanpura and Ramgarh* coalfields lie to the west of Bokaro and cover an area of about 1,522 sq km. The *North Karanpura* covers an area of 1,230 sq km having estimated reserves of 13,110.84 million tonnes upto a depth of 900 metre. The *South Karanpura* field covers an area of 194 sq km and possesses estimated reserves of 5,757.85 million tonnes down to 900 metres in depth. *Ramgarh coal field* situated about 9 km away from the Bokaro field covers an area of 98 sq km having 22 seams. The total coal reserves of this field are estimated to be at 1,059.20 m tonnes down to a depth of 900 m. The *Auranga* coalfield in Palamu occupies an area of 240 sq km with a seam of 3 m thickness. Coal taken out from this field is of inferior quality which is used in cement furnaces and brick kilns. The *Hutar* coalfield, about 19 km west of Auranga field in Palamu district, covers an area of about 200 sq km with 5 seams of inferior coal. The *Deltenganj* coalfield in Palamu district covers a total area of 51 sq km. This field has proved reserves of 84 million tonnes. There are 14 coal seams near *Rajhera* whose thickness varies from 15 cm to 1.5 m. *Devgarh* coalfields lie in Dumka district where Jayanti, Saharjuri and Kundithuraih are the main mining areas. These coalfields occupy the valleys of the Adjai and Barakar rivers and covers a total area of about 46 km. *Jayanti* coalfields have 5 seams while the remaining two have two seams each. The coal here is of inferior quality and has high ash content. The *Rajmahal* coalfield along the western side of the Rajmahal hills is in a scattered form which spreads over an area of about 182 sq km. It lies in the north of the river where Hura, Gilhuria Chuparhita, Pachwara, Mahuagahi and Brahmki are important mines. This coalfield has proved coal reserves of 1,913 million tonnes. However, coal available here is of inferior quality and is mostly used in brick kilns.

4. Madhya Pradesh. Madhya Pradesh has about 8.83 per cent of the coal reserves but contributes about 15.8 per cent of the total coal production of India. Currently Madhya Pradesh is the fourth largest coal producing state of India. *Singrauli* (Waidhian) coalfield in Sidhi and Shahdol districts is the largest coalfield of Madhya Pradesh. Spreading over an area of 2,337 sq km, this coalfield has 9,207 million tonnes of coal reserves. Jhingurda, Panipahari, Khadia, Purewa and Turra are important coal seams. Jhingurda with a total thickness of 131 m is the richest coal seam of the country. This field supplies coal to thermal power plants at *Singrauli* and *Obra*. *Pench-Kanhan-Tawa* in Chhindwara district is another important coalfield of Madhya Pradesh. It contains 1956 million tonnes of semi-coking and non-coking coal. Ghoravari seam in Kanhan field is 4.6 m thick and contains coking coal. *Sohagpur* coalfield in Shahdol district has 2,284 million tonnes reserves of coal. Here coal seams attain a thickness of 3-5 m and even more. *Umaria* coalfield is situated at a distance of 58 km to the south of Katni. It contains 6 coal seams of which 4 are important (thickness 2 to 4 m). The total reserves are estimated at about 58 million tonnes. However, the coal is of inferior quality with high percentage of moisture and ash.

5. Telangana. With only 6.85 per cent of the reserves, Telangana produces about 9 per cent of India's coal. Most of the coal reserves are in the Godavari valley spread over an area of 10,350 sq km in the districts of Adilabad, Karimnagar, Warangal and Khammam. The actual workable collieries are situated at Tandur, Singareni and Kothagudam. Almost the entire coal is of non-coking variety. The reserves of all types of coal in the Godavari valley upto a depth of 1210 metre have been estimated at 10,435.50 million tonnes. These are the southern most coalfields of India and a source of coal supply to most of south India.

6. Maharashtra. Though Maharashtra has only 4 per cent reserves, the state accounts for over 6 per cent of the production of coal in India. Most of the coal deposits are found in the Kamptee coalfields in Nagpur District; Wardha valley, Ghughus, Ballarpur and Warora in Chandrapur district and the Wun field in Yavatmal district. Coal has also been located in Umrer, Nand, Makardhokda and Bokhara areas.

7. West Bengal. Although West Bengal produces just over four per cent of India's coal, the state has over 10 per cent of the coal reserves of the country. Barddhaman, Bankura, Purulia, Birbhum, Darjeeling and Jalpaiguri are the chief producing districts. Raniganj is the largest coalfield of West Bengal. In fact, it is at Raniganj that coal-mining started in India in 1774. It covers an area of 1,500 sq km mainly in Barddhaman, Bankura and Purulia districts. Small part of this field is in Jharkhand state. This field produces mainly non-coking steam coal, which is mainly used by thermal power plants.

In Darjeeling district, coal reserves are found in Dalingkot coalfield; several seams occur near Tindharia where the best seam is about 3.3 metre thick. In Jalpaiguri district, a few seams are located near the Duars area.

8. Uttar Pradesh. Most parts of Uttar Pradesh are covered with sediment brought by rivers and do not possess coal reserves. But some of the coal seams of Madhya Pradesh project into the terrritory of this state. A small portion of the Singrauli field of Madhya Pradesh falls within Mirzapur district of Uttar Pradesh. A high grade coal seam, about 1 to 1.5 m thick occurs near Kotah.

Tertiary Coal-fields

Tertiary coal-fields mainly occur in association with limestone and slates of either Eocene or Oligocene-Miocene age. The statewise distribution of tertiary coal is as follows :

1. Assam. The major coalfields in Assam are the Makum, Nazira, Mikir Hills, Dilli-Jeypore and Lakhuni. Of these, the Makum coalfield in Sibsagar district is the most developed field. It is 28 km long and about 5 km wide. The total reserves of all types of coal in this field are estimated to be 511 million tonnes, down to a depth of 600 m. In the Mikir Hills coalfield, Koilajan, Langlor, Diogarang river areas are worth mentioning. Assam coals contain very low ash and high coking qualities but the sulphur content is high, as a result of which this coal is not suitable for metallurgical purposes. But these coals are best suited for hydrogenation process and are used for making liquid fuels.

2. Meghalaya. Garo, Khasi and Jaintia hills are believed to have deposits of tertiary coal belonging to lower Eocene. The total reserves of all types of coal in Meghalaya are estimated to be 576 million tonnes. The Garo hills have important coalfields near Darrangiri. In the Khasi and Jaintia hills, Siju, Cherrapunji, Liotryngew, Maolong and Langrin coalfields are important.

3. Arunachal Pradesh. The Upper Assam Coal belt extends eastwards as Namchick-Namrup coalfield in the Tirap district of Arunachal Pradesh. The seams of this coalfield are 4 to 19 metre thick. The coal is generally high in volatiles and in sulphur. The total reserves of all kinds of coal upto a depth of 330 m are estimated to be about 91 million tonnes.

The other tertiary coalfields include the Kalakot, Jangali, Chinkah, Metka, Maholgala and Ladde areas of Jammu and Kashmir and the Chamba district of Himachal Pradesh.

LIGNITE

As on 1 April, 2017, the total reserves of lignite in India have been estimated at 44,698.13 million tonnes. The main concentration is in Tamil Nadu (80.05%), Rajasthan (12.83%) and Gujarat (6.09%) (**Table 24.5**). Puducherry and Jammu and Kashmir also have some deposits of lignite.

1. Tamil Nadu. Tamil Nadu accounts for 80 per cent of the reserves and about 58 per cent of the production of lignite in India. This state produced 26.2 million tonnes of lignite out of a total of 45.23 million tonnes for the whole of India in 2016-17. The Neyveli Lignite fields of Cuddalore district, extending

TABLE 24.5. Reserves of Lignite in India (2017)

State	Reserves in million tonnes	Percentage of all India reserves
1. Tamil Nadu	35,782.38	80.05
2. Rajasthan	5,735.96	12.83
3. Gujarat	2,722.05	6.09
4. Puducherry	416.61	0.94
Others	41.13	0.09
Total	44,698.13	100.00

Source: Data computed from EnviStats, India, 2018, p. 135.

over an area of 480 sq km, have an estimated reserves of 4,150 million tonnes. However, Neyveli mines suffer from the artesian structure and constant pumping out of water is a formidable task. Lignite reserves at Neyveli are exploited by Neyveli Lignite Corporation Limited (NLC). Incorporated as a private limited company in 1956, NLC was wholly owned by the Government and converted into a public limited company with effect from 7 March, 1986. Over the years, it has acquired considerable expertise and has established itself as a premier organisation in the field of lignite mining and lignite based power generation. Geological reserves of about 1,168 million tonnes of lignite have been identified in Jayamkondacholapuram of Trichy district of Tamil Nadu. In Mannargudi and east of Veeranam, geological reserves of around 22,898 million tonnes and 1,342.45 million tonnes have been estimated respectively.

TABLE 24.6. Distribution of Lignite in India 2016-17

State	Production in million tonnes	Percentage of all India production
1. Tamil Nadu	26.20	57.93
2. Gujarat	10.55	23.32
3. Rajasthan	8.48	18.75
Total	45.23	100.00

Source : Data Computed from EnviStat, India, 2018, p. 158.

2. Gujarat. Lignite occurs in Kachchh district at Umarsar, Lefsi, Jhalrai and Baranda and also in Bharuch district. The lignite of all the places except that of Umarsar is of poor quality. The total reserves of the state are estimated at 2,722 million tonnes. In 2016-17 Gujarat produced over 10.55 million tonnes of lignite which amounted to over 23 per cent of the total lignite production of India.

3. Rajasthan. Lignite deposits occur over an area 4 km long and 0.8 km wide at Palana in Bikaner district. Most of lignite is found at depths varying from 61 to 78 metres in 6 to 12 metre thick seams. Huge lignite reserves to the tune of 1,640 million tonnes were discovered in Barmer district in the year 2003. This discovery has made the state richer at total reserves of 5,736 million tonnes. The 250 MW thermal plant at Bikaner wholly depends upon lignite as the basic fuel. Lignite is found in some other parts of the state also. In 2016-17, Rajasthan produced 8.5 million tonnes of lignite which amounted to 18.75 per cent of total lignite produced in the country in that year.

4. Jammu and Kashmir. Lignite deposits belonging to Pliocene or even a newer age have been found here in sufficient quantity. The main lignite fields occur in the Shaliganga river, continuing to the northwest upto Nichahom area in Handwara region of Baramula district. The lignite here is of poor quality. The total reserves are 90 million tonnes.

5. Kerala. Irregular seams of lignite occur near Pathiarpally in Alappuzha district, Warkala in Thiruvananthapuram district and Paravur in Kollam district. But the main deposit of lignite occurs as a band of 45 to 60 cm thick near Kannur and Kasargod taluka. It is also found in Kozhikhode district.

6. West Bengal. Good quality lignite is reported to occur on the western side of the Jainti river and in Buza hills north-west of Jainti. Darjeeling and Jalpaiguri districts have small pockets of lignite. Beds of lignite also occur in the Ganga delta.

7. Puducherry. The total reserves of lignite in Puducherry are estimated at 417 million tonnes. The main deposits are reported from near Bahur, Araganur and Kanniyankovil. Pudukaddai, Vadapuram, Kadvanur and Paliayam also have some deposits.

PEAT

Peat is confined to a few areas only. It occurs in Nilgiri hills at an elevation of over 1,800 m. In the Kashmir valley, peat occurs in the alluvium of the Jhelum and in swampy grounds in higher valleys. In West Bengal peat beds at depths ranging from 2 to 11 metre have been noted in Kolkata and its suburbs. In the Ganga delta, there are layers of peat which are composed of forest and rice plants.

Imports

In spite of phenomenal growth in coal production, India has to import huge quantities of coal as growth in production is not able to meet fast growth in demand (**Table 24.7**).

TABLE 24.7. Imports of Coal in India

Year	Quantity ('000 tonnes)	Value (₹ thousand)
2013-14	1,66,861	92,33,53,960
2014-15	2,12,106	1,04,53,00,819
2015-16	2,04,000	8,61,07,357

Source : EnviStats of India 2018, p. 125.

Problems of Coal Mining in India

Coal mining industry in India is facing a lot of problems. Some of the major problems confronting the coal mining are discussed as under :

1. Uneven distribution of coal. The distribution of coal is uneven. The major coal producing areas are confined to Jharkhand, Chhattisgarh, Madhya Pradesh, Odisha and West Bengal. Most of north plains and western parts of India are devoid of coal. This involves high transport cost to carry a heavy commodity like coal over long distances. Consequently, the coal consuming industries have to pay much higher prices for coal.

2. Inferior quality of coal. Indian coal has high ash content and low caloric value. The ash content varies from 20 to 30 per cent and sometimes exceeds even 40 per cent. This reduces the energy output of coal and complicates the problem of ash disposal.

3. Low Efficiency of work force. A large percentage of coal is taken out from underground mines where the productivity of labour and machinery is very low. This has stagnated at an output per man shift (OMS) of 0.55 tonnes for the last two decades despite massive investments made in modernisation of underground mines. The under-ground mines employ 80 per cent of the man power, but contribute only 30 per cent of the total output.

4. Fires in mines. There are heavy losses due to fires in the mines and at pit heads. This leads to hike in price of coal and sets off a vicious circle of price spiral in the economy.

5. Environmental problems. Mining and utilisation of coal leads to serious problem of environmental pollution. The open cast mining ravages the whole area converting it into a rugged and ravinous land. The coal dust in mines and near pit-heads creates health hazards to workers and their families. The burning of coal in thermal plants and factories releases several toxic gases in the atmosphere. Safety measures against environmental pollution caused by mining and utilisation of coal are very costly and complicated and are beyond the reach of ordinary entrepreneurs.

Conservation of Coal

The misuse of good quality coal for burning into transport and industries, the short life of metallurgical coal, selective mining leading to large scale wastage of raw coal, frequent fires in mines and unscientific method of extraction of coal have been identified as important indications for conservation of coal in India. Conservation of coal implies that every bit of energy that can be obtained from coal must be obtained and every bit of by-product that can be recovered must be recovered. Conservation of coal is an integral part of mine planning and operation. Following measures are suggested for conservation of coal in India.

1. Coking coal should be used for metallurgical industry only.
2. Low grade coal should be washed and blended with superior quality coal in requisite proportion and used in industries.
3. Selective mining should be discouraged and all possible coal from the mines should be taken out.
4. New reserves should be discovered and new techniques should be adopted.
5. Small and uneconomic collieries should be amalgamated and be made economically viable units.

PETROLEUM OR MINERAL OIL

The word 'petroleum' has been derived from two Latin words Petra (meaning rock) and Oleum (meaning oil). Thus petroleum is oil obtained from rocks; particularly sedimentary rocks of the earth. Therefore, it is also called ***mineral oil***. Technically speaking, petroleum is an inflammable liquid that is composed of hydrocarbons which constitute 90 to 95 per cent of petroleum and the remaining is chiefly organic compounds containing oxygen, nitrogen, sulphur and traces of organo-metallic compounds. Crude petroleum consists of a mixture of hydrocarbons—solid, liquid and gaseous. These include compounds belonging to the paraffin series and also some unsaturated hydrocarbons and small proportion belonging to the benzene group.

Utilisation of Petroleum

Petroleum and petroleum products are mainly used as motive power. It is a compact and convenient liquid fuel which has revolutionised transportation on land, in the air and on water. It can be easily transported from the producing areas to the consuming areas with the help of tankers and more conveniently, efficiently and economically by pipelines. It emits very little smoke and leaves no ash, (as is the case in coal utilisation) and can be used upto the last drop. It provides the most important lubricating agents and is used as an important raw material for various petro-chemical products.

Origin and Occurrence of Petroleum

Petroleum has an organic origin and is found in sedimentary basins, shallow depressions and in the seas (past and present). Most of the oil reserves in India are associated with anticlines and fault traps in the sedimentary rock formations of tertiary times, about 3 million years ago. Some recent sediments, less than one million years old also show evidence of incipient oil. Oil and natural gas originated from animal or vegetable matter contained in shallow marine sediments, such as sands, silts and clays deposited during the periods when land and aquatic life was abundant in various forms, especially the minor microscopic forms of flora and fauna. Conditions for oil formation were favourable especially in the lower and middle Tertiary period. Dense forests and sea organisms flourished in the gulfs, estuaries, deltas and the land surrounding them during this period. The decomposition of organic matter in the sedimentary rocks has led to the formation of oil. Though oil is mainly found in sedimentary rocks, all sedimentary rocks do not contain oil. An oil reservoir must have three pre-requisite conditions : (*i*) porosity so as to accommodate sufficiently large amounts of oil; (*ii*) permeability to discharge oil and/or gas when well has been drilled; (*iii*) the porous sand beds sandstone, conglomarates of fissured limestone containing oil should be capped by impervious beds so that oil does

TABLE 24.8. Proved and indicated balance Recoverable Reserves of Crude Oil and Natural Gas as on 14-2-2017

Area	Crude Oil (million tonnes)	Natural Gas (billion cubic metre)
India	604.10	1289.81
Onshore	**324.24**	**479.71**
Andhra Pradesh	1.52	0.93
Assam	159.66	158.57
Gujarat	118.61	62.28
Jharkhand	—	8.02
Madhya Pradesh	—	32.11
Nagaland	2.38	0.09
Rajasthan	24.55	34.86
Tamil Nadu	9.00	34.98
Tripura	0.07	36.10
West Bengal	—	66.45
Offshore	**279.86**	**810.10**
Western Offshore	239.20	302.35
Eastern offshore	40.67	507.76

Note : Totals may not tally due to rounding off.
Source : Indian Mineral Yearbook 2017, p. 22.3

not get dissipated by percolation in the surrounding rocks. Oil on a commercial scale is usually found where the sedimentary rock strata are inclined and folded; in a sort of chamber or reservoir, in the highest possible situation *e.g.* crests of anticlines. Normally, oil is associated with water. Being lighter than water (specific gravity of 0.8 to 0.98), it collects in the anticlines or fault traps above the surface of water. Gas is still lighter and occurs above oil. Thus on drilling an oil well, one finds gas followed by oil, although gas seepage is not always a sure indication of an oil reservoir.

As already mentioned, oil as well as natural gas in India occur in sedimentary rocks. About 14.1 lakh sq km or about 42 per cent of the total area of the country is covered with sedimentary rocks out of which about 10 lakh sq km form marine basins of Mesozoic and Tertiary times. Besides, the country has offshore areas having Mesozoic and Tertiary rocks of

TABLE 24.9. Production of Petroleum (Crude) in India (Million tonnes)

Year	1950-51	1960-61	1970-71	1980-81	1990-91	2000-01	2010-11	2015-16	2016-17	2017-18	2018-19	2019-20
On-shore	0.3	0.5	6.8	5.5	11.8	11.8	16.4	—	—	—	—	—
Off-shore	—	—	—	5.0	21.2	20.6	21.3	—	—	—	—	—
Total	**0.3**	**0.5**	**6.8**	**10.5**	**33.0**	**32.4**	**37.7**	**36.94**	**36.01**	—	**34.20**	**32.17**

Source : The Economic Survey, 2012-13, p. A-31 and Economic Survey 2013-14, Statistical Appendix, p. 27. Economic Survey 2020-21.
EnviStats India 2018, p. 139.

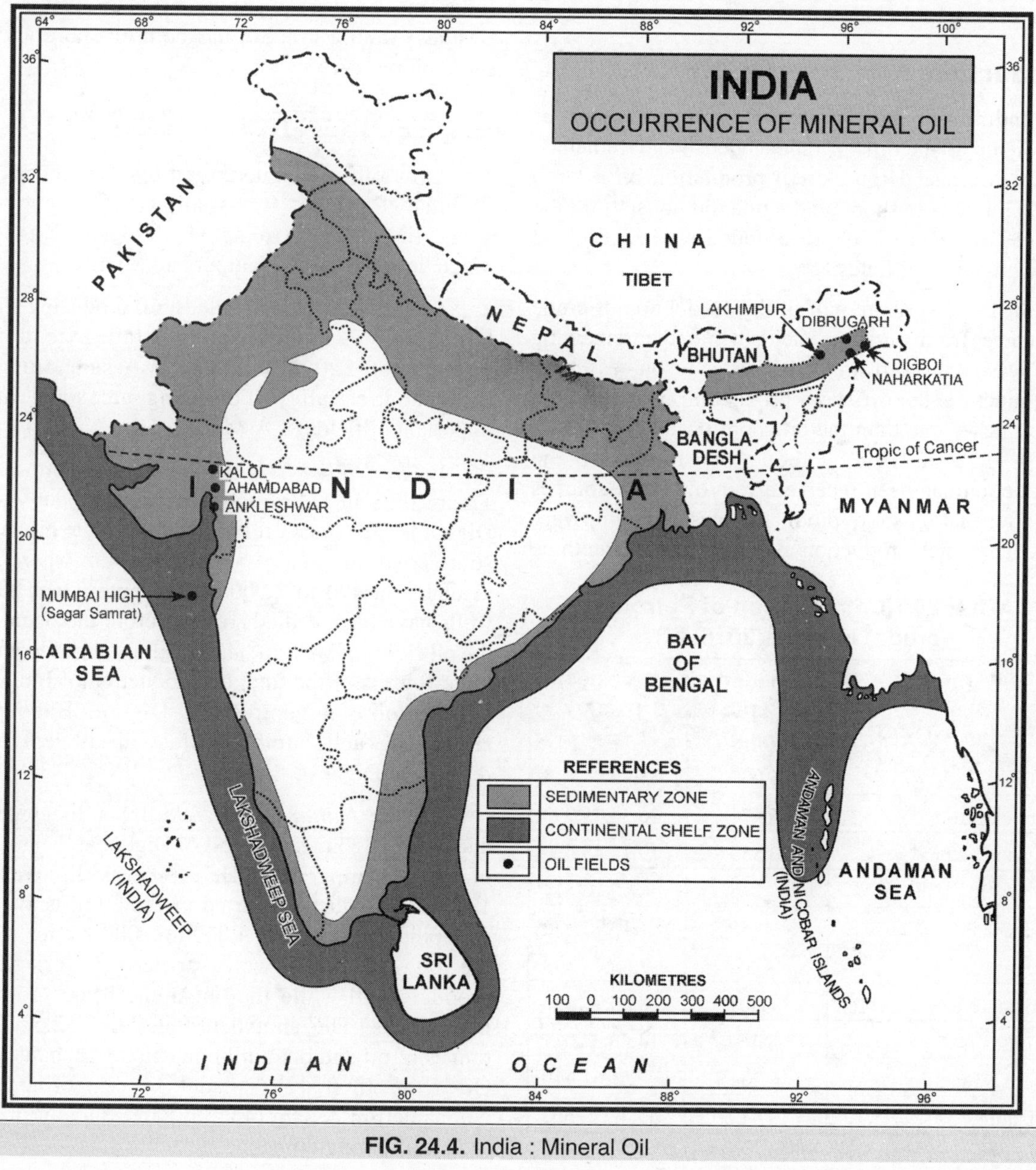

FIG. 24.4. India : Mineral Oil

marine origin covering an area of 2.5 lakh sq km upto a depth of 100 metre and another area of 0.7 lakh sq km upto a depth between 100 and 200 metre. Thus the total continental shelf of probable oil bearing rocks amounts to 3.2 lakh sq km (*see* **Fig. 24.4**) The total sedimentary area including both on shore and offshore comprises 27 basins. The geological and geophysical studies have been conducted in 14 basins while exploratory drilling has been done in 9 basins. Mumbai High, the Khambhat Gulf and the Assam are the most productive areas.

Production

India was a very insignificant producer of petroleum at the time of Independence and remained so till Mumbai High started production on a large scale. In fact, off-shore production did not start till the mid 1970s and the entire production was received from on-shore oil fields.

In 1980-81 about half of the production of crude oil came from on-shore fields while the remaining half was received from the off-shore resources. After that juncture, the off-shore production increased at a much faster rate than the on-shore production. From 1990-91 to 2010-11, about two-thirds of the production had been received from off-shore sources but after that on-shore production has picked up very fast. The total production recorded more than three

TABLE 24.10. Production of Petroleum (crude) in India (2016-17)

Area	Quantity ('000 tonnes)	Value (₹ 000)
India	36,009	65,44,94,413
Public Sector	25,476	46,30,45,993
Private Sector	10,533	19,14,45,420
Andhra Pradesh	276	50,16,514
Arunachal Pradesh	59	10,72,372
Assam	4,202	7,63,74,598
Gujarat	4,605	8,36,99,436
Rajasthan	8,164	14,83,87,011
Tamil Nadu	284	51,61,919
Offshore	18,419	33,47,79,563

Source : Indian Minerals Yearbook 2017, p. 22-5.

times increase after 1980-81. Crude oil production during 2019-20 was 32.17 million metre tonnes (**Table 24.9**).

As mentioned earlier, oil in India is obtained both from on-shore and off-shore areas, but off-shore areas made a major contribution.

On-shore Oil Production. On-shore oil fields are located in the Brahmaputra valley of north-east India, Barmer area of Rajasthan, Gujarat coast in western India and Cauvery on-shore basin in Tamil Nadu. Besides Andhra Pradesh has both on-shore and off-shore oil reserves.

OILFIELDS IN NORTH-EAST INDIA

The major oilfields in north-east India are those of the Brahmaputra valley in Assam and its neighbouring areas including Arunachal Pradesh, Nagaland, Meghalaya, Tripura, Manipur and Mizoram.

Assam. Assam is the oldest oil producing state in India. The main oil bearing strata extend for a distance of 320 km in upper Assam along the Brahmaputra valley. Following are some of the important oilfields of Assam :

1. ***The Digboi field.*** Located in the north-east of Tipam hills in Dibrugarh district of Upper Assam, Digboi is the oldest oil field of India. The oil bearing strata cover an area of about 13 sq km where oil is available at 400 to 2,000 metre depth. Over 800 oil wells have been drilled so far. Before the opening of the oil fields of west India, Digboi used to account for three-fourths of the total oil production of India. The most important centres are Digboi, Bappapang, Hassapang and Paintola. Most oil is sent to oil refinery at Digboi.

2. ***The Naharkatiya field.*** It is located at a distance of 32 km southwest of Digboi at the left bank of Burhi Dihing river. Here oil was discovered in 1953 and production started in 1954. Oil is available at depths varying from 4,000 to 5,000 metres. Out of the 60 successful wells drilled so far, 56 are producing oil while the remaining 4 are producing natural gas. The annual production is 2.5 million tonnes of oil and one million cubic metre natural gas. Oil from this area is sent to oil refineries at Noonamati in Assam (443 km) and Barauni in Bihar (724 km) through pipeline.

3. *The Moran-Hugrijan field.* It is located about 40 km south-west of Naharkatiya. Oil at Moran-Hugrijan field was discovered in 1953 and production started in 1956. Drilling has proved an oil bearing Barail horizon at a depth of 3,355 metre. Moran's potential may be estimated at one million tonnes per annum. As many as 20 wells have been drilled which yield oil as well as gas.

Other fields have been discovered at Rudrasagar, Sibsagar, Lakwa, Galeki, Badarpur, Barholla and Anguri.

Oilfields of Assam are relatively inaccessible and are distantly located from the main consuming areas. Oil from Assam is, therefore, refined mostly in the refineries located at Digboi, Guwahati, Bongaigaon, Barauni and Numaligarh.

Arunachal Pradesh has oil reserves at Manabhum, Kharsang and Charali. In ***Tripura***, promising oilfields have been discoverd at Mamunbhanga, Baramura-Deotamura Subhang, Manu, Ampi Bazar, Amarpur-Dambura areas. ***Nagaland*** also has some oil bearing rock strata.

ON-SHORE OIL FIELDS OF WESTERN INDIA

Gujarat. Explorations by Oil and Natural Gas Commission (ONGC) have yielded valuable findings of oil bearing rock strata over an area of about 15,360 sq km around the Gulf of Khambhat. The main oil belt extends from Surat to Amreli. Kachchh, Vadodara, Bharuch, Surat, Ahmedabad, Kheda, Mahesana, etc. are the main producing districts. In 2016-17, Gujarat produced 4,605 thousand tonnes of crude oil which accounted for over 12.7 per cent of the total oil production of India. Ankleshwar, Lunej, Kalol, Nawgam, Kosamba, Kathana, Barkol, Mahesana and Sanand are the important oilfields of this region.

1. *Ankleshwar.* The first major oil-find came in 1958 with the discovery of Ankleshwar field located about 80 km south of Vadodara and nearly 160 km south of Khambhat. Ankleshwar anticline is about 20 km long and 4 km wide. Oil is available at depths varying from 1,000 to 1,200 metres. It has a capacity of 2.8 million tonnes per annum. It is such a prolific oilfield that Pt. Jawahar Lal Nehru called it the *fountain of prosperity*. As many as 170 oil wells have been bored so far. It is estimated that 25 lakh tonnes per year of oil can be obtained from this field. Oil from this field is sent to refineries at Trombay and Koyali.

2. *Khambhat or Lunej field.* The oil and Natural Gas Commission drilled test wells in 1958 at Lunej near Ahmadabad and confirmed the occurrence of a commercially exploitable oil field. Oil was obtained on 4th Sept. 1959. Till 1969, a total of 62 wells were drilled out of which 19 yielded gas while 3 yielded oil. The annual production is 15 lakh tonnes of oil and 8-10 lakh cubic metres of gas. The total reserves are estimated at 3 crore tonnes.

3. *Ahmedabad and Kalol field.* It lies about 25 km north-west of Ahmedabad. This field and a part of Khambhat basin contain 'pools' of heavy crude trapped in chunks of coal. Nawgam, Kosamba, Mahesana, Sanand, Kathana, etc. are important producers.

Oil has also been struck in Olkad, Dholka, Kadi, Asjol, Sandkhurd, Siswas, Nandesan, Bandrat, Sobhasan and Vadesar areas.

Rajasthan. One of the largest inland oil discoveries was made in Barmer district of Rajasthan in 2004. The oil block covers an area of approximately 5,000 sq km. State-of-the-art technology with innovative geological modelling was used in discovering this oil field. Initial estimates of the oil in place of this discovery range from 63 to 153 million tonnes. Two important discoveries, *viz.*, Sarswati and Rajeshwari, with a total 35 million tonnes of inplace oil reserves were made earlier in 2002. The Sarswati discovery had found 14 million tonnes of in place oil reserves for which drilling upto 3,476 m was done. Mangala oil field discovered in 2004 is the largest oil discovery since 1985. It has nearly one billion barrels of recoverable oil. In the year 2016-17, Rajasthan produced 8,164 thousand tonnes of oil which accounted for over 22 per cent of the total production. Thus Rajasthan became the largest on shore oil producing state of India surpassing Assam and Gujarat in quick succession.

WESTERN COAST OFF-SHORE OILFIELDS

Extensive surveys have been conducted by ONGC in the offshore areas of Kachchh, Khambhat, Konkan,

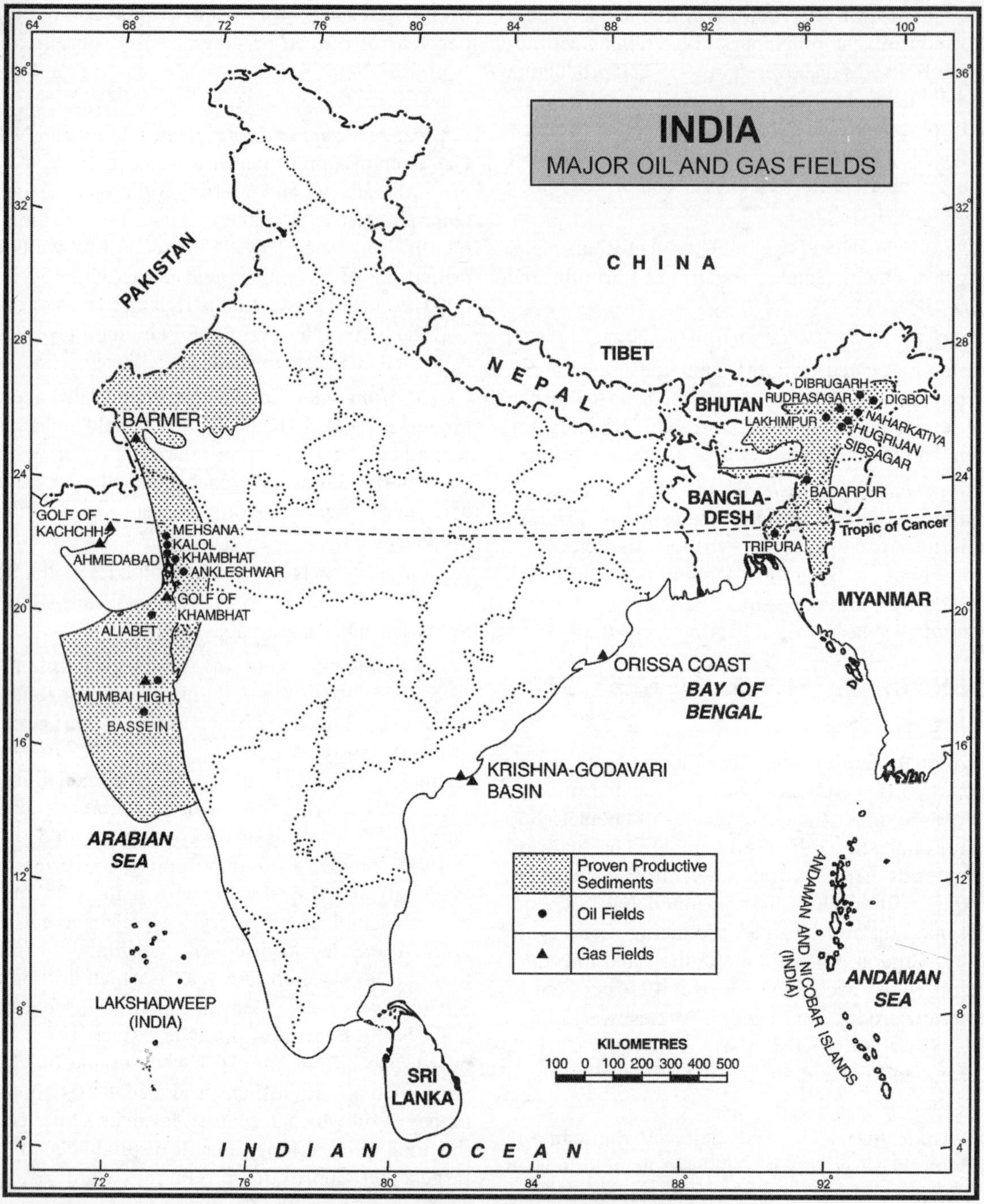

FIG. 24.5. India : Major Oil and Gas fields

Malabar and Coromandal coasts, Krishna-Godavari delta and Sunderbans. Success on commercial scale has been achieved at Mumbai High, Bassein and Aliabet.

1. Mumbai High. The greatest success achieved by the ONGC with respect to offshore surveys for oil was that of Mumbai High in 1974. It is located on the continental shelf off the coast of Maharashtra about

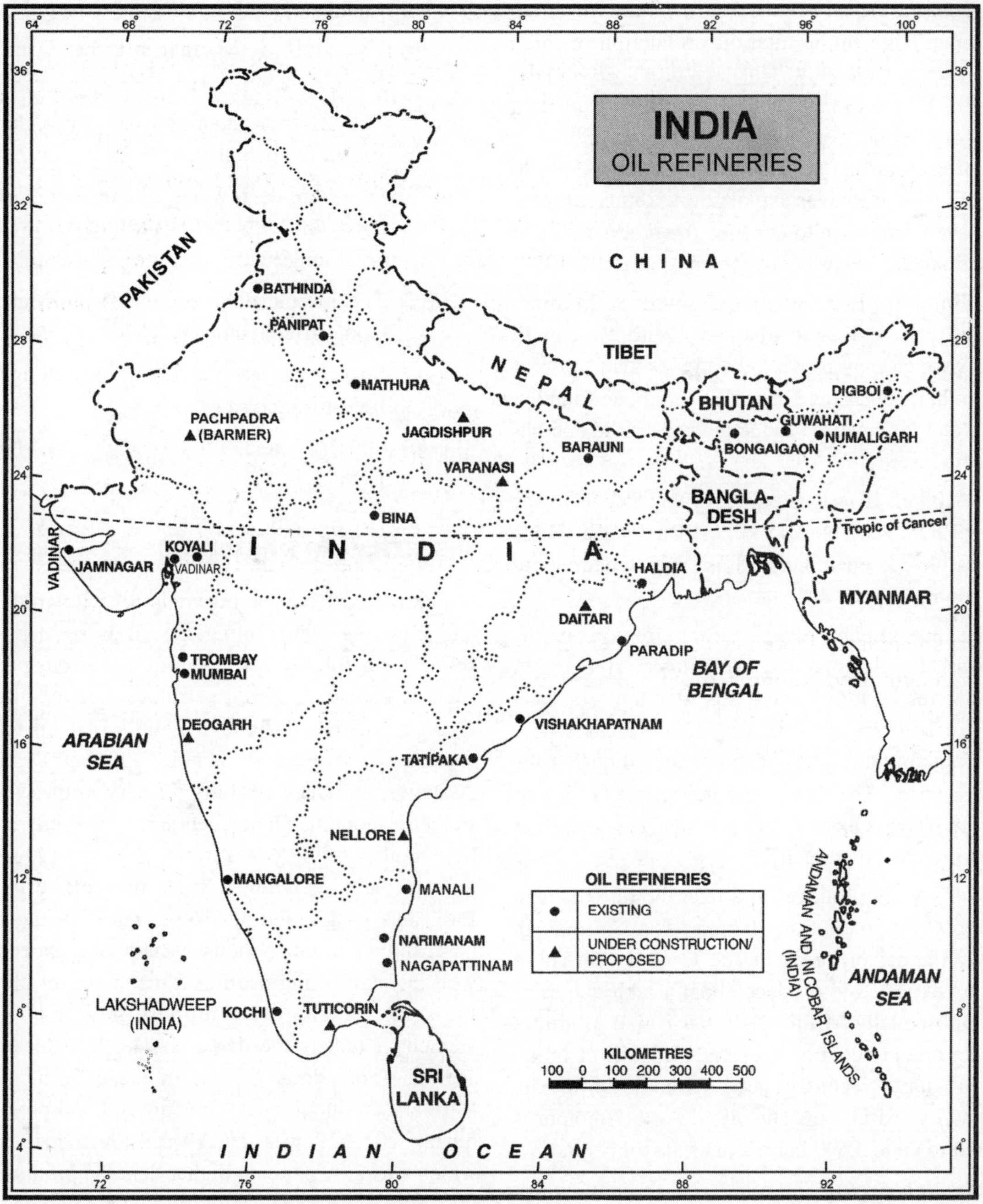

FIG. 24.6. India : Oil Refineries

176 km north-west of Mumbai. Here the rock strata of Miocene age covers an area of 2,500 sq km with estimated reserves of about 330 million tonnes of oil and 37,000 million cubic metres of natural gas. Production on commercial scale began in 1976. Oil is taken from a depth of over 1,400 metre with the help of a specially designed platform known as ***Sagar Samrat.*** The discovery of Mumbai High has

revolutionised the oil production in India. The share of Mumbai High in the total oil production of India has shot up considerably. This area produced 85 lakh tonnes of oil in 1982 which rose to over 189 lakh tonnes or over 62 per cent of all India in 1991-92. Production from this field declined between 1989-90 and 1993-94 due to over exploitation. Remedial measures have been taken to enhance the production and the declining trend has been reversed since 1994-95.

2. Bassein. Located to the south of Mumbai High, this is a recent discovery endowed with reserves which may prove to be higher than those of the Mumbai High. Huge reserves have been found at a depth of 1,900 metre. Production has started and has picked up fast.

3. Aliabet. It is located at Aliabet island in the Gulf of Khambhat about 45 km off Bhavnagar. Huge reserves have been found in this field. Commercial production is expected to start soon.

East Coast. The basin and delta regions of the Godawari, the Krishna and the Cauvery rivers hold great potential for oil and gas production. As such these are both on-shore and off-shore areas where extensive exploration has been conducted during the last few years. The Rawa field in Krishna-Godawari off-shore basin is expected to produce 1 to 3 million tonnes of crude oil annually.

Tamil Nadu produces less than the one per cent of the total oil production of India. The Narimanam and Kovilappal oilfields in the Cauvery on-shore basin are expected to produce about 4 lakh tonnes of crude oil annually. ***Andhra Pradesh*** also produces less than one per cent of the total crude oil of India. Oilfields have recently been discovered in the Krishna-Godavari basin. The oilfield near Amolpur is expected to yield 3,600 barrels of crude oil per day.

Probable Areas. There are vast possibilities of finding oil from about one lakh sq km area of sedimentary rocks in different parts of the country (see **Fig. 24.4**). Some of the outstanding areas which hold possibilities of oil are :

(*i*) Jawalamukhi, Nurpur, Dharamsala and Bilaspur in Himachal Pradesh.

(*ii*) Ludhiana, Hoshiarpur and Dasua in Punjab.

(*iii*) The Gulf of Mannar off the Tirunelveli coast.

(*iv*) The off-shore area between Point Calimere and Jaffna peninsula.

(*v*) Off-shore deep water area in Bay of Bengal between 12°N–16°N latitudes and 84°E–86°E longitudes.

(*vi*) The marine delta region of the Mahanadi, Godavari, Krishna and Cauvery rivers.

(*vii*) Stretch of sea between South Bengal and Baleshwar coast.

(*viii*) Off-shore area of the Anadaman and Nicobar Islands.

PETROLEUM REFINING

Oil extracted from the oil wells is in its crude form and contains many impurities. It is refined in oil refineries before use. After refining, various products such as kerosene, diesel, petrol, lubricants, bitumen, etc. are obtained. Although India's first oil refinery started working way back in 1901 at Digboi in Assam, it remained the only refinery in the whole of India for more than half a century. It was only in 1954 that another refinery at Tarapur (Mumbai) joined the lone refinery of Digboi. Since then oil refining in India has progressed at a rapid pace. In the recent past, Indian refining industry has done exceedingly well in establishing itself as a major player globally. India is emerging as a refinery hub and refining capacity exceeds the demand. The last decade has seen a tremendous growth in the refining sector. Refinery production (in terms of crude oil processed) during 2018-19 was 257.20 million metric tonnes which was 2.09 per cent higher than the production in 2017-18. Refinery capacity in India is projected to be nearly 400 million metric tonnes by the year 2030. India has sufficiently enough refining capacity comprising 23 refineries—18 under public sector, 3 under private sector and 2 in joint venture. The capacity-wise details of the refineries are given below :

TABLE 24.11. List of Oil Refineries in India (2018)

Sr. No.	Refinery	Oil Company	Sector	State	Location	Capacity 10^6 tonnes/ annum
1.	Jamnagar Refinery	Reliance Industries Limited	Private	Gujarat	Jamnagar SEZ	33
2.	Jamnagar Refinery	-do-	-do-	-do-	Jamnagar DTA	27
3.	Nayara Energy Refinery	Nayara Energy Ltd.	Private	Gujarat	Vadinar	20
4.	Kochi Refinery	Bharat Petroleum Corporation Ltd.	Public	Kerala	Kochi	15.5
5.	Mangalore Refinery and Petrochemicals Limited	Oil and Natural Gas Corporation	Public	Karnataka	Mangaluru	15
6.	Paradip Refinery	Indian Oil Corporation Ltd.	Public	Odisha	Paradip	15
7.	Panipat Refinery	Indian Oil Corporation Ltd.	Public	Haryana	Panipat	15
8.	Gujarat Refinery	Indian Oil Corporation Ltd.	Public	Gujarat	Koyali	13.7
9.	Mumbai Refinery	Bharat Petroleum Corporation Ltd.	Public	Maharashtra	Mumbai	12
10.	Mumbai Refinery	Hindustan Petroleum Corporation Ltd.	Public	-do-	-do-	7.5
11.	Guru Gobind Singh Refinery	HPCL Mittal Energy Ltd.	Joint	Punjab	Bathinda	11.3
12.	Manali Refinery	Chennai Petroleum Corporation Ltd.	Public	Tamil Nadu	Chennai	10.5
13.	Vishakhapatnam Refinery	Hindustan Petroleum Corporation Ltd.	Public	Andhra Pradesh	Vishakhapatna m	8.3
14.	Mathura Refinery	Indian Oil Corporation Ltd.	Public	Uttar Pradesh	Mathura	8.0
15.	Haldia Refinery	Indian Oil Corporation Ltd.	Public	West Bengal	Haldia	7.5
16.	Bina Refinery	Bharat Oman Refinery Ltd.	Joint	Madhya Pradesh	Bina	7.8
17.	Barauni Refinery	Indian Oil Corporation Ltd.	Public	Bihar	Barauni	6
18.	Numaligarh Refinery	• Bharat Petroleum Ltd. • Oil India • Govt. of Assam	Public	Assam	Namaligarh	9
19.	Bongaigaon Refinery	Indian Oil Corporation Ltd.	Public	Assam	Bongaigaon	2.735
20.	Guwahati Refinery	Indian Oil Corporation Ltd.	Public	Assam	Guwahati	1
21.	Nagapattnam Refinery	Chennai Petroleum Corporation Ltd.	Public	Tamil Nadu	Nagapattnam	1
22.	Digboi Refinery	Indian Oil Corporation Ltd.	Public	Assam	Digboi	0.65
23.	Tatipaka Refinery	Oil and Natural Gas Corporation	Public	Andhra Pradesh	Tatipaka	0.07

PIPELINES

Pipelines are most convenient, efficient and economical mode of transporting liquids like petroleum, petroleum products, natural gas, water, milk, etc. Even solids can also be transported through pipelines after converting them into slurry.

Transportation by pipelines is a new development in India. Pipelines have relieved the increasing pressure on the existing surface transport system (railways and roadways).

The country had a network of about 5,035 km long pipelines in 1980 which has now increased to over 10,909 km with a capacity of 75.55 million metric tonnes per annum (MMTPA) of crude oil and petroleum products and 96 million metric standard cubic metre per day (MMSCMD) of gas. Pipelines have both advantages and disadvantages as described below.

Advantages of Pipeline. Pipelines have the following advantages over other means of transport :

1. They are ideally suited to transport the liquids and gases.
2. Pipelines can be laid through difficult terrains as well as under water.
3. It involves very low energy consumption.
4. It needs very little maintenance.
5. Pipelines are safe, accident-free and environment friendly.

Disadvantages of Pipelines. Following are the main disadvantages of pipeline transport :

1. It is not flexible, *i.e.,* it can be used only for a few fixed points.
2. Its capacity cannot be increased once it is laid.
3. It is difficult to make security arrangements for pipelines.
4. Underground pipelines cannot be easily repaired and detection of leakage is also difficult.

Some of the important pipelines are briefly described as under :

1. Naharkatia-Nunmati-Barauni Pipeline. This was the first pipeline constructed in India to bring crude oil from Naharkatia oilfield to Nunmati. It was later extended to transport crude oil to refinery at Barauni in Bihar. It is 1,167 km long. It is now extended to Kanpur in U.P. The pipeline between Naharkatia and Nunmati became operative in 1962 and that between Nunmati and Barauni in 1964. Construction works on pipelines from Barauni to Kanpur and Haldia were completed in 1966. It has a number of pumping stations and subsidiary pipelines.

(*i*) Nunmati-Siliguri pipeline transports oil from Nunmati (Guwahati) in Assam to Siliguri in West Bengal.

(*ii*) Lakwa-Rudrasagar-Barauni pipeline has been constructed to carry crude oil from Lakwa and Rudrasagar to oil refinery at Barauni.

(*iii*) Barauni-Haldia pipeline, completed in 1966, transports refined petroleum products to Haldia port and bring back crude oil to Barauni refinery.

(*iv*) Barauni-Kanpur pipeline carries refined petroleum products from Barauni to Kanpur.

(*v*) Nunmati-Bongaigaon section of this pipe is used to transport raw materials for Bongaigaon petro-chemical complex.

(*vi*) Haldia-Rajbandh-Maurigram pipeline has been constructed to mect the requirements of southern part of West Bengal.

2. Mumbai High-Mumbai-Ankleshwar-Koyali Pipeline. This pipeline connects oilfields of Mumbai High and Gujarat with oil refinery at Koyali. A 210 km long double pipeline connects Mumbai with Mumbai High. It provides facilities for transporting crude oil and natural gas. Ankleshwar-Koyali pipeline was completed in 1965. It transports crude oil from Ankleshwar oilfield to Koyali refinery.

3. Salaya-Koyali-Mathura Pipeline. An important pipeline has been laid from Salaya in Gujarat to Mathura in U.P. via Viramgram. This is 1,256 km long pipeline which supplies crude oil to refineries at Koyali and Mathura. From Mathura, it has been extended to the oil refinery at Panipat in Haryana and further to Jalandhar in Punjab. It has an off-shore terminal for imported crude oil.

4. Hajira-Bijapur-Jagdishpur (HBJ) Gas Pipeline. This pipeline has been constructed by Gas Authority of India Limited (GAIL) to transport gas. It

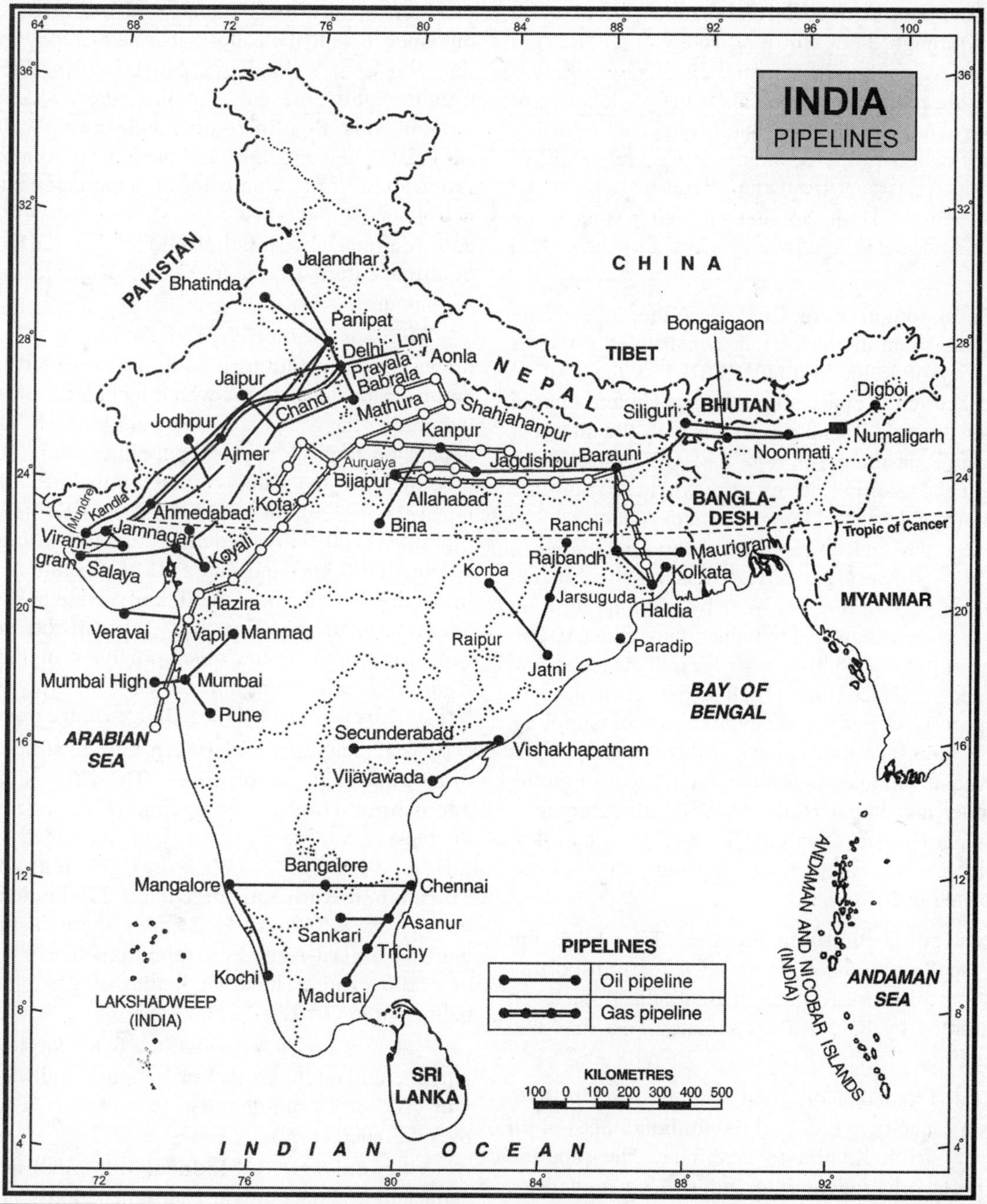

FIG. 24.7. India : Pipelines

is 1,750 km long and connects Hazira in Maharashtra to Bijapur in M.P. and Jagdishpur in U.P. It carries 18 million cubic metres of gas everyday to three power houses at Kawas (Gujarat), Anta (Rajasthan) and Auraiya (U.P.) and to six fertilizer plants at Bijapur, Sawai Madhopur. Jagdishpur, Shahjahanpur, Aonla and Babrala. Each of the fertilizer plants has a capacity of producing 1,350 tonnes of ammonia per day. The construction of this pipeline is a unique engineering feat and has been completed at an estimated cost of over ₹ 1,700 crore. The pipeline passes through 343.7 km long rocky area, 56.3 km long forest area, besides crossing 29 railway crossings and 75 big and small rivers. This is the world's largest underground pipeline and has brought about a big transformation in the economy of Gujarat, Madhya Pradesh, Rajasthan and Uttar Pradesh. It has been extended upto Delhi so that enough gas is made available to meet the growing demand of the capital city.

5. Jamnagar-Loni LPG Pipeline. This 1,269 km long pipeline has been constructed by Gas Authority of India Limited (GAIL) at the cost of ₹ 1,250 crore. It connects Jamnagar in Gujarat to Loni near Delhi in U.P. and passes through the states of Gujarat, Rajasthan, Haryana and U.P. This is the longest LPG pipeline of the world. It is like transporting 3.5 lakh LPG gas cylinders across 1,269 km every day and its capacity is being increased to 5.0 lakh cylinders per day. It has resulted in net saving of ₹ 500 crore per year by eliminating road tanker movement and led to reduction of about 10,000 tonnes of pollutant emission per year. This is the first time that cross-country pipeline has been used to transport LPG adding to availability of supplies, safety in transportation and wider distribution. LPG is received at various points along the route for bottling in Ajmer and Jaipur (Rajasthan), Piyala (Haryana), Madanpur Khadar (Delhi) and Loni (U.P.). Phase-I of the pipeline was completed in 2001 and Phase-II was completed in 2003.

6. Kandla-Bhatinda Pipeline. This 1,443 km long pipeline has been constructed for transporting crude oil to the oil refinery at Bhatinda. It has been constructed by IOC at the estimated cost of ₹ 690 crore.

It was started in 1993 and commissioned in 1996. It has a capacity to transport 6.0 million tonnes of oil per year from Kandla to Bathinda. The pipeline transports oil brought into India to consumers of North India which is land locked.

Apart from the above mentioned important pipelines, several other pipelines have also been laid in different parts of the country. Construction of some of them has already been completed while others are at different stages of completion. In Gujarat, a number of pipelines carry crude oil, gas and refined products to refineries and markets. Important pipelines include the Kalol-Sabarmati Crude Pipeline, the Nawgam-Kalol-Koyali Crude Pipeline, the Cambay-Dhiwaran Gas Pipeline, the Ankleshwar-Uttaran Gas Pipeline, the Ankleshwar-Vadodara Associated Gas Pipeline, and the Koyali-Ahmedabad Products Pipeline. Mumbai is an important centre for petroleum products. As such, it is joined with Pune and Manmad by pipelines. The Haldia-Kolkata pipeline caters to the needs of Kolkata and its neighbouring areas.

The Gas Authority of India Ltd. (GAIL) has drawn up an ambitious plan at the cost of ₹ 10,000 crore for the "near term" which includes projects for integrating the gas pipeline network through capacity expansion and adding new pipelines. Some of the pipelines are the Dahej-Vijaipur pipeline, 300 km north-south Gujarat pipeline, Dehej-Hazira-Uran pipeline (400 km), Kochi-Mangalore-Bangalore pipeline (900 km) and Phase III of pipeline network in Andhra Pradesh. A 600 km Vishakhapatnam-Secunderabad pipeline has 1.1 million tonnes capacity. Some of the other pipelines include the Kanpur-Bina, Mangalore-Chennai, Vijayawada-Vishakhapatnam, and Haldia-Budge Budge pipelines.

GAIL has also undertaken the construction of three important gas pipelines. The 2,050 km long **Jagdishpur-Haldia** gas pipeline is the longest and the most expensive pipeline. The approximate cost of this pipeline is ₹ 7,600 crore. The **Karanpur-Moradabad-Rudrapur** pipeline is 275 km long and its cost is ₹ 250 crore. Besides there is plan to upgrade **Bajera-Agra-Ferozabad** pipeline at the cost of ₹ 200 crore. The total length of gas pipelines belonging to GAIL is 14,800 km.

There is a proposal to build a 6,000 km long gas pipeline grid on the pattern of National Highways that will criss-cross the country. It is estimated to cost over ₹ 30,000 crores.

Chennai-Trichy-Madurai oil pipeline was dedicated to the nation on 26th June, 2006. Cairn has constructed 582 km pipeline from Barmer to Salaya. It has a capacity of 1,50,000 barrels per day. **Mundra-Delhi** oil pipeline is 1,054 km long with 18 inch (about 45 cm) diameter was dedicated to the nation on February 2, 2009. Built at a cost of ₹ 1,757 crore, it carries 5 MMTPA products which is expandable to 6 MMTPA.

IMPORTS

Consumption of oil and its products has always outstripped production in India. In 1950-51, India produced only 2.7 lakh tonnes of oil against consumption of 34 lakh tonnes. With the progress in industries and transport, need for oil has increased dramatically in the post-Independent era. Consequently, India has to lean heavily on the imports of oil and its products.

It is clear from **Table 24.12** that our oil bill has been rising rather disturbingly both in terms of quantity and value due to our increased demands for economic growth. Our oil import bill increased from less than 9 per cent of the total imports in 1960s to 30 per cent following the first oil crisis in 1973-74 and to 75 per cent in 1980s after the second oil crisis. Decline in indigenous production during 1989-90 and 1993-94 further worsened the balance between production and consumption. Our imports of oil are more than three times the indigenous production. This puts serious strain on our foreign exchange reserves. With increase in growth of industries and transport, demand for oil will increase further and make things still worse. In 2019-20, our oil bill accounted for 27.5 per cent of the total imports.

NATURAL GAS

Natural gas usually accompanies petroleum accumulations. Whenever a well for oil is drilled, it is natural gas which is available before oil is struck. Natural gas is fast becoming an important source of energy in India.

Discovery of gas made rapid strides after 1985. Oil strikes at Cauvery off-shore and at Nanda in Khambhat basin as also gas found at Talot in Jaisalmer Basin in Rajasthan were major discoveries during 1988-89. Production from South Bassein Gas Field started in September 1988. During 1989-90 oil/gas structures had been discovered in Adiyakkamanglam in Tamil Nadu, Andada in Gujarat, Khovaghat in Assam, Lingla in Andhra Pradesh, Mumbai off-shore and Kachchh offshore. Another survey conducted in 1997 in the Andamans has revealed 1,700 billion cubic feet of gas reserves which can meet the country's requirements for the next 30 years.

The face of the country's gas map has undergone a decisive change. The gas discovered by Reliance Industries in deep waters of the Krishna-Godavari off-shore basin has brought the sector into sharp focus since 2002. The reserves are estimated to 14 trillian cubic feet. The field is 200 km into the sea from Vishakhapatnam and has blocks spread over 200,000 sq km. The wells have been drilled at depths 1350-2700 metres. This is the deepest oil exploration in India. The finding, according to a number of energy analysts, is just the tip of a **'gasberg'**. However, the Krishna-Godawari basin is a very complex area because it is under the confluence of two big rivers of the peninsular India. Stratigraphy is now unraveling how sand and silt formations have taken place over the centuries. Undoubtedly the basin is rich in natural gas in its deep recesses. Hence it requires technology and expertise which goes beyond geology and geosciences and brings in engineering skill of high order. This discovery is believed to yield 60-80 million cubic metres of natural gas per day from a single field.

In June 2005, Oil and Natural Gas Corporation (ONGC) made a significant hydrocarbon find in the shallow waters of Krishna-Godawari basin, south-west of Ravva field discovered by ONGC in 1987. The new find is around 12 km from Amalpuram coast.

The above two discoveries have helped in saving the precious foreign exchange and also helped in transforming the economy of the Krishna-Godavari basin. The gas obtained from this basin is used for

TABLE 24.12. Imports of Petroleum, Oil and Lubricants

Year	1960-61	1970-71	1980-81	1990-91	2000-01	2020-21	2021-22
Quantity ('000 tonnes)	800.0	12,767	23,537.0	29,359.0	—	—	—
Value (₹ crore)	69	136	5,264	10,816	71,497	6,11,353	12,07,803

Source : Economic Survey, 2022-23. Statistical Appendix, pp. 112-14.

TABLE 24.13. Gross Production of Natural Gas in India

Year	1970-71	1980-81	1990-91	2000-01	2010-11	2015-16	2016-17
Production (in million cubic metres)	1,445	2,358	17,998	29,477	52,219	32,249	31,897

Source : EnviStats India, 2018, p. 151.

power generation, as fuel for fertilizer plants and even for running a transport system. Since the find is equivalent to 165 million tonnes of crude oil, it will help in replacing the more polluting fuels like naphtha, fuel oil and coal.

In yet another gas discovery, the Reliance Industries struck gas off the Odisha coast in Bay of Bengal in June, 2004. Here the reserves are estimated at 4 to 5 trillion cubic feet.

In Barmer district of Rajasthan also, gas was discovered along with crude oil in 2003. It has capacity to produce 7.3 million standard cubic feet of gas per day. Possibilities of findings gas exist in Gulf of Kachchh and Gulf of Khambhat as well as in Tripura.

About four-fifths of the total gas comes from Offshore sources. Over five per cent of the total gas of India is produced by Assam. The rest is obtained from Gujarat, Andhra Pradesh, Tamil Nadu, Tripura, and Rajasthan.

The production of natural gas during 2018-19 was 32.873 BCM which was 0.69 per cent higher than the production achieved in 2017-18.

Keeping in view the increasing significance of natural gas as an important source of energy, the Gas Authority of India Limited (GAIL) was incorporated in August, 1984 for transportation, processing and marketing of natural gas. The immediate task before GAIL was to complete HBJ Gas pipeline. Presently, GAIL is the largest natural gas processing, transmission and distribution company in India. It supplies gas to power plants for generation of about 5,500 MW of power, as feedstock for gas based fertilizer plants to produce about 10 million metric tonnes per annum (MMTPA) of urea. GAIL has seven plants for production of LPG with a total capacity of over one million tonnes per annum.

GAIL has taken major steps in Joint Venture companies, namely Mahanagar Gas Limited (with British Gas of U.K.) in Mumbai and Indraprastha Gas Limited (with Bharat Petroleum Corporation Limited) in Delhi for City Gas Distribution Schemes including CNG for transport sector. Besides GAIL has joint ventures with several other companies particularly in Gujarat and Kerala. On 29 November 2002, GAIL entered with a Joint Venture agreement with HPCL to distribute Piped Natural Gas, CNG and Auto LPG in cities of Andhra Pradesh.

TABLE 24.14. Distribution of Natural Gas (utilised) in India (2016-17)

State	Quantity (million cubic metres)	Percentage of total India production
Public Sector	24,356	78.91
Private Sector	6510	21.09
Andhra Pradesh	834	2.70
Arunachal Pradesh	12	0.04
Assam	2,914	9.44
Gujarat	1,519	4.92
Rajasthan	1,214	3.93
Tamil Nadu	967	3.13
Tripura	1,429	4.63
West Bengal	317	1.04
Offshore	21,660	70.17
Total	**30,866**	**100.00**

Source : Data computed from Minerals Yearbook, 2017.

Although India has made rapid strides in the discovery and production of natural gas, yet the production falls short of the demand which is increasing with each passing day. Gas in large quantities has to be imported to fill the widening gap between supply and demand. The Government of India has taken notable initiative in this direction.

Conservation of Petroleum Products

India is fast becoming a big producer, consumer and importer of petroleum products. The demand for petroleum products is increasing at an alarmingly accelerated rate. At current rate of consumption, our known oil reserves will last only for 30-40 years. This puts heavy strain on our economic resources and calls for an urgent need to conserve petroleum products. There is a vast scope of petroleum conservation in India provided there are technological improvements, financial incentives, policy initiatives and legislative measures for implementing conservation. Transport, industry, domestic/household and agriculture are four major sectors which consume bulk of petroluem. Awareness programmes for each sector need to be developed and then extensive publicity of the measures which would bring about savings need to be undertaken to bring about the desired results.

The transport sector is one of the largest consumers of petroleum products mainly petrol and high speed diesel. This sector accounts for 50 per cent consumption of petroleum in India. Road transport alone accounts for about 37 per cent of the total oil consumption. With fast expanding transport network this consumption level is bound to increase. Oil products constitute nearly 80 per cent of total commercial energy used in transport. Therefore, any energy conservation programme in transport sector is bound to strengthen our economy. This can be done largely by introducing more fuel efficient engines, operation of transport vehicles at optimum speed, improving road conditions and avoiding traffic bottlenecks, especially in urban areas. These measures can save about 30-35 per cent fuel. Provision of better roads alone can save 10 per cent petroleum. Extension of railway network has led to substantial savings in high speed diesel since a diesel loco is 7-8 times more efficient compared to diesel truck. It is, therefore, logical to switch all long distance freight movement to rail and confine road transport to short distances only. Railways have initiated several measures to achieve high degree of energy conservation. The introduction of Eastern and Western Dedicated Freight Corridors is a good initiative which will increase the freight carrying capacity of the railways and save a lot of fuel. Public transport system should be strengthened so that people are not forced to use private vehicles. This can save a lot of energy.

Industries consume about 16-20 per cent of the total oil products consumed in the country. Several industries have achieved commendable success in energy conservation by better management of energy. For example, iron and steel industry, petrochemicals, cement and paper industry have saved 21, 32, 28 and 25 per cent energy respectively between 2003 and 2018. Similar steps can be taken by other industries for saving oil and other sources of energy.

In agriculture sector, the main petroleum products consumed are high speed diesel and light diesel oil. The consumption of oil in this sector can be reduced by programmes for pump rectification, more efficient foot valves for lift irrigation, better farm machinery, use of agro residues and other non-commercial sources of energy. As regards use of petroleum for production of fertilizers, plants based on liquid oil products should be discouraged and these should be based on use of natural gas or coal as feedstocks of which the country has plentiful resources.

Kerosene and LPG are two major oil products used in the domestic sector for cooking and lighting in urban and semi-urban areas. Since there are no viable alternatives to these fuels and they are to be extended to the rural areas to check deforestation, it is imperative that their utilisation is carried out at the optimum efficiency level.

The need for conservation of petroleum products has been increasingly felt during the last few years. The Petroleum Conservation Research Association (PCRA) in 1978 under the Ministry of Petroleum and Natural Gas is doing a commendable job to promote conservation of petroleum products. PCRA's activities cover conservation of all energy sources, development, evaluation and commercialization of efficient equipment and additives, popularizing petro-crop cultivation and production of bio-fuels, environment protection etc. It has been conducting public awareness campaigns for promoting energy conservation in petroleum sector on a regular basis which includes various educational and practical sessions to a variety of consumer sectors. PCRA has also launched media campaign to promote efforts of conservation and is working in close association with Bureau of Energy Efficiency (BEE) for popularizing standards and labelling programmes for equipment using petro-based fuels such as domestic LPG stoves, diesel irrigation pump sets and diesel generating sets.

ELECTRICITY

Electricity plays a dominant role in the progress and prosperity of any country. Consumption of electricity is a barometer of a nation's economic well being and standard of living of its people. Availability of abundant electricity means unrestricted growth of industries, transport and agriculture which means freedom from hunger and poverty and the resultant economic prosperity of the masses. Although India set up its first power plant over a century ago and the electrification of Kolkata began within just a decade after that of London, power development could truly take off only after Independence. The power sector registered an impressive growth by about 200 times during 70 years from 1950-51 to 2020-21. The installed capacity rose from 2.4 thousand MW in 1950-51 to 459.2 thousand MW by 2020-21 and generation from 6.6 billion kWh to 1598.5 billion kWh during the same period. However, India still lags far behind regarding consumption of electricity. The per capita consumption in India was only 189 kWh in 2015-16 as against 3,481 kWh in Britain, 6,434 kWh in Sweden and 6,550 kWh in the USA and the world average of 1,000 kWh.

Following types of electricity are recognised depending upon the raw material used and mode of its production.

1. Hydroelectricity.
2. Thermal Electricity (including steam, gas and oil).
3. Nuclear Electricity.

Their relative importance and temporal growth has been shown in **Table 24.15**.

TABLE 24.15. Progress of Installed Capacity of Electricity (Utilities and Non-utilities) in Thousand MW

Year	Utilities					Non-utilities	Grand Total
	Hydro	Thermal	RES	Nuclear	Total		
1950-51	0.6	1.15	0.00	0	1.8	0.6	2.4
1960-61	1.9	2.74	0.00	0	4.6	1.0	5.6
1970-71	6.4	7.91	0.00	0.4	14.7	1.6	16.3
1980-81	11.8	17.56	0.00	0.9	30.3	3.1	33.4
1990-91	18.8	45.74	0.03	1.5	66.1	8.6	74.7
2000-01	25.1	72.34	1.27	2.9	101.6	16.2	117.8
2010-11	37.6	112.82	18.45	4.8	173.7	34.4	208.1
2015-16	42.8	210.68	45.92	5.8	305.2	48.3	353.5
2016-17	44.5	218.33	57.24	6.8	326.9	51.1	378.4
2017-18	45.3	222.91	69.02	6.8	344.0	54.9	398.9
2018-19	45.4	226.28	77.64	6.8	356.1	75.2	431.3
2019-20	45.7	230.60	87.03	6.8	370.1	76.2	446.3
2020-21	46.2	234.73	94.43	6.8	382.2	77.0	459.2
2021-22	46.7	236.1	109.9	6.8	399.5	82.7	482.2

RES = Renewable Energy Source

Source: Economic Survey 2021-22 and 2022-23, Statistical Appendix, p. 47.

HYDROELECTRICITY

The future prosperity of India depends to a great extent on our ability to produce and use hydroelectricity. The other two sources of energy, coal and petroleum, are exhaustible and will not be available to us for ever. Therefore, we should reduce our dependence on coal and petroleum and develop hydroelectricity as far as possible. Currently, hydroelectricity accounts for about 10 per cent of the total installed capacity. This has to be increased so that increasing demand for energy is met and at the same time, precious and scarce coal and petroleum resources are saved from over exploitation. Hydroelectricity is a renewable, cheap, clean and environmentally benign source of energy and will be available to us for all times to come. River water, if not properly used, will wastefully drain into the sea. India is blessed with huge water resources and there are vast possibilities of producing hydroelectricity. However, India has developed only a small percentage of the total potential available. India's exploitable hydro-electric potential in terms of installed capacity is estimated to be about 1,48,700 MW out of which a capacity of 45.4 thousand MW has been developed so far. This is due to certain geographical factors as well as because of developing stage of economy. Most of the river regimes in India are extremely erratic because they are fed by monsoon rains which are highly seasonal and whimsical. Further, many rivers do not have natural waterfalls and huge capital has to be invested for constructing dams.

Most of the sites suitable for generating hydroelectricity are located away from the consuming centres as a result of which a lot of energy is wasted in transmission. Under normal circumstances, there is loss of 8 per cent energy for transmitting it through a distance of 160 km and 21 per cent loss for 800 km. Thus if hydroelectricity generated at Bhakhra Nangal dam is to be consumed at Delhi, the average loss is about 15 per cent.

The hydroelectric power generation in India made a humble start at the end of the 19th century with the commissioning of a hydroelectric power plant in 1897 to supply electricity to Darjeeling. Another hydroelectric power plant was set up at Shivasamundram waterfall on the Cauvery river in Karnataka in the year 1902. At a later stage some hydroelectric power plants were erected in the Western Ghats. These were designed to meet the growing demands of Mumbai. In 1930s, a number of hydropower plants were commissioned in Himachal Pradesh, Uttar Pradesh, Tamil Nadu and Karnataka. The total generation capacity was 508 MW at the time of independence in 1947. Planned period started immediately after independence and several multipurpose projects were undertaken during the Five Year Plans. The National Hydroelectric Power Corporation (NHPC) was set up in 1975. Total installed capacity of hydroelectricity increased from 0.6 thousand MW in 1950-51 to 45.7 thousand MW in 2019-20 (**Table 24.15**).

Hydroelectric power can play a significant role in view of the energy crisis which India is currently facing. Indian rivers drain 1,677 billion cubic metres of water to the sea every year. The Central Water and Power Commission estimated the potential of hydroelectric power at about 40 million kW at 60% load factor from these rivers. Central Electricity Authority re-estimated this potential at 84,000 MW at 60% load factor. It is equivalent to about 450 billion units of annual energy generation. Basin-wise distribution of the potential is given in **Table 24.16**.

TABLE 24.16. India : Basin-wise Estimated Potential of Hydropower (Potential in thousand MW at 60 per cent load factor)

Basin	Potential	% of Total
Indus	20.0	23.8
Brahmaputra	35.0	41.7
Ganga	11.0	13.1
Central Indian basins	3.0	3.6
West flowing rivers	6.0	7.1
East flowering rivers	9.0	10.7
Total	**84.0**	**100.0**

Following factors influence the development of Hydroelectric Power (HEP) in India :

(*i*) There should be perennial flow of large volume of water which depends upon the amount of rainfall.

(*ii*) The water should fall from a sufficient height. This height may be in the form of a natural waterfall or a fall obtained by constructing a dam across the river. It may also be obtained by diverting the water from one river basin to another.

(*iii*) A readily available market is an essential requirement for generating HEP as electricity cannot be stored.

(*iv*) The generation of HEP requires huge capital investment as it is capital intensive activity.

(*v*) It also requires technological advancement because production, distribution and utilisation are closely related to the technological level of the concerned area.

The rivers originating from the northern mountainous region and the peninsular rivers differ markedly with respect to their suitability for hydroelectric production. Some outstanding facts are explained as under :

1. Northern Rivers. These rivers are very useful for hydroelectric generation due to a large number of factors. Major factors are :

(*i*) Himalayan rivers originate from the mountainous region and have their sources in glaciers and snowfields. Therefore, they receive water both from rain in rainy season and snowmelt in hot season and have enough flow of water throughout the year. As such they are known as perennial rivers and supply water for hydroelectric production all the year round.

(*ii*) Velocity of water flow is high because of dissected terrain and steep slope. This helps in generation of hydroelectric power.

(*iii*) Low competition for use of water for other purposes makes water available for HEP production. Water used in hydroelectric generation can be gainfully used for irrigation.

(*iv*) About three-fourth of the total potential is confined to the river basins originating from the northern mountainous region. The major rivers are the Indus, the Ganga and the Brahmaputra.

2. Peninsular Rivers. The peninsular rivers are comparatively poor with respect to hydroelectric power potential and production due to following reasons :

(*i*) The peninsular rivers are purely dependent on rainfall as a result of which flow of water in these rivers is very erratic. They have exceptionally high flow during the rainy season which is followed by a prolonged dry season of lean flow. They are thus not perennial rivers and are not much suited to hydroelectric production.

(*ii*) Storage of water is essential to regulate the flow.

(*iii*) The bulk of potential is confined to hilly regions.

However there are some factors which favour the development of hydroelectric power in the peninsular India.

TABLE 24.17. Progress of Electricity Generated (Utilities and Non-utilities) in Billion kWh

Year	Utilities					Non-utilities	Grand Total
	Hydro	Thermal	RES	Nuclear	Total		
1950-51	2.5	2.6	0.0	—	5.1	1.5	6.6
1960-61	7.8	9.1	0.0	—	16.9	3.2	20.1
1970-71	25.2	28.2	0.0	2.4	55.8	5.4	61.2
1980-81	56.5	61.3	0.0	3.0	120.8	8.4	129.2
1990-91	71.7	686.5	0.0	6.1	264.3	25.1	289.4
2000-01	74.5	408.4	1.6	16.9	501.4	55.0	556.4
2010-11	114.4	664.8	39.3	26.3	844.8	120.9	965.7
2015-16	121.4	943.0	65.8	37.4	1167.6	168.4	1336.0
2016-17	122.4	993.5	81.6	37.9	1235.4	172.0	1407.4
2017-18	126.1	1037.1	101.8	38.3	1303.3	179.8	1483.1
2018-19	134.9	1072.3	126.8	37.8	1371.8	213.1	1584.9
2019-20	155.8	1042.8	138.8	46.5	1383.4	239.6	1623.0
2020-21	150.3	1032.6	147.3	43.0	1373.2	224.8	1598
2021-22	151.6	1114.8	170.9	47.1	1484.4	235.0	1719.4

RES = Renewable Energy Resources

Source: Economic Survey 2022-23, Statistical Appendix, p. 48.

(*i*) The topographical features in upper reaches of the major rivers are seldom favourable for development of irrigation. Consequently, development of hydroelectric sites would not clash with other priority uses of water. The Western Ghats, north-western Karnataka, Nilgiri and Anamalai hills and upper Narmada basin are major areas of concentration of potential in peninsular India.

(*ii*) Most of the areas in the southern states, especially the western part of the peninsula, are far away from the coal deposits of north-eastern part of the peninsular plateau. As such they have to depend upon the hydroelectric power to meet their energy requirements.

Generation of hydroelectricity registered a gradual increase from 2.5 billion kWh in 1950-51 to 82.9 billion kWh in 1998-99. After that juncture, downward trend started and continued till 2002-03, when the electricity generation was 64 billion kWh only. This downward trend was reversed in 2004-05 and the electricity generated increased to 150.3 billion kWh in 2020-21.

Hydroelectric power plants are scattered in different parts of the country. **Table 24.18** gives the list of important hydroelectric power plants in different states.

Figure 24.8 shows the areal distribution of major hydel power stations in India.

Although hydroelectricity accounts for about 12 per cent of the total installed capacity of electricity in India, it is the single largest source of energy for some states. For example, Himachal Pradesh, Meghalaya, Nagaland, Sikkim and Uttarakhand wholly depend upon hydroelectricity for their energy requirements. All these states are located in hilly and mountainous areas where geographical conditions like terrain, drainage, etc. are congenial to hydroelectric generation. The other states where energy supply scene is dominated by hydroelectricity are Kerala, Mizoram, Odisha and the union terriroty of Jammu and Kashmir. So far as total installed capacity of hydroelectricity is concerned, Andhra Pradesh, Telangana, Karanataka, Kerala, Maharashtra, Odisha, Punjab, Rajasthan and Tamil Nadu are the outstanding states. Some of the important hydroelectric power projects are briefly discussed here.

1. Andhra Pradesh. Andhra Pradesh has about three thousand MW installed capacity of hydroelectricity. Nagarjuna Sagar (Guntur District), Lower Sileru, Upper Sileru and Polavaram are the main centres.

2. Telangana. Telangana's installed capacity is 815.6 MW. Nagarjuna Sagar (Nalgonda District), Srisailam, Pochampad, Nizam Sagar, Inchampalli, etc. are the main centres.

3. Karnataka. Mahatama Gandhi Project on Jog Falls of the Sharavati river, Shivasamudram and Krishnaraja Sagar on Cauvery, Shimsha on Shishma river (a tributary of the Cauvery) are some of the outstanding power projects of Karnataka. Karnataka's installed capacity is 2,944 MW. Hydroelectricity accounts for about 56.6 per cent of the total installed capacity of the state.

4. Tamil Nadu. Mettur, Pykara, Papanasam, Kadamparai, Pandiyar, Kodayar, Kundah and Periyar are the main hydroelectric producing stations in Tamil Nadu. Most of the power stations are located in hilly districts and have been connected by the electric grid system to regulate the supply of hydro power.

5. Punjab. Two power stations at Bhakhra dam on the Satluj along with Ganguwal and Kotla power projects are the main producers of hydroelectricity. Pong dam on the Beas and the Upper Bari Canal system are the other producers.

6. Uttar Pradesh. The Ganga Electric Grid System is very important for production and distribution of hydroelectricity. The Upper Ganga Canal forms 12 falls between Haridwar and Meerut and most of them have been used to produce hydroelectricity. The power stations are situated at Bahadurabad, Mohammadpur, Nirghni, Chitaura, Salwa, Bhola, Plara and Sumera. The stretch of the Upper Ganga Canal and the power stations in this area are shared by Uttarakhand and Uttar Pradesh. Matatila dam on Betwa river near Jhansi and Rihand project at Pipri in Mirzapur district are the other important producers of hydroelectricity.

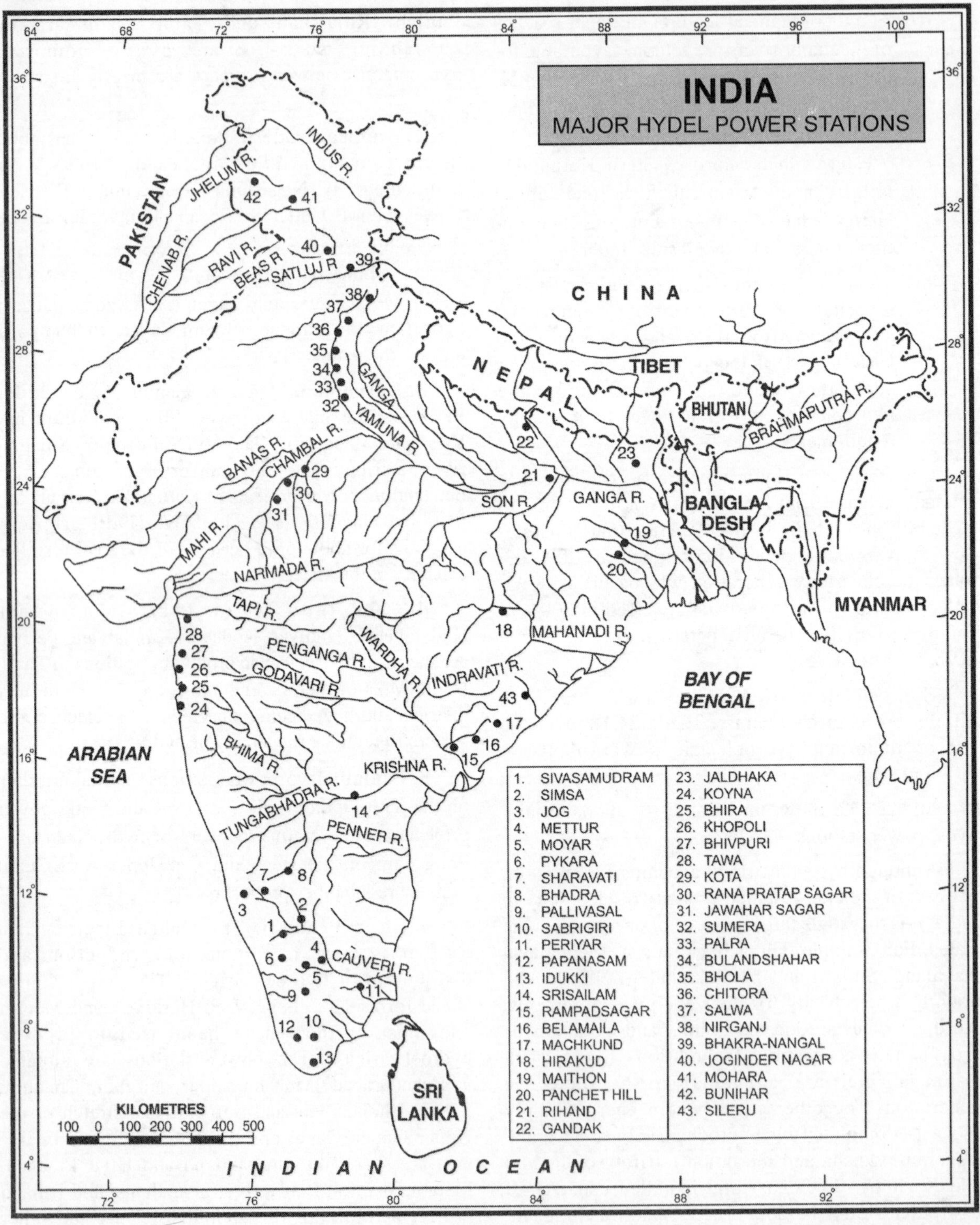

FIG. 24.8. India : Major Hydel Power Stations

TABLE 24.18. Important Hydroelectric Plants in Different States of India

States	Name of Hydroelectric Plants
Jammu and Kashmir	Lower Jhelum, Salal on Chenab, Dool Hasti, Karrah and Banihar.
Punjab and Himachal Pradesh	Bhakra-Nangal on Satluj, Dehar on Beas, Giri Bata, Andhra, Binwa, Rukti, Rongtong, Bhabanagar, Bassi, Baira Siul, Chamera, Nathpa-Jhakri on Satluj (biggest hydel power project in India).
Uttar Pradesh	Rihand, Khodri, Chibro on Tons.
Uttarakhand	Tehri dam on Bhagirathi.
Rajasthan	Ranapratap Sagar and Jawahar Sagar on Chambal.
Madhya Pradesh	Gandhi Sagar on Chambal, Pench, Bargi on Narmada, Bansagar-Tons.
Bihar	Kosi.
Jharkhand	Subarnarekha, Maithon, Panchet, Tilaiya (all three under DVC).
West Bengal	Panchet.
Odisha	Hirakud on Mahanadi, Balimela.
North-eastern states	Dikhu, Doyang (both in Nagaland), Gomuti (Tripura), Loktak (Manipur), Kopili (Assam), Khandong and Kyrdemkulai (Meghalaya), Serlui and Barabi (Mizoram), Ranganadi (Arunachal Pradesh).
Gujarat	Ukai (Tapi), Kadana (Mahi).
Maharashtra	Koyana, Bhivpuri (Tata Hydroelectric Works), Khopoli, Bhola, Bhira, Purna, Vaiterna, Paithon, Bhatnagar Beed.
Andhra Pradesh and Telangana	Lower Sileru, Upper Sileru, Machkund, Nizam Sagar, Nagarjun Sagar, Srisailam (Krishna).
Karnataka	Tungabhadra, Saravati, Kalinadi, Mahatma Gandhi (Jog fall), Bhadra, Sivasamudram (Kaveri), Shimasapura, Munirabad, Lungnamakki.
Kerala	Iddikki (Periyar), Sabarigiri, Kuttiaddy, Sholayar, Sengulum, Pallivasal, Kallada, Neriamangalam, Parambikulum Aliyar, Poringal, Ponniar.
Tamil Nadu	Pykara, Mettur, Kodayar, Sholayar, Aliayar, Sakarpathi, Moyar, Suruliyar, Papanasam.

7. Kerala. The state is located away from the main coal producing areas of the country and is blessed with several small but swift streams descending from the Western Ghats. Many of them form waterfalls which provide ideal sites for the generation of hydroelectricity. The state has 1,807 MW installed capacity. Idduki, Kuttiyadi, Sabargiri, Sholayar, Mananthavadi, Chalakudy, Puyankutty, Lower Periyar and Pallivasal are important projects.

8. Maharashtra. Most of the hydro-power potential is in the Sahyadris where geographical conditions are favourable for its production. Tata Hydro-electric grid at Lonavala comprises of three power houses at Khopoli (capacity 72 MW), Bivpuri (capacity 72 MW) and Bhira (capacity 137 MW). These stations supply electricity to Mumbai and its surroundings over a distance of 112 km. The Koyna project located about 240 km away from Mumbai on Koyna river (a tributary of the Krishna river) is another major landmark in Maharashtra. Purna and Vaitarna are other important projects.

9. Odisha. Over four-fifths of the total available capacity in the state is contributed by hydroelectricity. Hirakud, Bhimkud, Balimela and Rengali are the main projects producing hydroelectricity.

10. Himachal Pradesh. Mandi Project at Joginder Nagar, Pandoh project on Beas river and the Shimla project produce bulk of the hydroelectricity in the state. Installed capacity is 700 MW.

11. Jammu and Kashmir. The hill terrain of the state provides several suitable sites for hydroelectric generation. Chenani, Sind, Jhelum, Dool Hasti and Salal are the main producing projects.

THERMAL ELECTRICITY

This type of electricity is generated by using fossil fuels like coal, diesel and natural gas. Obviously, thermal power is influenced by the availability of these fuels.

Thermal electricity has recorded a much faster rate of growth, both with respect to installed capacity and actual generation, than its hydro and nuclear counterparts. In 1950-51 the installed capacity of thermal electricity was 1.1 thousand MW as compared to 0.6 thousand MW hydroelectricity. In 2020-21, the installed capacity of thermal electricity reached a level of 234.73 thousand MW against only 46.2 thousand MW installed capacity of hydroelectricity (**Table 24.15**). Generation of hydro and thermal electricity was almost about the same level in 1950-51. But thermal power generation reached a level of 1032.6 billion kWh in 2020-21 from an insignificant level of only 2.6 billion kWh in 1950-51 (**Table 24.17**). The thermal power generation recorded more than 454 times increase from 1950-51 to 2019-20 whereas hydroelectricity increased from 2.5 billion kWh in 1950-51 to 150.3 billion kWh in 2020-21 (**Table 24.17**) thus recording only 62-fold increase during the same period. In the year 2020-21, thermal power accounted for over 75 per cent of the total power generation as compared to only 51 per cent in 1950-51.

National Thermal Power Corporation (NTPC) was established in 1975, after which the share of thermal electricity increased considerably. At present, NTPC has 13 coal-based super thermal power projects and seven gas/liquid fuel based combined cycle projects. It has an installed capacity of 19,435 MW.

Thermal electricity has special significance in those areas where geographical conditions are not very favourable for generating hydroelectricity. It accounts for more than half of the installed capacity in 14 states. In the states of Assam, Bihar, Jharkhand, Chhattisgarh, Gujarat, Mizoram and West Bengal it accounts for over 90 per cent of the installed capacity. In Madhya Pradesh, Maharashtra, Manipur and Uttar Pradesh it accounts for about three-fourths of the total installed capacity. Statewise distribution of thermal electricity is briefly described as under :

1. Maharashtra. Maharashtra is the largest producer of thermal electricity and accounts for about 16 per cent of the total thermal electricity produced in India. Chola and Trombay steam power plants, Kolhapur Diesel Turbine, Uran Gas Turbine, Chandrapur, Bhusawal, Khaparkeda, Parli, Dhobal and Ujjaini are the main thermal power producing plants.

2. Gujarat. Gujarat does not have any major river flowing through its territory which can offer suitable sites for producing hydroelectricity in appreciable measure. Therefore, 91 per cent of the installed capacity of the state comprises thermal electricity. The main plants producing thermal electricity in the state are located at Banas, Gandhinagar, Kachchh, Sabarmati, Wanakbari, Kawas, Sikka, Mahuva, Utran, Shahpur, Porbandar, Kandla, Ahmedabad, Dhuvaran, etc.

3. Andhra Pradesh. Andhra Pradesh has some coal deposits as a result of which the state has developed thermal electricity. Thermal energy account's for about 90 per cent of the total energy produced in the state. The main thermal power stations in the state are at Ramagudam, Kothagudam, Nellore, Vijayawada, Bhadrachalam, Manuguru, etc.

4. West Bengal. West Bengal has large reserves of low grade steam coal which is quite suitable for generation of thermal power. Consequently, thermal electricity comprises 97.7 per cent of the total installed capacity of the state. The major power projects of the state are Bundel, Kolkata, Durgapur, Farakka, Murshidabad, Birbhum, Kalaghat, Titagarh, Mejia, Santaldih, etc.

5. Tamil Nadu. Thermal power accounts for about 72.1% of the installed capacity of Tamil Nadu. The state produces about 5 per cent of the total thermal electricity produced in India. Neyveli, Mettur, Ennore (Chennai), etc. are important projects of Tamil Nadu.

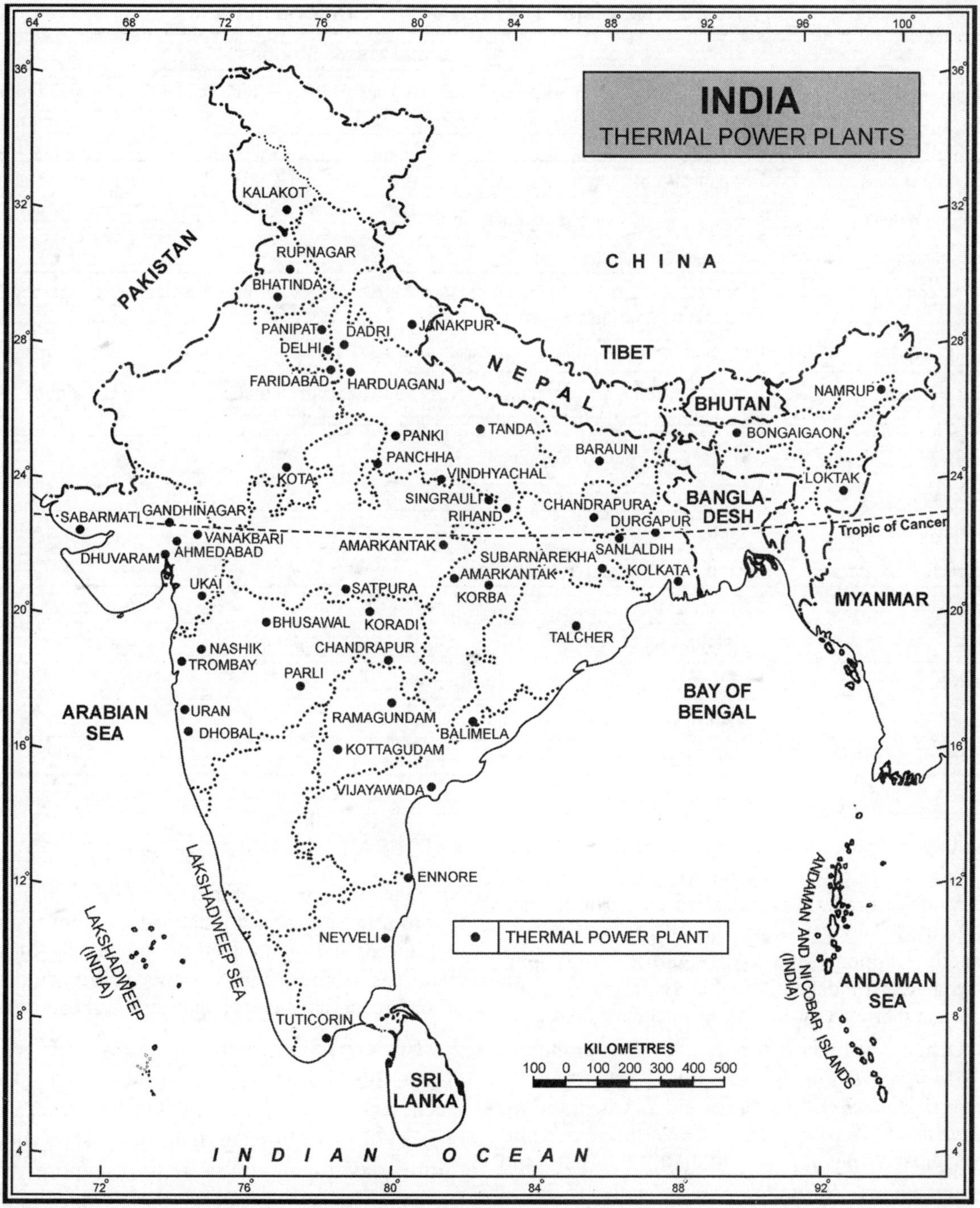

FIG. 24.9. Thermal Power Plants

6. Uttar Pradesh. Although Uttar Pradesh has mighty rivers flowing through its territory, the hydro-potential has not been properly harnessed. The result is that hydroelectricity forms only 6.4 per cent of the total installed capacity and over 93.6 per cent is constituted by thermal electricity. Most of the coal

TABLE 24.19. Major Thermal Power Stations in India

State	Thermal Power Stations
Maharashtra	Karadi, Nashik, Kaparkheda, Paras, Bhusawal, Parli, Uran, Ballarshah, Chola, Trombay, Kolhapur, Dhobal, Ujjaini.
Gujarat	Banas, Gandhinagar, Kachchh, Sabarmati, Wanskbori, Kawas, Sikka, Mahuva, Utran, Shapur, Porbandar, Kandla, Ahmedabad, Dhuvaram, Ukai.
Andhra Pradesh	Nellore, Vijayawada, Bhadrachalam, Manuguru.
Telangana	Ramagudam, Kottagudam
West Bengal	Bundel, Kolkata, Durgapur, Farakka, Murshidabad, Birbhum, Kalaghat, Titagarh, Mejia, Santaldih, Gauripur.
Tamil Nadu	Neyveli, Methur, Ennore, Tuticorin.
Uttar Pradesh	Obra, Harduaganj, Renusgar, Rosa, Jawaharpur, Unchohar, Rihand, Kanpur, Mau, Gorakhpur, Dohrighat, Moradabad, Tundla, Bahraich
Madhya Pradesh	Singrauli, Satpura, Amarkantak.
Jharkhand	Durgapur, Subarnarekha, Chandrapura.
Chhattisgarh	Korba.
Haryana	Faridabad, Panipat, Yamunanagar.
Punjab	Bhatinda, Roopnagar.
Delhi	Indraprastha, Rajghat, Badarpur. (All closed due to pollution)
Rajasthan	Kota, Polana, Sawai Modhiopur, Banswara, Anta.
Assam	Namrup, Bongaigaon, Chandarpur.
Odisha	Talcher, Balimela.
Bihar	Barauni, Kahalgaon.

used in generating thermal electricity is brought from the neighbouring states of Jharkhand, Chhattisgarh and Madhya Pradesh. Obra (Mirzapur), Harduaganj (Aligarh), Renusagar, Rosa, Jawaharpur, Unchahar, Rihand, Kanpur, Mau (Azamgarh), Gorakhpur, Dohrighat, Moradabad, Tundla, Bahraich, etc. are important thermal power stations in Uttar Pradesh.

Other producers. Among the other producers are *Madhya Pradesh* (Singrauli, Amarkantak, Satpura), *Jharkhand* (Subarnarekha, Chadrapura), *Chhattisgarh* (Korba), *Punjab* (Bhatinda and Rupnagar), *Haryana* (Faridabad, Panipat and Yamunanagar), *Rajasthan* (Kota, Palana, Sawai Madhopur, Banswara); *Karnataka* (Raichur), *Assam* (Namrup, Bongaigaon, Chandarpur), and *Odisha* (Ib Valley, Talcher).

Table 24.19 gives a list of major thermal power stations in India. Areal distribution of thermal power plants is shown in **Figure 24.9**.

NUCLEAR ENERGY

Nuclear energy is obtained from uranium and thorium. India has vast untapped uranium resources and there is urgent need to make use of these resources if India really wants to get out of the present scenario of power shortages and energy crisis.

Nuclear power contributes only 3.36 per cent of our total power generation at present. This is much less as compared to Sweden (98.2%) and France (78.1%). However, India has vast potential for future development. It requires quite higher technology to develop nuclear power which India has fortunately attained now. India is one of the few countries which have developed the capability of designing, constructing, commissioning and operating a nuclear power station without any help from outside.

Most of the nuclear power stations in India have been constructed near sources of water because it is required in great quantity for cooling purposes.

Nuclear power programme was initiated in 1940s when 'Tata Atomic Research Commission' was incorporated in August, 1948. However, the real progress was made only after the establishment of the Atomic Energy Institution at Trombay in 1954. This was renamed as the 'Bhabha Atomic Research Centre' (BARC) in 1967. The first nuclear power station with 320 MW capacity was set up at Tarapur near Mumbai in 1969. Later, atomic reactors were installed at Rawatbhata (300 MW) near Kota in Rajasthan, Kalpakkam (440 MW) in Tamil Nadu, and Narora in Uttar Pradesh. Kaiga in Karnataka and Kakarapara in Gujarat also have nuclear energy plants. Nuclear power plants are proposed to be set up at Kumharia (Haryana), Bargi (M.P.) Haripur (West Bengal), Jaitapur (Maharashtra) Mithi Virdi (Gujarat) and Kovvada (Andhra Pradesh).

Nuclear Power Programme

A three stage Nuclear Power Programme was drafted by the Department of Atomic Energy (DAE) in 1954. This programme aimed at meeting the growing demand for energy in the country by utilizing natural resources of uranium and thorium. The programme involves pressurized heavy water reactors (PHWREs) in the first stage, fast breeder reactors (FBRs) in the second stage and thorium-based advanced heavy water reactors (AHWRs) in the third stage. Development of technology relating to spent-fuel reprocessing, waste management, safety and environment were also included in this programme.

Nuclear Power Programme, Stage-I. This programme was initiated in 1960s with PHWR reactor as the initial choice. It was aimed at gaining experience with two boiling water reactors (BWRs) imported from the U.S.A. installed at Tarapur Atomic Power Station (TAPs) about 110 km north of Mumbai and commissioned in 1969. In 1972, the PHWR of Candou design nuclear power plant was commissioned at Rawatbhata near Kota in Rajasthan in technical collaboration with the Atomic Energy Canada Ltd. (AECL).

The U.S.A. and Canada ceased their nuclear collaboration with India following nuclear explosion made by India at Pokaran in 1974 and Indian engineers engaged themselves in developing indigenous technology. Thus the next reactor was established in 1980. The indigenous technology reached its commercial maturity with the commissioning of a 2X22 PHWR at Kakarapara (Gujarat) in 1992. The state-of-the-art 2X22. PHWRs were commissioned one each at Kakarapara, Kaiga (Karnataka) and Rawatbhata (Rajasthan) in 1995, 1999 and 2000 respectively.

The Nuclear Power Corporation of India (NPCIL) a public sector undertaking of the DAE is responsible for designing, constructing and operating the nuclear power stations in India. **Table 24.20** gives details of nuclear power plants run by the NPCIL.

TABLE 24.20. Details of nuclear power plants in India

Reactor	Type	Commercial Operation
Tarapur 1 and 2	BWR	1969
Tarapur 3 and 4	PHWR	2005-06
Kaiga 1 and 2	PHWR	1999-2000
Kaiga 3 and 4	PHWR	2007
Kakarapara 1 and 2	PMWR	1993-95
Kakarapara 3 and 4	PHWR	2012
Kalpakkam 1 and 2	PHWR	1984-86
Kalpakkam PFBR	FBR	2010
Narora 1 and 2	PHWR	1991-92
Rawatbhata 1	PHWR	1973
Rawatbhata 2	PHWR	1981
Rawatbhata 3 and 4	PHWR	1999-2000
Rawatbhata 5 and 6	PHWR	2007-08
Rawatbhata 7 and 7	PHWR	2012
Kudankulam 1 and 2	PWR	2007-08
Kudankulam 3 and 4	PWR	2012
Jaitapur 1 and 2	PWR	2012

Notes : BWR = Boiling Water Reactors.
PHWR = Pressurised Heavy Water Reactors
PWR = Pressurised Water Reactor,
FBR = Fast Breeder Reactor

***Source* :** World Nuclear Association.

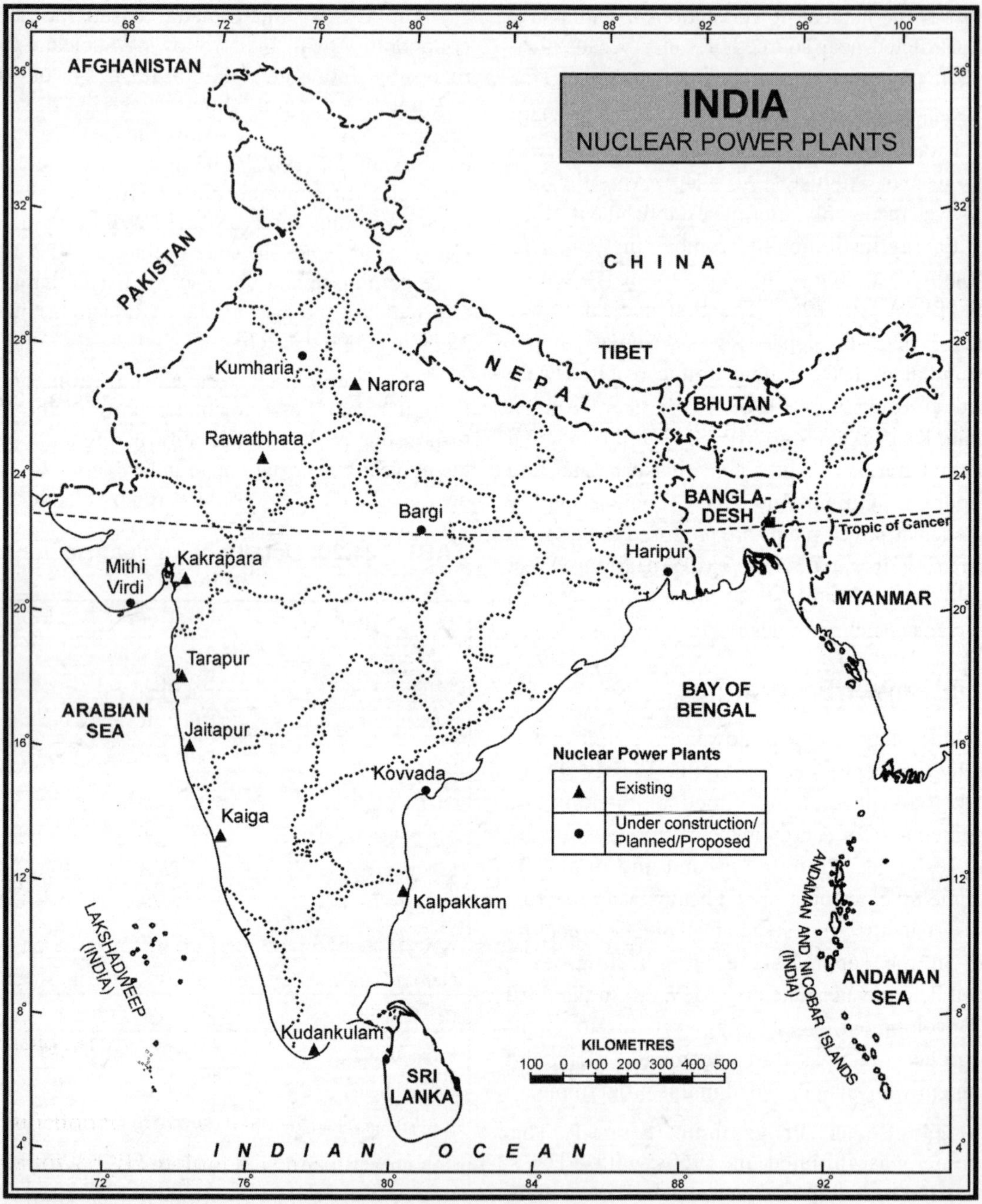

FIG. 24.10. India : Nuclear Power Plants

Nuclear fuel fabrication for nuclear reactors and research reactors is done at Nuclear Fuel Complex, Hyderabad and Bhabha Atomic Research Centre (BARC), Mumbai. The BARC and Indira Gandhi Centre for Atomic Research (ICAR) are also engaged in developing new fuels to be used in future.

Nuclear Power Programme Stage-II. This stage envisages setting up of Fast Breeder Reactors (FBRs) heated by reprocessing plants and plutonium based fuel fabrication plants. These fast breeder systems produce more fuel than consumed by them. FBRs can increase the utilization of fuel by about 60 times than PHWRs. India is vigorously pursuing the FBRs programme to tide over the current power problem faced by the country. To manage the FBRs programme, a new PSU named Bhartiya Nabhikiya Vidhyut Nigam (BHAVIN) was set-up in 2003.

Nuclear Power Programme, Stage-III. This stage is concerned with setting up of thorium-based reactors and associated fuel cycle facilities. Since thorium is only a fertile material, no reactor can be started using thorium directly as a fuel. Thorium is converted into uranium-233 through irradiation with neutrons in a reactor whether PHWR or FBR.

The above description brings us to the conclusion that there is vast scope for developing nuclear energy in India. However, nuclear energy development programme has to face very tough resistance from environmentalists and those who fear the occurrence of nuclear mishaps either due to natural disasters like earthquakes, tsunamis, cyclones etc. or because of human failure. Damage to Fukushima nuclear power plant in Japan caused by tsunami of March 11, 2011 is an eye opener. Several protests have been noticed against Jaitapur and Kudankulam nuclear power plants. These fears are not totally unfounded as several mishaps have taken place in different parts of the country. Some of the serious mishaps in nuclear power plants in India are detailed below.

NUCLEAR MISHAPS IN INDIA

- **May 4, 1987 (Kalpakkam).** Refueling accident damages reactor core.
- **September 10, 1989 (Tarapur).** Radioactive iodine leaks at above normal levels.
- **May 13, 1992 (Tarapur).** A malfunctioning tube causes plant to release 12 Curies of radioactivity.
- **March 31, 1993 (Narora).** Plant suffers a fire at its steam turbine blades. Shutdown lasted a year.
- **May 13, 1994 (Kaiga).** During construction, an inner containment dome meant to contain radiation collapses.
- **February 2, 1995 (Kota).** Plant leaks radioactive helium and heavy water into the Rana Pratap Sagar reservoir.
- **Dec. 26, 2004 (Kalpakkam).** During the tsunami, sea water enters intake tunnel of plant. Plant shut down.
- **November 25, 2009 (Kaiga).** Employees fall ill. Tests show tritium content in urine of 92 employees who drank water from cooler. A disgruntled staff had poured radioactive heavy water into it.

NON-CONVENTIONAL ENERGY SOURCES

With increasing demand for energy and with fast depleting conventional sources of energy such as coal, petroleum, natural gas, etc. the non-conventional sources of energy such as energy from sun, wind, biomass, tidal energy, geo-thermal energy and even energy from waste material are gaining importance. This energy is abundant, renewable, pollution free and eco-friendly. It can be more conveniently supplied to urban, rural and even remote areas. Thus it is capable of solving the twin problems of energy supply in a decentralised manner and helping in sustaining cleaner environment. *It is the energy of the future. No wonder, non-conventional energy is fast catching the imagination of the people in India.*

India has today one of the largest programmes for renewable energy. The activities cover all major renewable energy sources, such as biogas, biomass, solar, wind, small hydropower and other emerging technologies. Several renewable energy systems and devices are commercially available. The renewable energy programmes cover the entire gamut of technologies, including improved wood stoves, biogas plants, biomass gasifier, solar thermal and solar photovoltaic systems, wind mills, co-generation, small hydropower, energy recovery from urban/municipal and industrial wastes, geothermal energy, hydrogen energy, electric vehicles and bio-fuels, etc.

According to India 2022, A Reference Annual (p. 555), India has an estimated renewable energy potential of about 900 GW from commercially exploitable resources, *viz.* Wind—102 GW (at 80 metre mast height), small hydro—20 GW, bio-energy—25 GW and 750 GW solar power, assuming 3 per cent wasteland is made available.

During the period of ten years in from 2007 to 2017, the installation of renewable energy for electricity has grown at an annual rate of 25 per cent. It has reached 30,000 MW as of January 2017. During this period, wind power installation has grown ten times and solar energy has grown from nothing to 2,500 MW. Currently renewable energy accounts for 20 per cent of total electricity generation capacity and contributes about 10 per cent of the electricity produced in the country. Renewables, therefore,

TABLE 24.21. State-wise Renewable Energy Potential (in MW)

Sl. No.	States/Union Territories	Wind Power	Small Hydro Power	Bio-Energy			Solar	Total
				Biomass Power	Bagasse Cogene-ration	Waste to Energy		
1.	Andhra Pradesh	14,497	978	578	300	123	38,440	54,916
2.	Arunachal Pradesh	236	1,341	8	—	—	8,650	10,236
3.	Assam	112	239	212	—	8	13,760	14,330
4.	Bihar	144	223	619	300	73	11,200	12,559
5.	Chhattisgarh	314	1,107	236	—	24	18,270	19,951
6.	Goa	—	7	26	—	—	880	912
7.	Gujarat	35,071	202	1,221	350	112	35,770	72,726
8.	Haryana	93	110	1,333	350	24	4,560	6,470
9.	Himachal Pradesh	64	2,398	142	—	2	33,840	36,446
10.	Jammu & Kashmir	5,685	1,431	43	—	—	1,11,050	1,18,208
11.	Jharkhand	91	209	90	—	10	18,180	18,580
12.	Karnataka	13,593	4,141	1,131	450	—	24,700	44,015
13.	Kerala	837	704	1,044	—	36	6,110	8,732
14.	Madhya Pradesh	2,931	820	1,364	—	78	61,660	66,853
15.	Maharashtra	5,961	794	1,887	1250	287	64,320	74,500
16.	Manipur	56	109	13	—	2	10,630	10,811
17.	Meghalaya	82	230	11	—	2	5,860	6,185
18.	Mizoram	—	169	1	—	2	9,090	9,261
19.	Nagaland	16	197	10	—	—	7,290	7,513
20.	Odisha	1,384	295	246	—	22	25,780	27,728
21.	Punjab	—	441	3,172	300	45	2,810	6,768
22.	Rajasthan	5,050	57	1,039	—	62	1,42,310	1,48,518
23.	Sikkim	98	267	2	—	—	4,940	5,307
24.	Tamil Nadu	14,152	660	1,070	450	151	17,670	34,152
25.	Telangana	—	—	—	—	—	20,410	20,410
26.	Tripura	—	47	3	—	2	2,080	2,131
27.	Uttar Pradesh	1,260	461	1,617	1,250	176	22,830	27,593
28.	Uttarakhand	534	1,708	24	—	5	16,800	19,071
29.	West Bengal	22	396	396	—	148	6,260	7,222
30.	A. & N. Islands	365	8	—	—	—	0	373
31.	Chandigarh	—	—	—	—	6	0	6
32.	D. & N. Haveli	—	—	—	—	—	0	0
33.	Daman & Diu	4	—	—	—	—	-	4
34.	Delhi	—	—	—	—	131	2,050	2,181
35.	Lakshadweep	—	—	—	—	3	0	123
36.	Puducherry	120	—	—	—	3	0	123
	Others	—	—	—	—	1,022	790	1,812
	INDIA	**1,02,772**	**19,749**	**17,536**	**5,000**	**2,554**	**7,48,990**	**8,96,602**

Source : EnviStats India 2018, p. 155.

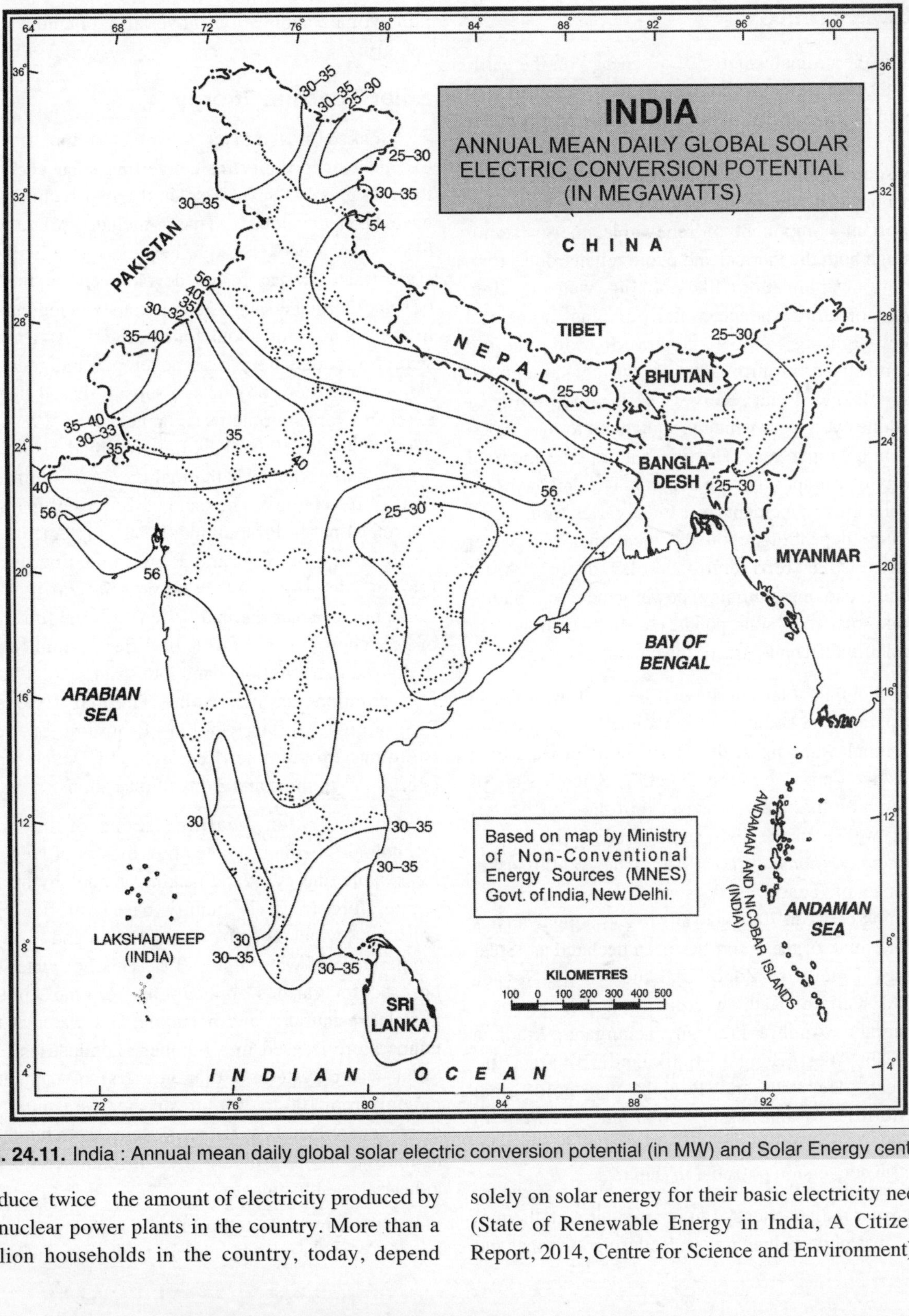

FIG. 24.11. India : Annual mean daily global solar electric conversion potential (in MW) and Solar Energy centres

produce twice the amount of electricity produced by all nuclear power plants in the country. More than a million households in the country, today, depend solely on solar energy for their basic electricity needs (State of Renewable Energy in India, A Citizen's Report, 2014, Centre for Science and Environment).

SOLAR ENERGY

Sun is the ultimate source of all energy on the earth. India, being a tropical country, is well endowed with plenty of solar energy. Most parts of the country have bright sun-shine throughout the year except a brief monsoon period. As our country is literally soaked in sunshine, exploitation of solar energy is an extremely important component of renewable energy sector through both the thermal and photovoltaic routes for a variety of applications like cooking, water heating, drying of farm produce, water pumping, home and street lighting, power generation for meeting decentralised requirements in villages, schools, hospitals, etc. India receives nearly 3,000 hours of sunshine every year which is equivalent to over 5,000 trillion kWh per year. This is far more than the total energy consumption of the country. The daily average of solar energy incident over India varies from 4 to 7 kWh/m^2 depending upon the location. Solar water heaters, solar refrigeration, solar drying, street lighting, cooking, pumping, power generation, photo-voltaic solar cells, solar ponds, etc. are becoming very popular in different parts of the country.

Although solar energy can be gainfully used in any part of the country except some higher areas in the Himalayan ranges, the Thar Desert of Rajasthan holds great promise in this direction. Scientists are of the opinion that the vast expanse of the Thar Desert could well earn the distinction of being the *biggest solar power house of the world*. The 35,000 sq km expanse of Thar Desert is sufficient to generate anything between 700 gigawatts to 2,100 gigawatts. A major chunk of the desert has been declared as 'Solar Energy Enterprise Zone' like the one in Nevada (USA). Parts of Kathiawar peninsula. Maharashtra, Karnataka, Andhra Pradesh, Telangana, Madhya Pradesh, West Bengal, Jharkhand, Bihar, Uttar Pradesh, Haryana and Punjab also hold great possibilities of harnessing solar energy. Map in **Figure 24.11** shows annual mean daily global solar electric conversion potential in India.

Solar radiant energy can be used through thermal as well as photovoltaic routes. Both solar, thermal and photovoltaic applications have large potential in the country.

Solar Thermal Energy

Soaked in abundant sunshine, India offers an excellent opportunity for converting solar energy to thermal energy. Several solar thermal technologies have been developed. These include solar water heaters, solar cookers, solar heaters, solar distillation systems, etc. Research and development in the field of solar thermal energy is continuously being pursued in the country for about four decades. As a result, several products have been developed indigenously. To promote these products, a subsidy-based thermal extension programme was launched in 1984 and continued upto 1993. This initiative had resulted in disseminating the solar thermal products in different parts of the country. The main objectives of the Solar Thermal Energy Programme, being implemented by the Ministry of New and Renewable Energy, are market development, commercialization and utilization of solar thermal systems for the fulfilment of heat energy requirements of different applications in domestic, institutional and industrial sectors. It has five components *viz.* Solar Thermal Extension Programme, Solar Cooker Programme, Solar Buildings Programme, Research and Development (R&D) Programme and Aditya Solar Shops.

Solar Water heating is one of the main technologies being promoted by Ministry of New and Renewable Energy. Water heating technology for low temperature range is mainly based on flat plate collectors, which absorb solar radiation and raise the temperature of water upto 80°C. This hot water can be used for various applications in homes, hotels, hostels, restaurants and hospitals. Hot water at this temperature is used in a number of industries also. Solar water systems (solar geysers) of capacities ranging from 100 to 300 litres per day are suited for domestic applications. Larger systems from hundreds to thousands of litres are used in commercial and industrial establishments.

Although the initial cost of solar water heating system is rather high, the system pays back the investment within 3 to 6 years depending on the fuels substituted. The technical potential of solar water heaters in the country has been estimated to be 140 million sq metres of collector area. There is an enormous possibility for harvesting solar energy through this technology. With the increasing acceptability in the residential sector, solar water heaters can be set up in multistoreyed residential flats for meeting the hot water requirement.

The use of solar water heaters saves electricity and contributes to a reduction in peak load demand. It has been estimated that the use of 1,000 domestic solar water heating systems of 100 litres capacity each can contribute to a peak load saving of 1 MW.

Solar air heaters and dryers can conveniently be used both in industry and agriculture. Already a number of solar drying systems have been installed in the country and these are helping to save significant amounts of conventional fuels. Among the industries using these are tea, food processing, *dal* mills and spice manufacturers. Solar air heaters are also being used for space heating in the cold regions.

Solar cooker is a simple device which cooks food with the help of solar energy and saves conventional fuels to a significant extent. On clear sunny days, it is possible to cook both noon and evening meals with a solar cooking device. Different types of solar cookers have been developed in the past, which include box solar cooker, steam cooker, solar meal maker with heat storage and concentrating type community cooker. The Ministry of Non-Conventional Energy Sources (MNES) had been promoting the box solar cooker in the country till 1993-94 due to its various advantages over the others. Thereafter, different designs of solar cooker have been propagated under its market-oriented and demonstration programmes. Currently two types of cookers, *viz,* box solar cooker and concentration type cooker are popular among the users. *Box solar* cooker can cook meals for a family of 4 to 5 members and saves 3 to 4 LPG cylinders in a year on full use. If provided with electrical back-up, it can be used during non-sunshine hours also within the kitchen with nominal consumption of electricity. *Concentrating* type solar cooker is of three broad types, *viz.,* dish solar cooker. community solar cooker and solar steam cooking system. Dish solar cooker is a fast cooking device which can cook food for 10 to 15 people under sun. It saves upto 10 LPG cylinders per year on full use in small establishments. Community solar cooker (Schefler) can cook food for around 40 people inside the kitchen and saves 35 LPG cylinders per year on full use in community kitchens. Solar steam cooking system can cook food for thousands of people using steam inside the kitchen in very short time. It is useful for installation at ashrams, temples, churches, gurudwaras, etc. A solar cooking system has been installed in Shirdi to cater to 3,000 devotees every day. World's largest system with a capacity to prepare food for 15,000 pilgrims was also set up by the Tirumala Tirupathi Devasthanam in October, 2002.

Solar Photovoltaic (SPV) technology enables direct conversion of sunlight into electricity without any moving parts and without causing pollution. Photovoltaic systems and power plants have emerged as viable power sources for applications such as lighting, water pumping and telecommunication and are being increasingly used for meeting the electrical energy needs in remote villages, hamlets and hospitals, besides households in the hilly, forest, and desert areas as well in islands. During the past few years, many organisations have started using SPV systems for a variety of applications on commercial basis as these are found to be economically viable as compared to other alternatives. Recently a programme on the deployment of SPV water pumping systems for agriculture and related uses has been implemented.

Efforts are being made to popularise the use of solar greenhouse for growing vegetables during off-season in cold and dry areas of Leh and Kargil. Solarised huts are being designed in cold areas of Jammu and Kashmir and Himachal Pradesh to keep the buildings warm.

Solar Energy Potential. About 5,000 trillion kWh per year energy is incident over India's land area with most parts receiving 4-7 kWh per square metre per day. National Institute of Solar Energy has assessed the country's solar potential of about 7489 W assuming 3% of the waste land area to be covered by solar PV module. Solar energy has taken central place in India's National Solar Mission as one of the key missions. National Solar Mission (NSM) was launched on 11th Jan. 2010. It is a major initiative of the Government of India with active participation of States to promote ecological sustainable growth while addressing Inda's energy security challenges. This is in line with India's intended Nationally Determined Contribution (INDCs) target to achieve capacity from non-fossil fuel based energy resources and to reduce emission intensity of its GDP by 33 to 35 per cent from 2005 level by 2030. Recently India achieved 5th global position in solar power deployment by surpassing Italy. The countries ahead of India are China, U.S.A., Japan and Germany. Solár power capacity has increased by more than 11 times in a span of five year from 2.6 GW in Mach 2014 to 30 GW in July 2019.

TABLE 24.22. Indias Solar Power Generation

Year	Solar Power Generation (TWh)
2016-17	12.09
2017-18	25.87
2018-19	39.27
2019-20	50.23

NEW SOLAR POLICY 2021

- Any person/developer/consumer can set up solar projects without any capacity ceiling.
- Existing ceiling of 50% of sanctioned load/constructed demand for setting up solar projects is removed.
- Residential consumers can set-up solar projects on their roof/premises or can give their/premises on lease to third party for generation and consumption of power in same premises.
- Consumers other than Residential can set-up solar projects on their roof/premises or can give their roof/premises on lease to third party for generation and consumption of power within or outside the premises.
- Surplus energy generated by any consumer shall be purchased by DISCOMs at rates specified by the policy.
- A group of consumers can set-up solar projects for self-consumption as a collective ownership project and consume the generated solar energy in the ratio of ownership.
- Security deposit to be furnished by developer to DISCOMs for PPA, reduced from ₹ 25 lakh per MW to ₹ 5 lakh per MW.
- New Solar Power Policy will be operative for five years i.e. upto 31-12-2025. Solar projects setup under this policy shall be eligible for benefits of this policy for a project life of 25 years.

Installed Capacity and Production. India's total installed capacity, including ground and roof mounted plants, was 47.7 GW as of 31 October, 2021.

WIND ENERGY

Wind is another important source of non-conventional energy. The cost inputs are only at the initial stage and the power generation starts immediately after commissioning. Once the generation starts, cost-free power is available for about 20 years because there is no recurring cost on fuel. India has vast wind potential and windfarms have emerged as a viable option with the advancement of wind technology in the country.

Since 1985, the MNES has carried out an extensive wind monitoring and mapping programme to identify better sites and to assess the resource potential. The programme has been implemented in a very systematic manner. The compiled data has been analysed and has been published with the ultimate objective of preparing a wind atlas of the country. This data has been utilised in identifying some excellent sites for windfarms. Map in **Figure 24.12** shows that coastal areas in Gujarat, Tamil Nadu, Andhra Pradesh, Odisha and West Bengal as well as vast areas of Maharashtra, Madhya Pradesh and Rajasthan are in a much better position with respect to wind power resources. The map shows wind resources in 10 states as detailed data are available for these states only.

TABLE 24.23. Distribution of wind potential

Sl. No.	State	Wind potential at 100 m (GW)	Wind potential at 120 m (GW)
1.	Gujarat	84.43	142.56
2.	Rajasthan	18.77	127.75
3.	Maharashtra	45.39	98.21
4.	Tamil Nadu	33.79	68.75
5.	Madhya Pradesh	10,.48	15.40
6.	Karnataka	55.85	124.35
7.	Andhra Pradesh	44.26	74.70
	Total 7 windy states	292.97	651.72
8.	Others	9.28	43.78
	Total	302.25	695.50

Potential of Wind Energy in India. The Government has installed over 800 wind monitoring stations in different parts of the country and issued wind potential mps at 50 m, 100 m and 120 m above the ground level. The recent assessment indicates a gross wind power potential of 302.25 MW in the country at 100 m and 695.5 MW at 120 m above the ground level. Most of it exists in seven windy states as is given in **Table 24.23**.

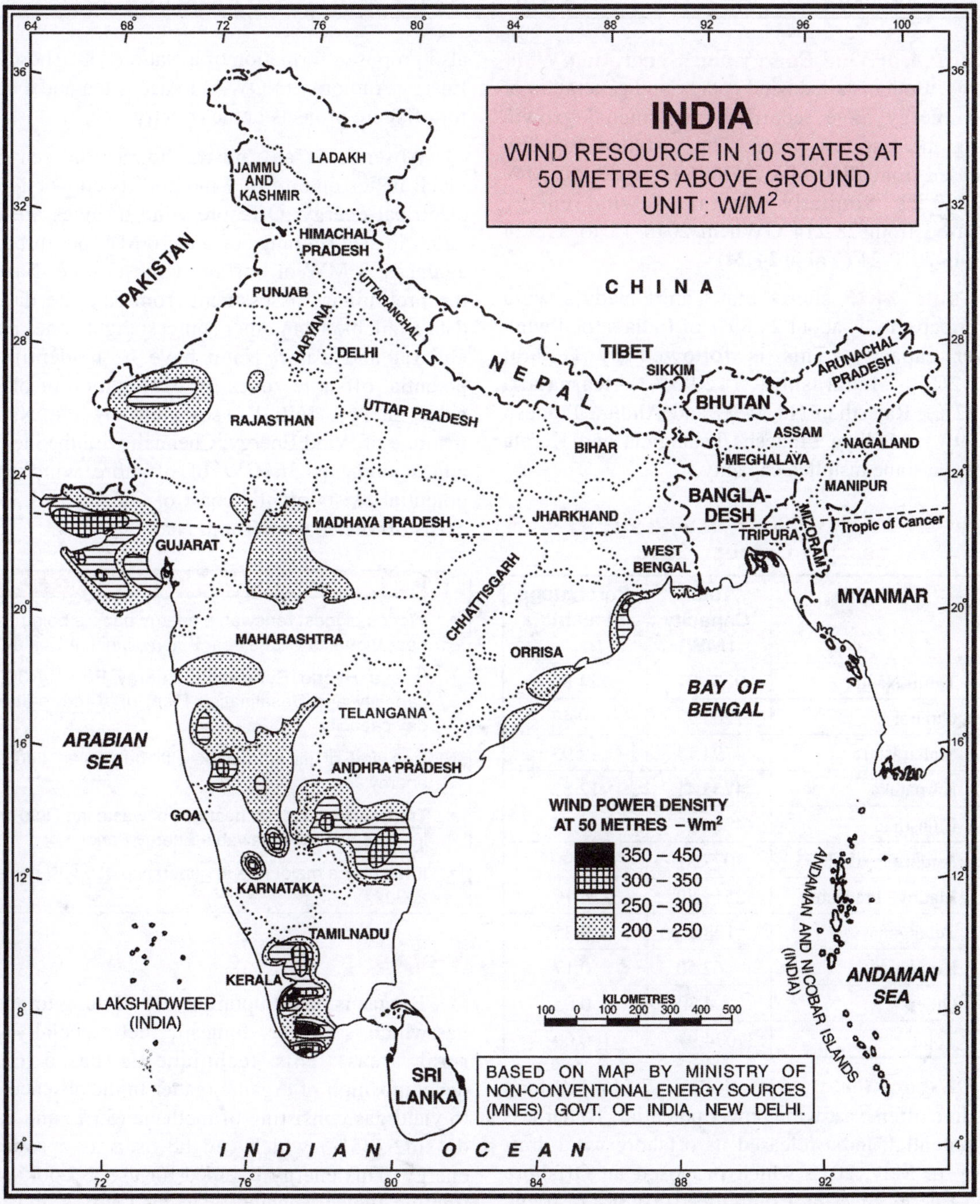

FIG. 24.12. India : Wind Resource in 10 states at 50 m above ground

TABLE 24.24. Installed wind power and generation of wind energy in India

Financial Year	2006-07	2010-11	2014-15	2015-16	2016-17	2017-18	2018-19	2019-20	2020-21
Installed Capacity (MW)	7,850	16,084	23,449	26,777	32,280	34,046	35,626	37,669	38,785
Generation (GWh)	—	—	28,214	28,604	46,011	52,036	64,036	64,485	59,824

Installed Wind Energy and Generation Wind Energy. Both installed wind energy and generation of wind energy have recorded phenomenal growth during the last few years. The installed capacity increased from 7850 MW in 2006-07 to 38,785 MW in 2020-21. Similarly general of wind energy increased from 28,214 GWh in 2014-15 to 59,824 GWh in 2020-21 (**Table 24.24**).

Table 24.25 shows that Tamil Nadu's wind power capacity is about 24.89% of India's total wind power capacity. This is followed by Gujarat (19.42%), Maharashtra (12.93%), Karnataka (12.82%), Rajasthan (11.54%) and Andhra Pradesh (10.99%), Madhya Pradesh, Telangana and Kerala also have some installed capacity.

TABLE 24.25. Installed wind capacity by state as of 31 October, 2019

Sl. No.	State	Total Capacity (MW)	Percentage of all India
1.	Tamil Nadu	9231.77	24.89
2.	Gujarat	7203.77	19.42
3.	Maharashtra	4794.13	12.93
4.	Karnataka	4753.41	12.82
5.	Rajasthan	4299.73	11.54
6.	Andhra Pradesh	4077.37	10.99
7.	Madhya Pradesh	2519.89	6.79
8.	Telangana	128.10	0.35
9.	Kerala	62.50	0.17
	Others	4.30	0.05
	Total	37074.97	100.00

Offshore Possibilities. India's 7,517 km long coastline offers vast wind power potential. To harness this potential, India released its offshore wind draft policy in May, 2013 which proposed an Offshore Wind Energy Steering Committee (OWSC) to frame the policy for offshore wind energy development. It also proposed formation of a National Offshore Wind Energy Authority (NOWA) to act as the nodal agency for offshore projects in the country.

Absence of any obstruction in the sea offers much better quality of wind and its conversion into electrical energy. Offshore wind turbines are much larger in size (in range of 5 to 10 MW per turbine) as against 2-3 MW of onshore wind turbine. Based on the preliminary assessment from satellite data and data available from other sources, eight zones each in Gujarat and Tamil Nadu have been identified as potential offshore zones for exploitation of offshore wind energy. Initial assessment by the National Institute of Wind Energy, Chennai, with the identified zones suggests 36 GW of offshore wind energy potential exists off the coast of Gujarat and 35 GW exists off the coast of Tamil Nadu.

WORLD'S BIGGEST RENEWABLE ENERGY PARK

- Worlds biggest renewable energy pack is being set-up near Vighakot village near Kutch district of Gujarat
- It is a Hybrid Renewable Energy Park of 30 GW capacity and Desalination Plant of 10 crore litre per day capacity
- It is first of its kind park with both wind and solar energy.
- Transforming 72,600 hectares of waste land into first of its wind Hybrid Renewable Energy Park.
- It will play a major role in generating 450 GW power by 2030.

Biogas

Biogas is based upon the use of dung to produce gas which is used as domestic fuel especially in the rural areas. This technique is based on the decomposition of organic matter in the absence of air to yield gas consisting of methane (55%) and carbon dioxide (45%) which can be used as a source of energy. This energy is piped for use as cooking and lighting fuel in specially designed stoves and lamps respectively. It can also be used for replacing diesel oil in dual fuel engines for generation of motive power and electricity. The left-over digested slurry serves as an enriched manure. Biogas technology is taking deep roots in rural India because of certain inherent advantages. Biogas has higher thermal efficiency when compared with kerosene, firewood, dung and charcoal. It is observed that the thermal efficiency of gobar gas is 60 per cent while dung, which is commonly used in villages for cooking, has only 11 per cent thermal efficiency. Thus the use of gobar gas fuel is advantageous from the point of view of not only fuel efficiency but also fuel saving.

Biogas technology has a bright future in India. It has been estimated that 1 to 1.5 billion tonnes of waste dung is available per annum in the country. If two-third of this quantity is used to produce biogas, it could yield 22,425 million cubic metre of biogas which could save 33,904 million litres of kerosene. Currently, it saves about 70 lakh tonnes of fuel wood annually. Further, it could yield manure equivalent to 14 million tonnes of nitrogen, 13 million tonnes of phosphate and 0.9 million tonnes of potash. The application of recommended dose of biogas slurry increases the yield by 12 per cent to 35 per cent of all types of crops, fruits and vegetables. About 1.2 million households are using biogas to meet their cooking energy needs. The success of biogas technology has brought about a sort of *Brown Revolution* in rural India.

Although biogas plants and improved *chullahs* have been disseminated in India since 1940s, they picked up only in early 1980s. The National Project on Biogas Development (NPBD) was taken up in Central Sector during 1981-82 on country wide basis. It seeks to provide clean and convenient fuel for cooking and lighting purposes in rural areas, produce enriched organic manure, improve sanitation and hygiene by way of linking household biogas plants with toilets and reduce the drudgery of women.

Besides family size biogas plants, the Ministry of Non-conventional Energy Sources is also promoting the setting up of community, institutional and nightsoil based biogas plants. The Community and Institutional Biogas Plants (CBP/IBP) Programme was initiated in 1982-83. Under this programme, the biogas is generally used for motive power and generation of electricity, besides meeting the cooking fuel requirements. A component on biogas plants linked with community toilet complexes was added in the year 1993-94 to facilitate on site treatment of human waste.

Production and Distribution of Biogas. The total production of biogas in 2014-15 was 20,757 lakh cubic metres. This is equivalent to 6.6 crore domestic LPG cylinders. Among the major producers Maharashtra tops the list and this state produced 3578 lakh cubic metres of biogas. This is followed by Andhra Pradesh Karnataka, Gujarat, Uttar Pradesh, West Bengal, Assam, Madhya Pradesh and Tamil Nadu (**Table 24.26**).

TABLE 24.26. Production and Distribution of Biogas in India (2014-15)

State	Production (Lakh cubic metre)
1. Maharashtra	3578
2. Andhra Pradesh	2165
3. Karnataka	1985
4. Gujarat	1965
5. Uttar Pradesh	1697
6. West Bengal	1460
7. Assam	1403
8. Madhya Pradesh	1370
9. Tamil Nadu	1137

Source: https//factly.in biogas production.

National Programme on Improved Chulhas

The National Programme on Improved Chulhas (NPIC) was initiated in 1986-87 with the following objectives :

- fuel wood conservation;
- elimination/reduction of smoke;
- reduction in drudgery of women and children from cooking in smoky kitchens and collection of fuel wood;
- environmental upgradation and check on deforestation; and
- employment generation in rural areas.

Improved chullahs that do not emit smoke and use less wood, are fast replacing traditional cookstoves in rural areas ushering in what is described as *smokeless revolution* in different parts of the country. While the traditional chullahs have thermal efficiency of 8-10 per cent, the improved chullahs have minimum efficiency of 20-25 per cent. An improved chullah saves on an average about 375 kg of bio-fuel and 3 litres of kerosene per year under field conditions. Besides, an improved chullah results in a saving of 45 minutes to one hour per family which would otherwise be spent on collecting and processing the fuel material, cleaning of utensils and cooking. The improved chullah also helps in making the environment inside the kitchen smoke-free, thereby reducing the incidences of eye and lung diseases amongst women and children. The NPIC is also generating employment in rural areas for women at the rate of 0.3 person day per chullah.

A pilot project has been launched to test the efficiency and marketability of improved cook-stoves, so that the consumption of fire wood is reduced. Under the National Business Cookstoves, initiative, a pilot scale project for deployment for 5,500 biomass community cookstoves in Anganwadis, Mid-day meal scheme in schools, Tribal Hostels, etc. and demonstration for 15,000 family sized/portable cookstoves has been taken up.

The world's largest system for cooking in community kitchen has been installed at Shirdi to cook food for 20,000 people per day and is saving around 60,000 kg of LPG every year. All institutions including large institutions with hostels, hospitals/ medical colleges military/paramilitary establishments, industrial organisations, academies wherever large number of meals is cooked, are the targets.

Biomass Power. Agricultural and agro-industrial residues are collectively known as biomass. Biomass gasification is a thermochemical process in which fuel gas is formed as a result of partial combustion of biomass, such as wood waste, crop residues, agro-industrial wastes, etc. (or any organic material). The main advantage of the gasification technique is that it enables solid biomass to be converted into a more convenient and versatile fuel. The produced gas could either be burnt directly for thermal applications or be used for replacing diesel oil in dual-fuel engines for mechanical and electrical applications. Its relevance in today's world is greater than in the past as it provides an option to reduce Green House gas emissions because biomass can be CO_2 neutral in terms of emission. Gasification has vast potential for rural lighting purposes. The social and environmental benefits of biomass power for long-term sustainability have been accepted. The gasifier programme in India was launched in 1986 by the Ministry of New and Renewable Energy (MNRE) based on 3.7 kW wood based gasifiers.

MNRE reckons that over 500 million tonnes of biomass generated every year in the country has energy potential equivalent to 175 million tonnes of oil. It is also estimated that 150-200 million tonnes of biomass goes waste. This waste can generate 15,000 to 25,000 MW of electrical power at typically prevalent plant factor load. A recent report from the Central Statistics Office estimates the potential of biomass power at 17,538 MW. MNRE also believes that more than 70,000 MW electricity can be generated from biomass grown on wastelands, road sides and on plantations along railway tracks. It has been estimated that vast stretches a wasteland would be able to produce 400 million tonnes of fuel wood per year equivalent to 60,000 MW of power. The total electricity generation potential from biomass in India had been estimated at about 1,00,000 MW. The MNRE has initiated biomass programme with a view to increase fast growing short rotation fuel wood species, suitable for plantation under the given set of agro-climatic conditions. This aims at increasing the productivity to about 40 tonnes per hectare per year as compared to the average forest tree production rate of 0.5 tonne per hectare per year. Biomass yields ranging from 12 to 37 tonnes per hectare per year have been achieved by established Biomass Research Centres.

Biomass gasifier systems of upto 500 kW capacity based on fuel-wood have been developed indigenously and are being manufactured in the country. Similarly, the technology for producing biomass briquettes from agricultural residues and forest litter at both household and industrial levels has been developed. The exploitable potential for power generation from agro residues, agro-industrial residues (excluding bagasse) and forestry residues has been estimated at 16,000 MW on a conservative basis.

Be it a small-scale unit or a remote island, the biomass gasifier has come as a saviour cutting down on the consumption of precious fossil fuels on the one hand and bringing about economic development on the other. Thanks to the initiatives and programmes of the MNES, techno-economic viability of biomass gasifiers has been demonstrated and gasifier systems for varied applications have been deployed in different parts of the country. In the process India has emerged a world leader in the development and deployment of gasifier technology and systems.

Gasifier plants have been set up in Gujarat, Uttar Pradesh, Andhra Pradesh, Tamil Nadu, West Bengal, Nagaland and some other states. In certain areas gasifier plants have played a vital role for meeting the energy requirements of the people living in remote and inaccessible areas. It is difficult to extend grid electricity to Chhottomallakhali Island in the Sunderbans area of South 24 Parganas district of West Bengal due to prohibitive cost involved in crossing various rivers and creeks. The switching on of the 4 × 125 KW power plant based on biomass gasifier on 29th June, 2001 has completely changed the lives of inhabitants of this remote island. The plant is catering to the electricity needs of domestic, commercial and industrial users, drinking water supply, hospital, ice factory, etc.

The MNRE has planned to initiate the National Bioenergy Mission in association with state governments and other stakeholders to promote biomass related projects in the country. The mission sets targets for scaling up biomass energy in the country. Its overall target is 20,000 MW biomass projects by 2022.

Cogeneration

Cogeneration is the simultaneous production of power either electrical or mechanical and useful thermal energy from a single fuel source. A cogeneration system is an integration of various components (energy conversion system, balance-of-plant system, heat source, heat pump, etc.) into a total system which provides the electrical and thermal requirements of a specific industrial process.

There have been very few studies undertaken to estimate the potential for cogeneration in India. However, some surveys conducted in different industries have revealed some interesting facts. For example, nearly 30 per cent of the electricity used in the pulp and paper industry is cogenerated with steam production in paper mills. In fertilizer industry, almost all the energy requirements of the ammonia and urea plants are met with steam generated through high pressure waste heat-boilers. According to a survey of 300 industrial units, conducted by Tata Energy Research Institute (TERI) in 1993, there exists a cogeneration potential of 7,574 MW in the country (*see* **Table 24.27**)

TABLE 24.27. Cogeneration Potential Estimates by TERI

Industry	Potential	Per cent
Alumia	59	0.8
Caustic soda	394	5.2
Cement	78	1.0
Cotton textile	506	6.7
Iron & steel	362	4.8
Man-made fibre	144	1.9
Paper	594	7.8
Refineries	232	3.1
Sugar	5,131	67.7
Sulfuric Acid	74	1.0
Total	**7,574**	**100**

Source : Planning for the Indian Power Sector (1995), TERI, p. 117.

It is clear from the table that over two-thirds of the cogeneration potential is accounted for by sugar industry alone. Bagasse based cogeneration programme, launched in January 1994, envisaged creation of 300 MW power generation capacity during the Eighth Plan. According to a report from the Central Statistics Office, 5,000 MW power can be obtained from bagasse based cogneration. The Tata Energy Research Institution has estimated the cogeneration power potential based on bagasse at 5,100 MW. The programme is targeted at around 420 existing and around 90 new sugar mills. Already ten mills have commissioned co-generation projects having total exploitable power to the tune of 29 MW. This brings out the extent of untapped potential. As of January 2014, the installed capacity of bagasse base cogeneration was 2,513 MW. The potential is also significant in paper and textile industries. Considering the gains from cogeneration, the Government of India is giving several incentives for its development.

Small Hydropower

There is no globally accepted definition of a small hydropower plant (SHP) because different countries use different capacity limits for defining SHP. For example developed countries have a range of 1.5 to 2.0 MW for defining SHP whereas in developing countries this limit varies from 25 to 50 MW. The capacity limit for SHP is 50 MW in China, 30 MW in Brazil, 20 MW in Australia and European Union, 10 MW in Norway, 5 MW in UK and only 1.5 MW in Sweden. In India, however, hydropower projects with a capacity upto 25 MW are considered as SHP. Interestingly, in India, hydropower plants with a capacity of less than 25 MW are considered renewable power source; there is no justification available why hydropower plants with more than 25 MW are not considered renewable. Small hydropower plants have further been classified into micro (100 MW or below), mini (101 KW-2 MW), and small hydro (2.25 MW).

Generation of electricity from small sized hydropower sources is a low cost, environment friendly and renewable source of energy. Small and mini hydel projects have the potential to provide energy in remote and hilly areas where extension of grid system is un-economical. Although SHPs have become very popular and an important source of hydropower in India, this is one of the oldest form of harnessing power in the country. The first SHP of 130 KW was installed at Darjeeling in 1897. This was followed by SHP of 2 MW at Mysore in 1902, 3 MW plant at Mussoorie in 1907 and 1.75 MW plant at Chaba and 50 MW plant at Jubbal (Near Shimla) in 1930. Some of the plants are still working and contributing to hydropower generation of the country.

MNRE has estimated the potential of SHP at about 21,135.37 MW in 7135 sites spread all over the country. Arunachal Pradesh, Chhattisgarh, Himachal Pradesh, Jammu and Kashmir, Karnataka and Uttarakhand have maximum potential. The other states with reasonable potential are Andhra Pradesh, Assam, Kerala, Madhya Pradesh, Maharashtra, Uttar Pradesh and West Bengal.

Geothermal Energy

There are vast possibilities of developing and exploiting geothermal energy in India. About 340 hot spring localities have been identified; many of them have temperature nearing boiling point. Extensive surveys have been conducted to develop geothermal energy for direct heat and power generation. Assessment of Geothermal energy potential of selected sites in Jammu and Kashmir, Himachal Pradesh, Uttarakhand, Jharkhand and Chhattisgarh has been/is being undertaken. A five kW geothermal pilot power plant has been commissioned at Manikaran in Kullu district of Himachal Pradesh. A potential of 4-5 MW geo-thermal power has been estimated in Puga Valley of Ladakh. Plan to exploit this potential has been finalised. Geothermal power plant at Tattapani geothermal fields in Chhattisgarh has been commissioned. A project on mushroom cultivation and poultry farming using geothermal power is under implementation at Regional Research Laboratory, Jammu.

Tidal Energy

It is estimated that India possesses 8,000-9,000 MW of tidal energy potential. The Gulf of Khambhat is the best suited area with about 7,000 MW potential of tidal energy. This is followed by Gulf of Kachchh (1,000 MW) and Sunderbans (100 MW). A 900 MW tidal power plant is proposed to be set up in the Gulf of Kachchh.

Wave Energy

Wave energy potential in India is estimated at about 40,000 MW. Coastal areas can get special benefit from this energy. One wave energy power plant of 150 kW (maximum) has been installed at Vzhinjam near Thiruvananthapuram. Another one MW wave energy plant has been set up in the Andaman and Nicobar Islands.

Ocean Thermal Energy Conversion

India's ocean thermal energy is estimated at 50,000 MW. The first ever plant for ocean thermal energy conversion with a capacity of 100 MW is proposed to be set up off the coast of Tamil Nadu.

Energy from Waste

Generation of waste is inevitable in many human activities. In nature, a balance is maintained as the waste generated by animals and human beings serve as nourishment for the plant life. So long as this natural cycle is maintained, the question of pollution due to waste does not arise. However, with increasing industrialisation, urbanisation and changes in the pattern of life, which are inevitable in any process of economic growth, waste causes damage to environment unless it is very carefully managed. According to 2011 census figures, nearly 377 million people live in urban areas and generate 110,000 – 150,000 tonnes per day (TPD) of municipal solid waste (MSW). In 2012 the Ministry of New and Renewal Energy (MNRE) estimated 4,000 MW of potential. However, not much headway has been made due to environmental issues and financial constraints.

Energy Crisis

India is passing through a critical phase of energy crisis as demand for energy has always outpaced its supply. Demand for energy is increasing at an accelerated rate due to rapid increase in population, industrialization, urbanization and unprecedented increase in means of transport. Rise in living standard of the common man has further complicated the matter. The prevailing scenario of energy crisis is reflected in frequent power failures, declared and undeclared power cuts, low and fluctuating voltage and closure of power plants in different parts of the country. Most the power plants are working at below the installed capacity due to short supply of fuels like coal and gas and mismanagement of the production process. Large scale electric thefts and inefficient distribution system are other causes of energy crisis.

The only way out to solve the problem of energy crisis is to increase production and conserve energy.

A number of schemes have been launched to provide electricity to every citizen of India. Some of the important schemes, launched recently, are described as under :

1. Deen Dayal Upadhyaya Gramin Jyoti Yojana was launched in December, 2014 to extend financial assistance for capital expenditure by distribution companies (DICOMS) for strengthening distribution infrastructure, including metering, in rural areas. The scheme is being implemented by states and their DISCOMS with support from Central Government to the tune of 60 per cent in 'General Category' states and 85 per cent in 'Special Category' state.

2. Ujwal DISCOM Assurance Yojana (UDAY) was launched on November 20, 2015. The scheme envisages reduction in interest burden, cost of power and Aggregate Technical and Commercial losses. It is a voluntary scheme for participation, 26 states and one union territory have joined it. The scheme is being monitored by an inter-ministerial committee and a state level committee.

UDAY is a path breaking reform for affordable and accessible 24 × 7 power for all. This is through four initiatives (*i*) improving operational efficiencies of DISCOMS; (*ii*) reduction of cost of power; (*iii*) reduction in interest cost of DISCOMS; (*iv*) enforcing financial discipline on DISCOMS through alignment with state finances.

3. Saubhagya (Pradhan Mantri Sahaj Bijli Har Ghar Yojana) was launched in September, 2017 to ensure electrification of all remaining willing households in the country in rural and urban areas with an outlay of ₹ 16320 crore. The scheme envisages electrification of around 4 crore households that do not have electricity connection by March, 2019. For unelectrified households located in remote and inaccessible areas, solar photo voltaic based standalone systems with power packs of 200-300 Watt with battery back-up are to be provided.

Energy Conservation. Energy conservation assumes great significance in the face of power shortage and fast increasing demand for power. Lighting itself accounts for about 20 per cent of the total electricity consumption in the country. Some of the initiatives taken by the Government in this direction are as under :

- National LED programme for promoting use of most efficient lighting technology at affordable rates was launched in January, 2015. The programme includes two components (*a*) Unnat Jyoti by Affordable LED for All (UJALA) providing LED bulbs to domestic consumers with a target to replace 77 crore incandescent bulbs with LED bulbs and (*b*) Street Lighting National Programme (SLNP) to replace 1.34 crore conventional street lights with smart and energy efficient LED strict lights by March, 2019.
- Bureau of Energy Conservation is simultaneously taking up a number of programmes for energy conservation including standardisation and labelling of appliances, buildings, passenger cars and heavy duty vehicles, etc.

REFERENCES

Athawale, Suneel B. (1994) : *Wind Energy*, National Book Trust, New Delhi.
Bhandari, Neena. : *A powerhouse on the sands*, The Hindustan Times, May 8, 1996.
Bhatia, B. and Natrajan B. (1993) : *Review of performance of power sector in India.*
Chanderi, M.R. (1970) : *Power Resources of India*, Oxford & IBH Publication Co. Calcutta.
Energy Management Centre (1994) : *Policy Guidelines for Energy Efficiency Promotion in Asia.*
Government of India (1991) : *Report on Utilisation and Conservation of Energy*, Inter Ministerial Working Group.
—— **Economic Survey,** 2022-23
—— **EnviStats India 2018**
—— India, 2024, *A Reference Annual.*
Gupta, Y.P. *Warm and user friendly : Sunshine*, The Hindustan Times, Aug. 31, 1994.
Hande, H. Harish and Jose G., Martin : "*Photovoltaics for Rural Electrification in India*", in the Solar Energy, Society of Canada's 19th Annual Conference on Energy Efficiency and Solar Energy Conference Proceedings, July 9-13, 1993. Ottawa, Canada.
Kayastha, S.L. (1981) : *Energy problems in India — Socio Economic Aspects* in R.P. Chaturvedi *et al.* (ed.), *Socio-Economic Aspects of Energy*, Catland, New York.
Krishnamurthy, V. : *Managing Energy Supply Vital for Economy*, Hindustan Times, Feb. 10, 1992.
Kymal Ramesh (2004) : *Wind Power—A viable energy option,* The Hindu Survey of Indian Industry 2004, Kasturi and Sons, Chennai, pp. 120-122.
Ministry of Non-conventional Energy Sources (MNES) : Lodhi Road, New Delhi, Website http ://mnes.nic.in
Narshimhan, S.V. (2004) : *Petroleum Products—Coping with deregulation,* The Hindu Survey of Indian Industry (2004), Kasturi & Sons, Chennai, pp. 181-184.
Natarajan, B. and Bhatia, B. : *Electricity Savings : Adopt DSM*, Economic Times, 8th April, 1994.
Padmanabhan, B.S. (2004) : *Renewable Energy—Promotion lacks vigour,* The Hndu Survey of Indian Industry, Kasturi and Sons, Chennai, 2004, p. 116-119.
Patel, S.M. (1988) : "*Low Cost and Quick Yielding Measure for Energy Conservation in agriculture Pumps*", Pacific and Asian Journal of Energy, Vol. 2. No. 1.
Planning for the Indian Power Sector : *Environmental and Development Considerations* (*June 1995*), Canadian Energy Research Institute, Alberta (Canada) and Tata Energy Research Institute (TERI), New Delhi.
Ramakrishnan, P. : *Energy Crisis and Possible Solutions*, The Hindustan Times, Feb. 16, 1992.
Ratnam, C. : *Planning Future energy needs.* H.T. Feb 10, 1992.
Raza Moonis (ed.) (1989) : *Renewable Resources for Regional Development*, New Delhi.
Razdan, Deepak : *Power—A Major issue before Indian economy*, The Hindustan Times, Jan. 29, 1996.
Rehman, Hifzm (1985) : *Trends in Energy Consumption in India*, The Geography, Vol. xxxii (I), pp. 36-51.
Rudra, P.K. : *Recycling Energy Waste*, The Hindustan Times, Feb. 10, 1992.
Sachidanand, N.N. (2004) : *Coal—Warrisome Supply days,* The Hindu Survey of Indian Industry, Kasturi and Sons, Chennai, 2004, p. 206-208.
Sankar, N. : *Awareness must for cutting Consumption*, The Hindustan Times, Feb. 10, 1992.
Sawhney, M. Dhruv : *Vast Scope for Conservation*, The Hindustan Times, Feb 10, 1992.
Sengupta, S. (1988) : *Problems and Potentials of Non-conventional Energy Resources in India*, NATMO, Monograph 8, Calcutta.
Singh, H.B. : *And now, Waste into Fuel Pellets.* The Hindustan Times, February 23, 1995.
Singh, R.M. (ed.) (1987) : *Coal Resources of* India, Department of Geology, BHU, Varanasi.
Singh, S.D. (2004) : *Renewable Energy—Lack of policy stalls growth,* The Hindu Survey of Indian Industry, Kasturi & Sons, Chennai, 2004, pp. 123-124.
Sridhar, K and Terway, A. (1991) : *Coal Use for Power Generation in India—Needs and Options*, in the International Conference on Coal, the Environment and Development : Technologies to Reduce Greenhouse Gas Emissions, Paris, France, International Energy Agency.
Srinivasan, Raghuvir (2004) : *Oil and Gas—It is turnaround time,* The Hindu Survey of Indian Industries, Kasturi and Sons, Chennai 2004, pp. 175-179.
State of Renewable Energy in India : A Citizen's Report (2014), Centre for Science and Environment, New Delhi.
The Hindu Survey of Industry, 2013.
Vasant, N.S. and Aggarwal, S.K. : *Rational Power Policy Needed.* The Hindustan Times, Jan 25, 1995.
Venkatasubramaniam, K.N. : *MRTS only solution to oil crisis.* The Hindustan Times, Feb. 10, 1992.
Wadia, Mehar D.N. (1994) : *Minerals of India*, National Book Trust, New Delhi, India.
World Bank : *"Energy Efficiency and Conservation in the Developing World" (1993).*

◆◇◆

CHAPTER 25

Manufacturing Industries

INTRODUCTION

Manufacturing is the processing of primary products into more refined and usable products. Many of the natural resources cannot be utilised directly without processing. It is because of this reason that we manufacture cloth from cotton, sugar from sugarcane, paper from wood pulp, and petro-chemicals from mineral oil. By doing so, we make the primary products more valuable and usable. In other words, *manufacturing means transformation of natural material endowments into commodities of utility by processing, assembling and repairing*.

Manufacturing is vital for our very existence. It is an activity that works as an *engine of economic growth,* helps in removing poverty, and unemployment and transforms a traditional society into a modern one. The economic strength of a country is judged by the development of its industries. All the developed countries of the world are highly industrialised.

FACTORS INFLUENCING THE LOCATION OF INDUSTRIES

Many important geographical factors involved in the location of individual industries are of relative significance, *e.g.*, availability of raw materials, power resources, water, labour, markets and the transport facilities. But besides such purely geographical factors influencing industrial location, there are factors of historical, human, political and economic nature which are now tending to surpass the force of geographical advantages. Consequently, the factors influencing the location of industry can be divided into two broad categories *i.e.*

(I) Geographical factors, and

(II) Non-geographical factors.

I. Geographical Factors

Following are the important geographical factors influencing the location of industries.

1. Raw Materials. The significance of raw materials in manufacturing industry is so fundamental that it needs no emphasising. Indeed, the location of industrial enterprises is sometimes determined simply by location of the raw materials. Modern industry is so complex that a wide range of raw materials is necessary for its growth. Further we should bear in mind that finished product of one industry may well be the raw material of another. For example, pig iron, produced by smelting industry, serves as the raw material for steel making industry. Industries which use heavy and bulky raw materials in their primary stage in large quantities are usually located near the supply of the raw materials. It is also true in the case of raw materials which lose weight in the process of manufacture or which cannot bear high transport cost or cannot be transported over long distances because of their perishable nature. This has been recognised since 1909 when **Alfred Weber** published his theory of location of industry. The jute mills in West Bengal, sugar mills in Uttar Pradesh, cotton textile mills in Maharashtra and Gujarat are concentrated close to the sources of raw materials for this very reason. Industries like iron and steel, which use very large quantities of coal and iron ore, losing lot of weight in the process of manufacture, are generally located near the sources of coal and iron ore.

Some of the industries, like watch and electronics industries, use very wide range of light raw materials and the attractive influence of each separate material diminishes. The result is that such industries are often located with no reference to raw materials and are sometimes referred to as **'footloose industries'** because a wide range of locations is possible within an area of sufficient population density.

2. Power. Regular supply of power is a pre-requisite for the localisation of industries. Coal, mineral oil and hydro-electricity are the three important conventional sources of power. Most of the industries tend to concentrate at the source of power. The iron and steel industry which mainly depends on large quantities of coking coal as source of power are frequently tied to coal fields. Others like the electro-metallurgical and electro-chemical industries, which are great users of cheap hydro-electric power, are generally found in the areas of hydro-power production, for instance, aluminium industry.

As petroleum can be easily piped and electricity can be transmitted over long distances by wires, it is possible to disperse the industry over a larger area. Industries moved to southern states only when hydro-power could be developed in these coal-deficient areas. Thus, more than all other factors affecting the location of large and heavy industries, quite often they are established at a point which has the best economic advantage in obtaining power and raw materials. Tata Iron and Steel Plant at Jamshedpur, the new aluminium producing units at Korba (Chhattisgarh) and Renukoot (Uttar Pradesh), the copper smelting plant at Khetri (Rajasthan) and the fertilizer factory at Nangal (Punjab) are near the sources of power and raw material deposits, although other factors have also played their role.

3. Labour. No one can deny that the prior existence of a labour force is attractive to industry unless there are strong reasons to the contrary. Labour supply is important in two respects (*a*) workers in large numbers are often required; (*b*) people with skill or technical expertise are needed. **Estall and Buchanan** showed in 1961 that labour costs can vary between 62 per cent in clothing and related industries to 29 per cent in the chemical industry; in the fabricated metal products industries they work out at 43 per cent.

In our country, modern industry still requires a large number of workers in spite of increasing mechanisation. There is no problem in securing unskilled labour by locating such industries in large urban centres. Although, the location of any industrial unit is determined after a careful balancing of all relevant factors, yet the light consumer goods and agro-based industries generally require a plentiful of labour supply.

4. Transport. Transport by land or water is necessary for the assembly of raw materials and for the marketing of the finished products. The development of railways in India, connecting the port towns with their hinterlands determined the location of many industries around Kolkata, Mumbai and Chennai. As industrial development also furthers the improvement of transport facilities, it is difficult to estimate how much a particular industry owes to original transport facilities available in a particular area.

5. Market. The entire process of manufacturing is useless until the finished goods reach the market. Nearness to market is essential for quick disposal of manufactured goods. It helps in reducing the transport cost and enables the consumer to get things at cheaper rates. It is becoming more and more true that industries are seeking locations as near as possible to their markets; it has been remarked that market attractions are now so great that a market location is being increasingly regarded as the normal one, and that a location elsewhere needs very strong justification. Ready market is most essential for perishable and heavy commodities. Sometimes, there is a considerable material increase in weight, bulk or fragility during the process of manufacture and in such cases industry tends to be market oriented.

6. Water. Water is another important requirement for industries. Many industries are established near rivers, canals and lakes, because of this reason. Iron and steel industry, textile industries and chemical industries require large quantities of water, for their proper functioning. Significance of water in industry is evident from **Table 25.1.** Also it requires 36,400 litres of water to produce one kwh of thermal electricity. Further, it is worth noting that water used in industries often gets polluted and is therefore not available for any other purpose.

TABLE 25.1. Requirement of Water in Industry

Name of the industry	Amount of water required in litres/tonne
Steel	300,000
Sulphite paper	290,000
Oil refining	25,600
Rayon	1,000,000
Paper from wood	173,000

Source : H.R. Jarrett, A Geography of Manufacturing (1977).

7. Site. Site requirements for industrial development are of considerable significance. Sites, generally, should be flat and well served by adequate transport facilities. Large areas are required to build factories. Now, there is a tendency to set up industries in rural areas because the cost of land has shot up in urban centres.

8. Climate. Climate plays an important role in the establishment of industries at a place. Harsh climate is not much suitable for the establishment of industries. There can be no industrial development in extremely hot, humid, dry or cold climate. The extreme type of climate of north-west India hinders the development of industries. In contrast to this, the moderate climate of west coastal area is quite congenial to the development of industries. Because of this reason, about 24 per cent of India's modern industries and 30 per cent of India's industrial labour is concentrated in Maharashtra-Gujarat region alone. Cotton textile industry requires humid climate because thread breaks in dry climate. Consequently, majority of cotton textile mills are concentrated in Maharashtra and Gujarat. Artificial humidifiers are used in dry areas these days, but it increases the cost of production.

II. Non-Geographical Factors

Now-a-days alternative raw materials are also being used because of modern scientific and technological developments. Availability of electric power supply over wider areas and the increasing mobility of labour have reduced the influence of geographical factors on the location of industries. The non-geographical factors are those including economic, political, historical and social factors. These factors influence our modern industries to a great extent. Following are some of the important non-geographical factors influencing the location of industries.

1. Capital. Modern industries are capital-intensive and require huge investments. Capital is usually available in urban centres. Big cities like Mumbai, Kolkata, Delhi, Chennai are big industrial centres, because the required capital is available in these cities.

2. Government Policies. Government activity in planning the future distribution of industries, for reducing regional disparities, elimination of pollution of air and water and for avoiding their heavy clustering in big cities, has become no less an important locational factor. There is an increasing trend to set up all types of industries in an area, where they derive common advantage of water and power and supply to each other the products they turn out. The latest example in our country is the establishment of a large number of industrial estates all over India even in the small-scale industrial sector.

It is of relevance to examine the influence of India's Five Year plans on industrial location in the country. The emergence of suitable industries in south India around new nuclei of public sector plants and their dispersal to backward potential areas has taken place due to Government policies.

The state policy of industrial location has a greater hand in the establishment of a number of fertiliser factories, iron and steel plants, engineering works and machine tool factories including railway, shipping, aircraft and defence installations and oil refineries in various parts in the new planning era in free India. We may conclude by noting that the traditional explanation of a location of industry at a geographically favourable point is no longer true. Location of oil refinery at Mathura, coach factory at Kapurthala and fertiliser plant at Jagdishpur are some of the results of government policies.

3. Industrial Inertia. Industries tend to develop at the place of their original establishment, though the original cause may have disappeared. This phenomenon is referred to as inertia, sometimes as *geographical inertia* and sometimes *industrial inertia*. The lock industry at Aligarh is such an example.

4. Efficient Organisation. Efficient and enterprising organisation and management is essential for running modern industry successfully. Bad management, sometimes squanders away the capital and puts the industry in financial trouble leading to industrial ruin. Bad management does not handle the labour force efficiently and tactfully, resulting in labour unrest. It is detrimental to the interest of the industry. Strikes and lock-outs lead to the closure of industries. Hence, there is an imperative need of effective management and organisation to run the industries.

5. Banking Facilities. Establishment of industries involves daily exchange of crores of rupees which is possible through banking facilities only. So the areas with better banking facilities are better suited to the establishment of industries.

6. Insurance. There is a constant fear of damage to man, machines and materials in industries for which insurance facilities are badly needed.

EVOLUTION OF INDUSTRIES IN INDIA

Indian industries were in a very backward state before Independence. The British rulers encouraged only those industries which were beneficial for them. In 1947, India got Independence but the Indian subcontinent was divided into India and Pakistan. This shattered the entire economic structure of the country and industries were badly affected. The Industrial Policy of 1948 gave some impetus to industries in India.

Planning era started immediately after Independence and more emphasis was laid on the industrial development. A brief account of industrial growth during the planning period is given below :

First Five-Year Plan (1951-56). This plan commenced just four years after Independence. More emphasis was laid on agriculture than on industries because India was facing acute shortage of foodgrains. Therefore it was decided that capacity of the then existing industries should be increased rather than establishing new industries. Cotton, woollen and jute textiles, sugar, *vanaspati*, cement, paper, news paper, power-looms, medicines, paints, chemical and engineering goods and transport equipments showed some progress. Although the main emphasis during this plan period was on agriculture, industries recorded 40% growth as against only 30% recorded by agriculture. In fact the foundation for industrial growth was laid in the First Five-year Plan itself.

Second Five-Year Plan (1956-61). Basic and heavy industries were given priority during this plan period. The second industrial policy was announced on 20th April 1956. The main thrust of industrial development was on iron and steel, heavy engineering, lignite projects and fertiliser industries. Moreover, there was emphasis on expansion of existing steel plants, like Jamshedpur, Kulti-Burnpur and Bhadrawali. Three new iron and steel plants were located at Durgapur, Rourkela and Bhilai. The Chittaranjan Locomotive Workshop, The Hindustan Ship-building Yard at Vishakhapatnam, the Sindri Fertiliser Factory and Hindustan Mechanic Tools (HMT) at Bangalore were expanded. Two new fertiliser plants were established at Nangal and Rourkela.

Third Five-Year Plan (1961-66). There was emphasis on the expansion of basic industries like iron and steel, fossil-fuel, power and machine building. Three more HMT units were established making a total of five units. Machine building, locomotive and railway coach making, ship-building, air-craft manufacturing, chemicals, drugs, and fertiliser industries also made steady progress.

Annual Plans. The situation created by Indo-Pakistan conflict in 1965, two successive years of severe drought devaluation of the currency, general rise in prices and erosion of resources available for plan purposes delayed the finalisation of the Fourth Five-year Plan. Instead, between 1966 and 1969, three Annual Plans were formulated within the framework of the draft outline of the Fourth Plan. Industrial growth index increased by 1.7 and 0.3%

during 1966-67 and 1967-68 respectively. Things improved a bit in 1968-69 and the index increased by 7%.

Fourth Five-Year Plan (1969-74). Economy was just trying to come out of the economic depression and saw many ups and downs. The highest growth of 7.3% was recorded in 1969-70 while the year 1973-74 saw zero growth. During this plan, there was much emphasis on the agro-based industries such as sugar, cotton, jute *vanaspati*, metal-based and chemical industries. It was during this plan when much progress was made in alloys, aluminum, automobile-tyres, electronic goods, machine tools, tractors, and special steel. Efforts were also made to accelerate the process of industrial dispersal.

Fifth Five-Year Plan (1974-79). The main stress in this plan was on rapid growth of steel plants, export-oriented articles, and goods of mass consumption. The annual growth rate was 8.2 per cent. The steel plan at Salem, Vijaynagar, and Vishakhapatnam were proposed to create additional capacity. The Steel Authority of India (SAIL) was constituted. Moreover, drug manufacturing, oil refining, chemical fetilisers, and heavy engineering industries made steady progress.

Sixth Five-Year Plan (1980-85). The main emphasis on the Sixth Five-Year Plan was on producing goods to exploit the domestic and international markets. To achieve this objective industries like aluminium, automobiles, electric equipments, thermostatics were given the priority. Production targets were achieved in industries like commercial vehicles, drugs, T.V. receivers, automobiles, cement, coal, jute industry, non-ferrous metals, textiles, railway wagons, sugar industry, etc.

Seventh Five-Year Plan (1985-90). The main thrust of the Seventh Five-Year Plan was on 'high tech' and electronic industries. Industrial dispersal, self employment, exploitation of local resources, and proper training were the preference areas of the plan.

Eighth Annual Plans. The Eighth Five-Year Plan could not take off due to fast-changing political situation at the centre. Consequently 1990-91 and 1991-92 were treated as separate Annual Plans during which not much industrial growth was reported.

Eighth Five-Year Plan (1992-1997). New Industrial Policy was declared in 1991 and the policy of liberalisation was adopted for investment by foreign multinationals. The plan aimed at an average industrial growth rate of about 7.5 per cent. Removal of restrictions on import, reduction of regional disparities and role of public sector, as well as improving prospects of employment were the main objectives of this plan. The growth rate was slow in the beginning but packed up later on. The main industries recorded a growth rate of 12% in 1995-96.

Ninth Five-Year Plan (1997-2002). The main emphasis during this plan was on cement, coal, crude oil, consumer goods, electricity, infrastructures, refinery, and quality steel products. Industrial growth slowed down. The growth rate of six basic industries *viz*. electricity, mineral oil, refining, coal, steel and cement, reduced from 9.1% in 1999-2000 to 5.3% in 2001-02. But software industry progressed at a rapid pact.

Tenth Five-Year Plan (2002-07). During the Tenth Five-Year Plan, the main emphasis was on :

(*i*) the modernisation, technology upgrading, reducing transaction costs, and increased export;

(*ii*) to enhance export and to increase global competitiveness; and

(*iii*) to achieve balanced regional development.

To achieve these objectives and reduce the regional inequalities Special Economic Zones (SEZs) policy was adopted. Moreover, industrial packages had been given to Himachal Pradesh, Jammu and Kashmir, and Uttarakhand. In addition to these, production of basic foods, capital goods, intermediate goods, and consumer goods had been encouraged.

Eleventh Five-year Plan (2007-12). This plan provided a comprehensive strategy for inclusive development, building on growing strength of economy, while also addressing the weaknesses that had surfaced. It act a target for 9 per cent growth in five-year period with acceleration during the period to reach 10 per cent by the end of the plan.

Index of Industrial Production IIP was adopted with 2004-05 as the base year. Main industries were included in this index. More important industries were mining, construction, electricity, capital goods and consumer goods. Along with other industries, these industries progressed a lot between 2007-08 and 2011-12. But the economic recession of 2008-09 caused heavy loss to a large number of industries. However, most of the industries soon recovered from the economic recession and recorded reasonably good growth rate.

Twelfth Five-Year Plan (2012-17). This plan recognised that the objective of development is broad-based improvement in economic and social conditions of our people. The Approach Paper to Twelfth Plan, had set a target of 9 per cent average growth of GDP over the Plan Period (2012-17). The plan envisaged that the current slow down in GDP growth can be reversed through strong corrective measure.

CLASSIFICATION OF INDUSTRIES

Industries can be classified into several groups. The following table gives an understanding about them.

I. On the Basis of Strength of Labour

1. Large Scale Industry. Industries which employ a large number of labourers in each unit are called large-scale industries. Cotton and jute textile industries are large scale industries.

2. Medium Scale Industries. The industries which employ neither very large nor very small number of labourers are put in the category of medium scale industries. Cycle industry, radio and television industries are some examples of medium scale industries

3. Small Scale Industries. Industries which are owned and run by individuals and which employ a small number of labourers are called small scale industries.

II. On the Basis of Raw-Material and Finished Goods

Industries classified on the basis of raw materials and finished goods are :

1. Heavy Industries. Industries which use heavy and bulky raw-materials and produce products of the same category are called ***heavy*** industries. Iron and steel industry presents a good example of heavy industries.

2. Light Industries. The light industries use light raw-materials and produce light finished products. Electric fans, sewing machines are light industries.

III. On the basis of Ownership

Since the start of the planned development of Indian economy in 1951, industries are divided in the following four classes :

1. Private Sector Industries. Industries owned by individuals or firms such as Bajaj Auto or TISCO situated at Jamshedpur are called private sector industries.

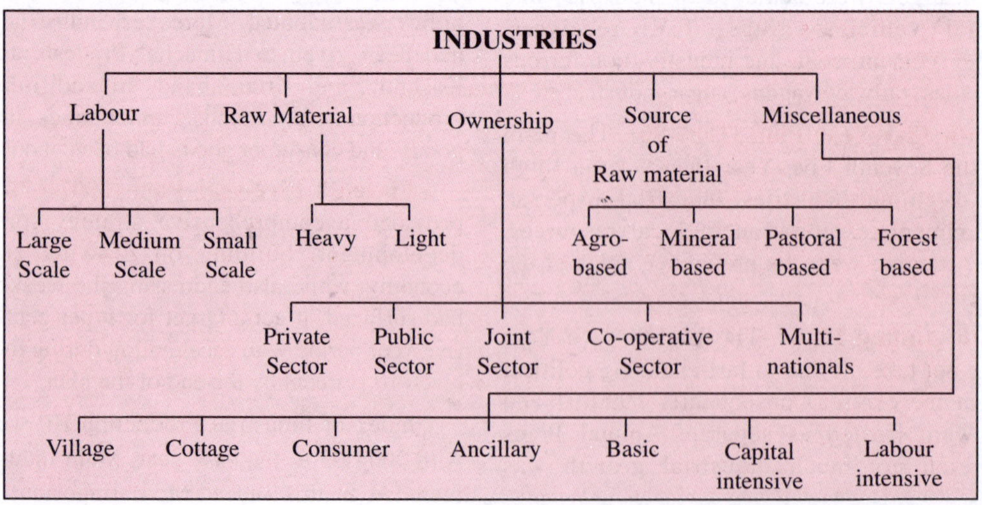

2. Public Sector Industries. Industries owned by the State and its agencies like Bharat Heavy Electricals Ltd., or Bhilai Steel Plant or Durgapur Steel Plant are public sector industries.

3. Joint Sector Industries. Industries owned jointly by the private firms and the state or its agencies such as Gujarat Alkalies Ltd., or Oil India Ltd. fall in the group of joint sector industries.

4. Co-operative Sector Industries. Industries owned and run co-operatively by a group of people who are generally producers of raw materials of the given industry such as a sugar mill owned and run by farmers are called co-operative sector industries.

5. Multinationals. Industries owned and seen by more than one country are called multinational industries.

IV. On the Basis of Source of Raw Material

On the basis of source of raw materials, industries are classified as under :

1. Agro Based Industries. Agro based industries are those industries which obtain raw-material from agriculture. Cotton textile, jute textile, sugar and vegetable oil are representative industries of agro-based group of industries

2. Mineral Based Industries. The industries that receive raw materials primarily from minerals such as iron and steel, aluminium and cement industries fall in this category.

3. Pastoral-Based Industries. These industries depend upon animals for their raw material. Hides, skins, bones, horns, shoes, dairy, etc. are some of the pastoral-based industries.

4. Forest Based Industries. Paper card-board, lac, rayon, resin, tanning of leather, leaf-utensils, basket industries are included in this type of industries.

V. Miscellaneous Industries

Industries are also classified into the following miscellaneous categories.

1. Village Industries. Village industries are located in villages and primarily cater to the needs of the rural people. They usually employ local machinery such as oil extraction, grain grinding and agricultural implements.

2. Cottage Industries. Industries which artisans set-up in their own houses, work with wood, cane, brass, stone, etc. are called cottage industries. Handloom, khadi and leather work at the artisans' home fall in this category.

3. Consumer Goods Industries. Consumer industries convert raw materials or primary products into commodities directly used by the people. Ready made garments, bakery products, sugar, etc. are some of the consumer goods industries.

4. Ancillary Industries. The industries which manufacture parts and components to be used by big industries for manufacturing heavy articles like trucks, buses, railway engines, tractors, etc. are called ancillary industries.

5. Basic Industries. Industries on which depend many other industries for their manufacturing processes are called basic industries. Iron and steel industry and power generating industry are included in this category.

6. Capital-Intensive Industries. Industries requiring huge investments are called capital-intensive industries. Iron and steel, cement and aluminium are outstanding examples of capital-intensive industries.

7. Labour-Intensive Industries. Industries which require huge labour force for running them are called labour-intensive industries. In these industries, labour is more important than capital. Shoe-making and bidi-manufacturing, etc. are included in these industries.

TEXTILE INDUSTRIES

Textile is a broad term which includes cotton, jute, wool, silk and synthetic fibre textiles. The Indian textile industry is the second largest manufacturer and exporter in the world. It contributes 7 per cent to manufacturing, 2.0 per cent to GDP and 15 per cent of the country's export earnings. The textile sector occupies a significant place in terms of employment generation. This sector is the biggest employer after agriculture, employing 4.5 crore people directly and another 6 crore people in allied sectors. Apparel also plays a critical role in improving social dynamics as mostly women are employed in this sector. About all the backward linkages of the sector to the rural economy give huge opportunity to millions of farmers, artisans, handloom and handicraft manufacturers. The sectors like handlooms, handicrafts, powerlooms and readymade garments are specially known for their employment potential. It is the only industry which is self reliant from raw material to the highest value added products *viz.* garments/made up. The sector is perfectly aligned with Government's key initiatives *viz.* Make in India, Skill India, Women Empowerment and Rural Youth Employment.

Table 25.2 gives details of production of various textiles in India.

COTTON TEXTILE INDUSTRY

Growth and Development

India held world monopoly in the manufacturing of cotton textiles for about 3,000 years from about B.C. 1500 to A.D. 1500. In the middle ages, Indian cotton textile products were in great demand in the Eastern and European markets. The muslins of Dhaka, chintzes of Masulipatnam, calicos of Calicut, baftas of Cambay and gold-wrought cotton piece goods of Burhanpur, Surat and Vadodara acquired a worldwide celebrity by virtue of their quality and design. This industry could not survive in the face of strong competition from the modern mill industry of Britain which provided cheap and better goods as a result of Industrial Revolution in that country. Moreover, the British textile industry enjoyed political advantage at that time.

The first modern cotton textile mill was set up in 1818 at Fort Glaster near Kolkata. But this mill could not survive and had to be closed down. The first successful modern cotton textile mill was established at Mumbai in 1854 by a local Parsi entrepreneur **C.N. Dewar**. Shahpur mill in 1861 and Calico mill in 1863 at Ahmedabad were other landmarks in the development of Indian cotton textile industry. The real expansion of cotton textile industry took place in 1870's. By 1875-76 the number of mills rose to 47 of which over 60 per cent were located in Mumbai city alone. The industry continued to progress till the outbreak of the First World War in 1914. The total number of mills reached 271 providing employment to about 2.6 lakh persons. The First World War, the Swadeshi Movement and the grant of fiscal protection favoured the growth of this industry at a rapid pace. Demand for cloth during the Second World War led to further progress of the industry. Consequently, the number of mills increased from 334 in 1926 to 389 in 1939 and 417 in 1945. Production of cloth also increased from 4,012 million yards in 1939-40 to 4,726 million yards in 1945-46.

The industry suffered a serious setback in 1947 when most of the long staple cotton growing areas went to Pakistan as a result of partition. However, most of the cotton mills remained in India. Under such circumstances, India faced a severe crisis of obtaining raw cotton. The country had, therefore, to resort to large-scale imports of long staple cotton which was an extremely difficult task in view of the limited foreign exchange reserves. The only solution to this problem was to increase production of long staple cotton within the country. This goal was achieved to a great extent in the post partition era.

TABLE 25.2. Estimated Production of Man-made Fibre, Filament Yarn, Spun Yarn and Cloth

Period	Man-made fibre (million kg.)	Man-made filament yarn (million kg.)	Cotton yarn (million kg.)	Blended and 100% Non-cotton yarn (million kg.)	Total Spun yarn (million kg.)	Cloth		
						Mill sector (million sq. mt.)	Decentralized sector (million sq. mt.)	Grand Total (million sq. mt.)
2015-16	1347	1164	4138	1527	5665	2315	62269	64584
2016-17	1364	1159	4055	1604	5659	2264	61216	63480
2017-18	1319	1187	4064	1616	5680	2157	6468	66845
2018-19	1443	1159	4182	1680	5862	2012	68034	70046

Source : Economic Survey 2019-20, p. 231.

Present Position

At present, cotton textile industry is the largest organised modern industry of India. There has been a phenomenal growth of this industry during the last five decades. About 16 per cent of the industrial capital, 14 per cent of indutrial production and over 20 per cent of the industrial labour of the country is engaged in this industry. There are at present 1,719 textile mills in the country, out of which 188 mills are in public sector, 147 in cooperative sector and 1,384 in private sector. About three-fourths are spinning mills and the remaining one-fourth composite mills. Apart from the mill sector, there are several thousand small factories comprising 5 to 10 looms. Some of them have just one loom. These are based on conventional handloom in the form of cottage industry and comprise decentralised sector of this industry.

Production. Cotton cloth is produced in three different sectors *viz*. 1. Mills, 2. Power-looms and 3. Handlooms.

1. Mills. The mill sector played a dominant role in cotton textile industry at the initial stage. But its importance was reduced drastically with the growth of powerlooms and handlooms. The share of mill sector in cotton cloth production came down from 80.69 per cent in 1950-51 to only 3.5 per cent in 2020-21.

In spite of steep fall in its relative significance during the recent years, the mill sector still plays a significant role in the overall scenario of the industry. This is an organised sector which has more than 3,400 big and small textile mills both in small scale industry and non-small scale industry sectors. The total installed capacity of spindles is the second largest in the world with more than 50 million spindles and 8,42,000 rotors.

2. Powerloom Sector. The decentralized powerloom sector is one of the most important segments of the textiles industry in terms of fabric production and employment generation. India had nearly 24.86 lakh powerlooms as on December 31, 2017. However, the number of modern shuttleless looms is only 1.25 lakh which accounts for only 5 per cent of the total powerlooms in the country. This sector of textile industry provides direct/indirect employment to 61.12 lakh persons and contributes about 57 per cent to the total cloth production. This sector produces almost all varieties of cloth including technical textiles. More than 60 per cent of the fabric meant for export is also sourced from powerlooms sector. The ready-made garments and home textile sectors are heavily dependent on this sector to meet their fabric requirements. The major powerloom clusters are at Bhiwandi, Erode, Salem, Madurai, Ichalkaranji, Solapur, Burchanpur, Bhilwara, Kishangarh, Ludhiana, Amritsar and Panipat. The Government is also developing four handloom clusters, *i.e.*, Bhiwandi, Erode, Bhilwara and Ichalkaranji. The Government has also set-up 44 Powerloom Service Centres in different clusters for solving the problems of the weavers and for improving the technology level, skill development, design development, etc. The main emphasis is on Technology Upgradation Fund Scheme (TUFS) which has given much needed impetus to this sector. With a view to enhance quality and productivity of the powerloom, the Govt. of India launched implementing the INSITU upgradation of plain powerloom as part of Power Tax India under which plain powerlooms are attached with process control equipment leading to higher productivity, better quality and more than 50 per additional value realisation.

In spite of the above mentioned various steps taken by the government, it pains to point out that 75 per cent of the shuttle rooms are obsolete and outdated with vintage of more than 15 years and have virtually no process or quality control devices/ attachments.

3. Handlooms. Handloom symbolizes rich heritage and ethos of vibrant Indian culture. Due to the uniqueness and exclusivity of designs, capability to produce small batch sizes and having eco-friendly fabric, handloom products are in the high demand both in the domestic and international markets. This sector has made significant progress in the recent past with respect to quantity, quality and variety of its products. With about 15 per cent share of handlooms in textile production and major contribution to export earnings, this sector provides employment to 43.31 lakh persons across the country. The fabric production which was witnessing downward trend due to economic slowdown in 2008 made a quick recovery in the post-slowdown period. The export of handloom fabric also staged a smart recovery in the same period. Five Indian Institutes of Handloom Technology (IIHTs) at Varanasi, Salem, Guwahati, Jodhpur and Bargarh and 25 Weavers' Service Centres (WSCs) across the country are working to impart technical knowledge to weavers.

Locational Factors

Several factors, like availability of raw cotton, market, transport, etc. play a key role in the localisation of cotton textile industry. The significance of raw cotton is evident from the fact that 80 per cent of the industry is coterminous with the cotton growing tracts of the country. Some of the important centres such as Ahmedabad, Solapur, Nagpur, Coimbatore and Indore are located in the areas of large scale cotton cultivation. Mumbai is also not far away from the cotton producing areas of Maharashtra and Gujarat which have contributed a good deal in the localisation and growth of cotton textile industry here. It is equally important to note that cotton is a **pure raw material**, in the sense that it does not lose much of its weight in the process of manufacturing and the slight loss in weight is more than compensated by the use of sizing materials. There is not much of difference between the cost of transporting raw cotton and finished cloth. Both can be transported with equal ease and without adding much to the total cost of production. Hence, this industry normally tends to be located at such centres which have favourable transport facilities with respect to market. In other words, it is now primarily a market oriented industry. With tropical and sub-tropical climate, all parts of India provide vast market potential for cotton textile industry. West Bengal, Bihar, Uttar Pradesh, Kerala and Odisha do not grow cotton and still have large number of big centres where cotton textile industry has flourished well. Thus although in earlier stages of industrialisation, cotton textile manufacturing was concentrated in Mumbai, it has witnessed great spatial spread and now covers almost the entire country. Since, it was a traditional cottage industry, cheap and skilled labour was readily available. The most notable feature of the distribution of the industry is that even within a state, the industry is localised within particular areas and regions, almost to the complete exclusion of others. Dispersal of industry from the old nuclei started after 1921 with railway lines penetrating into the peninsular region. New centres like Coimbatore, Madurai, Bengaluru, Nagpur, Indore, Solapur and Vadodara were more favourably located in respect to raw material, market and labour than places of original locations. This industry also reached some places with some additional advantages, such as nearness to coal (Nagpur), financial facilities (Kanpur) and wide market with port facilities (Kolkata).

Dispersal of cotton textile industry was further boosted with the development of hydroelectricity. The growth of this industry in Coimbatore, Madurai and Tirunelveli is largely due to the availability of hydroelectricity from Pykara dam. The industry also tended to shift from areas of high labour cost to those with low labour cost. The labour cost factor played a crucial role in establishing this industry at Madurai, Turunelveli and Coimbatore.

Distribution

Although cotton textile mills are located in over 100 towns and cities of India, yet its larger concentration is found in Maharashtra, Gujarat, West Bengal and Uttar Pradesh. **Figure 25.1** shows the spatial distribution of cotton textile industry in India.

Maharashtra. Maharashtra had been occupying the top position with respect to cotton textile industry for a pretty long time. This state used to produce nearly 40% mill cloth and about 11% yarn of India. About three lakh workers were engaged in this industry. Mumbai had been holding the prestigious position with as much as 65 mills out of a total of 122 mills in Maharashtra. It was rightly called the *Cottonpolis* of India. But it has lost much of its sheen as most the mills have been closed on account of various reasons including labour unrest, and requirement of space for more useful purposes. Today, Mumbai looks like a *grave yard* of cotton textile industry and is in a dormant position. Following factors had favoured the development of cotton textile industry in Mumbai and its contiguous areas.

(*i*) Mumbai enjoys humid climate which is helpful for this industry because thread does not break so frequently.

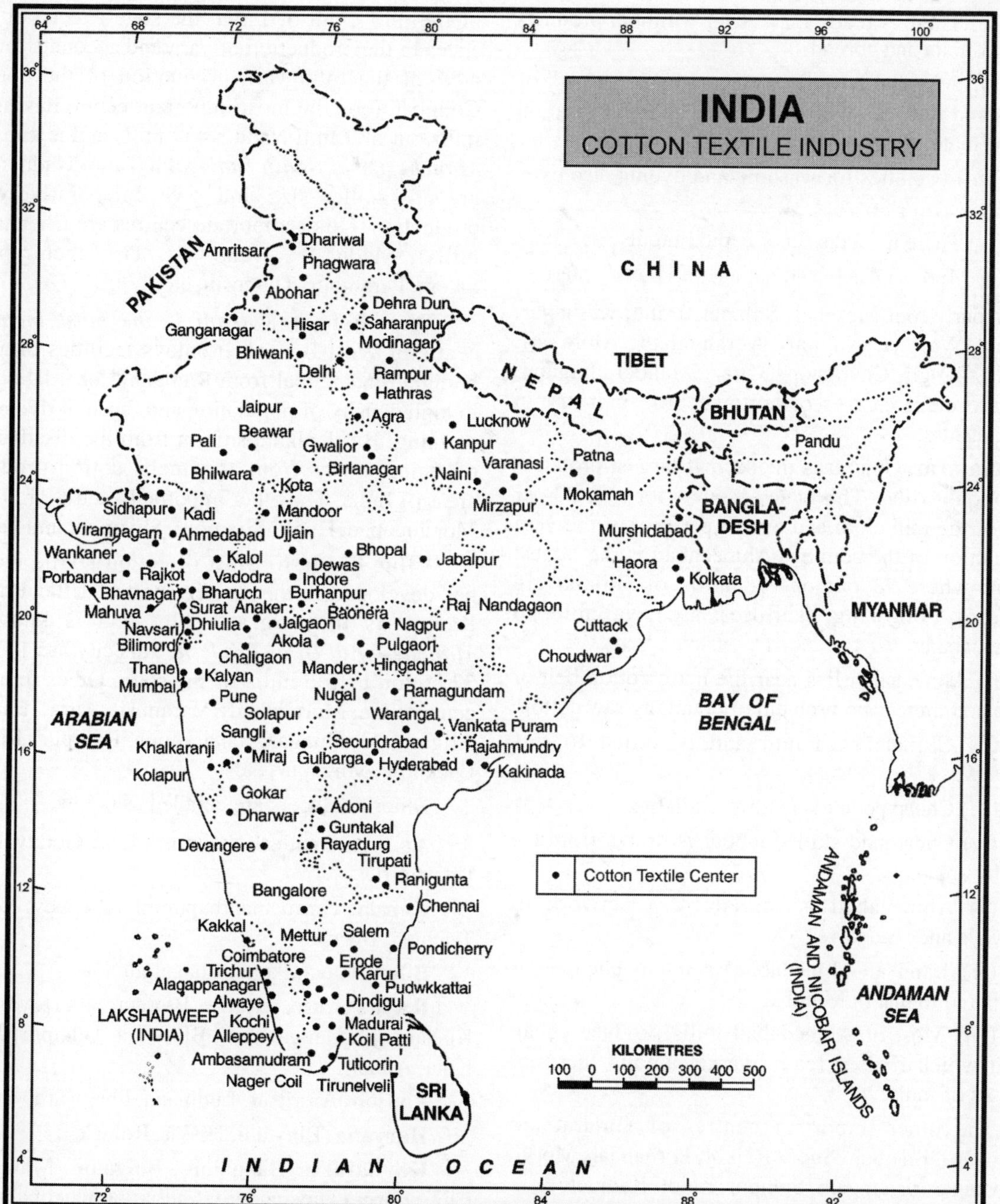

FIG. 25.1. India : Cotton Textile Industry

(*ii*) Mumbai is a very important port which helped in import of machinery and long staple cotton and export of cloth.

(*iii*) Cheap hydro-electricity is readily available from the nearby areas.

(*iv*) The black-cotton soil in the hinterland of Mumbai provides cotton as the basic raw material.

(*v*) Cheap labour could be drawn from the surrounding and far off areas.

(*vi*) There is ready market for Mumbai products both in India and abroad.

(*vii*) Mumbai is well-connected by a network of roads and railways which help in easy transportation of raw material and finished goods.

(*viii*) Facilities for washing and dyeing also exist here.

(*ix*) There is no dearth of capital inputs.

(*x*) Mumbai has the advantage of an early start.

Apart from Mumbai, Solapur, Pune, Kolhapur, Satara, Wardha, Nagpur, Aurangabad, Amravati, Akola, Sangli, Chaligaon, Miraz, Mander, Jalgaon, etc. are other centres of cotton textile industry in Maharashtra.

Gujarat. Gujarat is the second largest producer of cotton textiles. This state accounts for over 33 per cent of the mill cloth and over 8 per cent of the yarn production of the country. Ahmedabad is the largest centre where 73 out of 118 mills of Gujarat are located. Following facilities are available to Ahmedabad :

(*i*) Ahmedabad lies near the main cotton belt of India and there is no problem of obtaining raw cotton.

(*ii*) Climate is humid and is suited to this industry.

(*iii*) Cheap power is readily available.

(*iv*) Cheap and skilled labour is drawn from the nearby areas.

(*v*) Ahmedabad is served by a network of railways and roadways.

(*vi*) Land at Ahmedabad is much cheaper as compared to that in Mumbai.

(*vii*) Most of Ahmedabad mills produce cheap cloth which finds a ready market among the poor masses of India.

The other important centres of Gujarat are Vadodara, Bharuch, Surat, Rajkot, Porbandar, Morvi, Bhavnagar, Viramgam, Sidhpur, Kelot, Kadi, etc.

Madhya Pradesh. Cotton is locally grown. Coal provides necessary energy. Abundant cheap labour is available due to backward economy of the masses. Gwalior, Ujjain, Indore, Dewas, Ratlam, Jabalpur, Bhopal, etc. are important centres.

Tamil Nadu. Among the southern states, Tamil Nadu is an important cotton textile producer. Although Tamil Nadu produces only about 6 per cent of the mill cloth of India, the state excels all other states in the production of yarn and accounts for over 44% of the total yarn production of the country. Coimbatore is the most important centre having 200 mills out of Tamil Nadu's 439 mills and is known as ***Manchester of South India***. But Tamil Nadu's mills are of smaller size and give comparatively less production. Other important centres are Chennai (10 mills), Madurai, Tirunelveli, Tirucchchirappalli, Salem, Perambur, Tuticorin, etc.

West Bengal. *Kolkata is the most important* centre of West Bengal. It enjoys facilities of a port, humid climate, coal from Raniganj, local labour due to high density of population and those of dyeing and washing. But Kolkata suffers from the disadvantage of being away from the main cotton-producing areas of India. The other important centres are Haora, Murshidabad, Hugli, Sirampur, Shiampur and Panihar.

Uttar Pradesh. Most of cotton textile industry has developed in the western part of Uttar Pradesh. Kanpur is the largest centre and is known as ***Manchester of Uttar Pradesh***. This city has 10 out of 52 cotton textile mills of the state. Other important centres are Moradabad, Varanasi, Agra, Bareilly, Aligarh, Modinagar, Saharanpur, Rampur, Etawah, Lucknow, Mirzapur, etc.

Other producers are detailed as below:

Andhra Pradesh. Guntur, East Godavari and Udayagiri.

Kerala. Thiruvananthapuram, Alleppey, Kollam, Trichur, Alwaye.

Bihar. Gaya, Patna, Bhagalpur

Rajasthan. Pali, Beawar, Vijayanagar, Kishangarh, Ganganagar, Bhilwara, Udaipur, Jaipur, Kota, Ajmer.

Punjab. Amritsar, Ludhiana, Phagwara.

Haryana. Bhiwani, Hissar, Rohtak.

Karnataka. Bangalore, Belgaum, Mangalore, Chitradurga, Davanagere, Gulbarga, Chennapatnam, Mysore.

Telangana. Hyderabad, Secunderabad.

Apparel and Garmenting

The Indian textile sector includes a large apparel and garment sector which employs 12.3 million persons and produces 3.6 million tonnes of apparel and garments. The readymade garmenting segment contributes to 43 per cent of the Indian textile exports which include cotton garments and accessories, man made fibre garments and other textile clothing.

Problems of Cotton Textile Industry

Although cotton textile is one of the most important industries of India, it suffers from many problems. Some of the burning problems are briefly described as under :

1. Scarcity of Raw Cotton. Indian cotton textile industry suffered a lot as a result of partition because most of the long staple cotton growing areas went to Pakistan. Although much headway has been made to improve the production of raw cotton, its supply has always fallen short of the demand. Consequently, much of the long staple cotton requirements are met by resorting to imports.

2. Obsolete Machinery. Most of the textile mills are old with obsolete machinery. This results in low productivity and inferior quality. In the developed countries, the textile machinery installed even 10-15 years ago becomes outdated and obsolete, whereas in India about 60-75 per cent machinery is 25-30 years old. Only 18-20 per cent of the looms in India are automatic whereas percentage of such looms ranges from cent per cent in Hong Kong and the USA. to 99 per cent in Canada, 92 per cent in Sweden, 83 per cent in Norway, 76 per cent in Denmark, 70 per cent in Australia, 60 per cent in Pakistan and 45 per cent in China.

3. Erratic Power Supply. Power supply to most cotton textile mills is erratic and inadequate which adversely affects the production.

4. Low Productivity of Labour. Labour productivity in India is extremely low as compared to some of the advanced countries. On an average a worker in India handles about 2 looms as compared to 30 looms in Japan and 60 looms in the USA. If the productivity of an American worker is taken as 100, the corresponding figure is 51 for U.K. 33 for Japan and only 13 for India.

5. Strikes. Labour strikes are common in the industrial sector but cotton textile industry suffers a lot due to frequent strikes by a labour force. The long drawn strike in 1980 dealt a severe below to the organised sector.

6. Stiff Competition. Indian cotton mill industry has to face stiff competition from powerloom and handloom sector, synthetic fibres and from products of other countries.

7. Sick Mills. The above factors acting singly or in association with one another have resulted in many sick mills. As many as 177 mills have been declared as sick mills. The National Textile Corporation set up in 1968 has been striving to avoid sick mills and has taken over the administration of 125 sick mills. What is alarming is 483 mills have already been closed.

Technical Textiles

Technical textiles are materials and products used for their technical performance and functional properties. These products are used in different fields such as aerospace, shipping, sports, agriculture, defence, medicine/health, manufacturing, etc. This sector of textile industry has huge potential and a bright future. The global size of technical textiles is about $ 136 billion and is growing at an average annual rate of 3.5 per cent. Domestic market for these textiles was ₹ 70,151 crore in 2012-13 and is growing at an average annual rate of 11 per cent. The Government of India launched an ambitious scheme known as Scheme for Growth and Development of Technical Textiles (SGDTT) during 2007-08 to exploit the vast potential of Technical Textiles.

TABLE 25.3. Value of Cotton Textile Exports (₹ Crore)

Year	1960-61	1970-71	1980-81	1990-91	2000-01	2019-20	2020-21
Cotton Yarn, Fabrics, made ups, etc.	65	142	408	2,100	16,030	61,857	65,160
Readymade Garments	1	29	550	4,012	25,478	1,09,691	90,639
Total	**66**	**171**	**958**	**6,112**	**41,508**	**1,71,548**	**1,55,799**

Source : The Economic Survey 2021-22, Statistical Appendix, pp. 116-18.

Exports

India is a major exporter of cotton textiles. Cotton yarn, cloth and ready-made garments form important items of Indian exports. Indian garments are well known throughout the world for their quality and design and are readily accepted in the world of fashion. **Table 25.3** shows the export trends of cotton textile products from India. It is clear that export of ready-made garments has increased tremendously since 1960-61. In 2020-21, about 58 per cent of the total export of cotton textiles consisted of ready-made garments.

India's textiles including handlooms and handicrafts are exported to more than hundred countries. However, the USA and the European Union account for about two-thirds of India's textile exports. The other major textile destinations are China, U.A.E. Sri Lanka, Saudi Arabia, Republic of Korea, Bangladesh, Turkey, Pakistan, Brazil, Hong-Kong, Canada and Egypt.

JUTE TEXTILES

This is the second important textile industry of India after cotton textile industry. This industry occupies a significant place in our economy and is one of the major industries in the eastern region, particularly in West Bengal. Jute the *golden fibre* meets all the standards of safe packaging in view of being a natural, renewable, biodegradable and eco-friendly product. Globally, India is the largest producer of jute goods and this sector supports livelihood of about 40 lakh farm families and provides direct and indirect employment to 4 lakh workers.

Many of the sick and inefficient mills had to be closed down due to shortage of raw material. A relentless campaign to increase the production of raw jute by increasing area under jute cultivation in the Brahmaputra valley, West Bengal, Tarai and in East coastal areas and by increasing yield per hectare eased the situation to a great extent. The production of raw jute increased from 3.31 million bales (of 180 kg each) in 1950-51 to 10.14 million bales in 2017-18.

At present there are 83 composite jute mills in the country out of which 64 are in West Bengal, 7 in Andhra Pradesh, 3 each in Bihar and Uttar Pradesh, 2 each in Assam and Chhattisgarh and one each in Odisha and Tripura. As in January, 2012, total number of looms stood at 49,529 considing of 21,122 hessian looms, 26,663 sacking looms and 1,744 other looms.

Production and Distribution. India is the largest producer of jute textiles. Following are the main producing states.

West Bengal has the largest concentration of jute industry. This state has 64 jute mills and 39,623 looms which respectively account for 81 per cent and 79 per cent of all India installation. Over 84 per cent of jute goods production of India comes from West Bengal. Andhra Pradesh is a distant second producing only 10 per cent of the Indian jute goods. Most of the mills are within a distance of 64 kilometres from Kolkata along the Hugli river. As a matter of fact, there is a narrow belt of jute mills which is 100 km long and 3 km wide along both the banks of Hugli river. Jute mills in the Hoogly region are much larger than mills in other parts of the country and give much higher production as compared to mills of other areas. Apart from Kolkata, the other important centres of jute textile industry are Titagarh (9 mills), Jagatdal (8 mills), Budge Budge (8 mills), Haora (8 mills), Bhadreswar (6 mills), Bally, Agarpara, Rishra, Serampara, Shibpur, Shyamnagar, Bansbaria, Kankinara, Uluberia, Naihati, Baidyahati and many others (**Fig. 25.2**).

Following factors have been responsible for high concentration of jute mills in Hugli basin :

(*i*) The Ganga-Brahmaputra delta grows about three fourth of India's jute and provides raw material to jute mills here.

(*ii*) Coal is obtained from Raniganj fields which are hardly 200 km away.

(*iii*) Cheap water transportation is available. The area is also served by a network of roads and railways (**Fig. 25.2**).

(*iv*) Abundant water is available for processing, washing and dyeing jute.

(*v*) Humid climate is very convenient for spinning and weaving.

(*vi*) Kolkata is a big port which helps in the import of machinery and spare parts and in the export of finished jute products.

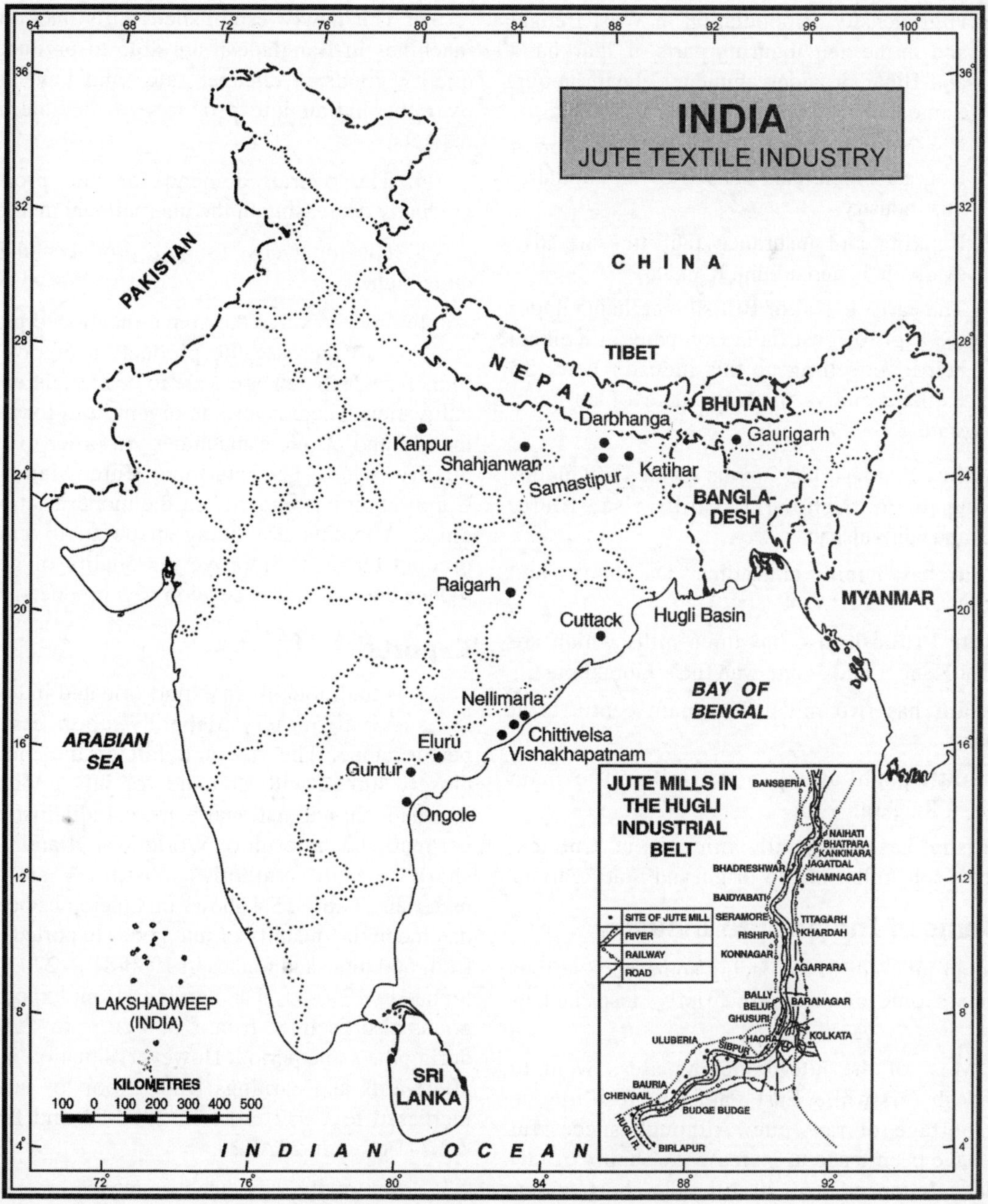

FIG. 25.2. India : Jute Textile Industry

TABLE 25.4. Exports of Jute Products

Year	1960-61	1970-71	1980-81	1990-91	2000-01	2019-20	2020-21
Quantity in thousand tonnes	790	560	660	220	—	—	—
Value in ₹ crore	135	190	330	298	932	2,428	2,741

Source : Economic Survey 2021-22, Statistical Appendix, pp. 116-18.

(*vii*) High density of population in West Bengal and in the neighbouring parts of Jharkhand and Bihar provides abundant cheap labour. Some labour comes from Uttar Pradesh also.

(*viii*) Big capitalists are living in and around Kolkata which makes easy flow of capital in this industry.

(*ix*) Banking and insurance facilities are also available in and around Kolkata.

(*x*) The early arrival of British merchants under the aegis of East India Company in Kolkata helped in setting up this industry here. So Kolkata enjoys the advantage of an early start.

Andhra Pradesh has 7 mills located at Guntur, Vishakhapatnam, Nelimarla, Chillivelsa, Eluru, Ongole, and Chitvalshah.

Bihar has 3 mills at Katihar, Darbhanga and Samastipur.

Uttar Pradesh also has three mills which are located at Kanpur and Shahjawan (near Gorakhpur).

Assam has two mills. The main centre is at Gauripur.

Chhattisgarh also has two mills. The main centre is at Raigarh.

Odisha has only mill which is at Cuttack. **Tripura's** sole mill is of new origin and is at Agartala.

Problems of Indian Jute Industry

Indian jute industry is facing some very serious problems. Some of these are briefly described as under:

1. Most of the jute-producing areas went to Bangladesh (erstwhile East Pakistan) resulting in acute shortage of raw jute. Although successful efforts have been made to increase the supply of raw jute since Independence, it still falls short of our current requirements.

2. Most of our customers could not get our jute products during World War II as a result of which several countries developed many substitutes to jute. Even today, our jute industry has to face a very tough competition from synthetic packing materials of the advanced countries of Europe and North America. As such the market for jute goods has shrunk considerably.

3. The newly established mills and improved machines in Bangladesh are able to produce better quality goods at cheaper rates and have an edge over the Indian jute products in the international markets.

4. The overall demand for jute products is gradually decreasing in the international market.

5. The input cost for jute products in India is quite high.

In order to solve the above-mentioned problems, we have to increase the production of raw jute in India for which new areas are to be brought under jute cultivation. There is also an urgent need for replacing the old and obsolete machinery in order to compete quality wise. The National Jute Manufactures Corporation has undertaken the modernisation of its units. Attempts are being made to diversify the product range, to improve the quality of goods, to reduce the cost and to develop new products.

Exports

It is traditionally an export oriented industry and its survival largely depends upon its export performance. The rise and fall of the industry is closely linked with demand for jute goods in the international and national market. India lost much of her market as a result of World War-II and owing to sharp rise in synthetic substitutes as packing materials. **Table 25.4** shows that there has been sharp decline in the quantity of jute goods exported by India from 660 thousand tonnes in 1980-81 to 220 thousand tonnes in 1990-91. The earnings from export of jute goods also declined from ₹ 330 crore to ₹ 298 crore during the same period. However, situation improved afterwards and earnings from export of jute goods increased to ₹ 932 crore in 2000-01 and further to ₹ 2,741 crore in 2020-21.

Some of the advanced countries are becoming conscious about the environmental degradation with the increasing use of non-biodegradable materials such as plastic bags for packing and are trying to discourage such a practice. This has once again turned the tide in favour of jute bags in some advanced countries and India's jute industry is poised for a big boost in the years to come. The main buyers of Indian jute products are the USA, Canada, Russia, UAR, Australia, U.K., Czech Republic, Slovakia, etc.

WOOLLEN TEXTILES

Growth and Development

Woollen textile is one of the oldest industries of India. There is evidence to prove that Aryans inhabiting the Indus Valley knew the use of wool as far back as 5000 B.C. Woollen textiles were quite popular as cottage industry in India during the historical times. However, the modern woollen textile industry started with the establishment of 'Lal Imli' at Kanpur in 1876. It was followed by setting up of woollen textile mills at Dhariwal (Punjab) in 1881, Mumbai in 1882 and Bengaluru in 1886. The industry could not make much headway till the Second World War. Rapid progress has been observed during the planning era. At present there are 621 big and small mills, 1,100 hosiery units and 155 yarn spinning units.

Distribution

The main concentration of woollen textile industry is found in Punjab, Maharashtra and Uttar Pradesh. These states account for about three-fourths of the total spindleage capacity. Gujarat, Karnataka, West Bengal and Jammu and Kashmir are next in order of importance.

Punjab. Punjab leads all other states in manufacturing woollen goods. There are currently 257 big and small mills and the state possesses about half of the total spindleage installed capacity. Dhariwal is the largest centre. The other centres are Amritsar, Ludhiana and Kharar. The industry enjoys the benefit of hydroelectricity from Bhakhra Nangal Dam and wool from Kashmir and Kumaon regions.

Maharashtra. Maharashtra is the second largest producer of woollen textiles. The state has 31 mills, most of which are located at Mumbai. Better quality woollens are manufactured from the imported wool. Being an important port, Mumbai facilitates the import of wool which mainly comes from Australia, Italy and U.K. Shoddy and inferior goods are manufactured from the indigenous wool.

Uttar Pradesh. Kanpur in Uttar Pradesh is the birth place of modern woollen textile industry in India and is the largest centre in the state. The other centres are Shahjahanpur, Mirzapur, Varanasi, Agra, Modinagar and Tanakpur. As many as 37 mills are engaged in manufacturing different varieties of woollen goods in the state.

Gujarat. Gujarat has 10 mills located at Jamnagar, Ahmedabad, Kalol and Vadodara. Jamnagar specializes in worsteds, blankets and shoddy goods. Inferior wool is obtained from Rajasthan and Kathiawar while superior wool is imported from abroad.

Others. ***Haryana*** has 160 small mills located mainly at Panipat, Faridabad, Gurugram and Bahadurgarh. ***Rajasthan's*** 72 mills are scattered over Bikaner, Alwar, Bhilwara, Sikar, Jaipur and Nagaur. Pushkar and Ajmer have handloom units producing mainly coarse blankets. Bengaluru and Bellary are important centres of ***Karnataka***. The north-eastern parts of the state produce large quantities of coarse wool which helps in establishing handloom industry. Superior wool is imported. Most of the woollen textile centres of ***West Bengal*** are concentrated in the Hugli basin especially in Kolkata and Haora. The state has 6 mills in all. Kolkata is famous for shoddy goods, blankets, serges and flannels. ***Jammu and Kashmir*** is an important producer of handloom and powerloom woollen goods. Kashmiri woollen products are prepared from fine quality wool which is locally available. They are very attractive and find ready market throughout the country. Srinagar has 2 large mills in public sector. Kulu in ***Himachal Pradesh*** is famous for shawls. In ***Tamil Nadu***, Chennai and Salem are important centres.

Woollen Carpets and Felts (Namdahs)

It is primarily a cottage industry and flourishes in areas where cheap skilled labour is readily available. There are about 240 units engaged in manufacturing woollen carpets and felts. The important carpet producing centres are Mirzapur, Gopiganj, Shahjahanpur, Agra, Bhadohi, Kanpur and Khamana in **Uttar Pradesh**; Jaipur, Jodhpur, Bikaner and Deogarh in **Rajasthan**, Srinagar in **Jammu and Kashmir**; Amritsar in **Punjab**; Panipat in **Haryana**; Eluru in **Andhra Pradesh**; Warangal in **Telangana**. Obra and Danapur in **Bihar**; Bengaluru, Mysuru and Bellary in **Karnataka**; Chennai in **Tamil Nadu** and Gwalior in **Madhya Pradesh**. About 90 per cent of the total production is exported mainly to the USA, Britain, Canada and Australia.

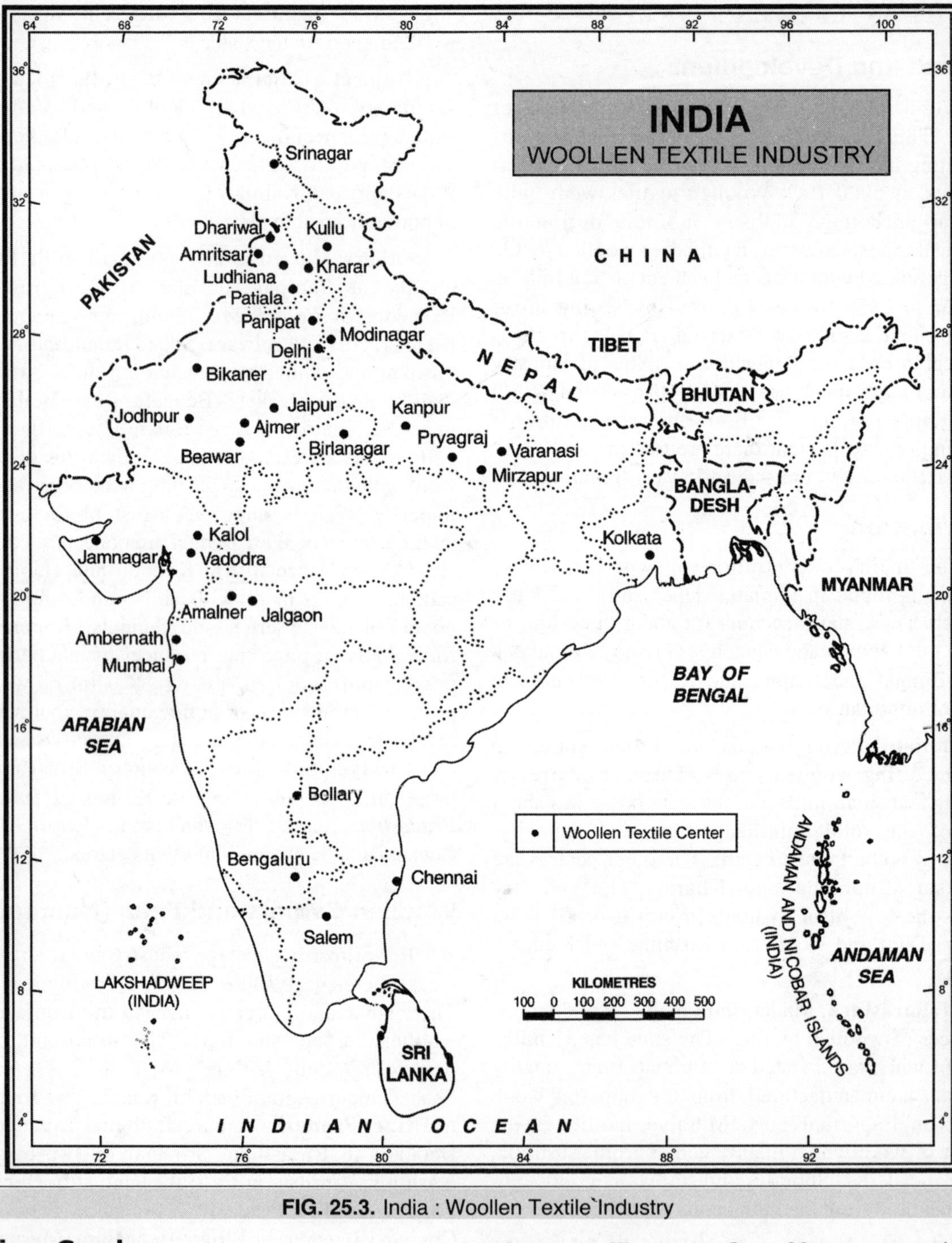

FIG. 25.3. India : Woollen Textile`Industry

Hosiery Goods

There are about 1,100 hosiery units manufacturing a variety of products like jerseys, sweaters, pullovers, shawls, cardigans, caps, socks, gloves, mufflers, etc. Over 90 per cent units are located in Punjab and Haryana. The rest are scattered over Uttar Pradesh, West Bengal, Delhi and Maharasthra. Ludhiana in Punjab is the largest centre of hosiery manufacturing in India.

Pashmina Wool Development Scheme. India produces in finest quality of pashmina wool which comes mainly from the Ladakh. This region has about 2.45 lakh pashmina goats which produce 40 to 50 tonnes of raw pashmina every year. To solve the problem of the pashmina goat rearing tribal people and to popularize this high quality wool in other non-traditional regions, the Government has taken up a pashmina development package in the 12th Five Year Plan (2012-17).

Trade (Imports and Exports)

India does not produce quality raw wool in sufficient quantity and has to import large quantities of fine wool mainly from Australia. But India is an important exporter of woollen goods. The important items of export are woollen fabrics, hosiery and knitwear, druggets, carpets, shawls, blankets, *lohis*, *nam dahs*, etc.

The USA., Russia, U.K., Canada, Australia, Germany, Denmark, Sweden, Chech Republic, Slovakia, France, Belgium, the Netherlands and some West Asian and African countries are the chief buyers of Indian woollen goods.

Problems of Woollen Textile Industry

1. Shortage of raw wool. As mentioned earlier, India does not produce sufficient quantity of fine quality raw wool. Also the productivity of Indian sheep is very low. On an average, an Indian sheep yields only 0.86 kg of wool per annum against 4.08 kg yielded by an Australian sheep. Large proportion of wool produced in India is of inferior quality and does not conform to international standards.

2. Lack of Market. Most parts of India have tropical and sub-tropical climate which restricts the demand for woollen clothes. The southern part of the country enjoys warm weather throughout the year and people do not require woollen clothes at all. Even in the northern parts of India, the winter season lasts only for four to five months in a year and it is only during this period that woollen clothes are required to some extent. The hot weather lasting for 7 to 8 months is the slack period during which production is carried out mainly for the Armed Forces and for export. For practical purposes, woollen textile industry is a seasonal phenomena in India. Moreover, low purchasing power of Indian masses also puts a limit to demand market.

3. Lack of Modern Equipment. Most of the equipment in woollen textile industry, like other textile industries, is obsolete and outdated as a result of which, its products are not able to cope with the everchanging designs and patterns, especially in the international market. There is an urgent need for mechanisation of the carpet industry keeping in view the rising demand to put a ban of the child labour. This will also help to increase the production and improve the quality.

4. Low Quality. Leaving aside a few exceptions, Indian woollen goods are considered to be of low quality in the international markets which results in lack of demand. Indian knitwear is often not shrink resistant, moth-proof and fast-coloured.

SILK INDUSTRY

India has been well-known for the production of silk since ancient times. Rulers of the medieval period encouraged this industry. India enjoys the distinction of being the only country producing all the five known commercial varieties of silk, *viz.,* Mulberry, Tropical Tasar, Oak Tasar, Eri and Muga (of which golden yellow muga silk is unique to India). India's total silk production amounting to about 17 per cent of the world production, ranks India as the second largest producer of raw silk, next only to China. Sericulture is a labour intensive industry and provides employment to nearly 7.56 million persons mainly to weaker and marginalized sections of society. Indian silk industry has improved manifold since Independence from raw silk production level of 1,437 metric tonnes during the first Five Year Plan to 23,060 metric tonnes at the end of the 11th Five Year Plan. Nearly 75 per cent of the country's total silk is mulberry silk which is produced mainly in the states of Karnataka, Andhra Pradesh, West Bengal, Tamil Nadu and Jammu and Kasmir. Tasar, eri and muga are collectively called ***vanya*** (wild) silk as these silks are mostly the products of the forests. These varieties are popular for their unique colour, lustre and are available in natural colours like golden brown, mauve, cream and beige. ***Tasar silk*** is largely produced in Jharkhand, Chhattisgrh, Madhya Pradesh, Bihar and parts of Odisha where it has great potential

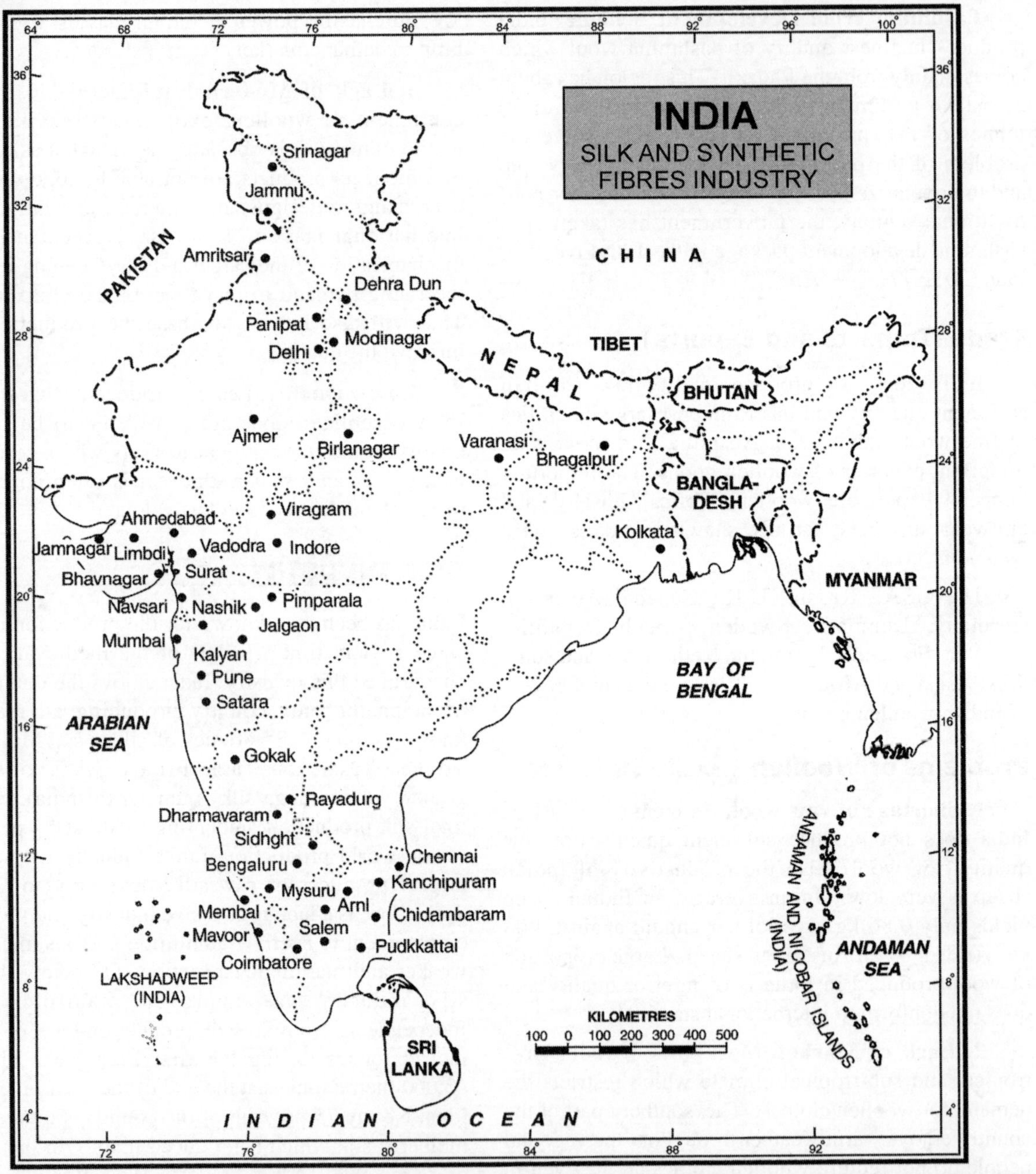

FIG. 25.4. India : Silk and Synthetic Fibres Industry

to enhance the livelihood of the tribal population. Idling women folk and unemployed youth are attracted towards tasar culture to a greater extent. The **oak tasar** culture is now practised in the sub-Himalayan states like Manipur, Himachal Pradesh, Uttar Pradesh, Assam, Meghalaya and Jammu and Kashmir. ***Eri silk*** ranks first among non-mulberry silk production and is produced mostly in hill tracks of the north-eastern states besides some parts of Bihar, West Bengal and Odisha. ***Mugas silk***, also known as golden silk is exclusively found in Assam and is widely distributed in the Brahmaputra valley.

Distribution

There are nearly 90 mills producing silk textiles. In addition, there are small and medium sized units also. It is also popular in the shape of handloom industry. About 98 per cent of the total production comes from Karnataka, West Bengal and Jammu and Kashmir.

Karnataka is the foremost silk producing state in India. It produces only mulberry silk and accounts for over 50 per cent of mulberry silk of the country. Mulberry trees grow over 8.5 thousand hectares. Mysuru, Bengaluru, Kolar, Mandya, Tumkur, Belgaum and Kodagu districts are the main producers. About 50 per cent of India's silk cloth is also manufactured in this state. The main centres of silk cloth manufacturing are Bengaluru, Kolar, Mysuru and Belgaum. ***West Bengal*** produces about 13 per cent of country's total silk, most of which is of mulberry variety. There are about five thousand handlooms mainly concentrated in Murshidabad, Bankura, North Twenty Four Parganas, South Twenty Four Parganas and Birbhum districts. ***Jammu and Kashmir*** has its main silk textile centres in Anantnag, Baramula, Jammu and Udhampur districts. Srinagar is a big centre for silk manufacturing where automatic machines are used for this purpose. ***Bihar*** is also an important producer. Patna, Gaya and Bhagalpur are the leading silk producing districts. In the neighbouring ***Jharkhand***, Ranchi, Palamu and Hazaribagh are important producers. ***Chhattisgarh*** produces about 2.6 per cent of India's silk, most of which comes from Bastar, Bilaspur and Surguja districts. The other areas of silk production are Mirzapur, Pratapgarh and Shahjahanpur in ***Uttar Pradesh***; Amritsar, Jalandhar, Ludhiana, Hoshiarpur and Gurdaspur in ***Punjab***; Goalpara, Kamrup and Nowgong in ***Assam***; Coimbatore, Tiruchirapalli, Dharmapur, Nilgiris, Salem, Thanjavur and Tirunelveli in ***Tamil Nadu***; Nagpur, Pune, Sangli, Chandrapur and Solapur in ***Maharashtra***; Ahmedabad, Surat, Bhavnagar, Porbandar in ***Gujarat***; Chittur, Kurnool, Vishakhapatnam and Anantpur in ***Andhra Pradesh,*** Warangal, Adilabad and Karimnagar in ***Telangana*** and Balaghat in ***Madhya Pradesh. Meghalaya, Manipur, Tripura and Himachal Pradesh*** also produce some quantity of silk.

Exports

The Indian silk is demanded in American, European as well as Asian markets. The USA, UK, Russia, Saudi Arabia, Kuwait and Singapore are the major importers of silk dress materials and scarves. The tough competition with Italy and Japan caused a setback to Indian silk industry. The synthetic fibres and the artificial silk being less expensive and easy to maintain, have been responsible for reducing the popularity of silk.

Man-Made Fibres

It is an important segment of textile industry. Man-made fibres have revolutionised the textile industry because they are in great demand due to their special qualities like strength, durability, dyeability and workability. Over and above, they are resistant to shrinkage.

Raw Materials. Man-made fibres are generally divided into two groups *viz.*, cellulose (rayon and acetate) and non-cellulose (nylon, polyster, etc.) The basic raw material for producing viscose rayon yarn is the cellulose pulp derived from bamboo, eucalyptus and other soft wood trees. The chemicals used for making these fibres include caustic soda, sodium sulphate, sulphuric acid, carbon disulphide and soda sulphate. For acetate yarn and staple fibre, the primary raw materials are alcohol and cotton linters. The other materials are acetic acid, acetic acetone and ethyl acetate. For the production of synthetic fibres like nylon, polyester, acrylic, etc. we require caprolactum, naphtha, polyster chips, ethylene, glycole, etc.

Growth and Development

The man made fibre weaving units were started between 1925 and 1935 when the cost of pure silk fabrics became exorbitant. The powerlooms and handlooms had been using imported fibre till 1950 when a factory was established by Travancore Rayons Ltd. at Rayapuram in Kerala to manufacture synthetic fibre. This was followed by National Rayon Co. at Mumbai and the Sirsilk Ltd. at Hyderabad. By 1960, Century Rayon Ltd. at Kalyan (Maharashtra), the Gwalior Rayon Silk Manufacturing Company at Nagda were also commissioned. Following units are making a significant contribution towards the manufacturing of synthetic fibres.

(*a*) Rayon Units. Kagaznagar (Telangana), Junagadh (Gujarat), Rayapuram (Kerala), Udhna (Gujarat), Birlagram (H.P.), Nagda (M.P.), Kalyan, Pimpri-Pune and Goregaon (Maharashtra), Kota (Rajasthan), Mettupalayam (Tamil Nadu), Kanpur (U.P.) and Tribeni (W. Bengal).

(*b*) Nylon Filament Yarn Units Kota, Pimpri, Pune, Bhonsari-Pune, Modinagar, Mumbai, Nagpur, Vadodara, Bengaluru, Chennai, Hyderabad, Thiruvananthapuram, Barauni, Kanpur, Ujjain and Kolkata.

(*c*) Nylon Staple Fibre Units. Kota, Mumbai.

(*d*) Nylon Tyre Cord Units. Kota, Mumbai, Chennai, Kalyan, Kanpur, Goregaon, New Delhi, Udhna.

(*e*) Polyster Staple Fibre Units. Thane, Ahmedabad, Vadodara, Ghaziabad, Mandi, Kota.

(*f*) Polyster Filament Yarn Units. Mumbai, Kota, Pimpri-Pune, Modinagar, Ujjain, Udhna, Vadodara.

India has to import large quantity of man-made fibres to meet the growing demand in the home market.

METALLURGICAL INDUSTRIES

Metallurgical industries are those industries which use metal as the basic raw material. It is a very wide term and includes several industries like iron and steel, aluminium, copper smelting, lead and zinc smelting, alloy, ferro-manganese, ferro-silicon, ferro-chrome, tungsten and a host of other industries. These industries form the *economic backbone* of a developing country. India has progressed a lot especially during the planning period and now occupies a place of pride with regard to the development and growth of metallurgical industries. Major metallurgical industries are briefly described in the following pages:

IRON AND STEEL INDUSTRY

We live essentially in an age of iron and steel. "Because of its hardness, strength and durability, because of the ease with which it can be cast and worked into any desired shape and because of its remarkable cheapness under modern methods of production, iron is the most important and widely used metal in the service of man". Iron and steel were the harbingers of industrial revolution in late 18th and early 19th century. Today this industry has proved to be the harbingers of globalisation. It is one of the very few industries that have assumed a global character with developments in one region affecting the industry almost everywhere else; and India is no exception. The proud machine civilization of modern age would not have existed without iron. The sturdy structure of modern industrial world is made of steel. Iron and steel is the **basic or key industry** and lays the foundation of a vibrant industrial economy. Most of the subsidiary industries such as automobiles, locomotives, rail tracks, ship-building, machine building, bridges, dams and a host of other industrial and commercial activities depend upon iron and steel industry. No wonder, per capita consumption of iron and steel is one of the most significant measures of the level of industrialisation and economic growth of a country.

Although Indians are known for their technique of smelting iron since early time, the first iron and steel unit on modern lines was established in 1830 at Porto Nova in Tamil Nadu. However, it could not succeed and was closed down in 1866. The other efforts made during the second half of the 19th century also met with the same fate.

The real beginning of modern iron and steel industry was made in 1907 only when Tata Iron and Steel Company (TISCO) was set up at Jamshedpur (Sakchi at that time). The Indian Iron and Steel Company (IISCO) was set up in 1919 at Burnpur followed by the setting up of Mysore Steel Works at Bhadravati (now Visveswaraya Iron and Steel Works) in 1923. Iron and steel Industry witnessed rapid growth after Independence. The development of iron and steel industry was envisaged during the first Five-Year Plan, but it was during the Second Five-Year Plan, that the three integrated steel projects were started at Bhilai, Rourkela and Durgapur. India is now the eighth largest producer of steel in the world. Recent developments have amply demonstrated the mettle of Indian steel industry to rise even further and become a major player in the world. However steel is known to be an industry witnessing periodic business cycles of upswings and downswings.

Steel Authority of India (SAIL) Established in 1973, SAIL is a government undertaking and is responsible for the management of steel plants at Bhilai, Durgapur, Rourkela, Bokaro and Burnpur and also the Alloy Steel Plant at Durgapur and Salem Steel Plant. The management of Indian Iron and Steel was taken over by Government on 14th July, 1976. SAIL also took over Maharashtra Elektrosmelt Limited, a mini steel plant, in January 1986. Visweswaraya Iron and Steel Limited was also taken over by SAIL in August 1989.

With the introduction of new liberalised industrial policy in 1991, some changes were visualised in the functioning of SAIL which had great impact on the performance of steel industry in the country. Over a period of five years it brought down its manpower by around 40,000 and substantially improved its techno-economic parameters to contain its cost of operation in spite of a steady rise in input prices.

India is the second largest producer of crude steel after China, surpassing Japan with a global share of 6 per cent. During 2019-20, the crude steels production stood at 111 million tonnes, as compared to 106.6 million tonnes in 2018-19.

India is the third largest consumer of finished steel after China and U.S.A. The per capita consumption of steel increased from 26 kg in 2001 to 74.7 kg in 2020. However, this is very small against the global average of 229 kg. With huge investments in infrastructure, construction and automobile sector, demand for steel and corresponding consumption is growing at an average rate of 7.4 per cent per annum. This will lead steel production to go up to 255 million tonnes by 2030 and per capita steel consumption to 160 kg.

TABLE 25.5. Per Capita Consumption of Steel in India

Year	Per capita consumption (in kg.)
2001	26
2005	31
2010	47
2015	61
2016	63
2017	66
2018	70
2019	74
2020	74.7

Source : Economic Suvey 2018-19, p. 206; 2019-20, p. 228; 2020-21, p. 282.

National Steel Policy. This policy was launched in 2017. Its aim was to ensure that the Indian steel sector is proposed to meet the growing requirements of modern India and to promote a healthy sustainable growth for the sector. The vision was to create a technologically advanced and globally competitive steel industry which promotes economic growth. Key features of this policy include establishing self-sufficiency in steel production by providing policy support and guidance to private manufacturers, Micro, Small and Medium Enterprises (MSME) steel producers and Central Public Sector Enterprises (CPSEs). It encourages capacity additions; development of globally competitive steel manufacturing capabilities as well as cost-efficient production by facilitating domestic availability of iron ore, cooking coal and natural gas; as well as overseas asset acquisitions of raw materials. It also aims at enhancing the domestic steel demand.

According to this Policy, it is projected that by 2030 crude steel capacity will reach 300 million tonnes and production will be about 225 million tonnes. The per capita consumption will reach 160 kg against the current 74.7 kg. The Policy also envisages 100 per cent indigenous fulfilment of demand for high grade automotive steel, electrical steel, special steels and alloys for strategic applications along with an increase in domestic availability of washed cooking coal so as to reduce dependence on cooking coal from about 85 per cent to around 65 per cent by 2030-31.

Locational Factors

Iron and steel industry uses large quantities of heavy and weight losing raw materials and its localisation is primarily controlled by the availability of raw materials. Coal and iron ore are the two basic raw materials used by iron and steel industry and on the basis of minimum transportation cost, most of the steel plants are located at three distinct places *viz.* (*i*) near coal fields, (*ii*) near iron ore mining centres and (*iii*) at places between areas of coal and iron ore production. Most of the iron and steel plants of India such as Jamshedpur, Burnpur, Durgapur, Rourkela, Bhilai and Bokaro are located in Jharkhand, West

Bengal, Odisha and Chhattisgarh. These states are very rich in coal and iron ore deposits and are important producers of these materials. Visveswaraya Iron and Steel Works at Bhadravati is a major exception which is located far away from the main coal producing areas of the country. Earlier, this centre was depending upon charcoal which was available locally. Now it uses hydroelectricity from the Sharavati Power Project. The other raw materials used in this industry are manganese, limestone, dolomite, chromite, silica, etc. These raw materials are used in small quantities and can be transported without much difficulty. Hence, they do not materially affect the localisation of this industry.

Another important factor influencing the localisation of iron and steel industry is the availability of market. Steel products of an integrated steel plant are quite bulky and it has been estimated that the transport cost per tonne-kilometre of steel product is about three times more than that of coal or iron ore. Thus, following the theory of *minimum transprotation cost* many centres of iron and steel production tend to be attracted by market. Moreover, recent technological developments in transport, the use of scrap as raw material and the agglomeration economics have made market oriented location more advantageous than ever before. With the increasing popularity of open hearth process, scrap has become a very important raw material in this industry *About half of the metal now melted in world's iron and steel furnaces is scrap*. Industrialized areas, specially with steel consuming industries, are the major sources of scrap iron. Thus, the market has double attraction, as the consumer of steel and as a source of raw materials. However, the use of scrap as raw material on a large scale is yet to pick up in India.

From the above discussion, it is clear that in the present day localization of iron and steel industry, each of the three factors *viz*., coal, iron ore and market has almost equal significance. The geographical coincidence of any two of the three factors would easily determine the site of the steel plant.

In another situation, when some ingredients are to be imported or finished steel is to be exported, sea port locations are preferred. This is exemplified by the establishment of the Vishakhapatnam Steel Plant at a sea port. A few more plants in the offing such as Mangaluru and Ratnagiri also favour seaboard location.

Centres of Production

At present there are 13 primary integrated plants and a large number of decentralised scondary units known as mini steel plants. Besides, there are several rolling and re-rolling mills and foundries which manufacture different items of steel using pig iron and ingot steel. There are about 10,000 foundries, 95 per cent of which are concentrated in the western states of Maharashtra and Gujarat and in the southern state of Tamil Nadu.

Large Integrated Iron and Steel Plants

1. Tata Iron and Steel Company (TISCO). This is the oldest iron and steel centre of India. It is a private sector enterprise. It was established in 1907 by Jamshedji Tata at Sakchi in Singhbhum district of Jharkhand. Later on, it was renamed as Jamshedpur after Jamshedji. It started producing pig iron in 1911 and steel in 1912. The plant initially had capacity of producing 1.21 million tonnes of pig iron and 1.1 million tonnes of steel per annum. This capacity has been enhanced to 3.9 million tonnes of pig iron, 2 million tonnes of ingot steel and 3 million tonnes of saleable steel. Currently it produces about 3 million tonnes of saleable steel. Following facilities are available to this centre :

(*i*) High grade haematite iron ore is available from Noamundi mines of Singhbhum in Jharkhand and Gurumahisani mines of Mayurbhanj in Odisha. These mines are located at a distance of 75-100 km from Jamshedpur.

(*ii*) Coal is available from Jharia and Raniganj coal mines located 160 to 200 km from Jamshedpur.

(*iii*) Manganese comes from Joda mines of Kendujhar district in Odisha.

(*iv*) Dolomite, limestone and fire clay used as flux material are available from Sundargarh district of Odisha.

(*v*) Kolkata, located at a distance of 250 km, provides port facilities and its industrialised hinderland provides market for the products.

(*vi*) Sufficient water for cooling purposes is obtained from Subarnrekha river. In addition to this, the storage dam on Kharkai river also provides water.

(*vii*) Jamshedpur is well connected with Kolkata, Mumbai and Chennai by road and rail and enjoys good transport facilities.

(*viii*) Densely populated regions of Jharkhand, Bihar and Odisha provide cheap labour. Major part of labour is drawn from tribal areas of Chota Nagpur plateau.

Dubari Steelworks. The shortage of space in Jamshedpur and some other allied factors persuaded TISCO to locate a second steelworks at Gopalpur in Odisha, about 170 kms south-east of Bhubaneshwar. The favourable factors are coastal location, proximity to a rich iron ore belt, availability of sparcely settled land for the project, a nearby source of fresh water in the form of a rivulet, proximity to the trunk rail line and a national highway and the presence of a minor deep water port which could be expanded. Tata Steel acquired land near Gopalpur to set up a shore based steel plant in 1997 but shelved the project later due to slow pace of development of Gopalpur port and a resistance movement by local people. Consequently the Tata Steel decided to build a six million tonne plant at Dubari. The infrastructure included development of the Dhamra port and a railway line between Dhamra and Bhadrakh. This plant will use the latest technology and will be cost competitive with the most efficient steel plants in the world.

2. Indian Iron and Steel Company (IISCO). Three plants at Kulti, Hirapur and Burnpur in West Bengal were set up in 1864, 1908 and 1937 respectively. These plants have been merged together and are known as Indian Iron and Steel Company (IISCO). It was brought under government control and management in July 1972. The three plants are linked by Kolkata-Asansol railway line. Hirapur plant produces pig iron which is sent to Kulti for making steel. The rolling mills are located at Burnpur. IISCO enjoys the following advantages :

(*i*) Iron ore is available from Guna mines in Singhbhum district of Jharkhand located at a distance of 285 km. Some iron ore is also obtained from Mayurbhanj area of Odisha.

(*ii*) It used to receive coal from Jharia, located at a distance of 137 km but now the power from the Damodar Valley Corporation is extensively used.

(*iii*) Dolomite and limestone are obtained from Sundargarh district of Odisha which is 327 km away. Limestone is also available from Gangpur and Paraghat areas of Odisha.

(*iv*) Rail and road links connect it to Kolkata which is just 200 km away.

(*v*) Cheap labour is readily available from the neighbouring areas.

IISCO has annual capacity of producing 10 lakh tonnes of steel. Currently it produces over 4 lakh tonnes of pig iron, more than 3.5 lakh tonnes of crude steel and around 3.8 lakh tonnes of saleable steel.

3. The Visweswaraya Iron and Steel Ltd. It was established as Mysore Iron and Steel Company (MISCO) in 1923 by the erstwhile state of Mysore. It is located at Bhadravati on the banks of river Bhadravati in Shimoga district of Karnataka. This plant was brought under Government control in 1962 and was renamed as Visveswaraya Iron and Steel Ltd. after the name of great engineer Dr. Visweswaraya. This plant has got a capacity of 1.38 lakh tonnes of steel. There are plans to raise its capacity to two lakh tonnes. This centre enjoys the following advantages.

(*i*) Bhadravati valley is 13 km wide as a result of which enough land is available.

(*ii*) High grade haematite iron ore is brought from Kemmangundi mines in Chikmaglur which is just 40 km away.

(*iii*) At the time of the setting up of the plant in 1923 the charcoal obtained from the forest-wood was used for smelting because coal was not available. Now it uses hydroelectric power obtained from Sharavati Power Project.

(*iv*) Limestone is available from Bhundiguda just 25 km away.

(*v*) Shimoga and Chitradurga supply manganese. These areas are just 50 km away.

(*vi*) Dolomite and chromite are also available within a radius of 45-50 km.

(*vii*) It lies on the main Biru-Shimoga railway line and makes use of railway facilities.

In order to increase the production of iron and steel, the Government of India established ***The Hindustan Steel Limited*** in public sector. Consequently, three plants under the public sector, *i.e.* Bhilai, Rourkela and Durgapur came into operation during the Second Five Year Plan. Capacity of each plant was fixed at 10 lakh tonnes of steel which was expanded during the Third Five Year Plan and a proposal of setting up a steel plant at Bokaro was also made.

4. Bhilai. Bhilai iron and steel centre was set up in Durg district of Chhattisgarh in 1957 with the technical and financial support of the then Soviet Union. It started production in 1959. Its initial capacity was 10 lakh tonnes which has been raised to 52 lakh tonnes. Durg happens to be a backward area and the purpose of setting this plant was to bring prosperity to this area. It enjoys following geographical advantages :

(*i*) It procures rich haematite iron ore from Dalli-Rajhara range which is 80 km south of Bhilai.

(*ii*) Coal is obtained from Korba and Kargali fields of Chhattisgarh located at 225 km away. Bokaro and Jharia (720 km) also supply coal.

(*iii*) Limestone comes from Nandini mines hardly 24 km away.

(*iv*) Bhandara of Maharashtra and Balaghat of Madhya Pradesh supply manganese.

(*v*) The Korba Thermal Power station is the main source of power.

(*vi*) It is connected with Kolkata-Nagpur railway line.

(*vii*) Dolomite comes from Bilaspur.

(*viii*) Cheap labour is available from the nearby areas.

5. Rourkela. Plant of Hindustan Steel Limited at Rourkela is situated in the Sundargarh district of Odisha. It was set up with the help of the then West German firm, Krupps and Demang, during the Second Five Year Plan (West Germany and East Germany have united to form one country now). It became operative in 1959. This plant has the following facilities for its successful operation :

(*i*) This plant uses iron ore obtained from Sundargarh and Keonjhar districts. These iron ore sources are located within a distance of 77 km from the site of the plant.

(*ii*) Coal is obtained from Jharia coalfields located at a distance of 225 km and Talcher, located at a distance of 169 km.

(*iii*) Hydro-electric power is obtained from Hirakud Power Project, located at a distance of 150 km.

(*iv*) The plant receives manganese from Barajmda, dolomite from Baradwar and limestone from Purnapani. These materials are located within a radius of 222 km in Odisha.

(*v*) It is located on the main Nagpur-Kolkata railway line and enjoys facilities of railway transport.

(*vi*) Kolkata provides the port facilities and its hinterland serves as market.

6. Durgapur. This plant of The Hindustan Steel Ltd. is located at Durgapur in Bardhaman district of West Bengal. It was set up in 1959 with the help of the United Kingdom. The production started in 1962. It has a total capacity of 35 lakh tonnes. It produced 12.45 lakh tonnes of crude steel, 10.93 lakh tonnes of saleable steel and 1.14 lakh tonnes of saleable pig iron in 2017-18. The Alloy Steel Plant at Durgapur has a capacity to produce 1.6 lakh tonnes of ingot steel which has been expanded to 2.6 lakh tonnes of crude steel. The following geographical factors favour its location and growth.

(*i*) Iron ore comes from Bolani mines. Mayurbhanj also supplies iron ore. These areas are located within a radius of 320 km.

(*ii*) Coal comes from Jharia and Raniganj.

(*iii*) Limestone is obtained from Birmitrapur in Sundargarh and manganese from Keonjhar district of Odisha.

(*iv*) Dolomite is supplied by Birmitrapur.

(*v*) Hydroelectricity is available from Damodar Valley Corporation.

(*vi*) Plenty of water is available from Durgapur Barrage built across Damodar river.

(*vii*) The Kolkata-Asansol railway line links it with other parts of the country.

(*viii*) Cheap labour is readily available from the surrounding areas.

7. Bokaro. A new public sector company, the Bokaro Steel Ltd. was formed in 1964 to erect a steel plant with the collaboration of the eartwhile Soviet Union at Bokaro near the confluence of the Bokaro and Damodar rivers in Hazaribagh district of Jharkhand. It is the second plant set up with the Soviet help. It started production in 1972. Its initial capacity was 10 lakh tonnes which was raised to 40 lakh tonnes. There are plans to raise its capacity to 100 lakh tonnes making it the largest iron and steel making centre in India. The achievement made by this plant has been made possible due to following few geographical factors :

(*i*) It receives iron ore from Kiriburu mine in Odisha.

(*ii*) Coal is obtained from Jharia coalfields located at a distance of 65 km.

(*iii*) Limestone comes from Palamu district of Jharkhand.

(*iv*) Hydroelectricity is obtained from Damodar Valley Corporation.

(*v*) Kolkata is just 300 km from here and provides port facilities.

Three more steel plants were planned during the Fourth Five-Year Plan in order to meet the growing requirement of steel. These plants are located at Salem in Tamil Nadu, Vishakhapatnam in Andhra Pradesh and Vijayanagar in Karnataka.

8. The Salem Steel Plant. The plant has been set up at Salem in the Salem district of Tamil Nadu. The plant has the advantage of rich iron ore and limestone, which is readily available in the adjoining areas. It also enjoys the facilities of cheap power, charcoal and vast market. The iron ore available here has low sulphur and phosphorus content and is suitable for producing special grade iron and steel.

The plant started commercial production in 1982. Its capacity was 32 thousand tonnes of stainless steel sheets in the beginning. This capacity was doubled in 1991 with the addition of another rolling mill. This capacity was further raised to 80 thousand tonnes of saleable steel in 1995-96. Today the Salem Steel Plant is a major producer of world class stainless steel and is in a position to export stainless steel to some of the advanced countries such as the USA, Mexico, Australia and some countries of South-East Asia. In order to cater to the growing demand for coinage of the Indian Government Mints, the management had also set up a blanking facility in 1993 with a capacity of 3,000 tonnes per annum. It also commissioned a hot rolling facility in November, 1995 which has state-of-the-art technology with high level of automation.

9. Vijayanagar Steel Plant. This plant has been set up at Tornagal near Hospet in Bellary district of Karnataka. It has the installed capacity of 32 lakh tonnes. It will be doubled in due course of time. The production of mild steel is its special feature. This plant enjoys the following facilities :

(*i*) Iron ore is obtained from Hospet region located in close proximity.

(*ii*) Coal comes from Kanhan valley in Chhattisgarh and Singareni coal fields in Andhra Pradesh.

(*iii*) Good quality limestone and dolomite is available at a distance of about 200 km.

(*iv*) Water and power requirements are met by the Tungabhadra hydel project located at a distance of about 36 km from the plant.

Another steel plant at *Paradwip* is fast coming up.

10. Vishakhapatnam Steel Plant (VSP). This integrated steel plant has a unique location on the sea port. In fact, it is the first shore based steel plant in the country. Although the foundation stone of the plant was laid in 1972, the construction work could not start in the real sense till February 1982 when *Rashtriya Ispat Nigam Limited* was incorporated as a public sector company to implement the construction of the plant.

The project has been completed in two stages : the first stage was completed by March 1992 and the second and final stage by July 1992. This is the most sophisticated modern integrated steel plant in the country. Though the production commenced in 1991-92, 1993-94 was the first full year of integrated

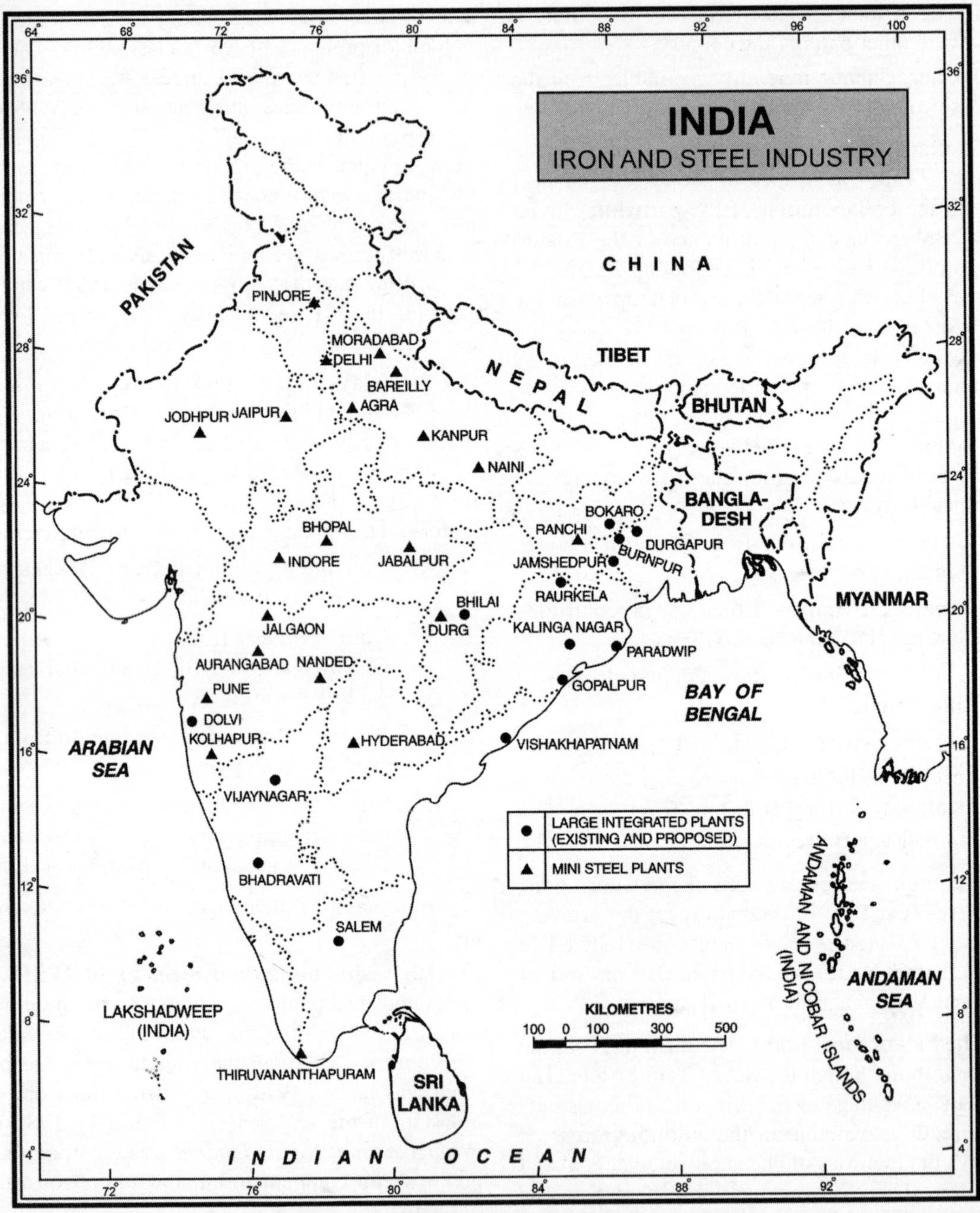

FIG. 25.5. India : Iron and Steel Industry

operation. It is a major export oriented steel plant and takes full advantage of its coastal location. Currently holding 67th rank among 80 largest steel makers on the globe, as certified by the Brussels-based International Iron and Steel Institute, VSP is smooth-sailing towards reaching its goal of turning into a 'world class company in steel industry'; as a result of the effective turn around strategy adopted by its management for the last couple of years. Buoyed by such a strong performance the VSP now intends to

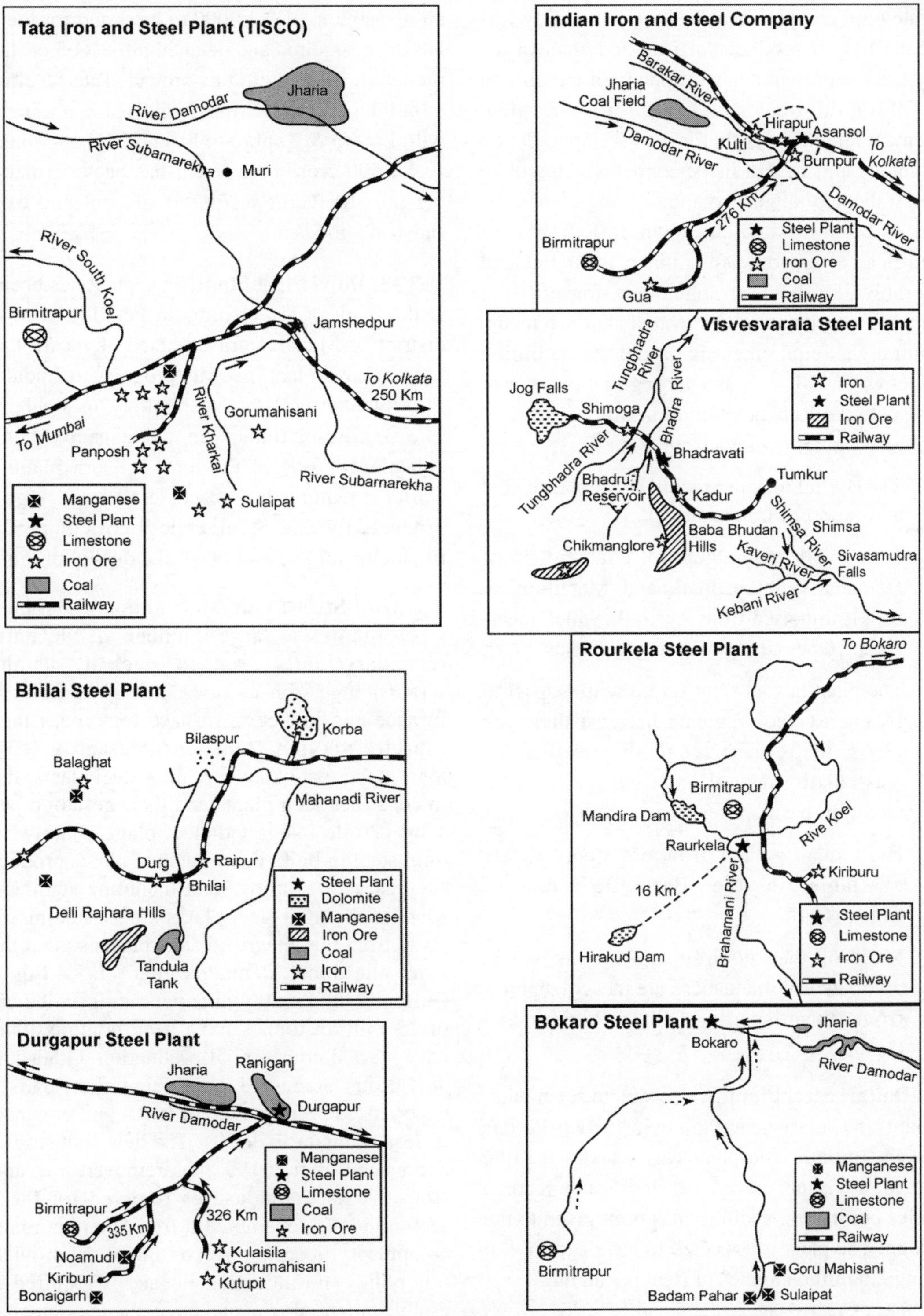

FIG. 25.6. Locational factors of major iron and steel plants in India.

accelerate on the expansion trajectory. Presently it is the second largest producer of iron and steel in the country and the present annual capacity of ten million tonnes of liquid steel can be raised. The plant management intends to go in for massive upgradation of technology and skill of its personnel which will be required if the natural gas from the Krishna-Godavari basin is to be utilised to cut down cost. Import of metallurgical coal from Australia can be reduced considerably if proper arrangements for utilising natural gas from Krishna-Godavari basin are made. The natural gas requirement is placed at one billion cubic metres (BCM) a year and negotiations are in progress with the Reliance Group in this connection. The plant has the following advantages :

(*i*) The coastal location facilitates import of coal and export of iron and steel.

(*ii*) It is well connected to coal fields of Damodar valley in Jharkhand. Metallurgical coal is imported from Australia which meets about 70 per cent power requirements.

(*iii*) The plant has a bright future with respect to its energy requirements because there are plans to replace coal imported from Australia by natural gas from the Krishna-Godavari basin.

(*iv*) High quality rich iron ore deposits are available in the Bailadila area of Chhattisgarh.

(*v*) Most of the requirements of limestone, dolomite and manganese are met by supplies from Chhattisgarh; Madhya Pradesh and Odisha.

11. Daitari Steel Plant. A decision to set another steel plant at Daitari near Paradwip in Odisha has been taken. Initially, the plant was scheduled to be built by joint venture of British and South Korean companies but its responsibility has been given to the Tata group. The plant is expected to have capacity of producing 2.6 million tonnes of steel per annum.

12. Tata Steel Kalinganagar. Tata Steel will set up a six million plant at Kalinganagar in Odisha with an investment of ₹ 15,000 crore. Land for the project has been acquired and detailed project report has been prepared. Along with this project, Tata Steel is going to build a port at Dharma in Odisha in a joint venture with Larsen & Toubro to handle 3,00,000 tonne ship. A ₹ 1,500 crore investment has been earmarked for the port to facilitate import of coal and export of finished goods.

13. Dolvi Steel Plant. This plant has been set up at the bank of river Amba in Pen Tehsil of Raigad district of Maharashtra. Located about 50 km from Mumbai this plant owes its progress to Jindal Steels. This plant uses latest technology in steel making. The 3 million tonnes annual capacity hot rolled coil plant is one of the most modern plants in the world. It requires less space, less energy, high labour productivity and involves less cost of production. Producing thin strips is a special quality of this plant.

Mini Steel Plants. In addition to the integrated steel plants, a large number of decentralised secondary units produce steel by using steel scrap/sponge iron as raw material and electric arc furnace and induction furnace for processing. With capacity varying from ten thousand to five lakh tonnes, these are known as mini steel plants. It is easy to construct such plants and their gestation period is short. While integrated steel plants mainly produce mild steel in bulk, the mini steel plants produce mild steel as well as alloy steel including stainless steel. Most of the mini steel plants are located in areas far away from the integrated steel plants so that they can meet the local demands there (*see* **Fig. 25.5**). Currently, about 200 units with an installed capacity of 12 million tonnes have been commissioned and have started commercial production. Other units are at various stages of implementation. This sector experienced rapid growth in 1970s but remained more or less stagnant in 1980s. The new Industrial Policy announced in July 1991 has removed iron and steel from the list of industries reserved for the public sector and also exempted it from the requirements of compulsory licensing. According to the provisions of this policy, no industrial licence is required for the establishment of iron and steel plants of any capacity in the private sector, except for locations within 25

TABLE 25.6. Imports of Iron and Steel

Year	1960-61	1970-71	1980-81	1990-91	2000-01	2019-20	2020-21	2021-22
Quantity in thousand tonnes	1325.2	683.4	2,031.1	1,920.5	1,613.6	14,288	10,857	10,460
Value in ₹ crore	123	147	852	2,113	3,569	75,855	61,288	94,053

Source : Economic Survey 2022-23, Statistical Appendix, pp. 112-14.

km of the city with a population of 10 lakh as per the 1991 census. Entrepreneurs are, therefore, free to set up steel plants of any capacity, subject to locational restrictions.

International Trade. India is both an importer and an exporter of iron and steel as is clear from the following brief description. However, our imports are much higher than the exports.

Imports. Production of iron and steel in the country falls short of our demands and India has to spend crores of rupees to import various items.

A look at **Table 25.6** reveals that imports of iron and steel by India have shown varying trends with respect to quantities imported and their value in terms of money. The production picked up quickly in the second Five Year Plan. Due to spurt in production and slackness in demand, the industry faced problem of stockpiles for a short duration from 1974 to 1977 and even exported some steel. But this boom could not survive for long time and soon the surplus was replaced by shortages, resulting in heavy imports.

Some quantity of steel is always needed to be imported, specially those grades and qualities which are required in small quantities and do not justify setting up of production capacities. Typical items of imports are rolled coils, cold rolled coils, semis and steel scrap.

Exports. Value addition in the Indian export basket has been a major trend with exports mainly consisting of hot rolled coils, cold rolled coils, colour coated sheets, galvanised plain (GP) sheets, galvanised corrugated (GC) sheets, and pig iron.

Problems of Indian Iron and Steel Industry

1. Capital. Iron and steel industry requires huge capital investment which a developing country like India cannot afford. Many of the public sector integrated steel plants have been established with the help of foreign aid.

2. Lack of Technology. Throughout the 1960s and upto the oil crisis in mid-1970s, Indian steel industry was characterised by a high degree of technological efficiency. This technology was mainly from abroad. But during the following two decades after the oil crisis, steep hike in energy costs and escalation of costs of other inputs, reduced the margin of profit of the steel plants. This resulted in lower levels of investment in technological developments. Consequently, the industry lost its technology edge and is now way behind the advanced countries in this regard. Material value productivity in India is still very low. In Japan and Korea, less than 1.1 tonnes (and in several developed countries 1.05 tonnes) of crude steel is required to produce a tonne of saleable steel. In India, the average is still high at 1.2 tonnes. Only about ₹ 200 crore is invested in R & D activities which is hardly 0.15 to 0.25 per cent of the sales turnover. Improvement in the yield at each stage of production, particularly for value added products will be more important in the coming years.

3. Low Labour Productivity. The per capita labour productivity in India is at 90-100 tonnes which is one of the lowest in the world. The labour productivity in Japan, Korea and some other major steel producing countries is about 600-700 tonnes per man per year. At Gallatin Steel a mini mill in the U.S. there are less than 300 employees to produce 1.2 million tonnes of hot rolled coils. A comparable facility in India employs 5,000 workers. Therefore, there is an urgent need to increase the productivity which requires retraining and redevelopment of the labour force.

4. Inefficiency of public sector units. Most of the public sector units are plagued by inefficiency caused by heavy investment on social overheads, poor labour relations, inefficient management, under-utilisation of capacity, etc. This hinders proper functioning of the steel plants and results in heavy losses.

5. Low potential utilisation. The potential utilisation in iron and steel is very low. Rarely the potential utilisation exceeds 80 per cent. For example, Durgapur steel plant utilises only 50 per cent of its potential. This is caused by several factors, like strikes, lockouts, scarcity of raw materials, energy crisis, inefficient administration, etc.

6. Heavy demand. Even at low per capita consumption rate, demand for iron and steel is increasing with each passing day and large quantities of iron and steel are to be imported for meeting the demands. Production has to be increased to save precious foreign exchange.

7. Shortage of metallurgical coal. Although India has huge deposits of high grade iron ore, her coal reserves, especially high grade coking coal for smelting iron are limited. Many steel plants are forced to import metallurgical coal. For example, steel plant at Vishakhapatnam has to import coal from Australia. Serious thought is now being given to replace imported coal by natural gas from Krishna-Godavari basin in this plant.

8. Inferior quality of products. Lack of modern technological and capital inputs and weak infrastructural facilities leads to a process of steel making which is more time consuming, expensive and yields inferior quality of goods. Such a situation forces us to import better quality steel from abroad. Thus there is urgent need to improve the situation and take the country out of desperate position.

ALUMINIUM SMELTING

Aluminium smelting is the second important metallurgical industry of India, next only to iron and steel industry. It plays a crucial role in the overall industrial development of the country. Its elasticity, flexibility, good conductivity of electricity and heat, and its capacity to be modulated into any desired shape has made aluminium a universally accepted metal. It is widely used in a large number of industries including generation and distribution of electricity, manufacturing of aeroplanes, railway coaches and bus bodies, building and architectural activities, defence and nuclear accessories, household utensils, packaging and for making coins. It is gaining popularity as a substitute to several other metals like steel, copper, zinc, lead, etc. in a large number of industries.

The per capita consumption of aluminium in India is barely 500 gram against 5.9 kilogram in the U.S. and 3.6 kilogram in Brazil.

Localisation

The production of one tonne of aluminium requires approximately 6 tonnes of bauxite, 0.26 tonnes of caustic soda, 0.09 tonnes of lime, small quantity of cryolite, aluminium flouride, calcium flouride, anthracite (calcined) and soda ash and 18,573 kWh of electricity. About 30-40 per cent of the production cost of aluminium is accounted for by electricity alone. This clearly indicates that the availability of bauxite and electricity are the two most significant factors which influence the localisation of this industry. Odisha and Gujarat are the major producers of bauxite in India.

Growth and Development

Manufacturing of aluminium metal commenced in 1886 and fabrication of utensils from imported metal started in 1929. Indian Aluminium Company started its production in 1938. Production of virgin aluminium from the indigenous bauxite ore made good progress during World War II. In fact, modern aluminium manufacturing is a war-born industry. The credit for this phenomenal growth goes to the Aluminium Corporation of India Ltd. It was formed in 1937 as a public limited company and its plant at Jaykaynagar in West Bengal started production of alumina in 1942 and that of aluminium in 1944. The Indian Aluminium Company Ltd. (INDAL) started fabrication of sheets from imported alumina in 1943 and production of alumina from indigenous bauxite in 1948. Its plant was set up at Alupuram in Kerala. These were the only two companies producing aluminium in the country till 1960. During the Second Five Year Plan, the demand for aluminium increased and two more plants at Hirakud in Odisha and Renukut in Uttar Pradesh were established by Indian Aluminium Company Ltd. (INDAL) and Hindustan Aluminium Corporation (HINDALCO) respectively.

Public sector has also contributed to the growth of aluminium industry. The Bharat Aluminium Company Ltd. (BALCO) came into being in 1965. Its first plant with a capacity of one lakh tonnes was set up at Korba in Chhattisgarh. It started production of alumina in 1973 and aluminium in 1975. The second plant by this company was established at Ratnagiri in Maharasthra. This plant has a capacity of 50,000 tonnes. Madras Aluminium Company Ltd. (MALCO) went into operation in 1965. Its plant is situated at Mettur. The National Aluminium Company Ltd. (NALCO) came into being in 1981. It has set up an integrated aluminium plant at Damanjodi near Jeypur in Koraput district of Odisha. This is the largest complex in India. The smelter with a capacity of about 2,18,000 tonnes a year has been set up at Angul in Dhenkanal district. Commercial production of aluminium started in 1988-89. As a result of expansion, the bauxite mining capacity of NALCO has doubled to 48 lakh tonnes a year and the alumina refinery at Damanjodi nearly doubled to 15.75 lakh tonnes. The new expansion plan has led to raising of the bauxite mining capacity to 63 lakh tonnes, alumina to 21 lakh tonnes, aluminium to 4.6 lakh tonnes and captive power generation to 1,200 MW. NALCO, the lowest cost producer of aluminium in the world, at present ranked among the top 10 global companies and is second to HINDALCO in India.

The Indian aluminium industry has taken giant strides following its decontrol in 1989 and launch of the economic liberalisation programme in 1991. The total production was 199 thousand tonnes which increased to 900 thousand tonnes in 2014-15. After this, the production declined primarily due to power shortages and interruptions in the supply of bauxite.

Imports. Earlier, India was almost self-sufficient with respect to supplies of aluminium. But the demand for aluminium has increased considerably in the past few years due to rapid progress in different spheres of economic activity. As such, India has started importing aluminium in a big way.

COPPER SMELTING

Though copper has been used for various purposes since time immemorial, the development of copper smelting industry took place only recently. Earlier, efforts to smelt copper proved abortive. Indian Copper Corporation was set up in 1924 and a plant was set up at Ghatsila in Singhbhum district of Jharkhand. The Hindustan Copper Ltd. came into being in 1967. It took over Indian Copper Corporation in 1972.

At present, copper is produced at only two centres. One is located at Maubhandar near Ghatshila in Singhbhum district (Jharkhand) and the other at Khetri in Jhunjhunu district (Rajasthan). The smelter at Maubhandar (Ghatshila) receives copper ore from Mosabani, Rakha, Dhobani Rajdah, Tamapahar and Turamdih. All these areas are located in Singhbhum district. It was the only producer of copper till 1971 and it produced about 9.3 thousand tonnes of copper annually. The smelting unit has been replaced by blister copper unit with a capacity of about thirty thousand tonnes. HCL plans to double its capacity at a marginal cost.

The Khetri Copper Complex at Khetri has been erected by Hindustan Copper Ltd. It is an integrated copper mining-cum-metallurgical plant. The smelter was commissioned for large scale production of electrolytic copper in 1974. It receives copper ore from Khetri, Kolihan (5 km south of Khetri), Chandmari (8 km south of Khetri) and Dariba in Alwar district about 150 km away. The copper smelter at Khetri has an installed capacity of 31,000 tonnes of copper metal per year. HCL plans to expand its capacity from 31,000 tonnes per annum to one lakh tonnes per annum at an estimated cost of ₹ 560 crore. The plant also has a production capacity of 2 lakh tonnes of super phosphates and 1.82 lakh tonnes of sulphuric acid per annum. The Malanjkhand mines in Balaghat district of Madhya Pradesh supplement the supply of copper ore to Khetri. For treatment of Malanjkhand concentrates the capacity of smelter and refinery plants at Khetri have been increased to 45,000 tonnes per annum. Agnigundala Copper-Lead project in Guntur district of Andhra Pradesh has started production.

Sterile Industries is a private sector company which uses imported copper concentrates to produce cathodes. Its smelting plant was commissioned at Tuticorin (Tamil Nadu) in 1997-98. The initial capacity of this plant was 60,000 tonnes which was raised to about 1.5 lakh tonnes in 2000-01.

Birla Copper Ltd. is a division of Indo-Gulf Fertilisers which has set-up a copper project at Dahej in Gujarat at an estimated cost of ₹ 1,850 crore. Its initial capacity was 1 lakh tonnes which was raised to 1.5 lakh tonnes later on. The smelter is based on imported copper concentrates.

Swil Copper Ltd. has set up a plant at Bharuch in Gujarat under the technical assistance from Boliden of Sweden. The plant has an annual capacity of 50,000 tonnes and is based on copper scrap. Most of the copper scrap is received from the USA, African countries and the Asia Pacific Region.

Production of copper in India increased substantially after 1970-71 when Khetri Copper Complex in Rajasthan became operative. Production of copper and copper products reached an all time high of 853.8 thousand tonnes in 2008-09 after which declining trends in production have been observed which is primarily due to low rate of productivity of our copper smelting plants. The red metal's usage is regarded as a reliable indicator of economic growth in the industrial community as it has a variety of sophisticated applications. However, the consumption rate is likely to increase with the overall industrial growth in the country. The surge in demand has attracted investments from the private sector.

Imports. Presently, India produces only half of her requirements of copper and the remaining half is imported. The main suppliers of copper to India are Zambia, Zaire, Chile, the USA, Canada and some West European countries.

LEAD AND ZINC SMELTING

Lead

The first lead smelting plant was set up at Tundoo near Dhanbad (Jharkhand) in 1942-43 by a private company named as the Metal Corporation of India. The commercial production by this plant commenced in 1945. It was taken over by Hindustan Zinc Ltd. (HZL), a public sector enterprise in 1965. The supply of lead cocnentrates is obtained from the Zawar and Rajpur-Dariba areas of Rajasthan. The plant has an installed capacity of 8,740 tonnes per annum. The HZL has set up another plant at Vishakhapatnam. It is primarily based upon the imported lead concentrates although some ore is obtained from Agnigundala also. Its installed capacity is 22,000 tonnes per annum. The lead-zinc complex at Chanderiya was constructed in 1991 under the British aid programme. It has an annual capacity of 35,000 tonnes of lead.

Indian Lead Ltd. (ILL) is a new plant with a rated capacity of 24,000 tonnes of lead per annum. The plant is based on lead scrap and concentrates. Varying trends are found in production of lead in India.

Imports. Production of lead always falls short of demand and large quantities of lead have to be imported every year to meet the growing demand.

Zinc

At present, there are four zinc smelters in the country, one each at Alwaye (Kerala), Debari and Chanderia (Rajasthan) and Vishakhapatnam (Andhra Pradesh). The Alwaye plant is wholly dependent upon imported supplies of zinc concentrates. It started production in 1967 and has an installed capacity of about fifty thousand tonnes per annum. It plans to raise its production from 20,000 to 30,000 tonnes. The Debari plant started production in 1968. It has installed capacity of smelting 18,000 tonnes of zinc annually which is being raised to 45,000 tonnes. Besides, it also produces 87,000 tonnes per annum of sulphuric acid, 190 tonnes of cadmium, 26,000 tonnes of phosphoric acid and 72,000 tonnes of high super phosphate annually. The ore is supplied by Balaria

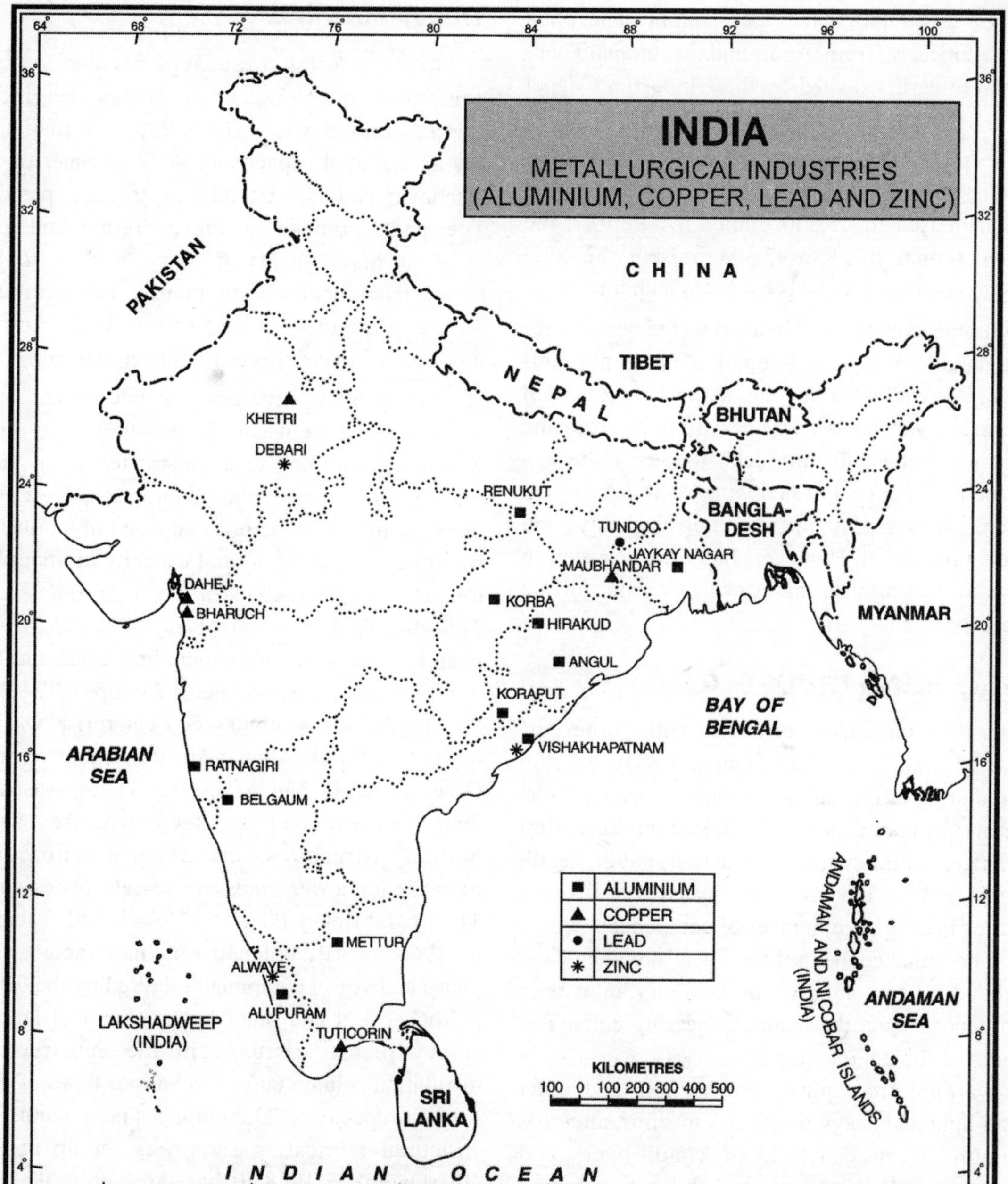

FIG. 25.7. India : Metallurgical Industries (Aluminium, Copper, Lead and Zinc)

and Rajpur-Dariba mines. However, the supplies of ore from these mines are not sufficient and half of the zinc concentrates have to be imported. The Chanderia smelting plant was set up in 1991. It is based upon the supplies of ore from Bhilwara and Chittaurgarh. It is an HZL enterprise and has an annual capacity to produce 70,000 tonnes of zinc, 35,000 tonnes of lead and 74 tonnes of silver. Vishakhapatnam produces over 2.8 thousand tonnes of zinc. This plant has an installed capacity of 30,000 tonnes which is being increased to 40,000 tonnes per annum. The plant receives zinc ore from Agnigundala mines. These mines were earlier owned by the Hindustan Copper Ltd.

The production of zinc in India started after 1967. The total production was 20,800 tonnes in 1973 which increased to 45,500 tonnes in 1980-81 and 1,70,000 tonnes in 1996-97. At present the total production of zinc in India is over two lakh tonnes.

With the expansion of industries, the demand for zinc is rapidly increasing. Keeping the present trends in mind, it is estimated that the demand for zinc will increase to over six lakh tonnes in 2020 from the present level of about three lakh tonnes. India can meet about 90 per cent of her demands if proper care is taken to augment the smelting capacity and increase production. However, India has resorted to imports of zinc to meet the growing demand for the commodity.

ENGINEERING INDUSTRIES

Engineering industries cover a wide range of industries and contribute substantially to the manufacturing of industrial machines, machine tools, transport, transmission and telecommunication equipments, building and construction, public health and sanitation and modern sophisticated goods. These industries need a large number of components which they do not produce themselves. Therefore, they have stimulated a large number of ancillary industries which depend upon the main engineering units. The products of the engineering industries cover a wide range varying from pins, screws, nuts and bolts through light and heavy machinery to ships, aircrafts, automobiles, rail coaches, air conditioners and computers, etc. Thus engineering industries are labour intensive and capital demanding and are located mainly in those areas where these major components are available. At present, engineering industries account for about one-third of the production capital in the organised sector, one-third of the value of output and nearly 30 per cent of the employment is in the public sector. Further, these industries contribute about 10 per cent of the total exports of the country.

Heavy Machinery

Manufacturing of heavy machinery made a beginning in 1958 when Heavy Engineering Corporation Ltd. was set up at Ranchi in Jharkhand. It has an installed capacity of 80,000 tonnes of heavy machinery and 25,000 tonnes of structures per annum. The plants produce castings, forgings and rolls to meet the requirements of heavy machine tools and heavy machine building plants. The Mining and Allied Machinery Corporation Ltd., Durgapur produces various items of mining equipment.

Structurals. Structurals include a wide range of items ranging from simple structures like ordinary warehouse building to sophisticated items such as railway bridges, steel plant buildings, etc. At present, there are over 130 factories engaged in fabrication of structural items with a total capacity of about 7 lakh tonnes. The Tungabhadra Steel Products Ltd. at Tungabhadra dam in Karnataka was set up in 1947. It manufactures structurals including gates and hoists, transmission towers and penstock pipes. The Triveni Structurals Ltd., a public sector enterprise, was set up in 1965 at Naini (Allahabad). It is a joint venture of Govt. of India and M/s. VOEST of Austria and manufactures complex steel structures such as building structures, crane construction, power transmission towers, pressure vessels, plate work, etc. The Bharat Heavy Plate and Vessels Ltd., established in 1966 at Vishakhapatnam, manufactures heavy plates and vessel equipment required by the fertilizer, petrochemical and other heavy chemical industries. Heavy plates are used in the construction of distillation column tanks and various types of vessels. It has a capacity of 23,000 tonnes per annum. A steel structural fabricating shop was set up at Ranchi (Jharkhand) in 1968. It has three units *viz.*, Heavy Machine Building Plant, Foundry Forge Plant and Heavy Machine Tools Plant. These plants produce castings, forgings and rolls to meet the requirements of heavy machine tools and heavy machine building plants. Another plant to manufacture high pressure boilers and boiler fittings has been set up at Tiruchchirappalli. Messrs Jessop and Co. Ltd. Kolkata, and Richardson & Cruddas Ltd., Mumbai produce machinery and heavy structurals. M/s. Larsen & Tubro Ltd., at Powai (Mumbai) produce machinery, heavy equipment for steel, under carriage parts, chemical, petro-chemical and cement industries. Drills for drilling holes in rocky areas are manufactured at Naroda near Ahmedabad.

Heavy machinery and structural industry has made much progress after Independence and the country is now in a position to export several items which were earlier imported from the western countries.

INDUSTRIAL MACHINERY MANUFACTURING

India today produces a wide spectrum of industrial machinery including textile, cement, sugar, paper, chemical, mining, agricultural machinery, pharmaceuticals, fertilizer, dairy, metallurgical, leather and food processing industries. These industries got a good start during the Second World War but the real progress has been achieved during the plans period.

***(i)* Textile Machinery.** India is an important producer of textile machinery. The country is in a position to export textile machinery after meeting the demand of the domestic market. Asian and African countries are the main buyers. Textile machinery manufacturing commenced with the Textile Machinery Corporation Ltd. (TEXMACO) at Mumbai in 1939. Later on, units were also started at Kolkata, Coimbatore, Ahmedabad, Ludhiana and Gwalior. The products include cording engines, ring frames, draw frames, fly frames, speed frames, sizing frames, open width bleaching plant, hot air stentors, mercerisers, polymerisers. The industry comprises over 1,446 machinery and components manufacturing with over 600 units producing complete machinery and other units mainly into the production of parts and accessories of textile machinery. To encourage the textile industry, the industry has been de-licenced and FDI upto 100 per cent under automatic route as well as technology collaboration is allowed freely.

***(ii)* Jute Mill Machinery.** Currently 63 units are engaged in manufacturing jute mill machinery. Majority of them are concentrated in the Kolkata-Haora region.

***(iii)* Sugar Mill Machinery.** The main centres of sugar mill machinery are at Kolkata, Chennai, Mumbai, Allahabad, Pimpri and Yamunanagar. About 12 units are engaged in manufacturing sugar mill machinery.

***(iv)* Cement Mill Machinery.** There are 13 units engaged in manufacturing cement mill machinery. These units are largely located in Tamil Nadu, West Bengal, Odisha and Karnataka. The main centres are Mumbai, Chennai, Pune, Delhi, Sahibabad and Durgapur.

***(v)* Paper Mill Machinery.** This machinery is manufactured by nearly 20 firms. These are spread over in West Bengal, Maharashtra, Jharkhand, Punjab and Odisha. The main centres are Titagarh, Jamshedpur and Rourkela.

***(vi)* Chemical and Pharmaceutical Machinery.** There are about 70 projects engaged in the manufacture of chemical and pharmaceutical mill machinery manufacturing a wide range of machinery such as sulphuric acid plants, superphosphate plants, water treatment plants, solvent extraction plants, heat exchangers, pressure vessels, crystallizers, evaporators and a host of other machines.

***(vii)* Agricultural Machinery.** Agricultural machinery covers a wide range of production including earth moving machinery, excavators, bulldozers, power tillers, steel discs, threshers, harvestors, cutters, tractors and many more implements and machines. Faridabad, Kolkata, Chennai, Vadodara, Hyderabad, Dehagaon and Sonepat are the main centres of producing agricultural machinery.

***(viii)* Coal Mining and Washery Machinery.** The Coal Mining Machinery Project at Durgapur produces grinding mills, rotary fans, rotary kilns, power shovels, coal cutters, loaders, conveyors, haulages, electrical winders, booster fans, axial fans, automatic cape keeps, safety hooks, shutter cars and mine locomotives. Drilling rigs and other equipments for oil exploration are also produced here. Kolkata and Jamshedpur are the other important producers. Other machines, such as power driven pumps, diesel engines, building and construction machinery, weighing machinery, dairy machinery, oil mill machinery, rubber machinery, etc. are also produced in India.

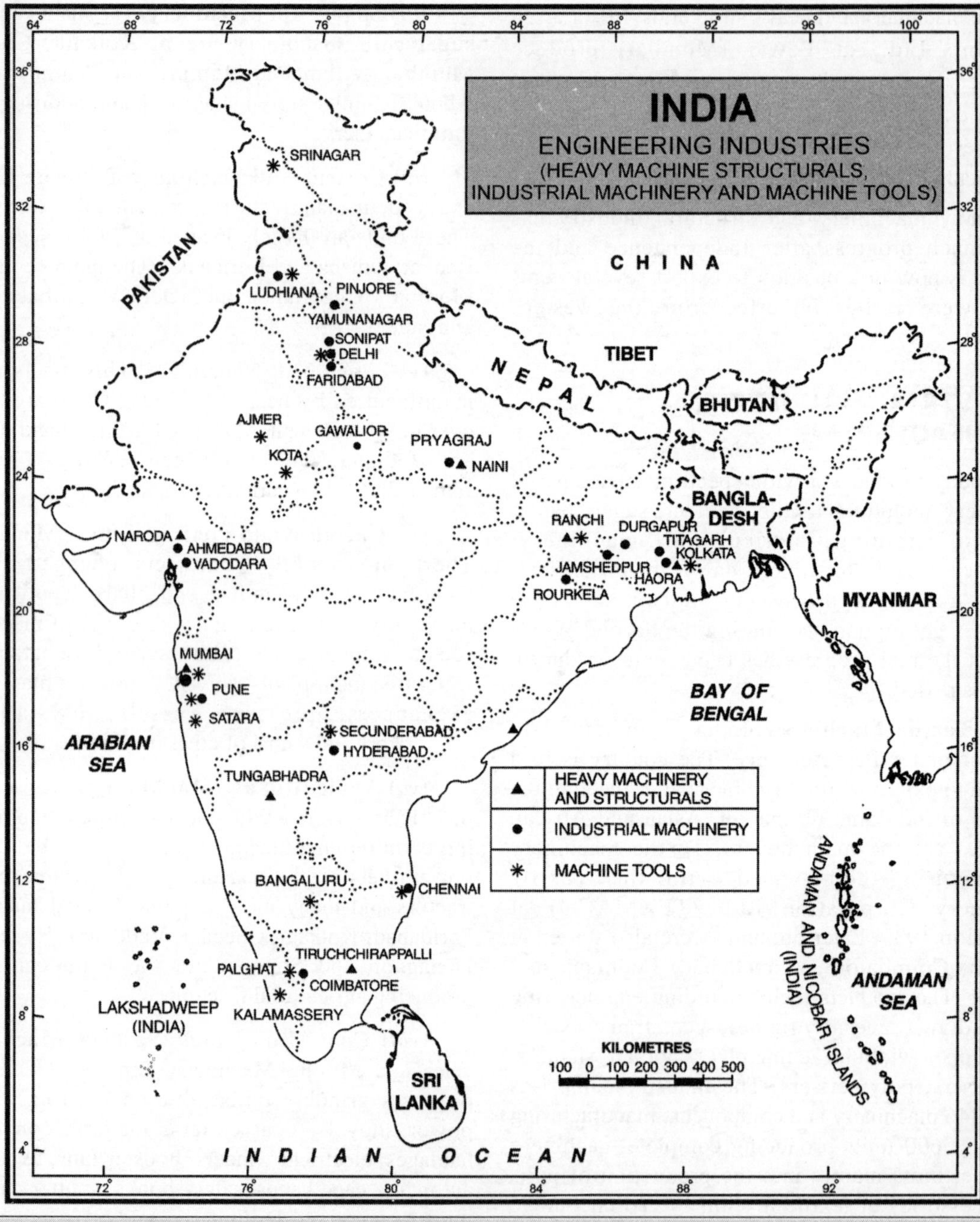

FIG. 25.8. India : Engineering Industries (Heavy machine, structurals, industrial machinery and machine tools)

MACHINE TOOLS

The machine tool industry is the core industry and forms the basis of developing engineering industry. The manufacturing of machine tools commenced in 1930s. Kirloskar Brothers Ltd. is the pioneer company in this regard. The industry got an impetus during the Second World War but rapid progress has been made during the plan period.

The Hindustan Machine Tools (H.M.T.) is the first large scale modern machine tool factory set up in public sector at Bengaluru in 1953, with Swiss collaboration. It is a multi-unit, multi-product company which is the largest manufacturer of machine tools in the country. It is a leading player in the machine tool industry and produces a wide variety of machine tools, wrist watches, tractors, printing machinery, high precision lathes, radial drilling machines, grinding machines, gears, shapers, gear hobbing machines, die castings, lamp making machinery, lamp and lamp components, etc. Two units of the company are located at Bangalore and the others are at Pinjore (Haryana), Kalamassery (Kerala), Hyderabad, Srinagar and Ajmer.

Besides H.M.T., there are other public and private companies which produce different types of machine and hand tools. ***The Heavy Machine Tools Plant*** at Ranchi started production in 1966. It was set up with Czech assistance. With a capacity of 10,000 tonnes, it produces axle turning, radial drilling machines, burnishing lathes, wheel lathes, central lathes, double column planing machines, etc. It also produces special machine tools for the railways. ***The Praga Tools Ltd.,*** another public sector enterprise at Secunderabad is mainly meant for defence equipment and stores. It also produces machine tools and accessories, precision tools, auto and diesel parts and railway components. ***The National Instruments Factory at Kolkata*** along with its Jadavpur unit produces precision instruments such as drawing instruments, office equipment, survey instruments, microscopes, binoculars, optical and vision sighting equipment and blood pressure equipment. ***The Instrumentation Ltd.*** has set up a precision plant at Kota and mechanical instruments plant at Palghat (Kerala). The Kota plant manufactures magnetic, electromagnetic and electronic instruments. The Palghat plant produces hydraulic and pneumatic instruments.

Currently there are around 200 machine tool manufacturers in the organized sector as also around 400 small scale units. They are dispersed all over the country with greater concentration at Mumbai, Kolkata, Delhi, Bangaluru, Chennai, Thane, Pune, Coimbatore, Satara, Ludhiana and Amritsar. Today 65 per cent of machine tool requirement in India is met by indigenous manufacturers. The industry lacks in design and engineering capability to undertake high procision machines. Due to technology gap in the field of metal cutting machine tools, metal forming technology as well as research and development initiatives are encouraged to bridge the gaps. The industry has been delicenced and FDI upto 100 per cent under automatic route as well as technology collaboration is allowed freely.

TRANSPORT EQUIPMENT INDUSTRY

This is vast and varied type of engineering industry which includes railway equipment (locomotives, coaches and wagons), auto-vehicles (trucks, buses, cars, three wheelers, two wheelers, shipbuilding, aircraft, and cycle manufacturing.)

RAILWAY EQUIPMENT

The manufacturing of railway equipment has made rapid strides since Independence. India today is not only self-sufficient with regard to production of railway equipment but is in a position to export substantial quantities of rails, locomotives, coaches, wagons, signalling equipment, etc. Rolling stock comprising the locomotives, coaches and wagons constitute the bulk of production.

LOCOMOTIVES

In the pre-Independence era, steam locomotives were assembled and partly manufactured at Ajmer in the B.B.&C.I. Railway workshop and at Kanchanpara in the workshops of Bengal-Assam Railway. Later on, three units started producing locomotives. They are : (*i*) Chittaranjan Locomotive Works, Chittaranjan (*ii*) The Diesel Locomotive Works at Varanasi and (*iii*) Tata Engineering and Locomotive Works, Jamshedpur.

The Chittaranjan Locomotive Works (CLW) at Chittaranjan has been set up in Burdwan district of West Bengal. It obtains iron from Asansol and Durgapur which are respectively 25 km and 67 km away from Chittaranjan. Raniganj Coalfields are also nearby. Hydroelectricity comes from DVC. Production started in 1950 and this locomotive works produced 2,351 steam locomotives till Dec., 1972. Since then, the production of steam locomotives has been discontinued and their place has been given to electric locomotives. In fact, the production of electric locomotives started in 1961 itself and since then their production has been increasing. There is no likelihood of resuming the production of steam locomotives because they use coal and emit huge quantity of smoke, thereby polluting the environment. Moreover, steam locomotives have less traction power and energy for a quick acceleration than that of electric or diesel locomotives.

The Diesel Locomotive Works at Varanasi started assembling locomotives from imported components in 1964. Its initial capacity was 150 broad gauge diesel locos per year. The first loco rolled out of the assembly lines in 1964 itself.

The Tata Engineering and Locomotive Works (TELCO) is a private sector unit and is located at Jamshedpur. It was set up in 1951 and it started production of steam locomotives in 1952. However, the production of steam locomotives was stopped in June, 1970 on the termination of company's commitment to the railways. It produced about 1200 steam locos during this period.

Bharat Heavy Electricals Limited (BHEL), **Bhopal** has developed capability to manufacture electric locomotives for the Indian railways. Diesel component works has been set up by the railways at **Patiala** for manufacturing and repairs of components of diesel locos and sub- assemblies.

A Wheel and Axle plant was set up at *Bengaluru* in 1984 to cut down imports in this field. The plant has performed exceedingly well and helped the railways to save valuable foreign exchange.

Other Railway Equipment. Rails and sleeper bars are manufactured in iron and steel works at Bhilai and Jamshedpur and wheels and axles at Durgapur, Jamshedpur and Rourkela. Coaches and wagons are manufactured both in public and private sectors.

COACHES

The Integral Coach Factory at Perambur near Chennai started production of railway coaches with Swiss collaboration in 1955. It now produces almost all types of coaches including air conditioned coaches, electric and diesel rail cars and electrical multiple units. It has an installed capacity of 1,150 coaches per annum.

The Bharat Earth Movers Limited (BEML) at Bengaluru manufactures coaches and electrical multiple units. It has an installed capacity of 400 broad gauge coaches per annum. A Rail Wheel Factory has also been set up at Bengaluru to cut down imports in this field.

Rail Coach Factory at Kapurthala in Punjab was set up in March, 1988. It has an installed capacity of 1000 coaches per annum. It is manufacturing AC 3-Tier coaches in addition to other coaches. Integral coach factory at **Rai Bareli** in Uttar Pradesh has also started functioning.

Some private companies are also manufacturing coaches and wagons. Jessop and Co. Ltd. at Kolkata has an installed capacity of 350 coaches and electrical multiple units. Textile Machinery Corporation (TEXMACO), Kolkata, Braithwaite Co. (India) Ltd., Kolkata, Burn and Co., Kolkata are some other companies which produce railway coaches.

WAGONS

Wagon manufacturing industry is fully geared to meet the growing demands of the railways. Most of the wagons are produced in private sector. There are 13 units with an installed capacity of 30,625 wagons (in terms of 4 wheelers) in private sector and three railway workshops with an annual capacity of about 4,000 units. About 60 per cent of wagons are produced in West Bengal and the rest come from Maharashtra, U.P., Punjab and Delhi.

AUTOMOBILE INDUSTRY

Automobile industry globally is one of the largest industries and is a key driver of economy. Owing to its deep linkages with several key segments of industry, automobile industry has a strong multiplier effect on the economy. A well developed Indian automobile industry ably fulfills the catalytic role by producing a wide variety of vehicles such as passenger cars, light, medium and heavy commercial vehicles, multi-utility vehicles, scooters, motor cycles, mopeds, three wheelers, etc.

Automobile industry did not exist in India in the real sense before Independence. Only assembly work was done from the imported parts. General Motors (India) Ltd. started assembling trucks and cars in 1928 in their factory at Mumbai. Ford Motor Co. (India) Ltd. started assembling of cars and trucks at Chennai in 1930 and at Mumbai in 1931. The real development of the industry began with the establishment of the Premier Automobiles Ltd. at Kurla (Mumbai) in 1947 and the Hindustan Motors Ltd. at Uttarpara (Kolkata) in 1948. Automobile industry in India has made considerable progress during the last three decades. Today, it is one of the most vibrant sectors of economy.

With gradual liberalisation of the automobile industry since 1991, more and more players have set up manufacturing facilities in India. At present there are 15 manufacturers of passenger cars and multi-utility vehicles, 9 manufacturers of commercial vehicles, 14 of two/three wheelers and 14 of tractors besides 5 manufacturers of engines. This industry currently employs 13.1 million people both directly and indirectly and contributes nearly 6 per cent to the national GDA. The industry is also making a hefty contribution of nearly 20 per cent to kitty of indirect taxes of the government.

Localisation

The automobile industry tends to be located near iron and steel producing centres because steel is the basic raw material used in this industry. The proximity of places producing tyres, tubes, storage batteries, paints and other ancillary industries is considered to be an added advantage. Port cities also find favour with this industry because of the import and export facilities offered by such places. Of late, automobile industry has become market oriented and prefers those locations which offer ready market for the manufactured vehicles. Under the Government plans for decentalization of industries, some locations in remote and industrially backward areas are given priority.

Production and Distribution

The Indian automobile sector is described as the *sun-rise* sector. During the last decade, the sector has been growing at approximately 12-15 per cent per annum. However, in 2008-09 the automobile sector was badly hit due to global economic slow down. But the industry soon recovered from this shock and bounced back on high growth track. Today, India is the seventh largest vehicle manufacturer of automobiles in the world, second largest manufacturer of two wheelers, largest manufacture of tractors and fifth largest manufacturer of commercial vehicles.

Mumbai, Chennai, Jamshedpur, Jabalpur and Kolkata are the chief centres producing automobiles. These centres produce almost all sorts of vehicles including trucks, buses, passenger cars, three wheelers and two wheelers. Motor cycles are also manufactured at Faridabad, Gurugram and Mysore. Scooters are also manufactured at Lucknow, Satara, Akurdi (Near Pune), Panki (near Kanpur) and Odhav (Ahmedabad dist.). Maruti Udyog Ltd. (MUL) at Gurgaon in Haryana started production of passenger cars in 1983. At present there are 38 units engaged in the production of automobiles producing four wheelers, three wheelers and two wheelers.

Commercial Vehicles. Commercial vehicles industry is divided into two broad segments, *viz.*, passenger and goods. The passenger segment is largely controlled by state owned transport undertakings (STUs) while goods vehicles are generally manufactured in private sector. The manufacture of commercial vehicles started in 1950s and the industry registered a rapid growth in the post-liberalisation period as a result of incentives given by the government. The production of commercial vehicles (including buses, trucks, tempos, 3 and 4 wheelers) increased from an insignificant of 8.6 thousand in 1950-51 and to 145.5 thousand in 1990-91 and 910.2 thousand in 2011-12.

Currently 7 companies are engaged in manufacturing buses and trucks. Tata Engineering and Locomotive Co. Ltd. (TELCO) is the leading producer of medium and heavy commercial vehicles and accounts for over 70 per cent of such vehicles produced in India. Four plants, each at Hyderabad, Pithampur (M.P.), Arson near Rupnagar (Punjab) and Surajpur in U.P. manufacture light commercial vehicles. Premier Automobiles and Mahendra and Mahendra are located in Mumbai, Ashoka Leyland Ltd. and Standard Motor Products of India Ltd. are located in Chennai, Hindustan Motors Ltd. is at Kolkata; and Bajaj Tempo Ltd. is located in Pune.

In addition to the above mentioned manufacturers, Shaktiman trucks are manufactured under the Ministry of Defence and Nissan Jeeps at Jabalpur in collaboration with the Nissan of Japan.

Passenger Cars. A number of companies are engaged in manufacturing passenger cars. Of these, Maruti Udyog Ltd. (MUL) is at the top. It is located at

Gurgaon in Haryana. It started production in 1983 with the collaboration of Suzuki Motor Corporation of Japan. Currently this company produces about four-fifths of total cars produced in India. It produces a variety of models of which Zen, Wagon R, Esteem and Gypsy, Swift desire, etc. are very popular. The company has been facing a lot of problems with respect to labour unrest and is trying to shift its manufacturing activities to Gujarat, where congenial atmosphere for industrial investors prevails Hindustan Motors (Kolkata and Chennai), the Premier Automobiles (Mumbai), Standard Motor Products (Chennai), and the Sunrise Industries (Bengaluru) are other important producers. Several new companies have entered the car manufacturing industry of India after liberalization in 1991. These include Hyuṇdai Motors India at Irrungattukottai (Tamil Nadu), Daewoo of Korea in 1995 at Surajpur (Uttar Pradesh), Telco at Pimpri (near Pune). Honda of Japan has set up a plant in Uttar Pradesh to manufacture 'City'. General Motors has launched Opel Astra. It has tie-up with Hindustan Motors. Ford in collaboration with Mahindra has introduced Ford. Hindustan Motors in collaboration with Mitsubishi of Japan has launched 'Lancer'. Mercedes Benz of Germany in collaboration with Telco is manufacturing E220 and 250D for upper strata of society. Premier Automobiles in collaboration with Fiat-India Auto Limited is manufacturing a number of models. Kia has recently set up a plant at Chennai.

Several factors have made it a buoyant industry in the recent past and the industry has a bright future. Reduction in excise duty on passenger cars has led to reduced car prices and created the potential for an increase in demand. This has helped in growth of the industry to a great extent. The government's Auto Policy also has the stated aim of making India an 'Asian hub' for the manufacture of small cars. The small car segment, which refers to the A and B segments, accounts for over 65 per cent of the market. Giving further push to this segment, therefore, adds impetus to a movement that is already underway. For India to become a large manufacturer of small cars, quality and price are the two basic deciding factors.

The potential for growth is good considering that passenger car penetration in India is a mere 28 car per 1,000 population as compared to 473 per 1000 in U.K., 816 per thousand in U.S.A., 837 in New Zealand and 866 in Iceland. Even Afghanistan (47), Bhutan (57), Indonesia (77) have better penetration than India. The market is projected to grow at a rate of around 7 per cent annually.

Loan facilities have also given impetus to this industry. There were times when a car loan meant tedious trips to bank, reams of paper work and long waiting periods for requisite approvals from the sanctioning authority. Today the scenario is quite different, with customers having fast access to very flexible financing options that suit their different needs. Attractive automobile financing schemes have definitely boosted sales of passenger cars. Today, almost 70 per cent of new cars are financed through auto-loans.

Jeeps. Almost the entire production of jeeps comes from Mahindra, Mumbai. It has a capacity to produce about 13,000 jeeps per annum.

Two Wheelers. Two wheeler industry mainly comprises of motor cycles, scooters, mopeds and scooterettes. The Indian two-wheeler industry made a humble beginning in the early 1950s when Automobile Products of India (API) started manufacturing scooters in the country. Until 1958, API and Enfield were the sole producers. In 1960, Bajaj Auto set up a plant in collaboration with Pioggio of Italy. Two wheeler industry has also made rapid strides. It came a long way from an insignificant production of 0.9 thousand units in 1960-61 to 18.4 million units in 2020-21. The two-wheeler market was opened to foreign competition in the mid-1980s. Practically, all the global giants have been present in India for quite some time. First to come was Suzuki Motor Corporation with TVS in 1984. Honda followed within a year, in a joint venture with Hero Group. Then Kawasaki and Yamaha entered into licence agreement with Bajaj Auto and Escorts respectively. Piaggio has joined up with LML which is planning to expand its capacity to six lakh vehicles per annum. Bajaj Auto is expanding its capacity to two million vehicles per annum. Hero Honda is expanding its Dharuhera plant capacity to over 2,52,000 vehicles a year and has set up a plant at Gurgaon with an investment of ₹ 160 crore. TVS Suzuki plans to invest ₹ 200 crore to expand its production capacity to one million vehicles per

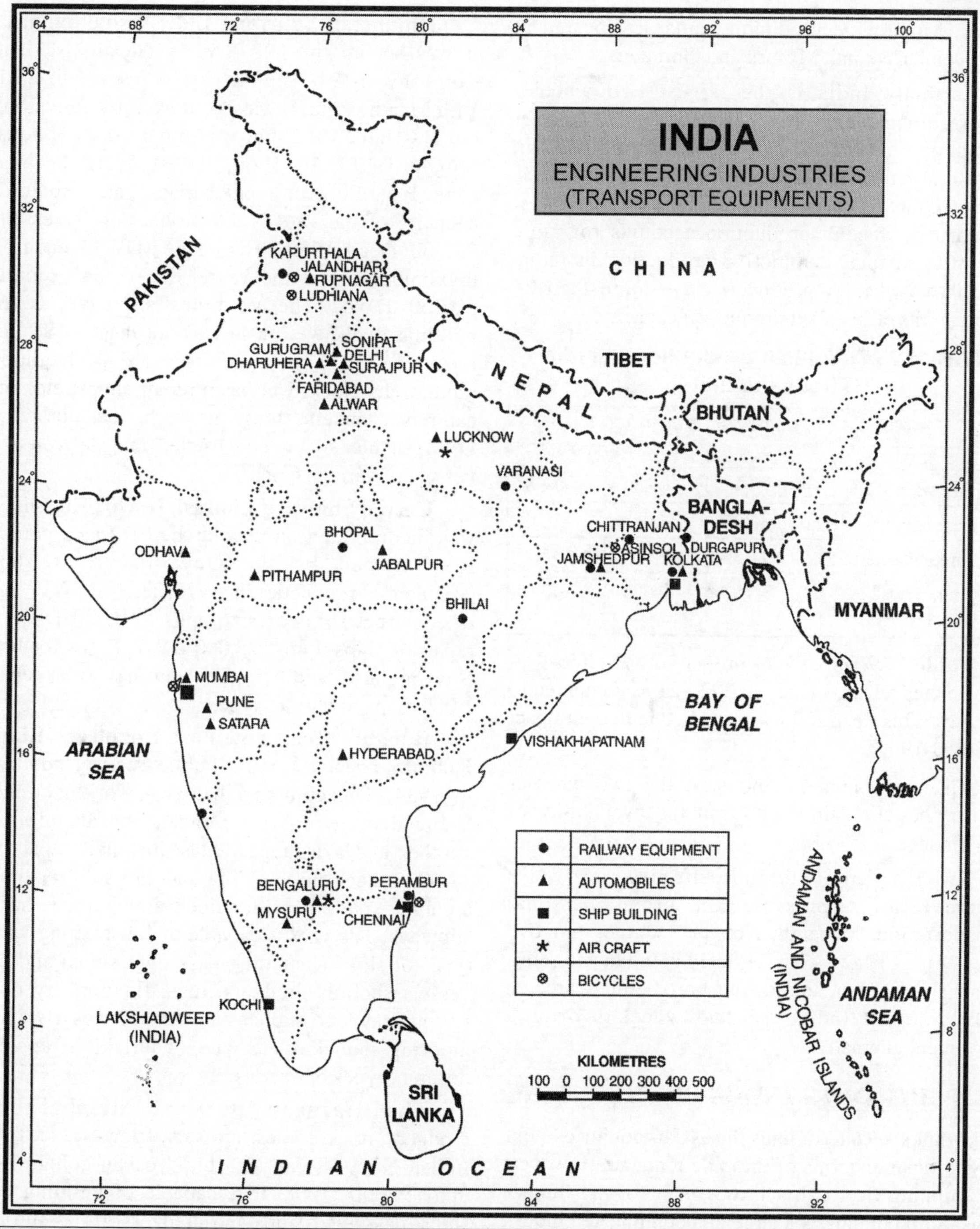

FIG. 25.9. India : Engineering Industry (Transport Equipments)

annum. Yamaha-Escorts, a joint venture has also announced plans to introduce new range of products and expansion facilities in its Surajpur plant. Mumbai, Pune, New Delhi and Kanpur are main centres of scooter manufacturing. Public sector units are located at Hyderabad, Bengaluru, Satara, Lucknow and

Alwar. Motorcycle producing units are located at New Delhi, Chennai, Mysuru and Gurgaon.

Currently India is the largest two wheeler manufacturing country in the world. The two wheeler sector contributes to 7% of manufacturing GDP in India. It also contributes to 2% of GST revenues. It is estimated that India has 190 million two-wheelers. At present two-wheeler production accounts for more than 80% of total automotive production in India. Indian two-wheeler producers have more than 25 million units of installed production capacity.

TABLE 25.7. Vehicle production in India (Thousand units)

Vehicle	2019-20	2020-21	Change (%)
1. Two wheelers	21,033	18,350	–12.8
2. Passenger vehicles	3,425	3,062	–10.6
3. Three wheelers	1,133	611	–46.1
4. Commercial vehicles	757	625	–17.6

Table 25.7 shows that production of almost all the vehicles fell drastically in 2020-21 as compared to 2019-20. This steep fall was mainly due to the impact of Covid-19.

Like passenger car industry, the two wheeler industry has also gained a lot from the availability of easy finance.

Two wheelers are the most effective safety valve which relieves pressure on urban personal transportation. More than 65 per cent of the two wheeler population is concentrated in urban and semi-urban areas. With public transport being scarce in most cities of India, the two wheelers offer a convenient alternative.

SHIP BUILDING INDUSTRY

India ranks second among the Asian countries next only to Japan in terms of shipping tonnage. However, her shipping fleet is much too small for her dimensions. Ship building is a large and complicated industry which requires huge capital investment. Harbours with large space are ideally suited for this industry.

At present, there are four main centres of ship building industry at Vishakhapatnam, Kolkata, Kochi and Mumbai, all in public sector.

Hindustan Shipyard Ltd., Vishakhapatnam. It was set up by M/s Scindia Steam Navigation Company in 1941 and the first ship was launched on 14th March, 1948. It was taken over by Government on 21st Jan., 1952 and was renamed as Hindustan Shipyard Ltd. In 1962, the shipyard became a central public sector enterprise. The shipbuilding capacity of the yard is 3.5 pioneer class vessels of 21,500 Dead Weight Tonnage (DWT) each. The maximum size of the vessel that could be built is 50,000 DWT. The yard has slipways, covered building dock, wet basin and outfit jetty. So far as repair of ships is concerned, the yard has facilities of modern dry dock, wet basin repair shops, etc. and it can repair tankers, ships and submarines upto 70,000 DWT. It has so far constructed and delivered 123 vessels of various types.

Cochin Shipyard Limited, Kochi. This shipyard was incorporated on 29 March, 1972 as a company fully owned by the Government. It started commercial production in 1976. The yard is designed to construct ships of size upto 1,10,000 DWT and repair of ships upto 1,25,000 DWT. Ships for Indian Navy are also built here. The yard has so far repaired 1,000 ships of various types.

Hooghly Dock and Port Engineers Limited Kolkata, became Central Public Sector Undertaking in 1984. The company has two working units in Haora district of West Bengal, one at Salkia and another at Nazirgunge. It has installed capacity of 11,000 tonnes shipbuilding and 125 ship repairs per annum. Apart from dry dock and jetty, it has six slipways. The yard is capable of constructing various types of ships (including passenger ships) and other vessels such as dredgers, tugs, floating dry docks, fishing trawlers, supply-cum-support vessels, barges mooring launches etc. and undertaking repairs of different types of vessels.

The Mazgaon Dock at Mumbai builds dredgers, dock cranes, cruisers, frigates, etc. for the Indian Navy. It can also build ocean-going vessels upto 27,000 DWT. It is capable of building cargo ships, passenger ships and dredgers. It has sub-units at Nhava and Mangalore. Recently it has started constructing submarines, missile boats, destroyers of the Navy and off-shore supply vessels etc for ONGC.

In addition to the above mentioned main centres, there are 33 smaller shipyards manufacturing vessels of small sizes meant for domestic purposes. *Goa Shipyard* Limited (GSL) at Vasco-da-Gama undertakes the manufacture of fibre glass boats, trawlers, dredgers and barges. It has undertaken construction/refitting of variety of vessels for Indian Navy and Coast Guard as well as for non-defence sector. Presently it is building advanced offshore petrol vessels of in-house design.

AIR CRAFT INDUSTRY

The first aircraft industry was set up at Bengaluru in 1940 under the name of Hindustan Aircraft Ltd. It was a private company and was taken over by the government in 1942. This factory was merged into Aeronautics India Ltd. in 1964 to form Hindustan Aeronautics Ltd. (HAL), Bengaluru. Different parts of aircrafts are manufactured at different places due to security reasons. The main divisions of the HAL are : (*i*) A three unit MIG complex comprising the Nashik Division where MIG airframe is manufactured. (*ii*) the Koraput division where the engine for MIG aircraft is manufactured and (*iii*) the Hyderated division where electronic equipment for the MIG is manufactured. Transport aircrafts are manufactured at Kanpur. Recently a factory was set up at Lucknow for producing equipment for aircraft. Among the other major products, mention may be made of Jaguars, Marut, Gnat Fighter Aircraft, Jet Trainer Aircraft, etc. and some helicopters. The first prototype of light combat aircraft (LCA) rolled out in November 1995.

BICYCLES

Bicycle is a poor man's drive and is much suited to the economic condition of Indian masses. It is the most convenient and popular mode of transportation among the lower and middle income groups. Presently, over sixty million bicycles are being used in the country. The first bicycle manufacturing factory was set up at Mumbai in 1940. At present, there are about 13 units manufacturing complete cycles and over 40 units manufacturing cycle parts. The main centres of bicycle production are Mumbai, Asansol, Sonipat, Delhi, Chennai, Jalandhar and Ludhiana. Ludhiana has emerged as the major bicycle production hub in the country. This industry has made rapid strides during the last four decades. Bicycle companies are now focusing on urban markets and are looking to expand their base in the professional and adventure categories. The production has increased from 99 thousand bicycles in 1950-51 to 125 million bicycles in 2019-20. This progress has made India the second largest producer after China. Hero, Atlas, TI and Avon account for 92 per cent of the production in the organised sector.

India has emerged as a major exporter of cycles to Pakistan, Afghanistan, Sri Lanka, Myanmar, Thailand, UAR, Turkey, Indonesia, Malayasia and Singapore. Indian exports of bicycles and components was around ₹ 205 crore in 2015-16. There has been a significant increase in exports to South America and African and European countries since 1991.

The bicycle industry has been de-licenced under the current industrial policy and this sector is qualified for 100 per cent FDI under automatic approval.

LIGHT MECHANICAL ENGINEERING GOODS

India produces a wide range of light mechanical engineering goods. The important items are sewing machines, office equipment, etc. The main centres of sewing machine production are Mumbai, Delhi, Secunderabad and Jalandhar. The production of sewing machines is showing varying trend.

ELECTRIC MACHINERY AND EQUIPMENT MAKING INDUSTRY

Electricity is an integral part of modern life. The increase in generation, distribution and utilisation of electricity has resulted in increased demand for electrical machinery and equipment in the recent past. Consequently, this industry has made rapid strides during the last two-three decades. This industry is divided into two components *viz*. (*i*) heavy electrical equipment industry and (*ii*) light electrical goods industry.

Heavy Electrical Equipment Industry

Heavy electrical industry comprises equipment used for the generation, transmission, distribution and utilisation of power. It includes items such as generators, boilers, turbines, transformers, switchgears, etc. The indigenous industry is equipped to meet the entire domestic requirement. The entire industry has been delicensed under the New Industrial Policy of the Government of India. Also under the New Policy, these items qualify for automatic approval as regards to foreign collaborations. Heavy electrical equipment is mainly manufactured by the government owned factories which have been set up with foreign financial assistance and technical skill. This industry took its birth in 1956 when Heavy Electricals (India) Ltd. was set up at Bhopal with British help. It started production in 1960. Another company known as the Bharat Heavy Electricals was formed in 1964. The two organisations have since been merged to form Bharat Heavy Electricals Ltd. (BHEL). BHEL is an important milestone in the development of heavy electrical equipment industry. In last about six decades since its inception, BHEL has taken the country from a position of total dependence on imports to self-reliance in power plant design, manufacture, installation and servicing. The sets manufactured and supplied by BHEL now account for 65.3 per cent of the country's installed power generation capacity. With an annual production capacity of 4,000 MW, BHEL ranks among the most important power equipment manufacturers in the world. To meet the demand for rapid growth in power generation, BHEL has steadily introduced higher capacity boilers—right upto 500 MW unit sizes. BHEL boilers and auxiliaries have also been exported to Libya, Malayasia and Egypt and boiler assemblies to China and the U.S.A. It has six units located at Bhopal, Tiruchchirapalli, Ramchandrapuram (Hyderabad), Jammu, Bengaluru and Haridwar.

Power Transformers. There are 33 units in the organised sector manufacturing power and distribution transformers. Transformers of upto 100 KVA capacity are mainly produced by small scale units. Most of the production comes from Mumbai, Chennai, Vishakhapatnam, Kolkata and Sonipat.

Electric Motors. India produces a wide range of electric motors to meet the requirements of industries, tube wells, pumping sets and electric traction. Heavy motors are manufactured by government factories while small motors are made by private factories. Mumbai, Chennai, Bengaluru, Kolkata, Pune, Patiala, Delhi, Coimbatore, etc. are the main centres of motor manufacturing.

Electric Wires and Cables. Wires and cables whether they are made of fibre, optics, iron or non-ferrous material (copper, aluminium, zinc), play a significant role in almost all areas of industrial and daily life. This is one of the earliest industries established in India in the field of electrical products. India produces a wide range of wires and cables which includes communication cables such as jelly fitted telephone cables, optic fibre cables, local area cables, switchboard cables, co-axial cables, electrical cables such as electrical wires, winding wires, automatic/battery cables, UPS cables, flexible wires, low voltage power cables, etc.

It is a fast developing industry as it has received a big boost from various government policies. The industry is de-licenced and is eligible for automatic approval for FDI upto 100 per cent.

Transmission Towers. Transmission towers support high voltage transmission lines which carry electricity over long distances. Demand for transmission towers has increased considerably with the rapid pace of electricity distribution over long distances from the electricity generating stations to the consuming areas. The country has sufficient capacity to cater to the domestic demands.

Cranes. Cranes and hoists are an important category of material handling equipment required by almost all sectors of industry. India produces a wide range of cranes which include Electric Overhead Travelling (EOT) cranes, mobile cranes, ladle cranes, hydraulic decks, crab cranes, floating cranes, controller cranes, etc. There is good potential for growth of this industry in India. The industry is de-licenced and is eligible for automatic approval for FDI upto 100 per cent.

Lifts and Escalators. During the recent past, Indian cities have tended to grow vertically due to fast growing urban population and also because of limited availability of land in urban areas. Consequently lifts and escalators have become the corner stone to support this development and lifeline for the buildings that constitute this development. Rapid urbanization and robust activity in the construction industry and corporalization of the real estate sector has led to a vary healthy growth of this industry. The use of lifts and escalators is increasing rapidly due to substantial investment in construction of multistoried housing complexes, large malls and supermarkets of international standards, modernization of airports and railway stations apart from industrial sectors. A wide range of lifts and escalators are manufactured in the country which include single speed, double speed, glearless, hydraulic, servo and Variable Voltage Variable Frequency (VVVF) elevators. The industry is de-licenced and is eligible for automatic approval for FDI upto 100 per cent.

Light Electrical Goods Industry

This industry covers a wide range of products which include white goods (refrigerators, washing machines, air conditioners, etc.), household electric appliances, electric fans, storage batteries, dry cells, wiring accessories, fittings, electric lamps, etc.

Refrigerators. In India, refrigerators have the second highest aspirational value, next only to television. It is on account of this fact that this industry has witnessed extremely high growth rate during the last few years. This industry has become highly competitive as a number of brands, both national and international, have entered the market. There are two basic designs of refrigerators which are currently manufactured in India. These are generally referred to as Direct Cool (DC) and Frost Free (FF). There has been a gradual consumer preference shift towards frost free segment. Increasing number of dual income householders are shifting the demand from the conventional 180L refrigerators to larger 220 L and higher capacity refrigerators with double doors. The industry is delicensed and its eligible for automatic approval for FDI upto 100 per cent.

Washing Machines. With growing number of working couples and higher aspirations as well as higher purchasing power of the urban middle class families, washing machines have become an integral part of domestic electrical appliances in most of the families. More efficient and energy saving washing machines are hitting the market every day as the technology advances. The washing machine market in the country can be divided into semi-automatic and fully automatic. There is a gradual shift towards higher capacity and fully automatic machines. The industry is de-licenced and eligible for automatic approval for FDI without any restriction.

Air Conditioners. Air conditioners are gradually being treated as a necessity in the changed socio-economic environment of the present day life in India. There are three types of ACs available in the market *viz.* window AC, split AC and central AC. Star rating of ACs has been introduced for energy efficiency. Market for ACs has grown considerably during the last few years. The industry is de-licenced and eligible for automatic approval for FDI upto 100 per cent.

Electric Fans. India is one of the foremost producers of electric fans in the world. A complete range of electric fans such as ceiling fans, table fans, pedestal fans, exhaust fans, cooler fans, railway carriage fans, cabin fans and air circulators are being produced in different parts of the country. Mumbai, Kolkata, Secunderabad, Chennai, Delhi, etc. are the leading centres for production of electric fans.

Electric Lamps. This industry was established in 1932 and has made tremendous progress, especially after Independence. The range of electric lamps includes mercury vapour lamps, automobile lamps, photoflash lamps, miniature lamps for torches and flourescent tubes. There is an urgent need for energy conservation which has encouraged the need for manufacture of energy efficient electric lamps. Compact Fluorescent Lamps (CFL) which consume about 20% of electricity for the same light output and last upto 8 times longer than the GLS are getting more popular. The industry is receiving adequate encouragement from the Government. A number of foreign collaborations have already been approved for manufacture of energy efficient lamps.

Dry and Storage Batteries. Established in 1926, this industry flourished well after the Second World War. Almost all types of dry and storage batteries are produced in the country. Storage batteries are required for automobile industry, train lighting, posts and telegraphs apparatus, power houses and for traction. New units are coming up and the process of upgradation of technology in this industry is continuously benefitting the producer and customer alike.

Electronics Industry

This industry covers a wide range of products including television sets, transistor sets, telephone exchanges, cellular telecom, computers, and varied equipments for posts and telegraph, defence, railway and meteorological departments. It is practically a post-independence phenomenon and has revolutionised the life style of the Indian masses in the recent past. Bengaluru is the largest centre of electronics goods production and is rightly termed as the ***Electronic Capital of India***. The other major producing centres are Hyderabad, Delhi, Mumbai, Chennai, Kolkata, Kanpur, Pune, Lucknow, Jaipur, Coimbatore, etc.

Information Technology

Two main components of Information Technology (IT) are software and hardware.

The software has emerged as the major industry in the field of electronics. This industry made a modest beginning in the 1970s and by mid-1980s, the forecasters, analysts and policy planners started understanding the potential of computer software applications. The industry achieved a major breakthrough in the 1990s and is now one of the important industries of India. The main cause of the rapid development of software industry is its vast reservoir of technically skilled manpower which has transformed India into a software super power. With a compound annual growth of about 52 per cent between 1991 and 1996, the Indian software sector has expanded almost twice as fast as the world's leading US software industry did, during the same period, although from a smaller base. There is now a critical mass of more than 500 software firms in the country and apart from these companies, there are an additional 1,000 start-up-companies. Today, India is one country that offers cost-effectiveness, great quality, high reliability, speedy deliveries and, above all, the use of state-of-the-art technologies in software industry. The year 1995-96 was a boom year for the Indian computer industry and the Information Technology (IT) industry of India really exploded in that year. Covid-19 during 2019-20 led to on line and work from home (WFH) which resulted in unprecedented increase in demand for computers and lap tops.

Hardware. The hardware segment of Information Technology (IT) industry is one of the fastest growing industry in terms of production, international trade and is characterised by innovation.

After liberalisation in 1991, India is fast catching up with the world in terms of penetration of IT, telecommunications and consumer electronics products. This can lead to a huge demand for these products as well as provide base for global competition.

"The Personal Computer (PC), not TV, will be the key information appliance of 21st century," said Andy Grove, co-founder of Intel, the world's largest chip-maker over a decade ago. He was proved right in the U.S. before the turn of the 20th century when for the first time PCs outsold TVs. This has not happened in India, and may not happen for another decade. But PCs are gaining more popularity in India as compared to TVs as the time passes.

The phenomenal increase in PC sales can be attributed to home segment which posted a growth of 48 per cent. Significant consumption by telecom, banking, manufacturing as well as business process outsourcing (BPO) and IT enabled services (ITES) segment also contributed to rise in PC sales. India has emerged as the most preferred destination for BPO, a key driver of growth of software industry and the service sector. Smaller cities and towns fueled the IT consumption with class C cities accounting for over 50 per cent of total PC sales.

The vision of IT policy is to use IT as a tool for raising the living standards of the common man and enriching their lives. Towards this end, the Department of Information Technology has taken up an ambitious programme of PC and Internet penetration to the rural and underserved urban areas. The Department has also announced a programme to establish State Wide Area Network (SWAN) up to the block level to provide connectivity for e-governance. The Department has also set up Community Information Centres (CICs) in hilly, far-flung areas of the North-East and Jammu and Kashmir to facilitate the spread of benefit of information and communication technology. It is also proposed to set up CICs in other hilly, far-flung areas of the country like Uttarakhand, Andaman & Nicobar and Lakshadweep. Distribution of Information Technology Industry (Software as Hardware) is shown in **Fig. 25.10**.

The television industry had also grown tremendously in the 1990s. The market size is about 2.5 million units with top four brands taking up over 80 per cent of the share. BPL and Videocon are on the top and accounted for 26 per cent each of the market

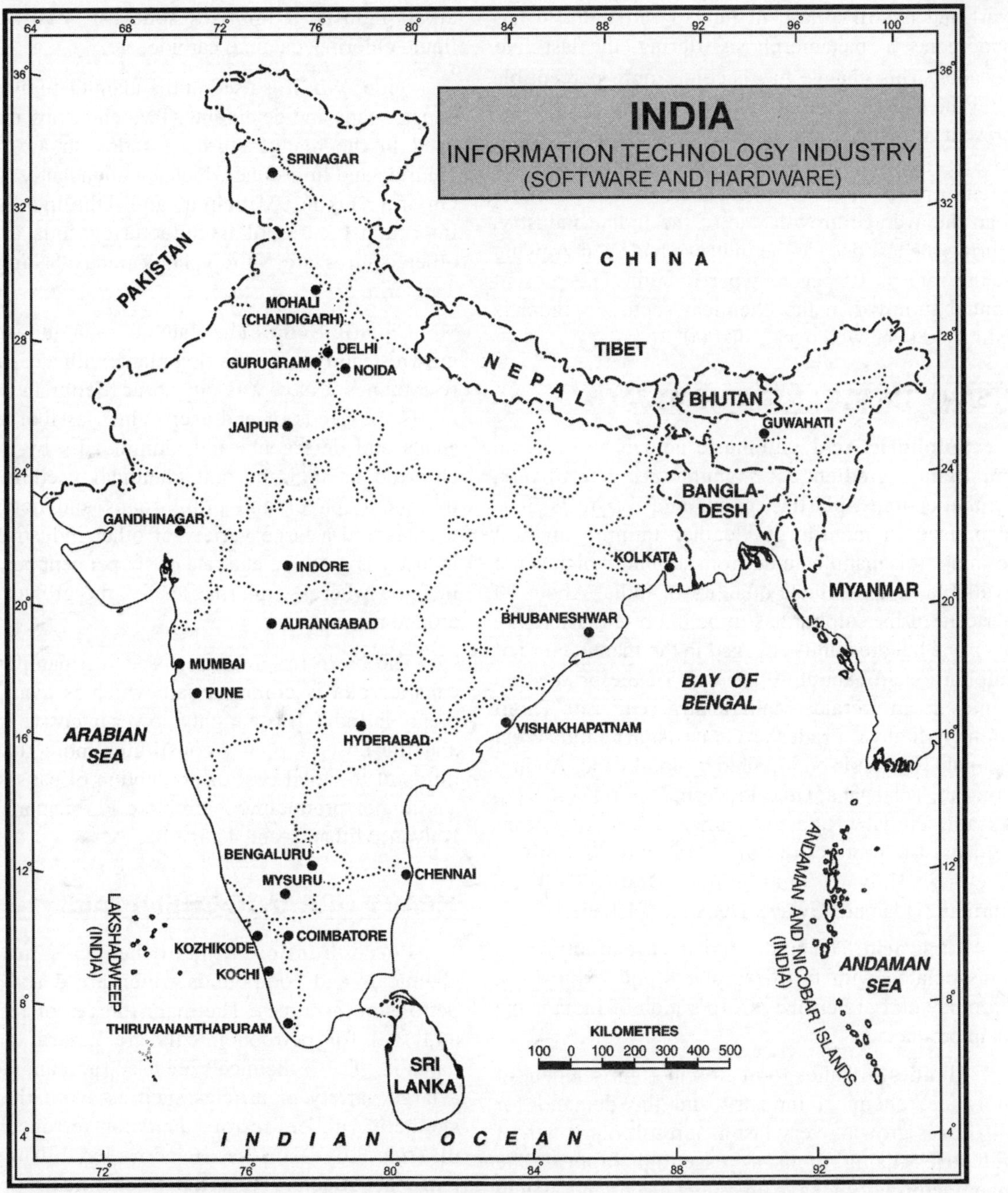

FIG. 25.10. India : Information Technology Industry (Software and Hardware)

shares. They are followed by Onida and Phillips. Among the international brands, Panasonic, Soni and Akai are famous. LG has also become quite popular It is now asserted that TV has entered the smallest of homes and the tiniest of villages and communities.

CHEMICAL AND ALLIED INDUSTRIES

Chemical and allied industries constitute one of the most vital and essential components of the country's economy. Although chemical industry made a late start and is still a nascent industry in India, it has undergone a metamorphosis during the last five decades. This change has become more perceptible after the liberalisation of industrial policy in 1991. At present, it is the fourth largest set of industries after textiles, iron and steel and engineering industries. The rate of growth of the chemical sector has been higher than the average growth rate of the Indian industry. During the last decade the industry has been growing at the rate of 10 per cent per annum. The present annual turnover in the chemical sector in India is estimated to be well over ₹ 40,000 crore.

HEAVY INORGANIC CHEMICALS

Sulphuric Acid. Sulphuric acid is used as an important ingredient for manufacturing fertilizers, synthetic fibres, plastics, paints and dyestuffs. It is also used in metallurgy, leather tanning and oil refining. It is manufactured from sulphur which is not available in appreciable quantity in India. About 90 per cent of the sulphur has to be imported. There are over one hundred units engaged in the manufacture of sulphuric acid. About 80 per cent of the production comes from Kerala, Maharashtra, Gujarat, Tamil Nadu, Madhya Pradesh, Chhattisgarh and West Bengal. The rest is contributed by Jharkhand, Andhra Pradesh, Telangana, Uttar Pradesh, Delhi, Karnataka, Assam, Punjab, Rajasthan and Odisha. Important centres of production are Mumbai, Trombay, Chennai, Kalyan, Jamshedpur, Delhi, Kolkata, Burnpur, Udaipur, Alwaye, Debari and Khetri.

Nitric acid. The production and use of nitric acid is associated with fertilizer plants and explosives. Trombay unit of Fertilizer Corporation of India is the main producer.

Alkalies. Alkalies form an important segment of inorganic chemical industry and the demand for alkalies is growing very fast as a result of growth of industries using them as essential inputs. The manufacturing of alkalies requires heavy and weight loosing raw materials like common salt (sodium chloride), limestone and coal. The final product has much less weight in comparison to the weight of the raw materials. This industry also requires cheap electricity in abundance. Thus the plants manufacturing alkalies tend to be located near the source of raw materials, electricity and market. The alkali industry comprises soda ash, caustic soda, liquid chlorine, calcium carbide, etc.

Soda Ash. It is used in the manufacture of glass, paper, soaps and detergents. Two chief raw materials used in the manufacture of soda ash are sodium chloride and limestone which are abundantly found in Gujarat. Okha, Mithapur and Dhrangadhra are important centres of manufacturing soda ash. The other centres are Sutrapada, Varanasi, Nangal and Tuticorin.

Caustic Soda. The caustic soda industry has grown steadily since the first plant, with a capacity of five tonnes a day, was put up at Mettur in 1936. It meets the needs of end-user industries like textiles, soaps and detergents and alumina. Its by-product, chlorine is an important chemical used in water treatment, paper and pulp, soaps and detergents, textiles and a large variety of other industries. The industry is growing at a rate of six per cent per annum in India whereas, internationally, the growth rate is around two per cent.

The basic raw material used for manufacturing caustic soda is common salt which is available in plenty in India. It is a highly power intensive industry and the cost of power constitutes more than two-thirds of the total cost of production of caustic soda. The major producing centres are Porbandar, Thane, Kalyan, Mithapur and Titagarh.

HEAVY ORGANIC CHEMICALS

Petrochemicals. Petrochemicals are those chemicals and compounds which are derived from petroleum resources. The main sources of feedstock and fuel for petrochemicals are natural gas and naphtha. These chemicals are used for manufacturing a large variety of articles such as synthetic fibres, synthetic rubber, ferrous and non-ferrous metals, plastics, dye-stuffs, insecticides, drugs and pharmaceuticals. Today, petrochemicals products permeate the entire spectrum of items of daily use, ranging across clothing, house construction, furniture, automobiles, household items, toys, agriculture, irrigation and packaging to medical appliances.

This is one of the fastest growing industries in the Indian economy. The demand for organic chemicals increased so rapidly that it became difficult to meet them by chemicals prepared from alcohol, calcium carbide, and coal. At the same time, petroleum refining industry expanded rapidly. Many things are derived from crude petroleum, which provide raw materials to several new industries; these are collectively known as petrochemical industries. This group of industries is divisible into four sub-groups : (*i*) polymers, (*ii*) synthetic fibres (*iii*) elastomers, and (*iv*) surfactant intermediates. Mumbai is the hub of petrochemical industries. Cracker units are also located in Auraiya (Uttar Pradesh), Jamnagar, Gandhar, Hazira (Gujarat) Nagothane, Ratnagiri (Maharashtra), Haldia (West Bengal) and Vishakhapatnam (Andhra Pradesh).

Three organisations are working in the petrochemical sector under the administrative control of the Department of Chemicals and Petrochemicals.

(*i*) Indian Petrochemical Corporation Limited (IPCL) is a public sector undertaking which is responsible for the manufacture and distribution of various petrochemicals like polymers, chemicals, fibres and fibre intermediates.

(*ii*) Petrofils Cooperative Limited (PCL) is a joint venture company of the Government of India and Weavers' cooperative societies. It produces polyester filament yarn and nylon filament yarn at its two plants located at Vadodara and Naldhari in Gujarat.

(*iii*) Central Institute of Plastics Engineering and Technology (CIPET) is involved in imparting training.

Synthetic Fibres

These are widely used for manufacturing a large variety of fabrics because of their special qualities like strength, durability, dyeability, workability, washability and resistant to wrinkles and shrinkage. These fabrics are very popular among the masses in India, both in urban and rural areas. Although nylon industry made a beginning immediately after the World War II, the real progress was made in 1960s.

The main cause of rapid progress in this period was the availability of feedstock from petroleum refineries. Units manufacturing nylon filament and polyester filament yarns are at Kota, Pimpri, Mumbai, Modinagar, Pune, Ujjain, Nagpur and Udhana. Acrylic staple fibre is manufactured at Kota and Vadodara. Plants manufacturing polyester staple fibre are at Thane, Ghaziabad, Manali, Kota and Vadodara.

Polymers

Polymers are made from ethylene and propylene. These materials are obtained in the process of refining crude oil in the refineries. Polymers provide the basic raw material for plastic industry. Among polymers, polyethylene is a widely used thermoplastic. Plastic is first converted into sheets, powder, resin and pellets, and then used for manufacturing plastic products. Plastic is a very useful industrial product in the modern day life.

Attempts to manufacture plastic polymers in India were made in late 1950s and early 1960s using other organic chemicals. The National Organic Chemical Industries Limited (NOCIL), instituted by the Mafatlal Group in 1961 started first naphtha based chemical industry at Mumbai. Later several other companies were formed. Mumbai, Barauni, Mettur, Pimpri and Rishra are major producers of plastic materials. The production of polymers (polyethylenes, polypropylene, poly-vinyl chloride and polystyrene) increased.

The Plastic Processing Industry is making available quality products at reasonable prices to a diverse range of consumer demand in the country. There are about 20,000 units engaged in manufacturing various plastic products. Nearly three-fourths of these units are in small-scale sector which account for 20-25 per cent of total polymer consumption. With a total investment of ₹ 15,000 crore, the industry gives the annual turnover of ₹ 25,000 crore. The present installed capacity of these units is of the order of 6-7 million tonnes. The industry has an average annual growth rate of 15 per cent and employs three million people directly. The industry processes about 3.6 million tonnes of prime or virgin polymers and 1.4 million tonnes (or 30 per cent of the total of reprocessed materials (recycled plastic).

FERTILIZERS

Indian soils are generally deficient in fertilizing elements namely nitrogen, phosphorus and potassium and do not give high yields. It is, therefore, essential to feed these soils with chemical fertilizers so that their productivity increases. The significant contribution made by the chemical fertilizers can be seen from the impact of the Green Revolution on Indian agriculture. The production of food grains increased from a miserably low of 50.82 million tonnes in 1950-51 to 308.6 million tonnes in 2020-21. The health and growth of the fertilizer industry is vital for increasing the growth of agricultural sector, to meet the foodgrain requirements of increasing population as well as increasing contribution to exports. We can ignore the significance of fertilizers for food production only at the cost of nation's

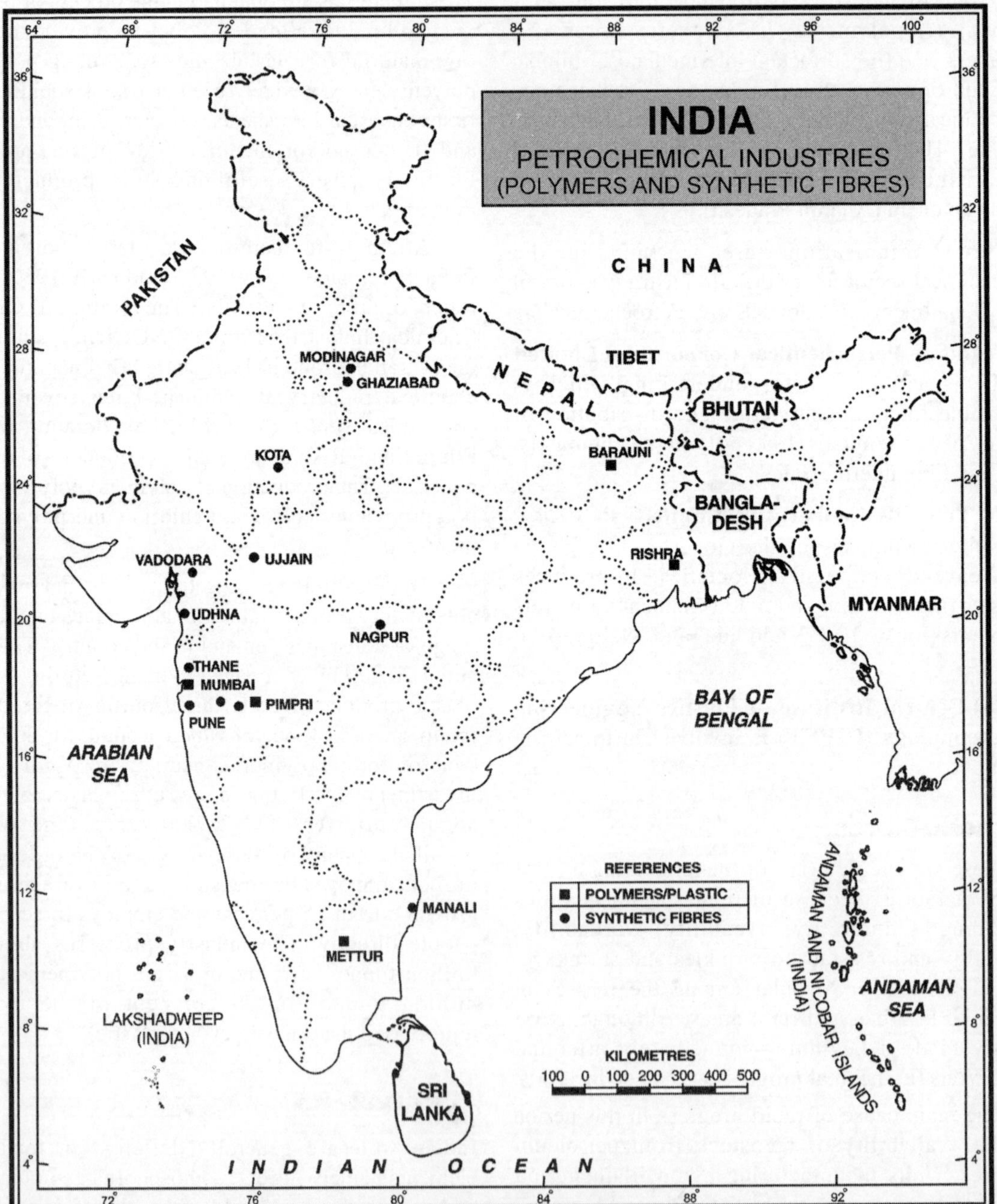

FIG. 25.11. India : Petrochemical Industries (Polymers and Synthetic Fibres)

welfare. The major role which fertilizers have played in raising the country's foodgrains production can be seen from the increase in fertilizer consumption from 0.55 kg per hectare in 1951-52 to 128.02 kg in 2017-18. This is much lower as compared to 427 kg/ha in Egypt, 250.7 kg/ha in Bangladesh, 281.5 kg/ha in Korea Republic and 229.9 kg/ha in Vietnam.

Localisation

The localisation of fertilizer industry is closely related to petrochemicals. About 70 per cent of the plants producing nitrogenous fertilizer use naphtha as the basic raw material. That is why, most of the fertilizer plants are located near the oil refineries. However, some fertilizer plants draw their feed stock from steel slag as well as coke and lignite. Phosphatic fertilizer plants are primarily dependent upon mineral phosphate which is available in Uttar Pradesh, Madhya Pradesh and Rajasthan. Some plants are based on imported phosphate mineral. Sulphur is another important mineral used for manufacturing fertilizers. This is available in Tamil Nadu. During the recent years transportation of Neptha or Gas through rail or pipelines has facilitated the widespread distribution of fertilizer plants with priority to seaboard location. The HBJ Gas pipeline has given birth to six gas based fertilizer plants at Vijaypur, Jagdishpur, Aonla, Gadepan, Babrala and Shahajahanpur.

Growth and Distribution

A modest beginning with respect to manufacturing of chemcial fertilizers was made in 1906 when the first super-phosphate factory was set up at Ranipet in Tamil Nadu. The actual growth of fertilizer indsutry is mainly a post-Independence phenomena. The setting up of the Sindri plant by the Fertilizer Corporation of India Ltd. (FCI) in 1961 was a turning point and this industry did not look back after that.

It is now one of the fastest growing basic industries which has made rapid strides in recent years. It produces a wide range of fertilizers to suit different soil and crop requirements in different parts of the country. India is now the third largest producer of nitrogenous fertilizers in the world only behind China and U.S.A. At present, there are 56 large size fertilizer units manufacturing a wide range of nitrogenous and complex fertilizers, including 29 units producing urea and 9 units producing ammonia sulphate as a by-product. Besides, there are about 72 medium and small scale units producing single superphosphate (SSP).

Gujarat, Tamil Nadu, Uttar Pradesh, Maharashtra, Andhra Pradesh, Punjab and Kerala are the main fertilizer producing states and account for about half of the total fertilizers produced in India. Odisha, Rajasthan, Bihar, Assam, West Bengal, Goa, Delhi, Madhya Pradesh and Karnataka are the other producers.

1. Gujarat. Gujarat is the largest producer of fertilizers in India and accounts for more than one-fourth of the total production of nitrogenous as well as phosphatic fertilizers of the country. This state has more than 14% of the country's total installed capacity. Units at Vadodara and Kalol produce both nitrogenous and phosphatic fertilizers while units at Bharuch, Udhna, Kandla and Bhavnagar produce phosphates only. Two new plants have been set up at Hazira and Surat. Ahmedabad and Navsari are also important producers.

2. Tamil Nadu. This state accounts for about 11 per cent of the country's installed capacity. At present Tamil Nadu is the second largest producer of phosphatic fertilizers (about 16%) and the fourth largest producer of nitrogenous fertilizers accounting for nearly 9% of India. Nearly 65 per cent of the capacity is for nitrogenous fertilizers. Neyveli, Ranipet, Tuticorin, Ennore, Coimbatore, Mettur Dam, Cuddalore, Avadi and Manali are the main producing centres.

3. Uttar Pradesh. This state accounts for about 9 per cent of the country's total installed capacity. About 17 per cent of nitrogenous and 3.4 per cent of phosphate fertilizers of India are produced in Uttar Pradesh. Important centres of fertilizer production in Uttar Pradesh are Kanpur, Phulpur, Gorakhpur, Varanasi and Magarwara. New plants have been set up at Babrala, Aonla, Shahjahanpur and Jagdishpur.

4. Maharashtra. Maharashtra accounts for over 11 per cent of the nitrogenous and about 7 per cent of the phosphatic fertilizers of the country. The state has six plants in all. Mumbai with three plants is the largest producer. The remaining three plants are located at Trombay, Ambarnath and Loni-Kalbhor.

5. Odisha. Odisha has 6 per cent of the total installed capacity and produces 11.6 per cent of the phosphate and 3 per cent of nitrogenous fertilizers of India. The industry is located at Rourkela, Talcher, and Paradwip.

6. Kerala. Kerala also accounts for 6 per cent of the total installed capacity and produces about 6 per cent phosphatic and 3.5 per cent nitrogenous fertilizers of India. Its fertilizer units are located at Alwaye, Kochi and Karimnangalam.

7. Rajasthan. This state has 5 per cent of the country's total installed capacity and accounts for about 2.3 per cent of nitrogenous and 1.6 per cent of phosphatic fertilizers of India. More than half of the state's production comes from Kota. Debari, Khetri, Saladipura (Sikar district) and Chittaurgarh are the other important centres.

Others. *Bihar*—Sindri and Barauni, *Jharkhand*—Jamshedpur and Dhanbad, *Punjab*—Nangal and Bhatinda, *West Bengal*—Burnpur, Durgapur, Rishra, Kharda, Haldia, *Chhattisgarh*—Bhilai, Kumhari and Korba, *Assam*—Namrup and Chadrapur, *Karnataka*—Mangaluru, Hubli, Mandya, Belagola and Munirchad, *Haryana*—Panipat, *Goa*—Sancoale (near Vasco), *Andhra Pradesh* (Vishakhapatnam, Kakinada, Tanukur Nidadavole, Tadepalli) *Telangana* (Maula Ali near Hyderabad, Ramagudam) are other centres.

Public Sector Undertakings

The public sector undertakings are playing a dominant role in manufacturing chemical fertilizers. At present, there are 11 public sector undertakings under the administrative control of Department of Fertilizers.

1. The Fertilizer Corporation of India (FCI) was incorporated in January, 1961. It has four units one each at Sindri (Jharkhand), Gorakhpur (U.P.), Talcher (Odisha) and Ramagundam (Telangana). Its total installed capacity is 8.06 lakh tonnes of nitrogen.

2. The National Fertilizers Limited (NFL) was established on 23 August 1974. It has six units, *viz.*, the Calcium Ammonium Nitrate (CAN) plant at Nangal and the Urea plants at Nangal, Bhatinda, Panipat and Vijaipur (two units). It has a total installed capacity of 15.66 lakh tonnes of nitrogen and is the largest producer of nitrogenous fertilizers in the country.

3. Fertilizers and Chemicals Travancore Limited (FACT). It was incorporated in 1943 and started production of ammonia sulphate in 1947 at Udyogmandal near Cochin (now Kochi) with an installed capacity of 50,000 MT. In 1980, FACT become a PSU and towards the end of 1982, the Government became a major shareholder. From a modest beginning, FACT has grown into a multi-division, multi-functional organisation with basic interest in manufacture and marketing of fertilizers and petro-chemicals, engineering consultancy in design and in fabrication and eraction of industrial equipments.

4. The Rashtriya Chemicals and Fertilizers Ltd. (RCF). It was incorporated as a separate company in 1978 as a result of reorganisation of the ertwhile Fertilizer Corporation of India Ltd. It is operating five fertilizer plants at Trombay and a large gas-based fertilizer plant at Thal which started production in 1985. The installed capacity of RCF plants is 9.55 lakh metric tonnes of nitrogen and 1.20 lakh metric tonnes of phosphate.

5. The Hindustan Fertilizer Corporation Limited (HFCL) has total annual installed capacity of 6.54 lakh tonnes of nitrogen. Out of a total of five units under its control, three are at Namrup (Assam) and one each at Durgapur (W. Bengal) and Barauni (Bihar). HFCL and Fertilizer Corporation of India (FCI) were declared sick in November 1992. After considering the rehabilitation proposals of these two PSUs, the Government decided to close down HFCL and FCI on 5th September, 2002, excepting its Jodhpur Mining Organisation which has been hived off into a new company.

6. The Madras Fertilizers Limited (MFL) is a joint venture of the Government of India and the National Iranian Oil Company. Its plant at Manali (Chennai) has an annual installed capacity of 3.67 lakh tonnes of nitrogen and 1.43 lakh tonnes of phosphate.

7. Pyrites, Phosphates and Chemicals Limited (PPCL) was set up in March 1960. It is engaged in exploration of pyrites deposits as well as production of single super phosphate at Amjhore (Bihar),

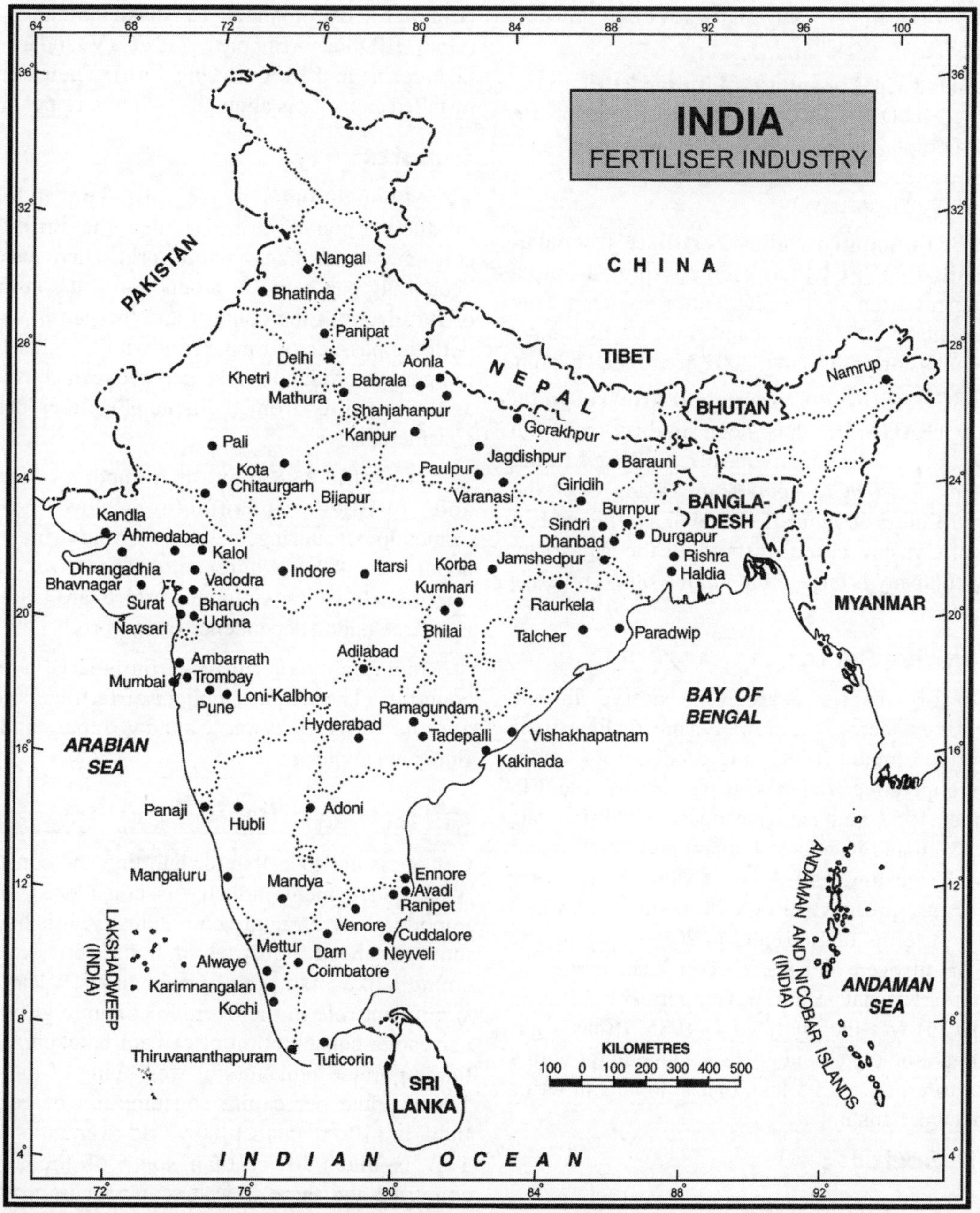

FIG. 25.12. India : Fertilizer Industry

exploration-cum-production mining of pyrites deposits at Saladipura in Rajasthan and mining of rock phosphate ore from the Mussoorie phosphorite deposits. The company was declared as a sick company in April 2000 and was closed down on November 20, 2000.

8. Project and Development India Limited (PDIL) formerly known as Fertilizer Planning and Development India Limited, is engaged in design, engineering, procurement and supervision of construction/commissioning of fertilizer and allied chemical plants. The company was declared as a sick company in December, 1992 and was closed down in April 2003.

9. Paradwip Phosphates Limited (PPL) was established in Dec. 1981 for setting up phosphatic fertilizer complex at Paradeep in Odisha. Phase I and II of this project went into commercial production in 1986 and 1992 respectively.

10. Brahmaputra Valley Fertilizer Corporation Limited (BVFCL) has been constituted into a new company from April 1, 2002 after hiving off the Namrup units of HFC. Units I and III of BVFCL have been commissioned in March 2002 after their revamp.

11. FCIL-Aravali Gypsum and Mineral India Limited (FAGMIL) has been incorporated on February 14, 2003 after hiving off Jodhpur Mining Organising from FCI. Apart from taking over the Jodhpur Mining Organisation which is engaged in mining of Gypsum in four districts of the Rajasthan, the new company is engaged in mining other minerals in Rajasthan.

Cooperative Sector

Two organisations are very active in the cooperative sector. Indian Farmers Fertilizer Cooperative Limited (IFFCO) is a federation of over 24,000 cooperative societies. It was incorporated in November 1967 and has five operating units, one each at Kalol and Kandla in Gujarat and Phulpur and Phulpur Extension, and Aonla Extension in Uttar Pradesh. At present, IFFCO's installed annual capacity in terms of nutrients is 20.16 lakh metric tonnes of nitrogen and 8.25 lakh metric tonnes of phosphate. Krishak Bharati Cooperative Limited (KRIBHCO) was incorporated in 1985. It has a gas based urea-ammonia plant at Hazira in Gujarat with a capacity to produce 7.95 lakh metric tonnes in terms of nitrogen per annum.

Private Sector

A number of private companies are also engaged in manufacturing fertilizers. Some of the important companies are IEL (Kanpur), SRC (Kota), GNFC (Bharuch), Coromandel, (Vishakhapatnam), ZACL (Goa), EID Pary (Ennore), GSFC (Vadodara), SPIC (Tuticorin) and MFC (Mangaluru). Their combined installed capacity is about 20 lakh tonnes per annum.

Imports

Although India is the third largest fertilizer producing country, she is also the third largest consumer of fertilizers in the world. There has always been a big gap between production and consumption of fertilizers in the country and the gap is widening with the passage of time. Even with a higher level of indigenous production, the gap between consumption and availability from indigenous sources increased considerably.

The **table 25.8** shows that imports of fertilizers and fertilizer manufactures has increased tremendously during in last few years. India meets about half of the requirements of fertilizers through imports. However, strenuous efforts are being made to reduce undue dependence on imports.

India imports fertilizers from a large number of countries. The major suppliers of fertilizers to India are the USA, Russia, Canada, Japan and some European countries.

CEMENT INDUSTRY

Cement is indispensible for building and construction work and cement industry is considered to be an important infrastructure core industry. It is one of the most advanced industries of India. In a developing country like India, the cement industry can play a significant role in the overall economic growth. The per capita consumption of cement is taken as one of the important indicators of well being of the people. The average per capita consumption of cement in India is 110 kg against the world average of 260 kg. This is much lower than some of the advanced countries and there is vast scope for improving the situation. India is currently on a fast track of economic growth and if the present growth trends

TABLE 25.8. Imports of Fertilizers and Fertilizer Manufactures in India

Year	1960-61	1970-71	1980-81	1990-91	2000-01	2019-20	2020-21	2021-22
Quantity (Thousand tonnes)	307	2,392.7	5,560.2	7,560.3	7,423.4	29,962	31,738	30,935
Value (₹ crore)	13	86	818	1,766	3,034	52,817	56,405	1,05,796

***Source*:** Economic Survey, 2022-23, Statistical Appendix, pp. 112-14.

continue, the per capita consumption is expected to touch 160 kg in 2025 even in the face of fast growing population.

Locational Factors

Manufacturing of cement requires heavy, low value and weight loosing materials and is primarily a raw material oriented indsutry. Limestone is the main raw material and comprises 60-65 per cent of the total product by weight. On an average 1.5 tonnes of limestone are required to produce one tonne of cement. Hence, the location of a cement plant is based on the limestone deposits. The other raw materials used are sea shells, slag from steel plants and fertilizer plants and these raw materials influence the localisation of cement industry in their own way. Silica (20-25%) and alumina (5-12%) are also important ingredients. Gypsum is necessary to regulate the setting time of cement. Power is used in raw material grinding, clinkerisation of limestone in the kiln operation and clinker grinding along with gypsum to form cement. The older plants required 120 to 130 units of electricity per tonne of cement produced. Modern energy efficient plants consume only 80 to 90 units of electricity per tonne. Coal is another major input along with electricity and forms 40 per cent of the total cost. Coal is used not only as fuel in the kiln but also to burn the limestone. The ash of the burnt coal combines with the limestone to form clinker. On an average 250 kg of coal is required to produce one tonne of cement. The quantum of other materials required to produce one tonne of cement are 4 kg of gypsum, 0.4 kg of bauxite and 0.2 kg of clay.

Cement and its raw materials are low value bulk materials and the transportation over long distance by rail and road transport involves huge costs. Some of the transportation cost on transporting limestone is reduced by beneficiating this mineral at the quarry heads. The transportation cost is also reduced if the manufacturing plant is located near the market. In fact ready market is the pre-requisite for the proper growth of an industry producing heavy commodity with low specific value like cement.

It is obvious from the above discussion that availability of raw materials, bulk transport facilities at reasonably low cost and market are the three main localisational factors, in that order, which favour the growth of cement industry in India.

Growth of Cement Industry

The first attempt to manufacture cement in India was made in 1904 when a mill based on sea shells as a source of limestone was established at Chennai. But this attempt proved abortive and a really successful attempt was made in 1912-13 when the Indian Cement Co. Ltd. set up a plant at Porbandar. World War I gave impetus to cement industry in India. Consequently, Katni Cement and Industrial Co. Ltd. at Katni (M.P.) started production in 1915 and Killick Nixon's Bundi Portland Cement Co. at Lakheri (Rajasthan) commenced in 1916. A number of companies came into existence to make use of the post war boom. Six new factories at Dwarka (Gujarat), Japla (Bihar), Banmore, Mehgaon, Kymore (M.P.) and Shahabad (Karnataka) were erected by 1922-23. Cement production virtually took off the ground only after tarrif protection was granted to this industry in 1924. A turning point came in 1934 when 10 out of 11 existing companies merged into the Associated Cement Co. Ltd. (ACC). The Dalmia Cement Group was also formed in 1937. This group set up factories at Dalmianagar (Bihar), Dalmiapuram (Tamil Nadu) and Dalmia (Charkhi) Dadri in Haryana. By 1947, there were 18 cement factories with a capacity of 21.15 lakh tonnes and production of 20.16 lakh tonnes. Rapid economic progress associated with massive building programmes during the plan period accelerated the demand for cement and provided stimulus to this industry. India achieved self-sufficiency in cement only in 1980s during the short five year period of partial decontrol. Prior to that Indian cement industry had seen days of total controls, partial decontrols and imports. The government took initiative to remove control on price and distribution of cement in 1989 and the industry was de-licensed in 1991 under the Industrial (Development and Regulation) Act, 1951. Since then, this industry has progressed at a fast rate both in capacity/production as well as in process technology. Today, in terms of quality, productivity and efficiency, the industry is second to none in the world. Its technology is state-of-the-art, its cost of production is one of the lowest in the world and its productivity is easily one of the highest.

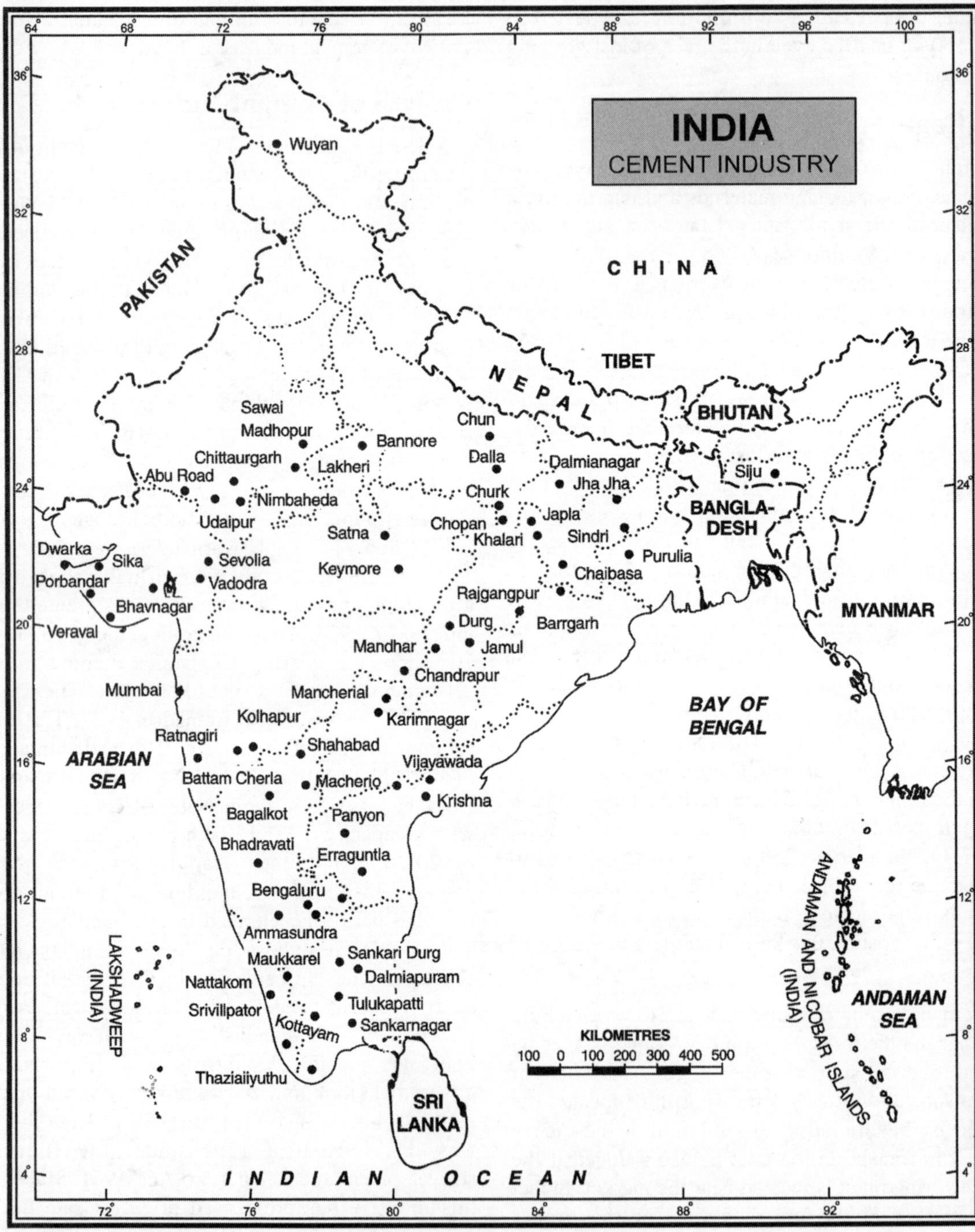

FIG. 25.13. India : Cement Industry

Currently, the Indian cement industry is the second largest in the world after that of China. The industry provides direct employment to 1.5 lakh persons and indirect employment to 1.2 million persons. There are 210 large cement plants producing 410 million tonnes of cement every year and 350 mini cement plants producing 92 million tonnes of cement per year. These plants are scattered in almost all parts of the country. The mini plants play a supplementary role. The concept of mini plants was accepted by the Government in 1979 to exploit smaller deposits of limestone scattered in remote and inaccessible areas. This concept was supported by incentives like 50 per cent reduction in excise duty. The main advantage of mini cement plants is that they provide employment opportunities to rural and remote areas and make cement easily available there. Further, they help in dispersal of production capacity and reduce strain on transportation infrastructure. Over 60 companies are engaged in the production of cement.

Given the enormous need for infrastructure and housing, which require large quantities of cement as a basic building material, the prospects of industry are bright. The Working Group on Cement Industry for the formulation of Tenth Five Year Plan and other studies on global competitiveness of the Indian cement industry highlight constraints such as high cost of power, high freight cost, inadequate infrastructure and poor quality of coal. In order to utilize the excess production capacity available with the cement industry, the Government has identified the following thrust areas for increasing demand :

- Further push to housing development programmes.
- Promotion of concrete highways.
- Use of ready-mix concrete in large infrastructure projects.
- Construction of concrete roads in rural areas under Prime Minister's Gram Sadak Yojana.

The Indian Cement industry today produces 11 varieties of cement including ordinary portland cement (71%), portland pozzolana cement (18%) and portland blast furnace slag cement (10%). The balance one per cent is of all special cements including white cement.

Distribution

A look at the distribution pattern of cement factories reveals that they are mainly concentrated along the Vindhayan ranges—running from eastern Rajasthan to Jharkhand—where abundant supply of good quality limestone is available. In fact limestone deposits have acted as big magnets for attracting cement factories and rarely a factory in this region is situated at a distance of over fifty kilometres from the limestone quarries. The vast northern plain, on the other hand, is devoid of limestone deposits and does not support cement factories to any appreicable extent. It is for this constraint of raw material that 86 per cent of the factories and 75 per cent of the production capacity is found in Andhra Pradesh, Telangana, Rajasthan, Madhya Pradesh, Chhattisgarh, Gujarat, Tamil Nadu, Karnataka and Jharkhand.

Andhra Pradesh and Telangana. With 15 per cent of the total installed capacity and about 18 per cent of the total production of India, Andhra Pradesh along with Telangana occupy first place among the cement producing states of India. Most of the 21 plants are concentrated in the Telangana belt. The location of plants along the trunk rail route skirting the plateau along its junction with the coastal plain offers the best advantages with respect to raw material, market and transport. Peddapalli is the biggest plant with an installed annual capacity of 7 lakh tonnes. The other important producers are Krishna, Karimnagar, Cementnagar, Vijaywada, Panyon, Macherla, Mancherial, Tandur, Vishakhapatnam, Vizianagram, Nadikundi, Erranguntla, Yerranguntala, Adilabad, etc. Several mini plants are also functioning.

Rajasthan has surpassed Tamil Nadu and is now the second largest cement producing state accounting for over 13 per cent of India. The major cement plants skirt the Aravali Range where plenty of limestone is available. The large scale conversion of metre gauge railway lines into broad gauge has given the much needed improved transport facilities and stimulate cement industry in this region. The state has 10 major plants and the main centres of production are Sawai Madhopur, Lakheri, Chittaurgarh, Udaipur, Nimbaheda and Sirohi. With an annual capacity of 8.5 lakh tonnes, the plant at Sawai Madhopur is the largest in Rajasthan.

Madhya Pradesh. Madhya Pradesh is the third largest producer of cement in India after Andhra Pradesh (including Telangana) and Rajasthan. Its major production units are in the limestone rich districts of Satna, Katni, Rewa and Neemuch. These are located at Kaimur, Katni, and Bammore (Katni), Satna, Maihar (Satna), Vikramnagar (Neemuch), Kharadia (Dhar), Itarsi (Hoshangabad) and Damoh.

This state supplies large quantities of cement to the neighbouring Uttar Pradesh.

Tamil Nadu. With an installed capacity of about 11 per cent of all India capacity, this state has 13 plants in different areas. Larger units are located at Thalaiyuthu (Tirunelveli), Alangulam, Tulukapatti (Ramnathpuram), Dalmiapuram, Poliyur, Alathiyur (Tiruchirapalli), Sankaridurg (Salem) and Madukkarai (Coimbatore).

Gujarat. Gujarat has an installed capacity of 13.93 million tonnes. This state had the advantage of an early start and the earliest successful attempt to manufacture cement was made at Porbandar. The industry enjoys the benefit of large deposits of limestone in the state. Besides, sea shells can also be used. Large market of Western India is readily available. Major plants of the state are located at Sika, Sevolia, Okha, Porbandar, Dwarka, Vadodara, Ranavav, Veraval and Bhavnagar.

Maharashtra. Maharashtra is perhaps the largest cement consuming state in India due to on going construction of a mega-structures like big malls, roads, bridges, etc. With an installed capacity of about 10.9 million tonnes, this state has its major cement factories at Chanderpur, Ratnagiri, Saweri, Mumbai, Battam and Kolhapur.

Karnataka. High quality cement grade limestone occurs in Bijapur, Gulburga, Chitradurga, Mysuru, Shimoga and Tumkur. The state has a total installed capacity of about 9.8 million tonnes. Bagalkot, Bhadravati, Shahabad, Ammasaudram, Wadi, Bengaluru and Krukunta are the main centres. The Shahabad unit is the largest and the only factory producing oil well cement in India currently. Two new units have come up at Hosdurga (Tumkur) and Tornagallu.

Chhattisgarh. This state has 6% of India's total limestone reserves and has an installed capacity of about 8.4 million tonnes. Chhattisgarh has 7 cement factories of which the larger ones are at Jamul (Durg), Mandhar and Modigram (Raipur), Lulli, Tilda and Akaltara (Bilaspur).

Jharkhand. Jharkhand has 5 cement plants with a total installed capacity of 5.9 million tonnes. These plants are located at Sindri, Khalari (Ranchi), Bhawnathpur (Palam), Chaibasa, Khodri and Jhikhani (West Singhbhum).

Uttar Pradesh. The main limestone deposits are found in the Upper Son Valley, where five cement manufacturing plants have been established. The major centres of production are Churk, Dalla, Chunar and Jhansi.

The other producers are *Himachal Pradesh* (Rajban in Sirmaur district and Gagal in Bilaspur district), *West Bengal* (Durgapur, Purulia and Madhukunda), *Odisha* (Rajgangpur and Bargash), *Bihar* (Banjari, Japla and Chopan), *Meghalaya* (Cherrapunji and Garo), *Nagaland* (Waziho in Tunsang and Garampani), and *Jammu and Kashmir* (Basholi in Kathua district and Kharan in Baramula district).

Exports

From a net importer, India has emerged as a major exporter of cement in the international market. She entered the world cement scenario after 1989 when the production of cement increased considerably as a result of decontrolling the cement industry. The main buyers of Indian cement are Sri Lanka, Bangladesh, Myanmar, Indonesia, Malaysia, Nepal, Pakistan, Middle East countries and South-East Asian countries. The exports to these countries are likely to increase as they do not have appreciable deposits of limestone and cannot develop cement industry on their own.

Problems

Cement industry in India is facing a large number of problems. Some of the more serious problems are briefly described as under :

1. Inadequate Transport Facilities. The main raw materials (limestone, coal, gypsum, etc.) and the end product of cement industry are cheap and bulky and cannot afford high cost and inadequacy of transportation. It is estimated that transport of raw materials and furnished product accounts for 25-30 per cent of the total cost of cement manufacturing and supplies to the consuming areas. For example, cement plant at Sawai Modhopur in Rajasthan uses coal from Dhanbad mines in Jharkhand about 1,300 km away and supplies cement to consumers in Srinagar about 1,100 km away. Cement has short shelf life and should be dispatched to the consuming centres without any loss of time. All these factors require quick, cheap and efficient means of transportation for cement immediately after it comes out of the manufacturing plant. Earlier, railways used to be the main source of transporting cement but shortages of wagons, delays in transit and frequent increase in freight made this mode of transportation unattractive to the cement manufacturers. Consequently, a shift was noticed in favour of road transportation after 1989 and 60 per cent of the current needs of cement industry are met by the road transport.

2. Shortage of Coal. As mentioned at the outset, it requires about 250 kg of coal to produce one tonne of cement. Coal in India is always in short supply and most of the Indian coal varieties are of poor quality. Further, most of the coal reserves are concentrated in the north-eastern part of the peninsular plateau area and it becomes difficult to supply coal to areas located at large distances from the main coal producing region.

3. Shortage of Power. Currently, India is a power starved nation and the country is facing a serious power crisis. Although the consumption of power has come down from 120-130 units to 80-90 units per tonne of cement by using energy-efficient methods originating from the new technology, yet the industry is facing perpetual problem of power shortage. Load shedding and frequent declared and undeclared power cuts is a routine matter. This problem can be solved by increasing the power production.

4. Low Per Capita Consumption. Although there had been a spurt in the total consumption of cement in the country, the per capita consumption of cement in the country is still at a miserably low level. Thus there is an overall shortage of demand for cement and the manufacturers do not get a ready market for their product. With new projects coming up very fast, it is hoped that there will be increase in consumption of cement and manufactures will not find it difficult to search for buyers of their product.

RUBBER GOODS INDUSTRY

Although nascent, rubber goods industry has emerged as an important industry in India. It covers a wide range of products such as tyres and tubes of all kinds, footwear, surgical gloves, prophylactics, conveyor and vee belts, hoses, instruments, sports goods, rubber coats and a host of other products.

The most important segment of rubber industry is constituted by tyres and tubes. Tyres play a very important role to ensure mobility of passengers and essential goods across urban and rural landscape of the country using all types of vehicles ranging from carts, tractors, trucks and buses to the latest generation of passenger cars that ply on the modern expressways.

India is one of the few countries which has a attained self sufficiency in manufacturing a wide range of tyres for all applications. Currently, Indian tyre industry consists of 39 companies with 60 tyres manufacturing plants. Large tyre companies are namely MRF Ltd., Apollo Tyres, JK Tyres, Ceat, Goodyear and Modi Rubber, etc. Three Indian companies namely MRF Ltd., Apollo Tyres and JK Tyres are in the list of top 25 global tyre companies. The annual turnover of the tyre industry was about ₹ 50,000 crore during 2018-19. Indian tyre market is growing rapidly and in order to meet the growing demand, major tyre companies have taken up the new expansion projects.

India has become a major tyre exporting country and has a track record of exporting to over 75 countries. India also imports tyres mainly from China and Thailand.

Rubber footwear is another important item of rubber goods industry. Among the hoses, radiator and vacuum hoses are the most important. Fan belt is another important item.

LEATHER AND LEATHER GOODS

India has the largest number of livestock in the world and has an old tradition of processing leather and manufacturing leather goods. This industry has been carried on in India at the village level since Vedic times. The two major sectors of this industry are tanning and goods manufacturing.

Tanning

Tanning of hides and skins is an old and a very important sector of leather industry. Raw leather has to be tanned before use so that it can be given proper colour, strength and shine. The first tannery was set up at Kanpur in 1867. At present, Kanpur, Chennai and Kolkata are three large tanning centres in the country. The other important tanning centres are Agra, Bengaluru, Belgaum, Bhopal, Mokameh (Bihar), Phulbani (Odisha), Sherbaug (Gujarat), Kapurthala (Punjab), Pallavaram, Trichirapalli, Perambur, Eluru, etc. Besides, there are about 300 small tanneries.

Tanning industry has become the target of severe criticism from environmentalists and several agencies because waste discharge from the tanneries has led to large-scale environmental pollution. This has called for the development and implementation of new and improved technologies for enabling the leather industry to comply with pollution control standards. The role of alternatives in the tanning sector is widely recognised. State-of-the-art technologies for leather chemicals and cleaner production methods are being developed and new technology is being absorbed at a rapid pace. Tanning without wasteful discharge of chromium, fashion dyed leather without the support of mineral tanning, reduction of salt, sulphide and many other environmentally disconcerting chemicals in tanning waste are now possible. The need of the hour is to adopt these technologies without any reservations and without further loss of time. The Central Leather Research Institute (CLRI) at Chennai is the largest of its kind in the world. It has developed a number of techniques for the benefit of the Indian leather industry.

Leather Goods

India produces a large variety of leather goods including footwear, suitcases, bags and portfolios, ladies handbags, belts and several industrial accessories. However, footwear is the most important of all the leather goods and far excels others.

Manufacturing of footwear has been tradition bound and India is now the second largest producer of footwear in the world. Vast raw material resources, skilled manpower, developed expertise, low labour costs and foreign collaboration have given fillip to Indian footwear industry. After 1991, this industry witnessed profound changes as a result of liberalisation policy and due to globalisation of economy. Consequently, the production increased tremendously, both of the Indian style and the western style footwear.

The Indian footwear industry can be divided into three segments—large organised units, small scale sector and tiny cottage industry. Under large units come Bata, Carona and Tata, small scale units include Liberty, Wasan, Aero, Bajaj, Tej and Lakhani. Large scale units contribute 7-8 per cent of the exports while the bulk is provided by the small scale units. The tiny cottage units contribute the bulk of production and offer their products to the domestic market. The main centres of producing leather shoes, shoe uppers and leather sandals are Chennai, Ranipet, Ambur, Vaniambadi, Mumbai, Kolkata, Kanpur, Agra, Kolhapur, Delhi, Jalandhar, Bengaluru, Hyderabad, Batanagar, Faridabad and Jaipur.

Exports

India has been a traditional exporter of raw hides and skins. But the scenario has undergone much change and now India is emerging as a strong contender of leather goods exports in the world market. India is the second largest exporter of leather garments and fifth largest exporter of leather goods. The industry has identified major export markets in the U.S.A, Germany, U.K., France, Japan, Australia and Russia. With the adoption of new technologies, Indian leather goods would be able to meet the international standards and exports are expected to increase substantially.

The value of exports increased from a petty amount of ₹ 28 crore in 1960-61 to a staggering amount of ₹ 32,971 crore in 2019-20 (**Table 25.9**).

TABLE 25.9. Value of Exports of Leather and Leather Manufacturers including Leather Footwear, Leather Travel Goods and Leather Garments

Year	1960-61	1970-71	1980-81	1990-91	2000-01	2019-20	2020-21	2021-22
Value (₹ crore)	28	80	390	2600	8914	32971	24403	32,670

Source : Economic Survey 2022-23, Statistical Appendix, 116-18.

PLASTICS

This industry made a humble beginning in 1945 but has registered a fast growth since then. The growth accelerated in the 1960s and with the liberalisation of Indian economy in 1991, it has shown tremendous growth and diversification and is now the star performer along with other fast growing industries like electronics, computers and white goods. Whereas the developed world has reported 3-4 per cent growth, the Indian plastic industry has reported double digit growth each year. It is now estimated that the plastic industry in India will grow at a rate of over 15 per cent per annum for several years to come.

The wonders of plastics cut across all sections of society. Due to its moulding capacity, its versatility, its non-corrosive and moisture resistant properties, its physical and chemical strength, its economic viability, its easy processability and also its attractiveness and durability in all weather conditions, it has become a material of choice and of universal use. The major uses of plastics are in the packaging industry, building industry and in the manufacturing of pipes, fittings, electrical accessories, consumer goods, houseware, toys, furniture and in a host of other spheres of life. A time has come when plastics industry has the capacity to influence the progress of all other sectors. Plastic industry is regarded as '*sunrise*' industry due to its increasing versatility and burgeoning worldwide demand. *No wonder we are living in the plastic age*.

At present, the plastic industry produces goods from virgin and reprocessed polymer products. This sector provides employment to over 10 lakh skilled people directly and indirectly. That is why, the Governement has accorded it a *thrust industry status*. High growth in user segments such as retail packaging and building construction as well as in pipes, bulk packaging and agricultural use are expected to increase polymer demand tremendously making India the third largest polymer consumer after the USA and China. The low per capita polymer consumption is a pointer to the sector's immense growth potential. Mumbai, Delhi, Kolkata, Chennai, Bengaluru, Vadodara, Vapi, Kanpur, Amritsar, Coimbatore, Bongaigaon, Barauni, Mettur, Durgapur, Pimpri (Pune), Rishra, etc. are some of the outstanding centres which are well known for the production of either plastic raw materials or finished plastic goods. The Government has set up the Central Institure of Plastics Engineering and Technology (CIPET) at Chennai with the primary objective to develop skilled manpower and provide tehnical services to the plastic industry. To meet the growing demand of trained personnel, CIPET has set up nine extension centres at Ahmedabad, Amritsar, Bhopal, Bhubaneshwar, Hyderabad, Imphal, Lucknow and Mysuru including the headquarter centre at Chennai. A service centre and a training centre of CIPET have also been opened at Goa and Patna respectively.

Exports

Though exports of plastics from India is a recent development, the country has made a big dent in the international market. Kuwait, Saudi Arabia, Egypt, Kenya, Nigeria, Iraq, Syria, U.K., Thailand, Sri Lanka, and a number of other countries are the regular buyers of Indian plastic goods.

Prospects and Problems

The plastic goods industry in India has bright prospects and serious problems stored for it in the time to come. As mentioned earlier, the plastic goods have universal acceptability and their production and use will grow unhindered. The per capita consumption of plastics in India at 11 kg is far below the global average of 22 kg, 32 kg in Brazil, 38 kg in China, 65 kg in European Union and 109 kg in U.S.A. Thus there is great scope for expanding this industry. India has 22,000 plastic processing units and 2000 exporters. India's plastic production was 311 million tonnes in 2014 which is likely to double over the next 20 years. This will create millions of unskilled and semi-skilled jobs and change the export profile from raw materials to value added plastic products. But the unprecedented growth in the use of plastics as a packing material has created a serious problem of solid waste disposal. Plastic is a non-biodegradable substance and creates problems of environmental pollution. Many advanced countries are seriously thinking of reverting back to conventional packing materials, thereby dealing big blow to this industry. However, Indian plastic industry may thrive on the recycling of the used plastic material for which ragpickers can make valuable contribution. Recycling solves the problem of waste disposal, lessens the burden on raw materials, lightens the pressure on the biosphere and provides jobs to millions. It may be mentioned here that this process needs technological development and upgradation.

DRUGS AND PHARMACEUTICALS

The growth of drugs and pharmaceuticals industry in India is primarily a post-Independence phenomenon. Before Independence, bulk of drugs were imported and only processing and formulations were done in the country. This industry has made tremendous progress during the last 50 years. The growth rate of this industry has been about ten per cent per annum, consistently for the last more than three decades. Today, India is in a position to meet 95 per cent of her requirements of bulk drugs and almost all the requirement of formulations. The present production covers a wide range of bulk drugs including antibiotics, vitamins, steroids and hormones, semi-synthetic penicillins, synthetic phytochemicals and biological products besides, practically the entire range of formulations required by the medical profession.

The industry is likely to grow rapidly in times to come.

Drug prices in India are some of the lowest in the world and India has one of the largest inventories of highly skilled pharmaceutical professionals. Modern drugs reach only 40 per cent of Indian population mainly in urban and semi-urban areas. Obviously the potential for growth in the domestic market is enormous.

There are about 250 units in the organised sector, five of these in the public sector (IDPL, HAL, SSPL, BCPL and BIL) and six in the joint sector. There are 21 MRTP companies and six FERA ones. About 5,000 units in the small scale sector are also engaged in the production of drugs. Of these, over 100 units are engaged in the production of bulk drugs.

Indian Drugs and Pharmaceuticals Limited (IDPL) a premier undertaking was incorporated on 5 April 1961. It has five plants located at Rishikesh for the manufacture of synthetic drugs, at Chennai for surgical instruments, at Gurgaon for formulations, Hyderabad (synthetic drugs plant) and at Muzaffarpur for drugs and chemical intermediates. Besides, IDPL has three subsidiary companies set up in association with the state governments. These are Rajasthan Drugs and Pharamaceuticals Ltd. (RDPL), Jaipur, Uttar Pradesh Drugs and Pharmaceuticals Ltd. (UPDPL), Lucknow and Odisha Drugs and Chemicals Ltd. (ODCL), Bhubaneshwar.

Hindustan Antibiotics Limited (HAL) was incorporated on 1 March, 1954. It is engaged in manufacturing Penicillin, Streptomycin, Amcillin, Hamycin, Gentamycin and Aureofungen, besides formulations. It has three subsidiary companies set up in collaboration with the state governments. They are (*i*) Maharashtra Antibiotics and Pharamceutical Limited (MAPL), Nagpur, (*ii*) Karnataka Antibiotics and Pharmaceuticals Limited (KAPL), Bengaluru and (*iii*) Manipur State Drugs and Pharmaceuticals Limited (MSDPL) Imphal.

The Government has also nationalised three sick units at Kolkata. These units have been converted into public sector companies. They are (*i*) Bengal Immunity Limited (BIL), (*ii*) Bengal Chemicals and Pharmaceuticals Limited (BCPL) and (*iii*) Smith Stanistreet Pharmaceuticals Ltd. (SSPL)

The leading companies in the private sector are Ciba, Sarabhai, Hoechst, Alembic, Glaxo, Unichem, Pfizer, Chemo Pharma and Warner Hindustan. These companies produce almost all types of drugs and formualtions.

Although small sector units are scattered all over the country, their heavier concentration is found at Mumbai, Chennai, Kolkata, Delhi, Vadodara, Kanpur and Hyderabad.

GLASS INDUSTRY

Indians are known to have acquired the knowledge of making glass since time immemorial. Glass industry came into being in India in 16th century when items like bangles, small bottles and flasks were made. By 17th century, enamalled glass was produced at a number of places in Karnataka and Uttar Pradesh. Although glass industry on modern lines was started in the 19th century, the real growth started only after 1932. The first successful organised glass factory was set up in 1941. The industry underwent large scale modernisation after Independence. At present, India is a major producer of glass and glass products.

Broadly speaking, glass industry covers seven items. These are (*i*) sheet and flat glass (including sheet, float, figured, wired, safety, mirror glass), (*ii*) glass fibre and glass wool, (*iii*) hollow glassware, (*iv*) laboratory glassware, (*v*) table and kitchen glassware, (*vi*) glass bangles, and (*vii*) other glassware.

There has been growing acceptability of Indian flat glass products in the global market.

Localisation

Glass industry requires a large number of raw materials. The most important raw material is silica sand which constitutes 75 per cent of the basic materials. This is a bulky material and cannot bear high transportation cost. Obviously the availability of silica sand influences the localisation of glass industry. The other raw materials used for manufacturing glass are soda ash, feldspar, limestone, dolomite, manganese dioxide, barium oxide, sulphur and copper. These raw materials affect the localisation of glass industry to some extent in their own way. All these raw materials except soda ash are abundantly available in the country. Sufficient supply of coal at cheap rates also influence this industry. But glass is a fragile commodity and easily breaks during transit. Therefore, this industry tends to be located near the market.

Distribution

West Bengal, Uttar Pradesh and Maharashtra are the main glass producing states and contribute the bulk of production.

West Bengal. West Bengal is the largest producer of glass and glass products. The state has 35 factories located at different places like Kolkata, Haora, Raniganj, Belgachiya, Belgharia, Belur, Sitarampur, Rishra, Durgapur and Asansol. Pure sand of high quality is available from the white Damudas sandstones at Mangalhat and Patharghata. Sand is also obtained from Bargarh and Lohagra. Good quality coal is obtained from the nearby coalfields at Jharia and Raniganj. Good market is readily available in the Hugli industrial region. Most of the items produced here are used as scientific instruments in the laboratories and as decoration prices.

Uttar Pradesh. This is the second largest glass producing state in India. There are 32 large factories and over 100 small units running as cottage industry. Silica sand is available in the Ganga, Yamuna and Chambal valleys. Limestone and sandstone are found in the southern part of the state. The state is well served by a close network of rail and road routes. Only coal has to be transported from Jharkhand and West Bengal. Firozabad is the largest centre. It is well known for making bangles, glass sheets, bulbs, articles of everyday use, and glassware used in scientific laboratories. Skilled glass workers known as ***shisgars*** of Firozabad have been engaged in the process of glass making for several generations and provide cheap and skilled labour for this purpose. The other major centres of glass production are Bahjoi, Naini, Hiranagau, Shikohabad, Hathras, Sasni, Pryagraj, Jaunpur, Kirtpur, Itawa, Ghaziabad, Ramnagar, etc.

Maharashtra. The state has 22 factories. Main centres of galss industry are Mumbai, Talegaon (Pune), Satara, Nagpur and Kolhapur. The industry specializes in bottles, shells, flasks, lampware, beakers and sheet glass.

The other producers are *Gujarat* (Bharuch, Vadodara, Morvi and Panch Mahals), *Tamil Nadu* (Salem, Chennai, Coimbatore), *Bihar* (Bhawaninagar, Patna), *Jharkhand* (Kandra, Jamshedpur, Kahalgaon), *Rajasthan* (Dhaulpur and Jaipur), *Haryana* (Ambala and Faridabad), *Telangana* (Warrangal and Hyderabad), *Delhi* (Shahdara), *Punjab* (Amritsar), *Kerala* (Alwaye), *Odisha* (Barang, Cuttack), *Madhya Pradesh* (Jabalpur, Gondia), *Assam* (Guwahati) and *Karnataka* (Bengaluru).

CERAMICS INDUSTRY

It is a very old industry and dates back to about 5000 years B.C. The excavations of Harappa and Mohen-jo-Daro have revealed that a wide variety of pottery goods and ceramics of different designs and colours was used when the Indus Valley civilisation was at its peak. The beginning of modern ceramics industry is associated with the discovery of deposits of China clay in Rajmahal hills in Jharkhand. The first ceramic factory was established at Patharghatta in 1860. In the same year, Burn and Co. set up a factory at Raniganj in West Bengal and the production of glazed tiles commenced. Further discoveries of China clay in different parts of the country added new dimensions to ceramic industry and a number of factories came up in West Bengal, Bihar, Madhya Pradesh,

Maharashtra, Gujarat, Karnataka and Kerala. World War II gave further impetus to this industry because metallic utensils could not be made due to ban on the import of metals. The industry has made rapid strides in the post-Independence era.

Today, ceramic industry produces a wide range of products, the outstanding being sanitary wares, porcelain wares, stoneware, enamel ware, tiles, crockery, insulators, etc. These products are used for a variety of purposes such as generation and transmission of power, construction of modern buildings, engineering, electronics, etc.

At present, there are above 150 units in the organised sector manufacturing different items. The important centres are Kolkata, Mumbai, Bengaluru, Delhi, Wankaner, Thanagarh, Ranipet, Rupnarainpur, Jabalpur, Nazarbagh, Gwalior, Jaipur, etc.

Ceramic tile industry is growing at a fast rate. India is the third largest producer as well as consumer of ceramic tiles in the world after China and Brazil. The industry gives direct and indirect employment to 2 lakh people. With majority of raw materials available in the country, the Indian tile industry has kept pace with the times embracing modern technology in designing and manufacturing and is capable of producing world class tiles.

Indian ceramic products are some of the best so far as their quality, shape, design and colour are concerned and are easily accepted in the international market. The main buyers of Indian ceramic goods are Iran, Iraq, Saudi Arabia, Kuwait, Kenya, Uganda, Tanzania, Zambia, Sudan, Mauritius, Sri Lanka, Myanmar, etc. India also imports ceramic products from some Asian and European countries like China, Japan, U.K., Germany, the Netherlands, Czech Republic and Slovakia.

FOREST BASED INDUSTRIES

Forests provide us with different types of materials which are used as raw materials for certain industries. Those industries which use forest products as raw materials are called forest based industries. Paper, match, lac, sports goods, plywood, etc. are such industries.

PAPER INDUSTRY

Paper is one of the core industries and is linked to the basic human needs. Paper is the pre-requisite for education and literacy and its use is an index of advancement in these two fields as well as the overall well being of the society.

Growth and Development

Paper manufacturing has been carried on in India since tenth century as a small cottage industry by the traditional craftsmen called ***kagzis***. They used gunny bags, rags, ropes, etc. for making paper. This industry could not survive the onslaught of the machine made paper and declined considerably. However, a part of it has managed to survive and even today, a large number of small units are producing handmade paper.

The beginning of modern paper industry goes back to 1816 when a factory was set up near Chennai. This venture proved abortive. Another papermill was set up in 1832 at Serampore on the bank of Hugli in West Bengal. This venture also failed and the first successful effort was made in 1870 with the setting up of the Royal Bengal Paper mills at Ballyganj near Kolkata. This is the principal seat of paper industry even today. Subsequent successful efforts were made at Lucknow in 1879, Titagarh in 1882, Pune in 1887, Raniganj in 1892, Kaukinra in 1892 and Naihati in 1918. The preferential treatment and tariff protection helped in solving the initial problems of this industry. World War II further infused life into this industry. But the progress was rather sluggish till Independence and it is only in the plan period that the industry has really made fast progress. The growing knowledge base coupled with synergistic contributions from flagship schemes of the government, namely Sarva Shiksha Abhiyan, Rashtriya Madhyamik Shiksha Abhiyan, Inclusive Education for Disabled at Secondary School, Adult Education and Right to Education assured a robust demand for paper and paper board.

The industry was delicenced in July, 1997. As per present policy, FDI upto 100 per cent is allowed on automatic route for pulp and paper sector.

Currently India is the largest growing paper market in the world and focus of paper manufacturing globally. Indian paper industry accounts for 5% of the world's production. The Indian pulp and paper industry provides direct employment to 5 lack people and around 15 lakh people indirectly. According to Central Pulp and Paper Research Institutes (CPPRI) data (2019-20), India has 861 paper mills with 526 operational with a total installed capacity of 27.15 million tonnes. The per capita consumption in India is around 15 kg. against the global average of 57 kg. The demand for paper is growing at 6-7 per cent per annum. The estimated output is 19 million tonnes per annum. The Indian paper industry is fragmented consisting of small mills (5-50 tonnes per day), medium mills (50-100 tonnes per day) and large mills (100-1,100 tonnes per day).

Raw Materials. Paper and paper board can be manufactured by using different types of raw materials. Raw materials account for 45-50 per cent of the total cost of production and form an important segment of manufacturing paper and paper board. Of the total installed capacity, 43 per cent is dependent on forest based raw materials, 28 per cent on agro based raw materials and the remaining 29 per cent on other materials including waste paper.

Bamboo. Generally speaking 2.3 to 2.4 tonnes of bamboo is required for producing one tonne of paper. The paper industry uses bamboo to the extent of 60-70 per cent of the total requirements of cellulosic raw materials. Bamboo has the advantage of possessing long fibre, dense stands and quick regeneration. It reaches maturity in 2-3 years and provides continuous flow of renewable source of raw material. However, there is danger of this source of important raw material being depleted if the rate of exploitation exceeds the rate of regeneration. The total supply of bamboo at the current rate is estimated at 20-30 lakh tonnes per annum. Assam, Odisha, Andhra Pradesh, Telangana Madhya Pradesh, Tamil Nadu, Karnataka and Maharashtra are important producers of bamboo.

Sabai Grass. This is another important raw material for manufacturing paper. It was the sole raw material before introduction of bamboo as a significant raw material, but its use has decreased considerably since then. It now constitutes 7 to 9 per cent of the total cellulosic raw material in the country. Although sabai grass has long fibre and requires low chemical consumption, it grows in tufts intermixed with other vegetation and it is often difficult to separate impurities from it. Moreover, its supplies are much less than those of bamboo. The annual supply of sabai grass along with other allied grasses is about one million tonnes. It mainly grows in the sub-Himalayan tracts of Shiwaliks and Tarai area.

Bagasse. It is a fibrous residue of the sugarcane stalk, mainly from the sugar mills, obtained after sucrose is extracted by crushing the sugarcane. On an average 50-60 lakh tonnes of bagasse is produced in the country, half of which is used for manufacturing paper.

Other materials. Paper is also manufactured by using materials other than those mentioned above. These include waste paper, rags, straw from rice and wheat, jute sticks and soft wood obtained from eucalyptus, pine-wood, wattle and mulberry trees.

Chemicals. In addition to the above mentioned cellulosic raw materials, certain chemicals are also used in the manufacture of pulp, paper and paper board. The important chemicals are caustic soda, soda ash, sodium sulphate, chlorine, sulphuric acid, sulphur, lime, ferric alumina, ammonium sulphate, resin and clay. A variety of sizing and colouring agents are also used. Coal is also an important input in paper industry. Normally, 3.5 to 4.1 tonnes of coal is required for producing one tonne of paper. Large quantity of soft water is also needed.

Localisation of the Industry

Paper and paper board manufacturing uses coarse, cheap and weight losing raw materials and seeks raw material oriented locations. Chemicals used in this industry are needed in small quantity and are easily transported even over long distances from the place of production to the place of consumption. Therefore, there is a strong tendency among the paper mills to be located near the forest tracts along the Western Ghats, the Eastern Ghats, central India and the Tarai-Bhabar area at the foot hills of the Himalayas. There are vast stands of temperate forests in the Himalayan region which can provide large quantities of soft woods. These soft woods can provide excellent cellulosic material for the paper industry. But because of the rugged terrain and poor means of transportation, these areas are inaccessible and the possibility of extraction of these soft woods and their supply to paper mills in the adjoining areas is remote. Moreover, the regeneration of wood pulp forests takes about 50-60 years, while bamboo forest is ready within 2-3 years. So far, no paper mill has been set up in the Himalayan region. Since the supply of cellulosic raw materials falls short of the demands of the paper industry, efforts are being made to bring more land under plantation of eucalyptus and other fast growing soft wood trees. However, plantations of eucalyptus has been discouraged during the last couple of decades in eastern parts of the country because this tree quickly exhausts the sub-soil water in large quantities. Some of the paper mills are located near the market where cheap labour is also available.

Distribution of Paper Industry

Maharashtra, Andhra Pradesh, Gujarat, Uttar Pradesh and West Bengal are the main producers and account for more than half of the paper and paper board production of India.

Maharashtra. Maharashtra has risen from fourth position in the mid-1980s to second position in 1993 and to first position in 1996 among the major paper producing states of India. The state has 63 mills, accounts for 16.52 per cent of the installed capacity and produces 18 per cent of the paper produced in India. Maharashtra is not very lucky with respect to the availability of raw material and coal. Most of the mills located in Maharashtra use rags, waste paper or pulp imported from Sweden and Canada as the principal cellulosic raw material. Hydroelectricity is used as a power resource in place of coal. Ballarpur has the largest paper mill. Sangli, Kalyan, Mumbai, Pune, Balarshah, Pimpri, Nagpur, Bhiwandi, Nandurbar, Tumur, Khopoli, Kamptee, Vikroli, Chinchwad, etc. are other major producers.

Andhra Pradesh. Although a late entrant in paper industry, Andhra Pradesh superceded West Bengal in early 1990s to become the top producer but conceded this position to Maharashtra in mid-1990s. At present, this state has 19 mills, accounts for 11.3 per cent of installed capacity and 13 per cent of India's total production of paper. Paper mills are mainly located at Rajahmundry, Tirupati, Kurnool, Srikakulum, Pallancheru, Nellore Bhadrachalam, Kakinada, Apidik, Bodhan, etc. Most of the mills use bamboo as the basic raw material. Bamboo grows widely in Andhra Pradesh and the recoverable reserves are estimated to be about 1.5 lakh tonnes per annum.

Telangana. In the neighbouring Telangana, Sirpur (Kagaznagar) is the most important centre of paper manufacturing. Sarapaka, Khammam are the other important centres of manufacturing paper.

Gujarat. Gujarat has improved its position considerably in paper manufacture. The share of Gujarat in the nation's total production of paper has gone up from a miserable 3 per cent in the mid-1980s to a respectable 10 per cent in the mid-1990s. The state has as many as 55 mills which largely depend upon bagasse and eucalyptus for supply of raw material. The main centres of production are Rajkot, Vadodara, Surat, Barjod, Bilmoria, Navsari, Songarh, Ahmedabad, Vapi, Bharuch, Dijandranagar, Limbdi, Gondal, Udvada and Bavla.

Uttar Pradesh. This state has the largest number of 68 mills, but the size of the mills being small, the installed capacity does not exceed 9 per cent. Saharanpur and Lalkuan have mills of large size. The other centres are at Meerut, Modinagar, Ghaziabad, Lucknow, Gorakhpur, Pipraich, Muzaffarnagar, Pryagraj (Naini), Varanasi, Kalpi, Budaun and Mainpuri.

West Bengal. West Bengal was the pioneer state in the paper industry at the initial stage and led the country till mid-1980s. The state has since slided down the ladder to sixth place with 22 mills, 7 per cent of the installed capacity and about the same percentage of production. West Bengal has the advantage of early start of this industry. The paper industry in West Bengal is based on bamboo which is available locally or is obtained from Assam, Odisha and Jharkhand, and sabai grass which is obtained from Chhattisgarh and Madhya Pradesh. Coal is abundantly available from Jharia and Raniganj. Ganga and other rivers provide sufficient water for processing. High density of population provides cheap labour. The highest concentration of paper mills is found along the Hugli river. Titagarh, Kankinara, Raniganj, Bansberia, Sheoraphuli, Chandrabati, Triveni, Naihati and Kolkata are some of the important centres of paper manufacturing in West Bengal.

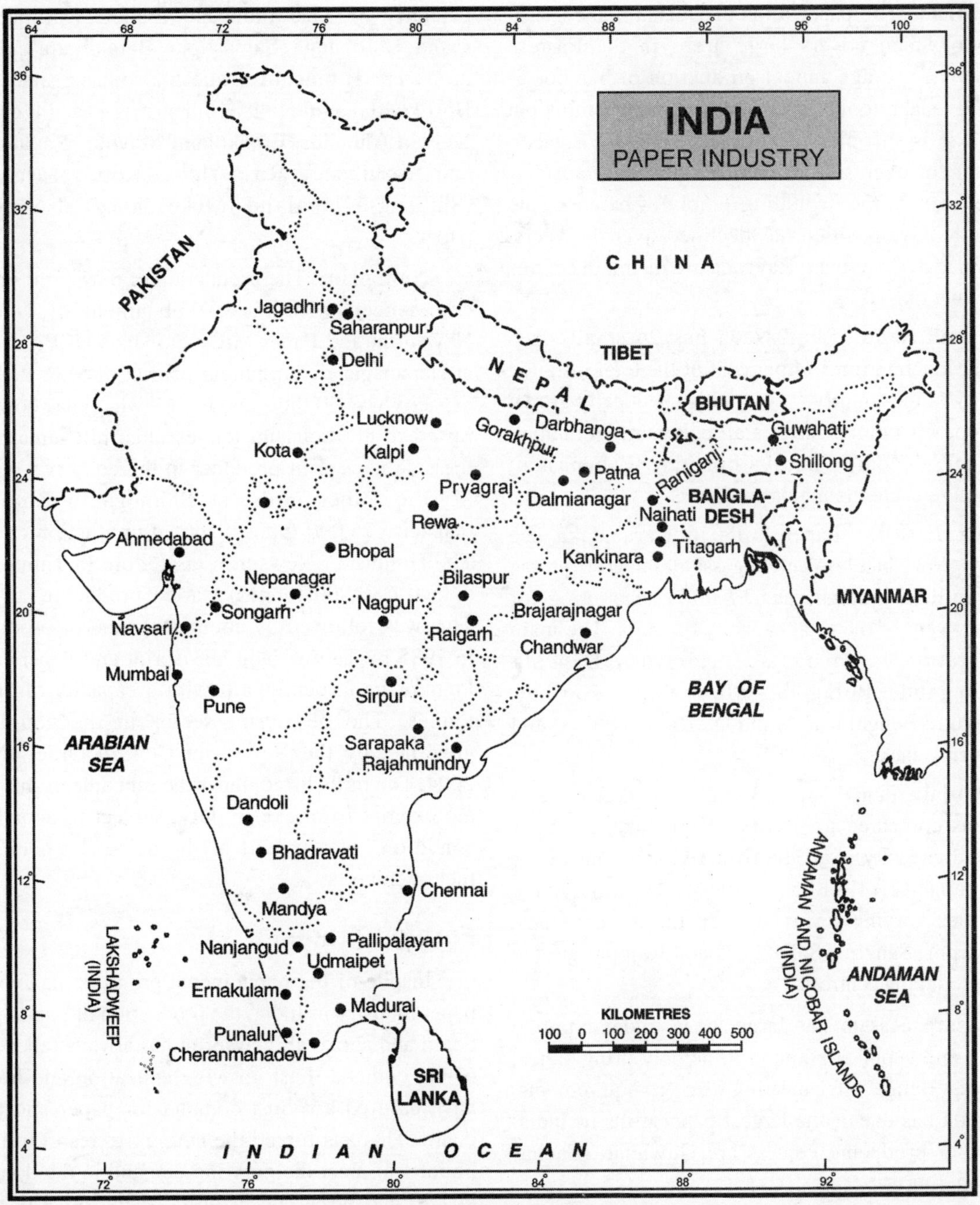

FIG. 25.14. India : Paper Industry

Madhya Pradesh. Madhya Pradesh has large tracts under cellulosic raw materials *viz*., bamboo, sabai grass, eucalyptus, etc. and provides solid base to paper industry. The state has 18 mills which account for 6.62 per cent of the total installed capacity of India. The main centres of production are Bhopal, Amlai, (Shahdol), Ratlam, Rajgarh, Vidisha, Abdullaganj, Rewa and Indore. Nepanagar is the birthplace of newsprint industry and is still an important centre of newsprint manufacturing in India.

Odisha. The paper mills of Odisha make use of bamboo which covers large areas in the Eastern Ghats. The average annual production of bamboo is about five lakh tonnes. The state has only 8 mills but their size is sufficiently large enabling the state to account for over six per cent of the total capacity. However, with the dwindling supply of bamboo, the state's relative position has declined over the years. Brijrajnagar, Chandwar, Raygada are the main centres of production.

Tamil Nadu. Tamil Nadu has 24 small sized mills accounting for 5.74 per cent of the total installed capacity of the country. These mills use locally grown bamboo. Cheranmahadevi, Pallipalayam, Udmalpet, Chennai, Salem, Amravathinagar, Pahanasam, Madurai, etc. are the major producers.

Karnataka. Paper mills in Karanataka use locally grown bamboo and bagasse obtained from the sugar mills. Karnataka has 17 mills accounting for 5.48 per cent of the total capacity of India. The main cocentration is found at Bhadravati, Dandoli, Nandangaud, Belagola, Munirabad, Harihar, Mundyud, Bengaluru, Mandya, Ramnagaram and Krishnarajsagar.

Punjab. Punjab has 23 mills but their size is small as compared to those of Karnataka and their installed capacity is a little over five per cent of the total capacity of India. Most of the mills have capacities varying from 10 to 20 thousand tonnes. Hoshiarpur, Sangrur, Sailakhurd and Rajpura are the chief producing centres.

Assam. Assam has large areas under bamboo growth and supplies bamboo to neighbouring states, like West Bengal after meeting its own requirements. Nowgaon has one of the largest paper mills of India. The other producing centres are Guwahati, Cachar and Lumding.

Haryana. Haryana's 18 mills mainly depend upon imported pulp and eucalyptus wood as raw material. Yamunanagar has the largest mill. Small units are functioning at Faridabad, Dharuhera and Jagadhari.

Others. Bihar has 6 mills with a total installed capacity of 87.6 thousand tonnes. Dalmianagar, Patna, Darbhanga, Samastipur, Barauni are the chief producers. **Himachal Pradesh** also has 6 small units. Borokwala and Kala Amb are the main centres. **Kerala** (Punalur, Ernakulam, Mavoor, Rayanpuram and Kozhikode), **Rajasthan** (Kota), **Meghalaya** (Shillong), **Nagaland** (Mokokchong) also produce paper.

Newsprint. The production of newsprint in India commenced with the establishment of National Newsprint and Paper Mills known as NEPA in 1956 at Nepanagar in Burhanpur district of M.P. Its initial capacity was 30 thousand tonnes which has now been raised to 88 thousand tonnes.This mill remained as the sole newsprint producer in the country till 1980. Mysore Paper mills, at Shimoga in Karnataka (capacity 75,000 tonnes) started production in 1981. The Hindustan Newsprint at Vellore in Tamil Nadu (capcity 80,000 tonnes) started producing in 1982. This was followed by commencement of production in 1985 by Tamil Nadu Newsprint and Paper Ltd. at Pugalur in Tiruchirapalli with a capacity of 50,000 tonnes. The newsprint sector in the country is governed by the Newsprint Control Order (NCO) 2004. The mills listed under the schedule of this order are exempt from excise duty, subject to actual user condition.. Nearly half of the newsprint demand is met by imports.

Foreign Trade

Inspite of the phenomenal progress made by the paper industry in India, the production of paper, paper board and above all newsprint has always fallen short of the demand. Fast growing education and literacy have assured a robust demand for paper and paper board. This has forced the country to resort to heavy imports to meet the growing demand in the domestic market and put a check on the rising price of paper and paper products. The present rate of consumption of about 15 kg per capita per annum is extremely low as compared to the world average of 57 kg and over 300 kg of the highly advanced countries. A little increase in the capita consumption can lead to serious crisis of paper availability in the coming years.

Table 25.10 shows that the amount spent on the import of raw materials, like pulp and waste paper as well as paper, paper board and manufactures as well as that of finished paper, paper board and manufactures thereof has been rising at an alarming rate.

The position with regard to newsprint is still worse. The current per capita consumption of newsprint in India is too meagre 600 grams as compared to the Asian (excluding Japan and China) average of 1.9 kg and the world average of 6 kg.

Pulp and waste paper are imported from Norway, Sweden, Canada and Holland and the main sources of paper, paper board and newsprint are Sweden, Poland, Canada, Czech Republic and Slovakia.

Problems and Prospects

Paper industry in India faces many serious problems and prospects do not seem very bright. The biggest problem faced by this industry is the scarcity of raw materials. Most of the materials used for manufacturing paper are derived from the forests. With the increasing degradation of forest and fast depletion of forest based raw materials, like bamboo, the paper industry is facing a severe raw material crisis. With the exploitation of forests based raw materials reaching its saturation, any substantial capacity expansion in the near future is practically ruled out and the growth of industry has come to a standstill. In order to survive and expand, the industry will have to look for unconventional raw materials. This will require new advanced technology which a developing country like India may not be able to afford. Another alternative is to increase the use of waste paper which will reduce dependence on the conventional raw materials. This will also reduce cost of energy used and other inputs. It may be mentioned here that only 15 per cent of the total output of paper and paper board is based on recycled material against the world average of 30-35 per cent. Thus there is a vast scope for using recycled material in paper industry.

Even now, large quantities of bagasse is just used as fuel in the sugar industry and is not made available to paper industry. If the sugar mills are encouraged to use coal fired boilers instead of those based on bagasse, this precious raw material can solve the problem of paper industry to great extent.

The growing consciousness for preservation of forests and maintenance of ecological balance and biodiversity during the last few years is further reducing the availability of raw materials to this industry. Environmentalists are also up in arms against this industry due to effluents released by the paper mills into open drains, rivulets and rivers thereby polluting the environment. Unless technology to solve the effluent problem is evolved and implemented with only marginal additional investment, many of the paper mills run the risk of economic losses and even closure.

TABLE 25.10. Import of Paper, Paper Board and Raw Materials

Year	Paper, paper board and manufactures thereof		Pulp and waste paper	
	Quantity ('000 tonnes)	Value (₹ crore)	Quantity ('000 tonnes)	Value (₹ crore)
1960-61	55.6	12	80.3	7
1970-71	159.0	25	71.7	12
1980-81	371.4	187	36.9	18
1990-91	286.4	456	678.2	458
2000-01	585.6	2,005	1,050.9	1,290
2020-21	—	20,095	—	6,303
2021-22	—	29,296	—	11,934

Source : Economic Survey, 2022-23, Statistical Appendix, pp. 112-14.

India is facing wide gap between supply and demand of paper even at the low rate of consumption. Currently, India has about 17.5 per cent of the world's population but consumes a little over 1 per cent of the world's paper and paper board. With the spread of education and literacy, demand for paper is bound to increase and is expected to be double the present demand within ten years. This will constraint India either to increase the indigenous production considerably or resort to large scale imports. As a result of Covid-19, there is a growing tendency of work from home, 'online work' and 'paperless work'. This may reduce demand for paper to some extent.

The average size of paper mills is abnormally low at annual production of less than 10,000 tonnes against 50,000 tonnes in South-east Asia and 85,000 tonnes in Asia Pacific. The small size of the manufacturing units makes them uneconomic, prohibits the induction of new technology, needs larger capital inputs in proportion to the final production and increases the cost of operations. Several mini plants are using old technologies and worn out machinery and badly lack competitive strength. The spiralling cost of imported wood pulp and waste paper coupled with unremunerative selling price of the finished paper has rendered many paper mills economically unviable and forced them to close down. Recently, the Department of Industrial Policy and Promotion had commissioned a study on 'Global

However, one encouraging aspect is that Indian paper industry has trained manpower whose skill can be gainfully used to adopt modern technology for manufacturing paper and paper board of international standard at relatively lower manpower cost inputs. The country has vast reservoir of unemployed rural labour which can be deployed for development of raw materials. Thus, we can overcome some of the weaknesses which have plagued the paper industry for a long time.

MATCH INDUSTRY

The first match factory was set up at Ahmedabad in 1921. The Western India Match Company (WIMCO) came into existence in 1923 and it set up five factories at Bareilly (U.P.), Kolkata, Chennai, Ambarnath (Mumbai) and Dhubri (Assam). These five factories of WIMCO along with Assam Match Co. (AMCO.) produce about 65 per cent matches in India. The remaining 35 per cent match is produced by de-centralised small scale units and cottage industry. There are about one thousand small scale units manufacturing match. The production has shown varying trends.

Localisation

The chief raw materials used in this industry are soft timber and paper for making sticks and boxes. The main chemicals used are phosphorus, potassium chlorate, paraffin, potash, etc. Earlier, most of the timber requirements were met by imports from Norway, Sweden, Finland and Japan. Now, locally available wood like *genwa*, *pipita*, *dhup*, *didu*, *bakoda*, *mango*, *semal* and *solai* is used and imports are drastically reduced. But most of the chemicals are still imported. Match factory requires cheap and skilled labour because one-third of the cost of manufacturing match is on labour.

Distribution

West Bengal, Tamil Nadu, Maharashtra and Gujarat are the main producers.

West Bengal has the highest concentration of match industry. Timber is obtained from Andaman and Nicobar Islands as well as from the Sunderbans. Some timber is imported from Sweden. Kolkata serves as a good port for importing chemicals. Coal is obtained from Jharia and Raniganj. The area is densely populated and there is no dearth of labour. 24 Parganas and Kolkata have the greatest concentration of match industry in this state. ***Tamil Nadu*** has nearly two-thirds of non-mechanised small scale units. The main centres are Ramnathpuram, Chingleput, Tirunelveli and Chennai. The other producers are ***Maharashtra*** (Pune, Thane, Chandrapur, Mumbai), ***Gujarat*** (Ahmedabad, Petlad, Ambarnath), ***Uttar Pradesh*** (Bareilly, Meerut, Pryagraj, Varanasi), ***Karnataka*** (Shimoga), ***Kerala*** (Thiruvananthapuram), ***Telangana*** (Hyderabad, Warangal), ***Assam*** (Dhubri), ***Rajasthan*** (Kota), ***Madhya Pradesh*** (Jabalpur) and ***Chhattisgarh*** (Bilaspur).

LAC INDUSTRY

Lac is obtained from an insect named Cerria lacca which secretes a resin. This insect lives on trees which grow in areas at an elevation of 300 metres above sea level and having 12°C temp. and 150 cm annual rainfall. Resin is known as sticklac in its crude form and shellac or lac in the refined form. Lac is used for a variety of purposes including gramophone records, French polish, electrical insulating materials, shellac moulded articles, micanite, hats, grinding wheels, adhesives, cements, wood turning, metal enamelling, printing ink, paints and varnishes, photographic equipment, bangles, toys and many more. This industry provides employment of about 10,000 people. In addition 60-70 lakh tribals are engaged in rearing the insect and obtaining resin from this insect.

The average annual production of lac in India is 20-25 thousand tonnes. Earlier, India was the largest producer of lac in the world and used to produce about 85 per cent of the world's total production of lac. But India's share in the world production of lac has fallen to 50-60 per cent due to increased production in Thailand, Myanmar and Indochina. Major part of the production comes from the **Chota Nagpur** plateau of Jharkhand which accounts for about 50 per cent of the total production. The remaining 50 per cent is produced by other areas in Madhya Pradesh, Chhattisgarh, West Bengal, Odisha, Meghalaya, Assam, Uttar Pradesh, Gujarat and Maharashtra. ***Madhya Pradesh*** produces 25 per cent of India's lac. The main producing districts are Balaghat, Chhindwara, Jabalpur and Shahdol. About 15 per cent lac of India comes from ***Chhattisgarh*** where Bilaspur, Raipur, Surguja and Koriya are the main producing districts.. In ***West Bengal***, Murshidabad, Malda and Bankura are the main lac producing districts. The other producers are ***Odisha*** (Mayurbhanj, Bolangir, Dhenkanal, Sambalpur and Keonjhar), ***Meghalaya*** (Garo, Khasi and Jaintia Hills), ***Assam*** (Nagaon, Kamrup and Sibsagar districts), ***Uttar Pradesh*** (Mirzapur), ***Gujarat*** (Panchmahals and Vadodara) and ***Maharashtra*** (Bhandara).

International Trade

India has been a traditional exporter of lac and is still the largest exporter in the world. About 85-90 per cent of the total product finds its way to foreign markets. The USA, UK, Germany and Russia are the main buyers and account for over 50 per cent of the country's total lac exports. The other buyers include Italy, France, Japan, Sweden, Australia, Brazil, Argentina, etc. India also imports raw lac from Thailand and re-exports it in the form of finished products.

SPORTS GOODS

Sports goods industry was first established at Sialkot (now in Pakistan) and Meerut in Uttar Pradesh. From a modest beginning the industry made rapid progress after Independence when several skilled workers shifted from Sialkot to various towns and cities of India including Meerut, Jalandhar, Delhi, Batala, Kolkata, etc. Later on this industry spread to Jammu, Srinagar, Patiala, Faridabad, Modinagar, Dehradun, Bulandshahr, Lucknow, Pryagraj, Amritsar, Bhopal, Katni, Chennai, Mumbai, Pune and many more cities. Meerut and Jalandhar dominate the production of sports goods. Meerut has 270 units out of a total of 700 units in the whole of the country. Jalandhar has made rapid progress during the last couple of decades and manufactures quality sports goods.

Today, India is in a position to manufacture goods for almost all the sports. The factors contributing to India's progress in sports goods are the availability of raw materials and skilled, cheap and abundant labour. The principal raw materials used in this industry are wood, leather, rubber, metal, etc.

Indian sports goods are of good quality and are in great demand throughout the world. U.K., Germany, Australia, Malaysia and Singapore are the main buyers of Indian sports goods.

FOOD AND ALLIED INDUSTRIES

This group includes those industries which are connected with food. Sugar, edible oils and vanaspati, flour milling, rice milling, alcohol, etc. are some of the examples of food and allied industries.

SUGAR INDUSTRY

Sugar can be produced from sugarcane, sugar-beet or any other crop having sugar content. But in India, sugarcane is the main source of sugar. At present, this is the second largest agro-based industry of India after cotton textile industry. India is the world's second largest producer of sugarcane after Brazil and second largest producer of sugar after Cuba. But India becomes the largest producer if *gur* and *khandsari* are also included. This industry provides employment to 2.4 lakh workers. In addition, 2.50 crore sugarcane growers also get benefit from this industry.

Growth and Development

India has a long tradition of manufacturing sugar. References of sugar making by the Indians are found even in the *Atharva Veda*. India is rightly called the *homeland of sugar*. But in ancient times, only *gur* and *khandsari* were made and modern sugar industry came on the Indian scene only in the middle of the 19th century, when it was introduced by the Dutch in North Bihar in about 1840. Unfortunately, this attempt could not succeed. The first successful attempt was made by the indigo planters at the initiative of Britishers in 1903 when Vacuum pan mills were started at Pursa, Pratabpur, Barachakia and Marhowrah and Rose in north-eastern U.P. and the adjoining Bihar. This happened when demand for indigo ceased to exist due to the introduction of synthetic blue in the market. In the early years of the 20th century, the industry grew rather sluggishly and there were only 18 mills in 1920-21 and 29 mills in 1930-31. The industry got a great fillip after the fiscal protection in 1932 and the number of mills rose to 137 in 1936-37. The production also shot up from 1.58 lakh tonnes to 9.19 lakh tonnes during the same period. The industry passed through an uncertain phase during and after the World War II and some stability was experienced only after 1950-51. There were 139 mills producing 11.34 lakh tonnes of sugar in 1950-51. After that, the plan period started and the industry made rapid strides. Currently there are 732 mills producing 339 lakh tonnes of sugar every year.

Localisation of Sugar Industry

Sugar industry in India is based on sugarcane which is a heavy, low value, weight losing and perishable raw material. Sugarcane cannot be stored for long as the loss of sucrose content is inevitable. Besides, it cannot be transported over long distances because any increase in transportation cost would raise the cost of production and the sugarcane may dry up on the way. It is estimated that 50 per cent cost of production is accounted for by sugarcane alone. Normally, it requires about 100 tonnes of sugarcane to produce 10-12 tonnes of sugar. Even today most of sugarcane is transported with the help of bullock carts and cannot be carried beyond 20-25 km. The introduction of tractor-trollies, trucks and even railway wagon have increased the distance covered by sugarcane to 70-75 kms. beyond which the transportation cost would increase exorbitantly. Therefore, the sugar industry is established in areas of sugarcane cultivation.

Distribution

Figure 25.15 makes it amply clear that sugar industry has two major areas of concentration. One comprises Uttar Pradesh, Bihar, Haryana and Punjab in the north and the other that of Maharashtra, Karnataka, Tamil Nadu and Andhra Pradesh in the south.

Maharashtra. Maharashtra has progressed a lot and captured first position from U.P. to emerge as the largest producer of sugar in India. Large production of sugarcane, higher rate of recovery (about 11.6%) and longer crushing period (about 162 days) are some of the factors which have helped the state to occupy this enviable position. The state has one-fourth of the total sugar mills and produces a little more than one-third of the total sugar of India. Sugar mills of Maharashtra are much larger as compared to the mills in other parts of the country. The major concentration of sugar mills is found in the river valleys in the western part of the Maharashtra Plateau. Ahmednagar is the largest centre. The other major centres are in the districts of Kolhapur, Solapur, Satara, Pune and Nashik.

Uttar Pradesh. Uttar Pradesh is the traditional producer of sugar and has been occupying the first rank among the major sugar producing states of India. However, its relative importance has been reduced during the last few years and the state has conceded the top position to Maharashtra and now occupies the second position. Uttar Pradesh has more mills than Maharashtra but they are of comapratively smaller size and yield less production. Presently, the state accounts for about 24 per cent of the total production of sugar in India. There are two distinct regions of sugar production in this state. One region consists of Gorakhpur, Deoria, Basti and Gonda in eastern Uttar Pradesh and the other lies in the upper

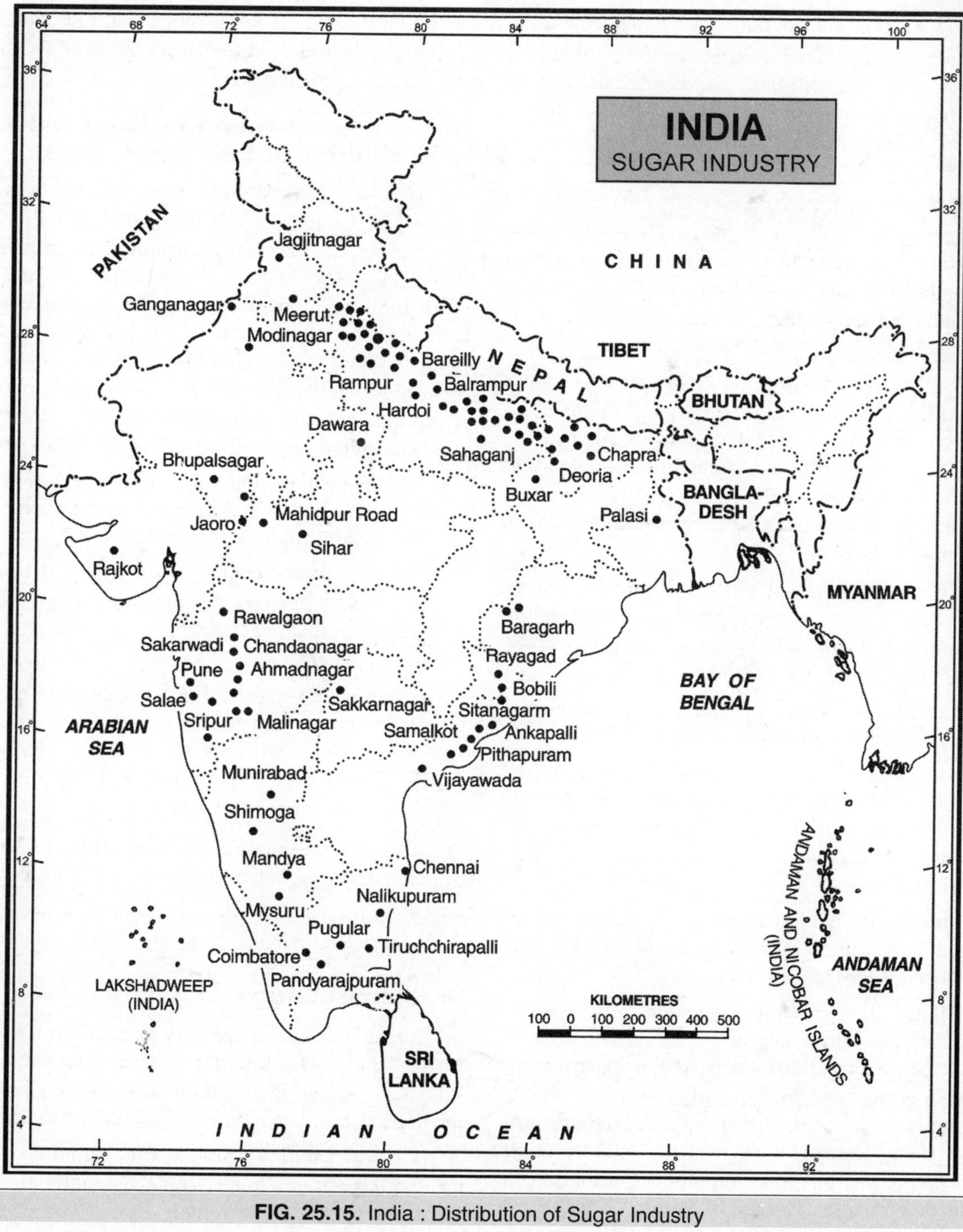

FIG. 25.15. India : Distribution of Sugar Industry

Ganga Plain consisting of Meerut, Saharanpur, Muzaffarnagar, Bijnore and Moradabad.

Tamil Nadu. Tamil Nadu has shown phenomenal progress with regard to sugar production during the last few years. High yield per hectare of sugarcane, higher sucrose content, high recovery rate and long crushing season have enabled Tamil Nadu to obtain one of the highest yield of 9.2 tonnes of sugar per hectare in the whole of India. As a result of these advantages, the state has emerged as the third largest producer of sugar, contributing over nine per cent of the total sugar production of India. Most of the 32 mills of the state are located in Coimbatore, Tiruppur, Karur and Tiruchchirapalli.

Karnataka. Karnataka has 30 mills producing over 6 per cent of the total sugar of India. Belgaum and Mandya districts have the highest concentration of sugar mills. Bijapur, Bellary, Shimoga and Chittradurga are the other districts where sugar mills are scattered.

Andhra Pradesh. Andhra Pradesh has more mills (35) than the neighbouring Karnataka but produces only 5.8 per cent of India's sugar. This means that the mills are comparatively smaller. Majority of the sugar mills are concentrated in East Godawari West Godavari, Krishna, Vishakhapatnam, and Chittoor districts.

Gujarat. Gujarat's 16 mills are scattered in Surat, Bhavnagar, Amreli, Banaskantha, Junagarh, Rajkot and Jamnagar districts. The state produces about 5.56 per cent of the total sugar produced in India.

Telangana. Most of the sugar mills are concentrated in Nizamabad and Medak districts.

Haryana. Haryana has only 8 mills but their large size enables the state to contribute 1.91 per cent of the total sugar production. Sugar mills are located in Rohtak, Ambala, Panipat, Sonipat, Karnal, Palwal and Hisar districts.

Punjab. Punjab has a total of 13 mills which are located in Amritsar, Jalandhar, Gurdaspur, Sangrur, Patiala and Rupnagar districts.

Bihar. Bihar was the second largest sugar producing state next only to Uttar Pradesh till mid-1960s. Since then the state has been experiencing sluggish growth and consequently lost its prestigious position to the peninsular states like Maharashtra, Tamil Nadu, Karnataka and Andhra Pradesh. Its 28 mills make an insignificant contribution to the production of sugar. The belt of eastern Uttar Pradesh extends further east in Bihar and the districts of Darbhanga, Saran, Champaran and Muzaffarpur are included in this belt.

Others. Among the other producers are Madhya Pradesh (8 mills in Morena, Gwalior and Shivpuri districts), Rajasthan (5 mills in Ganganagar, Udaipur, Chittaurgarh and Bundi districts), Kerala, Odisha, West Bengal and Assam.

Difference between the Sugar Industry of Northern and Peninsular India

There are marked differences between the sugar industry of the northern and the peninsular India. As a result of better conditions prevailing in the peninsular India, the sugar industry is gradually shifting from north India to the peninsular India. This is evident from the fact that previously north India used to produce about 90 per cent of India's sugar which is reduced to 35-40 per cent now. A brief description of differences between the sugar industry of the northern and peninsular India is given below :

1. Peninsular India has tropical climate which gives higher yield per unit area as compared to north India.
2. The sucrose content is also higher in tropical variety of sugarcane in the south.
3. The crushing season is also much longer in the south than in the north. For example, crushing season is of nearly four months only in the north from November to February, whereas it is of nearly 7-8 months in the south where it starts in October and continues till May and June.
4. The co-operative sugar mills are better managed in the south than in the north.
5. Most of the mills in the south are new which are equipped with modern machinery.

Problems of Sugar Industry

Sugar industry in India is plagued with several serious and complicated problems which call for immediate attention and rational solutions. Some of the burning problems are briefly described as under :

1. Low Yield of Sugarcane. Although India has the largest area under sugarcane cultivation, the yield per hectare is extremely low as compared to some of the major sugarcane producing countries of the world. For example, India's yield is only 70,394 kg/hectare as compared to 75,180 kg/ha in Brazil, 80,033 kg/ha in U.S.A., 86,399 kg/ha in Colombia and 1,29,049 kg/ha in Guatemala. The world average is 70,134 kg/hectare (data for 2016). This leads to low overall production and results in short supply of sugarcane to sugar mills. Efforts are being made to solve this problem through the introduction of high yielding, early maturing, frost resistant and high sucrose content varieties of sugarcane as well as by controlling diseases and pests which are harmful for sugarcane.

2. Short crushing season. Manufacturing of sugar is a seasonal phenomena with a short crushing season varying normally from 4 to 7 months in a year. The mills and its workers remain idle during the remaining period of the year, thus creating financial problems for the industry as a whole. One possible method to increase the crushing season is to sow and harvest sugarcane at proper intervals in different areas adjoining the sugar mill. This will increase the duration of supply of sugarcane to sugar mills.

3. Fluctuating Production Trends. Sugarcane has to compete with several other food and cash crops like cotton, oil seeds, rice, etc. Consequently, the land available to sugarcane cultivation is not the same and the total production of sugarcane fluctuates. This affects the supply of sugarcane to the mills and the production of sugar also varies from year to year.

4. Low rate of recovery. The average rate of recovery in India is less than ten per cent which is quite low as compared to other major sugar producing countries. For example recovery rate is as high as 14-16 per cent in Java, Hawaii and Australia.

5. High cost of Production. High cost of sugarcane, inefficient technology, uneconomic process of production and heavy excise duty result in high cost of manufacturing. The production cost of sugar in India is one of the highest in the world. Intense research is required to increase the sugarcane production in the agricultural field and to introduce new technology of production efficiency in the sugar mills. Production cost can also be reduced through proper utilisation of by-products of the industry. For example, bagasse can be used for manufacturing paper pulp, insulating board, plastic, carbon, coltex, etc. Molasses comprise another important by-product which can be gainfully used for the manufacture of power alcohol. This, in its turn, is useful in manufacturing DDT, acetate rayon, polythene, synthetic rubber, plastics, toilet preparations, etc. It can also be utilised for conversion into edible molasses and cattle feed. Press-mud can be used for extracting wax.

6. Small and uneconomic size of mills. Most of the sugar mills in India are of small size with a capacity of 1,000 to 1,500 tonnes per day. This makes large scale production uneconomic. Many of the mills are economically not viable.

7. Old and obsolete machinery. Most of the machinery used in Indian sugar mills, particularly those of Uttar Pradesh and Bihar is old and obsolete, being 50-60 years old and needs replacement. But low margin of profit prevents several mill owners from replacing the old machinery by the new one.

8. Competition with Khandsari and Gur. Khandsari and gur have been manufactured in rural India much before the advent of sugar industry in the organised sector. Since *khandsari* industry is free from excise duty, it can offer higher prices of cane to the cane growers. Further, cane growers themselves use cane for manufacturing *gur* and save on labour cost which is not possible in sugar industry. It is estimated that about 60 per cent of the cane grown in India is used for making *khandsari* and *gur* and the organised sugar industry is deprived of sufficient supply of this basic raw material.

9. Regional imbalances in distribution. Over half of sugar mills are located in Maharashtra and Uttar Pradesh and about 60 per cent of the production comes from these two states. On the other hand, there are several states in the north-east, Jammu and Kashmir and Odisha where there is no appreciable growth of this industry. This leads to regional imbalances which have their own implications.

10. Low per capita consumption. The per capita annual consumption of sugar in India is only 19.5 kg as against 48.8 kg in the USA., 53.6 kg in U.K., 57.1 kg in Australia and 78.2 kg in Cuba and the world average of about 21.1 kg. This results in low market demand and creates problems of sale of sugar.

VEGETABLE OIL INDUSTRY

Vegetable oil is an important item of Indian food as it is the major source of fat. Vegetable oil industry of India can be divided into three broad groups depending upon the technology used.

(*i*) ***Ghani*** is the main technology for expelling oil in the villages. Different oil seeds are used in different areas. For example, groundnut is used in Gujarat, coconut in Kerala and mustard seed in Uttar Pradesh.

(*ii*) Factories using intermediate level of technology are located in towns. Oil seeds used are region specific.

(*iii*) Large scale sophisticated mills are located in big cities and are oriented towards bigger market. They also procure oil seeds from a much larger area.

Vanaspati is 'hydrogenated' oil. The first vanaspati factory was established in 1930 which produced a meagre of 298 tonnes. The World War II and the levy of import duty on vanaspati gave a fillip to this industry and in 1951, there were 48 factories with a capacity of 3.3 lakh tonnes and a production of 155 thousand tonnes. The production showed a significant progress and stood at 1,445 thousand tonnes in 2000-11 after which varying trends in production have been observed.

Although vegetable oil industry has developed throughout India, Maharashtra has the largest number of vanaspati producing units. Other important vanaspati producing states are Uttar Pradesh, Gujarat, Punjab, Andhra Pradesh, West Bengal, Karnataka, Rajasthan, Tamil Nadu and Madhya Pradesh. Chennai, Akola, Modinagar, Kanpur, Ghaziabad, Indore and Vadodara are the main centres of vegetable oil industry.

Production of vegetable oils falls short of the domestic demand and the country has to import oil seeds as well as vegetable oils from other countries.

Table 25.11 shows that imports of edible oils and their value in terms of foreign exchange spent on these imports keeps on changing depending upon the availability of edible oils in the country. However, there had been steep rise in the imports of edible oils since 1990-91. In 2020-21, India had to spend a huge amount of ₹ 82,123 thousand to import 13,540 thousand tonnes of edible oils.

Cottage Industries

Industries which the artisans set-up in their houses and produce finished goods with the help of their family members by using locally available raw materials are known as cottage industries. In other words, cottage industry is a type of industry where creation of products and services is home based, rather than factory based. While the products and services created by cottage industry are often unique and distinctive given the fact that they are usually not mass-produced, producers in this sector often face numerous disadvantages when trying to compete with much larger factory based companies.

The artisan may sell the product directly to the consumes or to an entrepreneur who pays money to the artisan according to the quality and quantity of the finished product. This type of industry has a long history in India and was perhaps the first industry developed in the country at the village level. It involves minimum investment and modern technical input and can be found in all parts of the country. It provides employment to a large number of people, many of them are unskilled or semi-skilled and belong to SC/ST and other backward classes including women. *Bidi* making, sports goods, lac industry, basket making, extracting oil from oilseeds, manufacture of crackers, toys, embroidery, handloom, khadi, utensil making, idol making, leather work, etc. fall in the category of cottage industries. The main raw materials used by the artisans are wood, cane, leaves, leather, brass, stones of different varieties, etc. Some of the important industries have already been discussed in this chapter and some others are briefly described below.

Bidi Making. Bidi is made from *tandu* leaves or *Kachnal*. These two basic raw materials for bidi making are found in the forests of Chhattisgarh,

TABLE 25.11. Import of Edible oils

Year	1960-61	1970-71	1980-81	1990-91	2000-01	2020-21	2021-22
Quantity (Thousand tonnes)	31.1	84.7	1,663.3	525.8	4,267.9	13540	14278
Value (₹ crore)	4	23	677	326	6,093	82123	14152

Source : Economic Survey 2022-23, Statistical Appendix, pp. 112-14.

Madhya Pradesh and West Bengal as well as on the slopes of the Eastern Ghats. An important ingredient of *bidi* making is inferior quality tobacco which is available in Andhra Pradesh, Telangana, Bihar, Chhattisgarh, Madhya Pradesh and Uttar Pradesh. The annual production of *bidi* is more than ten billion numbers. The main areas of *bidi* production are Bastar and Jagatpur in Chhattisgarh, Belgaum and Mangaluru in Karnataka, Bhandura, Nagpur, Gondia, Kamptee, Pune and Nashik in Maharashtra, Jabalpur, Bhind and Murena in Madhya Pradesh, Kheda and Vadodra in Gujarat and Hyderabad in Telangana. As against cigarette, *bidi* is considered to be a cheap and light smoke and is relished by a large number of people belonging to lower strata of society in India. India is a net exporter of bidi to a large number of countries in Asia, Africa and Europe.

Basket Making. This is a typical cottage industry of hilly and mountain areas where baskets are prepared by using locally available raw materials from the forests. Bamboo, cane and willow are the chief raw materials used for basket making. The main areas of basket making are in the states of Sikkim, Arunachal Pradesh, Nagaland, Manipur, Mizoram, Meghalaya, Tripura, Jammu and Kashmir, Maharashtra, Madhya Pradesh, Uttar Pradesh, Odisha and West Bengal.

Others. Mirzapur district of Uttar Pradesh is famous for making *carpets* while Kashmir valley is well known for exclusive embroidery on shawls and other woolen garments. Lucknow's needle work on *kurtas* and other wearings is well known are over India. In Tamil Nadu, *Cracker* making is an important cottage industry. Almost each and every village in India has a cobbler who makes shoes for the villagers and for sale in the adjoining markets. Similarly there are persons who make utensils, particularly from brass.

INDUSTRIAL REGIONS

Industries are unevenly distributed in India because the factors affecting industrial location are not the same everywhere. Industries tend to concentrate in a few pockets because of certain favourable factors. The pockets having high concentration of industries are known as *industrial regions*.

Goegraphers and specialists in other allied fields have made several attempts to delineate industrial regions of India making use of different criteria. One of the first attempts was made by Trewartha and Burnet in 1944 using employment figures. Karan and Jenkins tried to demarcate industrial regions of India on empirical basis. These attempts have become outdated and have lost much of their relevance because industrial landscape in India has undergone drastic changes during the planning era. **J.E. Spencer** and **W.L. Thomas** have also discussed this subject in their book entitled *Asia, East by South*. **R.L. Singh** recognised ten regions based on empirical observation. They are (*i*) Hugli, (*ii*) Mumbai, (*iii*) Ahmedabad (*iv*) The north-western extending from Ghaziabad to Amritsar (*v*) Lucknow-Kanpur (*vi*) Bangalore-Madras (*vii*) Quilon-Coimbatore (*viii*) Madurai (*ix*) Dalmiapuram and (*x*) Digboi- Dibrugarh.

The Centre for Monitoring Indian Economy (CMIE) recognised the industrial centres on the basis of industrial employment exceeding 10,000 in 1971.

Dr. B.N. Sinha (1972) has classified industrial regions into following three categories :

(*i*) **Major Industrial Region** is identified on the basis of a minimum daily factory working force of 1.5 lakh

(*ii*) **Minor Industrial Region** must have a minimum of 25,000 working labour force.

(*iii*) **Manufacturing District** has a working labour force of less than 25,000.

Another basis of indentification of industrial regions is the number of manufacturing units located close to one another or the quantum of industrial environment. Major industrial regions of India have developed in the immediate hinterlands of the ports of Kolkata, Mumbai and Chennai. The Britishers developed these clusters because the hinterlands of these ports provided enough raw materials, power and market. With the development of good transportation, some power resources like coal could also be brought easily.

Major Industrial Regions

Following are the major industrial regions of India (**Fig. 25.16**) :

1. Mumbai-Pune Industrial Region.
2. Hugli Industrial Region.

3. Bengaluru-Tamil Nadu Industrial Region.
4. Gujarat Industrial Region.
5. Chotanagpur Industrial Region.
6. Vishakhapatnam-Guntur Industrial Region.
7. Gurgaon-Delhi-Meerut Industrial Region.
8. Kollam-Thiruvananthapuram Industrial Region.

1. Mumbai-Pune Industrial Region. This region extends from Thane to Pune and in adjoining districts of Nashik and Solapur. In addition, industries have grown at a rapid pace in Kolaba, Ahmednagar, Satara, Sangli and Jalgaon districts also. This region owes its origin to the British rule in India. The seeds of its growth were sown in 1774 when the island-site was obtained for construction of Mumbai port. The opening of the first railway track of 34 kms between Mumbai and Thane in 1853, opening of the Bhor and Thal Ghats respectively to Pune and Nashik and that of Suez canal in 1869 led to the development of Mumbai.

The growth of this industrial region is fully connected with the growth of cotton textile industry in India. As the coal was far removed, hydel power was developed in Western Ghats. Cotton was cultivated in the black cotton soil area of the Narmada and Tapi basins. Cheap labour-force came from the hinterland, the port facilities for export-import and communication links with the peninsular hinterland made Mumbai the *'Cottonopolis of India'*. With the development of cotton textile industry, the chemical industry developed too. Opening of the Mumbai High petroleum field and erection of nuclear energy plants added additional magnetic force to this region. Now the industrial centres have developed, from Mumbai to Kurla, Kolaba, Thane, Ghatkopar, Ville Parle, Jogeshwari, Andheri, Thane, Bhandup, Kalyan, Pimpri, Pune, Nashik, Manmad, Solapur, Ahmednagar, Satara and Sangli. In addition to cotton textile and chemical industries, engineering goods, leather, oil refineries; petrochemicals, synthetic and plastic goods, chemicals, drugs, fertilizers, electricals, electronics, software, ship-building, transport and food industries have also developed here.

The partition of the country in 1947 adversely affected this region because 81% of the total irrigated cotton area growing long staple cotton went to Pakistan. Mumbai, the nucleus of this industrial region, is facing the current limitation of space for the expansion of the industry. Dispersal of industries is essential to bring about decongestion.

2. The Hugli Industrial Region. Located in West Bengal, this region extends as a narrow belt running along the river Hugli for a distance of about 100 km from Bansbaria and Naihati in the north to Birlanagar in the south. Industries have also developed in Midinipur district in the west. The river Hugli offered the best site for the development of an inland river port as nucleus for the development of *Hugli industrial region*. The old trading centre of late 17th century has developed into the present industrial hub of Kolkata. Thus Kolkata-Haora form the nucleus of this region. It is very well-connected by the Ganga and its tributaries with the rich hinterland of Ganga-Brahmaputra plains. Besides navigable rivers, roads and the railways provided subsequent links to the great benefit of Kolkata port.

The discovery of coal and iron ore in Chotanagpur plateau, tea plantations in Assam and northern parts of West Bengal and the processing of deltaic Bengal's jute led to the industrial development in this region. Cheap labour could be found easily from the thickly populated states of Odisha, Bihar, Jharkhand and eastern part of U.P. Kolkata, having been designated capital city of the British India (1773-1912) attracted large scale British investment of capital. Establishment of first jute mill at Rishra in 1855 ushered in the era of modern industrial clustering in this region. A chain of jute mills and other factories could be established on either side of Hugli river with the help of Damodar valley coal. The port site was best-suited for export of raw materials to England and import of finished goods from that country. Kolkata's industries have established by drawing in the raw materials from adjoining regions and distributing the finished goods to consuming points. Thus, the role of transport and communication network has been as important as the favourable locational factors in the growth of this region. By 1921, Kolkata-Hugli region was responsible for two-thirds of factory employment in India.

Just after the partition of old Bengal province in 1947, the region faced, for some years, the problem of shortage of jute as most of the jute-growing areas went to East Pakistan (now Bangladesh). The problem was solved by gradually increasing home production of jute. Cotton textile industry also grew along with jute industry. Paper, engineering, textile machinery, electrical, chemical, pharmaceuticals, fertilizers and petrochemical industries have also developed in this region. Factory of the Hindustan Motors Limited at Konanagar and diesel engine factory at Chittaranjan are landmarks of this region. Location of petroleum refinery at Haldia has facilitated the development of a variety of industries. The major centres of this industrial region are Kolkata, Haora, Haldia, Serampur, Rishra, Shibpur, Naihati, Kakinara, Shamnagar, Titagarh, Sodepur, Budge Budge, Birlanagar, Bansbaria, Belgurriah, Triveni, Hugli, Belur, etc.

Alarming rate of silting of the Hugli river was a very serious problem. The depth of water in the channel from bay head to Kolkata docks must be kept at 9.2 metres for big ocean ships to come in. Dredging out of the silt rapidly filling up the water channel was very costly and not a permanent solution to save the life of Kolkata port. The construction of Farakka barrage about 300 kms upstream on Ganga and flushing of the channel were the only possible answers. The construction of Haldia port in the lower reaches of Hugli to the south of Kolkata is another landmark in relieving the great pressure of cargo ships on the port of Kolkata. However, the industrial growth of this region has slowed down as compared to the other regions. There are several reasons for this sluggish growth but decline in jute industry is said to be one of the main reasons.

3. Bengaluru-Tamil Nadu Industrial Region. Spread in two states of Karnataka and Tamil Nadu, this region experienced the fastest industrial growth in the post-independence era. Till 1960, industries were confined to Bengaluru district of Karnataka and Salem and Madurai districts of Tamil Nadu. But now they have spread over all the districts of Tamil Nadu except Viluppuram.

This region is a cotton-growing tract and is dominated by the cotton-textile industry. In fact cotton textile industry was the first to take roots in this region. But it has large number of silk-manufacturing units, sugar mills, leather industry, chemicals, rail wagons, diesel engines, radio, light engineering goods, rubber goods, medicines, aluminium, cement, glass, paper, cigarette, match box and machine tools, etc. This region is away from the main coal-producing areas of the country but cheap hydro-electric power is available from Mettur, Sivasamudram, Papanasam, Pykara and Sharavati dams. Cheap skilled labour and proximity to vast local market as well as good climate have also favoured the concentration of industries in this region. Coimbatore has grown rapidly mainly owing to its industrial growth based on Pykara power, local cotton, coffee mills, tanneries, oil presses and cement works. Coimbatore is known as *Manchester of Tamilnadu* because of its large-scale cotton textile industry. The establishment of public sector units at Bengaluru like Hindustan Aeronautics, Hindustan Machine Tools, Indian Telephone Industry and Bharat Electronics, etc. has further pushed up the growth of industries in the region. Madurai is known for its cotton textiles. Visvesvarayya Iron and Steel Works is located at Bhadravati. The other important centres of this region are Sivakasi, Tiruchirapalli, Madukottai, Mettur, Mysuru and Mandya. Petroleum refineries at Manali, Narimanam and Nagapattinam as well as iron and steal plant at Salem are recent developments.

4. Gujarat Industrial Region. The nucleus of this region lies between Ahmedabad and Vadodara as a result of which it is also known as **Ahmedabad-Vadodara** industrial region. However, this region extends upto Valsad and Surat in the south and Jamnagar in the west. The region corresponds to the cotton growing tracts of the Gujarat plains and the development of this region is associated with the location of textile industry since 1860s. This region became important textile region with the decline of cotton textile industry in Mumbai. Mumbai has the disadvantage of paying double freight charges for first bringing the raw cotton from the peninsular hinterland and then despatching the finished products to inland consuming points in India. But Ahmedabad is nearer the sources of raw material as well as the marketing centres of the Ganga and Satlui plains. Availability of cheap land, cheap skilled labour and other advantages helped the cotton textile industry to develop. This major industrial region of the country, mainly consisting of cotton textile industry, is expanding at a much faster rate in providing a greater factory employment.

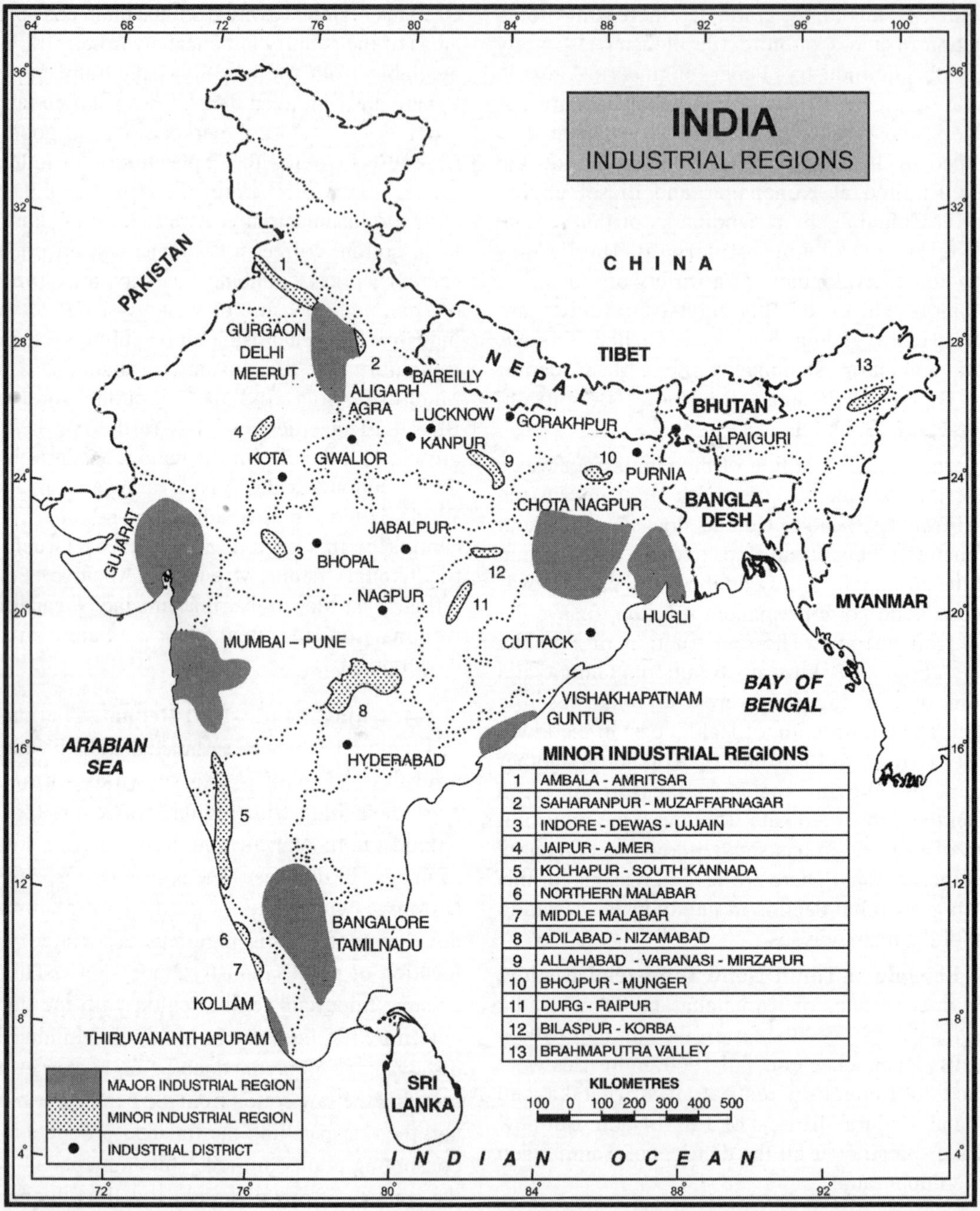

FIG. 25.16. India : Industrial Regions

The discovery and production of oil at a number of places in the Gulf of Khambhat area led to the establishment of petrochemical industries around Ankleshwar, Vadodara and Jamnagar. Petroleum refineries at Koyali and Jamnagar provide necessary raw materials for the proper growth of petrochemical industries. The Kandla port, which was developed immediately after independence, provides the basic infrastructure for imports and exports and helps in rapid growth of industries in this region. The region can now boast of diversified industries. Besides textiles (cotton, silk and synthetic fibres) and petrochemical industries, other industries are heavy and basic chemicals, dyes, pesticides, engineering, diesel engines, textile machinery, pharmaceuticals, dairy products and food processing. The main industrial centres of this region are Ahmedabad, Vadodara, Bharuch, Koyali, Anand, Khera, Surendranagar, Surat, Jamnagar, Rajkot and Valsad The region may become more important in the years to come.

5. Chotanagpur Industrial Region. As its name indicates, this region is located on the Chotanagpur plateau and extends over Jharkhand, Northern Odisha and Western part of West Bengal. The birth and growth of this region is linked with the discovery of coal in Damodar Valley and iron ore in the Jharkhand-Odisha mineral belt. As both are found in close proximity, the region is known as the '*Ruhr of India*'. Besides raw materials, power is available from the dam sites in the Damodar Valley and the thermal power stations based on the local coal. This region is surrounded by highly populated states of Jharkhand, Bihar, Odisha and West Bengal which provide cheap labour. The Kolkata region provides a large market for the goods produced in the Chotanagpur region. It also provides the port facility to the region. It has the advantages for developing ferrous metal industries. The Tata Iron and Steel Company at Jamshedpur, Indian Iron Steel Co., at Burnpur-Kulti, Hindustan Steel Limited at Durgapur, Rourkela and Bokaro are the important steel plants located in this region. Heavy engineering, machine tools, fertilizers, cement, paper, locomotives and heavy electricals are some of the other important industries in this region. Important nodal centres of this region are Ranchi, Dhanbad, Chaibasa, Sindri, Hazaribagh, Jamshedpur, Daltonganj, Garwa and Japla.

6. Vishakhapatnam-Guntur Industrial Region. This industrial region extends from Vishakhapatnam district in the north-eastern part of Andhra Pradesh to Kurnool and Prakasham districts in the south-east and covers most of the coastal Andhra Pradesh. The industrial development of this region mainly depends upon Vishakhapatnam and Machilipatnam ports. Developed agriculture and rich mineral resources in the hinterlands of these ports provide solid base to the industrial growth in this region. Coal fields of the Godavari basin are the main source of energy. Hindustan Shipyard Ltd. set up at Vishakhapatnam, in 1941 is the main focus. Petroleum refinery at Vishakhapatnam facilitated the growth of several petrochemical industries. Vishakhapatnam has the most modern iron and steel plant which has the distinction of being the only plant in India having coastal location. It uses high quality iron ore from Bailadila in Chhattisgarh. One lead-zinc smelter is functioning in Guntur district. The other industries of this region include sugar, textiles, paper, fertilizers, cement, aluminium and light engineering. The important industrial centres of this region are Vishakhapatnam, Vijaywada, Vijaynagar, Rajahmundry, Kurnool, Elum and Guntur. Recent discovery of natural gas in Krishna-Godavari basin is likely to provide much needed energy and help in accelerated growth of this industrial region.

7. Gurugram-Delhi-Meerut Industrial Region. This region was developed after independence, but is one of the fastest growing regions of India. It consists of two industrial belts adjoining Delhi. One belt extends over Agra-Mathura-Meerut and Saharanpur in U.P. and the other between Faridabad-Gurgaon-Ambala in Haryana. The region is located far away from the mineral and power resources, and therefore, the industries are light and market oriented. The region owes its development and growth to hydro-electricity from Bhakra-Nangal complex and thermal power from Harduaganj, Faridabad and Panipat. Sugar, agricultural implements, vanaspati, textiles, glass, chemicals, engineering, paper, electronics and cycle are some of the important industries of this region. Software industry is a recent addition, Agra and its environs have glass industry. Mathura has an oil refinery with its petro-chemical complex. One oil refinery has been set up at Panipat also. This has greatly helped to boost the industrial growth of this region. Gurugram has Maruti car factory as well as one unit of the IDPL. Faridabad has a number of engineering and electronic industries. Ghaziabad is a large centre of agro-industries. Saharanpur and Yamunanagar have paper mills. Modinagar, Sonipat, Panipat and Ballabhgarh are other important industrial nodes of this region.

8. Kollam-Thiruvananthapuram Industrial Region. This is comparatively small industrial region and spreads over Thiruvananthapuram, Kollam, Alwaye, Ernakulam and Allapuzha districts of south

Kerala. The region is located far away from the mineral belt of the country as a result of which the industrial scene here is dominated by agricultural products processing and market oriented light industries. Plantation agriculture and hydroelectricity provide the industrial base to this region. The main industries are textiles, sugar, rubber, match box, glass, chemical fertilizers, food and fish processing, paper, coconut coir products, aluminium and cement. Oil refinery set up in 1966 at Kochi provides solid base to petrochemical industries. Important industrial centres are Kollam, Thiruvananthapuram, Alluva, Kochi, Alappuzha and Punalur.

Besides the above mentioned eight major industrial regions, India has 13 minor industrial regions and 15 industrial districts. Their names are mentioned below :

Minor Industrial Regions

1. Ambala-Amritsar in Haryana-Punjab.
2. Saharanpur-Muzaffarnagar-Bijnaur in Uttar Pradesh.
3. Indore-Dewas-Ujjain in Madhya Pradesh.
4. Jaipur-Ajmer in Rajasthan.
5. Kolhapur-South Kannada in Maharashtra-Karnataka.
6. Northern Malabar in Kerala.
7. Middle Malabar in Kerala.
8. Adilabad-Nizamabad in Telangana.
9. Allahabad-Varanasi-Mirzapur in Uttar Pradesh.
10. Bhojpur-Munger in Bihar.
11. Durg-Raipur in Chhattisgarh.
12. Bilaspur-Korba in Chhattisgarh.
13. Brahmaputra Valley in Assam.

Industrial Districts

1. Kanpur, 2 Hyderabad, 3. Agra, 4. Nagpur, 5 Gwalior, 6. Bhopal, 7. Lucknow, 8. Jalpaiguri, 9. Cuttack, 10. Gorakhpur, 11. Aligarh, 12. Kota, 13. Purnia, 14. Jabalpur, 15. Bareilly.

INDUSTRIAL HOUSES IN INDIA

India has experienced rapid industrial growth, particularly after the new Industrial Policy of 1991 and India has emerged as one of the few industrial giants of the world. This is due to a variety of factors, important among which are vast and solid natural resource base, increasing interest of the common man in industries and incentives from the government. In addition to these factors, there is the major contribution made by some big industrial houses towards industrial growth. Although most of the industrial houses have emerged on the industrial scene in India after Independence, some of the industrial houses were functioning even before Independence. These industrial houses have laid solid foundation and given a new direction to industrial growth of India. They have made a commendable contribution to make India an industrial force to reckon with at the international level. In the year 1965, the Monopolies Inquiry Commission found that 75 industrial houses have absolute control over 1,536 companies. Industrial Licencing Policy Inquiry Committee accepted the concept of industrial houses and Dutt Committee identified 20 big industrial houses in 1963-64, each with an investment of ₹ 35 crores. Larger scale changes were observed in the functioning of these industrial houses after Industrial Policy and Policy of Liberalization in 1991. A brief description of some of the major industrial houses is given as under :

1. The Tata. This is one of the oldest industrial houses active for the last over a century in India. The credit of laying solid foundation of India's industrial growth goes to this house as Jamshedji Tata established India's first large scale integral Iron and Steel manufacturing house at Sakchi in Singhbhum district of Jharkhand way back in 1907. Also known as Tatanagar, this place was later on renamed as Jamshadpur after Jamshadji. Since then, this house has never looked back and established a large number of industries. The main industries established by this house include automobiles, cement, publication, pharmaceuticals, power, energy, hospitality (hotels), refrigeration, air-conditioning, telecommunication, soft drinks, finance, insurance, consumer non-durables, etc. Tata Consultancy Services is an export oriented software development company which is one of the largest companies of India. It has left an indelible impact in the international market. It launched the low cost Nano car in the small car segment in 2009 which hit the headlines all over the world. However this car could not come upto high expectations of the buyers.

2. The Birla. Like the Tatas, the Birlas are also one of the oldest industrial houses in India. Their major contribution since inception has been in textile, paper, cement, aluminum, paraffin and automobile industries. In the recent past, the Birlas have diverted their attention towards diversification of industries and ventured into some other industries like telecommunications, pharmaceuticals, machine tools, consumer durables and non-durables, etc.

3. The Modi. Set-up by Gujarmal Modi, this is also one of the oldest industrial houses of India. The famous industrial town of Modinagar in Uttar Pradesh owes its origin and growth to this industrial house. The main industries set-up by this industrial house are textiles, ready-made garments, carpets, sugar, pharmaceuticals, tyres and tubes, hospitality, etc.

4. The Bajaj. This is also a very old industrial house of India which is known all over the world for its popular product, the Bajaj Scooter. It has started manufacturing motor-cycles also. The major industries of this house are electric and home appliances and entertainment. They are also venturing into small car segment.

5. The Ambanis. Dhiru Bhai Ambani spent his early life in acute poverty and deprivation but built a strong industrial empire by dint of his hard work and innovative ideas. His life story is the story of **rags-to-riches**. His vast industrial empire was divided between his two sons, namely Mukesh Ambani and Anil Ambani. The *Reliance Company* of Mukesh Ambani is credited with the discovery and production of natural gas in the Krishna-Godavari basin. It is a great achievement because it has solved India's energy problem to a great extent and charted the path to India's industrial progress. The traditional industries of this house are paraffin and synthetic yarn. Today this house runs more than a dozen industries which includes power, oil refining, textiles, garments, software, hospitality, telecommunication, pharmaceuticals, banking, insurance, etc. The oil refinery at Jamnagar in Gujarat with a capacity of 33 million metric tonnes per annum is the largest in India and perhaps the largest in the world.

6. The Sighania. This is also one of the oldest industrial houses of India. The main industries run by this house are textiles, pharmaceuticals, machine tools, tyres and tubes and entertainment.

7. The Sriram. Being credited as one of the oldest industrial houses of India, this house has a wide variety of industries which include textiles, ready-made garments, fertilizers, electricity, home appliances, generator sets, pharmaceuticals, hospitality, etc.

8. The Goenka. This is also one of the oldest industrial houses of the country. The main industries of this house are power generation and distribution, mechanic tools, pharmaceuticals, textiles, entertainment, etc.

9. The Escorts. This house also enjoys the privilege of being one of the oldest industrial houses in India. Its most important traditional industry has been tractor manufacturing. This was followed by manufacturing of motor-cycles. Of late, this house has made its mark in medical treatment. The heart-treatment hospitals run by this house provide world class heart-treatment at reasonably low cost.

10. The Oberois. This industrial house is well known for its mark in the hospitality industry. It has built high class hotels in different parts of India and in different countries in he world.

11. The Godrej. This is one of the topmost, oldest and famous industrial house which has been popular for manufacturing high class locks, furniture including almirahs, detergents, refrigerators, air-conditioners, etc. The Godrej products are of high quality and meet international standards.

12. The Kirloskars. This is also one of the oldest and reputed industrial houses of India and is famous for manufacturing high class heavy and light machine tools. It has a place of pride in locomotion. The credit of producing India's first 'diesel pump set' goes to this industrial house. It has brought about a revolution in farm mechanization by manufacturing India's first tractor. Currently the company is producing a variety of farm implements. Its compressors are known for high quality performance. It is also producing cars in collaboration with Toyota.

13. The Jagjit Group. This industrial house is famous for textiles and sugar industries apart from distilleries.

14. The Sarabhais. The Sarabhais are mainly concerned with drugs and pharmaceutical industries for which they are renowned all over the world.

15. The Firodias. Being one of the oldest industrial houses the Firodias have attained specialization in the field of manufacturing automobiles (four wheelers and two wheelers) generator sets, etc.

16. The UB Group. This is considered to be one of the oldest industrial houses in India with manufacturing of alcoholic drink as its traditional industry. During the last few years, this industrial house has entered in fields of civil aviation, hospitality and structurals. Of late it has shown its presence in the real estate also.

17. The Thapars. This house is famous for paper, sugar and some allied industries. It has also made a significant contribution in the field of education.

18. The Infosys. The main industries of this house are electronics and software for which it is a well reputed house. Besides this house is making a significant contribution in the fields of education and social work.

19. The Wipro. The doyen of the sunrise industries, this industrial house of India is primarily involved in the field of software development and export including the production of computer peripherals. The company has earned world-wide acclaim in the development of software. This is also a leading company in the field of business process outsourcing (BPO). Like Infosys, this house is also seriously committed to social work.

20. The Bharti. The Bharti is famous for its 'Airtel' mobile phone for which it has sought collaboration with 'Airtel' company of America. Recently, this company has taken several steps to diversify its activities in the areas of software, internet, agriculture, hospitality and a host of others.

21. The Becon. Its main field of activity is Biotechnology. Although a comparatively new comer in the industrial field, it has carved a niche for itself.

22. The Ranbaxy. Its main specialization is the manufacturing of drugs and pharmaceuticals in which it has earned great reputation all over the world. In addition to catering to the needs of the domestic market, it exports its products to a number of countries. Many other companies have been inspired by the Ranbaxy.

23. Hamdard. This industrial house has done a pioneer work in the Unani tradition of medicines in India and has the high tradition of preserving, pursuing and propagating this stream of medicines. Its *Rooh Afza sharbat* is very useful to patients as well as to healthy people. This is very popular in a large number of countries and has a vast domestic and international market. Its child tonic called *Shinkara* is also very popular with the consumers.

24. The Dabur. Its primary activity is the manufacturing of Ayurvedic medicines for which it has earned a name for itself. Many of the medicines produced by the Dabur are used by allopathic doctors, some of which are useful even for the cancer patients. During the recent years, this industrial house has taken steps to diversify its products and is venturing in the areas of consumer non-durables like fruit-based drinks and toiletries.

25. The Baidhyanath. Like the Dabur, this house is also known to preserve and propagate the Ayurvedic medicines and is quite renowned in this field.

26. The Arya Vaidyashala. This is known as harbinger of the Ayurvedic tradition in India and also runs a hospital dealing with Ayurvedic treatment and naturopathy.

27. The Jaypee. This is one of the major real estate and construction companies of India which has earned the reputation of constructing major roads and bridges within the stipulated time.

28. Bombay Dying. This is one of the important companies engaged in textile industry and is well known for producing fine as well as cheap and durable bed-sheets and towels. Its products are well renowned in the international markets and are exported to a large number of European and American countries.

29. The Arvind Mills. This is one of the oldest textile mills in India which is engaged in producing and exporting some of the finest and cheapest cotton cloth and garments. This company is one of the largest producers of Jeans cloth and supplies this cloth to almost all the major brands in the world.

30. The Ansals. The Ansals are one of the largest and the oldest real estate developers and their activities can be seen in most of the cities of India.

31. The Pioneer. As the name goes, this company is doing a pioneer job in the development of high yielding varieties of seeds and is promoting hybrid seed-based farming in India. Thus it has played a significant role in the success of the Green Revolution of India. The company is also credited with producing different types of pesticides and herbicides.

32. The Pantnagar. This company is also producing better seeds and helps in the success of the Green Revolution. Apart form seeds, this company produces various pesticides and herbicides.

33. The ACC. This is the organisation of cement producing companies and is known as Associated Cement Company (ACC). It produces high quality cement which is in great demand in the market. The company has contributed a lot in boosting the construction activity in the country.

Industrial Complexes

Industries in India are unevenly distributed because geographical, economic and a host of other factors affecting these industries are not evenly distributed. This trend has led to the development of concept of industrial complexes and their growth in different parts of the country. ***An industrial complex consists of a set of specific industries with prominence of production, marketing and other interrelationships among them, strengthened by their togetherness and innovations.*** In simple language, an industrial complex may refer to a factory or collection of buildings relating to industrial production. These complexes have specific geographical locations.

The concept and growth of industrial complexes gained popularity during the communist regime in the erstwhile U.S.S.R. when vigorous industrialization and planning were pursued after taking clues from the planned industrial development in France. This concept is associated with the concept of **'the growth pole'** of Francois Perroux (1955) and 'the growth centre' of J.R. Boudeville (1966). The concept industrial complexes started developing when the governments of various countries started planned economies and paid attention to industrial growth in certain favoured regions. The idea of an industrial complex, as developed by **W. Isard** is a method of analysing the *linkages* between industries in an industrial complex with the objective of identifying the type of industrial complex which would be most suitable for a given development plan or region—it is based on inter-industry relations of *input-output analysis* and *comparative cost analysis*.

In an industrial complex, the finished product of one industry becomes raw material for another industry. For example, automobile industry uses a large number of components which are manufactured by anxillary industries. Thus there is large scale sale and purchase of goods among different industries and their bond becomes stronger. In this way industries of various types play a complementary role in the development of one another. The quality of products is assured from the local industries and transport cost is also minimized. Thus the industries tend to cluster in a particular area and the industries as well as industrial complexes grow fast via the '*spillover effect*' and the '*multiplier effect*' ultimately benefiting the whole economy through the '*trickle down effect*'.

Industrial Complexes in India

There are thousands of industrial complexes which have grown in all the major and minor industrial regions of India. Such complexes may develop in the vicinity of iron and steel plants, and are known as iron and steel complexes. Some of the other complexes are petro-chemical complexes, pharma complexes, hosiery complexes, garment complexes, electronic complexes, etc.

The major thrust for developing industrial complexes was given by the central and the state governments after the policy of liberalization and economic reforms was adopted as a result of the Industrial Policy of 1991. The Export Processing Zones (EPZs) and Export Oriented Units (EOUs) of the past and the Special Economic Zones (SEZs) of the present are some of the best examples of Industrial Complexes in India. Besides, Technology Parks, Hardware Parks, Biotechnology Parks, etc. are also considered as industrial complexes. These complexes will remain relevant to the industrial growth, till planning is relevant to the economic growth in India.

Problems of Industrial Complexes

In spite of their above mentioned advantages, the industrial complexes have to face several problems, some of which are briefly described as under :

1. It often becomes difficult to select an industry which can attract auxiliary industries so that the area may develop into an industrial complex.
2. India has a large number of such industries such as textile mills, paper mills, etc. This leads to break in the chain of *inter-linkage* which puts obstacles in the growth of industrial complexes.
3. Lack of infrastructural facilities, especially the power crisis has done untold damage to the growth of industrial complexes.
4. By and large, the market in India is underdeveloped and lopsided which is hampering the growth and development of the industrial complexes.
5. Indian industrial complexes have to face problems of proper support of external sector regarding imports-exports.
6. Lack of skill on the part of labour and entrepreneurship on the part of industrialists also act as big obstacles in the path of progress of industrial complexes.
7. Usually adequate finances are not available at the right time and industrial complexes have to suffer a great set-back.
8. Labour laws, generally prevailing in India are not conductive to the proper growth of industrial complexes.
9. Wrong selection of 'growth poles' and 'growth centres' often put hindrances in the growth of industrial complexes.

There is an urgent need to offer solutions to the above mentioned problems so that congenial atmosphere is provided for the proper growth of industrial complexes. Economic reforms were initiated in the policy of liberalization adopted in 1991 and some tangible results are visible in the industrial development. Experts are of the view that the industries will grow with the growth of economy and industrial complexes will also grow accordingly.

Public Sector Undertakings

A pubic sector undertaking is an organisation in which the finance, production, sale and management is governed by the Central or State Government. To put in simple words the Government owned corporations are termed as Public Sector Undertakings in India. In a Public Sector Undertaking, majority (51 per cent or more) of the paid-up share capital is held by the Central Government or a State Government or partly by the Central Government or partly by one or more State Governments. This is primarily a post-Independence phenomenon as there was virtually no Pubic Sector Undertaking before 1947. A few instances worth mentioning at that time were the Railways, the Post and Telegraphs, the Port Trusts, the Ordnance and Aircraft Factories and a few State managed undertakings like the government salt factories, quinine factories, etc. After Independence, the expansion of pubic sector was considered as an integral part of the 1956 Industrial Policy as this policy gave the public sector a strategic role in the Indian economy.

Immediately after Independence, the Indian economy was in a bad shape—basically an agrarian economy with a very weak industrial base, large scale unemployment, low level of savings and investment and near absence of infrastructural facilities. Under such circumstances, the Indian economy needed a big push which could come only from the Public Sector Undertakings because private sector was badly lacking in financial resources and managerial ability and was incapable of undertaking risks involved in large and long gestation period investments. At that time, the public sector was thought of as the engine for self-reliant economic growth to develop a sound agricultural and industrial base, diversify economy and overcome economic and social backwardness.

Central Government Enterprises

The Central Government has played a vital role with regard to public sector undertakings. The basic purpose was to facilitate the process of industrialization by establishing heavy and basic industries.

Objectives of Public Sector

Following objectives of setting up public sector enterprises have been concerned :

(*i*) to provide rapid economic development through creation and expansion of infrastructure

(*ii*) to generate financial resources for development
(*iii*) to promote redistribution of income and wealth
(*iv*) to create employment opportunities
(*v*) to promote balanced regional development
(*vi*) to encourage the development of small scale and ancillary industries, and
(*vii*) to promote exports on one side and import substitution on the other.

In addition to its major contribution to the industrial growth in the country, public sector has proved to be a great employment provider. Public sector has offered employment to much larger number of people than the private sector. Employment offered by public sector has never fallen below 62 per cent of the total employment offered by both public and private sector.

Strong Industrial Base by Pubic Sector

It is a well known fact that the pubic sector undertakings laid a solid industrial base in India in the first three decades after Independence. This solid base led to fast industrial growth rate in the country. The Industrial Policy Resolution reserved certain industries such as atomic energy, ammunition and armaments, aircraft, etc. with the government in the interest of national security. The state also took the responsibility for developing key industries like coal, iron and steel, aircraft, ship-building, etc. Setting of integrated iron and steel plants at Durgapur, Rourkela, Bhiali and Bokaro, The Hindustan Machine Tools, locomotives at Chittaranjan and Varanasi are the outstanding examples of contribution by the public sector undertakings towards establishing a strong industrial base in India. Rest of the industries were left to the private sector. But the experience of the first three Five Year Plans made it very clear that private sector was not capable of laying a sound industrial base in the country due some of its inherent handicaps such as lack of capital infrastructure and administrative skill. Simultaneously, the Planning Commission realised that a much more diversified development in the field of industries was necessary if the Indian economy was to become self-generating. Naturally, the government had to undertake the development of basic and strategic industries, capital goods industries and even some consumer goods industries in a big way to boost the Indian economy. Even after the introduction of economic reforms, private sector investment has not increased according to expectations and even now it is being suggested that the pubic sector should take the responsibility of infrastructure development. Consequently, the public sector has entered into a wide spectrum of industries and products. Its operations extend from basic and capital goods like steel, coal, copper, zinc and other minerals, heavy machinery on one hand and drugs, chemicals, fertilizers, consumer goods like textiles, hotel services, watches, bread, etc. on the other hand. Most of these industries are of great strategic importance in the Indian economy because these are industries of high linkage.

In highly critical areas such as copper, coal, petroleum products, hydro and steam turbines the share of public sector is 100 per cent. In quite a large number of products, it ranges between 50 to 95 per cent.

Reducing Regional Disparities

In addition to industrial growth, public sector undertakings have played a significant role in reducing regional disparities in industrial development. With the beginning of the plan period after Independence, care was taken to identify industrially backward regions and pubic enterprises of the Central Government were set-up in those regions. Good examples are setting up of three steel plants at Durgapur, Rurkela and Bhilai and Neyveli Project in Chennai. In certain cases, the State Governments could not raise sufficient resources for development of their regions. Under such circumstances, the only alternative available was the setting up of projects by the Central Government or to start enterprises which were financed by the centre. In the recent past, the Central Government had prepared an exhaustive plan to remove industrial backwardness of the north-eastern states and bring them in the main stream of the nation so far as industrial development is concerned.

Shortcomings of the Public Sector

Inspite of the fact that public sector undertakings have contributed a lot to the industrial development of India, there is lot of criticisms on account of the following shortcomings.

(i) **Mounting losses.** A review of the public sector enterprises reveals that most of them are running heavy recurrent accumulation of losses and very few are making little or no gain. This may be attributed to a large number of factors including subsidies.

(ii) **Political interference.** Often powerful politicians influence decisions regarding setting up of an industry in an area which is not suitable for that particular industry from the geographic and economic point of view. Such a situation results in high input, low output and considerable wastage of financial and human resources. A classic example of this type of political influence is the irrational approach in the decision of the Central Government to break up the MIG aircraft project into two parts to be located in two separate states. One of the project is at Nashik in Maharashtra and the other is at Koraput in Odisha. This was done to appease the political bosses of these two states.

(iii) **Delays in Completion of Projects.** Since the government machinery has its own way of working which is slow and inefficient, there are often prolonged delays in completion of the projects. This results in colossal loss of resources and escalates the cost of construction due to high rate of inflation.

(iv) **Over-capitalisation.** Most of the public sector projects are plagued with over-capitalisation which means that input-output ratio is unfavourable. Some of the public sector undertakings facing this problem are Heavy Engineering Corporation, Hindustan Aeronautics, Fertilizer Corporation etc. According to a report of the Study Team on Public Sector Undertakings (1967), "The cause leading to over-capitalisation can be traced to inadequate planning, delays and avoidable expenditure during construction, surplus machine capacity, tied aid resulting in the compulsion to purchase imported equipment on a non-competitive basis, expensive turn key contracts, bad location of projects and the provision of housing and other amenities on liberal scale."

(v) **Price Policy.** Prices of most of the commodities produced by public sector undertakings are fixed by the Government and these prices are fixed not based on the principle of profit maximisation, rather they are fixed keeping the public interest in mind. For example, fixing the prices of steel, oil, gas, coal, fertilizers and other essentials of everyday use will be suicidal for the overall growth of economy. Some times the prices of final products are fixed under public and political pressure.

(vi) **Excess of manpower.** In most of the public sector enterprises, recruitment of manpower is often in excess of actual requirement which is putting heavy strain on the input costs. On the other hand, low wages and lack of incentives are causing large scale flight of efficient and intelligent personnel from the public sector to the private sector.

(vii) **Under-Utilisation of Capacity.** There is a serious problem of under-utilisation of capacity in public sector enterprises. During 2005-06, out of 203 manufacturing/producing units, 103 units or 51% units had recorded capacity utilisation of more than 75 per cent. On the other hand, 33 public sector enterprises operated in the capacity utilisation range of 50 to 75 per cent and 67 functioned below 50 per cent of the rated capacity. There is urgent need to investigate into the causes of such a situation and find remedial measures.

(viii) **Inefficient management.** There is a serious crisis of inefficient management in almost all the public sector enterprises. This is largely due to strict government rules, lack of elasticity and concentration of power in the hands of a few top executives. Each and every worker should know his responsibility. Unfortunately, there is a general failure to define responsibilities and duties in public sector enterprises in India.

INDUSTRIAL POLICY

'Industrial policy' is a comprehensive concept which covers all those procedures, principles, policies, rules, and regulations that control the industrial undertakings in the country and shape the patterns of industrialisation. It incorporates fiscal and monetary policies, the tariff policy, labour policy and the Governments attitude not only towards external assistance but the pubic and private sectors also.

Industrial Policy Resolution, 1948

The first industrial policy was announced by he Government of India on August 6, 1948. It contemplated a mixed economy in which both public and private sectors were involved for the purpose of industrial development. In accordance with this approach, industries were divided into following four broad categories :

(i) **Exclusive monopoly of the Central Government.** This included manufacture of arms and ammunition, the production and control of atomic energy and ownership and management of railway transport.

***(ii)* State monopoly of New Units.** This category covered coal, iron and steel, aircraft manufacture, ship-building, manufacture of telephone, telegraph and wireless apparatus (excluding radio receiving sets) and mineral oil. New undertakings in this category could be undertaken only by the State.

***(iii)* State Regulation.** This category included industries of such basic importance like machine tools, chemicals, fertilizers, non-ferrous metals, rubber manufactures, cement, paper, newsprint, automobiles, electric engineering, etc. which the Central Government would feel necessary to plan and regulate.

***(iv)* Unregulated Private Enterprise.** This category of industries was left open to the private sector, individual as well as co-operative.

Industrial Policy Resolution, 1956

Significant development took place after the adoption of 1948 Industrial Resolution. The economic planning had proceeded on an organised basis and the First Five Year Plan had been completed. These important developments made it necessary to come out with fresh statements of industrial policy. Consequently a second Industrial Policy Resolution was adopted in April, 1956. Following were its important provisions.

***(i)* New Classification of industries.** Industries were divided into following three categories :

(a) Schedule A. These were exclusive responsibility of the state and included seventeen industries *viz.*, arms and ammunition, atomic energy, iron and steel, heavy castings and forgings of iron and steel, heavy machinery required for iron and steel production, for mining, for machine tool manufactures, etc., heavy electrical industries, coal, mineral oils, mining, iron ore and other important minerals like copper, lead, zinc; aircraft, air transport, railway transport, ship-building, telephone, telegraph and wireless equipment, generation and distribution of electricity.

(b) Schedule B. These were to be progressively state owned and in which the state would generally set-up new enterprises, but in which private enterprise would be expected only to supplement the efforts of the state. It included twelve industries *viz.* other mining industries, aluminium and non-ferrous metals not included in Schedule A, machine tools, ferro alloys and tool steels, the chemical industry, antibiotics and other essential drugs, fertilizers, synthetic rubber, carbonization of coal, chemical pulp, road transport and sea transport.

(c) Schedule C. It included all remaining industries and their future development, in general, would be left to the initiative and enterprise of the private sector.

***(ii)* Fair and non-discriminatory treatment for the private sector.** The state was supposed to facilitate and encourage the development of industries in the private sector by ensuring the development of transport, power, and other services and by appropriate fiscal and other measures without any discrimination.

***(iii)* Encouragement to village and small-scale enterprises.** The state would take appropriate steps to encourage village and small scale industries.

***(iv)* Removing regional disparities.** Special emphasis was laid on developing industries in industrially backward areas so that those area as well the whole country is benefited.

***(v)* Attitude towards foreign capital.** Keeping in view limited resources within the country, it was felt that there was need for securing the participation of foreign capital and enterprise particularly as regards to industrial technique and knowledge so as to foster the pace of industrialisation.

Industrial Policy Statement 1977

With the defeat of Congress Party in 1977, Janata Party came to power and announced its own New Industrial Policy in December, 1977. Following were the main elements of this policy :

1. Development of Small Scale Sector. It was felt that the policies gave priority to large scale industries at the cost of small-scale industries. Thus the main thrust of this policy was on effective promotion of cottage and small-scale industries which are widely dispersed in rural areas and in small towns. Small-scale sector was classified into the following three categories :

(*a*) Cottage and household industries which provide self-employment on a wide scale.

(*b*) Tiny sector incorporating investment in industrial units, in machinery and equipment upto ₹ 1 lakh.

(*c*) Small-scale industries comprising industrial units with an investment upto ₹ 10 lakhs and in case of ancillaries with an investment in fixed capital upto ₹ 15 lakhs.

2. Areas for Large-Scale Sector. Large-scale industries were divided into the following sectors for their proper development :

(*a*) basic industries such as steel, non-ferrous metals, cement and oil refineries.

(*b*) capital goods industries for meeting the machinery requirements of basic industries and small scale industries.

(*c*) high technology industries which required large-scale production and which were related to agricultural and small-scale industries development such as fertilizers, pesticides and petrochemicals, etc.

(*d*) other industries.

The other elements of this industrial policy were :

3. Rational approach towards large business houses.

4. Expanding role of public sector which includes development of wide range of ancillary industries and decentralization of production.

5. Restricted role of foreign collaboration.

Industrial Policy of 1980

The Janata Party Government at the centre was dislodged by the Congress Party Government which announced its own industrial policy in July, 1980. This policy suggested the following measures :

(*i*) Effective operational management of the Public Sector.

(*ii*) Integrating industrial development in private sector by promoting the concept of economic federalism.

(*iii*) Redefining small scale units as per following divisions :

(*a*) to increase the limit of investment in the case of tiny units from ₹ 1 lakh to ₹ 2 lakhs.

(*b*) to increase the limit of investment in case of small-scale units from ₹ 10 lakhs to ₹ 20 lakhs; and

(*c*) to increase the limit of investment in the case of ancillaries from ₹ 15 lakhs to ₹ 25 lakhs.

(*iv*) Regularisation of unauthorised excess capacity installed in private sector.

(*v*) Automatic expansion of all industries specified in the First Schedule of 1951 Industrial Development Regulation Act.

(*vi*) Revival of those sick industrial units which showed the requisite potential.

Industrial Policy of 1991

A major shift in the industrial policy was made by the Congress Government led by Mr. P.V. Narasimha Rao on July 24, 1991. The main aim of this policy was :

(*a*) to unsackle the Indian industrial economy from the cobwebs of unnecessary bureaucratic control.

(*b*) to introduce liberalisation with a view to integrate the Indian economy with the world economy,

(*c*) to remove restrictions on direct foreign investment as also to free the domestic entrepreneur from the restrictions of MRTP (Monopolies and Restrictive Practices Act) and

(*d*) the policy aimed to shed the load of the public enterprises which had shown a very low rate of return or were incurring losses over the years.

Keeping in view the above mentioned aims, the following initiatives were taken :

1. Industrial Licensing Policy

2. Foreign Investment
3. Foreign Technology Policy
4. Public Sector Policy
5. MRTP (Monopolies and Restrictive Trade Practices Act

1. Industrial Licensing Policy

The industrial licensing policy was framed with respect to the following :

(*a*) Industrial licensing to be abolished for all projects except for a short list of industries related to security and strategic concerns, social reasons, hazardous chemicals and overriding environmental reasons and items of elitist consumption. Industries reserved for the small-scale sector will continue to be so reserved.

(*b*) Areas where security and strategic concerns predominate, will continue to be reserved for public sector.

(*c*) In projects where imported capital goods are required, automatic clearance will be given in cases where foreign exchange availability is ensured through foreign equity. In other cases, imports of capital goods will require clearance from the Secretariat of Industrial Approvals (SIA) in accordance with the availability of foreign exchange resources.

(*d*) In locations other than cities of more than one million population, there would be no requirement for obtaining industrial approval from the central government except for industries subject to compulsory licensing. In respect of cities with population larger than one million, the industries other than non-polluting such as electronics, computer software, and printing will be located outside 25 km of the periphery, except in prior designated industrial areas.

2. Foreign Investment

For encouraging foreign investment in high priority industries, requiring large investments and advanced technology, it was decided to provide approval for direct foreign investment upto 51 per cent foreign equity in such industries.

3. Foreign Technology Policy

For improving the level of technology in Indian industry, the government proposed to provide automatic approval for technology agreements related to high priority industries within the specific parameters. No permission is required for hiring of foreign technicians, foreign testing of indigenously developed technologies.

4. Public Sector Policy

By and large public sector enterprises have given very small returns in proportion to the investment made in such enterprises largely due to their inefficient functioning and also due to take-over of sick private units. Thus they become a liability rather than an asset to the Government. The 1991 Industrial Policy adopted a new approach to public sector enterprises. The following areas were given priority for the future growth of public sector enterprises.

(*a*) Essential goods and services.

(*b*) Exploration and exploitation of oil and mineral resources.

(*c*) Technology development and building of manufacturing capabilities in areas which are crucial in the long-term development of the economy and where private sector investment is inadequate.

(*d*) Manufacture of products where strategic considerations predominate such as defence equipment.

5. MRTP (Monopolies and Restrictive Trade Practices) Act

With the growing complexity of the Indian industrial structure and increasing need for achieving economies of the scale for ensuring higher productivity and competitive advantage in the international market, the interference of the Government through the MRTP Act was to be restricted. To meet this end :

(*a*) The pre-entry scruitiny of investments or decisions by the so called MRTP companies will not be required. Instead, the emphasis will be on controlling and regulating monopolistic, restrictive and unfair trade practices rather than making it necessary for monopoly houses to obtain prior approval of the Central Government for expansion, establishment of new undertakings, merger, amalgamation and take over and appointment of certain directors.

(*b*) The thrust of the policy will be more on controlling unfair or restrictive business practices.

MULTINATIONAL COMPANIES (MNCs)

Induction and Definition

A multinational company/corporation is an organisation which has its headquarters in one country but has business in more than one country. In other words it is an organisation or enterprise carrying business not only in the country where it is registered but also in several other countries. It may also be called as international corporation, global giant and transnational corporation. According to the most widely accepted definition, an international company is that organisation which produces at least 25 per cent of its global investment outside its original country.

According to the United Nations, a multinational corporation is "an enterprise which owns or controls production or service facilities outside the county in which it is based." In words of W.H. Morcland, "Multinational Companies/Corporations are those enterprises whose management, ownership and controls are spread in more than one foreign country."

Characteristics of MNCs

1. Large Size. MNCs are generally of large size with assets worth billions of dollars and their annual sales turn over is usually more than the gross national product of many small countries.

2. Worldwide Operations. As mentioned earlier an MNC has business in more than one country and often it spreads over a large number of countries. For example, Coca Cola has its branches in more than seventy countries.

3. International Management. MNCs have management at the international level. They operate on the basis of best possible alternative available anywhere in the world. Its local subsidiaries are generally managed by the people of the host country. For example, the management of Hindustan Lever lies with Indians whereas the parent company Unilever is in the U.S.A.

4. Mobility of Resources. Operations of MNCs involve the mobility of capital, technology, entrepreneurship and other factors of production across the territories.

5. Integrated Activities. An MNC is usually a complete organisation comprising manufacturing, marketing, research and development (R&D) and other facilities.

6. Several Forms. An MNC may operate in host of countries in several forms *i.e.* branches, subsidiaries, franchise, joint ventures, etc.

Origin and Growth. MNCs started functioning even before the process of colonisation in the world by the European nations. East India Company was set-up in India before the British rule could establish itself in the country. Similarly the Dutch East Indonesian Company, the Royal African Company and the Hudson Bay Company started their business in early phase of the period of colonisation. Their main purpose was to find out new markets in different countries and exploit their potential. These companies started procuring raw materials from their markets to feed their industries at home. At the same time they found ready market for their furnished products in the countries under their control. Thus a totally new economic and political system was evolved and the era of colonisation began.

After the Second World War, most of the colonial countries became independent nations and started developing their own economies. During this period, the U.S.A. become the world's largest industrial nation and started spreading its influence in larger parts of the world through Direct Foreign Investment. So far so, the U.S.A. became a net investor in addition to being a major exporter.

According to estimates made by the U.N.O., about 35,000 corporates have nearly 1,70,000 companies doing business in countries other than their own country.

The quantum of business carried on by them can be judged from the fact that in 1998 about 200 big Multinational companies of the world accounted for

28.3 per cent of the world's GDP. Currently, their total sales are equivalent to the GDP of 182 countries of the world (excluding the U.S.A., Japan, Germany, France, Britain, Brazil, Canada, China and Italy).

Effects of MNCs. MNCs have both good and bad effects in India. For example, most industries were producing consumer goods at the time of Independence. There were only three iron and steel companies (TISCO, IISCO and MISCO) and capital goods were almost absent. At present, India has developed almost all types of Industries largely with the help of foreign finances, technology and expertise. In spite of all these achievements, MNCs are not flawless and have their own ill effects. In the year 1975-76, Dr. S.K. Goel studied 133 out of 171 companies and arrived at certain conclusions. Some of the conclusions can be summed up as follows :

1. Capital Investment. Some MNCs made some capital investment in the beginning to exploit the rich natural resources of India.

2. New Technology. Before Independence, most industries in India were using old and obsolete technology which gave low level of production of inferior goods. MNCs introduced new advanced technology to Indian industries which improved the quality of industrial products and also helped in increasing their production.

3. Damage to Cottage and Small-Scale Industries. India had a rich tradition of cottage and small-scale industries which were based on traditional methods of production. These industries could not compete with modern industries of MNCs which started producing better goods in larger quantities and at cheaper prices. Thus these industries suffered a great set-back at the hands of MNCs and may totally disappear in future if proper steps are not taken to safeguard the interest of these industries.

4. Low Level of Foreign Investment. Dr. S.K. Goel has rightly pointed out that most foreign subsidiaries have raised financial resources from within India by exploiting the natural resources of the country. Thus MNCs invested very little capital from their parent countries and total foreign investment has been only marginal.

5. Multi-Industry Pattern. Many MNCs have acquired multi-product and multi-industry pattern adding considerably to their original industry. For example, the Imperial (now India) Tobacco Company (ITC) has diversified its activities and ventured into hotel industry, constructing a chain of hotels in different parts of the country. This company has also entered into some allied industries such as salt, flour, biscuits, match boxes, agarbathi, etc.

6. Heavy Remittance Abroad. According to the Reserve Bank of India, 537 foreign companies have reported average rate of profit at 23.8 per cent during the period 1972 to 1993. This rate is reported to have increased to about 25 per cent now. This is a huge amount of money remitted outside the country causing heavy strain in India's economic resources.

7. Transfer of Technology—A Myth. According to Dr. S.K. Goel, 'the assumption that entry of multinational corporations would ensure transfer of latest sophisticated technology to developing countries has not been found valid in practice and has proved to be a myth.

LIBERALISATION

Liberalisation means to liberate the Indian industries from undue interference of the government agencies and to remove the bureaucratic control which have been hampering the proper growth of industries in the country. It was also aimed at enabling the Indian industries to successfully face the market forces and provide guidelines for their future growth. The process of liberalisation started on 24th July, 1991 when Dr. Manmohan Singh, the then Finance Minister in Mr. P.V. Narasimha Rao ministry introduced the new Industrial Policy. Following are the salient features of policy of liberalisation.

1. Under the policy of liberalisation, the bureaucratic control on the industries was removed as it was creating barriers in the industrial development in the country. As a matter of fact, this policy implies deregularisation of the industrial sector by cutting down the minimum administrative interference in operation, instead, letting the market forces operate through the profit motive of the producers and free competition among them to regulate and guide the future development of the industry sector.

2. The Industrial Policy of 1991 was slightly modified in March, 1993 according to which requirement of industrial licensing was abolished

except for 18 industries namely, (1) coal and lignite, (2) petroleum (other than crude) and its distillation products, (3) distillation and brewing of alcoholic drinks, (4) sugar, (5) animal fats and oils, (6) cigars and cigarettes of tobacco and manufactured tobacco substitutes, (7) asbestos and asbestos based products (8) plywood, decorative veneers and other wood based products like particle board, medium density fibre board, black board, (9) raw hides and skins, leather, chamois leather and patent leather, (10) tanned or dressed furskins, (11) motor car, (12) paper and newspaper except bagasse-based units, (13) electronic aerospace and defence equipment—all types, (14) industrial explosives, including detonating fuse, gun powder, nitrocellulose and matches, (15) hazardous chemicals, (16) entertainment electronics (VCRs), Colour TVs, CD players, tape recorders), (17) drugs and pharmaceuticals and (18) white goods (domestic refrigerators, domestic dish washing machines, programmable domestic washing machines, microwave ovens, airconditioners).

3. Gates for direct foreign investment (DFI) were opened and subsidies were reduced.

4. Restriction on export and import of a large number of items were abolished.

5. Import tariffs were substantially reduced and restrictions of foreign investment were totally removed.

6. The government share in industries was restricted to 51 per cent and the process of disinvestment in public sector was initiated. This led to reduction of delays in taking decisions at the government level and the industrialists become free to take their own-decisions without any loss of time.

7. Under the policy of liberalisation, the industrialists were encouraged to take active part in the fields of road, power, communication and petroleum so that centre and state governments could pay more attention to social and economic development programmes.

Impact of Liberalisation

Liberalisation has left both good and bad effects on Indian economy.

Good Effects

Indian economy in general and industries in particular have been benefited a lot by the policy of liberalisation as is clear from the following few points :

1. Increase in Foreign Direct Investment (FDI). The policy of liberalisation has provided congenial atmosphere for investment and given a boost to FDI. This provides great opportunities for development of industries and other spheres of economy. In 1991, FDI was at a very low level of ₹ 534.11 crore. It rose to US $ 92.51 billion in 2009-14 and US $ 163.87 in 2014-19 (Reserve Bank of India). Services, hotels and tourism, metallurgical industries, automobile industry, construction, drugs and pharmaceuticals, industrial machinery were the sectors that attracted maximum FDI inflows.

2. Increase in GDP (Gross Domestic Product). There had been significant increase in GDP after the policy of liberalisation was adopted by the government. In the year 1991, GDP was ₹ 6,92,871 crore which rose to ₹ 2.6 lakh crore USD in 2017.

3. Reduction in Industrial Recession. Before liberalisation, Indian industries were passing through the phase of recession. Foreign investment and technology started flowing in immediately after adoption of policy of liberalisation. Old industries were rejuvenated and new industries were established. All these developments resulted in reduction of recession. In 1991, Indian industries faced serious recession but its effect was reduced to a great extent by 1994. Industries like automobiles, autoparts, coal mining, consumer electronics, textiles, petrochemicals, software, sports-goods recorded a growth rate of 20 per cent in 1994. Besides, fertilizers, crude oil, tyres, tubes, etc. also recorded a 10 per cent growth at the same time. In the year 2007-08, the whole world was passing through a serious economic recession but India was one of the least effected nations in the world. However, some slowdown in the Indian industrial growth rate has been observed in the recent past.

4. Increase in employment. Liberalisation has opened the doors for domestic and foreign investment amounting to over ₹ three lakh crores. This will definitely strengthen the industrial base in the country and create 32 lakh jobs.

5. Development of Infrastructure. Before liberalisation, major components of infrastructure such as transport and electricity were in a bad shape and were adversely affecting the economic growth in general and industrial growth in particular. Liberalisation provided congenial atmosphere for heavy investment in infrastructure which facilitated the rapid economic and industrial growth in the country.

6. Rise in Exports. Liberalisation led to economic and industrial growth which in turn, boosted India's exports considerably. For example, India's total exports amounted to ₹ 32,553 crores in 1990-91 which rose to ₹ 21,59,045 in 2020-21.

Bad Effects

Although liberalisation has helped in developing economy and has led to industrial growth, it has its own bad effects, some of which are briefly described as under :

1. Increase in Regional Disparities in Industrial Growth. The policy of liberalisation and New Industrial Policy have increased the pre-existing regional disparities instead of reducing them. Both the Indian and the foreign investors invested money only in those regions which were already advanced from the industrial point of view. The maximum investment was made in industrially advanced states of Maharashtra, Gujarat, Rajasthan, West Bengal, Andhra Pradesh, Tamil Nadu and Karnataka and industrially backward states like Bihar, Himachal Pradesh, Uttarakhand, Jammu and Kashmir, Meghalaya, Mizoram, Tripura, Nagaland, Odisha, Uttar Pradesh, etc. are lagging behind. **Ghum** and **Ghuman** recognised four regions with respect to investment. These are : (*i*) Maharashtra-Gujarat, (*ii*) Andhra Pradesh, Tamil Nadu, Karnataka, Madhya Pradesh and Odisha, (*iii*) Haryana and Rajasthan and (*iv*) West Bengal. Uttar Pradesh, Himachal Pradesh, Uttarakhand and Bihar are the least benefited states. Increase in regional disparities has given birth to a large number of socio-economic and political problems. The Naxal Movement in a large number of states extending from Darjeeling in West Bengal to Kanniyakumari in Tamil Nadu. ULFA in Assam, and political turmoil in Jammu and Kashmir may be partly explained due to regional disparities in industrial growth.

2. Increase in Regional Disparities in Infrastructure. Along with regional disparities with respect to industrial growth, regional disparities in infrastructural development have also increased. For example, out of 33 sanction electricity projects (based on coal, gas lignite and water) in 1994, the state-wise sanction was 6 to Odisha, 5 to Karnataka, 3 each to Tamil Nadu, Andhra Pradesh and Maharashtra, 2 each to Haryana, West Bengal and Andhra Pradesh and the remaining seven were sanctioned to other states. Madhya Pradesh, Uttar Pradesh, Bihar, Mizoram, Nagaland, and Delhi were totally deprived of these projects.

3. Damage to Cottage and Small-scale Industries. A developing country like India has an old and rich tradition of cottage and small-scale industries. These industries form an important segment of our economy and provide employment to millions of artisans at the village level. Liberalisation opened the floodgates for investment by Indian as well as foreign companies in large scale industries. This gave much need boost to large-scale industries but cottage and small-scale industries suffered heavily because they could not compete with rich financial and technological resources of big companies. The revival of these industries requires help from the government in the form subsidies, technology, and proper facilities for exports.

4. Increase in Unemployment. Although liberalisation facilitated the smooth flow of foreign investment, it also encouraged the use of labour saving new work efficient technology. Thus in spite of heavy investment and rapid increase in industrial production, the employment opportunities were drastically reduced and unemployment increased.

5. Comparatively Little Direct Investment. The foreign investors are more interested in portfolio investment rather than in direct investment. The main reason for such a situation is portfolio investment can be withdrawn at will without much hurdles, thus adversely affecting the economy of the country.

6. Problem of Inflation. Since the adoption of policy of liberalisation, the rate of inflation has been almost consistently high, sometimes soaring to double digits. This often upsets the budget of a middle class family. Moreover, money is concentrated in the hands of a few rich persons and the gap between the poor and the rich widens. This type of situation creates a lot of economic and social problems often leading to social tension.

7. Excess of foreign money. Liberalisation has allowed large scale inflow of foreign money which could not be properly invested due to low level economy and poor base in the country. Investment for exports is limited and foreign trade is adversely affected.

8. Investment in Selected Industries. Most of the foreign investment comes to white-goods instead of wage-goods sector. Thus it may be useful in improving the high priority sector and bringing in the latest technology while ignoring the other sector. This can be counterproductive and is not in the interest of overall economic development.

9. Economic and Political Freedom at Stake. Liberalisation attracts more and more of foreign investment also involving a lot of foreign exchange. Thus there is a danger that the whole economy may be handed over to the multinationals putting our economic and political freedom of a great risk.

10. Agriculture is ignored. Liberalisation has given impetus to industries but has badly ignored agriculture. Agriculture is the **'back-bone'** of our economy which provides livelihood to more than 55 per cent of our population. If the trend of encouraging industries at the cost of agriculture continues unabated it will bring imbalance in the economy and the country will be confronted with a serious economic crisis.

SPECIAL ECONOMIC ZONE (SEZ)

Introduction and Definition

Special economic zone (SEZ) is a geographical region in which economic laws are more liberal than laws prevailing in other parts of a country and which is designed to export goods. SEZ may be exempt from laws regarding taxes, quotas, Foreign Direct Investment (FDI) loans, labour laws and other restrictive laws in order to make the goods manufactured in the SEZ at globally competitive prices. The SEZ includes Free Trade Zones (FTZ), Export Processing Zones (EPZ), Free Zones (FZ), or Free Economic Zones (FEZ), Industrial Parks or Industrial Estates, Free Parks, Bonded Logistics Parks and Urban Enterprise Zones. The operating definition of an Economic Zone is determined by each country's trade and customs administration. For a well organised SEZ there should be atleast 1000 hectares of land and a minimum investment of ₹ 10,000 crore.

Objectives. SEZ is established with the following objects in mind :

(*a*) Generation of additional economic activity.
(*b*) Promotion of exports of goods and services.
(*c*) Promotion of investment from domestic and foreign sources.
(*d*) Creation of employment opportunities.
(*e*) Development of infrastructure facilities.

Historical Perspective

In the opinion of Robert C. Hayward, Director, World Economic Processing Zones Association, the concept of free economic zones dates back to 300 B.C. He noted that such enclaves were found in the city of Tyre in the Greek island of Delos. The city became rich due to such policies and was considered as a challenge to the centralism of the Roman Empire. In the modern context, the most famous SEZs were established by the government of China under Deng Xiaoping in the early 1960s. This was followed by setting up of SEZs in a large number of countries including Brazil, India, Iran, Jordan, Kazakhstan, Pakistan, Philippines, Poland, Russia and Ukraine.

Special Economic Zones of India

India was one of the first countries in Asia to recognise the effectiveness of Export Processing Zone (EPZ) model in promoting exports with Asia's first EPZ set-up in Kandla in 1965. In order to overcome the shortcomings due to multiplicity of controls and clearances, absence of world class infrastructure, and an unstable fiscal regime and for attracting larger foreign investments in India, the Special Economic Zones (SEZs) Policy was announced in April, 2000.

In 2004, the Government of Gujarat, amended the Industrial Disputes Act to create special exemption for SEZs so that companies can terminate work with only one month's notice. This flexibility helped in growing manufacturing jobs in Gujarat by 60% from 2000 till 2012.

The SEZ Act, 2005, was a very important bill passed by the Government of India. This bill was passed with the aim to instil confidence in investors and signal the Government's commitment to a stable SEZ policy regime with a view to impart stability to SEZ regime thereby generating greater economic activity and employment through their establishment. This Act came into effect on February 10, 2006 and provided for drastic simplification of procedures and for single window clearance on matters relating to central as well as state governments. The remaining part of India, not covered by SEZ rules is known as *Domestic tariff area*.

INCENTIVES AND FACILITIES TO SEZ DEVELOPERS

- Exemption from customs/excise duties for authorised operations.
- Income Tax exemption on income derived from business of development of SEZ.
- Exemption from minimum alternative tax.
- Exemption from dividend distribution tax.
- Exemption from sales tax.
- Exemption from service tax.

INDUSTRIAL/ECONOMIC CORRIDORS

Industrial/economic are corridors along the main transport routes which have been selected for special thrust to be given for industrial/economic development in the country. The corridors involve multiple development projects including development in transport, socio-economic impact, urban development, environmental management, increase in exports, growth in employment opportunities and above all revolutionise the industrial growth. Currently following five industrial/economic corridors are proposed to be developed (**Fig. 25.17**).

1. Delhi-Mumbai Industrial Corridor (DMIC)
2. Amritsar-Delhi-Kolkata Industrial Corridor (ADKIC)
3. Chennai-Bengaluru Industrial Corridor (CBIC).
4. Bengaluru-Mumbai Industrial Corridor (BMIC)
5. Vizag-Chennai Industrial Corridor.

1. Delhi-Mumbai Industrial Corridor (DMIC). This is the most ambitious industrial development plan of the Government of India for which 'in principle' approval was accorded in August 2007. This 1483 km long industrial corridor will be developed along the *Western Dedicated Freight Corridor* of the Indian railways with view to using the high-capacity of this dedicated freight corridor as a backbone for creating a global manufacturing and investment destination. The project seeks to develop a series of futuristic infrastructure endowed smart industrial cities that can compete with the best international manufacturing and industrial regions. The master plan has a vision for 24 manufacturing cities. Potential production sectors include general manufacturing, IT/ITES components, electronics, agro and food processing, heavy engineering, pharmaceuticals, biotechnology, and services. The project involves a total investment of \$ 90 billion and will be completed with Japanese assistance.

The project will involve a total area of 4,36,486 sq km which is about 13.8 per cent of the total land area of India. It extends over seven states and two union territories *viz*. Delhi, Uttar Pradesh, Haryana, Rajasthan, Madhya Pradesh, Gujarat, Maharashtra, Daman and Diu and Dadra and Nagar Haveli (Delhi has been treated as a state in this context). The distribution of length of the corridor indicates that Rajasthan (39%) and Gujarat (38%) together constitute 77% of the total length of the freight corridor followed by Haryana and Maharashtra 10% each and Uttar Pradesh and National Capital of Delhi 1.5% each) About 17 per cent of India's total population will be affected. The project goals are to double employment potential in seven years, triple industrial output in nine years, quadruple exports from the region in 8-9 years and target 13-14 per cent growth per annum for the manufacturing sector on a sustainable basis. The project is expected to generate employment for 3 million persons. The Western Dedicated Freight Corridor will have nine junction stations along which other railroad networks will connect allowing the system to extend its reach across a wide swathe. Other infrastructure plans include

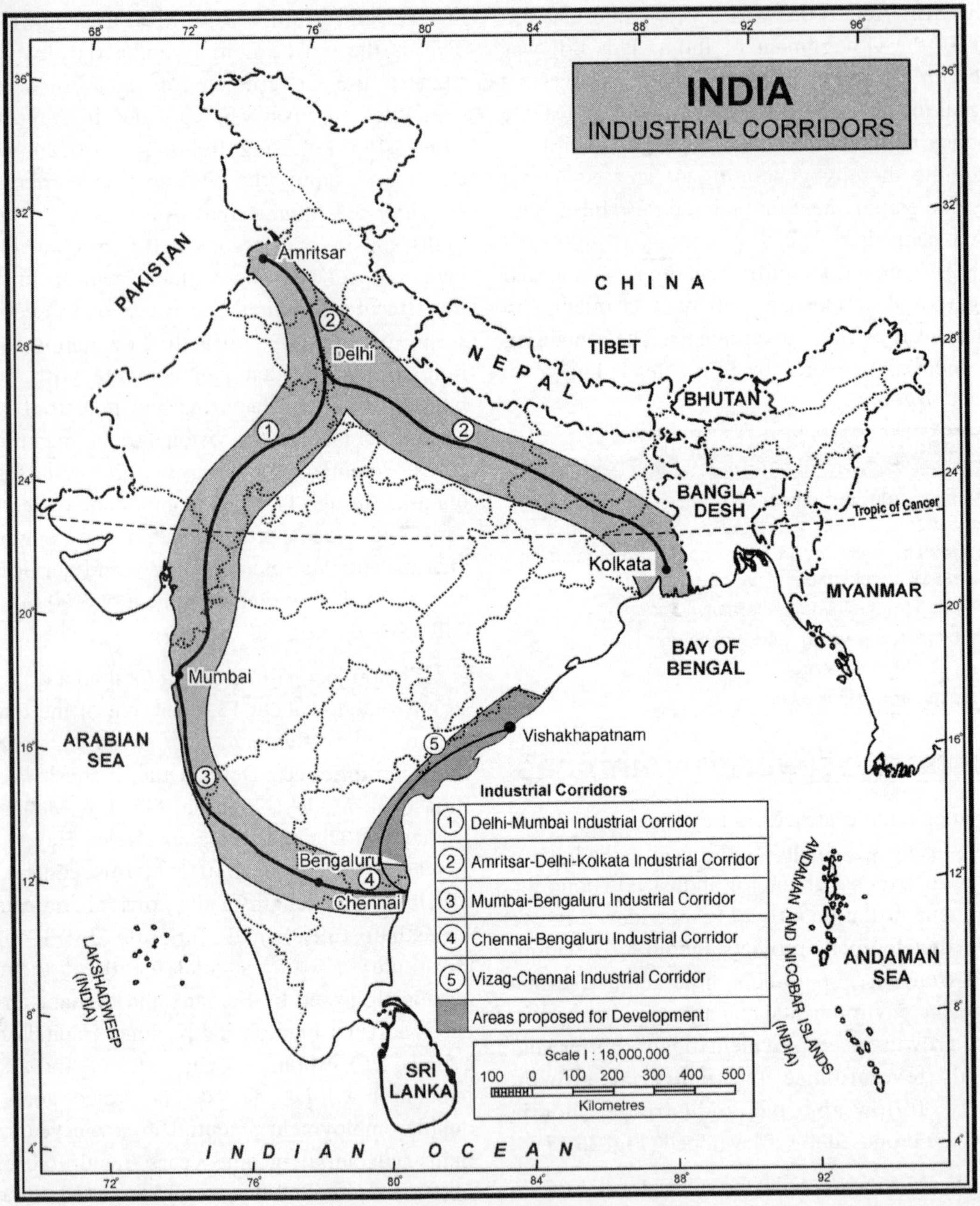

FIG 25.17. India : Industrial Corridors

logistic hubs, feeder roads, power generation facilities, up-gradation of existing ports and air-ports, developing greenfield ports, environment protection and social infrastructure. It will include three seaports and six airports in addition to connectivity with the existing ports. The major power inputs will come from six gas-based projects of around 1000-12000 MW each. Other power options include the use of renewable energy sources integrated through a smart grid.

The project seeks to upgrade the existing industrial clusters and also develop new industrial facilities. These will be developed on the concept of node-based development based on Investment Regions (IRs) and Industrial Areas (IAs). These are proposed as self-sustaining industrial townships with world class infrastructure. IRs will have a minimum area of 200 sq km and IAs will have 100 sq km. In all 24 IRs and IAs are planned.

Six mega investment regions are Dadri-Noida (U.P.), Manesar-Palwal (Haryana), Khushera-Bhiwadi-Neemra (Rajasthan), Pithampur-Dhar-Mhow (Madhya Pradesh), Bharuch-Dahej (Maharashtra) and Igatpuri-Nashik-Sinnar (Maharashtra). Five industrial areas are Meerut-Muzaffarnagar (Uttar Pradesh), Faridabad-Palwal (Haryana), Jaipur-Dausa (Rajasthan), Vadodara-Ankleshwar (Gujarat) and Alewari-Dighi Port (Maharashtra).

2. Amritsar-Delhi-Kolkata Industrial Corridor. This corridor will be structured around the Eastern Dedicated Freight Corridor (EDFC) and also along the existing highway transport route. It will also leverage the Inland Water Transport System being developed along the Ganga river from Allahabad to Haldia. It will cover the states of Punjab, Haryana, Uttar Pradesh, Uttarakhand, Bihar, Jharkhand and West Bengal. This is one of the most densely populated regions of the world and is the home for about 40% of India's population. It will cover the cities of Amritsar, Jalandhar, Ludhiana, Amabala, Saharanpur, Delhi, Roorkee, Moradabad, Bareilly, Aligarh, Kanpur, Lucknow, Pryagraj, Varanasi, Patna, Hazaribagh, Dhanbad, Asansol, Durgapur and Kolkata.

3. Chennai-Bengaluru Industrial Corridor (CBIC). This 560 km long corridor between Chennai-Bengaluru-Chitradurga will benefit the states of Karnataka, Andhra Pradesh and Tamil Nadu. Comprehensive plan for the CBIC was decided in December 2011, and it will be completed with Japanese financial help. Steel, cement, food processing, information technology, automobiles, readymade garments, petroleum, chemicals, and petrochemicals are some of the industries that would be benefited from this corridor. Karnataka government has proposed to extend this corridor upto Mangaluru and reposition this corridor as West seaport to East seaport connectivity corridor. A new proposal has been put forward where the corridor will have 2 seaports at the extreme ends and 3 international airports on the way. It will boost India's trade with both eastern and western countries.

4. Bengaluru-Mumbai Economic Corridor. Decision to develop this corridor was taken in February, 2013 and will be developed with the financial and technical help from the UK. This corridor is intended to facilitate development of a well-planned and efficient industrial base served by world-class connectivity infrastructure. It will also increase the possibility of private investments in manufacturing and industrial activities in Maharashtra and Karnataka.

5. Vizag-Chennai Industrial Corridor. The Vishakhapatnam-Chennai Industrial Corridor (VCIC) a key part of the East Coast Economic Corridor (ECEC). It is India's first coastal corridor. It is aligned with the Golden Quadrilateral and is poised to play a critical role in driving India's Act East Policy.

REFERENCES

Agarwala, A.K. (2004) : *Aluminium—Favourable Growth Factors,* The Hindu Survey of Indian Industry, Kasturi and Sons, Chennai, 2004, pp. 197-201.

Ahluwalia, Isher (1985) : *Industrial Growth in India—Stagnation since Mid 60's,* Oxford University Press.

Ahluwalia, I.J. (1994) : *Contribution of Planning to Indian Industrialisation* in T.J. Byres (ed.), *Development and Planning in India,* Oxford University Press, New Delhi.

Alexanderson, G. (1967) : *Geography of Manufacturing,* Prentice Hall, New Jersey.

Alexander, P.C. (1963) : *Industrial Estates in India,* Bombay.

Ashan Qamar and Mukhrjee, S. (1998) : *Indian Industry—A Geographical Perspective,* Transworld Media and Communication, Patna.

Balasubramaniam, S.V. (2004) : *Sugar—Time for total decontrol.* The Hindu Survey of Indian Industry, Kasturi & Sons, Chennai, 2004, pp. 297-301.

Biswanath, G. (1974) : *Changing Profile of India's Industrial Economy,* Calcutta.

Brush, John E. (1952) : *The Iron and Steel Industry in India,* The Geographical Review, Vol. XIII, No. 1.

Chander Shekhar, G. (2004) : *Constraints of growth.* The Hindu Survey of Indian Industry, Kasturi and Sons, Chennai, 2004, pp. 302-03.

Chaudhery, M.R. (1970) : *Indian Industries : Development and Location,* Oxford & I.B.H. Publishing Company, Calcutta.

—— (1982) : *The Public Sector in Indian Industries.* Calcutta.

—— (1982) : *Jute Industry in India,* Oxford Book and Stationery, Calcutta & New Delhi.

—— (1989) : *Industrial Development in India during the planning era,* Geographical Review of India, Vol. 51, 3. pp. 1-16.

De, Jayasree (1977) : *The Emergence of an Industrial Region—Gujarat,* Geographical Review of India, Vol. 39, No. 4, pp. 389-94.

Dharamakrishnan, A.V. and Moorthy Balaji, K. (2004) : *Cement—New thrust after pause,* The Hindu Survey of Indian Industry, Kasturi and Sons, Chennai, 2004, pp. 209-211.

Gadgil, D.R. (1971) : *Industrial Evolution of India in Recent Times (1860-1939),* Oxford University Press, Mumbai.

Gandhi, M.P. (1972) : *Major Industries of India,* 1970-71, Gandhi and Co, Bombay.

Ganguli, B.N. (1949) : *Bengal-Bihar Industrial Belt.* G.R.I. Vol. XI, No. 2.

Government of India (2024), *A Reference Annual.*

—— (2022-23) *Economic Survey.*

—— (Various years), *Five Year Plans* I-XII, Planning Commission.

—— (1996), *Handbook of Industrial Policy and Statistics.*

—— (Various years) *Industrial India Annual.*

Jarrett, H.R. (1977) : *A geography of manufacturing*, Macdonald and Evans, Plymouth.

Joshi, H.L. (1990) : *Industrial Geography of India,* Rawat Publications, Jaipur.

Karan, P.P. (1964) : *Changes of Indian Industrial Location*, Annals of the American Geographers (AAAG), Vol. 54.

Karnik, Kiran (2004) : *Software and Services—Riding on a boom.* The Hindu Survey of Indian Industry, Kasturi and Sons, Chennai, 2004, pp. 221-24.

Karunakaran, K. (2004) : *Fertilizers—Industry at the crossroads,* The Hindu Survey of Indian Industry, Kasturi and Sons, Chennai, 2004, pp. 275-78.

Kelkar, V.L. and Kumar, R. (1990) : *Industrial Growth in Eighties—Emerging Policy Issues,* Economic and Political Weekly, Vol. XXV.

Kennely, P. (1993) : *Preparing for Twenty First Century,* Indus, New Delhi.

Kuchchal, S.C. (1978) : *The Industrial Economy of India,* Chaitanya Publishing House, Allahabad.

Kulkarni, M.R. (2013) : *Industrial Development.* National Book Trust, India, New Delhi.

Kundu, A. and Raza, M. (1982) : *Indian Economy : The Regional Dimension,* CRD, JNU, New Delhi.

Kurian, G. (1962) : *An Analysis of the Spatial Distribution of Industry in India with special reference to Population,* Indian Geographical Journal, Vol. 37 (1), pp. 1-7.

Mehra, J. (2004) : *Steel-Strengthening global links,* The Hindustan Survey of Indian Industry, Kasturi and Sons, Chennai, 2004, pp. 192-194.

Mehta, M.M. (1961) : *Structure of Indian Industry,* Popular Book Depot., Mumbai.

Mehta, Vinnic (2004) : *Hardware—Brightening outlook.* The Hindu Survey of Indian Industry, Kasturi & Sons, Chennai, pp., 225-226.

Miller, E.W. (1962) : *A Geography of Manufacturing,* London.

Montanari Alberto (2004) : *Automobiles—Small segment to the fore,* The Hindu Survey of Indian Industry, Kasturi & Sons, Chennai, pp. 245-46.

Nair, M.D. (2004) : *Pharmaceuticals—Bid for global role.* The Hindu Survey of Indian Industry, Kasturi & Sons, Chennai, pp. 261-64.

Nair, Jharkhand (1991) : *Locational Analysis of Industries,* Rawat Publications, Jaipur.

Sampath, R. (2004) : *Aluminium—Poised for a global role.* The Hindu Survey of Indian Industry, Kasturi and Sons, Chennai, pp. 203-05.

Sampath, R. (2004) : *Vizag Steel Plant—Gains of turnaround,* The Hindu Survey of Indian Industry, Kasturi & Sons, Chennai, 2004, pp. 195-96.

Sengupta, P. (1959) : *The Indian Jute Belt,* Orient Longman, Calcutta.

Singh, Amar and Garg, Hari Shankar (1988) : *Industrial Planning and Development in India,* Aligarh, Chapter (NAGI).

Singh, M.B. (1985) : *Industrial Development in India—A Geographical Analysis,* Lotus Publications, Varanasi.

Sinha, B.N. (1972) : *Industrial Geography of India,* World Press Pvt. Ltd., Calcutta.

Sinha, N.K.P. and Singh, M.B. (1993) (Eds) : *Industrial Development in India,* Rawat Publications, Jaipur.

Smith, D.M. (1971) : *Industrial Location,* John Willey & Sons, New York.

Srinivasan Venu (2004) : *Two Wheelers—Changing market dynamics,* The Hindu Survey of Indian Industry, Kasturi and Sons, Chennai, 2004, pp. 251-55.

Taparia, M.P. (2004) : *Plastics—Energing opportunities,* The Hindu Survey of Indian Industry, Kasturi and Sons, Chennai, 2004, pp. 283-85.

Tata Services Ltd. (2012-13), Statistical Outline of India.

Tewari, R.T. (1988) : *Inter Regional Pattern of Industrialisation in India in R.T. Tewari and A. Josi (Eds.),* Development and Change in India, Ashish Publishing House, New Delhi.

The Hindu (2008, 09, 10, 11, 12) *Survey of Indian Industry.*

Venkataswamy, Vijay (2004) : *Textiles—Light end of the tunnel,* The Hindu Survey of Indian Industry, Kasturi and Sons, Chennai, 2004, pp. 292-95.

Vakil, C.N. (1953) : *Economic Consequences of Divided India, A study of Economy of India and Pakistan,* Mumbai.

◆◇◆

CHAPTER 26

Transport, Communication and Space

INTRODUCTION

Transport is a system in which passengers and goods are carried from one place to another. Development of cheap and efficient means of transport is necessary for the progress of a large and developing country like India. Transport routes are the basic economic arteries of the country. Transport system is regarded as the sinews of the national economy and provides a very important link between production and consumption. The amount of traffic moving in a country is a measure of its economic progress. In the modern age, even ideas and skills move from one place to another with the help of transport systems.

India is a vast country with long distances from Kashmir in the north to Kanniyakumari in the south and from Kandla in the west to Kohima in the east. She has vast natural resources of great diversity. In addition, India has great diversity in economic, social, cultural and ethnic structure. A well-knit transport system is essential to bring about unity in diversity in the Indian fabric.

MODES OF TRANSPORTATION

Three chief modes of transportation are land, water and air transport. Land transport consists of railways and roads. Water transport includes inland, coastal and ocean routes while air transport can be national or domestic and international. Pipelines can be constructed both over land and under water although most of the pipelines are over land.

All types of transport including railways, roadways, waterways, airways and pipelines are available in India.

RAILWAYS

Indian railway system is the main artery of the country's inland transport. Railways virtually form the life-line of the country, catering to its needs for large scale movement of traffic, both freight and passenger, thereby contributing to economic growth and also promoting national integration. In fact, railways constitute the backbone of surface transport system in India.

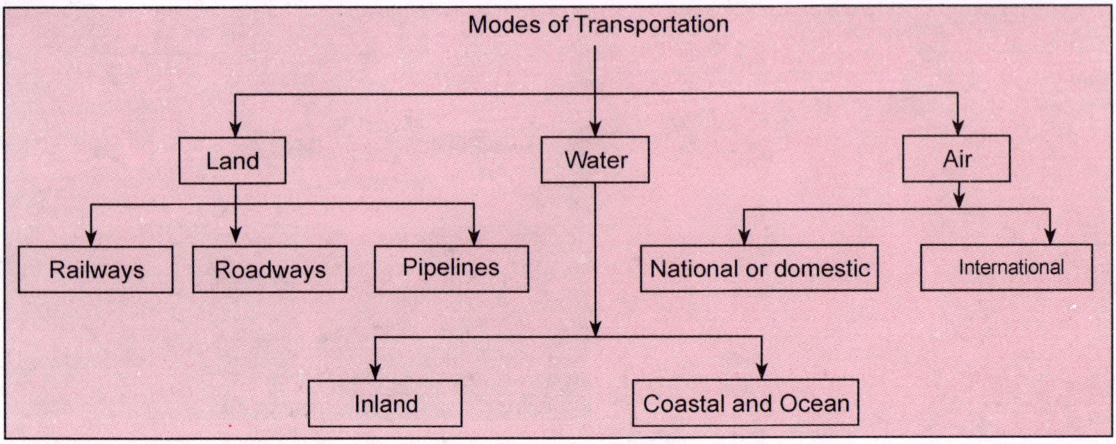

Development and Growth of Indian Railways

The first railway line in India was opened for public traffic on 16 April, 1853 between Mumbai and Thane over a distance of 34 km. This line was extended to Kalyan on 1 May, 1854 and to Khopoli on 12 May, 1856. The Khandala-Pune section was opened to traffic on 14 June, 1858.

Meanwhile, the construction of the railway lines was going on in eastern part of the country and the first section of the East Indian Railway, from Haora to Hugli, a distance of 37 km, was inaugurated on 15 August, 1854. The Haora-Hugli section was extended to Pundooah on 1 Sept, 1854 and to Raniganj Coal Mines on 3 Feb, 1855. The line from Kanpur to Pryagraj was opened in 1859 and the Haora-Khana-Rajmahal section was completed in 1860. Mughal Sarai (Deen Dayal Upadhyay) also appeared on the railway map of India in 1862. In 1860, the Kanpur-Etawah section was opened to traffic and between 1862 and 1866, all the gaps between Haora and Delhi were filled.

The southern part of the country did not lag behind and got its first 105 km long railway line from Royapuram to Arcot in 1856. This line was extended to Kadalundi (near Calicut) on the west coast in 1861. The Jolar Pettai-Bangalore Cantonment section was opened in 1864.

In 1870, the all-rail route between Kolkata and Mumbai started functioning and the main line from Mughal Sarai (Deen Dayal Upadhyay) to Lahore (now in Pakistan) was completed. In 1871, the Mumbai-Chennai route was also opened. Thus within a short span of 18 years from 1853 to 1871, most of the important cities of India were connected by rail. The total route kilometreage in 1870 was 7,680 km which rose to 39,834 km by the turn of 19th century. As on 15 August 1947, Indian Railways consisted of 65,217 km out of which 10,523 km went to Pakistan, leaving India with 54,694 km.

Phenomenal growth of Indian Railways has taken place in the post-Independence era as is clear from **table 26.1**.

Indian Railways with over 68.1 thousand route kms is the third largest network under a single management after U.S.A., Russia and China. But India tops world's leading countries with regard to passenger/kilometre carried. It is the largest public sector undertaking of the country comprising a vast network of 7,325 stations with a fleet of 12,729 locomotives, 67,956 passenger services vehicles, 6,372 other coaching vehicles and 2,93,077 vagons as on 31st March, 2021. The rolling stock comprised 39 steam, 5,898 diesel and 6,792 electric locomotives. The growth of Indian Railways since its inception in 1853 has been phenomenal. It has played a vital role in the economic, industrial and social development of the country.

In spite of progress made by India with respect to railway route length as shown in **Table 26.1**, Indian railway route length in growing at an abysmally low rate. Although, India has a large railway network, it is not sufficient considering the size and population of the country. In the past three decades the country's rail network has grown just 7.5% leaving many parts of the country without a rail line. Mountainous areas of Jammu & Kashmir, Himachal Pradesh and Uttarakhand, as well as hilly areas of most of North-Eastern States badly lack railway connectivity. Many people in remote areas have never boarded a train in their whole lifetime.

TABLE 26.1. Progress of Railways in India (Route length in thousand kilometres)

Year	Electrified	Non-electrified	Total
1950-51	0.4	53.2	53.6
1960-61	0.8	55.4	56.2
1970-71	3.7	56.1	59.8
1980-81	5.4	55.8	61.2
1990-91	10.0	52.4	62.4
2000-01	14.9	48.1	63.0
2010-11*	19.6	44.8	64.4
2015-16*	23.6	43.1	66.2
2016-17*	25.4	41.5	67.9
2017-18*	29.2	37.7	66.9
2018-19	34.3	33.1	67.4
2019-20	39.9	27.7	67.6
2020-21	44.8	23.3	68.1
2021-22	50.5	17.53	68.03

*Includes Metro Rail.

Source : Data computed from *Economic Survey, 2020-21*, p. A46 and Economic Survey 2021-22 and 2022-23 Statistical Appendix, p. 49.

Factors affecting Railways

The pattern of Indian railway network has been influenced by geographical, economic and political factors.

1. Geographical factors. The North Indian plain with its level land, high density of population and rich agriculture presents the most favourable conditions for the development of railways. However, the presence of large number of rivers makes it necessary to construct bridges which involves heavy expenditure. There are practically no railways in the flood plains of many rivers in Bihar and Assam. The plateau region of south India is not as much suitable for railways as the North Plain area. The Himalayan region in the north is almost entirely devoid of railways due to its rugged topography. Some railway terminals such as Hoshiarpur, Kotdwar, Dehra Dun, Kathgodam, etc. are found on the foothills. Some narrow gauge railway tracks are found in the Himalayan region. A railway link between Jammu and Kashmir valley has been executed at a very high cost. The sandy areas of Rajasthan are also not much favourable for railways. There was no railway line between Jodhpur and Jaisalmer till 1966. Similarly, forested areas of Madhya Pradesh and Odisha, deltaic swamps of West Bengal, marshy areas of Rann of Kachchh and hilly tract of Sahyadri are also unfavourable for the development of railways. Sahyadri can only be crossed through gaps like Thalghat, Bhorghat and Palghat to reach coastal rail heads like Mumbai, Vasco-de-Gama, Mangalore and Kochi. Obviously, the *railways tend to follow the path of least resistance*.

2. Economic factors. Railways develop more in the economically advanced areas where the need for railway network is felt more. Conversely, railways bring economic prosperity to the areas through which they pass. This is because of the economic linkages that we find the highest density of railways near big urban and industrial centres and in areas which are rich in mineral and agricultural resources.

3. Political and Administrative factors. The present railway system in India is the legacy of the British rule. The British administration planned the direction and pattern of the railway lines in such a way that they could exploit the valuable raw materials of India for the benefit of their industries and flood the Indian markets with the finished goods from Britain. Besides, the Britishers wanted to maintain their military supremacy, for which quick movement of troops, arms and ammunition was necessary and construction of railways became unavoidable. Thus, top priority was given to the big ports of Mumbai, Kolkata and Chennai. These ports were connected with their hinterlands by railway lines to facilitate imports and exports. It is from the ports that the railway network spread to the other parts of the country.

Distribution of Indian Railways

Fig. 26.1 shows the following distribution pattern of Indian railways :

1. The North Indian Plain. This region has a dense network of railways from Amritsar to Haora.

This is a plain area which is very much suitable for the construction of railways. This densely populated region has highly developed agriculture and industry. Large scale urbanisation has also helped in the development of the railways. The density of railway network is closely related to the agricultural and industrial development. There are a few focal points such as Delhi, Kanpur, Deen Dayal Upadhyay,

FIG. 26.1. India : Railways

Lucknow, Agra and Patna. However, Delhi is the main point from where railway lines radiate in all directions. For political, administrative and economic reasons, Delhi is connected with major ports like Mumbai, Kolkata, Haora and Chennai through superfast trains.

2. The Peninsular Plateau. The whole of Peninsular plateau has hilly and plateau terrain which hinders the development of railways. The population density is also moderate. For such reasons, excepting, Saurashtra and Tamil Nadu, a relatively open and more loose network has developed here. However, some trunk routes cross the peninsula and provide efficient rail service between Mumbai-Chennai, Chennai-Kochi, Chennai-Delhi, Mumbai-Kolkata, Chennai-Hyderabad and Mumbai-Thiruvananthapuram.

3. The Himalayan Region. Railways are conspicuous by their absence in the Hinmalayan region. The rugged terrain, hill and valley topography, backward economy and sparse population are the factors responsible for the sparse rail network in this region. There are only three narrow gauge railway lines in the Himalayan region. These are Kalka-Shimla, Pathankot-Kangra and Siliguri-Darjeeling. The Kalka-Shimla Railway, built in 1903 winds itself through picturesque country from Kalka to Shimla over a distance of 96.6 km. It has 103 tunnels, totalling 8 km in length, the longest tunnel is 1,144 metre. The railway track from Kalka to Shimla passes over 869 bridges. The Siliguri-Darjeeling Railway is 82 km long and was constructed in 1878. There is practically no railway line in the north-eastern states of Meghalaya, Tripura, Arunachal Pradesh, Mizoram, Manipur and Nagaland. These areas have rough terrain covered with thick forests. The population is sparse and the economy is in a backward state. Construction of railways under these conditions is a difficult and costly affair. However, plans are afoot to provide rail links to Meghalaya, Arunachal Pradesh and Tripura, although at a very heavy cost. A vital rail link to Kashmir valley has already made much headway.

4. The Coastal Plains. There is a distinct contrast in the rail network between eastern coastal plains and western coastal plains. The eastern coastal plain is quite wide and permits the construction of railways, as a result of which, there is a long trunk route along the east coast from Kolkata to Chennai. But such a route has been eluding the western coastal plain since long. This is due to the structure and relief of the area. The outcrops of the Western Ghats are very close to the coast, especially near Goa and make the construction of railway lines a difficult task. However, the completion of Konkan Railway Line

JAMMU-BARAMULA RAILWAY LINE

The Government of India has launched the most ambitious and at the same time most challenging programme to provide railway link connecting Baramula with Jammu-Tawi. When completed this 292 km long railway route will pass over 165 big and 650 small bridges. The bridge over the Chenab river is 1.3 km long and 359 metre tall which happens to be the tallest bridge in the world. Its 120.44 km stretch will pass through nearly 21 tunnels. The longest 11 km long tunnel is the PirPanjal tunnel which is the second largest tunnel of Asia after Wushaoling Tunnel (20 km) in Gansu, China. This tunnel has a maximum overburden (height of mountain strata above the tunnel) of 1,100 metres. This is 190 km away from Jammu and provides railway link between Qazigund in Kashmir valley and Jammu in the Jammu region. The work on the project started in 1984. Most of the work has already been done and the work on the remaining portion is likely to be completed in due course of time. Work on the prestigious railway line is often hampered by unwanted law and order position, terrorism, sabotages, land acquisition, difficult terrain, landslides, etc. This railway line is expected to usher into a new era of economic growth and national integration by bringing people of the Kashmir valley into the main stream of the country.

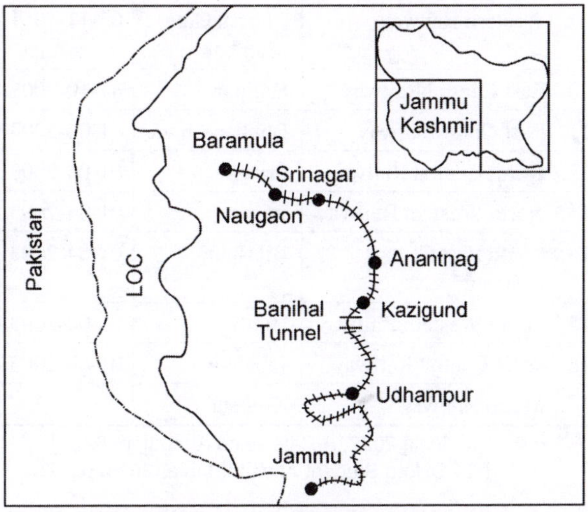

FIG. 26.2. Jammu-Baramula Railway

from Roha to Mangaluru is a dream come true. It passes through several tunnels and over numerous bridges. This line has one of the longest tunnels in the country 6.5 km long, about 23 km south of Ratnagiri. It has become the life line of the western coastal plain. The total saving in travel distances are Mangaluru-Mumbai (1,050 km), Mangaluru-Ahmedabad (1,218 km), Mangaluru-Delhi (707 km) and Kochi-Mumbai (437 km).

Railway Zones

The Indian railway network has been divided into 17 zones (**Table 26.2**) from the administration point of view.

TABLE 26.2. India : Railway Zones and their Headquarters

Zonal Railways	Headquarters	Date of establishment
1. Central Railway	Mumbai CST	05-11-1951
2. Eastern Railway	Kolkata	01-08-1955
3. Northern Railway	New Delhi	14-04-1952
4. North Eastern Railway	Gorakhpur	14-04-1952
5. Northeast Frontier Railway	Maligaon (Guwahati)	15-01-1958
6. Southern Railway	Chennai	14-04-1951
7. South Central Railway	Secunderabad	02-10-1966
8. South Eastern Railway	Kolkata	01-08-1955
9. Western Railway	Church Gate, Mumbai	05-11-1951
10. East Central Railway	Hajipur	01-10-2002
11. East Coast Railway	Bhubaneshwar	01-04-2003
12. North Central Railway	Pryagraj	01-04-2003
13. North Western Railway	Jaipur	01-04-2003
14. South East Central Railway	Bilaspur	01-04-2003
15. South Western Railway	Hubli	01-04-2003
16. West Central Railway	Jabalpur	01-04-2003
17. Metro Railway	Kolkata	

Source : (*i*) India 2022, A Reference Annual, p. 522.
(*ii*) Oxford Student Atlas for India, 2015, p. 47.

1. Railway Gauges. 'Gauge' is the name given to the distance between the inner faces of the pair of rails in the track. Indian railways comprise three gauges *viz*., broad gauge (1.675 metre), metre gauge (1.000 metre), and narrow gauge (0.762 metre and 0.610 metre).

Different gauges create serious hinderance in the smooth flow of traffic. Passengers have to change trains at the *break of gauge* station and are put to great inconvenience. Goods have to be transhipped which results in loss of time, increased cost of transportation, pilferage and damage to consignments. The Government of India have, therefore, adopted a policy of gauge conversion, mainly from metre gauge to broad gauge. *The unigauge system of railways assures larger capacity, higher speed and cheaper transportation*. The process of gauge conversion was initiated immediately after Independence but significant achievement has been recorded in recent years. Such a large scale gauge conversion is rightly called the *Operation Gauge Conversion or Operation Uni-Gauge*. At present, nearly 94.52% of total route length is broad gauge, 3.11% is metre gauge and only 2.37% is narrow gauge.

TABLE 26.3. Length of track gauge in India (2020)

Gauge	Track length (km)	Percentage of the total track
1.675 m broad gauge	63,950	94.52
1.000 m metre gauge	2,102	3.11
Two narrow gauges 0.762 m and 0.610 m	1,604	2.37
Total	67,656	100.00

Track Electrification. It has been estimated that use of electric locomotives increases the capacity by as much as 100 per cent. But the use of electric locomotives is possible only if the railway tracks are electrified. Track electrification is a major thrust area by virtue of which efficiency of the railways can be increased considerably. Track electrification was introduced in early 1920s and the first two sections from Victoria Terminus to Kurla and from Victoria Terminus to Bandra, totalling 16 route km. were electrified in 1925. Thus the Indian railways entered the push button era. In the first four decades from 1920-21 to 1960-61, the process of track electrification was rather slow and the length of electrified track stood at 388 km in 1950-51 and 748 km in 1960-61. After that the electrification of railway tracks picked up.

Under **Mission Electrification**, the Indian Railways has initiated a major electrification program for electrifying 100 per cent of its Board Gauge network. This would reduce the nation's dependence on imported diesel oil. **Table 26.1** gives an idea of race of electrification of the Indian railways. In 1950-51, only 0.7% of the total length of the railways was electrified which increased to 65.8% in 2020-21.

RAIL TRAFFIC

Rail traffic is broadly divided into two segments, *viz.*, (*i*) passenger traffic, and (*ii*) goods traffic.

Passenger Traffic

Railway journey particularly long journey is preferred because it is cheaper and more convenient. The number of passengers has risen from 1,284 million in 1950-51 to 8,086 million in 2019-20. This was less than 8439 million in 2018-19 perhaps due to Covid-19. Passenger kilometres represent the real indices of the volume of passengers handled. These are arrived at by multiplying the total number of passengers carried by the respective number of kilometres over which they are moved. The passenger kilometres increased from 66.5 billion in 1950-51 to 1,051 billion in 2019-20. This was also less than 1,157 billion in 2018-19 due to effect of Covid-19. For passenger service five types of trains are run by the Indian Railways, based on their speed and comfort levels— Ordinary Passenger trains, Express/Mail trains, superfast trains, Rajdhani Express, and Shatabdi and Jan Shatabdi trains. Sampark Kranti Express trains have been introduced to connect the national capital, Delhi, with the state capitals and other important places. Further, Indian Railways have introduced computer reservation system making it possible to get instant reservation between any two stations from any booking office.

The above developments have resulted in phenomenal growth in passenger earning by railways. The passenger earnings increased from ₹ 98.2 crore in 1950-51 to ₹ 50,669 crore in 2019-20. This also was less than ₹ 51,067 in 2018-19, perhaps due to effect of Covid-19.

There are plans to introduce high speed trains on following routes.

1. Delhi-Agra.
2. Delhi-Chandigarh
3. Delhi-Kanpur
4. Nagpur-Bilaspur
5. Mysuru-Bengaluru-Chennai
6. Mumbai-Goa
7. Mumbai-Ahmedabad
8. Chennai-Hyderabad
9. Nagpur-Secunderabad

Bullet Train

India's first bullet train got off the ground on September 14, 2017 when its foundation stone was laid at Sabarmati near Ahmedabad in Gujarat. The train will run for a distance of 509 km (156 km in Maharashtra, 351 km in Gujarat and 2 km in Daman and Diu). Most of the tract will be elevated to avoid the complications involved in the process of land acquisition. However, a small part of it will drive into a tunnel under sea after Thane towards Virar. The train will complete the 509 km long journey within 2 hours against 6 hours taken by the fastest trains at present. The project will create 20,000 construction jobs, 4,000 operations jobs and 20,000 indirect jobs. It will prove a big boost to urban, industrial and other

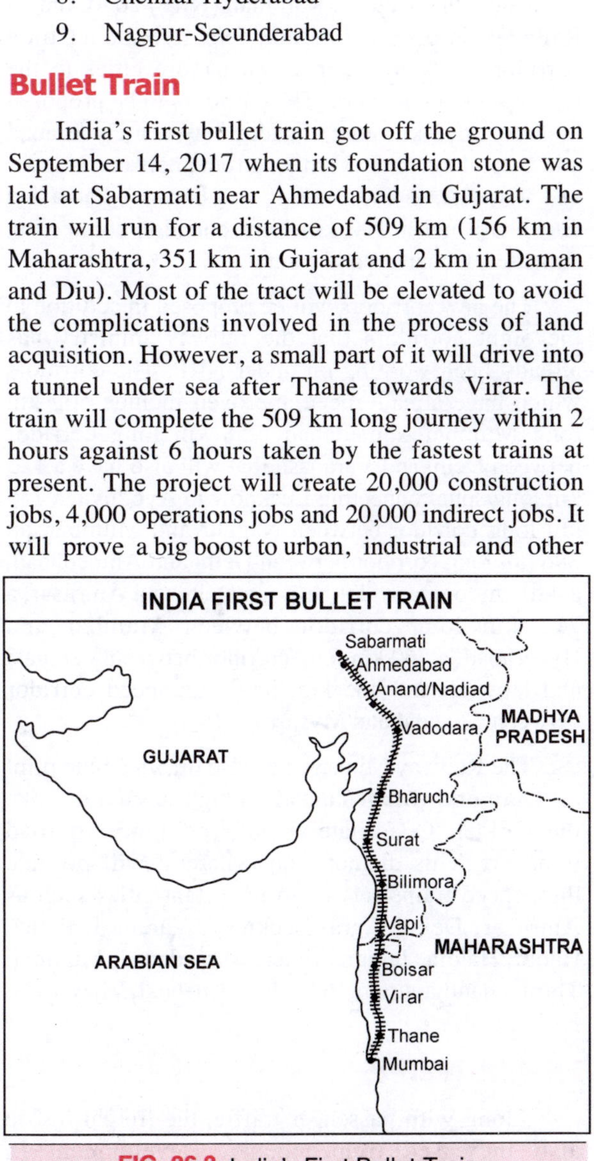

FIG. 26.3. India's First Bullet Train

infrastructure development along its route. Its cost would be nearly ₹ 1,08,000 crores. 81% of which will be met through soft loan from Japan at a nominal interest rate of 0.1% per annum.

The top speed, maximum operational speed and average speed of the train will be 350 kmph, 320 kmph and 250 kmph respectively. China's Shanghai Maglev with maximum operational speed of 430 kmph and average speed of 251 kmph is the fastest bullet train in the world.

Under the National Rail Plan (NRP) 2030, Indian Railways is planning to add four new bullet train corridors that will connect nine more cities to the high-speed rail network. The corridors to be proposed include a 618 km long corridor between Hyderabad and Bengaluru, an 855 km long route between Nagpur and Varanasi, an 850 km long way between Patna and Guwahati and a 190 km long corridor connecting Amritsar, Pathankot and Jammu.

The new corridors will be proposed in addition to the eight corridors that the railway ministry has already been working on under NRP. The corridors which have already been approved include 509 km long Mumbai-Ahmedabad, 958 km long corridor between Delhi and Varanasi, that will also have a 123 km long spur connecting Lucknow to Ayodhya, a 736 km long corridor between Nagpur and Mumbai, an 886 km long corridor between Delhi and Ahmedabad, a 480 km long corridor between Delhi and Amritsar, a 711 km long corridor between Mumbai and Hyderabad, a 760 km long corridor between Varanasi and Howrah and 435 km long high speed corridor between Channai and Mysuru.

The Railway Ministry has also drawn a blue print of **'Diamond Quadrilateral'** of high speed trains like the Golden Quadrilateral with reference to road transport. This diamond quadrilateral will provide high-speed trains service to important cities such as Amritsar, Delhi, Agra, Lucknow, Varanasi, Patna, Haora, Haldia, Jaipur, Ajmer, Ahmedabad, Mumbai, Thiruvananthapuram, Bengaluru, Chennai, Vijaywada, Hyderabad, etc.

Frieght Traffic

Along with passenger traffic, the freight traffic also increased tremendously. Development in industrial and agricultural sectors has generated high demand for rail transport. Major commodities transported by railways include coal, iron and steel, ores, petroleum products and such essential commodities as food grains, fertilizers, cement, sugar, salt, edible oils, etc. Consequently, freight traffic increased from 93.0 million tonnes in 1950-51 to 1,233.8 million tonnes in 2020-21. Transport effort measured in terms of net tonnes kilometres increased from 44.1 billion tonnes kilometres in 1950-51 to 720 billion tonnes kilometres in 2020-21. This was lower than 739 billion tonnes kilometer and this fall was caused by Covid-19. *Tonne kilometres are arrived at multiplying the total tonnage of goods carried by the number of kilometers over which they are moved.*

Table 26.4 shows that railways earnings from goods carried is always higher than the passenger earnings. Earning from freight traffic increased from ₹ 139.3 crore in 1950-51 to ₹ 1,11,472.3 crore in 2019-20. This was reduced from ₹ 1,22,580.3 crore in 2018-19 due to Covid-19.

Rail Tourism

Indian railways carry a large number of tourists to various tourist destinations cross the country. In addition, it has taken several steps to promote tourism such as operation of tourist trains/coach services on popular tourist circuits in different regions of the country, offering tour packages inclusive of rail travel, local transportation, accommodation, onboard services like catering, off board services such as conducted tours, etc. The important train/coach services offering package tours are : (*i*) Luxury Tourist trains, (*ii*) Semi-luxury trains, (*iii*) Buddhist special trains; (*iv*) Bharat Darshan trains, (*v*) Asha Circuit trains, (*vi*) Glass Top (Vistadome) Coaches; (*vii*) Steam Train, (*viii*) Value Added Tour Packages and (*ix*) Rail Tour Packages. Railways has been operating the luxury tourist train 'Palace on Wheels' since 1982. Four more trains namely Deccan Odyssey, 'Golden Chariot', 'Heritage Palace on Wheels' and 'Maharajas Express' on similar lines have been added over the years. All these trains are fully air conditioned and self sufficient to meet the requirements of the tourists for their mostly one week voyage but some shorter trips too. Each train is developed on a different theme based on the culture and heritage of India. The facilities provided in the trains include saloons of different types, multi cuisine

TABLE 26.4. Operations of Indian Railways

Year	1950-51	1960-61	1970-71	1980-81	1990-91	2000-01	2011-12	2017-18	2018-19	2019-20	2020-21	2021-22
Originating traffic (million tonnes)	93.0	156.2	196.5	220.0	341.4	504.2	975.2	1,162.60	1,225.3	1,212.2	1,233.8	1,418.8
Goods carried (billion tonnes km.)	44.1	87.7	127.4	158.5	242.7	315.5	668.6	692.9	738.5	707.7	719.8	871.8
Passengers originating (million)	1,284.0	1,594.0	2,431.0	3,613.0	3,858.0	4,833.0	8,224.4	8,286.0	8,439.0	8,086.0	1,250.0	3,519
Passengers kilometres (billion)	66.5	77.7	118.1	208.6	295.6	457.0	1,046.5	1,178.0	1,157.0	1,051.0	231.1	590.2
Earnings from goods carried (₹ crore)	139.3	280.5	600.7	1,550.9	8,247.0	23,045.4	67,761.4	1,13,523.5	1,22,580.3	1,11,472.3	1,15,738.4	1,39,287.3
Passengers earnings (₹ crore)	98.2	131.6	295.5	827.5	31,44.7	10,575.1	28,246.4	48,643.0	51,067.0	50,669.0	15,248	39,219

Source : (*i*) Economic Survey 2020-21, p. A-46; (*ii*) Economic Survey, 2021-22 and 2022-23 Statistical Appendix, p. 49.

restaurants, lounge, bar/mini bar, gym, spa, conference room, etc. Onboard facilities include individual temperature control, TV, WiFi, internet, telephone, piped music, CCTV, money changer laundary etc. The facilities may vary from train to train.

Dedicated Freight Corridor Project

A very ambitious plan of Dedicated Freight Corridors (DFC) was initiated in the year 2009 which aims at improving the freight carrying capacity of the Indian Railways, reducing the unit cost of transportation and improving service quality. It consists of two corridors *viz*. Eastern Dedicated Freight Corridor (EDFC) and Western Dedicated Freight Corridor (WDFC). The EDFC is 1,875 route kilometres (RKM) and extends from Dankuni near Kolkata to Ludhiana in Punjab while the WDFC is 1,506 route kilometre (RKM) and extends from Jawahar Lal Nehru port in Mumbai to Dadri in Uttar Pradesh. A special purpose vehicle, the Dedicated Freight Corridor Corporation of India Limited has been set-up to implement the project.

In addition to the above mentioned two Dedicated Freight Corridors (DFCs), the Railway Ministry intends construct three more DFCs for smooth, quick and safe movement of goods. The proposed corridors are mentioned below :

(*i*) 1868 km long East-west Sub Corridor which will connect Bhusawal, Nagpur, Kharagpur, Dakuni and Rajkharswan-Andal.

(*ii*) 1,115 km long East Coast Corridor connecting Kharagpur with Vijaywada.

(*iii*) 975 km North-South Sub-corridor connecting Vijaywada with Itarsi.

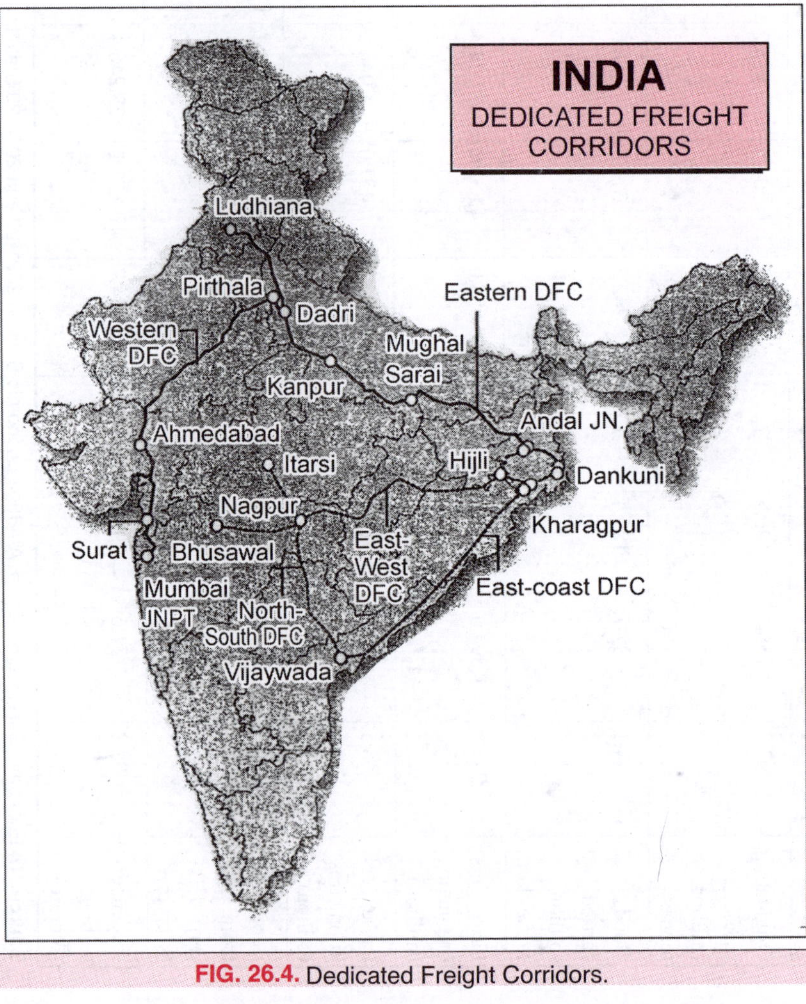

FIG. 26.4. Dedicated Freight Corridors.

Currently construction is underway for East and West DFCs. Work on the remaining three DFCs is likely to start soon. With this, the total length of DFCs will be 7,339 km.

Following are the salient features of this project.

- Only goods trains will be allowed to operate on these corridors. These trains will run at a speed of 100 kilometres per hour.
- These corridors are planned to run along the existing railway routes, but will provide services to railway junctions.
- With a view to keep minimum impact on social and environmental aspects, these corridors will have provision of bypasses for thickly populated big cities.
- Rail track on DFCs will be strengthened by providing heavier and stronger rails on concrete sleepers.
- Transport cost will be reduced by increasing work efficiency.
- Efforts will be made to deliver the goods at their destinations well in time and minimise losses due to delayed deliveries of goods.
- Give impetus to industrial growth by providing cheap and efficient transport.
- Bring freight carrying capacity at par with world's best. Currently, long distance carrying capacity of Indian Railways is only 5,000 tonnes as against 20,000 tonnes in China and 35,000 tonnes in Australia.

Road-railers

Road-railers are specially designed trailers which have wheels to operate both on road and railway track. A string of such trailers is used to form a rake, which is then pulled using a loco-engine to move it from the place of origin to their destination. After reaching the destination, these trailes can be separated from each other and driven off to respective areas of the city. This is a new system of transport which is likely to revolutionise the concept of transportation because it will add to speed, safety and economy of operations. It would be bimodal system in which trailers will move both on roads and railway tracks and would give the feeling that train is coming on our doorstep. Thus it would provide door-to-door transportation of goods because it does not require any handling or trans-shipment at rail terminals. It will be able to serve even in remote areas, where rail access is not possible.

The system is very much like roll-on, roll-off (RORO) in which trucks loaded with goods are transferred to open railway wagons and are driven to a particular railway station. From there, they are taken off from the railway open wagons and they travel to their respective destinations in the city.

Salient Features of Road-railers

- A road-railer (trailer) unit collects goods from designated consigner's place, a factory or a warehouse.
- All units loaded with cargo are then brought to a rail terminal to form a road-railer train.
- The rail-road wheel sets attached to the aft portion of the trailer are towered pneumalically, enabling them to move on the track.
- Delhi-Chennai is the first sector to be provided with road-railer transport service.
- Maximum speed limit is 100 kmph.
- One rake consists of nearly 50 trailers.

Benefits

- It provides door-to-door bimodal service involving road as well rail transport.
- It is a faster mode of transportation.
- It is cheaper mode of transport because it consumes less fuel.
- It helps in saving environmental pollution due to less consumption of fuel.
- It leads to decongestion of roads and highways.

Significance of the Indian Railways

1. Railways provide the cheapest and most convenient mode of passenger transport both for long distance and suburban traffic.
2. Railways have played a significant role in development and growth of industries. Growth of textile industry in Mumbai, jute industry in areas surrounding Kolkata, coal industry in Jharkhand, etc. is largely due to the development of railway network in these

areas. Railways help in supplying raw materials and other facilities to the factory sites and finished goods to the market.

3. Agriculture also owes its growth to railways to a great extent. Now farmers can sell their agricultural produce to distant places and even sell them in the world market at remunerative prices.
4. Railways are also helpful in removing isolation between cities and countryside and have played a significant role in disseminating innovations and new ideas.
5. Railways are particularly suited to long distance journey and provide a strong medium of national integration.
6. Railways play a vital role in mitigating the sufferings of the people in the event of natural calamities like droughts, floods, famines, earthquakes, epidemics, etc. This is done by carrying relief and rescue teams and essential items to the affected areas and save people from suffering and starvation.
7. Railways also help in facing man-made calamities like social, political, religious disturbances, insurgency, etc. It facilitates easy movement of police, troops, defence equipment, etc. The importance of railways to save the country's freedom and integrity from external aggression has been proved at several occasions.
8. Railways carry the British legacy and connect major ports to their hinterlands, thereby lending a helping hand to the overall prosperity of the coastal areas.
9. Introduction of superfast trains and container services in major cities of India have ensured quick movement of men and materials.
10. Railways are specially suited to long haulage of bulky materials like coal, petroleum and ores.

Swachh Rail

'Swachh Bharat' mission, launched on 2nd October, 2014, focuses on cleanliness and Indian Railways are paying due attention to this important aspect. Special emphasis is laid on bio-toilets in passenger coaches, mechanized cleaning, rag picking, dustbins, etc. The progress made in Swachh Rail, Swachh Bharat is given in **Table 26.5**.

In order to provide better amenities IR has embarked on providing Wi-Fi internet services at all stations (excluding halt stations). As on 5th December 2021, total 6,087 Railway Stations have been equipped with Wi-Fi facility. In addition to the above, projects connecting difficult terrain such as Rishikesh-Karnaprayag line as also the rail network to connect all capitals of north east states are ongoing. Further, a number of infrastructure development initiatives are envisaged in the National Rail Plan (NRP) being prepared by Indian Railways.

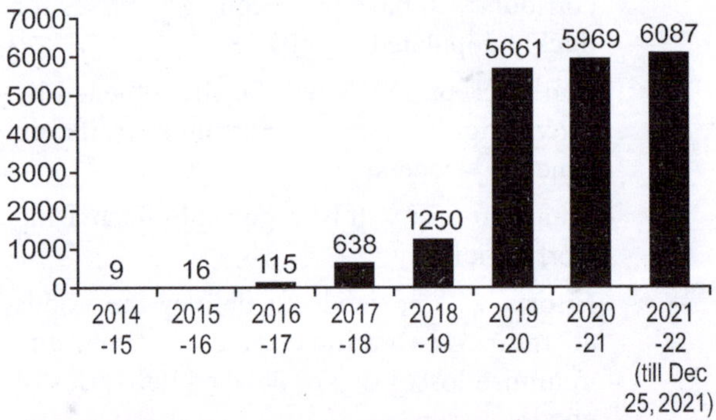

FIG. 26.5. Number of stations equipped with Wi-Fi

Source: Ministry of Railways in Economic Survey 2021-22, p. 293.

Problems of Indian Railways

Although Indian Railways have progressed a lot, both quantitatively and qualitatively, during the last few years, this system is still plagued by a number of problems which require immediate attention. A lot has been done, but a lot more is yet to be done. Some of the major problems faced by the Indian Railways are briefly discussed as under :

1. Safety. Indian Railways have been in the news albeit for wrong reasons. With the rapid increase in passenger and goods traffic, the frequency of train accidents had been increasing very fast. This has raised serious doubts in the public mind about safety of rail travel and the general health of the railway network. The credibility of an organisation with a long and proud history of nation building has been seriously eroded. Approximately 3000 passengers have lost their lives in ten years from 2003 to 2013.

TABLE 26.5. Progress of Swachh Rail, Swachh Bharat

Activity	Progress as on		
	March 2015	March 2020	October 2020
1. Bio-toilets in passenger coaches (numbers)	19,746	2,45,400	All
2. Mechanized cleaning contracts (at stations)	584	953	950
3. Plastic bottle crushing machines (number)	Nil	315	503
4. Plastic bottle crushing machines (at stations)	Nil	229	370
5. CCTV monitoring (at stations)	250	585	630
6. On board housekeeping service (pairs of trains)	525	1,060	1,100
7. Environment Management System certification to Railway stations	Nil	250	590
8. Funds allocated for station sanitation (₹ crores)	294	778	424

Source : Ministry of Railways in Economic Survey 2020-21, p. 296.

There are several factors which are responsible for increasing number of railway accidents; some outstanding being overaged tracks, wagons, coaches, bridges and signaling system. According to the Khanna Railways Safety Review Committee Report, nearly 25 per cent of the total railway track in India is overaged and is due for replacement. The tracks suffer from fatigue and wear and tear in due course of time, and their replacement should be carried on side by side. In several derailments poor condition of tracks had been found to be responsible. The condition of tracks becomes more significant when one looks at the other assets of the Railways. The Khanna Railways Safety Committee had reported that Indian Railways have 34,000 overaged wagons, 1,322 overaged coaches, and 1,560 stations with overaged signalling. Moreover, 262 bridges are listed "distressed." The white paper released by the Railway in April, 2003 acknowledges that over 51,000 bridges are of 19th century vintage. Out of a total of 1,27,154 bridges in India, 56,178 are more than 80 years old. Thus 44.17 per cent of the bridges have outlived their life. According to the review conducted by the Comptroller and Auditor General of India on various aspects of bridge management between 1997-98 to 2001-02, these old bridges include 339 important, 4,210 major and 51,629 minor bridges built before 1920. The Khanna Committee had further reported that 76 per cent of all rail accidents are due to derailments, 7 per cent due to collisions, 16 per cent take place at level crossings and 1 per cent are due to fires. Resource crunch is said to be the main cause of all these happenings in the Indian Railways.

Worried about the increasing rate of accidents and loss of life and property resulting from these accidents, the Indian Railways has come out with Anti-collision Devices (ACDs) or '*Raksha Kavach*' to get rid of such happenings. World's first Networked Anti-collision Devices (ACDs) *Raksha Kavach,* invented by Konkan Railway is a microprocessor system comprising of a central processing unit, a Global Position System (GPS) Receiver and a digital radio modem for communication between ACDs. When fitted to a guard van it becomes a Guard ACD. When fitted at stations it becomes a Station ACD and when fitted at level crossing gates, it becomes a Gate ACD. They all network among themselves, exchange information and take decision to prevent collision type of dangerous situations well in time automatically, without mannual inputs, forming *Raksha Kavach* against collisions.

Extensive field tests of *Raksha Kavach* have been conducted successfully on Indian Railways. Survey for implementation of ACDs has already been completed for 3,300 route km on Indian Railways and 760 km on Konkan Railway. About 1,770 ACDs on 1,736 km route length of North East Frontier Railways covering 183 stations have been installed at a cost of ₹ 50 crore.

Modern signalling like panel inter-locking, route relay inter-locking, centralised traffic control, automatic signalling and multi-aspect colour light signalling, are being progressively introduced.

As a result of the above mentioned precautionary measures all sorts of accidents, have either reduced or stayed constant over the years **(Table 26.6)**.

TABLE 26.6. Rate of occurrence of rail accidents

Type of Accidents	2016-17	2017-18	2018-19	2019-20
Collisions	5	3	0	5
Derailments	78	54	46	40
Manned Level Crossing Accidents	0	3	3	1
Unmanned Level Crossing Accidents	20	10	3	0
Fire in Trains	1	3	6	8
Miscellaneous	0	0	1	1
Total	**104**	**73**	**59**	**55**

Source : Ministry of Railways in Economic Survey, 2020-21, p. 296.

2. Cost and Revenue Problems. As is the case with most of the government organisations, Indian Railways face chronic financial crisis. The annual rate of increase in cost has overtaken that of revenues during the last few years. A study of Railways finances from 1998 to 2004 revealed that the revenues increased at an average annual rate of 8.7 per cent against the 9.65 per cent average annual growth in costs. While the Indian Railways' input costs increased by 10.6 per cent per annum between 2004-05 and 2010-11, passenger fairs remained unchanged or were even reduced in lower classes thereby constraining internal resource generation, essential for replacement/renewal of assets, operation and maintenance activities and critical safety and amenity works. In certain years, the revenue growth rate does exceed that of cost. But this position is achieved by providing inadequacy for replacements and severely controlling the costs. Such a situation has long term implications as it affects the internal generation of resources. In 2013-14, half of the 17 railway zones reported an operating ratio (rupee spent against every rupee earned) of more than 100 per cent which means that the railways are spending either equal or more than the money earned. Following are the main causes of costs and revenue problems.

(i) Low level of employee productivity. Indian Railways face a serious problem of low level of employee productivity. Transport output in terms of passengers and freight tonne kilometres per employee on Indian Railways is only 400 as compared to 500 for Chinese and 570 for French Railways. An estimated 30 per cent surplus workforce and operation of a number of lines with low traffic and assets not essential for the Railways are contributory factors.

(ii) Staff Wages. With the implementation of the recommendations of the Seventh Pay Commission, staff wages have increased tremendously and have put heavy strain on the financial resources of the Railways. With life expectancy going up and wage as well as pension escalations taking place periodically, the position will only worsen leaving little scope for development plans.

(iii) Increase in lease charges. Paucity of funds forces the Indian Railways to resort to market borrowings which results in increased lease charges. Market borrowings started in 1986 and the trend is increasing. At present payout of lease charges constitute about 8.5 per cent of the revenue.

3. Slowdown in Revenue Growth. With saturation of trunk routes and low quality of services and reliability, the revenue growth has registered a slowdown. The railways are increasingly becoming a transporter of bulk commodities for public sector (coal, iron ore, foodgrains, etc.) and are consistently loosing to roadways. Most of the national highways run parallel to railways and are consistently snatching revenues from the railways.

The share of Indian Railways in total freight movement of India has been declining over the years primarily due to non-competitive tariff structure. While the passenger fare had remained more or less flat, the freight fare has increased sharply over the years. To make rail transportation attractive and arrest the declining trend of rail share, various initiatives were taken in 2016-17 which include tariff rationalization, classification of new commodities, new policy guideline for station to station rates, expansion of freight basket through containerization, withdrawal of dual freight policy for export of iron ore, rationalization of coal tariff, policy guidelines for

Merry Go Round System, discount for loading of bagged consignment in open and flat wagons, new delivery models like Roll-on-Roll-off services, re-introduction of short load concession and reduction in minimum distance for charge, digital payment for freight business, Long Term Tariff Contract Policy (which provides tariff stability and attractive rebate in freight to customer), and Liberalised Automatic Freight Rebate scheme for tariff loaded in empty flow direction etc.

4. Social Burden. Indian Railways have to play a dual role of revenue earning as well as meeting the social obligations. The Expert Group, constituted in December 1998 to study the railway sector, termed it as the 'split personality'. On one hand, the Railways are seen as a commercial organisation and on the other hand, it is treated as a social organisation which must fulfil its social obligations. The two functions are diametrically opposite and difficult to reconcile. There are several social obligations on the railways which are always running below cost. Suburban passenger services, concessionary travel to certain section of travellers, concessional freight movement of certain commodities, particularly to remote and inaccessible areas like the North-east region, providing rail services to backward regions are some of the outstanding social obligations on the Indian Railways.

5. Other Problems. A large number of miscellaneous problems include late running of trains, lack of passenger facilities including lack of cleanliness at the railway stations, lack of security arrangement on the railways resulting in thefts and dacoities, etc. Political pressure and interference is a very big problem which the Indian Railways are facing with increasing impact. Several projects which are not economically viable have been initiated for political considerations.

METRO RAIL

Mero rail offers fast, cheap and comfortable journey in metropolitan cities of India. It helps in reducing pressure on the existing road transport and provides clean and eco-friendly transport at the local level. With the introduction of metro rail in the big cities, traffic jams on road crossings have reduced considerably. It is a part of rapid mass transport and is of recent origin in India. The first rapid transit system in India was the Kolkata Metro, which started operations in 1984. The Delhi Metro was India's first modern metro and third rapid transit system in India, after the Kolkata Metro and Chennai Mass Rapid Transit System. The Delhi Metro Rail started its operations in 2002 and is now providing transport facilities to most parts of the capital city. It also provides metro rail lines to most of the satellite towns like Gurugram, NOIDA, Faridabad, Bahadurgarh, Sonipat, Ghaziabad etc. Rapid Metro Rail Gurugram is India's first privately owned and operated metro rail system. It started its operations in November, 2013. After the grand success of Delhi Metro Rail, other cities are planning to have metro rails and in many cities this rail system is already in operation. The Government has planned to provide metro rail facilities in all cities of India having a population over two million.

As metro rail projects are highly capital intensive, it is difficult to fund metro rail projects from Government exchequer only. In this context, in order to create an ecosystem for proliferation of metro rail system in the country, the Government of India has notified Metro Rail Policy, 2017. The policy imbibes on the learnings from international examples and bridges the much needed gap for enhancing the feasibility of metro rail projects from economic, social and environmental perspective.

According to a report published in 2021, a total of 2,636 million people travel annually in metro system across India's thirteen major cities, placing the country as one of the busiest urban rapid transit hub in the world in terms of ridership. The combined length of 741.7 km of metro systems in India makes the fifth longest in the world with operation.

ROADWAYS

Roads have been existing in India for the last 5000 years. In early stages of Indian history, Ashoka and Chandragupta made efforts to construct roads. But the real progress was made during the Mughal period. A number of roads were laid during the Sultanate and Mughal periods. Most of the present trunk routes follow the Mughal routes. These routes were essential for strengthening and consolidating the empire. One such road was constructed by Sher Shah Suri which connected Peshawar to Kolkata. It was named as Grand Trunk (G.T.) Road and joined Amritsar with Kolkata after partition of India in 1947. Presently, it is known as 'Sher Shah Suri Marg'.

Importance of Roads

1. Roads play a very important role in the transportation of goods and passengers for short and medium distances.
2. It is comparatively easy and cheap to construct and maintain roads.
3. Road transport system establishes easy contact between farms, fields, factories and markets and provides door to door service.
4. Roads can negotiate higher gradients and sharp turns which railways cannot do. As such, roads can be constructed in hilly areas also.
5. Roads act as great feeders to railways. Without good and sufficient roads, railways cannot collect sufficient produce to make their operations possible.
6. Road transport is more flexible than the railway transport. Buses and trucks may be stopped anywhere and at any time on the road for loading and unloading passengers and goods whereas trains stop only at particular stations.
7. Perishable commodities like vegetables, fruits and milk are transported more easily and quickly by roads than by railways.

Due to above-mentioned advantages, the road transport has become very popular and its share in traffic is constantly increasing.

Growth and Development

Road transport in modern sense *i.e.* vehicles driven by internal combusion engines using petrol or diesel as fuel was practically negligible in India before World War II. Following plans have been drawn to develop roadways in India.

1. Nagpur Plan. First serious attempt to develop roadways was made in 1943 when ***Nagpur Plan*** was drawn. This plan envisaged increasing of the kilometreage of major roads to 1,96,800 km and of other roads to 3,32,800 km by 1953. The highlight of the plan was that no village in a developed agricultural region should be more than 8 km from a major road or 3 km away from any other road while the average distance of villages from a major road should be less than 3.2 km. In a non-agricultural region, these distances were fixed at 32, 8 and 10 km respectively. This plan could not be implemented immediately because the country was ruled by a number of princely states outside British India. Concerted efforts to achieve the objectives of this plan were made only after the reorganisation of the states. The targets of this plan were more or less achieved by 1961.

2. Twenty Year Plan. After achieving the objectives of the Nagpur Plan, another plan known as ***Twenty Year Road Plan*** was drawn in 1961. It aimed at increasing the road length from 6.56 lakh km to 10.60 lakh km and the density to 32 km of road per 100 sq km by 1981. The other objectives of the Twenty Year Road Plan were (*i*) to bring every village in a developed agricultural area within 6.4 km of a metalled road and 2.4 km of any other road, (*ii*) to bring every village in a semi-developed area within 12.8 km of a metalled road and (*iii*) to bring every village in an undeveloped and uncultivated area within 19.2 km of a metalled road and 8 km of any other road.

3. The Rural Development Plan includes construction of rural roads under Minimum Needs Programme (MNP), Rural Landless Employment Guarantee Programme (RLEGP), Jawahar Rojgar Yojana (JRY) and Command Area Development (CAD) programmes to connect all villages having a population of 1,500 or more with all weather roads and those having less than 1,500 population with link roads.

4. Build Operate Transfer (BOT) is a scheme under which private operators are invited to construct roads and bridges. They are allowed to collect toll tax from the vehicles using these roads and bridges for a specific period of time after which these assets are transferred to the government. The National Highways Act has been amended to facilitate private investment in road construction under BOT scheme.

5. Central Road Fund (CRF) is being raised for the betterment of roads by imposing additional excise/customs duty at the rate of ₹ 1.50 per litre on

TABLE 26.7. Share of road transport in total transport GVA in percentage

Mode of transport	Financial Year							
	2012	2013	2014	2015	2016	2017	2018	2019
1. Road	65.9	65.3	66.8	64.5	64.9	64.4	64.9	67.0
2. Rail	15.3	15.7	15.6	16.1	16.3	15.8	15.9	15.7
3. Water and Air	18.8	19.0	17.6	19.4	18.8	19.8	19.2	17.3
Total	**100.0**	**100.0**	**100.0**	**100.0**	**100.0**	**100.0**	**100.0**	**100.0**

***Source*:** Data computed from figure 25, p. 290 in Economic Survey, 2020-21.

petrol with effect from 2 June 1998 and on High Speed Diesel (HSD) with effect from February 28, 1999. The annual accrual through this source was to be about ₹ 5,500 crore. A part of this (₹ 0.4 per litre against sale of high speed diesel oil and ₹ 0.86 per litre against sale of petrol) goes to fund the NHDP. (National Highway Development Projects).

The Central Road Fund Act 2000 was enacted in December, 2000 with the primary objective of providing regular and adequate flow of funds for development of the road sector. This is a non-lapsable fund. The Act empowers the Centre to administer, manage and allocate the accrued amount to the following :

(*i*) Development of rural roads. About 43 per cent of the levy on diesel is to be spent on improving road connectivity,

(*ii*) Development and maintenance of National Highways,

(*iii*) Construction of road under/over bridges and safety works at unmanned railway crossings, and

(*iv*) Development and maintenance of State roads.

India runs on the road, be it the passenger or goods movement, road transport is the dominant mode of transportation in the country. **Table 26.7** highlights the significance of road transport *viz-a-viz* other modes of transport. The share of road transport in Gross Value Added (GVA) is about two-thirds of the total transport in the country. Although share of road transport in the total transport of the country has been found to be variable, it has risen from 64.4 per cent in 2017 to 67.0 per cent in 2019. In contrast, share of rail transport has fallen from 16.3 per cent in 2016 to 15.7 per cent in 2019.

Classification of Roads

The main significance of the Nagpur Plan lies in the fact that it classified roads into four categories on the functional basis. These are : (*i*) National Highways (*ii*) State Highways (*iii*) District Roads and (*iv*) Village Roads. To these may be added Border Roads, urban Roads and Project Roads. A brief description of each category is given as under :

1. National Highways

The main roads which are constructed and maintained by the Central Public Works Department (CPWD) are known as the National Highways. These roads are meant for inter-state and strategic defence movements and connect the state capitals, big cities, important ports, big railway junctions and link up with border roads. The length of National Highways increased from 19.8 thousand km in 1950-51 to more than 114 thousand km in 2017. National Highways form the lifeline of road transport and constitute the framework of road system in India. Although the percentage share of the National Highways to the total road length has decreased considerably from 4.95 per cent in 1951 to only 1.94 per cent in 2017, they carry nearly 40 per cent of the road traffic of India.

The National Highways have been classified on the basis of carriage way width of the highway. Generally, a lane has a width of 3.75 m in case of single lane and 3.5 m per lane in case of multi-lane.

The Government has embarked upon a massive National Highways Development Project (NHDP) in the country. The NHDP is the largest highway project ever undertaken in the country. The NHDP is being implemented mainly by NHAI in phases I to VII.

TABLE 26.8. Road Length in India, Category wise as on 31st March, 2017

Category of Road	Length in km	Share of total roads (per cent)
1. National Highways	1,14,158	1.94
2. State Highways	1,75,036	2.97
3. District Roads	5,86,181	9.94
4. Rural Roads	41,66,916	70.65
5. Urban Roads	5,26,483	8.93
6. Project Roads	3,28,897	5.57
Total	58,97,671	100.00

Source : Economic Survey 2018-19, p. 211.

- **NHDP Phase I & II:** Envisage 4/6 laning of about 14,000 km of National Highways, at an estimated cost of about ₹ 65,000 crore at 2004 prices. These two phases comprise Golden Quadrilateral (GQ), North-South and East-West corridors. (NSEW), Port Connectivity and Other Projects.
- **NHDP Phase III:** NHDP Phase-III involves 4-laning of 12,109 km of NHs having high density corridor connecting State capitals, important tourist places, economically important areas, etc. on PPP basis at an estimated cost of ₹ 80,628 crore.
- **NHDP Phase IV:** It involves upgradation/ strengthening of 20,000 km of single/intermediate/ two lane National Highways to two lanes with paved shoulders on BOT (Toll) and BOT (Annuity) basis.
- **NHDP Phase V:** It involves six laning of 6,500 km of NHs comprising 5,700 km of GQ and balance 800 km of other sections of NHs at a cost of ₹ 41,210 crore.
- **NHDP Phase VI:** It involves construction of 1,000 km of expressways with hill access control on new alignments at a cost of ₹ 16,680 crore.
- **NHDP Phase VII:** It involves construction of 700 km of ring roads of major towns and bypasses and construction of other stand-alone structures such as flyovers, elevated roads, tunnels, undespasses, grade separated interchanges etc. on National Highways at a cost of ₹ 16,680 crore.

Special Accelerated Road Development Programme for North Eastern region (SARDP-NE). It envisages improvement of road connectivity to the State Capitals with District Headquarters in the North Eastern region. The proposed programme includes improvement of 10,141 km of roads comprising National Highways (4,798 km) and state roads (5,343 km), to be implemented under Phase-A, Phase 'B' and Arunachal Pradesh Package for 'Road & Highways. Phase-A now consists improvement of 2,041 km of National Highways and 2,058 km of State Roads. Phase-B has now been modified to cover 2 laning of 1,285 km of NHs and 2 laning/ improvement of 2,438 km of State roads for preparation of Detailed Project Reports (DPRs). Arunachal Pradesh Package of roads and highways comprises 1,472 km of National Highways and 847 km of State Roads.

Development of Roads in Left Wing Extremism affected areas

The Government of India, approved the Road Requirement Plan (RRP) on 26 February, 2009 for upgrading 5,565 km long roads out of which 1,202 km are the national highways and 4,363 km are state roads. The upgradation is with respect to two-lane provision in 34 core districts affected by left-wing extremism (LWE). These districts are in the states of Andhra Pradesh, Bihar, Chhattisgarh, Jharkhand, Madhya Pradesh, Maharashtra, Odisha and Uttar Pradesh. The project is expected to cost ₹ 7,300 crores.

1. Golden Quadrilateral Super Highway. National Highways Development Project (NHDP) has taken up a massive programme of road building in the country. Launched on January 2, 1999, this is perhaps one of the largest programmes of road development ever taken up in the country. The project is being implemented by National Highways Authority of India (NHAI). NHDP has following two components.

(i) Phase I—Golden Quaderilateral. Comprising National Highways connecting Delhi-Mumbai-Chennai-Kolkata-Delhi by six-lane super highways. This component has a total length of 5,846 km. The four sides of the quadrilateral have varying length. The side of quadrilateral between Delhi and Mumbai is 1,419 km long, Mumbai to Chennai is 1,290 km long, Chennai to Kolkata is thelongest side which is

TABLE 26.9. Length of Various Sections of the Golden Quadrilateral

Name	Length of sides of the Golden Quadrilateral (Length in km)
Delhi-Mumbai	1,419
Mumbai-Chennai	1,290
Chennai-Kolkata	1,684
Kolkata-Delhi	1,453
Total	**5,846**
	Length of Corridors
North-South corridor Connecting Srinagar with Kanniyakumari	4,000
East-West corridor connecting Silchar with Porbandar	3,300
Total	**7,300**

1,684 km long. The side between Kolkata and Delhi is 1,453 km long.

(ii) Phase II—North-South Corridor. Comprising the National Highways connecting Srinagar to Kanniyakumari including Kochi-Salem Spur and *East-West corridor* comprising the National Highways connecting Silchar in Assam and Porbandar in Gujarat. The project has a total length of about 7,300 km, out of which the North-South corridor is 4,000 km and East-West corridor is 3,300 km long.Main components of the Golden Quadrilateral are shown in **Fig. 26.6**.

Year of Construction

The Ministry of Road Transport and Highways (MORTH) declared 2018-19 as the **'Year of Construction'** and has been making strenuous efforts to expand and upgrade the network of National Highways in the country. The pace at which the roads have been constructed has grown significantly from 12 km per day in financial year 2015 to 30 kms per day in financial year 2019 before it moderated to 28 kms per day in the financial year 2020. The decline in the construction of road at 22 km per day in the financial year 2021 is mostly due to COVID-19 shock (**Fig. 26.7**). With the unlocking of the economy, construction of roads is expected to return back to the high pace attained before COVID-19.

Major outcomes in Road sector in the recent past were construction of Eastern and Western Peripheral Expressways, Delhi-Meerut Expressway and Dhola-Sadiya Bridge. The Eastern and Western Expressways are aimed at decongesting the National Capital of Delhi by providing alternate routes to the traffic not destined for Delhi. Delhi-Meerut Expressway is the first national highway in the country with 14 lanes, dedicated cycle tracks and pedestrian paths with several eco-friendly features. Dhola-Sadiya Bridge connects Assam to Arunachal Pradesh and ensures 24 × 7 connectivity.

A WORLD RECORD

A 25.56 km long stretch of National Highway-6 between Amravati and Akola was built 78 hours in June 2022. This is a world record and is included in the Guinness Book of World Records. The earlier world record of 25.27 km was achieved in Doha, Qatar.

In Uttar Pradesh, an ambitious plan has been drawn for construction of a large number of Expressways (**Table 26.10**). Construction of some of them has already been completed.

TABLE 26.10. Expressways planned for construction in Uttar Pradesh

Expressway	Length in km.
1. Purvanchal Expressway	341
2. Bundelkhand Expressway	296
3. Delhi-Meerut Expressway	96
4. Eastern Peripheral Expressway	135
5. Gorakhpur Link Expressway	91.35
6. Ganga Expressway	594

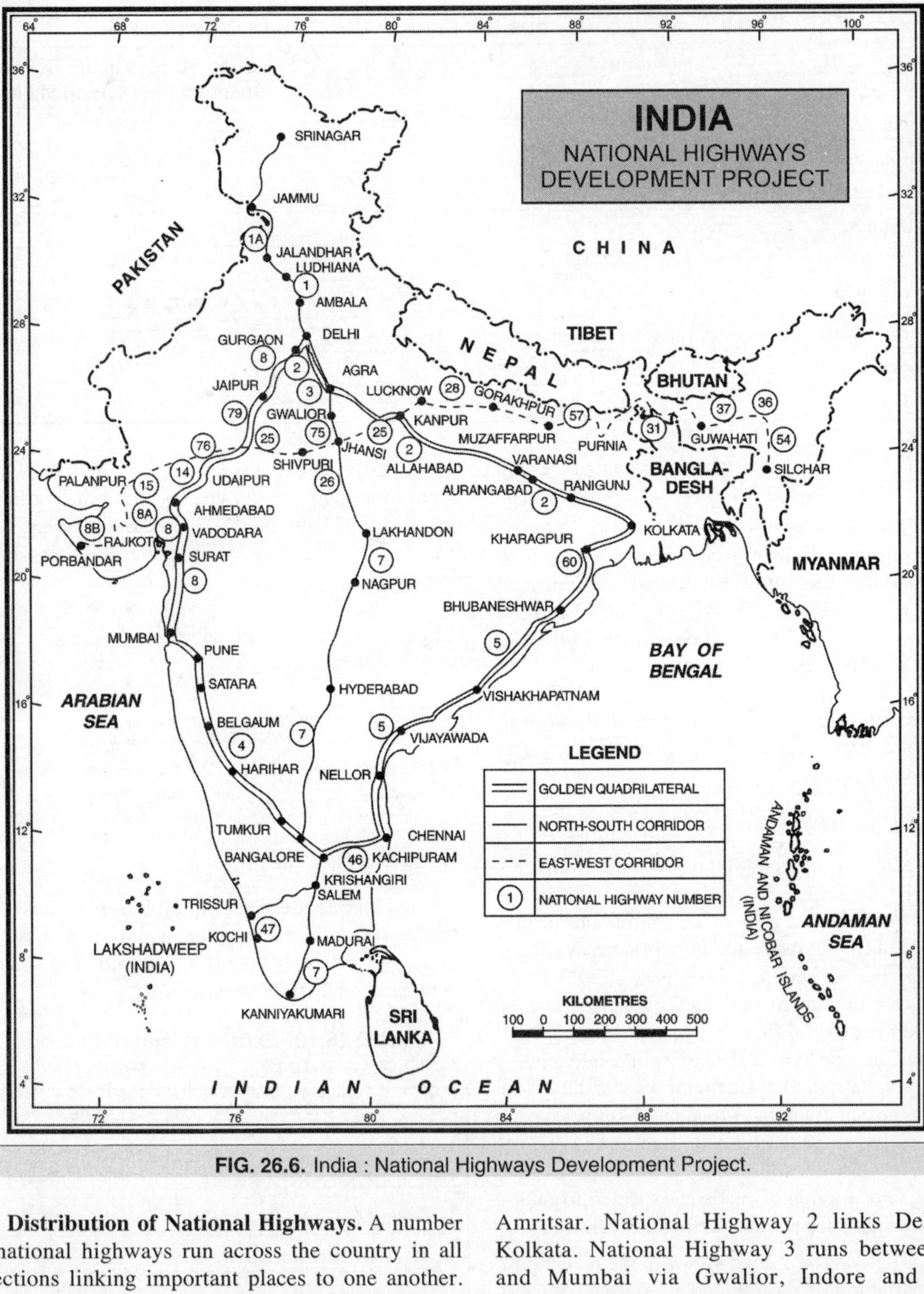

FIG. 26.6. India : National Highways Development Project.

Distribution of National Highways. A number of national highways run across the country in all directions linking important places to one another. The historically important Sher Shah Suri Marg is known as National Highway 1. It links Delhi and Amritsar. National Highway 2 links Delhi and Kolkata. National Highway 3 runs between Agra and Mumbai via Gwalior, Indore and Nasik. National Highway 7 is the longest one which links Varanasi with Kanniyakumari via Jabalpur, Nagpur,

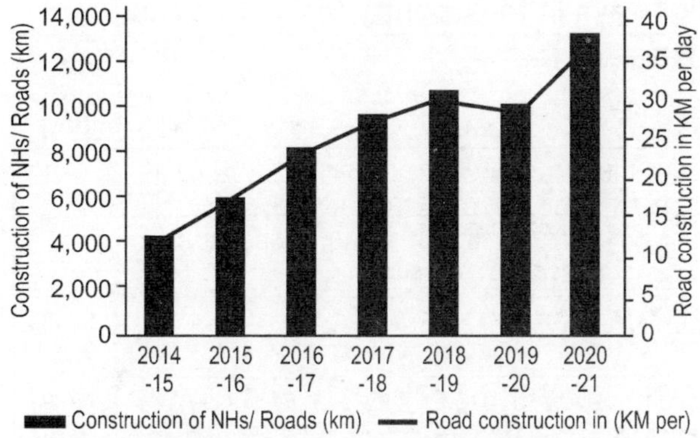

FIG. 26.7. Pace of road construction in India.

Source: Survey calculations based on data from M/o RTH. For year 2021-22.

Hyderabad, Bengaluru and Madurai. It traverses a distance of 2,325 km. National Highway 5 and 17 run along the eastern and western coasts respectively. National Highway 15 represents the border road in Rajasthan desert and runs through Kandla, Jaisalmer, Bikaner and joins the border road in the Punjab. **Fig. 26.8** shows the important national highways.

Table 26.11 depicts the state-wise distribution of National Highways. This table shows that Maharashtra has the longest length of National Highways covering a distance of 17,930 km. This is followed by Uttar Pradesh (11,830 km), Rajasthan (10,350 km), Madhya Pradesh (8,940 km) Karnataka (7,412 km) and Andhra Pradesh (7,340 km).

National Highway Corridors. Although India has over one lakh kilometres of national highways, yet there is no scientific road network pattern and drivers cannot take a straight road to reach from one place to the other. To address this, NHAI has prepared a grid of 27 corridors (12 vertical, north-south and 15 horizontal, east-west) at a distance of every 250 km criss-crossing the country. These stretches will be four lanes and will provide more road space for seamless transport. The total length of these corridors, including one such as Kanniyakumari to Srinagar, Porbandar to Kolkata, Surat to Paradip Port, Rameshwaram to Dehradun, and Mangaluru Port to Chennai Port, is about 36,600 km. Out of this about 30,100 km are already National Highways. But only 18,800 km of them are four lanes. The rest are either single or two lanes and have missing NH link of about 6,500 km, which are either state highways or major district roads. These stretches need to be converted into National Highways and widened to four lanes for smooth travel. This will cost about ₹ 25,000 crore.

The preparation of the grids will help in re-designating the NHs for easy identification. For example, all even number NH grids can be identified for roads connecting east to west and odd numbers for corridors joining northern end to south. This will also provide an integrated transport network. The existing NH system depicts unbalance network and even creates problem in numbering NHs systematically from the road users' perspective.

The proposed grid will not only improve connectivity in each region and state capital, but will also provide highway link to 12 major ports, 45 out of 53 million plus cities and 26 state capitals besides providing connectivity to major tourist and religious places.

Bharatmala Paryojana. This paryojana was established on July 31, 2015 with a view to develop the road connectivity to border areas, development of coastal roads including road connectivity for non-major ports, improvement in the efficiency of national corridors, development of economic corridors, inter and feeder routes along with integration with Sgarmala, etc. under Bharatmala Pariyojana. It envisages development of about 26,000 km length of economic corridors, which along with Golden Quadrilateral as well as North-South and East-West corridors are expected to carry majority of freight traffic on the roads. The total investment for 83,677 km committed new highways is estimated at ₹ 5.35 lakh crore, making it the single largest outlay for a government road construction scheme. Special emphasis will be given on providing connectivity to far-flung border and rural areas including tribal and backward areas. This project will interconnected 550 District Headquarters (up from current 300) through minimum four-lane highway by raising the number of corridors to 50 (up from current 6) and move 80% freight traffic to National Highways by interconnecting 24 logistics parks, 66 inter corridors of total 8,000 km, 116 feeder routes of total 7,500 km and 7 north-east Multi-Modal waterway ports. The programme envisages development of Ring Roads/Bypasses and elevated corridors to de-congest the traffic passing through cities and enhance logistic

TABLE 26.11. List of State-wise National Highways in the Country (as on 31-12-22)

Sl. No.	Name of States/UTs	National Highway No.	Total length (in km)
1.	Andhra Pradesh	16, 216, 216A, 716, 26, 326, 326A, 30, 40, 140, 340, 340C, 42, 44, 544D, 150A, 65, 165, 516D, 565, 765, 67, 167, 69, 71, 75, 167A, 516E, 167B, 365BB, 365BG, 544DD, 544E, 130CD, 716A, 716B, 516C, 167BG, 544F, 167K, 342, NE7, 440, 516B, 340B, 167AD, 167AG, 516F, 516W, 150C, 716G, 216E, 216H, 163G, 544G	8,683.15
2.	Arunachal Pradesh	13, 113, 313, 513, 713, 713A, 15, 115, 215, 315, 315A, 415, 515, 913	4,285.39
3.	Assam	2, 702, 702C, 702D, 6, 306, 8, 208A, 15, 115, 215, 315, 315A , 415, 515, 715, 715A, 17, 117, 117A, 217, 27, 127, 127A, 127B , 127C , 127D, 127E, 427, 627, 29, 129, 329, 329A, 37, 715K, 137, 137G	4,076.91
4.	Bihar	19, 119, 219, 319, 20, 120, 22, 122, 122A, 322, 722, 922, 27, 227, 227A, 327, 327A , 527, 527A, 527B, 527C, 527D, 727, 727A, 31, 131, 131A, 231, 331, 431, 531, 33, 133, 133B, 333, 333A, 333B, 139, 124C, 227F, 227J, 227L, 727AA, 133E, 122B, 333C, 527E, 327AD, 319A, 131B, 131G, 119A,119D, 139W, 727B	5,969.40
5.	Chandigarh	5, 1	15.28
6.	Chhattisgarh	30, 130, 130A, 130B, 130C, 130D, 930, 43, 343, 45, 49, 149B, 53, 153, 353, 63, 163, 163A, 130CD, 143B	3,620.45
7.	Delhi	9, 44, 48, 148A, 248BB, 709B, 344M, 148AE, 148NA, 344N, 344P, NE3	157.10
8.	Goa	748, 66, 366, 566, 748AA, 166S	299.30
9.	Gujarat	27, 927D, 41, 141, 341, 47, 147, 48, 848, 848A, 848B, 51, 151, 251, 351, 53, 753B, 953, 56, 58, 64, 68, 168, 168A, 756, 148M, 751D, 751DD, 351F, 147D, 751, 151A, 754A, NE1, NE4, 351K, 351G, 151K, 848K, 151AD, 927C, 927K, 168G	7,885.00
10.	Haryana	703, 5, 105, 7, 907, 9, 709, 709A, 11, 919, 334B, 44, 344, 444A, 48, 148A, 148B, 248A, 52, 152, 352, 352A, 54, 254, 248BB, 152A, 907G, 352R, 352W, 709AD, 334D, 152D, 148NA, 344N, 344P, NE2, NE5, NE4, 152G, 344GM	3,391.10
11.	Himachal Pradesh	3, 103, 303, 503, 503A, 5, 105, 205, 305, 505, 505A, 705, 7, 707, 907, 907A, 44, 154, 154A	2,606.88
12.	Jammu and Kashmir	1, 501, 701, 44, 244, 144, 144A, 444, 244A, 701A, NE5	1,752.16
13.	Jharkhand	114A, 18, 118, 19, 419, 20, 220, 320, 22, 522, 33, 133, 133A, 133B, 333, 333A, 39, 139, 43, 143, 143A, 343, 49, 143H, 143D, 320G, 143AG, 320D, 218, 143B, 320B	3,429.86
14.	Karnataka	44, 48, 648, 748, 948, 50, 150, 150A, 52, 160, 65, 66, 766, 766C, 67, 167, 367, 69, 169, 169A, 369, 73, 173, 75, 275, 181, 166E, 548B, 561A, 752K, 161A, 544DD, 544E, 548H, 748AA, 367A, 948A, 369E, 373, 275K, 766E, 766EE, NE7, 167N, 150C, 748A	8,037.18
15.	Kerala	544, 744, 66, 766, 966, 966A, 966B, 183, 183A, 85, 185	1,781.57
16.	Ladakh	1, 3, 301	806.45
17.	Madhya Pradesh	719, 27, 30, 34, 934, 135, 135B, 39, 339, 339B, 539, 43, 543, 943, 44, 45, 46, 146, 146B, 346, 47, 347, 347C, 347B, 547, 52, 552, 752B, 752C, 56, 548C, 752G, 161G, 347A, 753L, 147E, 135BB, 135BD, 135BG, 135C, 347BG, 552G, 752D, 753BE, NE4, 543K	9,104.64
18.	Maharashtra	130D, 930, 543, 44, 47, 347C, 547, 48, 348A, 348, 548, 848, 848A, 50, 150, 52, 53, 353C, 353D, 353E, 753, 753A, 753B, 953, 60, 160, 61, 161, 361, 63, 65, 965, 66, 166, 166A, 548C, 753E, 548E, 752G, 561, 753F, 548A, 166E, 266, 548B, 548CC, 161H,	18,459.25

		161G, 361H, 548D, 561A, 965C, 752I, 965G, 752K, 347A, 930D, 361B, 353B, 247, 161A, 361C, 161E, 353I, 753J, 753L, 353J, 353K, 752E, 752H, 753M, 548H, 160A, 160B, 753C, 965D, 753BB, 160D, 348B, 348BB, 753AB, 160C, 166H, 761, 753H, 166D, 652, 465, 647, 461B, 160H, 361F, 965DD, 166F, 166G, 548DD, NE4, 547E, 753BE, 548DG, 150C, 543K	
19.	Manipur	2, 102, 202, 102A , 102B , 102C, 29, 129A, 37, 137, 137A	1,840.34
20.	Meghalaya	6, 106, 206, 217, 127B	1,155.60
21.	Mizoram	2, 102B , 302, 502, 502A, 6, 306, 306A, 108	1,498.67
22.	Nagaland	2, 202, 702, 702A, 702B, 702D, 29, 129, 129A, 229, 329A, 202K	1,670.47
23.	Odisha	16, 316, 516, 18, 20, 220, 520, 26, 326, 326A, 130C, 143, 49, 149, 53, 153B, 353, 55, 57, 157, 59, 63, 126, 130CD, 316A, 516A, 157A, 126A, 655, 720, 143H, 320D	5,897.08
24.	Puducherry	32, 332	64.00
25.	Punjab	3, 503, 503A, 703, 703A, 5, 205, 205A, 7, 9, 44, 344, 344A, 344B, 148B, 52, 152, 54, 154, 154A, 254, 754, 62, 354, 148BB, 105B, 152A, 703B, 354E, 354B, 703AA, NE5, NE5A , 503D, 754A, 754AD, 205K, 205AG	4,239.32
26.	Rajasthan	709, 11, 919, 21, 23, 123, 25, 125, 325, 27, 927A, 44, 48, 148, 148B, 148D, 248, 248A, 448, 52, 552, 752, 54, 56, 156, 58, 158, 458, 758, 62, 162, 162A, 68, 168, 168A, 954, 311, 921, 70, 925, 925A, 911, 552G, 754K, 911A, 148C, 968, 752D, NE4, NE4C, 125A, 703	10,706.34
27.	Sikkim	10, 310, 310A, 310AG, 710, 510, 717A, 717B	709.07
28.	Tamil Nadu	16, 716, 32, 132, 332, 532, 36, 136, 336, 536, 38, 138, 40, 42, 44, 544, 944, 48, 648, 948, 66, 75, 77, 79, 81, 181, 381, 83, 183, 85, 87, 544H, 179A, 383, 381A, 381B, 785, 716A, 744A, 948A, 338, 136B, 179B, 132B, 179D, 332A, 844, 716B, NE7	7,000.20
29.	Telangana	30, 44, 150, 353C, 61, 161, 63, 163, 563, 65, 365, 365A, 365B, 363, 565, 765, 167, 353B, 161B, 365BB, 365BG, 765D, 161AA, 161BB, 167K, 765DG, 167N, 930P, 150C, 163G	4,925.76
30.	Tripura	8, 108, 108A, 208, 208A, 108B	888.61
31.	Uttar Pradesh	307, 9, 509, 709A, 19, 219, 519, 719, 21, 123, 24, 27, 227A, 727, 727A, 927, 28, 128, 30, 230, 330, 330B, 530, 330A , 730, 730A , 31, 731, 731A, 931, 931A, 34, 334, 334A, 334B, 334C, 534, 734, 234, 35, 135, 135B, 335, 39, 339, 539, 44, 344, 552, 709B, 135BB, 730H, 321, 731AG, 709AD, 319D, 124C, 727B, 727H, 727G, 128B, 128C, 328, 328A, 330D, 530B, 730B, 731K, 727BB, 730S, 730C, 334D, 128A, 135C, 135A, 124D, 321G, 334DD, 727AA, 731B, NE2, NE3, NE6, 227B, 344G, 344BG, 334GM, 148NA	12,270.23
32.	Uttarakhand	7, 107, 107A, 307, 507, 707, 707A, 9, 109, 109D, 309, 309A, 309B, 30, 34, 134, 334A, 534, 734, 344, 731K, 109K, 344BG, 107B, 134A	3,664.27
33.	West Bengal	10, 110, 12, 112, 512, 14, 114, 114A, 314, 16, 116, 116B, 17, 317, 317A, 517, 717, 717A, 18, 19, 419, 27, 327, 327B, 31, 131A, 33, 133A, 49, 316A, 116A, 327C, 312, 218	3,674.67
34.	A & N Islands	4	330.70
35.	Dadra and Nagar Havali	848A, NE4	37.00
36.	Daman & Diu	848B, 251	22.00
		Total	**1,44,955.40**

Source : Ministry of Road Transport & Highways http://morth.nic.in India 2024 : A Reference Annual, pp. 779-81.

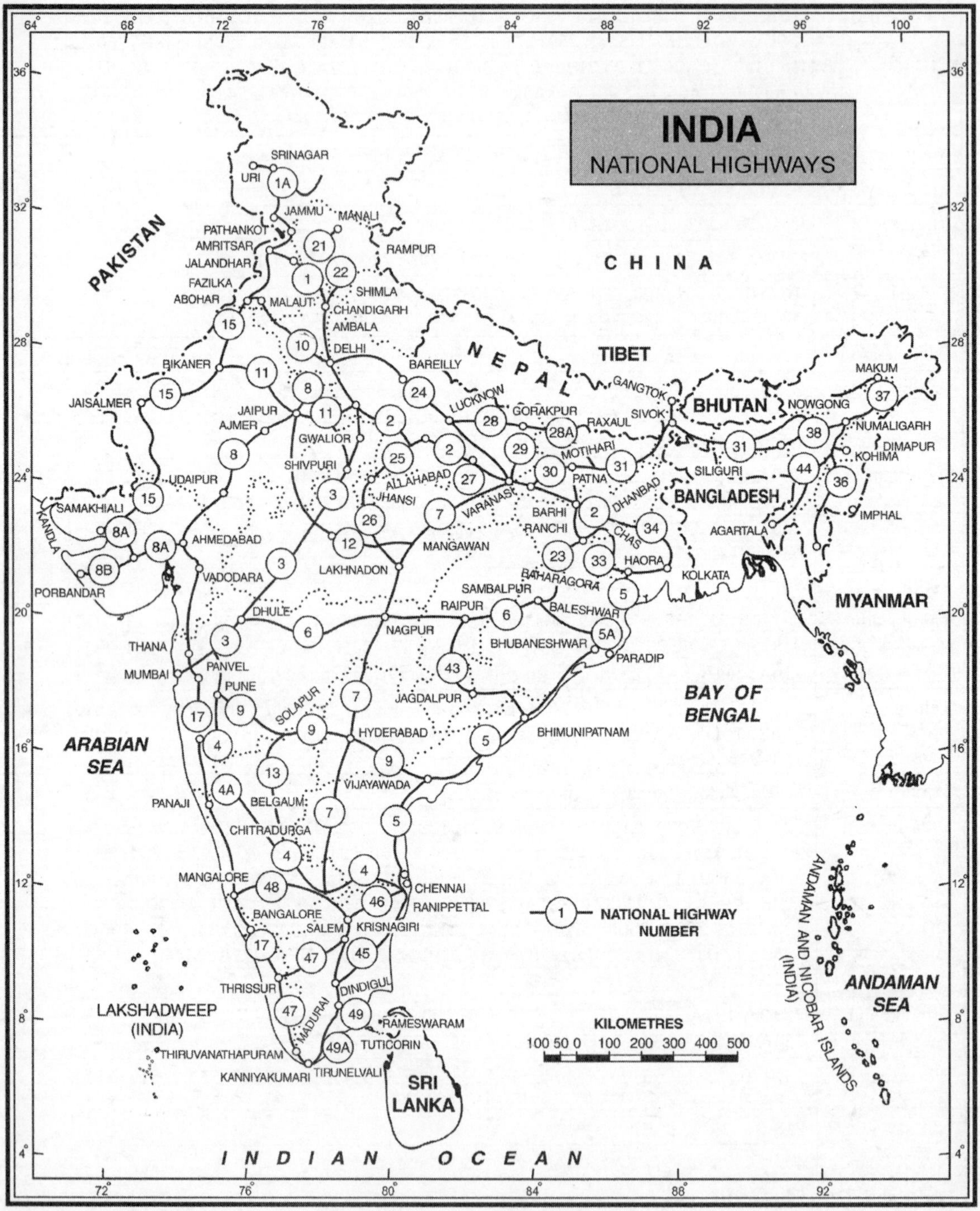

FIG. 26.8. India : National Highways

efficiency, 28 cities have been identified for Ring Roads; 125 chock points and 66 congestion points have been identified for their improvement. The main focus of the programme is improve speed of the traffic flow on key corridors. Due to poor road conditions, narrow width and congestion points, a truck in India covers an average 250-300 km daily in comparison to 700-800 km in developed countries.

Setu Bharatam Programme. Launched on 4th March, 2016, this programme aims at building bridges for safe and seamless travel on NHs. The main object of this programme is to make all NHs free of railway crossings. This is being done to prevent frequent accidents and loss of lives at level crossings. Under this programme, 208 road over bridges (ROBs) and road under bridges (RUBs) are to be built at a cost of ₹ 20,800 crore. The details of these bridges are as : Andhra Pradesh (33), Assam (12), Bihar (20), Chhattisgarh (5), Gujarat (8), Haryana (10), Himachal Pradesh (5), Jharkhand (11), Karnataka (17), Kerala (4), Madhya Pradesh (6), Maharashtra (12), Odisha (4), Punjab (10), Rajasthan (9), Tamil Nadu (9), Uttarakhand (2), Uttar Pradesh (9) and West Bengal (22). Besides, nearly 1,500 old and worn down bridges are to be improved by replacement/widening/strengthening in a phased manner at the cost of ₹ 30,000 crore.

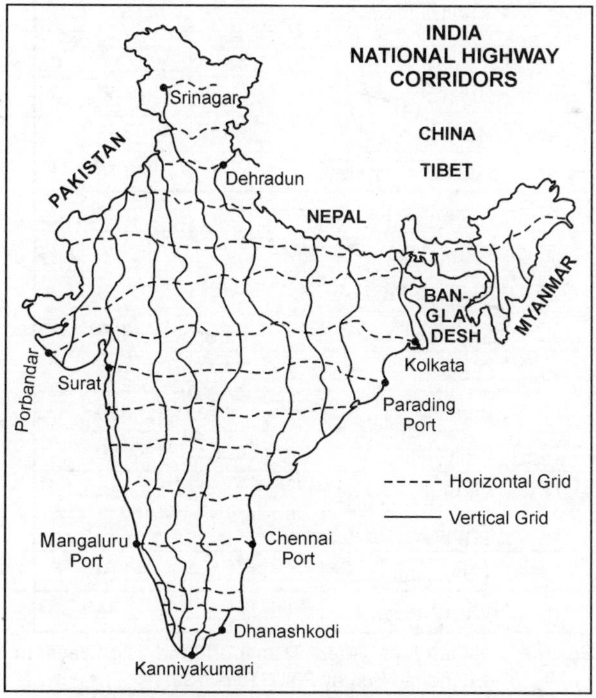

FIG. 26.9. National Highway Corridors

Delhi-Mumbai Expressway. NHAI has come out with an ambitious plan to construct an ultramodern expressway connecting Delhi with Mumbai. Conceived in March, 2018, the project is likely to be completed in January, 2023. It will be the longest signal free expressway in the country passing through five states of Haryana, Rajasthan, Madhya Pradesh, Gujarat and Maharashtra. About 12,000 hectares of land is being acquired for the project. The proposed expressway will be different from the existing highway in the sense that it is shorter, broader, signal free and less time consuming. The existing highway is 4/6 lane, 1450 km long and it takes about 25 hours to reach from one metropolis to another on this route. Compared to this, the proposed expressway will be 6/8 lane, 1260 km and it will take 12 hr 30 min to reach from one motropolis to another. Provision has been made for its expansion to 12 lanes to meet future traffic growth.

FIG. 26.10. Delhi-Mumbai Expressway

Another important 670 km long **Delhi-Amritsar Katra** expressway has been planned. It will extend from Delhi to Nakodar in Punjab from where it will split into Nakodar-Raja Sansi Airport (99 km) and Nakodar-Gurdaspur (92 km). Being developed by the

KEY FEATURES OF DELHI-MUMBAI EXPRESSWAY

- It is 1260 km long which is less than the length of existing 1450 km length of Delhi-Mumbai Highway.
- It will be 68 lane as against the existing 4/6 lane Highway.
- There will be 75 wayside amenities having restaurants, food courts, dormitories, convenience stores, fuel stations and helipds.
- Vehicles will run at 120 kmph.
- Rainwater harvesting at every 500 metres.
- Solar power lighting.
- It will save 320 million litres of fuel and reduce 850 million kg of CO_2 per year.

National Highway Authority of India (NHAI), it will bring down travel time from Delhi to Amritsar to barely four and a half hours and the journey time between Delhi and Katra (Jammu-Kashmir) will be reduced to six and a half-hours. Currently, the fastest train, Vande Bharat Express takes nearly eight hours to reach Katra from New Delhi.

2. State Highways

These are constructed and maintained by state governments and join the state capitals with district headquarters and other important towns. These roads are also connected to the national highways. The length of state roadways in India has more than trebled within a span of about four decades and has increased from 56.8 thousand km in 1970-71 to 186.5 thousand km in 2017-18. These roads constitute about 2.97 per cent of the total road length of India.

Although construction and maintenance of state highways is the responsibility of the concerned state governments, yet with the revamping of the Central Road Fund (CRF), 2000, the Centre provides financial assistance for development of state roads. Further, to promote inter-state facilities and also to assist the State Governments in their economic development through construction of roads and bridges, Central Government provides 100 per cent grant for inter-state connectivity and 50 per cent grant for projects of economic importance from CRF. Loan assistance from external funding agency is also taken by some states.

The distribution of State Highways is very uneven. Maharashtra has the maximum length of state highways. This is followed by Gujarat, Madhya Pradesh, Rajasthan, and Andhra Pradesh (including Telangana). Smaller states such as Goa, and states in hilly areas like Mizoram, Sikkim, Nagaland, Tripura, etc. have less than five hundred km length of State Highways each.

Conversion of State Highways to National Highways. The Government received proposals for declaration of more than 64,000 km of State roads as National Highways (NHs) from various State Governments, against which the Ministry has declared about 10000 km. of Roads/routes as new National Highways. Upto 2016-17 3180 km of State Highways had been converted to NHs. The State-wise conversion of State roads into National Highways is given in **Table 26.12**. Till 2018-19, 7300 km State Highways were notified as National Highways.

TABLE 26.12. Details of State Roads Converted to National Highways

Sl. No.	State/Union Territory	Length (in km)
1.	Andhra Pradesh	676
2.	Assam	9
3.	Bihar	160
4.	Haryana	395
5.	Himachal Pradesh	176
6.	Jammu & Kashmir	8
7.	Karnataka	70
8.	Madhya Pradesh	9
9.	Maharashtra	387
10.	Odisha	193
11.	Punjab	530
12.	Rajasthan	20
13.	Sikkim	154
14.	Tripura	228
15.	Telangana	119
16.	West Bengal	46

Source : Ministry of Road Transport and Highways in Economic Survey 2017-18 p. 132.

3. District Roadways

These roads join the district headquarters with the other places of the district. Development and maintenance of these roads fall within the purview of Zila Parishads. Currently there are about 5.86 lakh km long district roads in India and they account for about ten per cent of the total road length in India. Formerly most of the district roads were unsurfaced and lacked bridges and culverts. But now the situation has changed and most of these roads are surfaced. Such a situation has improved connectivity and paved way for economic development.

Maharashtra is at the top followed by Uttar Pradesh, Madhya Pradesh, Rajathan, Punjab, Karnataka, Assam, Himachal Pradesh, Haryana and Kerala.

4. Village Roads

The village roads are mainly the responsibility of village panchayats and connect the villages with the neighbouring villages, towns and cities. These are generally dusty tracks and are usable only during fair weather. They become muddy and unserviceable during the rainy season. Efforts have been made in the recent past to connect the villages with metalled roads. The length of these roads has increased by about 20 times from 2,06,408 km in 1951 to 45.2 lakh km in March 2019. These roads accounted for 71.4 per cent of the total road length of the country in March 2019. Still about 10 per cent of the villages having a population of 1,000 or more and 60 per cent of the villages with less than 1,000 people are not connected by all-weather roads. The network needs expansion and upgradation of existing roads to all-weather roads.

A new thrust was given to village roads when the Pradhan Mantri Gram Sadak Yojna (PMGSY) was launched in December 2000. This is a 100% Centrally Sponsored Scheme to provide rural connectivity to unconnected habitations with a population of 500 persons or more (250 persons in case of hilly, desert and tribal areas) in rural areas by the end of the Tenth Plan period. The scope of PMGSY has been expanded to include both construction of new links and upgradation of existing roads through routes associated with such link routes to form one complete sub-network for providing connectivity between the village and the market. A survey undertaken to identify the "core network" as part of PMGSY showed that over 1.70 lakh unconnected habitations needed to be undertaken under this programme. This would require new construction of 3.68 lakh kilometres of rural roads at a total cost of ₹ 1,33,000 crore.

Border Roads. Border Roads Organisation (BRO) Board was set up in May 1960 for accelerating economic development and strengthening defence preparedness through rapid and co-ordinated improvement of roads in the north and north-eastern border areas. This organisation has constructed world's highest road joining Chandigarh with Manali

TABLE 26.13. Length of road network in India (Thousand km)

Category of roads	1950-51	1960-61	1970-71	1980-81	1990-91	2000-01	2011-12	2012-13	2014-15	2015-16	2016-17	2017-18	2018-19	2019-20
Total Roads	399.9	524.5	914.9	1485.4	2327.4	3373.5	4865.4	5231.9	5472.1	5603.3	5897.7	6215.7	6371.8	NA
Total Surfaced roads	157.0	263.0	398.0	684.0	1090.0	1601.7	2698.6	3171.0	3341.0	3501.9	3729.8	3931.5	4116.4	NA
National Highways (Total)	19.8	23.8	23.8	31.7	33.7	57.7	76.8	79.1	98.0	101.0	114.2	126.8	132.5	NA
National Highways (Surfaced)	NA	21.0	23.3	31.5	33.4	57.7	76.8	79.1	98.0	101.0	114.2	126.3	132.5	NA
State Highways (Total)	NA	NA	56.8	94.4	127.3	132.1	164.4	169.2	167.1	176.1	174.1	186.9	179.5	NA
State Highways (Surfaced)	NA	NA	51.7	90.3	124.8	129.9	163.0	167.2	165.3	170.2	169.1	185.5	178.4	NA

NA : Data not available

Source : (*i*) Economic Survey 2019-20 p. A49. (*ii*) Economic Survey 2022-23, Statistical Appendix, p. 52.

in Himachal Pradesh and Leh in Ladakh. This road runs at an average altitude of 4,270 metres above sea level and negotiates four passes at heights ranging from 4,875 to 5,485 metres. It is a vital road link in the western Himalayas and has considerably reduced the distance between Chandigarh and Leh. The Border Roads Organisation has now spread its activities throughout the country and is presently working in states of Rajasthan, Himachal Pradesh, Maharashtra, Tamil Nadu, Andhra Pradesh, Uttar Pradesh, Sikkim, Assam, Meghalaya, Nagaland, Tripura, Manipur, Mizoram, Arunachal Pradesh, Bihar and Union Territories of Jammu and Kashmir and Andaman and Nicobar Islands. A prestigious project of developing the 160 km long Tamu-Kalemyo-Kalewa road in Myanmar was taken up in 1997 and completed in 2001. Another important work is the construction of Indo-Bangladesh Border (IBB) Roads and fencing of the border.

Urban Roads. A road within the limits of the area of municipality, military cantonment, port or railway authority is called an urban road. There has been a phenomenal growth in urban roads from a meagre 46,361 km in 1961 to 5,26,483 km in 2017 as a result of accelerated growth in urbanization.

Project Roads. A road within the limits of the area of a development project of a public authority for the exploitation of resources such as forests, irrigation, hydro-power, coal, sugarcane, etc. is called a project road. Various developmental projects have been undertaken as a result of which the length of project roads has increased from 1,30,893 km in 1971 to 3,28,897 km in 2017.

International Highways

The roads which are financed by the World Bank and connect India with neighbouring countries are called international highways. There are two categories of such highways. (*a*) the main arterial routes linking the capitals of neighbouring countries. Two important routes of this category are (*i*) the Lahore-Mandalay (Myanmar) route passing through Amritsar-Delhi-Agra-Kolkata-Golaghat-Imphal and (*ii*) Barhi-Kathmandu road. (*b*) Routes joining major cities, ports etc. with arterial network such as : (*i*) Agra-Mumbai road (*ii*) Delhi-Multan road (*iii*) Bengaluru-Chennai Road and (*iv*) Golaghat-Ledo road.

Geographical Distribution of Roads

The network of roads is more or less similar to that of railways, although former far excels the later with respect to kilometreage. There are great variations in the distribution pattern of roads in India. Uttar Pradesh, Rajasthan, Madhya Pradesh, Andhra Pradesh, Tamil Nadu, Maharashtra, Karnataka, Gujarat, Odisha and Bihar have much longer length of national highways. Incidently these are larger states with high density of population and comparatively advanced stage of industrial growth. However, length of National Highways passing through a state is not always a true index of economic well being of a state as they serve only the main routes. For example, Madhya Pradesh and Bihar have longer route length of the national highways and still have some of the most backward areas which are located far away from the main road routes.

The length of state highways could be a better index of road accessibility at the state level. The picture is slightly different with regard to state highways when compared to that of the national highways.

Maharashtra is the best served state by the state highways. Gujarat is a distant second with respect to length of state highways. The other states with sufficiently long route length of state highways are Madhya Pradesh, Rajasthan, Karnataka, Uttar Pradesh and Andhra Pradesh. The north-eastern states do not have sufficient route length of state highways.

The nature of roads, rather than their total length is more significant from the utility point of view. The largest concentration of roads is found in the Northern Plain, especially in West Bengal and in the Punjab-Haryana plain. But the ratio of surfaced road to the total road length is lower in the Northern Plain than the national average. For example, Bihar and West Bengal are the two large states in the plain but have only 42.32 per cent and 56.48 per cent of surfaced roads to the total length of roads respectively. Uttar Pradesh with 68.74 per cent of the surfaced roads is in a slightly better position. Punjab and Haryana are the two richest states of the country and have 100 per cent and 94.83 per cent of the surfaced roads respectively. The main cause of small proportion of surfaced roads in the Northern Plain is that it is made up of sand, silt and clay and there is shortage of stone for constructing surfaced roads. In contrast, the proportion of surfaced roads is much higher in the Peninsular plateau area because it is composed of hard rocks and stone for constructing roads is readily available here in plenty. Gujarat (93.95%), Maharashtra (79.68%), Goa (61.00%) are some of the states having higher than the national average of 60.33 per cent. Odisha presents an anomaly and is the poorest of all the states where surfaced roads account for less than one-third of the total length of roads. Among the Himalayan states Sikkim with 86.11 per cent is at the top and this is followed by 70.97 per cent in Jammu and Kashmir. All other Himalayan states have surfaced roads less than the national average. Among the union territories, Chandigarh, Dadra and Nagar Haveli, Daman and Diu. Delhi and Lakshadweep have 100 per cent surfaced roads. This is followed by 97.47 per cent in Andaman and Nicobar Islands and 77.09 per cent in Puducherry.

Density of Roads

A still better index of road accessibility is the *density of roads* which is defined as *the length of roads per 100 sq km of surface area*.

There are large scale spatial variations with regard to road density in different parts of the country. It varies from 10.0 km per 100 sq km in Jammu and Kashmir to 526.9 km per 100 sq km in Kerala with a national average of 96.5 km per 100 sq km (2013). Density of 200 to 400 km per 100 sq km is found in Assam, Goa, Tripura and West Bengal.

Andhra Pradesh, Bihar, Karnataka, Nagaland, Odisha, Tamil Nadu and Uttar Pradesh have road density varying from 100 to 200 km per 100 sq km. These two sets of states have moderate to high road density due to high level of urbanization and industrialisation. Level land of the Great Plain of North India provides ideal conditions for road construction but building material for road has to be transported from hilly region of the Himalayas or from the northern part of the plateau region as this plain is entirely made up of sand, silt and clay. On the other hand, the Himalayan region abounds in building material but rugged topography and rough terrain hampers the process of road building. This is the reason that most of the Himalayan states like Jammu

TABLE 26.14. State-wise Road Density (2013)

State	Length of roads km per 100 sq km
1. Andhra Pradesh (including Telangana)	125.4
2. Arunachal Pradesh	19.6
3. Assam	293.6
4. Bihar	127.5
5. Chhattisgarh	55.0
6. Goa	285.5
7. Gujarat	74.8
8. Haryana	67.2
9. Himachal Pradesh	65.2
10. Jammu & Kashmir	10.0
11. Jharkhand	21.9
12. Karnataka	133.2
13. Kerala	526.9
14. Madhya Pradesh	53.8
15. Maharashtra	72.6
16. Manipur	73.2
17. Meghalaya	43.9
18. Mizoram	29.2
19. Nagaland	134.5
20. Odisha	138.4
21. Punjab	89.7
22. Rajasthan	50.1
23. Sikkim	26.4
24. Tamil Nadu	139.3
25. Tripura	302.6
26. Uttar Pradesh	118.2
27. Uttarakhand	76.7
28. West Bengal	238.6
All States	**95.8**
Union Territories	
1. Andaman & Nicobar Islands	15.8
2. Chandigarh	1,857.9
3. Dadra & Nagar Haveli	128.7
4. Daman and Diu	200.0
5. Delhi	1,993.2
6. Lakshadweep	525.0
7. Puducherry	562.8
All Union Territories	**334.8**
All India	**96.5**

and Kashmir, Himachal Pradesh, Uttarakhand, Sikkim, Meghalaya, Manipur etc. have low density of roads. Most of the Union Territories are of small size and some of them are highly urbanized. As such they have moderate to high road density except Andaman and Nicobar Islands.

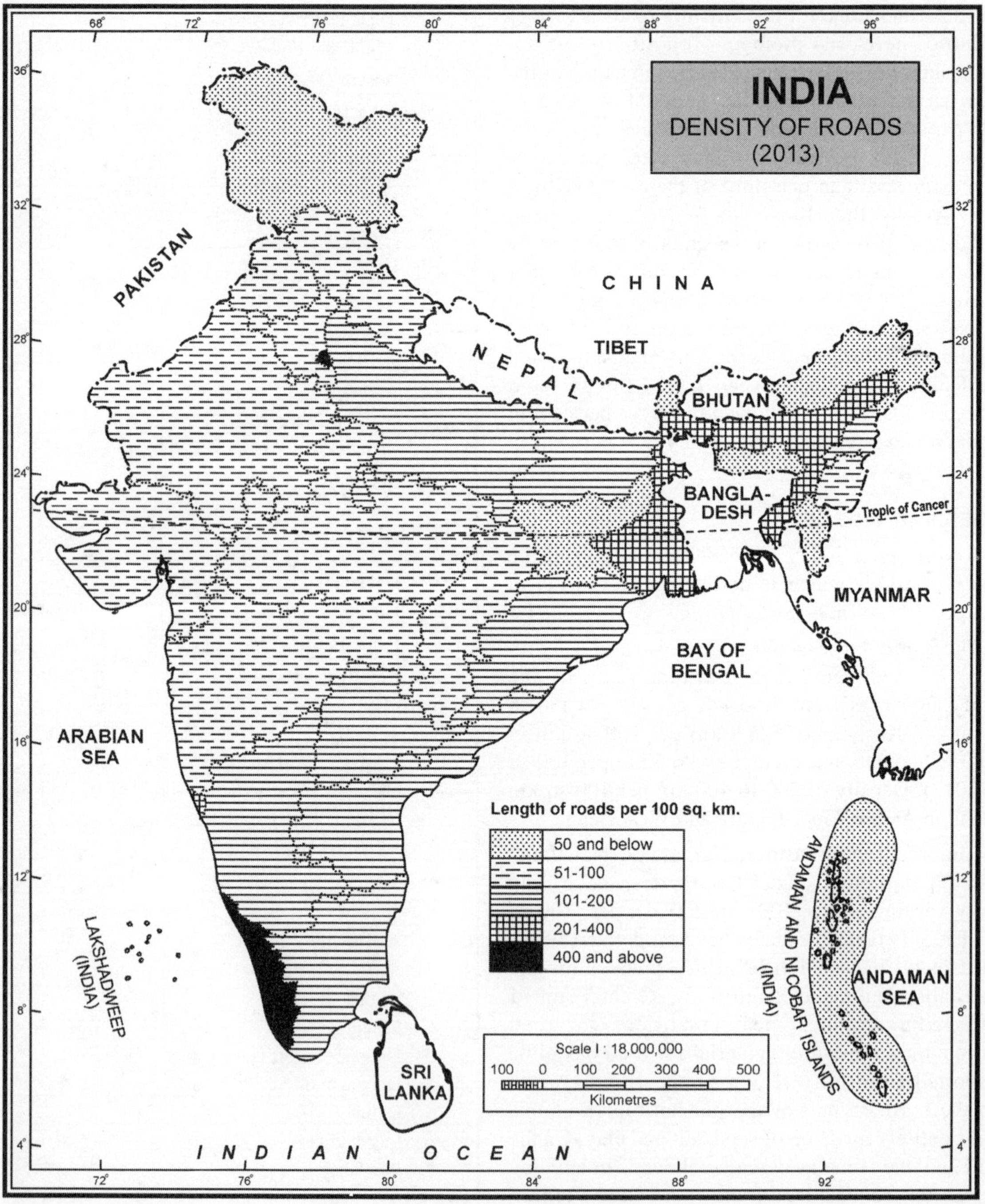

FIG. 26.11. India : Road Density (2013)

TABLE 26.15. Number of Registered Vehicles in India

(in thousands)

Year	1950-51	1960-61	1970-71	1980-81	1990-91	2000-01	2011-12	2012-13	2014-15	2015-16	2016-17	2017-18	2018-19	2019-20
All vehicles	306.0	665.0	1,865.0	5,391.0	21,374.0	54,991.0	1,59,491	1,76,044	2,10,023	2,30,031	2,53,311	2,72,587	2,95,772	3,26,299
Goods vehicles	82.0	168.0	343.0	554.0	1,356.0	2,948	7,658	8,307	9,344	10,516	12,256	12,773	13,766	14,288
Buses	34.0	57.0	94.0	162.0	331.0	634.0	1,677	1,814	1,971	1,757	1,864	1,943	2,049	2,196

Source : (*i*) The Economic Survey 2020-21, p. A-49. (*ii*) Economic Survey 2022-23, Statistical Appendix, p. 52.

Problems and Prospects

Road transportation in India faces a number of problems. Keeping in view the vastness of the dimensions of the country, her physiography, her unlimited natural resources and the fast growing passenger and freight traffic, inadequacy of road network is quite glaring. India's road length of 96.5 km per 100 sq km of area is desperately low as compared to 294.6 km in Japan, 131.2 km in Austria, 451.8 km in Belgium, 147.2 km in France and 172.2 km in Switzerland. Again, India has low road length of 379.3 km per one lakh population as against 893.6 in Japan, 497.2 in Malaysia, 1277.7 in Saudi Arabia, 1392.4 in Austria, 1556 in Hungary, 1572.4 in Sweden, 2494.5 in the USA, 3184.9 in Canada, 4635.4 in Australia and 2705.7 in New Zealand. Lakhs of villages in remote areas are still awaiting a road to reach them.

Another problem is that more than one-third of the roads (36.25%) are unsurfaced. They can be used only in fair weather and become muddy and unfit for transportation during the rainy season. Efforts need to be made to construct as many surfaced roads as is practically possible.

The national highway network will have to be improved to meet the growing traffic of men and materials. A large section has insufficient road pavement thickness. Other deficiencies are inadequate capacity; poor riding quality, weak and distressed bridges/culverts, congested city sections, too many railway level crossings, lack of wayside amenities and weak road safety measures. About 20 per cent of national highways need widening from single to double lanes and 70 per cent of two lane roads have to be strengthened and selected corridors on national highways need conversion into *expressways*. Thisis clearly an enormous task and needs huge capital investment which is beyond the capacity of the public sector. Consequently, National Highway Act was amended in 1995 for encouraging private sector to participate in the construction, maintenance and operation of roads on Build, Operate and Transfer (BOT) basis.

The future challenge in road sector revolves around building all-weather roads connecting each and every village to a State Highway or a National Highway. It is imperative to strengthen the road infrastructure for carrying rapidly increasing volumes of agricultural produces to the consuming centres in the near future. *Rural development is closely associated with road development*.

Another very important factor to be considered is the rapidly growing population of motor vehicles and increasing commerce. The number of registered vehicles increased from 306 thousand in 1950-51 to 2,95,772 thousand in 2018-19 (**Table 26.15**). However, carrying capacity of our roads has not been able to keep pace with the increase in the number of vehicles. This has led to traffic jams, delays and environmental pollution. The most disturbing factor is that population of motor vehicles is likely to increase at an accelerated rate in the near future. As such there is an urgent need to take steps to increase the capacity of roads.

Building adequate road infrastructure to accelerate the pace of economic progress is an uphill task and requires huge sums of money. The current estimates suggest that the cost of a four-lane highway works out at roughly ₹ 4.5 crore per kilometre and the cost of a protected access, six-lane express way works out at roughly ₹ 8.5 crore per kilometre. Funds on such a gigantic scale are managed by encouraging private participation, World Bank and imposing cess on fuel, etc.

AIR TRANSPORT

Air transport is the fastest mode of transport which has reduced distances and has led to drastic shrinking of the world. This mode of transport is indispensable when speed and time are the main considerations. One can easily cross and reach remote, inaccessible and hostile areas like lofty mountains, thick forests, marshy areas and sandy deserts by air transport which is almost impossible by other modes of transport. Air transport plays a vital role in times of emergency as well as in the event of natural and man-made calamities like floods, famines, epidemics and wars. Air transport is very essential for a vast country like India where distances are so long and the terrain and climatic conditions so diverse. The weather conditions in India are also quite congenial to air transport. Poor visibility due to clouds, fog and mist hinders air transport but India is lucky to have clear weather for most part of the year except for a short duration in rainy season and foggy winters particularly in North India.

Air transportation in India made a humble beginning in 1911 when air mail operation commenced over a very short distance of 10 km between Allahabad (Pryagraj) and Naini. The British, French and Dutch introduced air transport in 1929-30. Indian National Airways was formed in 1933 and it introduced air service between Karachi and Lahore. By the end of the World War II, major cities like Karachi, Mumbai, Delhi, Kolkata, Lahore and some other places were provided with air services. At the time of partition of the country in 1947, there were four companies namely Tata Sons Ltd./ Air India, Indian National Airways, Air Services of India and Deccan Airways. By 1951, four more companies *viz*. Bharat Airways, Himalayan Aviation Ltd., Airways India and the Kalinga Airlines also came up. In 1953, the air transport was nationalised and two corporations were formed: Air India International and the Indian Airlines. The face of Indian aviation is changing and is poised for a vibrant growth. India can definitely boast of a boom in air travel. Economic liberalisation has totally changed the outlook of a perspective air passenger. Now, he thinks in terms of time and gone are the days when he used to ponder over advantages and disadvantages of air travel.

The civil aviation is a potential sector in the country which can be a sunrise sector of growth. India has favourable conditions which are highly conducive for the sector's growth *i.e.* favourable demographics, a rising middle class population, highly disposable incomes, and faster economic growth.

Despite a strong home market for air travel, Indian (domestic) airlines have not captured the Indian market for international travel unlike many other countries. For example, Indian (domestic) airlines utilised only 38 per cent of its international Available Seat Kilometres (ASKs) in 2016 compared to 60.6 per cent for Netherlands, 49.1 per cent for China and 48.9 per cent for U.K. ASKs refer to *sum of products obtained by multiplying the number of seats available for sale on each flight by corresponding distance flown by flights*.

The civil aviation sector in India has recorded an unprecedented growth in the recent years. Currently there are large number of companies providing passenger transport and cargo handling services in the country. These are Air India Limited and its subsidiaries *viz*. Alliance Air, Air India Charters Ltd. (Air India Express etc.), Jet Airways (India) Ltd., Jetlite Airlines, Go Airlines (Indigo) Pvt. Ltd., Spice Jet Ltd., Inter Global Aviation Ltd. (Indigo) and Air Asia.

TABLE 26.16. Growth of Civil Aviation in India

(in thousands)

Year	1980-81	1990-91	1999-00	2011-12	2015-16	2016-17	2017-18	2018-19	2019-20	2020-21	2021-22	2022-23
Total fleet strength	63	80	119	355	448	493	584	691	669	716	691	714
Revenue tonne-kilometers (million)	1,380.4	2,040.3	2,767.4	11,945.6	14,787.2	17,141.6	20,258.7	22,779.5	20,945.7	7,458	7883.3	16778
Number of passengers carried	68.5	100.3	163.6	752.1	1,038.2	1,245.7	1,471.2	1,662.3	1,639.3	587.1	843	115601

Source : Economic Survey 2022-23. Statistical Appendix p. 53.

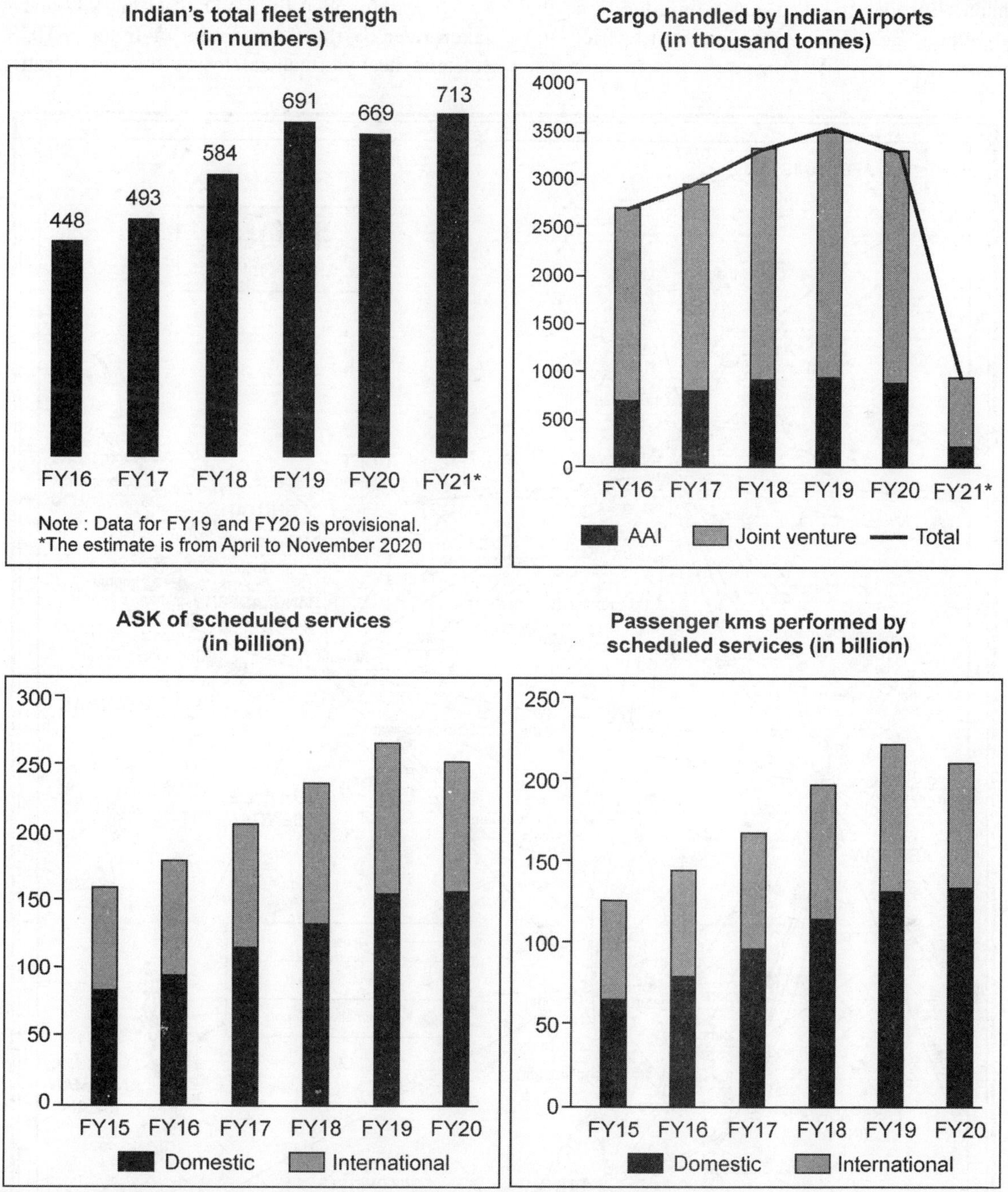

FIG. 26.12. Performance of Civil Aviation in India

* ASK: Available Seat Kilometres.

Source: Survey calculations based on DCCA data in Economic Survey 2020-21, p. 292.

1. Air India is responsible for international air services. It operates from international airports and connects India with almost all the continents of world through its services. Its main services are to the USA,

Canada, Europe, the Russian Confederation, the Gulf/Middle East, East Asia, Far East, Africa and Australia.

It was founded by J.R.D. Tata in 1932 and was taken over by the Government of India in 1953. It suffered heavy financial losses due to operational

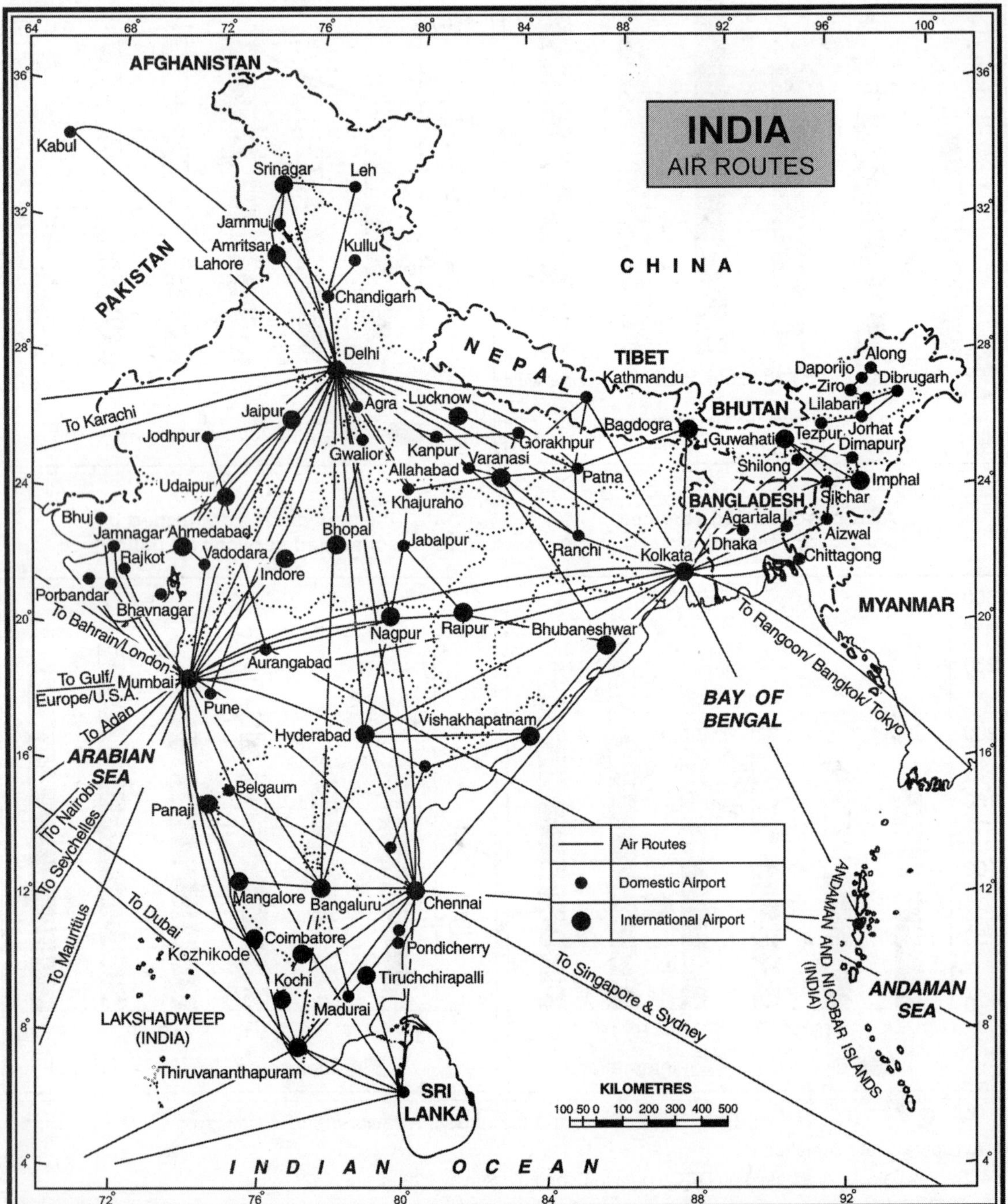

FIG. 26.13. India : Air Routes

TABLE 26.17. Growth of Civil Aviation in India

	1960-61	1970-71	1980-81	1990-91	1999-00	2005-06	2015-16	2016-17	2017-18	2018-19(P)	2019-20(P)
1. Total fleet strength (number)											
(i) Air India	13	11	18	24	26	38	—	—	—	—	—
(ii) Indian Airlines	88	73	45	56	44	55	—	—	—	—	—
(iii) Air India Ltd. (Erstwhile National Aviation Company of India Ltd.)	—	—	—	—	—	—	107	111	121	1256	127
(iv) Alliance Air	—	—	—	—	12	15	11	11	16	20	19
(v) Air India Express	—	—	—	—	—	4	18	23	23	25	25
(vi) Other Pvt. Scheduled Airlines	—	—	—	—	37	131	312	348	424	520	498
2. Revenue tonne-Kilometers (million)											
(i) Air India	109.4	275.2	980.1	1342.7	1456.5	2364.0	—	—	—	—	—
(ii) Indian Airlines	82.4	202.1	400.3	697.6	738.0	1589.0	—	—	—	—	—
(iii) Air India Ltd. (Erstwhile National Aviation Company of India Ltd.)	—	—	—	—	—	—	4363.7	4739.0	5395.7	5782.1	5848.1
(iv) Alliance Air	—	—	—	—	131.0	73.0	20.2	29.4	51.4	62.0	61.3
(v) Air India Expenses	—	—	—	—	—	113.0	672.2	836.2	999.1	1031.4	1192.8
(vi) Other Pvt. Scheduled Airlines	—	—	—	—	441.9	1789.0	9607.2	11404.6	13682.5	15779.6	21973.8
(vii) Dedicated Private Freighter	—	—	—	—	—	—	123.9	132.4	130.0	125.2	125.9
(viii) Total (Other Pvt Scheduled Airlines + Dedicated Private Freighter	—	—	—	—	—	—	9731.1	11537.0	13812.5	15904.8	22099.8
3. Number of passengers carried (lakh)											
(i) Air India	1.3	4.9	14.2	21.6	31.9	43.6	—	—	—	—	—
(ii) Indian Airlines	7.9	21.3	54.3	78.7	58.9	78.2	—	—	—	—	—
(iii) Air India Ltd. (Erstwhile National Aviation Company of India Ltd.)	—	—	—	—	—	—	184.8	196.0	211.7	222.8	224.8
(iv) Alliance Air	—	—	—	—	18.6	10.1	4.0	6.3	12.8	16.0	16.4
(v) Air India Express	—	—	—	—	—	4.9	27.8	34.2	38.9	43.4	48.3
(vi) Other Pvt. Scheduled Airlines	—	—	—	—	54.2	180.7	821.6	1009.2	1207.8	1380.1	1349.8

***Source*:** Directorate General of Civil aviation; Airport Authority of India in Economic Survey 2020-21 pp. A50-51

Notes: P : Provisional

inefficiencies of various types and became almost unviable economically in due course of time. Consequently it was returned back to Tata Sons on 27 January, 2022. It is hoped that with privatization, and work efficiency it will regain its past glory.

2. Indian Airlines handles domestic traffic and carries passengers, cargo and mail to different places in the country. It also provides services to 12 countries, *viz*., Pakistan, Maldives, Nepal, Sri Lanka, Malaysia, Bangladesh, Thailand, Singapore, U.A.E., Oman, Myanmar and Kuwait. Its operations cover 69 destinations including 15 abroad. Its fleet strength has shown varying trends, falling from 88 in 1960-61 to 73 in 1970-71 and 49 in 1980-81. It rose to 53 in 1999-2000. In 2005-06, its fleet stood at 64. The current projections indicate additional aircraft requirements for replacement of ageing aircraft as well as for growth.

3. Vayudoot was set up in 1981 to augment the air transport in the country. It provided links with remote and inaccessible areas which were not covered by Indian Airlines. It commenced its services in February, 1981. The Government had initially approved air services to 14 stations in the North-Eastern region. Subsequently, 23 stations outside this region were added for linking by Vayudoot. Its fleet rose from 1 in 1981 to 20 in 1988-89 but fell to 16 in 1992-92 after which it was merged in Indian Airlines. Its network which linked remote, inaccessible and thickly forested areas of the North-Eastern region has been taken over by the newly created Short Haul Operations Deptt. of Indian Airlines.

4. Pawan Hans Limited was established in 1985. It provides helicopter services to the petroleum sector including ONGC, Oil India Ltd. and Enron Oil and Gas, Mumbai High and connects remote and inaccessible areas. The company has a strong presence in North-East having its helicopters deployed in the states of Arunachal Pradesh, Meghalaya, Sikkim and Tripura. Apart from this it also provides services to certain state governments such as Punjab, M.P., Lakshadweep and public sector undertakings like NTPC, GAIL, and BSF.

Private Sector. Private air taxis started their services in 1990 and played feeder role to Indian Airlines. Only four parties ventured in this area at that time due to the restrictive environment. In a major policy change, the Government repealed the Air Corporation Act 1953 on March 1, 1994, thereby ending the monopoly of Indian Airlines and Air India on the scheduled operations. Private operators, who were hitherto operating as air taxis, have since been granted scheduled airlines status. The new policy on air taxi services provides for a route dispersal plan to ensure operation of a minimum number of services in the North-Eastern Region, Andaman and Nicobar Islands, Lakshadweep and Jammu and Kashmir. This policy has infused competitive environment between Indian Airlines Ltd., a public sector enterprise, on one hand and private operators on the other hand. The policy on domestic air transport service was approved in April 1997 according to which barriers to entry and exit from this sector have been removed.

Airports Authority of India (AAI)

International Airports Authority of India and National Airports Authority were merged on 1 April, 1995 to form Airports Authority of India (AAI). This authority is responsible for providing safe and efficient air traffic services and aeronautical communication services for effective control of air traffic in the Indian air space. It controls and manages the entire Indian space even beyond the territorial limits of the country in accordance with the norms set by International Civil Aviation Organisation (ICAO). It comprises of International Airports Division (IAD) and National Airports Division (NAD).

At present India has 487 air ports/air strips of which 35 are international. The distribution of international airports is given in **table 26.18**.

UDAN (Ude Desh ka Aam Nagrik)

This scheme was launched in October, 2016 to make flying accessible and affordable for the masses in the regionally important cities. This is a first-of-its-kind scheme globally to stimulate regional connectivity through a market based mechanism. It provides a few seats at affordable passenger fares for an hour long flight. Under UDAN, 70 airports and 128 routes are connected, and over 100 more unserved airports are to be connected in next rounds of bidding of routes.

The Government offers fiscal support through Viability Gap Funding (VGF) and infrastructural

TABLE 26.18. List of International Airports of India as on 31st March 2022

Sl. No.	Name of the Airport	City/State
1.	Rajiv Gandhi International Airport	Hyderabad/Telangana
2.	Sri Guru Ram Das Jee International Airport	Amritsar/Punjab
3.	Lokpriya Gopinath Bardoloi International Airport	Guwahati/Assam
4.	Biju Patnaik International Airport	Bhubaneswar/Odisha
5.	Gaya Airport	Gaya/Bihar
6.	Indira Gandhi International Airport	New Delhi/Delhi
7.	Veer Savarkar International Airport	Port Blair/Andaman Nicobar Islands
8.	Sardar Vallabhbhai Patel International Airport	Ahmedabad/Gujarat
9.	Kempegowda International Airport	Bengaluru/Karnataka
10.	Mangalore International Airport	Mangaluru/Karnataka
11.	Cochin International Airport	Kochi/Kerala
12.	Calicut International Airport	Kozhikod/Kerala
13.	Trivandrum International Airport	Thiruvananthapuram/Kerala
14.	Chhatrapati Shivaji Maharaj International Airport	Mumbai/Maharashtra
15.	Dr. Babasaheb Ambedkar International Airport	Nagpur/Maharashtra
16.	Jaipur International Airport	Jaipur/Rajasthan
17.	Chennai International Airport	Chennai/Tamil Nadu
18.	Tiruchirappalli International Airport	Tiruchirapalli/Tamil Nadu
19.	Chaudhary Charan Singh International Airport	Lucknow/Uttar Pradesh
20.	Lal Bahadur Shastri International Airport	Varanasi/Uttar Pradesh
21.	Netaji Subhas Chandra Bose International Airport	Kolkata/West Bengal
22.	Kannur International Airport	Kannur/Kerala
23.	Surat Airport	Surat/Gujarat
24.	Devi Ahilya Bai Holkar Airport	Indore/Madhya Pradesh
25.	Dabolim Airport	Dabolim/Goa
26.	Coimbatore International Airport	Coimbatore/Tamil Nadu
27.	Sheikh-ul-Aalam International Airport	Srinagar/J&K
28.	Imphal International Airport	Imphal/Manipur
29.	Madurai Airport	Madurai/Tamil Nadu
30.	Bagdogra International Airport	Siliguri/West Bengal
31.	Mangalore International Airport	Mangaluru/Karnataka
32.	Shahid Bhagat Singh International Airport	Chandigarh
33.	Nashik Airport	Nashik/Maharashtra
34.	Vadodara Airport	Vadodara/Gujarat
35.	Kushinagar Airport	Kushinagar/UP

Source : Airport Authority of India, 2022.

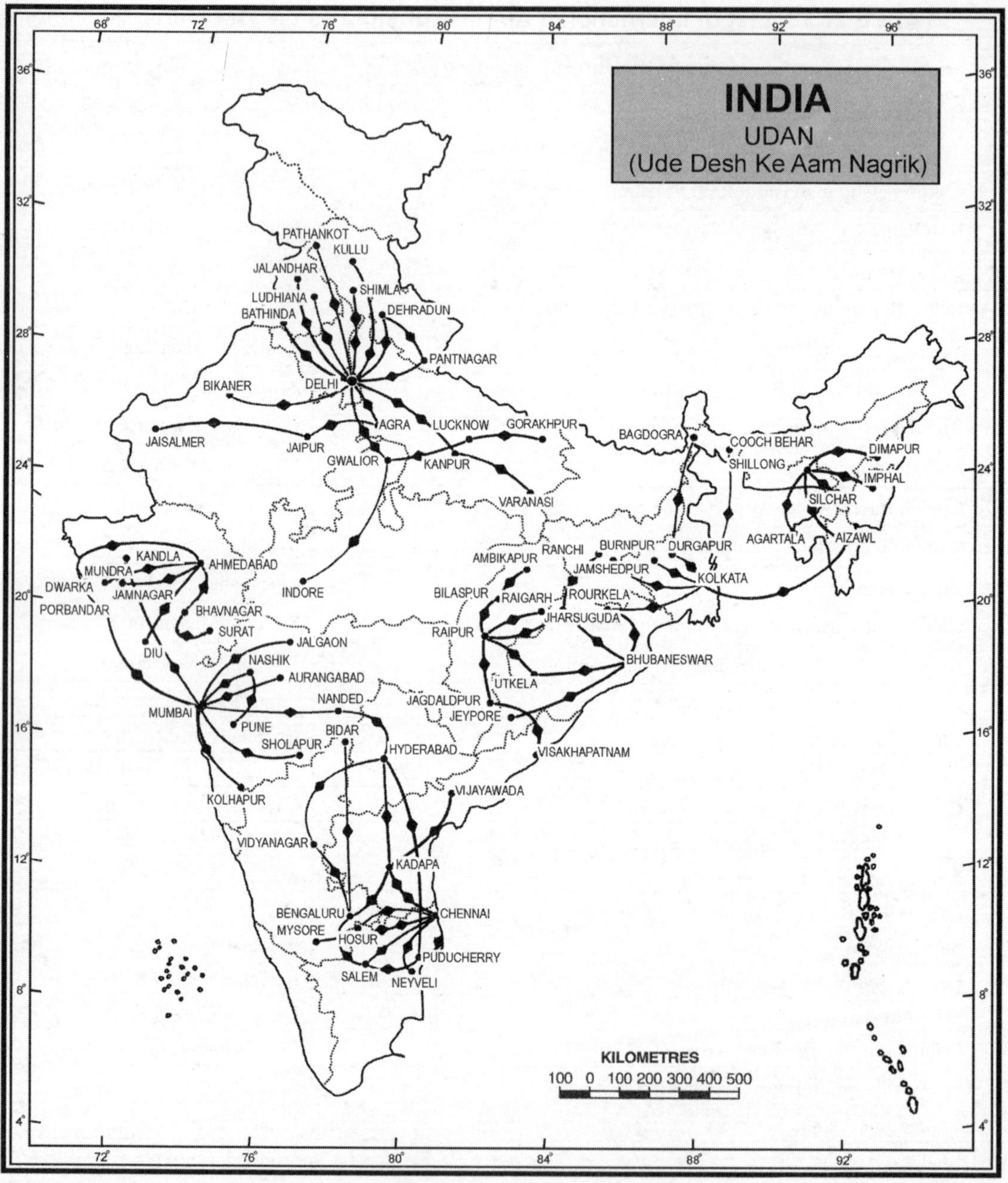

FIG. 26.14. India : Udan (Ude Desh ka Aam Nagrik)

development of under-utilized airport facilities to incentivize regional air traffic. UDAN ensures route profitability to airlines to sustain their operations through reducing operating costs by eliminating airport charges on UDAN routes, subsidizing ATF, providing market based subsidy for half of the seats, and guaranteeing three years exclusivity on routes. Under UDAN, 13 Regional Connectivity Scheme airports have been covered in the Eastern and North-Eastern region, 12 each in Northern and Western regions, and 8 in the Southern Region in the first round.

National Civil Aviation Policy, 2016

For the first time since Independence an Integrated Civil Aviation Policy was notified on 15 June, 2016, which aims at creating an ecosystem that will push the growth of the civil aviation sector, which in turn, would promote tourism, increase employment and lead to a balanced regional growth. The Policy aims to take flying to the masses by making it affordable and convenient, enhance ease of doing business through deregulation, simplified procedures and e-governance and promote the entire aviation sector chain in a harmonized manner covering cargo, general aviation, aerospace manufacturing and skill development. It aims at 50 crore domestic tickets and 20 crore international tickets by 2027.

WATER TRANSPORT

Inland waterways were the chief mode of transportation before the advent of railways. Waterways are the cheapest means of transport and are most suitable for carrying heavy and bulky materials having low specific value. Water transport is a fuel efficient and environment friendly mode of transportation which has vast employment generation potential. According to Govt. study conducted by RITEC, one litre of fuel moves 24 tonne-km on road, 95 tonne-km on rail and 215 tonne-km on inland waterways.

However, water transport suffered a great setback due to following reasons :

1. Boats sailing in water could not compete with fast moving railways and roadways. Since the traders wanted the delivery of their goods at the destination points at the earliest, they started using railways and roadways and discarding waterways.
2. Water transport was full of dangers as boats were trapped in plung poals or were looted by decoits on their way. Such problems did not exist in case of rail or road traffic. Consequently railways and roadways were given priority over waterways.
3. Rivers are often flooded in rainy season and are not fit for water transport. Such a problem is not faced in case of rail or road traffic.
4. In dry season, most of the rivers dry up and there is not enough water in the river to facilitate water transport.
5. Most rivers in South India flow through the rocky undulating terrain and form waterfalls at several places. Thus they are permanently unfit for water transport.
6. Most of the rails and roads run parallel to the waterways and offer stiff competition to water transport.

In spite of the above mentioned difficulties and drawbacks, India has one of the longest navigable inland water networks. However, cargo transport through these inland waterways is presently less than one per cent of the total cargo movement in the country.

Inland waterways Authority of India (IWAI) was constituted in 1986 for the development and regulation of Inland waterways for shipping and navigation. About 111 inland waterways have been declared as 'National Waterways' under the National Waterways Act 2016. Out of these 111 waterways, 106 waterways were created in 2016. The National Waterways network covers around 20,275.5 km. National Waterways 1, 2 and 3 are already operational. Cargo as well as passenger/cruise vessels are plying on these waterways. Detailed Project Report (DPR) for development of National Waterways 4 and 5 was completed in 2010. For newly declared 106 National Waterways, techno-economic feasibility studies have been initiated.

National Waterways in India handled 72 million tonnes and 100 million tonnes cargo in 2018-19 and 2021-22 respectively.

Details of some of the major waterways is given in **table 26.19**.

Ganga is the most important inland waterway in India. It is navigable by mechanised boats upto Patna and by ordinary boats upto Hardwar. It has been declared as National Waterway No. 1. The entire route has been divided into three parts for development purposes. These parts are Haldia- Farakka (560 km), Farakka-Patna (460 km) and Patna-Pryagraj 600 km). The National Waterways (Pryagraj-Haldia stretch of Ganga-Bhagirathi- Hooghly River system) Act, 1982 has the provision that the regulation and

TABLE 26.19. Details of National Waterways

Name	River (Stretch)	States	Length (km)
NW1	Pryagraj-Haldia stretch of the Ganga-Bhagirathi-Hoogly River System	U.P., Bihar, Jharkhand, West Bengal	1,620
NW2	Sadiya-Dhubri stretch of the Brahmaputra River	Assam	891
NW3	Kollam-Kozhikode stretch of West Coast Canal and Champakkara Canal and Udyogmandal Canal	Kerala	365
NW4	Kakinada to Puducherry (Buckingham Canal)	Andhra Pradesh, Tamil Nadu, Puducherry	1,095
NW5	(*a*) Talcher-Dhamra stretch of Brahmani River—Khasure River— Tasilighai River, Pendua Nala, Dudhai Nala—Kani (*b*) Goenkhali-Charbatia stretch of Coovum Canal (*c*) Harbatia-Dhamra stretch of Matai River and Mahanadi Delta Rivers	Odisha and West Bengal	623
NW6	Barak River	Assam	121

development of this waterway is the responsibility of the Central Government.

The 'Jal Marg Vikas Project' (on National Waterway-1; River Ganga) has been launched with the purpose of ensuring navigation of 1500 to 2000 tonne vessels by developing infrastructure and a fairway of 2.2 to 3 metres depth between Varanasi and Haldia covering a distance of 1380 km at an estimated cost of ₹ 5,369 crore. The project is being implemented by the Inland Waterways Authority of India.

India's first multimodal terminal (MMT) at Varanasi was inaugurated on 12 November 2018. Second multimodal terminal was inaugurated at Sahibganj (Jharkhand) on 11 September, 2019.

Brahmaputra is also navigable by steamers upto Dibrugarh for a distance of 1,384 km which is shared by India and Bangladesh. Its 891 km long stretch from Sadiya to Dhubri in Assam has been declared as a National Waterway and is being developed as an important inland waterway. Ro-Ro (Roll on-roll off) services have commenced between Dhubri and Hastingimari in July 2017. Roll on-roll off ships are vessels designed to carry wheeled cargo such as cars, trucks, semi-trailer trucks, trailers and rail-road cars that are driven on and off the ship on their own wheels or using platform vehicle such as self propelled modular transporter.

Rivers of South India are seasonal and are not much suited for navigation. However, the deltaic channels of the Godavari, the Krishna and the Mahanadi, lower reaches of the Narmada and the Tapi, back waters of Kerala, Mandovi and Juari rivers of Goa serve as waterways. The Godavari is navigable upto a distance of 300 km from its mouth. The Krishna is used as a waterway upto 60 km from its mouth.

There are some navigable canals also which serve as inland waterways. Buckingham canal in Andhra Pradesh and Tamil Nadu is one such canal which provides water transport for a distance of 413 km. It runs parallel to the eastern coast and joins all the coastal districts from Guntur to South Arcot. It has been made an integral part of NH4. The other navigable canals are Kurnool-Cuddapah Canal (116.8 km), Son Canal (326 km), Odisha Canal (272 km), Medinipur Canal (459.2 km), Damodar Canal (136 km). Some of the irrigation canals of Uttar Pradesh and Punjab are also utilised for local transport.

Regulatory amendment through the Inland Vessels Act 2021, replaced over 100 years old Inland Vessels Act, 1917 (1 of 1917) and ushered in a new era in inland water transport sector. The object of the Act is to promote economical safe transport and trade through inland waters. It will ensure transparency and accountability in administration of inland water transportation and promote the case of doing business.

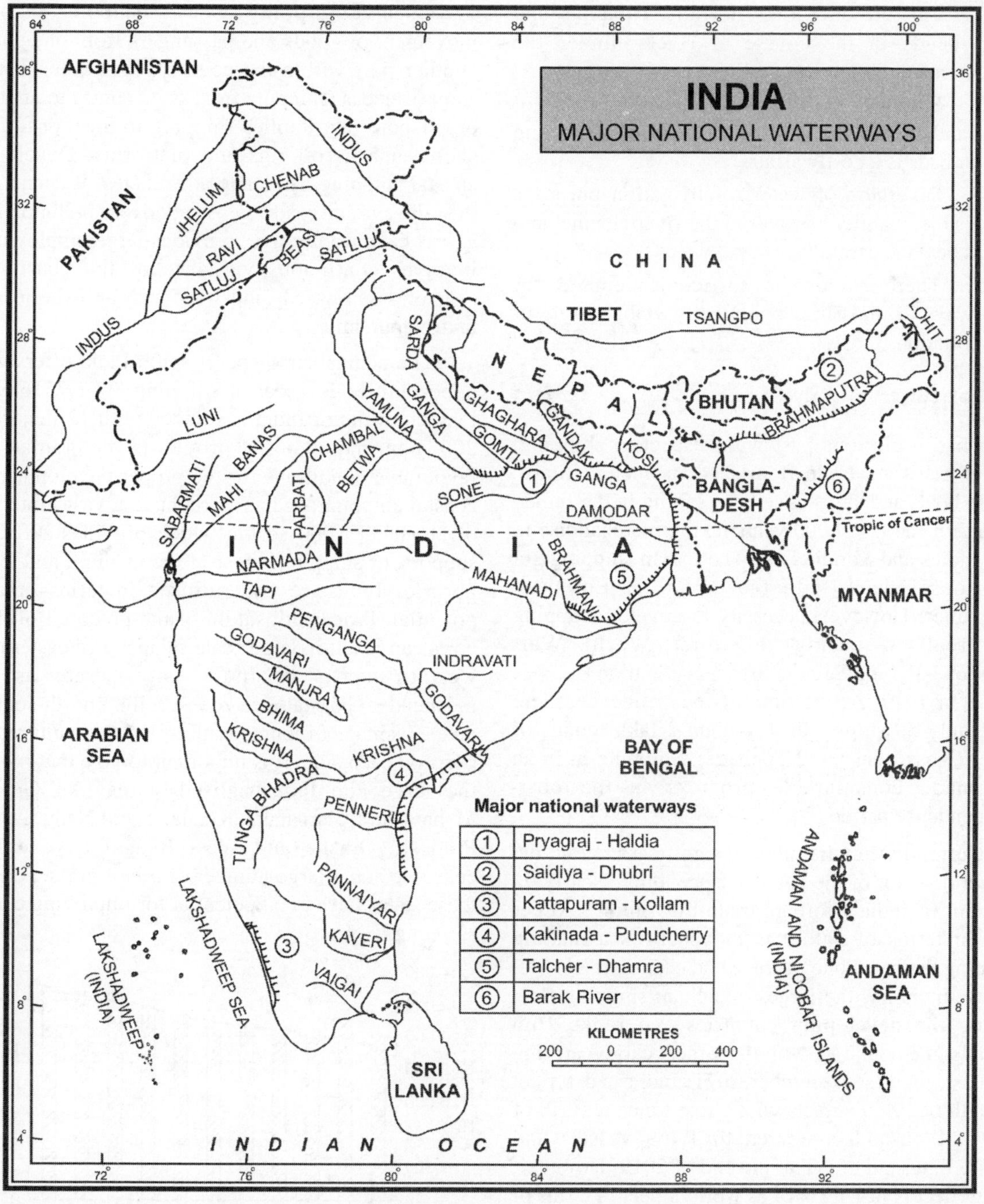

FIG. 26.15. Major National Waterways

In order to reduce the logistics cost of cargo and facilitate passenger movement between North East and mainland, MOUs have been signed with Bangladesh.

Factors affecting Inland Waterways

1. The rivers and canals should have regular flow of sufficient water.

2. The presence of waterfalls, cataracts and sharp bends in the course of rivers hinder the development of waterways.

3. Silting of the river bed reduces the depth of water and creates problems for navigation. Desilting of river beds is a costly affair.

4. Diversion of water for irrigation purposes reduces the quantity of water in the river channel and should be done carefully.

5. There should be sufficient demand for waterways to make it economically viable mode of transportation.

SHIPPING

India had a glorious past with respect to shipping. Indian maritime trade flourished in ancient times. Indian boats and ships have been sailing in the Indian Ocean for the last 4,000 years taking merchandise to East Indies and Middle East. The Indian shipping got a serious setback with the arrival of European companies. However, necessity to develop shipping was realised during the First World War. Consequently, the Scindia Steam Navigation Co. was set up in 1919. At the time of Independence, there were only 59 ships with less than 2 lakh tonnes of GRT (Gross Registered Tonnage). Shipping in India has made considerable progress in the post-Independence period.

Currently shipping plays a significant role in the transport sector of the country's economy. Nearly 95 per cent of India's foreign trade by volume (70 per cent in terms of value) is moved by sea making shipping the backbone of trade and economic growth. Today, India has the largest merchant shipping fleet among the developing countries and ranks 16th amongst the countries with the largest cargo carrying fleet. As on 31st December 2021, India had a fleet strength of 1,463 vessels with Gross tonnage (GT) of 13,011 thousand compared to 1,469 vessels and 12,746 thousand of GT at the end of 2019. However, Indian fleet is just 1.2% of as world's fleet in terms of capacity.

Thus it is clear that the present fleet strength is not adequate to support the trade flow in shipping sector. Therefore, there is an urgent need for augmenting the tonnage capacity in the coastal sector to meet the increasing demand.

Coastal Shipping. Coastal shipping involves movement of goods and passengers from one port to another port within the country. It is quite distinct from overseas (shipping from one country to another) or offshore that implies shipping to locations of the shore such as oil rigs and platforms. Developing coastal shipping has many advantages. It decongests the railways and roadways, is relatively pollution free, is less capital intensive, provides large employment, involves continuous vigilance of the coasts and promotes sea based industries such as fisheries and luxury tourism.

The peninsular shape of south India offers great opportunities for coastal shipping. The 7516.6 km long coastline of India is studded with 13 major and 200 non-major ports providing congenial and favourable conditions for the proper development of coastal shipping. Indian ports have a vast hinterland of about 3,80,000 sq km which provides sufficient support for shipping. Most of the medium and minor ports of India are underutilised in terms of their potential. Proper utilisation of the potential of these ports can result in large scale saving. The eastern and western coasts of India can operate as two independent coastal area zones, sailing north-south to take advantage of long coastline and the multitrade of ports on each side. It is important to note that most of the large and industrialised states like Gujarat, Maharashtra, Karnataka, Kerala, Tamil Nadu, Andhra Pradesh, Odisha and West Bengal have a long coastline and a large number of minor ports. As such these states have great potential for improving coastal shipping.

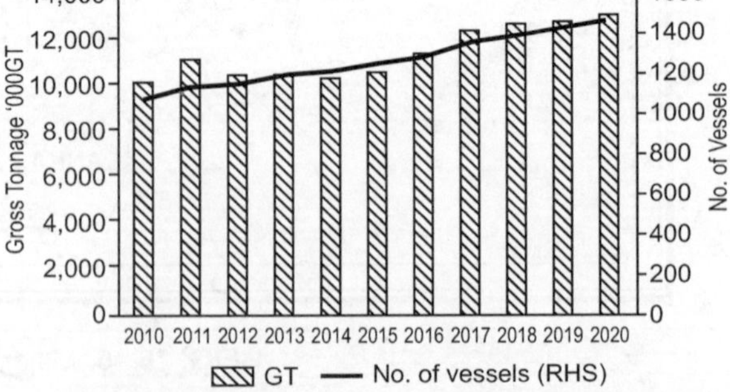

FIG. 26.16. Growth of Indian Shipping

***Source*:** Survey calculations based on data from Ministry of ports, shipping and waterways in Economic Survey 2021-22, p. 299.

Land transport involves huge capital expenditure —both initially and subsequently on maintenance— on roads and tracks, mobile equipment with high fuel consumption. A recent study of fuel costs involved in moving general or bulk cargo from Mumbai to Goa showed that sea mode was nine times less expensive than the road route.

Action plan for the development of coastal shipping is already on the anvil with the Central Government with a view to promote coastal shipping and sailing vessel industry, the home trade vessels and sailing vessels have been exempted from the payment of lighthouse dues. Meanwhile, a study has already been completed by the Tata Consultancy Services (TCS) to assess the potential of coastal shipping and the role of minor ports keeping in view the feasibility of routes and supporting environment needed for its development. Its report has been accepted by the Government in principle.

PORTS

There are 13 major and 200 medium and small ports in India. The major ports are under the supervision of the Central Government while the minor ones are managed by the concerned State Governments. The 13 major ports handle about 90% of our foreign trade. The major ports on the west coast are Mumbai, Jawaharlal Nehru, Kandla, Marmagao, Mangaluru and Kochi while on the east coast are Kolkata/Haldia, Paradwip, Vishakapatnam, Chennai, Ennor and Tuticorin. The number of cargo vessels handled at major ports is about 16,000 per annum.

The Major Ports in the country have an installed capacity of 1514.09 million tonnes per annum (MTPA) as in March, 2019 and handled traffic of 699.09 MT during 2018-19. While increasing the capacity of major ports, Ministry of Shipping has been striving to improve the operational efficiencies through mechanization, digitalization and process simplification. As a result key efficiency parameters have improved considerably. The traffic handled increased from 545.79 million metric tonnes in 2012-13 to 699.09 million metric tonnes in 2018-19. Similarly average turn around time decreased from 101.76 hours to 59.51 hours and average output per ship per berth day increased from 11800 tonnes to 16541 tonnes during the same period.

A brief description of major ports of India is given here :

1. Mumbai. It is a magnificent natural harbour on the West Coast of India. The deep 10-12 metre sea adjoining the harbour with no sand banks enables big ships to enter the port easily. It handles approximately one-fifth of India's foreign trade with predominance in dry cargo and mineral oil from the Gulf countries. It is the biggest port of India. It handles foreign trade with the Western countries and East African countries. The opening of the Suez Canal in 1869 brought it much closer to the European countries.

Mumbai has a vast hinterland covering the whole of Maharashtra and large part4s of Madhya Pradesh, Gujarat, Rajasthan and Delhi. This hinterland is very rich in agricultural and industrial resources. The entire hinterland has undergone large scale economic improvement which has helped in the rapid growth of this port. A dense network of roads and railways connects the port with its hinterland.

Mumbai is the gateway to India from the west and handles large scale trade of great variety. The major items of export are cotton textiles, leather, tobacco, manganese, machinery, chemical goods, etc. while the imports include crude oil, superior quality raw cotton, latest machines, instruments and drugs. This port is likely to progress further with the economic development of its hinterland.

2. Jawaharlal Nehru Port. Formerly known as *Nhava Sheva port*, this port was opened on 26th May, 1989. This port has been built at an island named Nhava Sheva across the famous Elepthanta caves, about 10 km from Mumbai. Built at the cost of ₹ 880 crore, this port is named Jawaharlal Nehru port as a tribute to the first prime minister of India. The main purpose of this port is to release pressure on the Mumbai port. The port is equipped with the most modern facilities having mechanised container berths for handling dry bulk cargo and service berth, etc. Most of the operations are conducted with the help of computers. The port is linked by road and rail to other railway routes and National Highways avoiding Mumbai city altogether. The initial capacity of the port was 5.9 million tonnes in 1995-96 which was raised to 9.9 million tonnes in 1997-98. The first private sector project Nhava Sheva International Container Terminal (NSICT) became operational in 2000. This terminal has already captured a substantial

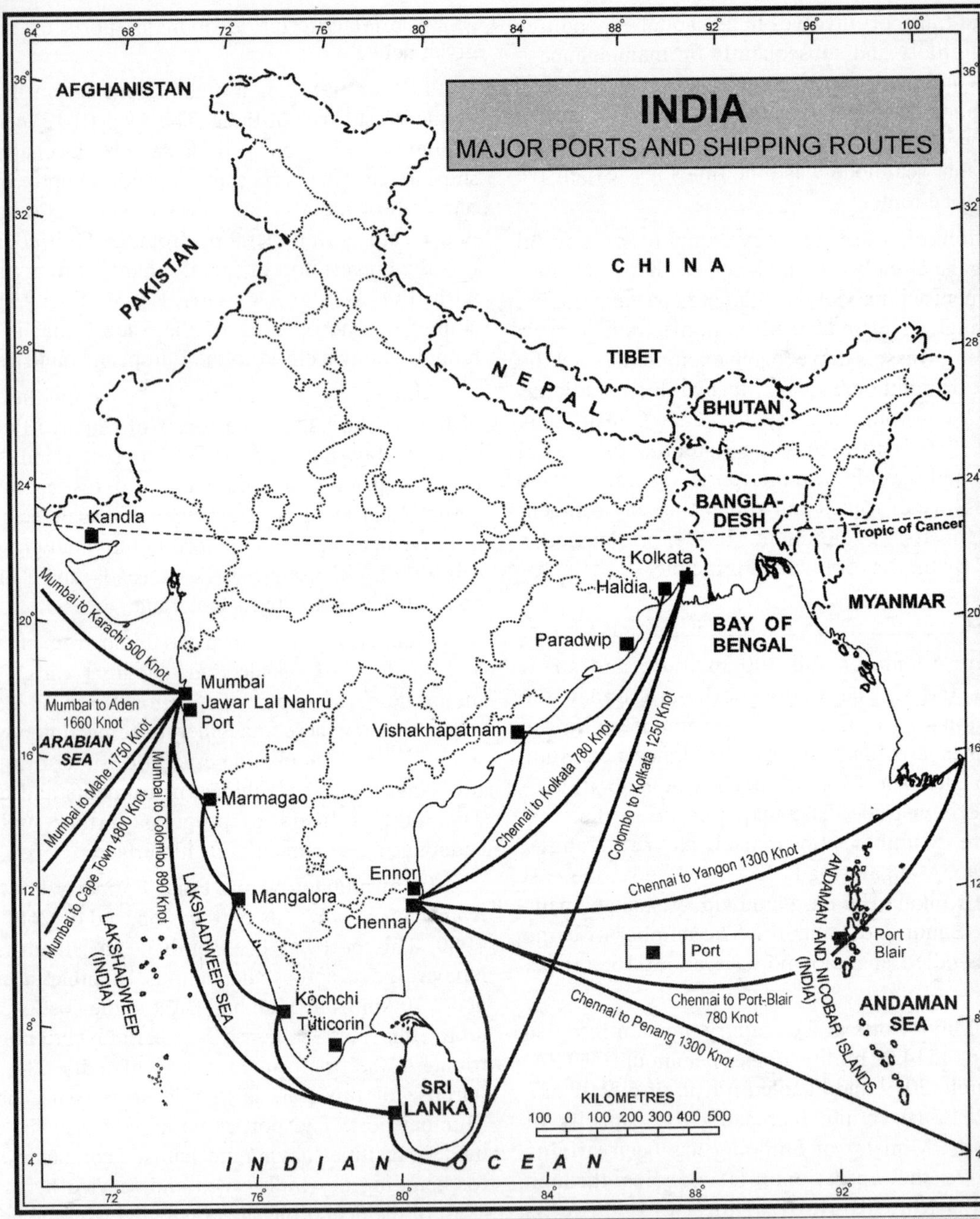

FIG. 26.17. India : Major Ports and Shipping Routes

share of container handling not only in the region but also among all the major ports of the country. The sea is quite deep near the port and there is no need for dredging in this port.

Jawaharlal Nehru Port Trust (JNPT) Navi Mumbai, signed an agreement with P&O Australia, for development of a two berth container terminal on 'Build, Operate and Transfer' (BOT) basis for a period of 30 years in July, 1997. The work was completed in April 1999 at the cost of ₹ 900 crore. Another agreement was signed on August 10, 2004 with Gateway Terminals India Pvt. Ltd. for

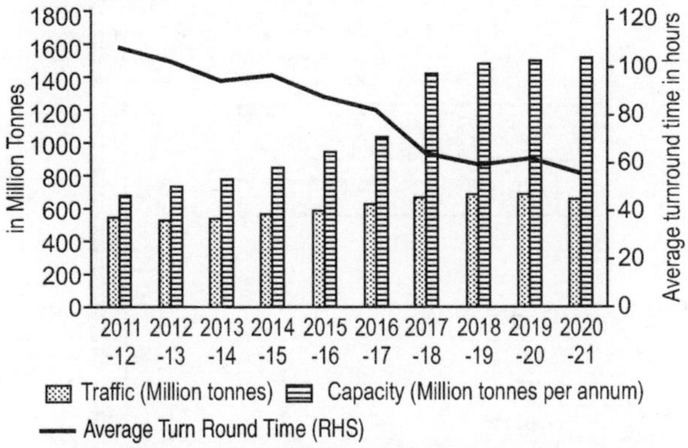

FIG. 26.18. Performance of Major ports

development of the third container terminal which will result in addition of 1.3 million TEUs (twenty feet equivalent units) container handling capacity of the port. JNPT is also taking preparatory action for setting up of fourth container terminal.

As a result of the above mentioned initiatives taken to enhance its capacity, Jawaharlal Nehru port has come to rank 28th among the world class container ports. Currently the port handles 55-60 per cent of the country's containerised cargo. This port handled vessels having draught upto 12.5 metres in 2011-12.

It is a twin port of Mumbai so far its hinterland is concerned.

3. Deendayal Port (Kandla). This port is located at the eastern end of Gulf of Kachchh about 48 km away from Bhuj. It is a natural sheltered harbour in Kandla creek with average depth of 10 m. The port is equipped with all the modern facilities and state-of-the-art technology. The traffic handled at Kandla consists of crude oil, petroleum products, fertilizers, food grains, salt, cotton, cement, sugar, edible oils and scrap. Kandla has a capacity of handling a total traffic of 23.3 million tonnes. This port has been developed after independence to relieve congestion on Mumbai port. With the loss of Karachi to Pakistan as a result of partition of the country in 1947, the necessity of constructing a port on the Gujarat coast was keenly felt. Consequently this port was constructed in 1951.

This port has a vast hinterland covering large parts of Gujarat, Rajasthan, Haryana, Punjab, Delhi, Himachal Pradesh, Jammu and Kashmir and Uttarakhand. The port is well connected by roads and railways and has a bright future.

4. Marmagao. It is an important port of Goa located at the entrance of Zuvari estuary and occupies fifth position in handling the traffic. Its harbour is protected and holds about 50 steamers in fair season and 15 steamers in rainy season. It has the capacity to handle 16.1 million tonnes of cargo traffic. For a long period, it handled the export of iron-ore from Goa. Currently the major items of exports are iron ore, manganese, coconut and other nuts, cotton, etc. Imports through this port are very few. It has a comparatively small hinterland covering the whole of Goa and parts of north Karnataka coastal region and southern Maharashtra.

With the opening of the Konkan railway, the importance of this port has increased significantly and it is fast emerging as a multi-commodity port. Four new harbours are being constructed in the Vasco Bay for handling container traffic and general cargo.

5. New Mangaluru. New Mangaluru port was declared as the 4th Major port on 4th May, 1974 and was formally inaugurated on 11th January 1975. This is an important port located at the southern tip of the Karnataka coast north of the Gurpur river. Initially it was designed for small ships. However, it was upgraded in the Fourth Five Year Plan to accommodate larger ships. A harbour was constructed so that business could be carried on throughout the year. Its hinterland lies in Karnataka and northern part of Kerala. Tea, coffee, rice, maize, wheat. cashew kernels, fish, rubber, oil products, iron ore pellets, granite, etc. are exported through this port. The major items of import through this port are crude oil, coal, iron ore fines, LPG, limestone, wooden logs, cement, liquid chemicals, fertilizers, edible oils etc. Its main importance lies in export of iron ore from the Kudremukh mines. The port is well linked through broad gauge rail line and NH-17 with Mumbai and Kanniyakumari.

6. Kochi. It is another natural harbour on the west coast of India and is located on the coast of Kerala. Kochi has a sheltered backwater bay. Located at Welllington island on the South-west coast of India about 930 km south of Mumbai and about 320 km north of Kanniyakumari. Kochi was given the status

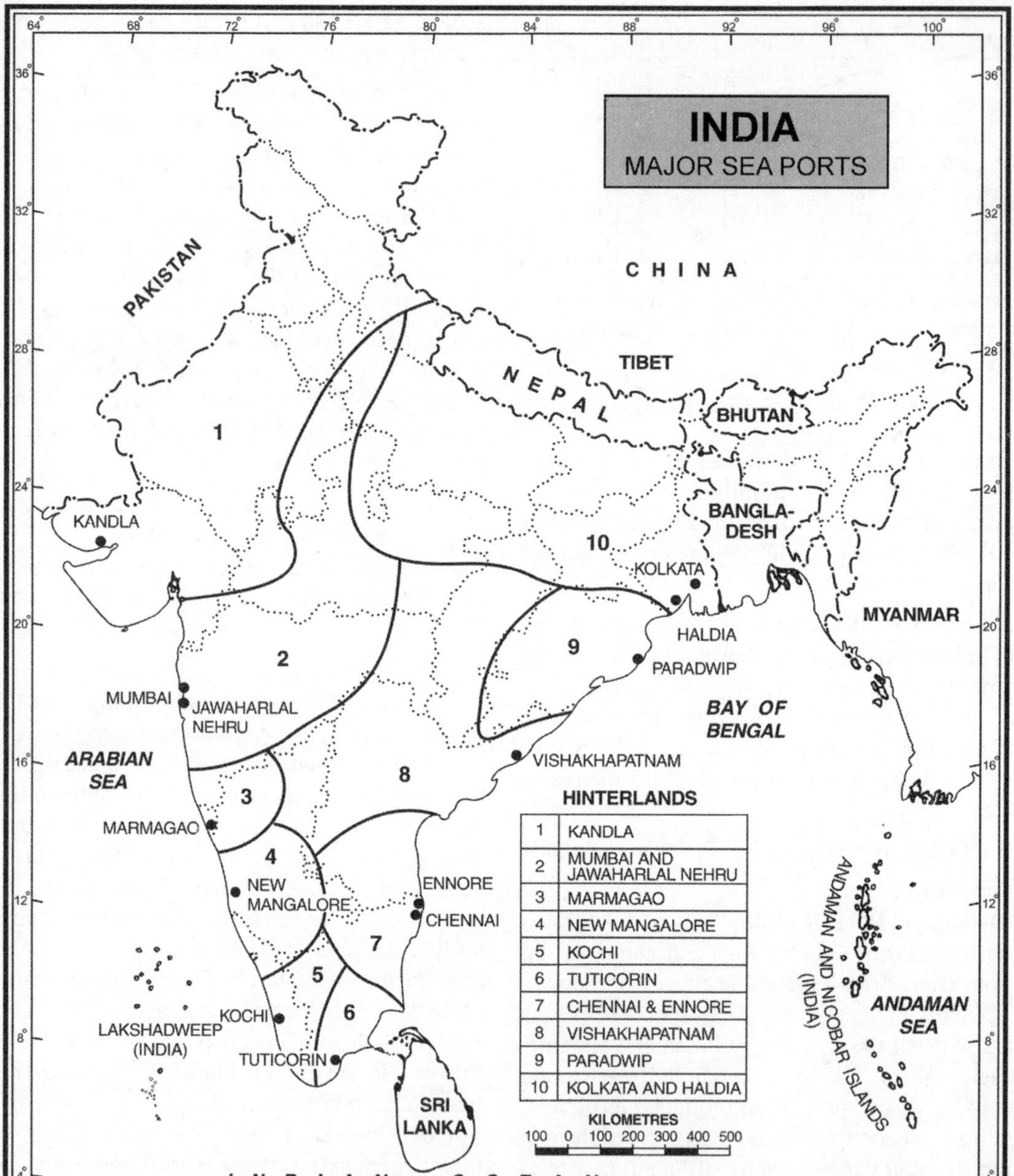

FIG. 26.19. India : Major Sea Ports

of Major Port in 1936. With its strategic location within the proximity of the international sea route between Europe in the west and the Far East and Australia in the east, it is capable of attracting a large number of container liners offering immense business opportunities. It handles the export of tea, coffee and spices and imports of mineral oil and chemical fertilizers. The Kochi Oil Refinery receives crude oil through this port. It is also a ship-building centre. Imports through this port far exceed the exports. In fact, imports are about five times more than the exports.

The hinterland of this port includes the whole of Kerala and parts of Tamil Nadu and Karnataka. About 97 per cent of the total volume of traffic is accounted for by Kerala alone. The port is the natural gateway to the vast industrial and agricultural market of south-west India.

7. Kolkata. It is a riverine port located on the left bank of river Hugli about 128 kms inland from the Bay of Bengal. Kolkata port handles goods coming from South-East Asian countries, Australia and New Zealand. Kolkata port is called the 'Gateway to Eastern India.' It is the world's most important centre of jute industry. Kolkata is the main port for exporting jute products, tea, coal, steel, iron ore, copper, leather and leather products, textiles, manganese and many more items. The imports consist mainly of machinery, crude oil, paper, fertilizers and chemical products.

Kolkata port suffers from a number of problems. It is located on the bank of the River Hugli, which suffers from the problem of silting as tidal bores enter this port frequently. Sandy bars and islands have been formed at several places. The river is in its old stage and bends at several places creating lot of problems for the ships because they do not find a straight passage from the coast to the port. To sum up *Kolkata port has a serious problem of 'bends', 'bars' and 'bores'*. The depth of water is gradually declining. This has necessitated constant dredging so that larger vessels are able to enter the port. Expert pilots are required to conduct the ships and the cost of mainitaining the port has become prohibitive.

To rid the Kolkata port of some of the problems mentioned above, a barrage has been constructed across the Ganga at Farakka. This barrage is designed to divert water along the Bhagirathi-Hugli. However, with the signing of water treaty between India and Bangladesh, adequate water is not available in the Hugli river. Only a change in the treaty can improve the situation.

Steps have been taken for modernising and replacing port craft; strengthening of cargo handling equipment; better utilisation of dry docks and deap drafted areas and promotion of ship-breaking activities.

Kolkata port has a vast hinterland. Almost whole of the eastern and north-eastern parts of the country are included in the hinterland of this port. The main areas comprising the hinterland of Kolkata are West Bengal, Bihar, Jharkhand, Uttar Pradesh, Uttarakhand, Sikkim, Assam, Arunachal Pradesh, Nagaland, Manipur, Mizoram, Meghalaya, Tripura and northern parts of Chhattisgarh and Madhya Pradesh. Two landlocked countries *viz*. Nepal and Bhutan are also parts of hinterland of this port.

8. Haldia port has recently been developed on the confluence of rivers Hugli and Haldi about 105 km downstream from Kolkata. Its main purpose is to relieve congestion at Kolkata. It receives larger vessels which otherwise would have to go to Kolkata. Some of the large vessels which cannot enter the Kolkata port can easily come upto Haldia. Haldia has an oil refinery and a fertilizer factory. A large integrated petro-chemical plant has also been set up here. An important rail link connects Haldia with Kharagpur. The main items of trade are mineral oil and petroleum products. Haldia-Dock complex has plans to develop a berth on BOT basis. The hinterland of Haldia covers the same territories as that of Kolkata although to a much lesser extent.

The Bhagirathi river has changed its course recently as a result of which Haldia's future has become uncertain. With the large-scale deposition of silt near the port (Nayachar), the entry of large ships has become difficult.

9. Paradwip. It is a deep water (depth 12 metres) and all weather port located on the Odisha coast about 100 km east of Cuttuck. Because of its great depth, this port is capable of handling bulk carriers of over 60,000 DWT. Construction of an exclusive oil jetty to handle about 6 to 8 million tonnes of petroleum products and crude tankers of 85,000 DWT was completed recently. Constructed in the Second Five Year Plan period, this handles iron-ore and coal along with some other dry cargo. Large quantity of iron ore is exported to Japan through this port. The imports through port are only half of its exports. The hinterland of Paradwip port is comparatively small and covers Odisha only.

10. Vishakhapatnam. It is the deepest protected port built at the coast of Andhra Pradesh. It was thrown open to commecial shiping on 7th October, 1933. Since then, this port has been consistently upgrading its infrastructure in tune with advancing technology. An outer harbour has been developed to handle the export of iron-ore. Elaborate arrangements have been made to handle crude oil and other petroleum products. It also handles fertilizers. Vishakhapatnam has a capacity of handling 16.7 million tonnes of cargo traffic. It also has the ship-building and ship-repair industry. The primary export items are iron ore (especially from Bailadila mines of Chhattisgarh to Japan), manganese ore, spices and wood. The imports comprise mainly of mineral oil, coal, luxury items and other industrial products.

The hinterland of Vishakhapatnam port commands an approximate area of 3.4 lakh sq km which is constituted by Andhra Pradesh, Telangana and the contiguous parts of Chhattisgarh, Madhya Pradesh, Maharashtra and Karnataka. This part of the country is very rich in mineral resources and agricultural produce.

11. Chennai. Chennai is the oldest artificial harbour on the East Coast of India established in 1875. It does not possess a natural harbour and an artificial harbour has been created in an area of 80 hectares near the coast. It mainly handles petroleum products, fertilizers, iron-ore and general cargo. The major items of exports are rice, textiles, leather and leather goods, tobacco, coffee, manganese ore, fish and fish products, coconut, copra etc. The imports consist of coal, crude oil, paper, cotton, vehicles, fertilizers, machinery, chemical products, etc. It has a capacity to handle a traffic of 21.37 million tonnes and can accommodate as many as 21 vessels inside the harbour. Chennai is often hit by cyclones in October and November and shipping becomes difficult during these months. It is ill-suited for large ships because of the lesser depth of water near the coast.

The hinterland of the Chennai port encompasses a large part of Tamil Nadu, southern part of Andhra Pradesh and some parts of Karnataka. The port is gaining importance due to increased significance of its hinterland.

12. Kamarajar Port Limited (Ennore). This port has recently been developed to reduce pressure of traffic on Chennai port. It was commissioned in 2001. Located slightly to the north of Chennai on the Tamil Nadu coast, this is the country's first corporate port. It envisages construction of two coal berths, one iron ore berth, one LNG berth, two POL/liquid chemicals berths and one crude oil berth for handling very large crude carriers. It has also a perspective plan to build three multi-purpose berths, five POL/liquid-chemicals berths and five container berths. The major items of traffic on the port are coal, iron ore, petroleum and its products, chemicals, etc. Its hinterland is a part of the hinterland of Chennai port.

13. V.O. Chidambarnar Port (Tuticorin). This port has also been recently developed on the Tamil Nadu coast about 8 km south-west of the old Tuticorin port. It has an artificial deep sea harbour which can accommodate vessels upto 8 metre draft in any season of the year. Two new berths are being developed. There is a plan to increase the depth of the port from the present 10.7 metres to 12.8 metres so that larger vessels could be accommodated. A long-term plan to build an outer harbour in four stages is also envisaged. The idea is to handle containers upto 6.3 million TEUs. The port handles the traffic of coal, salt, food grains, edible oils, sugar and petroleum products. Its main purpose is to carry on trade with Sri Lanka as it is very near to that country. Its hinterland is formed mainly by southern Tamil Nadu comprising districts of Madurai, Kanniyakumari, Ramnathpuram, Tirunelveli and southern part of Tiruchchirappally. It is well connected by railways and roads.

Minor Ports. There are 200 minor ports, with a pronounced accent on the west coast. These ports are located in Gujarat, Maharashtra , Goa, Daman and Diu, Karnataka, Kerala, Lakshdweep, Tamil Nadu, Punducherry, Andhra Pradesh, Odisha, West Bengal and Andaman & Nicobar.

Sagarmala Project

Originally conceived during former Prime Minister Atal Bihari Vajpaee's regime, this project was put in place in 2016 for industrial development along the coastline and develop smart cities near key ports. The project focuses on port modernization and capacity augmentation, efficient and speedy evacuation besides port-led industrial development of coastal communities. Following the Chinese model of growth driven by port-led development, the government has identified four industries which have strong linkages to ports—thermal coal, petroleum, oil and lubricants, steel and containers—to accelerate growth. These four industries contribute nearly 80% of the port traffic. In all 173 projects have been initially identified under four projects archetypes of Sagarmala. The project archetype and the number of projects identified are given in **Table 26.20**.

TABLE 26.20. Project under Sagarmala

Project	Number
1. Port modernisation	53
2. Port connectivity	83
3. Port let Industrialisation	29
4. Coastal community development	8
Total	173

Source : *India 2017 : A Reference Annual*, p. 734.

India is aiming at ₹ 11 lakh crore of investment in port sector, besides developing the port capacity by 2025. There is a strong pitch for turning the country's more than 7,500 km long coastline as an '*engine of growth*'. The aim is to increase the port capacity from 1400 million tonnes to 3000 million tonnes by 2025. It also targets to push coastal shipping by over three-fold and building rail and road networks for quicker evacuation of cargo traffic. There is a plan to integrate ports with Special Economic Zone (SEZ), smart cities, industrial parks and logistics parks. The project will boost export to $ 110 billion.

The Plan in Nutshell

- ₹ 1 lakh crore investment for port modernisation, including development of six new ports.
- ₹ 2,000 crore for coastal community development including skill development and fisheries.
- ₹ 1 lakh crore for infrastructure development for port led development.
- ₹ 2 lakh crore for improvement of port connectivity.

Major Targets

- Increasing port capacity from 1,400 million tonnes to 3,000 million tonnes.
- Integrated plan for 90% cargo movement to save ₹ 40,000 crore annually from logistic cost and time reduction.
- Road corridor from Mangalore to Chennai and Expressway from Hyderabad to Jawaharlal Nehru Port Terminal.

COMPETITIVE AND COMPLEMENTARY CHARACTER OF MEANS OF TRANSPORTATION IN INDIA

In India, there are four major modes of transportation *viz.* road, rail, air and water transport. To this may be added pipeline transportation. Air transport is the fastest and the costliest whereas water transport is the slowest and the cheapest mode of transportation. They serve entirely different purposes and have practically no competition and complementarity. Pipeline transport is of recent origin and is restricted to specific corridors only. Moreover pipelines are useful for transporting liquids and gases only and have a unique character of their own. Thus the main issue of competitiveness and complementarity is with respect to road, railways and water transport system. The problem of competition arises in a situation when optimum utilization of one mode of transportation is discarded and some other mode of transportation is given priority. As against this a situation of complementarity arises when one mode of transportation helps the other mode in such a way that both are used to their maximum capacity. Roadways and railways are complementary at certain places and under certain conditions whereas they are competitive with each other at some other places and in some other conditions.

Complementary Role of Different Modes of Transport

Generally speaking, roadways are useful for short distances whereas railways are more suitable for long distances. Roadways play an important role in carrying agricultural produce from fields, minerals from mining areas and industrial products from factories to the railway yards. Usually, these items are carried by trucks from fields, mines, and factories to the railway yards over distances varying from 100 to 150 km. It is for the railways to carry these goods in the wagons over longer distances. After reaching the destination railway yard, the goods are again transferred from the rail wagons to the trucks which carry the goods to their ultimate destination. Thus roads and railways play a complementary role for each other for short and long distances respectively.

Like freight, roads and railways play a complementary role for short and long distances with respect to passenger traffic also. Roadways are always given preference over railways for travelling over short distances. This is primarily due to flexibility possessed by the roadways. Vehicles can be stopped anywhere on the road-side. As against this, railways completely lack flexibility because railway trains stop at fixed stations only. India is par excellence, a country of villages and most villagers travel in buses to reach the nearest railway station and continue their journey to a distant railway station. After completing their rail journey they get down and board the buses to reach their ultimate destination.

It is because of the above mentioned complementary character of roadways and railways that the government has been laying more emphasis on construction of village roads. Pradhan Mantri Gram Sadak Yojna (PMGSY) was launched to provide single all-weater road connectivity to villages.

Normally, road transport is prefer for journey extending upto 200-250 km and railways are preferred for journey more than this distance. Much of the long distances journey by rail is performed at night because railway coaches are equipped with sleeper and toilet facilities. But people living in areas surrounding the metropolitan cities prefer to travel by rail because railways offer monthly, quarterly and half-yearly passes at concesional rates to commuters. But these commuters use road transport to reach the railway station from their place of residence. Thus rail and road transport play a complementary role with respect to commuters also.

Rail transport is almost completely absent in the Himalayan region (Jammu and Kashmir, Himachal Pradesh, Uttarakhand and north-eastern states). Under such a situation, the passengers travel long distances through the hilly areas to reach the rail head terminals at the foothills to continue their further journey by railway transport. Kalka, Hishiarpur, Dehra Dun, Kathgodam and Siliguri are such a railway terminals. People of the north-eastern states are almost completely deprived of the rail transport service. They have to travel long distances by road before they can reach any railway terminal from where they can reach their destinations situated at longer distances.

The Government of India has prepared an ambitious plan to improve inland waterways along major rivers of the country. The inland water transport system had been playing a vital role for transporting passengers and goods in early stage of development. But they lost much of their importance to road and rail transport because they could not compete with respect to speed. With growing pressure on road and rail transport and with crumbling infrastructure, inland water transport stands fair chances of reviving its old lost glory. As many as 111 inland waterways have been declared as National Waterways under the National Waterways Act, 2016. They are expected to play an important complementary role for both rail and road transport and relieve these modes of transport of much burden in the near future, particularly with respect to transportation of heavy and bulky commodities with low specific value. This is because water transport for heavy and bulky commodities is always much cheaper than roads and railways.

Roadways and railways play a major complimentary role to waterways with respect to international trade. The whole material for export is transported from the distant inner parts of the country to the ports by roads or railways from where it is exported to various countries by water transport. Similarly, imported goods imported through water transport are carried by roads and railways from the ports to different parts of the country. This is how ports develop their hinterlands. When the British rulers planned the development of rail routes in India, they paid due consideration to complimentarity that existed between ocean routes and railway routes. This is the reason that all major railway lines were constructed to serve the major ports of India such as Mumbai, Kolkata and Chennai. The purpose was to carry raw materials from different parts of India to ports by roads and railways and then to Britain and finished industrial goods from Britain to ports and then to different parts of the country by roads and railways.

Competitive Role of Different Modes of Transport. The Government of India introduced scheme of grouping national permit to road transporters in 1930s as a result of which trucks and buses started covering long distances across the inter-state borers. This initiated the period of competition between roads and railways. Road transport is now more preferred upto a distance of 250 km and there is a tough competition between roads and railways within this limit of the distance covered. But railways are still preferred for distances exceeding 400-500 kilometres. Whereas roadways offer more flexibility, over short distances, railways provide more comforts over long distances.

A casual look at the rail and road map of India reveals that in most parts of the country, railways and roadways run parallel to each other. This further adds to competition between these two modes of transportation as passengers and goods are divided between the two. Several surveys have indicated that railways are fast losing to roadways, particularly with respect to short distances upto about 250 km. For example most passengers from Delhi prefer to travel by road if they have to go to Chandigarh, Jaipur, Agra, Dehra Dun, Hardwar and *vice versa*. Similarly traders prefer their goods to be carried by trucks between Delhi and these destinations because trucks can provide door to door service.

However, railways have an edge over roadways with respect to travel to some places of tourist interest. For example tourists prefer rail travel over road travel from Kalka to Shimla from Shiliguri to Darjeeling and from Pathankot to Kangra although buses and taxies take much less travel time between these stations. This is because of the fact that railway routes follow meandering path through the hilly areas and offer attractive spots of natural beauty which almost mesmerize the tourists. A similar competition is found between railways and roadways for reaching Ooty in Tamil Nadu.

Even water transport offers tough competition to other modes of transportation in certain parts of India. **Figure 26.20** shows that road transport is cheaper for short distances, railway transport is cheaper for long distances and water transport is cheaper for very long distances. Any mode of transportation aims at recovering the initial cost as well as running or recurring cost from the passengers or goods transported by it. The road transport involves much less initial cost but the moving cost is very high. Therefore, road transport is the cheapest for short distance upto point A, beyond which rail transport is the cheapest upto point B. This is due to the fact that the initial cost in case of railways is much higher than that involved in roadways, but moving cost for railways is much lower than for roadways. Initial cost in case of water transport is extremely high because huge amount of money is required for building ships,

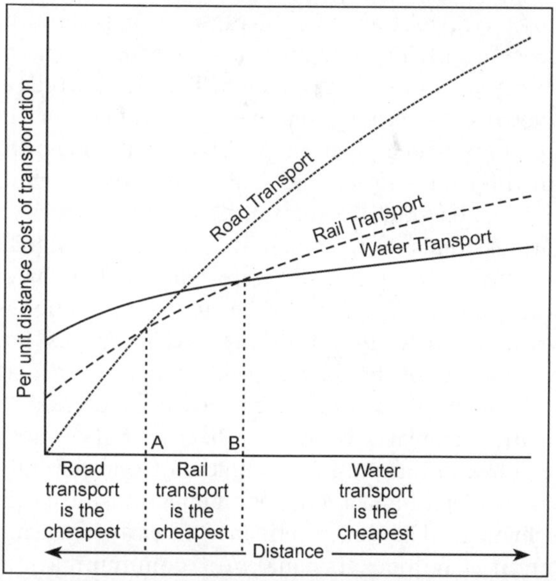

FIG. 26.20. Competition between road, rail and water transport

boats, steamers, ports and harbours. But moving cost is minimum in case of water transport because no expenditure is involved in maintaining water ways as against expenses for maintaining roads and railways tracks. Thus water transport is the cheapest for distances beyond point B.

Industries tend to be located in those areas where cost of transport is minimum. This is known as ***'Principle of Least Cost of Transport'***. This principle is not applicable to India because inland water transport is not properly developed here. However, some cases of competition between road and water transport can be found in some parts of the country. For example, there is tough competition between road and water transport in the Ganga-Brahmaputra delta region with respect to transportation of jute because most of the jute mills are located along the banks of the Hugli river. Similarly, lakes, lagoons and backwaters in the coastal region of Kerala have been joined with one another by canals which provide excellent facilities of inland water transport. These water ways offer stiff competition to road transport in this region.

COMMUNICATION

Communication is different from transport. Whereas transport involves the physical carriage of passengers and goods from one place to another, communication only involves transmission of word message and ideas. In early days the only way a message could be conveyed was by word of mouth. The use of words as a means of passing on messages relied on the proximity of the giver and receiver of the message, but in order to extend the range over which information could be transmitted, signal systems of various kinds were developed. The main signals through which messages were conveyed included drums, smoke signals or fly signals. The invention of writing helped in preserving the messages and communication became dependent upon transport. Written messages and letters could be carried by hand, by animals, by boat and later by rail or motor vehicles. In this way development of transportation had a deep impact on communications. But the technological advancements since the early nineteenth century helped in making communications independent of transport. Telecommunication (telegraph, telephone, radio, television, satellite, computer) are independent of transport network. But postal services still depend upon transport network.

Communication system contributes to the development of the economy, social relationships and also helps in promoting cultural unity. Internationally, it brings diverse people of the world close to one another. In the event of any impending calamity, accident or emergency, instant means of communication flash the news across the globe so that relief can be rushed to the spot immediately.

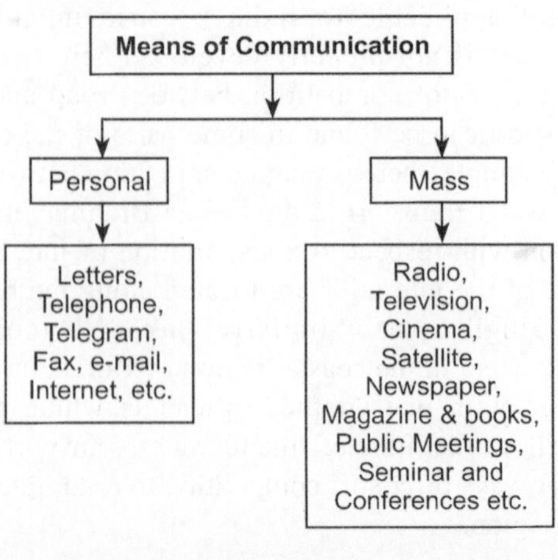

Communication is of two types *viz.* 1. personal communication and 2. mass communication. Personal communication takes place between individuals and is made through postal network, telecom (telephone, telegraph, telex etc.) supported by computers. Mass communication, on the other hand, involves communication with the masses. Radio, television, cinema, press and print media and satellites are the main media of mass communication.

Personal Communication

Following are the chief modes of personal communication.

POSTAL SERVICES

It is the most commonly used mode of communication in India, the potal services play a vital role in the rural areas of the country.

The modern postal system was established in India by Lord Clive in 1766 and was further developed by Warren Hastings in 1774.

The early postal system of India was solely used for official purposes and it was not until 1837 that the postal services were opened to the public. The first postage stamp was issued in Karachi in 1852, valid only in Sind province. In 1854, the Indian Post Office was reorganised as an institution, with a Director General-in-charge. 700 post offices were then already in existence. Since then, postal services have grown both in terms of the extent of the postal network and its scope and the variety of the services it offers.

Indian Postal network is the largest postal network in the world. As on 31st March, 2021 the country had 1,55,618 post offices, with 5,66,000 employees. At the time of Independence there were 23,344 post offices throughout the country. Of these 19,184 post offices were in the rural areas and 4,160 post offices in urban areas. The postal network has grown over six folds since then. On an average, a post office serves an area of 21.23 sq km and a population of 7,815. Post offices in rural areas are opened subject to satisfaction of norms regarding population, income and distance laid down by the Department for the purpose. The element of subsidy involved in opening of post offices in hilly, desert and inaccessible areas is to the extent of 85 per cent of the cost, whereas the subsidy in opening of post office in normal rural areas, is to the extent 66.6 per cent of the cost.

MAIL SYSTEM

First-class mail, *viz.,* post cards, inland letter cards and envelopes, are given air lift, without any surcharge, between stations connected by air. Second-class mail, *viz.,* book packets, registered newspapers and periodicals are carried by surface transport, *i.e.,* trains, buses and other means.

(*i*) (Postal Index Number) PIN has facilitated the prompt delivery of mail (*ii*) Speed post service has been introduced for fast and quick delivery of post (*iii*) Quick Mail Service (QMS) is another step in this direction.

Financial Services

In addition to delivering letters, parcels, and other important documents to the addresses, the Department of Posts is operating Small Savings Schemes of various types. Post Office Savings Bank has a customer base. As many as 1.56 lakh post offices are providing savings bank facilities to the common man. The Post Office saving bank operates savings accounts, Recurring Deposit (RD), Time Deposit (TD), Monthly Income Scheme (MIS), Public Provident Fund (PPF), National Saving Certificate (NSC), Kisan Vikas Patra (KUP), Senior Citizens Savings Scheme (SCSS), Sukanya Samridi account, etc.

Telecommunication

It is a device for the communication at individual and mass level. Telegraph, Telephone, Telex and Fax are the main means of telecommunication. The rise of telecommunication is directly linked with the advancement of electronic technology. Telecommunications brought about a revolution in communication system because of the high speed with which messages could be communicated. It became possible to send messages within a couple of minutes with the help of telegraphs which earlier took weeks together to reach their destinations. Further improvements made it possible to receive messages instantaneously. People on opposite sides of the globe can speak to each other directly with the help of telephone, radio, television, fax, internet, etc. Telecommunication services were introduced in India soon after invention of telegraphy and telephone.

1. Telegraph. First Telegraph line between Kolkata and Diamond Harbour was opened for traffic in 1851 just after seven years of invention. By March 1884, telegraph messages could be sent from Agra to Kolkata. By 1900, telegraph had started serving Indian Railways. Telegraph has now become outdated with the development of telephones, internet, e-mail, etc. In fact it is completely closed down in India now.

2. Telephone. Telephone was invented by Alexander Graham Bell in 1875. The invention of telephone made possible direct and instantaneous links from one part of the world to another. A close network of telephone wires and undersea cables was laid to provide facility of telephonic links in different parts of the world.

As in the case of telegraph, telephone service was also introduced in Kolkata in 1881-82, barely six years after invention of telephone. First automatic exchange was commissioned at Shimla in 1913-14 with a capacity of 700 lines.

The Department of Telecommunication (DoT) has improved significantly since Independence both in quality and quantity. Initially, the exchanges were of manual type, which subsequently were upgraded into automatic electro-mechanical type. In the last two decades, a significant qualitative improvement has been brought about by inducting Digital Electronic Exchanges in the network on a large scale. Today 100 per cent telphone exchanges in the country are of electronic type.

The telecom sector has been one of the fastest growing sectors in recent years. It is now the second largest telephone network in the world, after only China. A series of reform measures by the government, wireless technology, and active participation by the private sector played an important role in the exponential growth of the telecom sector in the country.

The total telephone subscriber base in India increased from 933.02 million in March, 2014 to 1200.88 million in March 2021. In the year 2021 about 45% subscribers were based in rural India and 55% subscribers were from urban areas (**Fig. 26.21**).

The wireless telephone now constitutes 98.27 per cent of all the subscription whereas the share of landline telephone now stands at only 1.73 per cent.

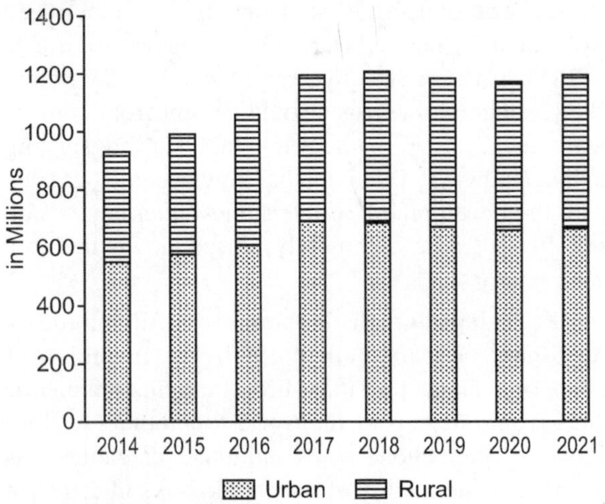

FIG. 26.21. Number and Composition of Telephone Subscribers (millions)

The overall tele-density (*i.e.* number telephones per 100 persons) is 88.51 per cent, the rural tele-density being 60.33 per cent and urban tele-density being 141.62 per cent at the end of July 2021.

Mass Communication

In a country of vest dimensions like India, mass communication plays a vital role in creating awareness among the masses, providing information and education as well as healthy entertainment. Electronic media (radio and television) and print media (newspapers and periodicals) are the main components of mass communication.

Prasar Bharati

Prasar Bharati is the public service broadcaster in the country, with All India Radio and Doordarshan as its two constituents. It was constituted on 23 November, 1997. The major objectives of the Prashar Bharati are :

(*i*) Upholding the unity and integrity of the country and the values enshrined in the constitution.

(*ii*) Promoting national integration.

(*iii*) Safeguarding citizen's rights to be informed on all matters of public interest and presenting a fair and balanced flow of information.

(*iv*) Paying special attention to the fields of education and spread of literacy, agriculture, rural development, health and family welfare and science and technology.

(*v*) Creating awareness about women's issues and other vulnerable sections of the society.

(*vi*) Providing adequate coverage to the diverse cultures, sports and games and youth affairs.

(*vii*) Promoting social justice, safeguarding rights of working classes, minorities and tribal communities.

(*viii*) Expanding broadcasting facilities and promoting research and development in broadcast technology.

Radio

Radio is a powerful medium which provides all sorts of useful information, news and variety of entertainment. Radio broadcasting began in India in early 1920's. The first programme was broadcast in 1923 by Radio Club of Bombay. This was followed by setting up of a Broadcasting Service in 1927 on experimental basis in Bombay (Mumbai) and Calcutta (Kolkata). The Government took over the transmitters and began operating them under the name Indian Broadcasting Service. It was changed to All India Radio (AIR) in 1936.

At the time of independence *i.e.* in 1947. AIR had a small network of six stations and 18 transmitters. The coverage was 2.5 per cent of the area and 11 per cent of the population of the country. Today AIR has a network of 225 broadcasting centres covering 91.42 per cent of the area almost reaching the entire population. Operating in a linguistically diverse country like India, AIR broadcasts in 24 languages and 146 dialects.

AIR operates its services through five channels namely : (*i*) Primary Channel, (*ii*) National Channel, (*iii*) Commercial Broadcasting Service (Vividh Bharti), (*iv*) FM Channels and (*v*) External Services Channels.

There has been a phenomenal progress in the news broadcasting by AIR. From 27 news bulletins in 1939-40. AIR now puts out 346 bulletins daily. News Headlines for every hour are broadcast on AIR FM, Delhi, AIR News on Phone was introduced in 1998.

Doordarshan

Doordarshan (DD), the national television service of India is one of the largest terrestrial networks in the world. Television broadcasting assumes tremendous importance in a developing country like India, which has low literacy rate and has varied cultures and multiple languages. Today, television is the most efficient and effective way to disseminate information and educate the masses. The country today has both terrestrial and satellite broadcasting services.

The first telecast in India orginated from a makeshift studio in the Akaswani Bhawan, New Delhi on 15 September, 1959. The regular service with a news bulletin became operational in 1965. After seven years, the second television channel commenced service in Mumbai, By 1975, TV service was available in Kolkata, Chennai, Srinagar, Amritsar and Lucknow.

The first experiment with satellite technology in India was conducted in 1975-76 under the programme Satellite Instructional Television Experiment (SITE). This was incidently, the first attempt in the world to use satellite broadcasting for social education. Colour transmission was introduced during the Asian Games held in New Delhi in 1982. The year 1982 also witnessed the introduction of a regular satellite link between Delhi and other transmitters, heralding the arrival of the National Network. Metro Channel was introduced in 1984.

Doordarshan presently operates 30 channels. It has three-tier programme service—National, Regional and Local. The National programmes are concerned with events and issues of the entire nation. The regional programmes cater to the interests of a particular state or region. The local programmes are area specific and cover local issues featuring local people. A new entertainment channel of Doordarshan, DD Bharati was launched on 26 January, 2002.

The programmes telecast by Doordarshan include news, current affairs, science, cultural magazines, documentaries, music, dance, drama, serials and feature films. Government policies, development programmes and current affairs are regularly telecast. It also transmits educational programmes for schools and universities. There are different channels for different types of programmes.

Several channels on television have been made available to serve entrepreneurs. This has promoted a keen and healthy competition to improve the quality of programmes. At present programmes of all types can be viewed at all the twenty-four hours of the day.

Cinema

Cinema is yet another powerful means of mass communication. It entertains millions of people everyday throughout the country. Feature films are being produced in India since 1912-13. The initial productions consisted of silent films. The era of silent films was overtaken by the talkie era in 1931 when Ardeshir Irani (1886-1969) produced *Alam Ara*. India now leads the world in the annual output of feature films. Before the advent of the TV era, cinema was the main source of entertainment for the masses but lost much of its viewership to TV. However, it has once again captured its viewers and regained the lost glory. In India, films can be publically exhibited only after they have been certified by the Central Board of Film Certification (CBFC).

Press and Print Media

Newspapers, periodicals and journals fall in the category of print media. According to the report of the Press Registrar, the total number of newspapers and periodicals being published in India was 65,032 in 2007. There were 7,131 dailies, 374 bi/triweeklies, 22,116 weeklies, 8,547 fortnighlies, 19,456 monthlies, 4,470 Quarterlies, 605 Annuals and 2,333 other periodicilies. Newspapers were published in as many as 123 languages and dialects during 2007. The largest number of newspapers (20,589) were published in Hindi followed by English (7,596) and Marathi (2,943). Daily newspapers were brought out in all the principal languages except Kashmiri.

Newspapers were published from all states and union territories. During 2007, the largest number of newspapers were published from Uttar Pradesh (8,397) followed by Delhi (6,926), Maharashtra (6,018) and Madhya Pradesh (3,555). Uttar pradesh continued to have the largest number of daily newspapers (841), followed by Maharashtra (573), and Karnataka (479). Bombay Samachar, a Gujarati daily published from Mumbai since 1822 is the oldest existing newspaper.

Computers

"Computer is an electronic device which is used for electronic data processing. It accepts the data, processes the data and converts the data into meaningful information. It is also used to perform mathematical and logical operations." This simple definition of computer amply proves that computers have a wide range of uses and play a dominant role in the sphere of communication system. Basically, a computer performs the following four functions :

(*i*) It accepts data as input.

(*ii*) It stores data, keeps it in its memory, and recalls the same as and when required.

(*iii*) It processes data as per instructions given to get required information.

(*iv*) It communicates the information as output.

The versatile use of computers in so many different fields is the outcome of its special capabilities in terms of speed, accuracy, consistency, storage capacity and automation. It plays an important role in the fields of education and transfer of knowledge by dint of its aforesaid qualities.

INDIAN SPACE PROGRAMME

India's space ventures date back to the ancient times when fireworks based on Chinese technology were first used in the country. Use of rockets by the Indians during the Mysore war against the British rulers inspired William Congreve to come out with Congreve rocket in 1804. Indian scientists and politicians recognised the significance of rocket technology and steps were taken immediately after Independence to develop this branch of science for defence, and for research and development in other fields. A young Indian British Veteran Sarabhai (1919-1971) founded a research laboratory for the study of cosmic rays. Another distinguished scientist, Homi Bhabha (1909-1966), also studied cosmic rays in 1940s. He realised the potential of satellites and Indian National Committee for Space Research was formed in 1962. The purpose of this committee was to identify the country's goals and priorities in the field of space research. Thumba near Thiruvananthapuram was chosen as an international facility for launching sound rockets and it became operational in 1963 for launching foreign rockets to study atmosphere. The main reason for selecting Thumba was its proximity to the magnetic equator of the earth.

The Indian Space Research Organisation (ISRO) prepared a programme to make sounding rockets and their propellants at Thumba. A Space Science and Technology Centre was also set-up at Thumba. Thumba Complex was renamed as Vikram Sarabhai Space Centre in honour of Sarabhai who made an outstanding contribution to the development of space science and technology in India.

The island of Sriharikota in Andhra Pradesh about 80 km north of Chennai, was chosen as a launch centre and it became operational in 1971 when Rohini-125, a sounding rocket was launched from here. Sriharikota has been renamed as Satish Dhawan Space Centre in honour of the late Prof. Dhawan who played a key role in realising the dream of Sarabhai. This centre has several tracking radars, including a Doppler Weather Radar (DWR) for detecting wind velocity in real time and for understanding severe weather conditions. Meanwhile, a space centre as set-up at Ahmedabad in 1967 to provide the necessary thrust for application of space technology in different fields of development. A satellite project as also established at Bengaluru in 1972 which was developed as ISRO Satellite Centre later on. This centre has the provision for design, development and launching of all spacecraft. Centres have also been established at Balasore in Odisha, Lucknow in U.P., Car Nicobar in Andaman and Nicobar Islands, Kavalur in Tamil Nadu.

Rohini-75. Rohini-75 (the number indicating the diameter of the rocket in millimetres) was the first successful rocket launched from Thumba in 1967. This was followed by Rohini-100; 125, 300 and 560. The main purpose of these rockets was to study the atmosphere and conduct tests for sub-systems for bigger launch vehicles. The success in sounding rockets led to the design and development of a Satellite Launch Vehicle (SLV). Till now over 1,000 sounding rockets have been launched. To begin with, only a single stage rocket weighing 100 kg. was launched and later on four stage rockets weighing 17,000 was reached.

Satellite Launch Vehicle (SLV) and Augmented Satellite Launch Vehicle (ASLV). The first SLV-3 was launched from Sriharikota in August 1979 which unfortunately proved a failure. The second attempt was made by launching 35 kg Rohini satellite in July, 1980 which was orbited by SLV3. This attempt was a successful one after which many successful attempts were made from 1981 to 1983.

ISRO started working on an ASLV which was designed to put a 150 kg satellite into near earth orbit. The attempts made in this regard in 1987 and 1988. could not meet much success. ASLV-D2 was launched in July 1988. In spite of many improvements, this attempt also failed. Further improvements were made in ASLV-D3 and its launch in May, 1992 was successful. The object of this mission was to operationalise the ASLV's capability of placing a 150 kg satellite—as a part of ISRO's Stretched Rohini Satellite Series (SROSS)—into a 400 km near circular orbit. This could achieve only partial success.

Geosynchronous Satellite Launch Vehicle (GSLV) is a three-stage vehicle for the launch of which an attempt was made in March 2001. It developed some technical problems which were removed at a later stage. The successful launch took place on 19 April, 2001. The second development flight of GSLV with GSAT-2 onboard weighing 1,825 kg was launched successfully in 2003. The rocket placed the geosynchronous satellite, EDUSAT, in correct orbit in September, 2004. The launch of GSLV-FO2 in July, 2006 was a complete failure as it crashed in the Bay of Bengal about 60 seconds after the take off.

APPLE (Ariane Passenger Payload Experiment) was launched from Kourou in French Guyana on 19 June, 1981. Its life was designed for two years but it worked for 27 months. It helped in conducting a number of experiments including those concerned with communication, TV, computer, etc.

INSAT (Indian National Satellite System) series. In 1980, the Government of India decided to purchase typical type of satellites which could help simultaneously in the fields of communication, TV and meteorology. The first in this series was INSAT-1 which had a capacity to provide 4,300 two-way telephone circuits through 12 transponders for national coverage. Transponders are devices which receive radio signals from the ground and transmit them. INSAT-1A launched on 10 April, 1982 from Cape Canaveral could not meet the desired goal due to a number of technical problems. INSAT-1B was an improved version and was launched from a US space shuttle in September 1983. It functioned well for over seven years, a duration which was longer than its designed life. INSAT-1C was launched in July, 1988 by an Ariane rocket. It met with only half success because it helped in operating only half the communication and television channels but the meteorological package worked fully. On 22 November, 1989, the spacecraft lost its contact with the earth. INSAT-1D was launched in June, 1990. All services of INSAT-1B were transferred to INSAT-1D.

Although India depended heavily on imports for the first generation of INSATs, the second generation of this series was indigenously developed and fabricated, INSAT-2A was launched on 2 July, 1992. This was followed by launch of INSAT-2B on 22 July, 1993. These two missions were successful to a great extent and provided increased capacity for telecommunications and broadcasting, better resolution for weather forecasting, etc.

INSAT-2C and INSAT-2D were designed without monitoring payload because there was sufficient facility for this purpose in the satellites which were working in 1995. These two satellites could not live their full life due to short circuit. INSAT-2E was an improved version of the earlier satellites. It was launched from Kourou in April, 1999. Its transponders were able to cover vast parts of South-east Asia, China, Europe and Australia. It was supposed to work for 12 years. The commissioning of INTELSAT-2E has strengthened the telecom services, TV broadcasting and it also carries an advanced meteorological payload.

The third generation of INSAT series started with the launch of INSAT-3B in March, 2000 from Kourou. It was a communication satellite launched ahead on INSAT-3A following the failure of INSAT-2D. This was followed by INSAT-2C in 2002 and INSAT-3A and 2003.

INSAT-4 series was started in December 2005 from Kourou. It has been used for direct to home television within the country and for communication within India as well as in countries located in south-east and north-west of India.

EDUSAT. Satellite links have become a strong tool in the field of education keeping in view the acute shortage of competent teachers at the school, college and university level as well as in the professional institutions. EDUSAT was launched by ISRO in 2004 to provide easy access to quality education. In addition to providing support to formal education, it is used for propagating information regarding health, hygiene, etc. among the rural masses and in remote areas. *Telemedicine through INSAT* has also become a reality and lakhs of people are benefited by this project.

Remote Sensing Satellites

Keeping in view the growing importance of Remote Sensing in everyday life, ISRO prepared a programme to launch remote sensing satellites in 1979 and consequently Bhaskara-I and Bhaskara-II were launched. Bhaskara-I was used to study agricultural land, dry deciduous forests within the tropics, dry temperate forests of the Himalayas and ice cover in the Himalayan region. Bhaskara-II was used to study land, ice cover, drought prone areas, geological structure and forest resources. Maps concerning land cover and land use were prepared for West Bengal and Bihar. Images of sedimentary cover over the Ganga plain were obtained. Maps of forest cover for Karnataka, West Bengal, Odisha and Kachchh were prepared. Besides, ice cover over parts of the Himalayas was also studied. Satellite Launch Vehicle (SLV)-3 was launched in April 1983 which helped in studying water and forest resources cloud and ice cover.

September 9, 2012 was the historic day for ISRO when it launched 100th mission Polar Satellite Launch Vehicle (PSLV-C21) from Sriharikota.

In the year 1972, National Remote Sensing Agency (NRSA) was set-up at Balanagar near Hyderabad. First remote sensing satellite known as Indian Remote Sensing Satellite (IRS-IA) was launched on 17th March, 1988. It was equipped with Linear Imaging Self Scanning Sensors (LISS) which help in giving better results.

The second generation **IRS-1C** remote sensing satellite was launched in 1995 which completed its ten year tenure in 2005. During those ten years it took over 6 lakh images while revolving around the earth for 60,000 times. **IRS-P3** was launched in 1995 for studying ice, clouds, moisture in crops and damages caused by floods. **IRS-1D** was launched in 1997 which took images of U.S.A., Germany, Japan, Dubai, Korea and Thailand in addition to those of India.

Oceansat was India's first satellite which aimed at studying fish resources, changes in sea level, speed of sedimentation, sea pollution, sea water temperature, air moisture over the sea surface and coral reefs. It was launched in 1999.

Technology Experiment Satellite (TES) was launched in 2001. It aimed at possibilities of introducing new technology in cartography.

Resource Set-I was an improved version of IRS-1C and IRS-1D and was meant to estimate the natural resources in the country. It was launched in 2003.

Cartosat-1 and Cartosat-2 were launched in 2005 and 2007 respectively. They are designed to up date topographical maps and prepare maps on large scale of 1 : 50,000. Maps prepared in this way are in great demand all over the world due to their accuracy and high quality. Normally, these satellites prepare maps for about 30 km in one day and complete map of the whole country within 100 days. At this speed, they can prepare a map for the whole world within one year.

Polar Satellite Launch Vehicle (PSLV) uses both gas and liquid fuel. **PSLV-D1 mission** was launched on 20 December 1993 from Sriharikota. It could not succeed much due to fault in its software. PSLV-D2 mission was launched on 15 October, 1994 again from Sriharikota. Third launch of PSLV was conducted in 1996. In the year 1997, PSLV-C1 launched IRS-1D. In the year 1999, PSLV-C2 or Oceansat was launched.

Chandrayan-1 Mission

Chandrayan which in Sanskrit means 'Moon Vehicle was the most ambitious and most talked about space project which thundered off into the space on 22 October, 2008 from Satish Dhawan Space Centre located in Sriharikota by means of the four stage space workhorse PSLV (Polar Satellite

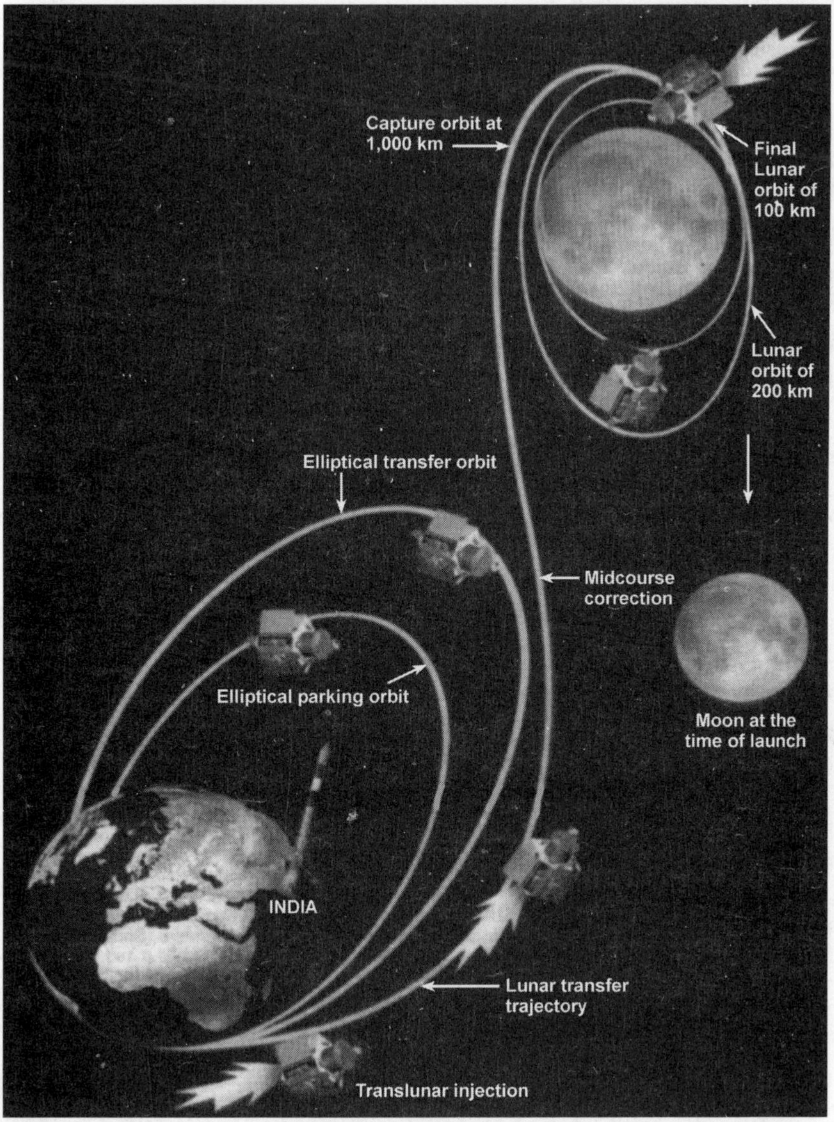

FIG. 26.22. Chandrayan-1

Launch Vehicle) featuring alternate liquid and solid fuel stages. It started revolving around the earth on an elliptical path being 240 km away in **perigee** and 24,000 km away in **apogee**. It started moving towards the moon after making a number of revolutions around the earth. The Chandrayan had a flawless take off and it was to be ignited in four stages. Soon after its separation from the fourth stage. Chandrayan was initially placed in an orbital with its farthest point from the earth being 22,860 km. On October 28, 2008 it was raised to 4,77,515 km and on October 26, 2008, it became the first Indian spacecraft to enter deep space crossing the 1,50,000 km mark. On October 29, 2008, it went up to 2,67,000 km and its terrain mapping camera was switched on. On 4 November, 2008, the lunar orbitor flew into lunar transfer trajectory which meant that the Chandrayan had successfully entered the main highway to the Moon.

November 8, 2008 was the day of nerve-wracking and nail-biting because lunar orbit insertion (LOI) was being eagerly awaited. This is the most critical exercise because nearly 30 per cent of the unmanned lunar satellites by the U.S.A. and the erstwhile U.S.S.R. had failed at this point. This is due to the fact that the gravity of the earth and the moon cancel each other at this point. Fortunately, this hurdle was crossed and the speed of the spacecraft was reduced to 366 metres per second so that the craft could enter the moon's gravitational field. Its farthest point from the moon at this stage was 75.02 km which was gradually reduced and on 12 November, 2008, it reached the operational orbit of 100 km by 100 km. Still one major milestone to be crossed was crash landing of the 29 kg indigenous Moon Improved Probe (MIP) about 32 km away from the Shackleton Crater in the Moon's South Pole. On 15 November, 2008, MIP slammed into the Moon's surface and India became the fifth country to show its presence on the Moon after the U.S., U.S.S.R., European Space Agency (17 countries) and Japan. The most outstand-ing feature of the Chandrayan was that its cost was only ₹ 386 crore which has been termed on the lowest expenditure according to international standards.

Immediately after landing on the moon's surface the Chandrayan-I started sending signals to ISRO Telemetry Tracking and Command Network (ISTRAC) at Bengaluru. The whole life of the Chandran was fixed for two years. But unfortunately on August 30, 2009, ISRO failed to restore the snapped communication link and contact with the spacecraft was permanently lost. Consequently the mission was terminated. Still the ISRO believed that it was a great success as the Chandrayan-I completed about 95 per cent of the mission. Its major achievement among others, was the discovery of the widespread presence of water molecules in the linear soil.

Chandrayan-II. Indian space scientists were not much discouraged by partial failure of Chandrayan-I and they moved ahead with their plan to launch Chandrayan-II for gathering more information about the moon. Chandrayan-II was launched on its mission to Moon from the second launch pad at Satish Dhawan Space Centre on July 22, 2019 at 2.43 pm (IST). The craft reached the Moon's orbit on August 20, 2019 and began orbital positioning manoeuvre for the landing of the Vikram lander. Vikram and rover were scheduled to land on the near side of the Moon, in the south polar region at latitude of about 70° south on September 6, 2019 and conduct scientific experiments for one lunar day (almost equal to two earth weeks). A successful soft landing would have made India fourth country in the world, after USSR, USA and China to achieve this destination.

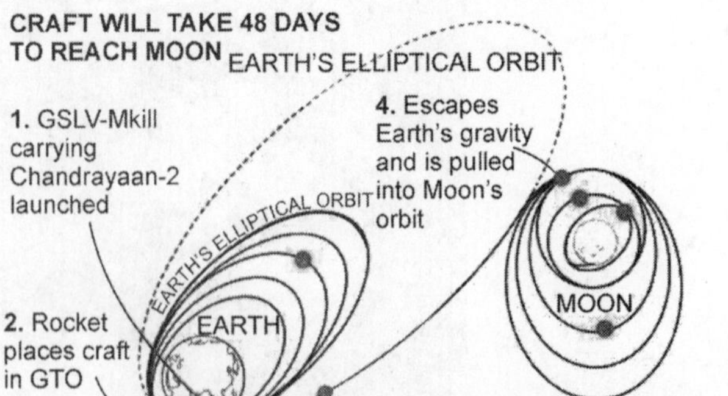

FIG. 26.23. Journey of Chandrayan-II

But unfortunately the lander deviated from its intended trajectory starting at 2.1 kilometres altitude and lost communication. It crash landed and was damaged. The Failure Analysis Committee, after investigation, conducted that the crash was caused by a software glitch.

Chandrayan-III

Following the partial failure of Chandrayan-II caused by last minute of glitch on soft landing, Indian Space Research Organisation (ISRO) launched Chandrayan-III. It consisted of a Vikram lunar lander and a Pragyan lunar rover as well as a propulsion module that carried the spacecraft from Earth to lunar orbit.

It was launched from Satish Dhawan Space Centre on 14th July, 2023 and the spacecraft entered lunar orbit on 5 August, 2023. The lander touched down near the lunar south pole on 23 August, 2023. This made India the fourth country to successfully land on the Moon and the first country to land near its south pole.

Mangalyan

Mangalyan which in Sanskrit means Mars craft is Indias Mars Orbiter Mission (MOM). It was launched from Satish Dhawan Space Centre, at Sriharikota, using a Polar Satellite Launch Vehicle (PSLV) on 5 November, 2013. The MOM probe spent about a month in the earth orbit, where it made a series of seven altitude-raising orbital maneuvers before trans-Mars journey on 30 November, 2013. It crossed half way to Mars on 9 April, 2014. Currently it is being monitored from the Spacecraft Control Centre at ISRO Telemetry Tracking and Command Network (ISTRAC) in Bengaluru with support from Indian Deep Space Network (IDSN). Three stages of MOM are depicted in **Fig. 26.24**. It is orbiting Mass since 24 September, 2014. It is India's first interplanetary mission and it made it the fourth s pace agency to achieve mass orbit after Roscosmos, NASA and European Space Agency. It made India the first Asian nation to reach Martian orbit and first nation in the world to do so on its maiden attempt.

Mangalyan-2

Mangalyan-2 is India's second interplanetary mission planned for launch to mars by the Indian Space Research Organisation in 2025. It will consist of an orbitor, and may include a lander and a rover.

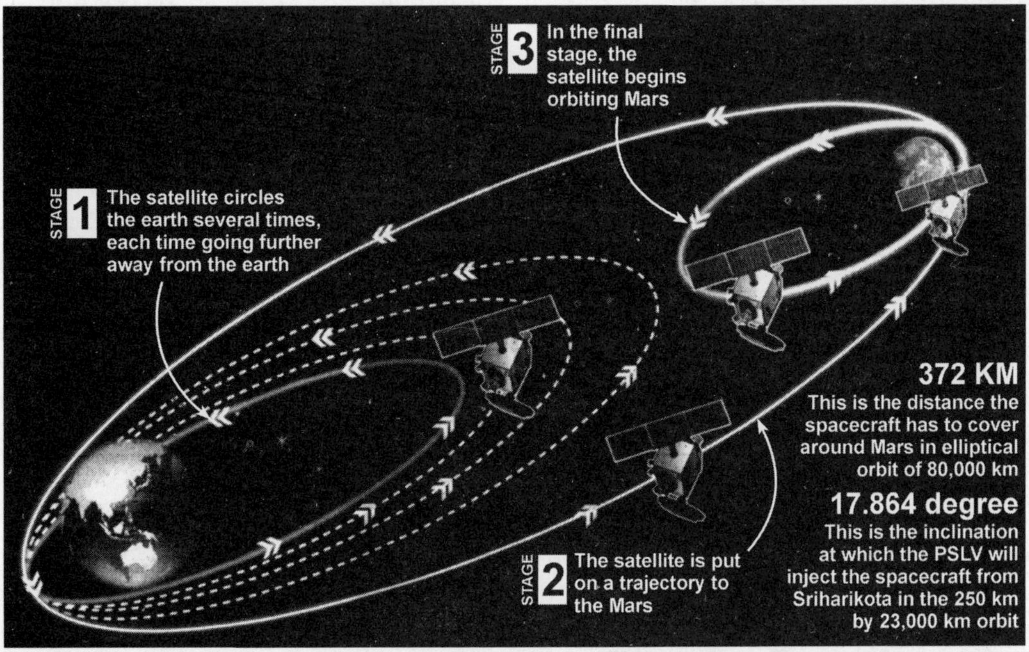

FIG. 26.24. Mangalyan's journey to Mars

Major Milestones in India's Space Programme

1962 : Department of Atomic Energy forms the Indian National Committee for Space Research Work on Establishing Thumba Equatorial. Rocket Launching Station (TERLS).

November 21, 1963 : India launches first sounding rocket from TERLS.

1965 : Space Science and Technology Centre established in Thumba.

1967 : Satellite Telecommunication Earth Station set-up at Ahmedabad.

August 15, 1969 : Indian Space Research Organisation (ISRO) formed under Department of Atomic Energy.

June 1, 1972 : Space Commission and Department of Space (DOS) set-up. ISRO brought under DOS.

April 19, 1975 : ISRO launches first Indian satellite Aryabhatta.

June 7, 1979 : An experimental satellite for earth observations, Bhaskara-1, launched.

August 10, 1979 : First experimental launch of SLV-3 with Rohini Technology Payload on board. Satellite can't be placed in orbit.

July 18, 1980 : Second experimental launch of SLV-3, Rohini satellite successfully placed in orbit.

1981 : First developmental launch of SLV-3, Bhasakara-II launched, Apple, an experimental geo-stationary communication satellite launched.

1982 : INSAT-1A launched on April 10, deactivated on September 6.

1983 : INAT-1B launched.

1984 : Indo-Soviet manned space mission.

1988 : Launch of first operational Indian Remote Sensing Satellite, IRS-1A.

1992 : First satellite of the indigenously-built second generation INSAT series, INSAT-2A launched.

1993 : First developmental launch of PSLV. Satellite couldn't be placed in orbit.

1994 : Second developmental launch of PSLV. Satellite successfully placed in orbit.

1999 : India launches two foreign satellites, one from Germany and one from Korea, for first time as co-passenger satellites.

2001 : First developmental launch of GSLV-D1.

2003 : Second developmental launch of GSLV-D2.

2008 : PSLV-C11 successfully launches CHANDRAYAAN 1.

2013 : PSLV-C25 successfully launches Mars Orbiter Mission spacecraft.

2019 : Chandrayan-II launched.

REFERENCES

Bhatnagar, L.P. (1970) : *Transport in Modern India*, Kishore Publishing House, Kanpur.

Budhiraja, Ravi Bhushan (2004) : *Jawaharlal Nehru Port—Modern Container facility,* The Hindu Survey of Indian Industry, Kasturi & Sons Ltd., Chennai, 2004, pp. 149-150.

Das, P.K. and Sinha, B.N. (1985) : *Inter-city airway connectivity in India,* Geographical Review of India, Vol. XLII (3), pp. 25-29.

Dheenkey, M.R. (1963) : *Air Transport in India : Growth and Problems*, Vora & Co., Mumbai.

Government of India, 2019 Basic Road Statistics.

Government of India, 2022-23, *Economic Survey.*

Government of India, 2024: *A Reference Annual.*

Jha, Mitu S. (2004) : *Shipping—Charting a steady course,* The Hindu Survey of Indian Industry, Kasturi & Sons Ltd., Chennai, 2004, pp. 154-155.

Kara, P.K., Tripathi, S.N. and Singh, D.K. (1982) : *A Comparative Analysis of Major Ports of India and their traffic characteristics.* NGJI, Vol. 28 (1&2), pp. 60-68.

Mohan Sundara Rajan (2008), *Space Today,* National Book Trust, N. Delhi.

Narain, Yogendra (2004) : *Road—Well-timed policy support,* The Hindu Survey of Indian Industry, Kasturi & Sons Ltd., Chennai, 2004, pp. 143-145.

National Institute of Science Communication and Information Research (NISCAR) : Science Reporter, August, 2004; March, 2008; January, 2009; October, 2009; November, 2009; February 2012.

Owen, W. (1968) : *Distance and Development Transport and Communication in India,* Washington, D.C.

Prasad, Amba (1971) : *Indian Railways,* Asia Publishing House, Mumbai.

Rao, M.A. (2012) : *Indian Railways,* National Book Trust, New Delhi, India.

Ravindra, M. (2004) : *Railways—Reorganisation a dire need,* The Hindu Survey of Indian Industry, Kasturi & Sons Ltd., Chennai, 2004, pp. 139-143.

Raza, Moonis and Yash Agarwal (1986) : *Transport Geography of India : Commodity Flows and The Regional Structure of Indian Economy,* Concept Publishing Company, New Delhi.

Sahni, S.N. (1953) : *Indian Railways : One Hundred Years :* 1853 to 1953, Manager of Publications, Delhi.

Singh, D.N. (1977) : *Transportation Geography in India : Survey of Research,* NGJI, Vol. 23 (1&2), pp. 95-114.

Sunder Ram, K.S. (2004) : *Coastal Shipping—A sleeping grant.* The Hindu Survey of Indian Industry, Kasturi and Sons Ltd., Chennai, 2004, pp. 154-155.

The Hindu Survey of Indian Industries, 2012.

Vasudeva, P.V. (2004) : *Railway Finances—New initiatives bring hope,* The Hindu Survey of Indian Industry, Kasturi & Sons, Chennai, 2004, pp. 139-141.

◆◇◆

CHAPTER 27

Foreign Trade

The exchange of goods or merchandise or their sale-purchase is known as trade. Trade is of three types (*i*) Local trade (*ii*) Regional trade and (*iii*) Foreign or International or Overseas trade. When one country exchanges goods or merchandise with another country, it is known as overseas trade or foreign trade. Advancement of international trade of a country leads to its economic prosperity. It is important both as a source of imports and as an outlet for exports. "International Trade of a country is rightly known as its *Economic Barometer*".

India has been maintaining trade links with China, Myanmar, Indonesia, Malaysia, Thailand, Iran, Iraq, Afghanistan, Saudi Arabia, Egypt, Greece, etc. since ancient times. With the establishment of British rule in India, our trade ties became stronger with Britain. India used to export raw materials such as cotton, iron-ore, etc. to Britain and import manufactured goods, machinery, medicines, etc. from that country. Tea and spices held a special position in Indian exports to Britain.

TRENDS IN INDIA'S FOREIGN TRADE

Table 27.1 shows the trends of foreign trade of India. There has been a rapid increase in the imports between 1950-51 and 1960-61. India had to import machinery and other items for industrial progress during that period. The value of exports increased considerably between 1960-61 and 1970-71 reducing the trade deficit from ₹ 480 crore in 1960-61 to a mere ₹ 99 crore in 1970-71. After that, both imports and exports have been increasing but imports have always been outpacing the exports. The rapid increase in India's overseas trade has been largely due to growth and diversification of Indian economy in the post-Independence era. However, inflationary trends and devaluation of the Indian currency have also contributed to increase in the volume of India's international trade :

Balance of Trade. All the countries of the world are importers as well as exporters of certain goods or services. *The difference between the value of a nation's exports and imports of all goods and services over a given period of time is called **balance of trade***. If the value of the total export is more than the value of the total import in a country, it has a ***positive balance*** of trade. A country with positive balance of trade is known as ***trade creditor nation***. If the value of the exports is lower than the value of imports, the country has a ***negative balance*** of trade. A country with negative balance of trade is known as a ***trade debtor nation***. **Table 27.1** shows that India's balance of trade has been invariably unfavourable. This is because of our increasing demands and growing economy. Imports, exports, total trade and trade deficit were reduced in 2019-20 and 2020-21 due to Covid-19.

In spite of the phenomenal growth of our foreign trade, India stands nowhere in the international market. As a matter of fact, India's share in the international trade has decreased with the passage of time.

As per press release of the World Trade Organisation (WTO) in April, 2013, India is the 19th largest exporter in the world with a share of 1.6 per cent and 10th largest importer with a share of 2.6 per cent in 2013. In commercial services, India is the 6th largest exporter with a share of 3.4 per cent and 7th largest importer with a share of 7.3 per cent.

Basis of Overseas Trade

Overseas trade depends upon several factors. Some important factors are discussed as under :

1. Difference in Natural Resources. The fundamental base of international trade is the difference in natural resources of different countries. There are variations in relief, structure, geology, climate and soil from one country to the other. These variations lead to variations in natural resources. Some countries produce certain things more than their requirement and export them while some other countries may be lacking them and import those commodities.

2. Marketable Surplus. Some countries are capable of producing certain things more than their internal consumption. In other words, these countries have marketable surplus which they trade with those countries having demand for such products. For example, tea from India, coffee from Brazil, jute from Bangladesh, paper and pulp from Norway and Sweden, wood from Australia and petroleum from the Middle East are available for export. Advanced countries like the United States of America, United Kingdom, Russia, Germany and Japan mostly export finished goods.

3. Scarcity of Goods. There is not even a single country in the world which does not have scarcity of one commodity or the other. Japan and Britain do not have raw materials. Hence, these countries have to import raw materials from a large number of

TABLE 27.1. India's Foreign Trade (in ₹ crore)

Year	Imports	Exports	Total value of trade	Balance of trade
1950-51	608	606	1,214	–2
1960-61	1,122	642	1,764	–480
1970-71	1,634	1,535	3,169	–99
1980-81	12,549	6,710	19,259	–5,839
1990-91	43,193	32,558	75,751	–10,635
2000-01	2,28,307	2,01,356	4,29,663	–26,950
2010-11	16,83,467	11,36,964	28,26,389	–5,46,503
2015-16	24,90,306	17,16,384	42,06,690	–7,73,921
2016-17	25,77,675	18,49,434	44,27,109	–7,28,242
2017-18	30,01,033	19,56,515	49,57,548	–10,44,519
2018-19	35,94,675	23,07,726	59,02,401	–12,86,694
2019-20	33,60,954	22,19,854	55,80,808	–11,41,100
2020-21	29,15,958	21,59,043	50,75,001	–7,56,915
2021-22	45,72,775	31,47,021	77,13,796	–14,25,753

Source : Economic Survey 2022-23. Statistical Appendix pp. 108-09.

countries to feed their industries. Thus, scarcity of goods also encourages international trade. Japan depends heavily on iron ore supplies from India.

4. Transport and Communication. Trade involves exchange of goods which requires proper arrangement for transportation and communication. Land, water and air transport have helped international trade to a great extent. Heavy commodities like coal and iron ore as well as light and perishable commodities like milk and milk products, meat, fruits and vegetables, etc. can be sent to all parts of the world by efficient means of transport and refrigeration. Countries with poor transport system have not been able to develop international trade. Realising the significant role played by transport system for promoting trade, India has launched an ambitious programme to improve surface transport. In addition, certain ports like Jawaharlal Nehru, Deendayal Port (Kandla), V.O. Chidambaranar Port Tuticorin, Kamraj Port (Ennore), Paradwip, Haldia, etc. have either been developed or improved.

5. Disparities in Economic Growth. There are disparities in the economic growth in different parts of the world. Some countries are still engaged in primary activities such as agriculture, mining, etc. These countries mainly export minerals and agricultural raw materials. India's export consisted of raw materials for a pretty long time even after Independence. It is only recently that India has been able to diversify its exports as a result of diversification of her economic activities.

6. Trade Policy. Free trade policy encourages international trade whereas restrictions on the trade discourage it. For example, India has restricted the export of oil-seeds to meet the domestic demand. Similarly, India has imposed heavy import duty on certain finished goods to encourage industries at home.

7. War and Peace. Peace is the most important condition for the development of international trade. International trade gets disrupted during the time of war.

8. Political Relations. Countries having cordial political relations have better exchange of goods which encourages international trade. For example, India and Russia have good political relations and trade between these two countries has increased. On the other hand, the U.S.A. and Russia have strained political relations and trade between these two big countries is at low level. India's trade with the neighbouring Pakistan remained at low level ever since that country came into being in 1947 primarily due to strained political relations between the two countries.

COMPOSITION OF EXPORT TRADE OF INDIA

India has been a traditional exporter of raw materials since time immemorial. But the complexion of India's export trade has undergone a world of change since Independence. At present, Indian exports cover a wide range of items of agricultural, industrial, handicrafts, handloom and cottage sectors. Exports of electronic, hardware and software goods have increased considerably since the beginning of 1990s, and more significantly, these items are mainly destined to the advanced countries. The export growth is gradually becoming broad based notwithstanding some deceleration in export growth of primary commodities. The major contributor to this increase has been the manufacturing sector in general, and engineering goods, chemicals and related products, leather and manufactures and gems and jewellery in particular.

1. Tea. India is a traditional exporter of tea in the world market. India has maintained its supremacy as an outstanding exporter of tea inspite of tough competition from the other tea producing countries. However, our percentage of tea and *mate* export in the world market has decreased considerably primarily, due to increased exports from other countries like Sri Lanka, China and Japan. In 2021-22 India exported 209 thousand tonnes of tea valued at about ₹ 5,597 crore. Britain has been the traditional buyer of our tea. Other important buyers are Russia, Poland, Iran, Egypt, Germany, Afghanistan, Netherlands, Australia and the USA.

2. Leather and Leather Goods. India has the largest population of livestock in the world which provide large quantity of leather in the shape of hides and skins. Previously, India used to export leather in its raw form but now the main export is that of leather goods. India exported leather and leather manufactures including leather footwear, leather travel goods and leather garments worth ₹ 28 crore in 1960-61 which rose to ₹ 298 crore in 1990-91 and to ₹ 32,670 crore in 2021-22. Footwear constitutes the most important item of leather exports. The USA, Russia, Germany, Britain, Yugoslavia, Italy, France and Japan are the main importers of our leather and leather goods. Recently, competition from China has grown considerably with respect to exports of leather and leather goods.

3. Ores and Minerals. India exports a large variety of ores and minerals which account for about one-fourth of our total exports. Higher international prices provided impetus to exports of metals and mineral ores.

(a) Precious stones, gems and jewellery have become the most important items of export from India in the recent past. They account for nearly 9.7 per cent of our total exports. India imports gems and precious stones in their raw form and exports them after proper cutting and polishing. India was an insignificant exporter of gems and jewellery and the exports were worth ₹ 7 crore only in 1960-61 which increased to ₹ 2,91,481 crore in 2021-22. This speaks volumes of increase in our gem and jewellery exports. The USA, Switzerland, Japan, Belgium, Netherlands, France, Hong Kong, Singapore and countries of the Middle East are our main buyers. Israel and Belgium have emerged as our main competitors in the recent years.

(b) Iron ore is the single largest metal exported by India. High quality magnetite and haematite ores are found in large quantities which are in great demand in the industrial countries. Japan is the largest buyer of our iron ore, purchasing more than two-thirds of our exports. The other customers are South Korea, Romania, Russia, Malaysia, and Germany. India exported 26 million tonnes of iron ore worth ₹ 24,145 crore in 2021-22. The recent trend has been to curtail the exports of iron ore and use it within the country to provide a solid base to iron and steel industry.

(c) Mica. India holds monopoly in the production of mica and is the largest exporter of mica in the world. In 2020-21, India exported 151 thousand tonnes of mica valued at ₹ 511 crore.

(d) Chemicals and allied products form an important segment of our exports. The main buyers are Russia, the USA, Britain, the Netherlands, France, Saudi Arabia, Bangladesh, UAE, Indonesia, Thailand, Hongkong and Kuwait. The value of chemicals and allied products was ₹ 3,06,618 crore in 2021-22.

4. Engineering Goods. As a result of industrial progress, India is in a position to export engineering goods even to some of the advanced countries. Saudi Arabia, Japan, Iran, Sri Lanka, Uganda, etc. are the main purchasers of Indian engineering goods. Besides, electronic goods are exported to advanced countries like the USA and Japan. Exports of engineering goods grew on the back of rising demand from countries in South-east Asia and China.

5. Cotton Textiles and Yarn. India now exports both inferior and superior quality cloth and yarn to different countries including the USA, Russia, Australia, New Zealand, the Netherlands, Britain, Germany, etc. While exports of manmade yarn, fabrics and made ups increased, those of cotton yarn and fabrics were stagnant. In the year 2021-22, India exported cotton yarn, fabrics made ups, etc. worth ₹ 1,02,118 crore.

6. Readymade Garments. Readymade garments have become a very important item of export during the last few years. Indian readymade garments have become very popular even in some of the advanced countries due to their attractive designs, new fashion and colour combinations. They are competitively cheap in the foreign markets because labour is quite cheap in India. The total earnings from the export of readymade garments were ₹ 1,067 crore in 1985-86 which increased to ₹ 1,19,474 crore in 2021-22.

7. Jute Manufactures. India was the largest producer and exporter of jute manufactures before partition in 1947. The partition of the country, introduction of substitutes and competition from other countries led to problems in the export of jute manufactures. In spite of these adversities, India has been doing well in the export of jute manufactures. India exported jute manufactures including twist and yarn valued at ₹ 3,786 crore in 2021-22. Eight countries purchase about two-thirds of our jute exports. These include the USA, Australia, Canada and European countries.

8. Marine Products. India is comparatively a new entrant in the export of marine products. The real progress in this field started after 1975. The total value of the exports of sea products was ₹132.15 crore in 1975 which rose to ₹ 18,833.06 crore in 2012-13. In 2020-21, India exported 1,168 thousand tonnes of fish and fish preparations worth ₹ 44,176 crore. The main items of marine exports are fish, dried prawn and shrimps. India exports these products to over 80 countries. Japan and the USA purchase about 90 per cent of our exports. Other important customers are the Netherlands, Kuwait and France.

9. Coffee. India exports coffee to the USA, Russia and some European countries. In 2021-22, 333 thousand tonnes of coffee worth ₹ 7,614 crore was exported.

10. Spices. India has been a traditional exporter of spices. It was a major item of export of India during the British period. But its relative importance was reduced after Independence as a result of increase in other exports. The total quantity of spices exported in 2020-21 was 1,428 thousand tonnes and the total earnings from the export of spices in that year was ₹ 29,039 crore.

COMPOSITION OF IMPORT TRADE OF INDIA

Like exports, Indian imports have also increased manifold. India used to import mainly the manufactured goods before Independence. At the time of Independence, roughly 70 per cent of imports into India consisted of either manufactured consumer goods or inputs for manufacturing industries. But after that, the import of manufactured goods decreased gradually and our imports saw a large variety of goods. The demand for petroleum and petroleum products increased tremendously which made it necessary for us to import large quantities of these items. The other major items of import are machinery, tools, cereals, fertilizers, edible oils, iron and steel, pearls and precious stones, superior quality cotton and paper, etc. Like exports, India's imports are also tending to become broad based. The growth in imports has been contributed by robust increase in imports of food and allied products (mainly edible oils), capital goods, raw materials and manufactured intermediate and consumer goods.

1. Petroleum and Petroleum Products. This is the largest single item of import by India these days. There is great demand for petroleum and petroleum products and this demand is increasing at an accelerated rate. Transport and industry are two major sectors of consumption. In 2021-22, petroleum, oil and lubricants worth ₹ 12,07,803 crore were imported which accounted for about 21% of our total imports. In mid 1970s when the world was facing oil crisis due to Arab-Israel war, oil accounted for about 75 per cent of our import bill. Gulf war and disintegration of the former USSR in early 1990s had very adversely affected our oil supplies. Efforts are being made to increase the home production of oil so as to reduce dependence on the imports. Iraq, Saudi Arabia, UAE, USA, Nigeria, Kuwait, Mexico, Oman, Russia and Brazil are the main sources of oil supply to India.

2. Machines. Special emphasis was laid on the industrial growth immediately after Independence which made it necessary to import machinery on a large scale. Now, most of the machines are manufactured in the country itself and only the machines involving high technology are imported. Textile machinery, electrical machines, farm implements and mining machines are the main items of import. These machines are mainly imported from the USA, Britain, Germany, Russia, France, Japan, Belgium, Poland, Italy, The Netherlands, Canada, Australia, etc.

3. Iron and Steel. Although there has been a considerable increase in the production of iron and steel in India, yet our production always falls short of our demand. This increase in demand has been largely due to the industrial growth in the country. India imported 10,460 thousand tonnes of iron and steel worth ₹ 94,053 crore in 2021-22. Our main suppliers are Japan, Germany, Belgium, Britain and S. Korea. Japan supplies about 20 per cent of our total imports.

4. Minerals. Apart from iron and steel, India imports a large variety of minerals including copper, lead, tin, zinc, aluminium, etc. These minerals are imported from the USA, Britain, Japan, Germany, Switzerland, Australia, Myanmar and Malaysia, etc.

5. Fertilizers. The increasing demand for fertilizers for agricultural growth has to be met by imports. India imported 30,925 thousand tonnes of fertilizers worth ₹ 1,05,796 crore in 2021-22. The USA, Germany and Japan are the main sources of fertilizers.

6. Pearls and Precious Stones. India imports a large quantity of pearls and precious stones in their raw form and exports them after cutting and polishing. The real increase in their imports has been seen after 1970. The import bill of pearls, precious and semi-precious stones, both worked and unworked was a meagre ₹ 1 crore in 1960-61 which shot up to ₹ 25 crore in 1970-71. ₹ 3,738 crore in 1990-91 and ₹ 2,31,279 crore in 2021-22.

7. Gold and Silver. Gold and silver have also become very important items of import. Gold and silver accounted for about 6.5 per cent of all imports of India in 2020-21.

8. Edible Oils. The production of edible oils has always been falling short of our demand which is met by imports. In 2021-22 about 14,278 thousand tonnes of edible oils costing ₹ 1,41,532 crore were imported. The USA, Brazil and Malaysia are the main sources of edible oils for India. These three countries supply about three-fourths of our edible oils.

9. Chemicals. India imports a large variety of chemicals including ammonia, sulphate, super phosphate, nitric acid, soda ash, bleaching powder and potash. India imported chemical elements and components worth ₹ 3,08,882 crore in 2021-22. The main sources of our chemicals are the USA, Japan, Germany, the Netherlands, Belgium, France, Britain, Italy, Kuwait and S. Korea.

10. Medicines. Large quantities of medicines are required to provide increasing medical treatment to the fast-growing population of India. A part of this requirement is met by importing medicines especially costly and life-saving drugs. India spent ₹ 67,545 crore in 2021-22 to import medicinal and pharmaceutical products. About half the imports are from Germany, Italy, China, Switzerland, Spain, Belgium and Poland.

11. Paper. With the increase in literacy and publication, India is finding it hard to meet her requirements from the indigenous production and has to import paper. The shortage of newsprint is badly felt. Pulp and scrap paper are also imported to manufacture paper. India imported pulp and waste paper worth ₹ 11,934 crore and paper board and manufactures thereof worth ₹ 29,296 crore in 2020-21. Russia, Sweden, Germany, Bangladesh, Brazil, China, Czech Republic, Slovakia and Korea are the main suppliers of these items.

12. Fibres. Although India is a major producer of fibres, she has to import different varieties of fibres to feed her textile industries which are growing at a fast pace. The main fibres imported are raw cotton, raw wool, raw jute and synthetic fibres.

(a) Raw Cotton. Although India is a big exporter of inferior quality cotton, she has to import superior quality cotton. Egypt, Uganda, the USA, Tanzania, Sudan and Peru are important suppliers. India imported 231 thousand tonnes of raw cotton worth ₹ 2,861 crore in 2021-22.

(b) Raw Wool. Large proportion of wool produced in India is of inferior quality and large quantity of superior wool is imported mainly from Australia. India had to spend ₹ 1,660 crore on the import of 108 thousand tonnes of raw wool in 2021-22.

(c) Raw Jute. India's jute supplies were drastically cut as a result of partition of the country in 1947, because most of the jute producing areas went to present Bangladesh. Strenuous efforts have been made to increase area and production of jute within the country, but demand of raw jute by jute textile industry far outstrips the supply. Therefore, India has to import raw jute primarily from Bangladesh. In 2021-22, India imported 29 thousand tonnes of raw jute at the cost of ₹ 179 crore.

(d) Synthetic Fibres. With the diversification of the textile industry and with trend in favour of synthetic fibres, the demand for these fibres is increasing rapidly; a large part of which has to be met by imports. India spent ₹ 3,592 crore to import synthetic fibres in 2021-22.

DIRECTION OF INDIA'S FOREIGN TRADE

Direction of foreign trade means those regions and countries with which India has trade relations. India imports from 180 countries and exports to about 135 countries. Obviously India has trade relations with almost all countries of the world. However, there are still some countries with whom we have more trade than other countries.

Britain was our most important trade partner before Independence and this condition prevailed for several years even after Independence. This was because Britain used India as a great source of raw materials and a huge market for her industrial products.

Planned development, Green Revolution and industrial growth added new dimensions to our foreign trade and there was conspicuous change in the composition and direction of our foreign trade.

For several years, the U.S.A. continued to be our largest trade partner but the global economic recession in 2007 dealt a severe blow to our trade with that country. In contrast, our trade with China increased tremendously although our political relations are not very cordial with that country. India has to import large quantities of oil and oil products to meet her fast growing domestic demand. This has enhanced our trade with a number of countries of the Middle East. Direction of India's exports and imports from different regions/sub-regions is outlined in **Table 27.2.**

TABLE 27.2. Direction of India's Foreign Trade (2022-23)

Region	Percentage Share	
	Imports	Exports
I. Europe	12.3	20.9
(*a*) European Union	7.8	15.9
(*b*) European Free Trade Association (EFTA)	2.6	0.4
(*c*) Other European Countries	2.0	4.6
II. Africa	7.2	11.2
(*a*) Southern African Customs Union (SACU)	1.6	2.2
(*b*) Other South African Countries	0.9	0.8
(*c*) West Africa	2.9	4.0
(*d*) Central Africa	0.1	0.4
(*e*) East Africa	0.4	2.3
(*f*) North Africa	1.0	1.6
III. America	11.4	24.2
(*a*) North America	8.1	19.9
(*b*) Latin America	3.3	4.3
IV. Asia	63.2	41.8
(*a*) East Asia (Oceania)	3.0	1.9
(*b*) ASEAN	12.3	9.9
(*c*) West Asia	18.8	11.3
(*d*) North-East Asia	22.5	8.9
(*e*) South Asia	0.8	6.4
V. CIS and Baltics	6.2	1.0
VI. Unspecified Region	0.1	0.9

Source : Economic Survey 2022-23, Statistical Appendix, pp. 120-139.

SALIENT FEATURES OF FOREIGN TRADE OF INDIA

1. Negative or Unfavourable Trade. India had to import various items like heavy machinery, agricultural implements, mineral oil and metals on a large scale after Independence for economic growth. But our exports could not keep pace with our imports which left us with negative or unfavourable trade balance. This situation still persists and is clear from **Table 27.1**.

2. Diversity in Exports. Previously, India used to export its traditional commodities only which included tea, jute, cotton textiles, leather, etc. But great diversity has been observed in India's export commodities during the last few years. India now exports over 7,500 commodities. Since 1991, India has emerged as a major exporter of computer software and that too to some of the advanced countries like the USA and Japan.

3. Worldwide Trade. India had trade links with Britain and a few selected countries only before Independence. But now India has trade links with almost all the regions of the world. India exports its goods to as many as 190 countries and imports from 140 countries.

4. Change in Imports. Earlier we used to import foodgrains and manufactured goods only. But now oil is the largest single commodity imported by India. Both the imports as well as exports of pearls and precious stones have increased considerably during the last few years. Our other important commodities of import are iron and steel, fertilizers, edible oils and paper.

5. Maritime Trade. About 95 per cent of our foreign trade is done through sea routes. Trade through land routes is possible with neighbouring countries only. But unfortunately, all our neighbouring countries including China, Nepal, Myanmar are cut off from India by lofty mountain ranges which makes trade by land routes rather difficult. We can have easy access through land routes with Pakistan only but the trade suffered heavily due to political differences between the two countries.

6. Trade through a few Selected Ports Only. We have only 13 major ports along the coast of India which handle the major part of overseas trade of India. Very small amount of foreign trade is handled by the remaining medium and small ports.

7. Insignificant Place of India in the World Overseas Trade. Although India has over 17.5 per cent of the world's population, her share in the world overseas trade is less than one per cent. This shows the insignificant place of India in the world's overseas trade. This is, however, partly due to very large internal trade, vast dimensions of the country provide a solid base for inter-state trade within the country. Europe is divided into a large number of smaller countries and the international trade is quite high (trade counted twice, first time as exports and second times as imports).

8. State Trading. Most of India's overseas trade is done in public sector by state agencies and very little trade is done by individuals or corporates.

REFERENCES

Balasubramanyam, V.N. (1984) : *The Economy of India*, London.

Bharadwaj, R. (1978) : *Structural Basis of India's Foreign Trade*, Allied Publishers, Mumbai.

Dagli, Vadilal (ed) (1973) : *India's Foreign Trade,* Vora and Co., Mumbai.

Dayal, E. : "The Changing Pattern of India's Foreign Trade", *Economic Geography,* Vol. 44, 1968, pp. 240-269.

Dutt, A.K. *et al.* (1972) : "Dimensions of India's Foreign Trade", *International Geography,* Vol. 1, Toronto, 1972.

Economic Survey, 2022-23.

Government of India (GOI), 2022, *A Reference Annual.*

Nayyar, D. (1976) : *India's Exports and Export Policies in the 1960's*, Cambridge.

Rafeeque Ahmed, M. (2004) : *Export Sector—Many Hurdles to Cross,* The Hindu Survey of Indian Industry, Kasturi & Sons, Chennai, 2004, pp. 28-29.

Robinson Francis *et al.* (Ed.) (1989) : *The Cambridge Encyclopedia of India, Pakistan, Bangladesh, Sri Lanka*, Cambridge University Press, Cambridge, pp. 310-313.

◆◇◆

CHAPTER 28

Tourism

INTRODUCTION AND DEFINITION

Tourism has different meanings in different languages and different regions and can be defined in different ways. The present day tourism is not the same as travels of the early periods of human history. In the Jewish language, the word '*Torah*' means study or search and 'tour' seems to have been derived from it. In Latin the original word '*Tornos*' stands close to it. 'Tornos' was a kind of round wheel-like tool hinting at the idea of a travel circuit or a package tour. In Sanskrit, '*Paryatan*' means leaving one's residence to travel for the sake of rest and for seeking knowledge. '*Deshatan*' is another word which means travelling for economic benefits. '*Tirthatan*' is the third equivalent which means travelling for religious purposes. All the three words convey the meaning and concept of tourism much more appropriately, as well as comprehensively.

Tourism as a modern term is applicable to both international and domestic tourists. It is the *temporary movement of people to destinations outside their normal place of work and residence*. Such a pleasure-seeking tourist is a traveller moving from place to place or visiting the same place time and again. Tourism includes all economic activities which are organised to fulfill the needs of such travellers. Any travel for holidaying, business or professional trips becomes a part of tourism if it is temporary, is undertaken voluntarily and does not aim at earning any livelihood.

According to **Ziffer** (1989), *"Tourism involves travelling to relatively undisturbed or uncontaminated natural areas with the specific object of studying, admiring and enjoying the scenery and its wild plants and animals, as well as any existing cultural aspects (both past and present) found in these areas"*. Tourism for some provides entertainment, for others it serves as a holiday and for yet others it is a means of understanding other peoples' ways of life, cultures and traditions. According to a Chinese proverb, "*Travelling a mile imparts more knowledge than reading a mile of written words.*" Tourism is now recognised as an industry generating a number of social and economic benefits. It promotes national integration and international understanding, helps in improving infrastructure, creates employment opportunities and augments foreign exchange earnings. Tourism is as much a part of socio- economic development as any other related acitvity. The advanced economies of the west and the emerging economies of south-east Asia underline the linkage between tourism and growth. In

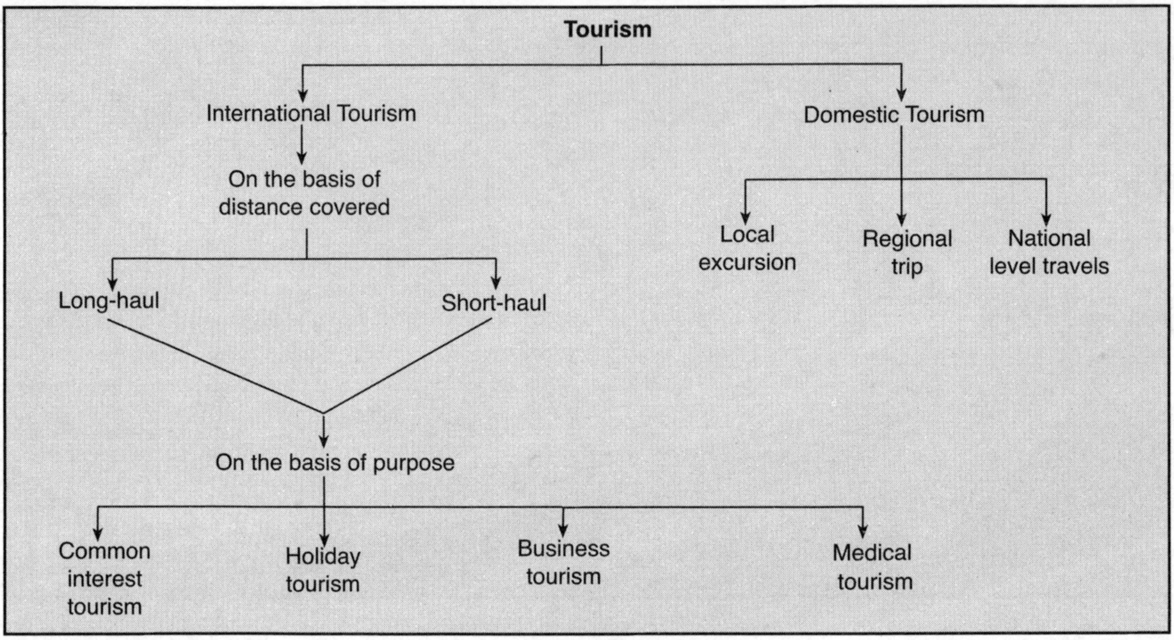

FIG. 28.1. Types of Tourism

any area where a place of tourist attraction is located, there exists a tourist economy besides the domestic area economy. It also lends a helping hand to local handicrafts and cultural activities. In the present day world, tourism is a major engine of economic growth, an important source of foreign exchange earnings and a generator of employment of diverse kinds in many countries including India. It contributes 6.88% to India's GDP and 12% to its jobs.

Types of Tourism

Different types of tourism can be recognised depending on length of stay, mode of transport used, distance travelled, purpose of trip and price paid by tourists. Broadly speaking, there are four major types of tourism namely : (*i*) international tourism, (*ii*) domestic tourism, (*iii*) long distance tourism, and (*v*) short distance tourism.

(*i*) **International Tourism.** An international tourist crosses the boundaries of one or more countries, uses different currencies, faces different languages and meets different types of people. Usually international tourism involves longer distances although crossing small countries or travelling in the neighbourhood of international borders may involve short distances.

(*ii*) **Domestic Tourism.** It is concerned with travelling within the country. It does not need a passport and visa or conversion of one currency into another. Domestic tourism has greater scope in countries of large dimensions such as India as compared to smaller countries. From a geographical viewpoint, domestic tourism may range from local excursions, regional trips to national level travels.

On the basis of purpose of tour or the motives of tourists, tourism is of four types, *viz.*, (*a*) common interest tourism, (*b*) holiday tourism (*c*) business tourism and (*d*) medical tourism. In case of *common interest tourism,* the purpose of visiting and the visited persons is common. Visiting friends and relatives in this type puts very low pressure on the provision of tourist facilities at the destination. *Holiday tourism* is the most popular type. A fine weather favourable for sightseeing, touring, recreation and going round different historical and cultural sites are sought after by incoming tourists of this category. *Business tourists* travel to boost their business, attend trade fairs and conferences pertaining to commerce or professions. They combine business with recreation using same facilities as provided for holiday tourists.

Medical Tourism combines medical treatment with tourism. Many foreign tourists come to India for cheap and efficient medical treatment and visit places of tourist interest after getting well.

Nature's Gift for Tourism

Nature is bountiful and provides vast opportunities in the form of gifts to enhance tourism.

1. Weather and Climate Resources. India is vast country and its different parts experience different types of climate. This extraordinary asset in the form of varied climates can be encashed to promote tourism in the country by attracting tourists from far and near. Regional variations in the climate of India are caused by five main factors, *viz*., (*i*) the vast size of the country, (*ii*) tapering of the peninsular India towards the south, (*iii*) the latitudinal extent astride the Tropic of Cancer, (*iv*) situation at the head of the Indian Ocean, and (*v*) role played by the Himalayas as a climatic barrier between India and central Asia, protecting the country from cold and dry winds from central Asia during winters. A tourist will find different types of climate in the Great Plain of North India, the snow covered Himalayas, in the peninsular plateau area and in the coastal regions. There are climatic variations from one state to another and even from one district to another. These climatic variations are of great advantage to engage tourists in a variety of activities during their stay in the country. However, strong influence of the monsoons along with their typical rhythm provides underlying unity to India's climatic diversity. This unity in diversity of the Indian climate is a potent factor which works to attract tourists. The visiting tourist can choose to stay in any type of climate and enjoy the changing status of weather while moving from one place to another. Temporal changes in weather are no less pronounced than the spatial changes. India is one of the few countries where there is change in weather after every two months. The spatial and temporal changes in India's climate and weather act as a great magnet to attract tourists and enable them to enjoy the gift of nature. Larger the diversity, greater is the potential of climate resources to develop tourism.

2. Landscape Resources. Geology and physiography are two major bases of landscape resources. These two aspects influence the landscape and determine the rocky peaks for climbing, cliffs or scarps for hand gliding, steep snowy slopes for skiing and caves for exploiting them for sightseeing. Rock formations exposed in areas like Deccan plateau or Ladakh attract tourists interested in trekking. There is ample scope for rural tourism also. Water bodies and waterpoints like riversides, gorges, waterfalls, springs, etc. provide spectacular views to the tourists. Sunbathing and adventurous raffling and rowing are associated with water tourism.

Although all sorts of landscape resources attract tourists, yet the degree to which different types of landscapes act as tourists attraction varies greatly. According a research report by A. Gilg of Exter University in U.K., mountains comprise the greatest attraction for the tourist. The other features, in descending order are bold hills, hilly country, plateau uplands and lowlands.

Forest landscape is found to be equally attractive, irrespective of topography. It is for this reason that National Parks, Sanctuaries as well as Biosphere Reserves are used as tourist spots. Seeing wildlife in its natural habitat rather than in a zoo or a safari-park is termed as Nature Tourism these days. Vegetation in wilderness and wildlife which depends on it constitute a very important element of tourist landscape.

3. Seascape Tourist Resources. Coastal waters of the mainland of India and that of the groups of islands (Andaman & Nicobar Islands in the Bay of Bengal and Lakshadweep in the Arabian Sea) comprise another major source of tourist attraction. Sandy beaches, coves, spits and lagoons, reefs and seaside cliffs are tourist hot spots. These are visited for sunbathing, swimming, boating and surfing. This type of tourism flourishes only in those areas where waves are gentle and tidal currents are not dangerous.

4. Historical and Cultural Resources. These are manmade features and are found in almost all parts of the country. Such resources are associated with historical, religious or cultural events. A rich historical and cultural heritage has developed during long course of history of India. These include statues, shrines, tombs, minarets, forts, palaces, ancient monuments as well as recent buildings and are famous for their architectural designs and structural beauty.

Classification of Tourist Resorts

There are several types of tourist resorts depending on their location in different geographic regions, diverse characteristics of their sites, and scope for a variety of tourists activities. A brief classification of tourist resorts is given as under :

1. Mountain and Hill Resorts. Almost all mountainous and hilly areas of India are dotted with places of tourist interest. In the north are the Himalayan ranges; blessed with scenic beauty which is beyond description. Besides, there are high hills in the northeast and in and around the Nilgiris in the south. There are hill ranges of medium to low heights in the Western Ghats, Vindhyas, Satpuras and the Aravalis in addition to isolated hillocks elsewhere. Hill resorts are divided into following three categories according to their altitude.

(*a*) Resorts at great heights (between 2,100—3,500 metres above mean sea level).

(*b*) Hill resorts at medium altitudes (between 1,200 and 2,100 metres above mean sea level).

(*c*) Hill resorts at low altitudes (between 800 and 1,200 metres above mean sea level).

Map in **Figure 28.2** shows that the maximum number of hill resorts are at medium heights followed by those located at low heights. Very few tourist resorts are at great heights.

2. Beach Resorts. The long coastline of India has several beaches which offer ample scope for promotion of beach tourism. A number of beautiful beaches of Goa and that of Kovalam in Kerala are quite popular with the tourists. There are vast stretches of shining golden sands of beaches at Chorwad and Ahmedpur-Mandvi. The Maharashtra coast has eight small beaches from Mumbai's famous Juhu to Murud at a distance of 220 km. Goa's 105 km long coastline has as many as 40 beaches, out of which 12 are very popular with the tourists. Karnataka's beaches are at Mangaluru and Karwar. The sun-swept Marina beach in Chennai (Tamil Nadu) is the world's second longest beach. It runs for a distance of 12 km from Chennai harbour in the north to Santhome (St. Thomas) Church in the south. Two beaches of Andhra Pradesh, namely Ramakrishna Mission beach and Rishi Konda beach are located in the close proximity of Vishakhapatnam. Odisha's Gopalpur beach is surrounded by sand dunes. Besides Puri and Konark, Chandipur near Balasore is known for its 55 km recession of the tides each day at its head. Digha beach of West Bengal is very close to the Ganga's mouth. With a total length of 6 km it is one of the widest in the world amidst gently rolling sea and thick casuarina forest on its two sides.

3. Cultural Centres (Heritage Tourism). India is proud of being very rich in heritage tourism. Her cultural centres are of varied types, the outstanding centres being temples or pilgrim centres (of all faiths), ancient ruins, forts, memorials, palaces, places of historical importance and gardens. These cultural centres are distributed all over the length and breadth of India.

4. Adventure Tourism (Off-beat Resorts). Adventure tourism includes a large variety of activities such as trekking, skiing, river rafting, water games, mountaineering, rock climbing, hang gliding, para gliding, hiking and camping in habitats of wild life. This type of tourism is not very popular as only about 7 per cent of the total tourists traffic opt for adventure tourism. However, there is great scope of making it popular if young people particularly in the age-group of 25-45 years are motivated for adventure tourism.

(i) Trekking. Trekking involves walking or hiking over long distance for fun and recreation. The real thrill of trekking comes by walking over rugged and remote terrain lacking in good means of transport, going up and down the hills, crossing passes and coping with extremes of variable weather in high altitudes. Some of the well defined trekking routes are marked in Ladakh, Uttarakhand, Sikkim, Arunachal Pradesh, etc.

(ii) Mountaineering. Lofty peaks in the Himalayan region have attracted mountaineers from all over the world. Mountaineering became very popular with the adventure lovers after the conquest of Mount Everest in 1953. The Himalayan region has the largest number of high peaks in the world. Most of the high peaks are under permanent snow and throw great challenge to mountaineers. In Jammu and Kashmir and Ladakh, there are a series of mountain peaks in Pir Panjal, Great Himalaya, Zaskar, Ladakh and Karakoram ranges. These are between 5,000 and 7,000 metres above sea level. High Himalayan ranges in Himachal Pradesh have about 150 peaks located at altitudes of over 5,400 metre. On an average, there is one high peak after every 20 km. In Uttarakhand, the source of the holy Ganga, a few kilometres above Gau Mukh ice cave, lies one of the best mountaineering areas of the world. Towards the east, Sikkim offers great opportunities to climbers. The

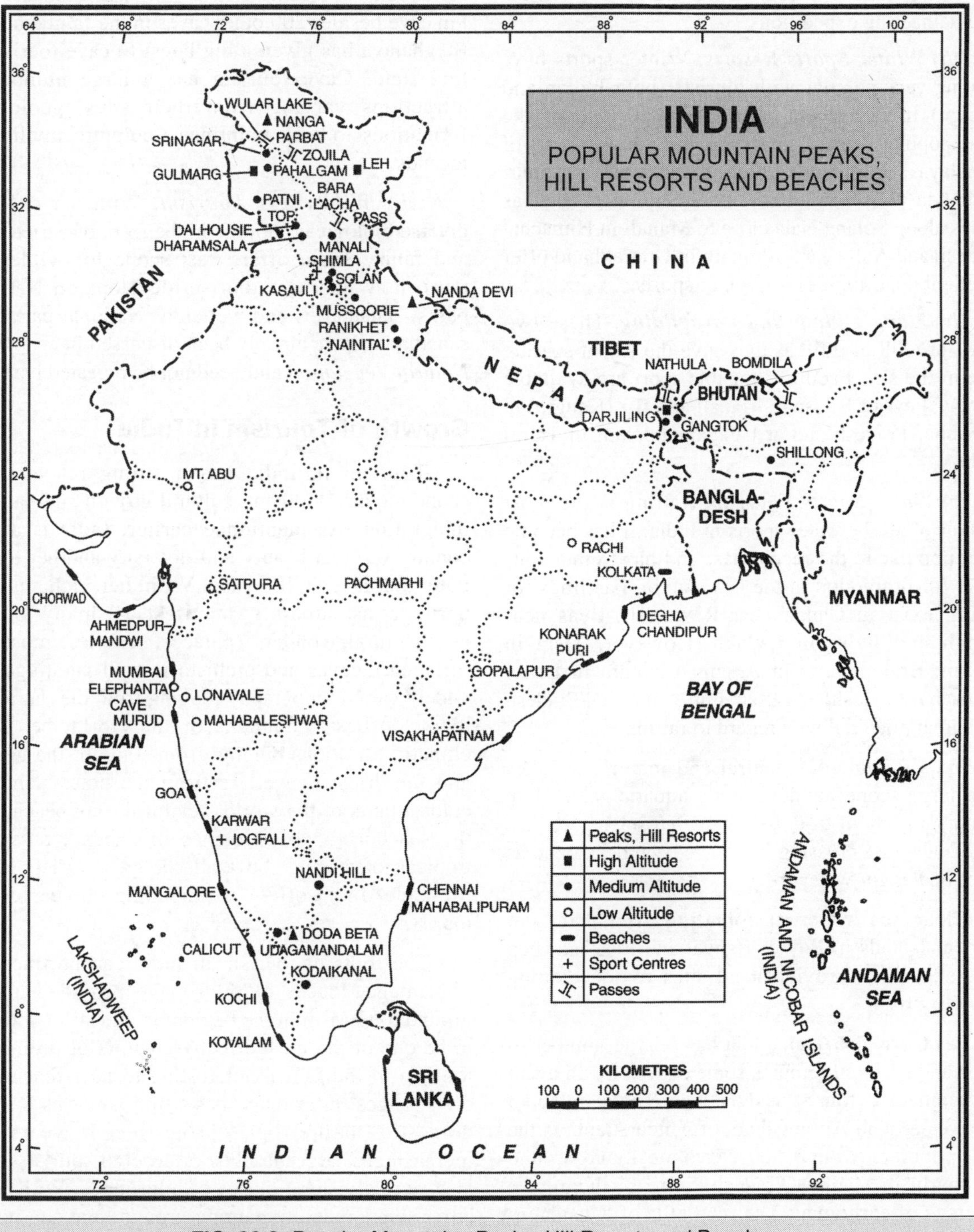

FIG. 28.2. Popular Mountains Peaks, Hill Resorts and Beaches

Indian Mountaineering Institutes at Manali, Darjeeling and Uttar Kashi and the Indian Mountaineering Foundation at Delhi render valuable help and guidance in organising and sponsoring the mountaineering expeditions.

(iii) Winter Sports Resorts. Winter sports have become very popular with tourists (both Indian and foreign) in the recent past. Skiing is one of the most popular winter sports on the snowy slopes of Himalayas. Gulmarg (2,730 m) in Jammu & Kashmir, Narkanda (2,700 m), Kufri near Shimla as well as slopes along Solang Nala close to Manali in Himachal Pradesh and Auli near Joshimath in Uttarakhand offer excellent opportunities for winter sports.

(iv) Hang gliding and Paragliding. These two sports are full of thrill as they give the joy of soaring high in the sky. Facilities for this sport are available in the valleys of Bilaspur, Manali and Bir (Kangra) in Himachal Pradesh and at Udagamandalam in Tamil Nadu.

(v) Water Sports Tourism. Rafting is the most popular of all the water sports in India. It has become quite popular in the recent past and has tremendous scope for promotion in the near future. Facilities for rafting exist in Ganga near Rishikesh, Beas near Manali and Indus in Ladakh. However, Tista in Sikkim, Brahmaputra in Assam, Chandra in Lahul (Himachal Pradesh) and Bharati in Arunachal Pradesh hold great potential with regard to rafting.

India's numerous natural and man-made lakes offer large scope for developing aquatic games like sailing, angling of trout fish and wind surfing. Rivers and canals criss crossing Goa have been developed as water bodies for water sports.

Clear sea water surrounding Andaman and Nicobar Islands in Bay of Bengal and Lakshadweep in Arabian Sea provide ideal sites for developing diving sports.

(vi) Cave Tourism. India has a large number of caves and rock-cut temples some of which still await development as tourist designations. There are about 30 caves around Aurangabad, of which Ajanta is the most popular. Around 500 caves are found around Pachmarhi in Madhya Pradesh. Some of them have paintings of early man. The twin hills of Khandagiri and Uraygiri near Bhubaneshwar have caves showing rock sculpture. Cherrapunji in the Garo hills of Meghalaya has a beautiful cave in limestone area. The cave has well developed stalactite and stalagmite formations. A recent discovery of Asia's longest 19.2 km cave besides 200 other caves in the Jaintia hills of Meghalaya has given a big boost to cave tourism in this state. Cave tourism has a large number of attractions varying from their sites, geological formations to rock paintings, sculpture myths and legends.

(vii) Wilderness Tourism. With her rich and diverse biological heritage in the form of varied flora and fauna, India offers vast scope for wilderness tourism. Wilderness tourism develops on National parks, sanctuaries and wetlands. National parks and sanctuaries have already been discussed in chapter on *Natural Vegetation* and need not be repeated here.

Growth of Tourism in India

Tourism in India has a strong relevance to economic development, cultural growth and national integration. As mentioned earlier, India is a vast country of great beauty and diversity and her tourist potential is equally vast. With her rich cultural heritage as superbly manifest in many of the architectural wonders (palaces, temples, mosques, forts, etc), caves and prehistoric wall paintings, her widely varied topography ranging from the flat plains to the loftiest mountains of the world, her large climatic variations ranging from some of the wettest and the driest as well as from the hottest and the coldest parts of the world, beautiful long beaches on the sea coast, vast stretches of sands, gregarious tropical forests and above all, the great variety of the life-style, India offers an unending choice for the tourist.

The roots of tourism in India can be traced to pilgrimage. Places of pilgrimage provided a firm ground to tourism in the beginning and still continues to be one of the most effective factors of promoting tourism in India. It is interesting to note that in the early stages, pilgrimage based tourism was only of domestic nature but during recent years, an increasingly large number of foreign tourists have also started visiting places of pilgrimage. With other attractions gradually gaining ground, one can notice detectable changes in the character of Indian tourism, with a tilt from pilgrimage to pleasure trips. Today, the majority of the tourists feel that travelling for sheer *pleasure* offers much wider scope than that offered by *pilgrimage* and thus the former is always more colourful, lively, intimate, soothing to mind and body and hence, more enjoyable. Sometimes, people combine pilgrimage with pleasure trips to draw the benefit of both.

Organised tourism in India began in 1950s with the genesis of planned development. Over the years, Indian tourism has grown considerably. Under normal circumstances, tourism sector is a major engine of economic growth that contributes significantly in terms of GDP, foreign exchange earnings and employment. However, the COVID-19 pandemic has had a debilitating impact on world travel and tourism and India was no exception. According to World Tourism Barometer of the United Nation's World Tourism Organisation (January 2021), International Tourist Arrivals (ITA) declined by 74 per cent globally in 2020 over the previous year. ITA had reached a total of 1.5 billion in 2019, and reduced to 381 million in 2020, leading to an estimated loss of US $ 1.3 trillion in export revenues. This weakness in international tourism continued in 2021 but is expected to recover in future as normalcy returns.

Before the spread of COVID-190 pandemic the tourism sector in India has been performing well with Foreign Tourist Arrivals (FTAs) growing at 14 per cent to 10.04 million and Foreign Exchange Earnings (FEEs) at 19.1 per cent to US $ 27.31 billion in 2017. However, the sector underwent a slowdown in 2018 and 2019 before declining sharply in 2020. The Foreign Tourist Arrivals (FTA) in 2019 stood at 10.93 million compared to 10.56 million in 2018. In terms of growth, the growth rate of FTAs declined from 14 per cent in 2017 to 5.2 per cent in 2018 and further to 3.5 per cent in 2019. Foreign Exchange Earnings (FEEs) from tourism stood at US $ 30.06 billion in 2019 as compared to US $ 28.59 billion in 2018. In terms of growth, the FEEs declined from 19.1 per cent in 2017 to 4.7 per cent in 2018, picking up slightly to 5.1 per cent in 2019.

India ranked 23rd in the world in terms of international tourist arrivals in 2019, falling slightly from 22nd position in 2018. The country accounts for 1.23 per cent of world's international tourist arrivals and ranked 12th in the world in terms of tourist foreign exchange, accounting for over 2 per cent of the world's tourism foreign exchange earnings (**Table 28.1**).

Tourists from the top countries visiting India—Bangladesh, U.S.A., U.K., Sri Lanka, Canada, Australia, Malaysia, China, Germany and Russia—accounted for 65 per cent of the total foreign tourist arrivals in India in 2018. Among the foreign tourists 62.4 per cent visited for leisure holiday and recreation, 16.3 per cent for business, and 13.5 per cent was Indian diaspora. Due to strained relations Pakistan does not send sufficient tourists to India, although there is a vast scope for it.

Outbound tourism has also picked up in recent years, with the number of departures of Indian nationals from India growing at 9.5 per cent during 2017 to 23.94 million from 21.87 million, in 2016. This is about double the foreign tourist arrivals in India.

Domestic Tourism

Domestic tourism is an important segment of the overall tourist scenario. It may be emphasised here that domestic tourism is no less pronounced than the foreign tourism. In fact, domestic tourists far outnumber foreign tourists. This is by far the largest segment of market with a potential of 20 crore local tourists travelling annually. It is domestic tourism which fosters a sense of unity in the otherwise diverse environment of the country and contributes to national integration. Even if 10 per cent of the population travels outside the native state, it involves a massive movement of over 14 crore people who develop the feeling that they are travelling within their own country.

Domestic tourism has increased considerably during the last couple of decades. The improved economy and greater exposure through mass media have developed increasing awareness among the people about tourism; a new dimension to their life style. Larger income, longer holidays coupled with certain incentives given by public and private organisations to their workers, have contributed a lot in infusing interest among the hitherto stay-at-home people, to look around for a place for an annual or biannual visit with family members.

TABLE 28.1. Foreign and International Tourist Arrivals and Tourism Receipts in India and the World

Indicator	2014	2015	2016	2017	2018	2019
Foreign Tourist Arrivals (FTA) in India (million)*	7.68	8.03	8.80	10.04	10.56	10.93
International Tourist Arrivals (ITAs) in India (million)**	13.11	13.76	15.03	16.81	17.42	17.91
International Tourist Arrivals around the world (million)	1137	1195	1241	1333	1408	1460
India's Share in World's ITAs (per cent)	1.15	1.15	1.21	1.26	1.24	1.23
India's Rank in World's ITAs	24th	24th	26th	26th	22nd	23rd
India's Share in World Tourist Receipts (per cent)	1.57	1.73	1.84	2.02	1.97	2.03
India's Rank in World Tourism Receipts	15th	14th	13th	13th	13th	12th

Source : Ministry of Tourism in Economic Survey 2019-20, p. 265, Economic Survey 2020-21 p. 315.

Note : * based on data from the Bureau of Immigration on tourists entering the country on a foreign passport.

** Tourists entering the country on a foreign passport plus non-resident citizens visiting the country International Tourist Arrivals is the sum of Foreign Tourist Arrivals and NRI arrivals in the country,

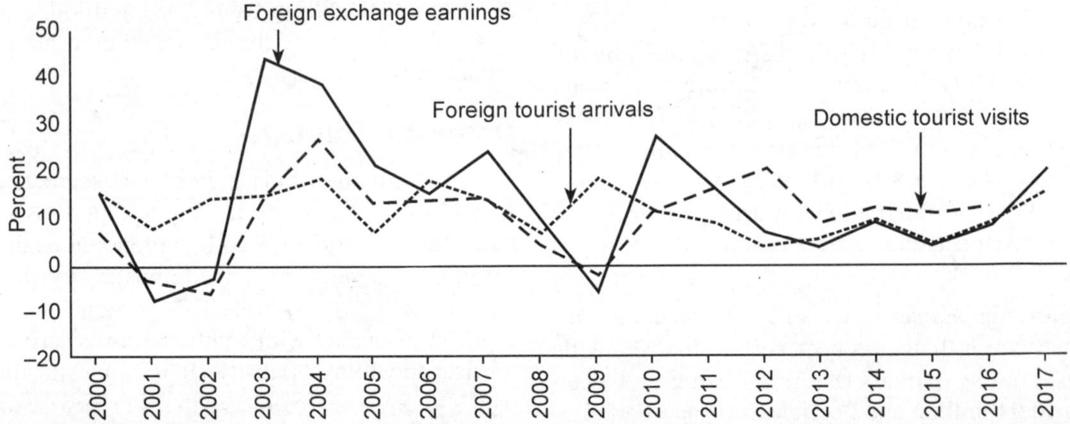

FIG. 28.3. Growth of Foreign Tourist Arrivals (FTA) Foreign Exchange Earnings (FEE) and Domestic Tourist Visits in India from 2000 to 2017

Source : Based on data from Ministry of Tourism data **(In Economic Survey, 2017-18)**

However, growth in domestic tourist visits slowed down from 12.7 per cent in 2016 to 2.4 per cent in 2017. In absolute numbers, there were 1652.5 million tourists in 2017 compared to 1615.4 million in 2016. Tamil Nadu (345.1 million), Uttar Pradesh (234 million), Karnataka (180 million), Andhra Pradesh (165.4 million) and Maharashtra (119.2 million) were the top five destination states, accounting for 63.2 per cent of total number of domestic tourist visits in 2017. Amongst the centrally protected ticketed monuments, Taj Mahal, Agra (5.66 million) was the most visited monument in 2017-18 for domestic visitors followed by Sun Temple Konark (3.22 million) and Red Fort Delhi (3.04 million).

The Ministry of Tourism launched the National Mission on Pilgrimage Rejuvenation and Spiritual, Heritage Augmentation Drive (PRASHAD) in January 2015 with the objective of pilgrimage infrastructure at selected important pilgrimage destinations. The total number of sites identified for development under the scheme at present is 41 spread over 25 states.

Under the **Swadesh Darshan Scheme,** the Ministry of Tourism has identified 15 thematic circuits for development, namely; Himalayan circuit, North-East Circuit, Krishna Circuit, Buddhist Circuit, Coastal Circuit, Desert Circuit, Tribal Circuit, Eco-circuit, Wildlife Circuit, Rural Circuit, Spiritual Circuit, Ramayana Circuit, Heritage Circuit, Tirthankar Circuit and Sufi Circuit. These circuits cover religious spiritual, cultural, natural, tribal sites in the country. Under the scheme, the Ministry is developing these thematic circuits on the principles of high tourist value, competitiveness and sustainability in an integrated manner with an objective of enriching tourist experience and enhance employment opportunities.

Various initiatives have been taken by the Government to promote tourism sector of the country. The most important step was the introduction of e-Tourist visa regime in September 2014 for 46 countries. Prior to the launch of the scheme, the e-Visa facility was available for only 12 countries. The government further liberalized the visa regime in 2016 renaming it to e-Visa scheme with five sub-categories *i.e.* 'e-Tourist Visa', 'e-Business Visa', 'e-Medical Visa', e-Conference Visa', and 'e-Medical Attendant Visa'. The e-Visa scheme is now available for 169 countries with valid entry through 28 designated airports and 5 designated seaports. With this, foreign tourist arrivals to India on e-Visa have increased from 4.45 lakh in 2015 to 23.69 lakh in 2018. The other initiatives are promotion of India as a 365 days destination, Swachhta Action Plan (SAP), Skill Development Initiative, launching of Multilingual Tourist Infoline, and Swachh Paryatan Mobile App launch of Global Media Campaign for 2017-18 on various Channels; launch of 'The Heritage Trail' to promote the World Heritage Sites in India, launch of International Media Campaign on various international TV channels; Celebration of 'Paryatan Parv' having 3 components namely 'Dekho Apna Desh' to encourage Indians to visit their own country, 'Tourism for All' with tourism events at sites across all states in the country, and 'Tourism and Governance' with interactive sessions and workshops with stakeholders on varied themes.

Business Tourism

Another healthy trend in the foreign tourism in India since 1991 has been a conspicuous increase in business travel with its spinoff effects in upgradation of accommodation and introduction of new technology in communications and other services. The metropolitan cities like Delhi, Mumbai, Chennai, Kolkata, Bengaluru, Hyderabad, etc. have experienced phenomenal growth in travel by business class people, some of whom combine business with pleasure. This trend has encouraged entrepreneurs and industrial houses to invest in hotel business apart from the existing chains to go in for large scale expansion and diversification of the hotel industry. On an average, a foreign tourist stays for about 27 days in India, but there are wide variations in the duration of stay by people from different parts of the world. The duration of stay of the foreign tourists in India is an important indicator of foreign exchange earned by the country.

Medical Tourism

Medical tourism is a recent trend which India is increasingly experiencing and is gaining a lot in terms of foreign exchange and goodwill—both in terms of medical treatment and tourist destination. India is fast emerging as a potential destination for medical tourism as foreign patients flee the chronic inefficiency of their own health services to more economical and effective alternatives in New Delhi and other Indian cities.

Patients in U.K. and U.S.A., South Africa, West Asia and South-East Asia are increasingly opting for private Indian hospitals. They are tempted by hospitality, swift treatment, economical pricing and world class quality of treatment. For example, a heart bypass costs £ 500 in India as against £ 5,000 in the U.K. British hip replacement surgery costs £ 6,600 in U.K. while in India it is available for just £ 860. In addition to cost factor, speed is a powerful incentive in India as there is no need for referrals or long waits. This important factor has led to Britain negotiating with some of the top hospitals in New Delhi to bring some of the patients to India who would otherwise have to wait for months or even for years to undergo routine operations on its National Health Service. Besides, patients in U.K. and the U.S.A. cannot avail of their insurance for not-critical treatment, such as knee replacement or cosmetic surgery. Under such circumstances, cost becomes an imperative factor and patients opt for cheapter and good quality treatment in India.

It is interesting to note that the main demand for medical tourism comes from about 20 million Indians living abroad, although the number of foreigners who are keen to avail of speedy, inexpensive and quality treatment in India is growing at a rapid pace. Two major factors influencing this trend are (*i*) world class private hospitals, and (*ii*) abundant experience of doctors, as many of them have already studied and practised medicine abroad.

The idea of medical tourism picked up fast after 1996 and the number of foreign patients seeking treatment or diagnosis in India, with holiday included, increased rapidly. For example, people from West Asia avoided visiting the U.S.A. and European countries for medical treatment after terrorist attack on the U.S.A. on 11th September, 2001 (known as 9/11 attack) and on London (U.K.) on 7th July, 2005 (known as 7/7 attack) and opted for India instead. Besides, value added services such as transfers from the airport, special food for foreigners and translators for patients not conversant with English have given a much needed boost to medical tourism. Realising the potential of health/medical tourism, top hospitals in Delhi and other major cities of India have initiated a new concept of treatment supplemented by vacation. Some hospitals have gone a step forward and are offering the patients and their relatives/friends stay in the hospital complex with all the luxuries of a five star hotel.

The Government of India has taken several steps to promote Health and Medical Tourism in India and India has emerged as a major Medical Tourism destination. The Ministry of Tourism has recognized Medical and Wellness Tourism including Ayurveda as a Niche product in order to overcome the aspect of **'seasonality'** and to promote India as 365 days destination and attract tourists with specific interests. In order to promote Medical Tourism, the Government offers financial support as Marketing Development Assistance for Publicity and for organizing Wellness and Medical Tourism promotion shows. Medical Tourism is regularly highlighted for promotion as part of the Incredible India campaign in print, electronic, online and outdoor media in India and abroad as well as the various travel trade exhibitions overseas. 'Medical visa' has been introduced, which can be given for specific purpose to foreign travellers coming to India for medical treatment. 'E-Medical Visa' has also been introduced for 169 countries. The arrival in India on Medical visa has registered a more than three-fold increase in a short duration from 2014 to 2017 (See **Table 28.2**)

TABLE 28.2. Arrivals in India on Medical Visa

Year	2014	2015	2016	2017
Number of arrivals	1,39,447	2,33,918	4,27,014	4,95,056

Source : Economic Survey 2018-19, p. 239

As a result of various developments in India, the market size of medical tourism rose from about US $ 2.17 billion in 2012 to about US $ 9 billion in 2020. Currently, India has about 18 per cent of global medical tourism market which is likely to increase further with the passage of time.

Employment Opportunities in Tourism. Tourism in India has vast employment potential, much of which still awaits exploitation. At present about 20.5 million persons are directly employed by hospitality services. This is about 5.6 per cent of the total work force of the country. In addition, the industry provides indirect employment to about 40 million persons. Further, it is interesting to note that the employment generation in proportion to investment is very high in tourist industry. According to one estimate, an investment of ₹ 10 lakh creates 89 jobs in hotel and restaurant sector as against 44.7 jobs in agriculture and 12.6 in manufacturing industries. The ratio further increases if one takes into account the ancillary services associated with hotels and restaurants. Further, it has been estimated that a spend of $ 1 million in this sector supports 407 jobs as compared to communication services (381 jobs), financial services (329 jobs) and manufacturing (315 jobs). Globally, every unit of investment in travel and tourism directly employs six times more than automotive manufacturing industry, five times more than chemical industry, four times more than mining industry and three times more than financial services industry. Another important aspect of employment in tourism is that it employs a large number of women, both educated and uneducated, as well as skilled and unskilled. In fact, women are greater in number than men in hotels, airline services, travel agencies, handicraft making and cultural activities, etc. Tourism has grown sharply due to entry of online-based service providers.

PLACES OF TOURIST INTEREST

Places of tourist interest are so numerous and of varied nature that it is not easy to describe these places comprehensively. These include places of archaeological and historical importance, pilgrimage centres, sanctuaries and national parks, hill resorts, sea beaches, etc. (**Fig. 28.4**). They are distributed throughout the length and breadth of the country. A brief description of the distribution of tourist places of India is as under :

1. The Himalayan Region. The Himalayan region is at present, one of the main tourist destinations, not only in India but in the whole of South Asia. Its lofty peaks, snow clad ranges, lush green valleys, turbulent rivers and varied fauna and flora offer a large variety of tourist attractions, games and skiing, hiking and trekking. The entire length of the Himalayas starting from Jammu and Kashmir in the west to Arunachal Pradesh and Meghalaya in the east is dotted with different types of tourist centres.

Srinagar, Gulmarg, Pahalgam, Patnitop, Jammu, Sonamarg, and Vaishno Devi are the important tourist centres of Jammu and Kashmir. The Union Territory of Ladakh has Leh and Amarnath Cave as important centres of tourist attraction. Pangong Tso a beautiful lake located on the Indo-China border has become a great source of attraction for the tourist in the recent past. In Himachal Pradesh, there are places like Shimla, Dalhousie, Kangra, Dharamsala, Kullu-Manali, Chamba, Kasauli, etc. which attract tourists from far and wide. In the adjoining Himalayan region of Uttarakhand are located famous religious places like Haridwar, Rishikesh, Kedarnath, Badrinath, Gangotri, Yamnotri, Gomukh, etc. Mussoorie, Almora and Nainital are hill stations. The Valley of Flowers presents a unique site in the higher Himalayas. Corbett has the world famous national park.

Towards the east are the famous hill resorts of Darjeeling, Mirik and Kalimpong in West Bengal, Gangtok in Sikkim and Shillong in Meghalaya. Some other places of tourist interest are Kohima in Nagaland, Imphal in Manipur, Aizawl in Mizoram and Agartala in Tripura. At the foothills in Assam are Mannas sanctuary, Soni-Rupa sancturary and Pabha sanctuary for protection of animals like one horned rhinoceros, wild buffaloe, etc.

2. The Great Plain of North India. Although monotonous from the view point of relief, this vast plain extends from the sandy stretches of Rajasthan in the west to the Sundarbans in the east and possesses a large number of tourist places. These places are of historical, archaeological, religious and industrial importance. It includes large parts of Rajasthan, Punjab, Haryana, Uttar Pradesh, Bihar, West Bengal and Assam, besides the union territories of Delhi and Chandigarh.

The whole of Rajasthan is famous for its palaces, forts, religious places and above all the vast stretches of sand dunes. More than half of the state to the west of the Aravali Range forms part of the Great Plain of North India. Here, Jaisalmer has become very popular during the recent years. Along with Jodhpur and Bikaner, it forms the famous **desert triangle**. In Punjab, Amritsar is an important seat of sikh religion. Jalandhar, Ludhiana, Patiala, Bathinda and Sirhind are other tourist centres of Punjab. Haryana Tourism has developed several places as tourist centres. They include Ambala, Kurukshetra, Karnal, Panipat, Pinjaur, Morni, Pehowa and Faridabad. Surajkund near Faridabad has become famous all over India due to its *annual crafts mela*. Delhi is the capital of the country and represents India in miniature. It has almost all sorts of tourist attractions including historical and administrative buildings, religious places, museums, parks and gardens, art galleries, etc. Chandigarh, the city beautiful, is one of the few planned cities of India. It is well known for its world famous rock garden, rose garden and Sukhna lake.

The plains of Uttar Pradesh present a grand mixture of Hindu and Islamic cultures. Mathura, Vrindaban, Pryagraj, Varanasi, Ayodhya, Sarnath are the sacred cities of Hindu religion. Lucknow, Agra and Aligarh represent Islamic religion, culture and history. Sarnath is an important Buddhist religious centre. Agra is known for its world famous Taj Mahal. Any foreign tourist who visits Delhi makes it a definite point to visit the Taj at Agra. As a matter of fact, Delhi, Agra and Jaipur form the world famous **Golden Triangle**. In the Bihar Plain are located Patna, Nalanda, Vaisali, Munger and Sitamarhi. In West Bengal, Kolkata is the most

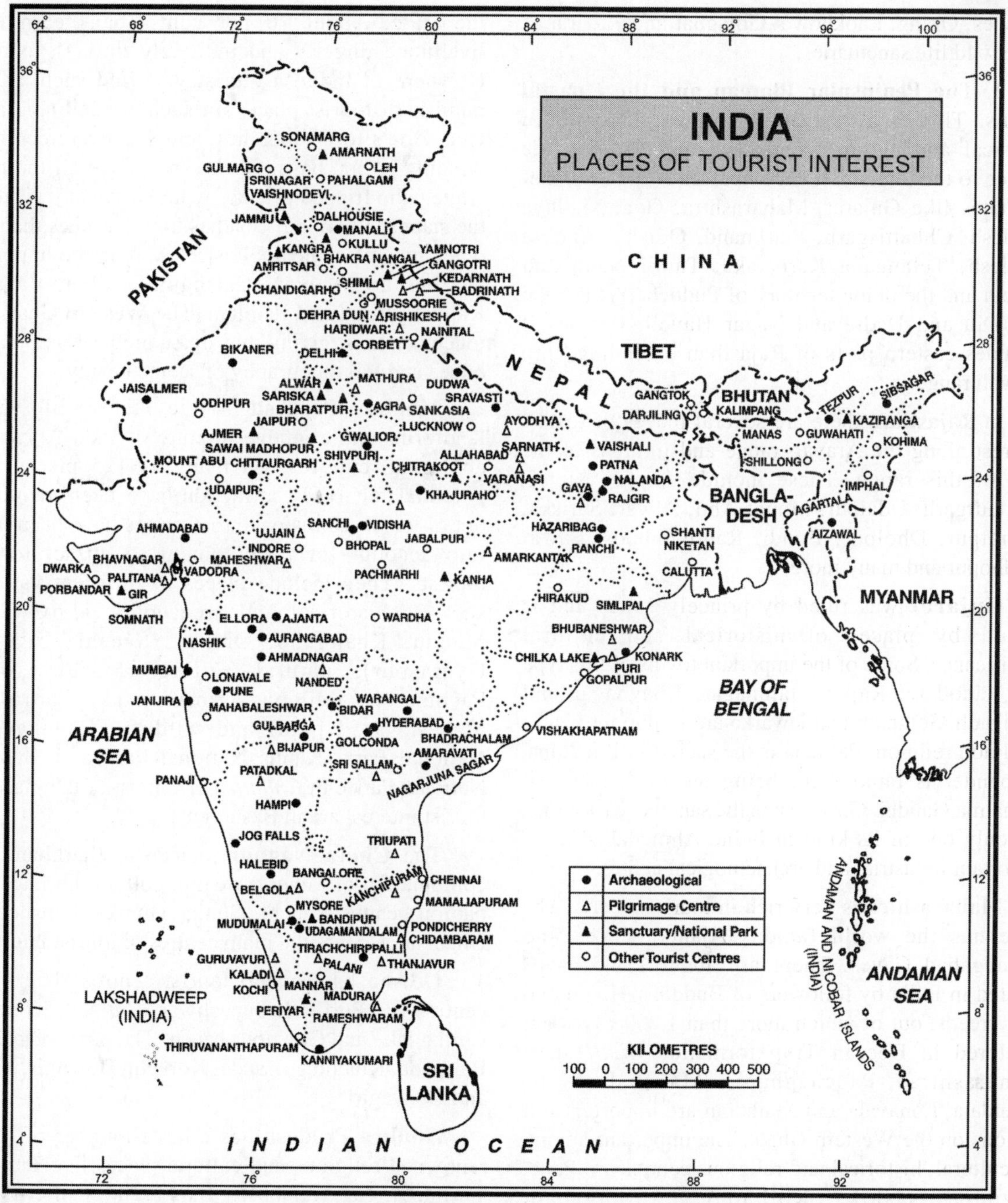

FIG. 28.4. India : Places of Tourist Interest

important place. This is the capital of West Bengal and is the third largest metropolitan city of India, after Mumbai, Delhi. Kolkata is often called the nerve centre of the entire eastern region. Shanti Niketan is famous for being the work place of Nobel laureate Gurudev Rabindranath Tagore. Sundarbans, the abode of the Royal Bengal Tigers offers much to the tourists. Murshidabad, Malda and Bishnupur are historical centres. In the Assam plain are located Barpeta, Hajo, Tejpur, Sibsagar and Guwahati.

Besides, Orang, Laokhowa, Golaghat and Kaziranga have wild life sanctuaries.

3. The Peninsular Plateau and the Coastal Plains. This is a vast area with varied types of physical and cultural landscape and offers wide choice to the tourists. It encompasses a large number of states like Gujarat, Maharashtra, Goa, Madhya Pradesh, Chhattisgarh, Jharkhand, Odisha. Andhra Pradesh, Telangana, Karnataka, Tamil Nadu and Kerala and the union territory of Puducherry, Daman and Diu and Dadra and Nagar Haveli. Besides it includes eastern parts of Rajasthan extending upto Aravali range.

In **Rajasthan** there are several places of tourist interest along the Aravali range and in areas to the east of this range. These include Jaipur, Ajmer, Chittaurgarh, Udaipur, Mount Abu, Alwar, Sariska, Bharatpur, Dholpur, Bundi, Ranthambhor, Sawai Madhopur and many more.

Gujarat was ruled by princely states and is dotted by places of historical and cultural significance. Some of the important towns of this type are Vadodara, Rajkot, Jamnagar, Bhavnagar and Junagadh. Somnath and Dwarka are important places of Hindu religion. Palitana is the sacred hill of Jains. Porbander is famous for being the birth place of Mahatma Gandhi. Gir forest is the sanctuary for lions, the only one of its kind in India. Ahmedabad is an important industrial and archaeological centre.

Maharashtra is very rich in tourist resorts. The state has the world famous Ajanta, Ellora and Aurangabad Caves. There are about 1,545 caves created in India by followers of Buddhist-Hindu and Jain creeds, out of which more than 1,200 caves are clustered in Deccan Trap formation of Western Maharashtra. Panchaghani, Mahabaleshwar, Khandala, Lonavala and Matheran are important hill stations on the Western Ghats. The important among the cultural, historical and religious centres are Pune, Nagpur, Nashik, Kolhapur, Aurangabad, Ahmadnagar, Shirdi, Nanded, Wardha and Melghat. Above all, Mumbai is the commercial capital of India and offers several opportunities to tourists.

Goa is a picturesque land full of scenic beauty and abounding places of tourist interest. Tourism forms a major segment of Goa's economy, contributing over 16 per cent of the total earnings of the state. About 20 per cent Goans earn their livelihood directly and indirectly through tourism. Nowhere in India, a tourist will find such a large number of tourist places, in such a small area as in Goa. Goa's forts, churches, temples, evergreen hills, winding rivers, bays, creeks and above all the pearl white palm fringed beaches would leave any visitor to the state spell bound. Goa has more beaches than any other state of comparable size. Among the important beaches are Anjuna, Calangute, Kolva, Vagator, Kandoli, Betul and Palolem. The Western Ghats just touch the coast giving this area a unique combination of sea and hill topography of scenic beauty.

Madhya Pradesh is the state with varied landforms and a unique mixture of Indo-Aryan and aboriginal cultures. As many as 125 places with potential for tourist attraction have been identified, out of which 16 major tourist places have been taken into account for development to attract foreign tourists. They include places of different varieties such as historical and archaeological (Gwalior, Mandu, Khajuraho, Orchha, Sanchi, Udaigiri, Gyaraspur), natural setting (Marble rocks, Pachmarhi), religious (Amarkantak, Chitrakoot, Maheshwar, Ujjain) and wildlife (Bandhavgarh National Park, Kanha National Park and Shivpuri National Park). In *Chhattisgarh*, most of the places of tourist interest are in Bastar area.

The **Chota Nagpur** plateau of Jharkhand is a combination of mineral wealth, cultural heritage and natural beauty. Ranchi, Gunha, Dumka, Jamshedpur, Dhanbad, etc. are the main centres of tourist interest.

Odisha has several tourist centres. The major centres are Puri, Bhubaneshwar, Cuttack, Konarka, Chilka lake and Gopalpur beach. The other places are Baripada, Khiching, Bauda, Koraput, Bolangir, Jajpur and Udayagiri.

Andhra Pradesh has a large number of world renowned religious and cultural places like Tirupati, Tirumala, Sri Kalahasti, Sankaram, Kolleru, etc. Nagarjuna Sagar Dam on the border with Telangana is a great centre of tourist attraction. Mukhalingam, Vishakhapatnam, Draksharamam, Vijayawada, Rajahmundry, Srikakulam, etc. are other centres.

Telangana has historical places like Hyderabad and Golconda Fort. Medak, Warangal, Bhadrachalam, Palampet, etc. are other places of tourist interest.

is a labour-intensive industry and is likely to offer more jobs in the coming years. And since most of the natural beauty and wildlife are to be found in non-urban areas, rural people could find employment as guides and transporters; with proper training. More jobs in rural areas would also help reduce continuous migration of people to towns.

The Approach Paper of 12th Five Year Plan highlights the need to adopt "pro-poor tourism" for increasing net benefit to the poor and ensuring that tourism growth contributes to poverty reduction. Tourism plays a key role in socio-economic progress through creation of jobs, entreprise, infrastructure and revenue earnings. The Planning Commission has identified tourism as the second largest sector in the country in providing employment opportunities for low skilled workers.

Hotel Accommodation. The hotel sector forms the key segment of tourism industry. Most of the foreign exchange earned by tourist industry is accounted for by hotel segment. Realising the importance of hotel segment, the Government has taken intiatives to encourage hotel indsutry by providing tax benefits and other incentives. The Industrial Policy has now placed the hotels and tourism related activities as a priority industry. Foreign investment and collaborations are now facilitated under the new economic policy.

The Department of Tourism classifies functioning hotels under the *star system* into various categories from one to five-star deluxe and Heritage (Heritage Classic, Grand and Heritage Renaissance) and Apartment Hotels from three star to five star deluxe. The Department also reclassifies these hotels after every five years to ensure that requisite standards are maintained by them. Heritage hotels have been introduced to cover functioning hotels in palaces, *havelies*, castles, forts and residences built prior to 1950. Since foreign tourists are crazy about Indian culture and heritage, care has been taken to preserve the original form of the buildings while developing them as heritage hotels. Industrial Policy has now placed hotels and tourism related activities as a priority industry. Foreign investment and collaborations are now facilitated under the new economic policy. Automatic approval is available for foreign direct investment (FDI) upto 100 per cent in hotel and tourism sector.

According to Ministry of Tourism, the number of approved hotels increased from 1,116 as on December, 2014 to 1961 as on December, 2018, registering a growth of 75.71 per cent during this period. Similarly the number of hotel rooms increased from 71,244 to 1,02,490 registering a growth of 43.85 per cent during the same period (**Table 28.3**). Of course, this is only a subset of the hotel market which

TABLE 28.3. Number of Approved Hotels and Hotel Rooms for Major Categories

Sl. No.	Category of Hotels	As on 31-12-2014		As on 31-12-2018	
		Hotels	Rooms	Hotels	Rooms
1.	One Star	41	1,193	9	348
2.	Two Star	80	1,902	37	990
3.	Three Star	554	22,724	535	18,889
4.	Four Star	134	7,969	322	16,451
5.	Five Star	92	11,744	181	22,673
6.	Five Star Delux	113	23,907	170	37,955
7.	Apartment Hotels	3	249	3	252
8.	Guest Houses	5	77	7	106
9.	Heritage Hotels	42	1,237	58	1,843
10.	B & B Estt.	52	242	639	2,983
	Total	1,116	71,244	1,961	1,02,490

Source : Economic Survey 2018-19, p. 240.

Karnataka is famous for its gardens, historical and religious places and natural beauty spots. The Brindavan Garden near Mysuru could be the envy of any beautiful garden in the world. Bengaluru, the garden city, is the capital of the state and is famous for many industries. Some of the important historical and religious centres are Mysuru, Bijapur, Gulbarga, Bidar, Nandi Hill, Shravan Belgola, Shrirangapattam, Chamundi, Belur, Halebidu, Chitradurga, etc. Ranganthitto bird sanctuary, Bandipur and Dendilio wildlife sanctuary, Jog Falls on Sharavati river along with Madog Falls and Lushington falls and the beaches of Murudeshwar, Maravanthe, Malpe, etc. are the other major tourist attractions.

Tamil Nadu is famous for its renowned temples at Rameshwaram, Madurai, Thanjavur, Tiruchchirappalli, Tiruvannamalai, Kanchipuram, etc. The state also boasts of famous hill stations like Udagamandalam (Ooty). Kodaikanal, Coonnoor and Yercaud. The capital city of Chennai with its Marina beach and Golden beach and several religious and historical buildings, is a great tourist attraction. Kanniyakumari is the southern-most tip of the mainland of the country where waters of Arabian Sea, Bay of Bengal and Indian Ocean merge with one another. It is world famous for its Rock Memorial. Mudumalai, Anaimalai and Mundanthurai have wildlife sanctuaries.

Kerala is not lagging behind with respect to tourism. The state is blessed with several places having buildings of historical and architectural importance, beautiful sea beaches, wildlife sanctuaries and some hill resorts. The main places of tourist interest are Thiruvananthapuram, Kochi, Alappuzha, Kannur, Thrissur, Kozhikode, Ernakulam, Malayatoor, Idukki, Periyar, Munnar, Thekkady, etc.

Among the union territories, Puducherry has the famous Aurobindo Ashram along with some other places of historical and religious importance. Daman and Diu on the Arabian Sea coast have a beach and some religious and historical places.

Islands. India has two groups of islands. They are having picturesque landscape and are fast becoming major tourist attractions. The Andaman and Nicobar Islands in the Bay of Bengal have beautiful beaches, lush green tropical rain forests and a vast variety of oceanic life in the blue seas surrounding them. Port Blair is the main tourist spot. Cellular Jail at Port Blair is a special attraction because it was used by the British rulers to punish the Indian Freedom Fighters and has been designated as National Memorial. The Lakshadweep group of islands is of coral origin and attracts tourists from far and wide.

PROBLEMS AND PROSPECTS

Although India has progressed a lot since 1950s with respect to tourism, she is still way behind the developed or even the developing countries. India's share in the tourist arrival has been growing at a snails pace from 0.23 per cent in 1975 to 0.28 per cent in 1980 and 0.42 per cent in 2004. Even now, it is only 1.23 per cent of the world tourist arrivals and is much less than he other South Asian countries like Singapore, Malaysia, Thailand, Hong Kong, Macao, etc. Even the diminutives like Maldives and Bhutan present an appreciable model of sustainable tourism by integrating environment and tourism. In India, tourism provides 5.6 per cent of the total jobs, China that just pulled away its iron curtain, is doing better with 7 per cent. Torn apart by civil war Sri Lanka still has 7.4 per cent of its jobs coming from the tourism sector. Travel is a happening business in Thailand, generating 11.2 per cent of the total jobs.

Still there are large areas which are untapped from the tourist point of view. Traditionally a popular destination, North India still draws about 49 per cent of the tourists; whereas only 4 per cent go to east. West is doing better getting 29 per cent of the tourist inflow. South, despite its beaches, temples, hills, etc. gets only 18 per cent of foreign and domestic tourists. Thus there is great potential for development of tourism, particularly in east and south, which require proper tapping with immediate effect.

Foreign tourists often fail to get suitable accommodation in hotels and go back as a dissatisfied lot. At the beginning of the millennium, India's shortage was estimated at 30,000 rooms. Today, the requirement is much larger, considering rapid increase in the inflow of foreign tourists after 2002. Today we need at least 90,000 more rooms (in five star segment) which requires a huge investment of ₹ 80,000 crore, just to fall in line with the internationally-benchmarked tourism models. The Common Wealth Games organised in Delhi in 2010 has once again highlighted our poor show from tourism point of view.

Another problem of recent origin is that Indian tourists have started travelling abroad rather than travelling within the country. Of late, Indian tourists are looking beyond Shimla, Srinagar, Ooty, Mumbai, Delhi and Goa, and are grabbing foreign holiday deals like never before. The main cause of outflow of Indian tourists is high aviation cost in the domestic sector. When a tourist in North India can get cheaper return tickets to Sri Lanka or Thailand than to Goa or Kerala, he will obviously go abroad. Our foreign exchange earnings from tourism are getting neutralised because of large number of Indians going abroad.

The unplanned and unrestricted growth of tourism has a detrimental effect on environment. The origin and growth of tourism in India as elsewhere in the world, is distinctly urban oriented rather than being spread over the countryside. This is due to the fact that urban centres provide the necessary infrastructure and wider choice of facilities and price ranges for accommodation, food, entertainment, etc. There is lack of appreciation of the holistic approach to tourism development which takes into account the linkages between environment and pressure of tourists. The result is that, almost all the popular tourist centres are groaning under the pressure of annual visitors and quite often facing difficult situation. The unplanned and unchecked growth of tourist centres deprives them of their original charm and ultimately proves disastrous. This is what is happening to some of the famous tourist centres like Shimla, Mussoorie, Darjeeling, etc. where sometimes even drinking water is not available to cater to the needs of the tourists. Sometimes, the area around a tourist spot is covered by commercial constructions and even by industries. Such unplanned growth tarnishes the image of the concerned spot. The Taj Mahal at Agra, the Meenakshi Temple at Madurai, the Lingaraja temple in Bhubaneshwar and the palace in Udaipur are examples of this nature where much of the original beauty and charm is lost due to unplanned constructions around them. Though many environments are endowed with great tourist potential, they are delicately balanced and large scale unplanned expansion of tourism in such areas often manifests itself in problems of serious environmental concern. Therefore, it is necessary that tourism should be developed in a planned way taking care of environment and infrastructural development. The emphasis should be on the *threshold* development keeping in view the carrying capacity of a given environment. In other words, growth and development of tourist industry should be *environmental friendly*. To be environmental friendly, the stress should be on what is now-a-days known as *soft tourism*. The tragedy that occurred in Uttarakhand on June 16-17, 2013, is an eye opener as it is considered to be largely due to unplanned growth of structures in the region.

With a view to improving tourism and at the same time taking care of the environment, the central and state governments are coming forward with new development plans. Although tourism is the responsibility of the state governments, the Central Government lends all help to develop tourism. India Tourism Development Corporation (ITDC) was established in October 1966. The Central Department of Tourism is extending financial assistance for improving tourism infrastructural facilities like *yatri niwases*, tourist lodges, wayside facilities, restaurants, cafeterias, tourist bungalows, etc. It also extends financial assistance for adventure sports activities, tourist transport, tented accomodation, etc. Travel Trade and Tourist Service Division of the Department of Tourism is helping the travel trade, both national and international. Department of Tourism has simplified and streamlined the rules from April 1992, regarding the grant of approval to travel agents, tour operators and tourist transport operators so that their services are standardised and are made broadbased, both for domestic and foreign tourists. The main thrust during the Eighth Five Year Plan had been, (*i*) development of selected tourist places/areas, (*ii*) diversification from culture-oriented tourism to holiday and liesure tourism, (*iii*) development of trekking, winter sports, wildlife and beach resort tourism, (*iv*) exploring new source markets in regions and countries having cultural affinity, (*v*) restoration of national heritage projects, (*vi*) launching of national image building and marketing plans in key markets, (*vii*) providing inexpensive accommodation in different tourist centres, (*viii*) improving service efficiency in public sector corporations and (*ix*) streamlining of facilitation procedures at airports.

Indian railways have also recognised its role in tourism development. The launching of Rajdhani and Shatabdi trains is a step in the right direction. **'Palace on Wheels'** is already operational in Rajasthan. Another special tourist train **'Orient Express'** has been introduced in Gujarat and one more covering temple towns of Tamil Nadu has become operational. Five more such luxury tourist trains with private sector participation connecting major tourist centres will soon be on the rails. Land for about 100 hotels at various railway stations has been set apart to augment accommodation for travellers. Considering the recent development, it is hoped that India will get her due share of tourism and at the same time steps will be taken to preserve the physical and social environment of the country.

REFERENCES

Adhikari, P.S. (1990) : *Mussoorie — A Tourist Resort of Garhwal Himalaya,* Geographical Observer (Meerut), Vol. 26, pp. 69-78.

Bhattacharya, B. (1992) : *Urban Tourism in the Himalaya in the context of Darjeeling and Sikkim,* TRR, Vol. 17, No. 2, pp. 79-83.

Bisht, Harashvanti (1994) : *Tourism in Garhwal Himalaya with special reference to mountaineering and trekking,* Indus Publishing House, Delhi.

Burkart, A.J. and Medlik, S. (1988) : *Tourism : Past, Present and Future,* ELBS Heinemann Professional Publishing, London.

Chopra, S. (1991) : *Tourism and Development in India,* Ashish Publishing House, New Delhi.

Economic Survey 2020-21 and 2021-22.

Gantzer, H. and Gantzer, C. (1983) : *Managing Tourists and Politicians in India,* Tourism Mangement, June, pp. 118-125.

Green Michael, J. (1993) : *Nature reserves of the Himalaya and the mountains of Central Asia :* IUNC. The World Conservative Union, Oxford University Press, Oxford, U.K.

India (1991) : *EIU* : International Tourism Reports No. 2 : 5-25.

India (2022) : *A Reference Annual,* Publication Division, Ministry of I&B, Govt. of India, New Delhi.

Kamath, U.D. (1993) : '*Development of Tourism, Tourism Development and Management of Coastal Areas* (Technical Papers), Institute of Town Planners, New Delhi, India.

Malkani, Ranjit (2004) : *Tourism—Key domestic segment,* The Hindu Survey of Indian Industry, Kasturi & Sons Ltd., Chennai, 2004, pp. 321-323.

Ministry of Civil Aviation and Tourism (MCAT) (1992) : *National Action Plan for Tourism,* New Delhi.

Pavoskar, M. (1982) : *Employment Effects of Tourism and the Indian Experience.* Journal of Travel Research, Vol. 21, No. 2, pp. 32-38.

Periasamy, Visalakshi (2010) : *Hotel Industry : Capacity additions imperative in Survey of Industry (2010).* The Hindu, Chennai.

Peter, M. (1969) : *International Tourism,* Hutchinson, London.

Pigram, J.J. (1990) : *Sustainable Tourism—Policy Consideration*; Journal of Tourism Studies, Vol. 1, No. 2, pp. 2-9.

Robinson, H. (1976) : *A Geography of Tourism,* Macdonald and Evans, London.

Seth, Rabindra (1996) : *Tourism : Strong linkage to growth in Survey of Indian Industry (1996),* The Hindu, Chennai.

Sharma, N.K. (1989) : *A Socio-economic study of Tourism Development in Himachal Pradesh in* S.C. Singh (ed). *Impact of Tourism on Mountain Environment,* Meerut, pp. 230-236.

Singh, Shalini and Singh, Tej Vir (1996) (ed.) : *Profiles of Indian Tourism,* A.P.H. Publishing Corporation, New Delhi.

Singh, Tej Vir (1975) : *Tourism and Tourist Industry,* New Heights, New Delhi.

—— (1989) : *On Developing Himalayan Tourism Ecologically* in Singh T.V. *et al.* (Eds), *Studies in Himalayan Ecology,* Himalayan Books, New Delhi.

—— (1990), (Ed.) : *Geography of the Mountains,* Heritage Publishers, New Delhi.

Stenigenberger, Consulting (1987) : *Tourism Development in Himachal Pradesh/India* GTZ Report.

Soundarajan, K.V. (1981) : *Cave Temples of the Deccan,* Archaeological Survey of India, New Delhi.

WTO/UNEP (1992) : *Tourism and Environment Guidelines : Development of national parks and protected areas for tourism.* World Tourism Organization, Madrid, Spain.

Ziffer, Karen A. (1989) : *Ecotourism : the uneasy alliances. Conservation International's series of working papers on ecotourism.*

◆◇◆